EPIGENETICS

Second Edition

OTHER TITLES FROM COLD SPRING HARBOR LABORATORY PRESS

Mammalian Development: Networks, Switches, and Morphogenetic Processes
Molecular Biology of the Gene, Seventh Edition
Protein Synthesis and Translational Control
RNA Worlds: From Life's Origins to Diversity in Gene Regulation
Signal Transduction: Principles, Pathways, and Processes

FROM *COLD SPRING HARBOR PERSPECTIVES IN BIOLOGY*

Glia
Innate Immunity and Inflammation
The Genetics and Biology of Sexual Conflict
The Origin and Evolution of Eukaryotes
Endocytosis
Mitochondria
Signaling by Receptor Tyrosine Kinases
DNA Repair, Mutagenesis, and Other Responses to DNA Damage
Cell Survival and Cell Death
Immune Tolerance
DNA Replication
Endoplasmic Reticulum
Wnt Signaling
Protein Synthesis and Translational Control
The Synapse
Extracellular Matrix Biology
Protein Homeostasis
Calcium Signaling
The Golgi
Germ Cells
The Mammary Gland as an Experimental Model
The Biology of Lipids: Trafficking, Regulation, and Function

EPIGENETICS

Second Edition

EDITORS

C. David Allis
The Rockefeller University, New York

Marie-Laure Caparros
London

Thomas Jenuwein
Max Planck Institute of Immunobiology and Epigenetics, Freiburg

Danny Reinberg
Howard Hughes Medical Institute
New York University School of Medicine-Smilow Research Center

ASSOCIATE EDITOR

Monika Lachner
Max Planck Institute of Immunobiology and Epigenetics, Freiburg

COLD SPRING HARBOR LABORATORY PRESS
Cold Spring Harbor, New York • www.cshlpress.org

Epigenetics, *Second Edition*

Chapters online at cshperspectives.org.

Printed in India

Publisher	John Inglis
Director of Editorial Development	Jan Argentine
Project Manager	Inez Sialiano
Permissions Coordinator	Carol Brown
Production Editor	Kathleen Bubbeo
Production Manager	Denise Weiss
Cover Designer	Michael Albano

Front cover artwork: Depicted is a schematic representation of the chromatin template. Epigenetic regulation affects and modulates this template through noncoding RNAs (ncRNAs) that associate with it, through covalent modification of histone tails (mod), methylation of DNA (Me), remodeling factors (blue oval), and nucleosomes that contain standard as well as variant histone proteins (the yellow nucleosome). In the background is a representation of several model organisms in which epigenetic control has been studied. From *top left* to *bottom right*: Pair of mouse chromosomes that may differ in their genomic imprint; a *Saccharomyces cerevisiae* colony, showing epigenetically inherited variegation of gene expression; anatomy of *Caenorhabditis elegans*; illustration of *Tetrahymena thermophila*, showing the large "active" macronucleus and the smaller "silent" micronucleus; *Drosophila melanogaster*, maize section with kernel color variegation; *Arabidopsis* flower.

Library of Congress Cataloging-in-Publication Data

Epigenetics (Allis)
Epigenetics / edited by C. David Allis, Marie-Laure Caparros, Thomas Jenuwein, Danny Reinberg ; associate editor, Monica Lachner. — Second edition.
p. cm.
Includes bibliographical references and index.
ISBN 978-1-936113-59-0 (hardcover)
I. Allis, C. David, editor. II. Caparros, Marie-Laure, editor. III. Jenuwein, Thomas, editor. IV. Reinberg, Danny, editor. V. Title.
[DNLM: 1. Epigenesis, Genetic. 2. Gene Expression Regulation. QU 475]
QH450
572.8′65--dc23

2014029474

10 9 8 7 6 5 4

All World Wide Web addresses are accurate to the best of our knowledge at the time of printing.

For a complete catalog of all Cold Spring Harbor Laboratory Press publications, visit our website at www.cshlpress.org.

Long before epigenetics changed from little more than a diverse collection of bizarre phenomena to a well-respected field covered by its own textbook, a talented group of foresighted molecular biologists laid a rich foundation upon which the modern era of chromatin biology and epigenetics is based. This group includes Vince Allfrey, Wolfram Hörz, Robert Simpson, Hal Weintraub, Jonathan Widom, Alan Wolffe, and Abe Worcel. This book is dedicated to their collective memory. Their passion and commitment to the study of chromatin biology inspired all of us who followed their work, and we now benefit from their many insights.

Contents

Preface

SINCE PUBLISHING THE FIRST Cold Spring Harbor Laboratory Press edition of *Epigenetics* in 2007, significant advances have been made by researchers worldwide working in multiple fields that touch on epigenetics. For those new to the field, we would point them first to the overview chapter (Chapter 3) for an introduction to the basic concepts and a synthesis of what the field encompasses. The editorial team acknowledges that, although numerous exciting findings could be cited here, several are particularly noteworthy and worth some expanded comments as we chart our course into this edition in 2014.

First, owing to dramatic innovation in sequencing technologies, often referred to as massively parallel or deep sequencing (e.g., genome-wide RNA-Seq or ChIP-Seq approaches), the textbook notion that the flow of genetic information is from DNA to protein via messenger RNA has undergone a remarkable paradigm shift. It is now widely accepted that RNA alone can perform many diverse roles and that a remarkably large fraction of the genome is transcribed, with some estimates as high as >90%. Interestingly only ~2% of these transcripts fall into the messenger RNA category, with a high percentage (~70%) accounting for divergently transcribed noncoding RNAs, be they long or short (see the essay by Rinn [Ch. 2]; also reviewed in Darnell 2011; Guttman and Rinn 2012). The function of these noncoding RNAs remains one of the intense areas of investigation, with emerging models suggesting that these RNAs may work to integrate or provide a scaffold for the chromatin-remodeling and -modifying enzyme complexes or to bring about critical changes in the nuclear architecture through *cis* or *trans* mechanisms, as well as to allow recruitment of factors that silence a chromatin domain (e.g., Polycomb) or facilitate transcription (e.g., eRNA recruiting mediator, elaborated in the Kim et al. essay). We also point out the intriguing links between defined histone modifications and the splicing of pre-messenger RNA (i.e., intronic and exonic definition) to underscore the concept that the RNA world is expanding and intimately linked to chromatin states (Huff et al. 2010).

Second, remarkable progress has been made in documenting the discovery and structures of chromatin-binding modules that "read" one or more histone modifications (see the new chapters in this book by Cheng, Patel, Marmorstein and Zhou, and Seto and Yoshida). How can one make sense of all of this staggering posttranslational modification complexity? Carrying epigenomics to a genome-wide scale, Zhong and co-workers (Xiao et al. 2012) have introduced what they refer to as "comparative epigenomics," wherein an impressive collection of epigenetic marks (histone modifications, genomic distributions of cytosine methylation, histone variants, transcription factors, etc.) have been mapped in human, mouse, and porcine cells, drawing upon evolution as a useful guide for highlighting the functional importance of various marks. Importantly, comparative epigenomics has revealed regulatory features of the genome that cannot be ascertained by sequence comparisons alone. Outside of better-known co-associating marks, such as those associated with bivalent domains (i.e., H3K4me3 and H3K27me3) at promoters of developmentally regulated genes, other highly conserved co-marks have been identified. For example, H3K27ac+H3K4me1/2 and H3K27ac+H3K4me2/3 mark active enhancer and promoter elements, respectively. The authors of these findings conclude that the general problem of "having too many epigenetic mark combinations and not knowing how to distinguish random versus functional co-localizations can be overcome by using evolutionary conservation." We applaud this study as it provides a fresh approach to the complexities of epigenomes, and we look forward to other studies that draw on the insights gleaned from using evolution as a guide.

Third, the fundamental question remains as to how any epigenetic marks are inherited, with our understanding being much more complete for how cytosine methylation marks in DNA are templated during replication. With respect to histone marks, an emerging literature suggests novel mechanisms that include allosteric regulation of the key histone-modifying enzyme complexes, wherein modifications on one histone tail, such as histone H2B ubiquitination (McGinty et al. 2008) or histone H3K27me3 (Margueron et al. 2009), can stimulate downstream activating (e.g., DOT1L [KMT4]) or inactivating (e.g., PRC2) histone methyltransferases (KMTs), respectively. Taken together, these groundbreaking studies suggest that new covalent modifications can be introduced to naïve chromatin templates, providing a potential mechanism of inheritance from unmodified (in some cases, newly synthesized histones) to newly modified states during replication and chromatin assembly, that can be passed on to future generations. We look forward to future studies along this line, especially when addressed by in vivo (i.e., mutants of histones and chromatin machinery; see Rando 2012a) and in vitro (i.e., the use of "designer chromatin" templates; see Fierz and Muir 2012) systems. Complexities of this "language" include elaborating the cross-talk relationships between histone marks, with the added complication of number and type of covalent modifications in both histone proteins (e.g., mono- vs. di- vs. trilysine methylation, lysine acetylation vs. crotonylation, arginine symmetric vs. asymmetric dimethylation) and DNA (e.g., methylation vs. hydroxymethylation on cytosine residues). There is no doubt that deciphering the links between histone modifications, DNA methylation, and noncoding RNAs promises to stimulate and challenge the next generation of scientists entering the general field of epigenetics.

Fourth, histone variants provide cells with the means to tailor chromatin assembly pathways to create distinct chromatin states at distinct genomic locations. We envision that the evolution of histone variants has given the cell a regulatory option to remodel the chromatin template, even outside of the classical notion of coupling histone synthesis to DNA replication during S phase (i.e., replication-independent histone deposition; see the chapter by Henikoff and Smith). It is not surprising then that histone variants, especially the replication-independent types, would require a dedicated machinery and energy to accomplish their task of "courting" and "escorting" their histones into place in the genome. Quite recently, a remarkable series of papers, spearheaded largely by physician scientists using exome sequencing, have identified mutations in "epigenetic regulators" in a remarkably wide range of human cancers. For example, DAXX, ATRX, and the H3.3 variant have been linked to tumorigenesis (pancreatic neuroendocrine tumors or panNETs for short; Jiao et al. 2011), strongly suggesting that DAXX-mediated, H3.3-specific chromatin assembly constitutes the tumor-suppressor function of the ATRX-DAXX complex, likely leading to chromosomal abnormalities that include dysfunctional telomeres. Perhaps the biggest surprise came with the finding that cancer-causing mutations exist in histone-encoding genes themselves (reviewed in Dawson and Kouzarides 2012; You and Jones 2012; Shen and Laird 2013). One of us (C.D.A.) has been known to say, "Every amino acid in histones matters," but this contention is difficult to test in organisms where histone genetics is not readily feasible. Given that oncogenic mutations have now been mapped to H3 amino termini at two "hot spots"—K27 and G34—in distinct groups of pediatric glioblastoma patients (interestingly those with stem vs. cortex tumors, respectively) (described in the Liu et al. essay [Ch. 2]), we look forward to insights here that will help to diagnose a devastating and lethal set of childhood cancers (see Rheinbay et al. 2012 for review and references). A high-frequency H3 mutation at K36 has also been linked to other pediatric cancers (e.g., chondroblastoma; Behjati et al. 2013), underscoring the functional importance of lysine-based covalent modifications in histone proteins. Examples are being uncovered in other components of the epigenetic machinery that have disease links that lie outside of cancer (e.g., pathways linked to neurological functions and mental retardation; see Schaefer et al. 2011; Lotsch et al. 2013). Cancer and other disease links (covered in the Baylin and Jones, Audia and Campbell, and Zoghbi and Beaudet chapters; see also the Qi, Schaefer, and Liu et al. chapters) promise to fuel the continued interest in epigenetics well into the future editions beyond this one.

Fifth, the idea that chromatin-remodeling pathways might provide therapeutically useful targets, which may permit mis-silenced or mis-activated genes to be reversed as the genes themselves are not altered by mutations, has led to the general acceptance that developing drugs against chromatin-based targets is a viable new route for treatment in clinical oncology. Specifically, the identification of altered DNA methylation and histone acetylase (HAT) activity in a range of human cancers, coupled with the use of histone deacetylase (HDAC) and DNA methylation inhibitors in the treatment of human cancer, make this a compelling argument, as do the well-documented genetic lesions in histone lysine methyltransferases such as EZH2 (KMT6A), MMSET, etc. Given the genetic links to these key epigenetic-based enzymes, small molecule inhibitors have been designed and tested with positive therapeutic outcomes. Some of these inhibitors are FDA-approved and in widespread use in clinical trials. It is clear that the regulatory signals provided by chromatin modifications will revolutionize our view of cancer as new models of "epigenetic carcinogenesis" are advanced (see also the chapter by Audia and Campbell).

Catalytic enzymes are not the only class of epigenetic regulators that have proven to be worth drugging. In late 2010, a pair of back-to-back papers (Filippakopoulos et al. 2010; Nicodeme et al. 2010) revealed that histone acetyllysine binding pockets, or bromodomains, are druggable by small molecules, with useful clinical outcomes (see the essays by Schaefer and by Qi [Ch. 2] and the chapters by Busslinger and Tarakhovsky and by Marmorstein and Zhou). Moreover, this work laid the foundation for an equally remarkable study wherein large-scale structural analysis of the human bromodomain family was performed, providing remarkable insights into the molecular discrimination by which the different histone acetyllysine reading modules discriminate different chromatin contexts

(Filippakopoulos et al. 2012). We look forward to extending these types of studies to other chromatin-reading "pockets," with added specificity holding promise for a new frontier for drug discovery (Arrowsmith et al. 2012). Last, with regard to potential therapeutic targets, we stress that *histones* are not the only physiologically relevant recipients of this covalent "language." Large cohorts of *nonhistone* proteins are now well known to be modified by what were originally described as histone-modifying enzymes (e.g., the acetylation and methylation of p53 by p300 [KAT3B] and Set7/9 [KMT7], respectively, originally reported by Gu and Roeder 1997; Chuikov et al. 2004). Histone "mimicry" has been well documented by Tarakhovsky and others (Sampath et al. 2007; Marazzi et al. 2012), suggesting that these mechanisms extend well beyond histone proteins (Sims and Reinberg 2008).

Finally, the very roots of epigenetics are grounded in problems of developmental biology, as articulated by Waddington and others (see the Felsenfeld chapter). The chromatin packaging system has evolved to make certain genes less or more accessible to transcription factors and other machinery that must engage the true genetic template (see the closing Pirrotta chapter). Although there can be little doubt that we are entering a "postgenomic" or "epigenomic" era, we acknowledge that transcriptional networks likely lie at the heart of reprogramming differentiated cell types from more pluripotent embryonic cell types. Nowhere is that better illustrated than with the generation of induced pluripotent stem cells (iPS cells or iPSCs) by Yamanaka and colleagues in 2006, wherein a set of master gene transcription factors encoded by key pluripotency genes (e.g., Oct-3/4, Sox2, c-Myc, and Klf4) were introduced into nonpluripotent cells, such as mouse adult fibroblasts, and shown to reprogram (or dedifferentiate) them backward to more pluripotent or totipotent states (Takahashi and Yamanaka 2006). These groundbreaking studies build nicely upon pioneering studies by Gurdon and others, who demonstrated early on that somatic, adult nuclei could be reprogrammed, provided that they were transplanted into an egg (oocyte) environment (Gurdon et al. 1958). Although the importance of transcription "master regulators" cannot be questioned, the low efficiency of reprogramming, the stability of the induced states, and the tendency for reprogrammed cells to take a turn toward a more neoplastic state suggest chromatin underpinnings or "barriers" to the reprogramming process have yet to be fully understood (Soufi et al. 2012; Chen et al. 2013). The editors are pleased that the discovery of induced pluripotency is described in the Takahashi essay (Ch. 2), and the general topic of "reprogramming" is covered in this textbook in the Hochedlinger and Jaenisch chapter. As well, topics that closely align epigenetics with developmental biology issues include the chapters by Grossniklaus and Paro, Kingston and Tamkun, and Reik and Surani.

In closing, this preface highlights only a few of the exciting areas that have come to light since the first edition was published. Our overview and the chapters that follow will not only develop these areas further, but also touch on many more. As well, a novel collection of short essays by junior scientists, who made important discoveries that have already set the field of epigenetics on new and exciting courses, appears in this edition. These essays touch on the history of how these discoveries were made. A quick comparison between the first and second edition of this book underscores the remarkable progress made by the field between these editions. Twelve new chapters have been added with a significant updating of all of the earlier chapters. For example, Figure 3 of Ch. 3 (also in the first edition) suggested that epigenetic alterations, as compared to true genetics, might not be stable or part of true germline inheritance. However, the long-standing debate contrasting the distinction between innate and acquired characteristics (Lamarckian theory) is being revisited in light of new research indicating that environmental factors can provide adaptive responses, via noncoding RNAs in somatic and germline lineages (Ashe et al. 2012; Lee et al. 2012; Rando 2012b). Clearly, new discoveries, likely fueled by readers of this edition, will form the foundation of other editions that take us beyond our current understanding. The developmental biologists of the past must be looking down with great pleasure.

Our goal here, as with the first edition, is to educate newcomers and seasoned veterans alike as to the key concepts that shape and guide the broad field of epigenetics. Words of others underscore the most general problem that our textbook hopes to address: "*We are more than the sum of our genes*" (Klar 1998); "*You can inherit something beyond the DNA sequence. That's where the real excitement is now*" (Watson 2003); or *Time* magazine's 2010 cover story headline "*Why your DNA isn't your destiny*" (Cloud 2010). The field of epigenetics does not appear to be slowing down; remarkably the slope of its citation index in the literature continues to climb. We hope that the readers of this textbook will share in our excitement and yet be inspired to tackle the many problems that remain unsolved or poorly understood. We remain grateful to all of those who have turned this edition into a long-awaited reality.

REFERENCES

Arrowsmith CH, Bountra C, Fish PV, Lee K, Schapira M. 2012. Epigenetic protein families: A new frontier for drug discovery. *Nat Rev Drug Discov* **11:** 384–400.

Ashe A, Sapetschnig A, Weick EM, Mitchell J, Bagijn MP, Cording AC, Doebley AL, Goldstein LD, Lehrbach NJ, Le Pen J, et al. 2012. piRNAs can trigger a multigenerational epigenetic memory in the germline of *C. elegans*. *Cell* **150:** 88–99.

Behjati S, Tarpey PS, Presneau N, Scheipl S, Pillay N, Van Loo P, Wedge DC, Cooke SL, Gundem G, Davies H, et al. 2013. Distinct H3F3A and H3F3B driver mutations define chondroblastoma and giant cell tumor of bone. *Nat Genet* **45:** 1479–1482.

Chen J, Liu H, Liu J, Qi J, Wei B, Yang J, Liang H, Chen Y, Chen J, Wu Y, et al. 2013. H3K9 methylation is a barrier during somatic cell reprogramming into iPSCs. *Nat Genet* **45:** 34–42.

Chuikov S, Kurash JK, Wilson JR, Xiao B, Justin N, Ivanov GS, McKinney K, Tempst P, Prives C, Gamblin SJ, et al. 2004. Regulation of p53 activity through lysine methylation. *Nature* **432:** 353–360.

Cloud J. 2010. Why genes aren't destiny. *Time* **175:** 48–53.

Darnell JE. 2011. *RNA: Life's indispensable molecule*. Cold Spring Harbor Laboratory Press, Cold Spring Harbor, NY.

Dawson MA, Kouzarides T. 2012. Cancer epigenetics: From mechanism to therapy. *Cell* **150:** 12–27.

Fierz B, Muir TW. 2012. Chromatin as an expansive canvas for chemical biology. *Nat Chem Biol* **8:** 417–427.

Filippakopoulos P, Qi J, Picaud S, Shen Y, Smith WB, Fedorov O, Morse EM, Keates T, Hickman TT, Felletar I, et al. 2010. Selective inhibition of BET bromodomains. *Nature* **468:** 1067–1073.

Filippakopoulos P, Picaud S, Mangos M, Keates T, Lambert JP, Barsyte-Lovejoy D, Felletar I, Volkmer R, Muller S, Pawson T, et al. 2012. Histone recognition and large-scale structural analysis of the human bromodomain family. *Cell* **149:** 214–231.

Gu W, Roeder RG. 1997. Activation of p53 sequence-specific DNA binding by acetylation of the p53 C-terminal domain. *Cell* **90:** 595–606.

Gurdon JB, Elsdale TR, Fischberg M. 1958. Sexually mature individuals of *Xenopus laevis* from the transplantation of single somatic nuclei. *Nature* **182:** 64–65.

Guttman M, Rinn JL. 2012. Modular regulatory principles of large non-coding RNAs. *Nature* **482:** 339–346.

Huff JT, Plocik AM, Guthrie C, Yamamoto KR. 2010. Reciprocal intronic and exonic histone modification regions in humans. *Nat Struct Mol Biol* **17:** 1495–1499.

Jiao Y, Shi C, Edil BH, de Wilde RF, Klimstra DS, Maitra A, Schulick RD, Tang LH, Wolfgang CL, Choti MA, et al. 2011. DAXX/ATRX, MEN1, and mTOR pathway genes are frequently altered in pancreatic neuroendocrine tumors. *Science* **331:** 1199–1203.

Klar AJ. 1998. Propagating epigenetic states through meiosis: Where Mendel's gene is more than a DNA moiety. *Trends Genet* **14:** 299–301.

Lee HC, Gu W, Shirayama M, Youngman E, Conte D Jr, Mello CC. 2012. *C. elegans* piRNAs mediate the genome-wide surveillance of germline transcripts. *Cell* **150:** 78–87.

Lotsch J, Schneider G, Reker D, Parnham MJ, Schneider P, Geisslinger G, Doehring A. 2013. Common non-epigenetic drugs as epigenetic modulators. *Trends Mol Med* **19:** 742–753.

Marazzi I, Ho JS, Kim J, Manicassamy B, Dewell S, Albrecht RA, Seibert CW, Schaefer U, Jeffrey KL, Prinjha RK, et al. 2012. Suppression of the antiviral response by an influenza histone mimic. *Nature* **483:** 428–433.

Margueron R, Justin N, Ohno K, Sharpe ML, Son J, Drury WJ 3rd, Voigt P, Martin SR, Taylor WR, De Marco V, et al. 2009. Role of the Polycomb protein EED in the propagation of repressive histone marks. *Nature* **461:** 762–767.

McGinty RK, Kim J, Chatterjee C, Roeder RG, Muir TW. 2008. Chemically ubiquitylated histone H2B stimulates hDot1 L-mediated intranucleosomal methylation. *Nature* **453:** 812–816.

Nicodeme E, Jeffrey KL, Schaefer U, Beinke S, Dewell S, Chung CW, Chandwani R, Marazzi I, Wilson P, Coste H, et al. 2010. Suppression of inflammation by a synthetic histone mimic. *Nature* **468:** 1119–1123.

Rando OJ. 2012a. Combinatorial complexity in chromatin structure and function: Revisiting the histone code. *Curr Opin Genet Dev* **22:** 148–155.

Rando OJ. 2012b. Daddy issues: Paternal effects on phenotype. *Cell* **151:** 702–708.

Rheinbay E, Louis DN, Bernstein BE, Suva ML. 2012. A tell-tail sign of chromatin: Histone mutations drive pediatric glioblastoma. *Cancer Cell* **21:** 329–331.

Sampath SC, Marazzi I, Yap KL, Sampath SC, Krutchinsky AN, Mecklenbrauker I, Viale A, Rudensky E, Zhou MM, Chait BT, et al. 2007. Methylation of a histone mimic within the histone methyltransferase G9a regulates protein complex assembly. *Mol Cell* **27:** 596–608.

Schaefer A, Tarakhovsky A, Greengard P. 2011. Epigenetic mechanisms of mental retardation. *Prog Drug Res* **67:** 125–146.

Shen H, Laird PW. 2013. Interplay between the cancer genome and epigenome. *Cell* **153:** 38–55.

Sims RJ III, Reinberg D. 2008. Is there a code embedded in proteins that is based on post-translational modifications? *Nat Rev Mol Cell Biol* **9:** 815–820.

Soufi A, Donahue G, Zaret KS. 2012. Facilitators and impediments of the pluripotency reprogramming factors' initial engagement with the genome. *Cell* **151:** 994–1004.

Takahashi K, Yamanaka S. 2006. Induction of pluripotent stem cells from mouse embryonic and adult fibroblast cultures by defined factors. *Cell* **126:** 663–676.

Watson JD. 2003. Celebrating the genetic jubilee: A conversation with James D. Watson. Interviewed by John Rennie. *Sci Am* **288:** 66–69.

Xiao S, Xie D, Cao X, Yu P, Xing X, Chen CC, Musselman M, Xie M, West FD, Lewin HA, et al. 2012. Comparative epigenomic annotation of regulatory DNA. *Cell* **149:** 1381–1392.

You JS, Jones PA. 2012. Cancer genetics and epigenetics: Two sides of the same coin? *Cancer Cell* **22:** 9–20.

Acknowledgments

As is likely with every major textbook undertaking, the project seems to grow beyond boundaries with many reasons why the book ever sees the light of day. Nowhere is this truer than with this second edition of the textbook *Epigenetics*. Here the boundaries grew: The number of chapters increased, as did the size of our overview and concepts chapter. Why is this? Here we can only suggest that part of it lies with all of the exciting science that is the collective field of epigenetics.

We are hugely grateful to all of the authors contributing to this second edition—some from the first edition of this book, but also many newcomers. Special thanks need to go to the junior scientists who stopped their benchwork to share with us, in highlighted essays contained in the second chapter, some of what was behind all of the exciting discoveries made firsthand by them and their colleagues, helping to make the field what it is today. It is the care and attention of all these authors that form the heart of this book; their knowledge and expertise makes this textbook what it we intended it to be—the latest detailed textbook on the exciting topic of epigenetics. With each chapter and essay, we have consulted with outside experts who weighed in with their constructive comments, making the written pieces as accurate and timely as possible. We thank them all.

As with the first edition, we are grateful for John Inglis's instigation and support for the second edition and to all of the staff at CSHL Press (Inez Sialiano, Kathy Bubbeo, Richard Sever, Jan Argentine, and Denise Weiss), who have been key players in this effort. With this second edition, the editors owe them special thanks as we pushed them to the breaking point with endless delays and behind-schedule demands. We also appreciate that CSHL Press will make each of the chapters of the textbook available online at *CSH Perspectives in Biology*, so that each contribution can be cited as a full publication. All of the editors' assistants (Marisa Cerio [C.D.A.], Marcela Mare [T.J.], and Michele Giunta [D.R.]) have also shown remarkable patience trying their best to arrange the countless calls, meetings, and calendars to make it all happen.

Special gratitude needs to be directed to two individuals, who, as much as anyone involved with this second edition, know every page, sentence, and word included in this textbook, inside and out. Put simply, Marie-Laure Caparros and Monika Lachner have made this textbook a reality. From text to tension, references to rants, figures to frustration, appendices to anxiety, they have done it all. Where does their patience come from? No one knows, but the three of us guess that some form of very special genetics and epigenetics are in both of them. Thank you beyond words for your amazing work, care, and attention to the fine details that make a good textbook, hopefully, a great one.

Finally, the three of us acknowledge that any mistakes or omissions in this textbook lie with us. Many folks have wondered what took this second edition so long to come out? What dark hole did those deadlines fall into? Slowness on our part has likely caused these delays. Please accept our apologies. Even so, we have enjoyed the challenge of trying to bring the many advances made in the field of epigenetics into a form where all of the readers of this textbook will join us in the genuine excitement that promises to come.

Financial support for this book was provided by CSHL Press (New York). Rockefeller University (New York), the NYU School of Medicine (New York), and the Max Planck Institute of Immunobiology and Epigenetics (Freiburg, Germany) provided further funds for the book's developmental editing.

C. David Allis, *The Rockefeller University, New York*

Thomas Jenuwein, *Max Planck Institute of Immunobiology and Epigenetics, Freiburg*

Danny Reinberg, *NYU School of Medicine Smilow Research Center*

September 15, 2014

(*Left to right*) Monika Lachner, Thomas Jenuwein, Danny Reinberg, Marie-Laure Caparros, and David Allis at an editorial meeting in New York.

CHAPTER 1

A Brief History of Epigenetics

Gary Felsenfeld

National Institute of Diabetes and Digestive and Kidney Diseases, National Institutes of Health, Bethesda, Maryland 20892-0540

Correspondence: garyf@intra.niddk.nih.gov

SUMMARY

The term "epigenetics" was originally used to denote the poorly understood processes by which a fertilized zygote developed into a mature, complex organism. With the understanding that all cells of an organism carry the same DNA, and with increased knowledge of mechanisms of gene expression, the definition was changed to focus on ways in which heritable traits can be associated not with changes in nucleotide sequence, but with chemical modifications of DNA, or of the structural and regulatory proteins bound to it. Recent discoveries about the role of these mechanisms in early development may make it desirable to return to the original definition of epigenetics.

Outline

Cite as *Cold Spring Harb Perspect Biol* doi: 10.1101/cshperspect.a018200

1 INTRODUCTION

The history of epigenetics is linked with the study of evolution and development. But during the past 50 years, the meaning of the term "epigenetics" has itself undergone an evolution that parallels our dramatically increased knowledge of the molecular mechanisms underlying regulation of gene expression in eukaryotes. Our present definitions of epigenetics reflect our understanding that although the complement of DNA is essentially the same in all of an organism's somatic cells, patterns of gene expression differ greatly among different cell types, and these patterns can be clonally inherited. This has led to a working definition of epigenetics as "the study of mitotically and/or meiotically heritable changes in gene function that cannot be explained by changes in DNA sequence" (Riggs et al. 1996; Riggs and Porter 1996). More recently added to this definition is the constraint that initiation of the new epigenetic state should involve a transient mechanism separate from the one required to maintain it (Berger et al. 2009). Until the 1950s, however, the word epigenetics was used more broadly (and less precisely) to categorize all of the developmental events leading from the fertilized zygote to the mature organism—that is, all of the regulated processes that, beginning with the genetic material, shape the final product (Waddington 1953). This concept had its origins in the much earlier studies in cell biology and embryology, beginning in the late 19th century, which laid the groundwork for our present understanding of the relationship between genes and development. There was a long debate among embryologists about the nature and location of the components responsible for carrying out the developmental plan of the organism. In trying to make sense of a large number of ingenious but ultimately confusing experiments involving the manipulation of cells and embryos, embryologists divided into two schools: Those who thought that each cell contained preformed elements that enlarged during development ("preformationism"), and those who thought the process involved chemical reactions among soluble components that executed a complex developmental plan ("epigenesis"). These views focused on the relative importance of the nucleus and cytoplasm in the developmental process. Although the definition that we choose for epigenetics has changed to accommodate our increasing knowledge, it is important to remember that the original problem was: How can a single fertilized egg give rise to a complex organism with cells of varied phenotypes?

Following Fleming's discovery of the existence of chromosomes in 1879, experiments by many investigators, including Wilson and Boveri, provided strong evidence that the developmental program resided in the chromosomes. Thomas Hunt Morgan (Morgan 1911) ultimately provided the most persuasive proof of this idea through his demonstration of the genetic linkage of several *Drosophila* genes to the X chromosome. From that point onward, rapid progress was made in creating linear chromosome maps in which individual genes were assigned to specific sites on the *Drosophila* chromosomes (Sturtevant 1913). Of course, the questions of classic "epigenesis" remained: What molecules within the chromosomes carried the genetic information, how did they direct the developmental program, and how was the information transmitted during cell division? It was understood that both nucleic acid and proteins were present in chromosomes, but their relative contributions were not obvious; certainly, no one believed that the nucleic acid alone could carry all of the developmental information. Furthermore, earlier questions persisted about the possible contribution of the cytoplasm to developmental events. Evidence from *Drosophila* genetics (see Section 2) suggested that heritable changes in phenotype could occur without corresponding changes in the "genes." This debate was dramatically altered by the identification of DNA as the primary carrier of genetic information. Ultimately, it became useful to redefine epigenetics so as to distinguish heritable changes that arise from sequence changes in DNA from those that do not. Further refinements of the definition have accompanied more detailed understanding of underlying mechanisms, but it may not be useful at this point to attempt ultimate precision in describing very complex regulatory processes in which "epigenetic" and "nonepigenetic" components are entwined.

2 CLUES FROM GENETICS AND DEVELOPMENT

Whatever the vagaries of the definition, the ideas and scientific data that underlie the present concept of epigenetics have been accumulating steadily since the early part of the 20th century. In 1930, H.J. Muller (Muller 1930) described a class of *Drosophila* mutations he called "eversporting displacements" ("eversporting" denoting the high rate of phenotypic change). These mutants involved chromosome translocations (displacements), but "even when all parts of the chromatin appeared to be represented in the right dosage—though abnormally arranged—the phenotypic result was not always normal." In some of these cases, Muller observed flies that had mottled eyes. He thought that this was probably attributable to a "genetic diversity of the different eye-forming cells," but further genetic analysis led him to connect the unusual properties with chromosomal rearrangement, and to conclude that "chromosome regions, affecting various characters at once, are somehow concerned, rather than individual genes or suppositious 'gene elements.'" Over the next 10 to 20 years, strong evidence provided by many laboratories (Hannah 1951) confirmed

Cite as *Cold Spring Harb Perspect Biol* doi: 10.1101/cshperspect.a018200

that this variegation arose when rearrangements juxtaposed the white gene with heterochromatic regions.

During that period, chromosomal rearrangements of all kinds were the object of a great deal of attention. It was apparent that genes were not completely independent entities; their function could be affected by their location within the genome—as amply shown by the many *Drosophila* mutants that led to variegation, as well as by other mutants involving translocation to euchromatic regions, in which more general (nonvariegating) position effects could be observed. The role of transposable elements in plant genetics also became clear, largely through the work of McClintock (McClintock 1965), although they are probably not involved in normal development.

A second line of reasoning came from the study of developmental processes. It was evident that during development there was a divergence of phenotypes among differentiating cells and tissues, and it appeared that such distinguishing features, once established, could be clonally inherited by the dividing cells. Although it was understood at this point that cell-specific programming existed, and that it could be transmitted to daughter cells, how this was done was less clear.

A number of mechanisms could be imagined, and were considered. For those with a biochemical point of view, a cell was defined by the multiple interdependent biochemical reactions that maintained its identity. For example, it was suggested in 1949 by Delbruck (quoted in Jablonka and Lamb 1995) that a simple pair of biochemical pathways, each of which produced as an intermediate an inhibitor of the other pathway, could establish a system that could switch between one of two stable states. Actual examples of such systems were found somewhat later in the lac operon of *Escherichia coli* (Novick and Weiner 1957), and in the lambda phage switch between lysogenic and lytic states (Ptashne 1992). Functionally equivalent models could be envisioned in eukaryotes: The kinds of self-stabilizing inhibitory and stimulatory mechanisms observed in lambda phage are in fact seen in greatly more complex form in higher organisms. In the sea urchin embryo, for example, development proceeds through the establishment and progression of a series of self-stabilizing regulatory networks. However, it is important to recognize an essential difference between the prokaryotic and eukaryotic systems: In the case of the sea urchin, each of the regulatory "modules" is not in a static state, but rather receives from and sends signals to other modules that give rise to the changing, time-dependent phenotype associated with a developing embryo. It should also be noted that although chromatin structure and biochemistry must certainly be involved in the implementation of this program (see Section 5), the system can be modeled entirely in terms of control of gene expression by specific binding of expressed factors to the regulatory regions of relevant genes.

The extent to which nucleus and cytoplasm each contributed to the transmission of a differentiated state in the developing embryo was of course a matter of intense interest and debate; a self-stabilizing biochemical pathway would presumably have to be maintained through cell division. A second kind of epigenetic transmission was clearly shown in *Paramecia* and other ciliates, in which the ciliary patterns may vary among individuals and are inherited clonally (Beisson and Sonneborn 1965). Altering the cortical pattern by microsurgery results in transmission of a new pattern to succeeding generations. It has been argued that related mechanisms are at work in metazoans, in which the organization of cellular components is influenced by localized cytoplasmic determinants in a way that can be transmitted during cell division (Grimes and Aufderheide 1991).

3 DNA IS THE SAME IN ALL SOMATIC CELLS OF AN ORGANISM

Although chromosome morphology indicated that all somatic cells possessed all of the chromosomes, it could not have been obvious that all somatic cells retained the full complement of DNA present in the fertilized egg. Nor was it even clear that a protein-free DNA molecule could carry genetic information until the work of Avery, MacLeod, and McCarty (Avery et al. 1944), and that of Hershey and Chase (Hershey and Chase 1952), a conclusion strongly reinforced by Watson and Crick's solution of the structure of DNA in 1953. Work by Briggs and King (Briggs and King 1952) in *Rana pipiens* and by Laskey and Gurdon (Laskey and Gurdon 1970) in *Xenopus* had shown that introduction of a nucleus from early embryonic cells into enucleated oocytes could result in development of an embryo. But as late as 1970, Laskey and Gurdon could state that: "It has yet to be proved that somatic cells of an adult animal possess genes other than those necessary for their own growth and differentiation." In the article containing this statement, they went on to show that to a first approximation, the DNA of a somatic cell nucleus was competent to direct embryogenesis when introduced into an enucleated egg. It was now clear that the program of development, and the specialization of the repertoire of expression seen in somatic cells, must involve signals that are not the result of some deletion or mutation in the germline DNA sequence when it is transmitted to somatic cells. At the same time, other experiments revealed that these signals could confer phenotypic stability over many generations of cell division. Even after undifferentiated *Drosophila* imaginal disc cells were transplanted and cultured in successive generations of adult flies, they maintained their disc-specific patterns of differentiation when transferred to larvae (Hadorn 1965; McClure and Schubiger 2007).

Of course, there are ways in which the DNA of somatic cells can come to differ from that of the germline with consequences for the cellular phenotype: For example, transposable elements can alter the pattern of expression in somatic cells as demonstrated by the work of Barbara McClintock and other plant geneticists. Similarly, the generation of antibody diversity involves DNA rearrangement in a somatic cell lineage. This rearrangement (or, more precisely, its consequences) can be considered a kind of epigenetic event, consistent with the early observations of position-effect variegation described by Muller (1930). However, much of the work on epigenetics in recent years has focused on systems in which no DNA rearrangements have occurred, and the emphasis has therefore been on modifications to the bases, and to the proteins that are complexed with DNA within the nucleus.

4 THE ROLE OF DNA METHYLATION

X-chromosome inactivation in the mouse provided an early model of this kind of epigenetic mechanism that involved no DNA rearrangement (Ohno et al. 1959; Lyon 1961). The silenced X chromosome was clearly chosen at random and then clonally inherited in somatic cells, and there was no evidence of changes in the DNA sequence itself. Riggs (1975) and Holliday and Pugh (1975) proposed that DNA methylation could act as an epigenetic mark, in part, to account for this kind of inactivation. The key elements in this model were the ideas that sites of DNA methylation were palindromic, and that distinct enzymes were responsible for methylation of unmodified DNA and DNA already methylated on one strand. It was postulated that the first DNA methylation event would be much more difficult than the second; once the first strand was modified, however, the complementary strand would quickly be modified at the same palindromic site. A DNA methylation mark present on a parental strand would be copied on the daughter strand following replication resulting in faithful transmission of the methylated state to the next generation. Shortly thereafter, Bird took advantage of the fact that the principal target of methylation in animals is the sequence CpG (Doskocil and Sorm 1962) to introduce the use of methylation-sensitive restriction enzymes as a way of detecting the DNA methylation state. Subsequent studies (Bird and Southern 1978; Bird 1978) then showed that endogenous CpG sites were either completely unmethylated or completely methylated. The predictions of the model were thus confirmed, establishing a mechanism for epigenetic transmission of the methylation mark through semiconservative propagation of the methylation pattern.

In the years following these discoveries, a great deal of attention has been focused on endogenous patterns of DNA methylation, the possible transmission of these patterns through the germline, the role of DNA methylation in silencing gene expression, possible mechanisms for initiation or inhibition of methylation at a fully unmethylated site, and the identification of the enzymes responsible for de novo methylation and for maintenance of methylation on already methylated sites. There also has been intense interest in possible mechanisms by which the methyl group at methylated cytosine residues (or 5mC itself) is removed, something that occurs early in development and in the germline. Evidently, the extent to which DNA methylation can be an epigenetic mark preserved through the germline is determined by which sites survive the demethylation events. Although much of the DNA methylation seen in vertebrates is associated with repetitive and retroviral sequences and may serve to maintain these sequences in a permanently silent state, there can be no question that in many cases this modification provides the basis for epigenetic transmission of the gene activity status. This is most clearly demonstrated at imprinted loci (Cattanach and Kirk 1985) such as the mouse or human *Igf2/H19* locus, in which one allele is marked by DNA methylation, which in turn controls expression from both genes (Bell and Felsenfeld 2000; Hark et al. 2000; Kanduri et al. 2000).

At the same time, it was clear that this could not be the only mechanism for epigenetic transmission of information. For example, as noted in Section 2, position-effect variegation had been observed many years earlier in *Drosophila,* an organism that has extremely low levels of DNA methylation. Furthermore, in subsequent years, *Drosophila* geneticists had identified the *Polycomb* and *Trithorax* groups of genes, which appeared to be involved in permanently "locking in" the state of activity, either off or on, respectively, of clusters of genes during development. The fact that these states were stably transmitted during cell division suggested an underlying epigenetic mechanism.

5 THE ROLE OF CHROMATIN

It had been recognized for many years that the proteins bound to DNA in the eukaryotic nucleus, especially the histones, might be involved in modifying the properties of DNA. Well before most of the work on DNA methylation began, Stedman and Stedman (1950) proposed that the histones could act as general repressors of gene expression. They argued that because all somatic cells of an organism had the same number of chromosomes, they had the same genetic complement (although this was not demonstrated until some years later, as noted in Section 3). Understanding the subtlety of histone modifications was far in the future though, so Stedman and Stedman operated on the assumption that different kinds of cells in an organism must have different kinds of histones to generate the observed differences in phenotype. Histones can indeed reduce levels of transcript far below those commonly observed for inactive genes in prokaryotes. Subsequent work addressed the capacity of chromatin to serve as a template for transcription, and asked whether that capacity was restricted in a cell-type-specific manner. Other results suggested that only a small fraction of DNA, packaged as chromatin, was accessible to enzyme probes (Cedar and Felsenfeld 1973). Nonetheless, there was a period in which it was commonly believed that the histones were suppressor proteins that passively silenced gene expression. In this view, activating a gene simply meant stripping off the histones; once that was done, it was thought transcription would proceed pretty much as it did in prokaryotes. There was, however, some evidence that extended regions of open DNA did not exist in eukaryotic cells (Clark and Felsenfeld 1971), focusing attention on promoters and other specific regulatory sites. Furthermore, even if the naked DNA model was correct, it was not clear how the decision would be made as to which histone-covered regions should be cleared.

The resolution of this problem began as early as 1964, when Allfrey and Mirsky (1964) speculated that histone acetylation might be correlated with gene activation, and that "active" chromatin might not necessarily be stripped of histones. In the ensuing decade, there was great interest in examining the relationship between histone modifications and gene expression. Modifications other than acetylation (methylation and phosphorylation) were identified, but their functional significance was unclear. It became much easier to address this problem after the discovery by Kornberg and Thomas (1974) that DNA was packaged into the nucleosome, the fundamental histone-containing chromatin subunit. The determination of the crystal structure of the nucleosome, first at 7-Å and then at 2.8-Å resolution, also provided important structural information, particularly evidence for the extension of the histone amino-terminal tails beyond the DNA–protein octamer core, making evident their accessibility to modification (Richmond et al. 1984; Luger et al. 1997). Beginning in 1980 and extending for some years, Grunstein and his collaborators (Wallis et al. 1980; Durrin et al. 1991), applying yeast genetic analysis, were able to show that the histone amino-terminal tails were essential for regulation of gene expression, and for the establishment of silent chromatin domains.

The ultimate connection to detailed mechanisms began with the critical demonstration by Allis and coworkers (Brownell et al. 1996) that a histone acetyltransferase from *Tetrahymena* was homologous to the yeast transcriptional regulatory protein Gcn5, providing direct evidence that histone acetylation was connected to control of gene expression. Complementary evidence came from a study showing that a mammalian histone deacetylase was related to the yeast repressive transcriptional regulator Rpd3p (Taunton et al. 1996). Since then there has been an explosion in the discovery of histone modifications, as well as a re-evaluation of the roles of those that were known previously.

This still did not answer the question of how the sites for modification were chosen in vivo. It had been shown, for example (Pazin et al. 1994), that Gal4-VP16 could activate transcription from a reconstituted chromatin template in an ATP-dependent manner. Activation was accompanied by repositioning of nucleosomes, and it was suggested that this was the critical event in making the promoter accessible. A fuller understanding of the significance of these findings required the identification of ATP-dependent nucleosome remodeling complexes such as SWI/SNF and NURF (Peterson and Herskowitz 1992; Tsukiyama and Wu 1995), and the realization that both histone modification and nucleosome remodeling were involved in preparing the chromatin template for transcription. More recent results from many laboratories reveal that individual sites vary in the order and identity of the enzymes that performed these steps. It should, however, be clear that the initial determinants of specific gene activity during normal development must involve regulatory factors that recognize and bind to particular DNA sequences at enhancers, promoters, and other control sites. These factors have typically been proteins with DNA sequence-specific binding domains coupled to domains that recruit cofactors, directly or indirectly affecting gene expression, including, in many cases, histone modification or nucleosome remodeling complexes. The first direct evidence for the primacy of DNA-binding factors came from the work of Weintraub (Davis et al. 1987; Tapscott et al. 1988; Weintraub et al. 1989) and his collaborators who showed that overexpression of the protein MyoD in fibroblasts and other tissues induced conversion to myoblasts. It is now understood that such sequence-specific binding events establish the initial states of regulation; the epigenetic mechanisms that follow provide ways of maintaining those initial states once they are established. Of course, disruption of such established epigenetic patterns can alter phenotypes.

Complexes that modify or remodel histones can be delivered in a site-specific way, not only by proteins, but also by RNA. Recently, it has become clear that certain kinds of noncoding RNAs are also capable of localized binding coupled to recruitment of regulatory complexes (Chu et al. 2011). For example, in the case of HOTAIR (Rinn et al. 2007) and Kcnq1ot1 (Pandey et al. 2008; Mohammad et al. 2010), the RNAs tend to associate with DNA fairly near their own sites of synthesis, and bring with them the histone modifying Polycomb complex PRC2 (see below), bound to a specific sequence on each RNA. Kcnq1ot1 can recruit a DNA methyltransferase as well (Mohammad et al. 2010). Both HOTAIR and the noncoding RNA telomerase RNA component have been shown to bind directly to distinct motifs on DNA, providing targeting specificity (Chu et al. 2012).

It was not clear how information about the state of activity could, using these mechanisms, be transmitted through cell division; their role in epigenetic transmission of information was thus unclear. The next important step came from the realization that modified histones could recruit proteins in a modification-specific way that could, in turn, affect the local structural and functional states of chromatin. It was found, for example, that methylation of histone H3 at lysine 9 resulted in the recruitment of the heterochromatin protein HP1 (Bannister et al. 2001; Lachner et al. 2001; Nakayama et al. 2001). Furthermore, HP1 could recruit the enzyme (Suv39 H1) that is responsible for that methylation. This led to a model for the propagation of the silenced chromatin state along a region, through a processive mechanism (Fig. 1A). Equally important, it provided a reasonable explanation of how that state could be transmitted and survive the DNA replication cycle (Fig. 1B). Recent attention has focused on the Polycomb group proteins (Margueron and Reinberg 2011), and particularly the PRC2 complex that contains a component protein, Ezh2/E(Z), that methylates histone H3 at residue K27, a mark associated with heterochromatin. Analogously to the H3 lysine 9-related mechanism, the PRC2 complex binds to H3K27me3 (Hansen et al. 2008). This involves another member of the PRC2 complex, Eed/ESC, that contains a domain that interacts with methylated H3K27, and this interaction in turn stimulates the methyltransferase activity of Ezh2 (Margueron et al. 2008; Margueron et al. 2009). The arrangement suggests the same kind of propagation mechanism proposed for H3K9 methylation as illustrated in Figure 1A. It still remains to be determined whether these mechanisms that can account for spreading of a histone modification down a polynucleosome chain also function during mitosis.

Despite these results, the issue of the role of histone modifications in epigenetic processes continues to be a source of confusion. It is clear that although the term "epigenetic modification" is frequently used, a given histone modification occurring at a given site in the genome may not necessarily be part of an epigenetic mechanism, but simply part of a biochemical process such as gene expression or DNA strand breakage repair.

A Propagation of an epigenetic mark

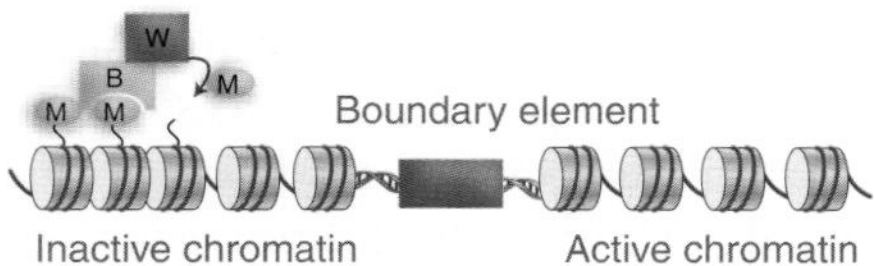

B Replication-dependent propagation of an epigenetic mark

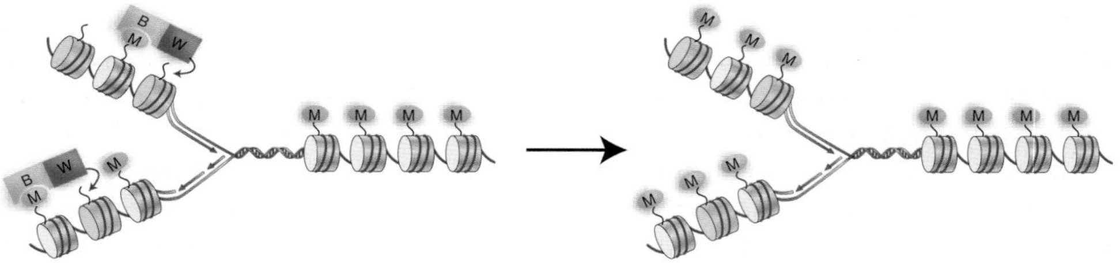

Figure 1. Propagation of epigenetic marks. (*A*) A general mechanism for propagating a histone modification such as H3K9 methylation typically found in heterochromatic regions. The modified histone tail (M) interacts with a protein binder (B) that has a binding site specific for that modification. B also has a specific interaction site with an enzyme "writer" (W) that carries out the same histone modification on an adjacent nucleosome (gray cylinder). Spreading of the histone mark will continue until the modifying machinery reaches a boundary element, delineating the boundary between heterochromatin and euchromatin. (*B*) A general mechanism for maintaining a histone modification during replication. Newly deposited nucleosomes (yellow), which may incorporate histone variants, are interspersed with parental nucleosomes (shaded in gray) following DNA replication. The modified histone tail (M) on the parental nucleosome interacts with a protein binder (B). As in *A*, B interacts with a "writer" (W), which catalyzes a histone modification on the histone tail from an adjacent daughter nucleosome.

 Cite as *Cold Spring Harb Perspect Biol* doi: 10.1101/cshperspect.a018200

Different kinds of propagation mechanisms have been suggested that depend on variant histones rather than modified histones (Ahmad and Henikoff 2002; McKittrick et al. 2004). Histone H3 is incorporated into chromatin only during DNA replication. In contrast, the histone variant H3.3, which differs from H3 by four amino acids, is incorporated into nucleosomes in a replication-independent manner, and it tends to accumulate in active chromatin, in which it is enriched in the "active" histone modifications (McKittrick et al. 2004). It has been proposed that the presence of H3.3 is sufficient to maintain the active state and that after replication, although it would be diluted twofold, enough H3.3 would remain to maintain the active state. The consequent transcription would result in replacement of H3-containing nucleosomes with H3.3, thus perpetuating the active state in the next generation. Results obtained by Ng and Gurdon (Ng and Gurdon 2005; Ng and Gurdon 2008a; Ng and Gurdon 2008b) strongly support such a model. In nuclear transplant experiments, they showed that when nuclei from cells expressing endodermal genes are transplanted into enucleated *Xenopus* eggs there is considerable expression of those genes in animal pole cells (which should not express them) from the resulting embryos. The extent of this epigenetic defect is controlled by the abundance of histone H3.3: Decreased H3.3 leads to a diminution in the fraction of cells expressing the endodermal genes, whereas increased H3.3 expression results in a larger fraction of animal pole cells with aberrant endodermal gene expression. Other histone variants can help confer stability on a silenced epigenetic state. The variant macroH2A (mH2A) is associated with the irreversible inactivation of the mouse X chromosome. Incorporation of this variant helps confer resistance to reprogramming of the inactive X in nuclear transfer experiments (Pasque et al. 2011).

It has been proposed that the definition of an epigenetic mechanism should include, in addition to the property of being maintained through cell division, a requirement for an initial signal, such as expression of a transcription factor, that is not needed once the new state is established (Berger et al. 2009). Behavior of this kind has been described for the glucocorticoid response element in which transient binding of glucocorticoid receptor (GR) to some sites leads to nucleosome remodeling that makes it possible for a modified estrogen receptor molecule to bind after the GR has departed (Voss et al. 2011). It should be kept in mind that most eukaryotic transcription factors do not have long residence times at their binding sites, but turn over rapidly. Certain kinds of chromatin modifications could, in principle, provide a mechanism to integrate signals from multiple transcription factors (Struhl 1999).

6 ALL MECHANISMS ARE INTERRELATED

These models finally begin to complete the connection between modified or variant histones, specific gene activation, and epigenetics, although there is much more to be done. Although we have some ideas about how the heterochromatic state may be maintained, they do not explain how silencing chromatin structures are established. Much of the evidence for such mechanisms has come from work on the silencing of mating-type locus and centromeric sequences in *Schizosaccharomyces pombe*. Formation of heterochromatin involves the production of RNA transcripts, particularly from repeated sequences, that are processed into small RNAs through the action of proteins such as Dicer, Argonaute, and RNA-dependent RNA polymerase (Zofall and Grewal 2006). These RNAs are subsequently recruited to the homologous DNA sites as part of complexes that will eventually include enzymes that deliver "silencing" histone modifications, thus initiating the formation of heterochromatin. There is also evidence that the same mechanisms are required for maintenance of at least some heterochromatic regions in plants and vertebrates.

We now know of countless examples of epigenetic mechanisms at work in the organism. In addition to the allele-specific and random X-chromosome inactivation described in Section 5 and similar allele-specific expression at many more imprinted loci, there are epigenetic phenomena involved in antibody expression in which the rearrangement of the immunoglobulin genes on one chromosome is selectively inhibited. In *Drosophila,* the Polycomb group genes are responsible for establishing a silenced chromatin domain that is maintained through all subsequent cell divisions (Ch. 17 [Grossniklaus and Paro 2014]). Epigenetic changes are also responsible for paramutation in plants, in which one allele can cause a heritable change in expression of the homologous allele (Brink 1956; Stam et al. 2002). This is an example of an epigenetic state that is inherited meiotically as well as mitotically, a phenomenon documented in plants and recently in animals (Rassoulzadegan et al. 2006). In addition, the condensed chromatin structure characteristic of centromeres in organisms as diverse as flies and humans has been shown to be transmissible through centromere-associated proteins rather than DNA sequence. In all of these cases, the DNA sequence remains intact, but its capacity for expression is suppressed. This is likely in all cases to be mediated by DNA methylation, histone modification, presence of a histone variant, or all three; in some cases, we already know that to be true. Perhaps the X chromosome, which inspired early ideas about the role of DNA methylation in epigenetic signaling, is the best example of how all of these mechanisms are interrelated and function together to achieve epigenetic regulation. Recent studies show that silencing of the inactive X chromosome involves, in addition to DNA methylation, specific silencing histone modifications, Polycomb group proteins, noncoding RNAs, and histone variants (Lee 2011). All of these are likely to be involved in transmission of the silenced state during cell division.

In recent years, the study of epigenetics has focused on defining mechanisms of transmission of information not encoded in DNA. Perhaps it is appropriate to reconsider the original use, 70 years ago, of the term epigenetics to describe the then poorly understood processes leading from fertilized zygote to organism. We now know a lot about these processes, thanks to recent results in embryonic stem cells, which show how expression of a few critical factors can establish a self-stabilizing pluripotent state. That state can be transmitted through cell division by what can be thought of as an epigenetic mechanism (according to present definitions). The state can also be perturbed, leading to different epigenetically maintained patterns of expression corresponding to different paths of differentiation into distinct cell types. In an elegant elaboration of these results and of the earlier nuclear transplant studies, somatic cells can be reprogrammed to pluripotency (Yamanaka and Blau 2010). There will be a proliferation of details for many years to come, but we now think we know more or less how this works.

Although this has been presented as a sequential story, it should more properly be viewed as a series of parallel and overlapping attempts to define and explain epigenetic phenomena. The definition of the term epigenetics has changed, but the questions about mechanisms of development raised by earlier generations of scientists are now once again the center of attention. Contemporary epigenetics still addresses those central questions. More than 80 years have passed since Muller described what is now called position-effect variegation. It is gratifying to trace the slow progress from observation of phenotypes through elegant genetic studies to the recent analysis and resolution of mechanisms at the molecular level, and especially to the synthesis of all this information in the analysis of the progression from pluripotent stem cell to individual differentiated states. With this knowledge has come the understanding that epigenetic mechanisms are, in fact, responsible for a considerable part of the phenotype of complex organisms.

ACKNOWLEDGMENTS

I am grateful to Dr. John Gurdon for illuminating exchanges, comments, and advice. This work was supported by the intramural research program of the National Institutes of Health, National Institute of Diabetes and Digestive and Kidney Diseases.

REFERENCES

Reference is also in this book.

Ahmad K, Henikoff S. 2002. The histone variant H3.3 marks active chromatin by replication-independent nucleosome assembly. *Mol Cell* **9:** 1191–1200.

Allfrey VG, Mirsky AE. 1964. Acetylation and methylation of histones and their possible role in the regulation of RNA synthesis. *Proc Natl Acad Sci* **51:** 786–794.

Avery OT, MacLeod CM, McCarty M. 1944. Studies on the chemical nature of the substance inducing transformation of pneumococcal types. *J Exp Med* **79:** 137–158.

Bannister AJ, Zegerman P, Partridge JF, Miska EA, Thomas JO, Allshire RC, Kouzarides T. 2001. Selective recognition of methylated lysine 9 on histone H3 by the HP1 chromo domain. *Nature* **410:** 120–124.

Beisson J, Sonneborn TM. 1965. Cytoplasmic inheritance of the organization of the cell cortex in *Paramecium aurelia*. *Proc Natl Acad Sci* **53:** 282.

Bell AC, Felsenfeld G. 2000. Methylation of a CTCF-dependent boundary controls imprinted expression of the Igf2 gene. *Nature* **405:** 482–485.

Berger SL, Kouzarides T, Shiekhattar R, Shilatifard A. 2009. An operational definition of epigenetics. *Genes Dev* **23:** 781–783.

Bird AP. 1978. Use of restriction enzymes to study eukaryotic DNA methylation: II. The symmetry of methylated sites supports semi-conservative copying of the methylation pattern. *J Mol Biol* **118:** 49–60.

Bird AP, Southern EM. 1978. Use of restriction enzymes to study eukaryotic DNA methylation: I. The methylation pattern in ribosomal DNA from *Xenopus laevis*. *J Mol Biol* **118:** 27–47.

Briggs R, King TJ. 1952. Transplantation of living nuclei from blastula cells into enucleated frogs' eggs. *Proc Natl Acad Sci* **38:** 455–463.

Brink RA. 1956. A genetic change associated with the R locus in maize which is directed and potentially reversible. *Genetics* **41:** 872–889.

Brownell JE, Zhou J, Ranalli T, Kobayashi R, Edmondson DG, Roth SY, Allis CD. 1996. Tetrahymena histone acetyltransferase A: A homolog to yeast Gcn5p linking histone acetylation to gene activation. *Cell* **84:** 843–851.

Cattanach BM, Kirk M. 1985. Differential activity of maternally and paternally derived chromosome regions in mice. *Nature* **315:** 496–498.

Cedar H, Felsenfeld G. 1973. Transcription of chromatin in vitro. *J Mol Biol* **77:** 237–254.

Chu C, Qu K, Zhong FL, Artandi SE, Chang HY. 2011. Genomic maps of long noncoding RNA occupancy reveal principles of RNA-chromatin interactions. *Mol Cell* **44:** 667–678.

Chu C, Quinn J, Chang HY. 2012. Chromatin isolation by RNA purification (ChIRP). *J Vis Exp* **61** pii: 3912. doi: 10.3791/3912.

Clark RJ, Felsenfeld G. 1971. Structure of chromatin. *Nat New Biol* **229:** 101–106.

Davis RL, Weintraub H, Lassar AB. 1987. Expression of a single transfected cDNA converts fibroblasts to myoblasts. *Cell* **51:** 987–1000.

Doskocil J, Sorm F. 1962. Distribution of 5-methylcytosine in pyrimidine sequences of deoxyribonucleic acids. *Biochim Biophys Acta* **55:** 953–959.

Durrin LK, Mann RK, Kayne PS, Grunstein M. 1991. Yeast histone H4 N-terminal sequence is required for promoter activation in vivo. *Cell* **65:** 1023–1031.

Grimes GW, Aufderheide KJ. 1991. Cellular aspects of pattern formation: The problem of assembly. *Monogr Dev Biol* **22:** 1–94.

* Grossniklaus U, Paro R. 2014. Transcriptional silencing by *Polycomb*-group proteins. *Cold Spring Harb Perspect Biol* **6:** a019331.

Hadorn E. 1965. Problems of determination and transdetermination. *Brookhaven Symp Biol* **18:** 148–161.

Hannah A. 1951. Localization and function of heterochromatin in *Drosophila melanogaster*. *Adv Genet* **4:** 87–125.

Hansen KH, Bracken AP, Pasini D, Dietrich N, Gehani SS, Monrad A, Rappsilber J, Lerdrup M, Helin K. 2008. A model for transmission of the H3K27me3 epigenetic mark. *Nat Cell Biol* **10:** 1291–1300.

Hark AT, Schoenherr CJ, Katz DJ, Ingram RS, Levorse JM, Tilghman SM. 2000. CTCF mediates methylation-sensitive enhancer-blocking activity at the H19/Igf2 locus. *Nature* **405:** 486–489.

Hershey AD, Chase M. 1952. Independent functions of viral protein and nucleic acid in growth of bacteriophage. *J Gen Physiol* **36:** 39–56.

Holliday R, Pugh JE. 1975. DNA modification mechanisms and gene activity during development. *Science* **187:** 226–232.

Jablonka E, Lamb MJ. 1995. *Epigenetic inheritance and evolution*. Oxford University Press, New York.

Kanduri C, Pant V, Loukinov D, Pugacheva E, Qi CF, Wolffe A, Ohlsson R, Lobanenkov VV. 2000. Functional association of CTCF with the insulator upstream of the H19 gene is parent of origin-specific and methylation-sensitive. *Curr Biol* **10:** 853–856.

Kornberg RD, Thomas JO. 1974. Chromatin structure; oligomers of the histones. *Science* **184:** 865–868.

Lachner M, O'Carroll D, Rea S, Mechtler K, Jenuwein T. 2001. Methylation of histone H3 lysine 9 creates a binding site for HP1 proteins. *Nature* **410:** 116–120.

Laskey RA, Gurdon JB. 1970. Genetic content of adult somatic cells tested by nuclear transplantation from cultured cells. *Nature* **228:** 1332–1334.

Lee JT. 2011. Gracefully ageing at 50, X-chromosome inactivation becomes a paradigm for RNA and chromatin control. *Nat Rev Mol Cell Biol* **12:** 815–826.

Luger K, Mader AW, Richmond RK, Sargent DF, Richmond TJ. 1997. Crystal structure of the nucleosome core particle at 2.8 A resolution. *Nature* **389:** 251–260.

Lyon MF. 1961. Gene action in the X-chromosome of the mouse. *Nature* **190:** 372–373.

Margueron R, Reinberg D. 2011. The Polycomb complex PRC2 and its mark in life. *Nature* **469:** 343–349.

Margueron R, Li G, Sarma K, Blais A, Zavadil J, Woodcock CL, Dynlacht BD, Reinberg D. 2008. Ezh1 and Ezh2 maintain repressive chromatin through different mechanisms. *Mol Cell* **32:** 503–518.

Margueron R, Justin N, Ohno K, Sharpe ML, Son J, Drury WJ 3rd, Voigt P, Martin SR, Taylor WR, De Marco V, et al. 2009. Role of the polycomb protein EED in the propagation of repressive histone marks. *Nature* **461:** 762–767.

McClintock B. 1965. The control of gene action in maize. *Brookhaven Symp Biol* **18:** 162–184.

McClure KD, Schubiger G. 2007. Transdetermination: *Drosophila* imaginal disc cells exhibit stem cell-like potency. *Int J Biochem Cell Biol* **39:** 1105–1118.

McKittrick E, Gafken PR, Ahmad K, Henikoff S. 2004. Histone H3.3 is enriched in covalent modifications associated with active chromatin. *Proc Natl Acad Sci* **101:** 1525–1530.

Mohammad F, Mondal T, Guseva N, Pandey GK, Kanduri C. 2010. Kcnq1ot1 noncoding RNA mediates transcriptional gene silencing by interacting with Dnmt1. *Development* **137:** 2493–2499.

Morgan TH. 1911. An attempt to analyze the constitution of the chromosomes on the basis of sex-linked inheritance in *Drosophila*. *J Exp Zool* **11:** 365–414.

Muller HJ. 1930. Types of visible variations induced by X-rays in *Drosophila*. *J Genet* **22:** 299–334.

Nakayama J, Rice JC, Strahl BD, Allis CD, Grewal SI. 2001. Role of histone H3 lysine 9 methylation in epigenetic control of heterochromatin assembly. *Science* **292:** 110–113.

Ng RK, Gurdon JB. 2005. Epigenetic memory of active gene transcription is inherited through somatic cell nuclear transfer. *Proc Natl Acad Sci* **102:** 1957–1962.

Ng RK, Gurdon JB. 2008a. Epigenetic inheritance of cell differentiation status. *Cell Cycle* **7:** 1173–1177.

Ng RK, Gurdon JB. 2008b. Epigenetic memory of an active gene state depends on histone H3.3 incorporation into chromatin in the absence of transcription. *Nat Cell Biol* **10:** 102–109.

Novick A, Weiner M. 1957. Enzyme induction as an all-or-none phenomenon. *Proc Natl Acad Sci* **43:** 553–566.

Ohno S, Kaplan WD, Kinosita R. 1959. Formation of the sex chromatin by a single X-chromosome in liver cells of *Rattus norvegicus*. *Exp Cell Res* **18:** 415–418.

Pandey RR, Mondal T, Mohammad F, Enroth S, Redrup L, Komorowski J, Nagano T, Mancini-Dinardo D, Kanduri C. 2008. Kcnq1ot1 antisense noncoding RNA mediates lineage-specific transcriptional silencing through chromatin-level regulation. *Mol Cell* **32:** 232–246.

Pasque V, Halley-Stott RP, Gillich A, Garrett N, Gurdon JB. 2011. Epigenetic stability of repressed states involving the histone variant macro-H2A revealed by nuclear transfer to *Xenopus* oocytes. *Nucleus* **2:** 533–539.

Pazin MJ, Kamakaka RT, Kadonaga JT. 1994. ATP-dependent nucleosome reconfiguration and transcriptional activation from preassembled chromatin templates. *Science* **266:** 2007–2011.

Peterson CL, Herskowitz I. 1992. Characterization of the yeast SWI1, SWI2, and SWI3 genes, which encode a global activator of transcription. *Cell* **68:** 573–583.

Ptashne M. 1992. *A genetic switch: Phage λ and higher organisms*, 2nd ed. Cell Press and Blackwell Scientific, Cambridge, MA.

Rassoulzadegan M, Grandjean V, Gounon P, Vincent S, Gillot I, Cuzin F. 2006. RNA-mediated non-Mendelian inheritance of an epigenetic change in the mouse. *Nature* **441:** 469–474.

Richmond TJ, Finch JT, Rushton B, Rhodes D, Klug A. 1984. Structure of the nucleosome core particle at 7 A resolution. *Nature* **311:** 532–537.

Riggs AD. 1975. X inactivation, differentiation, and DNA methylation. *Cytogenet Cell Genet* **14:** 9–25.

Riggs AD, Porter TN. 1996. Overview of epigenetic mechanisms. In *Epigenetic mechanisms of gene regulation* (ed. Russo VEA, Martienssen R, Riggs AD), pp. 29–45. Cold Spring Harbor Laboratory Press, Cold Spring Harbor, NY.

Riggs AD, Martienssen RA, Russo VEA. 1996. Introduction. In *Epigenetic mechanisms of gene regulation* (ed. Russo VEA, et al.), pp. 1–4. Cold Spring Harbor Laboratory Press, Cold Spring Harbor, NY.

Rinn JL, Kertesz M, Wang JK, Squazzo SL, Xu X, Brugmann SA, Goodnough LH, Helms JA, Farnham PJ, Segal E, et al. 2007. Functional demarcation of active and silent chromatin domains in human HOX loci by noncoding RNAs. *Cell* **129:** 1311–1323.

Stam M, Belele C, Dorweiler JE, Chandler VL. 2002. Differential chromatin structure within a tandem array 100 kb upstream of the maize b1 locus is associated with paramutation. *Genes Dev* **16:** 1906–1918.

Stedman E, Stedman E. 1950. Cell specificity of histones. *Nature* **166:** 780–781

Struhl K. 1999. Fundamentally different logic of gene regulation in eukaryotes and prokaryotes. *Cell* **98:** 1–4.

Sturtevant AH. 1913. The linear arrangement of six sex-linked factors in *Drosophila*, as shown by their mode of association. *J Exp Zool* **14:** 43–59.

Tapscott SJ, Davis RL, Thayer MJ, Cheng PF, Weintraub H, Lassar AB. 1988. MyoD1: A nuclear phosphoprotein requiring a Myc homology region to convert fibroblasts to myoblasts. *Science* **242:** 405–411.

Taunton J, Hassig CA, Schreiber SL. 1996. A mammalian histone deacetylase related to the yeast transcriptional regulator Rpd3p. *Science* **272:** 408–411.

Tsukiyama T, Wu C. 1995. Purification and properties of an ATP-dependent nucleosome remodeling factor. *Cell* **83:** 1011–1020.

Voss TC, Schiltz RL, Sung MH, Yen PM, Stamatoyannopoulos JA, Biddie SC, Johnson TA, Miranda TB, John S, Hager GL. 2011. Dynamic exchange at regulatory elements during chromatin remodeling underlies assisted loading mechanism. *Cell* **146:** 544–554.

Waddington CH. 1953. Epigenetics and evolution. *Symp Soc Exp Biol* **7:** 186–199.

Wallis JW, Hereford L, Grunstein M. 1980. Histone H2B genes of yeast encode two different proteins. *Cell* **22:** 799–805.

Weintraub H, Tapscott SJ, Davis RL, Thayer MJ, Adam MA, Lassar AB, Miller AD. 1989. Activation of muscle-specific genes in pigment, nerve, fat, liver, and fibroblast cell lines by forced expression of MyoD. *Proc Natl Acad Sci* **86:** 5434–5438.

Yamanaka S, Blau HM. 2010. Nuclear reprogramming to a pluripotent state by three approaches. *Nature* **465:** 704–712.

Zofall M, Grewal SI. 2006. RNAi-mediated heterochromatin assembly in fission yeast. *Cold Spring Harb Symp Quant Biol* **71:** 487–496.

 Cite as *Cold Spring Harb Perspect Biol* doi: 10.1101/cshperspect.a018200

CHAPTER 2

The Next Generation: Young Scientists Forging Exciting New Discoveries in Epigenetic Research

The editorial team recognizes that many, if not most, of the advances made in the field of epigenetics were brought about by the hard work, dedication, and creativity that many young scientists bring to the bench and the problem. It is they who carry out most of the experiments on which the findings described in our book chapters rest. In seeking to add another dimension to this edition, the editors have identified a collection of primary papers that we felt were groundbreaking in nature, that may well prove to be "game-changing" for the field in years to come, or that have already set the field in new directions (see below for details and references). Our charge to the original first authors (or co–first authors) of this important collection of essays was to write short retrospective historical reviews putting their spin on how the work took form—sometimes including details about what their mind-set was, what hurdles they faced, and how those hurdles had to be overcome. Here, we assemble the essays from these junior scientists, recognizing that many of them are already well immersed in epigenetic research, carrying on the torches of research in their own independent laboratories worldwide. We applaud their collective past discoveries, and we look forward to their continuing successes. Mostly, we hope that other junior scientists, students, and newcomers to the field will be inspired to follow in their footsteps, leading the charge to future discoveries that will help to advance the many fundamental questions that remain in epigenetics today. Although we have not been able to survey all of the exciting work that continues to be published, our goal has been to give our readers an insight into how many of these discoveries were made directly from those who spearheaded the experiments.

Essays

Histone Demethylases

The Discovery of Histone Demethylases, Yujiang Geno Shi, Yu-ichi Tsukada, pp 13–15

Primary papers:
Shi et al., *Cell* **119:** 941–953 (2004)
Tsukada et al., *Nature* **439:** 811–816 (2006)

Reprogramming

Cellular Reprogramming, Kazutoshi Takahashi, pp 16–17

Primary paper:
Takahashi and Yamanaka, *Cell* **126:** 663–676 (2006)

Functional Noncoding RNAs

lncRNAs: Linking RNA to Chromatin, John L. Rinn, pp 18–20

Primary paper:
Rinn et al., *Cell* **129:** 1311–1323 (2007)

Enhancer RNAs: A Class of Long Noncoding RNAs Synthesized at Enhancers, Tae-Kyung Kim, Martin Hemberg, and Jesse M. Gray, pp 21–23

Primary paper:
Kim et al., *Nature* **465:** 182–187 (2010)

DNA Demethylation

Expanding the Epigenetic Landscape: Novel Modifications of Cytosine in Genomic DNA, Skirmantas Kriaucionis and Mamta Tahiliani, pp 24–26

Primary papers:
Tahiliani et al., *Science* **324:** 930–935 (2009)
Kriaucionis and Heintz, *Science* **324:** 929–930 (2009)

Small Mobile RNAs

Plant Mobile Small RNAs, Patrice Dunoyer, Charles Melnyk, Attila Molnar, and R. Keith Slotkin, pp 27–29

Primary papers:
Molnar et al., *Science* **328:** 872–875 (2010)
Dunoyer et al., *Science* **328:** 912–916 (2010)
Slotkin et al., *Cell* **136:** 461–472 (2009)

CpG Island Chromatin

CpG Island Chromatin Is Shaped by Recruitment of ZF-CxxC Proteins, Neil P. Blackledge, John P. Thomson, and Peter J. Skene, pp 30–32

Primary papers:
Thomson et al., *Nature* **464:** 1082–1086 (2010)
Blackledge et al., *Mol Cell* **38:** 179–190 (2010)

BET Histone Readers as Targets for Epigenetic Therapy

Bromodomain and Extraterminal Domain Inhibitors (BETi) for Cancer Therapy: Chemical Modulation of Chromatin Structure, Jun Qi, pp 33–35

Primary paper:
Filippakopoulos et al., *Nature* **468:** 1067–1073 (2010)

Pharmacological Inhibition of Bromodomain-Containing Proteins in Inflammation, Uwe Schaefer, pp 36–39

Primary paper:
Nicodeme et al., *Nature* **468:** 1119–1123 (2010)

Cancer Driving Histone Mutations

Histone H3 Mutations in Pediatric Brain Tumors, Xiaoyang Liu, Troy A. McEachron, Jeremy Schwartzentruber, and Gang Wu, pp 40–43

Primary papers:
Wu et al., *Nat Genet* **44:** 251–253 (2012)
Schwartzentruber et al., *Nature* **482:** 226–231 (2012)

Long-Range Chromosome Interactions

Chromosome Folding: Driver or Passenger of Epigenetic State?, Tom Sexton and Eitan Yaffe, pp 44–46

Primary papers:
Sexton et al., *Cell* **148:** 458–472 (2012)
Lieberman-Aiden et al., *Science* **326:** 289–293 (2009)

The Discovery of Histone Demethylases

Yujiang Geno Shi[1] and Yu-ichi Tsukada[2]

[1]Harvard Medical School, and Endocrinology Division, Brigham and Women's Hospital, Boston, Massachusetts 02115;
[2]Research Center for Infectious Diseases, Medical Institute of Bioregulation, Kyushu University, 3-1-1 Maidashi, Higashi-ku, Fukuoka, Fukuoka 812-8582, Japan

Correspondence: yujiang_shi@hms.harvard.edu and ytsukada@bioreg.kyushu-u.ac.jp

Histone methylation is a key element of the eukaryotic epigenome. Since the discovery of the first histone demethylase (HDM) in 2004, more than 20 demethylases have been identified and characterized. They belong to either the LSD family or the JmjC family, demonstrating the reversibility of all methylation states at almost all major histone lysine methylation sites. These findings ended decades of debate about the reversibility of histone methylation, representing a major breakthrough that shifts our understanding of epigenetic inheritance and regulation of genome function. Here, we summarize the discovery of HDMs and more recent advances, challenges, and future prospects of HDM research.

Histone methylation at lysine and arginine residues are key covalent histone modifications in epigenetic regulation. Together with DNA methylation, they constitute hallmarks of epigenetic inheritance. Although other histone modifications such as acetylation and phosphorylation have been known to be reversible for some time, the reversibility of histone methylation was in question. It was not until 2004, with the discovery of the first histone demethylase (HDM), that this issue was resolved.

As is often the case, the first HDM, LSD1 (Gene ID: KDM1A), was discovered in an unexpected way. While studying how metabolic enzymes and their homologs and cofactors are involved in epigenetic gene regulation, Drs. Yang Shi and Yujiang Geno Shi became curious about how a metabolic enzyme homolog named nPAO (nuclear polyamine oxidase; also called KIAA0601/BHC110), which they had previously identified from the transcriptional CtBP complex, might function in epigenetic gene regulation (Shi et al. 2003). Based on nPAO's homology with known polyamine oxidases and given that polyamines are also minor components of chromatin, it was hypothesized that nPAO might regulate chromatin structure through a polyamine oxidation mechanism. Despite months of attempts, experiments aimed at detecting the putative polyamine oxidase activity of nPAO were unsuccessful. Although it is chemically possible that oxidation of methylated lysine could lead to demethylation of lysine through an amine oxidation reaction, such a reaction mechanism has never been discovered or reported. When the substrate was switched from polyamine to histone H3 dimethylated at lysine 4 (H3K4me2), Y.G. Shi was finally able to successfully detect the nPAO-mediated histone demethylation. Dr. Yang Shi's laboratory thus discovered the first lysine demethylase (KDM1) (Shi et al. 2004). nPAO was then renamed LSD1 for lysine-specific histone demethylase 1, and the LSD family of histone demethylases has since been defined. This discovery ended decades of debate about the reversibility of histone methylation, representing a major breakthrough and a paradigm shift in our understanding of the dynamics of histone methylation. The chemical reaction that LSD1/KDM1A catalyzes is an amine oxidation by oxidative cleavage of the α-carbon bond of the methylated lysine to form an imine intermediate, which is hydrolyzed to form formaldehyde, releasing one molecule of H_2O_2 and the demethylated lysine (see Fig. 1). Notably, because the formation of an imine intermediate requires protonated nitrogen, the LSD family of demethylases can only demethylate mono- and dimethylated (me1, me2), but not the trimethylated (me3) lysine residues. This raised the possibility of other classes of HDMs that were yet to be discovered.

Dr. Yi Zhang's laboratory took an unbiased activity-based biochemical purification approach to search for

Cite as *Cold Spring Harb Perspect Biol* doi: 10.1101/cshperspect.a017947

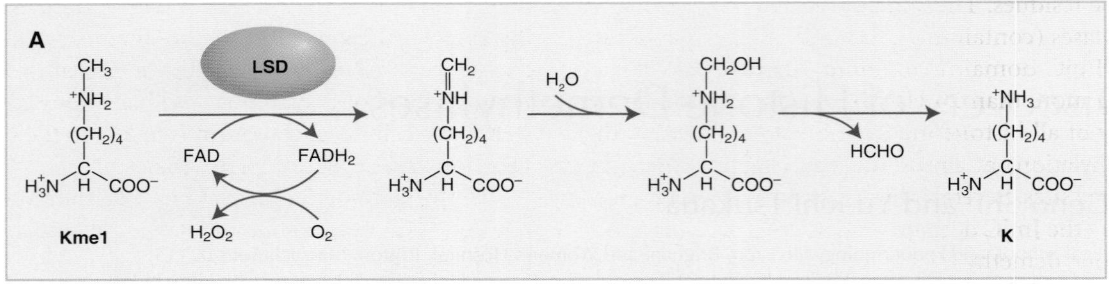

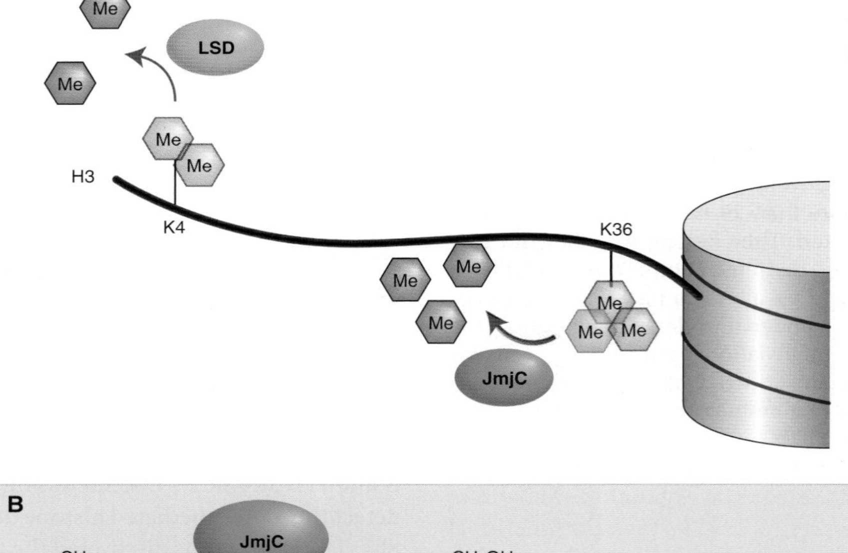

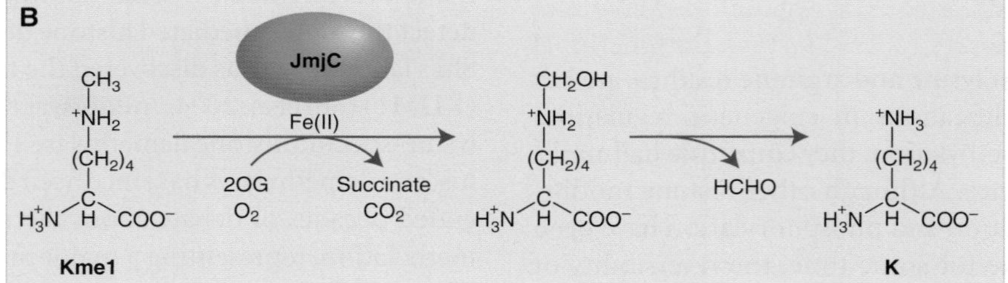

Figure 1. Histone demethylation mediated by (*A*) an LSD family demethylase through an FAD-dependent amine oxidase reaction, and (*B*) a JmjC domain family demethylase through a 2OG-Fe(II)-dependent dioxygenase reaction. LSD demethylases can demethylate the mono- and dimethylated states of histone lysine residues. The JmjC domain family demethylases can demethylate mono-, di-, and trimethylated histone lysine residues. For simplicity, only monomethyl-lysine is illustrated.

histone demethylases. The similarity between the chemistry of removing a methyl group from 1-meA or 3-meC of DNA by the AlkB family of DNA repair demethylases (Falnes et al. 2002; Trewick et al. 2002) to that of lysine demethylation prompted the hypothesis that a similar mechanism might be used for histone demethylation. Consequently, an assay measuring formaldehyde release using α-ketoglutarate (α-KG) and Fe(II) as cofactors was established, optimized, and used in histone demethylase purification through monitoring the enzymatic activities of column fractions. By 2005, Yu-ichi Tsukada, in Yi Zhang's laboratory, succeeded in identifying the first JmjC domain–containing HDM (JHDM1/KDM2), which demethylates histone H3K36 (Tsukada et al. 2006). It turns out that the mammalian genome has 30 different proteins that contain a JmjC domain. The chemical reaction catalyzed by the JmjC domain–containing HDMs is the oxygenation of a methyl group by a radical attack of a highly reactive oxoferryl species to form an unstable carbinolamine intermediate; subsequent release of formaldehyde from the carbinolamine produces demethylated lysine (see Fig. 1). Unlike the LSD1-mediated chemical reaction, this reaction is compatible with trimethylated (me3) lysine residues. Thus, in addition to the LSD family of demethylases (containing LSD1 and LSD2), the discovery of the JmjC domain–containing demethylase family (containing more than 20 HDMs) has further shown the reversibility of all histone methylation states (mono-, di-, and trimethylation) at almost all major histone methylation modified sites. Several other laboratories had also been investigating the JmjC domain–containing proteins as potential histone demethylase candidates and contributed to defining the JmjC family of demethylases.

Although it took more than 40 years to identify LSD1, the first HDM, in less than 4 years since the initial discovery we have seen the rapid progress in our knowledge of histone demethylation, such that it is now a major focus of epigenetic research. More than 20 HDMs actively catalyze the demethylation of almost all major histone lysine methylation sites and a number of arginine methylation sites. Yet, the following fundamental questions pertaining to enzymatic action, regulation, and biological function remain unaddressed: First, there is the possibility that a third class of HDMs is yet to be discovered. This is largely based on the premise that there is no known demethylase for H3K79 methylation. This modification is unique in that it is the only residue to be methylated by a non-SET domain histone methyltransferase, Dot1/Dot1L, and thus it is tempting to speculate that H3K79 demethylation may use a new and different class of HDMs. Also, new arginine demethylases may constitute a new class of HDMs. Second, the functional characterization of demethylases has been mostly limited to aspects of gene transcription and, in particular, promoter transcription initiation. This is a narrow focus that cannot fully explain the wide range of involvement of HDMs in biological and pathological processes. It will be worthwhile exploring how HDMs are involved in transcriptional elongation, cotranscriptional messenger RNA processing, DNA replication, and/or repair processes. Third, how HDMs are themselves regulated has yet to be explored. For example, it will be worthwhile knowing how they are specifically targeted to their functional sites. How do demethylases exert their functions in the context of larger complexes containing associated cofactors? How are their expression and activity regulated? And because the action of these enzymes requires metabolic cofactors, it will be important to understand how cellular metabolism is linked to the intracellular function and regulation of HDMs.

The impact of the discovery of histone demethylases has extended beyond the regulation of histone methylation, including spurring on the search for DNA demethylases, described in Kriaucionis and Tahiliani (2014). We expect to see more discoveries along this line that will greatly impact the field of epigenetics. Most importantly, HDMs are involved in many normal and pathological processes including gene transcription, stem cell self-renewal, development, and tumorigenesis. Now that we know that histone methylation is reversible, there is great hope that demethylases will be promising therapeutic targets in the future of "epigenetic medicines." Answers to the fundamental questions raised above are an absolute requisite not only for understanding the biological functions of histone demethylation, but for the progression of further clinical translational research on complex human diseases.

REFERENCES

Falnes PO, Johansen RF, Seeberg E. 2002. AlkB-mediated oxidative demethylation reverses DNA damage in *Escherichia coli*. *Nature* **419:** 178–182.

Kriaucionis S, Tahiliani M. 2014. Expanding the epigenetic landscape: Novel modifications of cytosine in genomic DNA. *Cold Spring Harb Perspect Biol* doi: 10.1101/cshperspect.a018630.

Shi YJ, Sawada J-I, Sui GC, Affar EB, Whetstine J, Lan F, Ogawa H, Luke MP-S, Nakatani Y, Shi Y. 2003. Coordinated histone modifications mediated by a CtBP co-repressor complex. *Nature* **422:** 735–738.

Shi Y, Lan F, Matson C, Mulligan P, Whetstine JR, Cole PA, Casero RA, Shi Y. 2004. Histone demethylation mediated by the nuclear amine oxidase homolog LSD1. *Cell* **119:** 941–953.

Trewick SC, Henshaw TF, Hausinger RP, Lindahl T, Sedgwick B. 2002. Oxidative demethylation by *Escherichia coli* AlkB directly reverts DNA base damage. *Nature* **419:** 174–178.

Tsukada Y, Fang J, Erdjument-Bromage H, Warren ME, Borchers CH, Tempst P, Zhang Y. 2006. Histone demethylation by a family of JmjC domain-containing proteins. *Nature* **439:** 811–816.

Cellular Reprogramming

Kazutoshi Takahashi

Center for iPS Cell Research and Application, Kyoto University, Kyoto, 606-8507, Japan

Correspondence: takahash@cira.kyoto-u.ac.jp

Nuclear reprogramming technology was first established more than 50 years ago. It can rejuvenate somatic cells by erasing the epigenetic memories and reconstructing a new pluripotent order. The recent discovery reviewed here that induced pluripotency can be achieved by a small set of transcription factors has opened up unprecedented opportunities in the pharmaceutical industry, the clinic, and laboratories. This technology allows us to access pathological studies by using patient-specific induced pluripotent stem (iPS) cells. In addition, iPS cells are also expected to be a rising star for regenerative medicine, as sources of transplantation therapy.

Differentiation has long been thought of as one-way traffic in which a cell can be depicted as a ball rolling down through a "developmental" landscape from an undifferentiated stem or progenitor cell state to a physiologically mature state, as described by Conrad Waddington (Fig. 1A) (Waddington 1957). In fact, all cells roll down this epigenetic landscape into deeper, inescapable valleys, representing the determination of cell fate during development. They continue rolling until they reach their final stable state at the lowest point, functionally equating to their final differentiated state. According to this metaphor, changes in cell fate would be strictly avoided by ridges that do not allow the movement from one valley to another. The discovery of an in vitro method we found for making pluripotent mammalian cells from differentiated somatic cells has added to the body of evidence that this dogma can be reversed. Even more importantly, it has provided an accessible technique to study reprogramming and epigenetics (Takahashi and Yamanaka 2006).

In the past, the inability to transmit genetic information from somatic cells to the next generation was commonly recognized as Weismann's barrier. Recent discoveries, however, show that cell fate now appears to be far more flexible than previously thought. Based on Waddington's landscape, rejuvenation refers to the process whereby cells travel back up their maturation path, through the epigenetic landscape, to become more immature cells, and eventually transform to the pluripotent state (Fig. 1B).

The concept of rejuvenation and cellular reprogramming was first proposed by John Gurdon with his landmark experiments producing clones from somatic cells in *Xenopus laevis* at roughly the same time as Waddington's doctrine was being advocated (Gurdon et al. 1958). Later, Ian Wilmut and his colleagues reported the successful cloning of a sheep, Dolly, showing that erasure of epigenetic memories that set somatic cell fate is possible even in mammals (Wilmut et al. 1997). The rejuvenation of a cell to the pluripotent state has also been shown by fusing somatic cells with pluripotent stem cells, such as embryonic stem (ES) cells (Tada et al. 2001). These two approaches suggested that fertilized eggs and pluripotent stem cells contain hidden "reprogramming factors" that are able to erase the somatic memories.

Other research countering Waddington's unidirectional epigenetic landscape model was the work on the conversion of cell fate by defined factors described 25 years ago (Davis et al. 1987). In their groundbreaking studies, Davis et al. performed subtraction experiments using complementary DNA (cDNA), and discovered the myogenic differentiation 1 (*MYOD1*) gene. Ectopic expression of MYOD1 alone is enough to induce the conversion of fibroblasts to myosin-expressing myoblasts. This pioneering work clearly showed that transcription factor(s) are crucial not only for the maintenance of cellular identity, but also for the determination of cell fate.

Cite as *Cold Spring Harb Perspect Biol* doi: 10.1101/cshperspect.a018606

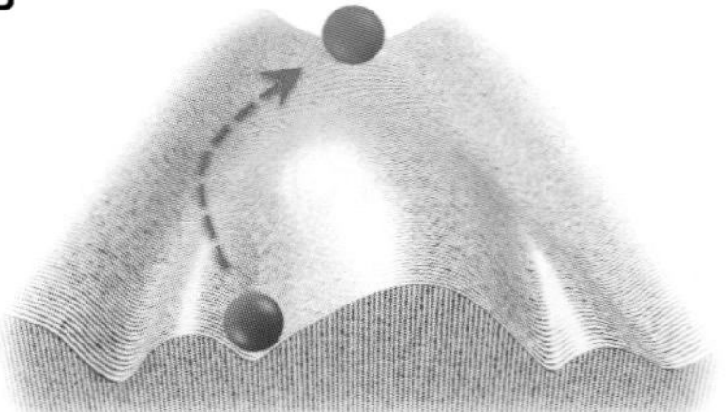

Figure 1. Cellular reprogramming depicted as a trajectory in Waddington's epigenetic landscape. (*A*) A cell's normal developmental trajectory can be traced starting from a pluripotent cell (green ball) at the top of the hill to its final differentiated state (blue ball), illustrating how epigenetics contributes to cell fate determination during development. (*B*) A terminally differentiated cell (blue ball) can be reprogrammed back to pluripotency when exposed to a cocktail of transcription factors.

Drawing encouragement from these studies, we reasoned and showed that latent pluripotency could be induced in differentiated somatic cells by using a defined cocktail of transcription factors without the need for transfer into an egg (Takahashi and Yamanaka 2006). The cocktail consisted of OCT3/4, SOX2, KLF4, and c-MYC, and was sufficient to revert differentiated somatic cells, including terminally differentiated cells such as T lymphocytes, to a pluripotent fate. The resulting dedifferentiated cells have been designated iPS cells, and they can theoretically be used to generate all cell types in the body, as well as ES cells. This discovery has confirmed the importance of transcription factor networks in cell fate determination, and has definitively affected our understanding of cellular reprogramming.

The efficiency of reprogramming from somatic cells to iPS cells is generally $<1\%$. This suggests that not only are the reprogramming factors important in triggering changes, but so are the subsequent stochastic events required to continue the reprogramming process. Because no major differences in the genomic sequences between the original cells and reprogrammed iPS cells have been observed, changes in the epigenetic status, such as by DNA methylation and histone modification, seem to be the critical events for reprogramming. In fact, the use of small-molecule compounds that inhibit histone deacetylase, thereby increasing overall chromatin acetylation levels, can improve the reprogramming efficiency. During reprogramming, silencing of somatic cell genes and reactivation of pluripotent stem cell–expressed genes have been observed. Although these changes are clearly linked to epigenetic statuses, their mechanisms and driving force are still unclear. The generation of iPS cells provides a good model and tool for understanding the epigenetic changes initiated by transcription factors. In addition, iPS cell technology is currently being used for research both as a source of stem cells for therapeutic use and as a tool for studying pathological processes.

After the landmark experiments that showed MYOD1 as a major commitment determinant for driving the closely related fibroblasts toward a muscle cell fate, the hunt for other factors directing particular cell fates has continued. Other relatively close conversions have been observed, such as from lymphoid to myeloid cells, and from glial cells to neurons, using defined transcription factors. Recent reports have also shown that it is possible to directly convert somatic cells to more distally related differentiated cell types, and even to transcend the germ layer origin (i.e., endoderm, mesoderm, or ectoderm), as illustrated by the conversion of fibroblasts into neurons, hematopoietic cells, cartilage, cardiomyocytes, and hepatocytes. All of these findings have lowered the hurdles of cell fate conversion. Cellular identities are clearly more flexible than previously thought and are largely defined by the epigenetic status of the cell.

REFERENCES

Davis R, Weintraub H, Lassar AB. 1987. Expression of a single transfected cDNA converts fibroblasts to myoblasts. *Cell* **51:** 987–1000.

Gurdon JB, Elsdale TR, Fischber M. 1958. Sexually mature individuals of *Xenopus laevis* from the transplantation of single somatic nuclei. *Nature* **182:** 64–65.

Tada M, Takahama Y, Abe K, Nakatsuji N, Tada T. 2001. Nuclear reprogramming of somatic cells by in vitro hybridization with ES cells. *Curr Biol* **11:** 1553–1558.

Takahashi K, Yamanaka S. 2006. Induction of pluripotent stem cells from mouse embryonic and adult fibroblast cultures by defined factors. *Cell* **126:** 663–676.

Waddington CH. 1957. *The strategy of the genes.* Allen & Unwin, London.

Wilmut I, Schnieke AE, McWhir J, Kind AJ, Campbell KH. 1997. Viable offspring derived from fetal and adult mammalian cells. *Nature* **385:** 810–813.

lncRNAs: Linking RNA to Chromatin

John L. Rinn

Harvard University, Department of Stem Cell and Regenerative Biology, Cambridge, Massachusetts 02138

Correspondence: john_rinn@harvard.edu

Numerous studies over the past decade have identified increasing numbers of long noncoding RNAs (lncRNAs) across many organisms. Research since has shown that lncRNAs constitute an important layer of genome regulation in diverse biological processes and disease. Here, we discuss the common emerging theme of lncRNAs interfacing with epigenetic machinery. This, in turn, modulates the activity and localization of the epigenetic machinery during cell fate specification.

Cellular identity is achieved through a complex choreography of DNA regulatory elements interacting with protein regulatory complexes to shape a myriad of unique epigenetic landscapes. The vast arrays of epigenetic landscapes are established by using ubiquitously expressed chromatin-modifying and -remodeling complexes, to give rise to countless unique combinations of histone posttranslational modifications and DNA methylation patterns. This raises the age-old question: How are the enzymatic complexes being guided to place these marks at a specific combination of sites under different cellular contexts? It has long been suspected that noncoding RNA molecules may provide some specificity to target these complexes to their sites of action. Indeed, it is becoming increasingly clear that a contingent of thousands of long noncoding RNAs (lncRNAs) represent a key layer of epigenetic control (see Fig. 24 of Ch. 3 [Allis et al.]).

A dramatic example of RNA-based epigenetic regulation that operates in mammalian dosage compensation has been known for more than 20 years (Ch. 25 [Brockdorff and Turner 2014]). Specifically, a long intergenic noncoding RNA (lincRNA) termed XIST (X inactive-specific transcript) is expressed from one female X chromosome resulting in the recruitment of Polycomb group complexes (PcG), such as PRC2, to this chromosome with the concomitant transcriptional silencing across a majority of the X chromosome. In other words, a single lncRNA gene is able to target and silence the majority of a chromosome *in cis*. This is a powerful precedent for RNA-mediated epigenetic regulation. However, the link between the RNA and PRC2 recruitment was, for a long time, elusive. A clue came from studying the epigenetic dynamics of a different lincRNA termed HOTAIR (Hox transcript antisense intergenic RNA; Rinn et al. 2007).

HOTAIR is expressed from one of the four clusters of HOX transcription factors (i.e., HOXC). The HOX family of proteins is a key regulator of body plan during development. HOTAIR was identified because of its distinctive pattern of expression in posterior and distal mouse tissues during development and adult human fibroblasts (Rinn et al. 2007). Intriguingly, HOTAIR expression demarcated a distinct epigenetic boundary between euchromatic and heterochromatic regions within the HOXC cluster. Moreover, the euchromatic-heterochromatic regions were inverted between anterior and posterior cell types; we termed these "diametric" chromatin domains. Collectively, the data led to the initial hypothesis that HOTAIR serves as an epigenetic boundary *in cis* within the HOXC cluster. To our surprise the HOXC chromatin boundary remained unchanged when HOTAIR RNA function was lost; however, the HOXD cluster, located on a separate chromosome, became active. Thus, similar to XIST, the expression of HOTAIR from the HOXC cluster results in the epigenetic silencing, yet the HOX cluster that it regulates (HOXD) is located on a different chromosome, i.e., is regulated *in trans*. This raised the question: How?

The answer was determined by a few critical experiments, which indicated that the HOTAIR lincRNA interacted with PRC2 and that this was required for proper

Cite as *Cold Spring Harb Perspect Biol* doi: 10.1101/cshperspect.a018614

localization of PRC2. The first experiment, using immunoprecipitation of the PRC2 complex, identified HOTAIR as being physically associated or coprecipitated with PRC2 (Rinn et al. 2007). The reciprocal experiment, which isolated proteins bound to the HOTAIR transcript, also revealed an association between HOTAIR and components of the PRC2 complex but not other chromatin regulatory factors. The final missing link was the demonstration that an RNA–protein interaction between HOTAIR and PRC2 is required for proper localization of PRC2 across the HOXD cluster. Indeed, depletion of HOTAIR resulted in the mislocalization of PRC2 from the HOXD cluster, *in trans*, and concomitant activation of HOXD genes. Together, these findings showed that the physical association between HOTAIR and PRC2 was required for the proper targeting of the chromatin regulatory machinery to their target sites. Thus, a novel layer of RNA-based modulation of epigenetic landscapes was unequivocally revealed.

Since unraveling the fact that PRC2 physically associates with HOTAIR or XIST, so targeting the complex to specific genomic regions, this mechanism has been shown to apply to numerous lncRNAs that physically associate with PRC2. Many of these associations are, in fact, required for the proper localization of the epigenetic regulatory machinery *in cis* and *trans* (see Fig. 1). Two independent studies, in particular, determined that in both human and mouse cells, hundreds of lncRNAs, accounting for up to 30% of the transcriptome, coprecipitate with PRC2. Moreover, many of the tested PRC2-bound lncRNAs were required for proper epigenetic and transcriptional regulation of PRC2 targets (Khalil et al. 2009; Zhao et al. 2010). It was further noted from these studies that one RNA could be bound to many different chromatin regulatory proteins, suggesting that it could function as an RNA bridge across multiple complexes (Khalil et al. 2009). Indeed, a detailed biochemical analysis of HOTAIR showed that in addition to binding PRC2, it also bound the histone demethylase LSD1 and NCOR. This pointed to a novel model for epigenetic regulation whereby a single lncRNA recruits several synergistic chromatin regulatory complexes to help guide, dock (NCOR), and facilitate heterochromatin formation (LSD1 and PRC2). Collectively, these studies led to the idea that an RNA scaffold can bridge numerous chromatin and additional regulatory complexes to impart genome target specificity (Fig.

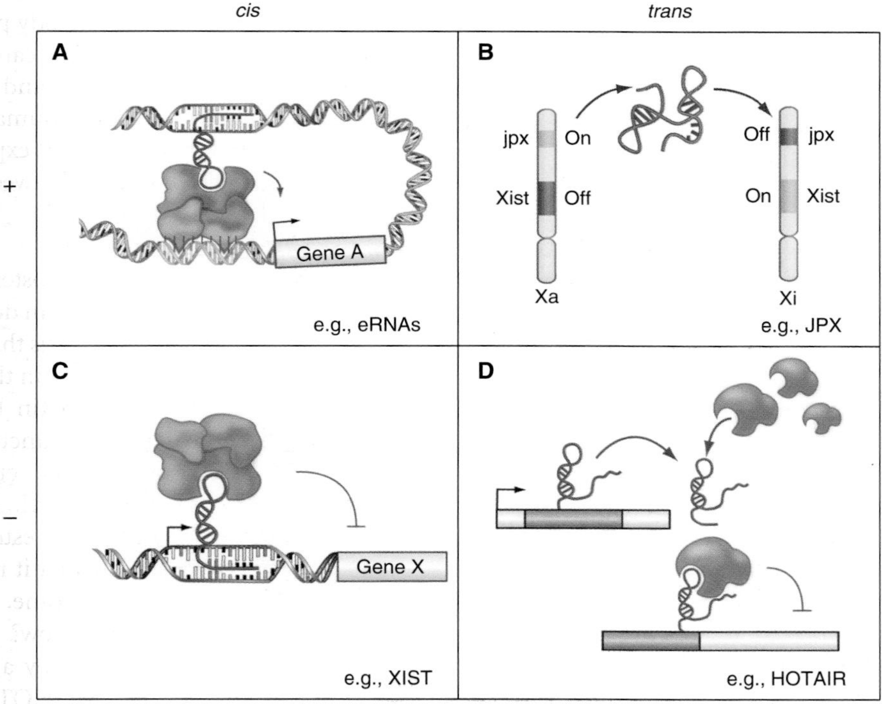

Figure 1. Models of how lncRNAs may function in the epigenetic control of gene expression, both activating and repressing transcription *in cis* and *in trans*. (*A*) Enhancer RNAs have been shown to play an RNA-mediated role in forming enhancer interactions resulting in long-range cis gene regulation by lncRNAs. (*B*) Activation by the JPX lncRNA as an example of a trans-acting lncRNA, which in this case facilitates Xist activation. (*C*) Xist as an example of an lncRNA that facilitates gene repression *in cis* across the majority of the X chromosome. (*D*) HOTAIR expression results *in trans* repression of HOX genes.

1D). The recent finding that HOTAIR overexpression is a hallmark in metastatic breast cancer (Gupta et al. 2010) has underscored the importance of HOTAIRs role in epigenetic regulation. In fact, HOTAIR serves as an "onco-lncRNA," inducing metastasis in breast cancer when overexpressed by remodeling the epithelial epigenome to resemble that of stromal cells. Cumulatively, the lessons learned from HOTAIR over the past five years have shown that lncRNAs play a critical role in interfacing with and modulating chromatin complexes during development and disease.

In the 50 years since RNA was identified as a central component in the flow of genetic information, it has become increasingly clear that RNA is more than a mere messenger and instead performs vast and diverse functions (Amaral et al. 2008). lncRNAs are, in fact, emerging as a critical layer of epigenetic regulation in which different lncRNAs are associated with distinctive epigenetic states, yet share a common mechanism; they physically associate with chromatin-modifying and -remodeling complexes, and guide them to specific genomic loci that are crucial for proper cellular function. However, this is only one facet of lncRNA biology; there is a diversity of other functional roles they play across numerous biological processes alluded to in Amaral et al. (2008).

REFERENCES

**Reference is also in this book.*

Amaral PP, Dinger ME, Mercer TR, Mattick JS. 2008. The eukaryotic genome as an RNA machine. *Science* **319:** 1787–1789.

* Brockdorff N, Turner BM. 2014. Dosage compensation in mammals. *Cold Spring Harb Perspect Biol* doi: 10.1101/cshperspect.a019406.

Gupta RA, Shah N, Wang KC, Kim J, Horlings HM, Wong DJ, Tsai MC, Hung T, Argani P, Rinn JL, et al. 2010. Long non-coding RNA *HOTAIR* reprograms chromatin state to promote cancer metastasis. *Nature* **464:** 1071–1076.

Khalil AM, Guttman M, Huarte M, Garber M, Raj A, Rivea Morales D, Thomas K, Presser A, Bernstein BE, van Oudenaarden A, et al. 2009. Many human large intergenic noncoding RNAs associate with chromatin-modifying complexes and affect gene expression. *Proc Natl Acad Sci* **106:** 11667–11672.

Rinn JL, Kertesz M, Wang JK, Squazzo SL, Xu X, Brugmann SA, Goodnough LH, Helms JA, Farnham PJ, Segal E, et al. 2007. Functional demarcation of active and silent chromatin domains in human *HOX* loci by noncoding RNAs. *Cell* **129:** 1311–1323.

Tsai MC, Manor O, Wan Y, Mosammaparast N, Wang JK, Lan F, Shi Y, Segal E, Chang HY. 2010. Long noncoding RNA as modular scaffold of histone modification complexes. *Science* **329:** 689–693.

Zhao J, Ohsumi TK, Kung JT, Ogawa Y, Grau DJ, Sarma K, Song JJ, Kingston RE, Borowsky M, Lee JT. 2010. Genome-wide identification of polycomb-associated RNAs by RIP-seq. *Mol Cell* **40:** 939–953.

Cite as *Cold Spring Harb Perspect Biol* doi: 10.1101/cshperspect.a018614

Enhancer RNAs: A Class of Long Noncoding RNAs Synthesized at Enhancers

Tae-Kyung Kim[1], Martin Hemberg[2], and Jesse M. Gray[3]

[1]The University of Texas Southwestern Medical Center, Department of Neuroscience, Dallas, Texas 75390-9111; [2]Boston Children's Hospital, Department of Ophthalmology, Boston, Massachusetts 02215; [3]Genetics Department, Harvard Medical School, Boston, Massachusetts 02115

Correspondence: taekyung.kim@utsouthwestern.edu

Recent studies have revealed that active enhancers are transcribed, producing a class of noncoding RNAs called enhancer RNAs (eRNAs). eRNAs are distinct from long noncoding RNAs (lncRNAs), but these two species of noncoding RNAs may share a similar role in the activation of mRNA transcription. Emerging studies, showing that eRNAs function in controlling mRNA transcription, challenge the idea that enhancers are merely sites of transcription factor assembly. Instead, communication between promoters and enhancers can be bidirectional with promoters required to activate enhancer transcription. Reciprocally, eRNAs may then facilitate enhancer–promoter interaction or activate promoter-driven transcription.

The functional contribution of enhancers to gene expression has been well shown over the past three decades. The mechanisms by which enhancers influence gene expression, however, remained poorly understood. Recent technological advances have made it possible to observe, on a genome-wide scale, the molecules and mechanisms that govern enhancer function. We know that enhancers recruit general coactivators, such as p300/CBP, and they show a common chromatin signature. This signature includes high levels of monomethylation at histone H3 lysine 4 (H3K4me1), but low levels of the promoter-specific H3K4me3 mark (Fig. 1). Using CBP and histone methylation patterns to identify neuronal enhancers, our study revealed that several thousand enhancers can recruit RNA polymerase II (Pol II) and transcribe noncoding RNAs upon neuronal activation (Kim et al. 2010). The transcripts, which we termed enhancer RNAs (eRNAs), have since been independently confirmed in many different cell types and species, suggesting that eRNA synthesis is not unique to neurons, but more likely a universal cellular mechanism involved in governing enhancer function.

Enhancer RNAs are clearly distinguishable from the canonical long noncoding RNAs (lncRNAs) whose functions have been better characterized. A first distinction is that although lncRNAs were broadly defined based on the presence of H3K4me3 at their promoters, eRNAs can be produced from enhancers without detectable levels of H3K4me3. This difference may stem from the 10- to 100-fold lower expression levels of eRNAs relative to lncRNAs, as H3K4me3 levels generally correlate with gene expression level. Second, unlike the promoters of lncRNAs and protein-coding genes, enhancers show little bias in the direction of transcription initiation. Third, whereas lncRNAs undergo maturation processes such as splicing and polyadenylation, eRNAs are shorter (<2 kb), with little evidence of being consistently spliced or polyadenylated. The lack of polyadenylation was inferred by the fact that eRNAs were first detected from analysis of total cellular RNA (using total RNA-seq) in neurons, but not observed in polyadenylated RNA (using mRNA-seq). Polyadenylated eRNAs, however, have been reported or implied from the analysis of other nonneuronal cell types. Despite some of these differences between stereotypical eRNAs and lncRNAs, we and others have observed a relatively small number of genomic loci that cannot be easily classified either as enhancers or promoters of lncRNAs because of the presence of both H3K4me3 and H3K4me1 marks. These loci could represent a distinct class of enhancers or may suggest that enhancer and promoter designations are only useful to a point. The distinction between enhancer

Cite as *Cold Spring Harb Perspect Biol* doi: 10.1101/cshperspect.a018622

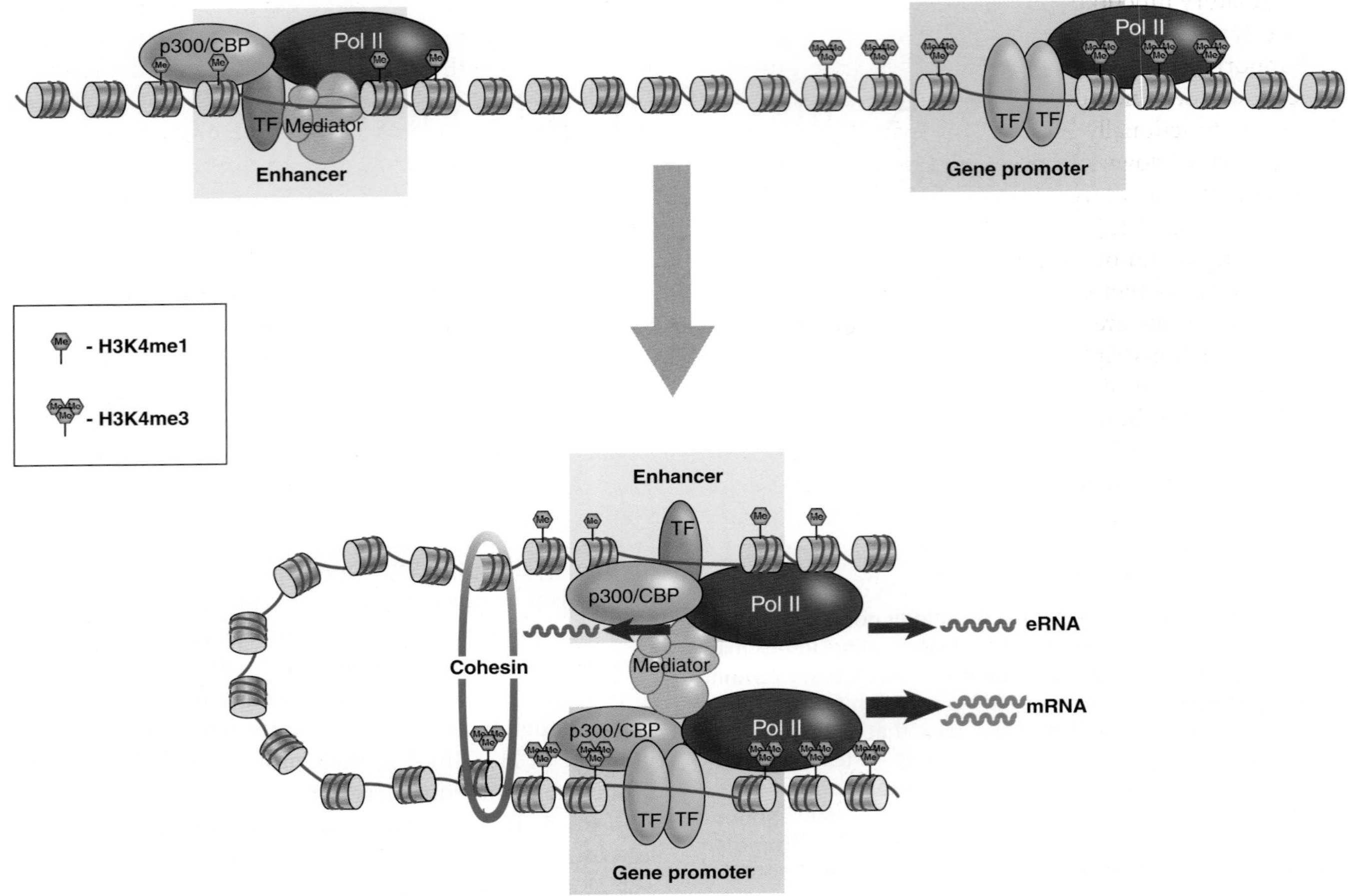

Figure 1. eRNA synthesis and function. During transcriptional activation, coactivator (e.g., p300/CBP) and RNA Pol II bind to a subset of enhancers and bidirectionally transcribe eRNAs. Chromatin looping between the enhancer and promoter will bring eRNAs near the target gene promoter to allow coordinate activation. Some eRNAs (e.g., eRNAs expressed from ER-α-bound enhancers in human breast cancer cells) facilitate and/or stabilize specific enhancer–promoter looping, in part by interacting with cohesin.

and promoter may simply be a quantitative one concerning the expression levels of transcripts. Indeed, protein-coding promoters have been reported to act as enhancers, regulating other nearby promoters.

The major question raised by the discovery of eRNAs relates to whether transcriptional activity at enhancers contributes to enhancer function. Several lines of circumstantial evidence suggest that eRNA synthesis is a regulated process and not mere transcriptional noise. Upon neuronal stimulation, only a subset of enhancers produces eRNAs, and this subset tends to be located near mRNAs that are strongly induced (Kim et al. 2010). Based on this observation, we propose that eRNA-producing enhancers are actively engaged in promoting the expression of target genes in response to stimulus-induced signaling. This hypothesis has been supported by correlative studies of eRNAs in various cell types. Moreover, kinetic analysis of eRNAs in lipopolysaccharide-activated macrophages showed that eRNA synthesis precedes transcription of adjacent protein-coding genes, suggesting an active role for eRNAs in the regulation of the target gene. Additional support comes from a recent siRNA-mediated knockdown study of human lncRNAs, inferring that they contribute to the activation of surrounding protein-coding mRNAs (Lai et al. 2013). Although it has not been clearly shown whether the lncRNAs with enhancer-like function are derived from enhancer regions, these GENCODE lncRNAs differ from eRNAs in that they are typically processed by splicing and polyadenylation, as well as having high levels of H3K4me3 at their promoters. Nonetheless, the combination of these two independent findings—eRNA derived from functionally defined enhancers and lncRNAs showing enhancer-like function—implies that non-protein-coding regions throughout the genome may produce transcripts with spe-

 Cite as *Cold Spring Harb Perspect Biol* doi: 10.1101/cshperspect.a018622

cific regulatory functions and have a more active role than previously anticipated in gene expression.

Supporting this notion, recent studies have provided more direct evidence demonstrating that at least some eRNAs are functionally important for target gene expression. The knockdown of several eRNAs caused decreased expression of nearby target genes (Lam et al. 2013; Li et al. 2013; Melo et al. 2013). In addition, the artificial tethering of eRNAs upstream of a minimal promoter in a plasmid-based reporter system enhances the reporter gene expression. These results are consistent with the proposed role of eRNAs in transcriptional activation. Intriguingly, the activating function of eRNAs appears to be sequence- or strand-specific, although the critical determinants for this specificity have not been identified (Lam et al. 2013). In other experiments in human breast cancer cells, eRNAs expressed from estrogen receptor α (ER-α)-bound enhancers increase the strength of specific enhancer–promoter looping, in part by interacting with cohesin (Li et al. 2013). The lncRNAs with enhancer-like function mentioned above were also shown to mediate chromatin looping, but through the interaction with a mediator complex (Lai et al. 2013).

Although these results collectively suggest that chromatin looping is an important regulatory step by which these distinct classes of activating lncRNAs commonly act (Fig. 1), there might be other functions of eRNAs. Our study, focusing on the neuronal *Arc* gene and enhancer, showed that eRNA synthesis requires the presence of an intact *Arc* gene promoter (Kim et al. 2010); that is, detectable levels of eRNAs are not synthesized when the *Arc* gene promoter is deleted, although RNA Pol II and transcription factors still bind to the enhancer. One possibility is that without its promoter, the *Arc* enhancer is missing an unknown factor required for its own transcription activity. Such an unknown factor would be present at the promoter and enable eRNA synthesis at the enhancer only when the enhancer is brought within close proximity with the promoter through a looping mechanism. These results challenge the standard model of unidirectional enhancer–promoter interactions in which enhancers act on promoters. Instead, enhancer and promoter activation may require feedback, with each contributing elements of the protein complement required for activation of the other. In this scenario, it is unlikely that the chromatin looping is facilitated entirely by eRNAs, as the eRNA synthesis would occur only after the enhancer–promoter looping. The chromatin looping would also keep nascent eRNAs in the proximity of target promoters, potentially providing an elegant way of preventing eRNAs from activating nonspecific target genes. A nascent eRNA transcript might then facilitate recruitment of activators to the promoter, acting as a scaffold for assembly of activating proteins. Because eRNAs are generally unstable, the specificity of eRNA function would come in part from its short half-life, preventing nonspecific activation away from its local site of synthesis, once synthesis was complete. It also needs to be pointed out that the act of eRNA transcription, in addition to the eRNA transcript itself, could have a specific biological function. For example, transcriptionally engaged RNA polymerase II could recruit chromatin modifiers to enhancers, stabilizing an enhancer domain in an active state.

The discovery and emerging functional roles of eRNAs certainly expand the growing regulatory capacity of noncoding RNAs. These findings not only illustrate a more complex role of *cis*-regulatory sequences than previously appreciated, but also provide an exciting avenue of future research in unraveling the intricate layers of gene regulation that are intertwined with lncRNAs, *cis*-regulatory sequences, epigenetic modifications, and three-dimensional chromatin configuration.

ACKNOWLEDGMENTS

We thank Drs. M.E. Greenberg and G. Kreiman for their guidance and support. This work was supported by the Whitehall Foundation (T-K.K.), and Welch Foundation (T-K.K.), and the Klingenstein Fund (T-K.K.).

REFERENCES

Kim TK, Hemberg M, Gray JM, Costa AM, Bear DM, Wu J, Harmin DA, Laptewicz M, Barbara-Haley K, Kuersten S, et al. 2010. Widespread transcription at neuronal activity-regulated enhancers. *Nature* **465:** 182–187.

Lai F, Orom UA, Cesaroni M, Beringer M, Taatjes DJ, Blobel GA, Shiekhattar R. 2013. Activating RNAs associate with Mediator to enhance chromatin architecture and transcription. *Nature* **494:** 497–501.

Lam MT, Cho H, Lesch HP, Gosselin D, Heinz S, Tanaka-Oishi Y, Benner C, Kaikkonen MU, Kim AS, Kosaka M, et al. 2013. Rev-Erbs repress macrophage gene expression by inhibiting enhancer-directed transcription. *Nature* **498:** 511–515.

Li W, Notani D, Ma Q, Tanasa B, Nunez E, Chen AY, Merkurjev D, Zhang J, Ohgi K, Song X, et al. 2013. Functional roles of enhancer RNAs for oestrogen-dependent transcriptional activation. *Nature* **498:** 516–520.

Melo CA, Drost J, Wijchers PJ, van de Werken H, de Wit E, Oude Vrielink JA, Elkon R, Melo SA, Leveille N, Kalluri R, et al. 2013. eRNAs are required for 53-dependent enhancer activity and gene transcription. *Mol Cell* **49:** 524–535.

Expanding the Epigenetic Landscape: Novel Modifications of Cytosine in Genomic DNA

Skirmantas Kriaucionis[1] and Mamta Tahiliani[2]

[1]Ludwig Institute for Cancer Research Ltd., University of Oxford, Nuffield Department of Clinical Medicine, Old Road Campus Research Building, Headington, Oxford OX3 7DQ, United Kingdom; [2]Skirball Institute/NYU School of Medicine, New York, New York 10016

Correspondence: mamta.tahiliani@med.nyu.edu

Methylation of the base cytosine in DNA is critical for silencing endogenous retroviruses, regulating gene expression, and establishing cellular identity, and has long been regarded as an indelible epigenetic mark. The recent discovery that the ten eleven translocation (TET) proteins can oxidize 5-methylcytosine (5mC) resulting in the formation of 5-hydroxymethylcytosine (5hmC) and other oxidized cytosine variants in the genome has triggered a paradigm shift in our understanding of how dynamic changes in DNA methylation regulate transcription and cellular differentiation, thus influencing normal development and disease.

Methylation of the base cytosine (termed 5-methylcytosine or 5mC) is an epigenetic mark often referred to as the fifth base, to underscore its heritability and importance in development. 5mC is considered an epigenetic mark because it directs biological function (i.e., transcriptional repression) without altering the protein coding capacity of the local DNA sequence dictated by the four conventional bases. 5mC is vital for processes including embryogenesis, parental imprinting, X inactivation, the silencing of endogenous retroviruses, and the regulation of gene expression and splicing. Cytosine methylation influences these processes by both modulating protein–DNA interactions and nucleating the formation of repressive heterochromatic structures. In 2009, 5-hydroxymethylcytosine (5hmC) was simultaneously identified by two research groups as a normal constituent of genomic DNA in mammalian neurons and embryonic stem (ES) cells (Kriaucionis and Heintz 2009; Tahiliani et al. 2009). This landmark finding has stimulated a tremendous amount of research focused on understanding how this modification exerts its influence on the regulation of the genome and how this modification ties into a 5mC demethylation pathway that was previously lacking in enzymatic players.

5hmC was serendipitously identified in Nathaniel Heintz's laboratory when Skirmantas Kriaucionis was elucidating the chromatin make-up of the strikingly euchromatic nuclei of cerebellar Purkinje neurons. Isolating Purkinje cell nuclei in itself was a technical achievement, requiring the use of transgenic mice with an eGFP labeled nucleolus (bacTRAP) and high-capacity fluorescence-activated cell sorting to get enough material for the assays. The goal was to compare 5mC abundance in Purkinje cells with granule cells using the classic "nearest neighbor" DNA composition analysis technique dating back to Kornberg's classic experiments of 1961 and used in Adrian Bird's pioneering experiments quantifying global levels of methylated CpGs. Unexpectedly, this sensitive, unbiased, and robust method revealed an additional signal, which was reproducibly enriched in Purkinje neurons and detectable in other neuronal cell types. The most exciting phase of these experiments was identifying the signal as 5hmC, a novel base modification in genomic DNA (Kriaucionis and Heintz 2009).

5hmC was concurrently discovered by Mamta Tahiliani in Anjana Rao's laboratory when her quest to identify a

Cite as *Cold Spring Harb Perspect Biol* doi: 10.1101/cshperspect.a018630

DNA demethylase took an unexpected twist. The search for such an enzyme was primarily motivated by the demonstration that DNA methylation is actively erased in the paternal genome immediately after fertilization. This seminal finding strongly suggested that resetting methylation patterns might be critical for epigenetic reprogramming (as illustrated in Fig. 3 of Ch. 15 [Li and Zhang 2014]). Mamta's bioinformatics collaborator L. Aravind predicted that the TET family of proteins were dioxygenases with a specificity for nucleic acids. Distantly related dioxygenases had recently been shown to remove methyl groups from both histones and damaged DNA bases. Therefore, the TET proteins were extremely attractive DNA demethylase candidates. In her initial experiments, Mamta found that overexpression of TET1 diminished levels of 5mC by immunofluorescence, suggesting tantalizingly that TET1 was acting as a true DNA demethylase. However, her attempts to confirm demethylation using thin-layer chromatography yielded puzzling results because the reduction in 5mC was not accompanied by the predicted increase in cytosine. However, when she adjusted the contrast on the scanned image, she noticed that what had appeared to be a faint smear under cytosine took on the shape of an independent spot suggesting that TET1 might be converting 5mC to a novel species. Because many dioxygenases initiate catalysis by hydroxylating their substrates, Mamta hypothesized and then confirmed that this nucleotide was 5hmC. The group also showed that 5hmC was present in the genome of ES cells, and that both TET1 and 5hmC levels decline when ES cells are differentiated. This suggested that 5hmC is a normal constituent of mammalian DNA, and that TET proteins and 5hmC play an important role in regulating gene expression and cell identity in ES cells (Tahiliani et al. 2009). Subsequent studies by multiple laboratories have established that each member of the TET family (TET1/TET2/TET3) is able to convert 5mC to 5hmC (Wu and Zhang 2011). However, studies in mice have shown that Tet3 is the only member of the TET family required in vivo for normal development.

The discovery that TET enzymes can oxidize 5mC to 5hmC led to the question of whether full DNA demethylation from 5hmC to C was passive (i.e., achieved by replication-dependent dilution), or actively catalyzed. TET enzymes have now been shown to successively oxidize 5hmC to 5-formylcytosine (5fC) and 5-carboxylcytosine (5acC); reviewed in Wu and Zhang 2011). The rapid loss of 5mC in the paternal genome coincides with the translocation of TET3 to the nucleus and the large-scale conversion of 5mC to 5hmC, 5fC, and 5caC (Wu and Zhang 2011). Immunostaining of metaphase chromatin further revealed that all three oxidized derivatives of 5mC are largely retained on the original strands of DNA and are passively diluted by replication during the early cleavage cycles, indicating that TET-mediated oxidation of 5mC can stimulate passive loss of 5mC oxidation products through replication. Alternatively or even concurrently, 5fC and 5caC can be removed by thymine DNA glycosylate (TDG) and replaced by cytosine via base excision repair (Fig. 1A) (Wu and Zhang 2011; Fig. 6 of Ch. 15 [Li and Zhang 2014]). When and where in the genome these mechanisms operate remains a topic of active research.

Understanding the biological function of 5hmC has required the development of innovative tools to detect it and distinguish it unequivocally from 5mC and C. It is clear now that bisulfite sequencing cannot distinguish 5hmC from 5mC, and also misinterprets 5fC and 5caC as cytosine (Pastor et al. 2013). Therefore, it is important to note that decades of bisulfite sequencing data must be interpreted with caution, as "methylation" could be either 5mC or 5hmC, whereas positions previously identified as cytosine could actually contain 5fC or 5caC. A number of techniques have now been developed to enrich for 5hmC-containing DNA and most recently to sequence it at single nucleotide resolution (Pastor et al. 2013).

Multiple lines of evidence indicate that 5hmC is not simply a demethylation intermediate, but rather a novel modification in DNA with an effector program of its own. 5hmC is present in a variety of mature cell types in adult organisms, and its levels range from 0.05% of all bases in some immune cells to as high as 0.6% in Purkinje cells. This leads to the question of whether readers of this mark exist to translate the presence of this modification into biological function, much as unmethylated cytosines can be read by CXXC domain-containing proteins (see Ch. 2 [Blackledge et al. 2013]), or methylated CpGs are recognized by MBD proteins. A number of proteins have already been identified that bind to 5hmC including MeCP2, MBD3, and Uhrf2, which are known to regulate transcription. 5fC- and 5caC-bound proteins include a number of DNA repair proteins, consistent with a role for these modifications as demethylation intermediates.

The cell type, developmental stage, and genomic locus specific distribution of 5hmC is beginning to suggest particular functions of this DNA modification. Techniques enriching for 5hmC as well as single nucleotide sequencing techniques have shown that in ES cells 5hmC levels are elevated at enhancers and CpG island (CGI)-containing promoters, which are free of methylation despite their high CpG content (Pastor et al. 2013). In neuronal cells, 5hmC is enriched in gene bodies (Fig. 1B,C) (Mellén et al. 2012; Pastor et al. 2013). Although gene body enrichment was also noted in ES cells, single nucleotide techniques have not verified this finding. It has been proposed that TET proteins and 5hmC play a role in keeping CGIs free of

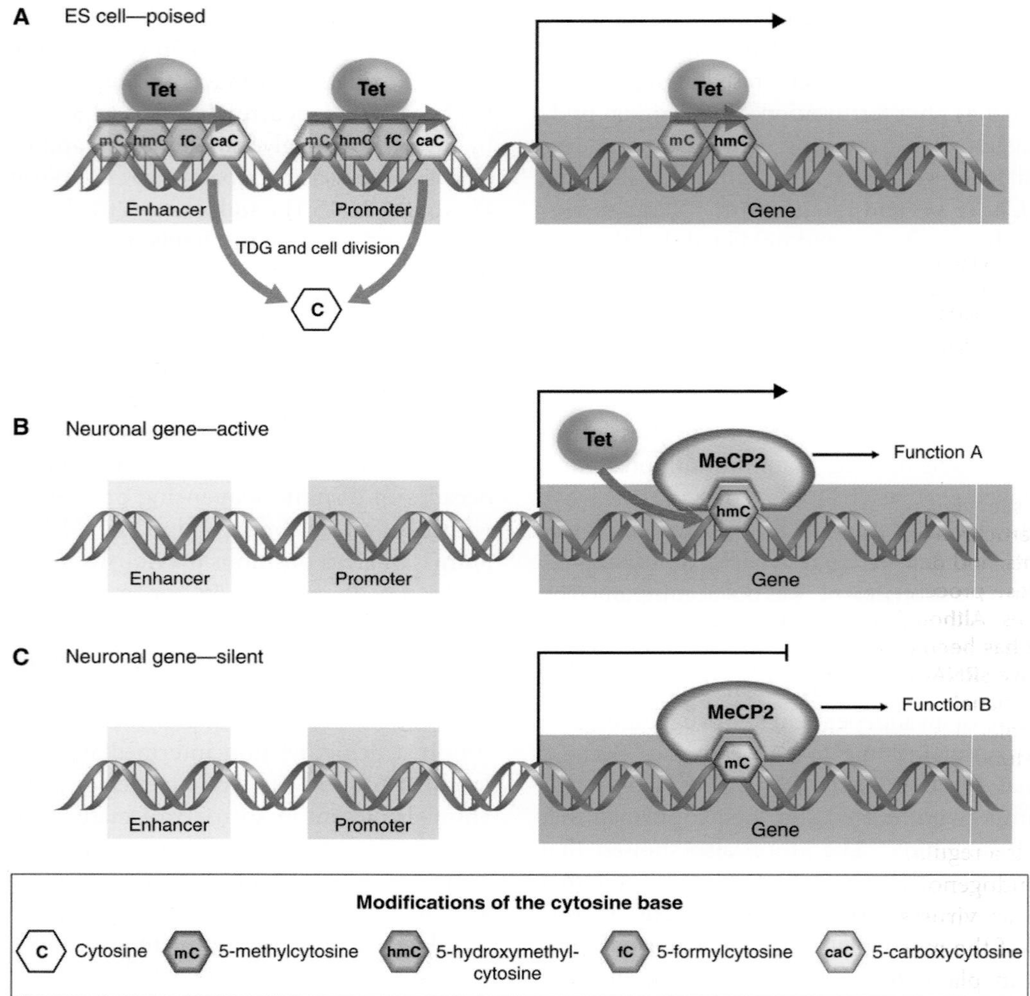

Figure 1. Distribution and metabolism of cytosine modifications within genes in ES cells and neurons. (*A*) TET-mediated oxidation of 5mC followed by base excision repair (BER)-mediated removal of 5caC keeps promoters and enhancers free of methylation in ES cells. It is also possible that oxidation of 5mC blocks maintenance methylation at these regions. MeCP2 binds both 5mC (*B*) and 5hmC (*C*) in neuronal gene bodies, where the cytosine modification state correlates with the level of expression.

methylation in ES cells, whereas the function of gene body 5hmC in neuronal cells is still unclear.

Future research will need to address the precise function of 5hmC in early development, hematopoiesis and neuronal function. It will be intriguing to know whether a single model can explain 5hmC function in all cell types or whether its function will vary for each cell type examined.

REFERENCES

*Reference is also in this book.

* Blackledge NP, Thomson JP, Skene PJ. 2013. CpG island chromatin is shaped by recruitment of ZF-CxxC proteins. *Cold Spring Harb Perspect Biol* **5:** a018648.

Kriaucionis S, Heintz N. 2009. The nuclear DNA base 5-hydroxymethylcytosine is present in Purkinje neurons and the brain. *Science* **324:** 929–930.

* Li E, Zhang Y. 2014. DNA methylation in mammals. *Cold Spring Harb Perspect Biol* **6:** a019133.

Mellén M, Ayata P, Dewell S, Kriaucionis S, Heintz N. 2012. MeCP2 binds to 5hmC enriched within active genes and accessible chromatin in the nervous system. *Cell* **151:** 1417–1430.

Pastor WA, Aravind L, Rao A. 2013. TETonic shift: Biological roles of TET proteins in DNA demethylation and transcription. *Nat Rev Mol Cell Biol* **14:** 341–356.

Tahiliani M, Koh KP, Shen Y, Pastor WA, Bandukwala H, Brudno Y, Agarwal S, Iyer LM, Liu DR, Aravind L, et al. 2009. Conversion of 5-methylcytosine to 5-hydroxymethylcytosine in mammalian DNA by MLL partner TET1. *Science* **324:** 930–935.

Wu H, Zhang Y. 2011. Mechanisms and functions of Tet protein-mediated 5-methylcytosine oxidation. *Genes Dev* **25:** 2436–2452.

Plant Mobile Small RNAs

Patrice Dunoyer,[1] Charles Melnyk,[2] Attila Molnar,[3] and R. Keith Slotkin[4]

[1]IBMP-CNRS, 67084 Strasbourg Cedex, France; [2]The Sainsbury Laboratory, University of Cambridge, Cambridge CB2 1LR, United Kingdom; [3]Department of Plant Sciences, University of Cambridge, Cambridge CB2 3EA, United Kingdom; [4]Department of Molecular Genetics and Center for RNA Biology, The Ohio State University, Columbus, Ohio 43210

Correspondence: patrice.dunoyer@ibmp-cnrs.unistra.fr

In plants, RNA silencing is a fundamental regulator of gene expression, heterochromatin formation, suppression of transposable elements, and defense against viruses. The sequence specificity of these processes relies on small noncoding RNA (sRNA) molecules. Although the spreading of RNA silencing across the plant has been recognized for nearly two decades, only recently have sRNAs been formally demonstrated as the mobile silencing signals. Here, we discuss the various types of mobile sRNA molecules, their short- and long-range movement, and their function in recipient cells.

RNA silencing is a regulatory mechanism that controls the expression of endogenous genes and exogenous molecular parasites such as viruses, transgenes, and transposable elements. One of the most fascinating aspects of RNA silencing found in plants and invertebrates is its mobile nature—in other words, its ability to spread from the cell where it has been initiated to neighboring cells. This phenomenon relies on the movement of small noncoding RNA molecules (sRNA, 21–24 nucleotides [nt] in length) that provide the sequence specificity of the silencing effects. In plants, there are two major classes of sRNAs: short interfering RNAs (siRNAs) and microRNAs (miRNAs). These sRNAs are generated by diverse and sometimes interacting biochemical pathways, which may influence their mobility. Movement of plant sRNAs falls into two main categories: cell-to-cell (short-range) and systemic (long-range) movement (Melnyk et al. 2011).

LONG-RANGE MOVEMENT OF SMALL RNA

The first hint of the mobile nature of RNA silencing in plants was provided by studies of tobacco exhibiting localized transgene silencing that was transmitted to new growth, following a pattern of vascular transport (see Fig. 1). The existence of a signaling molecule for RNA silencing was later confirmed by elegant plant grafting experiments (Melnyk et al. 2011). These studies demonstrated that a silencing signal emitted from the silenced transgenic rootstock travelled over long distances through the vasculature and could trigger de novo silencing of a homologous transgene in distant tissues of the plant. However, the identification of the systemic silencing signal took several years and was only recently conclusively addressed with the rise of high-throughput sequencing coupled to *Arabidopsis* micrografting experiments (Dunoyer et al. 2010a; Molnar et al. 2010). Detection of both transgenic and endogenous siRNAs in Dicer triple mutant grafted tissues (which are defective for siRNA biogenesis) provided strong support that siRNAs, as opposed to their precursor transcripts, are the mobile long-distance silencing signal. These experiments also demonstrated that mobile siRNAs are able to direct de novo methylation at target loci over long distances through a process of RNA-directed DNA methylation (Dunoyer et al. 2010a; Molnar et al. 2010). Intriguingly, only one third of the endogenous *Arabidopsis* siRNA loci generate mobile siRNAs, suggesting a complex channeling process that underlies mobility.

SHORT-DISTANCE CELL-TO-CELL MOVEMENT

Upon local induction, RNA silencing spreads from the site of initiation to the surrounding 10 to 15 neighboring cells (see Fig. 1). This cell-to-cell movement likely occurs through plasmodesmata (the pores interconnecting plant

Cite as *Cold Spring Harb Perspect Biol* doi: 10.1101/cshperspect.a017897

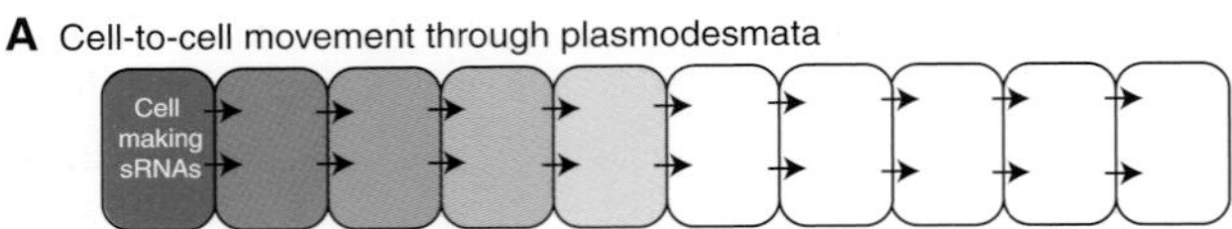

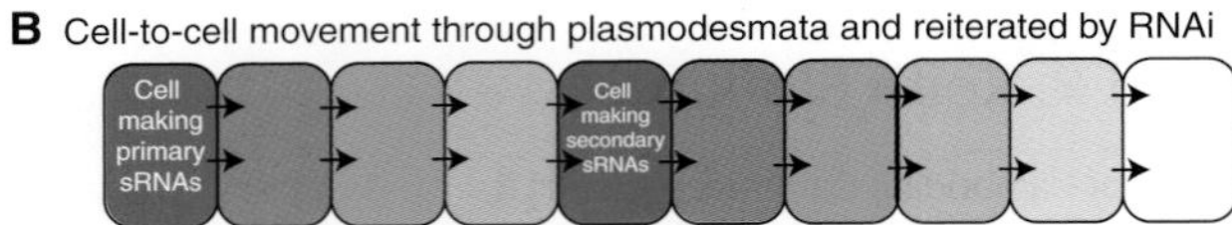

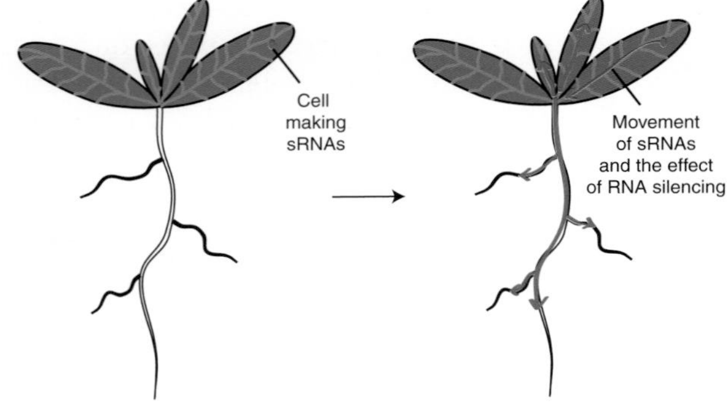

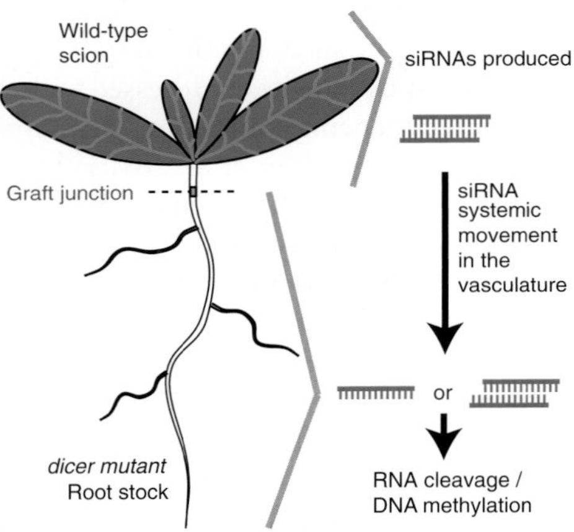

Figure 1. The movement of plant sRNAs, and therefore RNA silencing, can occur through two distinct mechanisms. (*A*) sRNAs can move through channels between plant cells called plasmodesmata. The diffusion of sRNAs from the cell in which they were made is seen in the red gradient. (*B*) Cell-to-cell movement can be extended beyond the range of diffusion through plasmodesmata by recipient cells using the primary sRNAs to initiate successive rounds of RNAi and produce secondary sRNAs. This amplification process is called transitivity. (*C*) Long-range movement from one organ to another is accomplished by loading the plant vascular system with sRNAs. Cells producing sRNAs and the movement of sRNAs are depicted in red. (*D*) In this critical experiment that defined sRNAs as the mobile factor in RNA silencing, a wild-type top of plant (scion) was grafted to a *dicer mutant* root, which is incapable of producing siRNAs. SiRNAs generated in the scion were transported into the root where they were identified by deep sequencing, and where they functioned in RNA cleavage and RNA-directed DNA methylation.

cells), as cells without plasmodesmata are resistant to mobile silencing signals. Occasionally, RNA silencing can be amplified through conversion of target RNAs into new double-stranded RNA (dsRNA) by the RNA-dependent RNA polymerase RDR6 and the production of secondary siRNAs (Melnyk et al. 2011). This amplification process, known as "transitivity" (see Fig. 1), results in a more extensive cell-to-cell spread of silencing through reiterated short-distance signaling events. Specific genetic screens were designed to identify the proteins required for biogenesis, movement, or perception of the cell-to-cell silencing signal (Melnyk et al. 2011). In these systems, Dicer-like4 (DCL4) was identified for this short-range silencing movement, suggesting a key role for DCL4-generated 21-nt siRNAs. Subsequently, these 21-nt siRNAs were directly identified as the cell-to-cell silencing signal by using a combination of transgenic reporter systems, viral suppressors of RNA silencing that specifically sequester 21-nt siRNAs, and the exogenous application (bombardment) of fluorescently labeled 21-nt siRNAs (Dunoyer et al. 2010b). In addition to DCL4, several mutations in the RNA-directed DNA methylation pathway compromised cell-to-cell silencing movement, but this mechanism remains ill defined. Of note, bombardment with RNA-directed DNA methylation pathway–associated 24-nt siRNAs moved to a similar extent as the 21-nt siRNAs, indicating that both size classes are mobile (Dunoyer et al. 2010a, 2010b).

FUNCTIONS OF SMALL RNA MOVEMENT

Developmental Patterning

Mobile small RNAs in plants are generated through several pathways from either endogenous or exogenous sources. During wild-type development, sRNA movement establishes gradients of target mRNA concentrations used in specifying developmental patterns. For example, movement of endogenous 21-nt siRNAs from the top of the leaves, where they are produced, to the bottom cell layers, generates a gradient that functions in the establishment of top-to-bottom leaf pattern formation (Melnyk et al. 2011). Certain miRNAs are also mobile. For instance, miR165/166 is produced in specific root cells and moves to adjacent cells, where it targets mRNAs encoding multiple transcription factors. The resulting gradient of target mRNA determines vascular cell types (Melnyk et al. 2011). Interestingly, some miRNAs are also thought to move long distances in plants. Upon phosphate starvation, one miRNA is translocated from the top of the plant to the root, where it degrades its target mRNA that encodes a suppressor of phosphate uptake (Melnyk et al. 2011).

Viral Resistance

Because viruses are potent triggers of RNA silencing in infected cells, they also represent huge sources of mobile small RNAs. These virus-derived siRNAs can move both cell-to-cell and systemically ahead of the infection front and prime the antiviral silencing response in naïve cells that are yet to be infected. This effect was shown using a movement-deficient virus carrying a fragment of an endogenous gene. Although this virus remained restricted to the infected lower leaves, it triggered silencing of the endogenous gene in and around the vasculature in distant parts of the plant. In addition, viral suppressors of RNA silencing that specifically sequester siRNAs are required for the successful infection of neighboring cells (Dunoyer et al. 2010b; Melnyk et al. 2011).

Epigenetic Changes

Work with viruses suggests that mobile silencing signals could direct epigenetic changes. Upon infection with an RNA virus, silencing can be established on a separate stably transformed nonviral homologous transgene. This silencing is manifested as DNA methylation at the transgene promoter and is inherited to subsequent generations in the absence of the virus, demonstrating heritable epigenetic silencing (Melnyk et al. 2011). Heritable epigenetic silencing is also known to occur at endogenous transposable elements (TEs). In *Arabidopsis*, TEs activate in the cells that neighbor the sperm cells and the embryo, becoming a source of siRNAs (Slotkin et al. 2009). The activation of TEs in these cells coincides with a corresponding increase in DNA methylation in the sperm cells and embryo, suggesting that the TE siRNAs may move into the neighboring gametes and subsequent embryo and reinforce RNA-directed DNA methylation and *trans*-generational TE silencing (Slotkin et al. 2009). However, the direct movement of endogenous TE siRNAs into gametes or the embryo has not been demonstrated.

MOVEMENT OF PLANT SMALL RNAS TO OTHER ORGANISMS

One interesting observation is the ability of mobile silencing signals to move outside of plants. RNA silencing transgenes expressed in a plant can silence complementary genes in fungal and invertebrate pathogens that feed on the plant (Melnyk et al. 2011). Transgenic RNA silencing signals can also pass between several plants through a parasitic plant intermediate (Melnyk et al. 2011). These examples represent important biotechnological applications for mobile RNA silencing and hint that similar sRNA movement may occur in nature from the plant to pathogens or symbionts.

REFERENCES

Dunoyer P, Brosnan CA, Schott G, Wang Y, Jay F, Alioua A, Himber C, Voinnet O. 2010a. An endogenous, systemic RNAi pathway in plants. *EMBO J* **29:** 1699–1712.

Dunoyer P, Schott G, Himber C, Meyer D, Takeda A, Carrington JC, Voinnet O. 2010b. Small RNA duplexes function as mobile silencing signals between plant cells. *Science* **328:** 912–916.

Melnyk CW, Molnar A, Baulcombe DC. 2011. Intercellular and systemic movement of RNA silencing. *EMBO J* **30:** 3553–3563.

Molnar A, Melnyk CW, Bassett A, Hardcastle TJ, Dunn R, Baulcombe DC. 2010. Small silencing RNAs in plants are mobile and direct epigenetic modification in recipient cells. *Science* **328:** 872–875.

Slotkin RK, Vaughn M, Borges F, Tanurdzić M, Becker JD, Feijó JA, Martienssen RA. 2009. Epigenetic reprogramming and small RNA silencing of transposable elements in pollen. *Cell* **136:** 461–472.

CpG Island Chromatin Is Shaped by Recruitment of ZF-CxxC Proteins

Neil P. Blackledge[1], John P. Thomson[2], and Peter J. Skene[3]

[1]Department of Biochemistry, University of Oxford, Oxford OX1 3QU, United Kingdom; [2]MRC Human Genetics Unit at the Institute of Genetics and Molecular Medicine, University of Edinburgh, Edinburgh EH4 2XU, United Kingdom; [3]Fred Hutchinson Cancer Research Center, Seattle, Washington 98109

Correspondence: neil.blackledge@bioch.ox.ac.uk

Most mammalian gene promoters are embedded within genomic regions called CpG islands, characterized by elevated levels of nonmethylated CpG dinucleotides. Here, we describe recent work demonstrating that CpG islands act as specific nucleation sites for the zinc finger CxxC domain–containing proteins CFP1 and KDM2A. Importantly, both CFP1 and KDM2A are associated with enzymatic activities that modulate specific histone lysine methylation marks. The action of these zinc finger CxxC domain proteins therefore imposes a defined chromatin architecture on CpG islands that distinguishes these important regulatory elements from the surrounding genome. The functional consequence of this CpG island–directed chromatin environment is discussed.

Approximately two-thirds of mammalian gene promoters are found within genomic regions known as CpG islands (CGIs). In contrast to bulk genomic DNA in which CpG dinucleotides are underrepresented and pervasively methylated, CGIs show a high density of CpGs and are refractory to DNA methylation. Despite more than 25 years of work aimed at understanding CGI function, it remains unclear how they contribute to the activity of gene promoters. However, of notable progress in the past few years is our demonstration that CGIs, through specific recruitment of proteins that bind nonmethylated DNA, can specifically alter the local chromatin environment of gene regulatory elements (Blackledge et al. 2010; Thomson et al. 2010).

We know that methylated CpGs act as nucleation sites for methyl-CpG binding domain (MBD) proteins that are generally associated with transcriptional repression. Based on the existence of MBD proteins, Skalnik and colleagues hypothesized that proteins may exist that specifically recognize nonmethylated CpGs. Through subsequent work, they identified the CpG binding protein (CGBP), later named CxxC finger protein 1 (CFP1), as such a factor (Voo et al. 2000). Importantly, this protein was found to bind to DNA through a ZF (zinc finger)-CxxC domain that specifically recognizes CpG dinucleotides in vitro.

The question remained: Does CFP1 recognize nonmethylated CGIs in vivo and thus impact on CGI function? This possibility was particularly intriguing, as CFP1 forms part of an extended family of ZF-CxxC domain–containing proteins that include chromatin-binding factors such as DNMT1, MLL1, and MBD1. With the emergence of massively parallel sequencing technologies coupled to chromatin immunoprecipitation (ChIP-seq), an exciting opportunity was presented to formally test the relationship between ZF-CxxC proteins and CGIs. To address this possibility, we performed independent studies, focusing on the ZF-CxxC proteins CFP1 and histone lysine demethylase 2A (KDM2A, also referred to as JHDM1a, FBXL11, or CXXC8). This unbiased approach showed a remarkable genome-wide association between sites of enrichment for both of these proteins and nonmethylated CGIs (Fig. 1A) (Blackledge et al. 2010; Thomson et al. 2010). Therefore, at least in the case of CFP1 and KDM2A, the ZF-CxxC domain appears to act as a CGI-targeting module. Significantly, these studies showed for the first time that CGIs are directly interpreted through recognition of nonmeth-

Cite as *Cold Spring Harb Perspect Biol* doi: 10.1101/cshperspect.a018648

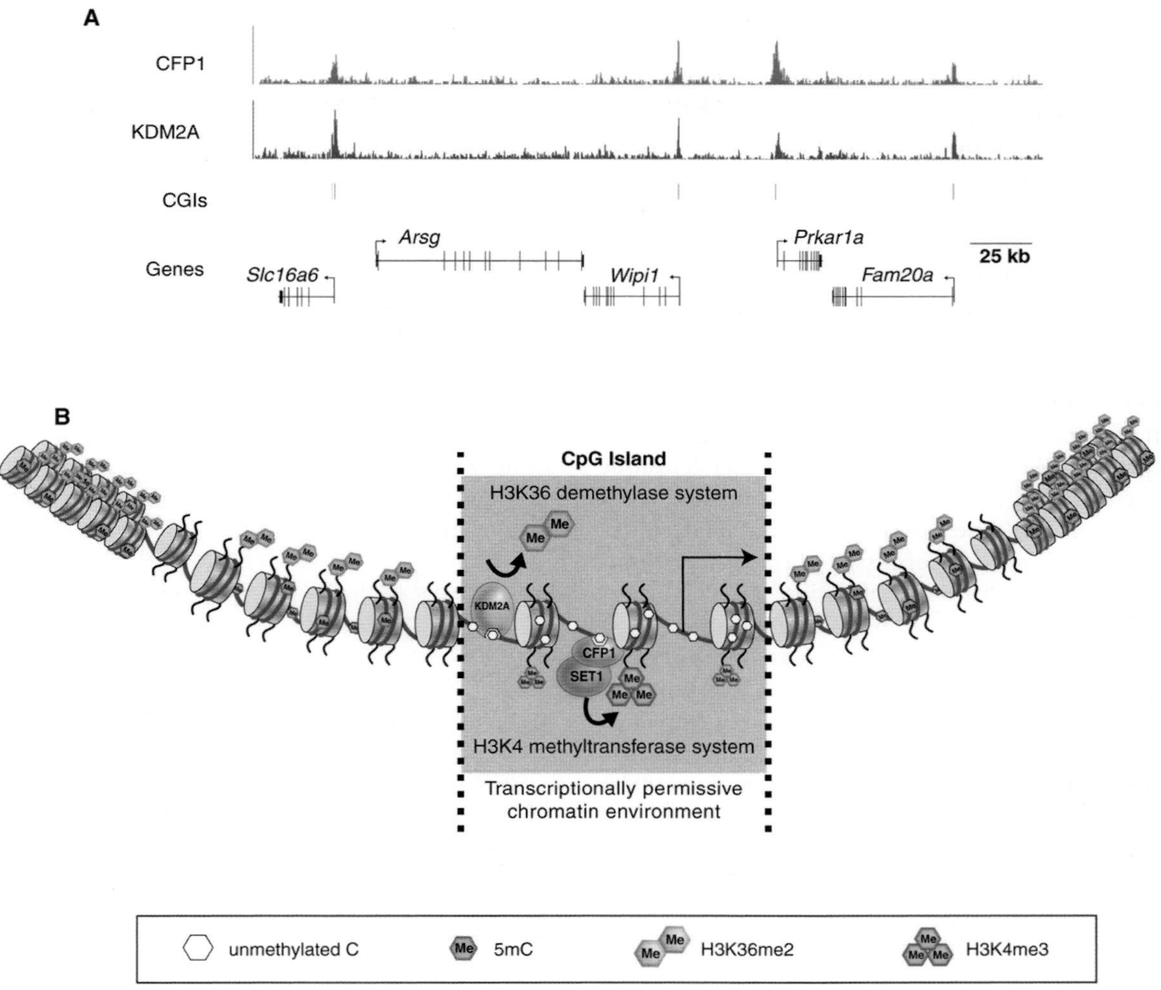

Figure 1. ZF-CxxC proteins at CpG islands. (*A*) CFP1 (*top* panel, red) and KDM2A (*bottom* panel, blue) ChIP-seq analysis for an ∼300 kb region of mouse chromosome 11 (data from Blackledge et al. 2010 and Thomson et al. 2010). Annotated genes and CGIs (green bars) are illustrated below the sequence traces. (*B*) CFP1 and KDM2A impose a specific chromatin environment at CGIs that is distinct from the rest of the genome.

ylated DNA, and that most, if not all, nonmethylated CGIs share common protein factors.

Interestingly, and of potential relevance to CGI function, most ZF-CxxC proteins are associated with chromatin-modifying activities. For example, CFP1 exists in a SETD1-containing methyltransferase complex that acts on the histone H3 lysine 4 (H3K4) residue, whereas KDM2A is a JmjC domain–containing demethylase enzyme that targets the histone H3 lysine 36 (H3K36) residue. In our studies, we observed that the histone-modifying activities associated with these ZF-CxxC proteins impose a defined chromatin environment at CGIs. Specifically, the action of CFP1 facilitates nucleation of histone H3K4 trimethylation (H3K4me3), a punctate mark generally associated with 5′ ends of genes, whereas KDM2A depletes H3K36 dimethylation (H3K36me2), a very abundant and broadly distributed modification adorning 30%–50% of total histone H3 (Blackledge et al. 2010; Thomson et al. 2010).

A surprising finding from these studies was that CFP1 and KDM2A bind at CGIs and modify CGI chromatin independently of transcriptional activity. A striking illustration of this was the demonstration that an exogenous CpG-rich sequence is sufficient to recruit CFP1 and nucleate H3K4me3 without concomitant recruitment of RNA polymerase II (RNA Pol II; Thomson et al. 2010). These bodies of work have therefore established a new paradigm, whereby the underlying DNA signal at CGIs (i.e., a high density of nonmethylated CpGs) is translated into a defined histone modification status (i.e., H3K4me3 enriched and H3K36me2 depleted). Therefore, via the ZF-CxxC system, nonmethylated CGIs have a "hard-wired" chromatin environment distinct from the rest of the genome (Fig. 1B).

The above studies prompted the question, within the context of CGI elements, what is the functional significance of ZF-CxxC protein binding and the chromatin modification states that they impose? Histone lysine methylation marks are thought to influence transcription by recruiting specific effector proteins via plant homeodomain (PHD) fingers or chromatin-modifier (chromo-) domains. In the case of H3K4me3, a number of studies suggest that this mark has the potential to recruit PHD finger proteins that support transcription, such as the core transcription factor TFIID, the NuRF chromatin remodeling complex, and ING4-containing histone acetyltransferase complexes. In contrast, studies in yeast suggest that H3K36me2 may repress transcription initiation by recruiting the chromodomain protein EAF3, a component of the RPD3S histone deacetylase complex. It is therefore possible that the combined effect of H3K4me3 enrichment and H3K36me2 depletion at CGIs creates a permissive chromatin environment that favors transcriptional initiation.

CGI promoters show a number of unique characteristics that may be attributable, at least in part, to the permissive chromatin environment created by ZF-CxxC proteins. For example, unlike classical TATA-box promoters, which use a defined transcriptional start point, CGI promoters tend to initiate transcription over a broad region of 100 bp or more. Furthermore, even in the absence of productive transcription, CGI promoters are enriched for RNA Pol II and show short, nonproductive, bidirectional transcripts (Core et al. 2008). Finally, inducible "primary response genes" that have CGI promoters can be rapidly activated by lipopolysaccharide stimulation without a requirement for chromatin-remodeling events (Ramirez-Carrozzi et al. 2009; illustrated in Fig. 9 of Ch. 29 [Busslinger and Tarakhovsky 2014]). This is in contrast to primary response genes with non-CGI promoters for which productive transcriptional output requires SWI/SNF-mediated chromatin remodeling. Further studies are required to determine whether ZF-CxxC proteins contribute to some or all of these CGI characteristics and ultimately to define the precise role that ZF-CxxC proteins play in CGI function.

As a final thought, it should be emphasized that protection from DNA methylation is at the crux of CGI existence, and is essential for ZF-CxxC protein nucleation at these regions. Although the mechanisms responsible for establishing and maintaining this DNA methylation-free state are poorly defined, one alluring possibility is that ZF-CxxC proteins themselves may contribute. For example, in vitro studies suggest that H3K4me3, the mark placed by the CFP1 complex, may block de novo DNA methylation, exemplified by the inhibitory effect this mark has on DNMT3L binding, which is part of the DNMT3A/3L de novo methylating complex (described in Ch. 6 [Cheng 2014]). Also, the ZF-CxxC protein TET1 is a hydroxylase enzyme able to convert methylcytosine into hydroxymethylcytosine, a reaction that has been implicated in DNA demethylation pathways (described in Ch. 15 [Li and Zhang 2014]). It is therefore tempting to speculate that at CGIs, ZF-CxxC proteins provide a self-reinforcing loop of nonmethylated CpG recognition and subsequent protection from DNA methylation.

REFERENCES

*Reference is also in this book.

Blackledge NP, Zhou JC, Tolstorukov MY, Farcas AM, Park PJ, Klose RJ. 2010. CpG islands recruit a histone H3 lysine 36 demethylase. *Mol Cell* **38:** 179–190.

* Busslinger M, Tarakhovsky A. 2014. Epigenetic control of immunity. *Cold Spring Harb Perspect Biol* **6:** a019307.

* Cheng X. 2014. Structural and functional coordination of DNA and histone methylation chromatin. *Cold Spring Harb Perspect Biol* **6:** a018747.

Core LJ, Waterfall JJ, Lis JT. 2008. Nascent RNA sequencing reveals widespread pausing and divergent initiation at human promoters. *Science* **322:** 1845–1848.

* Li E, Zhang Y. 2014. DNA methylation in mammals. *Cold Spring Harb Perspect Biol* **6:** a019133.

Ramirez-Carrozzi VR, Braas D, Bhatt DM, Cheng CS, Hong C, Doty KR, Black JC, Hoffmann A, Carey M, Smale ST. 2009. A unifying model for the selective regulation of inducible transcription by CpG islands and nucleosome remodeling. *Cell* **138:** 114–128.

Thomson JP, Skene PJ, Selfridge J, Clouaire T, Guy J, Webb S, Kerr AR, Deaton A, Andrews R, James KD, et al. 2010. CpG islands influence chromatin structure via the CpG-binding protein Cfp1. *Nature* **464:** 1082–1086.

Voo KS, Carlone DL, Jacobsen BM, Flodin A, Skalnik DG. 2000. Cloning of a mammalian transcriptional activator that binds unmethylated CpG motifs and shares a CXXC domain with DNA methyltransferase, human trithorax, and methyl-CpG binding domain protein 1. *Mol Cell Biol* **20:** 2108–2121.

 Cite as *Cold Spring Harb Perspect Biol* doi: 10.1101/cshperspect.a018648

Bromodomain and Extraterminal Domain Inhibitors (BETi) for Cancer Therapy: Chemical Modulation of Chromatin Structure

Jun Qi

Dana-Farber Cancer Institute, Boston, Massachusetts 02115

Correspondence: jun_qi@dfci.harvard.edu

In cancer, epigenetic proteins are intensely studied targets for therapeutic drug discovery, showing great promise. These proteins include the chromatin-modifying enzymes that "write" and "erase" histone posttranslational modifications (PTM), and those that "read" these marks through binding modules. In an effort to find a compound that could disrupt the protein–protein interactions between a PTM and reader, JQ1 has proven to be a first-in-class, drug-like inhibitor of the "bromodomain and extraterminal domain" epigenetic readers (BETs), which recognize histone lysine acetylation marks. JQ1 has facilitated the mechanistic study and therapeutic application in cancer of this kind of epigenetic inhibition. By using this chemical probe, we have discovered that the bromodomain inhibitors (BETi) have compelling activity in preclinical models of multiple myeloma and acute myeloid leukemia. In particular, BETi down-regulates the MYC, IL-7R, and E2F transcriptional programs. We are continuously integrating the transcriptional consequences of BETi with changes in the epigenomic landscapes of cancer cells to elucidate the mechanisms underlying response to BETi using chemical and genetic perturbations.

In cancer, epigenetic proteins are intensely studied targets for drug discovery owing to the general view that it is not just the DNA sequence that is altered in epigenetics-based diseases. Studies to date have indeed shown therapeutic promise; inhibitors of DNA methyltransferases (DNMTs) and histone deacetylases (HDACs), for example, have shown substantial clinical efficacy leading to regulatory approval for use in hematologic malignancies. These successes have stimulated broad-based efforts to develop other inhibitors of chromatin-modifying enzymes, so-called epigenetic "writers" and "erasers." Perhaps owing to the perception that it is difficult to interfere with protein–protein interactions, chromatin-binding modules or epigenetic "readers" have received comparatively little attention. Motivated by this challenge, we developed inhibitors of the BET (for bromodomain and extraterminal domain) epigenetic readers, using a tried and tested thienodiazepine core, which forms the core of a number of pharmaceutical drugs. Compounds identified showing potential were used as chemical probes to study the mechanistic details and its therapeutic application.

Readers are often regarded as functional "effector" proteins that can recognize particular posttranslational modifications (e.g., methylation or acetylation marks) on histone proteins or DNA, placed by "writers" as part of signaling-based transcriptional pathways (Fig. 1). Historically the bromodomain stands as the first well-characterized epigenetic reader, which recognizes the acetylated lysine side chain on histones. The binding affinity between a bromodomain protein reader and the histone tail is low—in the μM range. We designed and synthesized a thienodiazepine-based small molecule called JQ1, which shows excellent inhibition against the BET subfamily with low nanomolar binding potency, especially targeting the BET protein, BRD4. Typically, drug design seeks to achieve a high binding affinity, in the range of 0.1–10 nM with its target. The compound was selected by combining small-molecule library screening and

Cite as *Cold Spring Harb Perspect Biol* doi: 10.1101/cshperspect.a018663

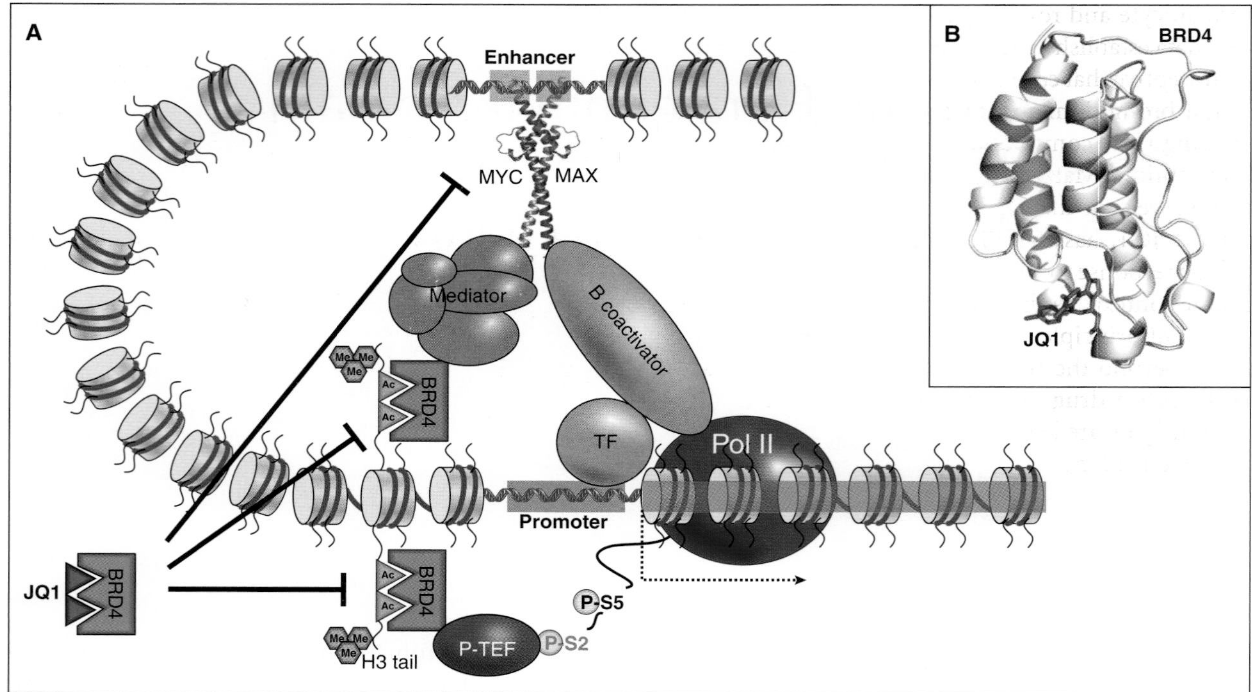

Figure 1. Model for JQ1 small-molecule inhibition of the BRD4 bromodomain in cancer. (*A*) The aberrant transcriptional activation of MYC target genes (indicated as red shaded regions of chromatin) is a common feature of many cancers. Transcriptional activation requires the bromodomain "reading" function of BRD4, which recognizes acetylation marks ("Ac"-labeled cyan triangles) on histone H3 tails at promoter-proximal target sequences. Acetyl-bound BRD4 interacts with both the MYC-MAX complex bound to enhancer sequences (via a mediator complex) and the PTEFb phosphorylase required for the release of RNA polymerase II (Pol II) during transcriptional elongation. The competitive binding of the JQ1 bromodomain inhibitor (red triangle) to BRD4 not only reduces transcription of the MYC gene (top inhibition arrow) but also its target genes by abrogating recruitment of enhancer complexes and PTEFb (middle and bottom inhibition arrows), possibly via chromatin looping. The active chromatin mark histone H3K4me3 is illustrated as triple green hexagons. (*B*) Crystal structure of the human BRD4 protein in complex with JQ1 (red).

structural information provided by crystallography. The cocrystal structure of JQ1 and the first bromodomain of BRD4 showed excellent structure complementarity between the small molecule and protein, explaining the high binding affinity. JQ1 also possessed good cell permeability, which allowed us to study the extent and effect of bromodomain inhibition in a BRD4-dependent cancer: the NUT-midline carcinoma, which arises from a BRD4-NUT translocation. The effect of the compound was to induce differentiation in patient-derived cancer cell lines in less than 24 hours after treatment. It also showed excellent efficacy in a murine xenograft model without obvious toxicity. Given the reasonable pharmacokinetic properties JQ1 was even entitled to be used in a human patient–derived xenograft model (Filippakopoulos et al. 2010). This prototype drug was further optimized to a preclinic candidate for BRD4-dependent cancers (M Mckewon, K Shaw, and J Qi, in prep.).

With a chemical probe in hand, the role of bromodomain epigenetic readers could be examined in transcriptional regulation. More recent studies showed that BET inhibitors (BETi) have compelling activity in preclinical models of multiple myeloma (Delmore et al. 2011) and acute myeloid leukemia (Zuber et al. 2011; Ott et al. 2012). Specifically, BETi down-regulates typically oncogenic MYC, IL7R, and E2F transcriptional programs. These observations have shown that BET bromodomain inhibition provides an efficient pathway to strategically target certain malignancies and other diseases that can be characterized by the pathologic activation of c-Myc.

The inhibitory effects of JQ1 on the testis-specific BRD member, BRDT, which is essential for chromatin remodeling during spermatogenesis, leads to a complete, but reversible contraceptive effect. As with BRD4, JQ1 occupies the BRDT acetyllysine binding pocket, preventing recognition of acetylated histone H4. In mice, this reduces the seminiferous tubule area, testis size, and spermatozoa number and motility without affecting hormone levels. JQ1-treated males show the inhibitory effects of JQ1 at

Cite as *Cold Spring Harb Perspect Biol* doi: 10.1101/cshperspect.a018663

the spermatocyte and round spermatid stages, in particular. These data establish a rationale for the development of a new contraceptive that can cross the blood–testis boundary and inhibit bromodomain activity during spermatogenesis, providing a lead compound targeting the male germ cell for contraception (Matzuk et al. 2012).

The discovery of the small-molecule inhibitor of bromodomains, JQ1, has shed light on the mechanisms underlying the response to BETi at a transcriptional level in its epigenomic context. Studies involving JQ1 have provided a clear "proof of principle" that other reader domains, both within and beyond the BET subfamily of bromodomains, may be attractive drug targets in years to come as more structural insights are learned from each class of chromatin-associated readers.

REFERENCES

Delmore JE, Issa GC, Lemieux ME, Rahl PB, Shi JW, Jacobs HM, Kastritis E, Gilpatrick T, Paranal RM, Qi J, et al. 2011. BET bromodomain inhibition as a therapeutic strategy to target c-Myc. *Cell* **146:** 904–917.

Filippakopoulos P, Qi J, Picaud S, Shen Y, Smith WB, Fedorov O, Morse EM, Keates T, Hickman TT, Felletar I, et al. 2010. Selective inhibition of BET bromodomains. *Nature* **468:** 1067–1073.

Matzuk MM, McKeown MR, Filippakopoulos P, Li Q, Ma L, Agno JE, Lemieux ME, Picaud S, Yu RN, Qi J, et al. 2012. Small-molecule inhibition of BRDT for male contraception. *Cell* **150:** 673–684.

Ott CJ, Kopp N, Bird L, Paranal RM, Qi J, Bowman T, Rodig SJ, Kung AL, Bradner JE, Weinstock DM. 2012. BET bromodomain inhibition targets both c-Myc and IL7R in high-risk acute lymphoblastic leukemia. *Blood* **120:** 2843–2852.

Zuber J, Shi J, Wang E, Rappaport AR, Herrmann H, Sison EA, Magoon D, Qi J, Blatt K, Wunderlich M, et al. 2011. RNAi screen identifies BRD4 as a therapeutic target in acute myeloid leukaemia. *Nature* **478:** 524–528.

Pharmacological Inhibition of Bromodomain-Containing Proteins in Inflammation

Uwe Schaefer

Laboratory of Immune Cell Epigenetics and Signaling, The Rockefeller University, New York, New York 10065

Correspondence: uschaefer@rockefeller.edu

Inflammation is associated with the activation of genes that contribute to immune defense and tissue repair. The bromodomain-containing proteins of the BET family, which recognize histone lysine acetylation, play a key role in the transcriptional control of inflammatory genes. Inhibition of BET proteins by the small-molecule inhibitor I-BET affects the expression of a particular subset of inflammatory genes—namely, ones that follow an "analog-like," but not "digital-like" activation pattern. This ability of I-BET to target genes based on the dynamic pattern of their activation may facilitate the further development of anti-inflammatory treatment protocols that are tuned to the individual or to disease-specific patterns of gene expression.

In humans as well as other organisms, tissue damage or interaction with pathogens leads to an inflammatory response associated with the production of numerous biologically active proteins and metabolites (see Nathan 2002). These mediators of the inflammatory response contribute to the elimination of pathogens as well as to the repair of tissue damage caused either by infection or other factors. The persistence of pathogens and environmental stresses makes inflammation an unavoidable part of life, yet the scope and timing of inflammation are crucial for the health of the individual. Recent research shows that chromatin plays an important role in the regulation of inflammation by activating cohorts of inflammatory genes (see Medzhitov and Horng 2009; Smale 2010). The Nicodeme et al. (2010) study describes a novel approach for the therapeutic intervention of inflammation, using a synthetic compound that targets bromodomain and extraterminal domain (BET) family epigenetic regulators of the transcriptional response. BET proteins are "readers" of the acetylated lysines found on the amino-terminal tails of histone proteins in chromatin. Acetylated histones are typically associated with transcriptionally active chromatin regions, and BET proteins play a role in achieving transcriptional competence as effector molecules, via the recruitment and association with other activating factors. Therefore, it follows that inhibitors of BET proteins may have a transcriptionally repressive effect by blocking the capacity of these epigenetic readers to recognize their acetylated lysine targets.

Excessive inflammatory responses are associated with numerous acute and chronic disorders that range from acute bacteria-induced sepsis to lingering chronic inflammatory conditions such as rheumatoid arthritis or Crohn's colitis (see Nathan 2002). Chronic inflammation has also been associated with cancer, as persistent repair of inflamed tissues may increase the likelihood of a cancerous mutation in dividing cells.

The magnitude and timing of inflammatory responses reflect tightly controlled and cell-type-specific gene expression patterns. Tissue-localized macrophages are the most immediate and perhaps the most potent cellular carriers of inflammatory responses on interaction with pathogens or tissue-derived proinflammatory triggers (see Medzhitov and Horng 2009). The patterns of gene expression associated with inflammation are likely to be trigger-specific. However, much of the general principles that govern inflammation can be discerned from the analysis of the

Cite as *Cold Spring Harb Perspect Biol* doi: 10.1101/cshperspect.a018671

macrophage response to Gram-negative bacteria-derived lipopolysaccharide (LPS), one of the best-studied triggers of inflammation in mice and man. Exposure of macrophages to LPS results in the activation of Toll-like receptor 4 followed by initiation of the signaling program that triggers the inflammatory response (see Medzhitov and Horng 2009; Smale 2010). At the gene level, this response has a well-defined pattern. Shortly after LPS triggering, macrophages up-regulate early response genes that can be divided into generic (e.g., c-Fos) and more inflammation-specific genes (e.g., TNF and Cxcl1/2). As the inflammatory response progresses, the expression of primary response genes is followed by the activation of secondary response genes that encode various regulators of the local and systemic inflammatory response. The stepwise recruitment of primary and secondary response genes into the inflammatory response reflects certain distinct features of these genes. These features include, but are not limited to, differences in AT content of gene promoters, dependence on chromatin-remodeling factors, and abundance of RNA polymerase II at gene promoters before cell activation by LPS (see Smale 2010; Ch. 29 [Busslinger and Tarakhovsky 2014]). To summarize, primary response genes appear to be activated according to a "digital"-like on/off response pattern, whereas secondary response genes follow an "analog"-like activation pattern that requires progression through discrete activation events to enable the initiation of transcription and elongation of mature RNA (see Fig. 1).

The apparent differences in activation of inflammatory genes have suggested that pharmacological regulation of inflammation could be based on a gene's dependence on common regulators of transcriptional responses rather than protein target specificity. Indeed, studies from 2009 showed that a deficiency in the BRD4 epigenetic transcriptional regulator (a BET family protein) resulted in greatly reduced LPS-driven expression of key inflammatory genes (Hargreaves et al. 2009). The natural ligands of BET proteins are the acetyl moieties at the amino-terminal portions of histones H3 and H4. Thus, our pharmacologically driven studies focused on determining the effects of inhibiting the BET family of proteins (BRD2, -3, and -4) implicated in the epigenetic regulation of gene expression during inflammation (Nicodeme et al. 2010). To our advantage was the availability of highly selective BET protein–specific ligands that affect the interaction between BET proteins and their natural ligands. The studies centered on the bromodomain inhibitor I-BET (initially called compound GSK525762A), which binds to BET proteins. We argued that I-BET would affect the ability of BET proteins to bind to their histone ligands, leading to a reduction in gene expression. Moreover, we speculated that the consequence of pharmacological intervention of BET protein–histone association is likely to be gene-specific and reflects the dependence of individual inflammatory genes on BET proteins.

BET proteins are associated with both primary and secondary response genes at relatively low but similar levels before macrophages are triggered with LPS. However, primary response genes have relatively higher levels of RNA polymerase II, and the "permissive" H3K4me3 and H3/H4Kac marks (see Nicodeme et al. 2010). These features suggested that primary genes are poised for immediate "digital" type transcriptional activation. Consequently, they may not require BET proteins as much as secondary response genes, which respond in a more "analog"-like way (see Fig. 1). In contrast, the incremental changes of secondary response genes during their activation require subsequent steps of chromatin remodeling, transcriptional initiation, and transcriptional elongation, making them more susceptible to I-BET. Indeed, we found that treatment of macrophages in vitro with I-BET resulted in strong and selective attenuation of secondary response gene expression while leaving the expression of primary response genes largely unaffected. Significantly, the selective effect of I-BET on secondary response genes holds true for both macrophage and fibroblast responses, not only to LPS, but to secondary mediators of the inflammatory responses such as TNF or IFN-β. We conclude that I-BET appears to "recognize" and target genes based on their epigenetic state.

This selective action of I-BET on secondary response gene activation has potential therapeutic advantages. First, most of the housekeeping genes do not rely on BET proteins for their expression. Second, the gene-state-specific effect of I-BET may enable accurate dosing of the drug within ranges that affect only a particular set of inflammatory genes. Thus, combined with the high rate of I-BET clearance in the organism (see Nicodeme et al. 2010), the duration and dose of I-BET could be computed for a highly selective treatment protocol.

The therapeutic potential of I-BET in in vivo inflammation has been validated by the successful treatment of bacteria- or LPS-induced sepsis in mice. The future of BET inhibitors, including I-BET and related compounds such as JQ1 (Filippakopoulos et al. 2010; Ch. 2 [Qi 2014]), for the treatment of inflammatory disorders will require the disease-tailored administration of I-BET alone or in combination with other drugs that control inflammatory responses. At this time, some of these questions are being addressed in the Phase I clinical trials of I-BET for the treatment of NUT midline carcinoma and also other cancers that are dependent on the continuous expression of MYC.

I-BET's ability to interfere with particular patterns of gene expression has set a strong precedent, validating further searches for synthetic ligands that can interfere with the binding of epigenetic regulators of gene expression to

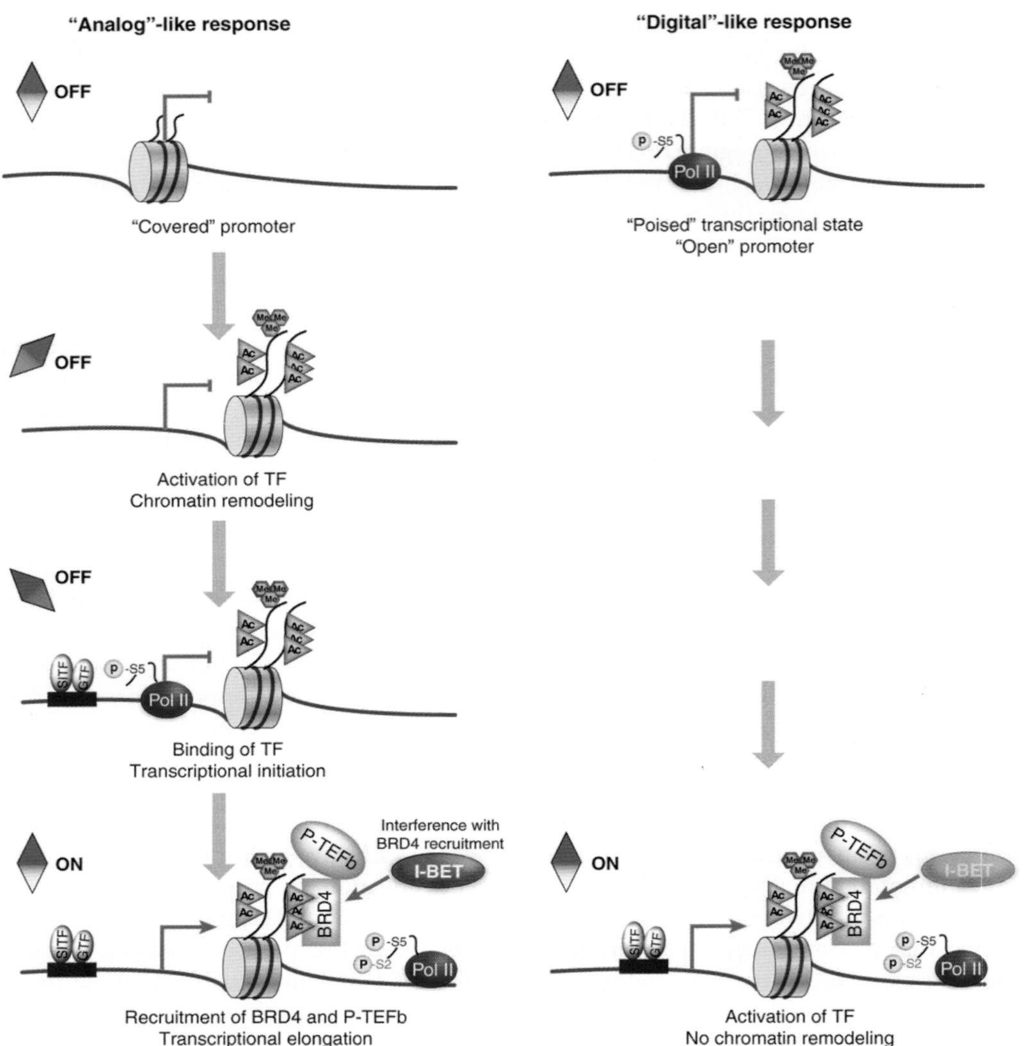

Figure 1. I-BET selectively inhibits genes that follow an "analog"-like activation pattern while not affecting the "digital"-like response. In response to an inflammatory stimulus, secondary response genes follow an "analog"-like activation pattern. This activation involves chromatin remodeling and exposure of the nucleosome-covered gene promoter. Next, transcriptional initiation commences with the binding of stimulus-induced transcription factors (SITF) and general transcription factors (GTF) to the accessible DNA. Induced acetylation of histones H3 and H4 (H3/H4Kac, illustrated as cyan triangles) recruits BRD4 and P-TEFb to chromatin. P-TEFb phosphorylates RNA polymerase II (Pol II) on serine 2 (S2) and allows pause-release of Pol II, resulting in elongation of mature RNA. Conversely, primary response genes follow a "digital"-like activation pattern in response to a stimulus. These genes already have relatively high levels of Pol II and the permissive histone marks, H3K4me3 (green hexagons) and H3/H4Kac, before stimulation, indicating a "poised" state that does not require chromatin remodeling. Stimulation results in TF binding and a H3/H4Kac-dependent recruitment of BRD4 and P-TEFb, allowing for the productive transcription of mature RNA. I-BET selectively prevents the transcription of genes that follow the "analog"-like, but not the "digital"-like activation pattern. This specificity suggests that "analog"-like secondary response genes are more dependent on BET protein function than "digital"-like primary response genes.

histone or DNA motifs or posttranslational modifications. Such synthetic histone or DNA mimics could represent a novel generation of drugs that target epigenetically defined gene states during inflammation and conceivably other diseases such as cancer (as discussed in Ch. 2 [Qi 2014]).

REFERENCES

*Reference is also in this book.

* Busslinger M, Tarakhovsky A. 2014. Epigenetic control of immunity. *Cold Spring Harb Perspect Biol* **6:** a019307.

 Cite as *Cold Spring Harb Perspect Biol* doi: 10.1101/cshperspect.a018671

Filippakopoulos P, Qi J, Picaud S, Shen Y, Smith WB, Fedorov O, Morse EM, Keates T, Hickman TT, Felletar I, et al. 2010. Selective inhibition of BET bromodomains. *Nature* **468:** 1067–1073.

Hargreaves DC, Horng T, Medzhitov R. 2009. Control of inducible gene expression by signal-dependent transcriptional elongation. *Cell* **138:** 129–145.

Medzhitov R, Horng T. 2009. Transcriptional control of the inflammatory response. *Nat Rev Immunol* **9:** 692–703.

Nathan C. 2002. Points of control in inflammation. *Nature* **420:** 846–852.

Nicodeme E, Jeffrey KL, Schaefer U, Beinke S, Dewell S, Chung C-W, Chandwani R, Marazzi I, Wilson P, Coste H, et al. 2010. Suppression of inflammation by a synthetic histone mimic. *Nature* **468:** 1119–1123.

* Qi J. 2014. Bromodomain and extraterminal domain inhibitors (BETi) for cancer therapy: Chemical modulation of chromatin structure. *Cold Spring Harb Perspect Biol* doi: 10.1101/cshperspect.a018663.

Smale ST. 2010. Selective transcription in response to an inflammatory stimulus. *Cell* **140:** 833–844.

Histone H3 Mutations in Pediatric Brain Tumors

Xiaoyang Liu[1], Troy A. McEachron[2], Jeremy Schwartzentruber[1], and Gang Wu[3]

[1]McGill University, Montreal, Quebec H3A 0G4, Canada; [2]Integrated Cancer Genomics Division, Translational Genomics Research Institute, Phoenix, Arizona 85004; [3]St. Jude Children's Research Hospital, Memphis, Tennessee 38105

Correspondence: tmceachron@tgen.org

Until recently, mutations in histones had not been described in any human disease. However, genome-wide sequencing of pediatric high-grade gliomas revealed somatic heterozygous mutations in the genes encoding histones H3.1 and H3.3, as well as mutations in the chromatin modifiers ATRX and DAXX. The functional significance and mechanistic details of how these mutations affect the tumors is currently under intensive investigation. The information gained from these studies will shed new light on normal brain development as well as increase our understanding of the tumorigenic processes that drive pediatric high-grade gliomas.

Histone proteins are among the most highly conserved proteins in eukaryotes, with the majority of the protein sequence being identical in organisms ranging from yeast to humans. Histones are involved in the basic packaging of DNA, allowing two meters of DNA to fit inside the nucleus of a single cell! There are four core histones: H2A, H2B, H3, and H4. DNA in the nucleus is wrapped around a histone octamer comprised of two of each of the core histones, forming a nucleosome. A string of nucleosomes is then further compacted to form chromatin. The amino-terminal tails of each of the core histones protrude from the nucleosome and receive a variety of posttranslational modifications (PTMs). Because essentially all cells in the body have the same genome but different sets of genes are expressed, it is the combination of PTMs of the histone tails—often referred to as the histone code—that largely determines the structure of the chromatin and whether genes will or will not be transcribed in each cell. This epigenetic regulation of gene expression is a key factor in cell determination and differentiation, and thus organismal development as a whole.

Until recently, there had not been any reports of histone mutations in any human disease. In January 2012, two studies simultaneously reported the first ever histone mutations in pediatric brain tumor patients. Both groups reported recurrent somatic heterozygous mutations in the gene encoding the histone variant H3.3 (i.e., *H3F3A*), in patients diagnosed with non-brainstem pediatric glioblastomas (non-BS-PGs) and diffuse intrinsic pontine gliomas (DIPGs; Schwartzentruber et al. 2012; Wu et al. 2012). One of the groups also reported heterozygous mutations in the gene encoding histone H3.1 (i.e., *HIST1H3B*) in a significant percentage of DIPGs (Wu et al. 2012). Strikingly, these mutually exclusive H3 mutations resulted in amino acid substitutions at two specific positions in the proteins: lysine-to-methionine substitutions at position 27 (K27M) in both H3.1 and H3.3 and either a glycine-to-arginine or glycine-to-valine substitution at position 34 (G34R, G34V) in H3.3. In fact, 78% of DIPG samples and 36% of non-BS-PG contained these histone mutations (Fig. 1B,C).

The histone H3.1 and H3.3 variants are structurally similar proteins that differ at only five amino acid positions. H3.1 is termed a replication-dependent histone because it is expressed and incorporated into nucleosomes during S phase of the cell cycle. Conversely, H3.3 is repli-

Cite as *Cold Spring Harb Perspect Biol* doi: 10.1101/cshperspect.a018689

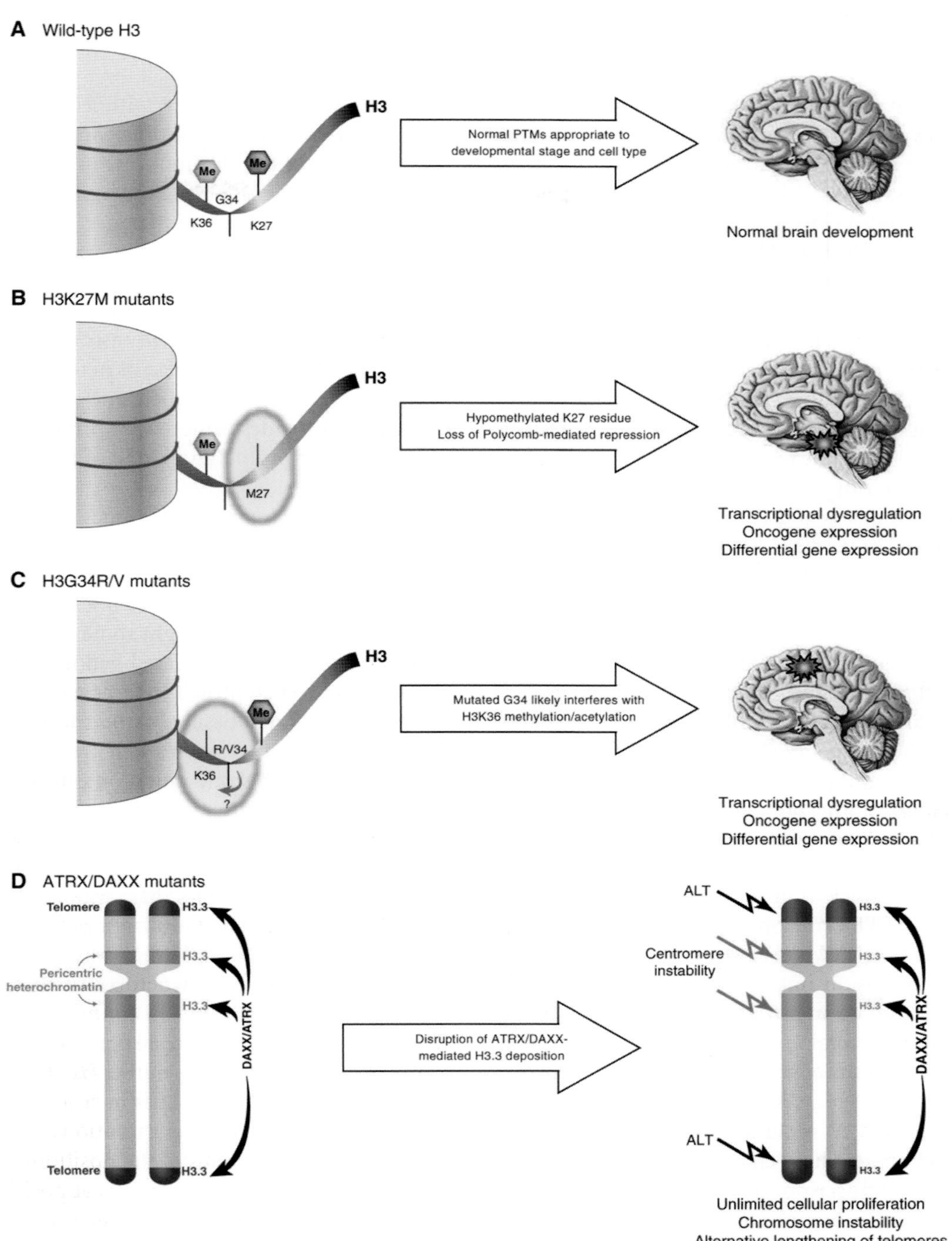

Figure 1. Potential effects of histone H3 and *ATRX/DAXX* mutations in the tumorigenic process of pediatric high-grade gliomas. (*A*) In the absence of mutations, the amino-terminal tail of histone H3 receives a myriad of cell-type- and developmental-stage-specific posttranslational modifications (PTMs) dictating the gene expression profiles required for normal brain development. (*B*) The H3K27M mutation results in a hypomethylated K27 residue (loss of red hexagon), preventing Polycomb-mediated repression of target genes. This likely results in oncogene expression and an altered gene expression profile suitable for the development of high-grade gliomas of the midline structures, most notably diffuse intrinsic pontine gliomas (blue/yellow star). (*C*) The H3G34R/V mutation diminishes the levels of H3K36 methylation (loss of yellow hexagon), which may impact transcriptional elongation in addition to generating a gene expression profile conducive to the formation of non-brainstem pediatric glioblastomas (blue/yellow star). The H3G34R/V may also impact on the acetylation of H3K36 (not shown for illustrative purposes). (*D*) Mutations in *ATRX/DAXX* alter the proper deposition of histone H3.3 at pericentric and telomeric heterochromatic loci, thus compromising chromatin structure and allowing for genomic instability and alternative lengthening of telomeres (ALT). H3K27me, red hexagon; H3K36me, yellow hexagon; high-grade glioma, blue/yellow star; telomeres, blue chromatin shading; pericentric heterochromatin, red chromatin shading.

cation-independent as it is expressed throughout the cell cycle and replaces existing nucleosomal histone H3 variants at a variety of loci along the genome (discussed in Ch. 20 [Henikoff and Smith 2014]). In the hundreds of brain tumor samples sequenced between the two studies, only residues K27 and G34 of histone H3 were affected. This begs the question: Why is there such an extreme selective pressure for mutations affecting these residues?

Lysine 27 of histone H3 (H3K27) is a critical residue that, when trimethylated (me3), is involved in transcriptional repression via Polycomb repressive complexes 1 and 2. The H3K27me3 modification regulates the expression of genes associated with lineage commitment, cellular differentiation, and anterior–posterior patterning (Faria et al. 2011; Ch. 7 [Grossniklaus and Paro 2014]). Thus, H3K27 has a role in normal brain development. Indeed, just a year after the discovery of these histone mutations, researchers are gaining some insight into the mechanistic details pertaining to the function of these mutations; namely, that the K27M mutation acts via a dominant-negative gain of function by competitively inhibiting the methyltransferase activity of EZH2 and thus abolishing Polycomb-mediated repression of numerous genes (Lewis et al. 2013).

The functional significance of the G34R/V mutation is less straightforward to interpret. Glycine 34 of histone H3 (H3G34) lies in close proximity to lysine 36 (H3K36), a residue that regulates transcriptional elongation. In fact, H3G34R/V mutant nucleosomes show reduced methylation of H3K36 by SETD2, the only human methyltransferase specific for H3K36 (Lewis et al. 2013). This suggests that the H3G34R/V mutation impacts the ability of histone-modifying complexes to methylate H3K36, thus altering the transcription of several target genes. Gene expression analyses revealed patterns of gene expression that were different in samples with the H3K27M mutation versus samples with the H3G34R/V mutation, both of which differ from the normal brain. These changes in gene expression could result in the transcription of oncogenes or microRNAs with oncogenic functions as well as prevent the expression of tumor-suppressor genes, promoting the growth of the respective tumors.

In addition to mutations in histone H3 genes, it was determined that there were frequent inactivating mutations in *ATRX* and *DAXX* in non-BS-PGs (Schwartzentruber et al. 2012). *ATRX* and *DAXX* encode chromatin-remodeling proteins responsible for the replication-independent incorporation of H3.3 at pericentric and telomeric heterochromatic loci. Inactivating mutations in these genes have indeed been previously found in pediatric and adult glioblastomas, neuroblastomas, and pancreatic neuroendocrine tumors and are among the many reported epigenetic regulators to be mutated in different cancers. Mutations in *ATRX/DAXX* may interfere with H3.3 incorporation at these heterochromatic loci, thus compromising the structural integrity of the chromosome (Fig. 1D). Interestingly, alternative lengthening of telomeres (ALT), a telomerase-independent telomere maintenance mechanism, was also frequently observed in tumors with *ATRX* mutations (Lovejoy et al. 2012; Schwartzentruber et al. 2012). Although it is not entirely clear how cancer-associated ALT operates, the genomic instability associated with mutations in *ATRX/DAXX* somehow results in telomeric dysfunction, allowing for ALT. This aberrant telomere lengthening mechanism provides tumor cells with the capacity for unlimited cellular proliferation, one of the hallmarks of cancer cells.

As there was a staggering frequency of histone H3 mutations, why were these mutations not found in earlier DNA sequencing studies? This is due in part to the new sequencing techniques now available. Rather than selecting and sequencing genes thought to be important, these two groups used different unbiased genome-wide sequencing methods to cover all protein-coding genes in samples from diseased and normal tissue. Furthermore, the clinical samples used in these two studies were from pediatric patients, not adults. These histone mutations are almost exclusively found in pediatric patients. This is of major significance as the human brain continues to develop postnatally. This then prompts the question: What is unique about the developing brain that enables these mutations to be tumorigenic in children and not adults? Surprisingly, the majority of tumors with H3K27M mutations were found in the thalamus or brainstem, structures in the midline of the brain. In contrast, H3G34R/V tumors arose in the cortex. The development of the normal brain is a very dynamic and complex process, involving numerous extracellular factors that are present at precise times in specific brain locations. It is possible that the different microenvironmental factors in these particular regions of the developing brain work in concert with the altered transcriptional profiles induced by the H3K27M and H3G34R/V mutations. Together, this may contribute to an increased transformation potential, resulting in the formation of these specific types of pediatric high-grade gliomas.

Often, the study of disease conditions leads to a deeper understanding of normal biological function. It is clear that more in-depth studies are needed to discern the functional roles of these mutations pertaining to tumor biology and normal brain development. The information and insight that will be gained from such studies will potentially provide clinically relevant diagnostic, prognostic, and/or therapeutic benefits for pediatric patients with this disease, as well as increasing our overall knowledge of epigenetics.

Cite as *Cold Spring Harb Perspect Biol* doi: 10.1101/cshperspect.a018689

REFERENCES

**Reference is also in this book.*

Faria CMC, Rutka JT, Smith C, Kongkham P. 2011. Epigenetic mechanisms regulating neural development and pediatric brain tumor formation. *J Neurosurg Pediatr* **8:** 119–132.

* Grossniklaus U, Paro R. 2014. Transcriptional silencing by *Polycomb*-group proteins. *Cold Spring Harb Perspect Biol* **6:** a019331.

* Henikoff S, Smith MM. 2014. Histone variants and epigenetics. *Cold Spring Harb Perspect Biol* doi: 10.1101/cshperspect.a019364.

Lewis PW, Müller MM, Koletsky MS, Cordero F, Lin S, Banaszynski LA, Garcia BA, Muir TW, Becher OJ, Allis CD. 2013. Inhibition of PRC2 activity by a gain-of-function H3 mutation found in pediatric glioblastoma. *Science* **340:** 857–861.

Lovejoy CA, Li W, Reisenweber S, Thongthip S, Bruno J, De Lange T, De S, Petrini JHJ, Sung PA, Jasin M, et al. 2012. Loss of ATRX, genome instability, and an altered DNA damage response are hallmarks of the alternative lengthening of telomeres pathway. *PLoS Genet* **8:** e1002772.

Schwartzentruber J, Korshunov A, Liu X-Y, Jones DTW, Pfaff E, Jacob K, Sturm D, Fontebasso AM, Quang D-AK, Tönjes M, et al. 2012. Driver mutations in histone H3.3 and chromatin remodelling genes in paediatric glioblastoma. *Nature* **482:** 226–231.

Wu G, Broniscer A, McEachron TA, Lu C, Paugh BS, Becksfort J, Qu C, Ding L, Huether R, Parker M, et al. 2012. Somatic histone H3 alterations in pediatric diffuse intrinsic pontine gliomas and non-brainstem glioblastomas. *Nat Genet* **44:** 251–253.

Chromosome Folding: Driver or Passenger of Epigenetic State?

Tom Sexton[1] and Eitan Yaffe[2]

[1]Institute of Human Genetics, Centre National de la Recherche Scientifique, 34396 Montpellier Cedex 5, France; [2]Department of Computer Science and Applied Mathematics and Department of Biological Regulation, Weizmann Institute of Science, Rehovot 76100, Israel

Correspondence: thomas.sexton@igh.cnrs.fr

Despite a growing understanding of how epigenetic marks such as histone modifications locally modify the activity of the chromatin with which they are associated, we know little about how marked regions on different parts of the genome are able to intercommunicate to effect regulation of gene expression programs. Recent advances in methods that systematically map pairwise chromatin interactions have uncovered important principles of chromosome folding, which are tightly linked to the epigenetic mark profiles and, hence, functional state of the underlying chromatin fiber.

Genes do not act independently of each other or of the genomic context in which they are present. As well as protection from the influences of surrounding repressive chromatin, many genes require control from long-range regulatory elements that may reside at megabase-scale distances in or close to unrelated genes for their appropriate expression. Furthermore, coordinately regulated genes from different chromosomes can come together at specific sites within the nucleus allowing them to share the same regulatory factors enriched at these foci. Despite our ever-growing comprehension of how epigenetic features such as histone modifications and DNA methylation determine the functional state of the chromatin they mark, such a "one-dimensional" view of the epigenome is unable to explain complex long-range control of gene expression programs. We also need to characterize the three-dimensional folding of the chromatin fiber to understand how epigenetic marks on both genes and their long-range regulatory elements intercommunicate.

Initially, chromatin interactions could only be studied in a low-throughput manner at a limited spatial resolution by light microscopy, but the field was revolutionized by the development of the 3C (chromosome conformation capture) method by Job Dekker in the Kleckner laboratory (Dekker et al. 2002). This molecular biology approach entails the fixation of chromatin in its native state, followed by restriction digestion, and religation. This generates hybrid DNA molecules from restriction fragments that may reside on different genomic locations, but are physically proximal at the time of fixation (see Fig. 5 of Ch. 19 [Dekker and Misteli 2014]). 3C allows the assessment of chromatin interactions at the resolution of individual restriction fragments and has been used to show chromatin loops bringing genes and their distal regulatory elements into direct physical proximity. Importantly, these loops tend to be specific to tissues in which the distal elements exert control, suggesting that they are functional and not just a consequence of folding the genome into a small nuclear volume. However, these studies were still anecdotal and gave little information on the global interrelationship of chromosome folding and gene expression control.

The next major advance came from using high-throughput sequencing to simultaneously detect all the pairwise chromatin interactions uncovered by one 3C experiment. This "Hi-C" method was first developed in the Dekker laboratory (Lieberman-Aiden et al. 2009), but has subsequently been performed at much higher sequencing depths to derive chromatin interaction maps of *Drosophila* (Sexton et al. 2012) and mammalian (Dixon et al. 2012)

Cite as *Cold Spring Harb Perspect Biol* doi: 10.1101/cshperspect.a018721

nuclei at very high resolution. These detailed maps uncovered conserved features of chromosome folding in metazoans, which are more tightly linked to the epigenomic features of the underlying chromatin fiber than previously appreciated.

A surprising and important finding was the conserved organization of the genome into distinctly folded modules or "topological domains"; represented as squares on the interaction map (see Fig. 1), interactions between genomic regions within a domain are very strong, but sharply depleted when crossing domain borders. Strikingly, nearly all epigenetic markers of chromatin activity (histone modifications, protein binding profiles, transcriptional output, DNase sensitivity, replication timing, etc.) correlated extremely well with topological domain organization—crossing the border from one domain to another is accompanied by a very sharp increase or decrease in the mark. In mammals, the domains are also associated with coordinately regulated genes, suggesting that the genome is functionally organized into distinct units, which are physically borne out as discretely folded chromatin modules (Fig. 1). This model implies that domains are physically segregated, but

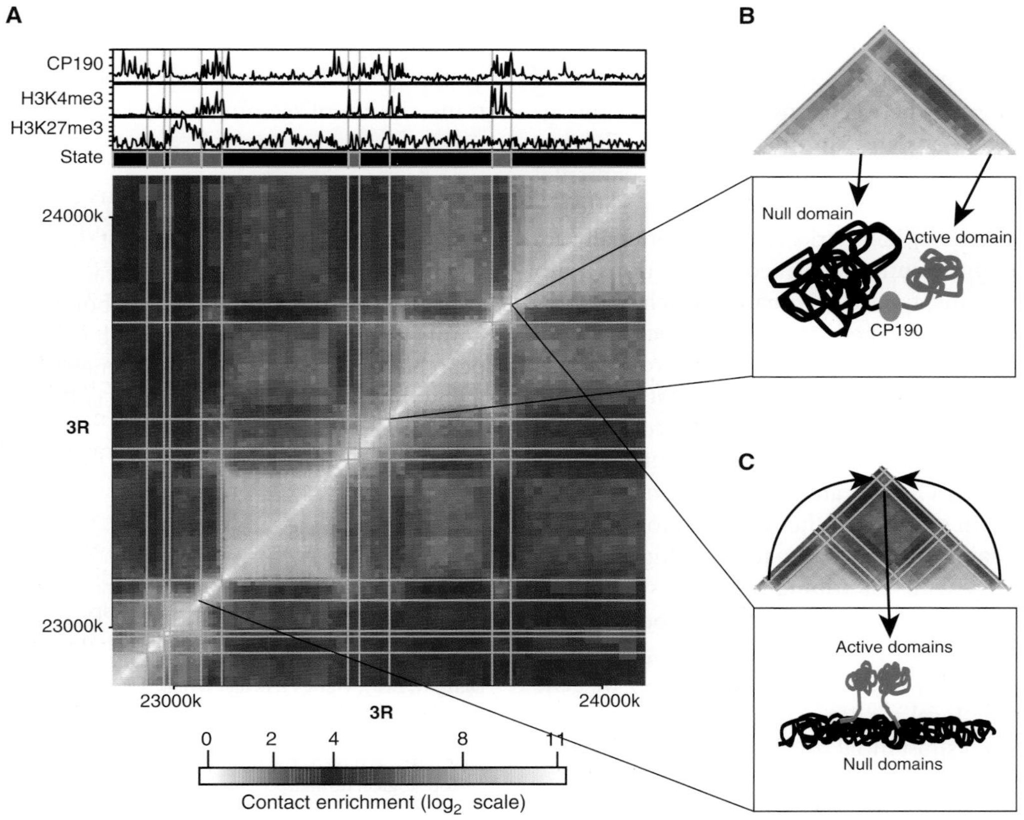

Figure 1. Spatial organization of the epigenome. (*A*) Chromatin interaction map for an ~1 Mb region of the genome in *Drosophila* embryos (data from the laboratories of Giacomo Cavalli and Amos Tanay). The interaction strength between two specific genomic regions is given by the heat map corresponding to their coordinates on the *x*- and *y*-axes. The presence of distinct topological domains is indicated by the patterns of squares on the heat map diagonal, with steep decreases in interaction strength when passing beyond the domain. The topological domain borders are denoted by gray lines. The linear profiles for binding of the insulator protein, CP190, and the histone modifications, H3K4me3 and H3K27me3, are shown above the map, along with a color-coding for the epigenetic "state" of each topological domain ("null" domains, devoid of known epigenetic marks are indicated in black; "active" domains, marked by H3K4me3, are indicated in green; domains repressed by the mark H3K27me3 are indicated in red). Key functional organization principles of the genome are shown in the insets. (*B*) A subset of the interaction map, rotated by 45°, is shown to highlight the topological domain organization of the genome. This inset shows a large null domain, physically segregated from a small active domain, with the border between them containing a binding site for the insulator protein CP190 (shown schematically below). (*C*) A specific interdomain interaction between two small active domains, kept separate from the surrounding repressive chromatin, is highlighted in a subset of the interaction map and schematically shown below the map.

also that this segregation prevents functional communication across domain borders. In support of this, the well-characterized regulatory elements acting over megabase distances have all been found to be present in the same topological domains as their target genes. Moreover, domain borders are enriched in the binding of insulator proteins (such as CTCF [CCCTC-binding factor] in mammals, or CP190 in *Drosophila*), which are factors that have been genetically characterized to shield genes from the influence of surrounding chromatin. It should be noted, however, that many insulator binding sites do not define topological domain borders; the genetic and epigenetic features distinguishing these sites from true domain borders will be an interesting topic of research in the near future.

In addition to a tight link between local folding of chromatin and its underlying epigenetic state, there is also a striking interplay between long-range, intertopological domain interactions and chromatin activity. The first Hi-C study in human cells identified two fundamental types of chromatin based on their interaction patterns: Type A, characterized by hallmarks of transcriptionally active chromatin, preferentially formed interactions with other type-A domains; type-B chromatin, which was transcriptionally repressed, formed strongest interactions with other type-B domains, and there were greatly reduced interactions between type-A and type-B chromatin. Such a model is consistent with observed clustering of coexpressed genes in shared transcriptional foci within the nucleus. The higher-resolution Hi-C maps are also consistent with the two-state chromatin model, but they extend it to show that topological domains form a basic unit of higher-order chromatin folding, with interdomain contacts predominantly occurring between active domains or between inactive domains, with little intermixing. Thus, although local epigenetic marks, long considered a proxy of gene activity, reflect the local topology of the chromatin fiber, they may also play a substantial role in global chromosome folding.

Thanks to these initial maps of genome configurations, the intimate link between genome structure, epigenetic state, and functional output is now clear, although the question of whether structure defines function or vice versa remains unresolved. Recent elegant experiments have provided evidence for chromatin loops as a cause of transcriptional activation rather than a consequence (Deng et al. 2012). Similar experiments may also be expected to assess the causal role of topological domains in regulating genome function. With ever-decreasing sequencing costs, it is reasonable to predict that more chromatin contact maps will be produced at increasing resolution and in a larger repertoire of cell types and experimental conditions. As a result, epigenomic profiling could become three-dimensional. Regions of histone modifications may no longer be viewed as tracks on conventional genome browsers, but could instead be viewed in the more physiological context of their nuclear neighbors. We anticipate that this will allow a much greater understanding of how entire gene expression programs are regulated—for example, allowing better in silico predictions of how mutations (e.g., previously uncharacterized disease-linked single-nucleotide polymorphisms) or transgenic insertions (e.g., during gene therapy) affect transcription of seemingly unrelated genes.

ACKNOWLEDGMENTS

We apologize to the investigators of the many works that we were unable to cite because of space constraints. The research of the investigators' host laboratories are funded by grants from the European Research Council and the European Network of Excellence EpiGeneSys. T.S. is funded by a fellowship by the Fondation pour la Recherche Médicale.

REFERENCES

*Reference is also in this book.

* Dekker J, Misteli T. 2014. Long-range chromatin interactions. *Cold Spring Harb Perspect Biol* doi: 10.1101/cshperspect.a019356.

Dekker J, Rippe K, Dekker M, Kleckner N. 2002. Capturing chromosome conformation. *Science* **295:** 1306–1311.

Deng W, Lee J, Wang H, Miller J, Reik A, Gregory PD, Dean A, Blobel GA. 2012. Controlling long-range genomic interactions at a native locus by targeted tethering of a looping factor. *Cell* **149:** 1233–1244.

Dixon JR, Selveraj S, Yue F, Kim A, Li Y, Shen Y, Hu M, Liu JS, Ren B. 2012. Topological domains in mammalian genomes identified by analysis of chromatin interactions. *Nature* **485:** 376–380.

Lieberman-Aiden E, van Berkum NL, Williams L, Imakaev M, Ragoczy T, Telling A, Amit I, Lajoie BR, Sabo PJ, Dorschner MO, et al. 2009. Comprehensive mapping of long-range interactions reveals folding principles of the human genome. *Science* **326:** 289–293.

Sexton T, Yaffe E, Kenigsberg E, Bantignies F, Leblanc B, Hoichman M, Parrinello H, Tanay A, Cavalli G. 2012. Three-dimensional folding and functional organization principles of the *Drosophila* genome. *Cell* **148:** 458–472.

CHAPTER 3

Overview and Concepts

C. David Allis[1], Marie-Laure Caparros[2], Thomas Jenuwein[3], Monika Lachner[3], and Danny Reinberg[4]

[1]The Rockefeller University, New York, New York 10065; [2]Halford Road, London SW6 1JZ, United Kingdom; [3]Max-Planck Institute of Immunobiology and Epigenetics, Freiburg 79108, Germany; [4]NYU Langone Medical Center, New York, New York 10016

Correspondence: alliscd@rockefeller.edu; danny.reinberg@nyumc.org; jenuwein@ie-freiburg.mpg.de

SUMMARY

Epigenetic traits are defined as stably heritable phenotypes, resulting from changes in chromatin without alterations in the DNA sequence. This overview discusses the fundamental concepts and general principles that explain how these epigenetic phenomena occur. Important aspects of modern epigenetic research, which include model organisms and molecular paradigms, basic mechanisms of epigenetic control, epigenetic response to the environment, epigenetic contribution to complex human diseases, and epigenetic inheritance, are discussed. This synthesis of the field of epigenetics not only provides a comprehensive summary of the current knowledge, but also exposes exciting and pressing questions to be addressed in the future.

Outline

OVERVIEW

Remarkably, nearly 15 years have passed since the first draft of our genetic blueprint became available in 2001—the tour de force culmination of the Human Genome Project (HGP). In many ways, the logic underlying the efforts of the HGP was in response to several widely held assumptions in biomedicine—genes determine disease and aging, and genetic analyses provide diagnosis and therapy. With more than a decade passing, we can ask: Is this the whole story? Why are genes alone not enough?

Despite providing the scientific community with a wealth of key information on an individual gene and whole-genome (genomics) basis, including the promise of "personalized medicine" as DNA sequencing costs decrease, important questions remain unanswered from the HGP. Why do so few genes, around 21,000 in humans, make us appear, by gene count alone, not much more complex than worms, flies, and fish? Where does individual variation come from, and can past experiences or our environment shape who we are, or skip generations, shaping our children's or grandchildren's lives? How can a cell remember what it is supposed to do for one, a few, or many generations? Are there diseases that cannot be explained by knowing our genetic "blueprints," based on the well-accepted rules of Mendelian inheritance that many of us learned in school? Although mistakes made in the DNA template, often referred to as "mutations," can clearly lead to the generation of disease states, such as cancer, might there also be groups of seemingly "classic mutations" that also affect other layers of regulatory information that are not based on changes in DNA sequence alone? What is the "missing heredity" underlying some human pathologies?

In humans, the genetic information (DNA) is organized into 23 chromosome pairs, containing approximately 21,000 genes. The DNA sequence of our genome is composed of 3.2×10^9 bases, abbreviated by four basic letters (or bases) A, C, G, and T within its sequence. If "genetics" is equal to words, "epigenetics" instructs how the words are read. Alternatively, if groups of genes can be thought of as computer hardware, epigenetic control can be compared with computer software. Understanding this additional regulatory layer is an extraordinary challenge. But understanding this "language" promises to reveal insights into how cellular and organismal events are coordinated during normal and abnormal development, leading to new therapies for complex human disorders in which mutations alone are not sufficient to explain the disease.

When summed across all chromosomes, the DNA molecule in higher eukaryotes would measure ~2-m long and, therefore, needs to be maximally condensed ~10,000-fold to fit into a nucleus, the compartment of the cell that stores our genetic material. The wrapping of DNA around "spools" of proteins, so-called histone proteins, provides an elegant solution to this packaging problem, giving rise to a repeating protein–DNA polymer known as "chromatin." However, in packaging DNA to better fit into a confined space, a problem develops, much like when one packs too many books onto library shelves: it becomes harder to find and read the book of choice. Thus, an indexing system is needed. Chromatin, as a genome-organizing platform, provides this indexing. Chromatin is not uniform in structure; it comes in different packaging designs from a highly condensed chromatin fiber (known as heterochromatin) to a less compacted type in which genes are typically transcribed (known as euchromatin). Variation can be introduced into the basic chromatin polymer through the introduction of unusual histone proteins (known as histone variants), chromatin-remodeling complexes that alter chromatin structures, and by the addition of chemical "flags" to the histones proteins themselves (known as covalent modifications). Moreover, addition of a methyl group directly to a cytosine (C) base in the DNA template (known as DNA methylation or subtle variations thereof, such as hydroxy-methylation) can act as docking sites for proteins, to alter the chromatin state or affect the covalent modification of resident histones. Recent evidence suggests that noncoding RNAs (ncRNAs) can "guide" specialized regions of the genome into more compacted or open chromatin states. Thus, chromatin should be viewed as a dynamic polymer that can index the genome and potentiate signals from the environment, ultimately determining which genes are expressed and which are not.

Together, these regulatory options provide chromatin with an organizing principle for genomes known as "epigenetics," the subject of this book. Epigenetic indexing patterns must be inherited through cell divisions, providing cellular "memory" that may extend the heritable information potential of the DNA. Epigenetics can, thus, be more narrowly defined as transmittable changes in gene "off–on" states through modulation of chromatin, which is not brought about by changes in the DNA sequence.

1 GENETICS VERSUS EPIGENETICS

In this overview, we attempt to explain the basic concepts of chromatin and epigenetics and discuss how epigenetic control may give us the clues to solve some long-standing mysteries, such as cellular identity, stem cell plasticity, regeneration, memory formation, aging, and to identify novel therapies for a better treatment of many forms of human disease. As readers comb through the chapters that follow, we encourage them to note the wide range of biological phenomena (see Fig. 1) uncovered in a diverse range of experimental models that seem to have an epigenetic (non-DNA alone) basis.

Determining that DNA was the "transforming" molecule and unraveling the structural details of the DNA double helix stand together as some of the landmark discoveries in all of biology. Without question, DNA is the primary macromolecule that stores genetic information and propagates this stored information from mother cell to daughter cell, and to the next generation through the germline.

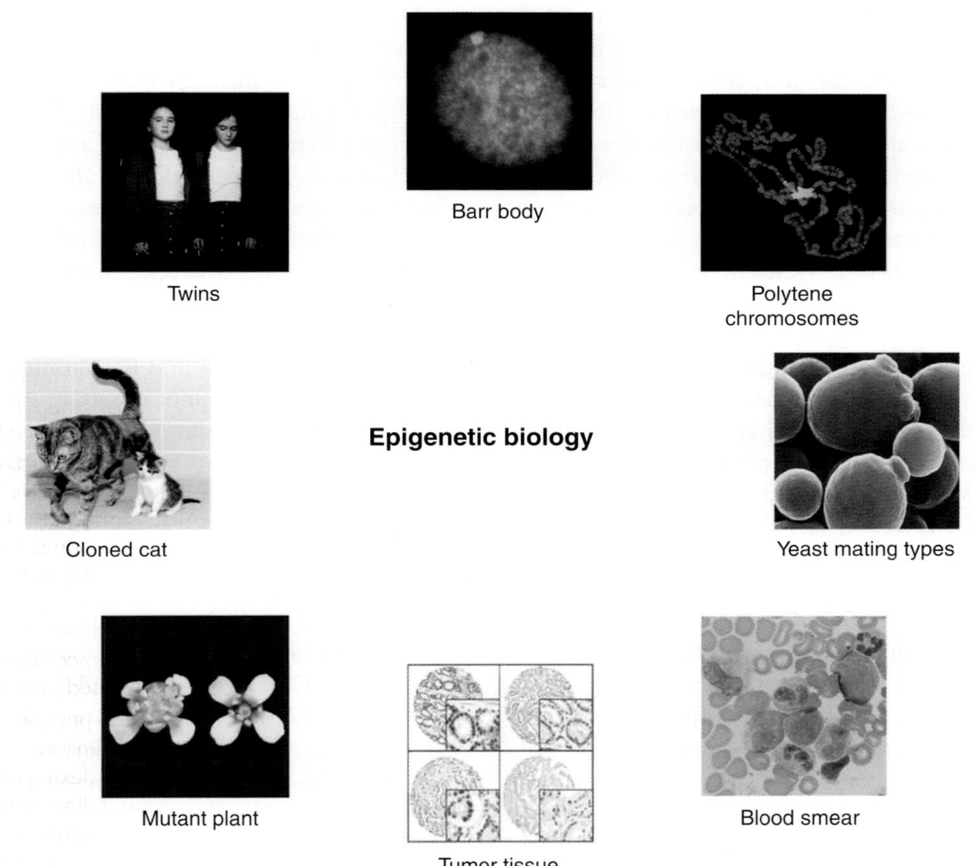

Figure 1. Biological examples of epigenetic phenotypes. Epigenetic phenotypes in a range of organisms and cell types are all attributable to "nongenetic" differences. Twins: Slight variations partially attributable to epigenetics. Barr body: The epigenetically silenced X chromosome in female mammalian cells, visible cytologically as condensed heterochromatin. Polytene chromosomes: Giant chromosomes in *Drosophila* salivary glands, ideally suited for correlating genes with epigenetic marks. Yeast mating type: Sex is determined by the active mating type (MAT) locus, whereas copies of both mating-type genes **a** and α are epigenetically silenced. Blood smear: Heterogeneous cells of the same genotype, but epigenetically determined to serve different functions. Tumor tissue: Metastatic cells (*left*) showing elevated levels of epigenetic marks in the tissue section. Mutant plant: *Arabidopsis* flower epiphenotypes, genetically identical, with epigenetically caused mutations. Cloned cat: Genetically identical, but with varying coat color phenotype. (Twins, © Randy Harris, New York; polytene chromosomes, reprinted from Schotta et al. 2003, © Springer; yeast mating type, © Alan Wheals, University of Bath; blood smear, courtesy Prof. Christian Sillaber; tumor tissue, reprinted, with permission, from Seligson et al. 2005, © Macmillan; mutant plant, reprinted, with permission, from Jackson et al. 2002, © Macmillan; cloned cat, reprinted, with permission, from Shin et al. 2002, © Macmillan.)

From this and other findings, the "central dogma" of modern biology emerged, including a clear paradigm for how gene units and groups of genes (operons) are regulated. This central view encapsulates the processes involved in maintaining and translating the genetic template: (1) self-propagation of DNA by semiconservative replication; (2) transcription in a unidirectional $5'-3'$ direction, templated by the DNA, to generate intermediary messenger RNAs (mRNAs); and (3) translation of mRNA to produce polypeptides consisting of amino- to carboxyl-terminal strings of amino acids that are colinear with the $5'-3'$ order of DNA. In simple terms, DNA $\leftrightarrow$ RNA $\rightarrow$ protein. The central dogma accommodates feedback from RNA to DNA by the process of reverse transcription, followed by integration into existing DNA (as shown by retroviruses and retrotransposons). However, this belief disavows feedback from protein to DNA, although a new twist to the genetic dogma is that rare proteins, known as prions, can be inherited in the absence of a DNA or RNA template. Thus, these specialized self-aggregating proteins have properties that resemble some properties of DNA itself, including a mechanism for replication and information storage (Aguzzi and Falsig 2012).

As well, emerging evidence suggests that a remarkably large fraction of our genome is transcribed into "noncoding" RNAs, so much so that some have stressed that RNA plays a central role as "life's indispensible molecule" (Darnell 2011; Mattick 2011). The function of these long and short noncoding RNAs (ncRNAs) (i.e., nonprotein encoding) is under active investigation, but RNA interference (RNAi) or RNAi-related pathways are now being linked to a remarkably broad number of key biological processes, including gene regulation, genome surveillance, and epigenetic "memory," which may even contribute to transgenerational effects that are potentially influenced by past events, such as environmental conditions, viral immunity, and longevity. The general process of RNAi stands as one of the major milestones in biology—one that impacts the core of epigenetic regulation. Like DNA methylation, which generally fits better into the category of being a "true carrier" of epigenetic information, RNA also provides a clear base-pairing mechanism for targeting epigenetic machinery that, in turn, can bring about stable and heritable silencing, although full details remain unclear.

The origin of epigenetics stems from long-standing studies of developmental biologists seeking to explain seemingly anomalous (i.e., non-Mendelian) and disparate patterns of inheritance in many organisms (see Ch. 1 [Felsenfeld 2014] for a historical overview; Carey 2012). Classical Mendelian inheritance of phenotypic traits (e.g., pea color, number of digits, or hemoglobin insufficiency) results from allelic differences caused by mutations of the DNA sequence. Collectively, mutations underlie the definition of phenotypic traits, which contributes to the determination of species boundaries. These boundaries are then shaped by the time-requiring pressures of natural selection, as explained by Darwin's theory of evolution. Such concepts place mutations and natural selection at the heart of classical genetics. In contrast, non-Mendelian inheritance (e.g., variation of embryonic growth, mosaic skin coloring, random X inactivation, plant paramutation) (see Fig. 1) can manifest, to take one example, through the expression of only one of two alleles within the same nuclear environment. Importantly, in these circumstances, the DNA sequence is not altered. This is distinct from another commonly referred to non-Mendelian inheritance pattern that arises from the maternal inheritance of mitochondria (Wallace and Chalkia 2013).

One challenge for epigenetic research is captured by the selective regulation of one allele within a nucleus. What then distinguishes two identical alleles, and how is this distinction mechanistically established and maintained through successive cell generations? These are questions addressed in the X inactivation chapter (Ch. 25 [Brockdorff and Turner 2014]), the genomic imprinting chapter (Ch. 26 [Barlow and Bartolomei 2014]), monoallelic expression in lymphocytes (Ch. 29 [Busslinger and Tarakhovsky 2014]), and olfactory receptor expression (Ch. 32 [Lomvardas and Maniatis 2014]). Also, what underlies differences observed in monozygotic ("identical") twins that make them not totally identical? Epigenetics is sometimes cited as one explanation for the differences in outward traits because it translates the influence of the environment, diet, and potentially other external sources to the expression of the genome (see Ch. 34 [Baylin and Jones 2014]; Ch. 30 [Berger and Sassone-Corsi 2014]; Ch. 33 [Zoghbi and Beaudet 2014]). Determining what components are affected at a molecular level, and how alterations in these components impact human biology and human disease, are major challenges for future studies.

Another related, key question is how important is the contribution of epigenetic information for normal development? And how do normal pathways become dysfunctional, leading to abnormal development and neoplastic transformation (i.e., cancer)? Because "identical" twins share the same DNA sequence, their phenotypic identity is often used to underscore the defining power of genetics. However, even identical twins can show outward phenotypic differences, likely imparted by epigenetic modifications that occur over the lifetime of the individuals (Fraga et al. 2005; Bell and Saffery 2012). Thus, the extent to which epigenetics is important in defining cell fate, identity, and phenotype remains to be fully understood. In the case of tissue regeneration and aging, it remains unclear whether these processes are dictated by alterations in the genetic program of cells or epigenetic modifications. The intensity of research on a global scale testifies to the recognition that the field of epigenetics is a critical new frontier in this "postgenomic" era.

2 MODEL SYSTEMS FOR THE STUDY OF EPIGENETICS

The study of epigenetics necessarily requires good experimental models, and as often is the case, these models seem at first sight far removed from studies using human or mammalian cells. The historical overview (Ch. 1 [Felsenfeld 2014]) makes reference to several important landmark discoveries that have emerged from early cytology, the growth of genetics, the birth of molecular biology, and relatively new advances in chromatin-mediated gene regulation. Many different model organisms have ultimately been pivotal in addressing and solving the various questions raised by epigenetic research (Fig. 2). Indeed, seemingly disparate epigenetic discoveries made in various model organisms have served to unite the research community by providing key insights into a highly conserved pathway or mechanism that would not have been easily uncovered using humans. Thus, in a very real way, the field of epigenetics owes much of its current glow to its rich history deeply seated in odd phenomena being pursued by talented investigators who saw the promise of unique biology presented in sometimes "offbeat" critters. As with the first edition, this section serves to highlight some of these major findings, which are discussed in more detail in the following chapters. As readers note these discoveries, they should focus on the fundamental principles that investigations using these model systems have exposed; their collective contributions point, more often, to common concepts than to diverging details.

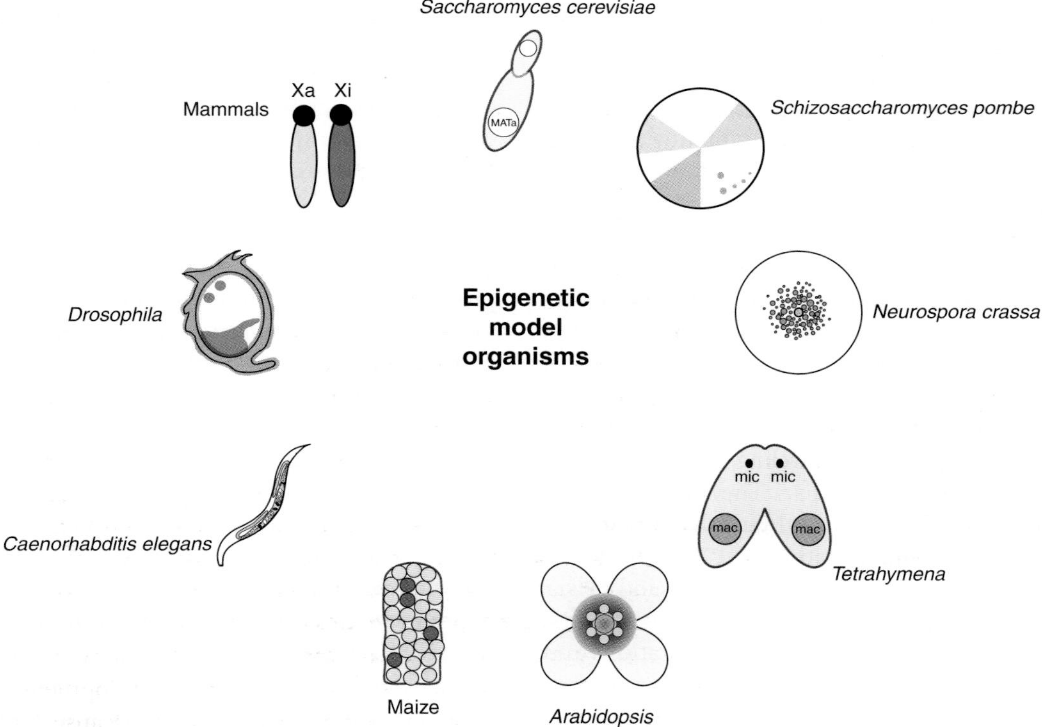

Figure 2. Model organisms used in epigenetic research. Schematic representation of model organisms used in epigenetic research. *Saccharomyces cerevisiae*: Mating-type switching to study epigenetic chromatin control. *Schizosaccharomyces pombe*: Variegated gene silencing manifests as colony sectoring. *Neurospora crassa*: Epigenetic genome defense systems include repeat-induced point mutation (RIP), quelling, and meiotic silencing of unpaired DNA, revealing interplay between RNAi pathways, DNA, and histone methylation. *Tetrahymena*: Chromatin in somatic and germline nuclei is distinguished by epigenetically regulated mechanisms. *Arabidopsis*: Model for repression by DNA, histone, and RNA-guided silencing mechanisms. Maize: Model for imprinting, paramutation, and transposon-induced gene silencing. *Caenorhabditis elegans*: Epigenetic regulation in the germline. *Drosophila*: Position effect variegation (PEV) manifest by clonal patches of expression and silencing of the white gene in the eye. Mammals: X-chromosome inactivation.

Unicellular eukaryotic organisms, *Saccharomyces cerevisiae*, *Schizosaccharomyces pombe*, and *Neurospora crassa*, allow for powerful genetic analyses, in part, facilitated by relatively small genome sizes and short haploid stages in their life cycles. Mating-type (MAT) switching that occurs in *S. cerevisiae* (Ch. 8 [Grunstein and Gasser 2013]) and *S. pombe* (Ch. 9 [Allshire and Ekwall 2014]) have provided remarkably instructive examples, demonstrating the importance of chromatin-mediated gene control. In the budding yeast, *S. cerevisiae*, the unique silent information regulator (SIR) proteins were shown to engage specific modified histones (see Sec. 12 and Ch. 8 [Grunstein and Gasser 2013]). This was preceded by elegant experiments using classical genetic approaches to document the active participation of histone proteins in gene regulation (Clark-Adams et al. 1988; Kayne et al. 1988). In the fission yeast, *S. pombe*, the patterns of histone modifications operating as activating and repressing signals are remarkably similar to those in metazoan organisms (see table on inside front cover of this book). In turn, genetic approaches in these organisms opened the door for powerful screens to look for gene products that suppress or enhance the silencing of genes. These studies led to a remarkably large number of genes that encode for universal eukaryotic epigenetic regulators. For example, mechanistic insights linking the RNAi machinery to the induction of histone modifications acting to repress gene expression was discovered in fission yeast (Hall et al. 2002; Volpe et al. 2002). The RNAi machinery was also implicated in gene silencing in the plant *Arabidopsis thaliana*, underscoring the potential importance of this regulation in a wide range of organisms (covered in Ch. 9 [Allshire and Ekwall 2014] and Ch. 16 [Martienssen and Moazed 2014]).

Other "offbeat" organisms have also made disproportionate contributions toward unraveling epigenetic pathways that, at first, seemed peculiar. The fungal species, *N. crassa*, revealed the unusual non-Mendelian phenomenon of repeat-induced point mutation (RIP) as a model for studying epigenetic control (Ch. 10 [Aramayo and Selker 2013]). Later, this organism was used to show the first functional connection between histone modifications and DNA methylation (Tamaru and Selker 2001), a finding later extended to plants (Jackson et al. 2002). Ciliated protozoa, such as *Tetrahymena* and *Paramecium*, although commonly used in biology laboratories as convenient microscopy specimens, facilitated important epigenetic discoveries owing to their unique nuclear dimorphism, as well as their remarkable set of programmed DNA elimination events that take place using some of the same themes and machinery described for RNAi-mediated gene silencing (Ch. 11 [Chalker et al. 2013]). Each ciliated cell carries two nuclei: a somatic macronucleus, which is transcriptionally active, and a germline micronucleus, which is transcriptionally inactive. Using macronuclei as an enriched starting source of "active" chromatin, the biochemical purification of the first nuclear histone-modifying enzyme, a histone acetyltransferase or HAT, was attained (Brownell et al. 1996). Ciliates' peculiar phenomenon of programmed DNA elimination during their sexual life cycle, triggered by small ncRNAs and histone modifications, may be considered the ultimate form of gene silencing as, here, varying portions of the germline genome itself are completely eliminated during a discrete window of the sexual pathway (Ch. 11 [Chalker et al. 2013]).

In multicellular organisms, genome size and organismal complexity generally increase from invertebrate (e.g., *Caenorhabditis elegans*, *Drosophila melanogaster*) or plant species (e.g., *Arabidopsis*) to vertebrate and to some, "more relevant" organisms (e.g., mammals). Plants, the masters of epigenetic control, have been providing a particularly rich source of epigenetic discoveries (Ch. 31 [Baulcombe and Dean 2014]; Ch. 13 [Pikaard and Mittelsten Scheid 2014]), ranging from transposable elements and paramutation (McClintock 1951) to the first description of ncRNAs involved in transcriptional silencing (Ratcliff et al. 1997). Crucial links between DNA methylation, histone modification, and components of the RNAi machinery came through plant studies. The discovery of plant epialleles with comic names, such as SUPERMAN and KRYPTONITE (e.g., Jackson et al. 2002), and several vernalizing genes (Bastow et al. 2004; Sung and Amasino 2004) have further provided the field with insight into understanding the developmental role of epigenetics and cellular memory. Plant meristem cells have also offered the opportunity to study crucial questions, such as somatic regeneration and stem cell plasticity (see Ch. 17 [Grossniklaus and Paro 2014]; Ch. 13 [Pikaard and Mittelsten Scheid 2014]).

With respect to the study of animal development, *Drosophila* has been an early and continuous genetic powerhouse. Based on the pioneering work of Muller (1930), many developmental mutations were generated, including the homeotic transformations and position effect variegation (PEV) mutants (see also Ch. 12 [Elgin and Reuter 2013]). The homeotic transformation mutants led to the idea that there could be regulatory mechanisms for establishing and maintaining cellular identity/memory, which was later shown to be regulated by the Polycomb and Trithorax family of genes (see Ch. 17 [Grossniklaus and Paro 2014] and Ch. 18 [Kingston and Tamkun 2014], respectively). For PEV, gene activity is dictated by the surrounding chromatin structure and not by primary DNA sequence. This system has been a particularly informative source for dissecting factors involved in epigenetic control (Ch. 12 [Elgin and Reuter 2013]). More than 100 suppressors of variegation (*Su(var)*) genes are predicted from the PEV screens and most of those are likely to encode components of heterochromatin. Without the foundation established by these landmark studies, the discovery of the first histone lysine methyltransferase, SUV39H1 (KMT1A) (Rea et al. 2000), and the resultant advances in histone lysine methylation would not have been possible. As is often the case in biology, comparable screens have been performed in fission yeast and plants, identifying silencing mutants with functional conservation with the *Drosophila Su(var)* genes (described in Ch. 9 [Allshire and Ekwall 2014] and Ch. 13 [Pikaard and Mittelsten Scheid 2014], respectively). From these studies, similar mammalian genes were identified and found to play crucial roles during development, cellular identity, and when mutated induce cellular transformation (see Ch. 14 [Blewitt and Whitelaw 2013]).

The use of reverse genetics via RNAi libraries in the nematode worm *C. elegans* has contributed to our understanding of epigenetic regulation in metazoan development. There, comprehensive cell-fate tracking studies, detailing all the developmental pathways of each cell, have highlighted the fact that Polycomb and Trithorax systems probably arose with the emergence of multicellularity. In particular, these mechanisms of epigenetic control are essential for gene regulation in the germline (see Ch. 23 [Strome et al. 2014]). More recently, *C. elegans* has emerged as a front-runner for the analyses of several related pathways that regulate gene function in the soma and germline, particularly, for double-stranded RNA (dsRNA)-mediated silencing (Fire et al. 1998) and epigenetic inheritance (Johnson and Spence 2011; Buckley et al. 2012).

The role of epigenetics in mammalian development has been mostly elucidated in the mouse, although a number of studies have been translated to diverse human cell lines and primary cell cultures. The advent of gene "knockout" and "knock-in" technologies has been instrumental for the functional dissection of key epigenetic regulators. For instance, the *Dnmt1* DNA methyltransferase mutant mouse provided functional insight for the role of DNA methylation in mammals (Li et al. 1992); it is embryonic lethal and shows impaired imprinting (see Ch. 15 [Li and Zhang 2014]). Disruption of DNA methylation has also been shown to cause genomic instability and reanimation of transposon activity, particularly, in germ cells (Walsh et al. 1998; Bourc'his and Bestor 2004). Over 100 chromatin-regulating factors (i.e., histone and DNA-modifying enzymes, components of nucleosome-remodeling complexes and RNAi machinery) have been characterized and a large number disrupted in the mouse. The mutant phenotypes affect cell proliferation, lineage commitment, stem cell plasticity, genomic stability, DNA repair, and chromosome segregation processes in both somatic and germ cell lineages. Not surprisingly, most of these mutants are also involved in disease development and cancer, many of which are listed in the appendices of Ch. 35 [Audia and Campbell 2014] (reviewed in Fodor et al. 2010; Dawson and Kouzarides 2012; You and Jones 2012; Shen and Laird 2013).

Thus, insight into "universal" epigenetic mechanisms took advantage of unique biological features showed by many, if not all, of the above-mentioned model organisms. Without these biological processes and the functional analyses (genetic and biochemical) that delved into them, many of the key advances in epigenetic control would have remained elusive. Finally, the new technology of next-generation sequencing (NGS) or massively parallel deep-sequencing of genomes (e.g., International Human Epigenome Consortium [IHEC], National Institutes of Health [NIH] roadmap, ENCODE [Encyclopedia of DNA Elements], and modEncode; see Appendix 1) has allowed the genomic mapping of histone modifications to search for common parameters that may underscore functionally important patterns within the epigenetic landscapes. Interestingly, the inherent complexities that plague epigenetic regulation become clearer when highly conserved elements are compared across a wide range of model organisms, pointing toward functionally important features (Beltrao et al. 2012; Xiao et al. 2012). So, it would appear that evolution is a very good guide in helping to decipher key features of epigenetic landscapes.

Ironically, few epigenetic pathways have been elucidated in mechanistic terms in humans, likely because of the limitations of obtaining and dealing with human samples, no less, from diseased tissue. Recently, however, high-resolution exome sequencing has identified mutations in a surprisingly large number of chromatin regulators in human cancers and other pathologies (Shen and Laird 2013). Quite remarkable are new findings from several sequencing consortia that have identified "hot spot" mutations in histone genes themselves (notably members of the histone H3 family), which exist in high frequency in a large cohort of pediatric gliomas (and other cancers) (described in Ch. 2 [Liu et al. 2014]; also see Schwartzentruber et al. 2012; Wu et al. 2012; Behjati et al. 2013). These groundbreaking studies suggest that human biology can also be a good "model" for epigenetics research, lending support to advances made in more traditional models for which residues in histones are functionally important, guided largely by histone genetics often performed in genetically tractable organisms, such as yeast and, now, *Drosophila* (Pengelly et al. 2013). Until these recent discoveries, the editorial team would never have envisioned discussing the subject of "onco-histones" in this overview. This fact underscores the exciting and fast-paced nature of current epigenetic research.

3 DEFINING EPIGENETICS

A common thread connecting diverse eukaryotic organisms with respect to fundamental epigenetic principles is that DNA is not "naked" in all organisms that maintain a true nucleus (eukaryotes). Instead, the DNA exists as an intimate complex with specialized histone and nonhistone proteins, which together comprise chromatin. Although initially regarded as a passive packaging molecule to wrap and organize eukaryotic DNA, it is now clear and widely accepted that distinctive forms of chromatin arise through nucleosome arrays carrying covalent and noncovalent modifications. This process encompasses a plethora of post-translational histone modifications (see Sec. 5), energy-dependent chromatin-remodeling steps that mobilize or alter nucleosome structures (see Sec. 6), the dynamic shuffling of histone variants in and out of nucleosomes (see Sec. 7), and the targeting role of small ncRNAs (Sec. 11). DNA itself can also be modified covalently in many higher eukaryotes by methylation at the cytosine (C) residue, usually, but not always, of CpG dinucleotides (Sec. 10). Together, these mechanisms provide a set of interrelated pathways that create variation in the chromatin polymer (Fig. 3).

In isolation, the modifications and changes in chromatin are reversible and, therefore, unlikely to be propagated through the germline. Yet, such transitory histone modifications can impose pivotal changes to the chromatin template in response to intrinsic and external stimuli (Badeaux and Shi 2013; Suganuma and Workman 2013) and, in doing so, regulate the access of the transcriptional machinery needed to "read" the underlying DNA template (Sims et al. 2004; Petesch and Lis 2012; Smith and Shilatifard 2013). Some histone modifications (like lysine methylation), methylated DNA regions, ncRNAs, and altered nucleosome structures can, however, be stable through several cell divisions. This stability contributes to the maintenance of "epigenetic states," perhaps, as a means of achieving cellular memory, which is poorly understood. Despite a lack of mechanistic understanding, chromatin "signatures" can be viewed as a highly organized system of information storage that can segregate distinct regions of the genome to accommodate a response to environmental signals that dictate gene expression programs, which, in some cases, may be heritable.

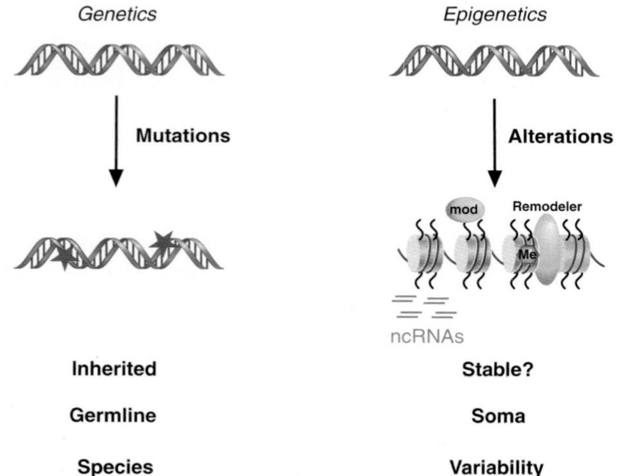

Figure 3. Genetics versus epigenetics. Genetics: Mutations (red stars) of the DNA template (green helix) are heritable somatically and through the germline. Epigenetics: Variations in chromatin structure modulate the use of the genome by (1) histone modifications (mod), (2) chromatin remodeling (remodeler), (3) histone variant composition (yellow nucleosome), (4) DNA methylation (Me), and (5) noncoding RNAs (ncRNAs). Marks on the chromatin template may be heritable through cell division and collectively contribute to determining cellular phenotype.

The significance of having a chromatin template that can potentiate the genetic information is that it provides multidimensional layers to the readout of DNA in keeping with the vast size and complexity of the eukaryotic genome, particularly, for multicellular organisms (see Sec. 13 for further details). In such organisms, a fertilized egg progresses through development, starting with a single genome that becomes epigenetically programmed to generate a remarkable multitude of distinct "epigenomes" in more than 200 different types of cells (Fig. 4).

The phenotypic alterations that occur from cell to cell during the course of development in a multicellular organism were originally described as an "epigenetic landscape" by the developmental biologist Conrad Waddington (Waddington 1957). This is essentially a contour map representing developmental potential, in which the combination of hills and valleys canalize the specification of cell type identities as development proceeds (an adaptation is illustrated in Fig. 1 of Ch. 2 [Takahashi 2014]). Yet, nearly all of the more than 200 cell types in humans share identical DNA sequences (with the exception of B and T cells that rearrange their antigen receptor loci), but differ remarkably in the profile of genes that they actually express. With this knowledge, "epigenetics" later came to be defined as "nuclear inheritance, which is not based on differences in DNA sequence" (Holliday 1994). In its more modern version, this sentence becomes molecularly (mechanistically) defined as the sum of the alterations to the chromatin template that collectively establish and propagate different patterns of gene expression (transcription) and silencing from the same genome.

4 THE CHROMATIN TEMPLATE

In its simplest form, the chromatin polymer is composed of repeating nucleosomal units, each consisting of an

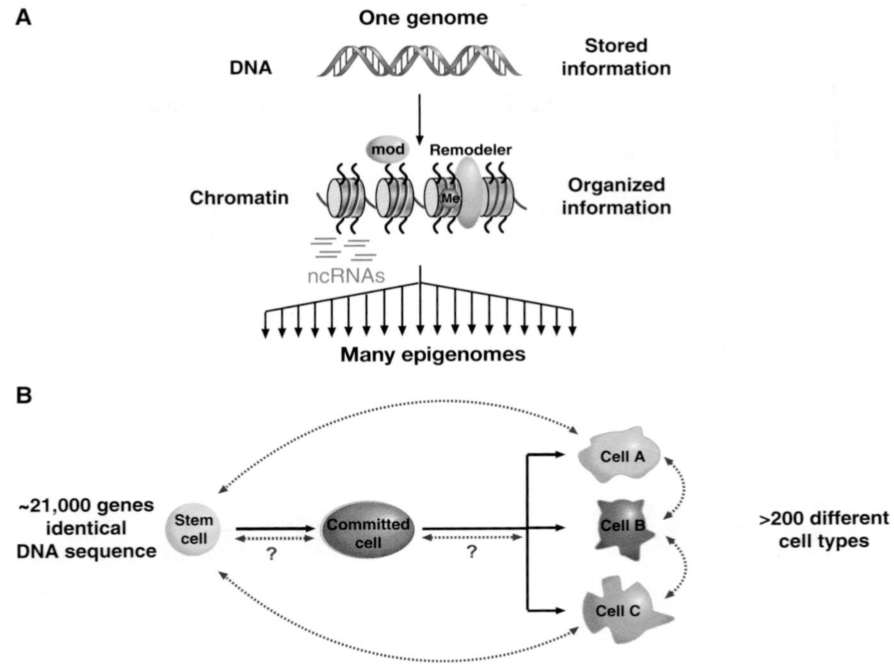

Figure 4. DNA versus chromatin. (*A*) The genome: Invariant DNA sequence (green double helix) of an individual. The epigenome: The overall chromatin composition, which indexes the entire genome in any given cell. It varies according to cell type and response to internal and external signals it receives. (*B*) Epigenome diversification occurs during development in multicellular organisms as differentiation proceeds from a single stem cell (the fertilized embryo) to more committed cells. Reversal of differentiation or change of cell type identities (blue dashed lines) requires the reprogramming of the epigenome of the individual cells.

octamer of histone proteins (i.e., two of each of the core histones H2A, H2B, H3, and H4) wrapped by 147 base pairs (bp) of DNA (Kornberg 1977). Repeating arrays of nucleosome particles had been seen early on in electron microscopic analyses of chromatin spreads, often described as the "bead-on-a-string" primary structure of chromatin, and represents a form of euchromatin. But, outside of the repeating and particulate nature of chromatin, details of nucleosomal organization were unclear. Considerable insights were gained into the nucleosome itself through elegant biochemical studies (Kornberg 1974), confirmed later by atomic resolution images of nucleosomes (Fig. 5) from X-ray crystallographic studies (Luger et al. 1997).

These landmark structures capture the disarming simplicity with which the nucleosomal unit is built; dimer sets of histone partners (H2A with H2B) and tetramers (H3 with H4) engage each other in what is known as the "handshake motif," forming an octamer (Arents et al. 1991). DNA itself organizes on the octamer surface leading to an overall symmetric particle with a defined dyad axis. Yet, these crystal structures do not accurately portray the well-studied histone tail domains that protrude from the histone DNA surface, giving rise to a flexible platform that carries many, but not all, of the posttranslational modifications (PTMs) that are described next.

5 HISTONE MODIFICATIONS: WRITERS AND ERASERS

The core histone proteins that make up the nucleosome are small and highly basic. They are composed of a globular domain and flexible (relatively unstructured) "histone tails," which protrude from the surface of the nucleosome (Fig. 5). Based on amino acid sequence, histone proteins are highly conserved from yeast to humans, supporting the general view that these proteins, even their unstructured tail domains, likely serve critical functions. The tails, particularly, of histones H3 and H4 hold important clues to nucleosomal and, hence, chromatin variability, as many of the residues are subject to extensive PTMs, as are some residues in the more structured globular core domains. The new studies of "human histone genetics," discussed above, wherein mutations in histones act as "onco-histones," underscore the importance of specific residues in histone H3 amino-terminal tails.

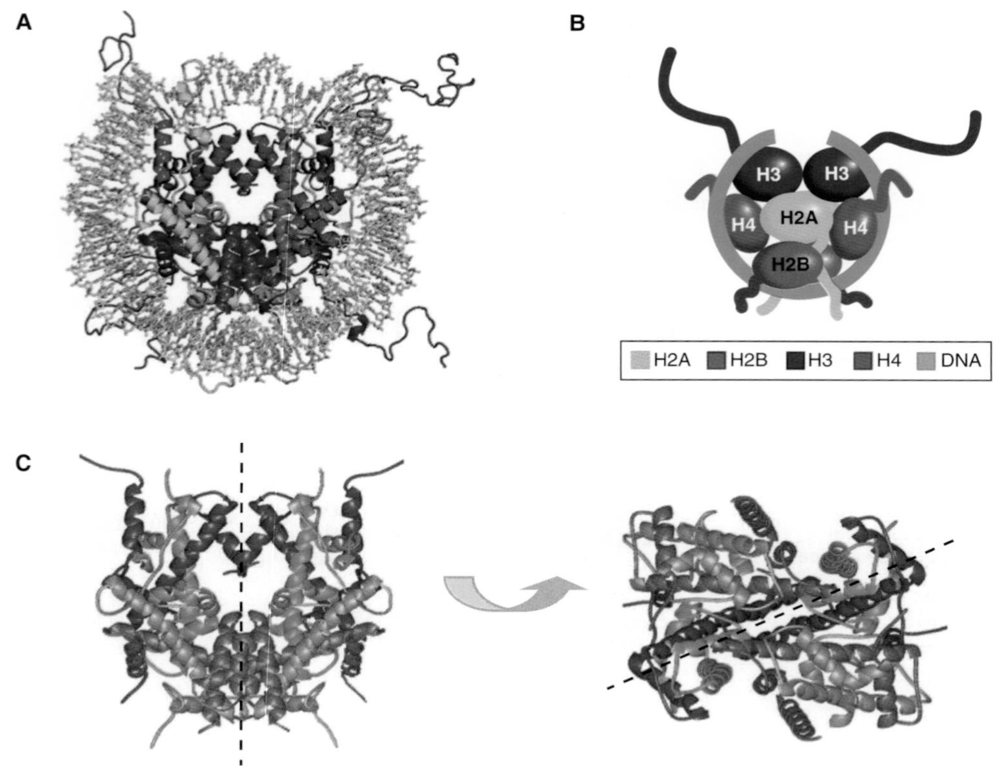

Figure 5. Nucleosome structure. (*A*) A 2.8-Å structure of a nucleosome. (*B*) A schematic representation of histone organization within the octamer core around which the DNA (light blue) is wrapped. Nucleosome formation occurs first through the deposition of an H3/H4 tetramer on the DNA, followed by two sets of H2A/H2B dimers. Unstructured amino-terminal histone tails extrude from the nucleosome core, which consists of structured globular domains of the eight histone proteins. (*C*) A top (*left*) and side (*right*) view of the octamer particle illustrates its twofold symmetry with a defined dyad axis. (*A,C*, Structural images, kindly provided by Karolin Luger, were adapted from Luger et al. 1997, with permission from Macmillan Publishers Ltd.)

Acetylation and methylation of core histones, notably H3 and H4, were among the first covalent modifications described and were long proposed to correlate with positive and negative changes in transcriptional activity (i.e., DNA accessibility or inaccessibility to the transcription machinery). Since the pioneering studies of Allfrey et al. (1964) and Paik and Kim (1971), many types of covalent histone modifications have been identified and characterized, including phosphorylation, ubiquitination, sumoylation, ADP-ribosylation, biotinylation, crotonylation, proline isomerization, among others. With the added sensitivities of modern approaches to mass spectrometry, others certainly await description (Sidoli et al. 2012). Many of these modifications occur at specific sites and residues, some of which are illustrated in Figure 6, and most that are presently known at this time are listed in App. 2 (Zhao and Garcia 2014). We stress that some of the new modifications reported are low in abundance with functional significance not known and with sites that may not be shown by every organism (listed in Sec. 2 of App. 2 [Zhao and Garcia 2014]).

How are histone modifications established ("written") or removed ("erased") in the first place? A wealth of work in the chromatin field has pointed to the catalytic action of chromatin-associated enzymatic systems. However, the identity of these enzymes eluded researchers for years. Over the last two decades, numerous biochemical and genetic studies have identified a remarkably large number of chromatin-modifying enzymes from many sources. These enzymes often reside in large multisubunit complexes that can catalyze the incorporation or removal of covalent modifications from histone and also nonhistone targets. Many of these enzymes catalyze their reactions with remarkable specificity to target residue and cellular context (i.e., dependent on external or intrinsic signals). For clarity, and by way of example, we discuss briefly the four major enzymatic systems that catalyze histone modifications, together with their counterpart enzymatic systems that reverse the modifications (Fig. 7). Together, these antagonistic activities govern the steady-state balance of each modification in question.

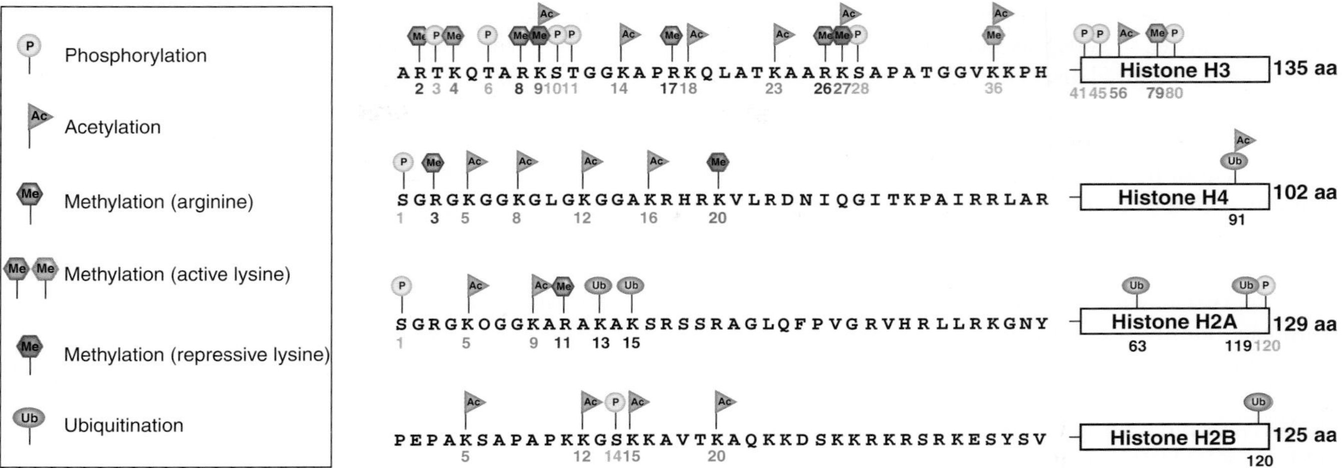

Figure 6. Sites of histone tail modifications. The amino-terminal tails of histones account for a quarter of the nucleosome mass. They host the vast majority of known covalent modification sites as illustrated. Modifications do occur in the globular domain (boxed), some of which are indicated. In general, active marks include acetylation (turquoise Ac flag), arginine methylation (purple Me hexagon), and some lysine methylation, such as H3K4 (green Me hexagon) and H3K36 (yellow Me hexagon). H3K79, found in globular domain, has an antisilencing function. Repressive marks include methylation at H3K9, H3K27, and H4K20 (red Me hexagon). In addition to the modifications shown here, many others exist (e.g., citrullination, ADP-ribosylation, sumoylation, *O*-GlcNAcylation), most of which are listed in App. 2 (Zhao and Garcia 2014).

HATs acetylate specific lysine residues in histone substrates (described in Ch. 4 [Marmorstein and Zhou 2014]) and are reversed (i.e., erased) by the action of opposing histone deacetylases (HDACs) (Ch. 5 [Seto and Yoshida 2014]). The histone kinase family of enzymes phosphorylates specific serine, threonine, or tyrosine residues, whereas the phosphatases remove phosphorylation marks. Particularly well known are the mitotic kinases, such as cyclin-dependent kinase or aurora kinase, which catalyze the phosphorylation of core (H3) and linker (H1) histones. Less clear in each case are the opposing phosphatases (PPTases) that act to reverse these phosphorylations as cells exit mitosis.

Two general classes of methylating enzymes have been described: the KMTs (histone lysine methyltransferases), which act on lysine residues, and the protein arginine methyltransferases, whose substrate is arginine (Ch. 6 [Cheng 2014]). Methylated lysine residues appear to be chemically more stable. Lysine methylation has been shown to be present in mono-, di-, or trimethylated states. Several trimethylated residues in the H3 and H4 amino termini appear to

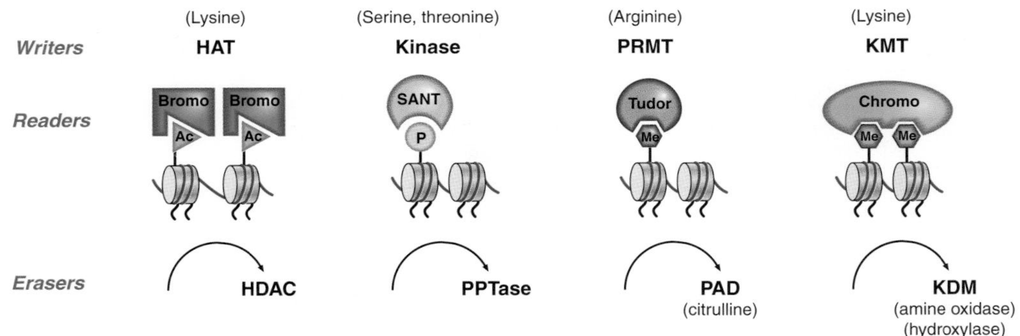

Figure 7. Histone-modifying enzymes. Covalent histone modifications are transduced by histone-modifying enzymes ("writers") and removed by antagonizing activities ("erasers"). They are classified into families according to the type of enzymatic action (e.g., acetylation, phosphorylation, or methylation). Protein domains with specific affinity for a histone tail modification are termed "readers." HAT, histone acetyltransferase; PRMT, protein arginine methyltransferase; KMT, lysine methyltransferase; HDAC, histone deacetylase; PPTase, protein phosphatase; PAD, peptidylarginine deiminase; KDM, lysine demethylase; Ac, acetylation; P, phosphorylation; Me, methylation.

have the potential to be stably propagated during cell divisions (Lachner et al. 2004), as well as the H4K20me1 mark in *Drosophila* imaginal discs (Karachentsev et al. 2005; Beck et al. 2012).

Arginine methylation is indirectly reversed by the action of protein arginine deiminases, which convert methylarginine (or arginine) to a citrulline residue (Wang and Wang 2013). The enzymes that remove methyl groups from lysine residues, the so-called histone lysine demethylases (KDMs), come in two flavors. The discovery of the first enzyme from each family is described in Ch. 2 (Shi and Tsukada 2013). One family comprises the amino oxidases represented by lysine-specific histone demethylase 1 (LSD1) (KDM1A) and lysine-specific histone demethylase 2 (LSD2) (KDM1B) that use FAD and oxygen as cofactors for demethylation and exclusively target mono- and dimethylation of H3K4 and H3K9 (Fig. 8). As part of the CoRest corepressor complex, LSD1 targets H3K4me1-2. LSD1 also associates with the androgen receptor, and, in this case, targets H3K9me1-2 during activation of transcription (Metzger et al. 2005). The other family of KDMs includes hydroxylases whose members all share the JMJC domain, which is the catalytically active site that mediates demethylation using 2-oxoglutarate and iron as cofactors (Black et al. 2012). This family of demethylases targets different residues on the H3 tail.

Considerable progress has been made in accessing the enzyme systems that govern the steady-state balance of these modifications. It remains a challenge to understand how enzyme complexes are regulated and how their physiologically relevant substrates and sites are targeted. As well, it remains unclear how covalent mechanisms impact on epigenetic phenomena. Interestingly, short stretches of "histone" sequences exist in nonhistones, leading to the

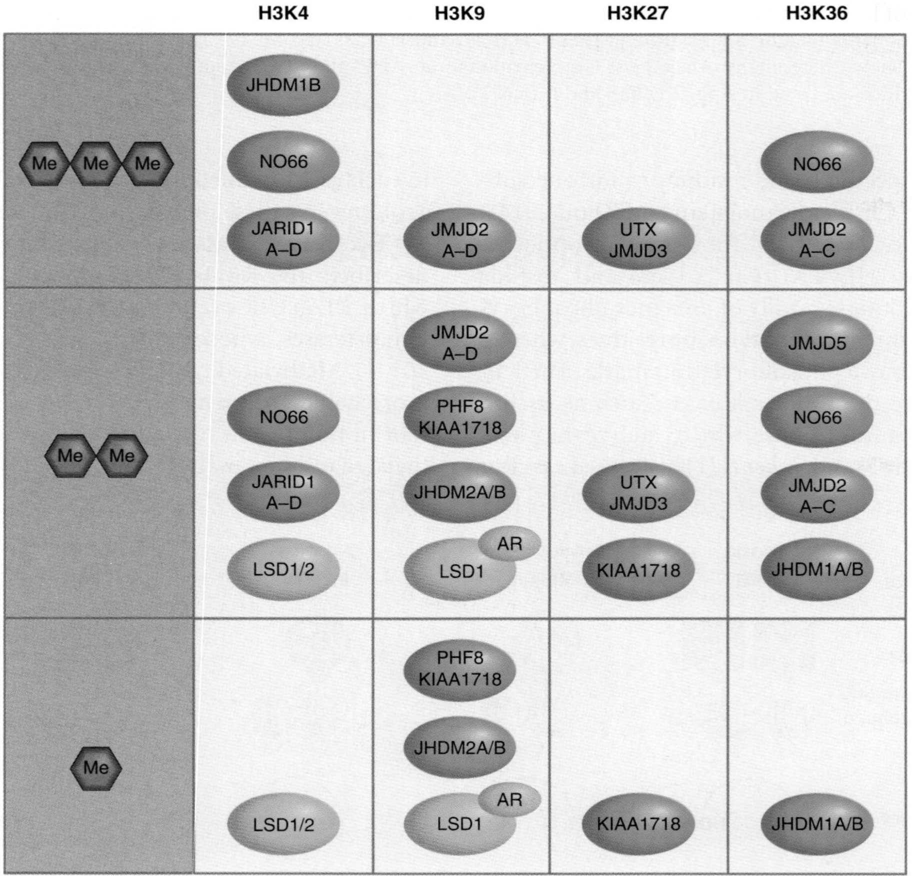

Figure 8. Histone lysine demethylases (KDMs). Histone lysine methylation can be removed by two distinct enzyme classes: amine oxidases (yellow) and hydroxylases (green). Because of their distinct catalytical mechanism, amine oxidases work only on mono- and dimethylation, whereas hydroxylases also convert trimethylation. The specificity of a subset of KDMs is shown for four prominent lysine positions within histone H3. Please note that the historical nomenclature for KDMs is used in the figure. A comprehensive listing and comparison between classical and new nomenclature for KDMs can be found in Black et al. (2012). AR, androgen receptor. (Data from Højfeldt et al. 2013.)

general concept of "histone mimicry" (Sampath et al. 2007; Marazzi et al. 2012). Histone mimicry is also reflected by the involvement of the same factors that write, erase, and read these modifications on histones. Thus, the phenomenon of histone modifications affecting chromatin structure and function must be broadened to include protein modifications more generally (Clarke 2013; Friedmann and Marmorstein 2013).

6 HISTONE READERS AND CHROMATIN REMODELERS

The chromatin template can be altered by *cis* and *trans* effects of covalently modified histone tails (Fig. 9). *Cis* effects are brought about by changes in the physical properties of modified histone tails, such as a modulation in the electrostatic charge or tail structure that, in turn, alters internucleosomal contacts and spacing. A well-known example, histone acetylation, has long been suspected of neutralizing the positive charges of highly basic histone tails, generating a localized expansion of the chromatin fiber, thereby enabling better access of the transcription machinery to the DNA double helix. Phosphorylation, through the addition of net negative charge, can generate "charge patches" (Dou and Gorovsky 2000), which are believed to alter nucleosome packaging or expose histone amino termini by altering the higher-order folded state of the chromatin polymer (Wei et al. 1999; Rossetto et al. 2012). In much the same way, linker histones (i.e., histone H1 family) that are incorporated at the dyad axis of nucleosomes (where the DNA enters and exits the nucleosome) are believed to promote the packaging of higher-order fibers by shielding the negative charge of linker DNA between adjacent nucleosomes (Izzo et al. 2008; Happel and Doenecke 2009). The addition of bulky adducts, such as ubiquitin (Wright et al. 2012), ADP-ribose (Messner and Hottiger 2011), and *O*-GlcNAcylation (Hanover et al. 2012), may also induce different arrangements of the histone tails and decondense nucleosome arrays. Similarly, PTMs in the globular histone fold domains, such as on H3K56, H3K64, or H3K122, are known to impact on chromatin structure and assembly (Tropberger and Schneider 2013).

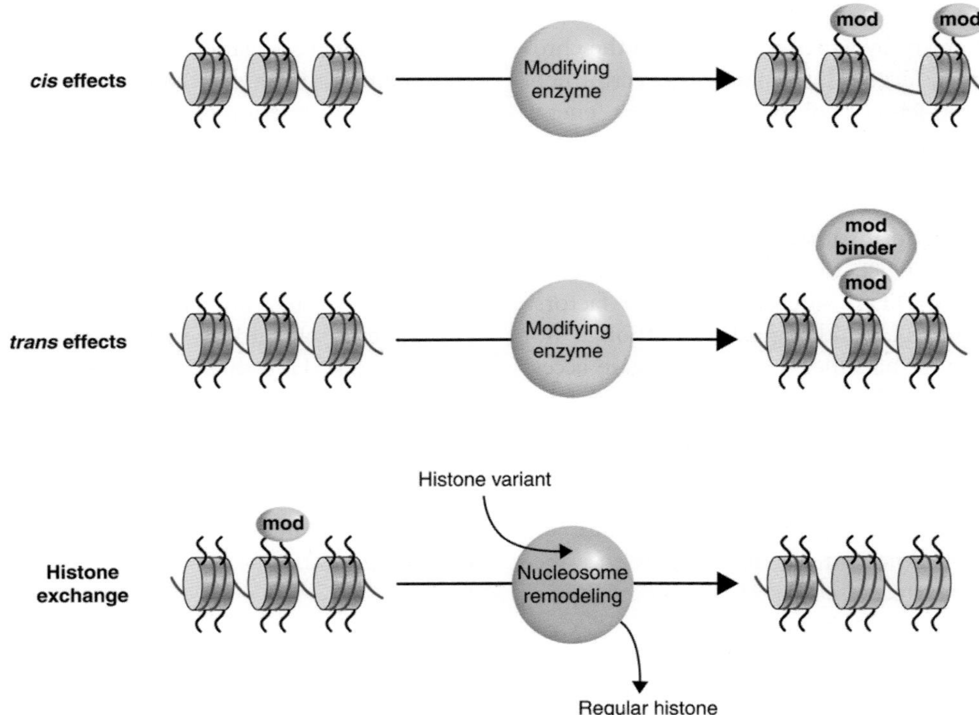

Figure 9. Transitions in the chromatin template. *cis* effects: A covalent modification of a histone tail residue (mod) results in an altered structure or charge that manifests as a change in chromatin organization. *trans* effects: The enzymatic modification of a histone tail residue (mod, e.g., H3K9 methylation) results in an affinity for a chromatin-associated protein reader (mod binder, e.g., heterochromatin protein 1, HP1). This, possibly with the association of protein complexes, causes downstream alterations in chromatin structure. Histone exchange: A covalent histone modification (or other stimulus) can signal the exchange of a core histone with a histone variant (indicated by yellow nucleosome) through a nucleosome-remodeling exchanger complex.

Histone modifications may also elicit "*trans* effects" on chromatin by recruiting modification-binding partners to the chromatin. This can be viewed as "reading" a particular covalent histone mark in a context-dependent fashion. Certain binding partners (bromodomain, chromodomain, Tudor domain) have a particular affinity for a certain histone modification and, hence, are known to "dock" onto specific modified histone tails (see Fig. 7) (see also Ch. 7 [Patel 2014]). When comprising much larger enzymatic complexes, these binding partners often serve as the chromatin "Velcro," engaging the entire complex with the chromatin polymer with subsequent consequences to the chromatin (Musselman et al. 2012). For instance, the bromodomain, a motif that recognizes acetylated histone residues (discussed in Ch. 4 [Marmorstein and Zhou 2014]), is often, but not always, part of a HAT enzyme that acetylates target histones as a portion of a larger chromatin-remodeling complex (Dhalluin et al. 1999; Jacobson et al. 2000). Similarly, methylated lysine residues embedded in histone tails can be read by "aromatic cages" present in chromodomains, or similar domains (e.g., MBT, Tudor) contained within complexes that facilitate downstream chromatin modulating events (see Ch. 7 [Patel 2014] for structural insights).

A major mechanism by which transitions in the chromatin template are induced is by signaling the recruitment of chromatin-"remodeling" complexes that use energy (ATP-hydrolysis) to change chromatin and nucleosome composition in a noncovalent manner. Nucleosomes, particularly when bound by repressive chromatin-associated factors, often impose an intrinsic inhibition to the transcription machinery. Hence, only some sequence-specific transcription factors (TFs), so-called "pioneer factors" (Zaret and Carroll 2011), and regulators (although not the basal transcription machinery) are able to gain access to their binding site(s). This problem of accessibility (discussed in Ch. 36 [Pirrotta 2014]) is solved, in part, by protein complexes that mobilize nucleosomes and/or alter nucleosomal structure. Chromatin-remodeling activities often work in concert with activating chromatin-modifying enzymes, but are also known to stabilize repressed rather than active chromatin states (see Ch. 21 [Becker and Workman 2013] for details on the different families).

Histone modifications of both the tail and globular core regions (see App. 2 [Zhao and Garcia 2014]) can engage ATP-dependent remodeling complexes to elicit a transition from a repressive chromatin state to an active one. This can be established by (1) the mobilization of nucleosomes, which can proceed by octamer sliding, (2) the alteration of nucleosome structure through DNA looping (see Ch. 21 [Becker and Workman 2013] for more detail), and (3) the replacement of specific core histones with histone variants (Ch. 20 [Henikoff and Smith 2014]). ATP-dependent chromatin remodelers (such as SWI/SNF, an historically important example) hydrolyze energy to bring about significant changes in histone:DNA contacts, resulting in looping, twisting, and sliding of nucleosomes. These noncovalent mechanisms are as critically important for gene regulatory events (Narlikar et al. 2013) as those involving covalent histone modifications, and evidence suggests that these two broad mechanisms are indeed linked.

Some of the ATP-hydrolyzing activities resemble "exchanger complexes" that are themselves dedicated to the replacement of conventional core histones with specialized histone "variant" proteins (elaborated in Sec. 7). This ATP-costing shuffle may actually be a means by which existing modified histones are replaced with a "clean slate" of variant histones (Schwartz and Ahmad 2005). Alternatively, recruitment of chromatin-remodeling complexes, such as SAGA (Spt-Ada-Gcn5-acetyltransferase) can also be enhanced by preexisting histone modifications to ensure transcriptional competence of targeted promoters (Grant et al. 1997; Hassan et al. 2002).

In addition to transcriptional initiation and establishing the primary contact with a promoter region, the passage of RNA polymerase (Pol) II (or RNA Pol I) during transcriptional elongation is further obstructed by the presence of nucleosomes. Mechanisms are therefore required to ensure the completion of nascent transcripts (particularly of long genes). A series of histone modifications and docking effectors act in concert with chromatin-remodeling complexes, such as SAGA and FACT (facilitate chromatin transcription) (Orphanides et al. 1998) to allow RNA Pol II passage through nucleosomal arrays (Reinberg and Sims 2006; Petesch and Lis 2012). These combined activities will, for example, induce increased nucleosomal mobility, displace H2A/H2B dimers, and promote the exchange of core histones with histone variants (see Fig. 9 in Ch. 20 [Henikoff and Smith 2014]). As such, they provide an excellent example of the close interplay between histone modifications, chromatin remodeling, and histone variant exchange to facilitate transcriptional initiation and elongation (Sims et al. 2004).

7 THE HISTONE VARIANTS

Histone proteins, as conserved as they are, have evolved to "specialize," through variant forms, in ways that allow for an alternative mechanism for introducing meaningful variation into the epigenome. This gives rise to differences in chromatin structure, nucleosomal dynamics, and functional properties. Variant isoforms exist for all of the core histones except H4 (mostly listed in Fig. 1 of Ch. 20 [Henikoff and Smith 2014]; Maze et al. 2014).

Using the histone H3 family as an example, two major H3 isoforms exist, H3.1 and H3.2, often referred to as "canonical" H3. Less widespread "variant" H3 isoforms also exist, including H3.3 and a centromere-specific isoform, CENP-A. In some cases, histone variants, such as H3.3, differ from their canonical histone counterpart by only a few amino acids, but genomic localization and functional studies hint that these differences do matter greatly. Histone modifications installed on canonical histones clearly are important epigenomic regulators (see Sec. 5). Histone variants also have their own pattern modifications (some listed in App. 2 [Zhao and Garcia 2014]). How histone variants and their modifications contribute to chromatin structure and function remains an active area of investigation (Bernstein and Hake 2006).

Transcriptionally active genes have general canonical histone H3 (H3.1 and H3.2), exchanged by the H3.3 variant in a transcription-coupled (replication-independent) mechanism (Ahmad and Henikoff 2002). Interestingly, H3.3 is also incorporated into repressed chromatin regions, such as pericentric and telomeric regions, by distinct histone chaperone complexes (see Fig. 10). Replacement of the canonical H3 with the H3.3 variant into active chromatin occurs via the action of the HIRA (histone regulator A) exchanger complex (Tagami et al. 2004), whereas incorporation of H3.3 into heterochromatic regions is mediated by the DAXX-ATRX complex (Goldberg et al. 2010). Thus, even a single histone variant (e.g. H3.3) can exhibit distinct, and seemingly opposing, genomic locations and biological functions (active vs. silent), adding increasingly complex "epigenetic signatures."

Behavior of histone variants also has a strong cell-cycle component. As is commonly held, H3.1 and H3.2, like canonical H4, are deposited into the bulk of the chromatin during DNA replication in S-phase by a dedicated complex known as chromatin assembly factor 1 (CAF1) (see Ch. 22 [Almouzni and Cedar 2014]). However, not all histones are synthesized and deposited during S-phase. Synthesis and substitution of histone variants, such as H3.3 and CENP-A, occur independently of DNA replication. Similarly, H2A is replaced by H2A.Z through the "on-demand" activity of the Swi2/Snf2-related ATPase 1 exchanger complex (Mizuguchi et al. 2004; illustrated in Fig. 9 of Ch. 20 [Henikoff and Smith 2014]). Hence, active replacement of generic histones by histone variants outside of S-phase provides a wealth of novel regulatory options (e.g., transcriptional activity or kinetochore tension during cell division), or "stress signals" (e.g., DNA damage or nutrient starvation), illustrated for the H3 family in Figure 10. Replacement of canonical H2A with the H2A.Z variant correlates with transcriptional activity and can mark the 5′ end of nucleosome-free promoters. Dual incorporation of H2A.Z and H3.3 into nucleosomes results in unstable chromatin, which closely correlates with enhancer elements via a more open chromatin fiber (Chen et al. 2013b). However, H2A.Z alone has also been associated with repressed chromatin (Rangasamy et al. 2003). CENP-A, the centromere-specific H3 variant (Fig. 10), is essential for centromeric function and, hence, chromosome segregation. H2A.X, together with other histone marks, is associated with sensing DNA damage and appears to signify a DNA lesion for recruitment of DNA repair complexes (illustrated in the context of cancer in Fig. 4 of Ch. 35 [Audia and Campbell 2014]). MacroH2A is a histone variant that specifically associates with the inactive X (Xi) chromosome in mammals. Taken together, these examples underscore the remarkable complexity of variant histone use with distinct biological

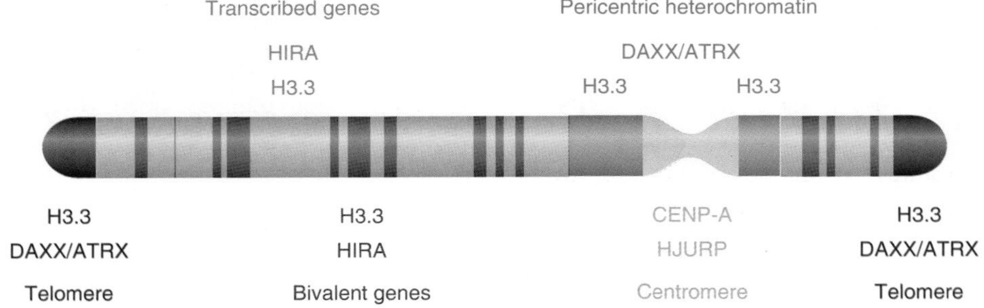

Figure 10. Genomic localization of histone H3 variants. The genomic localization of histone H3 variants, as determined by immunofluorescence and/or chromatin immunoprecipitation (ChIP)-sequencing, is schematically illustrated on a mitotic human chromosome. The centromere-specific variant CENP-A is deposited at centromeric chromatin by the HJURP chaperone. For the variant H3.3, there are two distinct histone–chaperone complexes: DAXX/ATRX deposits H3.3 at pericentric heterochromatin and telomeres, whereas HIRA regulates the incorporation at transcribed and also at bivalent genes. It is important to note that the deposition of these histone variants is replication independent. In contrast, the incorporation of canonical histone H3 (H3.1 and H3.2) is restricted to S phase.

readouts. Many important questions and challenges remain. How do histone variants know where to go in chromatin and how are they recognized by the machinery that deposits them where they need to be? The landmark recent findings that genes encoding the histone variant H3.3 (and to a lesser extent canonical H3.1) contain mutations in a remarkably high frequency in human patients with varying cancers, described in Ch. 2 [Liu et al. (2014]), promises to carry interest in histone variants to a new level (Schwartzentruber et al. 2012; Wu et al. 2012).

8 THE BIOLOGICAL OUTPUT OF HISTONE MODIFICATIONS

Modifications present in histones do not function in isolation; their establishment often requires the presence or absence of other modifications and their potential information is conveyed through their recognition by other factors (Fig. 11). Patterns of chromatin structure that correlate with bulk histone tail modifications have long suggested that specific epigenetic marks may provide "ON" (i.e., active) or "OFF" (inactive) signatures (summarized in the

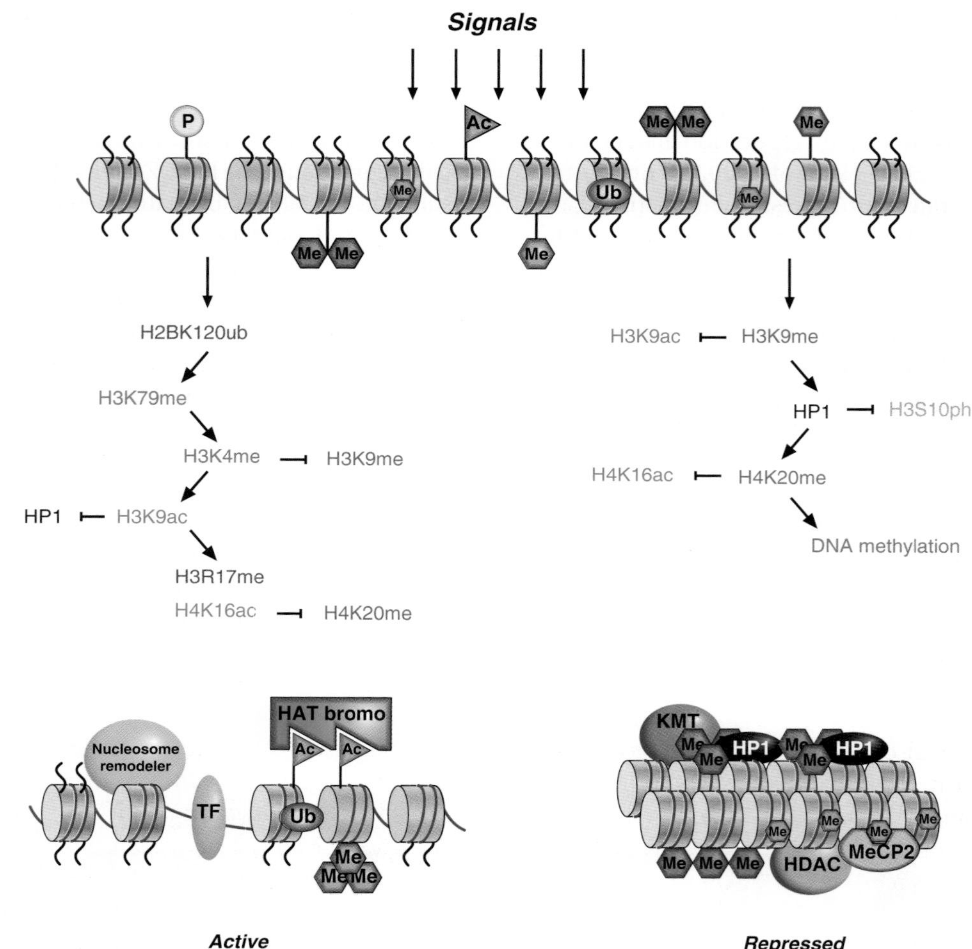

Figure 11. Coordinated modification of chromatin. The transition of a naïve chromatin template to active euchromatin (*left*) or the establishment of repressive heterochromatin (*right*) involves a series of coordinated chromatin modifications. In the case of transcriptional activation, this is accompanied by the action of nucleosome remodeling complexes to create a nucleosome-depleted region (NDR) at the promoter for TF binding and the replacement of core histones with histone variants (yellow, namely, H3.3). Also, euchromatin is typically enriched for histone acetylation (Ac), recruiting readers with a bromodomain, and contains ubiquitinated histones and H3K4me3 (green Me hexagons) at the transcriptional start site. Heterochromatin is typically depleted of histone acetylation via HDAC complexes and contains methylated DNA (pink Me hexagons), recruiting methyl-CpG-binding domain (MBD) readers, such as MeCP2. In addition, repressive histone methyllysine marks (red Me hexagons) are placed by KMTs and can recruit readers, such as HP1, via H3K9 methylation.

table on the inside front cover of this book). Notably, histone acetylation is often associated with active chromatin domains or regions that are generally permissive for transcription. In contrast, other modifications, such as certain phosphorylated histone residues, have been associated with condensed chromatin, which generally fails to support transcriptional activity. Figure 11 illustrates two examples of established hierarchies of histone modifications that seem to pattern chromatin active in transcription or, in contrast, heterochromatic domains.

Emerging studies also show that numerous combinations of "active" marks can function in a concerted fashion to simultaneously counteract "repressive" modifications and vice versa. Using the H3 amino-terminal tail as an example (Oliver and Denu 2011), Figure 12 illustrates some of the general principles underlying these complex interactions, which often use the orchestrated binding and interaction of writers, readers, and erasers. These concepts apply, in principle, to all of the canonical and linker histone tail domains and may be extendable to the histone core domains as well (see App. 2 [Zhao and Garcia 2014] for a compilation of the staggering complexity of histone modifications).

Figure 12A touches on some of the relationships that have been reported in a single H3 tail, termed "intrahistone" modification cross talk. The plant homeodomain (PHD) finger of the PHF8 (KDM7B) protein, a H3K9 demethylase, having been recruited by the active H3K4me3 mark, functions to remove repressive H3K9 methylation, in part, because the domain between the PHD finger and its catalytic domain is "bent," permitting it to act on nearby H3K9 (Horton et al. 2010). In contrast, PHF8 is unable to demethylate at H3K27, in part, because it cannot "reach" that residue from its H3K4me3 "anchor." Interestingly, the solved structure of the H3K27 demethylase KIAA1718 (KDM7A) has an "extended" linker region between its PHD finger, which also docks on H3K4me3, allowing this enzyme to reach across the H3 tail in an intrahistone mechanism to demethylate at H3K27 (Horton et al. 2010).

Figure 12A also depicts the general concept that "every amino acid in histones matters." In particular, residues adjacent to or very close to a modified lysine, such as H3K9, may become modified themselves in a cross talk pathway known as "phospho-methyl switching" (Fischle et al. 2003). In this case, a H3K9 methyl "reader," such as HP1, can bind its target lysine, but fails to do so when the adjacent serine (H3S10) is phosphorylated (Fischle et al. 2005). Experimental evidence supports phospho-methyl switching, in both histones and nonhistone proteins (Zhang et al. 2005; reviewed in Latham and Dent 2007; Suganuma and Workman 2011), although it remains unclear whether this mechanism will apply to other marks (e.g., acetyl-phospho switching), or if there will be histone readers that are immune to this mechanism, wherein adjacent or nearby phosphorylation events fail to disrupt the binding interactions.

Figure 12B illustrates "interhistone" or "trans-tail" cross talk, calling attention to the well-documented fact that histone H2B monoubiquitination located in its carboxy-terminal tail (H2BK120ub) is known to provide an upstream signal required for downstream H3K4 and H3K79 methylation (Briggs et al. 2002; Dover et al. 2002; Sun and Allis 2002; Kim et al. 2009). In vitro systems, in which homogeneously ubiquitinated H2BK120 mononucleosome particles (i.e., "designer chromatin") can stimulate intranucleosomal H3K79 methylation mediated by human DOT1L (McGinty et al. 2008). Conclusions about which subunit of the Set1 (KMT2) complex, the only H3K4 KMT in yeast, is responsible for "sensing" the H2B ubiquitination modification remain controversial, likely due, in part, to the inherent complexities of these intricate cross talk relationships.

X-ray structures solved for "monovalent" histone "readers" with their corresponding ligand (discussed above, and see Ch. 4 [Marmorstein and Zhou 2014]; Ch. 7 [Patel 2014]) provide structural insights in the discrimination of intra- and internucleosomal binding reactions. As well, Figure 12 illustrates "multivalent" binding considerations that come from having multiple modules in one polypeptide (Fig. 12C) or several polypeptides of a multisubunit effector complex (Fig. 12D) (Ruthenburg et al. 2007). Work with "designer chromatin" (Fierz and Muir 2012) has shown that the BPTF (bromodomain and PHD finger transcription factor) chromatin remodeler PHD finger docks on the H3K4me3 mark, whereas its bromodomain prefers H4K16ac, interestingly, within the context of a single, intranucleosomal particle (Ruthenburg et al. 2011). This multivalent interaction provides increased binding affinity and added specificity. In contrast, multivalent interactions can also occur between nucleosome on the same chromatin polymer or between separate polymers. Figure 12D shows the well-documented situation in which the chromodomain module in the heterochromatin-associated protein (HP1) dimerizes through chromo shadow domains to essentially "cross-link" chromatin strands between nucleosomes (internucleosomal binding) using chromodomain binding to separate H3K9me3 marks. In this case, chromatin compaction is facilitated through this internucleosomal binding reaction, which may be reversed by "ejection" of the chromodomains through phospho-methyl switching (Fig. 12A).

Finally, Figure 12E illustrates complex binding reactions between histone modifications and DNA methylation. In the left, Cfp1 recognizes unmethylated DNA via its CXXC finger and recruits the H3K4-specific KMT

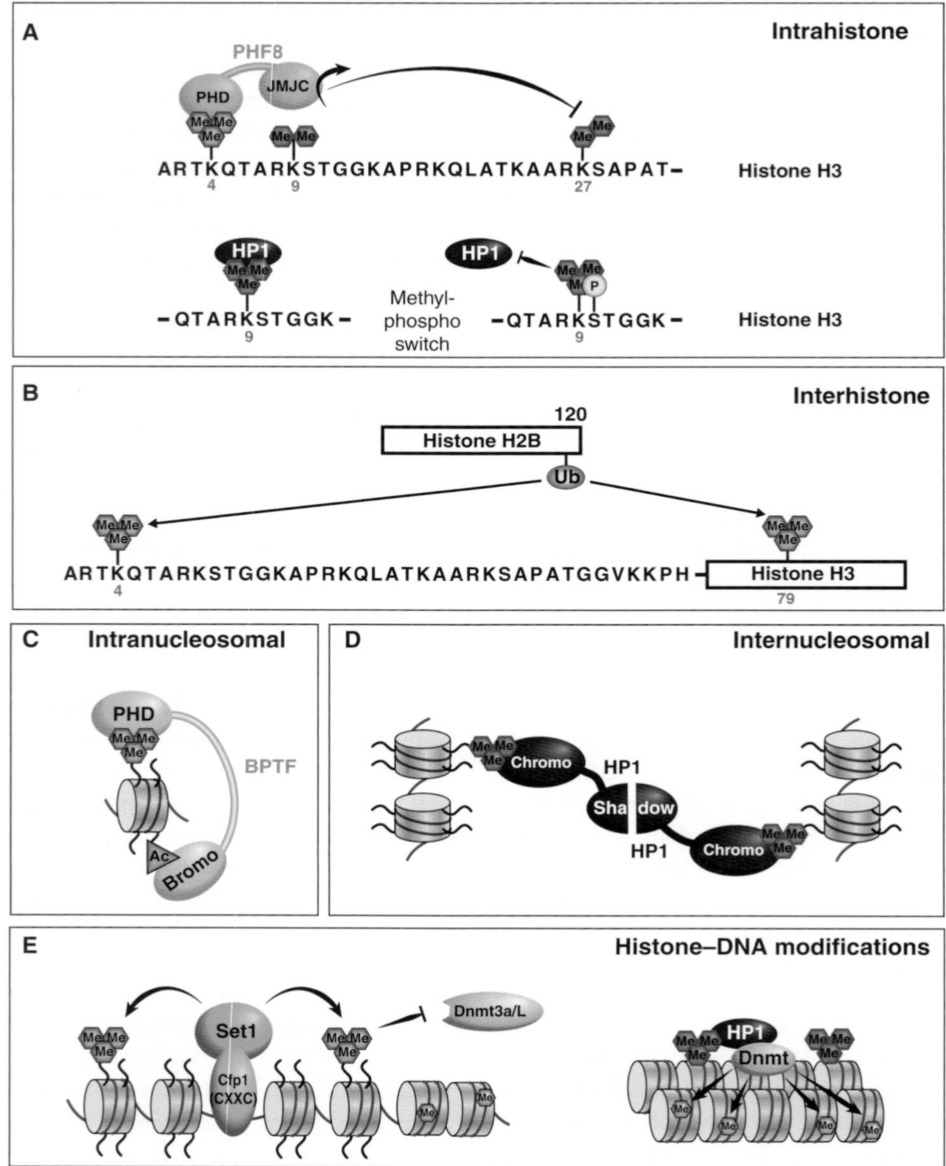

Figure 12. Combinatorial readout of chromatin modifications. (*A*) Intrahistone interplay: Histone modifications within one histone tail generate a specific downstream readout. For example, the position of the catalytic JMJC domain of the PHF8 KDM is structurally constrained to remove only H3K9me2, but not H3K27me2 because of its anchoring to chromatin via H3K4me3 binding to its PHD domain. The KIAA1718 KDM operates via a similar PHD domain-anchoring mechanism to H3K4me3, but because of different protein structure only acts on H3K27 and not H3K9 methyl marks. Another illustration is the ejection of HP1 bound to H3K9me3 when the neighboring H3S10 becomes phosphorylated, a mechanism known as "phospho-methyl" switch. (*B*) Interhistone cross talk: Modifications on two distinct histones influence each other. For example, H2BK120ub is a prerequisite for H3K4me3 and K3K79me3 to occur. (*C*) Intranucleosomal association: A chromatin reader is recruited via two distinct modifications within one nucleosome. For example, the PHD domain and bromodomain of BPTF (bromodomain and PHD finger transcription factor) bind to H3K4me3 and H4K16ac within one nucleosome. (*D*) Internucleosomal association: A chromatin reader interacts with histone modifications present on different nucleosomes. For example, the two chromodomains of HP1 dimers link H3K9me3-containing nucleosomes. (*E*) Histone-DNA modification cross talk: Modifications on histones and DNA influence each other. For example, recruitment of the Set1 KMT via Cfp1 bound to unmodified CpG-rich DNA results in H3K4me3 chromatin, which, in turn, inhibits association of Dnmt3a/L, thereby protecting these regions from DNA methylation. Conversely, recruitment of Dnmts to H3K9me3 chromatin via HP1 leads, subsequently, to DNA methylation.

Set1 (Thomson et al. 2010; see Ch. 2 [Blackledge et al. 2013]; Ch. 6 [Cheng 2014]). Loss of Cfp1 deficiency results in a strong decrease of H3K4me3 at expressed CpG island-associated genes, but also emergence of "ectopic" H3K4me3 peaks at regulatory elements. These results show that Cfp1 is a "specificity factor" that integrates multiple signals, including promoter CpG content and gene activity, to regulate genome-wide patterns of H3K4me3, a mark closely associated with active promoters (Clouaire et al. 2012). In these H3K4me3-rich regions, proteins containing the ADD (ATRX-DNMT3-DNMT3L) domain—namely, members of the DNMT3 family and several chromatin remodelers—are repelled by active marks, such as H3K4me3, but bind to unmethylated H3K4 heterochromatic regions (right) enriched in H3K9me3 (Ooi et al. 2007; Iwase et al. 2011). Collectively, the drawings shown in Figure 12 underscore the complex interactions between covalent modifications on histones and DNA with interacting proteins and complexes to bring about meaningful biological output.

In most cases involving the deposition of repressive methylation marks, the enzymes responsible for installing the histone modification also interact with factors that bind to it. Examples of such histone modifier/binder pairs are Suv39h1 and HP1 for H3K9me3, or Ezh2 and Eed for H3K27me3. In the latter case, the Polycomb repressive complex 2 (PRC2) provides a prime example of how the copying process could occur. PRC2, which "writes" the modification associated with repressive chromatin (i.e., di- and trimethylation of histone H3 on lysine 27 (H3K27me2-3)) becomes "allosterically activated" through the binding of the embryonic ectoderm development (Eed) subunit to H3K27me3, the product of its reaction (Fig. 13) (Margueron et al. 2009). This mechanism can explain the inheritance of a repressive chromatin state from parental to naïve octamers (i.e., during replication from mother to daughter cells [see Sec. 23]). The interaction between the modifier (writer) and binder (reader) can also be important as, in the case of constitutive heterochromatin, it allows the domain to expand and importantly, be maintained (reinforced) by a feedforward loop (Fig. 13). This can also result in an allosteric stimulation of the writer with important consequences for the transmission of the modification (and thus the chromatin domain it comprises) during replication and/or mitosis. An allosteric effect similarly operates, in the case of propagating cytosine methylation, during replication, catalyzed by DNMT1, explained in Section 10. Clearly, this is a key mode of establishing and propagating PTM-marked chromatin domains (Rando 2012).

Chromatin that is accessible to the transcription machinery is enriched in acetylated lysine residues. Transcriptionally active chromatin also correlates with the presence of H3K4me3 around the transcriptional start site (TSS) and H3K36me3 within the coding sequences. These modifications are established through the onset of transcription, as the carboxyl terminus of the largest RNA Pol II subunit (known as CTD) becomes phosphorylated in the heptapeptide repeat (Y-S_2-PT-S_5-PS), RNA Pol II consequently

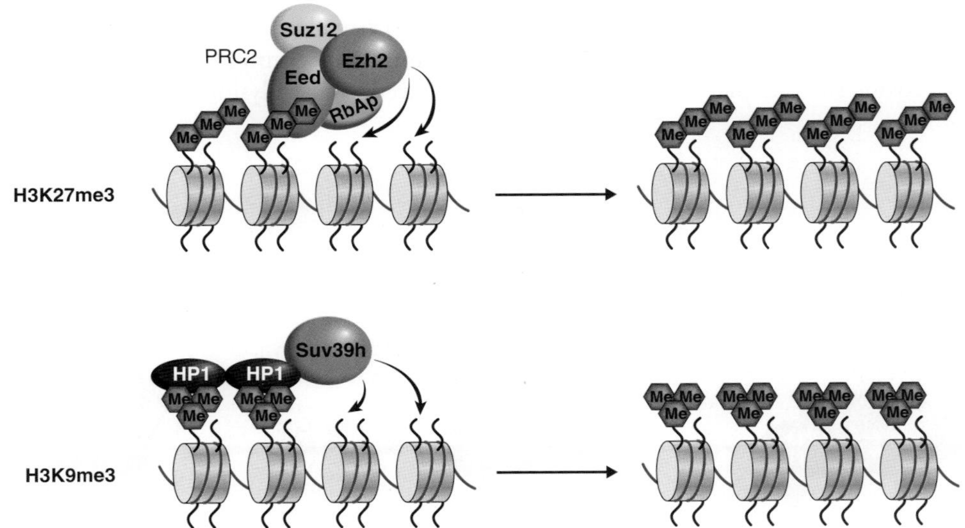

Figure 13. Chromatin domain extension by KMT complexes. The molecular mechanism for establishing repressive chromatin domains is illustrated for the PRC2 and Suv39h/HP1 systems. In addition to a catalytic component (Ezh2; Suv39h), both complexes contain a modification-specific binder (Eed; HP1), which recognizes the respective enzymatic product (H3K27me3; H3K9me3) and thus allows the progressive extension of the histone mark into the adjacent chromatin regions.

recruits the factors that set the histone modifications in place (Fig. 14) (Sims et al. 2004; Smith and Shilatifard 2013).

The process, in more detail, involves RNA Pol II being recruited to promoters, in the first place, by a family of factors known as the "general transcription factors" (GTFs), but in a phosphorylation-dependent manner; RNA Pol II is recruited to promoters via GTFs in the non-phosphorylated form and, then, recruits TFIIH, which contains the CDK7 kinase that phosphorylates serine 5 of the heptapeptide repeat. This phosphorylation disrupts interactions of RNA Pol II with most of the GTFs and allows the initiation of transcription. RNA Pol II, then, moves along the template, but soon after initiation, when the nascent RNA exits the catalytic channel, the polymerase ceases transcription to allow a series of regulatory steps, including the addition of the CAP structure. This highly regulated step results in the recruitment of different factors, including the capping enzyme through recognition of CTD, phosphorylated at serine 5 within the heptapeptide repeat. Once this is accomplished, a kinase that phosphorylates serine 2 of the CTD ensures that RNA Pol II is ready to escape the promoter and engage in elongation, a step known as promoter clearance.

The differential phosphorylation of the CTD (serine 2 and 5) recruits different histone-modifying factors, including KMTs, that act on H3K4 (serine-5 phosphorylation), such as Set 1 in yeast and the SET1 and MLL1 (KMT2) complexes in mammals, and KMTs that act on H3K36 (serine-2 phosphorylation), such as Set2 (KMT3) in yeast and related mammalian orthologs (Fig. 14) (Sims et al. 2004; Smith and Shilatifard 2013). What is the function of these modifications being deposited during transcription? In a simplistic way, they provide surfaces that are recognized by specific domains present in factors that regulate gene expression downstream from transcription initiation. For example, H3K4me3 is recognized by the chromodomains of mammalian CHD1 (Sims et al. 2005), which, in turn, recruits factors that affect transcription elongation, such as FACT, the PAF complex, and factors modulating splicing (Sims et al. 2007), among others. H3K4me3 is also "read" by the PHD finger present in BPTF (Wysocka et al. 2006), a subunit of the NURF (nucleosome-remodeling factor) nucleosome-remodeling complex (Barak et al. 2003), as described above and elaborated in Ch. 7 (Patel 2014). How a modification, such as H3K4me3, can be read by different factors is unknown, but they may bind to a particular modification at different stages of the transcription cycle or in a promoter-specific manner. The H3K36me3 modification recruits distinct factors, an example of which is the Sin3A complex, containing HDACs (Carrozza et al. 2005; Keogh et al. 2005). It is proposed that the deacetylases associated within the Sin3A complex function to overcome the acetylation required for transcription, promoting the reestablishment of nucleosomes and suppressing cryptic initiation in the open chromatin as a function of transcription.

9 HIGHER-ORDER CHROMATIN ORGANIZATION

Chromatin, the DNA-nucleosome polymer, is a dynamic molecule existing in many configurations. Historically, chromatin has been classified as either euchromatic or heterochromatic (elaborated further in Sec. 12), stemming from the nuclear staining patterns of dyes used by cytologists to visualize DNA. Euchromatin is a decondensed type of chromatin, exemplified by histone hyperacetylation (e.g., H4 acetylated on lysine 16, H4K16ac), and is, for the most

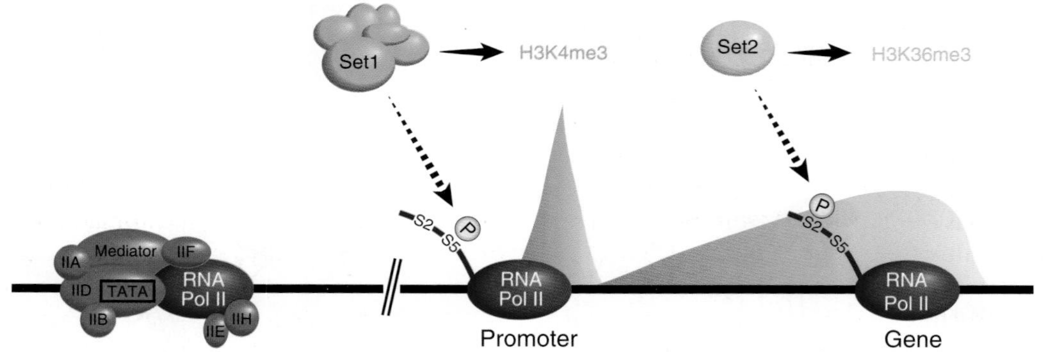

Figure 14. Modifications of the carboxy-terminal domain (CTD) of RNA Pol II. Unmodified RNA Pol II is recruited to promoters via the interaction with general transcription factors (purple). Serine 5 (S5) phosphorylation of the CTD initiates transcription, whereas serine 2 (S2) phosphorylation allows promoter clearance and, subsequently, transcriptional elongation. Importantly, these differential phosphorylation states of the CTD also mediate the recruitment of distinct chromatin-modifying enzymes: H3K4me3-specific KMT complexes (e.g., Set1) at promoters and an H3K36me3-specific KMT (Set2) over gene bodies.

part, transcriptionally active. Heterochromatin is, in contrast, highly compacted and silenced, existing either as permanently silent chromatin (constitutive heterochromatin), in which genes are rarely expressed in any cell type of the organism, or as facultative heterochromatin, in which genes can be derepressed during a specific cell cycle or developmental stage. Constitutive heterochromatin is present at pericentric and subtelomeric regions of the genome (see Fig. 10) and is more uniform in structure; facultative heterochromatin is more heterogeneous (elaborated further in Secs. 16 and 17). Thus, there is a spectrum of chromatin states whose "signatures" are being dissected genome-wide as more and more molecular markers are used. Moreover, these chromatin states are constantly in dynamic flux as physiologically relevant inputs signal to the chromatin fiber from upstream pathways. Another example illustrating the plasticity of chromatin is inferred by the extensive reprograming occurring in differentiated cells when induced to revert to a more "plastic" embryonic stem (ES) cell state (referred to as an induced pluripotent stem cell, or iPS, elaborated in Sec. 19).

Chromatin structure visualized microscopically as 11-nm "beads on a string" polymer represents a largely "unfolded" configuration of chromatin wherein DNA is periodically wrapped around repeating units of nucleosomes (Fig. 15). The chromatin fiber, however, is not always made up of regularly spaced nucleosomal arrays. Nucleosomes may be irregularly packed and fold into higher-order structures that are only beginning to be observed at atomic resolution (Luger et al. 2012; Song et al. 2014). Differential higher-order chromatin conformations occur along the length of the genome, changing more subtly during cell-fate specification and dramatically during distinct stages of the cell cycle (i.e., interphase vs. mitotic chromatin).

More compact and repressive higher-order chromatin structures (30 nm) are achieved partially through the recruitment of linker histone H1 (Robinson and Rhodes 2006; Li et al. 2010a) and/or modification-dependent "architectural" chromatin-associated factors, such as HP1 (Canzio et al. 2011) or Polycomb (Pc) (Francis et al. 2004). It is commonly held that compaction of nucleosomal chromatin (11 nm) into a 30-nm transcriptionally incompetent conformation is accomplished by the incorporation of linker histone H1 during interphase. However, the exact contribution that this fifth histone type makes to higher-order chromatin structures has proven difficult to firmly establish (Fan et al. 2005). In mammals, the histone H1 family occurs with as many as eight different isoforms, making detailed genetic analyses challenging; some H1 isoforms are redundant, whereas others hold tissue-specific functions (Izzo et al. 2008; Happel and Doeneke 2009). H1 itself can be covalently modified (listed in App. 2 [Zhao and

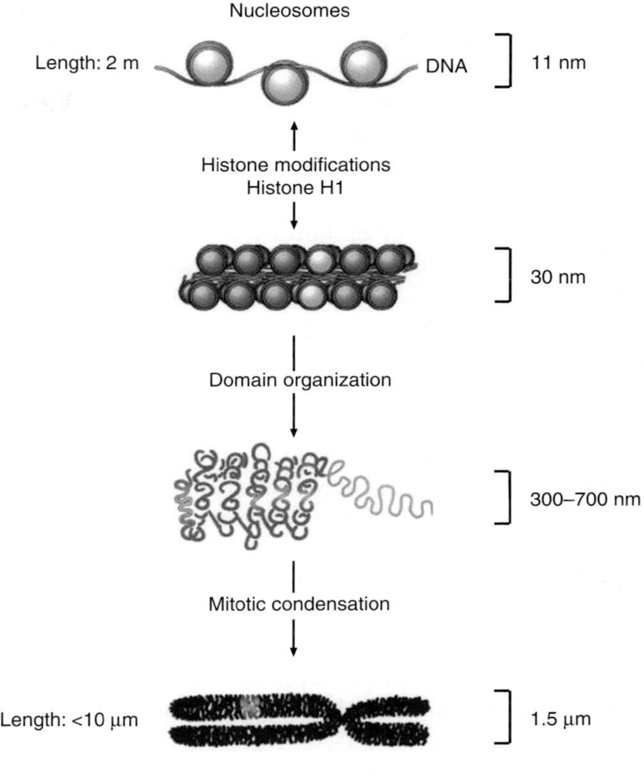

Figure 15. Higher-order structuring of chromatin. The 11-nm fiber represents DNA wrapped around nucleosomes. The 30-nm fiber is further compacted into an as-yet unconfirmed structure (illustrated as solenoid conformation, here), involving linker histone H1. The 300–700-nm fiber represents dynamic higher-order looping that occurs in both interphase and metaphase chromatin. The 1.5-μm condensed chromosome represents the most compacted form of chromatin, which occurs only during nuclear division (mitosis or meiosis). It is not yet clear how mitotic chromosome-banding patterns (i.e., G- or R-banding) correlate with particular chromatin structures. (Adapted from Felsenfeld and Groudine 2003; originally redrawn, with permission, from Alberts et al. 1998.)

Garcia 2014] as phosphorylated, methylated, poly(ADP) ribosylated, etc.), raising the possibility that *cis* and *trans* mechanisms, discussed above for core histones, likely extend to this important class of linker histones.

Details of how the 30-nm chromatin fiber forms and is organized is under considerable debate. In general, either "solenoid" (one-start helix) models, wherein the nucleosomes are gradually coiled around a central axis (6–8 nucleosomes/turn), or more open "zigzag" models, which adopt higher-order self-assemblies (two-start helix), have been described. Evidence, including that collected from X-ray structure using a defined template containing four nucleosomes, suggests a fiber arrangement more consistent with a two-start, zigzag arrangement of linker DNA connecting two stacks of nucleosome particles (Schalch et al. 2005; Li and Reinberg 2011; Song et al. 2014). Continued structural studies at the nucleosomal level, exploiting our continually improved methodology to engineer "designer chromatin" (Fierz and Muir 2012), will allow for better understanding of higher-order chromatin structures. This technology also permits systematic evaluation of the contribution of variants or modifications in a homogeneous population of nucleosomes. By reducing the complexity and heterogeneity of endogenous chromatin, insights will be gained as to how remodeling and histone-modifying enzymes operate to dynamic alterations in higher-order chromatin structures (Pepenella et al. 2014). In addition, single molecule studies, which are routinely used to analyze the stability and dynamics of single nucleosomes, are providing new insights into higher-order structures when applied to nucleosomal arrays (Killian et al. 2012). Because linker histone H1 is often not present in the current structures, considerably more levels of chromatin compaction remain to be understood.

Organization into larger looped chromatin domains (300–700 nm) occurs, perhaps, through anchoring the chromatin fiber to the nuclear periphery or other nuclear scaffolds via chromatin-associated proteins, such as nuclear lamins (Amendola and van Steensel 2014). The extent to which these associations give rise to meaningful functional "chromosome territories" remains unclear, but an emerging literature is beginning to provide compelling evidence for nonrandom, physiologically relevant positioning within nuclei (Bickmore 2013; Cavalli and Misteli 2013). For instance, clustering of multiple active chromatin sites to RNA Pol II TFs has been observed and such clustering correlates with the timing of replicating DNA and DNA polymerase as well. Clustering of "silent" heterochromatin (particularly, pericentric foci) and genes localized in *trans* have also been documented (see Ch. 8 [Grunstein and Gasser 2013]; Ch. 29 [Busslinger and Tarakhovsky 2014]). How these associations are controlled and the extent to which nuclear localization of chromatin domains affects genome regulation is not yet clear. There is, nonetheless, an increasing body of evidence showing correlations of an active or silent chromatin configuration with a particular nuclear territory (Cremer and Cremer 2010; see Ch. 19 [Dekker and Misteli 2014]). Certain techniques, including immunofluorescence and other microscopy methods, chromosome conformation capture (3C) and its derivatives, DNA adenine methyltransferase identification, and experiments engineering the tethering of a locus to defined nuclear domains are bringing great progress in our knowledge of nuclear organization (the topic of Ch. 19 [Dekker and Misteli 2014]; Ch. 2 [Sexton and Yaffe 2014]). These techniques are also beginning to attribute particular functions to nuclear domains, such as lamin-associated domains (LAD), which are predominantly heterochromatin-rich regions located at the nuclear periphery (Amendola and van Steensel 2014). Studies of particular biological processes, such as immunoglobulin recombination in B and T cells or olfactory receptor recombination in olfactory neurons, are revealing that the dynamic localization of relevant loci (immunoglobulins and olfactory receptors, respectively) is necessary to achieve monoallelic expression in a nuclear context (elaborated in Ch. 29 [Busslinger and Tarakhovsky 2014]; Ch. 32 [Lomvardas and Maniatis 2014]).

Perhaps, the most condensed chromatin structure is observed during the time of chromosome formation in metaphase stage of mitosis or meiosis. This compaction permits the faithful segregation of exact copies of our genome (one or two copies of each chromosome, depending on the division at hand) to each daughter cell. This condensation involves a dramatic restructuring of the DNA from an $\sim$2-m molecule when fully extended into discrete chromosomes measuring, on average, 1.5 μm in diameter (Fig. 15). This remarkable 10,000-fold compaction is achieved, in part, by the hyperphosphorylation of linker H1 and select phosphorylation sites in the amino terminus of histone H3 (e.g., serine 10 and 28). In addition, the ATP-dependent action of topoisomerase II, the condensin (Hirano 2012), and cohesin complexes (Nasmyth and Haering 2009) are absolutely required for higher-order structuring in mitosis and meiosis as there occurs very little chromatin condensation in their absence. Exactly how nonhistone complexes engage mitotic chromatin (or M-phase chromatin modifications), and what rules dictate their association and release from chromatin in a cell-cycle regulated fashion remain to be determined (Bernard et al. 2001; Watanabe et al. 2001). How the well-known mitotic phosphorylation of H3 and members of the H1 family contribute to the function of any of the enzymatic or structural complexes described above is not clear.

Specialized chromosomal domains, such as telomeres and centromeres, serve distinct functions dedicated to proper chromosomal dynamics. Telomeres act as chromosomal caps, preventing erosion of the chromosomal ends during subsequent cell divisions. Centromeres provide an attachment anchor for spindle microtubules during nuclear division. Both of these specialized domains have a fundamental role in the events that lead to faithful chromosome segregation. Interestingly, both telomeric and centromeric heterochromatin is distinguishable from euchromatin, and even other heterochromatic regions, by the presence of unique chromatin structures that are largely repressive for gene expression and recombination. Moving expressed genes from their normal positions in euchromatin to new positions at or near centromeric (or telomeric) heterochromatin can silence these genes, giving rise to powerful screens that sought to identify suppressors or enhancers of PEV (see Ch. 12 [Elgin and Reuter 2013]; Ch. 9 [Allshire and Ekwall 2014]). Centromeres and telomeres have molecular signatures that include, for example, hypoacetylated histones. Centromeres are also "marked" by the presence of the histone variant CENP-A, which plays an active role in chromosome segregation. Thus, the proper assembly and maintenance of distinct centromeric and pericentric heterochromatin is critical for the completion of mitosis or meiosis and, hence, cellular viability. Moreover, progress is being made regarding the mechanisms of epigenetic control of centromeric (and telomeric) "identity." Clever experiments have shown that "neocentromeres" (see Sec. 5 and Fig. 5 of Ch. 20 [Henikoff and Smith 2014]) can function in place of normal centromeres, demonstrating that DNA sequences do not dictate the identity of centromeres. Instead, epigenetic hallmarks, including centromere-specific modification patterns and histone variants, mark this specialized chromosomal domain. Considerable progress is also being made into how other coding, noncoding, and repetitive regions of chromatin contribute to these epigenetic signatures. How any of these mechanisms relate, if at all, to chromosomal banding patterns is not known, but remains an intriguing possibility. Achieving an understanding of the epigenetic regulation of these portions of unique chromosomal regions is needed, as highlighted by the numerous human cancers characterized by genomic instability, a hallmark of certain disease progression and neoplasia.

10 DNA METHYLATION

Since its discovery around 35 years ago (Razin and Riggs 1980), DNA methylation at cytosine residues is recognized as one of the prime epigenetic mechanisms to correlate with gene repression (see Ch. 34 [Baylin and Jones 2014]; Ch. 15 [Li and Zhang 2014]). This modification converts cytosine by the addition of a methyl group to 5-methylcytosine (5mC) in the DNA template. It is truly an epigenetic modification because it can be transmitted by both sister strands from mother to daughter cells during DNA replication and, thus, can be inherited through cell division. It is present, to varying degrees, in most multicellular organisms except worms and, arguably, *Drosophila* (see the table on the inside front cover of this book). It mainly occurs at CpG dinucleotides in mammals. Its distribution along the genome shows enrichment at noncoding regions (e.g., centromeric heterochromatin) and interspersed repetitive elements (e.g., retrotransposons). Conversely, its abundance is low in "CpG islands" present in the 5′ regulatory regions of many genes (Bird 1986); however, it is important to recognize that most exons and introns are highly DNA methylated (70%–80% of CpG sites).

DNA methyltransferases (DNMTs) are the "effectors" of DNA methylation and catalyze either de novo (DNMT3A, DNMT3B) or maintenance methylation (DNMT1) of hemimethylated DNA following DNA replication (see Fig. 2 of Ch. 15 [Li and Zhang 2014]). Maintenance methylation during DNA replication is regulated by an intrinsic autoregulatory loop of DNMT1 (Song et al. 2011); DNMT1, when bound to unmethylated CpGs at the replication fork via its CXXC motif, becomes inhibited. However, when it encounters a hemimethylated site, it becomes allosterically activated to, then, add a methyl group on the unmethylated sister strand (for details, see Sec. 2.1 of Ch. 6 [Cheng 2014]). Importantly, the interaction of DNMT1 with UHRF1 enhances the stability of DNMT1. UHRF1 is a protein that specifically interacts with H3K9me3-marked chromatin, providing a functional connection between DNA methylation and repressive histone methylation. De novo methylation is established by DNMT3A and DNMT3B enzymes, which associate with the catalytically inactive DNMT3L. As illustrated in Figure 12E, there is an antagonism between H3K4me3 and DNA methylation in which the presence of this histone mark inhibits binding of DNMT3L and the de novo DNMT3A and DNMT3B enzymes, thereby protecting CpG islands from DNA methylation (shown in more detail in Fig. 10A of Ch. 6 [Cheng 2014]). The functional interdependence between DNA methylation and histone modifications in mammalian cells has been reviewed (Cedar and Bergman 2009).

In *N. crassa* and plants, highly repetitive tandem repeat sequences of the genome (e.g., pericentric chromatin) rely on repressive H3K9 methylation marks to direct DNA methylation de novo (see Ch. 10 [Aramayo and Selker 2013]; Ch. 13 [Pikaard and Mittelsten Scheid 2014]). Interspersed repeats can also signal de novo DNA methylation, described in the context of RIP in *N. crassa* (Tamaru and Selker 2001) and retrotransposon silencing in the male germline of mammals. A mouse protein has been identified, Dnmt3L, and may function by scanning the genome to identify high levels of homology–heterology junctions, which signal the requirement for DNA methylation (Bourc'his and Bestor 2004). In plants, ncRNAs provide the signal for de novo DNA methylation through a unique mechanism termed RNA-dependent DNA methylation (detailed in Ch. 13 [Pikaard and Mittelsten Scheid 2014]).

Once established, the way in which DNA methylation functions to silence chromatin is not entirely clear, although evidence points to *trans*-regulation, by either providing specific binding motifs or inhibiting binding. Factors with affinity for methylated cytosines, called methyl-CpG-binding domain proteins (MBD) can be considered the DNA methylation equivalent to binders (or "readers") of modified histone motifs (Fig. 16). For example, the methylcytosine-binding protein (MeCP2) binds methylated CpGs and recruits HDACs to abrogate activating histone acetylation marks (see Ch. 15 [Li and Zhang 2014]). DNA methylation, on the other hand, is also known to disturb the recognition sites of transcriptional regulators, such as CTCF (see Ch. 26 [Barlow and Bartolomei 2014]).

The allelic difference in methylated DNA can occur at imprinted loci to silence either the maternal or paternal allele in plants and placental mammals, including the inactive X chromosome. This suggests that, in the course of evolution, they uniquely harnessed this epigenetic mechanism to stabilize silencing. Interestingly, in marsupials, there is a lack of DNA methylation at imprinted loci, indicating that its involvement in mammalian gene imprinting is a relatively recent evolutionary event (discussed in Ch. 26 [Barlow and Bartolomei 2014]; Ch. 25 [Brockdorff and Turner 2014]). Surprisingly, DNA methylation is present in low abundance in insects, such as honeybees and ants, and appears to be part of the imprinting process that stabilizes phenotypic polymorphism (Bonasio et al. 2012). Conversely, in dipteran insects, such as *Drosophila*, DNA methylation has largely been lost overall as a functional epigenetic mechanism (Krauss and Reuter 2011; Lyko and Maleszka 2011).

Highly repetitive regions of the mammalian genome, which are typically repressed by DNA methylation, become highly mutagenic when unmethylated to the extent of causing global genomic instability (Chen et al. 1998). Chromosomal abnormalities ensue, which is a major cause of many cancer progressions and diseases, such as ICF (immunodeficiency, centromeric instability, and facial abnormalities) (elaborated in Ch. 33 [Zoghbi and Beaudet 2014]). This underlines the crucial role that DNA methylation plays in genome integrity.

At the individual level of a methylated cytosine base, 5mC has a high propensity to spontaneously mutate. Thus, over time, C-T transitions occur through a deamination reaction. But, this characteristic is thought to be beneficial for protecting the host genome as it permanently deactivates parasitic DNA sequences, such as transposons. In a B- and T-cell context, this same chemical reaction, actively catalyzed by the activation-induced deaminase (AID), causes somatic hypermutation at the B- and T-cell antigen receptor loci. This is an important mechanism for expanding the repertoire of antigen receptors and, hence, strengthening the immunity of mammals (Pavri and Nussenzweig 2011). AID expression, also observed in early mammalian

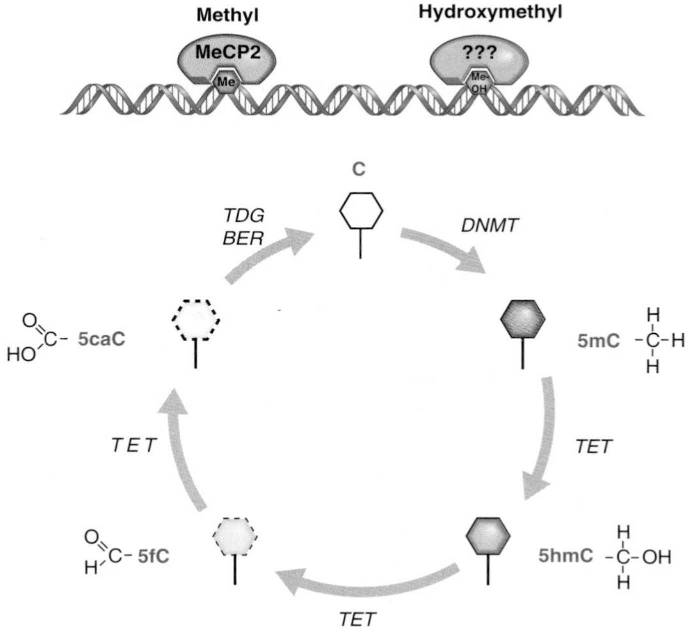

Figure 16. The DNA (de)methylation cycle. DNMTs convert unmodified cytosine nucleotides into 5mC, which can be further oxidized by the TET (ten eleven translocation) enzymes to generate 5-hydroxymethylcytosine (5hmC), 5-formylcytosine (5fC), and 5-carboxylcytosine (5caC). The combined action of thymine-DNA glycosylase (TDG) and the base excision repair (BER) machinery on 5caC sites results in unmodified cytosine. Although there are several readers for 5mC (e.g., MeCP2, MBDs), a "specific" binding protein for 5hmC remains unknown. (Y Zhang, pers. comm.)

development, has led to the suggestion that it may provide an alternative route to demethylating DNA (illustrated in Fig. 6 of Ch. 15 [Li and Zhang 2014]), although this would happen at the risk of increased point mutation rates.

A long-standing "big question" has been whether DNA methylation is reversible and could be actively removed by distinct enzymes. Studies performed after the first edition of this textbook clearly showed that DNA methylation is reversible (described in Kriaucionis and Tahiliani 2014). 5mC can initially be converted, by oxidative attack, through the action of the TET enzymes into 5hmC, followed by conversion to 5fC, and, finally, 5caC (see Fig. 16). To be strictly a reversible reaction, however, 5caC would need to be converted back to cytosine; this has been suggested to occur through a passive, repair type of mechanism, although it could also involve decarboxylation. An alternative mechanism based on DNA BER is also possible. Indeed, TDG has been shown to remove 5caC (He et al. 2011). Although the discoveries of enzymes that actively eliminate DNA methylation rank as a landmark finding in epigenetic research, more work is required to understand how and when DNA methyl marks are removed and whether the intermediates, such as 5hmC, 5fC, and 5caC, have functional significance themselves.

11 RNAi AND RNA-DIRECTED GENE SILENCING

RNAi is a host defense mechanism that breaks down dsRNA species into small RNA molecules (known as short interfering RNA or siRNA). This process ultimately leads to RNA degradation or the use of the small RNAs to inhibit translation, which is known as posttranscriptional gene silencing (PTGS). The more recently discovered transcriptional gene silencing (TGS) mechanism leads to heterochromatin formation by using the RNAi machinery to act in *cis* at sites of nascent transcription to recruit epigenetic machinery. The most convincing progress that tied together the RNAi and chromatin fields came from elegant studies in *S. pombe*, in which mutations in all known RNAi machinery components resulted in defects in chromosome segregation (Reinhart and Bartel 2002; Volpe et al. 2002; Hall et al. 2002). This was brought about by the inability to stabilize centromeric heterochromatin. It also highlighted the importance of heterochromatin at centromeres in maintaining genome integrity and, hence, cell viability. Evidence also suggests that siRNAs are required for defining other specialized regions of functional heterochromatin, such as telomeres.

The discovery that there was transcription from both DNA strands of *S. pombe* pericentric repeats and that processed siRNA derivatives could be detected was the start to unraveling the mechanism of TGS; it provided strong evidence that the dsRNA derivatives were the critical substrate to target the RITS (RNA-induced transcriptional silencing) complex to the centromeres for silencing (Fig. 17) (Reinhart and Bartel 2002; Volpe et al. 2002). Furthermore, *clr4* mutants (the *S. pombe* ortholog of mammalian Suv39h KMT) were unable to process dsRNA into siRNAs, strengthening the case for the functional connection between the RNAi machinery and heterochromatin assembly (Motamedi et al. 2004). Exactly how siRNAs, generated by the RNAi machinery (i.e., Dicer, Argonaute, RNA-dependent RNA polymerase [RdRP]), initiate heterochromatin assembly or "guide" it to appropriate genomic loci is still not fully understood. We know that RNA output is required for nucleation, but not extension of a heterochromatic domain in *S. pombe* (Buscaino et al. 2013). A model that has emerged proposes a complex interaction between the RNAi machinery, the RITS and SHREC (Snf2/HDAC-containing repressor complex) (Sugiyama et al. 2007) complexes, and centromeric repeats, which leads to a self-reinforcing cycle of heterochromatin formation, involving HDACs, Clr4, Swi6 (ortholog to mammalian HP1), probably via Argonaute-directed annealing of RNA:RNA hybrids to the nascent transcript (Fig. 17) (elaborated in Ch. 16 [Martienssen and Moazed 2014]). An alternative mechanism described in the chapter shows that heterochromatin assembly can also be triggered via an RNA exosome degradation pathway (Halic and Moazed 2010).

In *Tetrahymena*, a ciliated protozoan, a similar RNA-mediated targeting mechanism has been recognized to direct DNA elimination in the somatic macronucleus. In this case, transcription occurs from both strands of the internal eliminated segment sequences in the silent, germline micronucleus at the appropriate stage of the sexual pathway (Chalker and Yao 2001; Mochizuki et al. 2002). A "scan RNA" model was proposed to explain how DNA sequences in the parental macronucleus can control genomic alterations in the new macronucleus, involving small RNAs (for more detail, see Ch. 11 [Chalker et al. 2013]). These exciting results provide a compelling demonstration of an RNAi-like process directly altering a somatic genome and represent the ultimate form of gene silencing by excising the target DNA sequences. This raises the intriguing possibility that intergenic RNAs produced at the *V-DJ* locus (Bolland et al. 2004) may potentially direct DNA sequence elimination during V-DJ recombination of the *IgH* locus in B cells and the *TCR* loci in T cells (Ch. 29 [Busslinger and Tarakhovsky 2014]).

In plants, there are orthologs for many of the RNAi components, resulting in a number of RNA silencing pathways that can act with greater specificity for particular DNA sequences, although there is some redundancy between

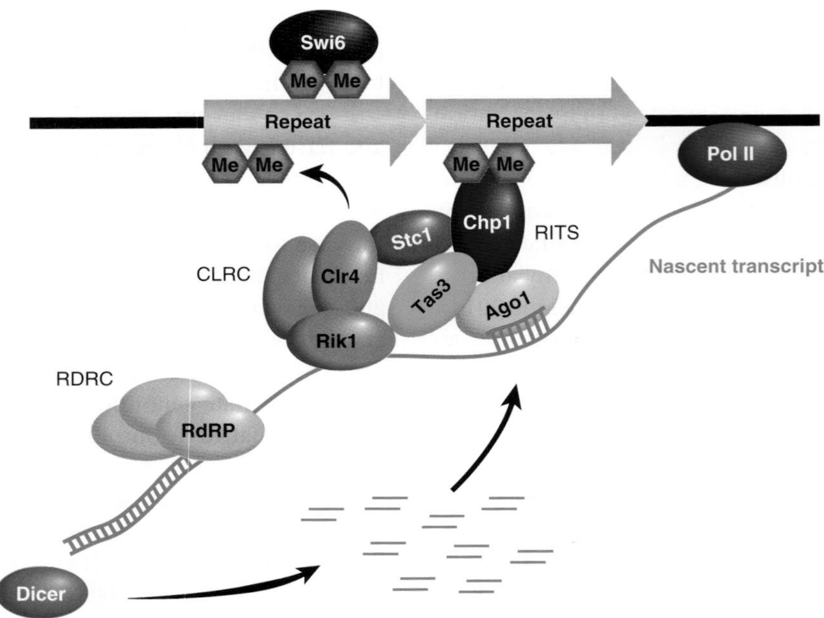

Figure 17. RNA-directed heterochromatin formation in *Schizosaccharomyces pombe*. Complementary dsRNA transcripts at pericentric regions, produced by either transcription of both strands or the folding back of inverted repeat transcripts, result in the generation of siRNAs through the action of Dicer. Incorporation of the siRNAs into the RITS complex through binding with the Argonaute protein (Ago1) activates the complex for targeting to complementary DNA or nascent RNA. The complex attracts Clr4 and the CLRC complex, which transduces histone H3K9me2. The modification-specific binder, Swi6, binds to these modified histones, facilitating the extension of a repressive chromatin domain. The action of RdRP amplifies the levels of siRNAs by using existing siRNAs as primers, reinforcing the targeting capacity of the RITS complex to specific regions of DNA.

factors. Studies of RNAi-mediated TGS in plants revealed a novel class of RNA polymerases, RNA Pol IV, that may transcribe DNA solely at heterochromatic regions (Herr et al. 2005; Pontier et al. 2005; see Ch. 31 [Baulcombe and Dean 2014]). Also unique to plants is the demonstration that RNAi pathways directly affect DNA methylation (Chan et al. 2004; explained in detail in Ch. 13 [Pikaard and Mittelsten Scheid 2014]).

RNAi-like chromatin effects have also been uncovered in *Drosophila* and mammals. In *Drosophila*, silencing of tandem arrays of the *mini-white* gene, subject to PEV, has been shown to be suppressed by *Aubergine* mutations (Pal-Bhadra et al. 2004; see Ch. 12 [Elgin and Reuter 2013]), which regulates processing of Piwi-associated small RNAs (piRNAs). In mammals, RNAse A treatment of permeabilized mouse cells rapidly removes heterochromatic H3K9me3 marks, suggesting that an RNA moiety may be a structural component of pericentric heterochromatin (Maison et al. 2002). Although mutations in siRNA processing factors, including Dicer, have been reported to impair vertebrate heterochromatin (Fukagawa et al. 2004; Kanellopoulou et al. 2005), recent studies indicate that expression of transcripts from major satellite repeats at pericentric regions (Casanova et al. 2013) and other repeat-associated ncRNAs in retrotransposon (LINE [long interspersed element]) repression (De Fazio et al. 2011; Fadloun et al. 2013) could even directly affect heterochromatin formation and gene silencing, particularly during early mouse embryogenesis.

Although there is no clear evidence for siRNA-mediated TGS in mammalian cells, another class of small RNAs, called piRNAsr, has been described in vertebrates (e.g., mouse) and invertebrates (e.g., *Drosophila*) (reviewed in Ghildiyal and Zamore 2009; Guzzardo et al. 2013; Peng and Lin 2013). *Drosophila* Piwi (P-element-induced wimpy testis), like Aubergine, belongs to a second clade of Argonaute proteins that are primarily expressed during spermatogenesis, in which they regulate the Dicer- and dsRNA-independent generation of piRNAs. piRNAs, in contrast to siRNAs (21–23 nucleotides), are between 24 and 30 nucleotides long, originate from repeat-rich clusters, and have a major function in silencing retrotransposons in male germ cells. Another intriguing characteristic of piRNAs is their "ping-pong" amplification loop, which does not require RdRP (Ghildiyal and Zamore 2009; Khurana and Theurkauf 2010). These fascinating connections suggest that piRNAs can provide a second system (in addition to Ago-siRNA-dependent TGS) for directing retrotransposon silencing, and the Piwi-piRNA pathway could also offer a plausible mechanism for the augmentation and transmission of epigenetic information through the male and even female (Ni et al. 2011) germline (see Sec. 22).

Collectively, these studies indicate a crucial, and probably, primary role for ncRNAs in triggering epigenetic transitions and maintaining, in a heritable manner, specific chromatin states of the template. In fact, these ncRNAs have provided the answer for how diverse repetitive sequences in different organisms achieve heterochromatinization through an RNA-mediated mechanism. In an effort to identify more targets of RNAi, the sequencing of small RNAs has revealed that they are largely transcribed from endogenous transposons and other repetitive sequences in organisms ranging from plants to *Drosophila* and mammals (see Ch. 16 [Martienssen and Moazed 2014]). Thus, although RNA silencing and RNA-mediated heterochromatin formation appear to have evolved to repress invading selfish DNA sequences, so maintaining genomic stability, they may even impart a much broader function in suppressing ncRNA transcription and serving as a genome-wide RNA surveillance mechanism (Aygün et al. 2010; Reyes-Turcu et al. 2011).

Together, the above examples indicate a striking variation to the central dogma of gene control that has emerged as follows: DNA ⇒ ncRNA ⇒ *chromatin* ⇒ gene function. The notion that ncRNAs would actively participate in RNAi-like mechanisms that also target locus-specific domains for chromatin remodeling and gene silencing could not have been anticipated.

12 THE DISTINCTION BETWEEN EUCHROMATIN AND HETEROCHROMATIN

For clarity, this overview has been divided into discussions of "euchromatin" and "heterochromatin," although we acknowledge that a spectrum of chromatin states exists. Euchromatin, or "active" chromatin, largely consists of coding and regulatory (e.g., promoters and enhancers) sequences of the genome. As explained earlier, extensive literature has suggested that euchromatin exists in an "open" (decompacted), more nuclease-sensitive configuration making it "poised" for gene expression, although not necessarily transcriptionally active. Some genes are ubiquitously expressed (housekeeping genes); others are developmentally regulated or stress induced in response to environmental cues. In some instances, an RNA Pol II complex has started transcription, but the transcript elongation is stalled. This promoter proximal RNA Pol II-pausing complex appears to be widely distributed among regulated genes; however, its function is currently unknown (Adelman and Lis 2012; Smith and Shilatifard 2013). In other genes, for example, those regulated by steroid hormones, a local compaction of the chromatin, involving a few nucleosomes around the TSS, is frequently observed and generated through nuclear receptor-mediated recruitment of HDAC complexes (e.g., NuRD and Sin3). This type of chromatin is frequently referred to as repressed euchromatin. In the presence of ligand, however, the DNA-bound nuclear receptor undergoes a conformational change, releasing the HDAC complex, concomitant with the recruitment of HATs, such as p300 or CBP (KAT3A), resulting in chromatin opening (Wiench et al. 2011).

Although descriptions of epigenetic regulation of transcription often involve modifications and nucleosome remodeling, *cis*-acting DNA sequences (promoters, enhancers, locus control regions, insulators, etc.), bound by combinations of *trans*-acting factors, trigger or restrict gene transcription in concert with RNA polymerase and associated factors (Sims et al. 2004; Voss and Hager 2014). These regulatory sequences have been highly selected during evolution and illustrate the overriding power of the DNA sequence to initiate and direct transcription programs that are then stabilized and propagated by associated chromatin marking. As such, chromatin provides an "indexing system" that memorizes when and potentiates where the transcription machinery had accessed its target sequences in the appropriate cell type.

At the DNA level, the A/T-rich composition of promoters is often devoid of nucleosomes, referred to as NDR, and may exist in a rigid noncanonical B-form DNA configuration, promoting TF occupancy (Sekinger et al. 2005; Brogaard et al. 2012; Struhl and Segal 2013). However, TF occupancy is not enough to ensure full-level transcription. The recruitment of nucleosome remodeling machines, through the induction of activating histone modifications (e.g., acetylation and H3K4 methylation), facilitates the engagement of the transcription machinery (see Ch. 21 [Becker and Workman 2013]). Also, the exchange of displaced histones with histone variants after the transcription machinery has unraveled and transcribed the chromatin fiber ensures integrity of the chromatin template (Weber and Henikoff 2014). Additionally, achieving fully matured mRNAs requires the prevention of cryptic and antisense transcripts (Venkatesh et al. 2012) and posttranscriptional processes involving splicing, polyadenylation, and nuclear export. Thus, the collective term "euchromatin" likely represents a complex chromatin state(s) that encompasses a dynamic and elaborate mixture of dedicated machines that interact together and closely with the chromatin fiber, to bring about the transcription and production of functional RNAs. The "rules" as to how TFs and the "activating machinery" interact with the chromatin template more strictly classify as transcription and chromatin dynamics studies and not as epigenetic control. However, some of the histone modifications associated with euchromatin, however, do participate in steps necessary to generate a functional mRNA. An example would be the discovery of the first nuclear HAT, GCN5, which was studied first as a TF (coactivator) and later shown to function as a chromatin-modifying enzyme (i.e., it possesses intrinsic HAT activity; Brownell et al. 1996).

What, then, defines "heterochromatin"? Heterochromatin plays a critically important role in the functional organization of genomes and secures accurate chromosome segregation. The great majority of the mammalian genome consists of noncoding sequences, and >50% are repeat-rich and/or contain repetitive elements (e.g., retrotransposons, LINEs, satellite repeats, etc.) that can adopt heterochromatic marks. Transcription across reiterated repeat elements has a high probability of generating dsRNA, which is subject to silencing by an RNAi mechanism in somatic cells. The production of such dsRNAs can act as an "alarm signal," reflecting the fact that the underlying DNA sequence cannot generate a functional product or has been invaded by RNA retrotransposons or other viruses. Using a variety of model systems (discussed in Ch. 9 [Allshire and Ekwall 2014]; Ch. 16 [Martienssen and Moazed 2014];

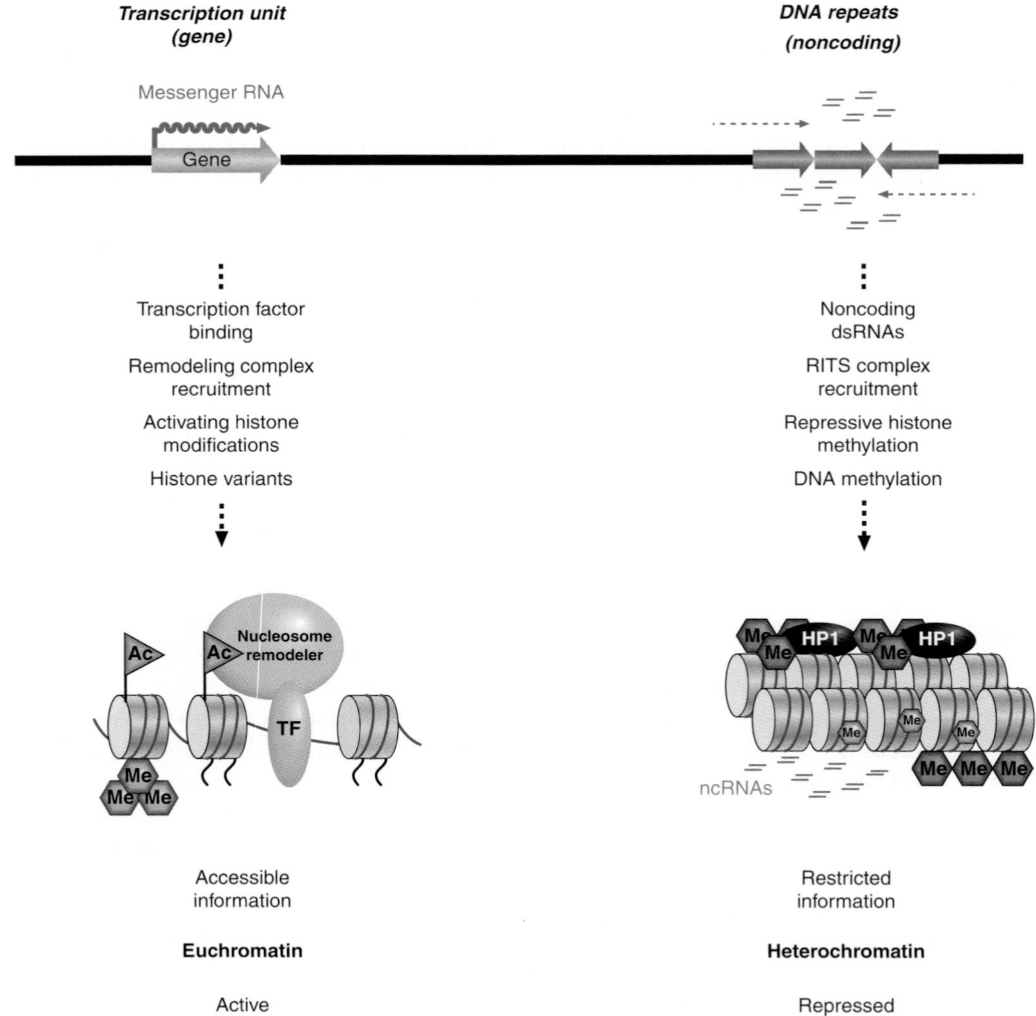

Figure 18. Distinction between euchromatic and heterochromatic domains. Summary of pathways leading to the formation of euchromatin or constitutive heterochromatin. Chromatin regions differ in the type of RNA transcripts produced from the underlying DNA sequence that can signal distinct recruitment of chromatin-associated proteins and complexes, covalent histone modifications, histone variant composition, and DNA methylation. See the text for detailed explanation.

Ch. 13 [Pikaard and Mittelsten Scheid 2014]; Ch. 23 [Strome et al. 2014]), a highly conserved pathway leading to a heterochromatin "locked-down" state has been dissected. Although the exact order and details may vary, this general silencing pathway typically involves histone tail deacetylation, methylation of specific histone lysine residues (particularly, H3K9), recruitment of heterochromatin-associated proteins (e.g., HP1), and establishment of DNA methylation in organisms that maintain active DNMTs. In addition, heterochromatin has also been shown to have reduced nucleosome mobility through HDAC-mediated suppression of histone turnover (see Fig. 18, right) (Aygün et al. 2013). *S. cerevisiae* has a distinct form of heterochromatin, lacking H3K9 and DNA methylation, and its assembly is initiated by hypoacetylation of histone tails and then "locked" by binding of acetylation-sensitive heterochromatin factors (i.e., the SIR proteins elaborated in Ch. 8 [Grunstein and Gasser 2013]). For all these examples, it is likely that sequestering of selective genomic regions to repressive nuclear environments (e.g., the nuclear lamina) or other territories may enhance heterochromatin formation (discussed in Sec. 9 and Ch. 19 [Dekker and Misteli 2014]). This is particularly evident for LADs, which have been described in *Drosophila* (Pickersgill et al. 2006), *C. elegans* (Towbin et al. 2012), and mammals (Zullo et al. 2012).

12.1 What Useful Functions Might Heterochromatin Serve?

Centromeres are heritable regions of constitutive heterochromatin, serving an essential nuclear function in chromosome segregation during mitosis and meiosis. Hence, there has been evolutionary pressure to maintain the largest clustering of repeats and repetitive elements on a chromosome to generate relatively stable heterochromatic domains marked by repressive "epigenetic signatures." It is noteworthy, thus, that centromeric repeats and the corresponding chromatin marks that associate with them have been duplicated and moved onto other chromosome arms to create "silencing domains," such as the mating-type loci of fission yeast (discussed in Ch. 9 [Allshire and Ekwall 2014]). This domain of heterochromatin acts to suppress gene transcription and genetic recombination, in contrast to the default state of being poised for transcription, typical in unicellular organisms. Constitutive heterochromatin at telomeres (the protective ends of chromosomes) ensures stability of the genome by serving as chromosomal "caps." Last, heterochromatin formation is a major defense mechanism against invading DNA, particularly in higher eukaryotes.

Although the broad functional distinction between euchromatin and heterochromatin tends to be operationally defined as active versus repressed chromatin, these states are much more dynamic than previously anticipated. For instance, in mammals, there is a distinction in the definition (e.g., levels of H3K9me3 and H3K9me2) and stability of association of heterochromatin proteins (e.g., HP1) between ES and more differentiated cells (Meshorer et al. 2006; Wen et al. 2009). Furthermore, components of constitutive heterochromatin can interchange with other subnuclear compartments of diverse function, including telomeres, imprinted regions, promyelocytic leukemia (stress) bodies, ribosomal DNA (rDNA) cluster in the nucleolus, and senescence-associated foci (see Fig. 19) (reviewed in Fodor et al. 2010), clearly illustrating the fact that chromatin is highly dynamic.

Despite all of the genetic screens and subnuclear localization analyses, no "magic" factor(s) has been identified that exclusively defines euchromatin or heterochromatin (Fodor et al. 2010). Moreover, TF binding sites are embedded in many repetitive elements, and TFs have been shown to be important constituents for heterochromatin formation (Bulut-Karslioglu et al. 2012). A model has been proposed in which the more random and reiterated distribution of TF binding sites in repeat elements versus their synergistic organization at promoters and enhancers significantly governs the RNA output from the underlying DNA sequence and the concomitant recruitment of silencing or activating chromatin-modifying enzymes (see Fig. 20). This general view is corroborated by studies in which the insertion of a strong promoter or enhancer, producing a productive RNA transcript, can override heterochromatin (Festenstein et al. 1996; Ahmad and Henikoff 2001)

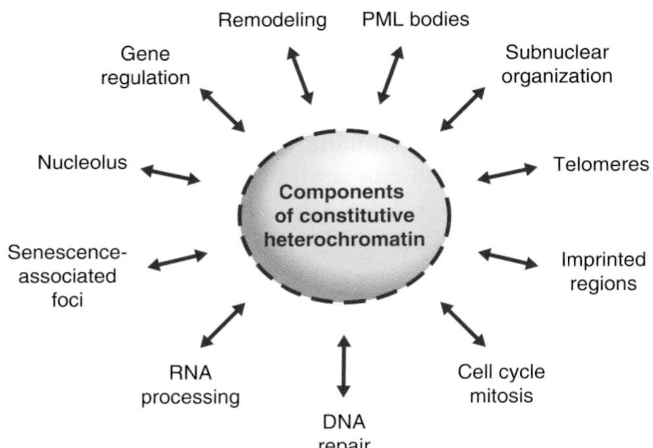

Figure 19. Interchange of heterochromatin components. Core components of constitutive heterochromatin are not restricted to DAPI-dense foci (blue circle), but frequently interchange with other subnuclear structures and are involved in many other chromatin-related processes. (Adapted from Fodor et al. 2010.)

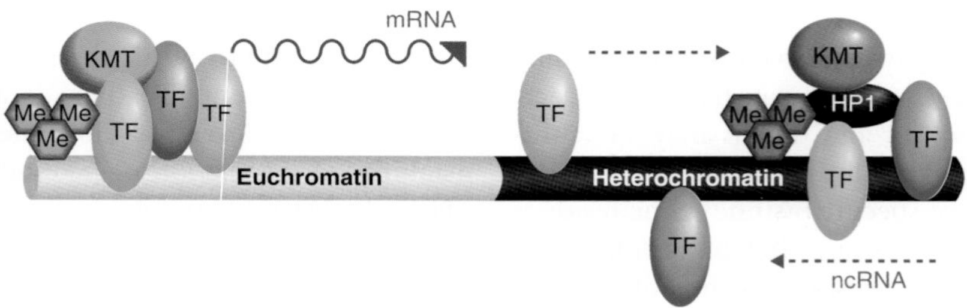

Figure 20. A transcription factor (TF) model to distinguish euchromatin and heterochromatin. This model proposes that the distinction of euchromatin and heterochromatin is based on a synergistic versus a more random organization of TF binding sites. Although euchromatic gene transcription and the generation of an mRNA is dependent on cooperative binding of TFs at fully functional promoters, repeat-rich heterochromatic regions display a more uncoordinated and reiterated arrangement of TF binding sites. This results in the occurrence of aberrant or misprocessed ncRNAs and the recruitment of repressive KMT complexes. (Adapted from Bulut-Karslioglu et al. 2012.)

or, vice versa, in which weakening of a promoter and misprocessing of a primary RNA transcript, particularly, if there is accumulation of dsRNA, can induce heterochromatic marks at a euchromatic position (Reyes-Turcu et al. 2011; Zofall et al. 2012).

With these insights, the distinction between euchromatin and heterochromatin can be attributed to three key principles. First, the nature of the underlying DNA sequence is important: for example, whether it contains AT-rich "rigid" DNA around promoters, whether there is synergy of TF binding sites versus more random TF binding sites at repetitive sequences, and/or whether other binding sites signal the recruitment of chromatin enzymes/factors. Second, the quality of the RNA produced during transcription determines whether it is fully processed into a mRNA that can be translated, or the RNA is aberrant and earmarked for silencing by the RNAi machinery or other mechanisms (e.g., exosome processing). And, third, the spatial organization within the nucleus into subnuclear compartments potentially plays a role in assembling, and more certainly in sequestering, repressed regions of chromatin into heterochromatic foci versus maintaining active chromatin in transcription hubs (Edelman and Fraser 2012; Wendt and Grosveld 2014; discussed in Ch. 8 [Grunstein and Gasser 2013]; Ch. 19 [Dekker and Misteli 2014]).

13 FROM UNICELLULAR TO MULTICELLULAR SYSTEMS

The 5000–6000 genes contained in the genomes of budding and fission yeasts are sufficient to regulate basic metabolic, repair, and cell division processes. There is, however, no requirement for cell differentiation, as these unicellular organisms are essentially clonal and, as such, repetitive "immortal" entities. In contrast, mammals code for ~21,000 genes required in more than 200 different cell types. Understanding how multicellular complexity is generated and coordinated from the same genetic template is a key question in epigenetic research.

A comparison of the genome sizes between eukaryotic organisms, including yeasts, flies, plants, and mammals, indicates that genome size significantly expands with the complexity of the respective organism. There is at least a 200-fold difference or more between the genome sizes of yeast and mammals, but only a modest, approximate fourfold increase in overall gene number (Fig. 21). This fact seems counterintuitive at first glance. But, in fact, the ratio of coding to noncoding and repetitive sequences is a better indicator of the complexity of the genome; the largely "open" genomes of unicellular fungi have relatively little noncoding DNA compared with the highly repeat-rich genomes of multicellular organisms. Mammals, in particular, have accumulated considerable repetitive elements and noncoding regions, which accounts for the majority of its DNA sequence (e.g., 54% noncoding and 45% repetitive DNA in the mouse genome).

"*What makes a fish a fish?*" is thus not determined by the number of genes, but how they are regulated and organized within the genome. Complex organisms have more complex genome organization, in which, for example, >95% of all transcripts are noncoding in the case of mammals (Mattick 2011). Evidence indicates that repetitive elements allow for genome evolution and plasticity and a certain degree of stochastic gene regulation. In the mammalian genome, this stochastic gene regulation is partly a result of the expansion of repetitive elements infiltrating the transcriptional units. This results in transcription units that are frequently much larger (30–200 kb), commonly containing multiple promoters and DNA repeats within untranslated introns. Interestingly, plants, with similarly large genomes to mammals, generally possess smaller transcription units with much shorter introns because they have evolved defense

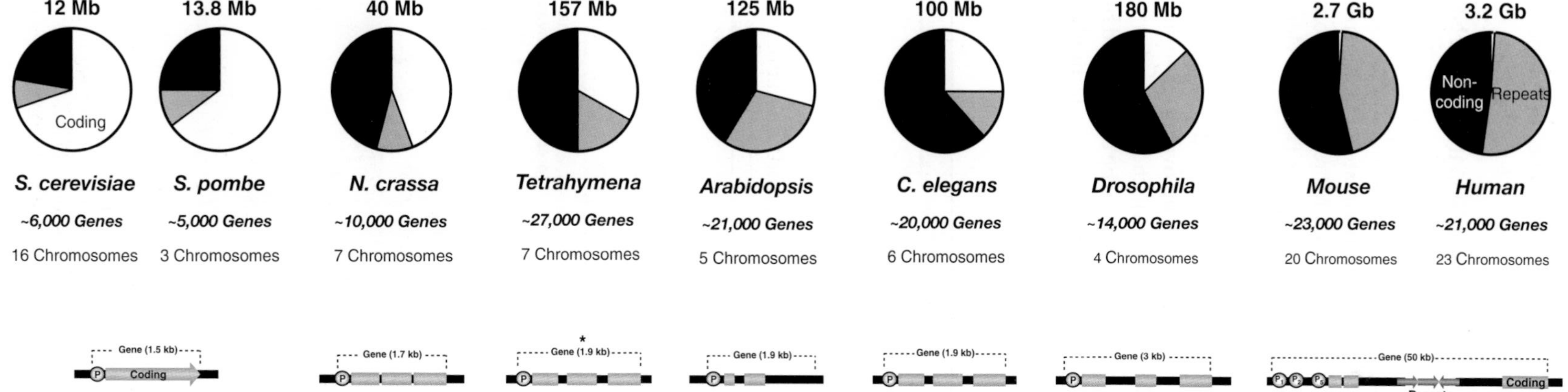

Figure 21. Pie charts of organismal genome organization. Genome sizes are indicated for the major model organisms used in epigenetic research at the top of each pie chart. The enlargement in genome size correlates with the increase of noncoding (i.e., intronic, some regulatory elements) and vast expansion of repetitive (i.e., retrotransposons [LINEs and ERVs (endogenous retroviruses) and other repeats [SINEs (short interspersed elements) and satellite repeats]) sequences in more complex multicellular organisms. This expansion is accompanied with an increase in the number of epigenetic mechanisms (particularly repressive) that regulate the genome (see the table on the inside front cover of this book). Expansion of the genome also correlates with an increase in size and complexity of transcription units, with the exception of plants; they have evolved mechanisms that are intolerant to insertions or duplications within the transcription unit. For *Tetrahymena*, the composition of the micronuclear genome is shown, whereas the common gene structure (asterisk) refers to the gene arrangement in the macronucleus, where gene expression occurs. P, promoter DNA element.

mechanisms to ensure that transposon insertion within transcription units is not tolerated.

There are important organismal differences in the types of epigenetic pathways used, despite high degrees of functional conservation for many mechanisms across species (see the table on the inside front cover of this book). Differences are, in part, related to genome size, wherein the vast expansion of noncoding and repetitive DNA in higher eukaryotes requires more extensive epigenetic silencing mechanisms. This correlates with the fact that mammals and plants use a full range of repressive histone lysine methylation, DNA methylation, and RNAi silencing mechanisms. Another challenge accompanying multicellularity is how to coordinate and maintain multiple cell types (cellular identity). This is partly governed by the delicate balance between Polycomb (PcG) and Trithorax (TrxG) group protein complexes (see Sec. 16; this is further covered in Ch. 17 [Grossniklaus and Paro 2014]; Ch. 18 [Kingston and Tamkun 2014]). The PcG proteins, in particular, correlate with the emergence of multicellularity.

14 EPIGENOMIC REFERENCE MAPS

A representative human genome sequence was first published in 2001, and current research estimates that it contains approximately 21,000 genes (Lander et al. 2001; Venter et al. 2001). More recent genome-wide association studies (GWAS), such as the 1000 Genomes project (http://www.1000genomes.org) and HapMap (http://hapmap.ncbi.nlm.nih.gov), have indicated considerable genetic variation, and revealed numerous regulatory elements and differences in overall genome organization between individuals, which suggests that only 15% of the human genome seems to be constrained by natural selection (Ponting and Hardison 2011). Technology has progressed from microarray, ChIP-chip, and chromosome mapping analyses to

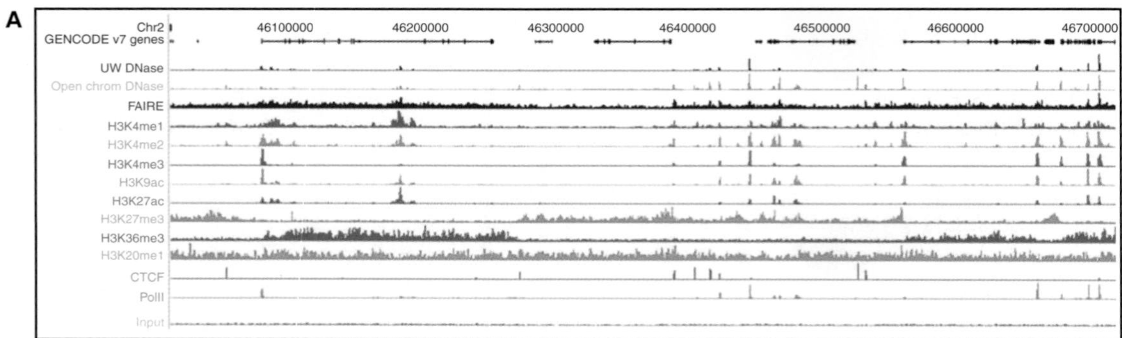

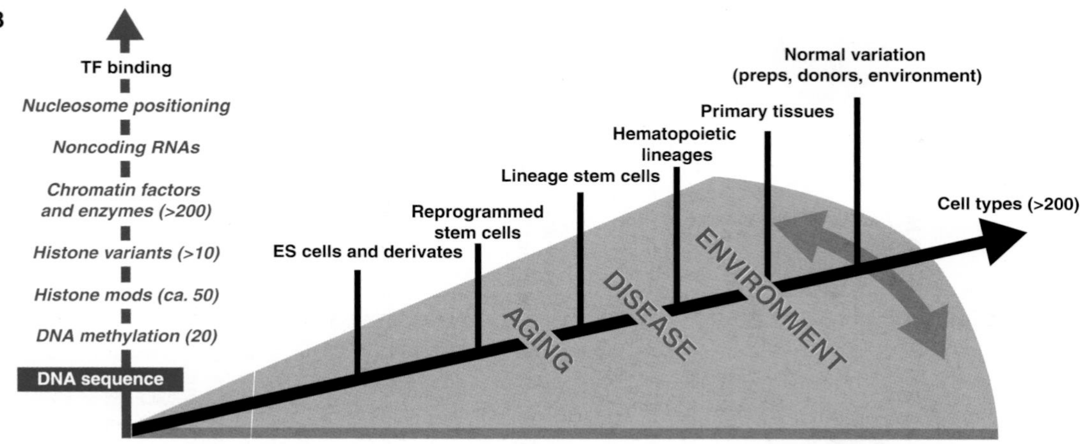

Figure 22. The epigenomic landscape is very large. (*A*) The epigenomic characterization of a 700-kb chromatin region in a lymphoblastoid human cell line (GM12878). Annotated genes are indicated on *top* and signal tracks for open chromatin (blue), histone modifications (red), and CTCF/RNA Pol II (green) are shown. (*B*) In addition to the DNA sequence, the definition of an "epigenome" requires analysis of the chromatin structure and RNA output (*y*-axis) in a given cell type (*x*-axis) under defined differentiation and environmental conditions (gray shaded area). Thus, the analysis of epigenomes is much more complex than the description of the DNA sequence in one genome. (*A*, Adapted from ENCODE Project Consortium 2012; *B*, IHEC Steering Committee, pers. comm.)

NGS or massively parallel sequencing, such that sequencing the entire human genome can be performed within 3 d and with < 10 ng of DNA harvested from a drop of blood. Thus, rather than one or two target genes being analyzed, the entire genome can be examined. Further progress is anticipated as technologies continue to be refined so that sequencing from the genome of a single cell will be possible. Also, detailed analyses of chromatin marking, including bisulfite mapping of DNA methylation (and hydroxymethylation), deep-sequencing of histone modifications, the determination of nucleosome positioning (Segal and Widom 2009), and the profiling of RNA transcripts produced from nearly all genomic sequences (Amaral et al. 2008) will be able to reveal the signatures of epigenetic control that manifest a specific cell type at a given point in time. Even allelic differences are beginning to be resolved by the clustering of long sequencing reads (Xie et al. 2012). Although all these new technologies offer a true revolution in genome and epigenome analysis, they also ask for significant bioinformatic expertise in data processing and interpretation.

Early epigenomic mapping studies, focused on characterizing histone modifications, described novel regions of "bivalent chromatin" (see below), as well as active and repressive histone modifications in several human cell types (Azuara et al. 2006; Bernstein et al. 2006; Barski et al. 2007; Mikkelsen et al. 2007). Later, five "colors" of chromatin were categorized in *Drosophila* by genome-wide mapping of histone modifications and TF binding (Filion et al. 2010). Now, more than 15 distinct patterns of chromatin and more than 50 chromatin signatures (Ernst et al. 2011; Kharchenko et al. 2011) have been described in diverse human cell types (similar to Fig. 22A) (data are accessible via the IHEC data portal; http://epigenomesportal.ca/ihec).

Characterizing the epigenome needs to incorporate at least three dimensions of input: (1) alterations of the chromatin structure, including patterns of DNA methylation, histone modifications, histone variants, nucleosome positioning, TF occupancy, and high-coverage transcriptome profiling for coding and ncRNA transcripts; (2) the definition and access to cell types, preferably primary cells, to establish reference epigenomes for cell type identities; and (3) careful consideration of distinct epigenetic states (i.e., by comparing how the epigenome of "genetically identical" cells responds to physiological and pathological signals from the environment [see Fig. 22B]). Considerable effort has been put into establishing guidelines and recommendations for standardizing the mapping of these reference epigenomes in the context of "time and space." This has largely been spearheaded by the NIH roadmap, the ENCODE program, the European Blueprint platform, and is being integrated by the IHEC, a worldwide alliance that connects several national activities (see Appendix 1 for URLs of the various organizations). A current IHEC canon has been defined as comprising DNA methylation, six instructive histone modifications (Fig. 23), nucleosome positioning, and the readout of the full transcriptome (coding and ncRNA transcripts). These initiatives have described epigenomes for around 60 human cell types (see also Appendix 1), with the goal of identifying 1000 epigenomes over the next 5 years (IHEC standards).

Given the enormous complexity of defining epigenomes accurately, other approaches have proven useful. For example, "comparative epigenomics" between organisms has re-

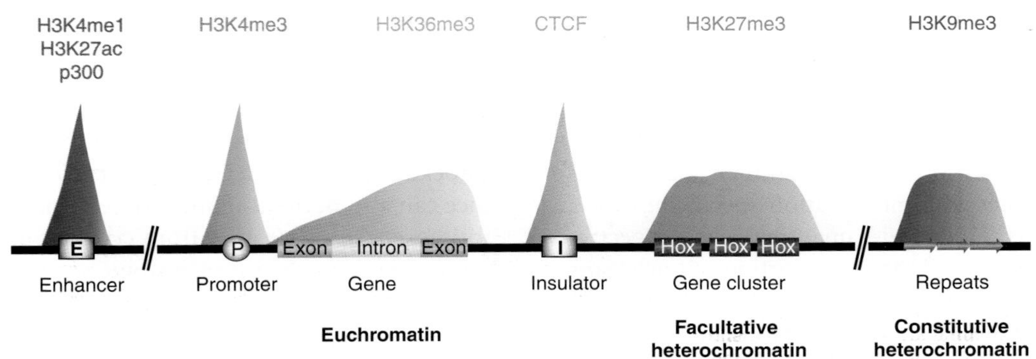

Figure 23. Instructive histone modifications mark functional genomic elements. Epigenomic mapping efforts identified instructive histone modifications and epigenetic factors that correlate with distinct genomic elements: H3K4me1, H3K27ac, and p300 (HAT) peak over enhancers (E), whereas CTCF (zinc-finger factor) accumulates at insulators (I). For active genes, there is an H3K4me3 peak at the promoter (P) and an extended tracking of H3K36me3, which reflects transcription of the elongating RNA Pol II across the gene body. Signatures for repressed chromatin regions are H3K27me3 (facultative heterochromatin, e.g., at Hox clusters) and H3K9me3 (constitutive heterochromatin, e.g., at repeats).

vealed functional genomic elements that would otherwise go undetected, an example being the characterization of enhancer elements (Xiao et al. 2012). Enhancers (see Fig. 23), typified by H3K4me1 and H3K27ac marks (Heintzman et al. 2009; Shen et al. 2012) and "superenhancers" (Whyte et al. 2013), show significant alterations of chromatin signatures in different cell types. Intriguingly, enhancer sequences are preferentially mutated in GWAS studies (Ernst et al. 2011), yet it remains to be seen to what extent genetic (i.e., mutated DNA sequence and single-nucleotide polymorphisms) versus perturbed epigenetic control (altered chromatin configuration) contribute to the development and manifestation of human disease.

In principle, it would seem that active (H3K4me3 and histone acetylation) and repressed (e.g., H3K27me3) chromatin signatures should not coexist in a given localized chromatin region. Nonetheless, a highly informative chromatin signature described as so-called "bivalent chromatin domains" presented a conundrum because they were enriched for both H3K4me3 and H3K27me3 marks. Such bivalent chromatin domains are found at developmentally regulated gene promoters, predominantly in ES cells, but also in other cell types (Bernstein et al. 2006; Mikkelsen et al. 2007; Fisher and Fisher 2011). Although this bivalent chromatin can be retained at some genes on differentiation, it is mostly resolved into active or repressed chromatin states, thereby allowing or preventing the expression of the associated gene loci (Fig. 24A).

The question of what exactly bivalent chromatin is at the single nucleosome level and how it is established has only recently been uncovered with the use of high-resolution biochemical analyses. This allowed the distinction of whether there is physical coexistence of the histone modifications on the same nucleosome and whether they occur on distinct or the same tail of "sister" histones within a single nucleosome particle. The approach needed not only to exclude averaged chromatin enrichments from a heterogeneous cell population, but also to distinguish between different alleles and separate signals from adjacent nucleosomes. The analysis implemented affinity purification of mononucleosomes with modification-specific histone antibodies and was coupled with quantitative mass spectrometry. The results showed that a proportion of nucleosomes are indeed asymmetrically modified for H3K4me3 and H3K27me3 in ES cells, mouse embryonic fibroblasts, and human HeLa cells (Fig. 24B) (Voigt et al. 2012). This discovery, together with mechanistic insight from previous studies, has direct implications for the establishment of bivalent domains; the presence of H3K4me3 (or H3K36me3) inhibits, in *cis*, the activity of the PRC2 complex that catalyzes H3K27me3. Thus, if H3K4me3 (or H3K36me3) is present on one H3 tail, only the sister H3 tail that lacks these modifications can be modified by PRC2, offering a route for the establishment of bivalent chromatin domains. Interestingly, deposition of the histone variant H3.3 via the HIRA-dependent chaperone system plays a role in PRC2 recruitment and H3K27 methylation, but not H3K4 methylation, at select developmental loci in mouse ES cells (Banaszynski et al. 2013). These findings support the emerging notion that H3.3 has multiple functions in distinct genomic locations that are not always correlated with an "active" chromatin state (see Sec. 7).

15 AN EXTENDED GENE CONCEPT AND ncRNAs

There has been a long-standing debate about what the exact nature of a gene is and what possible function the large amounts of noncoding DNA serve in our genomes. Detailed epigenomic maps have highlighted the numerous chromatin markings found at regions devoid of coding sequences or known regulatory regions (e.g., enhancers). This, together with data showing that the entire genome is transcribed (Carninci et al. 2005; Kapranov et al. 2007; Djebali et al. 2012), has been taken to suggest that there is much more functional importance in the (epi)genome than previously described. Indeed, chromatin can be seen as a genome-organizing device (Ridley 2003) and the genome as an "RNA machine" (Amaral et al. 2008). But, actually proving that noncoding DNA and RNA play a role in genome management still requires rigorous testing and ingenuity (Doolittle 2013; Bird 2013).

In the case of mammalian genomic imprinting or X-chromosome inactivation, RNA output is the chief mechanism that discriminates between two identical alleles. Intriguingly, this allelic choice is initially triggered and ultimately dependent on competing transcriptional activities between sense and antisense long noncoding RNAs (lncRNAs). For example, the lncRNA *Air* interferes and inhibits expression of the Igf2r imprinted cluster (Sleutels et al. 2002; Ch. 26 [Barlow and Bartolomei 2014]) and the antisense *Tsix* RNA blocks sense *Xist* transcription (Ch. 25 [Brockdorff and Turner 2014]). Transcriptional interference can also occur at nonimprinted genes, particularly if antisense transcripts arise from cryptic promoters, which can be present in exons or intronic sequences (Venkatesh et al. 2012; Smolle and Workman 2013).

The limited definition of a gene as being delineated by its single TSS and having a fixed length (e.g., ~20–50 kb in mammals) has been challenged by the fact that many mammalian genes have several promoters at their 5′ ends, which are used differentially in distinct cell types to generate alternative mRNAs, encoding a variety of gene product isoforms. Nascent RNA sequencing, by global nuclear run-on or Gro-sequencing, has also revealed paus-

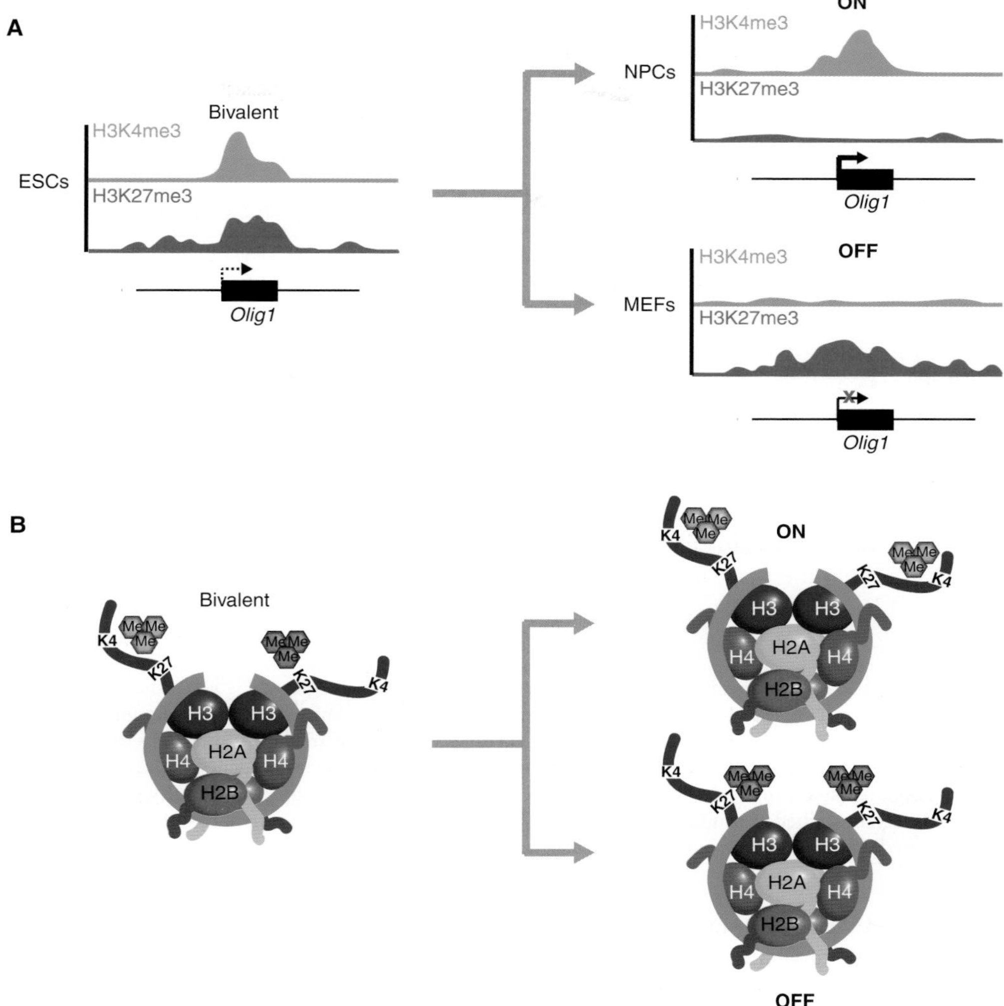

Figure 24. Bivalent chromatin. (*A*) Bivalent chromatin domains are characterized by the coexistence of active H3K4me3 and repressive H3K27me3 histone marks and are often found at promoter regions of developmentally regulated genes in pluripotent cells. Upon differentiation, this poised chromatin state is altered, which results in either gene activity or gene repression. This concept is illustrated for the "bivalent" *Olig1* gene. (*B*) Within a bivalent nucleosome, H3K4me3 and H3K27me3 do not occur on the same H3 tail, but are present on the two different H3 tails. This asymmetric distribution needs to be resolved to convert bivalent chromatin into an active (containing only H3K4me3) or a repressed (containing only H3K27me3) chromatin state. (*A*, Adapted from Mikkelsen et al. 2007.)

ing and divergent initiation at many human promoters (Core et al. 2008; Adelman and Lis 2012). In addition, enhancer RNAs (eRNAs) can modulate promoter-driven transcripts, although eRNAs may be located >100 kb from a promoter start site (see Ch. 2 [Kim et al. 2014]). The identification of so many new RNA transcripts has been facilitated by these new deep-sequencing technologies, able to obtain high RNA transcript coverage, because most of the noncoding or cryptic RNAs are significantly less abundant as compared with mRNAs. It is largely this progress that has extended the boundaries of our classic definition of a gene (Fig. 25). Therefore, a DNA region that demarcates an extended gene locus can span >300 kb, including regions in which newly identified RNA transcript signatures (particularly, the lncRNAs and eRNAs) have been observed. Such a locus probably uses complex regulatory mechanisms in which ncRNAs can have either a stimulatory or inhibitory function on the generation and processing of an mRNA.

One of the first descriptions of an lncRNA, *HOTAIR* (HOX transcript antisense intergenic RNA), described it as being associated with nearby gene repression (Rinn et al. 2007; Ch. 2 [Rinn 2014]), probably via recruitment of members of the Polycomb complex, some of which dis-

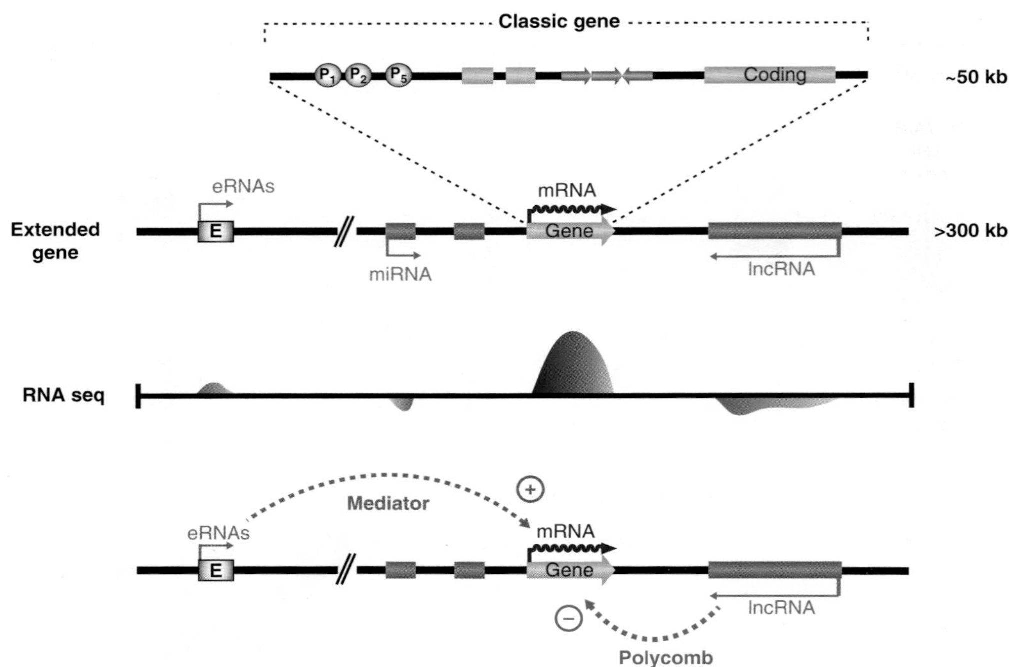

Figure 25. An extended gene concept. An average mammalian gene spans a region of ~50 kb and consists of several alternative promoters (P1, P2, P5), 7–8 exons, and 6–7 introns. Genome-wide RNA sequencing (RNA-seq) indicates that many "classic genes" are embedded within larger regions of transcriptional activity. In addition to the gene locus giving rise to the primary RNA transcript (mRNA), the "extended gene" concept integrates these neighboring regions (spanning >300 kb) that produce ncRNAs (e.g., eRNAs, miRNAs, lncRNAs). Several of these ncRNAs have been described to modulate the expression of the primary RNA transcript by either the recruitment of stimulating (mediator) or repressive protein complexes (Polycomb).

play affinity toward RNA (reviewed in Margueron and Reinberg 2011). Other ncRNAs can have stimulatory functions and facilitate enhancer–promoter communication through the assembly of the mediator complex, which is stabilized by the lncRNA (see Fig. 25) (Lai et al. 2013; Ch. 2 [Kim et al. 2014]).

ncRNAs have been classified based on size (i.e., lncRNAs of >200 bp and short ncRNAs of <200 bp) and function, summarized in Figure 26. Many of the ncRNAs can be very low in abundance (as low as 0.1%) when compared with fully processed mRNA, but clearly serve diverse regulatory functions. Some ncRNAs are compartmentalized in the nucleus, whereas others are exported to the cytoplasm. Of all the ncRNA transcripts that originate from the genome, only ~10% appear to be processed and exported to the cytoplasm. A special case is the germ-cell-specific assembly of piRNAs in the cytoplasmic vicinity of nuclear pores in a structure known as "nuage," which is important for their biogenesis and amplification (Klattenhoff et al. 2009). The many roles of ncRNAs, in addition to their function in TGS and PTGS (see Sec. 11) (Ch. 9 [Allshire and Ekwall 2014]; Ch. 31 [Baulcombe and Dean 2014]; Ch. 16 [Martienssen and Moazed 2014]; Ch. 13 [Pikaard and Mittelsten Scheid 2014]), comprise processing of precursors, gene regulation by enhancer–promoter communication, RNA–protein scaffolds, RNA guidance, recombination, transposon repression, and RNA modifications, particularly of small nucleolar RNAs (Lee 2012; Rinn and Chang 2012; Sabin et al. 2013). The ncRNA field is evolving very fast and a new class of RNAs, circular RNAs, was recently identified and proposed to have functions in PTGS (Memczak et al. 2013).

It is known that RNA can be covalently modified by nearly endless forms of chemical additions to both the sugar backbone and the nucleotide base (Cantara et al. 2011; Machnicka et al. 2013). Although modifications in the sugar backbone primarily protect RNA molecules from nucleolytic attack, chemical additions to the nucleotide base can impart novel regulatory functions. This is particularly evident for N6-methyl adenosine (m(6)A), which is the most abundant RNA modification in coding and ncRNA transcripts (Meyer et al. 2012). m(6)A is reversibly modified by the antagonizing activities of the METTL3 RNA methyltransferase complex and the FTO (fat-mass and obesity-associated) RNA demethylase, which belongs to the family of AlkB dioxygenases (Jia et al. 2013). Moreover, a selective binder (a YTH family protein) for m(6)A has been identi-

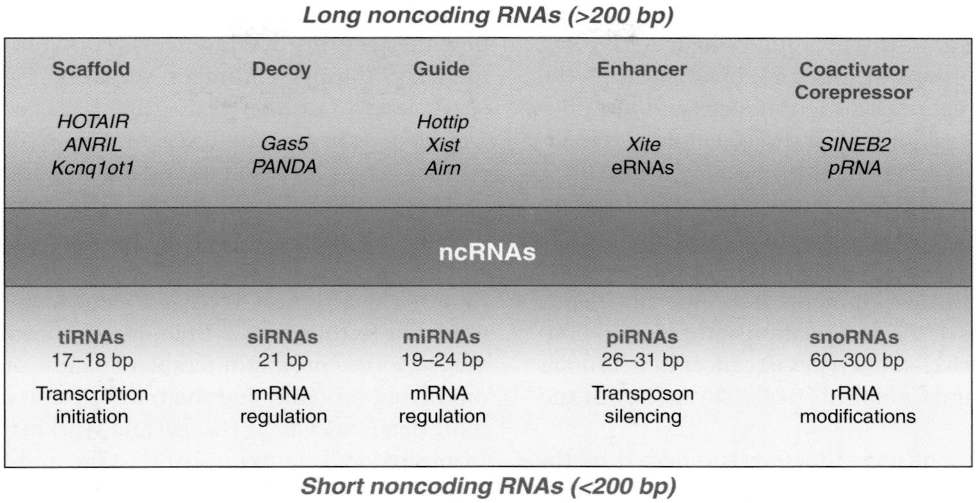

Long noncoding RNAs (>200 bp)

Scaffold	Decoy	Guide	Enhancer	Coactivator Corepressor
HOTAIR *ANRIL* *Kcnq1ot1*	*Gas5* *PANDA*	*Hottip* *Xist* *Airn*	*Xite* eRNAs	*SINEB2* *pRNA*

ncRNAs

tiRNAs	siRNAs	miRNAs	piRNAs	snoRNAs
17–18 bp	21 bp	19–24 bp	26–31 bp	60–300 bp
Transcription initiation	mRNA regulation	mRNA regulation	Transposon silencing	rRNA modifications

Short noncoding RNAs (<200 bp)

Figure 26. Function of noncoding RNAs (ncRNAs). Within the last years, a growing list of ncRNAs has been identified. They can be classified into short ncRNAs (<200 bp) and long ncRNAs (>200 bp). Examples and potential functions of long and short ncRNAs are shown and further described in the text.

fied (Wang et al. 2014), which targets m(6)A-containing RNAs to mRNA decay sites. The perturbed regulation of m(6)A, either through mutations of the obesity risk factor FTO or via experimental inhibition of METTL3, indicates profound roles of m(6)A RNA methylation for development (in particular, in the brain), metabolism, and fertility (Jia et al. 2013). Another example is cytosine methylation of transfer RNAs (tRNAs) by the DNMT homolog DNMT2 (Goll et al. 2006), which is relevant to protect tRNAs from stress-induced cleavage (Schaefer et al. 2010). Recent advances using high-throughput sequencing have made it possible to detect RNA cytosine methylation in many coding and ncRNAs (Squires et al. 2012; Khoddami and Cairns 2013). Together, these interesting findings suggest that RNA methylation may well represent a third type (besides DNA and histone methylation) of modification-based epigenetic control.

The physicochemical properties of RNA molecules to adopt a great variety of three-dimensional structures (e.g., stem loops and hairpins) and to base pair with complementary sequences would further extend their functions in interacting with and/or altering chromatin to regulate transcriptional output. In addition to forming dsRNAs, a nascent RNA can bind to the DNA template and induce a three-stranded nucleic acid structure, known as R-loop. R-loops originate naturally during transcription and have proposed roles in gene regulation (Skourti-Stathaki and Proudfoot 2014). Other than during R-loop formation, it is also plausible that *cis-* or *trans*-acting ncRNA molecules could associate with chromatin and invade the DNA template to form RNA:DNA hybrids or even triple helices (Felsenfeld and Rich 1957). Attractively, DNA (and/or RNA) cytosine methylation would favor Hoogsteen bonding (allowing a triple helix) rather than Watson–Crick base pairing at neutral pH (Lee et al. 1984) and could induce non B-form structures of nucleic acids. Although this is very difficult to directly prove on native chromatin, recent reports have described RNA:DNA hybrid or triple-helix formation in the repression of rDNA genes (Schmitz et al. 2010) and LINE retrotransposons (Fadloun et al. 2013), as well as for the stabilization of certain lncRNAs (Wilusz et al. 2012). A physiological role for RNA:DNA hybrids has also been identified during innate immunity, in which they are detected as novel pathogen-associated molecular patterns in the cytosol by Toll-like receptors (TLR; e.g., TLR9) (Rigby et al. 2014).

16 POLYCOMB AND TRITHORAX

Among some of the main effectors transducing signals to the chromatin template and participating in the maintenance of cellular identity (i.e., provide cellular memory) are members of the PcG and TrxG group of genes, which are the topics of Ch. 17 (Grossniklaus and Paro 2014) and Ch. 18 (Kingston and Tamkun 2014). These genes were discovered in *Drosophila* by virtue of their role in the developmental regulation of the *Hox* gene cluster and homeotic gene regulation. The studies showed that these groups of genes maintain a pattern of gene expression that is initially established by short-lived master regulators (TFs) during early development. PcG and TrxG have since been shown to be key regulators for cell proliferation, cellular identity, and the definition of primordial germ cells (PGCs) (see Ch. 27 [Reik and Surani 2014]) in multicellular eukaryotes. In addition, these groups of genes are involved in several signaling cascades that respond to mitogens and morphogens, regulate stem cell identity and proliferation, vernalization in plants (see Ch. 31 [Baulcombe and Dean 2014]), homeotic transformations and transdetermination, lineage commitment during B- and T-cell differentiation (see Ch. 29 [Busslinger and Tarakhovsky 2014]), and many other aspects of metazoan development. It is not surprising, then, that mutations or the deregulation of these genes contributes and even drives some types of cancer in mammals (see Ch. 35 [Audia and Campbell 2014]; Ch. 34 [Baylin and Jones 2014]).

The broad nature of their function is reflected by the variety of complexes comprising the PcG and TrxG components. The Polycomb and Trithorax group of proteins function antagonistically for the most part; the PcG family of proteins maintain a silenced chromatin state, whereas, in general, the TrxG family of proteins propagate gene activity. These complexes act by modifying chromatin through nucleosome alteration, chromatin compaction, or the placement of H3K4, H3K36, and H3K27 methylation, as well as by monoubiquitination of H2AK119. This section addresses what is known about the PcG and TrxG complexes and how they convert developmental cues into an "epigenetic memory" through chromatin structure.

At least 15 distinct TrxG genes have been identified in *Drosophila*. In general, TrxG proteins maintain an active state of gene expression at target genes and overcome (or prevent) PcG-mediated silencing. TrxG genes encode ATP-dependent chromatin-remodeling enzymes (e.g., Brahma), KMTs (e.g., Ash1, Trx, or Trr), and TFs (e.g., GAGA or Zeste). The action of TrxG ATP-dependent chromatin-remodeling factors, such as Brahma, is important for allowing access to the chromatin template bandy, other chromatin modifying proteins, and the transcription machinery (for more detail, see Ch. 21 [Becker and Workman 2013]; Ch. 18 [Kingston and Tamkun 2014]). Trx and related factors function as KMTs, catalyzing methylation of histone H3 at lysine 4. As described above (Fig. 23), H3K4me1 is associated with enhancers, whereas H3K4me3 demarcates the 5′ region of actively transcribed genes, and the two types of modifications are placed by different TrxG complexes (Fig. 27). This diversification of function in mammals likely correlates with genome complexity. However, how the mammalian complexes are targeted to their genomic location remains an open question. It is likely that MLL1, like the

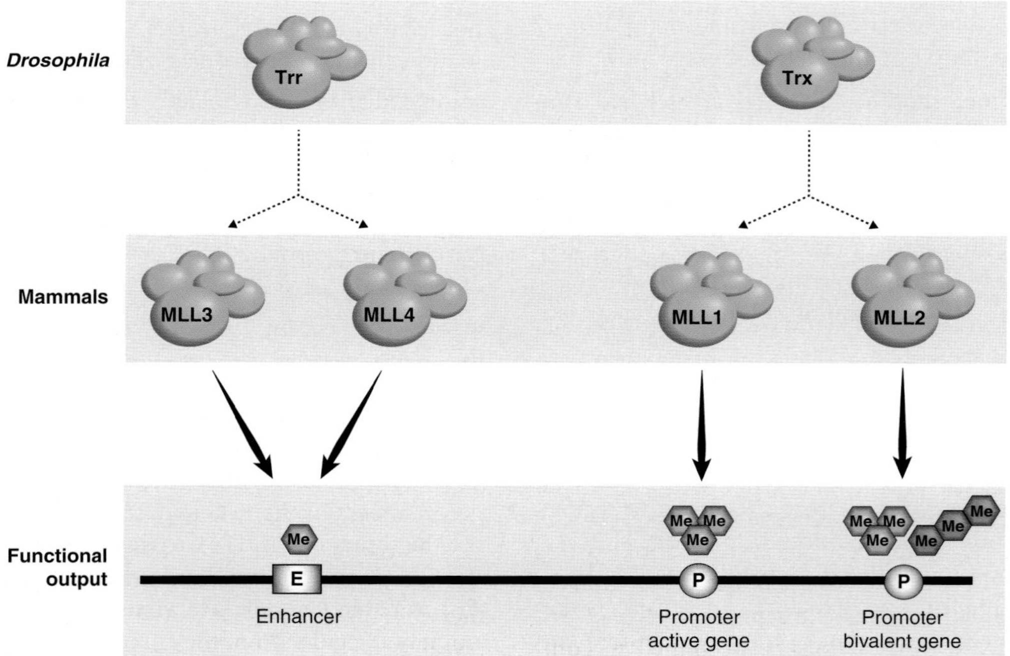

Figure 27. Trithorax complexes and their biological function. In *Drosophila*, there are two Trithorax complexes, Trx and Trx-related (Trr), with distinct biological functions. Trr and its mammalian orthologs, MLL3 and MLL4, specifically mediate H3K4me1 at enhancers. In contrast, Trx and its mammalian orthologs, MLL1 and MLL2, are recruited to promoters, in which they catalyze H3K4me3. In the mammalian system, MLL1 activity has been associated with active genes, whereas MLL2 seems to be specific for bivalent promoters.

Set1/Compass complex (Fig. 14), is directed to target regions by the transcriptional machinery.

Approximately 20 PcG genes have been identified in *Drosophila*. The molecular identification of the first protein in the group, Pc, known to stabilize patterns of gene repression over several cell generations, provided the first evidence for a molecular mechanism for "cellular" or "epigenetic" memory. As well, Pc provided an example of a chromodomain-containing protein with a high degree of similarity to the chromodomain of the heterochromatin-associated protein HP1 (Paro and Hogness 1991). The chromodomain forms an aromatic cage that has been well documented as a specific binding pocket to methylated histone residues (Fig. 7 and elaborated in Ch. 6 [Cheng 2014]).

Biochemical analyses in *Drosophila* point to the existence of two main PcG complexes: PRC1 and PRC2. PcG genes encode products that include DNA-binding proteins (e.g., Pho1), histone-modifying enzymes (e.g., Ring1B and Ezh), and other repressive chromatin-associated factors that contain a chromodomain with affinity for H3K27me3 (e.g., Pc). In *Drosophila*, the recruitment of PcG protein complexes to DNA is mediated via elements called PREs (Polycomb response elements). Equivalent sequences in mammals have remained elusive to date, although some reports (Sing et al. 2009; Woo et al. 2010) suggested the existence of PREs that target the PcG complex to specific loci, yet whether these are true PRE elements remains unresolved. Current models in *Drosophila* support PcG recruitment through interaction with DNA-binding proteins and the affinity of the Pc protein for H3K27me3 modified histones (Beisel and Paro 2011; Simon and Kingston 2013). It is unclear, then, how PcG protein complexes cause long-range silencing in a PRE-dependent manner, as, first, PREs in *Drosophila* are usually located kilobases from TSS of target genes and, second, PRE elements have reduced nucleosome density (Schwartz et al. 2005). These findings, coupled to the fact that *Drosophila* PcG complexes can associate in vitro with nucleosomes that lack histone tails (Francis et al. 2004), at first appear contradictory. The most logical explanation for some of these disparate observations is that PcG binding in vivo initially requires interaction with DNA-bound factors that is, then, stabilized by association with nucleosomes and modified H3K27me3 in the adjacent chromatin region, as illustrated in Figure 13. Clearly, more research is needed to link existing evidence of how PcG complexes are targeted to regions of chromatin and how they mediate repression. Newer evidence suggests distinct interactions between the PcG complexes, which affect their recruitment (described below). It should be noted that, although in some systems there is progress in mechanistic understanding, PcG and TrxG recruitment is likely to be organism dependent, as there is great heterogeneity in the PcG and TrxG complexes (see Ch. 17 [Grossniklaus and Paro 2014]; Ch. 18 [Kingston and Tamkun 2014]).

The mechanism of targeting and action of *Drosophila* PRC2 is illustrated in Figure 13. In mammals, PRC2 is more complex and the specific recruitment of PRC2 to target genes in mammals has remained elusive. Studies have documented that three of its subunits (i.e., Ezh2, Suz12, Jarid2) interact with RNA, suggesting that these interactions may participate in providing specificity in its recruitment to target genes. Once recruited, PRC2 then engages in a series of low-affinity interactions with chromatin (Fig. 28). For instance, we know that PRC2 complex binding to chromatin is aided by the RbAp histone-binding proteins. In ES cells, the complex is also associated with Jarid2, which contains a nucleosome binding activity that facilitates PRC2 recruitment to chromatin as well as a domain that binds with low affinity to C-G rich DNA sequences. As JARID2 is not expressed in differentiated cells, EZH1 (KMT6B) appears to compensate, showing inherent nucleosome-binding activity, in contrast to EZH2 (Son et al. 2013). The core E(z) gene, encoding the enzyme-catalyzing H3K27 methylation, has been duplicated in mammals, giving rise to EZH1 and EZH2, with EZH2 being considerably more potent in activity. The activity of either requires its association with EED and SUZ12, but EZH2 is present in actively dividing cells, whereas EZH1 is found in all cells. Clearly, more work is needed in characterizing the various PRC2 complexes in different mammalian cell types.

As with the extended diversity of the mammalian PRC2 complex(es), PRC1 is equally complex when compared with *Drosophila*. The fly PRC1 complex is composed of four polypeptides and this complex compacts chromatin, establishes H2AK119ub1, and includes a reader for H3K27me3 (Fig. 29). In mammals, this complex has evolved into at least six different types, which are categorized based on the presence of a unique PCGF (Polycomb group Ring finger) protein (Fig. 29A) (Gil and O'Loghlen 2014). The PCGF proteins interact with the enzyme that catalyzes H2AK119ub1 and Ring1B (or Ring 1A), and Ring1 interacts with the CBX proteins, which include a chromodomain that reads H3K27me3. Interestingly, two other proteins, RYBP and YAF2, also interact with Ring1, but in a competitive manner, as all three proteins (including the CBX family) bind to the same surface on Ring1B. This heterogeneity in mammalian PRC1 (Gao et al. 2012), depending on whether they contain CBX, RYBP, or YAF2 (Fig. 29B), has helped to reveal the modes of PRC1 complex recruitment.

The complex composed of PCGF1 includes the demethylase JMJD2 (KDM2B), which can bind to CpG islands via its CXXC domain (Blackledge et al. 2010; see Ch. 2 [Blackledge et al. 2013]). How the other complexes are recruited to their target genes remains unknown; however,

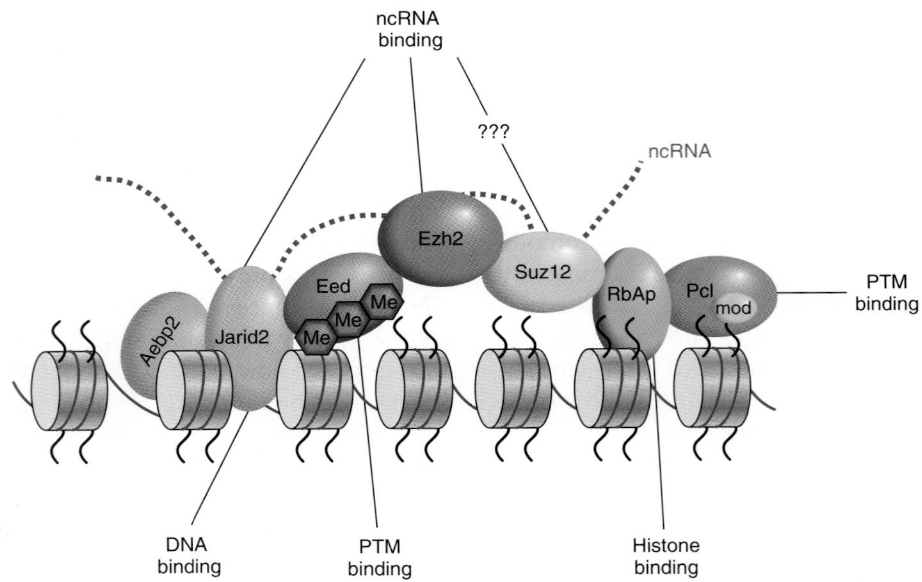

Figure 28. Low-affinity interactions of PRC2 with chromatin. The molecular mechanism of mammalian PRC2 recruitment to target genes remains under investigation (for details see the text). However, once recruited, the complex is stabilized by a multitude of low-affinity interactions of its various subunits with the chromatin template—that is, DNA binding (Jarid2), histone binding (RbAp), PTM binding (Pcl, Eed), ncRNA binding (Jarid2, Ezh2, and possibly Suz12). (Adapted from Margueron and Reinberg 2011.)

the finding that not all PRC1 complexes contain CBX proteins has opened new possibilities of recruitment that are independent of H3K27me3 recognition. Additionally, the diversity in PRC1 complexes is consistent with PRC1 not always establishing robust H2AK119ub1 at target genes. Indeed, chromatin bound by CBX-containing complexes are low in H2AK119ub1, a finding that is not surprising as RYBP stimulates the activity of Ring1b. Further, although initial studies suggested that PRC1 complexes function downstream from PRC2, recent studies established a very distinct mode of PRC1 recruitment, demonstrating that PRC1-catalyzed H2AK119ub1 could recruit PRC2 with resultant H3K27me2/3 catalysis (Blackledge et al. 2014; Cooper et al. 2014; Kalb et al. 2014). These latest findings suggest that PRC1/PRC2-mediated repressive domains can be established on newly replicated chromatin templates, with H3K27me2/3 and H2AK119ub1 fostering PRC1 and PRC2 targeting, respectively, in conjunction with PRC2-mediated propagation of its own modification through its EED component. It should be noted, however, that not all H3K27me2/3-decorated sites contain H2AK119ub1 and vice versa. As well, although the histone modifications catalyzed by PRC2 and PRC1 are associated with gene repression, it is important to note that the ability of these modifications alone to repress transcription has not been reported and is controversial. Nonetheless, these modifications are of paramount importance to the appropriate inheritance of gene expression profiles during development and into adulthood.

17 X INACTIVATION AND FACULTATIVE HETEROCHROMATIN

PcG-mediated gene silencing and X-chromosome inactivation are prime examples of developmentally regulated transitions from the active to inactive chromatin states (see Fig. 30), often referred to as facultative heterochromatin. Facultative heterochromatin occurs at coding regions of the genome, where gene silencing is dependent on developmental decisions that specify distinct cell fates and can be a reversible chromatin state. This is in contrast to constitutive heterochromatin (e.g., at pericentric domains), which may be, by default, induced at noncoding and highly repetitive regions (Fig. 18).

One of the better-studied examples of facultative heterochromatin formation is the inactive X chromosome in mammalian females. Only one of the two X chromosomes inactivates to equalize the dosage of X-linked gene expression with males that possess only one X (and a heteromorphic Y) chromosome. The chromosome-wide gene silencing of the Xi is visible as the Barr body, localized in the nuclear periphery of female mammalian cells (Fig. 30). How the two alleles of the mammalian X chromosomes are counted and how one particular X chromosome is chosen

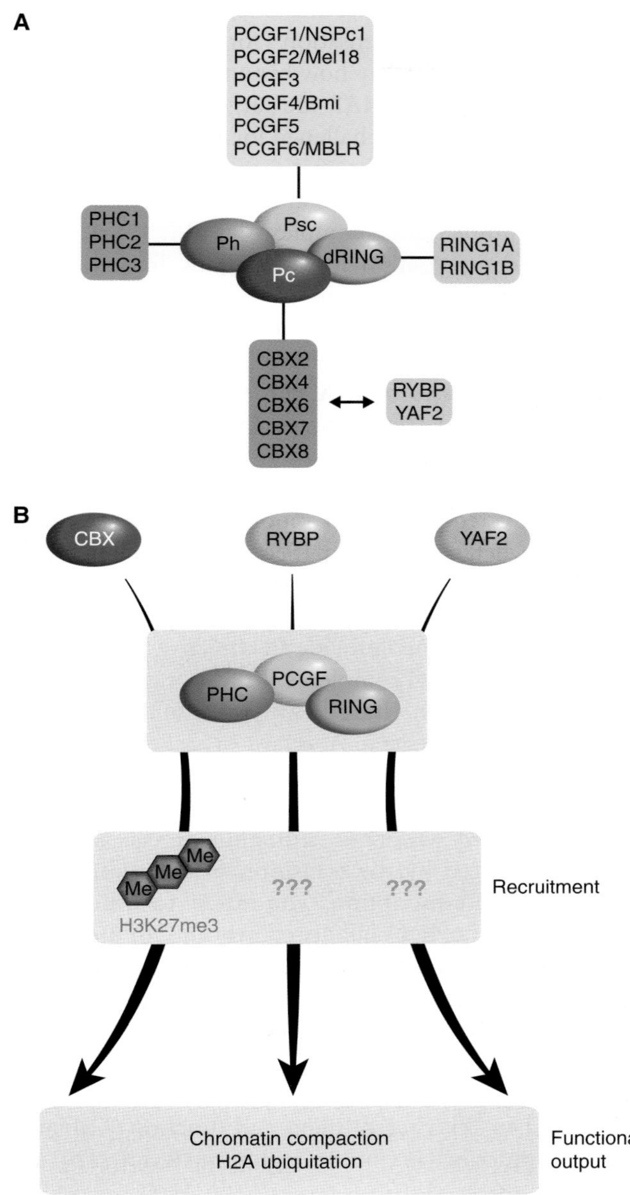

Figure 29. A variety of mammalian PRC1 complexes. (*A*) The *Drosophila* PRC1 complex consists of four core factors (Ph, Psc, Pc, dRing) that mediate chromatin compaction and H2A ubiquitination. In mammals, there are at least six distinct PRC1 complexes that differ by the incorporation of a unique Psc ortholog (PCGF1-6). Several mammalian orthologs for Ph, dRing, and Pc add to the complexity of PRC1. Interestingly, CBX proteins can also be replaced by two other factors, RYBP or YAF2. (*B*) This differential incorporation of CBX, RYBP, or YAF2 impacts on possible recruitment mechanisms and functional output. Although CBX-containing complexes bind to H3K27me3, the recruitment mechanisms for RYPB/YAF2-containing PRC1 remain unknown. Furthermore, CBX-containing PRC1 shows only limited H2A ubiquitination activity, whereas RYBP was shown to stimulate the enzymatic activity of RING.

for inactivation are challenging questions in today's epigenetic research and are addressed in Ch. 25 (Brockdorff and Turner 2014).

In mice, the large (~17 kb) ncRNA *Xist* appears to be the primary trigger for chromatin remodeling at the Xi (Brown et al. 1991; Brockdorff et al. 1992). We also know that the antisense transcript *Tsix* (expressed only "before" the onset of X inactivation) is involved in the initiation of X inactivation, but not through RNAi-dependent mechanisms (Lee and Lu 1999). The X inactivation center (XIC) and DNA "entry" or "docking" sites (postulated to be specialized repetitive DNA elements that are enriched on the X chromosomes) (Chow et al. 2010) are sites where *Xist* RNA associates for it to function as a scaffolding molecule, decorating the Xi in *cis*. A closer look at the XIC region that contains *Xist* and *Tsix* has also revealed other ncRNAs that play various activating and repressive roles involved in the process of establishing X inactivation. Examples include the Rnf12 ncRNA activator (Jonkers et al. 2009; Gontan et al. 2012) and *Jpx* ncRNA (Tian et al. 2010).

More recent studies suggest that YY1 (a zinc-finger TF) may act with dual specificity, on the one hand, binding to distinct DNA sequences at the Xi center and, on the other hand, interacting with a specific domain within *Xist* RNA (Jeon and Lee 2011). Although these findings need further clarification, they potentially could explain selective recruitment of *Xist* by YY1 to one of the X chromosomes and promote the recruitment and action of both PRC1 and PRC2 complexes (Fig. 30). The recruitment of PRC2 appears to be mediated by Jarid2 (da Rocha et al. 2014), which has been shown to bind to RNA (Kaneko et al. 2014), as well as to GC-rich DNA sequences (Li et al. 2010b). The distinction of the function of Jarid2 on the Xi is that it does not rely on PRC2 to bind. The mechanism of PRC1 recruitment appears to be dependent on the presence of H3K27me3, however, alternative mechanisms of PRC1 recruitment to the Xi are possible. Chromatin modifications, PcG complex binding, the subsequent incorporation of the histone variant macroH2A along the Xi, and extensive DNA methylation all contribute to generating a facultative heterochromatin structure along the entire inactivated X chromosome. Once a stable heterochromatic structure is established, *Xist* RNA is no longer required for its maintenance (Morey and Avner 2011; Gendrel and Heard 2014). A similar form of monoallelic silencing is genomic imprinting, which also uses a noncoding or antisense RNA to silence one allelic copy in a parent-of-origin-specific manner (Ch. 26 [Barlow and Bartolomei 2014]).

The general paradigm of dosage compensation, a classic epigenetically controlled mechanism, has also been addressed in other model organisms (reviewed in Ferrari et al. 2014), notably *C. elegans* (Ch. 23 [Strome et al.

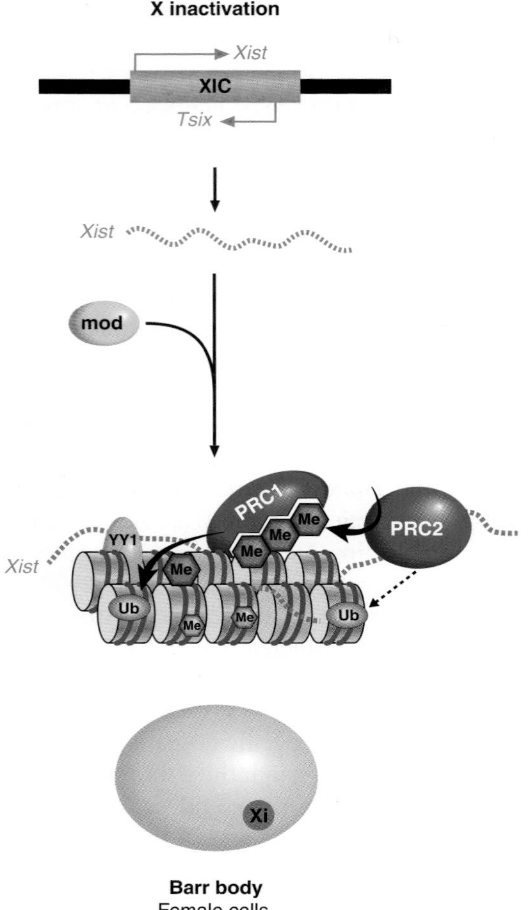

Figure 30. X inactivation. X-chromosome inactivation is triggered and then stabilized by the antagonistic expression of the antisense RNA *Tsix* and *Xist*. *Xist* has been shown to act as a scaffold (e.g., for the recruitment of PRC2), which allows for subsequent establishment of a collection of chromatin changes, including a combination of histone modifications (H3K27me3, H2Aub, and H4K20me1), the binding of repressive complexes (PRC1), DNA methylation, and the presence of histone variants (macroH2A). The inactive X chromosome is visible in the nucleus of female mammalian cells as the "Barr body."

2014]) and *Drosophila* (Ch. 24 [Lucchesi and Kuroda 2014]). It is not clear yet whether dosage compensation occurs in birds, despite the fact that they are heterogametic organisms. In *Drosophila*, dosage compensation between the sexes occurs not by X inactivation in the female, but by a twofold up-regulation from the single X chromosome in the male. Intriguingly, two ncRNAs, *roX1* and *roX2*, are known to be essential components and their expression is male specific (see Ch. 24 [Lucchesi and Kuroda 2014]). Although similar mechanistic details probably exist between flies and mammals, it is clear that activating chromatin remodeling and histone modifications, notably MOF (males absent on the first) (KAT8)-dependent H4K16 acetylation on the male X chromosome, plays a key role in *Drosophila* dosage compensation (for review, see Conrad and Akhtar 2011). Exactly how histone-modifying activities, such as the MOF HAT, are targeted to the male X chromosome remains a challenge for future studies. Further, ATP-dependent chromatin-remodeling activities, such as NURF, are thought to antagonize the activities of the dosage compensation complex (DCC).

Dosage compensation in the somatic lineages of *C. elegans* uses a hijacked version of the condensin complex, which is normally required for chromosome condensation and segregation during cell division. The condensin-related DCC down-regulates both X chromosomes in the hermaphrodite to equalize X dosage with the XO male (see Ch. 23 [Strome et al. 2014]).

There are intriguing parallels between the mechanisms for RNA-directed chromatin modifications occurring at constitutive heterochromatin (Sec. 11), the mammalian Xi, and possibly also for PcG-mediated gene silencing (see Sec. 16). One might postulate that an RNA moiety(s) or unpaired DNA would provide an attractive primary trigger for stabilizing PcG complexes at PREs or compromised promoter function, in which they may "sense" the quality of transcriptional processing. Aberrant or stalled elongation and/or splicing errors could stimulate the interaction between PRE-bound PcG and a promoter, resulting in transcriptional shutdown. Thus, initiation of PcG silencing would be induced by the transition from productive to nonproductive transcription. The extent to which TrxG complexes may use RNA quality control and/or processing of primary RNA transcripts in establishing or maintaining the transcriptional "ON" states is still uncertain. The recent description of the *DBE-T* lncRNA as being able to recruit the TrxG protein, ASH1L (KMT2H), in facioscapulohumeral muscular dystrophy suggests that ncRNAs may be involved in TrxG recruitment and function (Cabianca et al. 2012; described in Ch. 33 [Zoghbi and Beaudet 2014]).

18 REPROGRAMMING OF CELL FATES

The question of how cell fate can be altered or reversed has long intrigued scientists. The germ and early embryonic cells distinguish themselves from other cell compartments, as the "ultimate" stem cell, by their innate totipotency. Although cell-fate specification in mammals gives rise to around 200 different cell types, there are, in principle, two major differentiation transitions: from a stem (pluripotent) cell to a fully differentiated cell, and between a resting (quiescent or G_0) to a proliferating cell. There are, nonetheless, a multitude of different epigenomic "makeups" found in the various cell types during mammalian development.

Looking at embryogenesis, a dynamic increase in epigenetic modifications is detected during the transition from the fertilized oocyte to the blastocyst stage and, then, at implantation, gastrulation, organ development, and fetal growth (illustrated in Fig. 2 of Ch. 27 [Reik and Surani 2014]). Most of these modifications or imprints may be erased when a differentiated cell nucleus is transferred to the cytoplasm of an enucleated oocyte, a process called somatic cell nuclear transfer. However, some marks may persist, thereby restricting some aspects of normal development within cloned embryos, and a few could even be inherited as germline modifications (g-mod) (see Fig. 31A).

The reprogramming of somatic cells was first shown in cloned frogs (*Xenopus*) (Gurdon et al. 1958) and, more recently, by the generation of "Dolly," the first cloned mammal (Campbell et al. 1996). Interestingly, nuclear transfer (NT) of a tumor cell (mouse melanoma) nucleus into an enucleated oocyte alters the tumor spectrum of chimeric mice derived from the reprogrammed ES cells, but was not able to prevent tumorigenesis (Hochedlinger et al. 2004). In fact, tumorigenesis was more pervasive, with higher penetrance and shorter latency. This is an illuminating example showing that cancer is a genetic disease, although has a significant epigenetic contribution affecting the onset and progression of a given tumor type.

One of the more influential discoveries in the reprogramming field in recent years has been the demonstration that somatic cells (fibroblasts) can be induced to become pluripotent stem cells in tissue culture, without the need for NT, first shown in mouse (Takahashi and Yamanaka 2006) and then humans (Takahashi et al. 2007; described in Ch. 2 [Takahashi 2014]). This came on the shoulders of pioneering work on NT, heterokaryon and tetraploid fusion analyses, and the establishment of cell culture systems for ES and adult stem cells (Evans and Kaufman 1981; Martin 1981; Blau et al. 1983; Surani et al. 1986; Spangrude et al. 1988;

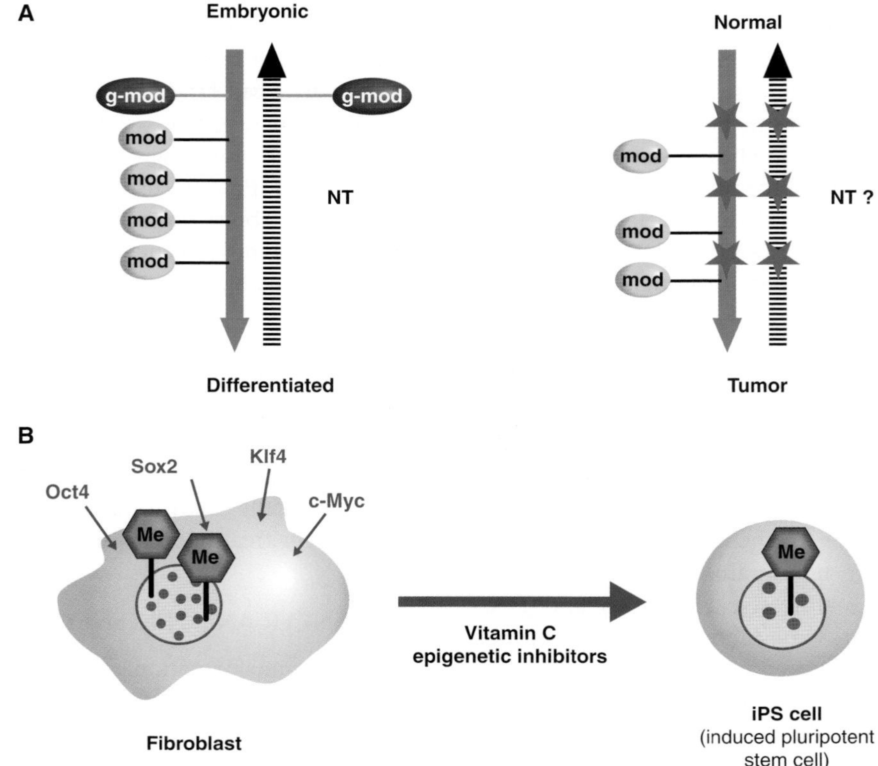

Figure 31. Reprogramming of somatic cells to pluripotency. (*A*) During the lifetime of an individual, epigenetic modifications (mod) are acquired in different cell lineages (*left*). Nuclear transfer (NT) of a somatic cell reverses the process of terminal differentiation, eradicating the majority of epigenetic marks (mod); however, some modifications that are also present in the germline (g-mod) may not be removed. (*Right*) During neoplastic transformation (from a normal to tumor cell), caused by a series of genetic mutations (red stars), aberrant epigenetic marks (mod) accumulate. These epigenetic alterations, but not the DNA mutations, can be erased through reprogramming on NT. (*B*) Reprogramming of somatic cells (fibroblast) into iPS cells by the transduction of four pluripotency factors (Oct4, Sox2, Klf4, c-Myc), also called "Yamanaka" factors. Treatment of somatic cells with epigenetic inhibitors or vitamin C enhances reprogramming efficiency. (*A*, R Jaenisch, pers. comm.)

Terranova et al. 2006). Ultimately, Yamanaka and coworkers showed that Oct4, Sox2, Klf4, and c-Myc, referred to as OSKM or "Yamanaka factors," are the key TFs expressed in ES, but not fibroblast cells. Although transfer of only one or two TFs was not sufficient to reprogram fibroblasts, the reintroduction of all four TFs resulted in the conversion of fibroblasts into so-called iPS cells (Fig. 31B). The mechanism by which these TFs induce reprogramming is detailed in Ch. 28 (Hochedlinger and Jaenisch 2014) and illustrated in their Figure 7. The concept of reprogramming cell type identities by the reintroduction or reactivation of master TFs or "pioneer" TFs (Zaret and Carroll 2011) has also been shown for other cell lineages and was, in fact, the inspiration for the groundbreaking discovery by Yamanaka. In particular, the work on ectopic expression of the master TF myoD, which resulted in the initiation of muscle-specific transcription in a variety of somatic cell types, illustrated this reprogramming potential (Lassar et al. 1986; Davis et al. 1987). Other examples of master TFs found to regulate lineage-specific cell differentiation and reprogramming have been described, particularly, in the hematopoietic system (see Fig. 2 of Ch. 29 [Busslinger and Tarakhovsky 2014]). iPS reprogramming and studies investigating cell lineage specification continue to be areas of intense research, which will hopefully soon be translated to the clinic in the form of viable stem cell therapies.

The conversion of somatic cells to iPS cells in a Petri dish has abrogated the need to use oocytes or NT and revolutionized the reprogramming field. Although the efficiency of reprogramming by the exogenous "Yamanaka factors" was initially very low ($<0.1\%$), technical advances in transducing these factors or altering the tissue culture conditions has greatly improved the generation of iPS cells (see Ch. 28 [Hochedlinger and Jaenisch 2014]). Three major obstacles to efficient somatic reprogramming and cloning in mammals have been identified and are detailed below.

The first major obstacle to efficient somatic reprogramming is the difficulty in reprogramming certain somatic epigenetic marks (e.g., repressive H3K9me3 and DNA methylation). These marks are stably transmitted through somatic cell divisions and even resist reprogramming in the oocyte. Changing the cell culture conditions to inhibit differentiation, such as with the use of 2i media (Ying et al. 2008), supplementing it with nutrients (e.g., vitamin C) that potentiate chromatin activating enzymes, or by using inhibitors of repressive chromatin complexes significantly increased the reprogramming potential (Fig. 31B). Vitamin C can stimulate the activity of jumonji KDMs that remove H3K9me3 marks (Chen et al. 2013a) and Tet enzymes that convert 5mC to 5hmC (Blaschke et al. 2013). Additional results from the inhibition of Mbd3/NuRD repressive complexes, for example, indicate that it may be possible to achieve a nearly synchronous conversion to iPS cells at a very high efficiency (Rais et al. 2013).

A second, closely related concern in somatic cell reprogramming is the fate of imprinted gene loci. For normal embryonic development to proceed, correct allelic expression at imprinted loci is required (see Ch. 26 [Barlow and Bartolomei 2014]). A parent-specific imprint must therefore be established in the germ cell following erasure of preexisting marks (Ch. 27 [Reik and Surani 2014]). Approximately 100 or more imprinted genes fall into the imprinted gene category, largely involved in functions dedicated to resource provision for embryonic and placental development (e.g., Igf2 growth factor). Intriguingly, there is evidence that imprinting may be perturbed during in vitro culture of iPS cells (Stadtfeld et al. 2010).

A third factor is the inability of a somatic cell nucleus to recapitulate the asymmetry of reprogramming that occurs in the fertilized embryo as a consequence of the differential epigenetic marks inherited by the male and female haploid genomes (see Ch. 26 [Barlow and Bartolomei 2014]; Ch. 27 [Reik and Surani 2014]). Although the in vitro systems show rapid but less extensive chromatin alterations, they display perturbed DNA methylation and histone methylation patterns.

With the major progress that has been achieved in the generation of iPS cells, using several human somatic cells, many of the above obstacles can be circumvented. This, coupled with novel gene modification and replacement technology, using for example, zinc finger nucleases or the CRISPR/Cas system (Cong et al. 2013), will hopefully allow the production of genetically compatible material for cellular therapy (see Ch. 28 [Hochedlinger and Jaenisch 2014]). Questions of immunoreactivity of autologous iPS cells, however, persist (Zhao et al. 2011), although personalized medicine by therapeutic cell replacement of reprogrammed adult stem cells remains a very promising approach for treating many human diseases.

19 CANCER AND EPIGENETIC THERAPY

Neoplastic transformation, or tumorigenesis, is regarded as the process whereby cells undergo a change involving uncontrolled cell proliferation. Other hallmarks of cancer include a loss of checkpoint control such that the cells tolerate the accumulation of chromosomal aberrations and genomic aneuploidies, misregulated differentiation, the evasion of cell death (apoptosis), and tissue invasiveness (Hanahan and Weinberg 2011). Cancer has commonly been thought of as being initiated by several genetic lesions, which could include a point mutation, a deletion or a translocation, disrupting either a tumor-suppressor gene, or activating a proto-oncogene. An accumulation of aberrant epigenetic modifications has now also been associated with tumor cells. The epigenetic changes involve altered DNA methylation patterns, histone modifications, and chromatin structure (detailed in the text and Fig. 1 of Ch. 34 [Baylin and Jones 2014]). Thus, neoplastic transformation is a complex multistep process, involving the silencing of tumor-suppressor genes and/or the random activation of proto-oncogenes, through a combination of genetic and epigenetic events (Fig. 32A). The finding that epigenetic dysfunction contributes to cancer has significantly extended the "Knudson two-hit" genetic theory to include "epigenetic" hits as well (Feinberg 2004; Feinberg and Tycko 2004). In a newer model, the epigenetic heterogeneity in a genetically identical cell population has been proposed to facilitate phase transitions (e.g., from a normal to a proneoplastic state) and would be consistent with higher "epigenetic noise" in cancer cells, as it is reflected by increased variability for DNA methylation patterns and altered chromatin signatures (Pujadas and Feinberg 2012).

DNA "hypomethylation" was the first type of epigenetic dysregulation to be associated with cancer (Feinberg and Vogelstein 1983). This has turned out to be a widespread phenotype of cancer cells. At the individual gene level, DNA hypomethylation can be neoplastic because of the activation of proto-oncogenes, causing aberrant cell function or the biallelic expression of imprinted genes (also termed loss of imprinting) (see Ch. 34 [Baylin and Jones 2014]; Ch. 33 [Zoghbi and Beaudet 2014]). On a more global genomic scale, broad DNA hypomethylation, particularly at regions of constitutive heterochromatin, predisposes cells to chromosomal translocations and aneuploidies and can result in the reactivation of transposons, all of which contribute to cancer progression.

DNA "hypermethylation," in many cancers, conversely, is concentrated at the promoter regions of tumor-suppressor genes. Silencing of tumor-suppressor genes through such aberrant DNA hypermethylation is particularly critical in cancer progression. Recent studies have further revealed that histone or chromatin changes may also contribute to tumorigenesis or cancer progression (Fig. 32B). In fact, there is considerable cross talk between chromatin modifications and DNA methylation, demonstrating that more than one epigenetic mechanism can be involved in the silencing of a tumor-suppressor gene (Dawson and Kouzarides 2012; You and Jones 2012; Shen and Laird 2013; see Ch. 34 [Baylin and Jones 2014]).

The reversibility of epigenetic modifications (i.e., "soft adaptation"), in contrast to the irreversibility of genetic mutations ("hard alteration"), has opened up a new horizon for cancer therapeutics, offering epigenetic therapy as a novel approach. The viability of this avenue was first shown in experiments using DNMT and HDAC inhibitors on tumor cell lines (elaborated in Sec. 9 of Ch. 34 [Baylin and Jones 2014]; Ch. 5 [Seto and Yoshida 2014]). Both of these

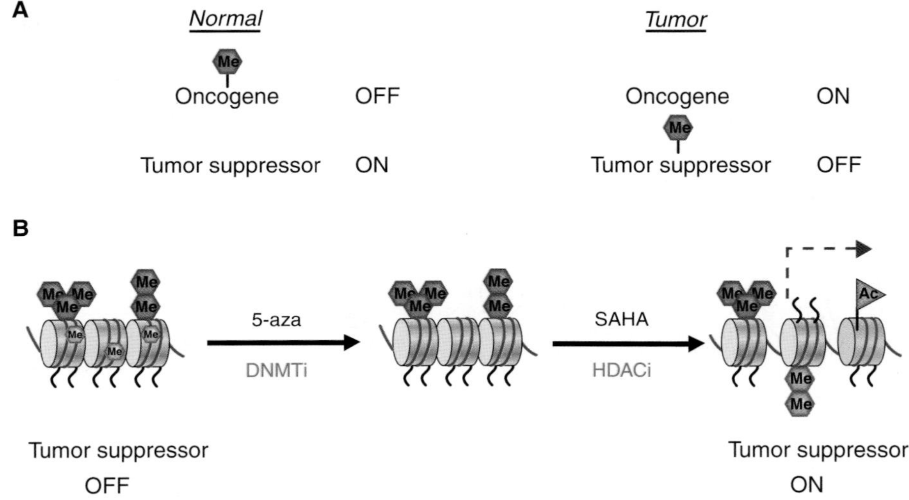

Figure 32. Epigenetic modifications in cancer. (*A*) Aberrant epigenetic marks at cancer-causing loci typically involve the derepression of oncogenes or silencing of tumor-suppressor genes. These epigenetic alterations often include perturbed DNA methylation and histone modifications. (*B*) The use of epigenetic agents for cancer therapy has consequences on the chromatin template, illustrated for a promoter of a tumor-suppressor locus. Exposure to DNMT inhibitors (DNMTi) results in a loss of DNA methylation and treatment with HDAC inhibitors (HDACi) increases histone acetylation, and subsequent downstream modifications, including active histone methylation marks, induction of a nucleosome-free region, and the incorporation of histone variants. These cumulative chromatin changes lead to gene reexpression.

classes of epigenetic inhibitors now have FDA-approved drugs that are in clinical use (e.g., 5-azacytidine and SAHA) (Fig. 32B). These drugs have proven particularly beneficial in cancer cells that have repressed tumor-suppressor genes (Batty et al. 2009; Marks 2010).

DNA methylation and histone modification patterns are proving to be valuable diagnostic markers of tumors. Highly proliferating and undifferentiated (stem) cells often, but not always, have more activating (e.g., histone acetylation) marks, whereas there is a progressive increase in inactivating modifications (repressive histone methylation and DNA methylation) when cells differentiate (Wen et al. 2009), senesce (Narita et al. 2003), or have an impaired nuclear lamina (Fig. 33A) (Shah et al. 2013). Global patterns of histone acetylation and methylation, in particular, are in fact, proving to be hallmarks for the progression of certain cancers (reviewed in Kurdistani 2011; Leroy et al. 2013).

The genetic mutation or deregulation of chromatin factors is indicated in many forms of cancer. These alterations ultimately feed into epigenomic changes, and many of the mutations for chromatin factors found in cancer are listed in the appendices in Ch. 35 (Audia and Campbell 2014). The fact that most chromatin factors are enzymes means they are ideally suited as drug targets for epigenetic therapy, particularly, if they are amplified or overexpressed in human cancer. Examples of such cases include the EZH2 KMT, GASC1 (KDM4C), and JMJD3 (KDM6B), among other enzymes. Many new, but also established, pharmaceutical and research companies are now significantly investing in epigenetic therapy (see Sec. 3 of Ch. 35 [Audia and Campbell 2014]). In addition, international drug discovery consortia, such as the Structural Genomics Consortium (Arrowsmith et al. 2012), are supporting the discovery of new epigenetic medicines through open access research. Inhibitor screens have been successful for HDACs, KMTs, DOT1L, LSD1, and the JMJC KDMs, and many more inhibitor compounds directed against other chromatin-modifying enzymes are currently being explored (see Figs. 5 and 6 of Ch. 35 [Audia and Campbell 2014]).

Although the specificity and efficacy of epigenetic inhibitors are mainly developed and tested in the treatment of cancer, great care is needed to detect possible side effects. For instance, EZH2 inhibitors may be valuable therapeutic

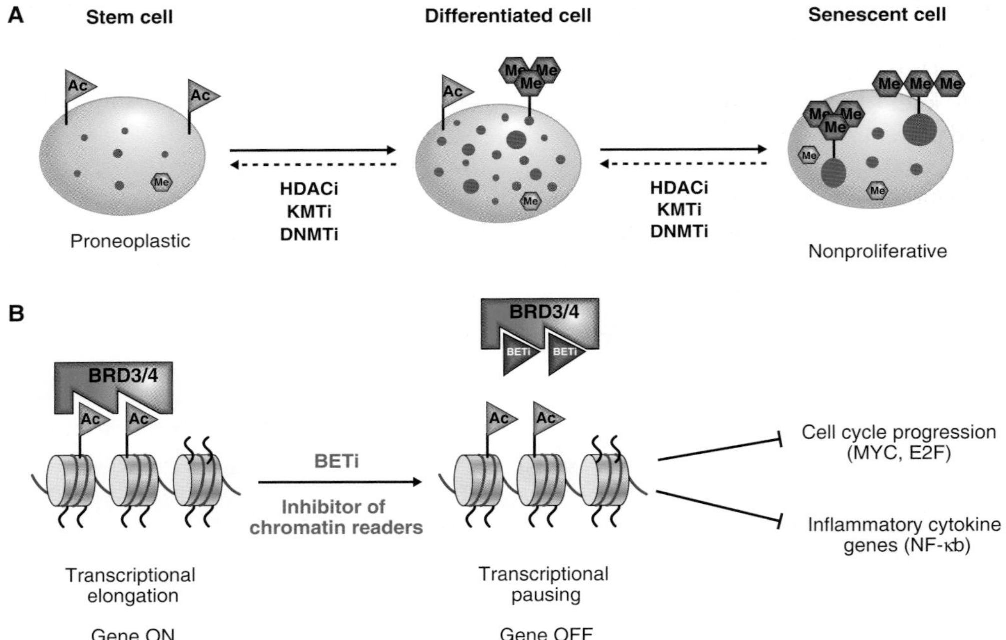

Figure 33. Epigenetic therapy. (*A*) The epigenome of a cell reflects cellular identity and differentiation state. Compared with differentiated cells, stem cells and proneoplastic cells are characterized by more open chromatin (high histone acetylation), whereas senescent and nonproliferative cells develop large heterochromatic foci (high histone and DNA methylation). The goal of epigenetic therapy is to reset the perturbed epigenomic profiles of tumor cells by the use of small chemical inhibitors of epigenetic enzymes (e.g., HDACi, KMTi, DNMTi). (*B*) In addition to chromatin-modifying enzymes, chromatin readers can also be targeted by epigenetic therapy. For example, BET (bromodomain and extraterminal) inhibitors (BETi) specifically block the binding of the BRD3/4 bromodomains to promoter-proximal histone acetylation marks, thereby preventing transcriptional elongation of key target genes (MYC, E2F, and NF-κB).

agents for leukemia and prostate cancer, however, EZH2 also has physiological functions in T-cell activation (Su et al. 2005), thereby strengthening an adaptive immune response (Ch. 29 [Busslinger and Tarakhovsky 2014]). Moreover, almost all histone-modifying enzymes also work on nonhistone substrates, such that establishing the precise mechanism(s) of action of an inhibitor presents a huge challenge. The considerable research on HDAC inhibitors, to date, has revealed that they operate at many levels (Bose et al. 2014) that go beyond transcriptional reactivation of distinct target genes. HDAC inhibitors can sensitize chromatin lesions, inhibit efficient DNA repair (elaborated in Ch. 35 [Audia and Campbell 2014]), and induce genomic instabilities that can trigger apoptosis in tumor cells. A major hurdle in achieving efficacy in chemotherapy treatment is the frequent occurrence of drug resistance thought to be predominantly caused by a lack of effect, possibly on dormant cancer stem cells that do not proliferate. Drug-tolerant states, however, are increasingly recognized to be reversible based on the underlying epigenetic mechanism; for example, drug resistance (as in the case of blocking hyperactive EGF-R [epidermal growth factor receptor] signaling) has been characterized to result from the up-regulation of JARID1A (KDM5A) (H3K4me3 demethylase), but interestingly showed sensitivity to HDAC inhibition (Sharma et al. 2010). Thus, a combination of HDAC inhibitors and JARID1A knockdown generated more fragile chromatin and could prevent the expansion of the dormant cancer cells. Judging from these and other results, it is conceivable that combination therapies using classic intervention (e.g., kinase inhibitors and γ-radiation), together with epigenetic modulation (e.g., HDAC, KMT, KDM or DNMT inhibitors), may be more selective in killing proneoplastic cells (probably even including dormant cancer stem cells) by driving them into information overflow and chromatin catastrophe. Attacking the chromatin makeup underscores the notion that cancer cells may indeed be more vulnerable than normal cells (reviewed in Dawson and Kouzarides 2012).

Epigenetic therapy has also been successful by interfering with the targeting of chromatin-modifying factors, rather than inhibiting their enzymatic activity. A small molecule inhibitor that disrupts the interaction between mixed lineage leukemia (MLL) and the chromatin binder, WDR5, is a recently described example (Senisterra et al. 2013). Another very powerful approach has been the development of inhibitors that can obstruct the binding of a chromatin "reader" to modified histones. This has been successful for a subset of BET factors, namely, BRD3 and BRD4, which read histone acetylation marks via their bromodomains. First-generation inhibitors I-BET and JQ1 are very efficient in blocking the binding of BRD3/BRD4 to promoter-proximal histone acetylation marks, so preventing transcriptional elongation of key cell-cycle genes, such as *myc* and *E2F*, thereby counteracting tumorigenesis. BETi can also stall the central NF-kB promoter and, so, reduce the production of inflammatory cytokines (Fig. 33B). (described in Ch. 29 [Busslinger and Tarakhovsky 2014]; Ch. 2 [Qi 2014]; Ch. 2 [Schaefer 2014]). BETi can, therefore, be used as both anticancer and immunomodulatory epigenetic drugs. BETi illustrate how epigenetic therapy can have a narrower window, enabling more selective gene repression than the inhibitors of more widespread chromatin-modifying enzymes. These exciting results provide an excellent precedent in the development of other antagonists and synthetic histone mimics for additional chromatin "readers" that, for example, contain methyl-binding pockets, such as chromo domains, PHD fingers, etc., as more X-ray structures with and without ligand become available (see Ch. 4 [Marmorstein and Zhou 2014]; Ch. 7 [Patel 2014]).

20 EPIGENETIC FACTORS IN HUMAN DISEASE

More than 10 years after the first description of an entire human DNA sequence, around 1200 genes can be associated with human disease (Lander 2011 and GWAS consortia such as the 1000 Genome Project or HapMap listed in App. 1). Some human diseases are caused by a single gene defect, whereas solid tumors, for example, can accumulate up to 10 gene lesions, and other more complex human disorders may be even more multigenic and have defied a clear genetic definition. Epigenetic mechanisms, with all their impact on packaging and interpreting the genome and their role in stabilizing gene expression patterns and controlling cellular identities, are increasingly recognized as being involved in human disease.

In line with a textbook overview, a word of clarification and a distinction between "genetic cause" of a human disease and "epigenetic contribution" to its onset, aggressiveness, and progression is needed: A DNA alteration (mutation) in a gene, even a gene encoding an epigenetic factor, is primarily a genetic and "not" an epigenetic disease. Despite this crucial difference, progress in both genetic and epigenetic research, together, have opened up new horizons to better combat human disease. For instance, there are greatly improved technologies for gene replacement and genomic engineering, such as via recombinant zing-finger nucleases or CRISPR/Cas (Cong et al. 2013). With these advances, the earlier work attempting to correct a single gene disorder in humans (e.g., *SCID*, Cavazzana-Calvo et al. 2000) and mice (e.g., *Rag2*, Rideout et al. 2002, or *MeCP2*, Guy et al. 2007) would have been greatly facilitated. Similarly, the discovery of somatic cell reprogramming to iPS cells and their use in therapeutic cell replacement (see Ch. 28 [Hochedlinger and Jaenisch 2014]), together with the promise of epigenetic therapy, allow unprecedented approaches for the treatment of human disease.

Diseases with distinct germline gene mutations classify as heritable (familial) disorders. This includes mutations in many epigenetic factors (reviewed in Fodor et al. 2010 and Dawson and Kouzarides 2012), prominent examples of which are highlighted in Figure 34. Well known are the frequent translocations in the *MLL* genes that abrogate the function of the catalytic SET domain in MLL and acute myeloid leukemia (Muntean and Hess 2012) and recurrent somatic mutations in the *UTX* (*KDM6A*) (van Haaften et al. 2009) and *EZH2* genes (Morin et al. 2010) in various forms of human cancer. Other classic cases include mutations in *DNMT3* genes, seen in ICF syndrome, and mutations in *ATRX* (α-thalassemia, mental retardation, X-linked) and *BRG1/SNF5* in pediatric cancers. Most of the identified epigenetic factors encode chromatin-modifying enzymes, in which the disease-associated mutation will abrogate the enzymatic activity. These factors decrease in all major classes of chromatin regulation: that is, histone methylation (KMTs and KDMs), histone acetylation (HATs and HDACs), nucleosome remodelers, signaling kinases, and DNA methylation (DNMTs and TET enzymes) (see Fig. 34). Intriguingly, even histone proteins themselves and their chaperones (DAXX/ATRX) are now proving to be "drivers" of human disease (see Ch. 2 [Liu et al. 2014]); somatic mutations in histone H3.3 (G34R/V) have been documented in pediatric glioblastoma (Schwartzentruber et al. 2012) and H3.3 (K36M) in chondroblastoma (Behjati et al. 2013). Even more surprising are the somatic H3.3 (K27M)-dominant mutations that inhibit PRC2 (EZH2) activity and result in reduced genome-wide H3K27me3 levels in human gliomas (Lewis et al. 2013).

In addition to perturbed protein function, there is also novel insight into the role of ncRNAs in human disease (Croce 2009; Esteller 2011). For instance, irregular profiles of microRNAs (miRNAs) between normal and cancer cells and in neurodegenerative disorders have been detected. Intriguingly, cancer-specific miRNAs are present even in plasma/serum (Chen et al. 2008), tumor exosomes (Taylor and Gercel-Taylor 2008), and the cerebrospinal fluid of patients with central nervous system lymphoma (Baraniskin et al. 2011). The more widespread distribution of

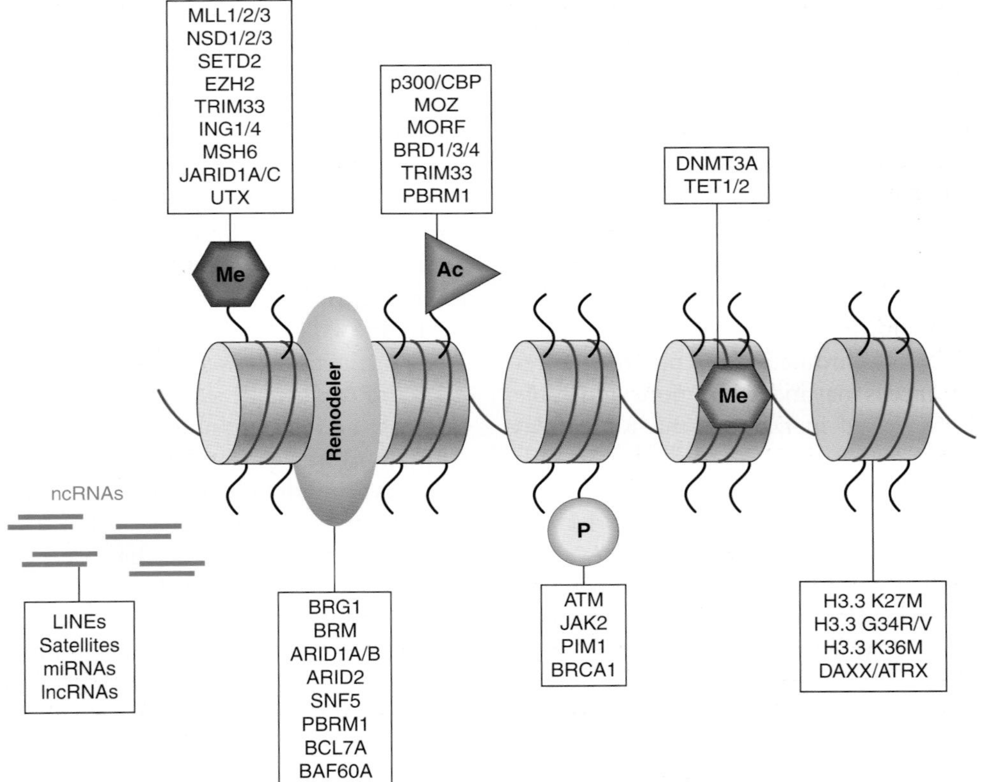

Figure 34. Mutations of the epigenetic machinery in cancer. Many genes encoding factors of the epigenetic machinery have been identified to be mutated in cancer and are summarized in this figure (based on review by Dawson and Kouzarides 2012).

miRNAs means they can potentially be used as surrogate diagnostic biomarkers instead of, or in addition to, biopsy profiling, although caution in the interpretation based on blood cell hemolysis has been noted (Pritchard et al. 2012). Other examples of ncRNAs implicated in human disease include the lncRNA *HOTAIR* (see Ch. 2 [Rinn 2014]), which is aberrantly overexpressed in advanced breast and colorectal cancer and mistargets Polycomb (PRC2) and LSD1 complexes (Li et al. 2013), resulting in perturbed chromatin signatures for H3K27me3 and H3K4me3. Also, dysregulation (i.e., aberrant overexpression) of heterochromatic satellite and other repeat transcripts (e.g., LINEs) in human and mouse cancer cells (Ting et al. 2011), and in *BRCA1*-deficient breast tumors (Zhu et al. 2011), has been correlated with disease progression.

Epigenetic mechanisms involved in the etiology, diagnosis, and progression of complex human disorders are not restricted to cancer, but include diabetes, cardiopulmonary diseases, neuropsychiatric disorders, imprinting disorders, inflammation, autoimmune diseases, and many others (discussed in Tollefsbol 2012; Ch. 33 [Zoghbi and Beaudet 2014]). Several common themes have emerged and some examples will be illustrated. In particular, for metabolic (see also Sec. 21) and neurodegenerative disorders, there is a conspicuous incidence of dysregulation in three classes of genes that encode central epigenetic enzymes (Fodor et al. 2010). These are the HDACs, including SIRTs (sirtuins, i.e., class III HDACs), the DNMT gene families, and genes encoding products that are involved in RNA processing (e.g., triplet repeat expansion). In most of the described cases, the chromatin regulator appears to have an aberrant expression profile without an identified DNA mutation in the gene locus. This malfunction would seem to be disease associated because several of the symptoms can be improved in animal (mouse and rat) models by small molecule inhibitors against HDAC, SIRT, and DNMT activity, demonstrating that epigenetic therapy extends well beyond cancer treatment (see Ch. 35 [Audia and Campbell 2014]).

Epigenetic therapy is hoped to have future prospects in treating psychiatric disorders, depression, and trauma (Holsboer 2008; Schmidt et al. 2011). For instance, epigenetic contributions to behavioral defects and habitual conditioning are starting to be uncovered; the Williams–Beuren syndrome is an example in which the activity of the amygdala, as the emotion and stress response center of the brain, is impaired. This manifests, among other defects, as heightened fear, a lack of social inhibition, but increased empathy (Meyer-Lindenberg et al. 2006). One of the 28 deleted genes encodes the Williams Syndrome Transcription Factor or BAZ1B. This protein is multifunctional, involved in transcription, DNA replication, DNA damage response; notably, it contains a bromodomain, a PHD domain, and a zinc finger, which interact with other factors as part of chromatin-remodeling complexes involved in maintaining heterochromatin (Barnett and Krebs 2011). Based on a hemizygous deletion, and when compared with region duplications, transcript levels of *BAZ1B* fluctuate and become dose sensitive to down-regulation by upstream heterochromatic repeat elements (Merla et al. 2010), thereby failing to fully activate amygdala activity. When considered together with research showing that normal adults are stress and fear conditioned depending on their life-style upbringing, which can be monitored by neuro-imaging of the amygdala (Lederbogen et al. 2011), a connection between genetic and epigenetic effects for age-related anxiety and fear conditioning, which is largely governed by amygdala function, is starting to be unveiled. In another study, childhood abuse has been correlated with increased suicidal risk through the epigenetic imprinting (increased DNA methylation) and down-regulation of glucocorticoid receptor expression in the hippocampus (McGowan et al. 2009).

Beyond the involvement of epigenetic factors in human disease, there is also evidence, in rat and mouse models, for their contribution in habitual functions, such as long-term potentiation (LTP) of learning, and memory and other forms of adaptive behavior. For example, increased Dnmt1 levels in the hippocampus facilitate fear conditioning and learning by silencing the Pp1 memory suppressor gene, whereas pharmacological inhibition of Dnmt1 blocks LTP (Miller and Sweatt 2007; Day and Sweatt 2010). In contrast, attenuation of Hdac2 function by small molecule inhibitors promotes synaptic plasticity and facilitates memory output (Fischer et al. 2007; Guan et al. 2009). Control of cognition and adaptive behaviors has also been shown to be mediated by G9a/Glp (KMT1C/KMT1D) KMTs (Schaefer et al. 2009). The brain is clearly one of the new frontiers for increased epigenetic research in years to come (reviewed in Sassone-Corsi and Christen 2012; Sweatt et al. 2012; Maze et al. 2013; Ch. 32 [Lomvardas and Maniatis 2014]).

In contrast to single gene lesions, complex and multi-genetic diseases have been difficult to dissect, despite GWAS (Manolio et al. 2009) and early EWAS (epigenome-wide association studies) analyses. Indeed, for many of the complex human disorders, making the distinction between purely heritable (genetic) diseases and more complex, non-Mendelian (epigenetic) disorders has been difficult. The problem of missing heritability in GWAS (Manolio et al. 2009) has been approached by using studies on genetically identical (monozygotic, MZ) twins that are discordant for a complex disease (Bell and Saffery 2012). For example, NGS for genetic, DNA methylation, and RNA expression differences in MZ twins (an EWAS study), which are discordant for multiple sclerosis, was performed (Baranzini et al. 2010). However, differences were minimal, with

$<10\%$ of RNA transcripts used as diagnostic markers for multiple sclerosis. Although only a small sample size was used, the data illustrates the dynamic and stochastic nature of gene expression, even among the same cell types of MZ twins, and highlights the difficulty in identifying disease-related gene expression changes or causality of disease-associated epigenetic variants (Bell and Saffery 2012). It should be noted, however, that the interpretation of results from analyses using MZ twins is complicated by the fact that they are not exempt from acquiring somatic mutations or triplet repeat expansions, and may be chromosomally mosaic, or have heteroplasmy for mitochondrial-encoded mutations and other DNA sequence alterations (Boomsma et al. 2002).

Despite the challenges presented in interpreting epigenomic mapping studies, mistakes in epigenetic programming do significantly contribute to the onset and progression of human disease. However, importantly, epigenomic changes are potentially reversible by drug treatments. Thus, to maximize the potential for intervention by epigenetic therapy and in advancing personalized medicine, it is critically important that more comprehensive characterization of epigenetic changes are documented during normal development, adult cell renewal, and disease. Also, the relationships between genetic and epigenetic variation and their impact on health needs further elucidation (Birney 2011; see also IHEC website). With the new advances in NGS (e.g., microfluidic technology), which may also allow single-cell epigenomic profiling (Shapiro et al. 2013) and the use of algorithms to better discriminate cell type heterogeneity in EWAS studies (Zou et al. 2014), this new, non-DNA-based knowledge is predicted to have a major impact for novel avenues in diagnosing and combating human disease.

21 EPIGENETIC RESPONSE TO METABOLISM AND THE ENVIRONMENT

There are around eight major signaling pathways in eukaryotic cells that are universally used in response to changing external and intrinsic signals (Pires-daSilva and Sommer 2003; Alberts et al. 2012). Almost all of the signaling pathways are activated by binding of an extracellular ligand to membrane-anchored receptors, which then transduce the signal via an amplification (often phosphorylation) cascade from the cytoplasm to the nucleus to initiate or redirect a distinct transcriptional program. Examples of extracellular signals include survival factors (e.g., IGF-1), death factors (e.g., Fasl), cytokines (e.g., interferon and interleukins), hormones (e.g., insulin, adrenaline), growth factors (e.g., TGF-α, EGF), Wnt/β-catenin signals (stem cell development), and hedgehog proteins (morphogens). Transduction of the signal results in transcription programs being stabilized and propagated, even in the absence of the initial signal, by alterations in the chromatin structure. This "signaling to chromatin" can involve activation or modulation of chromatin-modifying enzymes, chromatin-associated factors, and/or the modification of histones (Cheung et al. 2000; Schreiber and Bernstein 2002). Thus, it is now well recognized that chromatin is a signal integration platform (Badeaux and Shi 2013) and the intersection between metabolism and chromatin regulation is important for eliciting appropriate epigenetic control (Gut and Verdin 2013).

The environmental signals can either be physiological (e.g., diet, energy metabolism, circadian rhythms, etc.) or pathological (e.g., UV radiation, pathogen infection, etc.) (see examples in Fig. 35). A summary of chromatin factors known to be connected with signaling pathways has been described (Fodor et al. 2010). By way of illustration, the LSD1 and JMJD2C (KDM4C) KDMs can be recruited to target promoters in response to steroid hormone signaling, and the Brg1 (Smarca4) chromatin remodeler interacts with Smad transducers following TGF-β stimulation. In a recent finding, the binding of the Glp KMT to the RelA TF is abrogated by a phospho-methyl switch after NF-κB activation, thereby relieving target gene repression (Levy et al. 2011). The immune system represents a particularly rich area where chromatin responds to signaling pathways; for example, the TLR transduce an inflammation stimulus by establishing gene-specific chromatin signatures that cooperate to control transcription programs (Foster et al. 2007; see Ch. 29 [Busslinger and Tarakhovsky 2014]).

Signaling pathways and chromatin regulation also intersect with intermediary metabolism and the overall energy status of a cell. In a simplistic view, signaling pathways function from outside a cell to the nucleus, metabolic enzymes work in the cytoplasm, the mitochondria, and/or the nucleus, whereas chromatin factors associate with DNA and histones. Recent discoveries, however, have shown that this distinction is no longer as rigid as previously anticipated. Many chromatin factors have roles outside the nucleus and substrates beyond histones (Huang and Berger 2008). Even more surprisingly, metabolic enzymes can be recruited to chromatin (see below) or even bind RNAs (Hentze and Preiss 2010; Castello et al. 2012).

One of the prime examples for additional functions of a chromatin-modifying enzyme is the role of the Ezh2 KMT in promoting cytoplasmic actin polymerization during T-cell signaling (Su et al. 2005). Many more cases for nonnuclear and nonhistone substrates of chromatin-modifying enzymes exist and extend well beyond the documented modifications of the tumor-suppressor protein p53 (Glozak et al. 2005). Further, even histone-modifying enzymes themselves can become substrates for modification, such as

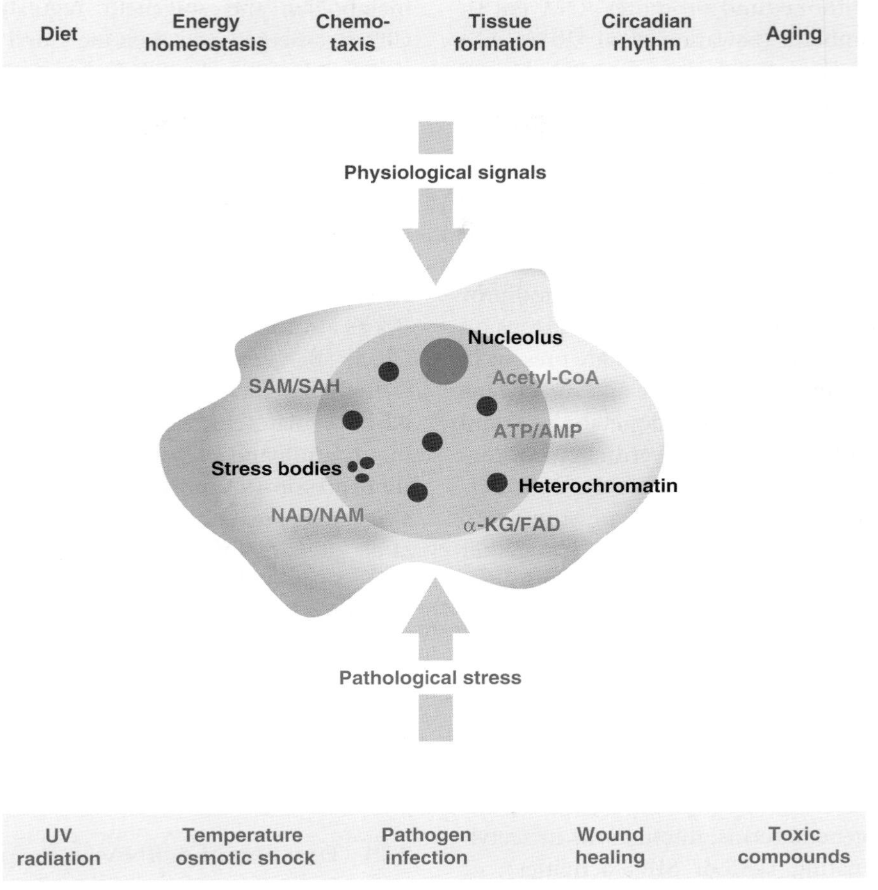

Figure 35. Epigenetic response to environmental signals and metabolism. Environmental conditions, such as physiological signals or pathological stress, are known to influence the epigenome. This is largely mediated via the availability of cofactors that are essential for the activity of chromatin-modifying enzymes. The distinct classes of relevant enzymes and their cofactors are illustrated: KMTs/DNMTs require *S*-adenosylmethionine (SAM); HATs require acetyl-coenzyme A (acetyl-CoA); sirtuins require nicotinamide adenine dinucleotide (NAD); kinases require ATP; KDMs require α-ketoglutarate (α-KG) or Flavin adenine dinucleotide (FAD). (Adapted from Fodor et al. 2010.)

the G9a KMT, which methylates itself in an autoregulatory loop to induce interaction with HP1 (Sampath et al. 2007). Another example is the stimulation of SUV39H1 KMT activity when K266Ac is deacetylated in its catalytic SET domain by SIRT1 (a NAD-dependent HDAC), which is itself upregulated on oxidative stress (caloric restriction) (Vaquero et al. 2007).

In a similar extension of their inherent functions, metabolic enzymes can become enriched in the nucleus and associate with chromatin. The nuclear salvage pathway enzyme NMNAT1 (nicotinamide nucleotide adenylyltransferase), which generates NAD, is recruited to chromatin by SIRT1 to increase local NAD levels, thereby securing SIRT1 activity at distinct promoter regions (Zhang et al. 2009). Another example is the AMP-kinase (AMPK), which senses reduced AMP levels and works as a central enzyme to control energy homeostasis in the cell. In response to stress (low energy), AMPK gets enriched in the nucleus and activates transcription through direct binding to chromatin and histone H2BS36 phosphorylation (Bungard et al. 2010). Energy-dependent adaptations and a switch from catabolic (release energy by glucose breakdown) to anabolic (store energy in fatty acids) metabolism have recently also been recognized as important mechanisms to safeguard function and fate of immune cells (Fox et al. 2005; Pearce and Pearce 2013). Well established is a change in metabolism to highly elevated (aerobic) glycolytic rates in tumor cells, known as the "Warburg effect" (Warburg 1956), that can be used as a diagnostic cancer marker.

Although not a metabolic enzyme, the heat shock factor Hsp90 has dual roles in the cytoplasm (as a chaperone for misfolded proteins) and at chromatin in the nucleus, where it can interact with Trithorax and promotes RNA Pol II, pausing at many promoters (Sawarkar et al. 2012). On environmental stimuli (e.g., the endotoxin LPS), Hsp90 is released from genomic targets, thereby allowing rapid transcriptional response.

The majority of epigenetic regulators are enzymes that require cofactors, such as ATP, acetyl-CoA, NAD, SAM, and others (see Fig. 35) (elaborated in Ch. 30 [Berger and Sassone-Corsi 2014]). In addition to the central energy sensor AMPK, the NAD-dependent sirtuins (HDACs) monitor nutrient availability and are involved in life-span extension and aging (Houtkooper et al. 2012). Sirtuins are present in the nucleus, cytoplasm, and mitochondria and direct gene repression, regulate glucose and fatty acid metabolism, and control energy homeostasis (Gut and Verdin 2013). Sirtuins are among the most versatile transducers of epigenetic control, removing acetylation marks from histones and many nonhistone proteins. In addition, SIRT1, together with the core regulator CLOCK (a HAT), are key enzymes for circadian rhythms (see Ch. 30 [Berger and Sassone-Corsi 2014]). In an attractive illustration of a direct transcriptional/coenzyme feedback loop, SIRT1 can bind and repress the promoter of nicotineamide phosphoribosyl transferase (which is a rate-limiting enzyme in the NAD salvage pathway), thereby modulating the synthesis of its own cofactor NAD (Nakahata et al. 2009). These results specify how, in more general terms, fluctuations in acetyl-CoA/NAD (for alternating HAT or SIRT activities), or SAM/α-KG (for alternating KMT or KDM activities), levels could account for periodic (day/night) gene expression. It has been estimated that ~15% of human genes oscillate according to distinct cell intrinsic clocks (Aguilar-Arnal and Sassone-Corsi 2013).

Catabolites, oncometabolites, and metabolic derivatives can also alter or divert normal epigenetic response to environmental stimuli and metabolism. The case of recurrent somatic mutations in isocitrate dehydrogenase (*IDH1* and *IDH2*) in brain (Turcan et al. 2012) and leukemic (Figueroa et al. 2010) tumors illustrates how a neomorphic mutation can produce a knockon effect on epigenetic dysfunction: the *IDH* mutations accumulate an excess of the abnormal 2-hydroxyglutarate, which is a competitive inhibitor of JMJC KDMs and TET enzymes (reviewed in Ward and Thompson 2012; see Fig. 4 of Ch. 30 [Berger and Sassone-Corsi 2014]). The resulting attenuation of these chromatin-modifying enzymes leads to elevated H3K9me3 marking and DNA hypermethylation and a differentiation block typical of many cancer types (discussed in Ch. 34 [Baylin and Jones 2014]).

SAM is the major methyl donor for most methyltransferases (DNA, RNA, histone, and nonhistone) in the cell, and SAM metabolism is a classic paradigm showing how metabolism and epigenetic regulation relate (Fig. 3 of Ch. 30 [Berger and Sassone-Corsi 2014]). The notion that SAM metabolism is linked to chromatin dates back to 1996 when it was reported that the gene-encoding SAM synthetase, *Su(z)5*, is a strong *Su(var)* modifier in *Drosophila* (Larsson et al. 1996). More evidence came from studies in *C. elegans*, showing that mutants in SAM-synthetase derepress heterochromatin and disrupt its localization to the nuclear periphery (Towbin et al. 2012). In humans, the MAT1 ortholog uses folate and vitamin B_{12} as a source from which to generate SAM (see Fig. 3 of Ch. 30 [Berger and Sassone-Corsi 2014]). In a mouse model, higher availability of methyl donors, which were provided in a folate-enriched diet to mothers, increased DNA methylation and silencing of repeat-associated gene (*agouti*) expression (Waterland et al. 2006); this, particularly, affected epigenetic variation in the offspring (Dolinoy et al. 2006; Dolinoy and Jirtle 2008). More recent studies using a paternal high-fat diet (HFD) in rats (Ng et al. 2010) or a low-protein diet in male mice (Carone et al. 2010) indicated (modestly) elevated levels of many genes in the pancreas and liver that direct lipid and cholesterol biosynthesis. The aspects of possible epigenetic transmission of diet-associated and other risk factors to the next generations are detailed in Section 22.

21.1 Function of ncRNAs in Response to Environmental Stimuli

ncRNAs and, in particular, miRNAs have, for example, been involved for the adaptation of synaptic function and memory formation in mammals (McNeill and Van Vactor 2012) and *C. elegans* (see below). Moreover, ncRNAs may also be relevant for adaptive behavior in eusocial insects (ants), in which caste-specific differences in the expression of miRNAs have recently been detected (Bonasio et al. 2010) that are further reflected by significant changes in DNA methylation patterns (Bonasio et al. 2012; Zhou et al. 2012). These interesting studies suggest that pheromones and ecdysteroids (cuticular hydrocarbons) can induce an altered RNA transcriptome in chemosensory neurons, thereby directing social behavior. This is similar to the paradigm of phenotypic polymorphism of genetically identical honeybee larvae, in which the discrimination to develop into the "queen" or "workers" is governed by differential feeding with "royal jelly," which has been shown to inhibit DNA methylation (Kucharski et al. 2008), and also contains HDAC inhibitors (Spannhoff et al. 2011). Another example is odor adaptation in the olfactory neurons in *C. elegans*. This organism is innately attracted to odors that require the activity of the *ODR-1* gene (guanyl cyclase). Odor adaptation involves down-regulation of *ODR-1* by endo-siRNAs that mediate binding of HP1-like (HPL-2) and the induction of repressive chromatin (Juang et al. 2013). Together, these striking illustrations for environment-driven adaptive behavior appear to manifest an epigenetic pathway that relies on ncRNAs and successive chromatin changes. Silencing mechanisms of olfactory genes to allow for allele-specific expression of odor receptors and epigenetic adaptations to brain function are detailed in Ch. 32 (Lomvardas and Maniatis 2014).

22 EPIGENETIC INHERITANCE

As discussed above, environmental exposure (e.g., diet or stress) can change the epigenetic signatures of chromatin. If these alterations could be stably propagated in the germline, then epigenetic modifications would offer a molecular explanation for the inheritance of acquired traits, as put forward by Lamarckian evolution theory. There have been significant new insights, across several model organisms, that are consistent with transgenerational epigenetic inheritance. Many questions are, nonetheless, unresolved and a critical debate toward its evolutionary benefits persists, particularly if acquired traits weaken the phenotype (Bird 2013; Grossniklaus et al. 2013; Heard and Martienssen 2014). The advances, however, allow a better understanding of some of the molecular mechanisms and chromatin alterations that manifest in transgenerational epigenetic inheritance. Current research points to three potential carriers that can transmit epigenetic information through the germline: chromatin state, DNA methylation, and ncRNAs (Grossniklaus et al. 2013). Here, we first discuss the observations of nongenetic transgenerational inheritance, followed by new insights in the mechanisms that may be involved in these processes.

There are well-known epidemiological studies in humans documenting paternal (the Överkalix study, see Kaati et al. 2002; Pembrey et al. 2006) or maternal (Dutch hunger winter, reviewed in Hales and Barker 2001) exposure to environmental stress factors, such as famine and undernutrition, that correlate with a higher vulnerability for metabolic and cardiovascular disorders in the next generations. The epigenetic mechanisms, which underpin such responses and their transmission, remained unclear (reviewed in Gluckman et al. 2009). Recent studies have been performed in animal models using NGS to detect possible differences in chromatin modifications (DNA methylation and histone marking) and RNA transcriptome profiles. For instance, a paternal HFD in rats (Ng et al. 2010) or low-protein diet in male mice (Carone et al. 2010) resulted in only in a modest deregulation of many genes (more than 1500) in the pancreas or livers of analyzed offspring, but indicated a distinct, although slight, increase in DNA methylation at a putative enhancer regulon (Pparα), which controls key genes in lipid and cholesterol biogenesis (Carone et al. 2010). Interestingly, diet-induced changes in RNA content (including many ncRNAs) and retention of nucleosomes and histones could be found in sperm. Comparable mouse work, looking at in utero undernourishment, to analyze response of imprinted gene expression identified locus-specific, rather than broad, perturbations of DNA methylation in sperm of second-generation offspring (F_2 descendants) that may be contributing to later-life metabolic syndrome (Radford et al. 2014).

Epigenetic inheritance is clearly indicated by the non-Mendelian segregation of a genotype as gauged by phenotype; epigenetically mutant mice can "remember" wild-type (wt) ancestry for more than 20 generations (see Ch. 14 [Blewitt and Whitelaw 2013]). Furthermore, a wt allele can be suppressed for several generations in *C. elegans* offspring when chromatin modifiers are mutated (Greer et al. 2011). Since the time of the seminal discovery of dsRNA-mediated gene silencing in *C. elegans* (Fire et al. 1998), this model organism has continued to provide illuminating insights into non-DNA-based inheritance (e.g., disabling the RNAi machinery abrogated multigenerational epigenetic transmission) and resulted in the germline losing its immortal character (Buckley et al. 2012). Plants, sometimes referred to as the "masters of epigenetic control," are well documented as using epigenetic inheritance for adapting to a changing environment (Henderson and Jacobsen 2007; Ch. 13 [Pikaard and Mittelsten Scheid 2014]). Recent findings showed that isogenic *Arabidopsis* lines stably segregate acquired traits, such as altered flowering time and root length, because epigenetic marks in the form of DNA methylation imprints were stably segregating for many generations (Cortijo et al. 2014). These examples of non-Mendelian and non-DNA-based inheritance still pose the old questions: How can offspring remember an environmental condition to which they were never exposed? How can a signal be transmitted from a somatic cell to germ cells? And, what could be the molecular mechanism(s) for this epigenetic information?

Long-established paradigms for the non-Mendelian propagation of altered gene expression in offspring comprise paramutation in maize (McClintock 1951) and PEV in *Drosophila* (Muller 1930). In these classic examples, variegation of an allele (the "epiallele") is associated with a DNA rearrangement that either involves insertion of retrotransposons in the proximity of the epiallele or its juxtaposition to repeat-rich, mostly heterochromatic chromatin regions (explained for *Drosophila* PEV in Ch. 12 [Elgin and Reuter 2013]). In maize, paramutation of the b1 locus (which regulates seed pigmentation) is dependent on several intracisternal A particle (IAP) retrotransposon repeats that have integrated ~100 kb upstream of the b1 locus, which then repress the activity of the linked b1 gene. This repressed b1 epiallele can also communicate in *trans* and instruct the naïve b1 allele to establish heritable gene silencing (Chandler and Alleman 2008). Mutants in RNAi and RNA processing, but not DNA methylation or chromatin modification, neutralize this paramutation (Arteaga-Vazquez and Chandler 2010). In mice, stochastic variation of epialleles, for example, *agouti* (*Avy*), is also associated with repeat insertions in proximity to the linked gene locus (see Ch. 14 [Blewitt and Whitelaw 2013]). These studies would suggest that not every gene locus qualifies as an epiallele, but juxtaposed repeat elements (mostly IAP and LINE retrotransposons) and their associated heterochromatic domains can frequently sensitize the linked gene for epigenetic silencing (see Fig. 36). In an intriguing extrapolation of this notion, it would not be inconceivable that repeat-rich genomes (e.g., mammals) could provide an adaptive benefit in facilitating epigenetic response and assisting a transgenerational memory of gene expression states. Indeed, in addition to their roles in genome evolution, retrotransposons (LINEs and ERVs) have been shown to modulate expression of associated genes (Han et al. 2004), and may contribute to germ or stem cell diversity (Lu et al. 2014), as is further suggested by their differential activity in iPS cells of chimpanzees or humans (Marchetto et al. 2013).

The mammalian oocyte genome is packaged by regular nucleosome-containing chromatin and germ-cell-specific histone variants that activate the paternal genome on fertilization (Ooi and Henikoff 2007; Shinagawa et al. 2014). In contrast, the sperm genome is largely made up of protamines, although there is a significant amount (~2%–10%) of canonical and variant histones (Hammoud et al. 2009; Erkek et al. 2013). This residual nucleosome retention occurs at distinct loci and over large gene-poor regions (Carone et al. 2014), thereby allowing chromatin imprints also in sperm. Transgenerational epigenetic inheritance could thus use nucleosome positioning, histone modifications, and DNA methylation in both the female and male germline. Although these mechanisms certainly contribute, RNA molecules are likely to be the central carriers of transgenerational epigenetic inheritance in most multicellular eukaryotes because not all model organisms have DNA methylation (see the table on the inside front cover of this book). The mutant analyses in maize and *C. elegans* (see above) indicate particular sensitivity toward defective RNAi and RNA processing, and only RNA molecules could confer sequence complementarity to a target DNA region.

There is, in fact, overwhelming evidence across many model organisms that RNA molecules are mobile epigenetic vectors (see Fig. 36). In plants, pathogen-infected leaves can send RNA signals to alarm uninfected leaves and initiate a defense program. These mobile RNAs can be short range (via cell-to-cell contact) or long range (via the vasculature) (see Ch. 2 [Dunoyer et al. 2013]; Ch. 31 [Baulcombe and Dean 2014]). *Arabidopsis* nurse cells transduce

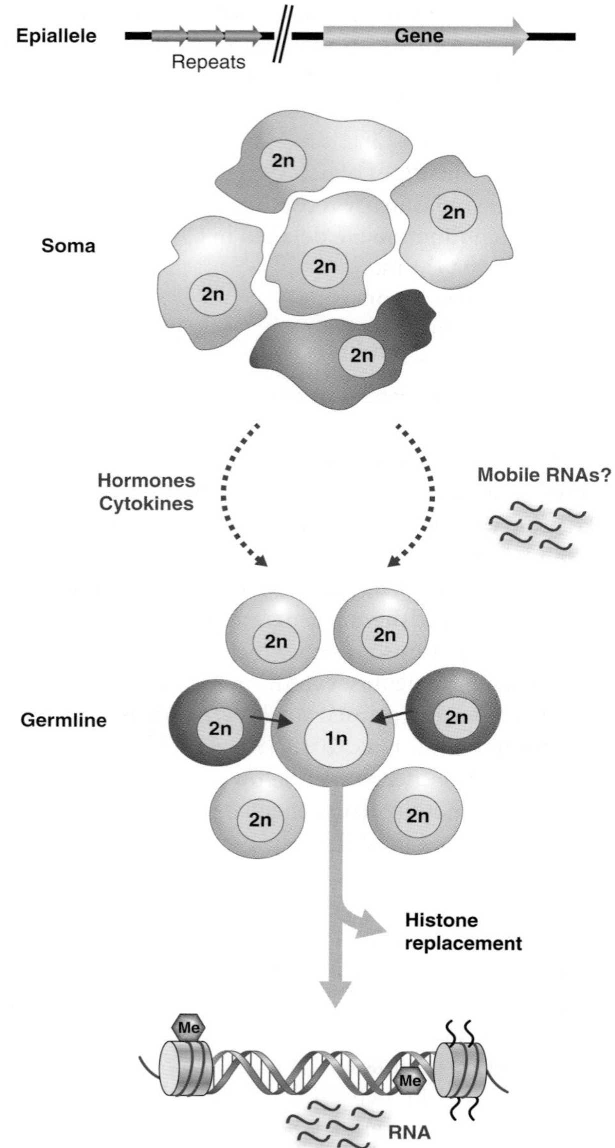

Figure 36. Epigenetic inheritance. The transcriptional response (e.g., silencing) of an epiallele can occur stochastically within a genetically identical population of somatic cells, as indicated by the single cell in dark gray shading. To inherit such an altered expression state, this epigenetic information must be transferred to the germ cells. Possible transducers for epigenetic information are mobile RNAs and hormones or cytokines. Within the germ cell epigenome, an epigenetic imprint (e.g., RNA association, histone modification, nucleosome position, and DNA methylation) can then be established and propagated to ensure transmission to the next generations.

ncRNAs to germ cells to protect them from retrotransposon activation, similar to the RNA scanning of repeat elements in the newly developing macronucleus in *Tetrahymena* (see Ch. 11 [Chalker et al. 2013]; Ch. 16 [Martienssen and Moazed 2014]). In *Drosophila*, a naïve transgenic locus can be converted into a strong *trans*-silencing element when piRNAs are up-regulated. This silencing capacity is induced and propagated by the transmission of Piwi-bound piRNAs in the maternal cytoplasm and can be transmitted over 50 generations (de Vanssay et al. 2012). Further, dsRNAs are diffusible effectors for gene silencing in *C. elegans* (Alcazar et al. 2008), and microinjection in germ cells of small RNAs complementary to the *fem-1* gene modulates feminization in *C. elegans* for many successive generations (Johnson and Spence 2011).

ncRNAs have been detected in human (Krawetz et al. 2011) and mouse sperm and display altered abundance in response to diverse environmental risk factors, such as smoking (Marczylo et al. 2012) or obesity (Fullston et al. 2013). Most instructive were experiments using RNA transfer into the early mouse embryo by microinjection of distinct RNA oligonucleotides into fertilized oocytes. Mice derived from these "primed" oocytes, as a result of the transcriptional modulation of sequence-identical target genes, displayed either altered skin color, heart development or overall body size (Rassoulzadegan et al. 2006; Grandjean et al. 2007; Wagner et al. 2008). These RNA-mediated "paramutations" can be backcrossed for up to four generations before they revert to the wt phenotype (Grandjean et al. 2007). Recent studies exploring trauma conditioning in mice extended this work and showed elevated levels of certain miRNAs and other ncRNAs in sperm of traumatized males. Strikingly, microinjection of sperm RNA from these traumatized males into wt fertilized oocytes transmitted the behavioral (trauma) and metabolic phenotypes to the first and second generation (Gapp et al. 2014). These results provide a captivating demonstration that sperm can deliver more than just the DNA genome.

Other signals that are rapidly distributed among many cells in the body (including germ cells) are hormones, cytokines, and odorants. Cytokine concentrations are particularly sensitive to environmental stimuli, as they can show a >1000-fold increase on trauma or infection. Hormone and olfactory receptors are also expressed in sperm (Goto et al. 2001) and could, therefore, directly respond to elevated hormone/cytokine levels in the blood and lymph vasculature. New work, using acetophenone (odorant) as a fear-conditioning agent in mice, results in reduced DNA methylation of the *Olfr151* (odorant receptor) gene in sperm. This sperm imprint can be transmitted, even by in vitro fertilization, to offspring that show increased *Olfr151* expression in olfactory neurons and enhanced sensitivity to acetophenone, although they have never been exposed to this stimulus (Dias and Ressler 2014).

Together, these examples represent exciting new findings, yet caution remains about whether mammalian germ cells are indeed this plastic and capable of integrating non-DNA-based information that can then be transmitted to the next generations. Much more work is still needed to strengthen the case for the molecular mechanism(s) that would allow an epigenetic signal to be passed on by the germ cells to the zygote and then be propagated, even through the significant epigenetic reprogramming that occurs in the developing embryo, to the newly forming PGCs (described in Ch. 27 [Reik and Surani 2014]), and also be able to mediate a distinct gene expression profile in somatic cells of the adult.

The most plausible mechanism for nongenetic transgenerational inheritance is via an RNA-based transmission of epigenetic information from germ cells to the early embryo. This is likely to work through the association of RNA to complementary DNA sequences, which could then induce a chromatin change (histone modifications and/or DNA methylation). This marked chromatin could then be further stabilized and propagated by a feedforward loop (e.g., see Fig. 13 in Sec. 8) in the developing "somatic" cells, even in the absence of the initial signal (described in the following section). However, for transmission to subsequent generations, newly formed PGCs must inherit the primary RNA trigger. This requires stability and protection of the RNA and its reiterative amplification. This is supported in plants and *C. elegans* via, for example, RdRPs (e.g., RDR6), which amplify an initial RNA signal by "transitivity" (explained in Ch. 2 [Dunoyer et al. 2013]) or reiterative generation of siRNAs, but has not been shown in mammals. However, an adaptive amplification loop for piRNAs in both mouse and fly germ cells has been identified (Aravin et al. 2007; Peng and Lin 2013), which is independent of RdRP. piRNAs are major components of ncRNA in developing germ cells and even sperm (Marczylo et al. 2012; Fullston et al. 2013), and their role for transgenerational epigenetic inheritance in mammalian gametes has recently been reviewed (Daxinger and Whitelaw 2012). Further, specification of PGCs is regulated by distinct pathways (see Ch. 27 [Reik and Surani 2014]), some of which directly connect RNA signals to chromatin alteration. For example, the maternal component Stella, which is important for germline differentiation, interacts with RNA, but can also bind to H3K9me2 to protect 5mC from conversion into 5hmeC (Nakamura et al. 2012).

In closing this section, we ask, then, does epigenetic control differ in a fundamental way from basic genetic principles? Although we may wish to view Waddington's epigenetic landscape as being demarcated patches of activating versus repressive domains along the continuum of the chromatin polymer, this notion could easily be overinterpreted. It is only in recent years that we have learned about the major enzymatic systems and the power of ncRNAs through which epigenetic control might be propagated. This has shaped our current thinking about the stability and, hence, the inheritance of certain epigenetic marks. Epigenetic modifications could reflect a minor and transient response to changes in the external environment or significantly contribute to phenotypic variation, which can then be maintained over many, but not indefinite, somatic cell divisions, and occasionally affect the germline. In view of our greatly improved knowledge on epigenetic mechanisms today, it is not inconceivable that some acquired traits could be transmitted in the sense of Lamarckian evolution. Even Darwin did not deny the evolutionary benefits of a plastic system: "*It is not the strongest, nor the most intelligent but the most adaptive organism that will survive.*"

23 WHAT DOES EPIGENETIC CONTROL ACTUALLY DO?

Approximately 10% of the protein pool encoded by the mammalian genome plays a role in transcription or chromatin regulation (UniProt database; www.uniprot.org). Given that the mammalian genome consists of 3×10^9 bp, it must accommodate $\sim 1 \times 10^7$ nucleosomes. This gives rise to an overwhelming array of possible regulatory messages, including DNA-binding interactions, histone modifications, histone variants, nucleosome remodeling, DNA methylation, and ncRNAs. Yet, the process of transcriptional regulation alone is quite intricate, often requiring the assembly of large multiprotein complexes (less than 100 proteins) to ensure initiation, elongation, and correct processing of mRNA from a single selected promoter. If DNA sequence-specific regulation is so elaborate, then one would expect the lower-affinity associations along the dynamic DNA–histone polymer to be even more so. Based on these considerations, there will rarely be one modification that correlates with one epigenetic state. More likely, and as experimental evidence suggests, it is the combination or cumulative effect of several (probably many) signals over an extended chromatin region that together stabilizes and propagates epigenetic states.

For the most part, TF binding is transient and lost in successive cell divisions. However, there are exceptions and DNAseI hypersensitive sites or certain "pioneer" TFs can remain at mitotic chromosomes (Martinez-Balbas et al. 1995; Caravaca et al. 2013). For persistent gene expression patterns, TFs are required at each subsequent cell division. Although local chromatin alterations can potentiate a primary trigger (e.g., TF binding and promoter stimulation), it is the establishment of a larger "chromatin domain" (e.g., locus control region, gene silencing, centromere definition, etc.) that facilitates transmission to successive cell generations and helps to maintain gene expression states, even in the absence of the initial signal (Fig. 37). Classic paradigms for this "cellular memory" are the PcG and TrxG systems (see Ch. 17 [Grossniklaus and Paro 2014]; Ch. 18 [Kingston and Tamkun 2014]; described in Sec. 16). For example, stimulation of a PRE element by a pulse of a single TF was shown to be sufficient to convert a silent into an activated chromatin region, which could then transmit histone H4 hyperacetylation over many cell divisions and even through the female germline (Cavalli and Paro 1999). Also in unicellular organisms, such as *S. pombe*, Swi6-dependent epigenetic variegation can be suppressed for many cell generations by a pulse of Swi6 (Grewal and Klar 1996). These findings were recently extended to mammalian cells, showing that a transient increase of HP1 (a Swi6 ortholog) leads to the induction of a broad H3K9me3 chromatin domain that can be stably propagated (Hathaway et al. 2012) (see also Fig. 13 in Sec. 8).

If histone modifications function together, an imprint may be left on the chromatin template that will help to mark nucleosomes, particularly if a signal is reestablished after DNA replication. For even more stable inheritance, collaboration between histone modifications, histone variant incorporation, and chromatin remodeling will convert an extended chromatin region into persistent structural alterations that can then be propagated over many cell divisions. Although explained in Figure 37 for the inheritance of transcriptional "ON" states, a similar synergy between repressive epigenetic mechanisms will more stably lock silenced chromatin regions, which could be further reinforced by additional DNA methylation.

The DNA double helix can be viewed, then, as a self-organizing polymer that, through its ordering into chromatin, can respond to epigenetic control and amplify a primary signal into a more long-term "memory." Understanding how this memory is cast into biologically relevant epigenetic signatures and how they are read, translated, and inherited, lies at the heart of current epigenetic research. It is, however, important to stress that epigenetic control requires an intricate balance between many factors and functional interaction is not always faithfully reestablished after each cell division. This is a functional contrast with genetics, which involves alteration of the DNA sequence, which is always stably propagated through mitosis and meiosis, if the mutation occurs in the germline.

An important question arising from the above considerations is how the information in the chromatin template is maintained from mother to daughter cells. In the case of DNA methylation, attractive mechanisms reveal how

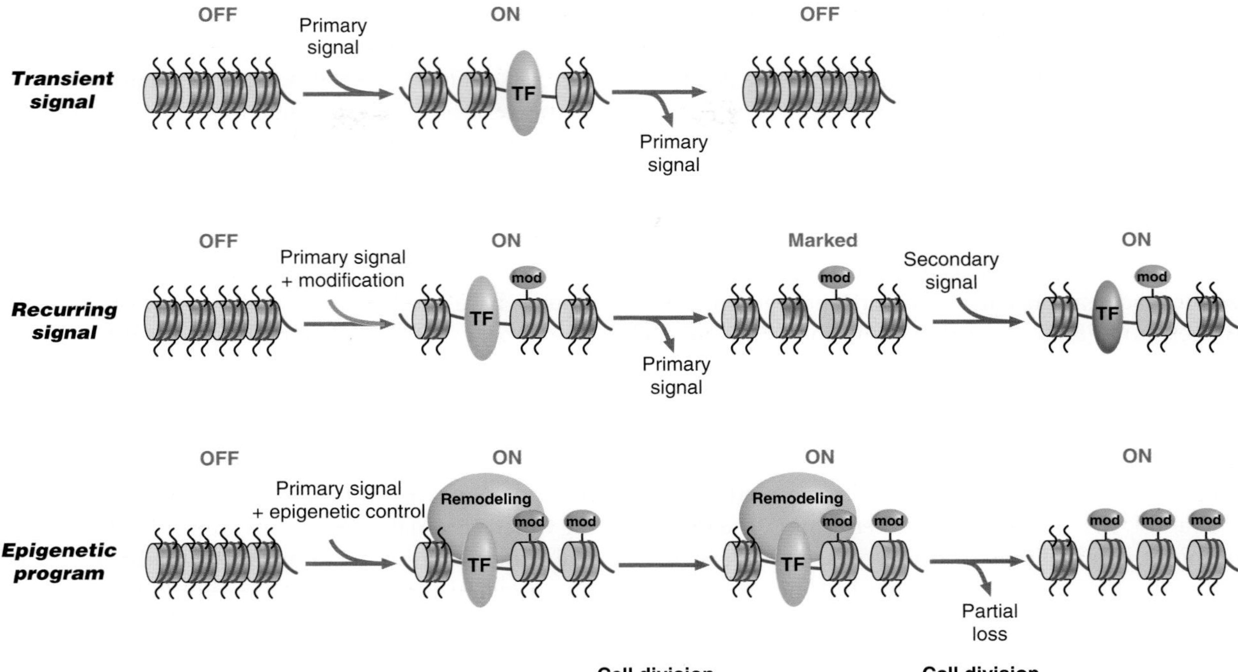

Figure 37. Epigenetic transmission of a primary signal. Classical genetics predicts that gene expression is dependent on the availability and binding of the appropriate panel of TF. Removal of such factors (i.e., a primary signal) results in the loss of gene expression and, thus, constitutes a transient activating signal (*top*). Chromatin structure contributes to gene expression, in which some conformations are repressive and others active. The activation of a locus may therefore occur through a primary signal and result in the downstream change in chromatin structure, involving activating histone marks (mod) and the replacement of core histones with variants (e.g., H3.3). Through cell division, this chromatin structure may only be reestablished in the presence of an activating signal (denoted "recurring signal"). Epigenetic memory results in the maintenance of a chromatin state through cell division, even in the absence of the primary or a recurring secondary signal. Such a memory system is not absolute, but involves multiple levels of epigenetic regulation for remodeling chromatin structure. The dynamic nature of chromatin means that although a chromatin state may be stable during mitosis, it is nonetheless prone to change, hence, affecting the longevity of epigenetic memory.

the epigenetic mark is templated to the replicating daughter strands via DNA hemimethyltransferases (see Ch. 6 [Cheng 2014]; Ch. 15 [Li and Zhang 2014]). Does a similar mechanism exist for histone modifications and is there an enzymatic system(s) that can copy a preexisting mark to another histone tail within the same nucleosome or to another nucleosome? To answer this question, we first have to consider how parental octamers are transmitted to the newly replicated DNA template (see Ch. 22 [Almouzni and Cedar 2014]). There are two prominent models that address octamer segregation during replication. The first model suggests that the H3/H4 tetramers split, such that each parental H3/H4 dimer mixes with naïve H3/H4 dimers synthesized and assembled during S phase; each of these newly arranged H3/H4 tetramers would then associate with the leading and lagging DNA strands. This model, however, has less supportive data than other possible mechanisms. The second model proposes that the entire octamer segregates in a stochastic manner to either the parental or daughter strand, behind the replication fork, interchanging with new "naïve" octamers synthesized and assembled during S phase. The epigenetic machinery would then copy information present in the parental octamers to the adjacent naïve octamers. These models await further experimental testing to delineate the molecular mechanism that allows newly synthesized daughter chromatin to retain crucial epigenetic information.

Chromatin has many distinct signatures that either reflect a transcriptional record, promote, or inhibit gene expression programs and also influence whether a chromatin region undergoes early versus late DNA replication. How these differences in transcriptional activity and replication timing could be incorporated into models of nucleosome segregation/propagation and how chromatin states are faithfully templated from one cell generation to the next, or through meiosis and germ cell formation, remains an active area of research. Current models and hypotheses are detailed in Ch. 22 (Almouzni and Cedar 2014), Ch. 20 (Henikoff and Smith 2014), and Ch. 36 (Pirrotta 2014).

24 BIG QUESTIONS IN EPIGENETIC RESEARCH

This textbook discusses the fundamental concepts and general principles that explain how epigenetic phenomena occur, as puzzling as they may seem. Our ultimate goal is to expose the reader to the current understanding of mechanisms that guide and shape these concepts, drawing on the rich biology from which they emerge. Since the first edition of this textbook, many exciting and remarkable insights have been advanced that are highlighted by some of the following breakthrough discoveries: description of enzymes that actively remove histone lysine methylation (KDMs) and DNA methylation (TET enzymes), the reprogramming of somatic cells into iPS cells, unprecedented functions of ncRNAs, cancer driving histone mutations, and development of novel inhibitors for epigenetic therapy (e.g., BETi) (see also Preface and Chapter 2 of this book). The field has also seen incredible technological progress in NGS that is accompanied by more sophisticated bioinformatic algorithms, the advent of epigenome analyses of single cells, genome engineering (TALEN and CRISPR/Cas), high-resolution mapping of chromatin domain interactions, and an overall explosion of publicly available genome-wide data sets. Despite these significant advances and our much-increased understanding of the underlying molecular mechanisms, the big questions in epigenetic research (see Fig. 38), as they were exposed in the first edition of this textbook, will continue to guide our creativity and aspiration to study and unravel these questions by rigorous experimentation.

There are plenty of reasons for renewed energy in research programs designed to gain molecular insights into epigenetic phenomena. Elegant biochemical and genetic studies have already successfully dissected many of the functional aspects of these pathways in an unprecedented manner. It could, therefore, be predicted that careful analysis of epigenetic transitions in different cell types (e.g., stem vs. differentiated; resting vs. proliferating, reprogrammed iPS cells) will uncover additional hallmarks of pluripotency. This will most likely be valuable in diagnosing which chromatin alterations are significant during normal differentiation as compared with disease states and tumorigenesis. For example, the revolution using NGS approaches and the new developments for single-cell sequencing (Shapiro et al. 2013) with normal, tumor, or ES cells to analyze the "epigenetic landscape" of the genome, and the knowledge generated, could be harnessed for novel therapeutic intervention approaches (see Appendix 1; e.g., NIH roadmap, ENCODE, Blueprint, IHEC websites). It is conceivable that differences in the abundance between distinct histone modifications, DNA methylation, and RNAi mechanisms in unicellular organisms (see the table on the

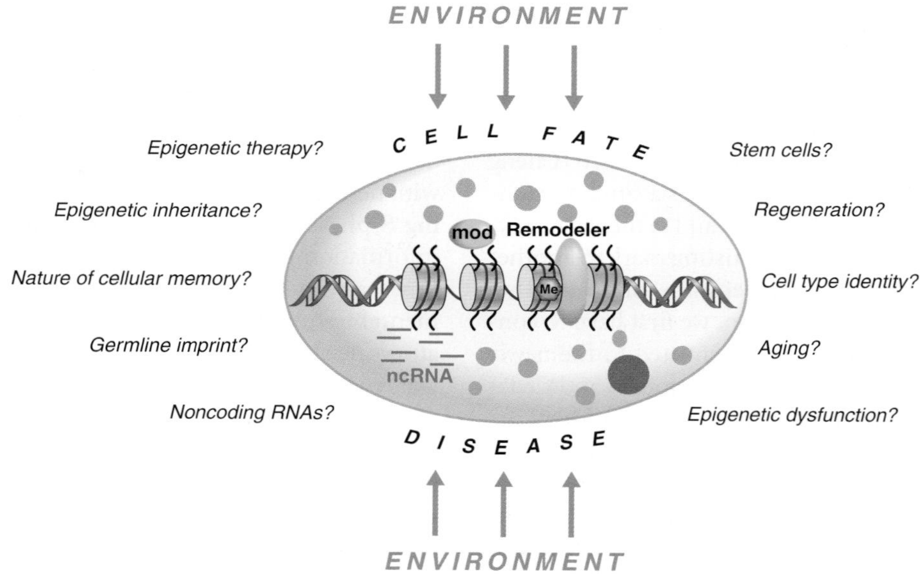

Figure 38. Big questions in epigenetic research. The many experimental systems used in epigenetic research have unveiled numerous pathways and novel insights into the mechanisms of epigenetic control. Many questions, as shown in the figure, still remain and require further elucidation or substantiation in new and existing model systems and methods.

inside front cover of this book), may reflect their greater proliferative and regenerative potential, as compared with the more restricted developmental programs of metazoan systems. Also, the functional links between the RNAi machinery, piRNAs, and histone and DNA methylation will continue to provide exciting surprises into the complex mechanisms required for cell-fate determination during development and for transgenerational epigenetic inheritance. Similarly, an enhanced understanding of the dynamics and specificity of nucleosome remodeling machines will contribute to this end. We predict that more "exotic" enzymatic activities beyond crotonylation, butyrylation, citrullination, etc., will be uncovered, catalyzing epigenetic transitions through modifications of histone and nonhistone substrates (currently known histone modifications are listed in App. 2 [Zhao and Garcia 2014]). It would appear that chromatin alterations, as induced by the above mechanisms, act largely as a response filter to the environment. Thus, it is hoped that this knowledge can ultimately be applied to enhanced therapeutic strategies for resetting some of a personalized epigenetic response that contributes to aging, disease, and cancer. This includes tissue regeneration, therapeutic cloning (using iPS cells and their derivatives), and adult stem cell therapy strategies. It is believed such strategies will extend cellular life span, modulate stress responses to external stimuli, reverse disease progression, and improve tissue engineering and personalized medicine. We predict that understanding the "chromatin basis" of pluripotency and totipotency will lie at the heart of understanding stem cell biology and its potential for therapeutic intervention.

Other fundamental epigenetic questions include: What distinguishes one chromatin strand from the other allele when both contain the same DNA sequence in the same nuclear environment? What defines the mechanisms conferring inheritance and propagation of epigenetic information? What is the molecular nature of cellular memory? Are there epigenetic imprints in the germline that serve to keep this genome in a totipotent state? If so, how are these marks erased during development? Alternatively, or in addition, are new imprints added during development that serve to "lock in" differentiated states? We look forward to the next generation of studies (and students) bold enough to tackle these questions with the heart and passion of previous generations of genetic and epigenetic researchers.

In summary, the genetic principles described by Mendel likely govern the vast majority of our development and outward phenotypes. But exceptions to the rule can sometimes reveal new principles and mechanisms, leading to inheritance, which has been underestimated and, in some cases, poorly understood previously. This second edition of the textbook *Epigenetics* hopes to inform and expose its readers to the newly appreciated basis of phenotypic variation—one that lies outside of DNA alteration. It is our hope that the systems and concepts described in this second edition will provide a useful resource for future generations of students and researchers alike who become intrigued by the curiosities of epigenetic phenomena.

REFERENCES

**Reference is also in this book.*

Adelman K, Lis JT. 2012. Promoter-proximal pausing of RNA polymerase II: Emerging roles in metazoans. *Nat Rev Genet* **13:** 720–731.

Aguilar-Arnal L, Sassone-Corsi P. 2013. The circadian epigenome: How metabolism talks to chromatin remodeling. *Curr Opin Cell Biol* **25:** 170–176.

Aguzzi A, Falsig J. 2012. Prion propagation, toxicity and degradation. *Nat Neurosci* **15:** 936–939.

Ahmad K, Henikoff S. 2001. Modulation of a transcription factor counteracts heterochromatic gene silencing in *Drosophila*. *Cell* **104:** 839–847.

Ahmad K, Henikoff S. 2002. The histone variant H3.3 marks active chromatin by replication-independent nucleosome assembly. *Mol Cell* **9:** 1191–1200.

Alberts B, Bray D, Johnson A, Lewis N, Raff M, Roberts K, Walter P. 1998. *Essential cell biology: An introduction to the molecular biology of the cell.* Garland, New York.

Alberts B, Johnson A, Lewis J. 2012. *Molecular biology of the cell.* Garland Science, New York.

Alcazar RM, Lin R, Fire AZ. 2008. Transmission dynamics of heritable silencing induced by double-stranded RNA in *Caenorhabditis elegans*. *Genetics* **180:** 1275–1288.

Allfrey VG, Faulkner R, Mirsky AE. 1964. Acetylation and methylation of histones and their possible role in the regulation of RNA synthesis. *Proc Natl Acad Sci* **51:** 786–794.

* Allshire RC, Ekwall K. 2014. Epigenetic regulation of chromatin states in *Schizosaccharomyces pombe*. *Cold Spring Harb Perspect Biol* doi: 10.1101/cshperspect.a018770.

* Almouzni G, Cedar H. 2014. Maintenance of epigenetic information. *Cold Spring Harb Perspect Biol* doi: 10.1101/cshperspect.a019372.

Amaral PP, Dinger ME, Mercer TR, Mattick JS. 2008. The eukaryotic genome as an RNA machine. *Science* **319:** 1787–1789.

Amendola M, van Steensel B. 2014. Mechanisms and dynamics of nuclear lamina-genome interactions. *Curr Opin Cell Biol* **28C:** 61–68.

* Aramayo R, Selker EU. 2013. *Neurospora crassa*, a model system for epigenetics research. *Cold Spring Harb Perspect Biol* 5: a017921.

Aravin AA, Hannon GJ, Brennecke J. 2007. The Piwi-piRNA pathway provides an adaptive defense in the transposon arms race. *Science* **318:** 761–764.

Arents G, Burlingame RW, Wang BC, Love WE, Moudrianakis EN. 1991. The nucleosomal core histone octamer at 3.1 A resolution: A tripartite protein assembly and a left-handed superhelix. *Proc Natl Acad Sci* **88:** 10148–10152.

Arrowsmith CH, Bountra C, Fish PV, Lee K, Schapira M. 2012. Epigenetic protein families: A new frontier for drug discovery. *Nat Rev Drug Discov* **11:** 384–400.

Arteaga-Vazquez MA, Chandler VL. 2010. Paramutation in maize: RNA mediated trans-generational gene silencing. *Curr Opin Genet Dev* **20:** 156–163.

* Audia JE, Campbell RM. 2014. Histone modifications and cancer. *Cold Spring Harb Perspect Biol* doi: 10.1101/cshperspect.a019521.

Azuara V, Perry P, Sauer S, Spivakov M, Jorgensen HF, John RM, Gouti M, Casanova M, Warnes G, Merkenschlager M, et al. 2006. Chromatin signatures of pluripotent cell lines. *Nat Cell Biol* **8:** 532–538.

Badeaux AI, Shi Y. 2013. Emerging roles for chromatin as a signal integration and storage platform. *Nat Rev Mol Cell Biol* **14:** 211–224.

Banaszynski LA, Wen D, Dewell S, Whitcomb SJ, Lin M, Diaz N, Elsasser SJ, Chapgier A, Goldberg AD, Canaani E, et al. 2013. Hira-dependent histone H3.3 deposition facilitates PRC2 recruitment at developmental loci in ES cells. *Cell* **155:** 107–120.

Barak O, Lazzaro MA, Lane WS, Speicher DW, Picketts DJ, Shiekhattar R. 2003. Isolation of human NURF: A regulator of Engrailed gene expression. *EMBO J* **22:** 6089–6100.

Baraniskin A, Kuhnhenn J, Schlegel U, Chan A, Deckert M, Gold R, Maghnouj A, Zollner H, Reinacher-Schick A, Schmiegel W, et al. 2011. Identification of microRNAs in the cerebrospinal fluid as marker for primary diffuse large B-cell lymphoma of the central nervous system. *Blood* **117:** 3140–3146.

Baranzini SE, Mudge J, van Velkinburgh JC, Khankhanian P, Khrebtukova I, Miller NA, Zhang L, Farmer AD, Bell CJ, Kim RW, et al. 2010. Genome, epigenome and RNA sequences of monozygotic twins discordant for multiple sclerosis. *Nature* **464:** 1351–1356.

* Barlow DP, Bartolomei MS. 2014. Genomic imprinting in mammals. *Cold Spring Harb Perspect Biol* **6:** a018382.

Barnett C, Krebs JE. 2011. WSTF does it all: A multifunctional protein in transcription, repair, and replication. *Biochem Cell Biol* **89:** 12–23.

Barski A, Cuddapah S, Cui K, Roh TY, Schones DE, Wang Z, Wei G, Chepelev I, Zhao K. 2007. High-resolution profiling of histone methylations in the human genome. *Cell* **129:** 823–837.

Bastow R, Mylne JS, Lister C, Lippman Z, Martienssen RA, Dean C. 2004. Vernalization requires epigenetic silencing of FLC by histone methylation. *Nature* **427:** 164–167.

Batty N, Malouf GG, Issa JP. 2009. Histone deacetylase inhibitors as antineoplastic agents. *Cnacer Lett* **280:** 192–200.

* Baulcombe DC, Dean C. 2014. Epigenetic regulation in plant responses to the environment. *Cold Spring Harb Perspect Biol* **6:** a019471.

* Baylin SB, Jones PA. 2014. Epigenetic determinants of cancer. *Cold Spring Harb Perspect Biol* doi: 10.1101/cshperspect.a019505.

Beck DB, Oda H, Shen SS, Reinberg D. 2012. PR-Set7 and H4K20me1: At the crossroads of genome integrity, cell cycle, chromosome condensation, and transcription. *Genes Dev* **26:** 325–337.

* Becker PB, Workman JL. 2013. Nucleosome remodeling and epigenetics. *Cold Spring Harb Perspect Biol* **5:** a017905.

Behjati S, Tarpey PS, Presneau N, Scheipl S, Pillay N, Van Loo P, Wedge DC, Cooke SL, Gundem G, Davies H, et al. 2013. Distinct H3F3A and H3F3B driver mutations define chondroblastoma and giant cell tumor of bone. *Nat Genet* **45:** 1479–1482.

Beisel C, Paro R. 2011. Silencing chromatin: Comparing modes and mechanisms. *Nat Rev Genet* **12:** 123–135.

Bell JT, Saffery R. 2012. The value of twins in epigenetic epidemiology. *Int J Epidemiol* **41:** 140–150.

Beltrao P, Albanese V, Kenner LR, Swaney DL, Burlingame A, Villen J, Lim WA, Fraser JS, Frydman J, Krogan NJ. 2012. Systematic functional prioritization of protein posttranslational modifications. *Cell* **150:** 413–425.

* Berger SL, Sassone-Corsi P. 2014. Metabolic signaling to chromatin. *Cold Spring Harb Perspect Biol* doi: 10.1101/cshperspect.a019463.

Bernard P, Maure JF, Partridge JF, Genier S, Javerzat JP, Allshire RC. 2001. Requirement of heterochromatin for cohesion at centromeres. *Science* **294:** 2539–2542.

Bernstein E, Hake SB. 2006. The nucleosome: A little variation goes a long way. *Biochem Cell Biol* **84:** 505–517.

Bernstein BE, Mikkelsen TS, Xie X, Kamal M, Huebert DJ, Cuff J, Fry B, Meissner A, Wernig M, Plath K, et al. 2006. A bivalent chromatin structure marks key developmental genes in embryonic stem cells. *Cell* **125:** 315–326.

Bickmore WA. 2013. The spatial organization of the human genome. *Annu Rev Genomics Hum Genet* **14:** 67–84.

Bird AP. 1986. CpG-rich islands and the function of DNA methylation. *Nature* **321:** 209–213.

Bird A. 2013. Genome biology: Not drowning but waving. *Cell* **154:** 951–952.

Birney E. 2011. Chromatin and heritability: How epigenetic studies can complement genetic approaches. *Trends Genet* **27:** 172–176.

Black JC, Van Rechem C, Whetstine JR. 2012. Histone lysine methylation dynamics: Establishment, regulation, and biological impact. *Mol Cell* **48:** 491–507.

Blackledge NP, Zhou JC, Tolstorukov MY, Farcas AM, Park PJ, Klose RJ. 2010. CpG islands recruit a histone H3 lysine 36 demethylase. *Mol Cell* **38:** 179–190.

* Blackledge N, Thomson JP, Skene PJ. 2013. CpG Islands chromatin is shaped by recruitment of ZF-CxxC proteins. *Cold Spring Harb Perspect Biol* **5:** a018648.

Blackledge NP, Farcas AM, Kondo T, King HW, McGouran JF, Hanssen LL, Ito S, Cooper S, Kondo K, Koseki Y, et al. 2014. Variant PRC1 Complex-dependent H2A ubiquitylation drives PRC2 recruitment and polycomb domain formation. *Cell* **157:** 1445–1459.

Blaschke K, Ebata KT, Karimi MM, Zepeda-Martinez JA, Goyal P, Mahapatra S, Tam A, Laird DJ, Hirst M, Rao A, et al. 2013. Vitamin C induces Tet-dependent DNA demethylation and a blastocyst-like state in ES cells. *Nature* **500:** 222–226.

Blau HM, Chiu CP, Webster C. 1983. Cytoplasmic activation of human nuclear genes in stable heterocaryons. *Cell* **32:** 1171–1180.

* Blewitt ME, Whitelaw E. 2013. The use of mouse models to study epigenetics. *Cold Spring Harb Perspect Biol* **5:** a017939.

Bolland DJ, Wood AL, Johnston CM, Bunting SF, Morgan G, Chakalova L, Fraser PJ, Corcoran AE. 2004. Antisense intergenic transcription in V(D)J recombination. *Nat Immunol* **5:** 630–637.

Bonasio R, Zhang G, Ye C, Mutti NS, Fang X, Qin N, Donahue G, Yang P, Li Q, Li C, et al. 2010. Genomic comparison of the ants *Camponotus floridanus* and *Harpegnathos saltator*. *Science* **329:** 1068–1071.

Bonasio R, Li Q, Lian J, Mutti NS, Jin L, Zhao H, Zhang P, Wen P, Xiang H, Ding Y, et al. 2012. Genome-wide and caste-specific DNA methylomes of the ants *Camponotus floridanus* and *Harpegnathos saltator*. *Curr Biol* **22:** 1755–1764.

Boomsma D, Busjahn A, Peltonen L. 2002. Classical twin studies and beyond. *Nat Rev Genet* **3:** 872–882.

Bose P, Dai Y, Grant S. 2014. Histone deacetylase inhibitor (HDACI) mechanisms of action: Emerging insights. *Pharmacol Ther* **143:** 323–336.

Bourc'his D, Bestor TH. 2004. Meiotic catastrophe and retrotransposon reactivation in male germ cells lacking Dnmt3L. *Nature* **431:** 96–99.

Briggs SD, Xiao T, Sun ZW, Caldwell JA, Shabanowitz J, Hunt DF, Allis CD, Strahl BD. 2002. Gene silencing: *Trans*-histone regulatory pathway in chromatin. *Nature* **418:** 498.

* Brockdorff N, Turner BM. 2014. Dosage compensation in mammals. *Cold Spring Harb Perspect Biol* doi: 10.1101/cshperspect.a019406.

Brockdorff N, Ashworth A, Kay GF, McCabe VM, Norris DP, Cooper PJ, Swift S, Rastan S. 1992. The product of the mouse Xist gene is a 15 kb inactive X-specific transcript containing no conserved ORF and located in the nucleus. *Cell* **71:** 515–526.

Brogaard K, Xi L, Wang JP, Widom J. 2012. A map of nucleosome positions in yeast at base-pair resolution. *Nature* **486:** 496–501.

Brown CJ, Ballabio A, Rupert JL, Lafreniere RG, Grompe M, Tonlorenzi R, Willard HF. 1991. A gene from the region of the human X inactivation centre is expressed exclusively from the inactive X chromosome. *Nature* **349:** 38–44.

Brownell JE, Zhou J, Ranalli T, Kobayashi R, Edmondson DG, Roth SY, Allis CD. 1996. Tetrahymena histone acetyltransferase A: A homolog to yeast Gcn5p linking histone acetylation to gene activation. *Cell* **84:** 843–851.

Buckley BA, Burkhart KB, Gu SG, Spracklin G, Kershner A, Fritz H, Kimble J, Fire A, Kennedy S. 2012. A nuclear Argonaute promotes multigenerational epigenetic inheritance and germline immortality. *Nature* **489:** 447–451.

Bulut-Karslioglu A, Perrera V, Scaranaro M, de la Rosa-Velazquez IA, van de Nobelen S, Shukeir N, Popow J, Gerle B, Opravil S, Pagani M, et al.

2012. A transcription factor-based mechanism for mouse heterochromatin formation. *Nat Struct Mol Biol* **19:** 1023–1030.

Bungard D, Fuerth BJ, Zeng PY, Faubert B, Maas NL, Viollet B, Carling D, Thompson CB, Jones RG, Berger SL. 2010. Signaling kinase AMPK activates stress-promoted transcription via histone H2B phosphorylation. *Science* **329:** 1201–1205.

Buscaino A, Lejeune E, Audergon P, Hamilton G, Pidoux A, Allshire RC. 2013. Distinct roles for Sir2 and RNAi in centromeric heterochromatin nucleation, spreading and maintenance. *EMBO J* **32:** 1250–1264.

* Busslinger M, Tarakhovsky A. 2014. Epigenetic control of immunity. *Cold Spring Harb Perspect Biol* **6:** a019307.

Cabianca DS, Casa V, Bodega B, Xynos A, Ginelli E, Tanaka Y, Gabellini D. 2012. A long ncRNA links copy number variation to a polycomb/trithorax epigenetic switch in FSHD muscular dystrophy. *Cell* **149:** 819–831.

Campbell KH, McWhir J, Ritchie WA, Wilmut I. 1996. Sheep cloned by nuclear transfer from a cultured cell line. *Nature* **380:** 64–66.

Cantara WA, Crain PF, Rozenski J, McCloskey JA, Harris KA, Zhang X, Vendeix FA, Fabris D, Agris PF. 2011. The RNA Modification Database, RNAMDB: 2011 update. *Nucleic Acids Res* **39:** D195–D201.

Canzio D, Chang EY, Shankar S, Kuchenbecker KM, Simon MD, Madhani HD, Narlikar GJ, Al-Sady B. 2011. Chromodomain-mediated oligomerization of HP1 suggests a nucleosome-bridging mechanism for heterochromatin assembly. *Mol Cell* **41:** 67–81.

Caravaca JM, Donahue G, Becker JS, He X, Vinson C, Zaret KS. 2013. Bookmarking by specific and nonspecific binding of FoxA1 pioneer factor to mitotic chromosomes. *Genes Dev* **27:** 251–260.

Carey N. 2012. *The epigenetics revolution: How modern biology is rewriting our understanding of genetics, disease, and inheritance.* Icon Books, London.

Carninci P, Kasukawa T, Katayama S, Gough J, Frith MC, Maeda N, Oyama R, Ravasi T, Lenhard B, Wells C, et al. 2005. The transcriptional landscape of the mammalian genome. *Science* **309:** 1559–1563.

Carone BR, Fauquier L, Habib N, Shea JM, Hart CE, Li R, Bock C, Li C, Gu H, Zamore PD, et al. 2010. Paternally induced transgenerational environmental reprogramming of metabolic gene expression in mammals. *Cell* **143:** 1084–1096.

Carone BR, Hung JH, Hainer SJ, Chou MT, Carone DM, Weng Z, Fazzio TG, Rando OJ. 2014. High-resolution mapping of chromatin packaging in mouse embryonic stem cells and sperm. *Dev Cell* **30:** 11–22.

Carrozza MJ, Li B, Florens L, Suganuma T, Swanson SK, Lee KK, Shia WJ, Anderson S, Yates J, Washburn MP, et al. 2005. Histone H3 methylation by Set2 directs deacetylation of coding regions by Rpd3S to suppress spurious intragenic transcription. *Cell* **123:** 581–592.

Casanova M, Pasternak M, El Marjou F, Le Baccon P, Probst AV, Almouzni G. 2013. Heterochromatin reorganization during early mouse development requires a single-stranded noncoding transcript. *Cell Rep* **4:** 1156–1167.

Castello A, Fischer B, Eichelbaum K, Horos R, Beckmann BM, Strein C, Davey NE, Humphreys DT, Preiss T, Steinmetz LM, et al. 2012. Insights into RNA biology from an atlas of mammalian mRNA-binding proteins. *Cell* **149:** 1393–1406.

Cavalli G, Misteli T. 2013. Functional implications of genome topology. *Nat Struct Mol Biol* **20:** 290–299.

Cavalli G, Paro R. 1999. Epigenetic inheritance of active chromatin after removal of the main transactivator. *Science* **286:** 955–958.

Cavazzana-Calvo M, Hacein-Bey S, de Saint Basile G, Gross F, Yvon E, Nusbaum P, Selz F, Hue C, Certain S, Casanova JL, et al. 2000. Gene therapy of human severe combined immunodeficiency (SCID)-X1 disease. *Science* **288:** 669–672.

Cedar H, Bergman Y. 2009. Linking DNA methylation and histone modification: Patterns and paradigms. *Nat Rev Genet* **10:** 295–304.

Chalker DL, Yao MC. 2001. Nongenic, bidirectional transcription precedes and may promote developmental DNA deletion in *Tetrahymena thermophila*. *Genes Dev* **15:** 1287–1298.

* Chalker DL, Meyer E, Mochizuki K. 2013. Epigenetics of ciliates. *Cold Spring Harb Perspect Biol* **5:** a017764.

Chan SW, Zilberman D, Xie Z, Johansen LK, Carrington JC, Jacobsen SE. 2004. RNA silencing genes control de novo DNA methylation. *Science* **303:** 1336.

Chandler V, Alleman M. 2008. Paramutation: Epigenetic instructions passed across generations. *Genetics* **178:** 1839–1844.

Chen RZ, Pettersson U, Beard C, Jackson-Grusby L, Jaenisch R. 1998. DNA hypomethylation leads to elevated mutation rates. *Nature* **395:** 89–93.

Chen X, Ba Y, Ma L, Cai X, Yin Y, Wang K, Guo J, Zhang Y, Chen J, Guo X, et al. 2008. Characterization of microRNAs in serum: A novel class of biomarkers for diagnosis of cancer and other diseases. *Cell Res* **18:** 997–1006.

Chen J, Liu H, Liu J, Qi J, Wei B, Yang J, Liang H, Chen Y, Chen J, Wu Y, et al. 2013a. H3K9 methylation is a barrier during somatic cell reprogramming into iPSCs. *Nat Genet* **45:** 34–42.

Chen P, Zhao J, Wang Y, Wang M, Long H, Liang D, Huang L, Wen Z, Li W, Li X, et al. 2013b. H3.3 actively marks enhancers and primes gene transcription via opening higher-ordered chromatin. *Genes Dev* **27:** 2109–2124.

* Cheng X. 2014. Structural and functional coordination of DNA and histone methylation. *Cold Spring Harb Perspect Biol* **6:** a018747.

Cheung P, Allis CD, Sassone-Corsi P. 2000. Signaling to chromatin through histone modifications. *Cell* **103:** 263–271.

Chow JC, Ciaudo C, Fazzari MJ, Mise N, Servant N, Glass JL, Attreed M, Avner P, Wutz A, Barillot E, et al. 2010. LINE-1 activity in facultative heterochromatin formation during X chromosome inactivation. *Cell* **141:** 956–969.

Clark-Adams CD, Norris D, Osley MA, Fassler JS, Winston F. 1988. Changes in histone gene dosage alter transcription in yeast. *Genes Dev* **2:** 150–159.

Clarke SG. 2013. Protein methylation at the surface and buried deep: Thinking outside the histone box. *Trends Biochem Sci* **38:** 243–252.

Clouaire T, Webb S, Skene P, Illingworth R, Kerr A, Andrews R, Lee JH, Skalnik D, Bird A. 2012. Cfp1 integrates both CpG content and gene activity for accurate H3K4me3 deposition in embryonic stem cells. *Genes Dev* **26:** 1714–1728.

Cong L, Ran FA, Cox D, Lin S, Barretto R, Habib N, Hsu PD, Wu X, Jiang W, Marraffini LA, et al. 2013. Multiplex genome engineering using CRISPR/Cas systems. *Science* **339:** 819–823.

Conrad T, Akhtar A. 2011. Dosage compensation in *Drosophila melanogaster*: Epigenetic fine-tuning of chromosome-wide transcription. *Nat Rev Genet* **13:** 123–134.

Cooper S, Dienstbier M, Hassan R, Schermelleh L, Sharif J, Blackledge NP, De Marco V, Elderkin S, Koseki H, Klose R, et al. 2014. Targeting polycomb to pericentric heterochromatin in embryonic stem cells reveals a role for H2AK119u1 in PRC2 recruitment. *Cell Rep* **7:** 1456–1470.

Core LJ, Waterfall JJ, Lis JT. 2008. Nascent RNA sequencing reveals widespread pausing and divergent initiation at human promoters. *Science* **322:** 1845–1848.

Cortijo S, Wardenaar R, Colome-Tatche M, Gilly A, Etcheverry M, Labadie K, Caillieux E, Hospital F, Aury JM, Wincker P, et al. 2014. Mapping the epigenetic basis of complex traits. *Science* **343:** 1145–1148.

Cremer T, Cremer M. 2010. Chromosome territories. *Cold Spring Harb Perspect Biol* **2:** a003889.

Croce CM. 2009. Causes and consequences of microRNA dysregulation in cancer. *Nat Rev Genet* **10:** 704–714.

Darnell JE. 2011. *RNA: Life's indispensable molecule.* Cold Spring Harbor Laboratory Press, Cold Spring Harbor, NY.

da Rocha ST, Boeva V, Escamilla-Del-Arenal M, Ancelin K, Granier C, Matias NR, Sanulli S, Chow J, Schulz E, Picard C, et al. 2014. Jarid2 is implicated in the initial Xist-induced targeting of PRC2 to the inactive X chromosome. *Mol Cell* **53:** 301–316.

Davis RL, Weintraub H, Lassar AB. 1987. Expression of a single transfected cDNA converts fibroblasts to myoblasts. *Cell* **51:** 987–1000.

Dawson MA, Kouzarides T. 2012. Cancer epigenetics: From mechanism to therapy. *Cell* **150:** 12–27.

Daxinger L, Whitelaw E. 2012. Understanding transgenerational epigenetic inheritance via the gametes in mammals. *Nat Rev Genet* **13:** 153–162.

Day JJ, Sweatt JD. 2010. DNA methylation and memory formation. *Nat Neurosci* **13:** 1319–1323.

De Fazio S, Bartonicek N, Di Giacomo M, Abreu-Goodger C, Sankar A, Funaya C, Antony C, Moreira PN, Enright AJ, O'Carroll D. 2011. The endonuclease activity of Mili fuels piRNA amplification that silences LINE1 elements. *Nature* **480:** 259–263.

* Dekker J, Misteli T. 2014. Long-range chromatin interactions. *Cold Spring Harb Perspect Biol* doi: 10.1101/cshperspect.a019356.

de Vanssay A, Bouge AL, Boivin A, Hermant C, Teysset L, Delmarre V, Antoniewski C, Ronsseray S. 2012. Paramutation in *Drosophila* linked to emergence of a piRNA-producing locus. *Nature* **490:** 112–115.

Dhalluin C, Carlson JE, Zeng L, He C, Aggarwal AK, Zhou MM. 1999. Structure and ligand of a histone acetyltransferase bromodomain. *Nature* **399:** 491–496.

Dias BG, Ressler KJ. 2014. Parental olfactory experience influences behavior and neural structure in subsequent generations. *Nat Neurosci* **17:** 89–96.

Djebali S, Davis CA, Merkel A, Dobin A, Lassmann T, Mortazavi A, Tanzer A, Lagarde J, Lin W, Schlesinger F, et al. 2012. Landscape of transcription in human cells. *Nature* **489:** 101–108.

Dolinoy DC, Jirtle RL. 2008. Environmental epigenomics in human health and disease. *Environ Mol Mutagen* **49:** 4–8.

Dolinoy DC, Weidman JR, Waterland RA, Jirtle RL. 2006. Maternal genistein alters coat color and protects Avy mouse offspring from obesity by modifying the fetal epigenome. *Environ Health Perspect* **114:** 567–572.

Doolittle WF. 2013. Is junk DNA bunk? A critique of ENCODE. *Proc Natl Acad Sci* **110:** 5294–5300.

Dou Y, Gorovsky MA. 2000. Phosphorylation of linker histone H1 regulates gene expression in vivo by creating a charge patch. *Mol Cell* **6:** 225–231.

Dover J, Schneider J, Tawiah-Boateng MA, Wood A, Dean K, Johnston M, Shilatifard A. 2002. Methylation of histone H3 by COMPASS requires ubiquitination of histone H2B by Rad6. *J Biol Chem* **277:** 28368–28371.

* Dunoyer P, Melnyk CW, Molnar A, Slotkin RK. 2013. Plant mobile small RNAs. *Cold Spring Harb Perspect Biol* **5:** a017897.

Edelman LB, Fraser P. 2012. Transcription factories: Genetic programming in three dimensions. *Curr Opin Genet Dev* **22:** 110–114.

* Elgin SCR, Reuter G. 2013. Position effect variegation, heterochromatin formation, and gene silencing in *Drosophila*. *Cold Spring Harb Perspect Biol* **5:** a017780.

ENCODE Project Consortium. 2012. An integrated encyclopedia of DNA elements in the human genome. *Nature* **489:** 57–74.

Erkek S, Hisano M, Liang CY, Gill M, Murr R, Dieker J, Schubeler D, van der Vlag J, Stadler MB, Peters AH. 2013. Molecular determinants of nucleosome retention at CpG-rich sequences in mouse spermatozoa. *Nat Struct Mol Biol* **20:** 868–875.

Ernst J, Kheradpour P, Mikkelsen TS, Shoresh N, Ward LD, Epstein CB, Zhang X, Wang L, Issner R, Coyne M, et al. 2011. Mapping and analysis of chromatin state dynamics in nine human cell types. *Nature* **473:** 43–49.

Esteller M. 2011. Non-coding RNAs in human disease. *Nat Rev Genet* **12:** 861–874.

Evans MJ, Kaufman MH. 1981. Establishment in culture of pluripotential cells from mouse embryos. *Nature* **292:** 154–156.

Fadloun A, Le Gras S, Jost B, Ziegler-Birling C, Takahashi H, Gorab E, Carninci P, Torres-Padilla ME. 2013. Chromatin signatures and retrotransposon profiling in mouse embryos reveal regulation of LINE-1 by RNA. *Nat Struct Mol Biol* **20:** 332–338.

Fan Y, Nikitina T, Zhao J, Fleury TJ, Bhattacharyya R, Bouhassira EE, Stein A, Woodcock CL, Skoultchi AI. 2005. Histone H1 depletion in mammals alters global chromatin structure but causes specific changes in gene regulation. *Cell* **123:** 1199–1212.

Feinberg AP. 2004. The epigenetics of cancer etiology. *Semin Cancer Biol* **14:** 427–432.

Feinberg AP, Tycko B. 2004. The history of cancer epigenetics. *Nat Rev Cancer* **4:** 143–153.

Feinberg AP, Vogelstein B. 1983. Hypomethylation distinguishes genes of some human cancers from their normal counterparts. *Nature* **301:** 89–92.

* Felsenfeld G. 2014. A brief history of epigenetics. *Cold Spring Harb Perspect Biol* **6:** a018220.

Felsenfeld G, Groudine M. 2003. Controlling the double helix. *Nature* **421:** 448–453.

Felsenfeld G, Rich A. 1957. Studies on the formation of two- and three-stranded polyribonucleotides. *Biochim Biophys Acta* **26:** 457–468.

Ferrari F, Alekseyenko AA, Park PJ, Kuroda MI. 2014. Transcriptional control of a whole chromosome: Emerging models for dosage compensation. *Nat Struct Mol Biol* **21:** 118–125.

Festenstein R, Tolaini M, Corbella P, Mamalaki C, Parrington J, Fox M, Miliou A, Jones M, Kioussis D. 1996. Locus control region function and heterochromatin-induced position effect variegation. *Science* **271:** 1123–1125.

Fierz B, Muir TW. 2012. Chromatin as an expansive canvas for chemical biology. *Nat Chem Biol* **8:** 417–427.

Figueroa ME, Abdel-Wahab O, Lu C, Ward PS, Patel J, Shih A, Li Y, Bhagwat N, Vasanthakumar A, Fernandez HF, et al. 2010. Leukemic IDH1 and IDH2 mutations result in a hypermethylation phenotype, disrupt TET2 function, and impair hematopoietic differentiation. *Cancer Cell* **18:** 553–567.

Filion GJ, van Bemmel JG, Braunschweig U, Talhout W, Kind J, Ward LD, Brugman W, de Castro IJ, Kerkhoven RM, Bussemaker HJ, et al. 2010. Systematic protein location mapping reveals five principal chromatin types in *Drosophila* cells. *Cell* **143:** 212–224.

Fire A, Xu S, Montgomery MK, Kostas SA, Driver SE, Mello CC. 1998. Potent and specific genetic interference by double-stranded RNA in *Caenorhabditis elegans*. *Nature* **391:** 806–811.

Fischer A, Sananbenesi F, Wang X, Dobbin M, Tsai LH. 2007. Recovery of learning and memory is associated with chromatin remodelling. *Nature* **447:** 178–182.

Fischle W, Wang Y, Allis CD. 2003. Binary switches and modification cassettes in histone biology and beyond. *Nature* **425:** 475–479.

Fischle W, Tseng BS, Dormann HL, Ueberheide BM, Garcia BA, Shabanowitz J, Hunt DF, Funabiki H, Allis CD. 2005. Regulation of HP1-chromatin binding by histone H3 methylation and phosphorylation. *Nature* **438:** 1116–1122.

Fisher CL, Fisher AG. 2011. Chromatin states in pluripotent, differentiated, and reprogrammed cells. *Curr Opin Genet Dev* **21:** 140–146.

Fodor BD, Shukeir N, Reuter G, Jenuwein T. 2010. Mammalian Su(var) genes in chromatin control. *Annu Rev Cell Dev Biol* **26:** 471–501.

Foster SL, Hargreaves DC, Medzhitov R. 2007. Gene-specific control of inflammation by TLR-induced chromatin modifications. *Nature* **447:** 972–978.

Fox CJ, Hammerman PS, Thompson CB. 2005. Fuel feeds function: Energy metabolism and the T-cell response. *Nat Rev Immunol* **5:** 844–852.

Fraga MF, Ballestar E, Paz MF, Ropero S, Setien F, Ballestar ML, Heine-Suner D, Cigudosa JC, Urioste M, Benitez J, et al. 2005. Epigenetic differences arise during the lifetime of monozygotic twins. *Proc Natl Acad Sci* **102:** 10604–10609.

Francis NJ, Kingston RE, Woodcock CL. 2004. Chromatin compaction by a polycomb group protein complex. *Science* **306:** 1574–1577.

Friedmann DR, Marmorstein R. 2013. Structure and mechanism of non-histone protein acetyltransferase enzymes. *FEBS J* **280:** 5570–5581.

Fukagawa T, Nogami M, Yoshikawa M, Ikeno M, Okazaki T, Takami Y, Nakayama T, Oshimura M. 2004. Dicer is essential for formation of the heterochromatin structure in vertebrate cells. *Nat Cell Biol* **6:** 784–791.

Fullston T, Ohlsson Teague EM, Palmer NO, DeBlasio MJ, Mitchell M, Corbett M, Print CG, Owens JA, Lane M. 2013. Paternal obesity initiates metabolic disturbances in two generations of mice with incomplete penetrance to the F2 generation and alters the transcriptional profile of testis and sperm microRNA content. *FASEB J* **27:** 4226–4243.

Gao Z, Zhang J, Bonasio R, Strino F, Sawai A, Parisi F, Kluger Y, Reinberg D. 2012. PCGF homologs, CBX proteins, and RYBP define functionally distinct PRC1 family complexes. *Mol Cell* **45:** 344–356.

Gapp K, Jawaid A, Sarkies P, Bohacek J, Pelczar P, Prados J, Farinelli L, Miska E, Mansuy IM. 2014. Implication of sperm RNAs in transgenerational inheritance of the effects of early trauma in mice. *Nat Neurosci* **17:** 667–669.

Gendrel AV, Heard E. 2014. Noncoding RNAs and epigenetic mechanisms during X-chromosome inactivation. *Annu Rev Cell Dev Biol* **30**. doi: 10.1146/annurev-cellbio-101512-122415.

Ghildiyal M, Zamore PD. 2009. Small silencing RNAs: An expanding universe. *Nat Rev Genet* **10:** 94–108.

Gil J, O'Loghlen A. 2014. PRC1 complex diversity: Where is it taking us? *Trends Cell Biol* pii: S0962-8924(14)00102-0.

Glozak MA, Sengupta N, Zhang X, Seto E. 2005. Acetylation and deacetylation of non-histone proteins. *Gene* **363:** 15–23.

Gluckman PD, Hanson MA, Buklijas T, Low FM, Beedle AS. 2009. Epigenetic mechanisms that underpin metabolic and cardiovascular diseases. *Nat Rev Endocrinol* **5:** 401–408.

Goldberg AD, Banaszynski LA, Noh KM, Lewis PW, Elsaesser SJ, Stadler S, Dewell S, Law M, Guo X, Li X, et al. 2010. Distinct factors control histone variant H3.3 localization at specific genomic regions. *Cell* **140:** 678–691.

Goll MG, Kirpekar F, Maggert KA, Yoder JA, Hsieh CL, Zhang X, Golic KG, Jacobsen SE, Bestor TH. 2006. Methylation of tRNAAsp by the DNA methyltransferase homolog Dnmt2. *Science* **311:** 395–398.

Gontan C, Achame EM, Demmers J, Barakat TS, Rentmeester E, van Ijcken W, Grootegoed JA, Gribnau J. 2012. RNF12 initiates X-chromosome inactivation by targeting REX1 for degradation. *Nature* **485:** 386–390.

Goto T, Salpekar A, Monk M. 2001. Expression of a testis-specific member of the olfactory receptor gene family in human primordial germ cells. *Mol Hum Reprod* **7:** 553–558.

Grandjean V, Yaman R, Cuzin F, Rassoulzadegan M. 2007. Inheritance of an epigenetic mark: The CpG DNA methyltransferase 1 is required for de novo establishment of a complex pattern of non-CpG methylation. *PloS One* **2:** e1136.

Grant PA, Duggan L, Cote J, Roberts SM, Brownell JE, Candau R, Ohba R, Owen-Hughes T, Allis CD, Winston F, et al. 1997. Yeast Gcn5 functions in two multisubunit complexes to acetylate nucleosomal histones: Characterization of an Ada complex and the SAGA (Spt/Ada) complex. *Genes Dev* **11:** 1640–1650.

Greer EL, Maures TJ, Ucar D, Hauswirth AG, Mancini E, Lim JP, Benayoun BA, Shi Y, Brunet A. 2011. Transgenerational epigenetic inheritance of longevity in *Caenorhabditis elegans*. *Nature* **479:** 365–371.

Grewal SI, Klar AJ. 1996. Chromosomal inheritance of epigenetic states in fission yeast during mitosis and meiosis. *Cell* **86:** 95–101.

* Grossniklaus U, Paro R. 2014. Transcriptional silencing by *Polycomb*-group proteins. *Cold Spring Harb Perspect Biol* **6:** a019331.

Grossniklaus U, Kelly WG, Ferguson-Smith AC, Pembrey M, Lindquist S. 2013. Transgenerational epigenetic inheritance: How important is it? *Nat Rev Genet* **14:** 228–235.

* Grunstein M, Gasser SM. 2013. Epigenetics in *Saccharomyces cerevisiae*. *Cold Spring Harb Perspect Biol* **5:** a017491.

Guan JS, Haggarty SJ, Giacometti E, Dannenberg JH, Joseph N, Gao J, Nieland TJ, Zhou Y, Wang X, Mazitschek R, et al. 2009. HDAC2 negatively regulates memory formation and synaptic plasticity. *Nature* **459:** 55–60.

Gurdon JB, Elsdale TR, Fischberg M. 1958. Sexually mature individuals of *Xenopus laevis* from the transplantation of single somatic nuclei. *Nature* **182:** 64–65.

Gut P, Verdin E. 2013. The nexus of chromatin regulation and intermediary metabolism. *Nature* **502:** 489–498.

Guy J, Gan J, Selfridge J, Cobb S, Bird A. 2007. Reversal of neurological defects in a mouse model of Rett syndrome. *Science* **315:** 1143–1147.

Guzzardo PM, Muerdter F, Hannon GJ. 2013. The piRNA pathway in flies: Highlights and future directions. *Curr Opin Genet Dev* **23:** 44–52.

Hales CN, Barker DJ. 2001. The thrifty phenotype hypothesis. *Br Med Bull* **60:** 5–20.

Halic M, Moazed D. 2010. Dicer-independent primal RNAs trigger RNAi and heterochromatin formation. *Cell* **140:** 504–516.

Hall IM, Shankaranarayana GD, Noma K, Ayoub N, Cohen A, Grewal SI. 2002. Establishment and maintenance of a heterochromatin domain. *Science* **297:** 2232–2237.

Hammoud SS, Nix DA, Zhang H, Purwar J, Carrell DT, Cairns BR. 2009. Distinctive chromatin in human sperm packages genes for embryo development. *Nature* **460:** 473–478.

Han JS, Szak ST, Boeke JD. 2004. Transcriptional disruption by the L1 retrotransposon and implications for mammalian transcriptomes. *Nature* **429:** 268–274.

Hanahan D, Weinberg RA. 2011. Hallmarks of cancer: The next generation. *Cell* **144:** 646–674.

Hanover JA, Krause MW, Love DC. 2012. Bittersweet memories: Linking metabolism to epigenetics through *O*-GlcNAcylation. *Nat Rev Mol Cell Biol* **13:** 312–321.

Happel N, Doenecke D. 2009. Histone H1 and its isoforms: Contribution to chromatin structure and function. *Gene* **431:** 1–12.

Hassan AH, Prochasson P, Neely KE, Galasinski SC, Chandy M, Carrozza MJ, Workman JL. 2002. Function and selectivity of bromodomains in anchoring chromatin-modifying complexes to promoter nucleosomes. *Cell* **111:** 369–379.

Hathaway NA, Bell O, Hodges C, Miller EL, Neel DS, Crabtree GR. 2012. Dynamics and memory of heterochromatin in living cells. *Cell* **149:** 1447–1460.

He YF, Li BZ, Li Z, Liu P, Wang Y, Tang Q, Ding J, Jia Y, Chen Z, Li L, et al. 2011. Tet-mediated formation of 5-carboxylcytosine and its excision by TDG in mammalian DNA. *Science* **333:** 1303–1307.

Heard E, Martienssen RA. 2014. Transgenerational epigenetic inheritance: Myths and mechanisms. *Cell* **157:** 95–109.

Heintzman ND, Hon GC, Hawkins RD, Kheradpour P, Stark A, Harp LF, Ye Z, Lee LK, Stuart RK, Ching CW, et al. 2009. Histone modifications at human enhancers reflect global cell-type-specific gene expression. *Nature* **459:** 108–112.

Henderson IR, Jacobsen SE. 2007. Epigenetic inheritance in plants. *Nature* **447:** 418–424.

* Henikoff S, Smith MM. 2014. Histone variants and epigenetics. *Cold Spring Harb Perspect Biol* doi: 10.1101/cshperspect.a019364.

Hentze MW, Preiss T. 2010. The REM phase of gene regulation. *Trends Biochem Sci* **35:** 423–426.

Herr AJ, Jensen MB, Dalmay T, Baulcombe DC. 2005. RNA polymerase IV directs silencing of endogenous DNA. *Science* **308:** 118–120.

Hirano T. 2012. Condensins: Universal organizers of chromosomes with diverse functions. *Genes Dev* **26:** 1659–1678.

* Hochedlinger K, Jaenisch R. 2014. Induced pluripotency and epigenetic reprogramming. *Cold Spring Harb Perspect Biol* doi: 10.1101/cshperspect.a019448.

Hochedlinger K, Blelloch R, Brennan C, Yamada Y, Kim M, Chin L, Jaenisch R. 2004. Reprogramming of a melanoma genome by nuclear transplantation. *Genes Dev* **18:** 1875–1885.

Højfeldt JW, Agger K, Helin K. 2013. Histone lysine demethylases as targets for anticancer therapy. *Nat Rev Drug Discov* **12:** 917–930.

Holliday R. 1994. Epigenetics: An overview. *Dev Genet* **15:** 453–457.

Holsboer F. 2008. How can we realize the promise of personalized antidepressant medicines? *Nat Rev Neurosci* **9:** 638–646.

Horton JR, Upadhyay AK, Qi HH, Zhang X, Shi Y, Cheng X. 2010. Enzymatic and structural insights for substrate specificity of a family of jumonji histone lysine demethylases. *Nat Struct Mol Biol* **17:** 38–43.

Houtkooper RH, Pirinen E, Auwerx J. 2012. Sirtuins as regulators of metabolism and healthspan. *Nat Rev Mol Cell Biol* **13:** 225–238.

Huang J, Berger SL. 2008. The emerging field of dynamic lysine methylation of non-histone proteins. *Curr Opin Genet Dev* **18:** 152–158.

Iwase S, Xiang B, Ghosh S, Ren T, Lewis PW, Cochrane JC, Allis CD, Picketts DJ, Patel DJ, Li H, et al. 2011. ATRX ADD domain links an atypical histone methylation recognition mechanism to human mental-retardation syndrome. *Nat Struct Mol Biol* **18:** 769–776.

Izzo A, Kamieniarz K, Schneider R. 2008. The histone H1 family: Specific members, specific functions? *Biol Chem* **389:** 333–343.

Jackson JP, Lindroth AM, Cao X, Jacobsen SE. 2002. Control of CpNpG DNA methylation by the KRYPTONITE histone H3 methyltransferase. *Nature* **416:** 556–560.

Jacobson RH, Ladurner AG, King DS, Tjian R. 2000. Structure and function of a human TAFII250 double bromodomain module. *Science* **288:** 1422–1425.

Jeon Y, Lee JT. 2011. YY1 tethers Xist RNA to the inactive X nucleation center. *Cell* **146:** 119–133.

Jia G, Fu Y, He C. 2013. Reversible RNA adenosine methylation in biological regulation. *Trends Genet* **29:** 108–115.

Johnson CL, Spence AM. 2011. Epigenetic licensing of germline gene expression by maternal RNA in *C. elegans*. *Science* **333:** 1311–1314.

Jonkers I, Barakat TS, Achame EM, Monkhorst K, Kenter A, Rentmeester E, Grosveld F, Grootegoed JA, Gribnau J. 2009. RNF12 is an X-Encoded dose-dependent activator of X chromosome inactivation. *Cell* **139:** 999–1011.

Juang BT, Gu C, Starnes L, Palladino F, Goga A, Kennedy S, L'Etoile ND. 2013. Endogenous nuclear RNAi mediates behavioral adaptation to odor. *Cell* **154:** 1010–1022.

Kaati G, Bygren LO, Edvinsson S. 2002. Cardiovascular and diabetes mortality determined by nutrition during parents' and grandparents' slow growth period. *Eur J Hum Genet* **10:** 682–688.

Kalb R, Latwiel S, Baymaz HI, Jansen PW, Muller CW, Vermeulen M, Muller J. 2014. Histone H2A monoubiquitination promotes histone H3 methylation in Polycomb repression. *Nat Struct Mol Biol* **21:** 569–571.

Kaneko S, Bonasio R, Saldana-Meyer R, Yoshida T, Son J, Nishino K, Umezawa A, Reinberg D. 2014. Interactions between JARID2 and noncoding RNAs regulate PRC2 recruitment to chromatin. *Mol Cell* **53:** 290–300.

Kanellopoulou C, Muljo SA, Kung AL, Ganesan S, Drapkin R, Jenuwein T, Livingston DM, Rajewsky K. 2005. Dicer-deficient mouse embryonic stem cells are defective in differentiation and centromeric silencing. *Genes Dev* **19:** 489–501.

Kapranov P, Cheng J, Dike S, Nix DA, Duttagupta R, Willingham AT, Stadler PF, Hertel J, Hackermuller J, Hofacker IL, et al. 2007. RNA maps reveal new RNA classes and a possible function for pervasive transcription. *Science* **316:** 1484–1488.

Karachentsev D, Sarma K, Reinberg D, Steward R. 2005. PR-Set7-dependent methylation of histone H4 Lys 20 functions in repression of gene expression and is essential for mitosis. *Genes Dev* **19:** 431–435.

Kayne PS, Kim UJ, Han M, Mullen JR, Yoshizaki F, Grunstein M. 1988. Extremely conserved histone H4 N terminus is dispensable for growth but essential for repressing the silent mating loci in yeast. *Cell* **55:** 27–39.

Keogh MC, Kurdistani SK, Morris SA, Ahn SH, Podolny V, Collins SR, Schuldiner M, Chin K, Punna T, Thompson NJ, et al. 2005. Cotranscriptional set2 methylation of histone H3 lysine 36 recruits a repressive Rpd3 complex. *Cell* **123:** 593–605.

Kharchenko PV, Alekseyenko AA, Schwartz YB, Minoda A, Riddle NC, Ernst J, Sabo PJ, Larschan E, Gorchakov AA, Gu T, et al. 2011. Comprehensive analysis of the chromatin landscape in *Drosophila melanogaster*. *Nature* **471:** 480–485.

Khoddami V, Cairns BR. 2013. Identification of direct targets and modified bases of RNA cytosine methyltransferases. *Nat Biotechnol* **31:** 458–464.

Khurana JS, Theurkauf W. 2010. piRNAs, transposon silencing, and *Drosophila* germline development. *J Cell Biol* **191:** 905–913.

Killian JL, Li M, Sheinin MY, Wang MD. 2012. Recent advances in single molecule studies of nucleosomes. *Curr Opin Struct Biol* **22:** 80–87.

Kim J, Guermah M, McGinty RK, Lee JS, Tang Z, Milne TA, Shilatifard A, Muir TW, Roeder RG. 2009. RAD6-Mediated transcription-coupled H2B ubiquitylation directly stimulates H3K4 methylation in human cells. *Cell* **137:** 459–471.

* Kim T-K, Hemberg M, Gray JM. 2014. Enhancer RNAs: A class of long noncoding RNAs synthesized at enhancers. *Cold Spring Harb Perspect Biol* doi: 10.1101/cshperspect.a018622.

* Kingston RE, Tamkun JW. 2014. Transcriptional regulation by trithorax-group proteins. *Cold Spring Harb Perspect Biol* **6:** a019349.

Klattenhoff C, Xi H, Li C, Lee S, Xu J, Khurana JS, Zhang F, Schultz N, Koppetsch BS, Nowosielska A, et al. 2009. The *Drosophila* HP1 homolog Rhino is required for transposon silencing and piRNA production by dual-strand clusters. *Cell* **138:** 1137–1149.

Kornberg RD. 1974. Chromatin structure: A repeating unit of histones and DNA. *Science* **184:** 868–871.

Kornberg RD. 1977. Structure of chromatin. *Annu Rev Biochem* **46:** 931–954.

Krauss V, Reuter G. 2011. DNA methylation in *Drosophila*—A critical evaluation. *Prog Mol Biol Transl Sci* **101:** 177–191.

Krawetz SA, Kruger A, Lalancette C, Tagett R, Anton E, Draghici S, Diamond MP. 2011. A survey of small RNAs in human sperm. *Hum Reprod* **26:** 3401–3412.

Kriaucionis S, Tahiliani M. 2014. Expanding the epigenetic landscape: Novel modifications of cytosine min genomic DNA. *Cold Spring Harb Biol* **6:** a018630.

Kucharski R, Maleszka J, Foret S, Maleszka R. 2008. Nutritional control of reproductive status in honeybees via DNA methylation. *Science* **319:** 1827–1830.

Kurdistani SK. 2011. Histone modifications in cancer biology and prognosis. *Prog Drug Res* **67:** 91–106.

Lachner M, Sengupta R, Schotta G, Jenuwein T. 2004. Trilogies of histone lysine methylation as epigenetic landmarks of the eukaryotic genome. *Cold Spring Harbor Symp Quant Biol* **69:** 209–218.

Lai F, Orom UA, Cesaroni M, Beringer M, Taatjes DJ, Blobel GA, Shiekhattar R. 2013. Activating RNAs associate with Mediator to enhance chromatin architecture and transcription. *Nature* **494:** 497–501.

Lander ES. 2011. Initial impact of the sequencing of the human genome. *Nature* **470:** 187–197.

Lander ES, Linton LM, Birren B, Nusbaum C, Zody MC, Baldwin J, Devon K, Dewar K, Doyle M, FitzHugh W, et al. 2001. Initial sequencing and analysis of the human genome. *Nature* **409:** 860–921.

Larsson J, Zhang J, Rasmuson-Lestander A. 1996. Mutations in the *Drosophila melanogaster* gene encoding S-adenosylmethionine synthetase [corrected] suppress position-effect variegation. *Genetics* **143:** 887–896.

Lassar AB, Paterson BM, Weintraub H. 1986. Transfection of a DNA locus that mediates the conversion of 10T1/2 fibroblasts to myoblasts. *Cell* **47:** 649–656.

Latham JA, Dent SY. 2007. Cross-regulation of histone modifications. *Nat Struct Mol Biol* **14:** 1017–1024.

Lederbogen F, Kirsch P, Haddad L, Streit F, Tost H, Schuch P, Wust S, Pruessner JC, Rietschel M, Deuschle M, et al. 2011. City living and urban upbringing affect neural social stress processing in humans. *Nature* **474:** 498–501.

Lee JT. 2012. Epigenetic regulation by long noncoding RNAs. *Science* **338:** 1435–1439.

Lee JT, Lu N. 1999. Targeted mutagenesis of Tsix leads to nonrandom X inactivation. *Cell* **99:** 47–57.

Lee JS, Woodsworth ML, Latimer LJ, Morgan AR. 1984. Poly(pyrimidine). poly(purine) synthetic DNAs containing 5-methylcytosine form stable triplexes at neutral pH. *Nucleic Acids Res* **12:** 6603–6614.

Leroy G, Dimaggio PA, Chan EY, Zee BM, Blanco MA, Bryant B, Flaniken IZ, Liu S, Kang Y, Trojer P, et al. 2013. A quantitative atlas of histone modification signatures from human cancer cells. *Epigenetics Chromatin* **6:** 20.

Levy D, Kuo AJ, Chang Y, Schaefer U, Kitson C, Cheung P, Espejo A, Zee BM, Liu CL, Tangsombatvisit S, et al. 2011. Lysine methylation of the NF-κB subunit RelA by SETD6 couples activity of the histone methyltransferase GLP at chromatin to tonic repression of NF-κB signaling. *Nat Immunol* **12:** 29–36.

Lewis PW, Muller MM, Koletsky MS, Cordero F, Lin S, Banaszynski LA, Garcia BA, Muir TW, Becher OJ, Allis CD. 2013. Inhibition of PRC2 activity by a gain-of-function H3 mutation found in pediatric glioblastoma. *Science* **340:** 857–861.

Li G, Reinberg D. 2011. Chromatin higher-order structures and gene regulation. *Curr Opin Genet Dev* **21:** 175–186.

⋆ Li E, Zhang Y. 2014. DNA methylation in mammals. *Cold Spring Harb Perspect Biol* **6:** a019133.

Li E, Bestor TH, Jaenisch R. 1992. Targeted mutation of the DNA methyltransferase gene results in embryonic lethality. *Cell* **69:** 915–926.

Li G, Margueron R, Hu G, Stokes D, Wang YH, Reinberg D. 2010a. Highly compacted chromatin formed in vitro reflects the dynamics of transcription activation in vivo. *Mol Cell* **38:** 41–53.

Li G, Margueron R, Ku M, Chambon P, Bernstein BE, Reinberg D. 2010b. Jarid2 and PRC2, partners in regulating gene expression. *Genes Dev* **24:** 368–380.

Li L, Liu B, Wapinski OL, Tsai MC, Qu K, Zhang J, Carlson JC, Lin M, Fang F, Gupta RA, et al. 2013. Targeted disruption of Hotair leads to homeotic transformation and gene derepression. *Cell Rep* **5:** 3–12.

⋆ Liu X, McEachron T, Schwartzentruber J, Wu G. 2014. Histone H3 mutations in pediatric brain tumors. *Cold Spring Harb Perspect Biol* **6:** a018689.

⋆ Lomvardas S, Maniatis T. 2014. Histone and DNA modifications as regulators of neuronal development and function. *Cold Spring Harb Perspect Biol* doi: 10.1101/cshperspect.a024208.

Lu X, Sachs F, Ramsay L, Jacques PE, Goke J, Bourque G, Ng HH. 2014. The retrovirus HERVH is a long noncoding RNA required for human embryonic stem cell identity. *Nat Struct Mol Biol* **21:** 423–425.

⋆ Lucchesi JC, Kuroda MI. 2014. Dosage compensation in *Drosophila*. *Cold Spring Harb Perspect Biol* doi: 10.1101/cshperspect.a019398.

Luger K, Mader AW, Richmond RK, Sargent DF, Richmond TJ. 1997. Crystal structure of the nucleosome core particle at 2.8 Å resolution. *Nature* **389:** 251–260.

Luger K, Dechassa ML, Tremethick DJ. 2012. New insights into nucleosome and chromatin structure: An ordered state or a disordered affair? *Nat Rev Mol Cell Biol* **13:** 436–447.

Lyko F, Maleszka R. 2011. Insects as innovative models for functional studies of DNA methylation. *Trends Genet* **27:** 127–131.

Machnicka MA, Milanowska K, Osman Oglou O, Purta E, Kurkowska M, Olchowik A, Januszewski W, Kalinowski S, Dunin-Horkawicz S, Rother KM, et al. 2013. MODOMICS: A database of RNA modification pathways—2013 update. *Nucleic Acids Res* **41:** D262–D267.

Maison C, Bailly D, Peters AH, Quivy JP, Roche D, Taddei A, Lachner M, Jenuwein T, Almouzni G. 2002. Higher-order structure in pericentric heterochromatin involves a distinct pattern of histone modification and an RNA component. *Nat Genet* **30:** 329–334.

Manolio TA, Collins FS, Cox NJ, Goldstein DB, Hindorff LA, Hunter DJ, McCarthy MI, Ramos EM, Cardon LR, Chakravarti A, et al. 2009. Finding the missing heritability of complex diseases. *Nature* **461:** 747–753.

Marazzi I, Ho JS, Kim J, Manicassamy B, Dewell S, Albrecht RA, Seibert CW, Schaefer U, Jeffrey KL, Prinjha RK, et al. 2012. Suppression of the antiviral response by an influenza histone mimic. *Nature* **483:** 428–433.

Marchetto MC, Narvaiza I, Denli AM, Benner C, Lazzarini TA, Nathanson JL, Paquola AC, Desai KN, Herai RH, Weitzman MD, et al. 2013. Differential L1 regulation in pluripotent stem cells of humans and apes. *Nature* **503:** 525–529.

Marczylo EL, Amoako AA, Konje JC, Gant TW, Marczylo TH. 2012. Smoking induces differential miRNA expression in human spermatozoa: A potential transgenerational epigenetic concern? *Epigenetics* **7:** 432–439.

Margueron R, Reinberg D. 2011. The Polycomb complex PRC2 and its mark in life. *Nature* **469:** 343–349.

Margueron R, Justin N, Ohno K, Sharpe ML, Son J, Drury WJ, 3rd, Voigt P, Martin SR, Taylor WR, De Marco V, et al. 2009. Role of the polycomb protein EED in the propagation of repressive histone marks. *Nature* **461:** 762–767.

Marks PA. 2010. The clinical development of histone deacetylase inhibitors as targeted anticancer drugs. *Exppert Opin Investig Drugs* **19:** 1049–1066.

⋆ Marmorstein R, Zhou M-M. 2014. Writers and readers of histone acetylation: Structure, mechanism, and inhibition. *Cold Spring Harb Perspect Biol* **6:** a018762.

⋆ Martienssen R, Moazed D. 2014. RNAi and heterochromatin assembly. *Cold Spring Harb Perspect Biol* doi: 10.1101/cshperspect.a019323.

Martin GR. 1981. Isolation of a pluripotent cell line from early mouse embryos cultured in medium conditioned by teratocarcinoma stem cells. *Proc Natl Acad Sci* **78:** 7634–7638.

Martinez-Balbas MA, Dey A, Rabindran SK, Ozato K, Wu C. 1995. Displacement of sequence-specific transcription factors from mitotic chromatin. *Cell* **83:** 29–38.

Mattick JS. 2011. The central role of RNA in human development and cognition. *FEBS Lett* **585:** 1600–1616.

Maze I, Noh KM, Allis CD. 2013. Histone regulation in the CNS: Basic principles of epigenetic plasticity. *Neuropsychopharmacology* **38:** 3–22.

Maze I, Noh KM, Soshnev AA, Allis CD. 2014. Every amino acid matters: Essential contributions of histone variants to mammalian development and disease. *Nat Rev Genet* **15:** 259–271.

McClintock B. 1951. Chromosome organization and genic expression. *Cold Spring Harbor Symp Quant Biol* **16:** 13–47.

McGinty RK, Kim J, Chatterjee C, Roeder RG, Muir TW. 2008. Chemically ubiquitylated histone H2B stimulates hDot1L-mediated intranucleosomal methylation. *Nature* **453:** 812–816.

McGowan PO, Sasaki A, D'Alessio AC, Dymov S, Labonte B, Szyf M, Turecki G, Meaney MJ. 2009. Epigenetic regulation of the glucocorticoid receptor in human brain associates with childhood abuse. *Nat Neurosci* **12:** 342–348.

McNeill E, Van Vactor D. 2012. MicroRNAs shape the neuronal landscape. *Neuron* **75:** 363–379.

Memczak S, Jens M, Elefsinioti A, Torti F, Krueger J, Rybak A, Maier L, Mackowiak SD, Gregersen LH, Munschauer M, et al. 2013. Circular RNAs are a large class of animal RNAs with regulatory potency. *Nature* **495:** 333–338.

Merla G, Brunetti-Pierri N, Micale L, Fusco C. 2010. Copy number variants at Williams-Beuren syndrome 7q11.23 region. *Hum Genet* **128:** 3–26.

Meshorer E, Yellajoshula D, George E, Scambler PJ, Brown DT, Misteli T. 2006. Hyperdynamic plasticity of chromatin proteins in pluripotent embryonic stem cells. *Dev Cell* **10:** 105–116.

Messner S, Hottiger MO. 2011. Histone ADP-ribosylation in DNA repair, replication and transcription. *Trends Cell Biol* **21:** 534–542.

Metzger E, Wissmann M, Yin N, Muller JM, Schneider R, Peters AH, Gunther T, Buettner R, Schule R. 2005. LSD1 demethylates repressive histone marks to promote androgen-receptor-dependent transcription. *Nature* **437:** 436–439.

Meyer KD, Saletore Y, Zumbo P, Elemento O, Mason CE, Jaffrey SR. 2012. Comprehensive analysis of mRNA methylation reveals enrichment in 3′ UTRs and near stop codons. *Cell* **149:** 1635–1646.

Meyer-Lindenberg A, Mervis CB, Berman KF. 2006. Neural mechanisms in Williams syndrome: A unique window to genetic influences on cognition and behaviour. *Nat Rev Neurosci* **7:** 380–393.

Mikkelsen TS, Ku M, Jaffe DB, Issac B, Lieberman E, Giannoukos G, Alvarez P, Brockman W, Kim TK, Koche RP, et al. 2007. Genome-wide maps of chromatin state in pluripotent and lineage-committed cells. *Nature* **448:** 553–560.

Miller CA, Sweatt JD. 2007. Covalent modification of DNA regulates memory formation. *Neuron* **53:** 857–869.

Mizuguchi G, Shen X, Landry J, Wu WH, Sen S, Wu C. 2004. ATP-driven exchange of histone H2AZ variant catalyzed by SWR1 chromatin remodeling complex. *Science* **303:** 343–348.

Mochizuki K, Fine NA, Fujisawa T, Gorovsky MA. 2002. Analysis of a piwi-related gene implicates small RNAs in genome rearrangement in tetrahymena. *Cell* **110:** 689–699.

Morey C, Avner P. 2011. The demoiselle of X-inactivation: 50 years old and as trendy and mesmerising as ever. *PLoS Genet* **7:** e1002212.

Morin RD, Johnson NA, Severson TM, Mungall AJ, An J, Goya R, Paul JE, Boyle M, Woolcock BW, Kuchenbauer F, et al. 2010. Somatic mutations altering EZH2 (Tyr641) in follicular and diffuse large B-cell lymphomas of germinal-center origin. *Nat Genet* **42:** 181–185.

Motamedi MR, Verdel A, Colmenares SU, Gerber SA, Gygi SP, Moazed D. 2004. Two RNAi complexes, RITS and RDRC, physically interact and localize to noncoding centromeric RNAs. *Cell* **119:** 789–802.

Muller HJ. 1930. Types of visible variations induced by x-rays in *Drosophila*. *J Genet* **22:** 299–U297.

Muntean AG, Hess JL. 2012. The pathogenesis of mixed-lineage leukemia. *Annu Rev Pathol* **7:** 283–301.

Musselman CA, Lalonde ME, Cote J, Kutateladze TG. 2012. Perceiving the epigenetic landscape through histone readers. *Nat Struct Mol Biol* **19:** 1218–1227.

Nakahata Y, Sahar S, Astarita G, Kaluzova M, Sassone-Corsi P. 2009. Circadian control of the NAD+ salvage pathway by CLOCK-SIRT1. *Science* **324:** 654–657.

Nakamura T, Liu YJ, Nakashima H, Umehara H, Inoue K, Matoba S, Tachibana M, Ogura A, Shinkai Y, Nakano T. 2012. PGC7 binds histone H3K9me2 to protect against conversion of 5mC to 5hmC in early embryos. *Nature* **486:** 415–419.

Narita M, Nunez S, Heard E, Narita M, Lin AW, Hearn SA, Spector DL, Hannon GJ, Lowe SW. 2003. Rb-mediated heterochromatin formation and silencing of E2F target genes during cellular senescence. *Cell* **113:** 703–716.

Narlikar GJ, Sundaramoorthy R, Owen-Hughes T. 2013. Mechanisms and functions of ATP-dependent chromatin-remodeling enzymes. *Cell* **154:** 490–503.

Nasmyth K, Haering CH. 2009. Cohesin: Its roles and mechanisms. *Annu Rev Genet* **43:** 525–558.

Ng SF, Lin RC, Laybutt DR, Barres R, Owens JA, Morris MJ. 2010. Chronic high-fat diet in fathers programs β-cell dysfunction in female rat offspring. *Nature* **467:** 963–966.

Ni JQ, Zhou R, Czech B, Liu LP, Holderbaum L, Yang-Zhou D, Shim HS, Tao R, Handler D, Karpowicz P, et al. 2011. A genome-scale shRNA resource for transgenic RNAi in *Drosophila*. *Nat Methods* **8:** 405–407.

Oliver SS, Denu JM. 2011. Dynamic interplay between histone H3 modifications and protein interpreters: Emerging evidence for a "histone language". *Chembiochem* **12:** 299–307.

Ooi SL, Henikoff S. 2007. Germline histone dynamics and epigenetics. *Curr Opin Cell Biol* **19:** 257–265.

Ooi SK, Qiu C, Bernstein E, Li K, Jia D, Yang Z, Erdjument-Bromage H, Tempst P, Lin SP, Allis CD, et al. 2007. DNMT3L connects unmethylated lysine 4 of histone H3 to de novo methylation of DNA. *Nature* **448:** 714–717.

Orphanides G, LeRoy G, Chang CH, Luse DS, Reinberg D. 1998. FACT, a factor that facilitates transcript elongation through nucleosomes. *Cell* **92:** 105–116.

Paik WK, Kim S. 1971. Protein methylation. *Science* **174:** 114–119.

Pal-Bhadra M, Leibovitch BA, Gandhi SG, Chikka MR, Bhadra U, Birchler JA, Elgin SC. 2004. Heterochromatic silencing and HP1 localization in *Drosophila* are dependent on the RNAi machinery. *Science* **303:** 669–672.

Paro R, Hogness DS. 1991. The Polycomb protein shares a homologous domain with a heterochromatin-associated protein of *Drosophila*. *Proc Natl Acad Sci* **88:** 263–267.

* Patel DJ. 2014. A structural perspective on readout of epigenetic histone and DNA methylation marks. *Cold Spring Harb Perspect Biol* doi: 10.1101/cshperspect.a018754.

Pavri R, Nussenzweig MC. 2011. AID targeting in antibody diversity. *Adv Immunol* **110:** 1–26.

Pearce EL, Pearce EJ. 2013. Metabolic pathways in immune cell activation and quiescence. *Immunity* **38:** 633–643.

Pembrey ME, Bygren LO, Kaati G, Edvinsson S, Northstone K, Sjostrom M, Golding J, Team AS. 2006. Sex-specific, male-line transgenerational responses in humans. *Eur J Hum Genet* **14:** 159–166.

Peng JC, Lin H. 2013. Beyond transposons: The epigenetic and somatic functions of the Piwi-piRNA mechanism. *Curr Opin Cell Biol* **25:** 190–194.

Pengelly AR, Copur O, Jackle H, Herzig A, Muller J. 2013. A histone mutant reproduces the phenotype caused by loss of histone-modifying factor Polycomb. *Science* **339:** 698–699.

Pepenella S, Murphy KJ, Hayes JJ. 2014. Intra- and inter-nucleosome interactions of the core histone tail domains in higher-order chromatin structure. *Chromosoma* **123:** 3–13.

Petesch SJ, Lis JT. 2012. Overcoming the nucleosome barrier during transcript elongation. *Trends Genet* **28:** 285–294.

Pickersgill H, Kalverda B, de Wit E, Talhout W, Fornerod M, van Steensel B. 2006. Characterization of the *Drosophila melanogaster* genome at the nuclear lamina. *Nat Genet* **38:** 1005–1014.

* Pikaard CS, Mittelsten Scheid O. 2014. Epigenetic regulation in plants. *Cold Spring Harb Perspect Biol* doi: 10.1101/cshperspect.a019315.

Pires-daSilva A, Sommer RJ. 2003. The evolution of signalling pathways in animal development. *Nat Rev Genet* **4:** 39–49.

* Pirrotta V. 2014. The necessity of chromatin: A view in perspective. *Cold Spring Harb Perspect Biol* doi: 10.1101/cshperspect.a019547.

Pontier D, Yahubyan G, Vega D, Bulski A, Saez-Vasquez J, Hakimi MA, Lerbs-Mache S, Colot V, Lagrange T. 2005. Reinforcement of silencing at transposons and highly repeated sequences requires the concerted action of two distinct RNA polymerases IV in *Arabidopsis*. *Genes Dev* **19:** 2030–2040.

Ponting CP, Hardison RC. 2011. What fraction of the human genome is functional? *Genome Res* **21:** 1769–1776.

Pritchard CC, Kroh E, Wood B, Arroyo JD, Dougherty KJ, Miyaji MM, Tait JF, Tewari M. 2012. Blood cell origin of circulating microRNAs: A cautionary note for cancer biomarker studies. *Cancer Prev Res* **5:** 492–497.

Pujadas E, Feinberg AP. 2012. Regulated noise in the epigenetic landscape of development and disease. *Cell* **148:** 1123–1131.

* Qi J. 2014. Bromodomain and extraterminal domain inhibitors (BETi) for cancer therapy: Chemical modulation of chromatin structure. *Cold Spring Harb Perspect Biol* doi: 10.1101/cshperspect.a018663.

Radford EJ, Ito M, Shi H, Corish JA, Yamazawa K, Isganaitis E, Seisenberger S, Hore TA, Reik W, Erkek S, et al. 2014. In utero undernourishment perturbs the adult sperm methylome and intergenerational metabolism. *Science* **345:** 1255903.

Rais Y, Zviran A, Geula S, Gafni O, Chomsky E, Viukov S, Mansour AA, Caspi I, Krupalnik V, Zerbib M, et al. 2013. Deterministic direct reprogramming of somatic cells to pluripotency. *Nature* **502:** 65–70.

Rando OJ. 2012. Combinatorial complexity in chromatin structure and function: Revisiting the histone code. *Curr Opin Genet Dev* **22:** 148–155.

Rangasamy D, Berven L, Ridgway P, Tremethick DJ. 2003. Pericentric heterochromatin becomes enriched with H2A.Z during early mammalian development. *EMBO J* **22:** 1599–1607.

Rassoulzadegan M, Grandjean V, Gounon P, Vincent S, Gillot I, Cuzin F. 2006. RNA-mediated non-mendelian inheritance of an epigenetic change in the mouse. *Nature* **441:** 469–474.

Ratcliff F, Harrison BD, Baulcombe DC. 1997. A similarity between viral defense and gene silencing in plants. *Science* **276:** 1558–1560.

Razin A, Riggs AD. 1980. DNA methylation and gene function. *Science* **210:** 604–610.

Rea S, Eisenhaber F, O'Carroll D, Strahl BD, Sun ZW, Schmid M, Opravil S, Mechtler K, Ponting CP, Allis CD, et al. 2000. Regulation of chromatin structure by site-specific histone H3 methyltransferases. *Nature* **406:** 593–599.

* Reik W, Surani MA. 2014. Germline and pluripotent stem cells. *Cold Spring Harb Perspect Biol* doi: 10.1101/cshperspect.a019422.

Reinberg D, Sims RJ 3rd. 2006. de FACTo nucleosome dynamics. *J Biol Chem* **281:** 23297–23301.

Reinhart BJ, Bartel DP. 2002. Small RNAs correspond to centromere heterochromatic repeats. *Science* **297:** 1831.

Reyes-Turcu FE, Zhang K, Zofall M, Chen E, Grewal SI. 2011. Defects in RNA quality control factors reveal RNAi-independent nucleation of heterochromatin. *Nat Struct Mol Biol* **18:** 1132–1138.

Rideout WM, 3rd, Hochedlinger K, Kyba M, Daley GQ, Jaenisch R. 2002. Correction of a genetic defect by nuclear transplantation and combined cell and gene therapy. *Cell* **109:** 17–27.

Ridley M. 2003. *Nature via nurture.* Harper, New York.

Rigby RE, Webb LM, Mackenzie KJ, Li Y, Leitch A, Reijns MA, Lundie RJ, Revuelta A, Davidson DJ, Diebold S, et al. 2014. RNA:DNA hybrids are a novel molecular pattern sensed by TLR9. *EMBO J* **33:** 542–558.

* Rinn JL. 2014. lncRNAs: Linking RNA to chromatin. *Cold Spring Harb Perspect Biol* **6:** a018614.

Rinn JL, Chang HY. 2012. Genome regulation by long noncoding RNAs. *Annu Rev Biochem* **81:** 145–166.

Rinn JL, Kertesz M, Wang JK, Squazzo SL, Xu X, Brugmann SA, Goodnough LH, Helms JA, Farnham PJ, Segal E, et al. 2007. Functional demarcation of active and silent chromatin domains in human HOX loci by noncoding RNAs. *Cell* **129:** 1311–1323.

Robinson PJ, Rhodes D. 2006. Structure of the '30 nm' chromatin fibre: A key role for the linker histone. *Curr Opin Struct Biol* **16:** 336–343.

Rossetto D, Avvakumov N, Cote J. 2012. Histone phosphorylation: A chromatin modification involved in diverse nuclear events. *Epigenetics* **7:** 1098–1108.

Ruthenburg AJ, Li H, Patel DJ, Allis CD. 2007. Multivalent engagement of chromatin modifications by linked binding modules. *Nat Rev Mol Cell Biol* **8:** 983–994.

Ruthenburg AJ, Li H, Milne TA, Dewell S, McGinty RK, Yuen M, Ueberheide B, Dou Y, Muir TW, Patel DJ, et al. 2011. Recognition of a mononucleosomal histone modification pattern by BPTF via multivalent interactions. *Cell* **145:** 692–706.

Sabin LR, Delas MJ, Hannon GJ. 2013. Dogma derailed: The many influences of RNA on the genome. *Mol Cell* **49:** 783–794.

Sampath SC, Marazzi I, Yap KL, Sampath SC, Krutchinsky AN, Mecklenbrauker I, Viale A, Rudensky E, Zhou MM, Chait BT, et al. 2007. Methylation of a histone mimic within the histone methyltransferase G9a regulates protein complex assembly. *Mol Cell* **27:** 596–608.

Sassone-Corsi P, Christen Y, eds. 2012. *Epigenetics, brain and behavior.* Springer, Heidelberg.

Sawarkar R, Sievers C, Paro R. 2012. Hsp90 globally targets paused RNA polymerase to regulate gene expression in response to environmental stimuli. *Cell* **149:** 807–818.

* Schaefer U. 2014. Pharmacological inhibition of bromodomain-containing proteins in inflammation. *Cold Spring Harb Perspect Biol* **6:** a018671.

Schaefer A, Sampath SC, Intrator A, Min A, Gertler TS, Surmeier DJ, Tarakhovsky A, Greengard P. 2009. Control of cognition and adaptive behavior by the GLP/G9a epigenetic suppressor complex. *Neuron* **64:** 678–691.

Schaefer M, Pollex T, Hanna K, Tuorto F, Meusburger M, Helm M, Lyko F. 2010. RNA methylation by Dnmt2 protects transfer RNAs against stress-induced cleavage. *Genes Dev* **24:** 1590–1595.

Schalch T, Duda S, Sargent DF, Richmond TJ. 2005. X-ray structure of a tetranucleosome and its implications for the chromatin fibre. *Nature* **436:** 138–141.

Schmidt U, Holsboer F, Rein T. 2011. Epigenetic aspects of posttraumatic stress disorder. *Dis Markers* **30:** 77–87.

Schmitz KM, Mayer C, Postepska A, Grummt I. 2010. Interaction of noncoding RNA with the rDNA promoter mediates recruitment of DNMT3b and silencing of rRNA genes. *Genes Dev* **24:** 2264–2269.

Schotta G, Ebert A, Reuter G. 2003. SU(VAR)3-9 is conserved key function in heterochromatic gene silencing. *Genetica* **117:** 149–158.

Schreiber SL, Bernstein BE. 2002. Signaling network model of chromatin. *Cell* **111:** 771–778.

Schwartz BE, Ahmad K. 2005. Transcriptional activation triggers deposition and removal of the histone variant H3.3. *Genes Dev* **19:** 804–814.

Schwartz YB, Kahn TG, Pirrotta V. 2005. Characteristic low density and shear sensitivity of cross-linked chromatin containing polycomb complexes. *Mol Cell Biol* **25:** 432–439.

Schwartzentruber J, Korshunov A, Liu XY, Jones DT, Pfaff E, Jacob K, Sturm D, Fontebasso AM, Quang DA, Tonjes M, et al. 2012. Driver mutations in histone H3.3 and chromatin remodelling genes in paediatric glioblastoma. *Nature* **482:** 226–231.

Segal E, Widom J. 2009. What controls nucleosome positions? *Trends Genet* **25:** 335–343.

Sekinger EA, Moqtaderi Z, Struhl K. 2005. Intrinsic histone-DNA interactions and low nucleosome density are important for preferential accessibility of promoter regions in yeast. *Mol Cell* **18:** 735–748.

Seligson DB, Horvath S, Shi T, Yu H, Tze S, Grunstein M, Kurdistani SK. 2005. Global histone modification patterns predict risk of prostate cancer recurrence. *Nature* **435:** 1261–1266.

Senisterra G, Wu H, Allali-Hassani A, Wasney GA, Barsyte-Lovejoy D, Dombrovski L, Dong A, Nguyen KT, Smil D, Bolshan Y, et al. 2013. Small-molecule inhibition of MLL activity by disruption of its interaction with WDR5. *Biochem J* **449:** 151–159.

* Seto E, Yoshida M. 2014. Erasers of histone acetylation: The histone deacetylase enzymes. *Cold Spring Harb Perspect Biol* **6:** a018721.

* Sexton T, Yaffe E. 2014. Chromosome folding: Driver or passenger of epigenetic state? *Cold Spring Harb Perspect Biol* doi: 10.1101/cshperspect.a018721.

Shah PP, Donahue G, Otte GL, Capell BC, Nelson DM, Cao K, Aggarwala V, Cruickshanks HA, Rai TS, McBryan T, et al. 2013. Lamin B1 depletion in senescent cells triggers large-scale changes in gene expression and the chromatin landscape. *Genes Dev* **27:** 1787–1799.

Shapiro E, Biezuner T, Linnarsson S. 2013. Single-cell sequencing-based technologies will revolutionize whole-organism science. *Nat Rev Genet* **14:** 618–630.

Sharma SV, Lee DY, Li B, Quinlan MP, Takahashi F, Maheswaran S, McDermott U, Azizian N, Zou L, Fischbach MA, et al. 2010. A chromatin-mediated reversible drug-tolerant state in cancer cell subpopulations. *Cell* **141:** 69–80.

Shen H, Laird PW. 2013. Interplay between the cancer genome and epigenome. *Cell* **153:** 38–55.

Shen Y, Yue F, McCleary DF, Ye Z, Edsall L, Kuan S, Wagner U, Dixon J, Lee L, Lobanenkov VV, et al. 2012. A map of the *cis*-regulatory sequences in the mouse genome. *Nature* **488:** 116–120.

* Shi YG, Tsukada Y. 2013. The discovery of histone demethylases. *Cold Spring Harb Perspect Biol* **5:** a017947.

Shin T, Kraemer D, Pryor J, Liu L, Rugila J, Howe L, Buck S, Murphy K, Lyons L, Westhusin M. 2002. A cat cloned by nuclear transplantation. *Nature* **415:** 859.

Shinagawa T, Takagi T, Tsukamoto D, Tomaru C, Huynh LM, Sivaraman P, Kumarevel T, Inoue K, Nakato R, Katou Y, et al. 2014. Histone variants enriched in oocytes enhance reprogramming to induced pluripotent stem cells. *Cell Stem Cell* **14:** 217–227.

Sidoli S, Cheng L, Jensen ON. 2012. Proteomics in chromatin biology and epigenetics: Elucidation of post-translational modifications of histone proteins by mass spectrometry. *J Proteomics* **75:** 3419–3433.

Simon JA, Kingston RE. 2013. Occupying chromatin: Polycomb mechanisms for getting to genomic targets, stopping transcriptional traffic, and staying put. *Mol Cell* **49:** 808–824.

Sims RJ 3rd, Belotserkovskaya R, Reinberg D. 2004. Elongation by RNA polymerase II: The short and long of it. *Genes Dev* **18:** 2437–2468.

Sims RJ 3rd, Chen CF, Santos-Rosa H, Kouzarides T, Patel SS, Reinberg D. 2005. Human but not yeast CHD1 binds directly and selectively to histone H3 methylated at lysine 4 via its tandem chromodomains. *J Biol Chem* **280:** 41789–41792.

Sims RJ 3rd, Millhouse S, Chen CF, Lewis BA, Erdjument-Bromage H, Tempst P, Manley JL, Reinberg D. 2007. Recognition of trimethylated histone H3 lysine 4 facilitates the recruitment of transcription post-initiation factors and pre-mRNA splicing. *Mol Cell* **28:** 665–676.

Sing A, Pannell D, Karaiskakis A, Sturgeon K, Djabali M, Ellis J, Lipshitz HD, Cordes SP. 2009. A vertebrate Polycomb response element governs segmentation of the posterior hindbrain. *Cell* **138:** 885–897.

Skourti-Stathaki K, Proudfoot NJ. 2014. A double-edged sword: R loops as threats to genome integrity and powerful regulators of gene expression. *Genes Dev* **28:** 1384–1396.

Sleutels F, Zwart R, Barlow DP. 2002. The non-coding Air RNA is required for silencing autosomal imprinted genes. *Nature* **415:** 810–813.

Smith E, Shilatifard A. 2013. Transcriptional elongation checkpoint control in development and disease. *Genes Dev* **27:** 1079–1088.

Smolle M, Workman JL. 2013. Transcription-associated histone modifications and cryptic transcription. *Biochim Biophys Acta* **1829:** 84–97.

Son J, Shen SS, Margueron R, Reinberg D. 2013. Nucleosome-binding activities within JARID2 and EZH1 regulate the function of PRC2 on chromatin. *Genes Dev* **27:** 2663–2677.

Song J, Rechkoblit O, Bestor TH, Patel DJ. 2011. Structure of DNMT1-DNA complex reveals a role for autoinhibition in maintenance DNA methylation. *Science* **331:** 1036–1040.

Song F, Chen P, Sun D, Wang M, Dong L, Liang D, Xu RM, Zhu P, Li G. 2014. Cryo-EM study of the chromatin fiber reveals a double helix twisted by tetranucleosomal units. *Science* **344:** 376–380.

Spangrude GJ, Heimfeld S, Weissman IL. 1988. Purification and characterization of mouse hematopoietic stem cells. *Science* **241:** 58–62.

Spannhoff A, Kim YK, Raynal NJ, Gharibyan V, Su MB, Zhou YY, Li J, Castellano S, Sbardella G, Issa JP, et al. 2011. Histone deacetylase inhibitor activity in royal jelly might facilitate caste switching in bees. *EMBO Rep* **12:** 238–243.

Squires JE, Patel HR, Nousch M, Sibbritt T, Humphreys DT, Parker BJ, Suter CM, Preiss T. 2012. Widespread occurrence of 5-methylcytosine in human coding and non-coding RNA. *Nucleic Acids Res* **40:** 5023–5033.

Stadtfeld M, Apostolou E, Akutsu H, Fukuda A, Follett P, Natesan S, Kono T, Shioda T, Hochedlinger K. 2010. Aberrant silencing of imprinted genes on chromosome 12qF1 in mouse induced pluripotent stem cells. *Nature* **465:** 175–181.

* Strome S, Kelly WG, Ercan S, Lieb JD. 2014. *Cold Spring Harb Perspect Biol* 6: a018366.

Struhl K, Segal E. 2013. Determinants of nucleosome positioning. *Nat Struct Mol Biol* **20:** 267–273.

Su IH, Dobenecker MW, Dickinson E, Oser M, Basavaraj A, Marqueron R, Viale A, Reinberg D, Wulfing C, Tarakhovsky A. 2005. Polycomb group protein ezh2 controls actin polymerization and cell signaling. *Cell* **121:** 425–436.

Suganuma T, Workman JL. 2011. Signals and combinatorial functions of histone modifications. *Annu Rev Biochem* **80:** 473–499.

Suganuma T, Workman JL. 2013. Chromatin and signaling. *Curr Opin Cell Biol* **25:** 322–326.

Sugiyama T, Cam HP, Sugiyama R, Noma K, Zofall M, Kobayashi R, Grewal SI. 2007. SHREC, an effector complex for heterochromatic transcriptional silencing. *Cell* **128:** 491–504.

Sun ZW, Allis CD. 2002. Ubiquitination of histone H2B regulates H3 methylation and gene silencing in yeast. *Nature* **418:** 104–108.

Sung S, Amasino RM. 2004. Vernalization in *Arabidopsis thaliana* is mediated by the PHD finger protein VIN3. *Nature* **427:** 159–164.

Surani MA, Barton SC, Norris ML. 1986. Nuclear transplantation in the mouse: Heritable differences between parental genomes after activation of the embryonic genome. *Cell* **45:** 127–136.

Sweatt JD, Meaney MJ, Nestler EJ, Akbarian S. 2012. *Epigenetic regulation in the nervous system: Basic mechanisms and clinical impact.* Academic, New York.

Tagami H, Ray-Gallet D, Almouzni G, Nakatani Y. 2004. Histone H3.1 and H3.3 complexes mediate nucleosome assembly pathways dependent or independent of DNA synthesis. *Cell* **116:** 51–61.

* Takahashi K. 2014. Cellular reprogramming. *Cold Spring Harb Perspect Biol* **6:** a018606.

Takahashi K, Yamanaka S. 2006. Induction of pluripotent stem cells from mouse embryonic and adult fibroblast cultures by defined factors. *Cell* **126:** 663–676.

Takahashi K, Tanabe K, Ohnuki M, Narita M, Ichisaka T, Tomoda K, Yamanaka S. 2007. Induction of pluripotent stem cells from adult human fibroblasts by defined factors. *Cell* **131:** 861–872.

Tamaru H, Selker EU. 2001. A histone H3 methyltransferase controls DNA methylation in *Neurospora crassa*. *Nature* **414:** 277–283.

Taylor DD, Gercel-Taylor C. 2008. MicroRNA signatures of tumor-derived exosomes as diagnostic biomarkers of ovarian cancer. *Gynecol Oncol* **110:** 13–21.

Terranova R, Pereira CF, Du Roure C, Merkenschlager M, Fisher AG. 2006. Acquisition and extinction of gene expression programs are separable events in heterokaryon reprogramming. *J Cell Sci* **119:** 2065–2072.

Thomson JP, Skene PJ, Selfridge J, Clouaire T, Guy J, Webb S, Kerr AR, Deaton A, Andrews R, James KD, et al. 2010. CpG islands influence chromatin structure via the CpG-binding protein Cfp1. *Nature* **464:** 1082–1086.

Tian D, Sun S, Lee JT. 2010. The long noncoding RNA, Jpx, is a molecular switch for X chromosome inactivation. *Cell* **143:** 390–403.

Ting DT, Lipson D, Paul S, Brannigan BW, Akhavanfard S, Coffman EJ, Contino G, Deshpande V, Iafrate AJ, Letovsky S, et al. 2011. Aberrant overexpression of satellite repeats in pancreatic and other epithelial cancers. *Science* **331:** 593–596.

Tollefsbol T. 2012. *Epigenetics in human disease.* Academic, New York.

Towbin BD, Gonzalez-Aguilera C, Sack R, Gaidatzis D, Kalck V, Meister P, Askjaer P, Gasser SM. 2012. Step-wise methylation of histone H3K9 positions heterochromatin at the nuclear periphery. *Cell* **150:** 934–947.

Tropberger P, Schneider R. 2013. Scratching the (lateral) surface of chromatin regulation by histone modifications. *Nat Struct Mol Biol* **20:** 657–661.

Turcan S, Rohle D, Goenka A, Walsh LA, Fang F, Yilmaz E, Campos C, Fabius AW, Lu C, Ward PS, et al. 2012. IDH1 mutation is sufficient to establish the glioma hypermethylator phenotype. *Nature* **483:** 479–483.

van Haaften G, Dalgliesh GL, Davies H, Chen L, Bignell G, Greenman C, Edkins S, Hardy C, O'Meara S, Teague J, et al. 2009. Somatic mutations of the histone H3K27 demethylase gene UTX in human cancer. *Nat Genet* **41:** 521–523.

Vaquero A, Scher M, Erdjument-Bromage H, Tempst P, Serrano L, Reinberg D. 2007. SIRT1 regulates the histone methyl-transferase SUV39H1 during heterochromatin formation. *Nature* **450:** 440–444.

Venkatesh S, Smolle M, Li H, Gogol MM, Saint M, Kumar S, Natarajan K, Workman JL. 2012. Set2 methylation of histone H3 lysine 36 suppresses histone exchange on transcribed genes. *Nature* **489:** 452–455.

Venter JC, Adams MD, Myers EW, Li PW, Mural RJ, Sutton GG, Smith HO, Yandell M, Evans CA, Holt RA, et al. 2001. The sequence of the human genome. *Science* **291:** 1304–1351.

Voigt P, LeRoy G, Drury WJ, 3rd, Zee BM, Son J, Beck DB, Young NL, Garcia BA, Reinberg D. 2012. Asymmetrically modified nucleosomes. *Cell* **151:** 181–193.

Volpe TA, Kidner C, Hall IM, Teng G, Grewal SI, Martienssen RA. 2002. Regulation of heterochromatic silencing and histone H3 lysine-9 methylation by RNAi. *Science* **297:** 1833–1837.

Voss TC, Hager GL. 2014. Dynamic regulation of transcriptional states by chromatin and transcription factors. *Nat Rev Genet* **15:** 69–81.

Waddington CH. 1957. *The strategy of the genes.* MacMillan, New York.

Wagner KD, Wagner N, Ghanbarian H, Grandjean V, Gounon P, Cuzin F, Rassoulzadegan M. 2008. RNA induction and inheritance of epigenetic cardiac hypertrophy in the mouse. *Dev Cell* **14:** 962–969.

Wallace DC, Chalkia D. 2013. Mitochondrial DNA genetics and the heteroplasmy conundrum in evolution and disease. *Cold Spring Harb Perspect Biol* **5:** a021220.

Walsh CP, Chaillet JR, Bestor TH. 1998. Transcription of IAP endogenous retroviruses is constrained by cytosine methylation. *Nat Genet* **20:** 116–117.

Wang S, Wang Y. 2013. Peptidylarginine deiminases in citrullination, gene regulation, health and pathogenesis. *Biochim Biophys Acta* **1829:** 1126–1135.

Wang X, Lu Z, Gomez A, Hon GC, Yue Y, Han D, Fu Y, Parisien M, Dai Q, Jia G, et al. 2014. N6-methyladenosine-dependent regulation of messenger RNA stability. *Nature* **505:** 117–120.

Warburg O. 1956. On respiratory impairment in cancer cells. *Science* **124:** 269–270.

Ward PS, Thompson CB. 2012. Metabolic reprogramming: A cancer hallmark even Warburg did not anticipate. *Cancer Cell* **21:** 297–308.

Watanabe Y, Yokobayashi S, Yamamoto M, Nurse P. 2001. Pre-meiotic S phase is linked to reductional chromosome segregation and recombination. *Nature* **409:** 359–363.

Waterland RA, Dolinoy DC, Lin JR, Smith CA, Shi X, Tahiliani KG. 2006. Maternal methyl supplements increase offspring DNA methylation at Axin Fused. *Genesis* **44:** 401–406.

Weber CM, Henikoff S. 2014. Histone variants: Dynamic punctuation in transcription. *Genes Dev* **28:** 672–682.

Wei Y, Yu L, Bowen J, Gorovsky MA, Allis CD. 1999. Phosphorylation of histone H3 is required for proper chromosome condensation and segregation. *Cell* **97:** 99–109.

Wen B, Wu H, Shinkai Y, Irizarry RA, Feinberg AP. 2009. Large histone H3 lysine 9 dimethylated chromatin blocks distinguish differentiated from embryonic stem cells. *Nat Genet* **41:** 246–250.

Wendt KS, Grosveld FG. 2014. Transcription in the context of the 3D nucleus. *Curr Opin Genet Dev* **25:** 62–67.

Whyte WA, Orlando DA, Hnisz D, Abraham BJ, Lin CY, Kagey MH, Rahl PB, Lee TI, Young RA. 2013. Master transcription factors and mediator establish super-enhancers at key cell identity genes. *Cell* **153:** 307–319.

Wiench M, Miranda TB, Hager GL. 2011. Control of nuclear receptor function by local chromatin structure. *FEBS J* **278:** 2211–2230.

Wilusz JE, JnBaptiste CK, Lu LY, Kuhn CD, Joshua-Tor L, Sharp PA. 2012. A triple helix stabilizes the 3′ ends of long noncoding RNAs that lack poly(A) tails. *Genes Dev* **26:** 2392–2407.

Woo CJ, Kharchenko PV, Daheron L, Park PJ, Kingston RE. 2010. A region of the human HOXD cluster that confers polycomb-group responsiveness. *Cell* **140:** 99–110.

Wright DE, Wang CY, Kao CF. 2012. Histone ubiquitylation and chromatin dynamics. *Front Biosci* **17:** 1051–1078.

Wu G, Broniscer A, McEachron TA, Lu C, Paugh BS, Becksfort J, Qu C, Ding L, Huether R, Parker M, et al. 2012. Somatic histone H3 alterations in pediatric diffuse intrinsic pontine gliomas and non-brainstem glioblastomas. *Nat Genet* **44:** 251–253.

Wysocka J, Swigut T, Xiao H, Milne TA, Kwon SY, Landry J, Kauer M, Tackett AJ, Chait BT, Badenhorst P, et al. 2006. A PHD finger of NURF couples histone H3 lysine 4 trimethylation with chromatin remodelling. *Nature* **442:** 86–90.

Xiao S, Xie D, Cao X, Yu P, Xing X, Chen CC, Musselman M, Xie M, West FD, Lewin HA, et al. 2012. Comparative epigenomic annotation of regulatory DNA. *Cell* **149:** 1381–1392.

Xie W, Barr CL, Kim A, Yue F, Lee AY, Eubanks J, Dempster EL, Ren B. 2012. Base-resolution analyses of sequence and parent-of-origin dependent DNA methylation in the mouse genome. *Cell* **148:** 816–831.

Ying QL, Wray J, Nichols J, Batlle-Morera L, Doble B, Woodgett J, Cohen P, Smith A. 2008. The ground state of embryonic stem cell self-renewal. *Nature* **453:** 519–523.

You JS, Jones PA. 2012. Cancer genetics and epigenetics: Two sides of the same coin? *Cancer Cell* **22:** 9–20.

Zaret KS, Carroll JS. 2011. Pioneer transcription factors: Establishing competence for gene expression. *Genes Dev* **25:** 2227–2241.

Zhang K, Lin W, Latham JA, Riefler GM, Schumacher JM, Chan C, Tatchell K, Hawke DH, Kobayashi R, Dent SY. 2005. The Set1 methyltransferase opposes Ipl1 aurora kinase functions in chromosome segregation. *Cell* **122:** 723–734.

Zhang T, Berrocal JG, Frizzell KM, Gamble MJ, DuMond ME, Krishnakumar R, Yang T, Sauve AA, Kraus WL. 2009. Enzymes in the NAD^+ salvage pathway regulate SIRT1 activity at target gene promoters. *J Biol Chem* **284:** 20408–20417.

* Zhao Y, Garcia BA. 2014. Comprehensive catalog of currently documented histone modifications. *Cold Spring Harb Biol* doi: 10.1101/cshperspect.a025064.

Zhao T, Zhang ZN, Rong Z, Xu Y. 2011. Immunogenicity of induced pluripotent stem cells. *Nature* **474:** 212–215.

Zhou X, Slone JD, Rokas A, Berger SL, Liebig J, Ray A, Reinberg D, Zwiebel LJ. 2012. Phylogenetic and transcriptomic analysis of chemosensory receptors in a pair of divergent ant species reveals sex-specific signatures of odor coding. *PLoS Genet* **8:** e1002930.

Zhu Q, Pao GM, Huynh AM, Suh H, Tonnu N, Nederlof PM, Gage FH, Verma IM. 2011. BRCA1 tumour suppression occurs via heterochromatin-mediated silencing. *Nature* **477:** 179–184.

Zofall M, Yamanaka S, Reyes-Turcu FE, Zhang K, Rubin C, Grewal SI. 2012. RNA elimination machinery targeting meiotic mRNAs promotes facultative heterochromatin formation. *Science* **335:** 96–100.

* Zoghbi HY, Beaudet AL. 2014. Epigenetics and human disease. *Cold Spring Harb Perspect Biol* doi: 10.1101/cshperspect.a019497.

Zou J, Lippert C, Heckerman D, Aryee M, Listgarten J. 2014. Epigenome-wide association studies without the need for cell-type composition. *Nat Methods* **11:** 309–311.

Zullo JM, Demarco IA, Pique-Regi R, Gaffney DJ, Epstein CB, Spooner CJ, Luperchio TR, Bernstein BE, Pritchard JK, Reddy KL, et al. 2012. DNA sequence-dependent compartmentalization and silencing of chromatin at the nuclear lamina. *Cell* **149:** 1474–1487.

WWW RESOURCES

http://www.1000genomes.org 1000 Genomes project
http://epigenomesportal.ca/ihec IHEC data portal
http://hapmap.ncbi.nlm.nih.gov HapMap

C H A P T E R 4

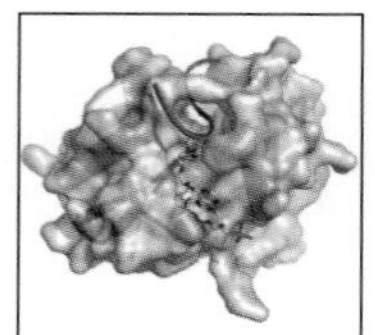

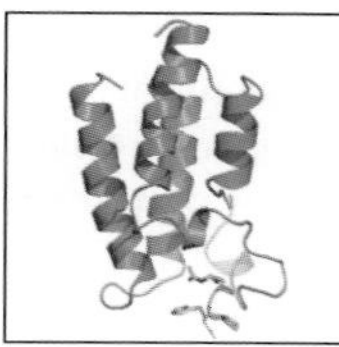

Writers and Readers of Histone Acetylation: Structure, Mechanism, and Inhibition

Ronen Marmorstein[1] and Ming-Ming Zhou[2]

[1]Program in Gene Expression and Regulation, Wistar Institute, and Department of Chemistry, University of Pennsylvania, Philadelphia, Pennsylvania, 19104; [2]Department of Structural and Chemical Biology, Icahn School of Medicine at Mount Sinai, New York, New York 10065

Correspondence: marmor@wistar.org

SUMMARY

Histone acetylation marks are written by histone acetyltransferases (HATs) and read by bromodomains (BrDs), and less commonly by other protein modules. These proteins regulate many transcription-mediated biological processes, and their aberrant activities are correlated with several human diseases. Consequently, small molecule HAT and BrD inhibitors with therapeutic potential have been developed. Structural and biochemical studies of HATs and BrDs have revealed that HATs fall into distinct subfamilies containing a structurally related core for cofactor binding, but divergent flanking regions for substrate-specific binding, catalysis, and autoregulation. BrDs adopt a conserved left-handed four-helix bundle to recognize acetyllysine; divergent loop residues contribute to substrate-specific acetyllysine recognition.

Outline

Chapter-opening figures: (*Right*) Adapted from Zeng et al. (2008b).

Cite as *Cold Spring Harb Perspect Biol* doi: 10.1101/cshperspect.a018762

OVERVIEW

Histone acetyltransferases (HATs), sometimes referred to as lysine acetyltransferases or KATs, form a superfamily of enzymes that acetylate the side-chain amino group of lysine residues on histones, and in some cases also other proteins. These enzymes contribute to several different transcription-mediated biological processes including cell-cycle progression, dosage compensation, and hormone signaling. Aberrant HAT function is correlated with several human diseases, including leukemic translocations, solid tumors, and metabolic disorders. In addition, protein acetylation reaches beyond histones and transcription-associated biological processes to other cellular processes, based on recent proteomic studies.

The acetylation marks on lysine residues are read by small protein modules called bromodomains (BrDs), sometimes referred to as "readers." These domains are conserved within many chromatin-associated proteins including some HATs, as well as other posttranslational modification enzymes (sometimes referred to as "writers") and ATP-dependent remodeling proteins. More recently, a PHD finger, previously shown to target methylated lysine residues, was also shown to bind acetyllysine (Kac), opening up the possibility that other types of domains may also read acetyllysine marks. Currently, many BrD-containing proteins do not have well-characterized functions, although some have been implicated in diseases such as inflammation, viral infection, solid tumors, and leukemias.

Structural and biochemical studies on HATs and BrDs have provided important mechanistic insights into the function of these writers and readers of histone acetylation. Five well-studied HAT subfamilies include Hat1 (or KAT1 according to the Allis et al. 2007 nomenclature), Gcn5/PCAF (KAT2A/KAT2B), MYST (KAT5), p300/CBP (KAT3B/KAT3A), and Rtt109 (KAT11). These HAT enzyme subfamily writers perform similar overall chemistry and have structurally related core regions that template substrates in an analogous fashion; however, they fall into subfamilies with very limited to no sequence homology. Consequently, they contain structurally divergent core flanking regions, which mediate divergent mechanisms of catalysis and possibly different modes of substrate recognition and regulation. Many HATs are regulated by autoacetylation. Inhibition of HAT enzymes by small molecule compounds is in the very early stages of development, but the prospects for exploiting HATs as therapeutic targets are strong.

The BrD readers adopt a conserved left-handed four-helix bundle and possess conserved residues within interhelical loops that recognize acetyllysine. Other residues flanking either side of the acetylated lysine contribute to binding specificity. Interestingly, many bromodomains come in multiples, and many have divergent functions such as binding two or more acetyllysine residues simultaneously or, in some cases, may have other functions distinct from acetyllysine recognition. Given the association of BrD-containing proteins with disease, there has been considerable interest in developing BrD inhibitors. Remarkably, several potent and selective inhibitors have already been developed that look promising for therapeutic applications.

 Cite as *Cold Spring Harb Perspect Biol* doi: 10.1101/cshperspect.a018762

1 INTRODUCTION TO WRITERS, ERASERS, AND READERS OF HISTONES

DNA within the eukaryotic nucleus is compacted into chromatin containing the histone proteins H1, H2A, H2B, H3, H4. The appropriate regulation of chromatin orchestrates all DNA-templated reactions such as DNA transcription, replication, repair, mitosis, and apoptosis (Williamson and Pinto 2012). The macromolecules that regulate chromatin fall into distinct classes of molecules. These include ATP-dependent remodeling proteins that mobilize the histones within chromatin (Ch. 21 [Becker and Workman 2013]), histone chaperones that insert and remove generic or variant histones into chromatin (covered in Ch. 22 [Almouzni and Cedar 2014]), posttranslational modification enzymes that add and remove chemical groups to the DNA or histone components of chromatin (Bannister and Kouzarides 2011), chromatin recognition proteins that specifically recognize DNA, histones or modified histone, or DNA (Yap and Zhou 2010; Glatt et al. 2011), and noncoding RNA molecules that bind and modulate chromatin regulatory proteins (Mattick and Makunin 2006; Kurth and Mochizuki 2009). These macromolecules work in a highly coordinated fashion to regulate distinct chromatin templated activities.

The posttranslational modification (PTM) enzymes include proteins that add chemical groups as well as those that remove them. The enzymes that mediate histone modification (i.e., writers) include acetyltransferases, methyltransferases, kinases, and ubiquitinases. The enzymes that remove these modifications (i.e., erasers) include deacetylases, phosphatases, demethylases, and deubiquitinases (Bannister and Kouzarides 2011). Protein domains have also been identified that can recognize specific histone modifications (i.e., readers), although there appears to be more flexibility than the enzymes that create the modifications (Yap and Zhou 2010; Glatt et al. 2011). For example, bromodomains selectively target acetyllysine residues, whereas many chromodomains bind methylated lysines, and tudor domains bind methylated arginines. However, methylated lysines are also recognized by PHD fingers, WD40 domains, and ankyrin repeats (Brent and Marmorstein 2008). Many of these protein domains recognize unmodified histones as well.

Of the enzymes that perform posttranslational modification on histones, the enzymes that mediate lysine acetylation and deacetylation were the first identified. In 1996, Allis and coworkers purified a histone acetyltransferase (HAT) from *Tetrahymena thermophila* that was orthologous to a previously identified transcriptional adaptor from yeast called Gcn5 and conserved from yeast to man (Brownell et al. 1996). During the same time, Sternglanz and coworkers (Kleff et al. 1995) and Gottschling and coworkers (Parthun et al. 1996) also identified a histone acetyltransferase called HAT1 that was initially proposed to be a cytoplasmic specific acetyltransferase and later shown to also harbor nuclear functions (Ruiz-Garcia et al. 1998; Ai and Parthun 2004; Poveda et al. 2004). In the same year, Schreiber and colleagues isolated a mammalian histone deacetylase (HDAC) that was highly homologous to a previously characterized transcriptional repressor Rpd3 (covered in Ch. 5 [Seto and Yoshida 2014]), also with conservation from yeast to man (Taunton et al. 1996). Subsequent to these groundbreaking studies, other HATs and HDACs were identified along with other types of enzymes that modify histones (Hodawadekar and Marmorstein 2007; Marmorstein and Trievel 2009). Many histone posttranslational modifications can be correlated with different DNA-templated activities. HAT and HDAC activities, however, are generally correlated with gene activation and repression/silencing, respectively. Aberrant HAT and HDAC activities are also correlated with diseases such as cancer and metabolic disorders (Keppler and Archer 2008a; Keppler and Archer 2008b). To date, the HATs and HDACs are, biochemically and structurally, the most well-characterized of the histone PTM enzymes. This chapter will cover what is known to date about the structure, mechanism of action, and inhibition of HAT enzymes. Readers are directed to the collection's chapter on HDACs dealing with the topic of histone lysine deacetylation (Ch. 5 [Seto and Yoshida 2014], and other excellent review articles therein).

Bromodomains were the first histone modification readers that were identified when Zhou, Aggarwal, and coworkers determined the three-dimensional solution structure of the PCAF bromodomain using nuclear magnetic resonance spectroscopy. They also performed related biochemical studies revealing that it is specifically capable of binding acetyllysine modifications in peptides derived from histones H3 and H4 (Dhalluin et al. 1999). Subsequent studies by others confirmed the acetyllysine-binding properties of bromodomains from other proteins along with the identification and characterization of other types of protein domains that specifically recognize other histone modifications (Yap and Zhou 2010). Interestingly, bromodomains are found in many different types of chromatin regulators as well as protein complexes that play important roles in the human biology of health and disease. Thus, recently they have become attractive therapeutic targets, illustrated by the discovery of JQ1 and I-BET described in Ch. 2 (Qi 2014; Schaefer 2014) (also discussed in Ch. 29 [Busslinger and Tarakhovsky 2014]). In addition to describing HAT structure and function, this chapter will also cover what is known to date about the structure, mechanism of action, and inhibition of bromodomains.

2 HISTONE ACETYLTRANSFERASES

2.1 Classification of HATs

Since the isolation of the Gcn5 HAT from *Tetrahymena* by Allis and coworkers (Brownell et al. 1996), and the identification of HAT1 by Sternglanz and coworkers (Kleff et al. 1995) and Gottschling and coworkers (Parthun et al. 1996) just more than a decade ago, many other HATs have been identified from yeast to man. Some of these HATs (e.g., PCAF and HAT1) show sequence conservation with Gcn5 within their catalytic domain, leading to their classification as Gcn5-related histone *N*-acetyltransferases (GNATs; Neuwald and Landsman 1997). Many other HATs, like CBP/p300, Rtt109, and the MYST proteins have extremely limited sequence conservation. Based on this sequence divergence within the HAT domain, HATs can be grouped into at least five different subfamilies (Table 1). This includes HAT1 (named histone acetyltransferase 1 as the founding member of the superfamily or KAT1 according to the Allis et al. 2007 nomenclature), Gcn5/PCAF (named for its founding member yeast Gcn5 and its human ortholog, PCAF, or KAT2a/KAT2B according to the alternative nomenclature), MYST (named for the founding members MOZ, Ybf2/ Sas3, Sas2, and TIP60, or KAT5), p300/CBP (named for the two human paralogs p300 and CBP, or KAT3B/KAT3A), and Rtt109 (named for its initial identification as a regulator of Ty1 transposition gene product 109, also referred to as KAT11). Although the Gcn5/PCAF, HAT1, and MYST subfamilies have homologs from yeast to man, p300/CBP is metazoan specific, and Rtt109 is fungal specific. Although other nuclear HAT subfamilies have been identified, such as the steroid receptor coactivators (ACTR/AIB1, SRC1) (Spencer et al. 1997), TAF250 (Mizzen et al. 1996), ATF-2 (Kawasaki et al. 2000), and CLOCK (Doi et al. 2006), their HAT activities have not been studied as extensively as the five major HAT classes and will not be further discussed here.

2.2 Overall HAT Structure

Representative structures of each of the five subfamilies of HAT proteins have been determined by X-ray crystallography, revealing the molecular characteristics of the enzymatic domains and molecular insights into catalysis and substrate acetylation.

Yeast histone acetyltransferase HAT1 (yHAT1) was the first HAT structure to be reported (Dutnall et al. 1998), setting the stage for structural analysis of this superfamily of enzymes. The structure of HAT1 bound to its AcCoA cofactor consists of an elongated α-β structure (Fig. 1A). The structure contains a conserved core region containing a three-stranded β-sheet and a long helix in parallel, span-

Table 1. The five major HAT families

Major HAT subfamilies	Prominent members	Key structural and biochemical properties
HAT1	yHat1	Member of the GNAT family Amino- and carboxy-terminal segments used for histone substrate binding Requires the yHat2 regulatory subunit for maximal catalytic activity
Gcn5/PCAF	yGcn5 hGCN5 hPCAF	Member of the GNAT family Uses a ternary complex catalytic mechanism Amino- and carboxy-terminal segments used for histone substrate binding
MYST	yEsa1 ySas2 ySas3 hMOZ dMof hMOF hTIP60 hHBO1	Uses a ping-pong catalytic mechanism Requires autoacetylation of a specific lysine at the active site for cognate histone acetylation
p300/CBP	hp300 hCBP	Metazoan-specific, but shows structural homology with yRtt109 Uses a ternary Theorell–Chance (hit-and-run) catalytic mechanism Contains a substrate-binding loop that participates in AcCoA and lysine binding Contains an autoacetylation loop that requires lysine autoacetylation for maximal catalytic activity
Rtt109	yR11109	Fungal-specific, but shows structural homology with p300 Contains a substrate-binding loop that participates in AcCoA and probably also lysine binding Requires autoacetylation of a lysine residue near the active site for maximal catalytic activity Requires one of two histone chaperone cofactors (Asf1 or Vps75) for maximal catalytic activity and histone substrate specificity

y, yeast; h, human; GNAT, Gcn5-related *N*-acetyltransferase.

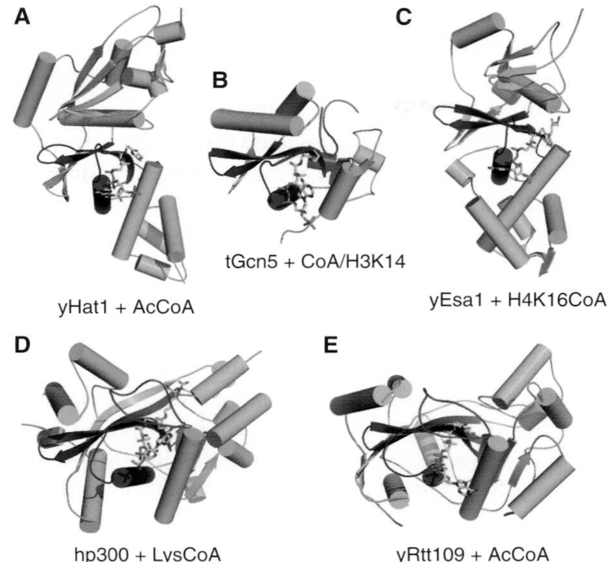

Figure 1. Overall structure of HAT proteins. Representative members of the five HAT subfamilies are illustrated as cartoons highlighting the structurally conserved core region (blue) and flanking amino- and carboxy-terminal regions (aqua). The cofactor is shown in stick figure in CPK coloring (carbon, yellow; oxygen, red; nitrogen, blue; phosphorous, orange; sulfur, brown): (*A*) yeast HAT1/AcCoA (PDB 1BOB), (*B*) *Tetrahymena* Gcn5/CoA/histone H3 (PDB code: 1PUA) with the histone H3 peptide shown in red, (*C*) yeast Esa1/H4K16-CoA (PDB code: 3TO6), (*D*) human p300/Lys-CoA (PDB code: 3BIY) with the substrate-binding loop shown in red, (*E*) yeast Rtt109/CoA (PDB code: 3D35) with the substrate-binding loop shown in red.

ning one side of the sheet (Neuwald and Landsman 1997). The AcCoA cofactor packs and makes interactions along one edge of this core region. The core region is flanked by a β-α-loop segment on one side (see top part of Fig. 1A) and an α-rich segment on the other (see bottom part of Fig. 1A) that together form a cleft over the central core domain for the histone substrate to bind and catalysis to occur.

To date, several Gcn5 crystal structures in various liganded forms (Rojas et al. 1999; Trievel et al. 1999; Poux et al. 2002; Clements et al. 2003; Poux and Marmorstein 2003), a Gcn5/CoA solution structure (Lin et al. 1999), and the human PCAF/AcCoA structure (Clements et al. 1999) have been reported. Each of the HAT domains shows a high degree of superposition. The *Tetrahymena* ternary structure of a Gcn5/CoA/histone H3 complex reveals a core region with structural conservation with yHat1 that make analogous AcCoA cofactor interactions, but structurally divergent amino- and carboxy-terminal regions flank the core region (Fig. 1B). The cleft over the core domain is deeper than yHat1 and accommodates the bound histone H3 peptide (red loop in Fig. 1B) that is contacted largely by residues from the amino- and carboxy-terminal segments.

Several structures have also been reported of HATs from the MYST subfamily from yeast to man, including *Drosophila* MOF bound to AcCoA (Akhtar and Becker 2001), hMOF in unliganded form (Yuan et al. 2012), hMOZ bound to AcCoA (Holbert et al. 2007), and yEsa1 in various liganded forms (Yan et al. 2000; Yan et al. 2002; Yuan et al. 2012). Each of the structures superimpose well and show a structurally conserved core region with HAT1 and Gcn5/PCAF, but divergent amino- and carboxy-terminal segments. The structure of yeast Esa1 bound to a linked peptide-CoA conjugate in which residues 11–22 of histone H4 is linked to CoA though lysine 16 (H4K16-CoA) reveals α/β amino- and carboxy-terminal domains, with the amino-terminal domain containing a TFIIIA zinc finger fold and the carboxy-terminal segment containing a helix-turn-helix domain typically found in DNA-binding proteins and proposed to participate in nucleosomal localization (Fig. 1C) (Holbert et al. 2007). The bound H4K16-CoA bisubstrate inhibitor provides important details about how both cosubstrates bind to yEsa1. The CoA portion is packed against the core domain similar to the Gcn5 and HAT complexes and the lysine portion of the inhibitor is located in the central cleft formed by the flanking amino- and carboxy-terminal segments. The rest of the H4 peptide is disordered in the crystal structure.

The structures of human p300 (hp300) bound to a Lys-CoA bisubstrate inhibitor (Liu et al. 2008), yeast Rtt109 (yRtt109) bound to CoA or AcCoA cofactors (Lin and Yuan 2008; Stavropoulos et al. 2008; Tang et al. 2008), or bound to cofactor and the Vps75 histone chaperone (Kolonko et al. 2010; Su et al. 2011; Tang et al. 2011) have been reported. Remarkably, the hp300 and yRtt109 structures show a high degree of superposition despite the absence of significant sequence homology (Fig. 1D,E). These structures reveal an elongated globular domain containing a central seven-strand β-sheet surrounded by nine α-helices and several loops. Roughly in the center of this domain is located the structurally conserved core region that is involved in cofactor interactions, whereas regions that flank this core diverge from the other HATs. A unique feature of these HATs relative to the others is a long, approximately 25 residue loop (colored in red in Fig. 1D,E), called the substrate-binding loop, that encapsulates the cofactor in yRtt109 and the Lys-CoA bisubstrate inhibitor in hp300.

Taken together, each of the 5 HAT subfamilies shares a conserved central core region that contributes to AcCoA cofactor binding, but divergent amino- and carboxy-terminal segments that flank this core also appear to contribute to histone substrate binding. The p300 and Rtt109 HAT subfamilies (Fig. 1D,E) are structurally conserved throughout the HAT domain and contain a unique, approximately 25 residue substrate-binding loop that appears to participate in the binding of both AcCoA and histone substrates.

2.3 Catalytic Mechanism

HATs transfer the acetyl group from the acetyl-CoA cofactor to the Nζ nitrogen of a lysine side chain within histones. Structural, biochemical, mutational, and enzymatic analyses have provided insights into the catalytic mechanism of these enzymes. A remarkable outcome of these studies is that each HAT subfamily uses a different catalytic strategy for acetyl transfer. This is unusual for a superfamily of enzymes that catalyze the same chemical reaction, but perhaps not so surprising for these enzymes because the transfer of an acetyl group from a thioester to an amine is not a chemically demanding reaction, thus allowing different HAT subfamilies to use different chemical strategies to mediate acetyl transfer.

Members of the Gcn5/PCAF subfamily of HATs were the first proteins of the superfamily whose enzymatic properties were characterized in detail in the context of the available crystal structures. Specifically, the crystal structures of Gcn5 and PCAF revealed the presence of a strictly conserved glutamate in the active site (Glu173 in yGcn5 and Glu570 in hPCAF) that is in position to act as a general base for catalysis through a well-ordered water molecule in the crystal structure (Fig. 2A). The glutamate is buried in a hydrophobic pocket, which likely raises its acid dissociation constant (pKa) for proton extraction. The importance of this glutamate residue was confirmed by mutational analysis (Tanner et al. 1999; Trievel et al. 1999) in which an E173Q mutant in yGcn5 showed a 360-fold decrease in enzymatic turnover (k_{cat}; Tanner et al. 1999). Denu and coworkers also showed that Gcn5 functions through a ternary complex mechanism in which both substrates (i.e., lysine and AcCoA) must be bound to the enzyme before catalysis can occur. This involves deprotonation of the lysine substrate by Glu173, facilitating the direct transfer of the acetyl group from AcCoA to the lysine side chain (Tanner et al. 1999). Denu and coworkers showed similar results for PCAF (Tanner et al. 2000), however, an acidic residue, if any, that protonates the CoA leaving group has not been identified.

The crystal structure of the yEsa1 member of the MYST subfamily of HATs revealed the presence of a glutamate residue (Glu338) that is strictly conserved within the MYST HATs. It overlays with the catalytic glutamate residue of Gcn5/PCAF, and a yEsa1 E338Q mutant only catalyzed background levels of acetylation (Yan et al. 2000). Interestingly, a subsequent structural analysis and enzymatic characterization revealed that an active site cysteine residue (Cys304), conserved within the MYST HAT subfamily, also plays an important catalytic role with C304S and C304A mutants shown to be highly defective (Fig. 2B) (Yan et al. 2002). Kinetic analysis with both substrates is consistent with a ping-pong catalytic mechanism, in which the enzyme first forms an acetylated intermediate involving Cys304 before the formation of an acetylated histone product. In this mechanism, Glu338 is proposed to serve as a general base, which deprotonates both Cys304 and the histone lysine side chain in preparation for both acetyl-transfer events (Yan et al. 2002). Paradoxically, a more recent report shows that Esa1 assembled within a physiologically relevant piccolo NuA4 complex does not show a strong dependence on Cys304 for catalysis, leading the investigators to conclude that yEsa1 within the piccolo NuA4 complex proceeds through a ternary catalytic mechanism similar to Gcn5/PCAF (Berndsen et al. 2007). This would suggest that the same HAT enzyme may use different catalytic mechanisms within different cellular contexts.

The p300 crystal structure reveals that there is no glutamate residue that is analogous to the key residues within the Gcn5/PCAF and MYST HAT subfamilies that function as general base residues for catalysis (Liu et al. 2008). Mutagenesis and kinetic analysis of the potential catalytic residues in the active site uncovered only two residues (Tyr1467 and Trp1436) that showed a significant effect on catalysis when mutated (Fig. 2C) (Liu et al. 2008); a Y1467F substitution showed about a 400-fold reduction in catalytic efficiency (k_{cat}/K_M), a W1436A substitution reduced the catalytic efficiency 50-fold, and a W1436F substitution showed a more modest effect on catalysis. Based on the position of these residues in the structure, Tyr1467 was proposed to play a role as a general acid for catalysis, whereas Trp1436 was proposed to help orient the target lysine into the active site. These residues are strictly conserved within the p300/CBP HAT subfamily, correlating with the notion that Tyr1467 and Trp1436 are important for catalysis. Thus, it appears that p300/CBP does not use a general base for catalysis, in contrast to the Gcn5/PCAF and MYST HAT subfamilies. Taken together with the fact that hp300 is inhibited potently by a more primitive Lys-CoA inhibitor, but poorly by bisubstrate inhibitors with longer peptide moieties (Lau et al. 2000) and the observation that longer peptides are better substrates for p300 than lysine, it was proposed that the p300/CBP subfamily of HATs use a "hit-and-run" or Theorell–Chance acetyl-transfer mechanism that is distinct from the catalytic mechanisms used by the Gcn5/PCAF and MYST HAT subfamilies.

The kinetic mechanism for acetylation by Rtt109 and Hat1 has been less well-characterized than the other HAT subfamilies. We know both enzymes require the association of other protein regulatory subunits for full activity. In the

 Cite as *Cold Spring Harb Perspect Biol* doi: 10.1101/cshperspect.a018762

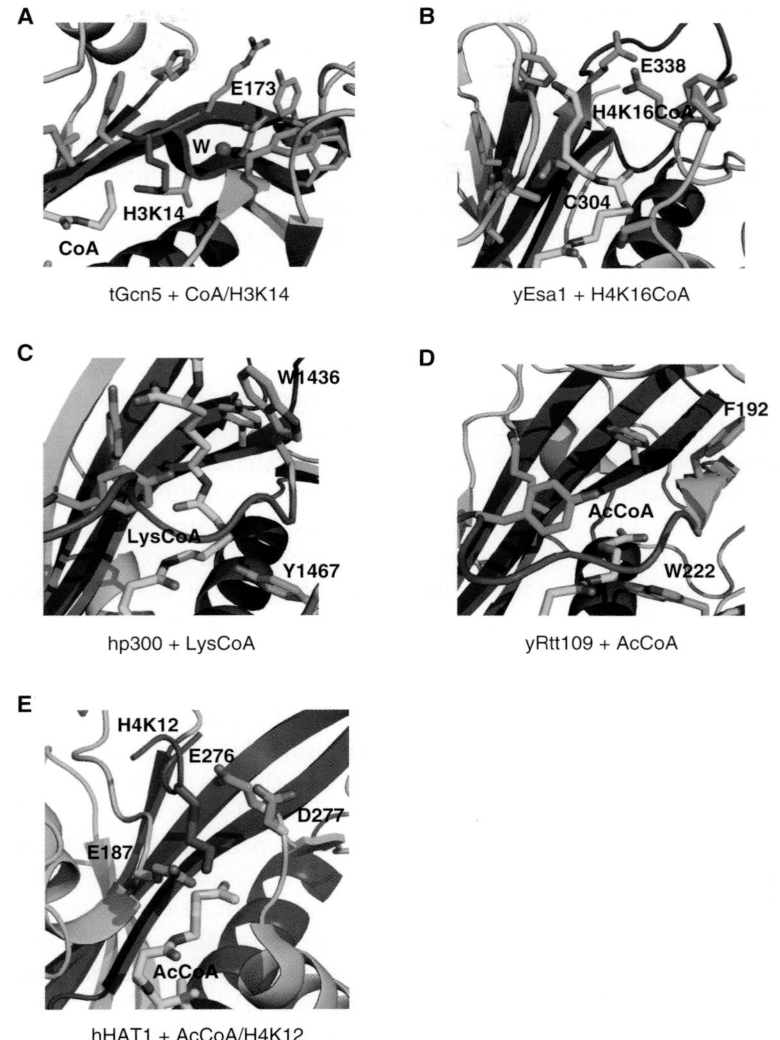

Figure 2. Catalytic mechanism of HAT proteins. Active sites of representative members of the HAT subfamilies are illustrated highlighting the relevant side chains on a backbone cartoon of the active site. (*A*) *Tetrahymena* Gcn5/CoA/histone H3. Key catalytic residues are labeled and hydrophobic residues of the active site that likely raise the pKa of Glu 173 are shown in stick figure in CPK coloring with carbon in green. A segment of the histone H3 peptide is shown in red. W indicates a well-ordered water molecule that participates in catalysis. The numbering is for yeast Gcn5. (*B*) Yeast Esa1 bound to the H4K16CoA bisubstrate inhibitor (stick figure and CPK coloring with carbon atoms in yellow). Key catalytic residues are labeled and hydrophobic residues of the active site that likely raise the pKa of Glu 338 are shown. Residues flanking K16 in the peptide are disordered in the structure. (*C*) Human p300 bound to the Lys-CoA bisubstrate inhibitor (stick figure and CPK coloring with carbon atoms in yellow). Residues shown to play catalytic roles are labeled with other potential catalytic residues shown in stick figure. The substrate-binding loop is shown in red. (*D*) Yeast Rtt109/CoA. Potential catalytic residues in the corresponding position of hp300 are shown. The CoA molecule is shown in stick figure in CPK coloring with carbon atoms in yellow. The substrate-binding loop is shown in red. (*E*) hHAT1/AcCoA/histone H4. The three general base candidate residues are represented as green stick figures and a segment of the histone H4 peptide is shown in red.

case of Rtt109, it harbors very low acetyltransferase activity on its own (Driscoll et al. 2007; Tsubota et al. 2007), but its activity is stimulated by association with either the Asf1 or Vps75 histone chaperone proteins (Han et al. 2007a; Han et al. 2007b; Tsubota et al. 2007; Berndsen et al. 2008; Albaugh et al. 2010). Although it is not clear how these histone chaperones enhance the catalytic activity of Rtt109, the crystal structures of Rtt109/Vps75 complexes (in both 2:2 and 2:1 stoichiometries) show no significant changes in the Rtt109 active site as a function of Vps75 binding, suggesting that the histone chaperones merely function to deliver histone substrates to Rtt109 for acetylation (Kolonko et al. 2010; Su et al. 2011; Tang et al. 2011). The crystal structure of Rtt109 alone revealed that despite its overall structural superposition with the hp300 HAT domain, the key catalytic residues of hp300 (Tyr1467 and Trp1436) are not conserved in Rtt109 (Fig. 2D). Instead, kinetic analysis showed that the Rtt109/Vps75 complex uses a sequential kinetic mechanism whereby the Rtt109–Vps75 complex, AcCoA, and histone H3 substrates form a complex before chemical catalysis (Albaugh et al. 2010). Structural, mutational, and kinetic analysis of Rtt109 points to the importance of Asp89 and Trp222 for catalysis. The D89N and W222F mutants showed about a 25-fold defect in catalytic rate; however, this was mostly because of a K_M effect for AcCoA binding (Tang et al. 2011), as was the 1000-fold reduced catalytic efficiency of a D288N mutant (Albaugh et al. 2010; Kolonko et al. 2010). Key residues involved in catalytic turnover, such as a general base or acid, have not been identified for Rtt109 and thus may not be present in this case.

Although a detailed enzymatic study of yHat1 has not been reported, a recent study provides important new insights (Wu et al. 2012). Specifically, the structure determination of human HAT1 (hHAT1) bound to AcCoA and a histone H4 peptide centered around K12 reveals that three residues, Glu187, Glu276, and Asp277, are in proximity to the Nζ nitrogen of H4K12 and thus could act as potential general base residues for catalysis (Fig. 2E) (Wu et al. 2012). A superposition of the hHAT1 active site with yGcn5 and yEsa1 (MYST subfamily) reveals that residues in these enzymes that function as a general base of catalysis (i.e., glutamates 173 and 338, respectively) superimpose with Glu276 of hHAT1 (Glu255 in yHat1; Yan et al. 2000; Wu et al. 2012), which is strictly conserved among Hat1 orthologs. In hHAT1, mutating Glu276 to Gln causes a 28.5-fold decrease in k_{cat}. E187Q and D277N mutations, however, cause a more modest ~15.3- and ~8.1-fold decrease in k_{cat}, respectively. E276Q and E187Q increase the pKa of ionization from 8.15 to 8.74 and 9.15, respectively, whereas D277N only increases the pKa slightly to 8.35 (Wu et al. 2012). Hat1, however, unlike MYST HATs, does not contain a cysteine residue in the active site of the enzyme. Taken together, these observations suggest that Glu276 of hHAT1, and probably also Glu187, play a role as general base(s) for catalysis, through a ternary complex mechanism similar to the Gcn5/PCAF HAT subfamily.

A comparison of the catalytic mechanisms of the different HAT subfamilies reveals remarkable diversity in the way each subfamily mediates acetyl transfer. This is likely to be a reflection of the relatively low chemical "cost" that is required to transfer an acetyl group from a thioester to an amine, as opposed to some other more demanding chemical reactions such as phosphorylation, and the relatively long evolutionary time that HATs had to evolve.

2.4 Histone Substrate Binding

To date, the only direct molecular insights into histone binding by HATs has come from structural studies of two HAT proteins; one study provided structural information on the hHAT1 HAT domain bound to AcCoA and a histone H4 peptide centered around Lys12 (Wu et al. 2012) whereas other studies have focused on the *Tetrehymena* Gcn5 (tGcn5) HAT domain bound to CoA and several cognate substrate peptides including histone H3 (centered around Lys14; Rojas et al. 1999; Clements et al. 2003; Poux and Marmorstein 2003), histone H4 (centered around Lys8), and p53 (centered around Lys320; Poux and Marmorstein 2003). These structures reveal that the histone peptide substrates bind across a groove formed by the central core region found at the base and the amino- and carboxy-terminal regions on flanking sides that mediate the majority of the interactions with the substrate peptide (Fig. 3A,B). A comparison of the structures of tGcn5 bound to different peptides revealed that 15 of the 19 residues of H3 are ordered in the structure rather than less than 10 residues for the H4 and p53 peptides, which is consistent with the >1000-fold greater catalytic efficiency of Gcn5 for H3 over H4 and p53 (Trievel et al. 2000). These structures also revealed a more ordered and extensive protein–peptide interaction in residues carboxy-terminal to the reactive lysine, arguing that substrate interactions carboxy-terminal to the target are more important for substrate binding by the Gcn5/PCAF HAT subfamily (Fig. 3A). These interactions involve mostly hydrogen bonds to backbone residues and van der Waals interactions with side chains. Not surprisingly, the residues of tGcn5 that make contact to the peptide substrate are highly conserved within the Gcn5/PCAF subfamily.

When bound to hHAT1, the H4 peptide adopts a well-defined conformation that harbors a β-turn at its amino terminus, which would otherwise be in an extended form in free H4 (Fig. 3B) (Wu et al. 2012). Two hHAT1 conserved hydrophobic residues (Trp199 and Tyr225) interact with Gly9 and Lys8 at the β-turn of the H4 peptide, respectively, to enforce the orientation of the substrate at the entrance of a groove (Wu et al. 2012). The carboxyl terminus of the H4 peptide contains two positively charged residues (Arg17 and Arg19) that make extensive hydrogen-bond and charge–charge interactions with invariant residues of hHAT1 (Glu64 and Asp62). The substrate-binding groove of hHAT1 narrows where H4 Lys12 binds allowing only Gly11 and Lys12 to be accommodated. The strictly conserved Glu276 of hHAT1 also makes hydrogen-bond

Cite as *Cold Spring Harb Perspect Biol* doi: 10.1101/cshperspect.a018762

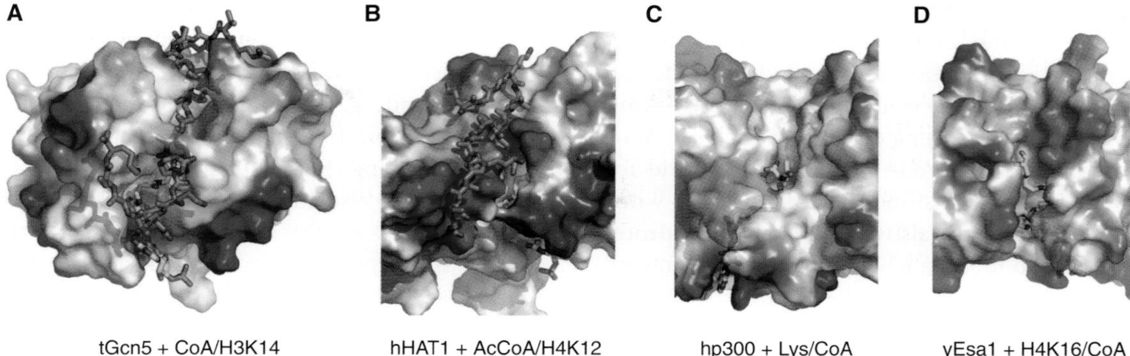

Figure 3. Histone substrate binding by HAT proteins. Close-up electrostatic view of HAT domain structures with histone peptide substrates or CoA-peptide bisubstrate inhibitors. Protein surfaces are colored according to electrostatic potential with the degree of red, blue, and white coloring correlating with electronegative, electropositive, and neutral charge, respectively. (*A*) Structure of tGcn5 bound to CoA (CPK coloring with carbon atoms shown in yellow) and a 19-residue histone H3 peptide (CPK coloring with carbon atoms shown in purple) centered around K14. (*B*) Structure of hHAT1 bound to AcCoA (CPK coloring with carbon atoms shown in yellow) and a 20-residue histone H4 peptide (CPK coloring with carbon atoms shown in purple) centered around K12. (*C*) Structure of the hp300/LysCoA complex. The LysCoA bisubstrate inhibitor is shown in CPK coloring with carbon atoms in yellow. (*D*) Structure of the yEsa1/H4K16CoA complex (only the lysine side chain of the H4K16 peptide component of the bisubstrate inhibitor is ordered in the crystal structure and shown in CPK coloring with carbon atoms in yellow).

interactions with H4 Gly11 and Lys12. Together, these specific interactions explain the preference for Lys12 of H4 as the acetylation target of hHAT1 (Wu et al. 2012).

The structures of human p300 (Liu et al. 2008) and the MYST HAT domain protein, yEsa1 (Yuan et al. 2012), crystallized with bisubstrate inhibitors has provided some information about histone substrate binding. hp300 was cocrystallized with a Lys-CoA bisubstrate inhibitor that shows an IC_{50} of ~400 nM, and is a more potent inhibitor than peptide-CoA inhibitors, consistent with its Theorell–Chance catalytic mechanism. The structure of the hp300/Lys-CoA complex revealed that the lysine portion of the bisubstrate inhibitor sits in a hydrophobic tunnel with the backbone of the lysine residue proximal to an electronegative groove on one side (Fig. 3C). This electronegative groove shown for hp300 also contains two pockets separated by a distance corresponding to about 3–4 amino acid residues. Correlating with this observation, an alignment of all known p300/CBP substrates reveals that they all contain a basic amino acid either three or four residues upstream of or downstream from the target lysine. Mutagenesis of residues that form these pockets increases the K_M for histone H3 substrate, highlighting the importance of this site for protein substrate binding by p300 (Liu et al. 2008). Taking this data together, the p300/CBP HAT subfamily have more promiscuous substrate-binding properties relative to the Hat1, Gcn5/PCAF, and MYST HATs.

The yEsa1 HAT domain was reported crystallized with an H4K16CoA bisubstrate inhibitor (Yuan et al. 2012). Although the yEsa1 HAT domain and the CoA, linker, and lysine portion of the bisubstrate inhibitor is well-resolved, the rest of the peptide portion of the bisubstrate inhibitor is not and is presumed to be disordered (Fig. 3C). Like the hp300/Lys-CoA structure, the yEsa1/H4K16CoA complex shows a groove for peptide binding proximal to the lysine portion of the bisubstrate inhibitor for histone peptide substrate to bind, although this groove is more apolar than that of the hp300 HAT domain. This likely reflects the greater degree of substrate selectivity of the MYST proteins over the p300/CBP HAT subfamily.

Taken together, the structures that are available of HAT domains bound to peptide substrates or CoA-peptide bisubstrate inhibitors provide important insights into the general mode of peptide binding although the molecular details that underlie substrate-specific binding is still unclear.

2.5 Regulation by Autoacetylation and Protein Cofactors

Acetyltransferase activity is regulated in at least two ways, through interaction with regulatory protein subunits and through autoacetylation. Many acetyltransferases function in the context of multiprotein complexes that modulate their catalytic activity and/or substrate specificity (Carrozza et al. 2003; Lee and Workman 2007). For example, the recombinant Gcn5 and PCAF proteins are active on free histones (acetylating H3K14 and, to a lesser extent, H4K8/K16) or histone peptides, but are much less active on nucleosomes. However, Gcn5/PCAF function in cells exclusively as multiprotein complexes and their assembly into these complexes facilitates nucleosomal acetylation and modulates intrinsic acetylation activity and substrate acetylation specificity (Carrozza et al. 2003; Lee and Workman 2007). Gcn5/PCAF-containing complexes such as SAGA/SLIK in yeast and TFTC/STAGA in human (Nagy and Tora 2007) acetylate lysines in histone H3, H4, and H2B within chromatin. MYST HATs are also assembled in multiprotein complexes in cells to modulate histone acetylation in chromatin (Sapountzi and Cote 2011). For example, Esa1 is part of NuA4 and piccolo/NuA4 complexes for chromatin acetylation. It is not clear yet what the molecular basis for how associated proteins within the HAT complexes modulate the HAT activity and specificity to their respective catalytic subunits.

Several HATs require the binding of cofactors for catalytic activity. For example, the Sas2 member of the MYST HATs requires the binding of Sas4 and Sas5 for acetyltransferase activity (Sutton et al. 2003; Shia et al. 2005). Hat2 and Hif1 elevate yHat1 acetyltransferase activity by about 10-fold when assembled into the NuB4 complex (Parthun et al. 1996). On its own, Rtt109 shows very low activity. The histone chaperones, Vps75 or Asf1, modulate Rtt109 acetylase activity and substrate specificity by several hundred-fold (Kolonko et al. 2010); the Rtt109/Vps75 complex selectively acetylates H3K9 and H3K27, whereas the Rtt109/Asf1 complex preferentially acetylates H3K56 near the H3 core region (Driscoll et al. 2007; Han et al. 2007a; Han et al. 2007b; Tsubota et al. 2007; Tang et al. 2011). Associated subunits of Hat1 (Ruiz-Garcia et al. 1998) and Rtt109 (Tang et al. 2008; Kolonko et al. 2010; Tang et al. 2011) may contribute to catalysis, at least in part, by facilitating productive binding to the respective cognate histone substrates, based on kinetic data, but more work is required.

Another mode of regulating HAT activity that has recently come to light is the autoacetylation of HAT proteins. In particular, three HAT subfamilies have now been shown to be regulated by autoacetylation: Rtt109, p300/CBP, and MYST.

p300 contains a highly basic loop of about 40 residues embedded in the middle of its HAT domain that was shown to undergo multiple autoacetylation through a intermolecular mechanism (Karanam et al. 2007) to regulate the catalytic activity of the protein, with the hyperacetylated forms correlated with the active and hypoacetylated forms with the inactive forms of the enzyme (Thompson et al. 2004). In addition, recombinant p300 protein with a cleaved "autoacetylation loop" has been shown to be constitutively active (Thompson et al. 2004). The molecular basis for how autoacetylation regulates p300 activity is not clear because the reported p300/Lys-CoA crystal structure does not contain an intact autoacetylation loop. However, Cole and colleagues proposed a model of regulation whereby the highly basic autoacetylation loop sits in the electronegative substrate-binding site to directly compete with substrate binding. The loop would then be released from the substrate-binding site on autoacetylation for cognate substrate binding (Fig. 4A) (Liu et al. 2008).

R1109 was also known to be autoacetylated but the molecular basis for this was not known until structures of Rtt109 were reported revealing a buried acetyllysine residue (Lys290; Fig. 4B) (Lin and Yuan 2008; Stavropoulos et al. 2008; Tang et al. 2008). The acetylation of Lys290 is required for full acetylation activity (Albaugh et al. 2011). This acetyllysine at 290 forms hydrogen bonds with Asp288, a mutationally sensitive Rtt109 residue. Mass spectrometry analysis showed that this acetylation occurs in vitro as well as in yeast cells (Tang et al. 2008). Reports addressing the functional importance of Lys290 modification have yielded contradictory results with regard to genotoxic agent sensitivity in vivo (Lin and Yuan 2008; Stavropoulos et al. 2008; Tang et al. 2008). A recent study by Denu and coworkers showed that Lys290 autoacetylation increases the overall K_{cat} and decreases the K_M for AcCoA binding (Albaugh et al. 2011), although the molecular basis for this is unknown.

Several recent reports have shown that the hMOF MYST protein is autoacetylated at an active site lysine (Lys274) and that this autoacetylation is required for cognate acetyltransferase activity both in vitro and in vivo (Lu et al. 2011; Sun et al. 2011; Yuan et al. 2012). The lysine that was found to be autoacetylated in hMOF is strictly conserved in MYST proteins, and Yuan et al. also showed that acetylation of this lysine residue occurs in yeast Esa1 and Sas2 and is required for acetylation of these proteins in vitro and for function of these enzymes in cells (Yuan et al. 2012). The structure of the yEsa1/H4K16CoA complex with this active site lysine in the acetylated form (Lys262Ac) reveals that it sits in the active site pocket with the acetyl CO group making hydrogen bonds to Tyr289 and Ser291, and the methyl group making van der Waals interactions with phenylalanines 271 and 273; the aliphatic region of Lys262Ac makes van der Waals interactions with the lysine of the H4K16CoA bisubstrate inhibitors (Fig. 4C) (Yuan et al. 2012). Each of the residues that contact the acetylated lysine is strictly or highly conserved among the entire MYST protein subfamily further arguing for the importance of this acetylated lysine in MYST function. The structure of the hMOF HAT domain has also been determined in unliganded form (Yuan et al. 2012). This structure reveals that Lys274 exists in two states. In one state it is acetylated and in the same conformation as Lys262Ac of yEsa1, making analogous intra-atomic interactions. In a second state, Lys274 is unacetylated and flipped by about 90° out of the

Cite as *Cold Spring Harb Perspect Biol* doi: 10.1101/cshperspect.a018762

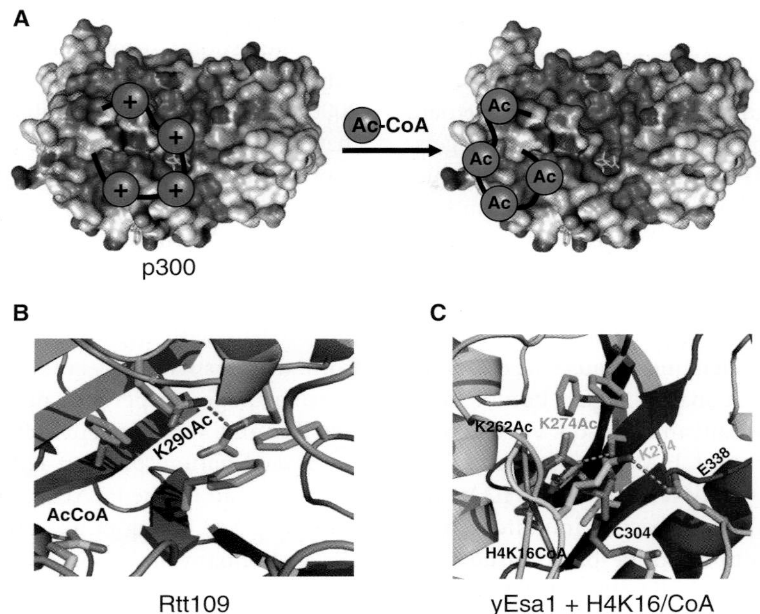

Figure 4. Autoacetylation regulation of HAT proteins. Close-up views of the autoacetylation site of HAT proteins. (*A*) Model for p300 activation by autoacetylation. The black loop and green acetylated lysine balls are modeled on the p300/Lys-CoA crystal structure. (*B*) Structure of the K290 autoacetylation site of Rtt109, highlighting the environment around acetylated K290. The acetylated lysine and other side chains that interact with the acetylated lysine are indicated in stick figure in CPK coloring with carbon in green and the hydrogen bond is shown as a dotted orange line. The AcCoA molecule is shown as a stick figure in CPK coloring with carbon atoms in yellow. (*C*) Structure of the yEsa1/H4K16CoA complex is shown, highlighting the environment around acetylated K262 (green). The corresponding K274 of hMOF is superimposed in the unacetylated (yellow) and acetylated (orange) conformations showing that the unacetylated conformation would clash with binding of the cognate substrate lysine (as represented by the lysine of the H4K16CoA bisubstrate inhibitor shown in purple).

active site, such that the lysine forms a long hydrogen bond with Glu350, the general base for catalysis. This position would block cognate lysine binding (Fig. 4C). This is consistent with biochemical studies that show that a K to R mutant of this lysine in hMOF is defective in cognate substrate binding (Yuan et al. 2012). Taken together, these structural, biochemical, and functional studies show that autoacetylation of MYST proteins in the active site is required for cognate substrate acetylation.

2.6 Inhibitors of Acetyltransferases and Implications for Drug Development

HATs mediate many different biological processes including cell-cycle progression, dosage compensation, repair of DNA damage, and hormone signaling. Aberrant HAT function is correlated with several human diseases including solid tumors, leukemias, inflammatory lung disease, viral infection, diabetes, fungal infection, and drug addiction (Heery and Fischer 2007; Renthal and Nestler 2009). The p300/CBP HAT has properties of both an oncoprotein and tumor-suppressor protein. As an oncoprotein, the CBP HAT forms translocation products with MLL (mixed lineage leukemia) and MOZ (monocytic leukemia zinc-finger protein), another HAT, in a subset of acute myeloid leukemias. When mutated, the p300 HAT is found in a subset of colorectal and gastric cancers making it a bonafide tumor suppressor. The activities of HATs and HDACs are also changed in asthma and chronic obstructive pulmonary disease because bronchial biopsies and alveolar macrophages from asthmatic patients show increased HAT and reduced HDAC activity (Barnes et al. 2005). p300-mediated acetylation of the HIV-1 viral protein, integrase, increases its activity in integrating the HIV-1 virus into the human genome (Cereseto et al. 2005). The leading diabetes drug, Metformin, was shown to act through p300/CBP inhibition, and heterozygous CBP knockout mice are noticeably lean with increased insulin sensitivity (He et al. 2009). The Rt109 HAT was also reported to be required for the pathogenesis of *Candida albicans*, the most prevalent cause of hospital-acquired fungal infections (Lopes da Rosa et al. 2010). Studies in drug addiction and the related disease of depression using animal models have also uncovered interesting correlations between stages of drug addiction and histone acetylation status (Renthal and Nestler 2009).

Because of the correlations between several human diseases and histone acetylation balance, proteins that mediate histone acetylation have become attractive drug targets. Although subnanomolar HDAC inhibitors are available, they show poor selectivity among the class I, II, and IV HDACs (Marsoni et al. 2008), and although potent sirtuin activators and inhibitors have been reported they also show modest selectivity (Sanders et al. 2009). HDAC-specific inhibitors are thus actively being pursued, but obtaining them may be particularly challenging because the class I, II, IV, and class III HDACs each share a highly homologous active site and catalytic mechanism (Marsoni et al. 2008).

The development of potent HAT inhibitors has not progressed nearly as far as the development of HDAC inhibitors. The most potent and specific inhibitors have come from the development of peptide-based bisubstrate inhibitors in which coenzyme A is directly linked to the Nζ nitrogen of the target lysine within histone peptides (Lau et al. 2000). Using this technology, Cole and coworkers developed submicromolar inhibitors that show selectivity between the Gcn5/PCAF and p300/CBP HATs (Lau et al. 2000). These bisubstrate inhibitors are a proof of principle that selective HAT inhibitors can be prepared, although peptide-based inhibitors do not generally have favorable pharmacokinetic properties (Heery and Fischer 2007). Nonetheless, structures of such bisubstrate inhibitors have been determined bound to Gcn5 (Poux et al. 2002) and p300 (Figs. 1D and 3B) (Liu et al. 2008). This has provided us with the knowledge of a molecular scaffold from which small molecule compounds with improved pharmacologic properties can be screened against (Hodawadekar and Marmorstein 2007; Wang et al. 2008).

Several natural product HAT inhibitors have been reported including the PCAF and p300 inhibitors, anacardic acid (Fig. 5A) (Sung et al. 2008), and garcinol (Fig. 5B) (Balasubramanyam et al. 2004a). Natural products that have been shown to inhibit p300 include curcumin (Fig. 5C) (Balasubramanyam et al. 2004b), Epigallocatechin-3-

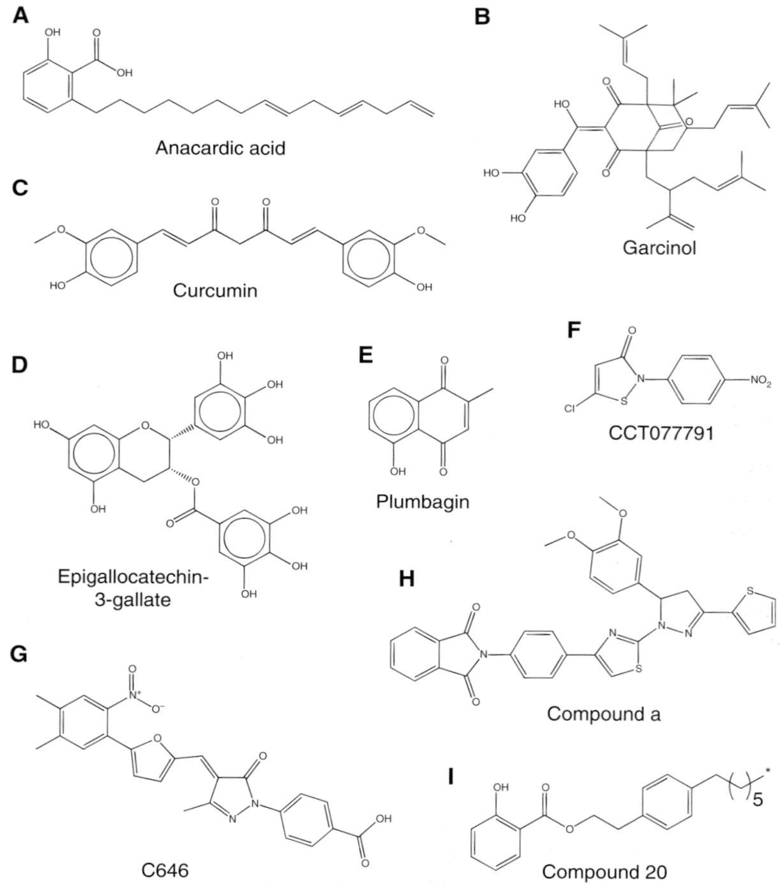

Figure 5. HAT inhibitors. Reported inhibitor specificities are as follows: (*A,B,F*) specific to p300 and PCAF; (*C–E*) specific to p300; (*G*) specific to p300; (*H,I*) specific to MYST.

Cite as *Cold Spring Harb Perspect Biol* doi: 10.1101/cshperspect.a018762

gallate (Fig. 5D) (Choi et al. 2009), and plumbagin (Fig. 5E) (Ravindra et al. 2009). Each of these inhibitors shows IC_{50} values in the low micromolar range. Although several derivatives of many of these inhibitors have been prepared and evaluated in cells (Heery and Fischer 2007; Furdas et al. 2012), they have not been rigorously biochemically evaluated in vitro and their structures bound to HAT proteins have not been determined, so their mode of action is still unclear.

Some high-throughput screens have been reported, identifying HAT inhibitors with similarly modest potencies to the compounds discussed in the previous paragraph. The high-throughput screening of a 70,000 compound library yielded a family of isothiazolones (CCT077791 and CCT077792) as PCAF inhibitors that were subsequently derivatized as low micromolar PCAF and p300 inhibitors (Fig. 5F) (Gorsuch et al. 2009). Cole and coworkers also used virtual ligand screening of about 50,000 compounds to identify a pyrazolone-containing small molecule p300 HAT inhibitor with $K_i \sim 400$ nM and selectivity against the Gcn5/PCAF, MYST, and Rtt109 HATs (Fig. 5G) (Bowers et al. 2010). More recently, there have been reports of small molecule inhibitors to MYST proteins. Zheng and coworkers performed an in silico inhibitor screen against the AcCoA-binding site of yEsa1 and then assayed the hits against the yEsa1 and hTIP60 MYST proteins, hp300 and yGcn5 (Wu et al. 2011). This resulted in the identification of four inhibitors that have IC_{50} values for hTIP60 ranging from 149 μM to 400 μM, with the most potent compound (compound **a**, Fig. 5H) shown to be competitive with AcCoA binding and with some selectivity over Gcn5, but not p300. Zheng and coworkers (Ghizzoni et al. 2012) prepared analogs of the known natural product HAT inhibitor, anacardic acid, and evaluated them as HAT inhibitors. They identified a series of 6-alkylsalicylates that showed selectivity for MYST proteins over Gcn5 and p300, with the most potent compound (**20**) competitively binding to AcCoA, with IC_{50} values of 74 μM and 47 μM for hTIP60 and hMOF, respectively (Fig. 5I). In summary, the HAT inhibitors that have been developed to date only have moderate potency and specificity, therefore motivating the need to develop more potent and selective HAT inhibitors.

2.7 Conclusions and Open Questions

Over the last decade, several HAT structures have been determined and their catalytic mechanisms have been characterized. These studies have revealed that HATs fall into distinct subfamilies that show remarkable diversity in amino acid sequence, but retain a structurally conserved core region that mediates AcCoA cofactor and substrate lysine binding. The structurally distinct flanking regions mediate different HAT-associated functions such as histone substrate binding and nucleosome targeting. Remarkably, the different HAT subfamilies use different chemical strategies to acetylate their substrates.

There is much that we still do not understand about HATs, setting the stage for future studies. Open questions include the following: (1) What are the exact molecular details of how autoacetylation regulates HAT function, and is autoacetylation indeed regulatory? It is clear that many HATs, including the p300/CBP, MYST, and Rtt109 subfamilies, are autoacetylated. We currently only have molecular insights into the mode of autoacetylation in MYST proteins. (2) How do associated HAT subunits in multiprotein complexes stimulate, modulate, or coordinate with HAT activity? The structure determination of relevant HAT complexes may address this. (3) What is the molecular basis of HAT substrate specificity? Different HATs appear to harbor different substrate preferences. For example, although the Gcn5/PCAF and Rtt109 HATs have a relatively restricted substrate preference, p300/CBP is considerably more promiscuous. The highly acidic substrate-binding site of p300/CBP has provided some insights into the promiscuity of this HAT subfamily, yet the molecular basis of substrate selectively of other HATs is poorly understood. (4) HAT inhibitors that have been identified to date have relatively modest potency and selectivity profiles. There is clearly significant interest in developing potent and selective small molecule HAT inhibitors. It is likely the structure determination of HAT/inhibitor complexes will be required to facilitate the development of more potent and selective inhibitors with possible therapeutic applications. (5) How much nonhistone acetyltransferases are similar or different to HATs is not currently known, yet is an area of significant interest. Recent proteomic studies on prokaryotic and eukaryotic cells have revealed that protein acetylation reaches beyond histones and transcription-associated biological processes (Choudhary et al. 2009; Zhang et al. 2009). In both eukaryotes and prokaryotes, thousands of acetylated sites have been identified, and protein acetylation occurs in cellular compartments outside of the nucleus, associated with most cellular processes in living organisms including protein translation, protein folding, DNA packaging, and mitochondrial metabolism (Smith and Workman 2009; Spange et al. 2009). Some of the proteins that mediate nonnuclear activities are just beginning to be identified (Ivanov et al. 2002; Akella et al. 2010), but it is likely that there are many other acetyltransferases, with sequence divergence from the currently known histone acetyltransferases, that are yet to be identified. (6) Zhao and coworkers have reported that lysine propionylation and butyrylation occur in cells and these reactions can be catalyzed by the p300 HAT (Chen et al. 2007). Whether known HATs or other acetyltransferases can indeed use propionyl-CoA and butyryl-CoA to catalyze such reactions, or whether they are catalyzed by other enzymes in cells to regulate distinct biological processes remains an open question.

3 ACETYLLYSINE READERS

3.1 Structure and Acetyllysine Recognition by Bromodomains

The bromodomain (BrD) is regarded as the first histone-binding module, whose function is to recognize or "read" acetylated lysine (Dhalluin et al. 1999; Sanchez and Zhou 2009). The bromodomain adopts a distinct structural fold involving a left-handed four-helix bundle (α_Z, α_A, α_B, and α_C), termed the "BrD fold." The interhelical α_Z-α_A (ZA) and α_B-α_C (BC) loops constitute a hydrophobic pocket that recognizes the acetylated lysine (Fig. 6A). Two conserved tyrosine residues in the interhelical loops (one in the ZA loop, the other at the carboxyl terminus of α_B) that contribute to the hydrophobic pocket are found in the majority of bromodomains (Sanchez and Zhou 2009). They are, however, not necessarily the determinants for acetyllysine recognition (Charlop-Powers et al. 2010). A highly conserved asparagine residue at the beginning of the BC loop (immediately following the second conserved tyrosine) forms a hydrogen bond with the acetyllysine carbonyl oxygen via its side-chain amide nitrogen (Fig. 6B). Together, these characteristics are critical for acetyllysine recognition. Bromodomain binding to acetyllysine in isolation is specific, but occurs with moderate affinity, with dissociation constants (K_d) typically in the range of tens-to-hundreds micromolar (Vandemark et al. 2007; Zhang et al. 2010).

The overall three-dimensional structure of the bromodomain is well-conserved (Dhalluin et al. 1999; Sanchez and Zhou 2009, and references therein), and changes little on acetyllysine peptide binding, with the exception of the conformational adjustment of the ZA and BC loops. Specificity by the bromodomains is dictated by the sequences within these loops interacting with both the acetylated lysine and up to three or more residues flanking either side of the acetylated lysine (Fig. 6A) (Zeng et al. 2008a; Zhang et al. 2010). The acetylated lysine ligand inserts into the pocket in a similar way as shown in different BrD structures, whether the acetylated lysine-containing ligand is a peptide derived from a histone tail, HIV-1 Tat, or p53 (Mujtaba et al. 2002; Mujtaba et al. 2004).

Bromodomains, like other histone-recognition modules, occur in multiples. The structure of the tandem bromodomains of TAF1 ($TAF_{II}250$) shows two bromodomains packed together to form a "U" shape (Fig. 6C) (Jacobson et al. 2000). The individual domains fold independently

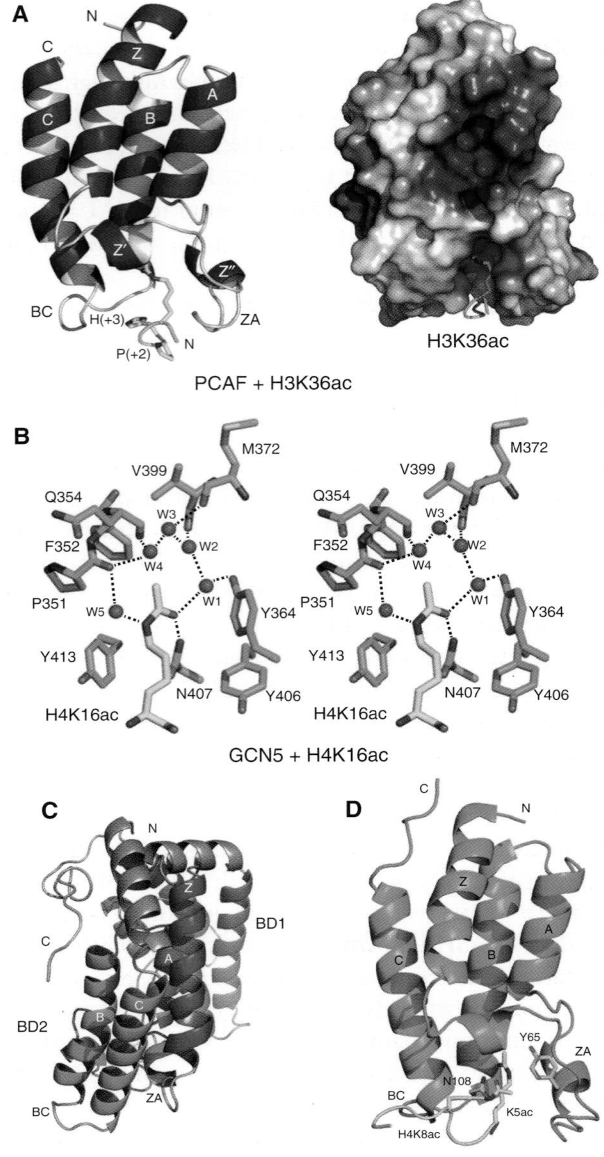

Figure 6. Bromodomains as acetyllysine-binding domains. In all structures, the histone peptide is in yellow and the main and side chains of the protein residues are color-coded by atom type. (*A*) The three-dimensional solution structure of the PCAF bromodomain bound to an H3K36ac peptide (PDB code: 2RNX) is illustrated as a ribbon diagram (*left*) and a surface electrostatic representation (*right*) of the protein with red and blue colors representing negatively or positively charged amino acid residues, respectively. (*B*) The acetyllysine-binding pocket is depicted from the crystal structure of the GCN5 bromodomain (green) in complex with an H4K16ac peptide (PDB code: 1E6I). This stick diagram shows key residues and bound water molecules (magenta spheres) contributing to acetyllysine recognition. Hydrogen bonding interactions are indicated by dotted lines. (*C*) The crystal structure of the tandem bromodomains of uncomplexed human TAF1 (PDB code: 1EQF). (*D*) The crystal structure of the first bromodomain of Brdt bound to an H4K5acK8ac peptide (PDB code: 2WP2).

and their acetyllysine-binding pockets are ~25 Å apart, equivalent to a span of 7–8 residues on a peptide. The TAF1 dual bromodomains bind with significantly higher affinity to peptides that are di- or tetra-acetylated at K5/K12, K8/K16, K5/K8/K12/K16 over a monoacetylated H4 peptide. This is consistent with the notion that each bromodomain binds one acetylated lysine on the same peptide. With the complex structure still elusive, however, this model remains to be confirmed.

A structural study of the tandem bromodomains in yeast Rsc4 showed that only the second bromodomain interacts with an acetylated H3K14 peptide, and this interaction is disrupted by phosphorylation at H3S10 (Vandemark et al. 2007). The two bromodomains fold like one autonomous unit with extensive contacts between the two bromodomains, and are more compact than the TAF1 structure, with the acetyllysine-binding sites 20 Å apart. Further, when a fused H3 peptide-Rsc4 dual bromodomain protein was acetylated by Gcn5, it was found that the amino terminus of Rsc4 was acetylated at a Gcn5-target consensus sequence. Acetylation of the Rsc4 amino-terminal sequence resulted in its binding to the first bromodomain, which precluded binding of the fused histone peptide to the second bromodomain. This suggests that Gcn5 provides an autoregulatory mechanism to control Rsc4 activity by acetylating both activating (i.e., H3 lysine acetylation) and inhibiting modifications (e.g., amino-terminal region of Rsc4).

The Polybromo (PB1) protein contains six bromodomains in tandem at its amino terminus and is involved in chromatin remodeling. It was hypothesized that the presence of multiple bromodomains might enable it to recognize specific nucleosomal acetylation patterns that would target PB1 and its parent PBAF complex to chromatin (Thompson 2009). Structures available for all but the fourth bromodomain confirm that secondary and tertiary structure is generally well-conserved among the bromodomains. All but the sixth bromodomain contain an additional two small helices within the ZA loop. However, sequence analysis indicates that these domains may be classified separately from one another, suggesting different ligands and affinities for each (Sanchez and Zhou 2009). Indeed, each bromodomain appears to have preferences for different acetylated lysines on H2A, H2B, H3, or H4 (Thompson 2009; Charlop-Powers et al. 2010) although the fifth and sixth bromodomains may serve as nonspecific-binding modules that stabilize PB1 binding to a specific acetylated histone sequence via the other four bromodomains. Notably, in the only PB1 bromodomain complex structure available, the H3K14ac peptide does not interact with the conserved Tyr residues that contribute to the hydrophobic pocket, but instead interacts with Leu and Val residues from the first additional helix within loop ZA and the amino terminus of α_C, respectively (Charlop-Powers et al. 2010). Because these residues are not conserved among other bromodomains within PB1, it is likely that PB1 uses multiple modes of recognition of their biological ligands.

Some bromodomains have been reported to interact with histones containing more than one acetylated lysine. For example, Brdt, a testis-specific bromodomain and extraterminal domain (BET) protein, recognizes and compacts hyperacetylated chromatin. Brdt contains two tandem bromodomains, the overall structures of which are similar, and their ligand-binding pockets are comparably large. Interestingly, however, the first bromodomain recognizes two acetylated lysines within its pocket (H4K5ac/K8ac), whereas the second recognizes a single acetylated lysine (H3K18ac; Fig. 6D) (Moriniere et al. 2009). Notably, H4K5ac recognition represents the canonical mode of acetyllysine binding by the bromodomain (i.e., interacting with the conserved Asn108 and Tyr65 in Brdt), whereas the acetylated side chain of H4K8ac is largely bound in a hydrophobic cavity on the protein surface consisting of the side chains of residues Trp49, Pro50, Leu60, and Ile114. Despite the general structural similarity of the two bromodomains in Brdt, the sequences in the ZA and BC loops differ and determine that only the first bromodomain is able to interact with the H4K5ac/K8ac peptide. Based on sequence conservation, the ligand-binding specificity of the two bromodomains in BET proteins is likely conserved for other BET proteins. Indeed, this mode of one domain interacting with two modifications within a single-binding pocket was also reported recently for BRD4 (Filippakopoulos et al. 2012), and also for BRD3, whose first bromodomain interacts with double acetylated lysines of a nonhistone protein—the hematopoietic transcription factor GATA1—that is important for erythroid target gene activation (Gamsjaeger et al. 2011; Lamonica et al. 2011).

3.2 The Human Bromodomain Proteins

The human genome encodes 42 bromodomain-containing proteins that contain a total of 56 unique individual bromodomains (Schultz et al. 2000; Sanchez and Zhou 2009). The diversity of the human bromodomain family can be illustrated by clustering the 56 bromodomain sequences into eight groups, each of which share similar sequence length and at least 30% sequence identity (Fig. 7) (Sanchez et al. 2000). Bromodomain proteins represent a wide variety of functionality in chromatin biology and gene transcription. Among the most notable bromodomain proteins are the HATs discussed in Section 2—PCAF, GCN5, and p300/CBP—which function as transcription coactivators.

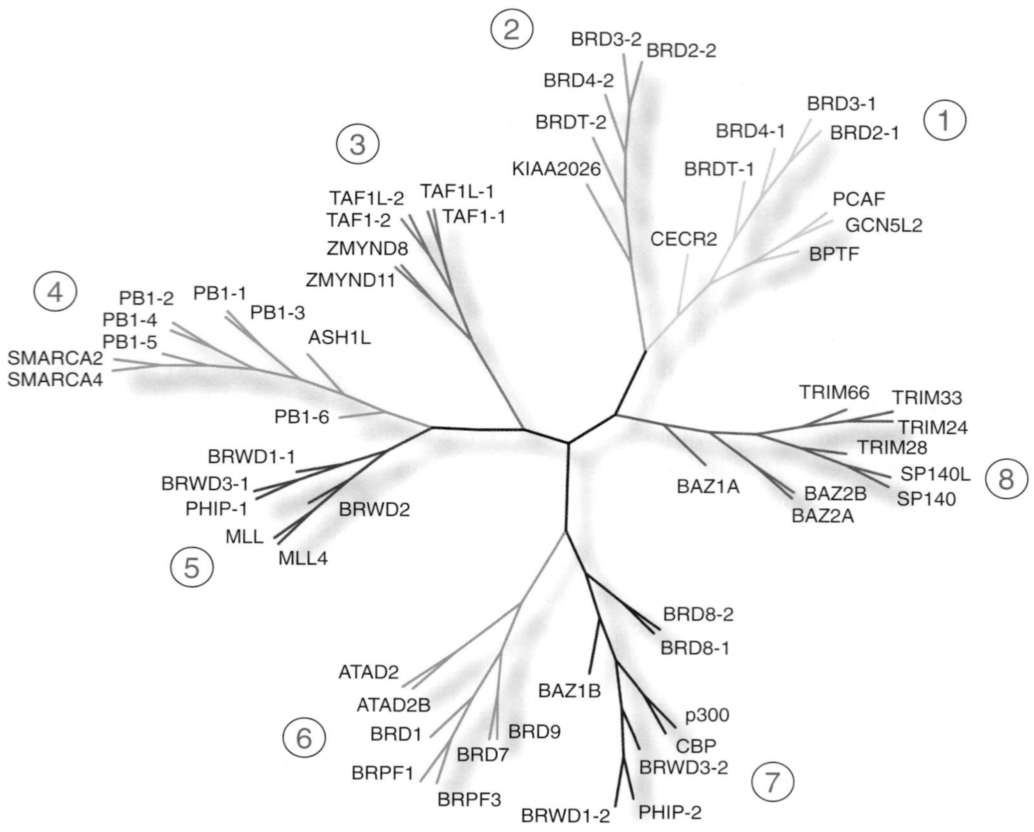

Figure 7. The phylogenetic tree of human bromodomains. Sequence similarity–based dendrogram of the human bromodomains was generated using the neighbor-joining method with MEGA (Kumar et al. 2004). Sequences of the human bromodomains were obtained from the SMART database (Letunic et al. 2004) and aligned with SMART bromodomains' hidden Markov models using Hmmalign (Sonnhammer et al. 1997). (Modified from Zhang et al. 2010.)

It has been suggested that the bromodomains in these nuclear HATs contribute to substrate recruitment and specificity involving histones and nonhistone proteins, thereby providing a functional link between lysine acetylation and acetylation-mediated protein–protein interactions in chromatin-mediated gene transcription (Sanchez and Zhou 2009). Bromodomains are also found in some histone lysine methyltransferases such as ASH1L and MLL, however, a detailed understanding of their function remains elusive.

Bromodomain proteins are involved in chromatin remodeling, the topic of Ch. 21 (Becker and Workman 2013). Bromodomain-containing remodelers include SMARC2 (also known as BRM, SNF2/SWI2) and SMARC4 (BRG1), and some with ATP-dependent helicase activity include ATAD2 (ATPase family AAA domain-containing protein 2; ANCCA) and ATAD2B. Further, double bromodomains are seen in many proteins including TAF1/TAF1L, the TFIID 250-kDa subunit of the transcription initiation complex, as well as the BET proteins of BRD2, BRD3, BRD4, and the testis-specific protein BRDT. The BET proteins play an important role in the assembly of the productive transcriptional activation complex through the recruitment of the p-TEFb complex (CDK9 and cyclin T1) to RNA polymerase II, which is required for transcriptional elongation (Chiang 2009).

3.3 The Association of Bromodomains with Other Chromatin Modules

Bromodomains are often present with a variety of other conserved protein modules of different functions within the same proteins (Basu et al. 2008; Basu et al. 2009). For instance, PCAF, GCN5, p300/CBP are histone lysine acetyltransferases, whereas HRX/ALL-1 is a histone lysine *N*-methyltransferase and SNF2L2 is an ATP-dependent helicase. More than 15 different domain types have been identified to occur within the same proteins as bromodo-

Cite as *Cold Spring Harb Perspect Biol* doi: 10.1101/cshperspect.a018762

mains, including the PHD, PWWP, B-box type zinc finger, ring finger, SAND, FY Rich, SET, TAZ zinc finger, helicase, ATPase, BAH (bromo-adjacent homolog) domain, WD40 repeat, and MBD (methyl-CpG-binding domain) (Schultz et al. 2000).

The modular domain that is most frequently associated with the bromodomain is the PHD (plant homeodomain) finger, which is a C4HC3 zinc-finger-like motif present in nuclear proteins (Sanchez and Zhou 2011). A PHD has been identified in 19 of the 42 human bromodomain-containing proteins. In 12 of these proteins the PHD and bromodomain are separated by a short amino acid sequence (less than 30 residues). The relative arrangement of the tandem domains vary based on the length and composition of the linker sequence, as well as the residues of the domains that may form an interacting surface. The PHD–bromodomain fragment of KAP1 (also known as TIF1β or TRIM28), a transcriptional corepressor for KRAB zinc-finger proteins (Zeng et al. 2008a), lacks several conserved residues within its bromodomain, believed to preclude its ability to directly bind to acetylated lysine. The tandem domains function cooperatively as a single unit, with helix α_Z forming a hydrophobic core between the two folds (Fig. 8A) (Zeng et al. 2008b). The intimate association between the domains enables the PHD finger, which is an intramolecular E3 sumoylation ligase, to sumoylate the bromodomain, which enables KAP1 to recruit SETDB1 (a histone H3 lysine 9 specific methyltransferase) to chromatin for gene silencing (Ivanov et al. 2007; Zeng et al. 2008a). Conversely, the PHD–bromodomain tandem module of BPTF, a subunit of a nucleosome-remodeling factor complex, represents two folded domains separated by a helical linker that

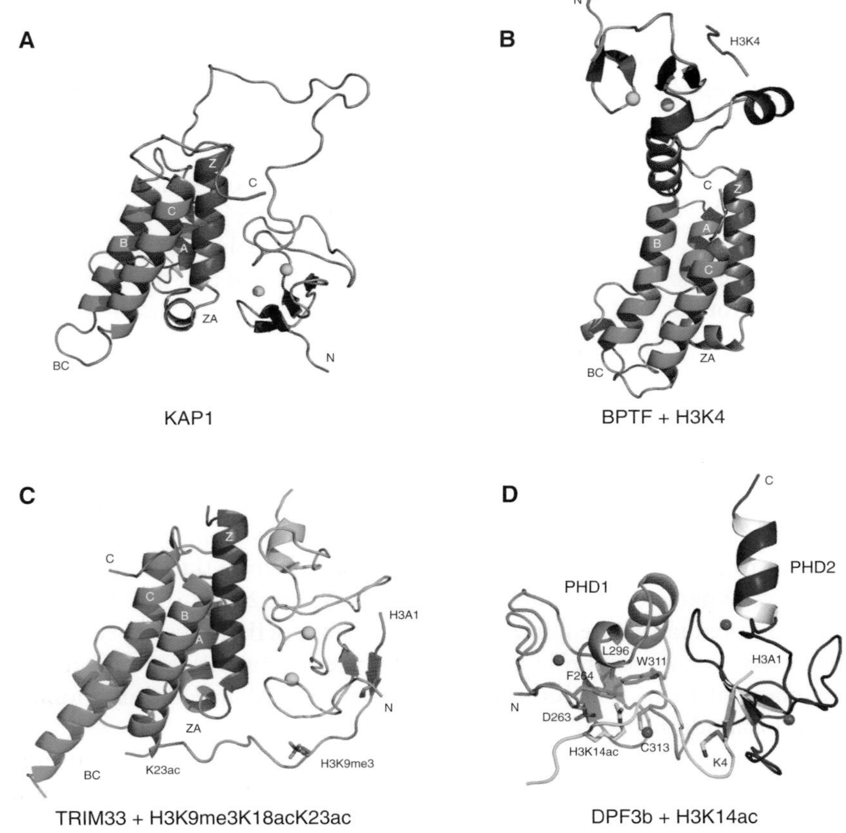

Figure 8. Interdomain interactions in tandem histone-binding modules. (*A*) The solution structure of the PHD (navy)–bromodomain (red and green) module of human KAP1 (PDB code: 2RO1). (*B*) The PHD–bromodomain module of human BPTF in complex with an H3K4 peptide (PDB code: 3QZV). (*C*) The crystal structure of the PHD–bromodomain module of human TRIM33 in complex with an H3K9me3K18acK23ac peptide (PDB code: 3U5P). Note that the second bromodomain in each of the above tandem modules are colored green, and each structure is oriented with respect to the α_Z helix (red) of this bromodomain. (*D*) The solution structure of the tandem PHD finger module of human DPF3b bound to an H3K14ac peptide (PDB code: 2KWJ). The zinc atoms are highlighted as red spheres, and the main and side chains of the protein residues involved in H3K14ac binding are color-coded by atom type with green, red, and blue for carbon, oxygen and nitrogen, respectively.

form no contacts between each other. Transcription proteins such as BPTF (Fig. 8B) (Li et al. 2006), MLL1 (Wang et al. 2010), TRIM24 (Tsai et al. 2010), or TRIM33 (Fig. 8C) (Xi et al. 2011) have tandem PHD domain modules. Several recent studies have shown that these are able to interact with histone H3 in both acetylation-dependent and methylation-sensitive manner, positively or negatively, thus highlighting a functional coordination between these two important histone modifications in gene transcriptional regulation.

The second most common domain associated with the bromodomain is another bromodomain; 11 of the 42 BRD-containing proteins contain two bromodomains. Polybromo is an exception, containing six bromodomains. In the transcription initiation factors TAF1 and the TFIID 210-kDa subunit, as well as in some of the Polybromo bromodomain pairs, the two bromodomains are separated by short amino acid sequences (less than 20 residues). The structure of the TAF1 bromodomains suggests that they form a tandem arrangement that binds selectively to multiple acetylated histone H4 peptides (Jacobson et al. 2000).

3.4 Functions of Human Bromodomain Proteins in Gene Expression

The complexity and variability of the domain composition in human bromodomain proteins and the influence of neighboring domains (such as the PHD finger) on the function of the bromodomain itself make it difficult to predict functions of bromodomain proteins based on sequence similarity alone. Indeed, growing evidence shows that in addition to histones, bromodomain-containing proteins bind acetylated lysine residues in nonhistone proteins that play an important role in control of gene transcription in chromatin. Many of the human bromodomain proteins do not have well-characterized functions, although some have been implicated in disease processes.

The BET protein BRD4 plays an important role in various biological processes, many of them nonchromatin-related, by means of its two bromodomains. BRD4 functions in the inflammatory response through interactions with both histone and nonhistone targets. It acts as a coactivator for the transcriptional activation of NF-κB via the binding of the bromodomains to the acetylated Lys310 on the RelA subunit of NF-κB (Huang et al. 2009). BRD4 also plays a cellular role by stimulating G_1 gene transcription and promoting cell-cycle progression to the S phase (Mochizuki et al. 2008). With regard to its role in chromatin-related processes, a recent study has reported that BRD4 binds, via its bromodomains, to H4K5ac, and is able to accelerate the dynamics of messenger RNA synthesis by decompacting chromatin and, hence, facilitating transcriptional reactivation (Zhao et al. 2011).

BRD4 can control the transcription of viral genes. For example, this protein has been shown to regulate HIV-1 transcription by inducing the phosphorylation of CDK9 (cyclin-dependent kinase 9) at Thr29 in the HIV transcription initiation complex, thereby inhibiting CDK9 kinase activity and leading to the inhibition of HIV transcription (Zhou et al. 2009). BRD4 is also involved in the inhibition of the proteasomal degradation of the papillomavirus E2 protein (Gagnon et al. 2009). Further, BRD4 associates with Kaposi's sarcoma-associated herpesvirus-encoded LANA-1 (latency-associated nuclear antigen) through molecular interactions involving the carboxy-terminal region (Ottinger et al. 2006) and the extraterminal domain (Lin et al. 2008) of BRD4. Additionally, both BRD4 and BRD2 proteins interact with the murine γ-herpesvirus 68 protein orf73, which is required to establish viral latency in vivo (Ottinger et al. 2009).

BRD4 plays a role in cancer as well. Its activation may predict the survival of patients with breast cancer (LeRoy et al. 2008). Crawford et al. proposed that the activation of BRD4 manipulates the response to the tumor microenvironment in vivo, resulting in a reduction of tumor growth and pulmonary metastasis in mice (Crawford et al. 2008). Microarray analysis of multiple human mammary tumor cell lines showed that the activation of BRD4 was predictive of a slower rate of cancer progression and/or improved survival. These results suggest that the dysregulation of BRD4-associated pathways likely plays a key role in breast cancer progression.

The coupling of histone acetylation to transcription in vivo by BRD2 and BRD3 was shown; in human 293 cells, these proteins preferentially associate with specific H4 modifications along the entire lengths of genes, allowing RNA polymerase II to transcribe through the nucleosomes (LeRoy et al. 2008). In addition, BRD2 also shows histone chaperone activity (LeRoy et al. 2008). BRD2 is essential for murine embryonic development (Shang et al. 2009) and is associated with juvenile myoclonic epilepsy in humans (Pal et al. 2003).

In mice, the bromodomain and WD-repeat-containing protein BRWD1 is required for normal spermiogenesis and the oocyte-embryo transition (Philipps et al. 2008). A mutation in BRWD1 leads to phenotypically normal, but infertile mice. The bromodomain of the HAT transcriptional coactivator p300 has been suggested to play a role in the IL-6 signaling pathway by mediating the interaction of the STAT3 amide-terminal domain with p300, thereby stabilizing enhanceosome assembly (Hou et al. 2008).

ATAD2, another bromodomain-containing protein, functions as an estrogen-regulated ATPase coactivator in

Cite as *Cold Spring Harb Perspect Biol* doi: 10.1101/cshperspect.a018762

both estrogen receptor α and androgen receptor signaling. This protein is required for the formation of transcriptional coregulator complexes in gene expression in chromatin (Zou et al. 2007). Chen and colleagues have suggested that ATAD2 also plays an important role in the development of prostate cancer by mediating specific androgen receptor functions during both cancer cell survival and proliferation (Zou et al. 2009).

3.5 Pharmacological Inhibition of Bromodomains in Gene Transcription

Owing to the functional importance of bromodomain-containing proteins in human biology, chemical modulation of bromodomain/acetyllysine binding, involved in chromatin-mediated gene transcription, is an attractive therapeutic strategy for a number of human diseases (Mujtaba et al. 2006; Prinjha et al. 2012). This strategy was first suggested for controlling HIV-1 transcriptional activation and replication in infected host cells (Zeng et al. 2005). Transcriptional activation of the integrated HIV provirus requires a molecular interaction between the HIV-1 Tat *trans*-activator, acetylated at lysine 50 and the bromodomain of the host transcriptional coactivator PCAF (Dorr et al. 2002; Mujtaba et al. 2002), suggesting that blocking this host–virus interaction could result in a reduction in Tat-mediated viral transcription. Indeed, small-molecule bromodomain inhibitors such as *N*1-aryl-propane-1,3-diamine compounds (Fig. 9A) were shown to be able to block PCAF bromodomain binding to K50-acetylated HIV-1 Tat in cells and effectively attenuate Tat-mediated HIV transcriptional activation (Zeng et al. 2005; Pan et al. 2007). These findings suggest that as a new antiviral strategy for intervening HIV-1 replication, targeting a host cell protein essential for viral reproduction rather than a viral protein may minimize the problem of drug resistance caused by mutations of the viral counterpart as observed with protease inhibitors (Ott et al. 2004).

Mounting evidence shows that many transcription factors undergo site-specific lysine acetylation, and the acetylated lysine then functions to recruit transcription or chromatin effector proteins to facilitate their target gene activation in the chromatin context. One such transcription factor is the human tumor suppressor p53, whose function in gene transcription is dependent on acetylation of several carboxy-terminal lysine residues by HAT coactivators such as CBP. It was shown that lysine acetylation of p53 is important for p53 recruitment of CBP via the CBP bromodomain binding to K382-acetylated p53 for p53 target gene expression (Mujtaba et al. 2004). Studies from the Zhou laboratory have shown that small-molecule chemical- or peptide-based inhibitors, selectively targeting the CBP bromodomain, inhibit p53 transcriptional activity in cells by blocking p53-K382ac binding to the CBP bromodomain, promoting p53 instability by changes in its posttranslational modification states (Mujtaba et al. 2006; Sachchidanand et al. 2006; Gerona-Navarro et al. 2011). Excessive p53 activity has been reported in numerous disease conditions. For instance, during myocardial ischemia, elevated p53 activity causes irreversible cellular injury and cardiomyocyte death. A small molecule named ischemin (Fig. 9A,B) has shown positive effects in reversing the detrimental consequences of myocardial ischemia. This molecule was developed to inhibit the acetyllysine-binding

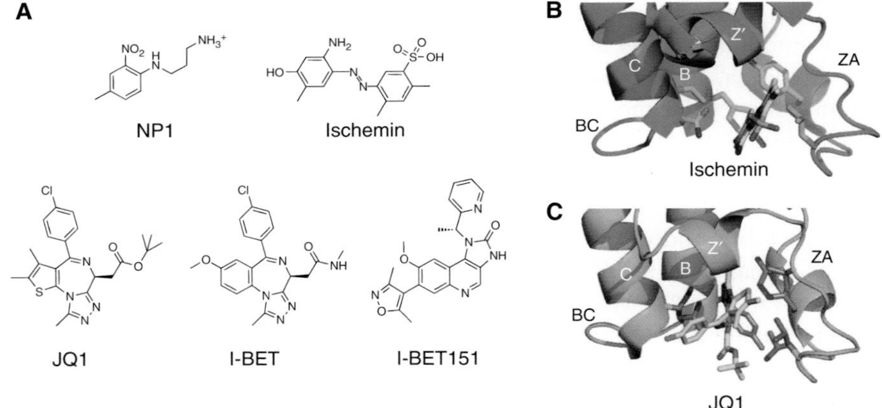

Figure 9. Small molecule inhibitors of bromodomains. (*A*) Chemical structures of representative small-molecule bromodomain inhibitors, including NP1 (for PCAF bromodomain), ischemin, JQ1, I-BET and I-BET151. (*B*) Ischemin, a small-molecule inhibitor developed for the CBP bromodomain, depicted in a complexed 3D structure bound to the protein (PDB code: 2L84). (*C*) JQ1, a BET bromodomain-specific inhibitor, shown when bound to the first bromodomain of BRD4 in the crystal structure (PDB code: 3MXF).

activity of the bromodomain of CBP. The study further showed that cells treated with ischemin have altered post-translational modifications on p53 and histones, inhibit the interaction between p53 and CBP, and reduce p53-mediated transcriptional activity in cells. Overall, this leads to the prevention of apoptosis in ischemic cardiomyocytes (Borah et al. 2011). These studies suggest that small-molecule modulation of acetylation-mediated interactions in gene transcription can serve as a new approach to therapeutic interventions of human disorders such as myocardial ischemia.

More recently, several highly selective and potent small-molecule inhibitors for the BET family bromodomain proteins have been developed (Fig. 9A,C) (Prinjha et al. 2012). For instance, the work by Bradner and colleagues first reported that a BET bromodomain-specific inhibitor JQ1 blocks the binding of BRD4 bromodomainsto lysine-acetylated histone H4 (Filippakopoulos et al. 2010). BRD4 is a known component of the recurrent chromosomal translocation product with the NUT protein in an aggressive human squamous carcinoma. Competitive binding by JQ1 displaces the BRD4-NUT fusion oncoprotein from chromatin, prompting squamous differentiation and specific antiproliferative effects in both BRD4-dependent cell lines and patient-derived xenograft models (Filippakopoulos et al. 2010). In another study, Vakoc and coworkers studied acute myeloid leukemia, an aggressive hematopoietic malignancy that is characteristic of altered epigenetic landscapes (Zuber et al. 2011). Suppression of *Brd4* by shRNA or inhibition of BRD4 by JQ1 led to dramatic antileukemic effects both in vivo and in vitro, and also led to myeloid differentiation and leukemia stem-cell depletion. Inhibition of BET protein recruitment to chromatin with small-molecule bromodomain inhibitors including I-BET and I-BET151 (Fig. 9A) has also been shown as an effective treatment for MLL-fusion leukemia (Dawson et al. 2011). In their study of multiple myeloma, a Myc oncoprotein-dependent hematologic malignancy, Delmore et al. have shown that the BET-specific inhibitor JQ1 generates a potent antiproliferative effect associated with cell cycle arrest and cellular senescence (Delmore et al. 2011). These studies suggest that inhibitors of BET bromodomains represent a new therapeutic strategy for malignancies that are characterized by pathologic activation of *c-Myc* (Delmore et al. 2011; Mertz et al. 2011). Finally, BET bromomdomain inhibition can reduce transcriptional activation of proinflammatory cytokines. For example, I-BET has been shown to be an effective means of down-regulating inflammatory gene expression in activated macrophages, conferring protection against lipopolysaccharide-induced endotoxic shock, and bacteria-induced sepsis (Nicodeme et al. 2010). Moreover, MS417, an improved thienodiazepine-based BET-specific bromodomain inhibitor (about five- to 10-fold more potent than JQ1) can attenuate HIV-infection triggered NF-κB transcriptional activity in activation of proinflammatory genes in mouse kidney cells, and also ameliorate HIV-induced kidney injury in HIV transgenic mice treated with MS417 (Zhang et al. 2012).

3.6 Other Acetyllysine Readers

Acetyllysine binding was regarded a unique molecular function exclusive to the bromodomains following its discovery in 1999 (Dhalluin et al. 1999). In 2008, Lange and coworkers reported that the tandem PHD fingers of human DPF3b, a component of the BAF chromatin remodeling complex, interacts with histone H3 in a manner sensitive to lysine acetylation and methylation (Lange et al. 2008). The PHD finger is a highly versatile protein module that has been shown to interact with histone H3 sequences in a modification-sensitive manner (Sanchez and Zhou 2011). A structural analysis of the tandem PHD fingers (PHD12) of human DPF3b provided a detailed molecular basis for the idea that its binding to histone H3 is modulated positively by lysine acetylation but negatively by lysine methylation (Zeng et al. 2010). Specifically, the tandem PHD fingers fold as one functionally cooperative unit, and interact with an unmodified H3 peptide with an affinity of K_d $\sim$2 μM; acetylation at K14 enhances H3 binding to a K_d of 0.5 μM, whereas methylation at H3K4 almost abolishes the interaction. The H3 peptide lies across a surface shared by the two domains with a β-strand from R2-K4 of H3 that contributes to the β-sheet of the second PHD finger (PHD2), a sharp kink at the middle of the peptide due to interactions between the first PHD finger (PHD1) and K9, and K14ac interacting with a hydrophobic pocket formed by PHD1 (Fig. 8D). Additionally, the H3K14ac acyl chain interacts with Arg289 and Phe264 of PHD1, and the acetyl amide group forms a hydrogen bond with the side chain of Asp263 of PHD1. The complex structure of DPF3b PHD12 bound to an amino-terminally acetylated H4 peptide confirms that the PHD1 interacts with the acetyl group with the same residues. The recognition of the acetyllysine by DPF3b PHD1, however, is structurally distinct from the pattern used by the bromodomains, and is completely different from PHD finger's mechanism for methyl-lysine recognition, which uses a surface on the opposite side of the fold.

3.7 Conclusions and Open Questions

The role of the bromodomain as the main protein domain known to recognize acetyllysine residues on proteins involved in gene transcription is much more complex and

Cite as *Cold Spring Harb Perspect Biol* doi: 10.1101/cshperspect.a018762

broad than initially envisioned (Dhalluin et al. 1999; Jacobson et al. 2000; Sanchez and Zhou 2009). Despite the large family size of the bromodomains, the acetyllysine recognition mechanism appears strictly conserved. However, it appears that our current knowledge on sequence-dependent recognition by individual bromodomains is very limited. This is in part because of the binding affinity of lysine-acetylated histones or nonhistone proteins by bromodomains is quote modest (in the range of tens-to-hundreds micromolars). Furthermore, the high degree of sequence variation and conformational flexibility of bromodomain ZA and BC loops, which comprise the acetyllysine-binding pockets, strongly indicates that more structural characterization of bromodomains in complex with their true biological ligands is required. Studies of individual bromodomains have identified varied ligand-binding specificities that are dependent not only on the characteristics of the bromodomain itself, but also on the other protein domains present in the same protein. Studies of bromodomain-containing proteins have highlighted the role of these domains in many important biological processes and their association with disease. The characterization of the multiplicity of molecular interactions mediated by bromodomains is therefore essential for deciphering the role of individual domains and proteins in chromatin-dependent gene transcription. This challenging task may be facilitated by the high structural coverage of the human bromodomain family, which presents a unique opportunity for the rational design of selective small molecules that could serve as tools to modulate and control gene expression in human biology. The more recent observation that the PHD fingers can also bind acetylated lysine residues also leaves open the possibility that still other types of domains may also be used for acetyllysine recognition.

4 PERSPECTIVES

Over the last decade we have learned that the writers and readers of lysine acetylation have many fascinating and unanticipated properties. The HAT enzymes that create the modification come in many different flavors, are regulated by autoacetylation, and interact with other protein subunits. They acetylate many different substrates—histone and nonhistone—to mediate diverse biological process and are attractive drug targets because of their association with diseases such as cancers and metabolic disorders. Acetylation marks are also recognized ("read") by dedicated protein modules such as bromodomains, PHD fingers, and possibly other types of domains, to mediate downstream biological signals. Interestingly, many of the properties listed for lysine acetyltransferases are common to the superfamily of kinase enzymes, including the recognition of phospho marks by protein segments such as SH2, PTB, and FHA domains (Taylor and Kornev 2011). However, because studies of kinases are more extensive and date back further, they are much better understood. Indeed, one could argue that the acetylation pathway for signal transduction and gene expression is in the same position phosphorylation was several decades ago. This raises the question: Does acetylation rival phosphorylation for mediating key signal transduction events in biology? Time will tell.

ACKNOWLEDGMENTS

The work was supported in part by the National Institutes of Health grants GM060293, GM098910, and AG031862 awarded to R.M., and grants CA87658 and HG004508 awarded to M.-M.Z.

REFERENCES

*Reference is also in this book.

Ai X, Parthun MR. 2004. The nuclear Hat1p/Hat2p complex: A molecular link between type B histone acetyltransferases and chromatin assembly. *Mol Cell* **14:** 195–205.

Akella JS, Wloga D, Kim J, Starostina NG, Lyons-Abbott S, Morrissette NS, Dougan ST, Kipreos ET, Gaertig J. 2010. MEC-17 is an α-tubulin acetyltransferase. *Nature* **467:** 218–222.

Akhtar A, Becker PB. 2001. The histone H4 acetyltransferase MOF uses a C_2HC zinc finger for substrate recognition. *EMBO Rep* **2:** 113–118.

Albaugh BN, Kolonko EM, Denu JM. 2010. Kinetic mechanism of the Rtt109–Vps75 histone acetyltransferase-chaperone complex. *Biochemistry* **49:** 6375–6385.

Albaugh BN, Arnold KM, Lee S, Denu JM. 2011. Autoacetylation of the histone acetyltransferase RTT109. *J Biol Chem* **286:** 24694–24701.

Allis CD, Berger SL, Cote J, Dent S, Jenuwien T, Kouzarides T, Pillus L, Reinberg D, Shi Y, Shiekhattar R, et al. 2007. New nomenclature for chromatin-modifying enzymes. *Cell* **131:** 633–636.

* Almouzni G, Cedar H. 2014. Maintenance of epigenetic information. *Cold Spring Harb Perspect Biol* doi: 10.1101/cshperspect.a019372.

Balasubramanyam K, Altaf M, Varier RA, Swaminathan V, Ravindran A, Sadhale PP, Kundu TK. 2004a. Polyisoprenylated benzophenone, garcinol, a natural histone acetyltransferase inhibitor, represses chromatin transcription and alters global gene expression. *J Biol Chem* **279:** 33716–33726.

Balasubramanyam K, Varier RA, Altaf M, Swaminathan V, Siddappa NB, Ranga U, Kundu TK. 2004b. Curcumin, a novel p300/CREB-binding protein-specific inhibitor of acetyltransferase, represses the acetylation of histone/nonhistone proteins and histone acetyltransferase-dependent chromatin transcription. *J Biol Chem* **279:** 51163–51171.

Bannister AJ, Kouzarides T. 2011. Regulation of chromatin by histone modifications. *Cell Res* **21:** 381–395.

Barnes PJ, Adcock IM, Ito K. 2005. Histone acetylation and deacetylation: Importance in inflammatory lung diseases. *Eur Respir J* **25:** 552–563.

Basu MK, Carmel L, Rogozin IB, Koonin EV. 2008. Evolution of protein domain promiscuity in eukaryotes. *Genome Res* **18:** 449–461.

Basu MK, Poliakov E, Rogozin IB. 2009. Domain mobility in proteins: Functional and evolutionary implications. *Brief Bioinform* **10:** 205–216.

* Becker PB, Workman JL. 2013. Nucleosome remodeling and epigenetics. *Cold Spring Harb Perspect Biol* **5:** a017905.

Berndsen CE, Albaugh BN, Tan S, Denu JM. 2007. Catalytic mechanism of a MYST family histone acetyltransferase. *Biochemistry* **46:** 623–629.

Berndsen CE, Tsubota T, Lindner SE, Lee S, Holton JM, Kaufman PD, Keck JL, Denu JM. 2008. Molecular functions of the histone acetyltransferase chaperone complex Rtt109–Vps75. *Nat Struct Mol Biol* **15:** 948–956.

Borah JC, Mujtaba S, Karakikes I, Zeng L, Muller M, Patel J, Moshkina N, Morohashi K, Zhang W, Gerona-Navarro G, et al. 2011. A small molecule binding to the coactivator CREB-binding protein blocks apoptosis in cardiomyocytes. *Chem Biol* **18:** 531–541.

Bowers EM, Yan G, Mukherjee C, Orry A, Wang L, Holbert MA, Crump NT, Hazzalin CA, Liszczak G, Yuan H, et al. 2010. Virtual ligand screening of the p300/CBP histone acetyltransferase: Identification of a selective small molecule inhibitor. *Chem Biol* **17:** 471–482.

Brent MM, Marmorstein R. 2008. Ankyrin for methylated lysines. *Nat Struct Mol Biol* **15:** 221–222.

Brownell JE, Zhou J, Ranalli T, Kobayashi R, Edmondson DG, Roth SY, Allis CD. 1996. Tetrahymena histone acetyltransferase A: A homolog of yeast Gcn5p linking histone acetylation to gene activation. *Cell* **84:** 843–851.

* Busslinger M, Tarakhovsky A. 2014. Epigenetic control of immunity. *Cold Spring Harb Perspect Biol* **6:** a019307.

Carrozza MJ, Utley RT, Workman JL, Cote J. 2003. The diverse functions of histone acetyltransferase complexes. *Trends Genet* **19:** 321–329.

Cereseto A, Manganaro L, Gutierrez MI, Terreni M, Fittipaldi A, Lusic M, Marcello A, Giacca M. 2005. Acetylation of HIV-1 integrase by p300 regulates viral integration. *EMBO J* **24:** 3070–3081.

Charlop-Powers Z, Zeng L, Zhang Q, Zhou MM. 2010. Structural insights into selective histone H3 recognition by the human Polybromo bromodomain 2. *Cell Res* **20:** 529–538.

Chen Y, Sprung R, Tang Y, Ball H, Sangras B, Kim SC, Falck JR, Peng J, Gu W, Zhao Y. 2007. Lysine propionylation and butyrylation are novel post-translational modifications in histones. *Mol Cell Proteomics* **6:** 812–819.

Chiang CM. 2009. Brd4 engagement from chromatin targeting to transcriptional regulation: Selective contact with acetylated histone H3 and H4. *F1000 Biol Rep* **1:** 98.

Choi KC, Jung MG, Lee YH, Yoon JC, Kwon SH, Kang HB, Kim MJ, Cha JH, Kim YJ, Jun WJ, et al. 2009. Epigallocatechin-3-gallate, a histone acetyltransferase inhibitor, inhibits EBV-induced B lymphocyte transformation via suppression of RelA acetylation. *Cancer Res* **69:** 583–592.

Choudhary C, Kumar C, Gnad F, Nielsen ML, Rehman M, Walther TC, Olsen JV, Mann M. 2009. Lysine acetylation targets protein complexes and co-regulates major cellular functions. *Science* **325:** 834–840.

Clements A, Rojas JR, Trievel RC, Wang L, Berger SL, Marmorstein R. 1999. Crystal structure of the histone acetyltransferase domain of the human PCAF transcriptional regulator bound to coenzyme A. *EMBO J* **18:** 3521–3532.

Clements A, Poux AN, Lo WS, Pillus L, Berger SL, Marmorstein R. 2003. Structural basis for histone and phosphohistone binding by the GCN5 histone acetyltransferase. *Mol Cell* **12:** 461–473.

Crawford NP, Alsarraj J, Lukes L, Walker RC, Officewala JS, Yang HH, Lee MP, Ozato K, Hunter KW. 2008. Bromodomain 4 activation predicts breast cancer survival. *Proc Natl Acad Sci* **105:** 6380–6385.

Dawson MA, Prinjha RK, Dittmann A, Giotopoulos G, Bantscheff M, Chan WI, Robson SC, Chung CW, Hopf C, Savitski MM, et al. 2011. Inhibition of BET recruitment to chromatin as an effective treatment for MLL-fusion leukaemia. *Nature* **478:** 529–533.

Delmore JE, Issa GC, Lemieux ME, Rahl PB, Shi J, Jacobs HM, Kastritis E, Gilpatrick T, Paranal RM, Qi J, et al. 2011. BET bromodomain inhibition as a therapeutic strategy to target c-Myc. *Cell* **146:** 904–917.

Dhalluin C, Carlson J, Zeng L, He C, Aggarwal K, Zhou MM. 1999. Structure and ligand of a histone acetyltransferase bromodomain. *Nature* **399:** 491–496.

Doi M, Hirayama J, Sassone-Corsi P. 2006. Circadian regulator CLOCK is a histone acetyltransferase. *Cell* **125:** 497–508.

Dorr A, Kiermer V, Pedal A, Rackwitz H, Henklein P, Schubert U, Zhou M, Verdin E, Ott M. 2002. Transcriptional synergy between Tat and PCAF is dependent on the binding of acetylated Tat to the PCAF bromodomain. *EMBO J* **21:** 2715–2723.

Driscoll R, Hudson A, Jackson SP. 2007. Yeast Rtt109 promotes genome stability by acetylating histone H3 on lysine 56. *Science* **315:** 649–652.

Dutnall RN, Tafrov ST, Sternglanz R, Ramakrishnan V. 1998. Structure of the histone acetyltransferase Hat1: A paradigm for the GCN5-related N-acetyltransferase superfamily. *Cell* **94:** 427–438.

Filippakopoulos P, Qi J, Picaud S, Shen Y, Smith WB, Fedorov O, Morse EM, Keates T, Hickman TT, Felletar I, et al. 2010. Selective inhibition of BET bromodomains. *Nature* **468:** 1067–1073.

Filippakopoulos P, Picaud S, Mangos M, Keates T, Lambert JP, Barsyte-Lovejoy D, Felletar I, Volkmer R, Muller S, Pawson T, et al. 2012. Histone recognition and large-scale structural analysis of the human bromodomain family. *Cell* **149:** 214–231.

Furdas SD, Kannan S, Sippl W, Jung M. 2012. Small molecule inhibitors of histone acetyltransferases as epigenetic tools and drug candidates. *Arch Pharm (Weinheim)* **345:** 7–21.

Gagnon D, Joubert S, Senechal H, Fradet-Turcotte A, Torre S, Archambault J. 2009. Proteasomal degradation of the papillomavirus E2 protein is inhibited by overexpression of bromodomain-containing protein 4. *J Virol* **83:** 4127–4139.

Gamsjaeger R, Webb SR, Lamonica JM, Billin A, Blobel GA, Mackay JP. 2011. Structural basis and specificity of acetylated transcription factor GATA1 recognition by BET family bromodomain protein Brd3. *Mol Cell Biol* **31:** 2632–2640.

Gerona-Navarro G, Yoel R, Mujtaba S, Frasca A, Patel J, Zeng L, Plotnikov AN, Osman R, Zhou MM. 2011. Rational design of cyclic peptide modulators of the transcriptional coactivator CBP: A new class of p53 inhibitors. *J Am Chem Soc* **133:** 2040–2043.

Ghizzoni M, Wu J, Gao T, Haisma HJ, Dekker FJ, Zheng YG. 2012. 6-alkylsalicylates are selective Tip60 inhibitors and target the acetyl-CoA binding site. *Eur J Med Chem* **47:** 337–344.

Glatt S, Alfieri C, Muller CW. 2011. Recognizing and remodeling the nucleosome. *Curr Opin Struct Biol* **21:** 335–341.

Gorsuch S, Bavetsias V, Rowlands MG, Aherne GW, Workman P, Jarman M, McDonald E. 2009. Synthesis of isothiazol-3-one derivatives as inhibitors of histone acetyltransferases (HATs). *Bioorg Med Chem* **17:** 467–474.

Han J, Zhou H, Li Z, Xu RM, Zhang Z. 2007a. Acetylation of lysine 56 of histone H3 catalyzed by RTT109 and regulated by ASF1 is required for replisome integrity. *J Biol Chem* **282:** 28587–28596.

Han J, Zhou H, Li Z, Xu RM, Zhang Z. 2007b. The Rtt109–Vps75 histone acetyltransferase complex acetylates non-nucleosomal histone H3. *J Biol Chem* **282:** 14158–14164.

He L, Sabet A, Djedjos S, Miller R, Sun X, Hussain MA, Radovick S, Wondisford FE. 2009. Metformin and insulin suppress hepatic gluconeogenesis through phosphorylation of CREB binding protein. *Cell* **137:** 635–646.

Heery DM, Fischer PM. 2007. Pharmacological targeting of lysine acetyltransferases in human disease: A progress report. *Drug Discov Today* **12:** 88–99.

Hodawadekar SC, Marmorstein R. 2007. Chemistry of acetyl transfer by histone modifying enzymes: Structure, mechanism and implications for effector design. *Oncogene* **26:** 5528–5540.

Holbert MA, Sikorski T, Carten J, Snowflack D, Hodawadekar S, Marmorstein R. 2007. The human monocytic leukemia zinc finger histone acetyltransferase domain contains DNA-binding activity implicated in chromatin targeting. *J Biol Chem* **282:** 36603–36613.

Hou T, Ray S, Lee C, Brasier AR. 2008. The STAT3 NH_2-terminal domain stabilizes enhanceosome assembly by interacting with the p300 bromodomain. *J Biol Chem* **283:** 30725–30734.

Huang B, Yang XD, Zhou MM, Ozato K, Chen LF. 2009. Brd4 coactivates transcriptional activation of NF-κB via specific binding to acetylated RelA. *Mol Cell Biol* **29:** 1375–1387.

Ivanov AV, Peng H, Yurchenko V, Yap KL, Negorev DG, Schultz DC, Psulkowski E, Fredericks WJ, White DE, Maul GG, et al. 2007. PHD domain-mediated E3 ligase activity directs intramolecular sumoylation of an adjacent bromodomain required for gene silencing. *Mol Cell* **28:** 823–837.

Ivanov D, Schleiffer A, Eisenhaber F, Mechtler K, Haering CH, Nasmyth K. 2002. Eco1 is a novel acetyltransferase that can acetylate proteins involved in cohesion. *Curr Biol* **12:** 323–328.

Jacobson RH, Ladurner AG, King DS, Tjian R. 2000. Structure and function of a human $TAF_{II}250$ double bromodomain module. *Science* **288:** 1422–1425.

Karanam B, Wang L, Wang D, Liu X, Marmorstein R, Cotter R, Cole PA. 2007. Multiple roles for acetylation in the interaction of p300 HAT with ATF-2. *Biochemistry* **46:** 8207–8216.

Kawasaki H, Schiltz L, Chiu R, Itakura K, Taira K, Nakatani Y, Yokoyama KK. 2000. ATF-2 has intrinsic histone acetyltransferase activity which is modulated by phosphorylation. *Nature* **405:** 195–200.

Keppler BR, Archer TK. 2008a. Chromatin-modifying enzymes as therapeutic targets–Part 1. *Expert Opin Ther Targets* **12:** 1301–1312.

Keppler BR, Archer TK. 2008b. Chromatin-modifying enzymes as therapeutic targets—Part 2. *Expert Opin Ther Targets* **12:** 1457–1467.

Kleff S, Andrulis ED, Anderson CW, Sternglanz R. 1995. Identification of a gene encoding a yeast histone H4 acetyltransferase. *J Biol Chem* **270:** 24674–24677.

Kolonko EM, Albaugh BN, Lindner SE, Chen Y, Satyshur KA, Arnold KM, Kaufman PD, Keck JL, Denu JM. 2010. Catalytic activation of histone acetyltransferase Rtt109 by a histone chaperone. *Proc Natl Acad Sci* **107:** 20275–20280.

Kumar S, Tamura K, Nei M. 2004. MEGA3: Integrated software for molecular evolutionary genetics analysis and sequence alignment. *Brief Bioinform* **5:** 150–163.

Kurth HM, Mochizuki K. 2009. Non-coding RNA: A bridge between small RNA and DNA. *RNA Biol* **6:** 138–140.

Lamonica JM, Deng W, Kadauke S, Campbell AE, Gamsjaeger R, Wang H, Cheng Y, Billin AN, Hardison RC, Mackay JP, et al. 2011. Bromodomain protein Brd3 associates with acetylated GATA1 to promote its chromatin occupancy at erythroid target genes. *Proc Natl Acad Sci* **108:** E159–E168.

Lange M, Kaynak B, Forster UB, Tonjes M, Fischer JJ, Grimm C, Schlesinger J, Just S, Dunkel I, Krueger T, et al. 2008. Regulation of muscle development by DPF3, a novel histone acetylation and methylation reader of the BAF chromatin remodeling complex. *Genes Dev* **22:** 2370–2384.

Lau OD, Kundu TK, Soccio RE, Ait-Si-Ali S, Khalil EM, Vassilev A, Wolffe AP, Nakatani Y, Roeder RG, Cole PA. 2000. HATs off: Selective synthetic inhibitors of the histone acetyltransferases p300 and PCAF. *Mol Cell* **5:** 539–595.

Lee KK, Workman JL. 2007. Histone acetyltransferase complexes: One size doesn't fit all. *Nat Rev Mol Cell Biol* **8:** 284–295.

LeRoy G, Rickards B, Flint SJ. 2008. The double bromodomain proteins Brd2 and Brd3 couple histone acetylation to transcription. *Mol Cell* **30:** 51–60.

Letunic I, Copley R, Schmidt S, Ciccarelli F, Doerks T, Schultz J, Ponting C, Bork P. 2004. SMART 4.0: Towards genomic data integration. *Nucleic Acids Res* **32:** D142–D144.

Li H, Ilin S, Wang W, Duncan EM, Wysocka J, Allis CD, Patel DJ. 2006. Molecular basis for site-specific read-out of histone H3K4me3 by the BPTF PHD finger of NURF. *Nature* **442:** 91–95.

Lin C, Yuan YA. 2008. Structural insights into histone H3 lysine 56 acetylation by Rtt109. *Structure* **16:** 1503–1510.

Lin Y, Fletcher CM, Zhou J, Allis CD, Wagner G. 1999. Solution structure of the catalytic domain of Tetrahymena GCN5 histone acetyltransferase in complex with coenzyme A. *Nature* **400:** 86–89.

Lin YJ, Umehara T, Inoue M, Saito K, Kigawa T, Jang MK, Ozato K, Yokoyama S, Padmanabhan B, Guntert P. 2008. Solution structure of the extraterminal domain of the bromodomain-containing protein BRD4. *Protein Sci* **17:** 2174–2179.

Liu X, Wang L, Zhao K, Thompson PR, Hwang Y, Marmorstein R, Cole PA. 2008. The structural basis of protein acetylation by the p300/CBP transcriptional coactivator. *Nature* **451:** 846–850.

Lopes da Rosa J, Boyartchuk VL, Zhu LJ, Kaufman PD. 2010. Histone acetyltransferase Rtt109 is required for *Candida albicans* pathogenesis. *Proc Natl Acad Sci* **107:** 1594–1599.

Lu L, Li L, Lv X, Wu XS, Liu DP, Liang CC. 2011. Modulations of hMOF autoacetylation by SIRT1 regulate hMOF recruitment and activities on the chromatin. *Cell Res* **21:** 1182–1185.

Marmorstein R, Trievel RC. 2009. Histone modifying enzymes: Structures, mechanisms, and specificities. *Biochim Biophys Acta* **1789:** 58–68.

Marsoni S, Damia G, Camboni G. 2008. A work in progress: The clinical development of histone deacetylase inhibitors. *Epigenetics* **3:** 164–171.

Mattick JS, Makunin IV. 2006. Non-coding RNA. *Hum Mol Genet* **15:** R17–R29.

Mertz JA, Conery AR, Bryant BM, Sandy P, Balasubramanian S, Mele DA, Bergeron L, Sims RJ 3rd. 2011. Targeting MYC dependence in cancer by inhibiting BET bromodomains. *Proc Natl Acad Sci* **108:** 16669–16674.

Mizzen CA, Yang X-J, Kokubo T, Brownell JE, Bannister AJ, Owen-Hughes T, Workman J, Wang L, Berger SL, Kouzarides T, et al. 1996. The $TAF_{II}250$ subunit of TFIID has histone acetyltransferase activity. *Cell* **87:** 1261–1270.

Mochizuki K, Nishiyama A, Jang MK, Dey A, Ghosh A, Tamura T, Natsume H, Yao H, Ozato K. 2008. The bromodomain protein Brd4 stimulates G_1 gene transcription and promotes progression to S phase. *J Biol Chem* **283:** 9040–9048.

Moriniere J, Rousseaux S, Steuerwald U, Soler-Lopez M, Curtet S, Vitte AL, Govin J, Gaucher J, Sadoul K, Hart DJ, et al. 2009. Cooperative binding of two acetylation marks on a histone tail by a single bromodomain. *Nature* **461:** 664–668.

Mujtaba S, He Y, Zeng L, Farooq A, Carlson J, Ott M, Verdin E, Zhou M. 2002. Structural basis of lysine-acetylated HIV-1 Tat recognition by PCAF bromodomain. *Mol Cell* **9:** 575–586.

Mujtaba S, He Y, Zeng L, Yan S, Plotnikova O, Sachchidanand, Sanchez R, Zeleznik-Le NJ, Ronai Z, Zhou MM. 2004. Structural mechanism of the bromodomain of the coactivator CBP in p53 transcriptional activation. *Mol Cell* **13:** 251–263.

Mujtaba S, Zeng L, Zhou MM. 2006. Modulating molecular functions of p53 with small molecules. *Cell Cycle* **5:** 2575–2578.

Nagy Z, Tora L. 2007. Distinct GCN5/PCAF-containing complexes function as co-activators and are involved in transcription factor and global histone acetylation. *Oncogene* **26:** 5341–5357.

Neuwald AF, Landsman D. 1997. GCN5-related histone *N*-acetyltransferases belong to a diverse superfamily that include the yeast SPT10 protein. *Trends Biochem Sci* **22:** 154–155.

Nicodeme E, Jeffrey KL, Schaefer U, Beinke S, Dewell S, Chung CW, Chandwani R, Marazzi I, Wilson P, Coste H, et al. 2010. Suppression of inflammation by a synthetic histone mimic. *Nature* **468:** 1119–1123.

Ott M, Dorr A, Hetzer-Egger C, Kaehlcke K, Schnolzer M, Henklein P, Cole P, Zhou M, Verdin E. 2004. Tat acetylation: A regulatory switch between early and late phases in HIV transcription elongation. *Novartis Found Symp* **259:** 182–193 discussion 193–196, 223–225.

Ottinger M, Christalla T, Nathan K, Brinkmann M, Viejo-Borbolla A, Schulz T. 2006. Kaposi's sarcoma-associated herpesvirus LANA-1 interacts with the short variant of BRD4 and releases cells from a BRD4- and BRD2/RING3-induced G_1 cell cycle arrest. *J Virol* **80:** 10772–10786.

Ottinger M, Pliquet D, Christalla T, Frank R, Stewart JP, Schulz TF. 2009. The interaction of the gammaherpesvirus 68 orf73 protein with cellular BET proteins affects the activation of cell cycle promoters. *J Virol* **83:** 4423–4434.

Pal DK, Evgrafov OV, Tabares P, Zhang F, Durner M, Greenberg DA. 2003. BRD2 (RING3) is a probable major susceptibility gene for common juvenile myoclonic epilepsy. *Am J Hum Genet* **73:** 261–270.

Pan C, Mezei M, Mujtaba S, Muller M, Zeng L, Li J, Wang Z, Zhou M. 2007. Structure-guided optimization of small molecules inhibiting human immunodeficiency virus 1 Tat association with the human coactivator p300/CREB binding protein-associated factor. *J Med Chem* **50:** 2285–2288.

Parthun MR, Widom J, Gottschling DE. 1996. The major cytoplasmic histone acetyltransferase in yeast: Links to chromatin replication and histone metabolism. *Cell* **87:** 85–94.

Philipps DL, Wigglesworth K, Hartford SA, Sun F, Pattabiraman S, Schimenti K, Handel M, Eppig JJ, Schimenti JC. 2008. The dual bromodomain and WD repeat-containing mouse protein BRWD1 is required for normal spermiogenesis and the oocyte-embryo transition. *Dev Biol* **317:** 72–82.

Poux AN, Marmorstein R. 2003. Molecular basis for Gcn5/PCAF histone acetyltransferase selectivity for histone and nonhistone substrates. *Biochemistry* **42:** 14366–14374.

Poux AN, Cebrat M, Kim CM, Cole PA, Marmorstein R. 2002. Structure of the GCN5 histone acetyltransferase bound to a bisubstrate inhibitor. *Proc Natl Acad Sci* **99:** 14065–14070.

Poveda A, Pamblanco M, Tafrov S, Tordera V, Sternglanz R, Sendra R. 2004. Hif1 is a component of yeast histone acetyltransferase B, a complex mainly localized in the nucleus. *J Biol Chem* **279:** 16033–16043.

Prinjha RK, Witherington J, Lee K. 2012. Place your BETs: The therapeutic potential of bromodomains. *Trends Pharmacol Sci* **33:** 146–153.

⋆ Qi J. 2014. Bromodomain and extraterminal domain inhibitors (BETi) for cancer therapy: Chemical modulation of chromatin structure. *Cold Spring Harb Perspect Biol* doi: 10.1101/cshperspect.a018663.

Ravindra KC, Selvi BR, Arif M, Reddy BA, Thanuja GR, Agrawal S, Pradhan SK, Nagashayana N, Dasgupta D, Kundu TK. 2009. Inhibition of lysine acetyltransferase KAT3B/p300 activity by a naturally occurring hydroxynaphthoquinone, plumbagin. *J Biol Chem* **284:** 24453–24464.

Renthal W, Nestler EJ. 2009. Histone acetylation in drug addiction. *Semin Cell Dev Biol* **20:** 387–394.

Rojas JR, Trievel RC, Zhou J, Mo Y, Li X, Berger SL, Allis CD, Marmorstein R. 1999. Structure of *Tetrahymena* GCN5 bound to coenzyme A and a histone H3 peptide. *Nature* **401:** 93–98.

Ruiz-Garcia AB, Sendra R, Galiana M, Pamblanco M, Perez-Ortin JE, Tordera V. 1998. HAT1 and HAT2 proteins are components of a yeast nuclear histone acetyltransferase enzyme specific for free histone H4. *J Biol Chem* **273:** 12599–12605.

Sachchidanand, Resnick-Silverman L, Yan S, Mutjaba S, Liu WJ, Zeng L, Manfredi JJ, Zhou MM. 2006. Target structure-based discovery of small molecules that block human p53 and CREB binding protein association. *Chem Biol* **13:** 81–90.

Sanchez R, Zhou MM. 2009. The role of human bromodomains in chromatin biology and gene transcription. *Curr Opin Drug Discov Devel* **12:** 659–665.

Sanchez R, Zhou MM. 2011. The PHD finger: A versatile epigenome reader. *Trends Biochem Sci* **36:** 364–372.

Sanchez R, Pieper U, Melo F, Eswar N, Marti-Renom M, Madhusudhan M, Mirkovic N, Sali A. 2000. Protein structure modeling for structural genomics. *Nat Struct Biol* (suppl.) **7:** 986–990.

Sanders BD, Jackson B, Brent M, Taylor AM, Dang W, Berger SL, Schreiber SL, Howitz K, Marmorstein R. 2009. Identification and characterization of novel sirtuin inhibitor scaffolds. *Bioorg Med Chem* **17:** 7031–7041.

Sapountzi V, Cote J. 2011. MYST-family histone acetyltransferases: Beyond chromatin. *Cell Mol Life Sci* **68:** 1147–1156.

⋆ Schaefer U. 2014. Pharmacological inhibition of bromodomain-containing proteins in inflammation. *Cold Spring Harb Perspect Biol* **6:** a018671.

Schultz J, Copley R, Doerks T, Ponting C, Bork P. 2000. SMART: A web-based tool for the study of genetically mobile domains. *Nucleic Acids Res* **28:** 231–234.

⋆ Seto E, Yoshida M. 2014. Erasers of histone acetylation: The histone deacetylase enzymes. *Cold Spring Harb Perspect Biol* **6:** a018713.

Shang E, Wang X, Wen D, Greenberg DA, Wolgemuth DJ. 2009. Double bromodomain-containing gene *Brd2* is essential for embryonic development in mouse. *Dev Dyn* **238:** 908–917.

Shia WJ, Osada S, Florens L, Swanson SK, Washburn MP, Workman JL. 2005. Characterization of the yeast trimeric-SAS acetyltransferase complex. *J Biol Chem* **280:** 11987–11994.

Smith KT, Workman JL. 2009. Introducing the acetylome. *Nat Biotechnol* **27:** 917–919.

Sonnhammer EL, Eddy SR, Durbin R. 1997. Pfam: A comprehensive database of protein domain families based on seed alignments. *Proteins* **28:** 405–420.

Spange S, Wagner T, Heinzel T, Kramer OH. 2009. Acetylation of non-histone proteins modulates cellular signalling at multiple levels. *Int J Biochem Cell Biol* **41:** 185–198.

Spencer TE, Jenster G, Burcin MM, Allis CD, Zhou JX, Mizzen CA, McKenna NJ, Onate SA, Tsai SY, Tsai MJ, et al. 1997. Steroid receptor coactivator-1 is a histone acetyltransferase. *Nature* **389:** 194–198.

Stavropoulos P, Nagy V, Blobel G, Hoelz A. 2008. Molecular basis for the autoregulation of the protein acetyl transferase Rtt109. *Proc Natl Acad Sci* **105:** 12236–12241.

Su D, Hu Q, Zhou H, Thompson JR, Xu RM, Zhang Z, Mer G. 2011. Structure and histone binding properties of the Vps75–Rtt109 chaperone-lysine acetyltransferase complex. *J Biol Chem* **286:** 15625–15629.

Sun B, Guo S, Tang Q, Li C, Zeng R, Xiong Z, Zhong C, Ding J. 2011. Regulation of the histone acetyltransferase activity of hMOF via autoacetylation of Lys274. *Cell Res* **21:** 1262–1266.

Sung B, Pandey MK, Ahn KS, Yi T, Chaturvedi MM, Liu M, Aggarwal BB. 2008. Anacardic acid (6-nonadecyl salicylic acid), an inhibitor of histone acetyltransferase, suppresses expression of nuclear factor-κB-regulated gene products involved in cell survival, proliferation, invasion, and inflammation through inhibition of the inhibitory subunit of nuclear factor-κBα kinase, leading to potentiation of apoptosis. *Blood* **111:** 4880–4891.

Sutton A, Shia WJ, Band D, Kaufman PD, Osada S, Workman JL, Sternglanz R. 2003. Sas4 and Sas5 are required for the histone acetyltransferase activity of Sas2 in the SAS complex. *J Biol Chem* **278:** 16887–16892.

Tang Y, Holbert MA, Wurtele H, Meeth K, Rocha W, Gharib M, Jiang E, Thibault P, Verreault A, Cole PA, et al. 2008. Fungal Rtt109 histone acetyltransferase is an unexpected structural homolog of metazoan p300/CBP. *Nat Struct Mol Biol* **15:** 738–745.

Tang Y, Holbert MA, Delgoshaie N, Wurtele H, Guillemette B, Meeth K, Yuan H, Drogaris P, Lee EH, Durette C, et al. 2011. Structure of the Rtt109-AcCoA/Vps75 complex and implications for chaperone-mediated histone acetylation. *Structure* **19:** 221–231.

Tanner KG, Trievel RC, Kuo MH, Howard RM, Berger SL, Allis CD, Marmorstein R, Denu JM. 1999. Catalytic mechanism and function of invariant glutamic acid 173 from the histone acetyltransferase GCN5 transcriptional coactivator. *J Biol Chem* **274:** 18157–18160.

Tanner KG, Langer MR, Denu JM. 2000. Kinetic mechanism of human histone acetyltransferase P/CAF. *Biochemistry* **39:** 15652.

Taunton J, Hassig CA, Schreiber SL. 1996. A mammalian histone deacetylase related to the yeast transcription regulator Rpd3p. *Science* **272:** 408–411.

Taylor SS, Kornev AP. 2011. Protein kinases: Evolution of dynamic regulatory proteins. *Trends Biochem Sci* **36:** 65–77.

Thompson M. 2009. Polybromo-1: The chromatin targeting subunit of the PBAF complex. *Biochimie* **91:** 309–319.

Thompson PR, Wang D, Wang L, Fulco M, Pediconi N, Zhang D, An W, Ge Q, Roeder RG, Wong J, et al. 2004. Regulation of the p300 HAT domain via a novel activation loop. *Nat Struct Mol Biol* **11:** 308–315.

Trievel RC, Rojas JR, Sterner DE, Venkataramani RN, Wang L, Zhou J, Allis CD, Berger SL, Marmorstein R. 1999. Crystal structure and mechanism of histone acetylation of the yeast GCN5 transcriptional coactivator. *Proc Natl Acad Sci* **96:** 8931–8936.

Trievel RC, Li FY, Marmorstein R. 2000. Application of a fluorescent histone acetyltransferase assay to probe the substrate specificity of the human p300/CBP-associated factor. *Anal Biochem* **287:** 319–328.

Tsai WW, Wang Z, Yiu TT, Akdemir KC, Xia W, Winter S, Tsai CY, Shi X, Schwarzer D, Plunkett W, et al. 2010. TRIM24 links a non-canonical histone signature to breast cancer. *Nature* **468:** 927–932.

Tsubota T, Berndsen CE, Erkmann JA, Smith CL, Yang L, Freitas MA, Denu JM, Kaufman PD. 2007. Histone H3-K56 acetylation is catalyzed by histone chaperone-dependent complexes. *Mol Cell* **25:** 703–712.

Vandemark AP, Kasten MM, Ferris E, Heroux A, Hill CP, Cairns BR. 2007. Autoregulation of the rsc4 tandem bromodomain by gcn5 acetylation. *Mol Cell* **27:** 817–828.

Wang L, Tang Y, Cole PA, Marmorstein R. 2008. Structure and chemistry of the p300/CBP and Rtt109 histone acetyltransferases: Implications for histone acetyltransferase evolution and function. *Curr Opin Struct Biol* **18:** 741–747.

Wang Z, Song J, Milne TA, Wang GG, Li H, Allis CD, Patel DJ. 2010. Pro isomerization in MLL1 PHD3-bromo cassette connects H3K4me readout to CyP33 and HDAC-mediated repression. *Cell* **141:** 1183–1194.

Williamson WD, Pinto I. 2012. Histones and genome integrity. *Front Biosci* **17:** 984–995.

Wu J, Wang J, Li M, Yang Y, Wang B, Zheng YG. 2011. Small molecule inhibitors of histone acetyltransferase Tip60. *Bioorg Chem* **39:** 53–58.

Wu H, Moshkina N, Min J, Zeng H, Joshua J, Zhou MM, Plotnikov AN. 2012. Structural basis for substrate specificity and catalysis of human histone acetyltransferase 1. *Proc Natl Acad Sci* **109:** 8925–8930.

Xi Q, Wang Z, Zaromytidou AI, Zhang XH, Chow-Tsang LF, Liu JX, Kim H, Barlas A, Manova-Todorova K, Kaartinen V, et al. 2011. A poised chromatin platform for TGF-β access to master regulators. *Cell* **147:** 1511–1524.

Yan Y, Barlev NA, Haley RH, Berger SL, Marmorstein R. 2000. Crystal structure of yeast Esa1 suggests a unified mechanism of catalysis and substrate binding by histone acetyltransferases. *Mol Cell* **6:** 1195–1205.

Yan Y, Harper S, Speicher DW, Marmorstein R. 2002. The catalytic mechanism of the ESA1 histone acetyltransferase involves a self-acetylated intermediate. *Nature Struct Biol* **9:** 862–869.

Yap KL, Zhou MM. 2010. Keeping it in the family: Diverse histone recognition by conserved structural folds. *Crit Rev Biochem Mol Biol* **45:** 488–505.

Yuan H, Rossetto D, Mellert H, Dang W, Srinivasan M, Johnson J, Hodawadekar S, Ding EC, Speicher K, Abshiru N, et al. 2012. MYST protein acetyltransferase activity requires active site lysine autoacetylation. *EMBO J* **31:** 58–70.

Zeng L, Li J, Muller M, Yan S, Mujtaba S, Pan C, Wang Z, Zhou M. 2005. Selective small molecules blocking HIV-1 Tat and coactivator PCAF association. *J Am Chem Soc* **127:** 2376–2377.

Zeng L, Yap KL, Ivanov AV, Wang X, Mujtaba S, Plotnikova O, Rauscher FJ 3rd, Zhou MM. 2008a. Structural insights into human KAP1 PHD finger-bromodomain and its role in gene silencing. *Nat Struct Mol Biol* **15:** 626–633.

Zeng L, Zhang Q, Gerona-Navarro G, Moshkina N, Zhou MM. 2008b. Structural basis of site-specific histone recognition by the bromodomains of human coactivators PCAF and CBP/p300. *Structure* **16:** 643–652.

Zeng L, Zhang Q, Li S, Plotnikov AN, Walsh MJ, Zhou MM. 2010. Mechanism and regulation of acetylated histone binding by the tandem PHD finger of DPF3b. *Nature* **466:** 258–262.

Zhang J, Sprung R, Pei J, Tan X, Kim S, Zhu H, Liu CF, Grishin NV, Zhao Y. 2009. Lysine acetylation is a highly abundant and evolutionarily conserved modification in *Escherichia coli*. *Mol Cell Proteomics* **8:** 215–225.

Zhang Q, Chakravarty S, Ghersi D, Zeng L, Plotnikov AN, Sanchez R, Zhou MM. 2010. Biochemical profiling of histone binding selectivity of the yeast bromodomain family. *PLoS One* **5:** e8903.

Zhang G, Liu R, Zhong Y, Plotnikov AN, Zhang W, Zeng L, Rusinova E, Gerona-Nevarro G, Moshkina N, Joshua J, et al. 2012. Down-regulation of NF-κB transcriptional activity in HIV-associated kidney disease by BRD4 inhibition. *J Biol Chem* **287:** 28840–28851.

Zhao R, Nakamura T, Fu Y, Lazar Z, Spector DL. 2011. Gene bookmarking accelerates the kinetics of post-mitotic transcriptional re-activation. *Nat Cell Biol* **13:** 1295–1304.

Zhou M, Huang K, Jung KJ, Cho WK, Klase Z, Kashanchi F, Pise-Masison CA, Brady JN. 2009. Bromodomain protein Brd4 regulates human immunodeficiency virus transcription through phosphorylation of CDK9 at threonine 29. *J Virol* **83:** 1036–1044.

Zou JX, Revenko AS, Li LB, Gemo AT, Chen HW. 2007. ANCCA, an estrogen-regulated AAA+ ATPase coactivator for ERα, is required for coregulator occupancy and chromatin modification. *Proc Natl Acad Sci* **104:** 18067–18072.

Zou JX, Guo L, Revenko AS, Tepper CG, Gemo AT, Kung HJ, Chen HW. 2009. Androgen-induced coactivator ANCCA mediates specific androgen receptor signaling in prostate cancer. *Cancer Res* **69:** 3339–3346.

Zuber J, Shi J, Wang E, Rappaport AR, Herrmann H, Sison EA, Magoon D, Qi J, Blatt K, Wunderlich M, et al. 2011. RNAi screen identifies Brd4 as a therapeutic target in acute myeloid leukaemia. *Nature* **478:** 524–528.

C H A P T E R 5

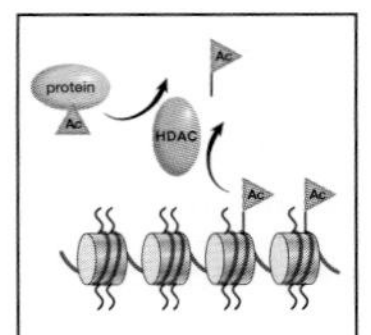

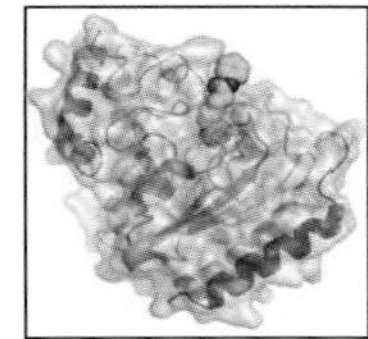

Erasers of Histone Acetylation: The Histone Deacetylase Enzymes

Edward Seto[1] and Minoru Yoshida[2]

[1]Department of Molecular Oncology, Moffitt Cancer Center and Research Institute, Tampa, Florida 33612;
[2]Chemical Genetics Laboratory, RIKEN, Wako, Saitama 351-0198, Japan

Correspondence: ed.seto@moffitt.org

SUMMARY

Histone deacetylases (HDACs) are enzymes that catalyze the removal of acetyl functional groups from the lysine residues of both histone and nonhistone proteins. In humans, there are 18 HDAC enzymes that use either zinc- or NAD^+-dependent mechanisms to deacetylate acetyllysine substrates. Although removal of histone acetyl epigenetic modification by HDACs regulates chromatin structure and transcription, deacetylation of nonhistones controls diverse cellular processes. HDAC inhibitors are already known potential anticancer agents and show promise for the treatment of many diseases.

Outline

Cite as *Cold Spring Harb Perspect Biol* doi: 10.1101/cshperspect.a018713

OVERVIEW

The posttranslational modification (PTM) of histones can cause a functional change in gene expression or chromatin structure—many epigenetic phenomena have testified to this. A common form of histone modification and, indeed, one of the first discovered is acetylation, which occurs at the ε-amino group of lysines, largely in the amino-terminal tail of histones. Results from many early studies suggest that histone acetylation regulates gene transcription. The exact number and combination of acetylated lysine residues that occur in histones that are self-perpetuating and heritable in the cell is unknown. It is clear, nonetheless, that histone acetylation is an abundant source of potential epigenetic information.

Histone lysine acetylation is highly reversible. A lysine residue becomes acetylated by the action of the histone/lysine acetyltransferase enzymes (HATs/KATs), and is removed by histone deacetylases (HDACs). In humans, there are 18 HDAC enzymes divided into four classes: the Class I Rpd3-like proteins (HDAC1, HDAC2, HDAC3, and HDAC8); the Class II Hda1-like proteins (HDAC4, HDAC5, HDAC6, HDAC7, HDAC9, and HDAC10); the Class III Sir2-like proteins (SIRT1, SIRT2, SIRT3, SIRT4, SIRT5, SIRT6, and SIRT7); and the Class IV protein (HDAC11). Like HATs, some HDACs possess substrate specificity. Accumulating evidence suggests that many, if not all, HDACs can also deacetylate nonhistone proteins. It is therefore important to take this fact into consideration when trying to ascertain an HDAC's function.

Structural comparisons among different Class I and II HDACs, as well as HDAC homologs from different species that share significant homology with human classical HDACs, reveal a conserved group of active site residues, suggesting a common mechanism for the metal-dependent hydrolysis of acetylated substrates. The Class III HDACs use NAD^+ as a reactant to deacetylate acetyllysine residues of protein substrates forming nicotinamide, the deacetylated product, and the metabolite 2′-*O*-acetyl-ADP-ribose.

By removing acetyl groups from ε-amino lysines of proteins, HDACs not only alter transcription, but also promote either the establishment or erasure of alternative posttranslational lysine modifications such as methylation, ubiquitination, and sumoylation. Additionally, they may change the dynamics of histone modification "cross talk." Like many important cellular enzymes, HDACs are subject to a variety of controlling mechanisms, including protein–protein interactions and posttranslational modifications. Abnormal HDACs play a key role in many human diseases. A thorough understanding of the functions and mechanisms of HDACs action is a prerequisite to further our understanding of how this family of enzymes impacts on human health and disease.

The availability of HDAC inhibitors has accelerated our understanding of HDAC functions and mechanism of actions. A number of compounds that inhibit HDAC activity have now been developed and characterized. They reportedly cause cell growth arrest, differentiation and/or apoptosis, and restrain tumor growth in animals. In parallel, research is increasingly showing that epigenetic abnormalities are tightly associated with a large number of human diseases, providing a rationale for the use of epigenetic-based therapies such as HDAC inhibitors.

Cite as *Cold Spring Harb Perspect Biol* doi: 10.1101/cshperspect.a018713

1 INTRODUCTION

An enzyme activity that catalyzes the removal of acetyl functional groups from histones was first discovered in a calf thymus extract in 1969 (Inoue and Fujimoto 1969). The deacetylase activity has preference toward the ε-amino acetyl groups of lysine residues compared to the α-amino acetyl groups of amino-terminal amino acids in histones. Treatment of deacetylation reactions with proteinases destroyed most of the enzymatic deacetylation activity. A number of studies followed in the early 1970s further dissecting the biochemical characterization of the histone deacetylase activity in various tissues. Many different properties of histone deacetylase activity were learned from these early chromatography studies, including the finding that the activity contains multiple enzymes. However, initial attempts to purify HDACs to homogeneity using conventional chromatography were unsuccessful. It was not until 1996 that the histone deacetylation field exploded, when the first bona fide histone deacetylase, HDAC1, was isolated and cloned (Taunton et al. 1996). Since that time, more than 15,000 papers have been published on this topic (compared to less than 100 papers published up to that time). There is now overwhelming support showing that HDACs play crucial roles in gene transcription and most likely in all eukaryotic biological processes that involve chromatin. Notably, recent discussions on eukaryotic transcriptional repression refer to some aspects of histone deacetylation. This chapter will focus on important discoveries in the HDAC field in the last two decades since the cloning of the first HDAC. Particular emphasis will be placed on highlighting the structures, functions, mechanisms of action, and regulation of HDACs that will be beneficial to readers interested in histone modifications, chromatin, or epigenetics, as well as scientists not currently working in this area.

2 HDAC FAMILIES AND CLASSES: TWO FAMILIES AND FOUR CLASSES

Each HDAC belongs to either the histone deacetylase family or the Sir2 regulator family. In humans, HDACs are traditionally divided into separate categories called classes based on sequence similarities (Table 1; illustrated in Fig. 1). The Class I proteins (HDAC1, HDAC2, HDAC3, and HDAC8) have sequence similarity to the yeast Rpd3 protein. The Class II proteins (HDAC4, HDAC5, HDAC6, HDAC7, HDAC9, and HDAC10) have sequence similarity to the yeast Hda1 protein. Three proteins in *Saccharomyces cerevisiae*—Hos1, Hos2, and Hos3—however, have 35%–49% identity to Rpd3, and 21%–28% identity to Hda1. Thus, mammalian Class I and II HDACs are also related to the yeast Hos proteins. The Class III proteins (SIRT1, SIRT2, SIRT3, SIRT4, SIRT5, SIRT6, and SIRT7) have sequence similarity to the yeast Sir2 protein. The Class IV protein (HDAC11) shares sequence similarity to both Class I and II proteins. It should be noted that the different classes of HDACs are not to be confused with "Class" in taxonomy classification, or the Structural Classification of Proteins hierarchy classification, in which all HDACs belong to the α and β proteins class.

The Class I, II, and IV HDACs are numbered according to their chronological order of discovery. For example, HDAC1 was first reported several months before HDAC2, both in 1996 (Taunton et al. 1996; Yang et al. 1996). HDAC3 was discovered the following year (Yang et al. 1997). HDAC4, 5, and 6 were first described in 1999 (Grozinger et al. 1999), HDAC7 in early 2000 (Kao et al. 2000), and so on. It is important to clarify the term HDAC isoforms (or isoenzymes) that are frequently used in the literature. Many different forms of an HDAC may arise, such as by single-nucleotide polymorphisms or by alternative splicings. For example, the HDAC9 transcript is alternatively spliced to generate multiple protein isoforms with distinct biological activities. Thus, HDAC9 isoforms (several different forms of the same HDAC9 protein) clearly exist. However, HDAC4, HDAC5, HDAC7, and HDAC9 each originate from a distinct gene and, therefore, although functionally related, they are technically not isoforms of each other.

2.1 Arginase/Deacetylase Superfamily and the Histone Deacetylase Family

The Class I, II, and IV HDACs belong to the arginase/deacetylase superfamily of proteins (Table 1). This superfamily contains the arginase-like amidino hydrolases, as well as the histone deacetylases. It was proposed that eukaryotic HDACs originated from a prokaryotic enzyme similar to the acetylpolyamine amidohydrolases. The prokaryotic enzyme targeted the reversible acetylation and deacetylation of the aminoalkyl group of a DNA binding molecule to achieve a gene regulatory effect (Leipe and Landsman 1997).

The histone deacetylase family of HDACs (sometimes referred to as the classical HDAC family) is made up of three classes of proteins: Class I, II, and IV HDACs. Proteins within each class descend from a common ancestor and have similar three-dimensional structures, functions, and significant sequence homology. A phylogenetic analysis of all histone deacetylase family proteins and all proteins related to this family from all fully sequenced free-living organisms has been performed, together with the analysis of gene duplication events. Results indicate that the common ancestor of metazoan organisms contained two Class I, two Class II, and a single Class IV HDAC (Gregoretti et al.

Table 1. HDAC classification

Superfamily	Family	Class	Protein (*S. cerevisiae*)	Subclass	Protein (human)
Arginase/deacetylase superfamily	Histone deacetylase family	Class I	Rpd3, Hos1, Hos2, Hos3		HDAC1, HDAC2, HDAC3, HDAC8
		Class II	Hda1	Class IIa	HDAC4, HDAC5, HDAC7, HDAC9
				Class IIb	HDAC6, HDAC10
		Class IV			HDAC11
Deoxyhypusine synthase like NAD/FAD-binding domain superfamily	Sir2 regulator family	Class III	Sir2, Hst1, Hst2, Hst3, Hst4	I	SIRT1, SIRT2, SIRT3
				II	SIRT4
				III	SIRT5
				IV	SIRT6, SIRT7

2004). Functional prediction reveals that self-association is common among this family of HDACs. Interestingly, all Class I, II, and IV HDACs precede the evolution of histone proteins suggesting that the primary substrates for this HDAC family may be nonhistones.

2.1.1 Class I (HDAC1, HDAC2, HDAC3, HDAC8)

In *S. cerevisiae*, the histone deacetylase-A 1 (Hda1) protein, which shares sequence similarity to the transcription regulator reduced potassium dependency 3 (Rpd3), is a subunit of a large histone deacetylase complex, Hda. Hda1 also shares similarity to three yeast proteins designated Hos1, Hos2, and Hos3. Another yeast histone deacetylase complex, Hdb, contains Rpd3 as an associated factor. Using a trapoxin (an inhibitor of histone deacetylase) affinity matrix, Stuart Schreiber purified and cloned a human 55-kDa protein related to the yeast protein Rpd3 (Taunton et al. 1996). Immunoprecipitation of this 55-kDa protein, HDAC1 (initially called HD1), showed that it contains his-

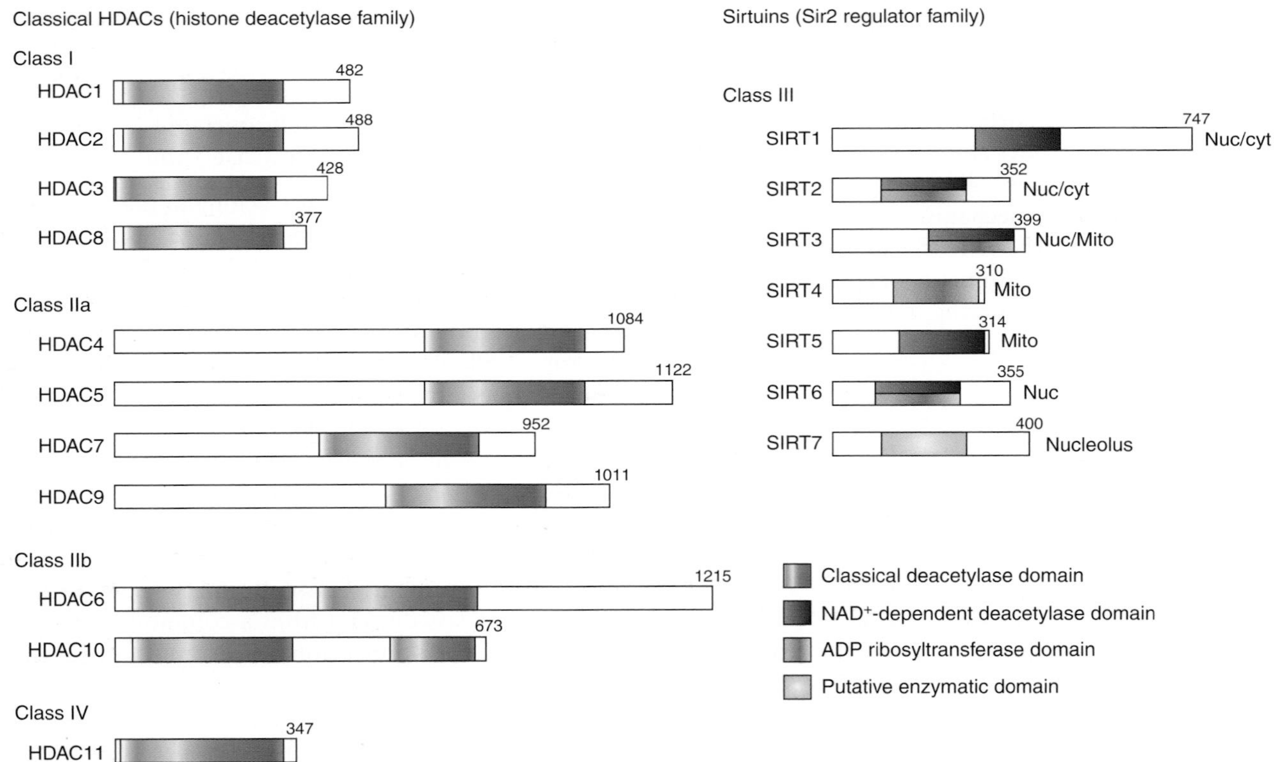

Figure 1. Domain organization of human HDACs. The total number of amino acid residues in each HDAC is shown on the right of each protein. Many HDACs have multiple isoforms and, for simplicity, only the longest isoform is shown. Enzymatic domains (or putative enzymatic domains) are shown in colors. Sirtuin localizations: Nuc, nuclear; cyt, cytoplasmic; Mito, mitochrondial.

 Cite as *Cold Spring Harb Perspect Biol* doi: 10.1101/cshperspect.a018713

tone deacetylase activity. A second human histone deacetylase protein, HDAC2 (initially called mRPD3), with high homology with yeast Rpd3 was independently identified as a transcription factor (Yang et al. 1996). HDAC2 negatively regulates transcription by being recruited to DNA as a corepressor. The third human Rpd3-related protein, HDAC3, was discovered by searching the GenBank database for DNA and protein sequences with homology to HDAC1 and HDAC2 (e.g., Yang et al. 1997). Like HDAC1 and HDAC2, HDAC3 represses transcription, binds to and is recruited by transcription factors, and is expressed in many different cell types. Similar to the identification of HDAC3, a search of the GenBank database for protein sequences similar to HDAC1, 2, and 3, led to the discovery of HDAC8 (e.g., Hu et al. 2000). The highly conserved deacetylase domains of Class I HDACs share extensive homology with each other, with 45%–94% amino acid sequence identity.

It is widely stated in the literature that Class I HDACs are located in the nucleus and are ubiquitously expressed. Results from more thorough studies reveal that HDAC3 expression is restrictive to certain tissues, and HDAC1, HDAC2, HDAC3, and HDAC8 can be localized to the cytoplasm or specialized cellular organelles. Therefore, the generalization that Class I HDACs are nuclear and ubiquitously expressed may be oversimplified, and it is fair to predict that Class I HDACs might possess yet to be discovered extranuclear or tissue-specific functions.

2.1.2 Class II (HDAC4, HDAC5, HDAC6, HDAC7, HDAC9, HDAC10)

HDAC4, HDAC5, and HDAC6 were discovered together after GenBank databases search for human HDACs with sequence similarity to yeast Hda1 (e.g., Grozinger et al. 1999). These proteins possess certain features present in the conserved catalytic domains of Class I human HDACs, but also contain additional sequence domains that have no similarity to Class I enzymes. The divergence of the Class I and II HDACs appears to have occurred relatively early in evolution. Like Class I HDACs, immunopurified recombinant HDAC4, HDAC5, and HDAC6 possess in vitro HDAC activity, although at a much lower level. HDAC6, interestingly, contains an internal duplication of two deacetylase catalytic domains, which appear to function independently of each other.

HDAC7 was first isolated as a protein that interacts with the transcription corepressor silencing mediator for retinoid or thyroid-hormone receptors (SMRT) (Kao et al. 2000). HDAC7 possesses three repression domains, two of which contain autonomous repressor functions that are independent of the third deacetylase repressor activity. Shortly after the discovery of HDAC7, Paul Marks reported the identification of a protein, HDAC-related protein (HDRP), that shares 50% identity in deduced amino acid sequence to the noncatalytic amino-terminal domain of HDAC4 and HDAC5 (Zhou et al. 2000). Subsequently, HDAC9 was identified by a homology database search using the human HDAC4 amino acid sequence. HDAC9 has multiple alternatively spliced isoforms. One of these isoforms, which is an amino-terminal splice variant, is the HDRP or myocyte enhancer–binding factor 2–interacting transcriptional repressor (MITR). Like all Class I and II HDACs, HDAC9 possesses a conserved deacetylase domain, which represses gene activity when recruited to a promoter through deacetylation of histones. HDAC4, 5, 7, and 9 make up the Class IIa HDACs and share 48%–57% overall identity.

HDAC10 was discovered independently by four different groups (e.g., Kao et al. 2002). Homology comparison indicates that HDAC10 is most similar to HDAC6 (55% overall identity), and both contain a unique, putative second catalytic domain not found in other HDACs. Therefore, HDAC6 and HDAC10 are subclassified as Class IIb. An interesting feature of both Class IIa and IIb HDACs is their subcellular localization; each member of these classes shows at least some cytoplasmic localization, suggesting a major cytoplasmic functional role for Class II HDACs. The conserved deacetylase domains of Class II HDACs share 23%–81% amino acid sequence identity to each other.

2.1.3 Class IV (HDAC11)

HDAC11 is the sole member of the Class IV HDAC. It uniquely shares sequence homology with the catalytic domains of both Class I and II HDACs, and was first discovered by Basic Local Alignment Search Tool searches of GenBank databases using the yeast Hos3 protein as the query sequence (Gao et al. 2002). HDAC11 regulates the protein stability of DNA replication factor CDT1 (Glozak and Seto 2009) and the expression of interleukin 10. HDAC11, together with HDAC10, are probably the least studied and most poorly understood HDACs in the classical HDAC family.

2.2 Deoxyhypusine Synthase–Like NAD/FAD-Binding Domain Superfamily and the Sir2 Regulator Family

The deoxyhypusine synthase (DHS)-like NAD/FAD-binding domain superfamily of proteins includes silent information regulator 2 (Sir2) proteins, as well as deoxyhypusine synthase, carboxy-terminal domain of the electron transfer flavoprotein α subunit, pyruvate oxidase and decarboxylase middle domain, transhydrogenase domain III, and the ACDE2-like families. The founding member of the Sir2 regulator family of proteins, the *S. cerevisiae* Sir2, was originally identified in a genetic screen for genes involved in controlling expression of silent mating type loci. In yeast, Sir2 is required for transcription silencing (see Ch. 8 [Grunstein and Gasser 2013] for extensive description). The Sir2 regulator family has only one class (i.e., Class III) of nicotinamide adenine dinucleotide (NAD^+)-dependent histone deacetylase, with seven Sir2-like proteins in humans (SIRT1, SIRT2, SIRT3, SIRT4, SIRT5, SIRT6, SIRT7). Sir2-like proteins (sirtuins) are phylogenetically conserved in eukaryotes, prokaryotes, and archaea, and based on phylogenetic relationships, they can be grouped into more than a dozen classes and subclasses. The first classification was organized into five major classes: I (SIRT1, SIRT2, SIRT3), II (SIRT4), III (SIRT5), IV (SIRT6, SIRT7), U (cobB in bacteria, no human homolog) (see Fig. 4 in Ch. 8 [Grunstein and Gasser 2013]). All sirtuins contain a conserved core domain with several sequence motifs.

2.2.1 Class III (SIRT1, SIRT2, SIRT3, SIRT4, SIRT5, SIRT6, SIRT7)

S. cerevisiae homologs of Sir2 (Hsts) and the conservation of this protein family from bacteria to humans were first described by Lorraine Pillus and Jef Boeke (Brachmann et al. 1995). Subsequently, five human sirtuins (SIRT1, SIRT2, SIRT3, SIRT4, SIRT5) were identified in the GenBank database using *S. cerevisiae* Sir2 amino acid sequence as the probe (Frye 1999). Two additional human sirtuins (SIRT6 and SIRT7) were similarly identified using human SIRT4 as the probe. The seven sirtuins share 22%–50% overall amino acid sequence identity, and 27%–88% iden-

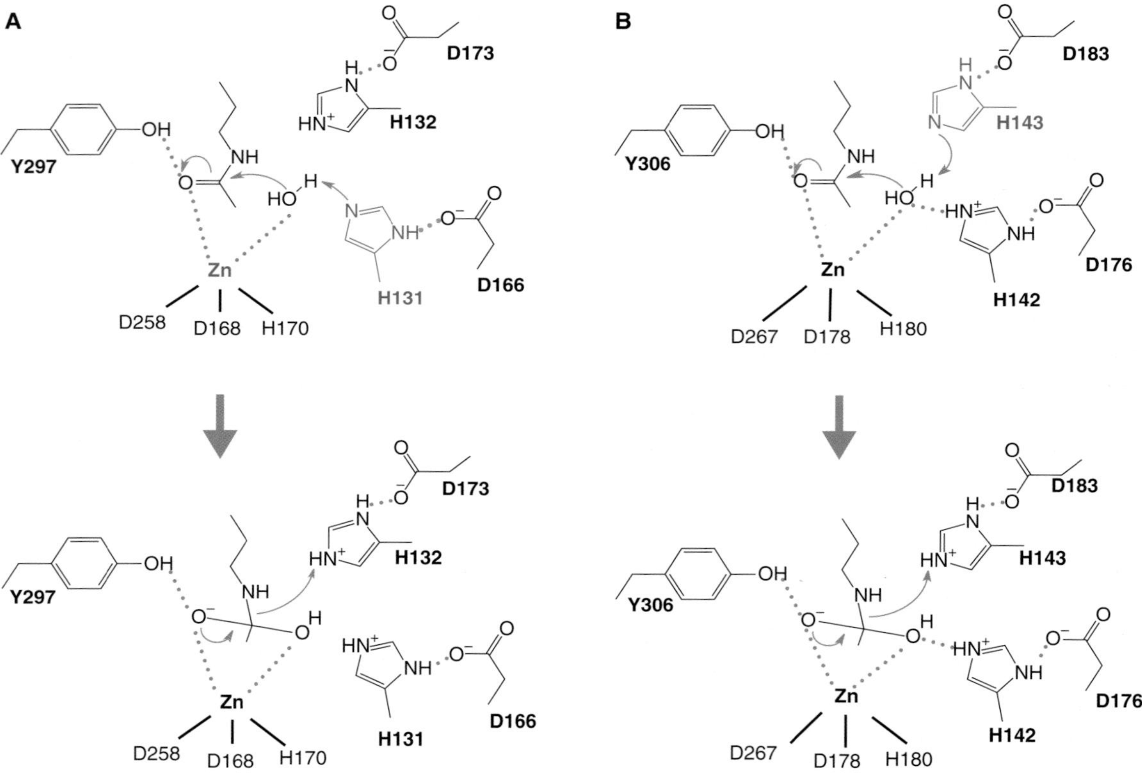

Figure 2. Catalytic mechanism of HDACs. Two models for the catalytic mechanism of the Zn-dependent HDAC reaction have been proposed. (*A*) A model proposed from the HDLP structure. The HDLP catalytic core consists of a tubular pocket, a zinc-binding site, and active-site residues (in bold) of a tyrosine (Y297) and two histidines (H131 and H132) that make hydrogen bonds to two aspartic acids (D166 and D173). One of these catalytic histidines (red) facilitates nucleophilic attack at the substrate carbonyl by activating a water molecule coordinated with the zinc ion. Initially, two tandem histidine residues (H131 and H132) were proposed to function as Asp-His charge relay systems, typical of serine proteases such as chymotrypsin and chymotrypsinogen in the enzyme reaction. The active site zinc ion is coordinated by three residues (two aspartic acids and one histidine). (*B*) A model proposed from the HDAC8 structure, in which the other histidine residue (red) plays an essential role in the electron transfer. Hydrogen bond interactions are drawn in dotted lines.

Cite as *Cold Spring Harb Perspect Biol* doi: 10.1101/cshperspect.a018713

tity in the conserved catalytic domains. Of the seven human sirtuins, SIRT1 is most similar to the yeast Sir2 protein, possesses the most robust histone deacetylase activity, and has been most extensively studied.

A remarkable feature of sirtuins is that they have two enzymatic activities: mono-ADP-ribosyltransferase and histone deacetylase. SIRT5 possesses additional protein lysine desuccinylase and demalonylase activity in vitro (Du et al. 2011). Another interesting characteristic of sirtuins is their localizations (see Fig. 1), with SIRT1 and SIRT2 found in the nucleus and cytoplasm, SIRT3 in the nucleus and mitochondria, SIRT4 and SIRT5 exclusively in the mitochondria, SIRT6 only in the nucleus, and SIRT7 in the nucleolus. Like the Class I, II, and IV HDACs, sirtuins also have nonhistone substrates, at least in eukaryotes.

3 CATALYTIC MECHANISMS AND STRUCTURES

3.1 Catalytic Mechanisms and Structures of Classical HDACs (Class I and II)

The classical HDAC family of enzymes (Class I, II, IV) share a common catalytic mechanism that requires a zinc ion (Fig. 2). Insights into the catalytic mechanisms for the metal-dependent hydrolysis of the acetamide bond in acetylated lysine have been provided by structural, biochemical, and mutational analysis.

The first X-ray crystal structure of a classical HDAC family protein was determined for the histone deacetylase-like protein (HDLP) from a hyperthermophilic bacterium *Aquifex aeolicus* (Fig. 3A). HDLP has the same topology as arginase containing an α/β fold and an 8-stranded parallel β-sheet (Finnin et al. 1999). This similarity to arginase (Fig. 3B), a metalloenzyme that catalyzes the hydrolysis of arginine to ornithine, suggests that it evolved from a common metalloprotein ancestor.

Structural studies of the HDLP catalytic core points to a catalytic reaction model illustrated in Figure 2A. Studies from X-ray crystallography of mammalian HDACs showed essentially the same catalytic domain structure as HDLP, in which the residues that make up the active site and contact inhibitors are conserved across the HDAC family. Structural analysis of HDAC8 (Fig. 3D) and its mutants, however, suggested a distinct model. This model proposes that one of the histidine residues acts as the general base (H143), whereas the other histidine (H142) serves as a general electrostatic catalyst (Fig. 2B). The HDAC8 H143A mutant has an almost complete loss of activity in contrast to the residual activity of an H142A mutant, in concordance with the proposed model of action (Gantt et al. 2010). Furthermore, quantum mechanical/molecular mechanical molecular dynamics simulations suggest that a neutral H143 first serves as the general base to accept a proton from the zinc-bound water molecule in the initial rate-determining nucleophilic attack step, and then shuttles it to the amide nitrogen atom to facilitate the cleavage of the amide bond (Wu et al. 2011). This model seems to be a more likely catalytic mechanism for HDAC-related enzymes, which is similar to the conventional metalloenzymes thermolysin and carboxypeptidase A.

All of the catalytic residues illustrated in Figure 2B are conserved in other HDACs except Class IIa enzymes, in which the tyrosine residue is replaced by a histidine residue. This tyrosine residue is positioned next to the zinc, opposite the histidine-aspartic acid residues, and is important for stabilizing the tetrahedral intermediate. Both the zinc and the tyrosine residue participate in polarizing the substrate carbonyl (C = O) for the nucleophilic attack. Thus the lack of the tyrosine residue in HDAC4 and other verte-

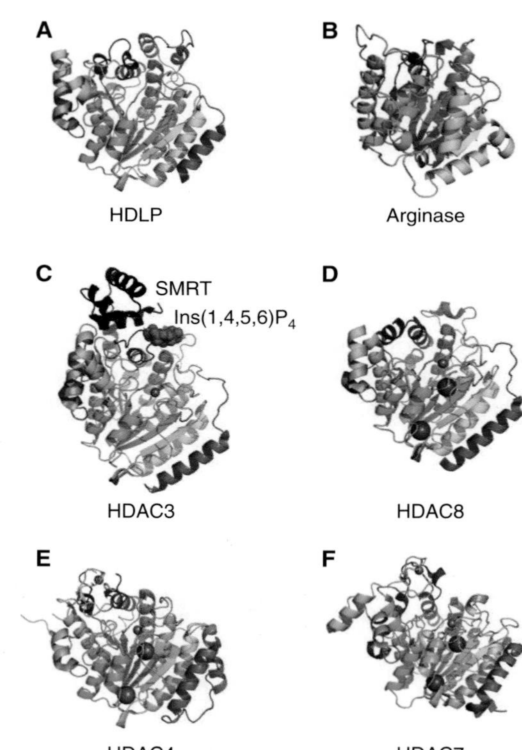

Figure 3. Structure of Class I and II HDACs. Crystal structures of the arginase/deacetylase superfamily of proteins are illustrated. Metal ions are represented as space filled spheres: red, gray, and violet spheres indicate Mn, Zn, and K atoms, respectively. (*A*) *Aquifex aeolicus* HDLP (1C3P), (*B*) rat arginase (PDB ID: 1RLA), (*C*) human Class I HDAC3 (4A69) in complex with Ins(1,4,5,6)P4 (highlighted by magenta space-filled spheres) and the deacetylase activation domain from the human SMRT corepressor (depicted with a ribbon model in black), (*D*) human Class I HDAC8 (3F07), (*E*) human Class IIa HDAC4 (2VQW), and (*F*) human Class IIa HDAC7 (3C0Y).

brate Class IIa HDACs may be responsible for the low catalytic activity. Indeed, catalytic activity of HDAC4 was restored by a single His to Tyr substitution, reaching levels comparable to that of Class I enzymes (Lahm et al. 2007).

Most recently, X-ray crystallographic structure of HDAC3 complexed with the deacetylase activation domain of the human SMRT (NCoR) corepressor was determined (Fig. 3C). Surprisingly, inositol (1,4,5,6)-tetrakisphosphate was present in the interface between HDAC3 and SMRT acting as an "intermolecular glue." This inositol phosphate molecule is essential for the interaction between the two proteins as well as the catalytic activity of HDAC3 (Watson et al. 2012).

3.2 Catalytic Mechanisms and Structures of Sirtuins (Class III)

Class III HDACs require NAD^+ as the cofactor for enzyme activity in contrast to the zinc-dependent catalysis by Class I, II, and IV enzymes (e.g., Imai et al. 2000). Structural studies of archaeal, yeast and human homologs of Sir2 have shown that the catalytic domain of sirtuins resides in a cleft formed between a large domain with a Rossmann-fold and a small zinc-binding domain (Fig. 4) (e.g., Finnin et al. 2001). The amino acid residues in the cleft are conserved in the sirtuin family and form a protein-tunnel in which the substrate interacts with NAD^+. Recently, the crystal structure of SIRT5 was solved, revealing a larger substrate-binding site that may accept a larger acyl group on the lysine residue (Schuetz et al. 2007). This agrees with SIRT5 acting as a protein lysine desuccinylase and demalonylase rather than a deacetylase (Du et al. 2011).

A proposed chemical mechanism for nicotinamide cleavage from NAD^+ and ADP-ribose transfer to an acetylated lysine is illustrated in Figure 5. This is based on detailed structural analysis of substrates and/or cofactor analogs complexed with Sir2 homologs (e.g., Avalos et al. 2004). The first step of the reaction involves nucleophilic addition of the acetamide oxygen to the C1′ position of the nicotinamide ribose to form a C1′-*O*-alkylamidate intermediate and free nicotinamide. The C1′-*O*-alkylamidate intermediate is then converted to a 1′,2′-cyclic intermediate from which lysine and 2′-*O*-acetyl-ADP ribose are eventually released. Nicotinamide, one of the byproducts, acts as an inhibitor of sirtuins.

Some sirtuin family members may also possess intrinsic mono-ADP-ribosyltransferase activity. The ADP-ribosyltransferase activity of sirtuins has been thought of as a low efficiency side-reaction caused by the partial uncoupling of intrinsic deacetylation and acetate transfer to ADP-ribose. Recently, however, mono-ADP-ribosyltransferase activity was found to be the main enzymatic activity of at least SIRT4, whereas SIRT2 and SIRT6 display both deacetylase and mono-ADP-ribosyltransferase activities (e.g., Frye 1999; Liszt et al. 2005). It is currently unclear whether NAD^+-dependent deacetylation and mono-ADP-ribosylation of proteins can occur simultaneously.

4 HDAC SUBSTRATES

4.1 Histone Substrates

Soon after the discovery of HATs, it became apparent that each may have a particular histone substrate specificity (discussed in Ch. 4 [Marmorstein and Zhou 2014]). The

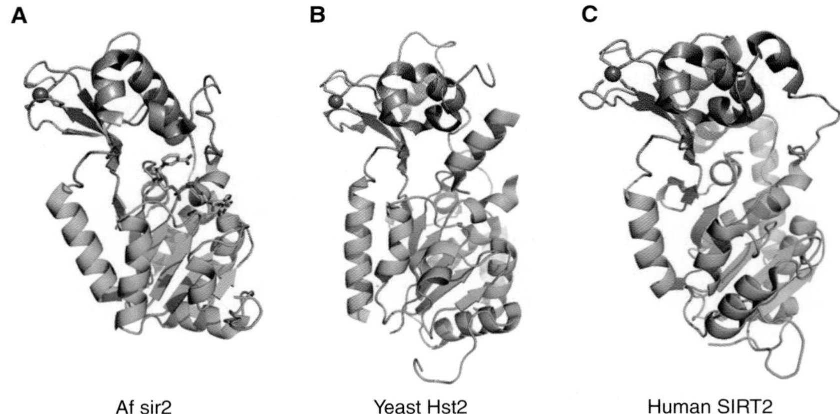

Figure 4. Overall structure of sirtuins (Class III). Crystal structures of the sirtuin family of proteins are illustrated as cartoons highlighting the large Rossmann-fold domains (cyan) and small zinc-binding domains (brown). (*A*) *Archaeoglobus fulgidus* sir2 (PDB ID: 1ICI). NAD^+ is drawn in stick model form in which yellow, blue, red, and orange represent C, N, O, and P atoms, respectively. (*B*) Yeast Hst2 (1Q14). (*C*) Human SIRT2 (1J8F). Zinc ions are represented as space-filled spheres in gray.

 Cite as *Cold Spring Harb Perspect Biol* doi: 10.1101/cshperspect.a018713

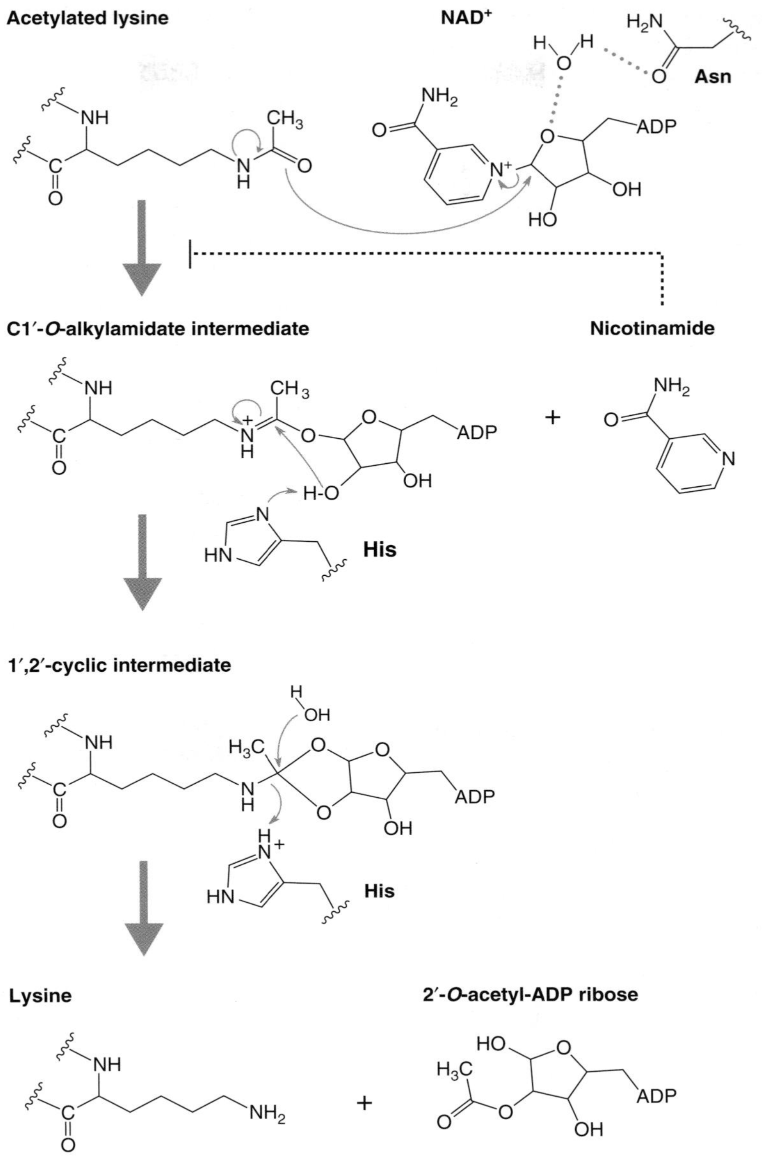

Acetylated lysine

NAD+

Asn

C1′-O-alkylamidate intermediate

Nicotinamide

His

1′,2′-cyclic intermediate

His

Lysine

2′-O-acetyl-ADP ribose

Figure 5. Catalytic mechanism of sirtuins (Class III). Proposed mechanism of the NAD^+-dependent deacetylase reaction. The first step of the reaction involves nucleophilic addition of the acetamide oxygen to the C1′ position of the nicotinamide ribose to form a C1′-O-alkylamidate intermediate and free nicotinamide. Next, the 2′-hydroxy group of the NAD^+ ribose is activated by an active site histidine residue that, in turn, attacks the C1′-O-alkylamidate to form the 1′,2′-cyclic intermediate. The 1′,2′-cyclic intermediate is then attacked by an activated water molecule resulting in the formation of deacetylated lysine and 2′-O-acetyl-ADP ribose. 2′-O-acetyl-ADP ribose can be readily converted to 3′-O-acetyl-ADP ribose in aqueous solution by nonenzymatic intramolecular transesterification. Thus, nicotinamide, the deacetylated peptide, and a mixture of 2′- and 3′-O-acetyl-ADP ribose are the final reaction products.

search for Class I, II, and IV HDAC histone substrate specificity has turned out to be much more difficult than for HATs. One obstacle to identifying HDAC substrate specificity is that most HDACs within this family possess very low measurable histone deacetylase activity when purified to homogeneity. Functional redundancy of many HDACs also contributes to the difficulty in deciphering substrate specificity. For example, knockdown of one classical HDAC can be compensated by the activity of another HDAC within the same class or even from a different class. Additionally, some Class I HDACs exist in several different complexes, with each complex potentially having different substrate preferences. For example, HDAC1 is present in at least three different stable protein complexes, with each complex possibly targeting different substrates (Fig. 6A). Finally, substrate specificity may differ depending on the source of the substrate, such as nucleosomal histones versus free histones.

Early studies indicated that purified HDAC1 alone or an HDAC1/2 complex deacetylated all four core histones. The same results for HDAC4, HDAC5, and HDAC6 suggested a lack of specificity (Grozinger et al. 1999). A later study, however, suggested that HDAC1 can deacetylate all four core histones at all lysines tested, but with varying efficiency (Johnson et al. 2002). Another study suggested that HDAC8 preferentially deacetylates histones H3 and H4 (Hu et al. 2000), whereas HDAC11 might specifically deacetylate H3K9 and H3K14.

The complexity in elucidating Class I, II, and IV HDAC histone substrate specificity can be illustrated by attempts to identify HDAC3 substrates. Using an immunoprecipitated HDAC3 complex and purified nucleosomes, it was

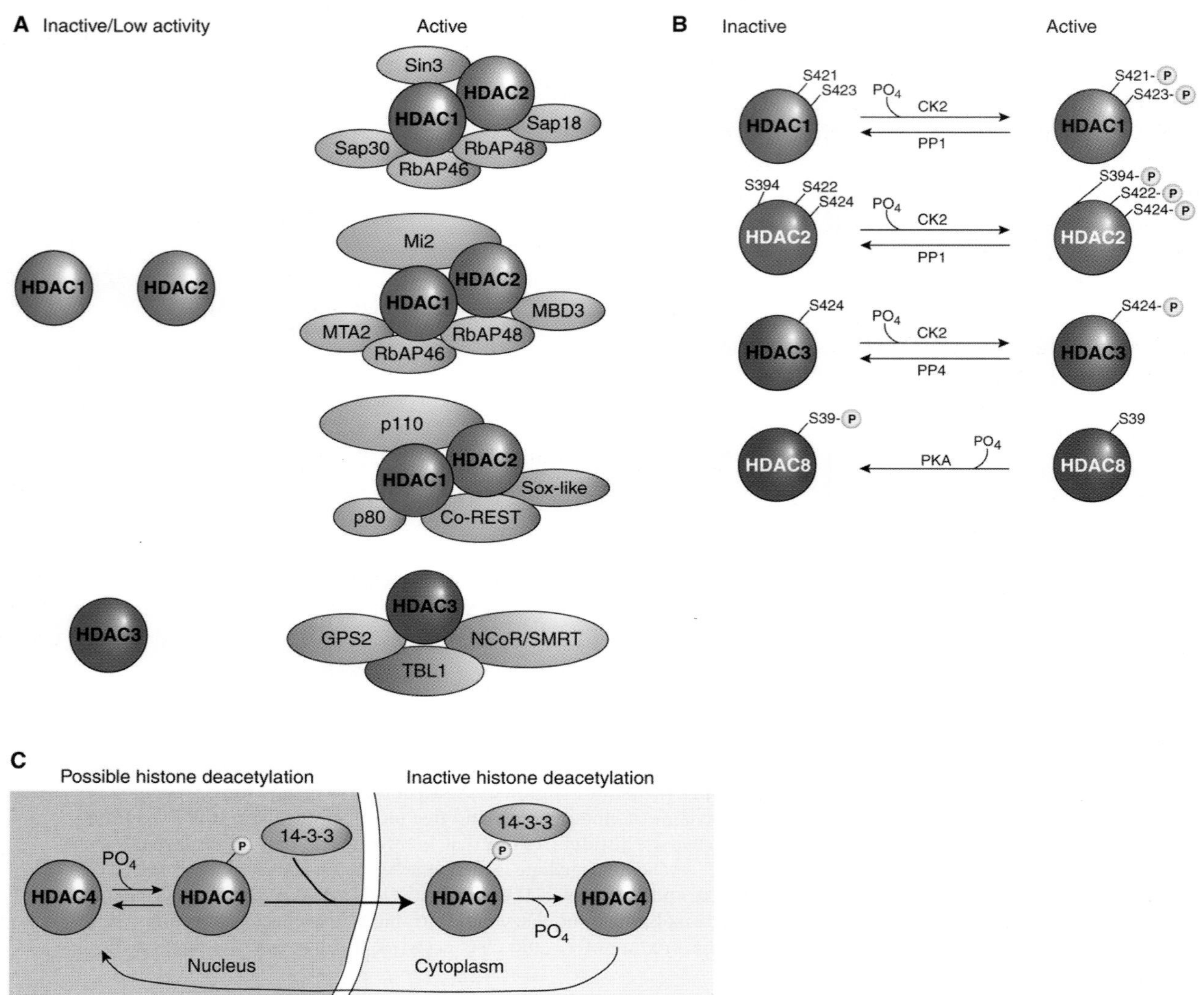

Figure 6. Examples of HDAC regulation. (*A*) Class I HDACs are commonly regulated by protein complex formations. HDAC1, HDAC2, and HDAC3 possess low enzymatic activities when in isolation, and their activities increase significantly when present in holoenzyme complexes. (*B*) The activities of Class I HDACs are modulated by phosphorylation and dephosphorylation. In general, phosphorylation activates HDAC1, HDAC2, and HDAC3, although it represses HDAC8 activities. (*C*) Phosphorylation of a Class II HDAC, HDAC4, promotes its interaction with the 14-3-3 protein and, subsequently, changes its localization. Multiple residues (S246, S467, and S632 on HDAC4, and corresponding conserved sites on HDAC5, HDAC7, and HDAC9) confer the HDAC–14-3-3 interactions.

shown that HDAC3 deacetylates histone H4 more efficiently than HDAC1. Using a similar approach, however, showed that although HDAC3 immunocomplexes completely deacetylated H2A, H4K5, and H4K12, it only partially deacetylated H3, H2B, H4K8, and H4K16 (Johnson et al. 2002). Interestingly, compared to HDAC1, HDAC3 preferentially deacetylates H4K5, H4K12, and H2AK5 in HeLa cells, based on chromatin immunoprecipitation (ChIP) experiments. Although these studies show that HDAC3 (and perhaps all Class I HDACs) has distinct substrate specificity, it is important to note that immunopurified HDAC3 complexes contain other proteins that themselves possess HDAC activity, thus complicating the interpretation of these results. In fact, in a study using homozygous HDAC3-deficient DT40 cells, the acetylation levels of H4K8 and H4K12 were similar to those detected in wild-type cells, suggesting that histone H4 might not be a major target of HDAC3 in vivo. Also, in an in vitro reconstituted chromatin system, an HDAC3-containing protein complex selectively deacetylated histone H3, compared to an HDAC1/2 complex that deacetylated both histones H3 and H4 (Vermeulen et al. 2004). In another study, there was no change in the overall acetylation status of histones H3 or H4 or the acetylation of H4K8 or H4K12 in HeLa cells depleted of HDAC3 (Zhang et al. 2004). Knockdown of HDAC3 by expression of HDAC3-specific siRNA, however, increased the acetylation of H3K9 and H3K18. This knockdown increased the acetylation of H3K9 to a greater extent than the HDAC1-specific siRNA did, whereas the reverse was true for H3K18. Although these incompatible results are quite puzzling, they do support the general conclusion that each Class I, II, IV HDAC might possess distinct histone substrate specificity. The possibility that each HDAC within the classical family targets specific lysines in certain histones, however, still requires more comprehensive investigations.

The case for sirtuins histone substrate specificity is less ambiguous (Table 2). Yeast Sir2 deacetylates H3K9, H3K14, and H4K16 (Imai et al. 2000). Biochemical studies revealed that SIRT1 deacetylates histones H4K16 and H3K9, and interacts with and deacetylates histone H1K26, mediating heterochromatin formation (Vaquero et al. 2004). Although there are reports that SIRT1 can deacetylate all four core histones in vitro, SIRT1's chief target site on histones is H4K16 (Vaquero et al. 2004). Also, SIRT1 associates with lysine-specific histone demethylase 1 (LSD1) and together play a concerted role in deacetylating histone H4K16 and demethylating H3K4 to repress gene expression (Mulligan et al. 2011).

Like SIRT1, SIRT2 and its yeast ortholog, Hst2, have a strong preference for histone H4K16 (Vaquero et al. 2006). Although SIRT2 is located mainly in the cytoplasm, it localizes to chromatin during the mammalian G2/M transition phase of the cell cycle and deacetylates H4K16. Further, SIRT3 is a predominant mitochondrial protein, but like SIRT2, it also deacetylates H4K16 when transported to the nucleus under certain conditions.

Table 2. Sirtuin histone substrates

Sirtuin	Histone substrate	Biological relevance
SIRT1	H3K9 H3K14 H3K56 H4K16 H1K26	Chromatin organization, DNA repair/genome stability, cancer
SIRT2	H4K16 H3K56	Chromatin condensation/mitosis, DNA repair, cancer
SIRT3	H4K16	Chromatin silencing, DNA repair, cellular stress
SIRT4	None	
SIRT5	None	
SIRT6	H3K9 H3K56	Telomeric chromatin/senescence, DNA repair/genome stability
SIRT7	H3K18	Cellular transformation

Both SIRT4 and SIRT5 are located exclusively in the mitochondria and, therefore, do not deacetylate histones. Initially, SIRT6 was determined to be a mono-ADP-ribosyltransferase, and not a histone deacetylase enzyme (Liszt et al. 2005). Later studies, however, reveal that SIRT6 deacetylates H3K9 and H3K56 (e.g., Michishita et al. 2008), although biochemical and structural analyses argue that SIRT6 has very low deacetylase activity compared to other sirtuins. Deacetylation of H3K9 by SIRT6 modulates telomeric chromatin. SIRT7 is a highly selective H3K18 deacetylase that plays a key role in cellular transformation (Barber et al. 2012).

4.2 Nonhistone Substrates

ε-amino lysine acetylation and deacetylation of nonhistone proteins were first reported for high-mobility group proteins 1 and 2 (HMG-1 and HMG-2; Sterner et al. 1979). It was shown that an enzyme preparation that deacetylates histone H4 can also deacetylate HMG-1 and HMG-2. In the last 15 years, a large number of acetylated nonhistone proteins have been discovered and shown to be deacetylated by HDACs. Their functional consequences have been a subject of intense research. Perhaps the best characterized nonhistone HDAC substrate is the tumor-suppressor protein p53. HDAC1 interacts with and deacetylates p53, promoting p53 ubiquitination and degradation (Luo et al. 2000). In addition, SIRT1 also binds to and deacetylates p53 (e.g., Vaziri et al. 2001). Deacetylation of p53 by SIRT1 decreases the ability of p53 to transcriptionally activate the cell cycle inhibitor p21, which causes cells to reenter into the cell cycle following DNA repair. Another well-characterized nonhistone HDAC substrate is the cytoskeletal protein α-tubulin. HDAC6 deacetylates lysine 40 of α-tubulin and regulates microtubule-dependent cell motility (Hubbert et al. 2002). These examples illustrate the ability of HDACs to regulate important biological processes without modifying histones.

More recently, using high-resolution mass spectrometry, more than 3600 acetylation sites on 1750 proteins were identified (Choudhary et al. 2009). The acetylation sites are present on nuclear, cytoplasmic, and mitochondrial proteins involved in many different cellular processes. The use of suberoylanilide hydroxamic acid (SAHA) and MS-275, two broad spectrum Class I, II, IV HDAC inhibitors, upregulated ~10% of all acetylation sites by at least a factor of 2, suggesting that many of these acetylations are regulated by classical HDACs. In a similar study, using a proteomics approach, 213 nuclear proteins were unambiguously identified to be hyperacetylated in the absence of Sirt1 (Peng et al. 2012). Again, these proteins have a range of diverse functions including DNA damage repair, apoptosis and survival, cell cycle, transcription, RNA processing, translation, metabolism, and chromatin structure. Also, high-throughput genetic interaction profiling revealed that HDACs functionally regulate nonhistone substrates that govern a wide array of biological processes (Lin et al. 2012). Profiling of protein acetylation in bacteria (which do not possess histones), likewise, has indicated that HDACs deacetylate a significant number of proteins in lower organisms, emphasizing the highly conserved nature of HDACs (Wang et al. 2010).

Using computational prediction tools, it is anticipated that many more ε-amino lysine acetylation sites on a wide variety of proteins are yet to be discovered. Thus, nonhistone lysine acetylation is prevalent and comparable with that of other major posttranslational modifications. Determining which of these acetylated proteins are functionally regulated by HDACs, therefore, is an important endeavor in future research. Considering the large number of potential nonhistone HDAC substrates, coupled with the fact that functional HDACs are present in cells devoid of histones, it is tempting to speculate that the chief function of HDACs is to regulate nonepigenetic phenomena. However, it is interesting that the activities of DNMT1, a key enzyme responsible for DNA methylation, and many histone modification enzymes (HATs, HDACs, HMTs) that are critical for heritable changes in gene expression, are regulated by HDACs (Choudhary et al. 2009; Peng et al. 2012). Therefore, although deacetylation of nonhistones in itself may not directly serve to erase epigenetic modifications or help establish an epigenetic off chromatin state, it could modify inheritance of chromatin states indirectly by deacetylation of other epigenetic-modifying enzymes.

5 REGULATION OF HDAC ACTIVITY

Like nearly all enzymes that are involved in critical cellular functions, the activities of HDACs are highly regulated. This regulation is achieved by a variety of different mechanisms at the transcription, posttranscription, translation, and posttranslational levels. The most well-defined mechanisms of HDAC regulation are protein–protein interactions and posttranslational modifications (PTMs). Less studied, but perhaps equally important, is the regulation of some HDACs by control of expression, alternative RNA splicing, availability of cofactors, subcellular localization, and proteolytic processing.

5.1 Protein Complexes

Many important molecular processes in the cell are performed by large multisubunit protein complexes. A protein may interact, sometimes transiently, with another protein to activate or repress the other protein's activities and functions. This is indeed a common mechanism used to regulate many HDACs (Fig. 6A). Early studies indicated that the isolation of HDAC1 or HDAC2 individually, without associated proteins, generally yielded very low enzymatic activity. Subsequent biochemical analyses revealed that human HDAC1 and HDAC2 exist together in at least three distinct multiprotein complexes called the Sin3, the NuRD, and the CoREST complexes (reviewed by Ayer 1999). Sin3 and NuRD complexes share a common core, comprised of four proteins: HDAC1, HDAC2, RbAp46, and RbAp48. Additionally, each complex contains unique polypeptides (Sin3, SAP18, and SAP30 in the Sin3 complex; Mi2, MTA-2, and MBD3 in the NuRD complex). In a NuRD reconstitution experiment using purified subunits, it was shown that HDAC activity of the core complex was severely compromised compared to the native NuRD holo-complex. The addition of certain cofactors to the core complex, however, was sufficient to direct the formation of an enzymatically active complex. In a different study, it was shown that a protein in the Sin3 corepressor complex augments enzymatic activity of HDAC1 in vivo. Similarly, in the CoREST complex, the association of HDAC1/2 with CoREST is essential for HDAC enzymatic activity.

Another well-defined example of HDAC regulation by protein–protein interaction came from studies of HDAC3. Early studies suggested that SMRT and the nuclear receptor corepressor (NCoR) function as platforms for recruitment of HDACs. Subsequent studies surprisingly revealed that

Cite as *Cold Spring Harb Perspect Biol* doi: 10.1101/cshperspect.a018713

the interaction between HDAC3 and SMRT/NCoR resulted in the stimulation of HDAC3 enzymatic activity (e.g., Wen et al. 2000). In contrast, SMRT/NCoR mutants that did not bind HDAC3 could not activate HDAC3. Activation of this deacetylase enzymatic activity by SMRT/NCoR is specific to HDAC3.

For Class III HDACs, several proteins have been reported to interact and regulate SIRT1. Of these, the best described is the association of SIRT1 with deleted in breast cancer 1 (DBC1; e.g., Zhao et al. 2008). DBC1 negatively regulates SIRT1 deacetylase activity, which leads to an increase in p53 acetylation and up-regulation of p53-mediated function. Depletion or down-regulation of DBC1 conversely stimulates SIRT1-mediated deacetylation of p53 and inhibits p53-dependent apoptosis. SIRT1 also positively self-regulates via a small region in the carboxy-terminal of the protein, which competes with DBC1 binding (Kang et al. 2011). Another nuclear protein called active regulator of SIRT1, or AROS, when bound, enhances SIRT1-mediated deacetylation of p53, and inhibits p53-mediated transcriptional activity (Kim et al. 2007). Depletion of AROS enhances p21 expression and increases both the G_0/G_1 population and apoptosis in response to DNA damage, whereas AROS overexpression improves cell survival. The activation of SIRT1 by protein–protein interaction with AROS is reminiscent of the activation of Sir2 deacetylase activity in *S. cerevisiae* by interaction with Sir4 (discussed in Ch. 8 [Grunstein and Gasser 2013]).

5.2 Posttranslational Modifications

HDACs can undergo a variety of posttranslational modifications including acetylation, glycosylation, *S*-nitrosylation, sumoylation, ubiquitination, and phosphorylation. The most extensively studied modification, phosphorylation, affects HDAC functionality (Fig. 6B,C). HDAC1 is phosphorylated by cAMP-dependent kinase PKA and protein kinase CK2 (Pflum et al. 2001). The two phosphoacceptor sites on the carboxyl-terminal of HDAC1, S421, and S423 are essential for enzymatic activity and when mutated to alanine result in a significant reduction in enzymatic activity. These mutations also disrupt protein complex formation of HDAC1 with RbAp48, MTA2, Sin3, and CoREST. HDAC2 is similarly phosphorylated at residues S394, S422, and S424 (corresponding to S393, S421, and S423 of HDAC1; Tsai and Seto 2002). Its phosphorylation also promotes enzymatic activity and affects protein complex formation with Sin3 and Mi2. Intriguingly, in cancer cells HDAC1 is phosphorylated at Y221 (which corresponds to Y222 of HDAC2; Rush et al. 2005). Because HDAC1-Y221 (HDAC2-Y222) is conserved in humans, mice, *Xenopus laevis*, and *Caenorhabditis elegans*, tyrosine phosphorylation may also be important in the regulation of HDAC1 and HDAC2 activities.

Phosphorylation of HDAC1 and HDAC2 is reversibly regulated by the protein phosphatase PP1 (e.g., Galasinski et al. 2002). Interestingly, mitotic arrest but not G_1/S arrested cells results in hyperphosphorylation of HDAC2 (without any change in HDAC1), suggesting that spindle checkpoint activation provides a physiological stimulus that leads to HDAC2 hyperphosphorylation.

Phosphorylation of HDAC3 by CK2 and DNA-PKcs significantly enhances HDAC3 activity (e.g., Zhang et al. 2005). HDAC3 can also be phosphorylated by GSK-3β, and inhibition of GSK-3β protects against HDAC3-induced neurotoxicity (Bardai and D'Mello 2011). HDAC3's phospho-acceptor site, S424, which is a nonconserved residue among the Class I HDACs, when mutated to alanine severely compromises enzymatic activity reminiscent of HDAC1 and HDAC2. Unlike HDAC1 and HDAC2, however, HDAC3 associates with the catalytic and regulatory subunits of the protein serine/threonine phosphatase 4 complex ($PP4_c/PP4_{R1}$), and dephosphorylation of HDAC3 by PP4 down-regulates HDAC3 enzymatic activity (Zhang et al. 2005).

HDAC8 regulation is quite different to other Class I enzymes. For one, phosphorylation inhibits, rather than increases its enzymatic activity (Lee et al. 2004). Mutation of S39 to alanine enhances HDAC8 activity, whereas an activator of HDAC8 phosphorylation causes a reduction in HDAC8 activity. Also, HDAC8 is refractory to phosphorylation by protein kinase CK2, but is instead phosphorylated by PKA. Crystal structures of HDAC8 revealed that S39 lies at the surface of HDAC8, roughly 20 Å from the opening to the HDAC8 active site (Somoza et al. 2004). Phosphorylation of S39 predictably leads to a major structural disruption of this region of the surface, which ultimately would negatively affect HDAC8's activity.

The phosphorylation state and modifying enzymes specific to Class IIa HDACs have been extensively studied and found to affect enzyme activity, partly through cellular localization. The subcellular localization is regulated by the binding of the 14-3-3 proteins to a phosphorylated HDAC4 (Fig. 6C). A working model proposes that phosphorylation of HDAC4 at residues S245, S467, and S632 induces 14-3-3 binding to HDACs, preventing access of importin to the nuclear localization signal on HDAC4. This results in cytoplasmic sequestration of HDACs. No less than six groups of kinases have been shown to phosphorylate the 14-3-3 binding sites (reviewed by Seto and Yang 2010): Ca^{2+}/calmodulin-dependent kinases, protein kinase D, microtubule affinity-regulating kinases, salt-inducible kinases, checkpoint kinase-1, and AMP-activated protein kinase (AMPK). In addition to nuclear export and cytoplasmic retention, phosphorylation of Class IIa HDACs may lead to their ubiquitination and proteasomal degradation.

Two phosphatases have been implicated in the regulation of Class IIa HDACs activities and functions: protein phosphatase 1β (PP1β), including myosin phosphatase targeting subunit 1 (MYPT1, a regulatory subunit of PP1), and protein phosphatase 2A (PP2A). In addition to 14-3-3 binding sites, PP2A also dephosphorylates S298 of HDAC4, which is required for nuclear import of HDAC4.

For Class IIb HDACs, global proteomic profiling of phosphopeptides, revealed that HDAC6 is phosphorylated at S22 and T30 (e.g., Beausoleil et al. 2004). However, the functional significance of phosphorylation on these two sites is unknown. Aurora A (AurA) phosphorylates HDAC6 to activate tubulin deacetylase activity (Pugacheva et al. 2007). At the basal body of cilia, phosphorylation and activation of HDAC6 promotes ciliary disassembly, and this pathway is both necessary and sufficient for ciliary resorption. In line with these observations, small molecule inhibitors of AurA and HDAC6 selectively stabilize cilia from regulated resorption cues. Other HDAC6 kinases include GSK3β, G protein–coupled receptor kinase 2, protein kinase CK2, and epidermal growth factor receptor.

For Class III HDACs, SIRT1 is a known phosphoprotein (Beausoleil et al. 2004) containing 13 residues that can be phosphorylated based on mass spectrometry data (Sasaki et al. 2008). Dephosphorylation of SIRT1 results in a decrease in SIRT1 deacetylase activity, arguing that SIRT1 activity is regulated by phosphorylation. Cell cycle–dependent kinase cyclin B/Cdk1 phosphorylates SIRT1, and mutations of two Cdk1 target sites, T530 and S540, disturbs normal cell cycle progression and fails to rescue proliferation defects in SIRT1-deficient cells. Mutation of the Cdk1 phosphorylation sites, however, did not lead to a reduction in deacetylase activity, suggesting that other phosphorylation sites might be more important for deacetylase activity. The protein levels of SIRT1 may be regulated by phosphorylation of SIRT1 at S27 (Ford et al. 2008). Cells depleted of c-JUN amino-terminal kinase 2 (JNK2), but not JNK1, in cells resulted in a SIRT1 protein with a reduced half-life and a lack of phosphorylation at S27. However, it is not clear in this study if the decrease in protein levels corresponds to a decrease in SIRT1 deacetylase activity in these cells. In contrast, a different study showed that JNK1 phosphorylates SIRT1 at S27, S47, and T530, and this phosphorylation of SIRT1 increased its nuclear localization and enzymatic activity (Nasrin et al. 2009). Like many Class I and II HDACs, SIRT1 is phosphorylated by CK2, and the target sites are S154, S649, S651, and S683 (Kang et al. 2009). In this case, phosphorylation of SIRT1 increases its deacetylation rate and its substrate-binding affinity. Also, CK2-mediated phosphorylation increases the ability of SIRT1 to deacetylate p53 and protects cells from apoptosis after DNA damage.

AMPK has been shown to increase intracellular NAD^+ levels, which in turn enhances SIRT1 deacetylation activity (e.g., Canto et al. 2009). However, in a later study, it was shown that activation of the cAMP signaling pathway induces rapid deacetylation of SIRT1 substrates independent of changes in NAD^+ levels (Gerhart-Hines et al. 2011). Activation of PKA phosphorylates S434, a residue located in the NAD^+ binding pocket of the catalytic domain of SIRT1 and is phylogenetically conserved across all Sir2 orthologs. Other kinases downstream of PKA might also be responsible for the phosphorylation of SIRT1.

In summary, HDAC activities are modulated by multiple mechanisms. Both protein–protein interactions and posttranslational modifications (particularly phosphorylation) fine-tune HDAC activities. Multisubunit complexes dictate the activity of many HDAC deacetylases and their substrate specificity. Likewise, multiple kinases and phosphatases, through a variety of signaling pathways, can up- or down-regulate HDAC activities. A thorough understanding of HDAC regulation will not only provide tremendous insights into histone and protein deacetylation but also potential diagnostic and therapeutic approaches for the treatment of diseases that result from abnormal acetylation/deacetylation.

6 BIOLOGICAL IMPORTANCE OF HDACs

At the molecular level, the most obvious biological importance of HDACs is to oppose the functions of HATs. HDACs are critical in maintaining a dynamic equilibrium of protein acetylation. HDACs also exert profound effects on other protein posttranslational modifications. Deacetylation of histones and nonhistones may change chromatin conformation or modify the activities of transcription factors leading to a change in gene expression. Significantly, the molecular changes induced by HDACs impact on human health and disease. Abnormal HDACs have been documented to play a key role in many human diseases including (but not limited to) cancer, neurological diseases, metabolic disorders, inflammatory diseases, cardiac diseases, and pulmonary diseases.

6.1 HDACs Indirectly Regulate Many Posttranslational Modifications

The lysine ε-amino group is prone to many different PTMs including acetylation, methylation, ubiquitination, sumoylation, neddylation, biotinylation, propionylation, butyrylation, and crotonylation (Tan et al. 2011). Introduction of

Cite as *Cold Spring Harb Perspect Biol* doi: 10.1101/cshperspect.a018713

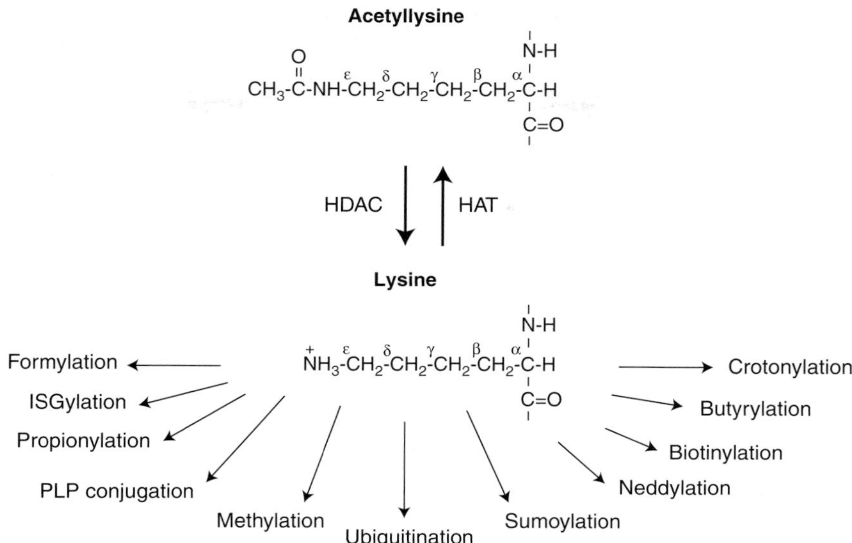

Figure 7. Examples of the many different potential posttranslational modifications on an ε-amino lysine subsequent to HDAC deacetylation.

an acetyl group, therefore, could exclude another modification on the same lysine residue within a protein (Fig. 7). An example of how lysine acetylation can interfere with other lysine modifications is the competition it often has with ubiquitination. Acetylation inhibits ubiquitination-dependent, proteasome-mediated protein degradation (Caron et al. 2005). Consequently, the stability of a number of proteins has been shown to increase after acetylation as a result of lysine site competition preventing ubiquitination. Conversely, removal of the acetyl group promotes lysine ubiquitination. Thus, one of the biological functions of HDACs is to accelerate protein degradation by exposing the ε-amino lysines for ubiquitination.

HDACs also play a key biological role in regulating histones (i.e., within a nucleosome) and chromatin (i.e., larger internucleosomal scale) cross talk. It has been fascinating to elucidate the many PTMs that communicate and interact with each other at multiple levels (e.g., see Fig. 10 in Ch. 3 [Allis et al.]). One of the best studied examples of histone modification cross talk occurs between acetylation and methylation (Fischle et al. 2003). Acetylation of histone H3K9 not only inhibits methylation of the same residue, but also promotes H3K4 methylation, which results in permissive chromatin and transcription activation. By logic, then, HDACs that deacetylate H3K9 will inhibit H3K4 methylation and ultimately repress transcription. Thus, HDACs not only provide access of previously acetylated lysines for another modification, but also promote histone and chromatin cross talk. Interestingly, several histone-modifying enzymes copurify with HDACs. For example, HDAC1, HDAC2, and G9a (a lysine methyltransferase) coordinate histone modifications from the same protein complex (Shi et al. 2003). Because thousands of combinations of histone modifications are possible, an abundance of regulatory potentials thus exist for HDACs. Cross-regulatory events between acetylation/deacetylation and other posttranslation modifications even extend to nonhistones (reviewed by Yang and Seto 2008). Furthermore, some HDAC proteins may possess other protein modification enzymatic activities in addition to deacetylase activity, suggesting that the effects of HDACs on protein modifications extend beyond the simple removal of acetyl groups from lysines.

6.2 HDACs Alter Gene Transcription

Early studies indicate that "relatively minor" histone modifications, particularly acetylation, could influence the rate of transcription (Allfrey et al. 1964). The thinking that emerged is that transcriptional activity is dictated, in part, by acetylation of nucleosomal histones. By conventional wisdom this would infer that the biological function of HDACs is to provide a reversible means of switching RNA synthesis off at different times and at different chromosomal loci. Mechanistically, it was believed that acetylation of core histones weakens its interaction with DNA, corroborated by the finding that acetylation markedly reduces the binding constant of the H4 tail to DNA in thermal denaturation studies (Hong et al. 1993). In contrast, deacetylation of histones by HDACs increases the positive charges on histones and, thereby, may strengthen histone–DNA interaction and repress transcription. Other, more recent evidence is accumulating in support of acetylation/

deacetylation generating specific docking surfaces for proteins. This, in turn, regulates transcription without a significant change in the overall electrostatic charge of histones. In other words, although hyperacetylated histones interact with transcription activators, HDACs may furnish deacetylated interaction sites for transcriptional repressors.

One of the first evidences that directly links HDACs to transcription repression came from the findings that Rpd3 is required for both full repression and full activation of transcription of target genes in yeast (Vidal and Gaber 1991). Subsequently, it was shown that a Gal4 DNA binding domain-HDAC2 (mRPD3) fusion protein strongly represses transcription from a promoter containing Gal4 binding sites (Yang et al. 1996). Similar observations were obtained using Gal4-HDAC1 (Yang et al. 1997), Gal4-HDAC3 (Yang et al. 1997), Gal4-HDAC7 (Kao et al. 2000), Gal4-HDAC9 (Zhou et al. 2000), and Gal4-HDAC10 (Kao et al. 2002). Based on these promoter-targeting reporter assay studies, it was suggested that the recruitment of HDAC enzymes is a very common, although maybe not universal, mechanism by which repressors and corepressors alter transcription. However, whether the deacetylase activity of these HDACs per se is necessary for this repression is unclear. In one study, HDAC5 and HDAC7 were shown to possess autonomous repressor functions that were independent of their deacetylase activity (Kao et al. 2000). Therefore, a "one-size-fits-all" model in which HDACs mediate transcriptional repression via deacetylation of histones may not always be correct. Using yeast Rpd3 mutants that lack detectable histone deacetylase activity, Kevin Struhl confirmed that histone deacetylase activity of Rpd3 is important, but not absolutely required, for transcriptional repression in vivo (Kadosh and Struhl 1998).

In a yeast genome-wide mapping study, using ChIP followed by analysis of the precipitated DNA on microarrays (ChIP-chip), Richard Young showed that histone acetylation is associated with transcription activity, and the modification occurs predominantly at the beginning of genes (Pokholok et al. 2005). In a set of complementary experiments, it was shown that in histone H4, mutation of K16 alone changed specific transcription, whereas mutations of K5, K8, and K12 change transcription nonspecifically in a cumulative manner (Dion et al. 2005). Consistent with this finding, it was reported that a single histone modification, acetylation of H4K16, modulates both higher-order chromatin structure and functional interactions between a nonhistone protein and the chromatin fiber (Shogren-Knaak et al. 2006). These results suggest that acetylation, at least for histone H4, is operated by two distinct mechanisms: one that is specific for H4K16 and a nonspecific mechanism for H4K5, H4K8, H4K12. By inference, then, HDACs may exert their effects on transcriptional repression depending on which particular lysine it deacetylates. That is, deacetylation of H4K16 may result in a significant global transcription repression, whereas deacetylation of H4K5, H4K8, or H4K12 alone has little outcome but together generates a cumulative effect on transcription.

It is commonly believed that HDACs are bound to repressed genes and are cyclically replaced by HATs on gene activation. However, in another genome-wide mapping study, HDACs were found bound to chromatin at active genes but not silent genes (Wang et al. 2009). HDAC1 and HDAC3 were mainly detected in promoters, whereas HDAC2 and HDAC6 localized to both promoter and gene body regions of active genes. The majority of these HDACs in the human genome function to reset chromatin by removing acetylation at active genes. Related to this unexpected observation is that HDACs do not always repress transcription. For example, although HDAC3 represses transcription when targeted to promoters and serves as a corepressor (Yang et al. 1997), paradoxically HDAC3 also is required for the transcriptional activation of at least one class of retinoic acid response elements (Jepsen et al. 2000). In cells derived from *Hdac3* knockout mice, both up-regulation and down-regulation of gene expression were detected (Bhaskara et al. 2008). Also, in gene expression profiling studies comparing cells treated and untreated with HDAC inhibitors, the number of genes down-regulated were comparable to up-regulated genes (e.g., LaBonte et al. 2009). One possibility is that HDACs may down-regulate transcription of transcriptional repressors, which leads to derepression of gene expression. Alternatively, HDACs may deacetylate and, consequently, activate transcription activators or inhibit the functions of transcription repressors independent of histone modifications. In summary, there is overwhelming evidence that a key biological function of HDACs is to modulate transcription, especially in repression. However, whether HDACs can directly activate transcription and the exact detailed mechanisms by which they regulate transcription still remain to be determined.

6.3 HDACs' Impact on Human Health and Diseases

One of the major motivations in studying HDACs is the expectation that an understanding of deacetylases will increase our understanding of histone modifications, chromatin biology, transcription regulation, and of course epigenetics. Equally important, however, is the need to understand the relevance of deacetylases in health and disease. Given the significant number of genes whose expression is regulated by HDACs, coupled with the finding that HDACs modulate the function of many proteins through nonhistone deacetylation, HDACs potentially play a role in

Cite as *Cold Spring Harb Perspect Biol* doi: 10.1101/cshperspect.a018713

nearly every aspect of health and disease. Numerous reports document the involvement of HDACs in cancer, neurodegenerative diseases, metabolic disorders, inflammatory diseases, immunological disorders, cardiac diseases, and pulmonary diseases. Because it is impractical to present a comprehensive discussion of HDACs in human diseases here, only a few examples are given.

The notion that HDACs are important in normal development, and that abnormal HDACs lead to disease is reinforced in HDAC knockout mice studies. For example, *Hdac2*-, *Hdac5*-, or *Hdac9*-null animals have cardiac defects (reviewed by Haberland et al. 2009). Conditional deletions of *Hdac3* reveal that HDAC3 is important in liver homeostasis and heart functions. *Hdac4* knockout mice expose the importance of HDAC4 in skeleton formation, and *Sirt1*-deficient mice have retinal, bone, and cardiac defects (e.g., McBurney et al. 2003).

Of the many roles HDACs play in human diseases, the most frequently discussed is cancer. Many correlative studies link cancer to epigenetic abnormalities. Somatic mutations in DNA and histone-modifying enzymes contribute to human malignancies (Dawson and Kouzarides 2012), and HDACs are no exception. For example, a frameshift mutation of HDAC2 has been found in sporadic carcinomas with microsatellite instability and in tumors arising in individuals with hereditary nonpolyposis colorectal cancer (Ropero et al. 2006). This mutation causes a loss of HDAC2 protein expression and enzymatic activity and renders these cells more resistant to the usual antiproliferative and proapoptotic effects of HDAC inhibitors. Loss of function HDAC2 mutations may derepress key genes from multiple cellular transformation pathways. HDACs can also be involved in cancer when recruited to specific loci in which repressive complexes are formed. This is the case for mutations in several oncogenic proteins generated by chromosome translocation such as PML-RARα and AML1-ETO fusion proteins that recruit HDACs (reviewed by Cress and Seto 2000).

Many studies have reported increased or decreased HDAC levels in various tumors compared to normal tissues. Although most of these studies focused on quantification of HDAC messenger RNA or proteins and not on HDAC enzymatic activities, conceivably any changes in HDACs can lead to changes in histone acetylation states, which in turn may lead to increased transcription of oncogenes or growth promoting factors, as well as a decrease in transcription of tumor suppressors or antiproliferative factors. It is widely known that HDACs can alter the expression of many cell cycle regulators. For example, early studies suggested that the cyclin-dependent kinase inhibitor p21 is an HDAC-response gene, and subsequent studies show that p21 expression is inversely correlated with HDAC2 in colorectal cancer cells (Huang et al. 2005). Some of the contributions that HDACs play in the development and progression of cancer may also be mediated through deacetylation of nonhistone proteins, including many oncogenes, tumor suppressors, and proteins that regulate tumor cell invasion and metastasis (reviewed by Glozak et al. 2005).

Like cancer, synaptic plasticity and cognition disorders can be manifestations of epigenetic dysregulation. HDAC inhibitors are therefore actively being pursued as potential therapeutics for neurodegenerative disorders. For example, SIRT2 deacetylates α-tubulin, and may promote α-synuclein toxicity and modified inclusion morphology in Parkinson's disease (Outeiro et al. 2007). Other HDACs are likely to be involved in these disease etiologies, as administration of two HDAC inhibitors, sodium butyrate and SAHA (which notably do not affect SIRT2 activity), also protect against α-synuclein-dependent neurotoxicity in a Parkinson's disease transgenic fly model. In the studies of Alzheimer's disease, mice overexpressing HDAC2 result in decreased synaptic plasticity, synapse number, and memory formation, and SAHA rescues the synaptic number and learning impairments in HDAC2-overexpressing mice. Conversely, Hdac2 deficiency results in increased synapse number and memory facilitation similar to chronic treatment with HDAC inhibitors in mice (e.g., Graff et al. 2012). HDAC6 protein levels in Alzheimer's disease brains is significantly increased, and it interacts with tau, a microtubule-associated protein that forms neurofibrillary tangles in Alzheimer's disease (Ding et al. 2008). Furthermore, SIRT1 is significantly reduced in the parietal cortex of Alzheimer's disease patients, and the accumulation of Aβ and tau in these patients may be associated with the loss of SIRT1 (Gao et al. 2010). HDACs may be linked to other neurological disorders, particularly psychiatric conditions including depression, anxiety disorders, and schizophrenia, as HDAC inhibitors show some potential promise in the treatments of these diseases.

There is much evidence to support the fact that heart diseases are linked to abnormal HDAC expression or activity (reviewed by Haberland et al. 2009). For example, HDAC9 is highly expressed in cardiac muscle, and one of the targets of HDAC9 is the transcription factor MEF2, which has been implicated in cardiac hypertrophy. In a series of elegant studies, Eric Olson showed that activation of the cardiac myocyte fetal gene program by a range of potent hypertrophic inducers could be blocked by expressing mutated HDAC9/MITR. Furthermore, mutant mice lacking HDAC9 are sensitized to hypertrophic signals and show stress-dependent cardiomegaly. Mice lacking HDAC5 display a similar cardiac phenotype and develop profoundly enlarged hearts in response to pressure overload, suggesting redundant functions of HDAC5 and HDAC9 in the control of cardiac development (Chang et al. 2004). Additionally, results from two independent studies clearly suggested that HDAC2 has an important role in heart biology (e.g., Trivedi et al. 2007).

HDACs also play an important role in inflammation and lung diseases. Perhaps this is best illustrated by the observation that HDAC2 expression and activity are reduced in lung macrophages, biopsies, and blood cells from patients with chronic obstructive pulmonary disease (COPD) and asthma (Ito et al. 2005). This decrease in expression and activity correlates with disease severity and the intensity of the inflammatory response. In patients with very severe COPD, the reduction in HDAC activity is accompanied by an increase in H4 acetylation at the *IL-8* promoter. These results, taken together, indicate that HDAC2 is a key regulator of *IL-8* gene transcription in inflammatory lung diseases. Results from a more recent study show that HDAC3 may be responsible for activation of almost half of the inflammatory gene expression under lipopolysaccharide-stimulation in macrophages, again underscoring the importance of HDACs in inflammation (see Ch. 29 [Busslinger and Tarakhovsky 2014] for more discussion on the epigenetics of the inflammatory response; Chen et al. 2012).

7 INHIBITORS

7.1 Discovery of HDAC Inhibitors

During biochemical analysis of HDAC activity in nuclei in the 1970s, it was found that millimolar concentrations of *n*-butyrate induced accumulation of acetylated histones in cells (Riggs et al. 1977). Soon after this finding, it was reported that *n*-butyrate inhibits deacetylation (e.g., Candido et al. 1978). Unfortunately, however, a causal relationship between the phenotypic consequence, such as cell cycle inhibition, and histone hyperacetylation induced by *n*-butyrate was doubted because of its nonspecific action on other enzymes and membranes. In 1990, potent HDAC inhibition by the natural product trichostatin A (TSA) was discovered. TSA was isolated from a *Streptomyces* strain and was originally identified as an antifungal antibiotic and a powerful inducer of murine erythroleukemia cell differentiation (Yoshida et al. 1987). TSA inhibited the activity of partially purified HDACs with a low nanomolar inhibition constant. Importantly, TSA-resistant mutant cells possessed a TSA-resistant HDAC enzyme providing genetic evidence that HDACs are the primary TSA target responsible for the cell cycle inhibition (Yoshida et al. 1990). TSA has a hydroxamic acid group, which can chelate a metal ion. Trapoxin, a fungal cyclic peptide that had been identified as an inducer of morphological change in transformed cells, was also found to strongly inhibit HDAC (Kijima et al. 1993). Unlike TSA, trapoxin irreversibly inhibits HDAC activity depending on its epoxyketone moiety. The potent ability of trapoxin to bind to HDACs was used for isolating the first HDAC protein (HDAC1) by means of trapoxin-affinity matrix (Taunton et al. 1996).

In 1998, two clinically important HDAC inhibitors were reported: SAHA, which had been designed and synthesized as a hybrid polar compound that strongly induces erythroid differentiation (Richon et al. 1998); and FK228 (romidepsin), an antitumor cyclic depsipeptide isolated from *Chromobacterium violaceum* (Nakajima et al. 1998). Like TSA, the zinc-interacting group of SAHA is its hydroxamic acid. On the other hand, there was no apparent zinc-interacting group in FK228. FK228 has an intramolecular disulfide bond, which is readily reduced in cells by the cellular reducing activity yielding a thiol side chain that coordinates to the active site zinc (Furumai et al. 2002). Although in vitro inhibition by FK228 was reversible, fluorescent live imaging of histone acetylation revealed that FK228 action in cells was sustained for several hours after removal, suggesting that efflux of the reduced form of FK228 is not efficient (Ito et al. 2011). Phase I clinical trials conducted by Susan Bates at the National Cancer Institute revealed that FK228 is effective for the therapy of cutaneous and peripheral T-cell lymphoma. This finding consequently accelerated the development of HDAC inhibitors as anticancer drugs. SAHA (Vorinostat) was the first HDAC inhibitor approved for cancer chemotherapy in 2006 (Bolden et al. 2006). Following SAHA, FK228 (Istodax) was also approved in 2009. Another class of clinically important HDAC inhibitors is benzamides, which includes CI-994 and MS-275 (Entinostat). MS-275, a synthetic benzamide derivative with HDAC inhibitory activity (Saito et al. 1999), showed moderate in vitro inhibitory activity, but the activity in cells was relatively strong because of the slow binding and time-dependent inhibition by the benzamide class of inhibitors (Bressi et al. 2010).

Figure 8 shows the large number of structurally diverse HDAC inhibitors. These have been discovered from natural sources or synthetically developed, and many are being studied clinically.

7.2 Mechanisms of HDAC Inhibition

HDAC inhibitors can be divided into four classes based on their chemical structures: hydroxamate, short-chain fatty acid (carboxylate), benzamide, and cyclic peptide (Fig. 8). The most studied class is hydroxamate inhibitors. Structural studies of TSA or SAHA (hydroxamate class inhibitors) cocrystallized with the HDLP HDAC showed that the inhibitors bind by inserting their long aliphatic chain into

 Cite as *Cold Spring Harb Perspect Biol* doi: 10.1101/cshperspect.a018713

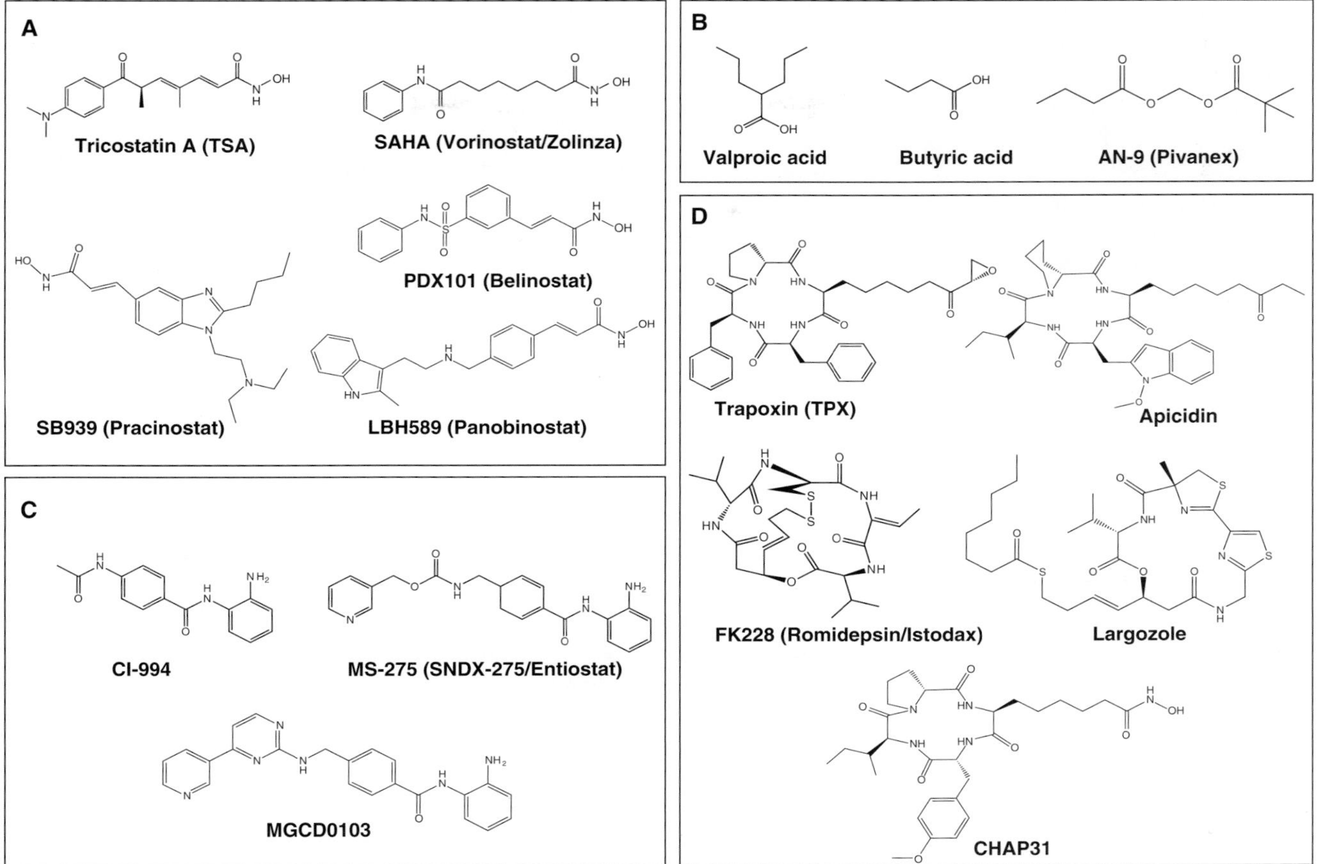

Figure 8. HDAC inhibitors. Four classes of HDAC inhibitors categorized according to their chemical structures are shown. (*A*) Hydroxamic acid, (*B*) short-chain fatty acid, (*C*) benzamide, and (*D*) cyclic peptide.

the HDLP pocket with multiple contacts to the tubular pocket. Also, the hydroxamic acid coordinates the zinc ion in a bidentate fashion through its carbonyl and hydroxyl groups (Fig. 9A). The hydroxamic acid also makes hydrogen bonds with both histidines and tyrosines located around the zinc, thereby replacing the zinc-bound water molecule with its hydroxy group. On the other hand, the aromatic ring group contacts residues at the rim of the pocket, which allows inhibitors to lock in the pocket.

The crystal structure of a benzamide inhibitor complexed with HDAC2 showed that the inhibitor coordinate (i.e., binds) to the catalytic zinc ion through both the carbonyl and amino groups to form a chelate complex (Fig. 9B). A simulation suggests that the transiently bound forms can be converted to the tightly bound, pseudoirreversible form over time, providing a basis for a time-dependent inhibition by this class of inhibitors (Bressi et al. 2010). Long-lasting effects on cellular histone acetylation levels after removal of MS-275 also suggest tight binding to the enzymes (Ito et al. 2011).

The cyclic peptide class contains different zinc-binding groups such as electrophilic ketones and thiols. Although crystal structure of the HDAC-FK228 (a cyclic peptide inhibitor) complex has not yet been solved, computer-modeling studies suggest that one of the thiol groups generated by reduction can coordinate to the active site zinc ion (Furumai et al. 2002). Recently, a similar inhibitory mechanism was proven by the crystal structure of HDAC8 with hydrolyzed largazole (Cole et al. 2011). Largazole has a thioester moiety, which can be hydrolyzed in cells to give an active thiol side chain (Fig. 8D). The HDAC8 structure complexed with the hydrolyzed largazole revealed that the thiol side chain coordinates to the catalytic zinc ion to inhibit the HDAC activity (Fig. 9C).

7.3 Target Enzyme Selectivity

TSA and SAHA, prototypical hydroxamate class inhibitors, are pan-HDAC inhibitors that block all Class I, II, and IV enzymes. The sensitivity to these inhibitors varies widely

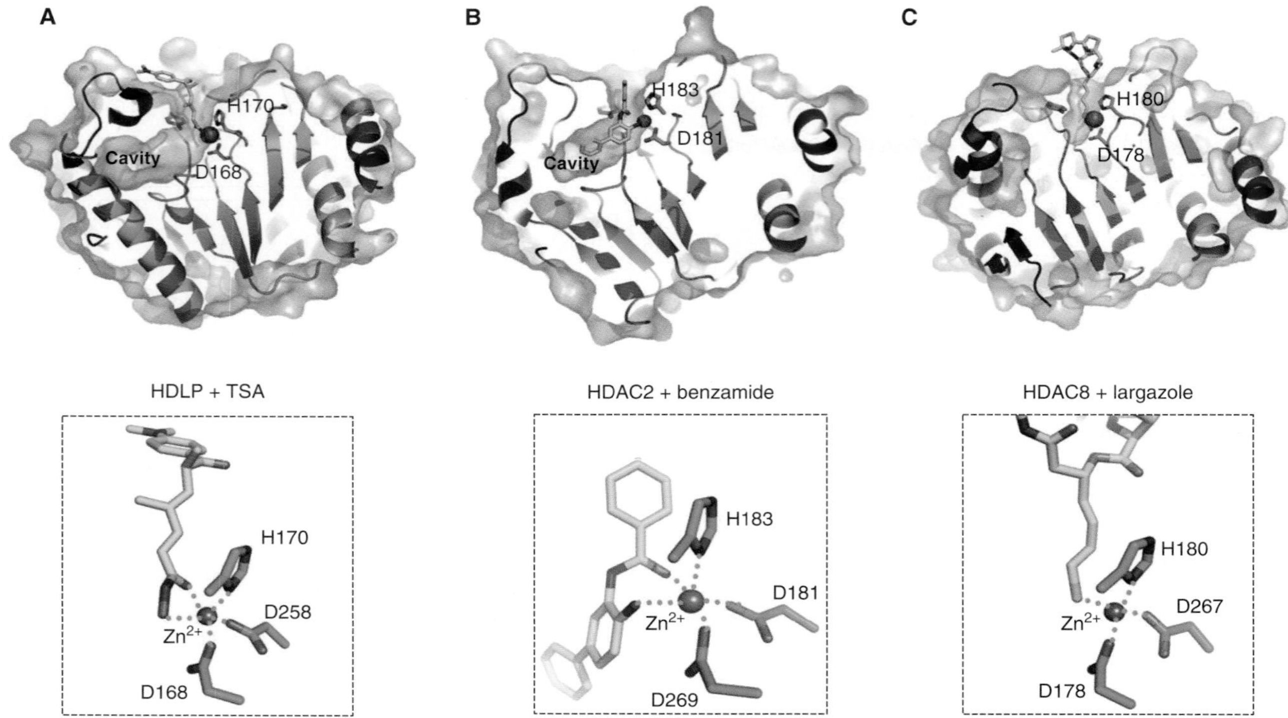

Figure 9. Crystal structure of HDAC proteins complexed with inhibitors. Molecular surfaces of the inhibitor-bound HDLP/HDACs cut at the level of the cavity are illustrated. (*A*) HDLP with TSA (PDB ID: 1C3R), (*B*) HDAC2 with benzamide (3MAX), and (*C*) HDAC8 with largazole (3RQD). The outer surfaces are depicted in light brown, and zinc atoms are shown in space-filled spheres in gray. Close-up views of the active sites are shown in the lower panels. Inhibitors are depicted in a stick model, in which yellow, blue, red, and gold represent C, N, O, and S atoms, respectively. Three Zn-coordinating residues (two aspartic acid and one histidine residue) are shown in a stick model. Hydrogen bond interactions are drawn in dotted lines.

among different HDACs, with some HDAC activities inhibited only above pharmacologically relevant concentrations (Bradner et al. 2010). On the other hand, *n*-butyrate, trapoxin, FK228, and MS-275 cannot inhibit HDAC6 (a Class IIb enzyme), suggesting a difference in structure of the catalytic pockets between Class IIb and other enzymes (Matsuyama et al. 2002). Indeed, the different sensitivity of HDAC6 to TSA and trapoxin was used for identifying α-tubulin as an important HDAC6 substrate (Hubbert et al. 2002; Matsuyama et al. 2002). Benzamide class inhibitors such as MS-275 and MGCD0103 preferentially inhibit Class I enzymes except HDAC8. The class selective inhibition by nonhydroxamate inhibitors may be because of subtle differences in the structure of the catalytic pocket or the internal cavity among the HDAC classes. Importantly, benzamide inhibitors can interact with the cavity residues (Fig. 9B). Although Class IIa enzymes such as HDAC4 showed weak intrinsic deacetylase activity toward canonical HDAC substrates containing acetyllysine, trifluoroacetyllysine was found to be a Class IIa-specific substrate in vitro (Lahm et al. 2007). Based on this finding, 2-trifluoroacetylthiophene derivatives were designed as selective inhibitors for Class IIa HDACs.

Development of selective inhibitors is very challenging because of the high homology in the active site structure and catalytic mechanism of HDACs within each class. The first selective HDAC inhibitor was tubacin, which was obtained from a high-throughput screening campaign for an HDAC6 inhibitor that increases tubulin acetylation but not histone acetylation (Haggarty et al. 2003). Another strategy comparing HDAC8 complexed with different inhibitors gave insights into the design of a specific HDAC8 inhibitor (Somoza et al. 2004). Based on the flexible structure of the surface around the active site, an HDAC8-specific inhibitor, PCI-34051, was synthesized, which induced apoptosis in T-cell lymphoma without increasing histone and α-tubulin acetylation (Balasubramanian et al. 2008). In another approach, novel biaryl derivatives of benzamide inhibitors (SHI-1:2), which were designed based on homology modeling of HDAC1 and HDAC3, showed HDAC1/HDAC2-selective inhibitory activity. The SHI-1:2 inhibitors may access the internal cavity with dif-

 Cite as *Cold Spring Harb Perspect Biol* doi: 10.1101/cshperspect.a018713

ferent shapes between HDAC1 and HDAC3 (Methot et al. 2008).

The first generation of clinical HDAC inhibitors was largely nonselective. Considering the diverse functions of Class I, II, and IV HDACs, it is obviously important to develop class- and HDAC-selective inhibitors for the next-generation of therapeutic HDAC inhibitors.

7.4 Biological Activity of HDAC Inhibitors

Treatment of tumor cells with anticancer HDAC inhibitors changes the expression of $\sim$10% of the total genes, including many important regulatory genes that control the cell cycle and apoptosis. The most remarkable one is the p53-independent increase in p21, a CDK inhibitory protein, and GADD45, a regulator of DNA repair and senescence. The induction of p21 might be responsible for the HDAC inhibitor-induced G_1 cell cycle arrest, with hypophosphorylation of pRb. In addition, repression of the cyclin D and cyclin A genes has been observed in a wide range of tumor cells on treatment, which is probably responsible in part for cell cycle inhibition. The therapeutic potential of HDAC inhibitors mainly comes from their capacity to selectively induce apoptosis in cancer cells. HDAC inhibitors activate expression of death receptors and their ligands in particular tumor cells. Furthermore, expression of proapoptotic BCL2 family proteins such as Bim or Bmf has been detected. Thus, both extrinsic and intrinsic apoptosis pathways, as well as reactive oxygen species production, may be involved in the HDAC-inhibitor-mediated apoptosis, although how important the activation of in vitro apoptosis pathways is for the in vivo therapeutic efficacy remains undefined (Bolden et al. 2006).

Tumor angiogenesis and cancer cell metastasis/invasion are also affected by HDAC inhibitors. Angiogenesis is an essential component of tumor growth and survival. The antiangiogenic activity of HDAC inhibitors has been associated with decreased expression of pro-angiogenic genes such as vascular endothelial growth factor, hypoxia-inducible factor-1α (HIFα), and chemokine receptor 4. Matrix metalloproteases (MMPs) also play a critical role in cancer metastasis/invasion. HDAC inhibitors can inhibit cancer cell invasion by up-regulating RECK, a tumor suppressor, to inhibit MMP-2 activation.

The increase in acetylation and the functional alteration of nonhistone substrates are likely to be involved in the biological activity of HDAC inhibitors. For example, the increase in p53 tumor-suppressor protein acetylation increases the stability and DNA-binding capacity, leading to the enhanced expression of p53 target genes and apoptosis. Acetylation of HIFs represses the function of HIF by either reducing the protein level or decreasing the transactivation activity (Chen and Sang 2011). Cortactin is an F-actin binding protein controlling cell motility and invasion by remodeling actin filaments in the cell cortex. Acetylation occurs in the F-actin binding repeat domain, which attenuates actin binding and cell migration activity. HDAC6 is responsible for this deacetylation and thus influences actin-dependent motility in cancer cells (Zhang et al. 2007). Conceivably, HDAC6 inhibitors might alter cancer cell migration through this pathway.

Compared to conventional anticancer drugs, adverse effects of clinically available HDAC inhibitors are marginal. Dose-limiting toxicities include cardiac arrhythmia, thrombocytopenia, nausea, and fatigue, which are clinically manageable. As cardiac abnormalities were observed in *Hdac5* null or *Hdac9* null mice (Chang et al. 2004), it seems possible that cardiac arrhythmia in patients treated with HDAC inhibitors is related to cardiac function of these Class II HDACs.

Owing to the varied functions of HDAC proteins themselves, untangling the mechanisms and targets by which HDAC inhibitors bring about selective improvement of certain human disease is complex.

7.5 Sirtuin Inhibitors

Sirtuins, the Class III HDACs, are involved in a number of cellular processes including insulin secretion, the cell cycle, and apoptosis. A great deal of effort has been put into developing sirtuin inhibitors as potential therapeutics. Nicotinamide is widely used as a global inhibitor of sirtuins, although high concentrations are required for inhibition of activity in cell culture. The nicotinamide molecule is, in fact, a byproduct of the sirtuin enzyme reaction, and thus a physiological inhibitor that decreases gene silencing, increases rDNA recombination, and accelerates aging in yeast. In search of a specific sirtuin inhibitor, there has been much focus on SIRT1 as it importantly regulates mitochondrial biogenesis and metabolic pathways, cellular redox, angiogenesis, and notch signaling. Malfunction of SIRT1 may contribute to diabetes, obesity, abnormal cancer metabolism, cancer stemness, and neurological disorders. These diseases represent a huge burden to our societies' health and healthcare systems, therefore sirtuin inhibitors represent hopeful therapeutics for many of these diseases.

The first synthetic sirtuin inhibitors, sirtinol and splitomicin, were identified through a screen based on telomere silencing in yeast (e.g., Bedalov et al. 2001). Following these, a number of compounds including cambinol, salermide, tenovin, EX-527, AGK2, etc. have been reported as sirtuin inhibitors (Fig. 10). The detailed mechanisms of how these compounds modulate sirtuin activity have

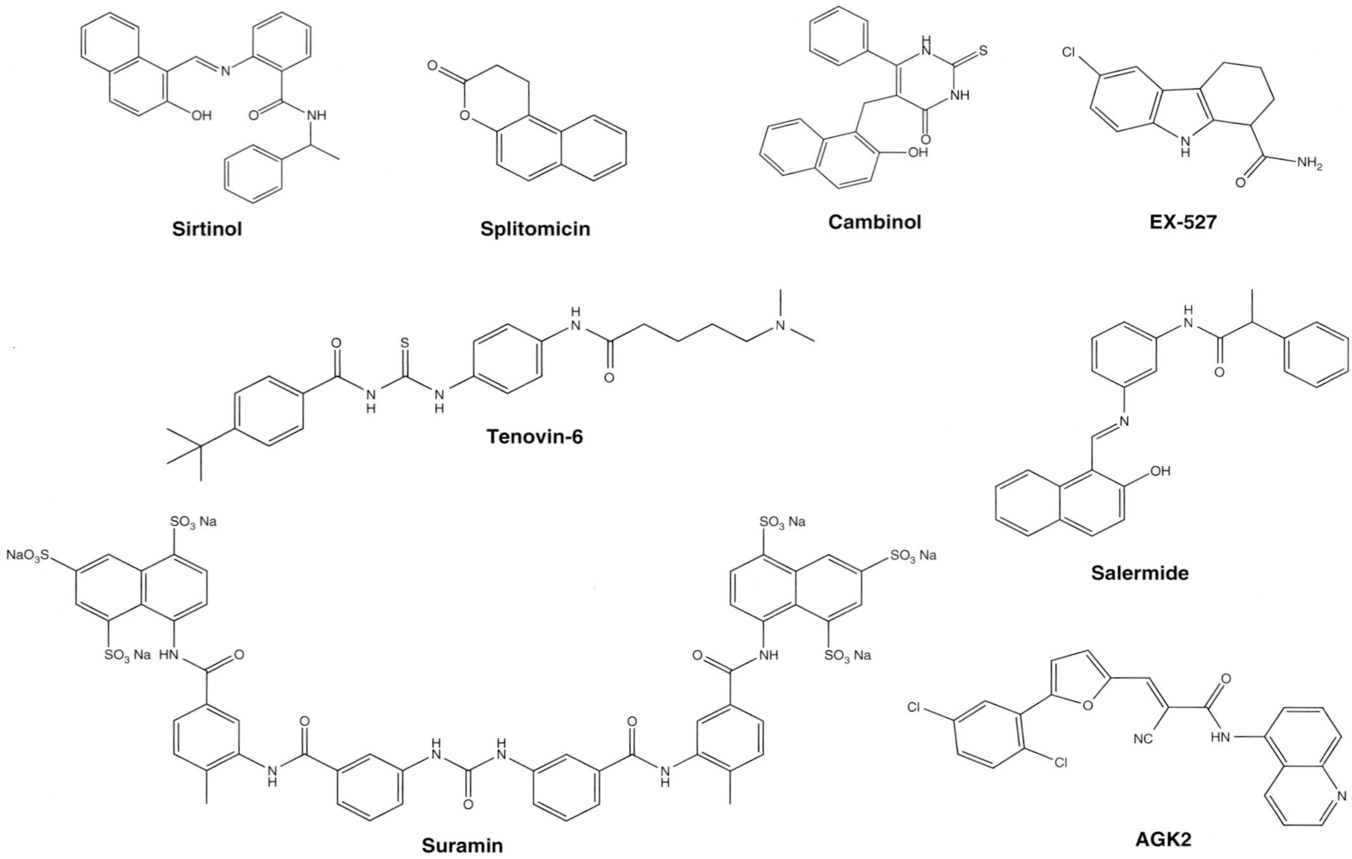

Sirtinol

Splitomicin

Cambinol

EX-527

Tenovin-6

Salermide

Suramin

AGK2

Figure 10. Sirtuin inhibitors. Sirtinol and splitomicin are identified as the first small molecule sirtuin inhibitors that affect telomere silencing in yeast. Cambinol is a splitomicin-related β-naphthol that is more stable than splitomicin and increases p53 acetylation showing antitumor activity in BCL6-expressing Burkitt's lymphoma. Salermide was designed based on the structure of sirtinol by molecular modeling and inhibits both SIRT1 and SIRT2 more effectively than sirtinol. Tenovin-1 and its water-soluble analog tenovin-6 induce p53 acetylation; their cellular targets were determined to be SIRT1 and SIRT2. A high-throughput screen revealed a number of indole compounds including EX-527, which selectively inhibits SIRT1 over SIRT2. Kinetic analysis suggests that EX-527 binds to the nicotinamide-binding site. AGK2, which was reported as a SIRT2 selective inhibitor, shows more than 10-fold selective inhibition relative to SIRT1 and SIRT3. Suramin, which was originally developed for treating trypanosomiasis and onchocerciasis, inhibits the NAD^+-dependent deacetylase activity of sirtuins by inducing sirtuin dimerization.

mostly not yet been determined because structural information about their binding sites is currently limited. Sirtuin inhibitors may be roughly divided into two groups, those interacting with the NAD^+ (nicotinamide and ADP-ribose) binding site and those interacting with the acetyllysine binding site. Suramin, for example, was shown to inhibit the NAD^+-dependent deacetylase activity of sirtuins (Fig. 10) (Schuetz et al. 2007). The crystal structure of suramin complexed with SIRT5 showed that the symmetrical suramin structure induces dimerization of SIRT5, mediated through interacting with both the NAD^+ and acetylated peptide binding sites.

Several acetyllysine analogs have been shown to inhibit sirtuins as a mechanism-based inhibitor. For instance, thioacetyl peptide acts as a tight-binding inhibitor that stalls reaction at the ADP-ribose-peptide enzyme intermediate by forming 1′-S-alkylamidate (Smith and Denu 2007). Also, an ethyl malonyllysine-based small molecule was shown to inhibit SIRT1 and produce a covalent conjugate that may occupy the binding sites for NAD^+ and acetyllysine (Asaba et al. 2009). A variety of these acetyllysine analogs have been useful for dissecting the catalytic mechanism of NAD^+-dependent protein deacetylases as the mechanistic probes.

In agreement with the diverse function of sirtuins, these inhibitors show a variety of biological activities. Tenovin-6 induces p53 acetylation, inhibits tumor growth, and eliminates cancer stem cells in chronic myelogenous leukemia

Cite as *Cold Spring Harb Perspect Biol* doi: 10.1101/cshperspect.a018713

(CML) in combination with Imatinib, a BCR-ABL tyrosine kinase inhibitor (Li et al. 2012). Importantly, SIRT2 inhibition by AGK2 rescued toxicity caused by α-synuclein (α-Syn) insoluble fibril aggregation seen in synucleinopathies such as Parkinson's disease. It did this by decreasing the number through increasing the size of aggregates in a cellular model of Parkinson's disease. Thus, development of small molecules modulators of sirtuin activity has become one of the most active areas in drug discovery.

8 SUMMARY

Today, we know that there are at least 18 human HDAC proteins. They contribute to the erasure of epigenetic modifications, help establish epigenetic off chromatin states, and regulate heritable changes in gene expression. How HDACs achieve these ends is a question of immediate interest to many working in this field. The finding that HDACs function as transcriptional repressors and corepressors has ushered a surge of interest in this subject, and reports of genes that are regulated by HDACs are continuously expanding. Global expression profiling experiments estimate that the transcription of 10% of genes are regulated by HDACs. However, questions of how each HDAC uniquely regulates a specific set of genes remain largely unexplored. Similarly, although we know that there are potentially more than a thousand HDAC substrates, the biological consequences of deacetylation of these substrates require intense research. The mechanistic details of how histone deacetylation promotes modification cross talk, and the global consequences of deacetylation of other histone and epigenetic-modifying enzymes are ripe for further research.

HDAC inhibitors show promise in the treatment of cancer, inflammation, and neurological diseases. Many laboratories are actively working on developing more effective HDAC inhibitors with minimal side effects in patients. Who would have thought that a humble beginning in a biochemistry curiosity on an enzymatic activity purified from calf thymus would lead to a better understanding of epigenetics and have such a potential for impacting the improvement of human health?

REFERENCES

*Reference is also in this book.

Allfrey VG, Faulkner R, Mirsky AE. 1964. Acetylation and methylation of histones and their possible role in the regulation of RNA synthesis. *Proc Natl Acad Sci* **51:** 786–794.

Asaba T, Suzuki T, Ueda R, Tsumoto H, Nakagawa H, Miyata N. 2009. Inhibition of human sirtuins by in situ generation of an acetylated lysine-ADP-ribose conjugate. *J Am Chem Soc* **131:** 6989–6996.

Avalos JL, Boeke JD, Wolberger C. 2004. Structural basis for the mechanism and regulation of Sir2 enzymes. *Mol Cell* **13:** 639–648.

Ayer DE. 1999. Histone deacetylases: Transcriptional repression with SINers and NuRDs. *Trends Cell Biol* **9:** 193–198.

Balasubramanian S, Ramos J, Luo W, Sirisawad M, Verner E, Buggy JJ. 2008. A novel histone deacetylase 8 (HDAC8)-specific inhibitor PCI-34051 induces apoptosis in T-cell lymphomas. *Leukemia* **22:** 1026–1034.

Barber MF, Michishita-Kioi E, Xi Y, Tasselli L, Kioi M, Moqtaderi Z, Tennen RI, Paredes S, Young NL, Chen K, et al. 2012. SIRT7 links H3K18 deacetylation to maintenance of oncogenic transformation. *Nature* **487:** 114–118.

Bardai FH, D'Mello SR. 2011. Selective toxicity by HDAC3 in neurons: Regulation by Akt and GSK3β. *J Neurosci* **31:** 1746–1751.

Beausoleil SA, Jedrychowski M, Schwartz D, Elias JE, Villen J, Li J, Cohn MA, Cantley LC, Gygi SP. 2004. Large-scale characterization of HeLa cell nuclear phosphoproteins. *Proc Natl Acad Sci* **101:** 12130–12135.

Bedalov A, Gatbonton T, Irvine WP, Gottschling DE, Simon JA. 2001. Identification of a small molecule inhibitor of Sir2p. *Proc Natl Acad Sci* **98:** 15113–15118.

Bhaskara S, Chyla BJ, Amann JM, Knutson SK, Cortez D, Sun ZW, Hiebert SW. 2008. Deletion of histone deacetylase 3 reveals critical roles in S phase progression and DNA damage control. *Mol Cell* **30:** 61–72.

Bolden JE, Peart MJ, Johnstone RW. 2006. Anticancer activities of histone deacetylase inhibitors. *Nat Rev Drug Discov* **5:** 769–784.

Brachmann CB, Sherman JM, Devine SE, Cameron EE, Pillus L, Boeke JD. 1995. The SIR2 gene family, conserved from bacteria to humans, functions in silencing, cell cycle progression, and chromosome stability. *Genes Dev* **9:** 2888–2902.

Bradner JE, West N, Grachan ML, Greenberg EF, Haggarty SJ, Warnow T, Mazitschek R. 2010. Chemical phylogenetics of histone deacetylases. *Nat Chem Biol* **6:** 238–243.

Bressi JC, de Jong R, Wu Y, Jennings AJ, Brown JW, O'Connell S, Tari LW, Skene RJ, Vu P, Navre M, et al. 2010. Benzimidazole and imidazole inhibitors of histone deacetylases: Synthesis and biological activity. *Bioorg Med Chem Lett* **20:** 3138–3141.

* Busslinger M, Tarakhovsky S. 2014. Epigenetic control of immunity. *Cold Spring Harb Perspect Biol* **6:** a019307.

Candido EP, Reeves R, Davie JR. 1978. Sodium butyrate inhibits histone deacetylation in cultured cells. *Cell* **14:** 105–113.

Canto C, Gerhart-Hines Z, Feige JN, Lagouge M, Noriega L, Milne JC, Elliott PJ, Puigserver P, Auwerx J. 2009. AMPK regulates energy expenditure by modulating NAD^+ metabolism and SIRT1 activity. *Nature* **458:** 1056–1060.

Caron C, Boyault C, Khochbin S. 2005. Regulatory cross-talk between lysine acetylation and ubiquitination: Role in the control of protein stability. *Bioessays* **27:** 408–415.

Chang S, McKinsey TA, Zhang CL, Richardson JA, Hill JA, Olson EN. 2004. Histone deacetylases 5 and 9 govern responsiveness of the heart to a subset of stress signals and play redundant roles in heart development. *Mol Cell Biol* **24:** 8467–8476.

Chen S, Sang N. 2011. Histone deacetylase inhibitors: The epigenetic therapeutics that repress hypoxia-inducible factors. *J Biomed Biotechnol* **2011:** 197946.

Chen X, Barozzi I, Termanini A, Prosperini E, Recchiuti A, Dalli J, Mietton F, Matteoli G, Hiebert S, Natoli G. 2012. Requirement for the histone deacetylase Hdac3 for the inflammatory gene expression program in macrophages. *Proc Natl Acad Sci* **109:** E2865–E2874.

Choudhary C, Kumar C, Gnad F, Nielsen ML, Rehman M, Walther TC, Olsen JV, Mann M. 2009. Lysine acetylation targets protein complexes and co-regulates major cellular functions. *Science* **325:** 834–840.

Cole KE, Dowling DP, Boone MA, Phillips AJ, Christianson DW. 2011. Structural basis of the antiproliferative activity of largazole, a depsipeptide inhibitor of the histone deacetylases. *J Am Chem Soc* **133:** 12474–12477.

Cress WD, Seto E. 2000. Histone deacetylases, transcriptional control, and cancer. *J Cell Physiol* **184:** 1–16.

Dawson MA, Kouzarides T. 2012. Cancer epigenetics: From mechanism to therapy. *Cell* **150:** 12–27.

Ding H, Dolan PJ, Johnson GV. 2008. Histone deacetylase 6 interacts with the microtubule-associated protein τ. *J Neurochem* **106:** 2119–2130.

Dion MF, Altschuler SJ, Wu LF, Rando OJ. 2005. Genomic characterization reveals a simple histone H4 acetylation code. *Proc Natl Acad Sci* **102:** 5501–5506.

Du J, Zhou Y, Su X, Yu JJ, Khan S, Jiang H, Kim J, Woo J, Kim JH, Choi BH, et al. 2011. Sirt5 is a NAD-dependent protein lysine demalonylase and desuccinylase. *Science* **334:** 806–809.

Finnin MS, Donigian JR, Cohen A, Richon VM., Rifkind RA, Marks PA, Breslow R, Pavletich NP. 1999. Structures of a histone deacetylase homologue bound to the TSA and SAHA inhibitors. *Nature* **401:** 188–193.

Finnin MS, Donigian JR, Pavletich NP. 2001. Structure of the histone deacetylase SIRT2. *Nat Struct Biol* **8:** 621–625.

Fischle W, Wang Y, Allis CD. 2003. Histone and chromatin cross-talk. *Curr Opin Cell Biol* **15:** 172–183.

Ford J, Ahmed S, Allison S, Jiang M, Milner J. 2008. JNK2-dependent regulation of SIRT1 protein stability. *Cell Cycle* **7:** 3091–3097.

Frye RA. 1999. Characterization of five human cDNAs with homology to the yeast SIR2 gene: Sir2-like proteins (sirtuins) metabolize NAD and may have protein ADP-ribosyltransferase activity. *Biochem Biophys Res Commun* **260:** 273–279.

Furumai R, Matsuyama A, Kobashi N, Lee KH, Nishiyama M, Nakajima H, Tanaka A, Komatsu Y, Nishino N, Yoshida M, et al. 2002. FK228 (depsipeptide) as a natural prodrug that inhibits class I histone deacetylases. *Cancer Res* **62:** 4916–4921.

Galasinski SC, Resing KA, Goodrich JA, Ahn NG. 2002. Phosphatase inhibition leads to histone deacetylases 1 and 2 phosphorylation and disruption of corepressor interactions. *J Biol Chem* **277:** 19618–19626.

Gantt SL, Joseph CG, Fierke CA. 2010. Activation and inhibition of histone deacetylase 8 by monovalent cations. *J Biol Chem* **285:** 6036–6043.

Gao L, Cueto MA, Asselbergs F, Atadja P. 2002. Cloning and functional characterization of HDAC11, a novel member of the human histone deacetylase family. *J Biol Chem* **277:** 25748–25755.

Gao J, Wang WY, Mao YW, Graff J, Guan JS, Pan L, Mak G, Kim D, Su SC, Tsai LH. 2010. A novel pathway regulates memory and plasticity via SIRT1 and miR-134. *Nature* **466:** 1105–1109.

Gerhart-Hines Z, Dominy JE Jr, Blattler SM, Jedrychowski MP, Banks AS, Lim JH, Chim H, Gygi SP, Puigserver P. 2011. The cAMP/PKA pathway rapidly activates SIRT1 to promote fatty acid oxidation independently of changes in NAD(+). *Mol Cell* **44:** 851–863.

Glozak MA, Seto E. 2009. Acetylation/deacetylation modulates the stability of DNA replication licensing factor Cdt1. *J Biol Chem* **284:** 11446–11453.

Glozak MA, Sengupta N, Zhang X, Seto E. 2005. Acetylation and deacetylation of non-histone proteins. *Gene* **363:** 15–23.

Graff J, Rei D, Guan JS, Wang WY, Seo J, Hennig KM, Nieland TJ, Fass DM, Kao PF, Kahn M, et al. 2012. An epigenetic blockade of cognitive functions in the neurodegenerating brain. *Nature* **483:** 222–226.

Gregoretti IV, Lee YM, Goodson HV. 2004. Molecular evolution of the histone deacetylase family: Functional implications of phylogenetic analysis. *J Mol Biol* **338:** 17–31.

Grozinger CM, Hassig CA, Schreiber SL. 1999. Three proteins define a class of human histone deacetylases related to yeast Hda1p. *Proc Natl Acad Sci* **96:** 4868–4873.

* Grunstein M, Gasser SM. 2013. Epigenetics in *Saccharomyces cerevisiae*. *Cold Spring Harb Perspect Biol* **5:** a017491.

Haberland M, Montgomery RL, Olson EN. 2009. The many roles of histone deacetylases in development and physiology: Implications for disease and therapy. *Nat Rev Genet* **10:** 32–42.

Haggarty SJ, Koeller KM, Wong JC, Grozinger CM, Schreiber SL. 2003. Domain-selective small-molecule inhibitor of histone deacetylase 6 (HDAC6)-mediated tubulin deacetylation. *Proc Natl Acad Sci* **100:** 4389–4394.

Hong L, Schroth GP, Matthews HR, Yau P, Bradbury EM. 1993. Studies of the DNA binding properties of histone H4 amino terminus. Thermal denaturation studies reveal that acetylation markedly reduces the binding constant of the H4 "tail" to DNA. *J Biol Chem* **268:** 305–314.

Hu E, Chen Z, Fredrickson T, Zhu Y, Kirkpatrick R, Zhang GF, Johanson K, Sung CM, Liu R, Winkler J. 2000. Cloning and characterization of a novel human class I histone deacetylase that functions as a transcription repressor. *J Biol Chem* **275:** 15254–15264.

Huang BH, Laban M, Leung CH, Lee L, Lee CK, Salto-Tellez M, Raju GC, Hooi SC. 2005. Inhibition of histone deacetylase 2 increases apoptosis and p21Cip1/WAF1 expression, independent of histone deacetylase 1. *Cell Death Differ* **12:** 395–404.

Hubbert C, Guardiola A, Shao R, Kawaguchi Y, Ito A, Nixon A, Yoshida M, Wang XF, Yao TP. 2002. HDAC6 is a microtubule-associated deacetylase. *Nature* **417:** 455–458.

Imai S, Armstrong CM, Kaeberlein M, Guarente L. 2000. Transcriptional silencing and longevity protein Sir2 is an NAD-dependent histone deacetylase. *Nature* **403:** 795–800.

Inoue A, Fujimoto D. 1969. Enzymatic deacetylation of histone. *Biochem Biophys Res Commun* **36:** 146–150.

Ito K, Ito M, Elliott WM, Cosio B, Caramori G, Kon OM, Barczyk A, Hayash S, Adcock IM, Hogg JC, et al. 2005. Decreased histone deacetylase activity in chronic obstructive pulmonary disease. *N Engl J Med* **352:** 1967–1976.

Ito T, Umehara T, Sasaki K, Nakamura Y, Nishino N, Terada T, Shirouzu M, Padmanabhan B, Yokoyama S, Ito A, et al. 2011. Real-time imaging of histone H4K12-specific acetylation determines the modes of action of histone deacetylase and bromodomain inhibitors. *Chem Biol* **18:** 495–507.

Jepsen K, Hermanson O, Onami TM, Gleiberman AS, Lunyak V, McEvilly RJ, Kurokawa R, Kumar V, Liu F, Seto E, et al. 2000. Combinatorial roles of the nuclear receptor corepressor in transcription and development. *Cell* **102:** 753–763.

Johnson CA, White DA, Lavender JS, O'Neill LP, Turner BM. 2002. Human class I histone deacetylase complexes show enhanced catalytic activity in the presence of ATP and co-immunoprecipitate with the ATP-dependent chaperone protein Hsp70. *J Biol Chem* **277:** 9590–9597.

Kadosh D, Struhl K. 1998. Histone deacetylase activity of Rpd3 is important for transcriptional repression in vivo. *Genes Dev* **12:** 797–805.

Kang H, Jung JW, Kim MK, Chung JH. 2009. CK2 is the regulator of SIRT1 substrate-binding affinity, deacetylase activity and cellular response to DNA-damage. *PLoS One* **4:** e6611.

Kang H, Suh JY, Jung YS, Jung JW, Kim MK, Chung JH. 2011. Peptide switch is essential for Sirt1 deacetylase activity. *Mol Cell* **44:** 203–213.

Kao HY, Downes M, Ordentlich P, Evans RM. 2000. Isolation of a novel histone deacetylase reveals that class I and class II deacetylases promote SMRT-mediated repression. *Genes Dev* **14:** 55–66.

Kao HY, Lee CH, Komarov A, Han CC, Evans RM. 2002. Isolation and characterization of mammalian HDAC10, a novel histone deacetylase. *J Biol Chem* **277:** 187–193.

Kijima M, Yoshida M, Sugita K, Horinouchi S, Beppu T. 1993. Trapoxin, an antitumor cyclic tetrapeptide, is an irreversible inhibitor of mammalian histone deacetylase. *J Biol Chem* **268:** 22429–22435.

Kim EJ, Kho JH, Kang MR, Um SJ. 2007. Active regulator of SIRT1 cooperates with SIRT1 and facilitates suppression of p53 activity. *Mol Cell* **28:** 277–290.

LaBonte MJ, Wilson PM, Fazzone W, Groshen S, Lenz HJ, Ladner RD. 2009. DNA microarray profiling of genes differentially regulated by the histone deacetylase inhibitors vorinostat and LBH589 in colon cancer cell lines. *BMC Med Genomics* **2:** 67.

Lahm A, Paolini C, Pallaoro M, Nardi MC, Jones P, Neddermann P, Sambucini S, Bottomley MJ, Lo Surdo P, Carfi A, et al. 2007. Unraveling the hidden catalytic activity of vertebrate class IIa histone deacetylases. *Proc Natl Acad Sci* **104:** 17335–17340.

Cite as *Cold Spring Harb Perspect Biol* doi: 10.1101/cshperspect.a018713

Lee H, Rezai-Zadeh N, Seto E. 2004. Negative regulation of histone deacetylase 8 activity by cyclic AMP-dependent protein kinase A. *Mol Cell Biol* **24:** 765–773.

Leipe DD, Landsman D. 1997. Histone deacetylases, acetoin utilization proteins and acetylpolyamine amidohydrolases are members of an ancient protein superfamily. *Nucleic Acids Res* **25:** 3693–3697.

Li L, Wang L, Li L, Wang Z, Ho Y, McDonald T, Holyoake TL, Chen W, Bhatia R. 2012. Activation of p53 by SIRT1 inhibition enhances elimination of CML leukemia stem cells in combination with imatinib. *Cancer Cell* **21:** 266–281.

Lin YY, Kiihl S, Suhail Y, Liu SY, Chou YH, Kuang Z, Lu JY, Khor CN, Lin CL, Bader JS, et al. 2012. Functional dissection of lysine deacetylases reveals that HDAC1 and p300 regulate AMPK. *Nature* **482:** 251–255.

Liszt G, Ford E, Kurtev M, Guarente L. 2005. Mouse Sir2 homolog SIRT6 is a nuclear ADP-ribosyltransferase. *J Biol Chem* **280:** 21313–21320.

Luo J, Su F, Chen D, Shiloh A, Gu W. 2000. Deacetylation of p53 modulates its effect on cell growth and apoptosis. *Nature* **408:** 377–381.

* Marmorstein R, Zhou M-M. 2014. Writers and readers of histone acetylation: Structure, mechanism, and inhibition. *Cold Spring Harb Perspect Biol* **6:** a018762.

Matsuyama A, Shimazu T, Sumida Y, Saito A, Yoshimatsu Y, Seigneurin-Berny D, Osada H, Komatsu Y, Nishino N, Khochbin S, et al. 2002. In vivo destabilization of dynamic microtubules by HDAC6-mediated deacetylation. *EMBO J* **21:** 6820–6831.

McBurney MW, Yang X, Jardine K, Hixon M, Boekelheide K, Webb JR, Lansdorp PM, Lemieux M. 2003. The mammalian SIR2alpha protein has a role in embryogenesis and gametogenesis. *Mol Cell Biol* **23:** 38–54.

Methot JL, Chakravarty PK, Chenard M, Close J, Cruz JC, Dahlberg WK, Fleming J, Hamblett CL, Hamill JE, Harrington P, et al. 2008. Exploration of the internal cavity of histone deacetylase (HDAC) with selective HDAC1/HDAC2 inhibitors (SHI-1:2). *Bioorg Med Chem Lett* **18:** 973–978.

Michishita E, McCord RA, Berber E, Kioi M, Padilla-Nash H, Damian M, Cheung P, Kusumoto R, Kawahara TL, Barrett JC, et al. 2008. SIRT6 is a histone H3 lysine 9 deacetylase that modulates telomeric chromatin. *Nature* **452:** 492–496.

Mulligan P, Yang F, Di Stefano L, Ji JY, Ouyang J, Nishikawa JL, Toiber D, Kulkarni M, Wang Q, Najafi-Shoushtari SH, et al. 2011. A SIRT1-LSD1 corepressor complex regulates Notch target gene expression and development. *Mol Cell* **42:** 689–699.

Nakajima H, Kim YB, Terano H, Yoshida M, Horinouchi S. 1998. FR901228, a potent antitumor antibiotic, is a novel histone deacetylase inhibitor. *Exp Cell Res* **241:** 126–133.

Nasrin N, Kaushik VK, Fortier E, Wall D, Pearson KJ, de Cabo R, Bordone L. 2009. JNK1 phosphorylates SIRT1 and promotes its enzymatic activity. *PLoS One* **4:** e8414.

Outeiro TF, Kontopoulos E, Altmann SM, Kufareva I, Strathearn KE, Amore AM, Volk CB, Maxwell MM, Rochet JC, McLean PJ, et al. 2007. Sirtuin 2 inhibitors rescue alpha-synuclein-mediated toxicity in models of Parkinson's disease. *Science* **317:** 516–519.

Peng L, Ling H, Yuan Z, Fang B, Bloom G, Fukasawa K, Koomen J, Chen J, Lane WS, Seto E. 2012. SIRT1 negatively regulates the activities, functions, and protein levels of hMOF and TIP60. *Mol Cell Biol* **32:** 2823–2836.

Pflum MK, Tong JK, Lane WS, Schreiber SL. 2001. Histone deacetylase 1 phosphorylation promotes enzymatic activity and complex formation. *J Biol Chem* **276:** 47733–47741.

Pokholok DK, Harbison CT, Levine S, Cole M, Hannett NM, Lee TI, Bell GW, Walker K, Rolfe PA, Herbolsheimer E, et al. 2005. Genome-wide map of nucleosome acetylation and methylation in yeast. *Cell* **122:** 517–527.

Pugacheva EN, Jablonski SA, Hartman TR, Henske EP, Golemis EA. 2007. HEF1-dependent Aurora A activation induces disassembly of the primary cilium. *Cell* **129:** 1351–1363.

Richon VM, Emiliani S, Verdin E, Webb Y, Breslow R, Rifkind RA, Marks PA. 1998. A class of hybrid polar inducers of transformed cell differentiation inhibits histone deacetylases. *Proc Natl Acad Sci* **95:** 3003–3007.

Riggs MG, Whittaker RG, Neumann JR, Ingram VM. 1977. n-Butyrate causes histone modification in HeLa and Friend erythroleukaemia cells. *Nature* **268:** 462–464.

Ropero S, Fraga MF, Ballestar E, Hamelin R, Yamamoto H, Boix-Chornet M, Caballero R, Alaminos M, Setien F, Paz MF, et al. 2006. A truncating mutation of HDAC2 in human cancers confers resistance to histone deacetylase inhibition. *Nat Genet* **38:** 566–569.

Rush J, Moritz A, Lee KA, Guo A, Goss VL, Spek EJ, Zhang H, Zha XM, Polakiewicz RD, Comb MJ. 2005. Immunoaffinity profiling of tyrosine phosphorylation in cancer cells. *Nat Biotechnol* **23:** 94–101.

Saito A, Yamashita T, Mariko Y, Nosaka Y, Tsuchiya K, Ando T, Suzuki T, Tsuruo T, Nakanishi O. 1999. A synthetic inhibitor of histone deacetylase, MS-27–275, with marked in vivo antitumor activity against human tumors. *Proc Natl Acad Sci* **96:** 4592–4597.

Sasaki T, Maier B, Koclega KD, Chruszcz M, Gluba W, Stukenberg PT, Minor W, Scrable H. 2008. Phosphorylation regulates SIRT1 function. *PLoS One* **3:** e4020.

Schuetz A, Min J, Antoshenko T, Wang CL, Allali-Hassani A, Dong A, Loppnau P, Vedadi M, Bochkarev A, Sternglanz R, et al. 2007. Structural basis of inhibition of the human NAD+-dependent deacetylase SIRT5 by suramin. *Structure* **15:** 377–389.

Seto E, Yang XJ. 2010. Regulation of histone deacetylase activities and functions by phosphorylation and dephosphorylation. In *Handbook of cell signaling*, 2nd ed. (ed Bradshaw RA, Dennis EA), pp. 2379–2388. Elsevier Science, Amsterdam.

Shi Y, Sawada J, Sui G, Affar el B, Whetstine JR, Lan F, Ogawa H, Luke MP, Nakatani Y. 2003. Coordinated histone modifications mediated by a CtBP co-repressor complex. *Nature* **422:** 735–738.

Shogren-Knaak M, Ishii H, Sun JM, Pazin MJ, Davie JR, Peterson CL. 2006. Histone H4-K16 acetylation controls chromatin structure and protein interactions. *Science* **311:** 844–847.

Smith BC, Denu JM. 2007. Mechanism-based inhibition of Sir2 deacetylases by thioacetyl-lysine peptide. *Biochemistry* **46:** 14478–14486.

Somoza JR, Skene RJ, Katz BA, Mol C, Ho JD, Jennings AJ, Luong C, Arvai A, Buggy JJ, Chi E, et al. 2004. Structural snapshots of human HDAC8 provide insights into the class I histone deacetylases. *Structure* **12:** 1325–1334.

Sterner R, Vidali G, Allfrey VG. 1979. Studies of acetylation and deacetylation in high mobility group proteins. Identification of the sites of acetylation in HMG-1. *J Biol Chem* **254:** 11577–11583.

Tan M, Luo H, Lee S, Jin F, Yang JS, Montellier E, Buchou T, Cheng Z, Rousseaux S, Rajagopal N, et al. 2011. Identification of 67 histone marks and histone lysine crotonylation as a new type of histone modification. *Cell* **146:** 1016–1028.

Taunton J, Hassig CA, Schreiber SL. 1996. A mammalian histone deacetylase related to the yeast transcriptional regulator Rpd3p. *Science* **272:** 408–411.

Trivedi CM, Luo Y, Yin Z, Zhang M, Zhu W, Wang T, Floss T, Goettlicher M, Noppinger PR, Wurst W, et al. 2007. Hdac2 regulates the cardiac hypertrophic response by modulating Gsk3 β activity. *Nat Med* **13:** 324–331.

Tsai SC, Seto E. 2002. Regulation of histone deacetylase 2 by protein kinase CK2. *J Biol Chem* **277:** 31826–31833.

Vaquero A, Scher M, Lee D, Erdjument-Bromage H, Tempst P, Reinberg D. 2004. Human SirT1 interacts with histone H1 and promotes formation of facultative heterochromatin. *Mol Cell* **16:** 93–105.

Vaquero A, Scher MB, Lee DH, Sutton A, Cheng HL, Alt FW, Serrano L, Sternglanz R, Reinberg D. 2006. SirT2 is a histone deacetylase with preference for histone H4 Lys 16 during mitosis. *Genes Dev* **20:** 1256–1261.

Vaziri H, Dessain SK, Ng Eaton E, Imai SI, Frye RA, Pandita TK, Guarente L, Weinberg RA. 2001. hSIR2(SIRT1) functions as an NAD-dependent p53 deacetylase. *Cell* **107:** 149–159.

Vermeulen M, Carrozza MJ, Lasonder E, Workman JL, Logie C, Stunnenberg HG. 2004. In vitro targeting reveals intrinsic histone tail specificity of the Sin3/histone deacetylase and N-CoR/SMRT corepressor complexes. *Mol Cell Biol* **24:** 2364–2372.

Vidal M, Gaber RF. 1991. RPD3 encodes a second factor required to achieve maximum positive and negative transcriptional states in *Saccharomyces cerevisiae*. *Mol Cell Biol* **11:** 6317–6327.

Wang Z, Zang C, Cui K, Schones DE, Barski A, Peng W, Zhao K. 2009. Genome-wide mapping of HATs and HDACs reveals distinct functions in active and inactive genes. *Cell* **138:** 1019–1031.

Wang Q, Zhang Y, Yang C, Xiong H, Lin Y, Yao J, Li H, Xie L, Zhao W, Yao Y, et al. 2010. Acetylation of metabolic enzymes coordinates carbon source utilization and metabolic flux. *Science* **327:** 1004–1007.

Watson PJ, Fairall L, Santos GM, Schwabe JW. 2012. Structure of HDAC3 bound to co-repressor and inositol tetraphosphate. *Nature* **481:** 335–340.

Wen YD, Perissi V, Staszewski LM, Yang WM, Krones A, Glass CK, Rosenfeld MG, Seto E. 2000. The histone deacetylase-3 complex contains nuclear receptor corepressors. *Proc Natl Acad Sci* **97:** 7202–7207.

Wu R, Lu Z, Cao Z, Zhang Y. 2011. Zinc chelation with hydroxamate in histone deacetylases modulated by water access to the linker binding channel. *J Am Chem Soc* **133:** 6110–6113.

Yang XJ, Seto E. 2008. Lysine acetylation: Codified crosstalk with other posttranslational modifications. *Mol Cell* **31:** 449–461.

Yang WM, Inouye C, Zeng Y, Bearss D, Seto E. 1996. Transcriptional repression by YY1 is mediated by interaction with a mammalian homolog of the yeast global regulator RPD3. *Proc Natl Acad Sci* **93:** 12845–12850.

Yang WM, Yao YL, Sun JM, Davie JR, Seto E. 1997. Isolation and characterization of cDNAs corresponding to an additional member of the human histone deacetylase gene family. *J Biol Chem* **272:** 28001–28007.

Yoshida M, Nomura S, Beppu T. 1987. Effects of trichostatins on differentiation of murine erythroleukemia cells. *Cancer Res* **47:** 3688–3691.

Yoshida M, Kijima M, Akita M, Beppu T. 1990. Potent and specific inhibition of mammalian histone deacetylase both in vivo and in vitro by trichostatin A. *J Biol Chem* **265:** 17174–17179.

Zhang X, Wharton W, Yuan Z, Tsai SC, Olashaw N, Seto E. 2004. Activation of the growth-differentiation factor 11 gene by the histone deacetylase (HDAC) inhibitor trichostatin A and repression by HDAC3. *Mol Cell Biol* **24:** 5106–5118.

Zhang X, Ozawa Y, Lee H, Wen YD, Tan TH, Wadzinski BE, Seto E. 2005. Histone deacetylase 3 (HDAC3) activity is regulated by interaction with protein serine/threonine phosphatase 4. *Gene Dev* **19:** 827–839.

Zhang X, Yuan Z, Zhang Y, Yong S, Salas-Burgos A, Koomen J, Olashaw N, Parsons JT, Yang XJ, Dent SR, et al. 2007. HDAC6 modulates cell motility by altering the acetylation level of cortactin. *Mol Cell* **27:** 197–213.

Zhao W, Kruse JP, Tang Y, Jung SY, Qin J, Gu W. 2008. Negative regulation of the deacetylase SIRT1 by DBC1. *Nature* **451:** 587–590.

Zhou X, Richon VM, Rifkind RA, Marks PA. 2000. Identification of a transcriptional repressor related to the noncatalytic domain of histone deacetylases 4 and 5. *Proc Natl Acad Sci* **97:** 1056–1061.

Cite as *Cold Spring Harb Perspect Biol* doi: 10.1101/cshperspect.a018713

CHAPTER 6

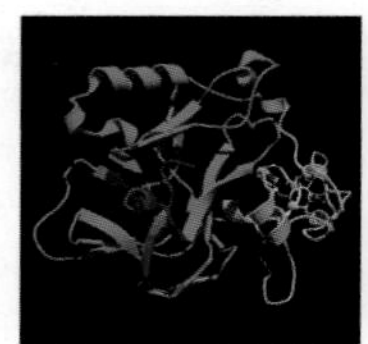

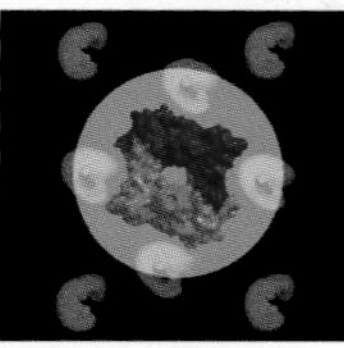

Structural and Functional Coordination of DNA and Histone Methylation

Xiaodong Cheng

Department of Biochemistry, Emory University School of Medicine, Atlanta, Georgia 30322

Correspondence: xcheng@emory.edu

SUMMARY

One of the most fundamental questions in the control of gene expression in mammals is how epigenetic methylation patterns of DNA and histones are established, erased, and recognized. This central process in controlling gene expression includes coordinated covalent modifications of DNA and its associated histones. This chapter focuses on structural aspects of enzymatic activities of histone (arginine and lysine) methylation and demethylation and functional links between the methylation status of the DNA and histones. An interconnected network of methyltransferases, demethylases, and accessory proteins is responsible for changing or maintaining the modification status of specific regions of chromatin.

Outline

Cite as *Cold Spring Harb Perspect Biol* doi: 10.1101/cshperspect.a018747

OVERVIEW

All cells face the problem of controlling the amounts and timing of expression of their various genes. In some cases, this control involves relatively long-term and heritable modifications to the chromatin, albeit nonpermanent. Such modifications that do not change the DNA sequence are referred to as "epigenetic." The resulting epigenetic effects maintain the various patterns of gene expression in different cell types. Epigenetic modifications include DNA methylation and histone posttranslational modifications (PTMs).

Nucleosomes consist of ~146 bp of DNA wrapped approximately 1.8 times around a histone octamer and are evolutionarily conserved across all eukaryotes. The combinatorial pattern of DNA and histone modifications constitutes an epigenetic code that shapes transcriptional patterns and genomic stability. The coding modification is "written" by sequence- and site-specific enzymes, and "interpreted" or "read" by effector (or reader) molecules that mediate the assembly of higher-order chromatin structures, involving remodeling complexes, histone variants, and noncoding RNAs (see Ch. 21 [Becker and Workman 2013]; Ch. 3 [Allis et al.]; Ch. 20 [Henikoff and Smith 2014]). Representative enzymes responsible for histone methylation (SET domain proteins, Dot1, and protein arginine methyltransferases), histone lysine demethylation (LSD1 and Jumonji proteins), and DNA methylation (Dnmt1 and Dnmt3) are discussed in this chapter. DNA methylation is an epigenetic modification that has long been known to repress transcription. Histone methylation, depending on the histone and residue of modification, contributes to either active or repressive chromatin configurations. The functional implications of the structural determination of enzymes in linking histone modifications to that of DNA in mammalian cells are also discussed. This chapter is complemented by subsequent chapters discussing the structure and function of proteins that read DNA and histone methylation (Ch. 7 [Patel 2014]), and previous chapters discussing histone acetylation writers (HATs), histone deacetylation (HDAC) "eraser" enzymes, and readers of these epigenetic marks (Ch. 4 [Marmorstein and Zhou 2014]; Ch. 5 [Seto and Yoshida 2014]).

Cite as *Cold Spring Harb Perspect Biol* doi: 10.1101/cshperspect.a018747

1 HISTONE METHYLATION AND DEMETHYLATION

1.1 Histone Lysine (K) Methyltransferases (HKMTs)

All known HKMTs (referred to in some of the literature as PKMTs or KMTs, as many of the substrates are nonhistone proteins) contain an evolutionarily conserved SET domain comprised of 130 amino acids (reviewed in Cheng et al. 2005), except Dot1 (see Sec. 1.2). The SET domain was first identified as a shared sequence motif in three *Drosophila* proteins: suppressor of variegation (Su(var)3-9), enhancer of zeste (E(z)), and homeobox gene regulator trithorax (Trx; Jenuwein et al. 1998). Mammalian homologs of the *Drosophila* Su(var)3-9 protein, SUV39H1 in humans, and Suv39h1 in mouse were the first characterized HKMTs involved in H3K9 methylation (Fig. 1A) (Rea et al. 2000). Since then, more than 50 SET domain-containing proteins have been identified in humans, with a proven or predicted enzymatic role in performing lysine methylation on histone tails (reviewed in Volkel and Angrand 2007).

The majority of the SET-containing HKMTs contain at least one additional protein module in their protein sequence (e.g., a chromodomain in SUV39H1; Fig. 1A). SET-containing HKMTs are grouped into six different subfamilies based on sequence homology within and around the catalytic SET domain, homology with other protein modules, and their structures. The six subfamilies include SET1, SET2, SUV39, EZH, SMYD, and PRDM (Volkel and Angrand 2007). A number of SET-containing HKMTs, however, do not fall into the above six subfamilies because of lacking sequences or conservation flanking their SET domains. Examples of such proteins include Set8 (also known as PR-Set7; Couture et al. 2005; Xiao et al. 2005), which monomethylates H4K20 (H4K20me1), and SUV4-20H1 and SUV4-20H2, which di- and trimethylate H4K20 (H4K20me2 and me3). Set7/9 can monomethylate H3K4 (H3K4me1; Xiao et al. 2003) and many other nonhistone substrates, whereas SetD6 monomethylates only the nonhistone substrate, RelA, a subunit of NF-κB (Levy et al. 2011).

Structures of many SET domains from different subfamilies have been solved in various combinations, including bound to peptide substrates and methyl donor (*S*-adenosyl-L-methionine, also known as AdoMet or SAM) or reaction product (*S*-adenosyl-L-homocysteine, also known as AdoHcy or SAH). The SET domain adopts a unique structure formed by a series of β-strands folded into three sheets surrounding a knot-like structure (Fig. 1B). The knot-like structure (or pseudoknot) is formed by the carboxyl-terminal segment of the SET domain, which passes through a loop formed by the preceding stretch of sequences. Formation of this pseudoknot structure brings the two conserved SET domain sequence motifs, III and IV (Fig. 1C), in close proximity to the AdoMet-binding region and peptide-binding channel (Fig. 1D).

Available crystal structures of the SUV39 HKMT subfamily, which methylates H3K9 (DIM-5, Clr4, GLP/EHMT1, G9a/EHMT2, and SUV39H2), show the presence of two closely packed cysteine-rich modules in both the pre-SET and post-SET (before and after the SET) domains (Fig. 1E). These two modules are important for maintaining structural stability (pre-SET) and forming part of the active site lysine channel (post-SET) (Zhang et al. 2002; Zhang et al. 2003). The pre-SET module contains nine conserved cysteines, which coordinate three Zn^{2+} atoms in a triangular geometry. The post-SET module contains three conserved cysteines, which along with a cysteine from the conserved motif III sequence, (R/H)F(I/V)NHxCxPN, tetrahedrally coordinate the fourth Zn^{2+} atom near the active site. Binding of the fourth Zn^{2+} at the active site is essential for the activity of the SUV39 subfamily (Zhang et al. 2003).

1.2 Dot1p: Non-SET Domain HKMT

Histone H3 Lys-79 (H3K79) is methylated by Dot1p (reviewed in Frederiks et al. 2011), a protein originally identified as a disruptor of telomeric silencing in *Saccharomyces cerevisiae* (Singer et al. 1998). Methylation of H3K79 in *S. cerevisiae* is important for the proper localization of the silent information regulator complex and DNA damage signaling (see Ch. 8 [Grunstein and Gasser 2013]).

Dot1p is a Class-I methyltransferase, as suggested by the presence of the AdoMet-binding sequence motifs (Dlakic 2001; Schubert et al. 2003), similar to those found in protein arginine methyltransferases and DNA methyltransferases (see Secs. 1.3 and 2). Class-I methyltransferases such as Dot1p are distinct from most other HKMTs because they do not contain the SET domain. Thus, they have an entirely different structural scaffolding and unrelated local active-site spatial arrangement that catalyzes AdoMet-dependent methyl transfer to a protein lysine side chain.

Yeast Dot1p contains a core region (indicated in Fig. 2A) conserved between Dot1p homologs in human, *Caenorhabditis elegans*, *Drosophila*, and the mosquito, *Anopheles gambiae*. The length of these Dot1 proteins varies from 582 amino acids in yeast to 2237 amino acids in *Drosophila*. The conserved Dot1p core is located at the carboxyl terminus in yeast, but is at the amino terminus in human, *C. elegans*, *Drosophila*, and *A. gambiae* Dot1p homologs. The Dot1p conserved core contains an amino-terminal helical domain and a seven-stranded catalytic domain that harbors the binding site for the methyl donor and an active-site pocket sided with conserved hydrophobic residues (Fig. 2B).

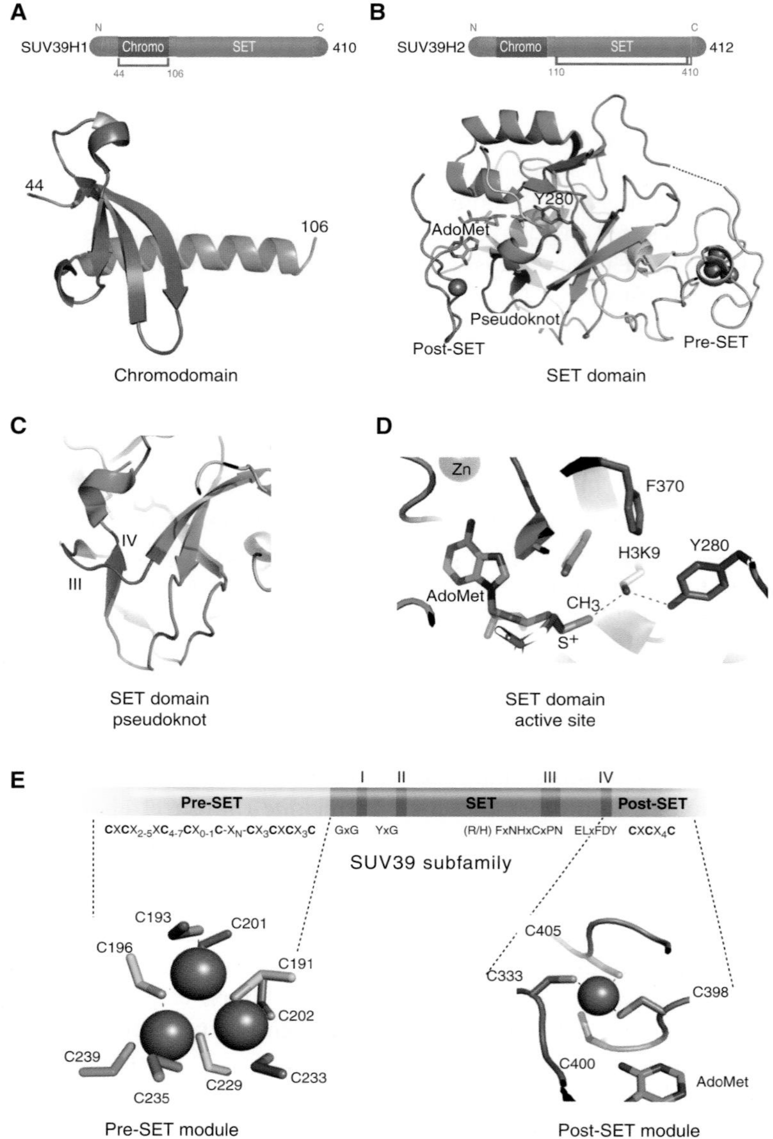

Figure 1. Structural features of SUV39H1 and SUV39H2. (*A*) Ribbon diagram of amino-terminal chromodomain of SUV39H1 (PDB 3MTS). (*B*) Carboxyl-terminal SET domain structure of SUV39H2 (PDB 2R3A). (*C*) Formation of the pseudoknot by motifs III and IV. (*D*) Formation of the active site showing the methyl donor (*S*-adenosyl-L-methionine [AdoMet]), target H3K9 lysine, catalytic Y280 residue, and F370 Phe/Tyr Switch (Collins et al. 2005). AdoMet-dependent methyltransferases (including HKMTs) share a reaction mechanism in which the nucleophile acceptor (NH_2) attacks the electrophilic carbon of AdoMet in an S_N2 displacement reaction. (*E*) SUV39H1 and H2 have a pre-SET segment containing nine invariant cysteines, the SET region containing four signature motifs, and the post-SET region containing three invariant cysteines. An enlargement of the pre-SET Zn_3Cys_9 triangular zinc cluster structure is illustrated on the *bottom left* and the post-SET zinc center on the *bottom right* panel.

Dot1p has several unique biochemical properties. Yeast Dot1p and its human homolog Dot1L methylate only nucleosomal substrates, but not free histone H3 protein (reviewed in Frederiks et al. 2011). A stretch of positively charged residues (i.e., Lys-rich) at the carboxyl-terminal end of the human Dot1L core or the amino-terminal end of the yeast Dot1p core were critical for nucleosome binding (Fig. 2C) and therefore for enzymatic activity (Min et al. 2003; Sawada et al. 2004; Oh et al. 2010). Given in *S. cerevisiae*, H3K79 methylation requires ubiquitination of H2B K123 in vivo (Briggs et al. 2002), and both histone residues are located on the same nucleosome disk surface

Cite as *Cold Spring Harb Perspect Biol* doi: 10.1101/cshperspect.a018747

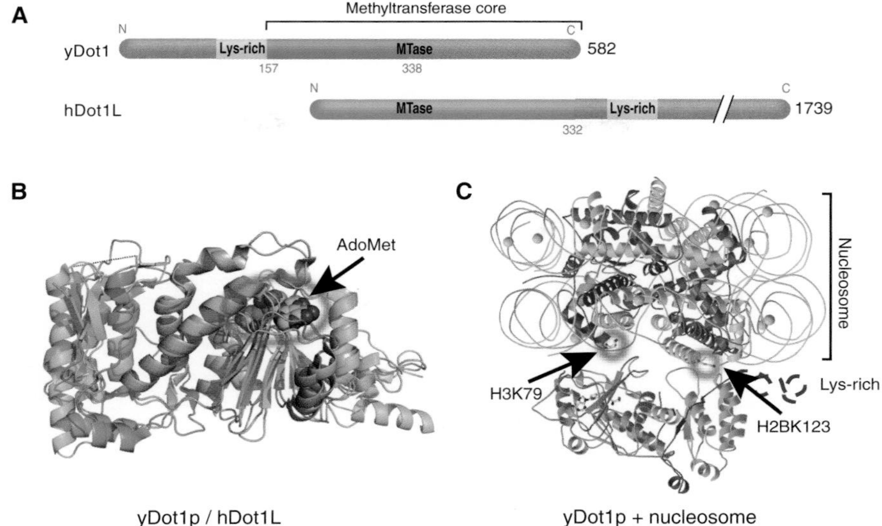

Figure 2. Dot1p family (non-SET HKMTs). (*A*) Schematic representation of Dot1 homologs from yeast and human, indicating the conserved methyltransferase core regions. (*B*) Superimposition of the conserved core regions of yeast Dot1p (residues 176-567; PDB 1U2Z), colored in green and brown, and human Dot1L (residues 5-332; PDB 1NW3), colored in cyan. The amino-terminal helical domains are shown on the *left* side of the panel in green for yDot1p or cyan for hDot1L. The carboxyl-terminal catalytic domain is shown on the *right* side of the panel with the bound methyl donor, AdoMet, as spheres (circled in red with an arrow). (*C*) A model of yDot1p docked with a nucleosome, adapted from Sawada et al. (2004). The structure of the nucleosome core particle is shown as ribbons (red, H3; green, H4; magenta, H2A; yellow, H2B; gray lines, DNA). The model was put together by aligning the target H3K79, located on the nucleosome disk surface, with the active site pocket of Dot1p.

~30 Å apart, Dot1p may interact specifically with nucleosomes containing ubiquitinated H2B (Fig. 2C) (Oh et al. 2010). Such an interaction could be significant in vivo because Dot1p could be recruited to specific high-order chromatin in which ubiquitinated histone H2B might serve as a spacer between adjacent nucleosome disk surfaces (Sun and Allis 2002), allowing Dot1p access to its target H3K79 lysine (see also Fig. 10 in Ch. 3 [Allis et al.]).

Mistargeted hDOT1L function has been implicated in the leukemogenesis of mixed lineage leukemia with the MLL-AF10 fusion (Okada et al. 2005). It does this by interacting with the AF10 protein and up-regulating genes such as Hox9a. Recently, the in vivo inhibition of hDot1L led to the increased survival of mice that had a xenograft model of mixed lineage leukemia (Daigle et al. 2011). This is, notably, the first example of selective inhibition of an HKMT that has efficacy in a cancer model.

1.3 Protein Arginine Methylation

Protein arginine methylation is a common posttranslational modification in eukaryotes. There are two major types of protein arginine (R) methyltransferases (PRMTs) that transfer the methyl group from AdoMet to the guanidino group of arginines in protein substrates (Lee et al. 1977), called type I and type II PRMTs (Fig. 3A). Both catalyze the formation of monomethylarginine (Rme1) as an intermediate, but type I PRMTs also form asymmetric dimethylarginine (Rme2a), whereas type II PRMTs form symmetric dimethylarginine (Rme2s). Among the nine canonical members of PRMT family (Herrmann et al. 2009; Table 1), only PRMT5 (also known as JBP1 for Jak-binding protein 1; Branscombe et al. 2001), and possibly PRMT7 and PRMT9 are type II PRMTs (Lee et al. 2005; Cook et al. 2006); they symmetrically dimethylate-specific arginines not only on histones, but also other proteins such as myelin basic protein (Kim et al. 1997), spliceosomal Sm proteins (Friesen et al. 2001), and Piwi proteins (Vagin et al. 2009). Highly relevant to the focus of this chapter, PRMT5-mediated methylation of histone H4R3 recruits DNMT3A, the de novo DNA methyltransferase, coupling histone arginine, and DNA methylation in gene silencing (Zhao et al. 2009).

The PRMT proteins vary in length from 353 amino acids for PRMT1 to 637 amino acids for PRMT5/JBP1, but they all contain a conserved core region of approximately 310 amino acids (Fig. 3B). The sequences beyond the conserved PRMT core region are all amino-terminal additions; however, PRMT4 also has a carboxyl-terminal addition. The size of the amino-terminal additions varies

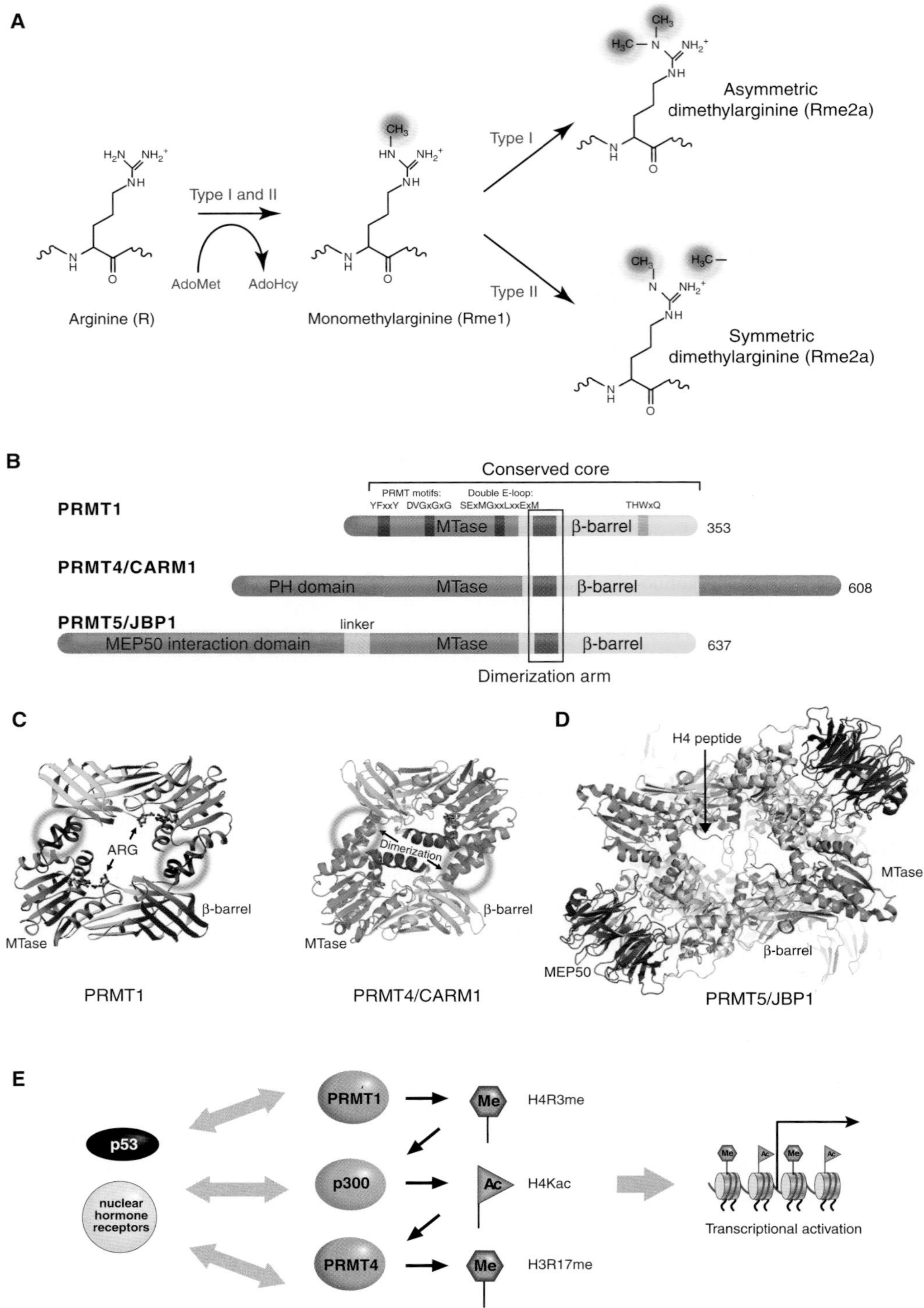

Figure 3. PRMT family. (*A*) Reactions catalyzed by the two major types of protein arginine methylation. (*B*) Representatives of PRMT members (type I: PRMT1 and PRMT4; type II: PRMT5). The conserved methyltransferase (MTase) domain is in green and the unique β-barrel domain in yellow. (*C*) Dimeric structures of PRMT1 (PDB 1OR8) and PRMT4 (PDB 3B3F). Dimerization arms are indicated by red circles. (*D*) Type II PRMT5-MEP50 tetramer complex (DPB 4GQB), formed by the stacking of two dimers. The second dimer is faded in the background. MEP50 (colored in brown) interacts with the amino-terminal domain (gray) of PRMT5. Bound H4 peptide is colored in red. (*E*) An example of three coactivators acting synergistically for p53-mediated transcription.

Table 1. Members of human PRMT family

Enzyme	Type	Activity	Chromosome	EST	Coding exon	Genomic size (kb)	Protein accession number	Protein size (residues)
PRMT1	I	+++	19q13	+++	9–10	10	CAA71764	361
PRMT2	I	-	21q22	++	10	30	P55345	433
PRMT3	I	+	11p15	+	13	50	AAH64831	531
PRMT4(CARM1)	I	+	19p13	++	16	50	NP_954592	608
PRMT5(JBP1)	II	+	14q11	++	17	8.5	AAF04502	637
PRMT6	I	+	1p13	+/−	1	2.5	AAK85733	375
PRMT7	II?	+	16q22	++	17	41	Q9NVM4	692
PRMT8	I	?	12p13	+	9	52	AAF91390	334
PRMT9	II?	?	4q31	+	10	40	AAH64403	845

between ~30 amino acids in PRMT1 to ~300 amino acids in PRMT5. The variation in amino termini allows each PRMT to be subject to a different mode of regulation. An interesting feature of PRMT7 and PRMT9 is that they seem to have arisen from a gene duplication event and contain two conserved core regions, each with a putative AdoMet-binding motif (Miranda et al. 2004; Cook et al. 2006).

Several crystal structures of type I PRMTs are currently available (Cheng et al. 2005; Troffer-Charlier et al. 2007; Yue et al. 2007). These structures reflect a striking structural conservation of the PRMT catalytic core (Fig. 3B). The overall monomeric structure of the PRMT core can be divided into three parts: a methyltransferase domain, a β-barrel, and a dimerization arm. The methyltransferase domain has a consensus fold conserved in Class-I AdoMet-dependent methyltransferases (like that of Dot1p) that harbor an AdoMet-binding site (Schubert et al. 2003). The β-barrel domain is unique to the PRMT family (Zhang et al. 2000). Dimer formation is a conserved feature in the type I PRMT family, validated by crystal structure studies, an example of which is illustrated in Figure 3C (Cheng et al. 2005). Dimerization may be required to correctly engage the residues in the AdoMet-binding site, so that they bind AdoMet and/or generate dimethylated (Rme2) products processively, akin to the spread of H3K9 trimethylation by DIM-5 (Zhang et al. 2003). Indeed, phosphorylation of PRMT4 at a conserved serine residue in the dimer interface results in inefficient AdoMet binding, and reduced histone methylation activity (Higashimoto et al. 2007). Similarly, an allosteric inhibitor that binds in the dimer interface of PRMT3 results in reduced binding of AdoMet and methylation activity (Siarheyeva et al. 2012).

The type II enzyme, PRMT5, functions as part of various high molecular weight protein complexes that also contain the WD repeat protein MEP50 (methylosome protein 50) (Friesen et al. 2002). The structure of the human PRMT5-MEP50 complex revealed that PRMT5 was organized into a tetramer through the stacking of two primary dimers and the amino-terminal domain of PRMT5 interacting with the seven-bladed β-propeller MEP50 (Fig. 3D) (Antonysamy et al. 2012).

Two well-studied enzymes, PRMT1 and PRMT4, methylate histones H2B, H3, and H4 (reviewed in Bedford and Clarke 2009), in addition to many nonhistone substrates including the carboxyl-terminal domain of RNA polymerase II (Sims et al. 2011). Histone arginine methylation is a component of the "histone code" that directs a variety of processes involving chromatin. For example, methylation of H4R3 by PRMT1 facilitates H4 acetylation and enhances transcriptional activation by nuclear hormone receptors. It acts synergistically with PRMT4 because PRMT4 prefers acetylated histone tails to generate methylated H3R17 (Wang et al. 2001; Daujat et al. 2002). Similarly, the synergistic action in vitro of PRMT1, PRMT4, and p300 for p53-mediated transcription is the greatest when all three coactivators are present, whether added sequentially or at the same time (Fig. 3E) (An et al. 2004). Even preincubating a chromatin template with p53 and PRMT1 significantly stimulated the histone acetyltransferase activity of p300, as did preincubation of the template with p53 and p300 to stimulate H3 arginine methylation by PRMT4. Indeed, PRMT4 was initially discovered as a transcriptional coactivator-associated arginine (R) methyltransferase 1 (CARM1; Chen et al. 1999). PRMT4/CARM1 also acts synergistically with the p160 coactivator to stimulate gene activation by nuclear receptors (Chen et al. 1999; Lee et al. 2002). These results provide compelling evidence that histones are relevant targets for PRMT4/CARM1, PRMT1, and p300, and that the resulting histone modifications are directly important for transcription.

1.4 Lysine Demethylation by Oxidation: LSD1

To the surprise of the research community, protein lysine methylation was finally shown to be a reversible posttranslational modification in 2004 (described in Ch. 2 [Shi and Tsukada 2013]). Before this, Bannister et al. had hypothesized that methyl groups from both lysine and arginine side

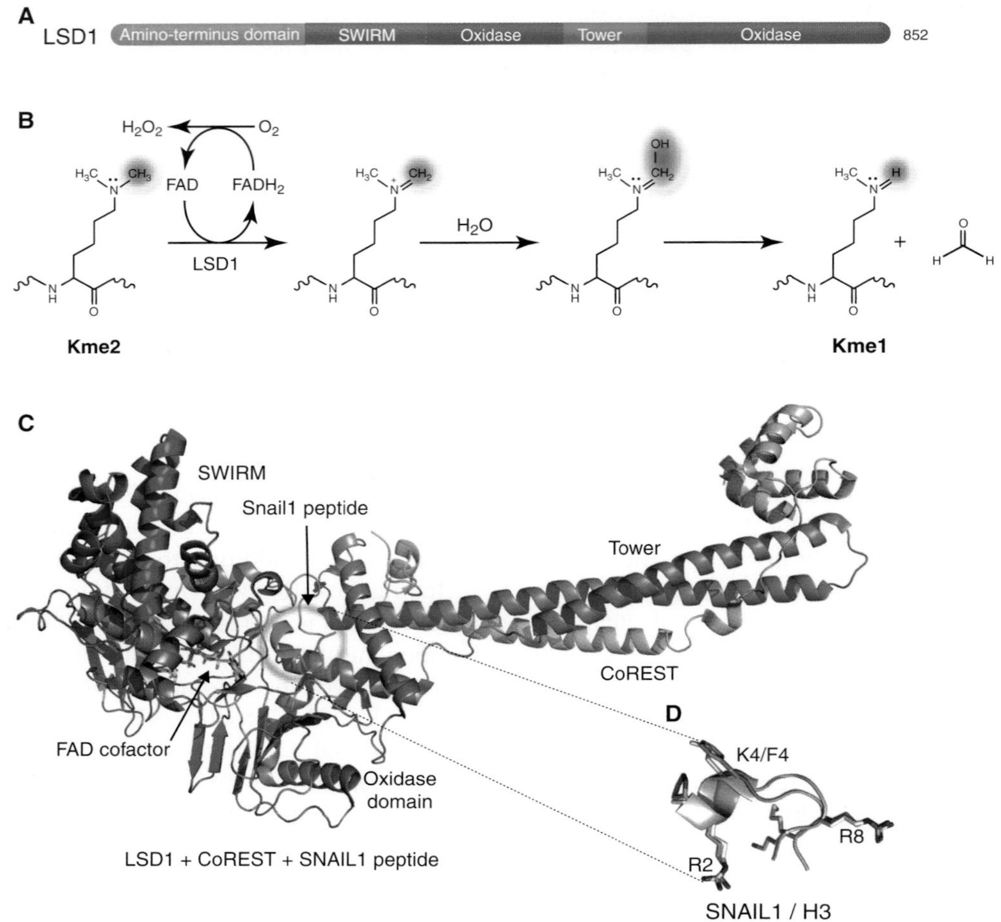

Figure 4. Histone demethylation by oxidation. (*A*) Schematic representation of human LSD1 domain organization: the amino-terminal putative nuclear localization signal, followed by a SWIRM (Swi3p, Rsc8p, and Moira) domain, and the catalytic oxidase domain. The oxidase domain contains an atypical insertion of the Tower domain not found in other oxidases. (*B*) Scheme of the demethylation reaction catalyzed by LSD1. (*C*) Crystal structure of LSD1-CoREST in complex with the SNAIL1 peptide (PDB 2Y48). LSD1 includes residues 171–836 in red, blue, and magenta. CoREST shows residues 308–440 in orange. The SNAIL1 peptide is in green, and the FAD cofactor is shown as a yellow ball-and-stick. (*D*) Superposition between SNAIL1 (orange) and histone H3 (gray) peptides. (Adapted from Baron et al. 2011).

chains could be oxidatively removed using a FAD (flavin adenine dinucleotide) cofactor as the electron acceptor (Bannister et al. 2002). Then came the discovery of the lysine-specific demethylase 1 (LSD1) protein (Fig. 4A) (Shi et al. 2004). LSD1 is a flavin-dependent amine oxidase that demethylates H3K4me2/me1 (Shi et al. 2004), H3K9me2/me1 in an androgen receptor-mediated pathway (Metzger et al. 2005), and even p53, a nonhistone protein (Huang et al. 2007). The closely related LSD2 demethylates H3K4me2/me1 (Karytinos et al. 2009). Both LSD1 and LSD2 demethylate methyl lysines by forming an imine intermediate that undergoes hydrolysis in aqueous buffer to complete the demethylation process (Fig. 4B). LSD1 and LSD2, however, cannot demethylate trimethylated lysines because of the mechanistic requirement for a protonated amine in this demethylation pathway.

Thus far, crystal structures of LSD1 in various configurations have been determined (Fig. 4C) (reviewed in Hou and Yu 2010). In one study, the first 16 residues of histone H3 was observed in a complex structure with LSD1-CoREST (Forneris et al. 2007), in perfect agreement with biochemical data that LSD1 is active on peptide substrates longer than 16 amino acids (Forneris et al. 2005). Interestingly, the amino-terminal extremity of the transcription factor SNAIL1 has sequence similarity with the amino-terminal tail of histone H3 (Lin et al. 2010), and binds to the LSD1 catalytic site in the same way as a histone H3 peptide substrate (Fig. 4D) (Baron et al. 2011). Specifically, the binding positions of the amino-terminal Arg2, Phe4, and Arg7 residues of the SNAIL1 peptide correspond to the amino-terminal Arg2, Lys4, and Arg8 residues of histone H3. Thus, the Snail1-LSD1-CoREST complex effectively inhibits LSD1 enzymatic activity (Baron et al. 2011), and is found in certain cancer cells. The fact that LSD1 recognizes the first amino-terminal amino group (a conserved positive charge), customizing the orientation of the fourth amino acid side chain (i.e., H3K4me2/me1) to point toward and be in direct contact with the flavin ring of the cofactor, raises the question of how LSD1 demethylates methyllysines further away from the amino terminus (e.g., H3K9me2/me1), in an androgen receptor-dependent manner (Metzger et al. 2005).

1.5 Lysine Demethylation by Hydroxylation: Jumonji-Containing Demethylases

In search of enzymes capable of reversing methylated lysines, Trewick et al. hypothesized that Jumonji domain-containing Fe(II)- and α-ketoglutarate-dependent dioxygenases can reverse lysine methylation via a similar mechanism as the bacterial AlkB family of DNA repair enzymes (Fig. 5A) (Trewick et al. 2005). This hypothesis was quickly verified by the discovery of JHDM1 as the Jumonji domain-containing histone demethylase 1, using a biochemical assay based on the detection of formaldehyde, one of the predicted reaction products (Tsukada et al. 2006; described in Ch. 2 [Shi and Tsukada 2013]). Jumonji-containing

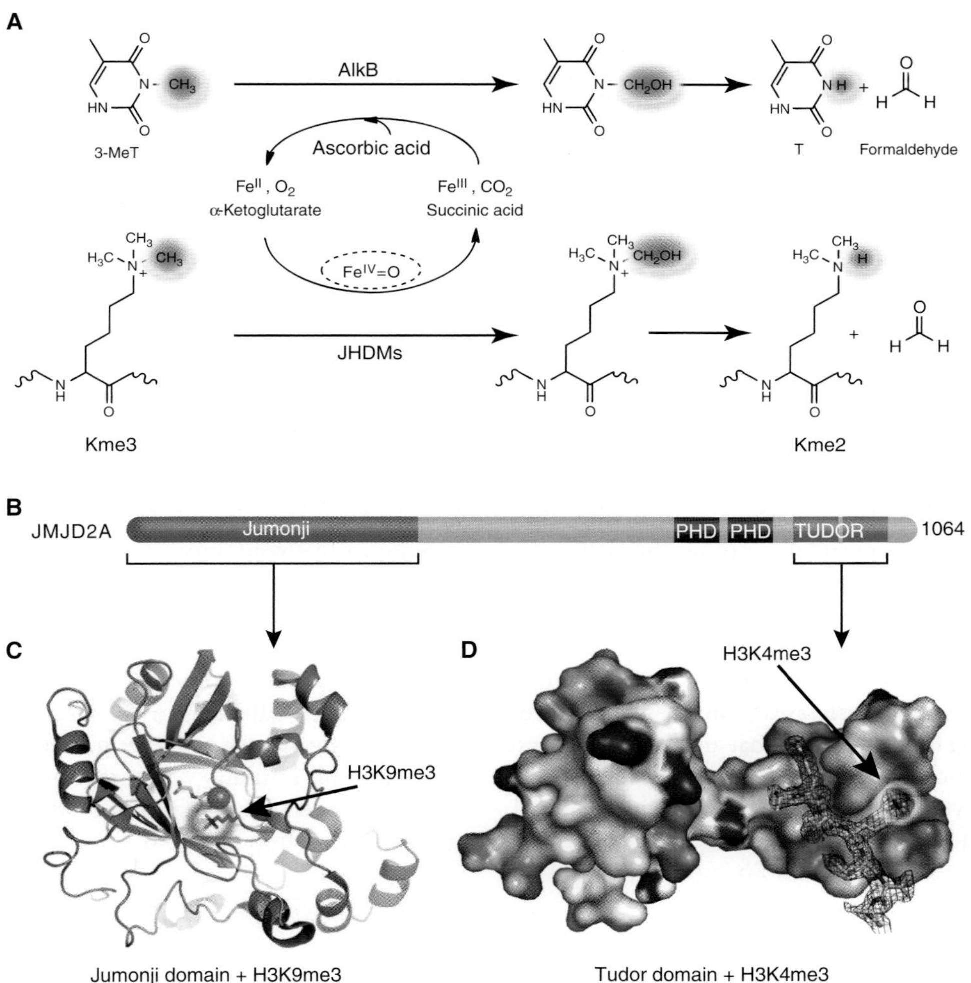

Figure 5. Demethylation by hydroxylation. (*A*) Mechanisms of demethylation of 3-methylthymine by AlkB (*top*) and of methyllysine by Jumonji-domain proteins (*bottom*). (*B*) Schematic representation of JMJD2A domain organization. (*C*) Structure of the amino-terminal Jumonji (ribbons) in complex with H3K9me3 (PDB 2OX0). (*D*) Structure of the carboxyl-terminal double Tudor domain (surface representation) in complex with H3K4me3 (PDB 2GFA). (Adapted, with permission, from Huang et al. 2006.)

proteins are members of the cupin super family (Clissold and Ponting 2001), including the Tet proteins involved in conversion of 5-methylcytosine to 5-hydroxymethylcytosine (Ch. 2 [Kriaucionis and Tahiliani 2014]; Ch. 15 [Li and Zhang 2014]). Demethylation reactions catalyzed by Jumonji enzymes follow a hydroxylation pathway involving a reactive Fe(IV) intermediate. As they do not require a lone pair of electrons on the target nitrogen atom, thereby they can demethylate mono-, di-, and trimethylated lysines (Fig. 5A) (Hoffart et al. 2006; Ozer and Bruick 2007).

JMJD2A contains carboxyl-terminal PHD and Tudor domains, which typically act as methyl-binding proteins, called readers, in addition to the amino-terminal Jumonji domain (Fig. 5B). The JMJD2A Jumonji domain alone is capable of demethylating H3K9me3/me2 and H3K36me3/me2. Structural studies revealed that the JMJD2A Jumonji domain predominantly recognizes the backbone of the histone peptides (unusual for a sequence-specific enzyme), which allows the enzyme to demethylate both H3K9 (Fig. 5C) and H3K36 (reviewed in Hou and Yu 2010; McDonough et al. 2010). The Tudor domain binds both H3K4me3 (Fig. 5D) and H4K20me3 (Huang et al. 2006; Lee et al. 2008). The functional connection between the methyl mark reader and eraser in JMJD2A, however, is currently not clear.

2 DNA METHYLATION

In mammals and other vertebrates, DNA methylation occurs at the C5 position of cytosine, generating 5-methylcytosine (5mC), mostly within CpG dinucleotides. This methylation, together with histone modifications, plays an important role in modulating chromatin structure, thus controlling gene expression and many other chromatin-dependent processes (Cheng and Blumenthal 2010; reviewed in Ch. 15 [Li and Zhang 2014]). The resulting epigenetic effects maintain the various patterns of gene expression in different cell types (reviewed in De Carvalho et al. 2010). In mammals, DNA methyltransferases (Dnmts) include three proteins belonging to two families that are structurally and functionally distinct. Dnmt3a and Dnmt3b establish the initial CpG methylation pattern de novo, whereas Dnmt1 maintains this pattern during chromosome replication and repair (see Fig. 2 of Ch. 15 [Li and Zhang 2014]).

2.1 Maintenance Methyltransferase Dnmt1

The so-called "maintenance" methyltransferase, Dnmt1, contains multiple functional domains (Fig. 6A) (Yoder et al. 1996). Structures are currently available for three amino-terminal deletions of mouse Dnmt1: an amino-terminal deletion of 350 residues (Δ350) in complex with AdoMet or its product AdoHcy (Takeshita et al. 2011), a larger deletion (Δ600) bound to DNA-containing an unmethylated CpG site (Song et al. 2011), and an even larger deletion (Δ730) bound to a hemimethylated CpG site (Song et al. 2012). When Dnmt1 is not bound to DNA, the amino-terminal replication focus targeting sequence (RFTS) domain of Δ350 is inserted into the DNA-binding surface cleft of the carboxyl-terminal MTase domain (Fig. 6B), indicating that this domain must be removed for methylation to occur. The RFTS domain is required for targeting Dnmt1 to replication foci, in which hemimethylated DNA is transiently generated. When the isolated RFTS domain is added in *trans* to an RFTS-lacking Dnmt1 protein, the RFTS domain acts as a DNA-competitive inhibitor of Dnmt1 (Syeda et al. 2011). In the structure of the Δ600 fragment, lacking the RFTS domain, the DNA bound CXXC domain (which specifically binds nonmethylated DNA) positions itself in the catalytic domain and prevents aberrant de novo methylation of CpG sequences (Fig. 6C). Only after physical removal of both the RFTS and CXXC domains can the carboxyl-terminal half of Dnmt1 (Δ730) bind to hemimethylated CpG DNA by flipping the target cytosine out of the double-stranded DNA helix into the active site (Fig. 6D). Thus, a multistep process, accompanied by structural changes, must occur during the targeting of full-length Dnmt1 to replication foci, which undergoes maintenance methylation of hemimethylated CpG DNA. This involves the removal of both RFTS and CXXC domains away from the catalytic center.

Dnmt1 alone, however, is insufficient for proper maintenance methylation. In vivo, maintenance methylation of hemimethylated CpG dinucleotides by Dnmt1 at DNA replication forks requires an accessory protein called UHRF1 (ubiquitin-like, containing PHD and RING finger domains 1; Bostick et al. 2007; Sharif et al. 2007). UHRF1, a multidomain protein (Fig. 6E), binds both hemimethylated CpG site (Fig. 6F), the substrate of Dnmt1, and histone H3 (reviewed in Hashimoto et al. 2009). Somehow Dnmt1 must displace UHRF1 from the site to allow methylation. In the coming years, it will be important to understand how these multiple binding events are coordinated and whether they are cooperative for faithful mitotic inheritance of genomic methylation patterns.

2.2 De Novo Methyltransferase Dnmt3 Family

The Dnmt3 family includes two active de novo Dnmts, Dnmt3a and Dnmt3b, and one regulatory factor, Dnmt3-Like protein (Dnmt3L; Fig. 7A) (Goll and Bestor 2005). Dnmt3a and Dnmt3b have similar domain arrangements: a variable region at the amino terminus, followed by a PWWP (Pro-Trp-Trp-Pro) domain that may be involved

 Cite as *Cold Spring Harb Perspect Biol* doi: 10.1101/cshperspect.a018747

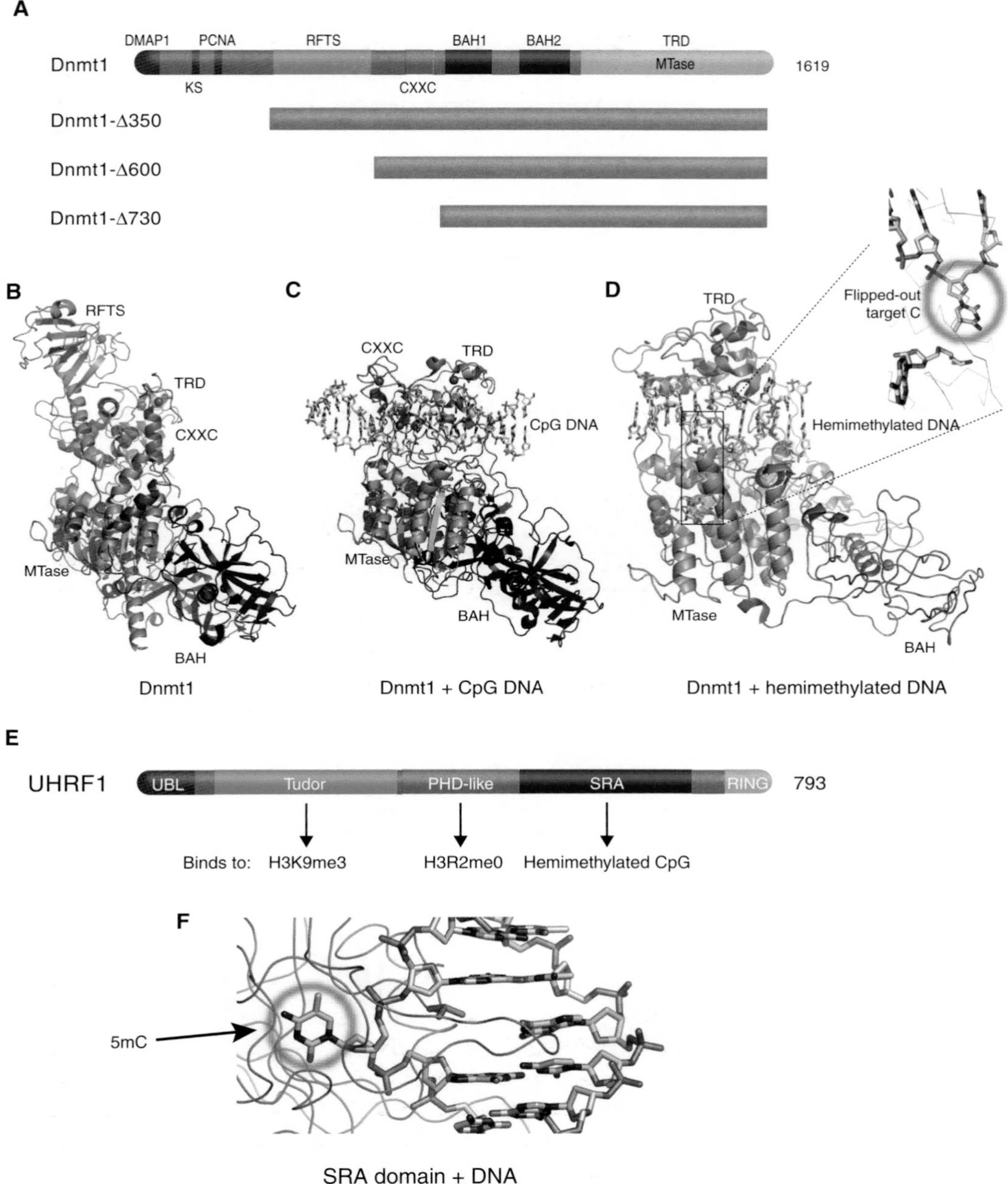

Figure 6. Structures of maintenance Dnmt1 and UHRF1. (*A*) Schematic representation of mouse Dnmt1 domain organization and available Dnmt1 amino-terminal deletion mutants. An amino-terminal region interacts with Dnmt1-associated protein(s) (DMAP1; Rountree et al. 2000). Then an adjacent lysine and serine are subject to a methylation and phosphorylation switch that determines Dnmt1 stability (Esteve et al. 2011), a PCNA (proliferating cell nuclear antigen) interacting sequence (Chuang et al. 1997), and an RFTS (Leonhardt et al. 1992) that interacts with the SET- and RING-associated (SRA) domain of UHRF1 (Achour et al. 2008). This is followed by a CpG-interacting CXXC domain (Song et al. 2011), a tandem BAH (bromo-adjacent homology) domain (Callebaut et al. 1999), and the catalytic DNA methyltransferase domain that includes the target-recognizing domain (Lauster et al. 1989) at the carboxyl terminus. (*B*) Structure of Dnmt1 in the absence of DNA (PDB 3AV4). (*C*) Structure of Dnmt1 in the presence of unmethylated CpG (PDB 3PT6). (*D*) Structure of Dnmt1 with hemimethylated CpG DNA oligonucleotides (PDB 4DA4). (*E*) UHRF1 harbors at least five recognizable functional domains: an ubiquitin-like domain at the amino terminus, followed by a tandem tudor domain recognizing H3K9me3 (Rothbart et al. 2012), a plant homeodomain (PHD) recognizing H3R2me0 (Rajakumara et al. 2011), an SRA domain recognizing hemimethylated CpG, and really interesting new gene (RING) domain at the carboxyl terminus that may endow UHRF1 with E3 ubiquitin ligase activity to histones (Citterio et al. 2004). (*F*) Structure of SRA-DNA complex illustrates 5mC flipped out from the DNA helix and bound in a cage-like pocket (circled in red; PDB 2ZO1). (Adapted from Hashimoto et al. 2009).

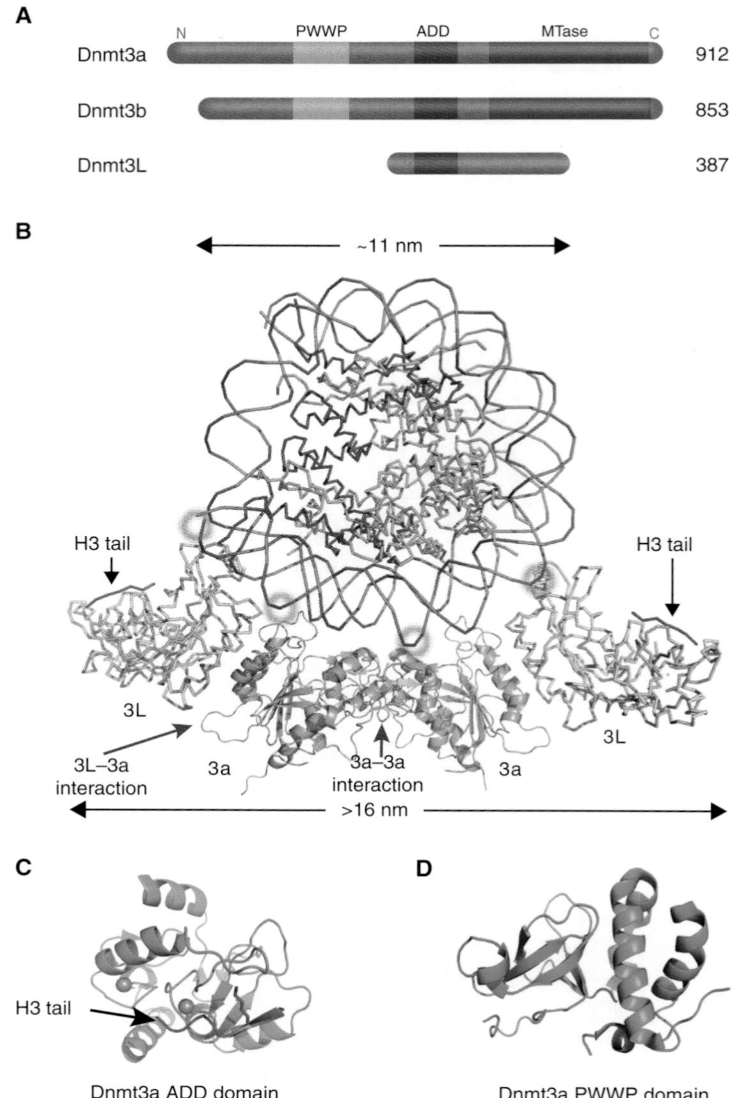

Figure 7. Structures of the de novo Dnmt3a-Dnmt3L complex. (*A*) Domain architecture of Dnmt3a, 3b, and 3L. (*B*) A nucleosome is shown docked to a Dnmt3L-3a-3a-3L tetramer (Dnmt3a is colored in green and Dnmt3L in gray; PDB 2QRV). (Adapted from Cheng and Blumenthal 2008). The position of the histone H3 amino-terminal tail (purple) bound to Dnmt3L is shown, taken from a cocrystal structure (PDB 2PVC). By wrapping the Dnmt3a/3L tetramer around the nucleosome, the two Dnmt3L molecules are able to bind both histone tails emanating from one nucleosome. The ~10 bp periodicity of binding to the major DNA groove is indicated by red circles. (*C*) Structure of the Dnmt3a ADD domain (PDB 3A1B) possibly interacting with histone tails from neighboring nucleosomes. (*D*) Structure of the Dnmt3a PWWP domain (PDB 3L1R).

in the nonspecific DNA binding for Dnmt3b (Qiu et al. 2002) or binding to histone H3K36me3 by Dnmt3a (Dhayalan et al. 2010), the Cys-rich 3-Zn-binding ADD (ATRX-DNMT3-DNMT3L) domain, and a carboxyl-terminal catalytic domain. The amino acid sequence of Dnmt3L is very similar to that of Dnmt3a and Dnmt3b in the ADD domain, but it lacks the conserved residues required for DNA methyltransferase activity in the carboxy-terminal domain. Structures are available for a complex between the carboxy-terminal domains of Dnmt3a and Dnmt3L (Jia et al. 2007) and the intact Dnmt3L in complex with a histone H3 amino-tail peptide (Fig. 7B) (Ooi et al. 2007). There is also structural data for an isolated ADD domain of Dnmt3a (Fig. 7C) (Otani et al. 2009), and isolated PWWP domains of Dnmt3a (Fig. 7D) and Dnmt3b (Qiu et al. 2002).

 Cite as *Cold Spring Harb Perspect Biol* doi: 10.1101/cshperspect.a018747

The phenotype of the Dnmt3L knockout mouse is indistinguishable from that of a Dnmt3a germ cell-specific conditional knockout—both have dispersed retrotransposons and aberrant germ cell patterns of de novo DNA methylation at loci that usually set parent-specific imprints (Bourc'his et al. 2001; Bourc'his and Bestor 2004; Kaneda et al. 2004; Webster et al. 2005). The interaction between Dnmt3L and Dnmt3a occurs in a minimal region located at the carboxy-terminal domain of both proteins, and is necessary for catalytic activity (Chedin et al. 2002; Suetake et al. 2004). The overall size of the Dnmt3a/Dnmt3L carboxy-terminal complex is ~16 nm long, which is greater than the diameter of an 11-nm core nucleosome (Fig. 7B). This complex contains two monomers of Dnmt3a and two of Dnmt3L, forming a tetramer with two 3L–3a interfaces and one 3a–3a interface (3L–3a–3a–3L). Substituting key, although noncatalytic residues at the Dnmt3a–3L or Dnmt3a–3a interfaces eliminates enzymatic activity, indicating that both interfaces are essential for catalysis (Jia et al. 2007).

The structure of the 3a–3a dimer interface suggests that the two active sites are located in adjacent DNA major grooves, facilitating the methylation by Dnmt3a of two CpGs separated by one helical turn, in one binding event (Fig. 7B). Methylation of CpG sites on long DNA substrates, by Dnmt3a, occurs with a periodicity of ~10 bp, suggesting a structural model in which Dnmt3a forms an oligomer docked to the DNA (Jia et al. 2007). This periodicity is also observed on maternally imprinted mouse genes (Jia et al. 2007). CpG methylation patterns on human chromosome 21 also correlate with a ~10-bp methylated CpG periodicity (Zhang et al. 2009). Interestingly, the ~10-bp methylation periodicity was evident in embryonic stem cells, however, often at non-CpG sites (which are substrates of Dnmt3a as well) that occur mostly in gene bodies as opposed to regulatory regions (Lister et al. 2009). Non-CG methylation is specific to the embryonic stem cell stage, as it disappears on induced differentiation and is restored in induced pluripotent stem cells. In the plant, *Arabidopsis thaliana* 10-bp periodic non-CpG DNA methylation by DRM2 (which is related to mammalian Dnmt3a) has similarly been observed (Cokus et al. 2008).

3 INTERPLAY BETWEEN DNA METHYLATION AND HISTONE MODIFICATION

3.1 Dnmt3L Connects Unmethylated H3K4 to De Novo DNA Methylation

Genome-scale DNA methylation profiles suggest that DNA methylation is correlated with histone methylation patterns (Meissner et al. 2008). Specifically, DNA methylation is correlated with the absence of H3K4 methylation and the presence of H3K9 methylation. Considering the inverse relationship between H3K4 methylation and DNA methylation, it is important to note that for the mammalian LSD histone demethylases, whose substrates include H3K4me2/me1, LSD1 is absolutely essential in maintaining global DNA methylation (Wang et al. 2009a) and LSD2 in establishing maternal DNA genomic imprints (Ciccone et al. 2009). Indeed, disruption of LSD1 results in earlier embryonic lethality and a more severe hypomethylation defect than disruption of the Dnmts themselves (Wang et al. 2009a).

The mammalian Dnmt3L-Dnmt3a de novo DNA methylation machinery can translate patterns of H3K4 methylation into heritable patterns of DNA methylation that mediate transcriptional silencing of the affected sequences (Ooi et al. 2007). Peptide interaction assays showed that Dnmt3L specifically interacts with the extreme amino terminus of histone H3; this interaction was strongly inhibited by H3K4 methylation, but was insensitive to modifications at other positions (Ooi et al. 2007). Cocrystallization of Dnmt3L with the amino tail of H3 showed this tail bound to the ADD domain of Dnmt3L (Fig. 7B), and substitution of key residues in the binding site eliminated the H3–Dnmt3L interaction. The main in vivo interaction partners of epitope-tagged Dnmt3L are Dnmt3a2, a shorter isoform of Dnmt3a (Chen et al. 2002), Dnmt3b, and the four core histones (Ooi et al. 2007). Given Dnmt3a and Dnmt3b bind nucleosomal DNA (Sharma et al. 2011), the data suggest that Dnmt3L is a probe of H3K4 methylation, and if the methylation is absent, then Dnmt3L induces de novo DNA methylation by docking activated Dnmt3a to the nucleosome.

Histone-Dnmt3L-Dnmt3a–DNA interactions have been studied in the budding yeast *S. cerevisiae* (Hu et al. 2009), which has no detectable DNA methylation (Proffitt et al. 1984) and lacks Dnmt orthologs. Introduction of the murine methyltransferases Dnmt1 or Dnmt3a leads to detectable, but extremely low levels of DNA methylation (Bulkowska et al. 2007). In contrast, a substantially higher level of de novo methylation can be achieved in yeast by coexpressing murine Dnmt3a and Dnmt3L (Hu et al. 2009). This induced DNA methylation was found preferentially in heterochromatic regions where H3K4 methylation is rare. When genes for components of the H3K4-methylating complex were disrupted in the context of Dnmt3a/3L overexpression, a greater level of genomic DNA methylation was observed. Deletions or targeted mutations in the ADD domain of Dnmt3L inhibited both global DNA methylation and the ability of Dnmt3L to associate with an H3K4me0 peptide. These same Dnmt3L mutants failed to restore normal DNA methylation to a specific promoter when introduced into embryonic stem cells from Dnmt3L$^{-/-}$ mice (Hu et al. 2009).

The above data has led to a model in which Dnmt3L binds to H3K4me0 (via its ADD domain) and recruits Dnmt3a to regions of chromatin where H3K4 is unmethylated. Such a model could explain part of the puzzle of how DNA methylation patterns are established de novo during embryonic and germ cell development, windows of time in which both proteins are expressed (Kato et al. 2007). However, whereas Dnmt3a and 3b expression is retained in somatic cells, Dnmt3L is expressed poorly, if at all, in differentiated cell types. This raises the question of how de novo DNA methylation is restricted in somatic cells, whether Dnmt3a and 3b alone are capable of discriminating H3K4 methylation status, and (if so) the structural basis for that discrimination. The key probably lies in the fact that, in vitro, the ADD domains of Dnmt3a or Dnmt3b possess the same H3 tail-binding specificity as that of Dnmt3L (Zhang et al. 2010), and a structure of the Dnmt3a ADD domain in complex with an amino-terminal-tail peptide from histone H3 indicates that the ADD domain is sufficient to recognize H3K4me0 (Fig. 7C) (Otani et al. 2009). Furthermore, Jeong et al. showed that in nuclei from HCT116 human colon cancer cells (which do not express DNMT3L) almost all of the cellular DNMT3a and 3b (but not DNMT1) was associated with nucleosomes (Jeong et al. 2009). Chromatin binding of DNMT3a and 3b required an intact nucleosomal structure, although no other chromatin factors, suggesting that DNMT3a and 3b alone are capable of direct interaction (via the ADD domain) with chromatin components (H3K4me0) in addition to DNA.

3.2 MLL1 Links H3K4 Methylation to Unmethylated CpGs

In humans there are at least eight HKMTs with specificity for H3K4. These include the mixed lineage leukemia (MLL) genes, MLL1–MLL5, hSET1a, hSET1b, and ASH1. MLL1/SET1-associated methyltransferase activity appears to be functional only in the context of multiprotein complexes; characterization of these reveals distinct multiprotein complexes for each with several shared components (reviewed in Cosgrove and Patel 2010). The MLL family plays an important role in embryonic development and is necessary for methylation of H3K4 at a subset of genes in the human and mouse genomes, particularly the HOX gene clusters (Ansari and Mandal 2010). Translocations involving MLL genes are involved in the etiology of myeloid and lymphoid leukemias. Considering the inverse relationship between H3K4 methylation and DNA methylation, it is interesting to note that disruption of the MLL1 gene in mice results in loss of H3K4 methylation and de novo DNA methylation at some *Hox* gene promoters (Milne et al. 2002; Terranova et al. 2006), suggesting that MLLs, directly or indirectly (through H3K4 methylation), prevent DNA methylation or perhaps stabilize unmethylated DNA. In fact, MLL proteins contain a CXXC domain, an evolutionarily conserved domain that mediates selective binding to unmethylated CpGs (Fig. 8A). The CXXC interaction with unmethylated CpGs was confirmed by a solution structure of an MLL1-CXXC domain complexed with unmethylated DNA (Fig. 8B) (Cierpicki et al. 2010) and an X-ray structure of DNMT1 in complex with unmethylated DNA (Fig. 6C) (Song et al. 2011). Structurally, the CXXC domain has a novel fold arranged in an elongated shape (Fig. 8B). The CXXC domain binds DNA in a clamp-like manner with the long axis of the structure linking the two Zn ions nearly perpendicular to the DNA axis (Fig. 8B) (Cierpicki et al. 2010).

The Set1 H3K4 methyltransferase also appears to interact with unmethylated DNA, although in this case it is via the Cfp1 accessory protein, which contains a CXXC domain (Fig. 8A) (Lee and Skalnik 2005; Lee et al. 2007). High throughput sequencing of Cfp1-bound chromatin identified a notable concordance between H3K4me3 and Cfp1 at unmethylated CpG islands in the mouse brain (Thomson et al. 2010). Also, Cfp1 binds specifically to the unmethylated allele of loci that are allele-specifically DNA methylated (e.g., imprinted loci, Xist gene). Depletion of Cfp1 results in a marked reduction in H3K4me3 genome-wide. The targeting of Cfp1 to CpG islands is independent of promoter activity as the insertion of an untranscribed, unmethylated CpG-dense construct into the genome of embryonic stem cells is sufficient to nucleate Cfp1 binding and H3K4me3. This suggests that unmethylated CpGs recruit Cfp1 and the associated methyltransferase Set1 creates new marks of H3K4me3 on the local chromatin (illustrated in Fig. 1 of Ch. 2 [Blackledge et al. 2013]).

CXXC domains are also found in Dnmt1 (Fig. 6), the methyl-CpG-binding protein MBD1 (Jorgensen et al. 2004), and Tet1, a Jumonji-like enzyme that catalyzes the conversion of 5mC to 5-hydroxymethylcytosine (Fig. 8A) (Tahiliani et al. 2009). Interestingly, the recurrent translocation, t(10;11)(q22;q23), has been described in acute myelogenous leukemias, and results in a fusion transcript that juxtaposes the first six exons of MLL1 (containing the AT hook and CXXC) to the carboxyl-terminal one third of TET1, thus "replacing" the TET1 CXXC with the MLL1 CXXC (labeled as the breakpoint in MLL1; Fig. 8A) (Ono et al. 2002; Lorsbach et al. 2003). Whether this leads to altered targeting of methyl hydroxylation remains to be determined.

Cite as *Cold Spring Harb Perspect Biol* doi: 10.1101/cshperspect.a018747

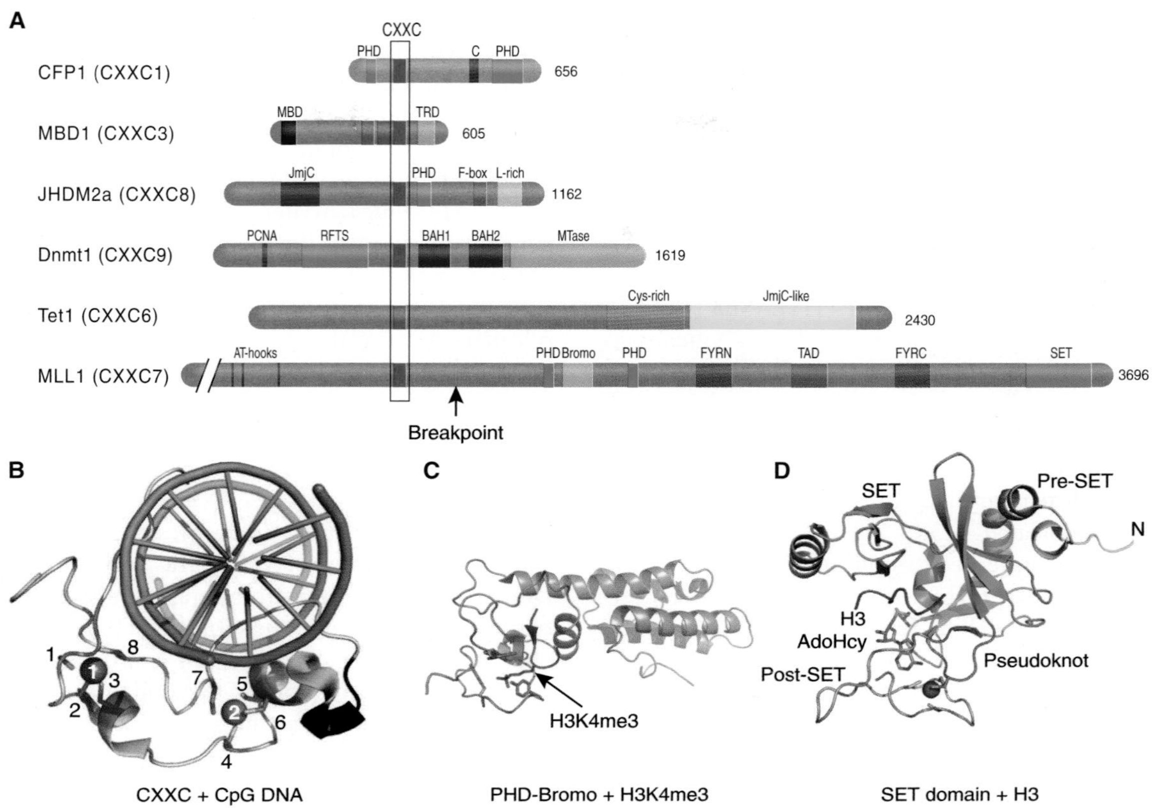

Figure 8. CpG-interacting proteins including MLL1. (*A*) Domain architecture of CXXC domain-containing proteins. (*B*–*D*) The structures of three isolated MLL1 domains in complex with interacting histones or DNA: (*B*) the amino-terminal CXXC domain in complex with CpG DNA (PDB 2KKF; Cierpicki et al. 2010), (*C*) the central PHD-bromodomain in complex with H3K4me3 peptide (PDB 3LQJ; Wang et al. 2010), and (*D*) the carboxyl-terminal SET domain (PDB 2W5Z; Southall et al. 2009).

3.3 JHDM1 Binds CpG DNA and Demethylates H3K36me2

Like the histone H3K4 methyltransferases of the MLL/SET1 family, the Jumonji domain-containing histone demethylases, JHDM1A (also known as CXXC8 or KDM2A; Fig. 8A) and JHDM1B (CXXC2 or KDM2B) have CXXC DNA-binding domains (Tsukada et al. 2006). Like the Set1-Cfp1 complex, JHDM1A is recruited to unmethylated CpG islands on a genome-wide scale via its CXXC domain (Blackledge et al. 2010). Like Cfp1, its localization to CpG islands is independent of promoter activity and gene-expression levels, and correlated with the selective depletion of H3K36me2/me1 within the CpG island, but not surrounding regions or the bodies of genes (see Fig. 1 of Ch. 2 [Blackledge et al. 2013]); knockdown of JHDM1A/KDM2A results in the selective accumulation of H3K36me2 in these regions. Consistent with the idea that DNA methylation restricts the localization of CXXC proteins, JHDM1A/KDM2A becomes mislocalized to DNA hypomethylated pericentric heterochromatin in DNA methyltransferase Dnmt1$^{-/-}$ mice. Although in vitro studies suggest that the CXXC domains can bind a single CpG site with micromolar affinity, both the Set1-Cfp1 and JHDM1A/KDM2A studies suggest that the targeting of CXXC proteins in vivo is dependent on CpG density as well as its methylation status. It could be possible that these proteins oligomerize and form nucleoprotein filaments on CpG-dense DNA, in a manner similar to that described for the DNA methyltransferase Dnmt3a-3L complex (Jurkowska et al. 2008). To reinforce the correlation between DNA methylation and H3K36 methylation, the PWWP domain of Dnmt3a is capable of binding H3K36me3 (Dhayalan et al. 2010) and directs DNA methylation (Chen et al. 2004).

3.4 Linkage between H3K9 Methylation and DNA Methylation

Methylation at H3K9 is positively correlated with DNA methylation, in contrast to its negative correlation with

H3K4 methylation. There is in vivo evidence that the H3K9-linked DNA methylations represent an evolutionarily conserved silencing pathway. Studies in *Neurospora* and *Arabidopsis* have shown a strict dependence of DNA methylation on the H3K9 methyltransferases Dim-5 and KRYPTONITE (KYP; Tamaru and Selker 2001; Jackson et al. 2002; Tamaru et al. 2003). The SRA domain of KYP (also known as SUVH4) binds directly to methylated CHG-containing oligonucleotides (Johnson et al. 2007), whereas a plant-specific DNA CHROMOMETHYLASE3 (CMT3), responsible for CHG methylation, binds H3K9me2-containing nucleosomes via its associated BAH and chromodomains within the same polypeptide (Du et al. 2012), resulting in a self-reinforcing loop between H3K9me2 and CHG methylation in plants.

With regard to mammals, G9a, and GLP, two related euchromatin-associated H3K9 methyltransferases form heterodimers and have been implicated in DNA methylation at various loci, including imprinting centers, retrotransposons and satellite repeats, a G9a/GLP target promoter, an Oct4 promoter, and a set of embryonic genes (reviewed in Collins and Cheng 2010; Shinkai and Tachibana 2011). Furthermore, G9a is required for de novo DNA methylation and the establishment of silencing of newly integrated proviruses in murine embryonic stem cells (Leung et al. 2011). The G9a/GLP heterodimer interacts with a chromodomain protein MPP8, which in turn interacts with DNA methyltransferase Dnmt3a and/or methylated H3K9 (Kokura et al. 2010; Chang et al. 2011). Together, these findings provide a molecular explanation, at least in part, for the co-occurrence of DNA methylation and H3K9 methylation in chromatin.

The functional relationships between DNA methylation and two other H3K9 methyltransferases in mammals are complex and context dependent, as deletion of Suv39h1/h2 (Lehnertz et al. 2003) or SETDB1 (Matsui et al. 2010) has only a minor impact on DNA methylation at constitutive heterochromatin or endogenous retroelements, respectively. A particularly interesting observation is that the methyl-CpG-binding domain protein MBD1 (Fig. 8A) forms a stable complex with SETDB1 (Sarraf and Stancheva 2004; Lyst et al. 2006) as well as the Suv39h1/HP1 complex (Fujita et al. 2003), constituting the heterochromatin-specific H3K9me3 writer and reader. The methyl-CpG-binding domain (MBD) is present in a family of proteins conserved throughout the eukaryotic lineage. This domain, in some but not all cases, confers the ability to bind fully methylated CpGs. Mammals have five well-characterized members of this family, each with unique biological characteristics (reviewed in Dhasarathy and Wade 2008; Ch. 15 [Li and Zhang 2014]). SETDB1 also contains an intrinsic putative MBD domain with two conserved DNA-interacting arginine residues known to make direct contact with DNA in the structures of the MBD domain (reviewed in Hashimoto et al. 2010). It remains to be seen whether the putative MBD domain of SETDB1 is similarly able to selectively bind methylated DNA. The intrinsic or associated coupling of a DNA methylation "reader" (MBD) with an H3K9me3 "writer" (SETDB1) implies an interdependent mechanism for the propagation or maintenance of these marks.

Finally, the identification of UHRF1 and its potential role in modulating the specificity of Dnmt1 for hemimethylated CpG sites and binding of histones provides another layer to the mechanism that ensures the faithful transmission of epigenetic information during DNA replication. Given that UHRF1 has the potential to interact with both hemimethylated CpGs (via the SRA domain) and H3K9me3 (via the Tudor domain), and is known to interact with a wide variety of epigenetic regulators, including Dnmt1, the H3K9 methyltransferase G9a and a histone acetyltransferase Tip60, it is possible that UHRF1 and the proteins in this larger complex play a more central role in coupling the transmission of DNA and histone methylation (H3K9, in particular) during mitotic cell division. It is interesting to know that, like plant CMT3, the mammalian Dnmt1 (and its homologs in *Neurospora* DIM2 and plant MET1) contains a BAH domain(s) within the same polypeptide (Fig. 6A). It remains to be seen whether the BAH domains of Dnmt1 are similarly able to bind methylated histones, either methylated H3K9, as recognized by the BAH domain of CMT3, or methylated H4K20, as recognized by the BAH domain of human ORC1 (origin of replication complex; Kuo et al. 2012).

3.5 PHF8 Binds H3K4me3 and Demethylates H3K9me2

The examples discussed so far provide a molecular explanation, at least in part, for the inverse correlation of DNA methylation and H3K4 methylation, as well as the co-occurrence of DNA methylation and H3K9 methylation in chromatin. The functional linkage of DNA methylation, H3K4 methylation, and H3K9 methylation is further illustrated by the finding that treatment with 5-aza-2′-deoxycytidine (5-aza), a DNA-demethylating drug (Yoo et al. 2007), leads to depletion of DNA methylation and H3K9 methylation, and a corresponding increase in H3K4 methylation (Nguyen et al. 2002). How is the inverse correlation of H3K4 and H3K9 methylation maintained?

PHF8 belongs to a small family of Jumonji proteins with three members in mice and human (PHF2, PHF8, and KIAA1718; Klose et al. 2006). Mutations in the PHF8 gene lead to X-linked mental retardation (Loenarz

 Cite as *Cold Spring Harb Perspect Biol* doi: 10.1101/cshperspect.a018747

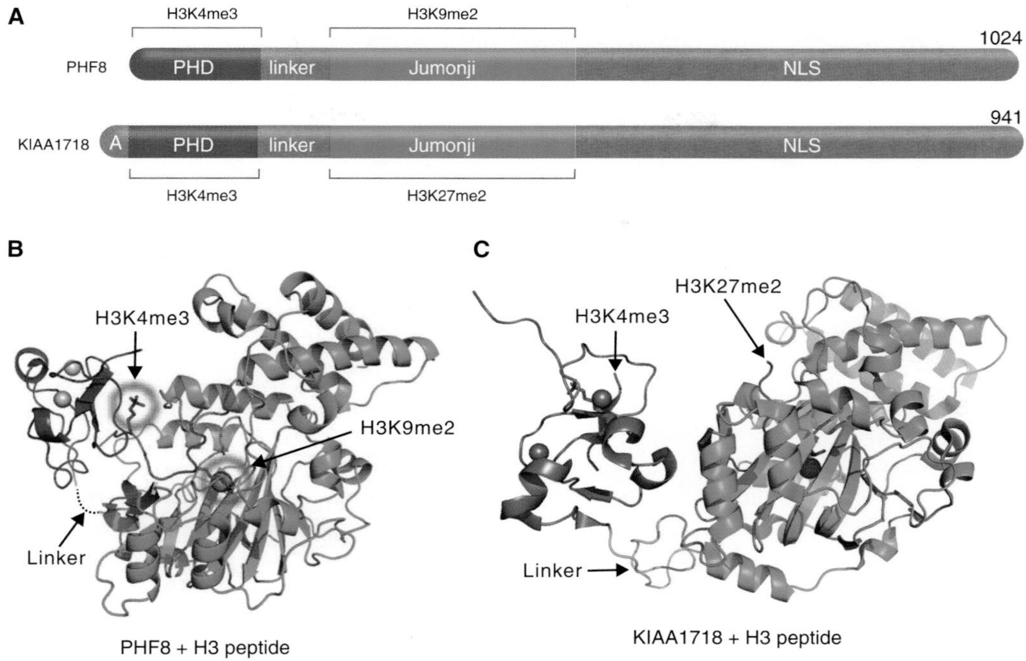

Figure 9. Coordinated methyllysine erasure between a Jumonji and a PHD. (*A*) Schematic representations of PHF8 and KIAA1718 domain structure. (*B*) A bent conformation of PHF8 bound to a histone H3 peptide (in red) containing K4me3 and K9me2 (circled in red; PDB 3KV4). (*C*) An extended conformation of KIAA1718 (PDB 3KV6). The position of the histone H3 peptide is taken from a cocrystal structure of *Caenorhabditis elegans* KIAA1718 (PDB 3N9P).

et al. 2010), and knockdown of KIAA1718 (also known as JHDM1D) and PHF8 homologs in zebrafish cause brain defects (Qi et al. 2010; Tsukada et al. 2010). These proteins harbor two domains in the amino-terminal half (Fig. 9A): a PHD domain that binds H3K4me3 and a Jumonji domain that demethylates H3K9me2, H3K27me2, or H4K20me1. However, the presence of H3K4me3 on the same peptide as H3K9me2 makes the doubly methylated peptide a significantly better substrate of PHF8 (Feng et al. 2010; Fortschegger et al. 2010; Horton et al. 2010; Kleine-Kohlbrecher et al. 2010). In contrast, the presence of H3K4me3 has the opposite effect in that it diminishes the H3K9me2 demethylase activity of KIAA1718 with no adverse effect on its H3K27me2 activity (Horton et al. 2010). Differences in substrate specificity between the two enzymes are explained by a bent conformation of PHF8 (Fig. 9B), allowing each of its domains to engage their respective targets, and an extended conformation of KIAA1718, which prevents the access to H3K9me2 by its Jumonji domain when its PHD domain engages H3K4me3 (Fig. 9C). Thus, the structural linkage between the PHD domain binding to H3K4me3 and the placement of the catalytic Jumonji domains relative to this active epigenetic mark determines which repressive marks (H3K9me2 or H3K27me2) are removed by these demethylases. Thus, the data indicate that the PHF8 and KIAA1718 Jumonji domains on their own are promiscuous enzymes; the PHD domains and linker—a determinant for the relative positioning of the two domains—are mainly responsible for substrate specificity.

Another structural study on *C. elegans* KIAA1718 suggested that the PHD and Jumonji domains might enable a *trans*-histone peptide-binding mechanism, in which the substrate peptide associated with the PHD domain and the peptide bound to the Jumonji domain could be coming from two separate histone molecules of the same nucleosome or two neighboring nucleosomes (Yang et al. 2010). The *trans*-binding mechanism is an attractive model for PHF8 and could explain the finding that PHF8 also functions in vivo as an H4K20me1 (histone H4 monomethylated at lysine 20) demethylase, whereas its PHD domain interacts with H3K4me3 in the context of nucleosome (Liu et al. 2010; Qi et al. 2010). But one has to explain why PHF8 is only active on monomethylated H4K20 (H4K20me1), whereas it is active on dimethylated H3K9 and H3K27. One possibility is that only H4K20me1 coexists with H3K4me3 in vivo.

4 SUMMARY

Combinatorial readout of multiple covalent chromatin modifications (including DNA methylation) is an explicit

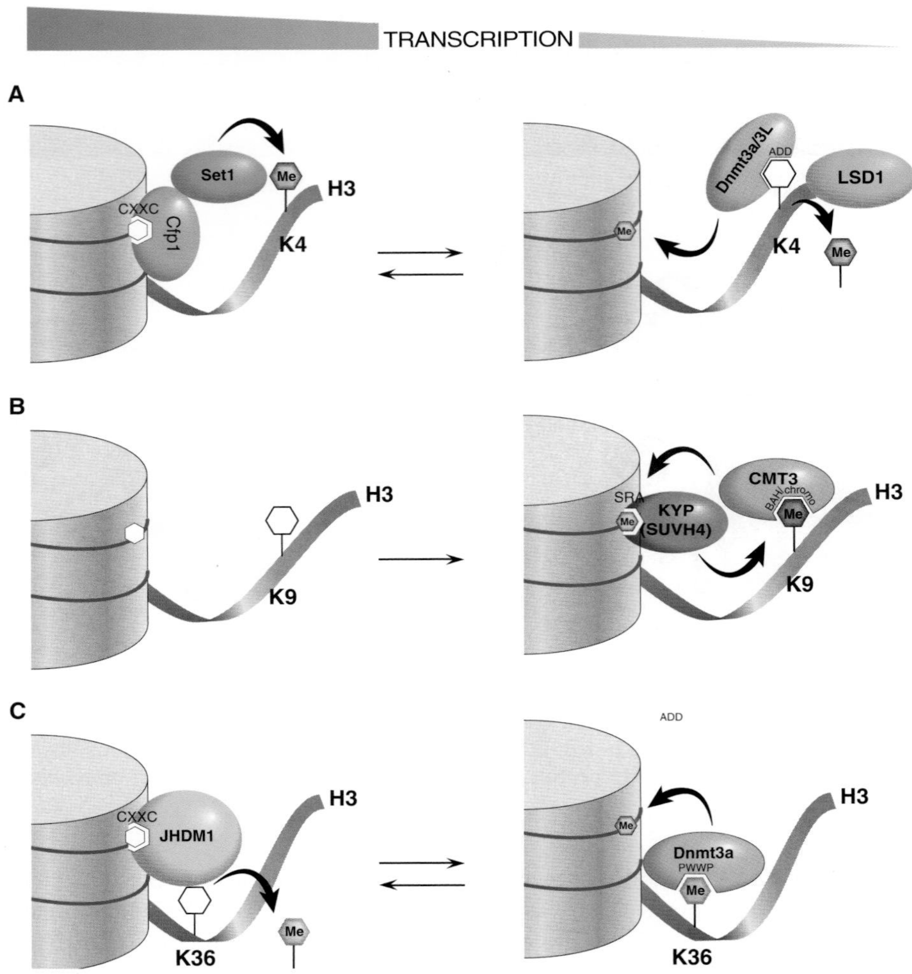

Figure 10. Cartoons of interactions that regulate DNA methylation and associated histone H3 modifications. There are three examples of PTM cross talk between histone and DNA methylation. Chromatin on the *left* represents transcriptionally active states, whereas chromatin on the *right* represents transcriptionally repressive states. The "Me"-labeled filled hexagons indicate one or more methyl groups in DNA (pink) or protein lysine residues (K). The catalytic action of methylation writers and erasers are indicated by curved black arrows. Methylation readers interact via specific labeled domains that fit in a lock-and-key fashion to methylated (filled hexagons) or unmethylated CpGs or lysines (unfilled hexagons). (*A*) Enzymatic reactions by the Dnmt3a-Dnmt3L complex and the SET1/CFP1 complex regulate the inverse correlation of DNA methylation and H3K4 methylation. (*B*) In plants, enzymatic reactions by CMT3 and KYP reinforce the correlation between DNA CHG methylation and H3K9 methylation. (*C*) Enzymatic reactions by Dnmt3a and JHDM1 positively regulate the association of DNA methylation with H3K36 methylation.

prediction of the "histone code" hypothesis (Strahl and Allis 2000; Jenuwein and Allis 2001; Turner 2007). Although it is well-accepted that DNA methylation patterns are replicated in a semiconservative fashion during cell division via a Dnmt1-dependent mechanism discussed (see also Ch. 15 [Li and Zhang 2014]), one of the fundamental unresolved questions is how, and indeed whether, histone modifications are similarly "inherited." Considering that the well-studied lysine methylation events reside on histone H3 (K4, K9, K27, K36, and K79) or H4 (K20), this evokes a model in which "old" histone methylation patterns may be retained (possibly by UHRF1) and copied onto newly deposited tetramers from neighboring parental nucleosomes. Indeed, many of the SET domain histone methyltransferases contain intrinsic or associated reader domains that recognize the same mark that they generate, allowing for the copying of these marks from old to new nucleosomes. For example, G9a/GLP catalyzes H3K9me1/2 and contains an ankyrin-repeat domain that binds H3K9me1/2 (Collins et al. 2008). Likewise, SUV39H1/2, the H3K9me3 writer, interacts with HP1, the H3K9me3 reader (reviewed in Grewal and Jia 2007). Similarly, yeast Clr4 methylates H3K9 and contains a chromodomain that binds H3K9me3 (Zhang et al. 2008). This interdomain cross talk provides a possible mechanism for propagating a methyl mark. Thus, higher organisms have evolved coordinated mechanisms of deposition and transmission of repressive chromatin marks to both DNA and histones. Enzymes that affect more complex cross talk include PHF8, which contains modules within the same polypeptide for both recognizing (PHD) and removing (Jumonji domain) two opposing methyl marks. This cross talk provides a possible mechanism for removing a repressive methyl mark (H3K9me2 or H4K20me1) based on an existing active methyl mark (H3K4me3). An even more complex situation is JARID1A, which contains multiple PHD domains for recognizing the substrate (H3K4me3) and product (H3K4me0) of its catalytic Jumonji domain, respectively (Wang et al. 2009b).

We have also discussed enzyme complexes that cross talk between DNA methylation (or lack thereof) and histone marks that are probably on the same nucleosome. These include the Dnmt3a-Dnmt3L complex, containing a reader domain for H3K4me0 coupled to DNA methyltransferase activity, whereas MLL1 (or Set1-Cfp1 complex) contains reader domains for DNA CpG and a SET domain for making methylated H3K4 (Fig. 10A). The mammalian Dnmt1-UHRF1 complex (Rothbart et al. 2012) and plant-specific CMT3 (Du et al. 2012) contain reader domains for H3K9me3/2 and DNA methyltransferase activity (Fig. 10B). The function of the Jumonji H3K36me3-specific demethylase JHDM1 is linked to the CXXC domain, which associates with unmethylated CpG DNA (Fig. 10C) (Blackledge et al. 2010).

Another intriguing observation involving DNA methylation is its mutually antagonistic relationship with histone variant H2A.Z (Zilberman et al. 2008; Conerly et al. 2010). How the exclusion is specifically established remains largely unknown. One possibility is that the histone variant H2A.Z is preferentially deposited by the remodeling ATPase complexes to regions lacking DNA methylation. Another scenario is that nucleosomes that contain the histone variant H2A.Z are no longer the substrates of DNA methyltransferases. Future experiments are needed to uncover the mechanisms of correct assembly of machinery required to accurately modify chromatin. Although the field still faces a number of critical questions, it is clear that structural analyses will continue to play a central and synergistic role, together with biochemical and genetic studies to address them.

ACKNOWLEDGMENTS

The author wishes to thank the former and current members of his laboratory, Xing Zhang for preparing Table 1, and support from the U.S. National Institutes of Health (GM049245).

REFERENCES

*Reference is also in this book.

Achour M, Jacq X, Ronde P, Alhosin M, Charlot C, Chataigneau T, Jeanblanc M, Macaluso M, Giordano A, Hughes AD, et al. 2008. The interaction of the SRA domain of ICBP90 with a novel domain of DNMT1 is involved in the regulation of VEGF gene expression. *Oncogene* **27:** 2187–2197.

An W, Kim J, Roeder RG. 2004. Ordered cooperative functions of PRMT1, p300, and CARM1 in transcriptional activation by p53. *Cell* **117:** 735–748.

Ansari KI, Mandal SS. 2010. Mixed lineage leukemia: Roles in gene expression, hormone signaling and mRNA processing. *FEBS J* **277:** 1790–1804.

Antonysamy S, Bonday Z, Campbell RM, Doyle B, Druzina Z, Gheyi T, Han B, Jungheim LN, Qian Y, Rauch C, et al. 2012. Crystal structure of the human PRMT5:MEP50 complex. *Proc Natl Acad Sci* **109:** 17960–17965.

Bannister AJ, Schneider R, Kouzarides T. 2002. Histone methylation: Dynamic or static? *Cell* **109:** 801–806.

Baron R, Binda C, Tortorici M, McCammon JA, Mattevi A. 2011. Molecular mimicry and ligand recognition in binding and catalysis by the histone demethylase LSD1-CoREST complex. *Structure* **19:** 212–220.

* Becker PB, Workman JL. 2013. Nucleosome remodeling and epigenetics. *Cold Spring Harb Perspect Biol* **5:** a017905.

Bedford MT, Clarke SG. 2009. Protein arginine methylation in mammals: Who, what, and why. *Mol Cell* **33:** 1–13.

Blackledge NP, Zhou JC, Tolstorukov MY, Farcas AM, Park PJ, Klose RJ. 2010. CpG islands recruit a histone H3 lysine 36 demethylase. *Mol Cell* **38:** 179–190.

* Blackledge N, Thomson JP, Skene PJ. 2013. CpG island chromatin is shaped by recruitment of ZF-CxxC proteins. *Cold Spring Harb Perspect Biol* **5:** a018648.

Bostick M, Kim JK, Esteve PO, Clark A, Pradhan S, Jacobsen SE. 2007. UHRF1 plays a role in maintaining DNA methylation in mammalian cells. *Science* **317:** 1760–1764.

Bourc'his D, Bestor TH. 2004. Meiotic catastrophe and retrotransposon reactivation in male germ cells lacking Dnmt3L. *Nature* **431:** 96–99.

Bourc'his D, Xu GL, Lin CS, Bollman B, Bestor TH. 2001. Dnmt3L and the establishment of maternal genomic imprints. *Science* **294:** 2536–2539.

Branscombe TL, Frankel A, Lee JH, Cook JR, Yang Z, Pestka S, Clarke S. 2001. PRMT5 (Janus kinase-binding protein 1) catalyzes the formation of symmetric dimethylarginine residues in proteins. *J Biol Chem* **276:** 32971–32976.

Briggs SD, Xiao T, Sun ZW, Caldwell JA, Shabanowitz J, Hunt DF, Allis CD, Strahl BD. 2002. Gene silencing: Trans-histone regulatory pathway in chromatin. *Nature* **418:** 498.

Bulkowska U, Ishikawa T, Kurlandzka A, Trzcinska-Danielewicz J, Derlacz R, Fronk J. 2007. Expression of murine DNA methyltransferases Dnmt1 and Dnmt3a in the yeast *Saccharomyces cerevisiae*. *Yeast* **24:** 871–882.

Callebaut I, Courvalin JC, Mornon JP. 1999. The BAH (bromo-adjacent homology) domain: A link between DNA methylation, replication and transcriptional regulation. *FEBS Lett* **446:** 189–193.

Chang Y, Sun L, Kokura K, Horton JR, Fukuda M, Espejo A, Izumi V, Koomen JM, Bedford MT, Zhang X, et al. 2011. MPP8 mediates the interactions between DNA methyltransferase Dnmt3a and H3K9 methyltransferase GLP/G9a. *Nat Commun* **2:** 533.

Chedin F, Lieber MR, Hsieh CL. 2002. The DNA methyltransferase-like protein DNMT3L stimulates de novo methylation by Dnmt3a. *Proc Natl Acad Sci* **99:** 16916–16921.

Chen D, Ma H, Hong H, Koh SS, Huang SM, Schurter BT, Aswad DW, Stallcup MR. 1999. Regulation of transcription by a protein methyltransferase. *Science* **284:** 2174–2177.

Chen T, Ueda Y, Xie S, Li E. 2002. A novel Dnmt3a isoform produced from an alternative promoter localizes to euchromatin and its expression correlates with active de novo methylation. *J Biol Chem* **277:** 38746–38754.

Chen T, Tsujimoto N, Li E. 2004. The PWWP domain of Dnmt3a and Dnmt3b is required for directing DNA methylation to the major satellite repeats at pericentric heterochromatin. *Mol Cell Biol* **24:** 9048–9058.

Cheng X, Blumenthal RM. 2008. Mammalian DNA methyltransferases: A structural perspective. *Structure* **16:** 341–350.

Cheng X, Blumenthal RM. 2010. Coordinated chromatin control: Structural and functional linkage of DNA and histone methylation. *Biochemistry* **49:** 2999–3008.

Cheng X, Collins RE, Zhang X. 2005. Structural and sequence motifs of protein (histone) methylation enzymes. *Annu Rev Biophys Biomol Struct* **34:** 267–294.

Chuang LS, Ian HI, Koh TW, Ng HH, Xu G, Li BF. 1997. Human DNA-(cytosine-5) methyltransferase-PCNA complex as a target for p21^{WAF1}. *Science* **277:** 1996–2000.

Ciccone DN, Su H, Hevi S, Gay F, Lei H, Bajko J, Xu G, Li E, Chen T. 2009. KDM1B is a histone H3K4 demethylase required to establish maternal genomic imprints. *Nature* **461:** 415–418.

Cierpicki T, Risner LE, Grembecka J, Lukasik SM, Popovic R, Omonkowska M, Shultis DD, Zeleznik-Le NJ, Bushweller JH. 2010. Structure of the MLL CXXC domain-DNA complex and its functional role in MLL-AF9 leukemia. *Nat Struct Mol Biol* **17:** 62–68.

Citterio E, Papait R, Nicassio F, Vecchi M, Gomiero P, Mantovani R, Di Fiore PP, Bonapace IM. 2004. Np95 is a histone-binding protein endowed with ubiquitin ligase activity. *Mol Cell Biol* **24:** 2526–2535.

Clissold PM, Ponting CP. 2001. JmjC: Cupin metalloenzyme-like domains in jumonji, hairless and phospholipase $A_2\beta$. *Trends Biochem Sci* **26:** 7–9.

Cokus SJ, Feng S, Zhang X, Chen Z, Merriman B, Haudenschild CD, Pradhan S, Nelson SF, Pellegrini M, Jacobsen SE. 2008. Shotgun bisulphite sequencing of the *Arabidopsis* genome reveals DNA methylation patterning. *Nature* **452:** 215–219.

Collins R, Cheng X. 2010. A case study in cross-talk: The histone lysine methyltransferases G9a and GLP. *Nucleic Acids Res* **38:** 3503–3511.

Collins RE, Tachibana M, Tamaru H, Smith KM, Jia D, Zhang X, Selker EU, Shinkai Y, Cheng X. 2005. In vitro and in vivo analyses of a Phe/Tyr switch controlling product specificity of histone lysine methyltransferases. *J Biol Chem* **280:** 5563–5570.

Collins RE, Northrop JP, Horton JR, Lee DY, Zhang X, Stallcup MR, Cheng X. 2008. The ankyrin repeats of G9a and GLP histone methyltransferases are mono- and dimethyllysine binding modules. *Nat Struct Mol Biol* **15:** 245–250.

Conerly ML, Teves SS, Diolaiti D, Ulrich M, Eisenman RN, Henikoff S. 2010. Changes in H2A.Z occupancy and DNA methylation during B-cell lymphomagenesis. *Genome Res* **20:** 1383–1390.

Cook JR, Lee JH, Yang ZH, Krause CD, Herth N, Hoffmann R, Pestka S. 2006. FBXO11/PRMT9, a new protein arginine methyltransferase, symmetrically dimethylates arginine residues. *Biochem Biophys Res Commun* **342:** 472–481.

Cosgrove MS, Patel A. 2010. Mixed lineage leukemia: A structure-function perspective of the MLL1 protein. *FEBS J* **277:** 1832–1842.

Couture JF, Collazo E, Brunzelle JS, Trievel RC. 2005. Structural and functional analysis of SET8, a histone H4 Lys-20 methyltransferase. *Genes Dev* **19:** 1455–1465.

Daigle SR, Olhava EJ, Therkelsen CA, Majer CR, Sneeringer CJ, Song J, Johnston LD, Scott MP, Smith JJ, Xiao Y, et al. 2011. Selective killing of mixed lineage leukemia cells by a potent small-molecule DOT1L inhibitor. *Cancer Cell* **20:** 53–65.

Daujat S, Bauer UM, Shah V, Turner B, Berger S, Kouzarides T. 2002. Crosstalk between CARM1 methylation and CBP acetylation on histone H3. *Curr Biol* **12:** 2090–2097.

De Carvalho DD, You JS, Jones PA. 2010. DNA methylation and cellular reprogramming. *Trends Cell Biol* **20:** 609–617.

Dhasarathy A, Wade PA. 2008. The MBD protein family-reading an epigenetic mark? *Mutat Res* **647:** 39–43.

Dhayalan A, Rajavelu A, Rathert P, Tamas R, Jurkowska RZ, Ragozin S, Jeltsch A. 2010. The Dnmt3a PWWP domain reads histone 3 lysine 36 trimethylation and guides DNA methylation. *J Biol Chem* **285:** 26114–26120.

Dlakic M. 2001. Chromatin silencing protein and pachytene checkpoint regulator Dot1p has a methyltransferase fold. *Trends Biochem Sci* **26:** 405–407.

Du J, Zhong X, Bernatavichute YV, Stroud H, Feng S, Caro E, Vashisht AA, Terragni J, Chin HG, Tu A, et al. 2012. Dual binding of chromomethylase domains to H3K9me2-containing nucleosomes directs DNA methylation in plants. *Cell* **151:** 167–180.

Esteve PO, Chang Y, Samaranayake M, Upadhyay AK, Horton JR, Feehery GR, Cheng X, Pradhan S. 2011. A methylation and phosphorylation switch between an adjacent lysine and serine determines human DNMT1 stability. *Nat Struct Mol Biol* **18:** 42–48.

Feng W, Yonezawa M, Ye J, Jenuwein T, Grummt I. 2010. PHF8 activates transcription of rRNA genes through H3K4me3 binding and H3K9me1/2 demethylation. *Nat Struct Mol Biol* **17:** 445–450.

Forneris F, Binda C, Vanoni MA, Battaglioli E, Mattevi A. 2005. Human histone demethylase LSD1 reads the histone code. *J Biol Chem* **280:** 41360–41365.

Forneris F, Binda C, Adamo A, Battaglioli E, Mattevi A. 2007. Structural basis of LSD1-CoREST selectivity in histone H3 recognition. *J Biol Chem* **282:** 20070–20074.

Fortschegger K, de Graaf P, Outchkourov NS, van Schaik FM, Timmers HT, Shiekhattar R. 2010. PHF8 targets histone methylation and RNA polymerase II to activate transcription. *Mol Cell Biol* **30:** 3286–3298.

Frederiks F, Stulemeijer IJ, Ovaa H, van Leeuwen F. 2011. A modified epigenetics toolbox to study histone modifications on the nucleosome core. *Chembiochem* **12:** 308–313.

Friesen WJ, Paushkin S, Wyce A, Massenet S, Pesiridis GS, Van Duyne G, Rappsilber J, Mann M, Dreyfuss G. 2001. The methylosome, a 20S complex containing JBP1 and pICln, produces dimethylarginine-modified Sm proteins. *Mol Cell Biol* **21:** 8289–8300.

Friesen WJ, Wyce A, Paushkin S, Abel L, Rappsilber J, Mann M, Dreyfuss G. 2002. A novel WD repeat protein component of the methylosome binds Sm proteins. *J Biol Chem* **277:** 8243–8247.

Fujita N, Watanabe S, Ichimura T, Tsuruzoe S, Shinkai Y, Tachibana M, Chiba T, Nakao M. 2003. Methyl-CpG binding domain 1 (MBD1) interacts with the Suv39h1-HP1 heterochromatic complex for DNA methylation-based transcriptional repression. *J Biol Chem* **278:** 24132–24138.

Goll MG, Bestor TH. 2005. Eukaryotic cytosine methyltransferases. *Annu Rev Biochem* **74:** 481–514.

Grewal SI, Jia S. 2007. Heterochromatin revisited. *Nat Rev Genet* **8:** 35–46.

* Grunstein M, Gasser SM. 2013. Epigenetics in *Saccharomyces cerevisiae*. *Cold Spring Harb Perspect Biol* **5:** a017491.

Hashimoto H, Horton JR, Zhang X, Cheng X. 2009. UHRF1, a modular multi-domain protein, regulates replication-coupled crosstalk between DNA methylation and histone modifications. *Epigenetics* **4:** 8–14.

Hashimoto H, Vertino PM, Cheng X. 2010. Molecular coupling of DNA methylation and histone methylation. *Epigenomics* **2:** 657–669.

* Henikoff S, Smith MM. 2014. Histone variants and epigenetics. *Cold Spring Harb Perspect Biol* doi: 10.1101/cshperspect.a019364.

Herrmann F, Pably P, Eckerich C, Bedford MT, Fackelmayer FO. 2009. Human protein arginine methyltransferases in vivo—Distinct properties of eight canonical members of the PRMT family. *J Cell Sci* **122:** 667–677.

Higashimoto K, Kuhn P, Desai D, Cheng X, Xu W. 2007. Phosphorylation-mediated inactivation of coactivator-associated arginine methyltransferase 1. *Proc Natl Acad Sci* **104:** 12318–12323.

Hoffart LM, Barr EW, Guyer RB, Bollinger JM Jr, Krebs C. 2006. Direct spectroscopic detection of a C-H-cleaving high-spin Fe^{IV} complex in a prolyl-4-hydroxylase. *Proc Natl Acad Sci* **103:** 14738–14743.

Horton JR, Upadhyay AK, Qi HH, Zhang X, Shi Y, Cheng X. 2010. Enzymatic and structural insights for substrate specificity of a family of jumonji histone lysine demethylases. *Nat Struct Mol Biol* **17:** 38–43.

Hou H, Yu H. 2010. Structural insights into histone lysine demethylation. *Curr Opin Struct Biol* **20:** 739–748.

Hu JL, Zhou BO, Zhang RR, Zhang KL, Zhou JQ, Xu GL. 2009. The N-terminus of histone H3 is required for de novo DNA methylation in chromatin. *Proc Natl Acad Sci* **106:** 22187–22192.

Huang Y, Fang J, Bedford MT, Zhang Y, Xu RM. 2006. Recognition of histone H3 lysine-4 methylation by the double tudor domain of JMJD2A. *Science* **312:** 748–751.

Huang J, Sengupta R, Espejo AB, Lee MG, Dorsey JA, Richter M, Opravil S, Shiekhattar R, Bedford MT, Jenuwein T, et al. 2007. p53 is regulated by the lysine demethylase LSD1. *Nature* **449:** 105–108.

Jackson JP, Lindroth AM, Cao X, Jacobsen SE. 2002. Control of CpNpG DNA methylation by the KRYPTONITE histone H3 methyltransferase. *Nature* **416:** 556–560.

Jenuwein T, Allis CD. 2001. Translating the histone code. *Science* **293:** 1074–1080.

Jenuwein T, Laible G, Dorn R, Reuter G. 1998. SET domain proteins modulate chromatin domains in eu- and heterochromatin. *Cell Mol Life Sci* **54:** 80–93.

Jeong S, Liang G, Sharma S, Lin JC, Choi SH, Han H, Yoo CB, Egger G, Yang AS, Jones PA. 2009. Selective anchoring of DNA methyltransferases 3A and 3B to nucleosomes containing methylated DNA. *Mol Cell Biol* **29:** 5366–5376.

Jia D, Jurkowska RZ, Zhang X, Jeltsch A, Cheng X. 2007. Structure of Dnmt3a bound to Dnmt3L suggests a model for de novo DNA methylation. *Nature* **449:** 248–251.

Johnson LM, Bostick M, Zhang X, Kraft E, Henderson I, Callis J, Jacobsen SE. 2007. The SRA methyl-cytosine-binding domain links DNA and histone methylation. *Curr Biol* **17:** 379–384.

Jorgensen HF, Ben-Porath I, Bird AP. 2004. Mbd1 is recruited to both methylated and nonmethylated CpGs via distinct DNA binding domains. *Mol Cell Biol* **24:** 3387–3395.

Jurkowska RZ, Anspach N, Urbanke C, Jia D, Reinhardt R, Nellen W, Cheng X, Jeltsch A. 2008. Formation of nucleoprotein filaments by mammalian DNA methyltransferase Dnmt3a in complex with regulator Dnmt3L. *Nucleic Acids Res* **36:** 6656–6663.

Kaneda M, Okano M, Hata K, Sado T, Tsujimoto N, Li E, Sasaki H. 2004. Essential role for de novo DNA methyltransferase Dnmt3a in paternal and maternal imprinting. *Nature* **429:** 900–903.

Karytinos A, Forneris F, Profumo A, Ciossani G, Battaglioli E, Binda C, Mattevi A. 2009. A novel mammalian flavin-dependent histone demethylase. *J Biol Chem* **284:** 17775–17782.

Kato Y, Kaneda M, Hata K, Kumaki K, Hisano M, Kohara Y, Okano M, Li E, Nozaki M, Sasaki H. 2007. Role of the Dnmt3 family in de novo methylation of imprinted and repetitive sequences during male germ cell development in the mouse. *Hum Mol Genet* **16:** 2272–2280.

Kim S, Lim IK, Park GH, Paik WK. 1997. Biological methylation of myelin basic protein: Enzymology and biological significance. *Int J Biochem Cell Biol* **29:** 743–751.

Kleine-Kohlbrecher D, Christensen J, Vandamme J, Abarrategui I, Bak M, Tommerup N, Shi X, Gozani O, Rappsilber J, Salcini AE, et al. 2010. A functional link between the histone demethylase PHF8 and the transcription factor ZNF711 in X-linked mental retardation. *Mol Cell* **38:** 165–178.

Klose RJ, Kallin EM, Zhang Y. 2006. JmjC-domain-containing proteins and histone demethylation. *Nat Rev Genet* **7:** 715–727.

Kokura K, Sun L, Bedford MT, Fang J. 2010. Methyl-H3K9-binding protein MPP8 mediates E-cadherin gene silencing and promotes tumour cell motility and invasion. *EMBO J* **29:** 3673–3687.

* Kriaucionis S, Tahiliani M. 2014. Expanding the epigenetic landscape: Novel modifications of cytosine in genomic DNA. *Cold Spring Harb Perspect Biol* **6:** a018630.

Kuo AJ, Song J, Cheung P, Ishibe-Murakami S, Yamazoe S, Chen JK, Patel DJ, Gozani O. 2012. The BAH domain of ORC1 links H4K20me2 to DNA replication licensing and Meier-Gorlin syndrome. *Nature* **484:** 115–119.

Lauster R, Trautner TA, Noyer-Weidner M. 1989. Cytosine-specific type II DNA methyltransferases. A conserved enzyme core with variable target-recognizing domains. *J Mol Biol* **206:** 305–312.

Lee JH, Skalnik DG. 2005. CpG-binding protein (CXXC finger protein 1) is a component of the mammalian Set1 histone H3-Lys4 methyltransferase complex, the analogue of the yeast Set1/COMPASS complex. *J Biol Chem* **280:** 41725–41731.

Lee HW, Kim S, Paik WK. 1977. *S*-adenosylmethionine: Protein-arginine methyltransferase. Purification and mechanism of the enzyme. *Biochemistry* **16:** 78–85.

Lee YH, Koh SS, Zhang X, Cheng X, Stallcup MR. 2002. Synergy among nuclear receptor coactivators: Selective requirement for protein methyltransferase and acetyltransferase activities. *Mol Cell Biol* **22:** 3621–3632.

Lee JH, Cook JR, Yang ZH, Mirochnitchenko O, Gunderson SI, Felix AM, Herth N, Hoffmann R, Pestka S. 2005. PRMT7, a new protein arginine methyltransferase that synthesizes symmetric dimethylarginine. *J Biol Chem* **280:** 3656–3664.

Lee JH, Tate CM, You JS, Skalnik DG. 2007. Identification and characterization of the human Set1B histone H3-Lys4 methyltransferase complex. *J Biol Chem* **282:** 13419–13428.

Lee J, Thompson JR, Botuyan MV, Mer G. 2008. Distinct binding modes specify the recognition of methylated histones H3K4 and H4K20 by JMJD2A-tudor. *Nat Struct Mol Biol* **15:** 109–111.

Lehnertz B, Ueda Y, Derijck AA, Braunschweig U, Perez-Burgos L, Kubicek S, Chen T, Li E, Jenuwein T, Peters AH. 2003. Suv39h-mediated histone H3 lysine 9 methylation directs DNA methylation to major satellite repeats at pericentric heterochromatin. *Curr Biol* **13:** 1192–1200.

Leonhardt H, Page AW, Weier HU, Bestor TH. 1992. A targeting sequence directs DNA methyltransferase to sites of DNA replication in mammalian nuclei. *Cell* **71:** 865–873.

Leung DC, Dong KB, Maksakova IA, Goyal P, Appanah R, Lee S, Tachibana M, Shinkai Y, Lehnertz B, Mager DL, et al. 2011. Lysine methyltransferase G9a is required for de novo DNA methylation and the establishment, but not the maintenance, of proviral silencing. *Proc Natl Acad Sci* **108:** 5718–5723.

Levy D, Kuo AJ, Chang Y, Schaefer U, Kitson C, Cheung P, Espejo A, Zee BM, Liu CL, Tangsombatvisit S, et al. 2011. Lysine methylation of the NF-κB subunit RelA by SETD6 couples activity of the histone methyltransferase GLP at chromatin to tonic repression of NF-κB signaling. *Nat Immunol* **12:** 29–36.

* Li E, Zhang Y. 2014. DNA methylation in mammals. *Cold Spring Harb Perspect Biol* **6:** a019133.

Lin Y, Wu Y, Li J, Dong C, Ye X, Chi YI, Evers BM, Zhou BP. 2010. The SNAG domain of Snail1 functions as a molecular hook for recruiting lysine-specific demethylase 1. *EMBO J* **29:** 1803–1816.

Lister R, Pelizzola M, Dowen RH, Hawkins RD, Hon G, Tonti-Filippini J, Nery JR, Lee L, Ye Z, Ngo QM, et al. 2009. Human DNA methylomes at base resolution show widespread epigenomic differences. *Nature* **462:** 315–322.

Liu W, Tanasa B, Tyurina OV, Zhou TY, Gassmann R, Liu WT, Ohgi KA, Benner C, Garcia-Bassets I, Aggarwal AK, et al. 2010. PHF8 mediates histone H4 lysine 20 demethylation events involved in cell cycle progression. *Nature* **466:** 508–512.

Loenarz C, Ge W, Coleman ML, Rose NR, Cooper CD, Klose RJ, Ratcliffe PJ, Schofield CJ. 2010. *PHF8*, a gene associated with cleft lip/palate and mental retardation, encodes for an N^{ε}-dimethyl lysine demethylase. *Hum Mol Genet* **19:** 217–222.

Lorsbach RB, Moore J, Mathew S, Raimondi SC, Mukatira ST, Downing JR. 2003. TET1, a member of a novel protein family, is fused to MLL in acute myeloid leukemia containing the t(10;11)(q22;q23). *Leukemia* **17:** 637–641.

Lyst MJ, Nan X, Stancheva I. 2006. Regulation of MBD1-mediated transcriptional repression by SUMO and PIAS proteins. *EMBO J* **25:** 5317–5328.

* Marmorstein R, Zhou M-M. 2014. Writers and readers of histone acetylation: Structure, mechanism, and inhibition. *Cold Spring Harb Perspect Biol* **6:** a018762.

Matsui T, Leung D, Miyashita H, Maksakova IA, Miyachi H, Kimura H, Tachibana M, Lorincz MC, Shinkai Y. 2010. Proviral silencing in embryonic stem cells requires the histone methyltransferase ESET. *Nature* **464:** 927–931.

McDonough MA, Loenarz C, Chowdhury R, Clifton IJ, Schofield CJ. 2010. Structural studies on human 2-oxoglutarate dependent oxygenases. *Curr Opin Struct Biol* **20:** 659–672.

Meissner A, Mikkelsen TS, Gu H, Wernig M, Hanna J, Sivachenko A, Zhang X, Bernstein BE, Nusbaum C, Jaffe DB, et al. 2008. Genome-scale DNA methylation maps of pluripotent and differentiated cells. *Nature* **454:** 766–770.

Metzger E, Wissmann M, Yin N, Muller JM, Schneider R, Peters AH, Gunther T, Buettner R, Schule R. 2005. LSD1 demethylates repressive histone marks to promote androgen-receptor-dependent transcription. *Nature* **437:** 436–439.

Milne TA, Briggs SD, Brock HW, Martin ME, Gibbs D, Allis CD, Hess JL. 2002. MLL targets SET domain methyltransferase activity to *Hox* gene promoters. *Mol Cell* **10:** 1107–1117.

Min J, Feng Q, Li Z, Zhang Y, Xu RM. 2003. Structure of the catalytic domain of human DOT1L, a non-SET domain nucleosomal histone methyltransferase. *Cell* **112:** 711–723.

Miranda TB, Miranda M, Frankel A, Clarke S. 2004. PRMT7 is a member of the protein arginine methyltransferase family with a distinct substrate specificity. *J Biol Chem* **279:** 22902–22907.

Nguyen CT, Weisenberger DJ, Velicescu M, Gonzales FA, Lin JC, Liang G, Jones PA. 2002. Histone H3-lysine 9 methylation is associated with aberrant gene silencing in cancer cells and is rapidly reversed by 5-aza-2′-deoxycytidine. *Cancer Res* **62:** 6456–6461.

Oh S, Jeong K, Kim H, Kwon CS, Lee D. 2010. A lysine-rich region in Dot1p is crucial for direct interaction with H2B ubiquitylation and high level methylation of H3K79. *Biochem Biophys Res Commun* **399:** 512–517.

Okada Y, Feng Q, Lin Y, Jiang Q, Li Y, Coffield VM, Su L, Xu G, Zhang Y. 2005. hDOT1L links histone methylation to leukemogenesis. *Cell* **121:** 167–178.

Ono R, Taki T, Taketani T, Taniwaki M, Kobayashi H, Hayashi Y. 2002. LCX, leukemia-associated protein with a CXXC domain, is fused to MLL in acute myeloid leukemia with trilineage dysplasia having t(10;11)(q22;q23). *Cancer Res* **62:** 4075–4080.

Ooi SKT, Qiu C, Bernstein E, Li K, Jia D, Yang Z, Erdjument-Bromage H, Tempst P, Lin S-P, Allis CD, et al. 2007. DNMT3L connects unmethylated lysine 4 of histone H3 to de novo methylation of DNA. *Nature* **448:** 714–717.

Otani J, Nankumo T, Arita K, Inamoto S, Ariyoshi M, Shirakawa M. 2009. Structural basis for recognition of H3K4 methylation status by the DNA methyltransferase 3A ATRX-DNMT3-DNMT3L domain. *EMBO Rep* **10:** 1235–1241.

Ozer A, Bruick RK. 2007. Non-heme dioxygenases: Cellular sensors and regulators jelly rolled into one? *Nat Chem Biol* **3:** 144–153.

* Patel DJ. 2014. A structural perspective on readout of epigenetic histone and DNA methylation marks. *Cold Spring Harb Perspect Biol* doi: 10.1101/cshperspect.a018754.

Proffitt JH, Davie JR, Swinton D, Hattman S. 1984. 5-Methylcytosine is not detectable in *Saccharomyces cerevisiae* DNA. *Mol Cell Biol* **4:** 985–988.

Qi HH, Sarkissian M, Hu GQ, Wang Z, Bhattacharjee A, Gordon DB, Gonzales M, Lan F, Ongusaha PP, Huarte M, et al. 2010. Histone H4K20/H3K9 demethylase PHF8 regulates zebrafish brain and craniofacial development. *Nature* **466:** 503–507.

Qiu C, Sawada K, Zhang X, Cheng X. 2002. The PWWP domain of mammalian DNA methyltransferase Dnmt3b defines a new family of DNA-binding folds. *Nat Struct Biol* **9:** 217–224.

Rajakumara E, Wang Z, Ma H, Hu L, Chen H, Lin Y, Guo R, Wu F, Li H, Lan F, et al. 2011. PHD finger recognition of unmodified histone H3R2 links UHRF1 to regulation of euchromatic gene expression. *Mol Cell* **43:** 275–284.

Rea S, Eisenhaber F, O'Carroll D, Strahl BD, Sun ZW, Schmid M, Opravil S, Mechtler K, Ponting CP, Allis CD, et al. 2000. Regulation of chromatin structure by site-specific histone H3 methyltransferases. *Nature* **406:** 593–599.

Rothbart SB, Krajewski K, Nady N, Tempel W, Xue S, Badeaux AI, Barsyte-Lovejoy D, Martinez JY, Bedford MT, Fuchs SM, et al. 2012. Association of UHRF1 with methylated H3K9 directs the maintenance of DNA methylation. *Nat Struct Mol Biol* **19:** 1155–1160.

Rountree MR, Bachman KE, Baylin SB. 2000. DNMT1 binds HDAC2 and a new co-repressor, DMAP1, to form a complex at replication foci. *Nat Genet* **25:** 269–277.

Sarraf SA, Stancheva I. 2004. Methyl-CpG binding protein MBD1 couples histone H3 methylation at lysine 9 by SETDB1 to DNA replication and chromatin assembly. *Mol Cell* **15:** 595–605.

Sawada K, Yang Z, Horton JR, Collins RE, Zhang X, Cheng X. 2004. Structure of the conserved core of the yeast Dot1p, a nucleosomal histone H3 lysine 79 methyltransferase. *J Biol Chem* **279:** 43296–43306.

Schubert HL, Blumenthal RM, Cheng X. 2003. Many paths to methyltransfer: A chronicle of convergence. *Trends Biochem Sci* **28:** 329–335.

* Seto E, Yoshida M. 2014. Erasers of histone acetylation: The histone deacetylase enzymes. *Cold Spring Harb Perspect Biol* **6:** a018713.

Sharif J, Muto M, Takebayashi S, Suetake I, Iwamatsu A, Endo TA, Shinga J, Mizutani-Koseki Y, Toyoda T, Okamura K, et al. 2007. The SRA protein Np95 mediates epigenetic inheritance by recruiting Dnmt1 to methylated DNA. *Nature* **450:** 908–912.

Sharma S, De Carvalho DD, Jeong S, Jones PA, Liang G. 2011. Nucleosomes containing methylated DNA stabilize DNA methyltransferases 3A/3B and ensure faithful epigenetic inheritance. *PLoS Genet* **7:** e1001286.

* Shi YG, Tsukada Y. 2013. The discovery of histone demethylases. *Cold Spring Harb Perspect Biol* **5:** a017947.

Shi Y, Lan F, Matson C, Mulligan P, Whetstine JR, Cole PA, Casero RA, Shi Y. 2004. Histone demethylation mediated by the nuclear amine oxidase homolog LSD1. *Cell* **119:** 941–953.

Shinkai Y, Tachibana M. 2011. H3K9 methyltransferase G9a and the related molecule GLP. *Genes Dev* **25:** 781–788.

Siarheyeva A, Senisterra G, Allali-Hassani A, Dong A, Dobrovetsky E, Wasney GA, Chau I, Marcellus R, Hajian T, Liu F, et al. 2012. An allosteric inhibitor of protein arginine methyltransferase 3. *Structure* **20:** 1425–1435.

Sims RJ III, Rojas LA, Beck D, Bonasio R, Schuller R, Drury WJ III, Eick D, Reinberg D. 2011. The C-terminal domain of RNA polymerase II is modified by site-specific methylation. *Science* **332:** 99–103.

Singer MS, Kahana A, Wolf AJ, Meisinger LL, Peterson SE, Goggin C, Mahowald M, Gottschling DE. 1998. Identification of high-copy disruptors of telomeric silencing in *Saccharomyces cerevisiae*. *Genetics* **150:** 613–632.

Song J, Rechkoblit O, Bestor TH, Patel DJ. 2011. Structure of DNMT1-DNA complex reveals a role for autoinhibition in maintenance DNA methylation. *Science* **331:** 1036–1040.

Song J, Teplova M, Ishibe-Murakami S, Patel DJ. 2012. Structure-based mechanistic insights into DNMT1-mediated maintenance DNA methylation. *Science* **335:** 709–712.

Southall SM, Wong PS, Odho Z, Roe SM, Wilson JR. 2009. Structural basis for the requirement of additional factors for MLL1 SET domain activity and recognition of epigenetic marks. *Mol Cell* **33:** 181–191.

Strahl BD, Allis CD. 2000. The language of covalent histone modifications. *Nature* **403:** 41–45.

Suetake I, Shinozaki F, Miyagawa J, Takeshima H, Tajima S. 2004. DNMT3L stimulates the DNA methylation activity of Dnmt3a and Dnmt3b through a direct interaction. *J Biol Chem* **279:** 27816–27823.

Sun ZW, Allis CD. 2002. Ubiquitination of histone H2B regulates H3 methylation and gene silencing in yeast. *Nature* **418:** 104–108.

Syeda F, Fagan RL, Wean M, Avvakumov GV, Walker JR, Xue S, Dhe-Paganon S, Brenner C. 2011. The replication focus targeting sequence (RFTS) domain is a DNA-competitive inhibitor of Dnmt1. *J Biol Chem* **286:** 15344–15351.

Tahiliani M, Koh KP, Shen Y, Pastor WA, Bandukwala H, Brudno Y, Agarwal S, Iyer LM, Liu DR, Aravind L, et al. 2009. Conversion of 5-methylcytosine to 5-hydroxymethylcytosine in mammalian DNA by MLL partner TET1. *Science* **324:** 930–935.

Takeshita K, Suetake I, Yamashita E, Suga M, Narita H, Nakagawa A, Tajima S. 2011. Structural insight into maintenance methylation by mouse DNA methyltransferase 1 (Dnmt1). *Proc Natl Acad Sci* **108:** 9055–9059.

Tamaru H, Selker EU. 2001. A histone H3 methyltransferase controls DNA methylation in *Neurospora crassa*. *Nature* **414:** 277–283.

Tamaru H, Zhang X, McMillen D, Singh PB, Nakayama J, Grewal SI, Allis CD, Cheng X, Selker EU. 2003. Trimethylated lysine 9 of histone H3 is a mark for DNA methylation in *Neurospora crassa*. *Nat Genet* **34:** 75–79.

Terranova R, Agherbi H, Boned A, Meresse S, Djabali M. 2006. Histone and DNA methylation defects at Hox genes in mice expressing a SET domain-truncated form of Mll. *Proc Natl Acad Sci* **103:** 6629–6634.

Thomson JP, Skene PJ, Selfridge J, Clouaire T, Guy J, Webb S, Kerr AR, Deaton A, Andrews R, James KD, et al. 2010. CpG islands influence chromatin structure via the CpG-binding protein Cfp1. *Nature* **464:** 1082–1086.

Trewick SC, McLaughlin PJ, Allshire RC. 2005. Methylation: Lost in hydroxylation? *EMBO Rep* **6:** 315–320.

Troffer-Charlier N, Cura V, Hassenboehler P, Moras D, Cavarelli J. 2007. Functional insights from structures of coactivator-associated arginine methyltransferase 1 domains. *EMBO J* **26:** 4391–4401.

Tsukada Y, Fang J, Erdjument-Bromage H, Warren ME, Borchers CH, Tempst P, Zhang Y. 2006. Histone demethylation by a family of JmjC domain-containing proteins. *Nature* **439:** 811–816.

Tsukada Y, Ishitani T, Nakayama KI. 2010. KDM7 is a dual demethylase for histone H3 Lys 9 and Lys 27 and functions in brain development. *Genes Dev* **24:** 432–437.

Turner BM. 2007. Defining an epigenetic code. *Nat Cell Biol* **9:** 2–6.

Vagin VV, Wohlschlegel J, Qu J, Jonsson Z, Huang X, Chuma S, Girard A, Sachidanandam R, Hannon GJ, Aravin AA. 2009. Proteomic analysis of murine Piwi proteins reveals a role for arginine methylation in specifying interaction with Tudor family members. *Genes Dev* **23:** 1749–1762.

Volkel P, Angrand PO. 2007. The control of histone lysine methylation in epigenetic regulation. *Biochimie* **89:** 1–20.

Wang H, Huang ZQ, Xia L, Feng Q, Erdjument-Bromage H, Strahl BD, Briggs SD, Allis CD, Wong J, Tempst P, et al. 2001. Methylation of histone H4 at arginine 3 facilitating transcriptional activation by nuclear hormone receptor. *Science* **293:** 853–857.

Wang J, Hevi S, Kurash JK, Lei H, Gay F, Bajko J, Su H, Sun W, Chang H, Xu G, et al. 2009a. The lysine demethylase LSD1 (KDM1) is required for maintenance of global DNA methylation. *Nat Genet* **41:** 125–129.

Wang GG, Song J, Wang Z, Dormann HL, Casadio F, Li H, Luo JL, Patel DJ, Allis CD. 2009b. Haematopoietic malignancies caused by dysregulation of a chromatin-binding PHD finger. *Nature* **459:** 847–851.

Wang Z, Song J, Milne TA, Wang GG, Li H, Allis CD, Patel DJ. 2010. Pro isomerization in MLL1 PHD3-bromo cassette connects H3K4me readout to CyP33 and HDAC-mediated repression. *Cell* **141:** 1183–1194.

Webster KE, O'Bryan MK, Fletcher S, Crewther PE, Aapola U, Craig J, Harrison DK, Aung H, Phutikanit N, Lyle R, et al. 2005. Meiotic and epigenetic defects in Dnmt3L-knockout mouse spermatogenesis. *Proc Natl Acad Sci* **102:** 4068–4073.

Xiao B, Jing C, Wilson JR, Walker PA, Vasisht N, Kelly G, Howell S, Taylor IA, Blackburn GM, Gamblin SJ. 2003. Structure and catalytic mechanism of the human histone methyltransferase SET7/9. *Nature* **421:** 652–656.

Xiao B, Jing C, Kelly G, Walker PA, Muskett FW, Frenkiel TA, Martin SR, Sarma K, Reinberg D, Gamblin SJ, et al. 2005. Specificity and mechanism of the histone methyltransferase Pr-Set7. *Genes Dev* **19:** 1444–1454.

Yang Y, Hu L, Wang P, Hou H, Lin Y, Liu Y, Li Z, Gong R, Feng X, Zhou L, et al. 2010. Structural insights into a dual-specificity histone demethylase ceKDM7A from *Caenorhabditis elegans*. *Cell Res* **20:** 886–898.

Yoder JA, Yen RW, Vertino PM, Bestor TH, Baylin SB. 1996. New 5′ regions of the murine and human genes for DNA (cytosine-5)-methyltransferase. *J Biol Chem* **271:** 31092–31097.

Yoo CB, Jeong S, Egger G, Liang G, Phiasivongsa P, Tang C, Redkar S, Jones PA. 2007. Delivery of 5-aza-2′-deoxycytidine to cells using oligodeoxynucleotides. *Cancer Res* **67:** 6400–6408.

Yue WW, Hassler M, Roe SM, Thompson-Vale V, Pearl LH. 2007. Insights into histone code syntax from structural and biochemical studies of CARM1 methyltransferase. *EMBO J* **26:** 4402–4412.

Zhang X, Zhou L, Cheng X. 2000. Crystal structure of the conserved core of protein arginine methyltransferase PRMT3. *EMBO J* **19:** 3509–3519.

Zhang X, Tamaru H, Khan SI, Horton JR, Keefe LJ, Selker EU, Cheng X. 2002. Structure of the *Neurospora* SET domain protein DIM-5, a histone H3 lysine methyltransferase. *Cell* **111:** 117–127.

Zhang X, Yang Z, Khan SI, Horton JR, Tamaru H, Selker EU, Cheng X. 2003. Structural basis for the product specificity of histone lysine methyltransferases. *Mol Cell* **12:** 177–185.

Zhang K, Mosch K, Fischle W, Grewal SI. 2008. Roles of the Clr4 methyltransferase complex in nucleation, spreading and maintenance of heterochromatin. *Nat Struct Mol Biol* **15:** 381–388.

Zhang Y, Rohde C, Tierling S, Jurkowski TP, Bock C, Santacruz D, Ragozin S, Reinhardt R, Groth M, Walter J, et al. 2009. DNA methylation analysis of chromosome 21 gene promoters at single base pair and single allele resolution. *PLoS Genet* **5:** e1000438.

Zhang Y, Jurkowska R, Soeroes S, Rajavelu A, Dhayalan A, Bock I, Rathert P, Brandt O, Reinhardt R, Fischle W, et al. 2010. Chromatin methylation activity of Dnmt3a and Dnmt3a/3L is guided by interaction of the ADD domain with the histone H3 tail. *Nucleic Acids Res* **38:** 4246–4253.

Zhao Q, Rank G, Tan YT, Li H, Moritz RL, Simpson RJ, Cerruti L, Curtis DJ, Patel DJ, Allis CD, et al. 2009. PRMT5-mediated methylation of histone H4R3 recruits DNMT3A, coupling histone and DNA methylation in gene silencing. *Nat Struct Mol Biol* **16:** 304–311.

Zilberman D, Coleman-Derr D, Ballinger T, Henikoff S. 2008. Histone H2A.Z and DNA methylation are mutually antagonistic chromatin marks. *Nature* **456:** 125–129.

C H A P T E R 7

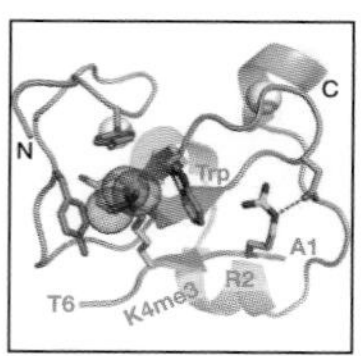

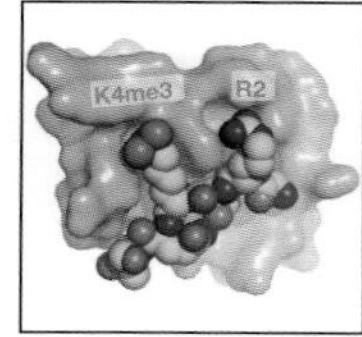

A Structural Perspective on Readout of Epigenetic Histone and DNA Methylation Marks

Dinshaw J. Patel

Structural Biology Department, Memorial Sloan-Kettering Cancer Center, New York, New York 10065

Correspondence: pateld@mskcc.org

SUMMARY

This chapter outlines the protein modules that target methylated lysine histone marks and 5mC DNA marks, and the molecular principles underlying recognition. The chapter focuses on the structural basis underlying readout of isolated marks by single reader molecules, as well as multivalent readout of multiple marks by linked reader cassettes at the histone tail and nucleosome level. Additional topics addressed include the role of histone mimics, cross talk between histone marks, technological developments at the genome-wide level, advances using chemical biology approaches, the linkage between histone and DNA methylation, the role for regulatory lncRNAs, and the promise of chromatin-based therapeutic modalities.

Outline

Cite as *Cold Spring Harb Perspect Biol* doi: 10.1101/cshperspect.a018754

OVERVIEW

This chapter focuses on the readout of histone and DNA methylation posttranslational modifications (PTMs) and their impact on chromatin structure and function. PTMs serve as docking sites for protein reader modules containing attached chromatin modifier and remodeling activities. The additional activities can alter noncovalent contacts within and between nucleosomes, thereby impacting on function. At a specific genomic site, there can be distinct combinations of methylation and other PTMs. The multivalent (more than a single mark) readout of these PTMs impacts on many DNA-templated processes ranging from gene transcription to DNA replication, recombination, and repair. Dysregulation of the readout due to mutated readers can result in aberrant gene expression patterns and/or genomic alterations, facilitating the onset of disease. A new generation of epigenetic drugs is being developed as a novel therapeutic approach to target these dysfunctions.

The chapter begins by introducing the landscape of histone and DNA methylation marks and then categorizes the various families of single and tandem reader modules that use an aromatic cage capture mechanism for readout of methyllysine (Kme) and methylarginine (Rme) marks. Next, the text highlights recent reader modules that target unmodified lysines and arginine marks, as well as reader cassettes involved as regulatory platforms for mediating functional output. The chapter also outlines the potential for cross talk between PTMs, whereby the binding of a reader module to a particular mark either sterically blocks an adjacent modification site or facilitates recruitment of additional modules to modify nearby residues. In addition, "histone mimics" are discussed as a distinct set of nonhistone proteins that are methylation targets, thereby expanding available methylated lysine recognition principles beyond the boundaries of direct chromatin regulation. The chapter next addresses DNA cytosine methylation (5mC) marks and their readout by 5mC-binding domains (MBDs) and zinc-finger-containing modules with the capacity to sequence specifically recognized 5mC-containing fully methylated CpG DNA sites. The chapter also highlights the contribution of 5mC-binding SRA (SET- and RING-associated) domains required for the establishment and/or maintenance of DNA methylation marks at hemimethyated CpG DNA sites in both mammals and plants.

The chapter ends by highlighting new initiatives and advances, as well as future challenges that promise to enhance our current mechanistic understanding of the readout of histone and DNA methylation marks. These include technological developments at the genome-wide level, chemical biology approaches to designer nucleosomes, and structural approaches to histone mark readout at the nucleosomal level. The chapter also outlines new developments related to readout of oxidative 5mC DNA adducts, the functional role for regulatory noncoding RNAs in epigenetic regulation, and the linkage between histone and DNA methylation. This chapter addresses the consequences of dysregulation of methylated lysine reader modules and long intergenic noncoding RNAs on epigenetic pathways resulting in the onset of disease states and outlines challenges toward identification and functional characterization of small molecules site-specifically targeted to aromatic-lined pockets involved in methyllysine readout.

Cite as *Cold Spring Harb Perspect Biol* doi: 10.1101/cshperspect.a018754

1 INTRODUCTION

The nucleosome core particle is composed of almost two turns of a DNA superhelix amounting to 147 bp wrapped around a compact histone octamer core containing four subunits labeled H2A, H2B, H3, and H4 (Luger et al. 1997). Nucleosomes are packaged into progressively higher-order folds to ultimately form chromosomes. Projecting from the four histone cores are amino-terminal tails that are subject to covalent posttranslational modifications (PTMs) (Allfrey et al. 1964), depositing marks such as methylation, acetylation, phosphorylation, and ubiquitination. Methylation of cytosines on DNA is also possible. More recently, with the advent of advanced mass spectroscopic and antibody-based techniques, PTMs have also been identified within the carboxy-terminal end of histone tails and even within the globular central histone fold. In addition, new covalent modifications have recently been identified such as sumoylation, ADP-ribosylation, proline isomerization, citrullination, and glycosylation (see App. 2 [Zhao and Garcia 2014]).

PTM marks are dynamic, being deposited and erased in the time frame of minutes. The recognition of a mark by a reader module that is part of a multidomain protein complex facilitates the recruitment and tethering of enzymatic activities intrinsic to other subunits to chromatin. Hence, histone and DNA covalent PTMs provide a scaffold for the assembly of activities that control the site- (e.g., lysine 4 of H3) and state-specific (e.g., mono-, di-, or trimethylated) readout of marks at the nucleosomal level. They also have the capacity to modulate higher-order chromatin structure and/or the ordered recruitment of nonhistone proteins and enzymes critical for DNA remodeling activities. Thus, PTMs serve as epigenetic information carriers that extend the message beyond that encoded in the DNA sequence. This capacity of chromatin to both store and transmit heritable information in the form of PTMs results in altered histone–DNA interactions. This leads to dynamic changes in chromatin-templated processes, including altered transcription rates reflecting the accessibility of particular segments of DNA.

The advent of histone mark-specific ChIP-chip (chromatin immunoprecipitation with DNA microarray analysis) and ChIP-seq (chromatin immunoprecipitation with next-generation sequence technology) has enabled the characterization of histone marks at the genome-wide level. This has provided unique insights into the distribution of marks to distinct segments of the genome. It has also identified global correlations of specific marks with downstream functional outcome. Genome-wide chromatin organization studies performed by the modENCODE consortium have, in addition, identified the linkage of PTMs with regulatory circuits at defined developmental stages in model organisms.

The methylation of certain histone residues and cytosine bases on the DNA are well-known and well-studied instances of PTMs. This chapter focuses on the structural aspects of how methylation mark readers interact with methylation marks, the functional consequence of these interactions, and how cross talk occurs with other histone PTMs.

1.1 Lysine and Arginine Methylation of Histone Tails

Lysine methylation is unique among known PTMs because of its relative stability, its multivalency (i.e., existence in the mono-, di-, and trivalent state), and its potential to engage in cross talk with other modifications. The hydrophobicity of lysine is increased on methylation with no net change in charge. This contrasts with the change in charge that occurs on lysine acetylation or serine phosphorylation.

The major lysine (K) methylation sites on histone H3 are at positions K4, K9, K27, K36, and K79 (Fig. 1A), together with a single site on H4 at K20 and one on H1 at K26. Of these, H3K4, H3K36, and H3K79 methylation marks are linked to the activation of transcription, whereas H3K9, H3K27, and H4K20 methylation marks are linked to transcriptional repression (Fig. 1A). Some sites such as H3K9 and H3K27 are embedded within a common A-R-K-S sequence context (Fig. 1B). Each site-specific mark is written by a specific enzyme classed as a histone lysine (K) methyltransferase, or KMT, and erased by other enzymes called histone lysine demethylases or KDMs. The catalytic activity of each writer and eraser enzyme is substrate specific—that is, effective on an individual or set of methylation states, such as mono-, di-, or trimethylated lysines (reviewed in Black et al. 2012; Greer and Shi 2012). Enrichment of particular lysine marks can be found at specific regions along the genome.

The major arginine (R) methylation sites are at R2, R8, R17, and R26 on histone H3 (Fig. 1A), at R3 on H4, and at R11 and R29 on H2A. The marks are written by protein arginine methyltransferase (PRMT) enzymes and erased by deiminases as a function of methylation state (mono-, symmetrical di-, and asymmetrical dimethylated arginines) context.

It has been proposed that multivalent readout of methylation together with other marks (e.g., acetylation, phosphorylation, ubiquitination) by reader modules regulates the transcriptional outcome. This occurs by directing the activity of chromatin-remodeling complexes (discussed in Ch. 21 [Becker and Workman 2013]) in a precisely con-

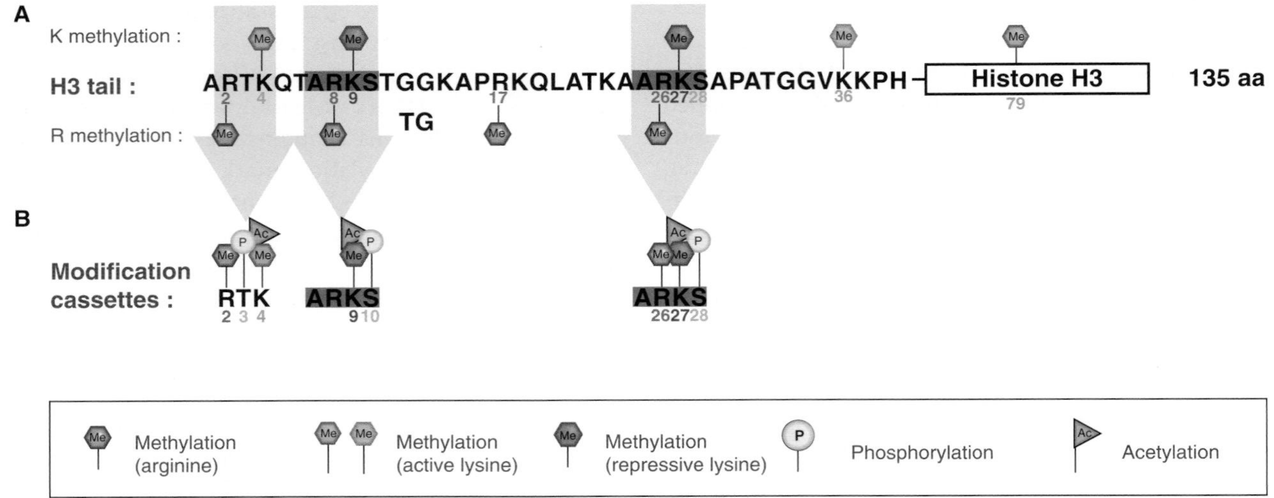

Figure 1. Sequence of H3 tail and distribution of PTMs. (*A*) Sequence of H3 tail and positioning of methyllysine (Kme) and methylarginine (Rme) marks. (*B*) Positioning of adjacent marks within the H3 tail at R2-T3-K4, A7-R8-K9-S10, and A25-R26-K27-S28 segments. Kme marks are partitioned between those that are activating and those that are repressive.

trolled manner according to the developmental stage and physiological state of cells, with transcription rates also susceptible to external stimuli (reviewed in Strahl and Allis 2000; Jenuwein and Allis 2001; Gardner et al. 2011). Meaningful variation in the chromatin polymer can also result from exchange of primary sequence histone variants such as H2A.Z in place of H2A or H3.3 instead of H3.1/2 (see Ch. 20 [Henikoff and Smith 2014]).

The binding pockets of reader modules recognize histone methylation marks in a sequence- and state-specific manner. The architecture of these binding pockets and principles of molecular recognition are of fundamental importance to our understanding of chromatin function. A salient question is whether recognition can occur by processes beyond the first identified mode involving aromatic cage capture of the methylated lysine side chain discussed in Section 2.1 and reviewed in Taverna et al. (2007). There is also interest as to whether known methyllysine-binding pockets can be engineered to recognize and distinguish between distinct (mono-, di-, and tri-) lysine methylation states.

An increasing number of human diseases ranging from autoimmune disorders to cancer are associated with aberrant writers, readers, and erasers of histone methylation marks (reviewed in Chi et al. 2010; Dawson and Kouzarides 2012). Research has revealed that many mutations in these molecules have affected the function of these epigenetic regulators and often entire chromatin-remodeling complexes. This has led to research looking for viable epigenetic therapies, including the identification of small molecules that selectively target Kme-binding pockets on reader modules.

There have been several excellent reviews covering earlier structure–function research on the readout of histone marks (Kouzarides 2007; Kouzarides and Berger 2007; Ruthenburg et al. 2007a,b; Taverna et al. 2007; see also more recent reviews by Yap and Zhou 2010; Bannister and Kouzarides 2011; Khorasanizadeh 2011; Musselman et al. 2012b). In this chapter, we provide a comprehensive up-to-date (until the end of December 2012) structural overview on the readout of histone methylation marks and place these results in a functional context. In companion chapters, a structure–function perspective is given on writers and erasers of histone lysine methylation marks (Ch. 6 [Cheng 2014]) and on writers, readers, and erasers of histone lysine acetylation marks (Ch. 4 [Marmorstein and Zhou 2014]; Ch. 5 [Seto and Yoshida 2014]).

1.2 Cytosine Methylation in DNA

Methylation of cytosines at CpG sites along the mammalian genomes constitutes an ancient evolutionary epigenetic mark. It operates as an epigenetic regulatory mark predominantly in mammals and plants (the latter of which is discussed in Ch. 13 [Pikaard and Mittelsten Scheid 2014]), although *Neurospora* and *Drosophila* contain this chromatin mark to a more limited extent. This epigenetic modification in mammals is associated with gene silencing and contributes to chromatin structure and genome stability (reviewed in Li and Bird 2007; Jones and Liang 2009; Law

Cite as *Cold Spring Harb Perspect Biol* doi: 10.1101/cshperspect.a018754

and Jacobsen 2010). The methyl mark is deposited at the 5-position of cytosines (5mC) in CpG steps by the de novo DNA methyltransferases, DNMT3A and DNMT3B, together with regulator DNMT3L, during embryonic development. The mark is faithfully maintained during multiple cell divisions by DNMT1, the maintenance DNA methyltransferase, thereby establishing a form of cellular memory through epigenetic marking of the genome (illustrated in Fig. 2 of Ch. 15 [Li and Zhang 2014]). The importance of DNA methylation is highlighted by the fact that targeted disruption of DNA methyltransferases results in lethality. Mechanistically, gene expression profiles can be altered by DNA methylation as a consequence of influencing the binding affinities of transcription factors. Alternatively, transcriptional silencing results from 5mC binding proteins recruiting repressor complexes to methylated promoter segments. Thus, DNA methylation plays a critical role in the establishment and maintenance of tissue-specific gene expression patterns at distinct stages of development. Critical developmental processes such as X-inactivation, monoallelic expression of imprinted genes, and suppression of transposable elements and proviral genomes use DNA methylation as a component of complex regulatory networks (further discussed in Ch. 26 [Barlow and Bartolomei 2014]; Ch. 25 [Brockdorff and Turner 2014]; Ch. 13 [Pikaard and Mittelsten Scheid 2014]).

Greater than 70% of CpG sites are methylated in the DNA of somatic tissues, but they are asymmetrically distributed along the genome within both CpG-rich and -poor regions. Pericentric heterochromatin, for example, is hypermethylated, yet overall CpG-poor. The global DNA methylation of such inaccessible heterochromatic regions requires the participation of SWI/SNF-like chromatin-remodeling proteins to allow DNA methyltransferase access. CpG islands, in contrast, are hypomethylated and immune to otherwise global DNA methylation. CpG islands are short CpG-rich sequences of $\sim$1 kb in length, making up $<$1% of genomic DNA, and mark the promoters and 5′ end of genes. Improved genome-scale mapping has provided more detailed insights into the distribution of DNA methylation patterns at transcription start sites, gene bodies, regulatory elements, and repeat sequences, with a relationship emerging between DNA methylation and transcription repression (see Ch. 15 [Li and Zhang 2014]; also reviewed in Jones 2012). DNA methylation is a dynamic mark, which is mostly deposited, removed, and reestablished during early mammalian development.

A growing number of human diseases ranging from imprinting disorders (e.g., Beckwith–Wiedemann, Prader–Willi, and Angelman syndromes) to repeat-instability diseases (e.g., fragile X syndrome and facioscapulohumeral muscular dystrophy) and cancer are associated with aberrant DNA methylation (as discussed in Robertson 2005; Baylin and Jones 2011; Ch. 33 [Zoghbi and Beaudet 2014]; Ch. 34 [Baylin and Jones 2014]). This may involve the improper establishment or maintenance of methylation, resulting in the alteration of chromatin states and/or nucleosome positioning (reviewed in Baylin and Jones 2011). Another feature of DNA methylation patterns that may be occasionally causal to disease is the susceptibility of 5mC to spontaneously deaminate to thymine. This instability at CpG sites is reflected by the fact that a third of point mutations are C-to-T transitions at CpG sequences, and explains the fivefold underrepresentation of CpG steps in mammalian genomes generated over evolutionary time. Mutations in components of the DNA methylation machinery can also lead to disease, such as the documented mutations in DNMT3B leading to immune deficiency and blood cancers, whereas mutations in MeCP2, a 5mCpG-binding protein, cause Rett syndrome, a severe neurological disorder.

Ch. 15 (Li and Zhang 2014), as well as several excellent reviews, covers the topic of genomic DNA methylation and its establishment and maintenance (reviewed in Klose and Bird 2006; Jones and Liang 2009; Law and Jacobsen 2010). In this chapter, we provide a comprehensive up-to-date (until the end of December 2012) structural overview on the readout of 5mCpG marks, placing these results in a functional context.

2 READOUT OF Kme MARKS BY PHD FINGER AND BAH MODULES

We start our description of reader modules targeting methylated lysine marks by focusing on PHD (plant homeodomain) fingers and BAH (bromo-adjacent homology) domains. The PHD finger is a very common module found among chromatin remodelers and is often positioned adjacent to other reader modules. Although it was initially thought to partake in protein–protein and protein–lipid interactions, in 2006, structure–function studies showed that PHD fingers are readers of methylated lysine marks on histone tails. This promoted a more structure-based perspective in the field of epigenetic regulation. The PHD reader often participates in multivalent readout (i.e., readout of combinations of histone PTMs with other reader modules) at the nucleosome level as elaborated in Sections 9.1 and 10.1. The BAH domain was also initially thought to be a protein–protein interaction module until it, too, was shown to be a reader of methylated lysine marks on histone tails in 2012. Both PHD finger and BAH domains are of great interest given that reports on their dysfunction results in the onset of disease, such as Meier–Gorlin primordial dwarfism syndrome in the case of specific mutants in the BAH domain of the ORC1 (origin of replication 1 protein) protein.

2.1 PHD Finger Domains

Lysine methylation of histone tails constitutes a key component of an epigenetic indexing system, demarcating the transcriptional activity of chromatin domains. The H3K4me3 mark, in particular, is associated with nucleosomes near the promoters and the 5′-ends of highly transcribed genes (Santos-Rosa et al. 2002; Bernstein et al. 2006). BPTF is a protein named after its two reader modules: bromodomain and PHD domain transcription factor. Functional studies of the PHD finger established its role in nucleosome remodeling factor (NURF) complex-mediated ATP-dependent chromatin remodeling, directly coupling the complex to H3K4me3 so as to maintain HOX gene expression patterns during development (Wysocka et al. 2006).

The PHD finger (50–80 residues) (reviewed in Bienz 2006) has limited secondary structure (a two-stranded β-sheet and a short α-helix) wherein cross-brace topology is stabilized by two zinc ions coordinated to a Cys_4-His-Cys_3-containing segment (Pascual et al. 2000). Simultaneous structure–function studies on the PHD fingers of BPTF (Li et al. 2006; Wysocka et al. 2006), ING2 (inhibitor of growth 2) (Pena et al. 2006; Shi et al. 2006), and YNG1 (Taverna et al. 2006) bound to higher methylated lysine states of H3K4 revealed the principles of molecular recognition occurring in a sequence and methylation-state-specific manner (reviewed in Ruthenburg et al. 2007a; Taverna et al. 2007). We focus below on the BPTF system because this PHD finger is part of a PHD finger-bromodomain cassette discussed at greater length in Sections 9 and 10, and because structure–function studies were undertaken both at the peptide (reviewed in Ruthenburg et al. 2007a) and nucleosomal (Ruthenburg et al. 2011) levels.

Unbiased pull-down assays using nuclear extracts initially identified the second PHD finger of BPTF as a specific reader of the H3K4me3 mark (Wysocka et al. 2006). This PHD finger bound to higher methylation states of H3K4 that are linked to transcriptional activation with dissociation constants (K_d) of 2.7 μM for H3K4me3- and 5.0 μM for H3K4me2-containing peptides, while discriminating against monomethylated (me1) and unmodified (me0) counterparts. The observed intermediate range (i.e., μM; high range, nM; and low range, mM) of dissociation constants reflects the required balance between binding and dissociation of these marks by reader and eraser modules during epigenetic regulation. The structure of the H3(1-15)K4me3 peptide bound to the BPTF PHD finger solved by both X ray (Fig. 2A) and nuclear magnetic resonance (NMR) analysis showed that the histone peptide adopts a β-conformation paired with the β-sheet of the PHD finger to form a three-stranded, antiparallel β-sheet stabilized by backbone intermolecular hydrogen bonds (Fig. 2B) on the surface of the PHD finger on complex formation (Li et al. 2006). The sequence specificity for H3K4me3 involves recognition of the amino terminus, the R2 side chain, and the K4me3 mark in the complex. This distinguishes the H3 A1-R2-T3-K4me3 sequence segment from other Kme marks such as H3K9 and H3K27 embedded within an A-R-Kme3-S sequence context (Fig. 1B), and H3K36 and H4K20. The K4me3 is positioned in a preformed pocket composed of four aromatic residues, referred to as an "aromatic cage" and stabilized by electrostatic cation–π (Ma and Dougherty 1997) and hydrophobic interactions (Fig. 2A,C). The long side chains of R2 and K4me3 are positioned in adjacent preformed "surface groove" recognition pockets separated by the indole group of an invariant Trp residue (Fig. 2D), with this stapling role contributing to the specificity of H3K4me3 recognition. The guanidinium group of Arg2 is restrained by the formation of intermolecular hydrogen bonds between backbone carbonyls and an acidic side chain, whereas amino-terminal recognition (through hydrogen-bonding to a backbone carbonyl) is also important for complex formation (Fig. 2B), reflected by the discrimination against peptides with amino-terminal extensions. The binding affinity decreases dramatically on mutation of the aromatic cage residues, especially of the Trp residue separating R2 and K4me3, consistent with developmental defects associated with this mutation (Wysocka et al. 2006). The observed modest preference for H3K4me3 (K_d of 2.7 μM) over H3K4me2 (K_d of 5 μM) can be reversed following replacement of one of the cage-lining aromatic amino acids by a Glu, thereby facilitating formation of a hydrogen bond between the dimethylammonium proton of Kme2 and the carboxylate of the Glu side chain (Fig. 2E) (Li et al. 2007a). The above structural results provide an explanation for why global loss of H3K4me3 results in the loss of chromatin association of BPTF (Wysocka et al. 2006).

Structure–function studies on additional PHD fingers have highlighted the plasticity (i.e., aromatic cages formed by both aromatic and nonaromatic amino acids) in their recognition of methylated lysine marks on histone tails (reviewed in Musselman and Kutateladze 2011; Sanchez and Zhou 2011). Thus, parallel structure–function studies on the PHD finger of ING2, a native subunit of the repressive mSin3a-HDAC1 histone deacetylase complex, explained why, in response to DNA damage, recognition of H3K4me3 by the PHD finger of ING2 stabilizes the mSin3a-HDAC1 complex at the promoters of proliferation genes (Pena et al. 2006; Shi et al. 2006). In the ING2 complex, the aromatic cage is composed of a Trp and a Tyr, with a Met side chain contributing to pocket formation. Disruption of binding interactions in the H3K4me3-ING2 PHD

Cite as *Cold Spring Harb Perspect Biol* doi: 10.1101/cshperspect.a018754

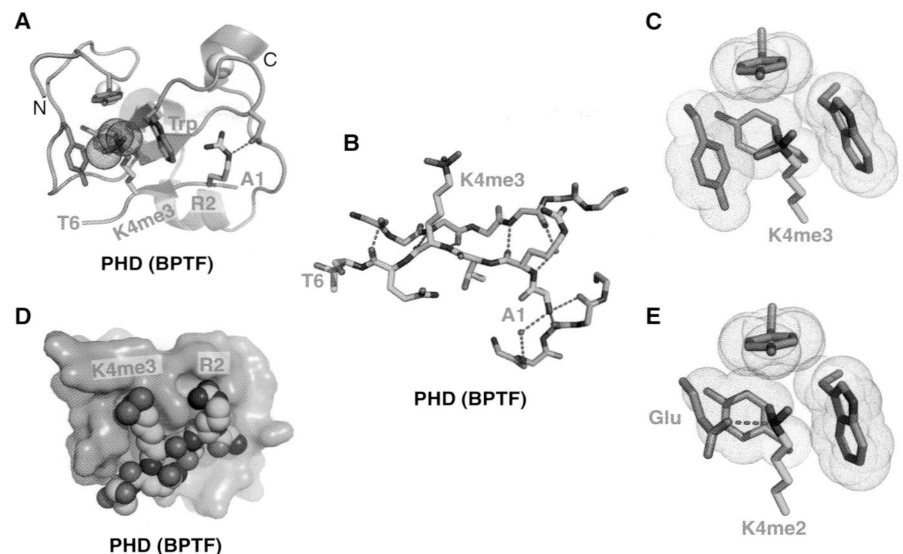

Figure 2. Structure of BPTF PHD finger bound to H3K4me3-containing peptide. (*A*) 2.0-Å crystal structure of the complex of the BPTF PHD finger bound to H3(1-15)K4me3 peptide (PDB: 2F6J). The PHD finger (as part of a PHD finger-bromo cassette) in a ribbon representation is in green, with two stabilizing bound Zn ions in silver balls. The bound peptide from A1 to T6 is shown in yellow with the trimethyl group of Kme3 shown by dotted balls in magenta. The residues forming the aromatic-lined cage are colored in orange. (*B*) Details showing the antiparallel alignment of the β-strands of the bound H3K4me3-containing peptide and PHD finger, resulting in formation of an antiparallel β-pleated sheet on complex formation. Note that the positively charged amino terminus is anchored in its own pocket. (*C*) Positioning of the K4me3 group within the aromatic-lined cage in the complex. (*D*) Positioning of R2 and K4me3 side chains in adjacent open surface pockets (surface groove mode), separated by the indole ring of an invariant Trp in the complex. The PHD finger and peptide are shown in surface- and space-filling representations, respectively. (*E*) Positioning of the K4me2 group into an engineered pocket, containing a Glu residue replacing the Tyr residue in *C* (PDB: 2RIJ).

finger complex impacts on function, impairing the ability of ING2 to induce apoptosis in vivo (Pena et al. 2006).

In another important structure–function study, the PHD finger of RAG2, an essential component of the RAG1/2V(D)J recombinase, that mediates antigen-receptor gene assembly (explained in Sec. 4 of Ch. 29 [Busslinger and Tarakhovsky 2014]) couples H3K4me3 readout with V(D)J recombination (Matthews et al. 2007; Ramon-Maiques et al. 2007). The investigators solved the structure of the RAG2 PHD finger bound to H3K4me3, evaluated the impact of Arg2 methylation, and then showed that mutations that abrogate intermolecular recognition severely impact on V(D)J recombination in vivo, as does depletion of the H3K4me3 mark. Interestingly, the Trp residue that is bracketed by the side chains of Arg2 and K4me3 in the RAG2 PHD finger-H3K4me3 peptide complex was found to be mutated in patients with immunodeficiency syndromes, highlighting the concept that disruption of the readout of a Kme mark can be the causal factor underpinning an inherited human disease (Matthews et al. 2007).

Histone mark dysregulation by PHD reader modules can impact on human cancers as reflected by the impact of somatic mutations in ING (inhibitor of growth) PHD fingers on solid tumors (reviewed in Chi et al. 2010).

2.2 BAH Domains

BAH domains are protein folds associated with epigenetic inheritance and gene regulation processes, having been identified in mammalian ORC1, MTA1 (a subunit of NuRD, a histone deacetylase and nucleosome remodeling complex), Ash1 (a SET domain-containing H3K36 methyltransferase), and the *Saccharomyces cerevisiae* Sir3 protein (part of the Sir2-Sir3-Sir4 complex required for silencing at the yeast mating-type loci). Tandem BAH domain pairs have been found in DNMT1 (the mammalian maintenance DNA methyltransferase) and the budding yeast RSC (a chromatin-remodeling complex) (reviewed in Callebaut et al. 1999; Goodwin and Nicolas 2001; Yang and Xu 2012).

The BAH domain adopts a conserved β-sheet core from which emerge loop and short helical segments shown in the structural determination of yeast ORC1p (Fig. 3A) (Zhang et al. 2002). The BAH domain (approximately 130 residues) was initially thought of as solely a protein–protein

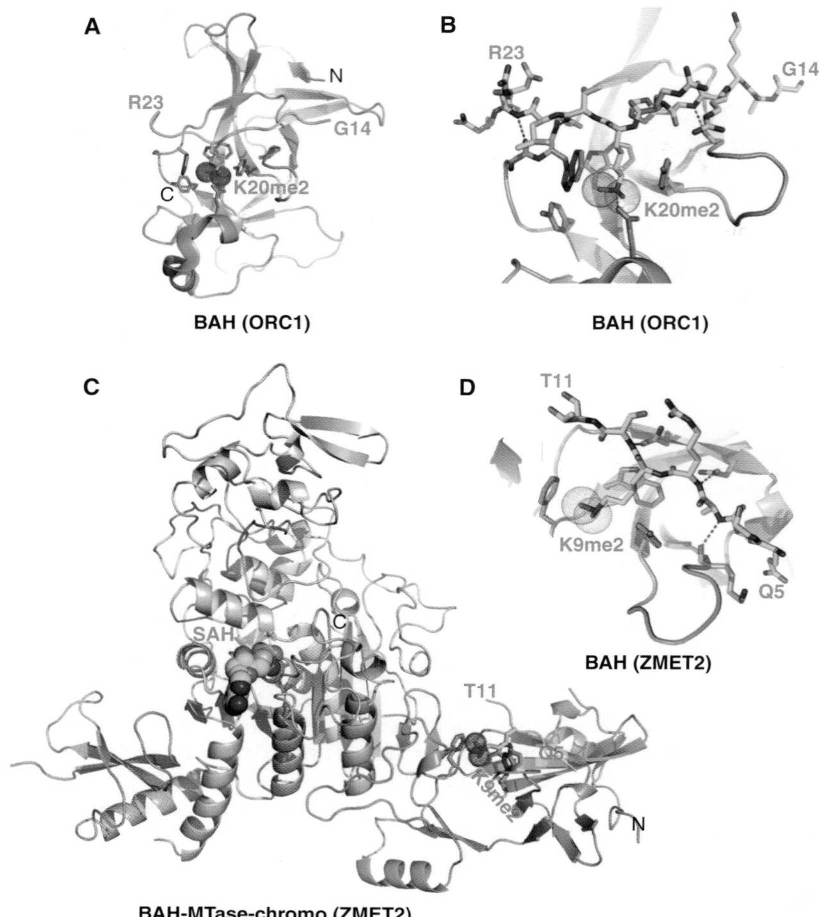

Figure 3. Structures of BAH domains of mammalian ORC1 and plant ZMET2 bound to methylated lysine histone peptides. (*A*) 1.95-Å crystal structure of the complex of mouse ORC1 BAH domain bound to H4(14-25)K20me2 peptide (PDB: 4DOW). The bound K20me2-containing H4 peptide can be traced from G14 to R23. (*B*) Enlargement of (*A*) showing details of the alignment of the K20me2-containing H4 peptide from G14 to R23 positioned on the mouse ORC1 BAH domain in the complex. The dimethylammonium group of H4K20 inserts into an aromatic-lined pocket in the BAH domain. (*C*) 2.7-Å crystal structure of the complex of maize ZMET2 BAH domain bound to H3(1-32)K9me2 peptide (PDB: 4FT4). The chromodomain, methyltransferase, and BAH domains are colored in pink, blue, and green, respectively. The bound K9me2-containing H3 peptide in yellow can be traced from Q5 to T11. (*D*) Enlargement showing details of the alignment of the K9me2-containing H3 peptide from Q5 to T11 positioned on the maize ZMET2 BAH domain in the complex. The dimethylammonium group of H3K9me2 inserts into an aromatic-lined pocket in the BAH domain.

interaction module of unknown function. The recent demonstration that BAH domains recognize methylated lysine as constituents of the mammalian ORC1 (Kuo et al. 2012) and *Arabidopsis thaliana* CMT3 (ZMET2 in maize) (Du et al. 2012) proteins at the peptide level, as well as recognition of unmodified mononucleosomes by yeast Sir3 (Armache et al. 2011), has now brought prominent attention to this reader module. The BAH domain is often flanked by reader modules of other histone marks, such as bromo and PHD domains, suggesting the potential for combinatorial readout (reviewed in Ruthenburg et al. 2007b).

A major advance in our understanding of the function of the BAH domain emerged following structure–function studies demonstrating that the mammalian ORC1 protein is a reader of the H4K20me2 mark, a property shared with diverse metazoans, but not the yeast ORC1 proteins (Kuo et al. 2012). Mammalian ORC1 constitutes the largest subunit of ORC (origin of replication complex) and it is the only ORC subunit of the six-subunit complex that contains a BAH domain. ORC1, which mediates pre-DNA replication licensing (Duncker et al. 2009), contributes both to ORC complex association and regulation of origin activity.

 Cite as *Cold Spring Harb Perspect Biol* doi: 10.1101/cshperspect.a018754

The structure of the ORC1 BAH domain bound to the H4K20me2 peptide (Fig. 3A) showed that the side chain of K20me2 inserts into an aromatic-lined cage (surface-groove recognition) within the BAH domain of ORC1, with the dimethylammonium group hydrogen-bonded to a Glu side chain and stabilized by hydrophobic and cation–π interactions (Fig. 3B). The side chains of Trp and Glu pocket-forming residues undergo conformational changes, thereby generating a fully walled aromatic cage pocket on complex formation. Mutation of the ORC1 BAH domain aromatic cage residues impairs ORC1 occupancy at replication origins, ORC chromatin loading, and cell-cycle progression (Kuo et al. 2012). The Meier–Gorlin syndrome, a form of primordial dwarfism (Klingseisen and Jackson 2011), has been linked to mutations in the ORC1 BAH domain (Bicknell et al. 2011). In fact, phenotypic rescue of growth retardation in *orc1*-depleted zebrafish morphants using wild-type, but not H4K20me2-binding mutants confirmed the central role the ORC1 BAH domain in this syndrome (Kuo et al. 2012). The above study thus identified the first direct link between histone methylation and the metazoan DNA replication machinery and linked the canonical histone H4K20me2 mark recognized by ORC1 with a primordial dwarfism syndrome.

More recently, studies have shown that the mammalian ORC1 BAH domain has several protein–protein interacting surfaces. Thus, in addition to the aromatic cage segment involved in H4K20me2 interaction, the BAH domain interacts with the protein kinase Cyclin E-CDK2 to inhibit the kinase activity involved in centrosome duplication (Hossain and Stillman 2012). The Meier–Gorlin syndrome mutants located within the BAH domain of ORC1 attenuated this inhibition activity. A mechanistic understanding of the inhibition characteristics of the kinase activity could emerge following attempts at structure determination of the complex between interacting partners.

The yeast ORC1 BAH domains are involved in the transcriptional silencing of the *HM* mating-type loci in budding yeast, which is an epigenetically regulated process (see Fig. 5 of Ch. 8 [Grunstein and Gasser 2013]). Structural studies of Sir1 (silencing information regulator 1) bound to the BAH domain of Orc1 revealed that the ORC1 BAH domain used a surface distinct from its aromatic cage-binding face to target Sir1 (Hou et al. 2005; Hsu et al. 2005), which contrasts with the mode of binding of higher eukaryotic ORC1 proteins.

In a groundbreaking study, structural details of the complex between the yeast Sir3 BAH domain (containing hypermorphic D205N mutant) bound to a nucleosome core particle (NCP) have emerged (Armache et al. 2011). Sir3 is part of the distinct Sir2-3-4 silent chromatin-forming structure used in *S. cerevisiae*, which contrasts with the more widely used H3K9me-HP1 rich heterochromatin found in higher eukaryotes. The Sir3-NCP structure is discussed briefly so as to highlight key features of the recognition process. Two Sir3 molecules are bound on either side of the pseudosymmetric NCP surface, contacting all four core histones on complex formation (schematically illustrated in Fig. 5 of Ch. 8 [Grunstein and Gasser 2013]). The complex is formed through a broad contiguous interaction between a surface on the NCP and a complementary surface on the BAH domain of Sir3 that is critical for transcriptional silencing. Complex formation results in the structural rearrangement of disordered segments on both the BAH domain and the nucleosome. The specific intermolecular contacts readily explain the numerous genetic mutations identified previously, as well as regulation of the silencing complex by modifications of H3K79 and H4K16 positions.

Tandem BAH domains in the structure of DNMT1 are separated by an α-helix, thereby ensuring a fixed separation and relative orientation between domains (Song et al. 2011). The DNMT1 BAH1 domain notably contains an aromatic cage along the same surface as mammalian ORC1. The BAH2 domain, however, lacks such an aromatic cage, but instead projects a very long loop, which interacts at its tip with the TRD (target recognition domain) of the methyltransferase domain, thereby holding the TRD in a retracted position away from the DNA in its complex with unmethylated DNA (illustrated in Fig. 6 of Ch. 6 [Cheng 2014]) and facilitating the formation of an autoinhibitory conformation. Importantly, both BAH1 and -2 of DNMT1 show large accessible surfaces available for further recognition, potentially with histone tails of nucleosomes and/or other interacting protein partners.

A novel example of recognition of methylated lysine histone marks by dual domains was highlighted in structure–function studies of the plant CMT3 protein (chromomethylase 3; ZMET2 in maize) (Du et al. 2012). CMT3 is a plant-specific DNA methyltransferase that methylates CpHpG (H stands for C, T, or G) sites in an H3K9me2-dependent manner (discussed in Sec. 2.1 of Ch. 13 [Pikaard and Mittelsten Scheid 2014]). CMT3 is composed of an amino-terminal BAH domain and a DNA methyltransferase domain, which has a chromodomain (for chromatin organization modifier) embedded within it. The structure of ZMET2 shows that the chromodomain and BAH domain are positioned at the corners of a triangular architecture, both binding H3K9me2-containing peptides through aromatic cage capture of the K9me2 side chain. The structure of the H3K9me2 peptide bound with directionality to the BAH domain of ZMET2 is shown in Fig. 3C, with details of the intermolecular interactions shown in Fig. 3D. Functional studies established a perfect correlation along the genome between methylated H3K9 and CMT3, and also showed that CMT3 is stably associated with H3K9me2-containing nucleosomes. Triple mutations of aromatic cage residues from the BAH domain or chromodomain disrupt CMT3 binding to nucleosomes and show a complete loss of CMT3 activity in vivo (Du et al. 2012). These studies definitively establish that DNA methylation in plants is directed by dual binding of the CMT3 BAH domain and chromodomain to H3K9me2-containing nucleosomes.

3 READOUT OF METHYLLYSINE MARKS BY SINGLE "ROYAL FAMILY" MODULES

The "Royal Family" of reader modules include the chromo, Tudor, PWWP (named after the conserved Pro-Trp-Trp-Pro motif), and MBT (malignant brain tumor) repeat domains (reviewed in Maurer-Stroh et al. 2003). Structural studies of these readers were among the first to elucidate the molecular principles behind methylated lysine recognition. In this section, we outline results on the structural principles by which single domains recognize lysine methylation marks and, in the next section, we look at tandem domains of the Royal Family.

3.1 Chromodomains

Structural studies of the chromodomains of HP1 (heterochromatin-associated protein 1) and Polycomb provided the first insights into how the aromatic cage pockets (illustrated in orange stick representation form in Fig. 4A) create a structural framework for methylated lysine recognition (reviewed in Yap and Zhou 2011). These proteins are known repressors that contribute to epigenetic silencing. More specifically, HP1 is a chromatin-associated protein first discovered in *Drosophila* and hallmark of condensed and highly repetitive heterochromatin (elaborated in Ch. 12 [Elgin and Reuter 2013]). Polycomb is another protein originally characterized in *Drosophila* shown to direct heritable changes in chromatin organization (see Ch. 17 [Grossniklaus and Paro 2014]).

Before structural studies, in situ immunofluorescence showed that H3K9me marks and HP1 colocalized to heterochromatic regions of *Drosophila* polytene chromosomes (illustrated in Figs. 3 and 4 of Ch. 12 [Elgin and Reuter 2013]; also see Jacobs et al. 2001). Next, X-ray and NMR studies (Jacobs and Khorasanizadeh 2002; Nielsen et al. 2002) independently corroborated the fact that the H3K9me3 mark was recognized by the HP1 chromodomain. We focus below on X-ray structural studies of the complex (Fig. 4A) that form with a $K_d = 2.5$ μM, illustrative of an intermediate affinity interaction. The H3 tail in an extended β-strand conformation (residues 5–10) binds to one face of the incomplete β-barrel architecture of HP1, where it aligns between β-strands of HP1 through an antiparallel-aligned-induced β-sandwich alignment. The K9me3 side chain inserts into a pocket lined by three conserved aromatic amino acids (i.e., the so-called aromatic cage pocket), where it is stabilized by cation–π interactions. The complex is stabilized by intermolecular contacts involving the ARKS (Ala-Arg-Lys-Ser) motif (Fig. 1), associated with two amino

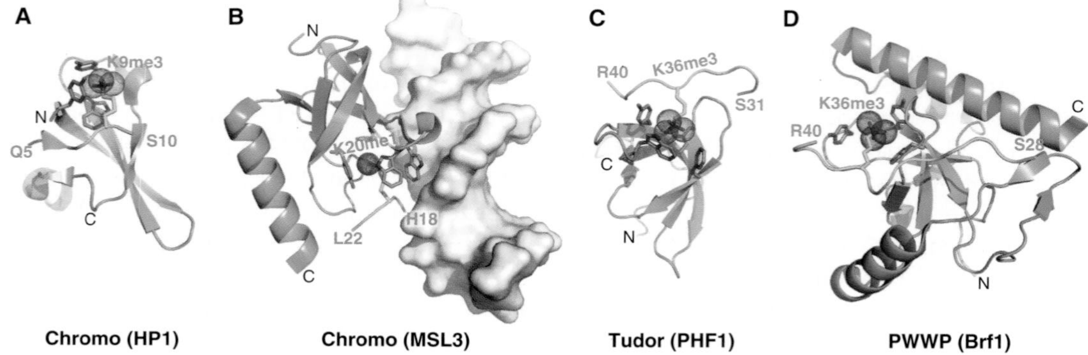

Figure 4. Structures of single Royal Family modules bound to methylated lysine histone peptides. (*A*) 2.4-Å crystal structure of the complex containing an HP1 chromodomain bound to H3(1-15)K9me3 peptide (PDB: 1KNE). The bound K9me3-containing H3 peptide can be traced from Q5 to S10. The HP1 residues in orange illustrate the aromatic cage that captures K9me3. (*B*) 2.35-Å crystal structure of the complex between the male-specific lethal (MSL)3 chromodomain bound to a H4(9-31)K20me1 peptide in the presence of duplex DNA (in surface representation) (PDB: 3OA6). The bound K20me1-containing H4 peptide can be traced from H18 to L22. (*C*) 1.85-Å crystal structure of the complex containing the PHF1 (a Polycomb-like protein) Tudor domain bound to H3(31-40)K36me3 peptide (PDB: 4HCZ). The bound K36me3-containing H3 peptide can be traced from S31 to R40. (*D*) 1.5-Å crystal structure of the complex of Brf1 PWWP domain bound to H3(22-42)K36me3 peptide complex (PDB: 2X4W). The bound K36me3-containing H3 peptide can be traced from S28 to R40.

acids preceding and one following the K9me3 mark, providing a sequence context for the readout of the H3K9me3 mark. Mutation of the conserved aromatic amino acids lining the cage of HP1 results in a substantial loss of binding affinity (approximately 20-fold), as does mutation of Ala7 and Arg8 on the H3 peptide, with a smaller loss (threefold) observed on mutation of Ser10 (Jacobs and Khorasanizadeh 2002).

Subsequent structural studies showed that cage-capture of higher methylation groups (i.e., Kme3/2) was also observed for other chromodomains. Notably, this was shown for the chromodomain of Polycomb bound to an H3K27me3 peptide (Min et al. 2003), CHD1 bound to H3K4me2/3 peptides (Sims et al. 2005), MGR15 bound to an H3K36me2 peptide (Zhang et al. 2006), Eaf3 bound to an H3K36me3 peptide (Sun et al. 2008; Xu et al. 2008), and Chp1 bound to an H3K9me3 peptide (Schalch et al. 2009). This reinforces the principle that sequences flanking the Kme3/2 mark must determine the specificity of recognition by chromodomains (reviewed in Brehm et al. 2004; Yap and Zhou 2010).

A recent structural study has taken our understanding of chromodomains to a new level; the MSL3 chromodomain can target the lower lysine methylation mark H4K20me1 only in the presence of DNA (Kim et al. 2010). This unexpected corecognition was specific for GA-rich DNA, with binding greater by two orders of magnitude over RNA. MSL3 is a subunit of the male-specific lethal (MSL) complex, which is necessary for dosage compensation on *Drosophila* male X chromosomes (the topic of Ch. 24 [Lucchesi and Kuroda 2014]). A preassembled complex contains the MSL3 chromodomain and DNA bound to H4K20me1 peptide with a K_d = 15 μM, with discrimination against unmodified and trimethylation states on H4K20. The crystal structure of this ternary complex is shown in Fig. 4B, with the chromodomain targeting the DNA minor groove and the K20me1 inserting into an adjacently positioned cage lined by four aromatic amino acids. Interestingly, the active H4K16ac mark antagonizes the DNA-mediated recognition of H4K20me1 by MSL3, suggesting that regulation of the MSL complex may be controlled by the readout of closely positioned marks. The investigators speculate that corecognition of the H4K20me1 mark and the DNA of two adjacent nucleosomes could contribute to the in vivo targeting of the MSL complex (Kim et al. 2010).

3.2 Tudor Domains

Polycomb group proteins are repressive chromatin modifiers essential for metazoan development, cellular differentiation, and the maintenance of cell fate. The Tudor fold, a domain within Polycomb, consists of a β-sheet core against which are packed one or more helical segments (Selenko et al. 2001). No fewer than four groups have recently undertaken functional studies of human Polycomb-like proteins PHF1 and PHF19, including the structural characterization of complex formation between its amino-terminal Tudor domain and H3K36me3-containing peptide (Ballare et al. 2012; Brien et al. 2012; Musselman et al. 2012a; Cai et al. 2013). X-ray and NMR studies of complex formation show that the H3K36me3 peptide in an extended conformation targets the five-stranded β-barrel of the Tudor domain (Fig. 4C). The trimethylammonium group of K36me3 inserts into an aromatic-lined pocket, whereas the flanking side chains of the peptide interact with a hydrophobic patch (bound by Pro38-His39 segment) and a shallow acidic groove (bound by Thr32-Gly33-Gly34 segment). Functionally, recognition of the H3K36me3 mark by polycomb-like proteins promotes the intrusion of Polycomb repressive complex 2 (PRC2) into active chromatin regions to then promote gene silencing, thereby impacting on the chromatin landscape during development.

3.3 PWWP Domains

The PWWP domain (containing a highly conserved Pro-Trp-Trp-Pro motif) along with chromodomain, MBT, and Tudor domains, belongs to the Royal Family of proteins that have the potential to recognize methylated lysine marks using an aromatic cage capture mechanism. It has been observed in many chromatin-associated proteins, often in combination with other domains such as SET modules, which are known writers of lysine methylation marks. PWWP-containing proteins are involved in diverse functions including transcriptional regulation, DNA repair, and DNA methylation (reviewed in Slater et al. 2003). The PWWP fold, as first determined by structural studies of the de novo methyltransferase DNMT3B, consists of a five-stranded β-barrel followed by an α-helical bundle with one of the α-helices packing against the β-barrel to generate a single structural motif (Qiu et al. 2002).

A recent structure of the PWWP domain of the Brf1 (bromodomain and plant homeodomain finger 1) protein bound to an H3K36me3 peptide (weak affinity of K_d = 2.7 mM) has validated the PWWP domain as a reader of methylated lysine marks (Fig. 4D) (Vezzoli et al. 2010). The peptide is positioned with directionality on a narrow surface groove with the K36me3 side chain positioned in an aromatic cage. Binding is abolished on mutation of any of the three conserved aromatic residues. The Brf1 protein associates specifically with the H3K36me3 mark and discriminates against other Kme3 marks on H3 and H4, in part, because of its ability to site-specifically accommodate Gly residues at positions 33 and 34. Functional studies in vivo indicate that Brf1 localizes to actively transcribed

Hox genes, in which its enrichment parallels that of the H3K36me3 mark (Vezzoli et al. 2010). It should be noted that the binding of methylated lysine peptides to the chromodomain and the PWWP domain use different surfaces of the β-barrel for recognition.

4 READOUT OF METHYLLYSINE MARKS BY TANDEM ROYAL FAMILY MODULES

In this section, we outline how tandem reader modules of the Royal Family are involved in the recognition of Kme marks. These examples highlight the diversity of the recognition process between tandem chromo, tandem Tudor, and MBT repeats, and also within the tandem Tudor family.

4.1 Tandem Chromodomains

Proteins have been identified with tandem chromodomains. CHD (chromo-ATPase/helicase-DNA-binding) is one such protein that is involved in regulating ATP-dependent nucleosome assembly and mobilization at sites of transcription activity. It contains amino-terminal tandem chromodomains that target the H3K4me3 mark with a $K_d = 5\ \mu M$. The structures of the human CHD1 tandem chromodomains have been solved both in the free state and bound to H3K4m3-containing peptides (Flanagan et al. 2005). Both chromodomains (1 and 2) adopt canonical chromodomain folds with a rigid helix-turn-helix motif connecting them such that the tandem chromodomains are juxtaposed together to form a continuous surface. One H3K4me3 peptide binds per CHD1 with the H3 peptide backbone positioned within an acidic surface between chromodomains 1 and 2 (Fig. 5A). The K4me3 side chain is positioned in a pocket composed of two Trp rings, with adjacent Arg2 also forming cation–π interactions with one of these Trp residues, thereby contributing to the specificity of recognition. Indeed, mutation of either Trp results in a substantial reduction in binding affinity. In addition, the binding affinity drops fourfold for complex formation with H3K4me3R2me2a, containing asymmetric dimethylation of Arg2 and a 25-fold reduction for complex formation with H3T3phK4me3-containing phosphorylation of adjacent Thr3 (Flanagan et al. 2005). It is important to note that chromodomains 1 and 2 of CHD1 do not use their canonical Kme-binding surfaces for peptide recognition partly because of sequence inserts in the CHD1 sequence that block canonical binding sites used by HP1 and Polycomb protein chromodomains.

4.2 Tandem Tudor Domains

Tandem Tudor domains have been identified as readers of methylated lysine marks and shown to use two distinct modes of recognition. The relative positioning of the pair of Tudor domains determines the binding mode; one class has a single linker that connects individually folded Tudor domains (examples include 53BP1, UHRF1, and SHH1) and another involves domain swapping and interdigitation of Tudor domains connected by a β-sheet whose β-strands are shared between domains (e.g., JMJD2A).

The tandem Tudor domain came into prominence following studies on 53BP1 (p53-binding protein). This study showed that histone lysine methylation marks facilitated the recruitment of 53BP1 and relocalized it to double-strand breaks on exposure to DNA-damaging agents. The recruitment element within 53BP1 was identified as a tandem Tudor domain (Huyen et al. 2004) and its mode of action emerged following the structural determination of its complex with H4K20me2 peptides ($K_d = 19.7\ \mu M$) (Botuyan et al. 2006). The structure of the complex is shown in Fig. 5B with the peptide positioned between the Tudor domains, but primarily interacting with Tudor 1 (in green). The K20me2 side chain inserts into an aromatic cage positioned within Tudor 1 that is lined by four aromatic residues and an Asp that hydrogen bonds to the dimethylammonium proton, with the dimensions of the pocket preventing insertion of a K20me3 group ($K_d = 1\ mM$) as a consequence of steric exclusion. The complex is also stabilized by cation–π interactions between the side chain of Arg19 of the peptide and a tyrosine ring of the protein. The importance of H4K20me2 recognition emerged following mutation of the aromatic residues lining the pocket, whereby binding was lost in vitro and also impacted on the targeting of 53BP1 to DNA double-strand breaks in vivo (Botuyan et al. 2006). A recent structural study has also reported that selective H3K4me3 recognition occurs by Spindlin1 Tudor domain 2 of this three-tandem Tudor domain-containing protein, with the Tudor domains aligned in a triangular-shaped architecture (Yang et al. 2012).

Structures have been solved for the tandem Tudor domains of UHRF1 (ubiquitin-like PHD and Ring finger 1) (Nady et al. 2011) and SHH1 (Sawadee homeodomain homolog 1) (Law et al. 2013) when bound to H3K9me2/3-containing peptides. These systems, like 53BP1, involve a single linker connecting individually folded Tudor domains, but unlike 53BP1, the bound H3 peptide interacts with both Tudor domains through positioning with directionality within a channel between domains. Importantly, the tandem Tudor domains within the UHRF1 and SHH1 proteins serve as dual lysine readers, probing for both unmethylated K4 (K4me0) and methylated K9 on the histone tail. Functional studies on the UHRF1 system show that Tudor domain mutants that can no longer bind H3K4me0K9me3 have reduced localization to heterochro-

Cite as *Cold Spring Harb Perspect Biol* doi: 10.1101/cshperspect.a018754

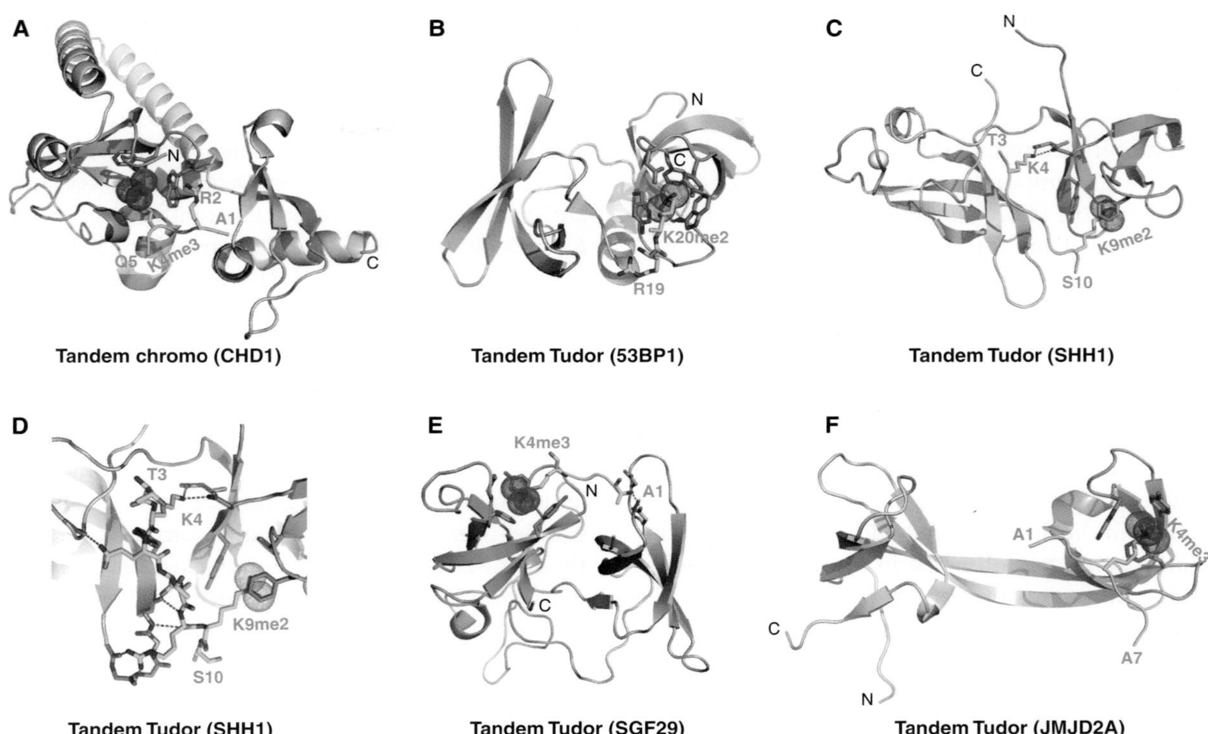

Figure 5. Structures of tandem Royal Family modules bound to methylated lysine histone peptides. (*A*) 2.4-Å crystal structure of the complex of the human CDH1 tandem chromodomains bound to H3(1-19)K4me3 peptide (PDB: 2B2W). Chromodomains 1 and 2 are colored in green and blue, respectively, with the connecting helix-turn-helix linker in pink. The bound K4me3-containing H3 peptide can be traced from A1 to Q5. (*B*) 1.7-Å crystal structure of the complex of 53BP1 tandem Tudor domains bound to H4(15-24)K20me2 peptide (PDB: 2IG0). Tudor domains 1 and 2 are colored in green and blue, respectively. The bound K20me2-containing H4 peptide can be traced for the R19-K20me2 step. (*C*) 2.7-Å crystal structure of the complex of tandem Tudor domains of *A. thalania* SHH1 protein bound to a H3(1-15)K9me2 peptide (PDB: 4IUT). A bound zinc ion is shown in a silver ball. Tudor domains 1 and 2 are colored in green and blue, respectively. The bound K9me2-containing H3 peptide can be traced from T3 to S10. (*D*) The enlargement shows details of the alignment of the K9me2-containing H3 peptide from T3 to S10 positioned on the *A. thaliana* SHH1 domain in the complex with intermolecular interactions formed with both Tudor domains. (*E*) 1.26-Å crystal structure of the complex of tandem Tudor domains of SGF29 bound to H3(1-11)K4me3 peptide (PDB: 3MEA). Tudor domains 1 and 2 are colored in blue and green, respectively. The bound K4me3-containing peptide can be traced from A1 to K4me3. (*F*) 2.1-Å crystal structure of the complex of tandem Tudor domains of JMJD2A bound to H3(1-10)K4me3 peptide (PDB: 2GFA). Individual Tudor domains are colored in green and blue, respectively. The bound K4me3-containing peptide can be traced from A1 to A7.

matic chromocenters and fail to silence expression of the $p16^{INK4A}$ gene (Nady et al. 2011).

The structure of an H3(1-15)K9me2 peptide bound to the tandem Tudor domains of SHH1 ($K_d = 1.9$ μM) is shown in Fig. 5C with elements involved in the recognition of unmodified K4me0 and K9me2 shown in the expanded view in Fig. 5D. No conformational change was observed in the tandem Tudor fold when complexed with an H3K9me2 peptide (Law et al. 2013). Functional studies on the SHH1 system in plants show that the protein acts upstream of the RNA-dependent DNA methylation (RdDM) pathway (described in Sec. 3.5 of Ch. 13 [Pikaard and Mittelsten Scheid 2014]) to enable small interfering RNA production from a large subset of the most active RdDM targets. SHH1 is required for polymerase-IV (pol-IV) occupancy at these same loci (Law et al. 2013). Further, key residues within both lysine-binding pockets of SHH1 are required to maintain DNA methylation in vivo, thereby providing the first insights into the mechanism by which SHH1 recruits RNA pol-IV to RdDM targets in plants.

SGF29, a component of the SAGA (Spt-Ada-Gcn5 acetyltransferase) complex, contains a pair of tandem Tudor domains that bind to H3K4me2/3 peptides. Binding occurs along one surface of the tightly packed face-to-face dimeric alignment (Fig. 5E) (Bian et al. 2011). The side chains of Ala1 insert into a pocket of the Tudor 1 domain (in blue), whereas the K4me2/3 inserts into an aromatic cage of the Tudor 2 domain (in green). Functionally, SGF29 targets active chromatin via H3K4me2/3, thereby recruiting the SAGA complex to mediate acetylation of H3 tails.

The jumonji histone KDM, JMJD2A, demethylates H3K9me2 via its jumonji domain (discussed in Sec. 1.5 of Ch. 6 [Cheng 2014]). Its tandem Tudor domains adopt a domain-swapped interdigitated topology with a two-stranded β-sheet serving as a linker between the well-separated domains (Huang et al. 2006; Lee et al. 2008). The structure of the interdigitated tandem Tudor domains bound to an H3K4me3 peptide is shown in Fig. 5F; the side chain of K4me3 is positioned within a three-aromatic-residue-lined cage associated primarily with Tudor 2 (in green), yet binding affinity is also attributed to the intermolecular contacts it makes with side chains from Tudor 1 (in blue) within the interdigitated Tudor scaffold. No conformational change was observed in the bilobal interdigitated Tudor motif on proceeding from the free to the H3K4me3-bound state. In a related structural study (Lee et al. 2008), it was shown that the JMJD2A interdigitated Tudor domains bind with similar affinities to H3(1-10)K4me3 ($K_d = 0.50\ \mu M$) and H4(16-25)K20me3 ($K_d = 0.40\ \mu M$) peptides, although the two peptides share no amino acid sequence similarity other than the trimethylated lysine. Strikingly, the two peptides are aligned in opposite orientations despite the Kme3 side chains of both peptides inserting into the same aromatic-lined pocket in Tudor 2. Details of the intermolecular contacts led Lee et al. (2008) to identify single point mutants that inhibited recognition of H4K20me3, but not H3K4me3 or vice versa.

4.3 Tandem MBT Repeats

The MBT repeats are approximately 70 residues long and aligned in tandem. MBT acts as a transcription repressor with its repeats often perturbed in hematopoietic malignances (Koga et al. 1999). At the functional level, MBT proteins impact on diverse processes, ranging from regulation of mitosis and tumor suppression to the maintenance of cellular identity and body pattern during development (Bonasio et al. 2010).

The MBT fold was first identified following structure determination of the three MBT repeats from human L3MBTL1 (Wang et al. 2003) and the two MBT repeat-containing *Drosophila* SCML2 (Sex comb on midleg-like 2) protein (Sathyamurthy et al. 2003). The MBT unit is composed of a four-stranded β-barrel core followed by an extended arm of helices. Interdigitation occurs between the extended arms and cores of adjacent MBT subunits. In the case of L3MBTL1, this results in the formation of a three-leaved propeller-like architecture with each MBT unit containing an aromatic-lined pocket located on the same face of the triangular architecture (Fig. 6A) (Wang et al. 2003). Binding studies with histone peptides established that L3MBTL1 shows a preference for lower (mono- and di-) lysine methylation states. However, binding to a particular histone Kme1/2 mark is somewhat promiscuous and of relatively low to intermediate affinity (in the range of $K_d = 5-40\ \mu M$). The Kme1/2 marks that L3MBTL1 binds to include H1.4K26me, H3K4me, H3K9me, H3K27me, H3K36me, and H4K20me, as measured at the peptide level by fluorescence polarization-based binding assays (Li et al. 2007a). Structural studies showed that the side chains of Kme1 and Kme2 insert deep into an aromatic cage pocket of the second MBT repeat of L3MBTL1 (Fig. 6B). Simultaneously, an adjacent L3MBTL1 protomer in the crystal lattice inserts its Pro ring from a carboxy-terminal tail Pro-Ser segment into a shallow aromatic cage pocket of the first MBT repeat (Fig. 5C). Pocket 2 is both deep and narrow and serves as a size-selective filter with the side chains that project from the gating and caging loops restricting access to the larger Kme3 group (Li et al. 2007a). Such a "cavity insertion" mode of methyllysine recognition (Fig. 6B) is distinct from the surface groove mode of methyllysine recognition (Fig. 2D) observed for other Royal Family members, PHD finger, and BAH domain reader complexes. A parallel structural study looking at the binding of the H4K20me2 peptide to L3MBTL1 also identified pocket 2 as a reader of lower lysine methylation marks (Min et al. 2007). In addition, it proposed an unanticipated mode of peptide-mediated dimerization leading to a model for chromatin compaction by L3MBTL1. More recent binding studies concluded that some MBT proteins showed sequence specificity, whereas others were promiscuous in the targeting of lower methylated lysine marks (Nady et al. 2012).

Reconstituted L3MBTL1-histone complexes analyzed by electron microscopy showed that L3MBTL1 does indeed compact nucleosomal arrays containing lower lysine methylation marks at H1K26 and H4K20. Thus, a combinatorial readout of methylated H1K26 and H4K20 by L3MBTL1 and methylated H3K9 by HP1γ results in chromatin condensation at Rb-regulated genes (Trojer et al. 2007; reviewed in Trojer and Reinberg 2008). Somewhat unexpectedly, recent functional studies indicate that the related protein L3MBTL2 can act in concert with PcG protein-mediated ubiquitination of H2A to establish a repressive chromatin structure without entailing the contribution of histone lysine methylation marks (Trojer et al. 2011).

Structural studies have also been undertaken on the four MBT repeats in human L3MBTL2 (Guo et al. 2009) and its *Drosophila* counterpart, dSfmbt (Grimm et al. 2009), bound to an H4K20me1 peptide. The four MBT

Cite as *Cold Spring Harb Perspect Biol* doi: 10.1101/cshperspect.a018754

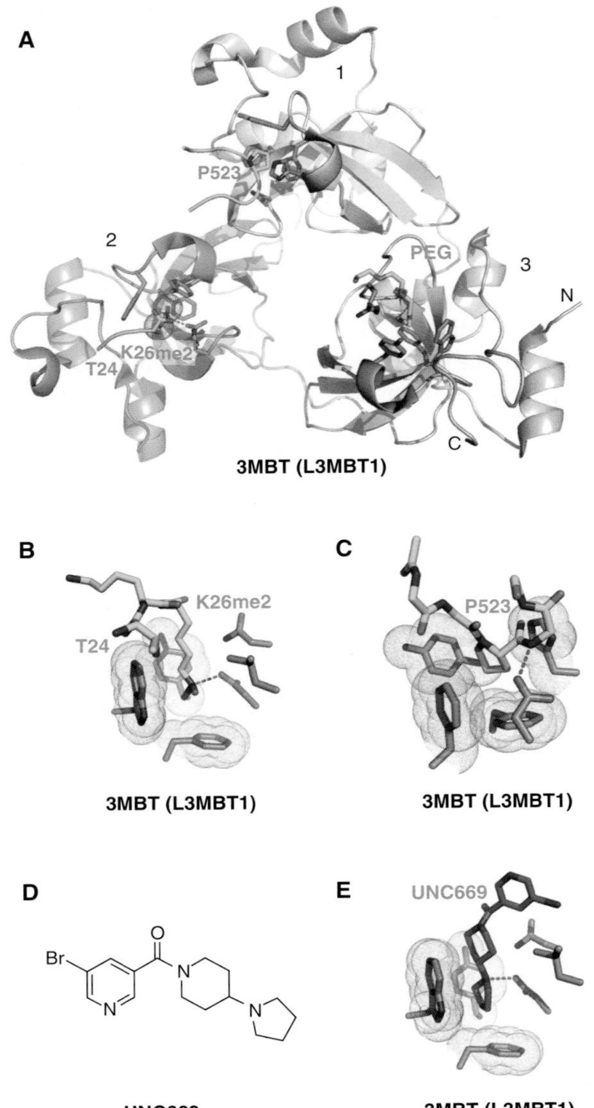

Figure 6. Structures of L3MBTL1 bound to methylated lysine histone peptides and an inhibitor. (*A*) 1.66-Å crystal structure of the complex-containing L3MBTL1 bound to H1(22-26)K26me2 peptide (PDB: 2RHI). A carboxy-terminal peptide from an adjacent L3MBTL1 in the crystal lattice inserts Pro523 into the aromatic-lined pocket of MBT domain 1 (in pink). The dimethylammonium group of bound K26me2 inserts into the aromatic-lined pocket of MBT domain 2 (in green) with the K26me2-containing H1 peptide traced from T24 to K26me2. A polyethylene glycol (PEG) molecule inserts into the aromatic-lined pocket of MBT domain 3 (in blue). (*B*) Details of how the H1K26me2 dimethylammonium group inserts into the aromatic-lined pocket of MBT domain 2. (*C*) Structural detail showing how the proline from an adjacent L3MBTL1 in the crystal lattice inserts into the aromatic-lined pocket of MBT domain 1. This pocket is shallower than the one shown in *B*. (*D*) Chemical formula of UNC669. (*E*) Details of how UNC669 inserts into the aromatic-lined pocket of MBT domain 2, based on the 2.55-Å crystal structure of L3MBTL1 bound to UNC669 (PDB: 3P8H).

repeats adopt an asymmetric rhomboid scaffold, with MBT repeats 2, 3, and 4 forming the triangular architecture and repeat 1 projecting off of it. The K20me1 side chain inserts into pocket 4 (corresponds to pocket 2 of L3MBTL1) with preferential recognition of lower methylation states. It was proposed that the observed lack of sequence specificity between bound histone peptides by MBT repeats could reflect the absence of distinct surface contours surrounding the methylated lysine-binding pocket (Guo et al. 2009). Functional experiments suggest that dSfmbt interacts with a related MBT repeat protein, Scm, and these two proteins together synergize in the repression of target genes crucial for Polycomb silencing (Grimm et al. 2009).

5 READOUT OF METHYLLYSINE MARKS BY EXPANDED AND PAIRED MODULES

We outline, below, two examples in which the readout of methylated lysine marks requires an expanded PHD finger module in one case and the modulation of PHD finger binding by a cofactor in the other case. The expanded PHD finger module example is most illuminating because the structural studies have identified a new principle for the recognition of Kme marks that is distinct from the surface groove (e.g., parts of PHD finger illustrated in Fig. 2D) and "insertion cavity" (e.g., Fig. 6B) modes of recognition already outlined in Sections 2.1 and 4.3. In addition, we outline recognition of Kme marks by ankyrin repeats.

5.1 GATA-1 and PHD Fingers of the ADD Domain

The ADD (ATRX-DNMT3A-DNMT3L) domain is found both in the ATRX protein, whose mutated form is associated with X-linked mental retardation (ATR-X) syndrome, and mammalian DNMT3A-DNMT3L required for de novo DNA methylation. ATRX is a large protein composed of adjacently positioned zinc-coordinated GATA and PHD fingers, referred to as the expanded PHD finger module. Half of the missense mutations in ATRX are associated with disease, many of them identified with pancreatic endocrine tumors. Further, about half are clustered toward the amino terminus within the Cys-rich ADD domain (Jiao et al. 2011); the other half cluster within the helicase/ATPase domain. The NMR solution structure of the ADD domain of ATRX in the free state showed that the GATA finger, PHD finger, and long carboxy-terminal α-helix form a single globular domain (Argentaro et al. 2007). Disease-causing mutations lie within this ADD domain dispersed between Zn-coordinating residues, those involved in packing, and others distributed along the outer surface.

The structure of H3K4me0-containing peptides bound to the ADD domain of DNMT3A has been solved (see

Sec. 3 and Fig. 10 of Ch. 6 [Cheng 2014] and Otani et al. 2009). More recently, two groups have solved the structures of the ATRX ADD domain bound to H3K9me3-containing peptides (Eustermann et al. 2011; Iwase et al. 2011). The binding is promoted by H3K9me3, but inhibited by H3K4me3. A crystal structure of the complex reported at 1.0-Å resolution has provided details of intermolecular contacts and bridging water molecules at the highest resolution (Fig. 7A,B) (Iwase et al. 2011). The structure explains the requirement for unmodified H3K4; this side chain ammonium group is hydrogen bonded to the carboxyate groups of acidic amino acids within the PHD component of the ADD domain, leaving no room to accommodate methylation marks (Fig. 7B). Strikingly, the K9me3 mark is positioned within an "interfacial composite pocket" composed of residues from both GATA-1 and PHD fingers, with the dimensions of the pocket exquisitely sculpted to perfectly allow accommodation of the bulky trimethyllysine group (Fig. 7C). Unlike classical trimethyllysine readers, which usually consist of an aromatic-lined cage for higher methylation state-specific readout (reviewed in Taverna et al. 2007), the interfacial composite pocket of the ADD domain that binds K9me3 involves favorable van der Waals contacts associated with a high degree of surface complementarity (Fig. 7D), supplemented by a set of carbon–oxygen hydrogen bonds with the K9me3 group (Fig. 7C) (Iwase et al. 2011). The K4 and K9me3 binding pockets adopt a rigid mutual orientation (Fig. 7B), thereby contributing to the combinatorial readout of dual marks as reflected in the large increase in affinity (sevenfold on comparing H3K4me0K9me3, $K_d = 0.5$ μM, with H3K9me0K9me0, $K_d = 3.6$ μM) and enthalpy (twofold on comparing H3K4me0K9me3, $\Delta H = -12.2$ Kcal mol^{-1}, with H3K9me0K9me0, $\Delta H = -6.1$ Kcal mol^{-1}) for their joint readouts compared with their individual mark counterparts (Eustermann et al. 2011; Iwase et al. 2011). The structure

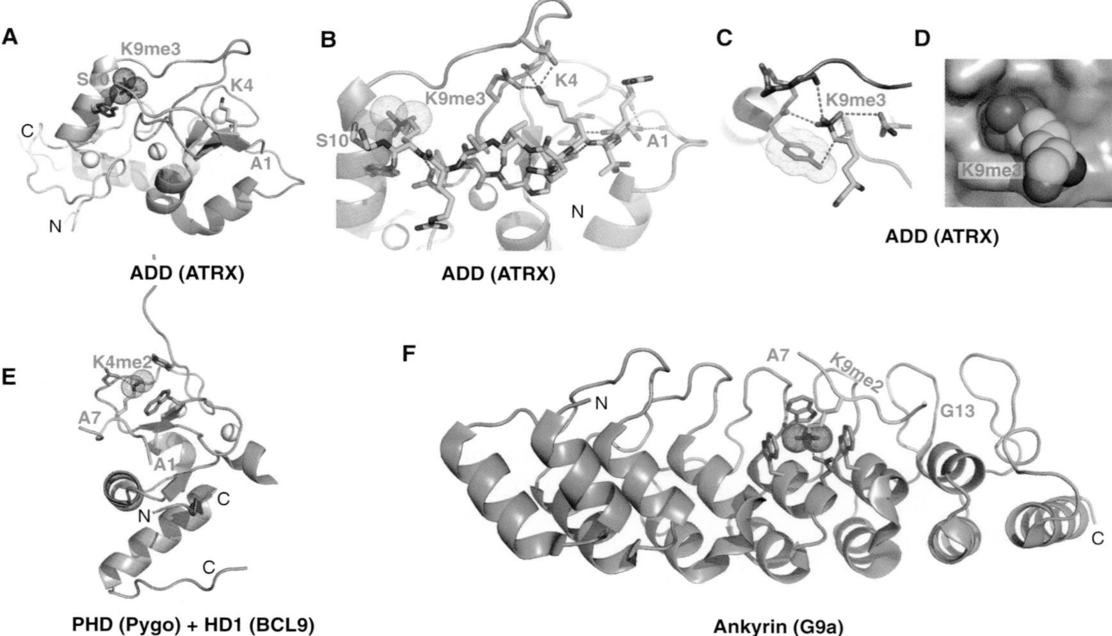

Figure 7. Structures of expanded and paired modules bound to methylated lysine histone marks. (*A*) 1.6-Å crystal structure of the complex containing the ADD domain of ATRX bound to the H3(1-15)K9me3 peptide (PDB: 3QLA). The ADD GATA-1 and PHD fingers are colored in blue and green, respectively. Bound Zn ions are shown as silver balls. The H3 peptide containing K9me3 is traced from A1 to S10. (*B*) Enlargement of *A* showing intermolecular contacts between the K9me3-containing H3 peptide, traced from A1 to S10, complexed with the ADD domain. The me3 are shown as magenta spheres. (*C*) Ribbon and stick representation of K9me3 positioned to interact with the GATA-1 and PHD finger domains in the complex. (*D*) Surface and space-filling representation of surface complementarity between K9me3 and the walls of the pocket lined by the GATA-1 and PHD finger domains in the complex. (*E*) 1.7-Å crystal structure of the ternary complex of the Pygo PHD finger (in green) bound to H3(1-7)K4me2 peptide (in yellow) in the presence of the HD1 domain of BCL9 (in pink) (PDB: 2VPE). (*F*) 2.99-Å crystal structure of the complex consisting of the G9a ankyrin repeats (green) bound to an H3(1-15)K9me2 peptide traced from A7 to G13 in yellow (PDB: 3B95). The K9me2 aromatic-lined binding pocket is positioned between the fourth and fifth ankyrin repeats of G9a.

Cite as *Cold Spring Harb Perspect Biol* doi: 10.1101/cshperspect.a018754

of the complex containing the H3K9me3 peptide bound to the ATRX ADD domain highlights the role of integrated modules and the unanticipated composite reader pockets in methyllysine recognition. Functionally, ATRX localization at pericentric heterochromatin is lost for mutations in the H3K9me3 binding pocket and ones identified from ATRX syndrome patients (Iwase et al. 2011). In addition, the in vivo studies of Eustermann et al. (2011) showed that the readout of H3 with unmodified K4 and K9me3 by the ADD domain of ATRX was facilitated by recruitment of HP1, which independently recognizes H3K9me3. Such tripartite recognition could have the potential for spanning neighboring nucleosomes.

5.2 Pygo PHD Finger and Its Cofactor BCL9

The Pygopus (Pygo) protein contains a PHD domain, which, together with its cofactor BCL9, operates during development via the Wnt signaling pathway by regulating β-catenin-mediated transcription through an interaction with methylated H3K4 marks. This regulatory function relies on the interaction between the carboxy-terminal PHD domain of Pygo and the homology domain 1 (HD1) of BCL9. Structural studies on the Pygo PHD finger (Nakamura et al. 2007), its binary complex with BCL9 HD1 (Fiedler et al. 2008), and its ternary complex with added H3K4me2 (Fig. 7E) (Fiedler et al. 2008; Miller et al. 2010) identified the principles underlying K4me2 mark recognition. The H3K4me2 mark and HD1 are positioned on opposite faces of the Pygo PHD finger, with efficient mark recognition requiring association of the Pygo PHD finger with HD1. HD1 binding to the PHD finger, in fact, triggers an allosteric transition, thereby facilitating optimal recognition of the H3K4me2 mark. The K4me2-binding pocket is composed of four aromatic residues and an Asp with a twofold preference for the Kme2 over the Kme3 state.

5.3 Ankyrin Repeats

G9a and GLP (G9a-like) are euchromatin-associated KMTs composed of amino-terminal ankyrin repeats and a carboxy-terminal methyltransferase SET domain. These enzymes repress transcription following SET domain-catalyzed writing of H3K9me1 and H3K9me2 marks. It was subsequent to identifying the enzymatic H3K9me writing activity that G9a and GLP were also shown to be readers of H3K9me marks, using the ankyrin repeats within the protein (Collins et al. 2008). Ankyrin repeats involve helix-turn-helix-β-turn modules, aligned through the stacking of helices, with the β-turns projecting outward at right angles. The H3K9me1 and H3K9me2 peptides bind to the ankyrin repeats of G9a with intermediate binding affinity ($K_d = 14$ μM and 6 μM, respectively), with a crystal structure solved for the complex of ankyrin repeats with the H3K9me2 peptide. The bound H3 peptide is sandwiched between β-turns and helices of the fourth and fifth ankyrin repeats, with the dimethylammonium group of K9me2 inserted into an aromatic pocket lined by three Trp residues and a Glu (Fig. 7F). Intermolecular recognition involves peptide residues 9 to 11, which includes K9me2. Mutation of these peptide or aromatic cage residues adversely impacts on complex formation (Collins et al. 2008). Perturbation of methyllysine recognition by the ankyrin repeats, however, has no effect on G9a's SET domain-mediated methyltransferase activity, indicating that the reading and writing domains function independently.

6 READOUT OF METHYLARGININE MARKS BY TUDOR MODULES

Tudor modules, as well as recognizing lysine methylation (Kme) marks, recognize methylated arginine (Rme) marks, and the latter can be subdivided into two classes: the canonical and the expanded Tudor reader modules. Tudor proteins play key roles in a range of cellular processes ranging from germ cell development to RNA metabolism, processing, and silencing, as well as DNA damage response and chromatin-remodeling (reviewed in Bedford and Clarke 2009; Siomi et al. 2010; Chen et al. 2011). The structural studies described in this section are relevant to chromatin biology not so much for directly reading arginine methylation marks contained on the histone proteins, but on proteins such as PIWI, which are part of the machinery involved in chromatin-templated processes such as RNA silencing. Although we describe the details of the binding structures of Tudor domains with nonchromatin targets, the details become relevant to epigenetic regulation in that they affect Piwi, a component of RNA interference (RNAi) pathways.

6.1 Canonical Tudor Domains

The first insights into the canonical Tudor domain fold emerged from structural studies of the SMN (survival of motor neuron) protein. When mutated, SMN causes spinal muscular atrophy, a degenerative motor neuron disease. The SMN protein contains a single highly conserved canonical Tudor domain required for uridine-rich small nuclear ribonucleoprotein complex assembly. The Tudor domain facilitates binding to the seven-membered hetero-oligomeric ring-shaped Sm proteins. The SMN Tudor domain forms a strongly bent five-stranded antiparallel β-sheet barrel-like fold (Selenko et al. 2001), with a cluster of aromatic amino acids forming a cage on one face of the

Tudor domain (Sprangers et al. 2003). This targets methylated arginines positioned within the carboxy-terminal Arg-Gly-rich tails of Sm proteins (Brahms et al. 2001). The Tudor-containing protein, TDRD3, is also a reader of methylated arginine marks on histone tails (H3R17me2a and H4R3me2a), acting as a transcriptional coactivator (Yang et al. 2010).

NMR-based solution structures have been solved for the Tudor domains of SMN and SFP30 bound to short symmetrical dimethylated arginine peptides corresponding to the carboxy-terminal Arg-Gly-rich tails of Sm proteins (Tripsianes et al. 2011). The binding specificity decreases in the order symmetrical dimethylated arginine (Rme2s) > asymmetrical dimethylated arginine (Rme2a) > monomethylated arginine (Rme1). Binding specificity for Rme2s marks was independent of residues within the peptide on either side of the mark. The dimethyl guanidinium moiety of the Rme2s mark, in an *anti–anti* alignment (*anti* orientation of both N-CH_3 bonds in relation to the side chain C-Nϵ bond), inserts into a cage formed by four aromatic residues and an Asn residue (Fig. 8A), in which it is stabilized by cation–π interactions. The E134K mutation associated with spinal muscular atrophy impairs a hydrogen bond between the Glu134 side chain and hydroxyl group of a Tyr residue lining the aromatic cage, highlighting the role of aromatic amino acids lining the cage for optimally orienting the dimethyl guanidinium group of the Rme2s, ensuring recognition.

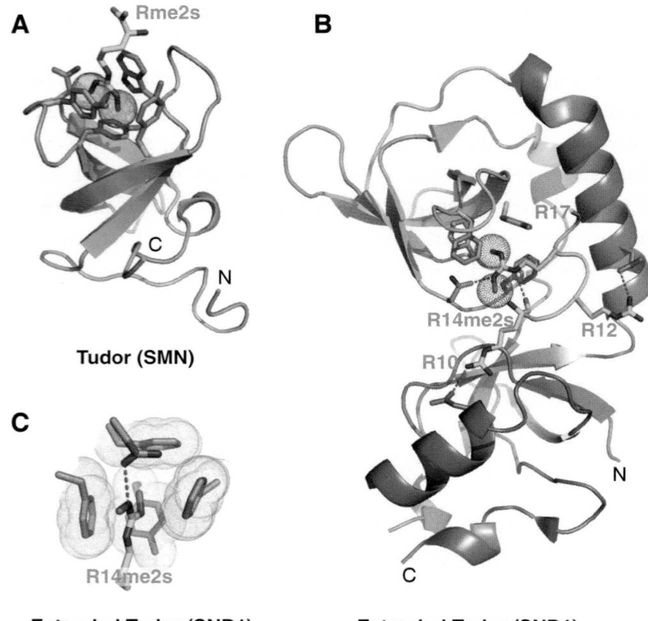

Figure 8. Structures of expanded and paired modules bound to methylated arginine histone marks. (*A*) NMR solution structure of the complex containing the SMN Tudor domain (green ribbon representation) bound to a symmetrical Rme2s-containing peptide (yellow) (PDB: 4A4E). The methyl groups are illustrated with magenta spheres and the aromatic-lined cage in orange. (*B*) 2.8-Å crystal structure of the complex containing the SND1 extended Tudor module bound to the amino-terminal PIWI peptide, traced from R10 to R17, with the R14me2s modification (PDB: 3NTI). The core fold of the Tudor domain is shown in green, whereas the extensions are shown in blue. (*C*) Enlargement of *B* showing the positioning of R14me2s in the aromatic-lined cage of the SND1 Tudor domain in the complex.

6.2 Expanded Tudor Domains

Certain proteins contain expanded multiple tandem repeats as a 180-residue module, which itself consists of a 60-residue canonical Tudor core domain and flanking amino- and carboxy-terminal conserved elements. Many of these Tudor proteins are germline-specific and exert their regulatory effect via the recognition of arginine methylation at the amino-terminal end of PIWI family proteins. PIWI proteins are themselves important regulators of the germline via the RNA silencing pathway. This pathway functions predominantly in silencing transposable elements in early development (reviewed in Siomi et al. 2010; Chen et al. 2011).

Several structure–function studies have been performed on complexes containing the expanded Tudor domains bound to methylated arginine peptides from the Arg-Gly/Arg-Ala-rich amino terminus of PIWI-family proteins. These studies have shed light on the principles underlying recognition of the symmetrical dimethylated arginine mark (Rme2s) (Liu et al. 2010a; Liu et al. 2010b; Mathioudakis et al. 2012). The structures of complexes containing other Tudor group proteins led to similar conclusions about the nature of expanded Tudor domain interactions with methylated arginines (Liu et al. 2010a; Liu et al. 2010b; Mathioudakis et al. 2012). The results of the complex between the SND1 (staphylococcal nuclease domain-containing 1) extended Tudor module and the R14me2s-containing amino-terminal PIWI peptide is discussed here (Liu et al. 2010b).

The SN-like domain of the SND1 protein is split into two segments by the canonical Tudor domain with an α-helical linker connecting the Tudor (in green) and SN-like (in blue) domains (Fig. 8B). This so-called extended Tudor module forms an OB (oligonucleotide and oligosaccharide-binding) fold. The R14me2s-containing amino-terminal PIWI peptide binds with directionality within a wide and negatively charged groove of the SND1 Tudor domain (Fig. 8B), in the process inserting the planar dimethyl-guanidinium group of R14me2s in an *anti–syn* orientation into an aromatic cage lined by four aromatic amino acids and an Asn residue (Fig. 8C). SND1 can also form a complex with the PIWI R4me2s-containing peptide and its structure has also been reported (Liu et al. 2010a).

Unlike canonical Tudor domains, binding affinity on complex formation is affected by proximal residues flanking the Rme2s mark. And both the canonical Tudor domain and flanking amino- and carboxy-terminal extensions are equally required for binding because deletion of either of the extensions resulted in loss of binding. Mutation of either individual aromatic amino acids or the Asn (involved in hydrogen bonding to the dimethyl guanidinium group) lining the pocket resulted in 11- to 22-fold loss in binding affinity. The SND1 extended Tudor module shows specificity for the R4me2s mark (K_d = 10 µM), with reduction in binding by fourfold for the R4me2a mark, twofold for the R4me1 mark, and a larger ninefold for the unmodified R4 mark. It is likely that monomethylation would reduce hydrophobic interactions with the aromatic pocket, and asymmetrical dimethylation could disrupt the intermolecular hydrogen bond with the Asn, accounting for the modest reduction in binding affinity. Interestingly, the R4me2s-containing amino-terminal PIWI peptide binds to the SND1 extended Tudor in the opposite orientation to the R14me2s peptide yet retains insertion of R4me2s into the aromatic cage pocket, thereby suggesting plasticity in the recognition process. Functional studies have identified an intricate interplay between writers and readers of the Rme2s mark and the mark itself and in regulating transposon silencing and germ cell development (reviewed in Chen et al. 2011). A remaining challenge in the field will be to identify potential readers of Rme2a and Rme1 marks and the principles underlying optimal selectivity as a function of methylation state of arginine.

7 READOUT OF UNMODIFIED LYSINE MARKS

Although the focus of this chapter is largely on reader modules that target methylated lysine marks, it should be noted that, as introduced in Sections 4.2 and 5.1, certain reader modules target unmodified lysines on histone tails. Such recognition events can then be blocked or weakened by methylation with functional consequences.

7.1 PHD Fingers

The first indication that unmethylated lysines contribute to molecular recognition emerged from the structure of the BHC80 PHD finger bound to the histone H3(1-10) peptide (K_d = 30 µM) (Lan et al. 2007). A high-resolution (1.4-Å) crystal structure study readily revealed details of the intermolecular contacts; the ammonium group of K4 hydrogen bonded to both an acidic side chain and backbone carbonyl group (Fig. 9A). There was no room to accommodate a methylated K4 in this complex, thereby accounting for the specificity of unmodified lysine recognition. The intermolecular interactions are stabilized by interdigitation of side chains involving the H3 peptide (R2 and K4) with the PHD finger (Met and Asp); binding is lost on mutation of these Met and Asp residues (Lan et al. 2007). RNAi knockdown of BHC80 results in derepression of LSD1 target genes, whereas ChIP studies show that BHC80 and LSD1

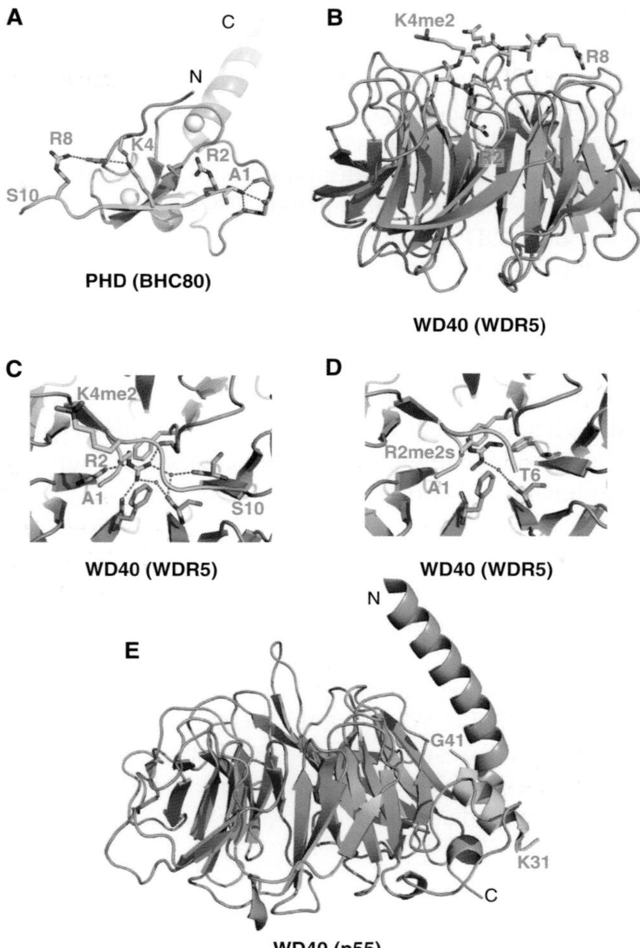

Figure 9. Structures of reader modules bound to lysines and arginines. (*A*) 1.43-Å crystal structure of the complex of the PHD finger of BHC80 bound to H3(1-10) peptide (in yellow) (PDB: 2PUY). The bound H3 peptide can be traced from A1 to S10. Zinc ions are shown as silver balls. (*B*) 1.5-Å crystal structure of the complex of the WD40 motif of WDR5 bound to H3(1-9)K4me2 peptide. The bound K9me2-containing peptide can be traced from A1 to R8 (PDB: 2H6N). (*C*) Insertion of R2 into the central channel of the WD40 motif in the H3(1-9)K9me2-WDR5 complex. (*D*) Insertion of R2me2s into the central channel of the WD40 motif in the H3(1-15)R2me2sK9me2-WDR5 complex solved at 1.9 Å (PDB: 4A7J). (*E*) 3.2-Å crystal structure of the complex of the WD40 motif of p55 bound to H4(15-41) peptide (PDB: 3C9C). The bound H4 peptide can be traced from K31 to G41.

mutually depend on each other to associate with chromatin. These results point to BHC80 function being coupled to that of LSD1, which is consistent with BHC80 having a role in LSD1-mediated gene repression via the recognition of unmodified H3K4 (Lan et al. 2007).

Other examples of reader modules that recognize unmodified lysines at H3K4 include DNMT3l (Ooi et al. 2007), the ADD domains of DNMT3A and ATRX (discussed in Sec. 5.1), PHD finger domains of TRIM24 and TRIM33 (to be discussed in Sec. 9), and tandem Tudor domains of UHRF1 and SSH1 (discussed in Sec. 4.2). Recognition of unmodified H3K4 and discrimination against methylation at this position is achieved by intermolecular hydrogen bonding between the ammonium group of lysine and acidic side chains, or the backbone carbonyl groups, or a combination of both.

8 READOUT OF UNMODIFIED ARGININE MARKS

The readout of unmodified lysines can solely be achieved by PHD fingers, as outlined in Section 7. In contrast, unmodified arginines can be read by several distinct reader modules including WD40 motifs, PHD fingers, and chromodomains. We outline below the distinct recognition principles by which each of these reader modules binds with unmodified arginine.

8.1 WD40 Motifs

WDR5 is a member of the WD-40 (for 40 amino acids) repeat family. Individual WD40 repeats form a toroidal B-propeller fold, much like the blades of a propeller. WDR5 is a common component of the SET1-family of KMT complexes that play an essential role in *Hox* gene activation and vertebrate development. Several groups simultaneously solved the structure of the H3K4me2 peptide-WDR5 complex (Couture et al. 2006; Han et al. 2006; Ruthenburg et al. 2006; Schuetz et al. 2006) with the anticipation that the structure would identify the principles underlying H3K4me2 recognition, given that WDR5 associates with and is essential for H3K4 methylation and vertebrate development (Wysocka et al. 2005). The H3K4me2 peptide binds on the surface above the central channel of the toroid-shaped fold of WDR5 (Fig. 9B), where it is stabilized by a set of extensive intermolecular contacts between the bound H3 peptide and WDR5. Unexpectedly, it is the side chain of unmodified Arg2 rather than K4me2 that inserts into the narrow central channel of the toroidal β-propeller fold (cavity insertion mode), in which it is stabilized by stacking with staggered Phe side chains and oriented through direct and water-mediated hydrogen bonds (Fig. 9C).

A further surprise came from a recent structure–function study demonstrating that the central channel of the WDR5 WD40 fold also accommodates the symmetrically dimethylated H3R2me2s mark (Fig. 9D), but not its asymmetrical H3R2me2a counterpart (Migliori et al. 2012). Comparison of the H3R2 and H3R2me2s complexes indicated that the guanidinium group forms direct and water-mediated hydrogen bonds involving two anchored water molecules in the H3R2 complex (Fig. 9C), with one of these waters displaced in the H3R2me2s complex, together with protrusion of the dimethylated side chain toward the Phe ring (Fig. 9D) to form enhanced hydrophobic interactions. Functionally, the H3R2me2s mark retains euchromatic genes in a poised state ready for transcriptional activation following transit from the cell cycle and differentiation.

Another study has investigated the structure of WRD5 complexed with a peptide from the KMT writer enzyme, MLL1 (mixed lineage leukemia 1) (Patel et al. 2008; Song and Kingston 2008). Unexpectedly, the histone H3-like MLL1 peptide bound in an almost identical orientation as an H3 peptide and, in addition, similarly inserted the Arg side chain into the central channel of the WD40 scaffold. It was also shown that H3K4me-containing peptides compete with the MLL1 peptide for WDR5 because disruption of WDR5-MLL1 binding was more effective by H3K4me1/me2 peptides over H3K4me0/me3 peptides. One interpretation of these observations is that there is a delicate interplay between components of the MLL1 complex, namely WDR5 (the reader), the SET domain of MLL1 (the writer), and the histone H3 tail (the substrate) (Song and Kingston 2008).

In another structural study, the WD40 motif of p55, a common component of several chromatin-modifying complexes, bound the first A-helix of H4 within a surface binding channel located on the side of the β-propeller scaffold (Fig. 9E) (Song et al. 2008). Because the first helix of H4 is buried within the canonical fold of the nucleosome, a substantial alteration of the conformation of this H4 segment was observed on complex formation with p55. Further, it was shown that the histone H4-binding pocket was important for the functional activities of p55-containing complexes involved in chromatin assembly (CAF1), remodeling (NURF), and deacetylation (NuRD) (Song et al. 2008).

8.2 PHD Fingers

The PHD finger is a reader module that can recognize unmodified arginines. A recent example includes the recognition of unmodified R2 on the H3 tail by the PHD finger of UHRF1, an important regulator of CpG methylation (discussed in Sec. 2 of Ch. 6 [Cheng 2014]; Rajakumara

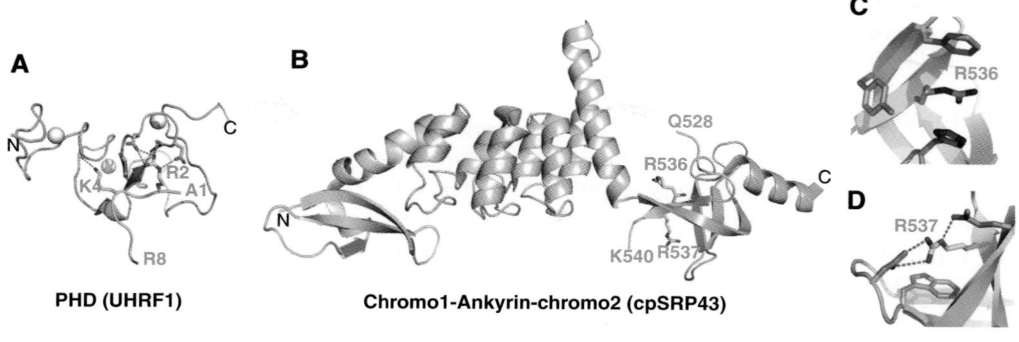

Figure 10. Structures of PHD finger and chromodomains bound to unmodified arginines. (*A*) 1.8-Å crystal structure of the complex of the PHD finger of UHRF1 bound to H3(1-9) peptide (PDB: 3SOU). The bound H3 peptide can be traced from A1 to R8. Zinc ions are shown by silver balls. (*B*) 3.18-Å crystal structure of the complex of the chromodomains and Ankyrin repeats of *A. thaliana* chloroplast signal recognition particle (cpSRP)43 bound to an RRKR (Arg-Arg-Lys-Arg)-containing peptide (yellow) (PDB: 3UI2). The side chains of R536 and R537 of the bound RRKR-containing peptide are positioned in adjacent pockets at the interface between the fourth Ankyrin repeat (purple) and the second chromodomain (green) in the complex. (*C*) Positioning of Arg536 of the RRKR-containing peptide within an aromatic-lined cage in the complex. (*D*) Positioning of Arg537 of the RRKR-containing peptide in a pocket lined by a Trp and two acidic side chains in the complex.

et al. 2011b; see also Hu et al. 2011; Wang et al. 2011; Xie et al. 2012). In this complex, the guanidinium group of R2 participates in an extensive intermolecular hydrogen bond network (Fig. 10A), and methylation of H3R2, but not H3K4 or H3K9, disrupts complex formation. Functionally, the ability of UHRF1 to suppress the expression of its target genes is dependent on the PHD finger of UHRF1 binding to unmodified H3R2, thereby linking this recognition event involving UHRF1 to regulation of euchromatic gene expression (Rajakumara et al. 2011b).

Another example of a PHD finger that recognizes the Rme0 mark is the MOZ (monocytic leukemia zinc-finger protein) PHD tandem cassette, detailed in Section 9.3.

8.3 Chromodomains

Recent research reported an unexpected interaction in the cytosol between cpSRP43 (composed of chromodomains and Ankyrin repeats) and a peptide containing the RRKR motif from cpSRP54. The latter is associated with the chloroplast signal recognition particle in *A. thaliana* (Holdermann et al. 2012). The structure of cpSRP43ΔCD3 bound to an RRKR-containing peptide of cpSRP54 is shown in Fig. 10B, with the peptide binding at the interface between the fourth Ankyrin repeat and the second chromodomain (CD2) ($K_d = 6.4\ \mu M$). In the complex, the RRKR-containing peptide adopts an extended conformation except for a type-II β-turn at its amino terminus, with recognition occurring through a β-completion mode within the hydrophobic binding groove of CD2. The specificity of molecular recognition resides in two adjacent arginines in the peptide (Arg536 and Arg537), which are positioned in adjacent pockets of cpSRP43, one of which is composed of a three-aromatic residue-lined cage (Arg536; Fig. 10C), whereas the other is composed of an aromatic amino acid and two acidic residues (Arg537; Fig. 10D), with replacement of either Arg on the peptide or Tyr and Trp residues lining the pockets, abolishing complex formation and its functional readout (targeting of Ankyrin repeats to light-harvesting chlorophyll *a*,*b* binding proteins). The investigators also propose that a similar recognition mechanism likely exists for the interaction between cpSRP43 and the carboxy-terminal tail of the membrane insertase Alb3 involved in recruiting cpSRP to the thylakoid membrane (Falk et al. 2010). This insightful contribution introduces the novel concept of readout of adjacent unmodified arginines by a twinned aromatic cage (Holdermann et al. 2012) to generate the cpSRP complex (Goforth et al. 2004) and defines an unexpected nonnuclear function for chromodomains in posttranslational targeting.

9 MULTIVALENT READOUT BY LINKED BINDING MODULES AT THE PEPTIDE LEVEL

Given the low affinity in the readout of single marks by their reader modules, it should not be surprising that the affinity of binding can be increased following multivalent readout of two or more marks by linked binding modules (reviewed in Ruthenburg et al. 2007b; Wang and Patel 2011). In this section, we outline recent results on histone combinatorial PTM readout at the peptide level and, in the following section, discuss results at the nucleosomal level.

9.1 PHD Finger-Bromo Cassettes

The adjacent positioning of PHD finger and bromodomain to form the PHD finger-bromodomain (designated PHD-bromo) cassette is the most frequently observed combination of dual reader modules impacting on epigenetic regulation (Ruthenburg et al. 2007b). The PHD finger and bromodomain read Kme and Kac marks, respectively, and different proteins with PHD-bromo cassettes show the potential for reading different combinations of these two marks. To date, structure–function studies have been undertaken on the PHD-bromo cassettes of BPTF (Li et al. 2006; Ruthenburg et al. 2011; Wysocka et al. 2006), MLL1 (Wang et al. 2010b), TRIM24 (Tsai et al. 2010b), TRIM33 (Xi et al. 2011), and KAP1 (Ivanov et al. 2007; Zeng et al. 2008). There are similarities and differences in structure–function aspects associated with these PHD-bromo cassettes at the peptide level. We outline their distinct roles in epigenetic regulation below.

The PHD-bromo cassette of the BPTF component of the NURF chromatin remodeler introduced in Section 2.1 provided the first insights into the structure and function of this cassette. The structure of the BPTF PHD-bromo cassette showed that the two domains are separated by an α-helical linker (Fig. 11A), resulting in a fixed distance between domains, with their histone mark-binding pockets defined by a fixed relative orientation (Li et al. 2006). The PHD finger binds the H3K4me3 mark (Li et al. 2006; Wysocka et al. 2006), whereas the bromodomain binds a range

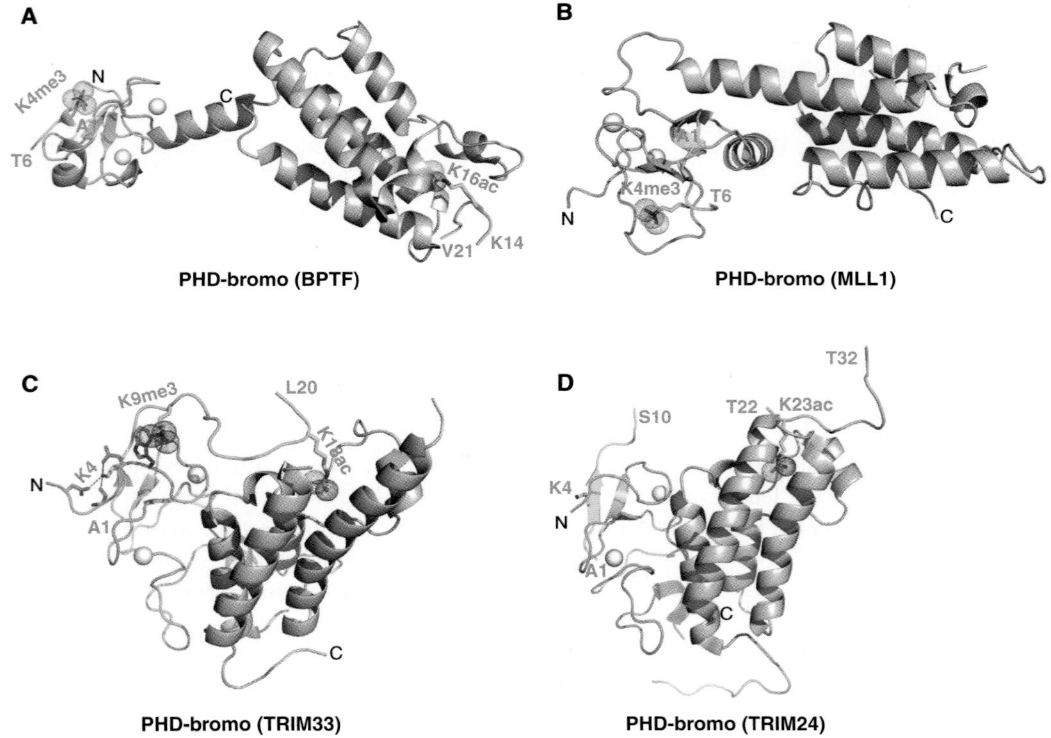

Figure 11. Structures of PHD-bromo cassettes involved in multivalent readout. (*A*) 2.0-Å crystal structure of the PHD-bromo cassette of BPTF bound to H3(1-15)K4me3 peptide (PDB: 2F6Z). A separate 1.8-Å crystal structure of the BPTF bromodomain bound to H4(12-21)K16ac peptide was also solved (PDB: 3QZS) and that information was superpositioned on the structure shown in this panel. The bound H3(1-15)K4me3-containing peptide can be traced from A1 to T6, whereas the bound H4(12-21)K16ac-containing peptide can be traced from K14 to V21 in the complexes. (*B*) 1.9-Å crystal structure of the MLL1 PHD-bromo cassette bound to H3(1-9)K4me3 peptide (PDB: 3LQJ). The bound H3(1-9)K4me3-containing peptide can be traced from A1 to T6 in the complex. (*C*) 2.7-Å crystal structure of the TRIM33 PHD-bromo cassette bound to H3(1-22)K9me3K18ac peptide (PDB: 3U5O). The bound H3(1-22)K9me3K18ac peptide can be traced from A1 to L20 in the complex. (*D*) The crystal structures of the TRIM24 PHD-bromo cassette bound to H3(1-10) peptide (2.0 Å) (PDB: 3O37) and bound to H3(13-32)K23ac peptide (1.9 Å) (PDB: 3O37). The structures were superpositioned to generate the composite structure shown in this panel. The bound H3(13-32)K23ac peptide can only be traced from T22 to T32 in the complex.

of Kac marks with a preference for H4K16ac (Ruthenburg et al. 2011). Thus, the BPTF PHD-bromo cassette shows capabilities for targeting distinct marks on different histone tails (Ruthenburg et al. 2011).

The leukemia-inducing MLL1 protein targets its SET domain methyltransferase activity to *HOX* gene promoters during hematopoiesis (Milne et al. 2002). This protein contains adjacently positioned PHD finger (PHD3) and bromodomain, which adopt the characteristic individuals folds of these motifs, with the connecting linker adopting a turn segment that leads into the extended αZ-helix of the bromodomain (Fig. 11B) (Wang et al. 2010b). This alignment results in interactions between the PHD and bromodomains that manifests as an increase in binding affinity for PHD finger recognition of the H3K4me3 mark as compared with the isolated PHD finger. The K4me3 side chain positions into an aromatic-lined pocket (Fig. 11B), and a conformational change closes an additional face of the aromatic-lined cage on complex formation. The bromodomain part of the cassette has lost its ability to target Kac-containing histone peptides. ChIP measurements established colocalization of MLL1 and H3K4me3 marks on *HOX* genes, suggesting that the binding of MLL1 PHD3 to the H3K4me3 mark contributes to MLL1 localization at target genes (Milne et al. 2010; Wang et al. 2010b).

Structure–function studies on the carboxy-terminal PHD-bromo cassettes of TRIM24 (Tsai et al. 2010b) and TRIM33 (Xi et al. 2011) provide further insights into combinatorial recognition of dual marks. For both TRIM24 and TRIM33, the adjacently positioned PHD finger and bromodomain folds interact extensively with each other to generate a single folded unit (Fig. 11C,D). Histone peptide binding studies on TRIM33 elucidated that the PHD finger recognized a combination of unmodified K4 (me0) and K9me3 on the same H3 tail, whereas the bromodomain recognized K18ac. Further, the binding affinity for H3(1-28)K9me3K18ac ($K_d = 0.06$ μM) dual mark was greater than the sum of binding to individual H3(1-28)K9me3 ($K_d = 0.20$ μM) and H3(1-28)K18ac ($K_d = 0.21$ μM) marks, consistent with combinatorial recognition occurring, even at the peptide level.

The crystal structure of the TRIM33 PHD finger-bromo cassette bound to H3(1-22)K9me3K18ac peptide has been solved (Fig. 11C) and provides a molecular explanation for the results obtained from binding studies. Key intermolecular contacts contributing to the specificity of recognition include interaction between the TRIM 33 amino terminus and H3 peptide backbone, between the unmodified K4 and three acidic side chains, as well as stacking between K9me3 and a Trp side chain, all involving the PHD finger. Additional specificity is associated with positioning of K18ac in an atypical binding pocket on the bromodomain (Xi et al. 2011). Functionally, nodal signaling-induced formation of TRIM33-Smad2/3 and Smad4-Smad2/3 complexes triggered the interaction between TRIM33-Smad2/3 and the H3K9me3 mark, thereby displacing the chromatin-compacting factor HP1γ, and making nodal response elements accessible to Smad4-Smad2/3 for RNA polymerase II recruitment. In essence, nodal TGF-β signals use the poised H3K9me3 mark as a platform to switch master regulators of stem cell differentiation from the poised to the active state (Xi et al. 2011).

The TRIM24 bromodomain, which is part of the PHD-bromo cassette, targets the H3K23ac mark (Tsai et al. 2010b). The difference in acetyllysine residue recognition between TRIM24 (H3K23ac recognition) and TRIM33 (H3K18ac recognition) is a reflection of distinct contributions of sequences that flank the Kac mark, translating into different intermolecular recognition capabilities of the two complexes. Crystal structures of two TRIM24 PHD finger-bromo cassette complexes, one bound to H3K9me3-, the other to H3K18ac-containing peptides, are superimposed and shown in Fig. 11D. The spatial alignment and extensive interaction between the paired reader domains generates a continuous binding surface, enhancing the combinatorial readout capacity (Tsai et al. 2010b). TRIM24 binds chromatin with H3K4me0 and H3K23ac marks and also to the estrogen receptor protein, which itself binds to estrogen response elements when signaled by estrogen to activate estrogen-dependent genes associated with cellular proliferation and tumor development. Thus, it is not surprising that the overexpression of TRIM24 frequently observed in breast cancer patients correlates with poor survival. The coexistence of estrogen response elements and distinct dual-marked H3 chromatin, thus, provides a distinct system conducive to regulating estrogen-dependent target genes (Tsai et al. 2010b).

9.2 The Tandem Tudor-PHD Finger Cassette

A striking pair of studies shows the multivalent readout (i.e., of multiple PTMs) on H3 tails by the tandem Tudor-PHD cassette, a linked binding module of UHRF1 (Arita et al. 2012; Cheng et al. 2013). The X-ray, NMR, small-angle X-ray scattering, and binding studies by Arita et al. (2012) show that a UHRF1-H3 peptide complex forms with a 1:1 stoichiometry and $K_d = 0.37$ μM. This strong binding affinity is suggestive of combinatorial recognition given that the tandem Tudor domain or PHD finger alone bind with affinities of $K_d = 1.75$ μM and $K_d = 1.47$ μM, respectively. The crystal structure of the complex is shown in Figure 12A, with the 17-residue linker between the tandem Tudor (in cyan and blue) and PHD finger (in green) domains essentially packed between the Tudor domains,

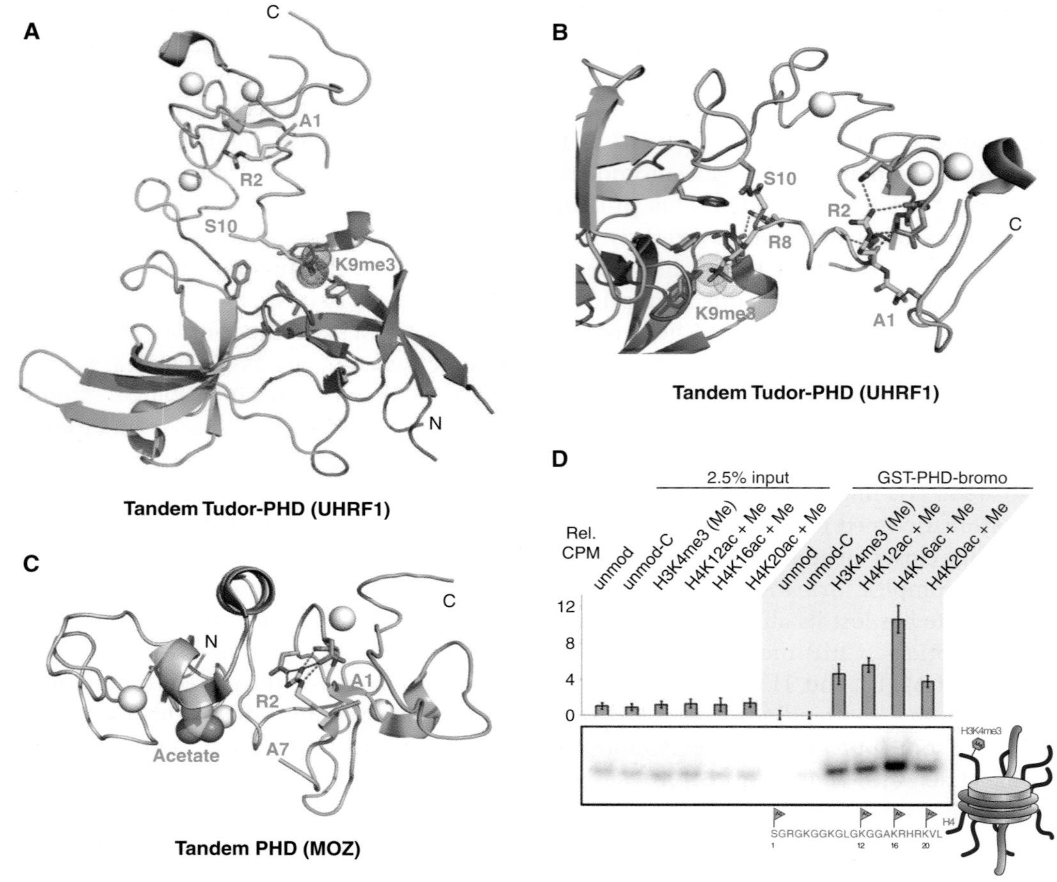

Figure 12. Structures of linked binding modules involved in multivalent readout at the peptide and nucleosomal levels. (*A*) 2.9-Å crystal structure of the UHRF1 tandem Tudor-PHD finger cassette bound to H3(1-13)K9me3 peptide (PDB: 3ASK). The tandem Tudor domains are shown in cyan and purple, whereas the PHD finger is shown in green. Zinc ions are shown as silver balls. The bound H3(1-13)K9me3-containing peptide can be traced from A1 to S10. (*B*) Enlargement of *A*, showing details of the intermolecular contacts between H3(1-13)K9me3-containing peptide (A1 to S10) and the tandem Tudor-PHD finger cassette of UHRF1. (*C*) 1.47-Å crystal structure of the complex of the tandem PHD finger cassette of MOZ bound to H3(1-18)K14ac peptide (PDB: 3V43). There is a bound acetate (in space-filling representation) from buffer bound in a pocket in the amino-terminal PHD finger (in blue). The bound H3(1-18)K14ac-containing peptide can be traced from A1 to A7, in which it is bound to the carboxy-terminal PHD finger (in green). (*D*) glutathione S-transferase (GST) pull-down of modified nucleosomes with semisynthetic histones produced by expressed protein ligation. Nucleosomes containing dual marks involving H4K12ac, H4K16ac, or H4K20ac in combination with H3K4me3 are pulled down with resin bound GST-BPTF PHD-bromo cassette and detected by autoradiography after native gel electrophoresis. (*D*, Reprinted from Ruthenburg et al. 2011.)

with minimal contacts between the Tudor domains and the PHD finger. There is a central hole in the overall architecture of the protein scaffold, which accommodates the compactly folded bound H3K9me3-containing peptide in the complex (expanded view in Fig. 12B). On the H3K9me3-containing peptide side, residues 1 to 4 participate in intermolecular β-sheet formation, with the guanidine group of unmodified Arg2 involved in a network of intermolecular hydrogen bonds such that methylation at this position results in reduced binding affinity (see Sec. 8.2). Peptide residues 5 to 8 unexpectedly adopt an α-helical conformation stabilized by a N-capping hydrogen bond, but otherwise do not form specific intermolecular contacts in the complex. The 9-10 peptide segment forms contacts with the first Tudor domain (in blue), thereby positioning the trimethylammonium group of K9me3 within an aromatic cage. The phosphorylation state of the adjacent Ser10 (and also Thr3) residue of the bound peptide significantly impacts on binding affinity. On the UHRF1 protein side, the importance of the linker element connecting the tandem Tudor and PHD finger was highlighted by the loss of binding following dual Arg to Ala mutations of adjacent Arg

 Cite as *Cold Spring Harb Perspect Biol* doi: 10.1101/cshperspect.a018754

residues within the linker segment. In addition, phosphorylation of a Ser (target of protein kinase A) within the linker segment resulted in a 30-fold loss in binding affinity. It is conceivable that modifications (such as phosphorylation) within the linker segment could act as a switch between potential regulatory pathways used by UHRF1 during its functional role in the maintenance of DNA methylation and transcription repression (detailed below in Sec. 12.1; Arita et al. 2012).

9.3 Tandem PHD Finger Cassettes

The histone acetyltransferase (HAT), MOZ, which contains a pair of tandem PHD fingers, is important for the expression of *HOX* genes and contributes to embryonic and postnatal development. Two groups have now shown that the tandem PHD fingers of MOZ target H3 peptides containing unmodified R2 and K14ac marks (Ali et al. 2012; Qiu et al. 2012). In the crystal structure of the complex bound to an H3(1-18)K14ac peptide (Fig. 12C), the R2 side chain forms a network of hydrogen bonds to the side chains of Asp residues on the second PHD finger (in green), whereas the binding site for K14ac on the first PHD finger (in blue) (by analogy with the related structure of the tandem PHD fingers of DPF3b bound to H3K14ac peptide; Zeng et al. 2010) was occluded by bound acetate from buffer and crystal packing interactions (Qiu et al. 2012). NMR-binding studies by both groups showed that methylation of either R2 or K4 significantly impacted on complex formation. Fluorescence microscopy studies showed that both PHD fingers are necessary for binding to H3K14ac in vivo and localization to chromatin (Ali et al. 2012). Further, ChIP studies showed that the tandem PHD finger facilitated the localization of MOZ to the promoter locus of the *HOXA9* gene, upregulating *HOXA9* messenger RNA (mRNA) levels through promotion of H3 acetylation (Qiu et al. 2012).

Recently, it has been shown that the tandem PHD fingers of CHD4, a catalytic subunit of the NuRD (nucleosome remodeling and deacetylase) complex, are required for NuRD complex transcriptional repression. The tandem PHD fingers concurrently bind both H3 tails within the same nucleosome, displacing HP1γ from pericentric sites (Musselman et al. 2012c). This interaction is modulated by PTMs on the H3 tail, with binding enhanced by H3K9me and H3K9ac marks, and abolished by H3K4me mark.

10 MULTIVALENT READOUT BY LINKED BINDING MODULES AT THE NUCLEOSOMAL LEVEL

Although the emphasis of structure–function studies investigating the readout of Kme marks has been perfomed at the peptide level, the major challenge has been to extend these studies to the nucleosomal level. Recent progress in this area centers on the BPTF PHD-bromo cassette described in Section 9.1 at the H3 peptide level, now outlined below at the nucleosomal level.

10.1 PHD Finger-Bromo Cassette

In a seminal contribution, Ruthenburg et al. (2011) extended studies of the readout of dual histone marks by the BPTF PHD-bromo cassette from the peptide to the nucleosomal level, thereby experimentally validating aspects of multivalent engagement of chromatin marks by linked binding modules (Ruthenburg et al. 2007b). Contrary to peptide level analysis, they showed that the bromodomain of BPTF displayed discrimination among different acetylated lysines on H4 (K12ac, K16ac, K20ac) with marked selectivity for H4K16ac in combination with H3K4me3 at the mononucleosomal level (Fig. 12D). This is corroborated by the observation that a significant pool of nucleosomes in vivo are doubly modified for H3K4me3 and H4K16ac and the demonstration that the BPTF PHD-bromo cassette colocalizes with the H3K4me3 and H4K16ac marks in the genome (Ruthenburg et al. 2011). Intriguingly, the writers of these marks—MLL1 for H3K4me3 and MOF for H4K16ac—interact with each other (Dou et al. 2005). Structure-based modeling suggests that the BPTF PHD-bromo cassette can potentially be snugly docked on a mononucleosome, with bound PHD finger and bromodomains simultaneously recognizing adjacently positioned H3 and H4 tails (Ruthenburg et al. 2007b).

11 ALTERNATE FUNCTIONAL ROLES FOR PHD-BROMO CASSETTES

Recent structure–function studies have identified new non-chromatin-associated roles for the PHD-bromo cassettes beyond their identities as multivalent readers of Kme and Kac marks at the peptide (Sec. 9.1; Li et al. 2006) and nucleosomal level (Sec. 10.1; Ruthenburg et al. 2011). Two examples are described here, outlining alternate mechanisms of functional synergy between the PHD finger and bromodomain in the PHD-bromo cassette.

11.1 PHD Finger as an E3 SUMO Ligase of Adjacent Bromodomain

The PHD-bromo cassette of the KAP1 (KRAB-associated protein 1) corepressor binds neither the expected Kme nor Kac marks under conditions in which both reader domains interact extensively with each other (Ivanov et al. 2007; Zeng et al. 2008). Rather, the PHD finger unexpectedly functions as an intramolecular E3 small ubiquitin-related modifier (SUMO) ligase, which through interaction with the E2 enzyme facilitates SUMOylation of the linked bromodomain. This recruits the NuRD complex via SUMO-interacting motifs to facilitate the establishment and stabilization of silent chromatin. This silencing occurs through the NuRD-catalyzed deacetylation of nucleosomes, which stimulates SETDB1 (an H3K9 KMT) to methylate H3K9. This mark is, in turn, targeted by HP1γ to induce the silenced state (reviewed in Peng and Wysocka 2008). It should be noted that site-specific phosphorylation of a Ser located near the bromodomain of KAP1 appears to inhibit KAP1 SUMOylation (Li et al. 2007b), implying a role for PTM-mediated cross talk in the regulation of KAP1 function.

11.2 PHD-Bromo Cassette as a Regulatory Platform

The *MLL1* gene is essential for embryonic development and hematopoiesis, as well as a frequent target for recurrent chromosomal translocations, resulting in transformation of hematopoietic precursors into leukemia stem cells. MLL1 has been shown to maintain *Hoxa9* expression in stem and progenitor cells, yet is also required for *Hoxa9* silencing during blood cell maturation. Failure to silence results in the expansion of self-renewing progenitor cells and the onset of leukemia (reviewed in Grow and Wysocka 2010). Indeed, induction of the epigenetically regulated switch during haematopoiesis depends on the PHD finger 3 of MLL1 (Xia et al. 2003), given its loss results in the immortalization of hematopoietic stem cells (Chen et al. 2008). The MLL1 PHD3-bromo cassette is targeted by the cyclophilin CyP33, composed of PPIase (peptidyl prolyl isomerase) and RRM (RNA recognition motif) domains. Functional studies have established that the MLL1-CyP33 interaction is required for the histone deacetylase-mediated repression of *HOX* target genes during blood cell maturation in vivo (Fair et al. 2001; Xia et al. 2003). Structural and biochemical studies show that the PPIase domain of CyP33 regulates the conformation of the MLL1 PHD-bromo cassette (Fig. 13A). It does this through *cis-trans* isomerization of a linker proline, thereby disrupting the PHD3-bromo interface and facilitating access of an otherwise occluded MLL1 PHD3 finger in the cassette to the Cyp33 RRM domain (Fig. 13B) (Wang et al. 2010b). This perspective is supported by an NMR-based structure of a segment of the MLL1 PHD finger bound to the Cyp33 RRM domain (Fig. 13C), which implies that the H3K4me3 mark and RRM domain target distinct surfaces of PHD3 and, further, can integrate distinct regulatory inputs by coexisting as a ternary complex. Collectively these results highlight the role of the MLL1 PHD3-bromo cassette both as a regulatory platform and a switch; the *cis-trans* proline isomerization within the linker segment connects H3K4me3 readout (see Fig. 11B) to CyP33 and HDAC-mediated repression (Wang et al. 2010b). Given that cellular RNAs can compete for the RRM motif in CyP33, unanswered questions remain as to the potential role of RNAs in the targeting, stabilization, and/or release of CyP33 from MLL1. Two additional studies on this system have focused solely on the binding of CyP33 to the MLL1 PHD3 finger in efforts to provide insights into regulation by CyP33 of MLL1-mediated activation and repression (Hom et al. 2010; Park et al. 2010).

A

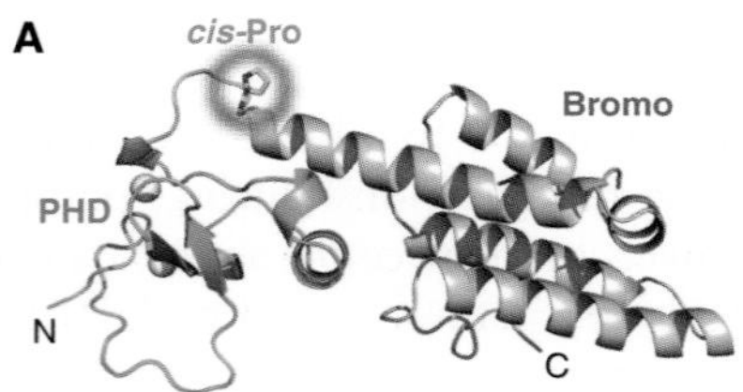

PHD-bromo (MLL1)

B

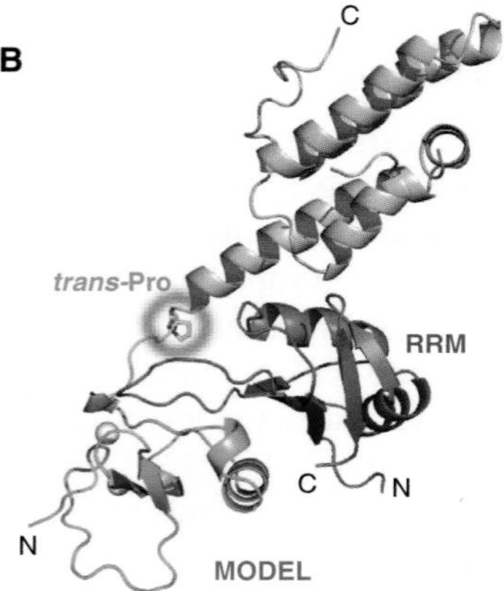

PHD-bromo (MLL1) + RRM (CyP33)

C

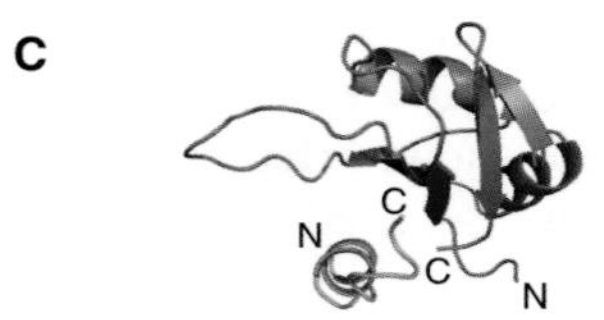

PHD (MLL1) + RRM (CyP33)

Figure 13. Structures of BPTF PHD-bromo cassette with *cis* and *trans* linker prolines. (*A*) 1.72-Å crystal structure of the MLL1 PHD-bromo cassette with a *cis* linker proline (circled in red) in the free state (PDB: 3LQH). The PHD finger and bromodomain are colored in green and blue, respectively. (*B*) Model of the MLL1 PHD-bromo cassette with a *trans* linker proline (circled in red) with this alignment stabilized by a bound RRM domain of CyP33 (magenta). (*C*) NMR solution structure of the complex containing the MLL1 PHD3 fragment (1603-1619, in green) and RRM domain of CyP33 (2-82, in magenta) (PDB: 2KU7).

12 CROSS TALK BETWEEN HISTONE MARKS

A notable feature of histone tails is their extreme density (adjacent or closely spaced) and variety of marks (see Fig. 6 of Ch. 3 [Allis et al.]). In some cases, a single amino acid such as lysine can be labeled by more than one kind of mark, including methylation, acetylation, or ubiquitination (e.g., H3K9). Such features have led to the concept of dynamic "binary switches" in which the readout of one mark is modulated by an adjacent/nearby second mark, thereby impacting on processes ranging from gene transcription, DNA replication, repair, and recombination (reviewed in Fischle et al. 2003a,b; Latham and Dent 2007; Garske et al. 2010; Oliver and Denu 2011). In this regard, the concept of "modification cassettes" was proposed (Fischle et al. 2003a), in which the segment of a histone tail consists of adjacent positions that are subject to different modifications, such as the R2(me)-T3(ph)-K4(me) and R26(me)-K27(me/ac)-S28(ph) segments on H3 (Fig. 1B). Thus, the readout of lysine methylation marks can be impacted by nearby serine/threonine phosphorylation, arginine methyation, lysine acetylation, and lysine ubiquitination marks. These can occur in *cis* within the same histone (intrahistone), in *trans* between histone pairs (interhistone, e.g., between H3 and H4), or even within (intranucleosomal) or across (internucleosomal) nucleosomes (see Fig. 10 of Ch. 3 [Allis et al.]). The binding of a reader module to a mark has the potential of sterically blocking an adjacent modification site or, conversely, may recruit an additional module to modify neighboring residues. Several examples of such cross talk are outlined below, with a complete listing of cross-regulation of histone marks found in comprehensive reviews (Latham and Dent 2007; Oliver and Denu 2011).

Genome-wide location analysis has uncovered the combinatorial patterns of histone marks and chromatin regulators along the genome in human cells (Ram et al. 2011). This study established that specific combination of chromatin marks and regulators colocalize in characteristic patterns with distinct genomic regions and chromatin environments. Interestingly, the chromatin regulators retain their modular and combinatorial associations even during redistribution to different loci as a function of cell type.

12.1 Kme-Sph Cross Talk

An examination of histone tail sequences indicates that lysines and serine/threonines are often proximally positioned in sequence. Examples on H3 include Thr3-Lys4, Lys9-Ser10, Thr22-Lys23, Lys27-Ser28, and Lys79-Thr80 (Fig. 1A). To take an example, Lys9 and Ser10 occupy adjacent positions on the histone H3 tail, with methylation of lysine a more stable mark than phosphorylation of serine. The H3K9me3 mark (written by the Suv39h KMT) serves as a recruitment site for HP1, involved in heterochromatin formation. Phosphorylation of adjacent Ser10 is written by the mitotic kinase Aurora B during mitosis, resulting in the ejection of HP1 from the adjacent H3K9me3 mark (Fischle et al. 2005; Hirota et al. 2005). Subsequent dephosphorylation of Ser10 at the end of mitosis reestablishes the association of HP1 with the H3K9me3 mark. These studies thus point to a "methyl/phospho switch" that dynamically controls the H3K9me3-HP1 interaction, and whose impairment impacts on chromosome alignment and segregation, spindle assembly, and cytokinesis.

The proposed methyl/phospho switch also plays a role in the recognition of the H3K4me3 mark by the PHD finger domains of BPTF and RAG2 (discussed in Sec. 2.1), and the tandem chromodomains of CHD1 (Sec. 4.1). Through the use of peptide microarrays containing a library of combinatorially modified histone peptides, the recognition of H3K4me3 by reader modules was found to be either blocked or attenuated when Thr3 or Thr6 were phosphorylated (Fuchs et al. 2011). The same conclusions were obtained by a parallel investigation, which also showed the structural basis for a 20-fold decrease in ING2 PHD finger binding affinity to H3K4me3, when Thr6 is phosphorylated and no longer able to be accommodated into the shallow pocket that targets the side chain hydroxyl of Thr6 (Garske et al. 2010).

The association of UHRF1 with the H3K9me3 mark through its tandem Tudor domain is insensitive to phosphorylation at the adjacent S10 site (Rothbart et al. 2012), unlike the cell-cycle restricted HP1 binding to H3K9me3. It appears that interactions between the side chains on the H3K9me3 reader modules and S10ph determine whether the reader module bound to H3K9me3 is retained or released on phosphorylation of S10. Thus, replacement of an Asn in the tandem Tudor domain of UHRF1 by Glu/Asp makes readout sensitive to phosphorylation of Ser10. Thus, given that the UHRF1 tandem Tudor domain is the only H3K9me3 reader that is insensitive to the K9me-S10ph switch, its function may be important in tethering DNMT1 to chromatin through mitosis in heterochromatic regions that require considerable maintenance of DNA methylation.

In another noteworthy study, the nonreceptor tyrosine kinase JAK2 was shown to phosphorylate Tyr41 on histone H3 in the nucleus of haematopoietic cells, thereby excluding HP1α from targeting an adjacent presumably lysine methylation site using its chromoshadow domain (Dawson et al. 2009). There are also examples of methyl/phospho switches in nonhistone proteins such as the *S. cerevisiae* Dam1 protein (Zhang et al. 2005).

12.2 Kme-Kac Cross Talk

A number of histone mark combinations have been shown to be synergistic or antagonistic. A few examples are described below.

The ING4 association with chromatin provides an example of synergistic Kme3-Kac cross talk. The PHD finger of ING4 (inhibitor of growth 4) targets the H3K4me3 mark. ING4 is a subunit of the HBO1 HAT complex, and acetylation activity augments the recognition of H3K4me3 on H3 tails in *cis*. This effectively activates ING4 target promoters and attenuates cellular transformation through its tumor-suppressor activity (Hung et al. 2009).

An antagonistic cross talk in *cis* (intrahistone) is observed between H4K20me3 and acetylation of lysines toward the amino terminus of H4; hyperacetylation of H4 antagonizes H4K20me3, and the reverse also holds, in which H4K20 methylation appears to inhibit acetylation of H4 (Sarg et al. 2004).

Another synergistic cross talk example occurs in *trans* (interhistone) in human cells; the H3K4 methylating MLL1 associates with the H4K16 acetylating MOF, thereby potentially linking these two marks during transcription (Dou et al. 2005).

12.3 Kme-Rme Cross Talk

The amino-terminal A1-R2-T3-K4 sequence of histone H3 has methylation sites on both R2 and K4 (Fig. 1A,B). The arginine methyltransferase PRMT6 deposits the asymmetrical R2me2a mark on H3 and this mark is distributed within the body and the 3′-ends of human genes (Guccione et al. 2007; Hyllus et al. 2007; Kirmizis et al. 2007). Antagonistic cross talk between the R2me2a and H3K4me3 mark seems to occur; that is, the R2me2a mark is prevented by the H3K4me3 mark and, conversely, the H3R3me2a mark prevents trimethylation of H3K4 by the MLL1 KMT and its associated factors (ASH2 and WDR5). It is conceivable that this mutual antagonism reflects the inability of WDR5 to recognize the H3R2me2a mark and, hence, is unable to recruit MLL1 necessary for H3K4 trimethylation (Guccione et al. 2007). Certainly, in *S. cerevisiae*, the H3R2me2a mark appears to occlude binding by the Set1 KMT via the Spp1 subunit PHD finger, necessary for H3K4 trimethylation (Kirmizis et al. 2007).

In another example, binding studies showed that recognition of unmodified H3K4 (me0) by the AIRE (autoimmune regulator) PHD finger is abrogated by H3R2me2 in vitro; this results in the reduction of AIRE target gene activation (Chignola et al. 2009).

Section 2.1 introduced the RAG2 PHD finger as being able to bind to the dual H3R2me2sK4me3 mark in vitro. In a follow-up study, high levels of the H3R2me2sK4me3 dual mark were found at antigen receptor gene segments poised for rearrangement (Yuan et al. 2012). Strikingly, this dual mark colocalizes at active promoters throughout the genome, with the implication that readout of the K4me3 mark on H3 can be modulated by the R2me2s mark.

12.4 Kme-Kub Cross Talk

Early studies established that ubiquitination (Ub) of histone H2B regulates H3 methylation and gene silencing (Ng et al. 2002; Sun and Allis 2002; Lee et al. 2007; reviewed in Shilatifard 2006). This has been elegantly investigated following the generation of designer mono- and dinucleosomes incorporating a site-specific ubiquitination mark on H2B at Lys120 (McGinty et al. 2008). Biochemical studies established that histone H2BK120-monoubiquitination (H2BK120ub1) stimulated intranucleosomal methylation of H3K79 (illustrated in Fig. 12 of Ch. 3 [Allis et al.]). This observed cross talk is mediated by the catalytic domain of hDot1, potentially through an allosteric mechanism (McGinty et al. 2008).

These Kme-Kub cross talk studies have been recently extended to the *Drosophila* MSL complex system, which regulates dosage compensation (the topic of Ch. 24 [Lucchesi and Kuroda 2014]); H2B ubiquitination by MSL1/2 directly regulates H3K4 and H3K79 methylation through intranucleosomal *trans*-tail cross talk both in vitro and in vivo (Wu et al. 2011b). Given that MSL1/2 activity contributes to transcriptional activation at *HOXA1* and *MEIS1* loci, the MSL complex shows two distinct chromatin-modifying activities: MSL1/2-mediated H2BK34 ubiquitination and MOF-mediated H4K16 acetylation (see Fig. 3 of Ch. 24 [Lucchesi and Kuroda 2014]). These studies provide insights into the underlying intricate network of interactions whereby chromatin-modifying enzymes function coordinately to program gene activation.

13 HISTONE MIMICS

In the above sections, we have largely outlined structure–function studies on the readout of methylation marks on histone tails by reader modules. This has raised the question of whether nonhistone chromatin proteins also use similar modification-recognition systems. In this regard, recent advances have highlighted the identification of "histone mimics" as a distinct set of nonhistone proteins that can have PTMs (Kme, Kac, Yph) written and read in a similar fashion to histone proteins. In this chapter, the focus will be on studies of histone mimics that are methylated, thereby expanding our insights of methylated lysine recognition principles beyond the boundaries of histones.

 Cite as *Cold Spring Harb Perspect Biol* doi: 10.1101/cshperspect.a018754

We outline, below, four examples of histone mimicry involving methylated lysine marks on nonhistone proteins functioning in chromatin-templated processes, including KMTs, tumor suppressors, RNA polymerases, and influenza viral proteins.

13.1 G9a Methyltransferase

Our first insights into histone mimics emerged from structure–function studies on a member of the Suvar(3-9) family of SET domain lysine methyltransferases called G9a. This KMT writes methylation marks on H3K9. In a striking result, G9a was shown to be a self-methylating KMT, trimethylating itself at K165 and di- and trimethylating K94 (Sampath et al. 2007). These Kme sites in G9a have marked sequence similarity to H3K9 sites and are in vivo binding targets for the chromodomain protein HP1. Recognition of the G9aK165me3 mark by HP1 can be reversed by concomitant phosphorylation of adjacent Thr166. Thus, G9a contains a lysine methylation cassette that mimics features of sequence context, reader module recognition, and regulation by phosphorylation (Sampath et al. 2007), which are characteristic features of histone lysine methylation.

13.2 p53 Tumor Suppressor

The transcriptional activity of the tumor suppressor p53 is modulated by multiple PTMs. Three lysine residues within its regulatory carboxy-terminal domain (CTD) undergo methylation by KMTs at Lys370, Lys372, and Lys382. The DNA repair factor 53BP1 (p53-binding protein 1) uses its tandem Tudor domain to recognize p53 during DNA damage via the K382me2 mark. Structural studies on the complex established that the p53K382me2 mark inserts into the aromatic-lined pocket of the tandem Tudor domain of 53BP1 in a similar manner to H4K20me2 (Roy et al. 2010). The adjacent His380 and Lys381, in a HKKme2 sequence context, contribute to the sequence specificity of molecular recognition ($K_d = 0.9\ \mu\text{M}$). ChIP and DNA repair assays suggest that recognition of the p53K382me2 mark by the tandem Tudor domains of p53BP1 may facilitate accumulation of p53 at DNA damage sites and promote repair.

p53K382me1 is generated by the SET8 KMT and recognized by the MBT repeats of L3MBTL1 (whose structure is illustrated in complex with Kme2-containing histone peptides in Fig. 6) (West et al. 2010). Functionally, the activation of p53 by DNA damage is coupled with a decrease in p53K382me1 levels and, hence, an abrogation of the p53-L3MBTL1 interaction, which causes the dissociation of L3MBTL1 from p53 target promoters. This study provides a mechanistic explanation for the basis by which the MBT repeats of L3MBTL1 link SET8-mediated p53 methylation at Lys382 to regulate p53 activity (West et al. 2010).

13.3 Carboxyl Terminus of RNA Polymerase II

The RNA polymerase II CTD can be modified by site-specific methylation of Arg1810. This modification, along with Ser2 and Ser5 phosphorylation, is essential for transcriptional initiation and elongation (Sims, III et al. 2011). Given that RNA polymerase II participates in chromatin-templated processes, the R1810me mark within the CTD mimics Arg methylation observed in histone tails.

Interestingly, the Tudor domain–containing protein TDRD3, but not SMN and SPF30, binds to the R1810 me2a-containing CTD peptide, but not to its mono- or symmetrically dimethylated counterparts. This is in addition to its binding to histone tail sites discussed in Section 6.1. Mutation of the aromatic cage residues of the TDRD3 reader results in a loss of binding to the R1810me2a-containing CTD peptide.

13.4 Carboxyl Terminus of Influenza Virus NS1 Protein

In a landmark study, Marazzi et al. (2012) describe the suppression of the antiviral response by an influenza histone mimic residing within the carboxyl terminus of the viral NS1 (nonstructural protein 1) protein. The NS1 protein of the H3N2 influenza virus subtype contains an ARSK (Ala-Arg-Ser-Lys) sequence that is similar to the amino-terminal ARKS sequence of histone H3 (Fig. 1A). The SET1 KMT, which methylates H3K4, an activation mark, also methylates the lysine in the ARSK sequence of NS1, supporting the notion of histone mimicry by NS1. Normally, hPAF1C (PAF1 transcription elongation complex) function is potentiated for the transcriptional elongation of inducible antiviral gene sets when methylated H3K4 tails bind to it. The binding of the methylated NS1 mimic to hPAF1C sequesters it and the methylating Set1 enzyme from acting at the hosts normal genomic targets, thus interfering with the gene transcription program of the host cell. Therefore, histone mimicry provides a selective advantage to viruses by inducing specific suppression of host antiviral gene transcription (also discussed in Sec. 7 and Fig. 12 of Ch. 29 [Busslinger and Tarakhovsky 2014]).

14 READOUT OF FULLY METHYLATED 5mCpG SITES ON DNA

DNA methylation may exert its repressive effect on transcription as a consequence of 5mC marks serving as docking sites for 5mC-binding proteins, which, in turn, recruit corepressors that can modify chromatin. In this section, we discuss structural studies involving 5mCpG-binding (MBDs) proteins and their zinc-finger counterparts, including Kaiso, bound to 5mC sites on duplex DNA.

14.1 Methylcytosine-Binding Proteins

Two MBD protein structural complexes are described below: first, the MBD1 protein and then MeCP2, bound to symmetrical 5mCpG/5mCpG sites. These structures illustrate the different principles by which recognition of 5mC marks occurs on duplex DNA.

MBD1 is a transcriptional regulator containing an amino-terminal 5mCpG-binding domain and a carboxy-terminal transcription repression domain (illustrated in Fig. 8 of Ch. 15 [Li and Zhang 2014]). The MBD domain targets methylated CpG islands of tumor suppressors and imprinted genes, thereby inhibiting their promoter activities in cooperation with the transcription repressor domain. An NMR-based solution structure has been solved for the complex of MBD1 with a DNA duplex containing a fully methylated 5mCpG/5mCpG step (1 protein bound per DNA duplex). In this structure, loops project from the protein core (an α/β sandwich composed of four β-strands and an α-helix) and interact with the major groove of the DNA duplex (Fig. 14A) (Ohki et al. 2001). Loop L1 adopts a defined hairpin-like structure on complex formation and interacts with one of the DNA strands in the major groove. Loop L2 and a segment of α-helix toward the amino terminus interact with the other DNA strand in the major groove, with the loop forming base-specific contacts, whereas the helical segment forms contacts with the sugar-phosphate backbone (schematically shown in Fig. 14B). Five protein residues form a continuous hydrophobic patch that interacts with the methyl groups of the 5mC residues (indicated by magenta arrows in Fig. 13B), with distinct interactions observed for the 5mC residues on partner strands. In addition, a pair of arginines and a tyrosine interact with guanines of the 5mCpG sites (indicated by blue arrows), with binding affinity lost following mutation of the Arg residues. The small observed protein–DNA interface that is essentially restricted to the 5mCpG/5mCpG binding site suggests that MBD1 should be able to access such sites in the major groove of nucleosomes most likely without encountering steric interference.

A second 5mC-binding protein, MeCP2, is of interest because of its role in the maintenance of neuronal function; mutations in the *MECP2* gene are responsible for the majority of cases of Rett syndrome, a late-onset autism spectrum disorder (discussed more extensively in Ch. 15 [Li and Zhang 2014] and Ch. 33 [Zoghbi and Beaudet 2014]; see also Amir et al. 1999). The crystal structure of MeCP2 bound to a 5mCpG/5mCpG-containing DNA duplex (Fig. 14C) established that the methyl groups of 5mC make contacts predominantly with a hydrophilic surface along the major groove, including C—H••O hydrogen bonds to tightly bound water molecules (Fig. 14D) (Ho et al. 2008). In addition, Arg residues are involved in 5mC–Arg-G interactions, whereby the Arg is positioned in the plane of the guanine base and locked in place by a salt bridge to the carboxylate of an Asp, with the guanidinium group of the Arg positioned directly above the methyl group of the 5mC. The most frequently mutated residue in Rett syndrome involves a Thr residue, and it was shown that this Thr plays an important role in maintaining a structural motif within the MeCP2 fold (Ho et al. 2008).

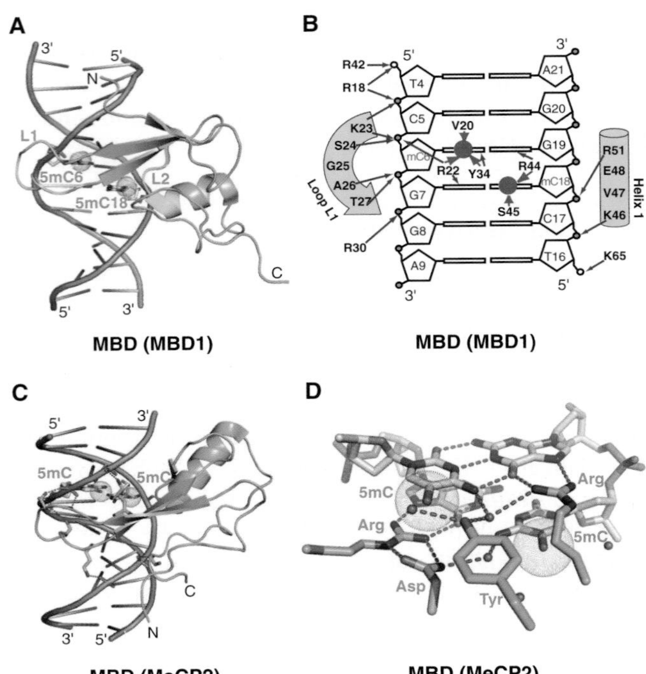

Figure 14. Structures of methylcytosine-binding proteins bound to fully methylated 5mCpG DNA. (*A*) NMR solution structure of MBD1 protein bound to fully methylated 5mCpG-containing DNA duplex (PDB: 1IG4). Two loops L1 and L2 are colored in yellow. The methyl groups of 5mC's are marked by magenta dotted circles. (*B*) Schematic of intermolecular contacts centered about the 5mCpG/5mCpG site involving loops L1 and L2, and the amino-terminal α-helix adjacent to L2. Methyl group of 5mC is represented by the magenta circle. Hydrophobic interactions between the 5mC residues and side chains of MBD1 are indicated by magenta colored arrows. (*C*) 2.5-Å crystal structure of MeCP2 protein bound to fully methylated 5mCpG-containing DNA duplex (PDB: 3C2I). The methyl groups of 5mCs are marked by magenta dotted circles. (*D*) Intermolecular contacts between hydrophilic amino acids of MeCP2 (green stick representation) and the 5mC groups (magenta dotted circles) in the major groove of the duplex, including C—H••O hydrogen bonds to tightly bound water molecules.

Cite as *Cold Spring Harb Perspect Biol* doi: 10.1101/cshperspect.a018754

14.2 Methylcytosine-Binding Zn-Finger Proteins

A subset of human zinc-finger-containing proteins have the capacity to specifically recognize 5mC-containing DNA (reviewed in Sasai et al. 2010). We outline the results obtained from structural studies of the Kaiso and Zpf57 zinc-finger proteins bound to symmetrical 5mCpG/5mCpG sites on duplex DNA.

Kaiso is a methylated DNA-binding factor involved in noncanonical Wnt signaling. It contains an amino-terminal BTB/POZ (BR-C, ttk and bab/Pox virus and Zn finger) domain involved in protein–protein interactions and three mCG-DNA-binding carboxy-terminal zinc fingers (Cys_2His_2 coordination) that repress transcription by recruiting chromatin-remodeling compression machinery to target genes (reviewed in Clouaire and Stancheva 2008). Recently, structural studies of the zinc-finger domains of Kaiso bound to a pair of sequential symmetrically methylated 5mCpG/5mCpG DNA sites from the E-cadherin promoter region have elucidated the details underlying the recognition process, in which 1 Kaiso molecule is bound per DNA duplex (Buck-Koehntop et al. 2012). The side chains of the first two zinc fingers target the major groove through base-specific recognition mediated by classical and C—H• •O hydrogen bonds, as well as phosphate backbone contacts. The carboxy-terminal extension following the third zinc finger (disordered in the free structure) targets the minor groove and contributes to high-affinity binding (Fig. 15A). The amino-terminal regions of the first two zinc fingers provide hydrophobic environments that accommodate the methyl groups of 5mC in the complex (Fig. 15B). Unexpectedly, the three zinc fingers of Kaiso span a total of 4–5 base pairs, contacting both the major and minor grooves, in contrast to most other triple zinc-finger proteins that solely target the major groove and span a total of 9–10 base pairs (Wolfe et al. 2000). Functionally, Kaiso DNA-binding sites are limited to target genes regulated by the Wnt signaling pathway that play key roles in early development and tumor progression. Kaiso targets and silences aberrantly methylated DNA repair and tumor-suppressor genes, acting oncogenically to contribute to tumor progression (Lopes et al. 2008).

The structure of tandem zinc fingers (C_2H_2 coordination) of the Zfp57 transcription factor and a DNA duplex containing a fully methylated 5mCpG/5mCpG-step within the T-G-C-5mC-G-C sequence element (1 Zfp57 bound per DNA duplex) has been solved at high resolution (Liu et al. 2012). Zfp57 is expressed during very early embryogenesis, and is responsible for the maintenance of paternal and maternal imprinted loci (the topic of Ch. 26 [Barlow and Bartolomei 2014]). Both zinc fingers target the DNA major groove on complex formation and together span 6 base pairs without perturbing the B-DNA conformation (Fig. 15C). The first zinc finger (depicted in green in the figure) contacts the 5′-T-G-C segment, whereas the second zinc finger (in blue) contacts the 5mC-G-C segment. The 5mC base is recognized asymmetrically: One case involves hydrophobic interactions with an Arg side chain (whose mutation results in loss in binding affinity) and a neighboring 3′-guanine (so called 5mC–Arg-G interaction) (Fig. 15D), whereas the other case is defined by a layer of ordered water molecules (Fig. 15E). The binding affinity of Zpf57 for its DNA target sites is reduced on formation of ten eleven translocation (Tet)-catalyzed oxidative products of 5mC such as 5-hydroxymethylcytosines (5hmC) (explained below in Sec. 16.4). Interestingly, in patients with transient neonatal diabetes, DNA-binding activity is abolished for two Zfp57 point mutations.

Future studies will have to elucidate the complex network of downstream protein–protein interactions through which 5mC-binding proteins regulate gene expression impacting on development and oncogenesis.

15 READOUT OF HEMIMETHYLATED 5mCpG SITES ON DNA

Proteins containing SRA 5mC-binding domains are required for the establishment and/or maintenance of DNA methylation in both mammals and plants. As we show below, the SRA domains of mammals and plants use both common and distinct recognition principles for targeting their DNA duplex sites. The SRA-containing mammalian UHRF1 protein plays a critical role in epigenetic inheritance of 5mC marks in a 5mCpG context. The SUVH (SU(VAR)3-9 homolog 5) family of proteins in plants, which contain the 5mC-binding SRA domains, have no obvious mammalian counterparts, and are more versatile, given that they target 5mCpG, 5mCpHpG, and 5mCpHpH sites.

15.1 Mammalian SRA Domains

Functional experiments initially showed that the protein UHRF1 plays a critical role in maintaining DNA methylation in mammalian cells by mediating epigenetic inheritance patterns through recruitment of DNMT1 to hemimethylated DNA at replication forks (Bostick et al. 2007; Sharif et al. 2007; reviewed in Cedar and Bergman 2009; Hashimoto et al. 2009). UHRF1 is composed of tandem Tudor, PHD-finger, and SRA domains, with the SRA domain involved in site-specific recognition of hemimethylated 5mCpG/CpG sites on duplex DNA (Fig. 6 of Ch. 6 [Cheng 2014]). The challenge has been to understand the molecular basis underlying the ability of mammalian SRA domains to target hemimethylated 5mCpG/CpG-

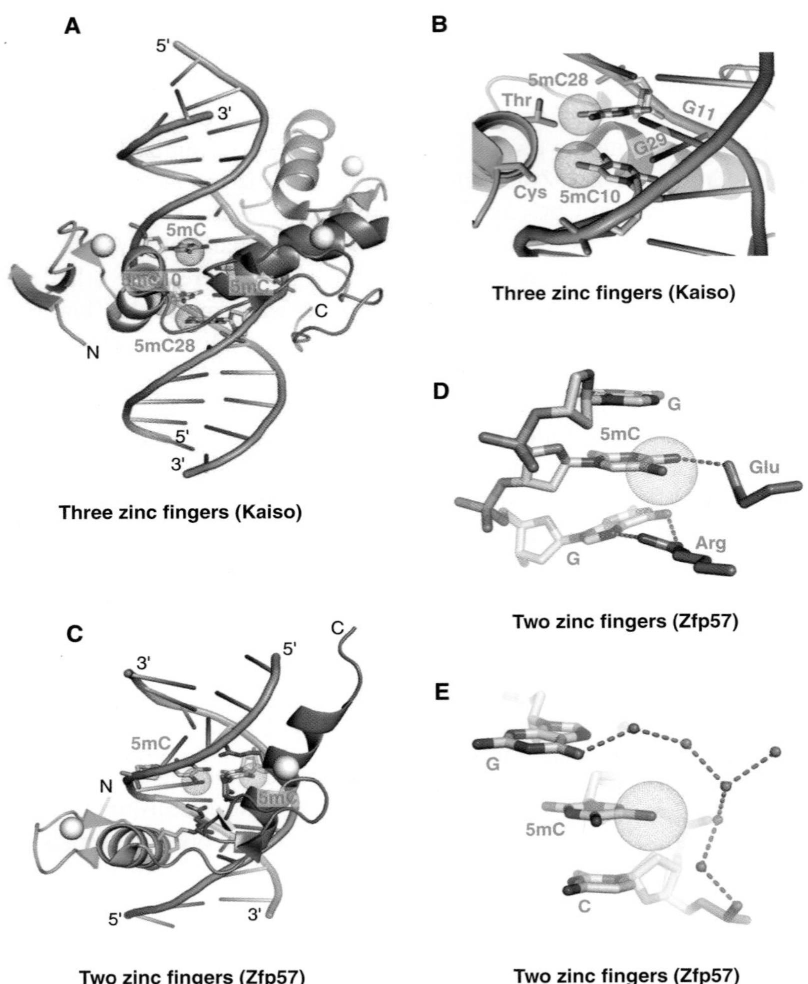

Figure 15. Structures of methylcytosine-binding zinc-finger proteins bound to fully methylated 5mCpG DNA. (*A*) 2.8-Å crystal structure of three zinc fingers of Kaiso protein bound to a pair of fully methylated 5mCpG-containing DNA duplex (PDB: 4F6N). The first, second, and third zinc fingers are colored in green, blue, and pink, respectively. Note that although the majority of the intermolecular contacts are with the major groove, involving zing fingers 1 (green) and 2 (blue), there are also contacts with the minor groove, involving zinc finger 3 and the carboxy-terminal extension of Kaiso. Zinc ions are shown as silver balls. The methyl groups of 5mC are marked by magenta dotted circles. (*B*) Intermolecular contacts between amino acids of the first zinc finger (in green) of Kaiso and 5mC groups in the major groove of the duplex. (*C*) 0.99-Å crystal structure of two zinc fingers of Zfp57 protein bound to a fully methylated 5mCpG-containing DNA duplex (PDB: 4GZN). Zinc ions are shown as silver balls. The methyl groups of 5mC are marked by magenta dotted circles. (*D*) One of the 5mC groups in the Zfp57-DNA complex is involved in hydrophobic interactions through positioning between an Arg side chain and a neighboring guanine. (*E*) The second 5mC in the Zfp57-DNA complex interacts with a layer of ordered water molecules (red circles).

containing DNA and discriminate against both fully methylated 5mCpG/5mCpG- and unmethylated CpG/CpG-containing DNA.

Three groups simultaneously solved the structure of mammalian (human or mouse) SRA domains of UHRF1 in the free state and when bound to 12-mer DNAs containing a central hemimethylated 5mCpG/CpG site (Arita et al. 2008; Avvakumov et al. 2008; Hashimoto et al. 2008). The SRA domain uses two loops (designated NKR [Asn-Lys-Arg]-containing loop and thumb) projecting from its highly conserved concave face to penetrate both major and minor grooves centered about the hemimethylated 5mCpG/CpG site (Fig. 16A), resulting in a 1:1 UHRF1 SRA:DNA duplex complex. Complex formation results in flipping out of the 5mC, thereby positioning it in a conserved binding pocket within the SRA domain, in which it is anchored through hydrogen-bonding along its Watson–Crick edge, as well as van der Waals and planar stacking interactions

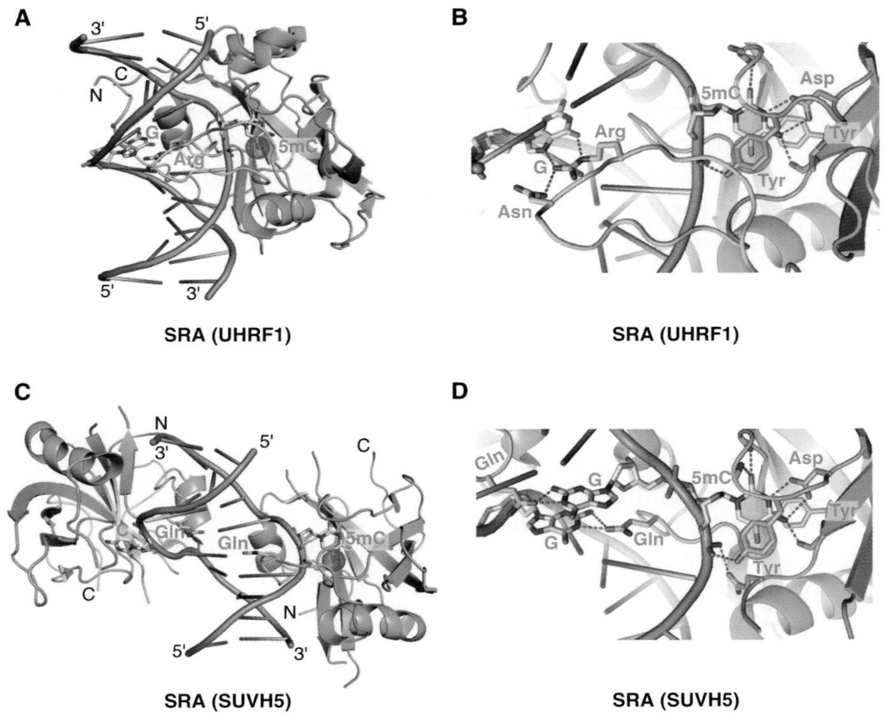

Figure 16. Structures of SRA domain-containing proteins bound to hemimethylated 5mCpG DNA. (*A*) 1.6-Å crystal structure of SRA domain of UHRF1 bound to a hemimethylated 5mCpG-containing DNA duplex (PDB: 2ZKD). The stoichiometry of the complex is one SRA domain per DNA duplex. The methyl groups of 5mC are marked by magenta dotted circles. (*B*) Alignment of the flipped-out 5mC (ring is shaded for clarity) in a pocket within the SRA domain of UHRF1. (*C*) 2.37-Å crystal structure of SRA domain of plant SUVH5 bound to a hemimethylated 5mCpG-containing DNA duplex (PDB: 3Q0D). The stoichiometry of the complex is two SRA domains per DNA duplex. 5mC's methyl groups are marked by magenta dotted circles. (*D*) Alignment of the flipped-out 5mC (ring is shaded for clarity) in a pocket within the SRA domain of plant SUVH5.

(Fig. 16B). The nature of the intermolecular contacts involving the flipped-out 5meC will discriminate against replacement by thymine, with specificity for 5mC over C conferred by the methyl group fitting precisely within an available hemispheric-like space. The orphan guanine remains stacked within the helix, with the DNA being both straight and adopting a B-DNA conformation, despite the flipping out of the 5mC on complex formation. An Arg side chain from the NKR-containing loop inserts from the major groove side into the cavity vacated by the flipped-out 5mC, in which it hydrogen bonds with both the orphan guanine and its adjacent cytosine of the CpG step on the unmodified partner strand. An Asn side chain interacts with and buttresses this inserted Arg, with both its main and side chain atoms of Asn positioned to clash with the C of the CG step, were it methylated. These results nicely explain why mammalian SRA domains bind exclusively to 5mCpG/CpG steps by recognizing both the flipped-out 5mC on the parental and opposing CpG step on the partner strand (Arita et al. 2008; Avvakumov et al. 2008; Hashimoto et al. 2008). Direct contacts involving protein and DNA are limited to the 5mCpG/CpG segment, indicative of binding and recognition restricted to this site and independent of flanking sequence.

15.2 Plant SRA Domains

The 5mC-binding capacity of several plant SRA domains has been characterized, and each has a preference for DNA methylation in different sequence contexts (Johnson et al. 2007). Binding studies established that the SRA domain of SUVH5 differs from other SRA domains in that it binds DNA to similar extents in all sequence contexts.

Unexpectedly, the structures of complexes containing the SUVH5 SRA domain bound to 5mC-containing DNA in either hemimethylated or fully methylated 5mCpG contexts, or a methylated 5mCpHpH context, show an SUVH5 SRA:DNA duplex stoichiometry of 2:1. These complexes revealed a dual flip-out mechanism in which both the 5mC from the parental strand and a base (5mC in

5mCpG/5mCpG context, C in 5mCpG/CpG, or G in 5mCpHpH/HpHpG context) from the partner strand are simultaneously extruded from the DNA duplex and positioned within individual binding pockets of SRA domains (Rajakumara et al. 2011a). The structure of the SUVH5 SRA domain bound to hemimethylated 5mCpG/CpG DNA duplex is shown in Fig. 16C. The 5mC and C bases on partner strands are simultaneously flipped out and positioned in conserved pockets on the symmetry-related SRA domains with the resulting gaps filled by Gln side chains projecting from the thumb loops into the minor groove (Fig. 16D). Interestingly, a second example of a dual flip out involving 5mC was recently reported for the recognition of this mark by the *Escherichia coli* methyl-specific McrBC (modified cytosine restriction BC) endonuclease (Sukackaite et al. 2012). Thus, mammalian and plant SRAs use different amino acids originating from different loops while inserting into the duplex from different grooves of the DNA (compare Figs. 16B and 16D). Complementary functional studies established that a functional SUVH5 SRA domain is required for both DNA methylation and accumulation of the H3K9me2 modification in vivo, suggesting a role for the SRA domain in recruitment of SUVH5 to genomic loci (Rajakumara et al. 2011a).

16 PERSPECTIVES AND FUTURE CHALLENGES

In this section, we outline new initiatives and advances, as well as future challenges, with the promise of enhancing our current mechanistic understanding of the readout of histone and DNA methylation marks. On the technology front, PTMs and their distribution have been identified and characterized at the genome-wide level using high-throughput and mass spectroscopic approaches. Chemical biological approaches have been developed to generate designer nucleosomes containing site-specifically incorporated histone lysine mark mimics, nonnatural and modified amino acids, and PTMs. The availability of designer nucleosomes should permit attempts at structural characterization of multivalent PTM readout at the nucleosomal level.

The contribution of DNA methylation in epigenetic regulation has been greatly expanded by the identification of oxidative adducts of 5mC (reviewed in Ch. 2 [Kriaucionis and Tahiliani 2014]), with current efforts aimed at understanding the role of these new PTMs as novel epigenetic states and/or intermediates in DNA demethylation. The potential role of long noncoding RNAs (lncRNAs) in mediating epigenetic regulation events has received increasing attention, but the available information from functional studies has yet to be matched at the structural level, heightening the need to understand the molecular basis for protein–RNA recognition facilitated by noncoding RNAs. Given the extensive information on both histone and DNA PTMs, attention has turned to the cross talk between these marks and the potential of structural and functional approaches, which can elucidate the interplay between these marks, preferably at the nucleosomal level.

Finally, dysregulation of chromatin is associated with autoimmune, neurological, and age-related disorders, as well as cancer. We consider the consequences caused by dysregulation of methylated lysine reader modules on epigenetic pathways and outline progress toward the identification of small molecules targeted to aromatic pockets involved in methyllysine readout. Each of these topics is addressed below in some detail so as to outline ongoing challenges and the future promise for further progress in the field.

16.1 Technological Developments at Genome-Wide Level

There have been several advances in microarray and next-generation sequencing technology that have enhanced approaches that precisely and comprehensively monitor the emerging patterns of PTM landscapes on a genome-wide basis. Such approaches have included PTM-modified peptide microarray platforms for proteome-wide identification of readers of histone marks (specific, or combinations thereof, discussed in Sec. 12.1) and the use of SILAC (stable isotope labeling by amino acids in cell culture)-based technology coupled with mass spectroscopy to identify the nucleosome-interacting proteins that read one or more PTMs incorporated into designer nucleosomes.

Modified histone peptide microarray platforms have been developed for high-throughput identification of chromatin reader modules, as well as assessing the epitope specificity of antibodies (Bua et al. 2009). The peptide arrays were built, for example, of 20-amino-acid-long H3 peptides that were unmodified or had one or a variety of PTMs. This approach led to the discovery that the BAH domain of mammalian ORC1 is a reader of the H4K20me2 mark, and was followed by the structural and functional characterization of this interaction and its implications for a dwarfism syndrome (Kuo et al. 2012). In another study, a large PTM-modified randomized combinatorial peptide library consisting only of amino-terminal H3 peptides was developed to investigate how the specificity of chromatin binding modules that read the H3K4me3 mark were affected by other PTMs (Garske et al. 2010). This approach showed that recognition of K4me3 was modulated by methylation of Arg2, as well as by phosphorylation of Thr3 and Thr6 on the H3 tail segment, discussed in Sections 12.1 and 12.3.

An alternate approach for monitoring the Kme3 mark involved a histone peptide pulldown approach coupled

Cite as *Cold Spring Harb Perspect Biol* doi: 10.1101/cshperspect.a018754

with quantitative SILAC proteomics technology to profile this mark on a genome-wide scale and identify readers of the Kme3 mark (Vermeulen et al. 2010). In an extension of this approach, SILAC nucleosome affinity purification technology based on designer nucleosomes labeled with specific modification states was used to identify nucleosome-interacting proteins regulated by DNA and histone methylation (Bartke et al. 2010). This study identified components of the ORC to be a methylation-sensitive nucleosome interactor based on cooperative recruitment by histone and DNA methylation. In another example, the recruitment of the jumonji-containing lysine demethylase KDM2A by methylated lysines on histone tails at the nucleosomal level was disrupted by DNA methylation.

More recently, an interspecies comparative epigenomics approach (using human, mouse, and pig pluripotent stem cells) has been undertaken to look at DNA and histone modifications as an approach for annotating the regulatory genome (Xiao et al. 2012). This approach identified conserved colocalization of different epigenetic marks with distinct regulatory functions during differentiation of embryoninc stem cells into mesendoderm cells.

These high throughput advances toward charting histone PTM modifications and the functional organization of mammalian genomes (reviewed in Zhou et al. 2011) have led to the development of a website titled the "structural genomics of histone tail recognition" (Wang et al. 2010a). Further advances are likely to involve both miniaturization and sensitivity enhancement of the assays so as to target individual cells in the future. In the longer term, the coupling of structural biology with these technological advances in the detection and distribution of PTMs should provide an improved molecular perspective of the multivalent readout of PTMs at the nucleosomal level.

16.2 Chemical Biology Approaches to Designer Nucleosomes

An emerging multidisciplinary challenge has been to harness the tools of chemical biology to address specific structure/function problems in chromatin biology, achievable through the preparation and manipulation of chemically defined chromatin (reviewed in Allis and Muir 2011; Voigt and Reinberg 2011; Fierz and Muir 2012). This type of research could result in an improved understanding of the underlying basic mechanisms that dictate the folding and inheritance of chromatin, the multivalent readout of chromatin marks at the nucleosomal level, as well as the principles underlying the spreading of PTMs within defined chromatin regions.

One chemical biology approach has been to use recombinant histones with lysines site-specifically replaced by cysteines. These sites can then be conjugated to introduce site-specific modifications of the cysteine that acts as a lysine methylation mimic within histone tails, even at the nucleosome level. Thus, mimics of methylated lysines have been introduced by *S*-alkylation through generation of *N*-methylated aminoethylcysteine residues (Simon et al. 2007) and this cysteine-based chemistry has also been applied to introduce ubiquitin modifications (Chatterjee et al. 2010) and acetyllysine analogs (Huang et al. 2010). Such approaches are limited by the standard constraints associated with mimics not being able to fully replicate their natural counterparts. Nevertheless, the simplicity of the cysteine conjugation approach has made it accessible to the nonchemist and could benefit from further efforts to mimic additional PTMs, as well as their incorporation in a multiple mark context.

In an alternate approach, a genetic code expansion methodology has been developed to introduce into recombinant proteins methylation, acetylation, and ubiquitination marks on Lys and phosphorylation marks on Ser/Tyr (reviewed in Davis and Chin 2012). The methodology incorporates nonnatural and modified amino acids site-specifically into recombinant proteins through the application of orthogonal aminoacyl-transfer RNA (tRNA) synthetase-tRNA pairs that direct incorporation in response to introduced amber stop codon placed in the gene of interest (reviewed in Liu and Schultz 2010; Davis and Chin 2012). Improvements in the methodology have resulted in the incorporation of a wider range of multiple unnatural amino acids (Neumann et al. 2010) and should be extendable to their multiple PTM counterparts.

The most promising approach for site-specifically incorporating PTMs at the nucleosomal level has been to use chemical ligation strategies, generating so-called "designer nucleosomes" containing the marks of interest. It is based on native chemical ligation chemistry in which a carboxy-terminal α-thioester-containing peptide reacts with an amino-terminal cysteine-containing peptide to generate a normal peptide linkage (reviewed in Dawson and Kent 2000). The corresponding expressed protein ligation (EPL) version can use a recombinant histone amino-terminal peptide with an α-thioester building block, which can be chemically modified to mimic a particular PTM (e.g., Kme1), ligated to the remainder of the unmodified core histone. This approach, when used with yet to be determined DNA templates that allow more uniform positioning of core histones on the DNA, has the potential for being able to incorporate standard PTMs positioned at multiple sites on histone tails into mononucleosomes in good yields, thereby generating designer chromatin for structural and functional studies (reviewed in Allis and Muir 2011; Frederiks et al. 2011; Voigt and Reinberg 2011; Fierz and Muir

2012). The potential exists for extending this EPL technology to PTM probes site-specifically incorporated into dinucleosomes and nucleosomal arrays by ligating the DNA of preformed nucleosomes. This approach has the potential to provide more useful and biologically relevant biochemical/biophysical probes. Another challenge that chemical approaches are trying to tackle is the generation of asymmetric nucleosomes, in which, for instance, the nucleosome contains a copy of a canonical H3.1 and variant H3.3. Such approaches could set the stage for a better understanding of the structural consequences of nucleosome asymmetry. The availability of designer nucleosomes could also assist toward elucidating a structure-based understanding of whether PTMs have an impact on nucleosome positioning, occupancy, and dynamics and, if so, the consequence of DNA damage on these processes.

Looking to the future, one anticipates the emergence of high-throughput biochemical approaches that couple libraries of PTM designer chromatin with "lab-on-a-chip" microfabrication methodology facilitated by microfluidic devices to screen reader modules at the nano scale (see review by Fierz and Muir 2012). Such studies could identify new systems for conducting structural studies of PTMs at the nucleosomal level.

16.3 Histone Mark Readout at the Nucleosomal Level

The field is moving toward understanding histone PTM readout at the nucleosomal and chromatin fiber level. It is very likely that histone modifications may alter structure at various levels. Not only might it affect the stability and dynamics of individual nucleosomes, but also the organization and compactness of nucleosome arrays involving 10-nm (primary chromatin structure) and 30-nm (secondary chromatin structure) fiber formation. Furthermore, it might affect fiber–fiber interactions leading to higher-order compaction (tertiary chromatin structure) observed in condensed chromatin (reviewed in Luger et al. 2012). That is to say, histone PTMs may impact on histone–DNA interactions at the mononucleosomal level, as well as on short-range and long-range internucleosomal contacts and the ability of readers targeted to these marks to stabilize particular chromatin conformations. The potential linkage of histone PTM patterns with chromatin accessibility, combined with information on nucleosomal positioning, transcription factor binding, and the role of noncoding RNAs, should thus provide us with a more complete understanding of chromatin structure and function.

Currently, there are several examples of proteins that bind to the unmodified nucleosome core particle, ranging from the Kaposi sarcoma herpes virus peptide LANA (Barbera et al. 2006) to the chromosome factor RCC1 (Makde et al. 2010) and BAH domain of Sir3 (Armache et al. 2011). These examples have provided insights into the interactions and common principles (i.e., that all proteins target the acidic patch on H2A/H2B) associated with protein–nucleosome recognition (reviewed in Tan and Davey 2011). Nevertheless, an important emerging challenge is the structural characterization of complexes between reader modules and their PTM targets at the nucleosomal, and in the longer term, nucleosomal array level. Potential candidate reader modules include isolated domains (PHD fingers and BAH domains) and cassettes (PHD-bromo dual domains). In addition, the chemical biology approaches mentioned in the previous section, capable of generating the required appropriate PTM-containing designer nucleosomes, should provide ideal substrates for such structural studies. It should also be possible to gain insights into the spreading of histone marks at a molecular level by structurally characterizing dinucleosome complexes in which the reader component of a dual reader/writer protein site-specifically reads a histone mark on one nucleosome and is positioned to write the same site-specific mark on an adjacent nucleosome. Such studies could determine whether there is a directional component to spreading and its molecular basis, as well as the role of antagonistic PTMs to spreading, potentially providing molecular insights into the role of boundary elements.

An even greater and longer term challenge is to structurally characterize PTM-modified designer nucleosomes bound to ATP-dependent chromatin-remodeling complexes (containing reader modules) that either facilitate movement of nucleosomes to different positions on the DNA (SNF2H family) or those that transiently unravel nucleosomes (SWI/SNF family) (reviewed in Clapier and Cairns 2009; also the topic of Ch. 21 [Becker and Workman 2013]). There has been promising initial progress toward this goal from X-ray studies on the chromatin-remodeling factor ISW1a bound to DNA (although the construct lacks the ATPase domain) and nucleosome-containing complexes of this system determined by cryo-electron microscopy (Yamada et al. 2011). Nevertheless, more needs to be performed toward the structural characterization of the complete system, including the use of PTM-modified nucleosomes.

Currently, there is an as yet unresolved controversy surrounding the structure of the compact 30-nm fiber: Does it consist of a one-start solenoid or two-start zig-zag model (reviewed in van Holde and Zlatanova 2007; Luger et al. 2012)? It is likely that the 30-nm structure and dynamics depend on several factors including the type and distribution of PTMs, linker length, the presence/absence of linker histones, and concentration of divalent cations. Thus, it has

Cite as *Cold Spring Harb Perspect Biol* doi: 10.1101/cshperspect.a018754

been shown that the charge change associated with the deposition of the H4K16ac mark disrupts formation of the 30-nm fiber and higher-order chromatin structures (Shogren-Knaak et al. 2006). This most likely reflects the disruption of internucleosomal interactions that promote chromatin folding via the interaction of H4K16 with an acidic patch on H2A on an adjacent nucleosome. The availability of designer nucleosomes with site-specifically incorporated fluorescence resonance energy transfer probes (see review by Allis and Muir 2011) could provide handles for investigating aspects of the architecture and dynamics of the 30-nm fiber.

Even less is known about higher-order tertiary folded states of chromatin, and the role of nuclear lamins and matrix proteins in the anchoring of looped-out chromatin domains to the nuclear periphery. Much work, however, has been performed to ascertain the likely genomic regions that interact with each other (see Ch. 19 [Dekker and Misteli 2014]). It should be possible to expand on earlier experiments involving cross-linking agents added in *trans* to probe higher-order tertiary chromatin structure (Kan et al. 2009) through involvement of designer nucleosomes containing site-specifically incorporated chemical cross-linkers (see review by Allis and Muir 2011). Given that H3 tails have been shown to contribute to interarray interactions, the next step would be to probe the effect of PTMs on higher-order tertiary folded states of chromatin, which could potentially be elucidated using small optical probes and fluorescence microscopy-based biophysical approaches.

16.4 Readout of Oxidative 5mC DNA Adducts

The field of DNA methylation received an unexpected boost with the discovery of 5hmC, a new PTM found on cytosines validated in a biological context (described in Ch. 2 [Kriaucionis and Tahiliani 2014]). One group focused on the enzymatic activity of Tet proteins, a common fusion partner of the MLL1 gene in acute leukemia; they showed Tet1 to be a 2-oxogulaterate and Fe^{2+}-dependent hydroxygenase, which catalyzes the conversion of 5mC to 5hmC, both in vitro and in cultured cells (Tahiliani et al. 2009). The other group identified 5hmC in neurons and the brain, constituting 0.6% of total nucleotides in Purkinje cells and 0.2% in granule cells (Kriaucionis and Heintz 2009). Both groups predicted a role for 5hmC in epigenetic regulation with a potential to influence chromatin structure and local transcription activity. The significance of these results was reinforced by the demonstration that Tet1 plays a role in embryonic stem (ES) cell self-renewal and inner cell mass specification (Ito et al. 2011; discussed in Ch. 27 [Reik and Surani 2014]).

Subsequent to identifying 5hmC, it was shown that 5hmC is converted by Tet to 5-formylcytosine (5fC) (Ito et al. 2011) and 5-carboxylcytosine (5caC) (He et al. 2011; Ito et al. 2011) in an enzymatic activity-dependent manner under in vitro and cell culture conditions. Further, 5hmC, 5fC, and 5caC can be detected in genomic DNA of mouse ES cells and organs with relative levels governed by the activity of the Tet proteins. Importantly, 5caC can be specifically recognized and excised by the enzyme TDG (thymine-DNA glycosylase); this provided the link for explaining how the conversion of 5mC to 5caC by Tet, when followed by the conversion of 5caC to C by TDG as part of the base-excision repair pathway, could constitute an active DNA demethylation pathway (He et al. 2011).

The identification of oxidative adducts of 5mC has led to the development of quantitative methods for 5mC and 5hmC detection. An oxidative bisulfite sequencing approach (oxidative conversion of 5hmC to 5fC, followed by bisulfite conversion of 5fC to uracil) has been developed to measure these adducts at single-base resolution (Booth et al. 2012). This research has identified high 5hmC levels at CpG islands associated with transcriptional regulators and in long interspersed nuclear elements, hinting that these regions may play a role in the epigenetic reprogramming of ES cells. Genome-wide studies of 5hmC and Tet1 have established that Tet1 controls 5mC and 5hmC levels in mouse ES cells by binding to CpG-rich regions thereby preventing unwanted DNA methyltransferase activity (Xu et al. 2011).

Functional studies have also identified a dual role of Tet1 in the transcriptional regulation of mouse ES cells, whereby Tet1 promotes transcription of pluripotency factors, as well as participates in the repression of Polycomb-targeted developmental regulators (Wu et al. 2011a). A regulatory link between histone PTMs and Tet has recently been established; maternal chromatin containing the H3K9me2 mark binds the maternal factor PGC7 (also known as STELLA), protecting 5mC from Tet3-mediated conversion to 5hmC (Nakamura et al. 2012; further explained in Sec. 3.2 of Ch. 27 [Reik and Surani 2014]).

The above studies suggest that 5hmC, 5fC, and 5caC may either represent new epigenetic states in genomic DNA and/or key intermediates in DNA demethylation pathways mediated by Tet and glycosylases/deaminases (reviewed in Bhutani et al. 2011; Wu and Zhang 2011; Branco et al. 2012). Also, it appears that Tet proteins are important regulators of cellular identity given that Tet2 is frequently mutated in haematopoietic tumors. It may do this by playing a key role in DNA methylation fidelity, with perturbation of Tet levels contributing to DNA hypermethylation phenotypes associated with certain cancers (reviewed in Williams et al. 2012; see Ch. 34 [Baylin and Jones 2014]).

At the molecular level, a structure has been solved for a 5caC adduct-containing DNA duplex bound to a human thymine DNA glycosylase (hTDG) (Zhang et al. 2012). In this complex, 5caC is specifically recognized through extrusion from the duplex and repositioning in the catalytic pocket of the enzyme, in which the flipped-out base is locked through polar interactions with pocket residues. This result is supportive of proposals that hTDG (together with Tet) plays a key role in mammalian 5-methylcytosine demethylation. A related study has reported the structure of the carboxy-terminal glycosylase domain of MBD4 (methyl-binding domain 4) bound to an extruded 5-hydroxylmethyl uridine, a deamination product of 5hmC (Hashimoto et al. 2012). Future challenges from the molecular perspective should focus on the identification and structural characterization of readers of oxidative 5mC marks, and whether such readers can selectively target and discriminate between 5mC, 5hmC, 5fC, and 5caC adducts, and whether this requires a base-flipping mechanism for adduct recognition (as illustrated in Fig. 6D of Ch. 6 [Cheng 2014]).

16.5 A Functional Role for Noncoding Regulatory RNAs

The vast majority of the human genome is transcribed, which means there is a much larger repertoire of RNAs beyond just protein-coding mRNAs (summarized in Fig. 6 of Ch. 3 [Allis et al.]). lncRNAs are generally greater than 200 nucleotide in length and, in many cases, are an integral component of chromatin. They serve a regulatory role by interacting with chromatin remodelers and modifiers, thereby changing the epigenetic status of target genes (reviewed in Guil and Esteller 2012; Guttman and Rinn 2012; Kugel and Goodrich 2012). It is believed that lncRNAs function at the interface of histone-modifying enzymatic complexes and the genome, thereby regulating chromatin states and epigenetic inheritance. The regulation can occur in *cis* or *trans* with the former controlling transcription at or in close proximity to the segment of the genome from which the lncRNA is transcribed, whereas the latter controls transcription at genomic sites removed from its site of transcription. From a different perspective, small RNAs in the 20- to 30-nucleotide range also play a role in directing effector complexes to target nascent chromatin-bound lncRNAs, thereby impacting on the recruitment of chromatin-modifying complexes to specific chromosomal regions (reviewed in Moazed 2009; see Ch. 16 [Martienssen and Moazed 2014]).

Long intergenic noncoding RNAs (lincRNAs) have been shown to play critical roles in imprinting, dosage compensation, and homeotic gene expression. One of the earliest identified lincRNAs acting in *cis* was *XIST,* expressed exclusively from the X chromosome and required for X inactivation in mammalian systems (the topic of Brown et al. 1991; Penny et al. 1996; Ch. 25 [Brockdorff and Turner 2014]). *XIST* RNA interacts directly with the Polycomb complex, facilitating condensation of chromatin and initiating transcriptional repression of the entire X chromosome (reviewed in Lee 2012; Ch. 25 [Brockdorff and Turner 2014]). From the perspective of this chapter, ~30% of lincRNAs are known to associate with chromatin-regulatory complexes that include writers, readers, and erasers of histone marks. It has been proposed that lincRNAs contribute regulatory specificity through localization of chromatin regulatory complexes to genomic DNA targets. Some examples of *trans*-acting lincRNAs that control gene transcription by mediating changes in chromatin structure are *HOTAIR* (*HOX* antisense intergenic RNA), *ANRIL* (antisense noncoding RNA in the *INK4* locus), and *HOTTIP* (*HOXA* transcript at the distal tip). We focus below on one such lincRNA, *HOTAIR*, whose discovery is described in Ch. 2 (Rinn 2014).

HOTAIR is one of the earliest and best-characterized lincRNAs (2.2 kb in length), expressed from the *HOXC* locus on chromosome 12 in human distal fibroblasts (Rinn et al. 2007) and metastatic breast tumors (Gupta et al. 2010). The 5′-domain of *HOTAIR* interacts with PRC2, whereas its 3′-counterpart interacts with the LSD1/CoREST/REST complex, thereby coordinating targeting of PRC2 and LSD1 to chromatin for coupled H3K27 methylation and H3K4me demethylation. Based on these observations, it has been proposed that *HOTAIR* and related lincRNAs could serve as scaffolds, providing binding surfaces for the assembly of select histone-modifying enzymes that can specify the pattern of histone modifications on target genes (Tsai et al. 2010a).

At the structural level, information is totally lacking both on the three-dimensional architecture of lincRNAs as well as the intermolecular contacts at protein–RNA interfaces, accounting for the specificity of the recognition process. It is likely that the inherent flexibility of lincRNAs will hinder attempts toward defining their structures in the free state. Nevertheless, this should be a tractable problem when considered at the complex level provided the lincRNAs adopt compact scaffolds in the presence of bound protein targets; such successful structural studies have been conducted on the riboswitch sensing domains bound to small cognate metabolites (reviewed in Serganov and Patel 2012). It is conceivable that lincRNAs may contain highly conserved surface patches involved in protein (and RNA and DNA) recognition and, hence, a tractable approach would require trimming of the RNA constructs down to a minimal functional size, preferably in the 200-nucleotide range (similar in size to the larger riboswitch sensing do-

Cite as *Cold Spring Harb Perspect Biol* doi: 10.1101/cshperspect.a018754

mains), as well as working with minimal protein constructs, so as to facilitate successful crystallization and structure determination of manageable complexes. Such an effort, premised on the functional modularity of lincRNAs, if successful, could provide insights into the RNA secondary structural elements involved in the recognition process and the diversity of such intermolecular interactions.

16.6 Linking Histone and DNA Methylation

There has been considerable interest in the potential cross talk linking histone and DNA methylation in mammals and the mechanisms underlying the relationships between these marks (reviewed in Cedar and Bergman 2009), given that DNA methylation correlates with histone methylation patterns in genome-scale DNA methylation maps of pluripotent and differentiated cells (Meissner et al. 2008). In this regard, DNA methylation is inversely correlated with H3K4 methylation and positively correlated with H3K9 methylation. Relevant questions relate to whether histone methylation serves as a template for directing DNA methylation and/or the converse. Other questions include how the relationship can be addressed on a mechanistic basis from a combined structural and functional perspective.

A notable example of cross talk is shown by the association of UHRF1 with H3K9me3 directing the maintenance of DNA methylation (Rothbart et al. 2012). UHRF1 is a multidomain protein composed of Ubl, tandem Tudor, PHD finger, SRA, and RING domains (see Fig. 6 of Ch. 6 [Cheng 2014]). DNMT1 is also a multidomain protein composed of RFD, CXXC, BAH1/2, and methyltransferase domains. Structures are available for the UHRF1 tandem Tudor domain bound to H3K9me2/3 (Nady et al. 2011), UHRF1 PHD finger bound to unmodified H3K4 (Rajakumara et al. 2011b), and UHRF1 SRA domain bound to hemimethylated CpG DNA (Arita et al. 2008; Avvakumov et al. 2008; Hashimoto et al. 2008), as well as for truncated DNMT1 constructs bound to unmodified CpG (Song et al. 2011) and hemimethylated 5mCpG (Song et al. 2012) sites on DNA. Recent functional studies have shown that UHRF1 association with the H3K9me2/3 histone mark via its tandem Tudor domain is required for the maintenance of DNA methylation, and this association is insensitive to phosphorylation at adjacent Ser10 (Rothbart et al. 2012). It was further established that mitotic binding of UHRF1 to methylated H3K9-containing chromatin stabilizes DNMT1 and contributes to maintenance DNA methylation through the cell cycle. It is not clear at this time whether UHRF1 and DNMT1 can simultaneously bind and be adjacently positioned on hemimethylated 5mCpG DNA given that UHRF1 targets unmodified H3K4 (through its PHD finger), H3K9me2/3 (tandem Tudor domain), and 5mCpG (SRA domain), whereas DNMT1 targets 5mCpG (through its methyltransferase domain) and potentially other histone PTMs (through its BAH1 and/or BAH2 domains). A future challenge will be to attempt making complexes of UHRF1 and DNMT1 on designer nucleosomes containing H3K9me2/3 marks and, if structurally tractable, to define the relative alignments and interactions between components of the complex.

Another example of cross talk between histone and DNA methylation has been noted for the maintenance of CpHpG DNA methylation in plants (reviewed in Law and Jacobsen 2010; Ch. 13 [Pikaard and Mittelsten Scheid 2014]). In this regard, KRYPTONITE (KYP) is a plant histone KMT that facilitates deposition of K9me2 marks on nucleosomal H3 tails (Jackson et al. 2002), whereas chromomethylase CMT3 is a plant DNA methyltransferase that facilitates deposition of 5mC methylation marks at CpHpG sites on nucleosomal DNA (Du et al. 2012). Genome-wide profiling of DNA and H3K9 methylation has established a high correlation between H3K9me2 and CpHpG methylation. In addition, loss of CMT3 or KYP results in a dramatic decrease in DNA methylation. It has been proposed that these two enzymes establish a feedback loop whereby methylated CpHpG DNA recruits KYP to maintain methylation at H3K9, whereas the H3K9me2 mark recruits CMT3 to methylate DNA, so completing the feedback loop. Structural information is available for the complex of ZMET2 (maize counterpart of *A. thaliana* CMT3) bound to H3K9me2 through its chromodomain and BAH domain (Fig. 2C; Du et al. 2012), but as yet no structural information is available for KYP (also designated SUVH4), which contains SRA, pre-SET, SET, and post-SET domains, either in the free state or when bound either to H3K9me2-containing peptide and/or methylated CpHpG-containing DNA (represented schematically in Fig. 10B of Ch. 6 [Cheng 2014]). There is also the additional challenge of elucidating whether direct interactions occur between histone (KYP) and DNA (CMT3) methyltransferases in this system and, if so, to elucidate the principles underlying recognition.

The take-home message is that a combined structure–function approach, as shown for the two examples above, should shed further light on mechanistic aspects of cross talk linking histone and DNA methylation, which can mediate epigenetic regulation in diverse systems (also discussed in Ch. 6 [Cheng 2014]).

16.7 Chromatin-Based Therapeutic Modalities

There is considerable interest in trying to determine the consequences of abnormal chromatin regulator expression and genome alterations and their impact in facilitating the onset of disease states. The types of diseases being looked at range from autoimmune to neurological disorders, and developmental abnormalities to cancer. A natural follow-up from such studies is research into the development of chromatin-based therapeutics, which holds promise given the potential for reversibility of epigenetic-based mutations (reviewed in Chi et al. 2010; Dawson and Kouzarides 2012).

For example, structure–function studies have highlighted that the dysregulation of a chromatin-binding PHD finger can give rise to hematological malignancies (Wang et al. 2009). This study examined the consequence of fusing the carboxy-terminal H3K4me3-binding PHD finger of the jumonji domain lysine demethylase JARID1A to nucleoporin-98 (NUP98), a common fusion partner, thereby generating potent oncoproteins that arrested haematopoietic differentiation and induced acute myeloid leukemia in murine models. The fused PHD finger-NUP98 cassette dominated over polycomb-mediated gene silencing to lock developmentally critical loci into a perpetually active chromatin state, essential for leukaemogenesis. Interestingly, the aromatic cage in the JARID1A PHD finger, which binds to H3K4me3 peptide, was composed of two orthogonally aligned Trp residues. The mutation of either Trp abrogated H3K4me3 binding, consequently abolishing leukaemic transformation.

Related studies have evaluated the consequences of somatic mutations in the ING PHD fingers on solid tumors (reviewed in Chi et al. 2010). Also, the impact of *INK4/ARF* pathway dysregulation has been investigated in aging and cancer (reviewed in Kim and Sharpless 2006).

There is very limited literature on inhibitors targeting the Kme-binding pockets of reader modules, in contrast to Kac-binding pockets (discussed in Sec. 6.3 of Ch. 29 [Busslinger and Tarakhovsky 2014]; Sec. 3.5 of Ch. 4 [Marmorstein and Zhou 2014]); Ch. 2 [Qi 2014; Schaefer 2014]; and reviewed in Arrowsmith et al. 2012). In earlier sections, we have outlined three general modes of molecular recognition of methylated lysine marks by reader modules. First, higher methylation state Kme3/2 marks are predominantly recognized within aromatic-lined cage-type pockets that involve a surface groove mode of recognition (Fig. 2D) (reviewed in Taverna et al. 2007). These are open and shallow pockets and, hence, are difficult to target, and as yet no inhibitors have been identified for such pockets.

Second, lower methylation state Kme1/2 marks can be positioned within the aromatic cage of MBT pocket 2 of L3MBTL1 (Li et al. 2007a; Min et al. 2007) that involve a cavity insertion mode of recognition (Fig. 6B) (reviewed in Taverna et al. 2007). Such pockets are both narrow and deep and, hence, a promising target for inhibitors. Indeed, a ligand and structure-guided design approach (Kireev et al. 2010) has identified UNC669 (Fig. 6D), a pyrrolidine-containing small molecule that targets L3MBTL1 ($K_d = 5$ μM) and shows fivefold increased affinity compared with cognate peptide (H4K20me1) binding, as well as selectivity against close homologs (L3MBTL3 and L3MBTL4) (Herold et al. 2011b). An X-ray structure of L3MBTL1 bound to UNC669 established that the ligand inserts its pyrrolidine ring system into the aromatic-lined pocket of MBT domain 2, with good shape complementarity between the inserted ligand and the walls of the pocket (Herold et al. 2011b) (Fig. 6E).

Third, Kme marks can be recognized by the recently identified interfacial composite pocket composed of residues from adjacently positioned reader domains, with recognition involving favorable van der Waals contacts associated with a high degree of surface complementarity, supplemented by a set of carbon–oxygen hydrogen bonds (Fig. 7C,D) (Iwase et al. 2011). It remains to be seen whether inhibitors can be designed to target such pockets.

An alternate approach would be to design linked small molecule inhibitors that can target closely positioned pockets such as, for instance, those for unmodified K4 and K9me3 on the same histone tail. Progress on development of small molecule inhibitors targeted to Kme reader modules is still in its infancy (reviewed in Herold et al. 2011a) and new approaches will need to be developed for further progress. In this regard, current drugs have rather broad specificity profiles, requiring the next-generation epigenetic drugs to target dysregulated processes with increased specificity.

Recently, it has been shown that *HOTAIR* lincRNA can reprogram chromatin states to promote cancer metastasis (Gupta et al. 2010). *HOTAIR* expression is increased in primary breast tumors, with expression levels in primary tumors a diagnostic predicator for eventual metastasis and death. In this regard, enforced expression of *HOTAIR* in epithelial cancer cells resulted in genome-wide retargeting of the PRC2 complex leading to altered H3K27 methylation, together with increased cancer invasiveness and metastasis. These results implicate an active role for *HOTAIR* in modulating the cancer epigenome, implying that lincRNAs could serve as potential targets for the diagnosis and therapy of cancer.

Because several protein–lincRNA complexes are involved in disease states, detailed structural information of intermolecular contacts could lead to directed functional studies toward identification and implementation of chromatin-based therapeutic modalities.

ACKNOWLEDGMENTS

I am grateful to Dr. Zhanxin Wang for his assistance in the preparation of this review.

Cite as *Cold Spring Harb Perspect Biol* doi: 10.1101/cshperspect.a018754

REFERENCES

*Reference is also in this book.

Ali M, Yan K, Lalonde ME, Degerny C, Rothbart SB, Strahl BD, Cote J, Yang XJ, Kutateladze TG. 2012. Tandem PHD fingers of MORF/MOZ acetyltransferases display selectivity for acetylated histone H3 and are required for the association with chromatin. *J Mol Biol* **424:** 328–338.

Allfrey VG, Faulkner R, Mirsky AE. 1964. Acetylation and methylation of histones and their possible role in the regulation of RNA synthesis. *Proc Natl Acad Sci* **51:** 786–794.

Allis CD, Muir TW. 2011. Spreading chromatin into chemical biology. *Chembiochem* **12:** 264–279.

Amir RE, Van den Veyver IB, Wan M, Tran CQ, Francke U, Zoghbi HY. 1999. Rett syndrome is caused by mutations in X-linked MECP2, encoding methyl-CpG-binding protein 2. *Nat Genet* **23:** 185–188.

Argentaro A, Yang JC, Chapman L, Kowalczyk MS, Gibbons RJ, Higgs DR, Neuhaus D, Rhodes D. 2007. Structural consequences of disease-causing mutations in the ATRX-DNMT3-DNMT3L (ADD) domain of the chromatin-associated protein ATRX. *Proc Natl Acad Sci* **104:** 11939–11944.

Arita K, Ariyoshi M, Tochio H, Nakamura Y, Shirakawa M. 2008. Recognition of hemi-methylated DNA by the SRA protein UHRF1 by a base-flipping mechanism. *Nature* **455:** 818–821.

Arita K, Isogai S, Oda T, Unoki M, Sugita K, Sekiyama N, Kuwata K, Hamamoto R, Tochio H, Sato M, et al. 2012. Recognition of modification status on a histone H3 tail by linked histone reader modules of the epigenetic regulator UHRF1. *Proc Natl Acad Sci* **109:** 12950–12955.

Armache KJ, Garlick JD, Canzio D, Narlikar GJ, Kingston RE. 2011. Structural basis of silencing: Sir3 BAH domain in complex with a nucleosome at 3.0 Å resolution. *Science* **334:** 977–982.

Arrowsmith CH, Bountra C, Fish PV, Lee K, Schapira M. 2012. Epigenetic protein families: A new frontier for drug discovery. *Nat Rev Drug Discov* **11:** 384–400.

Avvakumov GV, Walker JR, Xue S, Li Y, Duan S, Bronner C, Arrowsmith CH, Dhe-Paganon S. 2008. Structural basis for recognition of hemi-methylated DNA by the SRA domain of human UHRF1. *Nature* **455:** 822–825.

Ballare C, Lange M, Lapinaite A, Martin GM, Morey L, Pascual G, Liefke R, Simon B, Shi Y, Gozani O, et al. 2012. Phf19 links methylated Lys36 of histone H3 to regulation of Polycomb activity. *Nat Struct Mol Biol* **19:** 1257–1265.

Bannister AJ, Kouzarides T. 2011. Regulation of chromatin by histone modifications. *Cell Res* **21:** 381–395.

Barbera AJ, Chodaparambil JV, Kelley-Clarke B, Joukov V, Walter JC, Luger K, Kaye KM. 2006. The nucleosomal surface as a docking station for Kaposi's sarcoma herpesvirus LANA. *Science* **311:** 856–861.

* Barlow DP, Bartolomei MS. 2014. Genomic imprinting in mammals. *Cold Spring Harb Perspect Biol* **6:** a018382.

Bartke T, Vermeulen M, Xhemalce B, Robson SC, Mann M, Kouzarides T. 2010. Nucleosome-interacting proteins regulated by DNA and histone methylation. *Cell* **143:** 470–484.

Baylin SB, Jones PA. 2011. A decade of exploring the cancer epigenome—Biological and translational implications. *Nat Rev Cancer* **11:** 726–734.

* Baylin SB, Jones PA. 2014. Epigenetic determinants of cancer. *Cold Spring Harb Perspect Biol* doi: 10.1101/cshperspect.a019505.

* Becker PB, Workman JL. 2013. Nucleosome remodeling and epigenetics. *Cold Spring Harb Perspect Biol* **5:** a017905.

Bedford MT, Clarke SG. 2009. Protein arginine methylation in mammals: Who, what, and why. *Mol Cell* **33:** 1–13.

Bernstein BE, Mikkelsen TS, Xie X, Kamal M, Huebert DJ, Cuff J, Fry B, Meissner A, Wernig M, Plath K, et al. 2006. A bivalent chromatin structure marks key developmental genes in embryonic stem cells. *Cell* **125:** 315–326.

Bhutani N, Burns DM, Blau HM. 2011. DNA demethylation dynamics. *Cell* **146:** 866–872.

Bian C, Xu C, Ruan J, Lee KK, Burke TL, Tempel W, Barsyte D, Li J, Wu M, Zhou BO, et al. 2011. Sgf29 binds histone H3K4me2/3 and is required for SAGA complex recruitment and histone H3 acetylation. *EMBO J* **30:** 2829–2842.

Bicknell LS, Walker S, Klingseisen A, Stiff T, Leitch A, Kerzendorfer C, Martin CA, Yeyati P, Al Sanna N, Bober M, et al. 2011. Mutations in ORC1, encoding the largest subunit of the origin recognition complex, cause microcephalic primordial dwarfism resembling Meier–Gorlin syndrome. *Nat Genet* **43:** 350–355.

Bienz M. 2006. The PHD finger, a nuclear protein-interaction domain. *Trends Biochem Sci* **31:** 35–40.

Black JC, Van Rechem C, Whetstine JR. 2012. Histone lysine methylation dynamics: Establishment, regulation, and biological impact. *Mol Cell* **48:** 491–507.

Bonasio R, Lecona E, Reinberg D. 2010. MBT domain proteins in development and disease. *Semin Cell Dev Biol* **21:** 221–230.

Booth MJ, Branco MR, Ficz G, Oxley D, Krueger F, Reik W, Balasubramanian S. 2012. Quantitative sequencing of 5-methylcytosine and 5-hydroxymethylcytosine at single-base resolution. *Science* **336:** 934–937.

Bostick M, Kim JK, Esteve PO, Clark A, Pradhan S, Jacobsen SE. 2007. UHRF1 plays a role in maintaining DNA methylation in mammalian cells. *Science* **317:** 1760–1764.

Botuyan MV, Lee J, Ward IM, Kim JE, Thompson JR, Chen J, Mer G. 2006. Structural basis for the methylation state-specific recognition of histone H4-K20 by 53BP1 and Crb2 in DNA repair. *Cell* **127:** 1361–1373.

Brahms H, Meheus L, de Brabandere V, Fischer U, Luhrmann R. 2001. Symmetrical dimethylation of arginine residues in spliceosomal Sm protein B/B′ and the Sm-like protein LSm4, and their interaction with the SMN protein. *RNA* **7:** 1531–1542.

Branco MR, Ficz G, Reik W. 2012. Uncovering the role of 5-hydroxymethylcytosine in the epigenome. *Nat Rev Genet* **13:** 7–13.

Brehm A, Tufteland KR, Aasland R, Becker PB. 2004. The many colours of chromodomains. *Bioessays* **26:** 133–140.

Brien GL, Gambero G, O'Connell DJ, Jerman E, Turner SA, Egan CM, Dunne EJ, Jurgens MC, Wynne K, Piao L, et al. 2012. Polycomb PHF19 binds H3K36me3 and recruits PRC2 and demethylase NO66 to embryonic stem cell genes during differentiation. *Nat Struct Mol Biol* **19:** 1273–1281.

* Brockdorff N, Turner BM. 2014. Dosage compensation in mammals. *Cold Spring Harb Perspect Biol* 10.1101/cshperspect.a019406.

Brown CJ, Ballabio A, Rupert JL, Lafreniere RG, Grompe M, Tonlorenzi R, Willard HF. 1991. A gene from the region of the human X inactivation centre is expressed exclusively from the inactive X chromosome. *Nature* **349:** 38–44.

Bua DJ, Kuo AJ, Cheung P, Liu CL, Migliori V, Espejo A, Casadio F, Bassi C, Amati B, Bedford MT, et al. 2009. Epigenome microarray platform for proteome-wide dissection of chromatin-signaling networks. *PLoS One* **4:** e6789.

Buck-Koehntop BA, Stanfield RL, Ekiert DC, Martinez-Yamout MA, Dyson HJ, Wilson IA, Wright PE. 2012. Molecular basis for recognition of methylated and specific DNA sequences by the zinc finger protein Kaiso. *Proc Natl Acad Sci* **109:** 15229–15234.

* Busslinger M, Tarakhovsky A. 2014. Epigenetic control of immunity. *Cold Spring Harb Perspect Biol* **6:** a019307.

Cai L, Rothbart SB, Lu R, Xu B, Chen WY, Tripathy A, Rockowitz S, Zheng D, Patel DJ, Allis CD, Strahl BD, et al. 2013. An H3K36 methylation-engaging Tudor motif of Polycomb-like proteins mediates PRC2 complex targeting. *Mol Cell* **49:** 571–582.

Callebaut I, Courvalin JC, Mornon JP. 1999. The BAH (bromo-adjacent homology) domain: A link between DNA methylation, replication and transcriptional regulation. *FEBS Lett* **446:** 189–193.

Cedar H, Bergman Y. 2009. Linking DNA methylation and histone modification: Patterns and paradigms. *Nat Rev Genet* **10:** 295–304.

Chatterjee C, McGinty RK, Fierz B, Muir TW. 2010. Disulfide-directed histone ubiquitylation reveals plasticity in hDot1L activation. *Nat Chem Biol* **6:** 267–269.

Chen J, Santillan DA, Koonce M, Wei W, Luo R, Thirman MJ, Zeleznik-Le NJ, Diaz MO. 2008. Loss of MLL PHD finger 3 is necessary for MLL-ENL-induced hematopoietic stem cell immortalization. *Cancer Res* **68:** 6199–6207.

Chen C, Nott TJ, Jin J, Pawson T. 2011. Deciphering arginine methylation: Tudor tells the tale. *Nat Rev Mol Cell Biol* **12:** 629–642.

* Cheng X. 2014. Structural and functional coordination of DNA and histone methylation. *Cold Spring Harb Perspect Biol* **6:** a018747.

Cheng J, Yang Y, Fang J, Xiao J, Zhu T, Chen F, Wang P, Li Z, Yang H, Xu Y. 2013. Structural insight into coordinated recognition of H3K9me3 by the plant homeodomain (PHD) and tandem tudor domain (TTD) of the UHRF1 (ubiquitin-like, containing PHD and RING finger domains, 1) protein. *J Biol Chem* **288:** 1329–1339.

Chi P, Allis CD, Wang GG. 2010. Covalent histone modifications—Miswritten, misinterpreted and mis-erased in human cancers. *Nat Rev Cancer* **10:** 457–469.

Chignola F, Gaetani M, Rebane A, Org T, Mollica L, Zucchelli C, Spitaleri A, Mannella V, Peterson P, Musco G. 2009. The solution structure of the first PHD finger of autoimmune regulator in complex with non-modified histone H3 tail reveals the antagonistic role of H3R2 methylation. *Nucleic Acids Res* **37:** 2951–2961.

Clapier CR, Cairns BR. 2009. The biology of chromatin remodeling complexes. *Annu Rev Biochem* **78:** 273–304.

Clouaire T, Stancheva I. 2008. Methyl-CpG binding proteins: Specialized transcriptional repressors or structural components of chromatin? *Cell Mol Life Sci* **65:** 1509–1522.

Collins RE, Northrop JP, Horton JR, Lee DY, Zhang X, Stallcup MR, Cheng X. 2008. The ankyrin repeats of G9a and GLP histone methyltransferases are mono- and dimethyllysine binding modules. *Nat Struct Mol Biol* **15:** 245–250.

Couture JF, Collazo E, Trievel RC. 2006. Molecular recognition of histone H3 by the WD40 protein WDR5. *Nat Struct Mol Biol* **13:** 698–703.

Davis L, Chin JW. 2012. Designer proteins: Applications of genetic code expansion in cell biology. *Nat Rev Mol Cell Biol* **13:** 168–182.

Dawson PE, Kent SB. 2000. Synthesis of native proteins by chemical ligation. *Annu Rev Biochem* **69:** 923–960.

Dawson MA, Kouzarides T. 2012. Cancer epigenetics: From mechanism to therapy. *Cell* **150:** 12–27.

Dawson MA, Bannister AJ, Gottgens B, Foster SD, Bartke T, Green AR, Kouzarides T. 2009. JAK2 phosphorylates histone H3Y41 and excludes HP1alpha from chromatin. *Nature* **461:** 819–822.

* Dekker J, Misteli T. 2014. Long-range chromatin interactions. *Cold Spring Harb Perspect Biol* doi: 10.1101/cshperspect.a019356.

Dou Y, Milne TA, Tackett AJ, Smith ER, Fukuda A, Wysocka J, Allis CD, Chait BT, Hess JL, Roeder RG. 2005. Physical association and coordinate function of the H3 K4 methyltransferase MLL1 and the H4 K16 acetyltransferase MOF. *Cell* **121:** 873–885.

Du J, Zhong X, Bernatavichute YV, Stroud H, Feng S, Caro E, Vashisht AA, Terragni J, Chin HG, Tu A, et al. 2012. Dual binding of chromomethylase domains to H3K9me2-containing nucleosomes directs DNA methylation in plants. *Cell* **151:** 167–180.

Duncker BP, Chesnokov IN, McConkey BJ. 2009. The origin recognition complex protein family. *Genome Biol* **10:** 214.

* Elgin SC, Reuter G. 2013. Position-effect variegation, heterochromatin formation, and gene silencing in *Drosophila*. *Cold Spring Harb Perspect Biol* **5:** a017780.

Eustermann S, Yang JC, Law MJ, Amos R, Chapman LM, Jelinska C, Garrick D, Clynes D, Gibbons RJ, Rhodes D, et al. 2011. Combinatorial readout of histone H3 modifications specifies localization of ATRX to heterochromatin. *Nat Struct Mol Biol* **18:** 777–782.

Fair K, Anderson M, Bulanova E, Mi H, Tropschug M, Diaz MO. 2001. Protein interactions of the MLL PHD fingers modulate MLL target gene regulation in human cells. *Mol Cell Biol* **21:** 3589–3597.

Falk S, Ravaud S, Koch J, Sinning I. 2010. The C terminus of the Alb3 membrane insertase recruits cpSRP43 to the thylakoid membrane. *J Biol Chem* **285:** 5954–5962.

Fiedler M, Sanchez-Barrena MJ, Nekrasov M, Mieszczanek J, Rybin V, Muller J, Evans P, Bienz M. 2008. Decoding of methylated histone H3 tail by the Pygo-BCL9 Wnt signaling complex. *Mol Cell* **30:** 507–518.

Fierz B, Muir TW. 2012. Chromatin as an expansive canvas for chemical biology. *Nat Chem Biol* **8:** 417–427.

Fischle W, Wang Y, Allis CD. 2003a. Binary switches and modification cassettes in histone biology and beyond. *Nature* **425:** 475–479.

Fischle W, Wang Y, Allis CD. 2003b. Histone and chromatin cross-talk. *Curr Opin Cell Biol* **15:** 172–183.

Fischle W, Tseng BS, Dormann HL, Ueberheide BM, Garcia BA, Shabanowitz J, Hunt DF, Funabiki H, Allis CD. 2005. Regulation of HP1-chromatin binding by histone H3 methylation and phosphorylation. *Nature* **438:** 1116–1122.

Flanagan JF, Mi LZ, Chruszcz M, Cymborowski M, Clines KL, Kim Y, Minor W, Rastinejad F, Khorasanizadeh S. 2005. Double chromodomains cooperate to recognize the methylated histone H3 tail. *Nature* **438:** 1181–1185.

Frederiks F, Stulemeijer IJ, Ovaa H, van Leeuwen F. 2011. A modified epigenetics toolbox to study histone modifications on the nucleosome core. *Chembiochem* **12:** 308–313.

Fuchs SM, Krajewski K, Baker RW, Miller VL, Strahl BD. 2011. Influence of combinatorial histone modifications on antibody and effector protein recognition. *Curr Biol* **21:** 53–58.

Gardner KE, Allis CD, Strahl BD. 2011. Operating on chromatin, a colorful language where context matters. *J Mol Biol* **409:** 36–46.

Garske AL, Oliver SS, Wagner EK, Musselman CA, LeRoy G, Garcia BA, Kutateladze TG, Denu JM. 2010. Combinatorial profiling of chromatin binding modules reveals multisite discrimination. *Nat Chem Biol* **6:** 283–290.

Goforth RL, Peterson EC, Yuan J, Moore MJ, Kight AD, Lohse MB, Sakon J, Henry RL. 2004. Regulation of the GTPase cycle in post-translational signal recognition particle-based protein targeting involves cpSRP43. *J Biol Chem* **279:** 43077–43084.

Goodwin GH, Nicolas RH. 2001. The BAH domain, polybromo and the RSC chromatin remodelling complex. *Gene* **268:** 1–7.

Greer EL, Shi Y. 2012. Histone methylation: A dynamic mark in health, disease and inheritance. *Nat Rev Genet* **13:** 343–357.

Grimm C, Matos R, Ly-Hartig N, Steuerwald U, Lindner D, Rybin V, Muller J, Muller CW. 2009. Molecular recognition of histone lysine methylation by the Polycomb group repressor dSfmbt. *EMBO J* **28:** 1965–1977.

* Grossniklaus U, Paro R. 2014. Transcriptional silencing by *Polycomb*-group proteins. *Cold Spring Harb Perspect Biol* **6:** a019331.

Grow EJ, Wysocka J. 2010. Flipping MLL1's switch one proline at a time. *Cell* **141:** 1108–1110.

* Grunstein M, Gasser SM. 2013. Epigenetics in *Saccharomyces cerevisiae*. *Cold Spring Harb Perspect Biol* **5:** a017491.

Guccione E, Bassi C, Casadio F, Martinato F, Cesaroni M, Schuchlautz H, Luscher B, Amati B. 2007. Methylation of histone H3R2 by PRMT6 and H3K4 by an MLL complex are mutually exclusive. *Nature* **449:** 933–937.

Guil S, Esteller M. 2012. Cis-acting noncoding RNAs: Friends and foes. *Nat Struct Mol Biol* **19:** 1068–1075.

Guo Y, Nady N, Qi C, Allali-Hassani A, Zhu H, Pan P, Adams-Cioaba MA, Amaya MF, Dong A, Vedadi M, et al. 2009. Methylation-state-specific recognition of histones by the MBT repeat protein L3MBTL2. *Nucleic Acids Res* **37:** 2204–2210.

Gupta RA, Shah N, Wang KC, Kim J, Horlings HM, Wong DJ, Tsai MC, Hung T, Argani P, Rinn JL, et al. 2010. Long non-coding RNA HOTAIR reprograms chromatin state to promote cancer metastasis. *Nature* **464:** 1071–1076.

Guttman M, Rinn JL. 2012. Modular regulatory principles of large non-coding RNAs. *Nature* **482:** 339–346.

 Cite as *Cold Spring Harb Perspect Biol* doi: 10.1101/cshperspect.a018754

Han Z, Guo L, Wang H, Shen Y, Deng XW, Chai J. 2006. Structural basis for the specific recognition of methylated histone H3 lysine 4 by the WD-40 protein WDR5. *Mol Cell* **22:** 137–144.

Hashimoto H, Horton JR, Zhang X, Bostick M, Jacobsen SE, Cheng X. 2008. The SRA domain of UHRF1 flips 5-methylcytosine out of the DNA helix. *Nature* **455:** 826–829.

Hashimoto H, Horton JR, Zhang X, Cheng X. 2009. UHRF1, a modular multi-domain protein, regulates replication-coupled crosstalk between DNA methylation and histone modifications. *Epigenetics* **4:** 8–14.

Hashimoto H, Zhang X, Cheng X. 2012. Excision of thymine and 5-hydroxymethyluracil by the MBD4 DNA glycosylase domain: Structural basis and implications for active DNA demethylation. *Nucleic Acids Res* **40:** 8276–8284.

He YF, Li BZ, Li Z, Liu P, Wang Y, Tang Q, Ding J, Jia Y, Chen Z, Li L, et al. 2011. Tet-mediated formation of 5-carboxylcytosine and its excision by TDG in mammalian DNA. *Science* **333:** 1303–1307.

* Henikoff S, Smith MM. 2014. Histone variants and epigenetics. *Cold Spring Harb Perspect Biol* doi: 10.1101/cshperspect.a019364.

Herold JM, Ingerman LA, Gao C, Frye SV. 2011a. Drug discovery toward antagonists of methyl-lysine binding proteins. *Curr Chem Genomics* **5:** 51–61.

Herold JM, Wigle TJ, Norris JL, Lam R, Korboukh VK, Gao C, Ingerman LA, Kireev DB, Senisterra G, Vedadi M, et al. 2011b. Small-molecule ligands of methyl-lysine binding proteins. *J Med Chem* **54:** 2504–2511.

Hirota T, Lipp JJ, Toh BH, Peters JM. 2005. Histone H3 serine 10 phosphorylation by Aurora B causes HP1 dissociation from heterochromatin. *Nature* **438:** 1176–1180.

Ho KL, McNae IW, Schmiedeberg L, Klose RJ, Bird AP, Walkinshaw MD. 2008. MeCP2 binding to DNA depends upon hydration at methyl-CpG. *Mol Cell* **29:** 525–531.

Holdermann I, Meyer NH, Round A, Wild K, Sattler M, Sinning I. 2012. Chromodomains read the arginine code of post-translational targeting. *Nat Struct Mol Biol* **19:** 260–263.

Hom RA, Chang PY, Roy S, Musselman CA, Glass KC, Selezneva AI, Gozani O, Ismagilov RF, Cleary ML, Kutateladze TG. 2010. Molecular mechanism of MLL PHD3 and RNA recognition by the Cyp33 RRM domain. *J Mol Biol* **400:** 145–154.

Hossain M, Stillman B. 2012. Meier–Gorlin syndrome mutations disrupt an Orc1 CDK inhibitory domain and cause centrosome reduplication. *Genes Dev* **26:** 1797–1810.

Hou Z, Bernstein DA, Fox CA, Keck JL. 2005. Structural basis of the Sir1-origin recognition complex interaction in transcriptional silencing. *Proc Natl Acad Sci* **102:** 8489–8494.

Hsu HC, Stillman B, Xu RM. 2005. Structural basis for origin recognition complex 1 protein-silence information regulator 1 protein interaction in epigenetic silencing. *Proc Natl Acad Sci* **102:** 8519–8524.

Hu L, Li Z, Wang P, Lin Y, Xu Y. 2011. Crystal structure of PHD domain of UHRF1 and insights into recognition of unmodified histone H3 arginine residue 2. *Cell Res* **21:** 1374–1378.

Huang Y, Fang J, Bedford MT, Zhang Y, Xu RM. 2006. Recognition of histone H3 lysine-4 methylation by the double tudor domain of JMJD2A. *Science* **312:** 748–751.

Huang R, Holbert MA, Tarrant MK, Curtet S, Colquhoun DR, Dancy BM, Dancy BC, Hwang Y, Tang Y, Meeth K, et al. 2010. Site-specific introduction of an acetyl-lysine mimic into peptides and proteins by cysteine alkylation. *J Am Chem Soc* **132:** 9986–9987.

Hung T, Binda O, Champagne KS, Kuo AJ, Johnson K, Chang HY, Simon MD, Kutateladze TG, Gozani O. 2009. ING4 mediates crosstalk between histone H3 K4 trimethylation and H3 acetylation to attenuate cellular transformation. *Mol Cell* **33:** 248–256.

Huyen Y, Zgheib O, Ditullio RA Jr., Gorgoulis VG, Zacharatos P, Petty TJ, Sheston EA, Mellert HS, Stavridi ES, Halazonetis TD. 2004. Methylated lysine 79 of histone H3 targets 53BP1 to DNA double-strand breaks. *Nature* **432:** 406–411.

Hyllus D, Stein C, Schnabel K, Schiltz E, Imhof A, Dou Y, Hsieh J, Bauer UM. 2007. PRMT6-mediated methylation of R2 in histone H3 antagonizes H3 K4 trimethylation. *Genes Dev* **21:** 3369–3380.

Ito S, Shen L, Dai Q, Wu SC, Collins LB, Swenberg JA, He C, Zhang Y. 2011. Tet proteins can convert 5-methylcytosine to 5-formylcytosine and 5-carboxylcytosine. *Science* **333:** 1300–1303.

Ivanov AV, Peng H, Yurchenko V, Yap KL, Negorev DG, Schultz DC, Psulkowski E, Fredericks WJ, White DE, Maul GG, et al. 2007. PHD domain-mediated E3 ligase activity directs intramolecular sumoylation of an adjacent bromodomain required for gene silencing. *Mol Cell* **28:** 823–837.

Iwase S, Xiang B, Ghosh S, Ren T, Lewis PW, Cochrane JC, Allis CD, Picketts DJ, Patel DJ, Li H, et al. 2011. ATRX ADD domain links an atypical histone methylation recognition mechanism to human mental-retardation syndrome. *Nat Struct Mol Biol* **18:** 769–776.

Jackson JP, Lindroth AM, Cao X, Jacobsen SE. 2002. Control of CpNpG DNA methylation by the KRYPTONITE histone H3 methyltransferase. *Nature* **416:** 556–560.

Jacobs SA, Khorasanizadeh S. 2002. Structure of HP1 chromodomain bound to a lysine 9-methylated histone H3 tail. *Science* **295:** 2080–2083.

Jacobs SA, Taverna SD, Zhang Y, Briggs SD, Li J, Eissenberg JC, Allis CD, Khorasanizadeh S. 2001. Specificity of the HP1 chromo domain for the methylated N-terminus of histone H3. *EMBO J* **20:** 5232–5241.

Jenuwein T, Allis CD. 2001. Translating the histone code. *Science* **293:** 1074–1080.

Jiao Y, Shi C, Edil BH, de Wilde RF, Klimstra DS, Maitra A, Schulick RD, Tang LH, Wolfgang CL, Choti MA, et al. 2011. DAXX/ATRX, MEN1, and mTOR pathway genes are frequently altered in pancreatic neuroendocrine tumors. *Science* **331:** 1199–1203.

Johnson LM, Bostick M, Zhang X, Kraft E, Henderson I, Callis J, Jacobsen SE. 2007. The SRA methyl-cytosine-binding domain links DNA and histone methylation. *Curr Biol* **17:** 379–384.

Jones PA. 2012. Functions of DNA methylation: Islands, start sites, gene bodies and beyond. *Nat Rev Genet* **13:** 484–492.

Jones PA, Liang G. 2009. Rethinking how DNA methylation patterns are maintained. *Nat Rev Genet* **10:** 805–811.

Kan PY, Caterino TL, Hayes JJ. 2009. The H4 tail domain participates in intra- and internucleosome interactions with protein and DNA during folding and oligomerization of nucleosome arrays. *Mol Cell Biol* **29:** 538–546.

Khorasanizadeh S. 2011. Recognition of methylated histones: New twists and variations. *Curr Opin Struct Biol* **21:** 744–749.

Kim WY, Sharpless NE. 2006. The regulation of INK4/ARF in cancer and aging. *Cell* **127:** 265–275.

Kim D, Blus BJ, Chandra V, Huang P, Rastinejad F, Khorasanizadeh S. 2010. Corecognition of DNA and a methylated histone tail by the MSL3 chromodomain. *Nat Struct Mol Biol* **17:** 1027–1029.

Kireev D, Wigle TJ, Norris-Drouin J, Herold JM, Janzen WP, Frye SV. 2010. Identification of non-peptide malignant brain tumor (MBT) repeat antagonists by virtual screening of commercially available compounds. *J Med Chem* **53:** 7625–7631.

Kirmizis A, Santos-Rosa H, Penkett CJ, Singer MA, Vermeulen M, Mann M, Bahler J, Green RD, Kouzarides T. 2007. Arginine methylation at histone H3R2 controls deposition of H3K4 trimethylation. *Nature* **449:** 928–932.

Klingseisen A, Jackson AP. 2011. Mechanisms and pathways of growth failure in primordial dwarfism. *Genes Dev* **25:** 2011–2024.

Klose RJ, Bird AP. 2006. Genomic DNA methylation: The mark and its mediators. *Trends Biochem Sci* **31:** 89–97.

Koga H, Matsui S, Hirota T, Takebayashi S, Okumura K, Saya H. 1999. A human homolog of *Drosophila* lethal(3)malignant brain tumor (l(3)mbt) protein associates with condensed mitotic chromosomes. *Oncogene* **18:** 3799–3809.

Kouzarides T. 2007. Chromatin modifications and their function. *Cell* **128:** 693–705.

Kouzarides T, Berger SL. 2007. Chromatin modifications and their mechanism. In *Epigenetics* (ed. Allis CD, Jenuwen T, Reinberg D), pp. 191–209. Cold Spring Harbor Laboratory Press, Cold Spring Harbor, NY.

Kriaucionis S, Heintz N. 2009. The nuclear DNA base 5-hydroxymethylcytosine is present in Purkinje neurons and the brain. *Science* **324:** 929–930.

* Kriaucionis S, Tahiliani M. 2014. Expanding the epigenetic landscape: Novel modifications of cytosine in genomic DNA. *Cold Spring Harb Perspect Biol* **6:** a018630.

Kugel JF, Goodrich JA. 2012. Non-coding RNAs: Key regulators of mammalian transcription. *Trends Biochem Sci* **37:** 144–151.

Kuo AJ, Song J, Cheung P, Ishibe-Murakami S, Yamazoe S, Chen JK, Patel DJ, Gozani O. 2012. The BAH domain of ORC1 links H4K20me2 to DNA replication licensing and Meier–Gorlin syndrome. *Nature* **484:** 115–119.

Lan F, Collins RE, De Cegli R, Alpatov R, Horton JR, Shi X, Gozani O, Cheng X, Shi Y. 2007. Recognition of unmethylated histone H3 lysine 4 links BHC80 to LSD1-mediated gene repression. *Nature* **448:** 718–722.

Latham JA, Dent SY. 2007. Cross-regulation of histone modifications. *Nat Struct Mol Biol* **14:** 1017–1024.

Law JA, Jacobsen SE. 2010. Establishing, maintaining and modifying DNA methylation patterns in plants and animals. *Nat Rev Genet* **11:** 204–220.

Law JA, Du J, Hale CJ, Feng S, Krajewski K, Palanca AM, Strahl BD, Patel DJ, Jacobsen SE. 2013. Polymerase IV occupancy at RNA-directed DNA methylation sites requires SHH1. *Nature* **498:** 385–389.

Lee JT. 2012. Epigenetic regulation by long noncoding RNAs. *Science* **338:** 1435–1439.

Lee JS, Shukla A, Schneider J, Swanson SK, Washburn MP, Florens L, Bhaumik SR, Shilatifard A. 2007. Histone cross talk between H2B monoubiquitination and H3 methylation mediated by COMPASS. *Cell* **131:** 1084–1096.

Lee J, Thompson JR, Botuyan MV, Mer G. 2008. Distinct binding modes specify the recognition of methylated histones H3K4 and H4K20 by JMJD2A-tudor. *Nat Struct Mol Biol* **15:** 109–111.

Li E, Bird A. 2007. DNA methylation in mammals. In *Epigenetics* (ed. Allis CD, Jenuwen T, Reinberg D), pp. 341–356. Cold Spring Harbor Laboratory Press, Cold Spring Harbor, NY.

* Li E, Zhang Y. 2014. DNA methylation in mammals. *Cold Spring Harb Perspect Biol* **6:** a019133.

Li H, Ilin S, Wang W, Duncan EM, Wysocka J, Allis CD, Patel DJ. 2006. Molecular basis for site-specific read-out of histone H3K4me3 by the BPTF PHD finger of NURF. *Nature* **442:** 91–95.

Li H, Fischle W, Wang W, Duncan EM, Liang L, Murakami-Ishibe S, Allis CD, Patel DJ. 2007a. Structural basis for lower lysine methylation state-specific readout by MBT repeats of L3MBTL1 and an engineered PHD finger. *Mol Cell* **28:** 677–691.

Li X, Lee YK, Jeng JC, Yen Y, Schultz DC, Shih HM, Ann DK. 2007b. Role for KAP1 serine 824 phosphorylation and sumoylation/desumoylation switch in regulating KAP1-mediated transcriptional repression. *J Biol Chem* **282:** 36177–36189.

Liu CC, Schultz PG. 2010. Adding new chemistries to the genetic code. *Annu Rev Biochem* **79:** 413–444.

Liu H, Wang JY, Huang Y, Li Z, Gong W, Lehmann R, Xu RM. 2010a. Structural basis for methylarginine-dependent recognition of Aubergine by Tudor. *Genes Dev* **24:** 1876–1881.

Liu K, Chen C, Guo Y, Lam R, Bian C, Xu C, Zhao DY, Jin J, MacKenzie F, Pawson T, et al. 2010b. Structural basis for recognition of arginine methylated Piwi proteins by the extended Tudor domain. *Proc Natl Acad Sci* **107:** 18398–18403.

Liu Y, Toh H, Sasaki H, Zhang X, Cheng X. 2012. An atomic model of Zfp57 recognition of CpG methylation within a specific DNA sequence. *Genes Dev* **26:** 2374–2379.

Lopes EC, Valls E, Figueroa ME, Mazur A, Meng FG, Chiosis G, Laird PW, Schreiber-Agus N, Greally JM, Prokhortchouk E, et al. 2008. Kaiso contributes to DNA methylation-dependent silencing of tumor suppressor genes in colon cancer cell lines. *Cancer Res* **68:** 7258–7263.

* Lucchesi JC, Kuroda MI. 2014. Dosage compensation in *Drosophila*. *Cold Spring Harb Perspect Biol* doi: 10.1101/cshperspect.a019398.

Luger K, Mader AW, Richmond RK, Sargent DF, Richmond TJ. 1997. Crystal structure of the nucleosome core particle at 2.8 A resolution. *Nature* **389:** 251–260.

Luger K, Dechassa ML, Tremethick DJ. 2012. New insights into nucleosome and chromatin structure: An ordered state or a disordered affair? *Nat Rev Mol Cell Biol* **13:** 436–447.

Ma JC, Dougherty DA. 1997. The Cation–π interaction. *Chem Rev* **97:** 1303–1324.

Makde RD, England JR, Yennawar HP, Tan S. 2010. Structure of RCC1 chromatin factor bound to the nucleosome core particle. *Nature* **467:** 562–566.

Marazzi I, Ho JS, Kim J, Manicassamy B, Dewell S, Albrecht RA, Seibert CW, Schaefer U, Jeffrey KL, Prinjha RK, et al. 2012. Suppression of the antiviral response by an influenza histone mimic. *Nature* **483:** 428–433.

* Marmorstein R, Zhou M-M. 2014. Writers and readers of histone acetylation: Structure, mechanism, and inhibition. *Cold Spring Harb Perspect Biol* **6:** a018762.

* Martienssen R, Moazed D. 2014. RNAi and heterochromatin assembly. *Cold Spring Harb Perspect Biol* doi: 10.1101/cshperspect.a019323.

Mathioudakis N, Palencia A, Kadlec J, Round A, Tripsianes K, Sattler M, Pillai RS, Cusack S. 2012. The multiple Tudor domain-containing protein TDRD1 is a molecular scaffold for mouse Piwi proteins and piRNA biogenesis factors. *RNA* **18:** 2056–2072.

Matthews AG, Kuo AJ, Ramon-Maiques S, Han S, Champagne KS, Ivanov D, Gallardo M, Carney D, Cheung P, Ciccone DN, et al. 2007. RAG2 PHD finger couples histone H3 lysine 4 trimethylation with V(D)J recombination. *Nature* **450:** 1106–1110.

Maurer-Stroh S, Dickens NJ, Hughes-Davies L, Kouzarides T, Eisenhaber F, Ponting CP. 2003. The Tudor domain 'Royal Family': Tudor, plant Agenet, Chromo, PWWP and MBT domains. *Trends Biochem Sci* **28:** 69–74.

McGinty RK, Kim J, Chatterjee C, Roeder RG, Muir TW. 2008. Chemically ubiquitylated histone H2B stimulates hDot1L-mediated intranucleosomal methylation. *Nature* **453:** 812–816.

Meissner A, Mikkelsen TS, Gu H, Wernig M, Hanna J, Sivachenko A, Zhang X, Bernstein BE, Nusbaum C, Jaffe DB, et al. 2008. Genome-scale DNA methylation maps of pluripotent and differentiated cells. *Nature* **454:** 766–770.

Migliori V, Muller J, Phalke S, Low D, Bezzi M, Mok WC, Sahu SK, Gunaratne J, Capasso P, Bassi C, et al. 2012. Symmetric dimethylation of H3R2 is a newly identified histone mark that supports euchromatin maintenance. *Nat Struct Mol Biol* **19:** 136–144.

Miller TC, Rutherford TJ, Johnson CM, Fiedler M, Bienz M. 2010. Allosteric remodelling of the histone H3 binding pocket in the Pygo2 PHD finger triggered by its binding to the B9L/BCL9 co-factor. *J Mol Biol* **401:** 969–984.

Milne TA, Briggs SD, Brock HW, Martin ME, Gibbs D, Allis CD, Hess JL. 2002. MLL targets SET domain methyltransferase activity to Hox gene promoters. *Mol Cell* **10:** 1107–1117.

Milne TA, Kim J, Wang GG, Stadler SC, Basrur V, Whitcomb SJ, Wang Z, Ruthenburg AJ, Elenitoba-Johnson KS, Roeder RG, et al. 2010. Multiple interactions recruit MLL1 and MLL1 fusion proteins to the HOXA9 locus in leukemogenesis. *Mol Cell* **38:** 853–863.

Min J, Zhang Y, Xu RM. 2003. Structural basis for specific binding of Polycomb chromodomain to histone H3 methylated at Lys 27. *Genes Dev* **17:** 1823–1828.

Min J, Allali-Hassani A, Nady N, Qi C, Ouyang H, Liu Y, MacKenzie F, Vedadi M, Arrowsmith CH. 2007. L3MBTL1 recognition of mono- and dimethylated histones. *Nat Struct Mol Biol* **14:** 1229–1230.

Moazed D. 2009. Small RNAs in transcriptional gene silencing and genome defence. *Nature* **457:** 413–420.

Musselman CA, Kutateladze TG. 2011. Handpicking epigenetic marks with PHD fingers. *Nucleic Acids Res* **39:** 9061–9071.

Musselman CA, Avvakumov N, Watanabe R, Abraham CG, Lalonde ME, Hong Z, Allen C, Roy S, Nunez JK, Nickoloff J, et al. 2012a. Molecular basis for H3K36me3 recognition by the Tudor domain of PHF1. *Nat Struct Mol Biol* **19:** 1266–1272.

Musselman CA, Lalonde ME, Cote J, Kutateladze TG. 2012b. Perceiving the epigenetic landscape through histone readers. *Nat Struct Mol Biol* **19:** 1218–1227.

Musselman CA, Ramirez J, Sims JK, Mansfield RE, Oliver SS, Denu JM, Mackay JP, Wade PA, Hagman J, Kutateladze TG. 2012c. Bivalent recognition of nucleosomes by the tandem PHD fingers of the CHD4 ATPase is required for CHD4-mediated repression. *Proc Natl Acad Sci* **109:** 787–792.

Nady N, Lemak A, Walker JR, Avvakumov GV, Kareta MS, Achour M, Xue S, Duan S, Allali-Hassani A, Zuo X, et al. 2011. Recognition of multivalent histone states associated with heterochromatin by UHRF1 protein. *J Biol Chem* **286:** 24300–24311.

Nady N, Krichevsky L, Zhong N, Duan S, Tempel W, Amaya MF, Ravichandran M, Arrowsmith CH. 2012. Histone recognition by human malignant brain tumor domains. *J Mol Biol* **423:** 702–718.

Nakamura Y, Umehara T, Hamana H, Hayashizaki Y, Inoue M, Kigawa T, Shirouzu M, Terada T, Tanaka A, Padmanabhan B, et al. 2007. Crystal structure analysis of the PHD domain of the transcription co-activator Pygopus. *J Mol Biol* **370:** 80–92.

Nakamura T, Liu YJ, Nakashima H, Umehara H, Inoue K, Matoba S, Tachibana M, Ogura A, Shinkai Y, Nakano T. 2012. PGC7 binds histone H3K9me2 to protect against conversion of 5mC to 5hmC in early embryos. *Nature* **486:** 415–419.

Neumann H, Wang K, Davis L, Garcia-Alai M, Chin JW. 2010. Encoding multiple unnatural amino acids via evolution of a quadruplet-decoding ribosome. *Nature* **464:** 441–444.

Ng HH, Xu RM, Zhang Y, Struhl K. 2002. Ubiquitination of histone H2B by Rad6 is required for efficient Dot1-mediated methylation of histone H3 lysine 79. *J Biol Chem* **277:** 34655–34657.

Nielsen PR, Nietlispach D, Mott HR, Callaghan J, Bannister A, Kouzarides T, Murzin AG, Murzina NV, Laue ED. 2002. Structure of the HP1 chromodomain bound to histone H3 methylated at lysine 9. *Nature* **416:** 103–107.

Ohki I, Shimotake N, Fujita N, Jee J, Ikegami T, Nakao M, Shirakawa M. 2001. Solution structure of the methyl-CpG binding domain of human MBD1 in complex with methylated DNA. *Cell* **105:** 487–497.

Oliver SS, Denu JM. 2011. Dynamic interplay between histone H3 modifications and protein interpreters: Emerging evidence for a 'histone language'. *Chembiochem* **12:** 299–307.

Ooi SK, Qiu C, Bernstein E, Li K, Jia D, Yang Z, Erdjument-Bromage H, Tempst P, Lin SP, Allis CD, et al. 2007. DNMT3L connects unmethylated lysine 4 of histone H3 to de novo methylation of DNA. *Nature* **448:** 714–717.

Otani J, Nankumo T, Arita K, Inamoto S, Ariyoshi M, Shirakawa M. 2009. Structural basis for recognition of H3K4 methylation status by the DNA methyltransferase 3A ATRX-DNMT3-DNMT3L domain. *EMBO Rep* **10:** 1235–1241.

Park S, Osmers U, Raman G, Schwantes RH, Diaz MO, Bushweller JH. 2010. The PHD3 domain of MLL acts as a CYP33-regulated switch between MLL-mediated activation and repression. *Biochemistry* **49:** 6576–6586.

Pascual J, Martinez-Yamout M, Dyson HJ, Wright PE. 2000. Structure of the PHD zinc finger from human Williams–Beuren syndrome transcription factor. *J Mol Biol* **304:** 723–729.

Patel A, Dharmarajan V, Cosgrove MS. 2008. Structure of WDR5 bound to mixed lineage leukemia protein-1 peptide. *J Biol Chem* **283:** 32158–32161.

Pena PV, Davrazou F, Shi X, Walter KL, Verkhusha VV, Gozani O, Zhao R, Kutateladze TG. 2006. Molecular mechanism of histone H3K4me3 recognition by plant homeodomain of ING2. *Nature* **442:** 100–103.

Peng J, Wysocka J. 2008. It takes a PHD to SUMO. *Trends Biochem Sci* **33:** 191–194.

Penny GD, Kay GF, Sheardown SA, Rastan S, Brockdorff N. 1996. Requirement for Xist in X chromosome inactivation. *Nature* **379:** 131–137.

* Pikaard CS, Mittelsten Scheid O. 2014. Epigenetic regulation in plants. *Cold Spring Harb Perspect Biol* doi: 10.1101/cshperspect.a019315.

* Qi J. 2014. Bromodomain and extraterminal domain inhibitors (BETi) for cancer therapy: Chemical modulation of chromatin structure. *Cold Spring Harb Perspect Biol* doi: 10.1101/cshperspect.a018663.

Qiu C, Sawada K, Zhang X, Cheng X. 2002. The PWWP domain of mammalian DNA methyltransferase Dnmt3b defines a new family of DNA-binding folds. *Nat Struct Biol* **9:** 217–224.

Qiu Y, Liu L, Zhao C, Han C, Li F, Zhang J, Wang Y, Li G, Mei Y, Wu M, et al. 2012. Combinatorial readout of unmodified H3R2 and acetylated H3K14 by the tandem PHD finger of MOZ reveals a regulatory mechanism for HOXA9 transcription. *Genes Dev* **26:** 1376–1391.

Rajakumara E, Law JA, Simanshu DK, Voigt P, Johnson LM, Reinberg D, Patel DJ, Jacobsen SE. 2011a. A dual flip-out mechanism for 5mC recognition by the *Arabidopsis* SUVH5 SRA domain and its impact on DNA methylation and H3K9 dimethylation in vivo. *Genes Dev* **25:** 137–152.

Rajakumara E, Wang Z, Ma H, Hu L, Chen H, Lin Y, Guo R, Wu F, Li H, Lan F, et al. 2011b. PHD finger recognition of unmodified histone H3R2 links UHRF1 to regulation of euchromatic gene expression. *Mol Cell* **43:** 275–284.

Ram O, Goren A, Amit I, Shoresh N, Yosef N, Ernst J, Kellis M, Gymrek M, Issner R, Coyne M, et al. 2011. Combinatorial patterning of chromatin regulators uncovered by genome-wide location analysis in human cells. *Cell* **147:** 1628–1639.

Ramon-Maiques S, Kuo AJ, Carney D, Matthews AG, Oettinger MA, Gozani O, Yang W. 2007. The plant homeodomain finger of RAG2 recognizes histone H3 methylated at both lysine-4 and arginine-2. *Proc Natl Acad Sci* **104:** 18993–18998.

* Reik W, Surani MA. 2014. Germline and pluripotent stem cells. *Cold Spring Harb Perspect Biol* doi: 10.1101/cshperspect.a019422.

* Rinn JL. 2014. lncRNAs: Linking RNA to chromatin. *Cold Spring Harb Perspect Biol* **6:** a018614.

Rinn JL, Kertesz M, Wang JK, Squazzo SL, Xu X, Brugmann SA, Goodnough LH, Helms JA, Farnham PJ, Segal E, et al. 2007. Functional demarcation of active and silent chromatin domains in human HOX loci by noncoding RNAs. *Cell* **129:** 1311–1323.

Robertson KD. 2005. DNA methylation and human disease. *Nat Rev Genet* **6:** 597–610.

Rothbart SB, Krajewski K, Nady N, Tempel W, Xue S, Badeaux AI, Barsyte-Lovejoy D, Martinez JY, Bedford MT, Fuchs SM, et al. 2012. Association of UHRF1 with methylated H3K9 directs the maintenance of DNA methylation. *Nat Struct Mol Biol* **19:** 1155–1160.

Roy S, Musselman CA, Kachirskaia I, Hayashi R, Glass KC, Nix JC, Gozani O, Appella E, Kutateladze TG. 2010. Structural insight into p53 recognition by the 53BP1 tandem Tudor domain. *J Mol Biol* **398:** 489–496.

Ruthenburg AJ, Wang W, Graybosch DM, Li H, Allis CD, Patel DJ, Verdine GL. 2006. Histone H3 recognition and presentation by the WDR5 module of the MLL1 complex. *Nat Struct Mol Biol* **13:** 704–712.

Ruthenburg AJ, Allis CD, Wysocka J. 2007a. Methylation of lysine 4 on histone H3: Intricacy of writing and reading a single epigenetic mark. *Mol Cell* **25:** 15–30.

Ruthenburg AJ, Li H, Patel DJ, Allis CD. 2007b. Multivalent engagement of chromatin modifications by linked binding modules. *Nat Rev Mol Cell Biol* **8:** 983–994.

Ruthenburg AJ, Li H, Milne TA, Dewell S, McGinty RK, Yuen M, Ueberheide B, Dou Y, Muir TW, Patel DJ, et al. 2011. Recognition of a mononucleosomal histone modification pattern by BPTF via multivalent interactions. *Cell* **145:** 692–706.

Sampath SC, Marazzi I, Yap KL, Krutchinsky AN, Mecklenbrauker I, Viale A, Rudensky E, Zhou MM, Chait BT, Tarakhovsky A. 2007.

Methylation of a histone mimic within the histone methyltransferase G9a regulates protein complex assembly. *Mol Cell* **27:** 596–608.

Sanchez R, Zhou MM. 2011. The PHD finger: A versatile epigenome reader. *Trends Biochem Sci* **36:** 364–372.

Santos-Rosa H, Schneider R, Bannister AJ, Sherriff J, Bernstein BE, Emre NC, Schreiber SL, Mellor J, Kouzarides T. 2002. Active genes are tri-methylated at K4 of histone H3. *Nature* **419:** 407–411.

Sarg B, Helliger W, Talasz H, Koutzamani E, Lindner HH. 2004. Histone H4 hyperacetylation precludes histone H4 lysine 20 trimethylation. *J Biol Chem* **279:** 53458–53464.

Sasai N, Nakao M, Defossez PA. 2010. Sequence-specific recognition of methylated DNA by human zinc-finger proteins. *Nucleic Acids Res* **38:** 5015–5022.

Sathyamurthy A, Allen MD, Murzin AG, Bycroft M. 2003. Crystal structure of the malignant brain tumor (MBT) repeats in Sex Comb on Midleg-like 2 (SCML2). *J Biol Chem* **278:** 46968–46973.

* Schaefer U. 2014. Pharmacological inhibition of bromodomain-containing proteins in inflammation. *Cold Spring Harb Perspect Biol* **6:** a018671.

Schalch T, Job G, Noffsinger VJ, Shanker S, Kuscu C, Joshua-Tor L, Partridge JF. 2009. High-affinity binding of Chp1 chromodomain to K9 methylated histone H3 is required to establish centromeric heterochromatin. *Mol Cell* **34:** 36–46.

Schuetz A, Allali-Hassani A, Martin F, Loppnau P, Vedadi M, Bochkarev A, Plotnikov AN, Arrowsmith CH, Min J. 2006. Structural basis for molecular recognition and presentation of histone H3 by WDR5. *EMBO J* **25:** 4245–4252.

Selenko P, Sprangers R, Stier G, Buhler D, Fischer U, Sattler M. 2001. SMN tudor domain structure and its interaction with the Sm proteins. *Nat Struct Biol* **8:** 27–31.

Serganov A, Patel DJ. 2012. Metabolite recognition principles and molecular mechanisms underlying riboswitch function. *Annu Rev Biophys* **41:** 343–370.

* Seto E, Yoshida M. 2014. Erasers of histone acetylation: The histone deacetylase enzymes. *Cold Spring Harb Perspect Biol* **6:** a018713.

Sharif J, Muto M, Takebayashi S, Suetake I, Iwamatsu A, Endo TA, Shinga J, Mizutani-Koseki Y, Toyoda T, Okamura K, et al. 2007. The SRA protein Np95 mediates epigenetic inheritance by recruiting Dnmt1 to methylated DNA. *Nature* **450:** 908–912.

Shi X, Hong T, Walter KL, Ewalt M, Michishita E, Hung T, Carney D, Pena P, Lan F, Kaadige MR, et al. 2006. ING2 PHD domain links histone H3 lysine 4 methylation to active gene repression. *Nature* **442:** 96–99.

Shilatifard A. 2006. Chromatin modifications by methylation and ubiquitination: Implications in the regulation of gene expression. *Annu Rev Biochem* **75:** 243–269.

Shogren-Knaak M, Ishii H, Sun JM, Pazin MJ, Davie JR, Peterson CL. 2006. Histone H4-K16 acetylation controls chromatin structure and protein interactions. *Science* **311:** 844–847.

Simon MD, Chu F, Racki LR, de la Cruz CC, Burlingame AL, Panning B, Narlikar GJ, Shokat KM. 2007. The site-specific installation of methyl-lysine analogs into recombinant histones. *Cell* **128:** 1003–1012.

Sims RJ 3rd, Chen CF, Santos-Rosa H, Kouzarides T, Patel SS, Reinberg D. 2005. Human but not yeast CHD1 binds directly and selectively to histone H3 methylated at lysine 4 via its tandem chromodomains. *J Biol Chem* **280:** 41789–41792.

Sims RJ 3rd, Rojas LA, Beck D, Bonasio R, Schuller R, Drury WJ 3rd, Eick D, Reinberg D. 2011. The C-terminal domain of RNA polymerase II is modified by site-specific methylation. *Science* **332:** 99–103.

Siomi MC, Mannen T, Siomi H. 2010. How does the royal family of Tudor rule the PIWI-interacting RNA pathway? *Genes Dev* **24:** 636–646.

Slater LM, Allen MD, Bycroft M. 2003. Structural variation in PWWP domains. *J Mol Biol* **330:** 571–576.

Song JJ, Kingston RE. 2008. WDR5 interacts with mixed lineage leukemia (MLL) protein via the histone H3-binding pocket. *J Biol Chem* **283:** 35258–35264.

Song JJ, Garlick JD, Kingston RE. 2008. Structural basis of histone H4 recognition by p55. *Genes Dev* **22:** 1313–1318.

Song J, Rechkoblit O, Bestor TH, Patel DJ. 2011. Structure of DNMT1-DNA complex reveals a role for autoinhibition in maintenance DNA methylation. *Science* **331:** 1036–1040.

Song J, Teplova M, Ishibe-Murakami S, Patel DJ. 2012. Structure-based mechanistic insights into DNMT1-mediated maintenance DNA methylation. *Science* **335:** 709–712.

Sprangers R, Groves MR, Sinning I, Sattler M. 2003. High-resolution X-ray and NMR structures of the SMN Tudor domain: Conformational variation in the binding site for symmetrically dimethylated arginine residues. *J Mol Biol* **327:** 507–520.

Strahl BD, Allis CD. 2000. The language of covalent histone modifications. *Nature* **403:** 41–45.

Sukackaite R, Grazulis S, Tamulaitis G, Siksnys V. 2012. The recognition domain of the methyl-specific endonuclease McrBC flips out 5-methylcytosine. *Nucleic Acids Res* **40:** 7552–7562.

Sun ZW, Allis CD. 2002. Ubiquitination of histone H2B regulates H3 methylation and gene silencing in yeast. *Nature* **418:** 104–108.

Sun B, Hong J, Zhang P, Dong X, Shen X, Lin D, Ding J. 2008. Molecular basis of the interaction of *Saccharomyces cerevisiae* Eaf3 chromo domain with methylated H3K36. *J Biol Chem* **283:** 36504–36512.

Tahiliani M, Koh KP, Shen Y, Pastor WA, Bandukwala H, Brudno Y, Agarwal S, Iyer LM, Liu DR, Aravind L, et al. 2009. Conversion of 5-methylcytosine to 5-hydroxymethylcytosine in mammalian DNA by MLL partner TET1. *Science* **324:** 930–935.

Tan S, Davey CA. 2011. Nucleosome structural studies. *Curr Opin Struct Biol* **21:** 128–136.

Taverna SD, Ilin S, Rogers RS, Tanny JC, Lavender H, Li H, Baker L, Boyle J, Blair LP, Chait BT, et al. 2006. Yng1 PHD finger binding to H3 trimethylated at K4 promotes NuA3 HAT activity at K14 of H3 and transcription at a subset of targeted ORFs. *Mol Cell* **24:** 785–796.

Taverna SD, Li H, Ruthenburg AJ, Allis CD, Patel DJ. 2007. How chromatin-binding modules interpret histone modifications: Lessons from professional pocket pickers. *Nat Struct Mol Biol* **14:** 1025–1040.

Tripsianes K, Madl T, Machyna M, Fessas D, Englbrecht C, Fischer U, Neugebauer KM, Sattler M. 2011. Structural basis for dimethylarginine recognition by the Tudor domains of human SMN and SPF30 proteins. *Nat Struct Mol Biol* **18:** 1414–1420.

Trojer P, Reinberg D. 2008. Beyond histone methyl-lysine binding: How malignant brain tumor (MBT) protein L3MBTL1 impacts chromatin structure. *Cell Cycle* **7:** 578–585.

Trojer P, Li G, Sims RJ 3rd, Vaquero A, Kalakonda N, Boccuni P, Lee D, Erdjument-Bromage H, Tempst P, Nimer SD, et al. 2007. L3MBTL1, a histone-methylation-dependent chromatin lock. *Cell* **129:** 915–928.

Trojer P, Cao AR, Gao Z, Li Y, Zhang J, Xu X, Li G, Losson R, Erdjument-Bromage H, Tempst P, et al. 2011. L3MBTL2 protein acts in concert with PcG protein-mediated monoubiquitination of H2A to establish a repressive chromatin structure. *Mol Cell* **42:** 438–450.

Tsai MC, Manor O, Wan Y, Mosammaparast N, Wang JK, Lan F, Shi Y, Segal E, Chang HY. 2010a. Long noncoding RNA as modular scaffold of histone modification complexes. *Science* **329:** 689–693.

Tsai WW, Wang Z, Yiu TT, Akdemir KC, Xia W, Winter S, Tsai CY, Shi X, Schwarzer D, Plunkett W, et al. 2010b. TRIM24 links a non-canonical histone signature to breast cancer. *Nature* **468:** 927–932.

van Holde K, Zlatanova J. 2007. Chromatin fiber structure: Where is the problem now? *Semin Cell Dev Biol* **18:** 651–658.

Vermeulen M, Eberl HC, Matarese F, Marks H, Denissov S, Butter F, Lee KK, Olsen JV, Hyman AA, Stunnenberg HG, et al. 2010. Quantitative interaction proteomics and genome-wide profiling of epigenetic histone marks and their readers. *Cell* **142:** 967–980.

Vezzoli A, Bonadies N, Allen MD, Freund SM, Santiveri CM, Kvinlaug BT, Huntly BJ, Gottgens B, Bycroft M. 2010. Molecular basis of histone H3K36me3 recognition by the PWWP domain of Brpf1. *Nat Struct Mol Biol* **17:** 617–619.

Voigt P, Reinberg D. 2011. Histone tails: Ideal motifs for probing epigenetics through chemical biology approaches. *Chembiochem* **12:** 236–252.

Cite as *Cold Spring Harb Perspect Biol* doi: 10.1101/cshperspect.a018754

Wang Z, Patel DJ. 2011. Combinatorial readout of dual histone modifications by paired chromatin-associated modules. *J Biol Chem* **286:** 18363–18368.

Wang WK, Tereshko V, Boccuni P, MacGrogan D, Nimer SD, Patel DJ. 2003. Malignant brain tumor repeats: A three-leaved propeller architecture with ligand/peptide binding pockets. *Structure* **11:** 775–789.

Wang GG, Song J, Wang Z, Dormann HL, Casadio F, Li H, Luo JL, Patel DJ, Allis CD. 2009. Haematopoietic malignancies caused by dysregulation of a chromatin-binding PHD finger. *Nature* **459:** 847–851.

Wang M, Mok MW, Harper H, Lee WH, Min J, Knapp S, Oppermann U, Marsden B, Schapira M. 2010a. Structural genomics of histone tail recognition. *Bioinformatics* **26:** 2629–2630.

Wang Z, Song J, Milne TA, Wang GG, Li H, Allis CD, Patel DJ. 2010b. Pro isomerization in MLL1 PHD3-bromo cassette connects H3K4me readout to CyP33 and HDAC-mediated repression. *Cell* **141:** 1183–1194.

Wang C, Shen J, Yang Z, Chen P, Zhao B, Hu W, Lan W, Tong X, Wu H, Li G, et al. 2011. Structural basis for site-specific reading of unmodified R2 of histone H3 tail by UHRF1 PHD finger. *Cell Res* **21:** 1379–1382.

West LE, Roy S, Lachmi-Weiner K, Hayashi R, Shi X, Appella E, Kutateladze TG, Gozani O. 2010. The MBT repeats of L3MBTL1 link SET8-mediated p53 methylation at lysine 382 to target gene repression. *J Biol Chem* **285:** 37725–37732.

Williams K, Christensen J, Helin K. 2012. DNA methylation: TET proteins-guardians of CpG islands? *EMBO Rep* **13:** 28–35.

Wolfe SA, Nekludova L, Pabo CO. 2000. DNA recognition by Cys2His2 zinc finger proteins. *Annu Rev Biophys Biomol Struct* **29:** 183–212.

Wu H, Zhang Y. 2011. Mechanisms and functions of Tet protein-mediated 5-methylcytosine oxidation. *Genes Dev* **25:** 2436–2452.

Wu H, D'Alessio AC, Ito S, Xia K, Wang Z, Cui K, Zhao K, Sun YE, Zhang Y. 2011a. Dual functions of Tet1 in transcriptional regulation in mouse embryonic stem cells. *Nature* **473:** 389–393.

Wu L, Zee BM, Wang Y, Garcia BA, Dou Y. 2011b. The RING finger protein MSL2 in the MOF complex is an E3 ubiquitin ligase for H2B K34 and is involved in crosstalk with H3 K4 and K79 methylation. *Mol Cell* **43:** 132–144.

Wysocka J, Swigut T, Milne TA, Dou Y, Zhang X, Burlingame AL, Roeder RG, Brivanlou AH, Allis CD. 2005. WDR5 associates with histone H3 methylated at K4 and is essential for H3 K4 methylation and vertebrate development. *Cell* **121:** 859–872.

Wysocka J, Swigut T, Xiao H, Milne TA, Kwon SY, Landry J, Kauer M, Tackett AJ, Chait BT, Badenhorst P, et al. 2006. A PHD finger of NURF couples histone H3 lysine 4 trimethylation with chromatin remodelling. *Nature* **442:** 86–90.

Xi Q, Wang Z, Zaromytidou AI, Zhang XH, Chow-Tsang LF, Liu JX, Kim H, Barlas A, Manova-Todorova K, Kaartinen V, et al. 2011. A poised chromatin platform for TGF-β access to master regulators. *Cell* **147:** 1511–1524.

Xia ZB, Anderson M, Diaz MO, Zeleznik-Le NJ. 2003. MLL repression domain interacts with histone deacetylases, the polycomb group proteins HPC2 and BMI-1, and the corepressor C-terminal-binding protein. *Proc Natl Acad Sci* **100:** 8342–8347.

Xiao S, Xie D, Cao X, Yu P, Xing X, Chen CC, Musselman M, Xie M, West FD, Lewin HA, et al. 2012. Comparative epigenomic annotation of regulatory DNA. *Cell* **149:** 1381–1392.

Xie S, Jakoncic J, Qian C. 2012. UHRF1 double tudor domain and the adjacent PHD finger act together to recognize K9me3-containing histone H3 tail. *J Mol Biol* **415:** 318–328.

Xu C, Cui G, Botuyan MV, Mer G. 2008. Structural basis for the recognition of methylated histone H3K36 by the Eaf3 subunit of histone deacetylase complex Rpd3S. *Structure* **16:** 1740–1750.

Xu Y, Wu F, Tan L, Kong L, Xiong L, Deng J, Barbera AJ, Zheng L, Zhang H, Huang S, et al. 2011. Genome-wide regulation of 5hmC, 5mC, and gene expression by Tet1 hydroxylase in mouse embryonic stem cells. *Mol Cell* **42:** 451–464.

Yamada K, Frouws TD, Angst B, Fitzgerald DJ, DeLuca C, Schimmele K, Sargent DF, Richmond TJ. 2011. Structure and mechanism of the chromatin remodelling factor ISW1a. *Nature* **472:** 448–453.

Yang N, Xu RM. 2012. Structure and function of the BAH domain in chromatin biology. *Crit Rev Biochem Mol Biol* **48:** 211–221.

Yang Y, Lu Y, Espejo A, Wu J, Xu W, Liang S, Bedford MT. 2010. TDRD3 is an effector molecule for arginine-methylated histone marks. *Mol Cell* **40:** 1016–1023.

Yang N, Wang W, Wang Y, Wang M, Zhao Q, Rao Z, Zhu B, Xu RM. 2012. Distinct mode of methylated lysine-4 of histone H3 recognition by tandem tudor-like domains of Spindlin1. *Proc Natl Acad Sci* **109:** 17954–17959.

Yap KL, Zhou MM. 2010. Keeping it in the family: Diverse histone recognition by conserved structural folds. *Crit Rev Biochem Mol Biol* **45:** 488–505.

Yap KL, Zhou MM. 2011. Structure and mechanisms of lysine methylation recognition by the chromodomain in gene transcription. *Biochemistry* **50:** 1966–1980.

Yuan CC, Matthews AG, Jin Y, Chen CF, Chapman BA, Ohsumi TK, Glass KC, Kutateladze TG, Borowsky ML, Struhl K, et al. 2012. Histone H3R2 symmetric dimethylation and histone H3K4 trimethylation are tightly correlated in eukaryotic genomes. *Cell Rep* **1:** 83–90.

Zeng L, Yap KL, Ivanov AV, Wang X, Mujtaba S, Plotnikova O, Rauscher FJ 3rd, Zhou MM. 2008. Structural insights into human KAP1 PHD finger-bromodomain and its role in gene silencing. *Nat Struct Mol Biol* **15:** 626–633.

Zeng L, Zhang Q, Li S, Plotnikov AN, Walsh MJ, Zhou MM. 2010. Mechanism and regulation of acetylated histone binding by the tandem PHD finger of DPF3b. *Nature* **466:** 258–262.

Zhang Z, Hayashi MK, Merkel O, Stillman B, Xu RM. 2002. Structure and function of the BAH-containing domain of Orc1p in epigenetic silencing. *EMBO J* **21:** 4600–4611.

Zhang K, Lin W, Latham JA, Riefler GM, Schumacher JM, Chan C, Tatchell K, Hawke DH, Kobayashi R, Dent SY. 2005. The Set1 methyltransferase opposes Ipl1 aurora kinase functions in chromosome segregation. *Cell* **122:** 723–734.

Zhang P, Du J, Sun B, Dong X, Xu G, Zhou J, Huang Q, Liu Q, Hao Q, Ding J. 2006. Structure of human MRG15 chromo domain and its binding to Lys36-methylated histone H3. *Nucleic Acids Res* **34:** 6621–6628.

Zhang L, Lu X, Lu J, Liang H, Dai Q, Xu GL, Luo C, Jiang H, He C. 2012. Thymine DNA glycosylase specifically recognizes 5-carboxylcytosine-modified DNA. *Nat Chem Biol* **8:** 328–330.

* Zhao Y, Garcia BA. 2014. Comprehensive catalog of currently documented histone modifications. *Cold Spring Harb Perspect Biol* doi: 10.1101/cshperspect.a025064.

Zhou VW, Goren A, Bernstein BE. 2011. Charting histone modifications and the functional organization of mammalian genomes. *Nat Rev Genet* **12:** 7–18.

* Zoghbi HY, Beaudet AL. 2014. Epigenetics and human disease. *Cold Spring Harb Perspect Biol* doi: 10.1101/cshperspect.a019497.

CHAPTER 8

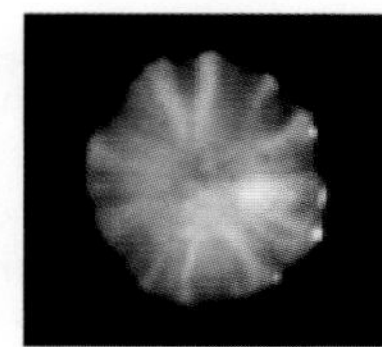
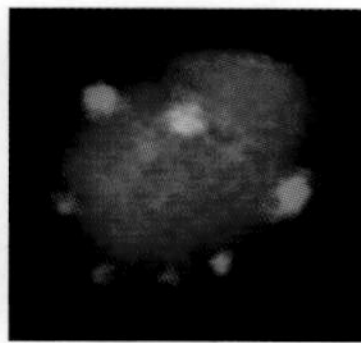

Epigenetics in *Saccharomyces cerevisiae*

Michael Grunstein[1] and Susan M. Gasser[2]

[1]University of California, Los Angeles, Los Angeles, California 90095; [2]Friedrich Miescher Institute for Biomedical Research, 4058 Basel, Switzerland

Correspondence: susan.gasser@fmi.ch and mg@mbi.ucla.edu

SUMMARY

Saccharomyces cerevisiae provides a well-studied model system for heritable silent chromatin, in which a nonhistone protein complex—the SIR complex—represses genes by spreading in a sequence-independent manner, much like heterochromatin in higher eukaryotes. The ability to study mutations in histones and to screen genome-wide for mutations that impair silencing has yielded an unparalleled depth of detail about this system. Recent advances in the biochemistry and structural biology of the SIR-chromatin complex bring us much closer to a molecular understanding of how Sir3 selectively recognizes the deacetylated histone H4 tail and demethylated histone H3 core. The existence of appropriate mutants has also shown how components of the silencing machinery affect physiological processes beyond transcriptional repression.

Outline

Cite as *Cold Spring Harb Perspect Biol* doi: 10.1101/cshperspect.a017491

OVERVIEW

The fraction of chromatin in a eukaryotic nucleus that bears active genes is termed euchromatin. This chromatin condenses in mitosis to allow chromosomal segregation and decondenses in interphase of the cell cycle to allow transcription to occur. However, some chromosomal domains were observed by cytological criteria to remain condensed in interphase, and this constitutively compacted chromatin was called heterochromatin. With the development of new techniques, molecular rather than cytological features have been used to define this portion of the genome, and heterochromatin, which is often found at centromeres and telomeres, was shown to contain many thousands of simple repeat sequences, particularly in higher eukaryotic organisms. The repeat-rich genomic DNA tends to replicate late in S phase of the cell cycle, is found clustered at the nuclear periphery or near the nucleolus, and is resistant to nuclease attack. Importantly, the characteristic chromatin structure that is formed on repeat DNA tends to spread and repress nearby genes. In the case of the fruit fly locus *white*, a gene that determines red eye color, epigenetic repression yields a red and white sectored eye through a phenomenon called position effect variegation (PEV). Mechanistically, PEV in flies reflects the recognition of methylated histone H3K9 by heterochromatin protein 1 (HP1), which can spread along the chromosomal arm. In *Saccharomyces cerevisiae*, also known as budding yeast, a distinct mechanism of heterochromatin formation has evolved, yet it achieves a very similar result.

S. cerevisiae is a microorganism commonly used for making beer and baking bread. However, unlike bacteria, it is a eukaryote. The chromosomes of budding yeast, like those of more complex eukaryotes, are bound by histones, enclosed in a nucleus and replicated from multiple origins during S phase. Still, the yeast genome is tiny with only 14 megabase pairs of genomic DNA divided among its 16 chromosomes, some not much larger than a bacteriophage genome. There are approximately 6000 genes in the yeast genome, closely packed along chromosomal arms, generally with less than 2 kb spacing between them. The vast majority of yeast genes are in an open chromatin state, meaning that they are either actively transcribed or can be rapidly induced. This, coupled with a very limited amount of simple repeat DNA, makes the detection of heterochromatin by cytological techniques very difficult in yeast.

Nonetheless, budding yeast has distinct heterochromatin-like regions adjacent to all 32 telomeres and at two silent mating loci on chromosome III (Chr III), shown using molecular tools. Transcriptional repression at telomeres and the silent mating loci can spread into adjacent DNA and repression of the silent mating loci is essential for maintaining a mating-competent haploid state. Both the subtelomeric regions and the silent mating type loci repress integrated reporter genes in a position-dependent, epigenetic manner; they replicate late in S phase and are present at the nuclear periphery. Thus, these loci bear most of the functional characteristics of heterochromatin, without having cytologically visible condensation in interphase. By exploiting the advantages afforded by the small genome of yeast and its powerful genetic and biochemical tools, many basic principles of chromatin-mediated repression that are relevant to heterochromatin in more complex organisms have been discovered. Nonetheless, silent chromatin in budding yeast is dependent on a unique set of nonhistone proteins that do not deposit nor recognize histone H3 lysine 9 methylation.

Cite as *Cold Spring Harb Perspect Biol* doi: 10.1101/cshperspect.a017491

1 THE GENETIC AND MOLECULAR TOOLS OF YEAST

Yeast provides a flexible and rapid genetic system for studying cellular events. With an approximate generation time of 90 min, colonies containing millions of cells are produced after just 2 d of growth. In addition, yeast can propagate in both haploid and diploid forms, greatly facilitating genetic analysis. Like bacteria, haploid yeast cells can be mutated to produce specific nutritional requirements or auxotrophic genetic phenotypes, and recessive lethal mutations can either be maintained in haploids as conditional lethal alleles (e.g., temperature-sensitive mutants), or in heterozygotic diploids, which carry both wild-type and mutant alleles.

Extremely useful is the efficient homologous recombination system of budding yeast, which allows the alteration of any chosen chromosomal sequence at will. In addition, portions of chromosomes can be manipulated and reintroduced on plasmids that are stably maintained through cell division, thanks to short sequences that provide centromere and replication origin function. Large linear plasmids, or minichromosomes, which carry telomeric repeats to cap their ends, also propagate stably in yeast.

Yeast also has a unique advantage in the genetic analysis of histones and their roles in gene regulation: Unlike mammalian cells, which have as many as 60–70 copies of the core histone coding genes (H2A, H2B, H3, and H4), yeast contains only two copies of each of these genes. Because these two copies are functionally redundant, this has enabled the production of cells containing single histone gene copies. Deletion analysis or mutation of defined histone amino acids in these genes has uncovered specific roles for histone residues in heterochromatin and other cellular functions. By searching for suppressors of these mutant phenotypes it has also been possible to identify heterochromatin proteins that interact with the histone sites in question. The ease of generating histone mutants in yeast has also led to the systematic mutagenesis of most amino acid residues in each of the core histones (http://www.ncbi.nlm.nih.gov/pmc/articles/PMC2666297/) and analysis of their effects on genome function (Huang et al. 2009).

Budding yeast also provides powerful cellular readouts for epigenetic gene regulation, conceptually similar to PEV in flies, in which the *white* gene provides a visible screen for variegated gene expression (see Ch. 12 [Elgin and Reuter 2013] for more detail). A parallel phenomenon called telomere position effect (TPE) occurs near yeast telomeres. The study of TPE has been aided analogously by the use of the *URA3* and *ADE2* reporter genes (Fig. 1). The *URA3* gene product is necessary for pyrimidine biosynthesis, and cells that do not express the Ura3 protein cannot grow on synthetic media lacking uracil. In addition, *URA3* allows for a counter-selection against expression in the presence of 5-fluoroorotic acid (5-FOA), because the Ura3 gene product converts 5-FOA to 5-fluorouracil, an inhibitor of DNA synthesis that causes cell death. Thus, when *URA3* is integrated near heterochromatin and the gene is repressed in some but not all cells, only the cells that silence *URA3* are able to grow in the presence of 5-FOA. Conversely, if the strain lacks the strong activator of *URA3*, Ppr1, positive selection is possible; only *URA3*-expressing cells can grow on uracil-deficient plates. Thus, the efficiency of *URA3* repression/expression can be scored accurately in serial dilution assays (over a $1-10^6$-fold range) on plates that either contain 5-FOA, or lack uracil (Fig. 1A). Because 5-FOA is mutagenic and puts strong pressure on cells to repress the reporter gene, some conditions or yeast backgrounds favor the use of uracil-drop out plates over counter-selection on 5-FOA.

Another useful assay for epigenetic repression is based on the insertion of the reporter gene *ADE2* near heterochromatin. When *ADE2* is repressed, a precursor in adenine biosynthesis accumulates, turning the cell red. When *ADE2* is expressed, cells are white (Fig. 1B). The epigenetic nature of repression of a subtelomeric *ADE2* gene is visible within a single colony of genetically identical cells because the variegated expression of *ADE2* generates white sectors in a red colony background, attributable to *ADE2* repression (Fig. 1B). This reporter avoids the stress of selective pressure against cells that fail to repress *ADE2*, and thus the sectored phenotype illustrates both the switching rate and inheritance of the epigenetic state through mitotic division, much like the sectored *white* phenotype in the *Drosophila melanogaster* eye.

Combined with these genetic approaches, biochemical techniques for mapping epigenetic modifiers are readily applied to yeast. Large yeast cultures can be grown either synchronously or asynchronously. A battery of molecular tools, which includes transcriptome analysis and chromatin immunoprecipitation (ChIP), can be combined with multiplexed next generation sequencing for efficient whole-genome coverage. Combining this with proteomic approaches that map protein networks enables a comprehensive and quantitative comparison of gene expression, transcription factor binding, histone modification, and protein–protein interactions. Finally, a technology developed in yeast, called chromosome conformation capture, scores protein-mediated contacts between unlinked chromosomal domains to monitor long-range chromatin interactions (for more detail, see Ch. 19 [Dekker and Misteli 2014]). With this broad range of tools, scientists have explored the mechanisms that regulate both the establishment of heterochromatin and its physiological roles in budding yeast for more than 30 years. Before describing

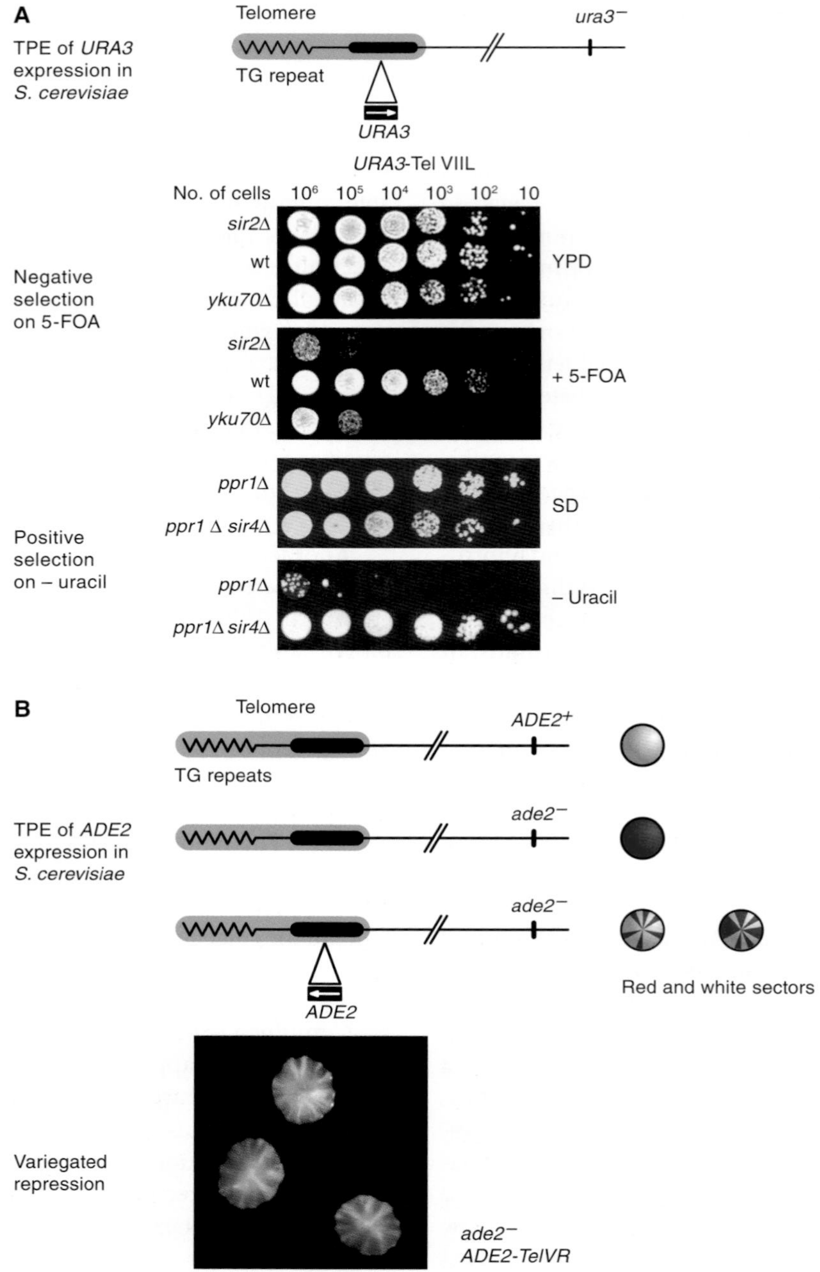

Figure 1. Silencing and TPE in yeast. (*A*) The *URA3* gene, inserted near the telomeric simple TG-rich repeat at the left arm of Chr VII, is silenced by telomeric heterochromatin in this yeast strain. In normal rich media (YPD) no growth difference can be detected between wild-type (wt) cells that repress the subtelomeric *URA3* gene, and silencing mutants that lose telomeric heterochromatin and express *URA3*. In media containing 5-FOA (*middle* panel), on the other hand, cells that repress *URA3* (e.g., wt cells) can grow, whereas cells that express it (*sir2*Δ and *yku70*Δ) cannot, because the *URA3* gene product converts 5-FOA to the toxic intermediate 5-flourouracil. The serial dilution/drop assay allows detection of silencing in as few as 1 in 10^6 cells. In cells deleted for the *URA3* activator, Ppr1 (*ppr1*Δ), one can screen for repression by plating on synthetic dextrose (SD) medium, lacking uracil. In this case, silencing the gene inhibits colony growth. (*B*) Cells containing the wild-type *ADE2* gene produce a colony that is white, whereas those containing mutant *ade2* appear red, because of the accumulation of a reddish intermediate in adenine biosynthesis. When the *ADE2* gene is inserted near the telomere at the right arm of Chr V it is silenced in an epigenetic manner. The silent *ADE2* state and the active *ADE2* state in genetically identical cells are both inherited creating red and white sectors in a colony.

 Cite as *Cold Spring Harb Perspect Biol* doi: 10.1101/cshperspect.a017491

these discoveries, however, we will first review the life cycle of yeast in more detail.

2 THE LIFE CYCLE OF YEAST

S. cerevisiae multiplies through mitotic division in either a haploid or a diploid state by producing a bud that enlarges and eventually separates from the mother cell (Fig. 2A). This is why it is called budding yeast. Haploid yeast cells can mate with each other (i.e., conjugate) because they exist in one of two mating types, termed **a** or α, reminiscent of the two sexes in mammals. Yeast cells of each mating type produce a distinct pheromone that attracts the cells of the opposite mating type: **a** cells produce a peptide of 12 amino acids (aa) called **a** factor, which binds to a membrane spanning **a**-factor receptor on the surface of an α cell. Conversely, α cells produce a 13 aa peptide (α-factor) that binds to the α-factor-receptor on the surface of **a** cells. These interactions result in the arrest of the two cell types in mid-to-late G_1 phase of the cell cycle. The arrested cells assume "shmoo"-like shapes (named after the pear-shaped Al Capp cartoon character; Fig. 2B). Shmoos of opposite mating type fuse at their tips to produce an **a**/α diploid cell.

In diploid cells the mating response is repressed, and cells propagate vegetatively (i.e., by mitotic division) unless they are exposed to starvation conditions. Nitrogen starvation induces a meiotic program in diploids that provokes the formation of an ascus containing four spores, two of each mating type. When nutrient levels are restored, these haploid spores grow into **a** or α cells that are again capable of mating to form a diploid, starting the life cycle over again.

Although haploid yeast cells in the laboratory are usually genetically constructed to be stable α or **a** cells, yeast in the wild switch their mating type nearly every cell cycle (Fig. 3A). Mating type switching is provoked by an endogenous cell-cycle-regulated endonuclease activity (HO) that induces a site-specific double-strand break at the *MAT* locus. A gene conversion event (in which the donor DNA sequence remains unchanged, but the recipient DNA is altered) transposes the opposite mating type information from one of two constitutively silent donor loci, *HML*α or *HMR***a**, to the *MAT* locus, in which the mating type–determining genes, **a**1 and **a**2, or α1 and α2, are expressed. Strains capable of mating type switching are called homothallic. This name reflects the fact that a single vegetative *MAT***a** cell can produce *MAT*α progeny, and vice versa, allowing offspring to mate with each other.

In the laboratory it is useful to have strains with stable mating types, and thus laboratory yeast generally contain a mutant HO endonuclease gene (*ho*$^-$). These cells fail to induce a double-strand cleavage at the *MAT* locus, and therefore cannot switch mating type. These heterothallic cells are stable haploid strains of either **a** or α mating type, unless placed in proximity of cells of the opposite mating type, in which case the two haploid cell types will mate to form a diploid.

Importantly, the two stable haploid cell types do not require silencing for viability, yet they must repress the genes at the homothallic mating type loci, *HML* and *HMR*, to retain their ability to mate. If repression fails and cells express both sets of mating type genes at once, then a haploid will behave as if it were diploid, suppressing mating competence and generating a sterile haploid strain (Fig. 3B). The mechanisms that repress the two homothallic mating type loci, *HML* and *HMR*, have become a classic system for the study of heterochromatin-mediated repression.

3 YEAST HETEROCHROMATIN IS PRESENT AT THE SILENT *HM* MATING LOCI AND AT TELOMERES

The three mating type loci, *HML*α, *MAT*, and *HMR***a**, are located on one small chromosome, Chr III, and contain the information that determines α or **a** mating type in yeast. The silent loci, *HML*α (~12 kb from the left telomere) and *HMR***a** (~23 kb from the right telomere), are situated between short DNA elements called E and I silencers (Fig. 3B,C). In a wild-type cell, the silent cassettes are active only once copied and integrated into the *MAT* locus, which lacks silencer elements. The transfer of *HML*α information into *MAT***a** results in an α mating type (*MAT*α) cell, whereas the transfer of *HMR***a** information into *MAT* results in the **a** mating type (*MAT***a**) (Fig. 3B). This shows that the promoters and genes at the *HM* loci are completely intact, and remain repressed because of their position between the E and I silencers. Deletion of the flanking silencer sequences indeed allows expression of the silent information, generating a nonmating state.

By scoring for haploid sterility, mutations that impair silencing at the *HM* loci were isolated (Rine et al. 1979). This allowed identification of the silent information regulatory proteins, Sir1, Sir2, Sir3, and Sir4, as being essential for the full repression of *HM* loci (Rine and Herskowitz 1987). Mutations in *sir2*, *sir3*, or *sir4* caused a complete loss of mating, which is attributable to a loss of *HM* repression. In *sir1* mutants, only a fraction of *MAT***a** cells were unable to mate. Taking advantage of the partial phenotype of *sir1* deficient cells, it could be shown that the two alternative states (mating and nonmating) are heritable through successive cell divisions in genetically identical cells (Pillus and Rine 1989). This provided a clear demonstration that mating type repression displays the hallmark characteristic of epigenetically controlled repression. Later genetic studies showed that the amino termini of histones H3 and H4,

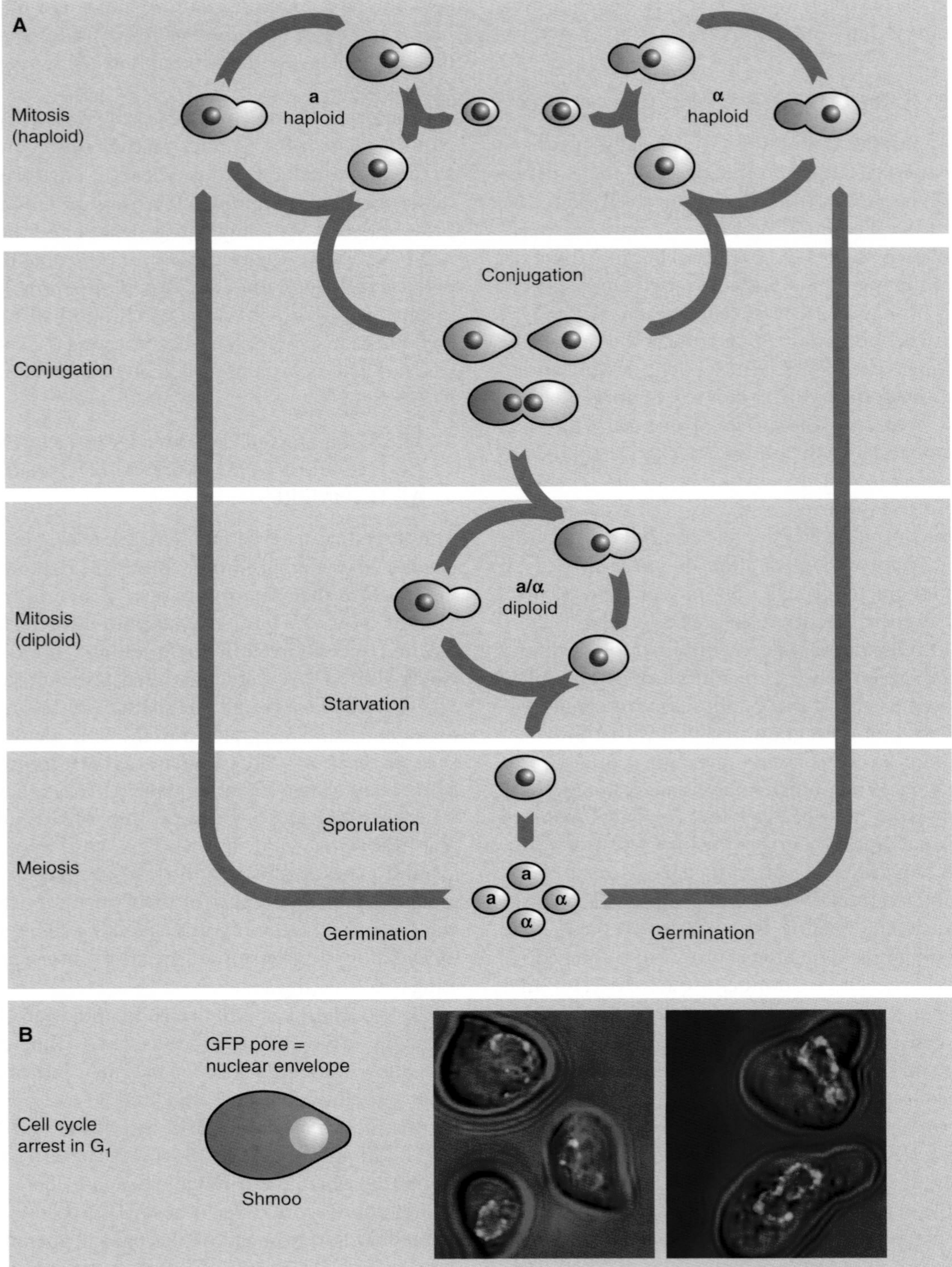

Figure 2. The life cycle of budding yeast. (*A*) Yeast cells divide mitotically in both haploid and diploid forms. Sporulation is induced in a diploid by starvation, whereas mating occurs spontaneously when haploids of opposite mating type are in the vicinity of each other. This occurs by pheromone secretion, which arrests the cell cycle in G_1 of a cell of the opposite mating type, and after sufficient exposure to pheromone the mating pathway is induced. The diploid state represses the mating pathway. (*B*) In response to pheromone, haploid cells distort toward cells of the opposite mating type. These are called shmoos. The nuclear envelope is shown in green fluorescence, showing distortions of the nucleus.

Cite as *Cold Spring Harb Perspect Biol* doi: 10.1101/cshperspect.a017491

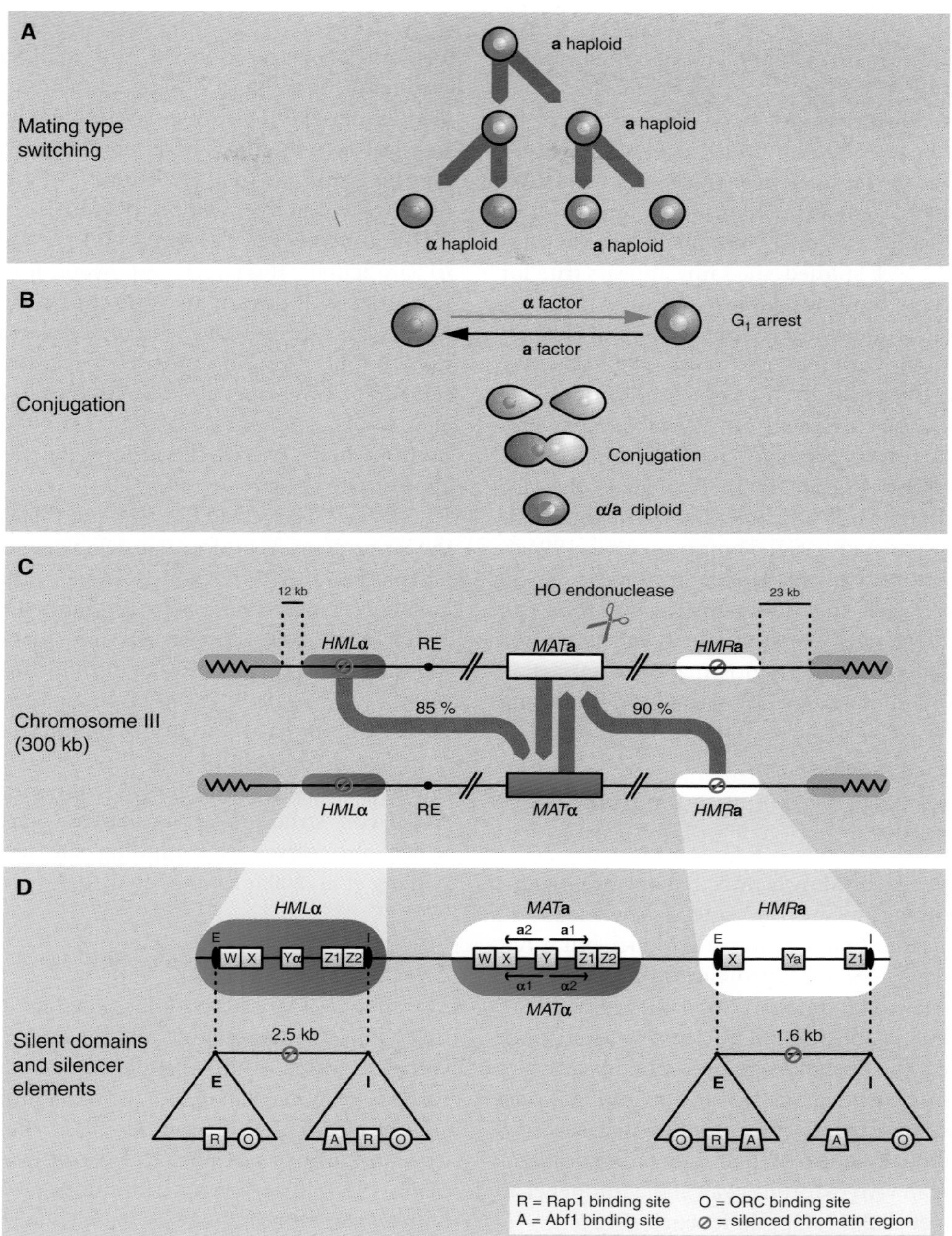

Figure 3. Mating type switching in yeast. (*A*) Homothallic yeast strains are able to switch mating type after one division cycle. The switch occurs before DNA replication so that both mother and daughter cells assume the new mating type. (*B*) In a wild-type population of yeast, this allows rapid conjugation between daughter cells to form a diploid. (*C*) The position of the silent and expressed mating type loci on Chr III are shown here. The active *MAT* locus is able to switch through gene conversion roughly once per cell cycle because of cleavage by the HO endonuclease. The percentages indicated show the frequency with which the gene conversion event replaced the *MAT* locus with the opposite mating type information. The directionality of switching is guaranteed by the recombination enhancer on the left arm of Chr III. (*D*) Repression at the silent mating type loci *HMR* and *HML* is mediated by two silencer DNA elements that flank the silent genes. These silencers are termed E (for essential) or I (for important) (Brand et al. 1987) and provide binding sites for Rap1 (R), Abf1 (A), and origin recognition complex (O). Artificial silencers can be created using various combinations of the redundant binding sites, although their efficiency is less than that of the native silencers. *HMLα* and *HMR***a** are 12 kb and 23 kb, respectively, from the telomeres of Chr III. Telomeric heterochromatin domains at Chr III are silenced independently from the *HM* loci in a process that is initiated at the telomeres through binding sites for Rap1.

repressor activator protein 1 (Rap1), and the origin recognition complex (ORC) are also components of silent mating locus heterochromatin (reviewed by Rusche et al. 2003). These latter two DNA binding factors have other essential functions in the nucleus, namely, the regulation of ribosomal protein gene expression or the initiation of DNA replication, and, thus, only "moonlight" as corepressors. Although less well studied, the same appears true for Abf1, a third silencer binding factor.

A similar position-dependent repression occurs immediately adjacent to the yeast telomeric repeat DNA ($C_{1-3}A$/TG_{1-3}) found at the ends of all yeast chromosomes. As mentioned above, the variegated but heritable repression of subtelomeric reporter genes such as *URA3* and *ADE2* is called TPE (Gottschling et al. 1990). TPE shares the *HM* requirement for Rap1, Sir2, Sir3, Sir4, and the histone amino termini (Aparicio et al. 1991; Thompson et al. 1994a), and the repression mechanisms have proven to be closely related. However, given that subtelomeric reporters can switch at detectable rates between silent and expressed states, unlike *HM* loci, telomere-proximal gene repression appears to be more similar to fly PEV (see Ch. 12 [Elgin and Reuter 2013]).

4 Sir PROTEIN STRUCTURE AND EVOLUTIONARY CONSERVATION

The known chromatin binding factors that are essential for Sir mediated silencing are Sir2, Sir3, and Sir4, whereas Sir1 enhances the efficiency of repression at *HM* loci, but is not found at telomeres. The Sir2-3-4 proteins work as a trimeric complex with 1:1:1 stoichiometry (Cubizolles et al. 2006). Both Sir3 and Sir4 are able to bind nucleosomes and DNA independently, yet the Sir holo-complex remains a trimer when bound to nucleosomes. Moreover, Sir3 and Sir4 each have homo- and heterodimerization motifs in their carboxyl termini. Mutation or deletion of their interaction domains disrupts silencing in vivo (Murphy et al. 2003; Rudner et al. 2005; Ehrentraut et al. 2011; Oppikofer et al. 2013).

Sir protein expression levels are tightly regulated and a single extra copy of the *SIR4* gene impairs repression, as does strong induction of *SIR2* (Cockell et al. 1998). On the other hand, increasing levels of Sir3 protein alone extends the spreading of Sir3 along nucleosomes, and with it, transcriptional repression (Renauld et al. 1993; Hecht et al. 1996). Balanced overexpression of all three proteins greatly improves silencing efficiency at telomeres, and allows repression of reporters located in euchromatic regions that are flanked by silencer elements, which at normal Sir protein levels are not repressed (Maillet et al. 1996).

Although Sir2, Sir3, and Sir4 are equally essential for the structural integrity of the Sir complex and therefore, for both the establishment and maintenance of silent chromatin, each has a different function. Sir2 provides a nicotinamide dinucleotide (NAD)-dependent histone deacetylase activity that is essential for repression in a wild-type background (Imai et al. 2000), whereas Sir3 and Sir4 fulfill structural roles without obvious enzymatic activities. Sir3 is a member of the AAA^+ ATPase family, which lacks ATPase activity. It is largely responsible for the specificity of Sir complex binding to nucleosomes because of its selective affinity for nucleosomes with unacetylated histone H4 lysine 16 (H4K16) and unmethylated histone H3 lysine 79 (H3K79) (Johnson et al. 1990; Altaf et al. 2007; Oppikofer et al. 2011). The sensitivity of Sir3 to these histone modifications helps restrict the binding of the Sir complex to appropriate sites.

Sir4 is the largest (152 kDa) and the least conserved of the Sir proteins, yet it forms a stable heterodimer with Sir2 (Moazed et al. 1997; Strahl-Bolsinger et al. 1997) and enhances Sir2 deacetylase activity (Tanny et al. 1999; Cubizolles et al. 2006). Structural information on the Sir4 interface indicates that most of the Sir2-interaction domain of Sir4 (aa 737–839) is buried in a pocket formed by the poorly conserved amino terminus and the carboxy-terminal catalytic domain of Sir2 (Hsu et al. 2013). The Sir3-binding domain within Sir4 is contained in the parallel coiled-coil structure at its extreme carboxyl terminus, which also serves as a binding site for other proteins (Chang et al. 2003; Rudner et al. 2005).

4.1 The "Scaffold" Role of Sir4

The affinity of Sir4 for both Sir2 and Sir3 already suggested that it might act as a scaffold for the assembly of the silencing complex. Much of this scaffolding role is achieved by the carboxy-terminal half of Sir4, which is sufficient for repression at *HM* loci (Kueng et al. 2012). Best studied is the extreme carboxy-terminal coiled-coil domain of Sir4 (aa 1257–1358), that forms a continuous parallel homodimer with two Sir3 binding sites on its outer surface (Chang et al. 2003). Mutations within the dimerization motif disrupt both Sir3 binding and silencing (Murphy et al. 2003). However, this same carboxy-terminal coiled-coil domain also binds yKu70 and Rap1, which recruit Sir4 to telomeric repeats or *HM* silencers (Moretti et al. 1994; Tsukamoto et al. 1997; Mishra and Shore 1999; Luo et al. 2002). Finally, Sir4 binds the second yKu subunit, called yKu80, through both yKu80's amino and carboxyl termini, and interacts with Esc1 (establishes silent chromatin 1) through its partitioning and anchoring domain (PAD; Sir4 aa 950–1262; Ansari and Gartenberg 1997; Andrulis et al. 2002). The yKu80 and Esc1 interactions serve to tether Sir4, and the silent chromatin it binds, to the nuclear periphery (Gartenberg et al. 2004; Taddei et al. 2004). The Sir4 PAD domain also binds ubiquitin binding protein 10 (Ubp10), a histone H2B deubiquitylase that reduces levels of H2B K123ub at telomeres (Gardner et al. 2005). Loss of H2B K123ub in turn reduces histone H3K79 methylation, which directly interferes in Sir3-nucleosome binding (Armache et al. 2011; Oppikofer et al. 2011).

The Sir4 amino terminus is necessary for silencing in subtelomeric domains, but not at *HM* loci. It appears to both regulate the efficiency of recruitment by binding yKu80 and provide linker DNA protection when bound to reconstituted nucleosomes (Kueng et al. 2012). The amino terminus is also heavily phosphorylated in vivo in a cell-cycle-dependent manner, allowing modification of repression by the cyclin-dependent kinase Cdc28 (Kueng et al. 2012). These observations highlight the role of Sir4 as a multifaceted scaffold that recruits, binds, and regulates the binding of various factors that impinge on Sir-mediated repression.

4.2 Evolutionary Conservation of Sir2 and Sir3

As mentioned above, the Sir2 deacetylase is well conserved, with homologs in all species extending from eubacteria and archaea to man (Fig. 4). Many species have multiple Sir2 family members, although some members are cytoplasmic and serve primarily to deacetylate nonhistone proteins (detailed in Ch. 5 [Seto and Yoshida 2014]). *S. cerevisiae* has five Sir2-related deacetylases (*SIR2* and *HST1–4*), but only Sir2 functions together with Sir3 and Sir4 in silent chromatin, in which it targets the amino-terminal tails of histones H3 and H4.

The Sir2 family is defined by a conserved catalytic domain in which deacetylation is coupled to the breakdown of NAD^+. The coupling of NAD hydrolysis with deacetylation produces *O*-acetyl-ADP-ribose, an intermediate that may have a function of its own (Tanner et al. 2000). The Sir2-like NAD-dependent histone deacetylases (HDACs) are implicated in transcriptional repression in many distant species such as fission yeast and flies, although they lack the other Sir proteins (reviewed in Chopra and Mishra 2005). Therefore, it is thought that an ancient Sir2 deacetylase evolved to acquire unique interaction interface with the species-specific factor Sir4, in budding yeast.

S. cerevisiae Sir2 plays an important role beyond TPE and *HM* locus silencing in that it suppresses nonreciprocal recombination in the highly repetitive rDNA locus (Gottlieb and Esposito, 1989). In this context Sir2 does not function as part of the Sir2-3-4 complex, but associates with an alternative group of factors that regulate exit from mitosis (the regulator of nucleolar silencing and telophase exit or RENT complex containing the phosphatase Cdc14, Net1/Cfi1, and the mitotic monopolin proteins, Lrs4 and Csm1; Mekhail et al. 2008; Chan et al. 2011). These proteins are also involved in the maintenance of rDNA repeat stability.

Sir3 contains several conserved domains, as the gene itself arose from an ancient version of *ORC1*, a subunit of the ORC that is found in all eukaryotes. The carboxy-terminal half of Sir3 contains a large AAA^+ ATPase domain, much like all ORC subunits and their loading protein, Cdc6 (Norris and Boeke 2010). AAA^+ domain proteins generally hydrolyze ATP to drive the assembly and disassembly of macromolecular complexes. Sir3, however, has an altered nucleotide binding pocket that precludes nucleoside binding (Ehrentraut et al. 2011). Sir3 and Orc1 further share a conserved amino-terminal BAH (bromo-adjacent homology) domain that binds nucleosomes (Armache et al. 2011), and a carboxy-terminal winged helix domain that mediates dimerization (Oppikofer et al. 2013). Interestingly, whereas the BAH domain of yeast Sir3 recognizes histone H4 deacetylated at K16, the BAH domain of evolutionarily related *Hs*ORC1 recognizes histone H4 dimethylated at K20 (H4K20me2), linking heterochromatin with origin function in vertebrates (Beck et al. 2012).

Whereas Orc1 is found in all eukaryotic species, Sir3 is only found in budding yeast species that underwent whole-genome duplication approximately 100 million years ago (Hickman et al. 2011). In very closely related budding yeast species that have both Sir3 and Orc1 orthologs, Sir3 mediates TPE and mating type repression. However, in *Kluyveromyces lactis*, which lacks Sir3, Orc1 appears to assume Sir3's role in repression. Indeed, the carboxy-terminal winged helix-turn-helix domain of Sir3 and Orc1 in these two yeasts has a similar dimerization function, which is lacking in Orc1 proteins from other species.

Sir1 and Sir4, in contrast to Sir2 and the Sir3/Orc1 family, are present only in species closely related to *S. cerevisiae* (i.e., the Saccharomycetaceae family). Intriguingly, despite the restricted evolutionary distribution of Sir1, it contains a functionally defined OIR (ORC-interacting region) domain that associates both with the Orc1 BAH domain and with Sir4 (Hickman et al. 2011).

5 SILENT CHROMATIN IS DISTINGUISHED BY A REPRESSIVE STRUCTURE THAT SPREADS THROUGH THE ENTIRE DOMAIN

Repression of gene activity in euchromatin often requires the presence of a repressive protein or complex that recognizes a specific sequence in the promoter of a gene, thus preventing the active engagement of the transcription machinery. In contrast, heterochromatic repression occurs

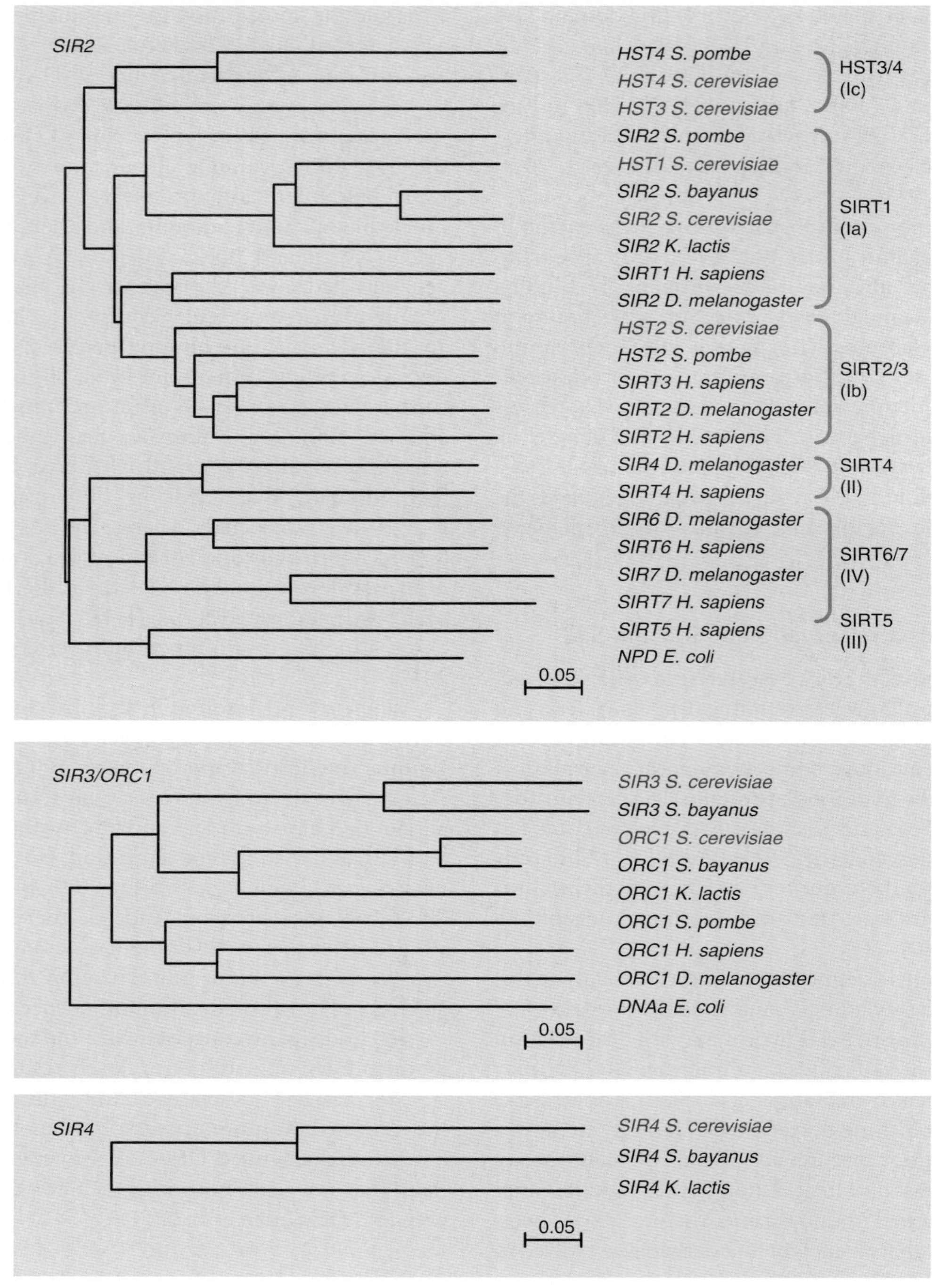

Figure 4. Sir protein family trees. Sir2 is the founding member of a large family of NAD-dependent deacetylases. The Sir2 family of proteins is highly conserved, found in multiple isoforms in organisms that range from bacteria to man. In the latter, there are both nuclear and cytoplasmic isoforms. Homologs of Sir2, Sir3, and Sir4 from *Saccharomyces bayanus*, *Kluyveromyces lactis*, *Schizosaccharomyces pombe*, *D. melanogaster*, and *Homo sapiens* were collected from UniProt and were aligned using ClustalW2 alignment. The phylogenetic tree was created using neighbor joining. Sir2 classification is according to Frye (2000). *K. lactis* has 4 *SIR2* orthologs (homologs to *S. cerevisiae* Sir2, Hst2, Hst3, and Hst4), but the *HST* homologs were omitted in the tree for clarity. For *S. bayanus*, only the Sir2 homolog is annotated to date. *S. cerevisiae* homologs are in red. Sir3 arose through a gene duplication of a gene encoding an ancient Orc1, and Sir4 is a rapidly evolving protein that is only found in related budding yeasts. The related proteins shown are not exhaustive, particularly for Sir2.

 Cite as *Cold Spring Harb Perspect Biol* doi: 10.1101/cshperspect.a017491

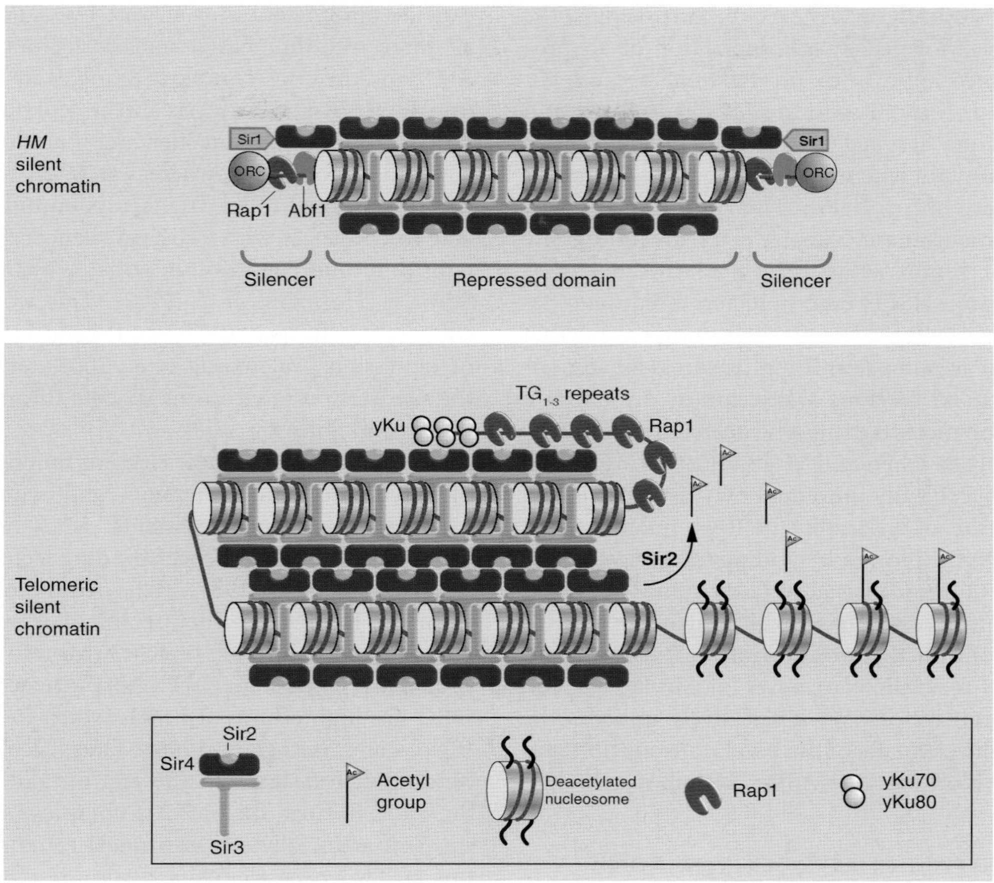

Figure 5. Model for yeast heterochromatin at telomeres and the *HM* loci. The telomere and HM silencer mechanisms for nucleating Sir complex spreading both use Rap1, Sir2, Sir3, and Sir4. Yet they differ in that telomeres also rely on yKu, whereas the *HM* silencer elements use the factors ORC, Abf1 and Sir1. Telomeric heterochromatin is thought to fold back onto itself to form a cap that protects the telomere from degradation and whose condensation and folding silences genes. In the case of *HM* heterochromatin, the repressed domain between the silencer elements consists of closely spaced nucleosomes that form a condensed structure. Both the telomeric and *HM* silent regions are inaccessible to a number of transcription factors and degradative enzymes.

through a different mechanism that is not promoter specific: Repression initiates at specific nucleation sites, yet spreads continuously throughout a domain, silencing any and all promoters in the region (Fig. 5) (Brand et al. 1985; Renauld et al. 1993). The correlation of transcriptional repression with Sir binding was confirmed by the use of ChIP, which showed that Sir2, Sir3, and Sir4 proteins interact physically with chromatin throughout the subtelomeric domain of silent chromatin and spread continuously inward from the chromosomal end (Strahl-Bolsinger et al. 1997). Evidence that this induces a repressive, less accessible chromatin structure in vivo comes from other approaches. For instance, it was shown that the DNA of silenced chromatin was not methylated efficiently in yeast cells that express a bacterial *dam* methylase, although the enzyme readily methylated sequences outside the silent region. This suggests that heterochromatin restricts access to macromolecules like *dam* methyltransferase (Gottschling, 1992). Similarly, the ~3 kb *HMR* locus in isolated nuclei is preferentially resistant to certain restriction endonucleases (Loo and Rine 1994), and nucleosomes were shown to be tightly positioned between two silencer elements creating nuclease resistant domains at silent, but not active, *HM* loci (Weiss and Simpson, 1998). The reduced accessibility of yeast silent chromatin to nucleolytic attack is also observed in vitro when Sir-nucleosome complexes are reconstituted from recombinant proteins (Martino et al. 2009).

The extent to which either yeast or metazoan heterochromatin is hypercondensed to hinder access to transcription factors sterically is less certain. Surprisingly, the repressive complex formed by the interaction of Sir proteins and histones appears to be dynamic because Sir proteins can be incorporated into *HM* silent chromatin even when cells are arrested at a stage in the cell cycle when heterochromatin assembly generally does not occur (Cheng and Gartenberg 2000). This may explain why Sir-bound heterochromatin can serve as a binding site for certain transcription factors (e.g., the heat shock transcriptional activator, HSF1) even in its repressed state (Sekinger and Gross 1999). Although such studies argue that heterochromatin does not simply hinder access for all nonhistone proteins, no obvious transcription occurs, and engaged RNA polymerases cannot be detected experimentally. Experiments by Chen and Widom (2005) argue that the step that is specifically prevented by yeast heterochromatin is formation of a complex of RNA polymerase II (RNA Pol II) with the promoter-binding transcription factors TFIIB and TFIIE. Consistently, a drop in RNA Pol II binding was seen in a system in which silencing was induced by the controlled expression of Sir3. Thus, silent yeast chromatin may allow turnover of Sir factors and some transcription factors, yet it selectively impedes the binding of specific elements of the basal transcription machinery, thereby blocking mRNA production.

6 DISTINCT STEPS IN HETEROCHROMATIN ASSEMBLY

The assembly of heterochromatin in budding yeast involves a series of molecular steps, starting with a site-specific nucleation step. This requires DNA recognition by a sequence-specific DNA binding factor. Next, heterochromatin spreads from the initiation site, limited by specific boundary mechanisms. A change in higher organization of the repressed chromatin then occurs, which is distinct from the simple binding of Sir factors. Finally, yeast silent chromatin is sequestered near the nuclear envelope, generating a subnuclear compartment that favors heterochromatin-mediated repression by promoting its duplication. Although the assembly of heterochromatin at telomeres varies in some aspects from its assembly at *HM* loci, both embody a very similar principle: the presence of specific DNA binding factors that nucleate the spread of general repressors. In both cases, the spreading requires active deacetylation by Sir2. These mechanisms are described in the following paragraphs.

6.1 *HM* Heterochromatin

The silent mating loci *HML* and *HMR* are bracketed by short DNA elements termed silencers (Fig. 3), which provide binding sites for at least two, and in most cases, all three multifunctional nuclear factors, namely Rap1, Abf1, and the ORC complex (Brand et al. 1987). The deletion of *HMR*-E, which has three recognition sites, has a much stronger effect on silencing than deletion of *HMR*-I, which has only two, whereas at *HML*, the two silencers have more equal roles. The factors bound to silencers are able to cooperate with a distant silencer through Sir proteins to promote repression, possibly by forming a looped domain (Hofmann et al. 1989). This would explain the cooperative effects of the E and I silencers on the initiation of repression (Valenzuela et al. 2008), and the effects of the silencers on nucleosome spacing throughout the locus (Weiss and Simpson 1998).

Redundancy of silencer element function is a hallmark of heterochromatic repression, and redundancy is also found within silencer elements: DNA binding sites for any two of the three silencer binding factors allow repression (Brand et al. 1987). This redundancy likely stems from the factors that they recruit. For example, Rap1 is able to recruit either Sir4 or Sir3 (Moretti et al. 1994; Luo et al. 2002; Chen et al. 2011), Abf1 interacts with Sir3, and ORC has high affinity for Sir1, which in turn binds Sir4 (Triolo and Sternglanz 1996). Thus, each of the silencer-binding factors leads to the recruitment of Sir4 and/or Sir3, and in turn, the Sir2-3-4 complex. Although targeting Sir2 artificially can also nucleate repression, none of the *HM* silencer elements nucleates repression by first recruiting Sir2. The redundancy among Rap1, Abf1, and ORC (at silencers) or the Ku heterodimer (at telomeres), thus reflects the ability of each nucleator to bind Sir3 or Sir4 and, in turn, to recruit the entire Sir complex to the silencer or telomeric repeat element. Thus, sequence-specific recognition is at the heart of position-dependent repression.

Sir1, which bridges from ORC to Sir4, is unique among the Sir factors. Unlike the others, Sir1 does not spread with the Sir complex beyond the silencers (Fig. 5) (Rusche et al. 2002). Moreover, once it helps establish silencing, Sir1 is no longer needed for the stable maintenance of the repressed state (Pillus and Rine 1989). This argues that Sir1 primarily serves in the establishment step of repression, most likely through its ability to bind the DNA-bound ORC and Sir4 (Triolo and Sternglanz, 1996). Its role in nucleation was shown by tethering the protein artificially through a Gal4 DNA binding domain to Gal4 binding sites, which replaced the *HMR*-E silencer. In this context, GBD-Sir1 could efficiently nucleate repression, rendering the silencer and its binding factors unnecessary (Chien et al. 1993), although the other Sir proteins and intact histone tails were still required for transcriptional repression.

Cite as *Cold Spring Harb Perspect Biol* doi: 10.1101/cshperspect.a017491

6.2 Telomeric Heterochromatin

At telomeres, an RNA-based enzyme called telomerase maintains a simple but irregular TG-rich repeat of 300–350 bp in length, which provides 16 to 20 consensus sites for Rap1 binding. This array of Rap1 binding sites forms a nonnucleosomal cap on the chromosomal end, and plays a critical role in telomere length maintenance (Kyrion et al. 1992; Marcand et al. 1997). Along the telomeric repeat, Rap1 binds its consensus through a core DNA binding domain and Sir4 through its carboxy-terminal domain, even in the absence of the other Sir proteins or the H4 amino terminus. Point mutations that disrupt the Rap1-Sir4 interaction disrupt TPE, although the effect on *HM* repression is very slight (Buck and Shore 1995). Rap1 also binds Sir3 through its carboxy-terminal domain, and mutation of this interface has similar effects on silencing (Chen et al. 2011). However, because the loss of Sir4 prevents other Sir proteins from binding to telomeric chromatin (Luo et al. 2002), Sir4 is apparently the crucial link between nucleation factors and the ensuing silent chromatin structure (Fig. 6).

Equally potent for the nucleation of repression at telomeres is the DNA end-binding complex yKu70/yKu80. The yKu heterodimer also recruits Sir4, and loss of yKu strongly derepresses TPE. Conversely, a targeted GBD-yKu fusion efficiently nucleates repression at silencer-compromised reporter genes. The requirement for yKu at telomeres can be bypassed by eliminating the Rap1-interacting factor, Rif1, which competes for the interaction of Sir4 with the Rap1 carboxy-terminal domain (Fig. 6) (Mishra and Shore 1999). This illustrates the redundancy between the two telomeric nucleation factors, yKu and Rap1, based on their affinity for Sir4.

There is a clear correlation between the amount of Rap1 bound at telomeres that is Rif1-free and the efficiency of silencing. Deletion of *RIF1* gene or lesions in Rap1 that block Rif1 binding leads to relatively stable increases in telomere length. Wild-type cells inheriting these longer telomeres show increased frequency of repression of reporter genes such as *URA3* or *ADE2* integrated at the telomeres on Chr VIIL or Chr VR (Kyrion et al. 1993). Moreover, in a diploid strain containing both elongated and wild-type telomeres on Chr VIIL, the elongated telomere with increased repression did not affect repression at a wild-type length telomere (Park and Lustig 2000). Thus, the effect of telomere length on the frequency of silencing occurs in *cis*. Moreover, the frequency of switching from a derepressed to a repressed state (white colonies to red in the case of subtelomeric *ADE2* gene) also depends on the length of its adjacent telomere; a longer telomeric repeat imposed a lower frequency of switching from derepressed to repressed states (Park and Lustig 2000). Thus, both Rap1 and yKu, which provide recruitment sites for Sir3 and Sir4, can be limiting for the nucleation of TPE.

7 THE CRUCIAL ROLE OF HISTONE H4K16 ACETYLATION AND ITS DEACETYLATION BY Sir2

The molecular interactions between the Sir proteins have been well characterized. Sir4 interacts strongly with Sir2 and separately with Sir3 in vivo and in vitro (Moazed et al. 1997; Strahl-Bolsinger et al. 1997; Hoppe et al. 2002). When coordinately expressed in insect cells, Sir2, Sir3, and Sir4 proteins can be isolated as a stable complex with a 1:1:1 stoichiometry (Cubizolles et al. 2006). Nonetheless, Sir3 has a special role in this process because Sir3 can form a stable extended multimer in vitro (Liou et al. 2005) and its overexpression extends the subtelomeric silent domain from its normal ~3 kb to ~15 kb from the telomeric end, coincident with the binding of Sir3 (Renauld et al. 1993; Hecht et al. 1996).

The platform on which the Sir complex spreads consists of nucleosomes with deacetylated histone H3 and H4 amino termini (Braunstein et al. 1996; Suka et al. 2001), and the manner in which Sir3 interacts with histones helps explain how spreading occurs (Fig. 7). Sir3 binds the deacetylated histone H4 amino terminus in a highly selective manner both in vitro and in vivo (Johnson et al. 1990; Johnson et al. 1992; Carmen et al. 2002; Yang et al. 2008a). In this regard, the most important histone region is contained in residues 16–29 of histone H4, and lysine 16 must be positively charged (unmodified or substituted by arginine) for Sir3 to bind (Johnson et al. 1990,1992).

A cocrystal structure of the Sir3 amino-terminal BAH domain explains this specificity quite well (Armache et al. 2011). Sixteen residues in the Sir3 BAH domain interact with H4 tail residues 13 to 23 which are held in a rigid conformation, primarily through electrostatic interactions with amino acid side chains. A negatively charged binding pocket of BAH Sir3 accommodates the side chains of unmodified H4K16 and H4H18. Indeed, acetylation of H4K16 could potentially disrupt most of the electrostatic contacts in this pocket. In contrast, the Sir2-Sir4 subcomplex binds with slight preference to nucleosomes bearing an acetylated H4K16 residue, at least in the absence of NAD^+ (Oppikofer et al. 2011). This is consistent with acetylated histone H4K16 being a preferred and crucial target for the Sir2 enzyme (Imai et al. 2000; Suka et al. 2002; Cubizolles et al. 2006). The AAA^+ domain of Sir3 also binds unmodified nucleosomes in vitro (Ehrentraut et al. 2011) and requires that all four H4 acetylation sites (K5, K8, K12, and K16) are deacetylated for optimal binding (Carmen et al. 2002) and for telomeric silencing

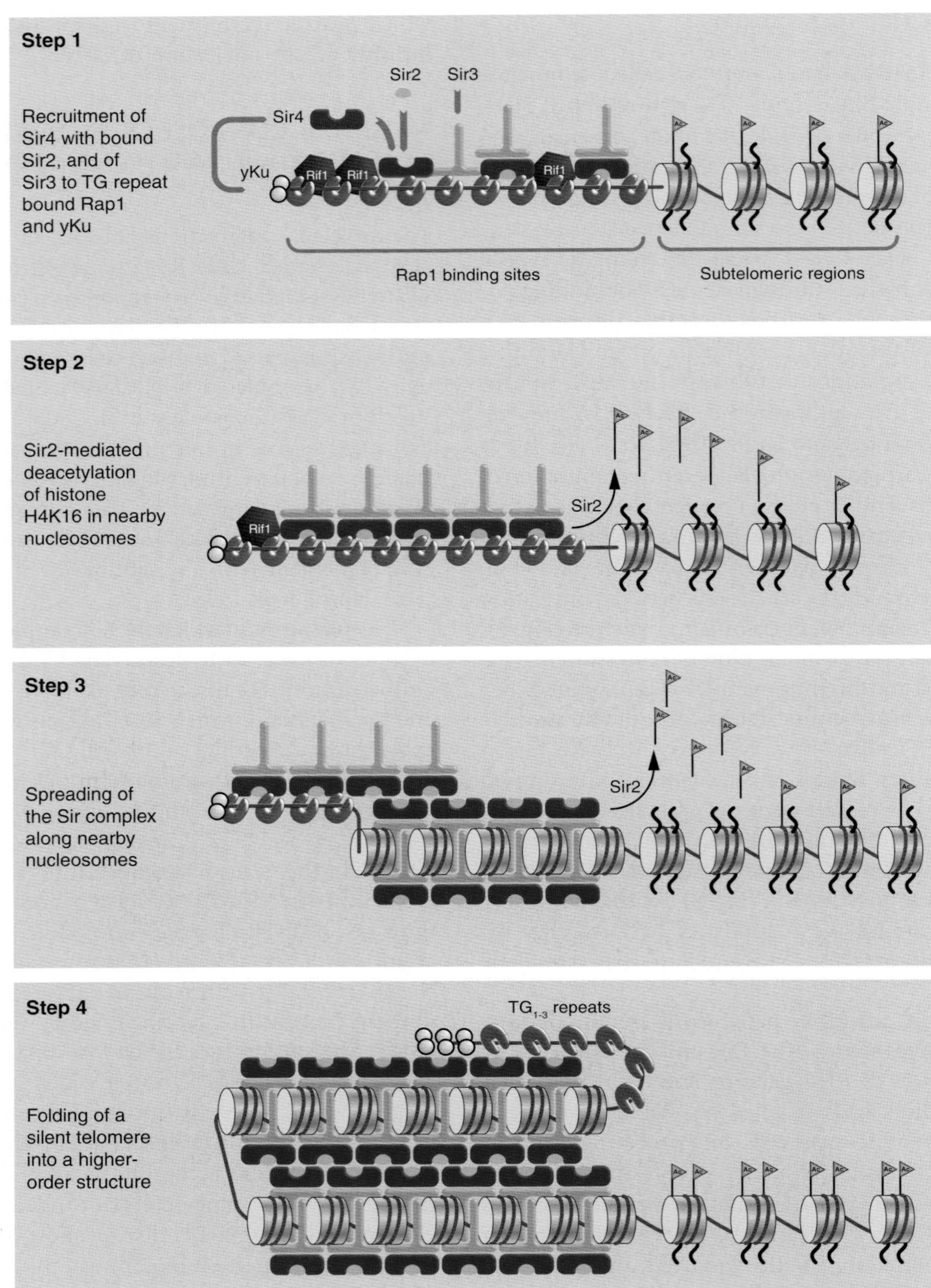

Figure 6. Model for stepwise assembly of heterochromatin in yeast. (*Step 1*) At telomeres, Rap1 and yKu recruit Sir4 even in the absence of Sir2 or Sir3. Only Sir4 can be recruited in the absence of the other Sir proteins, and its binding is antagonized by Rif1 and Rif2 (Mishra and Shore 1999). (*Step 2*) Sir4-Sir2 and Sir4-Sir3 interact strongly creating Sir complexes along the TG repeats. Sir2 NAD-dependent histone deacetylase activity is stimulated by complex formation and Sir2 deacetylates the acetylated histone H4K16 residue in nearby nucleosomes. (*Step 3*) Sir complexes spread along the nucleosomes, perhaps making use of the *O*-acetyl-ADP-ribose intermediate produced by NAD hydrolysis (Liou et al. 2005). Sir3 and Sir4 bind the deacetylated histone H4 tails. Although the deacetylated histone H3 amino-terminal tail also binds Sir3 and Sir4 proteins, it is not shown here. (*Step 4*) The silent chromatin "matures" at the end of M phase to create an inaccessible structure. This may entail higher-order folding and sequestering at the nuclear envelope.

Cite as *Cold Spring Harb Perspect Biol* doi: 10.1101/cshperspect.a017491

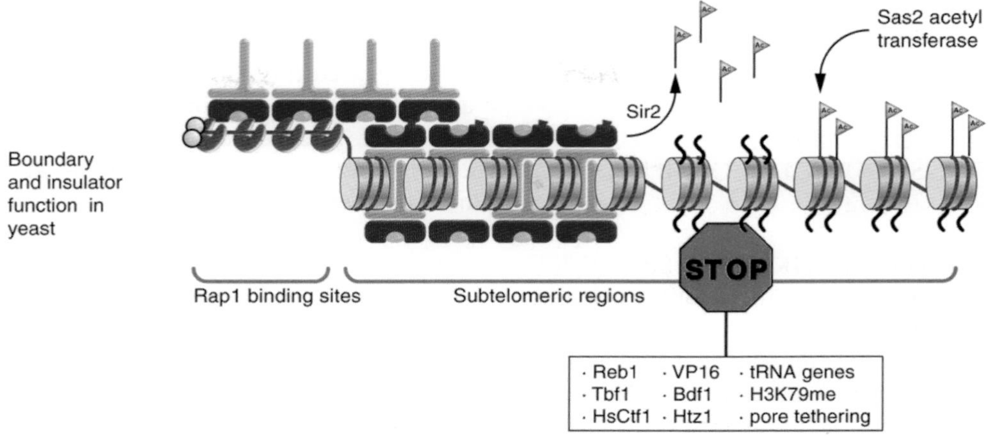

Figure 7. Heterochromatin boundary function in budding yeast. Spreading of heterochromatin through deacetylation of histone H4K16 by Sir2 is limited by the competing activity of Sas2 histone acetyltransferase, which acetylates H4K16 in adjacent euchromatin, thus preventing Sir3 binding. Methylation of K79 in histone H3 in adjacent euchromatin also affects the spreading of heterochromatin. In addition, factors such as Reb1, Tbf1, and mammalian or viral factors Ctf1 or VP16; nuclear pore tethering; and the presence of tRNA genes may also mediate boundary function. It is conceivable that several of these factors function through the recruitment of histone acetyltransferases, like Sas2.

(Thompson et al. 1994a). Because histones are naturally deacetylated throughout telomeric heterochromatin, the Sir3 carboxyl terminus may contribute to the stability of the Sir3 complex through its interaction with the fully deacetylated histone tail.

On addition of NAD^+, the Sir2-catalyzed deacetylation of H4K16ac generates a by-product called *O*-acetyl-ADP-ribose, as well as a deacetylated histone H4 tail (illustrated in Fig. 5 of Ch. 5 [Seto and Yoshida 2014]). Intriguingly, not only the generation of the high-affinity binding site for Sir3, but apparently also production of the intermediate metabolite *O*-acetyl-ADP-ribose, enhances the affinity of Sir2-3-4 complexes for chromatin (Johnson et al. 2009; Martino et al. 2009). The generation of *O*-acetyl-ADP-ribose enhances the interaction of Sir3 with Sir4-Sir2 in vitro (Liou et al. 2005), favors the oligomerization of Sir proteins on nucleosomal arrays (Onishi et al. 2007), and enhances protection of the linker DNA from micrococcal nuclease digestion (Oppikofer et al. 2011). This is consistent with genetic studies showing that any mutation of histone H4K16 disrupts telomere repression, even substitution of lysine by the similarly charged amino acid arginine, or glutamine, which mimics the uncharged nature of acetylated lysine.

Interestingly, the basic region of aa 16–24 within the H4 amino terminus also promotes nucleosomal array compaction in vitro, suggesting that the acetylation state of H4K16 may regulate the higher-order folding of the nucleosomal fiber (Shogren-Knaak et al. 2006). Thus, histone H4K16 deacetylation by Sir2 may actively promote silent chromatin formation in several ways: First, a conformational change may be triggered in the Sir complex by the by-product *O*-acetyl-ADP-ribose; second, the affinity of the Sir complex for chromatin increases thanks to generation of a high-affinity Sir3 site; and third, even when Sir complexes are not bound, nucleosomal arrays may compact because of contact between the H4 tail and the adjacent nucleosomal face. From this it was clear that control over Sir-mediated silencing would lie in the acetylation/deacetylation cycle of histone H4.

We summarize here the different steps for the initiation and spreading of heterochromatin in an environment enriched for acetylated histone H4, which is likely to be deposited immediately after replication (Fig. 6). At telomeres, Rap1 and yKu recruit Sir4, and Sir4 forms a dimer with Sir2 to deacetylate histone H4 and H3 amino-terminal tails of nearby nucleosomes. Sir3 is recruited by its affinity for Sir4, but also for Abf1 and Rap1. The deacetylation of the histone H4 tail produces a high affinity Sir3 binding site on the nucleosome, which favors assembly of a Sir2-3-4 complex. The interactions of Sir3 with Sir4, Sir3 with the H4 tail and the nucleosomal core, and Sir4 nonspecifically with linker DNA, all seem to contribute to the stable binding of the Sir complex to the nucleosomal fiber. The action of Sir2 on adjacent acetylated nucleosomes appears to trigger the spread of the complex along adjacent histone tails. Finally, the long-range folding of the chromatin fiber may stabilize the repressed state. Most of these events are likely to be very similar at *HM* loci, although there the initial recruitment of

Sir4 needs Rap1, Abf1, or ORC and Sir1. The question then arises, what causes Sir complex spreading to stop?

8 BARRIER FUNCTIONS: HISTONE MODIFICATIONS RESTRICT Sir COMPLEX SPREADING

Because the acetylated histone H4K16 binds Sir2-4 tightly, whereas its deacetylation by Sir2 is crucial for the spreading of heterochromatin, it is not surprising that interfering with the cycle of H4K16 acetylation/deacetylation impedes heterochromatin propagation (Kimura et al. 2002; Suka et al. 2002). The yeast histone acetyltransferase (HAT) Sas2, a member of the highly conserved MYST class of HATs, modifies K16 in bulk euchromatin. Therefore at the boundaries of silent chromatin, one expects to find acetylated histone H4K16. Accordingly, if the *SAS2* gene is deleted, or if H4K16 is changed to arginine to simulate the deacetylated state, Sir3, Sir4, and Sir2 spread at low levels inward from the telomeric repeats, approximately fivefold further than in a wild-type cell. This suggests that the spreading of subtelomeric heterochromatin is controlled, at least in part, by the opposing activities of Sir2 and Sas2 on lysine 16 of H4 (Fig. 7). Limiting the global amount of Sir2 (or of the Sir2-Sir4 complex) naturally limits spread by limiting deacetylation.

At the *HM* loci, restricting the spread of silent chromatin is perhaps even more critical than at telomeres because genes important for growth are found on Chr III, and it is known that silencing can spread bidirectionally from silencers into flanking DNA sequence. One boundary that prevents further spreading of silencing from *HMR*, is a tRNA gene (Donze and Kamakaka 2001). This boundary function is likely to require the HAT activity that is associated either with transcription or the transcriptional potential of this locus. It is significant that one of these HATs is Sas2, although the H3 HAT, Gcn5, can also affect boundary function of the tRNA gene. This suggests that transcriptional activators, in general, can restrict Sir complex propagation by recruiting HATs. Consistently, in subtelomeric euchromatin regions, boundary activity has been attributed to the general transcription factors, Reb1, Tbf1, and to the acidic *trans*-activating domain of VP16 (Fourel et al. 1999, 2001). These factors are likely to promote the hyperacetylation of histones (Fig. 7).

Another modification that antagonizes the spread of telomeric heterochromatin is histone H3K79 methylation, which is deposited by the lysine methyltransferase Dot1 (Van Leeuwen et al. 2002; Ng et al. 2002). This histone lysine methyltransferase (KMT) was discovered in a screen for factors whose overexpression caused loss of telomeric silencing (Singer et al. 1998). However, Dot1 does not methylate H3K79 in heterochromatin itself, but instead in adjacent euchromatin and at active genes. Indeed, the artificial targeting of Dot1 to telomeric heterochromatin derepresses silencing by Sir proteins (Stulemeijer et al. 2011), most likely by reducing the affinity of Sir3 for nucleosomes.

Surprisingly, the H4 tail is required for the bulk methylation of H3K79 by Dot1. An elegant set of experiments (Altaf et al. 2007; Fingerman et al. 2007) has generated a model that helps explain the interplay between the H4 tail and the demethylated state of K79 in heterochromatin. Namely, when H4K16 is deacetylated by Sir2, a high affinity binding site for Sir3 is generated by a charged patch in the H4 tail (K16, R17, H18, R19, K20). Interestingly, Dot1 and Sir3 compete for this charged patch, although Sir3 is sensitive to H4K16 acetylation, and Dot1 is not. Thus, in deacetylated heterochromatin, Sir3 is a potent inhibitor of Dot1 binding. However, in adjacent euchromatin, in which K16 is acetylated, Dot1 preferentially binds the charge patch and methylates histone H3K79. The Sir3N-nucleosome crystal structure confirmed genetic evidence, suggesting that Sir3 interacts with H3K79 and that it is in close proximity to H4K16 (Armache et al. 2011). Both the binding of Sir3 and that of the holo-SIR complex are, in turn, weakened by histone H3K79 methylation. Again, this could be reconstituted in vitro, in binding assays between Sir3 and reconstituted nucleosomes that bear either H4K16ac or H3K79me (Oppikofer et al. 2011). Thus, the weak binding of Sir3 to H4K16ac in euchromatin favors Dot1 binding to the H4 tail and subsequent H3K79 methylation, which, in turn, weakens interaction with Sir3. This provides a good example of interhistone interactions as discussed and illustrated in Figure 12 of Ch. 3 (Allis et al.).

In addition to the mechanism regulating H3K79 methylation and H4K16 acetylation near boundaries, it was reported that in euchromatin the presence of the variant histone H2A.Z and the RNA polymerase associated factor Bdf1 (Meneghini et al. 2003), as well as the tethering of DNA to nuclear pores (Ishii et al. 2002), generate boundaries that limit the spread of silent chromatin. Although the mechanisms by which these factors affect heterochromatin spreading are unknown, it is interesting to note that some inducible genes with H2A.Z-containing promoters, associate with nuclear pores on activation, and remain associated there in a manner that facilitates their re-induction (Ishii et al. 2002; Brickner and Walter 2004). Thus, boundary function may reflect a chromatin state that allows rapid recruitment of transcriptional activators, including HATs, KMTs like Dot1, or nucleosome remodelers that directly or indirectly disrupt histone interactions with heterochromatin proteins (Fig. 7).

 Cite as *Cold Spring Harb Perspect Biol* doi: 10.1101/cshperspect.a017491

9 A ROLE FOR THE H3 AMINO-TERMINAL TAIL IN HIGHER-ORDER CHROMATIN STRUCTURES

There is increasing evidence that the formation of heterochromatin involves a series of steps that include, but go beyond, the binding of Sir proteins. When Sir protein expression is induced artificially in G_1, Sir proteins spread from their initiation site by interacting with deacetylated H4 amino termini, but silencing is still defective (Kirchmaier and Rine 2006). Also, when the histone H3K56 acetylation site is mutated, Sir protein spreading occurs, but silencing is disrupted. Similarly, the presence of H3K56 acetylation leads to increased accessibility of Sir-bound nucleosomes to micrococcal nuclease in vitro without impairing Sir protein association (Oppikofer et al. 2011). Thus, the establishment of silencing requires not only Sir protein spreading, but also the deacetylation of H3K56, an event that enables nucleosomes to bind DNA more tightly to form a structure resistant to ectopically expressed bacterial *dam* methylase (Maas et al. 2006; Xu et al. 2007; Celic et al. 2008; Yang et al. 2008b).

Repression also involves the replacement of histone H3 methylated at K4 and K79 with unmethylated histone H3 (Katan-Khayakovich and Struhl 2005; Osborne et al. 2009). Whereas H3K56 deacetylation by Hst3 and Hst4, two homologs of Sir2, takes place every S phase, demethylation of H3K79 requires dilution by replication, and can take up to four cell divisions. These changes may parallel the formation of a compact, higher-order chromatin structure.

At first, the role of the H3 amino terminus in silencing appeared to be similar to that of the H4 amino terminus because the domain in each tail involved in silencing in vivo was shown to bind Sir3 and Sir4 in vitro (Hecht et al. 1995). However, although the H4 amino-terminal residues are required for the recruitment and spreading of Sir3 and the remaining Sir complex, the H3 tail was required neither for recruitment nor for spreading of the Sir proteins. Nonetheless, deletion of the H3 amino terminus or mutation of critical residues 11–15 (T–G–G–K–A) led to altered topology, increased accessibility to bacterial *dam* methylase expressed in yeast, and less tightly folded chromatin (Sperling and Grunstein 2009; Yu et al. 2011). Thus, although Sir proteins are clearly recruited by the H4 tail, they may subsequently interact with the H3 amino terminus to form compacted chromatin.

10 *TRANS*-INTERACTION OF TELOMERES, AND PERINUCLEAR ATTACHMENT OF HETEROCHROMATIN

In budding yeast, as in many lower eukaryotes, telomeres cluster together during interphase, in close association with the nuclear envelope. This clustering was initially observed as prominent foci of Rap1 and Sir proteins, which were detected above a diffuse nuclear background by immunostaining (Fig. 8). Disruption of silencing by histone H4K16 mutation, or interference in Rap1 or yKu function, led to the dispersion of the Sir proteins from these clusters (Hecht et al. 1995; Laroche et al. 1998). Later it was shown that not only telomeres, but also the silent *HML* and *HMR* loci are closely associated with telomeres at the nuclear envelope. Binding to the nuclear envelope is mediated by redundant pathways that depend either on the telomere-bound yKu factor, or on components of silent chromatin itself. Interestingly, the interactions that lead to telomere clustering can be genetically separated from telomere anchoring to the nuclear envelope, although both pathways involve Sir proteins (Ruault et al. 2011).

Within silent chromatin, the anchoring function has been assigned to the PAD of Sir4 (aa 950–1262) and its interaction with the nuclear envelope associated protein, Esc1 (Andrulis et al. 2002; Taddei et al. 2004). Sir4-Esc1 interactions tether the Sir-repressed chromatin domain at perinuclear sites distinct from pores. Even in the absence of a yKu anchoring pathway, the association of telomeres with the nuclear periphery can be achieved by Sir4-Esc1 interactions, as long as silent chromatin is formed (Hediger et al. 2002). Thus, rings of silent chromatin excised from their chromosomal context by recombination and lacking TG repeats, remain associated with perinuclear foci in a Sir-dependent manner (Gartenberg et al. 2004).

Initially, telomeres are recruited to the nuclear envelope by yKu, given that yKu-dependent tethering occurs even in the absence of silencing. This interaction with the nuclear envelope, together with interactions in *trans* between telomeres, generates a nuclear subcompartment that appears to sequester Sir proteins from the rest of the nucleoplasm (Fig. 9). yKu mediated anchoring is achieved either through yKu-Sir4 interaction, or through the interaction of yKu with telomerase (Schober et al. 2009). The Est1 subunit of telomerase binds specifically to an inner nuclear membrane-spanning protein, Mps3, which is a member of the conserved SUN domain family of inner nuclear envelope proteins. Intriguingly, this interaction is cell-cycle specific and mediates the link between yKu and the nuclear envelope only in S phase, possibly to maintain anchorage while subtelomeric chromatin is disrupted by the replication fork. In G_1 phase, there appears to be a secondary yKu anchoring pathway, and Sir4 may bind Mps3 through an intermediary to help tether telomeres (Bupp et al. 2007).

Both pathways of telomere anchoring, through Sir4-Esc1 and S-phase yKu-Est1-Mps3 pathway, are controlled by posttranslational modifications. Sir4 and both subunits of yKu are modified by SUMO, a ubiquitin-like moiety,

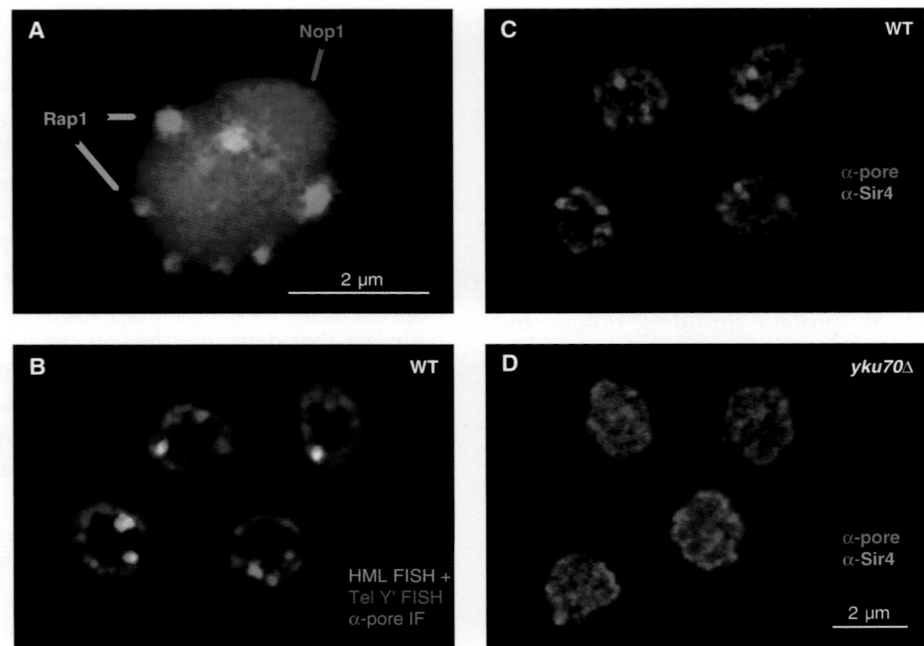

Figure 8. Sir proteins and Rap1 are found in foci at the nuclear periphery. (*A*) Rap1 (anti-Rap1, green) identifies seven clusters representing all 64 telomeres in this diploid yeast cell nucleus, in which DNA is stained red. Telomeres are either perinuclear or adjacent to the nucleolus (blue, anti-Nop1). (*B*) Telomeric repeat DNA (red) and *HML* (green) is identified by fluorescent in situ hybridization. The two colocalize in ~70% of the cases and both are adjacent to the nuclear envelope (anti-pore staining, blue). (*C*) The focal distribution of Sir4 (green) adjacent to the nuclear envelope (Mab414, red). (*D*) This pattern is lost in a yKu70 deletion strain, coincident with the loss of telomeric silencing (Laroche et al. 1998).

deposited specifically by the E3 SUMO ligase Siz2. Remarkably, loss of Siz2 led to the displacement of telomeres from the nuclear envelope (Ferreira et al. 2011). Silencing was reduced only slightly in *siz2* mutants, possibly because telomeres became abnormally long due to telomerase deregulation (see Section 12). Thus, the perinuclear compartment created by telomere anchoring and silent chromatin appears to regulate telomere functions beyond silencing.

The perinuclear clustering generates a subnuclear compartment that favors silencing (Fig. 9). Evidence supporting this conclusion includes the fact that silencer-flanked *HM* constructs repress less efficiently when they are integrated far from telomeres (Thompson et al. 1994b; Maillet et al. 1996), and this can be reversed by artificially tethering the domain at the nuclear envelope through a targeted transmembrane factor (Andrulis et al. 1998). Importantly, the ability to improve repression by peripheral tethering (or by being placed in telomere proximity) is lost when Sir3 and Sir4 are no longer sequestered in foci (Taddei et al. 2009). Similarly, coordinate overexpression of Sir3 and Sir4 proteins ablates the positive effects of tethering. Thus, an uneven distribution of Sir proteins is the relevant feature of telomere sequestration at the nuclear envelope for chromatin-mediated repression. Given that displaced Sir proteins can repress promiscuously (Taddei et al. 2009), the sequestration of Sir foci also positively reinforces active gene expression. It is proposed that the assembly of newly replicated DNA into heterochromatin is likely to be favored when DNA is replicated in a zone rich in silencing factors.

The crucial interaction that mediates telomere–telomere clustering, even in the absence of silencing, appears to be Sir3. Whether this is mediated by Sir3 itself or ligands of Sir3 remains to be determined, yet some form of clustering can occur even in the absence of Sir2 and Sir4 (Ruault et al. 2011). Other factors also affect telomere–telomere interaction, namely, the other Sir proteins, the Ku heterodimer, Asf1, Rtt109, Esc2, the Cohibin complex, and two factors involved in ribosome biogenesis, Ebp2 and Rrs1. However, because these factors also affect heterochromatin formation, one cannot rule out that they promote clustering by promoting Sir3 recruitment to telomeres.

11 TELOMERE LOOPING

A further long-range interaction may stem from the folding back of a single telomere on itself, which may allow silent chromatin to bypass subtelomeric boundary elements and stabilize repressed chromatin at subtelomeric genes (Figs. 5

 Cite as *Cold Spring Harb Perspect Biol* doi: 10.1101/cshperspect.a017491

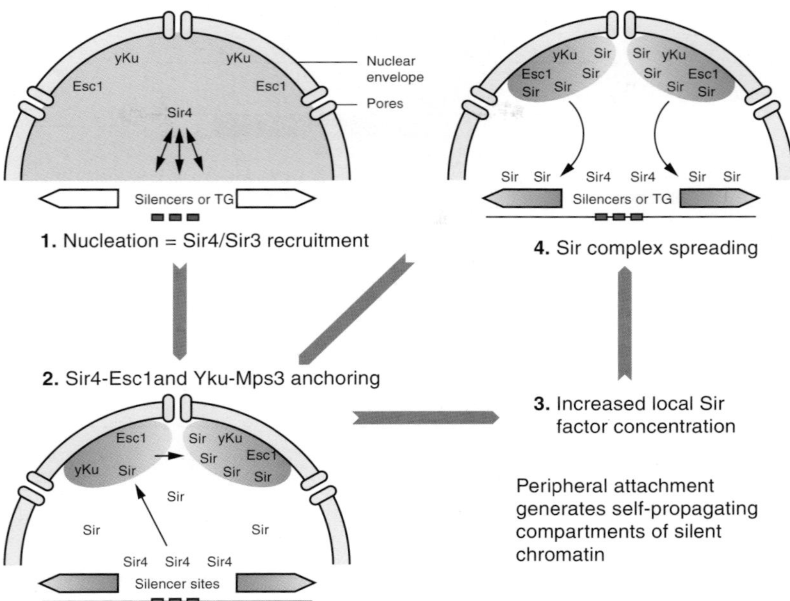

Figure 9. Spontaneous formation of silencing subcompartments. A simple model for the formation of subnuclear compartments is shown. (*1*) Sir4 is first recruited at the nucleation center by DNA binding proteins that can bind Sir4. These include Rap1, ORC, Abf1, and yKu. (*2*) The presence of Sir4 at the locus will then bring it to the nuclear periphery through one of the two Sir4 anchoring pathways (yKu or Esc1). (*3*) At the nuclear envelope, the high local concentrations of Sir proteins will help silencing complexes assemble and spread. (*4*) The ability of silent loci to remain attached at the periphery increases the local concentration of Sir proteins and reinforces the silencing of other loci within this region. Importantly, telomere-bound yKu can independently recruit telomeres to the nuclear envelope just as Sir4 recruits silencer sequences.

and 6). Although Rap1 binding sites are found only within the first ~300 bp of TG repeat DNA on the end of a telomere, ChIP showed that Rap1 is associated with nucleosomes as far as ~3 kb away from the TG repeat (Strahl-Bolsinger et al. 1997). Similarly, yKu is recovered for ~3 kb from the chromosomal end to which it binds (Martin et al. 1999). When silencing is disrupted by mutation of *SIR* genes, both Rap1 and yKu are lost exclusively from the more internal subtelomeric sequences and not from the terminal TG repeats (Hecht et al. 1996; Martin et al. 1999). This was interpreted as showing that the truncated telomere folds back, enabling TG-bound Rap1 and yKu to bind Sir proteins in *trans* (Figs. 5 and 6).

Evidence for telomere looping comes from the work of de Bruin et al. (2001), who have exploited the inability of yeast transcriptional activators, such as Gal4, to function from a site downstream of the targeted gene. Strains were constructed in which the Gal4 upstream activating sequence (UAS) element was placed beyond the 3′ termination site of a reporter gene, and the construct was inserted either at an internal chromosomal location or near a telomere. At an internal site, this construct could not be induced by activating Gal4, but in a subtelomeric context, the Gal4 UAS could activate the promoter from a site 1.9 kb downstream of the promoter. This was Sir3-dependent, arguing that the telomeric end can fold back in the presence of Sir3, but not in its absence, to allow the Gal4 UAS to position itself proximal to the transcription start site. In this way, silent chromatin appears to promote at least a transient folding of the chromosome end.

12 VARIABLE REPRESSION AT NATURAL SUBTELOMERIC DOMAINS

We have set forth here a simplistic view of continuous silent chromatin emanating from the telomeric Rap1 binding sites, yet the situation at native telomeres is significantly more complex, largely because of the presence of natural boundary elements found in subtelomeric repeat sequences. Generally, when reporter constructs for telomeric repression are integrated, the subtelomeric repeat elements called X and Y′ at telomeres are deleted, placing the reporter gene and unique sequence immediately adjacent to TG repeats. All native telomeres, on the other hand, contain a core subtelomeric repeat element, X, which is positioned between the TG repeat and the most telomere-proximal gene, and 50%–70% of native telomeres also contain at least one copy of a larger subtelomeric element called Y′ (Fig. 10).

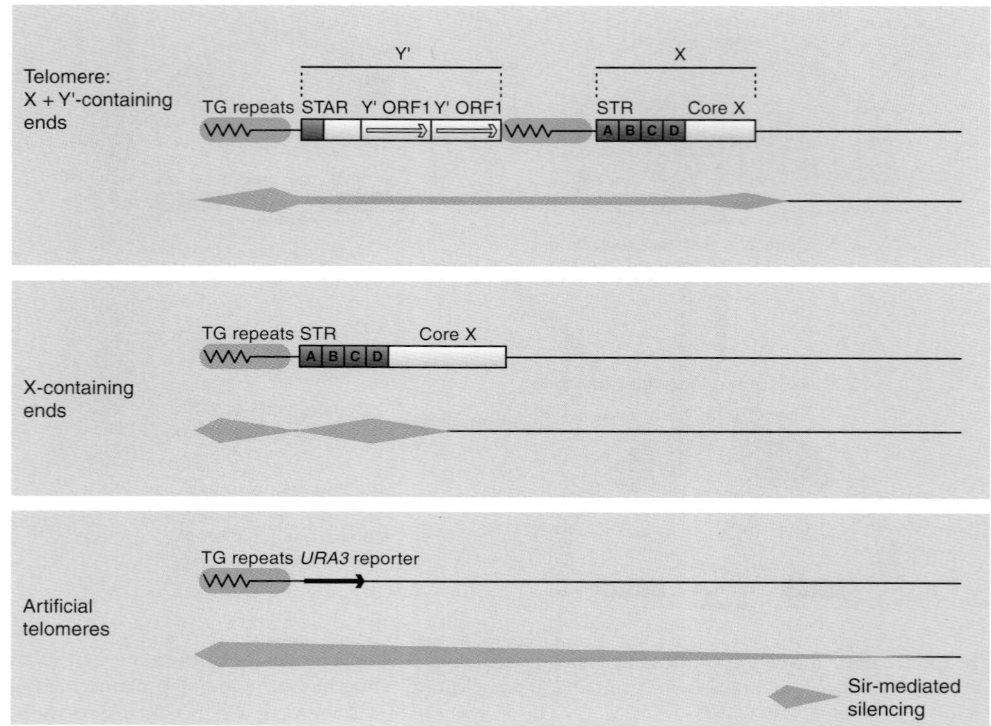

Figure 10. The organization of native telomeres and their silencing patterns. Subtelomeric elements are shown with their major protein binding sites. Telomeres fall into the two general classes: X-containing or X+Y′-containing ends. The STAR and STR elements block the propagation of repression and leave a region of reduced repression within the Y′ or X element. This is not the case at artificially truncated telomeres in which there is a gradient of repression that extends 3–4 kb from the TG repeat. Looping similar to that in Figures 5 and 6 is proposed for native telomeres, so that repressed regions contact each other leaving unrepressed chromatin in between areas of contact. (Adapted from Pryde and Louis 1999.)

Both X and Y′ elements contain binding sites for the transcriptional regulators Tbf1 and Reb1, which have been shown to reduce the spread of silent chromatin (Fourel et al. 1999). However, X elements also contain autonomously replicating sequence (ARS) consensus sequences, and binding sites for Abf1 and other transcription factors, which have the opposite effect: These re-initiate or boost the repression of reporters placed on the centromere proximal side of these elements. The result is one of discontinuity in silencing at many native telomeres. This adds a level of complexity to the model of continuous spreading outlined in Figure 6. Pryde and Louis (1999) have proposed that the unrepressed Y′ element loops out when it is found between two repressed domains, leading to discontinuity in silent domains without eliminating the need for nucleation and spreading from the TG repeats.

There is large variation in the efficiency of TPE at different native telomeres in budding yeast. If one inserts a reporter gene near a telomere without deleting the subtelomeric repeats, only about half of the telomeres appear to be subject to TPE (Pryde and Louis 1999). Empirically, it was shown that telomeres containing only X subtelomeric elements (rather than XY′ telomeres) are more likely to silence, possibly because the Y′ long terminal repeats bind factors that prevent Sir spreading, whereas various transcription factors in the X element (e.g., Reb1, Tbf1, and Abf1) contribute to Sir factor nucleation (Mak et al. 2009). These subtelomeric elements are particularly enriched for factors that regulate stress genes, whose effects on TPE may differ from their effects on transcription at nontelomeric loci. For example, Reb1 has boundary activity at telomeres (Fourel et al. 1999), but is a gene activator at internal sites. The binding of such factors can explain the discontinuity of Sir-mediated repression at native telomeres, yet they do not appear to affect the repression of classical reporters for TPE, which are integrated without X or Y′ elements (Renauld et al. 1993).

Interestingly, many of the genes found in subtelomeric domains are repressed by the HDAC Hda1 and the repressor Tup1 (Robyr et al. 2002), and are induced only on stress conditions (Ai et al. 2002). In general, the 267 genes that are within 20 kb of budding yeast telomeres produce roughly five times fewer mRNA molecules (average of 0.5/cell) than nontelomeric genes. Importantly, only 20 of these genes show derepression upon loss of Sir3. This leads to a logic of genome organization in which genes that are rarely or conditionally induced are found near telomeres, in which they are repressed by a non-Sir mechanism, but are adjacent to domains repressed by Sir proteins and, therefore, also tethered near the nuclear envelope.

13 INHERITANCE OF EPIGENETIC STATES

A universal characteristic of heterochromatin is that its silent state is passed from one generation to the next. This requires the reassembly of a repressive chromatin structure on daughter strands soon after replication of the DNA template. Pioneering work on the role of the cell cycle in the establishment or inheritance of the silent state was performed by Miller and Nasmyth, who studied the onset and loss of silencing with a temperature-sensitive $sir3^{ts}$ mutant (Miller and Nasmyth 1984). A shift from permissive temperature to nonpermissive temperature caused silencing to be lost immediately, indicating that *SIR3* was required for maintenance of the repressed state. However, in the reciprocal experiment, shifting from nonpermissive temperature to permissive temperature ($SIR3^+$) did not lead to immediate restoration of repression; passage through the cell cycle was required. They concluded that an event in S phase was required for establishment of heritably repressed chromatin. This requirement was later shown to involve events in both S and G_2/M phases (Lau et al. 2002).

Initially it was thought that origin firing from the silencer linked ARS elements might be a critical event in the establishment or inheritance of silent chromatin, but because there was no detectable initiation from the origins flanking the *HML* locus, this seemed an unlikely explanation. Indeed, an experiment showing that ORC can be efficiently replaced by a targeted GDB-Sir1 fusion protein put to rest the notion that origin firing is essential for the inheritance of silent chromatin. In addition, recent experiments have shown that establishment of repression can occur on DNA that does not replicate (Kirchmaier and Rine 2001; Li et al. 2001). Nonetheless, on rings of silent chromatin that are excised from the genome either with or without silencers, Sir complex association is in continual flux, arguing that nucleation and/or stabilization provided by silencer elements at *HM* loci is needed to actively suppress rapid decay (Cheng and Gartenberg 2000).

What occurs in S phase to enable the propagation of silent chromatin? One candidate event could be the deacetylation of H4K16ac or H3K56ac, or else the suppression of the enzymes that deposit these modifications (Xu et al. 2007; Neumann et al. 2009). Alternatively, chaperones necessary for histone deposition (CAF1) may be required for generating repressed chromatin after replication. The deacetylation of histone H3K56 can be achieved in vitro by Sir2 family members (Xu et al. 2007; Oppikofer et al. 2011) and in vivo during late S phase by Hst3 and Hst4 (Maas et al. 2006; Celic et al. 2008; Yang et al. 2008b). These two Sir2 paralogs are nuclear enzymes whose activities are required in S phase when newly synthesized DNA must be assembled into repressed chromatin, yet they are not structural components of heterochromatin. Importantly, disruption of the two genes weakens, but does not eliminate, Sir mediated silencing (Yang et al. 2008b).

Other studies have shown that robust silencing is not achieved until telophase, well beyond the S-phase window of nucleosome assembly. It appears that prevention of the metaphase degradation of the cohesin subunit, Scc1, inhibits stable repression (Lau et al. 2002). Propagation of repression thus depends both on a critical S-phase component, and a further event that entails continual recruitment and loading of Sir proteins.

Intriguingly, when the proteins of telomeric heterochromatin were examined by ChIP in the transcriptionally OFF and ON states, the major difference found between them was the presence of H3K79 methylation at telomeric chromatin in the ON state (Kitada et al. 2012). Because Dot1 is recruited during transcription (Shahbazian et al. 2005), this suggests a positive feedback loop in which the ON state is triggered, possibly by decreased telomere length, to initiate transcription and K79 methylation. Dot1 would then be responsible for maintaining the ON state through its promotion of K79 methylation.

14 OTHER FUNCTIONS OF Sir PROTEINS AND SILENT CHROMATIN

Although the standard function of heterochromatin is the silencing of adjacent genes, a closer examination of silencing factors has uncovered a plethora of new functions that correlate with silencing or require silencing factors. Particularly in organisms in which repetitive centromeric DNA plays a crucial role in centromere function, it is clear that heterochromatin contributes to centromere and kinetochores function. Budding yeast, on the other hand, does not depend on silent chromatin for centromere function. Nonetheless, a number of other roles have been identified for silent chromatin or silent chromatin factors, and these are described in this section.

14.1 Suppression of Recombination

In *Drosophila* highly active rDNA repeats are adjacent to centromeric heterochromatin, and in many higher eukaryotic species, nucleoli and condensed heterochromatin are spatially juxtaposed. It is significant, therefore, that yeast Sir2, independent of the other Sir proteins, is genetically and physically associated with rDNA repeats (Gotta et al. 1997). Importantly, the loss of Sir2 in budding yeast leads to a dramatic increase in rDNA recombination and reduction of the contiguous integrated array (Gottlieb and Esposito, 1989). Instability of the rDNA has been also correlated with an accumulation of extrachromosomal rDNA circles, which arise from unequal crossing over between sister chromatids (Kobayashi et al. 2004). These events that are normally suppressed by a complex called Cohibin, a V-shaped complex of two Lrs4 proteins and two Csm1 homodimers (Mekhail et al. 2008; Chan et al. 2011), which also mediates the association of rDNA repeats with the two nuclear envelope proteins, Heh1 (a human Man1 paralog) and Nur1. Loss of either Sir2 or the cohibin-Heh1 anchoring pathway leads to instability of the rDNA repeat, followed by cell-cycle arrest or premature senescence (Fig. 11A) (Sinclair and Guarente 1997; Kaeberlein et al. 1999). Although Sir2 can also silence RNA Pol II genes integrated in the rDNA (Smith and Boeke 1997), recent evidence separates this effect on transcription from the role of Sir2 in preventing rDNA recombination.

Exactly how the tethering of rDNA repeats to the nuclear envelope reduces recombination is still unknown. The simplest explanation may be that it imposes a steric hindrance on the binding of Rad52, a protein essential for homologous exchange in yeast. Indeed, Rad52 is excluded

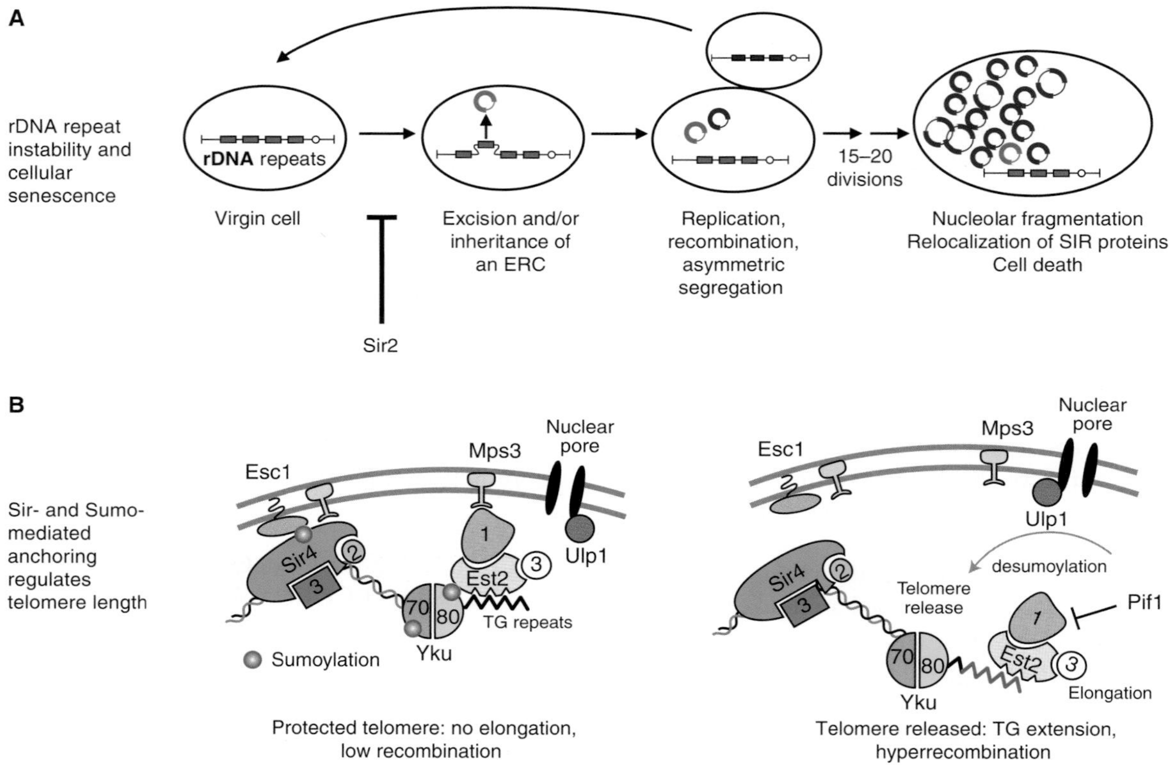

Figure 11. Secondary functions of Sir proteins and silent chromatin. (*A*) rDNA recombination leads to cellular senescence in yeast. The rDNA is organized in an array of 140–200 direct repeats of a 9.1 kb unit (red block). These encode the 18S, 5.8S, 25S, and 5S rRNAs, and contain two Sir2 responsive elements downstream of the 5S gene and within the 18S gene. The rDNA repeats tend to be excised in aging yeast cells, and the circles accumulate in the mother cell (Kaeberlein et al. 1999). This correlates with premature senescence and can be antagonized by Sir2, which helps suppress unequal recombination and ring excision. (*B*) Telomere anchoring and silent chromatin contribute to telomere homeostasis. Redundant pathways that tether yeast telomeres to the nuclear envelope include sumoylation targets, Sir4, yKu70, and yKu80 (Ferreira et al. 2011). The relevant Sumo E3 ligase is Siz2. Loss of Siz2, ablation of the Mps3 amino terminus, or deletion of Sir4 all lead to release of telomeres from the nuclear envelope and longer steady state telomere length. Loss of Mps3 amino terminus or yKu also increases telomere recombination. This suggests that sequestration at the nuclear envelope may limit access for both recombination and telomerase activation mechanisms, and that loss of anchoring increases both pathways. Regulated desumoylation by Ulp1 may play a role in releasing telomeres from the periphery allowing efficient elongation in late S phase. Siz2-mediated sumoylation is indicated by red circles.

 Cite as *Cold Spring Harb Perspect Biol* doi: 10.1101/cshperspect.a017491

from the nucleolus and when rDNA requires repair by recombination, the damage is extruded from the nucleolus. Intriguingly, other mutations that reduce the efficiency of rDNA excision, such as the elimination of the replication fork barrier protein Fob1, extend replicative lifespan of yeast cells, confirming that rDNA instability is indeed a key culprit for limiting cellular lifespan.

Because Sir2 is a NAD-dependent deacetylase, and because NAD levels act as a metabolic thermostat, it was proposed that the effect of yeast Sir2 on lifespan might be related to the extension of lifespan by calorie restriction, a conserved pathway that functions in many species. However, Sir2 and caloric restriction increase lifespan through independent pathways (Kaeberlein et al. 2004). It is also relevant to note that an accumulation of extrachromosomal rDNA rings has not been detected during aging in any other species. This is probably because of the unique budding mechanism through which yeast divides; the budding geometry and rapid kinetics of mitosis in yeast leads to an inevitable retention of noncentromere-bearing DNA elements in the mother cell (Gehlen et al. 2011).

14.2 Preventing Homologous Recombination on Chr III

Switching mating type involves the generation of a double strand break induced by the HO endonuclease at the *MAT* locus. That break is then repaired with homologous sequences that recombine from the heterochromatic silent *HM* loci. In vitro studies have shown that the holo-Sir complex and histone sequences involved in silencing prevent cleavage by the HO endonuclease at *HM* loci, and impair the early steps in strand invasion by the break induced at *MAT.* The nucleosome remodeling SWI/SNF complex, which displaces Sir3 from nucleosomes (Sinha et al. 2009), is needed to counteract the repression of strand invasion conferred by silent chromatin. This enables recombinational repair of *MAT* through an appropriate *HM* donor.

The deletion of *SIR3* enhanced general recombination rates throughout the genome (Palladino et al. 1993). More specifically the loss of telomeric anchoring correlates with increased rates of recombination between a telomeric sequence and internal sequences (Marvin et al. 2009). The latter involves the integrity of yKu, but probably does not simply reflect the loss of anchoring alone.

14.3 Chromosomal Cohesion

Cohesion of sister chromatids is made possible by a complex of proteins known as cohesin. Yeast heterochromatin, like that of more complex eukaryotes, is enriched in cohesin which holds these silent regions together in sister chromatids. When Sir proteins were tethered to chromosomal sites, in which pairing of sister chromatids could be seen by fluorescence microscopy, it was evident that Sir2 alone could mediate cohesion. Although this required the cohesin complex, it did not require the deacetylase activity of Sir2 indicating yet another function for Sir2 in chromosomal mechanics (Wu et al. 2011). It remains to be seen if this involves components of the Cohibin complex (Chan et al. 2011).

14.4 Telomere Length Regulation

In yeast, modification of histone H2A serine 129 by phosphorylation generates a modified form of H2A, called γH2A. In other species, this serine acceptor site is only present in an H2A variant called H2AX. Histone H2A phosphorylation is mediated by the checkpoint kinases, Tel1 or Mec1, and occurs when these central checkpoint kinases are recruited to DNA damage by a complex of replication protein A with ssDNA or by the Mre11 complex. Intriguingly, even without exogenously induced damage, γH2A was found by ChIP to be coextensive with and dependent on silent subtelomeric chromatin (Szilard et al. 2010; Kitada et al. 2011). The phosphorylation of H2A in telomeres is mediated by Tel1 kinase with weaker contributions from Mec1 kinase. This suggests that subtelomeric domains trigger a low-level checkpoint response, possibly during telomere replication. The persistence of γH2A in silent domains may stem from reduced histone turnover, given that the enzyme that dephosphorylates γH2A acts only on nonnucleosomal H2A (Keogh et al. 2006). Point mutations that remove the H2A phosphoacceptor residue render telomeres slightly shorter in certain genetic backgrounds, again linking telomeric chromatin structure to aspects of telomerase control (Kitada et al. 2011).

As discussed above, telomerase activity is increased when telomeres are released from the nuclear envelope because of the absence of Sir4 or the SUMO ligase Siz2 (Palladino et al. 1993; Ferreira et al. 2011). This further links telomere length homeostasis to subtelomeric chromatin status, implicating both the Sir proteins and the binding of the yKu complex to Mps3 (Fig. 11). Interestingly, activation of Mec1 kinase by double-strand breaks leads to a partial release of Sir proteins and displacement of telomeres (Martin et al. 1999), although it is unclear what Sir protein release achieves in this situation.

14.5 A Link with Replication Factors

Whereas passage through an event between early S and G_2/M is required for silencing, this critical event is not replication itself. Nevertheless, a number of replication factors are involved in silencing, most likely through structural roles that are independent of their roles in replication. ORC binding sites are found at each of the silencer E and I elements, and ORC directly recruits Sir1 in the absence of replication. Moreover, ORC binds not only at the silencer elements, but also in the region between E and I at *HMR* in a manner that is dependent on Sir proteins.

Like ORC, the helicase complex Mcm2-7 is part of the prereplicative complex (pre-RC) that assembles on replication origins before the initiation of replication in S phase. However, MCM proteins are found in abundance in yeast cells and may have functions in processes distinct from DNA replication, such as silencing. Indeed, Sir2 interacts indirectly with proteins of the Mcm2-7 complex outside of S phase through a carboxy-terminal protein bridge (53aa) found in another pre-RC component, Mcm10. Mutations in this bridge disrupt the binding of Mcm2-7 with Sir2 and weaken silencing efficiency, but they do not disrupt replication nor the association of Sir2 with chromatin (Liachko and Tye, 2009). In a speculative model, it is suggested that Mcm10, which itself forms a ring-shaped hexamer, binds Sir2 away from chromatin, ensuring that Sir2 undergoes a modification that renders it more competent for silencing. In the absence of Mcm10 and the MCM complex, a less competent Sir2 would be incorporated into chromatin, reducing repression. What this modification might be remains unknown, but it is noteworthy that in fruit flies, Mcm10 also interacts with Hp1, and that in yeast, Mcm10's silencing function can be genetically separated from its replication function.

Independent of Mcm10, there is a strong evolutionary relationship between silencing and certain replication factors, revealed by a phylogenetic analysis of Orc1, the largest subunit of the ORC complex and Sir3. *K. lactis* contains Orc1 that is found at replication origins and *HML*α. As mentioned above, *K. lactis* contains the Orc1 paralog of Sir3, but not Sir3. The BAH domain of Orc1, however, interacts with the deacetylated H4K16 just like *S. cerevisiae* Sir3, allowing Orc1 to spread along telomeric and *HML*α heterochromatin in a manner that is dependent on Sir2 and Sir4 (Hickman and Rusche 2010). Surprisingly, however, *HMR* in *K. lactis* is silenced by a different mechanism that involves neither Orc1 nor Sir4.

14.6 Regulating Replication Origin Choice

Replication origins are primed in G_1 phase of the cell cycle by the formation of a pre-RC that recognizes DNA replication origins. Interestingly, Sir2 inhibits pre-RC assembly at certain origins, including one that is found at *HMR*-E. This origin is sensitive to the presence of Sir2 and Sir3, arguing that the silencing mechanism itself influences origin choice. In contrast, certain origins that are not in heterochromatin are sensitive to the presence of Sir2 only, indicating that Sir2 has a unique function in origin function that is independent of transcriptional silencing. This function may be in nucleosome placement near the origins that prevents pre-RC assembly and is controlled by the deacetylation of histone H4K16 by Sir2 (Fox and Weinreich 2008). This is but one among many aspects of chromatin structure that affect origin function, particularly in higher eukaryotic species.

15 SUMMARY

Combined genetic, biochemical, and cytological techniques have been exploited in budding yeast to show fundamental principles at work during heterochromatin-mediated gene silencing. These principles include (1) the mechanism of initiation, spreading, and barriers to spreading of heterochromatin; (2) the balance of heterochromatin factors and their distribution within a subnuclear environment; (3) higher-order folding of heterochromatin; and (4) the effects of cell-cycle involvement in its formation. The in vivo and in vitro studies to date provide a strong mechanistic basis for our understanding of the assembly of heterochromatin from chromatin fibers in all eukaryotes. The in vitro systems currently being developed for the reconstitution of yeast heterochromatin promise to yield a structural reconstruction of Sir complexes bound to chromatin, enabling a first glimpse at higher-order chromatin folding.

REFERENCES

** Reference is also in this book.*

Ai W, Bertram PC, Tsang CK, Chan TF, Zheng XF. 2002. Regulation of subtelomeric silencing during stress response. *Mol Cell* **10:** 1295–1305.

Altaf M, Utley RT, Lacoste N, Tan S, Briggs SD, Côté J. 2007. Interplay of chromatin modifiers on a short basic patch of histone H4 tail defines the boundary of telomeric heterochromatin. *Mol Cell* **28:** 1002–1014.

Andrulis ED, Neiman AM, Zappulla DC, Sternglanz R. 1998. Perinuclear localization of chromatin facilitates transcriptional silencing. *Nature* **394:** 592–595.

Andrulis ED, Zappulla DC, Ansari A, Perrod S, Laiosa CV, Gartenberg MR, Sternglanz R. 2002. Esc1, a nuclear periphery protein required for Sir4-based plasmid anchoring and partitioning. *Mol Cell Biol* **22:** 8292–8301.

Ansari A, Gartenberg M. 1997. The yeast silent information regulator Sir4p anchors and partitions plasmids. *Mol Cell Biol* **17:** 7061–7068.

Aparicio OM, Billington BL, Gottschling DE. 1991. Modifiers of position effect are shared between telomeric and silent mating-type loci in *S. cerevisiae*. *Cell* **66:** 1279–1287.

Armache KJ, Garlick JD, Canzio D, Narlikar GJ, Kingston RE. 2011. Structural basis of silencing, Sir3 BAH domain in complex with a nucleosome at 3.0 Å resolution. *Science* **334:** 977–982.

 Cite as *Cold Spring Harb Perspect Biol* doi: 10.1101/cshperspect.a017491

Beck DB, Burton A, Oda H, Ziegler-Birling C, Torres-Padilla ME, Reinberg D. 2012. The role of PR-Set7 in replication licensing depends on Suv4-20h. *Genes Dev* **26:** 2580–2589.

Brand AH, Breeden L, Abraham J, Sternglanz R, Nasmyth K. 1985. Characterization of a "silencer" in yeast: A DNA sequence with properties opposite to those of a transcriptional enhancer. *Cell* **41:** 41–48.

Brand AH, Micklem G, Nasmyth K. 1987. A yeast silencer contains sequences that can promote autonomous plasmid replication and transcriptional activation. *Cell* **51:** 709–719.

Braunstein M, Sobel RE, Allis CD, Turner BM, Broach JR. 1996. Efficient transcriptional silencing in *Saccharomyces cerevisiae* requires a heterochromatin histone acetylation pattern. *Mol Cell Biol* **16:** 4349–4356.

Brickner JH, Walter P. 2004. Gene recruitment of the activated INO1 locus to the nuclear membrane. *PLoS Biol* **2:** e342.

Buck SW, Shore D. 1995. Action of a RAP1 carboxy-terminal silencing domain reveals an underlying competition between HMR and telomeres in yeast. *Genes Dev* **9:** 370–384.

Bupp JM, Martin AE, Stensrud ES, Jaspersen SJ. 2007. Telomere anchoring at the nuclear periphery requires the budding yeast Sad1-UNC-84 domain protein Mps3. *J Cell Biol* **179:** 845–854.

Carmen AA, Milne L, Grunstein M. 2002. Acetylation of the yeast histone H4 N terminus regulates its binding to heterochromatin protein SIR3. *J Biol Chem* **277:** 4778–4781.

Celic I, Verreault A, Boeke JD. 2008. Histone H3K56 hyperacetylation perturbs replisomes, causes DNA damage. *Genetics* **179:** 1769–1784.

Chan JN, Poon BP, Salvi J, Olsen JB, Emili A, Mekhail K. 2011. Perinuclear cohibin complexes maintain replicative life span via roles at distinct silent chromatin domains. *Dev Cell* **20:** 867–879.

Chang JF, Hall BE, Tanny JC, Moazed D, Filman D, Ellenberger T. 2003. Structure of the coiled-coil dimerization motif of Sir4 and its interaction with Sir3. *Structure* **11:** 637–649.

Chen Y, Rai R, Zhou ZR, Kanoh J, Ribeyre C, Yang Y, Zheng H, Damay P, Wang F, Tsujii H, et al. 2011. A conserved motif within RAP1 has diversified roles in telomere protection, regulation in different organisms. *Nat Struct Mol Biol* **18:** 213–221.

Chen L, Widom J. 2005. Mechanism of transcriptional silencing in yeast. *Cell* **120:** 37–48.

Cheng TH, Gartenberg MR. 2000. Yeast heterochromatin is a dynamic structure that requires silencers continuously. *Genes Dev* **14:** 452–463.

Chien CT, Buck S, Sternglanz R, Shore D. 1993. Targeting of SIR1 protein establishes transcriptional silencing at HM loci and telomeres in yeast. *Cell* **75:** 531–541.

Chopra VS, Mishra RK. 2005. To SIR with Polycomb: Linking silencing mechanisms. *Bioessays* **27:** 119–121.

Cockell M, Gotta M, Palladino F, Martin SG, Gasser SM. 1998. Targeting Sir proteins to sites of action: A general mechanism for regulated repression. *Cold Spring Harb Symp Quant Biol* **63:** 401–412.

Cubizolles F, Martino F, Perrod S, Gasser SM. 2006. A homotrimer-heterotrimer switch in Sir2 structure differentiates rDNA and telomeric silencing. *Mol Cell* **21:** 825–836.

de Bruin D, Zaman Z, Liberatore RA, Ptashne M. 2001. Telomere looping permits gene activation by a downstream UAS in yeast. *Nature* **409:** 109–113.

* Dekker J, Misteli T. 2014. Long-range chromatin interactions. *Cold Spring Harb Perspect Biol* doi: 10.1101/cshperspect.a019356.

Donze D, Kamakaka RT. 2001. RNA polymerase III and RNA polymerase II promoter complexes are heterochromatin barriers in *S. cerevisiae*. *EMBO J* **20:** 520–531.

Ehrentraut S, Hassler M, Oppikofer M, Kueng S, Weber JM, Mueller JW, Gasser SM, Ladurner AG, Ehrenhofer-Murray AE. 2011. Structural basis for the role of the Sir3 AAA$^+$ domain in silencing, interaction with Sir4 and unmethylated histone H3K79. *Genes Dev* **25:** 1835–1846.

* Elgin SCR, Reuter G. 2013. Position-effect variegation, heterochromatin formation, and gene silencing in *Drosophila*. *Cold Spring Harb Perspect Biol* **5:** a017780.

Ferreira HC, Luke B, Schober H, Kalck V, Lingner J, Gasser SM. 2011. The PIAS homologue Siz2 regulates perinuclear telomere position and telomerase activity in budding yeast. *Nat Cell Biol* **13:** 867–874.

Fingerman IM, Li HC, Briggs SD. 2007. A charge-based interaction between histone H4 and Dot1 is required for H3K79 methylation and telomere silencing: Identification of a new *trans*-histone pathway. *Genes Dev* **21:** 2018–2029.

Fourel G, Revardel E, Koering CE, Gilson E. 1999. Cohabitation of insulators, silencing elements in yeast subtelomeric regions. *EMBO J* **18:** 2522–2537.

Fourel G, Boscheron C, Revardel E, Lebrun E, Hu YF, Simmen KC, Muller K, Li R, Mermod N, Gilson E. 2001. An activation-independent role of transcription factors in insulator function. *EMBO Rep* **2:** 124–132.

Fox CA, Weinreich M. 2008. Beyond heterochromatin: SIR2 inhibits the initiation of DNA replication. *Cell Cycle* **7:** 3330–3334.

Frye RA. 2000. Phylogenetic classification of prokaryotic and eukaryotic Sir2-like proteins. *Biochem Biophys Res Commun* **273:** 793–798.

Gardner RG, Nelson ZW, Gottschling DE. 2005. Ubp10/Dot4p regulates the persistence of ubiquitinated histone H2B: Distinct roles in telomeric silencing and general chromatin. *Mol Cell Biol* **25:** 6123–6139.

Gartenberg MR, Neumann FR, Laroche T, Blaszczyk M, Gasser SM. 2004. Sir-mediated repression can occur independently of chromosomal and subnuclear contexts. *Cell* **119:** 955–967.

Gehlen LR, Nagai S, Shimada K, Meister P, Taddei A, Gasser SM. 2011. Nuclear geometry and rapid mitosis ensure asymmetric episome segregation in yeast. *Curr Biol* **21:** 25–33.

Gotta M, Strahl-Bolsinger S, Renauld H, Laroche T, Kennedy BK, Grunstein M, Gasser SM. 1997. Localization of Sir2p: The nucleolus as a compartment for silent information regulators. *EMBO J* **16:** 3243–3255.

Gottlieb S, Esposito RE. 1989. A new role for a yeast transcriptional silencer gene, SIR2, in regulation of recombination in ribosomal DNA. *Cell* **56:** 771–776.

Gottschling DE. 1992. Telomere-proximal DNA in *Saccharomyces cerevisiae* is refractory to methyltransferase activity in vivo. *Proc Natl Acad Sci* **89:** 4062–4065.

Gottschling DE, Aparicio OM, Billington BL, Zakian VA. 1990. Position effect at *S. cerevisiae* telomeres: Reversible repression of Pol II transcription. *Cell* **63:** 751–762.

Hecht A, Laroche T, Strahl-Bolsinger S, Gasser SM, Grunstein M. 1995. Histone H3, H4 N-termini interact with SIR3 and SIR4 proteins: A molecular model for the formation of heterochromatin in yeast. *Cell* **80:** 583–592.

Hecht A, Strahl-Bolsinger S, Grunstein M. 1996. Spreading of transcriptional repressor SIR3 from telomeric heterochromatin. *Nature* **383:** 92–96.

Hediger F, Neumann FR, Van Houwe G, Dubrana K, Gasser SM. 2002. Live imaging of telomeres. yKu and Sir proteins define redundant telomere-anchoring pathways in yeast. *Curr Biol* **12:** 2076–2089.

Hickman MA, Rusche LN. 2010. Transcriptional silencing functions of the yeast protein Orc1/Sir3 subfunctionalized after gene duplication. *Proc Natl Acad Sci* **107:** 19384–19389.

Hickman MA, Froyd CA, Rusche LN. 2011. Reinventing heterochromatin in budding yeasts, Sir2, the origin recognition complex take center stage. *Eukaryot Cell* **10:** 1183–1192.

Hofmann JF, Laroche T, Brand AH, Gasser SM. 1989. RAP-1 factor is necessary for DNA loop formation in vitro at the silent mating type locus HML. *Cell* **57:** 725–737.

Hoppe GJ, Tanny JC, Rudner AD, Gerber SA, Danaie S, Gygi SP, Moazed D. 2002. Steps in assembly of silent chromatin in yeast: Sir3-independent binding of a Sir2/Sir4 complex to silencers and role for Sir2-dependent deacetylation. *Mol Cell Biol* **12:** 4167–4180.

Hsu HC, Wang CL, Wang M, Yang N, Chen Z, Sternglanz R, Xu R-M. 2013. Structural basis for allosteric stimulation of Sir2 activity by Sir4 binding. *Genes Dev* **27:** 64–73.

Huang H, Maertens AM, Hyland EM, Dai J, Norris A, Boeke JD, Bader JS. 2009. HistoneHits: A database for histone mutations and their phenotypes. *Genome Res* **19:** 674–681.

Imai S, Armstrong CM, Kaeberlein M, Guarente L. 2000. Transcriptional silencing and longevity protein Sir2 is an NAD-dependent histone deacetylase. *Nature* **403:** 795–800.

Ishii K, Arib G, Lin C, Van Houwe G, Laemmli UK. 2002. Chromatin boundaries in budding yeast: The nuclear pore connection. *Cell* **109:** 551–562.

Johnson LM, Kayne PS, Kahn ES, Grunstein M. 1990. Genetic evidence for an interaction between SIR3 and histone H4 in the repression of the silent mating loci in *S. cerevisiae. Proc Natl Acad Sci* **87:** 6286–6290.

Johnson LM, Fisher-Adams G, Grunstein M. 1992. Identification of a non-basic domain in the histone H4 N-terminus required for repression of the yeast silent mating loci. *EMBO J* **11:** 2201–2209.

Johnson A, Li G, Sikorski TW, Buratowski S, Woodcock CL, Moazed D. 2009. Reconstitution of heterochromatin-dependent transcriptional gene silencing. *Mol Cell* **35:** 769–781.

Kaeberlein M, McVey M, Guarente L. 1999. The SIR2/3/4 complex, SIR2 alone promote longevity in *S. cerevisiae* by two different mechanisms. *Genes Dev* **13:** 2570–2580.

Kaeberlein M, Kirkland KT, Fields S, Kennedy BK. 2004. Sir2-independent life span extension by calorie restriction in yeast. *PLoS Biol* **2:** 296–307.

Katan-Khaykovich Y, Struhl K. 2005. Heterochromatin formation involves changes in histone modifications over multiple cell generations. *EMBO J* **24:** 2138–2149.

Keogh MC, Kim JA, Downey M, Fillingham J, Chowdhury D, Harrison JC, Onishi M, Datta N, Galicia S, Emili A, et al. 2006. A phosphatase complex that dephosphorylates γH2AX regulates DNA damage checkpoint recovery. *Nature* **439:** 497–501.

Kimura A, Umehara T, Horikoshi M. 2002. Chromosomal gradient of histone acetylation established by Sas2p and Sir2p functions as a shield against gene silencing. *Nat Genet* **3:** 370–377.

Kirchmaier AL, Rine J. 2001. DNA replication-independent silencing in *S. cerevisiae. Science* **291:** 646–650.

Kirchmaier AL, Rine J. 2006. Cell cycle requirements in assembling silent chromatin in *Saccharomyces cerevisiae. Mol Cell Biol* **26:** 852–862.

Kitada T, Schleker T, Sperling AS, Xie W, Gasser SM, Grunstein M. 2011. γH2A is a component of yeast heterochromatin required for telomere elongation. *Cell Cycle* **10:** 293–300.

Kitada T, Kuryan BG, Tran NN, Song C, Xue Y, Carey M, Grunstein M. 2012. Mechanism for epigenetic variegation of gene expression at yeast telomeric heterochromatin. *Genes Dev* **26:** 2443–2455.

Kobayashi T, Horiuchi T, Tongaonkar P, Vu L, Nomura M. 2004. Sir2 regulates recombination between different rDNA repeats, but not recombination within individual rRNA genes in yeast. *Cell* **117:** 441–453.

Kueng S, Tsai-Pflugfelder M, Oppikofer M, Ferreira HC, Roberts E, Tsai C, Roloff TC, Sack R, Gasser SM. 2012. Regulating repression: Roles for the sir4 N-terminus in linker DNA protection and stabilization of epigenetic states. *PLoS Genet* **8:** e1002727.

Kyrion G, Boakye KA, Lustig AJ. 1992. C-terminal truncation of RAP1 results in the deregulation of telomere size, stability, and function in *Saccharomyces cerevisiae. Mol Cell Biol* **12:** 5159–5173.

Kyrion G, Liu K, Liu C, Lustig AJ. 1993. RAP1 and telomere structure regulate telomere position effects in *Saccharomyces cerevisiae. Genes Dev* **7:** 1146–1159.

Laroche T, Martin SG, Gotta M, Gorham HC, Pryde FE, Louis EJ, Gasser SM. 1998. Mutation of yeast Ku genes disrupts the subnuclear organization of telomeres. *Curr Biol* **8:** 653–656.

Lau A, Blitzblau H, Bell SP. 2002. Cell-cycle control of the establishment of mating-type silencing in *S. cerevisiae. Genes Dev* **16:** 2935–2945.

Li YC, Cheng TH, Gartenberg MR. 2001. Establishment of transcriptional silencing in the absence of DNA replication. *Science* **291:** 650–653.

Liachko I, Tye BK. 2009. Mcm10 mediates the interaction between DNA replication and silencing machineries. *Genetics* **181:** 379–391.

Liou GG, Tanny JC, Kruger RG, Walz T, Moazed D. 2005. Assembly of the SIR complex and its regulation by *O*-acetyl-ADP-ribose, a product of NAD-dependent histone deacetylation. *Cell* **121:** 515–527.

Loo S, Rine J. 1994. Silencers and domains of generalized repression. *Science* **264:** 1768–1771.

Luo K, Vega-Palas MA, Grunstein M. 2002. Rap1-Sir4 binding independent of other Sir, yKu, or histone interactions initiates the assembly of telomeric heterochromatin in yeast. *Genes Dev* **12:** 1528–1539.

Maas NL, Miller KM, DeFazio LG, Toczyski DP. 2006. Cell cycle and checkpoint regulation of histone H3K56 acetylation by Hst3 and Hst4. *Mol Cell* **23:** 109–119.

Maillet L, Boscheron C, Gotta M, Marcand S, Gilson E, Gasser SM. 1996. Evidence for silencing compartments within the yeast nucleus: A role for telomere proximity and Sir protein concentration in silencer-mediated repression. *Genes Dev* **10:** 1796–1811.

Mak HC, Pillus L, Ideker T. 2009. Dynamic reprogramming of transcription factors to and from the subtelomere. *Genome Res* **19:** 1014–1025.

Marcand S, Gilson E, Shore D. 1997. A protein-counting mechanism for telomere length regulation in yeast. *Science* **275:** 986–990.

Martin SG, Laroche T, Suka N, Grunstein M, Gasser SM. 1999. Relocalization of telomeric Ku and SIR proteins in response to DNA strand breaks in yeast. *Cell* **97:** 621–633.

Martino F, Kueng S, Robinson P, Tsai-Pflugfelder M, van Leeuwen F, Ziegler M, Cubizolles F, Cockell MM, Rhodes D, Gasser SM. 2009. Reconstitution of yeast silent chromatin, multiple contact sites and O-AADPR binding load SIR complexes onto nucleosomes in vitro. *Mol Cell* **33:** 323–334.

Marvin ME, Becker MM, Noel P, Hardy S, Bertuch AA, Louis EJ. 2009. The association of yKu with subtelomeric core X sequences prevents recombination involving telomeric sequences. *Genetics* **183:** 453–467.

Mekhail K, Seebacher J, Gygi SP, Moazed D. 2008. Role for perinuclear chromosome tethering in maintenance of genome stability. *Nature* **456:** 667–670.

Meneghini MD, Wu M, Madhani HD. 2003. Conserved histone variant H2A.Z protects euchromatin from the ectopic spread of silent heterochromatin. *Cell* **112:** 725–736.

Miller AM, Nasmyth KA. 1984. Role of DNA replication in the repression of silent mating type loci in yeast. *Nature* **312:** 247–251.

Mishra K, Shore D. 1999. Yeast Ku protein plays a direct role in telomeric silencing, counteracts inhibition by Rif proteins. *Curr Biol* **9:** 1123–1126.

Moazed D, Kistler A, Axelrod A, Rine J, Johnson AD. 1997. Silent information regulator protein complexes in *S. cerevisiae*: A SIR2/SIR4 complex and evidence for a regulatory domain in SIR4 that inhibits its interaction with SIR3. *Proc Natl Acad Sci* **94:** 2186–2191.

Moretti P, Freeman K, Coodly L, Shore D. 1994. Evidence that a complex of SIR proteins interacts with the silencer and telomere-binding protein RAP1. *Genes Dev* **8:** 2257–2269.

Murphy GA, Spedale EJ, Powell ST, Pillus L, Schultz SC, Chen L. 2003. The Sir4 C-terminal coiled coil is required for telomeric and mating type silencing in *Saccharomyces cerevisiae. J Mol Biol* **334:** 769–780.

Neumann H, Hancock SM, Buning R, Routh A, Chapman L, Somers J, Owen-Hughes T, van Noort J, Rhodes D, Chin JW. 2009. A method for genetically installing site-specific acetylation in recombinant histones defines the effects of H3K56 acetylation. *Mol Cell* **36:** 153–163.

Ng HH, Feng Q, Wang H, Erdjument-Bromage H, Tempst P, Zhang Y, Struhl K. 2002. Lysine methylation within the globular domain of histone H3 by Dot1 is important for telomeric silencing and Sir protein association. *Genes Dev* **16:** 1518–1527.

Norris A, Boeke JD. 2010. Silent information regulator 3: The Goldilocks of the silencing complex. *Genes Dev* **24:** 115–122.

Onishi M, Liou GG, Buchberger JR, Walz T, Moazed D. 2007. Role of the conserved Sir3-BAH domain in nucleosome binding and silent chromatin assembly. *Mol Cell* **28:** 1015–1028.

Oppikofer M, Kueng S, Martino F, Soeroes S, Hancock SM, Chin JW, Fischle W, Gasser SM. 2011. A dual role of H4K16 acetylation in the establishment of yeast silent chromatin. *EMBO J* **30:** 2610–2621.

Oppikofer M, Kueng S, Keusch JJ, Hassler M, Ladurner AG, Gut H, Gasser SM. 2013. Dimerization through the Sir3 C-terminal winged helix domain is essential for yeast heterochromatin formation. *EMBO J* **32:** 437–449.

Osborne EA, Dudoit S, Rine J. 2009. The establishment of gene silencing at single-cell resolution. *Nat Genet* **41:** 800–806.

Palladino F, Laroche T, Gilson E, Axelrod A, Pillus L, Gasser SM. 1993. SIR3 and SIR4 proteins are required for the positioning and integrity of yeast telomeres. *Cell* **75:** 543–555.

Park Y, Lustig AL. 2000. Telomere structure regulates the heritability of repressed subtelomeric chromatin in *S. cerevisiae*. *Genetics* **154:** 587–598.

Pillus L, Rine J. 1989. Epigenetic inheritance of transcriptional states in *S. cerevisiae*. *Cell* **59:** 637–647.

Pryde FE, Louis EJ. 1999. Limitations of silencing at native yeast telomeres. *EMBO J* **18:** 2538–2550.

Renauld H, Aparicio OM, Zierath PD, Billington BL, Chhablani SK, Gottschling DE. 1993. Silent domains are assembled continuously from the telomere and are defined by promoter distance and strength, and by SIR3 dosage. *Genes Dev* **7:** 1133–1145.

Rine J, Herskowitz I. 1987. Four genes responsible for a position effect on expression from HML and HMR in *Saccharomyces cerevisiae*. *Genetics* **116:** 9–22.

Rine J, Strathern JN, Hicks JB, Herskowitz I. 1979. A suppressor of mating-type locus mutations in *S. cerevisiae*: Evidence for and identification of cryptic mating-type loci. *Genetics* **93:** 877–901.

Robyr D, Suka Y, Xenarios I, Kurdistani SK, Wang A, Suka N, Grunstein M. 2002. Microarray deacetylation maps determine genome-wide functions for yeast histone deacetylases. *Cell* **109:** 437–446.

Ruault M, De Meyer A, Loïodice I, Taddei A. 2011. Clustering heterochromatin: Sir3 promotes telomere clustering independently of silencing in yeast. *J Cell Biol* **192:** 417–431.

Rudner AD, Hall BE, Ellenberger T, Moazed D. 2005. A nonhistone protein-protein interaction required for assembly of the SIR complex and silent chromatin. *Mol Cell Biol* **25:** 4514–4528.

Rusche LN, Kirchmaier AL, Rine J. 2002. Ordered nucleation and spreading of silenced chromatin in *Saccharomyces cerevisiae*. *Mol Biol Cell* **7:** 2207–2222.

Rusche LN, Kirchmaier AL, Rine J. 2003. The establishment, inheritance, and function of silenced chromatin in *Saccharomyces cerevisiae*. *Annu Rev Biochem* **72:** 481–516.

Schober H, Ferreira H, Kalck V, Gehlen LR, Gasser SM. 2009. Yeast telomerase and the SUN domain protein Mps3 anchor telomeres and repress subtelomeric recombination. *Genes Dev* **23:** 928–938.

Sekinger EA, Gross DS. 1999. SIR repression of a yeast heat shock gene: UAS and TATA footprints persist within heterochromatin. *EMBO J* **18:** 7041–7055.

* Seto E, Yoshida M. 2014. Erasers of histone acetylation: The histone deacetylase enzymes. *Cold Spring Harb Perspect Biol* **6:** a018713.

Shahbazian MD, Zhang K, Grunstein M. 2005. Histone H2B ubiquitylation controls processive methylation but not monomethylation by Dot1 and Set1. *Mol Cell* **19:** 271–277.

Shogren-Knaak M, Ishii H, Sun J-M, Pazin MJ, Davie JR, Peterson CL. 2006. Histone H4K16 acetylation controls chromatin structure and protein interactions. *Science* **311:** 844–847.

Sinclair DA, Guarente L. 1997. Extrachromosomal rDNA circles—A cause of aging in yeast. *Cell* **91:** 1033–1042.

Singer MS, Kahana A, Wolf AJ, Meisinger LL, Peterson SE, Goggin C, Mahowald M, Gottschling DE. 1998. Identification of high-copy disruptors of telomeric silencing in *Saccharomyces cerevisiae*. *Genetics* **150:** 613–632.

Sinha M, Watanabe S, Johnson A, Moazed D, Peterson CL. 2009. Recombinational repair within heterochromatin requires ATP-dependent chromatin remodeling. *Cell* **138:** 1109–1121.

Smith JS, Boeke JD. 1997. An unusual form of transcriptional silencing in yeast ribosomal DNA. *Genes Dev* **11:** 241–254.

Sperling AS, Grunstein M. 2009. Histone H3 N-terminus regulates higher order structure of yeast heterochromatin. *Proc Natl Acad Sci* **106:** 13153–13159.

Strahl-Bolsinger S, Hecht A, Luo K, Grunstein M. 1997. SIR2 and SIR4 interactions differ in core and extended telomeric heterochromatin in yeast. *Genes Dev* **11:** 83–93.

Stulemeijer IJ, Pike BL, Faber AW, Verzijlbergen KF, van Welsem T, Frederiks F, Lenstra TL, Holstege FC, Gasser SM, van Leeuwen F. 2011. Dot1 binding induces chromatin rearrangements by histone methylation-dependent, and -independent mechanisms. *Epigenetics Chromatin* **4:** 2.

Suka N, Suka Y, Carmen AA, Wu J, Grunstein M. 2001. Highly specific antibodies determine histone acetylation site usage in yeast heterochromatin and euchromatin. *Mol Cell* **8:** 473–479.

Suka N, Luo K, Grunstein M. 2002. Sir2p and Sas2p opposingly regulate acetylation of yeast histone H4 lysine16 and spreading of heterochromatin. *Nat Genet* **3:** 378–383.

Szilard RK, Jacques PE, Laramée L, Cheng B, Galicia S, Bataille AR, Yeung M, Mendez M, Bergeron M, Robert F, et al. 2010. Systematic identification of fragile sites via genome-wide location analysis of γ-H2AX. *Nat Struct Mol Biol* **17:** 299–305.

Taddei A, Hediger F, Neumann FR, Bauer C, Gasser SM. 2004. Separation of silencing from perinuclear anchoring functions in yeast Ku80 Sir4 and Esc1 proteins. *EMBO J* **23:** 1301–1312.

Taddei A, Van Houwe G, Nagai S, Erb I, van Nimwegen E, Gasser SM. 2009. The functional importance of telomere clustering: Global changes in gene expression result from SIR factor dispersion. *Genome Res* **19:** 611–625.

Tanner KG, Landry J, Sternglanz R, Denu JM. 2000. Silent information regulator 2 family of NAD-dependent histone/protein deacetylases generates a unique product, 1-*O*-acetyl-ADP-ribose. *Proc Natl Acad Sci* **97:** 14178–14182.

Tanny JC, Dowd GJ, Huang J, Hilz H, Moazed D. 1999. An enzymatic activity in the yeast Sir2 protein that is essential for gene silencing. *Cell* **99:** 735–745.

Thompson JS, Ling X, Grunstein M. 1994a. Histone H3 amino terminus is required for telomeric and silent mating locus repression in yeast. *Nature* **369:** 245–247.

Thompson JS, Johnson LM, Grunstein M. 1994b. Specific repression of the yeast silent mating locus HMR by an adjacent telomere. *Mol Cell Biol* **14:** 446–455.

Triolo T, Sternglanz R. 1996. Role of interactions between the origin recognition complex and SIR1 in transcriptional silencing. *Nature* **381:** 251–253.

Tsukamoto Y, Kato J, Ikeda H. 1997. Silencing factors participate in DNA repair and recombination in *Saccharomyces cerevisiae*. *Nature* **388:** 900–903.

Valenzuela L, Dhillon N, Dubey RN, Gartenberg MR, Kamakaka RT. 2008. Long-range communication between the silencers of HMR. *Mol Cell Biol* **28:** 1924–1935.

van Leeuwen F, Gafken PR, Gottschling DE. 2002. Dot1p modulates silencing in yeast by methylation of the nucleosome core. *Cell* **109:** 745–756.

Weiss K, Simpson RT. 1998. High-resolution structural analysis of chromatin at specific loci: *Saccharomyces cerevisiae* silent mating type locus HMLα. *Mol Cell Biol* **18:** 5392–5403.

Wu CS, Chen YF, Gartenberg MR. 2011. Targeted sister chromatid cohesion by Sir2. *PLoS Genet* **7:** e1002000.

Xu F, Zhang Q, Zhang K, Xie W, Grunstein M. 2007. Sir2 deacetylates histone H3 lysine 56 to regulate telomeric heterochromatin structure in yeast. *Mol Cell* **27:** 890–900.

Yang B, Britton J, Kirchmaier AL. 2008a. Insights into the impact of histone acetylation and methylation on Sir protein recruitment, spreading, and silencing in *Saccharomyces cerevisiae*. *J Mol Biol* **381:** 826–844.

Yang B, Miller A, Kirchmaier AL. 2008b. HST3/HST4-dependent deacetylation of lysine 56 of histone H3 in silent chromatin. *Mol Biol Cell* **19:** 4993–5005.

Yu Q, Olsen L, Zhang X, Boeke JD, Bi X. 2011. Differential contributions of histone H3 and H4 residues to heterochromatin structure. *Genetics* **188:** 291–308.

C H A P T E R 9

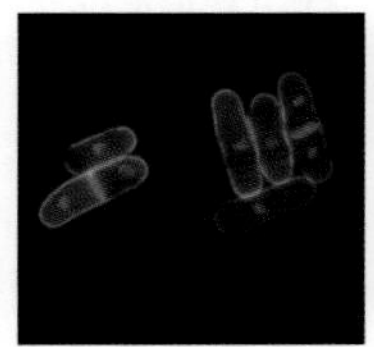

Epigenetic Regulation of Chromatin States in *Schizosaccharomyces pombe*

Robin C. Allshire[1] and Karl Ekwall[2]

[1]Wellcome Trust Centre for Cell Biology, The University of Edinburgh, Edinburgh EH9 3JR, Scotland, United Kingdom;
[2]Department of Biosciences and Nutrition, Karolinska Institutet, Center for Biosciences, NOVUM, S-141 83, Huddinge, Sweden

Correspondence: karl.ekwall@ki.se

SUMMARY

This chapter discusses the advances made in epigenetic research using the model organism fission yeast *Schizosaccharomyces pombe*. *S. pombe* has been used for epigenetic research since the discovery of position effect variegation (PEV). This is a phenomenon in which a transgene inserted within heterochromatin is variably expressed, but can be stably inherited in subsequent cell generations. PEV occurs at centromeres, telomeres, ribosomal DNA (rDNA) loci, and mating-type regions of *S. pombe* chromosomes. Heterochromatin assembly in these regions requires enzymes that modify histones and the RNA interference (RNAi) machinery. One of the key histone-modifying enzymes is the lysine methyltransferase Clr4, which methylates histone H3 on lysine 9 (H3K9), a classic hallmark of heterochromatin. The kinetochore is assembled on specialized chromatin in which histone H3 is replaced by the variant CENP-A. Studies in fission yeast have contributed to our understanding of the establishment and maintenance of CENP-A chromatin and the epigenetic activation and inactivation of centromeres.

Outline

Chapter-opening figures: (*Left*) Courtesy of Iain Hagan; (*right*) with permission from Johannes Hegemann and Ursula Fleig, Heinrich-Heine-Universität, Düsseldorf.

Cite as *Cold Spring Harb Perspect Biol* doi: 10.1101/cshperspect.a018770

OVERVIEW

This chapter discusses the advances made in epigenetic regulation using the fission yeast *Schizosaccharomyces pombe* as a model organism. *S. pombe* has been exploited for the investigation of epigenetically regulated processes since the discovery of position effect variegation (PEV), originally observed in *Drosophila*. This is a phenomenon in which a transgene is variably expressed, but often the expression pattern is then stably inherited in subsequent cell generations. This variable gene expression, however, occurs only when the transgene integrates in or near heterochromatic regions of the genome. The multitude of studies exploiting PEV or gene silencing in fission yeast have led to great progress not only in elucidating the mechanisms of heterochromatin assembly and maintenance, but understanding how heterochromatin creates a repressive environment. Because many of the mechanisms used to create repressive chromatin environments are common to higher eukaryotes including humans, *S. pombe* has provided a good model organism for epigenetic research.

Heterochromatin is found mainly at centromeres, telomeres, ribosomal DNA (rDNA) loci, and the mating-type regions of *S. pombe* chromosomes. Heterochromatin assembly requires the RNA interference (RNAi) machinery and enzymes that modify histones. One of the key modifying enzymes is the histone lysine (K) methyltransferase (HKMT or KMT) Clr4, which methylates histone H3 on lysine 9 (H3K9), now considered a hallmark of heterochromatin.

Centromeres consist of two types of chromatin: the pericentric heterochromatin regions and CENP-A (also referred to as CenH3 or Cnp1 in *S. pombe*)—containing chromatin. Pericentric heterochromatin plays a crucial function in mitotic and meiotic cell divisions, providing genome stability and sister-chromatid cohesion, partly by supporting the establishment of adjacent centromeric CENP-A-containing chromatin. CENP-A chromatin is required for the formation of a protein structure called the kinetochore, which interacts with spindle-microtubules to segregate chromosomes in mitosis and meiosis. CENP-A is a centromere-specific histone H3 variant, which replaces the canonical H3 histone in centromeric nucleosomes, and is targeted to centromeres by the combined action of chaperone proteins and loading factors. Heterochromatin in the mating-type region is essential for regulating gene expression, which directs the sexual life cycle of *S. pombe*. It provides a silencing mechanism to prevent the expression of inappropriate mating-type information in a haploid cell, yet allows for mating-type switching at the appropriate time. Heterochromatin at telomeres plays an important function in maintaining the integrity of linear chromosomes. Heterochromatin at rDNA repeats may suppress recombination, thus maintaining genome integrity.

In addition to heterochromatin research, *S. pombe* is used to study several other aspects of epigenetic control such as chromatin remodeling and nuclear organization. The relatively small size of its genome allows for convenient genomic analysis, especially when coupled to the simplicity of working with a unicellular organism in which one can exploit both the haploid and diploid phases of the life cycle. *S. pombe* has also become an important model organism for understanding how epigenetic mechanisms operate on a genome-wide scale, referred to as epigenomics.

 Cite as *Cold Spring Harb Perspect Biol* doi: 10.1101/cshperspect.a018770

1 *Schizosaccharomyces pombe* LIFE CYCLE

Fission yeast, *Schizosaccharomyces pombe*, is found in the fermentations involved in the production of beer in subtropical regions; "pombe" is the Swahili word for beer. *S. pombe* is primarily a haploid (1N) unicellular organism. In nutrient-rich media, wild-type cells undergo a mitotic division approximately every 2 h. However, a variety of conditions, or conditional mutants, can be used to block cells at distinct stages of the cell cycle or synchronize cell cultures in G_1/S, G_2, or at metaphase. This is particularly useful because the G_1 phase is very short in normally growing cultures and cells pass almost immediately into S phase following cytokinesis; the major portion of the cell cycle is spent in G_2 (Fig. 1) (Egel 2004).

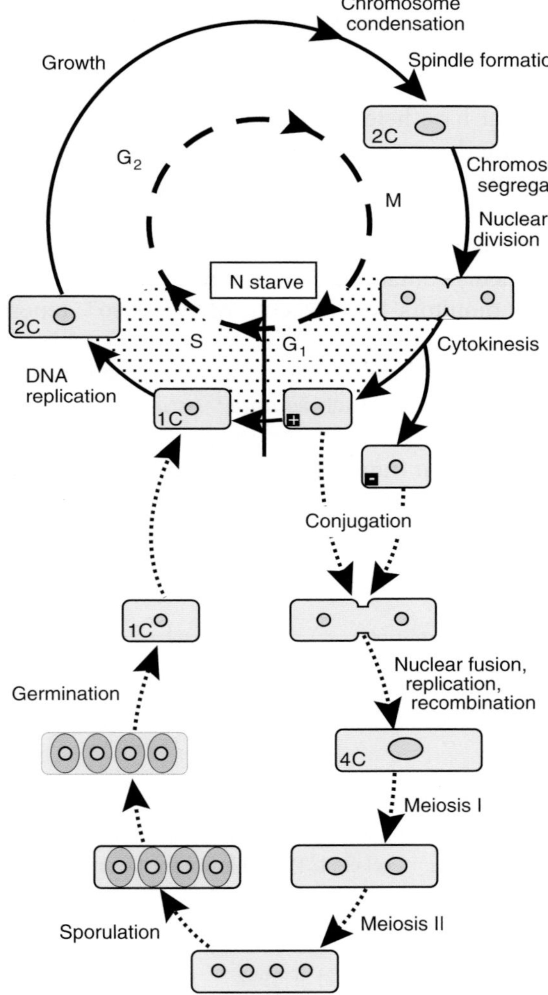

Figure 1. Life cycle of the fission yeast, *S. pombe*. Fission yeast has a short G_1 taking less than 10% of the cell cycle (stippled area is expanded to aid representation). In rich medium, G_1 cells proceed into S phase followed by a long G_2 (~70% of the cell cycle), mitosis, and cytokinesis. When starved of nitrogen, cells of opposite mating-type (+ and −) conjugate, after which nuclei fuse in a process known as karyogamy. Premeiotic replication and recombination allows meiosis I and II to proceed, resulting in four haploid nuclei that are separated into four spores in an ascus. Provision of rich medium allows germination of spores and resumption of the vegetative cell cycle.

S. pombe has two mating types, named plus (+) and minus (−), and, like *Saccharomyces cerevisiae,* can switch between opposite mating types. Mating types are equivalent to dimorphic sexes in higher eukaryotes, although they are haploid. The information for both mating types resides in the genome as epigenetically regulated silent cassettes: The (+) information cassette is found at the *mat2*-P locus and the (−) information cassette at the *mat3*-M locus. These silent loci provide the genetic template for mating-type identity, but mating type itself is determined by the information (+ or −) that resides at the active *mat1* locus. Switching of information at the active *mat1* locus occurs by recombination between one of the silent loci (*mat2*-P or *mat3*-M) and the *mat1* locus according to a strict pattern (Egel 2004). When starved of nitrogen, cells stop dividing and arrest in G_1, which promotes the sexual phase of the life cycle through conjugation of pairs of "+" and "−" cells to form diploid zygotes (Fig. 1). After mating and nuclear fusion, premeiotic replication occurs (increasing DNA content from 2N to 4N). This is followed by the pairing and recombination of homologous chromosomes, and ends with the reductional meiosis I division and equational meiosis II division. This produces four separate haploid nuclei (1N) that become encapsulated into spores enclosed in an ascus. The subsequent provision of a rich nutrient source allows germination and resumption of vegetative growth and mitotic cell division.

Nonswitching derivatives have been isolated or constructed in which all cells are either "+" or "−" mating type. This facilitates controlled mating between strains of distinct genotypes. Although *S. pombe* is normally haploid, it is possible to select for diploid strains. Such diploid cells can then divide by vegetative mitotic growth until starved of nitrogen when they, too, undergo meiosis and form "azygotic asci."

2 SCREENS TO IDENTIFY HETEROCHROMATIN COMPONENTS

The type of chromatin surrounding a gene can strongly influence its expression. This was originally shown in the fruit fly, *Drosophila melanogaster*, in which chromosomal rearrangements that move the *white* gene close to centromeric heterochromatin led to its variable expression in eye facets causing variegation in eye coloration (see Fig. 1 in Ch. 12 [Elgin and Reuter 2013]). Studies in *S. pombe* showed this through the variable silencing of a reporter gene when inserted into different centromeric regions. Scientists have subsequently studied silent chromatin (i.e., heterochromatin) as well as active chromatin (euchromatin) by exploiting PEV of reporter or translocated genes. In fact, in *S. pombe*, this approach has been fundamental in identifying many of the molecular components that define heterochromatin structures at centromeres and other regions of the genome.

The first genetic screens to identify modifiers of PEV were performed by selecting for mutants that allow expression of the normally silent *mat2-P* and *mat3-M* cassettes. Mutants were identified because of aberrant mating patterns or when selecting for expression of a $ura4^+$ transgene inserted in the silent mating-type region (Thon and Klar 1992; Ekwall and Ruusala 1994; Thon et al. 1994). These screens identified a set of genes called cryptic loci regulators (i.e., the $clr1^+$, $clr2^+$, $clr3^+$, $clr4^+$), as well as the $rik1^+$ and $swi6^+$ genes (for the gene products see Table 1). These proteins are the functional counterparts of the silent information regulator (Sir) proteins found in *S. cerevisiae* (described in Ch. 8 [Grunstein and Gasser 2013]). The link between heterochromatin formation occurring at centromeres and in the *mat2-mat3* region became clear when several of the Clr proteins were shown to be required for silencing at centromeres as well as the mating-type loci in *S. pombe* (Allshire et al. 1995).

To understand how the PEV screens have helped in identifying components involved in centromeric heterochromatin, it is useful to know the nature of the underlying DNA sequence at centromeres. At the DNA level, centromere regions in fission yeast are composed of outer repeats (subdivided into elements known as *dg* and *dh*), flanking a central domain. The central domain includes the innermost repeats (*imr*), and a central core (*cnt*; Fig. 2A). The three centromeres—*cen1*, *cen2*, and *cen3*—occupy ~40, ~60, and ~120 kb on chromosomes I, II, and III, respectively (Fig. 2B; reviewed in Egel 2004; Pidoux and Allshire 2004). The repetitive nature of fission yeast centromere DNA resembles the larger, more complex repeated structures associated with many metazoan centromeres, but they are more amenable to manipulation. Because repetitive DNA frequently correlates with the presence of heterochromatin in other eukaryotes (see also Ch. 16 [Martienssen and Moazed 2014]), this suggested that fission yeast centromeres might have heterochromatic properties such as the ability to hinder gene expression. Reporter gene silencing has been monitored in fission yeast by phenotypic assays similar to those used in *S. cerevisiae*. For example, when the $ura4^+$ reporter is silenced, colonies can form on counterselective media containing 5-fluoroorotic acid. An alternative assay monitors the silencing of the $ade6^+$ reporter, which results in red (repressed) rather than white (expressed) colony color (Fig. 2A). Thus, placement of a normally expressed gene, such as $ura4^+$ or $ade6^+$ within the centromere (as defined by the outer repeat and central domain) results in transcriptional silencing. Compared to the outer repeats, silencing of $ade6^+$ within the central domain is comparatively unstable, resulting in variegated

Table 1. Gene products identified in *S. pombe* forward PEV screens

Gene product	Molecular function	Reference(s)
Clr1 (Cryptic loci regulator 1)	Zinc ion binding protein, part of the SHREC complex	Thon and Klar 1992
Clr2 (Cryptic loci regulator 2)	Transcription-silencing protein Clr2; part of the SHREC complex	Ekwall and Ruusala 1994; Thon et al. 1994
Clr3 (Cryptic loci regulator 3)	Histone deacetylase class II (H3-K14-specific); part of the SHREC complex	Ekwall and Ruusala 1994; Thon et al. 1994; Grewal et al. 1998
Clr4 (Cryptic loci regulator 4)	Histone methyltransferase KMT activity (H3-K9-specific)	Ekwall and Ruusala 1994; Thon et al. 1994
Rik1	DNA-binding protein; part of the CLRC ubiquitin ligase complex	Egel et al. 1989; Ekwall and Ruusala 1994
Swi6	Chromo domain/shadow protein, heterochromatin protein 1 homolog	Ekwall and Ruusala 1994; Lorentz et al. 1994
Clr6 (Cryptic loci regulator 6)	Histone deacetylase class I	Grewal et al. 1998
Ago1 was identified as Csp9 (centromere: suppressor of position effect 9)	Argonaute protein, part of the RITS complex	Ekwall et al. 1999; Volpe et al. 2003
Rpb7 was identified as Csp3 (centromere: suppressor of position effect 3)	DNA-directed RNA polymerase II subunit	Ekwall et al. 1999; Djupedal et al. 2005
Rpb2	DNA-directed RNA polymerase II subunit	Kato et al. 2005
Epe1 (Enhanced position effect 1)	jmjC domain protein	Ayoub et al. 2003
Cwf10 was identified as Csp4	Splicing factor	Bayne et al. 2008; Ekwall et al. 1999
Prp39 was identified as Csp5	Splicing factor	Bayne et al. 2008; Ekwall et al. 1999

 Cite as *Cold Spring Harb Perspect Biol* doi: 10.1101/cshperspect.a018770

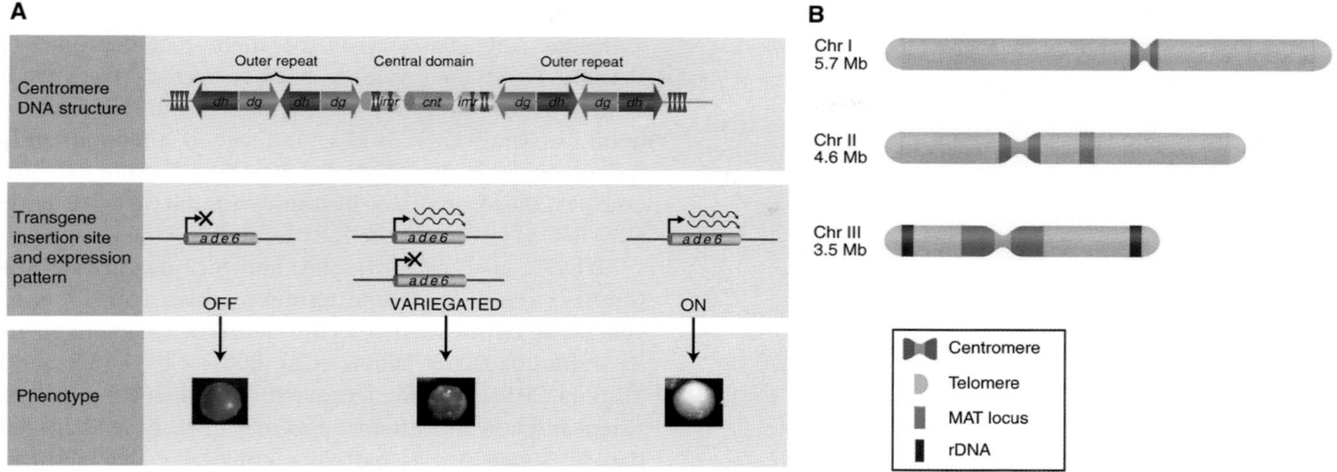

Figure 2. Distinct outer repeat heterochromatin and central kinetochore domains at fission yeast centromeres. (*A, top*) Representation of a fission yeast centromere. The central domain (pink, kinetochore) is composed of *imr* and *cnt* elements, the outer repeats contain transcribed *dg* and *dh* repeats (green, heterochromatin). All three centromeres have a similar overall arrangement; however, the number of outer repeats differs: *cen1* (40 kb) has two, *cen2* (65 kb) has three, and *cen3* (110 kb) has approximately 13. Clusters of transfer RNA (tRNA) genes (double arrowheads) occur in the *imr* region and at the extremities of all three centromeres. (*Middle*) Schematically shows transcription patterns of marker genes placed within the outer repeats, central domain, or beyond the centromere. (*Bottom*) Images showing the phenotype of *S. pombe* colonies of $ade6^+$ transgenics inserted at various sites within the centromere. Cells expressing $ade6^+$ from a transgene inserted in sequences outside the centromere form white colonies. When $ade6^+$ is inserted at sites within the outer repeats, expression is silenced and red colonies are formed. Expression of $ade6^+$ from the central domain is typically variegated, resulting in red, white, and sectored colonies. (*B*) A schematic representation of *S. pombe* chromosomes. The three chromosomes are depicted showing the four main regions of heterochromatin: centromere, telomere, *mat2/3*, and rDNA regions.

colonies (i.e., either red, white, or red–white sectored colonies). Moreover, just 1 kb distal to the outer repeats no silencing occurs (Allshire et al. 1994, 1995), indicating that transcriptional repression is confined to the centromere.

3 DIFFERENT TYPES OF HETEROCHROMATIN

Apart from centromeres, other repressive regions of the genome can silence transgene expression, notably the mating-type loci (*mat2-mat3*), the rDNA region, and telomeres (reviewed in Egel 2004). Together, these constitute the four regions of heterochromatin in *S. pombe* (Fig. 2B). With the possible exception of the rDNA region, it turns out that each region of silent chromatin has an essential function. At centromeres, heterochromatin ensures normal chromosome segregation (Allshire et al. 1995; Ekwall et al. 1995), whereas at the mating-type loci it facilitates and regulates mating-type switching (reviewed in Egel 2004). Silent chromatin formed adjacent to telomeres (Nimmo et al. 1994; Kanoh et al. 2005) plays a role in meiotic chromosome segregation (Nimmo et al. 1998), whereas the function of heterochromatin in the rDNA region of *S. pombe* has not been established yet (Thon and Verhein-Hansen 2000; Cam et al. 2005). It could be involved in maintaining rDNA stability by preventing recombination between rDNA repeat sequences just like in *S. cerevisiae* (described in Ch. 8 [Grunstein and Gasser 2013]).

The mechanisms that *S. pombe* uses to achieve heterochromatic silencing distinguish it from *S. cerevisiae*, making it more akin to *Neurospora crassa*, plants, and metazoa. So, although *S. pombe* and *S. cerevisiae* have silent chromatin at telomeres, the mating-type loci, and rDNA regions, only *S. pombe* shows silencing at centromeres. Silencing is mediated by distinct histone-modifying activities (in particular, histone H3K9 methylation), RNAi proteins, the RNA polymerase (RNA Pol) II machinery and homologs of heterochromatin protein 1 (HP1; see Table 1). These mechanisms are conserved in higher eukaryotes. Interestingly *S. pombe*, in contrast to *N. crassa* and many other eukaryotes, appears to lack any detectable DNA methylation (Wilkinson et al. 1995), a common mechanism used for silencing chromatin in higher eukaryotes (see Ch. 15 [Li and Zhang 2014]; Ch. 13 [Pikaard and Mittelsten Sheid 2014]); thus, silencing in fission yeast is mediated primarily by chromatin modifications and the RNAi machinery.

4 *S. pombe* CENTROMERES: A PARADIGM FOR STUDYING HETEROCHROMATIN

Heterochromatin formation at centromeres is a complex process involving chromatin factors, noncoding RNA, RNA Pol II, and its associated proteins. It is also strictly controlled by chromatin boundaries (see Sec. 4.5), which prevent the spreading of heterochromatin. It is important to note that chromatin boundaries need to be maintained throughout the cell cycle. This section discusses these factors and processes, which together control the formation of heterochromatin.

4.1 Chromatin Factors

In chromatin, the amino-terminal tails of histones H3 and H4 are subject to a range of posttranslational modifications, which generally correlate with active or repressed states (summarized in Ch. 3 [Allis et al.]). The centromeric heterochromatin domain at outer repeats in fission yeast contains di- and trimethylated histone H3 at lysine 9 (H3K9me2 and H3K9me3; Nakayama et al. 2001; Yamada et al. 2005). In most eukaryotic organisms, H3K9me2 and H3K9me3 are characteristic of silent heterochromatin. Because H3K9 methylation extends over a long stretch of heterochromatin, how does the process of heterochromatinization occur? Studies indicate it predominantly spreads through the concerted action of histone deacetylases (HDACs), the Clr4 histone lysine methyltransferase (KMT/HMTase), and the HP1 homolog, Swi6.

HDACs such as Clr3, Clr6, and Sir2 deacetylate histone H3 at various lysine residues, which contribute to generating repressive heterochromatin (discussed in Shankaranarayana 2003; Wiren et al. 2005; Ch. 5 [Seto and Yoshida 2014]). More recent studies, dissecting where, when, and how the different enzymes work, have identified that Clr3 operates as part of a complex called SHREC, which helps to deacetylate chromatin (Fig. 3) (Sugiyama et al. 2007). Clr4 is the key KMT/HKMT that methylates H3K9 over the centromeric outer repeats. Clr4 functions as part of a complex called CLRC, which, through methylation of H3K9, creates a specific binding site recognized by factors containing a chromodomain motif. Interestingly, Clr4 contains a chromodomain; thus, it not only methyl-

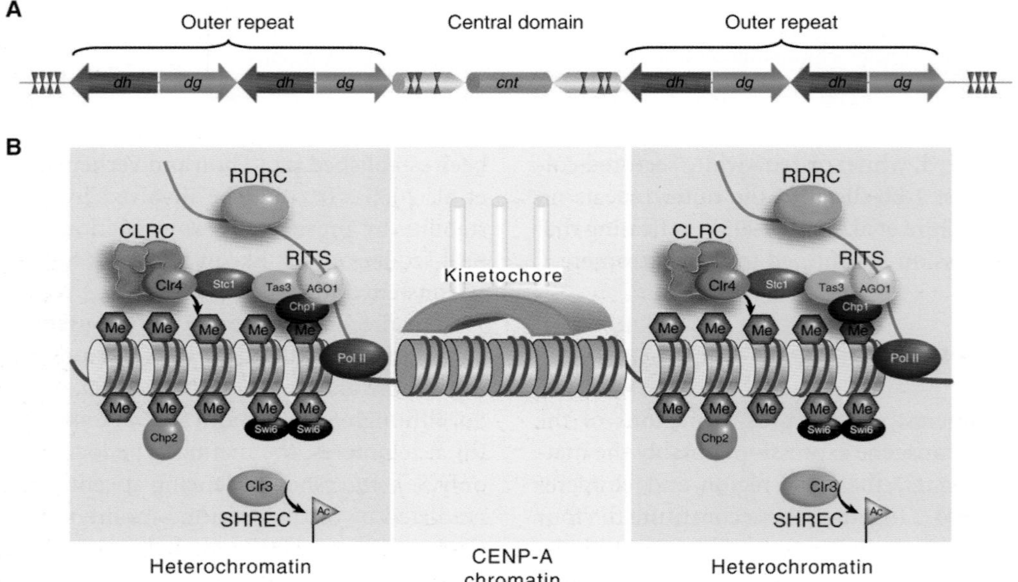

Figure 3. Centromeric chromatin domains in *S. pombe*. (*A*) A schematic representation of the symmetrical DNA arrangement in *S. pombe* centromeres. (*B*) Heterochromatin: outer repeats are packaged in nucleosomes, which are methylated on lysine 9 of histone H3 (H3K9me) by Clr4 as part of the CLRC complex. This allows the binding of chromodomain proteins Chp1 (a component of the RNAi RITS complex), Chp2, and Swi6. Collectively, these and other factors, including the SHREC complex-containing Clr3 histone deacetylase activity and the RDRC complex, act to assemble and propagate heterochromatin. Central "kinetochore" chromatin: CENP-A is found in the central domain where it probably replaces the majority of histone H3 to form specialized nucleosomes (coral colored). In addition to CENP-A, several proteins assemble at the central domain chromatin to form the inner and outer kinetochore multiprotein structures (coral arc). See Figure 6 for a description of the kinetochore.

Cite as *Cold Spring Harb Perspect Biol* doi: 10.1101/cshperspect.a018770

ates H3K9, but also binds H3K9me2/3, likely effecting adjacent H3K9 methylation through its catalytic domain in a sequential fashion (Zhang et al. 2008).

Swi6 is a chromodomain-containing HP1 homolog known to bind methylated H3K9 and contribute to the formation of silent chromatin over the outer repeats (Fig. 3) (Bannister et al. 2001; Nakayama et al. 2001). In addition to the chromodomain, Swi6 has another functional domain called the chromoshadow domain, which causes self-dimerization (Cowieson et al. 2000). This suggests an important role for nucleosome bridging in heterochromatin assembly (Canzio et al. 2011), presumably altering nucleosome repeat length in heterochromatin compared to euchromatin (Lantermann et al. 2010). Reporter gene insertions at the outer repeats (e.g., the *ura4*$^+$ gene) are notably enriched in H3K9me2 and Swi6 proteins, stressing how important Clr4-mediated H3K9 methylation and Swi6 heterochromatin spreading is not only within the centromere outer repeats but also into neighboring interposing sequences (Cowieson et al. 2000; Nakayama et al. 2001).

Chp1, another chromodomain protein, also associates with outer repeat chromatin at centromeres by binding methylated H3K9, as does Swi6. Chp1 is a member of the RNA-induced transcriptional silencing or RITS complex—another player in the heterochromatin-forming process. The RITS complex is part of the RNAi machinery, which is also required for the complete methylation of histone H3K9 over the outer repeats and inserted reporter genes (Partridge et al. 2002; Motamedi et al. 2004; Sadaie et al. 2004). The contribution of RNAi to heterochromatin formation is further explained in Section 4.2 and in Ch. 16 (Martienssen and Moazed 2014).

Altogether there are four chromodomain proteins in *S. pombe* (Chp1, Chp2, Swi6, and Clr4). Two of these, Clr4 and Chp1, have reduced affinities for methylated H3K9 when H3K4 is acetylated by the histone lysine acetyltransferase (HAT or sometimes abbreviated to KAT) enzyme, Mst1. This suggests that H3K4 methylation acts like a switch: After DNA replication, Clr4/Chp1 occupancy at H3K9me2/3 is promoted on newly deposited histones where H3K4 is unacetylated. On Mst1 acetylation of H3K4, occupancy switches to favor Chp2/Swi6 as the cell cycle progresses and heterochromatin is fully reassembled after DNA replication (Xhemalce and Kouzarides 2010). The concept of heterochromatin maintenance throughout the cell cycle is expanded in Section 4.4.

However, chromodomains are not the only interaction surface between these HP1-like proteins with chromosomes. Interestingly, Swi6 was recently shown to have a strong RNA-binding activity via a hinge region in the protein, and this RNA-binding domain has a specific function in the destruction of transcripts originating from heterochromatin regions (described in Sec. 4.2). Mutations in the hinge domain lead to reduced silencing of such transcripts, but the integrity of heterochromatin is not affected, indicating that Swi6 performs an effector function downstream from methylated H3K9 (Keller et al. 2012). Thus, chromatin factors including histone modifiers and HP1-like proteins cooperate to specify heterochromatin in *S. pombe.*

4.2 RNAi and the RNA Pol II Machinery in Heterochromatin Assembly

RNAi is an important mechanism contributing to heterochromatin formation in *S. pombe.* The phenomenon of RNAi was first discovered in *Caenorhabditis elegans* in which the expression of double-stranded RNA (dsRNA) abolished the expression of a homologous gene. It soon became apparent that this form of RNAi is related to the process of transcriptional gene silencing (TGS) described in plants (see Ch. 13 [Pikaard and Mittelsten Scheid 2014]) and quelling in *N. crassa* (Ch. 10 [Aramayo and Selker 2013]). These are processes of silencing that occur when a region is transcriptionally active, and transcripts that generate regions of dsRNA (e.g., through the self-annealing of inverted repeats) can be processed into small RNA fragments (termed small RNA biogenesis). These small RNAs are taken up by effector complexes and can trigger silent chromatin via targeting activities, which cause DNA methylation and histone modification (see Figs. 1 and 2 of Ch. 16 [Martienssen and Moazed 2014]). This process of silencing appears to be in operation from *S. pombe* to plants and metazoans, including mammals.

Studies of the components of the RNAi machinery in *S. pombe* have led to significant advances in our understanding of RNAi-mediated chromatin modification and silencing. Mutants in the RNAi machinery in *S. pombe* result in reduced H3K9me2 and loss of silencing over the outer repeats of centromeres (Volpe et al. 2002). Surprisingly at the time, these RNAi mutants revealed overlapping noncoding RNA (ncRNA) transcripts of a discrete size, originating from centromeric outer repeats. These ncRNAs were homologous to naturally occurring small dsRNAs called small interfering RNAs (siRNAs; $\sim$21 nt) that had been isolated and sequenced from *S. pombe* (Reinhart and Bartel 2002). We now know that these long noncoding double-stranded centromere repeat transcripts are cleaved by the Dicer (Dcr1) enzyme to generate siRNAs. These siRNAs then act to guide the RNAi machinery to homologous transcripts.

One of the initial questions in trying to elucidate how TGS worked in *S. pombe* was "Which RNA polymerase was responsible for the noncoding transcription of centromere repeats?" Mutation of either subunit of RNA Pol II (Rpb2 and Rpb7) results in defective centromere silencing (Djupedal et al. 2005; Kato et al. 2005) although these mutations display very different phenotypes. The *rpb7-1* mutant shows reduced levels of centromere repeat transcription, resulting in less ncRNA and, consequently, less siRNA production and a loss of silent chromatin. This implies that RNA Pol II is required for the transcription of centromere repeats, which then provides the primary substrate for RNAi. In contrast, centromeric transcripts in the *rpb2-m203* mutant are produced but not processed into siRNA, and H3K9 methylation at centromeres is reduced. These studies indicate that RNAi not only requires an RNA Pol II transcript but that, like other RNA-processing events, the production of centromeric siRNA may be coupled to transcription by interactions between the RNAi machinery, chromatin, histone-modifying enzymes, and RNA Pol II.

The RNAi machinery in *S. pombe* is complex and not yet fully understood. In addition to transcription of noncoding centromeric outer repeats by RNA Pol II and the processing of transcripts into siRNAs by Dcr1, the key activities of the RNAi machinery involve two complexes: RITS and RDRC (RNA-directed RNA polymerase complex; summarized in Fig. 4 of Ch. 16 [Martienssen and Moazed 2014]). The RITS complex incorporates siRNAs to then direct it to centromere outer repeats via sequence recognition as well as H3K9me2/3 recognition through the Chp1 chromodomain (see Sec. 4.3). RDRC is recruited to amplify the process of TGS by generating more long double-stranded ncRNAs through the action of Rdp1 (RNA-directed RNA polymerase 1). Rdp1 transcribes from transcripts primed with siRNAs presented by the RITS complex. The chromatin-modifying machineries that execute chromatin changes include the CLRC (for Clr4-Rik1-Cul4 complex) and the SHREC complex. CLRC is recruited by the RITS complex via the Stc1 protein (Bayne et al. 2010; see Sec. 4.4). Once recruited, Clr4 methylates H3K9 over the outer repeats and this allows it to bind directly to H3K9me2/3 via its chromodomain. The HP1 homologs Swi6 and Chp2 play further roles in establishing and maintaining heterochromatin.

Initiation of transcription, transcriptional elongation, and transcript processing are as important for heterochromatin assembly as they are for euchromatic gene expression. Several associated factors and activities are important for these different steps of the RNA Pol II transcription cycle, in addition to RNA Pol II itself. FACT, an RNA Pol II-associated chromatin assembly factor, and Spt6, another RNA Pol II-associated protein both colocalize to pericentric repeats. Spt6 is specifically required for facilitating trimethylation of H3K9, Swi6 binding, siRNA production, and recruitment of the HDAC enzyme Clr3 (Kiely et al. 2011). Mutations in the FACT component, Pob3, have a similar phenotype to *spt6* implicating FACT in the same processes (Lejeune et al. 2007).

Interestingly, mutations in chromatin-modifying activities or RNA-processing factors have been shown to suppress the need for Dcr1. For example, loss of Mst2 activity, a H3K14-specific acetyltransferase, completely suppresses *dcr1* mutants; that is, it eliminates the need for the RNAi machinery in heterochromatin maintenance, but not in the establishment of new heterochromatin (Reddy et al. 2011). This suggests that an important role of the RNAi directed heterochromatin process is to prevent Mst2 activity, which might interfere with CLRC recruitment.

Another example of bypassing the RNAi pathway is by knocking out the gene encoding Mlo3, involved in mRNP biogenesis and RNA quality control. This also suppresses the need for Dcr1 (Reyes-Turcu et al. 2011). It was suggested that in *mlo3 ago1* knockout cells there is an aberrant accumulation of centromeric transcripts, which results in the recruitment of the CLRC complex via Rik1 (Hong et al. 2005; Horn et al. 2005). In wild-type cells, CLRC acts downstream from the RNAi pathway and is required for heterochromatin silencing by recruiting and promoting the activity of the Clr4 enzyme. Thus, this RNAi-independent recruitment mechanism of CLRC allows the induction of heterochromatin assembly at repeat sequences in *mlo3 ago1* knockout cells (Reyes-Turcu et al. 2011).

Another RNA Pol II linked event is splicing of pre-mRNA by the spliceosome. There is not a general requirement for the spliceosome or the splicing process in heterochromatin formation, although some specific splicing factors are required for siRNA production (Bayne et al. 2008; Bernard et al. 2010). It remains to be determined how exactly these splicing factors contribute to RNAi, but their physical interaction with the RDRC suggests a direct role in RNAi. Findings in *C. elegans* and *Cryptococcus neoformans* corroborate the idea that splicing can be linked to RNAi. Clearly, heterochromatin formation is a complex process involving the RNA Pol II machinery, RNAi, splicing factors, and chromatin-modifying machinery.

4.3 Establishment of Heterochromatin

Detailed analysis of the genetic requirements for the initial establishment and the subsequent maintenance of heterochromatin indicate that these steps are distinct. Two models have been suggested to explain how heterochromatin is established over repeat sequences.

The first model suggests that RITS-associated siRNA may home in on nascent transcripts as they emerge from RNA Pol II engaged with the homologous locus. Consistent with this notion, artificial tethering of the RITS subunit

Cite as *Cold Spring Harb Perspect Biol* doi: 10.1101/cshperspect.a018770

Tas3 to the mRNA of the normally active *ura4*$^+$ locus can initiate heterochromatin formation over the *ura4*$^+$ gene (Buhler et al. 2006). But where do the initial siRNAs that trigger heterochromatin formation come from? The production of spurious small "primal" RNA by RNA degradation has been suggested to start off the heterochromatin establishment process (Halic and Moazed 2010). Alternatively, RNA Pol II transcripts could fold into self-structured double-stranded regions that are cleaved by Dcr1 to initiate heterochromatin formation (Djupedal et al. 2009; see Fig. 2 of Ch. 16 [Martienssen and Moazed 2014]). Once recognition has taken place, the RITS-siRNA complex might be stabilized on these transcripts resulting in the recruitment of chromatin-modifying activities such as Clr4 as part of the CLRC complex (see Fig. 4 of Ch. 16 [Martienssen and Moazed 2014]). Consistent with this idea, artificial tethering of Clr4 can completely bypass the need for RNAi during synthetic heterochromatin assembly (Kagansky et al. 2009).

A surprising result was that centromeric siRNAs were lost in cells lacking Clr4 activity (Noma et al. 2004; Hong et al. 2005). It is possible that the absence of Clr4 affects siRNA production by destabilizing associations between various components at the interface between transcription, RNAi, and chromatin modification (Motamedi et al. 2004). In some repeat regions, Clr4 can trigger RNAi independently of H3K9 methylation (Gerace et al. 2010) making it possible that Clr4 has additional targets for its HKMTase activity apart from histone H3, which might be important for siRNA amplification from repeat sequences.

A second model proposes that H3K9 methylation may be required to allow the generation of siRNAs in *cis* via the action of various RNA-processing activities (e.g., Rdp1) on primary centromeric transcripts (Noma et al. 2004; Sugiyama et al. 2005). Experiments with mutants of the RITS component Tas3, in which Tas3 is separated from the complex (so that Tas3 still interacts with Chp1 binding to H3K9me but not with Ago1), suggest that H3K9me is required before the production of small RNAs (Partridge et al. 2007). However, experiments showing that cells with H3K9 substitutions still produce small RNA argue against this model (Djupedal et al. 2009; Gerace et al. 2010). Clearly, more research is needed to understand what essential initiating steps versus reinforcing steps are required to generate heterochromatin.

4.4 Maintenance of Heterochromatin

Once heterochromatin is established, it must be maintained through cell division. The heterochromatin protein Swi6 is released from chromatin during mitosis because of the phosphorylation of the adjacent H3S10 residue (Fig. 4A) (Chen et al. 2008). Therefore, it is essential that Swi6 binding and the heterochromatic state are reestablished in S phase, when chromatin is replicated. Transcription of noncoding repeat sequences plays a crucial role in this maintenance mechanism, as it does for the establishment phase. The implication of RNA Pol II in heterochromatin assembly was initially thought to be a paradox because condensed heterochromatin was considered refractory to RNA polymerase activity. However, it appears that RNA Pol II is associated with the repeat sequences mainly during S phase of the cell cycle (Chen et al. 2008; Kloc et al. 2008). The decondensation of heterochromatin that occurs in this short window of time during DNA replication, before new heterochromatin is made, leads to a transient burst of RNA Pol II activity. This burst of transcription is thought to result in a boost of siRNA production leading to recruitment of heterochromatin assembly factors (Fig. 4A). As described in Section 4.1, other chromatin factors (i.e., HDACs) are also recruited at this stage to allow for reassembly of heterochromatin. A popular model is that siRNAs guide the RITS and RDRC complexes (i.e., RNAi machinery) to the chromosomes via the nascent noncoding RNA, providing a platform for the RNAi machinery (Motamedi et al. 2004). Interestingly, the LIM domain protein, Stc1, interacts both with the RITS Ago1 protein and CLRC; tethering of Stc1 is sufficient for heterochromatin assembly (Bayne et al. 2010). Thus, Stc1 seems to constitute the missing link between chromatin modification and the RNAi pathway, providing an explanation to why RNAi is needed to target H3K9 methylation.

In an alternative but not mutually exclusive model, RNAi plays a role in transcriptional termination leading to the avoidance of collisions between DNA polymerase and RNA Pol II (Fig. 4B) (Zaratiegui et al. 2011). In the absence of RNAi, experiments showed that such collisions cause stalled DNA replication forks that are repaired by homologous recombination (Zaratiegui et al. 2011). In this model, the HKMT activity is recruited via association with DNA polymerase epsilon interacting with the CLRC complex, linking the heterochromatin assembly process to lagging strand synthesis of DNA (Li et al. 2011). Thus, there is evidence for two distinct ways of CLRC recruitment to chromosomes in *S. pombe*. Clearly, further experiments are needed to determine their relative importance in the maintenance of heterochromatin.

4.5 Chromatin Boundaries: The Antisilencing Activity of Epe1 and Cullin Ligases

A key question is how distinct boundaries are set and maintained between heterochromatin and euchromatin. There are several types of heterochromatin boundary elements in

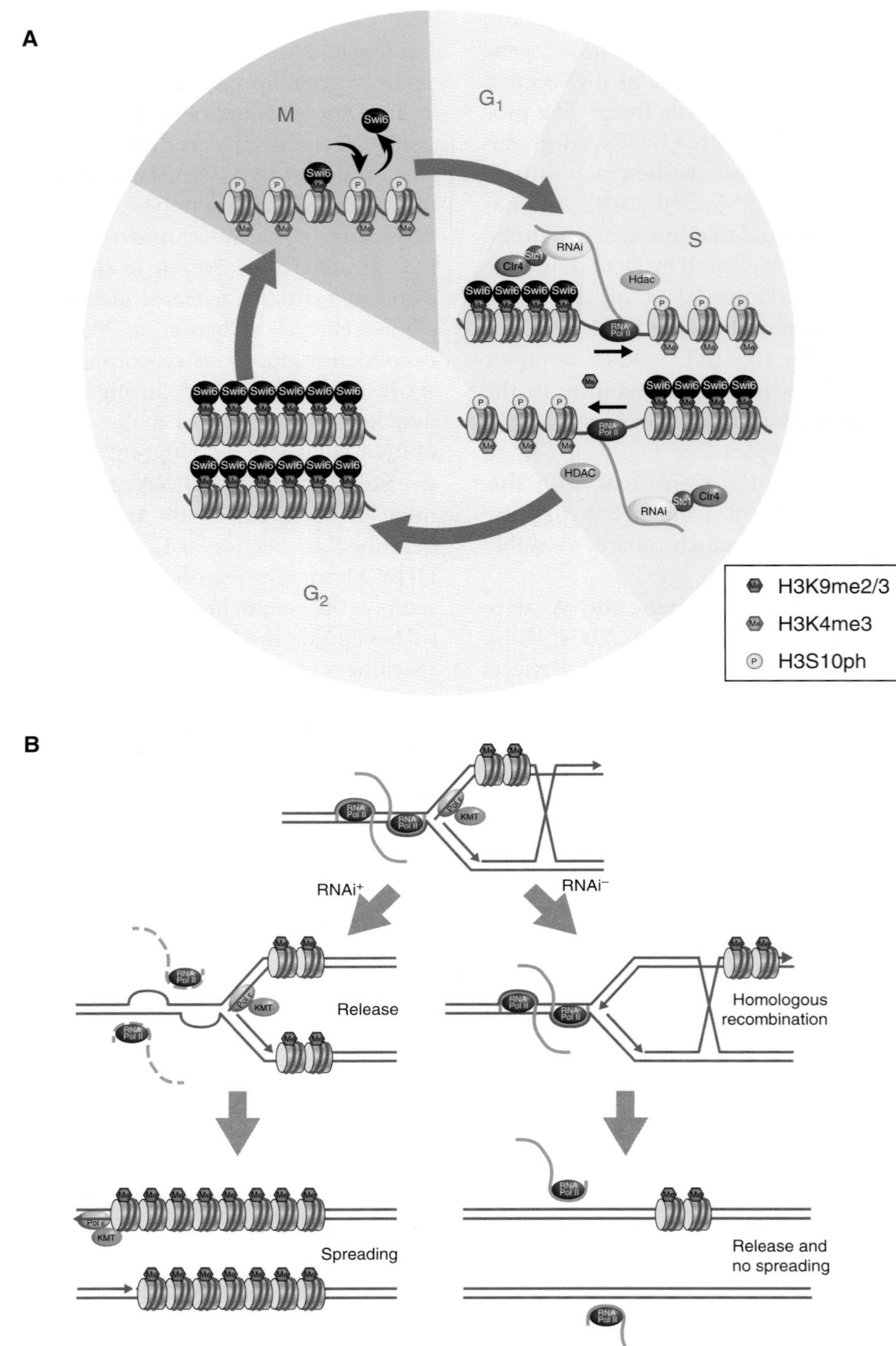

Figure 4. Cell-cycle regulation of centromere heterochromatin assembly. (*A*) Heterochromatin located at chromosomal centromeres becomes differentially methylated and phosphorylated on histones throughout the cell cycle as indicated. These modifications control the binding of the heterochromatin protein Swi6. During mitosis Swi6 is displaced by H3S10 phosphorylation. Swi6 binding is reestablished during subsequent DNA replication (S phase) when a more accessible chromatin structure permits RNA Pol II to transcribe centromeric DNA. This, in turn, recruits the RNAi machinery to direct H3K9me methylation. (*B*) Replication-coupled RNAi model (Li et al. 2011). This figure illustrates an alternative model for how RNAi works at centromeres. Here, RNAi serves to release RNA Pol II from chromatin to avoid collision with DNA replication machinery during S phase. See text for further details. (*A*, Adapted from Djupedal and Ekwall 2008.)

 Cite as *Cold Spring Harb Perspect Biol* doi: 10.1101/cshperspect.a018770

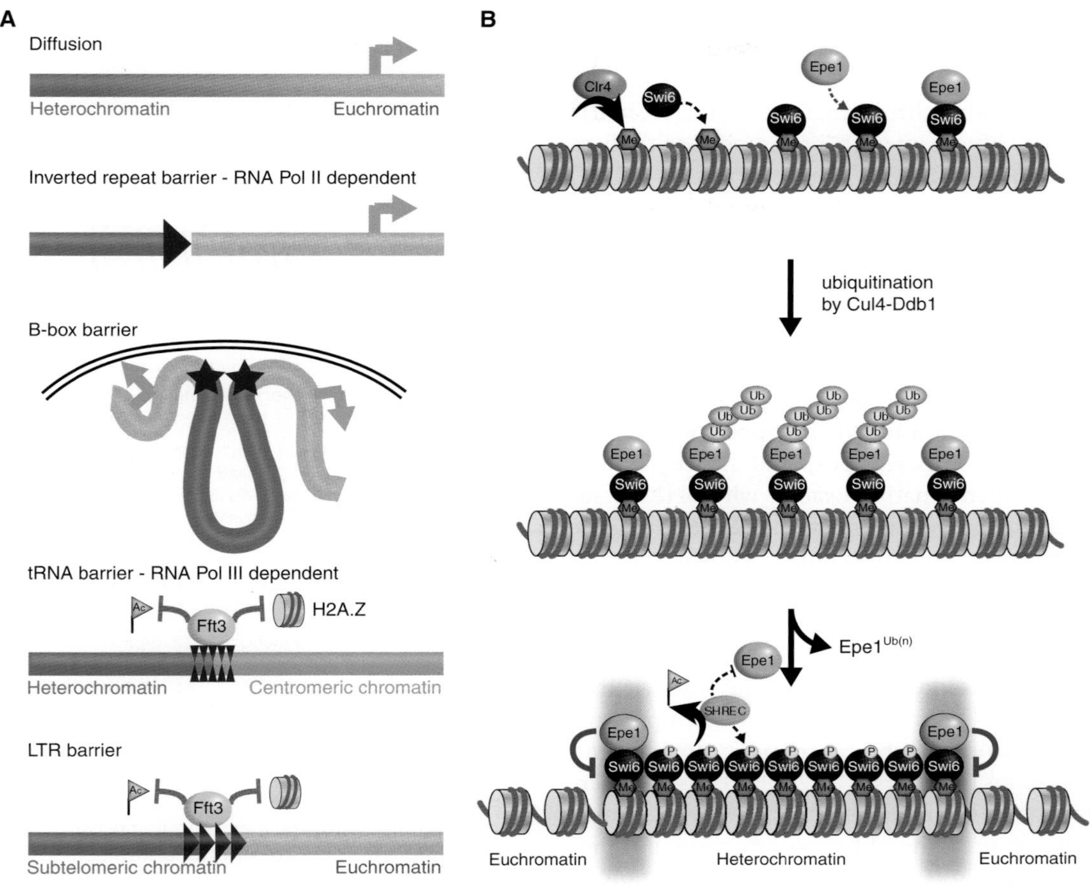

Figure 5. Chromatin boundaries and the boundary mechanism involving Epe1 (for enhancement of position effect) in *S. pombe*. (*A*) A schematic representation of different types of boundary elements in *S. pombe*. (*B*) The mechanism of boundary function by Epe1. Epe1 associates with Swi6, however, when the antisilencing factor is ubiquitinated by the Cul4-Ddb1 ligase and degraded in the heterochromatin region to allow for heterochromatin assembly. However, at boundaries, Epe1 is somehow protected from degradation, thus restricting the spreading of heterochromatin. Phosphorylation of Swi6 contributes to the dissociation of Epe1 at heterochromatin, while promoting the association with the HDAC complex SHREC in maintaining histone hypoacetylation. (*A*, Adapted from Scott et al. 2007.)

S. pombe: diffuse boundaries, RNA Pol II inverted repeat (IR) barriers, B-box barriers, RNA Pol III-dependent tRNA barriers, and RNA Pol III-independent LTR barriers (long terminal repeat sequences derived from retrotransposons) elements (Fig. 5A) (Scott et al. 2007; Stralfors et al. 2011). Several tRNA genes reside between the outer repeats and the central kinetochore domain (double arrowheads in Fig. 2A). These have been shown to act as a barrier preventing heterochromatin from encroaching into the central domain (Scott et al. 2006).

One of the best-studied boundary factors, Epe1, was first identified through a mutation that enhances heterochromatin formation (Ayoub et al. 2003). Epe1 interacts with Swi6 and somehow restricts H3K9 methylation to within heterochromatic regions. It also reduces siRNA production and promotes RNA Pol II occupancy in heterochromatin (Zofall and Grewal 2006; Isaac et al. 2007; Trewick et al. 2007), thereby limiting heterochromatin assembly beyond (IR) boundaries in the *mat2-mat3* region and pericentric regions of chromosomes 1 and 3 (Trewick et al. 2007; Braun et al. 2011). Based on its function in restricting heterochromatin domains from spreading into euchromatin, Epe1 has been termed an antisilencing factor. How, then, is the function of Epe1 restricted to chromatin boundary regions? It is thought to uniformly associate with heterochromatin through association with Swi6; however, the ubiquitinating action of the cullin-dependent ligase Cul4-Ddb1 degrades Epe1 in heterochromatin so that it becomes restricted to the heterochromatin/euchromatin boundary regions (Fig. 5B) (Braun et al. 2011). What prevents the ubiquitin ligase from acting on Epe1 at boundaries is currently an open question.

Determining and maintaining a particular chromatin structure is clearly a dynamic process involving the interaction and competition of many factors. Epe1 operates in the context of many other chromatin factors, and research is beginning to show how it interacts or competes with these other proteins and complexes. Epe1 counteracts HDAC activity by competing with the SHREC complex (containing the Clr3 HDAC, SNF2 remodeling factor, Mit1, and telomere end protection protein, Ccq1) for binding to heterochromatin (Sugiyama et al. 2007). Interestingly, phosphorylation of Swi6 by casein kinase 2 (CK2) promotes SHREC binding and heterochromatin spreading, which leads to reduced Epe1 levels (Shimada et al. 2009). Reduced phosphorylation of Swi6 leads to the opposite situation, with less SHREC and more Epe1, and hence reduced spreading of heterochromatin. Exactly why SHREC requires CK2-dependent phosphorylation of Swi6 to localize at the heterochromatic domain is not yet known.

The Fun30 remodeling factor Fft3, an ATPase helicase of the SNF2 family, also functions at boundaries (see Ch. 21 [Becker and Workman 2013] for details of how the SNF2 enzymes work). Fft3 is localized to the centromeric tRNA and subtelomeric LTR boundary regions, where it prevents euchromatin formation in the centromeric domain and subtelomeric regions (Stralfors et al. 2011). Thus, Fft3 performs a new type of boundary function at tRNAs, which does not limit the spreading of heterochromatin, but prevents euchromatin marks such as H3/H4 acetylation or the histone variant H2A.Z from invading the insulated regions. Fft3 works similarly at subtelomeric LTR boundaries preventing the formation of euchromatin in these regions (Buchanan et al. 2009). It is currently not known how chromatin remodeling at these boundaries leads to this type of insulator function, but one possibility is that this function involves changes in higher-order chromatin structure and nuclear organization (see Sec. 8 for a discussion of nuclear organization in *S. pombe*).

5 *S. pombe* CENTRAL DOMAIN CHROMATIN AND KINETOCHORES

The central domain is the site of kinetochore assembly and its chromatin is very distinct from the flanking outer repeat heterochromatin. The kinetochore is the protein structure that forms at centromeres to facilitate the separation of sister chromatids along spindle fibers during M phase. The variegated silencing of reporter genes in the ~10-kb central domain of *cen1* was indicative of a different chromatin configuration, which is essentially independent of Clr4 and therefore does not involve methylation of H3K9 (Fig. 3). The distinct chromatin composition was initially shown by micrococcal nuclease analysis (see Ch. 12 [Elgin and Reuter 2013] for an explanation of MNase digestion), which revealed a smear in contrast to the regular 150-bp ladder characteristic of flanking outer repeat chromatin (Polizzi and Clarke 1991; Takahashi et al. 1992). This distinct pattern differentiated central domain chromatin from both heterochromatin and euchromatin, reflecting its assembly into distinctive chromatin and the assembly of the kinetochore over it. Later research in various eukaryotes showed that all active centromere chromatin examined contained a histone H3-like protein known as CENP-A (or CenH3), which is critical for specifying the site of kinetochore assembly (Fig. 6).

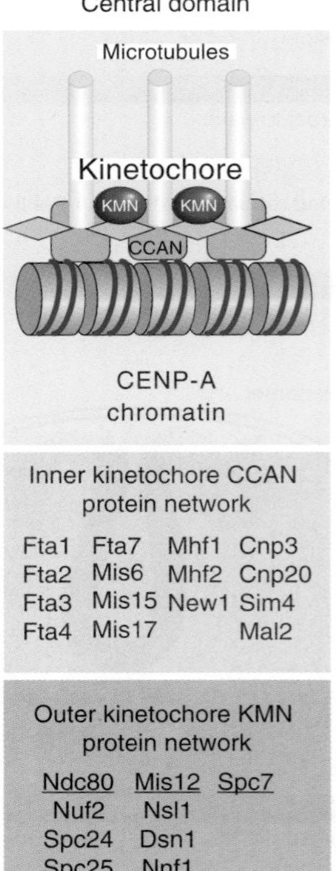

Figure 6. Centromeric chromatin and kinetochores in *S. pombe*. The central domain of *S. pombe* centromeres is composed of CENP-A-containing chromatin and a multiprotein network that makes up the kinetochore. The inner constitutive centromere-associated network (CCAN) and outer KMN kinetochore protein networks are depicted and the different protein components listed below. Kinetochore assembly within the central domain mediates attachment to microtubules on spindle formation and chromosome segregation.

5.1 The Central Domain Contains CENP-A

In the central core domain of fission yeast centromeres, the majority of histone H3 is replaced by the variant CENP-A ortholog known as Cnp1 (Takahashi et al. 2000). If CENP-A^{Cnp1} chromatin structure is disrupted (as in the case of kinetochore mutants), then the specific smeared micrococ-

Cite as *Cold Spring Harb Perspect Biol* doi: 10.1101/cshperspect.a018770

cal nuclease digestion pattern reverts to a pattern more typical of bulk chromatin (i.e., a nucleosomal ladder) without affecting adjacent heterochromatin or gene silencing in the outer repeats (Pidoux et al. 2003; Hayashi et al. 2004). Furthermore, the fact that CENP-A^{Cnp1} and other kinetochore proteins only associate with the central domain shows that the central kinetochore domain is structurally complex and functionally distinct from outer repeat silent heterochromatin (Fig. 6).

It has been estimated, based on Cnp1-GFP intensity, that there are 680 CENP-A^{Cnp1} molecules per cell in *S. pombe* (Coffman et al. 2011). Quantitation using single-molecule microscopy, however, concluded that there are 72–82 molecules of CENP-A^{Cnp1} per centromere cluster (containing all three centromeres) in G_2 cells. Estimates based on ChIP-seq (chromatin immunoprecipitation coupled to DNA sequencing) measurements yielded between 19 and 24 CENP-A nucleosomes per centromere; hence, a total of 64 molecules per haploid cell (Lando et al. 2012). These results indicate that there are ~10–20 CENP-A^{Cnp1} particles at each centromere and CENP-A is deposited in mid to late G_2 phase of the cell cycle.

Although not all known kinetochore domain proteins have been tested, thus far it appears that silencing within the central domain is a result of the steric hindrance caused from the assembly of an intact kinetochore. The inner kinetochore consists of the CCAN totaling 15 proteins. This includes the *S. pombe* homologs of several CENP proteins (Fig. 6) (reviewed in Perpelescu and Fukagawa 2011). Outside of CCAN, we find the "outer kinetochore" KMN complex, which is made up of another conserved network of proteins including the Ndc80 and Mis12 protein subcomplexes and Spc7 (reviewed in Buttrick and Millar 2011). Together, these protein networks bind centromeric CENP-A chromatin, helping to form stable kinetochore-microtubule attachments. This large complex of proteins presumably restricts access of RNA Pol II to reporter genes placed within this region and thereby impedes their transcription. In conditional temperature-sensitive kinetochore mutants, the kinetochore integrity is clearly only partially functional at the growth permissive temperature and this allows for increased transcription of reporter genes. An advantage of this is that a normally silent reporter gene that becomes active has been used to assay for defects in central core chromatin, leading to the identification of novel kinetochore proteins (Pidoux et al. 2003).

5.2 Epigenetic Inheritance of the Functional Centromere State

Evidence from many organisms has contributed to the hypothesis that CENP-A is the epigenetic mark that specifies centromere identity. Recent observations in *Drosophila* indicate that CENP-A chromatin is able to self-propagate. Studies in fission yeast have contributed to our understanding of CENP-A maintenance as well as the establishment of this unique chromatin state. In addition, elegant studies on formation of neocentromeres and inactivation of dicentric chromosomes highlight the importance of *S. pombe* in epigenetic research.

An interesting epigenetic phenomenon has been described with respect to the assembly of functional centromeres in fission yeast using plasmids containing minimal regions for centromere function. Plasmids that retain only part of an outer repeat and most of the central domain can assemble a functional centromere, albeit inefficiently. Once this functional centromere state has been established, however, it can be propagated through many mitotic divisions and even through meiosis (Steiner and Clarke 1994; Ngan and Clarke 1997). One interpretation is that the outer repeats provide an environment that is favorable for kinetochore assembly (Pidoux and Allshire 2005), but once assembled, CENP-A^{Cnp1} chromatin and thus the kinetochore is propagated at this position by a templating mechanism that may be coupled to replication. It is possible that heterochromatin somehow induces or aids the deposition of CENP-A^{Cnp1} in the central domain (Fig. 7A) such that only one block of heterochromatin is sufficient for kinetochore assembly. In fact, heterochromatin and RNAi are required for de novo assembly of CENP-A chromatin on naked DNA introduced into cells (Folco et al. 2008). Further evidence for the epigenetic nature of centromeric chromatin in *S. pombe* came from studies on neocentromeres (Ishii et al. 2008). Neocentromeres can be formed in subtelomeric regions, lacking any centromeric DNA sequences, on induced deletion of the centromere. This suggests that the formation of neocentromeres in the vicinity of subtelomeric heterochromatin probably reflects a key role for heterochromatin in centromere establishment (Castillo et al. 2013). Another illustration of epigenetic regulation in *S. pombe* is provided by the way that cells deal with dicentric chromosomes. One centromere in a dicentric chromosome can become inactivated because of disassembly of the kinetochore and the CENP-A chromatin and assembly of heterochromatin over the central domain (Sato et al. 2012). Even in the absence of heterochromatin components, centromeric shutdown can occur by the assembly of a hypoacetylated chromatin state. In addition, inactivated centromeres can become reactivated in certain situations. These observations serve to show the plasticity and epigenetic nature of centromeric chromatin; the same DNA can be assembled into a variety of different chromatin states, which have different functions.

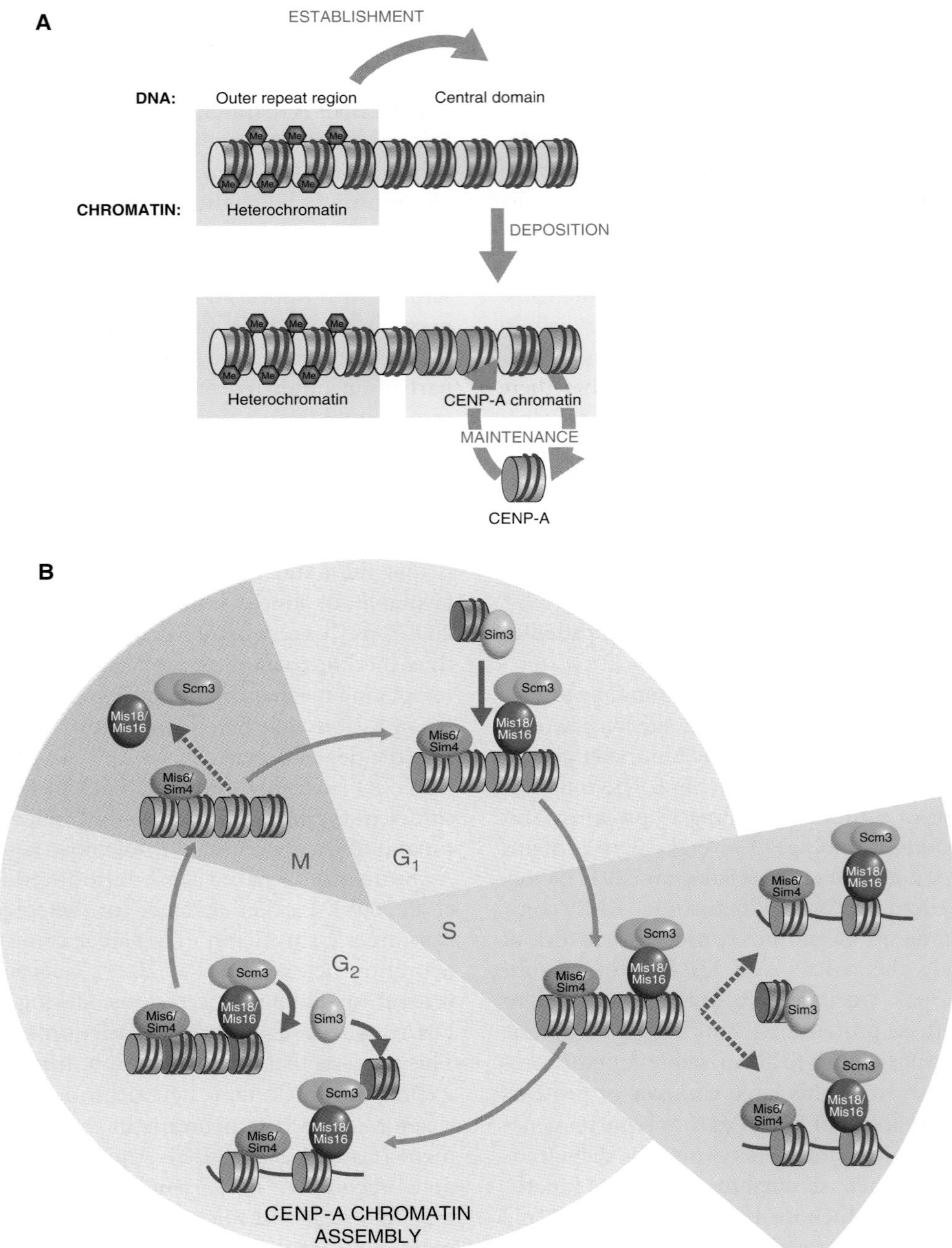

Figure 7. CENP-A chromatin establishment and propagation through the cell cycle. (*A*) Central domain DNA alone is unable to establish a functional centromere; outer repeats are required. Loss of heterochromatin from established centromeres does not affect CENP-A^{cnp1} or kinetochore maintenance in the central domain. This suggests that heterochromatin may, in some way, initially direct the site of CENP-A^{cnp1} chromatin and thus kinetochore assembly. (*B*) Cell-cycle dependency of CENP-A^{Cnp1} recruitment in *S. pombe*. In *S. pombe*, centromeric DNA is replicated and existing CENP-A^{Cnp1} is diluted by nucleosome segregation to sister chromatids during S phase. Recruitment of new CENP-A occurs during the G_2 phase, indicated by the pink nucleosomes. The Sim3 histone chaperone interacts with new CENP-A^{Cnp1} and delivers it to the centromere, where it is received by Scm3 and assembled into nucleosomes by unknown factors and mechanisms. Nucleosome gaps could be filled or H3 nucleosomes could be replaced. Scm3 is shown as a dimer interacting directly with Mis18. Scm3 recruitment at centromeres requires the Sim4/Mis6 and Mis16/Mis18 complexes. Mis16/Mis18 and Scm3 are removed from centromeres during mitosis and reassociate in G_1. (*B*, Adapted from Mellone et al. 2009.)

Cite as *Cold Spring Harb Perspect Biol* doi: 10.1101/cshperspect.a018770

Once CENP-A^{Cnp1} chromatin is established, heterochromatin is not required to propagate it in subsequent cell divisions. What is the mechanism of CENP-A propagation through the cell cycle? Unlike in *Drosophila* and mammals in which loading occurs in G_1, CENP-A^{Cnp1} deposition in *S. pombe* occurs mainly during the G_2 phase of the cell cycle (Fig. 7B) (Lando et al. 2012). Kinetochore proteins themselves govern the localization and assembly of CENP-A^{Cnp1} specifically within the central domain (Goshima et al. 1999; Takahashi et al. 2000; Pidoux et al. 2003). Several loading factors and chaperones for CENP-A^{Cnp1} have been identified. Mis16, an RpAp46/RpAp48-like protein, and Mis18/KNL are conserved centromeric proteins required for CENP-A loading and histone deacetylation in the central core region (Hayashi et al. 2004). They dissociate from centromeres, from early mitosis (prophase) until mid-anaphase, but remain associated for the rest of the cell cycle (Fujita et al. 2007). Sim3, a NASP-like protein, binds CENP-A^{Cnp1} and acts as a chaperone to deliver it to centromeres (Dunleavy et al. 2007). Scm3 (ortholog of HJURP) is a CENP-A^{Cnp1} chaperone required for assembly of CENP-A chromatin, and it dissociates from centromeres in mitosis similar to Mis16 and Mis18 (Pidoux et al. 2009; Williams et al. 2009). It is currently unknown what restricts CENP-A^{Cnp1} replenishment to the G_2 phase of the cell cycle in *S. pombe*.

In addition to the chaperones and loading factors, SNF2 family members of chromatin-remodeling factors are also required for CENP-A^{Cnp1} loading. The SNF2 remodeler Hrp1 (a Chd1 ortholog) is localized to centromeric chromatin where it is required to maintain high CENP-A^{Cnp1} levels (Walfridsson et al. 2005). Unstable noncoding transcripts that originate from the centromeric central core chromatin have been recently described in *S. pombe* (Choi et al. 2011). During the process of noncoding transcription in this region, nucleosome disassembly and reassembly reactions must occur, and it is conceivable that Hrp1 acts in these transcription-coupled chromatin remodeling and assembly activities to evict H3 and maintain CENP-A^{Cnp1} at centromeres. Consistent with this, loss of factors required for the reassembly and stabilization of H3 nucleosomes behind RNA Pol II, such as FACT and Clr6-complex II, promotes CENP-A assembly in place of histone H3 in noncentromeric regions of the genome (Choi et al. 2012). Moreover, some subunits of the Mediator complex are required for CENP-A^{Cnp1} localization, and the reduced levels of CENP-A^{Cnp1} in Mediator mutants can be suppressed by RNA Pol II inhibitors (Carlsten et al. 2012). This finding suggests that proper levels of centromeric transcription are critical to ensure inheritance of the functional centromere state.

Thus, initial establishment of CENP-A^{Cnp1} chromatin in *S. pombe* requires flanking heterochromatin. The subsequent maintenance of CENP-A chromatin is independent of heterochromatin, but depends on several loading factors and chaperones to facilitate G_2 stage–specific assembly of CENP-A chromatin. The chromatin-remodeling factor Hrp1 and appropriate levels of noncoding transcription help to maintain high CENP-A^{Cnp1} levels at centromeres.

5.3 Additional Chromatin Modifications Required for Mediating Sister-Centromere Cohesion, Condensation, and Normal Chromosome Segregation

How do outer repeat heterochromatin and CENP-A^{Cnp1} kinetochore chromatin affect the overall function of chromosome segregation? Experiments with naked plasmid DNA constructs have shown that outer repeats contribute to the assembly of functional centromeres (through unknown mechanisms), which allow these plasmids to segregate on mitotic and meiotic spindles. However, as already mentioned, neither the outer repeats nor the central domain alone are sufficient to assemble a functional centromere (Clarke and Baum 1990; Takahashi et al. 1992; Baum et al. 1994; Ngan and Clarke 1997; Pidoux and Allshire 2004).

Mutants that cause loss of silencing at the outer repeats (i.e., those defective in Clr4, RNAi components, or Swi6) have elevated rates of mitotic chromosome loss and a high incidence of lagging chromosomes on late anaphase spindles (Fig. 8A) (Allshire et al. 1995; Ekwall et al. 1996; Bernard et al. 2001; Nonaka et al. 2002; Hall et al. 2003; Volpe et al. 2003). Additionally, cells lacking Swi6 are defective in cohesion at centromeres, but retain cohesion along the chromosome arms (Bernard et al. 2001; Nonaka et al. 2002). One function of Swi6 is to recruit cohesin to outer repeat chromatin to mediate tight physical cohesion between sister centromeres. This ensures that sister kinetochores are tightly held together during the formation of a properly bioriented spindle in which sister kinetochores attach to microtubule fibers emanating from opposite spindle poles (Fig. 8B). Thus, one function of heterochromatin at centromeres is to mediate cohesion.

Cohesin is also strongly associated with telomeres and the *mat2-mat3* region (Bernard et al. 2001; Nonaka et al. 2002) and it is involved in the formation of subtelomeric heterochromatin (Dheur et al. 2011). In addition, cohesin is recruited to silent chromatin formed on a *ura4*$^+$ gene in response to an adjacent ectopic centromere repeat, underscoring the link between heterochromatin and cohesion (Partridge et al. 2002). Thus, the recruitment of cohesin seems to be a general property of Swi6-associated silent chromatin. How Swi6 chromatin brings about cohesin recruitment is not known, but Swi6 does interact with the

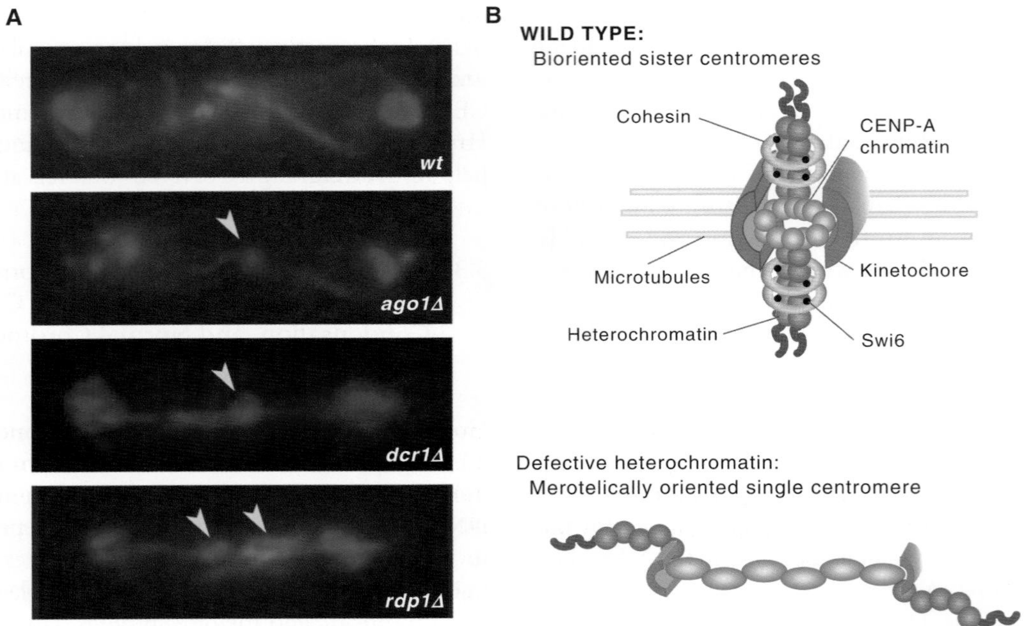

Figure 8. Defective heterochromatin leads to abnormal centromere structures. (*A*) Cells lacking RNAi or heterochromatin components display elevated rates of chromosome loss and lagging chromosomes (indicated by yellow arrows) on late anaphase spindles. Chromosomal DNA is stained by DAPI (blue) and mitotic spindle microtubules are labeled for immunofluorescence (IF) (red). (*B*) A schematic three-dimensional figure of a normal centromere illustrates the outer heterochromatin regions (green circles) decorated with Swi6 (black circles), which recruits cohesin to ensure sister-chromatid cohesion. The central domain consists of CENP-A-containing chromatin (red circles), associated with opposing kinetochores on each sister chromatid. Lagging chromosomes in cells with defective heterochromatin may be the result of disorganized kinetochores such that one centromere may attach to microtubules from opposite poles. Such merotelic orientation could persist into anaphase, in which the breakage of attachment with one pole or the other would lead to random segregation and result in chromosome loss/gain events.

cohesin subunit Psc3 (Nonaka et al. 2002). In addition, Dfp1, the regulatory subunit of the conserved kinase Hsk1 (Cdc7), interacts with Swi6 and is required to recruit cohesin to centromeres (Bailis et al. 2003). Cohesion at vertebrate centromeres from prophase to anaphase onset also depends on a protein known as Shugoshin. In *S. pombe*, Shugoshin associates with the centromere on phosphorylation of histone H2A at S121 by Bub1 to ensure proper chromosome segregation in mitosis and meiosis (Kawashima et al. 2010). Thus, cohesin and associated factors are not only required at centromeres, but along the length of chromosomes to mediate chromosome condensation and segregation functions during M phase.

The histone H2A variant H2A.Z was initially identified as a factor that when mutated led to defects in chromosome segregation (Carr et al. 1994). Histone H2A and its variant H2A.Z play important roles in chromosome segregation by acting as chromatin receptors for condensin recruitment both at kinetochores and along the entire chromosome to shape mitotic chromosomes during cell division. Recruitment of condensin to chromatin via H2A and H2A.Z only occurs in its phosphorylated state and is a cell-cycle regulated process (Tada et al. 2011). Acetylation of H2A.Z by the HAT enzyme Mst1 is also important for proper structure and separation of anaphase chromosomes (Kim et al. 2009). H2A.Z itself is not normally present at centromeres, but is required for expression of the centromere protein CENP-C^{Cnp3} (Fig. 6), and perturbs centromere function when mislocalized to centromeres in mutant conditions (Ahmed et al. 2007; Buchanan et al. 2009; Hou et al. 2010). Histone H2B is also implicated in centromere function, probably through ubiquitination, because mutations near the ubiquitination site cause anaphase defects and changes in centromeric chromatin structure (Maruyama et al. 2006). Thus, all core histones, together with the variants CENP-A^{Cnp1} (a histone H3 variant) and H2A.Z, are involved in orchestrating proper centromeric and chromosome-wide chromatin structure in *S. pombe* important for sister-centromere cohesion, condensation, and normal chromosome segregation.

6 HETEROCHROMATIN FORMATION AT OTHER SILENT REGIONS

Many of the molecular mechanisms controlling heterochromatin formation at centromeres also operate at other silenced chromatin regions such as the mating-type locus, telomeres, and rDNA (Fig. 2A). However, there are also additional pathways that are specific for these distinct regions; these are discussed below.

6.1 Assembly of Silent Chromatin at *cenH* Regions

At the DNA sequence level, most heterochromatic regions contain centromere homologous (*cenH*) DNA regions that are capable of assembling Clr4-dependent silent heterochromatin. *cenH* sequences were originally defined at the outer repeat regions of centromeres, but *cenH* DNA sequences can also be found at the silent mating-type loci (*mat2-mat3*; Noma et al. 2001) and at a subtelomeric region adjacent to the terminal telomeric repeats (consensus TTACAGG) synthesized by telomerase (Nimmo et al. 1994; Allshire et al. 1995; Nimmo et al. 1998; Kanoh et al. 2005). Seven of the 20-kb long *cenH* region that resides between *mat2* and *mat3* shares a high degree of sequence similarity with the centromeric outer repeats (Grewal and Klar 1997). In addition, a 0.5-kb DNA sequence located within the telomere-associated sequences (TASs) that occupy up to 40 kb proximal to the telomeres of chromosomes I and II shares >84% identity with *cenH* (Kanoh et al. 2005). This suggested that *cenH* repeats might act in *cis* to affect silent heterochromatin assembly.

Indeed, ectopic centromeric outer repeats (*dg*) or related *cenH* DNA sequences are sufficient to silence adjacent marker genes in a Clr4-dpendent manner (Ayoub et al. 2000; Partridge et al. 2002; Volpe et al. 2003). As already discussed, transcription of centromeric outer repeats (and *cenH* sequences) and the processing of resulting dsRNA by Dcr1 and the RNAi machinery provides a mechanism by which Clr4 is recruited to trigger the assembly of silent chromatin (Volpe et al. 2002).

6.2 Mechanisms of Heterochromatin Formation at the Mating-Type Loci

At the mating-type loci, two distinct mechanisms are involved in the establishment and maintenance of heterochromatin. This was revealed by experiments involving treatment of cells with trichostatin A, an HDAC inhibitor, in which derepression of silencing could be achieved only in RNAi mutant backgrounds (Hall et al. 2002). In addition to RNAi, an RNAi-independent silencing mechanism based on the DNA-binding proteins Atf1 and Pcr1 (which bind near *cenH*) and the HDAC Clr6 operates at the silent mating-type loci to maintain heterochromatin (Hall et al. 2002; Jia et al. 2004). RNAi acts to establish silent chromatin; however, its maintenance is achieved by the Atf1- and Pcr1-based silencing system that can maintain heterochromatin even in the absence of RNAi (Jia et al. 2004; Kim et al. 2004). Consistent with this notion, cells lacking components of the RNAi machinery and Atf1/Pcr1 show a complete loss of silent chromatin at the *mat2-mat3* region.

6.3 Mechanisms of Heterochromatin Formation at Telomeres

Overlapping mechanisms of silencing also function at telomeres. Terminal repeats are bound by the telomere repeat binding protein Taz1, which, in turn, recruits the Clr4 HKMT and thus Swi6 to form heterochromatin. However, RNAi also acts via the *cenH* region within the TAS elements to form an extended region of silent chromatin at telomeres (Nimmo et al. 1994; Allshire et al. 1995; Kanoh et al. 2005). Interestingly, heterochromatin formation contributes to the protection of chromosome ends even in the absence of telomerase, which typically acts to maintain telomere length. In a recent study, a subset of cells that survive telomerase ablation were found to maintain the integrity of chromosome ends through a mechanism of continual amplification and rearrangement of heterochromatic sequences, which, in turn, induces the recruitment of telomere end-protection factors Pot1 and Ccq1 (a component of SHREC). These strains were consequently named HAATI mutants (for heterochromatin amplification-mediated and telomerase-independent; Jain et al. 2010). Swi6-GFP localization in HAATI cells and the requirement of Clr4 for HAATI indicates that the amplified regions are indeed heterochromatic. Moreover, maintenance of linear chromosomes in HAATI cells depends on the Ccq1 component of the SHREC complex, which contains HDAC activity and plays a role in heterochromatin spreading (see Sec. 4.1). Thus, Ccq1 provides a molecular link between telomeric end-protecting protein complexes and histone-modifying activities involved in heterochromatin assembly.

6.4 Facultative Heterochromatin

In addition to the larger constitutive heterochromatin regions found at centromeres, the mating-type region, rDNA, and telomeres, there are also so-called facultative heterochromatin regions in *S. pombe* (Zofall et al. 2012). These heterochromatic "islands" are present in vegetative cells and contain clusters of genes that are expressed and, hence, euchromatic in meiotic cells. Epe1 and RNA degradation factors are important for the regulation of the meiotically induced genes in these islands. RNA degradation factors maintain silencing of the islands during vegetative growth, but Epe1 is needed to disassemble the heterochromatic islands during meiosis to allow for expression of the meiotic genes in response to nutritional signals. Some meiotically induced genes also cluster in the subtelomeric regions (Mata et al. 2002), and one such meiotic cluster resides in the *tel1L* (left subtelomere of chromosome 1) region. This cluster is hypoacetylated in mitotic cells and repression of the genes requires the HDAC Clr3, indicative of its facultative heterochromatin structure (Wiren et al. 2005).

Another type of transient heterochromatin is found in intergenic regions between some convergent gene pairs in *S. pombe* (Gullerova and Proudfoot 2008). This heterochromatin apparently depends on convergent transcription in the G_1 phase of the cell cycle, leading to local RNAi-dependent H3K9 methylation, which recruits Swi6 and cohesin. siRNA homologous to such intergenic regions have, so far, eluded detection. The accumulation of cohesin by interaction with Swi6 in these regions aids transcription termination in the G_2 phase at sites upstream of the termination sites used in G_1, so that alternative 3′ mRNA ends are generated. Interestingly, many of the genes that encode the RNAi machinery components are themselves autoregulated by this process (Gullerova et al. 2011).

7 NUCLEOSOME REMODELING IN *S. pombe*

Chromatin structure and function are not solely determined by histone modifications and mechanisms that facilitate this. Chromatin remodeling is also an important component in the regulation of chromatin structure and function. Although we have already alluded to a few chromatin remodeling proteins in our discussion of *S. pombe* heterochromatin formation and propagation, this section elaborates on the status of research in this field. Another important aspect that governs chromatin structure is nuclear architecture, which is the topic of Section 8.

SNF2 nucleosome-remodeling enzymes play key roles in epigenetic processes, such as the regulation of transcription and the loading or exchange of histones because of their ability to alter the positions of nucleosomes in the genome (Ch. 21 [Becker and Workman 2013]). *Homo sapiens* has 53 SNF2 enzymes, whereas *S. pombe* has only 20 (Flaus et al. 2006). The CHD-type SNF2 enzyme, Mit1, is part of the SHREC complex discussed in Section 4.5 and more extensively in Ch. 16 (Martienssen and Moazed 2014). It is required for silencing at the mating-type region, telomeres, and pericentric heterochromatin in *S. pombe*, but not at the 5′ end of genes (Sugiyama et al. 2007; Pointner et al. 2012).

The Chd1-like SNF2 remodelers, Hrp1 and Hrp3, are involved in silencing at pericentric heterochromatin and at the mating-type region. It is plausible that the role of Hrp1 at centromeres in CENP-A loading is mechanistically related to its ability, in conjunction with the Nap1 chaperone, to evict H3 at many gene promoters and its role in nucleosome assembly (Walfridsson et al. 2007; Pointner et al. 2012). Recently, Hrp1 and Hrp3 were implicated in the genome-wide suppression of cryptic transcription in gene-coding regions (Hennig et al. 2012; Pointner et al. 2012; Shim et al. 2012). The *hrp1* and *hrp3* mutants show reduced nucleosome arrays in coding regions, which allows for cryptic initiation of transcription by RNA Pol II. A mechanistic explanation for this phenomenon is the demonstration of a prominent nucleosome spacing activity for Hrp1 and Hrp3 in vitro (Pointner et al. 2012). Thus, Hrp1 and Hrp3 remodelers seem to have distinct activities that, at intergenic regions, favor disassembly versus coding regions, in which they favor the assembly and spacing of nucleosomes. The action of Hrp1 in intergenic regions is strongly correlated to topoisomerases I and II (Durand-Dubief et al. 2010). Topoisomerases I and II are generally required to maintain nucleosome-free regions both at gene promoters and in 3′ intergenic regions, most likely by removing negative supercoils that stabilize nucleosomes (Durand-Dubief et al. 2007, 2011). It is likely that this function of topoisomerases stimulates the nucleosome disassembly function of SNF2-remodeling enzymes including Hrp1. It is possible that the absence of Top I and II from coding regions alters the outcome of the remodeling process in favor of the spacing activity. Another example of SNF2 chromatin remodelers implicated in gene silencing is the Swr1 complex that replaces H2A with H2A.Z (Buchanan et al. 2009; see Fig. 9 of Ch. 20 [Henikoff and Smith 2014]). Thus, although the research on SNF2 enzymes in *S. pombe* is still in its budding stages, it is already clear that these enzymes play key roles in epigenetic regulating chromatin structure and control.

8 NUCLEAR ORGANIZATION IN *S. pombe*

Although the role of nuclear architecture in epigenetic control is well established in mammalian systems (see Ch. 19 [Dekker and Misteli 2014]), our knowledge in *S. pombe* is largely descriptive at present. The *S. pombe* interphase nucleus is spherical and consists of a chromatin-rich and an RNA-rich part, the latter of which contains the nucleolus (Fig. 9A). All heterochromatin is peripheral in the nucleus. The three centromeres are in fact attached to the nuclear periphery, adjacent to the spindle pole body (SPB) throughout interphase, but pericentric heterochromatin can be cytologically distinguished from centromeric chromatin by IF microscopy; kinetochores and centromeric CENP-A^{Cnp1} chromatin are visualized as being surrounded

Cite as *Cold Spring Harb Perspect Biol* doi: 10.1101/cshperspect.a018770

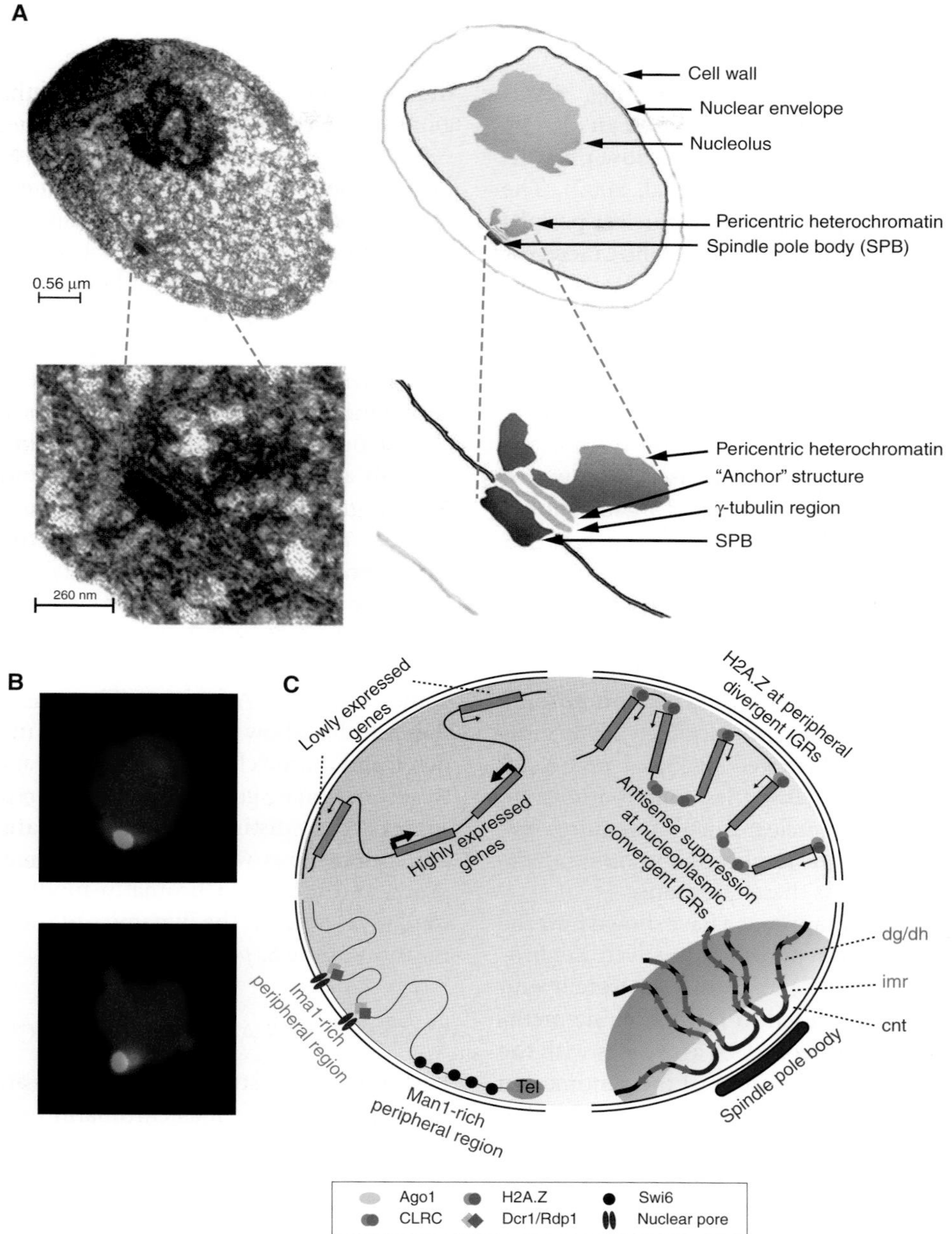

Figure 9. Nuclear organization in *S. pombe*. (*A*) Electron microscopy analysis of an *S. pombe* nucleus. (*Top*) Micrograph of a cross-section through a high-pressure fixed and Lowicryl-embedded interphase *S. pombe* cell. The cellular structures are indicated: cell wall, nuclear envelope, nucleolus, heterochromatin region, and SPB. (*Bottom*) A higher magnification of the same nucleus. The nuclear structures indicated are SPB, γ-tubulin region, anchor structure, and the centromeric heterochromatin. (*B*) Two interphase nuclei with heterochromatin (centromeres, telomeres, and the silent *mat2-mat3* loci) decorated by red fluorescent immunolocalization of Swi6, and kinetochore chromatin (centromeres only) decorated by green fluorescent immunolocalization of CENP-A^{Cnp1}. The red signals, not in close proximity to green, represent telomeres or the *mat2-mat3* loci. All centromeres are clustered at the nuclear periphery adjacent to the SPB. (*C*) A model for chromatin organization at the fission yeast nuclear periphery. (*Top*) Genes with low expression levels tend to associate with the nuclear periphery, whereas highly expressed genes tend to reside in the nuclear interior. (*Top*) Localization of divergent intergenic regions and H2A.Z at the nuclear envelope may present a mechanism for anchoring the promoters of convergent gene pairs at the periphery. (*Bottom left*) Differential localization of Ima1, nuclear pores, and Man1. The inner membrane proteins Ima1 and Man1 are not equally distributed at the nuclear periphery, but rather occupy distinct areas that interact with different chromosomal regions. The subtelomeric chromatin is associated with Man1-rich peripheral regions in which Swi6 is also located. Ima1 is colocalized with Dcr1 and Rdp1 at nuclear pores. (*Bottom right*) Organization of centromeric DNA at the SPB. Central domain *cnt* and *imr* regions are localized closer to the SPB than the heterochromatic *dg* and *dh* repeats. The two centromeric domains are shaded in colors symbolizing the different IF localization of Swi6 (*dg-dh* repeats) and CENP-A^{Cnp1} (*imr*/*cnt* regions). (*A*, Reprinted from Kniola et al. 2001; *C*, *bottom right*, Adapted from Takahashi et al. 1992.)

by a layer of pericentric heterochromatin (Fig. 9B,C) (Kniola et al. 2001). The mating-type region can also be found localized near the nuclear envelope in close proximity with the SPB (Alfredsson-Timmins et al. 2007). The telomeres, however, although found at the nuclear periphery, are not fixed in location, unlike the centromeres or mating-type regions.

Interestingly, specific transcriptionally active regions, such as tRNA and 5S rRNA genes transcribed by RNA Pol III, cluster and colocalize with centromeres (Iwasaki et al. 2010). It is hypothesized that this association with centromeres is mediated through interaction of the condensin complex with the RNA Pol III transcription machinery. Another interesting feature of nuclear organization is the distribution of retrotransposon elements such as Tf2 known to be silenced by HDACs; these elements cluster as discrete nuclear structures at the nuclear periphery in a CENP-B-dependent fashion, presumed to be a mechanism of host genome surveillance and defense (Cam et al. 2008). There is also evidence for the existence of chromosome territories in *S. pombe* (Molnar and Kleckner 2008; Tanizawa et al. 2010; concept illustrated in Fig. 1 of Ch. 19 [Dekker and Misteli 2014]). How these observations are interconnected and relate to nuclear function requires further investigation; however, the application of other methodology has begun to address these questions.

Studies using the DamID approach have shown how the nuclear envelope and nuclear pores are involved in chromosomal organization (Fig. 9C). In DamID, a protein with a known nuclear location is fused to DNA adenine methylase (Dam). When chromatin or DNA associates with the Dam-fusion protein, contacted regions of the genome can be mapped based on the presence of adenine methylation added by the Dam fusion in vivo. DamID thus allows transient associations to be detected. Using the nuclear envelope proteins Man1 and Ima1 as DamID fusion partners, repressed genes were frequently found to localize to the periphery (Steglich et al. 2012). This verifies IF data that show that Ima1 specifically binds to heterochromatic regions. However, Ima1 is dispensable for the tethering of centromeric DNA to the SPB region of the nuclear periphery (King et al. 2008; Hiraoka et al. 2011). Rather, the Ima1-interacting loci are enriched with the RNAi components, Dcr1 and Rdp1. The Ima1 protein is mainly localized around nuclear pores (Fig. 9C) where Dcr1 is involved in an RNA degradation mechanism that contributes to keeping stress-induced genes repressed (Woolcock et al. 2012). Man1 target loci are mostly within subtelomeric regions and bound by the heterochromatin protein Swi6 (Steglich et al. 2012). Aberrant expression of Man1, in contrast to Ima1, leads to the delocalization of telomeres from the nuclear periphery (Gonzalez et al. 2012). This implies that Man1 has a function in anchoring chromosomal regions to the inner membrane.

Little is known about the changes in nuclear organization that occur during gene activation in *S. pombe*. Clusters of inducible genes such as the meiosis-induced genes found at the subtelomere *1L* region (Mata et al. 2002) are spatially subject to repression by the HDAC enzyme Clr3 in mitotic cells—that is, their peripheral localization requires Clr3 activity (Hansen et al. 2005; Wiren et al. 2005). Other meiotic genes are posttranscriptionally repressed by an RNA elimination machinery that also directs assembly of heterochromatin islands during vegetative growth (see also Sec. 6.4). To allow for expression of the meiotic genes, some RNA degrading factors are inactivated in meiotic cells by the formation of a nuclear dot structure containing the Mei2 protein (Harigaya et al. 2006; Zofall et al. 2012). Evidence also suggests that nitrogen starvation can induce movement of stress-induced genes away from the nuclear membrane when a nitrogen-repressed gene cluster becomes activated (Alfredsson-Timmins et al. 2009). Thus, these studies show that the *S. pombe* nucleus is highly ordered and drastic changes in the nuclear organization seem to accommodate gene activation and repression. Chromosomes occupy distinct territories and are anchored via specific interactions with inner membrane proteins, nuclear pores, and the SPB. Chromatin-modifying activities seem to play key roles in the dynamics of the higher-order organization of the *S. pombe* nucleus.

9 GENOMICS AND EPIGENOMICS IN *S. pombe*

Historically, the study of chromatin and epigenetics has used techniques such as chromatin immunoprecipitation (ChIP) and IF that have focused necessarily either on particular loci or regions of the genome (ChIP) or provided low-resolution results (IF). Because *S. pombe* is such a tractable model system to study, great progress has nonetheless been made in characterizing regions of heterochromatin in this way. The entire DNA sequence, in itself, has been useful in studying the evolution of the underlying DNA sequences surrounding heterochromatic regions and has provided a foundation for the extensive epigenetic studies and facilitated the accurate mapping of histone modifications and proteins that associate with particular genomic regions. Recent epigenomic mapping and characterization studies have significantly accelerated progress by coupling our knowledge from the complete sequencing of *S. pombe* with the advent of high-throughput technology.

The DNA sequence has revealed a compact genome, 13.8 Mb in size, with only 4824 protein-coding genes (Wood et al. 2002). The small genome size of *S. pombe* makes it a convenient and inexpensive model organism

 Cite as *Cold Spring Harb Perspect Biol* doi: 10.1101/cshperspect.a018770

for genomics and epigenomics (see Fig. 19 of Ch. 3 [Allis et al.] for a comparison). The first epigenome maps produced in *S. pombe* were generated using spotted microarrays. Technology developments, including high-resolution tiling microarrays and parallel sequencing, have allowed even more comprehensive mapping of large and small RNA and nucleosome positions. Genome-wide nucleosome mapping in *S. pombe* revealed a surprisingly short linker length and positioning mechanisms that are distinct from those of *S. cerevisiae* (Lantermann et al. 2010). Examples of enzymatic functions subjected to epigenomic analysis in *S. pombe* have included HDAC, KAT and SNF2-remodeling enzymes, and the RNAi machinery (Wiren et al. 2005; Durand-Dubief et al. 2007; Nicolas et al. 2007; Walfridsson et al. 2007; Johnsson et al. 2009; Garcia et al. 2010; Halic and Moazed 2010; Hogan et al. 2010; Woolcock et al. 2011).

By comparing the *S. pombe* genome with the related fission yeasts, *Schizosaccharomyces octosporus* and *Schizosaccharomyces japonicus*, interesting insights into the evolution of centromeres, transposable elements, and RNAi were gained (Rhind et al. 2011). Centromeric DNA regions of all three fission yeasts consist of symmetric arrangements of repeat sequences and transposons, or transposon-like sequences. In *S. japonicus*, small RNAs are produced from transposon sequences, whereas in *S. pombe*, the majority of the small RNAs are produced from centromeric repeats indicating that the RNAi machinery has evolved from an ancestral role in transposon silencing to its role in *S. pombe* heterochromatin assembly. Genome-wide approaches have also recently been developed to study nuclear organization by combining chromosome conformation capture with parallel sequencing (Tanizawa et al. 2010), and Dam-ID with inner membrane proteins (Steglich et al. 2012) and nuclear pore proteins (Woolcock et al. 2012). These approaches are clearly paving the way for a holistic molecular understanding of epigenetics and cell biology in *S. pombe*.

10 CONCLUDING REMARKS

The fungus *S. pombe* has emerged as a very powerful system to discover and elucidate epigenetic phenomena. *S. pombe* epigenetics research began with the studies of mating type and centromeric silencing (i.e., the PEV phenomena). This led the community of epigenetic researchers to focus on heterochromatin assembly and centromere identity. Currently, new epigenomics tools are allowing us to explore other aspects of epigenetics in *S. pombe* including nucleosome remodeling and nuclear organization. Exciting discoveries related to epigenetic regulation will continue to be made in this excellent model organism as we strive to understand in detail the underlying mechanisms.

ACKNOWLEDGMENTS

We thank Monika Lachner for excellent help with figures and commenting on the manuscript. R.C.A thanks Alison L. Pidoux, Lakxmi Subramanian, and Sharon A. White for comments. Research in the Allshire laboratory is supported by the Wellcome Trust [095021], [065061], and EC FP7 "EpiGeneSys" Network of Excellence [HEALTH-F4-2010-257082], along with core funding from the Wellcome Trust to the Wellcome Trust Centre for Cell Biology [092076]. R.C.A. is a Wellcome Trust Principal Fellow. K.E. thanks Babett Steglich for proofreading and Figure 9C, and Jenna Persson for commenting on the manuscript. Research in the Ekwall laboratory is supported by grants from the Swedish Cancer Society and the Swedish Research Council (VR).

REFERENCES

**Reference is also in this book.*

Ahmed S, Dul B, Qiu X, Walworth NC. 2007. Msc1 acts through histone H2A.Z to promote chromosome stability in *Schizosaccharomyces pombe*. *Genetics* **177:** 1487–1497.

Alfredsson-Timmins J, Henningson F, Bjerling P. 2007. The Clr4 methyltransferase determines the subnuclear localization of the mating-type region in fission yeast. *J Cell Sci* **120:** 1935–1943.

Alfredsson-Timmins J, Kristell C, Henningson F, Lyckman S, Bjerling P. 2009. Reorganization of chromatin is an early response to nitrogen starvation in *Schizosaccharomyces pombe*. *Chromosoma* **118:** 99–112.

Allshire RC, Javerzat JP, Redhead NJ, Cranston G. 1994. Position effect variegation at fission yeast centromeres. *Cell* **76:** 157–169.

Allshire RC, Nimmo ER, Ekwall K, Javerzat JP, Cranston G. 1995. Mutations derepressing silent centromeric domains in fission yeast disrupt chromosome segregation. *Genes Dev* **9:** 218–233.

* Aramayo R, Selker EU. 2013. *Neurospora crassa*, a model system for epigenetics research. *Cold Spring Harb Perspect Biol* **5:** a017921.

Ayoub N, Goldshmidt I, Lyakhovetsky R, Cohen A. 2000. A fission yeast repression element cooperates with centromere-like sequences and defines a *mat* silent domain boundary. *Genetics* **156:** 983–994.

Ayoub N, Noma K, Isaac S, Kahan T, Grewal SI, Cohen A. 2003. A novel jmjC domain protein modulates heterochromatization in fission yeast. *Mol Cell Biol* **23:** 4356–4370.

Bailis JM, Bernard P, Antonelli R, Allshire RC, Forsburg SL. 2003. Hsk1-Dfp1 is required for heterochromatin-mediated cohesion at centromeres. *Nat Cell Biol* **5:** 1111–1116.

Bannister AJ, Zegerman P, Partridge JF, Miska EA, Thomas JO, Allshire RC, Kouzarides T. 2001. Selective recognition of methylated lysine 9 on histone H3 by the HP1 chromo domain. *Nature* **410:** 120–124.

Baum M, Ngan VK, Clarke L. 1994. The centromeric K-type repeat and the central core are together sufficient to establish a functional *Schizosaccharomyces pombe* centromere. *Mol Biol Cell* **5:** 747–761.

Bayne EH, Portoso M, Kagansky A, Kos-Braun IC, Urano T, Ekwall K, Alves F, Rappsilber J, Allshire RC. 2008. Splicing factors facilitate RNAi-directed silencing in fission yeast. *Science* **322:** 602–606.

Bayne EH, White SA, Kagansky A, Bijos DA, Sanchez-Pulido L, Hoe KL, Kim DU, Park HO, Ponting CP, Rappsilber J, et al. 2010. Stc1: A critical link between RNAi and chromatin modification required for heterochromatin integrity. *Cell* **140:** 666–677.

* Becker PB, Workman JL. 2013. Nucleosome remodeling and epigenetics. *Cold Spring Harb Perspect Biol* **5:** a017905.

Bernard P, Maure JF, Partridge JF, Genier S, Javerzat JP, Allshire RC. 2001. Requirement of heterochromatin for cohesion at centromeres. *Science* **294:** 2539–2542.

Bernard P, Drogat J, Dheur S, Genier S, Javerzat JP. 2010. Splicing factor Spf30 assists exosome-mediated gene silencing in fission yeast. *Mol Cell Biol* **30:** 1145–1157.

Braun S, Garcia JF, Rowley M, Rougemaille M, Shankar S, Madhani HD. 2011. The Cul4-Ddb1(Cdt)^2 ubiquitin ligase inhibits invasion of a boundary-associated antisilencing factor into heterochromatin. *Cell* **144:** 41–54.

Buchanan L, Durand-Dubief M, Roguev A, Sakalar C, Wilhelm B, Stralfors A, Shevchenko A, Aasland R, Ekwall K, Francis Stewart A. 2009. The *Schizosaccharomyces pombe* JmjC-protein, Msc1, prevents H2A.Z localization in centromeric and subtelomeric chromatin domains. *PLoS Genet* **5:** e1000726.

Buhler M, Verdel A, Moazed D. 2006. Tethering RITS to a nascent transcript initiates RNAi- and heterochromatin-dependent gene silencing. *Cell* **125:** 873–886.

Buttrick GJ, Millar JB. 2011. Ringing the changes: Emerging roles for DASH at the kinetochore-microtubule interface. *Chromosome Res* **19:** 393–407.

Cam HP, Sugiyama T, Chen ES, Chen X, FitzGerald PC, Grewal SI. 2005. Comprehensive analysis of heterochromatin- and RNAi-mediated epigenetic control of the fission yeast genome. *Nat Genet* **37:** 809–819.

Cam HP, Noma K, Ebina H, Levin HL, Grewal SI. 2008. Host genome surveillance for retrotransposons by transposon-derived proteins. *Nature* **451:** 431–436.

Canzio D, Chang EY, Shankar S, Kuchenbecker KM, Simon MD, Madhani HD, Narlikar GJ, Al-Sady B. 2011. Chromodomain-mediated oligomerization of HP1 suggests a nucleosome-bridging mechanism for heterochromatin assembly. *Mol Cell* **41:** 67–81.

Carlsten JO, Szilagyi Z, Liu B, Lopez MD, Szaszi E, Djupedal I, Nystrom T, Ekwall K, Gustafsson CM, Zhu X. 2012. Mediator promotes CENP-A incorporation at fission yeast centromeres. *Mol Cell Biol* **32:** 4035–4043.

Carr AM, Dorrington SM, Hindley J, Phear GA, Aves SJ, Nurse P. 1994. Analysis of a histone H2A variant from fission yeast: Evidence for a role in chromosome stability. *Mol Gen Genet* **245:** 628–635.

Castillo AG, Pidoux AL, Catania S, Durand-Dubief M, Choi ES, Hamilton G, Ekwall K, Allshire RC. 2013. Telomeric repeats facilitate CENP-A(Cnp1) incorporation via telomere binding proteins. *PLoS One* **8:** e69673.

Chen ES, Zhang K, Nicolas E, Cam HP, Zofall M, Grewal SI. 2008. Cell cycle control of centromeric repeat transcription and heterochromatin assembly. *Nature* **451:** 734–737.

Choi ES, Strålfors A, Castillo AG, Durand-Dubief M, Ekwall K, Allshire RC. 2011. Identification of noncoding transcripts from within CENP-A chromatin at fission yeast centromeres. *J Biol Chem* **286:** 23600–23607.

Choi ES, Strålfors A, Catania S, Castillo AG, Svensson JP, Pidoux AL, Ekwall K, Allshire RC. 2012. Factors that promote H3 chromatin integrity during transcription prevent promiscuous deposition of CENP-A^{Cnp1} in fission yeast. *PLoS Genet* **8:** e1002985.

Clarke L, Baum MP. 1990. Functional analysis of a centromere from fission yeast: A role for centromere-specific repeated DNA sequences. *Mol Cell Biol* **10:** 1863–1872.

Coffman VC, Wu P, Parthun MR, Wu JQ. 2011. CENP-A exceeds microtubule attachment sites in centromere clusters of both budding and fission yeast. *J Cell Biol* **195:** 563–572.

Cowieson NP, Partridge JF, Allshire RC, McLaughlin PJ. 2000. Dimerisation of a chromo shadow domain and distinctions from the chromodomain as revealed by structural analysis. *Curr Biol* **10:** 517–525.

* Dekker J, Misteli T. 2014. Long-range chromatin interactions. *Cold Spring Harb Perspect Biol* doi: 10.1101/cshperspect.a019356.

Dheur S, Saupe SJ, Genier S, Vazquez S, Javerzat JP. 2011. Role for cohesin in the formation of a heterochromatic domain at fission yeast subtelomeres. *Mol Cell Biol* **31:** 1088–1097.

Djupedal I, Ekwall K. 2008. Molecular biology. The paradox of silent heterochromatin. *Science* **320:** 624–625.

Djupedal I, Portoso M, Spahr H, Bonilla C, Gustafsson CM, Allshire RC, Ekwall K. 2005. RNA Pol II subunit Rpb7 promotes centromeric transcription and RNAi-directed chromatin silencing. *Genes Dev* **19:** 2301–2306.

Djupedal I, Kos-Braun IC, Mosher RA, Soderholm N, Simmer F, Hardcastle TJ, Fender A, Heidrich N, Kagansky A, Bayne E, et al. 2009. Analysis of small RNA in fission yeast; centromeric siRNAs are potentially generated through a structured RNA. *EMBO J* **28:** 3832–3844.

Dunleavy EM, Pidoux AL, Monet M, Bonilla C, Richardson W, Hamilton GL, Ekwall K, McLaughlin PJ, Allshire RC. 2007. A NASP (N1/N2)-related protein, Sim3, binds CENP-A and is required for its deposition at fission yeast centromeres. *Mol Cell* **28:** 1029–1044.

Durand-Dubief M, Sinha I, Fagerstrom-Billai F, Bonilla C, Wright A, Grunstein M, Ekwall K. 2007. Specific functions for the fission yeast Sirtuins Hst2 and Hst4 in gene regulation and retrotransposon silencing. *EMBO J* **26:** 2477–2488.

Durand-Dubief M, Persson J, Norman U, Hartsuiker E, Ekwall K. 2010. Topoisomerase I regulates open chromatin and controls gene expression in vivo. *EMBO J* **29:** 2126–2134.

Durand-Dubief M, Svensson JP, Persson J, Ekwall K. 2011. Topoisomerases, chromatin and transcription termination. *Transcription* **2:** 66–70.

Egel R. 2004. *The molecular biology of* Schizosaccharomyces pombe*: Genetics, genomics and beyond*. Springer, Berlin.

Egel R, Willer M, Nielsen O. 1989. Unblocking of meiotic crossing-over between the silent mating-type cassettes of fission yeast, conditioned by the recessive, pleiotropic mutant rik1. *Curr Genet* **15:** 407–410.

Ekwall K, Ruusala T. 1994. Mutations in *rik1, clr2, clr3* and *clr4* genes asymmetrically derepress the silent mating-type loci in fission yeast. *Genetics* **136:** 53–64.

Ekwall K, Javerzat JP, Lorentz A, Schmidt H, Cranston G, Allshire R. 1995. The chromodomain protein Swi6: A key component at fission yeast centromeres. *Science* **269:** 1429–1431.

Ekwall K, Nimmo ER, Javerzat JP, Borgstrom B, Egel R, Cranston G, Allshire R. 1996. Mutations in the fission yeast silencing factors clr4+ and rik1+ disrupt the localisation of the chromo domain protein Swi6p and impair centromere function. *J Cell Sci* **109:** 2637–2648.

Ekwall K, Cranston G, Allshire RC. 1999. Fission yeast mutants that alleviate transcriptional silencing in centromeric flanking repeats and disrupt chromosome segregation. *Genetics* **153:** 1153–1169.

* Elgin SCR, Reuter G. 2013. Position-effect variegation, heterochromatin formation, and gene silencing in *Drosophila*. *Cold Spring Harb Perspect Biol* **5:** a017780.

Flaus A, Martin DM, Barton GJ, Owen-Hughes T. 2006. Identification of multiple distinct Snf2 subfamilies with conserved structural motifs. *Nucleic Acids Res* **34:** 2887–2905.

Folco HD, Pidoux AL, Urano T, Allshire RC. 2008. Heterochromatin and RNAi are required to establish CENP-A chromatin at centromeres. *Science* **319:** 94–97.

Fujita Y, Hayashi T, Kiyomitsu T, Toyoda Y, Kokubu A, Obuse C, Yanagida M. 2007. Priming of centromere for CENP-A recruitment by human hMis18α, hMis18β, and M18BP1. *Dev Cell* **12:** 17–30.

Garcia JF, Dumesic PA, Hartley PD, El-Samad H, Madhani HD. 2010. Combinatorial, site-specific requirement for heterochromatic silencing factors in the elimination of nucleosome-free regions. *Genes Dev* **24:** 1758–1771.

Gerace EL, Halic M, Moazed D. 2010. The methyltransferase activity of Clr4Suv39h triggers RNAi independently of histone H3K9 methylation. *Mol Cell* **39:** 360–372.

Gonzalez Y, Saito A, Sazer S. 2012. Fission yeast Lem2 and Man1 perform fundamental functions of the animal cell nuclear lamina. *Nucleus* **3:** 60–76.

Goshima G, Saitoh S, Yanagida M. 1999. Proper metaphase spindle length is determined by centromere proteins Mis12 and Mis6 required for faithful chromosome segregation. *Genes Dev* **13:** 1664–1677.

Grewal SI, Klar AJ. 1997. A recombinationally repressed region between mat2 and mat3 loci shares homology with centromeric repeats and regulates directionality of mating-type switching in fission yeast. *Genetics* **146:** 1221–1238.

Grewal SI, Bonaduce MJ, Klar AJ. 1998. Histone deacetylase homologs regulate epigenetic inheritance of transcriptional silencing and chromosome segregation in fission yeast. *Genetics* **150:** 563–576.

* Grunstein M, Gasser SM. 2013. Epigenetics in *Saccharomyces cerevisiae*. *Cold Spring Harb Perspect Biol* **5:** a017491.

Gullerova M, Proudfoot NJ. 2008. Cohesin complex promotes transcriptional termination between convergent genes in *S. pombe*. *Cell* **132:** 983–995.

Gullerova M, Moazed D, Proudfoot NJ. 2011. Autoregulation of convergent RNAi genes in fission yeast. *Genes Dev* **25:** 556–568.

Halic M, Moazed D. 2010. Dicer-independent primal RNAs trigger RNAi and heterochromatin formation. *Cell* **140:** 504–516.

Hall IM, Shankaranarayana GD, Noma K, Ayoub N, Cohen A, Grewal SI. 2002. Establishment and maintenance of a heterochromatin domain. *Science* **297:** 2232–2237.

Hall IM, Noma K, Grewal SI. 2003. RNA interference machinery regulates chromosome dynamics during mitosis and meiosis in fission yeast. *Proc Natl Acad Sci* **100:** 193–198.

Hansen KR, Burns G, Mata J, Volpe TA, Martienssen RA, Bahler J, Thon G. 2005. Global effects on gene expression in fission yeast by silencing and RNA interference machineries. *Mol Cell Biol* **25:** 590–601.

Harigaya Y, Tanaka H, Yamanaka S, Tanaka K, Watanabe Y, Tsutsumi C, Chikashige Y, Hiraoka Y, Yamashita A, Yamamoto M. 2006. Selective elimination of messenger RNA prevents an incidence of untimely meiosis. *Nature* **442:** 45–50.

Hayashi T, Fujita Y, Iwasaki O, Adachi Y, Takahashi K, Yanagida M. 2004. Mis16 and Mis18 are required for CENP-A loading and histone deacetylation at centromeres. *Cell* **118:** 715–729.

* Henikoff S, Smith MM. 2014. Histone variants and epigenetics. *Cold Spring Harb Perspect Biol* doi: 10.1101/cshperspect.a019364.

Hennig BP, Bendrin K, Zhou Y, Fischer T. 2012. Chd1 chromatin remodelers maintain nucleosome organization and repress cryptic transcription. *EMBO Rep* **13:** 997–1003.

Hiraoka Y, Maekawa H, Asakawa H, Chikashige Y, Kojidani T, Osakada H, Matsuda A, Haraguchi T. 2011. Inner nuclear membrane protein Ima1 is dispensable for intranuclear positioning of centromeres. *Genes Cells* **16:** 1000–1011.

Hogan CJ, Aligianni S, Durand-Dubief M, Persson J, Will WR, Webster J, Wheeler L, Mathews CK, Elderkin S, Oxley D, et al. 2010. Fission yeast Iec1-Ino80-mediated nucleosome eviction regulates nucleotide and phosphate metabolism. *Mol Cell Biol* **30:** 657–674.

Hong EJ, Villen J, Gerace EL, Gygi S, Moazed D. 2005. A cullin E3 ubiquitin ligase complex associates with Rik1 and the Clr4 histone H3-K9 methyltransferase and is required for RNAi-mediated heterochromatin formation. *RNA Biol* **2:** 106–111.

Horn PJ, Bastie JN, Peterson CL. 2005. A Rik1-associated, cullin-dependent E3 ubiquitin ligase is essential for heterochromatin formation. *Genes Dev* **19:** 1705–1714.

Hou H, Wang Y, Kallgren SP, Thompson J, Yates JR III, Jia S. 2010. Histone variant H2A.Z regulates centromere silencing and chromosome segregation in fission yeast. *J Biol Chem* **285:** 1909–1918.

Isaac S, Walfridsson J, Zohar T, Lazar D, Kahan T, Ekwall K, Cohen A. 2007. Interaction of Epe1 with the heterochromatin assembly pathway in *Schizosaccharomyces pombe*. *Genetics* **175:** 1549–1560.

Ishii K, Ogiyama Y, Chikashige Y, Soejima S, Masuda F, Kakuma T, Hiraoka Y, Takahashi K. 2008. Heterochromatin integrity affects chromosome reorganization after centromere dysfunction. *Science* **321:** 1088–1091.

Iwasaki O, Tanaka A, Tanizawa H, Grewal SI, Noma K. 2010. Centromeric localization of dispersed Pol III genes in fission yeast. *Mol Biol Cell* **21:** 254–265.

Jain D, Hebden AK, Nakamura TM, Miller KM, Cooper JP. 2010. HAATI survivors replace canonical telomeres with blocks of generic heterochromatin. *Nature* **467:** 223–227.

Jia S, Noma K, Grewal SI. 2004. RNAi-independent heterochromatin nucleation by the stress-activated ATF/CREB family proteins. *Science* **304:** 1971–1976.

Johnsson A, Durand-Dubief M, Xue-Franzen Y, Ronnerblad M, Ekwall K, Wright A. 2009. HAT-HDAC interplay modulates global histone H3K14 acetylation in gene-coding regions during stress. *EMBO Rep* **10:** 1009–1014.

Kagansky A, Folco HD, Almeida R, Pidoux AL, Boukaba A, Simmer F, Urano T, Hamilton GL, Allshire RC. 2009. Synthetic heterochromatin bypasses RNAi and centromeric repeats to establish functional centromeres. *Science* **324:** 1716–1719.

Kanoh J, Sadaie M, Urano T, Ishikawa F. 2005. Telomere binding protein Taz1 establishes Swi6 heterochromatin independently of RNAi at telomeres. *Curr Biol* **15:** 1808–1819.

Kato H, Goto DB, Martienssen RA, Urano T, Furukawa K, Murakami Y. 2005. RNA polymerase II is required for RNAi-dependent heterochromatin assembly. *Science* **309:** 467–469.

Kawashima SA, Yamagishi Y, Honda T, Ishiguro K, Watanabe Y. 2010. Phosphorylation of H2A by Bub1 prevents chromosomal instability through localizing shugoshin. *Science* **327:** 172–177.

Keller C, Adaixo R, Stunnenberg R, Woolcock KJ, Hiller S, Buhler M. 2012. $HP1^{Swi6}$ mediates the recognition and destruction of heterochromatic RNA transcripts. *Mol Cell* **47:** 215–227.

Kiely CM, Marguerat S, Garcia JF, Madhani HD, Bahler J, Winston F. 2011. Spt6 is required for heterochromatic silencing in the fission yeast *Schizosaccharomyces pombe*. *Mol Cell Biol* **31:** 4193–4204.

Kim HS, Choi ES, Shin JA, Jang YK, Park SD. 2004. Regulation of Swi6/HP1-dependent heterochromatin assembly by cooperation of components of the mitogen-activated protein kinase pathway and a histone deacetylase Clr6. *J Biol Chem* **279:** 42850–42859.

Kim HS, Vanoosthuyse V, Fillingham J, Roguev A, Watt S, Kislinger T, Treyer A, Carpenter LR, Bennett CS, Emili A, et al. 2009. An acetylated form of histone H2A.Z regulates chromosome architecture in *Schizosaccharomyces pombe*. *Nat Struct Mol Biol* **16:** 1286–1293.

King MC, Drivas TG, Blobel G. 2008. A network of nuclear envelope membrane proteins linking centromeres to microtubules. *Cell* **134:** 427–438.

Kloc A, Zaratiegui M, Nora E, Martienssen R. 2008. RNA interference guides histone modification during the S phase of chromosomal replication. *Curr Biol* **18:** 490–495.

Kniola B, O"Toole E, McIntosh JR, Mellone B, Allshire R, Mengarelli S, Hultenby K, Ekwall K. 2001. The domain structure of centromeres is conserved from fission yeast to humans. *Mol Biol Cell* **12:** 2767–2775.

Lando D, Endesfelder U, Berger H, Subramanian L, Dunne PD, McColl J, Klenerman D, Carr AM, Sauer M, Allshire RC, et al. 2012. Quantitative single-molecule microscopy reveals that $CENP\text{-}A^{Cnp1}$ deposition occurs during G2 in fission yeast. *Open Biol* **2:** 120078.

Lantermann AB, Straub T, Stralfors A, Yuan GC, Ekwall K, Korber P. 2010. *Schizosaccharomyces pombe* genome-wide nucleosome mapping reveals positioning mechanisms distinct from those of *Saccharomyces cerevisiae*. *Nat Struct Mol Biol* **17:** 251–257.

Lejeune E, Bortfeld M, White SA, Pidoux AL, Ekwall K, Allshire RC, Ladurner AG. 2007. The chromatin-remodeling factor FACT contributes to centromeric heterochromatin independently of RNAi. *Curr Biol* **17:** 1219–1224.

* Li E, Zhang Y. 2014. DNA methylation in mammals. *Cold Spring Harb Perspect Biol* **6:** a019133.

Li F, Martienssen R, Cande WZ. 2011. Coordination of DNA replication and histone modification by the Rik1-Dos2 complex. *Nature* **475:** 244–248.

Lorentz A, Ostermann K, Fleck O, Schmidt H. 1994. Switching gene *swi6*, involved in repression of silent mating-type loci in fission yeast, encodes a homologue of chromatin-associated proteins from *Drosophila* and mammals. *Gene* **143:** 139–143.

* Martienssen R, Moazed D. 2014. RNAi and heterochromatin assembly. *Cold Spring Harb Perspect Biol* doi: 10.1101/cshperspect.a019323.

Maruyama T, Nakamura T, Hayashi T, Yanagida M. 2006. Histone H2B mutations in inner region affect ubiquitination, centromere function, silencing and chromosome segregation. *EMBO J* **25:** 2420–2431.

Mata J, Lyne R, Burns G, Bahler J. 2002. The transcriptional program of meiosis and sporulation in fission yeast. *Nat Genet* **32:** 143–147.

Mellone BG, Zhang W, Karpen GH. 2009. Frodos found: Behold the CENP-A "ring" bearers. *Cell* **137:** 409–412.

Molnar M, Kleckner N. 2008. Examination of interchromosomal interactions in vegetatively growing diploid *Schizosaccharomyces pombe* cells by Cre/loxP site-specific recombination. *Genetics* **178:** 99–112.

Motamedi MR, Verdel A, Colmenares SU, Gerber SA, Gygi SP, Moazed D. 2004. Two RNAi complexes, RITS and RDRC, physically interact and localize to noncoding centromeric RNAs. *Cell* **119:** 789–802.

Nakayama J, Rice JC, Strahl BD, Allis CD, Grewal SI. 2001. Role of histone H3 lysine 9 methylation in epigenetic control of heterochromatin assembly. *Science* **292:** 110–113.

Ngan VK, Clarke L. 1997. The centromere enhancer mediates centromere activation in *Schizosaccharomyces pombe*. *Mol Cell Biol* **17:** 3305–3314.

Nicolas E, Yamada T, Cam HP, Fitzgerald PC, Kobayashi R, Grewal SI. 2007. Distinct roles of HDAC complexes in promoter silencing, antisense suppression and DNA damage protection. *Nat Struct Mol Biol* **14:** 372–380.

Nimmo ER, Cranston G, Allshire RC. 1994. Telomere-associated chromosome breakage in fission yeast results in variegated expression of adjacent genes. *EMBO J* **13:** 3801–3811.

Nimmo ER, Pidoux AL, Perry PE, Allshire RC. 1998. Defective meiosis in telomere-silencing mutants of *Schizosaccharomyces pombe*. *Nature* **392:** 825–828.

Noma K, Allis CD, Grewal SI. 2001. Transitions in distinct histone H3 methylation patterns at the heterochromatin domain boundaries. *Science* **293:** 1150–1155.

Noma K, Sugiyama T, Cam H, Verdel A, Zofall M, Jia S, Moazed D, Grewal SI. 2004. RITS acts in cis to promote RNA interference-mediated transcriptional and post-transcriptional silencing. *Nat Genet* **36:** 1174–1180.

Nonaka N, Kitajima T, Yokobayashi S, Xiao G, Yamamoto M, Grewal SI, Watanabe Y. 2002. Recruitment of cohesin to heterochromatic regions by Swi6/HP1 in fission yeast. *Nat Cell Biol* **4:** 89–93.

Partridge JF, DeBeauchamp JL, Kosinski AM, Ulrich DL, Hadler MJ, Noffsinger VJ. 2007. Functional separation of the requirements for establishment and maintenance of centromeric heterochromatin. *Mol Cell* **26:** 593–602.

Partridge JF, Scott KS, Bannister AJ, Kouzarides T, Allshire RC. 2002. *cis*-acting DNA from fission yeast centromeres mediates histone H3 methylation and recruitment of silencing factors and cohesin to an ectopic site. *Curr Biol* **12:** 1652–1660.

Perpelescu M, Fukagawa T. 2011. The ABCs of CENPs. *Chromosoma* **120:** 425–446.

Pidoux AL, Allshire RC. 2004. Kinetochore and heterochromatin domains of the fission yeast centromere. *Chromosome Res* **12:** 521–534.

Pidoux AL, Allshire RC. 2005. The role of heterochromatin in centromere function. *Philos Trans R Soc Lond B Biol Sci* **360:** 569–579.

Pidoux AL, Richardson W, Allshire RC. 2003. Sim4: A novel fission yeast kinetochore protein required for centromeric silencing and chromosome segregation. *J Cell Biol* **161:** 295–307.

Pidoux AL, Choi ES, Abbott JK, Liu X, Kagansky A, Castillo AG, Hamilton GL, Richardson W, Rappsilber J, He X, et al. 2009. Fission yeast Scm3: A CENP-A receptor required for integrity of subkinetochore chromatin. *Mol Cell* **33:** 299–311.

* Pikaard CS, Mittelsten Sheid O. 2014. Epigenetic regulation in plants. *Cold Spring Harb Perspect Biol* doi: 10.1101/cshperspect.a019315.

Pointner J, Persson J, Prasad P, Norman-Axelsson U, Stralfors A, Khorosjutina O, Krietenstein N, Svensson JP, Ekwall K, Korber P. 2012. CHD1 remodelers regulate nucleosome spacing in vitro and align nucleosomal arrays over gene coding regions in *S. pombe*. *EMBO J* **31:** 4388–4403.

Polizzi C, Clarke L. 1991. The chromatin structure of centromeres from fission yeast: Differentiation of the central core that correlates with function. *J Cell Biol* **112:** 191–201.

Reddy BD, Wang Y, Niu L, Higuchi EC, Marguerat SB, Bahler J, Smith GR, Jia S. 2011. Elimination of a specific histone H3K14 acetyltransferase complex bypasses the RNAi pathway to regulate pericentric heterochromatin functions. *Genes Dev* **25:** 214–219.

Reinhart BJ, Bartel DP. 2002. Small RNAs correspond to centromere heterochromatic repeats. *Science* **297:** 1831.

Reyes-Turcu FE, Zhang K, Zofall M, Chen E, Grewal SI. 2011. Defects in RNA quality control factors reveal RNAi-independent nucleation of heterochromatin. *Nat Struct Mol Biol* **18:** 1132–1138.

Rhind N, Chen Z, Yassour M, Thompson DA, Haas BJ, Habib N, Wapinski I, Roy S, Lin MF, Heiman DI, et al. 2011. Comparative functional genomics of the fission yeasts. *Science* **332:** 930–936.

Sadaie M, Iida T, Urano T, Nakayama J. 2004. A chromodomain protein, Chp1, is required for the establishment of heterochromatin in fission yeast. *EMBO J* **23:** 3825–3835.

Sato H, Masuda F, Takayama Y, Takahashi K, Saitoh S. 2012. Epigenetic inactivation and subsequent heterochromatinization of a centromere stabilize dicentric chromosomes. *Curr Biol* **22:** 658–667.

Scott KC, Merrett SL, Willard HF. 2006. A heterochromatin barrier partitions the fission yeast centromere into discrete chromatin domains. *Curr Biol* **16:** 119–129.

Scott KC, White CV, Willard HF. 2007. An RNA polymerase III-dependent heterochromatin barrier at fission yeast centromere 1. *PLoS ONE* **2:** e1099.

* Seto E, Yoshida M. 2014. Erasers of histone acetylation: The histone deacetylase enzymes. *Cold Spring Harb Perspect Biol* **6:** a018713.

Shankaranarayana GD, Motamedi MR, Moazed D, Grewal SI. 2003. Sir2 regulates histone H3 lysine 9 methylation and heterochromatin assembly in fission yeast. *Curr Biol* **13:** 1240–1246.

Shim YS, Choi Y, Kang K, Cho K, Oh S, Lee J, Grewal SI, Lee D. 2012. Hrp3 controls nucleosome positioning to suppress non-coding transcription in eu- and heterochromatin. *EMBO J* **31:** 4375–4387.

Shimada A, Dohke K, Sadaie M, Shinmyozu K, Nakayama J, Urano T, Murakami Y. 2009. Phosphorylation of Swi6/HP1 regulates transcriptional gene silencing at heterochromatin. *Genes Dev* **23:** 18–23.

Steglich B, Filion G, van Steensel B, Ekwall K. 2012. The inner nuclear membrane proteins Man1 and Ima1 link to two different types of chromatin at the nuclear periphery in *S. pombe*. *Nucleus* **3:** 77–87.

Steiner NC, Clarke L. 1994. A novel epigenetic effect can alter centromere function in fission yeast. *Cell* **79:** 865–874.

Stralfors A, Walfridsson J, Bhuiyan H, Ekwall K. 2011. The FUN30 chromatin remodeler, Fft3, protects centromeric and subtelomeric domains from euchromatin formation. *PLoS Genet* **7:** e1001334.

Sugiyama T, Cam H, Verdel A, Moazed D, Grewal SI. 2005. RNA-dependent RNA polymerase is an essential component of a self-enforcing loop coupling heterochromatin assembly to siRNA production. *Proc Natl Acad Sci* **102:** 152–157.

Sugiyama T, Cam HP, Sugiyama R, Noma K, Zofall M, Kobayashi R, Grewal SI. 2007. SHREC, an effector complex for heterochromatic transcriptional silencing. *Cell* **128:** 491–504.

Tada K, Susumu H, Sakuno T, Watanabe Y. 2011. Condensin association with histone H2A shapes mitotic chromosomes. *Nature* **474:** 477–483.

Takahashi K, Chen ES, Yanagida M. 2000. Requirement of Mis6 centromere connector for localizing a CENP-A-like protein in fission yeast. *Science* **288:** 2215–2219.

Takahashi K, Murakami S, Chikashige Y, Funabiki H, Niwa O, Yanagida M. 1992. A low copy number central sequence with strict symmetry and unusual chromatin structure in fission yeast centromere. *Mol Biol Cell* **3:** 819–835.

Tanizawa H, Iwasaki O, Tanaka A, Capizzi JR, Wickramasinghe P, Lee M, Fu Z, Noma K. 2010. Mapping of long-range associations throughout the fission yeast genome reveals global genome organization linked to transcriptional regulation. *Nucleic Acids Res* **38:** 8164–8177.

Thon G, Klar AJ. 1992. The *clr1* locus regulates the expression of the cryptic mating-type loci of fission yeast. *Genetics* **131:** 287–296.

Thon G, Cohen A, Klar AJ. 1994. Three additional linkage groups that repress transcription and meiotic recombination in the mating-type region of *Schizosaccharomyces pombe*. *Genetics* **138:** 29–38.

Thon G, Verhein-Hansen J. 2000. Four chromo-domain proteins of *Schizosaccharomyces pombe* differentially repress transcription at various chromosomal locations. *Genetics* **155:** 551–568.

Trewick SC, Minc E, Antonelli R, Urano T, Allshire RC. 2007. The JmjC domain protein Epe1 prevents unregulated assembly and disassembly of heterochromatin. *EMBO J* **26:** 4670–4682.

Volpe TA, Kidner C, Hall IM, Teng G, Grewal SI, Martienssen RA. 2002. Regulation of heterochromatic silencing and histone H3 lysine-9 methylation by RNAi. *Science* **297:** 1833–1837.

Volpe T, Schramke V, Hamilton GL, White SA, Teng G, Martienssen RA, Allshire RC. 2003. RNA interference is required for normal centromere function in fission yeast. *Chromosome Res* **11:** 137–146.

Walfridsson J, Bjerling P, Thalen M, Yoo EJ, Park SD, Ekwall K. 2005. The CHD remodeling factor Hrp1 stimulates CENP-A loading to centromeres. *Nucleic Acids Res* **33:** 2868–2879.

Walfridsson J, Khorosjutina O, Matikainen P, Gustafsson CM, Ekwall K. 2007. A genome-wide role for CHD remodelling factors and Nap1 in nucleosome disassembly. *EMBO J* **26:** 2868–2879.

Wilkinson CR, Bartlett R, Nurse P, Bird AP. 1995. The fission yeast gene *pmt1+* encodes a DNA methyltransferase homologue. *Nucleic Acids Res* **23:** 203–210.

Williams JS, Hayashi T, Yanagida M, Russell P. 2009. Fission yeast Scm3 mediates stable assembly of Cnp1/CENP-A into centromeric chromatin. *Mol Cell* **33:** 287–298.

Wiren M, Silverstein RA, Sinha I, Walfridsson J, Lee HM, Laurenson P, Pillus L, Robyr D, Grunstein M, Ekwall K. 2005. Genomewide analysis of nucleosome density histone acetylation and HDAC function in fission yeast. *EMBO J* **24:** 2906–2918.

Wood V, Gwilliam R, Rajandream MA, Lyne M, Lyne R, Stewart A, Sgouros J, Peat N, Hayles J, Baker S, et al. 2002. The genome sequence of *Schizosaccharomyces pombe*. *Nature* **415:** 871–880.

Woolcock KJ, Gaidatzis D, Punga T, Buhler M. 2011. Dicer associates with chromatin to repress genome activity in *Schizosaccharomyces pombe*. *Nat Struct Mol Biol* **18:** 94–99.

Woolcock KJ, Stunnenberg R, Gaidatzis D, Hotz HR, Emmerth S, Barraud P, Buhler M. 2012. RNAi keeps Atf1-bound stress response genes in check at nuclear pores. *Genes Dev* **26:** 683–692.

Xhemalce B, Kouzarides T. 2010. A chromodomain switch mediated by histone H3 Lys 4 acetylation regulates heterochromatin assembly. *Genes Dev* **24:** 647–652.

Yamada T, Fischle W, Sugiyama T, Allis CD, Grewal SI. 2005. The nucleation and maintenance of heterochromatin by a histone deacetylase in fission yeast. *Mol Cell* **20:** 173–185.

Zaratiegui M, Castel SE, Irvine DV, Kloc A, Ren J, Li F, de Castro E, Marin L, Chang AY, Goto D, et al. 2011. RNAi promotes heterochromatic silencing through replication-coupled release of RNA Pol II. *Nature* **479:** 135–138.

Zhang K, Mosch K, Fischle W, Grewal SI. 2008. Roles of the Clr4 methyltransferase complex in nucleation, spreading and maintenance of heterochromatin. *Nat Struct Mol Biol* **15:** 381–388.

Zofall M, Grewal SI. 2006. Swi6/HP1 recruits a JmjC domain protein to facilitate transcription of heterochromatic repeats. *Mol Cell* **22:** 681–692.

Zofall M, Yamanaka S, Reyes-Turcu FE, Zhang K, Rubin C, Grewal SI. 2012. RNA elimination machinery targeting meiotic mRNAs promotes facultative heterochromatin formation. *Science* **335:** 96–100.

C H A P T E R 10

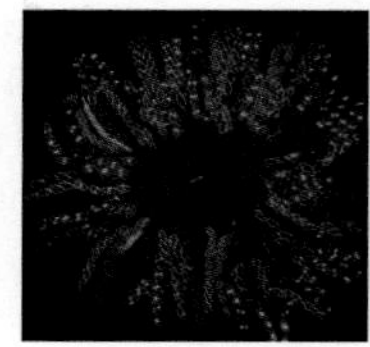
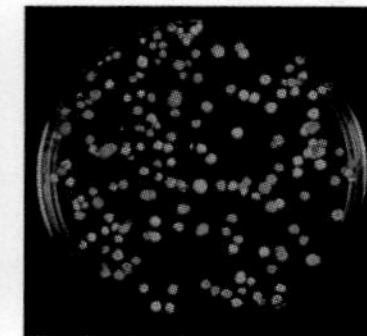

Neurospora crassa, a Model System for Epigenetics Research

Rodolfo Aramayo[1] and Eric U. Selker[2]

[1]Department of Biology, Texas A&M University, College Station, Texas 77843-3258; [2]Department of Biology and Institute of Molecular Biology, University of Oregon, Eugene, Oregon 97403-1229

Correspondence: selker@uoregon.edu

SUMMARY

The filamentous fungus *Neurospora crassa* has provided a rich source of knowledge on epigenetic phenomena that would have been difficult or impossible to gain from other systems. *Neurospora* sports features found in higher eukaryotes but absent in both budding and fission yeast, including DNA methylation and H3K27 methylation, and also has distinct RNA interference (RNAi)-based silencing mechanisms operating in mitotic and meiotic cells. This has provided an unexpected wealth of information on gene silencing systems. One silencing mechanism, named repeat-induced point mutation (RIP), has both epigenetic and genetic aspects and provided the first example of a homology-based genome defense system. A second silencing mechanism, named quelling, is an RNAi-based mechanism that results in silencing of transgenes and their native homologs. A third, named meiotic silencing, is also RNAi-based but is distinct from quelling in its time of action, targets, and apparent purpose.

Outline

Chapter-opening figures: (*Left*) From Freitag et al. (2004b).

Cite as *Cold Spring Harb Perspect Biol* doi: 10.1101/cshperspect.a017921

OVERVIEW

Fungi provide excellent models for understanding the structure and function of chromatin both in actively transcribed regions (euchromatin) and in transcriptionally silent regions (heterochromatin). The budding yeast, *Saccharomyces cerevisiae*, has been an invaluable eukaryotic model for studying chromatin structure associated with transcription at euchromatic regions and providing a paradigm for silent chromatin (Ch. 8 [Grunstein and Gasser 2013]). The fission yeast, *Schizosaccharomyces pombe*, has some epigenetic machinery that is absent from *S. cerevisiae* but common in higher organisms—most notably for RNA interference (RNAi) and for methylation of lysine 9 of histone H3 (H3K9me). As described in Ch. 9 (Allshire and Ekwall 2014), research using *S. pombe* has provided invaluable information on the structure and function of heterochromatin, principally found in regions of the centromeres, telomeres, and silent mating-type genes. This chapter focuses on a third model system, namely the filamentous fungus *Neurospora crassa*. Although not as commonly studied as the yeasts, *Neurospora* has proved to be a remarkably rich source of knowledge that would have been difficult or impossible to gain from other systems. *Neurospora* sports features found in higher eukaryotes, including DNA methylation and the H3K27 methylation ("Polycomb") system that both budding and fission yeasts lack, as well as RNAi and other epigenetic processes found in the yeasts. This has provided an unexpected wealth of information on gene silencing systems, some of which operate at distinct stages of its life cycle. The first such mechanism, named repeat-induced point mutation (RIP), has both epigenetic and genetic aspects and provided the first example of a homology-based genome defense system. The second, named quelling, is an RNAi-based mechanism that results in silencing of transgenes and their native homologs. The third, named meiotic silencing (or meiotic silencing by unpaired DNA), is also RNAi-based but is distinct from quelling in its time of action, targets, and apparent purpose. Although we are still in the early days of epigenetic studies in all organisms, it is already clear that yeasts and filamentous fungi such as *N. crassa* will continue to serve as rich sources of information on epigenetic mechanisms operative in a broad range of eukaryotes.

Cite as *Cold Spring Harb Perspect Biol* doi: 10.1101/cshperspect.a017921

1 *Neurospora crassa*: HISTORY AND FEATURES OF THE ORGANISM

The filamentous fungus *N. crassa* (see Figs. 1 and 2) was first developed as an experimental organism by Dodge in the late 1920s and about 10 years later was adopted by Beadle and Tatum for their famous "one gene–one protein" studies linking biochemistry and genetics (Davis and de Serres 1970). Beadle and Tatum selected *Neurospora*, in part, because this organism grows fast and is easy to propagate on defined growth media, and because genetic manipulations, such as mutagenesis, complementation tests, and mapping are simple. Although not as widely studied as some other model eukaryotes, *Neurospora* continues to attract researchers because of its moderate complexity and because it is well suited for a variety of genetic, biochemical, developmental, and subcellular studies (Borkovich et al. 2004). *Neurospora* has been especially useful for studies of photobiology, circadian rhythms, population biology, morphogenesis, mitochondrial import, DNA repair and recombination, DNA methylation, and other epigenetic processes (Borkovich et al. 2004).

N. crassa is commonly observed growing on burned wood after a forest fire (Fig. 1A). It comes in two mating types (*A* and *a*), which are morphologically indistinguishable from each other (Fig. 1B). The vegetative phase is initiated when either a sexual spore (ascospore) or an asexual spore (conidium) germinates, giving rise to multinucleate cells that form branched filaments (hyphae; Fig. 1C). In the wild, the heat of the fire provides the activation required for ascospore germination (Figs. 1D and 2). In contrast,

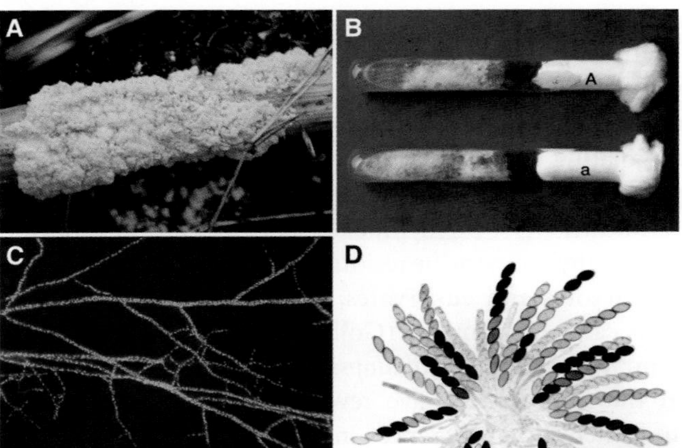

Figure 1. Images of *Neurospora crassa*. (*A*) Vegetative growth in the wild on sugarcane (photo by D. Jacobson, Stanford University). (*B*) Slants of vegetative cultures of *N. crassa* in the laboratory (photo by N.B. Raju, Stanford University). (*C*) Hyphae of *N. crassa* stained with 4′6-diamidino-2-phenylindole (DAPI) to show abundant nuclei (photo by M. Springer, Stanford University). (*D*) A rosette of maturing asci showing ascospores patterns (photo by N.B. Raju; reprinted, with permission, from Raju 1980, © Elsevier).

conidial cells germinate spontaneously. The hyphal system spreads out rapidly (linear growth >5 mm/h at 37°C) to form a "mycelium." After the mycelium is well established, aerial hyphae ("conidiophores") develop, leading to the production of the abundant orange conidia that are characteristic of the organism (Figs. 1A,B and 2). The conidia, which contain one to several nuclei each, can either establish new vegetative cultures or fertilize strains of the opposite mating type. If nutrients are limiting, *N. crassa* activates its sexual phase by producing nascent fruiting bodies ("protoperithecia"). When a specialized hypha ("trichogyne") projecting from the protoperithecium contacts tissue of the opposite mating type, a heterokaryon can form and an acquired "male" nucleus is transported back to the protoperithecium. Strains of either mating type can act as "females" or "males." The process of fertilization transforms a protoperithecium into a young perithecium. The sexual phase of *N. crassa* and other filamentous ascomycetes differs from that of yeasts in that the filamentous fungi have a prolonged heterokaryotic phase between fertilization and karyogamy (nuclear fusion). The heterokaryotic cells resulting from fertilization proliferate in the developing perithecium, which contains a mixture of both ascogenous (heterokaryotic) and maternal (homokaryotic) tissues. Development of ascogenous tissue involves a transition to strictly dikaryotic (binucleate) cells containing one nucleus of each mating type, which undergo synchronous nuclear divisions culminating in formation of the hook-shaped cells called "croziers" (Fig. 2). Croziers develop into three cells. Karyogamy, meiosis, and postmeiotic mitosis take place in the middle cell, also known as the ascus mother cell. It is noteworthy that the diploid nucleus formed by karyogamy immediately enters into meiosis. Thus, the diploid phase of the life cycle is brief (~24 h) and limited to a single developing cell. The eight nuclei that result from the first postmeiotic mitosis are compartmentalized, resulting in an ascus cell that contains eight haploid spores ("ascospores") arrayed in an order that reflects their lineage (Raju 1980, 1992). Ascospores are ejected from the beak of the perithecium and can germinate after exposure to high temperature to produce vegetative mycelia, completing the sexual cycle. One perithecium may contain up to 200 developing asci. Meiotic segregation and recombination can be studied in *Neurospora* by analyzing individual asci ("tetrads") or random spores ejected from numerous asci (Perkins 1966, 1988; Davis and de Serres 1970). Genetic analyses have indicated that, in general, all the asci of a perithecium derive from a single maternal nucleus and a single paternal nucleus.

The ~40-megabase *N. crassa* genome consists of seven chromosomes with approximately10,000 predicted protein coding genes (Galagan et al. 2003), and a total genetic map

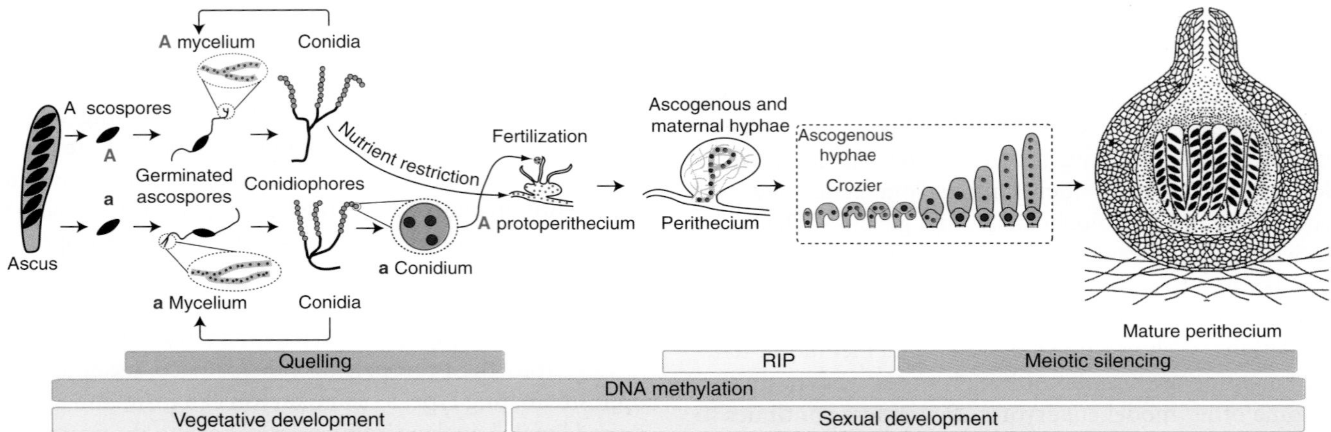

Figure 2. Life cycle of *N. crassa*. Half of the sexual spores (ascospores) are mating type *A* (red) and half are mating type *a* (blue). Sexual spores (ascospores) and vegetative spores (conidia) germinate and form mycelia, from which asexual fruiting bodies (conidiophores) emerge. Conidiophores form conidia, which are typically multinucleate. In response to nitrogen starvation, mycelia of either mating type form specialized female structures called protoperithecia. Vegetative tissue (e.g., a conidium) of the opposite mating type serves as the "male" to fertilize and initiate development of fruiting bodies (perithecia). After fertilization, male- and female-derived nuclei coexist in the same cytoplasm, where they undergo mitoses and eventually become organized into a dikaryotic tissue in which each cell has one nucleus of each mating type. The nuclei then pair and undergo a series of synchronous mitoses until the tip of the hyphal cell in which they reside bends to form a hook-shaped cell called a crozier. Fusion of haploid nuclei is immediately followed by meiosis and a mitotic division such that one crozier gives rise to one ascus containing eight ascospores. The approximate stages in which the epigenetic processes described in the text occur are indicated.

length of roughly 1000 map units (Perkins et al. 2001). Only ~9% of the genome consists of repetitive DNA and, aside from a tandem array of approximately 170 copies of the ~10-kb recombinant DNA (rDNA) unit encoding the three large ribosomal RNAs, most of the repetitive DNA consists of inactivated transposable elements. That most strains of *N. crassa* lack active transposons and have very few close paralogs almost certainly reflect the operation of RIP, the first homology-dependent genome defense system discovered in eukaryotes (Selker 1990a,b). We now know that *Neurospora* has at least three gene silencing processes that act to conserve the structure of the genome: RIP, quelling, and meiotic silencing (Borkovich et al. 2004). All of these processes have epigenetic aspects and have direct or indirect connections with DNA methylation, a basic epigenetic mechanism found in *Neurospora* and many other eukaryotes. We will discuss DNA methylation and then RIP, quelling, and meiotic silencing.

2 DNA METHYLATION IN *Neurospora*

Since its discovery decades ago, DNA methylation in eukaryotes has remained remarkably enigmatic. Basic questions are still debated, such as "What determines which chromosomal regions are methylated?" and "What is the function of DNA methylation?" *Neurospora* revealed itself to be an excellent system to study the control and function of DNA methylation. Some model eukaryotes, including the nematode *Caenorhabditis elegans* and the yeasts *S. cerevisiae* and *S. pombe*, lack detectable DNA methylation and isolated reports of DNA methylation in another model organism, *Drosophila melanogaster*, remain controversial. In some organisms such as mammals, DNA methylation is essential for viability, complicating certain analyses.

In *N. crassa* DNA, ~1.5% of the cytosines are methylated, but this methylation is dispensable, facilitating genetic studies. Although one must be cautious when extrapolating from one system to another, at least some aspects of DNA methylation appear conserved. For example, all known DNA methyltransferases (DMTs), the enzymes that methylate cytosine residues, including those from both prokaryotes and eukaryotes, show striking homology in their catalytic domains (Goll and Bestor 2005). Findings from *Neurospora*, *Arabidopsis*, mice, and other systems in the last decade have revealed important similarities and interesting differences in the control and function of DNA methylation, demonstrating the value of performing investigations in multiple model systems.

Discovery of DNA methylation in *Neurospora* initially attracted interest because it was not limited to symmetrical sites, such as CpG dinucleotides or CpNpG trinucleotides (see Ch. 15 [Li and Zhang 2014]; Ch. 13 [Pikaard and Mittelsten Scheid 2014]). Riggs, and Holliday and Pugh had

 Cite as *Cold Spring Harb Perspect Biol* doi: 10.1101/cshperspect.a017921

proposed an attractive model for the "inheritance" or "maintenance" of methylation patterns that relied on the symmetrical nature of methylated sites observed in animals (see Fig. 2 in Ch. 15 [Li and Zhang 2014]). Although results of a variety of in vitro and in vivo studies have supported the "maintenance methylase" model, mechanisms for maintenance methylation that do not rely on faithful copying at symmetrical sites can be imagined and may be operative in a variety of organisms (e.g., Selker 1990b; Selker et al. 2002). The possibility that the observed methylation at asymmetric sites represented "de novo methylation" was exciting because mechanisms that blindly propagate methylation patterns can complicate the determination of which sequences are methylated in the first place. Indeed, results of DNA-mediated transformation and methylation inhibitor studies with *Neurospora* showed reproducible de novo methylation (e.g., Singer et al. 1995). More recently, genomic experiments with methylation mutants revealed widespread and rapid de novo methylation after genetic reintroduction of the wild-type allele corresponding to the defective gene (Lewis et al. 2009). Additional studies defined, in part, the underlying signals for de novo methylation (e.g., see Tamaru and Selker 2003).

The first methylated patch characterized in detail was the 1.6-kb $\zeta-\eta$ (zeta–eta) region, which consists of a diverged tandem duplication of a 0.8-kb segment of DNA, including a 5S rRNA gene (Selker and Stevens 1985). Comparison of this region with the corresponding chromosomal region from strains lacking the duplication initially led to the idea that repeated sequences can somehow induce DNA methylation, and ultimately led to the discovery of the genome defense system named RIP (Fig. 3) (Selker 1990b). Elucidation of RIP revealed that repeated sequences do not directly trigger DNA methylation, at least in *Neurospora*; instead, repeats trigger RIP, which is closely tied to DNA methylation, as described below. Both the $\zeta-\eta$ region and the ψ63 (psi-63) region, the second methylated region discovered in *Neurospora*, are products of RIP. Moreover, subsequent genome-wide analyses of DNA methylation revealed that nearly all methylated regions in *Neurospora* are relics of transposons inactivated by RIP (Selker et al. 2003; Galagan and Selker 2004). Indeed, the only DNA methylation in *Neurospora* that may not have resulted from RIP is at the tandemly arranged rDNA genes (Perkins et al. 1986).

3 RIP, A GENOME DEFENSE SYSTEM WITH BOTH GENETIC AND EPIGENETIC ASPECTS

RIP was discovered as a result of a detailed analysis of progeny from crosses of *Neurospora* transformants (Selker 1990b). It was noticed that duplicated sequences, whether

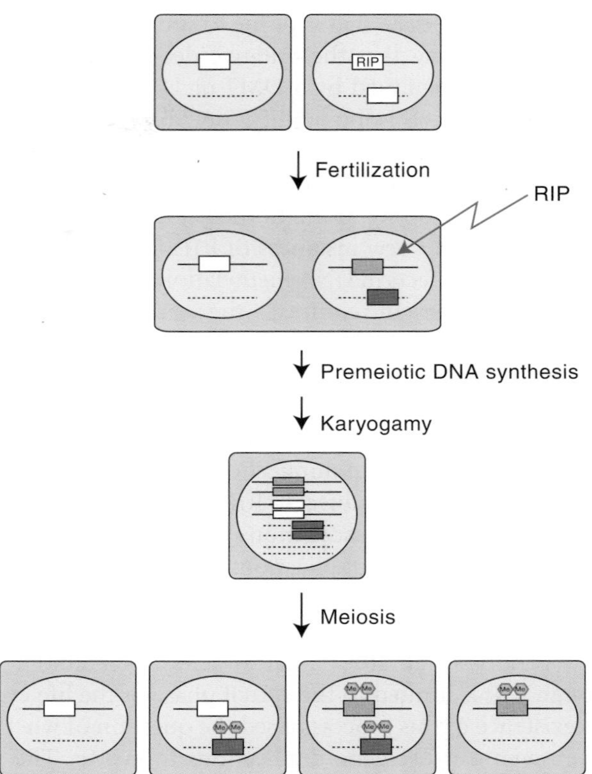

Figure 3. Repeat-induced point mutation (RIP). For clarity, only two chromosomes are illustrated. The open box represents a gene or chromosomal segment that was duplicated in one strain (*top, right*). Duplications are subject to RIP (symbolized by lightning bolt) between fertilization and karyogamy. Results of genetic experiments revealed that duplications can be repeatedly subjected to volleys of C to T transitions (symbolized by filled boxes) during this period of approximately 10 mitoses, right up to the final premeiotic DNA synthesis (Selker et al. 1987; Watters et al. 1999). The four possible combinations of chromosomes in progeny are indicated. Pink "Me" represents DNA methylation, which is frequently (although not always) associated with products of RIP.

native or foreign, and whether genetically linked or unlinked, were subjected to numerous polarized transition mutations (G:C to A:T) in the haploid genomes of the special heterokaryotic cells resulting from fertilization. When the stability of a gene was tested when it was unique in the genome or else combined with an unlinked homolog, it was found that RIP is not simply repeat-*associated*; it is truly repeat-*induced*. In a single passage through the sexual cycle, up to ~30% of the G:C pairs in duplicated sequences can be mutated. Frequently (but not invariably), the sequences that are altered by RIP become methylated de novo. It is likely that the mutations arising from RIP occur by enzymatic deamination of 5-methylcytosines (5mC) or by deamination of Cs followed by DNA replication (Selker 1990b). This was postulated partly because cytosine methylation involves a reaction intermediate that is prone to spontaneous deamination, suggesting that the putative deamination step of RIP might be catalyzed by a DMT or DMT-like enzyme. Consistent with this possibility, one of the two DMT homologs predicted from the *Neurospora* genome sequence—RID (RIP defective)—is involved in RIP (Freitag et al. 2002). Progeny from homozygous crosses of *rid* mutants show no new instances of RIP. *Rid* mutants have no noticeable defects in DNA methylation, fertility, growth, or development. In contrast, the second *Neurospora* DMT homolog (DIM-2), is necessary for all known DNA methylation, but is not required for RIP (Kouzminova and Selker 2001).

All indications are that every sizable duplication (>~400 bp for tandem duplication or ~1000 bp for unlinked duplication) is subject to RIP in some fraction of the special heterokaryotic ascogenous cells. Nevertheless, typically <1% of tandem duplications and ~50% of unlinked duplications escape RIP. Even duplications of chromosomal segments containing numerous genes are sensitive to RIP (Perkins et al. 1997; Bhat and Kasbekar 2001). Although RIP is limited to the sexual phase of the life cycle, the existence of this process raised the question of whether *Neurospora* can use gene duplications to evolve. The genome sequence revealed gene families, but tellingly, virtually all paralogs were found to be sufficiently divergent that they should not trigger RIP (Galagan and Selker 2004). Thus, RIP may indeed limit evolution through gene duplication in *Neurospora*. Interestingly, some fungi, such as *Ascobolus immersus*, show what appear to be milder genome defense systems that are similar to RIP. The most notable example is MIP (methylation induced premeiotically), a process that detects linked and unlinked sequence duplications during the period between fertilization and karyogamy, like RIP, but which relies exclusively on DNA methylation for inactivation; no evidence of mutations has been found in sequences inactivated by MIP (Rossignol and Faugeron 1994).

4 STUDIES OF RELICS OF RIP PROVIDED INSIGHTS INTO THE CONTROL OF DNA METHYLATION

4.1 Noncanonical Maintenance Methylation

The finding that a single DMT, DIM-2, is responsible for all detected DNA methylation was surprising. No previously identified DMT was known to methylate cytosines in a variety of sequence contexts. An obvious but important question was: Does methylation at nonsymmetrical sites necessarily reflect the potential of the corresponding sequences to induce methylation de novo? Early transformation experiments were consistent with this possibility; methylated sequences that were stripped of their methylation (e.g., by cloning) regained their normal methylation when reintroduced into vegetative cells. A surprise came when eight alleles of the *am* gene that were generated by RIP were tested for their capacity to induce methylation de novo (Singer et al. 1995). Consistent results were obtained from experiments performed in two ways: (1) Sequences were scored for remethylation after being stripped of methylation by treatment with a demethylating agent (5-azacytidine), and (2) unmethylated sequences introduced by transformation were scored for methylation de novo. Some products of RIP with relatively few mutations (Fig. 4; am^{RIP3} and am^{RIP4}) did not become remethylated, even at their normal locus, suggesting that the observed methylation represented propagation of methylation established earlier. Importantly, their methylation, like other observed methylation in *Neurospora*, was not limited to symmetrical sites, did not significantly spread with time, and was "heterogeneous" in the sense that the pattern of methylated residues was not invariant within a clonal population of cells. Thus this methylation, although dependent on preexisting methylation established in the sexual phase (perhaps by RIP), could not reflect the action of a "maintenance methylase" of the type envisioned in the original model for inheritance of methylation patterns.

The capacity of *Neurospora* to perform "maintenance methylation" was confirmed experimentally (Selker et al. 2002). Interestingly, propagation of methylation was found to be sequence-specific (i.e., it did not work on all sequences), adding a new dimension to the maintenance methylation concept. It is noteworthy that MIP in *Ascobolus* also provided evidence for propagation of DNA methylation in

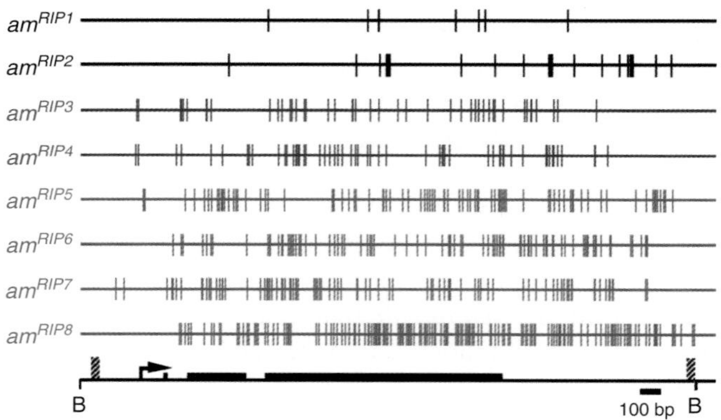

Figure 4. Mutations from RIP and methylation status of eight different *am* alleles (adapted from Singer et al. 1995). Vertical bars indicate mutations. Alleles shown in black were not methylated. Alleles in blue were initially methylated, but after loss of methylation induced by 5-azacytidine, or by cloning and gene replacement, did not become remethylated. Alleles shown in red were not only initially methylated, but also triggered methylation de novo.

 Cite as *Cold Spring Harb Perspect Biol* doi: 10.1101/cshperspect.a017921

fungi (Rossignol and Faugeron 1994). Although a number of potential schemes that would result in propagation of DNA methylation can be imagined, the actual mechanism operative in *Neurospora* remains unknown. In principle, maintenance of methylation at nonsymmetrical sites could depend on methylation of nearby symmetrical sites, but the observed heterogeneous methylation, including at CpG sites, renders this possibility unlikely. Feedback mechanisms involving proteins associated with the methylated DNA could result in methylation that depends on preexisting methylation (i.e., maintenance methylation). As discussed below, findings from *Neurospora* (and other organisms) implicate histone modifications in the control of DNA methylation, raising the possibility that histones play a role in the maintenance of DNA methylation.

4.2 Involvement of Histones in DNA Methylation

The first indication of a role of histones in DNA methylation came from the observation that blocking histone deacetylation in *Neurospora* reduced DNA methylation in some chromosomal regions (Selker 1998). This was performed by treatment with the histone deacetylase inhibitor trichostatin A (TSA). The selectivity of demethylation by TSA could reflect differential access to histone acetyltransferases (Smith et al. 2010), but has not been thoroughly investigated. Studies of the *dim-5* gene in *Neurospora* unambiguously tied chromatin to the control of DNA methylation. A *dim-5* mutant, like *dim-2* strains, shows a complete loss of DNA methylation, yet it is a SET domain protein that acts as a histone H3 lysine methyltransferase (HKMT), specifically trimethylating lysine 9 (Tamaru and Selker 2001; Tamaru et al. 2003). Confirmation that histone H3 is the physiologically relevant substrate of DIM-5 came from two demonstrations: (1) Replacement of lysine 9 in H3 with other amino acids caused loss of DNA methylation, and (2) trimethyl-lysine 9 (H3K9me3) was found specifically at DNA methylated chromosomal regions.

The discovery that histone methylation controls DNA methylation, at least in *Neurospora*, led to two important questions: (1) What instructs DIM-5 which nucleosomes to methylate? (2) What reads the trimethyl mark and transmits this information to the DMT, DIM-2? It has been easier to answer the second question than the first, in part because of information from other systems. In particular, knowledge that HP1, a protein first identified in *Drosophila*, binds H3K9me3 in vitro (discussed in Ch. 12 [Elgin and Reuter 2013]), motivated a search for an HP1 homolog in *Neurospora*. A likely homolog was found and its involvement in DNA methylation was tested by gene disruption (Freitag et al. 2004a). The gene, named *hpo* (HP one), was indeed essential for DNA methylation. As another test of whether *Neurospora* HP1 reads the mark generated by DIM-5, its subcellular localization was examined in wild type and *dim-5* strains. In wild type, HP1-GFP localized to heterochromatic foci, but this localization was lost in *dim-5*, confirming that *Neurospora* HP1 is recruited by the H3K9me3 mark generated by DIM-5. A yeast two-hybrid screen, and subsequent coimmunoprecipitation experiments, revealed that the chromoshadow domain of HP1 interacts directly with DIM-2 through PXVXL-related motifs in its amino-terminal region (Honda and Selker 2008).

The question of how DIM-5 is controlled has proved more difficult and is not yet fully answered. A combination of genetic, biochemical, and proteomic approaches has yielded insights, however. Results of such studies have culminated in the discovery that the localization and action of DIM-5 depends on a multiprotein complex, DCDC (DIM-5/-7/-9, CUL4/DDB1 complex; diagrammed in Fig. 5), which resembles an E3 ubiquitin ligase (Lewis et al. 2010a; Lewis et al. 2010b). Although all five core members of DCDC are essential for methylation of H3K9 and DNA, only DIM-7 is required to bring DIM-5 to heterochromatic regions. Interestingly, the *S. pombe* H3K9 MTase (Clr4) responsible for heterochromatin formation is in a similar complex, CLRC, comprised of Clr4, Cul4, Rik1, Raf1, Raf2, and Rad24 (Jia et al. 2005; Li et al. 2005; Thon et al. 2005), although CLRC and DCDC have significant structural and functional differences. No ubiquitination substrate of CLRC or DCDC has yet been identified, despite intensive efforts in multiple laboratories, raising the possibility that these H3K9 MTase complexes do not function as ubiquitin ligases in vivo.

DCDC may be controlled by one or more proteins that recognize products of RIP. It is noteworthy that most methylated sequences in the *Neurospora* genome are relics of RIP; most relics of RIP are methylated (Selker et al. 2003) and sequences resembling products of RIP are potent triggers of DNA methylation (Tamaru and Selker 2003). Indeed, analyses of the genomic distribution of 5mC, histone H3K9me3, HP1, and sequences showing evidence of RIP (high A + T content with unexpectedly high densities of TpA dinucleotides) in *Neurospora* revealed that these are all tightly correlated (Fig. 6). Extensive tests on synthetic and natural sequences that trigger DNA methylation led to the suggestion that an unidentified "A:T-hook"-type protein may mediate DNA methylation in *Neurospora*. Consistent with this idea, Distamycin A, an analog of the A:T-hook motif, interferes with de novo methylation in *Neurospora* (Tamaru and Selker 2003).

It is interesting to consider the possible implications of controlling DNA methylation through histones. First, the fact that DNA methylation patterns are relatively stable (i.e., they do not normally spread or shift significantly implies that the underlying histones, and the H3K9me3 mark, are similarly stable. Second, it raises the possibility that other histone modifications may play regulatory roles. Indeed, in vitro studies showed that DIM-5 is inhibited by phosphorylation of H3S10 and methylation of H3K4 (Adhvaryu et al. 2011). Thus, DIM-5 can integrate information relevant to whether DNA in a particular region should be methylated. This provides a possible explanation for the observation that TSA can inhibit DNA methylation in certain regions (Selker 1998).

Evidence that RNAi is important for heterochromatin formation and maintenance in *S. pombe* raised the question of whether the RNAi machinery of *Neurospora* is involved in HP1 localization and/or DNA methylation. *Neurospora* has homologs of a variety of genes implicated in RNAi (Galagan et al. 2003; Borkovich et al. 2004). Studies of mutants with null mutations in all three RNA-dependent RNA polymerase (RdRP) genes, in both Dicer genes, or in other presumptive RNAi genes revealed no evidence that RNAi is involved in methylation of H3K9, heterochromatin formation, or DNA methylation in *Neurospora* (Chicas et al. 2004; Freitag et al. 2004a; Lewis et al. 2009). However, as discussed in Sections 6 and 7, the *Neurospora* RNAi genes are involved in at least two other silencing mechanisms with epigenetic aspects, quelling, and meiotic silencing.

4.3 Modulators of DNA Methylation

An expectation is that DNA methylation would be subject to regulation. As mentioned above, there is already evidence from *Neurospora* that the HKMT underlying DNA methylation (DIM-5) is sensitive to histone modifications. It would not be surprising to find that other elements of the methylation machinery, such as HP1 and the DMT (DIM-2), would also be sensitive to histone modifications, and that DNA methylation would be subject to regulation in other ways. Indeed, we know that the extent of DNA methylation depends on environmental variables such as temperature and the composition of growth medium (Roberts and Selker 1995). Forward and reverse genetic studies have also identified proteins that modulate DNA methylation patterns. A notable example is DNA methylation modulator-1 (DMM-1) and its partner protein, DMM-2 (Honda et al. 2010). Mutants lacking either of these proteins show aberrant methylation of DNA and histone H3K9, with both epigenetic marks frequently spreading into genes adjacent to transposable elements. *Dmm-1* mutants grow poorly but growth can be restored by reduction or elimination of DNA methylation using the drug 5-azacytosine or mutation of the DNA methyltransferase gene, *dim*-2. The observation that *dmm-1*, *dim-2* double mutants

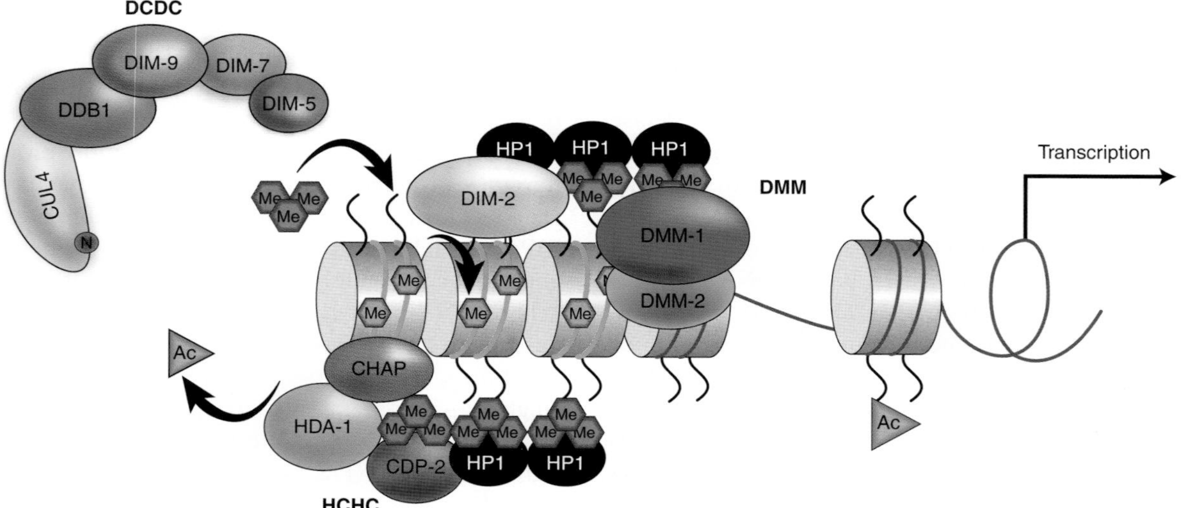

Figure 5. Basic components of the DNA methylation machine of *Neurospora*. Chromatin associated with DNA substantially mutated by RIP (orange spiral decorated with pink mC moieties) is subjected to methylation of H3K9 by the histone methyltransferase DIM-5, whose localization and action depends on a multiprotein complex, DCDC (DIM-5/-7/-9, CUL4/DDB1 complex; Lewis et al. 2010a,b). The CUL4 subunit of the DCDC complex is associated with the small protein Nedd (N), which resembles the E3 ubiquitin ligase complex. Trimethylated H3K9 (K9me3) is recognized by HP1, which is involved in at least three heterochromatin-associated complexes: (1) It recruits the DNA methyltransferase DIM-2 (Honda and Selker 2008); (2) it is required for localization and function of the HCHC silencing complex, which contains HP1, the chromodomain protein CDP-2, the histone deacetylase HDA-1, and a CDP-2/HDA-1-associated protein, CHAP (Honda et al. 2012); and (3) it is required to guide the DMM complex, which serves to block the spreading of heterochromatin into neighboring transcribed regions (Honda et al. 2010).

Cite as *Cold Spring Harb Perspect Biol* doi: 10.1101/cshperspect.a017921

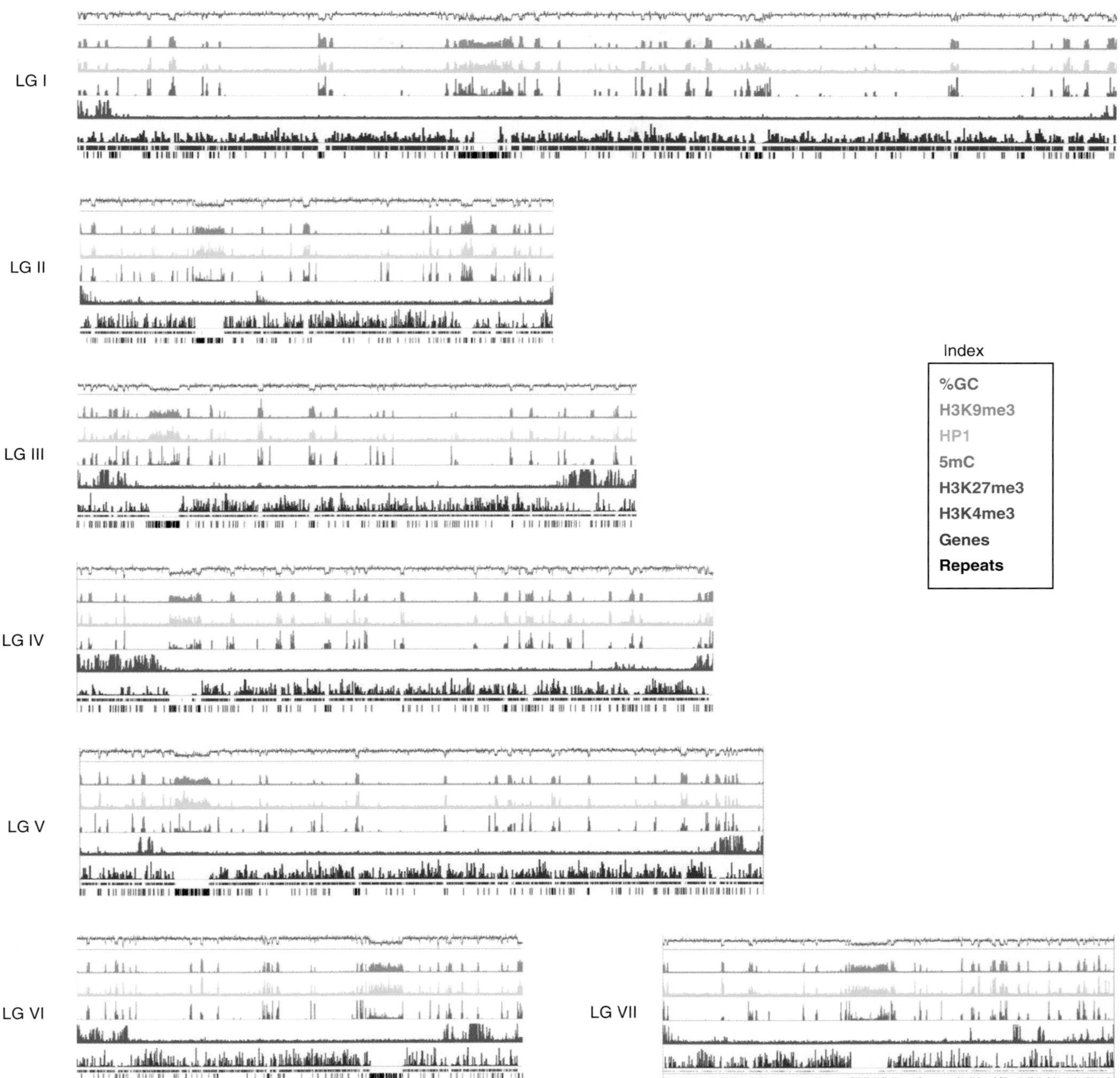

Figure 6. Epigenomic features of *N. crassa* genome. The genomic distributions of H3K9me3 (orange), HP1 (yellow), 5-methylcytosine (green), H3K27me3 (medium blue), and H3K4me3 (dark blue) are displayed for each of *N. crassa's* seven linkage groups (OR74A NC10 sequence assembly, http://www.broadinstitute.org/annotation/genome/neurospora/MultiHome.html) using the Integrative Genomics Viewer (http://www.broadinstitute.org/igv) (Jamieson et al. 2013; MR Rountree and EU Selker, unpubl.). Base composition is shown at the top of each linkage group as the moving average of %GC (red) calculated for 500-bp windows in 100-bp steps, whereas the positions of predicted genes (purple) and repeats (black) are indicated below. The predicted gene file was downloaded from The Broad Institute (http://www.broadinstitute.org/annotation/genome/neurospora) and repeats were determined using the RepeatMasker program (http://www.repeatmasker.org).

display normal H3K9me3 patterns implies that the spread of H3K9me3 involves DNA methylation. The DMM complex is preferentially localized to edges of methylated regions in an HP1-dependent manner, as cartooned in Fig. 5. A conserved residue within the JmjC domain of DMM-1 is essential for its function, raising the possibility that the complex functions as a histone demethylase.

Other proteins also affect the distribution of DNA methylation in *Neurospora*. For example, mutation of any of the genes encoding components of another HP1 protein complex, HCHC (HP1, CDP-2, HDA-1, and CHAP) cause hyperacetylation of centromeric histones, loss of centromeric silencing, increased accessibility of DIM-2 to centromere regions, and hypermethylation of the associated DNA. Interestingly, loss of HCHC also causes mislocalization of the DIM-5 H3K9 methyltransferase at a subset of interstitial methylated regions, leading to selective DNA hypomethylation (Honda et al. 2012). Figure 5 illustrates our current understanding of key known elements of the *Neurospora* DNA methylation/heterochromatin machinery.

5 HISTONE H3K27 METHYLATION

Trimethylated lysine 27 on histone H3 (H3K27me3) is present in metazoans such as *Arabidopsis*, *Drosophila*, and mammals, in which it is known to be involved with gene repression. H3K27me3 is absent from yeasts that have been examined, but is present in *Neurospora* (Smith et al. 2008), rendering the organism an attractive model to study basic aspects of this histone modification. H3K27me3 occupies ~7% of the *N. crassa* genome and is segregated into approximately 230 domains that are particularly enriched near telomeres, but are also found dispersed in the genome (Fig. 5). Approximately 700 predicted genes are covered by H3K27me3, all of which are normally silent. *Neurospora* possesses homologs of the four core components of the Polycomb repressive complex 2 (PRC2) but lacks clear homologs of members of the PRC1 complex found in *Drosophila*, mammals, and plants. Three of the PRC2 core components are required for H3K27 methylation, whereas the fourth, NPF (*Neurospora* homolog of *Drosophila* P55 and mammalian RbApP46/RbApP48), is not absolutely required for H3K27me3. Nevertheless, NPF is critical for H3K27me3 regionally, particularly in telomeric and subtelomeric domains (Jamieson et al. 2013). Loss of H3K27me3, caused by deletion of PRC2 genes, results in upregulation of a small subset of genes in both H3K27me3 and non-H3K27me3 regions. The emergence of *Neurospora* as a model system to explore the control and function of H3K27me3 promises to provide insights into this fascinating, but still largely mysterious, epigenetic mechanism. Important open questions include: What governs the distribution of H3K27me3? Do sequence elements akin to PREs (Polycomb response elements) in *Drosophila* regulate this epigenetic mark in *Neurospora* and other systems? To what extent are RNAs involved in H3K27me3 regulation? How much of the observed H3K27me3 reflects epigenetic "inheritance" and what is the actual mechanism of this process? What is the detailed function of H3K27me3 in gene silencing?

6 QUELLING

Soon after transformation techniques were established for *Neurospora*, researchers in several laboratories noticed that a sizable fraction (~30%) of *Neurospora* transformants showed silencing of transforming DNA, and more surprisingly, silencing of native sequences homologous to those of the transforming DNA. The latter form of vegetative phase silencing was named "quelling" by the Macino laboratory, which performed most of the pioneering research on this phenomenon (Pickford et al. 2002). Quelling is most apparent with visible markers such as the *albino* genes, which encode enzymes required for carotenoid biosynthesis (Fig. 7), and is comparable to "cosuppression" or PTGS (posttranscriptional gene silencing) in plants (Ch. 13 [Pikaard and Mittelsten Scheid 2014]). Interestingly, genes seem to vary in their sensitivity to quelling. For genes that are sensitive, quelling seems most common in transformants bearing multiple copies of transforming DNA in a tight array. Nuclei flow freely in hyphae of *Neurospora*, allowing for "heterokaryosis" in which genetically distinct nuclei share a common cytoplasm. Thus, it was easy to show that quelling is "dominant"—that is, a transformed nucleus can silence homologous sequences in nearby nuclei (Cogoni et al. 1996).

The ability to silence nearby nuclei implicated a cytoplasmic silencing factor, which we now know is RNA, or small RNAs, to be more precise. The roles of small RNAs in gene regulation, germ cell maintenance, and transposon silencing are widespread and are active areas of research. In *Neurospora*, the products of the *qde-1*, *qde-2*, and *qde-3* genes encode, respectively, an RNA/DNA-dependent RNA polymerase, an "Argonaute"-like protein, and a RecQ-like DNA helicase. Together, they have been implicated in the production of a new class of small RNAs called Qde-2-associated RNAs (or qiRNAs) (Lee et al. 2009). Unlike other *Neurospora* siRNAs, which are ~25 nucleotides long, qiRNAs are ~20–21 nucleotides long and have a strong preference for uridine at the 5′ end. In addition, they have been reported to originate mostly from the ribosomal DNA locus in response to DNA damage (Lee et al. 2009). The investigators hypothesized that the observed burst of

 Cite as *Cold Spring Harb Perspect Biol* doi: 10.1101/cshperspect.a017921

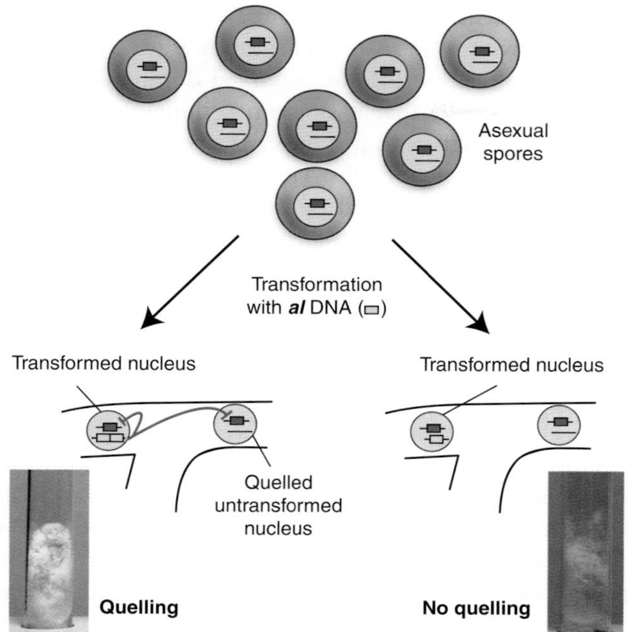

Figure 7. Quelling. For simplicity, only two of the seven chromosomes are diagrammed (straight line segments in gray circles representing nuclei). The native *albino* gene (*al*) is indicated by the dark orange rectangle on the top chromosome; rectangles on the lower chromosome (dark orange or yellow) represent ectopic *al* sequences introduced by transformation. Because transformed cells are often multinucleate, transformants are often heterokaryotic, as illustrated. Whether or not the transforming DNA includes the entire coding region, in some transformants it silences ("quells") the native al^+ gene in both transformed and nontransformed nuclei through an undefined *trans*-acting molecule (red lines emanating from the transforming DNA indicated by the yellow rectangle). This results in poorly pigmented or albino (Al^-) tissue in some transformants, as shown.

rDNA-related qiRNAs inhibits protein translation as a response to DNA damage. MicroRNA-like RNAs (milRNAs) and Dicer-independent small interfering RNAs (disiRNAs) have also been reported (Lee et al. 2010b). milRNAs are produced by at least four different mechanisms that use a distinct combination of factors, including Dicers, QDE-2, the exonuclease QIP, and an RNase III-like domain-containing protein, MRPL3. In contrast, disiRNAs do not require the known RNAi components as they originate from loci producing overlapping sense and antisense transcripts (Lee et al. 2010b). Notably, it was observed that the product of *qde-1*, QDE-1, has DNA- and RNA-dependent biochemical activities (Lee et al. 2010a). QDE-1 seems to play a central role generating the aberrant RNA required for RNAi. In vitro, QDE-1 produces dsRNA from ssDNA, a process that is strongly promoted by replication protein A (RPA). In vivo, this interaction probably occurs during DNA replication, as suggested by the observed interaction of QDE-1 with RPA and DNA helicases (Lee et al. 2010a). Importantly, although DNA methylation is frequently associated with transforming DNA, neither the DNA methyltransferase, DIM-2, nor the H3K9 methyltransferase, DIM-5, are required for quelling (Cogoni et al. 1996; Chicas et al. 2005).

Neurospora is proving to be a rich system for the study of the genesis and characterization of diverse pathways involved in the generation of small RNAs, thus shedding light on the diversity and evolutionary origins of eukaryotic small RNAs.

7 MEIOTIC SILENCING

The most recent addition to the list of known silencing mechanisms is meiotic silencing, which was originally called "meiotic transvection" and later referred to as "meiotic silencing by unpaired DNA" (MSUD; Aramayo and Metzenberg 1996; Aramayo et al. 1996; Shiu et al. 2001; Kelly and Aramayo 2007). As implied by its name, meiotic silencing operates only in meiosis, in which it evaluates the identity of the homologous chromosomes in two stages. First, regions located at equivalent locations on homologous chromosomes are evaluated by a process called *trans*-sensing. Second, regions identified as nonhomologous are silenced by a mechanism related to RNAi.

The discovery of meiotic *trans*-sensing and silencing was the result of a thorough characterization of an *Ascospore maturation-1* (*Asm-1*) deletion mutant of *Neurospora* generated by gene replacement (Aramayo et al. 1996). ASM-1 has a putative DNA binding domain, consistent with it being a transcription factor required for the expression of genes involved in ascospore maturation. The deletion mutant was unable to form aerial hyphae, and protoperithecia (see Fig. 2). Deletion strains carrying ectopically integrated DNA copies of the gene were normal. The ectopic copy could complement the vegetative defects. Interestingly, it could not rescue defects in the sexual phase. In contrast, strains carrying frameshift alleles of the gene ($asm\text{-}1^{fs}$) had the same mutant phenotype in vegetative cultures as those carrying the $Asm\text{-}1^{\triangle}$ deletion but showed different properties in sexual development (Fig. 8). Crosses between strains carrying functional ($asm\text{-}1^+$) and nonfunctional ($asm\text{-}1^{fs}$) alleles resulted in 4:4 segregation of mature and immature spores (Fig. 8; compare Panel A $asm\text{-}1^+ \times$ asm-1^+, with Panel B, $asm\text{-}1^+ \times$ asm-1^{fs}), suggesting that the product of $asm\text{-}1^+$ plays a critical role in ascospore development/maturation and indicating that the $asm\text{-}1^{fs}$ allele is recessive. Surprisingly, crosses heterozygous for a deletion allele of *Asm-1* (Fig. 8C; $asm\text{-}1^+ \times Asm\text{-}1^{\triangle}$), produced only white (inviable) spores, i.e., all spores within the ascus failed to develop, including the ones carrying the $asm\text{-}1^+$

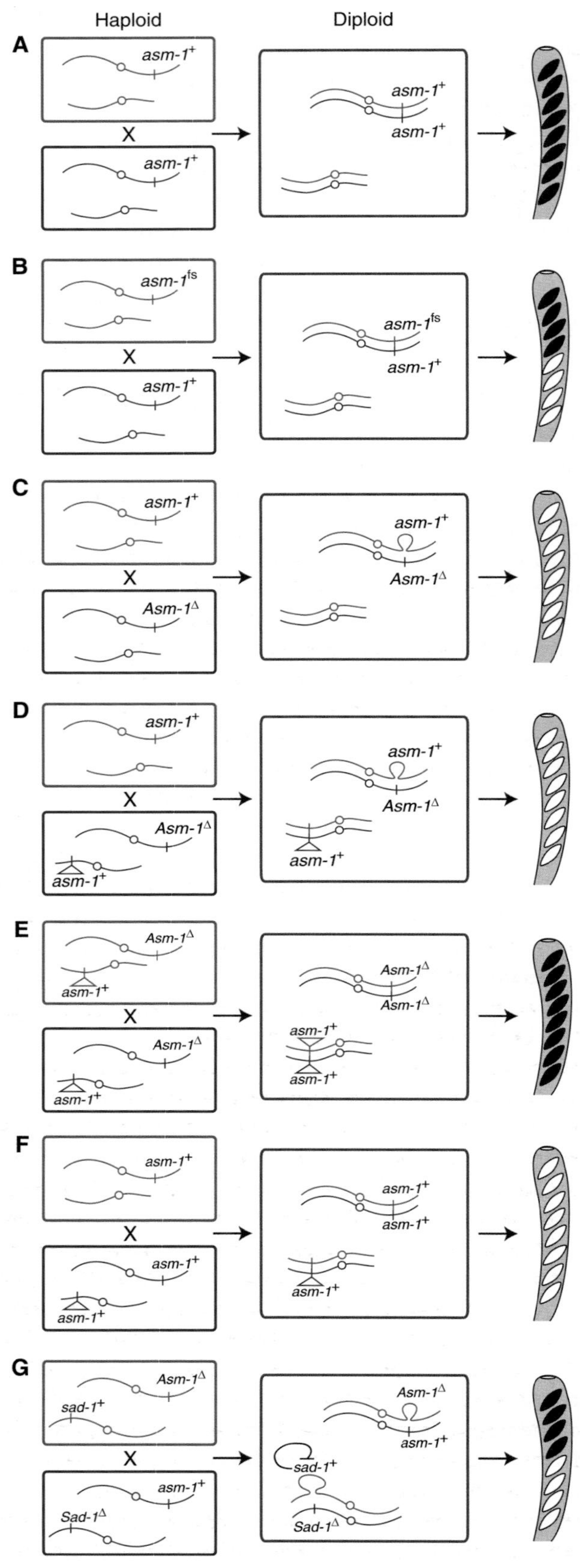

allele. This ascus-dominance of the *Asm-1*$^{\triangle}$ deletion allele contrasted with its recessive behavior in vegetative tissue.

One explanation for the observed dominance of the deletion allele was that a single functional allele of the gene was insufficient to produce adequate product in the diploid and/or meiotic cells. The possibility of such "haploinsufficiency" was tested by crossing wild-type strains with deletion strains carrying ectopic functional copies of the gene, that is, *asm-1*$^{+}$ × *Asm-1*$^{\triangle}$, *asm-1*$^{+}$ (Fig. 8D). The fully functional ectopic gene failed to correct the spore maturation defects (Aramayo and Metzenberg 1996). It was conceivable that the ectopic *asm-1*$^{+}$ failed to rescue the defect because of an unknown requirement for interactions between alleles at homologous chromosomal positions, reminiscent of the transvection phenomenon described in *D. melanogaster* (Wu and Morris 1999). This hypothesis was tested by crossing strains both carrying copies of the gene located at ectopic positions in an *Asm-1*$^{\triangle}$ background, i.e., *Asm-1*$^{\triangle}$, *asm-1*$^{+}$ (ectopic) × *Asm-1*$^{\triangle}$, *asm-1*$^{+}$ (ectopic), (Fig. 8E). Indeed, the two ectopic alleles rescued the ascus-dominant defect of *Asm-1* deletion alleles, supporting the idea that some form of *trans*-sensing was occurring, involving pairing.

To distinguish between the possibility that meiotic silencing is due to absence of pairing or due to an unpaired allele, crosses in which the meiotic nucleus had three copies of a gene: two wild-type alleles (which should pair) and an ectopic copy (which should be unpaired) were analyzed. Silencing was observed, implying that meiotic silencing results from the presence of unpaired alleles rather than from absence of paired ones [Fig. 8F; *asm-1*$^{+}$ × *asm-1*$^{+}$, *asm-1*$^{+(\text{ectopic})}$] (Shiu et al. 2001; Kutil et al. 2003; Lee et al. 2003; Lee et al. 2004).

Figure 8. Discovery and characterization of meiotic silencing. Key genetic experiments are illustrated using the *Ascospore maturation-1* (*Asm-1*) gene, as a reporter. For each cross, the relevant genotype of the haploid parents of *mating type A* (red boxes) or *mating type a* (blue boxes) is shown on the *left*, and cartoons showing the predicted chromosome pairing in the diploid cell (violet boxes) is shown on the *right*. The phenotypes of resulting asci are presented on the *far right*. Black represents mature (typically viable) ascospores and white represents immature (inviable) ascospores. (*A*) Wild-type cross. (*B*) 4:4 segregation of ascospores from a heterozygous cross of wild type and a frameshift mutant in which alleles can pair and no meiotic silencing occurs. (*C*) Cross of strains with wild-type and deletion alleles triggers meiotic silencing. (*D*) Meiotic silencing is not rescued by ectopic wild-type allele, indicating that the developmental defect is not due to haploinsufficiency. (*E*) Allelic (pairable) ectopic copies *asm-1*$^{+}$ in crossing partners rescue *Asm-1*$^{-}$ defect. (*F*) Presence of an unpaired allele triggers silencing of all *asm-1*$^{+}$ alleles (paired and unpaired) in meiosis. (*G*) Silencing of the suppressor of ascospore dominance (*Sad-1*), because of a *Sad-1* deletion in one parent, suppresses meiotic silencing.

A hunt for mutants defective in meiotic silencing resulted in the identification of a telling member of the silencing machinery. The *Sad-1* gene, encoding an RdRP was identified by selection for mutants that were able to pass through a cross in which *Asm-1* is not paired [Fig. 8G; $Sad\text{-}1^{\triangle}$, $asm\text{-}1^{+} \times sad\text{-}1^{+}$, $Asm\text{-}1^{\triangle}$]. This suggested that meiotic silencing is related to quelling in *Neurospora*, and to RNAi, generally. Further screens for meiotic silencing factors identified two RNAi-related genes in addition to *Sad-1*: an Argonaute-like protein, *suppressor of meiotic silencing-2* (*Sms-2*; Lee et al. 2003); and a Dicer-like protein, *suppressor of meiotic silencing-3* (*Dcl-1/Sms-3*; Alexander et al. 2008). In addition, the involvement of a putative helicase, *Sad-3*, has been reported (Hammond et al. 2011). Several other *Sms* mutants have also been identified (DW Lee and R Aramayo, pers. comm.). Although all of the genes shown to be required for meiotic silencing in *Neurospora* are also required for fertility, strains carrying loss-of-function mutations in the meiotic silencing pathway do not have a discernible vegetative phenotype, and as noted above, do not affect heterochromatin formation and DNA methylation. This contrasts with the situation in the fission yeast *S. pombe*, in which orthologs of *Sad-1*, *Sms-2*, and *Dcl-1/Sms-3* (Rdp1, Ago1, and Dcr1, respectively) are essential for normal chromosome biology, including heterochromatin formation (e.g., histone H3K9 methylation), and normal centromere and telomere functions (Martienssen et al. 2005).

The PTGS nature of meiotic silencing was confirmed using transgene reporters. Only regions containing homology with the reporter transcript result in silencing when unpaired (Lee et al. 2004). Intriguingly, all reported components of the meiotic silencing machinery identified so far have a perinuclear localization (Fig. 9) (Shiu et al. 2006). Our working model (Fig. 9) for the mechanism of meiotic silencing postulates two steps: (1) "sensing," in which pairing of homologous chromosomes reveals unpaired DNA, which then gives rise to an aberrant RNA (aRNA); and (2) "processing" of the aberrant RNA by perinuclear RNA silencing machinery (SAD-1, SAD-2, SMS-2, DCL-1/SMS-3, etc.). Silencing presumably results from degradation of normal RNAs targeted by siRNAs, generated in the cascade initiated by the aberrant RNA.

Exactly what constitutes "unpaired" DNA is an area of active investigation. Some of the quantitative and qualitative aspects of the "sensing" threshold have been addressed, however (Lee et al. 2004). The findings can be summarized as follows: (1) Given one small and one large loop of unpaired DNA, both carrying the same length of DNA homologous to a set of paired reporter genes, the large loop will silence more efficiently than the smaller one; (2) given two loops of identical size, but one carrying twice as much DNA homologous to a set of paired reporter genes, the loop carrying more homologous DNA will silence more efficiently than the one carrying less homologous DNA; (3) the silencing signal produced by an unpaired loop is confined to the unpaired region and does not "spread" to neighboring regions (e.g., paired reporter genes can be located next to a region of unpaired DNA without being significantly perturbed); (4) the canonical promoter of a gene need not be present in the loop of unpaired DNA for a gene to be silenced; and (5) meiotic silencing does not affect the ability of a promoter to direct transcription at a later developmental time (Kutil et al. 2003; Lee et al. 2003, 2004; Pratt et al. 2004).

In general, our understanding of homology sensing mechanisms—even those giving rise to homologous re-

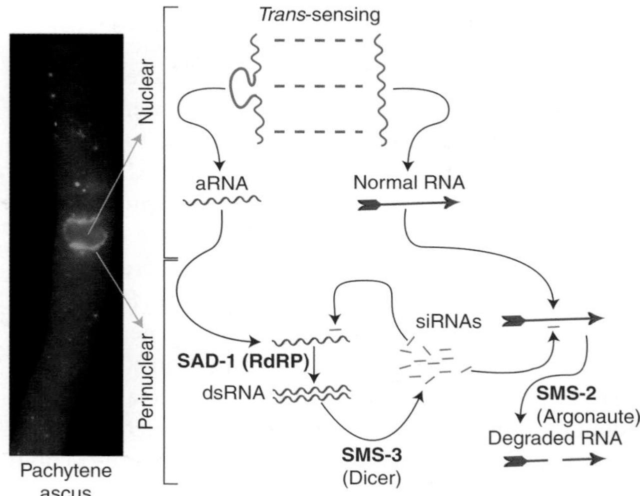

Figure 9. A model for meiotic silencing. An image of a developing ascus from a cross between parents engineered to contain paired copies of $sad\text{-}1^{+}$ fused to a reporter gene gfp^{+}(i.e., $sad\text{-}1^{+}::gfp^{+}$) at the Pachytene stage of meiosis I (*left*; DW Lee and R Aramayo, unpubl.). Inside this cell, the meiotic nucleus, delineated by its nuclear membrane, is surrounded by a perinuclear structure that supports the attachment of components of the meiotic silencing apparatus. Predicted nuclear and perinuclear steps in meiotic silencing are diagrammed (*right*). It is hypothesized that *trans*-sensing, a mechanism preceding silencing, identifies heterologous regions of interacting chromosomes. The degree of heterology determines the strength of the induction step, which presumably involves the synthesis of aRNA and its conversion to double-stranded RNA (dsRNA) by the SAD-1 RdRP, a perinuclear event. The presence of dsRNA triggers the initiation of the silencing process, which involves the conversion of the dsRNA trigger into siRNAs via the DCL-1/SMS-3 Dicer (initiation step), and use of these siRNAs primers and normal RNAs as templates, by SAD-1 RdRP to generate dsRNA (amplification cycle). The incorporation of the siRNAs, generated by both the initiation step and the amplification cycles, into the RNA-inducing silencing complex (RISC) directs the endonucleolytic cleavage of mRNA or ssRNA (single-stranded RNA).

combination (ectopic and standard) and RIP, as well as those behind meiotic silencing—is incomplete. Although details of the mechanism that detects unpaired DNA in meiotic silencing remain to be discovered, interesting features of this sensing mechanism have come to light. For example, it was found that quasi-homologous sequences (i.e., homeologous sequences), like those produced by RIP, can induce meiotic silencing. The degree of identity that is required to escape meiotic silencing was assessed by crossing strains carrying wild-type alleles of *Rsp* (*Round spore*) with strains carrying various alleles generated by RIP that differed in their density of mutations (Pratt et al. 2004). Some alleles (e.g., *Rsp*RIP93) conferred a dominant phenotype, comparable to that shown by a deletion of *Rsp*. As little as 6% sequence divergence (94% identity) could trigger silencing; 3% divergence (97% identity) did not. Interestingly, the methylation status of the RIP-mutated alleles shifted the sequence identity threshold—that is, methylated alleles triggered silencing more effectively than when methylation was prevented by a mutation in the DNA methyltransferase gene, *dim-2* (Pratt et al. 2004). This observation provided the first evidence that DIM-2 is functional in the sexual phase of *Neurospora* and suggested that either 5mCs contribute to heterology as a "fifth base," and/or that an indirect effect of DNA methylation (e.g., recruitment of a methylated DNA binding protein or an unknown effect on chromatin structure) impacts homology recognition.

In retrospect, it is not surprising that meiotic silencing escaped detection for many years of genetic studies with *Neurospora*. To be detected, a reporter gene must fulfill a series of strict requirements. Most unpaired genes (i.e., those whose gene product are involved in vegetative processes) would probably not impact meiotic and/or spore development. Thus, lack of pairing might only be evident for genes whose products are required for the completion of meiosis, the reestablishment of mitosis, cellularization, or maturation of ascospores and/or whose gene products are essential structural components of the ascus.

8 PROBABLE FUNCTIONS AND PRACTICAL USES OF RIP, QUELLING, AND MEIOTIC SILENCING

RIP seems custom-made to limit the expression of "selfish DNA" such as transposons that direct the production of copies of themselves in a genome. Consistent with this possibility, the vast majority of relics of RIP are recognizably similar to transposons known from other organisms, and most strains of *Neurospora* lack active transposons (Galagan et al. 2003; Selker et al. 2003; Galagan and Selker 2004). Nevertheless, because RIP is limited to the premeiotic dikaryotic cells, this process should neither prevent the spread of a new (e.g., horizontally acquired) transposon in vegetative cells nor prevent the duplication of a single-copy transposon in meiotic cells. Quelling and meiotic silencing should deal with such eventualities, however. Although quelling does not completely suppress the spread of transposons in vegetative cells, as evidenced by the proliferation of an introduced copy of the LINE (long interspersed element)-like transposon, Tad, it does appear to partially silence such transposons (Nolan et al. 2005). Information about the action of meiotic silencing suggests that this process will silence any transposed sequence in meiotic cells, even if it is only present as a single copy in the genome (Shiu et al. 2001; Kelly and Aramayo 2007). In addition to dealing with errant transposons in meiosis, some of the genes involved in meiotic silencing also appear to play an important role in the process of speciation, as shown by the observation that mutants defective in meiotic silencing relieve the sterility of strains bearing large duplications of chromosome segments and allow closely related species to mate with *N. crassa* (Shiu et al. 2001).

Although RIP, quelling, and meiotic silencing can all be a nuisance for some genetic experiments, all have been exploited for research purposes. RIP provided the first simple method to knock out genes in *Neurospora* and is still the preferred method for generating partial-function mutants. Quelling has also been used to reduce, if not eliminate, gene function, much as RNAi is exploited in a variety of organisms. And meiotic silencing provides a simple assay to test whether particular genes are required to function in (or immediately after) meiosis; if a gene is found to cause sterility when duplicated, or when at an ectopic location, and the sterility is rescued by a mutation blocking meiotic silencing, it is safe to assume that it plays an important function in meiosis.

In addition to the postulated evolutionary roles of RIP, quelling, and meiotic silencing, and to their utility in the laboratory, it is worth considering the possibility that these processes serve in other ways. For example, the fact that *Sad-1* function is required for full fertility suggests that meiotic silencing is directly or indirectly required for meiosis (Shiu et al. 2001). Surprisingly, however, not all genes shown to be required for meiotic silencing in *Neurospora* are required for fertility (R Pratt, DW Lee, and R Aramayo, unpubl.), indicating that despite their temporal and spatial colocalization, the connection(s) and/or interdependencies between the meiotic silencing pathway and meiosis are still poorly understood. In the case of RIP, although this process is nonessential, the distribution of products of RIP, which are concentrated in the genetically mapped centromeres of *Neurospora* (Fig. 5), suggest that junked transposons can serve the organism as substrates for kinetochore formation, much as repeated sequences do in *S. pombe* and other

 Cite as *Cold Spring Harb Perspect Biol* doi: 10.1101/cshperspect.a017921

organisms. Indeed, *Neurospora* centromere sequences consist primarily of relics of transposons heavily mutated by RIP and the normal distribution of kinetochore proteins depends on DIM-5 and HP1 (Smith et al. 2011). Relics of RIP are also found adjacent to telomere sequences of *Neurospora* (Smith et al. 2008). Interestingly, transposons and relics of transposons are also commonly found in heterochromatic sequences of other organisms, such as *Drosophila*, mammals, plants, and other fungi.

9 CONCLUDING REMARKS

The fungus *N. crassa* has emerged as a powerful system to discover and elucidate epigenetic phenomena. Because epigenetics is still a young field and studies of epigenetic processes have for the most part arisen from discoveries stemming from a variety of research programs, it is not surprising that our breadth and depth of understanding of epigenetic processes in *Neurospora*, yeasts, and other systems vary substantially. It is too early to know how general the various epigenetic mechanisms are. Nevertheless, it is already clear that various model eukaryotes have both important differences and striking similarities. For example, whereas *Neurospora* sports DNA methylation and H3K27 methylation and *S. pombe* does not, both of these fungi show histone H3K9 methylation and RNAi processes, neither of which are found in the budding yeast, *S. cereviseae.* It is also noteworthy that a given process may be functionally distinct in different organisms. For instance, in *Neurospora*, RNAi components have been implicated in quelling and meiotic silencing, but not heterochromatin formation, whereas in fission yeast, RNAi components have been implicated in heterochromatin formation but nothing more. Finally it is worth noting that even shared features, such as the association of heterochromatin with centromeres of both fission yeast and *Neurospora*, may have important differences. An important goal for the future is to discover the extent to which information gleaned from one organism is applicable to others. Continued exploitation of various model organisms, including *Neurospora*, should both provide this information and reveal features of epigenetic processes that are still unknown today. We anticipate that the richly diverse fungi will serve as useful systems for epigenetic research for many years.

ACKNOWLEDGMENTS

We thank N.B. Raju, Michael Rountree, and Shinji Honda for critical help generating Figures 2, 5, and 6, respectively. Research in the Selker laboratory was supported by U.S. Public Health Service Grants GM03569, GM093061, and S090064, and research in the Aramayo laboratory was supported by U.S. Public Health Service Grant GM58770.

REFERENCES

**Reference is also in this book.*

Adhvaryu KK, Berge E, Tamaru H, Freitag M, Selker EU. 2011. Substitutions in the amino-terminal tail of *Neurospora* histone H3 have varied effects on DNA methylation. *PLoS Genet* **7:** e1002423.

Alexander WG, Raju NB, Xiao H, Hammond TM, Perdue TD, Metzenberg RL, Pukkila PJ, Shiu PK. 2008. DCL-1 colocalizes with other components of the MSUD machinery and is required for silencing. *Fungal Genet Biol* **45:** 719–727.

* Allshire RC, Ekwall K. 2014. Epigenetic regulation of chromatin states in *Schizosaccharomyces pombe. Cold Spring Harb Perspect Biol* doi: 10.1101/cshperspect.a018770.

Aramayo R, Metzenberg RL. 1996. Meiotic transvection in fungi. *Cell* **86:** 103–113.

Aramayo R, Peleg Y, Addison R, Metzenberg R. 1996. *Asm-1*$^{+}$, a *Neurospora crassa* gene related to transcriptional regulators of fungal development. *Genetics* **144:** 991–1003.

Bhat A, Kasbekar DP. 2001. Escape from repeat-induced point mutation of a gene-sized duplication in *Neurospora crassa* crosses that are heterozygous for a larger chromosome segment duplication. *Genetics* **157:** 1581–1590.

Borkovich KA, Alex LA, Yarden O, Freitag M, Turner GE, Read ND, Seiler S, Bell-Pedersen D, Paietta J, Plesofsky N, et al. 2004. Lessons from the genome sequence of *Neurospora crassa*: Tracing the path from genomic blueprint to multicellular organism. *Microbiol Mol Biol Rev* **68:** 1–108.

Chicas A, Cogoni C, Macino G. 2004. RNAi-dependent and RNAi-independent mechanisms contribute to the silencing of RIPed sequences in *Neurospora crassa. Nucleic Acids Res* **32:** 4237–4243.

Chicas A, Forrest EC, Sepich S, Cogoni C, Macino G. 2005. Small interfering RNAs that trigger posttranscriptional gene silencing are not required for the histone H3 Lys9 methylation necessary for transgenic tandem repeat stabilization in *Neurospora crassa. Mol Cell Biol* **25:** 3793–3801.

Cogoni C, Irelan JT, Schumacher M, Schmidhauser TJ, Selker EU, Macino G. 1996. Transgene silencing of the *al-1* gene in vegetative cells of *Neurospora* is mediated by a cytoplasmic effector and does not depend on DNA-DNA interactions or DNA methylation. *EMBO J* **15:** 3153–3163.

Davis RH, de Serres FJ. 1970. Genetic and microbiological research techniques for *Neurospora crassa*. In *Metabolism of amino acids amines* (ed. Colowick SP, Kaplan NO), pp. 79–143. Academic, New York.

* Elgin SCR, Reuter G. 2013. Position-effect variegation, heterochromatin formation, and gene silencing in *Drosophila. Cold Spring Harb Perspect Biol* **5:** a017780.

Freitag M, Williams RL, Kothe GO, Selker EU. 2002. A cytosine methyltransferase homologue is essential for repeat-induced point mutation in *Neurospora crassa. Proc Natl Acad Sci* **99:** 8802–8807.

Freitag M, Hickey PC, Khlafallah TK, Read ND, Selker EU. 2004a. HP1 is essential for DNA methylation in *Neurospora. Mol Cell* **13:** 427–434.

Freitag M, Hickey PC, Raju NB, Selker EU, Read ND. 2004b. GFP as a tool to analyze the organization, dynamics and function of nuclei and microtubules in *Neurospora crassa. Fungal Genet Biol* **41:** 897–910.

Galagan JE, Calvo SE, Borkovich KA, Selker EU, Read ND, Jaffe D, FitzHugh W, Ma LJ, Smirnov S, Purcell S, et al. 2003. The genome sequence of the filamentous fungus *Neurospora crassa. Nature* **422:** 859–868.

Galagan JE, Selker EU. 2004. RIP: The evolutionary cost of genome defense. *Trends Genet* **20:** 417–423.

* Grunstein M, Gasser SM. 2013. Epigenetics in *Saccharomyces cerevisiae. Cold Spring Harb Perspect Biol* **5:** a017491.

Goll MG, Bestor TH. 2005. Eukaryotic cytosine methyltransferases. *Annu Rev Biochem* **74:** 481–514.

Hammond TM, Xiao H, Boone EC, Perdue TD, Pukkila PJ, Shiu PK. 2011. SAD-3, a putative helicase required for meiotic silencing by unpaired DNA, interacts with other components of the silencing machinery. *G3* **1:** 369–376.

Honda S, Selker EU. 2008. Direct interaction between DNA methyltransferase DIM-2 and HP1 is required for DNA methylation in *Neurospora crassa*. *Mol Cell Biol* **28:** 6044–6055.

Honda S, Lewis ZA, Huarte M, Cho LY, David LL, Shi Y, Selker EU. 2010. The DMM complex prevents spreading of DNA methylation from transposons to nearby genes in *Neurospora crassa*. *Genes Dev* **24:** 443–454.

Honda S, Lewis ZA, Shimada K, Fischle W, Sack R, Selker EU. 2012. HP1 forms distinct complexes to direct histone deacetylation and DNA methylation. *Nat Struct Mol Biol* **19:** 471–477.

Jamieson K, Rountree MR, Lewis ZA, Stajich JE, Selker EU. 2013. Regional control of histone H3 lysine 27 methylation in *Neurospora*. *Proc Natl Acad Sci* **110:** 6227–6232.

Jia S, Kobayashi R, Grewal SI. 2005. Ubiquitin ligase component Cul4 associates with Clr4 histone methyltransferase to assemble heterochromatin. *Nat Cell Biol* **7:** 1007–1013.

Kelly WG, Aramayo R. 2007. Meiotic silencing and the epigenetics of sex. *Chromosome Res* **15:** 633–651.

Kouzminova E, Selker EU. 2001. *dim-2* encodes a DNA methyltransferase responsible for all known cytosine methylation in *Neurospora*. *EMBO J* **20:** 4309–4323.

Kutil BL, Seong KY, Aramayo R. 2003. Unpaired genes do not silence their paired neighbors. *Curr Genet* **43:** 425–432.

Lee DW, Pratt RJ, McLaughlin M, Aramayo R. 2003. An argonaute-like protein is required for meiotic silencing. *Genetics* **164:** 821–828.

Lee DW, Seong KY, Pratt RJ, Baker K, Aramayo R. 2004. Properties of unpaired DNA required for efficient silencing in *Neurospora crassa*. *Genetics* **167:** 131–150.

Lee HC, Chang SS, Choudhary S, Aalto AP, Maiti M, Bamford DH, Liu Y. 2009. qiRNA is a new type of small interfering RNA induced by DNA damage. *Nature* **459:** 274–277.

Lee HC, Aalto AP, Yang Q, Chang SS, Huang G, Fisher D, Cha J, Poranen MM, Bamford DH, Liu Y. 2010a. The DNA/RNA-dependent RNA polymerase QDE-1 generates aberrant RNA and dsRNA for RNAi in a process requiring replication protein A and a DNA helicase. *PLoS Biol* **8:** e1000496.

Lee HC, Li L, Gu W, Xue Z, Crosthwaite SK, Pertsemlidis A, Lewis ZA, Freitag M, Selker EU, Mello CC, et al. 2010b. Diverse pathways generate microRNA-like RNAs and Dicer-independent small interfering RNAs in fungi. *Mol Cell* **38:** 803–814.

Lewis ZA, Honda S, Khlafallah TK, Jeffress JK, Freitag M, Mohn F, Schubeler D, Selker EU. 2009. Relics of repeat-induced point mutation direct heterochromatin formation in *Neurospora crassa*. *Genome Res* **19:** 427–437.

Lewis ZA, Adhvaryu KK, Honda S, Shiver AL, Knip M, Sack R, Selker EU. 2010a. DNA methylation and normal chromosome behavior in *Neurospora* depend on five components of a histone methyltransferase complex, DCDC. *PLoS Genet* **6:** e1001196.

Lewis ZA, Adhvaryu KK, Honda S, Shiver AL, Selker EU. 2010b. Identification of DIM-7, a protein required to target the DIM-5 H3 methyltransferase to chromatin. *Proc Natl Acad Sci* **107:** 8310–8315.

* Li E, Zhang Y. 2014. DNA methylation in mammals. *Cold Spring Harb Perspect Biol* **6:** a019133.

Li F, Goto DB, Zaratiegui M, Tang X, Martienssen R, Cande WZ. 2005. Two novel proteins, Dos1 and Dos2, interact with Rik1 to regulate heterochromatic RNA interference and histone modification. *Curr Biol* **15:** 1448–1457.

Martienssen RA, Zaratiegui M, Goto DB. 2005. RNA interference and heterochromatin in the fission yeast *Schizosaccharomyces pombe*. *Trends Genet* **21:** 450–456.

Nolan T, Braccini L, Azzalin G, De Toni A, Macino G, Cogoni C. 2005. The post-transcriptional gene silencing machinery functions independently of DNA methylation to repress a LINE1-like retrotransposon in *Neurospora crassa*. *Nucleic Acids Res* **33:** 1564–1573.

Perkins DD. 1966. Details for collection of asci as unordered groups of eight projected ascospores. *Neurospora Newslett* **9:** 11.

Perkins DD. 1988. Comments on Metzenberg's procedure for isolating unordered *Neurospora* asci. *Fungal Genet Newslett* **35:** 29.

Perkins DD, Metzenberg RL, Raju NB, Selker EU, Barry EG. 1986. Reversal of a *Neurospora* translocation by crossing over involving displaced rDNA, and methylation of the rDNA segments that result from recombination. *Genetics* **114:** 791–817.

Perkins DD, Margolin BS, Selker EU, Haedo SD. 1997. Occurrence of repeat induced point mutation in long segmental duplications of *Neurospora*. *Genetics* **147:** 125–136.

Perkins DD, Radford A, Sachs MS. 2001. *The Neurospora compendium: Chromosomal loci*. Academic, San Diego.

Pickford AS, Catalanotto C, Cogoni C, Macino G. 2002. Quelling in *Neurospora crassa*. *Adv Genet* **46:** 277–303.

* Pikaard C, Mittelsten Scheid O. 2014. Epigenetic regulation in plants. *Cold Spring Harb Perspect Biol* doi: 10.1101/cshperspect.a019315.

Pratt RJ, Lee DW, Aramayo R. 2004. DNA methylation affects meiotic *trans*-sensing, not meiotic silencing, in *Neurospora*. *Genetics* **168:** 1925–1935.

Raju NB. 1980. Meiosis and ascospore genesis in *Neurospora*. *Eur J Cell Biol* **23:** 208–223.

Raju NB. 1992. Genetic control of the sexual cycle in *Neurospora*. *Mycol Res* **96:** 241–262.

Roberts CJ, Selker EU. 1995. Mutations affecting the biosynthesis of *S*-adenosylmethionine cause reduction of DNA methylation in *Neurospora crassa*. *Nucleic Acids Res* **23:** 4818–4826.

Rossignol J-L, Faugeron G. 1994. Gene inactivation triggered by recognition between DNA repeats. *Experientia* **50:** 307–317.

Selker EU. 1990a. DNA methylation and chromatin structure: A view from below. *Trends Biochem Sci* **15:** 103–107.

Selker EU. 1990b. Premeiotic instability of repeated sequences in *Neurospora crassa*. *Annu Rev Genet* **24:** 579–613.

Selker EU. 1998. Trichostatin A causes selective loss of DNA methylation in *Neurospora*. *Proc Natl Acad Sci* **95:** 9430–9435.

Selker EU, Stevens JN. 1985. DNA methylation at asymmetric sites is associated with numerous transition mutations. *Proc Natl Acad Sci* **82:** 8114–8118.

Selker EU, Cambareri EB, Jensen BC, Haack KR. 1987. Rearrangement of duplicated DNA in specialized cells of *Neurospora*. *Cell* **51:** 741–752.

Selker EU, Freitag M, Kothe GO, Margolin BS, Rountree MR, Allis CD, Tamaru H. 2002. Induction and maintenance of nonsymmetrical DNA methylation in *Neurospora*. *Proc Natl Acad Sci* **99:** 16485–16490.

Selker EU, Tountas NA, Cross SH, Margolin BS, Murphy JG, Bird AP, Freitag M. 2003. The methylated component of the *Neurospora crassa* genome. *Nature* **422:** 893–897.

Shiu PK, Raju NB, Zickler D, Metzenberg RL. 2001. Meiotic silencing by unpaired DNA. *Cell* **107:** 905–916.

Shiu PK, Zickler D, Raju NB, Ruprich-Robert G, Metzenberg RL. 2006. SAD-2 is required for meiotic silencing by unpaired DNA and perinuclear localization of SAD-1 RNA-directed RNA polymerase. *Proc Natl Acad Sci* **103:** 2243–2248.

Singer MJ, Marcotte BA, Selker EU. 1995. DNA methylation associated with repeat-induced point mutation in *Neurospora crassa*. *Mol Cell Biol* **15:** 5586–5597.

Smith KM, Kothe GO, Matsen CB, Khlafallah TK, Adhvaryu KK, Hemphill M, Freitag M, Motamedi MR, Selker EU. 2008. The fungus *Neurospora crassa* displays telomeric silencing mediated by multiple sirtuins and by methylation of histone H3 lysine 9. *Epigenetics Chromatin* **1:** 5.

Smith KM, Dobosy JR, Reifsnyder JE, Rountree MR, Anderson DC, Green GR, Selker EU. 2010. H2B- and H3-specific histone deacetylases are required for DNA methylation in *Neurospora crassa*. *Genetics* **186:** 1207–1216.

Smith KM, Phatale PA, Sullivan CM, Pomraning KR, Freitag M. 2011. Heterochromatin is required for normal distribution of *Neurospora crassa* CenH3. *Mol Cell Biol* **31:** 2528–2542.

Tamaru H, Selker EU. 2001. A histone H3 methyltransferase controls DNA methylation in *Neurospora crassa*. *Nature* **414:** 277–283.

Tamaru H, Selker EU. 2003. Synthesis of signals for de novo DNA methylation in *Neurospora crassa*. *Mol Cell Biol* **23:** 2379–2394.

Tamaru H, Zhang X, McMillen D, Singh PB, Nakayama J, Grewal SI, Allis CD, Cheng X, Selker EU. 2003. Trimethylated lysine 9 of histone H3 is a mark for DNA methylation in *Neurospora crassa*. *Nat Genet* **34:** 75–79.

Thon G, Hansen KR, Altes SP, Sidhu D, Singh G, Verhein-Hansen J, Bonaduce MJ, Klar AJ. 2005. The Clr7 and Clr8 directionality factors and the Pcu4 cullin mediate heterochromatin formation in the fission yeast *Schizosaccharomyces pombe*. *Genetics* **171:** 1583–1595.

Watters MK, Randall TA, Margolin BS, Selker EU, Stadler DR. 1999. Action of repeat-induced point mutation on both strands of a duplex and on tandem duplications of various sizes in *Neurospora*. *Genetics* **153:** 705–714.

Wu CT, Morris JR. 1999. Transvection and other homology effects. *Curr Opin Genet Dev* **9:** 237–246.

WWW RESOURCES

http://www.broadinstitute.org/annotation/genome/neurospora/MultiHome.html A website housed by the Broad Institute that serves as the official repository of the information generated by, and associated with, the *Neurospora crassa* genome

http://www.broadinstitute.org/igv A website housed by the Broad Institute that serves as the official repository of the Integrative Genomics Viewer, a powerful Java-based high-performance visualization tool for interactive exploration of large, integrated genomic data sets

http://www.repeatmasker.org A website housed by the Institute for Systems Biology that serves as the official repository for RepeatMasker, an industry-standard program used for the screening of DNA sequences for interspersed repeats and low-complexity DNA sequences

C H A P T E R 11

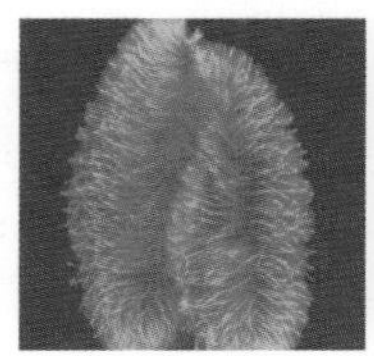
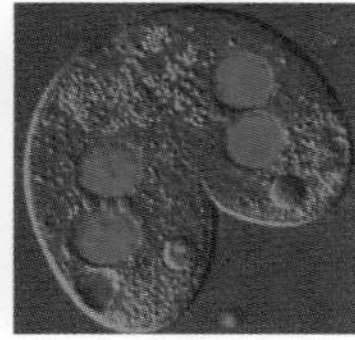

Epigenetics of Ciliates

Douglas L. Chalker[1], Eric Meyer[2], and Kazufumi Mochizuki[3]

[1]Department of Biology, Washington University, St. Louis, Missouri 63130; [2]Institut de Biologie de l'Ecole Normale Suprieure, CNRS UMR8197– INSERM U1024, 75005 Paris, France; [3]Institute of Molecular Biotechnology of the Austrian Academy of Sciences (IMBA), A-1030 Vienna, Austria

Correspondence: dchalker@biology2.wustl.edu

SUMMARY

Research using ciliates revealed early examples of epigenetic phenomena and continues to provide novel findings. These protozoans maintain separate germline and somatic nuclei that carry transcriptionally silent and active genomes, respectively. Examining the differences in chromatin within distinct nuclei of *Tetrahymena* identified histone variants and established that transcriptional regulators act by modifying histones. Formation of somatic nuclei requires both transcriptional activation of silent chromatin and large-scale DNA elimination. This somatic genome remodeling is directed by homologous RNAs, acting with an RNA interference (RNAi)-related machinery. Furthermore, the content of the parental somatic genome provides a homologous template to guide this genome restructuring. The mechanisms regulating ciliate DNA rearrangements reveal the surprising power of homologous RNAs to remodel the genome and transmit information transgenerationally.

Outline

Chapter-opening figures: (*Left*) Courtesy of Janine Beisson.

Cite as *Cold Spring Harb Perspect Biol* doi: 10.1101/cshperspect.a017764

OVERVIEW

Anyone watching ciliates under a microscope is sure to be fascinated by these complex little animals that use their hair-like cilia to swim, eat, and find a mate. Growing cells duplicate by simple binary fission; yet, periodically, ciliates will mate with a partner or, in some species, undergo self-fertilization, resulting in sexual progeny with a different genotype. What uniquely distinguishes these single-celled eukaryotes is that they maintain two functionally distinct genomes carried in separate nuclei within a common cytoplasm. The smaller of these, the micronucleus, contains the germline genome. It is transcriptionally silent during growth, but stores the genetic information that is passed to progeny at each sexual generation. The larger macronucleus performs somatic functions as it is responsible for all gene expression and, thus, governs the cell's phenotype. It is discarded at the end of each sexual generation when a new macronucleus differentiates from the germline.

The compartmentalization of gene expression of ciliates implies mechanisms exist that differentially regulate homologous sequences contained within the distinct nuclei. Early studies sought to elucidate the means by which the germline was kept silent and the somatic genome transcriptionally active. Researchers could readily correlate specific histones and their modifications with transcriptional activity or cell cycle stage. For instance, by comparing chromatin proteins from germline and somatic nuclei of *Tetrahymena thermophila*, some of the first histone variants were identified. Furthermore, new chromatin regulators, such as the first nuclear histone acetyltransferase (HAT), were identified in this ciliate, following up the observation that only the macronucleus contains acetylated histones.

Not only do germline and somatic nuclei have different transcriptional activities, but they have physically different genome organization. During somatic macronuclear development, massive DNA rearrangements generate a streamlined version of the genome. Much of its germline-derived DNA, including most repetitive sequences, is eliminated, whereas all the genes needed for the organism's survival throughout the life cycle are amplified to achieve a high ploidy level. Moreover, in some ciliate species, protein-encoding sequences that are "scrambled" in the micronucleus are assembled in proper order (unscrambled) in the macronucleus.

Many experiments have shown that DNA rearrangement patterns are not strictly genetically programmed, but are epigenetically controlled, at least in part, by preexisting rearrangements within the parental somatic genome. The implication is that the germline and somatic genomes are compared to each other during nuclear differentiation, a comparison that is likely mediated by homology-dependent interactions between germline and somatic RNAs. Recent studies have indicated that short RNAs produced by an RNA interference (RNAi)-related mechanism are used for the elimination of repetitive DNAs in *Paramecium* and *Tetrahymena*, whereas long RNAs play essential roles in the DNA unscrambling of *Oxytricha*.

In both *Paramecium* and *Tetrahymena*, short RNAs are generated from the germline genome during meiosis. The discovery of these small RNAs, together with the demonstration that Argonaute and Dicer homologs are required for DNA rearrangements in these ciliates, have led to the realization that an RNAi-like mechanism is involved. The small RNAs are thought to target homologous sequences for methylation at histone H3 at lysine 9 (H3K9me) and lysine 27 (H3K27me), marking them for elimination. Thus, ciliate DNA rearrangements are mechanistically similar to the more broadly used RNA-directed establishment of heterochromatin. The use of RNAi to eliminate transposable elements further underscores the importance of this pathway as a genome defense mechanism.

In contrast, another type of noncoding RNA, long macronuclear RNAs, is used to guide the DNA unscrambling event in *Oxytricha*, which is a ciliate distantly related to *Paramecium* and *Tetrahymena*. Transcription of the parental macronuclear genome occurs during early conjugation in *Oxytricha*, and RNAi knockdown of specific parental macronuclear long RNAs inhibits unscrambling in the corresponding loci in the new macronucleus. Moreover, the injection of artificial RNAs reprograms the unscrambling order. Therefore, long macronuclear RNAs epigenetically regulate DNA unscrambling probably by acting as templates to guide the rearrangements.

Fully understanding these two curious DNA rearrangement processes, small RNA-directed DNA elimination in *Paramecium* and *Tetrahymena* and long RNA-directed DNA unscrambling in *Oxytricha*, will undoubtedly provide new insights into the roles of RNA in the epigenetic programming of the genome.

 Cite as *Cold Spring Harb Perspect Biol* doi: 10.1101/cshperspect.a017764

1 CILIATES: SINGLE CELLS WITH TWO DIFFERENT GENOMES

Ciliates, which comprise a monophyletic lineage that is an estimated 1 billion years old (Philippe et al. 2000), were among the first unicellular eukaryotes to be used as genetic models. In the late 1930s, when T.M. Sonneborn discovered the mating types of *Paramecium aurelia* (Sonneborn 1937), the chromosome theory of inheritance elaborated by T.H. Morgan was still unsatisfying to many researchers, in particular embryologists (see Ch. 1 [Felsenfeld 2014] for historical detail). Unable to envision how such static entities as genes could be the sole basis of heredity, they believed that the cytoplasm had to be involved, if only to coordinate gene action (see Harwood 1985). Although mainstream geneticists largely focused on gene action, Sonneborn's early genetic analyses showed that the transmission of many heritable characteristics could not be fully explained by Mendel's laws. The study of ciliates has revealed some of the first examples of cytoplasmic inheritance and continues to provide new insights into epigenetic mechanisms.

A key biological attribute of ciliates that provides insight into epigenetic mechanisms is nuclear dimorphism: Each cell contains two kinds of nuclei that differ in structure and function. The diploid micronuclei are transcriptionally silent during vegetative growth, but contain the germline genome. These nuclei undergo meiosis to produce gametic nuclei that transmit the germline genome to the next sexual generation (Fig. 1). In contrast, the highly polyploid macronuclei are responsible for gene expression during vegetative growth and thus govern the cell's phenotype, but they are lost during sexual development and can therefore be considered the equivalent of the soma (Fig. 1). The numbers of nuclei of each type vary in different species. For example, *Paramecium tetraurelia* have two micronuclei and one macronucleus, whereas *T. thermophila* has just one of each.

Macro- and micronuclei divide by separate mechanisms during vegetative growth (Fig. 1A). Micronuclei divide via conventional closed mitosis. Macronuclei, in contrast, divide by a poorly understood amitotic mechanism that does not involve spindle formation or visible condensation of the centromere-less, somatic chromosomes. After DNA synthesis, the macronucleus simply splits into two roughly equal halves. There does not appear to be any mechanism to ensure equal segregation of macronuclear chromosomes to the two daughter cells. Instead, it is likely that the high ploidy level (~800C in *P. tetraurelia*, ~45C in *T. thermophila*) prevents lethal gene loss for a number of vegetative divisions. Most species have a finite vegetative life span, and clonal cell lines will eventually die if they do not engage in sexual reproduction before they become senescent.

2 CONJUGATION: DIFFERENTIATION OF THE GERMLINE AND SOMATIC GENOMES

Mature ciliate cells, which have reached an appropriate clonal age, will become sexually reactive on mild starvation and pair with cells of compatible mating types to initiate conjugation, a sexual process (Fig. 1B). If no compatible partner is available, some species will undergo a self-fertilization process called autogamy. In either case, nuclear reorganization ensues, starting with meiosis of micronuclei (i.e., containing the germline genome). The sequence of nuclear events is similar, with some variation, among all species and is depicted in Figure 1B for *T. thermophila* (Ray 1956; Sonneborn 1975; Martindale et al. 1982).

Postmeiotic development starts with the selection of a single haploid nucleus in each cell to pass on the genome. The selected nucleus undergoes an additional replicative division that produces two genetically identical gametic nuclei (Fig. 1B, step b). In the case of conjugation, the two mates exchange one of their two haploid nuclei, and subsequent karyogamy (i.e., the fusion of two haploid nuclei) therefore generates genetically identical zygotic nuclei in each conjugant (Fig. 1B, steps c,d). In ciliates that perform autogamy, two gametic nuclei within a single cell fuse to produce an entirely homozygous diploid genome. In either case, the resulting diploid zygotic nucleus (Fig. 1B, step d) divides twice more, and the four products differentiate, two into new micronuclei and two into new macronuclei (Fig. 1B, step e). On completion of nuclear development, cells return to vegetative growth with a special first caryonidal division (Fig. 1B, step f) to distribute new micro- and macronuclei to the two daughter cells of each conjugant, whereas the parental macronuclei are discarded (Davis et al. 1992).

3 MACRONUCLEI AND MICRONUCLEI: A MODEL FOR ACTIVE VERSUS SILENT CHROMATIN

The basic concept of epigenetics is that individual copies of a DNA sequence can possess different activities and such differential states can be stably maintained. The nuclear dimorphism of ciliates is a natural example of homologous sequences that are maintained in a common cytoplasm, yet possess opposite activity states. The macronucleus serves as a model for the transcriptionally active state, the micronucleus for the repressed or silent state (Fig. 2). Early biochemical and immunohistochemical studies, primarily in *Tetrahymena*, compared properties of these different nuclei. Both histone variants and chromatin modifications were found to correlate with these different activity states and revealed the importance of chromatin structure in epigenetic regulation.

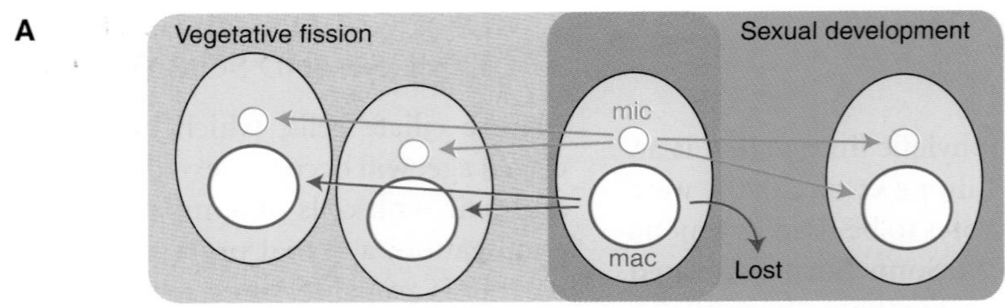

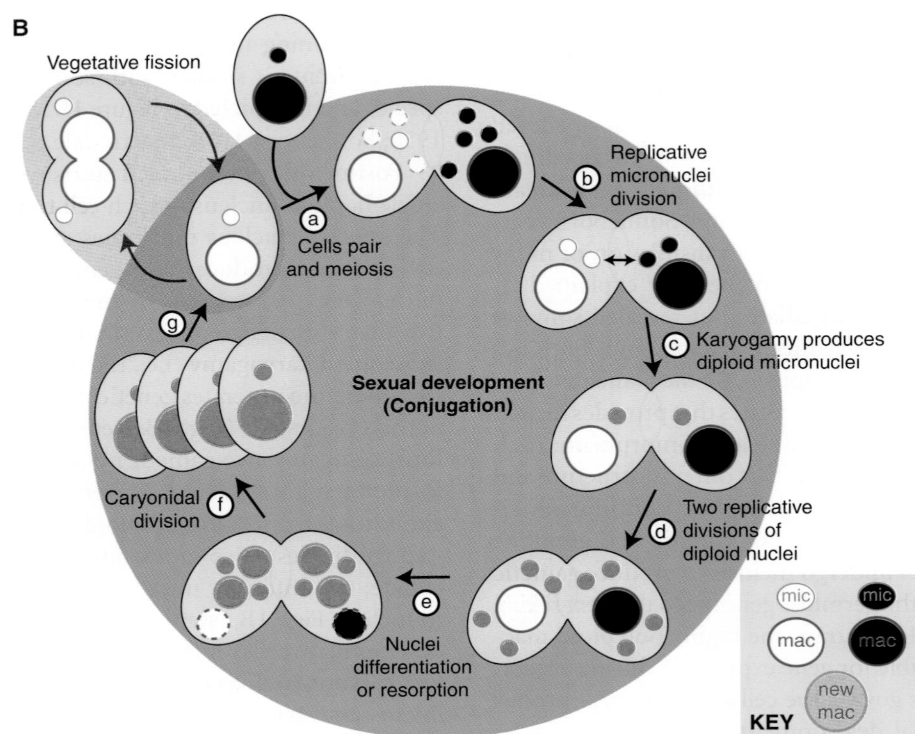

Figure 1. The ciliate life cycle and fates of their nuclei. (*A*) Vegetative cells multiply by binary fission, duplicating micro- and macronuclei. Sexual reproduction results in the loss of the parental somatic macronucleus and differentiation of new nuclei. (*B*) Conjugation: (a) cells pair inducing micronuclear meiosis that ends with selection of one of the haploid products as the gametic nucleus and degeneration of those remaining (dashed line nuclei); (b) an additional replicative division of the selected nucleus produces genetically identical haploid nuclei, and one from each mate is then transferred to its partner; (c) karyogamy produces a diploid zygotic (gray shading) nucleus in each mate; (d) two replicative divisions of these zygotic nuclei in each mate produce undifferentiated micro- and macronuclei; (e) nuclear differentiation: two nuclei become new micronuclei, whereas two begin differentiating into new macronuclei, and the parental macronuclei are resorbed; (f) pair-separation and caryonidal division; (g) resumption of vegetative growth. Nuclei with dashed outline are targets for destruction.

3.1 Macro- and Micronuclei Contain Distinct Histone Variants

Isolation of histone proteins separately from the macronucleus and micronucleus of *Tetrahymena* identified some of the first histone variants. Histone variant (hv) 1 and hv2, which correspond to the common variants H2A.Z and H3.3, respectively (see Ch. 20 [Henikoff and Smith 2014]), were detected exclusively within macronuclei, although the major histones are also present in both macro- and micronuclei (Fig. 2). Because of their macronucleus-specific localizations, these variants were suggested to be important for maintaining transcriptional activity (Allis et al. 1980; Hayashi et al. 1984). The hv2 (H3.3) variant and the more recently identified histone H3 variant H3.4 were shown to be constitutively expressed, a property critical for its DNA replication-independent deposition into chromatin, and indeed these histone H3 isoforms serve as

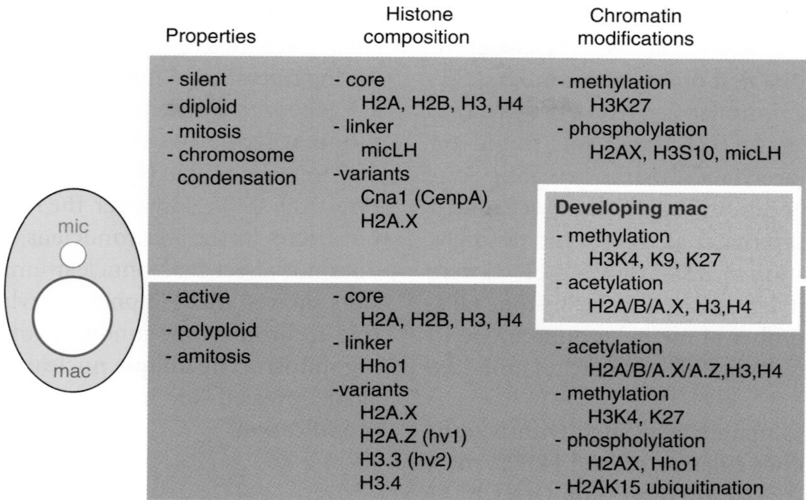

Figure 2. Nuclear dimorphism of ciliates. The germline micronucleus (mic), the developing macronucleus (mac), and the somatic macronucleus contain different histone complements and modifications. Those known to occur specifically in each or in the developing somatic genome are listed.

replacement histones after transcription-associated nucleosome removal (Yu and Gorovsky 1997; Cui et al. 2006). Consistent with this, studies with multicellular eukaryotes showed the association of H3.3 deposition with transcription (Ahmad and Henikoff 2002). Like many eukaryotes, *Tetrahymena* has the CenpA homolog Cnp1p, the centromere-specific histone H3 variant. Cnp1p only localizes to the micronucleus and is essential for the micronuclear chromosome segregation. Consistent with the fact that the macronucleus divides by amitosis, the macronucleus is devoid of Cnp1p (Cervantes et al. 2006; Cui and Gorovsky 2006).

In addition to the different complement of core histone variants, the macro- and micronucleus contain different linker histones, the macronuclear linker histone H1 (Hho1p) and the micronuclear linker histone (micLH; Allis et al. 1984; Wu et al. 1986; Hayashi et al. 1987). Neither linker histone gene is essential for cell viability, but individual gene knockouts result in an increase in volume of the nucleus in which each acts. Thus both are critical for full chromatin compaction, possibly by stabilizing higher order chromatin structure (Shen et al. 1995). Loss of the Hho1p also leads to changes in expression of specific genes implicating this linker histone in maintaining proper transcriptional regulation (Shen and Gorovsky 1996).

3.2 Chromatin Modifications Correlate with Activity States and Biological Processes

3.2.1 *Acetylation*

The hyperacetylation of histones in the macronucleus and lack of this modification from the micronucleus provided early evidence correlating this posttranslational modification with gene activation (Vavra et al. 1982). The enzymes that performed chromatin acetylation remained unknown in any organism until the mid-1990s when C. David Allis and coworkers purified the first type A (nuclear) HAT from *Tetrahymena* (Brownell and Allis 1995; Brownell et al. 1996). These researchers started with highly purified macronuclei to separate this activity from type B cytoplasmic HAT activity, and followed their purification using an in-gel assay in which purified histones were polymerized into the polyacrylamide gel matrix used to fractionate protein extracts. After electrophoresis, proteins were renatured and incubated with radiolabeled acetyl-CoA to reveal a polypeptide with an apparent molecular mass of 55 kDa that could incorporate the acetyl moiety into the histone matrix.

The real breakthrough came after microsequencing the purified protein and cloning the gene. This *Tetrahymena* HAT was found to be homologous to a well-characterized transcriptional regulator of baker's yeast, the Gcn5 protein. Before this discovery, transcriptional activators were primarily thought to act by recruiting RNA polymerase to promoters, but this work established that transcriptional activators also possess enzymatic activity, modifying chromatin, or other transcriptional regulators, thus changing the state of the template. Quickly thereafter, known regulators of many different eukaryotes were shown to act as HATs.

3.2.2 *Methylation*

Histone methylation patterns are specific for particular nuclei or developmental stage indicating the different modification states have different biological roles (Fig. 2).

Histone H3 lysine 4 (H3K4) methylation is restricted to the macronuclei in growing *Tetrahymena* cells (Strahl et al. 1999), which was one of the first observations linking this modification with transcriptional activity. Furthermore, the trimethylation of H3K4 (H3K4me3) was found on the same tail as multiple acetylation marks, revealing coupling between different modifications separately associated with active chromatin (Taverna et al. 2007). Interestingly, the mono- and dimethylation of H3K4 (H3K4me1/2) were not strongly linked to acetylation on the same histone tails. This indicates that the number of methyl groups added to the same lysine residue results in differential regulation (see Ch. 6 [Cheng 2014]).

H3K9me2/3, which is mainly associated with heterochromatin silencing in other eukaryotes, and H3K27me3, which is associated with polycomb repression (see Ch. 17 [Grossniklaus and Paro 2014]), are abundant in developing macronuclei. In particular, they are enriched on germline-limited sequences that are subsequently eliminated from the somatic macronuclear genome (Taverna et al. 2002; Liu et al. 2007). These germline-limited sequences include most repetitive sequences that other eukaryotes package as heterochromatin. The developmentally regulated establishment of these heterochromatin modifications provides a useful model with which to elucidate their targeting to specific sequences (described in detail in Sec. 8). H3K27me3 is also found in the macro- and micronuclei of vegetative cells (Liu et al. 2007; Taverna et al. 2007), but whether and how it contributes to the inactive state of chromatin in vegetative cells has not been explored.

3.2.3 Phosphorylation

Purification of ^{32}P-radiolabeled histones from macro- and micronuclei showed that linker histones and core histones H2A and H3 are highly phosphorylated in *Tetrahymena* (Allis and Gorovsky 1981). Multiple sites of Hho1 are phosphorylated and this modification was shown to participate in the regulation of specific gene transcription (Mizzen et al. 1999). Using mutational analysis, Dou and Gorovsky found that this requirement for phosphorylation could be mimicked by the addition of negatively charged amino acids into Hho1 (Dou et al. 1999). The charged residues, however, did not need to be present in the corresponding positions of the phosphorylated amino acid, but the complementary effect required a cluster of charged sites (Dou and Gorovsky 2000, 2002). These studies indicated that phosphorylation per se was not required, but that a critical charge density promotes proper transcription.

A single position, serine 10, is phosphorylated in histone H3 (H3S10ph; Wei et al. 1998). This modification is cell cycle–dependent and is correlated with mitosis in many eukaryotes. In *Tetrahymena*, it is restricted to micronuclei during mitosis and meiosis. Replacing the normal histone H3 gene with a mutant form containing an alanine substitution at serine 10 (S10A) causes defects in micronuclear division resulting in lagging chromosomes and aneuploidy (Wei et al. 1999). However, the H3S10ph modification does not occur in the macronucleus, thus the S10A mutation does not affect macronuclear amitotic division. These results showed that H3 phosphorylation plays an important role in chromosome condensation and/or segregation during mitosis. The unique nuclear dimorphism of the ciliate again revealed key insight into the role of a chromatin modification.

4 HOMOLOGY-DEPENDENT GENE SILENCING IN CILIATES

Homology-dependent, RNA-mediated silencing mechanisms are widely used in eukaryotes for epigenetic regulation (see Ch. 16 [Martienssen and Moazed 2014]). One adaptation of these mechanisms in ciliates leads to the ultimate means of gene silencing, DNA elimination (Sec. 6). The role of homologous RNAs in these genome rearrangements is covered in Section 8. More conventional silencing mechanisms were discovered in *Paramecium* when nonexpressed transgenes introduced into the vegetative macronucleus elicited a phenotype similar to a Mendelian mutant occurring in an endogenous gene homologous to the transgene. Feeding cells bacteria-expressing double-stranded RNA that is homologous to an endogenous gene results in its silencing, implicating RNAi-related mechanisms as mediators of homology-dependent silencing (Galvani and Sperling 2002).

4.1 Transgene-Induced Silencing

The *Paramecium* macronucleus is easy to transform by microinjection as any introduced DNA fragment will be maintained, replicating autonomously without the need for a specific origin. Transformation with high-copy, nonexpressed transgenes triggers posttranscriptional silencing of endogenous genes that possess sufficient sequence similarity (Ruiz et al. 1998; Galvani and Sperling 2001). Silencing is inefficient if the 3′UTR of the gene is present in the transgene, which suggests that improperly terminated transcripts are recognized as aberrant RNAs and directed to the RNAi machinery (Galvani and Sperling 2001). Silencing correlated with the accumulation of homologous short RNAs of approximately 23 nt in length (Garnier et al. 2004), indicating that an RNAi pathway is involved. The ~23-nt short RNAs appear to target degradation of homologous

Cite as *Cold Spring Harb Perspect Biol* doi: 10.1101/cshperspect.a017764

mRNAs and thus can be considered small interfering (si)RNAs. Endogenous siRNAs of 23–24 nt have been identified in growing *Tetrahymena* cells produced from loci that may be pseudogenes (Lee and Collins 2006).

This endogenous silencing pathway in ciliates requires an RNA-dependent RNA polymerase (encoded by *RDR3* in *Paramecium* and *RDR1* in *Tetrahymena*) for efficient siRNA production. In *Tetrahymena*, Rdr1p is associated with the Dicer ribonuclease Dcr2p, indicating that siRNAs generation is coupled with making double-stranded RNA (dsRNA; Lee and Collins 2007). In *Paramecium*, *RDR3* is required for the accumulation of endogenous siRNAs, some of which are derived from the regions between convergent genes. In *Schizosaccharomyces pombe*, regions of convergent transcription between adjacent loci are also regulated by RNA silencing mechanisms. In gene dense genomes as in yeast and ciliate macronuclei, it may be critical to abrogate possible readthrough transcription. These endogenous silencing pathways serve as a critical means to check the expression of aberrant RNAs.

4.2 Silencing Is Induced by dsRNA

dsRNA is likely to be the primary trigger for transgene-induced silencing observed in *Paramecium*. The silencing efficiency of transgenes correlates with the production of aberrant RNA molecules that correspond to both the sense and antisense strands of the injected sequence. dsRNA can also promote gene silencing when introduced by feeding *Paramecium* cells *Escherichia coli* transcribing both strands of a cloned gene (Galvani and Sperling 2002) using methodology developed for *Caenorhabditis elegans* (Timmons and Fire 1998; Timmons et al. 2001). Feeding of this dsRNA led to the accumulation of ~23-nt siRNAs, the same size as those observed on transgene-induced silencing (Nowacki et al. 2005), indicating that both phenomena rely on a common RNAi pathway. Feeding heat-killed *E. coli* to spirotrich species that normally feed on algae also promotes gene silencing (Paschka et al. 2003), suggesting that this mechanism in conserved throughout the ciliate lineage.

Silencing induced by feeding dsRNA to *Paramecium* can be reversed by replacing *E. coli* with the normal food bacterium in the culture medium; similarly, direct microinjection of dsRNA into the cytoplasm induces only transient silencing of the homologous genes, presumably because the injected dsRNA is rapidly diluted out during vegetative growth (Galvani and Sperling 2002). Thus, despite the RNA-dependent RNA polymerase(s) activity in vegetative *Paramecium* cells, dsRNA molecules cannot be sufficiently or continually amplified to establish a heritable silencing state, unlike the apparent fate of dsRNA in *C. elegans*. Furthermore, heritable silencing may require H3K9 methylation-mediated transcriptional gene silencing, which is apparently absent from the vegetative macronucleus, at least in *T. thermophila*.

5 GENOME-WIDE REARRANGEMENTS OCCUR DURING MACRONUCLEAR DEVELOPMENT

Chromosomes of the macronucleus and micronucleus do not just differ in transcriptional activity and chromatin modifications as described in Section 3 (Fig. 2), but show several fundamental differences is size and ploidy: (1) The size and complexity of the macronuclear genome is reduced relative to that found in the micronucleus; (2) macronuclear chromosome number (per haploid genome) is greater; but (3) macronuclear chromosomes are significantly shorter than micronuclear chromosomes; and (4) the arrangement of some gene segments is scrambled in the micronucleus relative to the "correct" order found in the macronucleus that is needed for proper gene expression. The first three differences are common among ciliates species, whereas the fourth has, so far, only been documented in spirotrichous ciliates, including *Stylonychia* and *Oxytricha*.

The structural differences between these nuclei result from programmed genome rearrangement events that transform the germline micronuclear genome into the somatic genome during macronuclear development. The reduction in genome size of the macronucleus, relative to the micronucleus, results primarily from DNA elimination (Sec. 6.1), whereas the increase in chromosome number and shortening of chromosomes is the product of chromosome fragmentation (Sec. 6.2). The scrambled order of gene segments in the micronucleus is corrected in the macronucleus by DNA elimination events, which illustrates the extreme degree of genome reorganization that occurs in some ciliates (Sec. 6.3). These DNA rearrangement events are described below to provide the background for the discussion of the epigenetic mechanisms that operate during differentiation of these nuclei.

5.1 DNA Elimination

Programmed DNA elimination (Fig. 3) creates a gene-condensed somatic genome in the macronucleus, reducing its size and sequence complexity significantly relative to the genome found in the germline micronucleus. For example, the macronuclear genome of *Tetrahymena* is only ~100 Mb, whereas its micronuclear genome is ~150 Mb (Eisen et al. 2006; Coyne et al. 2008). The amount of DNA eliminated varies greatly between ciliate species. Upward of 20%–30% of the micronuclear DNA is removed in the oligohymenophorean ciliates *Paramecium* and *Tetrahymena*, whereas more than 95% of the germline genome is

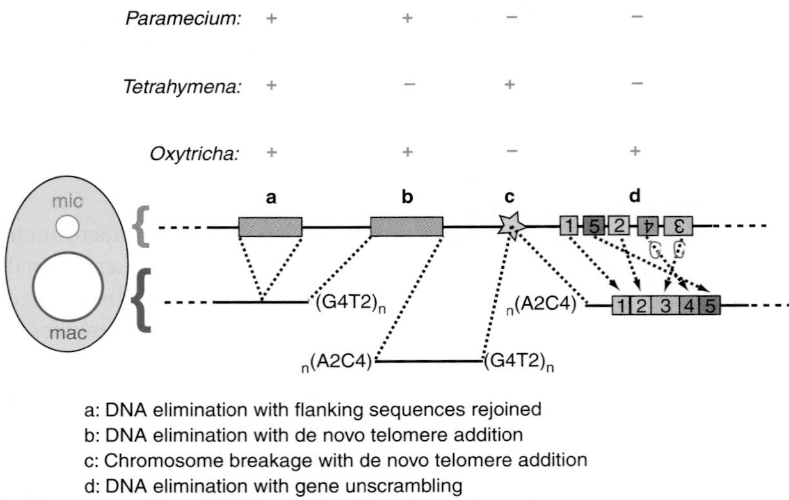

Figure 3. Ciliate DNA rearrangements. The four classes (a–d) of DNA rearrangements that occur during development of a new macronucleus are illustrated as listed. The presence or absence of each class is indicated for the ciliate species *Paramecium*, *Tetrahymena*, and *Oxytricha*. Colored bars are internal eliminated sequences (IESs) removed by precise or imprecise deletions or alternatively by fragmentation coupled with telomere $(G4T2)_n$ addition. The conserved, 15-bp chromosome breakage sequence is denoted by the star. DNA unscrambling reorganizes the "exons" (numbered boxes) misordered in the micronucleus.

eliminated from macronuclei of spirotrichous ciliates, such as *Euplotes*, *Stylonychia*, and *Oxytricha* (reviewed in Jahn and Klobutcher 2002). The sequences eliminated are enriched in transposon-related repetitive sequences, but also include many single-copy, nongenic sequences. As a result, the macronuclear genomes become highly gene rich: 1 gene per ~4 kb in *Tetrahymena*, ~2 kb in *Paramecium*, and ~2.5 kb in *Oxytricha*. This "streamlined" organization is consistent with genomes that are optimized for efficient gene expression.

DNA elimination can be classified as three different types based on the precision and outcome of each event: imprecise deletions of internal DNA segments healed by the rejoining of flanking sequences (Fig. 3, class a); imprecise deletions of DNA segments followed by de novo telomere formation (Fig. 3, class b); and precise deletions followed by the rejoining of flanking sequences (Fig. 3, class a). One or more of these events have been observed in the commonly studied ciliates. The first is the predominant type observed in *Tetrahymena*, whereas all three are observed in *Paramecium*. The first and second types may result from alternative healing outcomes of imprecise excision events in this ciliate. Although precise deletion events are not common in *Tetrahymena*, they are the major rearrangement found in many ciliates.

The DNA segments removed from the developing macronuclear genome are commonly called internal eliminated sequences (IESs). Double-strand DNA breaks are generated at the boundaries of each IES and the flanking macronuclear-destined sequences (MDSs). These are then healed by ligation (Fig. 3, class a) or by de novo telomere formation (Fig. 3, class b). In *Tetrahymena*, the removal of ~6000 IESs (per haploid genome) results in primarily imprecise deletion. In other words, the resulting macronuclear junctions formed by independent excision events can vary in position over tens of base pairs in different individuals and even for different copies of a chromosome in a same macronucleus. The IES/MDS boundaries often contain short (1–8 bp) direct repeats of variable sequence, one of which is retained in the macronuclear sequence. Although these repeats are not essential for excision, they may contribute to the precision of the joining process (Godiska et al. 1993).

Precise deletions of IESs occur at the same nucleotide positions in all copies of a macronuclear chromosome. This class of IESs is short (~25–900 bp in *Paramecium*), single-copy, noncoding DNA segments. There are >45,000 such IESs per haploid genome of *Paramecium*; they are abundant within coding sequences, but also distributed throughout intergenic or intronic regions of germline chromosomes (reviewed in Jahn and Klobutcher 2002; Arnaiz et al. 2012). These precisely excised IESs, the so-called "TA" IESs are identified as having invariable 5′-TA-3′ repeat at each boundary, one copy of which remains within the macronuclear locus after excision (see Betermier 2004). The few nucleotide positions internal to the TA dinucleotides form a loosely conserved consensus (5′-TAGYNR-3′) in *Paramecium* (Klobutcher and Herrick 1995).

Both precise and imprecise deletions occur reproducibly: The same DNA segments are eliminated in different individuals, generation after generation; however, alternative choices at boundaries are observed for some IESs of both types. Besides the short, variable direct repeats, no common sequence has been identified in or around the eliminated DNA segments. So, how do *Tetrahymena* cells identify ~6000 IESs? How do *Paramecium* cells precisely excise ~60,000 IESs? These questions are addressed in Sections 8 and 9.

5.2 Chromosome Fragmentation

Chromosome fragmentation occurs during macronuclear development resulting in macronuclei with more individual chromosomes than the micronuclei. Taking *Tetrahymena* as an example, each diploid micronucleus has five chromosome pairs whereas each macronucleus has nearly 200 different chromosomes (per haploid genome equivalent). Fragmentation is followed by the de novo addition of telomeric repeats to stabilize the shorten chromosomes. In many spirotrich ciliates, such as *Euplotes*, *Stylonychia*, and *Oxytricha*, chromosome fragmentations are so extensive that most macronuclear chromosomes are "gene-sized" nanochromosomes that possess a single open reading frame. In contrast, macronuclear chromosomes of *Tetrahymena* and *Paramecium* range in size from 20 kb to >1 Mb and contain many genes.

The recognition of chromosome fragmentation sites is not well understood. In *Tetrahymena*, a conserved 15-bp chromosome breakage sequence (CBS; Yao et al. 1987) is found at ~200 loci within micronuclear chromosomes (Cassidy-Hanley et al. 2005; Hamilton et al. 2005) and is both necessary and sufficient to direct fragmentation and new telomere addition (Fig. 3, class c) (Fan and Yao 1996). Similarly, a weak consensus sequence is found near chromosome fragmentation sites of *Euplotes* (Klobutcher 1999). The presence of CBS and CBS-like sequences probably indicates that the sites of chromosome fragmentation are genetically programmed in these ciliates. In contrast, in *Paramecium*, all chromosome fragmentation may result from imprecise DNA elimination that is healed by telomere addition instead of joining of dsDNA breaks (Fig. 3, class b) (Le Mouël et al. 2003). This process is epigenetically regulated and is discussed in Section 7.1.

5.3 DNA Unscrambling

An amazing variation to precise DNA elimination has been observed in some spirotrich ciliates in which IES removal occurs simultaneously with DNA "unscrambling" (Fig. 3, class d) (Prescott 1999). In the micronuclear versions of scrambled genes, coding regions are not only interrupted by IESs, but are also disordered relative to the linear arrangement found in the reorganized macronuclear sequence. Two MDSs that will be joined to form an expressed gene can be located far apart, sometimes found in an inverted orientation relative to each other, and can even be found at different chromosomal loci. It has been estimated that as much as 20%–30% of genes in *Stylonychia* and *Oxytricha* are scrambled in the micronucleus.

The precision of reordering appears to be guided, at least in part by relatively long (11 bp on average) homologous repeats, shared by cognate MDS ends, which are unrelated in sequence to those at other MDS ends. These long repeats have been found at all scrambled sites so far studied and thus are likely critical for accurate unscrambling. Yet, these sequences are not unique to MDS/IES junctions, and this has led to the proposal that a template specifying the correct order may be involved (Prescott et al. 2003). Indeed, it has been shown that long RNA templates derived from the parental macronucleus direct DNA unscrambling in *Oxytricha* (Nowacki et al. 2008). The presumed mechanism of RNA template-directed DNA unscrambling process is explained in Section 11.

5.4 Transposases Are Required for DNA Elimination

Many DNA segments eliminated from the new macronucleus have similarity to transposable elements (e.g., Herrick et al. 1985; Baird et al. 1989; Wuitschick et al. 2002; Fillingham et al. 2004), and the boundaries of "TA" IESs in *Paramecium* and *Euplotes* resemble the termini of Tc1/mariner transposons found in virtually all eukaryotes (Klobutcher and Herrick 1997). This has spurred the hypothesis that DNA elimination is derived from transposon excision. Indeed, it has been shown that transposases have been recruited to mediate DNA elimination in *Oxytricha*, *Paramecium*, and *Tetrahymena* (Baudry et al. 2009; Nowacki et al. 2009; Cheng et al. 2010).

In both *Paramecium* and *Tetrahymena*, the DNA excision step of DNA elimination is initiated by a DNA double-strand break (DSB), producing a 4-base 5′ overhang (Saveliev and Cox 1995; Saveliev and Cox 1996; Gratias and Betermier 2003). In *Paramecium*, the overhang is centered on the 5′-TA-3′, which is commonly observed at both ends of IESs. This structure is similar to that of DSBs produced by the piggyBac transposase: 4-base 5′ overhangs carrying the duplicated 5′-TTAA-3′ (Mitra et al. 2008). Both *Paramecium* and *Tetrahymena* have genes encoding piggyBac transposase-related proteins, Pgm and Tpb2p, respectively. RNAi knockdown of these genes indicated that they are required for DNA elimination (Baudry et al. 2009; Cheng et al. 2010). Like the piggyBac transposase, recombinant Tpb2p shows endonucleolytic cleavage activity producing DNA DSBs possessing 4-base 5′ protruding ends (Cheng et al. 2010). Although Pgm and Tpb2p are structurally similar to piggyBac transposases, the genes encoding these proteins are retained in the macronucleus. In contrast, most of the transposon-related sequences are limited to the micronuclear genome. Therefore, the genes encoding these transposase-related proteins have been domesticated within their host genomes during ciliate evolution.

In *Oxytricha*, the telomere-bearing element (TBE) family transposases are necessary for DNA elimination (Nowacki et al. 2009). In contrast to the piggyBac transposase-related proteins in *Paramecium* and *Tetrahymena*, TBE transposases are encoded in TBE transposons in the micronuclear-limited sequences in *Oxytricha*. Therefore, although *Paramecium* and *Tetrahymena* perform DNA elimination using domesticated transposases, *Oxytricha* uses transposases in the (potentially) active transposons to achieve DNA elimination. The TBE transposases and the piggyBac transposases are evolutionarily not directly related. Therefore, different classes of ciliates may have independently acquired their abilities to use transposon-derived enzymes for DNA elimination. This hypothesis suggests that the mechanisms of DNA elimination of oligohymenophorean (including *Paramecium* and *Tetrahymena*) and spirotrich (including *Oxytricha*) ciliates may be evolutionarily unrelated.

6 DNA ELIMINATION IS GUIDED BY HOMOLOGY-DEPENDENT MECHANISMS

Sequence-specific information can be communicated between macronuclear genomes of different generations to influence development. This comes from extensive experimental evidence showing that patterns of DNA elimination from a new macronucleus are epigenetically regulated by the DNA sequence of the preexisting, already reorganized genome of the parental macronucleus. In this section, we introduce classic examples of this regulation, which is followed in Section 8 with an explanation of how these epigenetic phenomena can be mediated by RNAi-related mechanisms.

6.1 Epigenetic Inheritance of Alternative Rearrangements

The first evidence that DNA rearrangements are subject to epigenetic regulation was uncovered when an aberrant DNA elimination pattern occurring in the previous sexual generation was shown to be inherited in new macronuclei of subsequent generations in *Paramecium* (Fig. 4). In wild-type macronuclei, the *A* surface antigen gene is located near a chromosome end, as chromosome fragmentation normally occurs downstream of this gene (Fig. 4A). A variant cell line, called d48, was found that failed to express the *A* gene because it lacked the gene itself, along with all downstream sequences in the macronucleus, as chromosome fragmentation placed the telomere at the 5′ end of the gene (Fig. 4B,left) (Forney and Blackburn 1988). This mutation was exclusive to the somatic macronucleus as the d48 strain micronucleus, when transplanted into a wild-type cell, gave rise to new macronuclei containing the full *A* gene after autogamy (Harumoto 1986; Kobayashi and Koizumi 1990). When the d48 strain was mated with a wild-type line, or allowed to undergo autogamy, the *A* gene of the macronucleus in the next generation was deleted (Fig. 4B, right) (Epstein and Forney 1984; Forney and Blackburn 1988). Reintroducing the *A* gene into the d48 macronucleus rescued the defect in *A* gene propagation during development (Fig. 4C). This restored, normal elimination pattern was maintained in the subsequent rounds of sexual reproduction (Koizumi and Kobayashi 1989; Jessop-Murray et al. 1991; You et al. 1991). These results indicated that the mere absence of the *A* gene from the parental macronucleus is sufficient to direct its future elimination from the new macronucleus. Similar experiments that focused on another telomere-proximal surface antigen gene, the *B* gene, showed that such epigenetic effects can be observed in other genomic regions (Scott et al. 1994).

The maternal macronuclear genome's influence on DNA rearrangements showed marked sequence specificity. The *A* and *B* gene coding sequences are 74% identical overall. Nevertheless, injection of the *A* gene sequences into the macronucleus of a cell line that carried macronuclear deletions of both genes could only prevent *A* gene deletion from the new macronucleus, and similarly, injection with the *B* gene could only prevent its own deletion (Scott et al. 1994). In addition, the macronuclear deletion of *A* gene in the d48 strain could not be reverted by introduction of the *G* gene from another species of *Paramecium*, *Paramecium primaurelia*, that shares 78% identity with the *A* gene. On the other hand, the macronuclear deletion of the *A* gene could be rescued by transformation with a different allele of the *A* gene having 97% identity (Forney et al. 1996). Thus, the maternal rescue of macronuclear deletions is a homology-dependent process that requires a minimum level of sequence identity.

These studies of inheritance of *Paramecium* surface antigen genes suggest that the germline-derived genome to be rearranged in a newly forming macronucleus is compared to the existing parental macronuclear genome that was rearranged in the previous sexual reproduction. The result

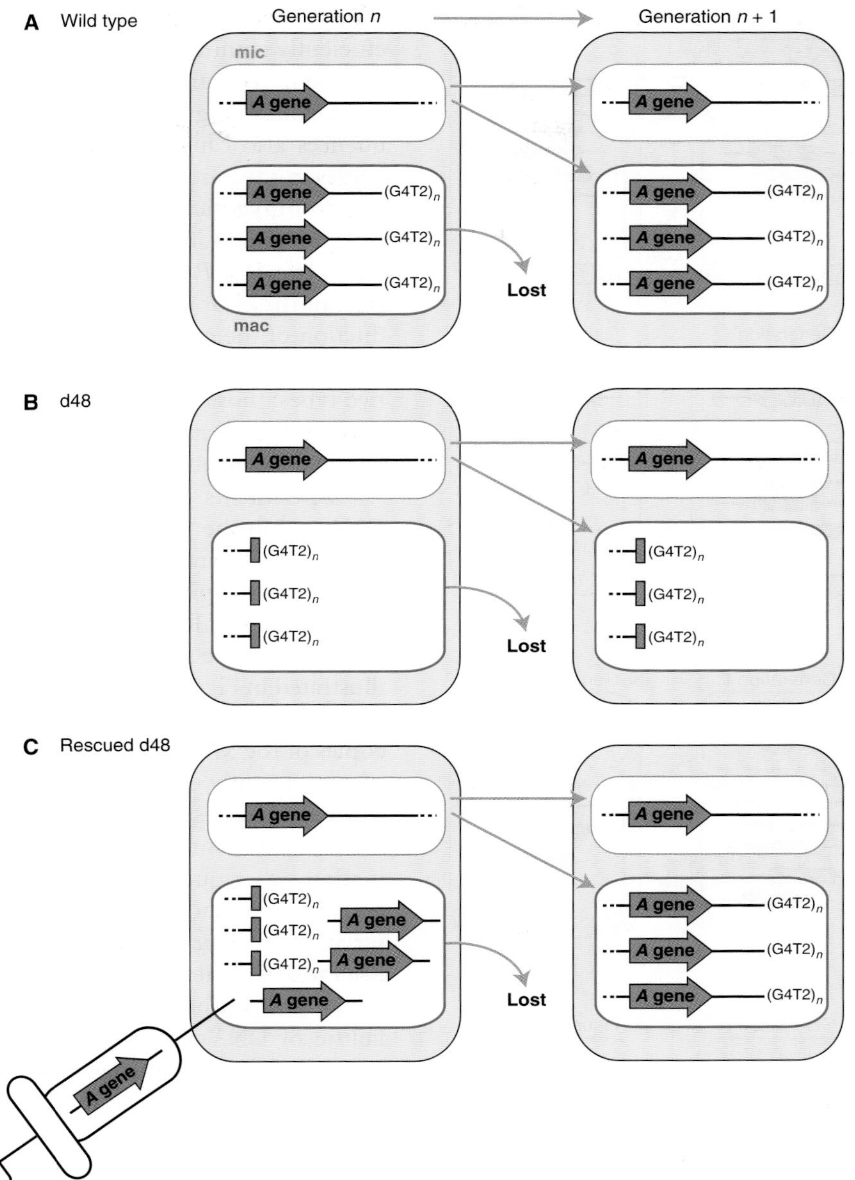

Figure 4. Epigenetic inheritance and experimental rescue of macronuclear *A* gene deletions. (*A*) Wild-type strain. (*B*) The d48 strain lacks the *A* gene in its macronucleus, but has a wild-type micronucleus. The *A* gene is reproducibly deleted during macronuclear development in each generation. (*C*) Transformation of the d48 strain macronucleus with *A* gene sequences restores amplification of the germline *A* gene in the developing macronucleus of sexual progeny.

of this comparison is that any sequences that are absent from the parental macronucleus are targeted for elimination in the next generation. Two further predictions can be made from these conclusions: (1) Excision of an IES could be perturbed in the next generation if a IES copy is introduced into the parental macronucleus, and (2) any DNA sequences introduced only into the micronucleus may be targeted for elimination from the new macronucleus. Both are indeed the case (see Secs. 7.2 and 7.3).

6.2 Homology-Dependent Inhibition of IES Elimination

As illustrated in Figure 5, when wild-type *Paramecium* cells were transformed by microinjection of DNA containing a fragment of the *G* coding sequence that includes an IES, the transformed clones produced progeny lines that retained the IES in their newly formed macronuclear chromosomes (compare Fig. 5A and B). In contrast the progeny of cells

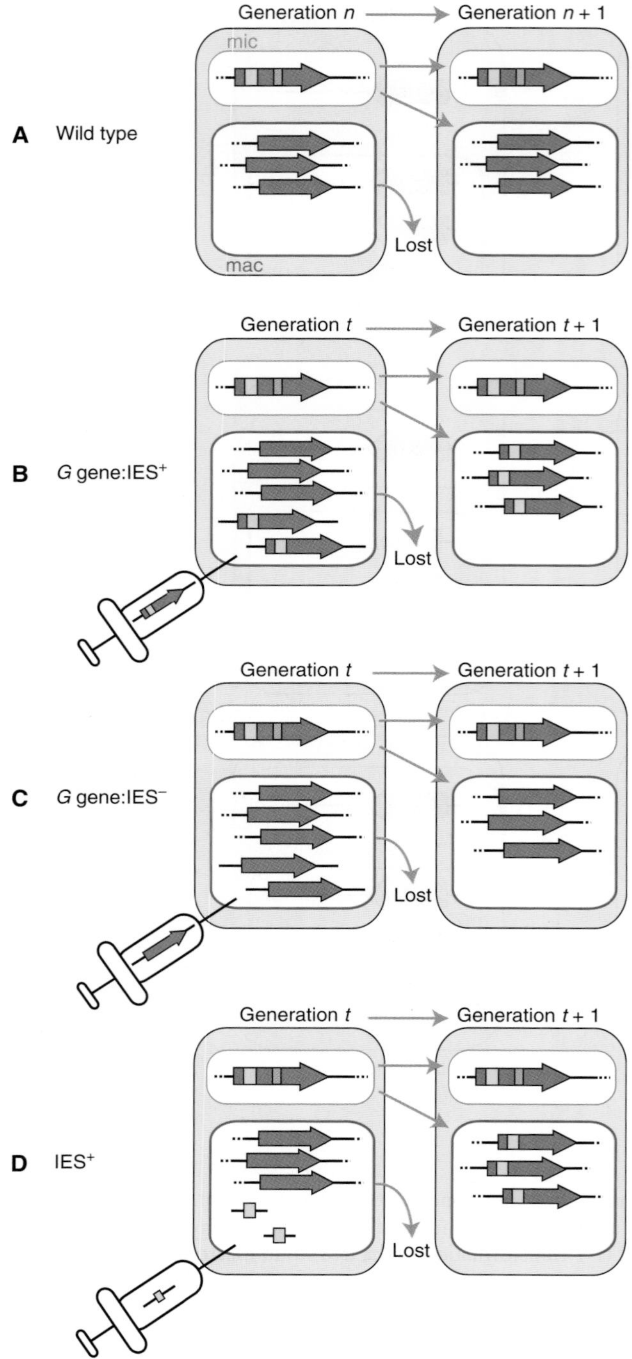

Figure 5. Homology-dependent inhibition of IES excision by the maternal macronucleus. (*A*) During development of wild type, IES (yellow and green bars) are excised efficiently. (*B*) Transformation of the maternal macronucleus with a high copy number of an IES (initial transformants = Generation *t*) can inhibit the elimination of the homologous IES during the subsequent (Generation $t + 1$, and future) rounds of new macronuclear differentiation. (*C*) Similar transformation with the IES$^-$ form of the gene does not change the wild-type rearrangement. (*D*) Transformation with IESs lacking flanking sequences results in inhibition of IES excision.

transformed with the *G* coding sequence without the IES efficiently eliminated the IES from the new macronucleus (Fig. 5C) (Duharcourt et al. 1995). Transformation with DNA containing only the IES without any flanking sequences also caused IES retention showing that the IES copies in the parental macronucleus alone inhibit its excision from zygotic macronuclei (Fig. 5D) (Duharcourt et al. 1995). However, not all IESs are controlled by similar maternal effects in *Paramecium*. Excision of only five out of 13 IESs in the *G* or *A* surface genes were inhibited by introduction of the corresponding IESs into the parental macronucleus (Duharcourt et al. 1998). Therefore, IESs are of two types: those for which elimination is maternally regulated and those not affected by the genome content of the parental macronucleus.

Experiments analogous to those above revealed that DNA within the *Tetrahymena* parental macronucleus can regulate the elimination of the homologous sequences from the developing macronuclear genome. The well-characterized M and R IESs were introduced into the macronuclei of wild-type strains. As per the results in *Paramecium* illustrated in Figure 5D, when these *Tetrahymena* cells were induced to conjugate, the progeny of the cells containing copies of the M or R IESs in the parental macronuclei failed to efficiently eliminate the chromosomal copies of the M or R IESs, respectively, during macronuclear development (Chalker and Yao 1996). The inhibition of DNA elimination was sequence-specific as significant inhibition of excision of nonhomologous IESs was not observed. As observed in *Paramecium*, sequences homologous to the IES itself were sufficient for this inhibition and the immediately flanking DNA had no effect. Importantly, this induced failure of DNA elimination was heritable as subsequent generations also retained genomic copies of the IES in their macronuclei. It is not clear whether *Tetrahymena* has any nonmaternally controlled IESs like *Paramecium* does.

Because the macronuclear and micronuclear chromosomes do not directly interact, the "signal" that mediates the inhibitory effect of the introduced IESs in the parental macronucleus must be transmitted to the new macronucleus through cytoplasm. *Tetrahymena* conjugating pairs extensively share cytoplasm. Even when one strain possessing the IES in its parental macronucleus was crossed to a wild-type partner, elimination of the IES in all new macronuclei, including those that differentiated within the cytoplasm of the wild-type partner, was blocked (compare Fig. 6A and 6B) (Chalker et al. 2005). Exchange of germline genomes (see Fig. 1B) is not required to transmit the inhibitory signal from one mating partner to another as blocking reciprocal exchange of micronuclei between pairs did not block the transmission of the inhibitory signal (Fig. 6C) (Chalker et al. 2005). Therefore, the influence of the

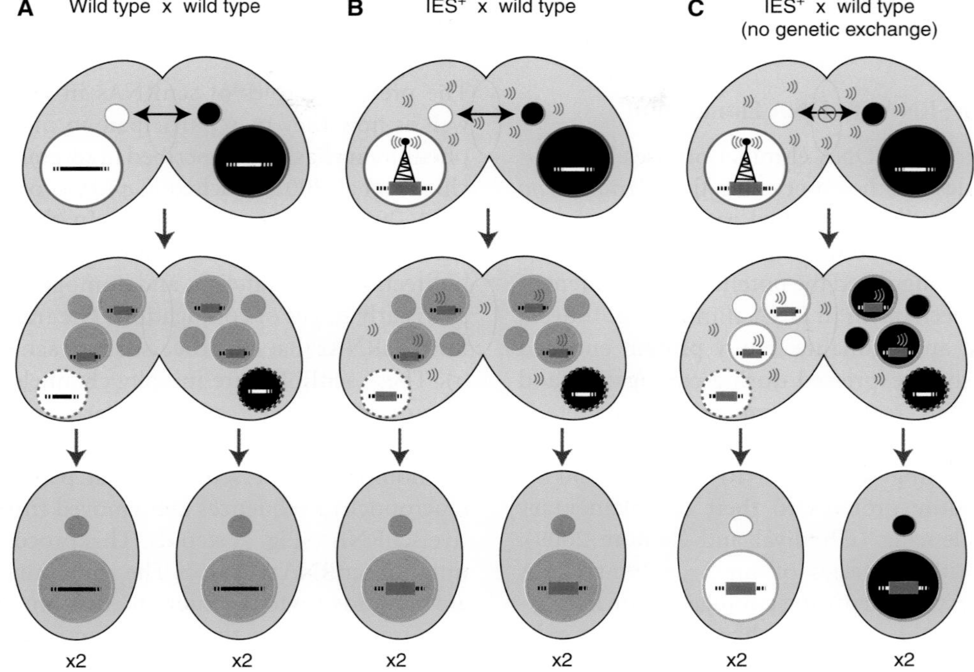

Figure 6. Inhibition of IES excision by the parental macronucleus is mediated by factors transmitted through the cytoplasm. (*A*) During macronuclear differentiation following mating of wild-type cells, IESs (e.g., the red bar) are excised efficiently; however, (*B*) the presence of an IES (IES^{+}) in the maternal macronucleus of one partner can signal its presence (depicted by radio tower emissions) to the developing macronuclei within both partners, inhibiting excision of the homologous IESs in all four resulting progeny. (*C*) The IES^{+} maternal macronucleus can inhibit IES excision in its wild-type mating partner even when genetic exchange is blocked.

parental macronuclear genome is transmitted through factors able to freely move through the cytoplasm.

6.3 "Spontaneous" Elimination of Foreign Sequences Introduced into the Micronuclear Genome

In *Tetrahymena*, when the *neo* gene is integrated into a micronuclear chromosome, it may be deleted from the genome of new macronuclei during successive rounds of conjugation (Yao et al. 2003; Liu et al. 2005; Howard-Till and Yao 2007). The *neo* gene is derived from the bacteria transposon Tn5, and thus is unlikely to contain by chance some specific signal to spontaneously induce its own elimination. Therefore, this result presents the possibility that any "foreign" sequence introduced into the micronuclear genome of *Tetrahymena* may be recognized as an IES.

Deletion of the *neo* gene is generally inefficient relative to the nearly 100% elimination observed for naturally occurring IESs. Furthermore, the efficiency of *neo* elimination from different micronuclear loci varied significantly (Liu et al. 2005; Howard-Till and Yao 2007). Not only did the genomic environment surrounding the *neo* gene influence elimination of this transgene, but multiple *neo* insertions into the micronuclear genome enhanced its DNA elimination (Liu et al. 2005; Howard-Till and Yao 2007). Thus, both the presence of a foreign sequence in the micronucleus as well as its repetitiveness in the genome affected DNA elimination. These position and copy number effects on transgene elimination will be discussed in Section 8 in the light of small RNA-directed regulation of DNA elimination mechanism. More generally, *neo* gene elimination suggests that IESs are recognized by more than sequence alone.

7 DNA ELIMINATION IS GUIDED BY SMALL RNA-DIRECTED TRANS-NUCLEAR COMPARISON OF WHOLE GENOMES

The homology-dependent effects described in Section 6 show that a cross talk occurs between the genomes of the parental macronucleus and the new macronucleus during nuclear differentiation, which can profoundly alter DNA elimination patterns and cellular phenotypes. The observations that only highly homologous sequences are affected suggest strongly that this cross talk is mediated by nucleic acids. As discussed in Sections 7.2 and 7.3, it is likely that the interplay between parental macronuclear transcripts and short RNAs derived from the meiotic micronucleus ultimately directs DNA elimination.

7.1 Linking Short RNAs to DNA Elimination

An initial indication that DNA elimination uses homologous RNA molecules was the observation that *Tetrahymena* IESs were bidirectionally transcribed early during conjugation (Chalker and Yao 2001). This bidirectional nature of transcription also suggested the potential involvement of an RNAi-related mechanism in DNA elimination. Indeed, it was shown that an Argonaute family protein encoded by *TWI1* is exclusively expressed during conjugation and is required for DNA rearrangement (Mochizuki et al. 2002). Argonaute proteins play key roles in RNAi-related processes by interacting with small RNAs (~20–30 nt) that target Argonaute proteins to their complementary RNAs for gene silencing (Ghildiyal and Zamore 2009). Twi1p interacts with a species of small (~29 nt) RNA that is exclusively expressed during conjugation in *Tetrahymena* (Mochizuki et al. 2002; Mochizuki and Gorovsky 2004). Disruption of the *TWI1* gene destabilizes these small RNAs and abolishes DNA elimination in the developing macronucleus (Mochizuki et al. 2002), indicating that these small RNAs play a pivotal role. The ~29-nt small RNAs are named scan (scn) RNAs because, as explained in Section 9.2, they "scan" the genomes for complementary sequences to direct DNA elimination.

Characterization of a gene, *DCL1*, encoding a Dicer-like ribonuclease further exposed the importance of these small RNAs in DNA elimination (Malone et al. 2005; Mochizuki and Gorovsky 2005). Dcl1p is expressed at high levels early in conjugation and localizes to premeiotic micronuclei, which indicated that generation of scnRNAs is temporally and spatially compartmentalized. Disruption of *DCL1* caused loss of scnRNA production, accumulation of the micronuclear transcripts, and ultimately failure of DNA elimination. These observations strongly support that scnRNAs are produced from the bidirectional micronuclear transcripts by Dcl1p and participate in DNA elimination.

In *Paramecium*, two Argonaute proteins, Ptiwi01 and Ptiwi09, and two Dicer proteins, Dcl2 and Dcl3, were identified as the functional counterparts of *Tetrahymena* Twi1p and Dcl1p, respectively (Lepere et al. 2009; Bouhouche et al. 2011). Double knockdown of the genes encoding these Argonaute or Dicer proteins causes both the loss of ~25-nt small RNAs and defects in DNA elimination. Thus, despite clear differences in IES size and structure between these ciliates, the ~25-nt small RNAs are *Paramecium* scnRNAs, and both *Paramecium* and *Tetrahymena* use an RNAi-related mechanism for DNA elimination.

7.2 Trans-Nuclear Comparison of Whole Genomes by Small RNAs

The presumed roles of scnRNAs in DNA elimination as well as how they may participate in mediating epigenetic phenomena has been described in several models (e.g., Mochizuki et al. 2002; Mochizuki and Gorovsky 2004; Aronica et al. 2008; Lepere et al. 2009). In the simplified model shown in Figure 7, the micronuclear genome is transcribed bidirectionally, randomly, and homogeneously (Fig. 7, step a) in early stages of conjugation. The transcripts produced form dsRNAs that are processed into scnRNAs (Fig. 7, step b). These scnRNAs are initially channeled to the parental macronucleus (Fig. 7, step c) in early–mid stages of conjugation. There they "scan" the existing rearranged genome for homology. The scnRNAs that pair with the parental macronuclear sequences are removed from the pool of active scnRNAs (Fig. 7, step d). This process is called "scanning" or scnRNA selection. The remaining scnRNAs, which are homologous to the micronuclear-limited sequences, are then transported to the developing macronucleus (Fig. 7, step e) where they target their complementary sequences for DNA elimination (Fig. 7, steps g and h) in late stages of conjugation.

This model is supported by the following observations: (1) In *Tetrahymena*, although scnRNAs complementary to both MDSs and the micronuclear-specific sequences (mostly IESs) are produced in the early conjugation stage, only the latter scnRNA population becomes gradually enriched in the mid-conjugation stages (Mochizuki and Gorovsky 2004; Aronica et al. 2008); (2) in the mid-conjugation stages, after the bulk of scnRNA production, Argonaute proteins (Ptiwi01 and Ptiwi09 in *Paramecium* and Twi1p in *Tetrahymena*) localize initially in the parental macronucleus, then later move to the new macronucleus (Mochizuki et al. 2002; Bouhouche et al. 2011); and (3) the RNA-binding protein Nowa1/2, which is required for the elimination of maternally regulated IESs, is also translocated from the parental to new macronucleus in *Paramecium* (Nowacki et al. 2005).

The model explains most of the epigenetic effects described in Section 8, but does not account for the position and the copy number effects associated with transgene elimination (Sec. 7.3). If transcription, which produces the scnRNA precursors, occurs homogeneously from the meiotic micronuclear genome, a transgene should be efficiently eliminated regardless of its genomic location or copy number in that nucleus; but if this micronuclear transcription occurs nonuniformly, the genomic loci harboring transgenes will influence the likelihood of being eliminated. In *Tetrahymena*, scnRNAs are preferentially produced from IESs (Schoeberl et al. 2012); future efforts to

 Cite as *Cold Spring Harb Perspect Biol* doi: 10.1101/cshperspect.a017764

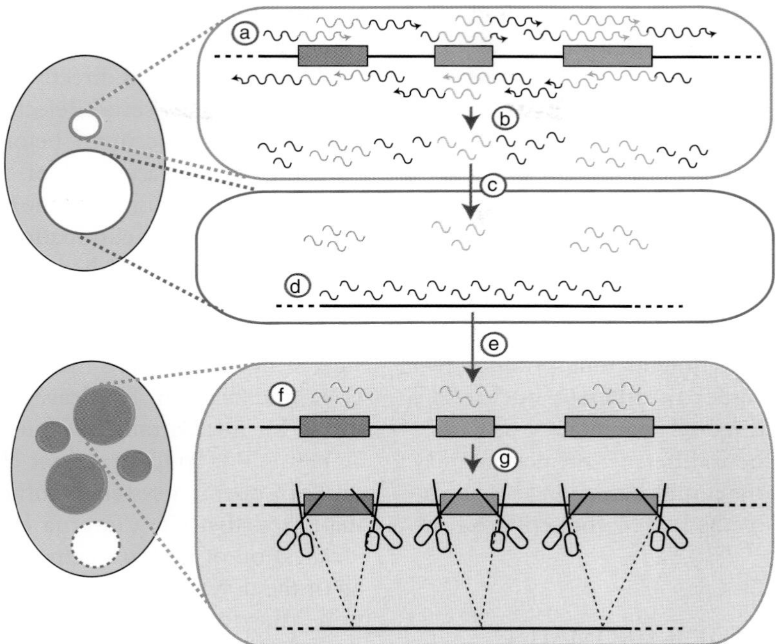

Figure 7. A template model for control of DNA deletion. Bidirectional transcription (a) of a large portion of the germline genome occurs early in development and leads to the production of scnRNAs (b). These are then transported into the maternal macronucleus (c) where any encounter homologous sequences (d), which triggers their removal from the active pool. The remaining micronucleus-specific RNAs are redirected to the developing macronucleus (e), where they mark homologous sequences (f), signaling their removal from the genome (g).

understand what underlies biased scnRNA production will likely show why some transgenes are eliminated and others are not. In *Paramecium*, although Ptiwi01 and Ptiwi09 are required for the excision of both maternally controlled and nonmaternally controlled IESs, Dcl2 and Dcl3 are necessary only for the elimination of former class of IESs (Bouhouche et al. 2011). Therefore, scnRNAs may have some fundamental role in DNA elimination besides its epigenetic regulation, at least in *Paramecium*.

7.3 Involvement of the Parental Macronuclear Transcripts in DNA Elimination

To transmit sequence-specific epigenetic information from the parental to the new macronuclear genome, scnRNAs have to interact with these genomes by base pairing. In both fission yeast and plants, small RNAs target genomic site by interacting with *cis*-transcripts emerging from the loci. Two reports, one of *Paramecium* and one of *Tetrahymena*, suggest that scnRNAs interact with parental and developing genomes through long noncoding RNAs (ncRNAs) (Aronica et al. 2008; Lepere et al. 2008). In *Tetrahymena*, a study of the Twi1p-interacting, putative RNA helicase called Ema1p suggested that nascent noncoding transcripts mediate the interaction between chromatin and scnRNA–Twi1p complexes in the parental macronucleus and that this interaction is dependent on Ema1p (Aronica et al. 2008). Moreover, Ema1p-depleted strains showed defects in scnRNA selection and DNA elimination (Aronica et al. 2008). The importance of the parental macronuclear transcripts for proper IES elimination has been shown more directly in *Paramecium* (Lepere et al. 2008): downregulation of noncoding transcripts from the parental macronucleus, by RNAi knockdown, blocked the scnRNA selection in the targeted regions and induced ectopic DNA elimination.

How scnRNA selection is achieved remains mostly speculation. One can hypothesize that the interaction between scnRNA and the parental macronuclear transcripts could remove the complementary scnRNAs from the pool by leading to their sequestration or degradation. Alternatively, scnRNAs that are not base paired with parental macronuclear transcripts could be selectively amplified. In *Tetrahymena*, northern blot hybridization studies of scnRNAs showed that those complementary to IESs were present at constant levels during early to late conjugation stages, whereas those complementary to MDSs were only detected in early stages (Aronica et al. 2008). These results suggest that the scnRNA selection process likely promotes scnRNA degradation, at least in *Tetrahymena*.

Although the interactions between scnRNAs and long noncoding RNAs within parental macronuclei result in scnRNA selection, similar interactions in the developing macronuclei likely induce DNA elimination. As we will explain in Section 10, these interactions effectively target heterochromatin formation on IESs, which appears to mark them for DNA elimination. The Twi1p-associated RNA helicase Ema1p is required for the interaction between Twi1p and nascent noncoding RNAs in the new macronucleus and the resulting heterochromatin formation (Aronica et al. 2008). It remains unclear how seemingly similar scnRNA–ncRNA interactions produce two dramatically different outcomes in different nuclei. The temporal (early vs. late) and spatial (parental vs. new macronucleus) separation of these different interactions may simply trigger alternative outcomes based on the downstream effectors recruited to the loci by the scnRNA–Argonaute complex.

8 scnRNA-INDUCED HETEROCHROMATIN FORMATION PRECEDES DNA EXCISION

In most eukaryotes, di- and trimethylated histone H3 lysine 9 (H3K9me2/3) is widely associated with transcriptionally repressed DNA that is partitioned in the nucleus as heterochromatin (see Sec. 10 of Ch. 3 [Allis et al.]). In *Tetrahymena*, H3K9me2/3 is not found in the transcriptionally silent micronucleus as one might presume, but is exclusively found in developing macronuclei, established immediately preceding DNA elimination, and particularly enriched on the histones associated with IESs (Taverna et al. 2002; Liu et al. 2007).

Different histone modifications are established on chromatin by specific "writers" and interact with different protein "readers" to enforce a regulatory state (see Ch. 6 [Cheng 2014]; Ch. 7 [Patel 2014]). In *Tetrahymena*, the writer for H3K9me2/3 appears to be the Polycomb group protein, enhancer-of-zeste-like (Ezl1p), a histone methyltransferase (Taverna et al. 2002; Liu et al. 2007). Ezl1p has been shown to be required for the accumulation of trimethylated H3K27 (H3K27me3) as well as H3K9me2/3 in the new macronucleus (Liu et al. 2007). The known readers for these modifications are the chromodomain-containing proteins Pdd1p and Pdd3p. Pdd1p binds to both H3K9me2/3 and H3K27me3, whereas Pdd3p only binds to H3K9me2/3 (Liu et al. 2007).

DNA elimination requires the function of Ezl1p and Pdd1p, in addition to the RNAi-related proteins described in Section 7.1 (Coyne et al. 1996; Liu et al. 2007). This shows that heterochromatin formation is a prerequisite of DNA elimination. Furthermore, *Tetrahymena* strains expressing a histone H3 mutant in which lysine 9 is replaced by glutamine (H3K9Q) are defective for DNA elimination (Liu et al. 2004). Although the requirement for H3K27me3 could not be assessed directly by K27Q, K27R, and K27A mutations because severe defects during meiosis and developmental arrest occurred before the appearance of new macronuclei, replacement of H3S28 to glutamic acid (H3S28E), which disrupted the H3K27me3-Pdd1p interaction, inhibited DNA elimination (Liu et al. 2007). Furthermore, this mutant abolished the accumulation of H3K9me3 (Liu et al. 2007), whereas conversely H3K27me3 accumulated normally in the H3K9Q mutant (Liu et al. 2004) suggesting that H3K27 methylation may be upstream of H3K9 methylation. This leaves open the possibility that an unknown methyltransferase, not Ezl1p, directs H3K9me. Tethering Pdd1p to an ectopic chromatin site using a LexA–LexA operator system is sufficient to trigger DNA elimination at that site (Taverna et al. 2002), indicating that localization of Pdd1p to chromatin is sufficient to recruit all of the downstream factors necessary for DNA elimination. Therefore, the structure of heterochromatin, but not the RNAi-related mechanism or H3K9/K27me, is the direct trigger for DNA elimination. The transposase, Tpb2p, discovered in *Tetrahymena* is localized to heterochromatic foci (Cheng et al. 2010) in which DNA elimination is believed to take place, suggesting that it may directly bind to some component of heterochromatin. Therefore, Tpb2p may directly recognize heterochromatin structure to catalyze DNA excision.

The RNAi-related pathway acts upstream of the establishment of heterochromatin modifications, as Dcl1p and Twi1p are required for the accumulation and/or targeting of H3K9me2/3 and H3K27me3, whereas the writer Ezl1p is not required for scnRNAs accumulation (Liu et al. 2004; Liu et al. 2007). It may be that Ezl1p is recruited to chromatin in which Twi1p–scnRNA complexes are interacting with nascent long ncRNA in the developing new macronucleus. Because the scnRNAs that are complementary to the micronucleus-limited sequences are selected in the parental macronucleus by the scnRNA selection (Section 8.2), the scnRNA–ncRNA interaction is expected to specifically recruit Ezl1p to induce the histone modifications only on IESs. The molecular mechanism underlying this recruitment is unknown because no direct interaction between Twi1p and Ezl1p has been detected. Nonetheless, we know that scnRNAs are the specificity factors targeting heterochromatin modifications to the correct genomic loci. The importance of heterochromatin in DNA elimination of other ciliates has yet to be established. It is interesting to note that many IESs in *Paramecium* are shorter than the length of DNA wrapped around a nucleosome, and thus a heterochromatin-independent mechanism is likely involved in the elimination of at least such short IESs.

 Cite as *Cold Spring Harb Perspect Biol* doi: 10.1101/cshperspect.a017764

9 DNA UNSCRAMBLING IS TEMPLATED BY MATERNAL RNAs

RNAs have roles in the DNA rearrangements of both oligohymenophorean and spirotrich ciliates; however, current evidence suggests that the mechanism(s) of RNA-directed DNA rearrangement of these two classes of ciliates may be fundamentally different. Although small, micronuclear RNAs direct DNA elimination in *Paramecium* and *Tetrahymena*; DNA elimination and DNA unscrambling (Figure 2D) of *Oxytricha* is regulated by long, macronuclear RNAs. Long RNA-guided DNA unscrambling was first proposed as a theoretical model (Fig. 8) (Prescott et al. 2003) and has since been shown experimentally (Fig. 9) (Nowacki et al. 2008). Transcription of the parental macronuclear genome occurs during early conjugation in *Oxytricha* (Nowacki et al. 2008). This transcription produces macronuclear long RNAs which likely span from telomere to telomere of "gene-sized" macronuclear nanochromosomes. RNAi knockdown of specific macronuclear long RNAs from a macronuclear nanochromosome, on injection of complementary dsRNAs, inhibited both DNA elimination and DNA unscrambling of the targeted loci in the new macronuclei of progeny cells (Fig. 9B). More strikingly, the injection of mutated artificial single-stranded RNAs was sufficient to reprogram the unscrambling order (Fig. 9C). Therefore, single-stranded long macronuclear RNAs can act as templates in DNA rearrangement in *Oxytricha*.

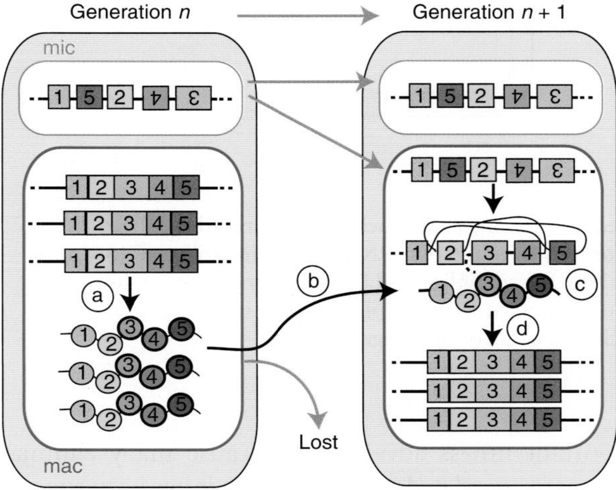

Figure 8. The template model of DNA unscrambling. The exons (numbered boxes) of scrambled genes from the micronucleus (mic) are faithfully reordered during nuclear differentiation. Transcripts produced from the maternal macronucleus (a) are transported to the developing macronucleus (b) where any encounter with a homologous germline-derived scrambled gene (c) and template correct reordering of the DNA in the somatic genome of the progeny (generation $n + 1$) (d).

Recently, it was reported that 27-nt small RNAs are accumulated during conjugation of *Oxytricha* (Fang et al. 2012; Zahler et al. 2012) and interact with the Argonaute protein Otiwi1 (Fang et al. 2012). Interestingly, these Otiwi1-bound small RNAs are produced from the parental macronucleus, not from the micronucleus as scnRNAs in *Tetrahymena* and *Paramecium*. Furthermore, injection of small RNAs complementary to an IES caused retention of the IES indicating that the Otiwi1-bound small RNAs in *Oxytricha* have an opposite role compared with scnRNAs in *Tetrahymena* and *Paramecium*: a protective role against DNA elimination (Fang et al. 2012). Although the division of labor between long macronuclear RNAs and Otiwi1-bound small RNAs in DNA rearrangement is currently unclear, *Oxytricha* clearly uses a fundamentally different strategy than *Tetrahymena* and *Paramecium* although these ciliates perform fundamentally similar DNA rearrangement processes.

The long macronuclear RNAs in *Oxytricha* have an additional surprising function in epigenetic genome regulation: These RNAs regulate not only the genome rearrangements, but also the chromosome copy numbers in the macronuclear development (Nowacki et al. 2010). In the macronucleus, chromosomes of the diploid zygotic nucleus are amplified by endoreplication. Although the copy number is the same for nearly all individual macronuclear chromosomes of *Paramecium* (~800C) and *Tetrahymena* (~45C), the copy numbers of different macronuclear chromosomes in *Oxytricha* can vary greatly, ranging from hundreds to tens of thousands. Because *Oxytricha* has more than 20,000 different macronuclear chromosomes per haploid genome (most containing a single gene), understanding the regulation of copy number has presented a challenge. It has been shown that the copy numbers of the new macronuclear chromosomes are influenced by the copy numbers of the corresponding chromosomes in the parental macronucleus. This copy number information is transmitted by the long macronuclear RNAs: More copies of a parental macronuclear chromosome produce more long macronuclear RNAs that correlate with the final copy number of that chromosome in the new macronucleus (Nowacki et al. 2010).

The mechanistic action of long macronuclear RNAs in directing DNA unscrambling and copy number regulation remains to be discovered. Their interactions with the developing somatic genome may facilitate the recruitment of TBE transposases to mediate DNA elimination/unscrambling. In turn, increasing the efficiency of DNA rearrangement may lead to increases in final chromosome copy number. Regardless of the actual molecular mechanism, the discovery that "RNA caches" participate in genome rearrangements undoubtedly broadens our horizons about the

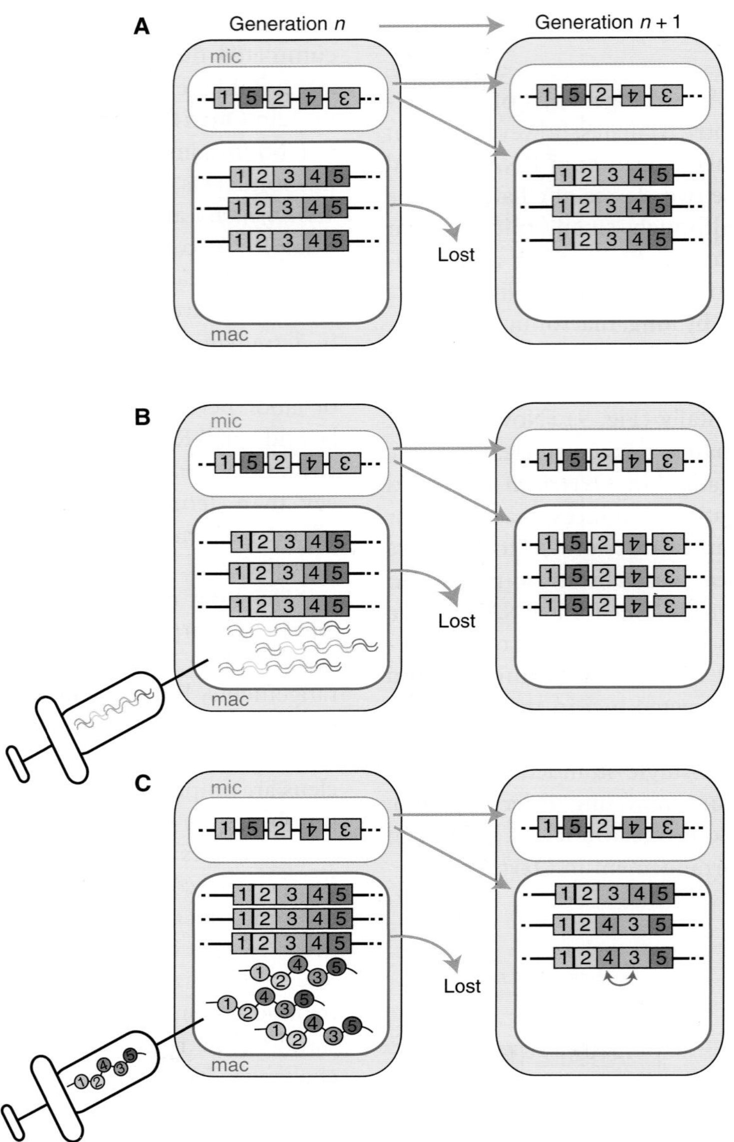

Figure 9. Maternal RNAs template unscrambling. (*A*) Scrambled genes in the micronucleus (mic) must have their exons (numbered boxes) reordered during nuclear differentiation. (*B*) Injection of dsRNA (RNAi) homologous to transcripts produced from the maternal macronucleus perturbs both unscrambling and DNA rearrangement in the macronuclei of the progeny generation ($n + 1$). (*C*) Injection of scrambled RNA templates results in the parental macronucleus directing misordering of the homologous nanochromosome of the progeny.

potential for RNAs to change the activity state of chromatin and even alter the physical structure of chromosomes.

10 BIOLOGICAL ROLES OF PROGRAMMED GENOME REARRANGEMENTS

DNA elimination occurs in all ciliates studied to date. Therefore, it must not be as wasteful a process as it may have seemed on its initial discovery, but likely offers important fitness advantages. Because many eliminated sequences are related to transposons, it has been suggested that DNA elimination is a defense mechanism against transposons (e.g., Yao et al. 2003; Fillingham et al. 2004). In this role, the epigenetic regulation of DNA elimination may act as a "genomic immune system" as discussed in Section 10.1. In addition, DNA elimination and DNA unscrambling may accelerate the emergence of new genes during the course of evolution, and even allow increase in the

 Cite as *Cold Spring Harb Perspect Biol* doi: 10.1101/cshperspect.a017764

coding potential from specific loci by alternative "DNA splicing" (Fass et al. 2011). Moreover, the epigenetic regulation of DNA elimination/unscrambling may allow inheritance of acquired characteristics of the soma to future sexual generations.

10.1 DNA Elimination as a Genome Defense Mechanism

The studies of RNAi and related mechanisms in many different eukaryotes support the hypothesis that this pathway evolved as a cellular defense mechanism controlling the proliferation of viruses and transposons by degrading mRNAs, inhibiting translation, and/or targeting the formation of heterochromatin on these genomic invaders (Matzke and Birchler 2005). Transposable elements of ciliates are largely eliminated from the somatic macronucleus during development, which will effectively limit their impact. The finding in *Tetrahymena* that histone H3K9 and H3K27 methylation is small RNA-dependent and marks genomic regions destined for the programmed DNA elimination revealed that the role of RNAi in ciliate nuclear differentiation is fundamentally similar to its role in establishing heterochromatin in other eukaryotes; ciliates just go one step further and eliminate heterochromatin from their somatic genome.

The specific sequences that are targeted by small RNAs for heterochromatin formation followed by DNA elimination are selected by a global comparison of maternal germline and somatic genomes. This mechanism would efficiently protect against the deleterious effects of transposition in the germline: Any new transposon integrating into the germline would be recognized as alien by comparison with the somatic genome during sexual reproduction, leading to its removal from the transcribed somatic genome of progeny, thereby limiting its future spread.

It has been suggested that RNAi is a primitive immune system against parasitic elements such as viruses and transposons (Waterhouse et al. 2001). The elimination of DNA, mediated by an RNAi-related mechanism, is conceptually more similar to the cellular immune system in vertebrates, both of which learn the distinction between self and nonself. A huge repertoire of lymphocytes expressing different antibodies is initially generated, but in early development all lymphocytes that recognize available antigens (likely self-antigens) are eliminated from the future pool (explained in Ch. 29 [Busslinger and Tarakhovsky 2014]). Once past this stage, the recognition of a cognate antigen (then likely to be alien) leads to the clonal expansion of the corresponding lymphocytes. By analogy, the DNA elimination mechanism can be considered to be a genomic immune system.

10.2 DNA Elimination and DNA Unscrambling as "DNA Splicing" Mechanism

DNA elimination occurs extensively from the coding regions. In this form, one can envisage DNA elimination removing "DNA introns" (= IESs) from the genome to assemble "DNA exons" (= MDSs), via "DNA splicing" events. Thus, like RNA splicing, IESs in the coding regions are removed with nucleotide level precisions. Because DNA unscrambling in *Oxytricha* is always accompanied with DNA elimination, this process is DNA splicing coupled with translocations and inversions of DNA exons. The biological consequences of DNA elimination and DNA unscrambling have obvious overlap with RNA splicing. RNA splicing offers two important advantages to cells. First, in evolutionary timescales, the presence of introns in DNA would allow genetic recombination to combine the exons of different genes and facilitate the emergence of new and useful proteins. Second, for individual organisms, the exon–intron organization of genes allows the same gene to produce multiple different proteins via alternative splicing. Thus, analogous to RNA splicing, DNA elimination/unscrambling has the potential to accelerate the emergence of useful genes over the course of evolution and allow individual loci to express multiple protein isoforms after alternative DNA elimination/rejoining.

The ability of DNA unscrambling, in which even translocations and inversions are possible, would greatly facilitate both evolutionary processes and alternative gene expression. Although alternative DNA elimination/unscrambling that produces different protein products has not been detected in ciliates, at least one case is known for which an IES in an exon results in alternative polyadenylation and splicing during development (Fass et al. 2011). Future large-scale genome sequence analyses will elucidate the extent of such outcomes.

The epigenetic regulation of DNA elimination and DNA unscrambling described in this chapter may also allow fixation of traits without alterations to the germline genome. As discussed above, DNA elimination/unscrambling can produce heritable variants of macronuclear chromosomes from a single micronuclear genome. Because the macronucleus is polyploid and its chromosomes are amitotically segregated during vegetative growth, variations in the macronuclear chromosomes can be selectively assorted based on positive or negative influences. So if a specific variant, or a combination of variants, is advantageous to the growth and/or survival of cells, it would dominate over other variants in the macronucleus during vegetative growth. Once a particular variant is fixed in the macronucleus, the pattern of DNA rearrangement in the next sexual generation can be biased toward this variant form by the epigenetic specification of DNA rearrangement patterns. This system could also work if an accidental rearrangement on macronuclear chromosomes during vegetative growth is beneficial to cells. In this way, acquired characters of somatic macronucleus can be inherited to the next sexual generations, and thus ciliates appear to naturally use a form of Lamarckian inheritance. Because the micronucleus retains the original genome, cells have a chance to return to an original character when the variant loses its advantage for survival owing to some environmental change.

11 EPIGENETIC REGULATION OF MATING-TYPE DETERMINATION

We finish this chapter by introducing yet another beautiful example of epigenetic inheritance in ciliates: the determination of mating type in *Paramecium* that was extensively studied by T.M. Sonneborn. Because the molecular mechanism behind this epigenetic inheritance has not been fully explained, we will let you imagine how it works.

The two complementary mating types of *P. tetraurelia*, which are called O and E, show a cytoplasmic pattern of inheritance (Fig. 10). The O and E traits are terminally differentiated phenotypes that are determined during development of the somatic macronucleus from the totipotent germline. After conjugation, a vegetative clone descended from the O parent is almost always mating-type O, whereas one arising from the E parent is almost always of type E (Fig. 10B), although both exconjugants develop from identical zygotic genomes (Fig. 10A). However, if a significant exchange of cytoplasm occurs between mating cells, all progeny typically develop as type E (Sonneborn 1977). Thus, a cytoplasmic factor must exist that directs development of the E type. One Mendelian mutation affecting mating-type determination has been shown to perturb IES excision at other loci (Meyer and Keller 1996). This supports the hypothesis that alternative genome rearrangements may determine mating type. If this is the case, then epigenetic regulation of the rearrangement may underlie the cytoplasmic inheritance of this trait. The E-determining cytoplasmic factor would have the capacity to direct an alternative rearrangement of the mating-type gene, resulting in a macronuclear form of the gene that specifies type E.

12 CONCLUSION

As future research illuminates the players (RNAs and proteins) mediating the many non-Mendelian phenomena that have been described in ciliates, more connections to known epigenetic mechanisms will become apparent. Clearly, further studies of DNA rearrangements will contribute new insights into RNA-based inheritance. The discovery of RNAi and other epigenetic mechanisms have largely come

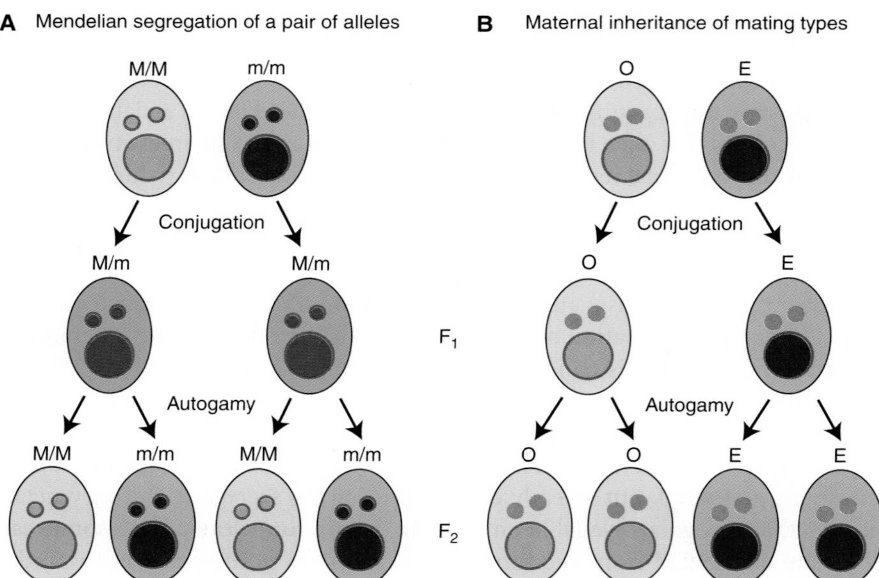

Figure 10. Epigenetic inheritance of mating-types in *Paramecium*. (*A*) Expected Mendelian segregation of genotypes and phenotypes. M and m alleles express yellow and blue phenotype, respectively. M/m F_1 heterozygote conjugation progeny shows green (intermediate) phenotype. F_2 homozygote progeny by autogamy shows either yellow or blue phenotype. (*B*) Mating type (O or E) is irreversibly determined during the development of the somatic macronucleus (large circle) from the totipotent germline micronucleus (small circle); however, the parental macronucleus directs differentiation of each exconjugant toward maintaining the existing mating type.

Cite as *Cold Spring Harb Perspect Biol* doi: 10.1101/cshperspect.a017764

from efforts of researchers working in many model organisms aiming to explain examples of unexpected phenotypes or inheritance patterns. Many homology-dependent effects, including meiotic silencing in fungi and paramutation in plants (see Ch. 9 [Allshire and Ekwall 2014]; Ch. 10 [Aramayo and Selker 2013]; Ch. 13 [Pikaard and Mittelsten Scheid 2014]), remain incompletely understood. It will be the combined insight provided by future experiments in all eukaryotes, including ciliates, which will expose the full scope of epigenetic processes.

REFERENCES

**Reference is also in this book.*

Ahmad K, Henikoff S. 2002. The histone variant H3.3 marks active chromatin by replication-independent nucleosome assembly. *Mol Cell* **9:** 1191–1200.

Allis CD, Gorovsky MA. 1981. Histone phosphorylation in macro- and micronuclei of *Tetrahymena thermophila. Biochemistry* **20:** 3828–3833.

Allis CD, Glover CV, Bowen JK, Gorovsky MA. 1980. Histone variants specific to the transcriptionally active, amitotically dividing macronucleus of the unicellular eucaryote, *Tetrahymena thermophila. Cell* **20:** 609–617.

Allis CD, Allen RL, Wiggins JC, Chicoine LG, Richman R. 1984. Proteolytic processing of h1-like histones in chromatin: A physiologically and developmentally regulated event in Tetrahymena micronuclei. *J Cell Biol* **99:** 1669–1677.

* Allshire R, Ekwall K. 2014. Epigenetics in *Saccharomyces pombe. Cold Spring Harb Perspect Biol* doi: 10.1101/cshperspect.a018770.

* Aramayo R, Selker EU. 2013. *Neurospora crassa*, a model system for epigenetics research. *Cold Spring Harb Perspect Biol* **5:** a017921.

Arnaiz O, Mathy N, Baudry C, Malinsky S, Aury JM, Wilkes CD, Garnier O, Labadie K, Lauderdale BE, Le Mouel A, et al. 2012. The *Paramecium* germline genome provides a niche for intragenic parasitic DNA: Evolutionary dynamics of internal eliminated sequences. *PLoS Genet* **8:** e1002984.

Aronica L, Bednenko J, Noto T, DeSouza LV, Siu KW, Loidl J, Pearlman RE, Gorovsky MA, Mochizuki K. 2008. Study of an RNA helicase implicates small RNA-noncoding RNA interactions in programmed DNA elimination in *Tetrahymena. Genes Dev* **22:** 2228–2241.

Baird SE, Fino GM, Tausta SL, Klobutcher LA. 1989. Micronuclear genome organization in *Euplotes crassus*: A transposonlike element is removed during macronuclear development. *Mol Cell Biol* **9:** 3793–3807.

Baudry C, Malinsky S, Restituito M, Kapusta A, Rosa S, Meyer E, Betermier M. 2009. PiggyMac, a domesticated piggyBac transposase involved in programmed genome rearrangements in the ciliate *Paramecium tetraurelia. Genes Dev* **23:** 2478–2483.

Betermier M. 2004. Large-scale genome remodelling by the developmentally programmed elimination of germ line sequences in the ciliate *Paramecium. Res Microbiol* **155:** 399–408.

Bouhouche K, Gout JF, Kapusta A, Betermier M, Meyer E. 2011. Functional specialization of Piwi proteins in *Paramecium tetraurelia* from post-transcriptional gene silencing to genome remodelling. *Nucleic Acids Res* **39:** 4249–4264.

Brownell JE, Allis CD. 1995. An activity gel assay detects a single, catalytically active histone acetyltransferase subunit in *Tetrahymena* macronuclei. *Proc Natl Acad Sci* **92:** 6364–6368.

Brownell JE, Zhou J, Ranalli T, Kobayashi R, Edmondson DG, Roth SY, Allis CD. 1996. *Tetrahymena* histone acetyltransferase: A homolog to yeast Gcn5p linking histone acetylation to gene activation. *Cell* **84:** 843–851.

* Busslinger M, Tarakhovsky A. 2014. Epigenetic control of immunity. *Cold Spring Harb Perspect Biol* **6:** a019307.

Cassidy-Hanley D, Bisharyan Y, Fridman V, Gerber J, Lin C, Orias E, Orias JD, Ryder H, Vong L, Hamilton EP. 2005. Genome-wide characterization of *Tetrahymena thermophila* chromosome breakage sites. II. Physical and genetic mapping. *Genetics* **170:** 1623–1631.

Cervantes MD, Xi X, Vermaak D, Yao MC, Malik HS. 2006. The CNA1 histone of the ciliate *Tetrahymena thermophila* is essential for chromosome segregation in the germline micronucleus. *Mol Biol Cell* **17:** 485–497.

Chalker DL, Yao MC. 1996. Non-Mendelian, heritable blocks to DNA rearrangement are induced by loading the somatic nucleus of *Tetrahymena thermophila* with germ line-limited DNA. *Mol Cell Biol* **16:** 3658–3667.

Chalker DL, Yao MC. 2001. Nongenic, bidirectional transcription precedes and may promote developmental DNA deletion in *Tetrahymena thermophila. Genes Dev* **15:** 1287–1298.

Chalker DL, Fuller P, Yao MC. 2005. Communication between parental and developing genomes during tetrahymena nuclear differentiation is likely mediated by homologous RNAs. *Genetics* **169:** 149–160.

Cheng CY, Vogt A, Mochizuki K, Yao MC. 2010. A domesticated piggyBac transposase plays key roles in heterochromatin dynamics and DNA cleavage during programmed DNA deletion in *Tetrahymena thermophila. Mol Biol Cell* **21:** 1753–1762.

* Cheng X. 2014. Structural and functional coordination of DNA and histone methylation. *Cold Spring Harb Perspect Biol* **6:** a018747.

Coyne RS, Chalker DL, Yao MC. 1996. Genome downsizing during ciliate development: Nuclear division of labor through chromosome restructuring. *Annu Rev Genet* **30:** 557–578.

Coyne RS, Thiagarajan M, Jones KM, Wortman JR, Tallon LJ, Haas BJ, Cassidy-Hanley DM, Wiley EA, Smith JJ, Collins K, et al. 2008. Refined annotation and assembly of the *Tetrahymena thermophila* genome sequence through EST analysis, comparative genomic hybridization, and targeted gap closure. *BMC Genomics* **9:** 562.

Cui B, Gorovsky MA. 2006. Centromeric histone H3 is essential for vegetative cell division and for DNA elimination during conjugation in *Tetrahymena thermophila. Mol Cell Biol* **26:** 4499–4510.

Cui B, Liu Y, Gorovsky MA. 2006. Deposition and function of histone H3 variants in *Tetrahymena thermophila. Mol Cell Biol* **26:** 7719–7730.

Davis MC, Ward JG, Herrick G, Allis CD. 1992. Programmed nuclear death: Apoptotic-like degradation of specific nuclei in conjugating *Tetrahymena. Dev Biol* **154:** 419–432.

Dou Y, Gorovsky MA. 2000. Phosphorylation of linker histone H1 regulates gene expression in vivo by creating a charge patch. *Mol Cell* **6:** 225–231.

Dou Y, Gorovsky MA. 2002. Regulation of transcription by H1 phosphorylation in *Tetrahymena* is position independent and requires clustered sites. *Proc Natl Acad Sci* **99:** 6142–6146.

Dou Y, Mizzen CA, Abrams M, Allis CD, Gorovsky MA. 1999. Phosphorylation of linker histone H1 regulates gene expression in vivo by mimicking H1 removal. *Mol Cell* **4:** 641–647.

Duharcourt S, Butler A, Meyer E. 1995. Epigenetic self-regulation of developmental excision of an internal eliminated sequence on *Paramecium tetraurelia. Genes Dev* **9:** 2065–2077.

Duharcourt S, Keller AM, Meyer E. 1998. Homology-dependent maternal inhibition of developmental excision of internal eliminated sequences in *Paramecium tetraurelia. Mol Cell Biol* **18:** 7075–7085.

Eisen JA, Coyne RS, Wu M, Wu D, Thiagarajan M, Wortman JR, Badger JH, Ren Q, Amedeo P, Jones KM, et al. 2006. Macronuclear genome sequence of the ciliate *Tetrahymena thermophila*, a model eukaryote. *PLoS Biol* **4:** e286.

Epstein LM, Forney JD. 1984. Mendelian and non-Mendelian mutations affecting surface antigen expression in *Paramecium tetraurelia. Mol Cell Biol* **4:** 1583–1590.

Fan Q, Yao M-C. 1996. New telomere formation coupled with site-specific chromosome breakage in *Tetrahymena thermophila*. *Mol Cell Biol* **16:** 1267–1274.

Fang W, Wang X, Bracht JR, Nowacki M, Landweber LF. 2012. Piwi-interacting RNAs protect DNA against loss during *Oxytricha* genome rearrangement. *Cell* **151:** 1243–1255.

Fass JN, Joshi NA, Couvillion MT, Bowen J, Gorovsky MA, Hamilton EP, Orias E, Hong K, Coyne RS, Eisen JA, et al. 2011. Genome-scale analysis of programmed DNA elimination sites in *Tetrahymena thermophila*. *G3 (Bethesda)* **1:** 515–522.

* Felsenfeld G. 2014. A brief history of epigenetics. *Cold Spring Harb Perspect Biol* doi: 10.1101/cshperspect.a018200.

Fillingham JS, Thing TA, Vythilingum N, Keuroghlian A, Bruno D, Golding GB, Pearlman RE. 2004. A non-long terminal repeat retrotransposon family is restricted to the germ line micronucleus of the ciliated protozoan *Tetrahymena thermophila*. *Eukaryot Cell* **3:** 157–169.

Forney JD, Blackburn EH. 1988. Developmentally controlled telomere addition in wild-type and mutant *Paramecia*. *Mol Cell Biol* **8:** 251–258.

Forney JD, Yantiri F, Mikami K. 1996. Developmentally controlled rearrangement of surface protein genes in *Paramecium tetraurelia*. *J Eukaryot Microbiol* **43:** 462–467.

Galvani A, Sperling L. 2001. Transgene-mediated post-transcriptional gene silencing is inhibited by 3' non-coding sequences in *Paramecium*. *Nucleic Acids Res* **29:** 4387–4394.

Galvani A, Sperling L. 2002. RNA interference by feeding in *Paramecium*. *Trends Genet* **18:** 11–12.

Garnier O, Serrano V, Duharcourt S, Meyer E. 2004. RNA-mediated programming of developmental genome rearrangements in *Paramecium tetraurelia*. *Mol Cell Biol* **24:** 7370–7379.

Ghildiyal M, Zamore PD. 2009. Small silencing RNAs: An expanding universe. *Nat Rev Genet* **10:** 94–108.

Godiska R, James C, Yao MC. 1993. A distant 10-bp sequence specifies the boundaries of a programmed DNA deletion in *Tetrahymena*. *Genes Dev* **7:** 2357–2365.

Gratias A, Betermier M. 2003. Processing of double-strand breaks is involved in the precise excision of paramecium internal eliminated sequences. *Mol Cell Biol* **23:** 7152–7162.

* Grossniklaus U, Paro R. 2014. Transcriptional silencing by *Polycomb*-group proteins. *Cold Spring Harb Perspect Biol* **6:** a019331.

Hamilton E, Bruns P, Lin C, Merriam V, Orias E, Vong L, Cassidy-Hanley D. 2005. Genome-wide characterization of *Tetrahymena thermophila* chromosome breakage sites. I. Cloning and identification of functional sites. *Genetics* **170:** 1611–1621.

Harumoto T. 1986. Induced change in a non-Mendelian determinant by transplantation of macronucleoplasm in *Paramecium tetraurelia*. *Mol Cell Biol* **6:** 3498–3501.

Harwood J. 1985. The erratic career of cytoplasmic inheritance. *Trends Genet* **1:** 298–300.

Hayashi T, Hayashi H, Fusauchi Y, Iwai K. 1984. *Tetrahymena* histone H3. Purification and two variant sequences. *J Biochem (Tokyo)* **95:** 1741–1749.

Hayashi T, Hayashi H, Iwai K. 1987. *Tetrahymena* histone H1. Isolation and amino acid sequence lacking the central hydrophobic domain conserved in other H1 histones. *J Biochem (Tokyo)* **102:** 369–376.

* Henikoff S, Smith M. 2014. Histone variants and epigenetics. *Cold Spring Harb Perspect Biol* doi: 10.1101/cshperspect.a019364.

Herrick G, Cartinhour S, Dawson D, Ang D, Sheets R, Lee A, Williams K. 1985. Mobile elements bounded by C4A4 telomeric repeats in *Oxytricha fallax*. *Cell* **43:** 759–768.

Howard-Till RA, Yao MC. 2007. Tudor nuclease genes and programmed DNA rearrangements in *Tetrahymena thermophila*. *Eukaryot Cell* **6:** 1795–1804.

Jahn CL, Klobutcher LA. 2002. Genome remodeling in ciliated protozoa. *Annu Rev Microbiol* **56:** 489–520.

Jessop-Murray H, Martin LD, Gilley D, Preer JR, Polisky B. 1991. Permanent rescue of a non-Mendelian mutation of *Paramecium* by microinjection of specific DNA sequences. *Genetics* **129:** 727–734.

Klobutcher LA. 1999. Characterization of in vivo developmental chromosome fragmentation intermediates in *E. crassus*. *Mol Cell* **4:** 695–704.

Klobutcher LA, Herrick G. 1995. Consensus inverted terminal repeat sequence of *Paramecium* IESs: Resemblance to termini of Tc1-related and *Euplotes* Tec transposons. *Nucl Acids Res* **23:** 2006–2013.

Klobutcher LA, Herrick G. 1997. Developmental genome reorganization in ciliated protozoa: The transposon link. *Prog Nucleic Acid Res Mol Biol* **56:** 1–62.

Kobayashi S, Koizumi S. 1990. Characterization of non-Mendelian and Mendelian mutant strains by micronuclear transplantation in *Paramecium tetraurelia*. *J Protozool* **37:** 489–492.

Koizumi S, Kobayashi S. 1989. Microinjection of plasmid DNA encoding the A surface antigen of *Paramecium tetraurelia* restores the ability to regenerate a wild-type macronucleus. *Mol Cell Biol* **9:** 4398–4401.

Lee SR, Collins K. 2006. Two classes of endogenous small RNAs in *Tetrahymena thermophila*. *Genes Dev* **20:** 28–33.

Lee SR, Collins K. 2007. Physical and functional coupling of RNA-dependent RNA polymerase and Dicer in the biogenesis of endogenous siRNAs. *Nat Struct Mol Biol* **14:** 604–610.

Le Mouël A, Butler A, Caron F, Meyer E. 2003. Developmentally regulated chromosome fragmentation linked to imprecise elimination of repeated sequences in *Paramecia*. *Eukaryotic Cell* **2:** 1076–1090.

Lepere G, Betermier M, Meyer E, Duharcourt S. 2008. Maternal noncoding transcripts antagonize the targeting of DNA elimination by scanRNAs in *Paramecium tetraurelia*. *Genes Dev* **22:** 1501–1512.

Lepere G, Nowacki M, Serrano V, Gout JF, Guglielmi G, Duharcourt S, Meyer E. 2009. Silencing-associated and meiosis-specific small RNA pathways in *Paramecium tetraurelia*. *Nucleic Acids Res* **37:** 903–915.

Liu Y, Mochizuki K, Gorovsky MA. 2004. Histone H3 lysine 9 methylation is required for DNA elimination in developing macronuclei in *Tetrahymena*. *Proc Natl Acad Sci* **101:** 1679–1684.

Liu Y, Song X, Gorovsky MA, Karrer KM. 2005. Elimination of foreign DNA during somatic differentiation in *Tetrahymena thermophila* shows position effect and is dosage dependent. *Eukaryot Cell* **4:** 421–431.

Liu Y, Taverna SD, Muratore TL, Shabanowitz J, Hunt DF, Allis CD. 2007. RNAi-dependent H3K27 methylation is required for heterochromatin formation and DNA elimination in *Tetrahymena*. *Genes Dev* **21:** 1530–1545.

Malone CD, Anderson AM, Motl JA, Rexer CH, Chalker DL. 2005. Germ line transcripts are processed by a Dicer-like protein that is essential for developmentally programmed genome rearrangements of *Tetrahymena thermophila*. *Mol Cell Biol* **25:** 9151–9164.

* Martienssen R, Moazed D. 2014. RNAi and heterochromatin assembly. *Cold Spring Harb Perspect Biol* doi: 10.1101/cshperspect.a019323.

Martindale DW, Allis CD, Bruns P. 1982. Conjugation in *Tetrahymena thermophila*: A temporal analysis of cytological stages. *Exp Cell Res* **140:** 227–236.

Matzke MA, Birchler JA. 2005. RNAi-mediated pathways in the nucleus. *Nat Rev Genet* **6:** 24–35.

Meyer E, Keller AM. 1996. A Mendelian mutation affecting mating-type determination also affects developmental genomic rearrangements in *Paramecium tetraurelia*. *Genetics* **143:** 191–202.

Mitra R, Fain-Thornton J, Craig NL. 2008. piggyBac can bypass DNA synthesis during cut and paste transposition. *Embo J* **27:** 1097–1109.

Mizzen CA, Dou Y, Liu Y, Cook RG, Gorovsky MA, Allis CD. 1999. Identification and mutation of phosphorylation sites in a linker histone. Phosphorylation of macronuclear H1 is not essential for viability in *Tetrahymena*. *J Biol Chem* **274:** 14533–14536.

Mochizuki K, Gorovsky MA. 2004. Small RNAs in genome rearrangement in *Tetrahymena*. *Curr Opin Genet Dev* **14:** 181–187.

Mochizuki K, Gorovsky MA. 2005. A Dicer-like protein in *Tetrahymena* has distinct functions in genome rearrangement, chromosome segregation, and meiotic prophase. *Genes Dev* **19:** 77–89.

Mochizuki K, Fine NA, Fujisawa T, Gorovsky MA. 2002. Analysis of a piwi-related gene implicates small RNAs in genome rearrangement in *Tetrahymena*. *Cell* **110:** 689–699.

Nowacki M, Zagorski-Ostoja W, Meyer E. 2005. Nowa1p and Nowa2p: Novel putative RNA binding proteins involved in trans-nuclear crosstalk in *Paramecium tetraurelia*. *Curr Biol* **15:** 1616–1628.

Nowacki M, Vijayan V, Zhou Y, Schotanus K, Doak TG, Landweber LF. 2008. RNA-mediated epigenetic programming of a genome-rearrangement pathway. *Nature* **451:** 153–158.

Nowacki M, Higgins BP, Maquilan GM, Swart EC, Doak TG, Landweber LF. 2009. A functional role for transposases in a large eukaryotic genome. *Science* **324:** 935–938.

Nowacki M, Haye JE, Fang W, Vijayan V, Landweber LF. 2010. RNA-mediated epigenetic regulation of DNA copy number. *Proc Natl Acad Sci* **107:** 22140–22144.

Paschka AG, Jönsson F, Maier V, Möllenbeck M, Paeschke K, Postberg J, Rupprecht S, Lipps HJ. 2003. The use of RNAi to analyze gene function in spirotichous ciliates. *Eur J Protistol* **39:** 449–454.

* Patel DJ. 2014. A structural perspective on readout of epigenetic histone and DNA methylation marks. *Cold Spring Harb Perspect Biol* doi: 10.1101/cshperspect.a018754.

Philippe H, Germot A, Moreira D. 2000. The new phylogeny of eukaryotes. *Curr Opin Genet Dev* **10:** 596–601.

* Pikaard C, Mittelsten Scheid O. 2014. Epigenetic regulation in plants. *Cold Spring Harb Perspect Biol* doi: 10.1101/cshperspect.a019315.

Prescott DM. 1999. The evolutionary scrambling and developmental unscrambling of germline genes in hypotrichous ciliates. *Nucleic Acids Res* **27:** 1243–1250.

Prescott DM, Ehrenfeucht A, Rozenberg G. 2003. Template-guided recombination for IES elimination and unscrambling of genes in stichotrichous ciliates. *J Theor Biol* **222:** 323–330.

Ray C. 1956. Meiosis and nuclear behavior in *Tetrahymena pyriformis**. *J Eukaryot Microbiol* **3:** 88–96.

Ruiz F, Vayssie L, Klotz C, Sperling L, Madeddu L. 1998. Homology-dependent gene silencing in *Paramecium*. *Mol Biol Cell* **9:** 931–943.

Saveliev SV, Cox MM. 1995. Transient DNA breaks associated with programmed genomic deletion events in conjugating cells of *Tetrahymena thermophila*. *Genes Dev* **9:** 248–255.

Saveliev SV, Cox MM. 1996. Developmentally programmed DNA deletion in *Tetrahymena thermophila* by a transposition-like reaction pathway. *EMBO J* **15:** 2858–2869.

Schoeberl UE, Kurth HM, Noto T, Mochizuki K. 2012. Biased transcription and selective degradation of small RNAs shape the pattern of DNA elimination in *Tetrahymena*. *Genes Dev* **26:** 1729–1742.

Scott JM, Mikami K, Leeck CL, Forney JD. 1994. Non-Mendelian inheritance of macronuclear mutations is gene specific in *Paramecium tetraurelia*. *Mol Cell Biol* **14:** 2479–2484.

Shen X, Gorovsky MA. 1996. Linker histone H1 regulates specific gene expression but not global transcription in vivo. *Cell* **86:** 475–483.

Shen X, Yu L, Weir JW, Gorovsky MA. 1995. Linker histones are not essential and affect chromatin condensation in vivo. *Cell* **82:** 47–56.

Sonneborn TM. 1937. Sex, sex inheritance and sex determination in *Paramecium aurelia*. *Proc Natl Acad Sci* **23:** 378–385.

Sonneborn TM. 1975. *Paramecium aurelia*. In *Handbook of genetics* (ed. King R), pp. 469–594. Plenum, New York.

Sonneborn TM. 1977. Genetics of cellular differentiation: stable nuclear differentiation in eucaryotic unicells. *Annu Rev Genet* **11:** 349–367.

Strahl BD, Ohba R, Cook RG, Allis CD. 1999. Methylation of histone H3 at lysine 4 is highly conserved and correlates with transcriptionally active nuclei in *Tetrahymena*. *Proc Natl Acad Sci* **96:** 14967–14972.

Taverna SD, Coyne RS, Allis CD. 2002. Methylation of histone h3 at lysine 9 targets programmed DNA elimination in *Tetrahymena*. *Cell* **110:** 701–711.

Taverna SD, Ueberheide BM, Liu Y, Tackett AJ, Diaz RL, Shabanowitz J, Chait BT, Hunt DF, Allis CD. 2007. Long-distance combinatorial linkage between methylation and acetylation on histone H3 N termini. *Proc Natl Acad Sci* **104:** 2086–2091.

Timmons L, Fire A. 1998. Specific interference by ingested dsRNA. *Nature* **395:** 854.

Timmons L, Court DL, Fire A. 2001. Ingestion of bacterially expressed dsRNAs can produce specific and potent genetic interference in *Caenorhabditis elegans*. *Gene* **263:** 103–112.

Vavra KJ, Colavito-Shepanski M, Gorovsky MA. 1982. Histone acetylation and the deoxyribonuclease I sensitivity of the *Tetrahymena* ribosomal gene. *Biochemistry* **21:** 1772–1781.

Waterhouse PM, Wang MB, Lough T. 2001. Gene silencing as an adaptive defence against viruses. *Nature* **411:** 834–842.

Wei Y, Mizzen CA, Cook RG, Gorovsky MA, Allis CD. 1998. Phosphorylation of histone H3 at serine 10 is correlated with chromosome condensation during mitosis and meiosis in *Tetrahymena*. *Proc Natl Acad Sci* **95:** 7480–7484.

Wei Y, Yu L, Bowen J, Gorovsky MA, Allis CD. 1999. Phosphorylation of histone H3 is required for proper chromosome condensation and segregation. *Cell* **97:** 99–109.

Wu M, Allis CD, Richman R, Cook RG, Gorovsky MA. 1986. An intervening sequence in an unusual histone H1 gene of *Tetrahymena thermophila*. *Proc Natl Acad Sci* **83:** 8674–8678.

Wuitschick JD, Gershan JA, Lochowicz AJ, Li S, Karrer KM. 2002. A novel family of mobile genetic elements is limited to the germline genome in *Tetrahymena thermophila*. *Nucleic Acids Res* **30:** 2524–2537.

Yao MC, Zheng K, Yao CH. 1987. A conserved nucleotide sequence at the sites of developmentally regulated chromosomal breakage in *Tetrahymena*. *Cell* **48:** 779–788.

Yao MC, Fuller P, Xi X. 2003. Programmed DNA deletion as an RNA-guided system of genome defense. *Science* **300:** 1581–1584.

You Y, Aufderheide K, Morand J, Rodkey K, Forney J. 1991. Macronuclear transformation with specific DNA fragments controls the content of the new macronuclear genome in *Paramecium tetraurelia*. *Mol Cell Biol* **11:** 1133–1137.

Yu L, Gorovsky MA. 1997. Constitutive expression, not a particular primary sequence, is the important feature of the H3 replacement variant hv2 in *Tetrahymena thermophila*. *Mol Cell Biol* **17:** 6303–6310.

Zahler AM, Neeb ZT, Lin A, Katzman S. 2012. Mating of the stichotrichous ciliate *Oxytricha trifallax* induces production of a class of 27 nt small RNAs derived from the parental macronucleus. *PLoS One* **7:** e42371.

CHAPTER 12

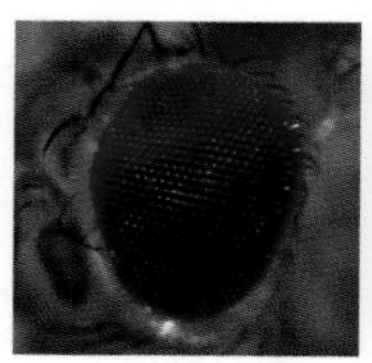

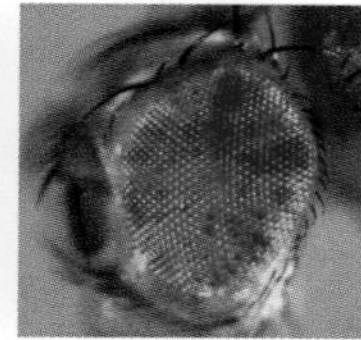

Position-Effect Variegation, Heterochromatin Formation, and Gene Silencing in *Drosophila*

Sarah C.R. Elgin[1] and Gunter Reuter[2]

[1]Department of Biology, CB-1137, Washington University, St. Louis, Missouri 63130; [2]Institute of Biology, Developmental Genetics, Martin Luther University Halle, D-06120 Halle, Germany

Correspondence: selgin@biology.wustl.edu

SUMMARY

Position-effect variegation (PEV) results when a gene normally in euchromatin is juxtaposed with heterochromatin by rearrangement or transposition. When heterochromatin packaging spreads across the heterochromatin/euchromatin border, it causes transcriptional silencing in a stochastic pattern. PEV is intensely studied in *Drosophila* using the *white* gene. Screens for dominant mutations that suppress or enhance *white* variegation have identified many conserved epigenetic factors, including the histone H3 lysine 9 methyltransferase SU(VAR)3-9. Heterochromatin protein HP1a binds H3K9me2/3 and interacts with SU(VAR)3-9, creating a core memory system. Genetic, molecular, and biochemical analysis of PEV in *Drosophila* has contributed many key findings concerning establishment and maintenance of heterochromatin with concomitant gene silencing.

Outline

Cite as *Cold Spring Harb Perspect Biol* doi: 10.1101/cshperspect.a017780

OVERVIEW

Genes that are abnormally juxtaposed with heterochromatin, either by rearrangement or transposition, show a variegating phenotype. This is a result of the gene being silenced in some of the cells in which it is normally active. Because the change is caused by a change in the position of the gene in the genome, rather than a change in the gene itself, this phenomenon is termed "position-effect variegation" (PEV). The silencing that occurs in PEV can be attributed to the packaging of the reporter gene in a heterochromatic form, indicating that endogenous heterochromatin formation, once initiated, can spread to encompass nearby genes. Genetic, cytological, and biochemical analyses are all possible in *Drosophila melanogaster*. In this chapter, we will show how these different approaches have converged to identify many contributors to this system, leading to characterization of both structural proteins and modifying enzymes that play key roles in establishing and maintaining heterochromatin.

Heterochromatin formation depends critically on methylation of histone H3 at lysine 9 (H3K9me2/3), with concomitant association of heterochromatin protein 1 (HP1a) and other interacting proteins, including H3K9 methyltransferases (HKMTs); the multiple interactions of these proteins are required for the spreading and maintenance of heterochromatin. Targeting of heterochromatin formation to particular regions of the genome appears to involve multiple mechanisms, from satellite DNA-specific binding proteins to utilization of the RNA interference (RNAi) machinery. Although heterochromatic regions (pericentric regions, telomeres, the Y chromosome, and the small fourth chromosome) share a common biochemistry, each is distinct, and each is complex in different ways. Heterochromatin in *Drosophila* is gene poor, but it is not devoid of genes, and counterintuitively, those genes that reside in heterochromatin are often dependent on this environment for full expression. A complete understanding of heterochromatin formation and maintenance (including targeting and spreading) will need to include an explanation for the varying responses of different genes to this chromatin environment.

Cite as *Cold Spring Harb Perspect Biol* doi: 10.1101/cshperspect.a017780

1 GENES ABNORMALLY JUXTAPOSED WITH HETEROCHROMATIN SHOW A VARIEGATING PHENOTYPE

Large segments of the eukaryotic genome are packaged in a permanently inactive form of chromatin termed constitutive heterochromatin. This chromatin fraction was originally identified as that portion of the genome that remains condensed and deeply staining (heteropycnotic) in interphase; such material is generally associated with the telomeres and pericentric regions of the chromosomes. Heterochromatic regions tend to be late replicating and show little or no meiotic recombination. These domains are dominated by repetitious DNA sequences (~30%–80%), both tandem repeats of short motifs (known as "satellite" DNA), and remnants of transposable elements (TEs), including DNA transposons and retroviruses. Although gene poor, these domains are not devoid of genes, and intriguingly, those genes that are present frequently are dependent on that environment for optimal expression. About one third of the *Drosophila* genome is considered heterochromatic, including the entire Y chromosome, most of the small fourth chromosome, the pericentric region that covers 40% of the X chromosome, and pericentric regions that cover 20% of the large autosomes (Smith et al. 2007). During the last few decades we have learned a great deal about the biochemistry of heterochromatin, and much of that understanding derives from our studies with *Drosophila* (see Schotta et al. 2003; Schulze and Wallrath 2007; Girton and Johansen 2008; Eissenberg and Reuter 2009 for prior reviews).

One of the first mutations identified in *D. melanogaster* was *white*, a mutation that results in a fly with a white eye, rather than the characteristic red pigmentation. Using X rays as a mutagen, Muller (1930) observed an unusual phenotype, in which the eye was variegating, with some patches of red and some patches of white facets (Fig. 1A). This phenotype suggested that the *white* gene itself was not damaged—after all, some facets remained red, and flies with entirely red eyes could be recovered as revertants, again using X rays as the mutagen. However, the *white* gene had clearly been silenced in some of the cells in which it is normally expressed. Subsequent examination of the polytene chromosomes indicated that such phenotypes are the consequence of an inversion or rearrangement with one breakpoint within the pericentric heterochromatin, and one breakpoint adjacent to the *white* gene (see Fig. 1A). Because the variegating phenotype is caused by a change in the position of the gene within the chromosome, this phenomenon is referred to as PEV. In *Drosophila*, virtually every gene that has been examined in an appropriate rearrangement has been shown to variegate, and rearrangements involving the pericentric heterochromatin of any chromosome can lead to PEV (Girton and Johansen, 2008). PEV has subsequently been observed in a variety of organisms, including yeasts (see Ch. 9 [Allshire and Ekwall 2014]), flies, and mammals (see Ch. 14 [Blewitt and Whitelaw 2013]; Ch. 25 [Brockdorff and Turner 2014]), but has been used as a tool to study heterochromatin formation primarily in *Drosophila*.

PEV indicates that such rearrangements allow packaging of the newly positioned gene into a heterochromatic configuration, and suggests that this is the consequence of heterochromatin "spreading" along the chromosome from the adjacent constitutive heterochromatin region. Apparently, the rearrangement has removed a normally existing barrier or buffer zone. The consequence is an altered packaging with concomitant silencing of genes normally packaged in a euchromatic form. Visual inspection of the polytene chromosomes of larvae carrying such a rearrangement shows that the region carrying the reporter gene is packaged in a dense block of heterochromatin only in the cells in which the gene is inactive (Zhimulev et al. 1986). Patterns of variegated expression, observed as a consequence of rearrangement of *white*, vary in the number of pigmented cells, the size of the pigmented patches, and the level of pigment in the two different cell types observed, one with a high level of expression, and one with a low level or no expression (Fig. 1A). In a system using an inducible *lac-Z* gene as a reporter, investigators observed that silencing occurs in early embryogenesis, just after heterochromatin is first observed cytologically, and is epigenetically inherited in both somatic and germline lineages; the mosaic phenotype is determined during differentiation by variegated relaxation of silencing in third instar larvae (Lu et al. 1996). However, not all variegating genes remain silent until after differentiation, and the balance of factors leading to the on/off decision no doubt differs for different genes. (See Ashburner et al. 2005, Chapter 28 for a more detailed discussion.)

A fly line showing a PEV phenotype can be used to screen for dominant second site mutations that are either suppressors or enhancers of PEV. These second site mutations can be induced by chemical mutagens that cause point mutations or small insertions/deletions, but do not impact the chromosome rearrangement responsible for the PEV phenotype. A suppressor (denoted "suppressor of variegation," *Su(var)*) results in a loss of silencing, whereas enhancers (denoted "Enhancer of variegation," *E(var)*) result in an increase in silencing (Fig. 1B). About 150 loci have been identified from such screens, and of these ~30 modifiers of PEV have been studied in detail. Where the gene has been cloned and the product characterized, one generally finds a chromosomal protein or a modifier of a chromosomal protein (see Table 1). A small subset of these loci cause both a haplo-abnormal and an opposite triplo-

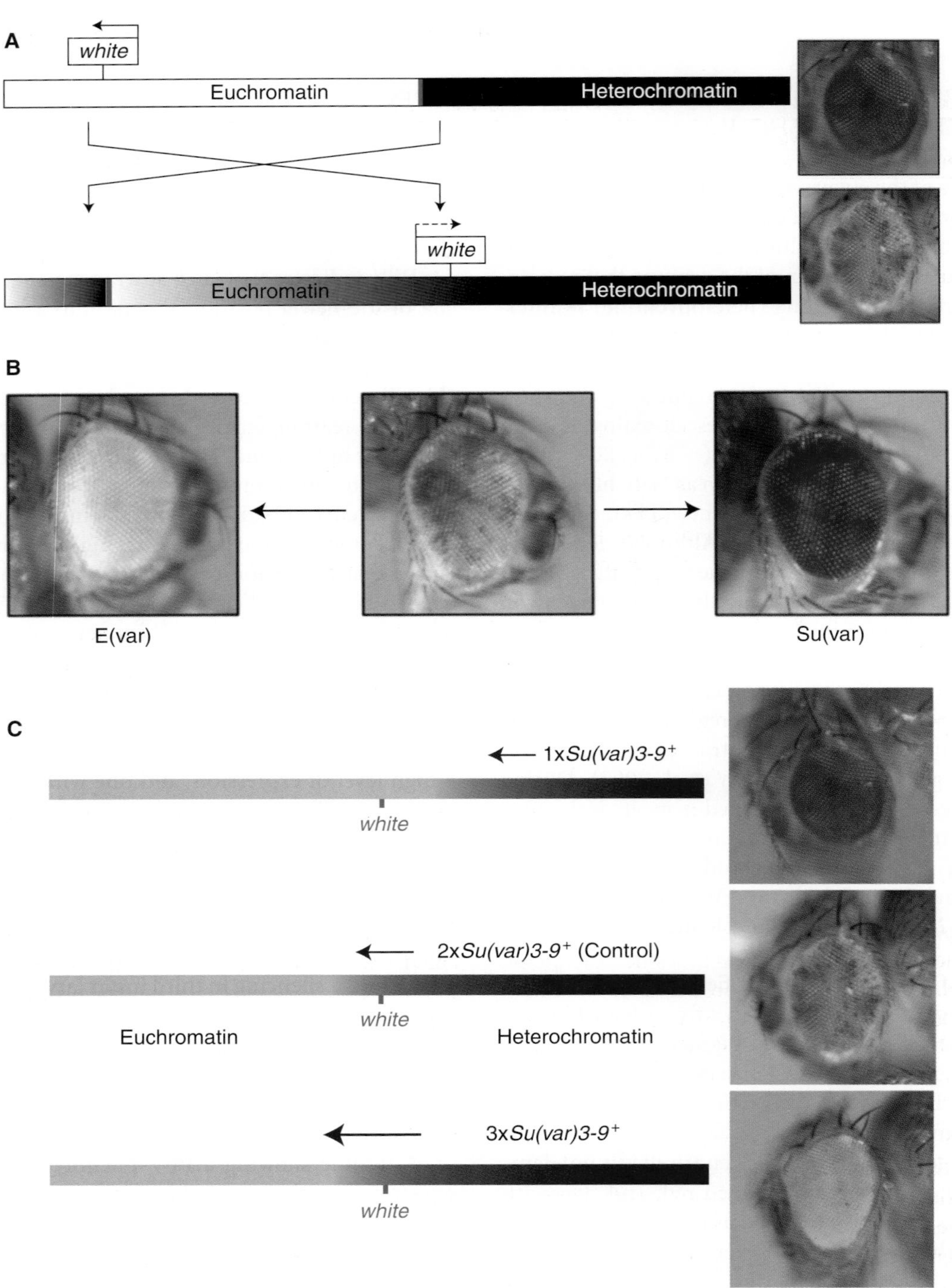

Figure 1. A schematic illustration of *white* variegation in the X chromosome inversion $In(1)w^{m4}$. (*A*) Rearrangement attributable to an X-ray-induced inversion places the *white* locus, normally located in the distal euchromatin (white bar) of the X chromosome (see *top* line), ~25 kb from a breakpoint in the pericentric heterochromatin (black bar; *bottom* rearranged line). Spreading of heterochromatin packaging into the euchromatic domain results in silencing (causing a white eye in this case); loss of silencing in some cells during differentiation results in a variegating phenotype (*bottom* line, *right*). (*B*) Given a variegating phenotype, screens for second site mutations can recover suppressors (*Su(var)*s) and enhancers (*E(var)*s) as described in the text. (*C*) Some *Su(var)* loci (e.g., *Su(var)3-9*, shown here) show an antipodal dosage-dependent effect, and are consequently thought to be structural proteins of heterochromatin. The presence of only one copy of the modifier gene results in less heterochromatin formation, and more expression from the reporter gene (suppression of PEV, *top* fly eye); conversely, the presence of three copies of such a modifier gene will drive more extensive heterochromatin formation, resulting in an enhancement of reporter gene silencing (enhancement of PEV, *bottom* fly eye).

Table 1. Genetically defined *Su(var)* and *E(var)* genes and their molecular function

Su(var)/E(var) gene	Cytological position	Molecular function, protein distribution, and phenotypic effects
Suv4-20 (*Su(var)*)	X; 1B13-14	Histone lysine methyltransferase (HKMT), histone H4K20 trimethylation
Su(z)5 (*Su(var)*)	2L; 21B2	*S*-adenosylmethionine synthetase
chm (*chameau*) (*Su(var)*)	2L; 27F3-4	Myst domain histone acetylase; suppresses PEV but enhances Polycomb-group mutations
Su(var)2-5 (HP1a)	2L; 28F2-3	Heterochromatin Protein 1 (HP1a), binding of di- and trimethyl H3K9; binding of SU(VAR)3-9
Su(var)2-HP2	2R; 51B6	Heterochromatin-associated protein, binds HP1a
Su(var)2-10	2R; 45A8-9	PIAS protein, negative regulators of JAK/STAT pathway
Su(var)3-64B (HDAC1 = RPD3)	3L; 64B12	Histone deacetylase HDAC1, deacetylation of H3K9
SuUR (*Su(var)*)	3L; 68A4	Suppresses heterochromatin underreplication; heterochromatin-associated protein
Su(var)3-1 (JIL1)	3L; 68A5-6	Antimorphic JIL1 mutations, carboxy-terminal protein truncations, blocking of heterochromatin spreading
Su(var)3-3	3L; 77A3	dLSD1, the H3K4me3 demethylase
Dom (*Domina*) (*Su(var)*)	3R; 86B1-2	Fork head winged-helix protein; heterochromatin-associated
Su(var)3-6	3R; 87B9-10	PP1 protein serine/threonine phosphatase
Su(var)3-7	3R; 87E3	Zinc-finger protein, heterochromatin-associated; interacts with HP1a and SU(VAR)3-9
Su(var)3-9	3R; 89E6-8	HKMT, histone H3-K9 methylation, heterochromatin-associated, interaction with HP1a
mod (*modulo*) (*Su(var)*)	3R; 100E3	DNA and RNA binding protein, phosphorylated Mod binds rRNA
E(var)3-64E/Ubp64^{Evar1}	3L; 64E5-6	Putative ubiquitin specific protease (Ubp46)
Trl (*trithorax-like*) (*E(var)*)	3L; 70F4	GAGA factor, binding of repetitive DNA sequences, abundant transcription factor
Mod(mdg4)/E(var)3-93D	3R; 93D7	Transcription regulator, more than 20 protein isoforms produced by *trans*-splicing
E(var)3-93E	3R; 93E9-F1	E2F transcription factor, haplo-enhancer, and triplo-suppressor

abnormal phenotype (i.e., if one copy of the gene results in suppression of PEV, three copies result in enhancement of PEV [Fig. 1C]). This suggests that the protein products of these genes play a structural role in heterochromatin, and that the spread of heterochromatic packaging can be driven by the dosage of these proteins in a stochastic manner (Locke et al. 1988). However, "spreading" is a complex process, not a simple linear continuum, and is most likely dependent on the organization of the region being silenced in addition to the adjacent heterochromatin mass (see Sec. 5).

The results observed on rearrangement of chromosomes suggest that a euchromatic gene inserted into a heterochromatic domain by transposition will also show a variegating phenotype. This has been found to be the case. The *P* element, a DNA transposon found in many strains of *Drosophila* in the wild, can be engineered for this purpose. A natural *P* element has distinctive inverted repeat sequences at each end, and codes for just one enzyme, the *P*-specific DNA transposase. Reporter constructs lacking the DNA transposase but containing other genes of interest can be inserted into the *Drosophila* genome in the presence of active transposase by coinjection into *Drosophila* embryos. A *P*-based TE, such as that shown (Fig. 2A) carrying an *hsp70*-driven copy of *white*, can be used in a fly with no endogenous copy of *white* to identify domains of heterochromatin. When the *P* element is inserted into euchromatin, the fly has a red eye. When this *P* is mobilized (by crossing in the gene encoding the transposase), approximately 1% of the lines recovered show a variegating eye phenotype. In situ hybridization shows that in these cases, the *P* element has jumped into the pericentric heterochromatin, the telomeres, the Y chromosome, or the small fourth chromosome (Wallrath and Elgin, 1995). This identification of heterochromatic domains is in agreement with earlier cytological studies.

The use of such *P* elements has allowed comparison of the packaging of the same reporter gene in heterochromatic and euchromatic environments. Heterochromatin is relatively resistant to cleavage by nucleases, whether nonspecific (e.g., DNase I) or specific (restriction enzymes), and is less accessible to other exogenous probes, such as *dam* methyltransferase. Analysis of the *hsp26* transgene (marked with a fragment of unique plant DNA; Fig. 2A) in both euchromatin and pericentric heterochromatin using micrococcal nuclease (MNase) reveals a shift to a more ordered nucleosome array, indicating that spacing of the nucleosomes in heterochromatin is more regular (Fig. 2B,C). The MNase cleavage fragments are well defined, suggesting a smaller MNase target than usual in the linker region. The ordered nucleosome array extends across the 5′ regulatory region of the gene, a shift that no doubt contributes to the observed loss of 5′ hypersensitive sites (HS sites) (Sun et al. 2001). Indeed, although the mechanism of silencing is as yet incompletely understood, there is abundant evidence of transcriptional repression in strongly variegating genes, which includes the

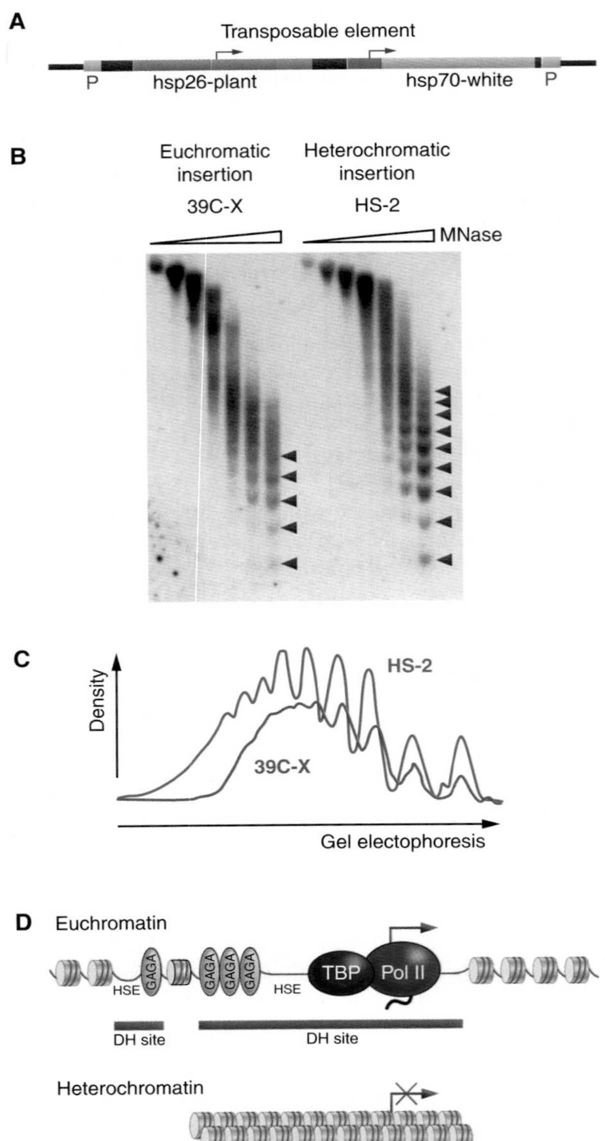

Figure 2. Heterochromatin is packaged into a regular nucleosome array. A TE such as that shown (*A*), carrying a marked copy of a heat shock gene for study and an *hsp70*-driven copy of *white* as a visual marker, can be used to study the same gene in different chromatin domains. (*B*) Nuclei from *Drosophila* lines carrying the transgene in a euchromatic domain (39C-X; red eye) and a heterochromatic domain (HS-2; variegating eye) were digested with increasing amounts of micrococcal nuclease (MNase), the DNA was purified and size-separated on an agarose gel, and the resulting Southern blot was hybridized with a probe unique to the transgene. Linker sites cleaved by MNase are marked with arrowheads. (*C*) Densitometer scans from the last lane of each sample are compared (*top to bottom* is left to right). An array of nine to 10 nucleosomes can be detected in heterochromatin (red line), compared with five to six in euchromatin (blue line), indicating more uniform spacing in the former case. (*D*) A diagrammatic representation of the results. DH site, deoxyribonuclease (DNase)-hypersensitive site; HSE, heat shock element. (*B*,*C*, Adapted from Sun et al. 2001, © American Society for Microbiology.)

loss of binding of TFIID and other transcription factors (Cryderman et al. 1999a). Heterochromatin structure apparently minimizes contact or slows down the processes of nuclear complexes that facilitate transcription, replication, recombination, etc. Heterochromatin packaging has also been found to be critical in maintaining genome integrity; *Su(var)* mutations can result in disorganized nucleoli, a substantial increase in extrachromosomal circular repeated DNAs, and other forms of DNA damage, potentially related to errors in replication (Peng and Karpen 2009). This packaging also appears to be important in minimizing productive transcription of TEs, helping to keep them immobilized, so protecting genome integrity (e.g., Wang and Elgin 2011).

2 SCREENS FOR SUPPRESSORS AND ENHANCERS OF PEV HAVE IDENTIFIED CHROMOSOMAL PROTEINS AND MODIFIERS OF CHROMOSOMAL PROTEINS

PEV can be modified by a variety of factors. The temperature of development and the amount of heterochromatin within the genome were the first factors shown to affect the extent of variegation. As a rule, an increase in developmental temperature (from 25°C to 29°C) results in suppression of variegation (loss of silencing), whereas lower temperatures (e.g., 18°C) cause enhancement of variegation (increase in silencing). Other changes in culture conditions that accelerate or slow the rate of development can have similar effects. Strong suppression is found in flies carrying an additional Y chromosome (XXY females and XYY males), whereas strong enhancement is shown in males without a Y chromosome (X0). In general, duplication of heterochromatic material suppresses whereas deletions of heterochromatic material enhance variegation. These observations suggest the titration of a fixed amount of key protein(s) required for heterochromatin packaging. One consequence is that Y chromosome polymorphisms that alter the amount of heterochromatic DNA in the genome can impact the expression of thousands of genes, presumably because of redistribution of key chromosomal proteins present in limited amounts. Interestingly, these Y polymorphisms have been shown to have a disproportionate influence on the expression of genes producing chromatin-associated protein products (Lemos et al. 2010).

The first mutations acting as a suppressor or enhancer of PEV were identified by Schultz (1950) and Spofford (1967). At present, approximately 150 such genes have been identified as causally connected with the onset and/or maintenance of the heterochromatic gene silencing seen in PEV. In most cases the modifying effect is dominant, and *Su(var)*/+ or *E(var)*/+ heterozygotes show a suppressed or enhanced PEV phenotype (Fig. 1B); not surprisingly, these mutations are often homozygous lethals. Efficient isolation and thorough genetic analysis of *Su(var)* and *E(var)* mutations has depended on the availability of an experimentally suitable PEV rearrangement. Of the many PEV rearrangements described (FlyBase 2012), one of the most useful lines for such experimental work is $In(1)w^{m4}$ (Muller 1930). This rearrangement variegates for *white*, a phenotype easily recognizable in the eye of adult flies, as shown in Figure 1. Penetrance of *white* variegation in appropriate w^{m4} stocks is 100%, so every fly in the starting stock shows an eye with a white variegated phenotype, although the degree of variegation can differ considerably from one individual to the next. Inactivation of the *white* gene does not affect viability or fertility, allowing unlimited work with flies homozygous for w^{m4}. Consequently, *white* has also been used as a reporter in *P* element constructs (such as that shown in Fig. 2A) used to examine the sensitivity of different heterochromatic domains to different modifiers.

In the w^{m4} rearrangement, the inversion results in juxtaposition of the *white* gene with heterochromatic material of the X chromosome, located at the distal border of the nucleolus organizer (Cooper 1959). This region contains tandem arrays of R1 type mobile elements; the heterochromatic breakpoint of $In(1)w^{m4}$ has been suggested to fall within an R1 repeat unit (Tartof et al. 1984). Phenotypic w^+ revertants of w^{m4} have been isolated after X-ray or EMS (ethane methyl sulfonate, a chemical mutagen) treatment. Analysis of a series of more than 50 of the w^+ revertant chromosomes indicates that all show reinversion or translocation of the *white* gene to a euchromatic neighborhood, suggesting that the heterochromatic material immediately flanking the breakpoint causes the inactivation of the *white* gene in w^{m4}. Most of the revertants show *white* variegation again if strong *E(var)* mutations are introduced, suggesting that some heterochromatic sequences remain associated with the *white* gene after relocation (Reuter et al. 1985), not surprising given that the breakpoint in the flanking DNA is randomly introduced. These studies implicate repetitious DNA (here R1, a retrotransposon) as a target for heterochromatin formation. In the fourth chromosome, *1360*, a remnant of a DNA transposon, has been implicated as a target (see Section 7). Available data suggest that many, but not all, TEs can be targets for heterochromatin formation (e.g., Riddle et al. 2008; Wang and Elgin 2011).

Most of the known PEV modifier mutations have been isolated using $In(1)w^{m4}$ or another reporter in a sensitized genetic background. For the isolation of dominant suppressor mutations, the test stock contains a dominant enhancer of PEV; consequently, the test stock has an almost entirely white eye, whereas the desired mutations (*Su(var)*s) result in a variegated or red eye. The converse is true for an *E(var)* screen; using a *Su(var)* mutation to generate a red eye in a variegating line, one screens for mutations that result in a variegated or white eye. More than 1 million flies have been inspected in different screens using this approach, and more than 140 *Su(var)* and 230 *E(var)* mutations have been isolated (Schotta et al. 2003). Mutations have been induced by EMS, by X-ray treatment, or by remobilization of *P* elements. Another set of *Su(var)* mutations has been isolated in a direct screen (no sensitization of the type described above) with w^{m4} (Sinclair et al. 1983). A screen using a *Df(1;f)* chromosome, which shows strong variegation for the *yellow* gene, a body color marker, resulted in the isolation of 70 PEV modifier mutations (Donaldson et al. 2002). In addition, screens for dominant modifiers of transposon reporter gene expression have identified several mutations with a *Su(var)* effect (Birchler et al. 1994). With modern technology, it is now possible to screen systematically for the impact of overproduction of proteins as well, and this strategy has identified many enhancers as well as suppressors of PEV (Schneiderman et al. 2010).

Altogether, these screens have identified approximately 500 dominant *Su(var)* and *E(var)* mutations. As noted above, based on the genetic analysis performed to date, the total number of *Su(var)* and *E(var)* genes can be estimated to be around 150. In naming these genes, the *Su(var)* and *E(var)* symbols are usually combined with numbers, indicating the chromosome where the mutation is located, the gene number, and the number of the allele. Thus $Su(var)3\text{-}9^{17}$ symbolizes allele 17 of the ninth *Su(var)* gene identified on the third chromosome. At present, only ca. 30 of the corresponding genes have been carefully mapped and alleles identified (Table 1). Dosage dependent effects (Fig. 1C) have been inferred for 15–20 of the 150 identified loci, either by using overlapping deficiencies and duplications (e.g., Schotta et al. 2003), or by using transgenes (e.g., Eissenberg et al. 1992); this suggests a structural role for these gene products in heterochromatin formation. As we have come to appreciate the modification of PEV as a means for identifying genes that code for chromosomal proteins, reverse screens have been increasingly used to test mutations in candidate genes for *Su(var)* or *E(var)* activity (e.g., Pal-Bhadra et al. 2004). Of course, a genetic test by itself does not tell one whether the effect of the mutation is direct or indirect; further characterization is always needed.

Analysis of the identified genes to date suggests that although a discrete set of proteins is required for heterochromatin formation and concomitant gene silencing (loci showing *Su(var)* mutations), the proteins involved in gene activation are used more broadly. For example, a subset of critical regulatory genes found in the euchromatic arms is maintained in a silent state by the Polycomb (*Pc*) group genes, and up-regulated by the trithorax group (*trxG*) genes (see Ch. 17 [Grossniklaus and Paro 2014]; Ch. 18 [Kingston and Tamkun 2014]). In direct tests, relatively few mutations in *Pc* group genes result in suppression of PEV (e.g., Sinclair et al. 1998). In contrast, many mutations in the *trxG* genes are enhancers of PEV (Dorn et al. 1993; Farkas et al. 1994). This indicates that the Pc and heterochromatin silencing mechanisms are distinct, although gene activation processes frequently share common components.

Three loci, *Su(var)2-5*, *Su(var)3-7*, and *Su(var)3-9*, can be used to describe some of the different approaches taken to investigate putative suppressors of PEV. *Su(var)2-5* was cloned by screening a copy DNA expression library with a monoclonal antibody that recognizes heterochromatin (Fig. 3A) (James and Elgin 1986). The encoded heterochromatin-associated protein was consequently designated HP1, heterochromatin protein 1 (now called HP1a). In situ hybridization analysis, using the isolated cloned DNA, iden-

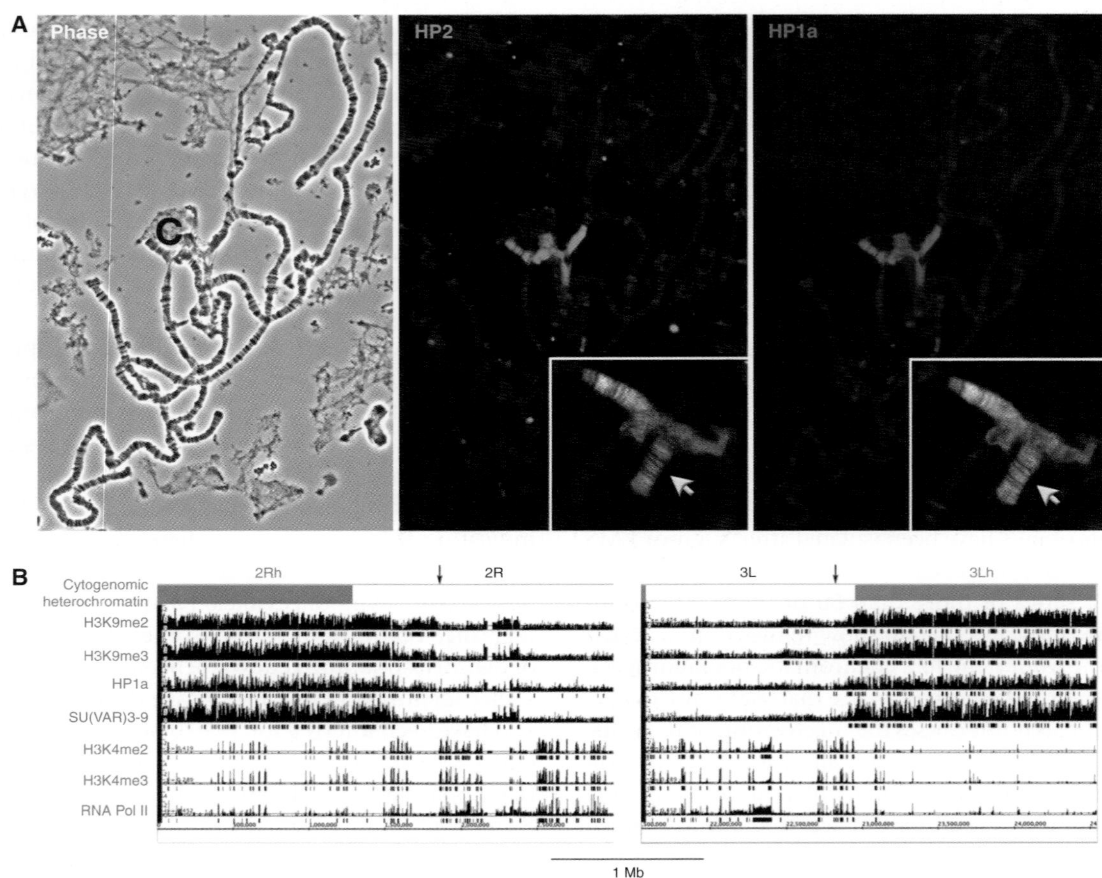

Figure 3. The distribution of chromosomal proteins and histone modifications defines different chromatin domains. (*A*) Immunofluorescent staining of the polytene chromosomes identifies proteins predominantly associated with heterochromatin. The polytene chromosomes, prepared by fixation and squashing of the larval salivary gland (shown by phase contrast microscopy, *left*; C = chromocenter) are "stained" by incubating first with antibodies specific for a given chromosomal protein, and then with a secondary antibody coupled to a fluorescent tag. HP1a (*right*) and HP2 (*center*) have similar distribution patterns, showing prominent association with the pericentric heterochromatin (found in the condensed chromocenter), small fourth chromosome (*inset*, arrow), and a small set of sites in the long euchromatin arms. (Adapted from Shaffer et al. 2002.) Note that the efficacy of any antibody can be affected by the choice of fixation protocol (see Stephens et al. 2003). (*B*) Chromatin marks define the epigenomic border between heterochromatin and euchromatin (indicated with an arrow). The border can be delineated based on chromatin immunoprecipitation (ChIP)-array data using antibodies to proteins known to be associated with heterochromatin (HP1a) or euchromatin (RNA polymerase II [RNA Pol II]), and to key histone modifications. Enrichment values are shown for the centromere-proximal 3 Mb of chromosome arms 2R and 3L (in BG3 cells). Boxes underneath the bar graphs indicate significant enrichment ($p < 0.001$). The cytologically defined heterochromatin is shown by the blue bar. The border, indicated here by the black arrow, is fairly well defined by the congruence of silencing marks. (*B*, Adapted from Riddle et al. 2011, © Cold Spring Harbor Laboratory Press.)

tified a gene in region 28–29 of the polytene chromosomes, where *Su(var)2-5* had been previously mapped by Sinclair et al. (1983). DNA sequence analysis of the mutant alleles confirmed that the *Su(var)2-5* locus at chromosome position 28F1-2 encodes HP1a (Eissenberg et al. 1990, 1992). HP1a contains two conserved domains, an amino-terminal chromo and a carboxy-terminal chromo-shadow domain (see Ch. 7 [Patel 2014] for further discussion of the chromodomain), and interacts with many other chromosomal proteins.

Su(var)3-7 was first cytogenetically mapped to region 87E1-4 in the third chromosome using a series of overlapping deletions and duplications. It was further assigned to within a 7.8-kb DNA fragment based on the triplo-enhancer effect it had on a variegating reporter (Reuter et al. 1990). *Su(var)3-7* encodes a protein with seven regularly spaced zinc fingers, domains that have been shown to function in DNA binding (Cleard and Spierer 2001).

Su(var)3-9 was cloned by *P* element transposon tagging (Tschiersch et al. 1994). The *Su(var)3-9* gene in *Drosophila* forms a bicistronic unit with the gene encoding eIF2γ, which can complicate genetic analysis. The SU(VAR)3-9 protein, like SU(VAR)2-5, contains a chromodomain in its amino-terminal region, but has a SET domain (identified first in the proteins SU(VAR)3-9, ENHANCER OF ZESTE [E(Z)], and TRITHORAX) at its carboxyl terminus. The SET domain allows this protein to function as a histone lysine methyltransferase (HKMT), specifically methylating histone H3 at lysine 9 (H3K9).

Immunocytological analyses using specific antibodies or transgene-expressed fusion proteins have shown that all three proteins—HP1a (encoded by *Su(var)2-5*), SU(VAR)3-7, AND SU(VAR)3-9—are preferentially associated with pericentric heterochromatin (see Fig. 3A for an example; James et al. 1989; Cleard et al. 1997; Schotta et al. 2002). Strong colocalization is particularly evident for HP1a and SU(VAR)3-9. Association of these proteins with each other has also been shown by coimmunoprecipitation (Delattre et al. 2000; Schotta et al. 2002). Thus, these proteins potentially form a core heterochromatin complex. Surprisingly, HP1a is also found at a number of euchromatic sites, and is involved in the positive regulation of a small set of euchromatic genes (Cryderman et al. 2005; Piacentini et al. 2009). Mutations in other genes encoding histone variants, chromatin-modifying enzymes, chromatin binders, or nucleosome remodeling factors often result in dominant PEV modifier effects (cf. Fodor et al. 2010). However, in the majority of these cases, causal analysis of the mutant effects on gene silencing in PEV is still missing, so it is possible that the effects are indirect. Despite intensive work on PEV modifiers, we still do not have a clear picture of the macromolecular assemblies in pericentric heterochromatin.

P element insertions carrying the w^+ reporter gene into other heterochromatic domains—telomeric regions, the Y chromosome, or the fourth chromosome—also show *white* variegation (e.g., Wallrath and Elgin 1995; Phalke et al. 2009). Genetic analysis of such reporters has revealed that although the different heterochromatic domains share some common features, they also can rely on different chromatin factors. For example, heterochromatin-like packaging is observed at TAS (telomere associated satellite) sequences, which are clusters of repetitious DNA elements just proximal to the HeT-A and TART retroviral elements that make up *Drosophila* telomeres (Cryderman et al. 1999b). Surprisingly, HP1a mutations do not show an effect on such reporters, although HP1a is important for telomere integrity (Fanti et al. 1998). Characteristics of this domain are sufficiently distinct that variegation here is referred to as telomere position effect (TPE). In another example, several groups have observed that silencing of reporters on the fourth chromosome is often sensitive to mutations in the gene for dSETDB1, and not to mutations in *Su(var)3-9*, although both encode an H3K9 HKMT (Seum et al. 2007; Tzeng et al. 2007; Brower-Toland et al. 2009). Investigations using a range of modifier loci to look at PEV reporters in different heterochromatin domains have shown that a unique complement of proteins is required for each domain to maintain silencing in somatic cells (Donaldson et al. 2002; Phalke et al. 2009).

3 DISTRIBUTION AND ASSOCIATION PATTERNS OF CHROMOSOMAL PROTEINS

One of the advantages of working with *Drosophila* is the ability to examine the polytene chromosomes, which provide a visual road map of the genome. Polyteny occurs during the larval stage when the chromosomes in many terminally differentiated cells are replicated, but do not go through mitosis; rather the chromatin strands remain paired, in perfect synapsis, with all copies aligned. The most extreme case is found in the salivary glands, where the euchromatic arms of the chromosomes have undergone 10 rounds of replication, generating ~1000 copies. Replication is not uniform, however; many repetitious sequences are underreplicated, and satellite DNA sequences are not replicated at all. All of the chromosome arms fuse in a common chromocenter (see Fig. 3A) (reviewed in Ashburner et al. 2005).

Polytene chromosomes provide an opportunity to determine the distribution pattern of chromosomal proteins by immunofluorescent staining, achieving much higher resolution than can be obtained using metaphase spreads (Silver and Elgin 1976). The approach has been used to discover heterochromatin-associated proteins (e.g., HP1a;

James and Elgin 1986) and to determine whether other candidates, identified genetically (by *Su(var)* phenotype) or by interaction with known heterochromatin proteins, show such localization (Fig. 3A) (e.g., HP2, identified by interaction with HP1a; Shaffer et al. 2006). Whereas genome-wide ChIP techniques described below have higher resolution, polytene chromosome staining remains a quick and inexpensive way (given specific antibodies in hand) to show distribution patterns. Approximately 10 heterochromatin-specific proteins have been identified using this approach; if mutations in the genes encoding these proteins are available, one often observes dominant suppression of PEV (e.g., Greil et al. 2007; reviewed in Ashburner et al. 2005, Chapter 28). These proteins are thus candidate structural components of heterochromatin.

With the availability of oligonucleotide arrays and high-throughput sequencing, it has become possible to map both chromosomal proteins and histone modifications across the sequenced genome by ChIP. In the most common approach, chromatin is cross-linked using formaldehyde, sonicated to obtain 500- to 1000-bp fragments, and the desired fragments pulled down using antibodies fixed to beads. The recovered fraction of genomic DNA is characterized either by qPCR (quantitative polymerase chain reaction, used if only a few loci are to be queried), hybridization to a genomic array (ChIP-chip), or deep sequencing (ChIP-seq). The validity of the results is critically dependent on the specificity of the antibodies used. As the commercial producers do not always provide sufficient quality control, the experimenters must do the controls themselves (Egelhofer et al. 2011). The technique depends on the ability to map the recovered DNA fragments to the assembled genome. In the case of *D. melanogaster*, only ~25% of the heterochromatic regions have been sequenced and assembled (not including satellite DNA, for example; Hoskins et al. 2007; Smith et al. 2007), and this limitation must be kept in mind in interpreting the results.

The modENCODE project, sponsored by the National Institutes of Health (USA), has reported genome-wide chromatin profiling in *Drosophila* for 25 histone modifications and many more chromosomal proteins and transcription factors, looking at several cell culture lines as well as several developmental stages. These data, as well as maps of transcription patterns, DNase I HS sites, and other chromatin features can be accessed through FlyBase (2012) or modMine (2011). Looking at the base of the chromosome arms in BG3 cells, for example, one observes a transition from a domain with abundant HP1a and SU(VAR)3-9 (pericentric heterochromatin) to a region where these proteins are only sporadically observed (euchromatin) (Riddle et al. 2011) (Fig. 3B). These results are in good agreement with the cytogenetically defined border between euchromatin and heterochromatin, but provide much higher resolution. The border clearly reflects the shift in genome organization around this point, with heterochromatic domains showing a decrease in gene density and increase in repeat density. However, the border can shift by hundreds of kilobases in S2 cells, suggesting that it is not fixed by a particular DNA sequence, but rather reflects the balance of chromosomal proteins or other factors specific to the cell type.

4 HISTONE MODIFICATION PLAYS A KEY ROLE IN HETEROCHROMATIN SILENCING

Analysis of SU(VAR)3-9 identified the key function of H3K9 methylation in heterochromatic gene silencing (Tschiersch et al. 1994). The protein contains a SET domain that enzymatically functions to methylate histone H3K9. That this protein is a histone methyltransferase (HKMT) targeting H3K9 was first shown by characterization of the human SUV39H1 homolog (Rea et al. 2000). In *Drosophila*, SU(VAR)3-9 is a major, but not the only, H3K9 HKMT (Schotta et al. 2002; Ebert et al. 2004). SU(VAR)3-9 contributes to di- and trimethylation of H3K9 (H3K9me2 and me3) in the bulk of the pericentromeric heterochromatin, but not in the majority of the fourth chromosome, telomeres, or euchromatic sites. The bulk of the dimethylation of these latter regions is independent of SU(VAR)3-9, as is monomethylation of H3K9 in pericentromeric heterochromatin (Ebert et al. 2004). dSETDB1 ("eggless") plays a major role in H3K9 methylation on the fourth chromosome (Seum et al. 2007; Tzeng et al. 2007; Brower-Toland et al. 2009; Riddle et al. 2012); G9a and potentially other HKMTs could also contribute, but the specifics are still unknown. The importance of H3K9 dimethylation in heterochromatic gene silencing is shown by the strong dosage-dependent effect of SU(VAR)3-9 on the PEV phenotype, as well as by the finding that suppression of gene silencing by *Su(var)3-9* mutations correlates with HKMT activity. The enzymatically hyperactive *Su(var)3-9*^{ptn} mutation is a strong enhancer of PEV and causes elevated H3K9 di- and trimethylation (H3K9me2 and H3K9me3) at the chromocenter, as well as generating prominent H3K9me2 and me3 signals at many euchromatic sites (ectopic heterochromatin) (Ebert et al. 2004). S-adenosylmethionine functions as the methyl donor for all of these methylation reactions; consequently, mutations in the gene encoding *S*-adenosylmethionine synthase, *Su(z)5*, are dominant suppressors of PEV (Larsson et al. 1996).

Studies using mutations in SU(VAR) genes have begun to reveal the sequence of molecular reactions required to establish heterochromatic domains. SU(VAR)3-9 binding at heterochromatic sequences depends on both its chromo and its SET domains (see Ch. 7 [Patel 2014] for details of protein structure; Schotta et al. 2002). How SU(VAR)3-9 binding is controlled is not yet understood. The act of methylating H3K9 by SU(VAR)3-9 establishes binding sites for HP1a. The HP1a chromodomain specifically binds H3K9me2 and H3K9me3 (Jacobs et al. 2001). That SU(VAR)3-9 also binds HP1a has been shown by yeast two-hybrid tests and by immunoprecipitation (Schotta et al. 2002). In fact, the region of SU(VAR)3-9 amino-terminal to its chromodomain interacts with the chromoshadow domain of HP1a, and this interaction stabilizes HP1a binding to H3K9me2/3 (Fig. 4A) (Eskeland et al. 2007). This region of SU(VAR)3-9 also interacts with the carboxy-terminal domain of SU(VAR)3-7. The SU(VAR)3-7 protein interacts at three different sites with the chromoshadow domain of HP1a (Delattre et al. 2000). This pattern of interactions suggests that the three proteins—HP1a, SU(VAR)3-7, and SU(VAR)3-9—physically associate in multimeric heterochromatin protein complexes.

Association of SU(VAR)3-9 and HP1a with pericentric heterochromatin is interdependent (Schotta et al. 2002). SU(VAR)3-9 causes H3K9 di- and trimethylation, which are specifically recognized by the chromodomain of HP1a (Jacobs et al. 2011). Consequently, in *Su(var)3-9* null larvae, HP1a binding to pericentric heterochromatin is impaired. As H3K9 dimethylation does not depend exclusively on SU(VAR)3-9 in the inner chromocenter, the fourth chromosome, telomeres and euchromatic sites, HP1a continues to be found at all of these sites in the mutant lines. Thus although SU(VAR)3-9 associates with these sites in wild-type cells, it appears to be relatively inactive.

Conversely, if HP1a is not present (having been depleted by mutations), SU(VAR)3-9 is no longer associated primarily with the pericentric heterochromatin, but is found along the euchromatic chromosome arms. It is now seen at almost all bands, where it causes ectopic mono- and dimethylation of H3K9 (H3K9me1 and H3K9me2) (Fig. 4B). Thus HP1a is essential for the restricted binding of SU(VAR)3-9 to pericentric heterochromatin. These data suggest a sequence of reactions starting with SU(VAR)3-9 association with heterochromatic domains and consequent generation of H3K9me2/3. This mark is recognized by the chromodomain of HP1a; binding of SU(VAR)3-9 to the HP1a chromoshadow domain ensures its association with heterochromatin (Fig. 4A). Interestingly, a chimeric HP1a-Pc protein has been generated in which the chromodomain of HP1a is replaced with the chromodomain of the Pc protein (Platero et al. 1996). The chromodomain of Pc binds strongly to H3K27me3 (Fischle et al. 2003), and the HP1a-Pc chimeric protein binds these sites in the euchromatic arms. In the presence of such a chimeric HP1a-Pc protein, the SU(VAR)3-9 protein is also found at Pc binding sites, demonstrating its strong association with the chromoshadow domain of HP1a (Schotta et al. 2002).

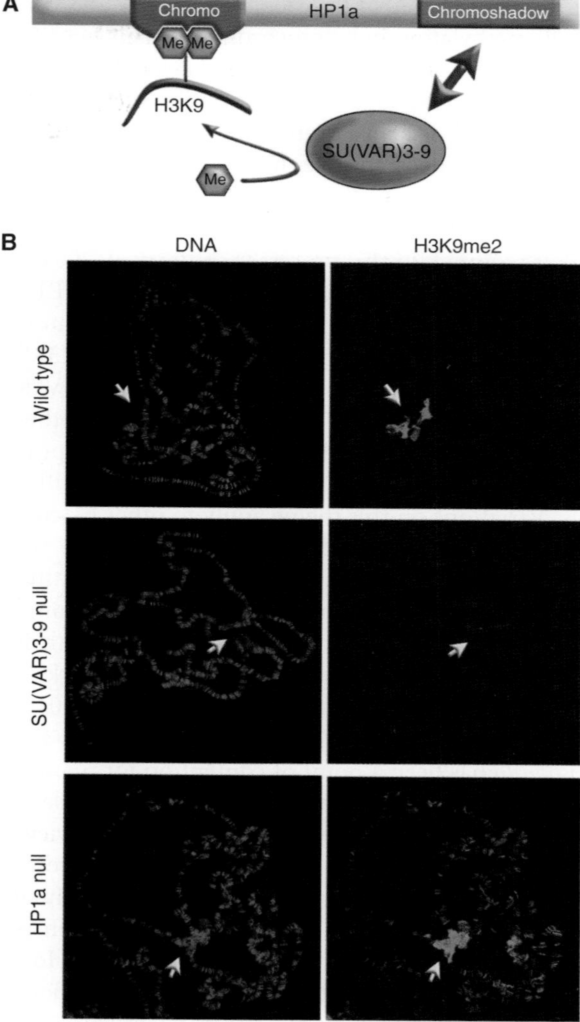

Figure 4. Interaction of SU(VAR)3-9 and HP1a in setting the distribution pattern of H3K9 methylation. (*A*) HP1a interacts with H3K9me2/3 through its chromodomain, and with SU(VAR)3-9 through its chromoshadow domain. By recognizing both the histone modification and the enzyme responsible for that modification, HP1a provides a mechanism for heterochromatin spreading and epigenetic inheritance. (*B*) SU(VAR)3-9 is responsible for much of the dimethylation of H3K9 (H3K9me2); loss of the enzyme results in loss of this modification in the pericentric heterochromatin, as shown by loss of antibody staining of the polytene chromosomes (compare *middle* panel with *top* panel). Loss of HP1a results in a loss of targeting of SU(VAR)3-9; high levels of H3K9me are consequently now seen throughout the chromosome arms (*bottom* panel).

In SU(VAR)3-9 null cells another heterochromatin-specific methylation mark, H4K20 trimethylation (H4K20-me3), is strongly reduced. The interdependence between H3K9 dimethylation and H4K20 trimethylation in heterochromatin has been shown to reflect an interaction between the SU(VAR)3-9, HP1a, and SUV4-20 proteins. SUV4-20 is a histone lysine methyl transferase (HKMT) that controls H4K20 methylation in heterochromatin. This heterochromatin-specific methylation mark is also strongly impaired in HP1a null cells, suggesting association of SU(VAR)3-9, HP1a, and SUV4-20 in a mutually dependent protein complex, although such a complex has not yet been isolated from flies. Mutations in the *Suv4-20* gene cause suppression of PEV-induced gene silencing, indicating that the H4K20me3 mark is required for this process (Schotta et al. 2004).

Taken together, the evidence argues that the HP1a protein has a central function in pericentric heterochromatin formation and associated gene silencing; it binds H3K9me2 and H3K9me3, and interacts directly with SU(VAR)3-9 (one of the H3K9 HKMTs) as well as several other key chromosomal proteins. The resulting complexes probably include several additional heterochromatin-specific proteins. Variations on this theme apply to other heterochromatic domains, such as the fourth chromosome (Riddle et al. 2012). However, given the number of identified *Su (var)* loci, the model is certain to become more complex!

In mammals and plants, histone H3K9 methylation and DNA methylation represent interrelated marks of repressed chromatin (Martienssen and Colot 2001; Bird 2002). Whether or not DNA methylation occurs at all in *Drosophila* has been a point of contention for many years. Recent reports showing low levels of DNA methylation in the early embryo have renewed this discussion (reviewed in Krauss and Reuter 2011). In *Drosophila* the only recognizable DNA methyltransferase present is Dnmt2. Mutations in this gene have a significant impact on retrotransposon silencing in somatic cells (Phalke et al. 2009). However, many inbred laboratory strains show only a very low level of *Dnmt2* expression; variation of this sort could explain conflicting results concerning DNA methylation in *Drosophila* (O Nickel, C Nickel, and G Reuter, unpubl.).

5 CHROMOSOMAL PROTEINS FORM MUTUALLY DEPENDENT COMPLEXES TO MAINTAIN AND SPREAD HETEROCHROMATIC STRUCTURE

PEV reflects a change in gene expression, specifically a loss in expression of a reporter gene in some of the cells in which it is normally active, as a consequence of a genetic rearrangement or transposition. Several different models, not all mutually exclusive, have been suggested to explain PEV. One possibility originally considered was the random loss of the gene, perhaps as a consequence of late replication (Karpen and Spradling 1990). Quantitative Southern blot analysis has shown that this explanation is not generally applicable; variegating genes are generally fully replicated in diploid tissue (Wallrath et al. 1996). A second test can be performed using a variegating reporter transgene (*white*) flanked by FRT sites; by inducing FLP recombinase, the gene is excised from the chromosome, forming an independent closed circle. A reporter gene subject to variegation (PEV) can be relieved of that silencing by excision from the chromosome in cells where it would otherwise be inactive (Ahmad and Golic, 1996). This shows not only the continuing presence of the gene, but that the heterochromatic state can be reversed once the reporter is removed from the heterochromatic environment.

Other models have focused on the association of the variegating gene with a heterochromatic compartment in the nucleus (see Sec. 6), and/or on the spreading of heterochromatic structure from the newly adjacent heterochromatin. The spreading model, which is based on extensive genetic and cytological data, explains gene silencing as a consequence of heterochromatin packaging spreading across the breakpoint into normally euchromatic domains. In normal chromosomes, euchromatic and heterochromatic regions appear to be differentiated by the higher density of repetitious sequences in heterochromatin, and are potentially insulated from each other by specific sequences or buffer zones. Because these "insulating sequences" (never well-defined in *Drosophila*) are not present at the euchromatic-heterochromatic junction in PEV rearrangements (see Fig. 1A), heterochromatinization of euchromatic sequences is variably induced. This heterochromatinization is cytologically visible in the polytene chromosomes as a shift from a banded to an amorphous structure at the base of the chromosome arms (Fig. 5A) (Hartmann-Goldstein 1967); the extent of this change can be modified by *Su(var)* and *E(var)* mutations (Reuter et al. 1982).

Inactivation of euchromatic genes over a distance along the chromosome can be genetically shown (Demerec and Slizynska 1937). The affected regions become associated with HP1a (Belyaeva et al. 1993; Vogel et al. 2009) and show dimethylation of histone H3 at lysine 9 (H3K9me2) (Ebert et al. 2004; Rudolph et al. 2007). The spreading model postulates a competition between packaging into euchromatin versus packaging into heterochromatin; the recovery of dosage-dependent modifiers, as discussed above, supports such a model (Locke et al. 1988; Henikoff 1996). However, spreading does not seem to be a simple matter of mass action, in which the regions closest to the heterochromatic mass show the greatest silencing. One can identify cells in which a marker closer to the pericentric heterochromatin in a rearrangement is active and one further away is silenced (Talbert and Henikoff 2000). Inspection using ChIP, followed by qPCR, shows that there is a gradient of the H3K9me2 heterochromatin mark declining with distance from the pericentriomeric heterochromatin (Fig. 5C) (Rudolph et al. 2007). Ironically, the regulatory region of the *white* gene appears to be particularly vul-

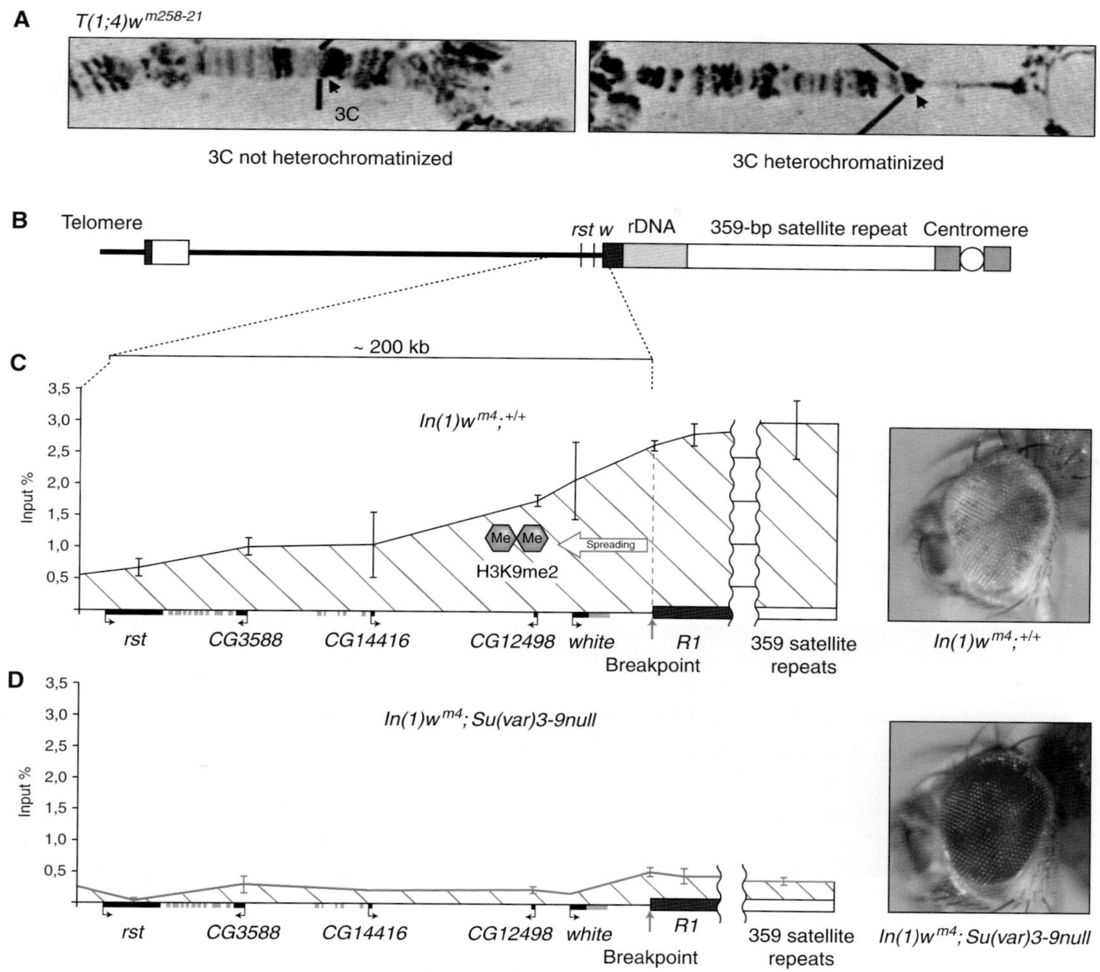

Figure 5. Spreading of histone H3K9me2 and cytological heterochromatinization at the *white* locus in PEV rearrangements. (*A*) In translocation $T(1;4)w^{m258-21}$ (with a breakpoint in the heterochromatin of chromosome 4), heterochromatinization becomes cytologically visible in the polytene larval salivary gland chromosomes as a loss of banding, the apparent consequence of condensation and underreplication (*right* portion of the *right* panel). (Reprinted from Reuter et al. 1982, with kind permission from Springer Science+Business Media.) (*B*) In the $In(1)w^{m4}$ chromosome, spreading of the heterochromatic chromatin state over ~200 kb of the adjacent euchromatic region is initiated by the segment of heterochromatin located distal to the rDNA cluster (dark gray box). (*C*) ChIP with an antibody specific for H3K9me2 detects spreading of this heterochromatic histone mark along the euchromatic region between the *roughest* (*rst*) gene and the breakpoint of w^{m4}. (*D*) Flies homozygous for a null mutation of the *Su(var)3-9* gene lose H3K9me2 in pericentric heterochromatin as well as in the *white* gene region, restoring wild-type activity of the *white* gene in $In(1)w^{m4}$ flies. (Adapted from Rudolph et al. 2007.)

nerable to the accumulation of silencing marks, shown in particular as an accumulation of HP1a in a DamID mapping study (method described below; Vogel et al. 2007). This regulatory region is present in reporter transgenes marked with *mini-white*, but not in those such as shown in Fig. 2, which use an *hsp70* promoter to drive *white* expression. Such differences should be kept in mind in evaluating studies using transgene reporters.

The heterochromatin spreading effect clearly depends on a series of molecular reactions within the euchromatic regions. Several histone modifications are now known to be mutually exclusive in defining alternative chromatin states. Acetylation of H3K9 (H3K9ac), di- and trimethylation of H3K4 (H3K4me2/3), and phosphorylation of H3S10 (H3S10ph) are typical marks of active euchromatin, whereas H3K9me2/3 and H4K20me3 are specific marks of silenced regions. Heterochromatinization of euchromatic regions therefore requires specific deacetylation, demethylation, and dephosphorylation reactions within euchromatin, as illustrated in Figure 6A. This transition depends

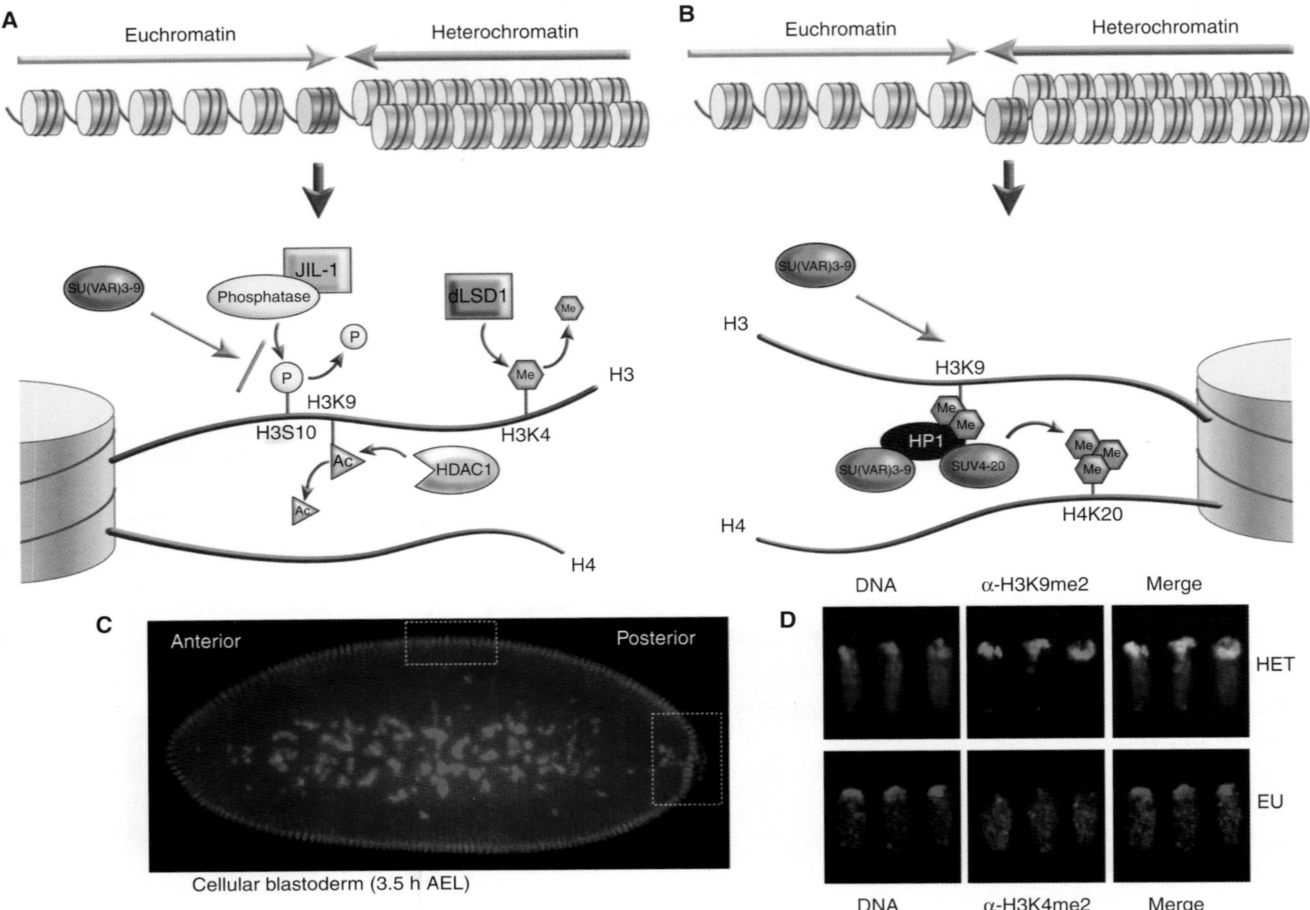

Figure 6. The transition from a euchromatic state to a heterochromatic state requires a series of changes in histone modification. (*A*) Active genes are marked by H3K4me2/3; if present, this mark must be removed by LSD1. H3K9 is normally acetylated in euchromatin; this mark must be removed by a histone deacetylase, HDAC1. Phosphorylation of H3S10 can interfere with methylation of H3K9; dephosphorylation appears to involve a phosphatase targeted by interaction with the carboxyl terminus of the JIL1 kinase. These transitions set the stage for acquisition of the modifications associated with silencing, shown in *B*, including methylation of H3K9 by SU(VAR)3-9, binding of HP1a, and subsequent methylation of H4K20 by SUV4-20, an enzyme potentially recruited by HP1a. (*C*) Differentiation of euchromatin and heterochromatin is initiated in early embryogenesis around cell cycle 10 and is completed when cellular blastoderm (*top* box) and primordial germline cells (*right-hand* box) are formed. (*D*) Blastoderm nuclei show an apicobasal polarity (Rabl conformation). Heterochromatin (H3K9me2 staining) is established at the apical site, whereas euchromatin (H3K4me2 staining) is organized toward the basal site. (Immunofluorescent images provided by Sandy Mietzsch.)

initially on H3K9 deacetylation by HDAC1. Mutations in the *rpd3* gene, encoding the histone H3K9-specific deacetylase HDAC1, are strong suppressors of PEV (Mottus et al. 2000), antagonizing the effect of SU(VAR)3-9 in gene silencing (Czermin et al. 2001). HDAC1 has been shown to be associated in vivo with the SU(VAR)3-9/HP1a complex; the two enzymes work cooperatively to methylate previously acetylated histones.

Spreading of heterochromatin into euchromatin is completely blocked in *Su(var)3-1* mutations (Ebert et al. 2004). *Su(var)3-1* mutations are frame shift mutations within the gene encoding JIL1 kinase that result in expression of a truncated JIL1 protein, lacking the carboxyl terminus region. The JIL1 protein contains two kinase domains, and catalyzes H3S10 phosphorylation in euchromatin. The $JIL1^{Su(var)3\text{-}1}$ mutations do not affect H3S10 phosphorylation, but probably impair dephosphorylation of H3S10, effectively inhibiting methylation of H3K9. This suggests involvement of a phosphatase. Whether the PP1 enzyme (that has been identified with *Su(var)3-6* mutations [Baksa et al. 1993]) is directly involved in this reaction is not yet known.

Cite as *Cold Spring Harb Perspect Biol* doi: 10.1101/cshperspect.a017780

Demethylation of H3K4 appears to be another prerequisite for heterochromatinization of euchromatic regions (Fig. 6A). Work in mammals has shown that the LSD1 amine oxidase functions as an H3K4 demethylase (Shi et al. 2005). Mutations in the *Drosophila* LSD1 homolog antagonize spreading of heterochromatin into euchromatic regions in all PEV rearrangements tested. In SU(VAR)3-3 null cells, lacking LSD1, the acquisition of H3K9 methylation in the euchromatin flanking a break point is eliminated, although constitutively heterochromatic regions are not affected (Rudolph et al. 2007). In the syncytial blastoderm, dLSD1 is concentrated in the nuclei at the boundary between heterochromatic and euchromatic domains (Fig. 6D). These findings show that the coordinated function of several enzymes is required to remove euchromatin-specific histone modification marks before the transition to heterochromatin packaging can take place (see Fig. 6A). It seems likely that the required enzymes will be found to form complexes with SU(VAR)3-9/HP1a, as has already been shown for HDAC1.

The above patterns of association among histone marks and chromosomal proteins suggest that the chromatin landscape could be described in terms of these patterns of coassociation, and indeed this has proved to be the case. Two different approaches have been taken for mapping genome-wide distribution patterns of chromosomal proteins: ChIP (described above) and DamID. In the latter case, DNA adenine methyltransferase (Dam, a bacterial specific protein) fused to a chromatin protein of interest deposits a stable adenine-methylation "footprint" in vivo at those sites where the chromatin protein associates. The modified DNA fragments are recovered and the pattern of association assessed using oligonucleotide microarrays. A study using 53 broadly selected chromosomal proteins in *Drosophila* Kc cells identified five major chromatin types: heterochromatin (enrichment of HP1a and H3K9me2), Pc silenced domains (enrichment in PC and H3K27me3), additional inactive regions, and two types of active domains, both associated with high levels of RNA polymerase but distinguished by molecular organization and H3K36 methylation levels. Principle component analysis shows that this classification can largely be achieved (85.5% agreement) using five proteins: histone H1, HP1a, PC, MRG15, and BRM (the latter two associated with nucleosome remodeling) (Filion et al. 2010).

The modENCODE study used the enrichment patterns of 18 different histone modifications to generate a model based on 30 alternative states, in contrast to the five categories by Filion et al. (2010) described above. Distribution patterns were determined by ChIP-chip experiments (with assessment of the captured DNA on oligonucleotide arrays); this approach, although requiring fixation, provides higher resolution. Much of the overall complexity is captured in nine prevalent combinatorial patterns (states), illustrated in Figure 7A (Kharchenko et al. 2011). The analysis shows the presence of correlated features, including those associated with the transcription start site (state 1), the body of the transcribed gene (state 2), and regulatory regions (states 3 and 4). Distinctive states associated with large domains include that found on the male X chromosome (presumably related to dosage compensation—see Ch. 24 [Lucchesi and Kuroda 2014]) (state 5), that associated with the Pc silencing complex (state 6; see Ch. 17 [Grossniklaus and Paro 2014]), and two associated with heterochromatin marks, one common to pericentric heterochromatin (state 7) and the other (with lower concentrations of H3K9me2/3) found in the euchromatic arms (state 8). (The pattern of state 8 domains is cell-type specific, suggesting "facultative" heterochromatin [Kharchenko et al. 2011].) Those domains with no distinctive features are grouped together in state 9 (see Fig. 7A).

Although the generation of this nine-state model is based solely on the map of histone modification marks, one observes distinctive patterns of enrichment and depletion for chromosomal proteins. For example, HP1a and SU(VAR)3-9 are greatly enriched in state 7, and moderately enriched in state 8, states that are depleted for proteins associated with gene expression. Mapping these states back across the genome provides both an overview of chromatin organization (Fig. 7B) and a detailed characterization of individual genes. The latter can be viewed on FlyBase (2012) by choosing the GBrowse option. The nine-state model allows us to see general patterns at the level of the chromosome or large domains, whereas going to more complex models (such as the 30-state model) can resolve more detail at the gene level.

Results from the two approaches, classifying by histone marks or by chromosomal proteins, are clearly in agreement in identifying alternative silencing domains (H3K9me2/HP1a vs. H3K27me3/Pc). However, the classification of active genes differs, with that based on histone modifications identifying different gene regions (1-transcription start site, 2-body of the transcribed region, 3/4-regulatory regions), whereas that based on chromosomal proteins identifies two groups of genes differing in remodeling strategies (Filion et al. 2010; Kharchenko et al. 2011). Analysis using key components identified by each study should provide a powerful approach in the future.

6 NOT ALL HETEROCHROMATIN IS IDENTICAL: SPATIAL ORGANIZATION MATTERS

In *D. melanogaster*, constitutive heterochromatin is arranged in large blocks that flank the centromeres, smaller

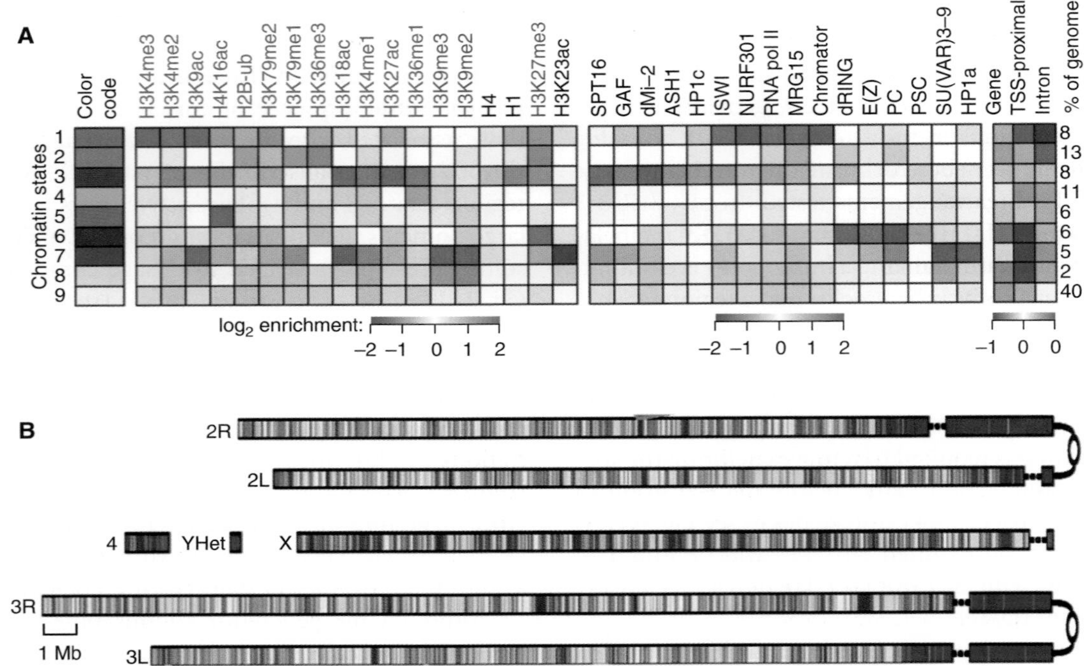

Figure 7. Chromatin annotation of the *D. melanogaster* genome. (*A*) A nine-state model of prevalent chromatin states was generated using data from S2 cells. Each chromatin state (row) is defined by a combinatorial pattern of enrichment (red) or depletion (blue) for specific histone modification marks. First (*left*) panel, color code for mapping; second panel, histone modification marks used (active marks labeled in green, repressive in blue, general in black); third panel, enrichment or depletion of chromosomal proteins found in that state; fourth (*right*) panel, fold over-/underrepresentation of genic and transcription start site (TSS)-proximal ($\pm$ 1 kb) regions relative to the entire tiled genome. (*B*) A genome-wide karyotype view of the domains defined by the nine-state model in S2 cells. Centromeres are shown as open circles; dashed lines span gaps in the genome assembly. Note the association of pericentric heterochromatin and the fourth chromosome distal arm with state 7 and the association of state 5 with the male X chromosome. (Adapted, by permission from Macmillan Publishers Ltd: NATURE, from Kharchenko et al. 2011, © 2011.)

blocks associated with the telomeres, the whole of the Y chromosome, and most of the small fourth chromosome. The centromeric regions are made up of large (0.2–1 Mb) blocks of satellite DNA interspersed with "islands" of complex sequences, primarily TEs (Le et al. 1995). Although gene-poor, these regions are not devoid of genes; a minimum of 230 protein-coding genes (conserved in other Drosophilids), as well as 32 pseudogenes and 13 noncoding RNAs reside in the pericentromeric heterochromatin (Smith et al. 2007). The telomeres of *Drosophila* do not have the typical G-rich repeats seen in other eukaryotes, but are composed of copies of HeT-A and TART retrotransposons. TAS, blocks of 10^2–10^3 nucleotide repeats, are found just proximal, and *white* transgene reporters inserted in these regions display the TPE variegating phenotype. Although the Y chromosome does carry the genes for a number of male fertility factors, the bulk of the chromosome is made up of satellite DNA, and it remains condensed in cells other than the male germline. Thus, although all of these domains are characterized by a high density of repeats, the type of repeats (and interspersion of different types) varies. The consequences have been seen in studies that look at the impact of different blocks of pericentric heterochromatin on expression from a reporter—one observes that the severity of the phenotype does not depend simply on the amount of heterochromatin in *cis*, but varies depending on the local heterochromatin environment (Howe et al. 1995). Heterochromatin-associated proteins that might play a role in specific subdomains include the AT-hook protein D1, preferentially associated with the 1.688 g/cm^3 satellite III (Aulner et al. 2002), and DDP1, a multi-KH-domain protein homologous to vigilin that binds the pyrimidine-rich C strand of the dodeca satellite (Cortes and Azorin 2000).

The small fourth chromosome is perhaps the most complex heterochromatic domain. It is on the order of 4.3 Mb in size, with ~3 Mb made up of satellite DNA. The distal 1.2 Mb can be considered euchromatic in that it is polytenized in the salivary gland (see Fig. 3A), but it appears heterochromatic by virtue of its late replication, its complete lack of meiotic exchange, and its association with HP1a, HP2, and H3K9me2/3 (Figs. 3A, 4B, and 7B). This region has a six- to sevenfold higher density of transposon fragments than found in the euchromatic arms, similar to regions at the junction of pericentric heterochromatin and euchromatin on the other chromosomes (Kaminker et al. 2002). Nonetheless, approximately 80 genes are present, a density similar to that of the long chromosome arms. An investigation of the fourth chromosome using the *white* reporter *P* element discussed above (Fig. 2) found primarily heterochromatic domains (resulting in a variegating phenotype), with a few permissive domains (resulting in a red eye phenotype) interspersed (Sun et al. 2004). Analysis using modENCODE data indicates that the permissive domains are under Pc regulation (Riddle et al. 2012).

The differences in DNA sequence organization described above are reflected in differences in the chromatin biochemistry and/or the enzymes used to achieve it. Examination of the impacts of mutations in 70 different modifiers on different variegating genes (including w^{m4}, bw^{D}, or *P* element reporters in pericentric heterochromatin or in a TAS array) showed that there is substantial overlap in the targets of modifiers, but there is also surprising complexity. This set of tests divided the modifiers into seven different groups in terms of their ability to impact silencing in a given compartment (Donaldson et al. 2002). The only modifier in this group to impact silencing in the TAS array was a new allele of *Su(var)3-9*. TAS silencing is also sensitive to alleles of *Psc* and *Su(z)2*, two Pc group genes (Cryderman et al. 1999b). Using a similar approach, Phalke et al. (2009) identified modifiers that distinguished pericentric heterochromatin, the fourth chromosome, retrotransposons, and TAS sequences from each other.

Heterochromatic masses are seen mostly at the nuclear periphery and around the nucleolus. In *Drosophila* embryos, this tendency is even more pronounced. Early development in *Drosophila* is syncytial until nuclear division cycle 14, when cell walls form between the nuclei, creating the typical blastula, a ball of cells. Heterochromatic masses are first seen in early embryogenesis, as the nuclei move to the periphery of the egg. The heterochromatic material (centromeres, chromosome four) is concentrated at one side of the nucleus, oriented to the exterior surface of the egg (Foe and Alberts 1985) (see Fig. 6C,D). Such spatial subdivision of the nucleus persists during development, leading to the concept of heterochromatin "compartments" within the nucleus (for more discussion of nuclear organization, see Ch. 19 [Dekker and Misteli 2014]). These compartments might maintain a high concentration of factors required for heterochromatin formation (such as HP1a and the H3K9 HKMTs), while being depleted in factors required for euchromatin assembly and gene expression (such as HATs and RNA Pol II). Indeed, proximity to heterochromatic masses, both in position along the chromosome and in three dimensions, has been shown to be a factor in PEV.

Chromosomal proximity to the mass of pericentric heterochromatin has been shown to have an impact on variegation both for euchromatic genes (of which *white* is an example), and for heterochromatic genes, the best studied examples being *light* and *rolled*. Genes that normally reside in heterochromatic domains can be observed to variegate when a rearrangement places them in juxtaposition with euchromatin; generally, they show the opposite dependencies, requiring normal levels of HP1a for full expression, and showing an enhancement of variegation when HP1a is depleted. Variegation of *light* depends not only on its juxtaposition to euchromatin, but also on the position of the breakpoint, relative to the distance along the chromosome arm from heterochromatin (Wakimoto and Hearn 1990). Similar results have been reported for *rolled*. Investigations of bw^{D}, a euchromatic gene induced to variegate by insertion of repetitious DNA, have shown that a shift in proximity of the locus to the pericentric heterochromatin can result in enhancement of silencing (if closer), or suppression of silencing (if further away) (Henikoff et al. 1995). Similarly, a reciprocal translocation that moves a fourth chromosome arm carrying a *white* reporter to the distal end of chromosome arm 2L or 2R results in a dramatic loss of silencing. The loss of silencing is correlated with a change in the position of the fourth chromosome fragment (now at the tip of the second chromosome arm) in the nucleus; it now frequently occupies sites in the salivary gland nucleus distant from the chromocenter (Cryderman et al. 1999b). These results suggest that proximity to a heterochromatic mass is necessary for effective silencing.

A recent study using high-resolution microscopy examined both gene activity (using antibodies specific for the product) and nuclear location of a reporter (using FISH, fluorescence in situ hybridization) in the same cell during the normal time frame of expression. A *white* variegating inversion, bw^{D}, and a variegating *lacZ* transgene were studied in differentiating eye discs or adult eyes. This investigation found a strong inverse correlation between the position of the reporter gene in the cell relative to pericentric heterochromatin and the level of expression, supporting the idea that a heterochromatic "compartment" exists, and that positioning within this compartment is correlated with gene silencing (Harmon and Sedat 2005). However, the correlation is not absolute. This is not surprising, given the stochastic nature of PEV. Further, studies with an *hsp70-white* reporter indicate the presence of both permissive and silencing domains interspersed on the small fourth chromosome (which is always close to the mass of pericentromeric heterochromatin in wild-type cells), indicating that local determinants also contribute to the decision to package the chromatin in one form or the other (Sun et al. 2004).

Genes normally residing in heterochromatin (*light* and *rolled*) function best in that domain, and show a loss of expression on depletion of HP1a. This is the opposite of what we see in PEV, in which our reporter genes (which normally function in euchromatin) show HP1a-dependent silencing. How do the genes in pericentric heterochromatin, or residing on the fourth chromosome, function in this presumably "hostile" environment? Examination of *light* using ChIP showed that although the region is generally enriched for H3K9me2, that mark is specifically depleted at the 5′ end of the gene (Yasuhara and Wakimoto 2008). The modENCODE project has allowed us to systematically examine the chromatin packaging for most pericentric and fourth chromosome genes to extend this analysis. There is indeed a conspicuous loss of silencing marks at the TSS of active genes in these domains (Fig. 8), although the usual heterochromatic marks (including H3K9me2) are still present upstream and across the body of the gene. As expected, the TSSs are occupied by RNA Pol II and are flanked downstream by nucleosomes with H3K4me2/3. Thus, these genes have "state 1" chromatin at their 5′ ends (around the TSSs), but state 7 chromatin over the body of the gene (Fig. 8). The presence of H3K9me3 and HP1a across the body of an active gene seems contradictory, but these marks are actually enriched there in preference to other sites (including intergenic spacer) on the fourth chromosome. The majority of the fourth chromosome genes show a loss of expression on HP1a depletion, showing a dependence on this chromatin structure (Riddle et al. 2012). How then is the gene expressed? POF (painting of fourth; Larsson et al. 2001), a protein found uniquely on the fourth chromosome, binds to nascent RNA and may play a role in transcript elongation (Johansson et al. 2007a; Johansson et al. 2012). And HP1a itself has been implicated in transcript elongation at some euchromatic sites (Piacentini et al. 2012). Although HP1a appears to bind to clusters of repeats on chromosome 4, as elsewhere in heterochromatin (utilizing H3K9me2/3 generated by SU(VAR)3-9), its association with fourth chromosome genes is dependent on POF (Johansson et al. 2007b; Riddle et al. 2012). This interaction may play a key role in facilitating transcription of fourth chromosome genes.

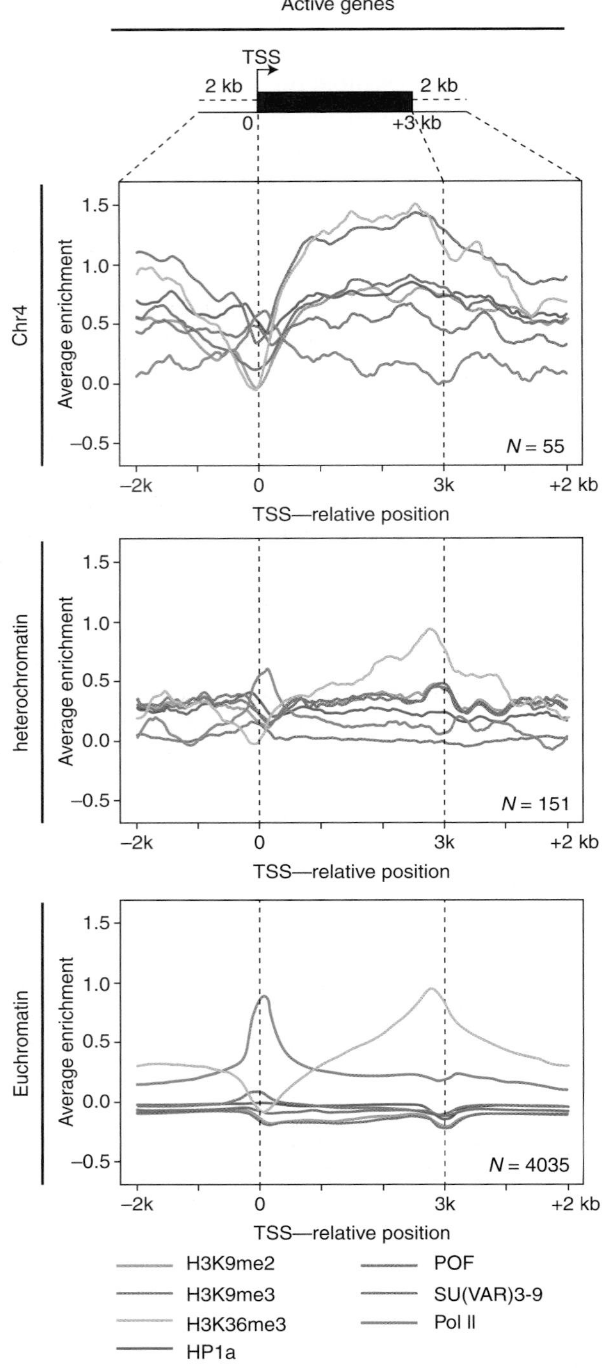

Figure 8. Packaging of active genes in chromosome 4, pericentric heterochromatin, and euchromatin. The plots show $\log_2$ enrichment (*y*-axis) for RNA Pol II (green), H3K36me3 (pink), H3K9me2 (yellow), H3K9me3 (purple), SU(VAR)3-9 (blue), POF (brown), and HP1a (red) for a scaled metagene and 2 kb flanking region created by averaging data for all active genes in a given compartment. Fourth chromosome (*top*) and pericentric (*middle*) genes show a similar depletion in silencing marks at the TSS; these marks reappear over the body of the gene only in the case of the fourth chromosome. As expected, euchromatic genes do not show association with any of the silencing marks (*bottom*). (Adapted from Riddle et al. 2012.)

7 HOW IS HETEROCHROMATIN FORMATION TARGETED IN *DROSOPHILA*?

Although we have learned a great deal about the biochemistry of heterochromatin structure, this leaves open the question of how heterochromatin formation is targeted to the desired regions of the genome in its normal configuration. The analysis of the fourth chromosome suggests that the presence of local elements in the DNA can signal the formation or stabilize the presence of heterochromatin. Genetic screens for a switch in phenotype (from red to variegating or vice versa) have shown that local deletions or duplications of 5–80 kb of DNA flanking a fourth chromosome transposon reporter can lead to the loss or acquisition of variegation, pointing to short-range *cis*-acting determinants for silencing in this domain. This silencing is dependent on HP1a, and correlates with a change in chromatin structure from an accessible (euchromatic) to a closed (heterochromatic) state based on changes in the nucleosome array, as shown by nuclease accessibility assays (Sun et al. 2001). Mapping data in one region of the fourth chromosome implicate the *1360* transposon (and other TEs) as a target for heterochromatin formation, and suggest that once heterochromatin formation is initiated at dispersed repetitive elements, it can spread along the fourth chromosome for ~10 kb, or until it encounters competition from a euchromatic determinant (Sun et al. 2004; Riddle et al. 2008). Short-range *cis*-acting determinants related to copy number are also implied by the observation that tandem or inverted repeats of reporter *P* elements result in heterochromatin formation and gene silencing (Dorer and Henikoff, 1994).

Such *cis*-acting elements in the DNA might function by sequence-specific binding of a protein capable of triggering heterochromatin formation. Proteins that bind specifically to some of the satellite DNAs have been identified, including D1, which is associated with the 1.672- and 1.688-g/cm^3 AT-rich satellite repeats (Aulner et al. 2002). Their importance has been inferred from the impact of satellite-specific DNA-binding drugs. For example, the P9 polyamide binds the X-chromosome 1.688-g/cm^3 satellite III, displacing the associated D1 and HP1a; this causes suppression of PEV in w^{m4h}, indicating a mechanistic link (Blattes et al. 2006). Other specific DNA binding proteins of this type could well be responsible for heterochromatin formation at satellite DNAs in the pericentric heterochromatin.

The findings in yeast and plants (see Ch. 9 [Allshire and Ekwall 2014]; Ch. 16 [Martienssen and Moazed 2014]; Ch. 13 [Pikaard and Mittelsten Scheid 2014]) suggest an additional model for dealing with TEs and their remnants by heterochromatization—one based on the RNAi system, able to recognize a diverse set of elements. Work from many laboratories has shown that the RNAi system is present in *Drosophila* and plays several important roles via posttranscriptional gene silencing. For example, there is a *Dicer-1* dependent miRNA system impacting developmental regulation by messenger RNA degradation or translational inhibition, and a *Dicer-2* dependent small interfering RNA (siRNA) system active in viral defense (reviewed in Kavi et al. 2008; Huisinga and Elgin 2009).

A Dicer-independent piRNA pathway exists, generating small piRNAs (24–30 nt) based on the cleavage activity of Piwi, Argonaute 3 (Ago3), and Aubergine (Aub), acting either alone or in a "ping-pong" mechanism to achieve a robust signal. rasiRNAs (repeat associated small interfering RNAs) generated by this pathway have been identified from 40% of the known TEs (including *1360*) and other repeated sequences (Aravin et al. 2003). To test the idea of targeted silencing, a *P* element carrying a *1360* copy immediately adjacent to an *hsp70*-driven *white* reporter was inserted at many different sites in the genome. The presence of a single copy of this TE was insufficient to induce silencing at most sites in the eukaryotic arms, but a variegating phenotype was observed when the *P* element inserted into a repeat-rich region at the base of chromosome arm 2L. Thus, formation of stable heterochromatin appears to be dependent on the nuclear location as well as a specific target; perhaps the spatial requirements are tied to the need for an abundant pool of heterochromatin proteins. The degree of silencing was dependent on the presence of the *1360* copy. This targeted silencing is dependent both on HP1a and SU(VAR)3-9, and on components of the rasiRNA pathway (Haynes et al. 2006). A larger screen identified many more *1360*-sensitive sites, some in euchromatic domains lying close to heterochromatic masses (base of chromosome 2L). In this type of insertion site, *1360* can drive heterochromatin formation (HP1a accumulation) at a normally euchromatic position. Use of a "landing pad" construct that allows one to replace the *1360* element with an altered copy led to the conclusion that the piRNA hot spots, but not the repetitious end sequences or putative TSSs, are critical for this *1360*-dependent silencing (Sentmanat and Elgin, 2012). This suggests a recognition event dependent on the piRNA system.

rasiRNAs (including piRNAs) are abundant in the female germline, where they clearly play a role in silencing TEs (Senti and Brennecke, 2010). Whether this role encompasses transcriptional silencing as well as posttranscriptional silencing is the question of interest here. In a genetic test, Pal-Bhadra et al. (2004) found that mutations in *piwi* (a member of the PAZ domain family) and *homeless* (a DEAD box helicase) suppress the PEV associated with tandem arrays of the *white* gene. Mutations in *piwi, aubergine,* and *homeless* (aka *spn-e*) suppress silencing of the *white* transgene *P*[*hsp70-w*] in pericentric heterochromatin or the fourth chromosome. The amount of rasiRNAs produced for a wide range of retroelements is significantly reduced in ovaries of flies carrying a mutation in *spn-E*, with concomitant depletion of HP1a at these TEs (Klenov et al. 2007). In the fly female germline, among the argonaute family of proteins (that bind rasiRNA), only Piwi is found to be a predominantly nuclear protein, and it has been reported to interact with HP1a (Brower-Toland et al. 2007). Specific depletion of HP1a in the female germline results in overexpression of some (but not all) TEs assayed, indicating a role for heterochromatin in silencing these elements (Fig. 9A). Germline depletion of Piwi also leads to a loss of silencing for this group of TEs, with concomitant loss of HP1a and H3K9me2 association. Piwi appears to function downstream of Aub here, suggesting that it is utilizing the products of the ping-pong rasiRNA system (Wang and Elgin 2011). A mutation that removes the nuclear localization signal of Piwi similarly leads to TE overexpression, with a shift in chromatin structure, demonstrating that Piwi is required to be in the nucleus for these functions (Klenov et al. 2011). These results support a model in which piRNAs, generated in the female germline and transported to the nucleus by Piwi, can promote HP1a deposition at target TEs (Fig. 9B). However, although *HeT-A*, *Blood*, *Bari*, and many other TEs are reported to be significantly affected, other TEs are not, including *Jockey* and *Roo*. It is apparent that there are multiple, redundant ways to silence TEs; it is not yet clear what determines the sensitivity of a given TE to a given mechanism.

Although a model using germline Piwi with associated rasiRNAs to direct silencing by heterochromatin formation at TEs is attractive, the evidence is as yet indirect. Any mutation that disrupts TE silencing (whether at the transcriptional or the posttranscriptional level) will result in TE mobilization and trigger a DNA damage response, which could lead to further destabilization of the genome (Khurana et al. 2011). The above experiments have focused on the germline, where the need for TE silencing is greatest, and Piwi is most abundant; similar effects are not seen in adult somatic cells (Klenov et al. 2007). However, the nurse cells deposit both RNA and protein into the oocyte, and could supply the needed materials for such Piwi-dependent targeted heterochromatin formation at blastoderm. Alternatively, an siRNA pathway has been suggested to operate in somatic cells based on the observation that expression of viral suppressor proteins results in suppression of PEV, with associated changes in chromatin structure (Fagegaltier et al. 2009).

8 PEV, HETEROCHROMATIN FORMATION, AND GENE SILENCING IN DIFFERENT ORGANISMS

The phenomenon of position-effect variegation was initially detected in *Drosophila*, simply because this was one of the first organisms for which X irradiation was used to induce mutations. X irradiation is much more likely than other commonly used mutagens to induce chromosomal rearrangements that can result in PEV. Similar mutations have been isolated from the mouse, in which variegating coat color indicates PEV (see Ch. 14 [Blewitt and Whitelaw 2013]). For example, the insertion of an autosomal region carrying a fur color gene onto the X chromosome results in variable silencing of the allele (Russel and Bangham 1961; Cattanach 1961). Variegation, however, is only observed in females carrying this insertion combined with a homozygous mutation in the original coat color gene. This is because the translocated wild-type allele becomes inactivated as a consequence of X inactivation by heterochromatinization (see Ch. 25 [Brockdorff and Turner 2014]). In plants the only unequivocal case of PEV that has been described was reported in *Oenothera blandina* (Catcheside 1939). In these cases, as is the case in *Drosophila*, PEV silencing of euchromatic genes is connected with placement of those genes into a new heterochromatic neighborhood.

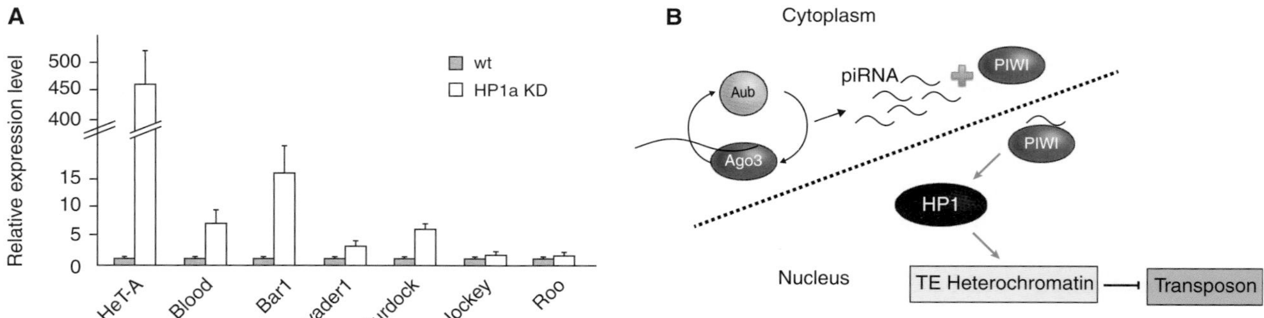

Figure 9. The piRNA system may drive silencing of some TEs through heterochromatin formation. (*A*) Depletion of HP1a (shown here), or of Piwi, results in overexpression of a subset of the TEs in the female germline. (*B*) This suggests that Piwi can promote transcriptional gene silencing, directly or indirectly, by targeting HP1a deposition at these TEs, presumably by a piRNA recognition event. (Adapted from Wang and Elgin 2011.)

Transcriptional gene silencing has also been observed for repeated sequences (repeat induced gene silencing), particularly in plants (Ch. 31 [Baulcombe and Dean 2014]). Analysis of the affected sequences has revealed the appearance of similar epigenetic marks (histone modifications and DNA methylation) as found in heterochromatin and in regions silenced by PEV. For example, if DNA fragments containing tandemly arranged luciferase genes are introduced into *Arabidopsis*, one observes either complete silencing or variegated luciferase expression. In the variegated line, the *Arabidopsis* SU(VAR)3-9 homolog SUVH2 shows a dosage dependent effect on silencing, suggesting that heterochromatin formation is responsible for the gene silencing observed and that the underlying molecular mechanisms are similar to those seen in other higher eukaryotic organisms (Naumann et al. 2005).

A central feature of pericentric heterochromatic gene silencing in *Drosophila* is the interaction of HP1a with H3K9me2/me3 and SU(VAR)3-9, a histone H3K9 HKMT. These three components of heterochromatin are highly conserved and found in the majority of eukaryotic taxa. HP1a is conserved from the fission yeast *Schizosaccharomyces pombe* to man, and is consistently associated with pericentric heterochromatin. The human HP1a genes can even be used to rescue the deficiency in *Drosophila* (Ma et al. 2001). However, a protein that binds to H3K9me2/me3, like HP1a, has not been identified in plants as such, and a clear homolog is lacking in some metazoans, such as *Caenorhabditis elegans*. SU(VAR)3-9 is even more widely represented, having been identified in fission yeast (clr4p), *Neurospora* (DIM5), *Arabidopsis* and mammals (SUV39h). All of these SU(VAR)3-9 homologs catalyze H3K9 methylation and function in heterochromatin formation. Also, as is the case for HP1a, a human SUV39H1 transgene can completely compensate for the loss of the endogenous *Drosophila* protein in mutant lines (Schotta et al. 2002). In higher plants (rice, *Arabidopsis*, and maize), there are in fact several SU(VAR)3-9 homologous proteins (SUVH) (Baumbusch et al. 2001). The high number of HKMTs may reflect the plasticity of plant development, or the need to respond to environmental factors (discussed further in Ch. 13 [Pikaard and Mittelsten Scheid 2014]).

Several other genes identified by *Drosophila Su(var)* mutations encode proteins with conserved functions. For instance, many of the key enzymes controlling histone modification are evolutionarily conserved, supporting the idea of a histone code (Jenuwein and Allis 2001). However, examination of the heterochromatin-specific histone modification marks observed in *Drosophila*, mammals, and plants (*Arabidopsis*) also identifies some genus-specific features (cf. Ebert et al. 2006), arguing that the histone code is not completely universal, but rather exists in different dialects.

9 SUMMING UP: THERE IS MUCH THAT WE DO NOT KNOW ABOUT HETEROCHROMATIN

Although PEV has provided us with an extraordinary opportunity to study heterochromatin formation and gene silencing, the development of the phenotype itself remains puzzling. Studies with an *hsp70*-driven *lacZ* reporter (that can be scored at any point in development and in most tissues) show that silencing begins in early embryogenesis, and is most extensive at cellular blastoderm, after which islands of inducible cells emerge (Lu et al. 1996). Events that occur during embryogenesis can impact the adult phenotype (Hartmann-Goldstein 1967). This persistence is of course implied by the concept of an epigenetic state: once formed at a specific site, heterochromatin will be maintained at that site through many rounds of mitosis by replication of the chromatin assembly. However, this "memory" has not yet been directly tested in *Drosophila*.

Many questions remain. Why do we observe a variegating pattern of silencing? What tips the balance, leading to a switch from the silent to the active state? When can that happen? What sets the patterns that indicate clonal inheritance, but are not easily related to the developmental process? Can faulty heterochromatin be restored? PEV is generally analyzed as a problem of maintaining the reporter gene "on" or "off," but in many instances (particularly when using *P* element based reporters) one observes red facets on a yellow or pale orange background, suggesting that gene expression has been reduced uniformly, but that that down-regulation has been lost in some cells. Careful analysis of such lines might lead to identification of chromatin states with an intermediate impact on gene expression. Although the data support a crude model for loss or maintenance of silencing based on mass action, the final model will be complex, involving numerous interacting proteins (see, e.g., the proposal by Henikoff, 1996). One is tempted to consider the nucleosome as a summation device, collecting modifications and displaying the results within a chromosomal domain in terms of both particular protein binding patterns, and facility for remodeling. The chromatin state might then reflect the results of competition for achieving different modifications. Such a model could be useful in sorting out the effects noted above. It is also compatible with observations demonstrating that the frequency of silencing of a GAL4-dependent reporter is sensitive to GAL4 levels (Ahmad and Henikoff 2001).

As might be anticipated for a system dependent on approximately 150 modifier loci, PEV phenotypes frequently differ in penetrance and expression of variegation in different inbred laboratory strains. For example, several genetically stable lines of $In(1)w^{m4}$ have been identified that differ significantly in expression of *white* variegation. In some cases the difference is linked to the heterochromatic breakpoint, suggesting that spontaneous alterations at the inactivating heterochromatic sequences are involved (Reuter et al. 1985). Variations in copy number of repeated sequences among strains could also have a significant effect on the degree of variegation caused by the titration of a fixed amount of key proteins required for heterochromatic packaging. Such changes may be more frequent in lines carrying *Su(var)* mutations, where heterochromatin structure is less stable (Peng and Karpen 2009).

It is standard technique to use inbred *Drosophila* lines with strong variegation when screening for or assessing putative *Su(var)* loci. However, fly genetics is not performed with homozygous lines and genetic background must always be carefully controlled. Different laboratories will make different choices of which reporter loci etc. to test, resulting in the presence of different chromosomes. It seems likely that the resulting differences in genetic background are the basic cause of some recent controversies in the field. The observed redundancies in the silencing system coupled with the absence of a recent transposon challenge in laboratory stocks appear to have allowed more variation in silencing system components than we might anticipate. This provides us with a very rich set of starting materials for investigating differences in penetrance and expressivity, a critical issue in understanding the health impact of such mutations in humans. We note that many basic features of the heterochromatin silencing system, first shown in *Drosophila*, have proven to be generally applicable.

The RNAi system provides a plausible mechanism for targeting heterochromatin formation to silence TEs in the germline, presumably by targeting a complex including HP1a, an H3K9 HKMT, or both. RNAi systems have repeatedly been found to play a role in genome surveillance and modification (see Ch. 11 [Chalker et al. 2013]). However, many questions remain for *Drosophila*. What is the source of the double-stranded RNA (dsRNA)? Must it be produced in *cis* (as implied by the results in *S. pombe*), or can it operate in *trans* (as suggested by results in plants) (i.e., can the production of dsRNA from one *1360* site result in targeting of all *1360* sites)? Must the target site be transcribed? Are all repetitious elements potential targets? The latter seems unlikely from the fourth chromosome analysis described above. If a subset of repetitious elements plays a key role, what determines that choice? The results obtained on the fourth chromosome indicate that the density and distribution of critical repetitious elements will impact expression of the genes in the vicinity. Thus, sequencing of an entire genome, not just the protein coding regions, is likely to be necessary when studying a novel organism.

How is spreading of heterochromatin accomplished, and what are the normal barriers to spreading? Note that there is no evidence for transitive RNAi, originally discovered in plants, in *Drosophila* (i.e., the spread of silencing to target genes that lie upstream of an introduced sequence that generates dsRNA [Celotto and Gravely 2002]). This is congruent with the lack of evidence for any RNA-dependent RNA polymerase in this system. An assembly system based on the interactions of HP1a, H3K9me2/3, and an HKMT might well account for the spread of heterochromatin for approximately 10 kb, as observed on the fourth chromosome; this type of spreading could be limited by a site of histone acetylation. But what about the spreading that occurs in rearrangements, which has been found to extend for hundreds of kb, as shown in Figure 5C? This form of spreading is not contiguous, but again appears to depend critically on chromatin proteins, notably JIL-1, in a role that does not depend on its kinase activity.

These and other questions remain unanswered. There is much to do to understand the heterochromatin system in *Drosophila*.

ACKNOWLEDGMENTS

We thank Gabriella Farkas for creating several of the original drawings used here, and the members of our research groups for a critical review of this chapter. Our work is supported by Deutsche Forschungsgemeinschaft (DFG) (G.R.) and by grants from the National Institutes of Health (S.C.R.E.).

REFERENCES

**Reference is also in this book.*

Ahmad K, Golic KG. 1996. Somatic reversion of chromosomal position effects in *Drosophila melanogaster*. *Genetics* **144:** 657–70.

Ahmad K, Henikoff S. 2001. Modulation of a transcription factor counteracts heterochromatin gene silencing in *Drosophila*. *Cell* **104:** 839–847.

* Allshire RC, Ekwall K. 2014. Epigenetic regulation of chromatin states in *Schizosaccharomyces pombe*. *Cold Spring Harb Perspect Biol* doi: 10.1101/cshperspect.a018770.

Aravin AA, Lagos-Quintana M, Yalcin A, Zavolan M, Marks D, Snyder B, Gaasterland T, Meyer J, Tuschl T. 2003. The small RNA profile during *Drosophila melanogaster* development. *Dev Cell* **5:** 337–350.

Ashburner M, Golic KG, Hawley RS. 2005. Chromosomes and position effect variegation. In *Drosophila: A laboratory handbook*, 2nd ed., Chaps. 4, 28. Cold Spring Harbor Laboratory Press, Cold Spring Harbor, NY.

Aulner N, Monod C, Mandicourt G, Jullien D, Cuvier O, Sall A, Janssen S, Laemmli UK, Kas E. 2002. The AT-hook protein D1 is essential for *Drosophila melanogaster* development and is implicated in position-effect variegation. *Mol Cell Biol* **22:** 1218–1232.

Baksa K, Morawietz H, Dombradi V, Axton M, Taubert H, Szabo G, Török I, Gyurkovics H, Szöör B, Gloover D, et al. 1993. Mutations in the phosphatase 1 gene at 87B can differentially affect suppression of position-effect variegation and mitosis in *Drosophila melanogaster*. *Genetics* **135:** 117–125.

* Baulcombe DC, Dean C. 2014. Epigenetic regulation in plant responses to the environment. *Cold Spring Harb Perspect Biol* **6:** a019471.

Baumbusch LO, Thorstensen T, Krauss V, Fischer A, Naumann K, Assalkhou R, Schulz I, Reuter G, Aalen R. 2001. The *Arabidopsis thaliana* genome contains at least 29 active genes encoding SET domain proteins that can be assigned to four evolutionary conserved classes. *Nucleic Acid Res* **29:** 4319–4333.

Belyaeva ES, Demakova OV, Umbetova GH, Zhimulev IF. 1993. Cytogenetic and molecular aspects of position-effect variegation in *Drosophila melanogaster*. V. Heterochromatin-associated protein HP1 appears in euchromatic chromosomal regions that are inactivated as a result of position-effect variegation. *Chromosoma* **102:** 583–590.

Birchler JA, Bhadra U, Rabinow L, Linsk R, Nguyen-Huyuh AT. 1994. *Weakener of white* (*Wow*), a gene that modifies the expression of white eye color locus and that suppresses position effect variegation in *Drosophila melanogaster*. *Genetics* **137:** 1057–1070.

Bird A. 2002. DNA methylation patterns and epigenetic memory. *Genes Dev* **16:** 6–21.

Blattes R, Monod C, Susbielle G, Cuvier O, Wu J-H, Hsieh T-S, Laemmli UK, Kas E. 2006. Displacement of D1, HP1 and topoisomerase II from satellite heterochromatin by a specific polyamide. *EMBO J* **25:** 2397–2408.

* Blewitt M, Whitelaw E. 2013. The use of mouse models to study epigenetics. *Cold Spring Harb Perspect Biol* **5:** a017939.

* Brockdorff N, Turner B. 2014. Dosage compensation in mammals. *Cold Spring Harb Perspect Biol* doi: 10.1101/cshperspect.a019406.

Brower-Toland B, Findley SD, Jiang L, Liu L, Yin H, Dus M, Zhou P, Elgin SCR, Lin H. 2007. *Drosophila* PIWI associates with chromatin and interacts directly with HP1a. *Genes Dev* **21:** 2300–2311.

Brower-Toland B, Riddle NC, Jiang H, Huisinga KL, Elgin SCR. 2009. Multiple SET methyltransferases are required to maintain normal heterochromatin domains in the genome of *Drosophila melanogaster*. *Genetics* **181:** 1303–1319.

Catcheside DG. 1939. A position effect in *Oenothera*. *J Genet* **38:** 345–352.

Cattanach BM. 1961. A chemically-induced variegated-type position effect in the mouse. *Z Vererbungsl* **92:** 165–182.

Celotto AM, Graveley BR. 2002. Exon-specific RNAi: A tool for dissecting the functional relevance of alternative splicing. *RNA* **8:** 718–724.

* Chalker DL, Meyer E, Mochizuki K. 2013. Epigenetics of ciliates. *Cold Spring Harb Perspect Biol* **5:** a017764.

Cleard F, Spierer P. 2001. Position-effect variegation in *Drosophila*: The modifier *Su(var)3-7* is a modular DNA-binding protein. *EMBO Rep* **21:** 1095–1100.

Cleard F, Delattre M, Spierer P. 1997. SU(VAR)3-7, a *Drosophila* heterochromatin-associated protein and companion of HP1 in the genomic silencing of position-effect variegation. *EMBO J* **16:** 5280–5288.

Cooper KW. 1959. Cytogenetic analysis of major heterochromatic elements (especially Xh and Y) in *Drosophila melanogaster* and the theory of "heterochromatin". *Chromosoma* **10:** 535–588.

Cortes A, Azorin F. 2000. DDP1, a heterochromatin-associated multi-KH-domain protein of *Drosophila melanogaster*, interacts specifically with centromeric satellite DNA sequences. *Mol Cell Biol* **20:** 3860–3869.

Cryderman DE, Tang H, Bell C, Gilmour DS, Wallrath LL. 1999a. Heterochromatic silencing of *Drosophila* heat shock genes acts at the level of promoter potentiation. *Nucleic Acids Res* **27:** 3364–3370.

Cryderman DE, Morris EJ, Biessmann H, Elgin SCR, Wallrath LL. 1999b. Silencing at *Drosophila* telomeres: Nuclear organization and chromatin structure play critical roles. *EMBO J* **18:** 3724–3735.

Cryderman DE, Grade SK, Li Y, Fanti L, Pimpinelli S, Wallrath LL. 2005. Role of HP1 in euchromatic gene expression. *Dev Dyn* **232:** 767–774.

Czermin B, Schotta G, Hülsmann BB, Brehm A, Becker PB, Reuter G, Imhof A. 2001. Physical and functional interaction of SU(VAR)3-9 and HDAC1 in *Drosophila*. *EMBO Rep* **2:** 915–919.

* Dekker J, Misteli T. 2014. Long-range chromatin interactions. *Cold Spring Harb Perspect Biol* doi: 10.1101/cshperspect.a019356.

Delattre M, Spierer A, Tonka CH, Spierer P. 2000. The genomic silencing of position-effect variegation in *Drosophila melanogaster*: Interaction between the heterochromatin-associated proteins Su(var)3-7 and HP1. *J Cell Sci* **113:** 4253–4261.

Demerec M, Slizynska H. 1937. Mottled white 258-18 of *Drosophila melanogaster*. *Genetics* **22:** 641–649.

Donaldson KM, Lui A, Karpen GH. 2002. Modifiers of terminal deficiency-associated position effect variegation in *Drosophila*. *Genetics* **160:** 995–1009.

Dorer DR, Henikoff S. 1994. Expansion of transgene repeats causes heterochromatin formation and gene silencing in *Drosophila*. *Cell* **77:** 993–1002.

Dorn R, Krauss V, Reuter G, Saumweber H. 1993. The enhancer of position-effect variegation *E(var)93D*, codes for a chromatin protein containing a conserved domain common to several transcriptional regulators. *Proc Natl Acad Sci* **90:** 11376–11380.

Ebert A, Schotta G, Lein S, Kubicek S, Krauss V, Jenuwein T, Reuter G. 2004. *Su(var)* genes regulate the balance between euchromatin and heterochromatin in *Drosophila*. *Genes Dev* **18:** 2973–2983.

Ebert A, Lein S, Schotta G, Reuter G. 2006. Histone modification and the control of heterochromatic gene silencing in *Drosophila*. *Chromosome Res* **14:** 377–392.

Egelhofer TA, Minoda A, Klugman S, Lee K, Kolasinska-Zwierz P, Alekseyenko AA, Cheung MS, Day DS, Gadel S, Gorchakov AA, et al. 2011. An assessment of histone-modification antibody quality. *Nat Struct Mol Biol* **18:** 91–93.

Eissenberg JC, Reuter G. 2009. Cellular mechanism for targeting heterochromatin formation in *Drosophila*. *Int Rev Cell Mol Biol* **273:** 1–47.

Eissenberg JC, James TC, Foster-Hartnett DM, Hartnett T, Ngan V, Elgin SCR. 1990. A mutation in a heterochromatin-specific chromosomal protein is associated with suppression of position effect variegation in *Drosophila melanogaster*. *Proc Natl Acad Sci* **87:** 9923–9927.

Eissenberg JC, Morris GD, Reuter G, Hartnett T. 1992. The hetero-chromatin-associated protein HP-1 is an essential protein in *Drosophila* with dosage-dependent effects on position-effect variegation. *Genetics* **131:** 345–352.

Eskeland R, Eberharter A, Imhof A. 2007. HP1 binding to chromatin methylated at H3K9 is enhanced by auxiliary factors. *Mol Cell Biol* **27:** 453–465.

Fagegaltier D, Bouge A-L, Berry B, Poisot E, Sismeiro O, Coppee J-Y, Theodore L, Voinnet O, Antoniewski C. 2009. The endogenous siRNA pathway is involved in heterochromatin formation in *Drosophila*. *Proc Natl Acad Sci* **106:** 21258–21263.

Fanti L, Pimpinelli S. 2004. Immunostaining of squash preparations of chromosomes of larval brains. *Methods Mol Biol* **247:** 353–361.

Fanti L, Giovinazzo G, Berloco M, Pimpinelli S. 1998. The heterochromatin protein 1 prevents telomere fusions in *Drosophila*. *Mol Cell* **2:** 527–538.

Farkas G, Gausz J, Galloni M, Reuter G, Gyurkovics H, Krach F. 1994. The *trithorax-like* gene encodes the *Drosophila* GAGA factor. *Nature* **371:** 806–808.

Filion GJ, van Bemmel JG, Braunschweig U, Talhout W, Kind J, Ward LD, Brugman W, de Castro IJ, Kerkhoven RM, Bussemaker HJ, et al. 2010. Systematic protein location mapping reveals five principal chromatin types in *Drosophila* cells. *Cell* **143:** 1–13.

Fischle W, Wang Y, Jacobs SA, Kim Y, Allis CD, Khorasanizadeh S. 2003. Molecular basis for the discrimination of repressive methyl-lysine marks in histone H3 by Polycomb and HP1 chromodomains. *Genes Dev* **17:** 1870–1881.

FlyBase. 2012. *The* Drosophila *database*. Available from the World Wide Web at the URLs http://morgan/harvard.edu and http://www.ebi.ac.uk/flybase/.

Fodor DB, Shukeir N, Reuter G, Jenuwein T. 2010. Mammalian *Su(var)* genes in chromatin control. *Annu Rev Cell Dev Biol* **26:** 471–501.

Foe VE, Alberts BM. 1985. Reversible chromosome condensation induced in *Drosophila* embryos by anoxia: Visualization of interphase nuclear organization. *J Cell Biol* **100:** 1623–1636.

Girton JR, Johansen KM. 2008. Chromatin structure and the regulation of gene expression: The lessons of PEV in *Drosophila*. *Adv Genet* **61:** 1–43.

Greil F, de Wit E, Bussemaker HJ, van Steensel B. 2007. HP1 controls genomic targeting of four novel heterochromatin proteins in *Drosophila*. *EMBO J* **26:** 741–751.

* Grossniklaus U, Paro R. 2014. Transcriptional silencing by *Polycomb*-group proteins. *Cold Spring Harb Perspect Biol* **6:** a019331.

Harmon B, Sedat J. 2005. Cell-by-cell dissection of gene expression and chromosomal interactions reveals consequences of nuclear reorganization. *PLoS Biol* **3:** e67.

Hartmann-Goldstein IJ. 1967. On the relationship between heterochromatization and variegation in *Drosophila*, with special reference to temperature sensitive periods. *Genet Res* **10:** 143–159.

Haynes KA, Caudy AA, Collins L, Elgin SCR. 2006. Element *1360* and RNAi components contribute to HP1-dependent silencing of a pericentric reporter. *Curr Biol* **16:** 2222–2227.

Henikoff S. 1996. Dosage-dependent modification of position-effect variegation in *Drosophila*. *Bioessays* **18:** 401–409.

Henikoff S, Jackson JM, Talbert PB. 1995. Distance and pairing effects on the $brown^{Dominant}$ heterochromatic element in *Drosophila*. *Genetics* **140:** 1007–1017.

Hoskins RA, Carlson JW, Kennedy C, Acevedo D, Evans-Holm M, Frise E, Wan KH, Park S, Mendez-Lago M, Rossi F, et al. 2007. Sequence finishing and mapping of *Drosophila melanogaster* heterochromatin. *Science* **316:** 1625–1628.

Howe M, Dimitri P, Berloco M, Wakimoto BT. 1995. *Cis*-effects of heterochromatin on heterochromatic and euchromatic gene activity in *Drosophila melanogaster*. *Genetics* **140:** 1033–1045.

Huisinga KL, Elgin SCR. 2009. Small RNA-directed heterochromatin formation in the context of development: What flies might learn from fission yeast. *Biochim Biophys Acta* **1789:** 3–16.

Jacobs SA, Taverna SD, Zhang Y, Briggs SD, Eissenberg JC, Allis CD, Khorasanizadeh S. 2001. Specificity of the HP1 chromo domain fort he methylated N-terminus of histone H3. *EMBO J* **20:** 5232–5241.

James TC, Elgin SCR. 1986. Identification of a nonhistone chromosomal protein associated with heterochromatin in *Drosophila melanogaster* and its gene. *Mol Cell Biol* **6:** 3862–3872.

James TC, Eissenberg JC, Craig C, Dietrich V, Hobson A, Elgin SCR. 1989. Distribution patterns of HP1, a heterochromatin-associated nonhistone chromosomal protein of *Drosophila*. *Eur J Cell Biol* **50:** 170–180.

Jenuwein T, Allis CD. 2001. Translating the histone code. *Science* **293:** 1074–1080.

Johansson AM, Stenberg P, Pettersson F, Larsson J. 2007a. POF and HP1 bind expressed exons, suggesting a balancing mechanism for gene regulation. *PLoS Genet* **3:** e209.

Johansson AM, Stenberg P, Bernhardsson C, Larsson J. 2007b. Painting of fourth and chromosome-wide regulation of the 4th chromosome in *Drosophila melanogaster*. *EMBO J* **26:** 2307–2316.

Johansson AM, Stenberg P, Allgardsson A, Larsson J. 2012. POF regulates the expression of genes on the fourth chromosome in *Drosophila melanogaster* by binding to nascent RNA. *Mol Cell Biol* **32:** 2121–2134.

Kaminker JS, Bergman CM, Kronmiller B, Carlson J, Svirskas R, Patel S, Frise E, Wheeler DA, Lewis SE, Rubin GM, et al. 2002. The transposable elements of the *Drosophila melanogaster* genome: A genomics perspective. *Genome Biol* **3:** RESEARCH0084.

Karpen GH, Spradling AC. 1990. Reduced DNA polytenization of a minichromosome region undergoing position-effect variegation in *Drosophila*. *Cell* **63:** 97–107.

Kavi HH, Fernandez H, Xie W, Birchler JA. 2008. Genetics and biochemistry of RNAi in *Drosophila*. *Curr Top Microbiol Immunol* **320:** 37–75.

Kharchenko PV, Alekseyenko AA, Schwartz YB, Minoda A, Riddle NC, Ernst J, Sabo PJ, Larschan E, Gorchakov AA, Gu T, et al. 2011. Comprehensive analysis of the chromatin landscape in *Drosophila melanogaster*. *Nature* **471:** 480–485.

Khurana JS, Wang J, Xu J, Koppetsch BS, Thomson TC, Nowosielska A, Li C, Zamore PD, Wenig Z, Theurkauf WE. 2011. Adaptation to *P* element transposon invasion in *Drosophila melanogaster*. *Cell* **147:** 1551–1563.

* Kingston R, Tamkun J. 2014. Transcriptional regulation by trithorax-group proteins. *Cold Spring Harb Perspect Biol* doi: 10.1101/cshperspect.a019349.

Klenov MS, Lavrov SA, Stolyarenko AD, Ryazansky SS, Aravin AA, Tuschl T, Gvozdev VA. 2007. Repeat-associated siRNAs cause chromatin silencing of retrotransposons in the *Drosophila melanogaster* germline. *Nucleic Acids Res* **35:** 5430–5438.

Klenov MS, Sokolova OA, Yakushev EY, Stolyarenko AD, Mikhaleva EA, Lavrov SA, Gvozdev VA. 2011. Separation of stem cell maintenance and transposon silencing functions of Piwi protein. *Proc Natl Acad Sci* **108:** 18760–18765.

Krauss V, Reuter G. 2011. DNA methylation in *Drosophila*—A critical evaluation. *Prog Mol Biol Transl Sci* **101:** 177–191.

Laible G, Wolf A, Dorn R, Reuter G, Nislow C, Lebesorger A, Popkin D, Pillus L, Jenuwein T. 1997. Mammalian homologs of *Enhancer of zeste* mediate position-effect variegation in *Drosophila* and restore telomeric silencing in *S. cerevisiae*. *EMBO J* **16:** 3219–3232.

Larsson J, Zhang J, Rasmuson-Lestander A. 1996. Mutations in the *Drosophila melanogaster S*-adenosylmethionine synthase suppress position-effect variegation. *Genetics* **143:** 887–896.

Larsson J, Chen JD, Rasheva V, Rasmuson-Lestander A, Pirrotta V. 2001. Painting of fourth, a chromosome-specific protein in *Drosophila*. *Proc Natl Acad Sci* **98:** 6273–6278.

Le MH, Duricka D, Karpen GH. 1995. Islands of complex DNA are widespread in *Drosophila* centric heterochromatin. *Genetics* **141:** 283–303.

Lemos B, Branco A, Hartl D. 2010. Epigenetic effects of polymorphic Y chromosomes modulate chromatin components, immune response, and sexual conflict. *Proc Natl Acad Sci* **107:** 15826–15831.

Locke J, Kotarski MA, Tartof KD. 1988. Dosage-dependent modifiers of position effect variegation in *Drosophila* and a mass action model that explains their effect. *Genetics* **120:** 181–198.

Lu BY, Bishop CP, Eissenberg JC. 1996. Developmental timing and tissue specificity of heterochromatin-mediated silencing. *EMBO J* **15:** 1323–1332.

* Lucchesi J, Kuroda M. 2014. Dosage compensation in *Drosophila*. *Cold Spring Harb Perspect Biol* doi: 10.1101/cshperspect.a019398.

Ma J, Hwang KK, Worman HJ, Courvalin JC, Eissenberg JC. 2001. Expression and functional analysis of three isoforms of human heterochromatin-associated protein HP1 in *Drosophila*. *Chromosoma* **109:** 536–544.

Martienssen RA, Colot V. 2001. DNA methylation and epigenetic inheritance in plants and filamentous fungi. *Science* **293:** 1070–1074. modMine. 2011. Available on the World Wide Web at the URL http://intermine.modencode.org/.

* Martienssen R, Moazed D. 2014. RNAi and heterochromatin assembly. *Cold Spring Harb Perspect Biol* doi: 10.1101/cshperspect.a019323.

Mottus R, Sobels RE, Grigliatti TA. 2000. Mutational analysis of a histone deacetylase in *Drosophila melanogaster*: Missence mutations suppress gene silencing associated with position effect variegation. *Genetics* **154:** 657–668.

Muller HJ. 1930. Types of visible variations induced by X-rays in *Drosophila*. *J Genet* **22:** 299–334.

Naumann K, Fischer A, Hofmann I, Krauss V, Phalke S, Irmler K, Hause G, Aurich AC, Dorn R, Jenuwein T, et al. 2005. Pivotal role of *At*SUVH2 in control of heterochromatic histone methylation and gene silencing in *Arabidopsis*. *EMBO J* **24:** 1418–1429.

Pal-Bhadra M, Leibovitch BA, Gandhi SG, Rao M, Bhadra U, Birchler JA, Elgin SCR. 2004. Heterochromatic silencing and HP1 localization in *Drosophila* are dependent on the RNAi machinery. *Science* **303:** 669–672.

* Patel DJ. 2014. A structural perspective on readout of epigenetic histone and DNA methylation marks. *Cold Spring Harb Perspect Biol* doi: 10.1101/cshperspect.a018754.

Peng J, Karpen G. 2009. Heterochromatic genome stability requires regulators of histone H3 K9 methylation. *PLoS Genet* **5:** e1000435.

Phalke S, Nickel O, Walluscheck D, Hortig F, Onorati MC, Reuter G. 2009. Epigenetic control of retrotransposon silencing and telomere integrity in somatic cells of *Drosophila* depends on the cytosine-5 methyltransferase DNMT2. *Nature Genet* **41:** 696–702.

Piacentini L, Fanti L, Negri R, Del Vescovo V, Fatica A, Altieri F, Pimpinelli S. 2009. Heterochromatin protein 1 (HP1a) positively regulates euchromatic gene expression through RNA transcript association and interaction with hnRNPs in *Drosophila*. *PLoS Genet* **5:** e1000670.

* Pikaard CS, Mittelsten Scheid O. 2014. Epigenetic regulation in plants. *Cold Spring Harb Perspect Biol* doi: 10.1101/cshperspect.a019315.

Platero JS, Sharp EJ, Adler PN, Eissenberg JC. 1996. In vivo assay for protein-protein interaction using *Drosophila* chromosomes. *Chromosoma* **104:** 393–404.

Rea S, Eisenhaber F, O'Carroll D, Strahl BD, Sun Z-W, Schmid M, Opravil S, Mechtler K, Ponting CP, Allis CD, et al. 2000. Regulation of chromatin structure by site-specific histone H3 methyltransferases. *Nature* **406:** 593–599.

Reuter G, Werner W, Hofmann HJ. 1982. Mutants affecting position-effect heterochromatinization in *Drosophila melanogaster*. *Chromosoma* **85:** 539–551.

Reuter G, Wolff I, Friede B. 1985. Functional properties of the heterochromatic sequences inducing w^{m4} position-effect variegation in *Drosophila melanogaster*. *Chromosoma* **93:** 132–139.

Reuter G, Giarre N, Farah J, Gausz J, Spierer A, Spierer P. 1990. Dependence of position-effect variegation in *Drosophila* on dose of a gene encoding an unusual zinc-finger protein. *Nature* **344:** 219–223.

Riddle NC, Leung W, Haynes KA, Granok H, Wuller J, Elgin SCR. 2008. An investigation of heterochromatin domains on the fourth chromosome of *Drosophila melanogaster*. *Genetics* **178:** 1177–1191.

Riddle NC, Minoda A, Kharchenko PV, Alekseyenko AA, Schwartz YB, Tolstorukov MY, Gorchakov AA, Kennedy C, Linder-Basso D, Jaffe JD, et al. 2010. Plasticity in patterns of histone modifications and chromosomal proteins in the *Drosophila* heterochromatin. *Genome Res* **21:** 147–163.

Riddle NC, Minoda A, Kharchenko PV, Alekseyenko AA, Schwartz YB, Tolstorukov MY, Gorchakov AA, Jaffe JD, Kennedy C, Linder-Basso D, et al. 2011. Plasticity in patterns of histone modifications and chromosomal proteins in *Drosophila* heterochromatin. *Genome Res* **21:** 147–163.

Riddle NC, Jung Y, Gu T, Alekseyenko AA, Asker D, Gui H, Kharchenko PV, Minoda A, Plachetka A, Schwartz YB, et al. 2012. Enrichment of HP1a on *Drosophila* chromosome 4 genes creates an alternate chromatin structure critical for regulation in this heterochromatic domain. *PLoS Genet* **8:** e1002954.

Rudolph T, Yonezawa M, Lein S, Heidrich K, Kubicek S, Schafer C, Phalke S, Walther M, Schmidt Q, Jenuwein T, et al. 2007. Heterochromatin formation in *Drosophila* is initiated through active removal of H3K4 methylation by the LSD1 homolog SU(VAR)3-3. *Mol Cell* **13:** 103–115.

Russel LB, Bangham JW. 1961. Variegated type position effects in the mouse. *Genetics* **46:** 509–525.

Schneiderman JI, Goldstein S, Ahmad K. 2010. Perturbation analysis of heterochromatin-mediated gene silencing and somatic inheritance. *PLoS Genet* **6:** e1001095.

Schotta G, Ebert A, Krauss V, Fischer A, Hoffmann J, Rea S, Jenuwein T, Reuter G. 2002. Central role of *Drosophila* SU(VAR)3-9 in histone H3-K9 methylation and heterochromatic gene silencing. *EMBO J* **21:** 1121–1131.

Schotta G, Ebert A, Dorn R, Reuter G. 2003. Position-effect variegation and the genetic dissection of chromatin regulation in *Drosophila*. *Semin Cell Dev Biol* **14:** 67–75.

Schotta G, Lachner M, Sarma K, Ebert A, Sengupta R, Reuter G, Reinberg D, Jenuwein T. 2004. A silencing pathway to induce H3-K9 and H4-K20 trimethylation at constitutive heterochromatin. *Genes Dev.* **18:** 1251–1262.

Schultz J. 1950. Interrelations of factors affecting heterochromatin-induced variegation in *Drosophila*. *Genetics* **35:** 134.

Schulze SR, Wallrath LL. 2007. Gene regulation by chromatin structure: Paradigms established in *Drosophila melanogaster*. *Annu Rev Entomol* **52:** 171–192.

Senti KA, Brennecke J. 2010. The piRNA pathway: A fly's perspective on the guardian of the genome. *Trends Genet* **26:** 499–509.

Sentmanat MF, Elgin SCR. 2012. Ectopic assembly of heterochromatin in *Drosophila melanogaster* triggered by transposable elements. *Proc Natl Acad Sci* **109:** 14104–14109.

Seum C, Pauli D, Delattre M, Jaquet Y, Spierer A, Spierer P. 2002. Isolation of *Su(var)3-7* mutations by homologous recombination in *Drosophila melanogaster*. *Genetics* **161:** 1125–1136.

Seum C, Reo E, Peng H, Rauscher FJ, Spierer P, Bontron S. 2007. *Drosophila* SETDB1 is required for chromosome 4 silencing. *PLoS Genet* **3:** e76.

Shaffer CD, Stephens GE, Thompson BA, Funches L, Bernat JA, Craig CA, Elgin SCR. 2002. Heterochromatin protein 2 (HP2), a partner of HP1 in *Drosophila* heterochromatin. *Proc Natl Acad Sci* **99:** 14322–14337.

Shaffer C, Cenci G, Thompson B, Stephens GE, Slawson E, Adu-Wusu K, Gatti M, Elgin SCR. 2006. The large isoform of *Drosophila melanogaster* heterochromatin protein 2 plays a critical role in gene silencing and chromosome structure. *Genetics* **174:** 1189–1204.

Shi Y, Lan F, Matson C, Mulligan P, Whetstine JR, Cole PA, Casero RA, Shi Y. 2005. Histone demethylation mediated by the nuclear amine oxidase homolog LSD1. *Cell* **119:** 941–953.

Silver LM, Elgin SCR. 1976. A method for determination of the in situ distribution of chromosomal proteins. *Proc Natl Acad Sci* **73:** 423–427.

Sinclair DAR, Mottus RC, Grigliatti TA. 1983. Genes which suppress position effect variegation in *Drosophila melanogaster* are clustered. *Mol Gen Genet* **191:** 326–333.

Sinclair DAR, Clegg NJ, Antonchuk J, Milner TA, Stankunas K, Ruse C, Grigliatti TA, Kassis J, Brock HW. 1998. Enhancer of Polycomb is a suppressor of position-effect variegation in *Drosophila melanogaster*. *Genetics* **148:** 211–220.

Smith CD, Shu S, Mungall CJ, Karpen GH. 2007. The Release 5.1 annotation of *Drosophila melanogaster* heterochromatin. *Science* **316:** 1586–1591.

Spofford JB. 1967. Single-locus modification of position-effect variegation in *Drosophila melanogaster*. I. White variegation. *Genetics* **57:** 751–766.

Stephens GE, Craig CA, Li Y, Wallrath LL, Elgin SCR. 2003. Immunofluorescent staining of polytene chromosomes: Exploiting genetic tools. *Methods Enzymol* **376:** 372–393.

Sun F-L, Cuaycong MH, Elgin SCR. 2001. Long-range nucleosome ordering is associated with gene silencing in *Drosophila melanogaster* pericentromeric heterochromatin. *Mol Cell Biol* **21:** 2867–2879.

Sun F-L, Haynes K, Simpson CL, Lee SD, Collins L, Wuller J, Eissenberg JC, Elgin SCR. 2004. *cis*-Acting determinants of heterochromatic formation on *Drosophila melanogaster* chromosome four. *Mol Cell Biol* **24:** 8210–8220.

Talbert PB, Henikoff S. 2000. A reexamination of spreading of position-effect variegation in the *white-roughest* region of *Drosophila melanogaster*. *Genetics* **154:** 259–272.

Tartof KD, Hobbs C, Jones M. 1984. A structural basis for variegating position effects. *Cell* **37:** 869–878.

Tschiersch B, Hofmann A, Krauss V, Dorn R, Korge G, Reuter G. 1994. The protein encoded by the *Drosophila* position effect variegation suppressor gene *Su(var)3-9* combines domains of antagonistic regulators of homeotic gene complexes. *EMBO J* **13:** 3822–3831.

Tzeng TY, Lee CH, Chan LW, Shen CK. 2007. Epigenetic regulation of the *Drosophila* chromosome 4 by the histone H3K9 methylatransferase dSETDB1. *Proc Natl Acad Sci* **104:** 12691–12696.

Vogel MJ, Peric-Hupkes D, van Steensel B. 2007. Detection of in vivo protein-DNA interactions using DamID in mammalian cells. *Nat Protoc* **2:** 1467–1478.

Vogel MJ, Pagie L, Talhout W, Nieuwland M, Kerkhoven RM, van Steensel B. 2009. High-resolution mapping of heterochromatin redistribution in a *Drosophila* position-effect variegation model. *Epigenetics Chromatin* **2:** 1.

Wakimoto BT, Hearn MG. 1990. The effects of chromosome rearrangements on the expression of heterochromatic genes in chromosome 2L of *Drosophila melanogaster. Genetics* **125:** 141–154.

Wallrath LL, Elgin SCR. 1995. Position effect variegation in *Drosophila* is associated with an altered chromatin structure. *Genes Dev* **9:** 1263–1277.

Wallrath LL, Gunter VP, Rosman LE, Elgin SCR. 1996. DNA representation of variegating heterochromatic P element inserts in diploid and polytene tissue of *Drosophila melanogaster. Chromosoma* **104:** 519–527.

Wang Y, Zhang W, Jin Y, Johansen J, Johansen KM. 2001. The JIL-1 tandem kinase mediates histone H3 phosphorylation and is required for maintenance of chromatin structure in *Drosophila. Cell* **105:** 433– 443.

Wang SH, Elgin SCR. 2011. *Drosophila* Piwi functions downstream of piRNA production mediating a chromatin-based transposon silencing mechanism in female germ line. *Proc Natl Acad Sci* **108:** 21164–21169.

Yasuhara JC, Wakimoto BT. 2008. Molecular landscape of modified histones in *Drosophila* heterochromatic genes and euchromatin-heterochromatin transition zones. *PLoS Genet* **4:** e16.

Zhimulev IF, Belyaeva ES, Fomina OV, Protopopov MO, Bolshkov VN. 1986. Cytogenetic and molecular aspects of position effect variegation in *Drosophila melanogaster. Chromosoma* **94:** 492–504.

 Cite as *Cold Spring Harb Perspect Biol* doi: 10.1101/cshperspect.a017780

C H A P T E R 13

Epigenetic Regulation in Plants

Craig S. Pikaard[1] and Ortrun Mittelsten Scheid[2]

[1]Department of Biology, Department of Molecular and Cellular Biochemistry, and Howard Hughes Medical Institute, Indiana University, Bloomington, Indiana 47405; [2]Gregor Mendel-Institute of Molecular Plant Biology, Austrian Academy of Sciences, 1030 Vienna, Austria

Correspondence: ortrun.mittelsten_scheid@gmi.oeaw.ac.at

SUMMARY

The study of epigenetics in plants has a long and rich history, from initial descriptions of non-Mendelian gene behaviors to seminal discoveries of chromatin-modifying proteins and RNAs that mediate gene silencing in most eukaryotes, including humans. Genetic screens in the model plant *Arabidopsis* have been particularly rewarding, identifying more than 130 epigenetic regulators thus far. The diversity of epigenetic pathways in plants is remarkable, presumably contributing to the phenotypic plasticity of plant postembryonic development and the ability to survive and reproduce in unpredictable environments.

Outline

Chapter-opening figures: (*Left, right*) Kindly provided by Enrico Coen.

Cite as *Cold Spring Harb Perspect Biol* doi: 10.1101/cshperspect.a019315

OVERVIEW

Plants are masters of epigenetic regulation. All of the major epigenetic mechanisms known to occur in eukaryotes are used by plants, with the responsible pathways elaborated to a degree that is unsurpassed in other taxa. DNA methylation occurs in CG, CHG, and CHH sequence contexts in plant genomes, in patterns that reflect a balance between enzyme activities that install, maintain, or remove methylation. As in other eukaryotes, histone-modifying enzymes influence epigenetic states in plants and these enzymes are encoded by comparatively large gene families, allowing for diversified as well as overlapping functions. RNA-mediated gene silencing is accomplished using multiple distinct pathways to combat viruses, tame transposons, orchestrate development, and help organize the genome. The interplay between DNA methylation, histone modification, and noncoding RNAs provides plants with a multilayered and robust epigenetic circuitry.

The prominence of epigenetic regulation in plants reflects their mode of development, lifestyle, and evolutionary history. Unlike growth in mammals, in which organ and tissue formation is largely specified during embryonic development, plants grow by continuously producing new organs from self-sustaining stem cell populations known as meristems. Consequently, postembryonic development in plants is a continuous process shaped by environmental influences resulting in a high degree of phenotypic plasticity. Because plants are unable to escape their surroundings, they are forced to cope with changeable and often unfavorable growth conditions. Epigenetic regulatory mechanisms can facilitate metastable changes in gene activity and fine-tune gene expression patterns, thus enabling plants to survive and reproduce successfully in unpredictable environments. Polyploidization, an increase in the number of sets of chromosomes, is common in plants, amplifying gene families and fostering the functional specialization of duplicated genes, including those involved in epigenetic regulation.

Understanding the epigenetic regulatory machinery of plants has come, in large part, from genetic screens, most notably in *Arabidopsis thaliana*, a member of the mustard family that is highly amenable to genetic analyses and was the first plant species to have its genome sequenced. Crop plants, particularly maize, have also contributed substantially to the discovery of epigenetic phenomena and epigenetic regulatory mechanisms. The study of plant epigenetics and epigenomics has a long and rich history and, in synergy with parallel studies in animal and fungal systems, is contributing significantly to our basic understanding of epigenetic regulation.

Cite as *Cold Spring Harb Perspect Biol* doi: 10.1101/cshperspect.a019315

1 PLANTS AS MODELS FOR EPIGENETIC RESEARCH

1.1 Overview

Plant studies have provided numerous seminal contributions to the field of epigenetics. Among them is the distinction between euchromatin and heterochromatin based on cytological analyses (Heitz 1929). The observation in tomato and maize of heritable changes in a gene's expression state on exposure to an allele with an alternative state, a phenomenon known as paramutation, was early evidence for non-Mendelian epigenetic inheritance (reviewed in Arteaga-Vazquez and Chandler 2010) and is now apparent in mammals and flies, as well as plants. Parental imprinting of individual genes, that is, the expression of a single allele of either maternal or paternal origin, which is a process whose misregulation is the basis for multiple genetic disorders in humans (see Ch. 33 [Zoghbi and Beaudet 2014]), was first observed in maize (reviewed in Alleman and Doctor 2000). The repeated occurrence of individuals with altered flower symmetry, first described by Carl von Linné in the 18th century as "peloria" (monster) individuals, is now known to be caused by the formation of a silenced epiallele whose DNA sequence is identical to expressed alleles of the gene (Cubas et al. 1999). Epialleles can affect developmental switches; the *FWA* gene in *Arabidopsis* is an example in which silencing in some natural ecotypes (strains) delays flowering (Soppe et al. 2000). Cytogenetic analyses of metaphase chromosomes in plant hybrids in the 1930s showed that secondary constrictions and other cytogenetic characteristics were reproducibly altered in certain hybrid combinations. This differential production of secondary constrictions was later shown to be caused by the reproducible silencing of one parental set of ribosomal RNA (rRNA) genes, and the selective expression of the other parental set at these loci, an epigenetic phenomenon widespread among eukaryotes and known as nucleolar dominance (reviewed in Preuss and Pikaard 2007). Only the active rRNA genes gives rise to secondary constrictions, which are less condensed than flanking regions because of the persistent binding of RNA polymerase I transcription factors throughout the cell cycle. The pioneering work on transposable elements in maize by Barbara McClintock and others in the 1940s revealed numerous links between genetic behavior and epigenetic regulation (reviewed in Lisch 2009). Indeed, extant transposons and degenerate remains of transposons provide a means of establishing new regulatory elements and epigenetic modifications in eukaryotic genomes.

As transgenic technology in plants such as tobacco, petunia, and *Arabidopsis* became routine in the late 1980s, it became evident that transgene expression was unpredictable (reviewed in Wassenegger 2002; Matzke and Matzke 2004). This led to the realization that homology-dependent gene silencing can occur when transgenes bearing identical sequences to genes already present in the genome are introduced, with an enhanced messenger RNA (mRNA) turnover (posttranscriptional gene silencing, PTGS) or repression of transcription (transcriptional gene silencing, TGS) being the molecular basis for silencing. A striking example of PTGS, termed "cosuppression," occurred when overexpression of the gene encoding chalcone synthase (*CHS*), a regulator of flower pigmentation, was attempted in petunia, this unexpectedly resulted in variegated or even completely white flowers rather than darker purple flowers. The lack of pigmentation in white sectors was because of coordinated gene silencing of both the *CHS* transgene and the endogenous *CHS* gene (reviewed in Jorgensen et al. 2006). PTGS is a form of RNA interference (RNAi), later discovered in *Caenorhabditis elegans* and other organisms (see Sec. 3).

By the mid-1990s, links between PTGS and virus resistance had been forged. Epigenetic mechanisms involving RNA molecules were shown to naturally protect plants from uncontrolled replication of viruses, which can be both inducers and targets of PTGS. It was soon realized that these mechanisms could be exploited to experimentally down-regulate plant genes by constructing viral vectors containing plant gene sequences, thus harnessing the virus-induced gene silencing pathway (VIGS, a topic discussed in more detail in Ch. 31 [Baulcombe and Dean 2014] and reviewed in Senthil-Kumar and Mysore 2011). Another mechanism discovered in viroid-infected plants is the RNA-guided specification of DNA methylation (RNA-directed DNA methylation, RdDM) and other heterochromatic marks at homologous DNA regions (Wassenegger et al. 1994). Transgene-induced RdDM is now widely used to methylate and transcriptionally silence promoters and is accomplished by intentionally generating double-stranded RNA (dsRNA) matching promoter sequences (see Sec. 3.4), a strategy that also works in animals (reviewed in Verdel et al. 2009).

Another pioneering contribution of plant epigenetic research was the application of deep sequencing to conduct genome-wide analyses (Zhang et al. 2006), including the first analysis of genomic methylation at single-base resolution (Cokus et al. 2008; Lister et al. 2008). These studies required the development of bioinformatics tools that have since been applied broadly for biomedical research.

1.2 The Epigenomes of Plants and Mammals Are Similar

Despite their obvious morphological differences and long evolutionary separation, the fundamental similarities between animals and plants are remarkable. These similarities include many aspects of genome and epigenome organization, especially between plants and mammals. For instance, genome size, genome complexity, and the ratio of heterochromatin to euchromatin in seed plants are generally comparable to mammals. Also, plants and mammals make similar use of DNA methylation and histone post-translational modifications (PTMs) for gene regulation. Overall, a comparison of genome organization and epigenetic regulation in different model systems reveals that there are more common features between plants and mammals than there are within the animal kingdom itself (App. 2 [Zhao and Garcia 2014]). Consequently, epigenetic mechanisms discovered in plants or mammals are generally relevant to both systems.

1.3 Unique Aspects of Plant Development Relevant to Epigenetic Regulation

In considering the similarities and differences between plants and mammals, it is important to consider the special life histories of plants. In mammals, fertilization is achieved by fusion of two haploid cells that are direct products of the preceding meiosis. In contrast, plants have a haploid (gametophyte) growth stage that follows meiosis and precedes fertilization (Fig. 1). The male and female gametophytes are the pollen and embryo sac, respectively, each composed of multiple cells that are produced by mitotic divisions of the initial haploid meiotic products. In haploid gametophytes, which are genetically and metabolically active, loss of genetic or epigenetic information cannot be compensated for by information on homologous chromosomes; hence, deleterious mutations in essential genes are selected against. Unlike mammals, there is no evidence for a massive erasure of epigenetic marks during plant gametogenesis. Instead, repressive epigenetic marks in plant sperm and egg cells appear to be reinforced by specific *trans*-silencing RNAs produced in neighboring nuclei. This might explain how epigenetic changes are often transmissible through meiosis in plants.

Another distinctive feature of plants relative to animals is the lack of a defined germline set aside early in embryogenesis. Instead, germ cells are produced late in plant development after stem cells in the shoot undergo a transition from producing progenitor cells of vegetative organs to producing daughter cells that give rise to the floral organs in which meiosis and gametogenesis occur (Fig. 1). Thus, epigenetic modifications acquired by meristem cells in response to the plant's interactions with its environment have the potential to be conveyed to the germ cells.

In addition to the clusters of stem cells at the apex of the shoot and root, known as apical meristems, plants have lateral meristems. These include the stem cells that give rise to new xylem and phloem each year in perennials, resulting in the characteristic growth rings of tree trunks, as well as buds that are located at the base of each leaf and that can sprout to form vegetative organs or flowers. Many plants also have specialized organs, such as underground rhizomes, tubers, or bulbs that have stem cells capable of giving rise to shoots that form independent, new plants. These mechanisms of vegetative or clonal propagation are common in plants and are often more successful than seed dissemination as a strategy for colonizing a favorable location. Importantly, mitotically transmissible epigenetic states can be perpetuated among clones resulting from vegetative propagation.

In tissue culture, some differentiated plant somatic cells can be reprogrammed to form somatic embryos that develop into plants without a requirement for fertilization. As a result, cloning via somatic embryogenesis is routine for many plant species. However, among what should be genetically uniform clones, a surprising amount of phenotypic variability is observed. This so-called "somaclonal variation" has a strong epigenetic basis and is potentially useful for plant breeding and selection of adaptive traits (reviewed in Miguel and Marum 2011).

Another plant-specific feature is the existence of plasmodesmata, which are cytoplasmatic connections between cells allowing metabolites, proteins, RNAs, and even viruses to pass. Plant shoots can be cut and grafted as scions onto genetically different rootstocks (Fig. 1). This results in chimeras in which roots and shoots are genetically different. Diffusible epigenetic signals move through plasmodesmata and the vascular system and can be transmitted between roots to shoots in grafted plants (see Sec. 3.6.2 and Ch. 2 [Dunoyer et al. 2013]). In this way, the epigenetic state of stem cells and gametes can potentially be modified by signals emanating from distant organs of the plant.

Plants have a high tolerance for polyploidy (the multiplication of whole chromosome complements) compared with mammals. In fact, most flowering plants have undergone polyploidization at least once in their evolutionary histories. Polyploid crops include wheat, cotton, potato, peanut, sugarcane, coffee, canola, and tobacco. The prevalence of polyploidy in plants suggests that it offers certain fitness advantages, such as built-in hybrid vigor or resistance to the effects of deleterious mutations, thereby allowing duplicated genes to potentially acquire beneficial mutations. The formation of polyploids is often associated with significant genomic and epigenetic changes (reviewed in Jackson and Chen 2010). Some of these changes occur within one or a few generations and can contribute to rapid adaptation and evolution in plants. Even diploid plants contain polyploid cells as a result of endoreplication,

Cite as *Cold Spring Harb Perspect Biol* doi: 10.1101/cshperspect.a019315

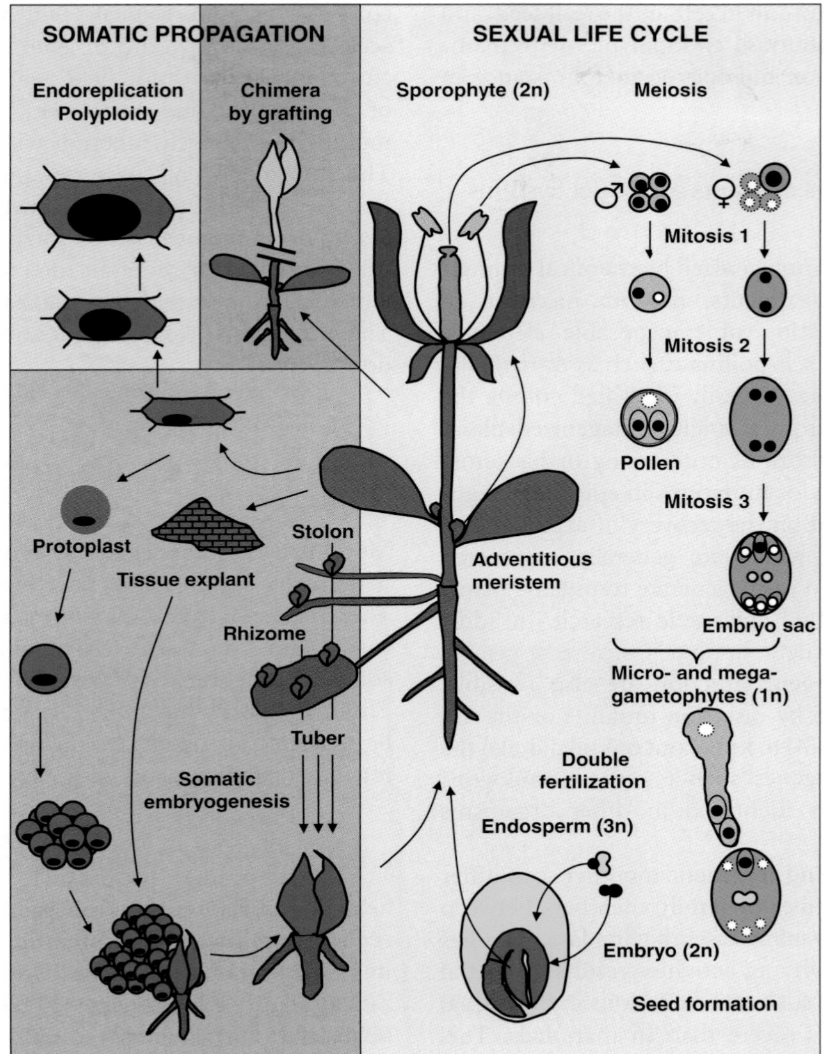

Figure 1. Unique aspects of the plant life cycle. Plants can propagate sexually (gametogenesis, fertilization, and seed formation; *right*) as well as somatically (vegetative explants, cell de- and re-differentiation, or somatic embryogenesis; *left*). The body of higher plants with roots, stem, leaves, and flowers is the diploid sporophyte. During meiosis, the chromosome number is reduced by half. Whereas in animals the meiotic products become the gametes without further division and fuse directly to produce the diploid embryo, plants form haploid male or female gametophytes via two or three mitotic divisions to form pollen or embryo sacs, respectively. The pollen grain ultimately contains one vegetative (white) and two generative (black) nuclei. The two generative nuclei fertilize the egg cell (black) and the central cell, the latter having a diploid nucleus derived from fusion of the two polar nuclei (yellow). This double fertilization gives rise to a diploid embryo and a triploid endosperm, the latter providing a nutrient source for the developing embryo. After seed germination, the embryo will grow into a new sporophyte. Most plants also have the potential for vegetative propagation through activation of lateral meristems, outgrowth of specialized organs (tubers, rhizomes or stolons), amplification in tissue culture, or regeneration from individual somatic cells after removal of the cell wall (protoplasts). Endoreduplication is frequent in plants, producing polyploid cells or tissues. Plants can be grafted to produce chimeras.

with endopolyploidy common in cells of the leaf, seed, and other plant organs. By contrast, endopolyploidy in mammalian cells is less common, but does occur, for instance, in muscle and liver cells.

1.4 Genetic Attributes of Plants as Model Systems for Epigenetics

Plants can be efficiently mutagenized by chemical or physical (e.g., radiation) treatments, random insertion of transgenes, or mobilization of transposable elements. Moreover, in plants that self-pollinate, such as *A. thaliana*, homozygous mutants can be easily identified among the thousands of progeny from a single mutagenized plant, without the need for laborious outcrossing or backcrossing procedures. Screens for mutations in epigenetic regulators are typically based on the recovery of expression of silenced marker genes, which are generally engineered transgenes. The ability to easily generate transgenic plants has thus greatly benefitted epigenetic research. In addition to such forward genetic approaches, reverse genetic approaches that ablate gene function are also possible. This has been facilitated by insertion mutants or the use of transgene-induced RNAi to knockout or knockdown the expression of candidate genes, such as genes homologous to epigenetic regulators identified in other organisms (Table 1).

Gene families encoding epigenetic modifiers can differ significantly in the number of family members between plants and mammals. In many cases, having large families of plant chromatin-modifying activities results in partial functional redundancies such that mutations in individual family members are less severe than in mammals. This can be useful if complete loss of function is lethal, thereby making mutants unavailable for study.

Nonessential genes in pathways that determine coloration of plant tissues have proven useful for identifying mutants affecting their epigenetic control (Fig. 2A–E). Other epigenetic modifier mutants cause morphological defects that are often tolerated in plants without lethal consequences (Fig. 2F). Thousands of individual plants can be screened for changes in the expression of endogenous or transgenic traits (Fig. 2G). Once specific epigenetic mutants are characterized, suppressor screens are often successful for identifying interacting components or alternative pathways, a strategy that has also been used in *Drosophila* (Ch. 12 [Elgin and Reuter 2013]) and mouse (see Ch. 14 [Blewitt and Whitelaw 2013]).

A. thaliana has emerged as a leading model system for plant epigenetics research because of the availability of comprehensive collections of insertion mutations in nearly every gene, facile mutagenesis, extensive natural variation, complete genome sequence information for numerous accessions, and a wealth of genome-wide expression and chromatin-modification data. *Arabidopsis* is thus the focus of this chapter, and Table 1 summarizes the epigenetic modifiers for which functional information is available. This information has been obtained through a combination of forward- and reverse-genetic screens, biochemical assays, or homologies with known epigenetic modifiers in other systems. For more detailed information about individual genes in *Arabidopsis*, the reader is referred to TAIR, The *Arabidopsis* Information Resource (http://www.arabidopsis.org).

2 MOLECULAR COMPONENTS OF CHROMATIN IN PLANTS

More than 130 genes encoding proteins involved in epigenetic regulation in plants have been identified (Table 1). The epigenetic modifiers known to date can roughly be sorted into five groups according to their proven or presumptive functions, as discussed in Sections 2.1–2.4 and 3. This list is clearly incomplete, as new activities continue to be identified in ongoing research, but provides a framework for understanding plant epigenetics.

2.1 Regulators of DNA Modification

5-Methylcytosine (5mC) is a hallmark of epigenetic gene silencing and heterochromatin in both plants and mammals (the latter of which is discussed in Ch. 15 [Li and Zhang 2014]; see also App. 2 [Zhao and Garcia 2014]; reviewed in Furner and Matzke 2011; Meyer 2011). Whereas in nuclei of differentiated mammalian cells 5mC is found nearly exclusively at CG sites (often referred to as CpG sites), plants methylate cytosines within CG, CHG, or CHH motifs (in which H is A, T, or C). Mammalian promoters are often present within methylation-free CG-rich regions known as CpG islands, but CpG islands are not easily distinguishable in plants. Nevertheless, cytosine methylation is nonrandomly distributed in plants, occurring primarily in repetitive regions of the genome that are enriched in transposable elements, centromeric repeats, or arrays of silent 5S or 45S rRNA gene repeats. Cytosine methylation also occurs at some differentially regulated promoters and within the protein-coding regions of highly expressed genes (Zilberman et al. 2007). The latter gene body methylation is evolutionarily conserved, occurring in animal species as diverse as humans and honeybees (Feng et al. 2010). The significance of gene body CG methylation is not yet clear, but its enrichment within exons suggests a potential role in pre-mRNA splicing.

Table 1. Components of epigenetic regulation in the model plant species *Arabidopsis thaliana*

Gene or mutant acronym[a]	Gene or mutant name	Confirmed or putative function of protein
DNA modification		
CMT3	Chromomethyltransferase	DNA methyltransferase (mainly CHG and CHH)
DME	Demeter	DNA glycosylase-domain protein, cytosine demethylation
DML2,3	Demeter-like proteins	DNA glycosylase-domain protein, cytosine demethylation
DNMT2	DNA methyltransferase	Interaction with HD2
DRM1	Domains-rearranged methyltransferase	DNA methyltransferase (CG, CHG, and CHH)
DRM2	Domains-rearranged methyltransferase	Major de novo DNA methyltransferase (CG, CHG, and CHH)
HOG1	Homology-dependent gene silencing	*S*-adenosyl-L-homocysteine hydrolase
MBD10	Methylcytosine-binding domain protein	Methylcytosine-binding protein
MBD6	Methylcytosine-binding domain protein	Methylcytosine-binding protein
MET1, DDM2	Methyltransferase, decreased DNA methylation	DNA methyltransferase (mainly CG)
ROS1	Repressor of silencing	DNA glycosylase-domain protein, cytosine demethylation
ROS3	Repressor of silencing	DNA glycosylase-domain protein, cytosine demethylation
VIM1,-2,-3	Variant in methylation	Methylcytosine-binding proteins
ZDP	Zinc finger DNA 3′-phosphoesterase	3′-Phosphatase
Histone modification		
ATX1	*Arabidopsis* homolog of trithorax	Histone methyltransferase
ATXR3/SDG2	*Arabidopsis* homolog of trithorax, SET domain group	Histone methyltransferase
ATXR5,-6,-7	*Arabidopsis* trithorax-related proteins	Histone methyltransferase (ATXR5,-6: H3K27me1)
EFS/SDG8/ASHH2	Early flowering in short days, SET domain group, ASH1 homolog	Histone methyltransferase
ELF6	Early flowering	Histone demethylase (H3K4me1,-2,-3)
FLD, LDL1,-2	Flowering locus D, LDS1-like	Histone demethylases
HAC1,-5,-12	Histone acetyltransferase CBP-like	Histone acetyltransferase
HAG1-,2	Histone acetyltransferase GCN5-like	Histone acetyltransferase (H3K14ac)
HAM1,-2	Histone acetyltransferase Myst-like	Histone acetyltransferase
HD2a-d; HDT1, -2,-3,-4	Histone deacetylase	Atypical histone deacetylase
HDA1	Histone deacetylase	Histone deacetylase
HDA6, SIL1, AXE1 RTS1	Histone deacetylase, modifier of silencing, auxin-gene repression, RNA-mediated transcriptional silencing	Histone deacetylase
HUB1,-2	E3 ubiquitin ligase	H2B monoubiquitination
IBM1	Increase in bonsai methylation	Histone demethylase
JMJ14	*Arabidopsis thaliana* jumonji	Histone demethylase (H3K4me1,-2,-3)
MEE27	Maternal effect embryo arrest	Histone demethylase
OTLD1	Otubain-like deubiquitinase	H2B deubiquitination
REF6	Relative of early flowering	Histone demethylase
SUP32/UBP26	Suppressor of ros, ubiquitin protease	H2B deubiquitination
SUVH2	Su(Var)3-9 homolog	H3K9 methyltransferase
SUVH4, KYP1	Su(Var)3-9 homolog, kryptonite	H3K9 methyltransferase
SUVH5,-6	Su(Var)3-9 homolog	Recognition of DNA methylation (SUVH5)
UBC1,-2	E2 ubiquitin-conjugating enzyme	H2B monoubiquitination
ULT1	Ultrapetala	Regulator of histone methylation, ATX1 interactor
Polycomb-group proteins and interacting components		
AtBMI1a,-b,-c	B cell-specific Mo-MLV integration site 1	PRC1 subunit
AtCYP71	Cyclophilin	LHP1 and H3 interaction
AtRING1a,-b	RING finger protein	PRC1 subunit
CLF	Curly leaf	Polycomb-group protein (E(z))
CUL4	Cullin protein	Subunit of E3 ligase complex, MSI1 interactor

Continued

Table 1. *Continued*

Gene or mutant acronym[a]	Gene or mutant name	Confirmed or putative function of protein
DDB1	Damaged DNA-binding	Subunit of E3 ligase complex, MSI1 interactor
EMF2	Embryonic flower	Polycomb-group protein (Su(z)12)
FIE, FIS3	Fertilization-independent endosperm, fertilization-independent seeds	Polycomb-group protein (Esc)
FIS2	Fertilization-independent seeds	Polycomb-group proteins (Su(z)12)
LHP1/TFL2	Like heterochromatin protein, terminal flower	PRC1 subunit, formation of repressive chromatin
LIF2	LHP1-interacting factor	RNA processing, LHP1 interaction
MEA, FIS1	Medea, fertilization-independent seeds	Polycomb-group protein (E(z))
MSI1	Multicopy suppressor of IRA homolog	Polycomb-group protein (p55)
MSI4/FVE	Multicopy suppressor of IRA homolog	Polycomb-group protein (p55), Cul4-DDB1 and PCR2 interactor
MSI5	Multicopy suppressor of IRA homolog	Polycomb-group protein (p55), interactor of HDA6, FLC silencing
RBR	Retinoblastoma-related protein	PRC2 regulator during development
SWN	Swinger	Polycomb-group protein (E(z))
VEL1	Vernalization-like	Homeodomain protein
VIL2	Vin3-like	Homeodomain protein
VIN3	Vernalization-insensitive	Homeodomain protein
VRN2	Vernalization	Polycomb-group protein (Su(z)12)
VRN5	Vernalization	Homeodomain protein
Chromatin formation or chromatin remodeling		
ARP4,-5	Actin-related protein	Subunits of INO80 complex
AtASF1a,-b	*Arabidopsis* antisilencing factor	Histone chaperone H3/H4
AtCHR12	*Arabidopsis* chromatin remodeling	SNF2/Brahma-type protein
AtNAP1 1–4	*Arabidopsis* nucleosome assembly protein	Histone chaperone H2A/H2B
AtSWI3 A-D	*Arabidopsis* homolog of SWI3	Subunits of SWITCH/SUCROSE NONFERMENTABLE (SWI/SNF)-remodeling complexes
AtSWP73 A,B	*Arabidopsis* homolog of SWP73	Subunit of SWI/SNF-remodeling complex
BRM	Brahma	SWI2/SNF2 ATPase family protein
BRU1/MGO3/TSK	Brushy, Mgoun, Tonsoku	Uncharacterized protein
BSH	Bushy	Subunit of SWI/SNF-remodeling complexes
CHR11	Chromatin-remodeling protein	ISWI-like chromatin remodeling protein
CLSY1	Classy	SWI2/SNF2 family ATPase
DDM1, SOM	Decreased DNA methylation, somniferous	SWI2/SNF2 family ATPase
DMS3/IDN1	Defective in meristem silencing, involved in de novo	Structural-maintenance-of-chromosomes protein; required for RNA Pol V transcription
DMS11	Defective in meristem silencing	GHKL ATPase, interaction with DMS3
DRD1	Defective in RNA-directed DNA methylation	SWI2/SNF2 ATPase; required for RNA Pol V transcription
FAS1	Fasciated	Chromatin assembly factor subunit H3/H4
FAS2	Fasciated	Chromatin assembly factor subunit H3/H4
HIRA	Histone regulator A	Histone chaperone H3/H4
INO80	Inositol-requiring	Subunits of INO80 complex
MGO1	Mgoun	DNA topoisomerase
MOM1	Morpheus' molecule	Genetic interaction with RNA Pol V
MSI1	Multicopy suppressor of IRA homolog	Chromatin assembly factor subunit, H3/H4
NRP1,-2	NAP (nucleosome assembly protein)-related protein	Histone chaperone H2A/H2B
PIE	Photoperiod-independent early flowering	ATPase subunit of SWR1
PKL	Pickle	CHD3 chromatin remodeling factor
RPA2	Replication protein A	Single-stranded DNA-binding protein
SEF/SWC6	Serrated and early flowering, SWR1 complex	Subunit of SWR1
SPD	Splayed	SWI2/SNF2 family ATPase
SPT16	Suppressor of Ty insertion-like	Histone chaperone H2A/H2B

Continued

Table 1. *Continued*

Gene or mutant acronym[a]	Gene or mutant name	Confirmed or putative function of protein
SSRP1	Structure-specific recognition protein	High mobility group (HMG) protein, histone chaperone subunit
SUF3/ESD1/ARP6	Suppressor of Frigida, early in short days, actin-related protein	Subunit of SWR2, H2A.Z deposition
TSL	Tousled	Phosphorylation of ASF1
RNA silencing		
ABH1/CBP80	ABA-hypersensitive, cap-binding complex	microRNAs (miRNA) processing
AGO1	Argonaute	PAZ-PIWI domain protein, translational repression
AGO10, PNH/ZLL	Argonaute, Pinhead, Zwille	PAZ-PIWI domain protein, translational repression
AGO4	Argonaute	PAZ-PIWI domain protein, siRNA-binding
AGO6	Argonaute	PAZ-PIWI domain protein, siRNA-binding
AGO7, ZIP	Argonaute, Zippy	PAZ-PIWI domain protein
AGO9	Argonaute	PAZ-PIWI domain protein, female gametophyte formation, siRNA binding
AtNUC-l1	Nucleolin-like	rRNA gene regulation
DCL1, CAF1 ,SIN1, EMB76, SUS1	Dicer-like, Carpel factory, short integuments, embryo-defective, suspensor	RNase III (dsRNase), miRNA and siRNA generation
DCL2	Dicer-like	RNase III (dsRNase), siRNA generation
DCL3	Dicer-like	RNase III (dsRNase), siRNA generation
DCL4, SMD	Dicer-like	RNase III (dsRNase), siRNA generation
DDL	Dawdle	FHA domain protein, miRNA processing
DRB1	Double-stranded RNA-binding protein	Dicer 1-interacting protein, miRNA strand selection
DRB2,-3,-4	Double-stranded RNA-binding protein	Dicer-interacting proteins, siRNA processing
ESD7	Early in short days	Subunit of DNA polymerase epsilon
FCA	Flowering time	RRM-domain protein
FDM1-5	Factor of DNA methylation	dsRNA-binding protein (FDM1)
FPA	Flowering time	RRM-domain protein
FRY1	Fiery	Nucleotidase
HEN1	HUA enhancer	dsRNA-binding, RNA methyltransferase
HST	Hasty	miRNA export receptor
HYL1	Hyponastic leaves	Nuclear dsRNA-binding protein
IDN2/RDM12	Involved in de novo, RNA-directed DNA methylation	dsRNA-binding protein
KTF1/RDM3/SPT5-l	KOW domain-containing transcription factor, suppressor of Ty insertion-like	Part of RdDM effector complex
NRP(A/B/C/D)5	Nuclear RNA polymerases I, II, III, IV	5th subunit common to Pols I–IV
NRP(B/D/E)3a, -3b	Nuclear RNA polymerases II, IV, and V	Alternative 3rd subunits of Pols II, IV, or V
NRP(B/D/E)9a, -9b	Nuclear RNA polymerases II, IV, and V	Alternative 9th subunits of Pols II, IV, and V
NRP(D/E)2, DRD2	Nuclear RNA polymerases IV and V; defective in RNA-directed DNA methylation 2	2nd subunit of Pol IV and PolV
NRP(D/E)4	Nuclear RNA polymerases IV and V	4th subunit of RNA Pol IV and RNA Pol V
NRP(D/E)5b	Nuclear RNA Pol IV and V	Alternative 5th subunit of RNA Pol IV (and RNA Pol V likely)
NRPD1, SDE4	Nuclear RNA polymerase IV, silencing defective 4	Pol IV largest subunit
NRPD7a	Nuclear RNA Pol IV	Major 7th subunit of RNA Pol IV
NRPE1, DRD3	Nuclear RNA Pol V, defective in RNA-directed DNA methylation	RNA Pol V largest subunit
NRPE5	Nuclear RNA Pol V	5th subunit of RNA Pol V
NRPE7/NRPD7b	Nuclear RNA Pol V	Major 7th subunit of RNA Pol V; alternative seventh subunit of RNA Pol IV
RDM1	RNA-directed DNA methylation	Component of DDR complex required for RNA Pol V transcription
RDM12	RNA-directed DNA methylation	Coiled-coil protein
RDM4/DMS4	RNA-directed DNA methylation, defective in meristem silencing	IWR1-like regulator of multisubunit RNA polymerase assembly
RDR1	RNA-dependent RNA polymerase	RNA-dependent RNA polymerase

Continued

Table 1. *Continued*

Gene or mutant acronym[a]	Gene or mutant name	Confirmed or putative function of protein
RDR2	RNA-dependent RNA polymerase	RNA-dependent RNA polymerase
RDR6, SDE1, SGS2	RNA-dependent RNA polymerase, silencing defective, suppressor of gene silencing	RNA-dependent RNA polymerase
SDE3,-5	Silencing-defective	RNA helicase
SE	Serrate	Zn finger protein, miRNA processing
SGS3	Suppressor of gene silencing	Coiled-coil protein
SHH1/DTF1	Sawadee homeodomain homolog, DNA-binding transcription factor	Homeodomain protein
SR45	Arginine/serine-rich	Splicing factor
WEX	Werner syndrome-like exonuclease	RNase D exonuclease
XRN2,-3	XRN homolog	Exoribonuclease
XRN4/EIN5	XRN homolog, ethylene insensitive	Exoribonuclease, small RNA processing

[a] Different acronyms in row one refer to the same gene.

2.1.1 DNA Methyltransferases

DNA methyltransferases catalyze cytosine methylation de novo at previously unmethylated cytosines or maintain preexisting cytosine methylation patterns (illustrated in Fig. 2 in Ch. 15 [Li and Zhang 2014]). In plants, de novo methylation is directed by small interfering RNAs (siRNAs) and occurs on cytosines present within CG, CHG, or CHH motifs (see Sec. 3.5). Maintenance methylation perpetuates preexisting methylation patterns, especially at CG or CHG motifs, which are known as symmetrical sites because each DNA strand reads CG or CHG in the 5′ to 3′ direction. This symmetry provides a basis for cytosine methylation patterns on mother strands to be imparted to daughter strands after each round of DNA replication. Maintenance methylation also occurs following DNA repair, allowing newly synthesized DNA to be methylated based on the methylation pattern of the nonrepaired strand.

Three families of DNA methyltransferase are conserved among eukaryotes and are present in plants: homologs of mammalian *Dnmt1*, *Dnmt2*, and *Dmnt3* (Fig. 4 in Ch. 15 [Li and Zhang 2014]). DNA METHYLTRANSFERASE 1 (*MET*1), the plant homolog of mammalian Dnmt1, is the major CG maintenance methyltransferase and may also contribute to CG de novo methylation. DNMT2 homologs are also present in plants and possess transfer RNA (tRNA) methylase activity, like mammalian DNMT2 (Goll et al. 2006); however, no catalytic activity on DNA has been observed for these enzymes. The plant DOMAINS-REARRANGED METHYLTRANSFERASEs (*DRM*) and their mammalian homologs, the Dnmt3 group, are predominantly de novo methyltransferases. DRM proteins have the domains of their amino- and carboxy-terminal halves arranged in reverse order when compared with Dnmt3 (domains VI–X are followed by I–V). DRM2 catalyzes methylation of cytosines in all sequence contexts and is the prominent cytosine methyltransferase in the RNA-directed DNA methylation pathway (see Sec. 3.5.2).

Unlike mammals, plants have a unique family of cytosine methyltransferases wherein the defining characteristic is the presence of a chromodomain that binds to methylated histones. CHROMOMETHYLASE 3 (CMT3) is the enzyme primarily responsible for CHG maintenance methylation. CMT3's chromodomain binds histone H3 that is dimethylated on lysine 9 (H3K9me2). CHG methylation, in turn, provides a binding site for the SET- AND RING-ASSOCIATED (SRA) domain of the H3K9 methyltransferase SUVH4, leading to H3K9 dimethylation. Thus, CMT3 and SUVH4 constitute a self-reinforcing loop in which repressive DNA methylation and histone modification marks specify one another to maintain an epigenetic state (reviewed in Law and Jacobsen 2010). CHROMOMETHYLASE 2 (CMT2) plays a role in maintaining CHH methylation in specific genomic contexts, such as the central regions of large transposable elements, presumably via cross talk with histone modifications like its paralog, CMT3 (Zemach et al. 2013).

In contrast to mammals in which *dnmt*1 and *dnmt*3 mutants die during embryonic development, or shortly after birth, *met*1, *cmt2*, *cmt*3, and *drm* mutants are viable, even in combination with one another. As a result, the role of DNA methylation in a variety of processes can be studied in plants, including vegetative and reproductive development, gametogenesis, fertilization, and the cross talk between DNA methylation and histone modification (as in the case of CMT3 described in the previous paragraph; reviewed in Furner and Matzke 2011; Meyer 2011).

 Cite as *Cold Spring Harb Perspect Biol* doi: 10.1101/cshperspect.a019315

Figure 2. Assays for epigenetic control in plants. Genes determining coloration of plant tissue allow easy and inexpensive gene expression readout in vivo. (*A*) Expression of the dihydroflavonol reductase (*DFR*) gene is required for dark purple petunia flowers, whereas silencing of the DFR promoter gives rise to variegated, light coloration. (*B*) Seeds from *Arabidopsis* expressing the chalcone synthase (*CHS*) gene have dark seed coats, whereas silencing of CHS upon expression of a homologous transgene results in yellow seeds. (*C*) Maize plants with the B-I gene have purple pigmentation, in contrast to green plants with a paramutagenic, inactive B′ allele whose DNA sequence is identical. (*D*) Maize ear, segregating a transposon insertion (Spm) in the B-Peru gene required for anthocyanin pigment. Purple kernels represent revertants in which the Spm element excised from the gene in the germline. The heavily spotted kernels contain the Spm element in the active form that induces frequent somatic excision sectors during kernel development. The kernels with rare, small purple sectors represent kernels in which the Spm element has been epigenetically silenced. (*E*) The dark color of soybeans (*middle*) is extinguished in cultivated varieties (*left*) because of natural posttranscriptional silencing of the *CHS* gene and can be partially reversed by infection of the parental plant with a virus possessing a PTGS suppressor protein, producing a mottled pattern (*right*). (*F*) Epigenetic regulation can become manifest also in plant morphology; reduced function of a chromatin assembly factor subunit leads to a "fasciated" *Arabidopsis* stem. (*G*) Release of silencing from a transgenic resistance marker in *Arabidopsis* can be scored by growth on selective medium. (*A*, Courtesy of Jan Kooter; *B*, courtesy of Ian Furner; *C*, reprinted from Chandler et al. 2000, with kind permission from Springer Science and Business Media; *D*, courtesy of Vicki Chandler; *E*, reprinted, with permission, from Senda et al. 2004, © American Society of Plant Biologists.)

2.1.2 Cytosine Demethylation and DNA Glycosylases

In spite of mechanisms allowing DNA methylation to be maintained, cytosine methylation can be lost. Passive loss occurs when methylation fails to be maintained during replication or following DNA repair. Active demethylation can also occur via enzymatic activity. Active demethylation in animals was shown decades ago, but the exact players and mechanisms are still debated (reviewed in Gehring et al. 2009; Niehrs 2009). In *Arabidopsis*, active demethylation is catalyzed by REPRESSOR OF SILENCING 1 (ROS1), DEMETER (DME), or DEMETER-LIKE proteins (DML2,-3). These are large proteins that contain DNA glycosylase domains. ROS1 displays nicking activity on methylated DNA, resulting in the removal and replacement of methylated cytosines via a pathway related to base excision repair (Agius et al. 2006). The demethylation process requires the ZINC FINGER DNA 3′PHOSPHOESTERASE (ZDP), which is thought to remove the 3′ phosphate at the nick site, thereby generating a 3′ hydroxyl that allows DNA ligase to seal the gap (Martínez-Macías et al. 2012). ROS1 is expressed constitutively, potentially contributing to loss of DNA methylation in nondividing cells at all stages of development. ROS1 is thought to be guided to its sites of action by its association with ROS3, an RNA-binding protein, suggesting that RNA may play a role in guiding demethylation as well as in guiding de novo methylation in the RNA-directed DNA methylation pathway (Zheng et al. 2008). The DNA glycosylase DME is especially important in the gametophytes, where it helps erase 5mC patterns that otherwise silence a specific subset of genes (Choi et al. 2002; Schoft et al. 2011). Mutations eliminating the CG methyltransferase MET1 suppress *dme* mutants, suggesting that DME is required primarily for demethylation of CG dinucleotides (Xiao et al. 2003). The DME paralogs, DML2 and -3, affect cytosine methylation in CG, CHG, and CHH contexts at specific genomic sites (Ortega-Galisteo et al. 2008).

Collectively, the functions of the multiple demethylating enzymes in *Arabidopsis* indicate that reversibility of DNA methylation is critical for the regulation of specific genes during development. There is also growing evidence that cytosine methylation can be rapidly lost in other ways (reviewed in Reinders and Paszkowski 2009). Spontaneous changes in cytosine methylation can occur sporadically among isogenic populations (Becker et al. 2011; Schmitz et al. 2011), thereby creating or derepressing epialleles that may influence fitness and natural selection (Roux et al. 2011).

2.1.3 *Methylcytosine-Binding Proteins*

METHYL-C-BINDING DOMAIN (MBD) proteins, which share sequence similarity with mammalian proteins such as MeCP2, are epigenetic "reader" modules thought to help transduce DNA methylation patterns into altered transcriptional activity (see Figs. 8 and 9 in Ch. 15 [Li and Zhang 2014]). Plants have more MBD-containing genes (13 in *Arabidopsis*) than mammals. However, outside the conserved MBD domain, there is little homology with mammalian MBDs (Springer and Kaeppler 2005). Only three members of the *Arabidopsis* family are known to bind specifically to methylated DNA, and these are missing in monocotyledonous plants such as maize, wheat, and rice (reviewed in Grafi et al. 2007). Other protein domains also confer 5mC binding. For instance, the SRA domain of SUVH4 (an H3K9 methyltransferase) binds to methylated CHG sites providing a mechanistic link between repressive histone and DNA modifications as mentioned in Section 2.1.1. Likewise, the SRA domains of VIM proteins (orthologs of UHRF1 in mammals) bind to hemimethylated DNA and recruit MET1/DNMT1 to modify the new daughter strand following replication (Woo et al. 2007, 2008).

2.1.4 *Proteins Required for Methyl Group Donor Synthesis*

Most methylating enzymes require the cofactor *S*-adenosyl-methionine (SAM, or AdoMet) as the methyl group donor. A key enzyme in SAM biosynthesis, S-ADENOSYL-L-HOMOCYSTEINE HYDROLASE, regulates SAM levels by removing substrate inhibition. Mutations in the *Arabidopsis* gene encoding this enzyme result in decreased DNA methylation and release of transcriptional silencing, especially from genes in pericentromeric heterochromatin (Rocha et al. 2005; Jordan et al. 2007). The mutations also affect histone methylation, further impacting epigenetic regulation (Baubec et al. 2010).

2.2 Histone-Modifying Enzymes and Histone Variants

As in other organisms, plants contain several histone variants and enzymes that posttranslationally modify histones and influence gene regulation (introduced in Ch. 3 [Allis et al.]). Application of chromatin immunoprecipitation followed by deep sequencing has given insight into the genome-wide distribution of histone variants and histones bearing different posttranslational modifications (reviewed in Roudier et al. 2009), including the coincidence or mutual exclusion of specific chromatin marks (Roudier et al. 2011). Histone-modifying enzymes are often encoded by large gene families in plants (reviewed in Berr et al. 2011; Deal and Henikoff 2011; Lauria and Rossi 2011).

2.2.1 *Histone Deacetylases and Histone Acetyltransferases*

Histone acetylation is an epigenetic mark generally associated with active chromatin and transcription, whereas transcriptionally nonactive sequences usually lack acetylation. The functions of histone acetyltransferase (abbreviated to HAT) "writer" activities can be reversed by histone deacetylase (HDAC) "eraser" enzymes that allow for the reversibility of acetylation as an epigenetic mark. The structure and function of these genes is extensively discussed in Ch. 4 (Marmorstein and Zhou 2014) and Ch. 5 (Seto and Yoshida 2014). Plants have multiple gene family members for both HATs and HDACs (Pandey et al. 2002; Chen and Tian 2007). The members of the HAT family can be classified into five different subfamilies based on their structure and different substrate specificities (Earley et al. 2007). The GCN5-homolog, HAG1, for instance specifically methyl-

Cite as *Cold Spring Harb Perspect Biol* doi: 10.1101/cshperspect.a019315

ates H3K14, as in yeast and mammals, and regulates several developmental processes in *Arabidopsis* (reviewed in Servet et al. 2010). We currently know of plant homologs to three of the five HAT subfamilies (see Table 1). Plants also have homologs of the histone deacetylases that are highly conserved throughout all eukaryotes, catalyzing the removal of histone acetylation to generate more repressive chromatin. There is a plant-specific family of putative histone deacetylases, named the HD2 or HDT family, which is implicated in gene silencing, but definitive HDAC activity has not yet been shown for recombinant HDT proteins. Genetic screens have so far identified only two HDACs that function as epigenetic regulators: *HDA*1 and *HDA*6 (Table 1). HDA6 plays a role in maintaining CG and CHG methylation, interacting with the DNA methyltransferase MET1, and is involved in transgene and transposon silencing, rRNA gene repression, and nucleolar dominance. Although deficiency mutants have only subtle morphological defects, HDA6 is involved in seed maturation, flowering time control, and stress responses (reviewed in Aufsatz et al. 2007; Kim et al. 2012). Pleiotropic morphological changes are apparent on reduced expression and/or overexpression of HDA1 and HD2A/HDT1, but their functions are not clear. Dissection of HAT and HDAC function is somewhat complicated by their redundancy and their potential operation in the context of large multiprotein complexes.

2.2.2 Histone Methyltransferases and Histone Demethylases

Like acetylation, histone methylation is a potentially reversible mark. Histone lysine methyltransferases (HKMTs) installing this modification are usually characterized by the presence of a SET domain (SU(VAR)/E(Z)/TRX (discussed extensively in Ch. 6 [Cheng 2014]). Depending on the specific histone lysine that is methylated, HKMTs can promote or inhibit transcription. Some SET domain proteins are members of the Polycomb group (PcG) or trithorax group (TrxG) that maintain transcriptionally repressed or active states, respectively, of specific genes during plant and animal development (see Sec. 2.3 and Ch. 17 [Grossniklaus and Paro 2014]). Other SET domain proteins of the Su(var)3-9 family participate in maintaining condensed heterochromatin, silencing transposons, or controlling DNA replication.

The *Arabidopsis* genome encodes 49 SET domain proteins, grouped into four conserved families: E(Z), ASH1, TRX, and SU(VAR)3-9-related proteins (Pontvianne et al. 2010). The latter and largest group is mostly, but not exclusively, responsible for H3K9 methylation. KRYPTONITE/SUPPRESSOR OF VARIEGATION 3-9 homolog 4 (*KYP/SUVH4*) and SUVH2, -5, and -6 catalyze H3K9me1 and -me2 installation. As described for SUVH4, their SRA domain, which binds 5mC, contributes to their cooperation with CMT3 to maintain CHG methylation and H3K9me2 at silenced loci. H3K9me3 is likely accomplished by the more distantly related SUVR proteins, and in contrast to animals in which it is a prominent heterochromatic mark, in plants it is found at expressed euchromatic genes. The TRX family member ARABIDOPSIS TRITHORAX 1 (ATX1) activates floral homeotic genes, presumably by means of its ability to catalyze histone H3 lysine 4 (H3K4) methylation. ATXR5 and ATXR6 monomethylate H3K27, a mark of repressed heterochromatin. Overreplication of heterochromatin occurs in *atxr5 atxr6* double mutants, pointing to a role in replication control (Jacob et al. 2010). Mutants lacking ATXR3 have reduced H3K4me2/3 and severe developmental defects, partly because of reduced cell size. ASH 1 HOMOLOG 2 (ASHH2), a member of the ASH1 group, catalyzes H3K36me2/3, thereby reinforcing active chromatin marks at target loci, including several pathogen resistance genes. Genes of the E(z) group are relevant to Polycomb regulation (as discussed in Sec. 2.3).

There are two classes of proteins that can demethylate histone tails at different positions. The lysine-specific histone demethylase-LIKE (LDL) proteins act via amine oxidation, and the JUMONJI-C-DOMAIN (JmjC) proteins act via hydroxylation (discussed extensively in Ch. 6 [Cheng 2014]; see also Ch. 2 [Shi and Tsukada 2013]). *Arabidopsis* has four LDL and 21 JmjC genes in several subgroups (reviewed in Chen and Tian 2007; Liu et al. 2010). The LDL proteins and some JmjC proteins are involved in flowering time control, have specific target genes, and can act in cooperation with transcription factors. The JmjC protein, INCREASE IN BONSAI METHYLATION (IBM1), counteracts H3K9 methylation and CHG DNA methylation. Developmental defects due to loss of IBM1 are consequently suppressed in double mutations with *suvh4* and *cmt3*. Several enzymes potentially changing histone methylation marks are still uncharacterized, and antagonism or cooperativity, as well as target specificity are likely to result in a complex pattern of writing and reading these marks.

2.2.3 Enzymes for Other Histone Modifications

One of the other important histone PTMs, in addition to acetylation and methylation, is phosphorylation (reviewed in Houben et al. 2007). Histone phosphorylation is involved in DNA repair (γH2AX) and the regulation of chromosome segregation and cell division (Aurora kinases). Histone phosphorylation can be influenced by other epigenetic histone marks, as is the case for phosphorylation of H3S10, which is influenced by the modification state of the adjacent lysine, H3K9. ADP-ribosylation of histones occurs in plants, but has not been studied extensively. Ubiquitination of histones H2A and/or H2B also has important regulatory functions in animals and plants (reviewed in Berr et al. 2011; Bycroft 2011) and can recruit or exclude other modifying enzymes. Proteins that install or remove ubiquitination have been characterized and influence cell cycle, development, and pathogen resistance.

2.2.4 *Histone Variants, Linker Histones, and Nonhistone Proteins*

Histones are among the most conserved proteins across eukaryotic taxa and are encoded by highly redundant gene families. Like animals, plants have developed structurally and functionally distinct classes of H2A and H3 histone variants (discussed in Ch. 20 [Henikoff and Smith 2014]). The physical properties of the variants play an important role in their dynamic association with DNA (reviewed in Ingouff and Berger 2010; Deal and Henikoff 2011).

Phosphorylation of the H2AX variant marks the sites of DNA damage and is believed to recruit DNA repair proteins. H2A.Z is a variant found mainly around the transcriptional start site of genes, likely regulating transcription, and is mutually exclusive with DNA methylation (Zilberman et al. 2008). Its incorporation requires the activity of a SWR1 chromatin-remodeling complex. On heat stress, H2A.Z-containing nucleosomes dissociate from the DNA, accompanied by changes in gene expression (Kumar and Wigge 2010). The CenH3 variant of histone H3 specifies the nucleosomes of centromeric regions and is required for kinetochore assembly, microtubule attachment, and chromosome segregation during cell division. Histone variant H3.3, which differs from the canonical H3 subunit at only a few amino acids, is found predominantly at regulatory regions and expressed genes. It is incorporated into chromatin in a replication-independent manner, involving special chaperones and remodeling complexes. However, the mechanisms responsible for histone variant replacement outside of replication are not well understood in plants.

In addition to the four histones that form nucleosome core particles, the compaction of DNA and its accessibility to interacting proteins is also determined by linker histones, particularly histone H1. Linker histones display substantial diversity, and their functional specialization is suggested by the expression of stress-inducible variants. Down-regulating specific linker histones is sometimes compensated for by the up-regulation of other variants, but can also result in DNA hypomethylation and pleiotropic phenotypic defects (reviewed in Jerzmanowski 2007).

As in other eukaryotes, plants have nonhistone chromosomal proteins that may contribute to epigenetic regulation, including HMG proteins. The best characterized and most diverse subgroup of these proteins in plants is the HMGB family, whose members differ in expression level, pattern, localization, and interaction with DNA and other proteins. Mutation and ectopic expression of individual family members indicate their partial subfunctionalization and a role in development and stress response (reviewed in Pedersen and Grasser 2010). One HMG protein, STRUCTURE-SPECIFIC RECOGNITION PROTEIN (SSRP1), is indirectly involved in DNA demethylation of imprinted genes in the central cell of the female gametophyte (Ikeda et al. 2011).

Cohesin complexes secure sister chromatid alignment before their separation during mitotic and meiotic anaphase, and are also involved in DNA repair, spindle attachment, chromosome condensation, and regulation of DNA accessibility. Cohesin structure, assembly, and removal appear to be highly conserved, but some family members in plants may have specialized functions (reviewed in Yuan et al. 2011). For instance, DEFECTIVE IN MERISTEM SILENCING3/INVOLVED IN DE NOVO (DMS3/IDN1) is a protein related to the hinge domain region of cohesins and condensins and is necessary for DNA-dependent RNA polymerase V (RNA Pol V) transcription in the establishment of RdDM (see Sec. 3.5.2).

Proteins such as REPLICATION PROTEIN A2 (RPA2) or REPLICATION FACTOR C1 have been identified in mutant screens for epigenetic regulators (Elmayan et al. 2005; Kapoor et al. 2005). Enhanced phenotypes of chromatin mutants when combined with loss-of-function mutants of the topoisomerase homolog MGOUN (MGO) also indicate a role for this protein, especially in connection with stem cell and meristem maintenance (Graf et al. 2010). It can be expected that many other nonhistone proteins interacting with DNA will also turn out to be direct or indirect epigenetic regulators.

2.3 Polycomb Proteins and Interacting Components

PcG group proteins were initially identified as master regulators and suppressors of homeotic genes in *Drosophila*. In balance with the activating TrxG proteins, PcGs determine cell proliferation and cell identity (see PcG proteins in Ch. 17 [Grossniklaus and Paro 2014] and TrxG proteins in Ch. 18 [Kingston and Tamkun 2014]).

POLYCOMB REPRESSIVE COMPLEX 2 (PRC2) in plants is the more conserved of two different types of PcG complexes and is responsible for trimethylation of histone H3 at lysine 27 (H3K27me3), as in animals. Each subunit in the *Drosophila* PRC2 complex has several paralogous equivalents in *Arabidopsis* (see Fig. 3 in Ch. 17

Cite as *Cold Spring Harb Perspect Biol* doi: 10.1101/cshperspect.a019315

[Grossniklaus and Paro 2014]). MEDEA (MEA), CURLY LEAF (CLF), and SWINGER (SWN) corresponding to the *Drosophila* protein E(Z); FERTILIZATION-INDEPENDENT SEEDS 2 (FIS2), VERNALIZATION 2 (VRN2), and EMBRYONIC FLOWER 2 (EMF2) corresponding to Su(z)12; FERTILIZATION-INDEPENDENT ENDOSPERM (FIE) corresponding to Esc; and MULTICOPY SUPPRESSOR OF IRA HOMOLOG 1–5 corresponding to p55 (reviewed in Koehler and Hennig 2010). This diversification of PRC2 component proteins is thought to be associated with the evolutionary expansion of land plants (reviewed in Butenko and Ohad 2011). Plant PRC2 components are required at different developmental stages and function at specific, but sometimes overlapping, subsets of genes. For many functions, plant PRC2 interacts with additional proteins or specific RNA transcripts, as in the control of flowering time (discussed in Ch. 31 [Baulcombe and Dean 2014]). PcG proteins themselves are under strict regulatory control, in part via DNA demethylation. At least two PcG genes (*MEA* and *FIS2*) are imprinted genes, meaning that they are differentially expressed depending on the parent from which they were inherited (reviewed in Raissig et al. 2011). Although imprinting evolved independently in plants and mammals (Table 1), DNA methylation and PcG proteins are key components in both cases.

PRC1, the other PcG protein complex, is more dissimilar in *Arabidopsis* compared with animals, but has related functions. Like the *Drosophila* protein POLYCOMB, the *Arabidopsis*-LIKE HETEROCHROMATIN PROTEIN 1 (LHP1) binds to H3K27me3 and might "read" and "translate" this modification into additional and more repressive chromatin configurations, such as by interaction with the H2A ubiquitinating enzymes, AtRING 1, -2, and AtBMI1, also homologs of PRC1 components.

Subunits of both PRC core complexes interact with other proteins that could modulate specificity, but the means by which PcG complexes are recruited to specific target genes in plants is still not clear. Importantly, the consensus sequences for Polycomb or Trithorax responsive elements determined in animals have not been identified in plants.

2.4 Nucleosome-Organizing Proteins

Replication, transcription, recombination, and repair require transient or lasting changes in the positioning of nucleosomes and their association with DNA. Therefore, dynamic processes at the chromatin level are not restricted to reversible DNA or protein modifications, but can include changes in nucleosome occupancy, nucleosome composition, and the accessibility of the DNA to other proteins.

2.4.1 Chromatin-Remodeling Complexes

Relocation or dissociation of nucleosomes can be accomplished by chromatin-remodeling ATPases such as the SWI/SNF complex first identified in yeast and named according to the processes affected in the mutants (see Ch. 21 [Becker and Workman 2013]). Plants have several such complexes (reviewed in Jerzmanowski 2007). Genetic screens have provided functional information for only a handful of putative chromatin remodelers, the first identified being DECREASE IN DNA METHYLATION 1 (DDM1).

Loss-of-DDM1-function results in genome-wide reduction of DNA methylation and H3K9me2, transcriptional activation of repetitive elements, and dysregulation of numerous genes. As a result, *ddm1* mutants display severe developmental and morphological defects that increase in severity with successive generations. The gradual reduced fitness of *ddm1* mutants is attributable to the accumulation of epimutations and insertional mutations caused by reactivated transposons. Part of the epigenetic information, in the form of DNA methylation, is irreversibly lost in *ddm1* mutants, yet backcrosses with wild-type plants can restore the pattern of modification at some loci as a result of de novo methylation (Teixeira et al. 2009). Like the SWI2/SNF2 ATPase protein, DDM1 displays ATP-dependent nucleosome repositioning activity in vitro (Brzeski and Jerzmanowski 2003). The loss of cytosine methylation that occurs in *ddm1* mutants does not occur in mutants that lack both DDM1 and linker histone H1 (Zemach et al. 2013) indicating that DDM1 is needed for the maintenance methylation machinery to be able to access nucleosomal DNA that is highly packaged, involving both core and linker histones. DDM1's ortholog in mammals, LYMPHOID-SPECIFIC HELICASE, is similarly important for global CpG methylation and development.

Two members of the SWI2/SNF2 family in *Arabidopsis*, DEFECTIVE IN RNA-DIRECTED DNA METHYLATION 1 (DRD1) and CLASSY 1 (CLSY1) are unique to the plant kingdom and have specialized roles in RNA-directed DNA methylation (see Sec. 3.5.2). Four other SWI2/SNF2 proteins, BRAHMA (BRM), SPLAYED (SPD), and MINUSCULE 1 and 2, are involved in the control of similar, but not identical, subsets of genes involved in hormone responses and stem cell maintenance (Sang et al. 2012).

Besides the ATPases, other core subunits of SWI/SNF remodelers are represented in plants, including several SWI3 family members (AtSWI3 A-D), one SNF5 homolog (BSH), and two SWP73 homologs (reviewed in Jerzmanowski 2007). Their roles in plants have not been studied extensively, but SWI3 has been shown to interact with RNA binding proteins involved in RNA-directed DNA methylation and to influence the spacing of nucleosomes at silenced loci (Zhu et al. 2012).

Subunits of other remodeling complexes include CHROMATIN-REMODELING PROTEIN 11 (CHR11; an ISWI complex protein), PICKLE (PKL; a CHD3 complex protein), and INOSITOL-REQUIRING 80 and ACTIN-RELATED PROTEINS 4 AND 5 (INO80, ARP4, ARP5; INO80 complex proteins). Mutants defective for these activities are characterized by developmental abnormalities and impaired DNA repair.

MORPHEUS' MOLECULE 1 (MOM1) is a plant-specific epigenetic regulator that is related to SWI2/SNF2 by virtue of a homologous, but incomplete, ATPase domain and is also related to CHD3. The lack of *mom1*, unlike *ddm1* or *pkl*, does not result in morphological defects. Its exact mode of action is unknown, yet *mom1* mutants are characterized by having an intermediate chromatin state at target genes that partially overlap with those targeted by the RdDM pathway (reviewed in Habu 2010).

The SWR1 complex C plays an important epigenetic role in the deposition of H2A.Z at transcription start sites (and the antagonistic exclusion of DNA methylation. The *Arabidopsis* SWR1 complex includes the ATPase subunit PHOTOPERIOD-INDEPENDENT EARLY FLOWERING (PIE), SWR1 COMPLEX 6 (SWC6), and ACTIN-RELATED PROTEIN 6 (ARP6). How this complex is directed to specific target genes is not clear, but its function is essential for developmental decisions and stress responsiveness.

2.4.2 Chromatin Assembly Factors

Whereas SWI/SNF and other remodeling complexes act on nucleosomes that are already associated with DNA, other activities are required for assembling core histones into new nucleosomes after replication, reestablishing chromatin after repair or recombination-associated DNA synthesis, or for the exchange of histones in connection with transcriptional processes. These functions are exerted by histone chaperones, which are mostly acidic proteins that interact with each other and specific canonical, or variant, histones (see Ch. 22 [Almouzni and Cedar 2014] for more detail).

Plants have three chaperones for loading H3/H4 tetramers and three for adding H2A/H2B dimers (reviewed in Zhu et al. 2011b). The CHROMATIN ASSEMBLY FACTOR 1 (CAF-1) complex helps to bring H3/H4 tetramers to the replication fork. Mutations in genes encoding the two larger CAF-1 subunits in *Arabidopsis* (*fas1*, *fas2*) cause characteristic morphological anomalies (fasciation, Fig. 2F), deficiencies in DNA repair, reductions in rRNA gene copy numbers, and derepression of repetitive elements. This indicates that correct nucleosome deposition is essential for development, genome stability, and epigenetic control. As would be expected, the lack of CAF-1 subunits does not interfere with maintenance of DNA methylation, but it can lead to the erasure of other histone-based epigenetic marks. Reduced levels of the third CAF-1 component, MSI1, does not cause fasciation, but lead to distorted seed development and several morphological changes, probably caused by to its multiple roles, including a role within the PRC2 complex. There are two other H3/H4 chaperones that function independently of replication, but also in cooperation with CAF-1: HISTONE REGULATOR A (HIRA), for which there is one *Arabidopsis* homolog and ANTISILENCING FUNCTION 1 (ASF1), which has two *Arabidopsis* homologs.

Histones H2A and H2B are assumed to be installed by NUCLEOSOME ASSEMBLY PROTEIN 1 (NAP1), NAP1-RELATED PROTEIN (NRP), and FACILITATES CHROMATIN TRANSCRIPTION (FACT) proteins. In *Arabidopsis*, there are a number of homologs to these proteins: four NAP1, two NRP, and two FACT homologs, each displaying similarity to the yeast or animal counterparts (reviewed in Zhu et al. 2011b).

Some chaperone subunits can be regulated by posttranslational modification, as indicated by the phenotype of TOUSLED (TSL) mutants, which lack an ASF1 kinase. A mutant of the *Arabidopsis* protein named BRUSHY (BRU) has a similar phenotype to plant CAF-1 *fas* mutants—that is, sensitivity to DNA damage, interference with TGS, and developmental defects. Although BRU has no homology with any known histone chaperones, the additional dysregulation of PcG-controlled genes at specific genomic clusters, likely indicates BRU's importance for maintaining different epigenetic states (Ohno et al. 2011).

It is striking that most mutations in genes for histone chaperones affect the organization of the meristems, the maintenance of stem cells, or processes in the differentiation zone adjacent to the stem cell niche within the meristem. This might reflect the requirement for correct histone and nucleosome deposition to perpetuate stem cell identities in plants.

3 MOLECULAR COMPONENTS OF RNA-MEDIATED GENE SILENCING PATHWAYS

As discussed in the preceding sections, cytosine methylation and posttranslational histone modifications are important aspects of epigenetic gene regulation in plants, helping to establish or maintain gene "on" or "off" states at the level of transcription. Epigenetic regulation also occurs posttranscriptionally through targeted mRNA degradation or translational inhibition. This PTGS can thus act to control the temporal or spatial distribution of develop-

 Cite as *Cold Spring Harb Perspect Biol* doi: 10.1101/cshperspect.a019315

mentally important mRNAs and serve as a potent defense against invaders that include viruses, microbial pathogens, and transgenes (Ruiz-Ferrer and Voinnet 2009; Vazquez et al. 2010).

A common feature of transcriptional and posttranscriptional silencing mechanisms in plants is the involvement of small RNAs, specifically miRNAs or siRNAs. The biogenesis of these small RNAs in plants shares similarities with their biogenesis in other eukaryotes, suggesting that RNA-based silencing mechanisms, involving mi- and siRNA, have a common ancient origin (Chapman and Carrington 2007; Shabalina and Koonin 2008). However, duplication and subfunctionalization of genes involved in miRNA- or siRNA-mediated processes in plants has led to the evolution of multiple pathways that are specialized to accomplish specific tasks (Herr 2005; Baulcombe 2006; Xie and Qi 2008; Chen 2009; Vazquez et al. 2010). These include (1) a pathway for the biogenesis of miRNAs that are complementary to target sequences, resulting in the down-regulation of individual transcribed mRNAs (Sec. 3.2); (2) a pathway in which a miRNA initiates the production of secondary *trans*-acting siRNAs that down-regulate multiple targets with no complementarity to the initiating miRNA; (3) a pathway for siRNA-mediated degradation of invading viral RNAs or transgene RNAs (Sec. 3.4); and (4) a pathway for siRNA-mediated DNA methylation and transcriptional silencing of transposons, viruses, and specific genes (Sec. 3.5). Collectively, these pathways provide plants with an arsenal of RNA-mediated silencing capabilities that is unrivaled by other eukaryotes.

3.1 Common Features of miRNA and siRNA Biogenesis and Silencing

miRNAs and siRNAs share a number of common features in plants, as in other eukaryotes (Bartel 2004; Voinnet 2009; Axtell et al. 2011). Both are produced from dsRNA precursors by the action of RNaseIII-related Dicer (DCL) endonucleases. Resulting small RNAs are then incorporated into a multiprotein RNA-induced silencing complex (RISC) that has, at its core, a member of the Argonaute (AGO) protein family (Czech and Hannon 2011). The AGO protein binds the 3′ end of the small RNA via its PAZ domain and uses the small RNA to base-pair with complementary target RNAs (Joshua-Tor 2006). As a consequence, the target RNA can be cleaved by the AGO protein's PIWI domain, or translation can be blocked without cleavage of the associated RNA, or chromatin-modifying machinery can be recruited to transcriptionally silence the locus. The different outcomes depend on both the class of small RNA involved and the particular AGO protein partner (see also Ch. 16 [Martienssen and Moazed 2014]).

The double-stranded precursors of miRNAs or siRNAs are generated in several ways. In the case of miRNAs, DNA-dependent RNA polymerase II (RNA Pol II) transcripts with extensive self-complementarity fold back on themselves to form stem-loop structures with imperfect double-stranded stems that can be diced by DCL1 (Fig. 3). In the case of siRNAs, the double-stranded precursors can be generated via convergent, bidirectional transcription by a DNA-dependent RNA polymerase such as RNA Pol II, thereby generating transcripts that overlap and base-pair. Alternatively, RNA transcripts can be used as templates for an RNA-dependent RNA polymerase (RdRP), to generate a complementary strand.

Diversification of the core machinery responsible for siRNA biogenesis and function underlies the evolution of the different small RNA silencing pathways in plants (Vazquez et al. 2010; Xie and Qi 2008). Like fission yeast (*S. pombe*) and nematodes (*C. elegans*), but unlike mammals or *Drosophila*, plants make use of RNA-dependent RNA polymerases for the production of dsRNA. Six distinct RdRPs are encoded by the *Arabidopsis* genome. Likewise, *Arabidopsis* has four distinct Dicer endonucleases, DICER-LIKE (DCL) 1 through 4, whereas mammals, fission yeast, and nematodes have only one Dicer, and fruit flies have two. The plant Dicers generate small RNAs with different sizes: miRNAs of (mostly) 21 nt (DCL1), or siRNAs that are 21 nt (DCL4), 22 nt (DCL2), or 23–24 nt (DCL3). The different sized siRNAs have distinct, yet partially overlapping functions, based on their associations with a diversified AGO protein family that has 10 members in *Arabidopsis* (Vaucheret 2008). The different RdRPs, Dicers, and AGO proteins are used in a variety of permutations to accomplish diverse small RNA-mediated silencing phenomena.

3.2 miRNA Biogenesis and Function

miRNAs are critical for plant and animal development, helping limit the functions of complementary mRNAs to a specific subset of cells by bringing about their posttranscriptional degradation or translational repression in adjacent cells (Carrington and Ambros 2003; Ambros 2004; Bartel 2009; Cuperus et al. 2011). miRNAs silence gene expression by base-pairing to target mRNAs in the context of a multiprotein RISC complex, facilitating mRNA cleavage, referred to as "slicing," or the inhibition of mRNA translation (Fig. 3). The importance of miRNAs in plant development is underscored by the fact that many of the genes involved in miRNA biogenesis and function were initially identified in genetic screens for mutants affecting development. Developmental processes requiring miRNAs include stem cell maintenance and differentiation,

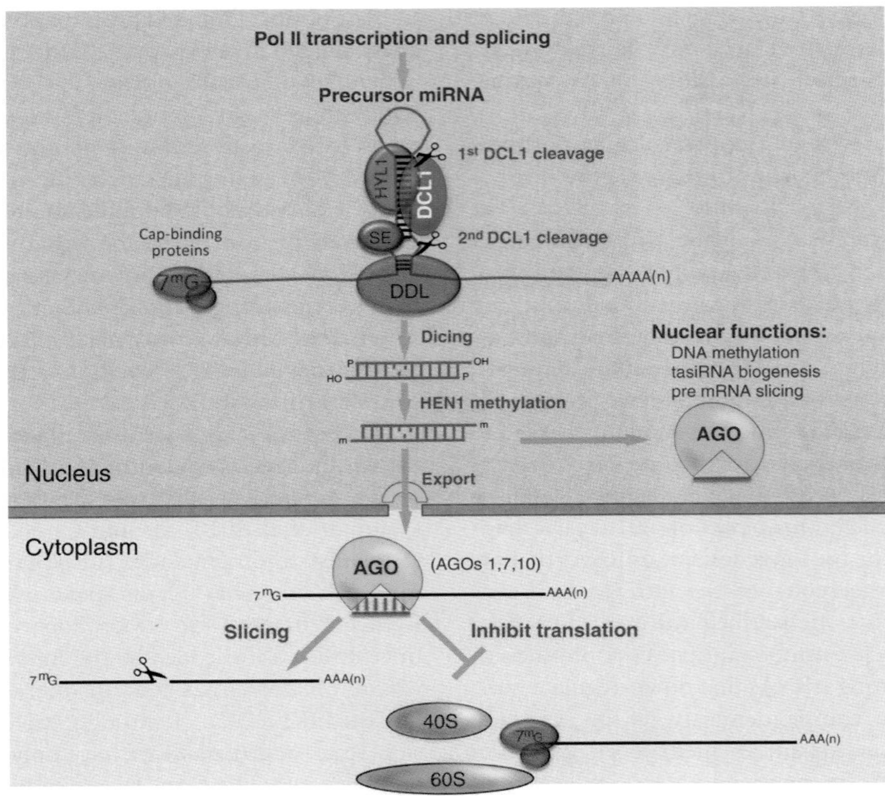

Figure 3. miRNA biogenesis and modes of action. Capped and polyadenylated transcripts of nuclear RNA Pol II that form imperfect hairpins serve as the precursors for miRNAs. DCL1 cleaves these precursors with the help of the HYL1, DDL1, and SE nucleic acid–binding proteins. Diced products are end-methylated by HEN1 and exported to the cytosplasm in a process involving the exportin 5 homolog, HST. In association with AGO1 or a related AGO protein, such as AGO10, the miRNA guides the cleavage or translational inhibition of complementary mRNAs. Nuclear functions in which miRNAs are implicated include specific cases of cytosine methylation, *trans*-acting siRNA (tasiRNA) production or pre-mRNA degradation.

specification of organ polarity, vascular development, floral patterning, hormone signaling, and responsiveness to environmental stresses (reviewed in Rubio-Somoza and Weigel 2011; Khraiwesh et al. 2012).

In plants, as in nematodes, loci encoding miRNAs were initially discovered as genes not encoding a protein, but required for the negative regulation of developmentally important mRNAs (Ambros 2004; Chen 2005). We now know that the transcripts of these noncoding RNA loci are processed into small (micro) RNAs with perfect or near-perfect complementarity to the down-regulated mRNAs. miRNAs with mismatches relative to their target mRNAs tend to repress translation by binding stably to the mRNA, hence interfering with the translation machinery. By contrast, miRNAs with perfect complementarity to their target mRNAs tend to induce mRNA cleavage through slicing by the PIWI domain of the associated AGO protein. In animals, in which miRNAs rarely match their target mRNAs perfectly, translation inhibition is the most prevalent form of miRNA action (Ambros 2004; Bartel 2004). In plants, perfect or near-perfect complementarity of plant miRNAs to their target mRNAs is the norm, thereby favoring mRNA slicing (Jones-Rhoades et al. 2006; Axtell et al. 2011; Cuperus et al. 2011). However, these are tendencies, not rules, such that some plant miRNAs and siRNAs block translation, and some mammalian miRNAs direct mRNA slicing.

miRNAs are transcribed by DNA-dependent RNA Pol II, sometimes as separate miRNA genes and sometimes as sequences present within long noncoding RNAs or in introns of protein-coding genes. As in animals, plant miRNAs originate from 70- to more than 600-bp-long precursor RNAs that fold back to form imperfect hairpin or stem-loop structures (Bartel 2004; Jones-Rhoades et al. 2006; Cuperus et al. 2011). In mammals, the RNase III-related endonuclease, Drosha, cleaves the initial pri-miRNA stem-loop precursor on the loop-proximal side of the stem, aided by the dsRNA binding protein, Pasha. Alternatively, in the case of intron-encoded miRNAs, or "mirtrons," splicing and lariat debranching enzymes generate the initial pre-miRNA (Okamura et al. 2007; Westholm and Lai 2011). The resulting partially processed pre-RNA is then transported to the cytoplasm by an Exportin 5-mediated pathway to undergo further processing. In the cytoplasm, Dicer cuts the stem at a site 22 nt from the Drosha cleavage site, thereby generating a duplex that includes the mature miRNA, or miR, and the complementary passenger strand, miR* (Du and Zamore 2005; Carthew and Sontheimer 2009). Plants, however, lack Drosha and Pasha orthologs and are thought to generate miRNAs entirely within the nucleus, with DCL1 responsible for both cleavages of the stem-loop precursor RNA, generating miR/miR* duplexes that are typically 21-nt long (Fig. 3). DCL1 processing is assisted by the RNA-binding proteins HYPONASTIC LEAVES 1 (HYL1; also known as DRB1) and DAWDLE (DDL), as well as a zinc-finger protein, SERRATE (SE). The miR/miR* duplex is then methylated by HUA ENHANCER 1 (HEN1) at the 3′-terminal ribose of each RNA strand, thereby protecting the small RNAs from uridylation and increasing their stability. Most methylated miR/miR* duplexes are then thought to be exported from the nucleus to the cytoplasm in a process involving HASTY (HST), a homolog of mammalian Exportin 5. However, the abundance of specific miRNAs is differentially affected in *hst* mutants, suggesting variation in the trafficking of different miRNAs. The miR strand is loaded onto AGO1, or a related AGO family member, including AGO10 or AGO7, and the resulting AGO-RISC complex carries out slicing or translational inhibition of mRNAs complementarity to the miRNA (reviewed in Chen 2009; Poethig 2009; Axtell et al. 2011; Cuperus et al. 2011).

Comparative analyses of angiosperms (flowering plants), gymnosperms, and more primitive plants, such as ferns and mosses, have identified a number of highly conserved miRNAs that are critical for meristem function and development (Axtell et al. 2007; Axtell and Bowman 2008). Although the coding sequences of the mRNAs targeted by these miRNAs have changed during evolution, the miRNA-binding sites within the mRNAs are nearly invariant, indicating that the need for such complementarity has constrained the evolution of the matching miRNA and mRNA sequences for at least 400 million years.

AGO1 is the founding member of the AGO protein family and is the principal mRNA slicer in *Arabidopsis* for DCL1-, DCL2-, and DCL4-generated small RNAs (Baumberger and Baulcombe 2005). AGO1 was identified as an *Arabidopsis* mutant defective in leaf development before the discovery of miRNAs. The name "Argonaute" was inspired by the phenotype of *ago1* mutants, whose narrow, filamentous leaves gave the plants the appearance of a small octopus of the genus *Argonauta*. *Ago1* hypomorphic mutants display severe shoot apical meristem defects, as do *ago10* mutants. The latter are also known as pinhead (*pnh*) mutants (see Table 1) because of the absence of lateral organ formation at the shoot apex. These dramatic apical meristem phenotypes derive from defects in stem cell maintenance and cell lineage specification (reviewed in Kidner and Martienssen 2005; Vaucheret 2008).

Approximately 50% of the known miRNA targets in *Arabidopsis* are transcription factors, many of which modulate meristem formation and identity. Other miRNAs target mRNAs encoding F-box proteins that are involved in the ubiquitin-mediated turnover of proteins in developmentally important signaling pathways. The biogenesis of some miRNAs is induced by specific environmental stresses, leading to down-regulation of complementary mRNAs during adaptive stress responses (Axtell and Bowman 2008; Poethig 2009; Cuperus et al. 2011).

An interesting regulatory strategy in plants is the use of decoys to sequester specific miRNAs from key targets. For instance, one function attributed to AGO10 is the sequestration of miR166/165 so as to make the miRNAs unavailable to AGO1 (Zhu et al. 2011a). In other cases, mRNAs with imperfect complementarity to miRNAs serve as decoys, or target mimics, that tie up miRNA-AGO RISC complexes such that other mRNA targets escape inactivation (Franco-Zorrilla et al. 2007). Interestingly, AGO1 and DCL1 are themselves targeted by miRNAs, allowing for negative feedback of miRNA biogenesis.

Mature miRNAs are detected in both the nucleus and cytoplasm, suggesting that miRNAs may have functions in both cellular compartments. Evidence that miRNAs can direct DNA cytosine methylation was among the earliest clues that miRNAs can regulate nuclear activities (Bao et al. 2004). There is also evidence that posttranscriptional silencing can be mediated by small RNA-guided cleavage of intronic sequences (Hoffer et al. 2011). Collectively, these observations suggest that pre-mRNAs have the potential to be targeted by miRNA or siRNA RISCs before mRNA maturation or export to the cytoplasm. Whether dedicated machinery exists for RISC trafficking from the cytoplasm to the nucleus, or whether RISC assembly can occur in both compartments, is not clear.

3.3 *Trans*-Acting siRNAs

Most miRNAs directly regulate specific mRNAs. However, plants have evolved another pathway in which miRNAs initiate the production of secondary siRNAs that then target mRNAs (Fig. 4). This so-called tasiRNA pathway gets its name because the secondary siRNAs act on loci that are distinct from the loci that give rise to the siRNAs (reviewed in Chen 2009; Chuck et al. 2009; Allen and Howell 2010).

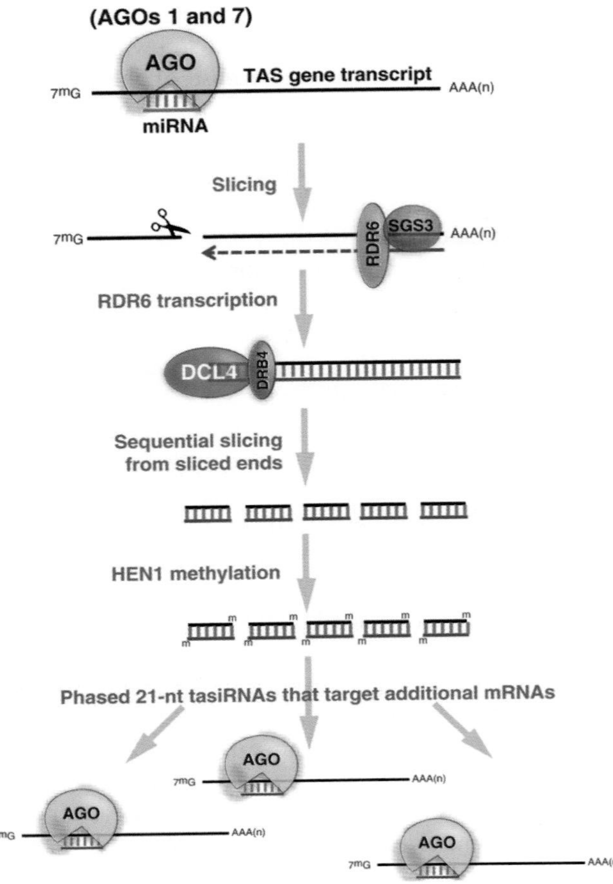

Figure 4. tasiRNA biogenesis and function. Specific miRNAs, associated with AGO1 or AGO7, target and cleave the transcripts of TAS gene loci, ultimately giving rise to tasiRNAs. In the case of AGO1, initiating miRNAs that are 22 nt in length induce the recruitment of the RNA-dependent RNA polymerase, RDR6 (and its partner SGS3), to the 3′ cleaved fragment resulting in transcription of the complementary strand to generate dsRNA. Dicing by DCL4, aided by the dsRNA-binding protein, DRB4, results in the production of 21-nt tasiRNAs that are phased with respect to the miRNA-cleaved end. These tasiRNAs, in turn, target specific complementary RNAs in association with AGO1, thus behaving like target-specific miRNAs.

In *Arabidopsis*, there are four families of TAS gene, loci that generate tasiRNAs: TAS 1–4. TAS1 and TAS2 are both targeted by miR173 in association with AGO1, TAS4 is targeted by miR828-AGO1, and TAS3 is targeted by miR390-AGO7 (Montgomery et al. 2008). TAS loci generate long noncoding RNAs that are cleaved by the miRNA-AGO-RISC complex, triggering the conversion of the 3′ cleavage product into dsRNA through the action of RNA-DEPENDENT RNA POLYMERASE 6 (RDR6) and its partner, SUPPRESSOR OF GENE SILENCING 3 (SGS3). The dsRNA is then diced by DCL4 into 21-nt tasiRNAs that are phased with respect to the original miRNA cleavage site, such that tasiRNA sequences have defined sequences, much like miRNAs, and function like miRNAs to direct the slicing of downstream complementary target mRNAs (Fig. 4). The tasiRNAs produced by the TAS3 locus target several Auxin response factor mRNAs, thereby affecting developmental transitions that are regulated by the plant hormone auxin. By modification of auxin responses, tasiRNAs contribute to the transition of the meristem from a juvenile phase, with vegetative organs only, to an adult phase capable of responding to signals promoting flowering (Poethig et al. 2006).

3.4 siRNA-Mediated Silencing as a Genome Defense Strategy

Multifunctional siRNA-AGO complexes and/or dicer endonucleases have evolved as useful components in the defense against viral and microbial pathogen attack (Ch. 16 [Martienssen and Moazed 2014]). Most plant viruses are RNA viruses whose dsRNA forms are subject to DCL-catalyzed dicing and degradation. For DNA viruses, such as geminiviruses or Cauliflower Mosaic Virus, the viral genome can be the target of siRNA-mediated DNA methylation, or siRNAs can facilitate the degradation or inactivation of viral transcripts, resulting in transcriptional silencing and decreased viral RNA synthesis. Defense against foreign or invasive nucleic acids thus provides the most likely explanation for the evolution of small RNA-mediated RNAi pathways in eukaryotes and for the diversification of RNA-silencing capabilities in plants (Ding 2010). These include the ability to silence genes at a transcriptional and/or posttranscriptional level and an ability to spread defensive small RNAs to cells distant from the site of initial infection, thereby arming the plant against a spreading virus, enabling it to mount a systemic defense (illustrated in Fig. 1 of Ch. 2 [Dunoyer et al. 2013]).

Most plant viruses are single-stranded RNA viruses whose replication by endogenous or viral encoded RNA-dependent RNA polymerase transcription produces dsRNA intermediates (Wassenegger and Krczal 2006) that can be diced into 21–24-nt siRNAs, primarily by DCLs 2, -3, and -4 (Fig. 5) (Vazquez et al. 2010). Other plant viruses have dsRNA genomes providing direct substrates for Dicer cleavage. Single-stranded DNA viruses, such as geminiviruses, replicate via double-stranded DNA intermediates and generate RNA transcripts from both strands of the DNA. These RNAs can overlap and base-pair resulting in dsRNA that can be diced. Alternatively, long viral RNA transcripts might fold back on themselves into structures with some dsRNA regions that might be diced. Thus, antiviral RNAi responses can be enabled by dsRNAs produced by the virus itself.

Cite as *Cold Spring Harb Perspect Biol* doi: 10.1101/cshperspect.a019315

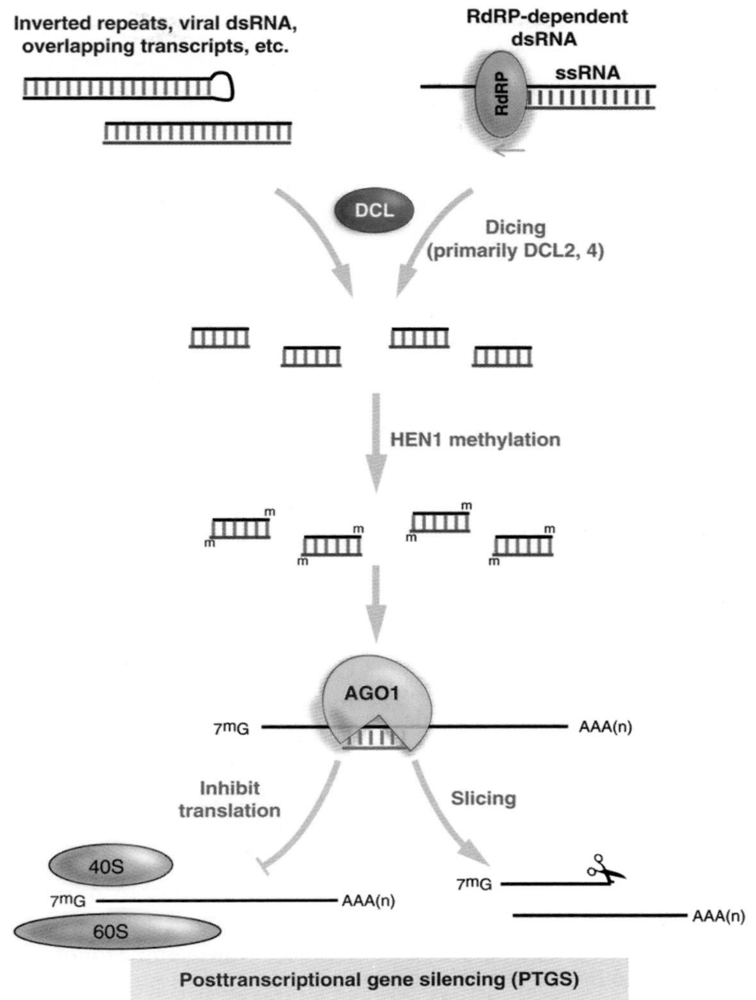

Figure 5. PTGS induced by viruses or transgenes. Double-stranded precursors of siRNAs that bring about PTGS can be dsRNA viral genomes or replication intermediates, dsRNAs formed by overlapping transcripts from adjacent transcription units, or inverted repeats resulting from tandem transgenes that integrate in opposite orientations. Alternatively, single-stranded RNAs can be made double-stranded by the action of RNA-dependent RNA polymerases, such as RDR6 or RDR1. Dicing can then occur by DCLs 2, -3 or -4, with 22-nt DCL2 products and 21-nt DCL4 products being primarily associated with PTGS. On association with AGO1, the siRNAs target complementary RNAs for degradation or translational inhibition.

Transgenes introduced into plant genomes can fall victim to silencing pathways that probably evolved for antiviral defense, as illustrated in Figure 5. This occurs because transgenes often integrate in multiple copies, and in inverted orientation relative to one another. Thus, read-through transcripts that extend from one transgene into an adjacent transgene integrated in the opposite orientation produce RNA with regions of self-complementarity, or inverted repeats, resulting in dsRNA that can be diced. A transgene might also integrate into a chromosome adjacent to a strong promoter pointed in the opposite orientation relative to the transgene's promoter, resulting in convergent bidirectional transcription and dsRNA formation.

PTGS of transgenes and VIGS involve the production of 21–24-nt siRNAs generated by DCLs 2, -3, and -4 (Fig. 5). In general, the 21- and 22-nt siRNAs generated by DCL4 are thought to guide mRNA cleavage in association with AGO1. Following endonucleolytic cleavage by a 21-nt siRNA-AGO1 complex, the severed 3′ segment of the mRNA lacking a 7-methylguanosine cap at the 5′ end is degraded in the 5′ to 3′ direction by the exonuclease, AtXRN4 (Rymarquis et al. 2011). The 5′ fragment is probably degraded by the exosome in a 3′ to 5′ direction. An intriguing property of 22-nt small RNAs, which are less abundant than 21-nt siRNAs, is that their association with AGO1 brings about mRNA cleavage that is coupled to the recruitment of RDR6, similar to the action of the 22-nt miRNAs that give rise to tasiRNAs. As a result, the 3′ AGO cleavage fragment is converted into a dsRNA, which is then diced by DCL4 to generate 21-nt secondary siRNAs (Chen et al. 2010; Cuperus et al. 2010; Manavella et al. 2012).

siRNAs that are generated by DCL3 associate primarily with AGO4, or the related AGO6 or AGO9 proteins (Havecker et al. 2010), and can direct RNA slicing of viral or transgene-derived RNAs. However, these 24-nt siRNA-AGO RISC complexes have the added property of directing epigenetic modifications to homologous DNA sequences (see Sec. 3.5.1). The combination of mRNA degradation, production of secondary siRNAs, and transcriptional silencing constitutes a potent response to an invading nucleic acid. However, it is not surprising that viruses have evolved counter-measures, encoding proteins that suppress RNA silencing (discussed in Ch. 31 [Baulcombe and Dean 2014]), in an evolutionary arms race between pathogen and host (reviewed in Ding and Voinnet 2007; Bivalkar-Mehla et al. 2011).

3.5 RNA-Directed DNA Methylation and Heterochromatin Formation

RNA-mediated transcriptional silencing of transposons, retroviruses, and other genomic repeats is well documented in diverse eukaryotes, including plants, mammals, fission yeast, and fruit flies. In fission yeast and fruit flies, which do not use DNA methylation as part of their gene silencing toolkit, small RNAs mediate changes in histone modification to bring about transcriptional silencing (detailed in Ch. 12 [Elgin and Reuter 2013] and Ch. 9 [Allshire and Ekwall 2014]). However, in plants and mammals, which methylate their DNA, cytosine methylation and repressive histone modifications are partners in the silencing of repetitive elements and the specification of heterochromatin. In plants, the silencing of facultative heterochromatin, which includes retrotransposons and other expressed repeats such as excess rRNA genes, involves 24-nt siRNAs that direct the de novo cytosine methylation of corresponding genomic sequences (Matzke et al. 2009; Law and Jacobsen 2010; Haag and Pikaard 2011; Zhang and Zhu 2011). In the male germline of mammals, small RNAs similarly associate with proteins of the PIWI subfamily of AGO proteins (piRNAs) and are implicated in directing histone modifications and de novo cytosine methylation of transposons (Klattenhoff and Theurkauf 2008; He et al. 2011; Pillai and Chuma 2012).

3.5.1 RdDM

RdDM was first observed in tobacco plants infected with viroids (Wassenegger et al. 1994). Viroids are plant pathogens consisting solely of a circular noncoding RNA only several hundred nucleotides in length. Replicating viroids were found to trigger de novo methylation of viroid cDNAs that had been integrated as transgenes into the tobacco genome. RNA viruses were similarly found to elicit the methylation of homologous DNA sequences in the nuclear genome. Moreover, by expressing transgenes engineered to express dsRNAs homologous to promoter sequences, cytosine methylation was shown to be directed to the promoter, resulting in homology-dependent transcriptional silencing of the corresponding gene (Mette et al. 2000).

RdDM is typified by the methylation of cytosines primarily within the region of RNA-DNA sequence identity, although RNA-mediated spreading to other sequences can occur (see Sec. 3.6). RdDM establishes methylation in all sequence contexts (see Sec. 2.1), but CHH methylation is a specific hallmark of RdDM. This is because symmetrical CG and CHG methylation can be perpetuated at every round of replication by maintenance methylation, but perpetuation of CHH methylation at many silenced loci requires continuous, RNA-guided de novo methylation in every cell cycle (Law and Jacobsen 2010).

Combined data from forward and reverse genetic screens have revealed many of the key molecular components needed for RNA-directed DNA methylation and TGS (see Secs. 2.1 and 2.2). These genetic screens have relied on reporter transgenes or endogenous genes having a visible phenotype when silenced, allowing mutants that release (or prevent) silencing to be readily identified.

3.5.2 Plant-Specific Machinery for RNA-Directed DNA Methylation

As explained previously (Sec. 2.1.1), de novo methylation during RdDM is catalyzed by the DRM class of DNA methyltransferases. The full details of how DRM2 is recruited to DNA are not yet clear. However, a number of key activities have been identified, revealing a critical role for noncoding RNAs and siRNAs to allow the DNA methylation machinery to hone in on its target sites (Fig. 6).

All known eukaryotes have three highly conserved nuclear multisubunit DNA-dependent RNA polymerases that are essential for viability: RNA Pol I, II, and III. Interestingly, plants have evolved two additional nuclear DNA-dependent RNA polymerases—RNA Pol IV (in the early literature referred to as RNA Pol IVa) and RNA Pol V (previously referred to as RNA Pol IVb)—that play important roles in RdDM in *Arabidopsis* (Haag and Pikaard 2011). It is clear that RNA Pols IV and V evolved as specialized forms

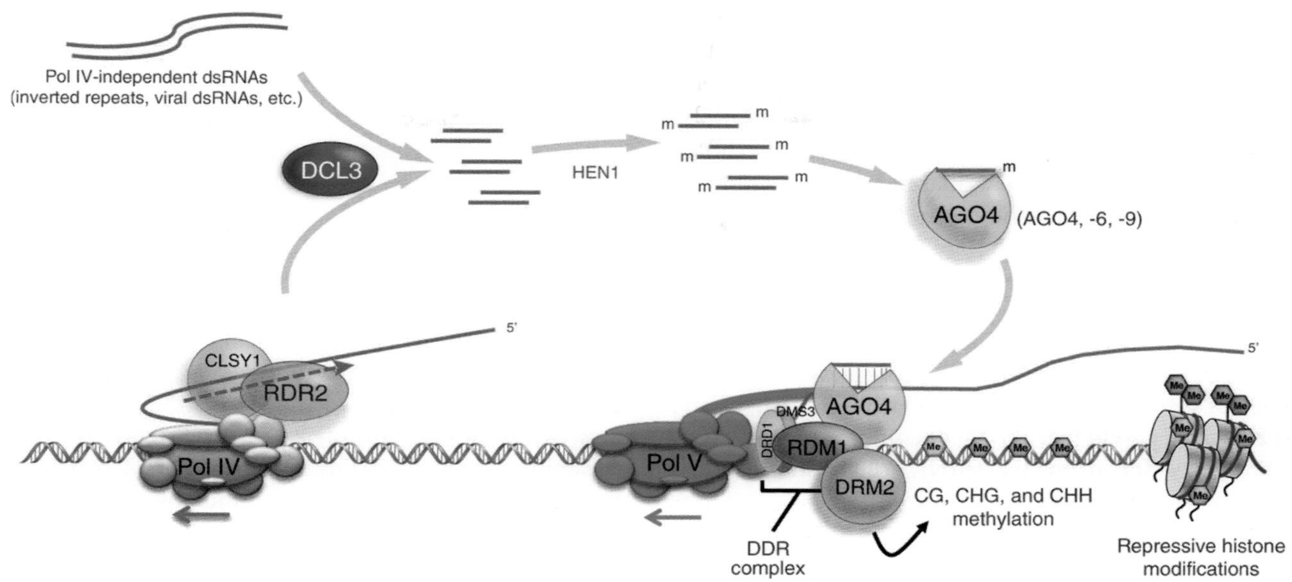

Figure 6. RdDM. Retrotransposons, viruses, transgenes, or repetitive genes are subject to silencing by RdDM. RNA Pol IV is thought to generate single-stranded RNAs (*bottom left*) that serve as templates for the RNA-dependent RNA polymerase, RDR2. Resulting dsRNAs are diced by DCL3 to generate 24-nt siRNA duplexes that are then methylated by HEN1 and loaded onto AGO4, or its closely related family members, AGO6 or AGO9. dsRNA derived from inverted repeat transgenes or viruses (*top left*) can bypass the need for RNA Pol IV and RDR2. AGO-siRNA complexes are recruited to their sites of action by binding to transcripts generated by RNA Pol V, as well as by physical interactions with the carboxy-terminal domain of the RNA Pol V largest subunit. At some loci, RNA Pol II is thought to substitute for RNA Pol V for the production of scaffold transcripts to which AGO-siRNA complexes bind. The DDR complex (DRD1, DMS3, and RDM1) enables RNA Pol V transcription. The RDM1 subunit of the DDR complex also interacts with AGO4 and the de novo cytosine methyltransferase, DRM2, thus potentially serving as a bridge that recruits DRM2 to sites of RNA Pol V transcription.

of RNA Pol II, a process that began in green algae before the emergence of land plants (Luo and Hall 2007; Tucker et al. 2010). *Arabidopsis* RNA Pols II, IV, and V each have 12 core subunits, approximately half of which are common to all three polymerases and encoded by the same genes (Ream et al. 2009). Subunits that are unique to RNA Pols IV or V are encoded by genes that arose via the duplication of RNA Pol II subunit genes, followed by subfunctionalization. For some subunits, there are two or more variants (Ream et al. 2009; Law et al. 2011). These alternative subunits can confer different capabilities to the enzymes, suggesting that there are multiple functional subtypes of RNA Pols II, IV, and V (Tan et al. 2012).

RNA Pol IV and Pol V have distinct roles in the RdDM pathway (reviewed in Matzke et al. 2009; Lahmy et al. 2010; Law and Jacobsen 2010; Haag and Pikaard 2011; Zhang and Zhu 2011). RNA Pol IV is thought to act early in the pathway because it colocalizes with genetic loci that give rise to abundant 24-nt siRNAs and is required for their biogenesis. Moreover, the loss of RNA Pol IV activity causes other proteins of the RdDM pathway to mislocalize. Genetic and biochemical evidence suggest that RNA Pol IV transcripts serve as templates for RNA-DEPENDENT RNA POLYMERASE 2 (RDR2), thereby generating dsRNAs. RNA Pol IV and RDR2 physically interact in vivo, with evidence that their activities are mechanistically coupled, allowing for the channeling of precursor RNAs for 24-nt siRNA biogenesis (Haag et al. 2012). CLSY1, a putative ATP-dependent chromatin remodeler in the SWI2/SNF2 helicase domain protein family, is thought to assist the RNA Pol IV–RDR2 transcription complex. As a result, nuclear localization patterns of RNA Pol IV and RDR2 are perturbed in *clsy1* mutants.

dsRNAs that result from the collaboration of RNA Pol IV with RDR2 are cleaved into 24-nt duplexes by DICER-LIKE 3 (DCL3), and HEN1 methylates the three ends of the siRNAs on their 2′ hydroxyl groups, helping to stabilize the siRNAs in the same way as for miRNAs (see Sec. 3.2). One strand of the duplex is then loaded into AGO4, or the closely related 24-nt siRNA-binding proteins, AGO6 or AGO9. Recruiting the resulting siRNA-AGO complexes is the function of RNA Pol V, whose RNA transcripts have been detected at a number of loci that are subjected to RdDM. These RNAs are dependent on the RNA Pol V active site and can be chemically crosslinked to RNA Pol V, suggesting that they are direct RNA Pol V transcripts. RNA Pol V transcription does not require siRNA biogenesis, but is dependent on the putative chromatin remodeler, DRD1, DMS3 (a protein that shares homology with the hinge domain regions of structural maintenance of chromosomes proteins, such as cohesins and condensins; Wierzbicki et al. 2008; Wierzbicki et al. 2009) and REQUIRED FOR DNA METHYLATION1 (RDM1), a single-stranded DNA-binding protein with a preference for binding methylated DNA in vitro. DRD1, DMS3, and RDM1 physically associate within a multifunctional complex (DDR) that copurifies with RNA Pol V (Law and Jacobsen 2010; Zhang and Zhu 2011). Whether the DDR complex recruits RNA Pol V to target loci or mediates RNA Pol V transcriptional elongation is still unclear.

Based on chemical crosslinking and immunoprecipitation studies, current models suggest that AGO4-RISC complexes are recruited to target loci by siRNA base-pairing to RNA Pol V transcripts (Wierzbicki et al. 2008; Wierzbicki et al. 2009). In this way, RNA Pol V transcripts are thought to serve as scaffolds for recruiting the chromatin-modifying machinery into proximity of the chromatin to be modified. AGO4 also binds tryptophan (W) and glycine (G) WG and GW repeats within the carboxy-terminal domain of the Pol V largest subunit, further contributing to RISC assembly at Pol V-transcribed loci (Lahmy et al. 2010).

Two AGO4-interacting proteins are important for RdDM: INVOLVED IN DE NOVO 2 (IDN2/RDM12), a protein that binds dsRNAs with 5′ overhangs, possibly siRNAs base-paired with RNA Pol V transcripts, and KOW DOMAIN-CONTAINING TRANSCRIPTION FACTOR (KTF1, also known as RDM3 or SPT5-LIKE) that shares similarity with the yeast RNA Pol II transcription factor, SPT5 (Lahmy et al. 2010; Law and Jacobsen 2010; Zhang and Zhu 2011). In vitro, KTF1 binds RNA and has a WG/GW-rich domain that facilitates interactions with AGO4. KTF1 might therefore play a role in the recruitment of AGO4 to RNA Pol V transcripts, promoting AGO4 slicing of RNA Pol V transcripts as well as recruitment of downstream silencing components.

At present, it is not clear how DRM2 DNA methyltransferase is recruited to Pol V-transcribed loci, but the RDM1 protein of the DDR complex might serve as a bridge between AGO4 and DRM2 (Zhang and Zhu 2011). RDM1 also interacts with RNA Pol II, which can substitute for RNA Pol V in the production of scaffold transcripts at some loci and is also implicated in the recruitment of RNA Pol IV and Pol V at some loci (Zheng et al. 2009). RDM1 might therefore be a key protein that can mediate DRM2 recruitment in both RNA Pol V- and Pol II-dependent pathways.

A classic epigenetic phenomena, first discovered in maize, is paramutation, the heritable alteration of gene activity that can occur when certain active and silenced alleles are brought together within the same nucleus. Maize genes required for paramutation include MAINTENANCE OF PARAMUTATION 1 (MOP1; the ortholog of *Arabidopsis* RDR2); REQUIRED TO MAINTAIN REPRESSION 6 (RMR6; the ortholog of the *Arabidopsis* RNA Pol IV largest subunit, NRPD1); RMR1, a putative chromatin remodeling ATPase related to CLSY1 and DRD1; and MOP2/RMR7, one of three homologs of the second-largest subunit of *Arabidopsis* RNA Pol IV and Pol V (Arteaga-Vazquez and Chandler 2010; Erhard and Hollick 2011). These findings clearly implicate proteins of the RdDM pathway in paramutation yet it remains unclear whether RNA-directed cytosine methylation occurs at paramutable and paramutagenic loci.

3.5.3 Repressive Histone Modifications Associated with Transcriptional Gene Silencing

At loci subjected to RdDM, the methylated DNA is wrapped by histones that bear posttranslational modifications typical of heterochromatin. Such a condensed chromatin state is refractive to transcription by RNA Pol I, II, or III, but enables transcription by Pol IV and Pol V. Several chromatin-modifying enzymes involved in establishing or maintaining heterochromatin (see Sec. 2.2) have been identified in genetic screens for mutants that interfere with RdDM. These include histone H3K9 and H3K27 methyltransferases, the broad-specificity histone deacetylase HDA6, UBIQUITIN PROTEASE 26 (UBP26 or SUP32), which is required to deubiquitinate histone H2B, and JMJ14, a JumonjiC domain-containing protein that demethylates H3K4me3, a modification typical of actively transcribed loci. Collectively, these activities are indicative of a silencing process that includes the establishment of chromatin marks typical of the silenced state (reviewed in Matzke et al. 2009; Lahmy et al. 2010; Law and Jacobsen 2010; Haag and Pikaard 2011; Zhang and Zhu 2011).

3.5.4 RNA-Mediated Silencing of Endogenous Genes

Most RNA-directed DNA methylation and TGS is focused on retrotransposons and repetitive genetic elements that are remnants of transposons, presumably as a means of transposon taming to keep their proliferation in check (Zaratiegui et al. 2007). One consequence is that a number of regulatory genes are affected by their proximity to transposons or repetitive elements. For example, transposon-derived repeats in the promoter of the *Arabidopsis*

Cite as *Cold Spring Harb Perspect Biol* doi: 10.1101/cshperspect.a019315

flowering time gene FWA are targeted by RdDM, thus silencing the gene. Because many plant genes have transposon insertions in the vicinity of promoters or in introns, this mode of regulation is likely to be common in the plant kingdom. Indeed, Barbara McClintock's hypothesis that transposons act as controlling elements regulating adjacent host genes (McClintock 1950) has been confirmed in multiple cases.

Not all repetitive sequences subjected to RdDM share a connection with transposons or are likely candidates for being targeted by machinery adapted for genome defense. Notable examples are the repetitive 5S rRNA genes transcribed by RNA Pol III and the 45S ribosomal RNA genes transcribed by RNA Pol I (Layat et al. 2012). These genes are essential for ribosome synthesis and are clustered in long tandem arrays at several loci, each having hundreds of gene copies. RdDM appears to help shut down excess rRNA genes such that their effective dosage can be regulated according to the cellular demand for ribosomes and protein synthesis (Preuss et al. 2008).

3.6 Amplification and Spreading of siRNA-Mediated Silencing

An important aspect of siRNA-mediated silencing in plants is that initial siRNA production can trigger additional, secondary siRNAs. Moreover, siRNAs can move from cell to cell or be transported to other organs, thereby amplifying siRNA-mediated silencing ultimately executed at the transcriptional or posttranscriptional levels (see also Ch. 2 [Dunoyer et al. 2013]; Ch. 31 [Baulcombe and Dean 2014]).

3.6.1 *Transitivity*

Transitivity is a term used to describe the generation of secondary siRNAs that are induced by a primary siRNA trigger. For instance, two adjacent genes that point toward one another can generate transcripts that overlap in their 3′ regions and can base-pair. Dicing of the resulting dsRNA in the region of overlap can generate primary siRNAs that can then base-pair with a complementary long mRNA (or other long RNA) sequence and prime transcription by RNA-dependent RNA polymerase, particularly RDR6. The result is a dsRNA that extends from the siRNA primer to the 5′ end of the RNA template. Subsequent dicing of these dsRNAs generates secondary RNAs corresponding to upstream (5′) regions (left side of Fig. 7).

Secondary siRNAs can also be generated in the region 3′ of a primary siRNA (or miRNA) trigger, as described for *trans*-acting siRNAs whose biogenesis is triggered by an miRNA. In this case, siRNA or miRNA-mediated cleavage of an RNA transcript facilitates the use of the uncapped 3′ fragment as a template for primer-independent RDR6 transcription (right side of Fig. 7). The mechanistic details of how this occurs are not yet clear, but RDR6 must initiate transcription from the 3′ end of the RNA fragment, a site that is distal from the siRNA or miRNA cleavage site. This process is specifically triggered by 22-nt siRNAs or 22-nt miRNAs that associate with AGO1 (Chen et al. 2010; Cuperus et al. 2010; Manavella et al. 2012), but not by 21-nt siRNA-AGO1 complexes. Although 22-nt miRNAs are not common, they can be produced by DCL1 from dsRNA hairpin precursors that have an extra nucleotide in the miR strand relative to the 21-nt miR* strand, forming a bulge that does not inhibit dicing. Presumably, a change in AGO1 conformation, or its association with other required proteins, occurs on binding 22-nt siRNAs, and results in the recruitment of RDR6. The resulting dsRNA is then diced by DCL4, with the help of the dsRNA binding protein DRB4, generating 21-nt secondary siRNAs that correspond to the region downstream from the primary siRNA or miRNA trigger.

Overall, the ability to produce secondary siRNAs both upstream of and downstream from the region complementary to the primary siRNA(s) magnifies the RNA silencing response, thus strengthening the plant's resistance to an invading virus or nucleic acid (Brodersen and Voinnet 2006).

3.6.2 *Non-Cell-Autonomous Silencing*

Small RNAs can bring about posttranscriptional or transcriptional silencing in neighboring cells or even in cells in distant organs. For short-range movement, RNAs travel from their cell of origin into neighboring cells through plasmodesmata. This conclusion is supported by the fact that guard cells, which form the openings (stomata) for gas exchange in leaves, lack plasmodesmata and are excluded from receiving silencing signals from adjacent cells (Voinnet et al. 1998). Long-range movement is a consequence of RNA loading into phloem cells, allowing their transport through the vascular system followed by their unloading and cell-to-cell spread in recipient tissues. This was shown in grafting experiments involving transgenes, mutants, and deep sequencing of small RNAs (see Ch. 2 [Dunoyer et al. 2013]).

Evidence that mobile RNAs are important transmitters of information between plant cells extends to the male and female gametophytes. In pollen, there is evidence that small RNAs produced by the vegetative cell can direct silencing in the two sperm cells. The hypothesis is that derepression of transposable elements in the vegetative cell allows for the biogenesis of siRNAs that then move to the sperm cells to reinforce the silencing of the corresponding transposons. In this way, epigenetic programming in sperm cells, in

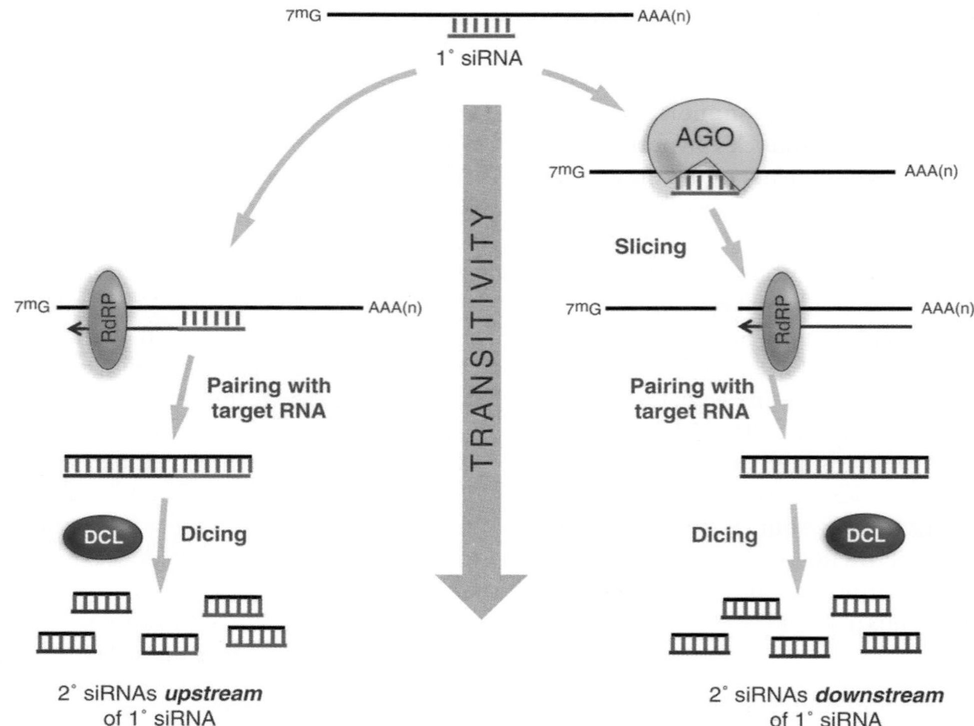

Figure 7. Transitivity: amplification and spreading of secondary siRNAs from the site of a primary siRNA target. Primary siRNAs are thought to prime RNA-dependent RNA polymerase activity, resulting in dsRNAs extending toward the 5′ end of the target RNA (*left*). Subsequent dicing gives rise to secondary siRNAs in the region upstream of the initiating primary siRNA. 22-nt siRNAs also have the property of recruiting RDR6 to the 3′ fragment following AGO slicing (*right*), resulting in dsRNA and secondary siRNA biogenesis in the region downstream from the initiating, primary siRNA.

which transposition would result in frequent deleterious mutations, is guaranteed via the activity of the vegetative cell, which does not contribute genomic information to the next generation through fertilization (Slotkin et al. 2009). Likewise, abundant RNA Pol IV–dependent siRNAs of maternal origin accumulate in embryos and developing seeds and may play roles in the epigenetic reprogramming of paternal chromosomes (Mosher et al. 2009).

4 OUTLOOK

The past decade has witnessed an explosion of information concerning the proteins, RNAs, and chemical modifications that contribute to epigenetic control. However, there is much to learn about the nature of epigenetic inheritance and the role of epigenetics as a source of variation contributing to fitness and natural selection. Much of what we know thus far has come from the analysis of mutants generated using harsh mutagenesis treatments and comparison of these mutants, under laboratory conditions, to nonmutagenized reference plants referred to as being "wild type." Breeders have long known that wild-type plants collected from different geographical origins are themselves a rich source of genetic diversity. Natural variation among ecotypes—strains or races of a species that are adapted to specific niches—reflects naturally occurring nucleotide sequence changes, recombination, or transposition events, gains, or losses of DNA sequences, or hybridization events. In plants with a high degree of inbreeding such as *A. thaliana*, individual ecotypes display a high degree of genetic homogeneity. However, the genetic diversity within the species, revealed by genomic comparisons of available ecotypes, is remarkable (http://1001genomes.org; Ossowski et al. 2010). It can be expected that the degree of epigenetic variation between ecotypes is also substantial, allowing an opportunity to explore whether epigenetic adaptations contribute to plant form, survival, and performance under different conditions (Becker et al. 2011; Richards 2011; Schmitz et al. 2011). Important agricultural challenges and questions, such as understanding the molecular basis for genotype by environment interactions or heterosis (hybrid vigor), are

 Cite as *Cold Spring Harb Perspect Biol* doi: 10.1101/cshperspect.a019315

likely to have answers rooted in epigenetic regulatory mechanisms.

Understanding what is genetic and what is epigenetic is not a simple matter. For instance, there is clear evidence for genetic changes that induce switches in epigenetic states (Durand et al. 2012), such as mutations that allow read-through transcripts to bring about the silencing of adjacent genes, including tumor-suppressor genes in humans. Conversely, epigenetically controlled accessibility of DNA is likely to influence the probability of genetic rearrangements caused by recombination or transposon mobilization (reviewed in Magori and Citovsky 2011; Mirouze and Paszkowski 2011). Understanding the degree to which genetically programmed expression of DNA-binding transcription factors combines with epigenetic mechanisms that limit enhancer or promoter specificity, RNA polymerase processivity, or splice-site selection, is a major challenge for the future.

Our understanding of how environmental conditions such as photoperiod or temperature provoke changes in RNA- or chromatin-based transcriptional regulation is still in its infancy. The majority of these changes lasts only as long as the trigger is present and are therefore not thought to be epigenetic. However, some environmental conditions induce altered chromatin and gene expression states that persist even after a return to the original environmental condition, as in the case of vernalization (Ch. 31 [Baulcombe and Dean 2014]) in which plants "remember" their experience of winter to flower the following spring. There is also the possibility that environmentally or pathogen-induced epigenetic states might be transmitted to progeny if the changes occur in meristems and can be maintained through meiosis. So far, there is only rudimentary evidence for transmission and inheritance of adaptive epigenetic states, as opposed to DNA sequence-based inheritance (Paszkowski and Grossniklaus 2011; Pecinka and Mittelsten Scheid 2012). Nevertheless, with our growing insight into epigenetic regulation and the transmission of mobile small RNAs influencing chromatin states, such neo-Lamarckian possibilities warrant careful consideration.

ACKNOWLEDGMENTS

We thank Marjori Matzke for her contribution to the first edition. O.M.S. acknowledges funding by the Austrian Academy of Sciences and the Austrian Science Fund (FWF). Pikaard lab research is supported by National Institutes of Health grants GM077590 and GM60380 and C.S.P.'s support as an Investigator of the Howard Hughes Medical Institute and the Gordon & Betty Moore Foundation.

REFERENCES

*Reference is also in this book.

Agius F, Kapoor A, Zhu J-K. 2006. Role of the *Arabidopsis* DNA glycosylase/lyase ROS1 in active DNA demethylation. *Proc Natl Acad Sci* **103:** 11796–11801.

Alleman M, Doctor J. 2000. Genomic imprinting in plants: Observations and evolutionary implications. *Plant Mol Biol* **43:** 147–161.

Allen E, Howell MD. 2010. miRNAs in the biogenesis of *trans*-acting siRNAs in higher plants. *Semin Cell Dev Biol* **21:** 798–804.

* Allshire RC, Ekwall K. 2014. Epigenetic regulation of chromatin states in *Schizosaccharomyces pombe*. *Cold Spring Harb Perspect Biol* doi: 10.1101/cshperspect.a018770.

* Almouzni G, Cedar H. 2014. Maintenance of epigenetic information. *Cold Spring Harb Perspect Biol* doi: 10.1101/cshperspect.a019372.

Ambros V. 2004. The functions of animal microRNAs. *Nature* **431:** 350–355.

Arteaga-Vazquez MA, Chandler VL. 2010. Paramutation in maize: RNA mediated trans-generational gene silencing. *Curr Opin Genet Dev* **20:** 156–163.

Aufsatz W, Stoiber T, Rakic B, Naumann K. 2007. *Arabidopsis* histone deacetylase 6: A green link to RNA silencing. *Oncogene* **26:** 5477–5488.

Axtell MJ, Bowman JL. 2008. Evolution of plant microRNAs and their targets. *Trends Plant Sci* **13:** 343–349.

Axtell MJ, Snyder JA, Bartel DP. 2007. Common functions for diverse small RNAs of land plants. *Plant Cell* **19:** 1750–1769.

Axtell MJ, Westholm JO, Lai EC. 2011. Vive la difference: Biogenesis and evolution of microRNAs in plants and animals. *Genome Biol* **12:** 221.

Bao N, Lye KW, Barton MK. 2004. MicroRNA binding sites in *Arabidopsis* class III HD-ZIP mRNAs are required for methylation of the template chromosome. *Dev Cell* **7:** 653–662.

Bartel DP. 2004. MicroRNAs: Genomics, biogenesis, mechanism, and function. *Cell* **116:** 281–297.

Bartel DP. 2009. MicroRNAs: Target recognition and regulatory functions. *Cell* **136:** 215–233.

Baubec T, Dinh HQ, Pecinka A, Rakic B, Rozhon W, Wohlrab B, von Haeseler A, Mittelsten Scheid O. 2010. Cooperation of multiple chromatin modifications can generate unanticipated stability of epigenetic states in *Arabidopsis*. *Plant Cell* **22:** 34–47.

Baulcombe DC. 2006. Short silencing RNA: The dark matter of genetics? *Cold Spring Harb Symp Quant Biol* **71:** 13–20.

* Baulcombe DC, Dean C. 2014. Epigenetic regulation in plant responses to the environment. *Cold Spring Harb Perspect Biol* **6:** a019471.

Baumberger N, Baulcombe DC. 2005. *Arabidopsis* ARGONAUTE1 is an RNA Slicer that selectively recruits microRNAs and short interfering RNAs. *Proc Natl Acad Sci* **102:** 11928–11933.

* Becker PB, Workman JL. 2013. Nucleosome remodeling and epigenetics. *Cold Spring Harb Perspect Biol* **5:** 017905.

Becker C, Hagmann J, Mueller J, Koenig D, Stegle O, Borgwardt K, Weigel D. 2011. Spontaneous epigenetic variation in the *Arabidopsis thaliana* methylome. *Nature* **480:** 245–249.

Berr A, Shafiq S, Shen W-H. 2011. Histone modifications in transcriptional activation during plant development. *Biochim Biophys Acta* **1809:** 567–576.

Bivalkar-Mehla S, Vakharia J, Mehla R, Abreha M, Kanwar JR, Tikoo A, Chauhan A. 2011. Viral RNA silencing suppressors (RSS): Novel strategy of viruses to ablate the host RNA interference (RNAi) defense system. *Virus Res* **155:** 1–9.

* Blewitt M, Whitelaw E. 2013. The use of mouse models to study epigenetics. *Cold Spring Harb Perspect Biol* **5:** 017939.

Brodersen P, Voinnet O. 2006. The diversity of RNA silencing pathways in plants. *Trends Genet* **22:** 268–280.

Brzeski J, Jerzmanowski A. 2003. Deficient in DNA methylation 1 (DDM1) defines a novel family of chromatin-remodeling factors. *J Biol Chem* **278:** 823–828.

Butenko Y, Ohad N. 2011. Polycomb-group mediated epigenetic mechanisms through plant evolution. *Biochim Biophys Acta* **1809:** 395–406.

Bycroft M. 2011. Recognition of non-methyl histone marks. *Curr Opin Struct Biol* **21:** 761–766.

Carrington JC, Ambros V. 2003. Role of microRNAs in plant and animal development. *Science* **301:** 336–338.

Carthew RW, Sontheimer EJ. 2009. Origins and mechanisms of miRNAs and siRNAs. *Cell* **136:** 642–655.

Chandler VL, Eggleston WB, Dorweiler JE. 2000. Paramutation in maize. *Plant Mol Biol* **43:** 121–145.

Chapman EJ, Carrington JC. 2007. Specialization and evolution of endogenous small RNA pathways. *Nat Rev Genet* **8:** 884–896.

Chen X. 2005. MicroRNA biogenesis and function in plants. *FEBS Lett* **579:** 5923–5931.

Chen X. 2009. Small RNAs and their roles in plant development. *Annu Rev Cell Dev Biol* **25:** 21–44.

Chen ZJ, Tian L. 2007. Roles of dynamic and reversible histone acetylation in plant development and polyploidy. *Biochim Biophys Acta* **1769:** 295–307.

Chen HM, Chen LT, Patel K, Li YH, Baulcombe DC, Wu SH. 2010. 22-Nucleotide RNAs trigger secondary siRNA biogenesis in plants. *Proc Natl Acad Sci* **107:** 15269–15274.

* Cheng X. 2014. Structural and functional coordination of DNA and histone methylation. *Cold Spring Harb Perspect Biol* **6:** a018747.

Choi YH, Gehring M, Johnson L, Hannon M, Harada JJ, Goldberg RB, Jacobsen SE, Fischer RL. 2002. DEMETER, a DNA glycosylase domain protein, is required for endosperm gene imprinting and seed viability in *Arabidopsis*. *Cell* **110:** 33–42.

Chuck G, Candela H, Hake S. 2009. Big impacts by small RNAs in plant development. *Curr Opin Plant Biol* **12:** 81–86.

Cokus SJ, Feng SH, Zhang XY, Chen ZG, Merriman B, Haudenschild CD, Pradhan S, Nelson SF, Pellegrini M, Jacobsen SE. 2008. Shotgun bisulphite sequencing of the *Arabidopsis* genome reveals DNA methylation patterning. *Nature* **452:** 215–219.

Cubas P, Vincent C, Coen E. 1999. An epigenetic mutation responsible for natural variation in floral symmetry. *Nature* **401:** 157–161.

Cuperus JT, Carbonell A, Fahlgren N, Garcia-Ruiz H, Burke RT, Takeda A, Sullivan CM, Gilbert SD, Montgomery TA, Carrington JC. 2010. Unique functionality of 22-nt miRNAs in triggering RDR6-dependent siRNA biogenesis from target transcripts in *Arabidopsis*. *Nat Struct Mol Biol* **17:** 997–1003.

Cuperus JT, Fahlgren N, Carrington JC. 2011. Evolution and functional diversification of *MIRNA* genes. *Plant Cell* **23:** 431–442.

Czech B, Hannon GJ. 2011. Small RNA sorting: Matchmaking for Argonautes. *Nat Rev Genet* **12:** 19–31.

Deal RB, Henikoff S. 2011. Histone variants and modifications in plant gene regulation. *Curr Opin Plant Biol* **14:** 116–122.

Ding SW. 2010. RNA-based antiviral immunity. *Nat Rev Immunol* **10:** 632–644.

Ding SW, Voinnet O. 2007. Antiviral immunity directed by small RNAs. *Cell* **130:** 413–426.

Du T, Zamore PD. 2005. microPrimer: The biogenesis and function of microRNA. *Development* **132:** 4645–4652.

* Dunoyer P, Melnyk CW, Molnar A, Slotkin RK. 2013. Plant mobile small RNAs. *Cold Spring Harb Perspect Biol* **5:** 017897.

Durand S, Bouche N, Perez Strand E, Loudet O, Camilleri C. 2012. Rapid establishment of genetic incompatibility through natural epigenetic variation. *Curr Biol* **22:** 326–331.

Earley KW, Shook MS, Brower-Toland B, Hicks L, Pikaard CS. 2007. In vitro specificities of *Arabidopsis* co-activator histone acetyltransferases: Implications for histone hyperacetylation in gene activation. *Plant J* **52:** 615–626.

* Elgin SCR, Reuter G. 2013. Position effect variegation, heterochromatin formation, and gene silencing in *Drosophila*. *Cold Spring Harb Perspect Biol* **5:** 017780.

Elmayan T, Proux F, Vaucheret H. 2005. *Arabidopsis* RPA2: A genetic link among transcriptional gene silencing, DNA repair, and DNA replication. *Curr Biol* **15:** 1919–1925.

Erhard KF Jr, Hollick JB. 2011. Paramutation: A process for acquiring trans-generational regulatory states. *Curr Opin Plant Biol* **14:** 210–216.

Feng S, Cokus SJ, Zhang X, Chen P-Y, Bostick M, Goll MG, Hetzel J, Jain J, Strauss SH, Halpern ME, et al. 2010. Conservation and divergence of methylation patterning in plants and animals. *Proc Natl Acad Sci* **107:** 8689–8694.

Franco-Zorrilla JM, Valli A, Todesco M, Mateos I, Puga MI, Rubio-Somoza I, Leyva A, Weigel D, Garcia JA, Paz-Ares J. 2007. Target mimicry provides a new mechanism for regulation of microRNA activity. *Nat Genet* **39:** 1033–1037.

Furner IJ, Matzke M. 2011. Methylation and demethylation of the *Arabidopsis* genome. *Curr Opin Plant Biol* **14:** 137–141.

Gehring M, Reik W, Henikoff S. 2009. DNA demethylation by DNA repair. *Trends Genet* **25:** 82–90.

Goll MG, Kirpekar F, Maggert KA, Yoder JA, Hsieh CL, Zhang XY, Golic KG, Jacobsen SE, Bestor TH. 2006. Methylation of tRNA(AsP) by the DNA methyltransferase homolog Dnmt2. *Science* **311:** 395–398.

Graf P, Dolzblasz A, Wuerschum T, Lenhard M, Pfreundt U, Laux T. 2010. MGOUN1 encodes an *Arabidopsis* Type IB DNA Topoisomerase required in stem cell regulation and to maintain developmentally regulated gene silencing. *Plant Cell* **22:** 716–728.

Grafi G, Zemach A, Pitto L. 2007. Methyl-CpG-binding domain (MBD) proteins in plants. *Biochim Biophys Acta* **1769:** 287–294.

* Grossniklaus U, Paro R. 2014. Transcriptional silencing by *Polycomb*-group proteins. *Cold Spring Harb Perspect Biol* **6:** a019331.

Haag JR, Pikaard CS. 2011. Multisubunit RNA polymerases IV and V: Purveyors of non-coding RNA for plant gene silencing. *Nat Rev Mol Cell Biol* **12:** 483–492.

Haag J, Ream T, Marasco M, Nicora C, Norbeck A, Pasa-Tolic L, Pikaard C. 2012. In vitro transcription activities of Pol IV, Pol V, and RDR2 reveal coupling of Pol IV and RDR2 for dsRNA synthesis in plant RNA silencing. *Mol Cell* **48:** 811–818.

Habu Y. 2010. Epigenetic silencing of endogenous repetitive sequences by MORPHEUS' MOLECULE1 in *Arabidopsis thaliana*. *Epigenetics* **5:** 562–565.

Havecker ER, Wallbridge LM, Hardcastle TJ, Bush MS, Kelly KA, Dunn RM, Schwach F, Doonan JH, Baulcombe DC. 2010. The *Arabidopsis* RNA-directed DNA methylation argonautes functionally diverge based on their expression and interaction with target loci. *Plant Cell* **22:** 321–334.

He XJ, Chen T, Zhu JK. 2011. Regulation and function of DNA methylation in plants and animals. *Cell Res* **21:** 442–465.

Heitz E. 1929. Heterochromatin, Chromocentren, Chromomeren. (Vorlaufige Mitteilung.) *Berichte der Deutschen Botanischen Gesellschaft* **47:** 274–284.

* Henikoff S, Smith MM. 2014. Histone variants and epigenetics. *Cold Spring Harb Perspect Biol* doi: 10.1101/cshperspect.a019364.

Herr AJ. 2005. Pathways through the small RNA world of plants. *FEBS Lett* **579:** 5879–5888.

Hoffer P, Ivashuta S, Pontes O, Vitins A, Pikaard C, Mroczka A, Wagner N, Voelker T. 2011. Posttranscriptional gene silencing in nuclei. *Proc Natl Acad Sci* **108:** 409–414.

Houben A, Demidov D, Caperta AD, Karimi R, Agueci F, Vlasenko L. 2007. Phosphorylation of histone H3 in plants—A dynamic affair. *Biochim Biophys Acta* **1769:** 308–315.

Ikeda Y, Kinoshita Y, Susaki D, Ikeda Y, Iwano M, Takayama S, Higashiyama T, Kakutani T, Kinoshita T. 2011. HMG domain containing SSRP1 is required for DNA demethylation and genomic imprinting in *Arabidopsis*. *Dev Cell* **21:** 589–596.

Ingouff M, Berger F. 2010. Histone3 variants in plants. *Chromosoma* **119:** 27–33.

Jackson S, Chen ZJ. 2010. Genomic and expression plasticity of polyploidy. *Curr Opin Plant Biol* **13:** 153–159.

Jacob Y, Stroud H, LeBlanc C, Feng S, Zhuo L, Caro E, Hassel C, Gutierrez C, Michaels SD, Jacobsen SE. 2010. Regulation of heterochromatic DNA replication by histone H3 lysine 27 methyltransferases. *Nature* **466:** 987–991.

Jerzmanowski A. 2007. SWI/SNF chromatin remodeling and linker histones in plants. *Biochim Biophys Acta* **1769:** 330–345.

Jones-Rhoades MW, Bartel DP, Bartel B. 2006. MicroRNAS and their regulatory roles in plants. *Annu Rev Plant Biol* **57:** 19–53.

Jordan ND, West JP, Bottley A, Sheikh M, Furner I. 2007. Transcript profiling of the hypomethylated hog1 mutant of *Arabidopsis*. *Plant Mol Biol* **65:** 571–586.

Jorgensen RA, Doetsch N, Muller A, Que Q, Gendler K, Napoli CA. 2006. A paragenetic perspective on integration of RNA silencing into the epigenome and its role in the biology of higher plants. *Cold Spring Harb Symp Quant Biol* **71:** 481–485.

Joshua-Tor L. 2006. The Argonautes. *Cold Spring Harb Symp Quant Biol* **71:** 67–72.

Kapoor A, Agarwal M, Andreucci A, Zheng X, Gong Z, Hasegawa PM, Bressan RA, Zhu JK. 2005. Mutations in a conserved replication protein suppress transcriptional gene silencing in a DNA-methylation-independent manner in *Arabidopsis*. *Curr Biol* **15:** 1912–1918.

Khraiwesh B, Zhu JK, Zhu J. 2012 Role of miRNAs and siRNAs in bioitic and abiotic stress responses of plants. *Biochim Biophys Acta* **1819:** 137–148.

Kidner CA, Martienssen RA. 2005. The developmental role of microRNA in plants. *Curr Opin Plant Biol* **8:** 38–44.

Kim JM, To TK, Seki M. 2012. An epigenetic integrator: New insights into genome regulation, environmental stress responses and developmental controls by HISTONE DEACETYLASE 6. *Plant Cell Physiol* **53:** 794–800.

* Kingston RE, Tamkun JW. 2014. Transcriptional regulation by trithorax-group proteins. *Cold Spring Harb Perspect Biol* **6:** a019349.

Klattenhoff C, Theurkauf W. 2008. Biogenesis and germline functions of piRNAs. *Development* **135:** 3–9.

Koehler C, Hennig L. 2010. Regulation of cell identity by plant Polycomb and trithorax group proteins. *Curr Opin Genet Dev* **20:** 541–547.

Kumar SV, Wigge PA. 2010. H2A.Z-containing nucleosomes mediate the thermosensory response in *Arabidopsis*. *Cell* **140:** 136–147.

Lahmy S, Bies-Etheve N, Lagrange T. 2010. Plant-specific multisubunit RNA polymerase in gene silencing. *Epigenetics* **5:** 4–8.

Lauria M, Rossi V. 2011. Epigenetic control of gene regulation in plants. *Biochim Biophys Acta* **1809:** 369–378.

Law JA, Jacobsen SE. 2010. Establishing, maintaining and modifying DNA methylation patterns in plants and animals. *Nat Rev Genet* **11:** 204–220.

Law JA, Vashisht AA, Wohlschlegel JA, Jacobsen SE. 2011. SHH1, a homeodomain protein required for DNA methylation, as well as RDR2, RDM4, and chromatin remodeling factors, associate with RNA polymerase IV. *PLoS Genet* **7:** e1002195.

Layat E, Saez-Vasquez J, Tourmente S. 2012. Regulation of Pol I-transcribed 45S rDNA and Pol III-transcribed 5S rDNA in *Arabidopsis*. *Plant Cell Physiol* **53:** 267–276.

* Li E, Zhang Y. 2014. DNA methylation in mammals. *Cold Spring Harb Perspect Biol* **6:** a019133.

Lisch D. 2009. Epigenetic regulation of transposable elements in plants. *Annu Rev Plant Biol* **60:** 43–66.

Lister R, O'Malley RC, Tonti-Filippini J, Gregory BD, Berry CC, Millar AH, Ecker JR. 2008. Highly integrated single-base resolution maps of the epigenome in *Arabidopsis*. *Cell* **133:** 523–536.

Liu C, Lu F, Cui X, Cao X. 2010. Histone methylation in higher plants. *Annu Rev Plant Biol* **61:** 395–420.

Luo J, Hall BD. 2007. A multistep process gave rise to RNA polymerase IV of land plants. *J Mol Evol* **64:** 101–112.

Magori S, Citovsky V. 2011. Epigenetic control of *Agrobacterium* T-DNA integration. *Biochim Biophys Acta* **1809:** 388–394.

Manavella PA, Koenig D, Weigel D. 2012. Plant secondary siRNA production determined by microRNA-duplex structure. *Proc Natl Acad Sci* **109:** 2461–2466.

* Marmorstein R, Zhou M-M. 2014. Writers and readers of histone acetylation: Structure, mechanism, and inhibition. *Cold Spring Harb Perspect Biol* **6:** a018762.

* Martienssen R, Moazed D. 2014. RNAi and heterochromatin assembly. *Cold Spring Harb Perspect Biol* doi: 10.1101/cshperspect.a019323.

Martínez-Macías M, Qian W, Miki D, Pontes O, Liu Y, Tang K, Liu R, Morales-Ruiz T, Ariza R, Roldán-Arjona T, et al. 2012. A DNA 3′ phosphatase functions in active DNA demethylation in *Arabidopsis*. *Mol Cell* **45:** 357–370.

Matzke MA, Matzke AJM. 2004. Planting the seeds of a new paradigm. *PloS Biol* **2:** 582–586.

Matzke M, Kanno T, Daxinger L, Huettel B, Matzke AJ. 2009. RNA-mediated chromatin-based silencing in plants. *Curr Opin Cell Biol* **21:** 367–376.

McClintock B. 1950. The origin and behavior of mutable loci in maize. *Proc Natl Acad Sci* **36:** 344–355.

Mette MF, Aufsatz W, van der Winden J, Matzke MA, Matzke AJ. 2000. Transcriptional silencing and promoter methylation triggered by double-stranded RNA. *EMBO J* **19:** 5194–5201.

Meyer P. 2011. DNA methylation systems and targets in plants. *FEBS Lett* **585:** 2008–2015.

Miguel C, Marum L. 2011. An epigenetic view of plant cells cultured in vitro: Somaclonal variation and beyond. *J Exp Bot* **62:** 3713–3725.

Mirouze M, Paszkowski J. 2011. Epigenetic contribution to stress adaptation in plants. *Curr Opin Plant Biol* **14:** 267–274.

Montgomery TA, Howell MD, Cuperus JT, Li D, Hansen JE, Alexander AL, Chapman EJ, Fahlgren N, Allen E, Carrington JC. 2008. Specificity of ARGONAUTE7-miR390 interaction and dual functionality in TAS3 trans-acting siRNA formation. *Cell* **133:** 128–141.

Mosher RA, Melnyk CW, Kelly KA, Dunn RM, Studholme DJ, Baulcombe DC. 2009. Uniparental expression of PolIV-dependent siRNAs in developing endosperm of *Arabidopsis*. *Nature* **460:** 283–286.

Niehrs C. 2009. Active DNA demethylation and DNA repair. *Differentiation* **77:** 1–11.

Ohno Y, Narangajavana J, Yamamoto A, Hattori T, Kagaya Y, Paszkowski J, Gruissem W, Hennig L, Takeda S. 2011. Ectopic gene expression and organogenesis in *Arabidopsis* mutants missing BRU1 required for genome maintenance. *Genetics* **189:** 83–95.

Okamura K, Hagen JW, Duan H, Tyler DM, Lai EC. 2007. The mirtron pathway generates microRNA-class regulatory RNAs in *Drosophila*. *Cell* **130:** 89–100.

Ortega-Galisteo AP, Morales-Ruiz T, Ariza RR, Roldan-Arjona T. 2008. *Arabidopsis* DEMETER-LIKE proteins DML2 and DML3 are required for appropriate distribution of DNA methylation marks. *Plant Mol Biol* **67:** 671–681.

Ossowski S, Schneeberger K, Lucas-Lledo JI, Warthmann N, Clark RM, Shaw RG, Weigel D, Lynch M. 2010. The rate and molecular spectrum of spontaneous mutations in *Arabidopsis thaliana*. *Science* **327:** 92–94.

Pandey R, Muller A, Napoli CA, Selinger DA, Pikaard CS, Richards EJ, Bender J, Mount DW, Jorgensen RA. 2002. Analysis of histone acetyltransferase and histone deacetylase families of *Arabidopsis thaliana* suggests functional diversification of chromatin modification among multicellular eukaryotes. *Nucleic Acids Res* **30:** 5036–5055.

Paszkowski J, Grossniklaus U. 2011. Selected aspects of transgenerational epigenetic inheritance and resetting in plants. *Curr Opin Plant Biol* **14:** 195–203.

Pecinka A, Mittelsten Scheid O. 2012. Stress-induced chromatin changes: A critical view on their heritability. *Plant Cell Physiol* **53:** 801–808.

Pedersen DS, Grasser KD. 2010. The role of chromosomal HMGB proteins in plants. *Biochim Biophys Acta* **1799:** 171–174.

Pillai RS, Chuma S. 2012. piRNAs and their involvement in male germline development in mice. *Dev Growth Differ* **54:** 78–92.

Poethig RS. 2009. Small RNAs and developmental timing in plants. *Curr Opin Genet Dev* **19:** 374–378.

Poethig RS, Peragine A, Yoshikawa M, Hunter C, Willmann M, Wu G. 2006. The function of RNAi in plant development. *Cold Spring Harb Symp Quant Biol* **71:** 165–170.

Pontvianne F, Blevins T, Pikaard CS. 2010. *Arabidopsis* histone lysine methyltransferases. *Adv Bot Res* **53:** 1–22.

Preuss S, Pikaard CS. 2007. RRNA gene silencing and nucleolar dominance: Insights into a chromosome-scale epigenetic on/off switch. *Biochim Biophys Acta* **1769:** 383–392.

Preuss SB, Costa-Nunes P, Tucker S, Pontes O, Lawrence RJ, Mosher R, Kasschau KD, Carrington JC, Baulcombe DC, Viegas W, et al. 2008. Multimegabase silencing in nucleolar dominance involves siRNA-directed DNA methylation and specific methylcytosine-binding proteins. *Mol Cell* **32:** 673–684.

Raissig MT, Baroux C, Grossniklaus U. 2011. Regulation and flexibility of genomic imprinting during seed development. *Plant Cell* **23:** 16–26.

Ream TS, Haag JR, Wierzbicki AT, Nicora CD, Norbeck AD, Zhu JK, Hagen G, Guilfoyle TJ, Pasa-Tolic L, Pikaard CS. 2009. Subunit compositions of the RNA-silencing enzymes Pol IV and Pol V reveal their origins as specialized forms of RNA polymerase II. *Mol Cell* **33:** 192–203.

Reinders J, Paszkowski J. 2009. Unlocking the *Arabidopsis* epigenome. *Epigenetics* **4:** 557–563.

Richards EJ. 2011. Natural epigenetic variation in plant species: A view from the field. *Curr Opin Plant Biol* **14:** 204–209.

Rocha PS, Sheikh M, Melchiorre R, Fagard M, Boutet S, Loach R, Moffatt B, Wagner C, Vaucheret H, Furner I. 2005. The *Arabidopsis HOMOLOGY-DEPENDENT GENE SILENCING1* gene codes for an *S*-adenosyl-L-homocysteine hydrolase required for DNA methylation-dependent gene silencing. *Plant Cell* **17:** 404–417.

Roudier F, Teixeira FK, Colot V. 2009. Chromatin indexing in *Arabidopsis*: An epigenomic tale of tails and more. *Trends Genet* **25:** 511–517.

Roudier F, Ahmed I, Berard C, Sarazin A, Mary-Huard T, Cortijo S, Bouyer D, Caillieux E, Duvernois-Berthet E, Al-Shikhley L, et al. 2011. Integrative epigenomic mapping defines four main chromatin states in *Arabidopsis*. *EMBO J* **30:** 1928–1938.

Roux F, Colome-Tatche M, Edelist C, Wardenaar R, Guerche P, Hospital F, Colot V, Jansen RC. 2011. Genome-wide epigenetic perturbation jump-starts patterns of heritable variation found in nature. *Genetics* **188:** 1015–1017.

Rubio-Somoza I, Weigel D. 2011. MicroRNA networks and developmental plasticity in plants. *Trends Plant Sci* **16:** 258–264.

Ruiz-Ferrer V, Voinnet O. 2009. Roles of plant small RNAs in biotic stress responses. *Annu Rev Plant Biol* **60:** 485–510.

Rymarquis LA, Souret FF, Green PJ. 2011. Evidence that XRN4, an *Arabidopsis* homolog of exoribonuclease XRN1, preferentially impacts transcripts with certain sequences or in particular functional categories. *RNA* **17:** 501–511.

Sang Y, Silva-Ortega CO, Wu S, Yamaguchi N, Wu MF, Pfluger J, Gillmor CS, Gallagher KL, Wagner D. 2012 Mutations in two non-canonical *Arabidopsis* SWI2/SNF2 chromatin remodeling ATPases cause embryogenesis and stem cell maintenance defects. *Plant J* **72:** 1000–1014.

Schmitz RJ, Schultz MD, Lewsey MG, O'Malley RC, Urich MA, Libiger O, Schork NJ, Ecker JR. 2011. Transgenerational epigenetic instability is a source of novel methylation variants. *Science* **334:** 369–373.

Schoft VK, Chumak N, Choi Y, Hannon M, Garcia-Aguilar M, Machlicova A, Slusarz L, Mosiolek M, Park J-S, Park GT, et al. 2011. Function of the DEMETER DNA glycosylase in the *Arabidopsis thaliana* male gametophyte. *Proc Natl Acad Sci* **108:** 8042–8047.

Senda M, Masuta C, Ohnishi S, Goto K, Kasai A, Sano T, Hong J-S, MacFarlane S. 2004. Patterning of virus-infected *Glycine max* seed coat is associated with suppression of endogenous silencing of chalcone synthase genes. *Plant Cell* **16:** 807–818.

Senthil-Kumar M, Mysore KS. 2011. New dimensions for VIGS in plant functional genomics. *Trends Plant Sci* **16:** 656–665.

Servet C, Conde e Silva N, Zhou D-X. 2010. Histone acetyltransferase AtGCN5/HAG1 is a versatile regulator of developmental and inducible gene expression in *Arabidopsis*. *Mol Plant* **3:** 670–677.

* Seto E, Yoshida M. 2014. Erasers of histone acetylation: The histone deacetylase enzymes. *Cold Spring Harb Perspect Biol* **6:** a018713.

Shabalina SA, Koonin EV. 2008. Origins and evolution of eukaryotic RNA interference. *Trends Ecol Evol* **23:** 578–587.

* Shi YG, Tsukada Y. 2013. The discovery of histone demethylases. *Cold Spring Harb Perspect Biol* **5:** 017947.

Slotkin RK, Vaughn M, Borges F, Tanurdzic M, Becker JD, Feijo JA, Martienssen RA. 2009. Epigenetic reprogramming and small RNA silencing of transposable elements in pollen. *Cell* **136:** 461–472.

Soppe WJJ, Jacobsen SE, Alonso-Blanco C, Jackson JP, Kakutani T, Koornneef M, Peeters AJM. 2000. The late flowering phenotype of fwa mutants is caused by gain-of-function epigenetic alleles of a homeodomain gene. *Mol Cell* **6:** 791–802.

Springer NM, Kaeppler SM. 2005. Evolutionary divergence of monocot and dicot methyl-CpG-binding domain proteins. *Plant Physiol* **138:** 92–104.

Tan EH, Blevins T, Ream T, Pikaard C. 2012. Functional consequences of subunit diversity in RNA polymerases II and V. *Cell Rep* **1:** 208–214.

Teixeira FK, Heredia F, Sarazin A, Roudier F, Boccara M, Ciaudo C, Cruaud C, Poulain J, Berdasco M, Fraga MF, et al. 2009. A role for RNAi in the selective correction of DNA methylation defects. *Science* **323:** 1600–1604.

Tucker SL, Reece J, Ream TS, Pikaard CS. 2010. Evolutionary history of plant multisubunit RNA polymerases IV and V: Subunit origins via genome-wide and segmental gene duplications, retrotransposition, and lineage-specific subfunctionalization. *Cold Spring Harb Symp Quant Biol* **75:** 285–297.

Vaucheret H. 2008. Plant ARGONAUTES. *Trends Plant Sci* **13:** 350–358.

Vazquez F, Legrand S, Windels D. 2010. The biosynthetic pathways and biological scopes of plant small RNAs. *Trends Plant Sci* **15:** 337–345.

Verdel A, Vavasseur A, Le Gorrec M, Touat-Todeschini L. 2009. Common themes in siRNA-mediated epigenetic silencing pathways. *Int J Dev Biol* **53:** 245–257.

Voinnet O. 2009. Origin, biogenesis, and activity of plant microRNAs. *Cell* **136:** 669–687.

Voinnet O, Vain P, Angell S, Baulcombe DC. 1998. Systemic spread of sequence-specific transgene RNA degradation in plants is initiated by localized introduction of ectopic promoterless DNA. *Cell* **95:** 177–187.

Wassenegger M. 2002. Gene silencing. *Int Rev Cytol* **219:** 61–113.

Wassenegger M, Krczal G. 2006. Nomenclature and functions of RNA-directed RNA polymerases. *Trends Plant Sci* **11:** 142–151.

Wassenegger M, Heimes S, Riedel L, Sanger HL. 1994. RNA-directed de novo methylation of genomic sequences in plants. *Cell* **76:** 567–576.

Westholm JO, Lai EC. 2011. Mirtrons: MicroRNA biogenesis via splicing. *Biochimie* **93:** 1897–1904.

Wierzbicki AT, Haag JR, Pikaard CS. 2008. Noncoding transcription by RNA polymerase Pol IVb/Pol V mediates transcriptional silencing of overlapping and adjacent genes. *Cell* **135:** 635–648.

Wierzbicki AT, Ream TS, Haag JR, Pikaard CS. 2009. RNA polymerase V transcription guides ARGONAUTE4 to chromatin. *Nat Genet* **41:** 630–634.

Woo HR, Pontes O, Pikaard CS, Richards EJ. 2007. VIM1, a methylcytosine-binding protein required for centromeric heterochromatinization. *Genes Dev* **21:** 267–277.

Woo HR, Dittmer TA, Richards EJ. 2008. Three SRA-domain methylcytosine-binding proteins cooperate to maintain global CpG methylation and epigenetic silencing in *Arabidopsis*. *PLoS Genet* **4:** e1000156.

Xiao WY, Gehring M, Choi Y, Margossian L, Pu H, Harada JJ, Goldberg RB, Pennell RI, Fischer RL. 2003. Imprinting of the *MEA* Polycomb gene is controlled by antagonism between MET1 methyltransferase and DME glycosylase. *Dev Cell* **5:** 891–901.

Xie Z, Qi X. 2008. Diverse small RNA-directed silencing pathways in plants. *Biochim Biophys Acta* **1779:** 720–724.

Yuan L, Yang X, Makaroff CA. 2011. Plant cohesins, common themes and unique roles. *Curr Prot Pept Sci* **12:** 93–104.

Zaratiegui M, Irvine DV, Martienssen RA. 2007. Noncoding RNAs and gene silencing. *Cell* **128:** 763–776.

Zemach A, Kim MY, Hsieh PH, Coleman-Derr D, Eshed-Williams L, Thao K, Harmer SL, Zilberman D. 2013. The *Arabidopsis* nucleosome remodeler DDM1 allows DNA methyltransferases to access H1-containing heterochromatin. *Cell* **153:** 193–205.

* Zhao Y, Garcia BA. 2014. Comprehensive catalog of currently documented histone modifications. *Cold Spring Harb Perspect Biol* doi: 10.1101/cshperspect.a025064.

Zhang H, Zhu JK. 2011. RNA-directed DNA methylation. *Curr Opin Plant Biol* **14:** 142–147.

Zhang X, Yazaki J, Sundaresan A, Cokus S, Chan SW, Chen H, Henderson IR, Shinn P, Pellegrini M, Jacobsen SE, et al. 2006. Genome-wide high-resolution mapping and functional analysis of DNA methylation in *Arabidopsis*. *Cell* **126:** 1189–1201.

Zheng X, Pontes O, Zhu J, Miki D, Zhang F, Li W-X, Iida K, Kapoor A, Pikaard CS, Zhu J-K. 2008. ROS3 is an RNA-binding protein required for DNA demethylation in *Arabidopsis*. *Nature* **455:** 1259–1262.

Zheng B, Wang Z, Li S, Yu B, Liu JY, Chen X. 2009. Intergenic transcription by RNA polymerase II coordinates Pol IV and Pol V in siRNA-directed transcriptional gene silencing in *Arabidopsis*. *Genes Dev* **23:** 2850–2860.

Zhu H, Hu F, Wang R, Zhou X, Sze SH, Liou LW, Barefoot A, Dickman M, Zhang X. 2011a. *Arabidopsis* Argonaute10 specifically sequesters miR166/165 to regulate shoot apical meristem development. *Cell* **145:** 242–256.

Zhu Y, Dong AW, Shen WH. 2011b. Histone variants and chromatin assembly in plant abiotic stress response. *Biochim Biophys Acta* **1819:** 343–348.

Zhu Y, Rowley MJ, Bohmdorfer G, Wierzbicki AT. 2012. A SWI/SNF chromatin-remodeling complex acts in noncoding RNA-mediated transcriptional silencing. *Mol Cell* **49:** 298–309.

Zilberman D, Gehring M, Tran RK, Ballinger T, Henikoff S. 2007. Genome-wide analysis of *Arabidopsis thaliana* DNA methylation uncovers an interdependence between methylation and transcription. *Nat Genet* **39:** 61–69.

Zilberman D, Coleman-Derr D, Ballinger T, Henikoff S. 2008. Histone H2A.Z and DNA methylation are mutually antagonistic chromatin marks. *Nature* **456:** 125–129.

* Zoghbi H, Beaudet A. 2014. Epigenetics and human disease. *Cold Spring Harb Perspect Biol* doi: 10.1101/cshperspect.a019497.

WWW RESOURCES

http://www.arabidopsis.leeds.ac.uk/act/coexpanalyser.php *Arabidopsis* coexpression mining

http://www.arabidopsis.org/index.jsp TAIR

http://asrp.cgrb.oregonstate.edu *Arabidopsis* small RNA

http://bbc.botany.utoronto.ca/efp/cgi-bin/efpWeb.cgi Gene expression in *Arabidopsis*

http://www.chromdb.org Chromatin genes

http://www.cymate.org/ Bisulfite sequence analysis

http://epigara.biologie.ens.fr/cgi-bin/gbrowse/a2e/ *Arabidopsis* epigenetics EPIGARA

http://epigenome.rutgers.edu/cgi-bin/gb2/gbrowse/Histone_modifications *Arabidopsis* histone modifications Rutgers

http://www.erapg.org/publicpage.m?key=everyone&trail=/everyone European plant genomic research

https://www.genevestigator.com/gv/ Gene expression in several organisms

http://genomes.mcdb.ucla.edu/AthBSseq/ *Arabidopsis* methylome of silencing mutants

http://katahdin.mssm.edu/kismeth/revpage.pl Bisulfite sequence analysis

https://www.mcdb.ucla.edu/Research/Jacobsen/LabWebSite/P_EpigenomicsData.shtml *Arabidopsis* histone modifications UCLA

http://mpss.udel.edu MPSS (Massively parallel signature sequencing)

http://neomorph.salk.edu/epigenome/epigenome.html *Arabidopsis* epigenome

https://www.plant-epigenome.org EPIC (Epigenomics of Plants International Consortium)

http://signal.salk.edu/cgi-bin/methylome *Arabidopsis* methylome

http://1001genomes.org *Arabidopsis* genetic variation

C H A P T E R 14

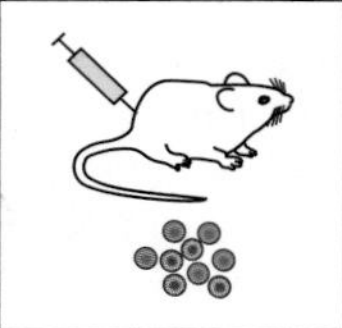

The Use of Mouse Models to Study Epigenetics

Marnie Blewitt[1] and Emma Whitelaw[2]

[1]Walter and Eliza Hall Institute, Melbourne, 3052 Victoria, Australia; [2]Queensland Institute of Medical Research, Brisbane, 4006 Queensland, Australia

Correspondence: blewitt@wehi.edu.au

SUMMARY

Much of what we know about the role of epigenetics in the determination of phenotype has come from studies of inbred mice. Some unusual expression patterns arising from endogenous and transgenic murine alleles, such as the *Agouti* coat color alleles, have allowed the study of variegation, variable expressivity, transgenerational epigenetic inheritance, parent-of-origin effects, and position effects. These phenomena have taught us much about gene silencing and the probabilistic nature of epigenetic processes. Based on some of these alleles, large-scale mutagenesis screens have broadened our knowledge of epigenetic control by identifying and characterizing novel genes involved in these processes.

Outline

Chapter-opening figures: (*Left*) From thefunmouse.com.

Cite as *Cold Spring Harb Perspect Biol* doi: 10.1101/cshperspect.a017939

OVERVIEW

The role of epigenetics in determining phenotype has been progressed through studies of inbred mice. Under laboratory conditions, the genome and the environment of mice are tightly controlled such that variance in phenotype or patterns of gene expression without a change in the underlying DNA sequence is, by definition, epigenetic. Interestingly, transgenes in mice appear to be particularly sensitive to epigenetic silencing, and as such provide a valuable model to study the underlying molecular mechanisms of epigenetic control. In fact, we have come to realize that there are some endogenous alleles, resulting from transposon insertions, which are similarly susceptible to epigenetic silencing. These alleles, termed metastable epialleles, display unusual expression and inheritance patterns: for example, variegated expression in a single cell type, variable expressivity between individuals, and transgenerational epigenetic inheritance. The study of these phenomena has revealed fundamental features of epigenetic control. In some cases, these recapitulate those found in other complex organisms, such as position-effect variegation (PEV) in *Drosophila* (see Ch. 12 [Elgin and Reuter 2013]) and paramutation in plants (Ch. 13 [Pikaard and Mittelsten 2014]), but in other cases, the phenomena are unique to mammals.

In addition to demonstrating many of the general features of epigenetic control, metastable epialleles and other reporter alleles have enabled random mutagenesis screens to be performed to find genes that are important in setting and resetting epigenetic marks at these loci. Similar screens for epigenetic regulators have been performed in lower organisms, using variegating phenotypes such as eye color in flies and pigmentation in maize. Mammalian screens have been important because certain epigenetic processes are specific to higher organisms: for example, the inactivation of the second X chromosome in females (see Ch. 25 [Brockdorff and Turner 2014]) and genomic imprinting (see Ch. 26 [Barlow and Bartolomei 2014]). Furthermore, by performing screens in mice, one immediately has the mutant mouse strains to study the effects of disruption to epigenetic processes on phenotypes relevant to humans. Two mouse mutagenesis screens have been specifically designed to identify genes involved in epigenetic control: the *Momme* and the X inactivation-choice screens. The *Momme* mutagenesis screen has used a green fluorescent protein (GFP) transgenic line that displays variegated expression equivalent to PEV in *Drosophila.* This screen has thus far revealed >30 modifiers of epigenetic regulation, some known but others entirely novel. Importantly, the novel players appear to be involved in mammalian specific processes, and these newly identified molecules expand our understanding of epigenetic control in the mammalian system.

Interestingly, mutation in one of the novel genes identified in the *Momme* screen has now been reported to be the underlying cause of a rare human disease. Studies in the mouse models generated by the *Momme* screen have been instrumental in helping us understand the molecular mechanisms of this disease. We anticipate the same will be true in other cases.

 Cite as *Cold Spring Harb Perspect Biol* doi: 10.1101/cshperspect.a017939

1 USING MOUSE MODELS TO IDENTIFY MODIFIERS OF EPIGENETIC REPROGRAMMING

Random mutagenesis screens performed in mice, which have isolated mutations in epigenetic regulators, are described in this section. First, we detail screens that were specifically designed for the purpose of identifying epigenetic regulators: the *Momme* and the X inactivation-choice screen (Secs. 1.1 and 1.3). Next, we briefly explain other screens that were primarily aimed at finding genes involved in embryonic development, hematopoiesis or immune function (Sec. 1.4), but have produced novel mutations in epigenetic regulators all the same.

1.1 A Screen for Modifiers of Murine Metastable Epialleles (*Mommes*)

The mutagenesis screens performed in yeast, plants, and flies used variegating or epigenetically controlled phenotypes as readouts of epigenetic state: for example, mating type switching in yeast (see Ch. 9 [Allshire and Ekwall 2014]), PEV in flies (Ch. 12 [Elgin and Reuter 2013]), and paramutation, RNAi, and RNA-directed DNA methylation in plants (Ch. 13 [Pikaard and Mittelsten Scheid 2014]). These screens have identified epigenetic modifiers critical for the phenotype being screened, but have also been a very useful tool in unraveling key molecular features of these unusual epigenetic processes. A similar screen has been performed in mice using a variegating metastable epiallele. A metastable epiallele has transcriptional activity that is less stable than expected and is associated with changes in epigenetic state (Rakyan et al. 2002). The screen was thus performed with the hope of finding novel epigenetic modifiers, creating useful new alleles of known modifiers, and helping us to understand more about the remarkable features of metastable epialleles. Briefly, the activity state of metastable epialleles varies among genetically identical individuals brought up in the same environment called variable expressivity, and is particularly sensitive to the epigenetic state of the locus. They also display variegation (i.e., different expression states within one tissue type). These phenomena are discussed in more detail in Section 2.

Several studies suggested that using a variegating metastable epiallele in a mutagenesis screen would be a good approach. First, both transgenes and endogenous metastable epialleles show strain specific differences in the extent of the variegation or variable expressivity, consistent with *trans*-acting genetic variants altering expression of these alleles (Wolff 1978; Belyaev et al. 1981; Allen et al. 1990; Weichman and Chaillet 1997; Sutherland et al. 2000; Chong et al. 2007). Second, the extent of variegation at metastable epialleles is shifted by altering the dose of some proteins known to be involved in epigenetic reprogramming such as Dnmt1 (Gaudet et al. 2004), HP1-β (Festenstein et al. 1999), and the polycomb group protein Mel18 (Blewitt et al. 2006). Together, these findings showed that metastable epialleles are particularly sensitive to alterations in genetic makeup, ideal for a mutagenesis screen.

A mouse line carrying a variegating green fluorescent protein (GFP) transgene directed to express in red blood cells was chosen (Fig. 1) (Preis et al. 2003). The advantages of using this transgenic line are many. First, the transgene is reproducibly expressed in ~55% of red blood cells in homozygous animals; the reproducibility of expression between isogenic littermates makes for a clean phenotype for screening with few false positives. Second, the transgene was produced and has been maintained on an inbred (FVB/N) genetic background, which simplifies later mapping of the ethylnitrosourea (ENU)-induced mutations. Third, the expression of GFP in red blood cells means that transgene expression was able to be efficiently and sensitively determined at a single cell level by flow cytometry. Fourth, by directing expression to red blood cells, analysis could be performed relatively simply (and without killing the animal) using a drop of blood taken from the tail of a mouse at weaning. Finally, alterations in the expression of the transgene itself do not inherently alter viability of the offspring.

1.1.1 The Dominant Screen

Males homozygous for the transgene were treated with the chemical mutagen ENU, which produces point mutations throughout the genome (Rinchik 1991). Mature germ cells are killed by the treatment, but point mutations are produced in the spermatogonial stem cells, so when treated males recover fertility, they can be bred and their G_1 offspring screened for dominant-acting mutations. In essence, the mice were screened for alterations in transgene silencing, assuming that any such alterations would be attributable to mutations in genes whose products are important in establishing epigenetic marks (see Fig. 1 for overview of the screen).

More than 4000 G_1 offspring have been screened and 40 strains have been isolated with transgene expression more than 2 standard deviations away from the mean of nonmutant transgenic offspring (E. Whitelaw, pers. comm.). Each of these strains possesses heritable dominant-acting mutations. This represents a dominant functional mutation rate of 1 in 100. These mutations have been named *Mommes* (Blewitt et al. 2005). The dominant mutations are called *MommeD1-40* and details of some of these are shown in Table 1. Identification of the mutations that segregate in 20 of the 30 lines has now been reported (Chong et al. 2007; Ashe et al. 2008; Blewitt et al. 2008; Daxinger et al.

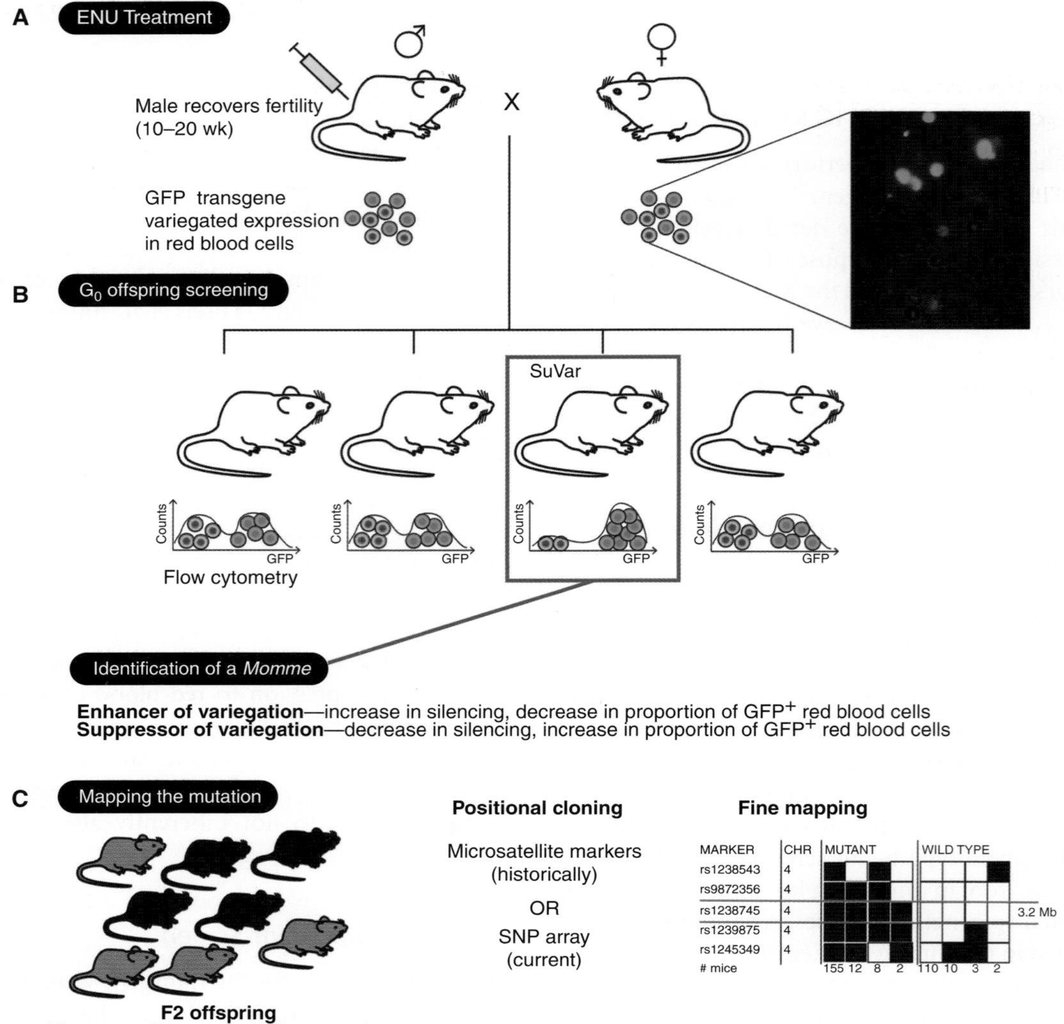

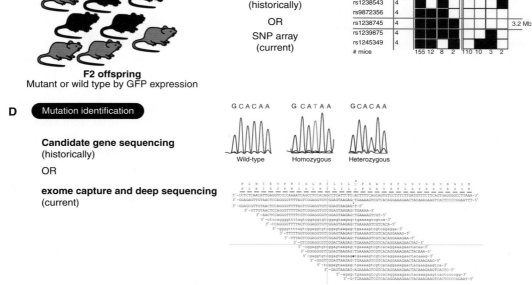

Figure 1. *Momme* screen for modifiers of epigenetic reprogramming. (*A*) GFP transgenic males are treated with ENU (now the G_0 generation), left to recover fertility, and then bred with GFP transgenic females to produce G_0 offspring. (*B*) A drop of blood is taken from all G_0 offspring at weaning and analyzed by flow cytometry to measure GFP expression in erythrocytes. Analysis is performed to look for variations in the extent of variegation of transgene expression. In this instance, an example of an individual with an enhancer of variegation phenotype is illustrated in the third mouse analyzed. (*C*) Animals with alterations in transgene variegation are backcrossed for two generations to allow mapping of the causative mutation. Mapping is performed with microsatellite markers or single-nucleotide polymorphism (SNP) arrays, and fine mapping followed up with large numbers of phenotypically mutant or wild-type animals, using additional SNPs or microsatellite markers. (*D*) The linked point mutation is then identified via exome capture (i.e., genomic DNA input selected using mouse exonic probes) followed by deep sequencing or by candidate gene sequencing.

Table 1. Summary of *Momme* mutants produced in dominant screen for modifiers of epigenetic reprogramming

Name	Effect on variegation	Homozygous lethality	Gene	Mutation	Chromosome	Reference(s)	Human homolog
MommeD1	Suppressor	Females E10, some male adults viable	*Smchd1*	C → T makes Stop	Chr 17	Blewitt et al. 2005, 2008	SMCHD1, mutated in FSHD2
MommeD2	Suppressor	E8–E9	*Dnmt1*	C → A in Exon 25, Thr → Lys	Chr 9	Chong et al. 2007	DNMT1
MommeD4	Enhancer	E17–E18	*Smarca5*	T → A in Exon 12, Trp → Arg	Chr 8	Blewitt et al. 2005; Chong et al. 2007	SMARCA5
MommeD5	Enhancer	E8–E9	*Hdac1*	7 bp Del in Exon 13, Frameshift	Chr 4	Blewitt et al. 2005; Ashe et al. 2008	HDAC1
MommeD6	Suppressor	E6–E8	*D14Abb1e*	T → C in Exon, Leu → His	Chr 14	Blewitt et al. 2005; Ashe et al. 2008; L Daxinger and E Whitelaw, pers. comm.	FAM208A
MommeD7	Enhancer	E18.5	*Hbb*	T → C in poly (A) signal	Chr 7	Brown et al. 2013	HBB β-thalassaemia
MommeD8	Enhancer	Some viable adults	*Rlf*	G → T Exon 8, Cys-Phe in Zn finger	Chr 4	Ashe et al. 2008; L Daxinger and E Whitelaw, pers. comm.	RLF
MommeD9	Enhancer	E6–E7	*Trim28*	T → C at splice site of Intron 13	Chr 7	Whitelaw et al. 2010a,b	TRIM28
MommeD10	Enhancer	Some viable adults	*Baz1b*	T → G in Exon 7, Leu → Arg	Chr 5	Ashe et al. 2008	BAZ1B Williams Syndrome
MommeD11	Suppressor	E14	*Klf1*	T → A in Exon 3, Cys → Stop,	Chr 8	E Whitelaw, pers. comm.	KLF1 Anaemia
MommeD12	Enhancer	E5–E7	*eIF3h*	T → A splice site −10 bp before Exon 5	Chr 15	Daxinger et al. 2012	EIF3H
MommeD13	Suppressor	E5–E8	*Setdb1*	A → G in Exon 20. Splicing defect	Chr 3	L Daxinger and E Whitelaw, pers. comm.	SETDB1 GWAS[a] for melanoma
MommeD14	Suppressor	Some viable adults	*Dnmt3b*	T → C in 3′ splice site of Exon 13. Exon 13 skipped	Chr 2	Youngson et al. 2013	DNMT3B ICF Syndrome
MommeD16	Enhancer	Some viable adults	*Baz1b*	C → T Exon 2, Leu → Pro	Chr 5	L Daxinger and E Whitelaw, pers. comm.	BAZ1B Williams Syndrome
MommeD17	Suppressor	Some viable adults	*Setdb1*	T → C in Exon 21, Val → Ala	Chr3	L Daxinger and E Whitelaw, pers. comm.	SETDB1 GWAS for melanoma
MommeD19	Suppressor	E5–E7	*Smarcc1*	T → G Intron 10 –Exon11 splice site	Chr 9	L Daxinger and E Whitelaw, pers. comm.	SMARCC1 Link to colon cancer
MommeD20	Suppressor	E6–E8	*D14Abb1e*	T → C at 5′ splice site of Intron 1	Chr 14	L Daxinger and E Whitelaw, pers. comm.	FAM208A
MommeD21	Suppressor		*Morc3*	T → A Exon 1, Met (start codon) → Lys	Chr 16	L Daxinger and E Whitelaw, pers. comm.	MORC3
MommeD23	Suppressor	Some viable adults	*Smchd1*	A → T Exon 12 Arg → Stop	Chr 17	L Daxinger and E Whitelaw, pers. comm.	SMCHD1, mutated in FSHD2
MommeD27	Suppressor		*Pbrm1*	A → G Exon 17, Tyr-Cys	Chr 14	L Daxinger and E Whitelaw, pers. comm.	PBRM1
MommeD28	Enhancer	Some viable adults	*Rlf*	A → G Intron 4, splicing defect	Chr 4	L Daxinger and E Whitelaw, pers. comm.	RLF
MommeD30	Enhancer	E10–E12	*Wiz*	Single base deletion Exon 5; frameshift	Chr 17	L Daxinger and E Whitelaw, pers. comm.	WIZ

Continued

Table 1. *Continued*

Name	Effect on variegation	Homozygous lethality	Gene	Mutation	Chromosome	Reference(s)	Human homolog
MommeD31	Enhancer	E6–E8	*Trim 28*	T → A Exon 3 Cys → Ser in Zn Finger	Chr 7	L Daxinger and E Whitelaw, pers. comm.	TRIM28
MommeD32	Suppressor	E8–E9	*Dnmt1*	T → C in Exon 29, Leu → Pro in BAH domain	Chr 9	L Daxinger and E Whitelaw, pers. comm.	DNMT1
MommeD33	Suppressor	Male hemizygotes live	*Suvar39h1*	A → G at Exon 1–Intron 1 splice site	Chr X	L Daxinger and E Whitelaw, pers. comm.	SUVAR39h1
MommeD34	Enhancer	Some adults viable	*Rlf*	C → A Exon 7, Cys → Stop, null allele	Chr 4	L Daxinger and E Whitelaw, pers. comm.	RLF
MommeD35	Enhancer		*Smarca5*	A → G Exon 9, Asn → Ser,	Chr 8	L Daxinger and E Whitelaw, pers. comm.	SMARCA5
MommeD36	Suppressor		*Smchd1*	G → A Exon 42, Glu → Stop	Chr 17	L Daxinger and E Whitelaw, pers. comm.	SMCHD1, mutated in FSHD2
MommeD37	Enhancer		*Smarca5*	T → C Exon 13, Leu → Pro	Chr 8	L Daxinger and E Whitelaw, pers. comm.	SMARCA5
MommeD38	Enhancer	No homozygotes at 3 wk	*eIF3h*	G → A Exon 7, Arg → Stop	Chr 15	Daxinger et al. 2012	EIF3H

[a]GWAS, genome-wide association studies.

2012; Youngson et al. 2013; L Daxinger and E Whitelaw, pers. comm.) and we will discuss some of these findings, along with the characterization of the strains, in Sections 1.1.2–1.1.5 and 1.2. Briefly, mutations have been identified in the genes coding for proteins known to play a role in epigenetic processes: DNA methyltransferases, Dnmt1 and Dnmt3b; histone deacetylase, Hdac1; chromatin-remodeling factors, Smarca5 (Snf2h), Smarcc1, Pbrm1, and Baz1b (WSTF); histone methyltransferases, Setdb1 and Suvar39h1; basal transcriptional machinery, Trim28 (KAP1); the transcription factor Klf1; and importantly in genes not previously identified to have a role in epigenetic silencing, such as Smchd1 and Rlf (see Fig. 2 and Daxinger et al. 2013).

1.1.2 *The Importance of* Mommes *in Development*

For each of the *MommeD* strains, heterozygous intercrosses were used to look for semidominant phenotypes and potential homozygous embryonic lethality. In all cases, the mutations were found to be semidominant with respect to transgene expression because either a third phenotypic class of transgene expression was observed at weaning (*MommeD1*, *MommeD8*, *MommeD10*, *MommeD14*, *MommeD16*, *MommeD17*, and *MommeD23*) or the ratio of mutant to wild-type offspring was consistent with embryonic lethality of homozygous mutants (Blewitt et al. 2005; Ashe et al. 2008). The semidominant nature of the observed phenotypes is consistent with epigenetic processes being dose-dependent as they are in lower organisms (Schotta et al. 2003).

For each of the *MommeD* strains, the timing of homozygous embryonic lethality has been established. Homozygous lethality is observed to some extent in all *MommeD* pedigrees (see Table 1). For *MommeD1*, *MommeD8*, *MommeD10*, *MommeD14*, *MommeD16*, *MommeD17*, *MommeD23*, *MommeD28*, *MommeD33*, and *MommeD45* some homozygotes survive to weaning, whereas for most other strains, it appears that all homozygotes die in utero or as neonates. The lethality of the majority of mice homozygous for *MommeD* mutations indicates the importance of the encoded proteins for normal development. In particular, *MommeD1* and *MommeD10* have provided substantial insights into mammalian developmental processes and are discussed in Sections 1.1.2.1 and 1.1.2.2.

1.1.2.1 **MommeD1.** *MommeD1* is unique among the *MommeD*s because it displays female-specific embryonic lethality; homozygous males are born in normal numbers, but only half survive to adulthood, whereas homozygous females die around E (embryonic day) 10.5 (Blewitt et al. 2005) caused by a failure of X inactivation (Blewitt

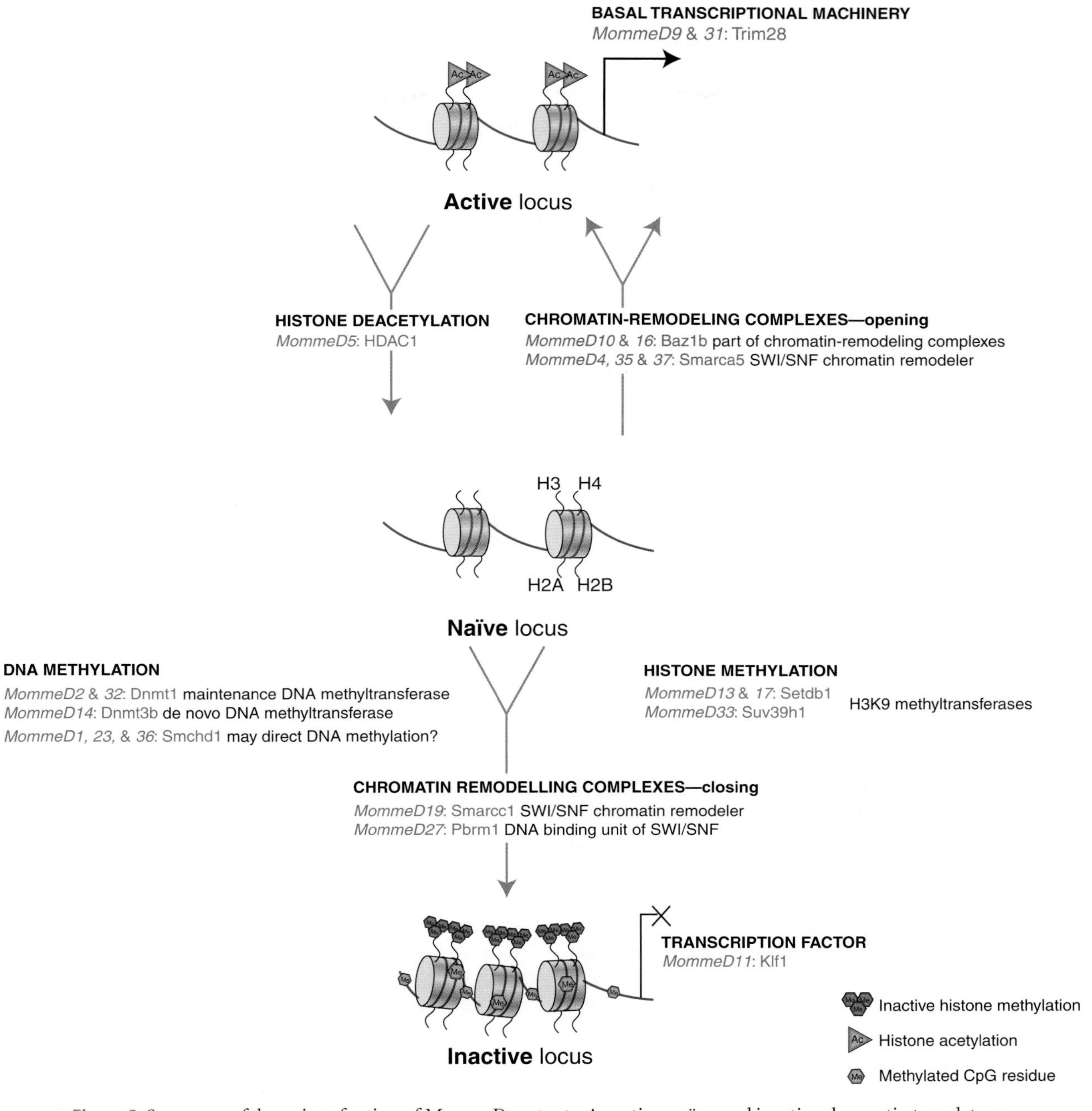

Figure 2. Summary of the point of action of *MommeD* mutants. An active, naïve, and inactive chromatin template are shown, and all of the *MommeD*s in Table 2 are indicated; red indicates that they are suppressor mutations and so the wild-type protein represses expression, whereas green indicates that they are enhancer mutations and so the wild-type protein activates expression.

et al. 2008). A nonsense mutation in a novel gene, called structural maintenance of chromosome hinge domain-containing 1 (*Smchd1*) linked to the alteration in transgene silencing. This nonsense mutation results in nonsense-mediated mRNA decay of the *Smchd1* transcript. Mutation of *Smchd1* was shown to cause both the alteration in transgene silencing and female-specific lethality through study of a genetrap null allele of *Smchd1*. Mutations in *Smchd1* have been identified in a further two *MommeD* pedigrees that display similar attributes (Table 1). $Smchd1^{MommeD1/MommeD1}$ mutant female embryos initiate X inactivation normally, as shown by normal *Xist* expression and H3K27me3 accumulation; however, they do not acquire DNA methylation at the CpG islands of genes on the inactive X, a characteristic of genes subject to X inactivation. Furthermore, both the embryo and the extraembryonic tissues display up-regulation of some genes normally subject to X inactivation (Blewitt et al 2008). The embryo proper normally undergoes random X inactivation, whereas the extraembryonic tissues undergo paternally imprinted X inactivation (elaborated in Ch. 25 [Brockdorff and Turner 2014]). So, these results indicate that Smchd1 is critical for both of these X inactivation processes. Although the mechanisms of X inactivation have been studied for decades, the process by which silencing occurs is still not completely understood, so the addition of a novel player in this field opens the door to many new functional studies.

Smchd1 was named because of the presence of a carboxy-terminal SMC hinge domain, which is normally found in the canonical SMC proteins, SMC1-6. These proteins form heterodimers that make up part of the cohesin and condensin complexes that are important for chromosome conformation during cell division and the SMC5/6 complex is involved in DNA repair. It is interesting to postulate how Smchd1 may function at the molecular level during X inactivation. It may belong to a new class of proteins that bridge the gap between epigenetic control and chromosome structure. The unbiased approach of the mutagenesis screen has enabled the identification of a novel protein involved in epigenetic gene silencing and critical for X inactivation. Because X inactivation is an epigenetic mechanism exclusive to higher organisms, this validates screening in the mammalian system. Very recently, Smchd1's role as a tumor suppressor has been described (Leong et al. 2013), and it is mutated in human facioscapulomuscular dystrophy type 2 (Lemmers et al. 2012). These results show the broader relevance to human health of novel epigenetic factors identified in screens.

1.1.2.2 MommeD10. The *Momme* screen has elucidated a new role for the known epigenetic modifier Baz1b. Baz1b is a mammalian bromodomain-containing protein, which is part of two different chromatin-remodeling complexes (WINAC SWI/SNF and WICH ISWI-containing) bound at promoters and replication foci (Bozhenok et al. 2002; Kitagawa et al. 2003). *Baz1b* is mutated in the *MommeD10* strain of mice, which displays subtle craniofacial abnormalities in the heterozygotes and more severe defects in the few surviving homozygotes (Ashe et al. 2008). The observed abnormalities are reminiscent of those seen in patients with Williams–Beuren syndrome (WBS) or Williams syndrome, and *Baz1b* is one of a linked group of 28 genes that are commonly heterozygously deleted in WBS patients. The *MommeD10* point mutation causes an amino acid substitution in a highly conserved region of Baz1b, which appears to destabilize the protein product. The $Baz1b^{MommeD10}$ allele is the first available mutant allele for *Baz1b*. Study of this strain shows that Baz1b has a hitherto uncharacterized role in craniofacial development, and suggests that reduction in the Baz1b protein contributes to the elfin facial features of WBS patients (Ashe et al. 2008). WBS patients also show hypersocial, anxious personalities that are associated with structural alterations in a number of parts of the brain (Jabbi et al. 2012). Whether haploinsufficiency for BAZ1B is involved in this phenotype is not known. The contribution of each of the 28 genes commonly deleted in WBS is controversial, so the Baz1b mouse model provides a tool to study some features of this syndrome. This study also reveals that at least some of the WBS features have an epigenetic component.

1.1.3 Mommes *Influence Expression at the* Agouti Viable Yellow *Allele*

One of the primary aims of the *Momme* screen was to use the mutants to learn more about the unusual features of metastable epialleles. Because a variegating transgene had been used for the original screen, some of the *MommeD* strains (*MommeD1-5*) were crossed with a strain carrying the *Agouti viable yellow* (A^{vy}) allele, a single copy endogenous metastable epiallele (Blewitt et al. 2005; Chong et al. 2007).

The A^{vy} allele is driven by an upstream retrotransposon and displays variable expressivity, subtle parent-of-origin effects, and transgenerational epigenetic inheritance. More specifically, the cryptic promoter within the LTR of the integrated IAP drives inappropriate expression of the *Agouti* gene (see Fig. 3A) (Michaud et al. 1994; Perry et al. 1994). Agouti is normally expressed for just a short period in the hair growth cycle causing a switch in pigment expression from brown or black to yellow. Normal Agouti expression results in a subapical yellow band on a dark hair, which en masse results in a brown appearance of the coat, termed agouti. When Agouti is expressed throughout the hair growth cycle, an entirely yellow hair shaft is produced. Mice carrying the A^{vy}, A^{hvy}, or A^{iapy} alleles all produce some mice with a variegated coat made up of agouti- and yellow-colored patches—a phenotype termed mottled (see Fig. 3B,C). The agouti-colored patches occur because sometimes the IAP LTR is epigenetically silenced and transcriptional control reverts to the normal promoter. Just like with X inactivation, the silent state is stable and heritable over hundreds of cell divisions, so one stochastic silencing event of the IAP LTR early in development can result in a patch of daughter cells, all showing normal Agouti expression.

Offspring were scored for coat color and phenotyped for expression of the GFP transgene (used to infer mutant or wild-type status for the *MommeD* in question when the mutation was not yet identified). In many cases, a shift was

Cite as *Cold Spring Harb Perspect Biol* doi: 10.1101/cshperspect.a017939

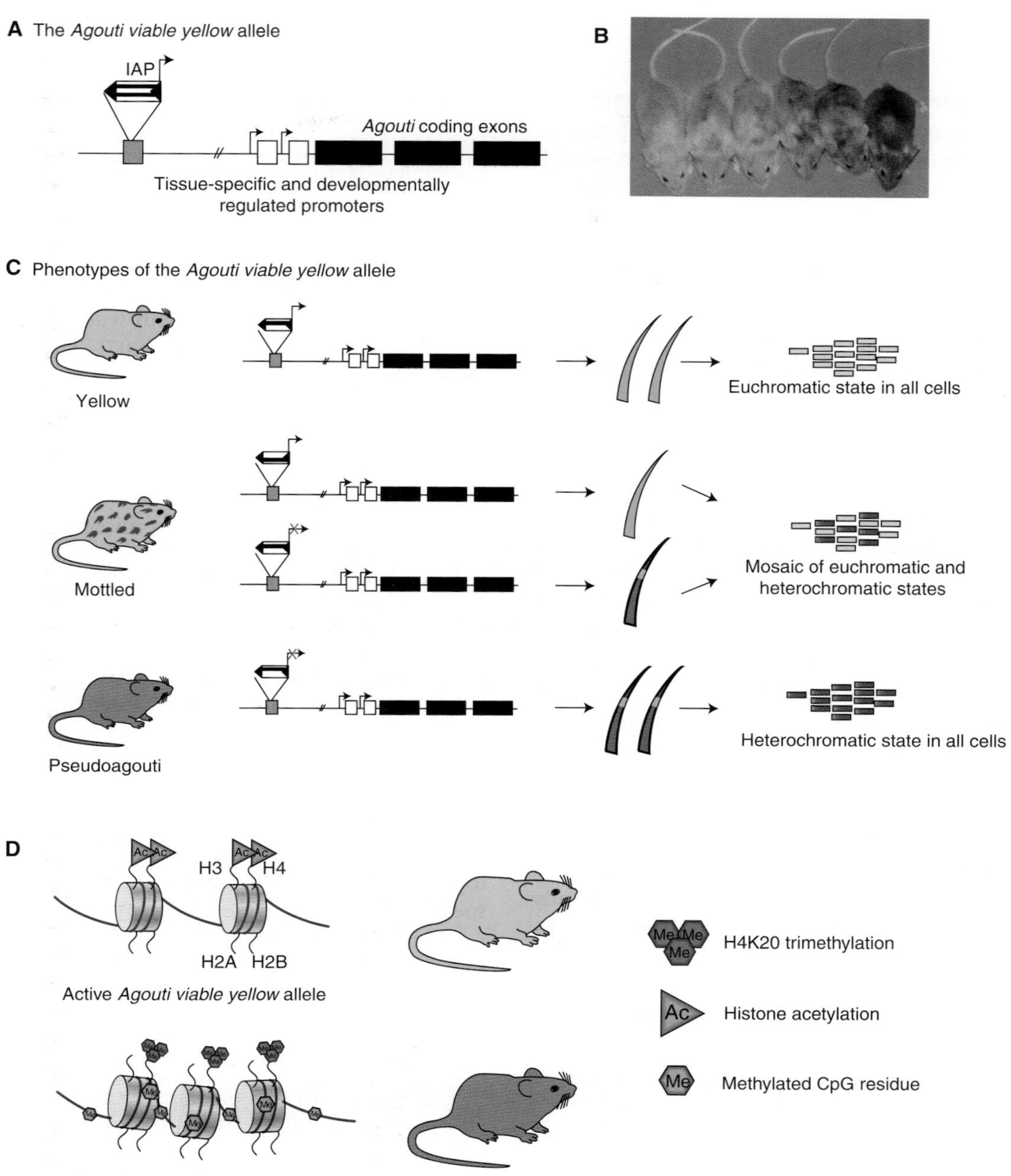

Figure 3. Features of the *Agouti viable yellow* allele. (*A*) The *Agouti viable yellow* allele (not to scale) has an intracisternal A particle (IAP) insertion (striped box) in pseudoexon 1a (gray box), ~100 kb upstream of the Agouti coding exons (black). IAP long terminal repeats (LTRs) are shown as arrowheads, and transcriptional start sites shown as arrows. (*B*) An agouti wild-type mouse with an A^+ allele has a brown coat color phenotype with a two-toned hair shaft, in which the base appears black in color, and closer to the tip shows a yellow color (see cartoon in *C*). This phenotype occurs because the *Agouti* gene, producing a yellow fur color, is only transiently expressed in the hair follicle microenvironment. The A^{vy} allele, if expressed ubiquitously from the IAP promoter, produces a phenotype with an entirely yellow hair shaft. There is a spectrum of coat color phenotypes observed, however, because of the A^{vy} allele ranging from completely yellow when the IAP LTR is active in all cells, to mottled due to patches of active and inactive cells, and finally animals with an agouti colored coat, called pseudoagouti, which are indistinguishable from wild-type agouti animals because of silenced IAP LTR in all cells. (Reproduced from Morgan et al. 1999.) (*C*) Mice with the *Agouti viable yellow* allele in a euchromatic state in all cells appear yellow, whereas those with the allele in a heterochromatic state in all cells appear agouti in color, termed pseudoagouti. Mice that are mosaic for the heterochromatic and euchromatic allele appear mottled. (*D*) Summary of the epigenetic marks found at an active or inactive *Agouti viable yellow* allele. The inactive allele is hypermethylated and enriched for H4K20 trimethylation. The active allele is hypomethylated and enriched for acetylated residues of histone H3 and H4 tails.

observed in the spectrum of A^{vy} coat color phenotypes seen in the mutant compared with the wild-type offspring. In general, the shift in variable expressivity was consistent with the alteration in expression of the GFP transgene; for example, if an increase in GFP expression was observed, then mutant animals more frequently displayed an active A^{vy} allele than wild-type animals. These shifts sometimes, but not always, had concomitant changes in DNA methylation at the LTR that controls expression at the A^{vy} locus (Blewitt et al. 2005). These results show that the mutant proteins (Smchd1, Dnmt1, Snf2h, Hdac1, and others unpublished) are important not only for transgene silencing but also for silencing at an endogenous metastable epiallele driven by a retrotransposon. This is reminiscent of the finding that many dominant modifiers of retrotransposon insertion alleles in the fly are also dominant modifiers of PEV (Fodor et al. 2010).

Some of *MommeDs* (*MommeD1-4*) showed complex alterations in the penetrance at A^{vy} including sex-specific effects. For example, female-specific effects were reported for both *MommeD1* and *MommeD2*. In both cases, mutant females showed a shift in penetrance that reflects a greater likelihood to have an active A^{vy} allele, compared with their littermate wild-type females (consistent with the role of *MommeD1* and *MommeD2* as suppressors of variegation), but this was not seen in the males (Blewitt et al. 2005). These female-specific effects are likely to occur because of the involvement of the mutated proteins, subsequently identified as Smchd1 and Dnmt1, respectively, in X inactivation (Sado et al. 2000; Blewitt et al. 2008). It was proposed that the *MommeD1* and *MommeD2* proteins bind the inactive X chromosome, and the inactive X acts as a sink for such repressor proteins, leaving fewer repressor molecules free to perform autosomal silencing compared with an XY male cell (Blewitt et al. 2005). In support of this argument, significantly more yellow females than yellow males are seen in the A^{vy} wild-type colony (Morgan 1999; Blewitt et al. 2005). It has been known for many years that even before sexual differentiation, female mammalian embryos are smaller than their male counterparts (Burgoyne et al. 1995; Ray et al. 1995). In combination, these findings suggest that an epigenetic difference driven by the presence or absence of an inactive X chromosome may be responsible for some of the phenotypic differences between the sexes. This idea has recently been extended using a different metastable epiallele, the hCD2 variegating transgene (see Table 2), and in mouse samples that have varying sex chromosome complements: XX, XY, XO, XXY (Wijchers et al. 2010). They found that several hundred autosomal genes are sensitive to X chromosome complement. It will be interesting to look at gene expression in humans with sex chromosome aneuploidies.

1.1.4 Paternal Effect Genes

A new and unusual phenomenon that may be the result of incomplete erasure of epigenetic marks between generations was detected when studying the coat colors of mice in *MommeD* colonies. For both $Dnmt1^{MommeD2}$ and $Smarca5^{MommeD4}$, the wild-type offspring of heterozygous mutant fathers have a different spectrum of A^{vy} coat color phenotypes when compared with wild-type offspring of wild-type parents (Chong et al. 2007). This is shown in a simple diagram in Figure 4. These offspring are genetically identical and differ only in the untransmitted genotype of the male parent. These paternal effects may be attributable to alterations in chromatin packaging or RNA populations in genetically wild-type sperm that is a result of their production in a genetically compromised male parent that then has an effect in *trans* on the A^{vy} allele inherited from the mother. Smarca5 and Dnmt1 are expressed in testes and in stage VII pachytene spermatids, which is before segregation of homologous chromosomes to produce haploid spermatids (La Salle et al. 2004; Chong et al. 2007). So, it is feasible that development of sperm in a Smarca5 or Dnmt1 depleted environment could result in epigenetic alterations to the wild-type set of chromosomes. Furthermore, the progeny of each spermatogonium remain connected by cytoplasmic bridges (channels connecting the cytoplasms of adjacent cells), which allows transcript sharing, providing another opportunity for wild-type haploid sperm to have depleted levels of Smarca5 or Dnmt1.

Although maternal effect genes have previously been reported in mammals and other organisms (Zheng and Liu 2012), this was the first report of paternal effect genes in mammals. Interestingly, the suppressors and enhancers of variegation identified in the *Drosophila* PEV screens also displayed paternal effects (Fitch et al. 1998). Our results alter the way we think about the transmission of phenotypic traits from parent to offspring. The laboratory mouse has provided an opportunity to observe these interesting effects.

1.1.5 Recessive Screen

As part of the *Momme* mutagenesis screen, 160 G_1 male progeny with no detectable change in transgene expression were screened for recessive mutations in modifiers of variegation. A pedigree was produced from each founder male by backcrossing the founder with four of his daughters to produce at least 32 offspring. Using this approach, >84% of recessive modifiers of variegation would theoretically be detected. Seven of the 160 pedigrees showed evidence of a recessive mutation, comprising both enhancers and suppressors of variegation (Blewitt 2004; Vickaryous 2005). These recessive mutants generally showed more

Cite as *Cold Spring Harb Perspect Biol* doi: 10.1101/cshperspect.a017939

Table 2. Table of metastable epialleles and reporter alleles

Allele	Type	Phenotype	Variegation	Variable expressivity	Parent-of-origin effects	Transgenerational epigenetic inheritance	Reference(s)
A^{vy}	Endogenous; IAP insertion	Coat color, obesity, diabetes	Yes	Yes	Yes	Yes	Morgan et al. 1999; Wolff 1978
A^{iapy}	Endogenous; IAP insertion	Coat color, obesity, diabetes	Yes	Yes	Yes	?	Michaud et al. 1994
A^{hvy}	Endogenous; IAP insertion	Coat color, obesity, diabetes	Yes	Yes	Yes	?	Argeson et al. 1996
$Axin^{fu}$	Endogenous; IAP insertion	Tail kinks	Yes	Yes	Yes	Yes	Rakyan et al. 2003; Reed 1937; Belyaev et al. 1981
Axial defects	Endogenous; Grhl2 up-regulation of unknown origin	Spina bifida	?	Yes	Yes	Yes?	Essien et al. 1990; Brouns et al. 2011
Disorganization	Endogenous; potential Gata4 disruption	Skeletal abnormalities	?	Yes	No	?	Hummel et al. 1959; White et al. 1995
$mCabp^{IAP}$	Endogenous; IAP insertion	*mCabp* expression	?	Yes	?	?	Druker et al. 2004
c^{m}	Endogenous; IAP insertion	Defective Tyrosinase expression; mottled coat color	Yes	?	?	?	Porter et al. 1991
C^{m1OR}	Endogenous; IAP insertion	Defective Tyrosinase expression; mottled coat color	Yes				Wu et al. 1997
239B	Transgene	LacZ-expressing red blood cells	Yes	Yes	Yes	No	Kearns et al. 2000
MTα#7	Transgene; insertion into L1 element	LacZ-expressing red blood cells	Yes	Yes	No	Yes	Sutherland et al. 2000
TKZ751	Transgene	LacZ-expressing somatic cells	?	Yes	Yes	No	Allen et al. 1990
RSVIgmyc	Transgene	Transgene expression in myocytes	?	Yes	Yes	No	Weichman and Chaillet 1997
BLG transgenic lines 7 and 45	Transgene; centromeric	β-lactoglobulin expression	Yes	Yes	?	?	Dobie et al. 1996
Tyr-SV40E	Transgene	SV40 expression in melanocytes	Yes	Yes	?	?	Bradl et al. 1991
hCD2-1.3b	Transgene; centromeric	Human CD2 expression in T cells	Yes	No	No	No	Festenstein et al. 1996
GFP1	Transgene	GFP-expressing red blood cells	Yes	No	No	No	Preis et al. 2003
GFP3	Transgene	GFP-expressing red blood cells	Yes	No	Yes	No	Preis et al. 2003

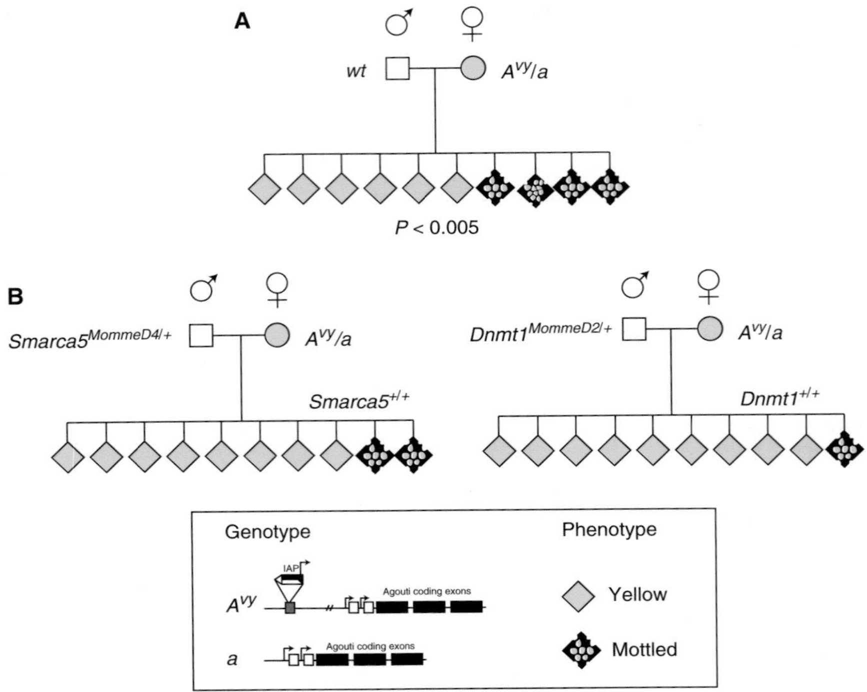

Figure 4. Paternal effects of *Smarca5*^{MommeD4} and *Dnmt1*^{MommeD2} on the A^{vy} allele. (*A*) For the A^{vy} progeny from a cross with a paternal wild-type mouse, the expected ratio of yellow versus mottled mice is 6:4. (*B*) Heterozygous males for *Momme* mutants of *Smarca5* (*left*) and *Dnmt1* (*right*) are bred with female yellow A^{vy} heterozygotes. Offspring are scored for their coat color and genotyped for the *MommeD* mutations and the A^{vy} allele. Only those offspring carrying the A^{vy} allele and the wild-type *Smarca5* or *Dnmt1* allele are shown for simplicity. The spectrum of phenotypes was compared with that observed for wild-type FVB/N males bred with yellow females (see *A*). Wild-type offspring of mutant males have a significantly different spectrum of coat color phenotypes than offspring of wild-type males, termed a paternal effect. Both the *Smarca5* and *Dnmt1* wild-type offspring produced from a *MommeD* heterozygous father show a skew toward less mottled mice, indicating that the haploinsufficiency of these genes in the spermatogonia (diploid sperm progenitors) is sufficient to affect the epigenomic programming of the A^{vy} locus in the next generation, and hence, the phenotype. (Adapted from Chong et al. 2007.)

subtle phenotypes than the dominant mutations, and were frequently sterile. Only one of the recessive mutations has been studied in detail (*MommeR1*).

MommeR1 was identified as a suppressor of variegation (i.e., mutants have increased expression of the GFP transgene). *MommeR1* has the interesting phenotype of premature ovarian failure in all female homozygotes, and ovarian teratomas in one-sixth of these animals. A missense mutation was identified in *Forkhead box protein 3a*, *Foxo3a*, which through complementation tests with a null allele of *Foxo3a*, was shown to be the causative mutation (Youngson et al. 2011). Foxo3a is a forkhead transcription factor, and so not traditionally viewed as an epigenetic modifier, although it does have a role in hematopoiesis. One possibility was that an altered blood cell compartment, in particular, an increased level of reticulocytes that display higher GFP transgene expression, could account for the elevation in transgene expression. Indeed, it is formally possible that any of the mutants identified are detected because of hematopoietic defects rather than epigenetic disruption per se. This was ruled out for *MommeR1*, as *Foxo3a*$^{MommeR1/MommeR1}$ animals had normal levels of reticulocytes setting them apart from the *Foxo3a*$^{-/-}$ animals. It is not yet clear whether Foxo3a has a direct or indirect effect on transgene variegation, but this new line of mice has linked premature ovarian failure with ovarian teratomas (Youngson et al. 2011), which have previously only been associated with late stage ovarian failure, and future studies should unravel the role of the protein in gene silencing.

1.2 ENU Mutagenesis Produces Hypomorphic Alleles

One of the advantages of ENU mutagenesis is that it predominantly produces point mutations. These point mu-

 Cite as *Cold Spring Harb Perspect Biol* doi: 10.1101/cshperspect.a017939

tations are more likely to produce hypomorphic alleles than standard knockout approaches, and can be informative about critical residues within a protein structure. This also provides an opportunity to create an allelic series, which can be useful when studying genes for which complete knockout is early embryonic lethal. The *Momme* screen has produced new mutations in nine known epigenetic modifiers for which null alleles already exist (*Dnmt1*, *Dnmt3b*, *Hdac1*, *Trim28*, *Smarca5*, *Setdb1*, *Smarcc1*, *Pbrm1*, and *Suvar39h1;* Chong et al. 2007; Ashe et al. 2008; Whitelaw et al. 2010a; Youngson et al. 2013; L Daxinger and E Whitelaw, pers. comm.). The missense mutations in *Dnmt1*, *Trim28*, and *Smarcc1* appear to destabilize the protein, whereas *Hdac1* has a small deletion and subsequent frameshift mutation that alters the very carboxyl terminus of the protein. The *Smarca5* missense mutation, the *Dnmt3b* splice-site mutation, and the *Setdb1* missense mutation appear to behave as hypomorphic alleles.

An independent ENU screen was performed to create a hypomorphic allele of *brahma related gene 1*, *Brg1* (Bultman et al. 2005). *Brg1* null mice fail to implant (Bultman et al. 2000), precluding the study of later stages of development. In this case, the investigators took advantage of a recessive phenotypic marker closely linked to Brg1—*curly whiskers* (*cw*). *cw/cw* males were treated with ENU, bred with wild-type females, and G_1 offspring bred with animals that had the *cw* marker in *cis* with the *Brg1* null allele (Fig. 5). 525 pedigrees were scored for the presence of mice with curly whiskers; one pedigree was found in which no such mice were found at weaning, suggesting that a mutation had occurred that failed to complement the *Brg1* null allele. The investigators went on to identify the mutation in the ATPase domain of Brg1, which uncoupled the ATPase activity from the chromatin-remodeling activity of Brg1. The mutant Brg1 protein was unable to remodel chromatin, although the ATPase activity itself was unchanged, and Brg1 still localized to chromatin and assembled into its usual SWI/SNF complex. Homozygous *Brg1* hypomorphs survived until mid-gestation, allowing many more studies than could be performed using the null mice. The observation of 1/525 pedigrees with a functional mutation in Brg1 is roughly consistent with the estimate of 1/700 animals with functional mutation (dominant or recessive) in a specific gene of interest

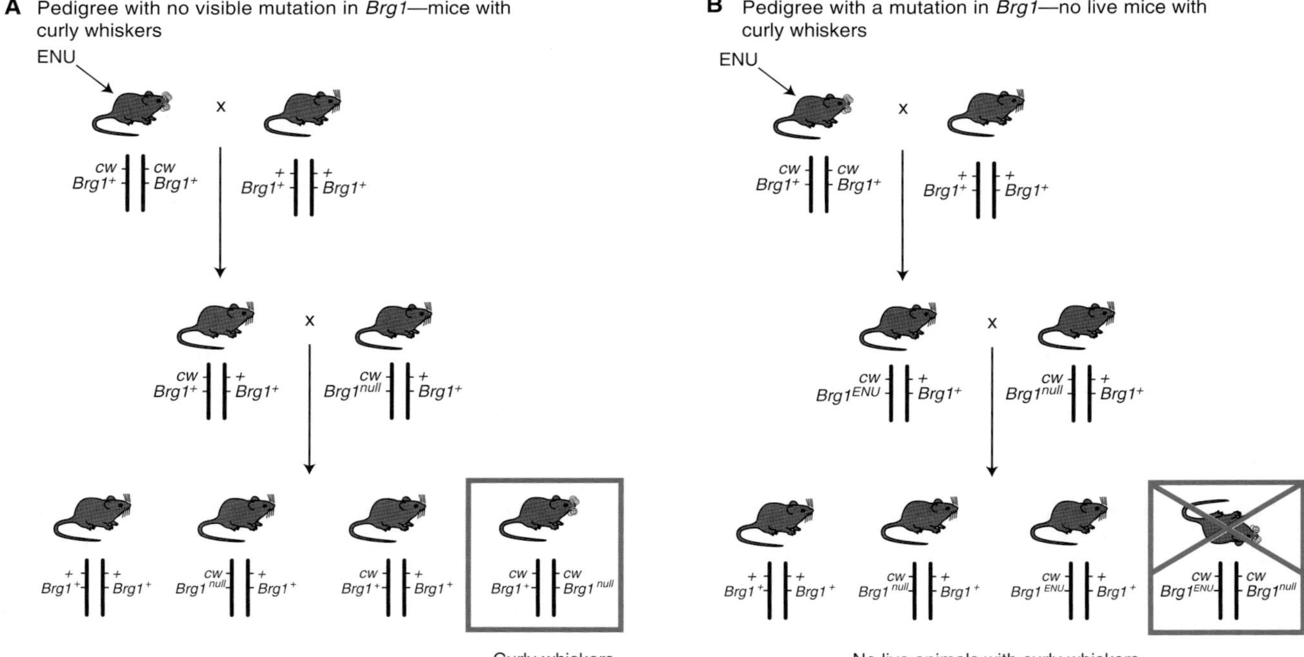

Figure 5. The breeding strategy for the *Brg1* hypomorph ENU screen. Only mouse chromosome 9 is shown indicating the *cw* locus and *Brg1* loci, which are 1 cM apart. *cw/cw* males were treated with ENU, and bred with wild-type females. Offspring were mated with *Brg1*^{null} heterozygotes, which carried the *cw* allele in *cis* with the null mutation. Resultant offspring were screened for the presence of pups with curly whiskers (in red). (*A*) Pups were observed with curly whiskers indicating no functional mutation in *Brg1*. (*B*) One pedigree was found in which no pups had curly whiskers indicating that a mutation had occurred that failed to complement the *Brg1*^{null} allele. A hypomorphic ENU-induced mutation in *Brg1* was later found. (Adapted from Bultman et al. 2005.)

(Hitotsumachi et al. 1985). This approach can be used to produce an allelic series without prior knowledge of the domain structures of the protein, and shows the power of ENU-induced point mutations over null alleles produced by homologous recombination.

1.3 A Screen for Genes Involved in X Inactivation Choice

Another ENU mutagenesis screen for epigenetic modifiers was established specifically to identify autosomal factors important in X inactivation choice (Percec et al. 2002; Percec et al. 2003). X inactivation choice is discussed in detail in Ch. 25 (Brockdorff and Turner 2014). Briefly, the choice of which X chromosome to inactivate happens in the embryo proper shortly after implantation. It has been known for some years that choice is governed by the Xce locus on the X chromosome; there are four different Xce alleles found in mice with differing likelihoods that the X chromosome bearing that allele will become the inactive X chromosome. Xce^{c} and Xce^{d} are found in wild mice such as *Mus castaneus*; Xce^{d} is the strongest allele and the least likely to be inactivated, whereas Xce^{c}, Xce^{b}, and Xce^{a} are of decreasing strength. Xce^{b} and Xce^{a} are found in various inbred *Mus musculus* mice. Xce heterozygotes display a predictable pattern of X inactivation skewing away from the 1:1 ratio seen in homozygotes and expected of a random process. The pattern of skewing, although predictable, displays considerable variance; for example, $Xce^{c/a}$ females show on average 25% of their cells with an active Xce^{a} carrying chromosome, but the range is from ~5% to 45%.

Little is known about the *trans*-acting factors that interact with the Xce or other elements on the X chromosome. It is clear that activation of *Xist* expression on just one of the two X chromosomes is required, and it is thought that pairing of the two X chromosomes immediately before the initiation of X inactivation may play some role in allowing cross talk and subsequent unequal distribution of *trans*-acting factors between the two X chromosomes. Both the chromatin insulator CTCF and pluripotency factor Oct4 play a role in this process. Elegant live-cell imaging experiments show that X chromosome pairing is frequently followed by up-regulation of a negative regulator of *Xist*. However, it is still unclear how the different *Xce* alleles influence this process (see Ch. 25 [Brockdorff and Turner 2014] for details).

The ENU screen performed by Percec and colleagues (Percec et al. 2002; Percec et al. 2003) took advantage of the predictable skewing of X inactivation present in $Xce^{a/c}$ and $Xce^{b/c}$ females. They screened for a skewing proportion more than 2 standard deviations away from the mean found in the control population by performing RT-PCR (reverse transcription polymerase chain reaction) followed by restriction fragment length polymorphism analysis of the X-linked gene *Pctk* in samples taken from females at weaning (Plenge et al. 2000). Females with aberrant skewing were further interrogated by allele specific qRT-PCR analysis of *Pctk*, *Pgk1*, and *Xist*. Three strains were identified and two have been followed in more detail. They have been shown to have one and two segregating mutations, respectively. The investigators have shown that, in each case, skewing is already present at E6.5 and E7.5 and is similar across all tissues. These results suggest that the causative mutations result in changes in primary choice, rather than skewing due to secondary nonrandom X inactivation (when cells die because of the choice of which X is inactivated). Furthermore, the mutations alter random X inactivation in the embryo proper, but not imprinted X inactivation in the extraembryonic tissues suggesting that the role of these factors, as least in heterozygotes, has some specificity for random X inactivation. Each of the mutations has been linked to specific large autosomal regions. These mutants have been called X inactivation autosomal factor 1, 2, and 3 (Xiafl, 2, and 3). It is anticipated that the underlying mutations will be identified soon.

1.4 Other Screens That Have Identified Mutations in Epigenetic Modifiers

Two ENU mutagenesis screens, performed to identify all genes within a defined chromosomal region, have identified homozygous embryonic lethal mutations in epigenetic modifiers. In each case, a long chromosomal deletion or inversion was used that covered the region of interest, and additionally covered a visible marker (the albino locus or the Rump White locus; Rinchik et al. 1990; Wilson et al. 2005), which allows for simple marking of the deleted or mutated chromosome, similar to the *curly whiskers* allele described above in Section 1.2.

The first of these screens isolated point mutations in the *Embryonic ectoderm development* (*Eed*) gene (Schumacher et al. 1996), which we now know is a core component of Polycomb repressive complex 2 (PRC2) in mammals. The null and hypomorphic alleles of *Eed* produced in this screen not only helped to clone the gene, but were also the only available mutant or knockout alleles of *Eed* available for many years. Study of these animals has informed us about the role of PRC2 in embryonic development, X inactivation, genomic imprinting, embryonic stem cell pluripotency, and differentiation and hematopoiesis (see Ch. 17 [Grossniklaus and Paro 2014]).

In the second of these screens (Wilson et al. 2005), a mutation was identified in the Promyelocytic zinc finger gene, *Plzf* (Ching et al. 2010). PLZF is a BTB/POZ domain-

 Cite as *Cold Spring Harb Perspect Biol* doi: 10.1101/cshperspect.a017939

containing zinc finger protein, which is involved in transcriptional repression. Although this is the third mutated allele of *Plzf*, the ENU induced mutant separates the functional requirements of PLZF in different biochemical pathways, and the missense mutation is potentially informative about BTB domain function.

Two mutations in epigenetic modifiers have been identified in one of the ENU mutagenesis screen run at the Walter and Eliza Hall Institute of Medical Research, Australia. This screen was designed as a sensitized suppressor screen, in which mice with thrombocytopenia and a hematopoietic stem cell (HSC) defect are subjected to ENU mutagenesis, and screens performed to identify genes that when mutated suppress these phenotypes. This screen has produced a point mutation in the histone acetyltransferase *Ep300* (Carpinelli et al. 2004) along with one in the PRC2 member *Suz12* (Majewski et al. 2008). Studies on the *Suz12* mutant line identified a novel role of PRC2 in the restriction of HSC function (Majewski et al. 2008).

Finally, a large-scale Australian ENU screen, run to identify genes important in immune function, has produced another mutation in *Eed*. This screen used flow cytometry-based screening of peripheral blood cell types and numbers (Jun et al. 2003), and identified that the *Eed* mutant, called Leukskywalker, had elevated leukocyte numbers. Although unpublished, this mutant is available to researchers who may be interested in such a strain.

Without knowing the identity of the majority of genes involved in epigenetic processes, it will be difficult to fully understand the molecular basis of epigenetic control. Therefore, screening approaches such as those described earlier in Section 1 will remain a valuable part of epigenetic research in the future.

2 EPIGENETIC PHENOMENA IN INBRED MOUSE COLONIES

Some endogenous murine alleles such as the *Agouti* coat color alleles can show unusual patterns of expression. These are termed metastable epialleles (Rakyan et al. 2002) because the transcriptional activity at these alleles is less stable than expected, and this is associated with changes in epigenetic state. This section summarizes the key features of the unusual expression patterns of, specifically, *Agouti viable yellow* (A^{vy}), an allele of *Axin*, *Axin fused* ($Axin^{fu}$), and some transgenic reporters (see Table 2). The behavior of these alleles will be used to define and describe variegation, variable expressivity, transgenerational epigenetic inheritance, parent-of-origin effects, and position effects. Together, these phenomena have taught us much about transcriptional gene silencing and the probabilistic nature of epigenetic processes.

2.1 Variegation

Variegation is the differential expression of a gene among cells of the same type. The variegated appearance of the leaves of some plants, and the coats of many dogs and cats, are good examples of variegation that we observe on a day-to-day basis. In some cases this variegation has a genetic origin (e.g., mosaic genetic states alter the color of corn kernels), but in many cases this is not the explanation. Because pigmentation patterns can be easily visualized, coat color gene mutations have been excellent models for the study of mosaicism in mammals. Since the 1970s, mouse geneticists, in particular Beatrice Mintz, became fascinated by this phenomenon (Mintz 1970). When discussing mottled mice she and her colleagues write (Bradl et al. 1991):

> Their patterns are apparently due to phenoclones—phenotypically different but genetically identical clones in which the same gene must be yielding more than one kind or amount of product in mitotic lineages of the same cell type. The phenomenon is presumably not limited to pigment cells.

At the level of gene expression, the existence of these phenoclones implies that there is some element of chance involved in whether a gene is expressed and that epigenetic control of gene expression is an inherently stochastic process.

2.1.1 *Variegation at Endogenous Alleles in the Mouse*

Historically, coat color variegation was first studied in female mice heterozygous for X-linked mutations of the mottled locus, later identified as *Atp7a*. Because hemizygous males never displayed variegation, and in fact were frequently embryonic lethal, the phenomenon was described as sex-linked variegation. The variant coat color patches in the case of the *Atp7a* mutations are determined by whether the wild-type or mutant X chromosome is subject to X inactivation in those particular cells. Analysis of the phenotype of these animals along with other critical studies led Mary Lyon to propose the Lyon hypothesis of X inactivation, in particular, that the process is random with respect to which of the two X chromosomes is inactivated in any particular cell (for further discussion see Ch. 25 [Brockdorff and Turner 2014]). Study of these animals has informed us about the timing of X inactivation in development, and the stable, heritable nature of that transcriptional silencing.

Some autosomal mutations in the mouse also result in variegated expression—for example, alleles of the *Agouti* gene: *Agouti viable yellow* (A^{vy}; Perry et al. 1994), *Agouti intracisternal A particle yellow* (A^{iapy}; Michaud et al. 1994), and *Agouti hypervariable yellow* (A^{hvy}; Argeson et al. 1996). Each of these alleles is characterized by a stable retrotransposon insertion. Similarly, a number of alleles at the albino locus, c^{m} and c^{m10R}, display coat color variegation that is the result of stable retrotransposon insertions in *cis* (Porter et al. 1991; Wu et al. 1997). Importantly, these autosomal alleles that display variegation show that loci throughout the genome can display this interesting characteristic of stochastic transcriptional silencing previously thought to be restricted to genes on the X chromosome.

2.1.2 *Variegation at Transgenes in the Mouse*

An unusually high proportion of transgenes also show variegated expression (Allen et al. 1990; Festenstein et al. 1996; Garrick et al. 1996; Weichman and Chaillet 1997; Kearns et al. 2000; Sutherland et al. 2000). The reason why transgenes have a tendency to variegate is not fully understood. There are several possible explanations, including high copy number within transgene arrays, integration of the transgenes next to heterochromatin (see Secs. 2.6 and 2.7), and in some cases, the foreign nature of sequences used within the transgene sequences (for review, see Martin and Whitelaw 1996). Transgenes composed of sequences with bacterial codon usage (e.g., LacZ) tend to show a greater degree of silencing than those with mammalian codon usage (Kearns et al. 2000; Sutherland et al. 2000; Preis et al. 2003). One possibility is that DNA sequences recognized as foreign are silenced by mechanisms normally reserved for integrated viruses and other invading transposable elements (Yoder et al. 1997), and in this light, it is interesting to note that all autosomal endogenous alleles that display variegation are associated with retrotransposon insertions (as mentioned for the *Agouti* and *Albino* alleles in Sec. 2.1.1).

Transgenes have also been shown to display additional peculiar phenotypes such as age-dependent silencing (Robertson et al. 1996) (i.e., transgene expression levels reduce with age). Because the promoters of tumor-suppressor genes are frequently silenced and show increased DNA methylation in cancer, and the greatest risk factor for cancer is increasing age, age-dependent silencing of transgenes raises the questions: Is this indicative of what occurs in the normal process of aging, and does it predispose to cancer development? It has now been shown that normal human prostate tissue (Kwabi-Addo et al. 2007) and normal mouse tissues (Maegawa et al. 2010) show widespread age-dependent DNA methylation changes. Furthermore, two groups (Rakyan et al. 2010; Teschendorff et al. 2010) have recently shown that specific DNA methylation changes occur in polycomb group protein target genes during aging, perhaps indicative of a preneoplastic state. Epigenetic deregulation appears to be a common feature of aging, first suggested by studies on transgene variegation in mice.

Analysis of variegated transgenes has allowed more detailed study of the epigenetic mechanisms contributing to variegation than was technically possible with variegated coat color alleles. One major advantage of the transgenic lines carrying transgenes directed to express in blood cells has been the ability to observe variegated gene expression at a single cell level rather than in cell populations (Robertson et al. 1995; Festenstein et al. 1996; Sutherland et al. 1997; Kearns et al. 2000; Sutherland et al. 2000; Preis et al. 2003). These studies have allowed consideration of different models of gene expression control (e.g., different models of enhancer action). Importantly, they have conclusively shown that gene expression can be controlled by probabilistic events rather than gradients of external factors (Sutherland et al. 1997). As techniques improve for transcriptome studies in single cells, it is possible that many loci throughout the genome will be shown to display similar variegated expression patterns, which may influence cellular behavior.

2.1.3 *Underlying Molecular Mechanisms*

The molecular mechanisms of transgene silencing or silencing of endogenous metastable epialleles such as *Agouti viable yellow* are still not completely understood; however, they show many features common to heterochromatin, including DNA methylation, chromatin packaging alterations including alterations in histone methylation, and acetylation (Fig. 3D) (Elliott et al. 1995; Garrick et al. 1996; Morgan et al. 1999; Sutherland et al. 2000; Blewitt et al. 2005; Blewitt et al. 2006; Dolinoy et al. 2010). Because these alleles show many of the same features as silencing at other sites throughout the genome, transgenes and endogenous alleles can be used as reporter alleles to test the effect of genetic or environmental alterations on epigenetic silencing. These sorts of studies have been performed using transgenic lines (Festenstein et al. 1999; Gaudet et al. 2004; Blewitt et al. 2005; Chong et al. 2007; Ashe et al. 2008; Whitelaw et al. 2010a; Youngson et al. 2011), and the *Agouti viable yellow* allele (Wolff 1978; Waterland and Jirtle 2003; Gaudet et al. 2004; Blewitt et al. 2005; Dolinoy et al. 2006; Chong et al. 2007; Kaminen-Ahola et al. 2010). Some of these studies were discussed in Section 1.

2.2 Variable Expressivity

In addition to variegation, many transgenic lines and endogenous metastable epialleles display variable expressivity (see Table 2), which can be defined as the differential expression of a gene between individuals. The term was originally used to describe a scenario, frequently observed in humans, in which patients with the same genetic aberration show variable severity of disease. In these cases, genetic

Cite as *Cold Spring Harb Perspect Biol* doi: 10.1101/cshperspect.a017939

heterogeneity at unlinked modifier loci (or quantitative trait loci [QTLs]) has been presumed (and in many cases, found) to be the cause of the phenotypic diversity. Variable expressivity at metastable epialleles occurs even when the organisms are inbred (ostensibly isogenic) and reared in controlled environments, suggesting that something in addition to genetic heterogeneity and environmental factors is involved.

Mice carrying the *Agouti viable yellow* allele show a spectrum of coat colors from normal agouti colored coats through various degrees of mottling to a completely yellow coat (Fig. 3B,C). The spectrum of phenotypes is associated with a range of Agouti expression; in the agouti mice the IAP LTR has been silenced in all cells, leaving the normal *Agouti* promoters and enhancers to drive expression of Agouti only during a short phase of the hair-growth cycle. In the yellow colored mice, the active IAP LTR drives constitutive expression of Agouti. Because Agouti is a signaling molecule in other pathways, these yellow mice also show other pleiotropic effects such as diabetes and obesity. The stochastic epigenetic silencing of the IAP LTR in some mice, but not others, results in phenotypes ranging from the extremes of obese and diabetic with a yellow coat through to indistinguishable from wild type (Fig. 3B).

A similar situation occurs for several other metastable epialleles that are also driven by IAP insertions: for example, *Axin fused* ($Axin^{fu}$) (Vasicek et al. 1997) and *murine CDK5 activator binding protein IAP* ($mCABP^{IAP}$; Druker et al. 2004). The IAP insertion at *Axin* is in intron 6 of the gene. When the IAP LTR is active, it produces a truncated transcript and protein, which correlates with a kinked backbone (Rakyan et al. 2003). Animals carrying the $Axin^{fu}$ allele range in phenotype from having a severely kinked backbone, most easily observed in the tail, to wild type in appearance. As for A^{vy}, this correlates with epigenetic state and activity of the fused IAP LTR (Rakyan et al. 2003). The IAP insertion at $mCABP^{IAP}$ was also found in intron 6 of the gene. In this case, no visible phenotype associated the allele has been reported; indeed, this allele is found in all C57BL/6 animals. Similar to $Axin^{fu}$, $mCABP^{IAP}$ produces an aberrant truncated transcript that initiates in the IAP LTR found in intron 6, in addition to the wild-type transcript. The expression of the truncated transcript varies between genetically identical littermates and is associated with hyper or hypomethylation of the IAP LTR (Druker et al. 2004). In the above cases, these transcriptional differences are thought to be the direct result of the different epigenetic marks found at the retrotransposable element in each case.

These findings raise the intriguing possibility that some proportion of the variable penetrance observed in humans is due to stochastic epigenetic silencing rather than effects of QTLs. Several approaches have been taken to identify loci that are epigenetically metastable in humans. These studies have primarily focused on studying genetically identical monozygotic twins (Bell and Spector 2011). Monozygotic twins show a high degree of discordance for many traits. In one case, a twin pair was identified discordant for a caudal duplication anomaly. Given the role of Axin in axis formation (Zeng et al. 1997), *AXIN1* was sequenced but no mutation was found (Kroes et al. 2002). Rather, the affected twin displayed DNA hypermethylation of the *AXIN1* promoter, compared with the unaffected twin (Oates et al 2006). This suggests that, at least in this case, stochastic epigenetic events that occur in the context of genetic identity can influence human disease (Oates et al. 2006). Further studies have looked for differences in DNA methylation within monozygotic twin pairs genome-wide. There is evidence for a small degree of metastability in DNA methylation between monozygotic twins (Fraga et al. 2005; Mill et al. 2006; Kaminsky et al. 2009) and some evidence that this increases with age (Fraga et al. 2005). These types of studies provide some support for the notion that epigenetic silencing may contribute to variable expressivity observed in humans, and possibly even to complex disease (Waterland et al. 2010; Pujadas and Feinberg 2012). In addition to genome-wide association studies, epigenome-wide association studies are now beginning to identify loci in which DNA methylation state might contribute to variable expressivity in humans (Rakyan et al. 2011).

2.3 Parent-of-Origin Effects

Metastable epialleles also frequently display subtle parent-of-origin effects (Rakyan et al. 2002). These differ from traditional parental imprinting because they are not strictly monoallelically expressed. Rather, the allele is preferentially in the active state when inherited from one parent, compared with the other. In cases in which transgenes display variegation, the degree of variegation is reduced by a small amount following transmission from the mother compared with the father (Reik et al. 1987; Sapienza et al. 1987; Preis et al. 2003; Williams et al. 2008). In cases in which the allele displays variable expressivity, the spectrum of phenotypes shifts slightly depending on the parent-of-origin. This has been observed for several transgenes (Allen et al. 1990; Weichman and Chaillet 1997; Kearns et al. 2000), along with all of the *Agouti* coat color alleles (Wolff 1978; Duhl et al. 1994; Argeson et al. 1996; Morgan et al. 1999) and the $Axin^{fu}$ allele (Reed 1937; Belyaev et al. 1981; Rakyan et al. 2003). For the *Axin fused* allele, the allele is 30% more likely to be active (producing a kinked tail) following paternal transmission, compared with maternal transmission. The converse is true for the *Agouti viable yellow* allele, which displays subtle paternal imprinting; there is a 15% greater likelihood

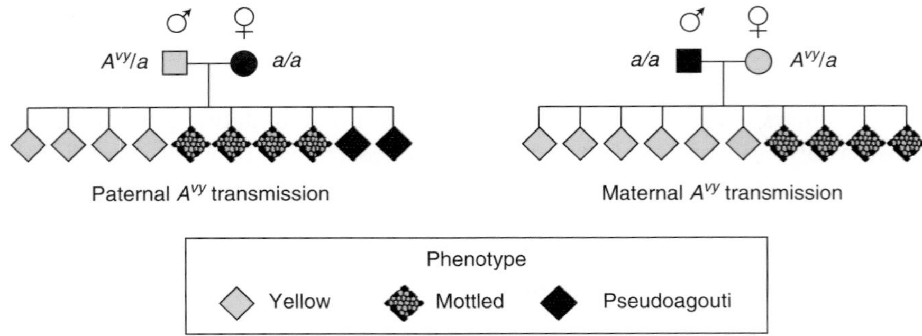

Figure 6. Subtle parent-of-origin effects at the *Agouti viable yellow* allele. Pedigrees showing the proportion of yellow, mottled, and pseudoagouti animals following paternal or maternal transmission of the allele. The *a/a* offspring have been excluded from these pedigrees for simplicity. More yellow offspring are observed following maternal transmission of the allele indicating a lower proportion of IAP LTR silencing. This illustrates what is termed a subtle parent-of-origin effect as incomplete erasure of the IAP LTR heterochromatic environment appears to occur during male transmission of the A^{vy} allele.

of expression when the allele is inherited from the mother, compared with the father (see Fig. 6).

These small changes in expression, dependent on parent-of-origin, were detectable because of the sensitivity of the assays used to measure transgene expression, or the dramatic effect on phenotype observed with endogenous metastable epialleles. The observation of this type of parental effect in mice suggested that that the maternal and paternal genomes might be differentially marked at more loci than only those subject to traditional parental imprinting (Pardo-Manuel de Villena et al. 2000; Ashe 2006). Indeed, very recent data using RNA-seq in the embryonic and adult mouse brain suggests that >1300 genes display parent-of-origin allelic effects (Gregg et al. 2010). Although it is not yet clear how widespread such effects are in humans, these results again show that epigenetic phenomena, first suggested by studies using reporter strains of mice, may be true more broadly throughout the genome.

2.4 Transgenerational Epigenetic Inheritance via the Gamete

Perhaps the most controversial of the effects observed at metastable epialleles is transgenerational epigenetic inheritance via the gametes. This term describes any situation in which the epigenetic state of the parent influences the epigenetic state of the offspring, but cannot be attributed to *cis*- or *trans*-acting genetic variation or mutations. This is a fascinating phenomenon that has intrigued biologists for many decades.

Transgenerational epigenetic inheritance has been reported following maternal transmission of transgenes (Allen et al. 1990; Kearns et al. 2000; Sutherland et al. 2000), maternal transmission of the A^{vy} allele (Wolff 1978; Morgan et al. 1999), and both maternal *and* paternal transmission of the $Axin^{fu}$ allele (Reed 1937; Belyaev et al. 1981; Rakyan et al. 2003). Therefore, transmission via either germline is capable of producing transgenerational epigenetic inheritance.

The A^{vy} allele is the most frequently studied case of transgenerational epigenetic inheritance. Yellow-colored females, which have an active A^{vy} allele, produce 60% yellow-coated offspring and 40% mottled offspring. Genetically identical agouti-colored females, with a silent A^{vy} allele, produce 40% yellow-colored, 40% mottled, and 20% agouti-colored offspring. The epigenotype of the mother at the A^{vy} IAP LTR alters the spectrum of epigenotypes observed in her offspring (see Fig. 7) (Morgan et al. 1999).

Using mice as a model, it has been possible to exclude confounding effects such as intrauterine environment or distant genetic alterations influencing epigenetic state. The A^{vy} allele displays transgenerational epigenetic inheritance, even in the inbred C57BL/6 background, which greatly reduces the likelihood of a genetic contribution to this effect (Morgan et al. 1999). Morgan and coworkers also showed that the intrauterine environment of the yellow female was not responsible. Because mice expressing Agouti constitutively (i.e., a yellow mouse) display pleiotropic effects, including obesity and diabetes, it was possible that development within the uterus of such a mouse may alter the epigenotype and, hence, phenotype of her offspring. To rule out this possibility, zygotes were transferred from yellow dams into congenic females that did not carry the A^{vy} allele. The transferred embryos developed the same spectrum of coat colors (60% yellow coats, 40% mottled coats) as those pups that were left in the uterus of their original yellow dams. This work showed that the uterine environment was not responsible for the unusual transgenerational epigenetic inheritance observed at the A^{vy} allele.

Cite as *Cold Spring Harb Perspect Biol* doi: 10.1101/cshperspect.a017939

Mothers with the same genotype but different epigenotype, hence phenotype

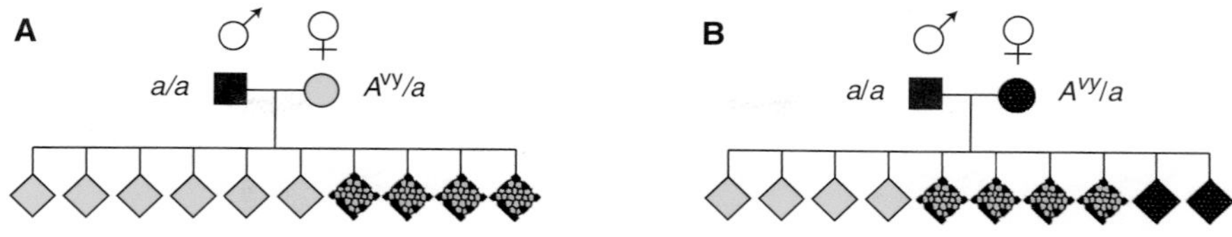

Distinct patterns of transgenerational inheritance:
a different spectrum of phenotypes are observed based on the epigenotype of the mother

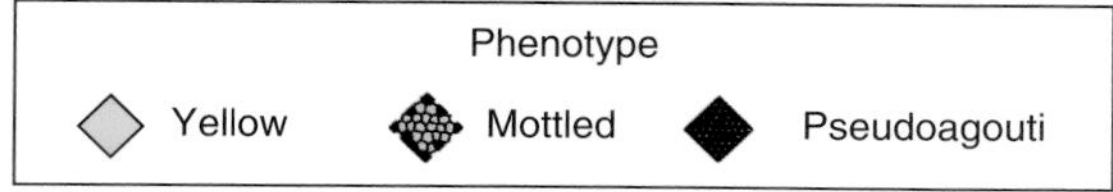

Figure 7. Transgenerational epigenetic inheritance at the *Agouti viable yellow* allele. Pedigrees showing the proportion of yellow, mottled, and pseudoagouti animals following transmission of the allele from an A^{vy} heterozygous mother with a yellow (*A*) or pseudoagouti (*B*) phenotype. The *a/a* offspring have been excluded from these pedigrees for simplicity. Yellow mothers, with an IAP promoter in an open chromatin configuration making the locus constitutively active in all cells, produce more yellow offspring (60%) than pseudoagouti mothers do (40%), in which the IAP promoter is in a repressive chromatin configuration and the promoter is consistently off. This difference is the visual result of transgenerational epigenetic inheritance, in which somehow the epigenetic configuration of the *Agouti* IAP LTR affects the ratio of phenotypes obtained in the next generation.

Epigenetic reprogramming (as discussed in Ch. 28 [Hochedlinger and Jaenisch 2014]) occurs between generations during both primordial germ cell development and preimplantation development. The instances of transgenerational epigenetic inheritance described above suggest that incomplete erasure of epigenetic marks between generations can sometimes occur. Working with mouse models, it is possible to study the transmission of epigenetic marks through germ cell development and embryonic development, and this has been performed for DNA methylation at the A^{vy} allele (Blewitt et al. 2006). The removal of the vast majority of histone proteins from sperm DNA made DNA methylation an ideal candidate for the epigenetic mark passed between generations.

The A^{vy} (Blewitt et al. 2006) and *Axin*^{fu} alleles (Rakyan et al. 2003) show similar levels of DNA methylation in mature gametes as in the somatic tissues suggesting that DNA methylation at the controlling IAP LTRs escape erasure during primordial germ cell development just as the bulk of IAPs do throughout the genome (Lane et al. 2003). However, DNA methylation was found to be completely cleared at the A^{vy} allele during preimplantation development (Blewitt et al. 2006). These studies suggest that DNA methylation is not the inherited mark at the A^{vy} allele, although they do not exclude the possibility that DNA methylation may transmit the heritable mark to other epigenetic modifications that are yet to be an analyzed.

This study also found that haploinsufficiency for Mel18, a polycomb repressive complex 1 (PRC1) component, produced epigenetic inheritance in cases in which it is not normally observed (Blewitt et al. 2006), raising the possibility that an altered makeup of PRC1 may decrease the efficiency of epigenetic reprogramming during early development, and implicate histone modifications in the process. The molecular processes underlying transgenerational epigenetic inheritance via the gametes remain unclear. Many reviews have been written about this with the most recent being Daxinger and Whitelaw (2012). We have summarized the incomplete erasure model in Figure 8A.

2.5 Paramutation-Like Effects in Mice

In a distinct, but related phenomenon, described as paramutation-like (see Ch. 13 [Pikaard and Mittelsten Scheid 2014] for paramutation in plants), RNA molecules in the sperm of mice have been implicated in the epigenetic events that are heritable across multiple generations (Rassoulzadegan et al. 2006). The initial study involved a mutant allele of the *Kit* locus. Kit is a receptor tyrosine kinase that is required for stem cells of various types. Reduced levels of Kit are associated with white spotting, particularly at the extremities, in many mammalian species including mice, and this is caused by reduced numbers of melanocyte precursors during development. Heterozygosity for the *Kit*^{tm1Alf} allele is associated with a white tail. The investigators found that following the cross of mice heterozygous for this null allele at the *Kit* locus, *Kit*^{tm1Alf}, with congenic wild-type littermates, some of the genetically wild-type offspring had white tails. Abnormal *Kit* mRNA transcripts were detected in sperm. The investigators injected microRNAs known to

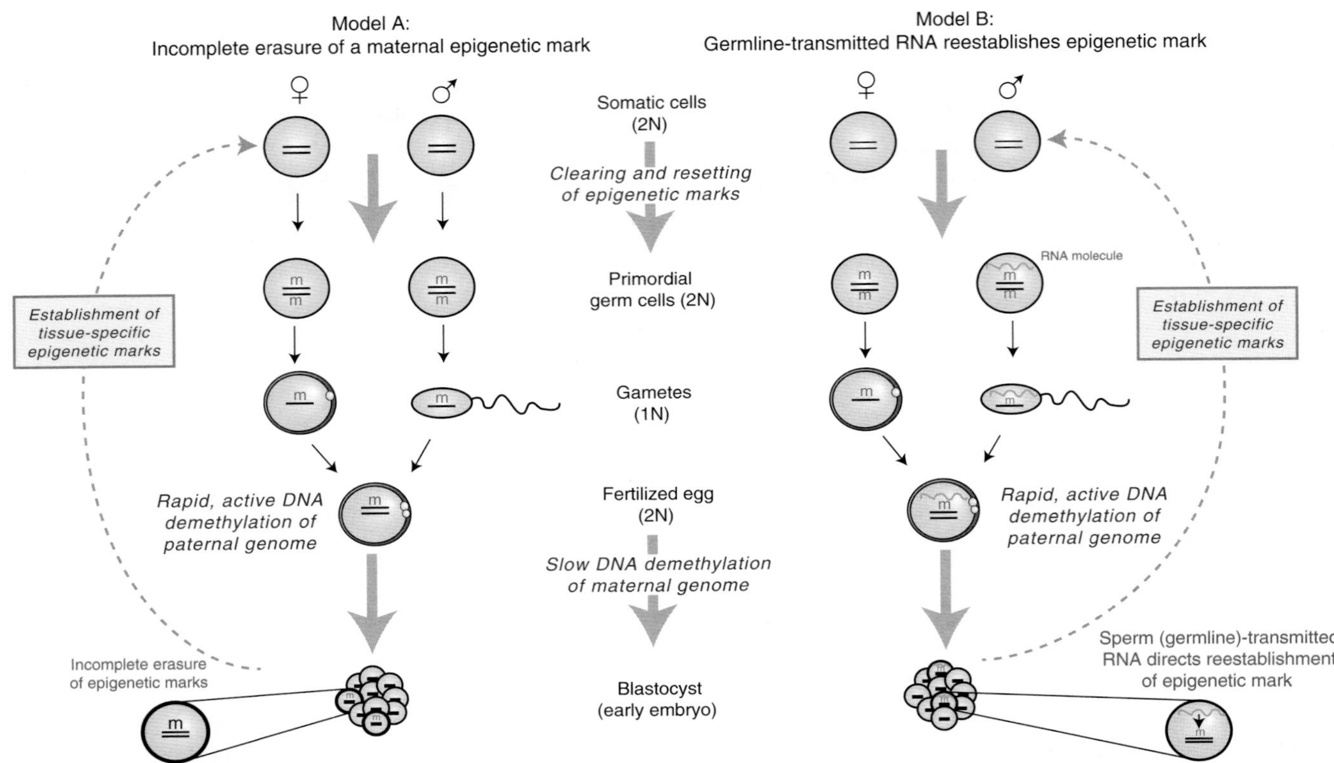

Figure 8. Models for transgenerational epigenetic inheritance. Only the gene showing transgenerational epigenetic inheritance is shown for simplicity. "m" denotes the epigenetic mark, in which pink shows the maternal mark and blue shows the paternal mark. Epigenetic reprogramming occurs in two phases (indicated by gray arrows): first, in primordial germ cell development and second, in preimplantation development. Model A: Incomplete erasure occurs of either a DNA methylation mark or histone mark. In a scenario in which transgenerational epigenetic inheritance is observed following maternal transmission (*left* gamete), some cells do not show complete removal of the epigenetic mark, and retain the inherited epigenetic mark when compared with other cells in the same blastocyst. In the instance in which there is complete erasure of epigenetic marks at a locus and transgenerational epigenetic inheritance does not occur all cells of the blastocyst would be unmarked (not shown). Model B: Germline RNA causes reestablishment of the inherited epigenetic mark. In a scenario in which transgenerational epigenetic inheritance is observed following paternal transmission, an RNA molecule is transmitted from the primordial germ cell to the mature sperm. This RNA is then transmitted to the fertilized egg and causes reestablishment of the paternal epigenetic mark in some cells that inherit the RNA molecule.

target *Kit* mRNA into zygotes and found white tails in the subsequent offspring. This implicates some long-lived effect of the RNA, presumably by epigenetic alteration of gene expression. Similar paramutation-like phenomena have been detected at some other loci in the mouse (Wagner et al. 2008; Grandjean et al. 2009). Given the heritability of this effect, these results suggest that the RNA content of sperm may play some role as an intermediary in transgenerational epigenetic inheritance. We have summarized that a model of transgenerational epigenetic inheritance occurs via an RNA molecule in Figure 8, Model B.

Phenotypic events reminiscent of transgenerational epigenetic inheritance have been reported in human populations (Lumey 1992; Pembrey et al. 2006); however, in these cases, there is no evidence that these are bona fide cases of transgenerational epigenetic inheritance via the gametes. In humans, it is impossible to adequately control for genetic makeup or confounding in utero effects, or to directly observe a failure to clear epigenetic marks in germ cell development and preimplantation development. Therefore, it remains to be established whether transgenerational epigenetic inheritance truly exists in humans. The existence of such an effect in humans would have profound implications for studies on the heritability of human phenotype and disease, and so there is considerable interest in this area, and it is probably only through mouse studies that we will be able to elucidate the mechanistic basis for such observations.

2.6 Position Effects

The term "position-effect variegation" (PEV) was coined for a phenomenon observed in *Drosophila* (see Ch. 12 [Elgin and Reuter 2013]) and yeast (see Ch. 8 [Grunstein and Gasser 2013]; Ch. 9 [Allshire and Ekwall 2014]). PEV has mostly been studied using translocations of the *white* gene in flies. Expression of the *white* gene was found to be dependent on its location with respect to pericentric heterochromatin. If the *white* gene (which produces red pigment in the eye) neighbored heterochromatin, then the *white* gene expression was frequently variegated, producing a mosaic red and white eye. This was attributed to incomplete spreading of neighboring heterochromatin through the *white* gene (Henikoff 1990). This interesting phenotype has been studied in classic mutagenesis screens to identify key components involved in silencing and spreading of heterochromatin (Schotta et al. 2003). Transgenes in mice are particularly susceptible to PEV, as described in Section 2.1; variegating transgenes have been located in repeats (Sutherland et al. 2000), near centromeres (Dobie et al. 1996), and near telomeres (Zhuma et al. 1999; Pedram et al. 2006; Gao et al. 2007), all of which are heterochromatic regions. The importance of this juxtaposition to heterochromatin is not known. Nevertheless, these studies show that heterochromatin has the capacity to spread in mammals, as it does in *Drosophila*.

If the aim of making a transgenic mouse is to achieve predictable expression of the gene under the control of a specific promoter, the integration site dependence of transgene expression is clearly a problem. Several groups have used transgenesis to identify the sequence elements required to produce integration-site-independent transgene expression (e.g., for the β-globin gene). In this case, there is interest in finding the controlling elements for gene therapy purposes. The sequence elements found to protect the transgene from PEV are known as locus control regions (LCRs). LCRs perform this function by somehow ensuring an open chromatin structure throughout the transgene, in a cell lineage-dependent manner (Kioussis and Festenstein 1997). There have been two main theories about how LCRs function: first, by direct contact between the LCR and the promoter, or second, by LCRs producing a general opening of chromatin in the region. With the development of new chromosome capture techniques (Carter et al. 2002), there is now some consensus that chromatin looping enables the β-globin LCR to interact with promoters. A recent study has found that such long-range DNA interactions are not detectable in every cell, and so at least for the β-globin locus, long-range interactions that occur in only a subset of cells could be responsible for variegation (Noordermeer et al. 2011).

2.7 Repeat-Induced Gene Silencing

The number of copies of a transgene that integrate as a concatemer at a single integration site also influences expression. Although there appears to be some interaction between the integration site and the copy number on transgene expression (Williams et al. 2008), in general, extremely high copy number transgenes show low or no expression per transgene copy. Garrick and colleagues (Garrick et al. 1998) investigated this effect using Cre-mediated deletion of two very low-expression, high-copy-number (>100) α-globin-driven LacZ transgenes. When these transgenes were reduced to five copies and one copy, respectively, the proportion of expressing red blood cells jumped from $<1\%$ to $>50\%$ with a concurrent decrease in DNA methylation and opening of the chromatin structure at the transgene. This finding was the first evidence that repeat-induced gene silencing, already reported in plants and flies (Henikoff 1998), exists in mammals.

Does repeat-induced gene silencing influence the expression of endogenous genes? There are many repeated gene families in mammals (e.g., rRNA genes, histone genes). Although often arranged as tandem repeats, similar to those seen at transgene arrays, these do not appear to be subject to repeat-induced gene silencing. Instead, there are specialist forms of epigenetic control that exist in these families (Keverne 2009). Interestingly, the D segment genes of the immunoglobulin heavy chain genes are also arranged as tandem repeats and show many of the hallmarks of regions subject to repeat-induced gene silencing found in lower organisms and at centromeric heterochromatin, such as antisense transcription and dimethylation of histone 3 at lysine 9 (H3K9me2; Chakraborty et al. 2007). The investigators propose that the D segments may be subject to repeat-induced silencing.

For transgene silencing and silencing at centromeres and telomeres, the repeats are clustered. In other circumstances, such as for endogenous retroviruses or retrotransposons, the repeats are spread throughout the genome with single copies in thousands of different locations. These dispersed repeats are frequently silenced throughout development, presumably to protect the genome from the mutagenic effects that result from active mobile elements, and additionally to protect neighboring regions from the strong promoters in these retroelements. The silencing in this case must occur in *trans* either by physical contact (e.g., looping) or by diffusible elements such as RNA. When this silencing is inefficient, it is possible for readthrough transcription to occur just as it does at the discussed endogenous metastable epialleles (e.g., A^{vy}, $Axin^{fu}$, and $mCabp^{IAP}$). The IAPs found at these metastable epialleles are all of a young class, in evolutionary terms, which may explain their different behavior compared with the rest of the complement of dispersed repeats. Most ancient dispersed retroviral repeats have accumulated mutations that make them nonfunctional.

2.8 The Effects of Environment on Metastable Epialleles

The finding that expression at metastable epialleles can be influenced by the environment has stimulated much research. In the case of A^{vy}, the decision about whether the locus will be "on" or "off" is made in the early postimplantation embryo (Blewitt et al 2006), and the percentage of yellow pups in a litter has been shown to decrease as a result of exposure of the female parent to a diet rich in methyl groups (folic acid, betaine, vitamin B12; Wolff et al. 1998; Cooney et al. 2002; Waterland and Jirtle 2003), genistein (Dolinoy et al. 2006), or ethanol (Kaminen-Ahola et al. 2010). It is interesting to compare these findings with those made decades ago at the variegating $w^{mottled}$ locus in *Drosophila*. Increasing the temperature during development has been shown to suppress variegation (reduce silencing) at this locus (see Ch. 12 [Elgin and Reuter 2013]). Another metastable epiallele, $Axin^{fu}$, has also been shown to be sensitive to maternal diet (Waterland et al. 2006). These findings emphasize the plasticity of the epigenome during early development, providing an opportunity for the environment to permanently influence an individual's phenotype. As such, and because it will be possible to document the epigenome in a particular tissue of an individual in fine detail, the epigenome may provide us with the capacity to infer certain facts about an individual's past and better predict disease risk for that individual (see Fig. 9). However, it remains unclear how much of the epigenome in any particular tissue will turn out to be both susceptible to environmental events and then stable for life, both of which are required for the epigenome to have predictive value.

2.9 "Intangible Variation"

Intangible variation, or developmental noise, is defined as the variation in phenotype that is unable to be explained by contributions from genotype and the environment alone (Falconer 1989). This type of variation is most easily observed when both genotype and environment are invariant (e.g., in inbred strains of mice [or other laboratory organisms] housed in standard conditions). Clearly, variable expressivity can be considered a case of intangible variation.

Intangible variation is more widespread than the few known instances of metastable epialleles. In inbred strains of mice, many quantitative traits conform to a bell-shaped curve despite the fixed environment and genetic background. Indeed, ~80% of the variance in body weight of inbred mice has been attributed to intangible variation rather than genetic or environmental differences (Gartner 1990). Gartner refers to this as "the third component", the first and second being genetics and environment, respectively. Other examples include incomplete penetrance of a phenotype associated with a particular genotype in inbred mice (Biben et al. 2000), and transmission ratio distortion owing to lethality of some but not all animals of a particular genotype in inbred mice (Carpinelli et al. 2004; Blewitt et al. 2005). Furthermore, regressing embryos are observed in isogenic lines of mice at far higher frequency than expected because of de novo lethal mutations, and independent of uterine position. All of these are examples of intangible variation at a tissue or organism level. The concept of intangible variation has emerged by studying inbred strains of mice.

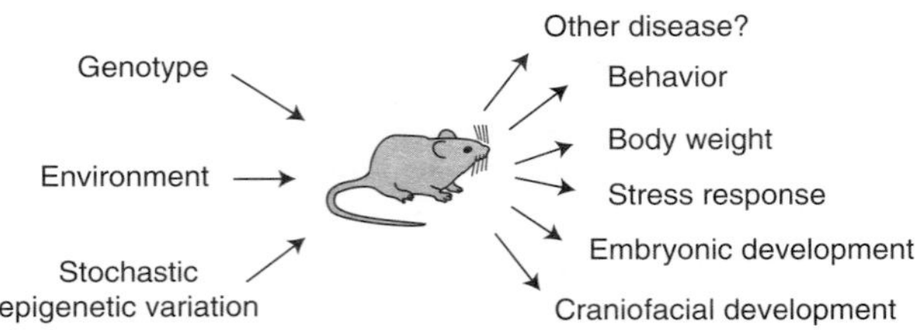

Figure 9. Contribution of stochastic epigenetic variation, genotype, and the environment to phenotype. Genotype, the environment (sometimes by altering epigenotype), and stochastic epigenetic variation can alter the phenotype of the mouse. Known and measurable differences that are a consequence of altered epigenetic control include body weight, behavior, stress responses, craniofacial development, and embryonic development.

What is the molecular basis for this intangible variation or phenotypic noise? Given the stochastic nature of gene expression that is responsible for the variable expressivity at a cellular level observed at many transgenes and endogenous metastable epialleles, one plausible explanation is that similar stochastic patterns of gene expression in early development subsequently maintained by epigenetic mechanisms is responsible for many cases of developmental noise (Blewitt et al. 2004). In addition to what is known about metastable epiallele expression, there is now ample evidence from single-cell analyses of gene expression in mammalian cells that gene expression has a stochastic element and inherent noise (e.g., Raj et al. 2006). Furthermore, increased variation at the cellular level in gene expression (Chi and Bernstein 2009) and at the level of the whole organism in quantitative traits is observed when epigenetic modifiers are depleted (Whitelaw et al. 2010b). This implicates epigenetic mechanisms in buffering transcriptional noise, and consequent phenotypic noise (see Fig. 10).

Cite as *Cold Spring Harb Perspect Biol* doi: 10.1101/cshperspect.a017939

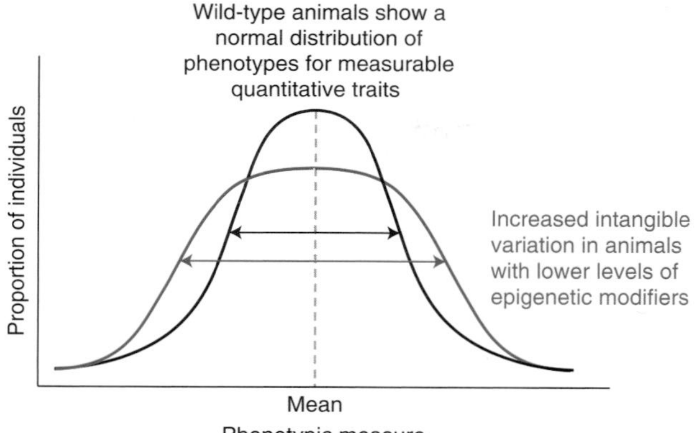

Figure 10. Intangible variation is increased in animals with decreased levels of epigenetic modifiers. Measurable phenotypes in wild-type animals conform to a normal distribution (bell-shaped curve) around a mean measurement (black line). Animals with lower levels of epigenetic modifiers display increased variance in such measurable phenotypes (red line).

The outcomes of stochastic gene expression and developmental noise can be both positive and negative. Noise can result in cell death or embryonic lethality, but the natural variance produced by stochastic gene expression can also be beneficial by providing plasticity (Eldar and Elowitz 2010). In development, it is commonly believed that external cues signal to homogeneous populations to begin differentiation. However, it is also possible that probabilistic events lead to differences in gene expression that result in phenotypic differences in what are considered homogeneous cell populations (Pujadas and Feinberg 2012). In the hematopoietic system, it has been shown that heterogeneity exists in clonal stem cell populations; importantly, this heterogeneity is not just for the expression of an individual gene, but rather represents differences that exist throughout the genome. In this case, the noise in gene expression appears to help determine lineage choice (Chang et al. 2008). Similar noise-dependent cell fate or developmental patterning decisions seem to exist in several scenarios—for example, neural crest cell fate (Shah et al. 1996) and inner cell mass lineage choice (Morris et al. 2010; Yamanaka et al. 2010). What we have learned about the stochastic mechanisms of gene expression from the study of variegating transgenes and metastable epialleles appears to hold true elsewhere in the genome, and may have important roles in development and differentiation (Pujadas and Feinberg 2012).

3 SUMMARY AND FUTURE DIRECTIONS

Over the last 20 years we have learned much about the epigenetic processes that are integral to development and differentiation in mammals, and much of this has emerged from studies in the laboratory mouse. This has been possible for a number of reasons: mice are relatively easy to rear in controlled environments, they have been maintained as inbred strains for decades, facilities such as the Jackson Laboratories have maintained thousands of different inbred strains for distribution to the research community, the fully annotated genome sequence of the C57BL/6 mouse has been available to all, and perhaps most importantly, their genomes can be manipulated by transgenesis, homologous recombination, and ENU mutagenesis.

Geneticists using mice to understand and model human health and disease did not fully embrace the field of epigenetics until the 1990s. Although gene silencing was being studied before this in the particular cases of X inactivation and parental imprinting, epigenetic processes at autosomal genes were not generally considered even among the developmental biologists. There were reports in the literature of the unusual behavior of some autosomal alleles (e.g., *Agouti viable yellow* and *Axin fused*)—that is, that they displayed variable phenotypes in inbred strains, but these were considered quirky. We now realize that epigenetic processes play an integral part of developmental processes in mammals and much of this has emerged from studying mice with mutations in the genes involved in laying down and reprogramming these marks. ENU mutagenesis screens have identified the genes and, possibly more importantly, have provided models that can be used to study the consequences of disruption of these processes to phenotype.

Studies in mice have stimulated those studying human health and disease to embrace the field of epigenetics and to apply this knowledge to their specific areas of interest. Some human diseases are now considered to be, at least in part, the consequence of mutations in modifiers of epigenetic reprogramming (e.g., Williams syndrome and ICF syndrome). One of the benefits of using the mouse to study epigenetic control is that one can study human-like phenotypes, such as behavior, memory, and learning. Colonies of animals provide opportunities to study reproductive fitness, germline epigenetic reprogramming, and the role of the epigenome in gene–environment interactions and we can look forward to much progress in this area in the near future.

REFERENCES

**Reference is also in this book.*

Allen ND, Norris ML, Surani MA. 1990. Epigenetic control of transgene expression and imprinting by genotype-specific modifiers. *Cell* **61:** 853–861.

* Allshire RC, Ekwall K. 2014. Epigenetic regulation of chromatin states in *Schizosaccharomyces pombe. Cold Spring Harb Perspect Biol* doi: 10.1101/cshperspect.a018770.

Argeson AC, Nelson KK, Siracusa LD. 1996. Molecular basis of the pleiotropic phenotype of mice carrying the *hypervariable yellow* (A^{hvy}) mutation at the *agouti* locus. *Genetics* **142:** 557–567.

Ashe A, Morgan DK, Whitelaw NC, Bruxner TJ, Vickaryous NK, Cox LL, Butterfield NC, Wicking C, Blewitt ME, Wilkins SJ, et al. 2008. A genome-wide screen for modifiers of transgene variegation identifies genes with critical roles in development. *Genome Biol* **9:** R182.

Ashe A, Whitelaw E. 2006. Imprinting: The hidden genome. In *The implicit genome* (ed. Caporale LH), pp. 282–293. Oxford University Press, Oxford.

* Barlow DP, Bartolomei MS. 2014. Genomic imprinting in mammals. *Cold Spring Harb Perspect Biol* **6:** a018382.

Bell JT, Spector TD. 2011. A twin approach to unraveling epigenetics. *Trends Genet* **27:** 116–125.

Belyaev DK, Ruvinsky AO, Borodin PM. 1981. Inheritance of alternative states of the fused gene in mice. *J Hered* **72:** 107–112.

Biben C, Weber R, Kesteven S, Stanley E, McDonald L, Elliott DA, Barnett L, Koentgen F, Robb L, Feneley M, et al. 2000. Cardiac septal and valvular dysmorphogenesis in mice heterozygous for mutations in the homeobox gene Nkx2-5. *Circ Res* **87:** 888–895.

Blewitt ME. 2004. "Epigenetic reprogramming in the mouse." PhD thesis, University of Sydney, Australia.

Blewitt ME, Chong S, Whitelaw E. 2004. How the mouse got its spots. *Trends Genet* **20:** 550–554.

Blewitt ME, Vickaryous NK, Hemley SJ, Ashe A, Bruxner TJ, Preis JI, Arkell R, Whitelaw E. 2005. An *N*-ethyl-*N*-nitrosourea screen for genes involved in variegation in the mouse. *Proc Natl Acad Sci* **102:** 7629–7634.

Blewitt ME, Vickaryous NK, Paldi A, Koseki H, Whitelaw E. 2006. Dynamic reprogramming of DNA methylation at an epigenetically sensitive allele in mice. *PLoS Genet* **2:** e49.

Blewitt ME, Gendrel AV, Pang Z, Sparrow DB, Whitelaw N, Craig JM, Apedaile A, Hilton DJ, Dunwoodie SL, Brockdorff N, et al. 2008. SmcHD1, containing a structural-maintenance-of-chromosomes hinge domain, has a critical role in X inactivation. *Nat Genet* **40:** 663–669.

Bozhenok L, Wade PA, Varga-Weisz P. 2002. WSTF-ISWI chromatin remodeling complex targets heterochromatic replication foci. *EMBO J* **21:** 2231–2241.

Bradl M, Larue L, Mintz B. 1991. Clonal coat color variation due to a transforming gene expressed in melanocytes of transgenic mice. *Proc Natl Acad Sci* **88:** 6447–6451.

* Brockdorff N, Turner B. 2014. Dosage compensation in mammals. *Cold Spring Harb Perspect Biol* doi: 10.1101/cshperspect.a019406.

Brouns MR, De Castro SC, Terwindt-Rouwenhorst EA, Massa V, Hekking JW, Hirst CS, Savery D, Munts C, Partridge D, Lamers W, et al. 2011. Over-expression of *Grhl2* causes spina bifida in the *Axial defects* mutant mouse. *Hum Mol Genet* **20:** 1536–1546.

Brown FC, Scott N, Rank G, Collinge JE, Vadolas J, Vickaryous N, Whitelaw N, Whitelaw E, Kile BT, Jane SM, Curtis DJ. 2013. ENU mutagenesis identifies the first mouse mutants reproducing human β-thalassemia at the genomic level. *Blood Cells Mol Dis* **50:** 86–92.

Bultman S, Gebuhr T, Yee D, La Mantia C, Nicholson J, Gilliam A, Randazzo F, Metzger D, Chambon P, Crabtree G, et al. 2000. A Brg1 null mutation in the mouse reveals functional differences among mammalian SWI/SNF complexes. *Mol Cell* **6:** 1287–1295.

Bultman SJ, Gebuhr TC, Magnuson T. 2005. A Brg1 mutation that uncouples ATPase activity from chromatin remodeling reveals an essential role for SWI/SNF-related complexes in β-globin expression and erythroid development. *Genes Dev* **19:** 2849–2861.

Burgoyne PS, Thornhill AR, Boudrean SK, Darling SM, Bishop CE, Evans EP. 1995. The genetic basis of XX-XY differences present before gonadal sex differentiation in the mouse. *Philos Trans R Soc Lond B Biol Sci* **350:** 253–260 [discussion 260–261].

Carpinelli MR, Hilton DJ, Metcalf D, Antonchuk JL, Hyland CD, Mifsud SL, Di Rago L, Hilton AA, Willson TA, Roberts AW, et al. 2004. Suppressor screen in $Mpl^{-/-}$ mice: *c-Myb* mutation causes supraphysiological production of platelets in the absence of thrombopoietin signaling. *Proc Natl Acad Sci* **101:** 6553–6558.

Carter D, Chakalova L, Osborne CS, Dai YF, Fraser P. 2002. Long-range chromatin regulatory interactions in vivo. *Nat Genet* **32:** 623–626.

Chakraborty T, Chowdhury D, Keyes A, Jani A, Subrahmanyam R, Ivanova I, Sen R. 2007. Repeat organization and epigenetic regulation of the DH-Cmu domain of the immunoglobulin heavy-chain gene locus. *Mol Cell* **27:** 842–850.

Chang HH, Hemberg M, Barahona M, Ingber DE, Huang S. 2008. Transcriptome-wide noise controls lineage choice in mammalian progenitor cells. *Nature* **453:** 544–547.

Chi AS, Bernstein BE. 2009. Developmental biology. Pluripotent chromatin state. *Science* **323:** 220–221.

Ching YH, Wilson LA, Schimenti JC. 2010. An allele separating skeletal patterning and spermatogonial renewal functions of PLZF. *BMC Dev Biol* **10:** 33.

Chong S, Vickaryous N, Ashe A, Zamudio N, Youngson N, Hemley S, Stopka T, Skoultchi A, Matthews J, Scott HS, et al. 2007. Modifiers of epigenetic reprogramming show paternal effects in the mouse. *Nat Genet* **39:** 614–622.

Cooney CA, Dave AA, Wolff GL. 2002. Maternal methyl supplements in mice affect epigenetic variation and DNA methylation of offspring. *J Nutr* **132:** 2393S–2400S.

Daxinger L, Whitelaw E. 2012. Understanding transgenerational epigenetic inheritance via the gametes in mammals. *Nat Rev Genet* **13:** 153–162.

Daxinger L, Oey H, Apedaile A, Sutton J, Ashe A, Whitelaw E. 2012. A forward genetic screen identifies eukaryotic translation initiation factor 3, subunit H (eIF3h), as an enhancer of variegation in the mouse. *G3* **2:** 1393–1396.

Daxinger L, Harten SK, Oey H, Epp T, Isbel L, Huang E, Whitelaw N, Apedaile A, Sorolla A, Yong J, et al. 2013. An ENU mutagenesis screen identifies novel and known genes involved in epigenetic processes in the mouse. *Genome Biol* **14:** R96.

Dobie KW, Lee M, Fantes JA, Graham E, Clark AJ, Springbett A, Lathe R, McClenaghan M. 1996. Variegated transgene expression in mouse mammary gland is determined by the transgene integration locus. *Proc Natl Acad Sci* **93:** 6659–6664.

Dolinoy DC, Weidman JR, Waterland RA, Jirtle RL. 2006. Maternal genistein alters coat color and protects A^{vy} mouse offspring from obesity by modifying the fetal epigenome. *Environ Health Perspect* **114:** 567–572.

Dolinoy DC, Weinhouse C, Jones TR, Rozek LS, Jirtle RL. 2010. Variable histone modifications at the A^{vy} metastable epiallele. *Epigenetics* **5:** 637–644.

Druker R, Bruxner TJ, Lehrbach NJ, Whitelaw E. 2004. Complex patterns of transcription at the insertion site of a retrotransposon in the mouse. *Nucleic Acids Res* **32:** 5800–5808.

Duhl DM, Vrieling H, Miller KA, Wolff GL, Barsh GS. 1994. Neomorphic agouti mutations in obese yellow mice. *Nat Genet* **8:** 59–65.

Eldar A, Elowitz MB. 2010. Functional roles for noise in genetic circuits. *Nature* **467:** 167–173.

* Elgin SCR, Reuter G. 2013. Position-effect variegation, heterochromatin formation, and gene silencing in *Drosophila*. *Cold Spring Harb Perspect* **5:** a17780.

Elliott JI, Festenstein R, Tolaini M, Kioussis D. 1995. Random activation of a transgene under the control of a hybrid hCD2 locus control region/Ig enhancer regulatory element. *EMBO J* **14:** 575–584.

Essien FB, Haviland MB, Naidoff AE. 1990. Expression of a new mutation (*Axd*) causing axial defects in mice correlates with maternal phenotype and age. *Teratology* **42:** 183–194.

Falconer DS. 1989. *Introduction to quantitative genetics*, 3rd ed. Wiley, New York.

Festenstein R, Tolaini M, Corbella P, Mamalaki C, Parrington J, Fox M, Miliou A, Jones M, Kioussis D. 1996. Locus control region function and heterochromatin-induced position effect variegation. *Science* **271:** 1123–1125.

Festenstein R, Sharghi-Namini S, Fox M, Roderick K, Tolaini M, Norton T, Saveliev A, Kioussis D, Singh P. 1999. Heterochromatin protein 1 modifies mammalian PEV in a dose- and chromosomal-context-dependent manner. *Nat Genet* **23:** 457–461.

Fitch KR, Yasuda GK, Owens KN, Wakimoto BT. 1998. Paternal effects in *Drosophila*: Implications for mechanisms of early development. *Curr Top Dev Biol* **38:** 1–34.

Fodor BD, Shukeir N, Reuter G, Jenuwein T. 2010. Mammalian *Su(var)* genes in chromatin control. *Annu Rev Cell Dev Biol* **26:** 471–501.

Fraga MF, Ballestar E, Paz MF, Ropero S, Setien F, Ballestar ML, Heine-Suner D, Cigudosa JC, Urioste M, Benitez J, et al. 2005. Epigenetic differences arise during the lifetime of monozygotic twins. *Proc Natl Acad Sci* **102:** 10604–10609.

Gao Q, Reynolds GE, Innes L, Pedram M, Jones E, Junabi M, Gao DW, Ricoul M, Sabatier L, Van Brocklin H, et al. 2007. Telomeric transgenes are silenced in adult mouse tissues and embryo fibroblasts but are expressed in embryonic stem cells. *Stem Cells* **25:** 3085–3092.

Garrick D, Sutherland H, Robertson G, Whitelaw E. 1996. Variegated expression of a globin transgene correlates with chromatin accessibility but not methylation status. *Nucleic Acids Res* **24:** 4902–4909.

Garrick D, Fiering S, Martin DI, Whitelaw E. 1998. Repeat-induced gene silencing in mammals. *Nat Genet* **18:** 56–59.

Gartner K. 1990. A third component causing random variability beside environment and genotype. A reason for the limited success of a 30 year long effort to standardize laboratory animals? *Lab Anim* **24:** 71–77.

Gaudet F, Rideout WM 3rd, Meissner A, Dausman J, Leonhardt H, Jaenisch R. 2004. Dnmt1 expression in pre- and postimplantation embryogenesis and the maintenance of IAP silencing. *Mol Cell Biol* **24:** 1640–1648.

Grandjean V, Gounon P, Wagner N, Martin L, Wagner KD, Bernex F, Cuzin F, Rassoulzadegan M. 2009. The *miR-124-Sox9* paramutation: RNA-mediated epigenetic control of embryonic and adult growth. *Development* **136:** 3647–3655.

Gregg C, Zhang J, Weissbourd B, Luo S, Schroth GP, Haig D, Dulac C. 2010. High-resolution analysis of parent-of-origin allelic expression in the mouse brain. *Science* **329:** 643–648.

* Grossniklaus U, Paro R. 2014. Transcriptional silencing by *Polycomb*-group proteins. *Cold Spring Harb Perspect Biol* **6:** a019331.

* Grunstein M, Gasser SM. 2013. Epigenetics in *Saccharomyces cerevisiae. Cold Spring Harb Perspect Biol* **5:** a017491.

Henikoff S. 1990. Position-effect variegation after 60 years. *Trends Genet* **6:** 422–426.

Henikoff S. 1998. Conspiracy of silence among repeated transgenes. *Bioessays* **20:** 532–535.

Hitotsumachi S, Carpenter DA, Russell WL. 1985. Dose-repetition increases the mutagenic effectiveness of *N*-ethyl-*N*-nitrosourea in mouse spermatogonia. *Proc Natl Acad Sci* **82:** 6619–6621.

* Hochedlinger K, Jaenisch R. 2014. Induced pluripotency and epigenetic reprogramming. *Cold Spring Harb Perspect Biol* doi: 10.1101/cshperspect.a019448.

Hummel KP. 1959. Developmental anomalies in mice resulting from action of the gene, *disorganization*, a semi-dominant lethal. *Pediatrics* **23:** 212–221.

Jabbi M, Kippenhan JS, Kohn P, Marenco S, Mervis CB, Morris CA, Meyer-Lindenberg A, Berman KF. 2012. The Williams syndrome chromosome 7q11.23 hemideletion confers hypersocial, anxious personality coupled with altered insula structure and function. *Proc Natl Acad Sci* **109:** E860–E866.

Jun JE, Wilson LE, Vinuesa CG, Lesage S, Blery M, Miosge LA, Cook MC, Kucharska EM, Hara H, Penninger JM, et al. 2003. Identifying the MAGUK protein Carma-1 as a central regulator of humoral immune responses and atopy by genome-wide mouse mutagenesis. *Immunity* **18:** 751–762.

Kaminen-Ahola N, Ahola A, Maga M, Mallitt KA, Fahey P, Cox TC, Whitelaw E, Chong S. 2010. Maternal ethanol consumption alters the epigenotype and the phenotype of offspring in a mouse model. *PLoS Genet* **6:** e1000811.

Kaminsky ZA, Tang T, Wang SC, Ptak C, Oh GH, Wong AH, Feldcamp LA, Virtanen C, Halfvarson J, Tysk C, et al. 2009. DNA methylation profiles in monozygotic and dizygotic twins. *Nat Genet* **41:** 240–245.

Kearns M, Preis J, McDonald M, Morris C, Whitelaw E. 2000. Complex patterns of inheritance of an imprinted murine transgene suggest incomplete germline erasure. *Nucleic Acids Res* **28:** 3301–3309.

Keverne B. 2009. Monoallelic gene expression and mammalian evolution. *Bioessays* **31:** 1318–1326.

Kioussis D, Festenstein R. 1997. Locus control regions: Overcoming heterochromatin-induced gene inactivation in mammals. *Curr Opin Genet Dev* **7:** 614–619.

Kitagawa H, Fujiki R, Yoshimura K, Mezaki Y, Uematsu Y, Matsui D, Ogawa S, Unno K, Okubo M, Tokita A, et al. 2003. The chromatin-remodeling complex WINAC targets a nuclear receptor to promoters and is impaired in Williams syndrome. *Cell* **113:** 905–917.

Kroes HY, Takahashi M, Zijlstra RJ, Baert JA, Kooi KA, Hofstra RM, van Essen AJ. 2002. Two cases of the caudal duplication anomaly including a discordant monozygotic twin. *Am J Med Genet* **112:** 390–393.

Kwabi-Addo B, Chung W, Shen L, Ittmann M, Wheeler T, Jelinek J, Issa JP. 2007. Age-related DNA methylation changes in normal human prostate tissues. *Clin Cancer Res* **13:** 3796–3802.

Lane N, Dean W, Erhardt S, Hajkova P, Surani A, Walter J, Reik W. 2003. Resistance of IAPs to methylation reprogramming may provide a mechanism for epigenetic inheritance in the mouse. *Genesis* **35:** 88–93.

La Salle S, Mertineit C, Taketo T, Moens PB, Bestor TH, Trasler JM. 2004. Windows for sex-specific methylation marked by DNA methyltransferase expression profiles in mouse germ cells. *Dev Biol* **268:** 403–415.

Lemmers RJ, Tawil R, Petek LM, Balog J, Block GJ, Santen GW, Amell AM, van der Vliet PJ, Almomani R, Straasheijm KR, et al. 2012. Digenic inheritance of an *SMCHD1* mutation and an FSHD-permissive D4Z4 allele causes facioscapulohumeral muscular dystrophy type 2. *Nat Genet* **44:** 1370–1374.

Leong HS, Chen K, Hu Y, Lee S, Corbin J, Pakusch M, Murphy JM, Majewski IJ, Smyth GK, Alexander WS, et al. 2013. Epigenetic regulator Smchd1 functions as a tumor suppressor. *Cancer Res* **73:** 1591–1599.

Lumey LH. 1992. Decreased birthweights in infants after maternal in utero exposure to the Dutch famine of 1944–1945. *Paediatr Perinat Epidemiol* **6:** 240–253.

Maegawa S, Hinkal G, Kim HS, Shen L, Zhang L, Zhang J, Zhang N, Liang S, Donehower LA, Issa JP. 2010. Widespread and tissue specific age-related DNA methylation changes in mice. *Genome Res* **20:** 332–340.

Majewski IJ, Blewitt ME, de Graaf CA, McManus EJ, Bahlo M, Hilton AA, Hyland CD, Smyth GK, Corbin JE, Metcalf D, et al. 2008. Polycomb repressive complex 2 (PRC2) restricts hematopoietic stem cell activity. *PLoS Biol* **6:** e93.

Martin DI, Whitelaw E. 1996. The vagaries of variegating transgenes. *Bioessays* **18:** 919–923.

Michaud EJ, van Vugt MJ, Bultman SJ, Sweet HO, Davisson MT, Woychik RP. 1994. Differential expression of a new dominant agouti allele (A^{iapy}) is correlated with methylation state and is influenced by parental lineage. *Genes Dev* **8:** 1463–1472.

Mill J, Dempster E, Caspi A, Williams B, Moffitt T, Craig I. 2006. Evidence for monozygotic twin (MZ) discordance in methylation level at two CpG sites in the promoter region of the catechol-*O*-methyltransferase (COMT) gene. *Am J Med Genet B Neuropsychiatr Genet* **141B:** 421–425.

Mintz B. 1970. Clonal expression in allophenic mice. *Symp Int Soc Cell Biol* **9:** 15.

Morgan HD. 1999. "Epigenetic inheritance at the murine agouti locus." PhD thesis, University of Sydney, Australia.

Morgan HD, Sutherland HG, Martin DI, Whitelaw E. 1999. Epigenetic inheritance at the agouti locus in the mouse. *Nat Genet* **23:** 314–318.

Morris SA, Teo RT, Li H, Robson P, Glover DM, Zernicka-Goetz M. 2010. Origin and formation of the first two distinct cell types of the inner cell mass in the mouse embryo. *Proc Natl Acad Sci* **107:** 6364–6369.

Noordermeer D, de Wit E, Klous P, van de Werken H, Simonis M, Lopez-Jones M, Eussen B, de Klein A, Singer RH, de Laat W. 2011. Variegated gene expression caused by cell-specific long-range DNA interactions. *Nat Cell Biol* **13:** 944–951.

Oates NA, van Vliet J, Duffy DL, Kroes HY, Martin NG, Boomsma DI, Campbell M, Coulthard MG, Whitelaw E, Chong S. 2006. Increased DNA methylation at the AXIN1 gene in a monozygotic twin from a pair discordant for a caudal duplication anomaly. *Am J Hum Genet* **79:** 155–162.

Pardo-Manuel de Villena F, de la Casa-Esperon E, Sapienza C. 2000. Natural selection and the function of genome imprinting: Beyond the silenced minority. *Trends Genet* **16:** 573–579.

Pedram M, Sprung CN, Gao Q, Lo AW, Reynolds GE, Murnane JP. 2006. Telomere position effect and silencing of transgenes near telomeres in the mouse. *Mol Cell Biol* **26:** 1865–1878.

Pembrey ME, Bygren LO, Kaati G, Edvinsson S, Northstone K, Sjostrom M, Golding J. 2006. Sex-specific, male-line transgenerational responses in humans. *Eur J Hum Genet* **14:** 159–166.

Percec I, Plenge RM, Nadeau JH, Bartolomei MS, Willard HF. 2002. Autosomal dominant mutations affecting X inactivation choice in the mouse. *Science* **296:** 1136–1139.

Percec I, Thorvaldsen JL, Plenge RM, Krapp CJ, Nadeau JH, Willard HF, Bartolomei MS. 2003. An *N*-ethyl-*N*-nitrosourea mutagenesis screen for epigenetic mutations in the mouse. *Genetics* **164:** 1481–1494.

Perry WL, Copeland NG, Jenkins NA. 1994. The molecular basis for dominant yellow agouti coat color mutations. *Bioessays* **16:** 705–707.

* Pikaard CS, Mittelsten Scheid O. 2014. Epigenetic regulation in plants. *Cold Spring Harb Perspect Biol* doi: 10.1101/cshperspect.a019315.

Plenge RM, Percec I, Nadeau JH, Willard HF. 2000. Expression-based assay of an X-linked gene to examine effects of the X-controlling element (*Xce*) locus. *Mamm Genome* **11:** 405–408.

Porter S, Larue L, Mintz B. 1991. Mosaicism of tyrosinase-locus transcription and chromatin structure in dark vs. light melanocyte clones of homozygous *chinchilla-mottled* mice. *Dev Genet* **12:** 393–402.

Preis JI, Downes M, Oates NA, Rasko JE, Whitelaw E. 2003. Sensitive flow cytometric analysis reveals a novel type of parent-of-origin effect in the mouse genome. *Curr Biol* **13:** 955–959.

Pujadas E, Feinberg AP. 2012. Regulated noise in the epigenetic landscape of development and disease. *Cell* **148:** 1123–1131.

Raj A, Peskin CS, Tranchina D, Vargas DY, Tyagi S. 2006. Stochastic mRNA synthesis in mammalian cells. *PLoS Biol* **4:** e309.

Rakyan VK, Blewitt ME, Druker R, Preis JI, Whitelaw E. 2002. Metastable epialleles in mammals. *Trends Genet* **18:** 348–351.

Rakyan VK, Chong S, Champ ME, Cuthbert PC, Morgan HD, Luu KV, Whitelaw E. 2003. Transgenerational inheritance of epigenetic states at the murine $Axin^{Fu}$ allele occurs after maternal and paternal transmission. *Proc Natl Acad Sci* **100:** 2538–2543.

Rakyan VK, Down TA, Maslau S, Andrew T, Yang TP, Beyan H, Whittaker P, McCann OT, Finer S, Valdes AM, et al. 2010. Human aging-associated DNA hypermethylation occurs preferentially at bivalent chromatin domains. *Genome Res* **20:** 434–439.

Rakyan VK, Down TA, Balding DJ, Beck S. 2011. Epigenome-wide association studies for common human diseases. *Nat Rev Genet* **12:** 529–541.

Rassoulzadegan M, Grandjean V, Gounon P, Vincent S, Gillot I, Cuzin F. 2006. RNA-mediated non-mendelian inheritance of an epigenetic change in the mouse. *Nature* **441:** 469–474.

Ray PF, Conaghan J, Winston RM, Handyside AH. 1995. Increased number of cells and metabolic activity in male human preimplantation embryos following in vitro fertilization. *J Reprod Fertil* **104:** 165–171.

Reed SC. 1937. The inheritance and expression of *fused*, a new mutation in the house mouse. *Genetics* **22:** 1–13.

Reik W, Collick A, Norris ML, Barton SC, Surani MA. 1987. Genomic imprinting determines methylation of parental alleles in transgenic mice. *Nature* **328:** 248–251.

Rinchik EM. 1991. Chemical mutagenesis and fine-structure functional analysis of the mouse genome. *Trends Genet* **7:** 15–21.

Rinchik EM, Carpenter DA, Selby PB. 1990. A strategy for fine-structure functional analysis of a 6- to 11-centimorgan region of mouse chromosome 7 by high-efficiency mutagenesis. *Proc Natl Acad Sci* **87:** 896–900.

Robertson G, Garrick D, Wu W, Kearns M, Martin D, Whitelaw E. 1995. Position-dependent variegation of globin transgene expression in mice. *Proc Natl Acad Sci* **92:** 5371–5375.

Robertson G, Garrick D, Wilson M, Martin DI, Whitelaw E. 1996. Age-dependent silencing of globin transgenes in the mouse. *Nucleic Acids Res* **24:** 1465–1471.

Sado T, Fenner MH, Tan SS, Tam P, Shioda T, Li E. 2000. X inactivation in the mouse embryo deficient for *Dnmt1*: Distinct effect of hypomethylation on imprinted and random X inactivation. *Dev Biol* **225:** 294–303.

Sapienza C, Peterson AC, Rossant J, Balling R. 1987. Degree of methylation of transgenes is dependent on gamete of origin. *Nature* **328:** 251–254.

Schotta G, Ebert A, Dorn R, Reuter G. 2003. Position-effect variegation and the genetic dissection of chromatin regulation in *Drosophila*. *Semin Cell Dev Biol* **14:** 67–75.

Schumacher A, Faust C, Magnuson T. 1996. Positional cloning of a global regulator of anterior–posterior patterning in mice. *Nature* **384:** 648.

Shah NM, Groves AK, Anderson DJ. 1996. Alternative neural crest cell fates are instructively promoted by TGFβ superfamily members. *Cell* **85:** 331–343.

Sutherland HG, Martin DI, Whitelaw E. 1997. A globin enhancer acts by increasing the proportion of erythrocytes expressing a linked transgene. *Mol Cell Biol* **17:** 1607–1614.

Sutherland HG, Kearns M, Morgan HD, Headley AP, Morris C, Martin DI, Whitelaw E. 2000. Reactivation of heritably silenced gene expression in mice. *Mamm Genome* **11:** 347–355.

Teschendorff AE, Menon U, Gentry-Maharaj A, Ramus SJ, Weisenberger DJ, Shen H, Campan M, Noushmehr H, Bell CG, Maxwell AP, et al. 2010. Age-dependent DNA methylation of genes that are suppressed in stem cells is a hallmark of cancer. *Genome Res* **20:** 440–446.

Vasicek TJ, Zeng L, Guan XJ, Zhang T, Costantini F, Tilghman SM. 1997. Two dominant mutations in the mouse fused gene are the result of transposon insertions. *Genetics* **147:** 777–786.

Vickaryous NK. 2005. "An ENU mutagenesis screen to identify modifiers of gene silencing in the mouse." PhD thesis, University of Sydney, Australia.

Wagner KD, Wagner N, Ghanbarian H, Grandjean V, Gounon P, Cuzin F, Rassoulzadegan M. 2008. RNA induction and inheritance of epigenetic cardiac hypertrophy in the mouse. *Dev Cell* **14:** 962–969.

Waterland RA, Jirtle RL. 2003. Transposable elements: Targets for early nutritional effects on epigenetic gene regulation. *Mol Cell Biol* **23:** 5293–5300.

Waterland RA, Dolinoy DC, Lin JR, Smith CA, Shi X, Tahiliani KG. 2006. Maternal methyl supplements increase offspring DNA methylation at *Axin fused*. *Genesis* **44:** 401–406.

Waterland RA, Kellermayer R, Laritsky E, Rayco-Solon P, Harris RA, Travisano M, Zhang W, Torskaya MS, Zhang J, Shen L, et al. 2010. Season of conception in rural Gambia affects DNA methylation at putative human metastable epialleles. *PLoS Genet* **6:** e1001252.

Weichman K, Chaillet JR. 1997. Phenotypic variation in a genetically identical population of mice. *Mol Cell Biol* **17:** 5269–5274.

White RA, Dowler LL, Pasztor LM, Gaston LL, Adkison LR, Angeloni SV, Wilson DB. 1995. Assignment of the transcription factor GATA4 gene to human chromosome 8 and mouse chromosome 14: *Gata4* is a candidate gene for *Ds* (disorganization). *Genomics* **27:** 20–26.

Whitelaw NC, Chong S, Morgan DK, Nestor C, Bruxner TJ, Ashe A, Lambley E, Meehan R, Whitelaw E. 2010a. Reduced levels of two modifiers of epigenetic gene silencing, Dnmt3a and Trim28, cause increased phenotypic noise. *Genome Biol* **11:** R111.

Whitelaw NC, Chong S, Whitelaw E. 2010b. Tuning in to noise: Epigenetics and intangible variation. *Dev Cell* **19:** 649–650.

Wijchers PJ, Yandim C, Panousopoulou E, Ahmad M, Harker N, Saveliev A, Burgoyne PS, Festenstein R. 2010. Sexual dimorphism in mammalian autosomal gene regulation is determined not only by Sry but by sex chromosome complement as well. *Dev Cell* **19:** 477–484.

Williams A, Harker N, Ktistaki E, Veiga-Fernandes H, Roderick K, Tolaini M, Norton T, Williams K, Kioussis D. 2008. Position effect variegation and imprinting of transgenes in lymphocytes. *Nucleic Acids Res* **36:** 2320–2329.

Wilson L, Ching YH, Farias M, Hartford SA, Howell G, Shao H, Bucan M, Schimenti JC. 2005. Random mutagenesis of proximal mouse chromosome 5 uncovers predominantly embryonic lethal mutations. *Genome Res* **15:** 1095–1105.

Wolff GL. 1978. Influence of maternal phenotype on metabolic differentiation of agouti locus mutants in the mouse. *Genetics* **88:** 529–539.

Wolff GL, Kodell RL, Moore SR, Cooney CA. 1998. Maternal epigenetics and methyl supplements affect agouti gene expression in A^{vy}/a mice. *FASEB J* **12:** 949–957.

Wu M, Rinchik EM, Wilkinson E, Johnson DK. 1997. Inherited somatic mosaicism caused by an intracisternal A particle insertion in the mouse tyrosinase gene. *Proc Natl Acad Sci* **94:** 890–894.

Yamanaka Y, Lanner F, Rossant J. 2010. FGF signal-dependent segregation of primitive endoderm and epiblast in the mouse blastocyst. *Development* **137:** 715–724.

Yoder JA, Walsh CP, Bestor TH. 1997. Cytosine methylation and the ecology of intragenomic parasites. *Trends Genet* **13:** 335–340.

Youngson NA, Vickaryous N, van der Horst A, Epp T, Harten S, Fleming JS, Khanna KK, de Kretser DM, Whitelaw E. 2011. A missense mutation in the transcription factor Foxo3a causes teratomas and oocyte abnormalities in mice. *Mamm Genome* **22:** 235–248.

Youngson NA, Epp T, Roberts AR, Daxinger L, Ashe A, Huang E, Lester KL, Harten SK, Kay GF, Cox T, et al. 2013. No evidence for cumulative effects in a *Dnmt3b* hypomorph across multiple generations. *Mamm Genome* **24:** 206–217.

Zeng L, Fagotto F, Zhang T, Hsu W, Vasicek TJ, Perry WL 3rd, Lee JJ, Tilghman SM, Gumbiner BM, Costantini F. 1997. The mouse *Fused* locus encodes Axin, an inhibitor of the Wnt signaling pathway that regulates embryonic axis formation. *Cell* **90:** 181–192.

Zheng W, Liu K. 2012. Maternal control of mouse preimplantation development. *Results Probl Cell Differ* **55:** 115–139.

Zhuma T, Tyrrell R, Sekkali B, Skavdis G, Saveliev A, Tolaini M, Roderick K, Norton T, Smerdon S, Sedgwick S, et al. 1999. Human HMG box transcription factor HBP1: A role in hCD2 LCR function. *EMBO J* **18:** 6396–6406.

C H A P T E R 15

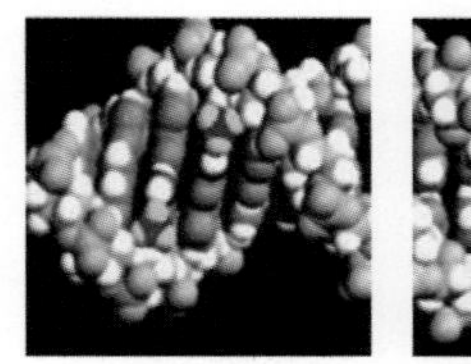
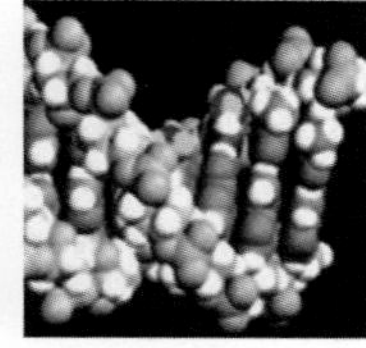

DNA Methylation in Mammals

En Li[1] and Yi Zhang[2]

[1]China Novartis Institutes for BioMedical Research, Pudong New Area, Shanghai 201203, China; [2]Boston Children's Hospital, Harvard Medical School, Boston, Massachusetts 02115

Correspondence: en.li@novartis.com

SUMMARY

DNA methylation is one of the best characterized epigenetic modifications. In mammals it is involved in various biological processes including the silencing of transposable elements, regulation of gene expression, genomic imprinting, and X-chromosome inactivation. This chapter describes how DNA methylation serves as a cellular memory system and how it is dynamically regulated through the action of the DNA methyltransferase (DNMT) and ten eleven translocation (TET) enzymes. Its role in the regulation of gene expression, through its interplay with histone modifications, is also described, and its implication in human diseases discussed. The exciting areas of investigation that will likely become the focus of research in the coming years are outlined in the summary.

Outline

Cite as *Cold Spring Harb Perspect Biol* doi: 10.1101/cshperspect.a019133

OVERVIEW

The DNA of vertebrate animals can be covalently modified by methylation of the cytosine base in the dinucleotide sequence 5′CpG3′. CpG is an abbreviation for cytosine and guanine separated by a phosphate, which links the two nucleotides together in DNA. In mammals, DNA methylation patterns are established during embryonic development by de novo methylating enzymes called Dnmt3a and Dnmt3b. They are maintained by a Dnmt1-mediated copying mechanism when cells divide. The heritability of DNA methylation patterns provides an epigenetic marking of the genome that is stable through multiple cell divisions and therefore constitutes a form of cellular memory. For this reason, historically, DNA methylation has represented the archetypal mechanism of epigenetic inheritance.

DNA methylation is found in some lower eukaryotes such as Neurospora and invertebrates (discussed in Ch. 10 [Aramayo and Selker 2013] and Ch. 12 [Elgin and Reuter 2013]). There is also quite an elaborate DNA methylation system in plants, involving many enzymes and specific binding proteins covered in depth in Ch. 13 (Pikaard and Mittelsten Scheid 2014).

Molecular and genetic studies in mammals have shown that DNA cytosine methylation (abbreviated to 5mC, for 5-methylcytosine) is associated with gene silencing. It also plays an important role in developmental processes such as X-chromosome inactivation and genomic imprinting. The methyl moiety of methylcytosine resides in the major groove of the DNA helix where many DNA-binding proteins make contact with DNA. The methylation, therefore, likely exerts its effect by attracting or repelling various DNA-binding proteins. A family of proteins, known as methyl-CpG binding domain proteins (or MBDs), are attracted to and bind DNA-containing methylated CpG dinucleotides and have been shown to recruit repressor complexes to methylated promoter regions, thereby contributing to transcriptional silencing. Conversely, regions of CpG methylation are known to prevent protein binding of certain transcription factors, thereby preventing transcription.

Certain regions of the genome contain clusters of CpG sequences, termed CpG islands, and are mostly found directly upstream of gene promoters. In general, CpG islands are DNA methylation-free. Certain transcription factors have been discovered to bind to nonmethylated CpG-containing DNA sequences via a CXXC binding domain motif, and contribute to creating a transcriptionally competent chromatin configuration, preventing DNA methylation from occurring at these regions.

Although DNA methylation patterns can be transmitted from cell to cell, they are not permanent. In fact, changes in DNA methylation patterns can occur throughout the life of an individual. Some changes can be a physiological response to environmental changes, whereas others might be associated with a pathological process such as oncogenic transformation or cellular aging. DNA methylation marks can be removed by either an active demethylation mechanism involving a family of DNA hydroxylases called Tet proteins or a passive demethylation process by inhibition of the maintenance methyltransferase, Dnmt1, during cell divisions. DNA methylation patterns fit into an epigenetic framework directly, but also indirectly through their intimate link to other epigenetic mechanisms such as histone lysine methylation and acetylation.

The clinical relevance of DNA methylation first became apparent in relation to cancer. Reduced levels of DNA methylation led to the suppression of some forms of tumors in mouse models of cancer through genetic manipulation or treatment with DNA methyltransferase inhibitors. Conversely, low levels of DNA methylation (referred to as DNA hypomethylation) can enhance the formation of certain tumor types as well. Several other human diseases have been linked to mutations of genes that encode critical components of the DNA methylation machinery. Mutations of the DNA methyltransferase Dnmt3b leads to immune deficiency, whereas mutations of the methyl-CpG binding protein, MeCP2, causes a severe neurological disorder known as Rett syndrome. It is apparent that the integrity of the DNA methylation system is of paramount importance for the health of mammals. Thus, the study of DNA methylation in human disease represents an important frontier in medicine and will contribute to our understanding of the impact of epigenetic modification on human life.

 Cite as *Cold Spring Harb Perspect Biol* doi: 10.1101/cshperspect.a019133

1 A MECHANISM OF CELL MEMORY

1.1 The Hypothesis

Cytosine methylation in mammalian cells occurs predominantly in CpG dinucleotides (Fig. 1). The idea that DNA methylation in animals could represent a mechanism of cell memory arose independently in two laboratories (Holliday and Pugh 1975; Riggs 1975). Recognizing that the CpG dinucleotide is self-complementary, both groups reasoned that patterns of methylated and nonmethylated CpG could be copied when cells divide. Immediately after replication of DNA, the parental DNA strand would maintain its pattern of modified cytosines, but the newly synthesized strand would be unmodified. To insure copying of the parental pattern onto the progeny strand, they postulated a "maintenance methyltransferase" that would exclusively methylate CpGs base-paired with a methylated parental CpG. Nonmethylated CpGs would not be substrates for the maintenance methyltransferase (see Fig. 2). The consequence of this simple mechanism is that patterns of DNA methylation would be replicated semiconservatively like the base sequence of DNA itself.

1.2 Maintenance of DNA Methylation Patterns

The mammalian DNA methyltransferase activity responsible for semiconservatively replicating DNA methylation patterns was detected early on in crude cellular extracts. It was finally purified as a 200-kDa protein (Bestor and Ingram 1983). This enzyme, called Dnmt1 for DNA methyltransferase 1, is specific to CpG and has significant activity against nonmethylated DNA in biochemical assays. Its preferred DNA substrate, however, is DNA methylated at CpG on one of the two strand (so-called hemimethylated DNA; see Fig. 2). Subsequent genetic studies indicated that inactivation of Dnmt1 in mouse embryonic stem cells (Table 1) led to a genome-wide loss of CpG methylation (Li et al. 1992). This evidence fit with the view that Dnmt1 maintains DNA methylation at CpGs by completing hemimethylated sites as originally postulated by Riggs (1975) and Holliday and Pugh (1975) (Fig. 2).

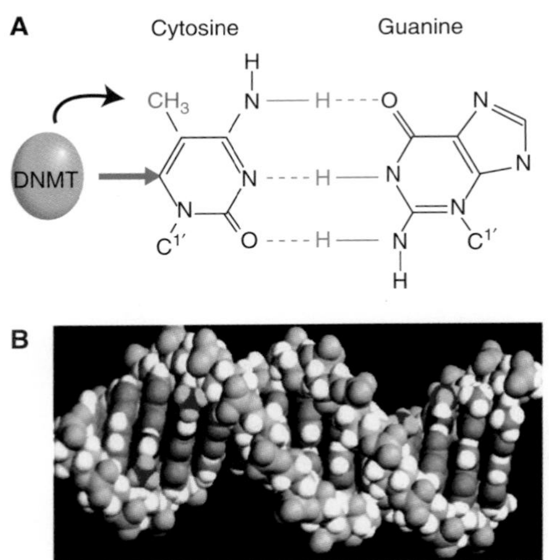

Figure 1. Cytosine methylation in DNA. (*A*) Addition of a methyl group, CH_3 (red), at the five position of the cytosine pyrimidine ring (black arrow) does not sterically interfere with GC base pairing (blue lines). DNA methyltransferases associate covalently with the carbon 6 position (straight green arrow) during methyl group transfer. (*B*) A model of B-form DNA methylated at cytosines in two self-complementary CpG sequences. The paired methyl moieties (magenta and yellow) lie in the major groove of the double helix.

2 THE ESTABLISHMENT OF DNA METHYLATION PATTERNS

The discovery of the maintenance DNA methyltransferase, Dnmt1, provided a mechanism by which DNA methylation patterns could be maintained through cell generations. This left the problem of determining when in development new DNA methylation patterns are established in an individual and how de novo methylation occurs. This is addressed in this section, as well as describing important work mapping the distribution and methylation patterns of CpG sequences in the mammalian genome and their functional significance.

2.1 De Novo Methylation of DNA in Early Embryos

In the life cycle of an individual, the genome undergoes dynamic changes in DNA methylation during early development. After fertilization, genome-wide demethylation takes place. De novo DNA methylation then occurs around the time of implantation when the inner cell mass cells start to differentiate to form the embryonic ectoderm (Fig. 3).

De novo DNA methylation was first detected when foreign DNA was introduced into a preimplantation embryo in an unmethylated state, which then became methylated. Reports of both retroviral DNA from infected mouse preimplantation embryos and DNA injected into mouse zygotes showed that the DNA became stably methylated in cells of the animal (Jahner et al. 1982). However, retroviral DNA did not become methylated if infection occurred in embryos at the later stage of gastrulation. This suggested that the process of de novo methylation of DNA is confined to pluripotent cells of early embryos. The hypothesis was further tested using mouse embryonal carcinoma cells and then embryonic stem (ES) cells. When these cells were infected by retroviruses, retroviral DNA became completely methylated and viral genes were consequently silenced (Stewart

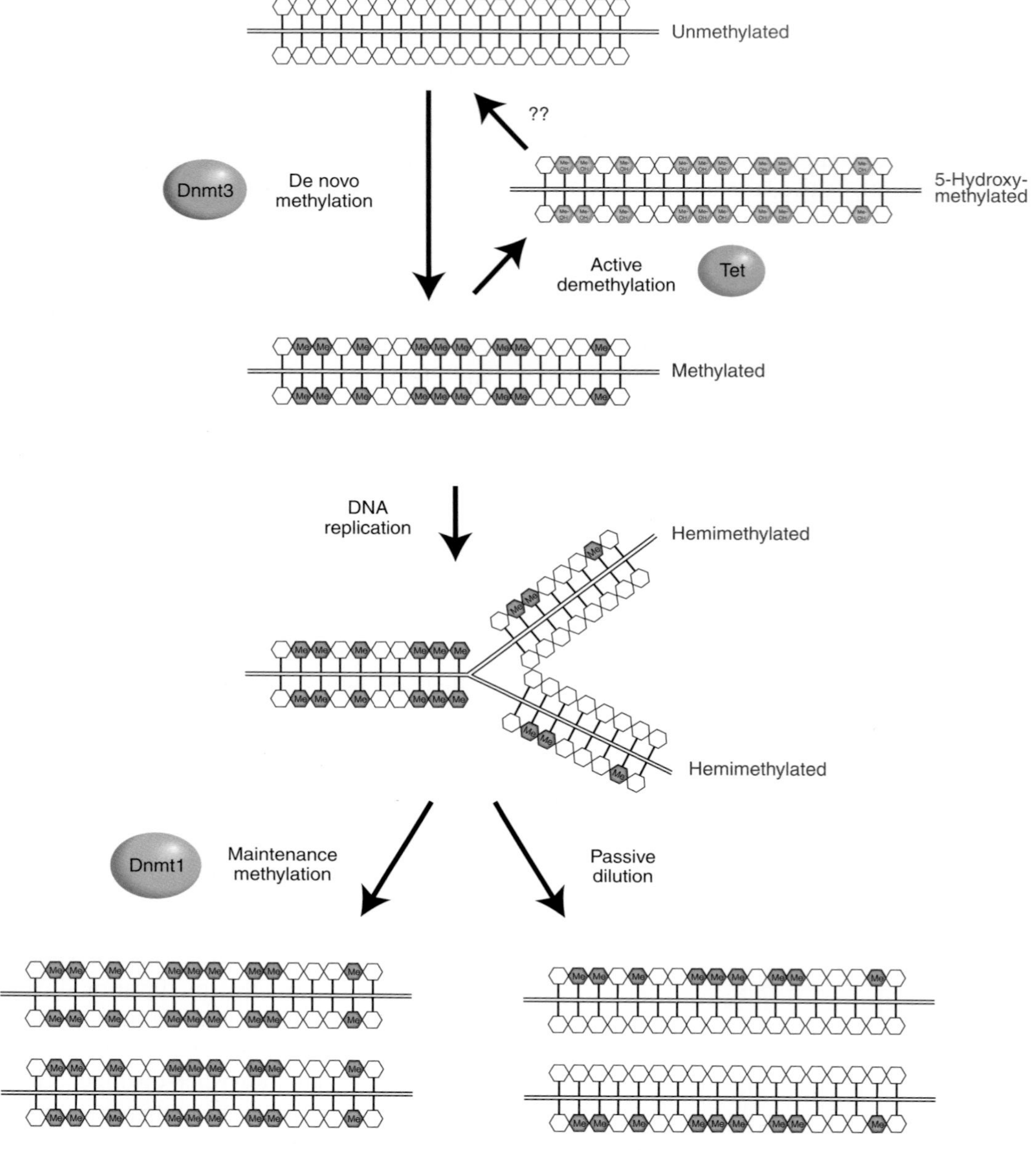

Figure 2. De novo methylation and maintenance methylation of DNA. A stretch of genomic DNA is shown as a line with self-complementary CpG pairs marked as vertical strokes. Unmethylated DNA (*top*) becomes methylated "de novo" by Dnmt3a and Dnmt3b to give symmetrical methylation at certain CpG pairs. On semiconservative DNA replication, a progeny DNA strand is base-paired with one of the methylated parental strands (the other replication product is not shown). Symmetry is restored by the maintenance DNA methyltransferase, Dnmt1, which completes half-methylated sites, but does not methylate unmodified CpGs.

et al. 1982). Even when the maintenance Dnmt1 gene was deleted, de novo methylation of retroviral DNA in ES cells still occurred proving that other DNA methyltransferases must be at work (Lei et al. 1996). Somatic cells that are infected with viral DNA, however, did not become methylated, again indicating that it is a process that occurs in early development (Stewart et al. 1982).

2.2 Discovery of De Novo Methyltransferases

De novo methyltransferases were discovered by searching for sequence homology with prokaryotic cytosine DNA methyltransferases using expressed sequence tag databases. Prokaryotic cytosine DNA methyltransferases share a set of conserved protein motifs (Posfai et al. 1989), and these

 Cite as *Cold Spring Harb Perspect Biol* doi: 10.1101/cshperspect.a019133

Table 1. Function of mammalian DNA methyltransferases

DNA methyltransferase	Species	Major activity	Major phenotypes of loss of function
Dnmt1	Mouse	Maintenance methylation of CpG	Genome-wide loss of DNA methylation, embryonic lethality at embryonic day 9.5 (E9.5), abnormal expression of imprinted genes, ectopic X-chromosome inactivation, activation of silent retrotransposon. In cancer cell lines, it leads to cell cycle arrest and mitotic defects.
Dnmt3a	Mouse	De novo methylation of CpG	Postnatal lethality at 4–8 wk, male sterility, and failure to establish methylation imprints in both male and female germ cells
Dnmt3b	Mouse	De novo methylation of CpG	Demethylation of minor satellite DNA, embryonic lethality around E14.5 days with vascular and liver defects. (Embryos lacking both Dnmt3a and Dnmt3b fail to initiate de novo methylation after implantation and die at E9.5.)
DNMT3B	Human	De novo methylation of CpG	ICF syndrome: immunodeficiency, centromeric instability, and facial anomalies. Loss of methylation in repetitive elements and pericentromeric heterochromatin.

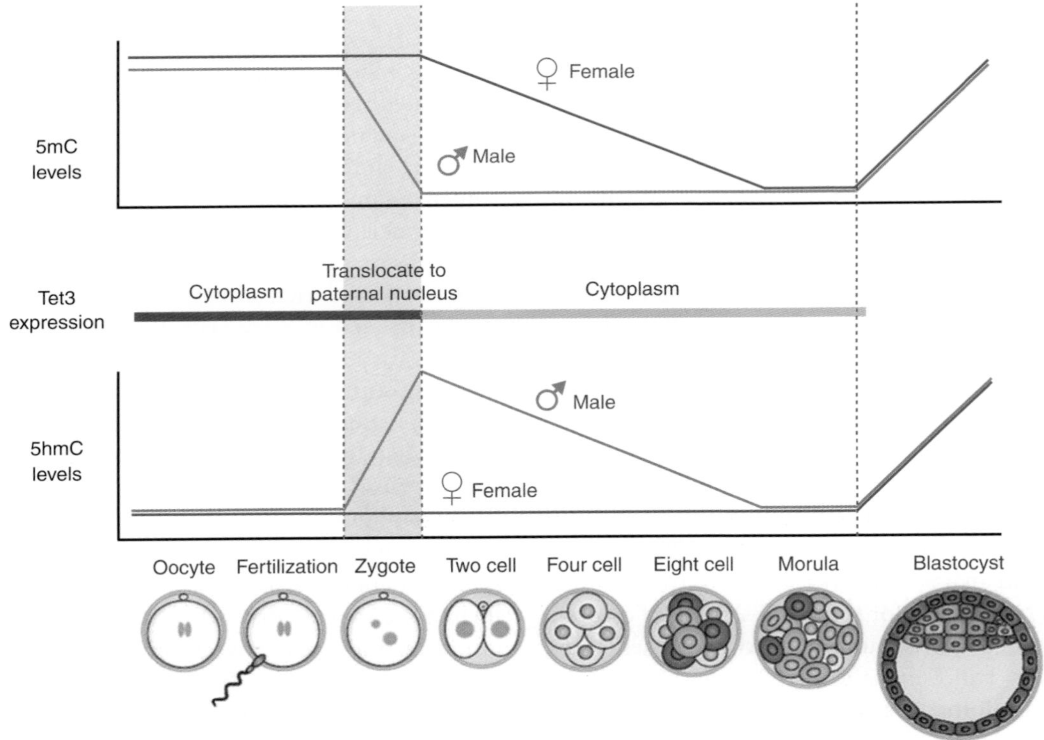

Figure 3. Dynamics of 5mC/5hmC/5fC/5caC in paternal and maternal genomes during preimplantation development. DNA demethylation of the zygote, gauged by 5mC levels, occurs by a passive mechanism in the female pronucleus, diluting the marks with the passage of every cell cycle. The male pronuclear genome becomes demethylated actively by the action of the Tet enzymes. Tet3 is expressed in the oocyte and zygote. After fertilization, Tet3 is relocated from the cytoplasm to the paternal nucleus to convert 5mC to 5hmC/5fC/5caC. Subsequently, paternal and maternal genomes undergo replication-dependent dilution of 5hmC/5fC/5caC in males and 5mC in females. It is possible that replication-independent active DNA demethylation may occur in a loci-specific manner in zygotes, but the exact mechanism is currently unclear. DNA methylation patterns in ICM are reestablished by de novo DNA methyltransferases Dnmt3a and Dnmt3b at the blastocyst stage.

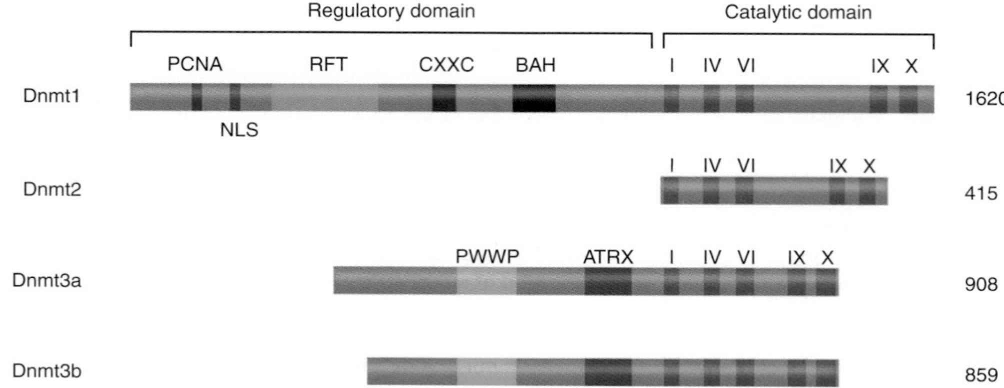

Figure 4. Mammalian DNA methyltransferases. The catalytic domains of Dnmt1, Dnmt2, and the Dnmt3 family members are conserved (the signature motifs, I, IV, VI, IX, and X, are most conserved in all cytosine methyltransferases), but there is little similarity among their amino-terminal regulatory domains. Domain abbreviations: PCNA, PCNA-interacting domain; NLS, nuclear localization signal; RFT, replication foci-targeting domain; CXXC, a cysteine-rich domain implicated in binding DNA sequences containing CpG dinucleotides; BAH, bromo-adjacent homology domain implicated in protein–protein interactions; PWWP, a domain containing a highly conserved "proline-tryptophan-tryptophan-proline" motif involved in heterochromatin association; ATRX, an ATRX-related cysteine-rich region containing a C2-C2 zinc finger and an atypical PHD domain implicated in protein–protein interactions.

features were also found in the mammalian maintenance DNA methyltransferase, Dnmt1. The homology search identified three genes that could potentially encode novel DNA methyltransferases (Fig. 4). One candidate protein, Dnmt2, has minimal DNA methyltransferase activity in vitro and its absence has no discernible effect on de novo or maintenance methylation of DNA (Okano et al. 1998b). The other two genes, Dnmt3a and Dnmt3b, encoded related catalytically active polypeptides that showed no preference for methylating hemimethylated DNA in vitro, unlike Dnmt1 (Okano et al. 1998a). Inactivation of both Dnmt3a and Dnmt3b by gene targeting in ES cells confirmed that these genes constituted the missing de novo DNA methyltransferases (Table 1)—ES cells and embryos lacking both proteins were unable to de novo methylate proviral genomes and repetitive elements (Okano et al. 1999). Moreover, Dnmt3a and an associated regulatory factor Dnmt3L were shown to be required for the establishment of distinct DNA methylation patterns found at imprinted genes (Hata et al. 2002; Table 1 in Kaneda et al. 2004; see also Ch. 26 [Barlow and Bartolomei 2014]).

2.3 CpG Islands and Patterns of DNA Methylation across the Genome

DNA across the genome of mammalian somatic tissues is methylated at 70%–80% of all CpG sites. With the continual advances in technology there has been a constant refinement in the level of detail with which we can measure DNA methylation at single CpG site resolution in any given cell type. Mapping studies (see Box 1) indicate that highly methylated sequences include satellite DNAs, repetitive elements (including transposons and their inert relics), nonrepetitive intergenic DNA, and exons of genes. Most sequences are methylated according to their frequency of CpG dinucleotides. Key exceptions to this global methylation of the mammalian genome are the CpG islands (CGIs).

CGIs are GC-rich sequences of ~1 kb in length that are nonmethylated in germ cells, the early embryo, and most somatic tissues (Fig. 5) (Bird et al. 1985). It was in early mapping studies of individual gene promoters that these GC-rich regions were identified (McKeon et al. 1982). It is now evident that most (if not all) CGIs mark the promoters and 5′ regions of genes. In fact, ~60% of human genes have CGI promoters.

Genes with CGI promoters that are expressed in a tissue-specific manner are usually expressed in early embryos and then in the soma. A distinct pattern of DNA methylation is found on the inactive X chromosome in females where, contrary to the norm, CGIs become de novo methylated in large numbers during the embryonic process of X-chromosome inactivation in female placental mammals (Wolf et al. 1984). This process is essential for the leak-proof silencing of genes on the inactivated chromosome necessary for dosage compensation, as DNA methylation-deficient mice or cells show frequent transcriptional reactivation of X-linked genes (see also Ch. 25 [Brockdorff and Turner 2014]).

Studies of DNA methylation patterns have focused on the question of how gene expression is regulated by DNA methylation. CGIs normally remain unmethylated and

Cite as *Cold Spring Harb Perspect Biol* doi: 10.1101/cshperspect.a019133

BOX 1 MAPPING DNA METHYLATION

To understand the functions of DNA methylation, it is first necessary to map methyl CpG in the genome and its dynamic changes during cell proliferation and differentiation or in development and disease. Several methods have been developed for quantitative analysis of DNA methylation at the genome scale as well as in a gene locus-specific manner.

- ***Bisulfite-sequencing (Frommer et al. 1992)*** This is the most reliable method for testing all cytosines within a region of the genome. It involves the "bisulfite modification" of single-stranded DNA, which leads to deamination of unmodified cytosines, whereas 5-methylcytosine is protected. As a result, cytosines that survive bisulfite treatment are identified as methylated. Because of its high resolution and positive identification of methylated cytosine, this is the method of choice for analyzing DNA methylation patterns. In combination with next-generation sequencing, the method is widely used for methylome analysis at the whole-genome scale. Several polymerase chain reaction (PCR)-based methods that depend on prior bisulfite treatment of DNA have also been developed for rapid analysis of methylation of genes of interest (Herman et al. 1996).
- ***MeDIP (Weber et al. 2005)*** Methylated DNA immunoprecipitation (MeDIP) is a versatile approach for unbiased detection of methylated DNA and can be applied to generate comprehensive DNA methylation profiles on a genome-wide scale. This method uses a monoclonal antibody that specifically recognizes 5-methylcytidine to enrich methylated genomic DNA fragments by immunoprecipitation. The methylation status is then determined by PCR for specific regions or by using DNA microarrays for the whole genome.
- ***Pyrosequencing (Tost and Gut 2007)*** Analysis of DNA methylation patterns by pyrosequencing yields reproducible and accurate measures of the degree of methylation at several CpGs in close proximity with high resolution. The method is highly sensitive and quantitative, and it is often applied to methylation analysis of specific regions.
- ***CHARM DNA methylation analysis (Irizarry et al. 2009)*** Comprehensive high-throughput array-based relative methylation (CHARM) analysis is a microarray-based method. It can be applied to custom-designed microarray covering the whole genome (usually nonrepetitive sequences) or specific regions (e.g., all CGIs). The method is quantitative and data analysis is straightforward. It has an advantage over other methods when it comes to comparing DNA methylation patterns from a large number of samples.
- ***A few commercially available technologies***
  Illumina methylation chip http://www.illumina.com/products/methylation_450_beadchipkits.ilmn
  Sequenom EpiTYPER http://www.sequenom.com/Sites/Genetic-Analysis/Applications/DNA-Methylation

multiple mechanisms are involved in protecting CGIs from de novo methylation by *Dnmt3a* and *Dnmt3b*. The discovery that two proteins containing a DNA-binding CXXC domain—Cfp1 and Kdm2a (Fig. 5C)—are able to bind specifically to unmethylated CpGs within CGIs (Blackledge et al. 2010, 2013 [Ch. 2]; Thomson et al. 2010) has provided an inroad to understanding how unmethylated DNA can contribute to creating a transcriptionally competent chromatin environment as well as protecting it from de novo DNA methylation (see Section 5.3 and Deaton and Bird 2011; Ch. 6 [Cheng 2014]).

The increasing resolution with which the methylome of particular cell types is being analyzed is beginning to reveal patterns of DNA methylation associated with normal development versus disease states such as cancer and aging. The detail has resulted in the categorization of different types of CGIs—that is, CGIs containing a transcriptional start site versus orphan CGIs that can be located intra- or intergenically. Studies have also revealed other differentially methylated regions termed shores (up to 2 kb away from a CGI) and shelves (located within 2–4 kb of a CGI; Fig. 5A) (Irizarry et al. 2009). One study proposes that the DNA methylation status of its first exon is a better indicator of transcriptional repression than its CGI (Brenet et al. 2011). The relevance these categories of CpG-containing regions of the genome have with the chromatin environment, transcriptional control and repression, and disease development or association requires further analyses.

3 DNA DEMETHYLATION

Developmental biologists have described the waves of genome-wide DNA demethylation that occur in the germline and in early embryogenesis (Fig. 3), however, the process by which DNA methylation is erased has been elusive. Recent discoveries, described in Sections 3.2 and 3.3, have advanced our understanding of the DNA demethylation process.

3.1 Active and Passive Demethylation during Development

The mammalian genome is reprogrammed during early development by both active and passive DNA demethyla-

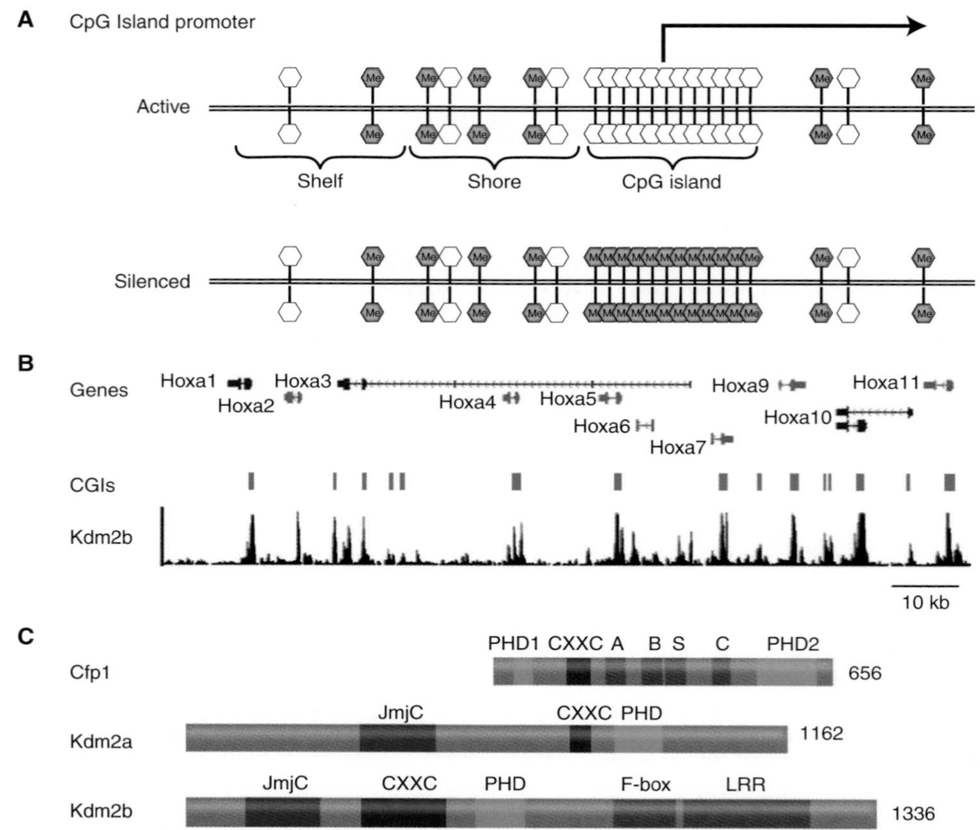

Figure 5. CpG islands. (*A*) CGIs are regions of high CpG density (>50%), usually 200 bp–2 kb in length that lack CpG methylation, found at promoters of most human genes. Long-term silencing of the gene can be insured by methylation of the CGI region. For example, genes on the inactive X chromosome and certain imprinted genes are silenced in this way. Also, in cancer cells certain genes are aberrantly silenced by CGI methylation. Shores are regions of the genome that reside up to 2 kb from CGIs, whereas shelves are found 2–4 kb away from CGIs. (*B*) Chromatin immunoprecipitation (IP) analysis of Kdm2b binding sites shows that Kdm2b is enriched at the CpGs of the Hox locus in which the unmethylated CGIs (green bars) are located. (*C*) Cfp1, Kdm2a, and Kdm2b proteins share a common CXXC domain that binds specifically to unmethylated CpG sites. Protein length is indicated to the *right* of each protein. Abbreviations for other domains include PHD, plant homeodomain; A, acidic domain; B, basic domain; S, Set1 interacting domain; C, coiled coil domain; LRR, leucine-rich repeat domain.

tion processes (Wu and Zhang 2010). Active DNA demethylation refers to an enzymatic process that results in the removal of the methyl-group from 5mC. In contrast, passive DNA demethylation refers to the lack of maintenance methylation during successive rounds of DNA replication either in the absence of Dnmt1 or because of its inhibition (Fig. 2). Immunostaining using an anti-5mC antibody initially showed that the 5mC levels of the maternal genome went through a replication-dependent dilution process (i.e., passive demethylation) during preimplantation development (Rougier et al. 1998). In contrast, the 5mC levels of the paternal genome dramatically decrease a few hours after fertilization (Fig. 3) (Mayer et al. 2000). Bisulfite sequencing confirmed that some of the repeat sequences, but not imprinted genes, of the paternal genomes were indeed demethylated (Oswald et al. 2000). Given that no DNA replication occurs during this period, the loss of 5mC in the paternal genome is considered "active."

Another place where global loss of 5mC is observed is in the primordial germ cells (PGCs). At embryonic stage E7.5, a subset of posterior epiblast cells is specified to become PGCs. At the beginning of their specification as well as during migration toward the genital ridge, PGCs are believed to have the same epigenetic marks as other epiblast cells. However, by the time they arrive at the genital ridge at E11.5 many of the epigenetic marks including DNA methylation have been erased (Hajkova et al. 2002; Yamazaki et al. 2003). Because PGCs have undergone several cell cycles in the presence of Dnmt1 during this process, the loss of DNA methylation is likely to be active. Also, because DNA

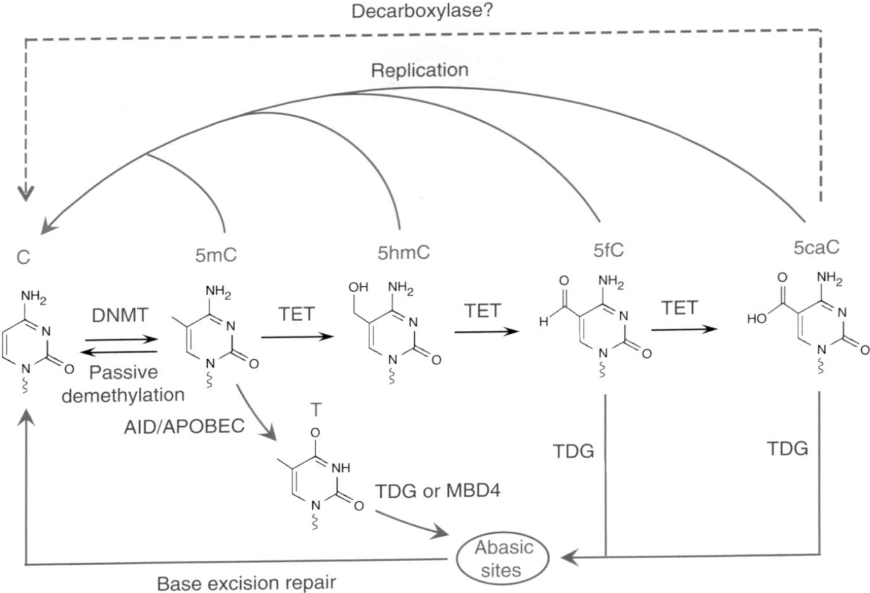

Figure 6. Model of Tet-initiated DNA demethylation pathways. DNA methylation (5mC) is established and maintained by DNMT. 5mC can by oxidized by Tet family of dioxygenases to generate 5hmC, 5fC, and 5caC. Because the oxidized 5mC derivatives cannot serve as substrates for DNMT1, they can be lost by replication-dependent passive demethylation. 5hmC can be deaminated by AID/APOBEC to become 5hmU, which together with 5fC and 5caC can be excised by glycosylases such as TDG, followed by DNA repair to generate C. Alternatively, a putative decarboxylase may convert 5caC to C.

methylation of the imprinted genes is also erased during this period, it is believed that DNA methylation patterns are reset during germ cell development (Sasaki and Matsui 2008).

Active DNA demethylation has been reported in somatic cells in a locus-specific manner. For example, activated T lymphocytes undergo active demethylation at the *interleukin-2* promoter-enhancer region in the absence of DNA replication within 20 min of stimulation (Bruniquel and Schwartz 2003). In addition, locus-specific demethylation occurs at the promoter of brain-derived neurotrophic factor (*Bdnf*) in depolarized neurons (Martinowich et al. 2003), and for other loci during nuclear hormone regulated gene activation (Kangaspeska et al. 2008; Metivier et al. 2008). Furthermore, locus-specific DNA demethylation occurs at the Oct4 and Nanog promoters when mouse ES cells are fused with human fibroblasts (Bhutani et al. 2010). Because no DNA replication takes place in the processes described above, active DNA demethylation is believed to be responsible for the loss of DNA methylation.

3.2 Tet-Mediated 5mC Oxidation

The observations described above have prompted the hunt for putative DNA demethylases. Most of the early studies were inconclusive in identifying the putative enzyme(s) or elucidating the mechanism of demethylation (Ooi and Bestor 2008; Wu and Zhang 2010). This situation changed with the identification of 5-hydroxymethylcytosine (5hmC) as a bona fide base of mammalian genomic DNA (Kriaucionis and Heintz 2009; Tahiliani et al. 2009) and the demonstration that Tet proteins are responsible for the conversion of 5mC to 5hmC (Fig. 6) (Tahiliani et al. 2009; Ito et al. 2010). This initial discovery is described in Ch. 2 (Kriaucionis and Tahiliani 2014).

Using thin layer chromatography and mass spectrometry assays, it was found that 5hmC is relatively abundant in the Purkinje neurons and mouse ES cells (Kriaucionis and Heintz 2009; Tahiliani et al. 2009). Importantly, the human TET1 protein was shown to be able to convert 5mC to 5hmC in an iron and 2-oxoglutarate-dependent manner (Tahiliani et al. 2009). This enzymatic activity is conserved in all three Tet family proteins (Fig. 7) (Ito et al. 2010). Based on the similar chemistry between Tet-catalyzed 5mC oxidation and thymine hydroxylase-catalyzed thymine oxidation, it was proposed that Tet-mediated 5mC oxidation should be able to proceed further to generate 5-formylcytosine (5fC) and 5-carboxylcytosine (5caC) (Fig. 6) (Wu and Zhang 2010). This proposed activity was experimentally shown both in vitro and in vivo (He et al. 2011; Ito et al. 2011). Mass spectrometry analysis has indicated that 5hmC and

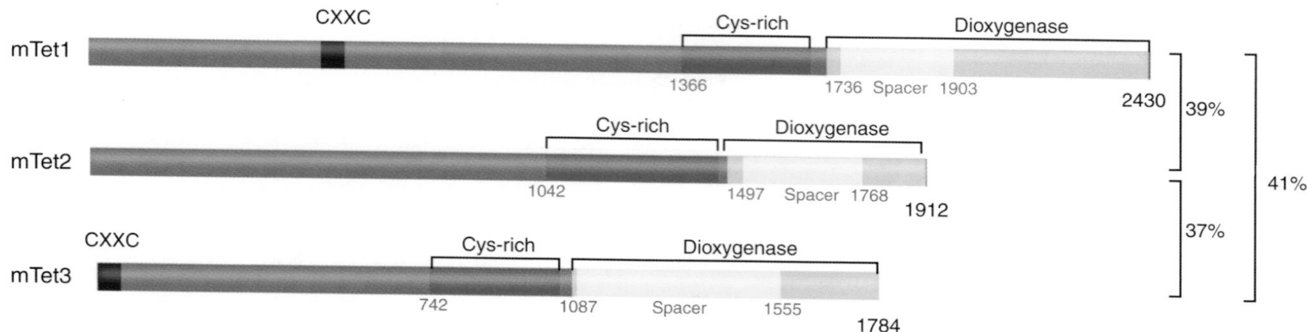

Figure 7. Domain structure of the mouse Tet family proteins. Schematic diagrams of predicted conserved domain structures in the three mouse Tet proteins. The conserved domains include CXXC zinc-binding domain, the cysteine-rich domain, and the double-stranded β-helix (DSBH) found in members of the dioxygenase superfamily proteins. Both the Cys-rich and the DSBH domains have been shown to be critical for enzymatic activity. Numbers indicate amino acid numbers.

5fC are broadly present in the genomic DNA of various tissues and cell types (Ito et al. 2011; Pfaffeneder et al. 2011), whereas the presence of 5caC appears to be more limited (He et al. 2011; Ito et al. 2011). Collectively, Tet proteins can oxidize 5mC not only to 5hmC, but also to 5fC and 5caC.

3.3 Tet-Mediated DNA Demethylation

Tet-mediated iterative oxidation of 5mC described in Sec. 3.2 (Fig. 6), may contribute to the dynamic changes in global or locus-specific levels of 5mC (Wu and Zhang 2011). The conversion of 5mC to 5hmC is expected to prohibit maintenance of existing DNA methylation patterns given that 5hmC is not recognized by Dnmt1 during DNA replication (Valinluck and Sowers 2007), leading to passive DNA demethylation during cell division. In addition, the 5mC oxidation products 5fC and 5caC may serve as intermediates for active DNA demethylation as both can be cleaved by the thymine-DNA glycosylase (TDG) (He et al. 2011; Maiti and Drohat 2011). These studies have raised the possibility that Tet-mediated 5mC oxidation followed by TDG-mediated excision of 5fC/5caC and base excision repair (BER) might be one of the pathways for active DNA demethylation (Fig. 6). Consistent with this, disruption of the Tdg gene in mouse results in increased DNA methylation at certain genomic loci (Cortazar et al. 2011; Cortellino et al. 2011). In plants, a similar glycosylase-initiated BER mechanism is responsible for DNA demethylation through members of the repressor of silencing 1 (ROS1)/Demeter family of 5mC glycosylases (Zhu 2009). More studies are required to ascertain which in vivo mechanisms of active or passive DNA demethylation are at work at the different loci and genomic regions, in different cell types, and at different stages of development. A start has been made in establishing the players involved in early embryonic DNA demethylation as described in Section 4.4.

3.4 Tet-Mediated Demethylation in Zygotes and Preimplantation Embryos

The discovery that Tet proteins are biochemically able to oxidize 5mC made it possible to test which protein may be responsible for the specific loss of 5mC in the paternal genome in zygotes. Immunostaining revealed that the decrease in 5mC staining in the paternal pronucleus in zygotes coincides with appearance of 5hmC/5fC/5caC (Gu et al. 2011; Inoue et al. 2011; Iqbal et al. 2011; Wossidlo et al. 2011). Studies showed that Tet3 is highly expressed in zygotes, and its small interfering RNA-mediated knockdown or targeted deletion abolishes 5mC oxidation, supporting the notion that Tet3 is responsible for 5mC oxidation in the zygotes (Gu et al. 2011; Wossidlo et al. 2011).

The asymmetric appearance of 5hmC in the sperm-derived chromosomes persists into two-cell-stage embryos, and gradually decreases until the morula stage (Inoue and Zhang 2011). This suggests that 5hmC is not rapidly removed at a genome-wide scale, but rather lost by "passive" replication-dependent dilution. Similar results have also been obtained for 5fC and 5caC (Inoue et al. 2011). Therefore, the dynamic changes in 5mC during preimplantation appear to involve both "passive" and "active" processes: DNA methylation in the maternal genome is "passively" diluted through replication, and 5mC in the paternal genome first goes through a Tet3-mediated oxidation process in zygotes followed by a "passive" replication-dependent dilution process (Fig. 3). The biological significance of the Tet3-catalyzed oxidation in zygotes is currently unknown, and it remains to be seen whether individual loci conform to this general trend.

Cite as *Cold Spring Harb Perspect Biol* doi: 10.1101/cshperspect.a019133

4 REGULATION OF GENE EXPRESSION BY DNA METHYLATION

DNA methylation of gene promoter regions is associated with transcription repression. Studies of the effects of DNA methylation in mammalian cells became possible with the discovery that the nucleoside analog 5-azacytidine could inhibit DNA methylation in living cells (Jones and Taylor 1980). 5-azacytidine is incorporated into DNA in place of cytidine and forms a covalent adduct with DNA methyltransferases preventing further DNA methylation. Silencing of several genes, including viral genomes (Harbers et al. 1981) and genes on the inactive X chromosome (Wolf et al. 1984), had previously been shown to correlate with their methylation. The ability of 5-azacytidine treatment to restore their expression (Mohandas et al. 1981) argued that DNA methylation played a causal role in their repression. This was later corroborated by genetic analysis using Dnmt1 knockout mice, demonstrating that inactivation of Dnmt1 resulted in genome-wide loss of DNA methylation and activation of inactive X silenced genes, viral genes, and imprinted genes such as H19 and IGF2 (Li et al. 1993).

4.1 Interference with Transcription Factor Binding

How does DNA methylation interfere with gene expression? One obvious possibility is that the presence of methyl groups in the major groove (see Fig. 1) interferes with the binding of transcription factors that activate transcription from a specific gene. A number of transcription factors recognize GC-rich sequence motifs that can contain CpG sequences. Several of these are unable to bind DNA when the CpG sequence is methylated (Watt and Molloy 1988). Proof that this mechanism operates in gene regulation comes from studies of the role of the CTCF protein in imprinting at the H19/Igf2 locus in mice (Bell and Felsenfeld 2000). CTCF is associated with transcriptional domain boundaries (Bell et al. 1999) and can insulate a promoter from the influence of remote enhancers. The maternally derived copy of the Igf2 gene is silent owing to the binding of CTCF between its promoter and a downstream enhancer. At the paternal locus, however, the CpG-rich CTCF binding sites are methylated, preventing CTCF binding and thereby allowing the downstream enhancer to activate Igf2 expression. Although there is evidence that H19/Igf2 imprinting involves additional processes, the role of CTCF represents a clear example of transcriptional regulation by DNA methylation (for more details, see Ch. 26 [Barlow and Bartolomei 2014]).

A recent study revealed that DNA-binding factors can also shape DNA methylation patterns. Whole genome bisulphite sequencing analysis of embryonic stem cells and neuronal progenitors revealed the existence of low-methylated regions (LMRs) at CpG-poor distal regulatory regions (Stadler et al. 2011). Interestingly, LMRs are occupied by transcription factors and their binding is both necessary and sufficient to generate LMRs, indicating transcription factors can influence local DNA methylation.

4.2 Recruitment of Methyl-CpG Binding Proteins and Repressor Complexes

The second mode of repression is opposite to the first as it involves proteins that are attracted to rather than repelled by methyl-CpG (Fig. 8). Evidence for such a mechanism came initially from the identification of the methyl-CpG binding protein complex, MeCP1, and subsequent purification and cloning of MeCP2 (Meehan et al. 1989). Proteins with DNA-binding motifs related to that of MeCP2 were identified using database searches, and designated the methyl-CpG binding domain (MBD) family comprising MeCP2, MBD1, MBD2, MBD3, and MBD4 (Bird and Wolffe 1999). Three of the MBD proteins, MBD1, MBD2, and MeCP2, have been implicated in methylation-dependent repression of transcription (Table 2) (Bird and Wolffe 1999). An unrelated protein, Kaiso, has also been shown to bind methylated DNA and bring about methylation-dependent repression in model systems (Table 2) (Prokhortchouk et al. 2001; Yoon et al. 2003).

MeCP2 associates with the mSin3a corepressor complex and depends on histone deacetylation for its action (Jones et al. 1998; Nan et al. 1998). This finding showed that DNA methylation can be read by MeCP2, and provides a signal to alter chromatin structure (Fig. 9). Each of the four methyl-CpG binding proteins has since been shown to associate with different corepressor complexes. Of particular interest is MBD1, which associates with the histone lysine methyltransferase SETDB1 only during DNA replication (Sarraf and Stancheva 2004). This may ensure continued histone H3K9 methylation at chromosomal MBD1 target sequences for stable silencing of the associated genes.

MBD2 is the DNA-binding component of MeCP1, which was initially implicated as a transcriptional repressor in cellular extracts (Meehan et al. 1989; Boyes and Bird 1991). MeCP1 is a large multiprotein complex that includes the NuRD (or Mi-2) corepressor complex and MBD2 (Wade et al. 1999; Feng and Zhang 2001). NuRD comprises histone deacetylases (HDAC) and a large chromatin remodeling protein (Mi-2) (Zhang et al. 1998). NuRD can be recruited to DNA by several DNA-binding proteins other than MBD2. Cells that lack MBD2 are unable to effectively repress methylated reporter constructs, despite the presence of other methyl-CpG binding proteins in these cells, arguing that it is an important component of the repression

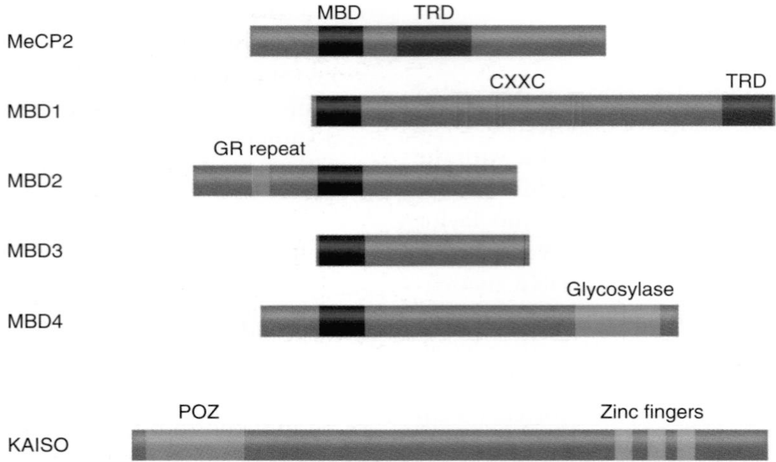

Figure 8. Proteins that bind methyl-CpG. Five members of the MBD protein family are aligned at their MBD domains (purple). Other domains are labeled and include TRD; CXXC domains, which are zinc fingers, some of which are implicated in binding to nonmethylated CpG; GR repeats that may bind; a T:G mismatch glycosylase domain that is involved in repair of 5mC deamination. Kaiso lacks the MBD domain, but binds methylated DNA via zinc fingers (orange) and possesses a POB/BTB domain that is shared with other transcriptional repressors. Domain abbreviations: MBD, methyl-CpG binding domain; TRD, transcriptional repression domain; POZ, poxvirus and zinc finger, a protein–protein interacting domain.

system (Hendrich et al. 2001). MBD2-deficient mice are viable and fertile, although they have a defect in maternal behavior (Hendrich et al. 2001), and careful examination has revealed aberrations in tissue-specific gene expression. For example, expression of the *interleukin-4* and *interferon-γ* genes during T-helper cell differentiation is significantly disrupted (Hutchins et al. 2002).

5 INTERACTIONS BETWEEN DNA METHYLATION AND HISTONE MODIFICATIONS

One of the important questions in the field is how DNA methylation coordinates with other modifications in chromatin to regulate gene expression and other chromatin-based processes. Given that DNA and histones are integral

Table 2. Functions of methyl-CpG binding proteins

MBP	Major activity	Species	Major phenotypes of loss-of-function mutations
MeCP2	Binds mCpG with adjacent run AT-rich run Transcriptional repressor	Mouse	Delayed onset neurological defects including inertia, hindlimb clasping, nonrhythmic breathing, and abnormal gait. Postnatal survival ~10 wk.
MECP2	Binds mCpG with adjacent AT run Transcriptional repressor	Human	Heterozygotes suffer from Rett syndrome, a profound neurological disorder characterized by apraxia, loss of purposeful hand use, breathing irregularities, and microcephaly
Mbd1	Binds mCpG via MBD; a major splice form is also able to bind CpG via a CxxC domain	Mouse	No overt phenotype, but subtle defects in neurogenesis detected
Mbd2	Binds mCpG Transcriptional repressor	Mouse	Viable and fertile, but show reduced maternal nurturing behavior. Defective gene regulation in T-helper cell differentiation leading to altered response to infection. Highly resistant to intestinal tumorigenesis.
Mbd3	Core component of NuRD corepressor complex Does not show strong binding to mCpG	Mouse	Early embryonic lethal
Mbd4	DNA repair protein that binds mCpG and T:G mismatches at mCpG sites Thymine DNA glycosylase that excises T from T:G mismatches	Mouse	Viable and fertile. three- to fourfold increase in mutations at CpG sites. Increased susceptibility to intestinal cancer correlates with C to T transitions within the *Apc* gene. Mbd4 functions to minimize the mutability of 5-methylcytosine.
Kaiso	Binds mCGmCG and CTGCNA Transcriptional repressor	Mouse	No overt phenotype. Small but significant delay in tumorigenesis on Min background.

Cite as *Cold Spring Harb Perspect Biol* doi: 10.1101/cshperspect.a019133

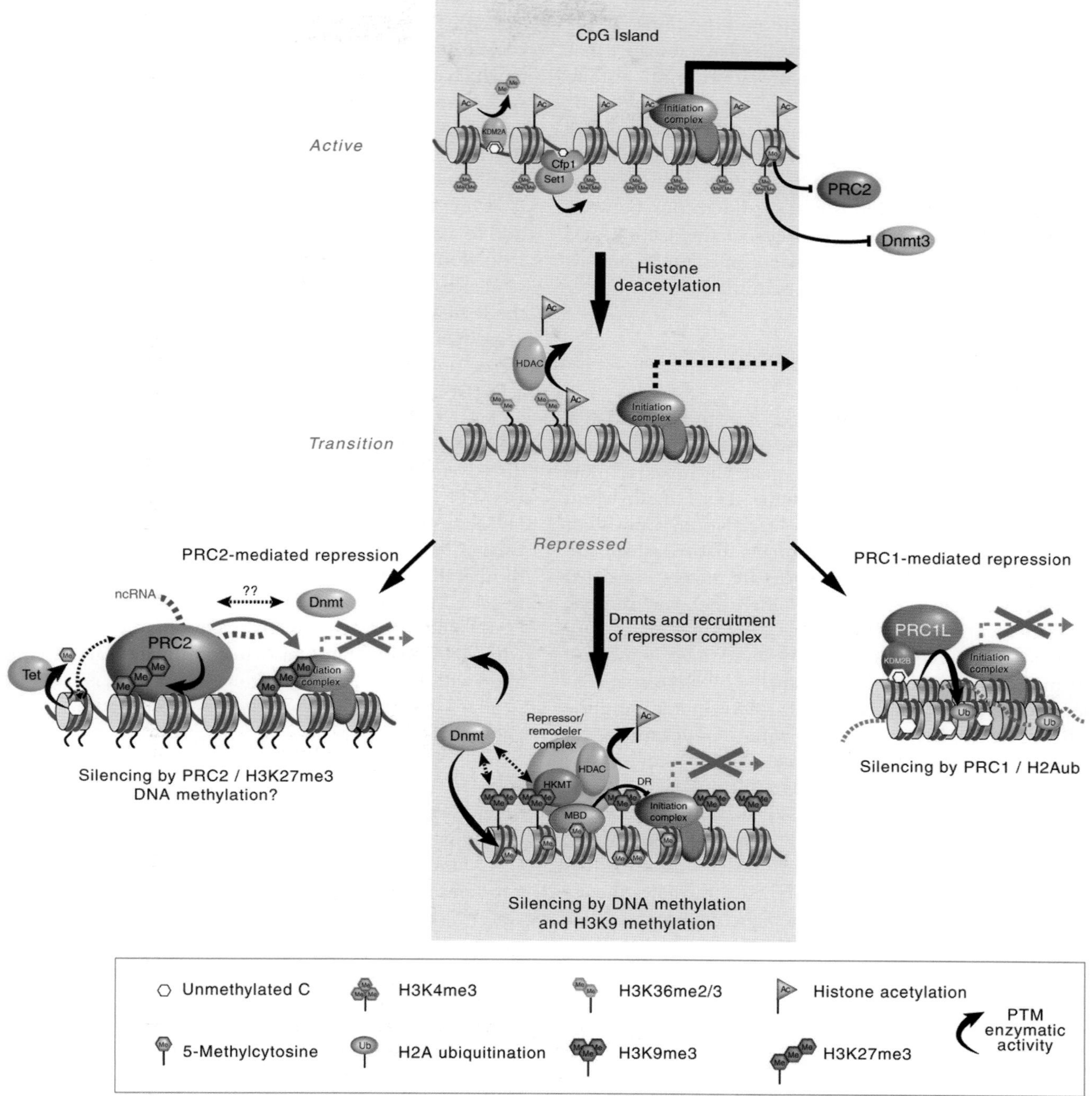

Figure 9. Recruitment of corepressors by methyl-CpG binding proteins. A hypothetical transition between an active, nonmethylated gene promoter and a repressed promoter whose silence is attributable to DNA methylation, as mediated by complexes containing an MBD protein such as MeCP2 (gray shading). The transition phase represents an intermediate step during which transcription is silenced and DNA methylation occurs. MeCP2 is envisaged to recruit the NCoR histone deacetylase (HDAC) complex and histone lysine methyltransferase (HKMT) activity to the methylated sites. In addition, there is some evidence that MeCP2 can directly repress (DR) transcription by contact with the transcription initiation complex. Other methyl-CpG binding proteins can also interact with and potentially recruit distinct corepressor complexes that include HKMT and/or HDAC activity. PRC1 and PRC2 are also involved in silencing gene expression through histone H3K27 methylation, catalyzed by PRC2 (*left* outcome). One of the mechanisms by which they function together to regulate the same set of target genes is through the recognition of the H3K27me3 mark by the chromodomain protein in the PRC1 complex. PRC1 can also be recruited by the CxxC domain–containing protein Kdm2b to effect gene silencing (*right* outcome). Proteins with known histone posttranslational modifying (PTM) activity are indicated.

components of chromatin, it is not surprising that DNA methylation interacts with histone modifications, as introduced in Section 4. Recent studies have revealed that interactions between DNA methylation and histone modifications play an important role in regulating chromatin dynamics and various biological processes (Cedar and Bergman 2009). Because histone modifications are highly conserved in evolution from yeast to mammals, whereas DNA methylation is not, the interaction between histone modifications, chromatin-associating factors, and DNA methylation may reveal new epigenetic mechanisms that are unique to vertebrates or mammals.

5.1 DNA Methylation and Histone Deacetylation

Previous studies have established that promoter DNA methylation is generally linked to transcriptional repression whereas histone acetylation is generally linked to transcriptional activation. Consequently, DNA methylation inversely correlates with histone acetylation. Consistent with this general trend, MBD proteins that recognize and bind to methylated DNA sequences are mostly associated with HDACs (Fig. 9) (Bird and Wolffe 1999). For example, the first identified MBD protein, MeCP2 associates with the Sin3A HDAC complex and is believed to be responsible for mediating transcriptional silencing by DNA methylation (Jones et al. 1998; Nan et al. 1998). In addition, the MBD protein, MBD2, associates with the nucleosome remodeling and histone deacetylase (NuRD) corepressor complex (Feng and Zhang 2001), and is believed to be the MeCP1 transcriptional repressor initially identified in cellular extracts (Meehan et al. 1989; Boyes and Bird 1991). Another example comes from a report demonstrating that interaction between MBD1 and HDAC3 mediates the establishment and maintenance of a silenced chromatin state directed by the chimeric PML-RARa gene, which functions as a constitutive transcriptional repressor in cancer (Villa et al. 2006).

5.2 DNA Methylation and H3K9 Methylation

Both DNA methylation and H3K9 methylation are associated with stable gene silencing, which is observed mainly in heterochromatin. Heterochromatin encompasses transcriptionally inert genomic regions that are enriched with repeat sequences and retrotransposons. It is believed that one of the major functions of DNA methylation is the silencing retrotransposons to maintain genome integrity (Bestor and Bourc'his 2004). Given that both H3K9 methylation and DNA methylation are features of heterochromatin, several studies have focused on the relationship between DNA methylation and H3K9 methylation. The first evidence that links DNA methylation to H3K9 methylation came from genetic studies in the fungus *Neurospora crassa*, in which deletion of the H3K9 methyltransferase Dim5 resulted in a complete loss of DNA methylation (Tamaru and Selker 2001; Ch. 10 [Aramayo and Selker 2013]). However, the relationship between DNA methylation and H3K9 methylation is more complex in mammals. Absence of the H3K9me3 methyltransferase, Suv39h1/Suv39h2, does not lead to a complete loss of DNA methylation, but results in reduction of DNA methylation in heterochromatic repeat sequences (Lehnertz et al. 2003). Other studies showed that the H3K9me2-specific methyltransferase G9a/GLP interacts directly with Dnmt1 (Esteve et al. 2006) and is required for de novo DNA methylation of retrotransposons in ES cells (Dong et al. 2008). The UHRF1 (ubiquitin-like PHD and RING finger domain-containing 1) protein further links DNA methylation to H3K9 methylation by binding directly to methylated H3K9 to maintain the stability of DNMT1 (Rothbart et al. 2012). Furthermore, MPP8, the H3K9me-binding protein, can mediate the interaction between G9a/GLP and Dnmt3a (Chang et al. 2011). Finally, a third H3K9 methyltransferase SETDB1 was reported to interact with Dnmt3a (Li et al. 2006). Therefore, multiple H3K9 methyltransferases can directly or indirectly associate with DNA methyltransferases to coordinate H3K9 methylation with DNA methylation.

5.3 H3K4 Methylation Inhibits Promoter CGI Methylation

A key question in the DNA methylation field has been how a CGI is protected from DNA methylation when most CpGs are methylated. The discovery of two proteins—Cfp1 and Kdm2a—that bind specifically to nonmethylated CpGs via the CXXC domain (Fig. 5) has brought us closer to understanding the mechanism. A feature of active gene promoters is its enrichment of the H3K4me3 mark (Guenther et al. 2007). An enzyme that writes this mark, Setd1, a member of the MLL family of H3K4 methyltransferases, has been shown to be recruited to CGIs in neuronal progenitor cells, through its interaction with the CpG binding protein Cfp1 (Thomson et al. 2010). Cfp1 binds only unmethylated CpG sequences found in CGIs through its CXXC domain. The enrichment of H3K4me3 at the CGIs, through the action of Setd1, is likely to be one of the reasons that prevent CGIs from being methylated. Conversely, unmethylated H3K4 enhances the access of the de novo DNA methylation enzymes, Dnm3a/Dnmt3b, to CGIs (Fig. 9). As Dnmt3a and Dnmt3b function in complex with Dnmt3L (Ooi et al. 2007), it was interesting to note that Dnmt3L selectively recognizes nucleosomes that lack H3K4

Cite as *Cold Spring Harb Perspect Biol* doi: 10.1101/cshperspect.a019133

methylation (elaborated in Ch. 6 [Cheng 2014]). Another factor contributing to the protection of CGIs from DNA methylation is an enrichment of Tet1 (Williams et al. 2011; Wu et al. 2011). Given that Tet1 has the capacity to oxidize 5mC to 5hmC and 5fC/5caC, and the latter can be processed by TDG followed by base excision repair (Fig. 6), any "accidental" methylation at CGIs could thus be potentially removed.

5.4 DNA Methylation and H3K27 Methylation

The first link between DNA methylation and H3K27 methylation was suggested by the demonstration that EZH2 can interact with all three DNMTs in vitro (Vire et al. 2006). However, subsequent studies in cancer cells revealed that gene silencing by H3K27me3 is independent of promoter DNA methylation (Kondo et al. 2008). Genome-wide location studies in ES cells have revealed that CpG-rich sequences are indeed enriched for the H3K27 methyltransferase PRC2 (Mendenhall et al. 2010) indicating that H3K27me3 enriched promoters are usually devoid of DNA methylation. However, many promoters marked by H3K27me3 in ES cells become silenced and DNA methylated during differentiation, suggesting a cell type–dependent cross talk between H3K27 methylation and DNA methylation (Fig. 9) (Mohn et al. 2008). In addition to cell types, the relationship between H3K27 methylation and DNA methylation can also be affected by the location of the modifications (Wu et al. 2010).

5.5 DNA Methylation and ATP-Dependent Chromatin Remodeling

In addition to histone modifications, ATP-dependent nucleosome remodelers have been linked to DNA methylation. Evidence that ATP-dependent remodeling factors are needed to ensure appropriate DNA methylation came initially from research in plants, in which the SNF2-like protein DDM1 was shown to be essential for full DNA methylation of the *Arabidopsis thaliana* genome (Jeddeloh et al. 1999). An equivalent dependence is seen in animals as mutations in human *ATRX* (Gibbons et al. 2000) and the mouse *Lsh2* genes (Dennis et al. 2001), both of which encode relatives of the chromatin remodeling protein SNF2, have significant effects on global DNA methylation patterns. Loss of the LSH2 protein, in particular, matches the phenotype of the *DDM1* mutation in *Arabidopsis*, in which methylation of highly repetitive DNA sequences is lost, although some DNA methylation is retained elsewhere in the genome. Perhaps efficient global DNA methylation of the genome requires conformational changes in chromatin structure by these chromatin remodeling factors so that DNMTs can gain access to the DNA. Collaboration between DNMTs and chromatin remodeling factors that allow them access may be particularly important in regions that are "heterochromatic" and inaccessible (Fig. 9).

6 DNA METHYLATION AND DISEASES

DNA methylation plays an important role in regulating tissue-specific gene expression and the repression of viral gene expression. It is also involved in the establishment and maintenance of genomic imprinting and X inactivation, and in the regulation of chromosome stability. Alterations in DNA methylation have been associated with cancer and several other diseases.

6.1 Methyl CpG as Hot Spots for Mutation

The disadvantage of having DNA methylation as an epigenetic system of cellular memory is the mutability of 5mC. Cytosine (C) deaminates spontaneously to give uracil (U), which is then mispaired with guanine. This potential mutation is recognized by uracil DNA glycosylases, which efficiently remove the inappropriate base and initiate repair to restore C in place of U. When 5mC deaminates, however, thymine (T) is formed. This also results in a mismatch, but the fact that T, unlike U, is a natural DNA base appears to interfere with the efficient repair of the lesion. As a result, the mutant thymine base can persist through DNA replication and is passed on to progeny cells as a C to T transition mutation. Mutations of this kind appear to be one of the most frequent single causes of genetic disease in humans, as approximately one third of all point mutations are C to T transitions at CpG sequences (Cooper and Youssoufian 1988). The instability of CpG over evolutionary time is further shown by the four- to fivefold underrepresentation of CpG in the mammalian genome (Bird 1980). The only exceptions are CGIs, within which CpGs are nonmethylated and therefore stable.

MBD4 is so far unique among methyl-CpG binding proteins in that it has enzymatic activity. The MBD4 carboxy-terminal domain is a thymine DNA glycosylase that can selectively remove T from a T-G mismatch in vitro (Hendrich et al. 1999). This activity would be expected of a DNA repair system that corrects 5mC deamination. Confirming this hypothesis, mice lacking MBD4 show enhanced mutability of methylated cytosine residues at a chromosomal reporter sequence (Millar et al. 2002). In addition, *Mbd4*-null mice acquire C to T transition mutations within the adenomatous polyposis coli gene and have an increased frequency of intestinal tumorigenesis

(Table 2). It is noteworthy that, in spite of the existence of a dedicated repair system, sites of cytosine methylation persist as hot spots for mutation.

6.2 Methylation of Promoter CGIs

Aberrant methylation of promoter CGIs can lead to silencing of gene expression and disease. Inactivation of growth inhibitory genes (and pathways) by methylation of promoter CGIs is the most common epigenetic mechanism contributing to cancer. What triggers aberrant methylation of promoter CGIs remains largely unknown (for further discussion, see Ch. 34 [Baylin and Jones 2014]).

Methylation of the CGI at the 5′ region of the FMR1 gene is shown to be the cause of fragile X syndrome (FXS). FXS is associated with triplet repeat expansion in the 5′ region of the FMR1 gene. Normal individuals have less than 50 CGG repeats, whereas the FXS patients have more than 200 CGG repeats. The genetic change caused by repeat expansion to more than 200 copies leads to de novo methylation of the CGIs and silencing of the FMR1 gene (Peprah 2012). In rare cases, individuals with more than 200 triplet repeats but an unmethylated CGI are completely normal. This indicates that promoter CGI methylation induced by CGG repeat expansion causes the disease and can be stably inherited (Peprah 2012).

6.3 DNA Hypomethylation and Chromosome Instability

Although DNA methylation is clearly mutagenic, there is evidence that its presence is beneficial with respect to chromosomal stability. Mice possessing ~10% of normal levels of DNA methylation, caused by a hypomorphic mutation of Dnmt1, acquire aggressive T cell lymphomas that often display trisomy of chromosome 15 (Gaudet et al. 2003). Mutations of DNMT3B in patients with ICF syndrome or inactivation of Dnmt3b in mice lead to various chromosomal aberrations including chromosome fusion, breakage, and aneuploidy (Ehrlich 2003; Dodge et al. 2005). These results are of interest because cancers often display reduced levels of DNA methylation, which may contribute to tumor initiation or progression (see also Ch. 34 [Baylin and Jones 2014]). One possible explanation for the result is that DNA methylation contributes to accurate chromosome segregation, and in its absence it is more frequent to have nondisjunction leading to chromosome aberrations. Alternatively, DNA methylation may suppress the expression and recombination of retrotransposons in the mammalian genome thereby protecting chromosomes from deleterious recombination. Indeed, DNA methylation has been shown to play a critical role in silencing the transcription of retrotransposons during embryonic development and spermatogenesis (Bestor and Bourc'his 2004).

6.4 Rett Syndrome

The existence of multiple methyl-CpG binding proteins with repressive properties argues that these may be important mediators of the methylation signal. This is illustrated most strikingly by the finding that mutations in the human *MECP2* gene are responsible for a severe neurological disorder called Rett syndrome (RTT). RTT affects females that are heterozygous for new mutations in the X-linked *MECP2* gene (Table 2) (Amir et al. 1999). Because of random X-chromosome inactivation, the patients are mosaic for expression of either the mutant or the wild-type gene. Affected girls develop apparently normally for 6–18 months, at which time they enter a crisis that leaves them with greatly impaired motor skills, repetitive hand movements, abnormal breathing, microcephaly, and other symptoms (Table 2). Males who are hemizygous for comparable mutations do not survive. Interestingly, duplication of the MECP2 gene also leads to a profound autism-like syndrome, suggesting that too much of this protein is also deleterious (Lubs et al. 1999; Meins et al. 2005). *Mecp2*-null mice are born and develop normally for several weeks, but they acquire neurological symptoms at about 6 weeks of age, leading to death at about 12 weeks. Several features of this phenotype recall human Rett syndrome, making the mouse a convincing model for the disorder (Guy et al. 2001). There is no evidence for increased brain cell death in MeCP2-deficient human patients or mice, raising the possibility that restoration of a functional *MECP2* gene could rescue the phenotype. Indeed, activation of the gene in severely affected male or female mice dramatically reverses neurological symptoms, raising the possibility that Rett syndrome in humans is curable (Guy et al. 2007).

Biochemical and immunocytochemical studies have established that MeCP2 is extremely abundant in the brain, particularly in neurons in which there are nearly 20 million molecules per nucleus (Skene et al. 2010). Accordingly, chromatin immunoprecipitation shows that MeCP2 is not targeted to specific genes, but coats the genome in a DNA methylation-dependent manner. Interestingly MeCP2 becomes phosphorylated at a specific serine residue on neuronal firing and this may play a role in modulating MeCP2 function (Zhou et al. 2006). The role of MeCP2 is not confined to neurons, however, as its presence is also required in glia (Lioy et al. 2011).

Given the potential role of MeCP2 as a transcriptional repressor, an attractive hypothesis to explain Rett syndrome

 Cite as *Cold Spring Harb Perspect Biol* doi: 10.1101/cshperspect.a019133

is that genes in the brain needing to be silenced escape repression in its absence. Although there is evidence for derepression of transcription from retrotransposons (Muotri et al. 2005) and repetitive elements (Skene et al. 2010) in the absence of MeCP2, results of genome-wide transcription analysis are complex. MeCP2-deficient mouse brains show modest over- and underexpression of many genes. Bdnf, for example, is consistently down-regulated in the MeCP2-deficient brain (Chen et al. 2003; Martinowich et al. 2003). Interestingly, overexpression of MeCP2 leads to increased expression of many genes that are down-regulated in *Mecp2*-null brains and vice versa. One potential explanation for this reciprocal behavior is that MeCP2 can activate as well as repress transcription (Chahrour et al. 2008). There is evidently more to learn about the precise role of MeCP2 in the brain.

7 FUTURE DIRECTIONS

Our understanding of the biological functions of DNA methylation in mammals has been growing steadily, but is far from complete. For instance, unlike genetic mutations, we know very little about the rate of changes in CpG methylation in mammals, or the intrinsic and environmental factors that induce changes in DNA methylation patterns. Accumulating evidence indicates that changes in DNA methylation and histone modifications contribute to the pathogenesis of many diseases such as cancer, diabetes, autoimmune, and neurological disorders as well as the aging process. Modulation of epigenetic states of the genome or gene expression is thus becoming a new therapeutic approach for the treatment of these diseases. Advances in the following areas will have significant impact on our understanding of the epigenetic mechanisms in disease and development of new medicine.

7.1 Profiling the Epigenome in Disease

Many complex diseases, such as type II diabetes, schizophrenia, autoimmune diseases, and some forms of cancer often cannot be explained by simple genetic alterations. The dynamic nature of epigenetic regulation is increasingly providing an alternative explanation for some of the features of complex diseases, which include late onset, gender effects, parent-of-origin effects, discordance of monozygotic twins, and fluctuation of symptoms (Gasser and Li 2011). Although growing evidence has linked aberrant DNA methylation and histone modifications to cancer, the role of epigenetic mechanisms in the etiology of many other complex diseases is largely unknown. Comparative studies of genome-wide DNA methylation patterns between normal and disease populations may provide insight into the epigenetic basis for various complex diseases (see also Ch. 33 [Zoghbi and Beaudet 2014]).

Recent advances in next generation sequencing technology has made it technically and financially possible to perform genome-wide analysis of DNA methylation patterns (i.e., the methylome) in normal and disease tissues, using bisulfite sequencing or MeDIP methods (Bibikova and Fan 2010) (Box 1). These approaches have been used to analyze DNA methylation changes, for example, during stem cell differentiation into neurons, and comparative analysis between normal and cancer tissues (Cortese et al. 2011; Hon et al. 2012). Such studies are beginning to shed light on how DNA methylation is regulated and its role in cell differentiation and disease. Two methods for the mapping of genome-wide 5hmC patterns at single base resolution have also been introduced recently (Booth et al. 2012; Yu et al. 2012). These technological breakthroughs mark a new era for understanding the role of DNA methylation and demethylation in normal development and disease.

7.2 Genetic and Environmental Factors That Induce Epigenetic Changes

It is well established that DNA methylation patterns in the mammalian genome are highly regulated during development. Genetic alterations such as CGG triplet repeat expansion in the FMR1 gene provide examples of genetic lesions that can lead to heritable changes in promoter CGI methylation and transcription repression. Mutations in genes encoding key epigenetic regulators such as MeCP2 and DNMT3B can also alter the epigenome and gene expression patterns, which lead to disease. An increasing number of other key epigenetic enzymes or chromatin-associated proteins, when mutated in cancer or other diseases, results in dramatic changes to the epigenome (You and Jones 2012).

How environmental factors may affect DNA methylation and gene expression is less well understood, although some recent studies are beginning to shed light on how they may induce epigenetic changes that can have long-lasting biological effects. One such example is the observation that rat maternal behavior produces stable alterations in DNA methylation in the offspring. Weaver and coworkers have reported that baby rats receiving different levels of maternal care have differences in DNA methylation in the promoter region of the glucocorticoid receptor (GR) gene, which are inversely correlated with GR expression, and these differences persist into adulthood (Weaver et al. 2004). Other factors, such as dietary supplements, drugs, and smoking can also impact the epigenome and lead to physiological or pathological changes.

7.3 Modulation of Epigenetic State and Reprogramming by Tet-Mediated 5mC Oxidation

One of the most exciting advances in the DNA methylation field is the demonstration that 5mC can be oxidized by Tet proteins to generate 5hmC, 5fC/5caC, and then later processed by the DNA repair mechanism enzyme TDG (Wu and Zhang 2011). Although there is clear evidence suggesting global loss of DNA methylation of the paternal genome is attributable to Tet-mediated oxidation (Wu and Zhang 2011), whether Tet proteins play a similar role in PGC reprogramming awaits to be determined (for more discussion, see Ch. 27 [Reik and Surani 2014]). Given the accumulation of 5hmC in certain tissues and cell types, 5hmC has also been proposed to serve as an epigenetic mark that may mediate a specific function. In this regard, a component of the NuRD complex has been shown to recognize and bind to 5hmC (Yildirim et al. 2011). Future studies along this line should reveal whether 5hmC signals as a new epigenetic mark or simply as an intermediate for DNA demethylation. Activation of pluripotency genes, such as Oct4 and Nanog, requires demethylation of their promoters. Thus, it is not surprising that factors implicated in DNA demethylation have been reported to facilitate somatic cell reprogramming (Bhutani et al. 2010). Whether Tet proteins contribute to Oct4 and Nanog activation during reprogramming should be determined in the near future.

7.4 Modulation of Reversible Epigenetic States

Most, if not all, epigenetic modifications are reversible, which makes modulation of epigenetic states a promising new therapeutic option for cancer and other diseases. A number of agents that alter patterns of DNA methylation or inhibit HDACs are in clinical use for the treatment of cancer (Baylin and Jones 2011). The demethylating agent 5-azacytidine, for instance, has been approved by the U.S. Food and Drug Administration for the treatment of myelodysplastic syndrome, a heterogeneous disease characterized by morphologic dysplasia of hematopoietic cells in 2004. The clinical use of 5-azacytidine and other nucleoside analogs, however, is limited by their toxicity, partly because these compounds are being incorporated into DNA. This has encouraged the search for agents that can inhibit DNA methyltransferases directly or target other epigenetic regulators that can modulate DNA methyltransferase activities or CGI methylation. Because DNA methylation is just one component of the complex epigenetic regulatory network, one approach to maximize therapeutic effects and minimize toxicity is combination therapy using DNA methyltransferase or HDAC inhibitors in combination with other anticancer therapeutics (see Ch. 34 [Baylin and Jones 2014] for more detail).

ACKNOWLEDGMENTS

We wish to thank Adrian Bird for his contribution to this chapter in the first edition, and critical reading and expert comments during the writing of this edition's chapter.

REFERENCES

*Reference is also in this book.

Amir RE, Van den Veyver IB, Wan M, Tran CQ, Francke U, Zoghbi HY. 1999. Rett syndrome is caused by mutations in X-linked MECP2, encoding methyl-CpG-binding protein 2. *Nat Genet* **23:** 185–188.

* Aramayo R, Selker EU. 2013. *Neurospora crassa,* a model system for epigenetics research. *Cold Spring Harb Perspect Biol* **5:** a017921.

* Barlow D, Bartolomei M. 2014. Genomic imprinting in mammals. *Cold Spring Harb Perspect Biol* **6:** a018382.

Baylin SB, Jones PA. 2011. A decade of exploring the cancer epigenome—Biological and translational implications. *Nat Rev Cancer* **11:** 726–734.

* Baylin S, Jones P. 2014. Epigenetics and cancer. *Cold Spring Harb Perspect Biol* doi: 10.1101/cshperspect.a019505.

Bell AC, Felsenfeld G. 2000. Methylation of a CTCF-dependent boundary controls imprinted expression of the *Igf2* gene. *Nature* **405:** 482–485.

Bell AC, West AG, Felsenfeld G. 1999. The protein CTCF is required for the enhancer blocking activity of vertebrate insulators. *Cell* **98:** 387–396.

Bestor TH, Bourc'his D. 2004. Transposon silencing and imprint establishment in mammalian germ cells. *Cold Spring Harb Symp Quant Biol* **69:** 381–387.

Bestor TH, Ingram VM. 1983. Two DNA methyltransferases from murine erythroleukemia cells: Purification, sequence specificity, and mode of interaction with DNA. *Proc Natl Acad Sci* **80:** 5559–5563.

Bhutani N, Brady JJ, Damian M, Sacco A, Corbel SY, Blau HM. 2010. Reprogramming towards pluripotency requires AID-dependent DNA demethylation. *Nature* **463:** 1042–1047.

Bibikova M, Fan JB. 2010. Genome-wide DNA methylation profiling. *Wiley Interdiscip Rev Systems Biol Med* **2:** 210–223.

Bird AP. 1980. DNA methylation and the frequency of CpG in animal DNA. *Nucleic Acids Res* **8:** 1499–1594.

Bird A, Wolffe AP. 1999. Methylation-induced repression—Belts, braces and chromatin. *Cell* **99:** 451–454.

Bird A, Taggart M, Frommer M, Miller OJ, Macleod D. 1985. A fraction of the mouse genome that is derived from islands of non-methylated, CpG-rich DNA. *Cell* **40:** 91–99.

Blackledge NP, Zhou JC, Tolstorukov MY, Farcas AM, Park PJ, Klose RJ. 2010. CpG islands recruit a histone H3 lysine 36 demethylase. *Mol Cell* **38:** 179–190.

* Blackledge NP, Thomson JP, Skene PJ. 2013. CpG island chromatin is shaped by recruitment of ZF-CxxC proteins. *Cold Spring Harb Perspect Biol* **5:** a018648.

Booth MJ, Branco MR, Ficz G, Oxley D, Krueger F, Reik W, Balasubramanian S. 2012. Quantitative sequencing of 5-methylcytosine and 5-hydroxymethylcytosine at single-base resolution. *Science* **336:** 934–937.

Boyes J, Bird A. 1991. DNA methylation inhibits transcription indirectly via a methyl-CpG binding protein. *Cell* **64:** 1123–1134.

Brenet F, Moh M, Funk P, Feierstein E, Viale AJ, Socci ND, Scandura JM. 2011. DNA methylation of the first exon is tightly linked to transcriptional silencing. *PLoS One* **6:** e14524.

* Brockdorff N, Turner B. 2014. Dosage compensation in mammals. *Cold Spring Harb Perspect Biol* doi: 10.1101/cshperspect.a019406.

Bruniquel D, Schwartz RH. 2003. Selective, stable demethylation of the interleukin-2 gene enhances transcription by an active process. *Nat Immunol* **4:** 235–240.

Cedar H, Bergman Y. 2009. Linking DNA methylation and histone modification: Patterns and paradigms. *Nat Rev Genet* **10:** 295–304.

Chahrour M, Jung SY, Shaw C, Zhou X, Wong ST, Qin J, Zoghbi HY. 2008. MeCP2, a key contributor to neurological disease, activates and represses transcription. *Science* **320:** 1224–1229.

Chang Y, Sun L, Kokura K, Horton JR, Fukuda M, Espejo A, Izumi V, Koomen JM, Bedford MT, Zhang X, et al. 2011. MPP8 mediates the interactions between DNA methyltransferase Dnmt3a and H3K9 methyltransferase GLP/G9a. *Nat Commun* **2:** 533.

Chen WG, Chang Q, Lin Y, Meissner A, West AE, Griffith EC, Jaenisch R, Greenberg ME. 2003. Derepression of BDNF transcription involves calcium-dependent phosphorylation of MeCP2. *Science* **302:** 885–889.

* Cheng X. 2014. Structural and functional coordination of DNA and histone methylation. *Cold Spring Harb Perspect Biol* **6:** a018747.

Cooper DN, Youssoufian H. 1988. The CpG dinucleotide and human genetic disease. *Hum Genet* **78:** 151–155.

Cortazar D, Kunz C, Selfridge J, Lettieri T, Saito Y, MacDougall E, Wirz A, Schuermann D, Jacobs AL, Siegrist F, et al. 2011. Embryonic lethal phenotype reveals a function of TDG in maintaining epigenetic stability. *Nature* **470:** 419–423.

Cortellino S, Xu J, Sannai M, Moore R, Caretti E, Cigliano A, Le Coz M, Devarajan K, Wessels A, Soprano D, et al. 2011. Thymine DNA glycosylase is essential for active DNA demethylation by linked deamination-base excision repair. *Cell* **146:** 67–79.

Cortese R, Lewin J, Backdahl L, Krispin M, Wasserkort R, Eckhardt F, Beck S. 2011. Genome-wide screen for differential DNA methylation associated with neural cell differentiation in mouse. *PLoS One* **6:** e26002.

Deaton AM, Bird A. 2011. CpG islands and the regulation of transcription. *Genes Dev* **25:** 1010–1022.

Dennis K, Fan T, Geiman T, Yan Q, Muegge K. 2001. Lsh, a member of the SNF2 family, is required for genome-wide methylation. *Genes Dev* **15:** 2940–2944.

Dodge JE, Okano M, Dick F, Tsujimoto N, Chen T, Wang S, Ueda Y, Dyson N, Li E. 2005. Inactivation of Dnmt3b in mouse embryonic fibroblasts results in DNA hypomethylation, chromosomal instability, and spontaneous immortalization. *J Biol Chem* **280:** 17986–17991.

Dong KB, Maksakova IA, Mohn F, Leung D, Appanah R, Lee S, Yang HW, Lam LL, Mager DL, Schubeler D, et al. 2008. DNA methylation in ES cells requires the lysine methyltransferase G9a but not its catalytic activity. *EMBO J* **27:** 2691–2701.

Ehrlich M. 2003. The ICF syndrome, a DNA methyltransferase 3B deficiency and immunodeficiency disease. *Clin Immunol* **109:** 17–28.

* Elgin S, Reuter G. 2013. Positive effect variegation, heterochromatin formation, and gene silencing in *Drosophila*. *Cold Spring Harb Perspect Biol* **5:** a017780.

Esteve PO, Chin HG, Smallwood A, Feehery GR, Gangisetty O, Karpf AR, Carey MF, Pradhan S. 2006. Direct interaction between DNMT1 and G9a coordinates DNA and histone methylation during replication. *Genes Dev* **20:** 3089–3103.

Feng Q, Zhang Y. 2001. The MeCP1 complex represses transcription through preferential binding, remodeling, and deacetylating methylated nucleosomes. *Genes Dev* **15:** 827–832.

Frommer M, McDonald LE, Millar DS, Collis CM, Watt F, Grigg GW, Molloy PL, Paul CL. 1992. A genomic sequencing protocol that yields a positive display of 5-methylcytosine residues in individual DNA strands. *Proc Natl Acad Sci* **89:** 1827–1831.

Gasser SM, Li E. 2011. Epigenetics and disease: Pharmaceutical opportunities. Preface. *Prog Drug Res* **67:** v–viii.

Gaudet F, Hodgson JG, Eden A, Jackson-Grusby L, Dausman J, Gray JW, Leonhardt H, Jaenisch R. 2003. Induction of tumors in mice by genomic hypomethylation. *Science* **300:** 489–492.

Gibbons RJ, McDowell TL, Raman S, O'Rourke DM, Garrick D, Ayyub H, Higgs DR. 2000. Mutations in ATRX, encoding a SWI/SNF-like protein, cause diverse changes in the pattern of DNA methylation. *Nat Genet* **24:** 368–371.

Gu TP, Guo F, Yang H, Wu HP, Xu GF, Liu W, Xie ZG, Shi L, He X, Jin SG, et al. 2011. The role of Tet3 DNA dioxygenase in epigenetic reprogramming by oocytes. *Nature* **477:** 606–610.

Guenther MG, Levine SS, Boyer LA, Jaenisch R, Young RA. 2007. A chromatin landmark and transcription initiation at most promoters in human cells. *Cell* **130:** 77–88.

Guy J, Hendrich B, Holmes M, Martin JE, Bird A. 2001. A mouse *Mecp2*-null mutation causes neurological symptoms that mimic Rett syndrome. *Nat Genet* **27:** 322–326.

Guy J, Gan J, Selfridge J, Cobb S, Bird A. 2007. Reversal of neurological defects in a mouse model of Rett syndrome. *Science* **315:** 1143–1147.

Hajkova P, Erhardt S, Lane N, Haaf T, El-Maarri O, Reik W, Walter J, Surani MA. 2002. Epigenetic reprogramming in mouse primordial germ cells. *Mech Dev* **117:** 15–23.

Harbers K, Schnieke H, Stuhlmann H, Jahner D, Jaenisch B. 1981. DNA methylation and gene expression; Endogenous retroviral genome becomes infectious after molecular cloning. *Proc Natl Acad Sci* **78:** 7609–7613.

Hata K, Okano M, Lei H, Li E. 2002. Dnmt3L cooperates with the Dnmt3 family of de novo DNA methyltransferases to establish maternal imprints in mice. *Development* **129:** 1983–1993.

He YF, Li BZ, Li Z, Liu P, Wang Y, Tang Q, Ding J, Jia Y, Chen Z, Li L, et al. 2011. Tet-mediated formation of 5-carboxylcytosine and its excision by TDG in mammalian DNA. *Science* **333:** 1303–1307.

Hendrich B, Hardeland U, Ng H-H, Jiricny J, Bird A. 1999. The thymine glycosylase MBD4 can bind to the product of deamination at methylated CpG sites. *Nature* **401:** 301–304.

Hendrich B, Guy J, Ramsahoye B, Wilson VA, Bird A. 2001. Closely related proteins Mbd2 and Mbd3 play distinctive but interacting roles in mouse development. *Genes Dev* **15:** 710–723.

Herman JG, Graff JR, Myohanen S, Nelkin BD, Baylin SB. 1996. Methylation-specific PCR: A novel PCR assay for methylation status of CpG islands. *Proc Natl Acad Sci* **93:** 9821–9826.

Holliday R, Pugh JE. 1975. DNA modification mechanisms and gene activity during development. *Science* **187:** 226–232.

Hon GC, Hawkins RD, Caballero OL, Lo C, Lister R, Pelizzola M, Valsesia A, Ye Z, Kuan S, Edsall LE, et al. 2012. Global DNA hypomethylation coupled to repressive chromatin domain formation and gene silencing in breast cancer. *Genome Res* **22:** 246–258.

Hutchins A, Mullen A, Lee H, Barner K, High F, Hendrich B, Bird A, Reiner S. 2002. Gene silencing quantitatively controls the function of a developmental trans-activator. *Mol Cell* **10:** 81–91.

Inoue A, Zhang Y. 2011. Replication-dependent loss of 5-hydroxymethylcytosine in mouse preimplantation embryos. *Science* **334:** 194.

Inoue A, Shen L, Dai Q, He C, Zhang Y. 2011. Generation and replication-dependent dilution of 5fC and 5caC during mouse preimplantation development. *Cell Res* **21:** 1670–1676.

Iqbal K, Jin SG, Pfeifer GP, Szabo PE. 2011. Reprogramming of the paternal genome upon fertilization involves genome-wide oxidation of 5-methylcytosine. *Proc Natl Acad Sci* **108:** 3642–3647.

Irizarry RA, Ladd-Acosta C, Wen B, Wu Z, Montano C, Onyango P, Cui H, Gabo K, Rongione M, Webster M, et al. 2009. The human colon cancer methylome shows similar hypo- and hypermethylation at conserved tissue-specific CpG island shores. *Nat Genet* **41:** 178–186.

Ito S, D'Alessio AC, Taranova OV, Hong K, Sowers LC, Zhang Y. 2010. Role of Tet proteins in 5mC to 5hmC conversion, ES-cell self-renewal and inner cell mass specification. *Nature* **466:** 1129–1133.

Ito S, Shen L, Dai Q, Wu SC, Collins LB, Swenberg JA, He C, Zhang Y. 2011. Tet proteins can convert 5-methylcytosine to 5-formylcytosine and 5-carboxylcytosine. *Science* **333:** 1300–1303.

Jahner D, Stuhlmann H, Stewart CL, Harbers K, Lohler J, Simon I, Jaenisch R. 1982. De novo methylation and expression of retroviral genomes during mouse embryogenesis. *Nature* **298:** 623–628.

Jeddeloh JA, Stokes TL, Richards EJ. 1999. Maintenance of genomic methylation requires a SW12/SNF2-like protein. *Nat Genet* **22:** 94–97.

Jones PA, Taylor SM. 1980. Cellular differentiation, cytidine analogues and DNA methylation. *Cell* **20:** 85–93.

Jones PL, Veenstra GJ, Wade PA, Vermaak D, Kass SU, Landsberger N, Strouboulis J, Wolffe AP. 1998. Methylated DNA and MeCP2 recruit histone deacetylase to repress transcription. *Nat Genet* **19:** 187–191.

Kaneda M, Sado T, Hata K, Okano M, Tsujimoto N, Li E, Sasaki H. 2004. Role of de novo DNA methyltransferases in initiation of genomic imprinting and X-chromosome inactivation. *Cold Spring Harb Symp Quant Biol* **69:** 125–129.

Kangaspeska S, Stride B, Metivier R, Polycarpou-Schwarz M, Ibberson D, Carmouche RP, Benes V, Gannon F, Reid G. 2008. Transient cyclical methylation of promoter DNA. *Nature* **452:** 112–115.

Kondo Y, Shen L, Cheng AS, Ahmed S, Boumber Y, Charo C, Yamochi T, Urano T, Furukawa K, Kwabi-Addo B, et al. 2008. Gene silencing in cancer by histone H3 lysine 27 trimethylation independent of promoter DNA methylation. *Nat Genet* **40:** 741–750.

Kriaucionis S, Heintz N. 2009. The nuclear DNA base 5-hydroxymethylcytosine is present in Purkinje neurons and the brain. *Science* **324:** 929–930.

Kriaucionis S, Tahiliani M. 2014. Expanding the epigenetic landscape: Novel modifications of cytosine in genomic DNA. *Cold Spring Harb Perspect Biol* **6:** a018630.

Lehnertz B, Ueda Y, Derijck AA, Braunschweig U, Perez-Burgos L, Kubicek S, Chen T, Li E, Jenuwein T, Peters AH. 2003. Suv39h-mediated histone H3 lysine 9 methylation directs DNA methylation to major satellite repeats at pericentric heterochromatin. *Curr Biol* **13:** 1192–1200.

Lei H, Oh SP, Okano M, Juttermann R, Gos KA, Jaenisch R, Li E. 1996. De novo DNA cytosine methyltransferase activities in mouse embryonic stem cells. *Development* **122:** 3195–3205.

Li E, Bestor TH, Jaenisch R. 1992. Targeted mutation of the DNA methyltransferase gene results in embryonic lethality. *Cell* **69:** 915–926.

Li E, Beard C, Jaenisch R. 1993. Role for DNA methylation in genomic imprinting. *Nature* **366:** 362–365.

Li H, Rauch T, Chen ZX, Szabo PE, Riggs AD, Pfeifer GP. 2006. The histone methyltransferase SETDB1 and the DNA methyltransferase DNMT3A interact directly and localize to promoters silenced in cancer cells. *J Biol Chem* **281:** 19489–19500.

Lioy DT, Garg SK, Monaghan CE, Raber J, Foust KD, Kaspar BK, Hirrlinger PG, Kirchhoff F, Bissonnette JM, Ballas N, et al. 2011. A role for glia in the progression of Rett's syndrome. *Nature* **475:** 497–500.

Lubs H, Abidi F, Bier JA, Abuelo D, Ouzts L, Voeller K, Fennell E, Stevenson RE, Schwartz CE, Arena F. 1999. XLMR syndrome characterized by multiple respiratory infections, hypertelorism, severe CNS deterioration and early death localizes to distal Xq28. *Am J Med Genet* **85:** 243–248.

Maiti A, Drohat AC. 2011. Thymine DNA glycosylase can rapidly excise 5-formylcytosine and 5-carboxylcytosine: Potential implications for active demethylation of CpG sites. *J Biol Chem* **286:** 35334–35338.

Martinowich K, Hattori D, Wu H, Fouse S, He F, Hu Y, Fan G, Sun YE. 2003. DNA methylation-related chromatin remodeling in activity-dependent BDNF gene regulation. *Science* **302:** 890–893.

Mayer W, Niveleau A, Walter J, Fundele R, Haaf T. 2000. Demethylation of the zygotic paternal genome. *Nature* **403:** 501–502.

McKeon C, Ohkubo H, Pastan I, de Crombrugghe B. 1982. Unusual methylation pattern of the α 2 (I) collagen gene. *Cell* **29:** 203–210.

Meehan RR, Lewis JD, McKay S, Kleiner EL, Bird AP. 1989. Identification of a mammalian protein that binds specifically to DNA containing methylated CpGs. *Cell* **58:** 499–507.

Meins M, Lehmann J, Gerresheim F, Herchenbach J, Hagedorn M, Hameister K, Epplen JT. 2005. Submicroscopic duplication in Xq28 causes increased expression of the *MECP2* gene in a boy with severe mental retardation and features of Rett syndrome. *J Med Genet* **42:** e12.

Mendenhall EM, Koche RP, Truong T, Zhou VW, Issac B, Chi AS, Ku M, Bernstein BE. 2010. GC-rich sequence elements recruit PRC2 in mammalian ES cells. *PLoS Genet* **6:** e1001244.

Metivier R, Gallais R, Tiffoche C, Le Peron C, Jurkowska RZ, Carmouche RP, Ibberson D, Barath P, Demay F, Reid G, et al. 2008. Cyclical DNA methylation of a transcriptionally active promoter. *Nature* **452:** 45–50.

Millar CB, Guy J, Sansom OJ, Selfridge J, MacDougall E, Hendrich B, Keightley PD, Bishop SM, Clarke AR, Bird A. 2002. Enhanced CpG mutability and tumorigenesis in MBD4-deficient mice. *Science* **297:** 403–405.

Mohandas T, Sparkes RS, Shapiro LJ. 1981. Reactivation of an inactive human X-chromosome: Evidence for X-inactivation by DNA methylation. *Science* **211:** 393–396.

Mohn F, Weber M, Rebhan M, Roloff TC, Richter J, Stadler MB, Bibel M, Schubeler D. 2008. Lineage-specific polycomb targets and de novo DNA methylation define restriction and potential of neuronal progenitors. *Mol Cell* **30:** 755–766.

Muotri AR, Chu VT, Marchetto MC, Deng W, Moran JV, Gage FH. 2005. Somatic mosaicism in neuronal precursor cells mediated by L1 retrotransposition. *Nature* **435:** 903–910.

Nan X, Ng H-H, Johnson CA, Laherty CD, Turner BM, Eisenman RN, Bird A. 1998. Transcriptional repression by the methyl-CpG-binding protein MeCP2 involves a histone deacetylase complex. *Nature* **393:** 386–389.

Okano M, Xie S, Li E. 1998a. Cloning and characterization of a family of novel mammalian DNA (cytosine-5) methyltransferases. *Nat Genet* **19:** 219–220.

Okano M, Xie S, Li E. 1998b. Dnmt2 is not required for de novo and maintenance methylation of viral DNA in embryonic stem cells. *Nucleic Acids Res* **26:** 2536–2540.

Okano M, Bell DW, Haber DA, Li E. 1999. DNA methyltransferases Dnmt3a and Dnmt3b are essential for de novo methylation and mammalian development. *Cell* **99:** 247–257.

Ooi SK, Bestor TH. 2008. The colorful history of active DNA demethylation. *Cell* **133:** 1145–1148.

Ooi SK, Qiu C, Bernstein E, Li K, Jia D, Yang Z, Erdjument-Bromage H, Tempst P, Lin SP, Allis CD, et al. 2007. DNMT3L connects unmethylated lysine 4 of histone H3 to de novo methylation of DNA. *Nature* **448:** 714–717.

Oswald J, Engemann S, Lane N, Mayer W, Olek A, Fundele R, Dean W, Reik W, Walter J. 2000. Active demethylation of the paternal genome in the mouse zygote. *Curr Biol* **10:** 475–478.

Peprah E. 2012. Fragile X syndrome: The FMR1 CGG repeat distribution among world populations. *Ann Hum Genet* **76:** 178–191.

Pfaffeneder T, Hackner B, Truss M, Munzel M, Muller M, Deiml CA, Hagemeier C, Carell T. 2011. The discovery of 5-formylcytosine in embryonic stem cell DNA. *Angew Chem Int Ed Engl* **50:** 7008–7012.

* Pikaard CJ, Mittelsten Scheid O. 2014. Epigenetic regulation in plants. *Cold Spring Harb Perspect Biol* doi: 10.1101/cshperspect.a019315.

Posfai J, Bhagwat AS, Posfai G, Roberts RJ. 1989. Predictive motifs derived from cytosine methyltransferases. *Nucleic Acids Res* **17:** 2421–2435.

Prokhortchouk A, Hendrich B, Jorgensen H, Ruzov A, Wilm M, Georgiev G, Bird A, Prokhortchouk E. 2001. The p120 catenin partner Kaiso is a DNA methylation-dependent transcriptional repressor. *Genes Dev* **15:** 1613–1618.

* Reik W, Surani A. 2014. Germline and pluripotent stem cells. *Cold Spring Harb Perspect Biol* doi: 10.1101/cshperspect.a019422.

Riggs AD. 1975. X inactivation, differentiation, and DNA methylation. *Cytogenet Cell Genet* **14:** 9–25.

Rothbart SB, Krajewski K, Nady N, Tempel W, Xue S, Badeaux AI, Barsyte-Lovejoy D, Martinez JY, Bedford MT, Fuchs SM, et al. 2012. Association of UHRF1 with methylated H3K9 directs the maintenance of DNA methylation. *Nat Struct Mol Biol* **19:** 1155–1160.

Rougier N, Bourc'his D, Gomes DM, Niveleau A, Plachot M, Paldi A, Viegas-Pequignot E. 1998. Chromosome methylation patterns during mammalian preimplantation development. *Genes Dev* **12:** 2108–2113.

Sarraf SA, Stancheva I. 2004. Methyl-CpG binding protein MBD1 couples histone H3 methylation at lysine 9 by SETDB1 to DNA replication and chromatin assembly. *Mol Cell* **15:** 595–605.

Sasaki H, Matsui Y. 2008. Epigenetic events in mammalian germ-cell development: Reprogramming and beyond. *Nat Rev Genet* **9:** 129–140.

Skene PJ, Illingworth RS, Webb S, Kerr AR, James KD, Turner DJ, Andrews R, Bird AP. 2010. Neuronal MeCP2 is expressed at near histone-octamer levels and globally alters the chromatin state. *Mol Cell* **37:** 457–468.

Stadler MB, Murr R, Burger L, Ivanek R, Lienert F, Scholer A, van Nimwegen E, Wirbelauer C, Oakeley EJ, Gaidatzis D, et al. 2011. DNA-binding factors shape the mouse methylome at distal regulatory regions. *Nature* **480:** 490–495.

Stewart CL, Stuhlmann H, Jahner D, Jaenisch R. 1982. De novo methylation and infectivity of retroviral genomes introduced into embryonal carcinoma cells. *Proc Natl Acad Sci* **79:** 4098–4102.

Tahiliani M, Koh KP, Shen Y, Pastor WA, Bandukwala H, Brudno Y, Agarwal S, Iyer LM, Liu DR, Aravind L, et al. 2009. Conversion of 5-methylcytosine to 5-hydroxymethylcytosine in mammalian DNA by MLL partner TET1. *Science* **324:** 930–935.

Tamaru H, Selker EU. 2001. A histone H3 methyltransferase controls DNA methylation in *Neurospora crassa*. *Nature* **414:** 277–283.

Thomson JP, Skene PJ, Selfridge J, Clouaire T, Guy J, Webb S, Kerr AR, Deaton A, Andrews R, James KD, et al. 2010. CpG islands influence chromatin structure via the CpG-binding protein Cfp1. *Nature* **464:** 1082–1086.

Tost J, Gut IG. 2007. DNA methylation analysis by pyrosequencing. *Nat Protoc* **2:** 2265–2275.

Valinluck V, Sowers LC. 2007. Endogenous cytosine damage products alter the site selectivity of human DNA maintenance methyltransferase DNMT1. *Cancer Res* **67:** 946–950.

Villa R, Morey L, Raker VA, Buschbeck M, Gutierrez A, De Santis F, Corsaro M, Varas F, Bossi D, Minucci S, et al. 2006. The methyl-CpG binding protein MBD1 is required for PML-RARα function. *Proc Natl Acad Sci* **103:** 1400–1405.

Vire E, Brenner C, Deplus R, Blanchon L, Fraga M, Didelot C, Morey L, Van Eynde A, Bernard D, Vanderwinden JM, et al. 2006. The Polycomb group protein EZH2 directly controls DNA methylation. *Nature* **439:** 871–874.

Wade PA, Gegonne A, Jones PL, Ballestar E, Aubry F, Wolffe AP. 1999. Mi-2 complex couples DNA methylation to chromatin remodelling and histone deacetylation. *Nat Genet* **23:** 62–66.

Watt F, Molloy PL. 1988. Cytosine methylation prevents binding to DNA of a HeLa cell transcription factor required for optimal expression of the adenovirus late promoter. *Genes Dev* **2:** 1136–1143.

Weaver IC, Cervoni N, Champagne FA, D'Alessio AC, Sharma S, Seckl JR, Dymov S, Szyf M, Meaney MJ. 2004. Epigenetic programming by maternal behavior. *Nat Neurosci* **7:** 847–854.

Weber M, Davies JJ, Wittig D, Oakeley EJ, Haase M, Lam WL, Schubeler D. 2005. Chromosome-wide and promoter-specific analyses identify sites of differential DNA methylation in normal and transformed human cells. *Nat Genet* **37:** 853–862.

Williams K, Christensen J, Pedersen MT, Johansen JV, Cloos PA, Rappsilber J, Helin K. 2011. TET1 and hydroxymethylcytosine in transcription and DNA methylation fidelity. *Nature* **473:** 343–348.

Wolf SF, Jolly DJ, Lunnen KD, Friedman T, Migeon BR. 1984. Methylation of the hypoxanthine phosphoribosyltransferase locus on the human X-chromosome: Implications for X-chromosome inactivation. *Proc Natl Acad Sci* **81:** 2806–2810.

Wossidlo M, Nakamura T, Lepikhov K, Marques CJ, Zakhartchenko V, Boiani M, Arand J, Nakano T, Reik W, Walter J. 2011. 5-Hydroxymethylcytosine in the mammalian zygote is linked with epigenetic reprogramming. *Nat Commun* **2:** 241.

Wu SC, Zhang Y. 2010. Active DNA demethylation: Many roads lead to Rome. *Nat Rev Mol Cell Biol* **11:** 607–620.

Wu H, Zhang Y. 2011. Mechanisms and functions of Tet protein-mediated 5-methylcytosine oxidation. *Genes Dev* **25:** 2436–2452.

Wu H, Coskun V, Tao J, Xie W, Ge W, Yoshikawa K, Li E, Zhang Y, Sun YE. 2010. Dnmt3a-dependent nonpromoter DNA methylation facilitates transcription of neurogenic genes. *Science* **329:** 444–448.

Wu H, D'Alessio AC, Ito S, Xia K, Wang Z, Cui K, Zhao K, Sun YE, Zhang Y. 2011. Dual functions of Tet1 in transcriptional regulation in mouse embryonic stem cells. *Nature* **473:** 389–393.

Yamazaki Y, Mann MR, Lee SS, Marh J, McCarrey JR, Yanagimachi R, Bartolomei MS. 2003. Reprogramming of primordial germ cells begins before migration into the genital ridge, making these cells inadequate donors for reproductive cloning. *Proc Natl Acad Sci* **100:** 12207–12212.

Yildirim O, Li R, Hung JH, Chen PB, Dong X, Ee LS, Weng Z, Rando OJ, Fazzio TG. 2011. Mbd3/NURD complex regulates expression of 5-hydroxymethylcytosine marked genes in embryonic stem cells. *Cell* **147:** 1498–1510.

Yoon HG, Chan DW, Reynolds AB, Qin J, Wong J. 2003. N-CoR mediates DNA methylation-dependent repression through a methyl CpG binding protein Kaiso. *Mol Cell* **12:** 723–734.

You JS, Jones PA. 2012. Cancer genetics and epigenetics: Two sides of the same coin? *Cancer Cell* **22:** 9–20.

Yu M, Hon GC, Szulwach KE, Song CX, Zhang L, Kim A, Li X, Dai Q, Shen Y, Park B, et al. 2012. Base-resolution analysis of 5-hydroxymethylcytosine in the mammalian genome. *Cell* **149:** 1368–1380.

Zhang Y, LeRoy G, Seelig HP, Lane WS, Reinberg D. 1998. The dermatomyositis-specific autoantigen Mi2 is a component of a complex containing histone deacetylase and nucleosome remodeling activities. *Cell* **95:** 279–289.

Zhou Z, Hong EJ, Cohen S, Zhao WN, Ho HY, Schmidt L, Chen WG, Lin Y, Savner E, Griffith EC, et al. 2006. Brain-specific phosphorylation of MeCP2 regulates activity-dependent *Bdnf* transcription, dendritic growth, and spine maturation. *Neuron* **52:** 255–269.

Zhu JK. 2009. Active DNA demethylation mediated by DNA glycosylases. *Annu Rev Genet* **43:** 143–166.

* Zoghbi H, Beaudet A. 2014. Epigenetics and human disease. *Cold Spring Harb Perspect Biol* doi: 10.1101/cshperspect.a019497.

WWW RESOURCES

http://www.illumina.com/products/methylation_450_beadchipkits.ilmn Illumina methylation chip

http://www.sequenom.com/Sites/Genetic-Analysis/Applications/DNA-Methylation Sequenom EpiTYPER

CHAPTER 16

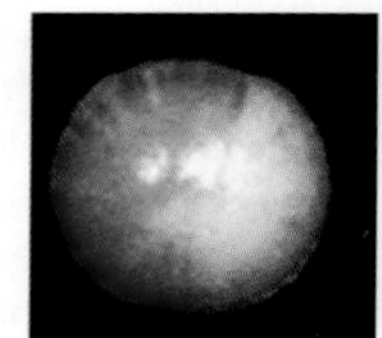
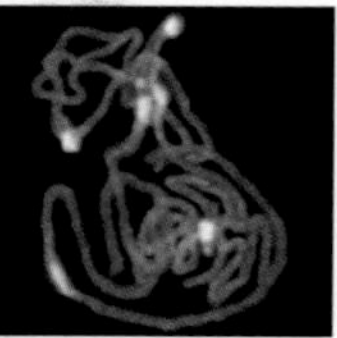

RNAi and Heterochromatin Assembly

Robert Martienssen[1] and Danesh Moazed[2]

[1]Cold Spring Harbor Laboratory, Cold Spring Harbor, New York 11724; [2]Department of Cell Biology, Harvard Medical School, Boston, Massachusetts 02115-5730

Correspondence: danesh@hms.harvard.edu

SUMMARY

The involvement of RNA interference (RNAi) in heterochromatin formation has become clear largely through studies in the fission yeast *Schizosaccharomyces pombe* and plants like *Arabidopsis thaliana*. This chapter discusses how heterochromatic small interfering RNAs are produced and how the RNAi machinery participates in the formation and function of heterochromatin.

Outline

Cite as *Cold Spring Harb Perspect Biol* doi: 10.1101/cshperspect.a019323

OVERVIEW

The intersection between RNA interference (RNAi) and heterochromatin formation has brought together two areas of gene regulation that had previously been thought to operate by different, perhaps even unrelated, mechanisms. Heterochromatin was originally defined nearly 80 years ago using cytological staining methods as those chromosome regions that retain a condensed appearance throughout the cell cycle. Early investigators studying the relationship between chromosome structure and gene expression noticed that certain chromosome rearrangements resulted in the spreading of heterochromatin into adjacent genes, which then became silent. But, the seemingly stochastic patterns of spreading gave rise to genetically identical populations of cells that had different phenotypes. This phenomenon, initially described in *Drosophila* as position-effect variegation, provides a striking example of epigenetic regulation. The term RNAi was first used to describe gene silencing that resulted from the introduction of homologous antisense or double-stranded RNA (dsRNA) into the nematode *Caenorhabditis elegans*. But, it was soon recognized that a related mechanism involving RNA accounted for posttranscriptional transgene silencing (PTGS) described earlier in petunia and tobacco. In contrast, heterochromatin was widely believed to operate directly at the chromatin level to cause transcriptional repression by a mechanism referred to as transcriptional gene silencing (TGS). This chapter focuses on the relationship between the RNAi pathway and the formation of epigenetically heritable heterochromatin at specific chromosome regions. It draws on recent examples that show this relationship in the fission yeast *Schizosaccharomyces pombe* and the mustard plant *Arabidopsis thaliana*.

The fission yeast nuclear genome is composed of three chromosomes that range in size from 3.5 to 5.7 Mb. Each chromosome contains large blocks of repetitive DNA, particularly at centromeres, which are packaged into heterochromatin. The mating-type loci (which control cell type) and subtelomeric DNA regions also contain repetitive sequences that are packaged into heterochromatin. We now know that the assembly of DNA into heterochromatin plays both regulatory and structural roles. In the case of the mating-type loci in yeast, regulation of gene transcription by heterochromatin is important for cell-type identity. In the case of telomeres and centromeres, heterochromatin plays a structural role that is important for proper chromosome segregation during cell division. Moreover, repetitive DNA sequences and transposable elements account for a large fraction, in some cases more than half, of the genomes of many eukaryotic cells. Heterochromatin and associated mechanisms play a critical role in regulating the activity of repeated sequences, thus maintaining genome stability.

Recent studies have uncovered a surprising requirement for components of the RNAi pathway in the process of heterochromatin formation in fission yeast and have provided insight into how these two pathways can work together at the chromatin level. Briefly, small interfering RNA (siRNA) molecules and their Argonaute-binding proteins assemble into the RNA-induced transcriptional silencing (RITS) complex and direct epigenetic chromatin modifications and heterochromatin formation at complementary chromosome regions. RITS uses siRNA-dependent base pairing to guide association with nascent RNA sequences at the target locus destined to be silenced, an association that is stabilized by direct binding to methylated histone H3 at lysine (K)9 (H3K9me). The presence of these two activities in RITS (i.e., siRNA base-pairing and association with chromatin via methylated H3K9) triggers heterochromatin formation in concert with well-known heterochromatin-associated factors, and RNA polymerase II (Pol II) directly linking RNA silencing to heterochromatin modification and silencing.

In *A. thaliana* and many other eukaryotes, repeat sequences such as retroelements and other transposons are targeted for inactivation at the chromatin level by mechanisms that couple small RNA-mediated targeting with histone H3K9, but also DNA methylation. Although the existence of a RITS complex is not always clear, components of the RNAi and related pathways are required for the initiation and maintenance of these repressive methylation events, along with Pol II and related polymerases. In this chapter, we will discuss how heterochromatic siRNAs are produced, and how they mediate DNA and/or chromatin modifications in fission yeast and *A. thaliana*.

Cite as *Cold Spring Harb Perspect Biol* doi: 10.1101/cshperspect.a019323

1 OVERVIEW OF THE RNAi PATHWAY

Although the term RNA interference (RNAi) was originally used to describe silencing that is mediated by exogenous double-stranded RNA (dsRNA) in *Caenorhabditis elegans* (Fire et al. 1998), it now broadly refers to gene silencing that is triggered by some kind of small RNA in association with a member of the Argonaute family of proteins (Fig. 1). In most cases, the small RNA is produced from dsRNA. However, new RNA-silencing pathways have recently been uncovered wherein the small RNA is produced from long single-stranded RNA (ssRNA) precursors. Despite the diversity of small RNA biogenesis pathways (briefly discussed below), the downstream steps use similar effector proteins and mechanisms; in all cases described so far, Argonaute-associated small RNAs target either messenger RNAs (mRNAs) posttranscriptionally (i.e., posttranscriptional transgene silencing, PTGS) or chromatin regions (i.e., transcriptional gene silencing, TGS) to effect silencing. Therefore, before introducing the components of the RNAi machinery specific to TGS—the topic of this chapter—the source of small RNAs that harness the RNAi machinery into action will be discussed.

1.1 Small RNA Biogenesis

Two classes of small RNA, small interfering RNA (siRNA) and microRNA (miRNA), are processed from longer dsRNA precursors. dsRNA may originate from the bidirectional transcription of repetitive DNA elements or transcription of RNA molecules that can base-pair internally to form dsRNA segments (see Fig. 2A,B, respectively). For example, transcription through inverted repeat regions produces RNA molecules that fold back on themselves to produce hairpin structures. dsRNAs are then cleaved by Dicer, an RNase III class ribonuclease, to generate siRNAs, or processed into miRNAs through a series of related steps (Bartel 2004; Filipowicz et al. 2005). The miRNA biogenesis pathway is distinct because miRNAs are produced from the introns of endogenous coding genes or from endogenous noncoding transcripts. The Dicer products are complementary duplexes, 21–24 nt in size, which have a characteristic 2-nt overhang at each 3′ end of the duplex (Hamilton and Baulcombe 1999; Zamore et al. 2000; Bernstein et al. 2001; Elbashir et al. 2001; Hannon 2002; Zamore 2002; Bartel 2004; Baulcombe, 2004). These duplexes are unwound into single-stranded siRNA (or miRNA) to act as guides through base-pairing interactions with complementary target sequences. miRNAs and siRNAs are therefore specificity factors and play a central role in all RNAi-mediated silencing mechanisms.

Although dsRNAs can form by the annealing of forward and reverse RNAs that result from bidirectional transcription or are present in hairpin structures, in some cells RNAi requires an additional enzyme to make dsRNA. This is the RNA-dependent RNA polymerase (RdRP) found in some viruses, many fungi, all plants, and *C. elegans* (Dalmay et al. 2000; Sijen et al. 2001). It is directed by siRNAs to generate more dsRNA, which can then be processed into additional siRNA by Dicer (Fig. 2C). The primary function of RdRP is thus thought to be in amplification of the RNAi response. Indeed, it seems to be involved in a process adapted for producing a better host defense response to the introduction of exogenous dsRNA. This idea is strengthened by the fact that RdRPs are not involved in the miRNA-

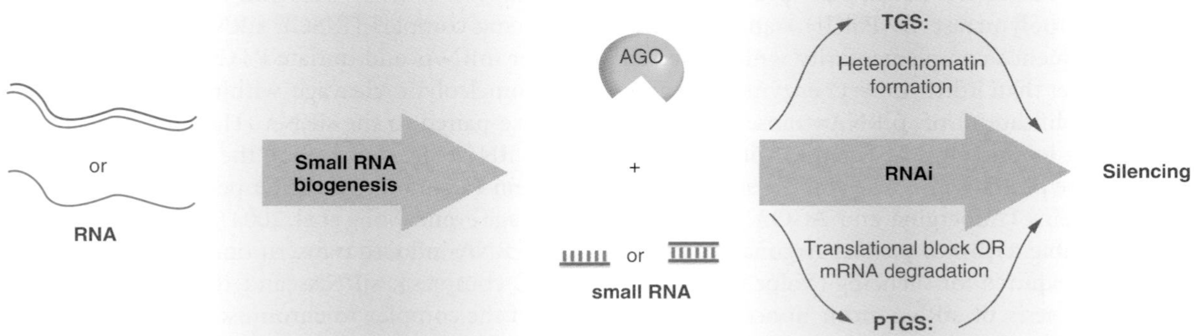

Figure 1. Overview of small RNA-silencing pathways. Silencing requires the biogenesis of a small RNA from either long ssRNA or dsRNA. The resulting small RNA is loaded onto effector complexes that contain a member of the Argonaute/Piwi (AGO) family of proteins, which bind to the small RNA (~22 to 28 nt in size) via their conserved middle (MID) and PIWI-Argonaute-Zwille (PAZ) domains. RNAi-mediated silencing occurs via multiple mechanisms. In the nucleus, RNAi promotes DNA and chromatin modification to induce heterochromatin formation and TGS. In addition, it cotranscriptionally degrades RNAs that are transcribed in heterochromatic domains by a process called cotranscriptional gene silencing (CTGS). In the cytoplasm, RNAi mediates the degradation of target mRNAs or their translational repression (PTGS).

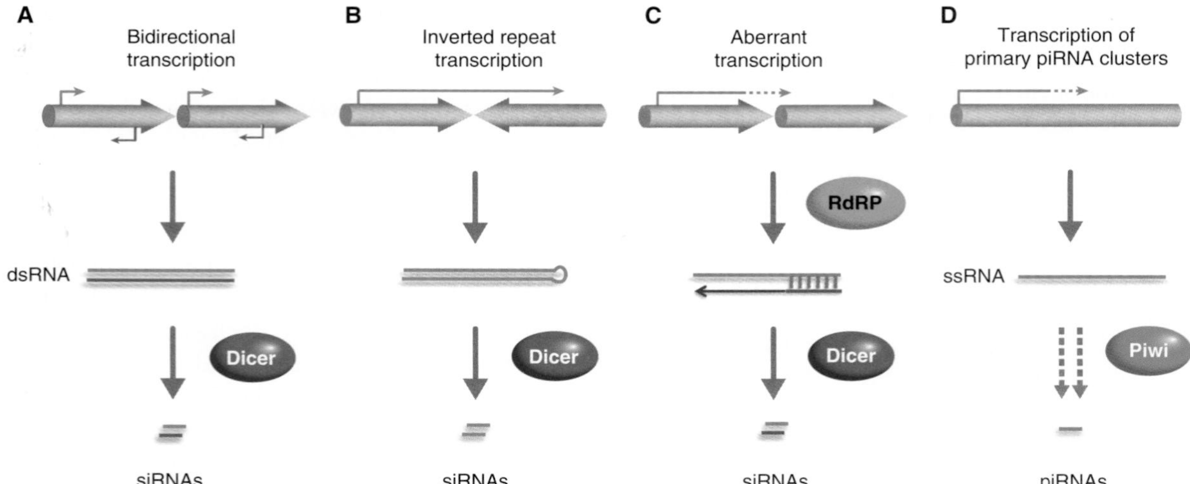

Figure 2. Pathways for the biogenesis of primary small RNAs that mediate silencing. (*A*) Bidirectional transcription has been observed at the *Schizosaccharomyces pombe* centromeric repeats and the *cenH* region of the silent mating-type locus and may provide a dsRNA substrate for the Dicer ribonuclease. (*B*) Transcription through inverted repeats found in many plant and animal cells can potentially produce dsRNA. (*C*) Transcription of aberrant RNAs that may lack proper processing signals may trigger dsRNA synthesis by RNA-dependent RNA polymerases (RdRPs). (*D*) Transcription from several driver loci gives rise to Piwi-associated small RNAs (piRNAs) that silence dispersed transposons. Piwi proteins together with other ribonucleases, which are not fully defined (represented by the gray dotted line), mediate primary piRNA generation.

silencing pathways (Sijen et al. 2001). Interestingly, insects (including *Drosophila*) and vertebrates (including mammals) lack recognizable RdRP-like sequences in their genome, but the possibility that other polymerases perform dsRNA synthesis in these organisms cannot be ruled out.

In the metazoan germline, a class of small RNAs called piRNA are generated from long single-stranded primary RNA transcripts, and function to silence transposons (Fig. 2d) (Aravin et al. 2007). piRNAs are produced from several long RNA polymerase II (Pol II) transcripts and have extensive sequence complementarity with dispersed transposons. Rather than RdRP or Dicer enzymes, the generation and amplification of piRNA transcripts from piRNA-producing loci involves association with the Piwi clade of Argonaute proteins and relies on the slicer activities of these proteins (Aubergine and AGO3 proteins in *Drosophila*; see Table 1). In *S. pombe*, Argonaute, Dicer, and RdRP are all required for silencing (Volpe et al. 2002) and for the biogenesis of siRNA from noncoding transcripts in centromeric repeats (Motamedi et al. 2004; Verdel et al. 2004). Dicer-independent small RNAs, called primal RNAs or priRNAs, have also been described and originate from single-stranded centromeric repeats as well as nearly the entire transcriptome (Halic and Moazed 2010). priRNAs have been proposed to initiate RdRP-dependent siRNA amplification by targeting long noncoding centromeric transcripts. The pathways that generate priRNAs are not fully understood. However, both siRNAs and priRNAs are trimmed to their mature size by a conserved 3–5′ exoribonuclease called Triman, which is required for efficient heterochromatin establishment (Marasovic et al. 2013).

1.2 RNA-Silencing Pathways

To date, two related complexes have been identified that incorporate siRNA: RISC and RITS. In the RNA-induced silencing complex (RISC), siRNAs or miRNAs recognize target mRNAs and initiate PTGS via their degradation by endonucleolytic cleavage within the mRNA region that is base-paired to the siRNA (Hannon 2002; Bartel 2004). The RNase H domain of the Argonaute/PIWI family protein (a subunit of RISC) performs this initial mRNA cleavage event (Song et al. 2004). In the nuclear RITS complex (RNA-induced transcriptional silencing; similar to the RISC complex), siRNAs and other protein components target the complex to chromosome regions for chromatin modification and silencing (i.e., TGS [Verdel et al. 2004; Bühler et al. 2006]) as well as cleavage of centromeric transcripts (Irvine et al. 2006). It is the RNAi-mediated TGS silencing pathway that is the focus of this chapter.

Argonaute and Dicer proteins are central components of virtually all RNA-silencing mechanisms including those involving siRNAs and miRNA, except, for example, the piRNA pathway (see Sec. 7 of Ch. 12 [Elgin and Reuter

 Cite as *Cold Spring Harb Perspect Biol* doi: 10.1101/cshperspect.a019323

Table 1. Conservation of RNAi and heterochromatin proteins

Schizosaccharomyces pombe	*Arabidopsis thaliana*	*Caenorhabditis elegans*	*Drosophila*	*Homo sapiens*
Dcr1	DCL1 to 4	Dcr-1	Dcr1 and 2	Dcr-1
Ago1	AGO1 to 10	Rde-1, Alg-1, and -2	Ago1 to 3, Piwi	Ago1 to Ago4
–	–	Prg-1 and 2, and 19 others	Aubergine/Sting	Piwi1 to Piwi4
Chp1[a]	CMT3	–	–	–
Tas3[b]	–	AIN-1	GW182	TNRC6
Rdp1	RDR1 to 6	Ego-1, Rrf-1 to -3	–	–
Hrr1	SGS2/SDE3[c]	ZK1067.2	GH20028p	KIAA1404
Cid12		Rde-3, Trf-4[c]	CG11265[c]	POLS[c]
Swi6	LHP1 (TFL2)	Hpl-1, Hpl-2, F32E10.6[d]	HP1a, b	HP1α, β, γ
Clr4	SUVH2 to 6		Su(var)3-9	Suv39h1 and 2
Rik1[e]	DDB1	M18.5	Ddb1	Ddb1
Cul4	CUL4	Cul4	Cul4	Cul4
Sir2	SIR2	Sir2–1	Sir2	SirT1
Clr3				
Clr6	HDA6	Hda-1	Rpd3	HDAC1
–	DDM1			
Eri1	ERI1	Eri-1	CG6393	THEX1

[a]An obvious ortholog of the chromodomain protein Chp1 has not been identified in the other model organisms listed here, but most eukaryotic cells contain multiple chromodomain proteins. CMT3 in *Arabidopsis* is a chromodomain DNA methyltransferase, which acts in the same pathway as AGO4 and may be analogous to Chp1.

[b]Tas3 is a GW motif protein. Members of this conserved family are found associated with Argonaute family members.

[c]Cid12 belongs to a large family of conserved proteins that share sequence similarity with the classical poly(A) polymerase as well as 2′–5′-oligoadenylate enzymes.

[d]*C. elegans* have about 20 SET domain proteins, but a histone H3 lysine (K) 9 methyltransferase (KMT) specific for H3K9 has not yet been identified in this organism.

[e]*Schizosaccharomyces pombe* contains another Rik1-like protein, Ddb1, which is involved in DNA damage repair. Metazoans and plants appear to contain only a single Rik1-like gene called Ddb1 involved in DNA damage repair; however, it is unknown whether it also participates in heterochromatin formation.

2013]). Like siRNAs, miRNAs are 21–24 nt in size and form part of the RISC complex via binding to the Argonaute protein to target specific mRNAs. This targeting can result in mRNA cleavage via the PIWI/RNase H domain and translational repression. This may be coupled to sequestration of the mRNA to cytoplasmic RNA-processing organelles known as P bodies (processing bodies). Thus, although at least two different dsRNA-processing pathways result in the generation of siRNA or miRNA (i.e., small RNA biogenesis), these small RNAs use a similar machinery to inactivate cognate mRNAs. The miRNA pathway distinguishes itself because the originating transcripts are largely developmentally regulated and, in turn, generally target and developmentally regulate the silencing of homologous genes.

As mentioned above (Overview and this section), nuclear small RNA-silencing mechanisms are widely conserved and play central roles in the regulation of gene expression and genome stability (through stable heterochromatin formation). The nuclear and cytoplasmic pathways are not separate and share common components. A striking example of the intersection of these pathways is represented by the recent finding that classical RNAi in *C. elegans*, first described as a purely PTGS mechanism, is coupled to histone H3K9 methylation and TGS that can be inherited in subsequent generations (Guang et al. 2010; Gu et al. 2012). Similarly, piRNAs, which associate with Piwi proteins to degrade transposon RNAs in the cytoplasm, can also act in the nucleus to promote histone H3K9 and DNA methylation. Finally, recent evidence in *C. elegans* indicates that small RNAs and Argonaute proteins form a sophisticated self–nonself recognition mechanism (Ashe et al. 2012; Lee et al. 2012; Shirayama et al. 2012). Small RNAs originating from foreign DNA elements load onto a specific Piwi protein (PRG-1), which, in turn, silences their transcription. This loading appears to involve a general scanning mechanism so that small RNAs from all transcripts are represented in PRG-1. However, the silencing of self-transcripts is prevented by the action of another Argonaute protein (CSR-1), which serves as a repository of small RNAs from all self-transcripts and prevents their silencing by PGR-1. This illustrates how RNA-silencing mechanisms can distinguish transcription of foreign DNA elements from self-transcription. Related mechanisms in the *Drosophila* germline can also detect transposons via the piRNA pathway resulting in hybrid dysgenesis (Aravin et al. 2007; Brennecke et al. 2007; discussed in Ch. 12 [Elgin and Reuter 2013]). RNA silencing clearly plays a central role in defense against transposons and RNA viruses at both the PTGS and TGS levels (Plasterk 2002; Li and Ding 2005; Ghildiyal and Zamore 2009).

2 EARLY EVIDENCE IMPLICATING RNA AS AN INTERMEDIATE IN TGS

Before discussing the better-understood examples of RNAi-based chromatin modifications in fission yeast and *Arabidopsis*, we will briefly discuss early experiments that suggested a role for RNA in mediating chromatin and DNA modifications. The earliest evidence for the role of an RNA intermediate in TGS came from studies of plant viroids. The potato spindle tuber viroid (PSTV) consists of a 359-nt RNA genome and replicates via an RNA–RNA pathway. The artificial introduction of PSTV transgenes into the tobacco genome resulted in DNA methylation (Wassenegger et al. 1994), but only in plants that supported viroid RNA replication. Thus, these experiments suggested the involvement of an RNA intermediate that directs DNA methylation to homologous sequences (Wassenegger et al. 1994). Further evidence in *Arabidopsis* came from the production of aberrant transcripts, which somehow resulted in DNA methylation of promoter regions homologous to the promoter of the aberrant transcript, thus causing TGS (Mette et al. 1999). Importantly, the silencing of viral genomes in plants leads to the production of small RNAs that are 22 nt in size, which were the first examples of small RNA (Hamilton and Baulcombe 1999). These observations as well as homology-dependent silencing of transgenes, which was first discovered in petunia and tobacco (discussed in Ch. 13 [Pikaard and Mittelsten Scheid 2014]), are now widely recognized as some of the earliest examples of silencing by RNA (Napoli et al. 1990). Classical silencing phenomena in maize, such as paramutation and transposon control, are early examples of transcriptional silencing that depend on RNAi (Slotkin and Martienssen 2007; Chandler 2010).

Further evidence for a link between RNAi and TGS comes from studies of repeat-induced gene silencing in *Drosophila*. The introduction of multiple tandem copies of a transgene results in the silencing of both the transgene and the endogenous copies (Pal-Bhadra et al. 1999). This silencing requires the chromodomain protein Polycomb, which is also involved in the packaging of homeotic regulatory genes into heterochromatin-like structures (Francis and Kingston 2001). In addition, this repeat-induced gene silencing requires Piwi, which associates with piRNAs in the nucleus (Pal-Bhadra et al. 2002; Aravin et al. 2007). However, Piwi and the piRNA amplification machinery appear to be exclusively localized in the germline and it remains to be determined how they can affect silencing of reporter genes in somatic cells in the above studies. In *Tetrahymena*, another Piwi protein family member Twi1 is required for small RNA accumulation and the massive DNA elimination that is observed in the somatic macronucleus of the protozoa (detailed in Ch. 11 [Chalker et al. 2013]). These and more recent results discussed in Section 4 suggest that the RNAi pathway is involved in the assembly of repressive chromatin structures in the fly germline and possibly somatic cells.

Other repeat-induced silencing mechanisms have been described in filamentous fungi (see Ch. 10 [Aramayo and Selker 2013]), including repeat-induced point mutation (RIP) in *Neurospora crassa* and methylation-induced premeiotically (MIP) in *Ascobolus immersus*, that do not appear to involve an RNA intermediate because they occur independently of the transcriptional state of the locus (Galagan and Selker 2004). Instead, RIP and MIP involve paired loci, in which, for example, two out of three gene copies are silenced, suggesting some kind of DNA–DNA interaction mechanism involving homologous loci for the induction of silencing. Conversely, meiotic silencing of unpaired DNA (MSUD), which occurs in *Neurospora*, requires the RNAi pathway (Shiu et al. 2001) and may have parallels in other organisms including *C. elegans* (Maine et al. 2005).

3 RNAi AND HETEROCHROMATIN ASSEMBLY IN *S. Pombe*

S. pombe chromosomes contain extensive heterochromatic regions that are associated with underlying repetitive DNA elements at the centromeres and the silent mating-type loci (*mat2/3*; Grewal 2000; Pidoux and Allshire 2004). The DNA sequence structure of fission yeast centromeres contains a unique central core region (*cnt*) that is flanked by two types of repeats called the "innermost" (*imr*) and "outermost" (*otr*) repeats (Fig. 3). The *otr* region itself is composed of *dh* and *dg* repeats.

Heterochromatin formation in *S. pombe* involves the concerted action of a number of *trans*-acting factors. These include histone deacetylases (HDACs), the histone H3 lysine (K) 9 methyltransferase (HKMT or KMT) called Clr4, and three chromodomain proteins that bind specifically to dimethylated (me2) or trimethylated (me3) histone H3K9 called Swi6, Chp2 (both HP1 homologs), and Chp1. It has been proposed that after their initial recruitment, Swi6 and Clr4 contribute to the spreading of H3K9 methylation and heterochromatin formation through sequential cycles of Clr4-catalyzed H3K9 methylation coupled to chromodomain-mediated spreading into adjacent nucleosomes (Grewal and Moazed 2003).

Fission yeast contains a single gene for each of the RNAi proteins: Dicer, Argonaute, and RdRP (*dcr1*$^+$, *ago1*$^+$, and *rdp1*$^+$, respectively). Mutations in the components of the

Cite as *Cold Spring Harb Perspect Biol* doi: 10.1101/cshperspect.a019323

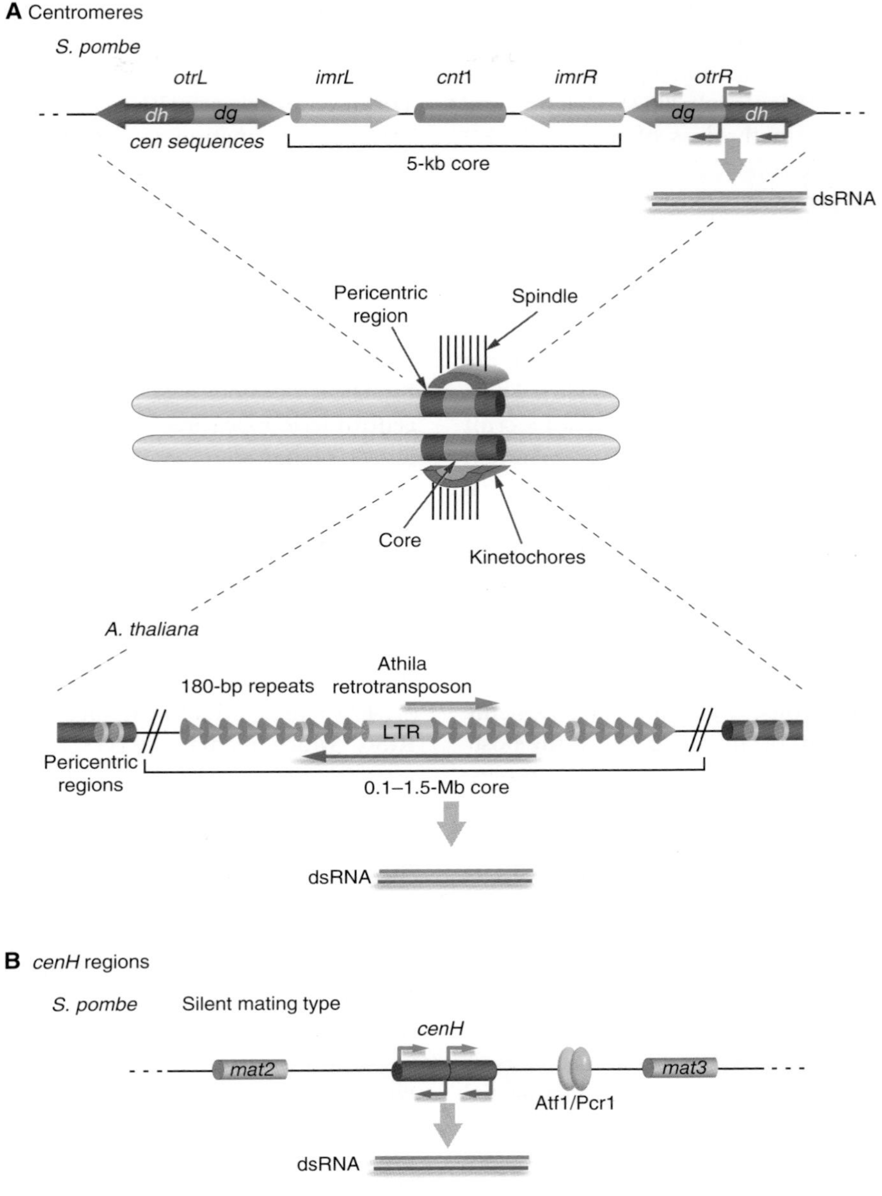

Figure 3. Organization of heterochromatic chromosome regions in *S. pombe* and *A. thaliana*. (*A*) The centromere of *S. pombe* chromosome 1 is shown as an example (*top* line), seen in the context of the whole chromosome below. The centromere core (orange) consists of the unique central core (*cnt1*) region flanked by innermost (*imrL* and *imrR*) and outermost (*otrL* and *otrR*) repeats. The pericentric *otr* region (green) is transcribed in both directions, giving rise to forward (blue) and reverse (red) transcripts. *A. thaliana* centromeres illustrated below are composed of 180-bp repeats (orange) interspersed with retrotransposable elements (yellow). Forward transcripts initiating within the long terminal repeat (LTR) of the retroelement and reverse transcripts initiating within the 180-bp repeats are indicated. (*B*) The region between the *mat2* and *mat3* genes contains a domain that is homologous to the centromeric repeats (*cenH*) and is also bidirectionally transcribed. Atf1 and Pcr1 are DNA-binding proteins that act in parallel with RNAi in mating-type silencing.

RNAi pathway surprisingly result in a loss of centromeric heterochromatin, and accumulation of noncoding forward and reverse transcripts from bidirectional promoters within each *dg* and *dh* repeat (Volpe et al. 2002). Deleting any of these genes also results in the loss of histone H3K9 methylation, and mutants display defects in chromosome segregation, which are generally associated with defects in heterochromatin assembly (Volpe et al. 2002, 2003). Moreover, sequencing of a fission yeast, small RNA library identified ~22-nt RNAs that mapped exclusively to centromeric repeat regions and ribosomal DNA repeats, suggesting that *cen* RNAs can produce dsRNAs that are processed into siRNAs (Reinhart and Bartel 2002). Thus, it was first suggested that the RNAi pathway could recruit Swi6 and Clr4 to chromatin to initiate and/or maintain heterochromatin formation at centromeric and ribosomal DNA repeat loci (Hall et al. 2002; Volpe et al. 2002).

Interestingly, both TGS and PTGS mechanisms appear to contribute to the down-regulation of *cen* RNAs. The forward-strand transcript is primarily silenced at the transcriptional level, as shown in RNAi mutants (Volpe et al. 2002). The reverse-strand *cen* transcripts, however, are not affected by Swi6 mutants (Volpe et al. 2002), and silencing of this *cen*-reverse transcript occurs primarily at the posttranscriptional level. Moreover, the TRAMP complex, containing the Cid14 poly(A) polymerase among other proteins, is required for efficient silencing of *dg*, *dh*, and centromeric transgenes (Bühler et al. 2007). TRAMP targets RNAs for degradation by a 3–5′ exonuclease complex called the Exosome (Houseley et al. 2006). Thus, efficient silencing of heterochromatic RNAs requires RNAi-dependent and -independent degradation mechanisms, which appear to act on the chromosome in *cis* and are thus referred to as *cis*-PTGS or -CTGS (co-TGS) mechanisms (Fig. 4A,B). The Swi6 protein, which binds to methylated H3K9, plays a major role in linking heterochromatic transcription with RNA degradation by promoting the interaction of heterochromatic transcripts with the RNAi and exosome pathways (see Fig. 4, dotted lines) (Motamedi et al. 2008; Reyes-Turcu et al. 2011; Hayashi et al. 2012; Keller et al. 2012; Rougemaille et al. 2012).

RNAi also plays a role in silencing the mating-type locus (*mat2/3*) but acts redundantly with other mechanisms (Hall et al. 2002). *mat2/3* is interrupted by a region of DNA that is highly homologous to centromeric repeats (called *cenH*, for *cenHomology*; Fig. 3). Like the *cen* repeats, the *cenH* region is divergently transcribed to produce forward and reverse RNA (Noma et al. 2004). These *cenH* transcripts accumulate to high levels in RNAi mutants in combination with mutations in Pcr1 and Atf1, site-specific DNA-binding proteins that can recruit the heterochromatin machinery independently of RNAi (Jia et al. 2004).

3.1 Small RNAs Initiate Heterochromatin Assembly in Association with an RNAi Effector Complex

The discovery that the RNAi pathway is involved in heterochromatin formation in fission yeast and TGS in other systems raised the question of how it could directly regulate chromatin structure. Purification of Chp1, a chromodomain protein that is a structural component of heterochromatin, led to the identification of the RITS complex (Verdel et al. 2004). RITS contains the fission yeast Ago1 protein and the Tas3 GW domain protein, in addition to Chp1. It also contains centromeric siRNAs, which are produced by the Dicer ribonuclease. Importantly, RITS associates with centromeric repeat regions in an siRNA-dependent fashion. RITS has therefore been proposed to use centromeric siRNAs to target specific chromosome regions for inactivation, and this provides a direct link between RNAi and heterochromatin assembly (Fig. 4A). The fission yeast Ago1 is present in a second complex called the Argonaute chaperone (ARC) complex, which may function in the specific delivery of duplex siRNAs to Ago1 before it is exchanged into the RITS complex (Fig. 4A) (Buker et al. 2007).

RITS uses siRNAs for target recognition, but unlike RISC, which mediates PTGS via mRNA inactivation, it associates with chromatin and initiates heterochromatin formation. How can siRNAs target specific chromosome regions? Two possible mechanisms have been proposed. In the first model, siRNAs bound to Ago1 in the RITS complex somehow base-pairs with an unwound DNA double helix. In the second model, RITS-associated siRNAs base-pair with noncoding RNA transcripts at the target locus (Fig. 4A). As discussed below (Sec. 3.3), the available evidence strongly supports the second model.

According to either model, the association of RITS with chromatin via siRNA results in the recruitment of the Clr4 KMT, with subsequent methylation of histone H3K9. However, Clr4 is also required for the association of RITS with chromatin, suggesting that it provides methylated H3K9 to which the RITS complex can bind, thereby stabilizing its association with chromatin (Fig. 4A). The chromodomain of Chp1 was already known to bind specifically to methylated H3K9 residues (Partridge et al. 2002), and mutations in Clr4 or the chromodomain of Chp1 that are involved in this interaction result in a loss of RITS binding to chromatin (Partridge et al. 2002; Noma et al. 2004). Moreover, RITS can also bind to chromatin domains that are coated with methylated H3K9 through the chromodomain of Chp1 at the *mat2/3* and telomeric regions in the absence of siRNAs (Noma et al. 2004; Petrie et al. 2005). In summary, the RITS complex shows affinity to chromatin via Chp1 binding to methylated H3K9 and through base-pairing of siRNA with either DNA or RNA transcripts. This

Cite as *Cold Spring Harb Perspect Biol* doi: 10.1101/cshperspect.a019323

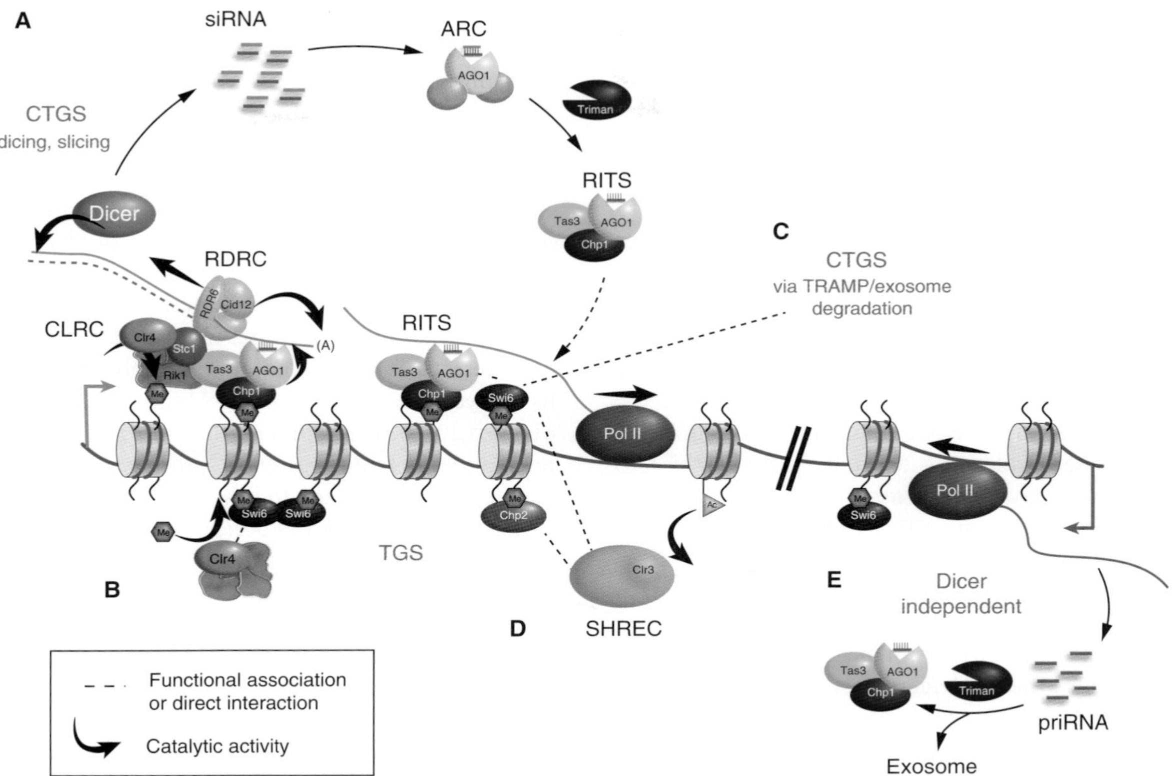

Figure 4. RNAi-mediated cotranscriptional assembly of heterochromatin in *S. pombe*. Transcription of pericentric repeats gives rise to long noncoding RNAs that are processed into primary small RNAs by Dicer-dependent and -independent pathways. (*A*) A small RNA loaded onto the RITS complex targets the nascent noncoding RNA by base-pairing interactions. This leads to the recruitment of the RDRC (RNA-directed RNA polymerase complex) and conversion of the targeted RNA into dsRNA, which is diced into siRNAs by Dicer. The resulting duplex siRNA is loaded onto the Argonaute chaperone (ARC) complex and converted into single-stranded siRNA after cleavage and release of the passenger strand in the RITS complex. The mature RITS complex containing single-stranded siRNA can now target additional noncoding RNAs completing a positive feedback loop. The RITS complex also recruits the CLRC H3K9 methyltransferase complex to chromatin via interactions with the Rik1 subunit of CLRC and Stc1, an adaptor protein. (*B*) H3K9 methylation stabilizes the association of RITS with chromatin and also provides binding sites for HP1 proteins (Swi6 and Chp2). Swi6 facilitates the recruitment of RDRC and degradation by the exosome (*C*), whereas Chp2 recruits the SHREC complex containing the Clr3 HDAC promotes TGS by mechanisms that remain to be defined (*D*). In addition to TGS, efficient silencing requires cotranscriptional RNA degradation (CTGS) by RNAi-dependent (*A*, dicing and slicing) and RNAi-independent (*C*, TRAMP/exosome degradation) mechanisms. Dicer-independent priRNAs contribute to low levels of H3K9 methylation (*E*) and may trigger siRNA amplification (*A*). The 3′ ends of priRNAs and siRNAs are trimmed by the Triman exonuclease (*A*,*E*). Black tapered arrows indicate enzymatic activity.

dual mode of RITS recruitment may provide an explanation for the epigenetic inheritance of heterochromatin by promoting the preferential establishment of heterochromatin at regions that inherit H3K9 methylation during the replication of chromatin (Moazed 2011).

The next step in the assembly of fully silenced heterochromatic domains involves the recruitment of HP1 proteins, Swi6 and Chp2, which associate the methylated H3K9. Swi6 appears to function mainly by stabilizing the RNAi and exosome complexes on the chromosome and thereby promotes efficient *cis*-RNA degradation (Fig. 4A,B) (Motamedi et al. 2008). Specifically, Swi6 interacts with an accessory factor called Ers1, which, in turn, interacts with the RDRC complex to promote dsRNA synthesis and RNA degradation (Hayashi et al. 2012; Rougemaille et al. 2012). In addition, Swi6 has nonspecific affinity for RNA and may help retain heterochromatin-transcribed RNAs on the chromosome until they are degraded by the exosome (Keller et al. 2012). Chp2, on the other hand, plays a critical role in recruitment of the SHREC histone deacetylase complex (Motamedi et al. 2008; Fischer et al. 2009). Deacetylation of histone H3 lysine 14 (H3K14) by the Clr3 subunit of SHREC is critical for shutting down transcription from the repeats. RNAi-mediated silencing acts in parallel to Clr3 to release Pol II from the centromeric repeats during S phase, preventing it from clashing with replicating DNA polymerase, which would lead to stalling of replication forks in centromeric repeats (Li et al. 2011; Zaratiegui et al. 2011). The consequent requirement for DNA repair in RNAi mutants has been proposed to underlie spreading of heterochromatic modifications along with the replication fork, as described below.

Recent evidence strongly suggests that RITS and siRNAs can initiate de novo heterochromatin assembly. Bühler et al. used a site-specific RNA-binding protein to artificially tether the RITS complex to the RNA transcript of the normally active *ura4*$^+$ gene (Bühler et al. 2006). Remarkably, this tethering results in the generation of *ura4*$^+$ siRNAs and silencing of the *ura4*$^+$ gene in a manner that requires both RNAi and heterochromatin components. In addition, this system allowed a direct evaluation of the ability of newly generated siRNAs to initiate H3K9 methylation and Swi6 binding, which are molecular markers for heterochromatin formation. Interestingly, the newly generated *ura4*$^+$ siRNAs were found to be largely restricted to the locus from which they were generated (*cis*-restricted). However, when the gene encoding Eri1, a conserved siRNA ribonuclease, is deleted, *ura4*$^+$ siRNAs are able to act in *trans* to silence a second copy of the *ura4*$^+$gene, which is inserted on a different chromosome in the same cell. This experiment shows that tethering of RNAi to a nascent transcript mediates heterochromatin formation. In addition, it indicates that siRNAs can act as specificity factors that direct RITS and heterochromatin assembly to a previously active region of the genome.

The ability of siRNAs to initiate silencing in *S. pombe* has also been examined using a different method that relies on the expression of a hairpin RNA to produce siRNAs homologous to *ura4*$^+$ or a green fluorescent protein transgene (Sigova et al. 2004; Iida et al. 2008; Simmer et al. 2010). Hairpin siRNAs can promote silencing at the PTGS, or both PTGS and TGS levels, depending on the locus, suggesting that properties of the targeted locus affect the ability of siRNAs to induce heterochromatin formation. For the *ura4*$^+$ locus, both preexisting H3K9 methylation and antisense transcription contribute to siRNA-mediated silencing (Iida et al. 2008). Thus, although siRNAs can initiate ectopic heterochromatin formation, their ability to do so is strongly affected by properties of the targeted locus, which have not yet been fully defined.

3.2 dsRNA Synthesis and siRNA Generation

Bidirectional transcription of centromeric DNA repeats could, in principle, provide the initial source of dsRNA in fission yeast (Volpe et al. 2002). dsRNA resulting from the annealing of forward and reverse transcripts could then be a substrate for the Dicer ribonuclease. However, RNA-directed RNA polymerase (Rdp1) and its associated cofactors, as well as the Clr4 KMT, are also required for siRNA production by Dicer (Hong et al. 2005; Li et al. 2005; Bühler et al. 2006). These observations indicate that the generation of heterochromatic siRNAs by Dicer is coupled to chromatin and Rdp1-dependent events (Fig. 4A). Moreover, recent high-throughput sequencing experiments, which are more sensitive than northern blots used in earlier experiments, have detected a class of Dicer-independent small RNAs in fission yeast, termed primal RNAs, or piRNAs (Halic and Moazed 2010). These small RNAs have the same size and 5′ nucleotide preference as Dicer-produced siRNAs and, not surprisingly, contribute to low levels of Dicer-independent H3K9 methylation at the pericentromeric repeats.

The Rdp1 enzyme resides in a multiprotein complex that also contains Hrr1, an RNA helicase, and Cid12, a member of the β family of DNA polymerases that includes poly(A) polymerase enzymes (Motamedi et al. 2004). This complex has been termed RDRC and all of its subunits are required for heterochromatin formation at centromeric DNA regions (Motamedi et al. 2004). As expected from the presence of Rdp1, RDRC has RNA-directed RNA polymerase activity in vitro, and mutations that abolish this activity also abolish RNAi-dependent silencing in vivo (Motamedi et al. 2004; Sugiyama et al. 2005). The in vitro RNA synthesis activity of RDRC does not require an siRNA primer (Motamedi et al. 2004). Therefore, in vivo, the RITS siRNA may just provide the specificity for recruiting RDRC to specific RNA templates. Consistent with this hypothesis, subunits of the RDRC complex are required for siRNA amplification, and RITS complexes purified from cells that lack any subunit of the RDRC complex contain only trace levels of small RNA (Motamedi et al. 2004; Li et al. 2005; Sugiyama et al. 2005; Bühler et al. 2006; Halic and Moazed 2010).

The presence of Cid12 in the RDRC complex is intriguing and raises the possibility that another polymerase activity participates in chromosome-associated RNA silencing. As some members of this family have poly(A) polymerase activity, one possibility is that adenylation may be important for either Rdp1-dependent dsRNA synthesis or further processing of Rdp1-produced dsRNA. The fission yeast Cid12 has monoadenylation activity and can adenylate template RNAs that are targeted by Rdp1, suggesting that it functions upstream of dsRNA synthesis (Halic and Moazed 2010). Interestingly, Cid12-like proteins are conserved throughout eukaryotes and appear to target small RNAs for exosome-mediated degradation (Table 1; Fig. 4A,B). In *C. elegans*, mutations in rde-3, a

 Cite as *Cold Spring Harb Perspect Biol* doi: 10.1101/cshperspect.a019323

member of this family, results in defective RNAi (Chen et al. 2005), corroborating a conserved role for these enzymes in the RNAi pathway.

There is evidence for dsRNA synthesis and processing associated with the generation of heterochromatic siRNAs occurring on the chromosome at sites of transcription of noncoding centromeric RNAs (Fig. 4). Evidence includes, first, that Rdp1 can be cross-linked to centromeric DNA repeats (Volpe et al. 2002; Sugiyama et al. 2005), and to the forward and reverse RNA transcripts that originate from *cen* regions (Motamedi et al. 2004). Cross-linking to centromeric RNAs requires Dicer and Clr4, and is therefore siRNA- and chromatin-dependent. Second, siRNA generation requires chromatin components, including Clr4, Swi6, and the HDAC Sir2 (Hong et al. 2005; Li et al. 2005; Bühler et al. 2006). Finally, the association of RDRC with RITS is dependent on siRNAs as well as Clr4 suggesting that it occurs on chromatin (Motamedi et al. 2004), although Clr4 also has nonhistone H3 targets that are critical for efficient siRNA generation (Gerace et al. 2010). Thus, the generation of dsRNA and heterochromatic siRNAs may involve the recruitment of RDRC to chromatin-associated nascent pre-mRNA transcripts (Martienssen et al. 2005; Verdel and Moazed 2005; Moazed 2011). The fact that transcription and siRNA generation are likely to occur simultaneously reinforces the difference between RNA-silencing mechanisms that mediate chromatin modifications and PTGS. However, this distinction is unlikely to be absolute. For example, in *C. elegans*, similar to *S. pombe*, mutations in several chromatin components result in defects in RNAi and transposon-induced RNA silencing (Sijen and Plasterk 2003; Grishok et al. 2005; Kim et al. 2005; see Table 1), raising the possibility that in some cases dsRNA synthesis and processing may occur on the chromosome regardless of whether silencing occurs at the TGS or PTGS level.

3.3 RNA–RNA versus RNA–DNA Recognition Models

As we discussed earlier, RITS and other nuclear Argonaute complexes are targeted to specific chromosomal regions by siRNA-mediated base-pairing interactions. Studies in *S. pombe* have provided support for base-pairing interactions between siRNAs and nascent RNA transcripts, giving rise to the general consensus that nuclear Argonaute complexes associate with chromosomes via siRNA-mediated base-pairing with nascent transcripts. The observation that tethering components of the RNAi machinery to the RNA transcript of a gene can initiate RNAi and heterochromatin-dependent gene silencing that is *cis*-restricted clearly shows that the process can initiate via initial interactions with nascent RNA transcripts (Bühler et al. 2006). Importantly, *cis*-restriction rules out the possibility that the initial events of dsRNA synthesis and siRNA generation occur on mature transcripts in which mRNA products from different alleles cannot be distinguished. Furthermore, a direct prediction of the RNA–RNA interaction model is that transcription at the target locus should be required for RNAi-mediated heterochromatin assembly. Although the requirement for transcription has not been directly tested, mutations in two different subunits of RNA Pol II, denoted Rpb2 and Rpb7, have specific defects in siRNA generation and heterochromatin assembly, but not on general transcription (Djupedal et al. 2005; Kato et al. 2005). This is reminiscent of Rbp1 mutants, which have defects in certain active histone modifications (i.e., H3K4 methylation and H2B ubiquitination), coupling it to the process of transcriptional elongation (Hampsey and Reinberg 2003). The Rbp1 paradigm provides a precedent for the hypothesis that RNAi-mediated H3K9 methylation and heterochromatin formation could be coupled to transcriptional elongation via the association of RNAi complexes with RNA Pol II. A notable development is the demonstration that transcription of the centromeric repeat regions and siRNA generation occur largely within a restricted window during S phase of the cell cycle (Chen et al. 2008; Kloc et al. 2008). Thus, both nascent transcripts and high levels of siRNA are present during the process of chromosome duplication when heterochromatin needs to be reestablished, ensuring heterochromatin maintenance (Fig. 4).

The RNA–RNA targeting model is also supported by the observation that components of both the RITS and RDRC complexes can be localized to noncoding centromeric RNAs using in vivo cross-linking experiments (Motamedi et al. 2004). This localization is siRNA-dependent, which suggests that it involves base-pairing interactions with the noncoding RNA. In addition, it requires the Clr4 KMT, suggesting that it is coupled to binding of RITS to methylated H3K9 and occurs on chromatin. Nonetheless, the possibility that siRNAs can also recognize DNA directly through base-pairing interactions cannot be ruled out. For example, in plants, siRNAs that are complementary to promoter regions that are (presumably) not transcribed still direct DNA methylation, another modification that takes place during heterochromatin formation within these regions (see Ch. 13 [Pikaard and Mittelsten Scheid 2014]).

3.4 How Does RNAi Recruit Chromatin-Modifying Enzymes?

The recruitment of Clr4 is a key step in initiating histone H3K9 methylation and heterochromatin assembly.

However, because RITS association with chromatin and Clr4-catalyzed histone H3K9 methylation are codependent processes, it has been difficult to determine the event that provides the initial trigger for RNAi-dependent heterochromatin assembly. Furthermore, low levels of H3K9 methylation are present at pericentromeric repeats in RNAi mutant cells, suggesting that other mechanisms also contribute to Clr4 recruitment. Nonetheless, as we discussed above, de novo generation of siRNAs, either by tethering RITS to RNA or from transcription of a long hairpin RNA, promotes silencing of a previously active copy of the *ura4*$^{+}$ gene, and that is coupled to H3K9 methylation and the recruitment of RITS and Swi6 to chromatin (Bühler et al. 2006; Iida et al. 2008; Simmer et al. 2010). These observations clearly suggest that siRNAs are capable of recruiting Clr4. However, the sensitivity of this capability to properties at the targeted locus suggests that unidentified factors may work together with an siRNA-programmed RITS to initiate H3K9 methylation.

Clr4 is a component of a multiprotein complex called CLRC, which contains the heterochromatin protein, Rik1, a Cullin E3 ubiquitin ligase, Cul4, and two variously named associated proteins, Raf1/Clr7 and Raf2/Clr8, which encodes a β propeller protein (Hong et al. 2005; Horn et al. 2005; Jia et al. 2005; Li et al. 2005). Consistent with a direct role for RNAi in recruitment of Clr4, subunits of the RITS and RDRC complex physically associate with the Clr4 complex (Bayne et al. 2010; Gerace et al. 2010). Interestingly, the efficient interaction of RITS/RDRC with Clr4 requires an adaptor protein called Stc1 and is diminished in cells that harbor a catalytically dead Dicer enzyme, and therefore lack siRNAs (Bayne et al. 2010; Gerace et al. 2010). These observations suggest that the interaction of these complexes may be stabilized by accessory proteins and siRNA-mediated mechanisms. In this regard, the Rik1 subunit of CLRC is a member of a large family of β propeller WD repeat proteins that have been implicated in RNA or DNA binding. Members of this protein family include the cleavage polyadenylation specificity factor A (CPSF-A) involved in pre-mRNA splicing, and the DNA damage binding 1 (Ddb1) protein involved in binding UV-damaged DNA. CPSF-A is of particular interest because Rik1 shares sequence similarity with its putative RNA-binding domain involved in the recognition of mRNA polyadenylation sequences (Barabino et al. 2000). The Ddb1 protein, like Rik1, is a component of a Cul4 E3 ubiquitin ligase complex and is involved in the recognition and repair of UV-damaged DNA (Higa et al. 2003; Zhong et al. 2003). An exciting possibility is that Rik1 acts in a fashion that is similar to CPSF-A and Ddb1, binding to an RNAi-generated product during heterochromatin assembly. Once recruited, the Rik1 complex is associated with the leading strand DNA polymerase, providing a mechanism for spreading of histone modification along with the replication fork. Because the repeats are transcribed during S phase, fork progression requires Pol II release by a mechanism that requires RNAi (Li et al. 2011; Zaratiegui et al. 2011). Failure to release Pol II invokes homologous recombination (HR) repair.

Studies in fission yeast have made enormous contributions to our understanding of RNAi-mediated heterochromatin formation, but important questions remain unanswered. Although progress has been made in defining the Dicer-dependent and -independent pathways of small RNA generation, the determinants that distinguish centromeric noncoding RNAs from other genomic transcripts as sources of abundant siRNAs as well as chromosome-bound scaffolds for siRNA-mediated heterochromatin formation remain to be determined. Furthermore, much remains to be learned about how RNAi contributes to the maintenance of genomic stability through the coordination of transcription, replication, and DNA repair within centromeric DNA repeats.

4 RNAi-MEDIATED CHROMATIN AND DNA MODIFICATIONS IN *Arabidopsis*

The mechanism by which RNAi guides heterochromatic modifications in plants resembles the mechanism in fission yeast, but there are also many differences. The most important difference is that plants have methylated DNA at most heterochromatin regions, resembling vertebrates in this respect, but differing from *S. pombe*, worms, and *Drosophila* (Fig. 5). Genetic screens for mutants that relieve RNA-mediated TGS have recovered mutants in H3K9-specific KMTs and RNAi components, but they have also uncovered the requirement for DNA methyltransferases, SWI/SNF remodeling complexes, and two novel RNA polymerases in heterochromatin formation. The genes are listed in Table 1 of Ch. 13 (Pikaard and Mittelsten Scheid 2014) and the findings from these screens are described in more detail in that chapter, but here we will briefly compare the mechanism in fission yeast and plants.

Many of the silencing mutant screens in *Arabidopsis* have used inverted repeats introduced in *trans* to induce the silencing of endogenous or transgenic reporter genes, driven by tissue-specific promoters or promoters that responded to epigenetic signals. In each case, the promoter was targeted for silencing through a TGS pathway involving local chromatin changes at the promoter. Genes of note in the RNAi and TGS *Arabidopsis* pathways (Fig. 5) include DNA methyltransferases, particularly those related to the mammalian DNMT3 encoded by *DRM1* and *DRM2*. Also, there were components of the RNAi apparatus responsible for biogenesis and usage of 24-nt siRNA, such as *DICER-*

 Cite as *Cold Spring Harb Perspect Biol* doi: 10.1101/cshperspect.a019323

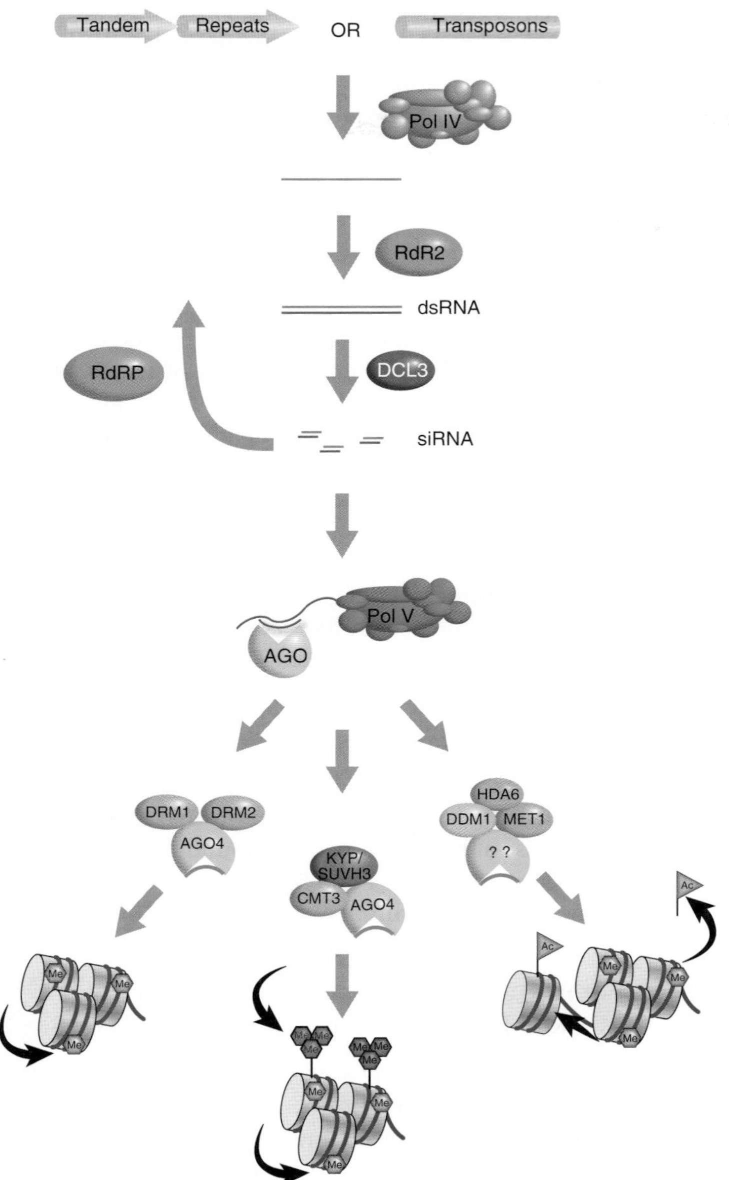

Figure 5. RNAi-mediated histone and DNA methylation in *Arabidopsis*. A summary of RNAi and chromatin proteins required for RNAi-mediated DNA and histone methylation in *Arabidopsis* are indicated. Synthesis of dsRNA from repeated DNA elements provides a substrate for Dicer-mediated cleavage and siRNA generation (DCL3 and other Dicers). RNA-directed RNA polymerases (RdRP, RdR2) and RNA polymerase IV (RNA Pol IV) may be directly involved in the synthesis of dsRNA or its amplification. siRNAs then load onto Argonaute proteins (e.g., AGO4), which is likely to help target cognate repeat sequences for DNA methylation (pink hexagons catalyzed by pink DNA methyltransferase enzymes) and H3K9 methylation (red hexagons via the red HKMT enzyme) in association with other factors, including chromatin remodeling proteins (pale green) and HDAC enzymes (cyan). See Ch. 13 (Pikaard and Mittelsten Scheid 2014) for more detail.

LIKE 3 (DCL3) and *ARGONAUTE 4 (AGO4)*. Yet, other mutants included those for the H3K9 methyltransferase genes *KYP/SUVH4*, *SUVH5* and *SUVH6*, and the chromodomain-containing DNA methyltransferase gene *CMT3*. Apart from the DNA methyltransferase genes, the parallels with fission yeast are striking, namely, that the *S. pombe* RITS complex contains both an Argonaute protein and the chromodomain protein Chp1, which depends on H3K9 methylation for its association with the chromosome. Unlike fission yeast, however, loss of heterochromatin-associated proteins that mediate TGS, like CMT3 or loss of the H3K9me2 mark does not result in loss of siRNA in *Arabidopsis* (Lippman et al. 2003), and it is not yet clear if these DNA/chromatin-modifying proteins form a complex with AGO4.

It turns out there is considerable redundancy in the proteins involved in *Arabidopsis* RNAi and TGS pathways, making it harder to dissect the detailed mechanisms. Furthermore, the pathway used to achieve heterochromatin formation depends on the underlying sequence—whether it is a transposon, retrotransposon, or tandem repeat—resulting in both RNAi and RNAi-independent mechanisms of silencing. Among the known players, SUVH4, -5, and -6, for example, all contain an SRA domain that binds methylated DNA, whereas DNA methyltransferase CMT3 has a chromodomain and a BAH domain that each recognize K9me2 (see Ch. 13 [Pikaard and Mittelsten Scheid 2014]). We know that, in some cases, DNA methylation and H3K9me2 in heterochromatic regions act to reinforce each other, bypassing the role of RNAi in heterochromatin maintenance. Interestingly, mutants in DNA polymerase epsilon and cullin 4, components of the Clr4-containing CLRC complex in *S. pombe*, relieve silencing of some RNAi-sensitive transgenes in *Arabidopsis*, implicating that it is a cell-cycle-restricted process in plants as well (Yin et al. 2009; del Olmo et al. 2010; Dumbliauskas et al. 2011; Pazhouhandeh et al. 2011). Furthermore, some of these mutants have elevated levels of HR repair, and Ago2 itself has been implicated in HR repair of double-strand breaks (Wei et al. 2012), reminiscent of the role of DNA replication and repair in heterochromatic silencing and spreading in *S. pombe* (Li et al. 2011; Zaratiegui et al. 2011). Thus, at least some aspects of the mechanism responsible for spreading of the Rik1/CLRC complex are conserved between fission yeast and *Arabidopsis*.

Transposons in *Arabidopsis* are a major source of 24-nt siRNA, unlike in *S. pombe*. However, *Arabidopsis* centromeric 180-bp satellite repeats, which are arranged in tens of thousands of tandem copies on either side of Athila LTR retroelements, are also transcribed and processed by RNAi as in fission yeast (Fig. 3). This processing depends on chromatin remodeling by the SWI2/SNF2 ATPase DDM1, dsRNA production from Pol IV transcripts by RDR2, and dicing by DCL3 (Fig. 5). Silencing also depends on H3K9me2 and the associated DNA methyltransferase CMT3. But silencing is more complex than in *S. pombe* as retrotransposon insertions into the repeats can silence the adjacent repeats, and this depends on other mechanisms including MET1, DDM1, and the histone deacetylase HDA6 (May et al. 2005). Interestingly, centromeres of the fission yeast *Styrax japonicus* also have multiple retrotransposon insertions, which generate siRNA-resembling plant pericentromeric regions in this respect (Rhind et al. 2011). DDM1 has an exquisite specificity for transposons and repeats, and must somehow recognize these as being different from genes, although the mechanism remains unclear. Loss of DNA methylation in *met1* and *ddm1* mutants is epigenetically inherited in crosses to wild-type plants, and only those transposons that retain siRNA can be remethylated (Teixeira et al. 2009). Thus, as in *S. pombe*, RNAi has a crucial role in the initiation of silencing in *Arabidopsis*.

As mentioned earlier, in fission yeast, subunits of RNA Pol II are required for silencing and siRNA production, supporting the idea that the RNAi and chromatin modification apparatuses are recruited to the chromosome by nascent transcripts (Fig. 4). In *Arabidopsis*, two novel RNA polymerases (Pol IV and Pol V) were recovered in some of the screens mentioned above. It is not yet known what template is used by Pol IV, but Pol V is responsible for intergenic transcription of repeats and recruits AGO4 via GW repeats in the carboxy-terminal domain and in accessory elongation factors. Only the largest subunits are unique to Pol IV and Pol V, which use many of the same small subunits as Pol II. Additional SWI2/SNF2 chromatin remodelers, namely, CLSY2, were recovered in these screens and they may alter local chromatin structure to facilitate processivity of RNA polymerases. It is therefore likely that they facilitate transcription by Pol IV (more detail can be found in Ch. 13 [Pikaard and Mittelsten Scheid 2014]). A similar role can be proposed for DDM1, although the requirement for DDM1 (also a chromatin remodeler) in silencing transposons is far more severe than that of Pol IV or the other SWI2/SNF2 proteins.

Thus, the role of RNAi-mediated heterochromatic silencing in plants is well documented, but more complicated than in *S. pombe*, largely because of DNA methylation. Parallels with fission yeast may become clearer in mutants that have lost RNAi-independent DNA methylation, such as *ddm1* (Teixeira et al. 2009) or in the germline. For example, in pollen grains, 24-nt siRNAs guide RNA-dependent DNA methylation independently of DDM1 (Calarco et al. 2012), whereas 21-nt siRNA are responsible for transposon silencing in sperm cells (Slotkin et al. 2009). Plants do not possess the piRNA pathway found in animal germlines (discussed in Secs. 1 and 5), and parallels with fission yeast may be more prominent as a result.

5 CONSERVATION OF RNAi-MEDIATED CHROMATIN MODIFICATIONS IN ANIMALS

Perhaps the most widely studied examples of epigenetic silencing are found in animals, including *Drosophila* and *C. elegans*, as well as in the mouse. The role of RNA and RNA interference in transcriptional silencing and heterochromatic modifications appears to be conserved in some model animals as well as in protists and plants. In *Drosophila*, both PIWI and the PIWI class Argonaute homolog, Aubergine (Sting), participate in epigenetic and heterochromatic silencing (Kawamura et al. 2008; Khurana et al. 2010; see also Ch. 12 [Elgin and Reuter 2013]). *Gypsy* retrotransposons are the target of silencing in ovary follicle cells and female gonads by PIWI and its associated piRNAs (Sarot et al. 2004). This is mediated by heterochromatic noncoding RNAs encoded by Flamenco, which give rise to the piRNAs that target complementary transposons for silencing. Cut-and-paste DNA transposons are also affected by RNAi. For example, certain telomeric P elements can suppress P activity elsewhere in the genome when inherited through the female germline, resulting in a strongly repressive "cytotype." This repression is completely dependent on the PIWI homolog, aubergine, as well as the Swi6 homolog HP1, for heterochromatin protein 1 (Reiss et al. 2004). However, not all P repressive cytotypes, such as those mediated by other nontelomeric P elements, are dependent on aubergine or HP1.

Unlinked transgenes in *Drosophila* are silenced posttranscriptionally when present in many copies (Pal-Bhadra et al. 1997, 2002). Silencing is associated with large amounts of 21-nt siRNA and depends on PIWI. Transgene fusions can also silence each other transcriptionally in a manner that requires the Polycomb chromatin repressor. This silencing is not associated with increased levels of siRNA from the transgene transcript, but is (largely) dependent on PIWI. Involvement of Polycomb in this example, and HP-1 in other examples of PIWI-dependent silencing, implicates the RNAi pathway and histone methylation in the silencing process. Tandem transgene arrays also show position-effect variegation in *Drosophila*, and this variegation is strongly suppressed by mutants in HP1 as well as in piwi, aubergine, and the putative RNA helicase Spindle-E (homeless; Pal-Bhadra et al. 2004). Transgenes inserted within centric heterochromatin are also affected, and heterochromatic levels of H3K9me2 are reduced in spindle-E mutant cells. These observations support a role for both chromatin proteins and components of the RNAi pathway in gene silencing within *Drosophila* heterochromatin.

In the *Drosophila* male germline, the heterochromatic suppressor of stellate repeats (Su(ste)), located on the Y chromosome, are transcribed first on the antisense strand and then on both strands during spermatocyte development, possibly following the insertion of a nearby transposon (Aravin et al. 2001). These nuclear transcripts are required to silence sense transcripts of the closely related X-linked *Stellate* gene, whose overexpression results in defects in spermatogenesis. Although heterochromatic sequences are involved, silencing in this case appears to be posttranscriptional, is associated with 25–27-nt piRNA, and depends on both *aubergine* and *Spindle-E*.

In *C. elegans*, examples of TGS in somatic cells have been reported. This depends on the RNAi pathway genes rde-1, dcr-1, rde-4, and rrf-1, a nuclear Argonaute and its associated factors, as well as HP1 homologs and the histone modification apparatus (Grishok et al. 2005; Guang et al. 2010; Buckley et al. 2012). An example of naturally occurring RNAi-dependent heterochromatic silencing has also been described in the germline (Sijen and Plasterk 2003). During meiosis, unpaired sequences, such as the X chromosome in males, are silenced via H3K9me2, and this silencing depends on RNA-dependent RNA polymerase (Maine et al. 2005; see Ch. 23 [Strome et al. 2014]), reminiscent of meiotic silencing of unpaired DNA in *Neurospora* (MSUD; Shiu et al. 2001; see Ch. 10 [Aramayo and Selker 2013]). It is possible therefore that the RNAi-dependent heterochromatin silencing pathway found in fission yeast is preserved in meiosis in higher organisms.

Finally, like *Drosophila*, mammalian cells lack genes related to RNA-dependent RNA polymerases found in plants, worms, and fungi. Nonetheless, antisense RNA has been implicated in the most widely studied epigenetic phenomena of all, imprinting and X inactivation (see Ch. 26 [Barlow and Bartolomei 2014] and Ch. 25 [Brockdorff and Turner 2014], respectively). In the case of X inactivation, a 17-kb spliced and polyadenylated noncoding RNA known as Xist is required to silence the inactive X chromosome from which it is expressed. Conversely, *Xist* itself is silenced on the active X chromosome, a process that depends in part on the antisense RNA *Tsix*. Silencing is accompanied by modification of histones associated with upstream chromatin regions, which are marked with H3K9me2 and H3K27me3 (Heard et al. 2001), and depends on the H3K27 methyltransferase Polycomb group gene Eed. Silencing of other imprinted loci in the mouse, including Igf2r and the Dlk1-Gtl2 region, is also maintained by antisense transcripts from the paternal and maternal allele, respectively. In the case of Dlk1-Gtl2, this noncoding RNA is specifically processed into miRNA that targets the antisense transcript from the paternal allele, encoding a sushi (gypsy) class retrotransposon (Davis et al. 2005). Perhaps the best example of RNAi-dependent DNA methylation in mammalian imprinting is at the *Rasgrf1* locus (Watanabe et al. 2011). At this locus, a long noncoding RNA is targeted by piRNA that match an embedded retrotransposon LTR. piRNA guides DNA methylation of the differentially methylated region required for silencing, so that imprinting in spermatocytes comes under the control of RNAi (Watanabe et al. 2011). In this case, clear parallels can be drawn with heterochromatic silencing in *S. pombe*, which also depends on the cleavage and targeting activities of Ago1 (Irvine et al. 2006; Buker et al. 2007), and with RNA-dependent DNA methylation in plants, which depends on catalytic activity of AGO4 (Qi et al. 2006).

6 CONCLUDING REMARKS

The possibility that genes may be regulated by small RNA molecules was suggested more than 40 years ago (Jacob and Monod 1961). An equally important hypothetical notion was that regulatory RNAs might be related to repeats (Britten and Davidson 1969). Since the identification of the λ and lac repressors as site-specific DNA-binding proteins in *Escherichia coli* or its infecting bacteriophage λ (Gilbert and Muller-Hill 1966; Ptashne 1967), studies of gene regulation have focused almost exclusively on the role of nucleic acid–binding proteins as specificity factors. The discovery of small RNA molecules as specificity agents in diverse RNA-silencing mechanisms now clearly establishes a role for RNA as a sequence-specific regulator of genes and their RNA products. Studies in fission yeast, *Arabidopsis*, and other model organisms have revealed a surprisingly direct role for small RNAs in mediating epigenetic modifications of the genome that direct gene silencing and contribute to heterochromatic domains necessary for genome stability and nuclear division. Many important mechanistic questions remain at large and future studies are likely to provide more surprises about how RNA regulates gene expression.

REFERENCES

**Reference is also in this book.*

* Aramayo R, Selker EU. 2013. *Neurospora crassa*, a model system for epigenetics research. *Cold Spring Harb Perspect Biol* **5:** a017921.

Aravin AA, Naumova NM, Tulin AV, Vagin VV, Rozovsky YM, Gvozdev VA. 2001. Double-stranded RNA-mediated silencing of genomic tandem repeats and transposable elements in the *D. melanogaster* germline. *Curr Biol* **11:** 1017–1027.

Aravin AA, Hannon GJ, Brennecke J. 2007. The Piwi-piRNA pathway provides an adaptive defense in the transposon arms race. *Science* **318:** 761–764.

Ashe A, Sapetschnig A, Weick EM, Mitchell J, Bagijn MP, Cording AC, Doebley AL, Goldstein LD, Lehrbach NJ, Le Pen J, et al. 2012. piRNAs can trigger a multigenerational epigenetic memory in the germline of *C. elegans*. *Cell* **150:** 88–99.

Barabino SM, Ohnacker M, Keller W. 2000. Distinct roles of two Yth1p domains in 3′-end cleavage and polyadenylation of yeast pre-mRNAs. *EMBO J* **19:** 3778–3787.

* Barlow DP, Bartolomei MS. 2014. Genomic imprinting in mammals. *Cold Spring Harb Perspect Biol* **6:** a018382.

Bartel DP. 2004. MicroRNAs: Genomics, biogenesis, mechanism, and function. *Cell* **116:** 281–297.

Baulcombe D. 2004. RNA silencing in plants. *Nature* **431:** 356–363.

Bayne EH, White SA, Kagansky A, Bijos DA, Sanchez-Pulido L, Hoe KL, Kim DU, Park HO, Ponting CP, Rappsilber J, et al. 2010. Stc1: A critical link between RNAi and chromatin modification required for heterochromatin integrity. *Cell* **140:** 666–677.

Bernstein E, Caudy AA, Hammond SM, Hannon GJ. 2001. Role for a bidentate ribonuclease in the initiation step of RNA interference. *Nature* **409:** 363–366.

Brennecke J, Aravin AA, Stark A, Dus M, Kellis M, Sachidanandam R, Hannon GJ. 2007. Discrete small RNA-generating loci as master regulators of transposon activity in *Drosophila*. *Cell* **128:** 1089–1103.

Britten RJ, Davidson EH. 1969. Gene regulation for higher cells: A theory. *Science* **165:** 349–357.

* Brockdorff N, Turner BM. 2014. Dosage compensation in mammals. *Cold Spring Harb Perspect Biol* doi: 10.1101/cshperspect.a019406.

Buckley BA, Burkhart KB, Gu SG, Spracklin G, Kershner A, Fritz H, Kimble J, Fire A, Kennedy S. 2012. A nuclear Argonaute promotes multigenerational epigenetic inheritance and germline immortality. *Nature* **489:** 447–451.

Bühler M, Verdel A, Moazed D. 2006. Tethering RITS to a nascent transcript initiates RNAi- and heterochromatin-dependent gene silencing. *Cell* **125:** 873–886.

Bühler M, Haas W, Gygi SP, Moazed D. 2007. RNAi-dependent and -independent RNA turnover mechanisms contribute to heterochromatic gene silencing. *Cell* **129:** 707–721.

Buker SM, Iida T, Bühler M, Villèn J, Gygi SP, Nakayama J, Moazed D. 2007. Two different Argonaute complexes are required for siRNA generation and heterochromatin assembly in fission yeast. *Nat Struct Mol Biol* **14:** 200–207.

Calarco JP, Borges F, Donoghue MT, Van Ex F, Jullien PE, Lopes T, Gardner R, Berger F, Feijo JA, Becker JD, et al. 2012. Reprogramming of DNA methylation in pollen guides epigenetic inheritance via small RNA. *Cell* **151:** 194–205.

* Chalker DL, Meyer E, Mochizuki K. 2013. Epigenetics of ciliates. *Cold Spring Harb Perspect Biol* **5:** a017764.

Chandler VL. 2010. Paramutation's properties and puzzles. *Science* **330:** 628–629.

Chen CC, Simard MJ, Tabara H, Brownell DR, McCollough JA, Mello CC. 2005. A member of the polymerase β nucleotidyltransferase superfamily is required for RNA interference in *C. elegans*. *Curr Biol* **15:** 378–383.

Chen ES, Zhang K, Nicolas E, Cam HP, Zofall M, Grewal SI. 2008. Cell cycle control of centromeric repeat transcription and heterochromatin assembly. *Nature* **451:** 734–737.

Dalmay T, Hamilton A, Mueller E, Baulcombe DC. 2000. Potato virus X amplicons in *Arabidopsis* mediate genetic and epigenetic gene silencing. *Plant Cell* **12:** 369–379.

Davis E, Caiment F, Tordoir X, Cavaille J, Ferguson-Smith A, Cockett N, Georges M, Charlier C. 2005. RNAi-mediated allelic trans-interaction at the imprinted Rtl1/Peg11 locus. *Curr Biol* **15:** 743–749.

del Olmo I, Lopez-Gonzalez L, Martin-Trillo MM, Martinez-Zapater JM, Pineiro M, Jarillo JA. 2010. EARLY IN SHORT DAYS 7 (ESD7) encodes the catalytic subunit of DNA polymerase epsilon and is required for flowering repression through a mechanism involving epigenetic gene silencing. *Plant J* **61:** 623–636.

Djupedal I, Portoso M, Spahr H, Bonilla C, Gustafsson CM, Allshire RC, Ekwall K. 2005. RNA Pol II subunit Rpb7 promotes centromeric transcription and RNAi-directed chromatin silencing. *Genes Dev* **19:** 2301–2306.

Dumbliauskas E, Lechner E, Jaciubek M, Berr A, Pazhouhandeh M, Alioua M, Cognat V, Brukhin V, Koncz C, Grossniklaus U, et al. 2011. The *Arabidopsis* CUL4-DDB1 complex interacts with MSI1 and is required to maintain MEDEA parental imprinting. *EMBO J* **30:** 731–743.

Elbashir SM, Lendeckel W, Tuschl T. 2001. RNA interference is mediated by 21- and 22-nucleotide RNAs. *Genes Dev* **15:** 188–200.

* Elgin SCR, Reuter G. 2013. Position-effect variegation, heterochromatin formation, and gene silencing in *Drosophila*. *Cold Spring Harb Perspect Biol* **5:** a017780.

Filipowicz W, Jaskiewicz L, Kolb FA, Pillai RS. 2005. Post-transcriptional gene silencing by siRNAs and miRNAs. *Curr Opin Struct Biol* **15:** 331–341.

Fire A, Xu S, Montgomery MK, Kostas SA, Driver SE, Mello CC. 1998. Potent and specific genetic interference by double-stranded RNA in *Caenorhabditis elegans*. *Nature* **391:** 806–811.

Fischer T, Cui B, Dhakshnamoorthy J, Zhou M, Rubin C, Zofall M, Veenstra TD, Grewal SI. 2009. Diverse roles of HP1 proteins in heterochromatin assembly and functions in fission yeast. *Proc Natl Acad Sci* **106:** 8998–9003.

Francis NJ, Kingston RE. 2001. Mechanisms of transcriptional memory. *Nat Rev Mol Biol* **2:** 409–421.

Galagan JE, Selker EU. 2004. RIP: The evolutionary cost of genome defense. *Trends Genet* **20:** 417–423.

Gerace EL, Halic M, Moazed D. 2010. The methyltransferase activity of Clr4Suv39h triggers RNAi independently of histone H3K9 methylation. *Mol Cell* **39:** 360–372.

Ghildiyal M, Zamore PD. 2009. Small silencing RNAs: An expanding universe. *Nat Rev Genet* **10:** 94–108.

Gilbert W, Muller-Hill B. 1966. Isolation of the Lac repressor. *Proc Natl Acad Sci* **56:** 1891–1898.

Grewal SI. 2000. Transcriptional silencing in fission yeast. *J Cell Physiol* **184:** 311–318.

Grewal SI, Moazed D. 2003. Heterochromatin and epigenetic control of gene expression. *Science* **301:** 798–802.

Grishok A, Sinskey JL, Sharp PA. 2005. Transcriptional silencing of a transgene by RNAi in the soma of *C. elegans*. *Genes Dev* **19:** 683–696.

Gu SG, Pak J, Guang S, Maniar JM, Kennedy S, Fire A. 2012. Amplification of siRNA in *Caenorhabditis elegans* generates a transgenerational sequence-targeted histone H3 lysine 9 methylation footprint. *Nat Genet* **44:** 157–164.

Guang S, Bochner AF, Burkhart KB, Burton N, Pavelec DM, Kennedy S. 2010. Small regulatory RNAs inhibit RNA polymerase II during the elongation phase of transcription. *Nature* **465:** 1097–1101.

Halic M, Moazed D. 2010. Dicer-independent primal RNAs trigger RNAi and heterochromatin formation. *Cell* **140:** 504–516.

Hall IM, Shankaranarayana GD, Noma K, Ayoub N, Cohen A, Grewal SI. 2002. Establishment and maintenance of a heterochromatin domain. *Science* **297:** 2232–2237.

Hamilton AJ, Baulcombe DC. 1999. A species of small antisense RNA in posttranscriptional gene silencing in plants. *Science* **286:** 950–952.

Hampsey M, Reinberg D. 2003. Tails of intrigue: Phosphorylation of RNA polymerase II mediates histone methylation. *Cell* **113:** 429–432.

Hannon GJ. 2002. RNA interference. *Nature* **418:** 244–251.

Hayashi A, Ishida M, Kawaguchi R, Urano T, Murakami Y, Nakayama J. 2012. Heterochromatin protein 1 homologue Swi6 acts in concert with Ers1 to regulate RNAi-directed heterochromatin assembly. *Proc Natl Acad Sci* **109:** 6159–6164.

Heard E, Rougeulle C, Arnaud D, Avner P, Allis CD, Spector DL. 2001. Methylation of histone H3 at Lys-9 is an early mark on the X chromosome during X inactivation. *Cell* **107:** 727–738.

Higa LA, Mihaylov IS, Banks DP, Zheng J, Zhang H. 2003. Radiation-mediated proteolysis of CDT1 by CUL4-ROC1 and CSN complexes constitutes a new checkpoint. *Nat Cell Biol* **5:** 1008–1015.

Hong EJ, Villèn J, Gerace EL, Gygi SP, Moazed D. 2005. A cullin E3 ubiquitin ligase complex associates with Rik1 and the Clr4 histone H3-K9 methyltransferase and is required for RNAi-mediated heterochromatin formation. *RNA Biol* **2:** 106–111.

Horn PJ, Bastie JN, Peterson CL. 2005. A Rik1-associated, cullin-dependent E3 ubiquitin ligase is essential for heterochromatin formation. *Genes Dev* **19:** 1705–1714.

Houseley J, LaCava J, Tollervey D. 2006. RNA-quality control by the exosome. *Nat Rev Mol Cell Biol* **7:** 529–539.

Iida T, Nakayama J, Moazed D. 2008. siRNA-mediated heterochromatin establishment requires HP1 and is associated with antisense transcription. *Mol Cell* **31:** 178–189.

Irvine DV, Zaratiegui M, Tolia NH, Goto DB, Chitwood DH, Vaughn MW, Joshua-Tor L, Martienssen RA. 2006. Argonaute slicing is required for heterochromatic silencing and spreading. *Science* **313:** 1134–1137.

Jacob F, Monod J. 1961. Genetic regulatory mechanisms in the synthesis of proteins. *J Mol Biol* **3:** 318–356.

Jia S, Noma K, Grewal SI. 2004. RNAi-independent heterochromatin nucleation by the stress-activated ATF/CREB family proteins. *Science* **304:** 1971–1976.

Jia S, Kobayashi R, Grewal SI. 2005. Ubiquitin ligase component Cul4 associates with Clr4 histone methyltransferase to assemble heterochromatin. *Nat Cell Biol* **7:** 1007–1013.

Kato H, Goto DB, Martienssen RA, Urano T, Furukawa K, Murakami Y. 2005. RNA polymerase II is required for RNAi-dependent heterochromatin assembly. *Science* **309:** 467–469.

Kawamura Y, Saito K, Kin T, Ono Y, Asai K, Sunohara T, Okada TN, Siomi MC, Siomi H. 2008. *Drosophila* endogenous small RNAs bind to Argonaute 2 in somatic cells. *Nature* **453:** 793–797.

Keller C, Adaixo R, Stunnenberg R, Woolcock KJ, Hiller S, Bühler M. 2012. $HP1^{Swi6}$ mediates the recognition and destruction of heterochromatic RNA transcripts. *Mol Cell* **47:** 215–227.

Khurana JS, Xu J, Weng Z, Theurkauf WE. 2010. Distinct functions for the *Drosophila* piRNA pathway in genome maintenance and telomere protection. *PLoS Genet* **6:** e1001246.

Kim JK, Gabel HW, Kamath RS, Tewari M, Pasquinelli A, Rual JF, Kennedy S, Dybbs M, Bertin N, Kaplan JM, et al. 2005. Functional genomic analysis of RNA interference in *C. elegans*. *Science* **308:** 1164–1167.

Kloc A, Zaratiegui M, Nora E, Martienssen R. 2008. RNA interference guides histone modification during the S phase of chromosomal replication. *Curr Biol* **18:** 490–495.

Lee HC, Gu W, Shirayama M, Youngman E, Conte D Jr, Mello CC. 2012. *C elegans* piRNAs mediate the genome-wide surveillance of germline transcripts. *Cell* **150:** 78–87.

Li HW, Ding SW. 2005. Antiviral silencing in animals. *FEBS Lett* **579:** 5965–5973.

Li F, Goto DB, Zaratiegui M, Tang X, Martienssen R, Cande WZ. 2005. Two novel proteins, Dos1 and Dos2, interact with Rik1 to regulate heterochromatic RNA interference and histone modification. *Curr Biol* **15:** 1448–1457.

Li F, Martienssen R, Cande WZ. 2011. Coordination of DNA replication and histone modification by the Rik1-Dos2 complex. *Nature* **475:** 244–248.

Lippman Z, May B, Yordan C, Singer T, Martienssen R. 2003. Distinct mechanisms determine transposon inheritance and methylation via small interfering RNA and histone modification. *PLoS Biol* **1:** E67.

Maine EM, Hauth J, Ratliff T, Vought VE, She X, Kelly WG. 2005. EGO-1, a putative RNA-dependent RNA polymerase, is required for heterochromatin assembly on unpaired DNA during *C. elegans* meiosis. *Curr Biol* **15:** 1972–1978.

Marasovic M, Zocco M, Halic M. 2013. Argonaute and Triman generate Dicer-independent priRNAs and mature siRNAs to initiate heterochromatin formation. *Mol Cell* **52:** 173–183.

Martienssen RA, Zaratiegui M, Goto DB. 2005. RNA interference and heterochromatin in the fission yeast *Schizosaccharomyces pombe*. *Trends Genet* **21:** 450–456.

May BP, Lippman ZB, Fang Y, Spector DL, Martienssen RA. 2005. Differential regulation of strand-specific transcripts from *Arabidopsis* centromeric satellite repeats. *PLoS Genet* **1:** e79.

Mette MF, van der Winden J, Matzke MA, Matzke AJ. 1999. Production of aberrant promoter transcripts contributes to methylation and silencing of unlinked homologous promoters in trans. *EMBO J* **18:** 241–248.

Moazed D. 2011. Mechanisms for the inheritance of chromatin states. *Cell* **146:** 510–518.

Motamedi MR, Verdel A, Colmenares SU, Gerber SA, Gygi SP, Moazed D. 2004. Two RNAi complexes, RITS and RDRC, physically interact and localize to noncoding centromeric RNAs. *Cell* **119:** 789–802.

Motamedi MR, Hong EJ, Li X, Gerber S, Denison C, Gygi SP, Moazed D. 2008. HP1 proteins form distinct complexes and mediate heterochromatic gene silencing by non-overlapping mechanisms. *Mol Cell* **32:** 778–790.

Napoli C, Lemieux C, Jorgensen R. 1990. Introduction of a chimeric chalcone synthase gene into petunia results in reversible co-suppression of homologous genes in *trans*. *Plant Cell* **2:** 279–289.

Noma K, Sugiyama T, Cam H, Verdel A, Zofall M, Jia S, Moazed D, Grewal SI. 2004. RITS acts in *cis* to promote RNA interference-mediated transcriptional and post-transcriptional silencing. *Nat Genet* **36:** 1174–1180.

Pal-Bhadra M, Bhadra U, Birchler JA. 1997. Cosuppression in *Drosophila*: Gene silencing of *Alcohol dehydrogenase* by *white-Adh* transgenes is *Polycomb* dependent. *Cell* **90:** 479–490.

Pal-Bhadra M, Bhadra U, Birchler JA. 1999. Cosuppression of nonhomologous transgenes in *Drosophila* involves mutually related endogenous sequences. *Cell* **99:** 35–46.

Pal-Bhadra M, Bhadra U, Birchler JA. 2002. RNAi related mechanisms affect both transcriptional and posttranscriptional transgene silencing in *Drosophila. Mol Cell* **9:** 315–327.

Pal-Bhadra M, Leibovitch BA, Gandhi SG, Chikka MR, Bhadra U, Birchler JA, Elgin SC. 2004. Heterochromatic silencing and HP1 localization in *Drosophila* are dependent on the RNAi machinery. *Science* **303:** 669–672.

Partridge JF, Scott KS, Bannister AJ, Kouzarides T, Allshire RC. 2002. *cis*-acting DNA from fission yeast centromeres mediates histone H3 methylation and recruitment of silencing factors and cohesin to an ectopic site. *Curr Biol* **12:** 1652–1660.

Pazhouhandeh M, Molinier J, Berr A, Genschik P. 2011. MSI4/FVE interacts with CUL4-DDB1 and a PRC2-like complex to control epigenetic regulation of flowering time in *Arabidopsis. Proc Natl Acad Sci* **108:** 3430–3435.

Petrie VJ, Wuitschick JD, Givens CD, Kosinski AM, Partridge JF. 2005. RNA interference (RNAi)-dependent and RNAi-independent association of the Chp1 chromodomain protein with distinct heterochromatic loci in fission yeast. *Mol Cell Biol* **25:** 2331–2346.

Pidoux AL, Allshire RC. 2004. Kinetochore and heterochromatin domains of the fission yeast centromere. *Chromosome Res* **12:** 521–534.

* Pikaard CS, Mittelsten Sheid O. 2014. Epigenetic regulation in plants. *Cold Spring Harb Perspect Biol* doi: 10.1101/cshperspect.a019315.

Plasterk RH. 2002. RNA silencing: The genome's immune system. *Science* **296:** 1263–1265.

Ptashne M. 1967. Specific binding of the lambda phage repressor to lambda DNA. *Nature* **214:** 232–234.

Qi Y, He X, Wang XJ, Kohany O, Jurka J, Hannon GJ. 2006. Distinct catalytic and non-catalytic roles of ARGONAUTE4 in RNA-directed DNA methylation. *Nature* **443:** 1008–1012.

Reinhart BJ, Bartel DP. 2002. Small RNAs correspond to centromere heterochromatic repeats. *Science* **297:** 1831.

Reiss D, Josse T, Anxolabehere D, Ronsseray S. 2004. *aubergine* mutations in *Drosophila melanogaster* impair P cytotype determination by telomeric *P* elements inserted in heterochromatin. *Mol Genet Genomics* **272:** 336–343.

Reyes-Turcu FE, Zhang K, Zofall M, Chen E, Grewal SI. 2011. Defects in RNA quality control factors reveal RNAi-independent nucleation of heterochromatin. *Nat Struct Mol Biol* **18:** 1132–1138.

Rhind N, Chen Z, Yassour M, Thompson DA, Haas BJ, Habib N, Wapinski I, Roy S, Lin MF, Heiman DI, et al. 2011. Comparative functional genomics of the fission yeasts. *Science* **332:** 930–936.

Rougemaille M, Braun S, Coyle S, Dumesic PA, Garcia JF, Isaac RS, Libri D, Narlikar GJ, Madhani HD. 2012. Ers1 links HP1 to RNAi. *Proc Natl Acad Sci* **109:** 11258–11263.

Sarot E, Payen-Groschene G, Bucheton A, Pelisson A. 2004. Evidence for a *piwi*-dependent RNA silencing of the *gypsy* endogenous retrovirus by the *Drosophila melanogaster flamenco* gene. *Genetics* **166:** 1313–1321.

Shirayama M, Seth M, Lee HC, Gu W, Ishidate T, Conte D Jr, Mello CC. 2012. piRNAs initiate an epigenetic memory of nonself RNA in the *C. elegans* germline. *Cell* **150:** 65–77.

Shiu PK, Raju NB, Zickler D, Metzenberg RL. 2001. Meiotic silencing by unpaired DNA. *Cell* **107:** 905–916.

Sigova A, Rhind N, Zamore PD. 2004. A single Argonaute protein mediates both transcriptional and posttranscriptional silencing in *Schizosaccharomyces pombe. Genes Dev* **18:** 2359–2367.

Sijen T, Plasterk RH. 2003. Transposon silencing in the *Caenorhabditis elegans* germ line by natural RNAi. *Nature* **426:** 310–314.

Sijen T, Fleenor J, Simmer F, Thijssen KL, Parrish S, Timmons L, Plasterk RH, Fire A. 2001. On the role of RNA amplification in dsRNA-triggered gene silencing. *Cell* **107:** 465–476.

Simmer F, Buscaino A, Kos-Braun IC, Kagansky A, Boukaba A, Urano T, Kerr AR, Allshire RC. 2010. Hairpin RNA induces secondary small interfering RNA synthesis and silencing in trans in fission yeast. *EMBO Rep* **11:** 112–118.

Slotkin RK, Martienssen R. 2007. Transposable elements and the epigenetic regulation of the genome. *Nat Rev Genet* **8:** 272–285.

Slotkin RK, Vaughn M, Borges F, Tanurdzic M, Becker JD, Feijo JA, Martienssen RA. 2009. Epigenetic reprogramming and small RNA silencing of transposable elements in pollen. *Cell* **136:** 461–472.

Song JJ, Smith SK, Hannon GJ, Joshua-Tor L. 2004. Crystal structure of Argonaute and its implications for RISC slicer activity. *Science* **305:** 1434–1437.

* Strome S, Kelly WG, Ercan S, Lieb JD. 2014. Regulation of the X chromosomes in *Caenorhabditis elegans. Cold Spring Harb Perspect Biol* **6:** a018366.

Sugiyama T, Cam H, Verdel A, Moazed D, Grewal SI. 2005. RNA-dependent RNA polymerase is an essential component of a self-enforcing loop coupling heterochromatin assembly to siRNA production. *Proc Natl Acad Sci* **102:** 152–157.

Teixeira FK, Heredia F, Sarazin A, Roudier F, Boccara M, Ciaudo C, Cruaud C, Poulain J, Berdasco M, Fraga MF, et al. 2009. A role for RNAi in the selective correction of DNA methylation defects. *Science* **323:** 1600–1604.

Verdel A, Moazed D. 2005. RNAi-directed assembly of heterochromatin in fission yeast. *FEBS Lett* **579:** 5872–5878.

Verdel A, Jia S, Gerber S, Sugiyama T, Gygi S, Grewal SI, Moazed D. 2004. RNAi-mediated targeting of heterochromatin by the RITS complex. *Science* **303:** 672–676.

Volpe TA, Kidner C, Hall IM, Teng G, Grewal SI, Martienssen RA. 2002. Regulation of heterochromatic silencing and histone H3 lysine-9 methylation by RNAi. *Science* **297:** 1833–1837.

Volpe T, Schramke V, Hamilton GL, White SA, Teng G, Martienssen RA, Allshire RC. 2003. RNA interference is required for normal centromere function in fission yeast. *Chromosome Res* **11:** 137–146.

Wassenegger M, Heimes S, Riedel L, Sanger HL. 1994. RNA-directed de novo methylation of genomic sequences in plants. *Cell* **76:** 567–576.

Watanabe T, Tomizawa S, Mitsuya K, Totoki Y, Yamamoto Y, Kuramochi-Miyagawa S, Iida N, Hoki Y, Murphy PJ, Toyoda A, et al. 2011. Role for piRNAs and noncoding RNA in de novo DNA methylation of the imprinted mouse Rasgrf1 locus. *Science* **332:** 848–852.

Wei W, Ba Z, Gao M, Wu Y, Ma Y, Amiard S, White CI, Rendtlew Danielsen JM, Yang YG, Qi Y. 2012. A role for small RNAs in DNA double-strand break repair. *Cell* **149:** 101–112.

Yin H, Zhang X, Liu J, Wang Y, He J, Yang T, Hong X, Yang Q, Gong Z. 2009. Epigenetic regulation, somatic homologous recombination, and abscisic acid signaling are influenced by DNA polymerase epsilon mutation in *Arabidopsis. Plant Cell* **21:** 386–402.

Zamore PD. 2002. Ancient pathways programmed by small RNAs. *Science* **296:** 1265–1269.

Zamore PD, Tuschl T, Sharp PA, Bartel DP. 2000. RNAi: Double-stranded RNA directs the ATP-dependent cleavage of mRNA at 21 to 23 nucleotide intervals. *Cell* **101:** 25–33.

Zaratiegui M, Castel SE, Irvine DV, Kloc A, Ren J, Li F, de Castro E, Marin L, Chang AY, Goto D, et al. 2011. RNAi promotes heterochromatic silencing through replication-coupled release of RNA Pol II. *Nature* **479:** 135–138.

Zhong W, Feng H, Santiago FE, Kipreos ET. 2003. CUL-4 ubiquitin ligase maintains genome stability by restraining DNA-replication licensing. *Nature* **423:** 885–889.

C H A P T E R 1 7

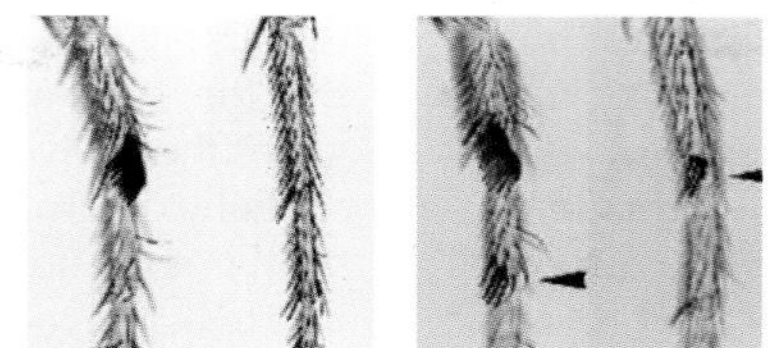

Transcriptional Silencing by *Polycomb*-Group Proteins

Ueli Grossniklaus[1] and Renato Paro[2]

[1]Institute of Plant Biology and Zürich-Basel Plant Science Center, University of Zürich, CH-8008 Zürich, Switzerland; [2]Department of Biosystems Science and Engineering, ETH Zürich, 4058 Basel, Switzerland

Correspondence: grossnik@botinst.uzh.ch

SUMMARY

Polycomb-group (PcG) genes encode chromatin proteins involved in stable and heritable transcriptional silencing. PcG proteins participate in distinct multimeric complexes that deposit, or bind to, specific histone modifications (e.g., H3K27me3 and H2AK119ub1) to prevent gene activation and maintain repressed chromatin domains. PcG proteins are evolutionary conserved and play a role in processes ranging from vernalization and seed development in plants, over X-chromosome inactivation in mammals, to the maintenance of stem cell identity. PcG silencing is medically relevant as it is often observed in human disorders, including cancer, and tissue regeneration, which involve the reprogramming of PcG-controlled target genes.

Outline

Chapter-opening figures: (*Left, right*) Reprinted from Yamamoto et al. (1997), *Development* **124:** 3385–3394, with permission from Company of Biologists Ltd.

Cite as *Cold Spring Harb Perspect Biol* doi: 10.1101/cshperspect.a019331

OVERVIEW

Organs of humans, animals, and plants are constructed from a large pool of distinct cell types, each performing a specialized physiological or structural function. With very few exceptions, all cell types contain the same genetic information encoded in their DNA. Thus, the distinctiveness of a given cell type is achieved through specific gene expression programs. As a consequence, cell lineages need to have these programs maintained during growth and cell division. This implies the existence of a memory system that ensures the faithful transmission of information (i.e., which gene is active or repressed) from mother to daughter cells. The existence of such a system is illustrated by the fact that cultured tissues of plants and animals usually maintain their differentiated characters even if grown in a foreign environment. By way of example, ivy plants regenerated after tissue culture produce the type of leaf corresponding to the phase of development from which the original tissue was taken (i.e., juvenile or adult leaf).

The major question to be addressed here and in Ch. 18 (Kingston and Tamkun 2014) concerns the molecular identity of factors contributing to the mechanism(s) of "cellular" or "transcriptional memory," which maintains a determined state over many cell divisions. Genetic analyses in *Drosophila melanogaster* have identified regulators crucial in maintaining the morphology of individual body segments that are determined by the action of the *HOX* genes. In *Drosophila* males, the first thoracic segment has legs with sex combs. Legs on the second and third thoracic segment lack these structures. In the 1940s, *Drosophila* mutants were identified (*Polycomb* and *extra sex combs*) wherein males had sex combs on all legs. These morphological alterations reflect homeotic transformations of the second and third leg identities into the first leg identity. Subsequent molecular studies showed that these mutations did not affect the products of the *HOX* genes themselves, but rather the way *HOX* gene activity was spatially controlled. Throughout the years, a large number of similar regulatory genes were identified, and were classified into two antagonistic groups: the *Polycomb* (PcG) and *Trithorax* (TrxG) group. Whereas the PcG proteins are required to maintain the silenced state of developmental regulators such as the *HOX* genes, the TrxG proteins are generally involved in maintaining the active state of gene expression. Thus PcG and TrxG proteins embody the molecular components of cellular memory.

Proteins of both groups form large multimeric protein complexes that act on their target genes by modulating chromatin structure. In this chapter, we will focus on the molecular nature and function of two of the major *Polycomb* repressive complexes, PRC1 and PRC2. The molecular nature of the TrxG complexes will be described in Ch. 18 (Kingston and Tamkun 2014). In *Drosophila*, it was shown that transcription factors recruit PcG complexes to a DNA sequence called a PcG response element (PRE). Once recruited, they establish a silent chromatin state that can be inherited over many cell divisions. Members of PRC2 are highly conserved between plants and animals, whereas PRC1 proteins are less well conserved. This implies conservation, but also diversity, in the basic building blocks of the cellular memory system. In addition to the function of PcG complexes in the maintenance of cell types, they may also play important roles in stem cell plasticity and regeneration. Also, their deregulation can lead to neoplastic transformation and cancer. Thus, PcG proteins play a crucial role in many fundamental processes of normal development and disease in multicellular eukaryotes.

Cite as *Cold Spring Harb Perspect Biol* doi: 10.1101/cshperspect.a019331

1 INTRODUCTION

All multicellular organisms start from a single cell, the zygote, which during development gives rise to a multitude of distinct cell types with specialized functions. This poses the problem of how, once determined, cell types can be maintained over many cell divisions occurring during growth phases.

1.1 The Concept of Cellular Memory

An adult animal has 200 to 300 structurally distinct cell types, whereas a plant has between 30 and 40. The identity and function of a given cell type is determined by its characteristic gene expression profile. During development and adult homeostasis, it is crucial to remember and faithfully reproduce this state after each cell division. This is particularly critical during the replication of genetic material (S-phase) and the separation of chromosomes during mitosis (M-phase; see Ch. 22 [Almouzni and Cedar 2014]). These are recurring events at each cell cycle that interrupt gene expression processes. Thus, how can differential gene expression patterns be inherited from one cell generation to the next, as illustrated in Figure 1?

We know from experiments performed in the 1960s and 1970s that plant and animal tissues remember a determined state even after prolonged passage in culture (Hadorn 1968; Hackett et al. 1987). Hadorn and colleagues showed that imaginal disc cells found in *Drosophila* larvae have an intrinsic memory, allowing them to remember determined states that are fixed in early embryogenesis. Imaginal discs are clusters of epithelial cells set aside in the developing embryo as precursors for the formation of specific external structures and appendages during metamorphosis. For instance, of the two pairs of imaginal discs in the second thoracic segment, one forms a midleg and the other a wing (see Fig. 2 in Ch. 18 [Kingston and Tamkun 2014]). Imaginal discs can be cultured by transplantation into the haemocoel of adult females, wherein they continue to proliferate, but do not differentiate. When transplanted back into a larva before metamorphosis, the disc will subsequently differentiate into the expected adult structures, even after successive passages in adult females. More recently, the PcG and TrxG proteins were shown to be required for the maintenance of the determined state of imaginal disc cells. Additionally, it was observed that in rare cases an imaginal disc could change its fate, a process called transdetermination. This process involves the down-regulation of PcG repression by the JNK signaling cascade in transdetermined cells (Lee et al. 2005). PcG mutants also have elevated frequencies of transdetermination, supporting a role for PcG proteins in maintaining imaginal disc cell fates (Katsuyama and Paro 2011). Thus, PcG proteins seem to play a crucial role in the maintenance and reprogramming of cellular fates during both normal development and regeneration.

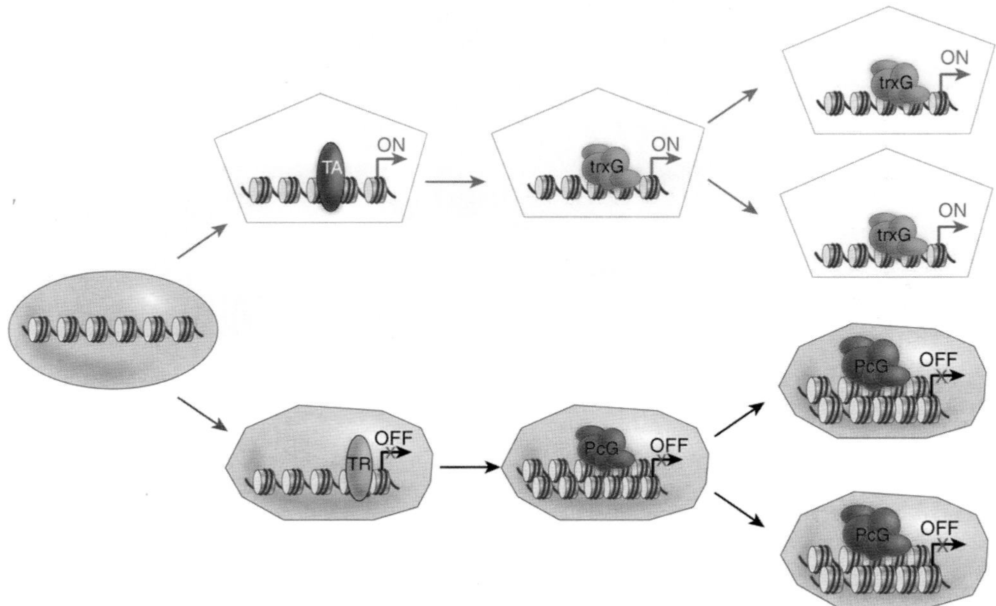

Figure 1. The concept of cellular memory. Schematic illustration of the involvement of PcG and TrxG complexes in the determination of active and repressed states of gene expression and, thereby, cellular differentiation, which is maintained over many cell divisions. TA, transcriptional activator; TR, transcriptional repressor.

1.2 The Genetic Identification of the *Polycomb* Group

In all metazoans, the anterior–posterior axis is specified through defined expression patterns of *HOX* genes (see Fig. 2 in Ch. 18 [Kingston and Tamkun 2014]). During *Drosophila* embryogenesis, the activity of maternally (i.e., inherited through the oocyte) and zygotically produced transcription factors generates a specific combination of *HOX* expression patterns that define the morphology of each body segment. This segment-specific profile of *HOX* gene expression is maintained throughout the development of the fly, long after the early transcriptional regulators have disappeared. When the function of *HOX* genes was genetically characterized, many *trans*-acting regulators were isolated. Among the first, *Polycomb* (*Pc*) was identified and genetically analyzed by Pam and Ed Lewis (Lewis 1978). Heterozygous *Pc* mutant males have additional sex combs on the second and third legs. Homozygous mutants are embryonic lethal, showing a transformation of all cuticular segments toward the most posterior abdominal segment (Fig. 2C,D). These classical PcG phenotypes are caused by the ectopic expression of *HOX* genes. Thus, *Pc* and other genes with similar phenotypes were defined as repressors of *HOX* gene activity. Detailed analyses subsequently uncovered the fact that the PcG proteins are only required for the maintenance of *HOX* repression, rather than the position-specific establishment of *HOX* activity. This latter task is performed by the transcription factors encoded by the early acting segmentation genes. Based on their repressing or activating influence on *HOX* expression, these newly identified *trans*-acting regulators were divided into two antagonistic classes, the PcG and TrxG, respectively (Kennison 1995).

The molecular isolation of *Drosophila* PcG genes made it possible to study the function of vertebrate orthologs in mice, which were subsequently also shown to be key regulators of *HOX* gene expression (van der Lugt et al. 1994; Core et al. 1997). In mammals, mutations in PcG genes typically lead to homeotic transformations of vertebrae (Fig. 2E,F). In addition, PcG genes play a crucial role in the control of cell proliferation, stem cell maintenance, and cancer (see Secs. 4.2 and 4.3).

In two other model organisms, namely, the worm *Caenorhabditis elegans* and the flowering plant *Arabidopsis thaliana*, the molecular characterization of mutants isolated in various genetic screens revealed the existence of other PcG protein orthologs in their genomes. In *C. elegans*, PcG members were identified in screens for maternal-effect sterile (*mes*) mutants and were shown to be involved in X-chromosome silencing in the hermaphrodite germline (Ch. 23 [Strome et al. 2014]).

In *Arabidopsis*, PcG genes were identified in several genetic screens investigating distinct developmental processes

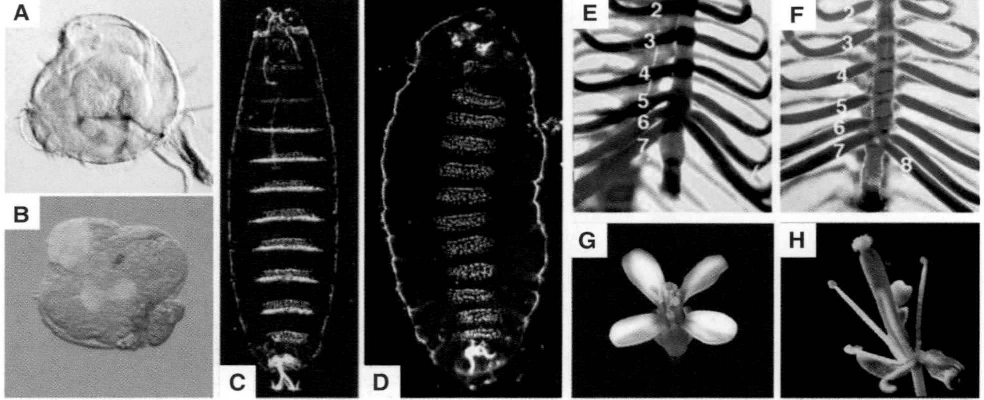

Figure 2. Homeotic transformations in PcG mutants of various species. (*A–D*) *Drosophila melanogaster*, (*E,F*) *Mus musculus*, (*G,H*) *Arabidopsis thaliana*. (*A,B*) Leg imaginal discs undergoing a transdetermination event as indicated by the expression of the wing-specific gene *vestigial* (marked by green fluorescent protein [GFP]). (*C,D*) Cuticles of a wild-type (*C*) and a *Su(z)12* mutant embryo (*D*). In the *Su(z)12* mutant embryo, all abdominal, thoracic, and several head segments (not all visible in this focal plane) are homeotically transformed into copies of the eighth abdominal segment because of misexpression of the *Abd-B* gene in every segment. (*E,F*) Axial skeleton of newborn wild-type (*E*) and *Ring1A*$^{-/-}$ mice (*F*). Views of the thoracic regions of cleared skeletons show bone (red) and cartilage (blue). The mutant displays anterior transformation of the eighth thoracic vertebra as indicated by the presence of an eighth (1–8) vertebrosternal rib, instead of seven (1–7) as in the wild type. (*G,H*) Wild-type (*G*) and *clf-2* mutant (*H*) flowers. The wild-type flower shows the normal arrangement of sepals, petals, stamens, and carpels. In the *clf-2* flower, petals are absent or reduced in number. (*A,B*, Courtesy of N. Lee and R. Paro; *C,D*, reprinted, with permission, from Birve et al. 2001, © Company of Biologists Ltd; *E,F*, reprinted, with permission, from Lorente et al. 2000, © Company of Biologists Ltd; *G,H*, courtesy of J. Goodrich.)

(Hsieh et al. 2003). The first PcG gene in plants, *CURLY LEAF* (*CLF*), was identified as a mutant with homeotic transformations of floral organs (Goodrich et al. 1997). Mutations in the *FERTILIZATION-INDEPENDENT SEED* (*FIS*) class of genes were found in screens for mutants showing maternal-effect seed abortion (Grossniklaus et al. 1998), or allowing aspects of seed development to occur in the absence of fertilization (Luo et al. 1999; Ohad et al. 1999). Finally, PcG genes were identified in screens for flowering time mutants, for example, mutants that flower directly after germination (Yoshida et al. 2001) or that disrupt the vernalization response (Gendall et al. 2001)—the process rendering plants competent to flower after prolonged exposure to cold (discussed in detail in Ch. 31 [Baulcombe and Dean 2014]).

The variety of processes regulated by PcG proteins illustrates the importance of maintaining the repressed state of key developmental regulators in different organisms. On the one hand, there is an amazing conservation of some biological functions from plants to mammals (e.g., the regulation of key developmental regulators such as homeotic genes or the involvement in the tight regulation of cell proliferation). On the other hand, PcG complexes appear to be versatile and dynamic molecular modules that have been used to control a large and diverse variety of developmental and cellular processes.

2 ESTABLISHING SILENCING MARKS ON CHROMATIN

PcG proteins are grouped into two major classes based on biochemical characterizations: members of the *Polycomb* repressive complex 1 or 2 (PRC1 and PRC2; Table 1). The two complexes are required for different steps in the repression of gene expression. PRC2 has histone modifying activity, namely, methylating H3K27 at genes targeted for silencing. PRC1 components can recognize and bind to this modification and induce appropriate structural changes in chromatin. Additionally, PRC1 is able to monoubiquitylate H2AK118/119 at target loci. Both complexes are widely conserved across metazoa (Whitcomb et al. 2007) and the plant kingdom (Köhler and Hennig 2010).

2.1 Components and Evolutionary Conservation of PRC2

Several variants of PRC2 have been purified from *Drosophila* embryos, but all of these complexes contain four core proteins: the SET domain histone lysine methyltransferase Enhancer of zeste (E(Z)), the WD40 protein ESC, the histone binding protein p55, and Suppressor of zeste 12 (SU(Z)12; Table 1 and Fig. 3).

The *E(z)* gene encodes a 760-amino acid protein, containing a SET domain that exerts KMT (lysine methyltransferase) activity to histones. The SET domain is preceded by a CXC or Pre-SET domain (Tschiersch et al. 1994), which contains nine conserved cysteines that bind three zinc ions and is thought to stabilize the SET domain (see Fig. 1 in Ch. 6 [Cheng 2014]). Such a structural role is supported by the fact that several temperature-sensitive *E(z)* alleles affect one of the conserved cysteines (Carrington and Jones 1996). In addition, E(Z) contains SANT domains implicated in histone binding, and a C5 domain required for the physical interaction with SU(Z)12. ESC is a short protein of 425 amino acids that contains five WD40 repeats, shown to form a β propeller structure. This serves as a platform for protein–protein interactions, hence, giving ESC a central role in PRC2 to physically interact with both E(z) and p55 in all model systems analyzed. The SU(Z)12 protein is 900 amino acids long and characterized by a C_2H_2-type zinc finger and a carboxy-terminal VEFS domain. The VEFS domain was identified as a conserved region between SU(Z)12 and its three homologs in plants: VERNALIZATION2 (VRN2), EMBRYONIC FLOWER2 (EMF2), and FIS2 (see Fig. 3). Several mutant *Su(z)12* alleles alter this domain, showing that it is required for the interaction with the C5 domain of E(Z) (Chanvivattana et al. 2004; Yamamoto et al. 2004).

The p55 protein was not identified as a PcG member by genetic approaches, possibly because it takes part in a multitude of other protein complexes associated with chromatin (Hennig et al. 2005). The p55 protein was, however, identified biochemically as part of PRC2. It is 430 amino acids long and contains six WD40 repeats, which physically interact with ESC or its orthologs in mammals and plants (Tie et al. 2001; Köhler et al. 2003a).

In addition to the core PRC2 proteins, some variants of the complex contain the RPD3 histone deacetylase (HDAC), or the Polycomb-like (PCL) protein. The interaction with RPD3 is noteworthy as histone deacetylation is correlated with a repressed state of gene expression (see Ch. 5 [Seto and Yoshida 2014]). The different compositions of PRC2 variants likely reflect both the dynamic changes that occur during development or cater for tissue-specific regulation. PRC2 is highly conserved in invertebrates, vertebrates, and plants (Fig. 3). In *C. elegans*, only homologs of E(Z) and ESC are present: MES-2 and MES-6. Together with a nonconserved protein, MES-3, they form a small complex of ~230 kDa required to repress the X-chromosome and somatically active genes in the hermaphrodite germline (see Ch. 23 [Strome et al. 2014]). In mammals and plants, all four core proteins of PRC2 are present. As in *Drosophila*, the mammalian complex is ~600 kDa and is not only involved in regulating homeotic gene expression,

Table 1. Core PcG proteins in model systems

Drosophila melanogaster			*Mus musculus*	*Arabidopsis thaliana*	*Caenorhabditis elegans*
PcG DNA binding proteins					
PHO	Pleiohomeotic	Zinc finger	YY1		
PHOL	Pleiohomeotic-like	Zinc finger			
PSQ	Pipsqueak	BTB-POZ domain			
DSP1	Dorsal switch protein 1	HMG domain protein	HMGB2		
PRC2 core proteins					
ESC	Extra sex combs	WD 40 repeats	EED	FIE	MES-6
E(Z)	Enhancer of zeste	SET domain	EZH1/ENX2 EZH2/ENX1	CLF MEA SWN	MES-2
SU(Z)12	Suppressor of zeste 12	Zinc finger VEFS box	SU(Z)12	FIS2 VRN2 EMF2	
p55	p55	Histone-binding domain	RBAP48 RBAP46	MSI1 (MSI2/3/4/5)	
PRC1 core proteins					
PC	Polycomb	Chromodomain	CBX2/M33 CBX4/MPC2 CBX6 CBX7 CBX8/MPC3		
PH	Polyhomeotic	Zinc finger SAM/SPM domain	EDR1/MPH1/RAE28 EDR2/MPH2 (EDR3)		SOP-2
PSC	Posterior sex combs	Zinc finger HTH domain	BMI1 MEL18/RNF110/ZFP144	AtBMI1A AtBMI1B AtBMI1C	MIG-32
SCE/dRING	Sex combs extra/dRing	RING zinc finger	RING1/RING1A RNF2/RING1B	AtRING1A AtRING1B	SPAT-3

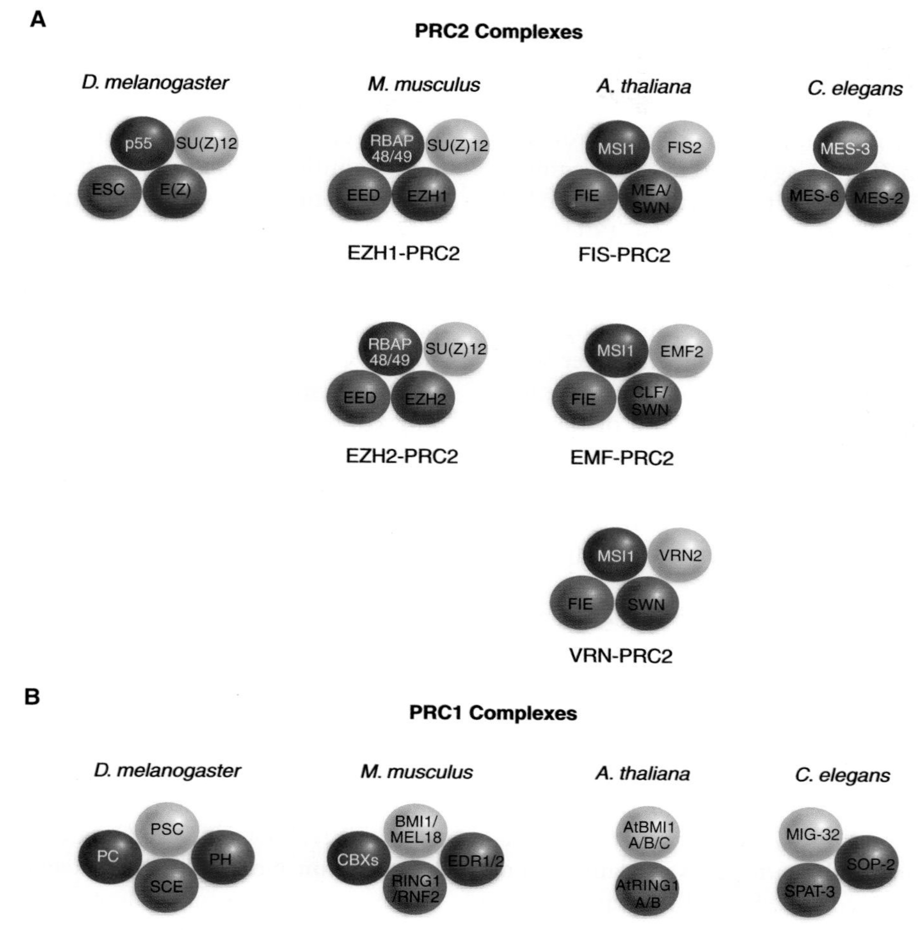

Figure 3. Conserved PRC2 core complexes. The conserved core proteins of PRC2 (*A*) and PRC1 (*B*) complexes in *Drosophila melanogaster, Mus musculus, Arabidopsis thaliana*, and *Caenorhabditis elegans* are shown. (*A*) In the mouse, PRC2 variants containing EZH1 or EZH2 have distinct functions, whereas in *Arabidopsis* the ancestral complex has diversified into at least three variants with discrete functions during development. In *C. elegans*, the PRC2 core complex contains only three proteins, with MES-3 not having homology with any other identified PRC2 protein. Apart from these core proteins, several other proteins, which are not shown here, interact with PRC2. For instance, mammalian complexes can contain the histone lysine demethylase JARID2, the Zn-finger protein AEBP2, and various homologs of the *Drosophila* PCL protein (PCL1/2/3). Proteins that share the plant homeodomain (PHD)-domain with PCL, but are otherwise not closely related, are also associated with the VRN-PRC2 complex in *Arabidopsis*. Homologous proteins are indicated by the same color. (*B*) The core proteins of PRC1 are less conserved than those of PRC2 across the four species. In mammals, all genes encoding the PRC1 core subunits have been expanded (see Table 1), such that a variety of complexes with different isoform composition can be formed. In addition to the core components, several additional proteins can be found in PRC1 that are, however, less well characterized and are not shown. In plants, only homologs of *Drosophila* PSC and SCE have been identified; these are encoded by small gene families. Homologous proteins are indicated by the same color. (Based on Reyes and Grossniklaus 2003, Chanvivattana et al. 2004, and Margueron and Reinberg 2011.)

but also in the control of cell proliferation, X-chromosome inactivation, and imprinted gene expression (for more detail, see Sec. 4; Ch. 26 [Barlow and Bartolomei 2014]; Ch. 25 [Brockdorff and Turner 2014]).

In plants, several genes encoding PRC2 components have undergone duplications such that now they are present as small gene families. In *Arabidopsis* there is only one homolog of ESC, FERTILIZATION-INDEPENDENT ENDOSPERM (FIE), but three homologs of E(Z), three homologs of SU(Z)12, and five homologs of p55 (referred to as MSI1-5; Table 1). Varying combinations of these proteins form at least three distinct complexes that control specific developmental processes, namely, the FIS-PRC2, EMF-PRC2, and VRN-PRC2 complexes (Fig. 3).

The first of these complexes that was studied in detail is formed by members encoded by the *FIS* class genes, which play a crucial role in the control of cell proliferation in the seed (Grossniklaus et al. 2001). The FIS-PRC2 contains MEDEA (MEA), FIE, FIS2, and MSI1. Using chromatin immunoprecipitation (ChIP) against one of its components, the FIS-PRC2 was found to directly regulate *MEA* itself, as well as *PHERES1* (*PHE1*) and *FUSCA3* (*FUS3*), coding for transcription factors of the MADS-domain and B3-domain class, respectively (Köhler et al. 2003b; Baroux et al. 2006; Makarevich et al. 2006). Interestingly, the paternal allele of *PHE1* is expressed at much higher levels than the maternal allele. This regulation of gene expression by genomic imprinting is under the control of the FIS-PRC2, which specifically represses the maternal allele (Köhler et al. 2005). Thus, as will be outlined in Section 4.1, the FIS-PRC2 shares with its mammalian counterpart functions in regulating cell proliferation as well as imprinted gene expression.

The EMF complex contains CLF and EMF2 (Chanvivattana et al. 2004). Mutations in either of them show weak homeotic transformations and an early flowering phenotype. The EMF-PRC2 is required to repress homeotic genes, whose combinatorial action determines the identity of floral organs (Goodrich et al. 1997). Thus, the EMF-PRC2 has a similar function in maintaining the repressed state of homeotic genes as PRC2 in *Drosophila* and vertebrates (Fig. 2). However, homeotic genes in plants do not encode homeodomain proteins, but rather transcription factors belonging to families containing the MADS-domain or the plant-specific AP2-domain. Strong *emf2* mutants, however, have more severe phenotypes, and produce flowers directly after germination, bypassing the vegetative phase of development (Yoshida et al. 2001). Thus, the EMF-PRC2 plays a role both early in development, in which it prevents immediate flowering, and later during floral organogenesis (Chanvivattana et al. 2004). The EMF-PRC2 directly represses *FLOWERING LOCUS T* (*FT*) and *SHOOTMERISTEMLESS* (*STM*), which are both required for the transition to flowering, and the floral homeotic MADS-box gene *AGAMOUS* (*AG*), which together with the homeobox gene *STM* regulates the development floral organs (Fig. 4) (Schubert et al. 2006; Jiang et al. 2008). The FIS class proteins FIE and MSI1 have also been implicated in the control of homeotic gene expression (Figs. 3 and 4). Because mutations in both cause maternal-effect embryo lethality, this function was only revealed when partial loss-of-function alleles could be studied at later stages of development (Kinoshita et al. 2001; Hennig et al. 2003).

Finally, the VRN-PRC plays a key role in a well-known process called vernalization. This epigenetic regulation governs the timing of flowering in winter annuals, induced by extended periods of exposure to low temperatures, but the effect is only seen after many cell divisions (Fig. 4D; see Fig. 1 in Ch. 31 [Baulcombe and Dean 2014] for detail). A plant cell will remember that it was vernalized for many months, or even years, after the cold period. This cellular memory is even maintained through passages in cell culture, but not from one generation to the next (Sung and Amasino 2004). The *VRN* genes mediate the response to vernalization. *VRN2* encodes a SU(Z)12 homolog (Gendall et al. 2001), which interacts with the plant E(Z) homologs CLF and SWINGER (SWN) in yeast two-hybrid assays (Chanvivattana et al. 2004). Floral induction is not only controlled by vernalization, but involves the perception of endogenous (developmental stage and age) as well as exogenous factors (day length, light conditions, temperature). Four pathways, two of which involve PcG factors, have been defined by genetic analyses (see Fig. 1B in Ch. 31 [Baulcombe and Dean 2014]): (1) the autonomous pathway, which constitutively represses flowering presumably via PcG-mediated H3K27 methylation; (2) the vernalization pathway, which induces flowering in response to a prolonged exposure to cold temperature; (3) the photoperiod pathway, which accelerates flowering under long days; and (4) the gibberellin pathway, a phytohormone that promotes flowering. The flowering time gene *FLOWERING LOCUS C* (*FLC*), encoding a MADS-domain transcription factor, is a key integrator of the flowering response as a repressor of flowering. Although the initial repression of *FLC* during vernalization is independent of the VRN-PRC2, the maintenance of *FLC* repression requires *VRN2* activity. Both the vernalization and the autonomous pathways reduce *FLC* expression, integrating diverse signals (Gendall et al. 2001; De Lucia et al. 2008; Jiang et al. 2008). The VRN-PRC2 contains the core subunits VRN2, SWN, FIE, and MSI1 (Fig. 3) and three associated PHD finger proteins (Wood et al., 2006; De Lucia et al. 2008). Interestingly, VRN2 interacts with the *FLC* locus independently of cold, and *FLC* silencing is triggered by the association of the VRN2-PRC2 with VRN5, a PHD finger protein that is cold-induced and shares limited similarity to PCL (see Sec. 2.2 and Fig. 2 in Ch. 31 [Baulcombe and Dean 2014]). In summary, the regulation of flowering time involves both the VRN-PRC2 and EMF-PRC2, which regulate *FLC* and *FT*, respectively (Fig. 4).

2.2 The Chromatin-Modifying Activity of PRC2

How does PRC2 mediate its repressive effect? In *Drosophila*, mammals, and plants, the hallmark histone modification H3K27me3 is produced by PRC2 (Cao et al. 2002; Czermin et al. 2002). This modification is generally thought to be

Cite as *Cold Spring Harb Perspect Biol* doi: 10.1101/cshperspect.a019331

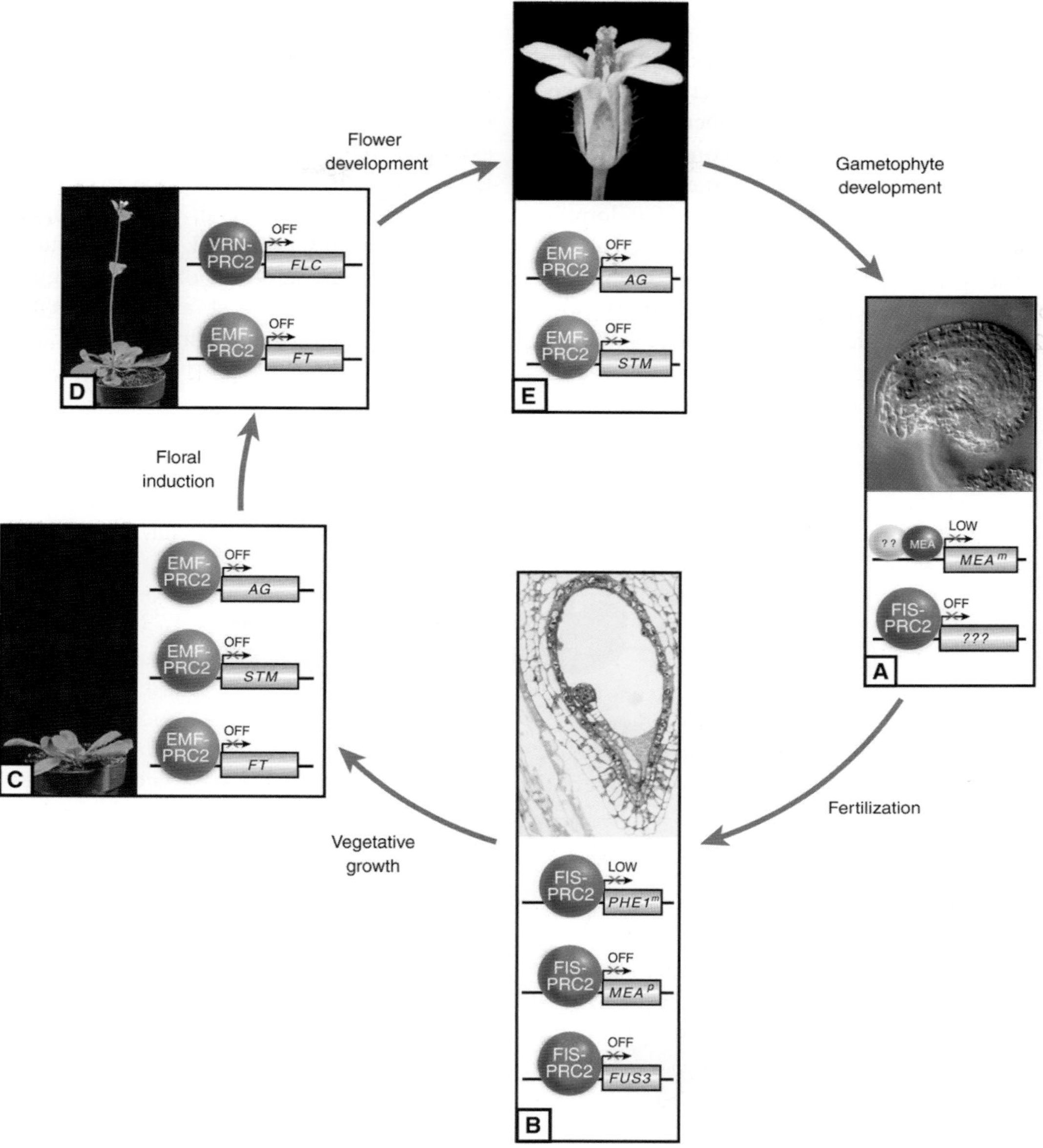

Figure 4. Involvement of distinct PRC2 complexes at various stages of plant development. During the plant life cycle, distinct variants of PRC2 (see Fig. 3) control developmental progression. (*A*) A cleared wild-type ovule harboring the female gametophyte in its center is represented. The FIS-PRC2 represses unknown target genes that control proliferation of the central cell; consequently, in all *fis* class mutants, this cell proliferates in the absence of fertilization. Around fertilization, MEA is also required to maintain expression of the maternal *MEA* allele (*MEA*^{m}) at a low level, but this activity is independent of other FIS-PRC2 components. (*B*) Section of a wild-type seed harboring embryo and endosperm, enclosed by the seed coat. After fertilization, the FIS-PRC2 is involved in the control of cell proliferation in embryo and endosperm. It maintains a low level of expression of the maternal *PHE1*^{m} allele and is involved in keeping the paternal *MEA*^{p} allele silent, although FIS-PRC2 only plays a minor part in its repression. Both parental alleles of *FUS3* are repressed by the FIS-PRC2. (*C*) Wild-type plant before flowering. The EMF-PRC2 prevents flowering by repressing *FT* and directly represses the floral genes *AG* and *STM*. (*D*) Wild-type plant after bolting—that is, floral induction induced by appropriate photoperiod and/or vernalization. The former relieves repression by EMF-PRC2 of *FT*, a promoter of flowering, whereas the latter leads to repression of the floral repressor *FLC*, thus inducing flowering. The maintenance of *FLC* repression depends on the VRN-PRC2. (*E*) Wild-type *Arabidopsis* flower. During flower organogenesis, the EMF complex regulates floral homeotic genes, such as *AG*, which determine the identity of floral organs, and *STM*, which is involved in floral organ development. (*A*, Courtesy of J.M. Moore and U. Grossniklaus; *B*, courtesy of J.-P. Vielle-Calzada and U. Grossniklaus; *C,D*, courtesy of D. Weigel; *E*, reprinted, with permission, from Page and Grossniklaus 2002, © Macmillan.)

crucial for PcG silencing because its distribution matches the binding sites of PcG components determined by genome-wide ChIP studies (Schuettengruber et al. 2009; Kharchenko et al. 2010). The PRC2 core complex contains E(Z), which catalytically adds up to three methyl groups at the target lysine residue K27 of H3 through its SET-domain (Fig. 5A). However, E(Z) alone seems to be inactive and needs to associate with the other PRC2 subunits, ESC and SU(Z)12, to provide the necessary catalytic activity (Cao and Zhang 2004; Pasini et al. 2004; Nekrasov et al. 2005). Although the mechanistic basis of this enhancement remains elusive, the process is conserved in mammals. Several more proteins and subcomplexes related to PRC2 exist. For example, the ESC-like gene encodes a protein similar to ESC and is able to fully replace it in its absence (Wang et al. 2006; Kurzhals et al. 2008). Also, a PRC2 variant additionally containing PCL has been found to specifically enhance the last addition of a methyl group to generate H3K27me3 (Nekrasov et al. 2007). Without PCL, the bulk H3K27me3 is reduced in embryonic and larval tissues,

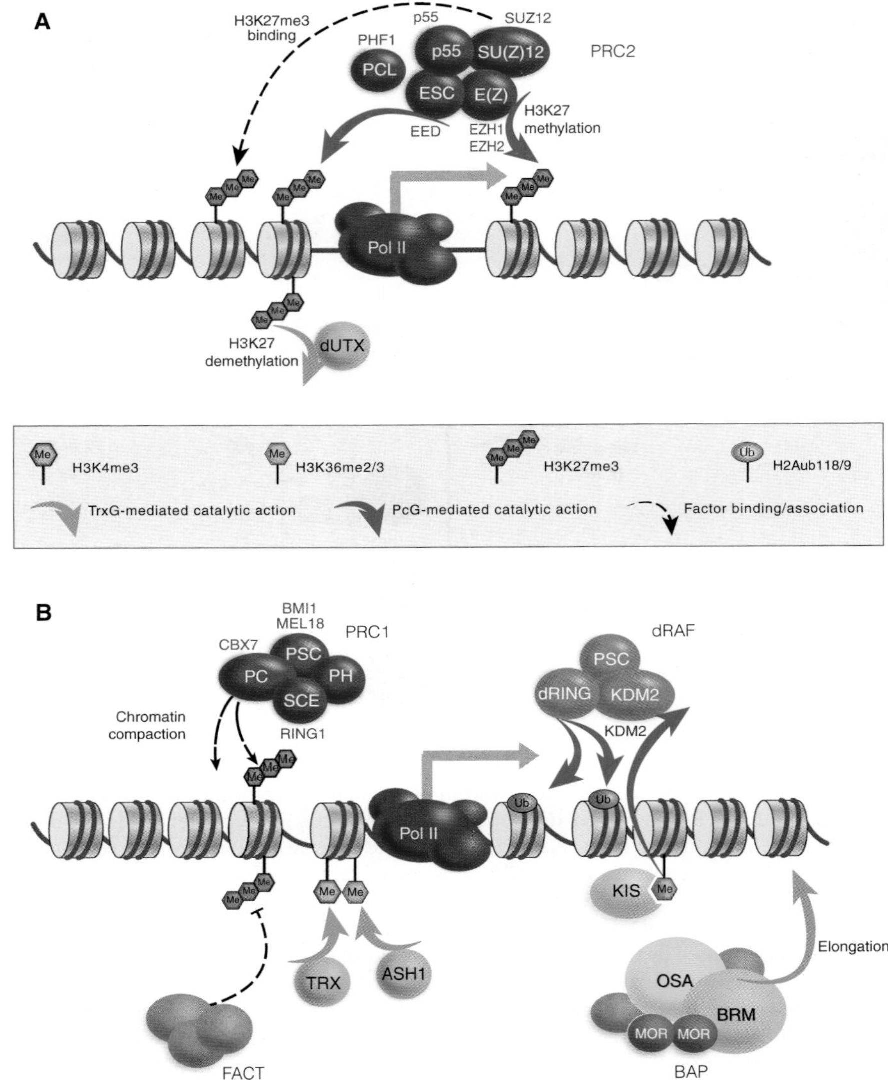

Figure 5. Schematic representation of the core PcG and TrxG protein complexes and their functions at promoters. *Drosophila* PcG proteins are depicted as red ovals with selected mammalian orthologs indicated in gray text. (*A*) Components and function of the PRC2 and counteracting activities of TrxG proteins (light green). (*B*) Components and functions of PRC1 and dRING-associated factor (dRAF) and the counteracting activities of the BAP SWI/SNF, facilitates chromatin transcription (FACT) remodeling complexes, and SET-domain histone KMTs TRX and ASH1. The TrxG protein Kismet-L is a member of the chromatin-helicase-DNA-binding (CHD) subfamily of chromatin-remodeling factors, stimulating elongation of Pol II. (Adapted from Enderle 2011.)

leading to the derepression of several target genes. Interestingly, a similar function and complex has also been described for the PHD finger protein PHF1 in mammals (Cao et al. 2008; Sarma et al. 2008).

The mammalian PRC2 has also been found to counteract H3K4 methylation, an active histone modification, by recruiting the H3K4me3 demethylase RETINOBLASTOMA BINDING PROTEIN 2 (RBP2) to target genes (Pasini et al. 2008) and control transcriptional elongation via JARID 2 (Landeira et al. 2010). The biological implications of these variations and additional activities, however, are not yet fully understood. Moreover, even the molecular function of H3K27me3 is still a matter of debate. There is no evidence yet that H3K27 methylation may directly alter nucleosomal structure to repress target genes. Rather, it seems to provide a binding platform for other PcG proteins; the PRC1 complex binds, albeit weakly, to H3K27me3 through the chromodomain of its PC subunit, and also mammalian PRC2 itself has been shown to bind H3K27me3 via the Embryonic ectoderm development protein (EED; Fischle et al. 2003; Margueron et al. 2009). Interestingly, this interaction seems to trigger the lysine methyltransferase activity of E(Z), providing a self-reinforcing positive-feedback loop (as illustrated in Fig. 13 of Ch. 3 [Allis et al.]), potentially contributing to the heredity of the PcG/TrxG system (Margueron et al. 2009). The counteracting force of the TrxG has also been described in mammals: UTX/KDM6A, the mammalian ortholog of *Drosophila* dUTX showing some genetic characteristics of a TrxG member (Smith et al. 2008), is able to demethlyate H3K27me3 in vivo and in vitro (Agger et al. 2007; Lee et al. 2007). This reveals a direct antagonistic relationship between PcG and TrxG proteins in the modification of their target chromatin (Fig. 5A).

2.3 The Dynamic Function of PRC2 during Development

As pointed out in Sections 2.1 and 2.4, the PRC1 and PRC2 core complexes are associated with distinct factors that may play a role in recruiting PcG complexes to tissue-specific target loci or in modulating target gene activity. PcG complexes may even differ between target genes in the same cell, suggesting a highly dynamic behavior at different developmental stages. Studies performed in mammals and plants clearly show that PcG complexes have distinct memberships in specific tissues and their composition changes during cellular differentiation. Similar to the situation in plants described in Section 2.1, some of the genes encoding PRC2 subunits have been duplicated in mammals. For instance, PRC2 complexes containing either EZH1 or EZH2 are functionally distinct (Fig. 3). EZH1-containing PRC2 has weak KMT activity and is abundant in nondividing cells of adult organs, whereas EZH2 confers high KMT activity and is expressed strongly in proliferating cells (Margueron et al. 2008). Furthermore, different isoforms of EED, which are derived from the same messenger RNA but different translational start sites, are able to methylate H3K27 and H1K26 (Kuzmichev et al. 2005).

In *Drosophila*, PcG proteins maintain repressed states of homeotic genes, established during early embryogenesis, thereby fixing developmental decisions. Once the silent state of a PcG target has been fixed, it will often remain in that state for the remainder of an individual's lifespan. In plants, a similar situation may occur with the VRN-PRC2; once vernalized, the target gene(s) will be permanently inactivated and only reset in the next generation (see Ch. 31 [Baulcombe and Dean 2014] for more detail). Other plant PRC2 variants, however, seem to respond quickly to developmental or environmental stimuli. For instance, one function of the FIS-PRC2 is to repress cell proliferation in the absence of fertilization. Upon fertilization, however, cell proliferation is rapidly induced, presumably through the derepression of PcG target genes. This indicates that PcG repression is the default state, which has to be overcome by some unknown mechanism to allow developmental progression to occur. Indeed, the major function of the various plant PRC2 variants seems to be the regulation of developmental transitions, such as those occurring at fertilization, during seed development, and after germination when the plant progresses from the juvenile to the adult, and finally the reproductive stage (reviewed in Holec and Berger 2012).

2.4 Components of the PRC1 Complex

The PRC1 core complex purified from *Drosophila* embryos contains stoichiometric amounts of PC, Polyhomeotic (PH), Posterior sex combs (PSC), and Sex combs extra (SCE/dRing 1) (Shao et al. 1999). Its mammalian counterpart comprises the same core components, but the genes encoding them have been amplified (Table 1). As mentioned before, the PC subunit is able to bind specifically to H3K27me3 in vitro. This does not necessarily mean that H3K27me3 is the primary recruiter of PRC1 (Fig. 5B) because it is experimentally challenging to distinguish recruitment from subsequent stabilization of local chromatin binding. However, experiments directly increasing the H3K27me3 levels in human cells clearly enhance PRC1 binding in vivo, demonstrating the importance of the PC subunit as a chromatin anchor point for the PRC1 (Fig. 5B) (Lee et al. 2007). Additionally, flies with a mutated H3K27 fail to repress transcription of PcG target genes depicting a similar phenotype as *Polycomb* mutants (Pengelly et al. 2013). The reconstituted core components of the mammalian PRC1 have been functionally tested on nucleosomal arrays and were found to inhibit chromatin remodeling by SWI/SNF and restrict access by RNA polymerase II (RNA Pol II) in vitro (Shao et al. 1999; King et al. 2002). Both the PRC1 subunits PSC and SU(Z)2 are crucial for chromatin accessibility, explaining the complete functional redundancy of PSC and SU(Z)2 (Lo et al. 2009). Another conserved hallmark function of PRC1 is its ability to monoubiquitinate H2A-K118/K119 (H2AK118/119ub1) by the E3 ubiquitin ligase SCE/dRing1 (Fig. 5B) (Wang et al. 2004b; Gutiérrez et al. 2012). This histone modification seems to be under a tight, dynamic control because PR-DUB, another PcG complex containing Calypso and Additional sex combs, actively removes this mark (Scheuermann et al. 2010). Moreover, there is cross talk with ubiquitination of histone H2B, extending the regulatory possibilities even further (reviewed in Weake and Workman 2008).

The function of H2AK118/119ub1 is not well understood, but somehow inhibits the recruitment of the FACT chromatin remodeling complex (Zhou et al. 2008). It was recently shown that the H2A ubiquitination activity of PRC1 is dispensable for target binding and its activity to compact chromatin at *HOX* loci, but is indispensable for efficient repression of target genes and thereby maintenance of embryonic stem (ES) cell identity (Endoh et al. 2012). The H2AK118/119ub1 mark in *Drosophila* is also set by the distinct and conserved dRAF complex (Fig. 5B) (Lagarou et al. 2008; Scheuermann et al. 2010). Interestingly, dRAF is able to demethylate H3K36 through its dKDM2 subunit, linking repression through H2A ubiquitination directly with the inhibition of transcriptional elongation by removing an activating mark.

The existence of a ubiquitously conserved PRC1 has been disputed because most of the core PRC1 subunits are not conserved in plants. However, although there is no clear PC homolog in plants, the *Arabidopsis* LIKE HETEROCHROMATIN PROTEIN1/TERMINAL FLOWER2 (LHP1/TFL2) protein serves as a functional counterpart to *Drosophila* PC. Like PC, LHP1/TFL2 binds H3K27me3 in vitro and is colocalized with this mark throughout the genome (Turck et al. 2007; Zhang et al. 2007). Also, *lhp1/tfl2* mutants show some of the phenotypes typical of PcG mutants. Furthermore, there are multiple homologs of PSC and SCE/dRing (Table 1), and double mutants for *Atbmi1a/1b* or *Atring1a/1b* result in phenotypes similar to those observed in mutants affecting PRC2. Indeed, the AtBMI1 homologous proteins have been shown to mediate H2A monoubiquitination in vivo (Bratzel et al. 2010). The *Arabidopsis* PSC/BMI 1 and SCE/dRING homologs interact with each other, the chromodomain protein LHP1, and EMF1, a plant-specific nucleoprotein. Thus, plants have a PRC1-like complex that contains some PRC1 homologs, but also plant-specific factors. However, this PRC1-like activity only regulates a subset of the PRC2 targets as it was also reported for *Drosophila* (Gutiérrez et al. 2012). The PRC2 target gene *AG*, for instance, is not up-regulated in either the *Atbmi1a/1b* or the *Atring1a/1b* double mutant (Xu and Shen 2008; Bratzel et al. 2010).

In spite of the already dazzling variety in PcG functions found, there is still scope for a better functional understanding of the role that proteins, other than the core subunits, play. There is a large group of loosely associated subunits such as TBP-associated factors in PRC1. This interaction might indicate a role in inhibiting the assembly of the preinitiation complex of RNA Pol II (Dellino et al. 2004). Several other enzymatic functions also seem to contribute to PcG silencing; PRC1 members associate with HDAC1 (Huang et al. 2002), indicating that histone-deacetylation may play a role in PcG-silencing. Furthermore, the PcG gene *super sex combs (sxc)* encodes an enzyme that posttranslationally modifies PH and RNA Pol II with β-O-linked *N*-acetylglucosamine residues and is necessary for the repression of several *HOX* genes (Gambetta et al. 2009; Sinclair et al. 2009). Yet another interesting link is the possibility of a switch-like mechanism through the acetylation of H3K27 by the acetyltransferase cAMP response element binding protein-binding protein (CBP), antagonizing its methylation (Tie et al. 2009). Most interestingly, in mammalian stem cells, PcG-based promoters marked by H3K27me3 frequently become DNA methylated during differentiation, suggesting that *Polycomb* repression and de novo DNA methylation are linked (Mohn et al. 2008). A direct physical interaction of PRC2 components and the MET1 DNA methyltransferase was recently also found in plants, indicating an evolutionarily old interaction between these major epigenetic pathways (Schmidt et al. 2013).

2.5 The PcG Connection to Paused RNA Pol II Promoters

The mechanism by which PcG complexes interact with the promoter via binding to specific *cis*-regulatory elements (PREs, detailed in Sec. 3.1) to prevent transcription in *Drosophila* has become clearer in recent years. The anchoring of paused RNA Pol II complexes at promoters, preventing initiation, has been attributed to PRE–PRC1 interactions described by the use of reporter constructs (Dellino et al. 2004). In mouse ES cells, Ring1-mediated ubiquitination of H2A was found to restrain paused RNA Pol II at PcG target genes (Fig. 6A) (Stock et al. 2007). Genome-wide ChIP-Seq profiling of *Drosophila* tissue culture cells uncovered a strong overlap between PRC1-binding sites and promoters with paused RNA Pol II (Enderle

 Cite as *Cold Spring Harb Perspect Biol* doi: 10.1101/cshperspect.a019331

et al. 2011). Indeed, this approach also found that many promoters of noncoding RNAs (ncRNAs) are targeted by PRC1. Among those, the promoters of primary transcripts for many micro RNAs stand out, suggesting that this important class of RNA regulators is under the control of the PcG system as well. Yet, the finding that paused Pol II at promoters is a major hallmark of PcG target genes indicates that there is a mechanistic link between PcG-mediated silencing and mechanisms of transcriptional elongation. In addition, PRC1 was shown to counteract remodeling of nucleosomes in vitro and to induce a compact chromatin structure. Thus, PRC1 potentially blocks the accessibility to DNA of transcription factors and other complexes required for transcription (Grau et al. 2011).

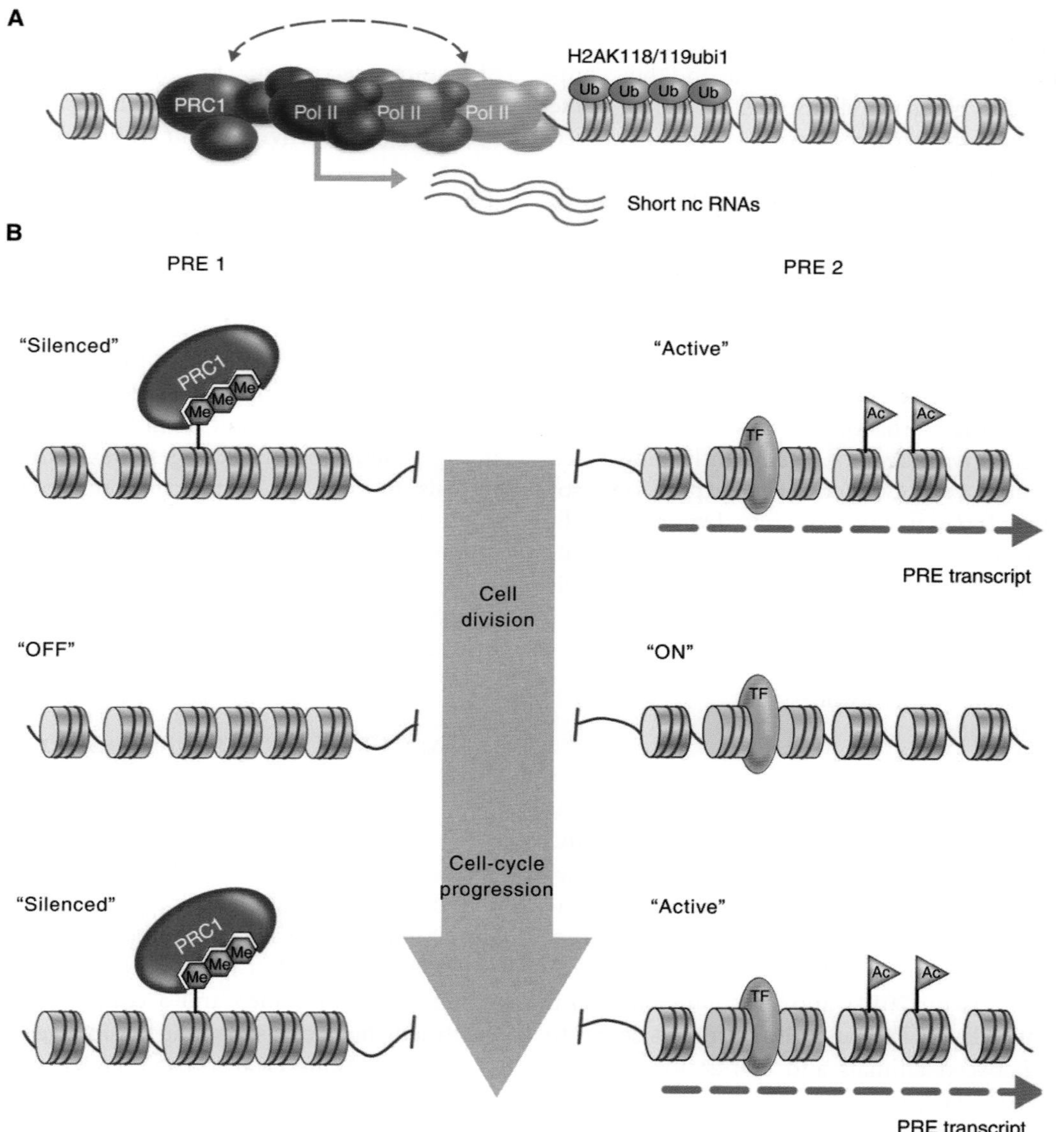

Figure 6. PRC1 at paused promoters and during cell division. (*A*) The PRC1 complex may repress target genes by stalling the elongation of RNA Pol II. This may be achieved by ubiquitination of histone H2A through the subunit SCE/dRING, compacting promoter proximal chromatin, or direct physical interaction with the transcriptional machinery (including the short RNAs produced by the paused RNA Pol II). (*B*) A possible model for how differential gene expression states can be inherited. The process of intergenic transcription places positive epigenetic marks (e.g., acetylated histone tails, histone variants) at PREs that control active genes (PRE 2). All other PREs are silenced by default (PRE 1). During DNA replication and mitosis, only the positive epigenetic signal needs to be transmitted to the daughter cells, ensuring that in the next interphase intergenic transcription is restarted at PRE 2 before default silencing is reestablished at all other PREs.

The stability of silencing complexes, as shown by anchoring via methylated histone tails, appears to be a major property of the long-term repressive function of PcG proteins. However, when analyzed in vivo at the cellular level, a remarkably dynamic behavior of the individual components is observed. PcG proteins cluster in PcG bodies, which vary in size and composition between cells (Bantignies and Cavalli 2011). Fluorescence recovery after photobleaching (FRAP) analyses of green fluorescent protein (GFP)-marked PC and PH proteins uncovered a very high rate of exchange between unbound proteins and their complexes at silenced target genes (Fonseca et al. 2012). These results suggest that long-term repression is primarily based on a chemical equilibrium between bound and unbound proteins rather than through high-affinity protection of DNA-binding sites. Additionally, a new method measuring nucleosomal turnover uncovered a rapid exchange over active gene bodies, epigenetic regulatory elements, and replication origins in *Drosophila* cells (Deal et al. 2010). Surprisingly, rapid nucleosome turnover is seen at many PcG- and TrxG-regulated elements. This finding questions whether PcG-associated histone marks can contribute to epigenetic stability. Indeed, it was recently shown that H3K4 methylation, a mark of active gene expression associated with the TrxG, is not essential; *Drosophila* cells completely lacking this histone mark show normal transcriptional activation in response to developmental signaling pathways (Hodl and Basler 2012).

2.6 Preventing Heritable Repression by Antisilencing

The binding of PRC1 complexes to PREs appears to be a default state, as many of the anchoring PcG components and DNA-binding proteins are expressed in all cells and transgenic constructs with PREs controlling reporter genes are globally silenced. The counteracting proteins of the TrxG do not, in fact, function as activators, but rather as antirepressors (Klymenko and Müller 2004; see Ch. 18 [Kingston and Tamkun 2014] and Fig. 7 therein). This antagonistic interplay of PcG and TrxG proteins seems to be conserved between animals and plants; for instance, several plant PRC2 targets such as the *AG* and *FLC* loci are similarly maintained in an active state through the activity of the homolog of *Drosophila* TRX, ATX1, which acts as a KMT specific to H3K4 (Alvarez-Venegas et al. 2003; Pien et al. 2008).

To maintain active transcription of a PRE-controlled gene, the silencing at that PRE, thus, has to be prevented in a tissue- and stage-specific manner. In *Drosophila*, for example, the early cascade of transcription factors encoded by the segmentation genes controls the activation of *HOX* genes. Interestingly, these factors do not only induce transcription of the *HOX* genes, but also of intergenic ncRNAs that are transcribed through the associated PREs often found upstream or downstream. It was shown that transcription through PREs is required to prevent silencing and to maintain the active state of a reporter gene using transgenic constructs (Schmitt et al. 2005). The process of transcription most probably remodels PRE chromatin to generate an active state that is characterized, for instance, by a lack of repressive histone methylation and the presence of histone acetylation. Thus, although the DNA binding proteins will attract PRC1 to this particular activated PRE, the histone environment will not allow anchoring of PC via H3K27me3, and no stable silencing would be established. Because silencing is induced by default in the PcG system, epigenetic inheritance of a differential gene expression pattern only requires the transmission of the active PRE state during DNA replication and mitosis (Fig. 6B). How this is achieved at the molecular level and which epigenetic mark(s) is responsible for maintaining an active PRE state is still an open question. It has been suggested that particular TrxG factors might act as "bookmarks" during epigenetic bottleneck stages, like DNA replication and mitosis, to mark a gene for continued expression (Blobel et al. 2009). Hence, finding the molecular constitution of chromatin components that self-template during DNA replication and carry over the signals for activity to the daughter cells may be key for advancing our understanding of epigenetic inheritance.

3 TARGETING PcG COMPLEXES TO SILENCED GENES

3.1 PcG Response Elements

A striking feature of the PRC1 and PRC2 core complexes is that they do not contain any obvious DNA sequence-binding activity, raising the question of how they might be targeted. Although PRC1 is bound to chromatin through its affinity for H3K27me3 and a consistent hierarchical recruitment has been shown at the *bxd* genomic region in *Drosophila* (Wang et al. 2004a), H3K27me3 alone is not sufficient to explain the targeting of the complex. First of all, PcG-binding sites are generally devoid of histones and a place of rapid nucleosome turnover (Mito et al. 2007; Deal et al. 2010). Furthermore, the generally broad distribution of H3K27me3 does not fit the localized binding of PRC1 (Fig. 7B) (Schuettengruber et al. 2009; Enderle et al. 2011). Also, removal of H3K27me3 does not lead to the immediate displacement of PRC1. Indeed, there are examples of PRC1 binding to sites without any apparent H3K27 methylation (Schoeftner et al. 2006; Tavares et al. 2012). Overall,

Cite as *Cold Spring Harb Perspect Biol* doi: 10.1101/cshperspect.a019331

H3K27me3 may contribute to several different low-affinity steps working together in recruitment or, more interestingly, allowing PRC1 to reach and modify histones distant to its initial binding site in a local domain.

The best characterized binding sites for PRC1 and PRC2 proteins were originally identified at the *bithorax* complex and subsequently termed PREs (Simon et al. 1993). PREs are thought to act as intergenic *cis*-regulatory elements, controlling gene expression by looping to the promoter regions of their target *HOX* genes. But mostly, PREs have been characterized by their ability to confer PcG silencing on reporter genes. PREs frequently contain binding sites for the zinc finger DNA-binding proteins Pleiohomeotic (PHO) and the related PHO-like (PHOL) protein, and these sites are essential for transgene and endogenous silencing functions (Fig. 7C). PHO forms a heterodimeric complex together with the Scm-related gene containing four mbt domains (SFMBT) protein, termed the Pho Repressive Complex (PhoRC) (Klymenko et al. 2006). The genome-wide distribution of PhoRC confirms its central role in the recruitment of PcG proteins: 45% of PHO-binding sites in larval and embryonic tissue are cooccupied by both PRC1 and PRC2. At the same time, the majority of PH-binding sites in embryos are enriched for PHO (Schuettengruber et al. 2009). Conversely, these data also show that PhoRC binding cannot be the only factor for targeting PRC1 because many loci bind PcG proteins without PhoRC being present. This is also reflected in PHO-binding sites being necessary, but not sufficient, for recruitment of PRC1 and PRC2. Several other proteins with DNA-binding abilities have been found as physical or genetic interactors of PcG proteins. Among them are Pipsqueak (PSQ) as subunit of the CHRASCH complex, Dorsal switch protein 1 (DSP1), Grainyhead (GRH), GAGA factor (GAF), and Sp1-like factor for pairing sensitive-silencing (SPSS), which is an Sp1/Klf protein family member (Fig. 7C). Despite this diversity, many PcG-binding sites do not contain any binding sites for the aforementioned transcription factors. Consistently, prediction algorithms based on their consensus binding sites are only able to predict a fraction of the many PRC1- and PRC2-binding sites found in a single cell type (Ringrose et al. 2003; Schwartz et al. 2006).

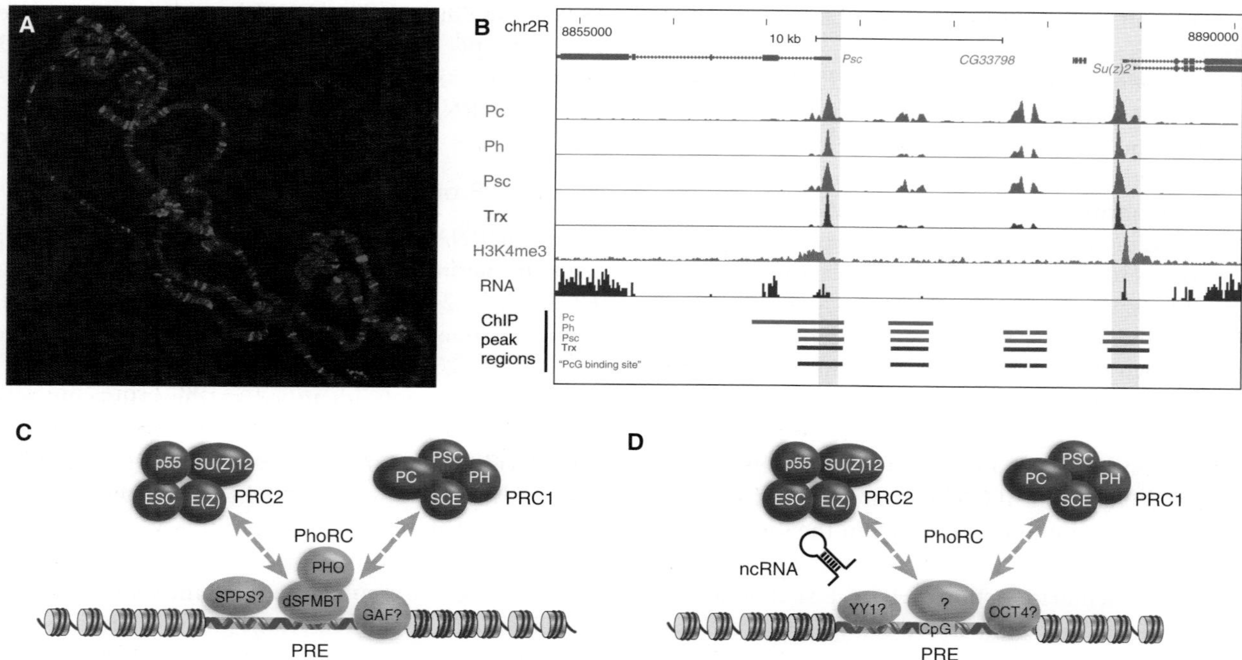

Figure 7. Chromosomal targeting of PRC1. (*A*) Immunostaining of *Drosophila* polytene chromosomes to visualize the distribution of the PC protein. (*B*) Genomic region encompassing the *Drosophila* PcG gene *Psc* and the *Su(z)2* gene. The genome browser section shows the result of a ChIP-Seq and RNA-Seq analysis of *Drosophila* S2 tissue culture cells. The distributions of PRC1 components (red) and the TRX protein (green) are shown (data from Enderle et al. 2011). (*C*) In *Drosophila*, the PhoRC is a key player in chromatin targeting of PRC1 and PRC2, but a number of other transcription factors also contribute to target gene specificity. (*D*) In mouse and human, several different anchoring factors have been proposed. These include the Pho ortholog Ying and Yang 1 (YY1), transcription factors like Jarid 2 and Oct4, long ncRNAs, and the CpG content of the target sequence. (Adapted from Enderle 2011.)

In contrast to *Drosophila*, PREs are ill-defined in plant and mammalian genomes, and only a few sequences have been described that, at least partially, fulfill the criteria for PRE function. According to Schwartz and Pirrotta (2008), the minimum criteria include (1) PREs attract H3K27me3, (2) they should form a new binding site for PcG proteins when inserted at a new location within the genome, and (3) they confer PcG-based repression to a reporter gene. Although no PREs have fulfilled all three criteria in plants, some sequences at well-studied PcG targets likely function as PREs. For instance, a transgene including sequences overlapping the promoter and parts of the coding sequence of the *AG* locus, a target of the EMF-PRC2, renders the reporter gene responsive to CLF, leading to H3K27me3 deposition (Schubert et al. 2006). More recently, the promoter of *LEAFY COTYLEDON2* (*LEC2*), which is regulated by EMF-PRC2, was shown to contain a repressive *LEC2* element (*RLE*), which is sufficient to trigger H3K27me3 deposition and silencing of a reporter gene in seedlings (Berger et al. 2011). Although the *RLE* is close to a CT-rich *cis*-regulatory element with similarities to the GAGA-box of *Drosophila* PREs, a possible function of the BASIC PENTACYSTEINE (BCP) proteins binding these CT-rich sequences in recruiting PRC2 has not yet been shown.

DNA sequences from the *FLC* and *MEA* loci can similarly confer PcG-dependent silencing to reporter genes, although in the case of *MEA* the FIS-PRC2 plays only a minor part in the repression of the paternal allele (Sheldon et al. 2002; Wöhrmann et al. 2012). For *AG*, *FLC*, or *MEA*, no DNA-binding factors are currently known that recruit PRC2 variants to these loci. Thus, other factors, such as a specific chromatin structure or long ncRNAs (lncRNAs), may be involved in PRC2 recruitment. Recent studies have, however, shown that DNA-binding proteins do play a role in PRC2 recruitment at the *WUSCHEL* (*WUS*) locus, whose repression is crucial for the appropriate termination of floral meristems. *WUS* is a target of the EMF-PRC2, and H3K27me3 levels at the *WUS* locus are reduced similarly in *ag*, *clf*, and *swn* mutants, which act in the same genetic pathway. As H3K27me3 levels at *WUS* increase rapidly after the experimental induction of *AG*, the MADS-domain protein AG likely plays a role in recruiting PcG proteins to this locus.

In the mouse and human genome, PcG proteins predominantly occupy regions around gene promoters (Boyer et al. 2006; Lee et al. 2006; Ku et al. 2008). Two intergenic PREs, however, were identified in mammalian genomes (Sing et al. 2009; Woo et al. 2010). An intergenic region of 1.8 kb, termed D11.12, was bound by PcG proteins in the human *HOX-D* complex. The other intergenic PRE, a 3-kb "PRE-kr" at the *MafB*/*Kreisler* locus in mouse, also recruited PcG proteins and was able to regulate its expression pattern. PcG protein recruitment is necessary for their potential to exert gene silencing. Importantly, the D11.12 element is able to maintain repression of a luciferase transgene throughout cell differentiation, providing the first example of a mammalian PRE sequence. Interestingly, both elements contain binding sites for the mammalian homolog of the *Drosophila* PhoRC, which contains SFMBT and YY1 (Fig. 7D). Still, binding of YY1 only accounts for a fraction of the genome-wide PRC2 sites in ES cells (Squazzo et al. 2006). Although there is substantial overlap with the pluripotency factors NANOG, OCT4, and SOX2, these three proteins have not been copurified with PcG complexes yet (Boyer et al. 2006; Lee et al. 2006). Also, the recent identification of the mammalian GAF homolog may provide new insights into PcG recruitment by transcription factors in the future (Matharu et al. 2010).

A surprising aspect was found in ES cells, in which nearly all PRC2 binding sites are found at CpG islands or other highly GC-enriched sequences (Ku et al. 2008). Indeed, GC-rich DNA from bacterial genomes is able to initiate recruitment of PRC2 (Mendenhall et al. 2010). This is especially interesting because Mixed lineage leukemia (MML), the mammalian TRX homolog, also has a preference for CpG dinucleotides, revealing a shared discriminant for targeting. Other interesting recruitment factors are specific lncRNAs, which are discussed in Section 3.2.

3.2 PcG Proteins Bind to Noncoding RNAs

Several ncRNAs have been proposed to recruit PcG proteins in mammals. The most prominent example may be *HOTAIR*, a 2.2-kb ncRNA from the human *HOX-C* cluster, acting in *trans* to mediate gene repression (described in Ch. 2 [Rinn 2014]; also Rinn et al. 2007). *HOTAIR* is one of many transcripts identified in the intergenic regions of human *HOX* clusters and its depletion leads to the loss of H3K27me3 in a large genomic region of *HOX-D*. Indeed, *HOTAIR* interacts with PRC2 components in vitro, suggesting that it may also recruit KMT activity to the *HOX-D* complex (Rinn et al. 2007). Another interaction between PRC2 components and an lncRNA is observed at the paternally imprinted *Kcnq1* locus in mouse (Fig. 8) (Wu and Bernstein 2008). Similar to *HOTAIR*, the 91-kb-long primary transcript of *Kcnq1 overlapping transcript 1* (*Kcnq1ot1*), coimmunoprecipitates with EZH2 and SUZ12, and additionally with the H3K9-specific KMT G9a (Kanduri et al. 2006; Pandey et al. 2008). The paternally transcribed ncRNA may facilitate silencing by recruitment of methyltransferases in *cis*, leading to the inactivation and compaction of genes at the locus (Terranova et al. 2008).

 Cite as *Cold Spring Harb Perspect Biol* doi: 10.1101/cshperspect.a019331

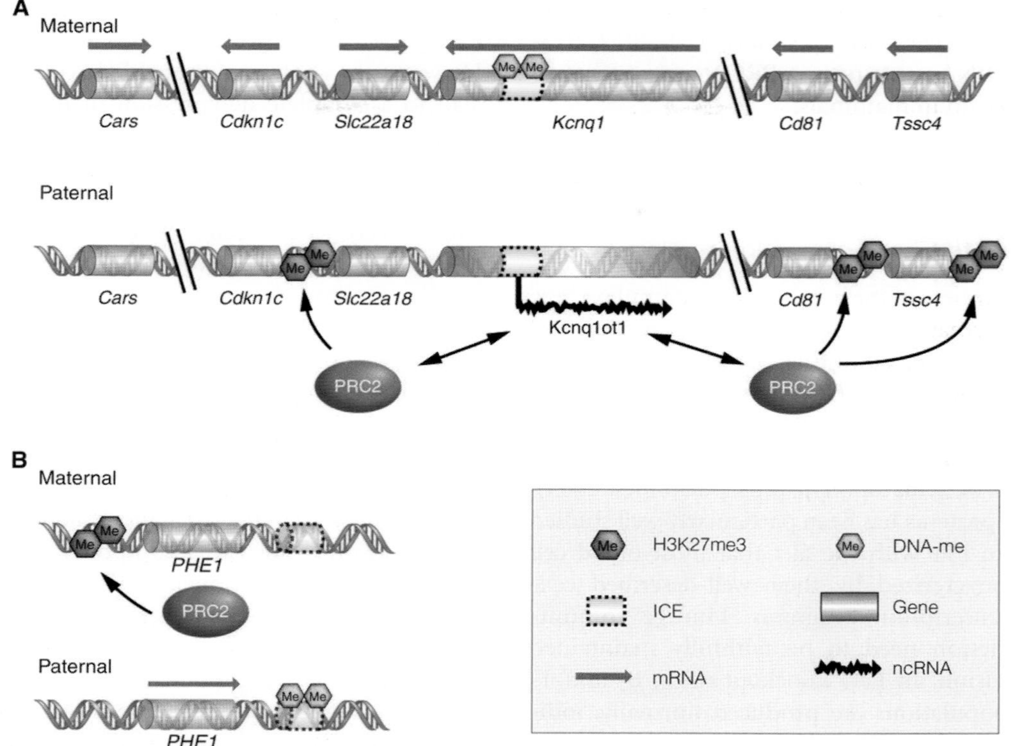

Figure 8. Interplay of PcG-mediated repression and DNA methylation regulates genomic imprinting in plants and mammals. (*A*) Regulation of genomic imprinting at the *Kcnq1* domain on distal chromosome 7. The imprinting control element (ICE) is maternally methylated and prevents the transcription of the lncRNA *Kcnq1ot1* from the maternal chromosome. The paternally expressed *Kcnq1ot1* associates with chromatin and recruits chromatin modifying complexes, such as PRC2, to mediate and maintain transcriptional silencing of several paternal, protein-coding alleles. (*B*) In *Arabidopsis* seeds, the paternally expressed *PHE1* gene is maternally repressed by the action of PRC2. A *cis*-regulatory element (shaded pink) downstream of the *PHE1* gene must be methylated for paternal expression, but demethylated for maternal repression.

Another *cis*-acting ncRNA that is a crucial component for inactivating one of the X chromosomes in female mammals is the 17-kb-long *X inactive-specific transcript* (*Xist*). It contains a 28-bp repeat element that interacts with EZH2 in vitro and in vivo (Zhao et al. 2010). This element folds into a double stem-loop structure and is necessary for X inactivation (covered in more detail in Ch. 25 [Brockdorff and Turner 2014]). Recently, similar stem-loop structures have been identified in a class of short promoter-proximal ncRNAs, which are transcribed from H3K27me3-marked genes (Kanhere et al. 2010). The small ncRNAs interact with PRC2 in vitro through secondary structures, revealing PRC2-bound and folded ncRNAs as a repeating theme in mammalian cells. However, details of the recruiting mechanism and a common RNA motif have yet to be described. Also, RNA-interacting PcG proteins are not restricted to members of the PRC2 complex. A recent example is *ANRIL*, an ncRNA at the murine *Ink4b*/*Arf*/*Ink4a* locus. This transcript has been shown to specifically associate with PRC1 members through the chromodomain of CBX7 (Yap et al. 2010). A local competition between *ANRIL* and H3K27me3 for the binding of CBX7 may therefore remove PRC1 from chromatin, leading to the derepression of the *Ink4b*/*Arf*/*Ink4a* locus.

An interaction between lncRNAs also seems to play a role in the regulation of the *Arabidopsis FLC* locus and, hence, vernalization (see Ch. 31 [Baulcombe and Dean 2014] for details). Expression of the sense ncRNA *COLDAIR* and the antisense ncRNA *COOLAIR* are induced by cold, and *COLDAIR* was shown to physically interact with CLF, indicating a possible role in VRN-PRC2 recruitment (Swiezewski et al. 2009). However, *FLC* transgenes without the *COLDAIR* promoter respond to cold, and also antisense *COOLAIR* is not required for vernalization-induced repression of *FLC*, such that the functional requirements for these ncRNAs are not clear (Sheldon et al. 2002; Helliwell et al. 2011). However, as the expression and processing of the *COOLAIR* ncRNA is affected by different genotypes and environments, and these correlate with changes in chromatin marks (reviewed in Ietswaart et al. 2012), it is possible that ncRNAs also recruit PRC2 components to target loci as they do in mammals.

4 PcG REPRESSION IN DEVELOPMENT AND DISEASE

4.1 From Gene to Chromosome Repression

Mutations in members of the murine PRC1 complex show homeotic transformations of the axial skeleton. This can cause the appearance of additional vertebrae as a consequence of the derepression of *HOX* genes (Fig. 2E,F) (Core et al. 1997). In addition, the mutant mice display severe combined immunodeficiencies caused by a lack of proliferative responses of hematopoietic cells (Raaphorst 2005). The role of PcG proteins has been particularly well studied in blood cells, in line with the fact that most blood cell lineages are characterized by their well-described cell-type-specific transcription programs. Lineage commitment and restriction need to be faithfully maintained through cell division. In PcG knockout mice, B- and T-cell precursor populations are produced normally, indicating that the establishment of lineage-specific gene expression patterns does not depend on PcG proteins. These proteins do, however, contribute to the irreversibility of the lineage choice, rather than the decision to follow a particular developmental pathway.

PcG proteins play a major role in controlling proliferation as well as the control of *HOX* genes, whose expression patterns characterize different blood cell lineages. The *Bmi1* gene, an ortholog of *Drosophila Psc* belonging to the PRC1 group, was initially identified as an oncogene that, in collaboration with *myc*, induces murine lymphomagenesis (van Lohuizen et al. 1991). The Bmi1 protein controls the cell-cycle regulators $p16^{INK4a}$ and $p19^{ARF}$ (Jacobs et al. 1999). Both Bmi1 and the related protein Mel-18 are negative regulators of the *Ink4c-Arf* locus required for normal lymphoid proliferation control. Misregulation of this important cell-cycle checkpoint affects apoptosis and senescence in mice.

Mammalian PcG proteins are also associated with X-chromosome inactivation as mentioned in Section 3.2 (see also Ch. 25 [Brockdorff and Turner 2014]). The inactivation of one X chromosome in XX female cells is accompanied by a series of chromatin modifications that involve PcG proteins. In particular, components of the PRC2 complex, like the ESC homolog EED, or the E(Z) homolog ENX1 (Table 1), play a major role in the establishment of histone marks associated with transcriptional silencing. Transient association of PRC2 with the X chromosome, coated by *Xist* RNA, is accompanied by H3K27 methylation. In contrast, *eed* mutant mouse embryos show no recruitment of the ENX1 KMT, and consequently no H3K27me3 is observed. However, the absence of these PRC2 components does not lead to a complete derepression of the entire inactive X chromosome; rather, some cells display sporadic reexpression of X-linked genes and an increase in epigenetic marks associated with an active state (H3K9ac and H3K4me3). This is likely because other, partially redundant, epigenetic mechanisms are in place to ensure the maintenance of one inactive X chromosome.

Recruitment of PRC2 to the inactive X chromosome is dependent on *Xist* RNA. As association of PRC2 to the inactive X is only transient, it appears that the complex is only required to set epigenetic marks (i.e., H3K27me3) for the maintenance of silencing. Currently, it is not known whether the PRC1 complex directly recognizes these marks. PRC1 is involved in the permanent silencing of the inactive X chromosome. The PRC2 component EED is required to recruit the PRC1 components MPH1 and MPH2, whereas RING1b, which can ubiquitinate H2A, is recruited independently of EED (Schoeftner et al. 2006). Thus, *Xist* RNA can recruit PRC1 components in both PRC2-dependent and -independent ways. In the absence of PRC2, *Xist*-dependent PRC1 recruitment is sufficient for PcG-based X-chromosome inactivation, which is further consolidated and maintained by DNA methylation.

PRC2 is involved in X-chromosome inactivation both in the embryo, in which an X chromosome is chosen at random for inactivation, and in extraembryonic tissues, wherein the paternally inherited X chromosome is systematically inactivated (imprinted X-chromosome inactivation). In addition, it was found that PRC2 is involved in the regulation of some autosomal imprinted genes. PRC2-mediated repression is, thus, a silencing mechanism that operates in addition to DNA methylation in the regulation of imprinted gene expression. By way of example, an analysis of 14 imprinted loci from six distinct imprinting clusters showed that four of these were biallelically expressed in *eed* mutant mice (Mager et al. 2003). At the *Kcnq1* imprinting cluster (Fig. 8), for instance, the predominantly maternally expressed genes *Cdk1*, *Cd81*, and *Tssc4* become biallelically expressed in *ezh2* mutant mice (Terranova et al. 2008). As similar results were also observed in mutants deficient for *Ring1b*, both PRC1 and PRC2 appear to be involved in regulating the monoallelic expression of some imprinted genes. Furthermore, it was shown that EZH2 is required for the association of the lncRNA *Kcnq1ot1* along the *Kcnq1* imprinting cluster (Terranova et al. 2008), confirming a link between PRC2 and ncRNAs in the regulation of imprinted genes (Fig. 8). Interestingly, all loci that lost imprinted expression were normally repressed when paternally inherited, whereas none of the maternally repressed loci were affected. As there appears to be a cross talk between PcG-based repression and gene silencing by DNA methylation, it is possible that PRC2 complex plays a role in the regulation of these imprinted genes via DNA methylation (see Ch. 15 [Li and Zhang 2014]).

An involvement of PRC2 in the regulation of imprinted gene expression has also been reported in *Arabidopsis*, in which the *PHE1* locus is expressed at much higher levels from the paternal allele (Köhler et al. 2005). In mutants affecting the *E(z)* homolog *MEA*, the maternal *PHE1* allele is specifically derepressed. *MEA* also regulates its own imprinted expression, noted by the strong repression of maternal *MEA* early in reproductive development in a *mea* mutant background. This effect, however, is independent of the other components of the FIS-PRC2 (Fig. 4) (Baroux et al. 2006). In contrast, later in development the FIS-PRC2 contributes to the stable repression of the paternal *MEA* allele (Baroux et al. 2006; Gehring et al. 2006; Jullien et al. 2006). In this latter case, the FIS-PRC2 is involved in the silencing of a paternally repressed imprinted allele similar to the situation in mammals. But *MEA* also has a role in keeping expression of the maternal *PHE1* and *MEA* alleles at low levels. Similar to the situation in mammals, regulation of imprinted expression at the *PHE1* and *MEA* loci involves both PRC2 and DNA methylation (Fig. 8). Although DNA methylation at the locus was suggested to regulate higher-order chromatin structure rather than directly distinguishing maternal and paternal alleles (Wöhrmann et al. 2012), both these epigenetic pathways seem to work together at the *PHE1* locus (Makarevich et al. 2008). The silencing of the maternal *PHE1* allele depends on a *cis*-regulatory region, which is differentially methylated. This regulatory element downstream from *PHE1* is methylated on the expressed paternal allele, but must not be methylated to mediate PRC2-dependent repression of the maternal allele (Fig. 8).

As PRC2 components are present in plants, invertebrates, and mammals, PRC2 represents an ancient molecular module suitable for gene repression that was already present in the unicellular ancestor of plants and animals, before the evolution of multicellularity. It was recently shown that DNA methylation at both the *MEA* and *PHE1* locus is affected in *mea* mutants, and PRC2 components directly interact with the DNA methyltransferase MET1, as found for the mammalian counterparts (Schmidt et al. 2012). Thus, although the interaction of these two major epigenetic pathways in the control of gene expression may have an ancient evolutionary origin, they were independently recruited for the regulation of imprinted genes in plants and mammals, the two lineages in which genomic imprinting evolved (Raissig et al. 2011).

4.2 Consequences of Aberrant Transcriptional Activation

The finding that *Bmi1* misregulation causes malignant lymphomas in mice raises the question of whether human BMI1 (a PRC1 component) itself contributes to the development of cancer in a similar fashion. There is now accumulating evidence that altered PcG gene expression is widespread in human malignant lymphomas (Shih et al. 2012). For instance, the level of BMI1 overexpression in B-cell lymphomas correlates with the degree of malignancy, suggesting that PRC1 components do play a role in the development of human cancers. However, the target genes of BMI1 in human cells appear to be different from those of mouse lymphocytes, as no obvious down-regulation of $p16^{INK4a}$ could be correlated with the overexpression of the oncogenes.

PcG gene overexpression is not only observed in hematological malignancies, but is also found in solid tumors, including meduloblastomas, and tumors originating from liver, colon, breast, lung, penis, and prostate (Fig. 9). The high expression of a PRC2 marker, EZH2, is often found in early stages of highly proliferative lung carcinomas. This suggests that the well-known cascade of PRC2 initiation and PRC1 maintenance might also accompany the development of a tumor cell lineage (for a review, see Sauvageau and Sauvageau 2010).

Interestingly, PRC2 components also play a crucial role in the control of cell proliferation in *Arabidopsis*. Although aberrant growth does not lead to cancer and death in plants, a strict control of cell proliferation is essential for normal development. In mutants of the *fis* class, the two fertilization products of flowering plants, the embryo and endosperm, overproliferate and the resulting seeds abort (Grossniklaus et al. 2001; Hsieh et al. 2003). Effects on cell proliferation are also observed in double mutants of *clf* and *swn*, two of the three plant *E(z)* homologs. Such plants undergo normal seed development, but produce a mass of proliferating, undifferentiated tissue (callus) rather than a differentiated shoot after germination (Chanvivattana et al. 2004).

Although it is currently not known how exactly PRC2 controls cell proliferation in plants, it is likely to involve interactions with RBR, the plant homolog of the Retinoblastoma (Rb) protein (Ebel et al. 2004; Mosquna et al. 2004). Mutants of the *fis* class do not only show proliferation defects during seed development after fertilization, but the *FIS* genes are also required to prevent proliferation of the endosperm in the absence of fertilization. This latter aspect of the phenotype is shared with *rbr* mutants and can be explained by the fact that RBR regulates the expression of genes encoding PRC2 components and *MET1* (Johnston et al. 2008). Remarkably, the Rb pathway also regulates the

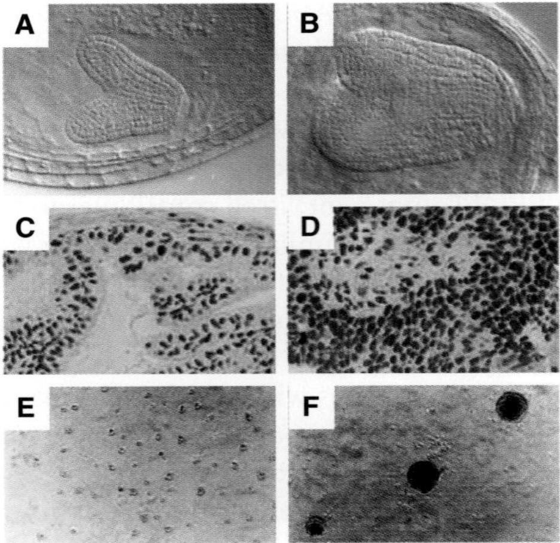

Figure 9. PRC2 regulates cell proliferation in mammals and plants. (*A,B*) Plant embryos derived from wild-type and *mea* mutant egg cells. *MEA* encodes a protein of the FIS-PRC2 and regulates cell proliferation. The *mea* embryo (*B*) is much larger than the corresponding wild-type embryo (*A*) at the same stage of development (late heart stage). Mutant embryos develop slower and have approximately twice the number of cell layers. (*C,D*) Normal and cancerous prostate epithelium of mice. In the cancerous epithelium, Ezh2 expression is highly increased (labeled with an anti-Ezh2 antibody). Thus, both loss of E(Z) function in plants and overexpression of E(Z) function in mice can lead to defects in cell proliferation. (*E,F*) Control and RING1 overexpressing rat 1a fibroblast cells. Overexpression of RING1 leads to anchorage-independent growth in soft agar, typical of neoplastically transformed cells. (*A,B*, Courtesy of J.-P. Vielle-Calzada and U. Grossniklaus; *C,D*, reprinted, with permission, from Kuzmichev et al. 2005, © National Academy of Sciences; *E,F*, reprinted, with permission, from Satijn and Otte 1999, © American Society for Microbiology.)

mammalian *Ezh2* and *Eed* genes encoding PRC2 subunits (Bracken et al. 2003), illustrating conserved regulatory networks between plants and animals.

4.3 Maintaining Stem Cell Fate

PcG regulation plays a very early role during oogenesis in the mouse for the formation of totipotent cell identities in the progeny (Posfai et al. 2012). Genetic ablation of the PRC1 components RING1 and RNF2 results in loss of chromatin-bound PRC1 in oocytes, induction of massive transcriptional misregulation during oocyte growth, and a developmental arrest at the two-cell stage of embryogenesis. These results indicate that PRC1 functions during oogenesis to specify maternal contributions in the cytoplasm as well as on maternal chromosomes, both of which contribute to the developmental competence of preimplantation embryos. Indeed, cultured mouse embryonic stem cells were used very effectively to study the role of PcG proteins in many aspects of cell proliferation and differentiation. The introduction of new technologies like ChIP-Seq allowed the correlation of PRC1/PRC2 components with many epigenetic marks, genetic regulatory elements of the murine genome, and the identification of functions related to ES cell pluripotency and plasticity during embryonic development (Boyer et al. 2006). Bivalent chromatin domains, characterized by the coexistence of the active H3K4me3 and repressive H3K27me3 histone marks, are resolved during differentiation (Mikkelsen et al. 2007). The cellular memory system takes a leading role in this process. Permanently repressed genes are tagged by the PcG system, together with DNA methylation, to establish stable silencing marks. Conversely, the TrxG system reiterates the H3K4me3 mark to keep the corresponding differentiation genes active.

Stem cells play an ever-increasing role in medicine. Their potential to provide progenitors for the healing of damaged tissue places them into a well treasured tool box of regenerative medicine. Not surprisingly, it is in the very well characterized blood cell lineage wherein we know most about the identity and location of stem cells. Hematopoietic stem cells (HSCs) maintain the pool of blood cells by self-renewing as well as producing daughter cells that differentiate into the lymphoid, myeloid, and erythroid lineages. The stem cell niche in the adult bone marrow provides the cells with specific external signals to maintain their fate. On the other hand, cell intrinsic cues for the maintenance of the "stemness" state seem to rely on the PcG system.

Mouse mutants affecting PRC1 genes (e.g., *bmi1*, *mel-18*, *mph1/rae28*, and *m33*; see Table 1) suffer from various defects in the hematopoietic system, such as hyperplasia in spleen and thymus, reduction in B and T cells, and an impaired proliferative response of lymphoid precursors to cytokines. The requirements for *Bmi1* and *Mel18* in stem cell self-renewal during different stages of development suggest a changing pool of target genes between embryonic and adult stem cells.

The PcG system is also required for neural stem cells (NSCs) as indicated by the neuronal defects observed in *bmi1* mouse mutants (Bruggeman et al. 2005; Zencak et al. 2005). In particular, these mice are depleted of cerebral NSCs postnatally, indicating an in vivo requirement for *Bmi1* in NSC renewal. It appears that embryonic NSC maintenance is thus under a different PcG network control than adult NSC self-renewal, similar to the regulation of the hematopoietic system.

External signals like the *Sonic hedgehog* (*Shh*) signaling cascade modulate the *Bmi1* response in NSCs and ensure a proliferative/self-renewal capacity (Leung et al. 2004). The identification of these external cues controlling PcG repression came through the analysis of the development of cerebellar granule neuron progenitors (CGNPs). A postnatal wave of proliferation is induced by the signaling factor Shh, secreted by the Purkinje cells. The *Shh* signal branches to control N-Myc and Bmi1 levels (Fig. 10). Thus, *Bmi1*-deficient CGNPs have a defective proliferative response upon Shh stimulation. The *Shh* signal is able to control proliferation of these stem cells ultimately by modulating both the downstream Rb pathway (via N-myc and Bmi1/p16^{INK4a}) and p53 pathway (via Bmi1/p19^{ARF}). This mechanism explains why hyperactivation of *Shh* signaling leads to the development of medulloblastomas. HSCs are regulated by a similar *Indian hedgehog*-controlled pathway. And, in NSCs, expression of the *Hoxd8*, *Hoxd9*, and *Hoxc9* loci is under the control of Bmi1. The appropriate *HOX* expression profile confers the necessary stem cell fate.

Indeed, as stem cells represent a defined and committed cellular state, it is not surprising that the PcG system maintains this particular fate in a mitotically heritable fashion. In the future, it will be interesting to identify the pool of targets of the PcG system in the different adult stem cell populations, and to learn how to influence the maintenance system to allow for the controlled reprogramming of stem cell fates. At the moment, little is known about the role of PcG genes in stem cell maintenance in plants. However, the reprogramming of plant cells, which are totipotent and have the potential to form a complete new organism under appropriate conditions, involves PcG regulation. Indeed, plants lacking the *E(z)* homologs *CLF* and *SWN* produce a mass of undifferentiated cells after germination, suggesting that PcG genes are required to maintain a differentiated state (Chanvivattana et al. 2004). Interestingly, the same PcG genes are required for the in vitro reprogramming of differentiated leaf cells into totipotent callus cells, possibly because PRC2 is required to repress leaf differentiation genes during this reprogramming process (He et al. 2012). In floral meristems, PRC2 plays a role to repress *WUS* (see Sec. 3.1), which itself is required for stem cell maintenance (Liu et al. 2011). Thus, although the molecular mechanisms differ greatly between animals and plants, PRC2 has been recruited to regulate cell stem identity and cell differentiation in both lineages.

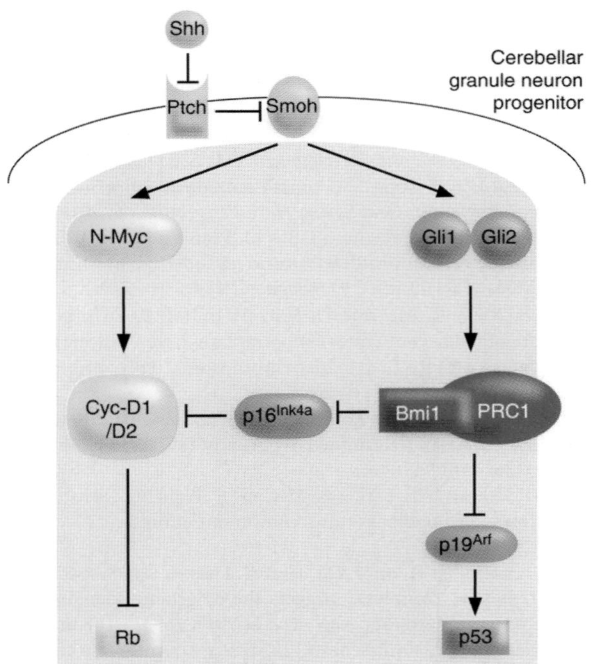

Figure 10. Sonic Hedgehog signaling maintains proliferation/self-renewal of cerebellar progenitor cells. The Shh signaling cascade regulates both the Rb pathway (which can be bound by the PRC2 RbAp48 protein) as well as the p53 pathway via Bmi1 control of the p16/p19 proliferation checkpoint. Inhibition of Smoothened (Smoh) by the Shh receptor Patched (Ptch) results in downstream signaling in the nucleus. One part of the signal induces N-Myc, Cyclin D1, and Cyclin D2, whereas the other part activates Bmi1 via the Gli effectors. (Adapted, with permission, from Valk-Lingbeek et al. 2004, © Elsevier.)

5 CONCLUSION AND OUTLOOK

It has been remarkable to follow the development of our understanding of PcG epigenetic regulation, starting with the initial genetic identification of a *Drosophila* mutant possessing additional sex combs on the second and third leg. This eventually led to the discovery of a new class of regulators found to be required for fundamental epigenetic processes such as vernalization in plants and silencing of the mammalian X chromosome. Control of genetic information is highly influenced by chromatin structure and composition of histones in their various modified forms. The proteins of the PcG are directly involved in generating epigenetic marks, for instance, H3K27me3 and H2AK118/119ub1, as a consequence of developmental decisions. The same group “reads” (i.e., shows high affinity to) these epigenetic marks through the action of the PRC1 proteins, and translates them into a stable, transcriptionally repressed state. In the model organism *Drosophila*, we have a relatively clear picture of how PcG complexes are anchored at PREs, for a defined group of target genes that are subject to long-term repression. However, to date very few PREs have been identified in other organisms. Although the basic functions of PcG proteins remain the same, it is not well understood how they are targeted to their site of action. Additionally, we need to get a better understanding of how an apparently dynamic group of proteins can impose a stable state of transcriptional repression through a chemical equilibrium.

The other major question in PcG research focuses on the heritability of the repressed state, the very essence of epigenetics. What is the identity of the molecular marks required to transmit a state of gene expression through DNA replication and mitosis? Do both active and repressed states need corresponding epigenetic marks, which are transmitted to daughter cells, or is only one sufficient, whereas the other represents the default state? The mechanism by which PcG proteins impose silencing on transcription during the interphase of the cell cycle has become increasingly clear. In the future, the focus of research will be on how the information regarding a state of gene expression endures the DNA replication process and is faithfully transmitted to the daughter cells following mitosis.

ACKNOWLEDGMENTS

R.P. thanks Daniel Enderle for help with the manuscript and design of some of the figures; U.G. is grateful to Michael Raissig and Heike Wöhrmann for their support in preparing figures, and to Hanspeter Schöb for help with the bibliography. Work on PcG-mediated silencing mechanisms in R.P.'s group is supported by the ETH Zürich and the Swiss National Science Foundation, and in U.G.'s laboratory by the University of Zürich, the Swiss National Science Foundation, and the European Research Council.

REFERENCES

**Reference is also in this book.*

Agger K, Cloos PA, Christensen J, Pasini D, Rose S, Rappsilber J, Issaeva I, Canaani E, Salcini AE, Helin K. 2007. UTX and JMJD3 are histone H3K27 demethylases involved in *HOX* gene regulation and development. *Nature* **449:** 731–734.

* Almouzni G, Cedar H. 2014. Maintenance of epigenetic information. *Cold Spring Harb Perspect Biol* doi: 10.1101/cshperspect.a019372.

Alvarez-Venegas R, Pien S, Sadder M, Witmer X, Grossniklaus U, Avramova Z. 2003. *ATX-1*, an *Arabidopsis* homolog of *trithorax*, activates flower homeotic genes. *Curr Biol* **13:** 627–637.

Bantignies F, Cavalli G. 2011. Polycomb group proteins: Repression in 3D. *Trends Genet* **27:** 454–474.

* Barlow DP, Bartolomei MS. 2014. Genomic imprinting in mammals. *Cold Spring Harb Perspect Biol* **6:** a018382.

Baroux C, Gagliardini V, Page DR, Grossniklaus U. 2006. Dynamic regulatory interactions of *Polycomb* group genes: *MEDEA* autoregulation is required for imprinted gene expression in *Arabidopsis*. *Genes Dev* **20:** 1081–1086.

* Baulcombe DC, Dean C. 2014. Epigenetic regulation in plant responses to the environment. *Cold Spring Harb Perspect Biol* **6:** a019471.

Berger N, Dubreucq B, Roudier F, Dubos C, Lepiniec L. 2011. Transcriptional regulation of *Arabidopsis LEAFY COTYLEDON2* involves *RLE*, a *cis*-element that regulates trimethylation of histone H3 at lysine-27. *Plant Cell* **23:** 4065–4078.

Birve A, Sengupta AK, Beuchle D, Larsson J, Kennison JA, Rasmuson-Lestander A, Müller J. 2001. *Su(z)12*, a novel *Drosophila* Polycomb group gene that is conserved in vertebrates and plants. *Development* **128:** 3371–3379.

Blobel GA, Kadauke S, Wang E, Lau AW, Zuber J, Chou MM, Vakoc CR. 2009. A reconfigured pattern of MLL occupancy within mitotic chromatin promotes rapid transcriptional reactivation following mitotic exit. *Mol Cell* **36:** 970–983.

Boyer LA, Plath K, Zeitlinger J, Brambrink T, Medeiros LA, Lee TI, Levine SS, Wernig M, Tajonar A, Ray MK, et al. 2006. Polycomb complexes repress developmental regulators in murine embryonic stem cells. *Nature* **441:** 349–353.

Bracken AP, Pasini D, Capra M, Prosperini E, Colli E, Helin K. 2003. EZH2 is downstream of the pRB-E2F pathway, essential for proliferation and amplified in cancer. *Embo J* **22:** 5323–5335.

Bratzel F, López-Torrejón G, Koch M, Del Pozo JC, Calonje M. 2010. Keeping cell identity in *Arabidopsis* requires PRC1 RING-finger homologs that catalyze H2A monoubiquitination. *Curr Biol* **20:** 1853–1859.

* Brockdorff N, Turner B. 2014. Dosage compensation in mammals. *Cold Spring Harb Perspect Biol* doi: 10.1101/cshperspect.a019406.

Bruggeman SW, Valk-Lingbeek ME, van der Stoop PP, Jacobs JJ, Kieboom K, Tanger E, Hulsman D, Leung C, Arsenijevic Y, Marino S, et al. 2005. *Ink4a* and *Arf* differentially affect cell proliferation and neural stem cell self-renewal in *Bmi1*-deficient mice. *Genes Dev* **19:** 1438–1443.

Cao R, Zhang Y. 2004. SUZ12 is required for both the histone methyltransferase activity and the silencing function of the EED-EZH2 complex. *Mol Cell* **15:** 57–67.

Cao R, Wang L, Wang H, Xia L, Erdjument Bromage H, Tempst P, Jones RS, Zhang Y. 2002. Role of histone H3 lysine 27 methylation in Polycomb-group silencing. *Science* **298:** 1039–1043.

Cao R, Wang H, He J, Erdjument-Bromage H, Tempst P, Zhang Y. 2008. Role of hPHF1 in H3K27 methylation and *Hox* gene silencing. *Mol Cell Biol* **28:** 1862–1872.

Carrington EA, Jones RS. 1996. The *Drosophila Enhancer of zeste* gene encodes a chromosomal protein: Examination of wild-type and mutant protein distribution. *Development* **122:** 4073–4083.

Chanvivattana Y, Bishopp A, Schubert D, Stock C, Moon YH, Sung ZR, Goodrich J. 2004. Interaction of Polycomb-group proteins controlling flowering in *Arabidopsis*. *Development* **131:** 5263–5276.

* Cheng X. 2014. Structural and functional coordination of DNA and histone methylation. *Cold Spring Harb Perspect Biol* **6:** a018747.

Core N, Bel S, Gaunt SJ, Aurrand Lions M, Pearce J, Fisher A, Djabali M. 1997. Altered cellular proliferation and mesoderm patterning in Polycomb-M33-deficient mice. *Development* **124:** 721–729.

Czermin B, Melfi R, McCabe D, Seitz V, Imhof A, Pirrotta V. 2002. *Drosophila* Enhancer of Zeste/ESC complexes have a histone H3 methyltransferase activity that marks chromosomal Polycomb sites. *Cell* **111:** 185–196.

Deal RB, Henikoff JG, Henikoff S. 2010. Genome-wide kinetics of nucleosome turnover determined by metabolic labeling of histones. *Science* **328:** 1161–1164.

Dellino GI, Schwartz YB, Farkas G, McCabe D, Elgin SC, Pirrotta V. 2004. Polycomb silencing blocks transcription initiation. *Mol Cell* **13:** 887–893.

De Lucia F, Crevillen P, Jones AM, Greb T, Dean C. 2008. A PHD-Polycomb repressive complex 2 triggers the epigenetic silencing of *FLC* during vernalization. *Proc Natl Acad Sci* **105:** 16831–16836.

Ebel C, Mariconti L, Gruissem W. 2004. Plant retinoblastoma homologues control nuclear proliferation in the female gametophyte. *Nature* **429:** 776–780.

Enderle D. 2011. Non-coding RNAs at intergenic Polycomb group binding sites. PhD thesis. University of Heidelberg, Heidelberg.

Enderle D, Beisel C, Stadler MB, Gerstung M, Athri P, Paro R. 2011. Polycomb preferentially targets stalled promoters of coding and non-coding transcripts. *Genome Res* **21:** 216–226.

Endoh M, Endo TA, Endoh T, Isono K, Sharif J, Ohara O, Toyoda T, Ito T, Eskeland R, Bickmore WA, et al. 2012. Histone H2A mono-ubiquitination is a crucial step to mediate PRC1-dependent repression of developmental genes to maintain ES cell identity. *PLoS Genet* **8:** e1002774.

 Cite as *Cold Spring Harb Perspect Biol* doi: 10.1101/cshperspect.a019331

Fischle W, Wang Y, Jacobs SA, Kim Y, Allis CD, Khorasanizadeh S. 2003. Molecular basis for the discrimination of repressive methyl-lysine marks in histone H3 by Polycomb and HP1 chromodomains. *Genes Dev* **17:** 1870–1881.

Fonseca JP, Steffen PA, Müller S, Lu J, Sawicka A, Seiser C, Ringrose L. 2012. In vivo Polycomb kinetics and mitotic chromatin binding distinguish stem cells from differentiated cells. *Genes Dev* **26:** 857–871.

Gambetta M, Oktaba K, Müller J. 2009. Essential role of the glycosyltransferase sxc/Ogt in Polycomb repression. *Science* **325:** 93–96.

Gehring M, Huh JH, Hsieh T-F, Penterman J, Choi Y, Harada JJ, Goldberg RB, Fischer RL. 2006. DEMETER DNA glycosylase establishes *MEDEA* Polycomb gene self-imprinting by allele-specific demethylation. *Cell* **124:** 495–506.

Gendall AR, Levy YY, Wilson A, Dean C. 2001. The *VERNALIZATION 2* gene mediates the epigenetic regulation of vernalization in *Arabidopsis*. *Cell* **107:** 525–535.

Goodrich J, Puangsomlee P, Martin M, Long D, Meyerowitz EM, Coupland G. 1997. A Polycomb-group gene regulates homeotic gene expression in *Arabidopsis*. *Nature* **386:** 44–51.

Grau DJ, Chapman BA, Garlick JD, Borowsky M, Francis NJ, Kingston RE. 2011. Compaction of chromatin by diverse Polycomb group proteins requires localized regions of high charge. *Genes Dev* **25:** 2210–2221.

Grossniklaus U, Vielle Calzada JP, Hoeppner MA, Gagliano WB. 1998. Maternal control of embryogenesis by *MEDEA*, a *Polycomb* group gene in *Arabidopsis*. *Science* **280:** 446–450.

Grossniklaus U, Spillane C, Page DR, Köhler C. 2001. Genomic imprinting and seed development: Endosperm formation with and without sex. *Curr Opin Plant Biol* **4:** 21–27.

Gutiérrez L, Oktaba K, Scheuermann JC, Gambetta MC, Ly-Hartig N, Müller J. 2012. The role of the histone H2A ubiquitinase Sce in Polycomb repression. *Development* **139:** 117–127.

Hackett WP, Cordero RE, Sinivasan C. 1987. *Apical meristem characteristics and activity in relation to juvenility in Hedera*. Butterworth, London, UK.

Hadorn E. 1968. Transdetermination in cells. *Sci Am* **219:** 110–114.

He C, Chen X, Huang H, Xu L. 2012. Reprogramming of H3K27me3 is critical for acquisition of pluripotency from cultured *Arabidopsis* tissues. *PLoS Genet* **8:** e1002911.

Helliwell CA, Robertson M, Finnegan EJ, Buzas DM, Dennis ES. 2011. Vernalization-repression of *Arabidopsis FLC* requires promoter sequences but not antisense transcripts. *PloS One* **6:** e21513.

Hennig L, Taranto P, Walser M, Schonrock N, Gruissem W. 2003. *Arabidopsis MSI1* is required for epigenetic maintenance of reproductive development. *Development* **130:** 2555–2565.

Hennig L, Bouveret R, Gruissem W. 2005. MSI1-like proteins: An escort service for chromatin assembly and remodeling complexes. *Trends Cell Biol* **15:** 295–302.

Hodl M, Basler K. 2012. Transcription in the absence of histone H3.2 and H3K4 methylation. *Curr Biol* **22:** 1–5.

Holec S, Berger F. 2012. Polycomb group complexes mediate developmental transitions in plants. *Plant Phys* **158:** 35–43.

Hsieh TF, Hakim O, Ohad N, Fischer RL. 2003. From flour to flower: How Polycomb group proteins influence multiple aspects of plant development. *Trends Plant Sci* **8:** 439–445.

Huang DH, Chang YL, Yang CC, Pan IC, King B. 2002. *pipsqueak* encodes a factor essential for sequence-specific targeting of a Polycomb group protein complex. *Mol Cell Biol* **22:** 6261–6271.

Ietswaart R, Wu Z, Dean C. 2012. Flowering time control: Another window to the connection between antisense RNA and chromatin. *Trends Genet* **28:** 445–53.

Jacobs JJ, Scheijen B, Voncken JW, Kieboom K, Berns A, van Lohuizen M. 1999. Bmi-1 collaborates with c-Myc in tumorigenesis by inhibiting c-Myc-induced apoptosis via INK4a/ARF. *Genes Dev* **13:** 2678–2690.

Jiang D, Wang Y, He Y. 2008. Repression of *FLOWERING LOCUS C* and *FLOWERING LOCUS T* by the *Arabidopsis* Polycomb repressive complex 2 components. *PLoS One* **3:** e3404.

Johnston AJ, Matveeva E, Kirioukhova O, Grossniklaus U, Gruissem W. 2008. A dynamic reciprocal RBR-PRC2 regulatory circuit controls *Arabidopsis* gametophyte development. *Curr Biol* **18:** 1680–1686.

Jullien PE, Katz A, Oliva M, Ohad N, Berger F. 2006. Polycomb group complexes self-regulate imprinting of the Polycomb group gene *MEDEA* in *Arabidopsis*. *Curr Biol* **16:** 486–492.

Kanduri C, Thakur N, Pandey RR. 2006. The length of the transcript encoded from the *Kcnq1ot1* antisense promoter determines the degree of silencing. *EMBO J* **25:** 2096–2106.

Kanhere A, Viiri K, Araújo CC, Rasaiyaah J, Bouwman RD, Whyte WA, Pereira CF, Brookes E, Walker K, Bell GW, et al. 2010. Short RNAs are transcribed from repressed Polycomb target genes and interact with Polycomb repressive complex-2. *Mol Cell* **38:** 675–688.

Katsuyama T, Paro R. 2011. Epigenetic reprogramming during tissue regeneration. *FEBS Lett* **585:** 1617–1624.

Kennison JA. 1995. The Polycomb and trithorax group proteins of *Drosophila*: Trans-regulators of homeotic gene function. *Annu Rev Genet* **29:** 289–303.

Kharchenko PV, Alekseyenko AA, Schwartz YB, Minoda A, Riddle NC, Ernst J, Sabo PJ, Larschan E, Gorchakov AA, Gu T, et al. 2010. Comprehensive analysis of the chromatin landscape in *Drosophila melanogaster*. *Nature* **471:** 480–485.

King IF, Francis NJ, Kingston RE. 2002. Native and recombinant Polycomb group complexes establish a selective block to template accessibility to repress transcription in vitro. *Mol Cell Biol* **22:** 7919–7928.

* Kingston RE, Tamkun JW. 2014. Transcriptional regulation by trithorax-group proteins. *Cold Spring Harb Perspect Biol* **6:** a019349.

Kinoshita T, Harada JJ, Goldberg RB, Fischer RL. 2001. Polycomb repression of flowering during early plant development. *Proc Natl Acad Sci* **98:** 14156–14161.

Klymenko T, Müller J. 2004. The histone methyltransferases Trithorax and Ash1 prevent transcriptional silencing by Polycomb group proteins. *EMBO Rep* **5:** 373–377.

Klymenko T, Papp B, Fischle W, Kocher T, Schelder M, Fritsch C, Wild B, Wilm M, Müller J. 2006. A Polycomb group protein complex with sequence-specific DNA-binding and selective methyl-lysine-binding activities. *Genes Dev* **20:** 1110–1122.

Köhler C, Hennig L. 2010. Regulation of cell identity by plant Polycomb and *trithorax* group proteins. *Curr Opin Genet Dev* **20:** 541–547.

Köhler C, Hennig L, Bouveret R, Gheyselinck J, Grossniklaus U, Gruissem W. 2003a. *Arabidopsis* MSI1 is a component of the MEA/FIE *Polycomb* group complex and required for seed development. *EMBO J* **22:** 4804–4814.

Köhler C, Hennig L, Spillane C, Pien S, Gruissem W, Grossniklaus U. 2003b. The *Polycomb*-group protein MEDEA regulates seed development by controlling expression of the MADS-box gene *PHERES1*. *Genes Dev* **17:** 1540–1553.

Köhler C, Page DR, Gagliardini V, Grossniklaus U. 2005. The *Arabidopsis thaliana* MEDEA Polycomb group protein controls expression of PHERES1 by parental imprinting. *Nat Genet* **37:** 28–30.

Ku M, Koche RP, Rheinbay E, Mendenhall EM, Endoh M, Mikkelsen TS, Presser A, Nusbaum C, Xie X, Chi AS, et al. 2008. Genomewide analysis of PRC1 and PRC2 occupancy identifies two classes of bivalent domains. *PLoS Genet* **4:** e1000242.

Kurzhals RL, Tie F, Stratton CA, Harte PJ. 2008. *Drosophila* ESC-like can substitute for ESC and becomes required for Polycomb silencing if ESC is absent. *Dev Biol* **313:** 293–306.

Kuzmichev A, Margueron R, Vaquero A, Preissner TS, Scher M, Kirmizis A, Ouyang X, Brockdorff N, Abate-Shen C, Farnham P, Reinberg D. 2005. Composition and histone substrates of Polycomb repressive group complexes change during cellular differentiation. *Proc Natl Acad Sci* **102:** 1859–1864.

Lagarou A, Mohd-Sarip A, Moshkin YM, Chalkley GE, Bezstarosti K, Demmers JA, Verrijzer CP. 2008. dKDM2 couples histone H2A ubiquitylation to histone H3 demethylation during Polycomb group silencing. *Genes Dev* **22:** 2799–2810.

Landeira D, Sauer S, Poot R, Dvorkina M, Mazzarella L, Jørgensen HF, Pereira CF, Leleu M, Piccolo FM, Spivakov M, et al. 2010. Jarid2 is a PRC2 component in embryonic stem cells required for multi-lineage differentiation and recruitment of PRC1 and RNA Polymerase II to developmental regulators. *Nature* **12:** 618–624.

Lee N, Maurange C, Ringrose L, Paro R. 2005. Suppression of Polycomb group proteins by JNK signalling induces transdetermination in *Drosophila* imaginal discs. *Nature* **438:** 234–237.

Lee TI, Jenner RG, Boyer LA, Guenther MG, Levine SS, Kumar RM, Chevalier B, Johnstone SE, Cole MF, Isono K, et al. 2006. Control of developmental regulators by Polycomb in human embryonic stem cells. *Cell* **125:** 301–313.

Lee MG, Villa R, Trojer P, Norman J, Yan K-P, Reinberg D, Di Croce L, Shiekhattar R. 2007. Demethylation of H3K27 regulates Polycomb recruitment and H2A ubiquitination. *Science* **318:** 447–450.

Leung C, Lingbeek M, Shakhova O, Liu J, Tanger E, Saremaslani P, van Lohuizen M, Marino S. 2004. Bmi1 is essential for cerebellar development and is overexpressed in human medulloblastomas. *Nature* **428:** 337–341.

Lewis EB. 1978. A gene complex controlling segmentation in *Drosophila*. *Nature* **276:** 565–570.

* Li E, Zhang Y. 2014. DNA methylation in mammals. *Cold Spring Harb Perspect Biol* **6:** a019133.

Liu X, Kim YJ, Müller R, Yumul RE, Liu C, Pan Y, Cao X, Goodrich J, Chen X. 2011. *AGAMOUS* terminates floral stem cell maintenance in *Arabidopsis* by directly repressing *WUSCHEL* through recruitment of Polycomb Group proteins. *Plant Cell* **23:** 3654–3670.

Lo SM, Ahuja NK, Francis NJ. 2009. Polycomb group protein Suppressor 2 of zeste is a functional homolog of Posterior Sex Combs. *Mol Cell Biol* **29:** 515–525.

Lorente MD, Marcos-Gutierrez C, Perez C, Schoorlemmer J, Ramirez A, Magin T, Vidal M. 2000. Loss- and gain-of-function mutations show a Polycomb group function for Ring1A in mice. *Development* **127:** 5093–5100.

Luo M, Bilodeau P, Koltunow A, Dennis ES, Peacock WJ, Chaudhury AM. 1999. Genes controlling fertilization-independent seed development in *Arabidopsis thaliana*. *Proc Natl Acad Sci* **96:** 296–301.

Mager J, Montgomery ND, de Villena FP, Magnuson T. 2003. Genome imprinting regulated by the mouse Polycomb group protein Eed. *Nat Genet* **33:** 502–507.

Makarevich G, Leroy O, Akinci U, Schubert D, Clarenz O, Goodrich J, Grossniklaus U, Köhler C. 2006. Different *Polycomb* group complexes regulate common target genes in *Arabidopsis*. *EMBO Rep* **7:** 947–952.

Makarevich G, Villar CB, Erilova A, Köhler C. 2008. Mechanism of *PHERES1* imprinting in *Arabidopsis*. *J Cell Sci* **121:** 906–912.

Margueron R, Reinberg D. 2011. The Polycomb complex PRC2 and its mark in life. *Nature* **469:** 343–349.

Margueron R, Li G, Sarma K, Blais A, Zavadil J, Woodcock CL, Dynlacht BD, Reinberg D. 2008. Ezh1 and Ezh2 maintain repressive chromatin through different mechanisms. *Mol Cell* **32:** 503–518.

Margueron R, Justin N, Ohno K, Sharpe ML, Son J, Drury WJ 3rd, Voigt P, Martin SR, Taylor WR, De Marco V, et al. 2009. Role of the Polycomb protein EED in the propagation of repressive histone marks. *Nature* **461:** 762–767.

Matharu NK, Hussain T, Sankaranarayanan R, Mishra RK. 2010. Vertebrate homologue of *Drosophila* GAGA factor. *J. Mol Biol* **400:** 434–447.

Mendenhall EM, Koche RP, Truong T, Zhou VW, Issac B, Chi AS, Ku M, Bernstein BE. 2010. GC-rich sequence elements recruit PRC2 in mammalian ES cells. *PLoS Genet* **6:** e1001244.

Mikkelsen TS, Ku M, Jaffe DB, Issac B, Lieberman E, Giannoukos G, Alvarez P, Brockman W, Kim T-K, Koche RP, et al. 2007. Genome-wide maps of chromatin state in pluripotent and lineage-committed cells. *Nature* **448:** 553–560.

Mito Y, Henikoff JG, Henikoff S. 2007. Histone replacement marks the boundaries of cis-regulatory domains. *Science* **315:** 1408–1411.

Mohn F, Weber M, Rebhan M, Roloff TC, Richter J, Stadler MB, Bibel M, Schübeler D. 2008. Lineage-specific Polycomb targets and de novo DNA methylation define restriction and potential of neuronal progenitors. *Mol Cell* **30:** 755–766.

Mosquna A, Katz A, Shochat S, Grafi G, Ohad N. 2004. Interaction of FIE, a Polycomb protein, with pRb: A possible mechanism regulating endosperm development. *Mol Genet Genomics* **271:** 651–657.

Nekrasov M, Wild B, Müller J. 2005. Nucleosome binding and histone methyltransferase activity of *Drosophila* PRC2. *EMBO Rep* **6:** 348–353.

Nekrasov M, Klymenko T, Fraterman S, Papp B, Oktaba K, Kocher T, Cohen A, Stunnenberg HG, Wilm M, Müller J. 2007. Pcl-PRC2 is needed to generate high levels of H3-K27 trimethylation at Polycomb target genes. *EMBO J* **26:** 4078–4088.

Ohad N, Yadegari R, Margossian L, Hannon M, Michaeli D, Harada JJ, Goldberg RB, Fischer RL. 1999. Mutations in FIE, a WD polycomb group gene, allow endosperm development without fertilization. *Plant Cell* **11:** 407–416.

Page DR, Grossniklaus U. 2002. The art and design of genetic screens: *Arabidopsis thaliana*. *Nat Rev Genet* **3:** 124–136.

Pandey RR, Mondal T, Mohammad F, Enroth S, Redrup L, Komorowski J, Nagano T, Mancini-Dinardo D, Kanduri C. 2008. *Kcnq1ot1* antisense noncoding RNA mediates lineage-specific transcriptional silencing through chromatin-level regulation. *Mol Cell* **32:** 232–246.

Pasini D, Bracken AP, Jensen MR, Denchi EL, Helin K. 2004. Suz12 is essential for mouse development and for EZH2 histone methyltransferase activity. *EMBO J* **23:** 4061–4071.

Pasini D, Hansen KH, Christensen J, Agger K, Cloos PA, Helin K. 2008. Coordinated regulation of transcriptional repression by the RBP2 H3K4 demethylase and Polycomb-Repressive Complex 2. *Genes Dev* **22:** 1345–1355.

Pengelly AR, Copur Ö, Jäckle H, Herzig A, Müller J. 2013. A histone mutant reproduces the phenotype caused by loss of histone-modifying factor Polycomb. *Science* **339:** 698–699.

Pien S, Fleury D, Mylne JS, Crevillen P, Inze D, Avramova Z, Dean C, Grossniklaus U. 2008. ARABIDOPSIS TRITHORAX1 dynamically regulates *FLOWERING LOCUS C* activation via histone 3 lysine 4 trimethylation. *Plant Cell* **20:** 580–588.

Posfai E, Kunzmann R, Brochard V, Salvaing J, Cabuy E, Roloff TC, Liu Z, Tardat M, van Lohuizen M, Vidal M, et al. 2012. Polycomb function during oogenesis is required for mouse embryonic development. *Genes Dev* **26:** 920–932.

Raaphorst FM. 2005. Deregulated expression of Polycomb-group oncogenes in human malignant lymphomas and epithelial tumors. *Hum Mol Genet* **14 Spec No 1:** R93–R100.

Raissing RT, Baroux C, Grossniklaus U. 2011. Regulation and flexibility of genomic imprinting during seed development. *Plant Cell* **23:** 16–26.

Reyes JC, Grossniklaus U. 2003. Diverse functions of *Polycomb* group proteins during plant development. *Semin Cell Dev Biol* **14:** 77–84.

Ringrose L, Rehmsmeier M, Dura JM, Paro R. 2003. Genome-wide prediction of Polycomb/Trithorax response elements in *Drosophila melanogaster*. *Dev Cell* **5:** 759–771.

* Rinn JL. 2014. lncRNAs: Linking RNA to chromatin. *Cold Spring Harb Perspect Biol* **6:** a018614.

Rinn JL, Kertesz M, Wang JK, Squazzo SL, Xu X, Brugmann SA, Goodnough LH, Helms JA, Farnham PJ, Segal E, et al. 2007. Functional demarcation of active and silent chromatin domains in human *HOX* loci by noncoding RNAs. *Cell* **129:** 1311–1323.

Sarma K, Margueron R, Ivanov A, Pirrotta V, Reinberg D. 2008. Ezh2 requires PHF1 to efficiently catalyze H3 lysine 27 trimethylation in vivo. *Mol Cell Biol* **28:** 2718–2731.

Sauvageau M, Sauvageau G. 2010. Polycomb group proteins: Multi-faceted regulators of somatic stem cells and cancer. *Cell Stem Cell* **7:** 299–313.

Scheuermann JC, Gaytán de Ayala Alonso A, Oktaba K, Ly-Hartig N, McGinty RK, Fraterman S, Wilm M, Muir TW, Müller J. 2010.

Histone H2A deubiquitinase activity of the Polycomb repressive complex PR-DUB. *Nature* **465:** 243–247.

Schmidt A, Wöhrmann HJ, Raissig MT, Arand J, Gheyselinck J, Gagliardini V, Heichinger C, Walter J, Grossniklaus U. 2013. The *Polycomb* group protein MEDEA and the DNA methyltransferase MET1 interact to repress autonomous endosperm development in *Arabidopsis*. *Plant J* **73:** 776–787.

Schmitt S, Prestel M, Paro R. 2005. Intergenic transcription through a Polycomb group response element counteracts silencing. *Genes Dev* **19:** 697–708.

Schoeftner S, Sengupta AK, Kubicek S, Mechtler K, Spahn L, Koseki H, Jenuwein T, Wutz A. 2006. Recruitment of PRC1 function at the initiation of X inactivation independent of PRC2 and silencing. *EMBO J* **25:** 3110–3122.

Schubert D, Primavesi L, Bishopp A, Roberts G, Doonan J, Jenuwein T, Goodrich J. 2006. Silencing by plant Polycomb-group genes requires dispersed trimethylation of histone H3 at lysine 27. *EMBO J* **25:** 4638–4649.

Schuettengruber B, Ganapathi M, Leblanc B, Portoso M, Jaschek R, Tolhuis B, van Lohuizen M, Tanay A, Cavalli G. 2009. Functional anatomy of Polycomb and trithorax chromatin landscapes in *Drosophila* embryos. *PLoS Biol* **7:** e13.

Schwartz YB, Pirrotta V. 2008. Polycomb complexes and epigenetic states. *Curr Opin Cell Biol* **20:** 266–273.

Schwartz YB, Kahn TG, Nix DA, Li XY, Bourgon R, Biggin M, Pirrotta V. 2006. Genome-wide analysis of Polycomb targets in *Drosophila melanogaster*. *Nat Genet* **38:** 700–705.

* Seto E, Yoshida M. 2014. Erasers of histone acetylation: The histone deacetylase enzymes. *Cold Spring Harb Perspect Biol* **6:** a018713.

Shao Z, Raible F, Mollaaghababa R, Guyon JR, Wu CT, Bender W, Kingston RE. 1999. Stabilization of chromatin structure by PRC1, a Polycomb complex. *Cell* **98:** 37–46.

Sheldon CC, Conn AB, Dennis ES, Peacock WJ. 2002. Different regulatory regions are required for the vernalization-induced repression of *FLOWERING LOCUS C* and for the epigenetic maintenance of repression. *Plant Cell* **14:** 2527–2537.

Shih AH, Abdel-Wahab O, Patel JP, Levine RL. 2012. The role of mutations in epigenetic regulators in myeloid malignancies. *Nat Rev Can* **12:** 599–612.

Simon J, Chiang A, Bender W, Shimell MJ, O'Connor M. 1993. Elements of the *Drosophila bithorax* complex that mediate repression by *Polycomb* group products. *Dev Biol* **158:** 131–144.

Sinclair DA, Syrzycka M, Macauley MS, Rastgardani T, Komljenovic I, Vocadlo DJ, Brock HW, Honda BM. 2009. *Drosophila O*-GlcNAc transferase (OGT) is encoded by the *Polycomb* group (PcG) gene, *super sex combs* (*sxc*). *Proc Natl Acad Sci* **106:** 13427–13432.

Sing A, Pannell D, Karaiskakis A, Sturgeon K, Djabali M, Ellis J, Lipshitz HD, Cordes SP. 2009. A vertebrate Polycomb response element governs segmentation of the posterior hindbrain. *Cell* **138:** 885–897.

Smith ER, Lee MG, Winter B, Droz NM, Eissenberg JC, Shiekhattar R, Shilatifard A. 2008. *Drosophila* UTX is a histone H3 Lys27 demethylase that colocalizes with the elongating form of RNA polymerase II. *Mol Cell Biol* **28:** 1041–1046.

Squazzo SL, O'Geen H, Komashko VM, Krig SR, Jin VX, Jang SW, Margueron R, Reinberg D, Green R, Farnham PJ. 2006. Suz12 binds to silenced regions of the genome in a cell-type-specific manner. *Genome Res* **16:** 890–900.

Stock JK, Giadrossi S, Casanova M, Brookes E, Vidal M, Koseki H, Brockdorff N, Fisher AG, Pombo A. 2007. Ring1-mediated ubiquitination of H2A restrains poised RNA polymerase II at bivalent genes in mouse ES cells. *Nat Cell Biol* **9:** 1428–1435.

* Strome S, Kelly WG, Ercan S, Lieb JD. 2014. Regulation of the X chromosomes in *C. elegans*. *Cold Spring Harb Perspect Biol* **6:** a018366.

Sung S, Amasino RM. 2004. Vernalization in *Arabidopsis thaliana* is mediated by the PHD finger protein VIN3. *Nature* **427:** 159–164.

Swiezewski S, Liu F, Magusin A, Dean C. 2009. Cold-induced silencing by long antisense transcripts of an *Arabidopsis* Polycomb target. *Nature* **462:** 799–802.

Tavares L, Dimitrova E, Oxley D, Webster J, Poot R, Demmers J, Bezstarosti K, Taylor S, Ura H, Koide H, et al. 2012. RYBP-PRC1 complexes mediate H2A ubiquitylation at Polycomb target sites independently of PRC2 and H3K27me3. *Cell* **148:** 664–678.

Terranova R, Yokobayashi S, Stadler MB, Otte AP, van Lohuizen M, Orkin SH, Peters AH. 2008. Polycomb group proteins Ezh2 and Rnf2 direct genomic contraction and imprinted repression in early mouse embryos. *Dev Cell* **15:** 668–679.

Tie F, Furuyama T, Prasad Sinha J, Jane E, Harte PJ. 2001. The *Drosophila* Polycomb group proteins ESC and E(Z) are present in a complex containing the histone-binding protein p55 and the histone deacetylase RPD3. *Development* **128:** 275–286.

Tie F, Banerjee R, Stratton CA, Prasad-Sinha J, Stepanik V, Zlobin A, Diaz MO, Scacheri PC, Harte PJ. 2009. CBP-mediated acetylation of histone H3 lysine 27 antagonizes *Drosophila* Polycomb silencing. *Development* **136:** 3131–3141.

Tschiersch B, Hofmann A, Krauss V, Dorn R, Korge G, Reuter G. 1994. The protein encoded by the *Drosophila* position-effect variegation suppressor gene *Su(var)3-9* combines domains of antagonistic regulators of homeotic gene complexes. *EMBO J* **13:** 3822–3831.

Turck F, Roudier F, Farrona S, Martin-Magniette M-L, Guillaume E, Buisine N, Gagnot S, Martienssen RA, Coupland G, Colot V. 2007. *Arabidopsis* TFL2/LHP1 specifically associates with genes marked by trimethylation of histone H3 lysine 27. *PLoS Genet* **3:** e86.

Valk-Lingbeek ME, Bruggeman SW, van Lohuizen M. 2004. Stem cells and cancer; the Polycomb connection. *Cell* **118:** 409–418.

van der Lugt NM, Domen J, Linders K, van Roon M, Robanus Maandag E, te Riele H, van der Valk M, Deschamps J, Sofroniew M, van Lohuizen M, et al. 1994. Posterior transformation, neurological abnormalities, and severe hematopoietic defects in mice with a targeted deletion of the *bmi-1* proto-oncogene. *Genes Dev* **8:** 757–769.

van Lohuizen M, Frasch M, Wientjens E, Berns A. 1991. Sequence similarity between the mammalian *bmi-1* protooncogene and the *Drosophila* regulatory genes *Psc* and *Su(z)2*. *Nature* **353:** 353–355.

Wang L, Brown JL, Cao R, Zhang Y, Kassis JA, Jones RS. 2004a. Hierarchical recruitment of Polycomb group silencing complexes. *Mol Cell* **14:** 637–646.

Wang H, Wang L, Erdjument-Bromage H, Vidal M, Tempst P, Jones RS, Zhang Y. 2004b. Role of histone H2A ubiquitination in Polycomb silencing. *Nature* **431:** 873–878.

Wang L, Jahren N, Vargas ML, Andersen EF, Benes J, Zhang J, Miller EL, Jones RS, Simon JA. 2006. Alternative ESC and ESC-like subunits of a Polycomb group histone methyltransferase complex are differentially deployed during *Drosophila* development. *Mol Cell Biol* **26:** 2637–2647.

Weake VM, Workman JL. 2008. Histone ubiquitination: Triggering gene activity. *Mol Cell* **29:** 653–663.

Whitcomb SJ, Basu A, Allis CD, Bernstein E. 2007. Polycomb group proteins: An evolutionary perspective. *Trends Genet* **23:** 494–502.

Wöhrmann HJ, Gagliardini V, Raissig MT, Wehrle W, Arand J, Schmidt A, Tierling S, Page DR, Schöb H, Walter J, et al. 2012. Identification of a DNA methylation-independent imprinting control region at the *Arabidopsis MEDEA* locus. *Genes Dev* **26:** 1837–1850.

Woo CJ, Kharchenko PV, Daheron L, Park PJ, Kingston RE. 2010. A region of the human HOXD cluster that confers Polycomb-group responsiveness. *Cell* **140:** 99–110.

Wood CC, Robertson M, Tanner G, Peacock WJ, Dennis ES, Helliwell CA. 2006. The *Arabidopsis thaliana* vernalization response requires a Polycomb-like protein complex that also includes VERNALIZATION INSENSITIVE-3. *Proc Natl Acad Sci* **103:** 14631–14636.

Wu HA, Bernstein E. 2008. Partners in imprinting: Noncoding RNA and Polycomb group proteins. *Dev Cell* **15:** 637–638.

Xu L, Shen W-H. 2008. Polycomb silencing of *KNOX* genes confines shoot stem cell niches in *Arabidopsis*. *Curr Biol* **18:** 1966–1971.

Yamamoto K, Sonoda M, Inokuchi J, Shirasawa S, Sasazuki T. 2004. Polycomb group suppressor of zeste 12 links heterochromatin protein 1α and enhancer of zeste 2. *J Biol Chem* **279:** 401–406.

Yap KL, Li S, Muñoz-Cabello AM, Raguz S, Zeng L, Mujtaba S, Gil J, Walsh MJ, Zhou M-M. 2010. Molecular interplay of the non-coding RNA *ANRIL* and methylated histone H3 lysine 27 by Polycomb CBX7 in transcriptional silencing of *INK4a*. *Mol Cell* **38:** 662–674.

Yoshida N, Yanai Y, Chen L, Kato Y, Hiratsuka J, Miwa T, Sung ZR, Takahashi S. 2001. EMBRYONIC FLOWER2, a novel Polycomb group protein homolog, mediates shoot development and flowering in *Arabidopsis*. *Plant Cell* **13:** 2471–2481.

Zencak D, Lingbeek M, Kostic C, Tekaya M, Tanger E, Hornfeld D, Jaquet M, Munier FL, Schorderet DF, van Lohuizen M, et al. 2005. *Bmi1* loss produces an increase in astroglial cells and a decrease in neural stem cell population and proliferation. *J Neurosci* **25:** 5774–5783.

Zhang X, Germann S, Blus BJ, Khorasanizadeh S, Gaudin V, Jacobsen SE. 2007. The *Arabidopsis* LHP1 protein colocalizes with histone H3 Lys27 trimethylation. *Nat Struct Mol Biol* **14:** 869–871.

Zhao J, Ohsumi TK, Kung JT, Ogawa Y, Grau DJ, Sarma K, Song JJ, Kingston RE, Borowsky M, Lee JT. 2010. Genome-wide identification of Polycomb-associated RNAs by RIP-seq. *Mol Cell* **40:** 939–953.

Zhou W, Zhu P, Wang J, Pascual G, Ohgi KA, Lozach J, Glass CK, Rosenfeld MG. 2008. Histone H2A monoubiquitination represses transcription by inhibiting RNA polymerase II transcriptional elongation. *Mol Cell* **29:** 69–80.

CHAPTER 18

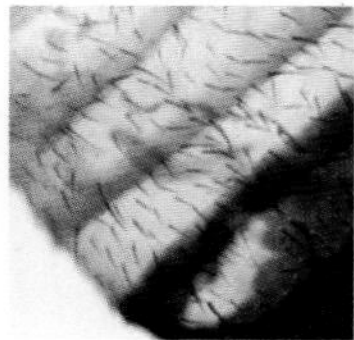

Transcriptional Regulation by Trithorax-Group Proteins

Robert E. Kingston[1] and John W. Tamkun[2]

[1]Department of Molecular Biology, Massachusetts General Hospital, Boston, Massachusetts 02114; [2]Department of Molecular, Cell and Developmental Biology, University of California, Santa Cruz, Santa Cruz, California 95064

Correspondence: kingston@molbio.mgh.harvard.edu

SUMMARY

The trithorax group of genes (trxG) was identified in mutational screens that examined developmental phenotypes and suppression of *Polycomb* mutant phenotypes. The protein products of these genes are primarily involved in gene activation, although some can also have repressive effects. There is no central function for these proteins. Some move nucleosomes about on the genome in an ATP-dependent manner, some covalently modify histones such as methylating lysine 4 of histone H3, and some directly interact with the transcription machinery or are a part of that machinery. It is interesting to consider why these specific members of large families of functionally related proteins have strong developmental phenotypes.

Outline

Chapter-opening figures: (*Left*) From the Biology Project, Department of Biochemistry and Molecular Biophysics, University of Arizona, photo by Ed Lewis; (*right*) from Daubresse et al. (1999).

Cite as *Cold Spring Harb Perspect Biol* doi: 10.1101/cshperspect.a019349

OVERVIEW

All cells in an organism must be able to "remember" what type of cell they are meant to be. This process, referred to as "cellular memory" or "transcriptional memory," requires two basic classes of mechanisms. The first class, discussed in Ch. 17 (Grossniklaus and Paro 2014), functions to maintain an "off" state for genes that, if turned on, would specify an inappropriate cell type. The Polycomb-group (PcG) proteins have as their primary function a repressive role in cellular memory. The second class of mechanisms is composed of those that are required to maintain key genes in an "on" state. Any cell type requires the expression of master regulatory proteins that direct the specific functions required for that cell type. The genes that encode these master regulatory proteins must be maintained in an "on" state throughout the lifetime of an organism to maintain the proper cell types within that organism.

The proteins that are involved in maintaining the "on" state are called trithorax-group (trxG) proteins in honor of the trithorax gene, the founding member of this group of regulatory proteins. A large group of proteins with diverse functions make up the trxG. The roles these proteins play in the epigenetic mechanisms that maintain the "on" state appear more complex at this juncture than the roles for PcG proteins in repression. The first complexity is that a very large number of proteins and mechanisms are needed to actively transcribe RNA from any gene. Thus, in contrast to repression, which might be accomplished by comparatively simple mechanisms that block access of all proteins, activation of a gene requires numerous steps, any of which might play a role in maintaining an "on" state. Thus, there are numerous possible stages in which a trxG protein might work.

A second complexity in thinking about trxG proteins is that proteins that function in activation can also, in different contexts, function in repression. This might appear counterintuitive, but depending on the precise architecture of a gene, the same protein performing its function might in one case help a gene become activated, and in another case help a different gene become repressed. At this time it does not appear that trxG proteins are dedicated solely to the maintenance of gene expression, but that these proteins can also play multiple roles in the cell. These complexities make for several interesting unanswered questions. Why are only some of the proteins needed to activate transcription also critical for maintenance of transcription? Do these proteins have functions that are uniquely suited to maintaining the active state? Or are some of these proteins needed for maintenance solely because an evolutionary accident that made them key regulators of gene(s) particularly important to development?

As we will see when we discuss mechanisms of action, some of the trxG proteins are involved in regulating chromatin structure in opposition to the mechanisms used by the PcG proteins. trxG proteins can place covalent posttranslational modifications (PTMs) on chromatin or can alter chromatin by changing the structure and position of the nucleosomes that are the building blocks of chromatin. Other trxG proteins function as part of the transcription machinery. Thus, these proteins are found in a wider variety of complexes than the PcG proteins and are likely to play more complicated roles in epigenetic mechanisms.

Cite as *Cold Spring Harb Perspect Biol* doi: 10.1101/cshperspect.a019349

1 INTRODUCTION

Numerous developmental decisions—including the determination of cell fates—are made in response to transient positional information in the early embryo. These decisions are dependent on changes in gene expression. This allows cells with identical genetic blueprints to acquire unique identities and follow distinct pathways of differentiation. The changes in gene expression underlying the determination of cell fates are heritable; a cell's fate rarely changes once it is determined, even after numerous cell divisions and lengthy periods of developmental time (Fig. 1). Understanding the molecular mechanisms underlying the maintenance of the determined state has long been a goal of developmental and molecular biologists.

Many of the regulatory proteins involved in the maintenance of heritable states of gene expression were identified in studies of *Drosophila* homeotic (Hox) genes. Hox genes encode homeodomain transcription factors that regulate the transcription of batteries of downstream target genes, which in turn specify the identities of body segments (Gellon and McGinnis 1998). In *Drosophila*, Hox genes are found in two gene complexes: the Antennapedia complex (ANT-C), which contains the Hox genes *labial* (*lab*), *Deformed*, *Sex combs reduced* (*Scr*), and *Antennapedia* (*Antp*); and the bithorax complex, which contains the Hox genes *Ultrabithorax* (*Ubx*), *abdominalA* (*abdA*), and *AbdominalB* (*AbdB*; Duncan 1987; Kaufman et al. 1990). Each Hox gene specifies the identity of a particular segment, or group of segments, along the anterior–posterior axis of the developing fly. For example, *Antp* specifies the identity of the second thoracic segment, including the second pair of legs, whereas *Ubx* specifies the identity of the third thoracic segment, including the balancer organs located behind the wings. Thus, the transcription factors encoded by Hox genes function as master regulatory switches that direct the choice between alternative pathways of development.

The transcription of Hox genes must be regulated precisely because dramatic alterations in cell fates can result from their inappropriate expression (Simon 1995; Simon and Tamkun 2002). For example, the derepression of *Antp* in head segments transforms antennae into legs, whereas the inactivation of *Ubx* in thoracic segments transforms balancer organs into wings. In *Drosophila*, the initial patterns of Hox transcription are established early in embryogenesis by transcription factors encoded by segmentation genes. The proteins encoded by segmentation genes—including the gap, pair-rule, and segment polarity genes—subdivide the early embryo into 14 identical segments. These proteins also establish the initial patterns of Hox transcription, the first step toward the development of segments with distinct identities and morphology. The majority of segmentation genes, however, are transiently expressed during early development. Once established, the segmentally restricted patterns of Hox transcription must

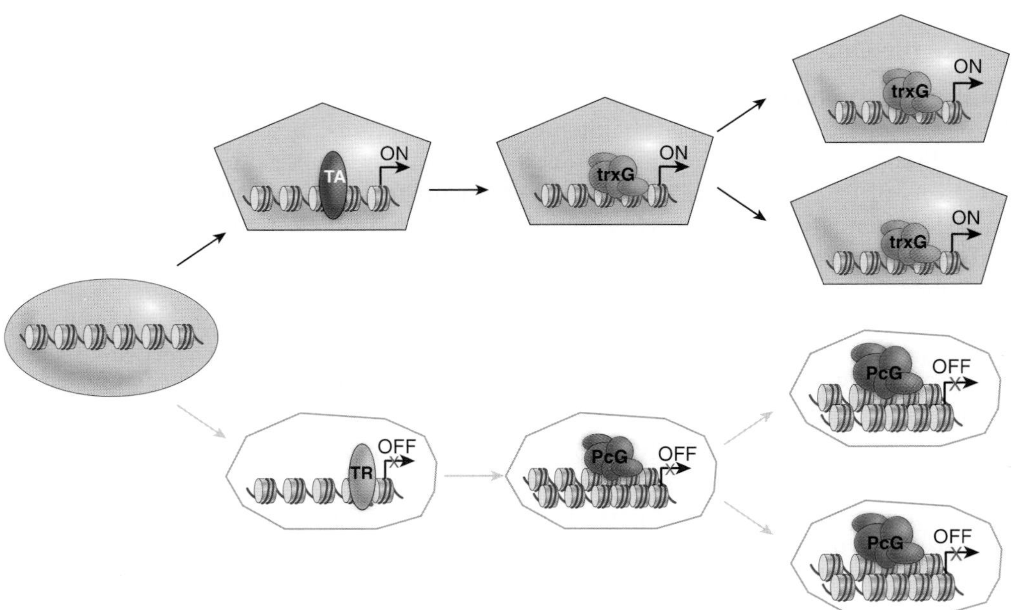

Figure 1. The concept of cellular memory. Schematic illustration highlighting the role of trxG complexes in maintaining heritable states of active gene expression in contrast to heritable silencing by PcG complexes, as defined originally for the *Drosophila* homeotic (Hox) gene cluster.

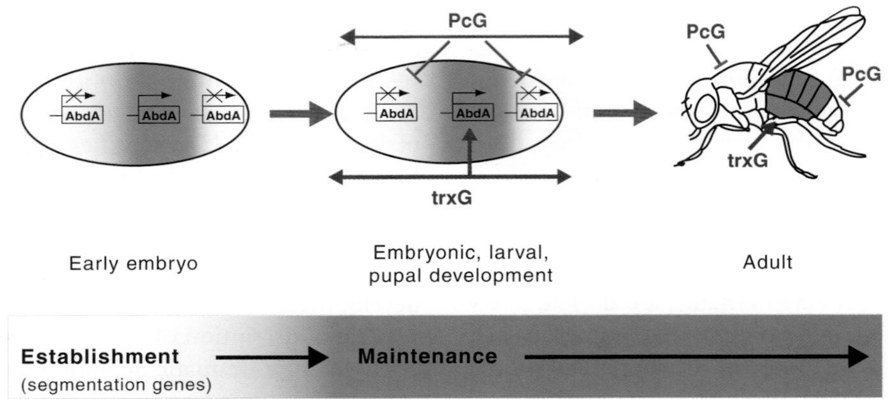

Figure 2. Regulation of Hox transcription. The boundaries of *abd-A* transcription and other Hox genes are established by segmentation proteins. These include the products of gap and pair-rule genes, which subdivide the embryo into 14 identical segments. During subsequent development, the "off" or "on" states of Hox transcription are maintained by the ubiquitously expressed members of the trxG of activators and the PcG of repressors via mechanisms that remain poorly understood.

be maintained throughout subsequent embryonic, larval, and pupal stages to maintain the identities of the individual body segments. This function is performed by two other groups of regulatory proteins: the Polycomb group of repressors (PcG) and the trithorax group of transcriptional regulators (trxG). The regulation of Hox transcription therefore consists of at least two distinct phases: establishment (by segmentation genes) and maintenance (by PcG and trxG genes; Fig. 2).

1.1 Identification of Genes Involved in the Maintenance of the Determined State

Because of their roles in the maintenance of cell fates, *Drosophila* PcG and trxG genes have been the subject of intense study for decades. As discussed in Ch. 17 (Grossniklaus and Paro 2014), the majority of PcG genes were identified by mutations that cause homeotic transformations due to the failure to maintain repressed states of Hox transcription. A classic example of a phenotype associated with PcG mutations is the transformation of second and third legs to first legs. This homeotic transformation results from the derepression of the ANT-C gene, *Scr*, and is manifested by the appearance of first leg bristles known as sex comb teeth on the second and third legs of the adult. This "Polycomb" or "extra sex combs" phenotype—together with other homeotic transformations resulting from the failure to maintain repression of Hox genes—led to the identification of more than a dozen PcG genes in *Drosophila*. The majority of PcG genes encode subunits of two complexes involved in transcriptional repression: Polycomb repressive complex (PRC) 1 and PRC2 (Levine et al. 2004). PRC1 and PRC2 are targeted to the vicinity of Hox (and other) promoters via *cis*-regulatory elements known as Polycomb-response elements (PREs) in the *Drosophila* genome (see Ch. 17 [Grossniklaus and Paro 2014]). A large body of evidence suggests that PcG complexes repress transcription by modulating chromatin structure (Francis and Kingston 2001; Ringrose and Paro 2004).

Members of the trxG were initially identified by mutations that mimic loss-of-function Hox mutations in *Drosophila* (Fig. 3) (Kennison 1995). Genes identified by this phenotype include *trithorax* (*trx*); *absent, small, or homeotic 1* (*ash1*), *absent, small, or homeotic 2* (*ash2*), and *female-sterile homeotic* (*fsh*). Mutations in *trx*—the founding member of the trxG—for example, cause the partial transformation of halteres to wings (because of decreased *Ubx* transcription), first legs to second legs (because of decreased *Scr* transcription), and posterior abdominal segments to more anterior identities (because of decreased *abdA* and *AbdB* transcription). Numerous other trxG members were identified in screens for extragenic suppressors of *Pc* (*Su(Pc)*) mutations (Kennison and Tamkun 1988). The rationale behind these genetic screens was that a reduction in the level of a protein that maintains an active state should compensate for a reduction in the level of a PcG repressor (Fig. 4). *brahma* (*brm*) and numerous other *Su(Pc)* loci were identified using this approach, bringing the total number of trxG members to more than 16 (Table 1). Many other proteins have been classified as trxG members based on other, less stringent criteria, including sequence homology with known trxG proteins, physical association with trxG proteins, biochemical activity, or effects on Hox transcription in vitro or in vivo.

The functional relationship between members of the trxG is complicated, as is the mechanistic connection be-

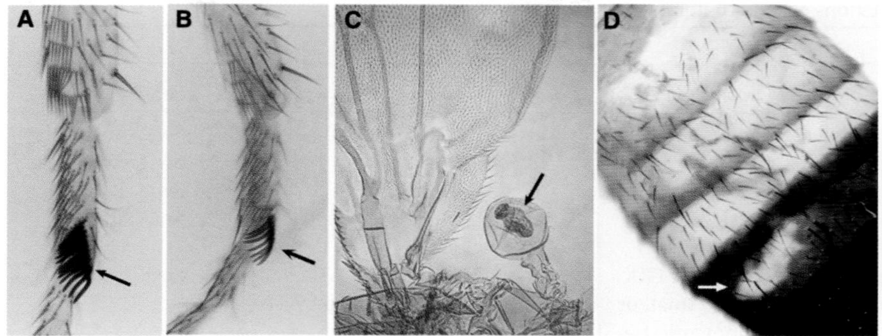

Figure 3. Examples of developmental cell fate transformations associated with mutations in *Drosophila* trxG genes. (*A*) Wild-type first leg. The sex comb, unique to the first leg, is marked by an arrow. (*B*) A patch of *kis* mutant tissue (marked by an arrow) is partially transformed from the first leg to the second leg because of decreased *Scr* transcription, albeit incomplete, as evidenced by a reduction in the number of sex comb teeth. (*C*) A patch of *mor* mutant tissue (marked by an arrow) displays the partial transformation from balancer organ to wing because of decreased Ubx expression. (*D*) A patch of *kis* mutant tissue (marked by an arrow) in the fifth abdominal segment is partially transformed to a more anterior identity because of decreased *Abd-B* expression, as evidenced by the loss of the dark pigmentation characteristic of this segment. (*A,B,D*, Reprinted from Daubresse et al. 1999.)

tween trxG function and the maintenance of cell fate. There are numerous mechanisms via which a protein might maintain an appropriately high level of expression of a homeotic gene (the genetic definition of a trxG protein) without being a devoted transcriptional activator, or a protein devoted to epigenetic control. Formal possibilities for trxG function (in addition to the ability to directly activate transcription) include the ability to increase function of

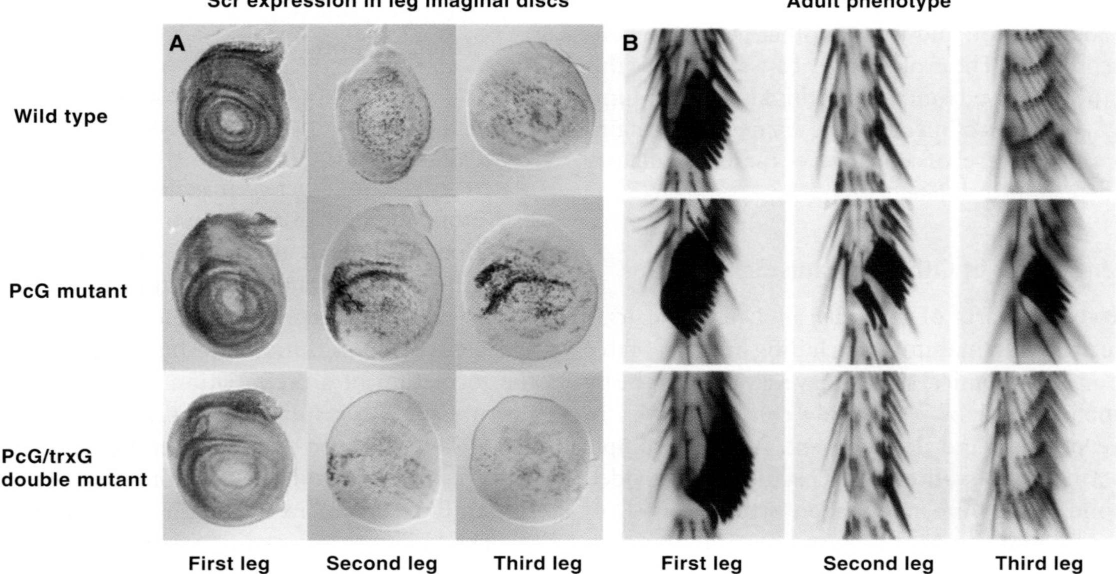

Figure 4. trxG mutations block the derepression of Hox genes in PcG mutants. (*A*) Leg imaginal discs stained with antibodies against the protein encoded by the Hox gene, *Scr*, which specifies the identity of the labial and first thoracic segments, including the first leg. (*B*) Basitarsal segments of the legs of wild-type and mutant adults. Note the presence of sex comb teeth on the first leg, but not the second and third legs of wild-type adults. The *Scr* gene is partially derepressed in the second and third leg discs, in which it is normally silent, in individuals heterozygous for mutations in PcG genes leading to the appearance of ectopic sex comb teeth on the second and third legs. These phenotypes are suppressed by mutations in *brm* and many other trxG genes. (*A*, Reprinted, with permission, from Tamkun et al. 1992, © Elsevier; *B*, portion modified, with permission, from Kennison 2003, © Elsevier.)

Table 1. Biochemical functions of trxG proteins

Known function	Organism			Complexed with non-trxG proteins?
	Drosophila	Human	Yeast	
ATP-dependent chromatin remodeling	BRM	BRG1/HBRM	Swi2/Snf2, Sth1	Yes (5–10)[a]
	OSA	BAF250	Swi1/Adr6	Yes (5–10)
	MOR	BAF155, BAF170	Swi3, Rsc8	Yes (5–10)
	SNR1	hSNF5/INI1	Snf5, Sfh1	Yes (5–10)
	Kismet (KIS)	CHD7	–	NK
Histone methyltransferases	Trithorax (TRX)	MLL1, MLL2, MLL3	Set1	Yes (5–20)
	Absent, small or homeotic 1(ASH1)	MlLL4, hSET1 hASH1	–	NK
Mediator subunits	Kohtalo (KTO)	TRAP230	Srb8	Yes (13–24)
	Skuld (SKD)	TRAP240	Srb9	Yes (13–24)
Cohesin subunit	Verthandi (VTD)	Rad21	Scc1/Rad21	Yes (>3)
Transcription factor	Trithorax-like (TRL)	BTBD14B	–	No
Growth factor receptor	Breathless (BTL)	FGFR3	–	NK
Other	Sallimus (SLS)	Titin		NK
	ASH2	hASH2L[b]	Bre2	Yes (5–20)

[a]BRM, OSA, MOR, and SNR1 can all be found in stable association with each other in a single complex.
[b]Relatively low sequence similarity to ASH2.
NK, not known.

direct activators, to block function of PcG repressors, and to create a "permissive" chromatin state that facilitates the function of numerous other regulatory complexes. Furthermore, as discussed in Section 2, some trxG proteins play complicated mechanistic roles that on some genes contribute to activation and that on other genes can contribute to repression. The evolutionary conservation of this family and the conserved functions of this family offer hints concerning what types of mechanisms are needed to maintain the appropriate level of activation of master regulatory genes that determine cell fate.

1.2 trxG Members in Other Organisms

Functional counterparts of virtually all *Drosophila* trxG proteins are present in mammals, including humans (Table 1). Genetic and biochemical studies have shown that the fly and mammalian proteins play highly conserved roles in both gene expression and development. A good example of the functional conservation of trxG proteins is provided by MLL1, one of the four mammalian orthologs of *Drosophila trx*. Mutations in MLL1 cause homeotic transformations of the axial skeleton of mice because of the failure to maintain active transcription of Hox genes (Yu et al. 1995; Yu et al. 1998). Both mixed lineage leukemia (MLL) proteins and trx function as histone lysine (K) methyltransferases (KMTs), and direct evidence of functional homology between the two proteins was provided by the use of human MLL to partially rescue developmental defects in flies resulting from the loss of *trx* function (Muyrers-Chen et al. 2004). Thus, the mechanisms underlying the maintenance of the determined state have been highly conserved during evolution.

1.3 trxG Proteins Play Diverse Roles in Eukaryotic Transcription

The trxG of activators is a large and functionally diverse group of regulatory proteins. This may reflect the complexity of eukaryotic transcription, which involves highly regulated interactions between gene-specific transcriptional activators, the numerous components of the general transcription machinery, and the DNA template that is transcribed. Transcriptional activation involves the binding of sequence-specific activating proteins, the recruitment of the general transcription machinery by those proteins, the formation of a preinitiation complex in which RNA polymerase II is bound to the promoter, the opening of the DNA helix near the promoter, the efficient escape of RNA polymerase from the promoter, and efficient elongation of RNA polymerase through the gene.

The ability to maintain an active transcriptional state might involve any of the numerous steps required for activation, as on any given gene, different steps might play a rate-determining role for transcriptional activity. The packaging of eukaryotic DNA into chromatin also provides another level at which trxG proteins can regulate transcription. Nucleosomes and other components of chromatin tend to inhibit the binding of general and gene-specific transcription factors to DNA, as well as inhibiting the elon-

 Cite as *Cold Spring Harb Perspect Biol* doi: 10.1101/cshperspect.a019349

gation of RNA polymerase. Alterations in chromatin structure, including changes in the structure or positioning of nucleosomes, can influence virtually every step in the process of transcription.

Any protein that is required for transcription is required for the maintenance of the active state. Indeed, some trxG proteins play relatively general roles in transcription and are not dedicated solely to the maintenance of the determined state. Other trxG proteins, however, may play specialized roles in this process either by directly counteracting PcG repression or maintaining heritable states of gene activity through DNA replication and mitosis. The latter class of trxG proteins is of particular interest to developmental biologists.

2 CONNECTIONS BETWEEN trxG PROTEINS AND CHROMATIN

Genetic studies indicating that trxG genes play key roles in transcription and development has stimulated significant work to understand the biochemical function of their products. Many of these experiments have used, as their conceptual basis, the hypothesis that chromatin will be the biologically relevant substrate of trxG proteins. All genes are packaged into chromatin and that packaging can create a compacted and inaccessible state, or can be in an open and permissive state. Both the permissive and inaccessible states may conceivably be heritable. These considerations led to the simple hypothesis that trxG proteins might modulate chromatin structure to affect regulation. Furthermore, as trxG genes were cloned and sequenced it became apparent that some of their products are related to proteins involved in ATP-dependent chromatin remodeling (covered in detail in Ch. 21 [Becker and Workman 2013]) or the covalent modification of nucleosomal histones in other organisms, including the yeast *Saccharomyces cerevisiae*. Thus, although yeast lack either Hox genes or PcG repressors, this organism has provided valuable clues about potential roles for trxG proteins in eukaryotic transcription.

One of the first connections between trxG proteins and chromatin was provided by the discovery that the *Drosophila* trxG gene *brahma* (*brm*) is highly related to yeast SWI2/SNF2 (Tamkun et al. 1992). SWI2/SNF2 was identified in screens for genes involved in mating-type switching (*switch* [*swi*] genes) and sucrose-fermentation (*sucrose nonfermenting* [*snf*] genes). It was subsequently shown to be required for the activation of numerous inducible yeast genes (Holstege et al. 1998; Sudarsanam et al. 2000). The transcription defects observed in *swi2/snf2* mutants are suppressed by mutations in nucleosomal histones, an early observation that first suggested that SWI/2/SNF2 activates transcription by counteracting chromatin repression (Kruger et al. 1995). Biochemical studies conducted in the early 1990s confirmed this hypothesis; SWI2/SNF2 and many other proteins identified in the *swi/snf* screens function as subunits of a large protein complex called SWI/SNF that uses the energy of ATP hydrolysis to increase the ability of proteins to bind to nucleosomal DNA (Cote et al. 1994; Imbalzano et al. 1994; Kwon et al. 1994). SWI2/SNF2 has ATPase activity and thus functions as the "engine" of this chromatin-remodeling machine; other subunits of the SWI/SNF complex mediate interactions with regulatory proteins or its chromatin substrate (Phelan et al. 1999).

Another connection between trxG and chromatin was suggested by the presence of SET domains in the trxG proteins, Trithorax (TRX) and Absent, small or homeotic (ASH1). The SET domain was originally defined by a stretch of amino acids that shows homology between Su(var)3-9, Enhancer of Zeste and TRX, the latter two proteins being, respectively, PcG and trxG members. In the late 1990s the SET family of proteins was shown to have lysine methyltransferase (KMT) activity. Su(var)3-9 methylates histone H3K9 (H3K9me), whereas Enhancer of Zeste methylates H3K27 (H3K27me; Rea et al. 2000; Levine et al. 2004; Ringrose and Paro 2004). As discussed elsewhere, H3K9 methylation promotes heterochromatin assembly, whereas H3K27 methylation appears to be required for PcG repression (see Ch. 12 [Elgin and Reuter 2013] and Ch. 17 [Grossniklaus and Paro 2014], respectively, for more detailed discussion). The presence of SET domains in trxG proteins suggested that the methylation of histone tails might also be important for the maintenance of active transcriptional states.

These findings, together with the growing realization that chromatin-remodeling and -modifying enzymes play key roles in transcriptional activation, motivated biochemists to identify protein complexes that contain trxG proteins and examine the effect of these complexes on chromatin structure in vitro. Other experiments tested the hypothesis that trxG proteins might interact directly with the transcriptional machinery, another well established method of affecting regulation. As described below, these studies revealed that some trxG proteins impact regulation by modifying chromatin structure whereas others function via direct interactions with components of the transcription machinery.

2.1 trxG Proteins Involved in ATP-Dependent Chromatin Remodeling

Chromatin-remodeling complexes have been implicated in a wide variety of biological processes including transcriptional repression and activation, chromatin assembly, the regulation of higher-order chromatin structure, and cellular differentiation (see Ch. 21 [Becker and Workman 2013]). Two trxG members, BRM and KIS, have ATPase domains and members of evolutionarily conserved families of proteins with chromatin-remodeling activity (Table 1). BRM is part of the SWI2/SNF2 subfamily of remodelers, whereas KIS is a member of the CHD7 subfamily. These two subfamilies are both able to move nucleosomes (see Ch. 21 [Becker and Workman 2013]), but differ in mode of action; CHD7 requires naked DNA adjacent to the nucleosome, and thus loosely spaced nucleosomes, to function, whereas the SWI/SNF complexes do not require adjacent naked DNA and are able to function on tightly packed nucleosomes (Bouazoune and Kingston 2012).

The most extensively studied trxG proteins involved in chromatin remodeling are BRM and its human counterparts, BRG1 and HBRM. As predicted, these proteins function as the ATPase subunits of complexes that are highly related to yeast SWI/SNF (Kwon et al. 1994; Wang et al. 1996). SWI/SNF complexes contain between eight and 15 subunits and have been highly conserved during evolution (Fig. 5). A second *Drosophila* trxG gene identified in the screens for suppressors of Pc mutations, *moira* (*mor*), encodes another key member of this ATP-dependent remodeling complex. Homologs of BRM and MOR interact directly to form a functional core of SWI/SNF in humans (Phelan et al. 1999). SWI/SNF complexes are abundant in higher eukaryotes. For example, each mammalian nucleus contains about 25,000 copies of SWI/SNF-subfamily complexes. They are potent remodelers in that they are able to efficiently create access to sites at the center of a mononucleosome, which is energetically difficult because sites at the center of the nucleosome have ~70 bp of constrained nucleosomal DNA on both sides.

Each species studied has at least two distinct SWI/SNF complexes, all of which contain BRM or a highly related chromatin-remodeling ATPase. Another trxG protein, OSA, provides distinction between the complexes, in that one class of complexes (Brahma associate proteins complex [BAP]) contains OSA and another evolutionarily con-

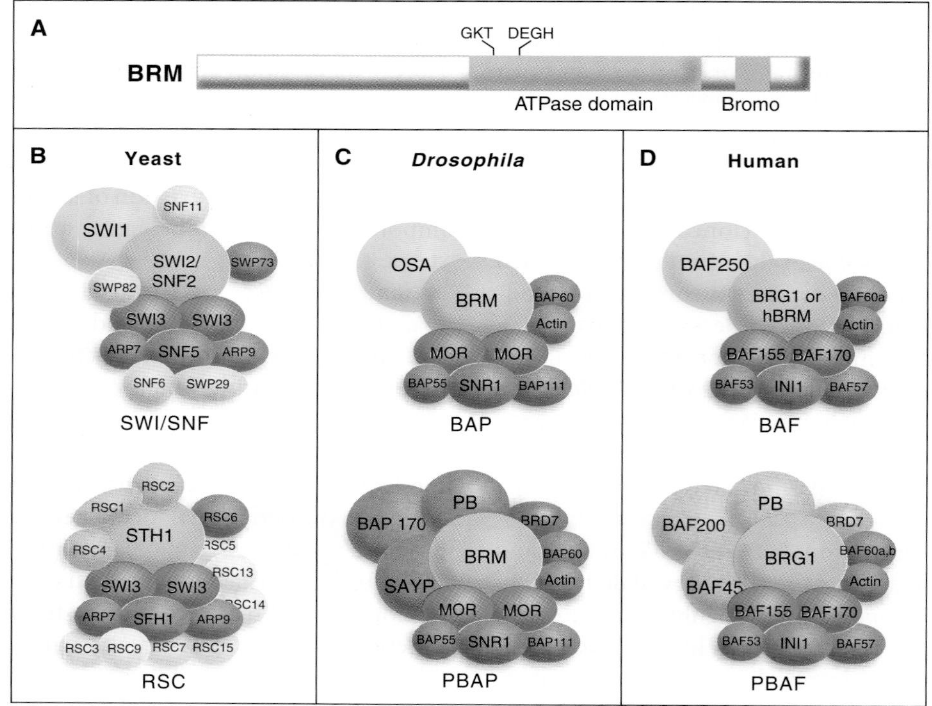

Figure 5. The SWI/SNF family of remodeling complexes. Each complex contains a member of SNF2/SWI2 family of ATPases and at least eight other subunits. (*A*) Schematic diagram of the BRM protein showing the location of the ATPase domain and carboxy-terminal bromodomain (which shows affinity to acetylated lysine residues in histone tails), which are conserved in all SNF2/SWI2 family members. SWI/SNF complexes in yeast (*B*), *Drosophila* (*C*), and humans (*D*) are shown. *Drosophila* trxG proteins (BRM, MOR, and OSA) and their counterparts in other organisms are shown in color. Further information about these complexes and their subunits may be found in Mohrmann and Verrijzer (2005).

Cite as *Cold Spring Harb Perspect Biol* doi: 10.1101/cshperspect.a019349

served complex (polybromo-containing BAP complex [PBAP]) contains a Polybromo-domain protein (Fig. 5) (Mohrmann and Verrijzer 2005). Further evidence distinguishing these complexes comes from fractionation studies that show that several other proteins, including SAYP and BAP170, are PBAP-specific subunits (Chalkley et al. 2008). SAYP has trxG properties, implying that the PBAP complex is directly involved in trxG function. There are many related SWI/SNF family subcomplexes in flies and mammals, and an interesting question, not yet fully resolved, is how each of these contributes to differentiation and trxG functions.

The identification of three different members of the SWI/SNF subfamily of complexes in the trxG screens highlights the importance of this diverse family of large complexes during developmental processes. It is believed that this primarily reflects the role for this complex in activating transcription, although this family of complexes can also be involved in repressive mechanisms (Ch. 21 [Becker and Workman 2013]), SWI/SNF complexes have been implicated in transcriptional activation in every species examined. They can be targeted to genes by interactions with transcriptional activators, remodel nucleosomes to assist in the initial binding of general transcription factors and RNA polymerase II (RNA Pol II), and become targeted later in the activation process to assist with transcriptional elongation. Thus, SWI/SNF complexes appear to function at every step in the process of transcriptional activation, although there appears to be an emphasis on function at the early steps that lead to loading of RNA Pol II.

The trxG gene *kis*, like *brm*, *mor*, and *osa*, was identified in a screen for extragenic suppressors of *Pc*. This suggested that it acts antagonistically to PcG proteins to maintain active states of Hox transcription (Kennison and Tamkun 1988). Genetic studies revealed that *kis* is required for both segmentation and maintenance of Hox transcription during *Drosophila* development (Daubresse et al. 1999). The molecular analysis of *kis* revealed that it encodes several large proteins, including a ~575 kDa isoform (KIS-L), which contains an ATPase domain characteristic of chromatin-remodeling factors (Daubresse et al. 1999; Therrien et al. 2000). Other conserved domains outside the ATPase domain (e.g., bromodomains and chromodomains) contribute to the functional specificity of chromatin-remodeling factors by mediating interactions with nucleosomes or other proteins. BRM and all ATPase subunits of the SWI/SNF subfamily contain a single bromodomain—a protein motif associated with the binding of certain acetylated histones (Fig. 5A). KIS-L contains two chromodomains (protein motifs that bind certain methylated histones are covered in Ch. 6 [Cheng 2014] and Ch. 7 [Patel 2014]) and is therefore more similar to Mi2 and other members of the CHD family of chromatin-remodeling factors. The large size of KIS-L (~575 kDa) has made it difficult to analyze this protein biochemically. Its sequence implies that it is an ATP-dependent nucleosome-remodeling protein, and, as discussed below, its apparent human homolog has been shown to have remodeling activity.

KIS-L is not physically associated with BRM and may function as a monomer or in a distinct protein complex (Srinivasan et al. 2005). The two proteins overlap extensively with each other and RNA Pol II on polytene chromosomes, however, suggesting that both play relatively global roles in transcription (Fig. 6) (Armstrong et al. 2002; Srinivasan et al. 2005). Loss of BRM function in the larval salivary gland blocks a relatively early step in transcription (Armstrong et al. 2002), whereas the loss of KIS-L function leads to a decrease in the level of elongating, but not initiating forms of RNA Pol II (Srinivasan et al. 2005, 2008). These findings suggest that BRM and KIS-L facilitate distinct steps in transcription by RNA Pol II by catalyzing ATP-dependent alterations in nucleosome structure or spacing.

Recent studies on the CHD7 gene in mammals, which is homologous to *kis*, highlight the importance of this protein in development and show that its product functions in a distinct manner to the BRM family of remodelers. Mutation of human CHD7 leads to CHARGE syndrome and other related developmental syndromes that impact approximately one in 10,000 live births (Vissers et al. 2004). Infants with CHARGE syndrome show serious defects in multiple tissues and organs, including the central nervous system, eye, ear, nose, heart, and genitalia (Jongmans et al. 2006). These phenotypes are due to altered gene expression in neural crest cells, where CHD7 and other human trxG

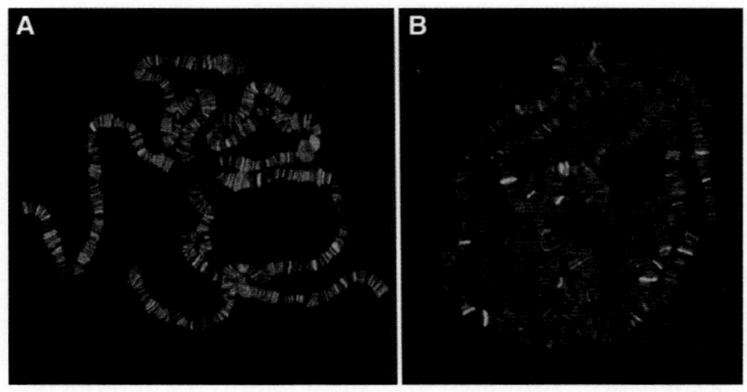

Figure 6. Chromosomal distribution of trxG proteins. The genome-wide distribution of trxG proteins was examined by staining *Drosophila* salivary gland polytene chromosomes with antibodies against BRM (*A*) or TRX (*B*). Consistent with a relatively global role in transcriptional activation, BRM is associated with hundreds of sites in a pattern that overlaps extensively with RNA Pol II. In contrast, strong TRX signals are detected at a much smaller number of sites on polytene chromosomes.

proteins synergistically regulate gene-expression programs critical for development (Bajpai et al. 2010). Mutations in CHD7 also lead to a variety of other human developmental defects and more than 500 distinct mutations in CHD7 have been identified in human patients, emphasizing the importance of this trxG protein to human development and human pathologies (Janssen et al. 2012).

These dramatic developmental impacts are likely to be related to defects in the ability of CHD7 to remodel nucleosomes. The CHD7 protein is a potent ATP-dependent remodeler that does not require other proteins for function in vitro. As expected by its difference in domain structure from the BRM-related proteins, and its distinct impacts on gene expression, this protein shows several functional differences in remodeling behavior from the SWI/SNF subfamily of remodelers (Bouazoune and Kingston 2012). CHD7 requires excess DNA to remodel a nucleosome, whereas BRM family proteins do not. CHD7 is also less adept at opening sites internal to the nucleosome and "slides" nucleosomes in a different manner than BRM family proteins. Many of the mutations found in human patients with CHARGE syndrome truncate the CHD7 protein in a manner that is expected to impair or eliminate remodeling activity. Missense CHD7 mutations in patients were shown to affect remodeling activity, suggesting it is needed for appropriate human development. The comparison between CHD7 and SWI/SNF function shows that trxG phenotypes can be caused by mutations that affect two distinct classes of remodeling machinery operating through different mechanisms, implying that both of these remodeling functions are required for appropriate developmental regulation.

An important goal for future research is to unravel the role that ATP-dependent remodeling plays in the maintenance of the activated state. It is intriguing that four trxG members (BRM, MOR, OSA, and KIS) are known members of two distinct ATP-dependent chromatin-remodeling complexes, but that none of the other numerous ATP-dependent remodeling complexes have been identified in genetic screens for *Drosophila* trxG proteins. Two predominant hypotheses that could explain this, which are not mutually exclusive, are (1) the BRM and KIS chromatin-remodeling complexes are targeted to genes important for developmental progression, and (2) they have special remodeling characteristics that are uniquely required for maintenance. Although the BRM family of remodeling complexes shows distinct mechanism from other families, consistent with the second possibility, CHD7 and, presumably, KIS function similarly to remodeling proteins in the ISWI family, yet generate distinct phenotypes when mutated from those generated by ISWI mutations, which do not impact maintenance. Thus, it is possible that generic ATP-dependent remodeling is required for all active states, and maintenance of the active state of developmentally important genes happens to require these trxG members because they are targeted to these genes.

It is also intriguing to think about the mechanisms that remodelers might use to contribute to epigenetic regulation of the active state. At least three classes of mechanisms can be envisioned that might apply. First, remodeling functions might be required in a somewhat indirect manner to facilitate the binding (or rebinding following replication) of gene-specific activating proteins that are needed to maintain active transcription. In this case, the remodelers would not be the "brains" of the epigenetic mechanism, but instead would act as a necessary tool to allow the proteins required to function efficiently. Second, remodelers could work alone or with histone chaperones to evict nucleosomes from a region, and this lack of occupancy by nucleosomes would hypothetically cause the region to remain nonnucleosomal following replication. As mentioned above, the ability of the replication/nucleosome deposition machinery to accurately recapitulate nucleosome modification or location is an important unanswered issue in epigenetics (covered in more detail in Ch. 22 [Almouzni and Cedar 2014]). Finally, remodeling machineries could reposition nucleosomes to create a structure that is amenable to activation. This latter mechanism has experimental support from studies of the albumin gene (Chaya et al. 2001; Cirillo et al. 2002). Several DNA-binding factors are required to maintain activity of this key gene in the liver. One of these factors, FoxA, binds directly to a nucleosome and the specific nucleosome-FoxA architecture is key to maintaining the active state of the albumin gene. Although it is not clear whether there is a required role for ATP-dependent remodeling to position this specific nucleosome in the liver, this example shows the potential for specific nucleosome positioning to play a key epigenetic role.

2.2 trxG Proteins That Covalently Modify Nucleosomal Histones

A second common method of regulating gene expression involves covalent PTM of the amino-terminal tails of the core histones that comprise the protein component of the nucleosome. These tails, which protrude from the surface of the nucleosome, can mediate interactions with other nucleosomes as well as with a wide variety of structural and regulatory proteins (illustrated in Fig. 12 of Ch. 3 [Allis et al.]). The covalent modification of histone tails by acetylation, methylation, or phosphorylation can help target regulatory complexes to chromatin and also directly change the ability of nucleosomes to compact into repressive structures by changing the charge on the tails. Covalent modification might also provide a mark to help maintain

Cite as *Cold Spring Harb Perspect Biol* doi: 10.1101/cshperspect.a019349

a specific regulated state, as covalently modified histones have the potential to divide to the two daughter strands and thereby propagate the information contained in the covalent mark to both mother and daughter cells following replication. The involvement of covalent modification in maintenance remains controversial, as the time frame for turnover of histones in vivo is under investigation and might be too fast for this type of mechanism to function, and the mechanisms that regulate transfer of histones to the daughter strand are also not understood (see also Ch. 22 [Almouzni and Cedar 2014]).

Several trxG proteins are able to covalently modify histone tails, and these proteins are frequently found in complexes that are able to perform more than one type of modification reaction. For example, *Drosophila* TRX and its counterparts in other organisms methylate histone H3 at lysine 4 (H3K4); this covalent mark is tightly associated with active genes in a wide variety of organisms including yeast, flies, and humans (Fig. 7A). The yeast homolog of *Drosophila* TRX, Set1, is a subunit of large complex (COMPASS or Set1C) that methylates H3K4. COMPASS contains six other subunits, including the yeast counterpart of a second trxG protein, ASH2 (Miller et al. 2001; Roguev et al. 2001). Six KMTs related to *Drosophila* TRX are present in humans, including hSET1A, hSET1B, and the MLL1-4 proteins. hSET1A and hSET1B are the histone KMT subunits of human COMPASS, whereas MLL1-4 are subunits of related complexes (Shilatifard 2012). By promoting H3K4 trimethylation (H3K4me3) in the vicinity of promoters, COMPASS is thought to promote the transition of RNA Pol II to active elongation (Ardehali et al. 2011). H3K4 methylation has also been implicated in the maintenance of active gene expression in yeast, based on the timing of its appearance and removal at active genes (Santos-Rosa et al. 2002; Pokholok et al. 2005).

It has widely been assumed that TRX maintains active states of transcription by methylating H3K4 in the vicinity of promoters, based on the above observations. The dSet1 KMT (a subunit of *Drosophila* COMPASS) is responsible for the bulk of H3K4me3 in flies, however, and the loss of TRX function does not significantly alter H3K4me3 (Ardehali et al. 2011; Hallson et al. 2012). These findings suggest that the histone KMT activity of TRX is not as important for the maintenance of the active state as initially suspected.

The proteolytic cleavage of TRX leads to the production of two proteins: TRX-C and TRX-N. TRX-C, which retains the SET domain, is associated with PREs and promoters (Schuettengruber et al. 2009; Schwartz et al. 2010). TRX-N, which lacks histone KMT activity, is associated with broad regions of active genes (Schuettengruber et al. 2009; Schwartz et al. 2010), suggesting that it activates transcription via mechanisms other than histone methylation. There is some evidence that TRX/MLL complexes are also involved in the acetylation of histones. In flies, TRX is associated with dCBP, a histone lysine (K) acetyltransferase (frequently abbreviated to HAT or KAT) with broad specificity that is involved in activation (Petruk et al. 2001; Tie et al. 2009). Acetylation is also known to prevent the methylation of residues such as H3K9 and H3K27 that direct repression of the template, providing yet another mechanism for maintaining active states (Fig. 7A,B).

The trxG protein ASH1 also has histone KMT activity. ASH1 methylates H3K4, H3K9, H4K20, and H3K36 in vitro (Beisel et al. 2002; Byrd and Shearn 2003; Tanaka et al. 2007), but its H3K36 methyltransferase activity appears to be particularly important for the maintenance of active states (Yuan et al. 2011; Dorighi and Tamkun 2013). ASH1 colocalizes with TRX-N over broad domains of derepressed PcG targets (Schwartz et al. 2010), but it is not understood how the activities of ASH1 and TRX are coordinated. ASH1 has been seen to colocalize and associate with the CBP family of acetyltransferases (Bantignies et al. 2000), once again suggesting that methylation and acetylation go hand in hand.

A distinct type of modification, most normally found on the carboxy-terminal tail of histones H2A and H2B, is monoubiquitination. As opposed to the covalent modifications discussed above, ubiquitin is a large peptide that not only impacts charge but has the potential for significant steric effects. The OSA subunit of the PBAP family of complexes has been implicated in promoting ubiquitination of histone H2B on lysine 120 (Li et al. 2010), a modification associated with activated transcription. The mechanism via which H2B ubiquitination functions during transcriptional activation is not understood, although it appears to be an ancient function as this role is conserved from *S. cerevisiae* through humans.

There are numerous fascinating questions concerning how covalent modification of histones might contribute to trxG function. What functional role do the marks play? Covalent modification can contribute to epigenetic regulation via a wide spectrum of mechanisms. Methylation and acetylation marks might serve to directly alter chromatin compaction (sometimes termed *cis*-effects, as in Fig. 14 of Ch. 3 [Allis et al.]). The ability of chromatin to enter a compacted state, which is generally assumed to be repressive for transcription, is influenced by the charge distribution on the histone tails. Modifications that occur on lysine (e.g., acetylation) can eliminate the positive charge normally found with this residue, and therefore might directly decrease the ability of nucleosomes to form compacted structures, thus increasing the ability of the template to be transcribed (discussed in Ch. 4 [Marmorstein and Zhou 2014]).

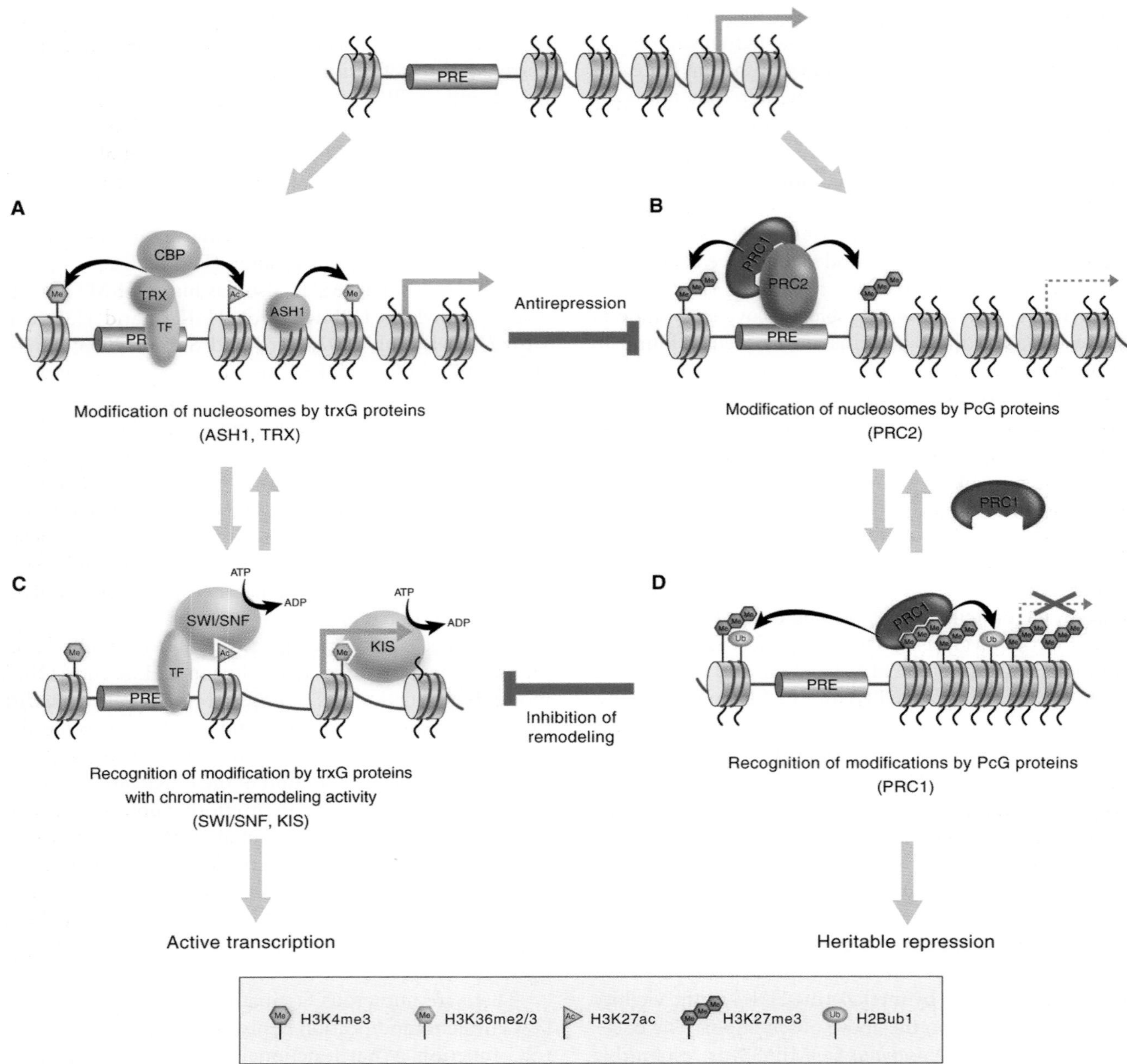

Figure 7. trxG and PcG functions and interactions. Both trxG and PcG families include proteins that covalently modify histones and those that noncovalently modify chromatin. Covalent modifications on histones can promote or block the binding or activity of trxG complexes (e.g., SWI/SNF and KIS), PcG complexes (e.g., PRC1 and PRC2), or other factors involved in the maintenance of active or repressed states. Binding by these latter complexes has the potential to lead to further covalent modification, thus leading to iterative cycles of covalent modification and recognition of the covalent marks.

Histone PTMs have been proposed to create strong binding sites for complexes that direct transcriptional activation. These covalent modifications are able to create specific "knobs" on the surface of the nucleosome that fit into pockets on the complexes that promote activation, thus increasing binding energy and function of these complexes. For example, acetylation of histone tails increases binding by homologs of the BRM protein, thus promoting ATP-dependent remodeling of acetylated templates (Fig. 7C) (Hassan et al. 2001). This type of mechanism, frequently referred to as the "histone code" or *trans*-effects of covalent histone modifications, has the potential to be a central

 Cite as *Cold Spring Harb Perspect Biol* doi: 10.1101/cshperspect.a019349

epigenetic function. Further studies are needed to determine which marks created by trxG proteins enhance binding of which complexes, and to determine the extent to which the energy of binding to a single modified residue can influence function and targeting. More research is also needed to determine the temporal order of addition of the marks and whether they are maintained across mitosis.

The flip side of this mechanism is that the marks could inhibit binding by repressive complexes. A covalent mark on a key residue required for optimal binding by a repressive complex could strongly inhibit binding by the repressive complex. For example, it is known that binding by repressive complexes is increased by methylation of histone H3 at K9 and K27 (Khorasanizadeh 2004). Acetylation of these residues would both block methylation and create an ill-shaped knob on the histone that impairs binding by the repressive complex. As discussed in Section 6, the ability of modifications to influence function of other complexes can cut in both directions, increasing the potency of this potential mode of epigenetic regulation.

These mechanisms by which PTMs achieve an effect are not mutually exclusive but, in fact, are likely to work together to help maintain an active state. Marks that chemically increase the ability to form a compacted state (a *cis*-effect) might also increase the ability of complexes to bind (a *trans*-effect), further promoting a compacted state. Conversely, marks that chemically decrease compaction might increase binding of complexes that also decompact nucleosomes. This mechanistically parsimonious use of covalent marks to alter several characteristics of chromatin structure and the ability of regulatory complexes to bind could create a powerful means of maintaining an active state.

3 CONNECTIONS BETWEEN trxG PROTEINS AND THE GENERAL TRANSCRIPTION MACHINERY

The theme that trxG proteins frequently are found in the same complex is continued with the *skuld* (*skd*) and *kohtalo* (*kto*) proteins. These two *Drosophila* proteins are homologs of the mammalian proteins identified biochemically as TRAP240 (Skuld) and TRAP230 (Kohtalo), which are both members of the Mediator complex (Janody et al. 2003). The Mediator complex is a large complex that functions at the interface between gene-specific activator proteins and formation of the preinitiation complex that contains RNA Pol II (Lewis and Reinberg 2003). Thus, these proteins are involved in general activation processes, much in the same way as SWI/SNF-family remodelers are involved in general activation. SKD and KTO might have some special function involved in maintenance, as other components of the Mediator complex were not identified in screens for trxG genes. The observation that SKD and KTO interact with each other, and *skd kto* double mutants have the same phenotype as either single mutant, has led to the hypothesis that the two proteins together form a functional module that somehow alters Mediator action (Janody et al. 2003).

4 CONNECTIONS BETWEEN trxG PROTEINS AND COHESIN

trxG proteins can influence levels of chromatin organization above the level of the nucleosome, as evidenced by the discovery that the *Drosophila* trxG gene, *verthandi* (*vtd*), encodes the Rad21 subunit of cohesin (Hallson et al. 2008). The four core subunits of cohesin, Rad21 (Scc1/Mcd1), Smc1, Smc3, and stromalin (SA/Scc3/Stag2), form a ring that encircles DNA. Cohesin is essential for sister chromatid cohesion during mitosis and meiosis (Haering et al. 2008; Nasmyth and Haering 2009). Cohesin also plays important roles in transcriptional activation and repression in interphase cells by promoting interactions between insulators, enhancers, and promoters via the formation of DNA loops (illustrated in Fig. 1 of Ch. 2 [Kim et al. 2014]; also Wood et al. 2010; Dorsett 2011; Fay et al. 2011; Seitan and Merkenschlager 2012). These and other findings raise the intriguing possibility that the stabilization of long-range chromosomal interactions by cohesin contributes to the maintenance of the active state (Cunningham et al. 2012; Schaaf et al. 2013).

5 BIOCHEMICAL FUNCTIONS OF OTHER trxG PROTEINS

The biochemical activities of the majority of other trxG proteins remain relatively mysterious. Brd2/Ring3, a human counterpart of the *Drosophila* trxG gene *fsh*, encodes a nuclear protein kinase with two bromodomains, implicated in cell-cycle progression and leukemogenesis, but the substrates of the kinase are currently unknown (Denis and Green 1996). The trxG gene *Tonalli* (*Tna*) encodes a protein related to SP-RING finger proteins involved in sumoylation, suggesting that it may regulate transcription via the covalent modification of proteins other than histones (Gutierrez et al. 2003; Monribot-Villanueva et al. 2013). The trxG gene *sallimus* (*sls*) was identified in a screen for extragenic suppressors of *Pc* and subsequently found to encode *Drosophila* Titin (Machado and Andrew 2000). Like its vertebrate counterpart, *Drosophila* Titin helps maintain the integrity and elasticity of the muscle sarcomere. In addition, Titin is a chromosomal protein that is required for chromosome condensation and segregation (Machado and Andrew 2000). These findings provide additional evidence of potential roles for trxG proteins in the regulation of higher-order chromatin structure.

6 FUNCTIONAL INTERACTIONS BETWEEN trxG AND PcG PROTEINS

As the basic biochemical activities of many trxG and PcG members have been identified, attention has shifted toward looking at how their activities are coordinated in regulating transcription and maintaining the active state. In spite of the lack of in vitro systems for studying the maintenance of the determined state, good progress has been made toward addressing this issue. One popular hypothesis is that the trxG and PcG members facilitate a sequence of dependent events required for the maintenance of the active or repressed state.

Support for this idea has come from studies of the PcG complexes PRC1 and PRC2; by methylating H3K27, the E(z) histone KMT subunit of PRC2 creates a covalent mark that is directly recognized by the chromodomain of the Pc subunit of PRC1 (Jacobs and Khorasanizadeh 2002; Min et al. 2003). Thus, one PcG complex appears to directly promote the binding of another PcG complex to chromatin. This simple model has been complicated by recent studies showing that not all complexes in the PRC1 family contain Pc and that methylation of H3K27, although involved in targeting PRC1, is neither necessary nor sufficient for targeting (Margueron and Reinberg 2011; Tavares et al. 2012). The basic concept that one complex can help target the function of another complex nonetheless remains an important paradigm.

By analogy with the PcG complex paradigm described in the previous paragraph, it is possible that trxG-mediated covalent modification of nucleosomes by members with histone KMT or KAT activities (e.g., TRX or ASH1) could directly regulate the activities of trxG members involved in ATP-dependent chromatin remodeling (e.g., the BRM or KIS proteins) (Fig. 7). Consistent with this possibility, BRM and other subunits of SWI/SNF complexes contain bromodomains that can directly interact with acetylated histone tails. KIS and its human homolog CHD7 contain chromodomains that may directly interact with methylated histone tails. Indeed global ChIP analyses have detected colocalization of CHD7 with methylated H3K4, leading to the possibility that there could be a direct interaction between CHD7 and this mark (Schnetz et al. 2009, 2010). This model, which is supported by studies of chromatin-remodeling factors in both yeast and mammals (Agalioti et al. 2000; Hassan et al. 2001), is particularly attractive because it provides a mechanism by which a heritable histone modification could perpetuate a constitutively "open" chromatin configuration that is permissive for active transcription.

Another important issue concerns the functional relationship between PcG repressors and trxG activators. Do these regulatory proteins have independent roles in activation and repression or do they act in direct opposition to maintain the heritable state? One interesting property of PcG proteins is that they are capable of repressing transcription when tethered near virtually any gene transcribed by RNA Pol II. trxG members that play global roles in transcription, including BRM, KIS, and other trxG members involved in chromatin remodeling, are thus excellent candidates for direct targets of PcG repressors (Fig. 7). PRC1 does apparently block access of the SWI/SNF-family complex to the chromatin template (Francis et al. 2001). This is consistent with the notion that a mechanism to achieve PcG repression is to prevent ATP-dependent remodeling by trxG members. The Brahma complex and PRC1 are further connected by the fact that both directly interact with the Zeste protein, which plays a complicated role in the regulation of gene expression in *Drosophila* by somehow helping to direct cross-talk between the two complexes.

A second protein that connects PcG proteins and trxG proteins is the GAGA factor, which is encoded by the *Trithorax-like (Trl)* gene and is thus a trxG member (Farkas et al. 1994). This protein can function as a sequence-specific activator protein at some promoters, but also is a prominent member of the proteins that bind to the PRE. PRE sequences direct PcG function, and at least one PRE can act as a memory module when affixed to a reporter construct, emphasizing the importance of these sequences. PRE sequences bind the GAGA factor and play an important role in PRE function; they tether the GAGA protein to DNA, which has been proposed to enhance binding and function of the PRC1 complex (Mahmoudi and Verrijzer 2001). Thus, the GAGA factor may play key roles in the maintenance of activation (via its transcriptional activating properties) and repression (via interactions with the PcG proteins). An important issue for future research is to understand the rationale for why proteins such as GAGA and Zeste appear to interact with both the activating and repressing machineries of maintenance.

Genetic studies have shown that removal of PcG complexes will reactivate genes even in the absence of TRX and ASH1 (Klymenko and Muller 2004), suggesting that some trxG proteins might use their activity to function as PcG antirepressors, as opposed to direct activators (Fig. 7). How do TRX and ASH1 then counteract PcG repression? trxG proteins may directly antagonize PcG repression by interfering with the binding of PRC2 to nucleosomes via methylated H3K27, inhibiting its histone KMT activity, directly blocking H3K27 methylation, or removing this repressive mark. Recent studies have shown that trxG proteins use each of these mechanisms to antagonize PcG repression. For example, a complex containing TRX and the histone acetyltransferase CBP acetylates H3K27, which prevents

 Cite as *Cold Spring Harb Perspect Biol* doi: 10.1101/cshperspect.a019349

H3K27 methylation by PRC2 (Fig. 7) (Tie et al. 2009). H3K4me3, another histone modification catalyzed by trxG proteins, blocks the interaction of PRC2 with its nucleosome substrate (Schmitges et al. 2011). Finally, the methylation of H3K36, a modification catalyzed by the trxG protein ASH1, inhibits the H3K27 methyltransferase activity of PRC2 (Tanaka et al. 2007; Schmitges et al. 2011; Yuan et al. 2011). These findings explain why repressive H3K27 methylation does not spread over actively transcribed genes and provides a satisfying molecular explanation for the functional antagonism between PcG and trxG proteins that has fascinated geneticists for decades.

7 NONCODING RNAs AND THE trxG

Recent studies suggest that long noncoding RNAs (lncRNAs) might be involved in modulating trxG targeting and function. This proposed mechanism would function in addition to targeting mechanisms that use covalently modified histones and those described in the voluminous literature on sequence-specific DNA-binding factors targeting different members of the trxG family, such as Mediator and SWI/SNF. The use of lncRNAs in targeting factors involved in maintenance is attractive because these RNAs would be expected to segregate to daughter cells and thus help faithfully maintain a regulated state.

Examples of lncRNAs that have been proposed to modulate trxG function include HOTTIP and DBE-T. HOTTIP is expressed from the HOXA cluster and has been proposed to regulate methylation of H3K4 via the interaction with a component (WDR5) of several MLL complexes (Wang et al. 2011). DBE-T has been proposed to directly target Ash1L to the facioscapulohumeral dystrophy (FSHD) locus to guide methylation of H3K36 and thereby assist in derepressing that locus in muscular dystrophy (Cabianca et al. 2012). Interesting areas for future work on these and other candidate lncRNAs include the identification and characterization of the domains that they bind to in trxG complexes, how they function to modulate the activity of trxG complexes, and whether they are able to target complexes in *trans* via direct interaction with targeting sequences in the DNA.

8 trxG PROTEINS AND HUMAN DISEASE

Cancer and other human diseases can result from the failure to maintain heritable states of gene expression. Not surprisingly, many human PcG and trxG genes function as protooncogenes or tumor-suppressor genes. For example, the human trxG gene *MLL*, was originally identified by 11q23 chromosome translocations associated with acute lymphoblastic or myeloid leukemia. Mutations in other mammalian trxG genes are also associated with a variety of cancers (Roberts and Orkin 2004; see Ch. 33 [Zoghbi and Beaudet 2014] for more detail). For example, BRG1, the human counterpart of *Drosophila brm*, physically interacts with the retinoblastoma tumor-suppressor protein; disruption of this interaction leads to increased cell division and malignant transformation in certain human tumor cell lines (Dunaief et al. 1994; Strober et al. 1996). Consistent with a role of BRG1 in tumor suppression, mice heterozygous for mutations in this gene are prone to develop a variety of tumors (Bultman et al. 2000). Mutations in INI1, the human counterpart of the *Drosophila* trxG gene *SNF5-related gene 1* (*SNR1*), also predispose individuals to cancers and have been identified in a large percentage of malignant rhabdoid tumors, an aggressive cancer of children (Versteege et al. 1998; Wilson et al. 2010). Drugs that alter the expression or function of trxG proteins might therefore prove to be useful therapeutic agents in treating human cancer.

9 CONCLUSION AND OUTLOOK

Two of the major questions surrounding the function of trxG proteins remain a matter for conjecture. First, what is the rationale for why a relatively small subset of proteins required for transcriptional activation score genetically as being important for maintenance of the active state? Is this because these proteins play global roles in transcription but are expressed in limiting quantities, or happen, by evolutionary serendipity, to be especially important for developmentally important genes? Second, how can the active state be maintained across replication and mitosis? Replication will create two daughter strands that must both be regulated identically, and mitosis requires condensation and thereby inhibition of transcription of most genes in a cell. What mechanisms create the epigenetic mark(s) that ensure reactivation of a gene on both daughter strands following mitosis?

The majority of trxG proteins are part of complexes that are broadly used in gene expression and most of these complexes also contain many other proteins not in the trxG (see Table 1). This raises the important question as to whether there are special functions that are used for maintenance of active gene expression. It is possible that SWI/SNF remodelers are able to perform a special remodeling function, H3K4 methylation targets special complexes and/or chromatin conformations, and Skuld/Kohtalo alter function of Mediator in a specific manner important for maintenance. Alternatively, it is possible that each of these proteins performs a reaction that is normally used in activation of all types of genes, and these complexes are among those that have emerged as being important for maintenance for a relatively uninteresting reason (e.g., because even relatively subtle changes in the expression of *Drosophila* Hox genes cause homeotic transformations). To resolve these issues, considerably more information is needed about the precise mechanisms that each of these proteins use in activation. For example, do the SWI/SNF complexes harness the energy of ATP hydrolysis in the same manner as other ATP-dependent remodeling complexes, or do they differ in an important way in how this energy is used to alter nucleosome structure? Structural techniques including crystallography, biophysical techniques such as single-molecule analysis and FRET (fluorescence resonance energy transfer), and detailed imaging in vivo might help to shed light on whether there are mechanisms specially designed for epigenetic maintenance of activation. The initial functional studies that have been performed with trxG complexes on simple model templates are just the beginning of the process for answering these important questions.

The epigenetic mechanisms that might maintain an active state are even less well understood. Are covalent marks distributed to help create an active mark? Are nucleosome positions maintained following replication to create open stretches of chromatin, or specially positioned nucleosomes, that increase binding of activators? Does trxG function cause active genes to compartmentalize within the nucleus to regions that favor active transcription? These are all viable hypotheses; more hypotheses exist, and others have not yet even been envisioned. The incredible complexity of the machinery that transcribes DNA offers numerous possibilities for regulation and the development of mechanisms that allow an epigenetic maintenance of active transcription. This intersection of two fields rich in intellectual history, transcriptional activation and epigenetic mechanism, will provide fertile ground for experimentalists for many years.

REFERENCES

*Reference is also in this book.

Agalioti T, Lomvardas S, Parekh B, Yie J, Maniatis T, Thanos D. 2000. Ordered recruitment of chromatin modifying and general transcription factors to the IFN-β promoter. *Cell* **103:** 667–678.

* Almouzni G, Cedar H. 2014. Maintenance of epigenetic information. *Cold Spring Harb Perspect Biol* doi: 10.1101/cshperspect.a019372.

Ardehali MB, Mei A, Zobeck KL, Caron M, Lis JT, Kusch T. 2011. *Drosophila* Set1 is the major histone H3 lysine 4 trimethyltransferase with role in transcription. *EMBO J* **30:** 2817–2828.

Armstrong JA, Papoulas O, Daubresse G, Sperling AS, Lis JT, Scott MP, Tamkun JW. 2002. The *Drosophila* BRM complex facilitates global transcription by RNA polymerase II. *EMBO J* **21:** 5245–5254.

Bajpai R, Chen DA, Rada-Iglesias A, Zhang J, Xiong Y, Helms J, Chang CP, Zhao Y, Swigut T, Wysocka J. 2010. CHD7 cooperates with PBAF to control multipotent neural crest formation. *Nature* **463:** 958–962.

Bantignies F, Goodman RH, Smolik SM. 2000. Functional interaction between the coactivator *Drosophila* CREB-binding protein and ASH1, a member of the trithorax group of chromatin modifiers. *Mol Cell Biol* **20:** 9317–9330.

* Becker PB, Workman JL. 2013. Nucleosome remodeling and epigenetics. *Cold Spring Harb Perspect Biol* **5:** a017905.

Beisel C, Imhof A, Greene J, Kremmer E, Sauer F. 2002. Histone methylation by the *Drosophila* epigenetic transcriptional regulator Ash1. *Nature* **419:** 857–862.

Bouazoune K, Kingston RE. 2012. Chromatin remodeling by the CHD7 protein is impaired by mutations that cause human developmental disorders. *Proc Natl Acad Sci* **109:** 19238–19243.

Bultman S, Gebuhr T, Yee D, La Mantia C, Nicholson J, Gilliam A, Randazzo F, Metzger D, Chambon P, Crabtree G, et al. 2000. A Brg1 null mutation in the mouse reveals functional differences among mammalian SWI/SNF complexes. *Mol Cell* **6:** 1287–1295.

Byrd KN, Shearn A. 2003. ASH1, a *Drosophila* trithorax group protein, is required for methylation of lysine 4 residues on histone H3. *Proc Natl Acad Sci* **100:** 11535–11540.

Cabianca DS, Casa V, Bodega B, Xynos A, Ginelli E, Tanaka Y, Gabellini D. 2012. A long ncRNA links copy number variation to a polycomb/trithorax epigenetic switch in FSHD muscular dystrophy. *Cell* **149:** 819–831.

Chalkley GE, Moshkin YM, Langenberg K, Bezstarosti K, Blastyak A, Gyurkovics H, Demmers JA, Verrijzer CP. 2008. The transcriptional coactivator SAYP is a trithorax group signature subunit of the PBAP chromatin remodeling complex. *Mol Cell Biol* **28:** 2920–2929.

Chaya D, Hayamizu T, Bustin M, Zaret KS. 2001. Transcription factor FoxA (HNF3) on a nucleosome at an enhancer complex in liver chromatin. *J Biol Chem* **276:** 44385–44389.

* Cheng X. 2014. Structural and functional coordination of DNA and histone methylation. *Cold Spring Harb Perspect Biol* **6:** a018747.

Cirillo LA, Lin FR, Cuesta I, Friedman D, Jarnik M, Zaret KS. 2002. Opening of compacted chromatin by early developmental transcription factors HNF3 (FoxA) and GATA-4. *Mol Cell* **9:** 279–289.

Cote J, Quinn J, Workman JL, Peterson CL. 1994. Stimulation of GAL4 derivative binding to nucleosomal DNA by the yeast SWI/SNF complex. *Science* **265:** 53–60.

Cunningham MD, Gause M, Cheng Y, Noyes A, Dorsett D, Kennison JA, Kassis JA. 2012. Wapl antagonizes cohesin binding and promotes Polycomb-group silencing in *Drosophila*. *Development* **139:** 4172–4179.

Daubresse G, Deuring R, Moore L, Papoulas O, Zakrajsek I, Waldrip WR, Scott MP, Kennison JA, Tamkun JW. 1999. The *Drosophila kismet* gene is related to chromatin-remodeling factors and is required for both segmentation and segment identity. *Development* **126:** 1175–1187.

Denis GV, Green MR. 1996. A novel, mitogen-activated nuclear kinase is related to a *Drosophila* developmental regulator. *Genes Dev* **10:** 261–271.

Dorighi KM, Tamkun JW. 2013. The trithorax group proteins Kismet and ASH1 promote H3K36 dimethylation to counteract Polycomb group repression in *Drosophila*. *Development* **140:** 4182–4192.

Dorsett D. 2011. Cohesin: Genomic insights into controlling gene transcription and development. *Curr Opin Genet Dev* **21:** 199–206.

Dunaief JL, Strober BE, Guha S, Khavari PA, Alin K, Luban J, Begemann M, Crabtree GR, Goff SP. 1994. The retinoblastoma protein and BRG1 form a complex and cooperate to induce cell cycle arrest. *Cell* **79:** 119–130.

Duncan I. 1987. The bithorax complex. *Annu Rev Genet* **21:** 285–319.

* Elgin SCR, Reuter G. 2013. Position-effect variegation, heterochromatin formation, and gene silencing in *Drosophila*. *Cold Spring Harb Perspect Biol* **5:** a017780.

Farkas G, Gausz J, Galloni M, Reuter G, Gyurkovics H, Karch F. 1994. The Trithorax-like gene encodes the *Drosophila* GAGA factor. *Nature* **371:** 806–808.

Fay A, Misulovin Z, Li J, Schaaf CA, Gause M, Gilmour DS, Dorsett D. 2011. Cohesin selectively binds and regulates genes with paused RNA polymerase. *Curr Biol* **21:** 1624–1634.

Francis NJ, Kingston RE. 2001. Mechanisms of transcriptional memory. *Nat Rev Mol Cell Biol* **2:** 409–421.

Francis NJ, Saurin AJ, Shao Z, Kingston RE. 2001. Reconstitution of a functional core polycomb repressive complex. *Mol Cell* **8:** 545–556.

Gellon G, McGinnis W. 1998. Shaping animal body plans in development and evolution by modulation of Hox expression patterns. *Bioessays* **20:** 116–125.

* Grossniklaus U, Paro R. 2014. Transcriptional silencing by *Polycomb*-group proteins. *Cold Spring Harb Perspect Biol* **6:** a019331.

Gutierrez L, Zurita M, Kennison JA, Vazquez M. 2003. The *Drosophila* trithorax group gene *tonalli* (*tna*) interacts genetically with the Brahma remodeling complex and encodes an SP-RING finger protein. *Development* **130:** 343–354.

Haering CH, Farcas AM, Arumugam P, Metson J, Nasmyth K. 2008. The cohesin ring concatenates sister DNA molecules. *Nature* **454:** 297–301.

Hallson G, Syrzycka M, Beck SA, Kennison JA, Dorsett D, Page SL, Hunter SM, Keall R, Warren WD, Brock HW, et al. 2008. The *Drosophila* cohesin subunit Rad21 is a trithorax group (trxG) protein. *Proc Natl Acad Sci* **105:** 12405–12410.

Hallson G, Hollebakken RE, Li T, Syrzycka M, Kim I, Cotsworth S, Fitzpatrick KA, Sinclair DA, Honda BM. 2012. dSet1 is the main H3K4 di- and tri-methyltransferase throughout *Drosophila* development. *Genetics* **190:** 91–100.

Hassan AH, Neely KE, Workman JL. 2001. Histone acetyltransferase complexes stabilize SWI/SNF binding to promoter nucleosomes. *Cell* **104:** 817–827.

Holstege FC, Jennings EG, Wyrick JJ, Lee TI, Hengartner CJ, Green MR, Golub TR, Lander ES, Young RA. 1998. Dissecting the regulatory circuitry of a eukaryotic genome. *Cell* **95:** 717–728.

Imbalzano AN, Kwon H, Green MR, Kingston RE. 1994. Facilitated binding of TATA-binding protein to nucleosomal DNA. *Nature* **370:** 481–485.

Jacobs SA, Khorasanizadeh S. 2002. Structure of HP1 chromodomain bound to a lysine 9-methylated histone H3 tail. *Science* **295:** 2080–2083.

Janody F, Martirosyan Z, Benlali A, Treisman JE. 2003. Two subunits of the *Drosophila* mediator complex act together to control cell affinity. *Development* **130:** 3691–3701.

Janssen N, Bergman JE, Swertz MA, Tranebjaerg L, Lodahl M, Schoots J, Hofstra RM, van Ravenswaaij-Arts CM, Hoefsloot LH. 2012. Mutation update on the *CHD7* gene involved in CHARGE syndrome. *Hum Mutat* **33:** 1149–1160.

Jongmans MC, Admiraal RJ, van der Donk KP, Vissers LE, Baas AF, Kapusta L, van Hagen JM, Donnai D, de Ravel TJ, Veltman JA, et al. 2006. CHARGE syndrome: The phenotypic spectrum of mutations in the *CHD7* gene. *J Med Genet* **43:** 306–314.

Kaufman TC, Seeger MA, Olsen G. 1990. Molecular and genetic organization of the antennapedia gene complex of *Drosophila melanogaster*. *Adv Genet* **27:** 309–362.

Kennison JA. 1995. The Polycomb and trithorax group proteins of *Drosophila*: Trans-regulators of homeotic gene function. *Annu Rev Genet* **29:** 289–303.

Kennison JA. 2003. Introduction to Trx-G and Pc-G genes. *Meth Enzymol* **377:** 61–70.

Kennison JA, Tamkun JW. 1988. Dosage-dependent modifiers of Polycomb and Antennapedia mutations in *Drosophila*. *Proc Natl Acad Sci* **85:** 8136–8140.

Khorasanizadeh S. 2004. The nucleosome: From genomic organization to genomic regulation. *Cell* **116:** 259–272.

* Kim T-K, Hemberg M, Gray JM. 2014. Enhancer RNAs: A class of long noncoding RNAs synthesized at enhancers. *Cold Spring Harb Perspect Biol* doi: 10.1101/cshperspect.a018622.

Klymenko T, Muller J. 2004. The histone methyltransferases Trithorax and Ash1 prevent transcriptional silencing by Polycomb group proteins. *EMBO Rep* **5:** 373–377.

Kruger W, Peterson CL, Sil A, Coburn C, Arents G, Moudrianakis EN, Herskowitz I. 1995. Amino acid substitutions in the structured domains of histones H3 and H4 partially relieve the requirement of the yeast SWI/SNF complex for transcription. *Genes Dev* **9:** 2770–2779.

Kwon H, Imbalzano AN, Khavari PA, Kingston RE, Green MR. 1994. Nucleosome disruption and enhancement of activator binding by a human SW1/SNF complex. *Nature* **370:** 477–481.

Levine SS, King IF, Kingston RE. 2004. Division of labor in Polycomb group repression. *Trends Biochem Sci* **29:** 478–485.

Lewis BA, Reinberg D. 2003. The mediator coactivator complex: Functional and physical roles in transcriptional regulation. *J Cell Sci* **116:** 3667–3675.

Li XS, Trojer P, Matsumura T, Treisman JE, Tanese N. 2010. Mammalian SWI/SNF—A subunit BAF250/ARID1 is an E3 ubiquitin ligase that targets histone H2B. *Mol Cell Biol* **30:** 1673–1688.

Machado C, Andrew DJ. 2000. D-Titin: A giant protein with dual roles in chromosomes and muscles. *J Cell Biol* **151:** 639–652.

Mahmoudi T, Verrijzer CP. 2001. Chromatin silencing and activation by Polycomb and trithorax group proteins. *Oncogene* **20:** 3055–3066.

Margueron R, Reinberg D. 2011. The Polycomb complex PRC2 and its mark in life. *Nature* **469:** 343–349.

* Marmorstein R, Zhou M-M. 2014. Writers and readers of histone acetylation: Structure, mechanism, and inhibition. *Cold Spring Harb Perspect Biol* **6:** a018762.

Miller T, Krogan NJ, Dover J, Erdjument-Bromage H, Tempst P, Johnston M, Greenblatt JF, Shilatifard A. 2001. COMPASS: A complex of proteins associated with a trithorax-related SET domain protein. *Proc Natl Acad Sci* **98:** 12902–12907.

Min J, Zhang Y, Xu RM. 2003. Structural basis for specific binding of Polycomb chromodomain to histone H3 methylated at Lys 27. *Genes Dev* **17:** 1823–1828.

Mohrmann L, Verrijzer CP. 2005. Composition and functional specificity of SWI2/SNF2 class chromatin remodeling complexes. *Biochim Biophys Acta* **1681:** 59–73.

Monribot-Villanueva J, Juarez-Uribe RA, Palomera-Sanchez Z, Gutierrez-Aguiar L, Zurita M, Kennison JA, Vazquez M. 2013. TnaA, an SP-RING protein, interacts with Osa, a subunit of the chromatin remodeling complex BRAHMA and with the SUMOylation pathway in *Drosophila melanogaster*. *PloS One* **8:** e62251.

Muyrers-Chen I, Rozovskaia T, Lee N, Kersey JH, Nakamura T, Canaani E, Paro R. 2004. Expression of leukemic MLL fusion proteins in *Drosophila* affects cell cycle control and chromosome morphology. *Oncogene* **23:** 8639–8648.

Nasmyth K, Haering CH. 2009. Cohesin: Its roles and mechanisms. *Annu Rev Genet* **43:** 525–558.

* Patel DJ. 2014. A structural perspective on readout of epigenetic histone and DNA methylation marks. *Cold Spring Harb Perspect Biol* doi: 10.1101/cshperspect.a018754.

Petruk S, Sedkov Y, Smith S, Tillib S, Kraevski V, Nakamura T, Canaani E, Croce CM, Mazo A. 2001. Trithorax and dCBP acting in a complex to maintain expression of a homeotic gene. *Science* **294:** 1331–1334.

Phelan ML, Sif S, Narlikar GJ, Kingston RE. 1999. Reconstitution of a core chromatin remodeling complex from SWI/SNF subunits. *Mol Cell* **3:** 247–253.

Pokholok DK, Harbison CT, Levine S, Cole M, Hannett NM, Lee TI, Bell GW, Walker K, Rolfe PA, Herbolsheimer E, et al. 2005. Genome-wide map of nucleosome acetylation and methylation in yeast. *Cell* **122:** 517–527.

Rea S, Eisenhaber F, O'Carroll D, Strahl BD, Sun ZW, Schmid M, Opravil S, Mechtler K, Ponting CP, Allis CD, et al. 2000. Regulation of chromatin structure by site-specific histone H3 methyltransferases. *Nature* **406:** 593–599.

Ringrose L, Paro R. 2004. Epigenetic regulation of cellular memory by the Polycomb and trithorax group proteins. *Annu Rev Genet* **38:** 413–443.

Roberts CW, Orkin SH. 2004. The SWI/SNF complex—Chromatin and cancer. *Nat Rev Cancer* **4:** 133–142.

Roguev A, Schaft D, Shevchenko A, Pijnappel WW, Wilm M, Aasland R, Stewart AF. 2001. The *Saccharomyces cerevisiae* Set1 complex includes an Ash2 homologue and methylates histone 3 lysine 4. *EMBO J* **20:** 7137–7148.

Santos-Rosa H, Schneider R, Bannister AJ, Sherriff J, Bernstein BE, Emre NC, Schreiber SL, Mellor J, Kouzarides T. 2002. Active genes are tri-methylated at K4 of histone H3. *Nature* **419:** 407–411.

Schaaf CA, Misulovin Z, Gause M, Koenig A, Gohara DW, Watson A, Dorsett D. 2013. Cohesin and polycomb proteins functionally interact to control transcription at silenced and active genes. *PLoS Genet* **9:** e1003560.

Schmitges FW, Prusty AB, Faty M, Stutzer A, Lingaraju GM, Aiwazian J, Sack R, Hess D, Li L, Zhou S, et al. 2011. Histone methylation by PRC2 is inhibited by active chromatin marks. *Mol Cell* **42:** 330–341.

Schnetz MP, Bartels CF, Shastri K, Balasubramanian D, Zentner GE, Balaji R, Zhang X, Song L, Wang Z, Laframboise T, et al. 2009. Genomic distribution of CHD7 on chromatin tracks H3K4 methylation patterns. *Genome Res* **19:** 590–601.

Schnetz MP, Handoko L, Akhtar-Zaidi B, Bartels CF, Pereira CF, Fisher AG, Adams DJ, Flicek P, Crawford GE, Laframboise T, et al. 2010. CHD7 targets active gene enhancer elements to modulate ES cell-specific gene expression. *PLoS Genet* **6:** e1001023.

Schuettengruber B, Ganapathi M, Leblanc B, Portoso M, Jaschek R, Tolhuis B, van Lohuizen M, Tanay A, Cavalli G. 2009. Functional anatomy of polycomb and trithorax chromatin landscapes in *Drosophila* embryos. *PLoS Biol* **7:** e13.

Schwartz YB, Kahn TG, Stenberg P, Ohno K, Bourgon R, Pirrotta V. 2010. Alternative epigenetic chromatin states of Polycomb target genes. *PLoS Genet* **6:** e1000805.

Seitan VC, Merkenschlager M. 2012. Cohesin and chromatin organisation. *Curr Opin Genet Dev* **22:** 93–100.

Shilatifard A. 2012. The COMPASS family of histone H3K4 methylases: Mechanisms of regulation in development and disease pathogenesis. *Annu Rev Biochem* **81:** 65–95.

Simon J. 1995. Locking in stable states of gene expression: Transcriptional control during *Drosophila* development. *Curr Opin Cell Biol* **7:** 376–385.

Simon JA, Tamkun JW. 2002. Programming off and on states in chromatin: Mechanisms of Polycomb and trithorax group complexes. *Curr Opin Genet Dev* **12:** 210–218.

Srinivasan S, Armstrong JA, Deuring R, Dahlsveen IK, McNeill H, Tamkun JW. 2005. The *Drosophila* trithorax group protein Kismet facilitates an early step in transcriptional elongation by RNA Polymerase II. *Development* **132:** 1623–1635.

Srinivasan S, Dorighi KM, Tamkun JW. 2008. *Drosophila* Kismet regulates histone H3 lysine 27 methylation and early elongation by RNA polymerase II. *PLoS Genet* **4:** e1000217.

Strober BE, Dunaief JL, Guha, Goff SP. 1996. Functional interactions between the hBRM/hBRG1 transcriptional activators and the pRB family of proteins. *Mol Cell Biol* **16:** 1576–1583.

Sudarsanam P, Iyer VR, Brown PO, Winston F. 2000. Whole-genome expression analysis of SNF/SWI mutants of *Saccharomyces cerevisiae*. *Proc Natl Acad Sci* **97:** 3364–3369.

Tamkun JW, Deuring R, Scott MP, Kissinger M, Pattatucci AM, Kaufman TC, Kennison JA. 1992. *brahma*: A regulator of *Drosophila* homeotic genes structurally related to the yeast transcriptional activator SNF2/SWI2. *Cell* **68:** 561–572.

Tanaka Y, Katagiri Z, Kawahashi K, Kioussis D, Kitajima S. 2007. Trithorax-group protein ASH1 methylates histone H3 lysine 36. *Gene* **397:** 161–168.

Tavares L, Dimitrova E, Oxley D, Webster J, Poot R, Demmers J, Bezstarosti K, Taylor S, Ura H, Koide H, et al. 2012. RYBP-PRC1 complexes mediate H2A ubiquitylation at Polycomb target sites independently of PRC2 and H3K27me3. *Cell* **148:** 664–678.

Therrien M, Morrison DK, Wong AM, Rubin GM. 2000. A genetic screen for modifiers of a kinase suppressor of Ras-dependent rough eye phenotype in *Drosophila*. *Genetics* **156:** 1231–1242.

Tie F, Banerjee R, Stratton CA, Prasad-Sinha J, Stepanik V, Zlobin A, Diaz MO, Scacheri PC, Harte PJ. 2009. CBP-mediated acetylation of histone H3 lysine 27 antagonizes *Drosophila* Polycomb silencing. *Development* **136:** 3131–3141.

Versteege I, Sevenet N, Lange J, Rousseau-Merck MF, Ambros P, Handgretinger R, Aurias A, Delattre O. 1998. Truncating mutations of hSNF5/INI1 in aggressive paediatric cancer. *Nature* **394:** 203–206.

Vissers LELM, van Ravenswaaij CMA, Admiraal R, Hurst JA, de Vries BBA, Janssen IM, van der Vliet WA, Huys EHLPG, de Jong PJ, Hamel BC, et al. 2004. Mutations in a new member of the chromodomain gene family cause CHARGE syndrome. *Nat Genet* **36:** 955–957.

Wang W, Cote J, Xue Y, Zhou S, Khavari PA, Biggar SR, Muchardt C, Kalpana GV, Goff SP, Yaniv M, et al. 1996. Purification and biochemical heterogeneity of the mammalian SWI-SNF complex. *EMBO J* **15:** 5370–5382.

Wang KC, Yang YW, Liu B, Sanyal A, Corces-Zimmerman R, Chen Y, Lajoie BR, Protacio A, Flynn RA, Gupta RA, et al. 2011. A long noncoding RNA maintains active chromatin to coordinate homeotic gene expression. *Nature* **472:** 120–124.

Wilson BG, Wang X, Shen X, McKenna ES, Lemieux ME, Cho YJ, Koellhoffer EC, Pomeroy SL, Orkin SH, Roberts CW. 2010. Epigenetic antagonism between polycomb and SWI/SNF complexes during oncogenic transformation. *Cancer Cell* **18:** 316–328.

Wood AJ, Severson AF, Meyer BJ. 2010. Condensin and cohesin complexity: The expanding repertoire of functions. *Nat Rev Genet* **11:** 391–404.

Yu BD, Hess JL, Horning SE, Brown GA, Korsmeyer SJ. 1995. Altered Hox expression and segmental identity in Mll-mutant mice. *Nature* **378:** 505–508.

Yu BD, Hanson RD, Hess JL, Horning SE, Korsmeyer SJ. 1998. MLL, a mammalian trithorax-group gene, functions as a transcriptional maintenance factor in morphogenesis. *Proc Natl Acad Sci* **95:** 10632–10636.

Yuan W, Xu M, Huang C, Liu N, Chen S, Zhu B. 2011. H3K36 methylation antagonizes PRC2-mediated H3K27 methylation. *J Biol Chem* **286:** 7983–7989.

* Zoghbi H, Beaudet A. 2014. Epigenetics and human disease. *Cold Spring Harb Perspect Biol* doi: 10.1101/cshperspect.a019497.

C H A P T E R 19

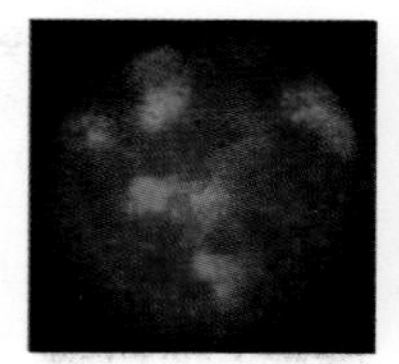

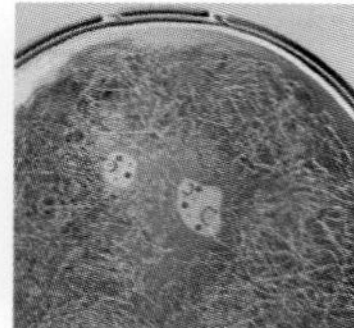

Long-Range Chromatin Interactions

Job Dekker[1] and Tom Misteli[2]

[1]University of Massachusetts Medical School, Worcester, Massachusetts 01605; [2]National Cancer Institute, National Institutes of Health, Bethesda, Maryland 20892

Correspondence: mistelit@mail.nih.gov

SUMMARY

To accommodate genomes in the limited space of the cell nucleus and ensure the correct execution of gene expression programs, genomes are packaged in complex fashion in the three-dimensional cell nucleus. As a consequence of the extensive higher-order organization of chromosomes, distantly located genomic regions on the same or distinct chromosomes undergo long-range interactions. This chapter discusses the nature of long interactions, mechanisms of their formation, and their emerging functional roles in gene regulation and genome maintenance.

Outline

Cite as *Cold Spring Harb Perspect Biol* doi: 10.1101/cshperspect.a019356

OVERVIEW

The higher-order packing of DNA in the cell nucleus is both driven by and leading to physical interactions between genomic regions. The formation of physical interactions between loci, even distantly located ones, is not surprising in itself. The question of whether these interactions are functionally meaningful or merely a by-product of the tight packing of DNA in a limited space is what needs deciphering. Although some interactions are likely purely because of the random proximity of the genome regions in the nucleus, two facts argue very strongly against mere haphazardness and for the existence of recurrent and functionally relevant interactions. The first is that if interactions were functionally irrelevant, then one would predict that they would occur largely randomly in a population. Based on unbiased mapping studies this is not the case—specific, high-frequency interactions can be mapped in a large range of organisms (Splinter and de Laat 2011). The second argument stems from the numerous examples of long-range interactions, both intra- and interchromosomal, that have been molecularly dissected and shown to have significant functional consequences. Among the best characterized are the globin genes, the Hox genes, imprinted loci, and interaction involved in X-inactivation (see discussion below).

Long-range interactions can occur intrachromosomally between regions located on the same chromosome or interchromosomally between regions on distinct chromosomes (Deng and Blobel 2010; Dean 2011). Intrachromosomal interactions have been reported between promoters and terminators over several kilobases, as have interactions between promoters and enhancers located tens of kilobases up to megabases away (Deng and Blobel 2010; Dean 2011). Another type of intrachromosomal interaction is insulator-mediated contacts, which appear to contribute to the organization of the genome into functionally distinct regions by separating differentially regulated regions from each other (Phillips and Corces 2009).

Compared with intrachromosomal interactions, many fewer interchromosomal interactions are known. Interchromosomal interactions are mostly involved in promoting the formation of chromatin domains such as centromere clusters. There are some intriguing examples suggesting that these types of interactions may also be involved in gene regulation, for example, in the control of olfactory receptors, interferon-responsive genes, and also more globally in the establishment of X-inactivation (discussed below and also the topic of Ch. 25 [Brockdorff and Turner 2014]; Ch. 29 [Busslinger and Tarakhovsky 2014]; Ch. 32 [Lomvardas and Maniatis 2014]). Beyond gene regulation, interchromosomal interactions are also now appreciated as playing a particularly critical role in the formation of chromosomal translocations because the rejoining of broken chromosomes requires their physical interaction (Misteli and Soutoglou 2009).

Traditionally, genomic interactions have been detected using fluorescence in situ hybridization (FISH). In this approach, cells are chemically fixed, chromatin denatured, and fluorescently labeled probes hybridized to their targets. This enables the visualization of specific genes, genomic regions, or entire chromosomes by fluorescence microscopy. In these approaches genomic regions are detected as bright fluorescent signals and use of multicolor labels allows detection of a few loci simultaneously. Distances between fluorescent spots can be measured to provide information regarding the proximity of the investigated regions. For example, this approach was used to investigate the organization of the immune *IgH* gene during VDJ recombination (discussed in Ch. 29 [Busslinger and Tarakhovsky 2014]). The advantage of FISH is the ability to interrogate interactions at the single cell level, thus allowing determination of the fraction of cells in a population that harbors a particular interaction. However, FISH methods are limited in the resolution with which they can determine spatial proximity of genomic regions and they do not provide information about actual physical association of two loci. Furthermore, FISH can only be applied in candidate approaches and is not easily amenable to the unbiased discovery of novel genome interactions.

More recently, biochemical approaches that map physical genome interactions have been developed (Sanyal et al. 2011; Splinter and de Laat 2011). These so-called "C" techniques (3C, 4C, 5C, and Hi-C) involve chemical cross-linking of the genomes, fragmenting chromatin (e.g., by digestion with restriction enzymes), religating cross-linked regions, and identifying the interacting regions either by polymerase chain reaction (PCR) or DNA sequencing (Sanyal et al. 2011; Splinter and de Laat 2011; Hakim and Misteli 2012). Most of these methods allow the unbiased and genome-wide mapping of interactions without prior knowledge of partners, thus permitting the discovery of new interactions. Their downside is that these are population-based approaches that only generate averaged data and cannot interrogate the behavior of single cells or subpopulations. A combination of the two methods is a powerful approach to solving the question of how genomes are spatially organized.

Cite as *Cold Spring Harb Perspect Biol* doi: 10.1101/cshperspect.a019356

1 INTRODUCTION: THE CHALLENGE OF STORING DNA IN VIVO

Genomes are organized in complex ways in the 3D space of the cell nucleus (Misteli 2007; Rajapakse and Groudine 2011; Cavalli and Misteli 2013). A typical higher eukaryotic cell contains ~2 m of DNA that must be packed into a nucleus ~10 μm in diameter. Assuming approximately five trillion cells in the human body, the total amount of DNA in an individual is on the order of 10 trillion meters—the equivalent of spanning the distance from Earth to Sun more than 100 times! Clearly, cells face a tremendous challenge to ensure the safekeeping and accurate propagation of DNA during replication and cell division, at the same time allowing accessibility of regulatory factors to genes at the right time and in the right place.

To accommodate the immense length of DNA in the nucleus and to ensure functionality, the genetic material is wrapped into higher-order chromatin fibers culminating in the organization of chromatin into chromosome territories (Fig. 1A,C). A chromosome territory is the physical space taken up by a given chromosome in the interphase nucleus (Cremer et al. 1982; Lanctot et al. 2007). Despite the extraordinary length of DNA in a cell, it is noteworthy that DNA-containing chromatin only takes up an estimated 15% of the nuclear volume.

Obviously, to achieve the necessary compaction of DNA, the chromatin fiber must be looped onto itself and

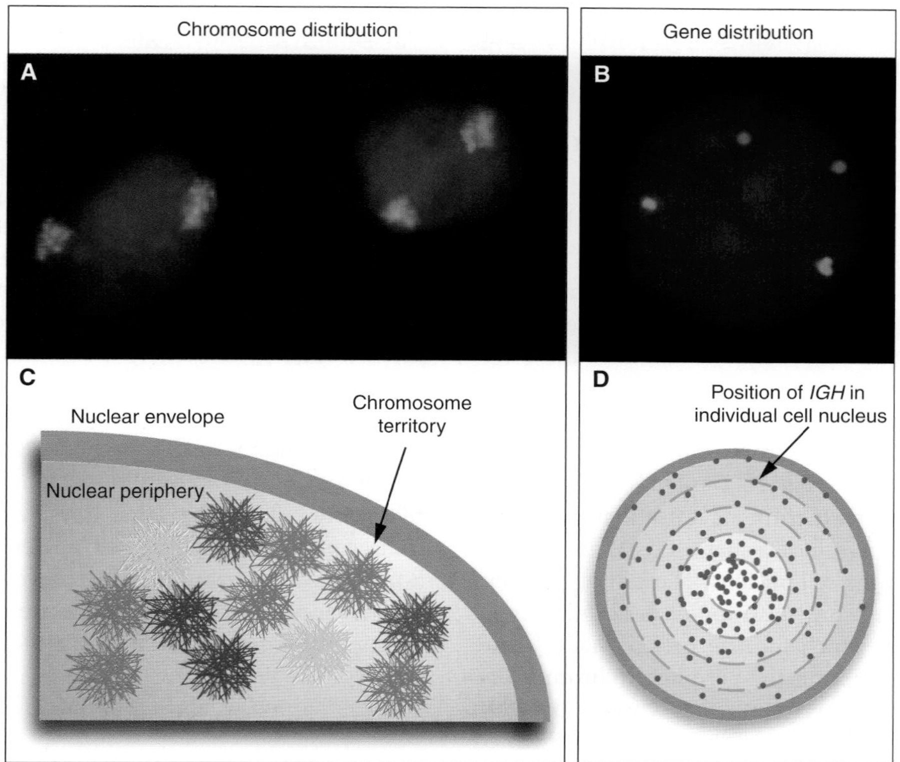

Figure 1. Chromosome territories and genes in three-dimensional (3D) space. Fluorescence in situ hybridization (FISH) visualizes the spatial organization of the genome. (*A*) Chromosomes exist in the form of chromosome territories in the interphase nucleus. The DNA of each chromosome occupies a spatially well-defined fraction of the nuclear volume, typically about 1–2 μm in diameter. Chromosome 11 (green) in the nucleus (blue) of MCF10A breast cancer cells is shown. (*B*) Individual genes appear as distinct spots. *MYC* (red) and *TGFBR2* (green) in MCF10A cells are shown. (*C*) The position of a chromosome or a gene can be expressed as the distance from the center of the nucleus or relative to other genes. The distribution of chromosomes and genes is nonrandom with some chromosomes (red) preferentially occupying internal positions and others (green) occupying peripheral positions. The nonrandom radial positioning also gives rise to nonrandom genome neighborhoods. (*D*) The distribution of a gene is probabilistic. Mapping of the position of the *IGH* gene in human lymphocytes in several hundred individual cells shows that its distribution is distinct from a random distribution; however, the *IGH* locus can be found in variable locations in individual cells. Each red dot represents the position of an *IGH* allele in a single cell. (*C*, Modified from Meaburn and Misteli 2007.)

condensed. How precisely this occurs in vivo is one of the enigmas in modern cell and molecular biology. The traditional view holds that DNA is hierarchically folded into higher-order fibers of 10 nm, 30 nm, 100 nm, and beyond (Felsenfeld and Groudine 2003; Li and Reinberg 2011). Although there is compelling evidence for the 10-nm fiber, several recent studies have questioned the existence under physiological conditions of the regular structure beyond the 30-nm fiber (Fussner et al. 2010). Higher-order regimes of organization are even less well defined and the physical properties of higher-order chromatin are currently largely unknown. The reason for this considerable uncertainty is the inability to accurately visualize higher-order chromatin fibers at the required level of resolution, thus limiting our ability to ascertain its structure. Despite the difficulty in assessing the path of the chromatin fiber in single cells, important insights into the higher-order spatial organization of chromatin have been obtained by improved imaging approaches and 3C-based technologies.

2 LONG-RANGE INTERACTIONS IN THE CONTEXT OF NUCLEAR ARCHITECTURE

2.1 Genomes Are Nonrandomly Organized in the Cell Nucleus

A fundamental property of genomes in higher eukaryotes is their nonrandom spatial organization in 3D space (Fig. 1) (Misteli and Soutoglou 2009; Rajapakse and Groudine 2011; Cavalli and Misteli 2013). The nonrandomness of a genome's spatial arrangement is easily shown in FISH experiments by determining the position of a chromosome or a gene relative to the center of the nucleus (Fig. 1C,D). This is referred to as its radial position. These experiments reveal that each chromosome, and many genes, have a characteristic distribution of positioning within the cell nucleus. The prototypical examples of radial chromosome positioning are human chromosomes 18 and 19, which localize preferentially to the center (chromosome 19) and the periphery (chromosome 18) of the nucleus in human lymphocytes (Croft et al. 1999). Similar preferential location patterns have been documented for all chromosomes in the human genome (Boyle et al. 2001). The radial position of genes and chromosomes is cell-type- and tissue-specific. For example, X chromosomes localize more peripherally in liver cells compared to kidney cells (Parada et al. 2004). Along the same lines, chromosome position often differs in disease cells. For example, in pancreatic cancer, chromosome 8 shifts to a more peripheral location (Wiech et al. 2005). Similarly, chromosomes 18 and 19 change nuclear location in several cancers types, including cervical and colon cancer (Cremer et al. 2003).

It is important to point out that the spatial position of a chromosome or a gene is a statistical property, and not an absolute indicator of its location (Parada et al. 2003; Rajapakse and Groudine 2011). A gene's location may be highly variable between individual cells (Fig. 1D). In other words, although a gene may, for example, preferentially be located toward the center of the nucleus, any given allele may be found at the nuclear periphery in a subpopulation of cells. The position of a gene is thus most accurately reflected by the statistical distribution of its positions in a population of cells.

The radial arrangement of chromosomes and genes in the nuclear space has an important consequence: Distinct genome neighborhoods arise through the preference for a genomic region to either be located toward the center or the periphery of the nucleus (Fig. 1C) (Meaburn and Misteli 2007). For example, more internal chromosomes or genes have a low probability of associating with genomic regions that are enriched at the nuclear periphery, thus physically segregating genomic regions within the nuclear space. Prominent examples of nonrandom spatial genome neighborhoods are the nucleolus, which harbors ribosomal RNA gene clusters located on human chromosomes 13, 14, 15, 21, and 22 or the congregation of chromosomes 12, 14, 15, which associate with each other with high frequency in mouse lymphocytes and are involved in chromosomal translocations in lymphoma (Parada et al. 2002).

For many genes the precise radial positioning is generally not sufficient to determine gene activity, nor is it predictive of gene activity. For instance, in a systematic study of 20 genes in a breast cancer model, no consistent correlation between gene activity and radial position was found (Meaburn and Misteli 2008). Active genes were found both internally and peripherally, as were inactive genes. Along the same lines, visualization of active transcription sites in the cell nucleus reveals a relatively homogenous distribution of active genes without any preferential accumulations. It can be concluded that the radial position of a gene alone is not sufficient to determine its activity.

2.2 Physical Interaction with the Nuclear Periphery as a Regulatory Mechanism

Although the precise location of a gene within the cell nucleus is not likely to be a key determinant of its activity, its positioning with respect to the nuclear envelope can have regulatory function (Kind and van Steensel 2010; Egecioglu and Brickner 2011). In most higher eukaryotes, the nuclear periphery is enriched in condensed heterochromatin, generally associated with the transcriptional repression. Furthermore, in budding yeast *Saccharomyces cerevisiae*, silent telomeric regions are preferentially found at the nuclear periphery (Egecioglu and Brickner 2011).

Cite as *Cold Spring Harb Perspect Biol* doi: 10.1101/cshperspect.a019356

Large-scale mapping (e.g., using targeted DNA methylation approaches [DamID] and antibody pull-down [chromatin immunoprecipitation, ChIP]) has identified genomic regions that preferentially associate with the nuclear lamina (Pickersgill et al. 2006; Guelen et al. 2008; McCord et al. 2013), a meshwork of the intermediate filament proteins lamin A, B, and C and associated proteins (Dechat et al. 2010). In humans, there are more than 1000 lamina-associated domains (LADs) and these regions are typically 0.1–1 Mb in size and generally gene poor, and the contained genes are either silent or expressed at low levels (Pickersgill et al. 2006; Guelen et al. 2008; Kind and van Steensel 2010). LADs are likely organized into higher-order structures as their boundaries are frequent binding sites for the insulator protein CTCF. Not surprisingly, because they are enriched in inactive genes, histone marks associated with active genes (e.g., H3K4me3) are depleted in LADs (Kind and van Steensel 2010).

The presence of heterochromatin and inactive genes suggests that the periphery is a transcriptionally repressive environment (Egecioglu and Brickner 2011). But is interaction with the nuclear periphery a gene regulatory mechanism? A functional role is suggested by the observation of widespread gene misregulation in cells lacking major components of the nuclear lamina such as lamin A or lamin C (Dechat et al. 2010). Similar effects are seen upon overexpression of lamins or dominant mutants of lamin A, many of which are associated with human diseases (Scaffidi and Misteli 2008). Furthermore, experimental repositioning of active genes to the nuclear periphery by artificial tethering to a component of the nuclear lamina results in silencing of at least some genes, suggesting that interaction of genes with the nuclear lamina can be sufficient to alter gene activity (Fig. 2A) (Finlan et al. 2008; Kumaran and Spector 2008; Reddy et al. 2008).

Gene repositioning relative to the nuclear envelope is likely a physiologically relevant regulatory mechanism because physical relocation away from the nuclear envelope has been implicated in several gene activation events during differentiation. For instance, the IgH and IgK loci dissociate from their location at the nuclear periphery in hematopoietic progenitor cells and assume a more internal location upon stimulation into pro-B-cells (Fig. 2B) (Kosak et al. 2002). Similarly, the CFTR gene detaches from the nuclear envelope upon its activation (Zink et al. 2004). These types of relocation events correlated with activity changes appear to be universal and can also be seen in diverse organisms such as *Caenorhabditis elegans* (Meister et al. 2010).

The mechanisms by which mammalian genes are regulated through interaction with the nuclear periphery are poorly understood, but likely involve specific sets of his-

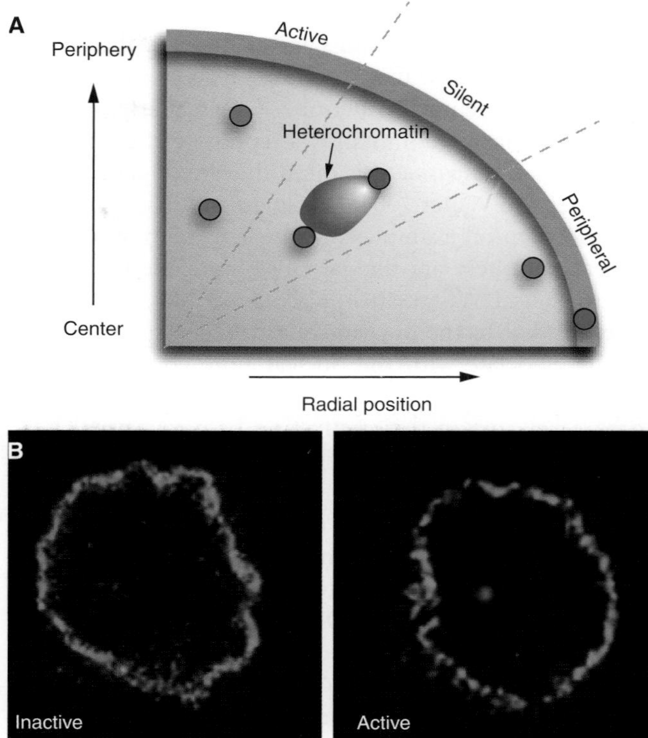

Figure 2. The role of the nuclear periphery in gene regulation. (*A*) Active genes (green) show a large range of radial positions; the precise radial position of a locus does not correlate with its activity level. Inactive genes (red) may associate with heterochromatin blocks at various radial positions. In contrast, physical association with the nuclear periphery is often linked to silencing. Genes that are in close proximity to the nuclear envelope but do not physically interact with it may be active. (*B*) Association of a gene locus with the nuclear edge often correlates with activity. The IgH locus (red) associates with the nuclear lamina when it is inactive in hematopoietic progenitor cells, but dissociates on activation in activated pro B cells. (*A*, Redrawn from Takizawa et al. 2008; *B*, reproduced, with permission, from Kosak et al. 2002, © AAAS.)

tone modifications (Egecioglu and Brickner 2011). A hint as to the molecular mechanisms involved in silencing peripheral genes comes from analysis of their histone modification patterns. Nuclear envelope–associated genes are generally hypoacetylated and treatment with histone deacetylase inhibitors reverses silencing of genes tethered to the periphery (Finlan et al. 2008; Guelen et al. 2008). In addition, peripherally located genes are enriched in H3K9me2 (Yokochi et al. 2009). Interestingly, dimethylation of H3K9 (H3K9me2) is mediated by the G9A histone methyltransferase, which physically interacts with BAF, one of the major lamina-interacting proteins (Montes de Oca et al. 2011). Elimination of G9A results in derepression of predominantly late-replicating genes typically associated with heterochromatin (Yokochi et al. 2009). A model therefore emerges in which genes may be marked as silent by specific histone modifications and tethered to the periphery by virtue of these modifications.

Although the nuclear periphery, particularly in higher eukaryotes, has predominantly a repressive effect on gene activity, in *S. cerevisiae* it is often the site of gene activity (Brickner and Walter 2004; Egecioglu and Brickner 2011). The prototypical examples of gene activation at the yeast nuclear periphery are the galactose-inducible GAL genes (Brickner and Walter 2004; Casolari et al. 2004). These genes localize to the nuclear interior when repressed, but rapidly associate with the nuclear pore complex (NPC) upon stimulation. The nuclear pore interaction is likely to be a key step in their activation because several NPC components and NPC-associated factors are required for activation of these genes. Targeting of these genes to the NPC requires their promoter region, but is independent of sequences in the gene body or the 3′ untranslated region (Ahmed et al. 2010). Importantly, mutational analysis of the promoter region of INO1, an NPC-targeted gene in which transcription is ablated, indicates that targeting to the pore can be uncoupled from transcription (Light et al. 2010). The INO1 promoter regions that target the gene to the pore are also distinct from the regions that control its activity. These so-called gene recruitment sequences are also sufficient to reposition an adjacent reporter gene to the pore (Ahmed et al. 2010; Light et al. 2010). It remains to be seen to what extent active genes are also found at the nuclear periphery in higher eukaryotes. In genome-wide mapping studies in *Drosophila*, hundreds of genomic sites containing active genes have been identified that physically interact with the nuclear pore complex (Kalverda et al. 2010).

2.3 Spatial Gene Clustering

The nonrandom positioning of genes in the nucleus raises the possibility that functionally related genes are spatially clustered in 3D space. Clustering of genes may be caused by sharing of overlapping activating or repressing protein complexes, and may provide a mechanism to coregulate groups of genes through their association with subnuclear compartments enriched in specific regulatory proteins and increase efficiency of transcription and RNA processing. There are several examples of spatial gene clustering.

2.3.1 The Nucleolus

The nucleolus is the site of ribosomal RNA (rRNA) synthesis and processing (Pederson 2010). The structure contains three morphologically distinct regions: the fibrillar center, which contains the rRNA genes; the dense fibrillar component, which contains newly synthesized rRNA transcripts at early stages of processing; and the granular component, which contains mature rRNA and early ribosomal assembly intermediates. Eukaryotic cells contain multiple copies of rDNA genes, typically several hundred in number, organized as tandem repeats located on multiple chromosomes. The rDNA repeat regions of the genome are referred to as nucleolar organizer regions (NORs). In humans, five chromosomes (chromosomes 13, 14, 15, 21, 22) contain NORs, each ~3 Mb in size, made up of approximately 80 copies of the 43-kb rDNA repeat. In contrast to the five NOR clusters, typically only two to three nucleolar foci are observed in human cells. FISH analysis shows that the reason for the numerical discrepancy between NORs and nucleoli is that rDNA regions from multiple chromosomes physically aggregate to form an individual nucleolus. A typical nucleolus occupies up to 25% of the nuclear volume and represents the most dramatic example of gene clustering (Pederson 2010).

The formation and maintenance of the nucleolus is driven by the transcriptional activity of the ribosomal genes (Dundr and Misteli 2010). Nucleoli form in late M phase/early G_1 as transcription of rDNA genes resumes. Importantly, nucleoli only form around active, not inactive, NORs. Initially, small prenucleolar bodies (PNBs) form in late M phase, which serve as precursors of mature nucleoli. PNBs likely contain partially processed rRNA molecules and, upon resumption of full rDNA transcription, mature into nucleoli (McStay and Grummt 2008). Evidence for a requirement for transcription as a driving force in formation of nucleoli is the observation that inhibition of rDNA transcription leads to disassembly of the nucleolar structure (Karpen et al. 1988). On the other hand, introduction into *Drosophila* cells of mini-NORs on plasmids is sufficient to nucleate structures, which, by morphological criteria, are indistinguishable from endogenous nucleoli (Karpen et al. 1988).

Although the formation of the nucleolus per se is strictly dependent on transcriptional activity, the association of rDNA genes is not. This was indicated in elegant hybrid experiments exploiting the fact that rDNA transcription is species specific. When a human chromosome containing a NOR is introduced into mouse cells, the human rDNA genes are not transcribed (Sullivan et al. 2001). However, the human chromosome still physically associates with preexisting mouse nucleoli, demonstrating that physical interaction of rDNA gene clusters is independent of their transcription (Sullivan et al. 2001).

Why do rDNA genes cluster? They are unusual in that they are expressed at exceedingly high levels. Although most RNA Pol II–transcribed genes contain a few polymerases at any given time, there are typically hundreds of RNA

Cite as *Cold Spring Harb Perspect Biol* doi: 10.1101/cshperspect.a019356

Pol I complexes associated with an rDNA gene. Live-cell imaging experiments have estimated that individual polymerases are separated by <100 nt and a new initiation event occurs every few seconds (Beyer et al. 1980; Dundr et al. 2002). As a result, thousands of rRNA transcripts are generated from the several hundred rDNA genes every minute, all of which need to undergo extensive processing. The spatial clustering of the rDNA genes and formation of a dedicated nuclear subcompartment, in which all necessary processing factors are enriched, may serve to facilitate efficient rRNA processing and promote ribosome assembly.

2.3.2 tRNA Gene Clustering

The transfer RNAs (tRNA), which act as adaptor molecules during translation, are encoded by multiple tRNA genes; humans contain ~500, *S. cerevisiae* 274, and *C. elegans* 620 tRNA genes, distributed over multiple chromosomes (Phizicky and Hopper 2010). Analysis of *S. cerevisiae* tRNA genes has provided a compelling case for gene clustering. When localized by FISH analysis, the tRNA genes, which are located on all 16 *S. cerevisiae* chromosomes, are found in a large spatial gene cluster near the nucleolus (Thompson et al. 2003). The proximity of the tRNA cluster to the nucleolus may be of functional relevance because tRNAs and rRNAs are often coregulated in response to environmental conditions. Clustering of tRNA genes persists through all phases of the cell cycle, including cell division when transcription ceases, further supporting the notion that gene clustering can be independent of transcriptional activity (Thompson et al. 2003). Clustering of tRNA genes has to date only been observed in *S. cerevisiae* and it remains to be seen whether higher eukaryotes organize their tRNA genes in a similar fashion.

The formation of tRNA clusters near the nucleolus appears to be a two-step process (Haeusler et al. 2008). The clustering of individual tRNA genes requires condensin subunits. In temperature-sensitive mutants of the condensin subunits, *smc2-8*, *smc4-1*, *ycg1-2*, *ycs4-1*, and *brn1-9* tRNA genes lose their clustered appearance and are dispersed throughout the nucleus. The condensin effect may reflect a global role of the complex on higher-order genome organization. But, interestingly, the fact that condensin specifically interacts with tRNA genes via the TFIIIB/TFIIIC complex and does not require the RNA Pol III complex that transcribes tRNAs, points to a direct role of condensin in the spatial clustering of tRNA genes. The second step in the clustering process is the position of the cluster near the nucleolus. This step is dependent on microtubules as indicated by the fact that disruption of microtubules results in displacement of the tRNA cluster from the nucleolus (Haeusler et al. 2008).

2.3.3 Transcription Factories

The notion of gene clustering, as observed in specialized classes of genes such as rRNA and tRNA genes, has been generalized in the concept of transcription factories (Fig. 3) (Jackson et al. 1993; Eskiw et al. 2011). These are defined as nuclear structures containing multiple active genes and multiple copies of the RNA Pol II transcription complex. In particular, they are characterized by the presence of hyperphosphorylated RNA Pol II, representing the actively elongating form of the polymerase. The existence of transcriptional hot spots, which are shared by multiple genes, is suggested by quantitative analysis of transcription sites. Although expression profiling indicates that typically several thousand genes are expressed in a given cell type, visualization of transcription using imaging methods reveals a much lower number of sites of nuclear transcription. Quantitative analysis in HeLa cells suggests that on average approximately eight polymerases and eight active genes cluster in a transcription factory (Jackson et al. 1993; Schoenfelder et al. 2010a).

In contrast to the traditional view of the transcription machinery being recruited to a gene, in the transcription factory model, genes are recruited to these centralized hot spots of transcription. Association of genes does not involve their physical interaction with each other, but their association with the protein components of the transcription factory or with adjacent splicing factor compartments (Brown et al. 2006). A key question then is whether transcription factories are specialized—that is, whether the genes they contain are coregulated or whether association of genes is largely random. Several observations raise the possibility of preferential clustering of coregulated genes. For example, in erythrocytes, the cell-type-specific Hba, Hbb, and Xpo7 genes can be found in a triplet cluster in a transcription factory at frequencies higher than expected based on random association (Schoenfelder et al. 2010b). Furthermore, pairwise clustering of Hba and Hbb with several other erythroid-specific genes has been observed at frequencies higher than expected. These genes are regulated by the transcription factor Krüppel-like factor 1, and their association with the transcription factory appears to be mediated by Klf1 because its loss interferes with clustering (Schoenfelder et al. 2010b). Another example of recruitment of coordinately regulated genes to a shared transcription factory is TNF-responsive genes (Papantonis et al. 2011). Upon stimulation, TNF-responsive genes, located on both the same and different chromosomes, spatially associate with each other in a coordinated fashion closely following their temporal expression pattern, suggesting that their physical clustering is directly related to their activity (Papantonis and Cook 2010). Further exam-

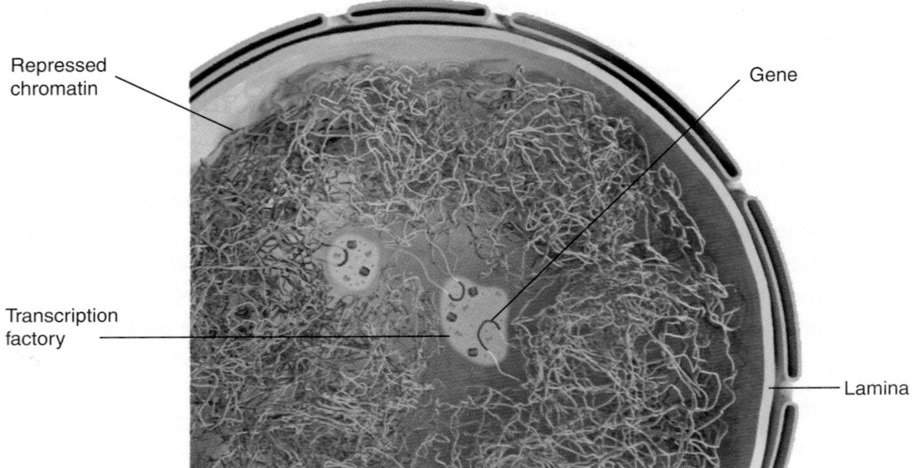

Figure 3. A view into the nuclear interior. Genomes exist in vivo in the form of chromosome territories made up of chromatin folded into still poorly understood higher-order fibers (multicolored areas). Chromatin fibers occupy the entire nuclear space, yet despite the presence of up to 2 m of DNA in a mammalian nucleus, there is considerable open nucleoplasmic space. Active genes (red) may loop out from chromosome territories and aggregate in centers of transcription (shaded areas), which contain appropriate transcription factors (blue, green). It has been suggested that coregulated genes cluster in shared transcription sites. (Reproduced from Misteli 2011; illustration by Anatomy Blue.)

ples include the juxtaposition of estrogen-responsive genes (Fullwood et al. 2009) and the physical association of androgen-responsive genes (Lin et al. 2009). However, these studies also show that expressed genes are not exclusively clustering with coordinately regulated genes. For instance, the hbb locus not only interacts with a number of Klf4-regulated genes, but also with hundreds of apparently unrelated but coexpressed genes (Schoenfelder et al. 2010b). Clearly, many parameters, ranging from general transcription status to binding of specific sets of transcriptions factors, contribute to determining the composition of clusters of active genes.

These observations suggest that spatial clustering of genes is not limited to high copy number genes such as rRNA and tRNA genes, but may also play a role in expression of RNA Pol II transcribed genes. An unanswered question is how globally applicable functional clustering of genes is. Even in the most prominent examples, such as the clustering of erythroid-specific genes, coalescence of multiple genes was only observed in 5%–10% of cells in a population, suggesting that clustering is not a prerequisite for proper gene expression. An additional related question is how many distinct transcription factories a nucleus contains. There is currently no strong evidence to suggest that a transcription factory caters exclusively to a specific set of genes and it seems more likely that most transcription factories contain genes that are regulated by multiple, distinct mechanisms.

2.4 Proximity in the Formation of Chromosomal Translocations

Translocations are a common hallmark of cancer cells. These genomic rearrangements are formed by the illegitimate joining of DNA ends on distinct or the same chromosomes. Translocations may represent a downstream bystander effect due to enhanced genomic instability and DNA repair defects in cancer cells. Translocations can be causal in carcinogenesis, as in the case of fusion between the BCR-ABL genes in chronic myelogenous leukemia or the PML-RAR genes in promyelocytic leukemia. The nonrandom arrangement of the genome in 3D space and the physical interaction of genomic regions located on different chromosomes can have a critical impact on the frequency of formation of chromosomal translocations (Fig. 4) (Misteli and Soutoglou 2009).

A prerequisite for the formation of a chromosome translocation is the physical association of the two translocating chromosomes in the nuclear space. Live-cell imaging experiments have shown that in mammalian cells the free chromosome ends generated by double-strand breaks are largely immobile in the cell nucleus (Soutoglou et al. 2007). As a consequence, translocations occur preferentially between genomic regions that are often in close spatial proximity because of the nonrandom arrangement of the genome in the interphase nucleus. In support, FISH experiments and, more recently, biochemical mapping approaches of fre-

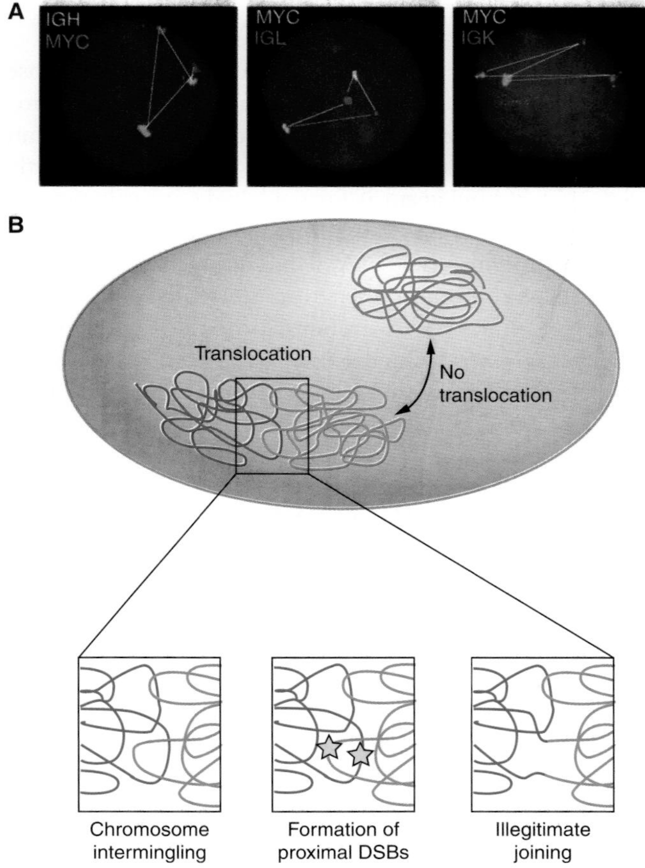

Figure 4. Roles of genome organization in determining chromosome translocations. The nonrandom organization of genes and chromosomes contributes to the formation of cancer translocations. (*A*) The physical distance of MYC to its translocation partners IGH, IGL, and IGK correlates with their translocations frequency (MYC-IGH > MYC-IGL > MYC-IGK). (*B*) Translocations preferentially occur between proximally positioned chromosomes (red, green), and only rarely between distally located chromosomes (blue). Closely juxtaposed double-strand breaks (yellow stars) occurring at the interface between chromosomes create free chromosome ends that may recombine to form a chromosome translocation by illegitimate joining. (*A*, Adapted from Roix et al. 2003; *B*, reproduced, with permission, from Misteli 2010, © Cold Spring Harbor Laboratory Press.)

quent translocation partners have revealed juxtaposition of translocating regions in the intact cell nucleus (Neves et al. 1999; Roix et al. 2003; Osborne et al. 2007; Kind and van Steensel 2010; Chiarle et al. 2011). For example, the MYC and IGH genes, whose translocation gives rise to Burkitt's lymphoma, are frequently found interacting or located in close proximity in B-cells (Roix et al. 2003; Osborne et al. 2007). Similar correlations between translocation frequency and physical proximity have been found for numerous translocation partners in mouse and human cells (Mani et al. 2009; Mathas et al. 2009). The notion that spatial interactions between translocating regions are a key determinant in translocation formation is also supported by observations on prostate cancer translocations. In prostate cancer, translocations between the TMPRSS2 gene and either ETV1 or ERG are frequent (Lin et al. 2009; Mani et al. 2009). In normal prostate cells, these genes do not physically associate, yet stimulation with androgen leads to their physical association and, when challenged with UV-irradiation, their rapid translocation. The importance of spatial positioning in the formation of chromosomal translocations is further shown by unbiased approaches that are able to map physical interactions of chromatin fibers at a genome-wide scale (see below). Clearly, analyses of naturally occurring and experimentally inducible random translocations have revealed a strong correlation between translocation frequency and spatial proximity (Chiarle et al. 2011; Klein et al. 2011; Hakim et al. 2012; Zhang et al. 2012).

Further support for the notion that the nonrandom spatial proximity of chromosomes is key in determining translocation frequency is the finding that adjacent chromosomes intermingle their DNA (Branco and Pombo 2006). Careful visualization and mapping indicates that chromosomes invade the body of their adjacent neighbors, thus creating a zone of chromatin intermingling. It appears likely that translocations preferentially occur when DNA damage occurs in this region creating double-strand breaks, which are immobile and prone to undergo illegitimate joining with proximal breaks. This view is supported by the finding that translocation frequencies correspond closely to the degree of chromosome overlap (Branco and Pombo 2006) (Fig. 4).

The finding that translocations occur preferentially between proximally positioned regions means that the nonrandom spatial organization of the genome is a key factor in determining translocation partners (Misteli and Soutoglou 2009). The tissue-specific arrangement of chromosomes is a significant contributor to the tissue-specific appearance of translocations because we know genomes are arranged differently in different cell types and tissues. In agreement with this notion, chromosomes 5 and 6, which are frequently translocated in mouse hepatoma, are often neighbors in normal liver cells, whereas in mouse lymphocytes it is translocation partners chromosomes 12 and 15 that are frequent neighbors (Parada et al. 2004). These observations support a role for the nonrandom spatial genome interactions in the formation of chromosomal translocations.

3 ANALYSIS OF CHROMATIN INTERACTIONS AND GENE REGULATION

3.1 Introduction: Gene Elements and Methodologies

Perhaps the most direct and specific role of long-range chromatin interactions and spatial chromatin organization is in controlling gene expression. In the large genomes of metazoans, only a small portion of the genome is protein coding. The enormous noncoding fraction of the genome, including both intragenic and intergenic DNA, is thought to represent a large reservoir of gene regulatory elements. A wide range of genome-wide approaches are being applied to identify the full set of functional elements in the human genome and initial estimates indicate the presence of hundreds of thousands of regulatory elements scattered throughout the genome (ENCODE Project Consortium 2011, 2012). Gene regulatory elements include enhancers, repressors, insulators, and possibly other classes of yet-to-be-discovered elements. These elements may be located at significant genomic distances from the nearest gene, suggesting that they are able to communicate over considerable distances in the linear genome. Consistently, studies using transgenes and reporter constructs have revealed that significant stretches of flanking DNA (e.g., hundreds of kilobases) are required to recapitulate the normal regulation of a gene.

One obvious mechanism by which long-range gene control can be achieved is through direct physical associations between regulatory elements and gene promoters (e.g., through the 3D folding and looping of the chromatin fiber). Technological innovations have facilitated detection of chromatin loops and indicate that they indeed play critical roles in gene regulation by bringing widely spaced genes and regulatory elements in direct physical contact.

Microscopic approaches can be used for the analysis of the relative spatial positioning of loci within individual nuclei, as illustrated above, but they generally lack in resolution and comprehensiveness. On the other hand, crystallographic studies can be used to determine chromatin structure at the level of individual nucleosomes. However, both these techniques do not determine the folding and structure of the chromatin fiber. This resolution gap has, for a long time, been an impediment to studying the role of chromatin structure in mediating long-range gene regulation often involving regulatory elements located tens to hundreds of kilobases from their target genes.

Over the last several years, an expanding set of related molecular methodologies based on the chromosome conformation capture (3C) technique have been developed to allow the quantification of chromatin interactions at increasing resolution (kb) and scale (whole genome; Dekker et al. 2002; van Steensel and Dekker 2011; Hakim and Misteli 2012). These 3C-based approaches are particularly powerful for the identification of chromatin looping interactions, and increasingly for determining whole-chromosome and whole-genome folding at the resolution of several kilobases.

3.2 3C

The basic concept of the 3C method is that loci in close spatial proximity (e.g., they directly interact to form a chromatin loop or a contact between two chromosomes) can be chemically cross-linked (Fig. 5). The most frequently used cross-linking agent is formaldehyde; it easily permeates cells and induces cross-links between proteins, and between proteins and DNA. Thus, physically interacting segments of chromatin fibers throughout the genome become cross-linked with protein complexes that associate with them. Next, chromatin is fragmented using a restriction enzyme. Cross-linked chromatin segments are then ligated under dilute conditions to strongly favor intramolecular ligations over random intermolecular ligation events. Finally, cross-links are broken by simply heating the chromatin to 95°C, and DNA is purified to obtain a library of 3C ligation products. Effectively, 3C converts pairs of loci that are spatially proximal into unique hybrid DNA molecules that can then be detected and quantified using any of a wide range of standard DNA detection methods (Fig. 5).

In a classical 3C experiment, a genome-wide 3C library is generated that can be used as DNA template in PCR assays using locus-specific primers to determine the presence and relative abundance of specific ligation products of interest (Dekker et al. 2002). Thus, interaction frequencies are determined one at a time using a candidate approach, which puts practical limits on the number of interactions that can reasonably be tested. Classical 3C is often used to determine interaction frequencies among up to several dozen individual genomic restriction fragments (e.g., to test whether a given gene promoter of interest is frequently interacting with a particular regulatory element). One of the first applications of 3C was to detect chromatin looping in the β-globin locus between the active globin genes and the upstream locus control region (LCR; Tolhuis et al. 2002). Since then, many other studies have used 3C to identify similar looping interactions in other loci.

The 3C technology has revolutionized the study of chromatin folding, particularly because it has spawned the development of a number of 3C-based technologies with increased resolution, sensitivity, and, most importantly, throughput. It has achieved this by changing the way 3C ligation product libraries are interrogated (van Steensel and Dekker 2011). In Sections 3.3 and 3.4, we will outline these detection methods in more detail and then describe how their application is starting to uncover the principles that determine the spatial organization of individual gene loci, chromosomes, and whole genomes, and how these different layers of genome organization impact gene expression.

Cite as *Cold Spring Harb Perspect Biol* doi: 10.1101/cshperspect.a019356

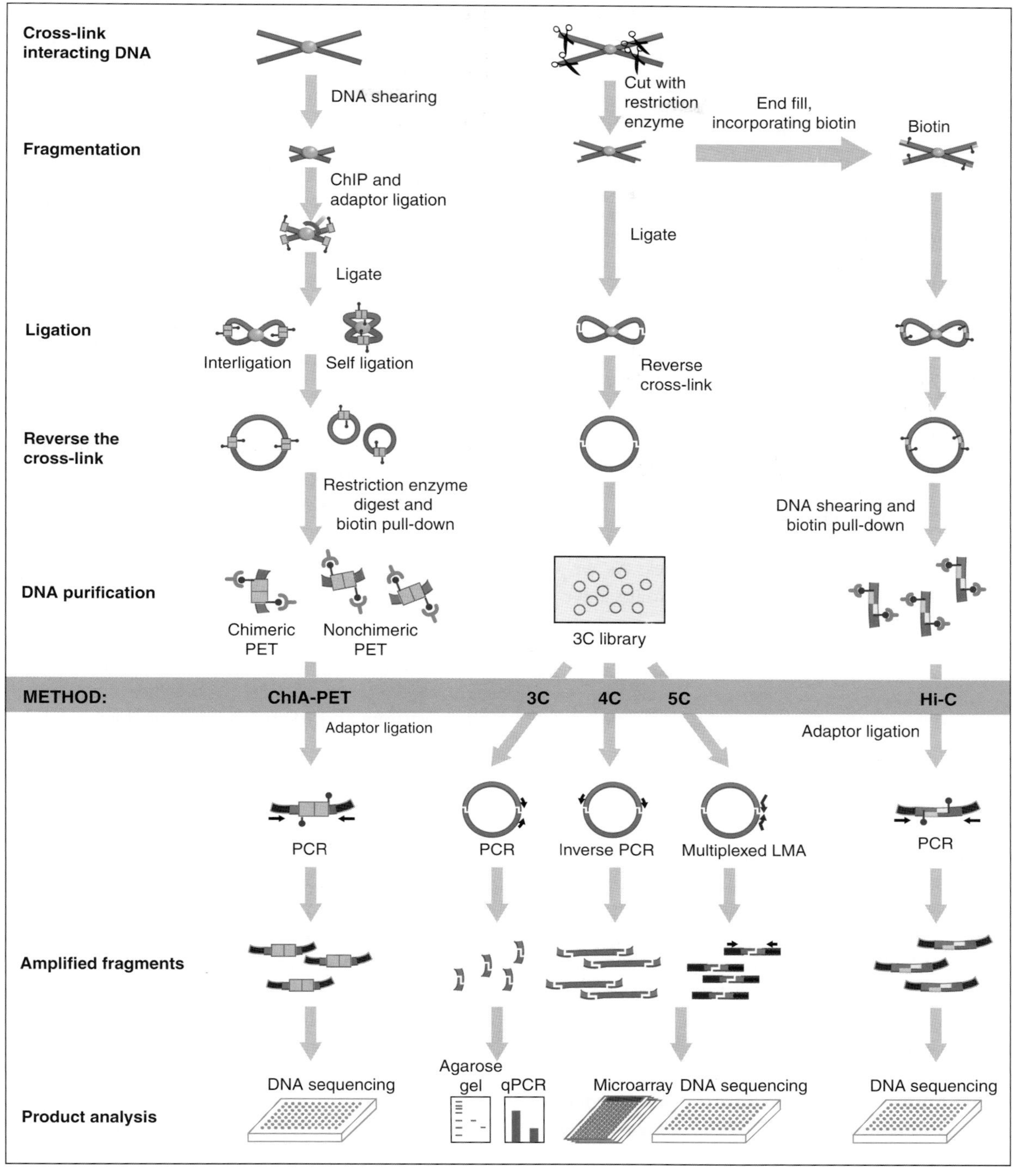

Figure 5. Overview of 3C-based methods. All 3C-based methods rely on covalently linking spatially proximal chromatin segments. Through a series of steps chromatin is then fragmented and religated and ligation products are detected using PCR, microarrays, or deep sequencing. In ChIA-PET (chromatin interaction analysis with paired-end tag sequencing), chromatin fragmentation is achieved by shearing and ligation junctions are marked by adaptors that contain recognition sites for type I restriction enzymes. Redigestion of ligation products with such enzymes yields small ligation, junction-containing molecules that can be analyzed by deep sequencing. 3C, 4C, and 5C use restriction enzyme digestion to fragment cross-linked chromatin. Religation of DNA then produces ligation products that can be directly analyzed by PCR (3C), inverse PCR (4C), or ligation-mediated amplification (LMA; 5C). Hi-C is as 3C, but includes a step to incorporate biotinylated nucleotides before religation. This facilitates purification of ligation junctions that are then analyzed by deep sequencing. (Adapted from Sanyal et al. 2011.)

3.3 Global Analysis of Complete 3C Libraries

The recent development of high-throughput DNA sequencing platforms has greatly facilitated comprehensive analysis of chromatin interactions. Improvements in 3C technology have culminated in the development of the Hi-C method, the first genome-wide 3C method, which includes a step to label the ends of restriction fragments with biotinylated nucleotides before DNA ligation, thereby labeling ligation junctions with biotin (Fig. 5) (Lieberman-Aiden et al. 2009). This facilitates purification and sequencing of DNA molecules only containing ligation junctions, greatly improving the number of informative DNA sequence reads. Moreover, more recently, it has become feasible, and no longer cost-prohibitive, to directly sequence a 3C library to obtain an unbiased genome-wide chromatin interaction map. Indeed, direct sequencing of 3C libraries of *Drosophila* embryos has yielded genome-wide maps of long-range interactions at a resolution of several kilobases (Sexton et al. 2012).

Genome-wide interaction maps provide a global overview of the spatial organization of genomes. As we will discuss in more detail in Section 4, these maps have revealed several novel features of higher-order chromosome structure and nuclear organization (Cavalli and Misteli 2013).

3.4 Comprehensive Analysis of Parts of the 3C Library

When relatively small genomes are analyzed (e.g., that of yeast), comprehensive interaction detection methods such as Hi-C can provide structural information at a resolution of a few kilobases. However, for large metazoan genomes, these global methods generally yield much lower resolution, typically 0.1–1 Mb. The reason for this is that the complexity of 3C libraries obtained for large genomes is truly enormous. For instance, for the human genome a 3C library generated with the HindIII enzyme will contain as many as 10^{12} unique ligation products, and the abundance of these products can vary over several orders of magnitude. Thus, to obtain a reliable chromatin interaction data set at a resolution of single restriction fragments, one needs to sequence far more molecules than is currently feasible.

To overcome the resolution limit of unbiased global detection methods, several targeted detection methods have been developed. These methods use different strategies to comprehensively sequence and analyze only a subset of the 3C library (Fig. 5). By targeting only a selected portion of the 3C library, one increases the sequencing coverage per interaction and thereby the resolution with which long-range interactions can be identified and quantified. This is particularly relevant for the analysis of long-range gene regulation that involves specific functional elements that are usually smaller than a kilobase.

3.4.1 4C Technology

In many studies, one is interested in identifying all genomic loci that interact with a particular gene or element (e.g., to find putative long-range regulatory elements that loop to and regulate a gene of interest). Thus, instead of probing the entire 3C library in an unbiased fashion, one would like to analyze a subfraction of it at high resolution. To comprehensively identify all fragments that are ligated to a single restriction fragment of interest (the "bait" or "anchor" fragment), the 4C technology combines 3C with inverse PCR detection of 3C ligation products (Fig. 5) (Simonis et al. 2006; Zhao et al. 2006). Briefly, many of the 3C ligation products are circular, or can be circularized by redigestion and re-ligation of 3C ligation products. Inverse PCR with primers pointing outward from the bait fragment can therefore amplify all DNA fragments ligated to the bait. 4C PCR products can then be detected and quantified by hybridization to a genome-wide DNA tiling array or direct DNA sequencing. A 4C experiment provides a genome-wide interaction profile of a bait locus to reveal the positions and identities of other loci at kilobase resolution on the same chromosome or on other chromosomes that the locus frequently interacts with.

3.4.2 5C Technology

The 5C (abbreviation for 3C-carbon copy) technology is used specifically to determine interaction frequencies between two selected sets of loci (e.g., between a set of promoters and a set of enhancers [Fig. 5] [Dostie et al. 2006]). 5C differs from 3C only by the way ligation products are detected. In 5C, LMA is used with pairs of locus-specific forward and reverse primers, termed 5C primers. These are designed to anneal directly across predicted 3C ligation junctions. Only when a ligation product is present in the 3C library can the two primers anneal and become ligated by a nick-specific DNA ligase. Ligated 5C primer pairs are subsequently PCR amplified using primers that recognize their universal tails. The advantage of 5C is that LMA can be performed at high levels of multiplexing, in contrast to PCR. For instance, one can design 5C reverse primers for thousands of genes and 5C forward primers for thousands of putative distal regulatory elements, and interrogate the millions of pairwise gene-element interactions in a single reaction. The resulting 5C library represents a "carbon copy" of a selected fraction of the 3C library and can be analyzed by direct DNA sequencing. In many respects, 5C is similar to other enrichment methods such as hybrid capture approaches for selectively purifying parts of the genome (e.g., all exons) before targeted sequencing (Mamanova et al. 2010).

3.4.3 ChIP-Loop and ChIA-PET Technologies

Another approach to selectively enrich for specific interactions is to combine 3C with ChIP to isolate cross-linked complexes that contain a protein of interest (Fig. 5). For instance, by immunoprecipitating chromatin complexes one can enrich for interactions that involve specific proteins of interest. The ChIP-loop assay and ChIA-PET assays use PCR and deep sequencing, respectively, to analyze the purified ligation junctions (Cai et al. 2006; Fullwood et al. 2009; Li et al. 2010).

3.5 3C-Based Data Reveal Cellular Heterogeneity in Genome Folding

The 3C procedure and all its high-throughput variants described in Sections 3.3 and 3.4 generate a highly complex mixture of hybrid DNA molecules in which each individual DNA molecule represents a single chromatin interaction event in a single cell. It is critical to appreciate that the 3C procedure captures interactions across a large population of cells, and thus the abundance of a single ligation product in the library represents the fraction of cells in which the two corresponding loci are sufficiently proximal to become cross-linked. When sufficient numbers of cells are studied, a given genomic locus can be found contacting almost any other locus in the genome, but with contact probabilities ranging over several orders of magnitude. Thus, 3C-based data can only be understood in terms of enormous cell-to-cell variability in chromosome folding and nuclear organization (Kalhor et al. 2011). Of course, this is exactly what has been observed by direct imaging of loci, revealing preferences in subnuclear localizations, but no determined and reproducible positions. 3C libraries thus reflect the contact probability landscapes of loci, revealing trends in spatial organization of the genome in a large cell population. These data sets must be carefully analyzed to determine patterns in contact probabilities of sets for loci, indicating the presence of looping interactions and other chromatin structural features (Bulger and Groudine 1999, 2011; Dekker 2006; Cavalli and Misteli 2013). Similarly, any model indicating the role of higher-order chromatin structure in gene regulation must take into consideration the significant cell-to-cell variability in chromatin folding.

4 MULTIPLE CATEGORIES OF CHROMATIN LOOPING INTERACTIONS

Studies of local and global patterns of chromatin organization suggest that looping and long-range interactions, in general, reflect different levels of chromatin organization with distinct roles in gene regulation. At the finest scale, we can discern frequent, precise, and specific looping interactions between gene promoters and their distal regulatory elements. At a more global scale, specific sets of loci that can be located on different chromosomes associate with each other. Finally, zones of interactions between and within large megabase-sized chromosomal regions indicate the presence of distinct higher-order structural domains, reflecting nuclear organization and functional compartmentalization, in general. This last category involves infrequent, broad, and rather nonspecific interactions. These different categories of chromatin associations are described in Sections 4.1–4.6 in more detail in relation to their impact on gene regulation (Cavalli and Misteli 2013).

4.1 Looping between Genes and Their Regulatory Elements

Enhancers are elements that can control expression over very large genomic distances. One of the best studied enhancers is the LCR in the β-globin locus, located 40–80 kb upstream of its cluster of target genes (Fig. 6). An even more dramatic example is the *sonic hedgehog* gene regulated by an enhancer located more than 1 Mb away. Many genetic experiments have shown that long-range regulation of gene expression is a general phenomenon in many complex genomes. Over the years, several mechanisms have been proposed by which these elements can activate their target genes. Most of these models propose the formation of a direct physical contact between the element and the target promoter, but they differ in the mechanisms by which these contacts are established (Bulger and Groudine 1999, 2011).

The application of 3C-based assays has led to the identification of long-range interactions between gene regulatory elements and genes. This has confirmed the idea that chromatin looping plays a major role in gene regulation. We will describe some of the best-characterized examples of long-range looping interactions.

4.2 The α- and β-Globin Loci

The α- and β-globin loci express high levels of α- and β-type globin proteins that combine to form hemoglobin. Expression of these loci has been studied in great detail and they provide a paradigm for long-range gene regulation by looping (Fig. 6).

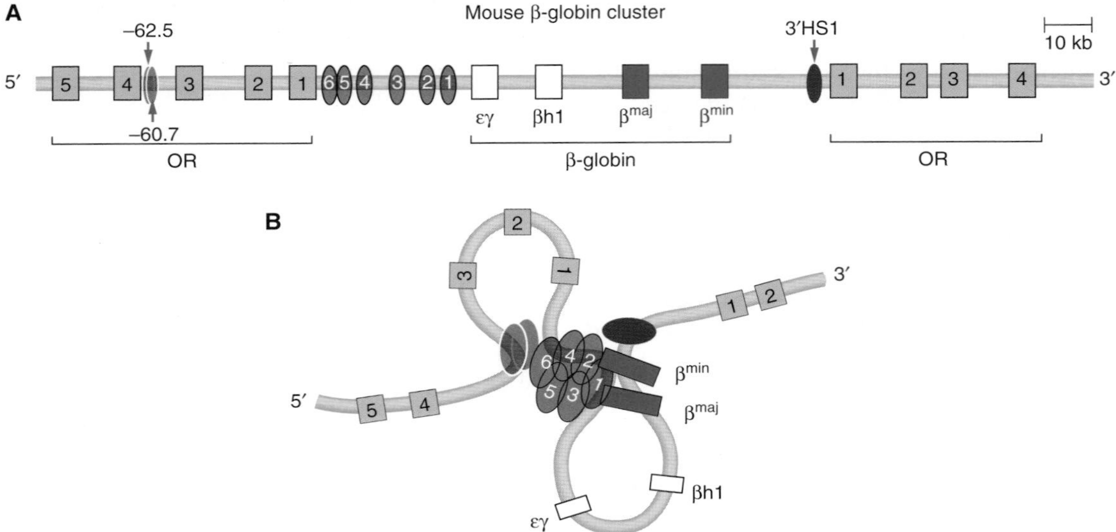

Figure 6. The mouse β-globin locus. (*A*) Schematic representation of the β-globin locus. (*B*) Looping between the LCR, the globin genes, and up- and downstream elements bound by CTCF (HS-62.5/-60.7 and 3′HS1) is observed in globin-expressing erythroid cells. (Redrawn from de Laat and Grosveld 2003, with permission from Springer Science and Business Media.)

Both loci are composed of clusters of related genes that are developmentally regulated. The order of the genes in the cluster corresponds to the order in which they are expressed during development. The β-globin gene cluster is activated by an extraordinarily strong composite element, the LCR. The α-globin genes are activated by a strong enhancer located 40–60 kb upstream of the gene cluster. 3C studies of the β-globin locus were among the first to show that the LCR directly touches the globin genes, resulting in the formation of a chromatin loop that is 40–80 kb in size (Tolhuis et al. 2002). This looping interaction is strictly correlated with gene expression; it is not observed in cells that do not express the locus, but appears in cells that express it. Further, during development, the LCR interacts sequentially with the appropriate gene in the cluster that is expressed at the corresponding stage (Palstra et al. 2003). 3C studies of the α-globin locus show, similarly, that the α-globin genes directly interact with their enhancer located 40 kb upstream of the genes, and again this interaction is only observed in cells that express the genes (Vernimmen et al. 2007; Bau et al. 2010). Since these initial studies, many additional examples of looping between genes and enhancers have been found, including in other complex gene loci such as the HoxD locus (Montavon et al. 2011), the immunoglobulin heavy chain locus (Guo et al. 2011), and the Th2 Interleukin cluster (Spilianakis et al. 2005).

Looping interactions are not limited to very highly expressed gene clusters or developmentally controlled genes, but are also observed in single gene loci encoding proteins involved in a large variety of biological processes. For instance, looping interactions with distal enhancers have been observed for CFTR (Ott et al. 2009; Gheldof et al. 2010), c-MYC (Wright et al. 2010), FoxL2 (D'Haene et al. 2009), and many other genes. A recent comprehensive analysis of more than 600 gene promoters in the human genome identified several looping interactions per gene with elements resembling enhancers, promoters, and insulator elements (bound by the CTCF protein; see Sec. 5.1) located within up to hundreds of kilobases from the promoter (Sanyal et al. 2012).

Long-range interactions are not only involved in gene activation, but can also lead to gene silencing. In particular, Polycomb complexes repress genes by compacting chromatin (discussed in Ch. 17 [Grossniklaus and Paro 2014]) and mediating looping between silencing elements and genes. For example, in the *Drosophila* bithorax complex looping interactions have been identified between Polycomb response elements and silenced hox genes (Lanzuolo et al. 2007). Extrapolating from these studies, one can infer that most genes are engaged in highly specific long-range interactions with regulatory elements located up to hundreds of kilobases around the promoter, leading to either activation or repression.

4.3 Looping between Regulatory Elements

Many genes are regulated by more than one enhancer. For instance, the HoxD locus is regulated by numerous elements spread out over a large gene desert that act in concert to ensure proper control of the gene cluster (Montavon et al. 2011). 3C and 4C studies show that the hoxD genes not only interact with each of these elements at the appropriate time and location during development, but that these elements also associate with each other. One interpretation is that a highly complex multilooped structure is formed. Similar observations have been made in the β-globin locus in which the globin genes interact with the LCR, but both also associate with elements bound by the CTCF protein located upstream of and downstream from the locus (Tolhuis et al. 2002). These observations have led to models in which complex higher-order structures are formed that may act as singular units to integrate regulatory input from multiple elements. Although appealing, there is currently no firm evidence that all pairwise interactions detected using 3C-based methods occur simultaneously in a cell to form stable and reproducible structures, leaving open the possibility that these interactions and the assembly of multilooped structures are dynamic and/or vary among cells in the population.

4.4 Interchromosomal Interactions between Specific Loci

It has long been known from genetic studies in *Drosophila* that regulatory elements on one chromosome can affect gene expression on the homologous chromosome. This phenomenon, called transvection, is thought to be made possible by close pairing of homologs chromosomes (e.g., in polytene chromosomes), which leads to close spatial proximity of homologous regions located on the two chromosomes. In that context, a regulatory element can activate or repress the target gene it is linked to in *cis*, as well as the target gene located on the nearby homolog in *trans*. One well-studied example is the yellow locus, in which the presence of an enhancer on one homolog can activate the yellow target gene on the homologous chromosome lacking the enhancer, in a process that depends on close interactions between the homologs (Morris et al. 1999; Ou et al. 2009). Thus, the mechanism of transvection may be related to other long-range gene regulation phenomena by also involving direct physical interactions.

Several tantalizing examples have been described in mammalian cells suggesting that interchromosomal interactions between regulatory elements on one chromosome can affect gene activity on other chromosomes including nonhomologous chromosomes. The best-characterized case involves the pair of X chromosomes in female mouse cells during the initiation of the X-chromosome inactivation process, which is the topic of Ch. 25 (Brockdorff and Turner 2014) (Fig. 7). In early embryonic development one of the two X chromosomes is chosen to become inactivated. This process is directed by a large and complex locus called the X inactivation center (Xic). This locus encodes the Xist noncoding RNA, which is activated and spreads in *cis* along the X chromosome from the site of transcription from the chromosome that becomes inactive. The Xist RNA recruits silencing factors such as Polycomb complexes that will then silence expression of most genes all along the X chromosome. The process of X inactivation ensures that one and only one of the two X chromosomes will express Xist and become silenced. Both DNA FISH and 3C studies show that during X inactivation the two Xic loci on the two X chromosomes briefly pair and directly interact (Bacher et al. 2006; Xu et al. 2006). The Xic pairing process requires specific CTCF-bound loci around the Xist gene, as well as a larger region located several hundred kilobases from the Xic (Augui et al. 2007; Donohoe et al. 2009). It is thought that this encounter somehow establishes that one of the two X chromosomes will activate the Xist gene, whereas the other will not. Recently, this was studied by live-cell imaging of tagged Xic loci, followed by RNA FISH analysis, demonstrating that pairing is followed by asymmetric expression of Tsix, the Xist antisense transcript and regulator, leading ultimately to monoallelic Xist up-regulation (Masui et al. 2011). Thus, the physical association between the Xics may directly coordinate the establishment of opposite epigenetic states on the X chromosomes. Indeed, mutants that affect pairing of Xics affect appropriate regulation of X-chromosome inactivation.

Other examples of interchromosomal interactions have been described that may be involved in similar coordinated events, but their relevance is highly debated. In olfactory neurons, only one of hundreds of olfactory receptor genes is expressed. 3C studies and imaging have shown that the mouse genome contains a single enhancer (the H-enhancer) that associates with a single olfactory gene either in *cis* or in *trans* in a given cell, leading to the activation of that gene, whereas the other gene copies remain silent (Lomvardas et al. 2006). Although this certainly is a very intriguing finding, subsequent deletion of the H-enhancer found no effect on olfactory genes located on other chromosomes that do not have the H-enhancer, putting in doubt the functional relevance of the interchromosomal interactions (Fuss et al. 2007). Another example of interchromosomal interaction is that of the Th2 interleukin cluster on mouse chromosome 11 and the interferon γ locus on chromosome 10 in naïve T cells (Spilianakis et al. 2005; Noordermeer et al. 2011). This interaction is lost when the cells differentiate into either Th1 or Th2 cells that express the interleukin cluster or interferon-γ gene, respectively. The interaction between the loci before commitment may keep the loci in a poised state, ready to coordinate activation and

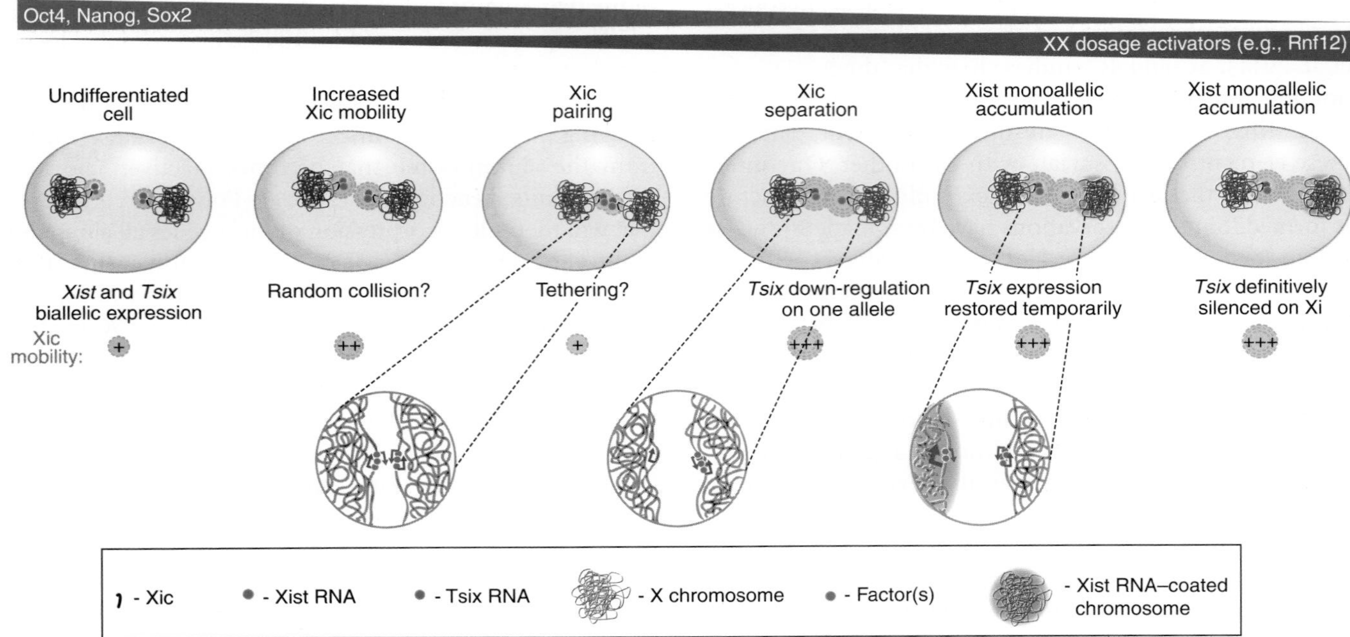

Figure 7. Interchromosomal interactions between Xics during early embryonic development. During early stages of embryonic stem (ES) cell differentiation, the two X chromosomes are highly mobile, possibly to allow pairing of the two Xics. During this pairing stage, through a process still largely unknown, one of the two X chromosomes will down-regulate expression of Tsix, a negative regulator of the Xist RNA gene. On that chromosome, Xist will be expressed and Xist RNA will accumulate resulting in X-inactivation. The other X chromosome will continue to express Tsix and the chromosome will remain active. (Redrawn, with permission, from Masui et al. 2011, © Elsevier.)

repression of the loci during subsequent differentiation. However, whether these interactions cause the poised state, or simply reflect it, remains an open question.

A direct test of the occurrence and relevance of interchromosomal interactions was performed by analyzing the effect of an ectopically inserted LCR on the expression of the endogenous β-globin genes (Noordermeer et al. 2011). 4C analyses showed that an ectopic LCR can directly associate with the endogenous globin locus, but only in a very small minority of cells. Interestingly, combined RNA and DNA FISH experiments showed that in the rare cases in which the ectopic LCR interacted with the globin locus, the globin genes were up-regulated. This important study shows that interchromosomal interactions occur very infrequently, even for a very strong enhancer, but that these rare events can have functional consequences for the interacting genes in the few cells in which the interaction occurs.

Despite these examples of potential functional interchromosomal interactions between specific loci and regulatory elements, the relevance of such interactions for gene regulation is currently far from established. In fact, many of these interchromosomal interactions may be of a different nature than interactions typically involved in intrachromosomal looping for gene regulation via bona fide gene elements. Interchromosomal associations may be related to overall nuclear organization of gene expression in which groups of genes are found gathered at subnuclear sites enriched in transcription machinery or silencing factors. These less-specific and typically low-frequency associations may be the result of gene activity, rather than the cause, and are discussed next.

4.5 Associations between Groups of Active and Inactive Loci

Although active genes can be found throughout the genome, within the nucleus transcription is highly nonhomogeneous and occurs most prominently at multiple foci or subnuclear regions enriched in transcribing RNA polymerases and splicing machineries. These sites can be visualized microscopically in fixed cells using antibodies against active RNA Pol II, splicing proteins, or by BrU labeling of nascent RNA. One interpretation is that groups of active genes, including those located on different chromosomes, cluster together in the nucleus at these transcription foci that are sometimes referred to as transcription factories. Indeed, both imaging and 3C-based studies show that colocalization and direct physical contacts between active

 Cite as *Cold Spring Harb Perspect Biol* doi: 10.1101/cshperspect.a019356

genes are readily detected (Osborne et al. 2004; Simonis et al. 2006; Lieberman-Aiden et al. 2009). In general, there appears to be limited specificity in these interactions; 4C experiments in erythroid cells detect interactions between the β-globin locus and many other active genes throughout the genome (Simonis et al. 2006). These interactions involve larger chromatin segments (e.g., entire genes instead of precise gene regulatory elements) and occur significantly less frequently in the cell population as compared with the chromatin looping interactions described in Sections 4.1–4.3 that drive gene regulation.

Hi-C analysis has shown that active chromosomal domains, typically hundreds of kilobases up to several megabases in size, preferentially associate with any other active chromosomal domains throughout the genome (Lieberman-Aiden et al. 2009). Similarly, inactive domains associate with any other inactive domain. There seems to be limited specificity in these interactions, although there are some exceptions. In *Drosophila*, loci repressed by the Polycomb complex are found clustered at a limited number of Polycomb bodies. In one case, it was found that disrupting the association of a locus with other Polycomb-repressed loci can affect gene regulation of that locus, albeit weakly and only after several generations, although still suggesting that clustering of loci directly impacts on gene expression (Bantignies et al. 2011).

These experiments highlight a recurring theme for these types of chromosomal interactions; active genes are found with other active genes, and inactive or silenced genes are found at other subnuclear sites (Cavalli and Misteli 2013) as indicated by direct imaging, although particular interactions are highly variable and nonspecific—that is, the precise subset of interacting genes and loci differs between cells, even in an otherwise homogeneous cell population. This does not rule out that some preferential clustering of loci regulated by overlapping transcriptional regulators may occur over and above the general tendency of colocalization.

4.6 Topologically Associating Domains: Building Blocks of Chromosomes

Both high-resolution, genome-wide Hi-C analysis and targeted 5C analysis of the X-chromosome inactivation center has led to the discovery that chromosomes are composed of series of chromosomal domains that are several tens of kilobases (in flies), or several hundreds of kilobases (mouse and human) in size (Dixon et al. 2012; Nora et al. 2012; Sexton et al. 2012; Cavalli and Misteli 2013). These domains are characterized by relatively high interaction frequencies between loci located within one domain, but lower interaction frequencies between loci located in different domains. Therefore, these domains have been dubbed "topologically associating domains" (TADs; Nora et al. 2012), or "topological domains" (Dixon et al. 2012). TADs are separated by genetically defined boundary elements (Nora et al. 2012), but the precise mechanism of their formation is currently unknown. The much larger active and inactive chromatin domains described above that interact with other large active and inactive domains genome-wide, respectively, are often composed of multiple smaller TADs.

Initial analyses have indicated that groups of genes located within the same TAD are correlated in their expression during differentiation (Nora et al. 2012), but how general this is across the genome is still an open question. Chromatin looping interactions between genes and their regulatory elements occurs mostly within TADs, which is perhaps not surprising given that chromatin interactions between TADs, in general, are of lower frequency (Sanyal et al. 2012; Shen et al. 2012). TADs may therefore represent not only structural building blocks of chromosomes, but also functional units of gene regulation. It has been proposed that TADs are a critical structural and functional level in the hierarchy of higher-order chromosome folding (Gibcus and Dekker 2013), but much is still unknown about this level of chromosome organization (Cavalli and Misteli 2013).

5 BUILDING CHROMATIN LOOPS

Up to now, we have described the various types of chromatin interactions that can be observed in cells. But what mediates these interactions? We will focus on the class of precise and specific looping interactions between gene promoters and long-range acting gene regulatory elements because these have been characterized in most detail, although many details driving these phenomena remain to be discovered.

Enhancer and promoter elements are bound by large protein complexes containing a variety of different types of proteins, ranging from specific DNA binding transcription factors to cofactors with a variety of enzymatic activities such as histone acetylation and methylation and noncoding RNAs. Long-range interactions between gene regulatory elements and promoters likely involve direct protein–protein interactions and may also require specific bridging complexes. The role of specific proteins in mediating looping has been confirmed for several cases. For example, the looping between the LCR and β-globin genes requires binding of the GATA1 and EKLF1 transcription factor to both the enhancer and the target gene (Drissen et al. 2004; Vakoc et al. 2005). Besides transcription factors that directly bind DNA, other complexes have been implicated in chromatin loop formation. For instance, the chromatin-remodeling enzyme Brg1 is involved in looping in the α- and β-globin loci (Kim et al. 2009a; 2009b). Brg1 directly binds GATA1 and EKLF complexes and may be required for remodeling chromatin to facilitate binding of other factors.

The presence of specific combinations of proteins on pairs of interacting elements may explain, at least in part, the specificity of looping interactions between promoters and their regulatory elements. Interestingly, there are also general factors that may contribute to long-range interactions. The Mediator complex is a transcriptional regulator that interacts with both the basal transcription machinery on the promoter and transcriptional activators, making this an ideal bridging factor between complexes bound at promoters and distal regulatory elements. Indeed, 3C analysis in mouse ES cells identified Mediator-dependent looping between promoters of pluripotency gene promoters and their enhancers (Fig. 8). These interactions also required the presence of the cohesin complex at these sites (Kagey et al. 2010). The cohesin complex mediates interactions between sister chromatids, but is increasingly implicated in mediating other types of long-range interactions, including looping between gene regulatory elements (Hadjur et al. 2009) and often in collaboration with the CTCF protein.

5.1 The General Role of the CTCF Protein in Looping and Chromosome Organization

The CTCF protein binds tens of thousands of sites throughout the mouse and human genomes. The protein is particularly enriched near genes (Kim et al. 2007). Initial studies found that the protein acts as an insulator or boundary complex to block long-range gene regulation by enhancers, at least in reporter constructs. Recently, it has become clear that the protein plays a particularly prominent role in building higher-order chromatin architectures by facilitating intra- and interchromosomal looping interactions (Phillips and Corces 2009).

As described in Sections 4.2–4.4, active gene promoters often interact with distal sites bound by the CTCF protein (e.g., for example, in the α- and β-globin loci) but also in many other cases. CTCF-bound elements also interact with each other, which has led to a model that these loci play general roles in organizing chromosomes (Phillips and Corces 2009). CTCF-bound sites appear to cluster in the nucleus, which will greatly constrain the folding of chromosomes (Yusufzai et al. 2004). How assembly of these higher-order chromatin structures plays roles in regulating genes is far from clear. CTCF-binding to the H19 imprinting control region affects the pattern of looping on the imprinted Igf2 locus, elaborated in Ch. 26 (Barlow and Bartolomei 2014), consistent with a role in controlling other looping interactions in a chromosomal locus. However, other studies are not consistent with such an insulator-related role. Deletion of single CTCF-binding elements in the β-globin locus does not affect globin gene regulation or promoter–enhancer looping, possibly because they are redundant (Bender et al. 2006; Splinter et al. 2006). This observation is not consistent with a simple enhancer-blocking model in which CTCF-bound sites determine which enhancer–promoter looping interactions can occur. One additional complicating factor, however, is the possibility that CTCF-bound sites include a variety of different types of elements depending on additional protein and RNA factors that are recruited to these sites (Yao et al. 2010).

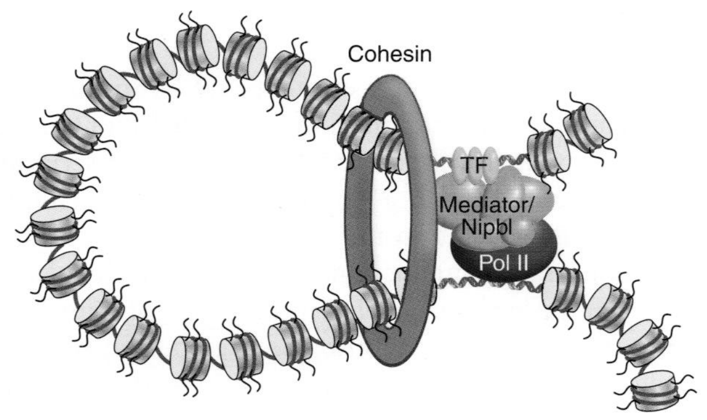

Figure 8. Cohesin and Mediator complexes mediate long-range interactions between promoters and enhancers. The Cohesin and Mediator have been shown to act together to form and stabilize chromatin looping interactions between gene promoters and distal enhancers (Kagey et al. 2010). A model is shown of how these complexes might mediate chromatin looping, although molecular details are currently unknown. (Redrawn, with permission, from Young 2011, © Elsevier.)

6 LOOPING INTERACTIONS AND GENE REGULATION

Although the occurrence of looping interactions between promoters and regulatory elements is now well established, how these interactions contribute to modulating expression levels of the target genes is less well understood. Several observations indicate that direct interactions between promoters and distal enhancers facilitate the recruitment of critical transcription regulators to the gene. For instance, the α-globin enhancer has been shown to recruit RNA Pol II first, which may then be transferred to the target promoter through looping (Vernimmen et al. 2007). Other protein complexes can be recruited in a similar manner. Also, the LCR influences the transition of RNA polymerase from its initiating phase to transcriptional elongation (Sawado et al. 2003), suggesting that the LCR-promoter interaction can help recruit elongation factors to the promoter.

Cite as *Cold Spring Harb Perspect Biol* doi: 10.1101/cshperspect.a019356

7 CONCLUSIONS

The statement, "genomes exist in space and time in the cell nucleus" is a trivial one, but one that has long been ignored in our studies of gene function. During the decades of pioneering work that have led to a now detailed molecular understanding of transcription, the consequences of spatial and temporal aspects of genome organization were largely ignored. This was necessary to reduce the complexity and to make the question amenable to experimental interrogation. The last decade, however, has brought the realization that gene expression is not merely controlled by the information contained in the DNA sequence, but that higher-order chromatin structure features such as nucleosome positioning, chromatin fiber formation, intra- and interchromosomal chromatin interactions, and even chromosome and gene localization contribute to genome function. Not only have we realized that these aspects of genome organization are critical, we are in the fortunate position to have experimental tools available to test their contribution. Although imaging-based methods have long been used for these studies, the emergence of biochemical methods to test physical genome interactions, coupled with affordable next-generation sequencing, has revolutionized our ability to now test their functional relevance. Using these methods, we can now fairly routinely generate complete maps of genome interactions.

Mapping of genome interactions is only the very first step in understanding the functional relevance of spatial and temporal genome organization. What needs to be done? A key step will be the generation of a large number of genome-wide interaction maps for various cell types, tissues, developmental stages, and diseases. Several international consortia, such as ENCODE and IHEC, are aiming to generate such data. Once this information is gathered, physical interaction profiles must be compared with gene expression profiles, and histone modification and DNA methylation patterns. These correlative studies should provide some hints as to the functional relationship between genome organization and function. To truly test their relevance, genome organization patterns will need to be manipulated experimentally. This can be achieved by knockdown and overexpression approaches. This should reveal, on the one hand, how genome organization affects function and, on the other hand, what cellular factors determine genome organization.

Mapping and understanding genome organization has important implications for disease. For instance, we know that the nonrandom genome organization plays a key role in determining cancer translocations and 3C-based methods are being used to detect translocations. It is intriguing, and not utopian anymore, to speculate that we will be able to use genome organization for diagnostic and prognostic purposes. Aberrant chromatin organization is increasingly recognized as a hallmark of various diseases ranging from common cancer to rare diseases. Of particular interest is the possibility of using genome organization as a marker for early stage detection of disease, taking advantage of the fact that changes in chromatin and genome organization often precede genetic changes.

In hindsight, it seems obvious that considering the spatial and temporal organization of the genome would be an integral part of epigenetic and genetic function, and needs to be taken into account when attempting to unravel the mysteries of the genome. For the first time, we are now in a position to explore this key aspect of genome function experimentally; there is no doubt that what we will find will enrich our understanding and appreciation of the complexity and intricacy of genome function.

REFERENCES

*Reference is also in this book.

Ahmed S, Brickner DG, Light WH, Cajigas I, McDonough M, Froyshteter AB, Volpe T, Brickner JH. 2010. DNA zip codes control an ancient mechanism for gene targeting to the nuclear periphery. *Nat Cell Biol* **12:** 111–118.

Augui S, Filion GJ, Huart S, Nora E, Guggiari M, Maresca M, Stewart AF, Heard E. 2007. Sensing X chromosome pairs before X inactivation via a novel X-pairing region of the Xic. *Science* **318:** 1632–1636.

Bacher CP, Guggiari M, Brors B, Augui S, Clerc P, Avner P, Eils R, Heard E. 2006. Transient colocalization of X-inactivation centres accompanies the initiation of X inactivation. *Nat Cell Biol* **8:** 293–299.

Bantignies F, Roure V, Comet I, Leblanc B, Schuettengruber B, Bonnet J, Tixier V, Mas A, Cavalli G. 2011. Polycomb-dependent regulatory contacts between distant Hox loci in *Drosophila*. *Cell* **144:** 214–226.

* Barlow DP, Bartolomei MS. 2014. Genomic imprinting in mammals. *Cold Spring Harb Perspect Biol* **6:** a018382.

Bau D, Sanyal A, Lajoie BR, Capriotti E, Byron M, Lawrence JB, Dekker J, Marti-Renom MA. 2010. The three-dimensional folding of the α-globin gene domain reveals formation of chromatin globules. *Nat Struct Mol Biol* **18:** 107–114.

Bender MA, Byron R, Ragoczy T, Telling A, Bulger M, Groudine M. 2006. Flanking HS-62.5 and 3′ HS1, and regions upstream of the LCR, are not required for β-globin transcription. *Blood* **108:** 1395–1401.

Beyer AL, Miller OL, McKnight SL. 1980. Ribonucleoprotein structure in nascent hnRNA is nonrandom and sequence-dependent. *Cell* **20:** 75–84.

Boyle S, Gilchrist S, Bridger JM, Mahy NL, Ellis JA, Bickmore WA. 2001. The spatial organization of human chromosomes within the nuclei of normal and emerin-mutant cells. *Hum Mol Genet* **10:** 211–219.

Branco MR, Pombo A. 2006. Intermingling of chromosome territories in interphase suggests role in translocations and transcription-dependent associations. *PLoS Biol* **4:** e138.

Brickner JH, Walter P. 2004. Gene recruitment of the activated INO1 locus to the nuclear membrane. *PLoS Biol* **2:** e342.

* Brockdorff N, Turner BM. 2014. Dosage compensation in mammals. *Cold Spring Harb Perspect Biol* doi: 10.1101/cshperspect.a019406.

Brown JM, Leach J, Reittie JE, Atzberger A, Lee-Prudhoe J, Wood WG, Higgs DR, Iborra FJ, Buckle VJ. 2006. Coregulated human globin genes are frequently in spatial proximity when active. *J Cell Biol* **172:** 177–187.

Bulger M, Groudine M. 1999. Looping versus linking: Toward a model for long-distance gene activation. *Genes Dev* **13:** 2465–2477.

Bulger M, Groudine M. 2011. Functional and mechanistic diversity of distal transcription enhancers. *Cell* **144:** 327–339.

* Busslinger M, Tarakhovsky A. 2014. Epigenetic control of immunity. *Cold Spring Harb Perspect Biol* **6:** a019307.

Cai S, Lee CC, Kohwi-Shigematsu T. 2006. SATB1 packages densely looped, transcriptionally active chromatin for coordinated expression of cytokine genes. *Nat Genet* **38:** 1278–1288.

Casolari JM, Brown CR, Komili S, West J, Hieronymus H, Silver PA. 2004. Genome-wide localization of the nuclear transport machinery couples transcriptional status and nuclear organization. *Cell* **117:** 427–439.

Cavalli G, Misteli T. 2013. Functional implications of genome topology. *Nat Struct Mol Biol* **20:** 290–299.

Chiarle R, Zhang Y, Frock RL, Lewis SM, Molinie B, Ho YJ, Myers DR, Choi VW, Compagno M, Malkin DJ, et al. 2011. Genome-wide translocation sequencing reveals mechanisms of chromosome breaks and rearrangements in B cells. *Cell* **147:** 107–119.

Cremer T, Cremer C, Schneider T, Baumann H, Hens L, Kirsch-Volders M. 1982. Analysis of chromosome positions in the interphase nucleus of Chinese hamster cells by laser-UV-microirradiation experiments. *Hum Genet* **62:** 201–209.

Cremer M, Kupper K, Wagler B, Wizelman L, von Hase J J, Weiland Y, Kreja L, Diebold J, Speicher MR, Cremer T. 2003. Inheritance of gene density-related higher order chromatin arrangements in normal and tumor cell nuclei. *J Cell Biol* **162:** 809–820.

Croft JA, Bridger JM, Boyle S, Perry P, Teague P, Bickmore WA. 1999. Differences in the localization and morphology of chromosomes in the human nucleus. *J Cell Biol* **145:** 1119–1131.

Dean A. 2011. In the loop: Long range chromatin interactions and gene regulation. *Brief Funct Genomics* **10:** 3–10.

Dechat T, Adam SA, Taimen P, Shimi T, Goldman RD. 2010. Nuclear lamins. *Cold Spring Harb Perspect Biol* **2:** a000547.

Dekker J. 2006. The three 'C's of chromosome conformation capture: Controls, controls, controls. *Nat Methods* **3:** 17–21.

Dekker J, Rippe K, Dekker M, Kleckner N. 2002. Capturing chromosome conformation. *Science* **295:** 1306–1311.

de Laat W, Grosveld F. 2003. Spatial organization of gene expression: The active chromatin hub. *Chromosome Res* **11:** 447–459.

Deng W, Blobel GA. 2010. Do chromatin loops provide epigenetic gene expression states? *Curr Opin Genet Dev* **20:** 548–554.

D'Haene B, Attanasio C, Beysen D, Dostie J, Lemire E, Bouchard P, Field M, Jones K, Lorenz B, Menten B, et al. 2009. Disease-causing 7.4 kb *cis*-regulatory deletion disrupting conserved non-coding sequences and their interaction with the FOXL2 promotor: Implications for mutation screening. *PLoS Genet* **5:** e1000522.

Dixon JR, Selvaraj S, Yue F, Kim A, Li Y, Shen Y, Hu M, Liu JS, Ren B. 2012. Topological domains in mammalian genomes identified by analysis of chromatin interactions. *Nature* **485:** 376–380.

Donohoe ME, Silva SS, Pinter SF, Xu N, Lee JT. 2009. The pluripotency factor Oct4 interacts with Ctcf and also controls X-chromosome pairing and counting. *Nature* **460:** 128–132.

Dostie J, Richmond TA, Arnaout RA, Selzer RR, Lee WL, Honan TA, Rubio ED, Krumm A, Lamb J, Nusbaum C, et al. 2006. Chromosome Conformation Capture Carbon Copy (5C): A massively parallel solution for mapping interactions between genomic elements. *Genome Res* **16:** 1299–1309.

Drissen R, Palstra RJ, Gillemans N, Splinter E, Grosveld F, Philipsen S, de Laat W. 2004. The active spatial organization of the β-globin locus requires the transcription factor EKLF. *Genes Dev* **18:** 2485–2490.

Dundr M, Misteli T. 2010. Biogenesis of nuclear bodies. *Cold Spring Harb Perspect Biol* **2:** a000711.

Dundr M, Hoffmann-Rohrer U, Hu Q, Grummt I, Rothblum LI, Phair RD, Misteli T. 2002. A kinetic framework for a mammalian RNA polymerase in vivo. *Science* **298:** 1623–1626.

Egecioglu D, Brickner JH. 2011. Gene positioning and expression. *Curr Opin Cell Biol* **23:** 338–345.

ENCODE Project Consortium. 2011. A user's guide to the Encyclopedia of DNA Elements. *PLoS Biol* **9:** e1001046.

ENCODE Project Consortium, Bernstein BE, Birney E, Dunham I, Green ED, Gunter C, Snyder M. 2012. An integrated encyclopedia of DNA elements in the human genome. *Nature* **489:** 57–74.

Eskiw CH, Cope NF, Clay I, Schoenfelder S, Nagano T, Fraser P. 2011. Transcription factories and nuclear organization of the genome. *Cold Spring Harb Symp Quant Biol* **75:** 501–506.

Felsenfeld G, Groudine M. 2003. Controlling the double helix. *Nature* **421:** 448–453.

Finlan LE, Sproul D, Thomson I, Boyle S, Kerr E, Perry P, Ylstra B, Chubb JR, Bickmore WA. 2008. Recruitment to the nuclear periphery can alter expression of genes in human cells. *PLoS Genet* **4:** e1000039.

Fullwood MJ, Liu MH, Pan YF, Liu J, Xu H, Mohamed YB, Orlov YL, Velkov S, Ho A, Mei PH, et al. 2009. An oestrogen-receptor-α-bound human chromatin interactome. *Nature* **462:** 58–64.

Fuss SH, Omura M, Mombaerts P. 2007. Local and *cis* effects of the H element on expression of odorant receptor genes in mouse. *Cell* **130:** 373–384.

Fussner E, Ching RW, Bazett-Jones DP. 2010. Living without 30 nm chromatin fibers. *Trends Biochem Sci* **36:** 1–6.

Gheldof N, Smith EM, Tabuchi TM, Koch CM, Dunham I, Stamatoyannopoulos JA, Dekker J. 2010. Cell-type-specific long-range looping interactions identify distant regulatory elements of the *CFTR* gene. *Nucleic Acids Res* **38:** 4325–4336.

Gibcus JH, Dekker J. 2013. The hierarchy of the 3D genome. *Mol Cell* **49:** 773–782.

* Grossniklaus U, Paro R. 2014. Transcriptional silencing by *Polycomb*-group proteins. *Cold Spring Harb Perspect Biol* **6:** a019331.

Guelen L, Pagie L, Brasset E, Meuleman W, Faza MB, Talhout W, Eussen BH, de Klein A, Wessels L, de Laat W, et al. 2008. Domain organization of human chromosomes revealed by mapping of nuclear lamina interactions. *Nature* **453:** 948–951.

Guo C, Gerasimova T, Hao H, Ivanova I, Chakraborty T, Selimyan R, Oltz EM, Sen R. 2011. Two forms of loops generate the chromatin conformation of the immunoglobulin heavy-chain gene locus. *Cell* **147:** 332–343.

Hadjur S, Williams LM, Ryan NK, Cobb BS, Sexton T, Fraser P, Fisher AG, Merkenschlager M. 2009. Cohesins form chromosomal *cis*-interactions at the developmentally regulated *IFNG* locus. *Nature* **460:** 410–413.

Haeusler RA, Pratt-Hyatt M, Good PD, Gipson TA, Engelke DR. 2008. Clustering of yeast tRNA genes is mediated by specific association of condensin with tRNA gene transcription complexes. *Genes Dev* **22:** 2204–2214.

Hakim O, Misteli T. 2012. Chromosome conformation techniques. *Cell* **148:** 1068.

Hakim O, Resch W, Yamane A, Klein I, Kieffer-Kwon KR, Jankovic M, Oliveira T, Bothmer A, Voss TC, Ansarah-Sobrinho C, et al. 2012. DNA damage defines sites of recurrent chromosomal translocations in B lymphocytes. *Nature* **484:** 69–74.

Jackson DA, Hassan AB, Errington RJ, Cook PR. 1993. Visualization of focal sites of transcription within human nuclei. *EMBO J* **12:** 1059–1065.

Kagey MH, Newman JJ, Bilodeau S, Zhan Y, Orlando DA, van Berkum NL, Ebmeier CC, Goossens J, Rahl PB, Levine SS, et al. 2010. Mediator and cohesin connect gene expression and chromatin architecture. *Nature* **467:** 430–435.

Kalhor R, Tjong H, Jayahilaka N, Alber F, Chen L. 2011. Genome architectures revealed by tethered chromosome conformation capture and population-based modeling. *Nat Biotechnol* **30:** 90–98.

Kalverda B, Pickersgill H, Shloma VV, Fornerod M. 2010. Nucleoporins directly stimulate expression of developmental and cell-cycle genes inside the nucleoplasm. *Cell* **140:** 360–371.

 Cite as *Cold Spring Harb Perspect Biol* doi: 10.1101/cshperspect.a019356

Karpen GH, Schaefer JE, Laird CD. 1988. A *Drosophila* rRNA gene located in euchromatin is active in transcription and nucleolus formation. *Genes Dev* **2:** 1745–1763.

Kim TH, Abdullaev ZK, Smith AD, Ching KA, Loukinov DI, Green RD, Zhang MQ, Lobanenkov VV, Ren B. 2007. Analysis of the vertebrate insulator protein CTCF-binding sites in the human genome. *Cell* **128:** 1231–1245.

Kim SI, Bresnick EH, Bultman SJ. 2009a. BRG1 directly regulates nucleosome structure and chromatin looping of the α globin locus to activate transcription. *Nucleic Acids Res* **37:** 6019–6027.

Kim SI, Bultman SJ, Kiefer CM, Dean A, Bresnick EH. 2009b. BRG1 requirement for long-range interaction of a locus control region with a downstream promoter. *Proc Natl Acad Sci* **106:** 2259–2264.

Kind J, van Steensel B. 2010. Genome-nuclear lamina interactions and gene regulation. *Curr Opin Cell Biol* **22:** 320–325.

Klein IA, Resch W, Jankovic M, Oliveira T, Yamane A, Nakahashi H, Di Virgilio M, Bothmer A, Nussenzweig A, Robbiani DF, et al. 2011. Translocation-capture sequencing reveals the extent and nature of chromosomal rearrangements in B lymphocytes. *Cell* **147:** 95–106.

Kosak ST, Skok JA, Medina KL, Riblet R, Le Beau MM, Fisher AG, Singh H. 2002. Subnuclear compartmentalization of immunoglobulin loci during lymphocyte development. *Science* **296:** 158–162.

Kumaran RI, Spector DL. 2008. A genetic locus targeted to the nuclear periphery in living cells maintains its transcriptional competence. *J Cell Biol* **180:** 51–65.

Lanctot C, Cheutin T, Cremer M, Cavalli G, Cremer T. 2007. Dynamic genome architecture in the nuclear space: Regulation of gene expression in three dimensions. *Nat Rev Genet* **8:** 104–115.

Lanzuolo C, Roure V, Dekker J, Bantignies F, Orlando V. 2007. Polycomb response elements mediate the formation of chromosome higher-order structures in the bithorax complex. *Nat Cell Biol* **9:** 1167–1174.

Li G, Reinberg D. 2011. Chromatin higher-order structures and gene regulation. *Curr Opin Genet Dev* **21:** 175–186.

Li G, Fullwood MJ, Xu H, Mulawadi FH, Velkov S, Vega V, Ariyaratne PN, Mohamed YB, Ooi HS, Tennakoon C, et al. 2010. ChIA-PET tool for comprehensive chromatin interaction analysis with paired-end tag sequencing. *Genome Biol* **11:** R22.

Lieberman-Aiden E, van Berkum NL, Williams L, Imakaev M, Ragoczy T, Telling A, Amit I, Lajoie BR, Sabo PJ, Dorschner MO, et al. 2009. Comprehensive mapping of long-range interactions reveals folding principles of the human genome. *Science* **326:** 289–293.

Light WH, Brickner DG, Brand VR, Brickner JH. 2010. Interaction of a DNA zip code with the nuclear pore complex promotes H2A.Z incorporation and INO1 transcriptional memory. *Mol Cell* **40:** 112–125.

Lin C, Yang L, Tanasa B, Hutt K, Ju BG, Ohgi K, Zhang J, Rose DW, Fu XD, Glass CK, et al. 2009. Nuclear receptor-induced chromosomal proximity and DNA breaks underlie specific translocations in cancer. *Cell* **139:** 1069–1083.

* Lomvardas S, Maniatis T. 2014. Epigenetics regulation in neuronal development. *Cold Spring Harb Perspect Biol* doi: 10.1101/cshperspect.a019489.

Lomvardas S, Barnea G, Pisapia DJ, Mendelsohn M, Kirkland J, Axel R. 2006. Interchromosomal interactions and olfactory receptor choice. *Cell* **126:** 403–413.

Mamanova L, Coffey AJ, Scott CE, Kozarewa I, Turner EH, Kumar A, Howard E, Shendure J, Turner DJ. 2010. Target-enrichment strategies for next-generation sequencing. *Nat Methods* **7:** 111–118.

Mani RS, Tomlins SA, Callahan K, Ghosh A, Nyati MK, Varambally S, Palanisamy N, Chinnaiyan AM. 2009. Induced chromosomal proximity and gene fusions in prostate cancer. *Science* **326:** 1230.

Masui O, Bonnet I, Le Baccon P, Brito I, Pollex T, Murphy N, Hupe P, Barillot E, Belmont AS, Heard E. 2011. Live-cell chromosome dynamics and outcome of X chromosome pairing events during ES cell differentiation. *Cell* **145:** 447–458.

Mathas S, Kreher S, Meaburn KJ, Johrens K, Lamprecht B, Assaf C, Sterry W, Kadin ME, Daibata M, Joos S, et al. 2009. Gene deregulation and spatial genome reorganization near breakpoints prior to formation of translocations in anaplastic large cell lymphoma. *Proc Natl Acad Sci* **106:** 5831–5836.

McCord RP, Nazario-Toole A, Zhang H, Chines PS, Zhan Y, Erdos MR, Collins FS, Dekker J, Cao K. 2013. Correlated alterations in genome organization, histone methylation, and DNA-lamin A/C interactions in Hutchinson-Gilford progeria syndrome. *Genome Res* **23:** 260–269.

McStay B, Grummt I. 2008. The epigenetics of rRNA genes: From molecular to chromosome biology. *Annu Rev Cell Dev Biol* **24:** 131–157.

Meaburn KJ, Misteli T. 2007. Cell biology: Chromosome territories. *Nature* **445:** 379–781.

Meaburn KJ, Misteli T. 2008. Locus-specific and activity-independent gene repositioning during early tumorigenesis. *J Cell Biol* **180:** 39–50.

Meister P, Towbin BD, Pike BL, Ponti A, Gasser SM. 2010. The spatial dynamics of tissue-specific promoters during *C. elegans* development. *Genes Dev* **24:** 766–782.

Misteli T. 2007. Beyond the sequence: Cellular organization of genome function. *Cell* **128:** 787–800.

Misteli T. 2010. Higher-order genome organization in human disease. *Cold Spring Harb Perspect Biol* **2:** a000794.

Misteli T. 2011. The inner life of the genome. *Sci Am* **304:** 66–73.

Misteli T, Soutoglou E. 2009. The emerging role of nuclear architecture in DNA repair and genome maintenance. *Nat Rev Mol Cell Biol* **10:** 243–254.

Montavon T, Soshnikova N, Mascrez B, Joye E, Thevenet L, Splinter E, de Laat W, Spitz F, Duboule D. 2011. A regulatory archipelago controls *Hox* genes transcription in digits. *Cell* **147:** 1132–1145.

Montes de Oca R, Andreassen PR, Wilson KL. 2011. Barrier-to-autointegration factor influences specific histone modifications. *Nucleus* **2:** 580–590.

Morris JR, Geyer PK, Wu CT. 1999. Core promoter elements can regulate transcription on a separate chromosome in trans. *Genes Dev* **13:** 253–258.

Neves H, Ramos C, da Silva MG, Parreira A, Parreira L. 1999. The nuclear topography of ABL, BCR, PML, and RARalpha genes: Evidence for gene proximity in specific phases of the cell cycle and stages of hematopoietic differentiation. *Blood* **93:** 1197–1207.

Noordermeer D, de Wit E, Klous P, van de Werken H, Simonis M, Lopez-Jones M, Eussen B, de Klein A, Singer RH, de Laat W. 2011. Variegated gene expression caused by cell-specific long-range DNA interactions. *Nat Cell Biol* **13:** 944–951.

Nora EP, Lajoie BR, Schulz EG, Giorgetti L, Okamoto I, Servant N, Piolot T, van Berkum NL, Meisig J, Sedat J, et al. 2012. Spatial partitioning of the regulatory landscape of the X-inactivation centre. *Nature* **485:** 381–385.

Osborne CS, Chakalova L, Brown KE, Carter D, Horton A, Debrand E, Goyenechea B, Mitchell JA, Lopes S, Reik W, et al. 2004. Active genes dynamcially colocalize to shared sites of ongoing transcription. *Nat Genet* **36:** 1065–1071.

Osborne CS, Chakalova L, Mitchell JA, Horton A, Wood AL, Bolland DJ, Corcoran AE, Fraser P. 2007. Myc dynamically and preferentially relocates to a transcription factory occupied by Igh. *PLoS Biol* **5:** e192.

Ott CJ, Blackledge NP, Kerschner JL, Leir SH, Crawford GE, Cotton CU, Harris A. 2009. Intronic enhancers coordinate epithelial-specific looping of the active *CFTR* locus. *Proc Natl Acad Sci* **106:** 19934–19939.

Ou SA, Chang E, Lee S, So K, Wu CT, Morris JR. 2009. Effects of chromosomal rearrangements on transvection at the *yellow* gene of *Drosophila melanogaster. Genetics* **183:** 483–496.

Palstra RJ, Tolhuis B, Splinter E, Nijmeijer R, Grosveld F, de Laat W. 2003. The β-globin nuclear compartment in development and erythroid differentiation. *Nat Genet* **35:** 190–194.

Papantonis A, Cook PR. 2010. Genome architecture and the role of transcription. *Curr Opin Cell Biol* **22:** 271–276.

Papantonis A, Larkin JD, Wada Y, Ohta Y, Ihara S, Kodama T, Cook PR. 2011. Active RNA polymerases: Mobile or immobile molecular machines? *PLoS Biol* **8:** e1000419.

Parada L, McQueen P, Munson P, Misteli T. 2002. Conservation of relative chromosome positioning in normal and cancer cells. *Curr Biol* **12:** 1692.

Parada LA, Roix JJ, Misteli T. 2003. An uncertainty principle in chromosome positioning. *Trends Cell Biol* **13:** 393–396.

Parada L, McQueen P, Misteli T. 2004. Tissue-specific spatial organization of genomes. *Genome Biol* **7:** R44.

Pederson T. 2010. The nucleolus. *Cold Spring Harb Perspect Biol* **3:** a000638.

Phillips JE, Corces VG. 2009. CTCF: Master weaver of the genome. *Cell* **137:** 1194–1211.

Phizicky EM, Hopper AK. 2010. tRNA biology charges to the front. *Genes Dev* **24:** 1832–1860.

Pickersgill H, Kalverda B, de Wit E, Talhout W, Fornerod M, van Steensel B. 2006. Characterization of the *Drosophila melanogaster* genome at the nuclear lamina. *Nat Genet* **38:** 1005–1014.

Rajapakse I, Groudine M. 2011. On emerging nuclear order. *J Cell Biol* **192:** 711–721.

Reddy KL, Zullo JM, Bertolino E, Singh H. 2008. Transcriptional repression mediated by repositioning of genes to the nuclear lamina. *Nature* **452:** 243–247.

Roix JJ, McQueen PG, Munson PJ, Parada LA, Misteli T. 2003. Spatial proximity of translocation-prone gene loci in human lymphomas. *Nat Genet* **34:** 287–291.

Sanyal A, Bau D, Marti-Renom MA, Dekker J. 2011. Chromatin globules: A common motif of higher order chromosome structure? *Curr Opin Cell Biol* **23:** 325–331.

Sanyal A, Lajoie BR, Jain G, Dekker J. 2012. The long-range interaction landscape of gene promoters. *Nature* **489:** 109–113.

Sawado T, Halow J, Bender MA, Groudine M. 2003. The β-globin locus control region (LCR) functions primarily by enhancing the transition from transcription initiation to elongation. *Genes Dev* **17:** 1009–1018.

Scaffidi P, Misteli T. 2008. Lamin A-dependent misregulation of adult stem cells associated with accelerated ageing. *Nat Cell Biol* **10:** 452–459.

Schoenfelder S, Clay I, Fraser P. 2010a. The transcriptional interactome: Gene expression in 3D. *Curr Opin Genet Dev* **20:** 127–133.

Schoenfelder S, Sexton T, Chakalova L, Cope NF, Horton A, Andrews S, Kurukuti S, Mitchell JA, Umlauf D, Dimitrova DS, et al. 2010b. Preferential associations between co-regulated genes reveal a transcriptional interactome in erythroid cells. *Nat Genet* **42:** 53–61.

Sexton T, Yaffe E, Kenigsberg E, Bantignies F, Leblanc B, Hoichman M, Parrinello H, Tanay A, Cavalli G. 2012. Three-dimensional folding and functional organization principles of the *Drosophila* genome. *Cell* **148:** 458–472.

Shen Y, Yue F, McCleary DF, Ye Z, Edsall L, Kuan S, Wagner U, Dixon J, Lee L, Lobanenkov VV, et al. 2012. A map of the *cis*-regulatory sequences in the mouse genome. *Nature* **488:** 116–120.

Simonis M, Klous P, Splinter E, Moshkin Y, Willemsen R, de Wit E, van Steensel B, de Laat W. 2006. Nuclear organization of active and inactive chromatin domains uncovered by chromosome conformation capture-on-chip (4C). *Nat Genet* **38:** 1348–1354.

Soutoglou E, Dorn JF, Sengupta K, Jasin M, Nussenzweig A, Ried T, Danuser G, Misteli T. 2007. Positional stability of single double-strand breaks in mammalian cells. *Nat Cell Biol* **9:** 675–682.

Spilianakis CG, Lalioti MD, Town T, Lee GR, Flavell RA. 2005. Interchromosomal associations between alternatively expressed loci. *Nature* **435:** 637–645.

Splinter E, de Laat W. 2011. The complex transcription regulatory landscape of our genome: Control in three dimensions. *EMBO J* **30:** 4345–4355.

Splinter E, Heath H, Kooren J, Palstra RJ, Klous P, Grosveld F, Galjart N, de Laat W. 2006. CTCF mediates long-range chromatin looping and local histone modification in the β-globin locus. *Genes Dev* **20:** 2349–2354.

Sullivan GJ, Bridger JM, Cuthbert AP, Newbold RF, Bickmore WA, McStay B. 2001. Human acrocentric chromosomes with transcriptionally silent nucleolar organizer regions associate with nucleoli. *EMBO J* **20:** 2867–2874.

Takizawa T, Meaburn KJ, Misteli T. 2008. The meaning of gene positioning. *Cell* **135:** 9–13.

Thompson M, Haeusler RA, Good PD, Engelke DR. 2003. Nucleolar clustering of dispersed tRNA genes. *Science* **302:** 1399–1401.

Tolhuis B, Palstra RJ, Splinter E, Grosveld F, de Laat W. 2002. Looping and interaction between hypersensitive sites in the active β-globin locus. *Mol Cell* **10:** 1453–1465.

Vakoc CR, Letting DL, Gheldof N, Sawado T, Bender MA, Groudine M, Weiss MJ, Dekker J, Blobel GA. 2005. Proximity among distant regulatory elements at the β-globin locus requires GATA-1 and FOG-1. *Mol Cell* **17:** 453–462.

van Steensel B, Dekker J. 2011. Genomics tools for unraveling chromosome architecture. *Nat Biotechnol* **28:** 1089–1095.

Vernimmen D, De Gobbi M, Sloane-Stanley JA, Wood WG, Higgs DR. 2007. Long-range chromosomal interactions regulate the timing of the transition between poised and active gene expression. *EMBO J* **26:** 2041–2051.

Wiech T, Timme S, Riede F, Stein S, Schuricke M, Cremer C, Werner M, Hausmann M, Walch A. 2005. Human archival tissues provide a valuable source for the analysis of spatial genome organization. *Histochem Cell Biol* **123:** 229–238.

Wright JB, Brown SJ, Cole MD. 2010. Upregulation of c-*MYC* in *cis* through a large chromatin loop linked to a cancer risk-associated single-nucleotide polymorphism in colorectal cancer cells. *Mol Cell Biol* **30:** 1411–1420.

Xu N, Tsai CL, Lee JT. 2006. Transient homologous chromosome pairing marks the onset of X inactivation. *Science* **311:** 1149–1152.

Yao H, Brick K, Evrard Y, Xiao T, Camerini-Otero RD, Felsenfeld G. 2010. Mediation of CTCF transcriptional insulation by DEAD-box RNA-binding protein p68 and steroid receptor RNA activator SRA. *Genes Dev* **24:** 2543–2555.

Yokochi T, Poduch K, Ryba T, Lu J, Hiratani I, Tachibana M, Shinkai Y, Gilbert DM. 2009. G9a selectively represses a class of late-replicating genes at the nuclear periphery. *Proc Natl Acad Sci* **106:** 19363–19368.

Young RA. 2011. Control of the embryonic stem cell state. *Cell* **144:** 940–954.

Yusufzai TM, Tagami H, Nakatani Y, Felsenfeld G. 2004. CTCF tethers an insulator to subnuclear sites, suggesting shared insulator mechanisms across species. *Mol Cell* **13:** 291–298.

Zhang Y, McCord RP, Ho YJ, Lajoie BR, Hildebrand DG, Simon AC, Becker MS, Alt FW, Dekker J. 2012. Spatial organization of the mouse genome and its role in recurrent chromosomal translocations. *Cell* **148:** 908–921.

Zhao Z, Tavoosidana G, Sjolinder M, Gondor A, Mariano P, Wang S, Kanduri C, Lezcano M, Sandhu KS, Singh U, et al. 2006. Circular chromosome conformation capture (4C) uncovers extensive networks of epigenetically regulated intra- and interchromosomal interactions. *Nat Genet* **38:** 1341–1347.

Zink D, Amaral MD, Englmann A, Land S, Clarke LA, Rudolph C, Alt F, Luther K, Braz C, Sadoni N, et al. 2004. Transcription-dependent spatial arrangement of CFTR and adjacent genes in human cell nuclei. *J Cell Biol* **166:** 815–825.

Cite as *Cold Spring Harb Perspect Biol* doi: 10.1101/cshperspect.a019356

C H A P T E R 20

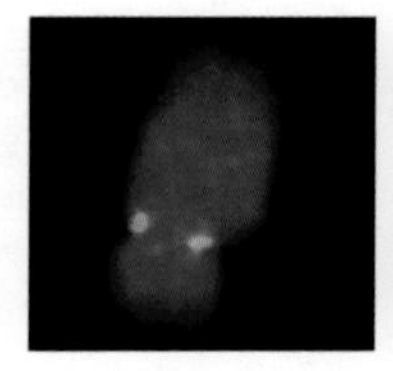
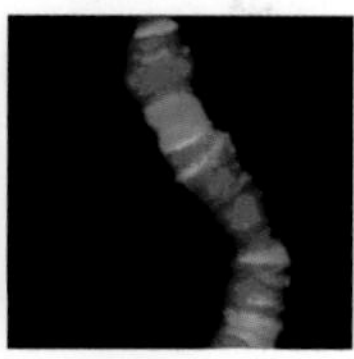

Histone Variants and Epigenetics

Steven Henikoff[1] and M. Mitchell Smith[2]

[1]Howard Hughes Medical Institute, Fred Hutchinson Cancer Research Center, Seattle, Washington 98109-1024;
[2]Department of Microbiology, University of Virginia, Charlottesville, Virginia 22908

Correspondence: steveh@fhcrc.org

SUMMARY

Histones package and compact DNA by assembling into nucleosome core particles. Most histones are synthesized at S phase for rapid deposition behind replication forks. In addition, the replacement of histones deposited during S phase by variants that can be deposited independently of replication provide the most fundamental level of chromatin differentiation. Alternative mechanisms for depositing different variants can potentially establish and maintain epigenetic states. Variants have also evolved crucial roles in chromosome segregation, transcriptional regulation, DNA repair, and other processes. Investigations into the evolution, structure, and metabolism of histone variants provide a foundation for understanding the participation of chromatin in important cellular processes and in epigenetic memory.

Outline

Chapter-opening figures: (*Left*) Courtesy of Peter Warburton; (*right*) from Schwartz and Ahmad (2005).

Cite as *Cold Spring Harb Perspect Biol* doi: 10.1101/cshperspect.a019364

OVERVIEW

Histones package DNA by assembling into nucleosome core particles, whereas the double helix wraps around. Over evolutionary time, histone-fold domain proteins have diversified from archaeal ancestors into the four distinct subunits that comprise the familiar octamer of the eukaryotic nucleosome. Further diversification of histones into variants results in differentiation of chromatin that can have epigenetic consequences. Investigations into the evolution, structure, and metabolism of histone variants provides a foundation for understanding the participation of chromatin in important cellular processes and in epigenetic memory.

Most histones are synthesized at S phase for rapid deposition behind replication forks to fill in gaps resulting from the distribution of preexisting histones. In addition, the replacement of canonical S-phase histones by variants, independent of replication, can potentially differentiate chromatin. The replacement of a canonical histone by a noncanonical variant is a dynamic process that changes the composition of chromatin.

The differentiation of chromatin by a histone variant is especially conspicuous at centromeres, in which the H3 variant, CENP-A, is assembled into specialized nucleosomes that form the foundation for kinetochore assembly. A centromeric H3 (cenH3) counterpart of CENP-A is found in all eukaryotes. In plants and animals, the faithful assembly of cenH3-containing nucleosomes at centromeres does not appear to require centromeric DNA sequences, a spectacular example of epigenetic inheritance. Some cenH3s have evolved adaptively in regions that contact DNA, which suggests that centromeres compete with each other, and cenH3s and other centromere-specific DNA-binding proteins have adapted in response. This process could account for the large size and complexity of centromeres in plants and animals.

Chromatin can also be differentiated outside of centromeres by incorporation of a constitutively expressed form of H3, called H3.3, which is the substrate for replication-independent nucleosome assembly. Replacement with H3.3 occurs at active genes, a dynamic process with potential epigenetic consequences. Differences between H3 and H3.3 in their complement of covalent modifications might underlie changes in the properties of chromatin at actively transcribed loci.

Several H2A variants can also differentiate or regulate chromatin. H2A.X is defined as a variant by a four-amino-acid carboxy-terminal motif whose serine residue is the site for phosphorylation at sites of DNA double-stranded breaks. Phosphorylation of H2A.X is an early event in double-strand break repair, in which it is thought to concentrate components of the repair machinery. H2A.X phosphorylation also marks the inactive XY bivalent during mammalian spermatogenesis and is required for condensation, pairing, and fertility.

H2A.Z is a structurally diverged variant that has long presented an enigma. Studies in yeast have implicated H2A.Z in establishing transcriptional competence and in counteracting heterochromatic silencing. The biochemical complex that replaces H2A with H2A.Z in nucleosomes is an ATP-dependent nucleosome remodeler, providing the first example of a specific function for a member of this diverse class of chromatin-associated machines.

Two vertebrate-specific variants, macroH2A and H2A.B (also called H2A.Bbd), display contrasting features when packaged into nucleosomes in vitro, with macroH2A impeding and H2A.B facilitating transcription. These features are consistent with their localization patterns on the epigenetically inactivated mammalian X chromosome: macroH2A showing enrichment and H2A.B showing depletion.

The emerging view from these studies is that histone variants and the processes that deposit them into nucleosomes provide a primary differentiation of chromatin that might serve as the basis for epigenetic processes.

 Cite as *Cold Spring Harb Perspect Biol* doi: 10.1101/cshperspect.a019364

1 DNA IS PACKAGED BY ARCHITECTURAL PROTEINS IN ALL ORGANISMS

The enormous length of the DNA double helix relative to the size of the chromosome that contains it requires tight packaging, and architectural proteins have evolved for this purpose. The first level of packaging shortens the double helix and protects it from damage while still allowing DNA polymerase to gain full access to each base pair every cell cycle. In addition, these architectural proteins facilitate higher-order folding to further reduce the length of a chromosome. Perhaps because of stringent requirements for packaging DNA, only two structural classes of architectural proteins are found in nearly all cellular life-forms (Talbert and Henikoff 2010): HU proteins that package bacterial DNA, and histones that package eukaryotic DNA. Archaeal DNA is packaged by either HU proteins or histones.

Histones package DNA into nucleosome particles, and this architectural role can account for the fact that histones comprise half of the mass of a eukaryotic chromosome. However, histones have also been found to play diverse roles in gene expression, chromosome segregation, DNA repair, and other basic chromosomal processes in eukaryotes. Specific requirements of these chromosomal processes have led to the evolution of distinct histone variants. The incorporation of a variant histone into a nucleosome represents a potentially profound alteration of chromatin. Indeed, some histone variants are deposited by distinct nucleosome assembly complexes, which suggests that chromatin is diversified, at least in part, by the incorporation and replacement of histone variants.

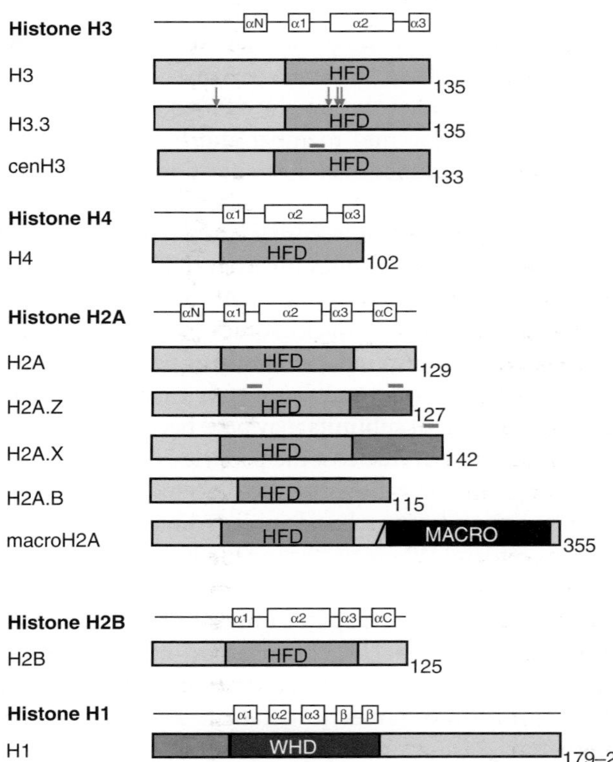

Figure 1. Histone variants. Protein domain structure for the core histones (H3, H4, H2A, and H2B), linker histone H1, and variants of histones H3 and H2A. The histone-fold domain (HFD) is where histone dimerization occurs. Regions of sequence variation in histone variants are indicated in red. WHD, winged-helix domain.

The four core histones, H2A, H2B, H3, and H4, differ with respect to their propensity to diversify into variants. For example, humans have only one H4 isotype but several H2A paralogs with different properties and functions. Evidently, the different positions of the core histones within the nucleosome particle have subjected them to different evolutionary forces, leading to important diversifications of H2A and H3 but not to H2B and H4 (Fig. 1). The availability of genomic sequences from a wide variety of eukaryotes allows us to conclude that these diversifications have occurred at various times during eukaryotic evolution. However, the evident diversification of an ancestral histone-fold protein into the familiar four core histones must have occurred early in the evolution of the eukaryotic nucleus or perhaps before. By considering these ancient events, we gain insight into the forces that have resulted in subsequent diversification into present-day variants.

2 EUKARYOTIC CORE HISTONES EVOLVED FROM ARCHAEAL HISTONES

The eukaryotic nucleosome is a complex structure, consisting of an octamer of four core histones wrapped nearly twice by DNA, with histone tails and linker histones mediating a variety of packaging interactions outside the core particle (Arents et al. 1991; Wolffe 1992; Luger et al. 1997). Archaeal nucleosomes are much simpler, and it is evident that they resemble the ancestral particle from which eukaryotic nucleosomes evolved (Malik and Henikoff 2003). An archaeal nucleosome consists of histone-fold domain proteins that lack tails and form a tetrameric particle that is wrapped only once by DNA. The genomes of all ancient archaeal lineages encode histones (Fig. 2), which implies that the eukaryotic nucleosome evolved from an as-yet unidentified archaeal ancestor. The kinship between archaeal and eukaryotic nucleosomes can be seen by comparing their structures: The backbone of the archeal tetramer nearly superimposes over that of the $(H3\text{-}H4)_2$ tetramer. When archaeal nucleosomes are reconstituted to form chromatin, the resulting fiber behaves similarly to "tetrasomes" of $(H3\text{-}H4)_2$, and when mapped in vivo,

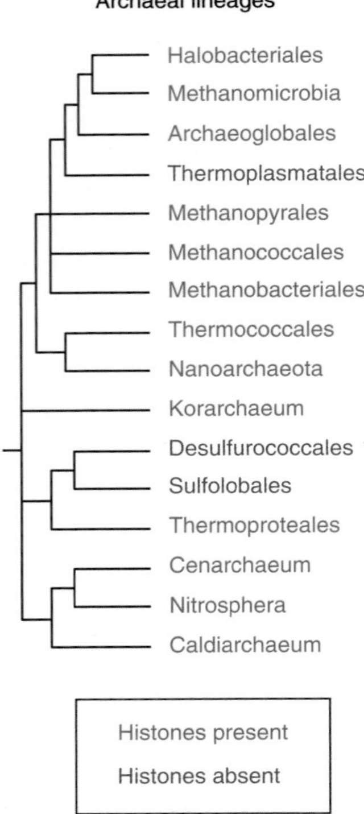

Figure 2. Archaeal cladogram indicating the presence of histones in all ancestral clades. Losses are attributable to horizontal transfer of HU proteins from bacteria. (Modified, with permission, from Brochier-Armanet et al. 2011, © Elsevier.)

they show phasing downstream from transcriptional start sites analogous to what is seen for eukaryotic nucleosomes (Ammar et al. 2012). Therefore, it is thought that eukaryotic nucleosomes evolved from an archaeal ancestor by doubling the number of subunits to allow for a second DNA wrap, and by acquisition of histone tails. In addition, DNA wraps into a right-handed superhelix around archaeal cores, but into a left-handed superhelix around eukaryotic cores.

Further insight into the origin of the eukaryotic nucleosomes comes from examination of the subunit structures of archaeal nucleosomes. Whereas most archaeal histones are undifferentiated monomers or are differentiated into structurally interchangeable variants that come together to form a tetramer, some are head-to-tail dimeric fusions that come together to form a dimer of fused dimers. When two of these fused dimers assemble into a nucleosome particle, each member of the fused pair is in a structurally distinguishable position. By occupying distinct positions in the particle, each member of the archaeal fused dimer will evolve independently, allowing it to adapt to a single position in the nucleosome particle. In contrast, monomers that occupy interchangeable positions are not free to adapt to particular positions. Indeed, the two members of archaeal dimers have diverged from one another in both independent lineages in which they are found. This process provides a possible scenario for the differentiation of an ancestral histone-fold domain protein into four distinct subunits that occupy distinct positions in the eukaryotic nucleosome. Like their presumed archaeal ancestors, eukaryotic histones form dimers, where H2A dimerizes with H2B, and H3 with H4 (which also stably tetramerizes in solution). The structural backbone of an archaeal histone dimer superimposes with those of H2A-H2B and H3-H4 at 2-Å resolution, with the first member of the dimeric repeat superimposing on H2A or H3 and the second member superimposing on H2B or H4. So, although all four eukaryotic histones lack significant sequence similarity to one another and to archaeal histones, the striking structural superposition of dimeric units suggests that eukaryotic histones evolved and differentiated from simpler archaeal ancestors.

The asymmetry of H2A-H2B and H3-H4 dimers, which appears to have originated from archaeal tandem dimers, could have led the way to subsequent diversification of eukaryotic histone variants. Both H2A and H3 correspond to the first member of archaeal tandem histone dimers and both have subsequently diversified multiple times in eukaryotic evolution. In contrast, H2B and H4 correspond to the second member and have shown little (H2B) or no (H4) functional diversification. Both H3 and H2A make homodimeric contacts in the octamer (Fig. 3), whereas H4 and H2B only contact other histones. As a result, changes in the residues involved in homodimerization of either H2A or H3 can potentially resist formation of mixed octamers, allowing nucleosomes containing an H2A or H3 variant to evolve independently of parental nucleosomes. In general, structural features that facilitated independent evolution of subunits may have been prerequisites for diversification of nucleosome particles.

Although we can rationalize the descent of the eukaryotic core histones from archaeal tandem dimers, other basic questions remain. Where did histone tails come from? Did $(H3\text{-}H4)_2$ tetrasomes evolve before acquiring flanking H2A-H2B dimers, or perhaps the four core histones evolved first as (H2A-H2B-H4-H3) "hemisomes" before doubling to form the octamer, or was there some other evolutionary progression from tetramer to octamer? Did these events occur before, during, or after the evolution of the eukaryotic nucleus? Did the emergence of octameric nucleosomes with two DNA wraps allow for the tight packaging of mitotic chromosomes, a eukaryotic-specific invention? Perhaps the sequences of more archaea or of primitive eukaryotes will reveal intermediate forms that can answer these questions.

Cite as *Cold Spring Harb Perspect Biol* doi: 10.1101/cshperspect.a019364

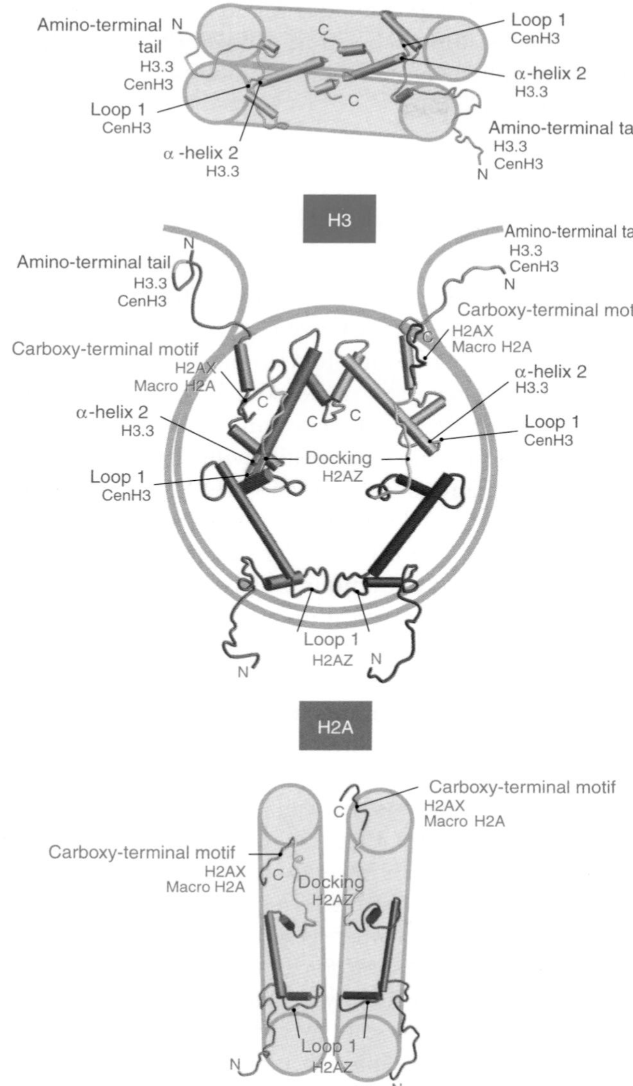

Figure 3. Location of histones H3 (blue) and H2A (brown) in the nucleosome core particle. Differences between variants are highlighted in yellow. (Reprinted, with permission, from Henikoff and Ahmad 2005.)

3 BULK HISTONES ARE DEPOSITED AFTER DNA REPLICATION

The packaging of essentially all DNA in a eukaryotic cell into nucleosomes requires that chromatin is duplicated when DNA replicates. Thus, canonical histones are produced during the DNA synthesis (S) phase of the cell cycle. S-phase coupling of histone synthesis to DNA synthesis is under tight cell-cycle control (Marzluff and Duronio 2002). This is especially evident in animals, in which special processing of histone transcripts by the U7 small nuclear ribonuclear protein complex, and messenger RNA (mRNA) stabilization by the stem-loop-binding protein (SLBP) contributes to the tight coordination of histone synthesis with DNA replication. The need for rapid and massive production of histones during S phase is very likely responsible for the fact that replication-coupled (RC) histones in animals are encoded in clusters that comprise many histone genes. For example, there are 14 H4 genes in the human genome, most of which are found in two major clusters, where these H4 genes are interspersed with other RC histone genes (Marzluff et al. 2002). In animals, RC histones are recognizable by the presence of a 26-bp 3′ sequence that forms a stem-loop for recognition by SLBP when transcribed into histone mRNA. Canonical plant histones are also encoded by multiple genes and are deposited during S phase, although plant histone transcripts are polyadenylated and there does not appear to be a counterpart to SLBP.

To the extent that epigenetic inheritance results from inheritance of a chromatin "state," the process of RC nucleosome assembly has been of intense interest. The biochemistry of the process was elucidated with the development of in vitro systems that could assemble nucleosomes onto replicating DNA. These studies revealed that a three-subunit complex, CAF-1 (chromatin assembly factor 1), acts as a histone chaperone that facilitates the incorporation of H3-H4 as a first step in nucleosome assembly (reviewed in Loyola and Almouzni 2004). CAF-1 was shown to interact with the replication processivity clamp, PCNA, which implies that DNA replication and RC assembly occur in close proximity (Fig. 4). Work in budding yeast revealed that none of the subunits of complexes involved in RC assembly in vitro is essential for growth, suggesting that in vivo, there are redundant mechanisms for RC assembly. The fact that much of yeast chromatin is assembled in a replication-independent (RI) manner (Altheim and Schultz 1999) provides a rationale for this evident redundancy. As we shall see, histone variants are typically deposited by RI nucleosome assembly.

RC assembly is not completely redundant in budding yeast. An intriguing finding is that absence of the large CAF-1 subunit leads to loss of epigenetic silencing at telomeres (Loyola and Almouzni 2004). Moreover, in human cells, depletion of CAF-1 results in the deposition of H3.3 at sites of DNA replication (Ray-Gallet et al. 2011). The connection between RC assembly and epigenetic silencing has been extended to *Arabidopsis*, in which loss of CAF-1 subunits results in a variety of defects attributable to loss of epigenetic memory (Kaya et al. 2001). Although the mechanistic basis for these observations is unknown, it seems clear that the proper deposition of new nucleosomes behind the replication fork is important for maintaining an epigenetically silenced state. An example of the importance of replication-coupled assembly for maintaining a devel-

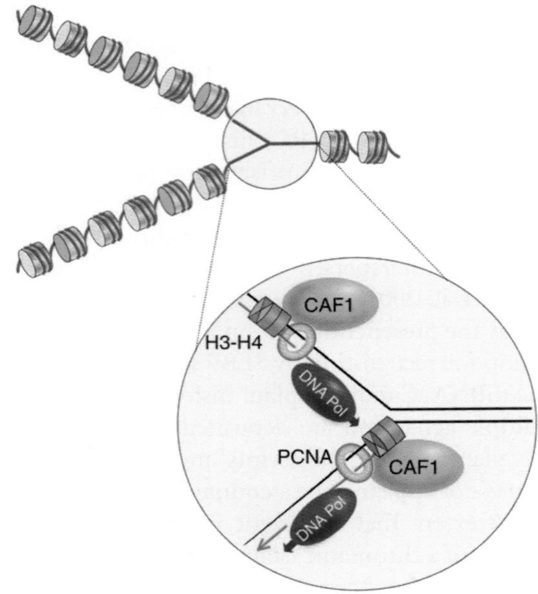

Figure 4. Distribution of old and new nucleosomes at a replication fork. Old nucleosomes (gray disks) are randomly distributed behind the replication fork and new nucleosomes (cyan disks) are deposited in the gaps. CAF-1-mediated nucleosome assembly is depicted on the leading and lagging strand in magnification. DNA polymerase (green); replication processivity clamp, PCNA (gray ring); histone H3-H4 tetramers (cyan); newly synthesized DNA (red lines).

opmental state is the observation that the identity of one of two sister neurons in *Caenorhabditis elegans* is disrupted by mutating either the H3-H3 dimerization interface or CAF-1 (Nakano et al. 2011).

A prerequisite for epigenetic inheritance of a nucleosome state is that preexisting nucleosomes must be distributed to daughter chromatids following replication. Indeed this is the case; classical studies have shown that old nucleosomes are inherited intact and evidently at random to daughter chromatids (Annunziato 2005). However, more recent studies of particular cell types and particular loci have challenged this dogma. Germline stem cells in *Drosophila* male early embryos show asymmetric inheritance of RC H3, but not RI H3.3, whereby old nucleosomes remain in the stem cell and new nucleosomes segregate to the differentiating daughter cell (Tran et al. 2012). Moreover, although RC H3 was not found to split at replication (Xu et al. 2010), the small amount of RI H3.3 that splits was shown to be enriched at active human genes and cell-type-specific enhancers (Huang et al. 2013). Taken together, these two findings suggest that RC nucleosome assembly is profoundly involved in initiating developmental decisions and RI assembly in epigenetic inheritance, resuscitating ideas proposed long before these pathways were elucidated (Weintraub et al. 1976). More work is needed to test these exciting possibilities and to explore the importance of nucleosome deposition pathways in establishing and maintaining epigenetic states (Jenuwein 2001; Henikoff and Ahmad 2005).

4 VARIANT HISTONES ARE DEPOSITED THROUGHOUT THE CELL CYCLE

As we have seen, core histones can be classified based on their ancestral sequence and position in the nucleosome. Linker histones are characterized by a winged-helix domain, rather than a histone-fold domain, and bind to the linker DNA that separates nucleosomes (Wolffe 1992). Although minor variants of these canonical histones exist, they appear to be interchangeable with the major form. For example, mammalian H3.1 and H3.2 differ by a single amino acid that is not known to impart different functional properties to the two isoforms. The existence of multiple genes that produce large amounts of canonical histones for S-phase deposition is typical of eukaryotic genomes. The near ubiquity and overwhelming abundance of canonical S-phase histones has resulted in relatively little attention being paid to histone variants until recently.

The renaissance of interest in histone variants came in part from the realization that they differ from canonical S-phase histones in ways that can lead to profound differentiation of chromatin. One way that they differ is in their mode of incorporation into chromatin. RC assembly incorporates new nucleosomes into gaps between old nucleosomes genome-wide, whereas RI assembly involves local replacement of an existing nucleosome or subunit (Marzluff et al. 2002). RI assembly therefore has the potential of switching a chromatin state by replacing a canonical histone with a variant. Replacing one histone with another also could erase or alter the pattern of posttranslational modifications. Therefore, RI assembly can potentially reset epigenetic states that are thought to be mediated by histones and their modifications. Recent progress in studying histone variants and the processes by which they are deposited has led to new insights into the basis for epigenetic inheritance and remodeling. Below we discuss features of particular histone variants that contribute to chromatin differentiation and might be involved in propagating epigenetic information.

5 CENTROMERES ARE IDENTIFIED BY A SPECIAL H3 VARIANT

A defining feature of the eukaryotic chromosome is the centromere, which is the site of attachment of spindle microtubules at mitosis. The first centromeres to be described in molecular detail were those of budding yeast (*Saccharo-*

myces cerevisiae), in which a 125-bp sequence is necessary and sufficient for centromere formation (Amor et al. 2004b). However, centromeres of plants and animals are very different, typically consisting of megabase arrays of short tandem repeats. Unlike the situation for budding yeast, the role of DNA sequence at these complex centromeres is uncertain because fully functional human neocentromeres are known to form spontaneously at ectopic sites that entirely lack sequences resembling centromeric repeats (Fig. 5A). These and other observations argue against a direct role of DNA sequence in determining the location of centromeres (see Ch. 9 [Allshire and Ekwall 2014]).

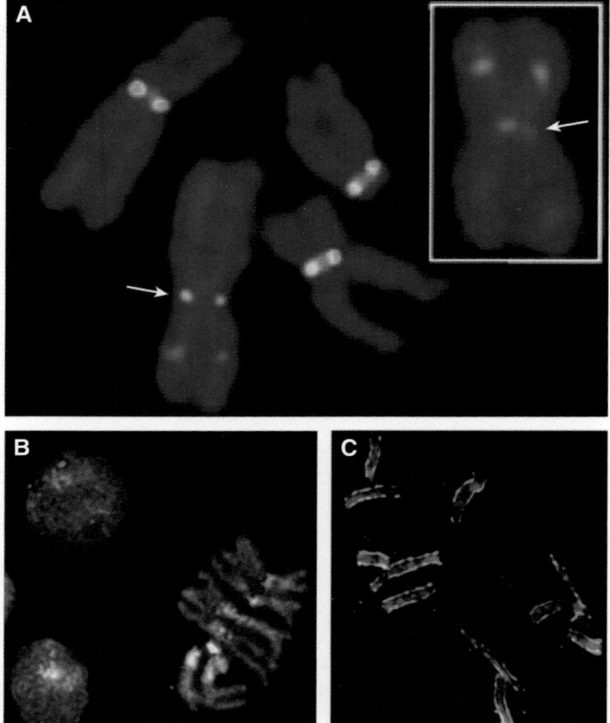

Figure 5. cenH3s at centromeres of eukaryotes. (*A*) Human neocentromeres (indicated by an arrow) lack centromeric α-satellite DNA, but have CENP-A and heterochromatin. Anti-CENP-A staining in green and Anti-CENP-B staining in red (which marks α-satellite DNA) identifies a Chromosome 4 neocentromere that lacks α-satellite (main panel). This Chromosome 4 is otherwise normal, having been transmitted for at least three meiotic generations in normal individuals. Inset shows anti-HP1 staining, which indicates that despite the lack of satellite DNA, heterochromatin forms around active neocentromeres (indicated by arrow). (Reprinted, with permission, from Amor et al. 2004a, © National Academy of Sciences.) (*B*) *Drosophila melanogaster* anti-cenH3 antibody (red) stains centromeres in metaphase chromosomes and throughout interphase. (Image courtesy of Suso Platero.) (*C*) *C. elegans* anti-cenH3 antibody (green) stains the end-to-end holocentromeres of prophase chromosomes (red). (Image courtesy of Landon Moore.)

A key insight into the basis for centromere identity and inheritance came from the identification of a histone H3 variant, CENP-A which was found to localize specifically to centromeres and be incorporated into nucleosomal particles in place of H3 itself (Palmer et al. 1991). Remarkably, CENP-A remains associated with centromeres during the transition from histones to protamines during spermatogenesis, when essentially all other histones are lost (Palmer et al. 1990). This early observation in the study of CENP-A suggested that CENP-A contributes to centromere identity of the male genome. The generality of this insight was not fully appreciated until it was realized that CENP-A is a much better marker for centromeres than DNA sequence (Fig. 5) (Amor et al. 2004b), and that counterparts of CENP-A (cenH3s) can be found in the genomes of all eukaryotes (Talbert and Henikoff 2010). So, although budding yeast centromeres are determined by a 125-bp consensus sequence, this is also the site of a centromeric nucleosome that contains the cenH3 variant. In fission yeast (*Schizosaccharomyces pombe*), an array of cenH3-containing nucleosomes occupies the central core region of the centromere flanked by H3-containing nucleosomes that display heterochromatic features (see Fig. 3 of Ch. 9 [Allshire and Ekwall 2014]; Amor et al. 2004b). In flies and vertebrates, cenH3s are present in arrays that alternate with H3-containing arrays that display a unique pattern of histone modifications (Sullivan and Karpen 2004). Alternation can account for the fact that centromeres occupy only the outside edge of the centromeric constriction of metaphase chromosomes. This is consistent with the observation that in worm "holokinetic" chromosomes, microtubules attach throughout the length of each anaphase chromosome, and cenH3 occupies the leading edge all along its length (Fig. 5C) (Malik and Henikoff 2003). Indeed a unique cenH3 variant is found to precisely mark the centromere in nearly all eukaryotes (Fig. 6A). This apparent ubiquity, and the presence of centromeres to perform mitosis in all eukaryotes, raises the possibility that the first canonical H3 evolved from a cenH3.

Genetic experiments in a variety of eukaryotes have confirmed the essentiality of cenH3 for formation of the kinetochore and for chromosome segregation (Amor et al. 2004b). Because they remain in place throughout the cell cycle, cenH3-containing nucleosomes form the foundation for assembly of other kinetochore proteins during mitosis and meiosis (see Ch. 9 [Allshire and Ekwall 2014]). An outstanding question in chromosome research is just how these proteins interact to provide a linkage between the centromere and spindle microtubules that can hold up to the strong pulling forces exerted on kinetochores at anaphase. Several dozen kinetochore-specific proteins have been identified in yeast (for more detail, see Ch. 9 [Allshire and Ekwall 2014]), although exactly how they in-

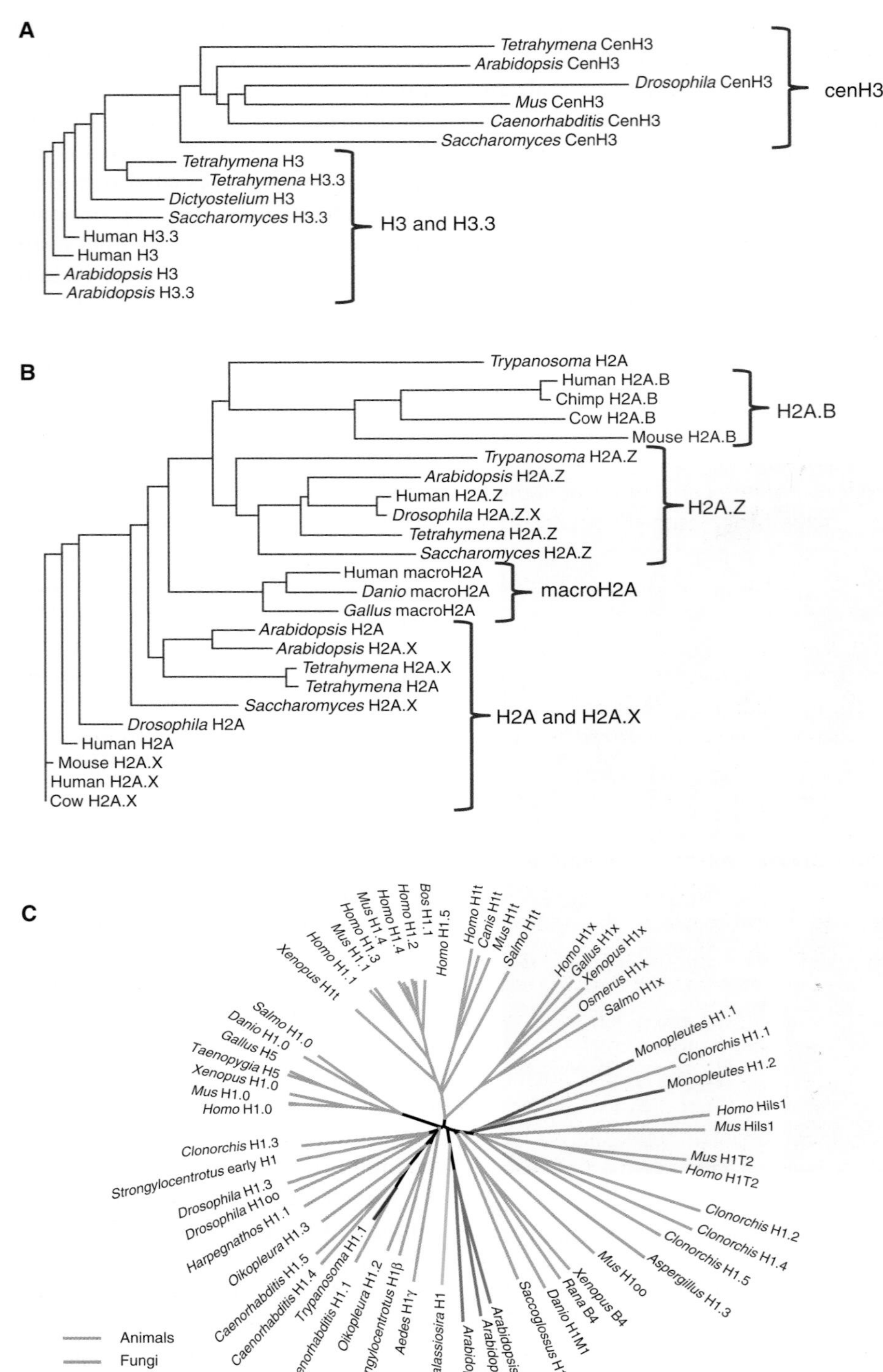

Figure 6. Histone variant phylogenies. Histone sequences from selected species were multiply aligned and neighbor-joining trees were produced using the EBI server (http://www.ebi.ac.uk/Tools/phylogeny). (*A*) Histone H3s. (*B*) Histone H2As. Note that there are no clear phylogenetic distinctions between RC H3 and RI H3.3, and between RC H2A and RI H2A.X. (*C*) H1 variants from diverse eukaryotes show a "star" phylogeny, which suggests that they are functionally interchangeable. (*C*, Modified from Talbert et al. 2012.)

 Cite as *Cold Spring Harb Perspect Biol* doi: 10.1101/cshperspect.a019364

teract with cenH3-containing nucleosomes and other foundation proteins, such as CENP-C, remains unclear.

The evolution of cenH3s is unlike that of any other histone class. Whereas histone H3 is almost invariant in sequence, which reflects extraordinarily strong purifying selection on every residue, cenH3s are evolving rapidly, especially in plant and animal lineages (Talbert and Henikoff 2010). This is most evident from the amino-terminal tails, which differ in length and sequence to such an extent that they cannot be aligned between the cenH3s of different taxonomic groups. Even the histone-fold domain of cenH3 is evolving orders of magnitude faster than that of H3. What is the reason for this striking evolutionary difference between an H3 that functions at centromeres and an H3 that functions everywhere else?

Rapidly evolving regions of *Drosophila* and *Arabidopsis* cenH3 genes display an excess of replacement nucleotide substitutions over what would be expected from the rate of synonymous substitutions (Malik and Henikoff 2009). This excess is a hallmark of adaptive evolution. Adaptive evolution in plants and animals is also seen for another major centromere foundation protein, CENP-C (Malik and Henikoff 2009). Although adaptive evolution is well documented for genes involved in genetic conflicts such as arms races between hosts and parasites, these are the only known essential single-copy genes that are adaptively evolving in any organism. In the case of cenH3 and CENP-C, the regions of adaptive evolution correspond to regions of DNA binding and targeting. This suggests that the major centromere-binding proteins are adapting to the evolving centromeric DNA, thus allowing centromeric chromatin to interact with the conserved kinetochore machinery that connects the centromere to spindle microtubules. It has been proposed that centromeres compete during female meiosis to be included in the egg nucleus rather than being lost as polar bodies (Malik and Henikoff 2009). An arms race would develop leading to expansion of centromeres, probably by unequal crossing-over between sister chromatids. Host suppression of this meiotic drive process by cenH3 and CENP-C would lead to an excess of replacement changes in regions that interact with DNA. Organisms in which there is no opportunity for centromeres to compete, such as budding yeast, would not undergo centromere drive, and this might account for the fact that they have small centromeres, and their cenH3 and CENP-C proteins are under strong purifying selection.

Thus, we see that a special region of the genome, the centromere, is distinguished by a single histone variant class, whose sequences reveal remnants of an arms race that may have led to the extraordinary complexity of centromeres. The RI assembly process that targets new cenH3-containing nucleosomes to centromeres every cell cycle has been elucidated by the discovery and detailed characterization of the related Scm3 (yeast) and HJURP (mammalian) cenH3-specific chaperones (Stoler et al. 2007; Dunleavy et al. 2009; Foltz et al. 2009). Detailed biochemical and structural characterization of Scm3/HJURP complexes (Shuaib et al. 2010; Cho and Harrison 2011; Hu et al. 2011) indicates a role in CenH3 nucleosome assembly that parallels that of other nucleosome assembly complexes, discussed further in Ch. 22 (Almouzni and Cedar 2014). Centromeric nucleosomes show a remarkable lack of sequence specificity in that they not only can faithfully localize to neocentromeres that are completely unlike native centromeres (Fig. 5A), but also the yeast homolog Cse4 can functionally replace human CENP-A (Wieland et al. 2004). It is extraordinary that our centromeres have remained in the same positions for tens of millions of years without any evident sequence determinants involved in the process that maintains them. To the extent that epigenetics refers to inheritance that does not depend on DNA sequence, the inheritance of centromeres on a geological timescale is the most extreme form imaginable. Yet, we are still seeking a mechanism to explain how they have maintained themselves for even a single cell cycle.

There is now general agreement in the centromere field that the cenH3 nucleosome is the key to understanding the epigenetic inheritance of centromeres (Black and Cleveland 2011; Henikoff and Furuyama 2012). It is not only necessary for recruitment of the other structural components of the centromere, in some experimental systems it is also sufficient (Guse et al. 2011; Mendiburo et al. 2011). However, its molecular structure has been the subject of controversy for several years. In vivo evidence from flies, humans, and yeast is most consistent with a right-handed hemisome (Henikoff and Furuyama 2012), whereas several groups have shown that reconstitution of cenH3-containing particles generally results in the formation of a partially unwrapped left-handed octameric nucleosome (Black and Cleveland 2011). Indeed, since 2007, the authors of this chapter coauthored the studies that provided the first evidence for nonoctameric cenH3 particles (Dalal et al. 2007; Mizuguchi et al. 2007), but the composition and structure of these proposed particles were completely different! In light of the continuing controversy, we leave a final resolution of this important issue to the future.

6 THE REPLACEMENT HISTONE VARIANT H3.3 IS FOUND AT ACTIVE CHROMATIN

Like centromeres, transcriptionally active chromatin is thought to be maintained epigenetically and is enriched in an H3 variant, H3.3, which is the substrate for RI deposition (Filipescu et al. 2013). H3.3 is very similar in sequence to the canonical forms of H3, differing by only four amino acids. With so few differences, it might have been assumed that these two forms are interchangeable. However, H3.3 is deposited exclusively by RI nucleosome assembly, whereas H3 is deposited only at replication foci in an RC manner. This difference between the two variants is encoded in the protein itself, with three of the four differences between H3 and H3.3 evidently involved in preventing H3 from being deposited by an RI pathway (illustrated in the α-helix 2 of Fig. 3). Purification of soluble nucleosome assembly complexes confirmed that these two forms participate in distinct assembly processes: H3.1 copurifies with CAF-1 for RC assembly, and H3.3 copurifies with other components, including the HirA and Daxx histone chaperones, and participates in RI assembly.

Although the four-amino-acid difference might seem practically insignificant, when one considers that humans, flies, and clams have precisely the same H3.3 sequence, these differences from H3 stand out. Phylogenetic analysis reveals that the H3/H3.3 pair evolved at least four separate times during eukaryotic evolution, in plants, animals/fungi, ciliates, and apicomplexans (Fig. 6A) (Talbert and Henikoff 2010). Despite having a separate origin from animals and fungi, the animal H3/H3.3 pair and the pair from plants (called H3.1 [RC] and H3.2 [RI]—to avoid confusion, we will refer to all RC isoforms as H3 and all RI isoforms as H3.3) are strikingly similar. The same cluster of amino acids (positions 87–90) that prevents RI deposition of H3 in *Drosophila* are found to differ in plants, and the remaining difference in animals (position 31 is Ala for H3 and either Ser or Thr for H3.3) is also found in plants. Fungi are especially interesting. Ancestrally, they have both H3 and H3.3; however, ascomycetes, which include yeasts and molds, have lost the H3 form. Thus, the obligate RC form of histone 3 that has received the most attention in animals is not even present in yeast.

Studies of H3.3 in bulk chromatin showed that it is enriched in transcriptionally active chromatin fractions (Filipescu et al. 2013). However, various factors contributed to the obscurity of this potential "mark" of active chromatin during a time of great excitement in the chromatin field when it was realized that histone modifications can distinguish active from silent chromatin. For one thing, no antibodies were available that could effectively distinguish H3 from H3.3 in chromatin (positions 87–90 are blocked by the DNA gyres in the nucleosome), whereas excellent antibodies against many different posttranslational modifications were readily available. Also, the seemingly slight sequence differences between H3 and H3.3 did not suggest any fundamental distinctions in chromatin, whereas histone modifications were mostly on tail lysines that were known to affect chromatin interactions or to bind chromatin-associated proteins. This perception that the two histone 3 forms should be interchangeable was confirmed by the finding in *Tetrahymena* and *Drosophila* that the S-phase form can in general substitute for its replacement counterpart. Finally, the influential "histone code" hypothesis envisioned nucleosomes as fixed targets of modification enzymes during chromatin differentiation (Jenuwein and Allis 2001). However, it has become increasingly evident that chromatin is highly dynamic, and even heterochromatin-associated proteins bind with residence times of a minute or less (Phair et al. 2004). It appears that the chromatin of actively transcribed genes is in constant flux, characterized by continual histone replacement (Dion et al. 2007). The three core amino acid differences that distinguish H3 and H3.3 make H3.3-H4 dimers the substrate for RI assembly, and RI assembly itself profoundly changes chromatin. As a result of this process, actively transcribed regions become marked by H3.3 (Fig. 7), and evidence for this process comes from the observation of RI replacement of H3 methylated on lysine-9 (H3K9me) with tagged H3.3 at RNA polymerase I and II (RNA Pol I and II) transcribed loci (Schwartz and Ahmad 2005).

The dynamic nature of chromatin at active loci results in the erasure of preexisting histone modifications, and yet histone modification states persist through multiple rounds of cell division. Therefore, the enzymes responsible for modifying histones must be targeted to their sites of action. For histone modifications that are typically associated with transcriptionally active chromatin, this is achieved by association with the carboxy-terminal domain (CTD) of RNA Pol II, a tandem array of YSPTSPS

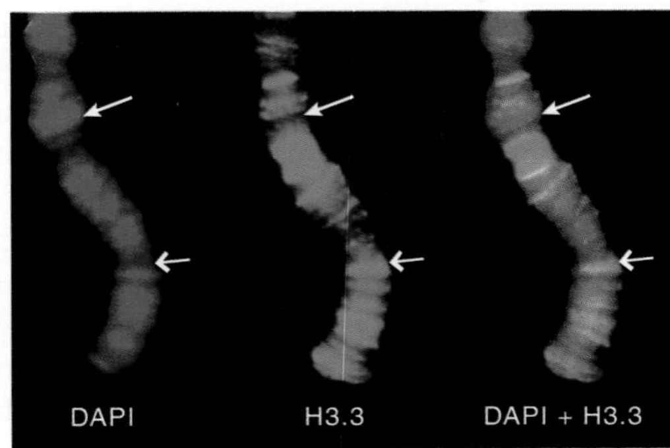

Figure 7. H3.3 preferentially localizes to actively transcribed regions of *Drosophila* polytene chromosomes. DAPI staining (red) shows the DNA banding pattern (*left*), and H3.3-GFP (green) localizes to interbands (*middle*), which are sites of RNA Pol II localization. The merge (Schwartz and Ahmad 2005) is shown on the *right*. In each image, the shorter arrow points to a decondensed interband that is enriched in H3.3, and the longer arrow points to a condensed band that lacks H3.3.

Cite as *Cold Spring Harb Perspect Biol* doi: 10.1101/cshperspect.a019364

heptamers. For example, the Set1 H3K4 methyltransferase associates with the CTD when it is heavily phosphorylated on Serine-5 during transcriptional initiation, and so encounters its substrate primarily near initiation sites. Similarly, the Set2 H3K36 methyltransferase associates with the CTD when it becomes heavily phosphorylated on Serine-2 during transcriptional elongation and encounters its substrate within gene bodies. When a nucleosome is evicted and replaced with unmodified histones during transcriptional elongation, the newly deposited H3.3 is therefore modified appropriately.

It appears that an analogous process maintains histone modifications that are typically associated with silent chromatin. Nucleosomes that are lost at sites of short-period tandem repeats, such as occur at mammalian telomeres and pericentric regions, are replaced by the Daxx H3.3-specific histone chaperone complex and the ATRX ATP-dependent nucleosome remodeling protein (Fig. 8) (Drane et al. 2010; Goldberg et al. 2010). ATRX has a bifunctional histone tail recognition domain with high affinity for an H3 tail that is both unmethylated at K4 and trimethylated at K9 (Eustermann et al. 2011), and so likely is recruited to telomeric sites that are enriched for H3K9me and lack H3K4me. Telomeres are also enriched for heterochromatin-associated protein 1 (HP1), which recruits the Su(var)3-9 H3K9 methyltransferase, and binds its H3K9 methylated product (Hines et al. 2009), and so an enzyme that methylates the tails of replacement H3.3 at telomeres is present at a high local concentration in which new H3.3 is incorporated. This implies that all of the components necessary for maintaining H3K9 methylation are present at telomeres: the enzyme that performs the modification, the modification-specific-binding module on the machine that uses ATP to provide energy for the replacement process, and the fresh unmodified H3.3 substrate that becomes incorporated into the new nucleosome (Fig. 8). It seems likely that a similar process occurs at other sites of short-period tandem repeats in DNA where nucleosomes frequently turn over, insofar as ATRX is also abundant in mammals at C + G-rich sites (Law et al. 2010), which are found at most promoters, and the *Drosophila* ortholog of ATRX, XNP, is abundantly present at a single site of a $(GATA)_n$ repeat where H3.3 is actively incorporated (Schneiderman et al. 2009). The incorporation of H3.3 nucleosomes at sites of telomeric heterochromatin belies the common notion that H3.3 is a "mark" of active chromatin. As the general substrate for replacement of nucleosomes wherever they are lost, the finding that H3.3 is mostly incorporated at active genes rather implies that these are sites where nucleosome turnover is most intense. This generic replacement function of H3.3 has important relevance to human disease as described in Section 15.

7 H3.3 FUNCTIONS IN THE GERMLINE

When cells exit the cell cycle and differentiate, they no longer produce or incorporate S-phase histones, and H3.3 accumulates as a result. For example, H3.3 accumulates in rat brains to a level of 87% of the histone 3 by the time that rats are 400 days old (Pina and Suau 1987). This classical observation suggested that replacement by H3.3 has no functional significance except to prevent holes in the

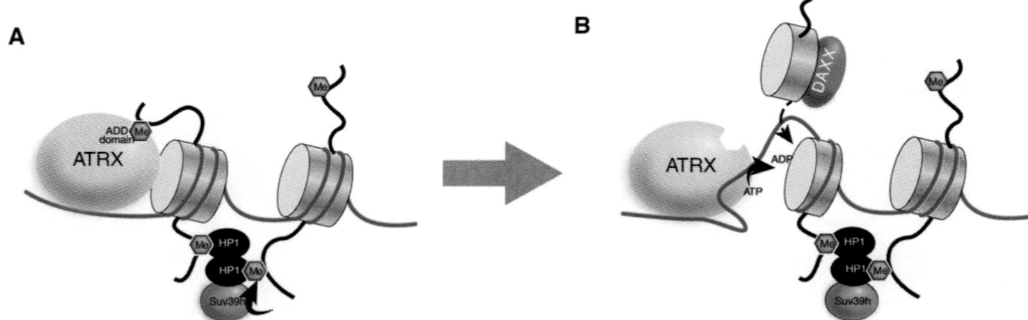

Figure 8. Model for maintenance of histone modifications by the concerted action of multiple chromatin regulators via RI replacement with H3.3. We address the question of how a histone modification can be inherited when a nucleosome is lost and replaced. (*A*) The Suv39h H3K9 methyltransferase (an ortholog of fly Su(var)3-9) is recruited by HP-1 protein, which binds specifically to methylated H3K9. To perpetuate this mark when the nucleosome turns over, we speculate that the ATRX ATPase is recruited to the site via its ATRX-DNMT3-DNMT3L (ADD) domain, which binds with high specificity to methylated H3K9 on tails that entirely lack H3K4 methylation (because there are no H3K4 methyltransferases in this region of the genome). (*B*) ATRX provides the energy of ATP and works together with the H3.3-specific DAXX histone chaperone complex to incorporate the new nucleosome (Goldberg et al. 2010), or half-nucleosome in the case of partial eviction (Xu et al. 2010). The high local concentration of Suv39h results in a new nucleosome with the same H3K9 methylation as the nucleosome that was lost.

nucleosomal landscape. Consistent with this view, H3.3 has been found to be nonessential for *Drosophila* development, as flies lacking both H3.3 genes develop normally to the pupal stage, with occasional adult escapers that die shortly after eclosion but show no specific morphological defects (Hodl and Basler 2009; Sakai et al. 2009). Moreover, H3.3 can functionally substitute for H3 and allow at least some developmental decisions to be made in *Drosophila* embryos, further suggesting that RC and RI substrates are largely interchangeable (Hodl and Basler 2012). It appears that histone replacement in H3.3 null proliferating cells can be accomplished by incorporating H3 nucleosomes using the RI pathway. However, there is no germline development in flies lacking key RI pathway components, such as HirA (e.g., Fig. 9). Females that lack the ChD1 ATP-dependent nucleosome remodeler protein are also sterile, evidently because ChD1 is required in the zygote for sperm nuclei decondensation and the replacement of protamines by maternally encoded RI histones (Orsi et al. 2009). This essential germline function of the RI pathway is conserved in mammals, in which H3.3 is required for remodeling of both maternal and paternal gametes (Santenard et al. 2010; Akiyama et al. 2011). Similar RI processes have been documented in both *C. elegans* and *Arabidopsis*, in which maternal H3.3 is incorporated into the paternal genome of the zygote (Ooi et al. 2006; Ingouff et al. 2007). Therefore, germline remodeling via RI assembly of H3.3 nucleosomes is a universal process that has evolved in both animals and plants, most likely to "reset" the chromatin landscape to a totipotent state.

Remodeling events in the zygote, where RI assembly of H3.3 nucleosomes plays a key role, is analogous to nuclear reprogramming, which can be artificially induced in *Xenopus* eggs, mouse embryonic stem cells, and induced pleuripotent cells. In *Xenopus*, transfer of an embryonic nucleus into an enucleated egg can result in the production of mostly normal embryos that nevertheless sometimes (mis)express genes that were active in the differentiated donor nucleus (Ng and Gurdon 2008). The lack of observable gene expression during the intervening 12 embryonic

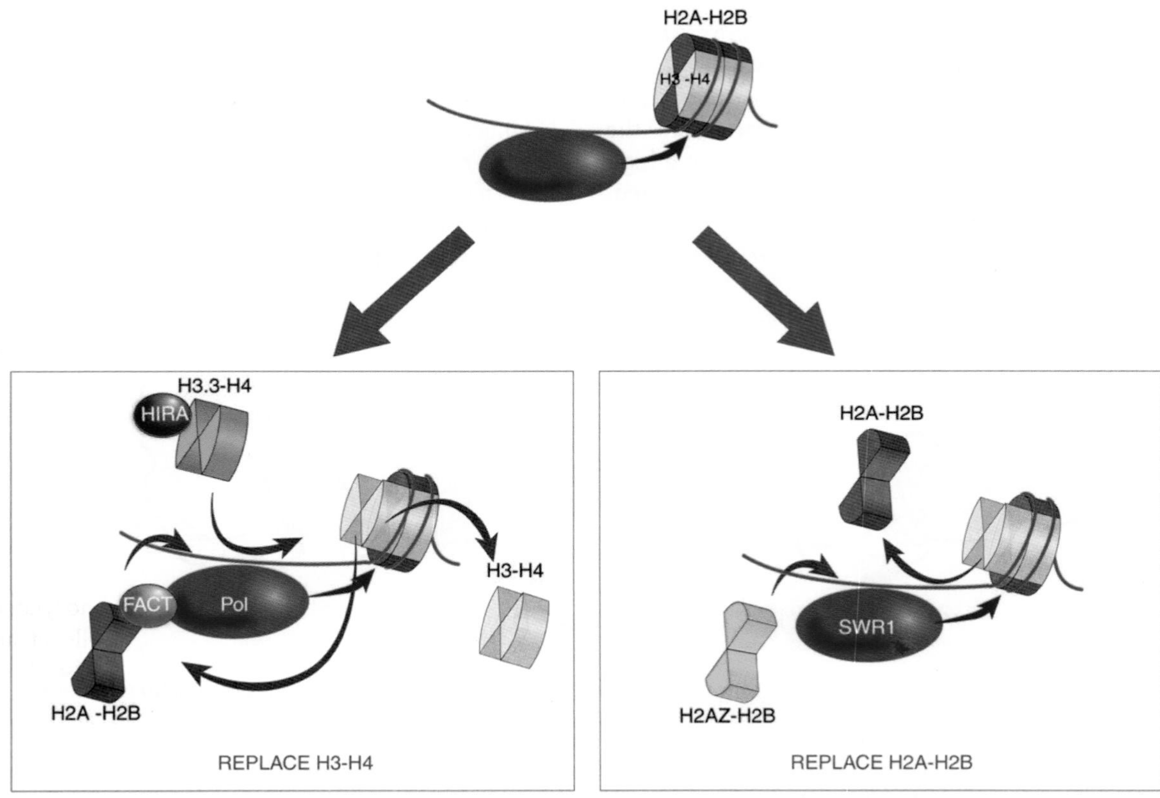

Figure 9. Models for RI replacement or exchange. A large molecular machine (either the SWR1 complex or RNA polymerase) partially or completely unravels a nucleosome during transit. The result is either retention of heterodimeric subunits, such as the FACT-facilitated transfer of H2A-H2B from in front of RNA polymerase to behind (Formosa et al. 2002; Belotserkovskaya et al. 2003) or loss of a heterodimer. In the latter case, chromatin repair replaces the lost heterodimer with either H3.3-H4 (*left*) or H2A.Z-H2B (*right*).

 Cite as *Cold Spring Harb Perspect Biol* doi: 10.1101/cshperspect.a019364

divisions implies that the persistence of an epigenetic mark maintains the memory of prior gene activity. Overexpression of H3.3 in developing embryos improved epigenetic memory and mutation of H3.3K4 to glutamine erased memory of the active state, whereas no effect was seen using a general DNA methyltransferase inhibitor. Further evidence for the importance of H3.3 in nuclear reprogramming comes from the finding that the shift from somatic to oocyte transcription does not require replication, but does require transcription and the H3.3-specific chaperone, HirA (Jullien et al. 2012). It is attractive to think of H3.3 and its histone partners as general mediators of totipotency both in the zygote and during nuclear reprogramming, insofar as histones are likely to be most accessible to the enzymes that posttranslationally modify their tails during nucleosome assembly (e.g., Fig. 8).

8 PHOSPHORYLATION OF H2A.X FUNCTIONS IN DNA DOUBLE-STRAND BREAK REPAIR

The H2A histones also comprise a family of distinct variants found throughout eukaryotes (Fig. 6B). The H2A.X variant is defined by the presence of a carboxy-terminal amino acid sequence motif, SQ(E or D)Ø, in which Ø indicates a hydrophobic amino acid. The serine in this sequence motif is the site of phosphorylation, producing a modified protein designated "γH2A.X." The dynamic nature of chromatin, and H2A.X phosphorylation, is especially evident when double-strand breaks (DSBs) occur in DNA (Morrison and Shen 2005). The lethality of even a single double-stranded (ds) break requires immediate action to repair the lesion and restore the continuity of the double helix. The detection of a ds break normally occurs within a minute or so of its formation and this, in turn, triggers the rapid phosphorylation of H2A.X in the immediate vicinity of a break site. This phosphorylation is performed by members of the phosphoinositol 3-kinase-like kinase family. Following this initial event, H2A.X phosphorylation then spreads quickly along the chromosome marking a relatively large chromatin domain surrounding the break. Finally, the ds break is eventually repaired by either homologous recombination or nonhomologous end-joining and the phosphorylation mark is removed.

Phosphorylation of H2A.X is not essential for detection or repair of DSBs because deletion of the gene or mutation of the target serine residue does not abolish repair. However, H2A.X is not just a marker of damage because such mutants have reduced efficiency of repair and are hypersensitive to radiation damage and genotoxic agents. Currently, H2A.X is thought to function in ds break repair in at least two ways. First, it may help recruit or retain proteins required for repair at the site of the break (Morrison and Shen 2005). Second, it may stabilize the chromosome surrounding the broken ends through the recruitment of cohesin, the protein complex responsible for keeping sister chromatids together (Lowndes and Toh 2005).

The evolution of H2A.X is unlike that of other histone variants. Although a gene for H2A.X is found in nearly all eukaryotes, it has had multiple relatively recent divergences from H2A (Fig. 6B) (Malik and Henikoff 2003; Talbert and Henikoff 2010). For example, the version of H2A.X found in *Drosophila* is different from that found in another dipteran insect, *Anopheles*. Some organisms, such as yeasts, have an H2A.X but lack an H2A, consistent with the possibility that most present-day canonical H2As have evolved from ancestral H2A.Xs (Talbert and Henikoff 2010). Presumably, the ability to evolve either a canonical H2A from an H2A.X or vice versa is a consequence of the simplicity of the SQE motif. Depending on selective constraints, the loss or gain of such a simple motif at the carboxyl terminus of a protein might occur repeatedly over evolutionary time. Occasional loss of an existing H2A.X with a newly minted version might be fueled by the need for H2A.X to be very uniformly distributed because DSBs can occur anywhere in the genome. If mutations occur in an existing H2A.X gene that reduce its similarity to the canonical H2A in such a way that its assembly becomes less efficient or uniform, then there will be strong selection to replace it with a version that is more similar to canonical H2A. This rationale could help account for the exceptional case of *Drosophila* H2A.X, which unlike other eukaryotes, is not derived from its canonical H2A, but rather from the distant H2A.Z variant lineage (Baldi and Becker 2013). If all that is necessary to be an H2A.X is to be in the H2A position in a nucleosome and to have the carboxy-terminal motif for phosphorylation, then an H2A.Z can evolve this capability.

DSB repair is clearly the universal function of H2A.X phosphorylation, and there would seem to be no stable epigenetic aspect to this process. However, H2A.X null mice are sterile, and cytological examination of mammalian spermatogenesis has revealed a striking epigenetic feature in which H2A.X is specifically phosphorylated on the XY bivalent (Fig. 10) (Fernandez-Capetillo et al. 2003). This chromosome pair occupies a distinct "sex body" during meiotic prophase, which has been implicated in silencing of sex-linked genes during male meiosis. H2A.X phosphorylation is essential for normal sex-body formation, and H2A.X-deficient spermatocytes fail to pair or condense and fail to inactivate X and Y genes during meiosis. H2A.X phosphorylation of the XY bivalent is distinct from the process that occurs at DSBs. XY phosphorylation in the sex body does not require breaks, but rather occurs

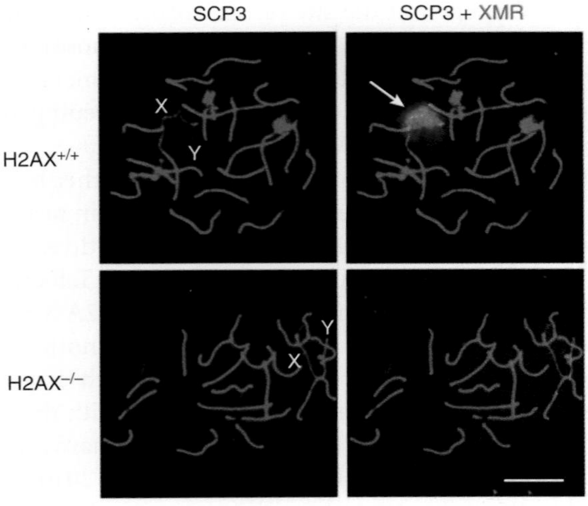

Figure 10. Pachytene stage of spermatogenesis showing the dependence of sex-body formation on H2A.X. In normal mammalian spermatocytes, a nuclear structure, the sex body (arrow, green, in *right* panels), is seen to encompass the unpaired XY bivalent (labeled in *left* panels). The synaptonemal complex, which aligns paired chromosomes, is stained red. H2A.X is normally enriched in the sex body (H2A.$X^{+/+}$). In H2A.$X^{-/-}$ spermatocytes, the sex body does not form and a sex-body epitope becomes dispersed (*lower right*). Scale bar, 10 μm. (Images courtesy of Shantha Mahadevaiah and Paul Burgoyne; Fernandez-Capetillo et al. 2003.)

most conspicuously at unpaired regions of the chromosomes. The mechanisms whereby H2A.X phosphorylation is targeted to unpaired chromosomes and how this event leads to condensation, pairing, and silencing are currently unknown. However, it is interesting to speculate that this role may be related to its ability to interact with and recruit cohesin.

9 H2A.Z PLAYS DIVERSE ROLES IN CHROMATIN REGULATION

Histone variant H2A.Z is found throughout most eukaryotic lineages and it has been intensely studied for its structure and function in chromatin biology (Zlatanova and Thakar 2008; Draker and Cheung 2009; Marques et al. 2010; Talbert et al. 2012). H2A.Z diverged from an ancestral H2A early in evolution and shares only ~60% similarity with its major histone H2A counterpart. Consistent with this separate lineage, genetic experiments in yeast, plants, flies, and mammals have shown that histones H2A and H2A.Z have evolved to play separate nonoverlapping functions. H2A.Z is an essential histone in most organisms, from ciliated protozoans to mammals. However, in budding and fission yeasts, cells that carry a deletion of the single-copy H2A.Z gene are viable, although the null mutants show a variety of conditional lethal phenotypes. Plants, as exemplified by *Arabidopsis thaliana*, have three closely related H2A.Z genes, HTA8, HTA9, and HTA11, with roughly 90% identity and a more distantly related HTA4 gene (Talbert and Henikoff 2010). There appears to be functional redundancy among these genes as organisms with single deletions of either HTA8, HTA9, or HTA11 are normal, whereas the hta9 hta11 double mutant shows developmental defects. Interestingly, vertebrates have evolved two closely related H2A.Z variants, H2A.Z.1 and H2A.Z.2, which differ at only three amino-acid residues (Dryhurst et al. 2009; Matsuda et al. 2010; Mehta et al. 2010). These two variants show different patterns of chromatin distribution (Dryhurst et al. 2009) and are apparently nonredundant as the single deletion of H2A.Z.1 in the mouse is lethal (Faast et al. 2001).

Evidence for functional diversification of H2A.Z comes from the discovery of an alternatively spliced form of human H2A.Z.2 mRNA that results in a protein with a reduced CTD. This shorter H2A.Z isoform destabilizes nucleosomes and is most enriched in brain (Bonisch et al. 2012), in which H2A.Z, like H3.3, is known to be especially abundant (Pina and Suau 1987). Although this particular isoform appears to be limited to primates, evidence for a longer alternatively spliced isoform with potential for nucleosome destabilization was reported in carp brain (Simonet et al. 2013). With the increasing popularity of RNA-seq for identification of alternatively processed variants, we expect that other examples of potential H2A.Z functional diversification will be discovered.

The high-resolution structure of an H2A.Z-containing nucleosome reveals several unique properties of the variant (Suto et al. 2000). Compared with H2A nucleosomes, H2A.Z presents an extended acidic patch domain on the surface of the nucleosome and mutational studies have shown this to have functional significance. The acidic patch is then part of a larger "docking domain," an essential part of the protein necessary for interaction with H3 in the nucleosome. Like other histones, H2A.Z is subject to a variety of posttranslational modifications, including acetylation, ubiquitylation, and sumoylation. There is strong genetic and biochemical evidence that these modifications affect the localization, dynamics, and function of H2A.Z nucleosomes (Talbert and Henikoff 2010).

H2A.Z has been linked to a wide variety of different and sometimes contradictory nuclear functions, including transcriptional activation, transcriptional repression, RNA Pol II elongation, heterochromatin, antisilencing, cell-cycle control, DNA replication, DNA damage repair, chromosome segregation, and genome integrity (Zlatanova and Thakar 2008; Altaf et al. 2009; Marques et al. 2010; Talbert and Henikoff 2010; Xu et al. 2012; Adkins et al.

2013). It likely has direct mechanistic roles in transcription initiation, elongation, antisilencing, and DNA damage repair. In some other cases, evidence suggests that H2A.Z function is manifest indirectly through its involvement in transcription. For example, delay in the G_1-S transition observed in budding yeast deleted for H2A.Z is likely due to the misregulation of cyclin gene expression and not defects in DNA replication initiation (Dhillon et al. 2006). In *S. pombe*, and perhaps other organisms, H2A.Z cooperates with heterochromatin-silencing factors (Clr4/SUV39H) to enforce RNA-processing fidelity and prevent deleterious antisense transcription (Zofall et al. 2009). Loss of this enforcement may account for the genome instability and chromosome segregation defects observed in H2A.Z deletion mutants. Indeed, at least part of the chromosome segregation defects observed in H2A.Z mutants of *S. pombe* might be caused by decreased transcriptional expression of the centromere protein, CENP-C (Hou et al. 2010). Nevertheless, such examples of indirect function are relatively rare. Indeed, ruling out indirect activities for any chromatin regulator is a challenging problem, particularly for H2A.Z in organisms for which it is essential.

10 H3.3 AND H2A.Z OCCUPY DISCRETE CHROMATIN LOCATIONS

Much of our understanding of histone variant function is inferred from its patterns of genomic chromatin occupancy. H3.3 makes up ~15%–25% of total H3 protein and H2A.Z makes up ~5%–10% of the total H2A protein in most organisms examined to date. Abundances increase when cells exit from the cell cycle and no longer replicate their DNA, such as during development (Pina and Suau 1987). These variants are widely, but not uniformly, distributed throughout the genome. High-resolution chromatin immunoprecipitation experiments in a number of model organisms have revealed that both H3.3 and H2A.Z preferentially occupy nucleosomes that flank gene promoters and both are particularly enriched at the +1 nucleosome bordering transcriptional start sites (TSSs). They are often enriched at the −1 or −2 nucleosome as well and thus flank a nucleosome-depleted region at the TSS (Talbert and Henikoff 2010). In animals, the H3.3 over gene bodies correlates with transcriptional levels, suggesting that it replaces nucleosomes that are occasionally lost during transcription (illustrated in the left panel of Fig. 9). Direct evidence for this interpretation comes from measuring nucleosome turnover by metabolic labeling, which showed nucleosome turnover patterns closely matching H3.3 patterns genome-wide (Deal et al. 2010).

H2A.Z has been mapped genome-wide in a variety of eukaryotes. In budding yeast, nematodes, and plants, H2A.Z occupancy around the promoter is correlated with nontranscribing genes "poised" for activation (Zhang et al. 2005; Mavrich et al. 2008; Whittle et al. 2008; Kumar and Wigge 2010). However, in flies and mammals, promoter H2A.Z occupancy appears to correlate more with actively transcribing genes (Barski et al. 2007; Mavrich et al. 2008; Hardy et al. 2009; Hardy and Robert 2010; Kelly et al. 2010), similar to the situation for H3.3. Although preferentially found at promoters and regulatory sites, H2A.Z nucleosomes can also be found at lower frequency in gene bodies and elsewhere (Hardy et al. 2009; Weber et al. 2010; Santisteban et al. 2011). Enrichment of H2A.Z over gene bodies closely corresponds to that of chromatin that is extracted with low salt, suggesting that H2A.Z changes the physical properties of nucleosomes (Weber et al. 2010).

H2A.Z is also specifically deposited near or within heterochromatin. In budding yeast, H2A.Z is enriched near telomeres where it serves as an antisilencing factor. Deletion of the H2A.Z gene results in extended spreading of silent chromatin inward from the telomeres and this defect can be suppressed by the additional deletion of genes encoding the silencing factors themselves (see Ch. 8 [Grunstein and Gasser 2013] for more detail). Indeed, this function may act globally, in parallel with the Set1 histone H3 methyltransferase, to prevent large-scale aberrant distribution of silencing factors (Venkatasubrahmanyam et al. 2007). In metazoans, H2A.Z is also localized in facultative and constitutive heterochromatin, the inactive X chromosome, transposable elements, and pericentric heterochromatin (Greaves et al. 2007; Draker and Cheung 2009; Boyarchuk et al. 2011; Zhang and Pugh 2011).

In contrast to the chromosomal features that are correlated with H2A.Z, there is a remarkable anticorrelation between histone H2A.Z nucleosome occupancy and DNA methylation (Zilberman et al. 2008; Kobor and Lorincz 2009; March-Diaz and Reyes 2009; Conerly et al. 2010; Edwards et al. 2010; Zemach et al. 2010; see Ch. 15 [Li and Zhang 2014] for a discussion of DNA methylation). There is strong evidence that this mutual antagonism is causal and not simply correlative. Mutants in *A. thaliana* with decreased DNA methylation show an increase in H2A.Z occupancy at loci where it is normally not found, independent of transcriptional activity. Conversely, mutants defective in H2A.Z deposition show increased DNA methylation over gene bodies normally occupied by H2A.Z nucleosomes. While the precise molecular pathways that account for this mutual exclusion remain to be elucidated, this functional relationship has important implications for development and carcinogenesis. For example, the stochastic or environmentally effected loss of H2A.Z from the promoter of a tumor-suppressor gene could well contribute to locally increased DNA methylation and heritable epigenetic repression.

11 H2A.Z NUCLEOSOME OCCUPANCY IS DYNAMIC AND CHANGES THE PROPERTIES OF CHROMATIN

The dynamic exchange of H2A.Z nucleosomes in chromatin appears to be an important part of its function. (See Ch. 21 [Becker and Workman 2013] for an in-depth discussion of histone exchange.) Unlike the major core histones, H2A.Z expression is not restricted to S phase and it can be incorporated into chromatin independent of DNA replication. The deposition of H2A.Z into nucleosomes is performed by multisubunit protein complexes, which have been conserved throughout the eukaryotic kingdom (Lu et al. 2009; March-Diaz and Reyes 2009; Morrison and Shen 2009). First identified in budding yeast, the SWR1 complexes contain, as their catalytic subunits, homologs of the protein Swr1, a member of the SWI/SNF family of ATP-dependent chromatin remodelers. The substrate for SWR1 is an H2A.Z-H2B dimer, which is used to replace one of the existing H2A-H2B dimers in the nucleosome in an ATP-dependent exchange reaction (Fig. 9). This reaction is stepwise and unidirectional, in vitro, resulting in the complete replacement of H2A-H2B dimers with H2A.Z-H2B dimers (Luk et al. 2010). In vivo, unidirectional replacement of H2A with H2A.Z by SWR1 is enforced by acetylation of H3K56, which allows the reverse reaction to occur, resulting in local reduction in H2A.Z incorporation, thus modulating transcription (Watanabe et al. 2013). SWR1 is likely dedicated to the task of replacing H2A with H2A.Z because the effects of eliminating SWR1 function are similar to the effects of deleting the gene encoding H2A.Z itself. Indeed, the activity of the SWR1 complex is actually deleterious to the cell in the absence of its H2A.Z-H2B substrate (Halley et al. 2010; Morillo-Huesca et al. 2010).

The removal of H2A.Z from chromatin proceeds by at least two pathways. Nucleosome exchange and eviction occurs in many contexts such as the remodeling and eviction of complete nucleosomes that can happen at promoters. Any H2A.Z that is a part of those nucleosomes will be removed as well. However, there is also evidence that H2A.Z-H2B dimers may be specifically removed from nucleosomes by the INO80 complex, a close relative of SWR1 (Morrison and Shen 2009; Papamichos-Chronakis et al. 2011). In budding yeast, the elimination of INO80 results in the global mislocalization of H2A.Z and a decrease in its apparent exchange rate. Additional mutational results are consistent with this interpretation. In vitro, purified INO80 is reported to catalyze the replacement of a nucleosomal H2A.Z-H2B dimer with a canonical H2A-H2B dimer, that is, the reverse of the SWR1 reaction (Luk et al. 2010; Papamichos-Chronakis et al. 2011), and this reaction may also be regulated by acetylation of H3K56 (Watanabe et al. 2013).

The factors that determine where H2A.Z is deposited are incompletely understood. At present, there is little evidence that H2A.Z templates its own deposition epigenetically (Viens et al. 2006). In budding yeast, a DNA sequence related to the binding site of transcription factor Reb1 is able to target the enrichment of H2A.Z at ectopic sites, independent of Reb1 (Raisner et al. 2005). In other cases, transcription factors themselves are implicated in targeting H2A.Z deposition (Updike and Mango 2006; Zacharioudakis et al. 2007; Gevry et al. 2009). The unifying theme underlying the function of these various factors appears to be the creation of a nucleosome-depleted region necessary for H2A.Z deposition, although how this recruits H2A.Z is not currently known (Hartley and Madhani 2009). Interestingly, SWR1 complexes often contain homologs of yeast Brd1, a protein containing dual bromodomain motifs capable of binding acetylated lysines. Thus, SWR1 may be recruited to, or stabilized at, chromatin neighborhoods rich in acetylated histones. H2A.Z deposition may also be blocked at specific loci. The *S. pombe* SWR1 complex contains a regulatory subunit, Msc1, which is dispensable for the loading of H2A.Z at promoter nucleosomes, but which is required to prevent H2A.Z deposition within the chromatin of the inner centromere and subtelomeric regions (Buchanan et al. 2009; Zofall et al. 2009). It has also been suggested that an additional pathway directing localized occupancy by H2A.Z may involve its random deposition and then specific eviction, perhaps as a consequence of transcription (Hardy and Robert 2010).

In any case, the consequence of H2A.Z deposition and replacement can be complex. Because of the exchange reactions, nucleosomes in cellular chromatin may be "ZZ," "ZA," or "AA," containing two, one, or zero H2A.Z-H2B dimers, respectively (Luk et al. 2010; Weber et al. 2010). In *Drosophila*, the pattern of ZZ and ZA nucleosome occupancy is different, and homotypic H2A.Z nucleosomes are enriched over the bodies of active genes, perhaps as a consequence of transcription elongation. In mouse trophoblast cells, there are distinct changes in the promoter content of ZZ and ZA nucleosomes during G_1, S, and M phase, independent of actual transcription activity, suggesting a major remodeling pathway dependent on the cell-division cycle (Nekrasov et al. 2012). Differential posttranslational modifications further enrich the situation as H2A.Z acetylation has been linked with both SWR1 and INO80 functions (Millar et al. 2006; Papamichos-Chronakis et al. 2011).

Cite as *Cold Spring Harb Perspect Biol* doi: 10.1101/cshperspect.a019364

A quantitative evaluation of the +1 nucleosome in yeast revealed that at steady state the relative abundance of ZZ, ZA, and AA nucleosomes is roughly 32%, 24%, and 44% respectively (Luk et al. 2010). How can H2A.Z drive chromatin function with this level of heterogeneity? One explanation is that H2A.Z nucleosomes are in dynamic exchange at these sites. In budding yeast, nucleosomes with rapid turnover were identified by the kinetic incorporation of newly synthesized histone H3. These "hot" nucleosomes preferentially map to promoter regions, including the nucleosomes at the TSS, which are enriched in H2A.Z histones (Dion et al. 2007). Remarkably, in budding yeast H2A.Z appears to increase the global turnover of nucleosomes and not simply those into which it is most abundantly incorporated (Dion et al. 2007; Santisteban et al. 2011). The mechanistic basis for this influence is currently unknown.

The stability of H2A.Z-containing nucleosomes in vitro has been examined in many studies with contrasting results (Zlatanova and Thakar 2008; Talbert and Henikoff 2010). In vivo it is clear that not all H2A.Z nucleosomes are created equal and that the lability of H2A.Z nucleosomes is affected by posttranslational modifications and the presence of other variant histones. Nucleosomes containing both H2A.Z and H3.3, especially H2A.Z/H2A nucleosomes with H3.3, are particularly unstable and sites occupied by these double-variant nucleosomes may be erroneously scored as entirely free of nucleosomes, depending on how the chromatin is isolated (Jin et al. 2009; Nekrasov et al. 2012). H2A.Z acetylation generally destabilizes nucleosomes in vitro and in vivo is associated with gene activation (Tanabe et al. 2008; Wan et al. 2009; Halley et al. 2010).

A particularly striking example of the differential properties of H2A.Z nucleosomes is the transcriptional response of cells to temperature. Plants possess a signaling pathway that senses ambient temperature and regulates gene expression (see Ch. 31 [Baulcombe and Dean 2014] for a discussion of plant responses to environmental factors). The developmental program for flowering, for example, is accelerated at higher temperature. To identify factors that regulate this response, *A. thaliana* was screened for mutants with a constitutive high temperature expression pattern and the mutations turned out to be in ARP6, which encodes one of the conserved subunits of the SWR1 complex (Kumar and Wigge 2010). Indeed, H2A.Z occupancy at the promoters of heat responsive genes was found to decrease with increasing temperature and this loss was independent of transcriptional activity per se. These findings have potentially broad implications for agriculture in a globally warming environment, insofar as depletion of H2A.Z in a model cereal phenocopies the ambient temperature response and impacts grain yield (Boden et al. 2013).

12 H2A.Z FUNCTIONS IN EPIGENETIC INHERITANCE

There is increasing evidence that H2A.Z participates in the heritable specification of chromatin function. Chromosomes undergo extensive remodeling each mitotic generation and the specification of silent and active chromatin domains across division cycles is essential for normal development. In budding yeast, H2A.Z is transiently lost from a subset of genes during telophase, as measured by chromatin immunoprecipitation and this displacement is required for the establishment of heterochromatin in an inducible model (Martins-Taylor et al. 2011). In mammalian cells most active transcription is repressed during mitosis and must be reactivated in the next cell cycle. The +1 nucleosome containing H2A.Z is a prime candidate for mediating this control and its translational positioning at TSSs has been examined in G_0/G_1-phase, S-phase, and M-phase cells (Kelly et al. 2010; Nekrasov et al. 2012). Indeed, the positions of these nucleosomes were found to be shifted during mitosis, altering the size of the nucleosome-depleted region and remodeling chromatin at the TSS. Together these features may help establish epigenetic memory during mitosis and rapid gene reactivation following cell division. The pathways regulating these cell-cycle behaviors of H2A.Z are currently unknown.

The location of a gene within the nucleus can be important for its expression. Positioning at the nuclear periphery is often associated with gene inactivation with the exception of localization at nuclear pores, which is a positive factor in expression (reviewed in Akhtar and Gasser 2007). Several genes in budding yeast, such as *GAL1* and *INO1*, have been found to move to the nuclear periphery on activation and remain there for multiple mitotic cell cycles even after repression. It has been argued that this localization contributes to "transcriptional memory," which is the observation that recently expressed genes are more rapidly reactivated than genes that have experienced long-term repression (Brickner 2009). The molecular mechanisms responsible for transcriptional memory are currently controversial (Zacharioudakis et al. 2007; Halley et al. 2010; Kundu and Peterson 2010). Nevertheless, H2A.Z appears to have a role in maintaining the localization of recently repressed genes at the nuclear periphery for multiple mitotic generations. In the case of *INO1*, this pathway requires a *cis*-acting DNA element in the *INO1* promoter and interaction with the nuclear pore protein NUP100 to direct the loading of H2A.Z (Light et al. 2010). If the H2A.Z gene is deleted, recently repressed *INO1* fails to remain at the nuclear periphery and becomes nucleoplasmic (Brickner 2009).

H2A.Z is also linked to the epigenetic specification of cell fates in mammalian stem cells (Creyghton et al. 2008;

see Ch. 29 [Reik and Surani 2014] for more on stem cell epigenetics). Whole genome-mapping studies in mouse embryonic stem cells revealed that H2A.Z preferentially occupies the promoters of genes that are poised to direct development and differentiation when activated (Ku et al. 2012; Li et al. 2012; Hu et al. 2013). Indeed, the distribution of H2A.Z is coincident with that of Suz12, a component of the Polycomb group complexes, which have primary roles in development (see Ch. 17 [Grossniklaus and Paro 2014] for a detailed discussion of Polycomb group proteins). Inhibiting H2A.Z expression results in increased and more stable nucleosome occupancy at regulatory regions, decreased methylation of H3K4 and H3K27 at promoters and enhancers, the derepression of developmental target genes, and aberrant differentiation of embryoid bodies. These results are consistent with a model in which H2A.Z variant nucleosomes impart a dynamic instability to chromatin structure increasing access to chromatin-modification factors such as mixed lineage leukemia and Polycomb repressive complex 2 (PRC2), and to lineage-specific factors such as the transcription factor FoxA2 for endoderm/hepatic differentiation (Li et al. 2012) or RARα for neuronal development (Hu et al. 2013).

In theory, stem cells are at a higher risk of mutation because of their continued proliferation during the life of the organism. Cairns first proposed that stem cells might protect against this risk by ensuring the inheritance of the "oldest" DNA strands exclusively to the stem cell daughter as opposed to the differentiating daughter cell (Cairns 1975). There is experimental support for this "immortal-strand model" in a number of systems in which asymmetric self-renewal can be experimentally manipulated. Interestingly, histone H2A.Z expression is a biomarker for this immortal-strand inheritance (Huh and Sherley 2011). First, H2A.Z mRNA is down-regulated in the differentiating sister cell during asymmetric cell division. This observation is consistent with the fact that the H2A.Z gene is a target of Oct4 regulation and is down-regulated following differentiation of both mouse and human stem cell populations (Du et al. 2001; Shaw et al. 2009; Huh and Sherley 2011). More striking, however, is the apparent asymmetric distribution of H2A.Z entirely to the immortal DNA strands segregating to the stem cell sister, as detected by immunofluorescence (Huh and Sherley 2011). This turns out to be because of the "cloaking" of the H2A.Z histones on the mortal-strand chromatids inherited by the differentiating sister cell. That is, H2A.Z is actually present on both sets of chromosomes, but it is only accessible to detection by the anti-H2A.Z antibody on the set segregating to the stem cell sister. Interestingly, mild acid treatment before immunostaining removes whatever is blocking detection and reveals H2A.Z on both sets of chromosomes. The nature of the cloaking mechanism is currently unknown, but these observations place H2A.Z quite proximal to the mechanism of immortal-strand inheritance.

13 OTHER H2A VARIANTS DIFFERENTIATE CHROMATIN, BUT THEIR FUNCTIONS ARE AS YET UNKNOWN

Further diversification of H2A has occurred in vertebrates. In mammals, macroH2A and H2A.B represent unique lineages that appear to play roles in the epigenetic phenomenon of dosage compensation (discussed in detail in Ch. 25 [Brockdorff and Turner 2014]). macroH2A is so-called because, in addition to the histone-fold domain and amino- and carboxy-terminal tails, it contains a more than 200-amino-acid carboxy-terminal globular domain (Ladurner 2003). macroH2A is enriched in discrete regions of the facultatively inactive X chromosome of human females that alternate with regions of constitutive heterochromatin (Fig. 11A) (Chadwick and Willard 2004). However, the role of macroH2A must be more diverse, insofar as macroH2A orthologs are present in many nonmammalian clades, and appear to be ancestral in the animal kingdom (Talbert and Henikoff 2010). In vitro, macroH2A reduces transcription factor access and excludes histone H1. The macro domain itself binds to the enzyme PolyADP ribose polymerase 1 and inhibits its activity in vitro (Nusinow et al. 2007), although exactly what is occurring in vivo to mediate gene repression is unclear.

In contrast to macroH2A, H2A.B seems to be undetectable on the Barr body, but otherwise ubiquitous through-

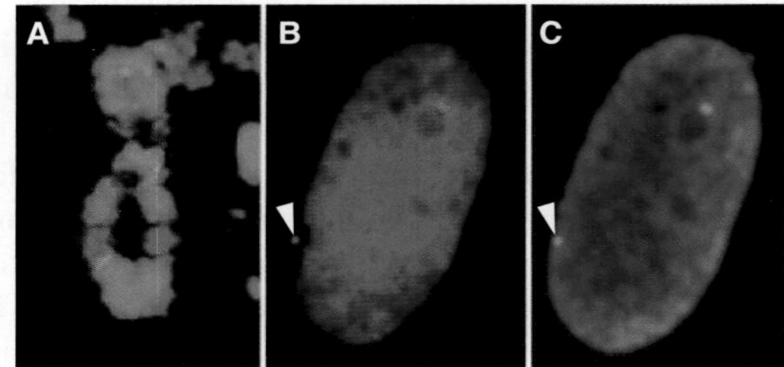

Figure 11. H2A variants and the inactive X chromosome of human females. (*A*) macroH2A (red) stains discrete regions of the inactive X chromosome that alternate with a marker for heterochromatin (histone H3K9me3). (*B*) H2A.B (green) is excluded from the inactive X chromosome (red dot with arrowhead pointing to it). (*C*) Same nucleus as in *B*, but stained with DAPI to show chromatin. (*A*, Reprinted, with permission, from Chadwick and Willard 2004, © National Academy of Sciences; *B*,*C*, reprinted, with permission, from Chadwick and Willard 2001, © 2001 The Rockefeller University Press. Originally published in *Journal of Cell Biology* **152:** 375–384. doi: 10.1083/jcb.152.2.375.)

out the nucleus (Fig. 11B) (Chadwick and Willard 2001). The short wrap of DNA around H2A.B nucleosomes (Bao et al. 2004) and its enrichment at TSSs (Soboleva et al. 2012) is consistent with it playing a role in facilitating transcription. H2A.B and its closest relative, H2A.L (Govin et al. 2007), are rapidly evolving relative to other H2A variant classes, perhaps related to the fact that they are both testes-specific variants (Talbert and Henikoff 2010). The roles that H2A.B and H2A.L play in mammalian male germ cell development remain to be elucidated.

14 MANY HISTONES HAVE EVOLVED TO MORE TIGHTLY PACKAGE DNA

When it is no longer necessary to gain access to DNA for replication and transcription, chromatin typically becomes further condensed, and this often involves replacement of canonical histones. This is obviously the case for sperm, and in some lineages, histone paralogs have evolved specialized packaging roles. For example, sea urchin sperm contains H1 and H2B variants with repeated tail motifs that bind to the minor grooves of DNA (Talbert and Henikoff 2010), presumably an adaptation to tightly package chromosomes for inclusion into sperm heads. A similar adaptation is found in pollen-specific H2A variants in flowering plants. In vertebrates, a sperm-specific H2B specialized histone variant is found in mammalian testes, including an H2B paralog (subH2Bv) that localizes to the acrosome and a testes-specific H3 variant (Witt et al. 1996).

The replacement of histones during sperm maturation by protamines and other proteins provides a potential means of erasing epigenetic information in the male germline. However, evidence for transgenerational inheritance (Rakyan and Whitelaw 2003; discussed in more detail in Ch. 14 [Blewitt and Whitelaw 2013]), especially in animals that lack DNA methylation, raises the possibility that a subset of nucleosomal histones survive this transition and transmit epigenetic information. As already pointed out, this is just what occurs for CENP-A at centromeres (Palmer et al. 1990), and it is possible that a small fraction of other variants, such as H3.3, remain with sperm for epigenetic inheritance of gene-expression information. Although our understanding of the process that replaces histones during sperm development is rudimentary, we expect that much more can be learned by understanding how CENP-A survives this transition.

Increased compaction also occurs in somatic cells that have finished dividing and undergo differentiation. In some cases, compaction involves quantitative and qualitative changes in linker histones. The stoichiometry of histone H1 relative to nucleosomes determines the average spacing within nucleosome arrays in vivo (Fan et al. 2003). In addition, the presence of H1 in chromatin promotes higher-order chromatin structure that generally inhibits transcription (Wolffe 1992). Linker histones are much more mobile than core histones in vivo. Residence times for H2A and H2B are hours in length, and cannot even be measured for H3 and H4, whereas the residence time of H1 is a few minutes (Phair et al. 2004). As a result, the incorporation of variant linker histones is unlikely to differentiate chromatin in a heritable manner. Rather, the role of H1 variants is thought to change the bulk properties of chromatin that can affect overall compaction (Wolffe 1992).

H1 variants share with core histones a distinction between RC and RI forms (Marzluff et al. 2002). RC variant forms of H1 appear to be interchangeable with one another, based on the fact that knockout mice lacking one or two of the five RC H1 variants are phenotypically normal (Fan et al. 2003). The functional interchangeability of histone variants is implicit in the "star" phylogeny of the H1 family, in which there is little if any evidence for evolutionary conservation of H1 branches in the tree (Fig. 6C). Rather, the greater divergence of H1 relative to core histone variants is likely to reflect weaker structural constraints, and the existence of multiple H1 genes in many lineages might represent an adaptation for regulating levels of linker histone. For example, in birds, the H1.0 linker histone variant (previously referred to as H5) is deposited during erythrocyte maturation, which accompanies extreme compaction of the nucleus. The mammalian H1.0 ortholog (previously referred to as H1°) is deposited at high levels in nondividing cells. Overexpression of H1.0 renders chromatin less accessible to nucleases than similar overexpression of a canonical form. The natural accumulation of H1.0 in nondividing cells might be a general mechanism for chromatin compaction as cells become quiescent.

15 HISTONE VARIANTS AND HUMAN DISEASE

The diverse roles of histone variants in basic epigenetic processes leads to the expectation that their loss or misexpression can result in disease, and indeed some examples of this have recently come to light. For example, ATRX, which is the DNA translocase component of the ATRX-Daxx-H3.3 pathway (Fig. 8), was originally identified as causing α-thalassemia mental retardation on the X (ATRX) syndrome. In this syndrome, loss of ATRX results in defects at the CpG island of the α-globin gene promoter, and can now be understood as resulting from frequent nucleosome loss at CpG islands (Law et al. 2010). Importantly, mutations causing the ATRX syndrome are most frequently found within the ADD histone tail-binding domain (Eustermann et al. 2011), strongly implicating H3/H3.3 tail interactions as being key causal factors in the syndrome.

The study of ATRX syndrome has also led to the realization that another histone variant, macroH2A, is also depleted from specific sites, including the α-globin CpG island (Ratnakumar et al. 2012).

A remarkable relationship has recently emerged between the ATRX-Daxx-H3.3 nucleosome assembly process and cancer that provides important insights into both the process itself and oncogenesis. Genomic sequencing of human pancreatic neuroendocrine tumors (PanNETs) revealed that ~40% harbored likely loss-of-function mutations in either Daxx or ATRX (Jiao et al. 2011). Notably, all such tumors displayed a distinct phenotype called "ALT" (alternative lengthening of telomeres) in which telomeres show massive lengthening without induction of telomerase (see Fig. 1 of Ch. 15 [Liu et al. 2014]; Heaphy et al. 2011a). Based on examination of more than 6000 human tumors, 3.73% showed ALT occurring in cancers of all different types (Heaphy et al. 2011b). ATRX or Daxx are mutated in the vast majority of cases of ALT examined by sequencing. Although the precise molecular basis for telomere lengthening in ALT is not well understood, lengthening is likely initiated by recombinational strand invasion between telomeres, which suggests that failure to replace a nucleosome at a simple sequence array can result in DNA breakage and strand invasion. In the case of PanNETs, ALT predicts a more favorable outcome than telomerase induction, as if the ALT pathway is less efficient than the telomerase pathway in preventing the senescence that occurs when telomeres erode (Jiao et al. 2011).

Sequencing of tumors has also revealed the very surprising presence of specific mutations in H3.3 itself in a large fraction of pediatric glioblastomas (the topic of Ch. 15 [Liu et al. 2014]; also Schwartzentruber et al. 2012; Wu et al. 2012). One of the two H3.3 genes in these tumors is found to encode either K27M or G34R/V, and is most likely to be a gain-of-function mutation that drives accelerated tumorigenesis (see Fig. 1 of Ch. 15 [Liu et al. 2014]). G34R/V is associated with ATRX loss and probably enhances ALT, whereas K27M can occur on an H3 as well, so would appear to be independent of the ATRX-Daxx-H3.3 pathway. Rather, K27 is a key substrate for methylation and acetylation involved in Polycomb silencing, and the presence of a nonmodifiable residue at position 27 of the H3/H3.3 tail causes gain-of-function inhibition of methylation by the EZH2 subunit of PRC2 (Lewis et al. 2013). It would then appear that the resulting global reduction in H3K27 trimethylation contributes to this aggressive pediatric tumor.

Misregulation of histone variant expression has also been implicated in cancer. In malignant melanoma cells, macroH2A levels are sharply reduced and restoration of macroH2A levels suppresses metastasis (Kapoor et al. 2010). Reduced levels of macroH2A result in up-regulation of many genes, including the CDK8 oncogene. macroH2A is also reduced in more rapidly proliferating lung cancers, and is elevated in cells undergoing senescence, which suggests that reduced senescence of tumor cells is a mechanism whereby loss of macroH2A contributes to oncogenesis (Sporn et al. 2009). It is possible that other defects in the expression of histone variants and in pathways that deposit them can contribute to cancer and other human diseases, but there is as yet insufficient evidence of causality as opposed to mere correlation. For example, excessively high levels of H2A.Z are found to correlate with poor outcomes in estrogen-positive breast cancer (Hua et al. 2008). In addition, CENP-A is found to be overexpressed and ectopically localized in a variety of cancers, suggesting that transient or permanent neocentromere formation might result in some of the aneuploidies that are hallmarks of cancer cells (Dalal 2009). We anticipate that as genome-wide technologies improve, other possible roles of histone variants incorporation pathways in human diseases will be uncovered.

16 CONCLUSIONS AND FUTURE RESEARCH

Histone variants provide the most fundamental level of differentiation of chromatin, and alternative mechanisms for depositing different variants can potentially establish and maintain epigenetic states. Histones H2A, H2B, H3, and H4 occupy distinct positions in the core particle as a result of an evolutionary process that began before the last common ancestor of eukaryotes. Key evolutionary innovations remain uncertain, including the emergence of an octamer from an ancestral tetramer, and we look forward to the sequencing of more archaeal and primitive eukaryotic genomes that might provide missing links. Subsequent elaborations of the four core histones into distinct variants have provided the basis for epigenetic processes, including development and chromosome segregation. For a full understanding of epigenetic inheritance, we need a better understanding of the processes that incorporate variants by replacing canonical histones. An important recent development is the initial characterization of RI assembly pathways dedicated to particular variants.

Centromeres are the most conspicuous examples of profoundly different chromatin that is attributable to special properties of a histone variant. Although it is clear that cenH3-containing nucleosomes form the foundation of the centromere, just how they are deposited in the same place every cell generation without any hint of sequence specificity is a major challenge for future research.

It is becoming evident that histone variants are also involved in epigenetic properties of active genes. Both H3.3 and H2A.Z are enriched at transcriptionally active loci, and understanding the assembly processes that are responsible for their enrichment is an exciting area of current research. The dynamic behavior of chromatin leads to the realization that transcription, chromatin remodeling, and histone modification might be coupled to nucleosome assembly and disassembly. The study of dynamic processes coupled to histone turnover is only at an early stage, and we look forward to technological advances in molecular biology, cytogenetics, biochemistry, and structural biology that can be harnessed to better understand the dynamic nature of chromatin.

In addition to these universal processes, histone variants are also involved in particular epigenetic phenomena. In the case of the mammalian X chromosome, three different H2A variants, phospho-H2A.X, macroH2A, and H2A.B have been recruited to participate in silencing or activation of genes for purposes of germline inactivation or dosage compensation. Understanding the function of these variants in epigenetic processes remains a major challenge for the future.

The availability of the first high-resolution structure of the nucleosome core particle (Luger et al. 1997) was a seminal advance in elucidating the properties of chromatin. By elaborating this basic structure in a way that has biological consequences, histone variants provide an opportunity to deepen our understanding of how these fascinating architectural proteins have evolved to play diverse roles in epigenetic processes.

REFERENCES

*Reference is also in this book.

Adkins NL, Niu H, Sung P, Peterson CL. 2013. Nucleosome dynamics regulates DNA processing. *Nat Struct Mol Biol* **20:** 836–842.

Akhtar A, Gasser SM. 2007. The nuclear envelope and transcriptional control. *Nat Rev Genet* **8:** 507–517.

Akiyama T, Suzuki O, Matsuda J, Aoki F. 2011. Dynamic replacement of histone h3 variants reprograms epigenetic marks in early mouse embryos. *PLoS Genet* **7:** e1002279.

* Allshire RC, Ekwall K. 2014. Epigenetic regulation of chromatin states in *Schizosaccharomyces pombe*. *Cold Spring Harb Perspect Biol* doi: 10.1101/cshperspect.a018770.

* Almouzni G, Cedar H. 2014. Maintenance of epigenetic information. *Cold Spring Harb Perspect Biol* doi: 10.1101/cshperspect.a019372.

Altaf M, Auger A, Covic M, Cote J. 2009. Connection between histone H2A variants and chromatin remodeling complexes. *Biochem Cell Biol* **87:** 35–50.

Altheim BA, Schultz MC. 1999. Histone modification governs the cell cycle regulation of a replication-independent chromatin assembly pathway in *Saccharomyces cerevisiae*. *Proc Natl Acad Sci* **96:** 1345–1350.

Ammar R, Torti D, Tsui K, Gebbia M, Durbic T, Bader GD, Giaever G, Nislow C. 2012. Chromatin is an ancient innovation conserved between Archaea and Eukarya. *eLife* **1:** e00078.

Amor DJ, Bentley K, Ryan J, Perry J, Wong L, Slater H, Choo KH. 2004a. Human centromere repositioning 'in progress'. *Proc Natl Acad Sci* **101:** 6542–6547.

Amor DJ, Kalitsis P, Sumer H, Choo KH. 2004b. Building the centromere: From foundation proteins to 3D organization. *Trends Cell Biol* **14:** 359–368.

Annunziato AT. 2005. Split decision: What happens to nucleosomes during DNA replication? *J Biol Chem* **280:** 12065–12068.

Arents G, Burlingame RW, Wang BC, Love WE, Moudrianakis EN. 1991. The nucleosomal core histone octamer at 3.1 A resolution: A tripartite protein assembly and a left-handed superhelix. *Proc Natl Acad Sci* **88:** 10148–10152.

Baldi S, Becker PB. 2013. The variant histone H2A.V of *Drosophila*—Three roles, two guises. *Chromosoma* **122:** 245–258.

Bao Y, Konesky K, Park YJ, Rosu S, Dyer PN, Rangasamy D, Tremethick DJ, Laybourn PJ, Luger K. 2004. Nucleosomes containing the histone variant H2A.Bbd organize only 118 base pairs of DNA. *EMBO J* **23:** 3314–3324.

Barski A, Cuddapah S, Cui K, Roh TY, Schones DE, Wang Z, Wei G, Chepelev I, Zhao K. 2007. High-resolution profiling of histone methylations in the human genome. *Cell* **129:** 823–837.

* Baulcombe D, Dean C. 2014. Epigenetic regulation in plant responses to the environment. *Cold Spring Harb Perspect Biol* **6:** a019471.

* Becker PB, Workman JL. 2013. Nucleosome remodeling and epigenetics. *Cold Spring Harb Perspect Biol* **5:** 017905.

Belotserkovskaya R, Oh S, Bondarenko VA, Orphanides G, Studitsky VM, Reinberg D. 2003. FACT facilitates transcription-dependent nucleosome alteration. *Science* **301:** 1090–1093.

Black BE, Cleveland DW. 2011. Epigenetic centromere propagation and the nature of CENP-A nucleosomes. *Cell* **144:** 471–479.

* Blewitt M, Whitelaw E. 2013. The use of mouse models to study epigenetics. *Cold Spring Harb Perspect Biol* **5:** 017939.

Boden SA, Kavanova M, Finnegan J, Wigge PA. 2013. Thermal stress effects on grain yield in *Brachypodium distachyon* occur via H2A.Z-nucleosomes. *Genome Biol* **14:** R65.

Bonisch C, Schneider K, Punzeler S, Wiedemann SM, Bielmeier C, Bocola M, Eberl HC, Kuegel W, Neumann J, Kremmer E, et al. 2012. H2A.Z.2.2 is an alternatively spliced histone H2A.Z variant that causes severe nucleosome destabilization. *Nucleic Acids Res* **40:** 5951–5964.

Boyarchuk E, Montes de Oca R, Almouzni G. 2011. Cell cycle dynamics of histone variants at the centromere, a model for chromosomal landmarks. *Curr Opin Cell Biol* **23:** 266–276.

Brickner JH. 2009. Transcriptional memory at the nuclear periphery. *Curr Opin Cell Biol* **21:** 127–133.

Brochier-Armanet C, Forterre P, Gribaldo S. 2011. Phylogeny and evolution of the Archaea: One hundred genomes later. *Curr Opin Microbiol* **14:** 274–281.

* Brockdorff N, Turner B. 2014. Dosage compensation in mammals. *Cold Spring Harb Perspect Biol* doi: 10.1101/cshperspect.a019406.

Buchanan L, Durand-Dubief M, Roguev A, Sakalar C, Wilhelm B, Strålfors A, Shevchenko A, Aasland R, Shevchenko A, Ekwall K et al. 2009. The *Schizosaccharomyces pombe* JmjC-protein, Msc1, prevents H2A.Z localization in centromeric and subtelomeric chromatin domains. *PLoS Genet* **5:** e1000726.

Cairns J. 1975. Mutation selection and the natural history of cancer. *Nature* **255:** 197–200.

Chadwick BP, Willard HF. 2001. A novel chromatin protein, distantly related to histone H2A, is largely excluded from the inactive X chromosome. *J Cell Biol* **152:** 375–384.

Chadwick BP, Willard HF. 2004. Multiple spatially distinct types of facultative heterochromatin on the human inactive X chromosome. *Proc Natl Acad Sci* **101:** 17450–17455.

Cho US, Harrison SC. 2011. Recognition of the centromere-specific histone Cse4 by the chaperone Scm3. *Proc Natl Acad Sci* **108:** 9367–9371.

Conerly ML, Teves SS, Diolaiti D, Ulrich M, Eisenman RN, Henikoff S. 2010. Changes in H2A.Z occupancy and DNA methylation during B-cell lymphomagenesis. *Genome Res* **20:** 1383–1390.

Creyghton MP, Markoulaki S, Levine SS, Hanna J, Lodato MA, Sha K, Young RA, Jaenisch R, Boyer LA. 2008. H2AZ is enriched at Polycomb complex target genes in ES cells and is necessary for lineage commitment. *Cell* **135:** 649–661.

Dalal Y. 2009. Epigenetic specification of centromeres. *Biochem Cell Biol* **87:** 273–282.

Dalal Y, Wang H, Lindsay S, Henikoff S. 2007. Tetrameric structure of centromeric nucleosomes in interphase *Drosophila* cells. *PLoS Biol* **5:** e218.

Deal RB, Henikoff JG, Henikoff S. 2010. Genome-wide kinetics of nucleosome turnover determined by metabolic labeling of histones. *Science* **328:** 1161–1164.

Dhillon N, Oki M, Szyjka SJ, Aparicio OM, Kamakaka RT. 2006. H2A.Z functions to regulate progression through the cell cycle. *Mol Cell Biol* **26:** 489–501.

Dion M, Kaplan T, Friedman N, Rando OJ. 2007. Dynamics of replication-independent histone turnover in budding yeast. *Science* **315:** 1405–1408.

Draker R, Cheung P. 2009. Transcriptional and epigenetic functions of histone variant H2A.Z. *Biochem Cell Biol* **87:** 19–25.

Drane P, Ouararhni K, Depaux A, Shuaib M, Hamiche A. 2010. The death-associated protein DAXX is a novel histone chaperone involved in the replication-independent deposition of H3.3. *Genes Dev* **24:** 1253–1265.

Dryhurst D, Ishibashi T, Rose KL, Eirin-Lopez JM, McDonald D, Silva-Moreno B, Veldhoen N, Helbing CC, Hendzel MJ, Shabanowitz J, et al. 2009. Characterization of the histone H2A.Z-1 and H2A.Z-2 isoforms in vertebrates. *BMC Biol* **7:** 86.

Du Z, Cong H, Yao Z. 2001. Identification of putative downstream genes of Oct-4 by suppression-subtractive hybridization. *Biochem Biophys Res Commun* **282:** 701–706.

Dunleavy EM, Roche D, Tagami H, Lacoste N, Ray-Gallet D, Nakamura Y, Daigo Y, Nakatani Y, Almouzni-Pettinotti G. 2009. HJURP is a cell-cycle-dependent maintenance and deposition factor of CENP-A at centromeres. *Cell* **137:** 485–497.

Edwards JR, O'Donnell AH, Rollins RA, Peckham HE, Lee C, Milekic MH, Chanrion B, Fu Y, Su T, Hibshoosh H, et al. 2010. Chromatin and sequence features that define the fine and gross structure of genomic methylation patterns. *Genome Res* **20:** 972–980.

Eustermann S, Yang JC, Law MJ, Amos R, Chapman LM, Jelinska C, Garrick D, Clynes D, Gibbons RJ, Rhodes D, et al. 2011. Combinatorial readout of histone H3 modifications specifies localization of ATRX to heterochromatin. *Nat Struct Mol Biol* **18:** 777–782.

Faast R, Thonglairoam V, Schulz TC, Beall J, Wells JR, Taylor H, Matthaei K, Rathjen PD, Tremethick DJ, Lyons I. 2001. Histone variant H2A.Z is required for early mammalian development. *Curr Biol* **11:** 1183–1187.

Fan Y, Nikitina T, Morin-Kensicki EM, Zhao J, Magnuson TR, Woodcock CL, Skoultchi AI. 2003. H1 linker histones are essential for mouse development and affect nucleosome spacing in vivo. *MolCell Biol* **23:** 4559–4572.

Fernandez-Capetillo O, Mahadevaiah SK, Celeste A, Romanienko PJ, Camerini-Otero RD, Bonner WM, Manova K, Burgoyne P, Nussenzweig A. 2003. H2AX is required for chromatin remodeling and inactivation of sex chromosomes in male mouse meiosis. *Dev Cell* **4:** 497–508.

Filipescu D, Szenker E, Almouzni G. 2013. Developmental roles of histone H3 variants and their chaperones. *Trends Genet* **29:** 630–640.

Foltz DR, Jansen LE, Bailey AO, Yates JR 3rd, Bassett EA, Wood S, Black BE, Cleveland DW. 2009. Centromere-specific assembly of CENP-A nucleosomes is mediated by HJURP. *Cell* **137:** 472–484.

Formosa T, Ruone S, Adams MD, Olsen AE, Eriksson P, Yu Y, Roades AR, Kaufman PD, Stillman DJ. 2002. Defects in SPT16 or POB3 (yFACT) in *Saccharomyces cerevisiae* cause dependence on the Hir/Hpc pathway: Polymerase passage may degrade chromatin structure. *Genetics* **162:** 1557–1571.

Gevry N, Hardy S, Jacques PE, Laflamme L, Svotelis A, Robert F, Gaudreau L. 2009. Histone H2A.Z is essential for estrogen receptor signaling. *Genes Dev* **23:** 1522–1533.

Goldberg AD, Banaszynski LA, Noh KM, Lewis PW, Elsaesser SJ, Stadler S, Dewell S, Law M, Guo X, Li X, et al. 2010. Distinct factors control histone variant H3.3 localization at specific genomic regions. *Cell* **140:** 678–691.

Govin J, Escoffier E, Rousseaux S, Kuhn L, Ferro M, Thevenon J, Catena R, Davidson I, Garin J, Khochbin S, et al. 2007. Pericentric heterochromatin reprogramming by new histone variants during mouse spermiogenesis. *J Cell Biol* **176:** 283–294.

Greaves IK, Rangasamy D, Ridgway P, Tremethick DJ. 2007. H2A.Z contributes to the unique 3D structure of the centromere. *Proc Natl Acad Sci* **104:** 525–530.

* Grossniklaus U, Paro R. 2014. Transcriptional silencing by *Polycomb*-group proteins. *Cold Spring Harb Perspect Biol* **6:** a019331.

* Grunstein M, Gasser SM. 2013. Epigenetics in *Saccharomyces cerevisiae*. *Cold Spring Harb Perspect Biol* **5:** 017491.

Guse A, Carroll CW, Moree B, Fuller CJ, Straight AF. 2011. In vitro centromere and kinetochore assembly on defined chromatin templates. *Nature* **477:** 354–358.

Halley JE, Kaplan T, Wang AY, Kobor MS, Rine J. 2010. Roles for H2A.Z and its acetylation in GAL1 transcription and gene induction, but not GAL1-transcriptional memory. *PLoS Biol* **8:** e1000401.

Hardy S, Robert F. 2010. Random deposition of histone variants: A cellular mistake or a novel regulatory mechanism? *Epigenetics* **5:** 368–372.

Hardy S, Jacques PE, Gevry N, Forest A, Fortin ME, Laflamme L, Gaudreau L, Robert F. 2009. The euchromatic and heterochromatic landscapes are shaped by antagonizing effects of transcription on H2A.Z deposition. *PLoS Genet* **5:** e1000687.

Hartley PD, Madhani HD. 2009. Mechanisms that specify promoter nucleosome location and identity. *Cell* **137:** 445–458.

Heaphy CM, de Wilde RF, Jiao Y, Klein AP, Edil BH, Shi C, Bettegowda C, Rodriguez FJ, Eberhart CG, Hebbar S, et al. 2011a. Altered telomeres in tumors with ATRX and DAXX mutations. *Science* **333:** 425.

Heaphy CM, Subhawong AP, Hong SM, Goggins MG, Montgomery EA, Gabrielson E, Netto GJ, Epstein JI, Lotan TL, Westra WH, et al. 2011b. Prevalence of the alternative lengthening of telomeres telomere maintenance mechanism in human cancer subtypes. *Am J Pathol* **179:** 1608–1615.

Henikoff S, Ahmad K. 2005. Assembly of variant histones into chromatin. *Ann Rev Cell Dev Biol* **21:** 133–153.

Henikoff S, Furuyama T. 2012. The unconventional structure of centromeric nucleosomes. *Chromosoma* **121:** 341–352.

Hines KA, Cryderman DE, Flannery KM, Yang H, Vitalini MW, Hazelrigg T, Mizzen CA, Wallrath LL. 2009. Domains of heterochromatin protein 1 required for *Drosophila melanogaster* heterochromatin spreading. *Genetics* **182:** 967–977.

Hodl M, Basler K. 2009. Transcription in the absence of histone H3.3. *Curr Biol* **19:** 1221–1226.

Hodl M, Basler K. 2012. Transcription in the absence of histone H3.2 and H3K4 methylation. *Curr Biol* **22:** 2253–2257.

Hou H, Wang Y, Kallgren SP, Thompson J, Yates JR 3rd, Jia S. 2010. Histone variant H2A.Z regulates centromere silencing and chromosome segregation in fission yeast. *J Biol Chem* **285:** 1909–1918.

Hu H, Liu Y, Wang M, Fang J, Huang H, Yang N, Li Y, Wang J, Yao X, Shi Y, et al. 2011. Structure of a CENP-A-histone H4 heterodimer in complex with chaperone HJURP. *Genes Dev* **25:** 901–906.

Hu G, Cui K, Northrup D, Liu C, Wang C, Tang Q, Ge K, Levens D, Crane-Robinson C, Zhao K. 2013. H2A.Z facilitates access of active and repressive complexes to chromatin in embryonic stem cell self-renewal and differentiation. *Cell Stem Cell* **12:** 180–192.

Hua S, Kallen CB, Dhar R, Baquero MT, Mason CE, Russell BA, Shah PK, Liu J, Khramtsov A, Tretiakova MS, et al. 2008. Genomic analysis of estrogen cascade reveals histone variant H2A.Z associated with breast cancer progression. *Mol Syst Biol* **4:** 188.

Huang C, Zhang Z, Xu M, Li Y, Li Z, Ma Y, Cai T, Zhu B. 2013. H3.3-h4 tetramer splitting events feature cell-type specific enhancers. *PLoS Genet* **9:** e1003558.

Huh YH, Sherley JL. 2011. Molecular cloaking of H2A.Z on mortal DNA chromosomes during nonrandom segregation. *Stem Cells* **29:** 1620–1627.

Ingouff M, Hamamura Y, Gourgues M, Higashiyama T, Berger F. 2007. Distinct dynamics of HISTONE3 variants between the two fertilization products in plants. *Curr Biol* **17:** 1032–1037.

Jenuwein T. 2001. Re-SET-ting heterochromatin by histone methyltransferases. *Trends Cell Biol* **11:** 266–273.

Jenuwein T, Allis CD. 2001. Translating the histone code. *Science* **293:** 1074–1080.

Jiao Y, Shi C, Edil BH, de Wilde RF, Klimstra DS, Maitra A, Schulick RD, Tang LH, Wolfgang CL, Choti MA, et al. 2011. DAXX/ATRX, MEN1, and mTOR pathway genes are frequently altered in pancreatic neuroendocrine tumors. *Science* **331:** 1199–1203.

Jin C, Zang C, Wei G, Cui K, Peng W, Zhao K, Felsenfeld G. 2009. H3.3/H2A.Z double variant-containing nucleosomes mark 'nucleosome-free regions' of active promoters and other regulatory regions. *Nat Genet* **41:** 941–945.

Jullien J, Astrand C, Szenker E, Garrett N, Almouzni G, Gurdon JB. 2012. HIRA dependent H3.3 deposition is required for transcriptional reprogramming following nuclear transfer to *Xenopus* oocytes. *Epigenetics Chromatin* **5:** 17.

Kapoor A, Goldberg MS, Cumberland LK, Ratnakumar K, Segura MF, Emanuel PO, Menendez S, Vardabasso C, Leroy G, Vidal CI, et al. 2010. The histone variant macroH2A suppresses melanoma progression through regulation of CDK8. *Nature* **468:** 1105–1109.

Kaya H, Shibahara KI, Taoka KI, Iwabuchi M, Stillman B, Araki T. 2001. FASCIATA genes for chromatin assembly factor-1 in *Arabidopsis* maintain the cellular organization of apical meristems. *Cell* **104:** 131–142.

Kelly TK, Miranda TB, Liang G, Berman BP, Lin JC, Tanay A, Jones PA. 2010. H2A.Z maintenance during mitosis reveals nucleosome shifting on mitotically silenced genes. *Mol Cell* **39:** 901–911.

Kobor MS, Lorincz MC. 2009. H2A.Z and DNA methylation: Irreconcilable differences. *Trends Biochem Sci* **34:** 158–161.

Ku M, Jaffe JD, Koche RP, Rheinbay E, Endoh M, Koseki H, Carr SA, Bernstein BE. 2012. H2A.Z landscapes and dual modifications in pluripotent and multipotent stem cells underlie complex genome regulatory functions. *Genome Biol* **13:** R85.

Kumar SV, Wigge PA. 2010. H2A.Z-containing nucleosomes mediate the thermosensory response in *Arabidopsis*. *Cell* **140:** 136–147.

Kundu S, Peterson CL. 2010. Dominant role for signal transduction in the transcriptional memory of yeast GAL genes. *Mol Cell Biol* **30:** 2330–2340.

Ladurner AG. 2003. Inactivating chromosomes: A macro domain that minimizes transcription. *Mol Cell* **12:** 1–3.

Law MJ, Lower KM, Voon HP, Hughes JR, Garrick D, Viprakasit V, Mitson M, De Gobbi M, Marra M, Morris A, et al. 2010. ATR-X syndrome protein targets tandem repeats and influences allele-specific expression in a size-dependent manner. *Cell* **143:** 367–378.

Lewis PW, Muller MM, Koletsky MS, Cordero F, Lin S, Banaszynski LA, Garcia BA, Muir TW, Becher OJ, Allis CD. 2013. Inhibition of PRC2 activity by a gain-of-function H3 mutation found in pediatric glioblastoma. *Science* **340:** 857–861.

* Li E, Zhang Y. 2014. DNA methylation in mammals. *Cold Spring Harb Perspect Biol* **6:** a019133.

Li Z, Gadue P, Chen K, Jiao Y, Tuteja G, Schug J, Li W, Kaestner KH. 2012. Foxa2 and H2A.Z mediate nucleosome depletion during embryonic stem cell differentiation. *Cell* **151:** 1608–1616.

Light WH, Brickner DG, Brand VR, Brickner JH. 2010. Interaction of a DNA zip code with the nuclear pore complex promotes H2A.Z incorporation and INO1 transcriptional memory. *Mol Cell* **40:** 112–125.

* Liu X, McEachron T, Schwartzentruber J, Wu G. 2014. Histone H3 mutations in pediatric brain tumors. *Cold Spring Harb Perspect Biol* **6:** a018689.

Lowndes NF, Toh GW. 2005. DNA repair: The importance of phosphorylating histone H2AX. *Curr Biol* **15:** R99–R102.

Loyola A, Almouzni G. 2004. Histone chaperones, a supporting role in the limelight. *Biochim Biophys Acta* **1677:** 3–11.

Lu PY, Levesque N, Kobor MS. 2009. NuA4 and SWR1-C: Two chromatin-modifying complexes with overlapping functions and components. *Biochem Cell Biol* **87:** 799–815.

Luger K, Mader AW, Richmond RK, Sargent DF, Richmond TJ. 1997. Crystal structure of the nucleosome core particle at 2.8 A resolution. *Nature* **389:** 251–260.

Luk E, Ranjan A, Fitzgerald PC, Mizuguchi G, Huang Y, Wei D, Wu C. 2010. Stepwise histone replacement by SWR1 requires dual activation with histone H2A.Z and canonical nucleosome. *Cell* **143:** 725–736.

Malik HS, Henikoff S. 2003. Phylogenomics of the nucleosome. *Nat Struct Biol* **10:** 882–891.

Malik HS, Henikoff S. 2009. Major evolutionary transitions in centromere complexity. *Cell* **138:** 1067–1082.

March-Diaz R, Reyes JC. 2009. The beauty of being a variant: H2A.Z and the SWR1 complex in plants. *Mol Plant* **2:** 565–577.

Marques M, Laflamme L, Gervais AL, Gaudreau L. 2010. Reconciling the positive and negative roles of histone H2A.Z in gene transcription. *Epigenetics* **5:** 267–272.

Martins-Taylor K, Sharma U, Rozario T, Holmes SG. 2011. H2A.Z (Htz1) controls the cell-cycle-dependent establishment of transcriptional silencing at *Saccharomyces cerevisiae* telomeres. *Genetics* **187:** 89–104.

Marzluff WF, Duronio RJ. 2002. Histone mRNA expression: Multiple levels of cell cycle regulation and important developmental consequences. *Curr Opin Cell Biol* **14:** 692–699.

Marzluff WF, Gongidi P, Woods KR, Jin J, Maltais LJ. 2002. The human and mouse replication-dependent histone genes. *Genomics* **80:** 487–498.

Matsuda R, Hori T, Kitamura H, Takeuchi K, Fukagawa T, Harata M. 2010. Identification and characterization of the two isoforms of the vertebrate H2A.Z histone variant. *Nucleic Acids Res* **38:** 4263–4273.

Mavrich TN, Jiang C, Ioshikhes IP, Li X, Venters BJ, Zanton SJ, Tomsho LP, Qi J, Glaser RL, Schuster SC, et al. 2008. Nucleosome organization in the *Drosophila* genome. *Nature* **453:** 358–362.

Mehta M, Braberg H, Wang S, Lozsa A, Shales M, Solache A, Krogan NJ, Keogh MC. 2010. Individual lysine acetylations on the N terminus of *Saccharomyces cerevisiae* H2A.Z are highly but not differentially regulated. *J Biol Chem* **285:** 39855–39865.

Mendiburo MJ, Padeken J, Fulop S, Schepers A, Heun P. 2011. *Drosophila* CENH3 is sufficient for centromere formation. *Science* **334:** 686–690.

Millar CB, Xu F, Zhang K, Grunstein M. 2006. Acetylation of H2AZ Lys 14 is associated with genome-wide gene activity in yeast. *Genes Dev* **20:** 711–722.

Mizuguchi G, Xiao H, Wisniewski J, Smith MM, Wu C. 2007. Nonhistone Scm3 and histones CenH3-H4 assemble the core of centromere-specific nucleosomes. *Cell* **129:** 1153–1164.

Morillo-Huesca M, Clemente-Ruiz M, Andujar E, Prado F. 2010. The SWR1 histone replacement complex causes genetic instability and genome-wide transcription misregulation in the absence of H2A.Z. *PLoS One* **5:** e12143.

Morrison AJ, Shen X. 2005. DNA repair in the context of chromatin. *Cell Cycle* **4:** 568–571.

Morrison AJ, Shen X. 2009. Chromatin remodelling beyond transcription: The INO80 and SWR1 complexes. *Nat Rev Mol Cell Biol* **10:** 373–384.

Nakano S, Stillman B, Horvitz HR. 2011. Replication-coupled chromatin assembly generates a neuronal bilateral asymmetry in *C. elegans*. *Cell* **147:** 1525–1536.

Nekrasov M, Amrichova J, Parker BJ, Soboleva TA, Jack C, Williams R, Huttley GA, Tremethick DJ. 2012. Histone H2A.Z inheritance during the cell cycle and its impact on promoter organization and dynamics. *Nat Struct Mol Biol* **19:** 1076–1083.

Ng RK, Gurdon JB. 2008. Epigenetic memory of an active gene state depends on histone H3.3 incorporation into chromatin in the absence of transcription. *Nat Cell Biol* **10:** 102–109.

Nusinow DA, Hernandez-Munoz I, Fazzio TG, Shah GM, Kraus WL, Panning B. 2007. Poly(ADP-ribose) polymerase 1 is inhibited by a histone H2A variant, MacroH2A, and contributes to silencing of the inactive X chromosome. *J Biol Chem* **282:** 12851–12859.

Ooi S, Priess J, Henikoff S. 2006. Histone H3.3 variant dynamics in the germline of *Caenorhabditis elegans*. *PLoS Genet* **2:** e97.

Orsi GA, Couble P, Loppin B. 2009. Epigenetic and replacement roles of histone variant H3.3 in reproduction and development. *Int J Dev Biol* **53:** 231–243.

Palmer DK, O'Day K, Margolis RL. 1990. The centromere specific histone CENP-A is selectively retained in discrete foci in mammalian sperm nuclei. *Chromosoma* **100:** 32–36.

Palmer DK, O'Day K, Trong HL, Charbonneau H, Margolis RL. 1991. Purification of the centromere-specific protein CENP-A and demonstration that it is a distinctive histone. *Proc Natl Acad Sci* **88:** 3734–3738.

Papamichos-Chronakis M, Watanabe S, Rando OJ, Peterson CL. 2011. Global regulation of H2A.Z localization by the INO80 chromatin-remodeling enzyme is essential for genome integrity. *Cell* **144:** 200–213.

Phair RD, Scaffidi P, Elbi C, Vecerova J, Dey A, Ozato K, Brown DT, Hager G, Bustin M, Misteli T. 2004. Global nature of dynamic protein-chromatin interactions in vivo: Three-dimensional genome scanning and dynamic interaction networks of chromatin proteins. *Mol Cell Biol* **24:** 6393–6402.

Pina B, Suau P. 1987. Changes in histones H2A and H3 variant composition in differentiating and mature rat brain cortical neurons. *Dev Biol* **123:** 51–58.

Raisner RM, Hartley PD, Meneghini MD, Bao MZ, Liu CL, Schreiber SL, Rando OJ, Madhani HD. 2005. Histone variant H2A.Z marks the 5′ ends of both active and inactive genes in euchromatin. *Cell* **123:** 233–248.

Rakyan V, Whitelaw E. 2003. Transgenerational epigenetic inheritance. *Curr Biol* **13:** R6.

Ratnakumar K, Duarte LF, LeRoy G, Hasson D, Smeets D, Vardabasso C, Bonisch C, Zeng T, Xiang B, Zhang DY, et al. 2012. ATRX-mediated chromatin association of histone variant macroH2A1 regulates α-globin expression. *Genes Dev* **26:** 433–438.

Ray-Gallet D, Woolfe A, Vassias I, Pellentz C, Lacoste N, Puri A, Schultz DC, Pchelintsev NA, Adams PD, Jansen LE, et al. 2011. Dynamics of histone H3 deposition in vivo reveal a nucleosome gap-filling mechanism for H3.3 to maintain chromatin integrity. *Mol Cell* **44:** 928–941.

* Reik W, Surani A. 2014. Germline and pluripotent stem cells. *Cold Spring Harb Perspect Biol* doi: 10.1101/cshperspect.a019422.

Sakai A, Schwartz BE, Goldstein S, Ahmad K. 2009. Transcriptional and developmental functions of the H3.3 histone variant in *Drosophila*. *Curr Biol* **19:** 1816–1820.

Santenard A, Ziegler-Birling C, Koch M, Tora L, Bannister AJ, Torres-Padilla ME. 2010. Heterochromatin formation in the mouse embryo requires critical residues of the histone variant H3.3. *Nat Cell Biol* **12:** 853–862.

Santisteban MS, Hang M, Smith MM. 2011. Histone variant H2A.Z and RNA polymerase II transcription elongation. *Mol Cell Biol* **31:** 1848–1860.

Schneiderman JI, Sakai A, Goldstein S, Ahmad K. 2009. The XNP remodeler targets dynamic chromatin in *Drosophila*. *Proc Natl Acad Sci* **106:** 14472–14477.

Schwartz BE, Ahmad K. 2005. Transcriptional activation triggers deposition and removal of the histone variant H3.3. *Genes Dev* **19:** 804–814.

Schwartzentruber J, Korshunov A, Liu XY, Jones DT, Pfaff E, Jacob K, Sturm D, Fontebasso AM, Quang DA, Tonjes M, et al. 2012. Driver mutations in histone H3.3 and chromatin remodelling genes in paediatric glioblastoma. *Nature* **482:** 226–231.

Shaw ML, Williams EJ, Hawes S, Saffery R. 2009. Characterisation of histone variant distribution in human embryonic stem cells by transfection of in vitro transcribed mRNA. *Mol Reprod Dev* **76:** 1128–1142.

Shuaib M, Ouararhni K, Dimitrov S, Hamiche A. 2010. HJURP binds CENP-A via a highly conserved N-terminal domain and mediates its deposition at centromeres. *Proc Natl Acad Sci* **107:** 1349–1354.

Simonet NG, Reyes M, Nardocci G, Molina A, Alvarez M. 2013. Epigenetic regulation of the ribosomal cistron seasonally modulates enrichment of H2A.Z and H2A.Zub in response to different environmental inputs in carp (*Cyprinus carpio*). *Epigenetics Chromatin* **6:** 22.

Soboleva TA, Nekrasov M, Pahwa A, Williams R, Huttley GA, Tremethick DJ. 2012. A unique H2A histone variant occupies the transcriptional start site of active genes. *Nat Struct Mol Biol* **19:** 25–30.

Sporn JC, Kustatscher G, Hothorn T, Collado M, Serrano M, Muley T, Schnabel P, Ladurner AG. 2009. Histone macroH2A isoforms predict the risk of lung cancer recurrence. *Oncogene* **28:** 3423–3428.

Stoler S, Rogers K, Weitze S, Morey L, Fitzgerald-Hayes M, Baker RE. 2007. Scm3, an essential *Saccharomyces cerevisiae* centromere protein required for G2/M progression and Cse4 localization. *Proc Natl Acad Sci* **104:** 10571–10576.

Sullivan BA, Karpen GH. 2004. Centromeric chromatin exhibits a histone modification pattern that is distinct from both euchromatin and heterochromatin. *Nat Struct Mol Biol* **11:** 1076–1083.

Suto RK, Clarkson MJ, Tremethick DJ, Luger K. 2000. Crystal structure of a nucleosome core particle containing the variant histone H2A.Z. *Nat Struct Biol* **7:** 1121–1124.

Talbert PB, Henikoff S. 2010. Histone variants—Ancient wrap artists of the epigenome. *Nat Rev Mol Cell Biol* **11:** 264–275.

Talbert PB, Ahmad K, Almouzni G, Ausio J, Berger F, Bhalla PL, Bonner WM, Cande WZ, Chadwick BP, Chan SW, et al. 2012. A unified phylogeny-based nomenclature for histone variants. *Epigenetics Chromatin* **5:** 7.

Tanabe M, Kouzmenko AP, Ito S, Sawatsubashi S, Suzuki E, Fujiyama S, Yamagata K, Zhao Y, Kimura S, Ueda T, et al. 2008. Activation of facultatively silenced *Drosophila* loci associates with increased acetylation of histone H2AvD. *Genes Cells* **13:** 1279–1288.

Tran V, Lim C, Xie J, Chen X. 2012. Asymmetric division of *Drosophila* male germline stem cell shows asymmetric histone distribution. *Science* **338:** 679–682.

Updike DL, Mango SE. 2006. Temporal regulation of foregut development by HTZ-1/H2A.Z and PHA-4/FoxA. *PLoS Genet* **2:** e161.

Venkatasubrahmanyam S, Hwang WW, Meneghini MD, Tong AH, Madhani HD. 2007. Genome-wide, as opposed to local, antisilencing is mediated redundantly by the euchromatic factors Set1 and H2A.Z. *Proc Natl Acad Sci* **104:** 16609–16614.

Viens A, Mechold U, Brouillard F, Gilbert C, Leclerc P, Ogryzko V. 2006. Analysis of human histone H2AZ deposition in vivo argues against its direct role in epigenetic templating mechanisms. *Mol Cell Biol* **26:** 5325–5335.

Wan Y, Saleem RA, Ratushny AV, Roda O, Smith JJ, Lin CH, Chiang JH, Aitchison JD. 2009. Role of the histone variant H2A.Z/Htz1p in TBP recruitment, chromatin dynamics, and regulated expression of oleate-responsive genes. *Mol Cell Biol* **29:** 2346–2358.

Watanabe S, Radman-Livaja M, Rando OJ, Peterson CL. 2013. A histone acetylation switch regulates H2A.Z deposition by the SWR-C remodeling enzyme. *Science* **340:** 195–199.

Weber CM, Henikoff JG, Henikoff S. 2010. H2A.Z nucleosomes enriched over active genes are homotypic. *Nat Struct Mol Biol* **17:** 1500–1507.

Weintraub H, Worcel A, Alberts B. 1976. A model for chromatin based upon two symmetrically paired half-nucleosomes. *Cell* **9:** 409–417.

Whittle CM, McClinic KN, Ercan S, Zhang X, Green RD, Kelly WG, Lieb JD. 2008. The genomic distribution and function of histone variant HTZ-1 during *C. elegans* embryogenesis. *PLoS Genet* **4:** e1000187.

Wieland G, Orthaus S, Ohndorf S, Diekmann S, Hemmerich P. 2004. Functional complementation of human centromere protein A (CENP-A) by Cse4p from *Saccharomyces cerevisiae*. *Mol Cell Biol* **24:** 6620–6630.

Witt O, Albig W, Doenecke D. 1996. Testis-specific expression of a novel human H3 histone gene. *Exp Cell Res* **229:** 301–306.

Wolffe AP. 1992. *Chromatin: Structure and function*. Academic, San Diego.

Wu G, Broniscer A, McEachron TA, Lu C, Paugh BS, Becksfort J, Qu C, Ding L, Huether R, Parker M, et al. 2012. Somatic histone H3 alterations in pediatric diffuse intrinsic pontine gliomas and non-brainstem glioblastomas. *Nat Genet* **44:** 251–253.

Xu M, Long C, Chen X, Huang C, Chen S, Zhu B. 2010. Partitioning of histone H3-H4 tetramers during DNA replication-dependent chromatin assembly. *Science* **328:** 94–98.

Xu Y, Ayrapetov MK, Xu C, Gursoy-Yuzugullu O, Hu Y, Price BD. 2012. Histone H2A.Z controls a critical chromatin remodeling step required for DNA double-strand break repair. *Mol Cell* **48:** 723–733.

Zacharioudakis I, Gligoris T, Tzamarias D. 2007. A yeast catabolic enzyme controls transcriptional memory. *Curr Biol* **17:** 2041–2046.

Zemach A, McDaniel IE, Silva P, Zilberman D. 2010. Genome-wide evolutionary analysis of eukaryotic DNA methylation. *Science* **328:** 916–919.

Zhang Z, Pugh BF. 2011. Genomic organization of H2Av containing nucleosomes in *Drosophila* heterochromatin. *PLoS One* **6:** e20511.

Zhang H, Roberts DN, Cairns BR. 2005. Genome-wide dynamics of Htz1, a histone H2A variant that poises repressed/basal promoters for activation through histone loss. *Cell* **123:** 219–231.

Zilberman D, Coleman-Derr D, Ballinger T, Henikoff S. 2008. Histone H2A.Z and DNA methylation are mutually antagonistic chromatin marks. *Nature* **456:** 125–129.

Zlatanova J, Thakar A. 2008. H2A.Z: View from the top. *Structure* **16:** 166–179.

Zofall M, Fischer T, Zhang K, Zhou M, Cui B, Veenstra TD, Grewal SI. 2009. Histone H2A.Z cooperates with RNAi and heterochromatin factors to suppress antisense RNAs. *Nature* **461:** 419–422.

WWW RESOURCES

http://www.ebi.ac.uk/Tools/phylogeny EBI server

CHAPTER 21

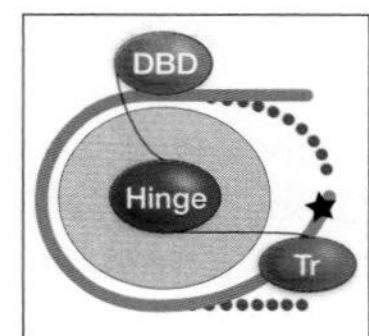

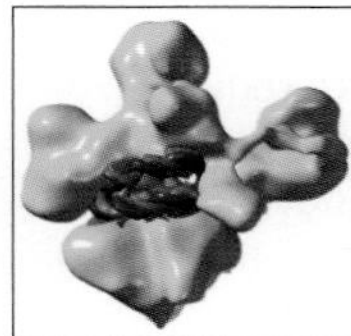

Nucleosome Remodeling and Epigenetics

Peter B. Becker[1] and Jerry L. Workman[2]

[1]BioMedical Center, Ludwig-Maximilians-University, D-80336 Munich, Germany; [2]Stowers Institute for Medical Research, Kansas City, Missouri 64110

Correspondence: pbecker@med.uni-muenchen.de; jlw@stowers.org

SUMMARY

Eukaryotic chromatin is kept flexible and dynamic to respond to environmental, metabolic, and developmental cues through the action of a family of so-called "nucleosome remodeling" ATPases. Consistent with their helicase ancestry, these enzymes experience conformation changes as they bind and hydrolyze ATP. At the same time they interact with DNA and histones, which alters histone–DNA interactions in target nucleosomes. Their action may lead to complete or partial disassembly of nucleosomes, the exchange of histones for variants, the assembly of nucleosomes, or the movement of histone octamers on DNA. "Remodeling" may render DNA sequences accessible to interacting proteins or, conversely, promote packing into tightly folded structures. Remodeling processes participate in every aspect of genome function. Remodeling activities are commonly integrated with other mechanisms such as histone modifications or RNA metabolism to assemble stable, epigenetic states.

Outline

Chapter-opening figures: (*Left*) Adapted from Clapier and Cairns (2009), © Annual Reviews; (*right*) courtesy Francisco Arturias, Scripps Research Institute, La Jolla.

Cite as *Cold Spring Harb Perspect Biol* doi: 10.1101/cshperspect.a017905

OVERVIEW

An unavoidable side effect of the organization of eukaryotic genomes into chromatin is the occlusion of DNA sequence by histones and nonhistone chromatin components. Making use of the genetic information, be it as part of a developmental program or in response to environmental cues, for faithful propagation of chromatin during replication or the repair of damage, necessitates that regulatory factors and complex machineries gain access to DNA sequence. The most fundamental issue of chromatin biology is, therefore, how to ensure access to DNA despite the compact and protective chromatin organization. By its very nature, this organization generates a default state of inaccessibility and, hence, inactivity of the DNA that is subject to it. There are several reasons for this. First, proteins cannot easily associate with DNA sequence that touches the nucleosomal histone surface. Second, nucleosomal DNA is strongly bent during its path around the histone octamer, and many DNA binding proteins find their target sequence distorted and unrecognizable. Finally, nonhistone chromatin components may associate with nucleosomes bearing chemical modifications (i.e., histone posttranslational modifications) and fold the nucleosomal fiber into "higher-order" structures that may be even less accessible.

The inevitable inaccessibility of DNA in chromatin may appear as a problem at first sight, but evolution has turned it into a strong asset by inventing enzymes that are able to "remodel" nucleosomes. This allows access to DNA to be regulated locally and differentially as needed. Nucleosome remodeling involves changing histone–DNA interactions as a means of disrupting, assembling or moving nucleosomes. Nucleosome remodeling enzymes may liberate segments of DNA by complete or partial disassembly of nucleosomes, may alter the composition of nucleosomes with respect to histone variants and, more indirectly, may also affect the folding of the nucleosomal fiber (Workman and Kingston 1998; Kingston and Narlikar 1999; Becker and Horz 2002; Clapier and Cairns 2009; Hargreaves and Crabtree 2011).

Nucleosomes are rather stable entities attributable to the cumulative effect of many weak histone–DNA interactions (Luger and Richmond 1998). It therefore comes as no surprise that nucleosome remodeling reactions require biochemical coupling to ATP hydrolysis. Like all biochemical reactions they are, in principle, reversible and the outcome of remodeling reactions—be it the disruption of histone–DNA interactions or their formation, the sliding of a histone octamer on or off a particular DNA sequence, or alteration of the histone variant composition of a nucleosome—are largely determined by the specification of remodeling enzymes and the involvement of cofactors.

Nucleosome remodeling ATPases are involved in each and every aspect of genome utilization, be it the regulated execution of developmental gene expression programs (Chioda and Becker 2010; Ho and Crabtree 2010) or the fast transcriptional response to environmental signals (Vicent et al. 2010). It also includes their involvement in the scheduled replication of the genome (Falbo and Shen 2006; Neves-Costa and Varga-Weisz 2006; Morettini et al. 2008) or the surveillance of the genome for DNA damage and its repair by a range of strategies, including the recombination of chromosomal segments (Altaf et al. 2007; Bao and Shen 2007; Downs et al. 2007). Defects in genes coding for nucleosome remodeling enzymes can have subtle or rather dramatic consequences, depending on the processes affected and the functional redundancy of the system. Failure of nucleosome remodeling during development may compromise viability or cause morphological defects (Chioda and Becker 2010; Ho and Crabtree 2010). In other cases, failure of remodeling systems may render cells unable to cope with DNA damage and lead to genome instability and cancer (Cairns 2001; Weissman and Knudsen 2009; Hargreaves and Crabtree 2011).

In this chapter we describe the discovery of nucleosome remodelers through a combination of genetics and biochemistry, their mechanism of action, and the outcomes of remodeling events. We outline the different families of remodeling enzymes, the protein domains that distinguish them, and the diversity of complex machineries they reside in. We describe the roles of remodelers in chromatin assembly, transcriptional regulation, and development. Finally, we discuss the functional cross talk of nucleosome remodeling with histone variants and posttranslational modifications of both histones and the remodelers themselves.

 Cite as *Cold Spring Harb Perspect Biol* doi: 10.1101/cshperspect.a017905

1 THE DISCOVERY OF NUCLEOSOME REMODELING: A HISTORICAL PERSPECTIVE

Monitoring the differential accessibility of a complex genome in a eukaryotic nucleus is a first step to understanding its function. Toward this end, probing the accessibility of sequences in intact nuclei with small, nonspecific deoxyribonucleases (DNases), such as DNase I, has been extremely successful (Elgin 1981; Becker and Horz 2002; see Ch. 12 [Elgin and Reuter 2013]). In these experiments, intact nuclei are treated mildly with nucleases that digest any DNA they can access within the time of the experiment. In these experiments, chromatin domains harboring active gene loci are an order of magnitude more sensitive toward digestion than inactive domains. Active regulatory elements, such as promoters, enhancers, and replication origins were found to be in even more open chromatin configurations, characterized by hypersensitivity to DNase. DNase I-hypersensitive sites (DHS) have been characterized as regions relatively depleted of canonical nucleosomes and, in turn, are usually occupied by regulatory DNA-binding proteins. The rapid—within minutes—new appearance of a DHS at a regulatory element on induction of transcription by hormones highlighted the dynamic nature of chromatin that enabled fast and very local structural transitions and rendered DNA accessible (Fig. 1). Conversely, the removal of the inducer led to a fast closing of chromatin on dissociation of factors (Reik et al. 1991). These findings motivated the search for the active molecules and processes underlying these transitions, which led to the identification of ATP-dependent nucleosome remodeling factors. These remodeling factors modulate nucleosome organization in a variety of ways (Fig. 2), which will be discussed in this chapter.

The best known of the remodeling complexes, Swi/Snf, was originally identified genetically in yeast. Its subunits are encoded by several genes required for the expression of the SUC2 invertase gene and the HO endonuclease gene. SUC2 is required for sucrose fermentation, therefore SNF (sucrose nonfermenting) and HO expression is required for mating-type switching, hence SWI (switch; reviewed in Winston and Carlson 1992). The SWI/SNF genes were later shown to be involved in the regulation of a broad set of genes in yeast (Hargreaves and Crabtree 2011). One of the Swi/Snf complex subunits, Swi2/Snf2, bears similarity to helicases and was found to have DNA-stimulated ATPase activity. Further genetic studies suggested that SWI/SNF genes functioned interdependently in positively regulating gene expression, leading to the suggestion that the proteins they encode might function in a multiprotein complex. A connection was made between SWI/SNF function and chromatin when several suppressor mutations of swi/snf phenotypes (called SIN for switch independent) mapped

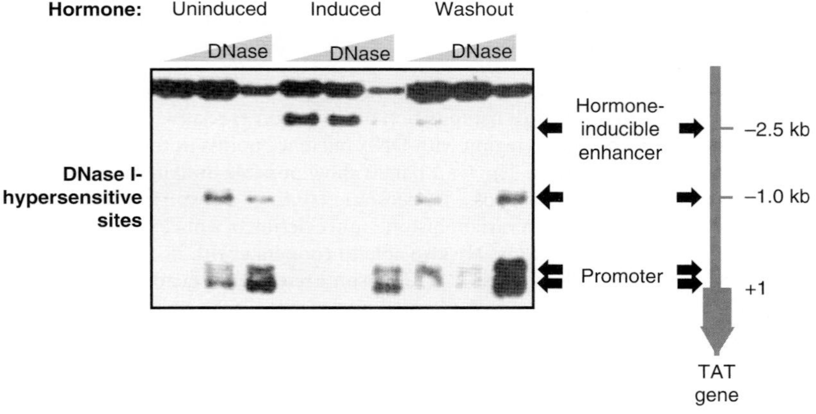

Figure 1. DNase I-hypersensitivity (DH) analysis reveals rapid and reversible local nucleosome remodeling in vivo. The figure shows primary data from a classical DH analysis (Reik et al. 1991). The chromatin organization at the glucocorticoid-responsive enhancer element, 2.5 kb upstream of the promoter of the tyrosine aminotransferase gene, was probed in rat liver cells. Isolated nuclei of cells are digested with increasing amounts of DNase I. Digested genomic DNA is purified, cleaved with a restriction enzyme, resolved by agarose gel electrophoresis, and subjected to Southern blotting. The DH sites are revealed by indirect end-labeling of restriction fragments through hybridization of a small radioactive probe). They are marked with arrows. In the silent, uninduced state, there are two DH sites at the promoter and one at −1 kb upstream. When the gene is activated on hormone induction with corticosterone, nucleosomes are remodeled at the enhancer within 15 min. A new DH site appears 2.5 kb upstream of the transcriptional start site, caused by chromatin remodeling (see "induced" columns). This correlates with the binding of glucocorticoid receptors and a complex set of remodeling factors. On removal of the hormone ("washout"), the factors dissociate and canonical nucleosomes reform within 15 min and the −2.5-kb enhancer DH disappears. The enhanced cleavage at the promoter reflects the transcriptional status of the gene.

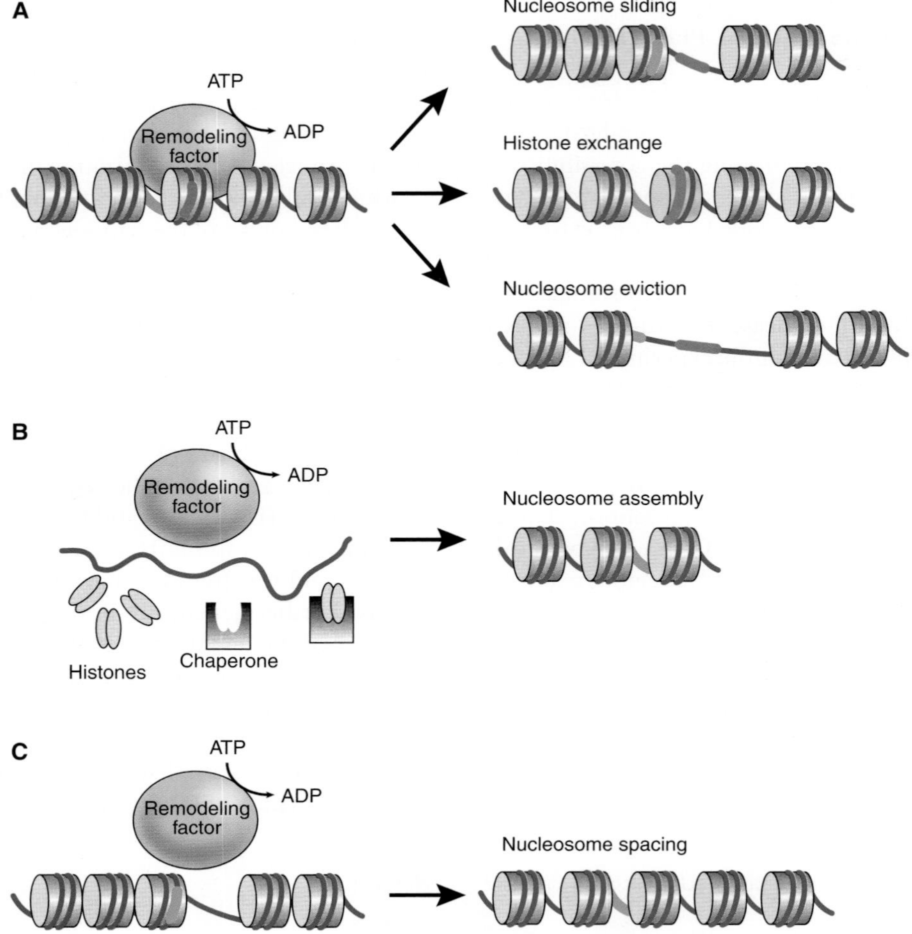

Figure 2. Consequences of ATP-dependent nucleosome remodeling. (*A*) Models for nucleosome remodeling are illustrated by showing the change in position or composition of nucleosomes relative to the DNA wrapped around it. The *left* panel indicates a starting chromatin region with DNA reference points in the linker DNA or on the nucleosome shown in blue and pink, respectively. The *right* panels show possible outcomes of the remodeling reaction (from *top* to *bottom*): translational movement of a nucleosome (sliding) to expose a region that was previously occluded, exchange of a standard histone for a variant histone, and eviction of a nucleosome to expose the associated DNA. (*B*) Some nucleosome remodeling factors are also able to cooperate with histone chaperones to wrap DNA around histone octamers to generate nucleosomes. (*C*) Nucleosome remodeling factors may equilibrate the distances between nucleosomes in irregular arrays in a process termed nucleosome "spacing."

to genes encoding histones or other chromatin components (Vignali et al. 2000; Fry and Peterson 2001). The connection became clearer when the complex was purified from yeast and mammalian cells. Functionally, purified Swi/Snf complex could disrupt nucleosome structure in an ATP-dependent manner and stimulate the binding of transcription factors to nucleosomal DNA in vitro (Fig. 2). Furthermore, these activities generated DNase-hypersensitive sites at transcription factor binding sites within nucleosome arrays (Vignali et al. 2000; Fry and Peterson 2001). Several versions of Swi/Snf-type remodeling complexes have since been found in eukaryotic cells. For example, a second essential and more abundant complex in yeast, RSC (remodels the structure of chromatin), contains many orthologs to Swi/Snf subunits (Clapier and Cairns 2009). In *Drosophila melanogaster*, the Swi2/Snf2 homolog, Brahma (brm), was identified in genetic screens for suppressors of the transcriptional repressor Polycomb (see Ch. 18 [Kingston and Tamkun 2014]). Brm is also part of multiple versions of Swi/Snf-type complexes in mammalian cells, which play important roles in development and cellular homeostasis (Hargreaves and Crabtree 2011, and see Section 10).

An alternative biochemical strategy led to the identification of remodelers of the ISWI-type (imitation switch) from *Drosophila*. Extracts from preblastoderm *Drosophila* embryos provided a powerful in vitro chromatin assembly

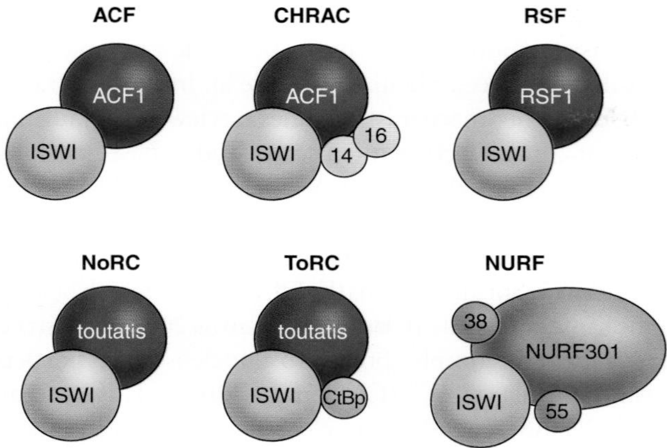

Figure 3. The ISWI ATPase resides in several remodeling factors. The known ISWI-containing remodeling complexes in *Drosophila* are schematically shown. The functions of ACF, CHRAC, RSF, and NURF are described in the text. In mammals, the NoRC remodeler is involved in regulating the activity of ribosomal RNA genes (Li et al. 2006). NoRC is defined by the signature factor Tip5. The homologous protein in flies, toutatis, also interacts with ISWI (Vanolst et al. 2005). NoRC interacts with CtBP to form ToRC, which is involved in transcription regulation and nucleosome assembly outside of the nucleolus (Emelyanov et al. 2012). In mammals, additional complexes are known (Bao and Shen 2011; Kasten et al. 2011; Sims and Wade 2011; Yadon and Tsukiyama 2011) and it is likely that further assemblies will be discovered in flies as well.

system, as they contain abundant histones chaperone and remodeling factors. The reconstitution of chromatin in the test tube, with physiological packaging and repression properties, provided an opportunity to search for factors that were able to increase the accessibility of DNA in an ATP-dependent manner (Becker and Wu 1992; Pazin et al. 1994; Tsukiyama et al. 1994; Varga-Weisz et al. 1995). The activities that were discovered by such strategies were protein complexes containing the ATPase ISWI, including the nucleosome remodeling factor (NURF), the ATP-dependent nucleosome assembly and remodeling factor (ACF), and the chromatin accessibility complex (CHRAC) (Fig. 3) (Tsukiyama et al. 1995; Varga-Weisz et al. 1997; Ito et al. 1999). ISWI had, in fact, been identified earlier because of its sequence similarity to the yeast and *Drosophila* SWI2/SNF2 proteins (Elfring et al. 1994). The convergence of genetic and biochemical discovery strategies revealed the existence of a novel family of enzymes, the nucleosome remodeling ATPases, dedicated to regulating the access to DNA in chromatin (Flaus et al. 2006; Cairns 2009).

Nucleosome remodeling ATPases related to Swi2/Snf2 (the SNF2 family; Fig. 4) are found in all eukaryotes from yeast to man. They have been refined during evolution, just like the basic chromatin organization itself and can be identified because of their sequence similarity. A comprehensive survey of all known and potential nucleosome remodeling ATPases lists some 1300 SNF2 family members (approximately 30 alone in humans) that can be grouped in no less than 23 subfamilies depending on their domain organization (Fig. 4) (Durr et al. 2006; Flaus et al. 2006; Ryan and Owen-Hughes 2011).

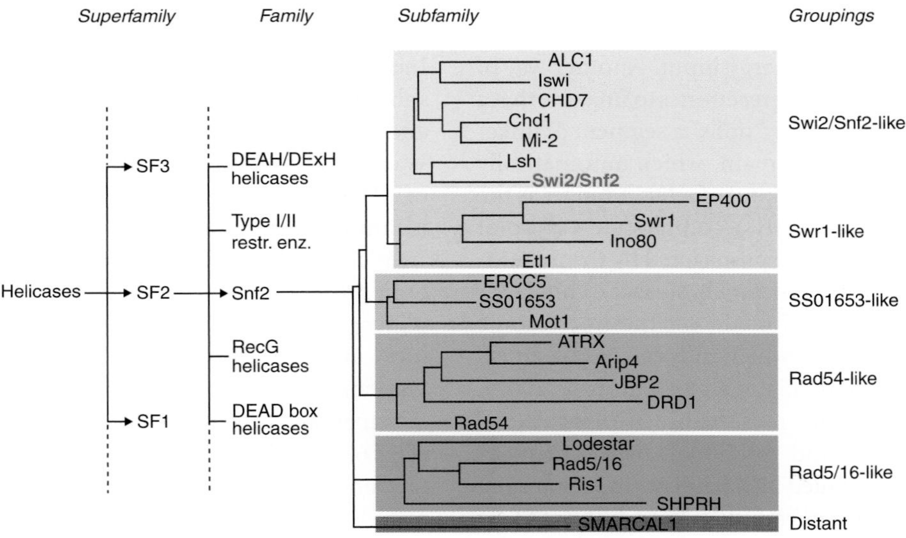

Figure 4. Sequence relationships for the Snf2 family. Cladogram showing the relationship of the Snf2 family to other helicase-like proteins of superfamily 2 (SF2) (Fairman-Williams et al. 2010). Subfamily relationships within the Snf2 family are indicated based on alignments of the helicase-like region from 1306 members (Flaus et al. 2006). Branch lengths are not to scale. Swi2/Snf2 is the founding member of the Snf2 family of remodelers. (Adapted from Flaus et al. 2006, by permission of Oxford University Press.)

2 THE NUTS AND BOLTS OF NUCLEOSOME REMODELING

The ATPase domains of all known nucleosome remodelers are characterized by a number of short sequence motifs (I–VI) that reveal their relatedness to the much larger superfamily of nucleic acid helicases and has led to six groupings of Snf2 subfamily proteins (Fig. 4) (Flaus et al. 2006; Flaus and Owen-Hughes 2011). Snf2 family helicases commonly bind a double-stranded nucleic acid and move along one strand in a defined direction, thereby separating the two strands. Detailed mechanistic studies on a small number of selected enzymes revealed that nucleosome remodelers are also DNA translocases—that is, they move along one strand of nucleosomal DNA, yet without separating the two strands (Saha et al. 2006; Gangaraju and Bartholomew 2007). Remodelers engage in other defined contacts with histones and linker DNA, at the same time, which position the ATPase domain at a strategic site within the nucleosomal DNA, ~2 helical turns off the dyad axis. According to the prevailing model, this anchoring combined with the translocation of the ATPase domain on nucleosomal DNA, leads to the detachment of DNA segments from the histone octamer surface (Saha et al. 2006; Gangaraju and Bartholomew 2007; Racki and Narlikar 2008; Flaus and Owen-Hughes 2011). Cycles of ATP binding, hydrolysis, and product release define a succession of conformation changes of the enzyme, which propel the movement of the enzyme on DNA. This automatically changes the position of DNA relative to the histone surface (see Fig. 5). Once a segment of DNA has been detached, the dislocation of the resulting DNA "bulge" or "loop" on the nucleosomal surface may not require much additional energy input. Another way of illustrating the basic remodeling reaction is to imagine that a remodeling enzyme "pushes" or "pulls" a segment of linker DNA into the nucleosomal domain, which automatically leads to the formation of some kind of a DNA bulge, as the histone octamer surface can only accommodate 147 bp of DNA. This bulge would then be propagated by the translocase activity over the histone octamer surface. This DNA translocation of the ATPase domain on the nucleosome causes a twisting of the DNA, which induces a local superhelical strain into the DNA (Lia et al. 2006; Cairns 2007). Nucleosome remodeling thus presumably involves a combination of translational and rotational displacement of DNA. A much more detailed mechanical understanding of the remodeling action is currently lacking, but it is likely that the precise mechanism will differ for each of the individual remodeling ATPases, depending on enzyme structure and geometry, the arrangement, affinities, and selectivity of histone and DNA interaction domains, including the translocase domain itself. Unfortunately, it has so far been very difficult to determine the structures of ATPase/translocase domains of remodeling enzymes. The very few notable structures that have been obtained provide an important framework for further mechanistic studies (reviewed in Hauk and Bowman 2011). Electron microscopy studies have generated structures of the Swi/Snf complex and the related RSC complex. The yeast RSC resembles the Swi/Snf complex in subunit composition and overall architecture. Instead of the ATPase Swi2/Snf2 it contains the Sth1 ATPase, which belongs to the same subfamily (Clapier and Cairns 2009). The structure of the RSC complex bound to a nucleosome is particularly insightful (Fig. 5C) (Chaban et al. 2008). This structure reveals that RSC engulfs the nucleosome within a central cavity. Importantly, ATP-independent binding of RSC to the nucleosome appears to alter histone–DNA interactions perhaps to facilitate ATP-dependent remodeling. Nucleosomal DNA within the structure appears relatively unconstrained by the complex and may be able to undergo movement of the kinds described above (Chaban et al. 2008). RSC-nucleosome complexes have been found in vivo: at the yeast UASg locus, RSC positions an apparently partially unwound nucleosome to facilitate transcription factor binding to nearby sites (Floer et al. 2010).

The current classification of the many SNF2-type ATPases into subfamilies rests on sequence features within the ATPase domain (Flaus and Owen-Hughes 2011), but the more closely related ATPases also share particular domains and sequence signatures outside of the ATPase domains. A notable feature of the members of the Ino80 or SWR1 subfamilies is a major insertion of some 300 amino acids in front of helicase motif IV (Fig. 6). The best-studied remodeling ATPases—Swi2/Snf2, ISWI, Chd1, Mi-2, Ino80, and Swr1—are prototypic enzymes that represent subfamilies of enzymes (Fig. 4) (Bao and Shen 2011; Flaus and Owen-Hughes 2011; Kasten et al. 2011; Sims and Wade 2011; Yadon and Tsukiyama 2011). The hallmark of SWI2-type ATPases (synonymous with SWI2/SNF2) is, among other features, a bromodomain. This domain "reads" acetylated lysine marks of histone and nonhistone proteins (for more structural information, see Ch. 6 [Cheng 2014]; Ch. 7 [Patel 2014]). ISWI-type enzymes feature carboxy-terminal SANT-SLIDE domains and CHD-type enzymes bear amino-terminal chromodomains, which recognize and bind methylated lysine residues of histone H3 (Fig. 6). It is likely that this classification lessens our appreciation of more fundamental similarities between remodelers. For example, the resolution of the structure of the carboxyl terminus of CHD1 revealed that it contains a domain that resembles the "SLIDE" domain, a DNA binding domain that so far was thought to be characteristic of enzymes belonging to the ISWI subfamily. This common structural motif had not been found by simple sequence comparison

 Cite as *Cold Spring Harb Perspect Biol* doi: 10.1101/cshperspect.a017905

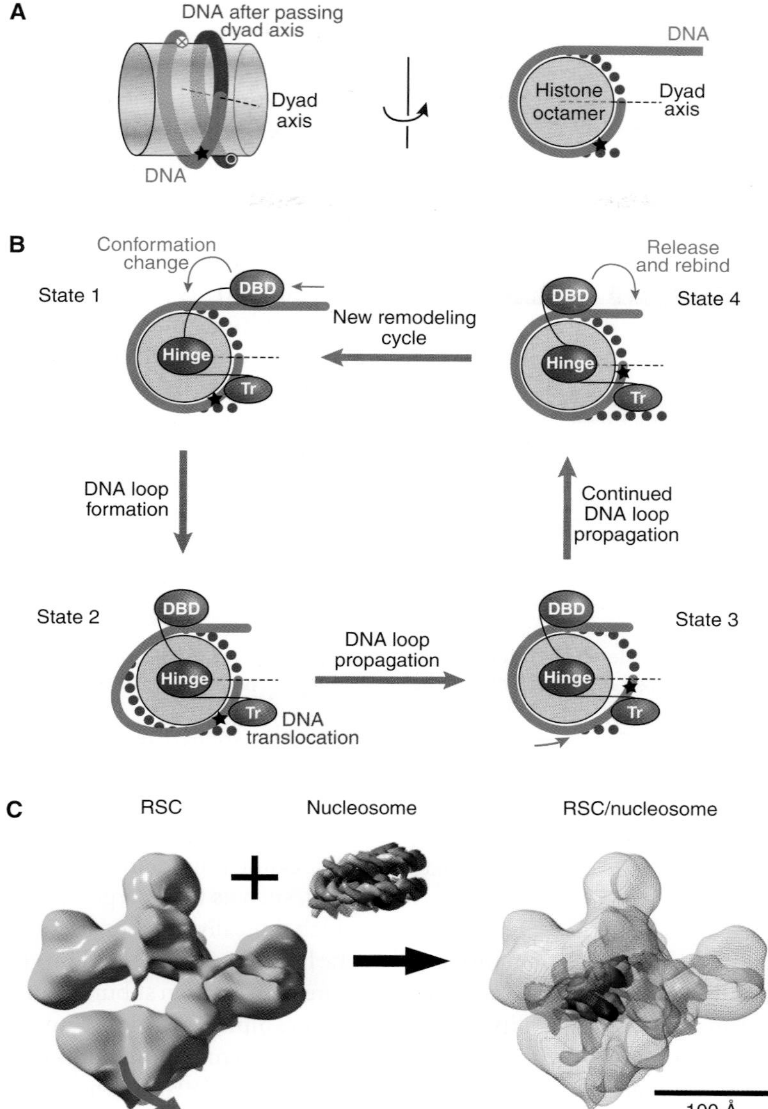

Figure 5. Mechanism of nucleosome remodeling. (*A*) Nucleosome view emphasizing the left hand wrapping of DNA. (*Left*) Side view of the nucleosome core with the histone octamer represented as a gray transparent cylinder and the DNA in orange (before the dyad axis) and red (after the dyad axis). (*Right*) Top view of the nucleosome (rotated 90°) in which the DNA after the dyad axis is represented by red dots. The star represents a reference point on the DNA sequence. (*B*) Model for DNA movement across the histone octamer during a remodeling event by the ISWI-type enzymes. Successive steps in a remodeling event are represented by States I–IV. In State I the DNA binding domain (DBD) is bound to the linker DNA and the translocase (Tr) domain is bound near the nucleosome dyad. A hypothetical "hinge" mediates the changes in conformation. In State II a conformational change between the DBD and the Tr "pulls in" DNA, which becomes visible as a bulge on the histone octamer surface. The Tr activity propagates this bulge across the surface of the histone octamer beyond the dyad axis (State III). The DNA loop continues to diffuse across the octamer surface and is released into the distal linker DNA (State IV). Loop diffusion thus effectively repositions the histone octamer relative to the DNA sequence (i.e., the star has moved closer to the dyad axis). A further conformational change triggered by aspects of the ATPase cycle lead to a resetting of the remodeler relative to the histone octamer (compare in States IV and I). The remodeler now engages with a different segment of linker DNA to start another cycle of remodeling. (*A*,*B*, Adapted, with permission, from Clapier and Cairns 2009, © Annual Reviews. *C*) Cryo-electron microscopy (EM) analysis of the RSC structure and nucleosome interaction. The yeast RSC resembles the Swi/Snf complex in subunit composition and overall architecture. Instead of the ATPase Swi2/Snf2, it contains the Sth1 ATPase, which belongs to the same subfamily (Clapier and Cairns 2009). A 25-Å cryo-EM map of RSC (*left*) shows a central cavity that closely matches the shape and dimensions of a nucleosome core particle. Movement (indicated by the red arrow) of the bottom RSC domain appears to control access to the central cavity. Incubation of RSC with nucleosome core particles (NCPs) results in formation of a RSC-NCP complex (*right* panel) in which NCP density is apparent in the central RSC cavity. Interestingly, interaction with RSC in the absence of any ATP hydrolysis appears to result in extensive changes in NCP organization. Histone density can be identified, but nucleosomal DNA appears disordered (semitransparent blue density). This loosening of DNA may facilitate DNA translocation during remodeling. (Image and interpretation provided by Francisco Asturias, Scripps Research Institute, La Jolla.)

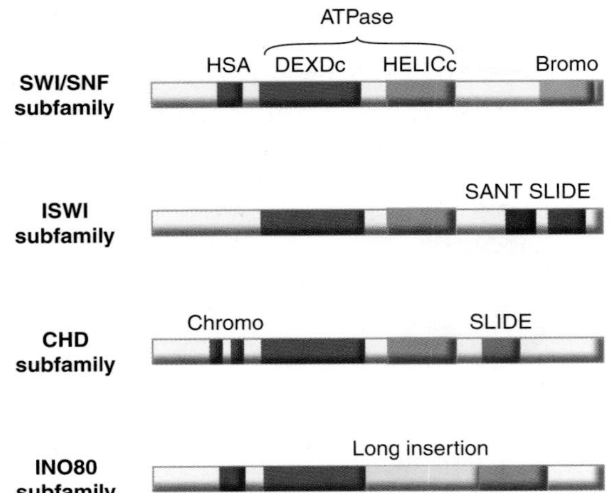

Figure 6. Four ATPase subfamilies: Signatures of SNF2, ISWI, CHD, and Ino80. The grouping of the remodeling ATPase of the Snf2 family is defined by signature motifs within the ATPase domain, whereas additional domains define the subfamilies. Members of the INO80 (and Swr1) subfamily of ATPases have a longer insertion between the two ATPase subdomains than other remodelers (Fig. 5). These subfamilies also contain a HSA (helicase-SANT) domain. The SWI/SNF family of ATPases contains a HSA domain, but is further defined by a carboxy-terminal bromodomain (capable of binding acetylated lysine residues). The ISWI and CHD family of ATPases each have SANT-SLIDE modules (blue) whereas only the CHD family has tandem chromodomains. (Adapted, with permission, from Clapier and Cairns 2009, © Annual Reviews.)

(Ryan et al. 2011). Also, some of the associated domains may be involved in auto-regulation of the ATPase, and may thus render the ATPase conditional on substrate availability. Mechanistically this may occur through allosteric conformational changes on substrate binding, or through direct competition between a regulatory domain and the substrate for access to the active site (Hauk et al. 2010; Flaus and Owen-Hughes 2011; Hauk and Bowman 2011).

In addition to the signature domains mentioned above, illustrated in Figure 6, other structures can be described that are found in ATPases of different subfamilies. For example, the HSA (helicase-SANT-associated) domain, which binds actin-related proteins (ARPs) is present in the ATPases of the SWI/SNF complex, the related RSC complex, the SWR1 complex, and the Ino80 complex (Fig. 6) (Dion et al. 2010).

Conceivably, the specific outcome of a basic remodeling reaction depends not only on the geometry and enzymatic parameters of the ATPases themselves, but also on associated proteins. The unpeeling of DNA segments from the histone surface may lead to a delocalization of nucleosomes (their "sliding") if the detached DNA segment is propagated around the octamer and expelled on the other side. However, it is conceivable that a very similar remodeling reaction may lead to the removal or replacement of histones, or to the complete eviction of a nucleosome (Fig. 2). An interesting variation of the theme has been suggested, whereby a nucleosome is used as a "wedge" to destabilize an adjacent particle. If nucleosomes are moved to collide with one another, DNA may be peeled off one or both of them, yielding unconventional remodeling intermediates (Chaban et al. 2008; Engeholm et al. 2009; Dechassa et al. 2010). This possibility is consistent with the fact that RSC engulfs the nucleosome to which it is bound (Fig. 5C), as it may be difficult for a remodeler to displace the histone octamer under those circumstances (Chaban et al. 2008). Whether histone octamers are moved intact or whether partial disassembly occurs may be largely determined by the cooperation of remodeling ATPases with associated histone chaperones that may either scavenge the histones as they are liberated from the toroidal DNA supercoil or perhaps deliver variant histones for exchange.

3 DIVERSITY OF NUCLEOSOME REMODELING COMPLEXES

Although isolated remodeling ATPases can alter histone–DNA interactions in vitro, they usually associate with other proteins to form defined multisubunit complexes. These complexes are commonly referred to as "remodeling factors." Some remodeling factors are made of only a few subunits, such as those organized around ISWI-type ATPases, which typically only consist of two to four subunits (Fig. 3). The large Ino80- and SWI2- type complexes, which contain more than a dozen subunits, mark the other end of the scale (for examples, see Figs. 7 and 8). The associated subunits frequently contribute additional domains (e.g., plant homeodomain [PHD] fingers, bromodomains) that bind modified epitopes on histones and nonhistone proteins. In most cases, however, the precise role of the ATPase-associated subunits is unknown, but is likely to target the remodeling ATPase to specific sites of action, to regulate their activity, and the precise outcome of the remodeling reaction as well as to integrate the remodeling reaction into the physiological context of a nuclear process, such as transcription, replication, or DNA repair.

Some ATPase-associated subunits appear as evolutionary conserved "recurring themes," as they are found in several remodeling factors. For example, the large SWI/SNF, RSC, Ino80, and SWR1 complexes contain actin-related proteins (ARPs) and even nuclear actin itself (Figs. 7 and 8). Different ARPs contribute to complex assembly and remodeling activity in various ways. In some cases, ARPs and nuclear actin may bind particularly modified histones or variants and thus mediate association of remodeling complexes to chromatin. In other cases, they may provide interfaces to the transcription machinery or to nuclear pores (Dion et al. 2010). Another example of a recurring theme is the hexameric rings of

 Cite as *Cold Spring Harb Perspect Biol* doi: 10.1101/cshperspect.a017905

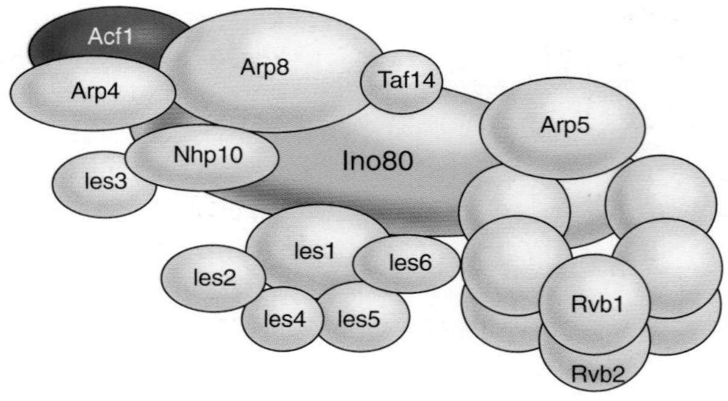

Figure 7. Example of a complex remodeling machine: INO80. The *Saccharomyces cerevisiae* INO80 complex provides an example of the subunit composition of a complex nucleosome-remodeling machine. INO80 subunits include the core ATPase, Ino80 (INOsitol requiring), Rvb1 (RuVB-like), Rvb2, Act1 (actin), Arp4 (actin-related protein), Arp5, Arp8, Nhp10 (nonhistone protein), Taf14 (TATA-binding protein-associated factor), Ies1 (Ino eighty subunit), Ies2, Ies3, Ies4, Ies5, and Ies6. (Adapted from Bao and Shen 2011, with permission from Elsevier.)

the RVB1 and RVB2 ATPases from the AAA family, which are important functional components of SWR1 and Ino80 subfamily of complexes (Fig. 7). These enzymes resemble the bacterial RuvB helicase that is involved in the resolution of holiday junctions during homologous recombination (Jha and Dutta 2009). The assembly of SWR1 and Ino80 complexes is compromised in the absence of the RVB hexamers, which suggest a scaffolding function. Nonetheless, the fact that these rings themselves have ATPase activity suggests more interesting, dynamic activities.

An important concept in the study of chromatin remodelers is that one particular ATPase may associate with alternative sets of proteins to form distinct types of remodeling complexes. For example, the ATPase ISWI in *Drosophila* can serve as the "engine" for at least five different remodeling factors with nuclear functions ranging from transcription control to DNA repair and chromosome organization (Fig. 3) (Yadon and Tsukiyama 2011).

The complexity of remodeling factors has increased dramatically during metazoan evolution and multiple "variations on a theme" have been described for the mammalian complexes, including subunit isoforms, variants, and post-translational modifications. This is best illustrated for the 11-subunit yeast SWI/SNF complex. SWI/SNF is mainly involved in regulating transcription initiation and it is targeted to promoters through direct interactions with sequence-specific DNA binding factors (see Section 4). The *Drosophila* homolog of the SWI2/SNF2 ATPase is called Brm (Brahma) for historical reasons (see Ch. 18 [Kingston and Tamkun 2014]). Brm associates with seven to eight brahma-associated proteins (BAPs) to form complexes that loosely resemble the yeast SWI/SNF complex. Interestingly, two related complexes, BAP and PBAP, share six subunits but differ in the signature subunits: Osa, dD4, and TTH for BAP; PB, BAP170, SAYP, and dBRD7 for PBAP (Fig. 8A) (Moshkin et al. 2007; Moshkin et al. 2012). The 'core complex" in the absence of these signature subunits is not functional in vivo. At the same time, they specify the preferential recruitment to distinct sets of target genes. BRM is a global regulator of transcription in *Drosophila* as the majority of transcriptionally active genes was bound by BRM in polytene chromosomes and required the ATPase for activity (Hargreaves and Crabtree 2011). BAP subunits function as *trithorax* gene products and oppose the function of polycomb proteins in many developmental pathways (see Ch. 18 [Kingston and Tamkun 2014]).

The mammalian orthologs of Brahma are the highly related ATPases BRM and BRG1. The proteins that associate with them, termed "BRG/BRM-associated factors" (BAFs), form a family of related remodeling machineries (Hargreaves and Crabtree 2011). The mammalian BAF complexes resemble the yeast SWI/SNF complexes vaguely: Eight subunits of both complexes are clearly evolutionary related, but yeast and mammalian complexes also contain several species-specific subunits (Ho and Crabtree 2010). A hallmark of these complexes is their diversity, as several subunits have cell-type-specific isoforms, which associate with BAF complexes in mutually exclusive and combinatorial ways (Fig. 8B). This explains why BAF complexes are involved in regulating very different gene expression programs in different tissues, from the maintenance of the pluripotent state of embryonic stem (ES) cells to the highly differentiated profile of postmitotic neurons (see Sec. 10).

4 NUCLEOSOME REMODELING FACTORS AS REGULATORS OF TRANSCRIPTION

Promoters can be tentatively sorted into different classes according to their chromatin architecture and arrangement of binding sites. The transcription start sites of many constitutively expressed "housekeeping" promoters tend to be depleted of nucleosomes and so depend less on nucleosome remodeling. Tightly regulated genes, in contrast, rely on remodeling factors to clear their promoters of repressive nucleosomes for expression (Cairns 2009; Rach et al. 2011). Promoter architecture is nonetheless more complex and diverse than this, and nucleosome remodeling factors are equally involved in setting up promoter organization as well as in its subsequent remodeling.

Nucleosome remodelers play important roles in regulating the initiation and elongation of transcription. They are often recruited to target genes through interactions with sequence-specific transcription factors to serve as

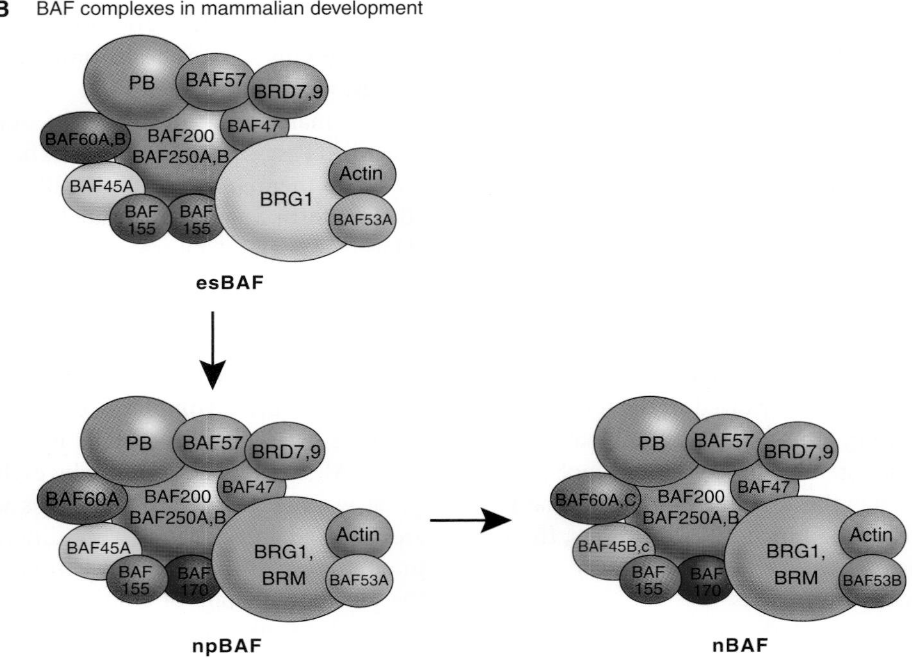

Figure 8. Diversity of Swi/Snf complexes in metazoa. (*A*) BAP and PBAP complexes in *Drosophila*. There are two distinct Swi/Snf-type complexes in *Drosophila*: the BRM-associated proteins complex (BAP) and the polybromo-containing BAP complex (PBAP). Although these complexes share multiple subunits including the ATPase BRM, they each have distinct subunits. The OSA, dD4, and TTH subunits are only found in the BAP complex and not in the PBAP complex. In contrast, Polybromo, BAF170, dBRD7, and SAYP are subunits of the PBAP complex but not the BAP complex. (Adapted from Ho and Crabtree 2010, with updates from Moshkin et al. 2012.) (*B*) Cell- and tissue-specific versions of BAF complexes in mammals. These complexes contain either the BRG1 or BRM ATPases. They may also contain polybromo (PB) and BAF200 (PBAF complexes) or BAF250A/B (BAF complexes). Shown here are composite representations of those variants. This figure serves to illustrate tissue-specific assemblies of BAF complexes, which have distinct functions in specific cell types. The subunits present in BAF complexes for each tissue are indicated (e.g., BAF60A or C). PB is colored the same as in *A*. The other subunits shaded with color are those whose presence varies in the different tissue, and define the tissue-specific complexes. (Adapted from Ho and Crabtree 2010.)

coactivators or corepressors (Clapier and Cairns 2009). In yeast, for example, many transcription factors are able to recruit the Swi/Snf complex to target genes by direct interaction (Fig. 9) (Vignali et al. 2000; Fry and Peterson 2001). The yISW2 complex is also targeted to meiotic genes by Ume6, to participate in repression (Clapier and Cairns 2009). In an interesting variation of the theme, the Hir1 and Hir2 corepressors recruit Swi/Snf to histone gene loci, but it is required there for subsequent transcription activation (Dimova et al. 1999).

In mammals, Swi/Snf interacts with a number of transcription factors, including steroid receptors, tumor suppressors, and oncogenes like RB, BRCA-1, c-Myc, and MLL (Hargreaves and Crabtree 2011). In *Drosophila*, the NURF

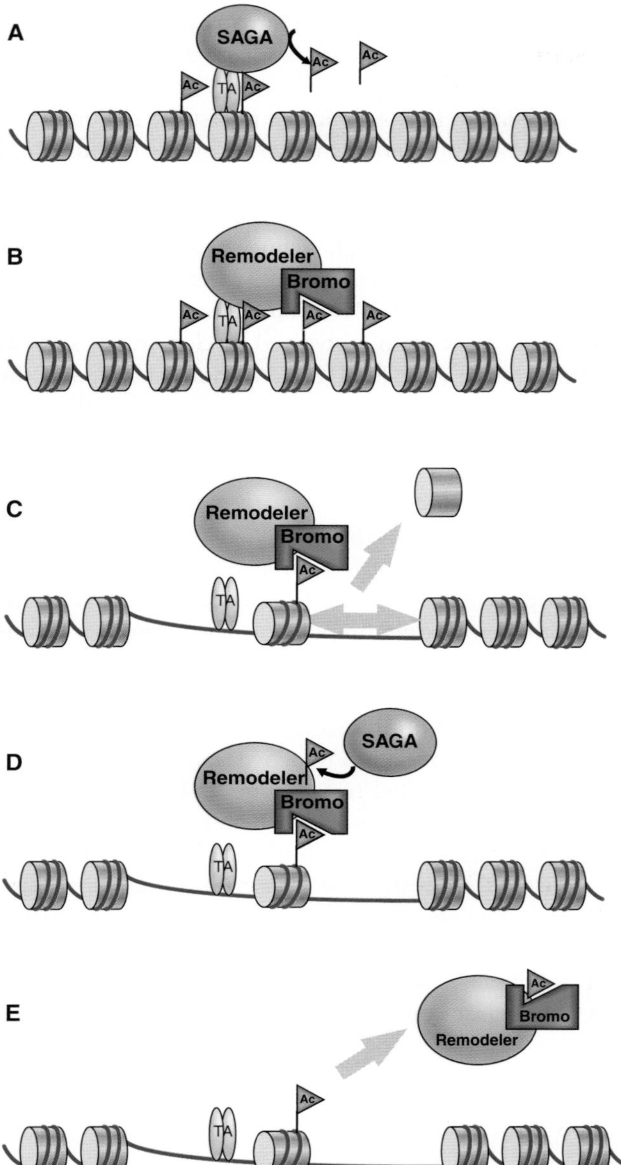

Figure 9. Model for the action of Swi/Snf-type nucleosome remodelers at promoters. and their regulation by acetylation. (*A*) SAGA or other histone acetyltransferase (HAT) complexes can be recruited to gene promoters by interacting with sequence-specific DNA-binding transcription activators (TA). Once recruited these HATs acetylate (blue Ac flag) nucleosomes in proximity of the activator recognition site. (*B*) The Swi/Snf or RSC nucleosome remodeling complexes (remodeler) can be recruited to promoters by interactions with transcription activators. Bromodomains (bromo) within subunits of these complexes then interact with the acetylated (Ac) nucleosomes at the promoter. (*C*) ATP-dependent remodeling and/or displacement (gray arrows) are preferentially directed at the acetylated nucleosomes bound by the bromodomains in the remodeling complex. (*D*) SAGA or other Gcn5-containing complexes acetylate specific lysines within subunits of the remodeling complex. These acetylated lysines compete for interaction with the bromodomains, which (*E*) leads to dissociation of the remodeler from the acetylated nucleosomes (gray arrow). (Adapted from Suganuma and Workman 2011.)

complex interacts with several transcription factors including GAGA factor, heat shock factor, ecdysone receptor, and the dKen repressor through the NURF301 subunit (Alkhatib and Landry 2011).

Once recruited to target gene promoters, chromatin remodelers alter the local chromatin organization by nucleosome movement or displacement, which can facilitate either gene activation (Li et al. 2007) or repression. In combination with histone chaperones, complexes like Swi/Snf and RSC are able to displace nucleosomes in *trans* either by forcing the histone octamer onto another piece of DNA or by moving the histones onto chaperones (Workman 2006). Nucleosome displacement by Swi/Snf is facilitated by histone acetylation (see Fig. 9, and Sec. 7) and by interactions of the complexes with transcription factors (Gutierrez et al. 2007).

Nucleosome remodeling complexes are involved in both the activation and repression of gene expression. Mutation of their subunits in yeast often leads to as many genes showing an increase in expression as those whose expression decreases (Hargreaves and Crabtree 2011). Although some of these effects will be indirect, it is not difficult to imagine a chromatin remodeler activating or repressing genes through the same mechanism (for example, moving nucleosomes on or off a promoter). Nonetheless, most chromatin remodelers have been characterized based on their activity in either activating or repressing gene expression.

In yeast, both Swi/Snf and the related RSC complex are required for expression of nonoverlapping subsets of genes (Hargreaves and Crabtree 2011). Whereas Swi/Snf regulates many inducible genes, RSC is more involved in controlling essential, constitutively expressed genes, such as genes encoding ribosomal protein subunits (Hargreaves and Crabtree 2011). Importantly, only a small minority of yeast genes absolutely requires Swi/Snf or RSC for expression. This does not negate that Swi/Snf or RSC may still function at many other genes. RSC was, in fact, found to bind to $>$700 target genes in the yeast genome, including histone genes, genes regulated by stress, and several genes transcribed by RNA polymerase III (Hargreaves and Crabtree 2011). Moreover, 20% of genes in yeast grown in glucose require Snf2 to keep their promoters nucleosome-depleted (Tolkunov et al. 2011). These data suggest that Swi/Snf and RSC function more broadly than initially indicated by gene expression analysis alone.

Other nucleosome remodeling factors also function in gene activation (Clapier and Cairns 2009). The yeast INO80 complex activates genes regulated by the inositol/choline response element and is required for activation of the PHO84 gene. In the absence of INO80, the expression levels of 150 genes changed (approximately half up and half down; Morrison and Shen 2009). The mammalian INO80 complex is recruited to target genes by the YY1 transcription factor (Cai et al. 2007).

Several nucleosome remodeling complexes appear to function primarily in repression of transcription, as noted above, be it directly or indirectly. Chromatin immunoprecipitation studies showed that Swi/Snf physically occupies some genes where it functions as a repressor (Martens and Winston 2003; Hargreaves and Crabtree 2011). Similarly, most Brg1 binding in ES cells occurs at enhancers and intragenic regulatory elements where it functions as a repressor (Hargreaves and Crabtree 2011). An instructive example of an indirect repressive effect is provided by action of the Swi/Snf on the SER3 gene in yeast. Swi/Snf activates the upstream SRG1 promoter, which leads to transcription of a noncoding RNA. The transcription extends through the downstream SER3 gene promoter and interferes with its transcription initiation (Martens et al. 2005).

The yeast Ume6 repressor recruits the ISW2 complex to meiotic promoters in which it organizes repressive chromatin. Isw2 collaborates with the Rpd3 deacetylase complexes to repress numerous Ume6-dependent and -independent genes (Fazzio et al. 2001). This was corroborated by a widespread analysis in which loss of ISWI function in *Drosophila* 3rd instar larvae indicated that 75% of the 500 genes with altered expression profiles showed increased expression (out of 15,000 genes tested; Hargreaves and Crabtree 2011). The Mi2 ATPase of higher eukaryotes functions as part of the repressive nucleosome remodeler/deacetylase complexes (NuRD). NuRD represents a heterogeneous group of factors of variable, cell-specific subunit composition (Bowen et al. 2004), and is part of the MeCP1 complex that connects chromatin remodeling, histone deacetylation, and DNA methylation. More recently, Mi2 was also found as part of a distinct complex, dMec, which functions as a SUMO-dependent corepressor (Kunert and Brehm 2009).

In addition to regulating transcription initiation by activation or repression, chromatin remodelers also affect the process of transcription elongation. Elongating RNA polymerase II has to confront nucleosomes along the entire gene body and a number of elongation factors, histone chaperones, histone modifications, and chromatin remodelers have been implicated in facilitating its progress (Li et al. 2007; Selth et al. 2010). In general, histones are acetylated and methylated during transcription elongation and all or part of the histones in the octamer are chaperoned around the elongating polymerase to be reassembled and deacetylated behind the polymerase (Li et al. 2007; Clapier and Cairns 2009). Remodelers then might play a role in disrupting nucleosomes in front of the polymerase and in reassembling and spacing the nucleosomes in its wake. Examples for such action come from all subfamilies of remodelers and several species. Human Swi/Snf was shown early on to help RNA polymerase pass a nucleosome-induced transcriptional pause in vitro (Vignali et al. 2000). Yeast Swi/Snf has been implicated in histone eviction during polymerase elongation in vivo (Schwabish and Struhl 2007). The yeast RSC complex can associate with RNA polymerase II (Soutourina et al. 2006) in vitro and stimulate transcription through a nucleosome in a manner enhanced by histone acetylation (Li et al. 2007). Yeast Chd1 colocalizes with RNA polymerase II and interacts with a number of factors involved in elongation, such as the PAF complex, the Spt4/5 elongation factors, and the FACT histone chaperone complex (Simic et al. 2003). The *Drosophila* CHD family remodeler Kismet associates with actively transcribed regions on polytene chromosomes and apparently functions in an early step in transcription initiation or elongation (Srinivasan et al. 2005; Murawska et al. 2008). Mutants in Kismet lead to reduced levels of elongating RNA polymerase II and reduced levels of the histone chaperone Spt6 on polytene chromosomes (Srinivasan et al. 2005).

To summarize, chromatin remodelers are involved in the entire transcription process from promoting or blocking transcription initiation to active transcription elongation. They are recruited to target genes by DNA-binding transcription factors, RNA polymerases, and elongation factors. They serve in disrupting nucleosomes to facilitate initiation and elongation of transcription, and reassembling and spacing nucleosomes in the wake of passing RNA polymerase. They are the engines that both deny RNA polymerase access to a gene (repression) and help RNA polymerase contend with nucleosomes (activation).

5 NUCLEOSOME REMODELING IN CHROMATIN ASSEMBLY AND ORGANIZATION

The previous paragraphs emphasized the roles of nucleosome remodeling factors in nucleosome disassembly as a means to access the genetic information. Less intuitive, but equally important is the involvement of nucleosome remodeling in the assembly and organization of chromatin with repressive properties as well as for the diversification of structures that are collectively called "the epigenome." Nucleosome remodeling is effective at all levels of chromatin organization: the de novo assembly of nucleosomes (Fig. 2B), their positioning relative to the underlying DNA, and the generation of fibers with defined distances between the nucleosomes ("nucleosome spacing"; Fig. 2C), which profoundly affects the folding of chromatin. Finally, nucleosome remodeling also affects the association of linker histones and perhaps even the interaction of nonhistone proteins with the nucleosome fiber.

 Cite as *Cold Spring Harb Perspect Biol* doi: 10.1101/cshperspect.a017905

The dual capacity of remodeling factors to detach segments of DNA from the histone surface and, conversely, to wrap DNA around histones during nucleosome assembly is best illustrated by the process of histone variant exchange. For example, nucleosome remodeling factors of the SWR1 family remove a "canonical" H2A/H2B dimer from a nucleosome and then replace it by a dimer containing the H2A.Z variant (for more detail, see Sec. 6 and Ch. 20 [Henikoff and Smith 2014]).

Although nucleosomes form spontaneously by mixing DNA and histone in high salt solutions and gradual removal of the salt, nucleosome assembly under physiological conditions requires the cooperation of histone chaperones and remodeling activities (Fig. 2B). Although histone chaperones keep the highly charged histones soluble and assure their ordered deposition onto DNA (see also Ch. 22 [Almouzni and Cedar 2014]), the example of ACF reveals that remodelers can catalyze the supercoiling of DNA around histone octamers (Torigoe et al. 2011). Indeed, nucleosome remodeling factors of the ISWI family appear to be particularly involved in de novo nucleosome assembly. In vitro, ACF and RSF are able to promote nucleosome formation, but whereas ACF requires the assistance of histone chaperones, RSF profits from "built-in" chaperone activity (Loyola et al. 2003; Lusser et al. 2005). In *Drosophila*, genetic analysis of the respective signature subunits of these remodeling factors (i.e., ACF1 and RSF1 [Fig. 3]), confirms their roles in chromatin assembly in vivo. Those animals that survive an ACF1 deficiency are characterized by "sloppy," irregular chromatin, and defects in heterochromatin specification as well as in chromatin-mediated gene silencing (Chioda and Becker 2010). In contrast, an RSF1 deficiency does not affect the basal chromatin organization, but leads to a more specific defect in the incorporation of the histone variant H2A.V and selective impairment of chromatin structures that depend on this variant, like heterochromatin (Hanai et al. 2008).

The bulk of nucleosome assembly occurs during S phase and accordingly a number of remodeling factors are targeted to sites of active DNA replication (Morettini et al. 2008; Rowbotham et al. 2011). Their physiological roles are difficult to assess because of their dual potential in opening chromatin to facilitate DNA synthesis, and in subsequent nucleosome assembly to reconstitute the integrity of the nucleosome fiber. Nucleosome assembly also occurs in interphase during DNA damage repair, and many different remodelers are recruited to sites of DNA damage repair (Altaf et al. 2007; Bao and Shen 2007; Downs et al. 2007; Lan et al. 2010).

The generation of repressive chromatin may, in the simplest case, involve the positioning of nucleosomes onto regulatory DNA sequences. The energy input of ATP hydrolysis allows the remodelers to move nucleosomes onto DNA sequences that are intrinsically unfavorable. For example, the yeast ISWI-type remodeler isw2 can slide nucleosomes onto a promoter to repress its activity (Whitehouse and Tsukiyama 2006). Genome-wide chromatin remodeling by ISW2 near promoters and in coding regions increases nucleosome density and enforces the accuracy of transcription (Yadon et al. 2010). More specifically, the isw2 and isw1 complexes contribute to the positioning of nucleosomes in intergenic regions or in the middle of genes, respectively, which suppresses cryptic, antisense transcription that would otherwise pose a risk of interfering with sense transcription (Whitehouse et al. 2007; Tirosh et al. 2010). The positioning and phasing of nucleosomes around transcription start sites in yeast, which frequently carry information in the form of histone variants and specific modifications, are brought about by the RSC complex in *S. cerevisiae* (Wippo et al. 2011; Zhang et al. 2011). In fission yeast the CHD-type remodeler, Mit1, has been shown to be involved in aligning nucleosomal arrays with respect to promoter boundaries, a process termed nucleosome "phasing" (Lantermann et al. 2010).

The nucleosome assembly activity of remodeling factors improves the integrity of the chromatin fiber by filling gaps in the succession of nucleosomes. The sliding of nucleosomes enables remodelers to improve the regularity of nucleosomal fibers by equalizing the inter-nucleosome distances (i.e., their "spacing") (Fig. 2C) (Becker and Horz 2002). The more regular a nucleosome array, the better it will fold into a variety of "next level" chromatin fibers characterized by diameters of ~30 nm. The folding of nucleosomal arrays into fibers is further promoted by the association of linker histones. The presence of linker histones affects the activity of remodeling factors differentially; some enzymes are unable to remodel nucleosomes that are "locked in" by H1 association, but others, notably the ISWI-containing ACF, are able to move nucleosomes even with H1 bound. ACF can also facilitate the incorporation of H1 into chromatin in vitro (Lusser et al. 2005; Torigoe et al. 2011). These biochemical observations may be physiologically significant as ISWI-containing remodelers are suggested to modulate the steady-state association of H1 with chromosomes (Chioda and Becker 2010). In summary, nucleosome remodeling factors are likely to regulate the folding of chromatin by modulating the integrity, regularity, and spacing parameters of nucleosomal arrays, thereby laying the foundations for higher-order chromatin structures (Korber and Becker 2010). This hypothesis is supported by observations in yeast and flies. In *Schizosaccharomyces pombe* the formation of silenced heterochromatin correlates with the regularity of nucleosome positioning, which is promoted by ATPase Mit1 in the context of the SHREC complex

(Sugiyama et al. 2007). *Drosophila* lacking the nucleosome assembly and spacing factor, ACF, show defects in heterochromatin formation and polycomb-dependent silencing (Chioda and Becker 2010). These examples show that nucleosome remodeling factors not only serve to catalyze fast, local, and reversible changes of nucleosome organization, but that their action may also promote the assembly of stable and epigenetically silenced chromatin domains.

6 RECOGNITION OF HISTONE MODIFICATIONS BY CHROMATIN REMODELERS

Most remodeling complexes contain domains that mediate their specific association with modified histones, in addition to interaction with sequence-specific DNA-bound transcription factors. For example, CHD family members often contain two tandem chromodomains (Fig. 6) (Brehm et al. 2004). The tandem chromodomains of human CHD1 allows it to specifically recognize a methylated lysine 4 residue (H3K4me) of a histone H3 tail (Hargreaves and Crabtree 2011). Because of their inherent weakness, interactions with modified histones are not likely to be primary targeting determinants, but may modulate the activity of the enzyme once it has been recruited by other means. As H3K4 methylation is an active mark at the 5′ end of genes, chromodomain-histone interactions are believed to modulate the activity of CHD1 at promoters. In agreement with this idea, the chromodomains of yeast and *Drosophila* CHD1 appear to be important for their ATPase and nucleosome remodeling activities, but did not significantly affect their chromosomal localization (Hauk et al. 2010; Morettini et al. 2011). Thus, the CHD1 chromodomain interaction with methylated H3K4 may play a regulatory role of the remodeling activity beyond its recruitment to active genes.

The plant homeodomain (PHD domain) is a versatile zinc finger domain that reads histone modifications along the amino-terminal tail of histone H3 (Sanchez and Zhou 2011). A PHD domain in the BPTF subunit of the NURF complex recognizes H3K4me3 and couples this modification to nucleosome remodeling (Hargreaves and Crabtree 2011). The CHD4 subunit of the NuRD complex contains two PHD domains, the second of which recognizes H3K9me3, and the first one interacts preferentially with the amino terminus of H3 if unmodified at K4 (Mansfield et al. 2011). Remarkably, DPF3b, a factor associated with human BAF, recognizes H3K14 acetylation through tandem PHD fingers (Zeng et al. 2010). Previously, recognition of acetylated histones was primarily attributed to bromodomains.

Bromodomains have the potential to facilitate the recruitment, retention, and/or activity of chromatin remodelers on acetylated nucleosomes. The presence of acetyllysine-binding bromodomains in several remodeling ATPases and associated subunits suggests that some remodelers are sensitive to the acetylation state of target nucleosomes (Horn and Peterson 2001; Workman 2006). All Snf2-like ATPases contain at least one bromodomain (Fig. 6) and additional ones may be found on other subunits of the Swi/Snf type complexes (Clapier and Cairns 2009).

7 REGULATION OF REMODELERS BY POSTTRANSLATIONAL MODIFICATION

There is increasing evidence that chromatin remodelers themselves are modified and regulated by enzymes that are better known for modifying histones. Gcn5, the acetyltransferase subunit of the SAGA complex, has long been linked to the function of chromatin remodelers like Swi/Snf. Thus, SAGA affects remodeling activity indirectly through acetylating histones and directly through remodeler acetylation (Fig. 9) (Clapier and Cairns 2009). The Rsc4 subunit of the SWI/SNF-type remodeler RSC contains tandem bromodomains one of which binds H3K14ac. Remarkably, the adjacent bromodomain is able to bind acetylated lysine 25 of Rsc4 itself (VanDemark et al. 2007). Binding of this bromodomain to K25ac of Rsc4 inhibits interaction of the other bromodomain with H3K14ac, thereby hampering the interaction of RSC with acetylated nucleosomes. Gcn5 is responsible for acetylation of both H3K14 and Rsc4K25, and hence, has the ability to promote RSC association with nucleosomes through H3 acetylation or inhibit RSC association with nucleosomes through Rsc4 acetylation (VanDemark et al. 2007). These opposing reactions may serve to fine-tune the local activity of remodeling enzyme. An analogous, but somewhat different mechanism has been described for the Swi/Snf complex. The bromodomain-containing Snf2 ATPase is acetylated by Gcn5 (Kim et al. 2010). The acetylated sites on the ATPase itself are in competition with similar epitopes on the histone substrates, a fact that suggests complex regulatory interactions (Fig. 9) (Kim et al. 2010). The mammalian Swi/Snf ATPase subunit Brm is also regulated through acetylation by the acetyltransferase PCAF (the metazoan Gcn5) (Bourachot et al. 2003). Moreover, a proteomic screen identified three sites of acetylation on the other mammalian Snf2-like ATPase, Brg1 (SMARCA4) (Choudhary et al. 2009). Regulation of chromatin remodelers by acetylation is not limited to the Swi/Snf type complexes; the *Drosophila* ISWI ATPase can be acetylated by the Gcn5 and p300 acetyltransferases on lysine 753, a region of the protein that bears similarity to the H3 tail (Ferreira et al. 2007). These examples show that acetylation may serve to regulate the properties of both components of nucleosome remodeling reaction, the nucleosome substrate, and the remodeling enzyme.

 Cite as *Cold Spring Harb Perspect Biol* doi: 10.1101/cshperspect.a017905

PARP-1 (poly[adenosine diphosphate-ribose] polymerase-1) has emerged as a major regulator of genome function. PARP-1 also modifies histones and proteins involved in chromatin modification, transcription, and DNA repair (Krishnakumar and Kraus 2010), most notably the ISWI ATPase. PARylation of ISWI inhibits its ATPase activity and nucleosome binding. This is corroborated by PARP-1 counteracting the function of ISWI in flies (Krishnakumar and Kraus 2010).

Phosphorylation is a modification that is widely used to coordinate processes during cell cycle progression and during the DNA damage response. In this context it is not surprising that Swi/Snf-type remodeling complexes are also regulated by phosphorylation in diverse ways, which is illustrated by the following, selected examples (Vignali et al. 2000). The Sfh1 subunit of the yeast RSC complex is phosphorylated in G_1 phase of the cell cycle, and temperature-sensitive alleles of Sfh1 arrest at the G_2/M transition. In humans, the hbrm and Brg1 proteins (Snf2 homologs) are phosphorylated during mitosis, which leads to their dissociation from mitotic chromatin. The *Drosophila* Mi-2 subunit of the NuRD complex is constitutively phosphorylated by the CK2 kinase increasing its affinity for nucleosomes and its nucleosome remodeling activity (Bouazoune and Brehm 2005). The Ies4 subunit of the INO80 complex is phosphorylated in response to DNA damage by the Mec1/Tel1 kinases (ataxia telangiectasia mutated [ATM]/ATM and Rad3-related [ATR] homologs) in yeast. Phosphorylation of Ies4 does not appear to be required for DNA repair, but is required for the DNA damage checkpoint responses (Morrison and Shen 2009). As multiple chromatin remodelers have been implicated in double-strand break repair (Polo and Jackson 2011), it would not be surprising if additional complexes are found to be regulated by phosphorylation during DNA repair.

In summary, known modifications involved in the regulation of chromatin remodeling complexes include acetylation, PARylation, and phosphorylation. Modification of remodeler subunits can activate or inhibit the activity of the complexes and their interactions with nucleosomes. Such modifications appear to "fine tune" the regulation of remodeling complexes.

8 INTERACTION OF CHROMATIN REMODELERS WITH DNA METHYLATION

DNA methylation has long been implicated in gene silencing and as it has a clear potential to be inherited through cell divisions, it is a prime component of epigenetic regulation (Clouaire and Stancheva 2008; Ch. 15 [Li and Zhang 2014]). In eukaryotes, DNA methylation occurs exclusively at CG dinucleotides and its effects on gene silencing are manifested through meCG binding proteins (MBDs). MBD proteins associate with a number of transcriptional corepressor complexes including histone methylases, histone deacetylases, and chromatin remodelers (Clouaire and Stancheva 2008). The chromatin remodeling complex with the most studied connection to DNA methylation is the Mi-2/NuRD complex. NuRD contains MBD2 or MBD3 as core subunits and preferentially remodels and deacetylates nucleosomes containing methylated DNA. MBD2 and MBD3 are in distinct versions of the NuRD complex and as MBD3 does not actually bind methylated DNA presumably only the MBD2 form of NuRD contributes to DNA methylation-mediated repression (Clouaire and Stancheva 2008).

9 CHROMATIN REMODELERS AND HISTONE VARIANTS

Nucleosome remodelers are commonly involved in incorporating histone variants into chromatin. In turn, the nucleosome remodeling activity can be affected by the presence of histone variants in nucleosomal substrates. The most prominent example is the function of the SWR1 complex in the incorporation of variant histone Htz1 (yeast H2A.Z) into chromatin. H2A.Z is a highly conserved variant that is implicated in gene activation, protection from heterochromatin spreading and chromosome segregation (Altaf et al. 2009; Morrison and Shen 2009). Htz1 is inserted into chromatin by the SWR1 complex in yeast and by the SRCAP or p400/Tip60 complexes in humans (Billon and Cote 2011). Swr1, SRCAP, and p400 are the ATPase subunits of the respective complexes and catalyze the ATP-dependent replacement of H2A with H2A.Z. Htz1 is inserted at promoters of silent genes, but assists in their subsequent activation (Workman 2006). The SWR complex contains a bromodomain protein, Bdf1, which may bring SWR to acetylated nucleosomes at promoters for the insertion of Htz1. In mammals, incorporation of H2A.Z at some promoters requires SRCAP whereas others require p400. p400 is part of the TIP60 complex in flies that can directly interact with transcription factors at promoters.

The histone variant H2A.X has been suggested to poise chromatin for remodeling at sites of DNA breaks (Talbert and Henikoff 2010). H2A.X is phosphorylated on a specific serine in the carboxy-terminal tail of the protein by the ATM kinase in response to DNA breaks. Phosphorylated H2A.X (γ-H2AX) is important for recruiting or retaining a number of chromatin modifying and repair proteins to the site of the double-strand break (Talbert and Henikoff 2010). γ-H2AX is bound by Arp4, a subunit of the NuA4 HAT complex as well as the SWR1 and INO80 chromatin remodeling complexes, and is thought to play a role in the recruitment of all three of these complexes to DNA breaks, where they contribute to the repair process (Altaf et al. 2009).

Other H2A variants also affect the function of nucleosome remodeling complexes. MacroH2A, the largest of the histone variants, contains a 30-kDa carboxy-terminal globular tail called "macrodomain" (Gamble and Kraus 2010). Macrodomains bind metabolites of NAD generated by PARP or Sirtuin-type deacetylases. MacroH2A is most characterized in its involvement in facultative heterochromatin and transcriptional repression at the inactive X chromosome. More recently, its presence in nucleosomes was shown to inhibit recruitment of or remodeling by the Swi/Snf complex (Gamble and Kraus 2010). In contrast to macroH2A, histone H2A.Bbd (Barr body deficient) appears to associate with active acetylated chromatin and is excluded from the inactive X chromosome (Gonzalez-Romero et al. 2008). This short mammalian-specific histone variant forms nucleosomes that are less stable than those containing canonical H2A (Gonzalez-Romero et al. 2008). Surprisingly, the presence of H2A.Bbd in nucleosomes reduced their ability to be remodeled/mobilized by the Swi/Snf and ACF complexes in vitro. At the same time, H2A.Bbd-containing nucleosome arrays were more active templates in p300-activated transcription (Gonzalez-Romero et al. 2008). Thus, the contribution of H2A.Bbd to gene expression most likely is attributable to inherently destabilizing nucleosomes rather than making nucleosomes more sensitive to nucleosome remodeling.

10 NUCLEOSOME REMODELING DURING DEVELOPMENT

Because of their versatile activities nucleosome remodeling factors affect every aspect of genome function, even those that are characterized by epigenetic stability. Nucleosome remodeling may affect the assembly of such stable states directly by laying the foundations of higher-order chromatin organization or by strategic placement of histone variants as described in Section 9. Alternatively, they may help to establish and maintain stable, cell lineage-specific gene expression programs by serving as essential coregulators of the transcription machinery. A case in point is the versatile BAF complex, involved in orchestrating a number of rather diverse gene expression programs ranging from embryonic stem cells to postmitotic neurons (Yoo and Crabtree 2009; Ho and Crabtree 2010; Hargreaves and Crabtree 2011).

The explanation for the extraordinary versatility of the BAF complexes lies in the diversity of their subunit composition. The BAF remodelers of different cells all share a "core complex" of identical subunits that may be considered the engine that drives the fundamental, BAF-specific remodeling reaction. In addition to these core subunits, BAF complexes purified from different cells contain additional, cell-specific subunits (Fig. 8B). For example, the BAF complex purified from ES cells (esBAF) never contains the ATPase isoform BRM or the associated BAF170 subunit, but features BAF53a and BAF45a subunits. In contrast, the BAF complex isolated from postmitotic neurons (nBAF), may contain BRM in place of BRG1, BAF170 instead of BAF155, and is characterized by the presence of the variants BAF53b and BAF45b/c. The closely related BAF45 variants differ in their amino-terminal Krüppel-like domain. Remarkably, the proliferating and self-renewing neural progenitor cells contain BAF complexes of intermediate subunit composition (npBAF) (Fig. 8B) (Yoo and Crabtree 2009; Ho and Crabtree 2010; Hargreaves and Crabtree 2011). The most likely scenario is that the specificity subunits mediate interactions with cell-type-specific transcription factors that recruit the remodeling coactivators to alternative sets of genes. The switch between the different BAF complexes is brought about by repression of BAF53a through microRNAs that are specifically activated in postmitotic neurons and suppressed by the action of the transcription repressor REST in precursor cells. Continued expression of the progenitor signature subunits BAF45a and BAF53a prevents neuronal differentiation (Yoo and Crabtree 2009; Ho and Crabtree 2010; Hargreaves and Crabtree 2011). In agreement with their roles as regulators of differentiation programs, many BAF proteins function as tumor suppressors and their expression is lost or reduced in several forms of cancer (Hargreaves and Crabtree 2011). In conclusion the BAF complexes provide a striking example of a novel mechanism regulating patterns of gene expression during development. Developmental patterns of gene expression are not only dictated by the presence or absence of tissue specific transcription factors and epigenetic marks on chromatin but also by the subunit composition of a nucleosome remodeling complex.

11 CONCLUSIONS

Nucleosome remodeling enzymes, despite the dynamic nature of the chromatin transitions they catalyze, are involved in the assembly and propagation of stable and lasting—epigenetic—chromatin states. Canonical roles for nucleosome remodeling factors involve local nucleosome remodeling at regulatory elements to affect specific gene expression programs, as well as assuring the integrity of the chromatin fiber by nucleosome assembly and spacing, and finally their role in the exchange of histone variants. Remodeling factors may have other, less explored functions as well. We know of isolated examples in which remodelers use their ATP-dependent DNA translocase activities to modulate the chromatin

 Cite as *Cold Spring Harb Perspect Biol* doi: 10.1101/cshperspect.a017905

association of nonhistone substrates (Kia et al. 2008; Wollmann et al. 2011). We also contemplate their impact on chromatin organization independent of their enzymatic remodeling reactions, given the large number of remodeling machines associated with chromatin (Varga-Weisz and Becker 2006). A lot remains to be discovered.

ACKNOWLEDGMENTS

We thank Joanne Chatfield for assistance in preparing this chapter. Work in the laboratory of J.L.W. is funded by National Institute of General Medical Sciences (NIGMS) grants GM047867 and GM99945 and the Stowers Institute for Medical Research. Work on nucleosome remodeling factors in the laboratory of P.B. was funded by Deutsche Forschungsgemeinschaft SFB594, SPP1356, and the European Union through the Network of Excellence "The Epigenome" (FP6-503433).

REFERENCES

*Reference is also in this book.

Alkhatib SG, Landry JW. 2011. The nucleosome remodeling factor. *FEBS Lett* **585:** 3197–3207.

* Almouzni G, Cedar H. 2014. Maintenance of epigenetic information. *Cold Spring Harb Perspect Biol* doi: 10.1101/cshperspect.a019372.

Altaf M, Auger A, Covic M, Cote J. 2009. Connection between histone H2A variants and chromatin remodeling complexes. *Biochem Cell Biol* **87:** 35–50.

Altaf M, Saksouk N, Cote J. 2007. Histone modifications in response to DNA damage. *Mutat Res* **618:** 81–90.

Bao Y, Shen X. 2007. Chromatin remodeling in DNA double-strand break repair. *Curr Opin Genet Dev* **17:** 126–131.

Bao Y, Shen X. 2011. SnapShot: Chromatin remodeling: INO80 and SWR1. *Cell* **144:** 158–158. e2.

Becker PB, Horz W. 2002. ATP-dependent nucleosome remodeling. *Annu Rev Biochem* **71:** 247–273.

Becker PB, Wu C. 1992. Cell-free system for assembly of transcriptionally repressed chromatin from *Drosophila* embryos. *Mol Cell Biol* **12:** 2241–2249.

Billon P, Cote J. 2011. Precise deposition of histone H2A.Z in chromatin for genome expression and maintenance. *Biochim Biophys Acta* **1819:** 290–302.

Bouazoune K, Brehm A. 2005. dMi-2 chromatin binding and remodeling activities are regulated by dCK2 phosphorylation. *J Biol Chem* **280:** 41912–41920.

Bourachot B, Yaniv M, Muchardt C. 2003. Growth inhibition by the mammalian SWI-SNF subunit Brm is regulated by acetylation. *EMBO J* **22:** 6505–6515.

Bowen NJ, Fujita N, Kajita M, Wade PA. 2004. Mi-2/NuRD: Multiple complexes for many purposes. *Biochim Biophys Acta* **1677:** 52–57.

Brehm A, Tufteland KR, Aasland R, Becker PB. 2004. The many colours of chromodomains. *Bioessays* **26:** 133–140.

Cai Y, Jin J, Yao T, Gottschalk AJ, Swanson SK, Wu S, Shi Y, Washburn MP, Florens L, Conaway RC, et al. 2007. YY1 functions with INO80 to activate transcription. *Nat Struct Mol Biol* **14:** 872–874.

Cairns BR. 2001. Emerging roles for chromatin remodeling in cancer biology. *Trends Cell Biol* **11:** S15–21.

Cairns BR. 2007. Chromatin remodeling: Insights and intrigue from single-molecule studies. *Nat Struct Mol Biol* **14:** 989–996.

Cairns BR. 2009. The logic of chromatin architecture and remodelling at promoters. *Nature* **461:** 193–198.

Chaban Y, Ezeokonkwo C, Chung WH, Zhang F, Kornberg RD, Maier-Davis B, Lorch Y, Asturias FJ. 2008. Structure of a RSC-nucleosome complex and insights into chromatin remodeling. *Nat Struct Mol Biol* **15:** 1272–1277.

* Cheng X. 2014. Structural and functional coordination of DNA and histone methylation. *Cold Spring Harb Perspect Biol* **6:** a018747.

Chioda M, Becker PB. 2010. Soft skills turned into hard facts: Nucleosome remodelling at developmental switches. *Heredity* **105:** 71–79.

Choudhary C, Kumar C, Gnad F, Nielsen ML, Rehman M, Walther TC, Olsen JV, Mann M. 2009. Lysine acetylation targets protein complexes and co-regulates major cellular functions. *Science* **325:** 834–840.

Clapier CR, Cairns BR. 2009. The biology of chromatin remodeling complexes. *Annu Rev Biochem* **78:** 273–304.

Clouaire T, Stancheva I. 2008. Methyl-CpG binding proteins: Specialized transcriptional repressors or structural components of chromatin? *Cell Mol Life Sci* **65:** 1509–1522.

Dechassa ML, Sabri A, Pondugula S, Kassabov SR, Chatterjee N, Kladde MP, Bartholomew B. 2010. SWI/SNF has intrinsic nucleosome disassembly activity that is dependent on adjacent nucleosomes. *Mol Cell* **38:** 590–602.

Dimova D, Nackerdien Z, Furgeson S, Eguchi S, Osley MA. 1999. A role for transcriptional repressors in targeting the yeast Swi/Snf complex. *Mol Cell* **4:** 75–83.

Dion V, Shimada K, Gasser SM. 2010. Actin-related proteins in the nucleus: Life beyond chromatin remodelers. *Curr Opin Cell Biol* **22:** 383–391.

Downs JA, Nussenzweig MC, Nussenzweig A. 2007. Chromatin dynamics and the preservation of genetic information. *Nature* **447:** 951–958.

Durr H, Flaus A, Owen-Hughes T, Hopfner KP. 2006. Snf2 family ATPases and DExx box helicases: Differences and unifying concepts from high-resolution crystal structures. *Nucleic Acids Res* **34:** 4160–4167.

Elfring LK, Deuring R, McCallum CM, Peterson CL, Tamkun JW. 1994. Identification and characterization of *Drosophila* relatives of the yeast transcriptional activator SNF2/SWI2. *Mol Cell Biol* **14:** 2225–2234.

Elgin SC. 1981. DNAase I-hypersensitive sites of chromatin. *Cell* **27:** 413–415.

* Elgin SCR, Reuter G. 2013. Position-effect variegation, heterochromatin formation, and gene silencing in *Drosophila*. *Cold Spring Harb Perspect Biol* **5:** a017780.

Emelyanov AV, Vershilova E, Ignatyeva MA, Pokrovsky DK, Lu X, Konev AY, Fyodorov DV. 2012. Identification and characterization of ToRC, a novel ISWI-containing ATP-dependent chromatin assembly complex. *Genes Dev* **26:** 603–614.

Engeholm M, de Jager M, Flaus A, Brenk R, van Noort J, Owen-Hughes T. 2009. Nucleosomes can invade DNA territories occupied by their neighbors. *Nat Struct Mol Biol* **16:** 151–158.

Fairman-Williams ME, Guenther UP, Jankowsky E. 2010. SF1 and SF2 helicases: Family matters. *Curr Opin Struct Biol* **20:** 313–324.

Falbo KB, Shen X. 2006. Chromatin remodeling in DNA replication. *J Cell Biochem* **97:** 684–689.

Fazzio TG, Kooperberg C, Goldmark JP, Neal C, Basom R, Delrow J, Tsukiyama T. 2001. Widespread collaboration of Isw2 and Sin3-Rpd3 chromatin remodeling complexes in transcriptional repression. *Mol Cell Biol* **21:** 6450–6460.

Ferreira R, Eberharter A, Bonaldi T, Chioda M, Imhof A, Becker PB. 2007. Site-specific acetylation of ISWI by GCN5. *BMC Mol Biol* **8:** 73.

Flaus A, Martin DM, Barton GJ, Owen-Hughes T. 2006. Identification of multiple distinct Snf2 subfamilies with conserved structural motifs. *Nucleic Acids Res* **34:** 2887–2905.

Flaus A, Owen-Hughes T. 2011. Mechanisms for ATP-dependent chromatin remodelling: The means to the end. *FEBS J* **278:** 3579–3595.

Floer M, Wang X, Prabhu V, Berrozpe G, Narayan S, Spagna D, Alvarez D, Kendall J, Krasnitz A, Stepansky A, et al. 2010. A RSC/nucleosome complex determines chromatin architecture and facilitates activator binding. *Cell* **141:** 407–418.

Fry CJ, Peterson CL. 2001. Chromatin remodeling enzymes: Who's on first? *Curr Biol* **11:** R185–R197.

Gamble MJ, Kraus WL. 2010. Multiple facets of the unique histone variant macroH2A: From genomics to cell biology. *Cell Cycle* **9:** 2568–2574.

Gangaraju VK, Bartholomew B. 2007. Mechanisms of ATP dependent chromatin remodeling. *Mutat Res* **618:** 3–17.

Gonzalez-Romero R, Mendez J, Ausio J, Eirin-Lopez JM. 2008. Quickly evolving histones, nucleosome stability and chromatin folding: All about histone H2A.Bbd. *Gene* **413:** 1–7.

Gutierrez JL, Chandy M, Carrozza MJ, Workman JL. 2007. Activation domains drive nucleosome eviction by SWI/SNF. *EMBO J* **26:** 730–740.

Hanai K, Furuhashi H, Yamamoto T, Akasaka K, Hirose S. 2008. RSF governs silent chromatin formation via histone H2Av replacement. *PLoS Genet* **4:** e1000011.

Hargreaves DC, Crabtree GR. 2011. ATP-dependent chromatin remodeling: Genetics, genomics and mechanisms. *Cell Res* **21:** 396–420.

Hauk G, Bowman GD. 2011. Structural insights into regulation and action of SWI2/SNF2 ATPases. *Curr Opin Struct Biol* **21:** 719–727.

Hauk G, McKnight JN, Nodelman IM, Bowman GD. 2010. The chromodomains of the Chd1 chromatin remodeler regulate DNA access to the ATPase motor. *Mol Cell* **39:** 711–723.

* Henikoff S, Smith MM. 2014. Histone variants and epigenetics. *Cold Spring Harb Perspect Biol* doi: 10.1101/cshperspect.a019364.

Ho L, Crabtree GR. 2010. Chromatin remodelling during development. *Nature* **463:** 474–484.

Horn PJ, Peterson CL. 2001. The bromodomain: A regulator of ATP-dependent chromatin remodeling? *Front Biosci* **6:** D1019–D1023.

Ito T, Levenstein ME, Fyodorov DV, Kutach AK, Kobayashi R, Kadonaga JT. 1999. ACF consists of two subunits, Acf1 and ISWI, that function cooperatively in the ATP-dependent catalysis of chromatin assembly. *Genes Dev* **13:** 1529–1539.

Jha S, Dutta A. 2009. RVB1/RVB2: Running rings around molecular biology. *Mol Cell* **34:** 521–533.

Kasten MM, Clapier CR, Cairns BR. 2011. SnapShot: Chromatin remodeling: SWI/SNF. *Cell* **144:** 310.e1.

Kia SK, Gorski MM, Giannakopoulos S, Verrijzer CP. 2008. SWI/SNF mediates polycomb eviction and epigenetic reprogramming of the INK4b-ARF-INK4a locus. *Mol Cell Biol* **28:** 3457–3464.

Kim JH, Saraf A, Florens L, Washburn M, Workman JL. 2010. Gcn5 regulates the dissociation of SWI/SNF from chromatin by acetylation of Swi2/Snf2. *Genes Dev* **24:** 2766–2771.

Kingston RE, Narlikar GJ. 1999. ATP-dependent remodeling and acetylation as regulators of chromatin fluidity. *Genes Dev* **13:** 2339–2352.

* Kingston RE, Tamkun JW. 2014. Transcriptional regulation by trithorax-group proteins. *Cold Spring Harb Perspect Biol* **6:** a019349.

Korber P, Becker PB. 2010. Nucleosome dynamics and epigenetic stability. *Essays Biochem* **48:** 63–74.

Krishnakumar R, Kraus WL. 2010. The PARP side of the nucleus: Molecular actions, physiological outcomes, and clinical targets. *Mol Cell* **39:** 8–24.

Kunert N, Brehm A. 2009. Novel Mi-2 related ATP-dependent chromatin remodelers. *Epigenetics* **4:** 209–211.

Lan L, Ui A, Nakajima S, Hatakeyama K, Hoshi M, Watanabe R, Janicki SM, Ogiwara H, Kohno T, Kanno S, et al. 2010. The ACF1 complex is required for DNA double-strand break repair in human cells. *Mol Cell* **40:** 976–987.

Lantermann AB, Straub T, Stralfors A, Yuan GC, Ekwall K, Korber P. 2010. *Schizosaccharomyces pombe* genome-wide nucleosome mapping reveals positioning mechanisms distinct from those of *Saccharomyces cerevisiae*. *Nat Struct Mol Biol* **17:** 251–257.

* Li E, Zhang Y. 2014. DNA methylation in mammals. *Cold Spring Harb Perspect Biol* **6:** a019133.

Li B, Carey M, Workman JL. 2007. The role of chromatin during transcription. *Cell* **128:** 707–719.

Li J, Langst G, Grummt I. 2006. NoRC-dependent nucleosome positioning silences rRNA genes. *EMBO J* **25:** 5735–5741.

Lia G, Praly E, Ferreira H, Stockdale C, Tse-Dinh YC, Dunlap D, Croquette V, Bensimon D, Owen-Hughes T. 2006. Direct observation of DNA distortion by the RSC complex. *Mol Cell* **21:** 417–425.

Loyola A, Huang JY, LeRoy G, Hu S, Wang YH, Donnelly RJ, Lane WS, Lee SC, Reinberg D. 2003. Functional analysis of the subunits of the chromatin assembly factor RSF. *Mol Cell Biol* **23:** 6759–6768.

Luger K, Richmond TJ. 1998. DNA binding within the nucleosome core. *Curr Opin Struct Biol* **8:** 33–40.

Lusser A, Urwin DL, Kadonaga JT. 2005. Distinct activities of CHD1 and ACF in ATP-dependent chromatin assembly. *Nat Struct Mol Biol* **12:** 160–166.

Mansfield RE, Musselman CA, Kwan AH, Oliver SS, Garske AL, Davrazou F, Denu JM, Kutateladze TG, Mackay JP. 2011. Plant homeodomain (PHD) fingers of CHD4 are histone H3-binding modules with preference for unmodified H3K4 and methylated H3K9. *J Biol Chem* **286:** 11779–11791.

Martens JA, Winston F. 2003. Recent advances in understanding chromatin remodeling by Swi/Snf complexes. *Curr Opin Genet Dev* **13:** 136–142.

Martens JA, Wu PY, Winston F. 2005. Regulation of an intergenic transcript controls adjacent gene transcription in *Saccharomyces cerevisiae*. *Genes Dev* **19:** 2695–2704.

Morettini S, Podhraski V, Lusser A. 2008. ATP-dependent chromatin remodeling enzymes and their various roles in cell cycle control. *Front Biosci* **13:** 5522–5532.

Morettini S, Tribus M, Zeilner A, Sebald J, Campo-Fernandez B, Scheran G, Worle H, Podhraski V, Fyodorov DV, Lusser A. 2011. The chromodomains of CHD1 are critical for enzymatic activity but less important for chromatin localization. *Nucleic Acids Res* **39:** 3103–3115.

Morrison AJ, Shen X. 2009. Chromatin remodelling beyond transcription: The INO80 and SWR1 complexes. *Nat Rev Mol Cell Biol* **10:** 373–384.

Moshkin YM, Chalkley GE, Kan TW, Reddy BA, Ozgur Z, van Ijcken WF, Dekkers DH, Demmers JA, Travers AA, Verrijzer CP. 2012. Remodelers organize cellular chromatin by counteracting intrinsic histone-DNA sequence preferences in a class-specific manner. *Mol Cell Biol* **32:** 675–688.

Moshkin YM, Mohrmann L, van Ijcken WF, Verrijzer CP. 2007. Functional differentiation of SWI/SNF remodelers in transcription and cell cycle control. *Mol Cell Biol* **27:** 651–661.

Murawska M, Kunert N, van Vugt J, Langst G, Kremmer E, Logie C, Brehm A. 2008. dCHD3, a novel ATP-dependent chromatin remodeler associated with sites of active transcription. *Mol Cell Biol* **28:** 2745–2757.

Neves-Costa A, Varga-Weisz P. 2006. The roles of chromatin remodelling factors in replication. *Results Probl Cell Differ* **41:** 91–107.

* Patel DJ. 2014. A structural perspective on readout of epigenetic histone and DNA methylation marks. *Cold Spring Harb Perspect Biol* doi: 10.1101/cshperspect.a018754.

Pazin MJ, Kamakaka RT, Kadonaga JT. 1994. ATP-dependent nucleosome reconfiguration and transcriptional activation from preassembled chromatin templates. *Science* **266:** 2007–2011.

Polo SE, Jackson SP. 2011. Dynamics of DNA damage response proteins at DNA breaks: A focus on protein modifications. *Genes Dev* **25:** 409–433.

Rach EA, Winter DR, Benjamin AM, Corcoran DL, Ni T, Zhu J, Ohler U. 2011. Transcription initiation patterns indicate divergent strategies for gene regulation at the chromatin level. *PLoS Genet* **7:** e1001274.

Racki LR, Narlikar GJ. 2008. ATP-dependent chromatin remodeling enzymes: Two heads are not better, just different. *Curr Opin Genet Dev* **18:** 137–144.

Reik A, Schutz G, Stewart AF. 1991. Glucocorticoids are required for establishment and maintenance of an alteration in chromatin structure: Induction leads to a reversible disruption of nucleosomes over an enhancer. *EMBO J* **10:** 2569–2576.

Rowbotham SP, Barki L, Neves-Costa A, Santos F, Dean W, Hawkes N, Choudhary P, Will WR, Webster J, Oxley D, et al. 2011. Maintenance of silent chromatin through replication requires SWI/SNF-like chromatin remodeler SMARCAD1. *Mol Cell* **42:** 285–296.

Ryan DP, Owen-Hughes T. 2011. Snf2-family proteins: Chromatin remodellers for any occasion. *Curr Opin Chem Biol* **15:** 649–656.

Ryan DP, Sundaramoorthy R, Martin D, Singh V, Owen-Hughes T. 2011. The DNA-binding domain of the Chd1 chromatin-remodelling enzyme contains SANT and SLIDE domains. *EMBO J* **30:** 2596–2609.

Saha A, Wittmeyer J, Cairns BR. 2006. Chromatin remodelling: The industrial revolution of DNA around histones. *Nat Rev Mol Cell Biol* **7:** 437–447.

Sanchez R, Zhou MM. 2011. The PHD finger: A versatile epigenome reader. *Trends Biochem Sci* **36:** 364–372.

Schwabish MA, Struhl K. 2007. The Swi/Snf complex is important for histone eviction during transcriptional activation and RNA polymerase II elongation in vivo. *Mol Cell Biol* **27:** 6987–6995.

Selth LA, Sigurdsson S, Svejstrup JQ. 2010. Transcript elongation by RNA polymerase II. *Annu Rev Biochem* **79:** 271–293.

Simic R, Lindstrom DL, Tran HG, Roinick KL, Costa PJ, Johnson AD, Hartzog GA, Arndt KM. 2003. Chromatin remodeling protein Chd1 interacts with transcription elongation factors and localizes to transcribed genes. *EMBO J* **22:** 1846–1856.

Sims JK, Wade PA. 2011. SnapShot: Chromatin remodeling: CHD. *Cell* **144:** 626–626. e1.

Soutourina J, Bordas-Le Floch V, Gendrel G, Flores A, Ducrot C, Dumay-Odelot H, Soularue P, Navarro F, Cairns BR, Lefebvre O, et al. 2006. Rsc4 connects the chromatin remodeler RSC to RNA polymerases. *Mol Cell Biol* **26:** 4920–4933.

Srinivasan S, Armstrong JA, Deuring R, Dahlsveen IK, McNeill H, Tamkun JW. 2005. The *Drosophila* trithorax group protein Kismet facilitates an early step in transcriptional elongation by RNA Polymerase II. *Development* **132:** 1623–1635.

Suganuma T, Workman JL. 2011. Signals and combinatorial functions of histone modifications. *Annu Rev Biochem* **80:** 473–499.

Sugiyama T, Cam HP, Sugiyama R, Noma K, Zofall M, Kobayashi R, Grewal SI. 2007. SHREC, an effector complex for heterochromatic transcriptional silencing. *Cell* **128:** 491–504.

Talbert PB, Henikoff S. 2010. Histone variants—Ancient wrap artists of the epigenome. *Nat Rev Mol Cell Biol* **11:** 264–275.

Tirosh I, Sigal N, Barkai N. 2010. Widespread remodeling of mid-coding sequence nucleosomes by Isw1. *Genome Biol* **11:** R49.

Tolkunov D, Zawadzki KA, Singer C, Elfving N, Morozov AV, Broach JR. 2011. Chromatin remodelers clear nucleosomes from intrinsically unfavorable sites to establish nucleosome-depleted regions at promoters. *Mol Biol Cell* **22:** 2106–2118.

Torigoe SE, Urwin DL, Ishii H, Smith DE, Kadonaga JT. 2011. Identification of a rapidly formed nonnucleosomal histone-DNA intermediate that is converted into chromatin by ACF. *Mol Cell* **43:** 638–648.

Tsukiyama T, Becker PB, Wu C. 1994. ATP-dependent nucleosome disruption at a heat-shock promoter mediated by binding of GAGA transcription factor. *Nature* **367:** 525–532.

Tsukiyama T, Daniel C, Tamkun J, Wu C. 1995. ISWI, a member of the SWI2/SNF2 ATPase family, encodes the 140 kDa subunit of the nucleosome remodeling factor. *Cell* **83:** 1021–1026.

VanDemark AP, Kasten MM, Ferris E, Heroux A, Hill CP, Cairns BR. 2007. Autoregulation of the rsc4 tandem bromodomain by gcn5 acetylation. *Mol Cell* **27:** 817–828.

Vanolst L, Fromental-Ramain C, Ramain P. 2005. Toutatis, a TIP5-related protein, positively regulates Pannier function during *Drosophila* neural development. *Development* **132:** 4327–4338.

Varga-Weisz PD, Becker PB. 2006. Regulation of higher-order chromatin structures by nucleosome-remodelling factors. *Curr Opin Genet Dev* **16:** 151–156.

Varga-Weisz PD, Blank TA, Becker PB. 1995. Energy-dependent chromatin accessibility and nucleosome mobility in a cell-free system. *EMBO J* **14:** 2209–2216.

Varga-Weisz PD, Wilm M, Bonte E, Dumas K, Mann M, Becker PB. 1997. Chromatin-remodelling factor CHRAC contains the ATPases ISWI and topoisomerase II. *Nature* **388:** 598–602.

Vicent GP, Nacht AS, Zaurin R, Ballare C, Clausell J, Beato M. 2010. Minireview: Role of kinases and chromatin remodeling in progesterone signaling to chromatin. *Mol Endocrinol* **24:** 2088–2098.

Vignali M, Hassan AH, Neely KE, Workman JL. 2000. ATP-dependent chromatin-remodeling complexes. *Mol Cell Biol* **20:** 1899–1910.

Weissman B, Knudsen KE. 2009. Hijacking the chromatin remodeling machinery: Impact of SWI/SNF perturbations in cancer. *Cancer Res* **69:** 8223–8230.

Whitehouse I, Rando OJ, Delrow J, Tsukiyama T. 2007. Chromatin remodelling at promoters suppresses antisense transcription. *Nature* **450:** 1031–1035.

Whitehouse I, Tsukiyama T. 2006. Antagonistic forces that position nucleosomes in vivo. *Nat Struct Mol Biol* **13:** 633–640.

Winston F, Carlson M. 1992. Yeast SNF/SWI transcriptional activators and the SPT/SIN chromatin connection. *Trends Genet* **8:** 387–391.

Wippo CJ, Israel L, Watanabe S, Hochheimer A, Peterson CL, Korber P. 2011. The RSC chromatin remodelling enzyme has a unique role in directing the accurate positioning of nucleosomes. *EMBO J* **30:** 1277–1288.

Wollmann P, Cui S, Viswanathan R, Berninghausen O, Wells MN, Moldt M, Witte G, Butryn A, Wendler P, Beckmann R, et al. 2011. Structure and mechanism of the Swi2/Snf2 remodeller Mot1 in complex with its substrate TBP. *Nature* **475:** 403–407.

Workman JL. 2006. Nucleosome displacement in transcription. *Genes Dev* **20:** 2009–2017.

Workman JL, Kingston RE. 1998. Alteration of nucleosome structure as a mechanism of transcriptional regulation. *Annu Rev Biochem* **67:** 545–579.

Yadon AN, Tsukiyama T. 2011. SnapShot: Chromatin remodeling: ISWI. *Cell* **144:** 453–453.e1.

Yadon AN, Van de Mark D, Basom R, Delrow J, Whitehouse I, Tsukiyama T. 2010. Chromatin remodeling around nucleosome-free regions leads to repression of noncoding RNA transcription. *Mol Cell Biol* **30:** 5110–5122.

Yoo AS, Crabtree GR. 2009. ATP-dependent chromatin remodeling in neural development. *Curr Opin Neurobiol* **19:** 120–126.

Zeng L, Zhang Q, Li S, Plotnikov AN, Walsh MJ, Zhou MM. 2010. Mechanism and regulation of acetylated histone binding by the tandem PHD finger of DPF3b. *Nature* **466:** 258–262.

Zhang Z, Wippo CJ, Wal M, Ward E, Korber P, Pugh BF. 2011. A packing mechanism for nucleosome organization reconstituted across a eukaryotic genome. *Science* **332:** 977–980.

C H A P T E R 22

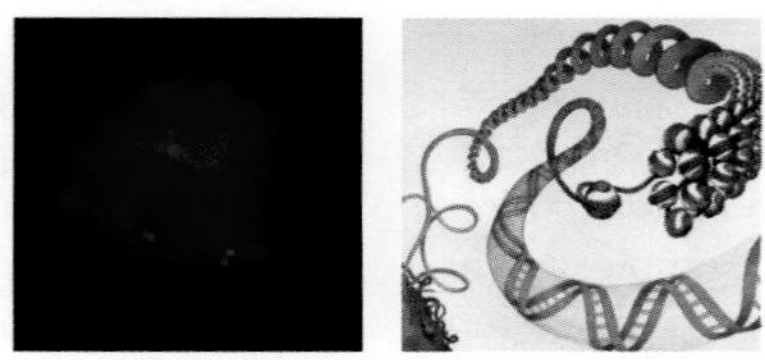

Maintenance of Epigenetic Information

Geneviève Almouzni[1] and Howard Cedar[2]

[1]Department of Nuclear Dynamics and Genome Plasticity, Institut Curie, Section de recherche, 75231 Paris Cedex 05, France; [2]Department of Developmental Biology and Cancer Research, Institute for Medical Research Israel–Canada, Hebrew University Medical School, Ein Kerem, Jerusalem, Israel 91120

Correspondence: genevieve.almouzni@curie.fr

SUMMARY

The genome is subject to a diverse array of epigenetic modifications from DNA methylation to histone posttranslational changes. Many of these marks are somatically stable through cell division. This chapter focuses on our knowledge of the mechanisms governing the inheritance of epigenetic marks, particularly, repressive ones, when the DNA and chromatin template are duplicated in S phase. This involves the action of histone chaperones, nucleosome-remodeling enzymes, histone and DNA methylation binding proteins, and chromatin-modifying enzymes. Last, the timing of DNA replication is discussed, including the question of whether this constitutes an epigenetic mark that facilitates the propagation of epigenetic marks.

Outline

Chapter-opening figures: (*Right*) Art by N. Bouvier.

Cite as *Cold Spring Harb Perspect Biol* doi: 10.1101/cshperspect.a019372

OVERVIEW

Every living organism has the ability to regulate the use of its genetic information. The most basic and usual mechanism for this process uses local regulatory sequences to bind transcription factors that can affect RNA polymerase initiation or elongation of nearby genes. Proteins or protein complexes can operate to either increase RNA synthesis or cause repression. This mechanism is transient in dividing cells because progression of the DNA replication fork usually disrupts these protein–DNA interactions, which then need to be reestablished in the resulting daughter cells. Because this system is based on the recognition of specific sequence elements in the DNA, however, the exact same expression pattern will be reconstructed following division.

The control of expression states using a sequence recognition-based method is perpetually dependent on the dynamic concentration of binding factors in progeny cells. As a result, it represents a nonautonomous mechanism that may not be completely suited for the long-term maintenance and stability of gene expression patterns. Furthermore, >50% of the genome in animals is repressed in any individual cell. These observations strongly suggest that higher organisms utilize a different strategy for gene regulation, using more global mechanisms to provide extended stability of gene expression through cell division and even development. In this chapter, we discuss three different mechanisms that may contribute to long-term epigenetic programming: DNA methylation, the inheritance of chromatin structure through DNA replication, and DNA replication timing.

Cite as *Cold Spring Harb Perspect Biol* doi: 10.1101/cshperspect.a019372

1 DNA METHYLATION

DNA methylation represents a major mechanism for stable gene repression. DNA methylation refers to the addition of a methyl group to the 5′ position of cytosine residues within a CpG dinucleotide sequence context in mammals. In general, methylation has a bimodal pattern of distribution in the genome; most regions are highly methylated (85%–100%), whereas CpG islands are unmethylated (0%–5%) (Straussman et al. 2009; Laurent et al. 2010). Many genes, including those only expressed in specific tissues, are located in the methylated fraction, whereas genes with CpG island promoters (mostly with housekeeping function) are constitutively unmodified.

A great deal is known about how DNA methylation patterns are maintained in vivo. Originally, it was shown that when in vitro–methylated DNA templates are introduced into somatic cells in culture, they retain the exact methylation pattern of the original substrate regardless of sequence, and even after many cell divisions (Pollack et al. 1980; Wigler et al. 1981). This suggested that there must be a mechanism for actually copying the position of methyl moieties during the process of replication. The basis for this lies in the symmetry of CpG dinucleotides—every CpG on one strand has a CpG complementary to it on the opposite strand, and methylated sites are almost always modified on both strands of the DNA. During the process of replication, however, synthesis of the new strand generates a hemimethylated site. This is, then, specifically recognized by the enzyme Dnmt1 (DNA methyltransferase 1) (Li et al. 1992), which then methylates the new CpG, thereby copying the methyl group from the native strand in a semiconservative manner (see Fig. 2 of Ch. 15 [Li and Zhang 2014]). Because the Dnmt1 enzyme has a high preference for hemimethylated sites, CpG sites that are not methylated on the parent strand do not serve as good substrates, thus, preserving their unmodified state on the newly synthesized DNA (Gruenbaum et al. 1982). It is now known that the specificity for this important reaction is not only dependent on the properties of Dnmt1 itself, but aided by additional proteins associated with the replication fork (elaborated in Ch. 6 [Cheng 2014]). As expected, knockdown of either Dnmt1 or other proteins in the complex will lead to overall, nonspecific demethylation in dividing cells (Gruenbaum et al. 1981; Lande-Diner et al. 2007).

If the maintenance of DNA methylation patterns is so stable using the Dnmt1-catalyzed process, the question of how patterns can be modified at all during development or ontogeny are intriguing. We know that, in the early morula, most DNA methylation gets erased, but the bimodal pattern is then generated anew at the time of implantation (Monk et al. 1987). This process occurs through a wave of generalized de novo methylation with CpG islands being protected by virtue of inherent *cis*-acting sequences, probably related to transcription (Brandeis et al. 1994; Straussman et al. 2009). Following this key stage in development, cells largely lose their ability to perform global de novo DNA methylation, but the original bimodal pattern is preserved through every cell division by the autonomous Dnmt-1-mediated maintenance mechanism (Siegfried et al. 1999). As a result, global methylation patterns seen in individual tissues throughout the organism actually reflect events that occurred at the time of implantation (Fig. 1) (Straussman et al. 2009; Laurent et al. 2010).

The mechanism of copying DNA methylation and histone posttranslational modification (PTM) patterns following DNA synthesis probably plays an important role in gene regulation. Passage of the DNA polymerase complex during replication disrupts nucleosome placement. The original chromatin structure must then be reconstructed on the newly synthesized daughter DNA molecules (Lucchini and Sogo 1995). Because DNA methylation takes part in creating inaccessible chromatin conformations and setting histone modification patterns (Eden et al. 1998; Jones et al. 1998; Nan et al. 1998; Hashimshony et al. 2003), the existence of an autonomous covalent mechanism for preserving DNA methylation patterns greatly helps in this reassembly process. Taken together, this system serves as a global, long-term repression pathway. In this scheme, most regions of the DNA, which are predominantly methylated at CpGs, are automatically put in a relatively closed conformation, whereas CpG islands are kept open. Global repression is thus achieved without the need to recognize specific sequence elements at each individual gene. It should be noted that although this overall repression state surely plays a role in reducing transcription, this represents only one link in the multilayer process of gene regulation (Lande-Diner et al. 2007).

One of the most important observations confirming the idea that methylation maintenance is based on strand symmetry came from studies of plant DNA. Although methylation in animal DNA was shown to be restricted to CpG residues, nearest-neighbor analysis revealed that DNA from plant sources can be methylated at all four C-containing dinucleotides. By analyzing further sequence context, however, it turned out that almost all of these methylated C moieties were actually included in the trinucleotide consensus sequence CpXpG (Gruenbaum et al. 1981). This suggested that, in addition to methylation at symmetrical dinucleotide CpG residues, plant DNA has the ability to maintain methylation patterns at trinucleotide sequences that have symmetrical C residues located two nucleotides away on the opposite strand. Further studies have indeed confirmed the existence of maintenance enzymes specific

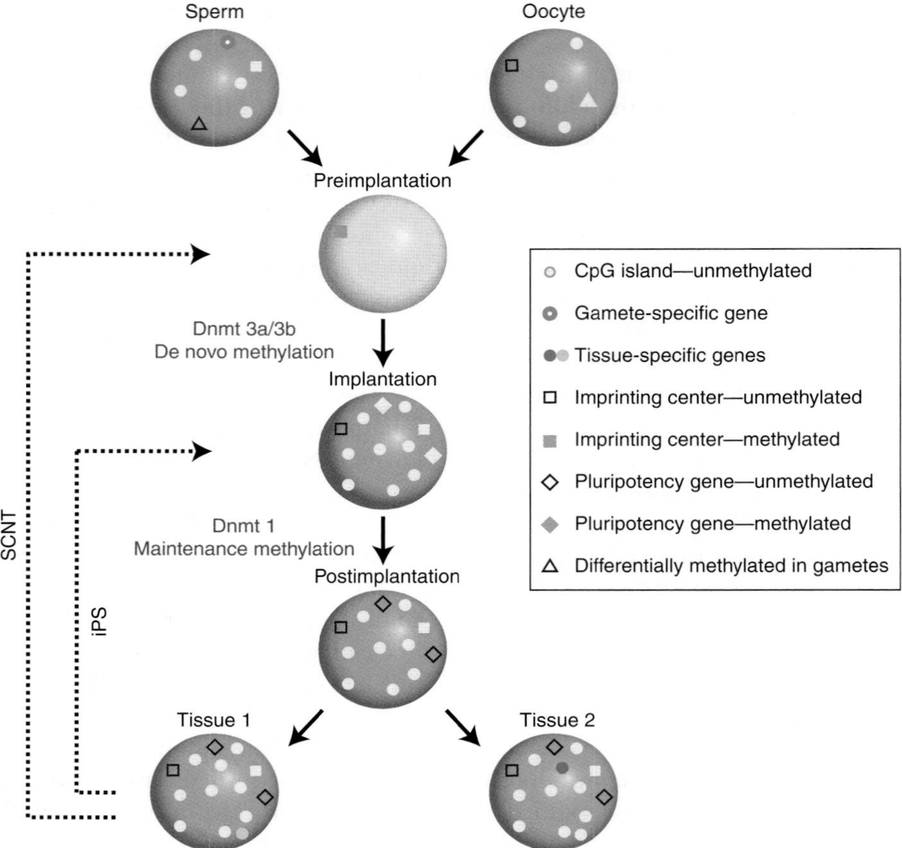

Figure 1. Generation of DNA methylation patterns during development. Gametic cells have a bimodal pattern of methylation with most regions methylated and CpG islands unmethylated (gray circles). Imprinting centers are methylated in one gamete (pink square), but not the other (white square). Gamete-specific genes (blue ring) are unmethylated. Some genes (triangles) are specifically unmethylated (gray) in one gamete. Almost all methylation in the gametes is erased (gray) in the preimplantation embryo, but imprinting centers retain methylation on one allele (pink square). At the time of implantation, the entire genome gets methylated (pink), with CpG islands being protected (gray circles). Postimplantation, pluripotency genes are de novo methylated (pink diamond). Tissue-specific genes undergo demethylation (yellow in Tissue 1, green in Tissue 2) in their cell type of expression. Imprinting centers remain differentially methylated throughout development. Somatic cell reprogramming by iPS or fusion resets the methylation pattern to the stage of implantation, whereas somatic cell nuclear transplantation (SCNT) resets to the preimplantation state.

for this type of modification site (elaborated in Sec. 2.1 of Ch. 13 [Pikaard and Mittelsten Scheid 2014]).

Proof of a DNA methylation maintenance model in vivo has been obtained from reverse epigenetics experiments in transgenic mice. A loxP-flanked CpG island element juxtaposed to a non-CpG island sequence upstream of a reporter gene protects the reporter sequence from de novo methylation at the time of implantation and, subsequently, in all tissues of the mouse (Siegfried et al. 1999; Goren et al. 2006). In contrast, if the CpG island element is removed by early Cre expression, this same reporter will be methylated throughout the animal. Interestingly, when the element is removed by Cre induction in the adult, the reporter still retains its unmethylated state. This confirms that the pattern of methylation is established in the early embryo and maintained for many cell generations, even if the original *cis*-acting elements responsible for this pattern are subsequently removed.

Although the basic bimodal DNA methylation pattern of the organism is formed without the need for recognizing specific sequences, all postimplantation changes in methylation take place in a gene- or tissue-specific manner (Fig. 1). By their very nature, these events are primarily driven by interactions between highly specific protein factors and *cis*-acting recognition signals on the DNA. These epigenetic alterations usually accompany changes in gene expression

 Cite as *Cold Spring Harb Perspect Biol* doi: 10.1101/cshperspect.a019372

and are involved in either demethylation or de novo methylation. Thus, for example, the pluripotency gene, Oct-3/4, undergoes de novo methylation in association with its repression following implantation, whereas many tissue-specific genes undergo demethylation together with their activation during development. In many cases, it is possible that this change in methylation is actually secondary to transcriptional activation or repression (Feldman et al. 2006). Nonetheless, these epigenetic alterations play an important role, at least, by bringing about a new stable state that can be maintained over long periods of time, even in cells that are continuously dividing. The devastating effects of cancer frequently accompanied by the aberrant acquisition or loss of DNA methylation of key genes illustrate the importance of this repressive maintenance mechanism in controlling appropriate gene expression programs (Ch. 34 [Baylin and Jones 2014]). In general, once a programmed change in methylation occurs, it is unlikely to return to its previous state because somatic cells mostly lack the ability to reprogram DNA methylation.

The most striking example of how DNA methylation acts as a *cis*-acting autonomous marker for gene repression can be seen at imprinted gene regions. In each of these domains, there is always a central imprinting center that dictates differential expression patterns in an allele-specific manner (Ch. 26 [Barlow and Bartolomei 2014]). These centers acquire DNA methylation on one allele during gametogenesis, whereas the other adopts an unmethylated profile. This differential state is then maintained through early embryogenesis (illustrated in Fig. 7 of Ch. 26 [Barlow and Bartolomei 2014]) and into the developing organism, and it is this pattern that ultimately dictates imprinted expression. Indeed, if DNA methylation is erased in the early embryo, imprinting is lost (Li et al. 1993). Similarly, genetic defects that affect methylation of imprinting centers also disrupt allele-specific expression profiles. Allelic methylation is also found on the X chromosomes in female cells (Ch. 25 [Brockdorff and Turner 2014]), in which it is known to play a role in the maintenance of X inactivation.

Despite the long-term stability afforded by DNA methylation during normal development, we now know that changes that occur as part of the differentiation process can actually be reversed by exposing somatic cells to an embryonic environment. This occurs, for example, when differentiated cell types are fused to embryonic stem (ES) cells in culture or altered by the introduction of pluripotency factors (Ch. 28 [Hochedlinger and Jaenisch 2014]). In this latter case, it appears that these exogenous genes serve to activate endogenous stem-cell master genes, which then turn on the specialized methylation machinery required for setting up a new basal methylation pattern. This occurs, for the most part, in the same manner as in vivo at the time of implantation.

2 MIXING PARENTAL AND NEW HISTONES

The organization of DNA into chromatin not only ensures its compaction, but also serves as a versatile structure that offers a range of possibilities for regulating genome function. The chromatin dynamics during DNA replication in eukaryotic cells enables the replication machinery access to compact structures while ensuring maintenance of its chromatin organization. Thus, because of the genome-wide alterations in chromatin structure that occur during replication, S phase has been considered a unique window of opportunity for cells to potentially modify chromatin structures that influence gene expression patterns and, thus, cell fate. Chromatin dynamics during replication, therefore, has to meet the dual challenge of maintaining the epigenetic fabric of the genome in a given lineage, but also orchestrating changes in chromatin structure that could promote a switch during cell differentiation and development (see G_2 panel of Fig. 2A).

During DNA replication, at the nucleosomal level, two fundamentally distinct processes proceed in parallel: (1) the transient disruption of histone–DNA interaction at preexisting nucleosomes located ahead of replication forks coupled to histone transfer/recycling onto nascent DNA, a reaction known as parental histone segregation; and (2) the deposition of newly synthesized histones through a pathway known as replication-dependent de novo nucleosome deposition (Groth et al. 2007b). Both the segregation of parental histone and de novo chromatin assembly after the replication fork affects the whole genome during each passage through S phase and, most likely, during active transcription and DNA repair at other stages of the cell cycle. Therefore, these two processes, if uncontrolled, can potentially have a widespread and profound impact on the ability of proliferating cells to propagate or modify epigenetic states that depend on specific chromatin structures.

To date, three main categories of factors have been considered during the duplication of chromatin (Fig. 2B): (1) chromatin remodelers, defined as large multiprotein complexes using ATP hydrolysis to slide nucleosomes or remove histones from DNA (Ch. 21 [Becker and Workman 2013]); (2) histone chaperones, which are escort factors that associate with histones and stimulate a reaction involving histone transfer without necessarily being part of the final product (Gurard-Levin et al. 2014); and (3) chromatin modifiers, which are involved in the establishment or removal of PTM marks on histones or, in the case of DNA, methylation of cytosines (Ch. 3 [Allis et al.]). These players, when acting in the vicinity of the replication fork, couple the duplication of DNA with its repackaging into chromatin. We will discuss how histone chaperones and chromatin modifications can possibly contribute to the maintenance

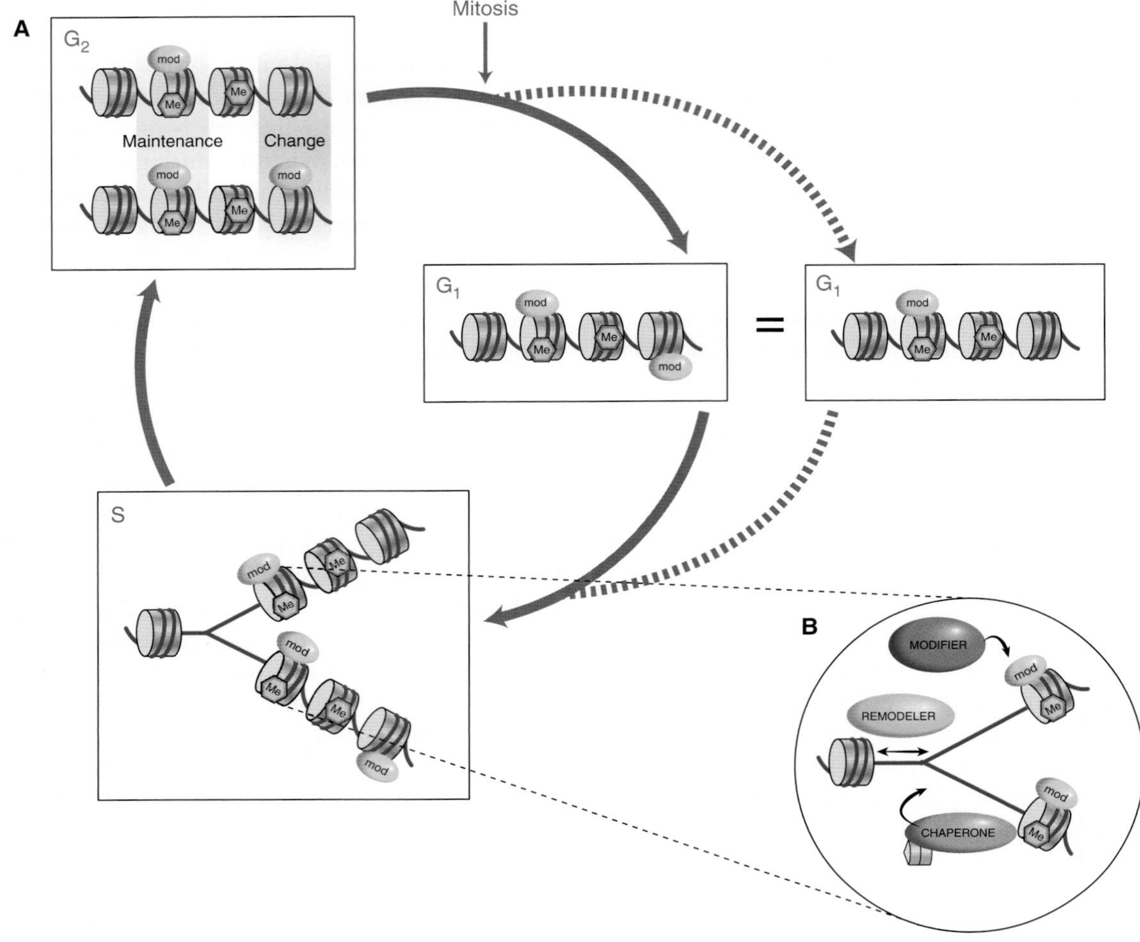

Figure 2. Duplicating chromatin in a cell-cycle-regulated manner. (*A*) In G_2, following DNA replication in S phase, duplicated chromatin, for the most part, maintains its epigenetic marks (blue shading), and although the opportunity for changes may occur at S phase, manifests as differences in marks between duplicated chromatin at a locus or region (purple shading). (*B*) The three types of factors believed to be involved, at the replication fork, in ensuring the propagation of epigenetic marks are chromatin-modifying enzymes (modifiers), nucleosome remodelers, and histone chaperones.

of particular epigenetic marks and, thereby, possibly help to perpetuate cellular memory.

2.1 Provision of Histones

2.1.1 Cell-Cycle-Regulated Synthesis of Histones

In mammals, if just considering the histone H3, there are four major variants: the canonical replicative variants H3.1 and H3.2, replacement variant H3.3, and CENP-A (centromeric protein A), a variant specifically present at centromeres (Loyola and Almouzni 2007; Ch. 20 [Henikoff and Smith 2014]). Replicative histones H3.1 and H3.2 are produced during S phase (Fig. 3) and ensure the provision of new histones to fulfill the requirement for nucleosome assembly on the two daughter strands in the wake of the replication fork (reviewed in Marzluff et al. 2008). Regulatory systems exist to control histone levels through transcriptional and posttranscriptional mechanisms (Gunjan et al. 2005; Marzluff et al. 2008). This is crucial for matching the demand for histones throughout S phase so that all new DNA can efficiently be packaged into nucleosomes while avoiding the deleterious effects of excess histones that can cause impaired replication or chromosome loss (Meeks-Wagner and Hartwell 1986; Gunjan and Verreault 2003; Groth et al. 2007a).

In contrast with the tight cell-cycle regulation observed for canonical histones, replacement (also referred to as replication-independent) histone variants do not show an S-phase regulation during the cell cycle. For example, H3.3

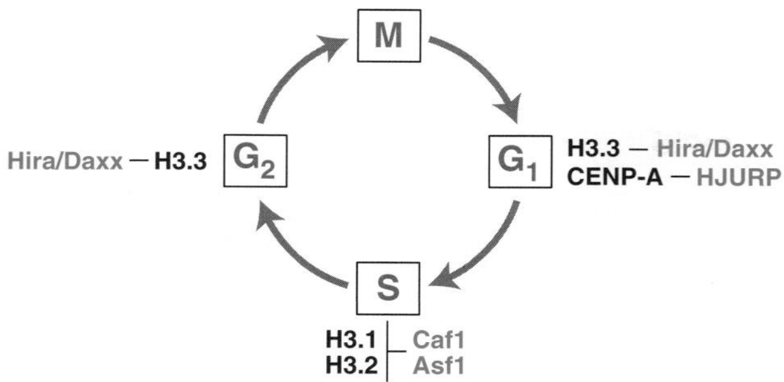

Figure 3. The cell-cycle-regulated provision of histone H3 variants. Canonical histones H3.1 and H3.2 are expressed mostly during S phase to ensure the supply of new histone subunits for major de novo chromatin assembly involved during DNA replication. Histone variants show significant incorporation patterns at distinct phases of the cell cycle. Chaperones (cyan) known to function specifically with particular types of histone (dark blue) are indicated.

is predominant in quiescent G_1 and G_2 phases (Fig. 3) (Wu et al. 1982). CENP-A, the most divergent H3 variant found at centromeric regions, shows a peak of expression in the G_2 phase before nuclear division (Shelby et al. 2000). Thus, when considering histone provision for nucleosome formation, one has to consider both the S-phase-dependent assembly of canonical histones and a replication-independent deposition of replacement histone variants, such as H3.3 and CENP-A (elaborated in Ch. 20 [Henikoff and Smith 2014]).

2.1.2 Modification of Preassembled Histones

An important question for understanding chromatin assembly at the replication fork is whether newly synthesized histones have preexisting modifications added before being assembled into chromatin. If so, a follow-up question is: Are these PTMs removed or not on chromatin assembly? Preexisting modifications on histones could influence the final modified state as a given histone-modifying enzyme may or may not work, depending on the histone substrate (Loyola and Almouzni 2007). Another important issue is to understand when and how PTMs are placed or removed, and whether they can account for the selective incorporation of histone variants at specific chromatin locations.

Acetylation is a major modification found on histones before their deposition on chromatin. Histone H4 acetylation, found on the amino-terminal tail at lysine 5 and lysine 12, is a characteristic mark for newly synthesized histones during S phase in most eukaryotes (Sobel et al. 1995; Loyola et al. 2006). Histones H4 diacetylated at K5 and K12 are found in a complex with the histone chaperones CAF-1 (chromatin assembly factor 1) and Asf1 (antisilencing function 1; discussed further in Sec. 2.2.1), and the K5-, K12-specific histone acetyltransferase, HAT1 (KAT1), before their deposition on chromatin (Fig. 4) (Tagami et al. 2004; Loyola et al. 2006; Parthun 2007). Intriguingly, however, histones H3–H4, which lack their amino-terminal tails, can still be efficiently deposited by CAF-1 onto chromatin in a SV40 DNA replication assay (Shibahara et al. 2000), indicating that histone acetylation–deacetylation is not a limiting step for histone deposition.

Histone H3 acetylation shows variability between organisms, with a peak in acetylation at K9 and K56 observed during S phase in *Saccharomyces cerevisiae*, and lower levels in G_2/M phases. In mammals, given the limited amount of H3K56ac reported so far in human cells (Garcia et al. 2007; Das et al. 2009; Xie et al. 2009), this mark is unlikely to occur on the bulk of newly synthesized histones, unless it is very rapidly removed after deposition. The role of H3K56 acetylation and chromatin assembly in mammals, thus, awaits further investigation.

Transient acetylation of newly synthesized histones H3 and H4 can even be observed in heterochromatin, which is largely underacetylated. Histone deacetylation at heterochromatic regions is, in fact, essential for the maintenance of the silent state of heterochromatic regions (Fig. 4) (Agalioti et al. 2002), proper chromosome segregation, and HP1 (heterochromatin protein 1) binding (Ekwall et al. 1997; Taddei et al. 2001). Failure to remove acetyl groups can have deleterious consequences, as shown by the inactivation of HDAC3 (histone deacetylase 3), which leads to impaired S-phase progression (Bhaskara et al. 2008). In addition, improper chromatin maturation on daughter strands could lead to DNA damage.

Loyola et al. (2006) addressed the question of whether or not histone methylation is placed before deposition. Predeposition H3.1 and H3.3 histones show a general lack of methylation marks, with the exception of lysine 9 (Loyola et al. 2006). H3.1 showed mainly H3K9 monomethylation (36%), whereas H3.3 can be monomethylated (17%) or dimethylated (4%), underscoring that K9 methylation marks may be imposed before deposition on chromatin (Fig. 4). Histone methylation is considered a relatively stable repressive mark at H3K9, found mostly at heterochromatin, although several histone lysine demethylases have recently been uncovered (Cloos et al. 2008). Nonnucleosomal H3K9me1 largely depends on the histone methyltransferase SetDB1 (Set domain bifurcated 1, or ESET/KMT1E) in complex with CAF-1 (Fig. 4) (Loyola et al. 2009). Intriguingly, Suv39h, an H3K9me2/3 histone methyltransferase, operates efficiently on monomethylated, but not dimethylated, H3K9 (Loyola et al. 2006). Thus, a predeposition PTM (i.e., H3K9me1) can potentiate further modification by enzymes (e.g., Suv39h-catalyzed conversion of H3K9me1 to me3) to impact on final chromatin states (Fig. 4). The combined action of histone-mod-

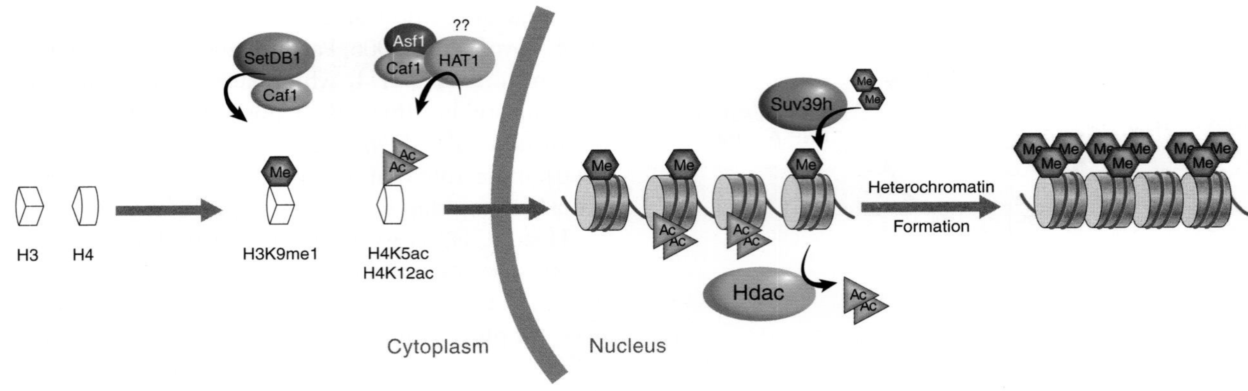

Figure 4. PTM of de novo synthesized histones prechromatin assembly. Histones H3 and H4 may have epigenetic marks added before chromatin assembly, which occurs after the replication fork. It is thought that H3K9 monomethylation and H4K5 and H4K12 acetylation are commonly added by modifying enzymes SetDB1 and HAT1, respectively, in complex with histone chaperones (e.g., Caf1 or Asf1). Once chromatin has been assembled, further modification may occur as illustrated in the case of heterochromatin formation.

ifying enzymes before deposition together with the local action of another one during deposition on chromatin will therefore determine the final signature of histones on chromatin. Where and when the modifications occur, and whether, as it has been proposed for the H3K56 and CAF-1 interaction in yeast, certain modifications can facilitate specific interactions with histone chaperones will be important to determine. This may also help to understand whether the modifications can contribute to the specificity of interactions between histone chaperones and particular histone variants discussed in the following section.

2.2 Histone Dynamics during DNA Replication

A key question in the research of how epigenetic marks are replicated is how the histone octamer becomes disassembled, transferred, and reassembled on replicated daughter strands of DNA. Also, how is this performed in conjunction with the incorporation of newly synthesized histones? This requires an understanding of the dynamics of chromatin assembly at replication forks, knowledge of the factors involved, and elucidation of how this is coordinated with any propagation or change in histone PTMs. We have, thus far, discussed the cell-cycle-regulated provision of histones and existence of histone PTMs prechromatin assembly. We will now take a closer look at the factors and dynamics involved at the replication fork.

2.2.1 Histone Chaperones

We now know that chromatin assembly relies on the activity of histone chaperones, which are histone-binding proteins. Indeed, chaperones are involved in functions as diverse as histone storage, transport, deposition, and eviction. Our understanding of the mechanisms and factors involved in histone deposition began with the development of pioneering cell-free systems consisting of *Xenopus laevis* egg extracts, which enabled chromatin assembly (Laskey et al. 1977). In 1986, Stillman (1986) biochemically identified human CAF-1, providing the first link between a histone chaperone and the capacity to deposit histones H3-H4 onto replicating DNA (Smith and Stillman 1989). The CAF-1 complex comprises three subunits (p150, p60, and RbAp48 in mammals) (Kaufman et al. 1995; Verreault et al. 1996). It is targeted in a phosphorylation-dependent manner to replication forks through an interaction with proliferating cell nuclear antigen (PCNA), a ring-shaped homotrimeric protein that serves as a processivity factor for the DNA polymerases (Fig. 5A) (Shibahara and Stillman 1999; Moggs et al. 2000; Gerard et al. 2006). A function of CAF-1 may be to ensure a tight coordination between histone deposition and ongoing DNA replication. The in vivo importance of CAF-1 is underscored by depletion studies in which loss of CAF-1 results in the loss of viability during development in mouse (Houlard et al. 2006), *Xenopus* (Quivy et al. 2001), and *Drosophila* (Song et al. 2007; Klapholz et al. 2009) and impaired S-phase progression in human cells (Hoek and Stillman 2003). More detail of its precise roles is elaborated in the following section.

Asf1 is another H3-H4 histone chaperone, identified initially in yeast, in a screen for silencing defects on overexpression (Le et al. 1997). Like CAF-1, it facilitates chromatin assembly coupled to DNA synthesis in vitro (Fig. 5) (Tyler et al. 1999; Mello et al. 2002). However, the sole addition of Asf1 to human cell extracts, or in HIRA (histone cell cycle regulator homolog A) and Asf1-depleted

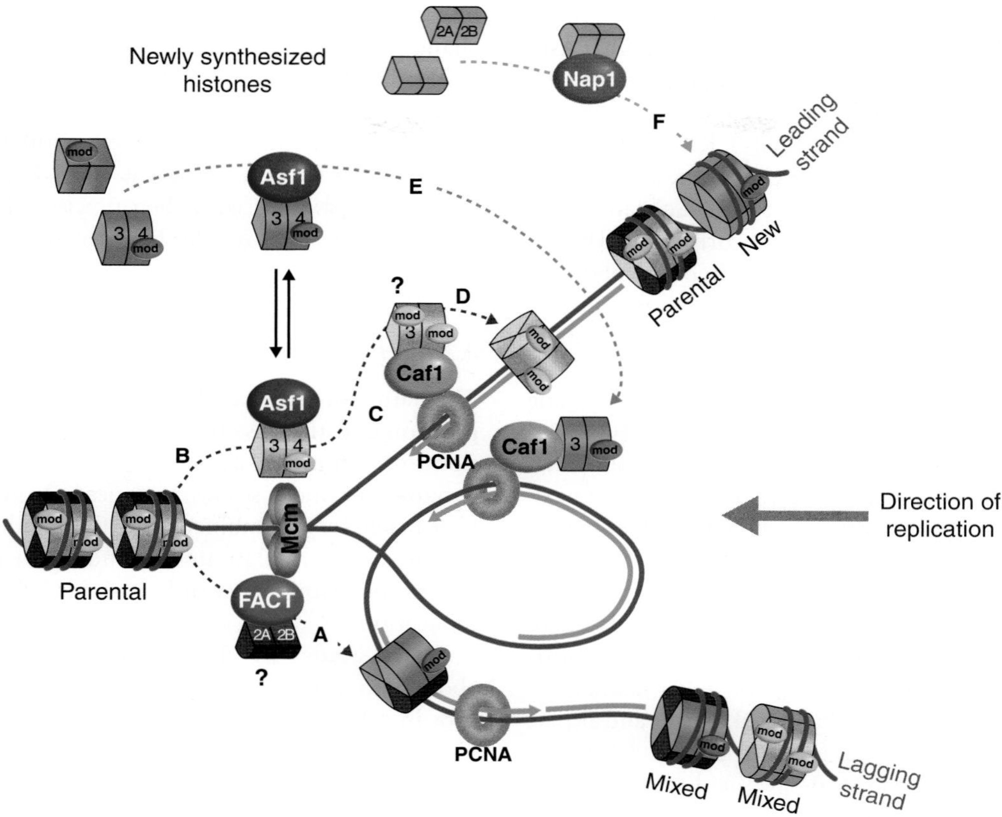

Figure 5. Chaperones and histone dynamics at the replication fork in eukaryotic cells. DNA replication proceeds in an asymmetric manner with continuous synthesis on the leading strand and discontinuous synthesis on the lagging strand. Folding of the two strands in space ensures coupling of replication between the two strands. Two fundamental processes affect the basic unit of chromatin during replication: "nucleosome removal" in front of the replication fork and "nucleosome deposition" on the two daughter strands behind the fork. Disruption of parental nucleosomes into two H2A-H2B dimers and an $(H3\text{-}H4)_2$ tetramer (or two H3-H4 dimers?) and their transfer/recycling onto the newly synthesized daughter strands provides a first source of histones (stages *A–D*). De novo assembly of new histones (H3-H4 as dimers in complex with histone chaperones) is necessary to restore a full nucleosomal density on the duplicated material (stages *E* and *F*). During S phase, this pool is provided by synthesis of replicative histone H3.1 variants in mammals. Although new histones carry a typical diacetylation K5, K12 modification on histone H4 (red "mod"), parental histone PTMs (yellow mod) potentially preserved during transfer can be used as a blueprint to reproduce marks on newly incorporated histones, which could be a means for epigenetic inheritance. How such events function in coordination with progression of the replication fork remains an open issue. The MCM (minichromosome maintenance) 2–7 is thought to mediate DNA unwinding in front of the replication fork. The histone binding activity of Mcm2 could aid the disruption, possibly in conjunction with chromatin remodelers and/or histone modifiers. An interaction with the MCM2–7 complex could favor targeting of the histone chaperones Asf1 and FACT (facilitates chromatin transcription), which would handle, respectively, parental H3-H4 (*B*) and H2A-H2B (*A*). In addition, Asf1 could pass parental histones onto CAF-1 (*C*). Because Asf1 interacts with H3-H4 in the form of dimers, parental tetramers (with their own marks) could potentially split and redistribute as dimers in a semiconservative fashion onto daughter strands. The reassembly on nascent DNA, in a stepwise fashion, would proceed via recruitment of CAF-1 to PCNA, which mediates the deposition of H3-H4 dimers provided by Asf1 acting as a histone donor (*E*). Having Asf1 handling both new and parental histones (*C* and *E*) would provide a means to coordinate histone supply with replication fork progression. With dynamics of H2A-H2B being relatively important throughout the cell cycle, it may be that assembly of these histones could simply use the NAP1 histone chaperone for H2A-H2B without a particular need to have a direct connection with the replication fork. NAP1 would bring in new histones H2A-H2B, and possibly old H2A-H2B, made available from transcriptional exchange (*C*). (Adapted, with permission, from MacAlpine and Almouzni 2013, © Cold Spring Harbor Laboratory Press.)

X. laevis egg extracts, is not sufficient to promote chromatin assembly or histone deposition. This indicates that Asf1 is unlikely to play a direct role in either replication-coupled or replication-independent chromatin assembly pathways (Mello et al. 2002; Ray-Gallet et al. 2007). Rather, it may act as a histone donor for the histone chaperone CAF-1 during DNA replication or repair (Fig. 5A,B), a collaboration conserved in various organisms. This is corroborated by the fact that Asf1 interacts with the B-domain of the p60 subunit of CAF-1 through a conserved hydrophobic groove (Tyler et al. 2001; Mello et al. 2002; Sanematsu et al. 2006; Tang et al. 2006; Malay et al. 2008) at a site opposite to that of its interaction with H3-H4 (English et al. 2006; Natsume et al. 2007). Formation of a ternary complex (CAF-1–Asf1–H3-H4) could thus be an intermediate, enabling histones to be handed over from one chaperone to the next. Hence, such histone transfer from Asf1 to CAF-1, as part of an "assembly line," would ensure an efficient histone deposition coupled to DNA replication (Fig. 5A or 5B). Other histone chaperones also have to be considered as players in histone dynamics. For example, NASP (nuclear autoantigenic sperm protein) is a chaperone for linker histone H1 (Finn et al. 2008), H3-H4 (Osakabe et al. 2010), and as part of a multichaperone complex (Tagami et al. 2004; Groth et al. 2005). There are also H2A-B-specific chaperones, such as Nap1 and FACT, not elaborated here, but illustrated in Figure 5D,E.

Stability of the histone $(H3\text{-}H4)_2$ tetramer when it is free of DNA in solution (Baxevanis et al. 1991) led to the long-standing assumption that these histones were deposited directly as a tetramer entity. Recent data have now challenged this view; histones H3-H4 were found as dimers together with histone chaperones within predeposition complexes in human cells (Fig. 5) (Tagami et al. 2004). Moreover, investigation of CenH3 (centromeric-specific H3) nucleosome properties identified hemisomes containing one copy of CenH3, H2A, H2B, and H4 each (Dalal et al. 2007). Together, these data suggest that histones H3-H4 or CenH3-H4 can first be provided as dimers. Two H3-H4 dimers next associate during deposition onto replicating DNA to form tetramers (Fig. 5C). The crystal structure of Asf1 interacting with a dimer of histones H3-H4 revealed that Asf1 physically blocks the formation of a $(H3\text{-}H4)_2$ tetramer (English et al. 2006; Natsume et al. 2007). However, whether both new histone H3-H4 dimers are provided by Asf1 and then deposited by CAF-1 onto DNA or additional chaperones are involved remains a possibility.

The NAP-1 chaperone adds histones H2A-H2B following the delivery of two dimers of H3-H4 onto newly synthesized DNA (Zlatanova et al. 2007), forming a complete nucleosomal core particle (Fig. 5D). Incorporation of new H2A-H2B, however, does not necessarily have to be tightly linked to DNA replication, as significant H2A-H2B exchange also occurs outside replication (Kimura and Cook 2001). The FACT complex also acts as an H2A-H2B chaperone in transcription and DNA repair, as well as in DNA replication (Fig. 5E). The apparent similarity of substrate and locality of action of the FACT and NAP-1 chaperones begs the question of what their precise functions are, and whether there is cross talk between them (Krogan et al. 2006).

2.2.2 *Histone Dynamics at the Replication Fork*

The nature of parental histones, in terms of variant composition and the ensemble of PTMs, represents information that can modulate genome function. Elucidating the dynamics of nucleosomes/histones in front of the replication machinery, combined with the handling of parental histones, is essential for understanding the parameters that govern the maintenance or loss of this information during replication.

Early studies by electron microscopy indicated that about 1 to 2 nucleosomes are destabilized in front of the fork in cell culture (Sogo et al. 1986; Gasser et al. 1996). However, determining whether destabilization involves a stepwise disassembly of the histone core octamer or, rather, a coordinated transfer of the octamer on the daughter strands is not yet clear. The fact that replication of SV40 mini-chromosomes with cross-linked nucleosomes can occur in human cell extracts, although at a reduced speed, indicates that dissociation of the histone octamer is not an absolute requirement for the passage of the replication fork and transfer of the parental nucleosomes (Vestner et al. 2000). In vivo metabolic labeling studies, however, led to the view that the core histone octamer is disrupted into two H2A-H2B dimers and a histone $(H3\text{-}H4)_2$ tetramer (Fig. 6A) (Annunziato 2005). More recently, the use of isotope labeling, combined with mass spectrometry analysis of histone content, has shed light on whether the histone $(H3\text{-}H4)_2$ tetramer remains intact during transfer (Xu et al. 2010); the vast majority of H3.1–H4 tetramers do not split, but the investigators did observe a significant number of splitting events for the H3.3-containing tetramers (Fig. 6B). Together with the provision of newly synthesized histones as dimers (Fig. 6C) (Tagami et al. 2004), ultimately, three modes of H3-H4 partitioning during nucleosome assembly at the replication fork can be considered: the transfer of parental histones onto replicated DNA, deposition of mixed old and new, and assembly of only newly synthesized histones (Fig. 6). It will be important to determine whether mixing events are exclusively variant-specific or if they reflect particular histone dynamics associated with specific chromatin regions.

Cite as *Cold Spring Harb Perspect Biol* doi: 10.1101/cshperspect.a019372

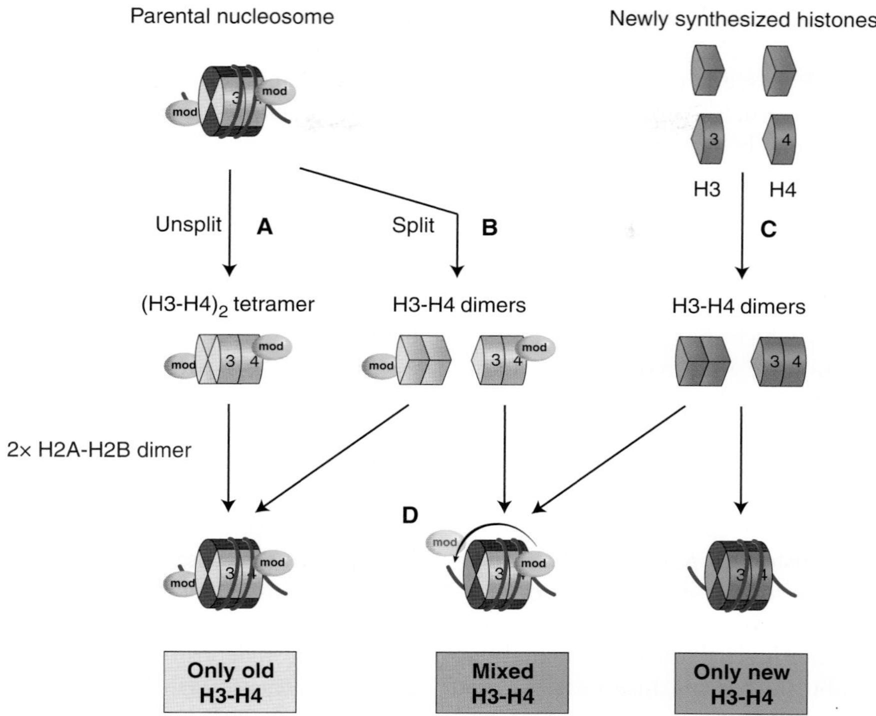

Figure 6. H3-H4 partitioning during nucleosome assembly. Upon nucleosome disruption during replication, parental $(H3\text{-}H4)_2$ tetramers can either remain intact (Unsplit) or broken up into two H3-H4 dimers (Split). Old nucleosomes will form either by inheritance of a stable $(H3\text{-}H4)_2$ tetramer (*A*) or by self-reassociation of two old recycled H3-H4 dimers (*B*). On the other hand, new nucleosomes result from de novo assembly of two newly synthesized H3-H4 dimers (*C*). Mixed particles can form on daughter strands by mixing an old H3-H4 dimer together with a new H3-H4 dimer (*D*). In all cases, association of two H2A-H2B dimers is necessary to complete the nucleosome. A modification on H4 is illustrated on the parental nucleosome to show the concept of intraparticle propagation of the mark between parental and new histones (*D*). (Adapted, with permission, from Nakatani et al. 2004; also MacAlpine and Almouzni 2013, © Cold Spring Harbor Laboratory Press.)

Disruption of the nucleosome during DNA replication may be achieved by the driving force of the moving fork, for example, through progression of the replicative helicase (Ramsperger and Stahl 1995). Interestingly, the large T antigen viral helicase and the MCM2–7 helicase bind histone H3 (Ramsperger and Stahl 1995; Ishimi et al. 1998) and, as ATP-driven machines, these helicases can provide the energy for the disruptive process. Interestingly, human cells depleted of Asf1 show defects in DNA unwinding, implicating its loss as impairing the removal of the nucleosome, so creating a barrier for replication fork progression (Groth et al. 2007a). This is corroborated by the fact that histones carrying typical parental PTMs can be detected in association with Asf1 on replication stress, suggesting that Asf1 can function as an acceptor of recycled parental dimers (Fig. 5F), as well as the previously mentioned role in chaperoning de novo synthesized H3-H4 dimers (Sec. 2.2.1) (Groth et al. 2007a; Jasencakova et al. 2010). Collectively, these results suggest that nucleosome disruption is closely coordinated with DNA unwinding in close proximity of the replication fork. However, additional factors acting at the histone level may also facilitate this process (see Sec. 2.2.3). Consequent to disruption, the consensus is that parental histones are randomly segregated on both sides of the replication fork (i.e., parental, mixed, or new), as illustrated in Figure 5 (Annunziato 2005). However, one can imagine that alternative mechanisms may operate at specific loci and in particular cell types, such as stem cells, which may harbor particular properties. Thus, whether the general principle of random segregation applies genome wide and during all stages of development requires further investigation.

In summary, three crucial steps (disruption, transfer, and de novo deposition) must take place for the duplication of chromatin organization during DNA replication. These events should not be considered independently but rather as a coordinated continuum between disruption, transfer, and de novo deposition, surrounding replication fork progression (Fig. 5), which is essential for preserving genetic stability and chromatin organization.

2.2.3 Propagating Histone Modifications

DNA methylation, histone PTMs, and even the incorporation of histone variants into nucleosomes provide extra layers of information that may be termed "epigenetic" if stably inherited throughout cell generations. The question of whether they are perpetuated during DNA and chromatin duplication or their stable propagation through multiple cell divisions occurs independently of DNA replication continues to be assessed. Our current understanding of the models and mechanisms by which these marks are propagated during DNA replication are discussed here. The manner in which DNA methylation is inherited requires that the preexisting methylation pattern on CpG is copied onto the new DNA strand after DNA replication via Dnmt1 (Sec. 1). One can envision that histone PTMs could also be perpetuated during DNA replication to maintain cell identity.

For the maintenance of histone PTMs, parental histones might be used as a template for the modification of new histones. Current models describing the random distribution of parental and newly synthesized histones on daughter strands suggest that neighboring parental histones serve as a template to spread a given modification to a de novo assembled nucleosome (for reviews, see Nakatani et al. 2004; Probst et al. 2009). Accordingly, the parental nucleosome would remain intact and no splitting of H3-H4 tetramers would be required (Fig. 6A) (Annunziato 2005). Such a mechanism can be effective in repetitive regions in which long arrays of nucleosomes carry the same marks, such as in pericentric heterochromatin in which H3K9me3 is bound by HP1 proteins (Bannister et al. 2001; Lachner et al. 2001). However, it cannot apply to regions in which particular marks are restricted to only one or two nucleosomes. In these regions, the splitting of tetramers, as recently observed with H3.3 variants (e.g., Fig. 6B) (Xu et al. 2010), could lead to the transmission of histone marks through an intraparticle mechanism (Fig. 6D).

To take the example of pericentric heterochromatin, in which the replication of not only DNA, but epigenetic information, such as histone marks and spatial organization, needs to occur, propagation represents a challenge. The maintenance of DNA and histone methylation patterns following DNA replication is, in fact, crucial for the stability of the genome and cell division. The local concentration of HP1 proteins, together with DNA methylation and H3K9me3, all contribute to the formation of these constitutive heterochromatic regions next to centromeres via a spreading mechanism that may partly rely on the templating of marked parental nucleosomes distributed randomly after DNA replication (Fig. 7A).

A common theme in the transmission of DNA methylation patterns or histone PTMs is the coupling of the enzymatic activity with the replication process. This may be facilitated via the interaction of several factors implicated in the duplication of epigenetic marks with PCNA, which is a major component of the replication fork machinery. Known interacting proteins include Dnmt1 (Fig. 7B), which is the principal enzyme responsible for propagating cytosine methylation inheritance owing to its preferred affinity for hemimethylated CpG motifs (Pradhan et al. 1999), CAF-1, introduced as a histone chaperone in the previous section, and Pr-Set7, discussed below.

Mouse CAF-1 p150 is required for replication of pericentromeric heterochromatic regions (Quivy et al. 2008). The largest subunit of CAF-1, p150, binds HP1 (Murzina et al. 1999) and this interaction is essential to promote the redistribution of HP1 during replication (Fig. 7C) (Quivy et al. 2004). An attractive hypothesis is that, by handling HP1 proteins, mouse CAF-1 p150 would relieve physical constraints impeding replication fork progression in HP1-rich pericentromeric heterochromatic regions and ensure an effective redeposition of HP1 molecules behind the fork on recycled H3K9me3-marked parental histones (Quivy et al. 2008; Loyola et al. 2009). A self-reinforcing loop involving the histone H3K9 methyltransferase Suv39h recruited by HP1 (Aagaard et al. 1999) would then create additional binding sites for HP1 (Fig. 7A). This would explain the propagation of HP1 binding in pericentric heterochromatin. The identification of a multimeric complex composed of the H3K9 histone methyltransferases Suv39h1, G9a, GLP, and SetDB1 is consistent with this view, implicating all of the KMTs (lysine methyltransferases) in the process of heterochromatin formation (Fig. 7) (Fritsch et al. 2010).

CAF-1 is also found in a complex together with MBD1 (methyl-CpG-binding protein 1), a protein that binds to methylated CpG motifs, and the histone H3K9 methylase SetDB1 (Reese et al. 2003; Sarraf and Stancheva 2004). A current hypothesis is that passage of the replication fork may displace MBD1, facilitating its interaction with the CAF-1 p150 subunit (Reese et al. 2003) to form a SetDB1–MBD1–CAF1 S-phase-specific complex (Fig. 7D) (Sarraf and Stancheva 2004). This epigenetic memory would ensure the coupling of postreplicative transmission of DNA and H3K9 methylation with nucleosomal assembly. In addition, MBD1 can interact with the histone methyltransferase Suv39h1 and HP1 (Fujita et al. 2003), which could provide another means to perpetuate H3K9 methylation states and chromatin compaction (Fig. 7A).

In mammals, the HDAC responsible for H4K5,K12 deacetylation remains unknown, yet deacetylation is observed within 20–60 min of chromatin assembly (Taddei et al. 1999). In *Drosophila*, deacetylation of H4K5,K12 depends

 Cite as *Cold Spring Harb Perspect Biol* doi: 10.1101/cshperspect.a019372

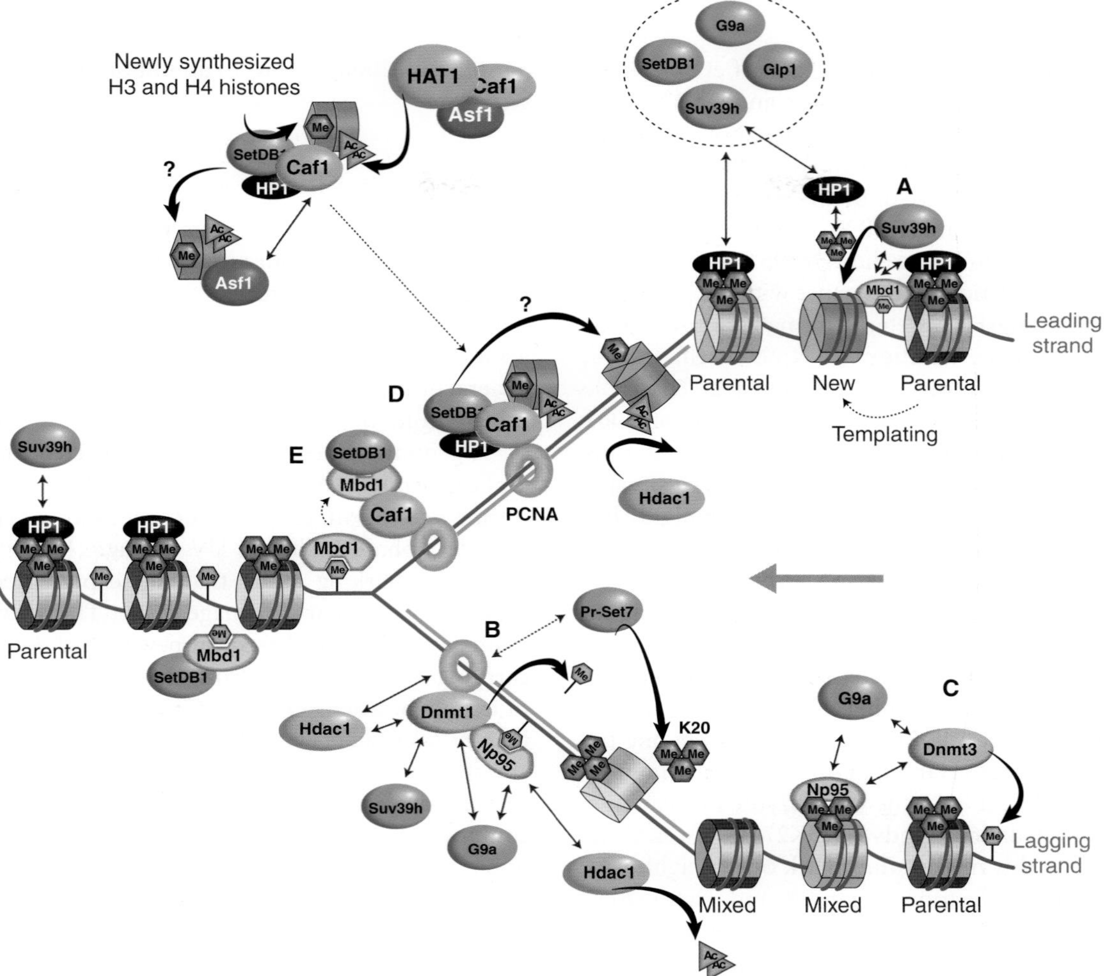

Figure 7. Model for the maintenance of constitutive pericentromeric heterochromatin during chromatin replication in mammals. (*A*) The maintenance of histone PTMs can be envisioned according to the following mechanism: A parental mark is recognized by a chromatin-binding protein, or reader protein (HP1), which, in turn, recruits a chromatin modifier or writer protein (Suv39h). This writer protein then imposes the parental modification on neighboring new histones (with respect to H3-H4 histones only illustrated in turquoise). This model has been suggested for the feedback loop in the maintenance of HP1 at pericentric heterochromatin or the repressive mark H3K27me3. HP1 binds to H3K9me3 (triangle of red hexagons) through its chromodomain and, in turn, recruits more of the H3K9 methyltransferase Suv39h (KMT1A). Suv39h could then further methylate the de novo assembled H3K9me1-marked nucleosomes into H3K9me3. The latter would provide additional binding sites for HP1α in pericentric heterochromatin. For H3K27me3 maintenance, both the reader and writer modules are part of the same protein, which is PRC2 (Polycomb repressive complex 2). Whether these modifications are immediately imposed on new histones after replication, or this happens at later stages, remains to be investigated. (*B*) DNA methylation (pink hexagon), histone hypoacetylation, H3K9me3 and H4K20me3 methylation, as well as the enrichment of HP1, propagate during replication of pericentric heterochromatin by exploiting a complex network of histone chaperones and modifiers still under investigation, in which PCNA functions as a central hub. Dnmt1, which is targeted to hemimethylated or H3K9me3 sites at replication foci via its interactions with PCNA and Np95, as described in the text, ensures maintenance of DNA methylation. Histone-modifying enzymes G9a and Suv39h can be recruited via their interaction with Dnmt1. (*C*) In addition, recent data suggest that Dnmt3A/3B could also be involved in maintenance of DNA methylation in highly methylated regions. These enzymes could be targeted by an interaction with the nucleosomes, or indirectly through Np95 or G9a. In this manner, maintenance of DNA and histone methylation can be coordinated. Formation of a multimeric complex between the four histone H3K9 methyltransferases, G9a, Suv39h, GLP (G9a-like protein 1), and SetDB1, could further ensure maintenance of H3K9 methylation states. (*D*) CAF-1, which can be found in distinct complexes with either H3.1 or HP1α, is also targeted via PCNA to the replication fork. The dual interaction ensures the handling of both histones and HP1 proteins in a successive manner. This could be advantageous to promote redistribution of HP1 proteins at pericentric heterochromatin. (*E*) Furthermore, the CAF-1 connection with an MBD1–SetDB1 and HP1α-SetDB1 complex could promote H3K9 methylation. Asf1 could participate in this assembly line by docking onto CAF-1 while monomethylation of H3K9 is imposed. As a current working hypothesis, this model should help to refine the precise mechanisms and protein interactions involved. Tapered arrows represent catalytic action. Straight arrows signify interactions (dashed when possible interactions). (Adapted from Corpet and Almouzni 2009, with permission from Elsevier.)

on the monomethylation of H4K20 (Scharf et al. 2009b). Cross talk between histone deacetylation and DNA methylation is also suggested in mouse cells, based on the hyperacetylation of H4K5,K12 observed following Np95 depletion. Np95 is an SRA domain–containing protein (also known as UHRF1 and ICPB90) that binds methylated DNA (Papait et al. 2007). Somehow, Np95 may function in linking DNA methylation maintenance with the proper deacetylation of histone H4 at heterochromatic regions. Other HDAC interactors, in addition to Np95 itself, include Suv39h1 (KMT1A), Dnmt1, and PCNA (Fig. 7B) (Fuks et al. 2000; Robertson et al. 2000; Milutinovic et al. 2002; Vaute et al. 2002; Unoki et al. 2004). HDACs would, thus, ensure the maintenance of a deacetylated state of chromatin, together with H3K9me3, at replication forks in highly histone-methylated heterochromatin regions. On the other hand, in regions in which HDACs are not tightly targeted, it may be easier to maintain acetylated states following the passage of the fork.

With respect to facultative heterochromatin regions, the process proposed for maintaining the H3K27me3 histone modification through cell division resembles the one put forward for H3K9me3 (Hansen and Helin 2009; Margueron et al. 2009). Here, PRC2, the very enzyme that catalyzes this modification, directly binds to H3K27me3. Accordingly, it could copy this methylation mark onto neighboring newly incorporated histones (Hansen and Helin 2009; Margueron et al. 2009). PRC1 (Polycomb repressive complex 1) complex, which remains associated with DNA during replication in vitro, could also potentially participate in the maintenance of transcriptional silent states through cell division (Francis et al. 2009). However, whether PRC1 remains in direct contact with DNA during DNA replication or its transfer involves component of the replication machinery remains to be elucidated.

Studies have provided a glimpse of how H4K20 monomethylation could be maintained during DNA replication. Pr-Set7 (also known as Set8 or KMT5A), an H4K20 histone monomethyltransferase, is found at replication sites and required for the progression through S phase (Jorgensen et al. 2007; Tardat et al. 2007; Huen et al. 2008). The targeting to replication forks could be by direct interaction with PCNA via a "PIP box" (PCNA-interacting protein box) in its amino terminus (Jorgensen et al. 2007; Huen et al. 2008). However, the reported low levels of Pr-Set7 during S phase challenge this view (Oda et al. 2009). Thus, further investigation is needed to reproduce Pr-Set7-catalyzed H4K20 monomethylation patterns on daughter strands via its interaction with PCNA. Nevertheless, the essential role of Pr-Set7 for mouse development (Huen et al. 2008; Oda et al. 2009), associated with its functions in cell division and genome stability (Jorgensen et al. 2007; Tardat et al. 2007), points to an important function of this enzyme.

2.2.4 *Histone Variants*

Given that histone variants can mark particular chromatin states, a challenging question is to understand how their incorporation is propagated through replication. And, if maintenance is ensured over the cell cycle, how is this achieved? Here, we will use H3 variants to illustrate how general principles can be derived and potentially applied to other chromatin marks.

A genome-wide distribution is expected for canonical replicative histone variants H3.1 or H3.2, mainly deposited during replication. However, incorporation of H3.1 occurs outside S phase at sites of UV damage (Polo et al. 2006), therefore, marking a site that has experienced damage, much as a scar marks damaged skin. This may provide a memory of damage, based on newly synthesized histones carrying different PTMs from the original ones. Furthermore, oligonucleosomes containing H3.1 are more prominently associated with HP1α and MBD1, suggesting an enrichment of H3.1 in constitutive pericentric heterochromatin (Loyola et al. 2006). These domains of concentration could reflect a default state if no other variant gets incorporated outside S phase. This is consistent with the fact that replacement variant H3.3 accumulates in actively transcribed chromatin regions, as first shown in *Drosophila* (Ahmad and Henikoff 2002) and within chromatin enriched in "active" marks when compared with the replicative variants (McKittrick et al. 2004; Hake et al. 2006; Loyola et al. 2006). Histone H3.3 accumulation at promoters of active genes or regulatory elements may exploit a replication-independent mechanism involving the histone chaperone HIRA (Ch. 20 [Henikoff and Smith 2014]). Although, during replication, the dilution of active marks on parental histones occurs, the amount provided may be sufficient to maintain a permissive state for transcription, which, in turn, would add more active marks. The presence of some H3.3 along with active marks would thus act as a seeding event. According to this view, the memory of an active transcriptional state could therefore involve both the choice of the H3.3 variant combined with an active mark, such as H3K4 methylation (Ng and Gurdon 2008; Muramoto et al. 2010). However, not only is H3.3 confined to sites of active transcription, but it can also be enriched in other genomic regions depending on the developmental context. At the time of fertilization, a massive and global accumulation of H3.3 occurs on sperm-derived DNA (Loppin et al. 2005) and, in ES cells, H3.3 accumulates at telomeres (illustrated in Fig. 10 of Ch. 3 [Allis et al.]) (Goldberg et al. 2010; Wong et al. 2010). How these events

 Cite as *Cold Spring Harb Perspect Biol* doi: 10.1101/cshperspect.a019372

are controlled and which factors are involved are beginning to be unraveled (for review, see Maze et al. 2014). It is amazing, nevertheless, to bear in mind that H3.1, H3.2, and H3.3 show very little sequence difference; thus, how specificity in their deposition is achieved remains a fascinating but unresolved issue.

In contrast to H3.3, CENP-A is highly divergent (Wolffe and Pruss 1996). It provides the best example of a histone H3 variant that specifies a functional locus: the site of centromere identity (Warburton et al. 1997), which serves as a platform for kinetochore assembly (elaborated in Ch. 9 [Allshire and Ekwall 2014]). During replication of centromeric chromatin, CENP-A nucleosomes become diluted to half the initial concentration on daughter chromatin (Shelby et al. 2000; Jansen et al. 2007). It is not until the next G_1 phase that new CENP-A gets incorporated again (Fig. 3) (Jansen et al. 2007; Schuh et al. 2007). This case illustrates a situation in which the disruption during replication is clearly separated from the reassembly event outside S phase. HJURP (Holliday junction recognition protein) was discovered to be a CENP-A chaperone, localized at centromeres precisely from late telophase to early G_1, promoting the specific targeting/incorporation and maintenance of CENP-A at centromeres (Fig. 3) (Dunleavy et al. 2009; Foltz et al. 2009).

An alternative view of the cell-cycle-regulated assembly of CENP-A chromatin is that incorporation of CENP-A in G_1 may be programmed in anticipation of the disruptive event during replication, rather than being a restoration of half the pool of CENP-A. Whichever way one looks at this issue, it provides a general conceptual framework for the mechanism by which chromatin marks can be dealt with during the cell cycle. Thus, it is informative to consider how and when histone PTMs or histone variants are actually imposed at particular domains, given that parental nucleosomes experience the disruption caused by passage of the replication fork.

Late steps in chromatin restoration may await reinitiation of transcription on daughter strands as recently proposed for H3K9me2 at heterochromatic repeats in *Schizosaccharomyces pombe* (Chen et al. 2008; Kloc and Martienssen 2008). Similarly, in human cells, di- or trimethylation of H3K27 and H4K20 has been detected during the next G_1 phase (Scharf et al. 2009a). This suggests that restoration of global levels of these histone modifications would occur mostly just before the next S phase, which could be regarded, again, as an anticipation of the dilution event occurring during replication. Thus, our understanding of the transmission of chromatin-based information in a replication-independent manner, concerning the histone H3 variants and specific PTMs, represents a general theme that may also help to understand the fate of other marks, like association of nuclear RNA, Polycomb proteins, or other chromatin proteins.

2.3 Summary

Thus, in eukaryotic cells, chromatin duplication entails a series of complex and coordinated events. These include nucleosomal disruption, histone transfer, and deposition, along with dynamic histone modifications that occur in a manner that maintains particular marks at defined chromatin regions. Chromatin remodelers and histone modifiers and chaperones are critical for chromatin duplication, both at the level of histone dynamics and in the context of reestablishing marks that define distinct chromatin states. It should be noted that replication has also been considered as a window of opportunity important to induce changes in chromatin states (Weintraub 1974). In addition, nonreplicative chromatin dynamics can also promote the maintenance of particular marks, as exemplified with the case of the histone variants H3.3 and CENP-A. To finely tune the balance between maintenance and plasticity during development, these processes are tightly controlled. A better understanding of the fine-tuning of this system and its network of partners not only will improve our basic knowledge of chromatin duplication, but also will be informative with respect to functional conservation between species and will help to gain insight into developmental diseases, as well as cancer.

3 REPLICATION TIMING

3.1 Patterns of Replication Timing

One of the least understood chromosomal epigenetic marks is DNA replication timing. The entire genome is replicated according to a predetermined program: Some regions replicate early in S, others during the middle of S, and, yet, other defined loci undergo replication only in late S phase. This pattern can be observed in dividing cells by labeling them with BrdU (bromodeoxyuridine) for fixed-time intervals and then visualizing the incorporated nucleotides in metaphase chromosomes using specific antibodies to BrdU (Hand 1978). The appearance of a clear-cut banding pattern on each individual metaphase chromosome nicely shows that this property is regulated at the regional level. Furthermore, comparative studies indicate that these replication zones actually correspond almost one-to-one with the Giemsa staining pattern in which dark G-bands represent the late-replicating regions (Holmquist 1988). Thus, it appears that chromosomes are organized into distinct subunits, each with its own structure and replication-timing properties.

Replication timing of individual genes can be measured by using this same labeling approach. After separating cells at different stages of S phase, it is possible to use polymerase chain reaction to detect specific genes in BrdU-labeled DNA. These studies have shown that there is a clear-cut correlation between gene repression and late replication. Active genes tend to replicate early in S phase, whereas inactive genes replicate late. Also, many developmentally regulated and tissue-specific genes are early replicating in the tissue of expression, but late replicating in all other cell types (Holmquist 1987). Although these experiments were originally performed on a small number of individual gene regions, recent genome-wide techniques have confirmed that this is, indeed, a general phenomenon (Farkash-Amar and Simon 2010).

Although there is, as yet, no direct proof that replication timing itself can influence gene expression, the fact that this mark is present from the earliest stages of development, as well as its regional nature, both suggest that it is set up independently of transcription and can be maintained in a stable manner. To gain some insight into the mechanisms involved in controlling replication timing, it is necessary to understand the overall scheme of this temporal epigenetic device. Replication is initiated at small DNA regions that are referred to as origins and, in general, DNA synthesis proceeds in a bidirectional manner that radiates from these sites. The determination of when in S phase each origin is fired appears to be under the control of nearby *cis*-acting sequences (Raghuraman et al. 1997; Ferguson et al. 1991; Ofir et al. 1999). In animal cells that have large genomes, the initiation of replication often occurs in a regional manner with multiple origins making up a cluster that is programmed to fire simultaneously in a coordinated fashion (Huberman and Riggs 1966, 1968), thus generating a large replication time zone (or band).

3.2 The Regulation of Replication Timing

The regulation of this replication-timing process must involve multiple molecular factors. Experiments in both yeast and animal cells indicate that the replication time of each zone is set up during early G_1 phase soon after nuclear reorganization (Raghuraman et al. 1997; Dimitrova and Gilbert 1999), and this is evidently accomplished by the recognition of *cis*-acting elements, which then inform each origin of when in S it should be fired. The human β-globin locus provides a good example for how this might work. The β-globin gene, together with its neighboring developmental paralogs, are imbedded in a gene region of ~100 kb that replicates late in almost all cell types, but becomes early replicating in erythroid lineage cells (Dhar et al. 1988). Both early and late replication are initiated by a single bidirectional origin located adjacent to the β-globin gene promoter (Kitsberg et al. 1993b; Aladjem et al. 1998).

Using a large bacterial artificial chromosome (BAC) transgene containing the full human β-globin domain, it has been shown that the locus control region (LCR) positioned upstream of the globin gene array plays a *cis*-acting role in mediating replication timing (Simon et al. 2001). When the full LCR is present, this entire gene region replicates late in most tissues, but early in erythroid tissues, regardless of its site of integration in the genome. Mutations in the regulatory unit, however, can severely affect this pattern (Forrester et al. 1990), thereby causing the transgene to lose its dominance and, thus, come under replication-timing control of surrounding endogenous sequences. These studies suggest that this multifunctional *cis*-acting domain affects both early and late time settings of the single origin sequence controlling this region. It is very likely that time zones, in general, are set up by long-range *cis*-acting sequences of this nature, which serve as recognition sites for cell-type-specific *trans*-acting factors.

Although little is known about the protein factors that participate in this process, a number of different studies have highlighted their existence. Fusion of lymphoid with erythroid cells, for example, causes the globin domain in this nonerythroid nucleus to undergo a dramatic switch from late to early replication, probably because of its exposure to a new set of factors that reset the timing mechanism (Forrester et al. 1990). Another striking example of the role of *trans*-acting factors is the reprogramming of replication timing that takes place during SCNT (somatic cell nuclear transfer). A number of specific gene regions have been shown to be early replicating in the early embryo, but become late replicating in all somatic cells following implantation (Hiratani et al. 2004; Perry et al. 2004). Interestingly, when a somatic cell nucleus is transplanted to fertilized oocyte cells, these same regions rapidly revert back to the early replication pattern, and this occurs almost immediately during the first division cycle. Other gene regions switch replication from early to late S phase following nuclear transplantation (Shufaro et al. 2010). These experiments clearly show that replication timing is dependent on the presence of local *trans*-acting factors, which apparently have a role in setting up the proper time slot for DNA synthesis during each passage through the cell cycle.

What is the mechanism by which these *cis*-acting control centers influence the firing of local origins? Very elegant studies in yeast were the first to show that the time of replication initiation may be influenced by histone acetylation at the origin sequence (Fig. 8A) (Vogelauer et al. 2002). When histones located at a late-replicating origin were forcibly acetylated, this had the effect of changing the firing time to an early slot. In a similar manner, removal

 Cite as *Cold Spring Harb Perspect Biol* doi: 10.1101/cshperspect.a019372

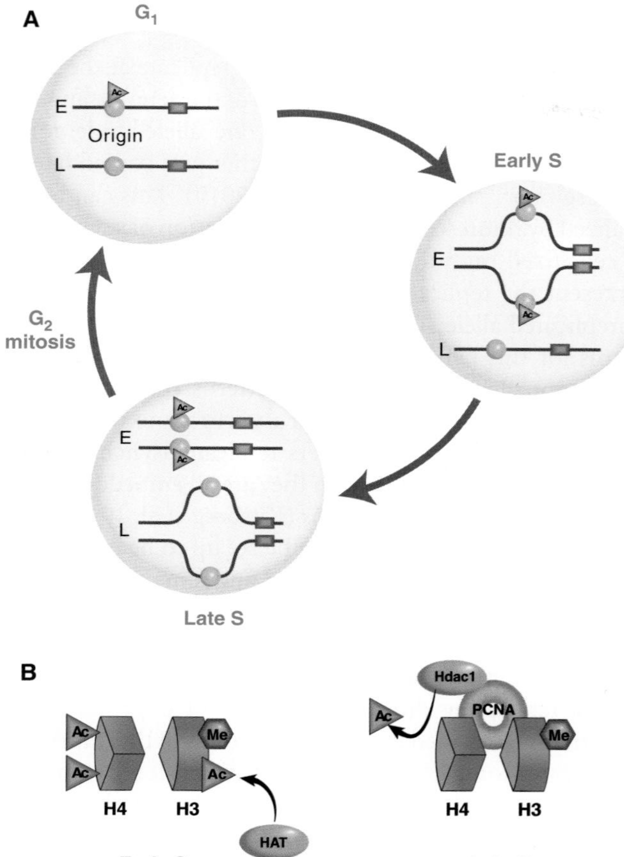

Figure 8. Regulation of replication timing. (*A*) The diagram shows two different DNA regions, one that replicates early (E) in S phase and one that replicates late (L). In G_1, histones packaging the early-replicating origin (orange circle) become acetylated because *trans*-acting factors in the cell recognize specific "early" *cis*-acting sequences (purple rectangle). In contrast, nucleosomes packaging the late-replicating origin are prevented from becoming acetylated by virtue of different *cis*-acting motifs (green). During early S phase, protein factors specific for this stage in the cell cycle recognize acetylated origins and initiate the process of replication. Unacetylated origins remain unreplicated. Finally, factors specific for late S phase recognize and initiate firing of unacetylated origins, bringing about replication of late-replicating DNA regions. (*B*) Replication origins that are marked with acetylated histones H3 and H4 replicate early in S phase, whereas the deacetylation of origin regions primes them for late replication in S phase.

of acetylation at an origin caused it to initiate replication late. More recent studies have shown that this same mechanism may also play a role in the control of replication timing in animal cells (Goren et al. 2008). The human globin domain origin, for example, is packaged with acetylated histones in erythroid cells in which it replicates early, whereas the exact same origin is deacetylated in nonerythroid cells. This pattern is recapitulated on a transgene containing the full human β-globin locus. Strikingly, tethering a histone acetylase to the origin region caused this domain to switch from late to early replication in nonerythroid cells of the mouse, and, in a parallel manner, forced deacetylation in vivo caused the globin domain to become abnormally late replicating in erythroid cells (Fig. 8B).

3.3 Cause or Effect?

Despite the clear-cut correlation between early replication timing and gene expression, little is known about the mechanistic connection between these two parameters. It does not appear that replication timing comes about as a direct result of local expression patterns because changes in expression do not automatically bring about replication-timing shifts (Cimbora et al. 2000; Simon et al. 2001; Farkash-Amar et al. 2008). Given that replication time zones are very large, often encompassing many gene sequences, it follows that replication timing is unlikely to be dictated by gene expression. In contrast, there does seem to be some evidence for the opposite idea that replication timing itself may influence histone modification and gene expression patterns.

An experiment testing the influence of replication timing on histone marks and gene expression involved the nuclear microinjection of reporter plasmids directly into early S-phase nuclei. Plasmids injected into early S-phase nuclei become packaged in an acetylated histone structure, whereas these same templates get assembled with deacetylated histones when introduced into cells during late S phase (Zhang et al. 2002). Furthermore, DNA that is already in a deacetylated chromatin structure becomes acetylated when it undergoes subsequent replication in early S, whereas templates carrying acetylated histones switch to a deacetylated state after replication in late S (Rountree et al. 2000; Lande-Diner et al. 2009). During the normal DNA replication cycle in vivo, preexisting histones on the original template DNA segregate randomly to both sister molecules, and new histones are then added to regenerate the original nucleosome-packing density (Lucchini and Sogo 1995). These new nucleosomes may be assembled in either an acetylated or a deacetylated form. The experiments described above strongly suggest that different regions of the DNA get repackaged, depending on their time of replication. Early-replicating DNA appears to incorporate nucleosomes that adopt an acetylated form, whereas late-replicating regions get repackaged with deacetylated nucleosomes, thus, transiently generating large chromosome domains that have a relatively uniform histone modification pattern (Fig. 8A).

Because these events take place through the action of histone modification and demodification enzymes that are closely associated with the dynamically progressing replication fork machinery, this does not constitute a reliable mechanism for maintaining these acetylation states throughout the cycle, and it is likely that this overall pattern may undergo further adjustments through the action of locally acting histone-marking enzymes recruited to specific sites on the DNA. It, thus, appears that zonal replication-timing control provides a crude system for setting up general background structure following DNA synthesis so that active chromosomal bands are assembled with acetylated histones, whereas more inactive zones are initially packaged with deacetylated histones. This mechanism would help make the task of postreplication reassembly much more efficient. According to this scheme, replication timing represents a type of ground-state maintenance mechanism that helps recreate the original chromatin structure following DNA synthesis.

Because the replication-banding pattern, as observed on mitotic chromosomes, appears to be extremely stable throughout development, it is likely that this epigenetic-marking system may have an autonomous maintenance system, which insures that the original timing pattern after each round of replication is restored. The histone acetylation pattern of nucleosomes at origin sequences may provide a simple system for such a maintenance scheme. As noted above, origins marked by histone acetylation fire early in S phase, whereas those that are packaged with deacetylated histones are instructed to skip the early times and only fire in late S phase (Fig. 8). Following activation of the origin, this same region then undergoes partial reassembly to fill in missing nucleosomes. Because of differences in the packaging machinery present at the replication fork in early as opposed to late S phase means that early origins will continue to be in an acetylated conformation, whereas those that fire late will once again acquire histones that are deacetylated. This clearly represents an elegant maintenance mechanism for autonomously preserving replication-timing patterns from one cell generation to the next (Fig. 8A).

Replication time slots are thought to be set up in the G_1 phase of each cycle (Raghuraman et al. 1997; Dimitrova and Gilbert 1999), probably through interactions of *trans*-acting factors with *cis*-acting control sequences that ultimately lead to the setting up of appropriate acetylation patterns over adjacent origins that are under their command (Vogelauer et al. 2002; Goren et al. 2008). Furthermore, it appears that changes in these factors can act very quickly to reset the timing pattern (Forrester et al. 1990; Shufaro et al. 2010). With this in mind, it may be best to view replication time maintenance as being governed by two separate parameters; the autonomous *cis*-acting acetylated system is backed up by a *trans*-acting mechanism that can, if necessary, reset the timing switches to their proper mode.

3.4 Asynchronous Replication Timing

Although most chromosome regions replicate at the same time on both parental alleles, >10% of the genome appears to replicate asynchronously with one allele undergoing DNA synthesis early, whereas the other gets copied late in S phase (Farkash-Amar and Simon 2010). This can be visualized by fluorescence in situ hybridization of such a gene in cells fixed midway through S phase in which two dots represent the replicated allele and one dot represents the unreplicated allele. There appear to be two different paradigms for this behavior, and both of them are associated with monoallelic gene expression patterns. Imprinted genes are examples of one paradigm for asynchronous replication; the alleles of such gene clusters replicate asynchronously in every somatic cell, although they are all embedded within large chromosomal domains (Kitsberg et al. 1993a). For example, a paternal allele could be set up to be early replicating during late spermatogenesis, and remains this way in the early preimplantation embryo and then throughout development (Simon et al. 1999). In a complementary manner, the maternal, often repressed, allele is set up to be late replicating during final stages of oogenesis. This pattern is maintained in all cells regardless of the imprinted gene expression patterns on the two alleles. It is commonly assumed that allele-specific methylation patterns in somatic cells are responsible for preserving the identity of each allele separately. However, parental identity can also be maintained in the absence of any DNA methylation marks, both during late gametogenesis (Perk et al. 2002; Reik and Walter 2001) and in the very early embryo (Birger et al. 1999). It is possible that asynchronous replication timing may play a role in this process, specifically during these stages of development.

Another paradigm for asynchronous replication occurs at a large number of chromosomal domains that are not imprinted. In these cases, one parental allele in each cell is early, whereas the other replicates late. In 50% of the cells, it is the maternal allele that replicates early, whereas, in other cells, it is the paternal allele that is copied first in S phase (Goren and Cedar 2003). These loci appear to initially replicate synchronously in the preimplantation embryo, but then become asynchronous at about the time of implantation when the embryo begins to undergo differentiation (Mostoslavsky et al. 2001). Once the decision is made as to which allele will be early, this pattern is then maintained in a clonal manner (Singh et al. 2003). This form of asynchronous replication is often associated with allelic choice. Antigen receptor arrays, like those at the κ, μ, or T-cell receptor (TCR)β loci, for example, are all positioned within asynchronous replicating domains (Mostoslavsky et al. 2001). During the process of antigen receptor selec-

 Cite as *Cold Spring Harb Perspect Biol* doi: 10.1101/cshperspect.a019372

tion, the primary rearrangement event always occurs preferentially on the early replicating allele, and this evidently serves as a first-line basis for allelic exclusion, a process that insures that only a single receptor is selected in each individual cell (see Ch. 29 [Busslinger and Tarakhovsky 2014] for more detail). Although there is no evidence showing that replication timing itself is directly involved in this selection process, there is no question that it represents an early marker for predicting allelic choice.

Olfactory receptors are also organized in a similar manner, with each cluster in the genome undergoing asynchronous replication, and here, too, there is a mechanism of allelic exclusion that restricts receptor choice to a single allele in each olfactory neuron (elaborated in Ch. 32 [Lomvardas and Maniatis 2014]). Because, in both of these systems, the activation of individual gene segments within each cluster is stochastic, it is allelic exclusion that actually facilitates the selection of only a single gene in each cell. In this sense, asynchronous replication appears to be part of the regulatory machinery that promotes gene diversity in the organism's ability to interact with its environment. Other examples of asynchronous replication and allelic exclusion include the recombinant DNA loci, which are only expressed from a single allele in each cell (Schlesinger et al. 2009) and, of course, the X chromosome in females, in which inactivation of one entire chromosome is associated with late replication timing (Ch. 25 [Brockdorff and Turner 2014]).

The finding that asynchronous replication timing at certain loci in the genome can be inherited in a clonal manner strongly suggests that there must be an autonomous mechanism for maintaining replication-timing patterns. This is especially poignant in light of the fact that each allele is capable of independently maintaining its differential timing profile for many cell generations. It should be noted that numerous asynchronous replication domains are located on each chromosome where they are dispersed among large intervening regions that replicate both alleles synchronously. Strikingly, on any individual chromosome, all of these loci appear to replicate in a coordinated manner (Singh et al. 2003; Ensminger and Chess 2004). On chromosome 6 in the mouse, for example, in cells that have the κ locus early replicating on one allele, the TCRβ locus on the same chromosome is also early replicating, and the same is true for additional asynchronous markers in *cis*. This suggests that asynchronous replication timing is actually controlled by central regulatory units that operate chromosome wide. Thus, any maintenance mechanism must be focused on these master regulatory sequences, and it is very possible that this mechanism may be based on other epigenetic markers, such as DNA methylation, which is known to have its own autonomous maintenance scheme.

4 CONCLUSIONS

Unravelling how epigenetic marks are inherited from one somatic generation to the next has been a key endeavor in epigenetic research for decades. We now know the many molecules involved in the process. The specifics, however, of how the inheritance of marks is orchestrated through the cell cycle still has many unanswered questions. This chapter first looks at how DNA methylation provides an autonomous mechanism for the propagation of a mark that is largely involved in gene repression. As research has delved more into investigating how the marks placed on histones in a chromatin context are inherited, we have discovered the involvement of histone chaperones, readers, nucleosome remodelers, and chromatin-modifying enzymes, which are inextricably linked with the DNA replication machinery. We have gained an understanding of how histone synthesis is cell-cycle-regulated, depending on the type of histone, in order to cater for the demand of new histones required in chromatin assembly. Their PTM before chromatin assembly may aid in assembling context appropriate chromatin. Concerning the dynamics of chromatin duplication, we present current models consistent with the available data on histone recycling or de novo introduction during the process of nucleosome disruption, transfer, and deposition at the replication fork. Likewise, the mechanisms of histone mark propagation on the chromatin template at other stages of the cell cycle are discussed. Last, our knowledge of how replication timing during S phase seems to present another layer in the epigenetic toolbox of mechanisms used to distinguish or even specify and ensure the propagation of chromatin states at the level of large genomic domains is discussed. Further research is still needed to understand how this intersects with other epigenetic marks in the maintenance or change of epigenomic specification through the cell cycle.

ACKNOWLEDGMENTS

This work was supported by la Ligue Nationale contre le Cancer (G.A.), the European Commission Network of Excellence EpiGeneSys (G.A.), the European Research Council (H.C. and G.A.), the Agence Nationalc de la Recherche (ANR) (G.A.), the Israel Science Foundation (H.C.), Lew Sanders (H.C.), and Norton Herrick (H.C.).

REFERENCES

**Reference is also in this book.*

Aagaard L, Laible G, Selenko P, Schmid M, Dorn R, Schotta G, Kuhfittig S, Wolf A, Lebersorger A, Singh PB, et al. 1999. Functional mammalian homologues of the *Drosophila* PEV-modifier Su(var)3-9 encode centromere-associated proteins which complex with the heterochromatin component M31. *EMBO J* **18:** 1923–1938.

Agalioti T, Chen G, Thanos D. 2002. Deciphering the transcriptional histone acetylation code for a human gene. *Cell* **111:** 381–392.

Ahmad K, Henikoff S. 2002. The histone variant H3.3 marks active chromatin by replication-independent nucleosome assembly. *Mol Cell* **9:** 1191–1200.

Aladjem MI, Rodewald LW, Kolman JL, Wahl GM. 1998. Genetic dissection of a mammalian replicator in the human β-globin locus. *Science* **281:** 1005–1009.

* Allshire RC, Ekwall K. 2014. Epigenetic regulation of chromatin states in *Schizosaccharomyces pombe*. *Cold Spring Harb Perspect Biol* doi: 10.1101/cshperspect.a018770.

Annunziato AT. 2005. Split decision: What happens to nucleosomes during DNA replication? *J Biol Chem* **280:** 12065–12068.

Bannister AJ, Zegerman P, Partridge JF, Miska EA, Thomas JO, Allshire RC, Kouzarides T. 2001. Selective recognition of methylated lysine 9 on histone H3 by the HP1 chromo domain. *Nature* **410:** 120–124.

* Barlow DP, Bartolomei MS. 2014. Genomic imprinting in mammals. *Cold Spring Harb Perspect Biol* **6:** a018382.

* Baylin SB, Jones PA. 2014. Epigenetic determinants of cancer. *Cold Spring Harb Perspect Biol* doi: 10.1101/cshperspect.a019505.

Baxevanis AD, Godfrey JE, Moudrianakis EN. 1991. Associative behavior of the histone (H3-H4)2 tetramer: Dependence on ionic environment. *Biochemistry* **30:** 8817–8823.

* Becker PB, Workman JL. 2013. Nucleosome remodeling and epigenetics. *Cold Spring Harb Perspect Biol* **5:** a017905.

Bhaskara S, Chyla BJ, Amann JM, Knutson SK, Cortez D, Sun ZW, Hiebert SW. 2008. Deletion of histone deacetylase 3 reveals critical roles in S phase progression and DNA damage control. *Mol Cell* **30:** 61–72.

Birger Y, Shemer R, Perk J, Razin A. 1999. The imprinting box of the mouse Igf2r gene. *Nature* **397:** 84–88.

Brandeis M, Frank D, Keshet I, Siegfried Z, Mendelsohn M, Nemes A, Temper V, Razin A, Cedar H. 1994. Sp1 elements protect a CpG island from de novo methylation. *Nature* **371:** 435–438.

* Brockdorff N, Turner BM. 2014. Dosage compensation in mammals. *Cold Spring Harb Perspect Biol* doi: 10.1101/cshperspect.a019406.

* Busslinger M, Tarakhovsky A. 2014. Epigenetic control of immunity. *Cold Spring Harb Perspect Biol* **6:** a019307.

Chen ES, Zhang K, Nicolas E, Cam HP, Zofall M, Grewal SI. 2008. Cell cycle control of centromeric repeat transcription and heterochromatin assembly. *Nature* **451:** 734–737.

* Cheng X. 2014. Structural and functional coordination of DNA and histone methylation. *Cold Spring Harb Perspect Biol* **6:** a018747.

Cimbora DM, Schubeler D, Reik A, Hamilton J, Francastel C, Epner EM, Groudine M. 2000. Long-distance control of origin choice and replication timing in the human β-globin locus are independent of the locus control region. *Mol Cell Biol* **20:** 5581–5591.

Cloos PA, Christensen J, Agger K, Helin K. 2008. Erasing the methyl mark: Histone demethylases at the center of cellular differentiation and disease. *Genes Dev* **22:** 1115–1140.

Corpet A, Almouzni G. 2009. Making copies of chromatin: The challenge of nucleosomal organization and epigenetic information. *Trends Cell Biol* **19:** 29–41.

Dalal Y, Wang H, Lindsay S, Henikoff S. 2007. Tetrameric structure of centromeric nucleosomes in interphase *Drosophila* cells. *PLoS Biol* **5:** e218.

Das C, Lucia MS, Hansen KC, Tyler JK. 2009. CBP/p300-mediated acetylation of histone H3 on lysine 56. *Nature* **459:** 113–117.

Dhar V, Mager D, Iqbal A, Schildkraut CL. 1988. The coordinate replication of the human β-globin gene domain reflects its transcriptional activity and nuclease hypersensitivity. *Mol Cell Biol* **8:** 4958–4965.

Dimitrova DS, Gilbert DM. 1999. The spatial position and replication timing of chromosomal domains are both established in early G_1 phase. *Mol Cell* **4:** 983–993.

Dunleavy EM, Roche D, Tagami H, Lacoste N, Ray-Gallet D, Nakamura Y, Daigo Y, Nakatani Y, Almouzni-Pettinotti G. 2009. HJURP is a cell-cycle-dependent maintenance and deposition factor of CENP-A at centromeres. *Cell* **137:** 485–497.

Eden S, Hashimshony T, Keshet I, Thorne AW, Cedar H. 1998. DNA methylation models histone acetylation. *Nature* **394:** 842–843.

Ekwall K, Olsson T, Turner BM, Cranston G, Allshire RC. 1997. Transient inhibition of histone deacetylation alters the structural and functional imprint at fission yeast centromeres. *Cell* **91:** 1021–1032.

English CM, Adkins MW, Carson JJ, Churchill ME, Tyler JK. 2006. Structural basis for the histone chaperone activity of Asf1. *Cell* **127:** 495–508.

Ensminger AW, Chess A. 2004. Coordinated replication timing of monoallelically expressed genes along human autosomes. *Hum Mol Genet* **13:** 651–658.

Farkash-Amar S, Simon I. 2010. Genome-wide analysis of the replication program in mammals. *Chromosome Res* **18:** 115–125.

Farkash-Amar S, Lipson D, Polten A, Goren A, Helmstetter C, Yakhini Z, Simon I. 2008. Global organization of replication time zones of the mouse genome. *Genome Res* **18:** 1562–1570.

Feldman N, Gerson A, Fang J, Li E, Zhang Y, Shinkai Y, Cedar H, Bergman Y. 2006. G9a-mediated irreversible epigenetic inactivation of Oct-3/4 during early embryogenesis. *Nat Cell Biol* **8:** 188–194.

Ferguson BM, Brewer BJ, Reynolds AE, Fangman WL. 1991. A yeast origin of replication is activated late in S phase. *Cell* **65:** 507–515.

Finn RM, Browne K, Hodgson KC, Ausio J. 2008. sNASP, a histone H1-specific eukaryotic chaperone dimer that facilitates chromatin assembly. *Biophys J* **95:** 1314–1325.

Foltz DR, Jansen LE, Bailey AO, Yates JR 3rd, Bassett EA, Wood S, Black BE, Cleveland DW. 2009. Centromere-specific assembly of CENP-a nucleosomes is mediated by HJURP. *Cell* **137:** 472–484.

Forrester WC, Epner E, Driscoll MC, Enver T, Brice M, Papayannopoulou T, Groudine M. 1990. A deletion of the human β-globin locus activation region causes a major alteration in chromatin structure and replication across the entire β-globin locus. *Genes Dev* **4:** 1637–1649.

Francis NJ, Follmer NE, Simon MD, Aghia G, Butler JD. 2009. Polycomb proteins remain bound to chromatin and DNA during DNA replication in vitro. *Cell* **137:** 110–122.

Fritsch L, Robin P, Mathieu JR, Souidi M, Hinaux H, Rougeulle C, Harel-Bellan A, Ameyar-Zazoua M, Ait-Si-Ali S. 2010. A subset of the histone H3 lysine 9 methyltransferases Suv39h1, G9a, GLP, and SETDB1 participate in a multimeric complex. *Mol Cell* **37:** 46–56.

Fujita N, Watanabe S, Ichimura T, Tsuruzoe S, Shinkai Y, Tachibana M, Chiba T, Nakao M. 2003. Methyl-CpG binding domain 1 (MBD1) interacts with the Suv39h1-HP1 heterochromatic complex for DNA methylation-based transcriptional repression. *J Biol Chem* **278:** 24132–24138.

Fuks F, Burgers WA, Brehm A, Hughes-Davies L, Kouzarides T. 2000. DNA methyltransferase Dnmt1 associates with histone deacetylase activity. *Nat Genet* **24:** 88–91.

Garcia BA, Hake SB, Diaz RL, Kauer M, Morris SA, Recht J, Shabanowitz J, Mishra N, Strahl BD, Allis CD, et al. 2007. Organismal differences in post-translational modifications in histones H3 and H4. *J Biol Chem* **282:** 7641–7655.

Gasser R, Koller T, Sogo JM. 1996. The stability of nucleosomes at the replication fork. *J Mol Biol* **258:** 224–239.

Gerard A, Koundrioukoff S, Ramillon V, Sergere JC, Mailand N, Quivy JP, Almouzni G. 2006. The replication kinase Cdc7-Dbf4 promotes the interaction of the p150 subunit of chromatin assembly factor 1 with proliferating cell nuclear antigen. *EMBO Rep* **7:** 817–823.

Goldberg AD, Banaszynski LA, Noh KM, Lewis PW, Elsaesser SJ, Stadler S, Dewell S, Law M, Guo X, Wen D, et al. 2010. Distinct factors control histone variant H3.3 localization at specific genomic regions. *Cell* **140:** 678–691.

Goren A, Cedar H. 2003. Replicating by the clock. *Nat Rev Mol Cell Biol* **4:** 25–32.

Goren A, Simchen G, Fibach E, Szabo PE, Tanimoto K, Chakalova L, Pfeifer GP, Fraser PJ, Engel JD, Cedar H. 2006. Fine tuning of globin gene expression by DNA methylation. *PLoS ONE* **1:** e46.

Goren A, Tabib A, Hecht M, Cedar H. 2008. DNA replication timing of the human β-globin domain is controlled by histone modification at the origin. *Genes Dev* **22:** 1319–1324.

Groth A, Ray-Gallet D, Quivy JP, Lukas J, Bartek J, Almouzni G. 2005. Human Asf1 regulates the flow of S phase histones during replicational stress. *Mol Cell* **17:** 301–311.

Groth A, Corpet A, Cook AJ, Roche D, Bartek J, Lukas J, Almouzni G. 2007a. Regulation of replication fork progression through histone supply and demand. *Science* **318:** 1928–1931.

Groth A, Rocha W, Verreault A, Almouzni G. 2007b. Chromatin challenges during DNA replication and repair. *Cell* **128:** 721–733.

Gruenbaum Y, Naveh-Many T, Cedar H, Razin A. 1981. Sequence specificity of methylation in higher plant DNA. *Nature* **292:** 860–862.

Gruenbaum Y, Cedar H, Razin A. 1982. Substrate and sequence specificity of a eukaryotic DNA methylase. *Nature* **295:** 620–622.

Gunjan A, Verreault A. 2003. A Rad53 kinase-dependent surveillance mechanism that regulates histone protein levels in *S. cerevisiae*. *Cell* **115:** 537–549.

Gunjan A, Paik J, Verreault A. 2005. Regulation of histone synthesis and nucleosome assembly. *Biochimie* **87:** 625–635.

Gurard-Levin ZA, Quivy JP, Almouzni G. 2014. Histone chaperones: Assisting histone traffic and nucleosome dynamics. *Annu Rev Biochem* **83:** 487–517.

Hake SB, Garcia BA, Duncan EM, Kauer M, Dellaire G, Shabanowitz J, Bazett-Jones DP, Allis CD, Hunt DF. 2006. Expression patterns and post-translational modifications associated with mammalian histone H3 variants. *J Biol Chem* **281:** 559–568.

Hand R. 1978. Eucaryotic DNA: Organization of the genome for replication. *Cell* **15:** 317–325.

Hansen KH, Helin K. 2009. Epigenetic inheritance through self-recruitment of the polycomb repressive complex 2. *Epigenetics* **4:** 133–138.

Hashimshony T, Zhang J, Keshet I, Bustin M, Cedar H. 2003. The role of DNA methylation in setting up chromatin structure during development. *Nat Genet* **34:** 187–192.

* Henikoff S, Smith MM. 2014. Histone variants and epigenetics. *Cold Spring Harb Perspect Biol* doi: 10.1101/cshperspect.a019364.

Hiratani I, Leskovar A, Gilbert DM. 2004. Differentiation-induced replication-timing changes are restricted to AT-rich/long interspersed nuclear element (LINE)-rich isochores. *Proc Natl Acad Sci* **101:** 16861–16866.

* Hochedlinger K, Jaenisch R. 2014. Induced pluripotency and epigenetic reprogramming. *Cold Spring Harb Perspect Biol* doi: 10.1101/cshperspect.a019448.

Hoek M, Stillman B. 2003. Chromatin assembly factor 1 is essential and couples chromatin assembly to DNA replication in vivo. *Proc Natl Acad Sci* **100:** 12183–12188.

Holmquist GP. 1987. Role of replication time in the control of tissue specific gene expression. *Am J Hum Genet* **40:** 151–173.

Holmquist G. 1988. DNA sequences in G-bands and R-bands. In *Chromosomes and chromatin* (ed. Adolph KW), pp. 75–121. CRC Press, Boca Raton, FL.

Houlard M, Berlivet S, Probst AV, Quivy JP, Hery P, Almouzni G, Gerard M. 2006. CAF-1 is essential for heterochromatin organization in pluripotent embryonic cells. *PLoS Genet* **2:** e181.

Huberman JA, Riggs AD. 1966. Autoradiography of chromosomal DNA fibers from Chinese hamster cells. *Proc Natl Acad Sci* **96:** 11434–11439.

Huberman JA, Riggs AD. 1968. On the mechanism of DNA replication in mammalian chromosomes. *J Mol Biol* **32:** 327–341.

Huen MS, Sy SM, van Deursen JM, Chen J. 2008. Direct interaction between SET8 and proliferating cell nuclear antigen couples H4-K20 methylation with DNA replication. *J Biol Chem* **283:** 11073–11077.

Ishimi Y, Komamura Y, You Z, Kimura H. 1998. Biochemical function of mouse minichromosome maintenance 2 protein. *J Biol Chem* **273:** 8369–8375.

Jansen LE, Black BE, Foltz DR, Cleveland DW. 2007. Propagation of centromeric chromatin requires exit from mitosis. *J Cell Biol* **176:** 795–805.

Jasencakova Z, Scharf AND, Ask K, Corpet A, Imhof A, Almouzni G, Groth A. 2010. Replication stress interferes with histone recycling and predisposition marking of new histones. *Mol Cell* **37:** 736–743.

Jones PL, Veenstra GJC, Wade PA, Vermaak D, Kass SU, Landsberg N, Strouboulis J, Wolffe AP. 1998. Methylated DNA and MeCP2 recruit histone deacetylase to repress transcription. *Nat Genet* **19:** 187–191.

Jorgensen S, Elvers I, Trelle MB, Menzel T, Eskildsen M, Jensen ON, Helleday T, Helin K, Sorensen CS. 2007. The histone methyltransferase SET8 is required for S-phase progression. *J Cell Biol* **179:** 1337–1345.

Kaufman PD, Kobayashi R, Kessler N, Stillman B. 1995. The p150 and p60 subunits of chromatin assembly factor I: A molecular link between newly synthesized histones and DNA replication. *Cell* **81:** 1105–1114.

Kimura H, Cook PR. 2001. Kinetics of core histones in living human cells: Little exchange of H3 and H4 and some rapid exchange of H2B. *J Cell Biol* **153:** 1341–1353.

Kitsberg D, Selig S, Brandeis M, Simon I, Keshet I, Driscoll DJ, Nicholls RD, Cedar H. 1993a. Allele-specific replication timing of imprinted gene regions. *Nature* **364:** 459–463.

Kitsberg D, Selig S, Keshet I, Cedar H. 1993b. Replication structure of the human β- globin gene domain. *Nature* **366:** 588–590.

Klapholz B, Dietrich BH, Schaffner C, Heredia F, Quivy JP, Almouzni G, Dostatni N. 2009. CAF-1 is required for efficient replication of euchromatic DNA in *Drosophila* larval endocycling cells. *Chromosoma* **118:** 235–248.

Kloc A, Martienssen R. 2008. RNAi, heterochromatin and the cell cycle. *Trends Genet* **24:** 511–517.

Krogan NJ, Cagney G, Zhong G, Guo X, Ignatchenko A, Datta N, Tikuisis AP, Punna T, Peregrin-Alvarez JM, Shales M, et al. 2006. Global landscape of protein complexes in the yeast *Saccharomyces cerevisiae*. *Nature* **440:** 637–643.

Lachner M, O'Carroll D, Rea S, Mechtler K, Jenuwein T. 2001. Methylation of histone H3 lysine 9 creates a binding site for HP1 proteins. *Nature* **410:** 116–120.

Lande-Diner L, Zhang J, Ben-Porath I, Amariglio N, Keshet I, Hecht M, Azuara V, Fisher AG, Rechavi G, Cedar H. 2007. Role of DNA methylation in stable gene repression. *J Biol Chem* **282:** 12194–12200.

Lande-Diner L, Zhang J, Cedar H. 2009. Shifts in replication timing actively affect histone acetylation during nucleosome reassembly. *Mol Cell* **34:** 767–774.

Laskey RA, Mills AD, Morris NR. 1977. Assembly of SV40 chromatin in a cell-free system from *Xenopus* eggs. *Cell* **10:** 237–243.

Laurent L, Wong E, Li G, Huynh T, Tsirigos A, Ong CT, Low HM, Kin Sung KW, Rigoutsos I, Loring J, et al. 2010. Dynamic changes in the human methylome during differentiation. *Genome Res* **20:** 320–331.

Le S, Davis C, Konopka JB, Sternglanz R. 1997. Two new S-phase-specific genes from *Saccharomyces cerevisiae*. *Yeast* **13:** 1029–1042.

* Li E, Zhang Y. 2014. DNA methylation in mammals. *Cold Spring Harb Perspect Biol* **6:** a019133.

Li E, Bestor TH, Jaenisch R. 1992. Targeted mutation of the DNA methyltransferase gene results in embryonic lethality. *Cell* **69:** 915–926.

Li E, Beard C, Jaenisch R. 1993. Role for DNA methylation in genomic imprinting. *Nature* **366:** 362–365.

* Lomvardas S, Maniatis T. 2014. Histone and DNA modifications as regulators of neuronal development and function. *Cold Spring Harb Perspect Biol* doi: 10.1101/cshperspect.a019489.

Loppin B, Bonnefoy E, Anselme C, Laurencon A, Karr TL, Couble P. 2005. The histone H3.3 chaperone HIRA is essential for chromatin assembly in the male pronucleus. *Nature* **437:** 1386–1390.

Loyola A, Almouzni G. 2007. Marking histone H3 variants: How, when and why? *Trends Biochem Sci* **32:** 425–433.

Loyola A, Bonaldi T, Roche D, Imhof A, Almouzni G. 2006. PTMs on H3 variants before chromatin assembly potentiate their final epigenetic state. *Mol Cell* **24:** 309–316.

Loyola A, Tagami H, Bonaldi T, Roche D, Quivy JP, Imhof A, Nakatani Y, Dent SY, Almouzni G. 2009. The HP1α-CAF1-SetDB1-containing complex provides H3K9me1 for Suv39-mediated K9me3 in pericentric heterochromatin. *EMBO Rep* **10:** 769–775.

Lucchini R, Sogo JM. 1995. Replication of transcriptionally active chromatin. *Nature* **374:** 276–280.

MacAlpine DM, Almouzni G. 2013. Chromatin and DNA replication. *Cold Spring Harb Perspect Biol* **5:** a010207.

Malay AD, Umehara T, Matsubara-Malay K, Padmanabhan B, Yokoyama S. 2008. Crystal structures of fission yeast histone chaperone Asf1 complexed with the Hip1 B-domain or the Cac2 C terminus. *J Biol Chem* **283:** 14022–14031.

Margueron R, Justin N, Ohno K, Sharpe ML, Son J, Drury WJ, Voigt P, Martin SR, Taylor WR, De Marco V, et al. 2009. Role of the polycomb protein EED in the propagation of repressive histone marks. *Nature* **461:** 762–767.

Marzluff WF, Wagner EJ, Duronio RJ. 2008. Metabolism and regulation of canonical histone mRNAs: Life without a poly(A) tail. *Nat Rev Genet* **9:** 843–854.

Maze I, Noh KM, Soshnev AA, Allis CD. 2014. Every amino acid matters: Essential contributions of histone variants to mammalian development and disease. *Nat Rev Genet* **15:** 259–271.

McKittrick E, Gafken PR, Ahmad K, Henikoff S. 2004. Histone H3.3 is enriched in covalent modifications associated with active chromatin. *Proc Natl Acad Sci* **101:** 1525–1530.

Meeks-Wagner D, Hartwell LH. 1986. Normal stoichiometry of histone dimer sets is necessary for high fidelity of mitotic chromosome transmission. *Cell* **44:** 43–52.

Mello JA, Sillje HH, Roche DM, Kirschner DB, Nigg EA, Almouzni G. 2002. Human Asf1 and CAF-1 interact and synergize in a repair-coupled nucleosome assembly pathway. *EMBO Rep* **3:** 329–334.

Milutinovic S, Zhuang Q, Szyf M. 2002. Proliferating cell nuclear antigen associates with histone deacetylase activity, integrating DNA replication and chromatin modification. *J Biol Chem* **277:** 20974–20978.

Moggs JG, Grandi P, Quivy JP, Jonsson ZO, Hubscher U, Becker PB, Almouzni G. 2000. A CAF-1-PCNA-mediated chromatin assembly pathway triggered by sensing DNA damage. *Mol Cell Biol* **20:** 1206–1218.

Monk M, Boubelik M, Lehnert S. 1987. Temporal and regional changes in DNA methylation in the embryonic, extraembryonic and germ cell lineages during mouse embryo development. *Development* **99:** 371–382.

Mostoslavsky R, Singh N, Tenzen T, Goldmit M, Gabay C, Elizur S, Qi P, Reubinoff BE, Chess A, Cedar H, et al. 2001. Asynchronous replication and allelic exclusion in the immune system. *Nature* **414:** 221–225.

Muramoto T, Muller I, Thomas G, Melvin A, Chubb JR. 2010. Methylation of H3K4 is required for inheritance of active transcriptional states. *Curr Biol* **20:** 397–406.

Murzina N, Verreault A, Laue E, Stillman B. 1999. Heterochromatin dynamics in mouse cells: Interaction between chromatin assembly factor 1 and HP1 proteins. *Mol Cell* **4:** 529–540.

Nakatani Y, Ray-Gallet D, Quivy JP, Tagami H, Almouzni G. 2004. Two distinct nucleosome assembly pathways: Dependent or independent of DNA synthesis promoted by histone H3.1 and H3.3 complexes. *Cold Spring Harb Symp Quant Biol* **69:** 273–280.

Nan X, Ng HH, Johnson CA, Laherty CD, Turner BM, Eisenman RN, Bird A. 1998. Transcriptional repression by the methyl-CpG-binding protein MeCP2 involves a histone deacetylase complex. *Nature* **393:** 386–389.

Natsume R, Eitoku M, Akai Y, Sano N, Horikoshi M, Senda T. 2007. Structure and function of the histone chaperone CIA/ASF1 complexed with histones H3 and H4. *Nature* **446:** 338–341.

Ng RK, Gurdon JB. 2008. Epigenetic memory of an active gene state depends on histone H3.3 incorporation into chromatin in the absence of transcription. *Nat Cell Biol* **10:** 102–109.

Oda H, Okamoto I, Murphy N, Chu J, Price SM, Shen MM, Torres-Padilla ME, Heard E, Reinberg D. 2009. Monomethylation of histone H4-lysine 20 is involved in chromosome structure and stability and is essential for mouse development. *Mol Cell Biol* **29:** 2278–2295.

Ofir R, Wong AC, McDermid HE, Skorecki KL, Selig S. 1999. Position effect of human telomeric repeats on replication timing. *Proc Natl Acad Sci* **96:** 11434–11439.

Osakabe A, Tachiwana H, Matsunaga T, Shiga T, Nozawa RS, Obuse C, Kurumizaka H. 2010. Nucleosome formation activity of human somatic nuclear autoantigenic sperm protein (sNASP). *J Biol Chem* **285:** 11913–11921.

Papait R, Pistore C, Negri D, Pecoraro D, Cantarini L, Bonapace IM. 2007. Np95 is implicated in pericentromeric heterochromatin replication and in major satellite silencing. *Mol Biol Cell* **18:** 1098–1106.

Parthun MR. 2007. Hat1: The emerging cellular roles of a type B histone acetyltransferase. *Oncogene* **26:** 5319–5328.

Perk J, Makedonski K, Lande L, Cedar H, Razin A, Shemer R. 2002. The imprinting mechanism of the Prader-Willi/Angelman regional control center. *EMBO J* **21:** 5807–5814.

Perry P, Sauer S, Billon N, Richardson WD, Spivakov M, Warnes G, Livesey FJ, Merkenschlager M, Fisher AG, Azuara V. 2004. A dynamic switch in the replication timing of key regulator genes in embryonic stem cells upon neural induction. *Cell Cycle* **3:** 1645–1650.

* Pikaard CS, Mittelsten Scheid O. 2014. Epigenetic regulation in plants. *Cold Spring Harb Perspect Biol* doi: 10.1101/cshperspect.a019315.

Pollack Y, Stein R, Razin A, Cedar H. 1980. Methylation of foreign DNA sequences in eukaryotic cells. *Proc Natl Acad Sci* **77:** 6463–6467.

Polo SE, Roche D, Almouzni G. 2006. New histone incorporation marks sites of UV repair in human cells. *Cell* **127:** 481–493.

Pradhan S, Bacolla A, Wells RD, Roberts RJ. 1999. Recombinant human DNA (cytosine-5) methyltransferase. I. Expression, purification, and comparison of de novo and maintenance methylation. *J Biol Chem* **274:** 33002–33010.

Probst AV, Dunleavy E, Almouzni G. 2009. Epigenetic inheritance during the cell cycle. *Nat Rev Mol Cell Biol* **10:** 192–206.

Quivy JP, Grandi P, Almouzni G. 2001. Dimerization of the largest subunit of chromatin assembly factor 1: Importance in vitro and during *Xenopus* early development. *EMBO J* **20:** 2015–2027.

Quivy JP, Roche D, Kirschner D, Tagami H, Nakatani Y, Almouzni G. 2004. A CAF-1 dependent pool of HP1 during heterochromatin duplication. *EMBO J* **23:** 3516–3526.

Quivy JP, Gerard A, Cook AJ, Roche D, Almouzni G. 2008. The HP1-p150/CAF-1 interaction is required for pericentric heterochromatin replication and S-phase progression in mouse cells. *Nat Struct Mol Biol* **15:** 972–979.

Raghuraman MK, Brewer BJ, Fangman WL. 1997. Cell cycle-dependent establishment of a late replication program. *Science* **276:** 806–809.

Ramsperger U, Stahl H. 1995. Unwinding of chromatin by the SV40 large T antigen DNA helicase. *EMBO J* **14:** 3215–3225.

Ray-Gallet D, Quivy JP, Sillje HW, Nigg EA, Almouzni G. 2007. The histone chaperone Asf1 is dispensable for direct de novo histone deposition in *Xenopus* egg extracts. *Chromosoma* **19:** 19.

Reese BE, Bachman KE, Baylin SB, Rountree MR. 2003. The methyl-CpG binding protein MBD1 interacts with the p150 subunit of chromatin assembly factor 1. *Mol Cell Biol* **23:** 3226–3236.

Reik W, Walter J. 2001. Genomic imprinting: Parental influence on the genome. *Nat Rev Genet* **2:** 21–32.

Robertson KD, Ait-Si-Ali S, Yokochi T, Wade PA, Jones PL, Wolffe AP. 2000. DNMT1 forms a complex with Rb, E2F1 and HDAC1 and represses transcription from E2F-responsive promoters. *Nat Genet* **25:** 338–342.

Rountree MR, Bachman KE, Baylin SB. 2000. DNMT1 binds HDAC2 and a new corepressor, DMAP1, to form a complex at replication foci. *Nat Genet* **25:** 269–277.

Sanematsu F, Takami Y, Barman HK, Fukagawa T, Ono T, Shibahara K, Nakayama T. 2006. Asf1 is required for viability and chromatin assembly during DNA replication in vertebrate cells. *J Biol Chem* **281:** 13817–13827.

Sarraf SA, Stancheva I. 2004. Methyl-CpG binding protein MBD1 couples histone H3 methylation at lysine 9 by SETDB1 to DNA replication and chromatin assembly. *Mol Cell* **15:** 595–605.

Scharf AN, Barth TK, Imhof A. 2009a. Establishment of histone modifications after chromatin assembly. *Nucleic Acids Res* **37:** 5032–5040.

Scharf AN, Meier K, Seitz V, Kremmer E, Brehm A, Imhof A. 2009b. Monomethylation of lysine 20 on histone H4 facilitates chromatin maturation. *Mol Cell Biol* **29:** 57–67.

Schlesinger S, Selig S, Bergman Y, Cedar H. 2009. Allelic inactivation of rDNA loci. *Genes Dev* **23:** 2437–2447.

Schuh M, Lehner CF, Heidmann S. 2007. Incorporation of *Drosophila* CID/CENP-A and CENP-C into centromeres during early embryonic anaphase. *Curr Biol* **17:** 237–243.

Shelby RD, Monier K, Sullivan KF. 2000. Chromatin assembly at kinetochores is uncoupled from DNA replication. *J Cell Biol* **151:** 1113–1118.

Shibahara K, Stillman B. 1999. Replication-dependent marking of DNA by PCNA facilitates CAF-1-coupled inheritance of chromatin. *Cell* **96:** 575–585.

Shibahara K, Verreault A, Stillman B. 2000. The N-terminal domains of histones H3 and H4 are not necessary for chromatin assembly factor-1- mediated nucleosome assembly onto replicated DNA in vitro. *Proc Natl Acad Sci* **97:** 7766–7771.

Shufaro Y, Lacham-Kaplan O, Tzuberi BZ, McLaughlin J, Trounson A, Cedar H, Reubinoff BE. 2010. Reprogramming of DNA replication timing. *Stem Cells* **28:** 443–449.

Siegfried Z, Eden S, Mendelsohn M, Feng X, Tzubari B, Cedar H. 1999. DNA methylation represses transcription in vivo. *Nat Genet* **22:** 203–206.

Simon I, Tenzen T, Reubinoff BE, Hillman D, McCarrey JR, Cedar H. 1999. Asynchronous replication of imprinted genes is established in the gametes and maintained during development. *Nature* **401:** 929–932.

Simon I, Tenzen T, Mostoslavsky R, Fibach E, Lande L, Milot E, Gribnau J, Grosveld F, Fraser P, Cedar H. 2001. Developmental regulation of DNA replication timing at the human β globin locus. *EMBO J* **20:** 6150–6157.

Singh N, Ebrahimi FA, Gimelbrant AA, Ensminger AW, Tackett MR, Qi P, Gribnau J, Chess A. 2003. Coordination of the random asynchronous replication of autosomal loci. *Nat Genet* **33:** 339–341.

Smith S, Stillman B. 1989. Purification and characterization of CAF-I, a human cell factor required for chromatin assembly during DNA replication in vitro. *Cell* **58:** 15–25.

Sobel RE, Cook RG, Perry CA, Annunziato AT, Allis CD. 1995. Conservation of deposition-related acetylation sites in newly synthesized histones H3 and H4. *Proc Natl Acad Sci* **92:** 1237–1241.

Sogo JM, Stahl H, Koller T, Knippers R. 1986. Structure of replicating simian virus 40 minichromosomes. The replication fork, core histone segregation and terminal structures. *J Mol Biol* **189:** 189–204.

Song Y, He F, Xie G, Guo X, Xu Y, Chen Y, Liang X, Stagljar I, Egli D, Ma J, et al. 2007. CAF-1 is essential for *Drosophila* development and involved in the maintenance of epigenetic memory. *Dev Biol* **311:** 213–222.

Stillman B. 1986. Chromatin assembly during SV40 DNA replication in vitro. *Cell* **45:** 555–565.

Straussman R, Nejman D, Roberts D, Steinfeld I, Blum B, Benvenisty N, Simon I, Yakhini Z, Cedar H. 2009. Developmental programming of CpG island methylation profiles in the human genome. *Nat Struct Mol Biol* **16:** 564–571.

Taddei A, Roche D, Sibarita JB, Turner BM, Almouzni G. 1999. Duplication and maintenance of heterochromatin domains. *J Cell Biol* **147:** 1153–1166.

Taddei A, Maison C, Roche D, Almouzni G. 2001. Reversible disruption of pericentric heterochromatin and centromere function by inhibiting deacetylases. *Nat Cell Biol* **3:** 114–120.

Tagami H, Ray-Gallet D, Almouzni G, Nakatani Y. 2004. Histone H3.1 and H3.3 complexes mediate nucleosome assembly pathways dependent or independent of DNA synthesis. *Cell* **116:** 51–61.

Tang Y, Poustovoitov MV, Zhao K, Garfinkel M, Canutescu A, Dunbrack R, Adams PD, Marmorstein R. 2006. Structure of a human ASF1a-HIRA complex and insights into specificity of histone chaperone complex assembly. *Nat Struct Mol Biol* **13:** 921–929.

Tardat M, Murr R, Herceg Z, Sardet C, Julien E. 2007. PR-Set7-dependent lysine methylation ensures genome replication and stability through S phase. *J Cell Biol* **179:** 1413–1426.

Tyler JK, Adams CR, Chen SR, Kobayashi R, Kamakaka RT, Kadonaga JT. 1999. The RCAF complex mediates chromatin assembly during DNA replication and repair. *Nature* **402:** 555–560.

Tyler JK, Collins KA, Prasad-Sinha J, Amiott E, Bulger M, Harte PJ, Kobayashi R, Kadonaga JT. 2001. Interaction between the *Drosophila* CAF-1 and ASF1 chromatin assembly factors. *Mol Cell Biol* **21:** 6574–6584.

Unoki M, Nishidate T, Nakamura Y. 2004. ICBP90, an E2F-1 target, recruits HDAC1 and binds to methyl-CpG through its SRA domain. *Oncogene* **23:** 7601–7610.

Vaute O, Nicolas E, Vandel L, Trouche D. 2002. Functional and physical interaction between the histone methyl transferase Suv39H1 and histone deacetylases. *Nucleic Acids Res* **30:** 475–481.

Verreault A, Kaufman PD, Kobayashi R, Stillman B. 1996. Nucleosome assembly by a complex of CAF-1 and acetylated histones H3/H4. *Cell* **87:** 95–104.

Vestner B, Waldmann T, Gruss C. 2000. Histone octamer dissociation is not required for in vitro replication of simian virus 40 minichromosomes. *J Biol Chem* **275:** 8190–8195.

Vogelauer M, Rubbi L, Lucas I, Brewere BJ, Grunstein M. 2002. Histone acetylation regulates the time of replication origin firing. *Mol Cell* **10:** 1223–1233.

Warburton PE, Cooke CA, Bourassa S, Vafa O, Sullivan BA, Stetten G, Gimelli G, Warburton D, Tyler-Smith C, Sullivan KF, et al. 1997. Immunolocalization of CENP-A suggests a distinct nucleosome structure at the inner kinetochore plate of active centromeres. *Curr Biol* **7:** 901–904.

Weintraub H. 1974. The assembly of newly replicated DNA into chromatin. *Cold Spring Harb Symp Quant Biol* **38:** 247–256.

Wigler M, Levy D, Perucho M. 1981. The somatic replication of DNA methylation. *Cell* **24:** 33–40.

Wolffe AP, Pruss D. 1996. Deviant nucleosomes: The functional specialization of chromatin. *Trends Genet* **12:** 58–62.

Wong LH, McGhie JD, Sim M, Anderson MA, Ahn S, Hannan RD, George AJ, Morgan KA, Mann JR, Choo KH. 2010. ATRX interacts with H3.3 in maintaining telomere structural integrity in pluripotent embryonic stem cells. *Genome Res* **20:** 351–360.

Wu RS, Tsai S, Bonner WM. 1982. Patterns of histone variant synthesis can distinguish G0 from G1 cells. *Cell* **31:** 367–374.

Xie W, Song C, Young NL, Sperling AS, Xu F, Sridharan R, Conway AE, Garcia BA, Plath K, Clark AT, et al. 2009. Histone h3 lysine 56 acetylation is linked to the core transcriptional network in human embryonic stem cells. *Mol Cell* **33:** 417–427.

Xu M, Long C, Chen X, Huang C, Chen S, Zhu B. 2010. Partitioning of histone H3-H4 tetramers during DNA replication-dependent chromatin assembly. *Science* **328:** 94–98.

Zhang J, Feng X, Hashimshony T, Keshet I, Cedar H. 2002. Establishment of transcriptional competence in early and late S-phase. *Nature* **420:** 198–202.

Zlatanova J, Seebart C, Tomschik M. 2007. Nap1: Taking a closer look at a juggler protein of extraordinary skills. *FASEB J* **21:** 1294–1310.

C H A P T E R 23

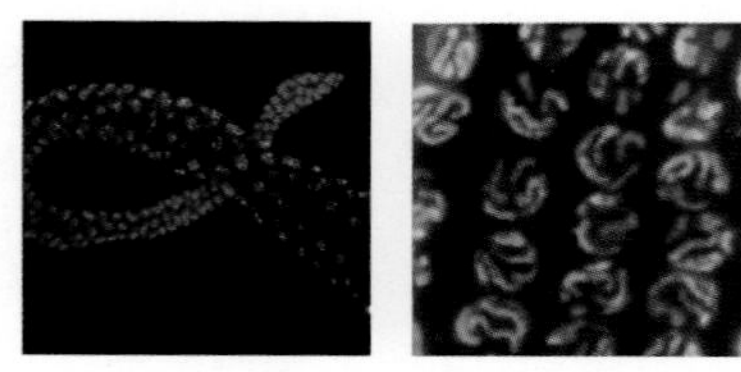

Regulation of the X Chromosomes in *Caenorhabditis elegans*

Susan Strome[1], William G. Kelly[2], Sevinc Ercan[3], and Jason D. Lieb[4]

[1]Department of Molecular, Cell and Developmental Biology, University of California, Santa Cruz, Santa Cruz, California 95064; [2]Department of Biology, Emory University, Atlanta, Georgia 30322; [3]Department of Biology and Center for Genomics and Systems Biology, New York University, New York, New York 10003; [4]Department of Biology and Carolina Center for Genome Sciences, The University of North Carolina at Chapel Hill, Chapel Hill, North Carolina 27599

Correspondence: sstrome@ucsc.edu

SUMMARY

Dosage compensation, which regulates the expression of genes residing on the sex chromosomes, has provided valuable insights into chromatin-based mechanisms of gene regulation. The nematode *Caenorhabditis elegans* has adopted various strategies to down-regulate and even nearly silence the X chromosomes. This chapter discusses the different chromatin-based strategies used in somatic tissues and in the germline to modulate gene expression from the *C. elegans* X chromosomes and compares these strategies to those used by other organisms to cope with similar X-chromosome dosage differences.

Outline

Cite as *Cold Spring Harb Perspect Biol* doi: 10.1101/cshperspect.a018366

OVERVIEW

Dosage compensation, which regulates the expression of genes residing on the sex chromosomes, has provided valuable insights into epigenetic mechanisms used to regulate whole chromosomes. The nematode *Caenorhabditis elegans* has adopted various strategies to down-regulate and even silence the X chromosomes. This chapter discusses the different chromatin-based strategies used in somatic tissues and in the germline to modulate gene expression from the *C. elegans* X chromosomes, and compares these strategies to those used by other organisms to cope with similar X-chromosome dosage differences.

Why is dosage compensation necessary? The mechanism of sex determination in *C. elegans* relies on the ratio of X chromosomes to sets of autosomes (X:A ratio). Diploid animals with two X chromosomes (X:A ratio of 1) develop as hermaphrodites, whereas those with one X chromosome (X:A ratio of 0.5) develop as males. The difference in X-chromosome dosage between the sexes, if not rectified, leads to imbalanced expression of X-linked genes and to lethality in one sex. Thus, equalizing expression of X-linked genes between the sexes is one function of dosage compensation.

In the somatic tissues of worms, dosage compensation occurs by down-regulating transcription approximately twofold from each of the two X chromosomes in the XX sex. This down-regulation is accomplished by the dosage compensation complex (DCC), a constellation of proteins assembled specifically on the X chromosomes of XX animals. The protein components of the DCC are homologous to members of the condensin complex, which is required for chromosome condensation and segregation during mitosis and meiosis. The similarity between the DCC and condensin complexes has led to the hypothesis that the DCC achieves repression of X-linked genes by partially condensing the X chromosomes. Long-standing questions surrounding dosage compensation in *C. elegans* include how the X:A ratio is assessed, how the DCC is assembled uniquely in XX animals, how the DCC is targeted to the X chromosomes, and the mechanism by which it accomplishes twofold down-regulation of X-linked gene expression. This chapter summarizes the substantial progress that has been made in answering these key questions.

The dosage compensation strategy used by *C. elegans* differs from the strategies used by mammals and fruit flies (*Drosophila*). Mammals achieve dosage compensation by globally silencing one X in the female (XX) sex. Fruit flies up-regulate gene expression from the single X in the male (XY) sex. This diversity of strategies probably reflects the co-option of different preexisting general chromatin mechanisms for the specialized role of equalizing X gene expression between sexes.

In addition to mechanisms that equalize gene dosage between sexes, there is strong evidence that these organisms also correct the chromosome dosage imbalance within cells, which is caused by a single dose of X-linked expression versus biallelic autosomal expression. In *C. elegans* and mammals, this is achieved by up-regulation of expressed X-linked genes in both sexes. Thus, in *C. elegans* hermaphrodites and mammalian females, down-regulation of the X via dosage compensation mechanisms is superimposed on this general up-regulation.

In the germline tissue (i.e., the reproductive cells) of worms, a more extreme modulation of X-linked gene expression occurs. Transcription from the single X in males and both Xs in hermaphrodites is significantly repressed in proliferating and meiotic germ cells. In germ cells of both sexes, the X chromosomes lack histone modifications that are associated with actively expressed chromatin. This is regulated at least in part by the MES proteins, chromatin modifiers whose loss leads to sterility. Furthermore, in males, the single X chromosome in each germ nucleus acquires histone modifications that are associated with heterochromatic silencing. This silencing depends on the unpaired status of the X in male meiosis. Genes expressed in the germline are strikingly underrepresented on the X chromosome. A favored view is that heterochromatinization of the unpaired X in male meiotic germ cells led to selection against X linkage of genes required for essential germline functions. In addition, the equalization of X-linked gene expression between the sexes may also be required in germ cells, and the most straightforward way to achieve this may have been to repress both X chromosomes in the XX sex. Thus, the *C. elegans* germline continues to be a critical system to explore how chromosome imbalances between the sexes have led to epigenetic regulation of chromosome states and gene expression.

Cite as *Cold Spring Harb Perspect Biol* doi: 10.1101/cshperspect.a018366

1 SEX CHROMOSOME IMBALANCE IN *C. elegans*

C. elegans exists as two sexes that are genetically distinguished by their X-chromosome complement. XX worms are hermaphrodites and XO worms are males. There is no sex-specific chromosome, such as a Y chromosome. Hermaphrodites and males display numerous sex-specific anatomical features and have different germline developmental programs (Fig. 1). These dramatic differences between the sexes are initiated in the early embryo, and result from counting and properly responding to the number of X chromosomes relative to autosomes (Nigon 1951; reviewed in Meyer 2000). How can a simple difference in sex chromosome number translate into such dramatically different developmental programs? An important concept is that *C. elegans* cells must assess not only the number of X chromosomes but also the number of sets of autosomes. It is the ratio of these—the X:A ratio—that determines sex. Diploid animals with two X chromosomes (X:A ratio of 1) develop as hermaphrodites, whereas those with one X chromosome (X:A ratio of 0.5) develop as males (Fig. 2). Many of the mechanistic details of appropriately responding to the X:A ratio have been elegantly dissected and are summarized in Sections 3–5.

The difference in X-chromosome dosage between the sexes leads, if uncorrected, to a difference in the levels of X-linked gene expression. Indeed, the double dose of X genes is lethal to hermaphrodites if not corrected. Intriguingly, somatic cells and germ cells have evolved different mechanisms to deal with this X-dosage challenge (Fig. 3). The germline and somatic lineages are fully separated from each other by the 24-cell stage of embryogenesis. Starting at about the 30-cell stage, the somatic lineages initiate a process termed "dosage compensation" whereby genes that reside on the X chromosomes of XX animals are down-regulated approximately twofold. In contrast, as discussed in Ch. 25 (Brockdorff and Turner 2014) and Ch. 24 (Lucchesi and Kuroda 2014), mammals implement dosage compensation by globally silencing one X in the XX sex, and fruit flies implement dosage compensation by up-regulating expression from the single X in the XY sex. In *C. elegans* germline tissue, a more extreme adjustment of X-linked gene expression occurs: The single X in males and both Xs in hermaphrodites are globally repressed. The chromatin-based mechanisms that accomplish dosage compensation in the soma and X-chromosome repression in the germline are the subjects of this chapter.

2 ASSESSING THE X:A RATIO

How do worm cells count Xs and autosomes so that dosage compensation is implemented when the X:A ratio is 1? Four small regions of the X, termed X signal elements (XSEs), have been identified as contributing to the numerator portion of the X:A ratio. By mutagenesis, four responsible X-linked genes have been identified within the XSE regions: *sex-1* (sex for signal element on X), *fox-1* (fox for feminizing gene on X), *ceh-39*, and *sex-2* (Carmi et al. 1998; Skipper et al. 1999; Gladden and Meyer 2007; Gladden et al. 2007). These four genes repress expression of *xol-1*, the most upstream gene in the sex determination and dosage compensation pathway, during a critical window of embryonic

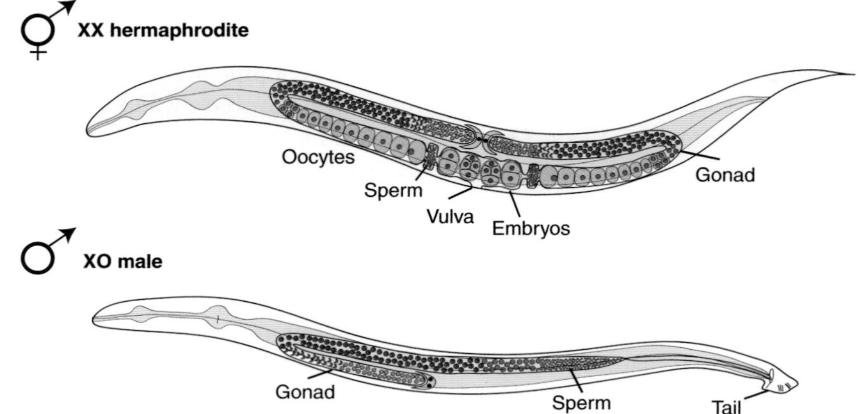

Figure 1. *C. elegans* hermaphrodite and male anatomy. *C. elegans* naturally exists as two sexes: XX hermaphrodites and XO males. Hermaphrodites and males display several sex-specific anatomical features, most notably a male tail designed for mating and a vulva on the ventral surface of hermaphrodites for reception of male sperm and for egg-laying. Their germline programs also differ. The two-armed gonad in hermaphrodites produces sperm initially and then oocytes throughout adulthood. The one-armed gonad in males produces sperm continuously. (Adapted, with permission, from Hansen et al. 2004, © Elsevier.)

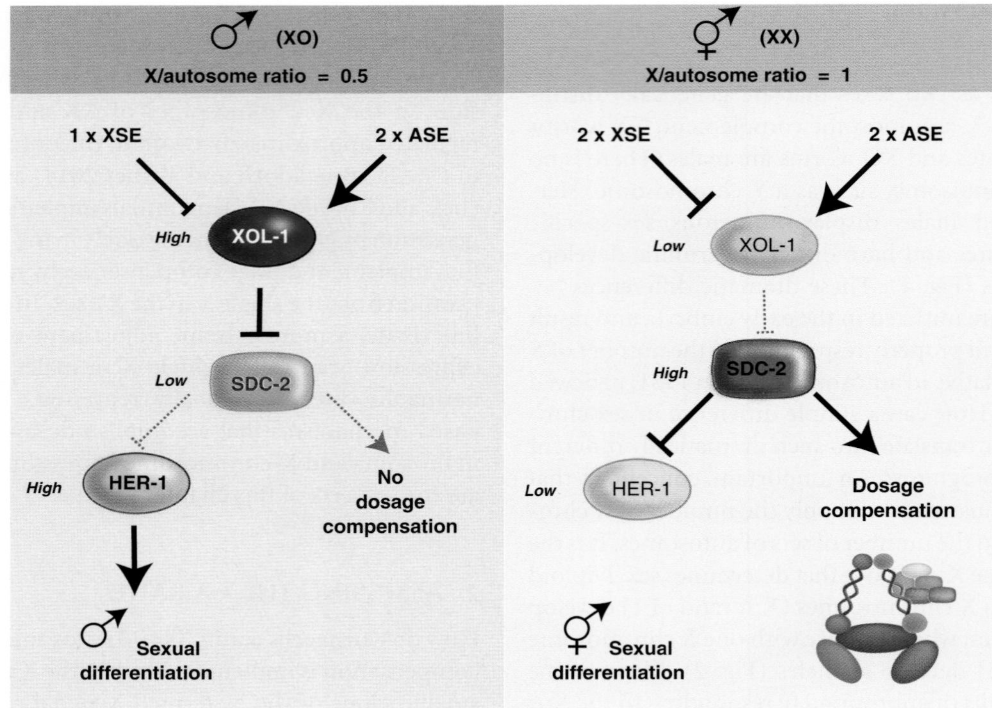

Figure 2. Pathway of sex determination and dosage compensation in *C. elegans*. The proposed roles of X signal elements (XSEs) and autosomal signal elements (ASEs) in regulating XOL-1 levels and the subsequent sexual differentiation and assembly of the dosage compensation complex (DCC), which represses X-linked gene expression about twofold in XX hermaphrodites, are highlighted.

development (Fig. 2). Because XX embryos produce approximately twice as much SEX-1, SEX-2, CEH-39, and FOX-1 as XO embryos, *xol-1* expression is much lower in XX than in XO embryos. One of the XSE genes, SEX-1, is a nuclear hormone receptor, which represses transcription of *xol-1* (Carmi et al. 1998). FOX-1 is an RNA-binding protein that reduces the level of XOL-1 protein through unknown posttranscriptional mechanisms (Nicoll et al. 1997; Skipper et al. 1999; Gladden et al. 2007). CEH-39 has a predicted DNA-binding domain that is similar to the ONECUT family of homeobox-containing transcription factors. CEH-39 does not contain sequence similarity to ONECUT proteins outside the DNA-binding domain (Gladden and Meyer 2007). Reducing the level of individual XSEs has a small effect on sex determination. Combined reductions have a larger effect, suggesting that the XSEs act cooperatively (Gladden and Meyer 2007). The functional readout of X-chromosome dosage is therefore equivalent to the dosage of X-linked factors expressed from XSEs that act to repress expression of *xol-1*. Thus far, only one autosomal signal element (ASE) has been identified as contributing to the denominator portion of the X:A ratio (Powell et al. 2005). The identified gene, *sea-1*, encodes a T-box transcription factor, which activates transcription of *xol-1*.

Autosomal dosage (ASE dosage) therefore counterbalances X dosage (XSE dosage) effects through antagonistic action on the master regulator switch, *xol-1* (Fig. 2). The working hypothesis for how this plays out in both sexes is as follows. Diploid XX embryos produce a double dose of XSE repressors of *xol-1*, which override the activating influence of ASEs. This keeps XOL-1 levels low and leads to hermaphrodite development and implementation of dosage compensation. Conversely, XO embryos produce a single dose of XSE repressors, which is insufficient to counteract the activating influence of ASEs. High XOL-1 levels lead to male development and failure to implement dosage compensation. In this way, the XSEs and ASEs translate an X:A ratio of either 0.5 or 1 into a switch for sex determination, and for deciding whether or not to implement dosage compensation.

3 THE DOSAGE COMPENSATION COMPLEX (DCC) RESEMBLES THE CONDENSIN COMPLEX

Understanding the assembly and composition of the DCC on the X chromosomes requires a brief introduction to the first few genes in the pathway that regulates both sex determination and dosage compensation (Fig. 2) (reviewed in

 Cite as *Cold Spring Harb Perspect Biol* doi: 10.1101/cshperspect.a018366

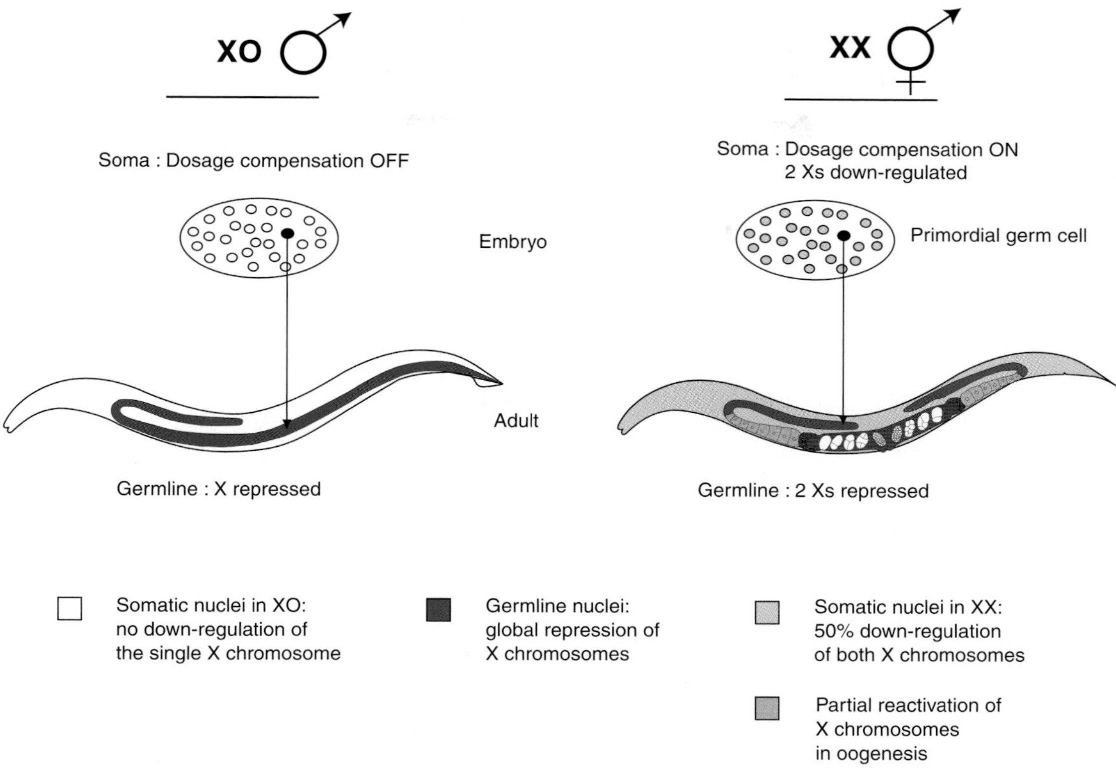

Figure 3. Overview of X-chromosome regulation. Dosage compensation occurs in somatic tissues only in XX hermaphrodites. Repression of the Xs in the germline occurs in both XO males and XX hermaphrodites. Hermaphrodites display late and partial activation of X-linked genes during late pachytene of oogenesis. The black cell highlights the single primordial germ cell in the embryo that generates the germline in the adult gonad.

Meyer 1997). *xol-1* (xol for XO-lethal), the first gene in the pathway, is considered a master switch gene because its activity is determined by the X:A ratio and it, in turn, dictates whether the pathway leads to male or hermaphrodite development. XOL-1 is a negative regulator of *sdc-2* (sdc for sex determination and dosage compensation defective) gene. The *sdc* genes, *sdc-1*, *sdc-2*, and *sdc-3*, encode components of the DCC and also regulate the *her-1* (her for hermaphroditization of XO animals) sex determination gene. In XO embryos, an X:A ratio of 0.5 leads to high XOL-1 protein levels and low SDC-2 protein levels; the DCC is not assembled, dosage compensation is not implemented, and the sex determination gene *her-1* is expressed, leading to male sexual development. In XX embryos, an X:A ratio of 1 results in low XOL-1 and high SDC-2 levels; the DCC is assembled, dosage compensation is implemented, and *her-1* is repressed, leading to hermaphrodite sexual development.

In addition to the SDC proteins (SDC-1, SDC-2, and SDC-3), the DCC also contains a set of DPY (DPY for dumpy) proteins (DPY-21, DPY-26, DPY-27, DPY-28, and DPY-30), MIX-1 (MIX for mitosis and X-associated), and the more recently identified CAPG-1 protein (Csankovszki et al. 2009) (see Table 1; reviewed in Meyer 2005). Significant insights into the mechanism of dosage compensation in worms came from the discovery that a portion of the *C. elegans* DCC resembles the 13S condensin I complex (Table 1 and Fig. 4). The condensin complex is conserved among all eukaryotes and is essential for proper chromosome compaction and segregation during mitosis and meiosis (reviewed in Hirano 2002). The core of the DCC includes DPY-26, DPY-27, DPY-28, MIX-1, and CAPG-1, which are homologous to members of the condensin complex. In fact, in addition to their functions in the DCC, all of the core proteins except DPY-27 function in a true condensin complex with SMC-4 to form the canonical five-subunit condensin I that acts in mitosis and meiosis. The SDC proteins DPY-21 and DPY-30, however, do not resemble known condensin subunits. The current view is that the DCC complex was derived from an ancestral condensin complex for targeted down-regulation of genes on the X chromosome, likely through a mechanism involving some degree of chromatin condensation.

Why is the DCC assembled only in XX embryos? Surprisingly, most of the DCC components are maternally

Table 1. Components that regulate gene expression from the X chromosome

Protein	Complex	Homolog and/or conserved domains
SDC-1	DCC	C2H2 zinc-finger domain
SDC-2	DCC	Novel protein
SDC-3	DCC	C2H2 zinc-finger domain and myosin-like ATP-binding domain
DPY-21	DCC	Conserved protein; no recognizable motifs
DPY-27	DCC	Condensin subunit homolog SMC-4/XCAP-C
DPY-30	DCC	Subunit of the MLL/COMPASS complex
DPY-26	DCC, condensin I	Condensin subunit XCAP-H
DPY-28	DCC, condensin I	Condensin subunit XCAP-D2/Cnd1/Ycs4p
CAPG-1	DCC, condensin I	Condensin subunit XCAP-G
MIX-1	DCC, condensin I, II	Condensin subunit SMC-2/XCAP-E
MES-2	MES-2/3/6	PRC2 subunit E(Z)/EZH2; SET domain
MES-3	MES-2/3/6	Novel protein
MES-6	MES-2/3/6	PRC2 subunit ESC/EED; WD40 domains
MES-4	Unknown	NSD proteins; PHD fingers and SET domain

supplied via the oocyte to both XX and XO embryos. The key regulator of DCC assembly is SDC-2 (Dawes et al. 1999). SDC-2 is not maternally supplied and is produced only in XX embryos (Fig. 2), in which it, along with SDC-3 and DPY-30, recruits the remaining DCC complex subunits to the X chromosomes (Fig. 5). In fact, driving expression of SDC-2 in XO embryos is sufficient to cause assembly of the DCC on the single X chromosome and to trigger dosage compensation, which kills the XO embryos. SDC-2 thus directs the specific recruitment of other DCC components, most of which have other cellular roles, and co-opts their activities for dosage compensation and sex determination.

4 RECRUITMENT AND SPREADING OF THE DCC

Several studies have focused on identifying features of the X chromosome involved in recruiting the DCC. An elegant approach was used to investigate whether the DCC is recruited independently to many sites or just a few sites, after which complexes spread into adjoining chromosomal regions (Lieb et al. 2000; Csankovszki et al. 2004; Meyer 2005). Worm strains containing duplications of different regions of the X chromosome were stained for DCC components. Association of the DCC with a duplication was interpreted to mean that the duplicated region contained a DCC recruitment site. Lack of DCC association with the duplication, but association of DCC with the corresponding region of the intact X, was interpreted to mean that the duplicated X region lacks a DCC recruitment site and instead acquires DCC at its endogenous locus by spreading from adjoining regions. These experiments identified at least 13 regions that can independently recruit the DCC, and provided evidence for DCC spreading along the X (Fig. 5A) (Csankovszki et al. 2004). Some regions recruit strongly and others weakly, suggesting either the presence of varying numbers of recruitment sites or of sites with varying capacity to recruit and/or promote spreading along the X (Csankovszki et al. 2004).

Finer mapping of the X-chromosome regions that are capable of recruiting the DCC identified smaller regions, called *rex* sites (rex for recruitment elements on the X) (McDonel et al. 2006). To test whether *rex* recruitment sites have a common DNA sequence, it was necessary to identify many other sites of DCC recruitment to the X. The key technology for these experiments was chromatin immunoprecipitation (ChIP) using two DCC subunits (DPY-27 and SDC-3) coupled to microarray analysis (ChIP-chip). This provided maps of DCC binding sites across the X and resulted in the identification of a 10-bp DNA sequence motif enriched at sites of high DCC binding (Ercan et al. 2007). The 10-bp motif was extended to 12 bp by a combination of ChIP-chip and extrachromosomal recruitment assays (Jans et al. 2009). The 12-bp DNA sequence motif (Fig. 5B) is necessary for recruiting the DCC to *rex* sites. It is estimated that there are 100–300 *rex* sites distributed along the ~17 Mb X chromosome (Jans et al. 2009). Although enriched on the X and more clustered on the X (Ercan et al. 2007), many copies of the 12-bp DNA sequence motif are present on the autosomes, but do not recruit the DCC. Thus, the motif is important, but not sufficient for DCC recruitment. The DNA or chromatin context of the motif may contribute to DCC recruitment, and clustering may be important for increasing the affinity for recruitment. SDC-2 is the first protein to localize to the X, and it recruits the other DCC members to the X chromosome, but whether it is SDC-2 or another factor that directly recognizes the DNA sequence motif at *rex* sites is not known. Regardless, it is clear that the initial recruitment of the DCC to the X in XX hermaphrodite embryos has at least two components: the DNA sequence motif at *rex* sites and the SDC-2 protein. It is interesting to note that in mammals and fruit flies, the dosage compensation machinery is targeted to the X chromosome by a combination of X-specific noncoding RNAs and X-chromosome sequence elements (see Ch. 25 [Brockdorff and Turner 2014]; Ch. 24 [Lucchesi and Kuroda 2014]). There is no evidence to date, however, that noncoding RNAs function in *C. elegans* dosage compensation.

Cite as *Cold Spring Harb Perspect Biol* doi: 10.1101/cshperspect.a018366

Vertebrates

Condensin I (13S)

SMC-2 SMC-4 CAP-H CAP-D2 CAP-G

Resolves and condenses mitotic and meiotic chromosomes

Condensin II

SMC-2 SMC-4 CAP-H2 CAP-D3 CAP-G2

Resolves and condenses mitotic and meiotic chromosomes

Subunits	Vertebrates
Core subunits (common to I & II)	
ATPase	CAP-E/SMC2
ATPase	CAP-C/SMC4
I-specific subunits	
HEAT	CAP-D2
HEAT	CAP-G
Kleisin	CAP-H
II-specific subunits	
HEAT	CAP-D3
HEAT	CAP-G2
Kleisin	CAP-H2/nessy

C. elegans

Condensin I

MIX-1 SMC-4 DPY-26 DPY-28 CAPG-1

Resolves and condenses mitotic and meiotic chromosomes

Condensin II

MIX-1 SMC-4 KLE-2 HCP-6 CAPG-2

Resolves and condenses mitotic and meiotic chromosomes

Condensin I^{DC}

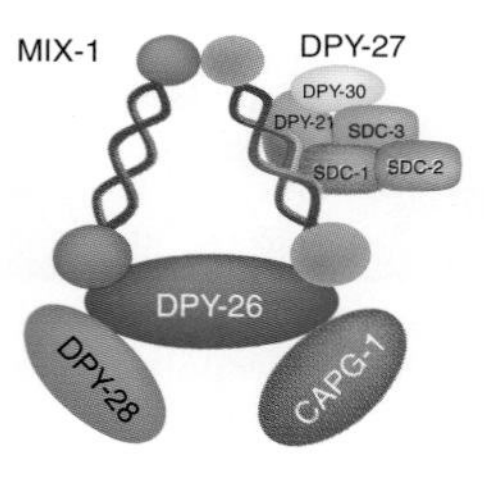

Reduces X-linked gene expression

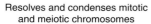

Figure 4. The DCC and condensin complexes. The worm DCC resembles the condensin complex, which functions in condensing chromosomes during nuclear division. In particular, the DCC contains several subunits that are homologous to the XCAP (XCAP for *Xenopus* chromosome-associated polypeptide) subunits of the 13S condensin complex I, originally characterized in *Xenopus*. There are two condensin complexes in most metazoans and three complexes in *C. elegans*. MIX-1 is present in all three *C. elegans* condensin complexes. Three additional DCC subunits (DPY-26, DPY-28, and CAPG-1) are present in both condensin I and condensin I^{DC}. The SDC proteins, DPY-21 and DPY-30, do not resemble known condensin subunits; they instead function in localizing condensin I^{DC} to the X chromosome. (Adapted from Meyer 2005 and Csankovszki et al. 2009.)

A critical unanswered question is how DCC spreading occurs from the initial recruitment sites (Fig. 5B). Most of the tested sites on the X chromosome are not able to recruit the DCC by themselves, yet the DCC is bound to them in the context of the natural chromosome. The DCC accumulates through a spreading mechanism, especially at the promoters of actively transcribed genes (Ercan et al. 2007). In experiments using X-to-autosome end-to-end fusion chromosomes, the DCC was also shown to spread from the X to juxtaposed autosomal sequences, indicating that unlike recruitment, DCC spreading is not dependent on any particular property of X-linked DNA sequences (Ercan et al. 2009). Furthermore, DCC spreading onto juxtaposed autosomal DNA is concentrated on the promoter regions of actively transcribed genes, just as it is on the natural X chromosome.

In summary, spreading of the DCC onto active promoters is governed by a mechanism that is not specific to the X chromosome, but DCC spreading is largely restricted to the X via recruitment of the DCC to *rex* sites (Fig. 5). What property of active promoters does the DCC recognize? One candidate is a histone variant H2A.Z because upon depletion of H2A.Z, DCC immunostaining is no longer as sharply restricted to the X chromosome as in wild type (Petty et al. 2009). Spreading may also be mediated by cooperative interactions between DCC complexes, physical interaction with the transcription machinery, or local modification of chromatin into a structure that facilitates more DCC binding in a self-reinforcing loop as shown for the spread of heterochromatin in *Schizosaccharomyces pombe* (see Ch. 16 [Martienssen and Moazed 2014]).

5 EFFECTS OF THE DCC: DOWN-REGULATION OF X-LINKED GENES AND THE AUTOSOMAL GENE *her-1*

Mammals, fruit flies, and worms appear to have co-opted different preexisting chromatin complexes to serve the spe-

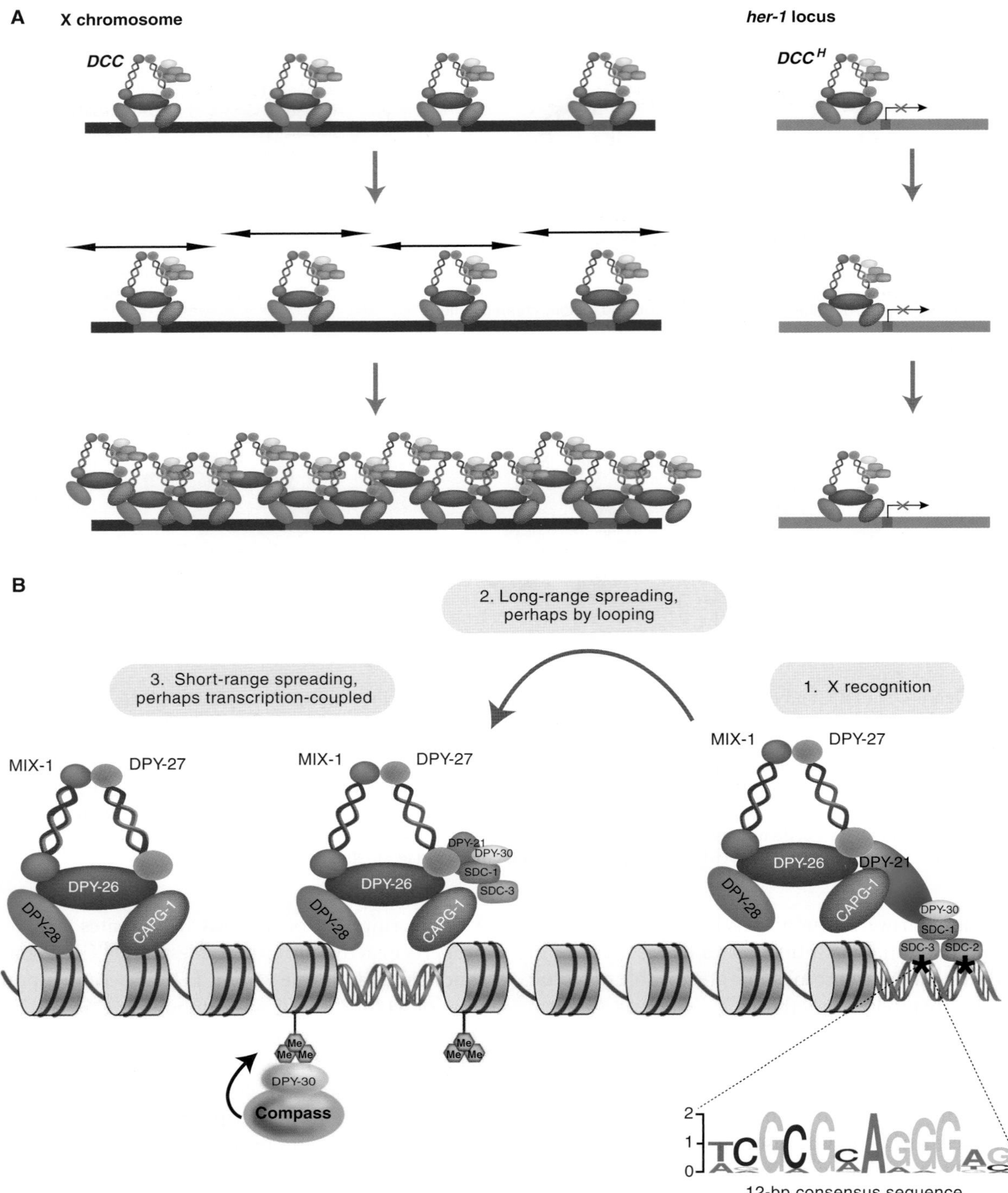

Figure 5. DCC recruitment and spreading on the X chromosome. (*A*) The DCC is recruited to approximately 100 to 300 recruitment elements and spreads along both X chromosomes. It also binds to the upstream region of the autosomal gene *her-1* and reduces expression by 20-fold, denoted DCC^H to indicate that the complex that binds to *her-1* lacks the DPY-21 protein. (Adapted from Alekseyenko and Kuroda 2004.) (*B*) From initial recruitment sites, the DCC spreads and accumulates preferentially at promoters by an unknown mechanism. The recruitment sites on the X chromosome, indicated by asterisks, contain multiple 12-bp DNA sequence motifs (shown below the DNA) that are important for recruitment. DCC binding is dynamic during development and is tuned to the transcriptional activity of individual genes. DCC accumulation is high in the body of highly expressed genes. (Adapted from Ercan and Lieb 2009.)

cialized role of modulating gene expression from the X chromosome. Mammals have used heterochromatin-based silencing to inactivate one of the two Xs in the XX sex. Fruit flies have used chromatin-modifying machinery to alter the state of the single X in XY animals, leading to up-regulation of gene expression. *C. elegans* has adapted a chromosome condensation mechanism typically used for mitosis and meiosis to down-regulate both Xs in the XX sex. Regulating chromatin structure to achieve modest twofold effects on X-linked gene expression, as occurs in worms and fruit flies, seems mechanistically challenging. The precise mechanism of down-regulation in worms and how down-regulation is limited to approximately twofold are critical questions.

Microarray analysis of RNA levels in XX wild type, XX DCC mutants, and XO embryos that are phenotypically hermaphrodite provided insight into *C. elegans* X-chromosome dosage compensation (Jans et al. 2009). In XX DCC mutant embryos, ~40% of expressed X-linked genes showed an increase in transcript levels, whereas only 2.5% showed a decrease. This is consistent with the expectation that the DCC represses the X chromosome. However, there was not a uniform twofold effect. The range of change was from 1.5-fold (the lowest ratio allowed for a "significant change") to 10-fold. Nevertheless, the *average* effect across X-linked genes whose expression significantly increased in DCC mutants was approximately twofold.

The mechanism by which the DCC achieves an average twofold reduction in transcript levels remains unknown. Microarray experiments revealed a disconnect between DCC binding and transcriptional response upon loss of DCC function. At the gene level, genes that were bound by the DCC were equally likely to be dosage compensated or not. This was interpreted to mean that the DCC does not directly regulate the genes to which it binds, but rather that it regulates "at a distance" (Fig. 6A) (Jans et al. 2009). At the genome level, although the DCC binds primarily to the X chromosome, loss of DCC function causes misexpression of some autosomal genes: repression of ~25% and activation of 7%. Jans et al. hypothesized that the DCC may repel a rate-limiting general transcription factor from the X, and that in the absence of the DCC this factor would be re-distributed on all chromosomes (Fig. 6B) (Jans et al.

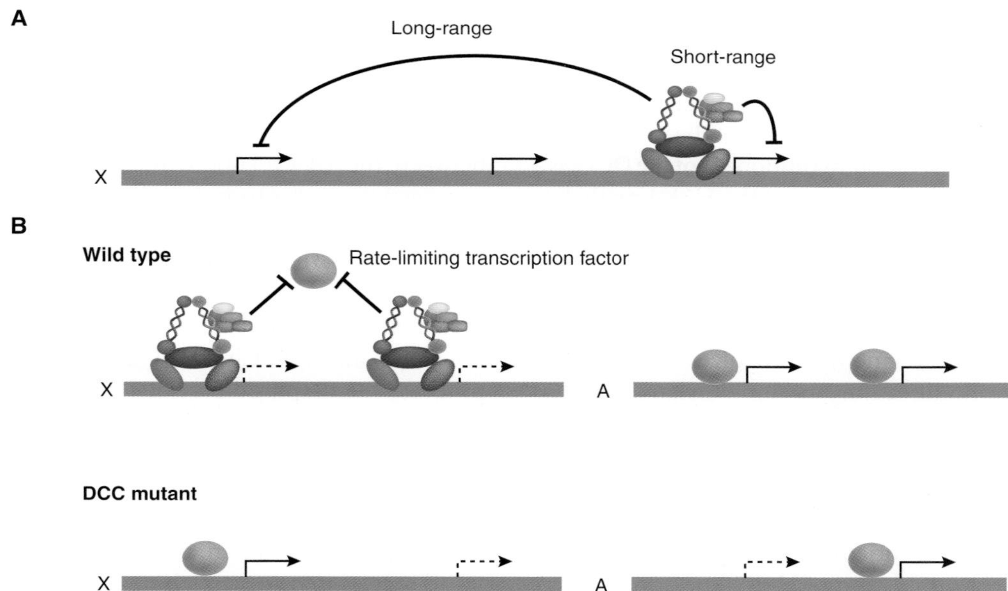

Figure 6. Models for how DCC concentrated on the X chromosomes may regulate gene expression. Upon loss of DCC function, most of the gene expression changes on the X chromosome are an increase in transcription, consistent with the DCC repressing X-chromosome transcription. However, only about half of DCC-bound genes increase in expression and many genes whose expression increases are not bound by the DCC. (*A*) The DCC may repress genes locally or cause structural changes that affect distant loci. (*B*) On the autosomes, the majority of expression changes due to loss of DCC function are a decrease in transcription. This could be explained by a model in which the DCC repels an activator from the X chromosomes. In the absence of the DCC, the activator is more evenly distributed between the X chromosomes and autosomes, resulting in increased gene expression from the X and decreased gene expression from the autosomes. An alternative model (not shown) is that the DCC may repress genes locally, and in the absence of the DCC additional effects on the X and autosomes result from the increase in transcription of hundreds of genes encoded from the X. The possibilities discussed above are not mutually exclusive.

2009). An alternative interpretation to explain the disconnect between binding and regulation at both the gene level and genome-wide level is that the DCC does indeed regulate genes directly at its sites of binding, but that steady-state measurements of mRNA levels in XX DCC mutants are dominated by secondary effects, rather than direct transcriptional responses to dosage compensation defects. Even in yeast, in which transcription factors bind very near to the genes they regulate, overlap between genes bound by a transcription factor and genes misregulated in the absence of that factor is often small (Gao et al. 2004; Chua et al. 2006; Hu et al. 2007). The issue of whether the DCC acts locally and/or at a distance is therefore unresolved. A high priority should be to determine whether the mechanism of DCC action is local, affecting genes near where it is bound, or whether the mechanism acts at a distance to affect genes across the entire X.

Histone modifications are often involved in establishing more "active" or "repressed" chromatin regions. In fruit flies, up-regulation of X gene expression in males is associated with H4K16 acetylation (H4K16ac) and loss of linker histone from the X (see Ch. 24 [Lucchesi and Kuroda 2014]). H4K16ac is enriched on the X chromosome by the action of MOF acetyltransferase, a subunit of the fly DCC, reducing compaction of nucleosomes. Its effect on transcription has been shown to be on transcription initiation (Conrad et al. 2012) and elongation (Larschan et al. 2011). To date, no histone modification activity has been attributed to the *C. elegans* DCC. DPY-30, a component of the DCC, is a noncatalytic structural component of the H3K4 methylating MLL/COMPASS complex (Fig. 5B) (Pferdehirt et al. 2011), but whether DPY-30 regulates H3K4 methylation specifically on the X chromosome has not been tested.

Although the DCC has no known direct histone-modifying activity, there is a newly recognized involvement of the histone modification monomethylated H4K20 (H4K20me1) in dosage compensation. H4K20me1, which is associated with gene repression in other species (Karachentsev et al. 2005; Yang and Mizzen 2009), becomes enriched on the X chromosomes in XX hermaphrodites (Liu et al. 2011). H4K20me1 enrichment is dependent on functional DCC, and accumulation of H4K20me1 occurs after dosage compensation is established (Vielle et al. 2012). Immunofluorescence analysis of H4K20me1 and H4K16ac in wild-type adult somatic nuclei showed that the X chromosomes have higher levels of H4K20me1 and lower levels of H4K16ac compared to the autosomes (Wells et al. 2012). Reducing SET-1 enzyme, which monomethylates H4K20, resulted in increased levels of H4K16ac on the Xs. Reducing the SIR-2.1 enzyme, which is responsible for depletion of H4K16ac on the X, however, did not affect H4K20me1 enrichment on the Xs (Wells et al. 2012). Therefore, it appears that enrichment of H4K20me1 on the X chromosomes results in a reduction of H4K16ac levels on the X (Vielle et al. 2012; Wells et al. 2012). It is not clear whether H4K20me1 exerts all of its effect on transcription through H4K16ac. In *C. elegans* as well as other organisms, H4K20me1 is greatly increased on chromosomes during mitosis, and human condensin II binding to chromosomes is mediated in part by H4K20me1 (Liu et al. 2010). It is possible that enrichment of H4K20me1 on the *C. elegans* X chromosomes has a mechanistic connection to condensed chromosome structure.

The DCC may function to restrict access of RNA polymerase or transcription factors to promoter regions, impede progression of RNA polymerase through transcription units, or slow reinitiation of transcription at each gene. These possibilities are consistent with the finding that loss of DCC function leads to increased RNA polymerase II on the X chromosome (Pferdehirt et al. 2011). Given the similarity of the DCC to the 13S condensin I complex, a likely mechanism for the down-regulation of gene expression is DCC-mediated condensation of chromatin. It is not known, however, whether the X chromosome is more condensed compared to the autosomes. X-chromosome promoters have higher nucleosome occupancy than autosomal promoters, but this higher nucleosome occupancy is not dependent on the DCC and is likely due to the higher GC content of X-linked promoters (Ercan et al. 2011). Future studies are required to dissect the functional connection between the DCC, H4K20 methylation, X-chromosome compaction, nucleosome organization, and transcription.

Clues about the mechanism of repression may also be found at *her-1*, the only autosomal target at which the DCC is known to function, and which displays 20-fold DCC-mediated down-regulation of gene expression (Fig. 5A) (Dawes et al. 1999; Chu et al. 2002). Repression of *her-1* promotes hermaphrodite sexual development (Fig. 2). An interesting question is how the DCC achieves 10-fold greater repression at *her-1* than it achieves on the X chromosome. A few differences between X-chromosome-associated DCC and *her-1*-associated DCC have been noted. First, although the DCC is recruited to the X chromosomes by SDC-2, the DCC is recruited to *her-1* by SDC-3 (Dawes et al. 1999; Yonker et al. 2003). Second, a different sequence motif is responsible for recruitment of the DCC to *her-1* than to the Xs (Chu et al. 2002). Third, DPY-21 is present on the X chromosomes, but not at the *her-1* locus (Yonker and Meyer 2003). DPY-21 is a conserved protein with a proline-rich amino terminus, but comparisons to other proteins do not reveal an apparent function. It could be that DPY-21 modulates the function of the DCC to weakly repress transcription of many genes on the X, whereas the DCC without DPY-21 strongly represses transcription at the *her-1* locus.

 Cite as *Cold Spring Harb Perspect Biol* doi: 10.1101/cshperspect.a018366

Dissecting how the DCC at *her-1* functions differently from the DCC localized to X may reveal important aspects of DCC mechanism.

6 COMPENSATORY UP-REGULATION OF X-LINKED GENES

Somatic down-regulation of the X chromosomes by the DCC compensates for the X-chromosome dosage difference between males and females. A second dosage compensation mechanism is thought to equalize overall transcript levels between X and autosomes. In animals with X chromosomes, cells in males contain a single X chromosome compared to two copies of each autosome (i.e., XY or XO, AA). This causes a potential haploinsufficiency for X-linked genes in males that is independent of the X-chromosome dosage imbalance between sexes. Susumu Ohno first hypothesized that during the evolution of X chromosomes, transcription of X-linked genes was doubled from the single X to match the transcript output from a diploid complement of autosomes; dosage compensation in females evolved to repress the inappropriate increase in X transcription in XX animals (Ohno 1967). This combination of up- and down-regulation effectively creates a diploid dose of X transcripts in both sexes (Fig. 7). Although one study claimed that compensatory up-regulation of X-linked genes does not occur in *C. elegans* (Xiong et al. 2010), subsequent studies that took into account the proliferation of germ cells during development showed that X up-regulation does occur in somatic cells (Deng et al. 2011; Kharchenko et al. 2011b). The clearest demonstration of this phenomenon was a microarray experiment showing that the mean and median X:A expression ratio in XO animals was 0.98, very close to the ratio of ~1 that occurs in XX animals, indicating that X-linked genes are up-regulated in XO animals (Deng et al. 2011). Thus, up-regulation of X-linked genes in both sexes requires a compensating mechanism that specifically down-regulates X-linked genes in animals with two X chromosomes. The molecular mechanism by which transcriptional up-regulation occurs on the X and how it is restricted to the X in both sexes are not known and are important areas for future investigation.

	Female or hermaphrodite		Male	
Human	↑ Xx	AA	↑ XY	AA
Fly	XX	AA	↑ XY	AA
Worm	↑ xx	AA	↑ X	AA
Effective dose	XX	AA	XX	AA

Figure 7. Coordinated up-regulation and DCC-mediated down-regulation of X-linked gene expression. Dosage compensation between the sexes is accomplished differently in the three well-studied systems shown. In mammals, one of the X chromosomes in females is inactivated (small x). In fruit flies, transcription from the single X in males is increased by a factor of 2 (larger X). In worms, transcription from both of the X chromosomes in hermaphrodites is decreased by a factor of 2 (smaller Xs). In all three systems, transcript levels from the X chromosome(s) and autosomes are similar in both females/hermaphrodites and males, suggesting that there is a mechanism to increase X transcription approximately twofold in both sexes (red arrows).

7 GERMLINE DEVELOPMENT AND GLOBAL REPRESSION OF THE X CHROMOSOMES

In *C. elegans*, the somatic mode of X-chromosome regulation by the DCC is inoperative in germ cells. Early evidence for this was the observation that some DCC components are not expressed in germ cells (e.g., SDC-2, DPY-27) and other DCC components are expressed in germ cells but are localized to all chromosomes (e.g., MIX-1, DPY-26, and DPY-28). MIX-1, DPY-26, and DPY-28 are, in fact, components of canonical condensin I in *C. elegans* (Fig. 4). In the germline, those DCC proteins are engaged in more general roles in mitotic and meiotic chromosome organization and segregation.

The absence of DCC members in the germline raises the question of whether any mode of dosage compensation occurs in this tissue. Recall that one function of dosage compensation is to achieve equivalent expression of X-linked genes between the sexes for viability and normal development. Current evidence suggests that in contrast to the twofold down-regulation of both X chromosomes in XX somatic cells, the X chromosomes are globally repressed during most stages of germline development in both XX and XO animals (Fig. 3). The specific mechanisms that regulate the X chromosomes in germ cells bear little resemblance in either form or function to the somatic mechanisms already described in this chapter.

The adult gonads of both sexes in *C. elegans* contain an orderly progression of germ cell stages. Germ cells proliferate in the distal region, enter meiosis in the middle region, and complete gametogenesis in the proximal region (Figs. 1 and 8) (reviewed in Schedl 1997). In XO males, this progression occurs in a single tubular testis that continuously produces sperm. In XX hermaphrodites, two tubular gonad arms initially produce sperm during a late larval stage and then switch to the production of oocytes in adulthood. Mature oocytes are pushed into the spermathecae and are fertilized by sperm residing there before further extrusion into the uterus. The production of some sperm for a limited

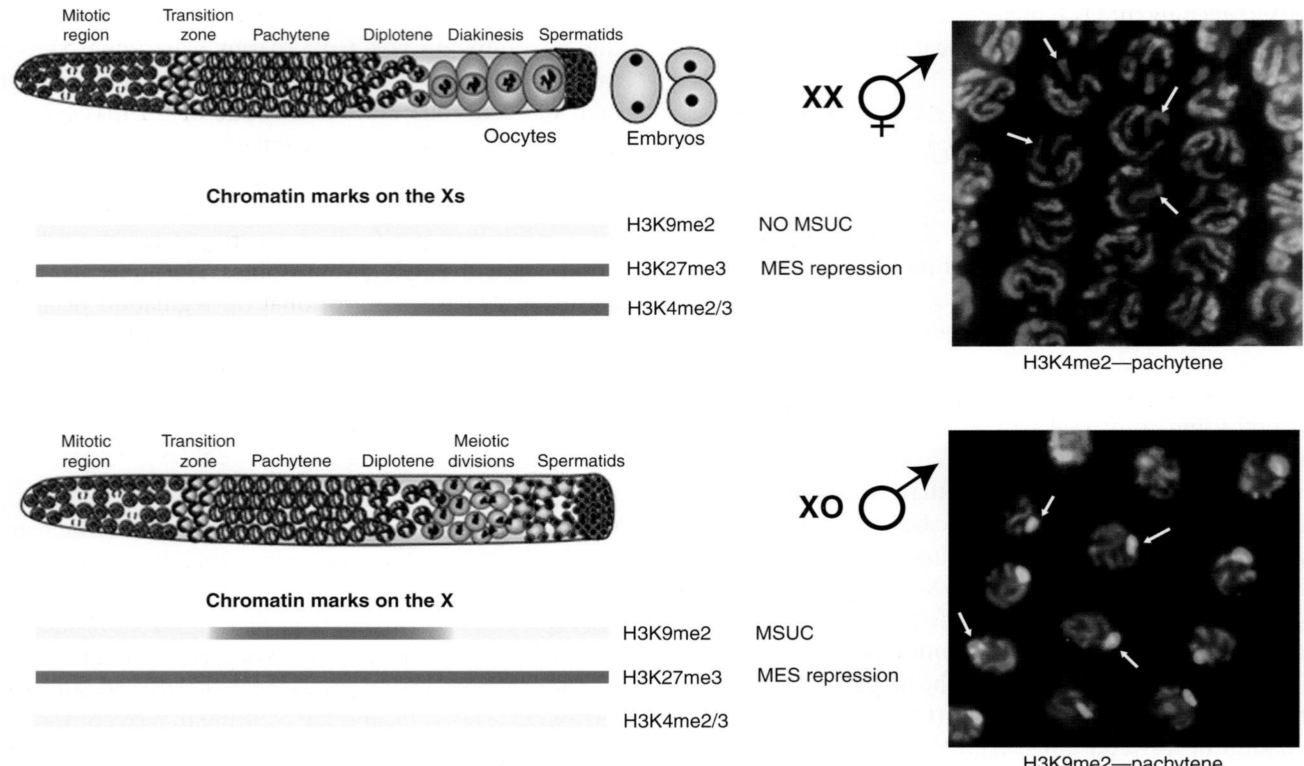

Figure 8. Epigenetic regulation of the X chromosomes during germ cell development. In both sexes, germ cells progress through mitosis (*left*), enter meiosis in the transition zone, and progress through meiosis I prophase. Cells destined to form sperm in both sexes complete the meiotic divisions in the gonad. In hermaphrodites, cells destined to form oocytes progress through meiotic prophase in the gonad and complete the meiotic divisions after ovulation and fertilization. The presence of various histone modifications on the X chromosome(s) in germ cells is shown by red bars (for repressive modifications) and green bars (for activating modifications). As shown on the *right*, antibodies to particular histone modifications reveal that the X chromosomes in germ nuclei are marked differently than the autosomes and are repressed. H3K4me2 (green), a mark of actively expressed chromatin, is excluded from the Xs in XX pachytene nuclei. H3K9me2 (green), a mark of heterochromatin, is concentrated on the X in XO pachytene nuclei as part of MSUC (meiotic silencing of unsynapsed chromatin). DNA is stained red. Arrows indicate representative X chromosomes in each image.

time in development by XX worms thus allows the "hermaphroditic" mode of reproduction in the adult. However, the ovary and certain other somatic tissues in adult XX animals can be considered female in identity and function.

The X chromosomes in the germline are initially transcriptionally repressed both in XX hermaphrodites and in XO males. The evidence for X repression comes from immunofluorescence analysis and genome-wide transcript profiling. Immunostaining of germlines revealed that the X chromosomes have higher levels of repressive histone modifications (H3K27 methylation and H3K9 methylation) than the autosomes (Fig. 8) (Kelly et al. 2002; Bender et al. 2004). The X chromosomes also lack several marks of actively expressed chromatin: H4 acetylation on K8, K12, and K16, H3K4 methylation, and the H3 variant H3.3 (Kelly et al. 2002; Ooi et al. 2006; Arico et al. 2011). Marks of active chromatin are absent from the X chromosomes throughout all stages of germline development in XO males. In XX hermaphrodites, such marks are absent from the X chromosomes in proliferating and early meiotic germ cells, but appear on the Xs during oogenesis (Kelly et al. 2002). These findings suggested that the X chromosomes are transcriptionally silent during all stages of germ cell development except oogenesis.

Genome-wide transcript profiling performed in worms that contain or lack germ cells showed that there is a striking asymmetry in the location of genes with germline-enriched expression; genes expressed in both the male and hermaphroditic germline, including those with enriched expression during spermatogenesis, are severely underrepresented on the X chromosome (Reinke et al. 2000; Reinke et al. 2004). Genes with enriched expression in oogenic germlines are also underrepresented on the X, but not to the same degree. Consequently, most of the genes that are required for the viability and function of germ cells are located on the autosomes. Most of the oogenesis-associated genes that reside on the X chromosome are expressed late in meiosis, which correlates with the accumulation of active histone modifications on the X chromosomes during oogenesis (Fig. 8). Thus, the bias against X-linkage is most stringent for genes that are common to both sexes and that function in the early stages of germ cell maturation. Consistent with this, primordial germ cells isolated from embryos express fewer genes from their X chromosomes than somatic cells at the same developmental stage, similar to the bias observed in adults (Spencer et al. 2011). Thus, preferential expression of autosomal genes is a feature of germ cells at all stages.

Recent transcript profiling from dissected adult hermaphrodite germlines has refined these results and revealed that the X is not as "silent" as previously thought (Wang et al. 2009; Tabuchi et al. 2011; Gaydos et al. 2012). Interestingly, genes that are transcribed in both soma and germline are observed to have significantly lower transcript levels in the germline relative to somatic tissues (Wang et al. 2009). It has been speculated that the X-linked transcripts detected in dissected full-length germlines come from oogenic germ cells, in which the X chromosomes acquire marks of active chromatin, as discussed in the preceding paragraphs. However, two observations instead suggest that there is some transcription from the X in germ cells at all stages and argue against oogenic germ cells being the sole source of X transcripts. First, isolated primordial germ cells express some X-linked genes (Spencer et al. 2011). Second, adult germlines that were dissected to include proliferating and early meiotic germ cells, but not oogenic germ cells, transcribe ~15% of the genes on the X (Tabuchi et al. 2011).

In summary, in the *C. elegans* germline there is chromosome-wide and significant dampening of transcription from the X. As a consequence, germline development relies predominantly on expression of genes located on the autosomes. The X chromosome does house its fair share of "housekeeping genes" that are expressed in all cells, but they are expressed at lower levels in germ cells than in somatic cells. The predominant reliance of *C. elegans* germ cells on autosomal genes raises the question of how germline-expressed genes became preferentially located on the autosomes. Interestingly, a number of gene duplications with X/autosome paralogs have been identified in which the autosomal copy is uniquely required for germ cell function and the X-linked copy functions in somatic lineages (Maciejowski et al. 2005). Gene duplication followed by germline dependence on the autosomal copy offers a potential mechanism for exclusion of essential germline genes from the inhospitable environment of the X chromosome. Genes that are only required during late oogenesis (i.e., only required in late stage female germ cells) were likely not subject to the selective pressures that excluded early germline-required genes from the X chromosome. The selective pressures likely included the evolved absence of a pairing partner for the X in male meiosis, as discussed in Section 8. Similar forces appear to be acting in other species including fruit flies and mammals (Wu and Xu 2003).

8 MEIOTIC SILENCING OF THE SINGLE X IN MALES

The earliest suggestion that the X chromosome differs from the autosomes in the germline came from cytological observations. The single X in male worms hypercondenses during pachytene in meiotic prophase, forming a ball-like structure reminiscent of the XY "sex body" seen during male meiosis in mammals (Fig. 8) (Goldstein and Slaton 1982; Handel 2004). The autosomes condense later in meiotic prophase near the onset of spermatogenesis. Premature X condensation is also seen during sperm meiosis in XX hermaphrodites and in sexually transformed XX males, suggesting that premature X condensation is in response to germ cell sex and not to X-chromosome ploidy or pairing status. Accordingly, in XX hermaphrodites, germ cells destined for oogenesis do not show premature X condensation.

In addition to premature condensation, a separate mechanism that depends on unpaired DNA in meiosis causes the single X chromosome in XO males to transiently accumulate a striking enrichment of H3K9me2, which appears during pachytene and disappears by diakinesis (Kelly et al. 2002; Bessler et al. 2010) (Fig. 8). This X enrichment of H3K9me2 does not occur during spermatogenesis in either XX hermaphrodites or sexually transformed XX males, but does occur during pachytene in sexually transformed XO hermaphrodites and in XX animals in which the X chromosomes do not pair (e.g., *him-8* mutants; Bean et al. 2004; Bessler et al. 2010). The specific acquisition of a heterochromatin mark on the X in XO meiosis thus appears to be a consequence of its unpaired status, and not the sex of the germline through which it is passing. Targeting of H3K9me2 to unpaired DNA is not limited to X-chromosome sequences; it is also found on unpaired autosomal fragments and translocations (Bean et al. 2004). H3K9me2 on unpaired X chromosomes is generated by the SETDB1 homolog MET-2 and its pattern is regulated by the chromatin protein HIM-17 (Reddy and Villeneuve 2004; Bessler et al. 2010). The finding that *met-2* mutant males do not display defects in transmission of X chromosomes during meiosis suggests that concentrating H3K9me2 on the X is not essential for proper meiotic chromosome segregation. Whether loss of H3K9me2 affects X silencing in males, however, has not yet been assessed.

Targeted repression of unpaired DNA in meiosis is not unique to *C. elegans*. Similar recognition and repression of unpaired DNA occurs in other organisms during meiosis, including *Neurospora* and mice. This is referred to as MSUC (meiotic silencing of unsynapsed chromatin; also referred to as MSUD, for meiotic silencing of unpaired DNA; Shiu et al. 2001; Baarends et al. 2005; Turner 2005; Kelly and Aramayo 2007). In mouse, for example, the poorly synapsed XY "sex body" is similarly enriched in H3K9me2 during male meiosis, and this is a consequence of its unsynapsed status (Cowell et al. 2002; Turner et al. 2005). Meiotic silencing in *Neurospora* requires the activity of proteins with conserved roles in RNA interference (see Ch. 10 [Aramayo and Selker 2013]). These include an RNA-dependent RNA polymerase (RdRP), an Argonaute-related protein (a conserved component of RNA-induced silencing complexes), and the Dicer nuclease (reviewed in Kelly and Aramayo 2007; see Ch. 16 [Martienssen and Moazed 2014]). In *C. elegans*, the enrichment of H3K9me2 on unpaired DNA requires the RdRP EGO-1 and the Argonaute protein CSR-1, but does not require Dicer (Maine et al. 2005; She et al. 2009). This suggests that although there is conservation of meiotic silencing (repression of unpaired DNA), the mechanism by which this is achieved likely evolved differently in different organisms.

In contrast to what happens in XO male meiosis, the X chromosomes in either XX spermatogonia or XX oogonia are not enriched in H3K9me2. This difference is likely because of the complete synapsis of the Xs in hermaphrodite meiosis. Why would silencing of unpaired DNA be a conserved feature of sexual reproduction? In many organisms, homolog pairing is unique to meiosis, and during synapsis novel insertions unique to one homolog would be exposed as regions of unpaired DNA. It has been proposed that recognition and silencing of unpaired sequences during meiosis provides a mechanism for self-scanning of a diploid genome, and could provide protection against invasion by, or expansion of, transposable elements. This protection would be needed only in the germline because somatic transposon insertions or expansions are not passed to the next generation. As a consequence, genes required during meiosis would encounter a strong selection against residence on an unpaired chromosome (such as the male X). Such selection may have led to the unique genetic profile of the X that is observed in *C. elegans*, as discussed in Section 7. It is interesting to speculate that the genomic warfare between transposons and their hosts, and the silencing mechanisms that have evolved as weapons in this battle, have shaped genomes and led to some of the X-autosome differences discussed in Section 7.

9 REGULATION OF X-CHROMOSOME REPRESSION BY THE MES HISTONE MODIFIERS

Section 8 describes repression via heterochromatin formation that is specific to the single X in XO males and restricted to the pachytene stage of meiosis due to MSUC. How are the X chromosomes maintained in a repressed chromatin state during other stages of male germline development and in the germlines of XX hermaphrodites? Genetic screens for maternal-effect sterile (*mes*) mutants identified a set of four *mes* genes that participate in X-chromosome repression in XX animals and likely in XO animals as well. A combination of genetic and molecular analyses have shown that the encoded MES proteins effect repression through regulating part of the spectrum of histone modifications found on the X chromosomes in germ cells. Their functions are essential for the survival and development of germ cells.

The MES proteins generate two opposing modifications on the tail of histone H3: methylation of H3K27, which is associated with gene repression, and methylation of H3K36, which is associated with active gene expression. Methylation of H3K27 is catalyzed by MES-2 in association with MES-3 and MES-6 (Fig. 9A). This trimeric complex resembles the Polycomb repressive complex PRC2 in fruit flies and vertebrates (Xu et al. 2001; Bender et al. 2004; Ketel et al. 2005; see Ch. 17 [Grossniklaus and Paro 2014]). MES-2 and MES-6 are the worm orthologs of two PRC2 subunits, E(Z) (enhancer of zeste) and ESC (extra sex combs; Table 1); MES-3 is a novel protein. MES-2's SET domain is responsible for its histone lysine methyltransferase (HKMT) activity, whereas MES-6 and MES-3 appear to be required either for substrate binding or to boost catalytic activity (Ketel et al. 2005).

MES-2, MES-3, and MES-6 are responsible for all detectable H3K27me2 and H3K27me3 in the germline and in early embryos, but another as-yet undiscovered HKMT contributes to H3K27 methylation in the primordial germ cells and in larval and adult somatic tissues. Importantly, within the germline H3K27me3 is enriched on the X chromosomes (Fig. 8) (Bender et al. 2004). Accordingly, one consequence of abolishing H3K27 methylation in the germline is activation of genes on the X. In progeny of *mes-2*, *mes-3*, or *mes-6* mutant mothers, the X chromosomes in the germline lack H3K27me2/3, acquire marks of active chromatin (e.g., H3K4me and H4K12ac), and become decorated with the transcriptionally active form of RNA polymerase II (Fong et al. 2002; Bender et al. 2004). These findings suggest that the MES-2/3/6 complex participates,

 Cite as *Cold Spring Harb Perspect Biol* doi: 10.1101/cshperspect.a018366

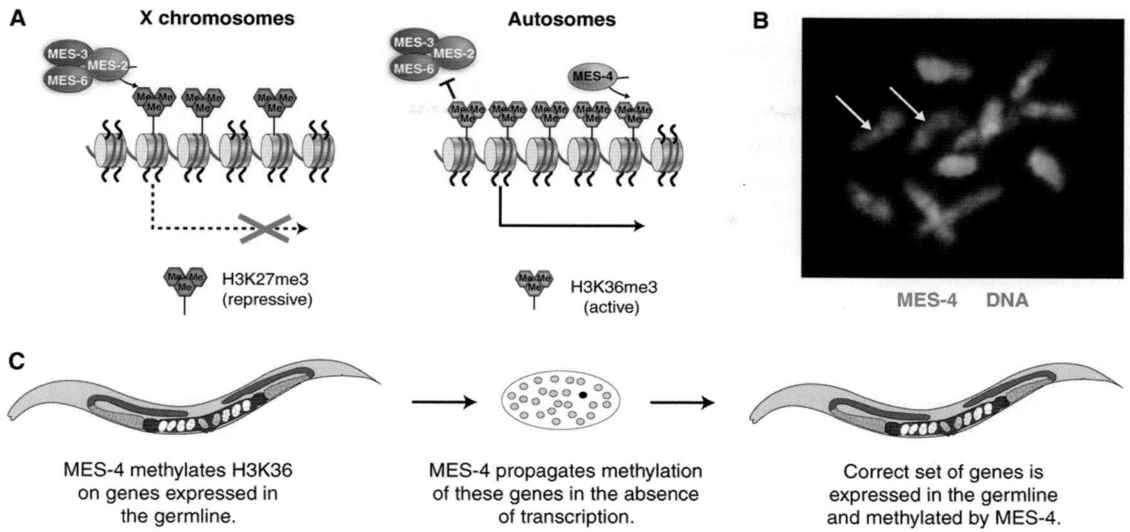

Figure 9. Model for the transgenerational role of MES-4 and how MES-4 and MES-2/3/6 participate in X repression in the germline. (*A*) MES-2/3/6-generated H3K27me3, a repressive histone modification, is concentrated on the X chromosomes. MES-4 and H3K36 methylation on autosomal genes repel the MES-2/3/6 complex, helping to concentrate its repressive action on the X chromosomes. (*B*) MES-4 is concentrated on the autosomes. MES-4 immunostaining is in green. DNA is stained red. Arrows mark the two X chromosomes, which lack MES-4 staining. (*C*) MES-4 associates with genes expressed in the maternal germline and marks them with H3K36me2/3. MES-4 propagation of H3K36 methyl marks in the absence of transcription enables MES-4 to transmit the memory of germline gene expression across generations.

perhaps directly, in repressing the X chromosomes in the germline of worms. Indeed, derepression of the two Xs is proposed to be the cause of germline degeneration observed in XX progeny of *mes* mutant mothers because XO progeny of those same mothers can be fertile, perhaps because derepression of a single X is tolerated during germline development, or more likely because another mechanism, such as heterochromatinization of the single X through MSUC, maintains repression in the XO germline (Garvin et al. 1998). It is interesting to note that, similar to the involvement of MES-2 and MES-6 in germline X repression in *C. elegans*, the vertebrate homologs of MES-2 and MES-6 are involved in somatic X inactivation in XX mammals (see Ch. 25 [Brockdorff and Turner 2014]; Ch. 17 [Grossniklaus and Paro 2014]).

The fourth MES protein involved in X repression in the germline is MES-4. Its distribution by immunostaining is novel among chromosome-associated proteins and exactly opposite to what might be expected. In contrast to the other MES proteins, MES-4 associates with the five autosomes in a banded pattern and is strikingly absent from most of the length of the X chromosome (Fig. 9B) (Fong et al. 2002). MES-4, like MES-2, contains a SET domain and also has HKMT activity (Bender et al. 2006). It is responsible for H3K36me2 and contributes to H3K36me3 in the germline and embryos. As predicted by the autosomal concentration of MES-4, H3K36 methylation is also dramatically concentrated on the autosomes. Recent analysis of the distributions of MES-4, H3K36 methyl marks, and RNA polymerase II by chromatin immunoprecipitation in embryos provided insights into how MES-4 is targeted to the autosomes and what role it serves there (Furuhashi et al. 2010; Rechtsteiner et al. 2010). In embryos, MES-4 and H3K36me2/3 reside in the body of genes that share the property of having been expressed in the maternal germline. Those genes include ubiquitously expressed genes, and germline-specific genes that were expressed in the maternal germline but are not expressed during embryogenesis and thus lack RNA polymerase II in embryos. The latter category revealed that MES-4 can associate with genes independently of RNA polymerase II, which distinguishes it from other H3K36 HKMTs. The picture that has emerged is that MES-4 is a maintenance HKMT whose function is to propagate the memory of germline gene expression from the parental germline to the germline in progeny (Fig. 9C). Thus, MES-4 serves a truly epigenetic role, by transmitting a chromatin mark (H3K36me) across generations. This function is critical for germ cell survival. The concentration of MES-4 on the autosomes is likely explained by MES-4's association with germline-expressed genes, which are concentrated on the autosomes and nearly absent from the X, as discussed in Section 7.

MES-4's participation in repressing the X chromosome emerged from transcript profiling of germlines dissected from fertile *mes-4* mutant mothers (Bender et al. 2006; Gaydos et al. 2012). The primary change in gene expression was up-regulation of genes on the X. How does MES-4, which is concentrated on the autosomes, participate in repressing genes on the X? The current model is that MES-4-mediated H3K36 methylation of germline-expressed genes on the autosomes repels MES-2/3/6 from those genes and helps concentrate its repressive activity on other regions of the genome, including the X chromosomes (Fig. 9A). The main evidence for this model is the finding that loss of MES-4 from germline genes causes H3K27me3 to spread to germline genes and results in reduced H3K27me3 on the X chromosome (Gaydos et al. 2012). In fact, in genome-wide chromatin immunoprecipitation studies in diverse species, H3K36me3 and H3K27me3 generally occupy nonoverlapping regions of the genome, consistent with those marks excluding each other and defining distinct genomic regions (Kharchenko et al. 2011a; Liu et al. 2011; Gaydos et al. 2012). Furthermore, the notion that H3K36 methylation antagonizes H3K27 methylation is supported by the finding that prior methylation of H3K36 prevents methylation of K27 on the same histone tails in vitro (Schmitges et al. 2011; Yuan et al. 2011). The model presented for MES-4 participation in X repression in worms (Fig. 9A) is similar to the model for Dot1 participation in telomeric silencing in *Saccharomyces cerevisiae* (van Leeuwen and Gottschling 2002). Dot1-mediated methylation of H3K79 along chromosomes is thought to repel the Sir repressors and help focus their action on the telomeres. Loss of Dot1 allows spreading of the Sirs from telomeres and results in telomeric desilencing. MES-4 and Dot1 illustrate how histone modifiers can contribute to proper distributions of repressors by antagonizing promiscuous repressor binding (van Leeuwen and Gottschling 2002).

The MES proteins function epigenetically in two intertwined but conceptually distinct manners: regulation of histone modifications and chromatin states, as described in the preceding paragraphs, and maternal-effect regulation. A maternal-effect mutation is defined as one whose mutant phenotype is not revealed in first-generation homozygous mutants, but instead shows up in their progeny. In the case of first-generation *mes/mes* mutants, the presence of some wild-type MES product produced by the *mes/+* mother and packaged into the oocyte is sufficient for the proper expansion of the two primordial germ cells into over a thousand functional germ cells in adults. The germ cells in these fertile *mes/mes* worms, however, cannot produce functional MES product for their offspring. As a result, in these offspring, the primordial germ cells undergo little proliferation and degenerate. The HKMT activities encoded by *mes-2* and *mes-4* must establish a heritable chromatin state that is properly maintained in the many descendants of the two initial primordial germ cells. MES-2, MES-3, and MES-6 operate in a complex to concentrate a repressive chromatin modification (H3K27me3) on the X chromosomes in the germline, and participate, probably directly, in X repression. MES-4 participates in X repression from a distance by methylating H3K36 on germline-expressed genes on the autosomes, which repels and helps focus the activity of the MES-2/3/6 complex and H3K27me3 on the X. The MES system is thought to act epigenetically in the mother's germline and in early embryos to establish chromatin domains that are properly marked for subsequent expression (autosomal regions) or repression (the X chromosomes) during germline development in larvae (see Fig. 9A,C). Loss of the MES system leads to germline death and sterility, likely due, at least in part, to derepression of the X chromosomes.

The discussion thus far has focused on factors and mechanisms involved in repressing expression of X-linked genes in the germline. Intriguingly, several members of the *C. elegans* DRM complex appear to be involved in promoting what little expression of X-linked genes does occur in the germline. The DRM complex includes the worm homologs of Retinoblastoma, E2F, DP, and LIN-54/Mip120. The complex is named for DP, Retinoblastoma, and Myb-MuvB and is related to the mammalian DREAM complex. Loss of any of the *C. elegans* DRM proteins causes genes on the X to be down-regulated in the germline (Tabuchi et al. 2011). Like MES-4, DRM components appear, by immunostaining, to be concentrated on the autosomes in germ cells, leading to the model that like MES-4, DRM indirectly regulates transcription of genes on the X. An interesting possibility is that MES-4 and DRM antagonistically influence transcription of germline-expressed genes on the autosomes, and as a result have opposite effects on transcription of genes on the X (Tabuchi et al. 2011).

10 PATERNAL X INACTIVATION IN EARLY EMBRYOS

The genomes contributed by the different gametes arrive in the zygote with vastly different epigenetic histories (Figs. 8 and 10). Although the X chromosome is inactive in early germ cell stages in both sexes (e.g., Fig. 10—nascent germline and mitosis stages), the X becomes transcriptionally active during the late pachytene stage of oogenesis (e.g., Figs. 8 and 10—oogenesis) (Fong et al. 2002; Kelly et al. 2002). In contrast, during spermatogenesis in both hermaphrodites and males the X is never activated (e.g., Figs. 8 and 10—spermatogenesis), undergoes premature condensation, and in XO meiosis is additionally enriched in

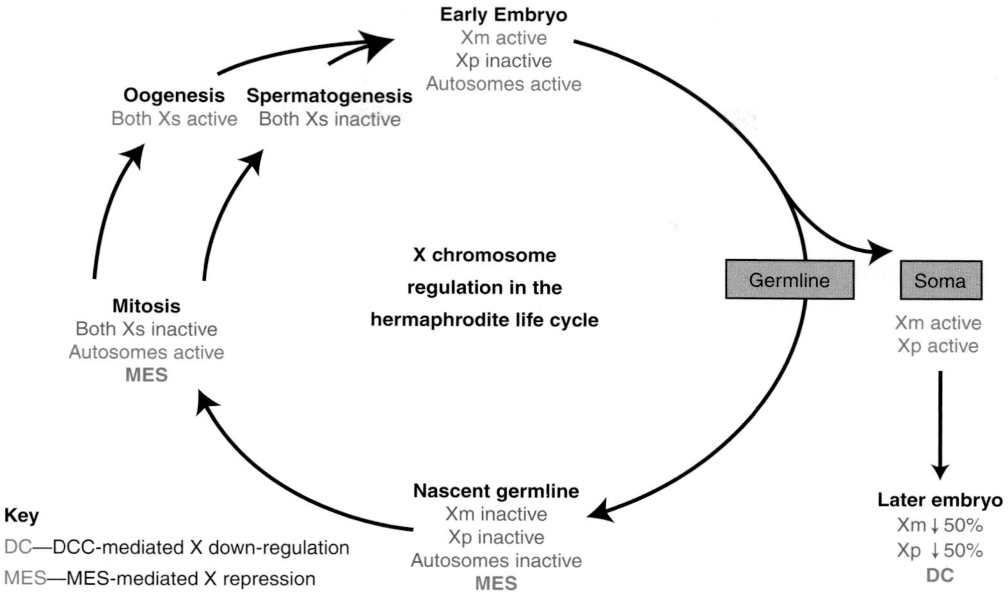

Figure 10. Regulation of the X chromosomes through the life cycle of an XX hermaphrodite. The X chromosomes are regulated by different mechanisms at different stages and in different tissues: paternal X inactivation in the early embryo, dosage compensation (DC) in the somatic tissues of 30-cell and later-stage embryos and worms, and MES-mediated repression in the germline. In XO male meiosis (not depicted), meiotic silencing of the unsynapsed X by MSUC additionally occurs in the germline.

H3K9me2 (Fig. 8) (Kelly et al. 2002; Bean et al. 2004). During sperm formation, although some canonical histones are replaced by specialized, sperm-specific histone variants and protamine-like basic proteins (Chu et al. 2006), the sperm genome possesses and enters the egg with easily detectable levels of histone H3 and several histone modifications, including H3K4me2, on the autosomes (Arico et al. 2011). The oocyte chromatin also continues to show significant levels of most activating histone modifications even during chromosome condensation in diakinesis (Fig. 8). RNA polymerase II levels associated with DNA drop precipitously in diakinesis-stage oocytes, suggesting that the histone modifications retained by the chromatin reflect recent transcriptional history instead of ongoing transcriptional activity (Kelly et al. 2002). Thus, the zygote inherits two epigenetically different genomes and, in particular, two X chromosomes with very different transcriptional histories—a recently active, oocyte-derived Xm and a sperm-derived Xp with little or no recent transcriptional activity.

After entry into the oocyte, the sperm DNA displays histone modifications such as H3K4me2 (Arico et al. 2011) as it begins to decondense to form the sperm pronucleus. In striking contrast to the autosomes, the decondensing Xp lacks H3K4me2 and histone H3 acetylation (Bean et al. 2004). There is, however, no difference in histone H4 modifications between the Xp and the autosomes. In the oocyte pronucleus, all chromosomes, including the Xm, are similarly modified and remain so throughout embryogenesis. Intriguingly, the Xp-specific absence of H3 modifications is maintained after DNA replication and survives multiple rounds of DNA replication and cell division, and thus has been termed an "epigenetic imprint." However, the imprint is gradually lost; that is, H3-specific modifications become increasingly detectable on the Xp until there are no obvious differences in H3K4me2 levels between the Xp and other chromosomes. An attractive scenario is that the absence of H3K4me2 and other marks of active chromatin from the Xp in early embryos reflects the transcriptionally silent state of the Xp during spermatogenesis (Figs. 8 and 10), and the Xp acquires marks of active chromatin as transcription from the zygotic genome is activated. It is noteworthy that maternally supplied H3.3, an H3 variant associated with active gene expression, exchanges onto the Xp, Xm, and the autosomes upon fertilization, yet it fails to acquire active histone modifications specifically on Xp (Ooi et al. 2006; Arico et al. 2011).

The Xp imprint in XX embryos is detected in both cross-progeny (from XO-derived sperm) and self-progeny (from XX-derived sperm). The pairing status of the X chromosome during spermatogenesis, and thus H3K9me2 targeting and enrichment on the X, therefore does not play an obvious role in establishment of the Xp imprint. However, the stability of the imprint—that is, the number of cell divisions after which it can still be readily observed—is significantly increased in offspring from XO-derived sperm relative to XX-derived sperm (Bean et al. 2004). Therefore, heterochromatin assembly on unpaired DNA during meiosis has effects that persist through early embryonic stages. Importantly, both pairing-based meiotic silencing and imprinted Xp inactivation are also observed in mammals, as discussed in Ch. 25 (Brockdorff and Turner 2014).

Genetic imprinting (covered in detail in Ch. 26 [Barlow and Bartolomei 2014]) has long been thought to be absent in *C. elegans*, as animals with uniparental inheritance of any single chromosome are viable and fertile (Hodgkin et al. 1979; Haack and Hodgkin 1991). In particular, there are no reported deleterious consequences of patriclinous inheritance of the X (i.e., XpXp animals). The unique genetic composition of the X chromosome, however, may help to explain why uniparental inheritance is not detrimental in *C. elegans*. In addition to a paucity of germline-expressed loci on the X chromosome, genes encoding early zygotic transcripts and those required for early embryonic development are also strikingly underrepresented on the X (Piano et al. 2000; Baugh et al. 2003). Thus, most genes whose products are essential for the very early stages during which the Xp is inactive are unlikely to reside on the X, rendering X-specific uniparental inheritance inconsequential in genetic tests.

Xp inactivation does, however, suggest a reason why somatic dosage compensation is not fully active in early embryos. The assembly of DCC components on the X chromosomes in XX embryos is not detectable by antibody staining until approximately the 30-cell stage. This is shortly after the Xp becomes fully decorated with H3 modifications, and is thus presumably fully activated. Activation of somatic dosage compensation is responsive to levels of the XSEs, the X-linked products that comprise the X portion of the X:A ratio, as discussed in Section 2. Full or partial repression of these elements on the Xp may render the early embryo functionally XO (i.e., the X:A is interpreted as 0.5). As the Xp reactivates with increased rounds of cell division, the level of X transcription may finally reach the critical threshold at which a two X-chromosome dosage is sensed and X:A equals 1, triggering the dosage compensation cascade. One might consider this a switch from maternal/paternal control of dosage compensation to zygotic control.

11 CONCLUDING REMARKS

C. elegans is one among many animals that uses a chromosome-based mechanism of sex determination in which the two sexes have a different number of X chromosomes. Interestingly, the germline and soma in worms have evolved distinct mechanisms to deal with this difference in X ploidy. In the somatic tissues, a conserved complex that is normally used for chromatin condensation and segregation during mitosis and meiosis appears to have been co-opted and adapted to achieve twofold repression of both X chromosomes in XX animals. This complex is recruited to the X chromosome in part by specific DNA sequences, and spreads along the X via a sequence-independent mechanism. Although it is conceptually attractive to envisage that an "inefficient" condensin complex might serve to decrease transcriptional efficiency, whether the complex acts locally at the level of individual genes, or at a distance across the entire X, and how repression is limited to twofold are currently not understood. This somatic dosage compensation mechanism appears to have been superimposed on a separate genome-wide dosage balancing process that compensates for X aneuploidy by up-regulating all expressed X-linked genes in both sexes by a factor of 2. The mechanism by which this second process operates is not understood.

In the germline, the X chromosomes are globally repressed during early stages of germ cell development in XX animals and during all stages of germ cell development in XO animals. Silencing of the single X chromosome in males, a mechanism to protect the integrity of unpaired chromosomal segments, likely created strong evolutionary pressure to move genes required for germ cell maturation off of the X chromosome, and also to repress the two X chromosomes in hermaphrodites. Global X-chromosome repression in the germline of XO and XX animals, in fact, is also one mechanism to compensate for the X-dosage difference between the sexes. If some X-linked genes that operate in both sexes escape repression, then the question arises as to whether the germline equalizes their expression in XX versus XO animals. The answer to this is not known. Equalization in the germline would involve a mechanism other than the DCC that operates in somatic tissues.

The germline and soma differ in many fundamental ways. This chapter highlights the different mechanisms of X-chromosome regulation in the germline and soma. One interesting theme that has emerged is co-option of different preexisting mechanisms, including utilization of a condensin-related complex to subtly down-regulate X expression in the soma, and utilization of a PRC2-related complex to repress the Xs in the germline. The heterochromatinization of the single X in males is thought to have dramatically altered the representation of genes on the X, such that genes required for general germline functions and early embryonic development are significantly underrepresented on this chromosome. The X chromosome and its regulation in *C. elegans* provide a window into chromosome-wide gene regulation mechanisms and genome evolution.

Cite as *Cold Spring Harb Perspect Biol* doi: 10.1101/cshperspect.a018366

REFERENCES

*Reference is also in this book.

Alekseyenko AA, Kuroda MI. 2004. Molecular biology. Filling gaps in genome organization. *Science* **303:** 1148–1149.

* Aramayo R, Selker EU. 2013. *Neurospora crassa*, a model system for epigenetics research. *Cold Spring Harb Perspect Biol* **5:** a017921.

Arico JK, Katz DJ, van der Vlag J, Kelly WG. 2011. Epigenetic patterns maintained in early *Caenorhabditis elegans* embryos can be established by gene activity in the parental germ cells. *PLoS Genet* **7:** e1001391.

Baarends WM, Wassenaar E, van der Laan R, Hoogerbrugge J, Sleddens-Linkels E, Hoeijmakers JH, de Boer P, Grootegoed JA. 2005. Silencing of unpaired chromatin and histone H2A ubiquitination in mammalian meiosis. *Mol Cell Biol* **25:** 1041–1053.

* Barlow DP, Bartolomei MS. 2014. Genomic imprinting in mammals. *Cold Spring Harb Perspect Biol* **6:** a018382.

Baugh LR, Hill AA, Slonim DK, Brown EL, Hunter CP. 2003. Composition and dynamics of the *Caenorhabditis elegans* early embryonic transcriptome. *Development* **130:** 889–900.

Bean CJ, Schaner CE, Kelly WG. 2004. Meiotic pairing and imprinted X chromatin assembly in *Caenorhabditis elegans. Nat Genet* **36:** 100–105.

Bender LB, Cao R, Zhang Y, Strome S. 2004. The MES-2/MES-3/MES-6 complex and regulation of histone H3 methylation in *C. elegans. Curr Biol* **14:** 1639–1643.

Bender LB, Suh J, Carroll CR, Fong Y, Fingerman IM, Briggs SD, Cao R, Zhang Y, Reinke V, Strome S. 2006. MES-4: An autosome-associated histone methyltransferase that participates in silencing the X chromosomes in the *C. elegans* germline. *Development* **133:** 3907–3917.

Bessler JB, Andersen EC, Villeneuve AM. 2010. Differential localization and independent acquisition of the H3K9me2 and H3K9me3 chromatin modifications in the *Caenorhabditis elegans* adult germline. *PLoS Genet* **6:** e1000830.

* Brockdorff N, Turner BM. 2014. Dosage compensation in mammals. *Cold Spring Harb Perspect Biol* doi: 10.1101/cshperspect.a019406.

Carmi I, Kopczynski JB, Meyer BJ. 1998. The nuclear hormone receptor SEX-1 is an X-chromosome signal that determines nematode sex. *Nature* **396:** 168–173.

Chu DS, Dawes HE, Lieb JD, Chan RC, Kuo AF, Meyer BJ. 2002. A molecular link between gene-specific and chromosome-wide transcriptional repression. *Genes Dev* **16:** 796–805.

Chu DS, Liu H, Nix P, Wu TF, Ralston EJ, Yates JR III, Meyer BJ. 2006. Sperm chromatin proteomics identifies evolutionarily conserved fertility factors. *Nature* **443:** 101–105.

Chua G, Morris QD, Sopko R, Robinson MD, Ryan O, Chan ET, Frey BJ, Andrews BJ, Boone C, Hughes TR. 2006. Identifying transcription factor functions and targets by phenotypic activation. *Proc Natl Acad Sci* **103:** 12045–12050.

Conrad T, Cavalli FM, Vaquerizas JM, Luscombe NM, Akhtar A. 2012. *Drosophila* dosage compensation involves enhanced Pol II recruitment to male X-linked promoters. *Science* **337:** 742–746.

Cowell IG, Aucott R, Mahadevaiah SK, Burgoyne PS, Huskisson N, Bongiorni S, Prantera G, Fanti L, Pimpinelli S, Wu R, et al. 2002. Heterochromatin, HP1 and methylation at lysine 9 of histone H3 in animals. *Chromosoma* **111:** 22–36.

Csankovszki G, McDonel P, Meyer BJ. 2004. Recruitment and spreading of the *C. elegans* dosage compensation complex along X chromosomes. *Science* **303:** 1182–1185.

Csankovszki G, Collette K, Spahl K, Carey J, Snyder M, Petty E, Patel U, Tabuchi T, Liu H, McLeod I, et al. 2009. Three distinct condensin complexes control *C. elegans* chromosome dynamics. *Curr Biol* **19:** 9–19.

Dawes HE, Berlin DS, Lapidus DM, Nusbaum C, Davis TL, Meyer BJ. 1999. Dosage compensation proteins targeted to X chromosomes by a determinant of hermaphrodite fate. *Science* **284:** 1800–1804.

Deng X, Hiatt JB, Nguyen DK, Ercan S, Sturgill D, Hillier LW, Schlesinger F, Davis CA, Reinke VJ, Gingeras TR, et al. 2011. Evidence for compensatory upregulation of expressed X-linked genes in mammals, *Caenorhabditis elegans* and *Drosophila melanogaster. Nat Genet* **43:** 1179–1185.

Ercan S, Lieb JD. 2009. *C elegans* dosage compensation: A window into mechanisms of domain-scale gene regulation. *Chromosome Res* **17:** 215–227.

Ercan S, Giresi PG, Whittle CM, Zhang X, Green RD, Lieb JD. 2007. X chromosome repression by localization of the *C. elegans* dosage compensation machinery to sites of transcription initiation. *Nat Genet* **39:** 403–408.

Ercan S, Dick LL, Lieb JD. 2009. The *C. elegans* dosage compensation complex propagates dynamically and independently of X chromosome sequence. *Curr Biol* **19:** 1777–1787.

Ercan S, Lubling Y, Segal E, Lieb JD. 2011. High nucleosome occupancy is encoded at X-linked gene promoters in *C. elegans. Genome Res* **21:** 237–244.

Fong Y, Bender L, Wang W, Strome S. 2002. Regulation of the different chromatin states of autosomes and X chromosomes in the germline of *C. elegans. Science* **296:** 2235–2238.

Furuhashi H, Takasaki T, Rechtsteiner A, Li T, Kimura H, Checchi PM, Strome S, Kelly WG. 2010. Trans-generational epigenetic regulation of *C. elegans* primordial germ cells. *Epigenetics Chromatin* **3:** 15.

Gao F, Foat BC, Bussemaker HJ. 2004. Defining transcriptional networks through integrative modeling of mRNA expression and transcription factor binding data. *BMC Bioinformatics* **5:** 31.

Garvin C, Holdeman R, Strome S. 1998. The phenotype of mes-2, mes-3, mes-4 and mes-6, maternal-effect genes required for survival of the germline in *Caenorhabditis elegans*, is sensitive to chromosome dosage. *Genetics* **148:** 167–185.

Gaydos L, Rechtsteiner A, Egelhofer T, Carroll C, Strome S. 2012. Antagonism between MES-4 and Polycomb repressive complex 2 promotes appropriate gene expression in *C. elegans* germ cells. *Cell Rep* **2:** 1169–1177.

Gladden JM, Meyer BJ. 2007. A ONECUT homeodomain protein communicates X chromosome dose to specify *Caenorhabditis elegans* sexual fate by repressing a sex switch gene. *Genetics* **177:** 1621–1637.

Gladden JM, Farboud B, Meyer BJ. 2007. Revisiting the X:A signal that specifies *Caenorhabditis elegans* sexual fate. *Genetics* **177:** 1639–1654.

Goldstein P, Slaton DE. 1982. The synaptonemal complexes of *Caenorhabditis elegans*: Comparison of wild-type and mutant strains and pachytene karyotype analysis of wild-type. *Chromosoma* **84:** 585–597.

* Grossniklaus U, Paro R. 2014. Transcriptional silencing by *Polycomb*-group proteins. *Cold Spring Harb Perspect Biol* **6:** a019331.

Haack H, Hodgkin J. 1991. Tests for parental imprinting in the nematode *Caenorhabditis elegans. Mol Gen Genet* **228:** 482–485.

Handel MA. 2004. The XY body: A specialized meiotic chromatin domain. *Exp Cell Res* **296:** 57–63.

Hansen D, Hubbard EJ, Schedl T. 2004. Multi-pathway control of the proliferation versus meiotic development decision in the *Caenorhabditis elegans* germline. *Dev Biol* **268:** 342–357.

Hirano T. 2002. The ABCs of SMC proteins: Two-armed ATPases for chromosome condensation, cohesion, and repair. *Genes Dev* **16:** 399–414.

Hodgkin J, Horvitz HR, Brenner S. 1979. Nondisjunction mutants of the nematode *Caenorhabditis elegans. Genetics* **91:** 67–94.

Hu Z, Killion PJ, Iyer VR. 2007. Genetic reconstruction of a functional transcriptional regulatory network. *Nat Genet* **39:** 683–687.

Jans J, Gladden JM, Ralston EJ, Pickle CS, Michel AH, Pferdehirt RR, Eisen MB, Meyer BJ. 2009. A condensin-like dosage compensation complex acts at a distance to control expression throughout the genome. *Genes Dev* **23:** 602–618.

Karachentsev D, Sarma K, Reinberg D, Steward R. 2005. PR-Set7-dependent methylation of histone H4 Lys 20 functions in repression of gene expression and is essential for mitosis. *Genes Dev* **19:** 431–435.

Kelly WG, Aramayo R. 2007. Meiotic silencing and the epigenetics of sex. *Chromosome Res* **15:** 633–651.

Kelly WG, Schaner CE, Dernburg AF, Lee M-H, Kim SK, Villeneuve AM, Reinke V. 2002. X-chromosome silencing in the germline of *C. elegans*. *Development* **129:** 479–492.

Ketel CS, Andersen EF, Vargas ML, Suh J, Strome S, Simon JA. 2005. Subunit contributions to histone methyltransferase activities of fly and worm polycomb group complexes. *Mol Cell Biol* **25:** 6857–6868.

Kharchenko PV, Alekseyenko AA, Schwartz YB, Minoda A, Riddle NC, Ernst J, Sabo PJ, Larschan E, Gorchakov AA, Gu T, et al. 2011a. Comprehensive analysis of the chromatin landscape in *Drosophila melanogaster*. *Nature* **471:** 480–485.

Kharchenko PV, Xi R, Park PJ. 2011b. Evidence for dosage compensation between the X chromosome and autosomes in mammals. *Nat Genet* **43:** 1167–1169; author reply 1171–1172.

Larschan E, Bishop EP, Kharchenko PV, Core LJ, Lis JT, Park PJ, Kuroda MI. 2011. X chromosome dosage compensation via enhanced transcriptional elongation in *Drosophila*. *Nature* **471:** 115–118.

Lieb JD, de Solorzano CO, Rodriguez EG, Jones A, Angelo M, Lockett S, Meyer BJ. 2000. The *Caenorhabditis elegans* dosage compensation machinery is recruited to X chromosome DNA attached to an autosome. *Genetics* **156:** 1603–1621.

Liu W, Tanasa B, Tyurina OV, Zhou TY, Gassmann R, Liu WT, Ohgi KA, Benner C, Garcia-Bassets I, Aggarwal AK, et al. 2010. PHF8 mediates histone H4 lysine 20 demethylation events involved in cell cycle progression. *Nature* **466:** 508–512.

Liu T, Rechtsteiner A, Egelhofer TA, Vielle A, Latorre I, Cheung MS, Ercan S, Ikegami K, Jensen M, Kolasinska-Zwierz P, et al. 2011. Broad chromosomal domains of histone modification patterns in *C. elegans*. *Genome Res* **21:** 227–236.

* Lucchesi JC, Kuroda MI. 2014. Dosage compensation in *Drosophila*. *Cold Spring Harb Perspect Biol* doi: 10.1101/cshperspect.a019398.

Maciejowski J, Ahn JH, Cipriani PG, Killian DJ, Chaudhary AL, Lee JI, Voutev R, Johnsen RC, Baillie DL, Gunsalus KC, et al. 2005. Autosomal genes of autosomal/X-linked duplicated gene pairs and germ-line proliferation in *Caenorhabditis elegans*. *Genetics* **169:** 1997–2011.

Maine EM, Hauth J, Ratliff T, Vought VE, She X, Kelly WG. 2005. EGO-1, a putative RNA-dependent RNA polymerase, is required for heterochromatin assembly on unpaired DNA during *C. elegans* meiosis. *Curr Biol* **15:** 1972–1978.

* Martienssen R, Moazed D. 2014. RNAi and heterochromatin assembly. *Cold Spring Harb Perspect Biol* doi: 10.1101/cshperspect.a019323.

McDonel P, Jans J, Peterson BK, Meyer BJ. 2006. Clustered DNA motifs mark X chromosomes for repression by a dosage compensation complex. *Nature* **444:** 614–618.

Meyer BJ. 1997. Sex determination and X chromosome dosage compensation. In *C. elegans II* (ed. Riddle DL, et al.), pp. 209–240. Cold Spring Harbor Laboratory Press, Cold Spring Harbor, NY.

Meyer BJ. 2000. Sex in the wormcounting and compensating X-chromosome dose. *Trends Genet* **16:** 247–253.

Meyer BJ. 2005. X-chromosome dosage compensation. In *WormBook*, ed. The *C. elegans* Research Community, Wormbook, doi/10.1895/wormbook.1.7.1, http://www.wormbook.org.

Nicoll M, Akerib CC, Meyer BJ. 1997. X-chromosome-counting mechanisms that determine nematode sex. *Nature* **388:** 200–204.

Nigon V. 1951. Polyploidie experimentale chez un nematode libre, *Rhaditis elegans* maupas. *Bull Biol Fr Belg* **85:** 187–255.

Ohno S. 1967. *Sex chromosomes and sex-linked genes*. Springer, Berlin.

Ooi SL, Priess JR, Henikoff S. 2006. Histone H3.3 variant dynamics in the germline of *Caenorhabditis elegans*. *PLoS Genet* **2:** e97.

Petty EL, Collette KS, Cohen AJ, Snyder MJ, Csankovszki G. 2009. Restricting dosage compensation complex binding to the X chromosomes by H2A.Z/HTZ-1. *PLoS Genet* **5:** e1000699.

Pferdehirt RR, Kruesi WS, Meyer BJ. 2011. An MLL/COMPASS subunit functions in the *C. elegans* dosage compensation complex to target X chromosomes for transcriptional regulation of gene expression. *Genes Dev* **25:** 499–515.

Piano F, Schetter AJ, Mangone M, Stein L, Kemphues KJ. 2000. RNAi analysis of genes expressed in the ovary of *Caenorhabditis elegans*. *Curr Biol* **10:** 1619–1622.

Powell JR, Jow MM, Meyer BJ. 2005. The T-box transcription factor SEA-1 is an autosomal element of the X:A signal that determines *C. elegans* sex. *Dev Cell* **9:** 339–349.

Rechtsteiner A, Ercan S, Takasaki T, Phippen TM, Egelhofer TA, Wang W, Kimura H, Lieb JD, Strome S. 2010. The histone H3K36 methyltransferase MES-4 acts epigenetically to transmit the memory of germline gene expression to progeny. *PLoS Genet* **6:** e1001091.

Reddy KC, Villeneuve AM. 2004. *C elegans* HIM-17 links chromatin modification and competence for initiation of meiotic recombination. *Cell* **118:** 439–452.

Reinke V, Smith HE, Nance J, Wang J, Van Doren C, Begley R, Jones SJM, Davis EB, Scherer S, Ward S, et al. 2000. A global profile of germline gene expression in *C. elegans*. *Mol Cell* **6:** 605–616.

Reinke V, Gil IS, Ward S, Kazmer K. 2004. Genome-wide germline-enriched and sex-biased expression profiles in *Caenorhabditis elegans*. *Development* **131:** 311–323.

Schedl T. 1997. Developmental genetics of the germline. In *C elegans II* (ed. Riddle DL, et al.), pp. 241–269. Cold Spring Harbor Laboratory Press, Cold Spring Harbor, NY.

Schmitges FW, Prusty AB, Faty M, Stutzer A, Lingaraju GM, Aiwazian J, Sack R, Hess D, Li L, Zhou S, et al. 2011. Histone methylation by PRC2 is inhibited by active chromatin marks. *Mol Cell* **42:** 330–341.

She X, Xu X, Fedotov A, Kelly WG, Maine EM. 2009. Regulation of heterochromatin assembly on unpaired chromosomes during *Caenorhabditis elegans* meiosis by components of a small RNA-mediated pathway. *PLoS Genet* **5:** e1000624.

Shiu PK, Raju NB, Zickler D, Metzenberg RL. 2001. Meiotic silencing by unpaired DNA. *Cell* **107:** 905–916.

Skipper M, Milne CA, Hodgkin J. 1999. Genetic and molecular analysis of fox-1, a numerator element involved in *Caenorhabditis elegans* primary sex determination. *Genetics* **151:** 617–631.

Spencer WC, Zeller G, Watson JD, Henz SR, Watkins KL, McWhirter RD, Petersen S, Sreedharan VT, Widmer C, Jo J, et al. 2011. A spatial and temporal map of *C. elegans* gene expression. *Genome Res* **21:** 325–341.

Tabuchi TM, Deplancke B, Osato N, Zhu LJ, Barrasa MI, Harrison MM, Horvitz HR, Walhout AJ, Hagstrom KA. 2011. Chromosome-biased binding and gene regulation by the *Caenorhabditis elegans* DRM complex. *PLoS Genet* **7:** e1002074.

Turner JM. 2005. Sex chromosomes make their mark. *Chromosoma* **114:** 300–306.

Turner JM, Mahadevaiah SK, Fernandez-Capetillo O, Nussenzweig A, Xu X, Deng CX, Burgoyne PS. 2005. Silencing of unsynapsed meiotic chromosomes in the mouse. *Nat Genet* **37:** 41–47.

van Leeuwen F, Gottschling DE. 2002. Genome-wide histone modifications: Gaining specificity by preventing promiscuity. *Curr Opin Cell Biol* **14:** 756–762.

Vielle A, Lang J, Dong Y, Ercan S, Kotwaliwale C, Rechtsteiner A, Appert A, Chen Q, Dose A, Egelhofer T, et al. 2012. H4K20me1 contributes to downregulation of X-linked genes for *C. elegans* dosage compensation. *PLoS Genet* **8:** e1002933.

Wang X, Zhao Y, Wong K, Ehlers P, Kohara Y, Jones SJ, Marra MA, Holt RA, Moerman DG, Hansen D. 2009. Identification of genes expressed in the hermaphrodite germline of *C. elegans* using SAGE. *BMC Genomics* **10:** 213.

Wells MB, Snyder MJ, Custer LM, Csankovszki G. 2012. *Caenorhabditis elegans* dosage compensation regulates histone H4 chromatin state on X chromosomes. *Mol Cell Biol* **32:** 1710–1719.

Wu CI, Xu EY. 2003. Sexual antagonism and X inactivation—The SAXI hypothesis. *Trends Genet* **19:** 243–247.

Xiong Y, Chen X, Chen Z, Wang X, Shi S, Wang X, Zhang J, He X. 2010. RNA sequencing shows no dosage compensation of the active X-chromosome. *Nat Genet* **42:** 1043–1047.

Xu L, Fong Y, Strome S. 2001. The *Caenorhabditis elegans* maternal-effect sterile proteins, MES-2, MES- 3, and MES-6, are associated in a complex in embryos. *Proc Natl Acad Sci* **98:** 5061–5066.

Yang H, Mizzen CA. 2009. The multiple facets of histone H4-lysine 20 methylation. *Biochem Cell Biol* **87:** 151–161.

Yonker SA, Meyer BJ. 2003. Recruitment of *C. elegans* dosage compensation proteins for gene-specific versus chromosome-wide repression. *Development* **130:** 6519–6532.

Yuan W, Xu M, Huang C, Liu N, Chen S, Zhu B. 2011. H3K36 methylation antagonizes PRC2-mediated H3K27 methylation. *J Biol Chem* **286:** 7983–7989.

WWW RESOURCE

http://www.wormbook.org/chapters/www_dosagecomp/dosagecomp.html Meyer BJ. 2005. X-chromosome dosage compensation. In *WormBook*.

C H A P T E R 24

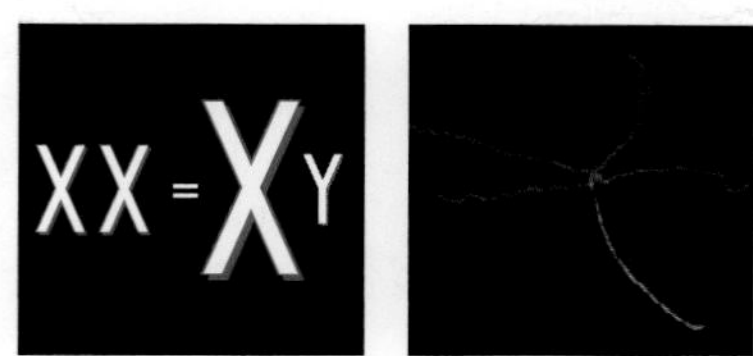

Dosage Compensation in *Drosophila*

John C. Lucchesi[1] and Mitzi I. Kuroda[2]

[1]Department of Biology, O. W. Rollins Research Center, Emory University, Atlanta, Georgia 30322; [2]Department of Genetics, Harvard Medical School, Boston, Massachusetts 02115

Correspondence: jclucch@emory.edu

SUMMARY

Dosage compensation in *Drosophila* increases the transcription of genes on the single X chromosome in males to equal that of both X chromosomes in females. Site-specific histone acetylation by the male-specific lethal (MSL) complex is thought to play a fundamental role in the increased transcriptional output of the male X. Nucleation and sequence-independent spreading of the complex to active genes serves as a model for understanding the targeting and function of epigenetic chromatin-modifying complexes. Interestingly, two noncoding RNAs are key for MSL assembly and spreading to active genes along the length of the X chromosome.

Outline

Cite as *Cold Spring Harb Perspect Biol* doi: 10.1101/cshperspect.a019398

OVERVIEW

In recent years, the long-standing dogma that cellular differentiation and development require the coordinate regulation of different sets of genes in time and space has led to the search for regulatory signals that would affect the activity of groups of functionally related genes. An example of coordinate regulation had been described in *Drosophila* long before these investigations were initiated. This involved a group of genes whose activity was regulated in unison, yet were not related by function. Rather, they shared a common location in the genetic material—the X chromosome. The purpose of the regulation was to ensure that females with two X chromosomes and males with only one X would have equal levels of gene products, in other words, to compensate for differences in the doses of X-linked genes between the sexes, commonly referred to as "dosage compensation." In studying this level of regulation, the question "How are groups of unrelated genes coordinately regulated?" became "What are the mechanisms that can regulate the activity of a whole chromosome?"

The study of dosage compensation in *Drosophila*, a mechanism that enhances the transcription of most of the genes on the single X chromosome in males, reveals the involvement of site-specific histone acetylation, X-specific noncoding RNAs (called roX1 and roX2), and chromosome-wide targeting of an evolutionarily conserved chromatin-modifying machine (called the MSL complex, for male-specific lethal complex).

 Cite as *Cold Spring Harb Perspect Biol* doi: 10.1101/cshperspect.a019398

1 THE PHENOMENON OF DOSAGE COMPENSATION WAS DISCOVERED IN *DROSOPHILA*

The karyotypes (i.e., ensemble of chromosomes) of many organisms include a pair of sex chromosomes. In *Drosophila*, females have two sex chromosomes called the X chromosomes that are identical in shape and genetic content; both X chromosomes are active in all somatic cells. Males have one X and a Y chromosome that differs from the X in morphology and genetic information that it contains. On the sex chromosomes there are genes that are responsible for sex determination and sexual differentiation. The Y chromosome is male specific, but the X chromosome carries many genes involved in basic cellular housekeeping functions or developmental pathways. Females with two X chromosomes have twice the number of these genes; males with a single X have only one dose. Yet, the level of the products of most of these genes is the same in the two sexes. In the early 1930s, this paradox was first noticed in *Drosophila* by H.J. Muller while he was studying the eye pigment level of individuals carrying partial loss-of-function X-linked mutations (Muller 1932). Muller reasoned that there must be a regulatory mechanism that helps flies to compensate for the difference in dosage of X-linked genes in males and females by equalizing the level of X-linked gene products between the two sexes. He called this hypothetical regulatory mechanism "dosage compensation" (Fig. 1).

Following its discovery in *Drosophila*, the phenomenon of dosage compensation was observed in additional species. We now know that in organisms belonging to distantly related groups—from round worms to mammals—transcriptional regulation leading to equal products of X-linked genes in males and females has been achieved in different ways: by decreasing the level of transcription of the two doses of X-linked genes in hermaphrodites relative to males (*Caenorhabditis elegans*) or by hypertranscribing the X chromosome in both males and females and then shutting down one of the two X chromosomes throughout most of its length in the somatic cells of females (mammals). The mechanisms underlying dosage compensation in these forms are described in Ch. 23 (Strome et al. 2014) and Ch. 25 (Brockdorff and Turner 2014).

The first evidence that dosage compensation in *Drosophila* is achieved by regulating the transcription of X-linked genes was obtained more than 30 years after Muller's seminal observations, by A.S. Mukherjee and W. Beermann (Mukherjee and Beermann 1965). Using transcription autoradiography of the giant polytenic chromosomes of larval salivary glands, a molecular technique that represented the state of the art at that time, these investigators observed that the level of [^{3}H]uridine incorporation by the single X in males and both Xs in females was equivalent. It appeared, therefore, that the rate of RNA synthesis by the single X chromosome in males was approximately twice the rate of each of the two Xs in females. The next experimental breakthrough consisted of the genetic identification by J. Belote and J. Lucchesi of four genes: *msl1*, *msl2*, *msl3*, and *mle*, with loss-of-function mutations that appeared inconsequential in females but lethal in males; notably, the mutant males showed approximately

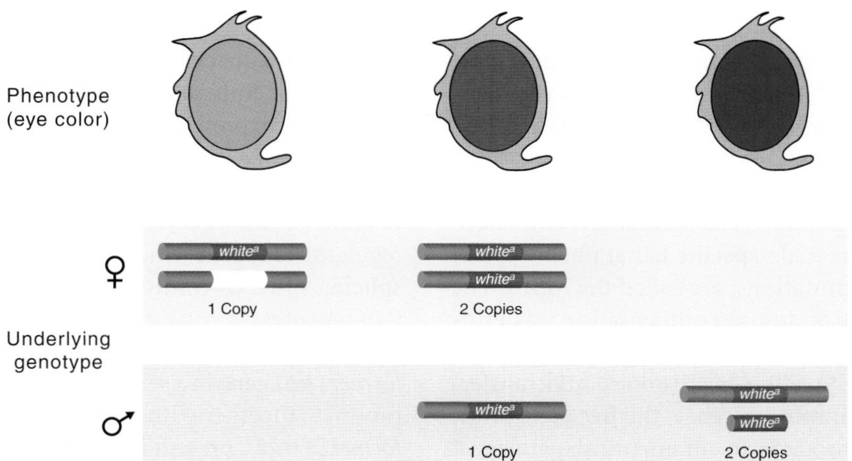

Figure 1. Diagrammatic representation of the results that led H.J. Muller to formulate the hypothesis of dosage compensation. The mutant allele of the X-linked *white* gene (w^a) is a hypomorph and allows partial eye-pigment synthesis; its presence on the X chromosomes is indicated. The level of pigmentation is directly proportional to the dosage of the w^a allele within each sex; yet, males with one dose and females with two doses have comparable amounts of pigment because of dosage compensation.

half of the normal level of [^{3}H]uridine incorporation by their X chromosome (Belote and Lucchesi 1980a,b). Furthermore, the X chromosome had lost its normal paler and somewhat puffed appearance that had been interpreted as an indication of an enhanced level of transcriptional activity in relation to each of the two X chromosomes in females. These results suggested that the equalization of X-linked gene products was achieved by doubling, on average, the transcriptional activity of the X chromosome in males rather than by halving the transcriptional activity of each X in females.

An alternate hypothesis was proposed based on an "inverse dosage effect," in which the activity of all chromosomes is set by general transcriptional regulators (reviewed in Birchler et al. 2011). In males, because of the absence of one X chromosome, a greater concentration of these regulators would be available than in females, driving the activity of all chromosomes to higher levels. For appropriate compensation to occur, the products of the *msl* loci would sequester some of these regulators away from the autosomes in males, thus leaving only the X chromosome with increased expression. In this model, *msl* gene mutations result in elevation of the expression of autosomal genes rather than a reduction of X-linked gene expression. However, a number of experimental results are incompatible with the inverse hypothesis (Arkhipova et al. 1997; Hamada et al. 2005; Straub et al. 2005; Deng et al. 2011). Particularly compelling is the recent observation that ectopic MSL complex on autosomes leads to a localized increase in transcription and suppression of phenotypes caused by haplo-insufficient mutants in the same region (Park et al. 2010).

Among the four genes introduced above, two were newly discovered (male-specific lethal 1, *msl1*; and male-specific lethal 2, *msl2*), whereas the other two (maleless, *mle*; and male-specific lethal 3, *msl3*) had been previously identified by other investigators in natural populations (specific references to this early phase of the study of dosage compensation can be found in Lucchesi and Manning 1987). For ease of reference, all of the gene products identified to date, on the basis of the male-specific lethal phenotype of their loss-of-function mutations, are called the MSLs. The next phase in the study of dosage compensation was initiated with the cloning of *mle* and the three *msl* genes, and the discovery and cloning of the *mof* histone acetyltransferase gene. By cytoimmunofluorescence, the five gene products were found to associate in an identical pattern at numerous sites along the polytene X chromosome in males (reviewed in Gelbart and Kuroda 2009). This observation and the interdependence of the different gene products for X-chromosome binding suggested that they form a complex. It is crucial for viability that the complex is present in males (XY) and absent in females (XX); therefore, the first step in dosage compensation is to establish this sex-specificity.

2 REGULATORS OF DOSAGE COMPENSATION

2.1 Regulation of Dosage Compensation Starts with Counting the Number of X Chromosomes

Each embryo needs to count its X chromosomes to make the critical decision whether or not to implement dosage compensation. An incorrect decision, such as failure to up-regulate the single male X or aberrant up-regulation of both female XXs, results in lethality. In *Drosophila*, the X-counting process is coordinated with the sex determination decision (reviewed in Cline and Meyer 1996). Phenotypic sex is determined by the number of X chromosomes per nucleus, such that XX embryos are females and XY embryos are male. The Y chromosome is required for male fertility, but unlike in mammals, it plays no role in phenotypic sex. Formally it is the X:autosome ratio that controls both sex and dosage compensation, as the X counting mechanism is sensitive to the number of sets of autosomes. This becomes apparent in 2X:3A triploids, which have an intermediate X:A ratio between XY:2A males and XX:2A females. 2X:3A triploids differentiate as intersexes with a mixture of both male and female cells.

The X:A ratio controls both sex determination and dosage compensation by regulating a critical binary switch gene, *Sex lethal* (*Sxl*). *Sxl* encodes a female-specific RNA-binding protein that regulates splicing and translation of key messenger RNAs (mRNAs) in the sex determination and dosage compensation pathways respectively (Fig. 2). The *Sex lethal* gene resides on the X chromosome and is positively regulated by transcription factors encoded by the X, such that embryos with two X chromosomes are able to initiate *Sxl* expression from an early, regulated promoter, P_e, whereas embryos with a single X per nucleus fail to express *Sxl* from P_e. This initial transient difference in activation of *Sxl* in early embryos is stabilized by an autoregulatory loop in which SXL protein positively regulates splicing of its own mRNA from a maintenance promoter that is expressed constitutively. SXL initiates differentiation in the female mode by regulating the splicing of the *transformer* (*tra*) gene in a sex-specific manner. In turn, this gene product (together with the product of another gene, *transformer2* (*tra2*), present in both sexes) directs the splicing of the *doublesex* (*dsx*) primary transcript to yield a regulatory protein that acts to repress genes required for male development, thus achieving female sexual differentiation. In male embryos, an alternate mode of splicing of the *dsx* transcripts occurs by default and leads to a product that

 Cite as *Cold Spring Harb Perspect Biol* doi: 10.1101/cshperspect.a019398

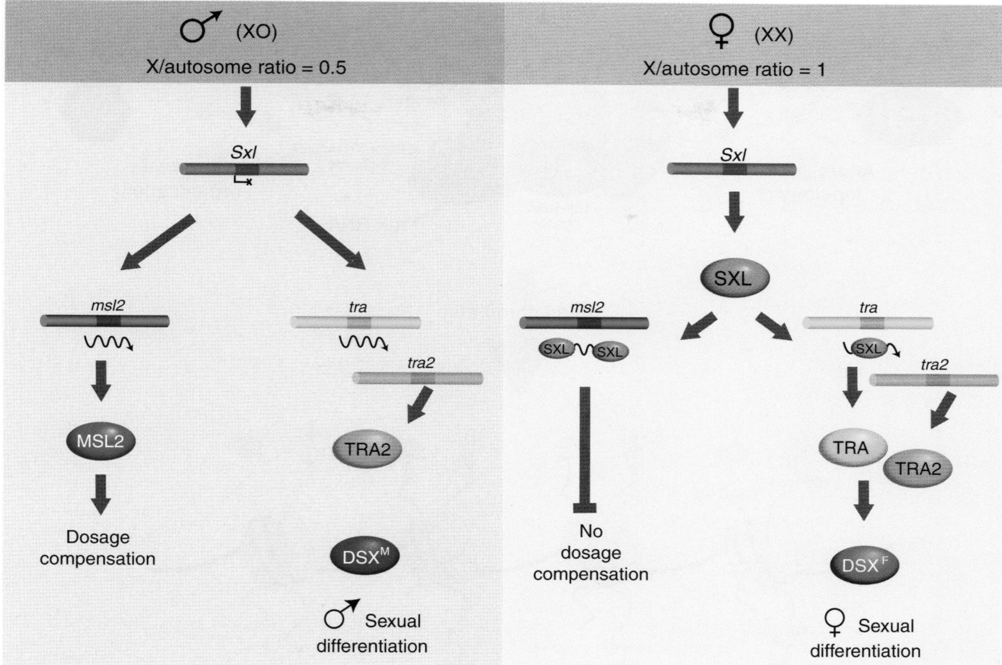

Figure 2. Diagram of the control of sex determination and dosage compensation. If the X/A ratio is equal to 1, a regulatory cascade leads to female sexual development. In females, the presence of the *Sxl* gene product prevents the translation of the *msl2* message and the assembly of the MSL complex. If the X/A ratio is only 0.5, absence of the cascade leads by default to male sexual development and to the formation of the MSL complex.

represses genes required for female development, resulting in male sexual differentiation.

2.2 The SXL Protein Prevents Formation of the MSL Complex in Females

The key target of SXL in the dosage compensation pathway is *msl2* mRNA (Bashaw and Baker 1997; Kelley et al. 1997). SXL binding sites are located in both the 5′ and 3′ untranslated regions (UTRs) of *msl2* mRNA. SXL is normally present only in females, in which it represses translation of the msl2 mRNA through association with its UTRs (see Fig. 2). If SXL is absent in females, dosage compensation is aberrantly turned on and these females die. Conversely, if SXL is ectopically expressed in males, dosage compensation is turned off and males die. Ectopic expression of MSL2 in females is sufficient to assemble MSL complexes on both female X chromosomes, indicating that all other MSL components are either turned on or stabilized by expression of MSL2.

In summary, dosage compensation must respond to the number of X chromosomes in the nucleus, and these are counted early in embryonic development. Females repress MSL2 translation, preventing inappropriate dosage compensation when two X chromosomes are present. In the absence of SXL-mediated repression, males express MSL2 protein and this leads to the assembly of a functional MSL complex. What are the components of the complex and how do they function together to regulate dosage compensation?

3 ASSEMBLY OF THE CHROMATIN-REMODELING COMPLEX RESPONSIBLE FOR COMPENSATION

The MSL complex, which is essential for male viability, consists of five known protein subunits and two noncoding RNAs (ncRNAs). The unifying function of the individual components of the MSL complex appears to be the targeting of histone H4K16 acetylation and additional chromatin-modifying activities to active X-chromosomal genes (Fig. 3). Association of MSL1 and MSL2 is essential for binding of the MSL complex to chromatin because these are the only subunits that associate with the X chromosome in the absence of the other MSL subunits (Copps et al. 1998; Gu et al. 1998). Whether or not there is a direct interaction with DNA is not known, as neither of these proteins contains an identifiable DNA-binding domain. Two MSL2 subunits interact with an MSL1 dimer and may initiate the assembly of the MSL complex (Hallacli et al. 2012). The direct interaction of MSL2 with MSL1 maps near the RING finger of MSL2 (a C_3HC_4 zinc-binding domain) and

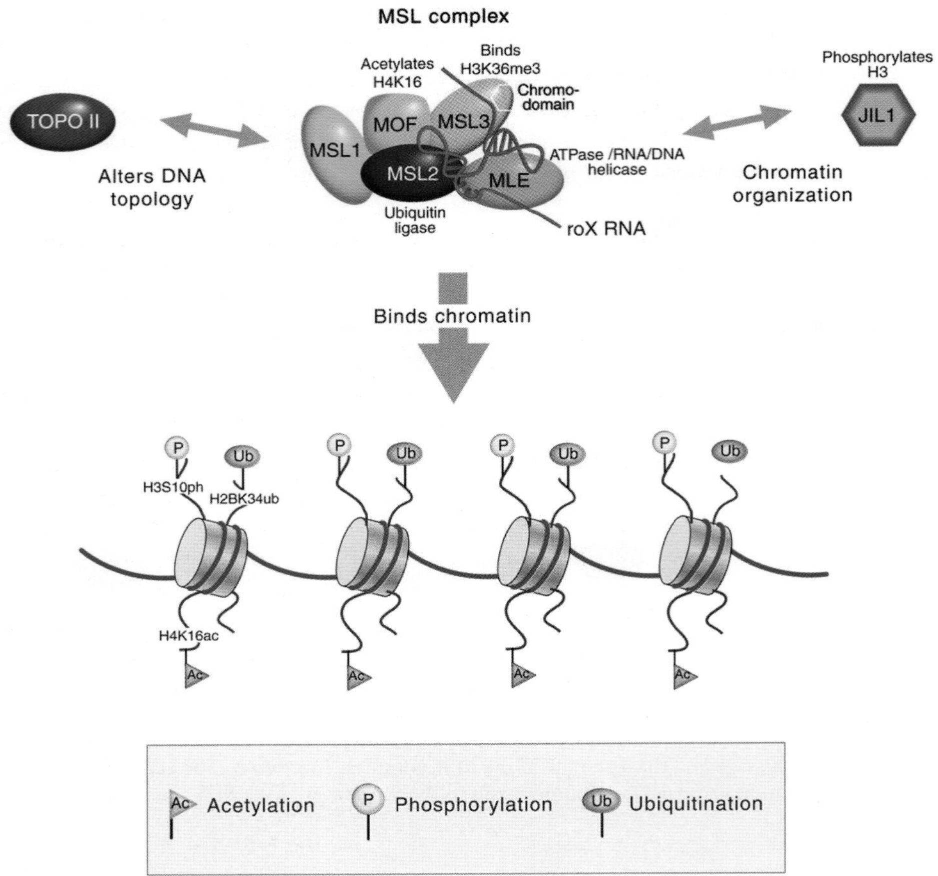

Figure 3. The various components of the MSL complex. Known or proposed functions of MSL components include the acetylation of histone H4K16 by MOF and the ubiquitination of H2BK34 by MSL2; MLE has ATPase and RNA/DNA helicase activity, and JIL-1 phosphorylates histone H3. The male-specific complex promotes enrichment of the general factors JIL-1 and topoisomerase II to the male X chromosome.

to an amino-terminal coiled coil domain in MSL1 (Scott et al. 2000). RING fingers are also associated with E3 ubiquitin ligase activity in many proteins, and both human and fly MSL2 show ubiquitin ligase activity directed toward histone H2B in vitro, dependent on interaction with MSL1 (Wu et al. 2011). MSL2 also ubiquitinates itself and other members of the MSL complex components, including MSL1, MSL3, and MOF, but not MLE in vitro (Villa et al. 2012). Current experiments are directed toward understanding the physiological role for this activity in vivo.

In addition to its essential role in chromatin binding, MSL1 forms a scaffold for interaction with MSL3 and MOF via adjacent conserved carboxy-terminal domains (Morales et al. 2004; Kadlec et al. 2011). In the absence of interaction with MSL2, MSL1 is destabilized, thus leading to failure of MSL complex formation (Chang and Kuroda 1998).

MSL3 belongs to a group of proteins that may have coevolved with the chromodomain-bearing histone acetyltransferases (HATs; Pannuti and Lucchesi 2000; see Ch. 4 [Marmorstein and Zhou 2014] for more on HAT structure and function). Yeast Eaf3, a member of the MRG15/MSL3 protein family, interacts via the chromodomain with methylated H3K36 and recruits the histone deacetylase complex Rpd3S to protect active genes from spurious transcription initiation within the coding regions (Carrozza et al. 2005; Joshi and Struhl 2005; Keogh et al. 2005). A similar interaction between the MSL3 chromodomain and active chromatin marks such as H3K36me3 may help the MSL complex to locate target genes (Fig. 3) (Larschan et al. 2007; Bell et al. 2008; Sural et al. 2008). Of particular interest is the existence of a human complex related to the MSL complex that includes human homologs of MOF, MSL1, MSL2, and MSL3 (Smith et al. 2005; Taipale et al. 2005). This complex, which specifically acetylates histone H4 at lysine 16 and is responsible for the majority of this histone isoform in human cells, can include one of three different versions of a *Drosophila* MSL3 homolog encoded by two different genes (Smith et al. 2005).

Cite as *Cold Spring Harb Perspect Biol* doi: 10.1101/cshperspect.a019398

Dosage compensation in *Drosophila* provided an early and compelling argument for a link between site-specific histone modifications and the regulation of gene expression. In 1992, Turner and colleagues made the seminal discovery that antibodies recognizing site-specific acetylation of histone H4 revealed distinct patterns on the *Drosophila* polytene chromosomes, with acetylation at position 16 (H4K16ac) showing strong enrichment on the male X chromosome (Turner et al. 1992). This enrichment requires the function of the *msl* genes (Bone et al. 1994), and the causal link between dosage compensation and chromatin modification was firmly established by the discovery that *mof* (males absent on the first) encodes a founding member of a new family of HATs (Hilfiker et al. 1997).

MOF is a member of the MYST subfamily of HATs. This subfamily, characterized by the presence of a chromodomain, can be further subdivided into enzymes that specifically acetylate lysine 16 in vivo (MOF and human MOF; Smith et al. 2005) and those such as Esa1p (essential SAS-related acetyl transferase 1) in yeast that acetylate all four terminal lysines of H4. Another MYST family member, SAS2, specifically acetylates lysine 16 in yeast, but lacks a chromodomain. Because the *mof* gene resides on the X chromosome, the discovery and characterization of mutants as male-specific lethals required a special genetic scheme to determine that an X-linked mutation lethal in males was viable in homozygous condition in females (Hilfiker et al. 1997).

mof encodes the MSL activity that is best implicated in gene regulation; therefore, one of the principal roles of the rest of the complex may be to localize MOF to its targets on the X chromosome. MOF recruitment is particularly important as MOF also participates in the nonspecific lethal (NSL) or MBD-R2 complex in both sexes (Raja et al. 2010; Feller et al. 2012; Lam et al. 2012). The NSL complex is found at 5′ ends of most active genes, coincident with MSL complex-independent H4K16 acetylation, and is essential for viability in both sexes. Because females are viable in the absence of MOF (albeit with low fertility), it is currently not clear how central MOF is to NSL function (Hilfiker et al. 1997; Gelbart et al. 2009).

MLE shows RNA/DNA helicase, adenosine triphosphatase (ATPase), and single-stranded RNA/single-stranded DNA binding activities in vitro (Lee et al. 1997), foreshadowing a potential role for RNA in MSL function. Mutants that retain the ATPase function but lack the helicase activity can still enhance transcription, but fail to support spreading of the complex along the X chromosome (Morra et al. 2008). Orthologs of MLE, which include human RNA helicase A (RHA), belong to the DEXH RNA helicase subfamily and are characterized by an additional domain implicated in double-stranded RNA binding (Pannuti and Lucchesi 2000). RHA is an abundant, essential protein in mammals, involved in numerous biological processes (Lee et al. 1998). It is likely that MLE performs its function in dosage compensation by interacting with or altering RNA structure, in particular the roX RNAs (see Sec. 4).

In addition to male-specific factors, it is likely that general factors involved in chromatin organization and transcription in both sexes also participate in dosage compensation. JIL-1, a tandem kinase, is found along all chromosomes in both males and females, but is more highly concentrated on the male X chromosome. This enrichment is dependent on the MSL complex. JIL-1 mediates histone H3 phosphorylation at Serine 10 and maintains open chromatin structure in transcriptionally active regions of the genome (Wang et al. 2001). However, despite its localization on active genes, definition of a direct role in transcription remains elusive (Ivaldi et al. 2007; Cai et al. 2008; Regnard et al. 2011).

In summary, the MSL proteins and JIL1 kinase have the ability to modify nucleosomes specifically on the male X chromosome. A central question is how these chromatin-modifying activities become targeted to a single chromosome. ncRNAs, DNA sequences, and active chromatin marks have all been implicated in the targeting process.

4 NONCODING roX RNAs FACILITATE ASSEMBLY AND TARGETING OF THE MSL COMPLEX ON THE X CHROMOSOME

One of the most intriguing and mysterious aspects of dosage compensation in both mammals and *Drosophila* is the role of ncRNAs in targeting compensation to the correct chromosome (reviewed in Gelbart and Kuroda 2009; also see Ch. 25 [Brockdorff and Turner 2014]). Two ncRNAs, called RNA on X (roX), are dissimilar in size and sequence, and yet function redundantly to target the MSL complex to the male X chromosome in *Drosophila* (Meller and Rattner 2002). Traditional mutant screens usually do not reveal the existence of genes that encode products with redundant functions such as the roX RNAs. Rather, roX RNAs were discovered by serendipity as male-specific RNAs in the adult brain (Amrein and Axel 1997; Meller et al. 1997). On closer examination, both RNAs displayed a lack of significant open reading frames and colocalization with the MSL complex along the length of the X. roX RNA function was not revealed until an X-chromosome mutant for both roX1 and roX2 was isolated. Most double mutant males die, with severely mislocalized MSL complexes, whereas single mutant males have no known phenotype (Meller and Rattner 2002). This is surprising in view of the fact that the two roX RNAs are very different in size (3.7 kb vs. 0.5–1.4 kb) and share little sequence similarity. A likely explanation for the apparent flexibility and sequence divergence of the roX RNAs comes from the discovery of a degenerate sequence, termed the roX box, which is found in multiple copies in each RNA and can participate in conserved secondary structures (Fig. 4) (Park et al. 2007; Kelley et al. 2008). An RNA, comprised primarily of tandem copies of these secondary structures, is capable of stimulating the H4K16 acetylation activity of the MSL complex in vivo (Park et al. 2007).

roX RNAs are recovered after coimmunoprecipitation of MSL proteins, demonstrating physical association of the RNAs with the complex (Meller et al. 2000; Smith et al. 2000). Partial purification of the complex suggests the presence of a tight core consisting of MSL1, MSL2, MSL3, and MOF proteins, with roX RNA and the MLE helicase lost except under very low salt concentrations (Smith et al. 2000). The minimal protein core complex lacking roX RNAs can still specifically acetylate histone H4 on lysine 16 within nucleosomes in vitro (Morales et al. 2004), and overexpression of MSL proteins can partially overcome the lack of roX RNAs, suggesting that the proteins possess all of the essential functions of dosage compensation but require the RNAs to stimulate assembly and spreading (Oh et al. 2003). Assembly of the complex is initiated when the secondary structure of the roX RNAs is modified, allowing the binding of MSL2 and providing the core for the full recruitment of the other MSL subunits (Ilik et al. 2013; Maenner et al. 2013).

5 HIGH-RESOLUTION ANALYSIS OF MSL BINDING ON THE X CHROMOSOME

Genomic analyses of MSL-binding sites have provided valuable insights into both the targeting principles and mechanism of action of the MSL complex. Chromatin immunoprecipitation (ChIP), enriching for MSL-associated DNA fragments, has been coupled to microarray or high-

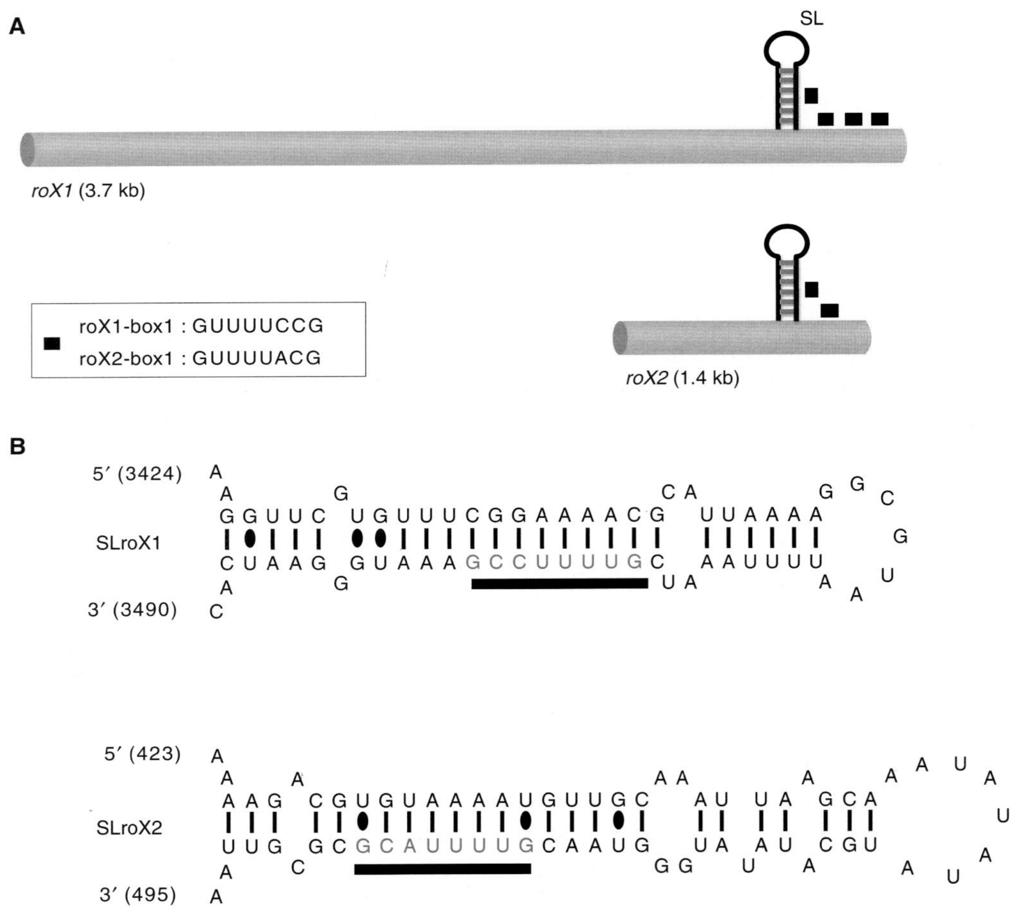

Figure 4. Evolutionarily conserved regions of *roX* RNAs. (*A*) Gene structure of *roX1* and *roX2* RNA showing location of stem-loop region (SL) and GUUNUACG roX boxes. (*B*) Stem-loop structures of the RNAs. (Modified from Maenner et al. 2013, © Elsevier.)

throughput DNA sequencing (ChIP-chip or ChIP-seq technology). Results show strong enrichment on the X chromosome, consistent with the previous cytological analyses. With the greatly improved resolution of ChIP, it was discovered that the MSL complex is enriched over the bodies of active X-linked genes, rather than binding primarily at the promoter or upstream intergenic regions as would be expected of a typical transcription factor (Fig. 5A,B) (Alekseyenko et al. 2006; Gilfillan et al. 2006). In fact, MSL binding is biased toward the 3′ ends of active genes, suggesting that it might act downstream from initiation, potentially at the level of elongation or the recycling of RNA polymerase II back to the promoter for reinitiation (Smith et al. 2001).

The full binding pattern does not reveal a simple solution to the targeting question. How might this pattern arise solely on the X chromosome? One can envisage two very general models for regulating a whole chromosome. A single site or a very limited number of sites might control the chromosome in *cis*, as is the case in mammalian X inactivation via the region called the X inactivation center (see Ch. 25 [Brockdorff and Turner 2014]). This mechanism requires either compartmentalization of the complex to a specific place in the nucleus or regulation over very long distances through the spreading of factors from the central control region to the rest of the chromosome. On the other extreme, a chromosome could have unique identifying sequences at every regulated gene along its entire

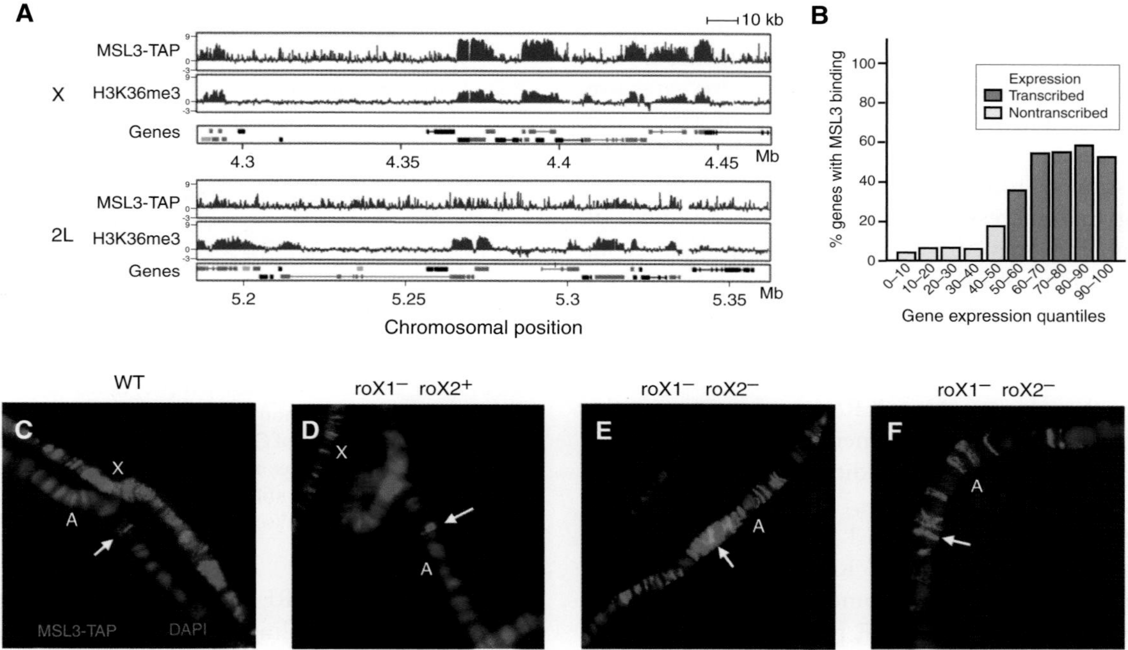

Figure 5. Localization of the MSL complex. (*A*) High-resolution ChIP-chip analysis of MSL3 binding and H3K36me3 along 180-kb sections of chromosomes X and 2L in males, illustrated on a logarithmic scale. MSL3 was tagged (MSL-TAP) to purify it using the tandem affinity purification (TAP) technique during ChIP analysis. Concomitant H3K36me3 ChIP-chip analysis shows that the MSL complex colocalizes with H3K36me3 on the middle and 3′ ends of transcribed genes in SL2 cells. Genes on the *top* line represent those expressed from left to right and those shown below are genes expressed from right to left. Rectangles represent exons, connected by lines that represent introns. Red genes are expressed, whereas black genes are not expressed. (*B*) Comparison of gene expression state and MSL binding. Genes were divided into quantiles by increasing Affymetrix expression values and graphed to show the percent of genes in each quantile that were clearly bound by MSL complex in ChIP-chip analysis. (*C–F*) Maximal autosomal spreading is achieved when a *roX* transgene is the only source of roX RNA in the cell. (*C*) Chromosomes from a male with a wild-type (WT) X; the presence of an autosomal *roX* transgene is indicated by the arrow, showing a narrow MSL band (red). X, X chromosome; A, autosome. (*D*) A male with only one active X-linked *roX* gene. The MSL complex spreads slightly more from the autosomal roX transgene than in wild type, but binding is reduced on the X chromosome. (*E,F*) Extensive spreading of the MSL complex in two roX transgenic male lines, in which the X chromosome has both *roX* genes deleted. (*A*, Adapted, with permission, from Alekseyenko et al. 2006, © Cold Spring Harbor Laboratory Press; and, with permission, from Larschan et al. 2007, © Elsevier; *B*, adapted, with permission, from Alekseyenko et al. 2006; *C–F*, adapted from Park et al. 2002, © AAAS.)

length. In this case, any segment of the chromosome could be regulated autonomously. The targeting model for the MSL complex appears to share characteristics of both of these possibilities, with a set of initiation sites potentially dispersing the complex in *cis* to its full set of targets along the chromosome.

roX RNAs are normally encoded by the X chromosome; the *roX1* gene is near the tip and the *roX2* gene is around the middle of the euchromatic part of the X. Like the *Xist* gene in mammals, the *roX* genes may reside on the X to target MSL complex assembly to this chromosome. When *roX* genes are moved to the autosomes as transgenes, they potently attract MSL proteins to their novel insertion sites (Fig. 5C–F), where the complex appears to spread in *cis*, variably into flanking sequences (Kelley et al. 1999). Under specific genetic conditions (e.g., when there are no competing endogenous *roX* genes on the X chromosome), extensive spreading from autosomal roX transgenes is consistently seen (Fig. 5E,F) (Park et al. 2002). This extensive spreading is augmented by overexpression of MSL1 and MSL2, the key limiting MSL proteins, and diminished by overexpression of roX RNA from competing transgenes, suggesting that successful cotranscriptional assembly of MSL complexes may drive local spreading (Oh et al. 2003). The ability of *roX* genes to direct MSL complexes to the wrong chromosome is one of the most intriguing aspects of the dosage compensation mechanism. However, in all cases in which *roX* genes direct spreading in *cis* on autosomes, they also provide roX RNA in *trans* to cover the X chromosome (Meller and Rattner 2002). Therefore, it is clear the X chromosome has additional targeting signals beyond the two known *roX* genes.

In the absence of either MSL1 or MSL2, none of the remaining MSL proteins or roX RNAs appears to retain specific recognition for the X chromosome. However, in the absence of MLE, MSL3, or MOF, partial MSL complexes bind a subset of approximately 35–70 sites by cytological mapping, including the two *roX* genes (Fig. 6A). Termed high-affinity sites (HASs) or chromatin entry sites (CESs), these were postulated to be nucleation sites that might enable the MSL complex to access the X chromosome. Again, a genomic approach has provided key insights into how targeting might occur. Results from ChIP of MSL2 in *msl3* mutant embryos, or MSL1 or MSL2 in suboptimal crosslinking conditions, identified a set of 130–150 candidate entry sites by their high affinity (Alekseyenko et al. 2008; Straub et al. 2008). These sites included the previously characterized entry sites in the *roX1* and *roX2* genes (Park et al. 2003). Motif searches yielded a 21-bp GA-rich (or TC-rich) common sequence motif, named the MSL recognition element or MRE (Fig. 6B). In functional assays, CES sequences attract the MSL complex when moved to autosomes. Conversely, MRE mutations abolish MSL recruitment, whereas scrambling the surrounding sequences has no effect (Alekseyenko et al. 2008). Therefore, MREs sprinkled along the length of the X are likely to play a key role in MSL recognition of the X chromosome. Recently, a zinc finger protein, CLAMP (chromatin-linked adaptor for MSL proteins), was found to bind directly to the MREs, with particular affinity for those MREs present within the CES sites; in the absence of CLAMP, the MSL complex was depleted along the entire X chromosome, suggesting that it functions in recruiting the complex to its initial binding sites (Larschan et al. 2012; Soruco et al. 2013).

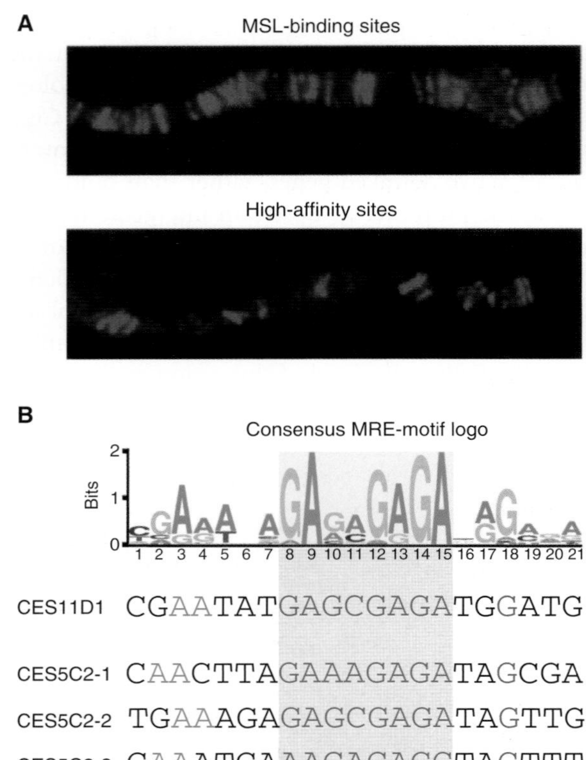

Figure 6. Targeting of the MSL complex. (*A*) The MSL complex is found associated with numerous sites along the X chromosome in males (*upper* panel); a mutant or incomplete complex containing at least MSL1 + MSL2 is found at fewer sites called CESs or HASs. (*B*) A GA-rich motif is a common feature in MSL CESs. The motif logo is shown with two examples of CESs presented below. This motif occurs once in CES11D1 and three times in CES5C2. The GA-rich core is highlighted in red. (*A*, Modified from Gu et al. 2000; *B*, reprinted, with permission, from Alekseyenko et al. 2008, © Elsevier.)

The identification of a functional DNA sequence element is a significant step toward understanding how the MSL complex initiates binding to the X chromosome. However, the MRE is less than twofold enriched on the X, and like the sites for most sequence-specific binding factors, there is an excess of nonused sites that match the consensus motif. Although it is enriched on the X chromosome, the CLAMP protein is present at the 5′ end of active genes on all chromosomes in both sexes. Therefore, there must be additional features, such as the local chromatin context (Alekseyenko et al. 2012), that help distinguish the MREs on the X that are used in vivo.

6 TRANSITION FROM INITIATION SITES TO TARGET GENES

If entry sites enable sequence-specific binding of the MSL complex to the X chromosome, how does the complex reach the majority of its target sites (i.e., transcriptionally active genes on the X chromosome)? Experiments to date support a model in which initial recruitment of the MSL complex to CESs generates high local concentrations that drive spreading to nearby sites of lower affinity. Recent evidence, based on high-shear ChIP-seq, suggests that MSL2 and MLE contact the HASs directly and provide a platform for the indirect association of the other subunits and roX RNA at these sites (Straub et al. 2013). Movement of the complex to active genes is facilitated by the activity of MOF and MLE (Gu et al. 2000; Morra et al. 2008) and is stabilized by the binding of MSL3 to H3K36me3 (Larschan et al. 2007).

It is not known whether spreading is a linear process, whereby the complex scans along chromatin and is only stabilized at active genes, or whether spreading is discontinuous, sampling chromatin in three-dimensional space. In either case, the majority of the MSL complex is targeted within 10 kb of a CES (Sural et al. 2008). Interestingly, the MSL-dependent H4K16ac mark is more widely distributed along the X chromosome than documented MSL binding, suggesting the transient interaction of the complex with a larger fraction of the X than that observed to have stable binding (Fig. 7) (Gelbart et al. 2009; Conrad et al. 2012; Straub et al. 2013).

The mechanism for recognizing active genes may involve additional sequence elements, features of transcription, or both. Ectopic MSL binding appears at regulated transgenes inserted on the X only upon induction of expression (Fig. 8) (Sass et al. 2003). The localization of the MSL complex to active genes on autosomes when ectopically recruited by a roX transgene suggests that the critical targeting features are not restricted to X-linked genes (Kelley et al. 1999). One clue suggesting the identity of one of these features came from the observation that the distribution of MSL complex along genes is strongly coincident with the pattern of H3K36me3, a histone modification associated with active genes on all chromosomes (Fig. 5A). This interaction is the hallmark of coding regions and is less pronounced in long intronic sequences in which H3K36me3 is less abundant (Straub et al. 2013). SET2, the enzyme responsible for H3K36 trimethylation, is attracted to gene bodies by the elongating form of RNA polymerase II. In the absence of SET2 activity, the MSL complex shows diminished binding to target genes on the X chromosome, suggesting a role for H3K36me3 in spreading (Larschan et al. 2007). The fact that binding is decreased, but not eliminated, in a *set2* mutant suggests that transcriptionally active genes are recognized by additional, partially redundant mechanisms. For example, roX RNAs participate in spreading by an unknown mechanism.

An 80-kb segment of autosomal DNA inserted on the X is capable of MSL recruitment to its active genes with no evidence for skipping, suggesting that there is not an obligate requirement for an X-specific sequence within active genes (Gorchakov et al. 2009). However, deletion mapping of X-linked genes inserted on autosomes has revealed specific gene segments that function in conjunction with transcription to attract the MSL complex (Kind and Akhtar 2007). As yet, not enough of these examples have been described to deduce common sequence characteristics

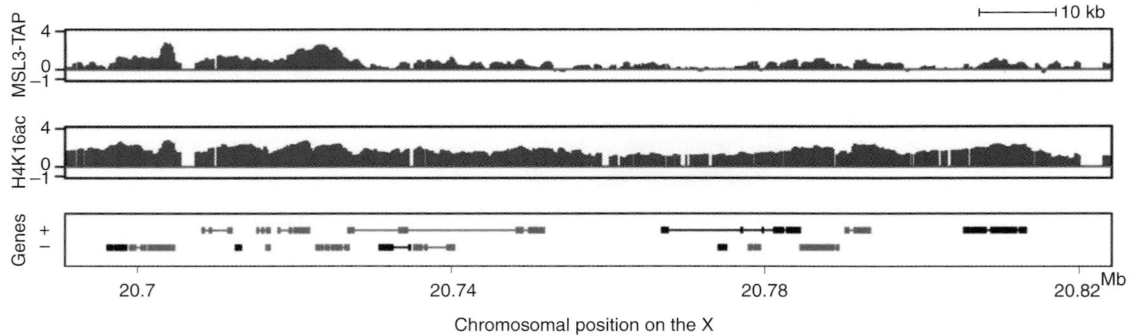

Figure 7. Correlation of H4K16 acetylation and MSL complex binding on the male X chromosomes. The distribution of H4K16ac on the male X chromosome is broader than MSL complex; active genes that lack stable MSL binding are nonetheless associated with H4K16ac. See Figure 5 for explanation of gene representation. (Adapted from Gelbart et al. 2009.)

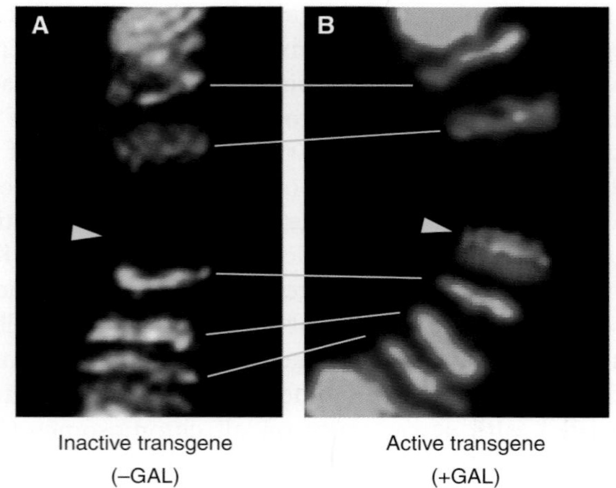

Figure 8. The MSL complex targets activated genes. (*A*) A construct containing a promoter under the control of the *trans*-activator GAL4 has been inserted at a site on the X indicated by the yellow arrowhead. This region is normally devoid of the MSL complex in larval salivary gland chromosomes. (*B*) When GAL4 is introduced, it binds to the construct (red), activates it, and recruits the MSL complex (blue). (Adapted, with permission, from Sass et al. 2003, © National Academy of Sciences.)

that can be tested experimentally by mutagenesis. In stable translocation stocks, spreading of MSL complexes from the X into contiguous autosomal sequences is not evident (Fagegaltier and Baker 2004; Oh et al. 2004). Therefore, even if spreading of MSL complexes is a major mechanism for covering the X chromosome, there is very likely an additional characteristic of the X that causes MSL complex to strongly favor X over autosomal binding. Possibilities include a distinct overall sequence composition, including enrichment for simple repeats that can be detected by principal component analysis (Stenberg and Larsson, 2011), and the distinct three-dimensional organization of each chromosome into its own "territory" within the nucleus (see Sec. 9).

How can these various observations be accommodated into an integrated scheme for X-chromosome targeting of the MSL complex? A model that is the best fit for existing data is depicted in Figure 9. In this model, MSL complexes assemble at CESs, in particular, the sites of roX RNA transcription, and subsequently access flanking and distant sites on the X based on their X-linkage and transcriptional activity.

7 CHROMATIN MODIFICATIONS ASSOCIATED WITH DOSAGE COMPENSATION

A key modification that is correlated with the association of the MSL complex with the X chromosome in males is the presence of a high level of histone H4 acetylated at lysine 16 (Turner et al. 1992; Bone et al. 1994). This chromatin mark occurs throughout active transcriptional units with a bias toward the middle and the 3′ end (Fig. 7) (Kind et al. 2008; Gelbart et al. 2009). In yeast, this particular covalent modification of histone H4 plays a key role in maintaining the boundary between silent and active chromatin; loss of function of Sas2, the HAT responsible for H4K16ac, allows the spreading of telomeric heterochromatin into adjacent subtelomeric chromatin (Suka et al. 2002; for more detail, see Ch. 8 [Grunstein and Gasser 2013]). Structural studies have indicated that a key internucleosomal interaction may occur between an acidic patch of the histone H2A-H2B dimer on one nucleosome and a positively charged segment of the histone H4 tail (residues 16–26) extending from a neighboring nucleosome (Luger et al. 1997; Schalch et al. 2005). When lysine 16 is acetylated, its positive charge becomes neutral suggesting that weakening a repressive internucleosomal structure could play a key role in dosage

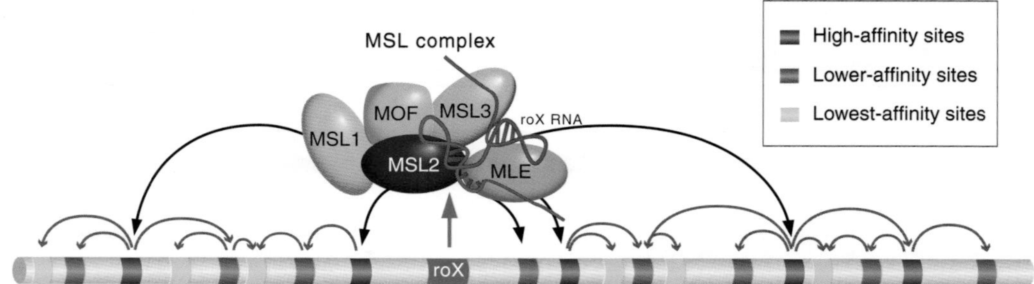

Figure 9. Model for the targeting of the MSL complex to the X chromosome. The dosage compensation complex in *Drosophila* is proposed to implement at least three targeting principles: (i) interaction with the sites of noncoding roX RNA synthesis, (ii) chromatin context–dependent binding to degenerate DNA sequences of varying affinities, and (iii) DNA sequence–independent movement from initiation sites to chromatin marks signaling gene expression (reviewed in Gelbart and Kuroda 2009). This movement has been characterized as "spreading" based on its apparent restriction in *cis* to the chromosome of origin, but the underlying molecular mechanism remains to be understood.

compensation. This contention was supported by the demonstration that reconstituted nucleosomal arrays acetylated at lysine 16 of histone H4 cannot achieve the level of salt-induced condensation of nonacetylated arrays (Shogren-Knaak et al. 2006; Robinson et al. 2008) and that this acetylation also weakens the self-association of reconstituted single nucleosome particles, reflecting the specific role of H4K16 in nucleosome–nucleosome stacking (Liu et al. 2011). Using molecular force spectroscopy, the acetylation of H4K16 was observed to weaken nucleosome packing in reconstituted chromatin fibers and to result in a more disordered architecture (Dunlap et al. 2012). Whether, in vivo, it is the formation of the 30-nm fiber (intramolecular compaction) or the higher-order 100- to 400-nm fibers (intermolecular compaction) that are affected by the presence of H4K16ac is not known (Shogren-Knaak et al. 2006). In either case, the presence of H4K16ac renders the chromatin of dosage compensated genes more accessible to factors or complexes. This is evidenced by the significantly greater accessibility of the compensated male X chromosome to an extrinsic DNA-binding protein (e.g., the bacterial DNA methyltransferase). The elevated accessibility of this protein follows the distribution of H4K16ac along the X (Bell et al. 2010). Given the long-standing correlation between active chromatin and early DNA replication (Hiratani and Gilbert 2009; Schubeler et al. 2002), it is not surprising that another feature of the compensated X chromosome is that it initiates replication earlier in S phase than the rest of the genome (Lakhotia and Mukherjee 1970; Bell et al. 2010).

Another finding is that the X chromosome is more susceptible to mechanical shearing than the autosomes in both males and females, indicating that it has a more open chromatin structure. In fact, the histone marks associated with active gene transcription (H3K4me2 and H3S10ph), as well as those specifically enriched on the dosage compensated X (H4K16ac) in males, are also slightly enriched on the X chromosomes of females. These findings suggest that the evolution of the distinct chromatin structure responsible for dosage compensation in males has affected the female X (Zhang and Oliver 2010).

In addition to the effect of H4K16 acetylation, there is mounting evidence that dosage compensation involves changes in the torsional stress of X-linked genes. Reducing the level of supercoiling factor, a protein known to associate at the 5′ end of active genes (Ogasawara et al. 2007), preferentially affects male viability because of a sex-specific decrease in the transcription of X-linked genes (Furuhashi et al. 2006). Compensated chromatin is topologically different from noncompensated chromatin. The difference requires the function of topoisomerase II, which associates with the MSL complex and is recruited to compensated genes in excess of the amount present on autosomal genes with similar transcription levels (Cugusi et al. 2013).

The rate of histone variant H3.3 incorporation into the X chromosome in male cells is enhanced in relation to the autosomes (Mito et al. 2005). This is to be expected, as the replication-independent nucleosome deposition of H3.3 occurs in transcriptionally active regions of chromatin and involves the replacement of histone H3 with the variant H3.3 (see Ch. 20 [Henikoff and Smith 2014]). However, contrary to the expectation that this increased level of H3.3 on the X may contribute to the mechanism of dosage compensation is the observation that the absence of the two genes that encode H3.3, although causing sterility in both sexes, has no effect on viability of mutant flies (Hodl and Basler 2009).

8 A MODEL FOR THE MECHANISM OF COMPENSATION

Gene expression can be regulated at multiple steps, particularly during transcription initiation, release from pausing, or elongation. A number of considerations have suggested that the transcriptional enhancement of X-linked genes responsible for dosage compensation occurs at the elongation step of transcription rather than at initiation. Foremost was the observation that the high level of H4K16 acetylation mediated by the MSL complex occurs throughout the length of transcriptional units with a bias that favors their 3′ ends rather than the promoter regions (Smith et al. 2001). Furthermore, genes with "weak" promoters and genes with "strong" promoters coexist on the X chromosome; in males, the activity of both types of genes is enhanced approximately twofold by the dosage compensation mechanism.

Recently, supporting evidence for the hypothesis that the MSL complex enhances transcription during elongation was obtained using GRO-seq (global run-on sequencing; Core et al. 2008). The experiments, measuring the relative density of RNA polymerase in male nuclei, found an increase in density specifically on X-linked gene bodies compared to autosomal genes, which was no longer evident after RNAi depletion of the MSL complex (Larschan et al. 2011). The nature of the mechanism underlying this experimental observation is not yet fully understood. In particular, it is not known whether increased density of RNA polymerase in gene bodies reflects a corresponding increase in initiation not evident in the GRO-seq assay, improved processivity (i.e., a decrease in premature termination), or both. However, the localization of the MSL complex and increased H4K16 acetylation is consistent with a direct effect on elongation. As modeled in vitro, the H4K16 acetylation of nucleosomes, widely spread along dosage compensated transcriptional units, might diminish internucleosomal interactions in vivo, thereby facilitating nucleosome eviction and RNAPII progression. MSL2 is an E3 ubiquitin ligase that ubiquitinates histone H2B in vitro; its human ortholog is an E3 ubiquitin ligase that ubiquitinates histone H2B and facilitates the methylation of H3K4 and H3K79 (Wu et al. 2011). Ubiquitinated H2B and methylated H3K79 represent histone modifications that are important for transcription elongation (Minsky et al. 2008). A similar function for MSL2 in *Drosophila* would bolster the role of the MSL complex in facilitating transcription elongation.

Clearly, an enhanced rate of elongation of RNAPII alone is not sufficient to generate the increase in steady-state level of X-linked gene transcripts necessary for dosage compensation. One possibility is that transcription along the length of genes is not always successful, and that the processivity of RNA polymerase II can be improved to increase final RNA production. If the dosage compensation mechanism is, in fact, based on enhancing the rate of elongation, then it is clear that to achieve an increase in the steady-state level of X-linked gene transcripts, there must occur a concomitant increase in the frequency of recruitment of polymerase or release from pausing. Given the tight relationship between these two phases of transcription, distinguishing the primary mechanism for up-regulation may require additional experimentation and perhaps technical advances not yet realized.

9 DOSAGE COMPENSATION AND NUCLEAR ORGANIZATION

During interphase, chromosomes are seen to occupy individual territories rather than an intermingling of unraveled chromatin strands (Cremer and Cremer 2010; see Ch. 19 [Dekker and Misteli 2014]). This organization is particularly evident in cells of *Drosophila* males in which the X chromosome can be identified by the presence of the MSL complex (Strukov et al. 2011). Within this compartment, the chromatin modifications that underlie the mechanism of dosage compensation appear to induce a particular higher-order topography to the X chromosome. Throughout development, X-linked sites that are separated by approximately a dozen megabases are located much closer in male than in female nuclei (Fig. 10) (Grimaud and Becker 2009). This difference is dependent on the presence of the MSL1-MSL2 chromatin-recognition component of the complex and is not affected by the absence of the other three MSL proteins. Because, in the absence of any one of the latter, the partial complex that includes MSL1-MSL2 is found only at HASs, the proximity of X-linked loci in male cells must be mediated by their clustering (Grimaud and Becker 2009).

It is interesting to note that the MSL proteins copurify with the nuclear pore complex proteins Nup153 and Megator (Mtor). Regions of the genome at the nuclear periphery that are proximal to the nuclear pore complex contain

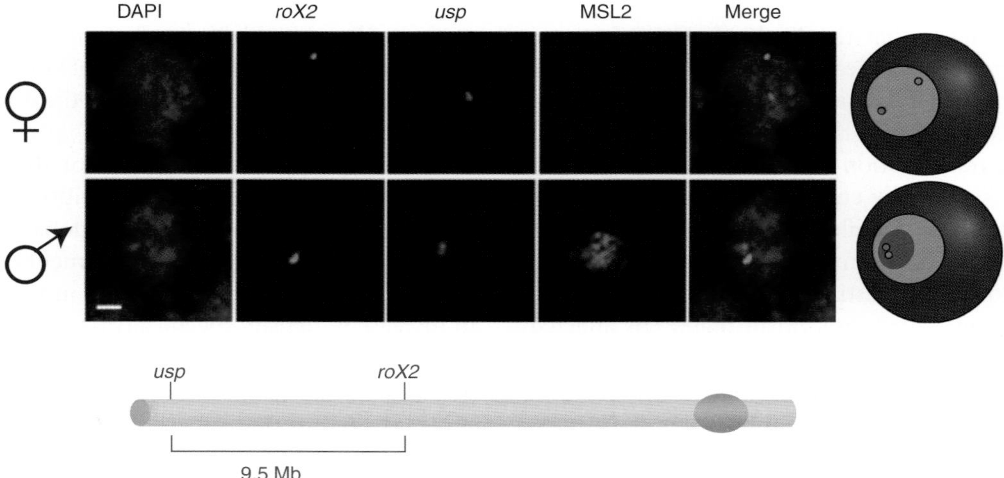

Figure 10. Male-specific conformation of the dosage-compensated X chromosome. A pair of high-affinity chromosomal sites (*roX2* and *usp*) were visualized by two-color FISH (fluorescence in situ hybridization) in female or male embryos. DNA was stained with DAPI (blue) and the X-chromosome territory (magenta) was painted with an antibody against MSL2 in male nuclei (there is no MSL2 in female nuclei). A merge of the channels reveals the proximity of the HASs and their residence relative to the MSL2 territory in male nuclei, clearly summarized in the cartoon on the right. The schematic diagram showing part of the X chromosome below indicates the distances separating the different HASs. (Modified, with permission, from Grimaud and Becker 2009, © Cold Spring Harbor Laboratory Press.)

 Cite as *Cold Spring Harb Perspect Biol* doi: 10.1101/cshperspect.a019398

groups of active genes, suggesting this compartment may have a regulatory effect on transcription (Vaquerizas et al. 2010). Depletion of the Nup153 and Mtor nucleoporins leads to the loss of dosage compensation (Mendjan et al. 2006), although this analysis could be complicated by general viability issues (Grimaud and Becker 2009).

10 INFLUENCE OF GLOBAL CHROMATIN FACTORS ON THE MALE X CHROMOSOME

The male polytene X chromosome shows special sensitivity to changes in the dosage or activity of several chromatin regulators that are thought to be general, non-X-chromosome-specific factors. For example, a functional interaction between ISWI-bearing complexes and the MSL complex was brought to light by the observation that loss-of-function mutations of *Iswi* (imitation switch) lead to a global structural defect on the X chromosome in salivary gland preparations of males (Deuring et al. 2000). ISWI is an ATPase found in four chromatin-remodeling complexes of *Drosophila*: NURF (nucleosome remodeling factor), ACF (ATP-dependent chromatin assembly and remodeling factor), CHRAC (chromatin accessibility complex; Hargreaves and Crabtree 2011), and RSF (remodeling and spacing factor; Hanai et al. 2008). In vivo, ACF and CHRAC behave as assembly factors, promoting the formation of chromatin, particularly repressive chromatin states. Nevertheless, high-resolution mapping of nucleosomes in *Iswi* mutant males and females indicates that the chromosome condensation defects that they show, especially of the X chromosome in males, are not correlated with global ISWI-dependent nucleosome spacing changes (Sala et al. 2011). A more convincing explanation is provided by the observation that ISWI facilitates the association of linker histone H1 and that a reduction of the H1 leads to a chromosomal phenotype similar to the one described above for ISWI loss of function (Siriaco et al. 2009). Somewhat surprisingly, loss of function of the RSF complex does not appear to alter chromosome structure (Hanai et al. 2008).

Early functional studies of the NURF complex indicated that it might be involved in enhancing transcription in some cases or repressing it in others (Badenhorst et al. 2002). Mutants in *nurf301* show the abnormally decondensed male X seen in *Iswi* mutants (Badenhorst et al. 2002). NURF also has a specific effect on *roX1* and *roX2* transcription; wild-type NURF negatively regulates these genes in females and reduces the level of transcription of roX2 in males by approximately one half (Bai et al. 2007).

The X-chromosome defect visible in *Iswi* and *nurf* mutant male salivary glands does not occur when the MSL complex is inactive (i.e., in the absence of H4K16ac). In contrast, the absence of either *roX* gene reduces the abnormal puffing of the X in the region of the roX mutation. This observation highlights the localized nature of opposing activities that are responsible for normal chromatin organization (Bai et al. 2007).

The down-regulation of certain structural components of heterochromatin, such as Su(var)3-7 (suppressor of variegation 3-7) and HP1 (heterochromatin protein 1) leads to a polytene X-chromosome phenotype that is similar to the one resulting from ISWI knockdown (Delattre et al. 2004; Spierer et al. 2005). In these cases as well, the bloated appearance requires the presence of an active MSL complex.

In wild-type males, the distribution of heterochromatin factors along the X chromosome may be regulated by the action of JIL-1, a histone H3 serine 10 kinase. JIL-1 localizes to active gene bodies with a 3′ bias, is approximately twofold more abundant on the X chromosome than the autosomes in males (Jin et al. 1999, 2000) and is necessary for proper dosage compensation of eye pigmentation (Lerach et al. 2005). Loss-of-function alleles result in global changes in the morphology of polytene chromosomes in both males and females, whereas the male X is once again shorter, fatter, and without any evidence of banding (Deng et al. 2005). JIL-1 loss-of-function alleles allow the spreading of H3 dimethylated at lysine 9 (H3K9me2), the modification that attracts HP1 to inactive chromatin, suggesting that JIL-1 normally marks and preserves the limits of euchromatic domains (Ebert et al. 2004; Zhang et al. 2006).

11 HOW DID DOSAGE COMPENSATION EVOLVE?

In *Drosophila*, the evolution of the regulatory mechanism of dosage compensation has been correlated to the evolution of heteromorphic sex chromosomes represented by an X chromosome that is similar to the major autosomes in the density of transcribing genes per unit length and a Y that is largely heterochromatic. The sex chromosomes are thought to have originated from the occurrence of an autosomal mutation that determined one of two mating types. In *Drosophila* this mating type became the heterogametic male sex. Over time, the autosome bearing the initial mutation degenerated by the concomitant reduction of recombination and the random occurrence of deleterious mutations that were retained if they were linked to an occasional beneficial mutation. In many organisms, the difference in levels of expression of genes present in two doses in the homogametic sex and in a single dose in the heterogametic sex appears to be tolerated and specific adjustments regulate the transcription of critical individual genes in the two sexes. In flies, mammals, and worms, however, the degeneration of the genetic content of the incipient Y provided the selective pressure necessary for the evolution of a chromosome-wide compensatory mechanism that increases the expression of the alleles present on its homolog, the X chromosome (Charlesworth 1978; Lucchesi 1978). Evidence for the occurrence of this evolutionary phenomenon can be found in several *Drosophila* species that have undergone whole chromosome arm fusions (reviewed in Charlesworth 1978). The wealth of classical cytological information, as well as the complete genomic sequencing of numerous *Drosophila* species, provides an invaluable window into the evolution of sex chromosomes and the concomitant mechanism of dosage compensation.

12 OUTLOOK

Dosage compensation is an epigenetic regulatory mechanism that presents several unique characteristics: It is a paramount example of the control of the expression of individual genes at the level of an entire chromosome, and it is mediated by a multiprotein complex (MSL) that includes not only enzymatic subunits with known chromatin-modifying functions, but also two long ncRNAs that are required for its assembly. Very substantial progress has been accomplished in the past few years in elucidating various aspects of the chromatin modifications effected by the MSL complex and in identifying some of the parameters that regulate its targeting. Further understanding of the targeting process will be facilitated by the molecular analysis of evolving sex chromosomes in different *Drosophila* species with fully sequenced genomes (see www.FlyBase.org). An understanding of the function of the roX RNAs in spreading of the MSL complex will make use of new RNA, rather than protein-based ChIP mapping techniques (Chu et al. 2011; Simon et al. 2011). The role of histone modifications, namely H4K16 acetylation and H2B ubiquitination, will be assessed by replacing canonical histones with multiple copies of modified histone transgenes (Gunesdogan et al. 2010). Topological characteristics of dosage compensated chromatin units will be determined by biophysical experiments (Kruithof et al. 2009; Allahverdi et al. 2011). An important insight into the function of the MSL complex will be to determine its status during the cell cycle and how it manages to survive through DNA replication and mitosis (Lavender et al. 1994; Strukov et al. 2011). In summary, a full understanding of all of these aspects of the regulatory mechanism will contribute key information regarding the organization of chromatin into transcriptional domains and the role that chromosomal organization and nuclear topology play in gene regulation (the topic of Ch. 19 [Dekker and Misteli 2014]); it will also provide important insights into molecular mechanisms that fine-tune the heritable expression of genes within precise ranges.

REFERENCES

*Reference is also in this book.

Alekseyenko AA, Larschan E, Lai WR, Park PJ, Kuroda MI. 2006. High-resolution ChIP-chip analysis reveals that the *Drosophila* MSL complex selectively identifies active genes on the male X chromosome. *Genes Dev* **20:** 848–857.

Alekseyenko AA, Peng S, Larschan E, Gorchakov AA, Lee OK, Kharchenko P, McGrath SD, Wang CI, Mardis ER, Park PJ, et al. 2008. A sequence motif within chromatin entry sites directs MSL establishment on the *Drosophila* X chromosome. *Cell* **134:** 599–609.

Alekseyenko AA, Ho JWK, Peng S, Gelbart M, Tolstorukov MY, Plachetka A, Kharchenko PV, Jung YL, Gorchakov AA, Larschan E, et al. 2012. Sequence-specific targeting of dosage compensation in *Drosophila* favors an active chromatin context. *PLoS Genet* **8:** e1002646.

Allahverdi A, Yang R, Korolev N, Fan Y, Davey CA, Liu CF, Nordenskiold L. 2011. The effects of histone H4 tail acetylations on cation-induced chromatin folding and self-association. *Nucleic Acids Res* **39:** 1680–1691.

Amrein H, Axel R. 1997. Genes expressed in neurons of adult male *Drosophila. Cell* **88:** 459–469.

Arkhipova I, Li J, Meselson M. 1997. On the mode of gene-dosage compensation in *Drosophila. Genetics* **145:** 729–736.

Badenhorst P, Voas M, Rebay I, Wu C. 2002. Biological functions of the ISWI chromatin remodeling complex NURF. *Genes Dev* **16:** 3186–3198.

Bai X, Larschan E, Kwon SY, Badenhorst P, Kuroda MI. 2007. Regional control of chromatin organization by noncoding roX RNAs and the NURF remodeling complex in *Drosophila melanogaster. Genetics* **176:** 1491–1499.

Bashaw GJ, Baker BS. 1997. The regulation of the *Drosophila msl-2* gene reveals a function for *Sex-lethal* in translational control. *Cell* **89:** 789–798.

Bell O, Conrad T, Kind J, Wirbelauer C, Akhtar A, Schubeler D. 2008. Transcription-coupled methylation of histone H3 at lysine 36 regulates dosage compensation by enhancing recruitment of the MSL complex in *Drosophila melanogaster. Mol Cell Biol* **28:** 3401–3409.

Bell O, Schwaiger M, Oakeley EJ, Lienert F, Beisel C, Stadler MB, Schubeler D. 2010. Accessibility of the *Drosophila* genome discriminates PcG repression, H4K16 acetylation and replication timing. *Nat Struct Mol Biol* **17:** 894–900.

Belote JM, Lucchesi JC. 1980a. Control of X chromosome transcription by the *maleless* gene in *Drosophila. Nature* **285:** 573–575.

Belote JM, Lucchesi JC. 1980b. Male-specific lethal mutations of *Drosophila* melanogaster. *Genetics* **96:** 165–186.

Birchler J, Sun L, Fernandez H, Donohue R, Xie W, Sanyal A. 2011. Re-evaluation of the function of the male specific lethal complex in *Drosophila. J Genet Genomics* **38:** 327–332.

Bone JR, Lavender J, Richman R, Palmer MJ, Turner BM, Kuroda MI. 1994. Acetylated histone H4 on the male X chromosome is associated with dosage compensation in *Drosophila. Genes Dev* **8:** 96–104.

* Brockdorff N, Turner BM. 2014. Dosage compensation in mammals. *Cold Spring Harb Perspect Biol* doi: 10.1101/cshperspect.a019406.

Cai W, Bao X, Deng H, Jin Y, Girton J, Johansen J, Johansen KM. 2008. RNA polymerase II–mediated transcription at active loci does not require histone H3S10 phosphorylation in *Drosophila. Development* **135:** 2917–2925.

Carrozza MJ, Li B, Florens L, Suganuma T, Swanson SK, Lee KK, Shia WJ, Anderson S, Yates J, Washburn MP, et al. 2005. Histone H3 methylation by Set2 directs deacetylation of coding regions by Rpd3S to suppress spurious intragenic transcription. *Cell* **123:** 581–592.

Chang KA, Kuroda MI. 1998. Modulation of MSL1 abundance in female *Drosophila* contributes to the sex specificity of dosage compensation. *Genetics* **150:** 699–709.

Charlesworth B. 1978. Model for evolution of Y chromosomes and dosage compensation. *Proc Natl Acad Sci* **75:** 5618–5622.

Cite as *Cold Spring Harb Perspect Biol* doi: 10.1101/cshperspect.a019398

Chu C, Qu K, Zhong FL, Artandi SE, Chang HY. 2011. Genomic maps of long noncoding RNA occupancy reveal principles of RNA-chromatin interactions. *Mol Cell* **44:** 667–678.

Cline TW, Meyer BJ. 1996. Vive la difference: Males vs females in flies vs worms. *Annu Rev Genet* **30:** 637–702.

Conrad T, Cavalli FM, Holz H, Hallacli E, Kind J, Ilik I, Vaquerizas JM, Luscombe NM, Akhtar A. 2012. The MOF chromobarrel domain controls genome-wide H4K16 acetylation and spreading of the MSL complex. *Dev Cell* **22:** 610–624.

Copps K, Richman R, Lyman LM, Chang KA, Rampersad-Ammons J, Kuroda MI. 1998. Complex formation by the *Drosophila* MSL proteins: Role of the MSL2 RING finger in protein complex assembly. *EMBO J* **17:** 5409–5417.

Core LJ, Waterfall JJ, Lis JT. 2008. Nascent RNA sequencing reveals widespread pausing and divergent initiation at human promoters. *Science* **322:** 1845–1848.

Cremer T, Cremer M. 2010. Chromosome territories. *Cold Spring Harb Perspect Biol* **2:** a003889.

Cugusi S, Ramos E, Ling H, Yokoyama R, Luk KM, Lucchesi JC. 2013. Topoisomerase II plays a role in dosage compensation in *Drosophila*. *Transcription* **4.** doi: 10.4161/trns.26185.

* Dekker J, Misteli T. 2014. Long-range chromatin interactions. *Cold Spring Harb Perspect Biol* doi: 10.1101/cshperspect.a019356.

Delattre M, Spierer A, Jaquet Y, Spierer P. 2004. Increased expression of *Drosophila* Su(var)3-7 triggers Su(var)3-9-dependent heterochromatin formation. *J Cell Sci* **117:** 6239–6247.

Deng H, Zhang W, Bao X, Martin JN, Girton J, Johansen J, Johansen KM. 2005. The JIL-1 kinase regulates the structure of *Drosophila* polytene chromosomes. *Chromosoma* **114:** 173–182.

Deng X, Hiatt JB, Nguyen DK, Ercan S, Sturgill D, Hillier LW, Schlesinger F, Davis CA, Reinke VJ, Gingeras TR, et al. 2011. Evidence for compensatory upregulation of expressed X-linked genes in mammals, *Caenorhabditis elegans* and *Drosophila melanogaster*. *Nat Genet* **43:** 1179–1185.

Deuring R, Fanti L, Armstrong JA, Sarte M, Papoulas O, Prestel M, Daubresse G, Verardo M, Moseley SL, Berloco M, et al. 2000. The ISWI chromatin-remodeling protein is required for gene expression and the maintenance of higher order chromatin structure in vivo. *Mol Cell* **5:** 355–365.

Dunlap D, Yokoyama R, Ling H, Sun HY, McGill K, Cugusi S, Lucchesi JC. 2012. Distinct contributions of MSL complex subunits to the transcriptional enhancement responsible for dosage compensation in *Drosophila*. *Nucleic Acids Res* **40:** 11281–11291.

Ebert A, Schotta G, Lein S, Kubicek S, Krauss V, Jenuwein T, Reuter G. 2004. Su(var) genes regulate the balance between euchromatin and heterochromatin in *Drosophila*. *Genes Dev* **18:** 2973–2983.

Fagegaltier D, Baker BS. 2004. X chromosome sites autonomously recruit the dosage compensation complex in *Drosophila* males. *PLoS Biol* **2:** e341.

Feller C, Prestel M, Hartmann H, Straub T, Söding J, Becker PB. 2012. The MOF-containing NSL complex associates globally with housekeeping genes, but activates only a defined subset. *Nucleic Acids Res* **40:** 1509–1522.

Furuhashi H, Nakajima M, Hirose S. 2006. DNA supercoiling factor contributes to dosage compensation in *Drosophila*. *Development* **133:** 4475–4483.

Gelbart ME, Kuroda MI. 2009. *Drosophila* dosage compensation: A complex voyage to the X chromosome. *Development* **136:** 1399–1410.

Gelbart ME, Larschan E, Peng S, Park PJ, Kuroda MI. 2009. *Drosophila* MSL complex globally acetylates H4K16 on the male X chromosome for dosage compensation. *Nat Struct Mol Biol* **16:** 825–832.

Gilfillan GD, Straub T, de Wit E, Greil F, Lamm R, van Steensel B, Becker PB. 2006. Chromosome-wide gene-specific targeting of the *Drosophila* dosage compensation complex. *Genes Dev* **20:** 858–870.

Gorchakov AA, Alekseyenko AA, Kharchenko P, Park PJ, Kuroda MI. 2009. Long-range spreading of dosage compensation in *Drosophila* captures transcribed autosomal genes inserted on X. *Genes Dev* **23:** 2266–2271.

Grimaud C, Becker PB. 2009. The dosage compensation complex shapes the conformation of the X chromosome in *Drosophila*. *Genes Dev* **23:** 2490–2495.

* Grunstein M, Gasser SM. 2013. Epigenetics in *Saccharomyces cerevisiae*. *Cold Spring Harb Perspect Biol* **5:** a017491.

Gu W, Szauter P, Lucchesi JC. 1998. Targeting of MOF, a putative histone acetyl transferase, to the X chromosome of *Drosophila melanogaster*. *Dev Genet* **22:** 56–64.

Gu W, Wei X, Pannuti A, Lucchesi JC. 2000. Targeting the chromatin-remodeling MSL complex of *Drosophila* to its sites of action on the X chromosome requires both acetyl transferase and ATPase activities. *EMBO J* **19:** 5202–5211.

Gunesdogan U, Jackle H, Herzig A. 2010. A genetic system to assess in vivo the functions of histones and histone modifications in higher eukaryotes. *EMBO Rep* **11:** 772–776.

Hallacli E, Lipp M, Georgiev P, Spielman C, Cusack S, Akhtar A, Kadlec J. 2012. Msl1-mediated dimerization of the dosage compensation complex is essential for male X-chromosome regulation in *Drosophila*. *Mol Cell* **48:** 587–600.

Hamada FN, Park PJ, Gordadze PR, Kuroda MI. 2005. Global regulation of X chromosomal genes by the MSL complex in *Drosophila melanogaster*. *Genes Dev* **19:** 2289–2294.

Hanai K, Furuhashi H, Yamamoto T, Akasaka K, Hirose S. 2008. RSF governs silent chromatin formation via histone H2Av replacement. *PLoS Genet* **4:** e1000011.

Hargreaves DC, Crabtree GR. 2011. ATP-dependent chromatin remodeling: Genetics, genomics and mechanisms. *Cell Res* **21:** 396–420.

* Henikoff S, Smith MM. 2014. Histone variants and epigenetics. *Cold Spring Harb Perspect Biol* doi: 10.1101/cshperspect.a019364.

Hilfiker A, Hilfiker-Kleiner D, Pannuti A, Lucchesi JC. 1997. *mof*, a putative acetyl transferase gene related to the Tip60 and MOZ human genes and to the SAS genes of yeast, is required for dosage compensation in *Drosophila*. *EMBO J* **16:** 2054–2060.

Hiratani I, Gilbert DM. 2009. Replication timing as an epigenetic mark. *Epigenetics* **4:** 93–97.

Hodl M, Basler K. 2009. Transcription in the absence of histone H3.3. *Curr Biol* **19:** 1221–1226.

Ilik IA, Quinn JJ, Georgiev P, Tavares-Cadete F, Maticzka D, Toscano S, Wan Y, Spitale RC, Luscombe N, Backofen R, et al. 2013. Tandem stem-loops in roX RNAs act together to mediate X chromosome dosage compensation in *Drosophila*. *Mol Cell* **51:** 156–173.

Ivaldi MS, Karam CS, Corces VG. 2007. Phosphorylation of histone H3 at Ser10 facilitates RNA polymerase II release from promoter-proximal pausing in *Drosophila*. *Genes Dev* **21:** 2818–2831.

Jin Y, Wang Y, Walker DL, Dong H, Conley C, Johansen J, Johansen KM. 1999. JIL-1: A novel chromosomal tandem kinase implicated in transcriptional regulation in *Drosophila*. *Mol Cell* **4:** 129–135.

Jin Y, Wang Y, Johansen J, Johansen KM. 2000. JIL-1, a chromosomal kinase implicated in regulation of chromatin structure, associates with the male specific lethal (MSL) dosage compensation complex. *J Cell Biol* **149:** 1005–1010.

Joshi AA, Struhl K. 2005. Eaf3 chromodomain interaction with methylated H3-K36 links histone deacetylation to Pol II elongation. *Mol Cell* **20:** 971–978.

Kadlec J, Hallacli E, Lipp M, Holz H, Sanchez-Weatherby J, Cusack S, Akhtar A. 2011. Structural basis for MOF and MSL3 recruitment into the dosage compensation complex by MSL1. *Nat Struct Mol Biol* **18:** 142–149.

Kelley RL, Wang J, Bell L, Kuroda MI. 1997. Sex lethal controls dosage compensation in *Drosophila* by a non-splicing mechanism. *Nature* **387:** 195–199.

Kelley RL, Meller VH, Gordadze PR, Roman G, Davis RL, Kuroda MI. 1999. Epigenetic spreading of the *Drosophila* dosage compensation complex from *roX* RNA genes into flanking chromatin. *Cell* **98:** 513–522.

Kelley RL, Lee OK, Shim YK. 2008. Transcription rate of noncoding roX1 RNA controls local spreading of the *Drosophila* MSL chromatin remodeling complex. *Mech Dev* **125:** 1009–1019.

Keogh MC, Kurdistani SK, Morris SA, Ahn SH, Podolny V, Collins SR, Schuldiner M, Chin K, Punna T, Thompson NJ, et al. 2005. Cotranscriptional set2 methylation of histone H3 lysine 36 recruits a repressive Rpd3 complex. *Cell* **123:** 593–605.

Kind J, Akhtar A. 2007. Cotranscriptional recruitment of the dosage compensation complex to X-linked target genes. *Genes Dev* **21:** 2030–2040.

Kind J, Vaquerizas JM, Gebhardt P, Gentzel M, Luscombe NM, Bertone P, Akhtar A. 2008. Genome-wide analysis reveals MOF as a key regulator of dosage compensation and gene expression in *Drosophila*. *Cell* **133:** 813–828.

Kruithof M, Chien FT, Routh A, Logie C, Rhodes D, van Noort J. 2009. Single-molecule force spectroscopy reveals a highly compliant helical folding for the 30-nm chromatin fiber. *Nat Struct Mol Biol* **16:** 534–540.

Lakhotia SC, Mukherjee AS. 1970. Chromosomal basis of dosage compensation in *Drosophila*. 3. Early completion of replication by the polytene X-chromosome in male: Further evidence and its implications. *J Cell Biol* **47:** 18–33.

Lam KC, Mühlpfordt F, Vaquerizas JM, Raja SJ, Holz H, Luscombe NM, Manke T, Akhtar A. 2012. The NSL complex regulates housekeeping genes in *Drosophila*. *PLoS Genet* **8:** e1002736.

Larschan E, Alekseyenko AA, Gortchakov AA, Peng S, Li B, Yang P, Workman JL, Park PJ, Kuroda MI. 2007. MSL complex is attracted to genes marked by H3K36 trimethylation using a sequence-independent mechanism. *Mol Cell* **28:** 121–133.

Larschan E, Bishop EP, Kharchenko PV, Core LJ, Lis JT, Park PJ, Kuroda MI. 2011. X chromosome dosage compensation via enhanced transcriptional elongation in *Drosophila*. *Nature* **471:** 115–118.

Larschan E, Soruco MM, Lee OK, Peng S, Bishop E, Chery J, Goebel K, Feng J, Park PJ, Kuroda MI. 2012. Identification of chromatin-associated regulators of MSL complex targeting in *Drosophila* dosage compensation. *PLoS Genet* **8:** e1002830.

Lavender JS, Birley AJ, Palmer MJ, Kuroda MI, Turner BM. 1994. Histone H4 acetylated at lysine 16 and proteins of the *Drosophila* dosage compensation pathway co-localize on the male X chromosome through mitosis. *Chromosome Res* **2:** 398–404.

Lee CG, Chang KA, Kuroda MI, Hurwitz J. 1997. The NTPase/helicase activities of *Drosophila* maleless, an essential factor in dosage compensation. *EMBO J* **16:** 2671–2681.

Lee CG, da Costa Soares V, Newberger C, Manova K, Lacy E, Hurwitz J. 1998. RNA helicase A is essential for normal gastrulation. *Proc Natl Acad Sci* **95:** 13709–13713.

Lerach S, Zhang W, Deng H, Bao X, Girton J, Johansen J, Johansen KM. 2005. JIL-1 kinase, a member of the male-specific lethal (MSL) complex, is necessary for proper dosage compensation of eye pigmentation in *Drosophila*. *Genesis* **43:** 213–215.

Liu Y, Lu C, Yang Y, Fan Y, Yang R, Liu CF, Korolev N, Nordenskiold L. 2011. Influence of histone tails and h4 tail acetylations on nucleosome-nucleosome interactions. *J Mol Biol* **414:** 749–764.

Lucchesi JC. 1978. Gene dosage compensation and the evolution of sex chromosomes. *Science* **202:** 711–716.

Lucchesi JC, Manning JE. 1987. Gene dosage compensation in *Drosophila melanogaster*. *Adv Genet* **24:** 371–429.

Luger K, Mader AW, Richmond RK, Sargent DF, Richmond TJ. 1997. Crystal structure of the nucleosome core particle at 2.8 A resolution. *Nature* **389:** 251–260.

Maenner S, Müller M, Fröhlich J, Langer D, Becker PB. 2013. ATP-dependent roX RNA remodeling by the helicase maleless enables specific association of MSL proteins. *Mol Cell* **51:** 174–184.

* Marmorstein R, Zhou M-M. 2014. Writers and readers of histone acetylation: Structure, mechanism, and inhibition. *Cold Spring Harb Perspect Biol* **6:** a018762.

Meller VH, Rattner BP. 2002. The *roX* genes encode redundant male-specific lethal transcripts required for targeting of the MSL complex. *EMBO J* **21:** 1084–1091.

Meller VH, Wu KH, Roman G, Kuroda MI, Davis RL. 1997. *roX1* RNA paints the X chromosome of male *Drosophila* and is regulated by the dosage compensation system. *Cell* **88:** 445–457.

Meller VH, Gordadze PR, Park Y, Chu X, Stuckenholz C, Kelley RL, Kuroda MI. 2000. Ordered assembly of *roX* RNAs into MSL complexes on the dosage-compensated X chromosome in *Drosophila*. *Curr Biol* **10:** 136–143.

Mendjan S, Taipale M, Kind J, Holz H, Gebhardt P, Schelder M, Vermeulen M, Buscaino A, Duncan K, Mueller J, et al. 2006. Nuclear pore components are involved in the transcriptional regulation of dosage compensation in *Drosophila*. *Mol Cell* **21:** 811–823.

Minsky N, Shema E, Field Y, Schuster M, Segal E, Oren M. 2008. Monoubiquitinated H2B is associated with the transcribed region of highly expressed genes in human cells. *Nat Cell Biol* **10:** 483–488.

Mito Y, Henikoff JG, Henikoff S. 2005. Genome-scale profiling of histone H3.3 replacement patterns. *Nat Genet* **37:** 1090–1097.

Morales V, Straub T, Neumann MF, Mengus G, Akhtar A, Becker PB. 2004. Functional integration of the histone acetyltransferase MOF into the dosage compensation complex. *EMBO J* **23:** 2258–2268.

Morra R, Smith ER, Yokoyama R, Lucchesi JC. 2008. The MLE subunit of the *Drosophila* MSL complex uses its ATPase activity for dosage compensation and its helicase activity for targeting. *Mol Cell Biol* **28:** 958–966.

Mukherjee AS, Beermann W. 1965. Synthesis of ribonucleic acid by the X-chromosomes of *Drosophila melanogaster* and the problem of dosage compensation. *Nature* **207:** 785–786.

Muller HJ. 1932. Further studies on the *Nature* and causes of gene mutations. *Proc 6th Int Congr Genet* **1:** 213–255.

Ogasawara Y, Furuhashi H, Hirose S. 2007. DNA supercoiling factor positively regulates expression of the homeotic gene *Abdominal-B* in *Drosophila melanogaster*. *Genes Cells* **12:** 1347–1355.

Oh H, Park Y, Kuroda MI. 2003. Local spreading of MSL complexes from roX genes on the *Drosophila* X chromosome. *Genes Dev* **17:** 1334–1339.

Oh H, Bone JR, Kuroda MI. 2004. Multiple classes of MSL binding sites target dosage compensation to the X chromosome of *Drosophila*. *Curr Biol* **14:** 481–487.

Pannuti A, Lucchesi JC. 2000. Recycling to remodel: Evolution of dosage-compensation complexes. *Curr Opin Genet Dev* **10:** 644–650.

Park Y, Kelley RL, Oh H, Kuroda MI, Meller VH. 2002. Extent of chromatin spreading determined by *roX* RNA recruitment of MSL proteins. *Science* **298:** 1620–1623.

Park Y, Mengus G, Bai X, Kageyama Y, Meller VH, Becker PB, Kuroda MI. 2003. Sequence-specific targeting of *Drosophila roX* genes by the MSL dosage compensation complex. *Mol Cell* **11:** 977–986.

Park SW, Kang Y, Sypula JG, Choi J, Oh H, Park Y. 2007. An evolutionarily conserved domain of *roX2* RNA is sufficient for induction of H4-Lys16 acetylation on the *Drosophila* X chromosome. *Genetics* **177:** 1429–1437.

Park SW, Oh H, Lin YR, Park Y. 2010. MSL cis-spreading from *roX* gene up-regulates the neighboring genes. *Biochem Biophys Res Commun* **399:** 227–231.

Raja SJ, Charapitsa I, Conrad T, Vaquerizas JM, Gebhardt P, Holz H, Kadlec J, Fraterman S, Luscombe NM, Akhtar A. 2010. The nonspecific lethal complex is a transcriptional regulator in *Drosophila*. *Mol Cell* **38:** 827–841.

Regnard C, Straub T, Mitterweger A, Dahlsveen IK, Fabian V, Becker PB. 2011. Global analysis of the relationship between JIL-1 kinase and transcription. *PLoS Genet* **7:** e1001327.

Robinson PJ, An W, Routh A, Martino F, Chapman L, Roeder RG, Rhodes D. 2008. 30 nm chromatin fibre decompaction requires both H4-K16 acetylation and linker histone eviction. *J Mol Biol* **381:** 816–825.

Sala A, Toto M, Pinello L, Gabriele A, Di Benedetto V, Ingrassia AM, Lo Bosco G, Di Gesu V, Giancarlo R, Corona DF. 2011. Genome-wide characterization of chromatin binding and nucleosome spacing activity of the nucleosome remodelling ATPase ISWI. *EMBO J* **30:** 1766–1777.

Sass GL, Pannuti A, Lucchesi JC. 2003. Male-specific lethal complex of *Drosophila* targets activated regions of the X chromosome for chromatin remodeling. *Proc Natl Acad Sci* **100:** 8287–8291.

Schalch T, Duda S, Sargent DF, Richmond TJ. 2005. X-ray structure of a tetranucleosome and its implications for the chromatin fibre. *Nature* **436:** 138–141.

Schubeler D, Scalzo D, Kooperberg C, van Steensel B, Delrow J, Groudine M. 2002. Genome-wide DNA replication profile for *Drosophila melanogaster*: A link between transcription and replication timing. *Nat Genet* **32:** 438–442.

Scott MJ, Pan LL, Cleland SB, Knox AL, Heinrich J. 2000. MSL1 plays a central role in assembly of the MSL complex, essential for dosage compensation in *Drosophila*. *EMBO J* **19:** 144–155.

Shogren-Knaak M, Ishii H, Sun JM, Pazin MJ, Davie JR, Peterson CL. 2006. Histone H4-K16 acetylation controls chromatin structure and protein interactions. *Science* **311:** 844–847.

Simon MD, Wang CI, Kharchenko PV, West JA, Chapman BA, Alekseyenko AA, Borowsky ML, Kuroda MI, Kingston RE. 2011. The genomic binding sites of a noncoding RNA. *Proc Natl Acad Sci* **108:** 20497–20502.

Siriaco G, Deuring R, Chioda M, Becker PB, Tamkun JW. 2009. *Drosophila* ISWI regulates the association of histone H1 with interphase chromosomes in vivo. *Genetics* **182**, 661–669.

Smith ER, Pannuti A, Gu W, Steurnagel A, Cook RG, Allis CD, Lucchesi JC. 2000. The *Drosophila* MSL complex acetylates histone H4 at lysine 16, a chromatin modification linked to dosage compensation. *Mol Cell Biol* **20:** 312–318.

Smith ER, Allis CD, Lucchesi JC. 2001. Linking global histone acetylation to the transcription enhancement of X-chromosomal genes in *Drosophila* males. *J Biol Chem* **276:** 31483–31486.

Smith ER, Cayrou C, Huang R, Lane WS, Cote J, Lucchesi JC. 2005. A human protein complex homologous to the *Drosophila* MSL complex is responsible for the majority of histone H4 acetylation at lysine 16. *Mol Cell Biol* **25:** 9175–9188.

Soruco MM, Chery J, Bishop EP, Siggers T, Tolstorukov MY, Leydon AR, Sugden AU, Goebel K, Feng J, Xia P, et al. 2013. The CLAMP protein links the MSL complex to the X chromosome during *Drosophila* dosage compensation. *Genes Dev* **27:** 1551–1556.

Spierer A, Seum C, Delattre M, Spierer P. 2005. Loss of the modifiers of variegation Su(var)3-7 or HP1 impacts male X polytene chromosome morphology and dosage compensation. *J Cell Sci* **118:** 5047–5057.

Stenberg P, Larsson J. 2011. Buffering and the evolution of chromosome-wide gene regulation. *Chromosoma* **120:** 213–225.

Straub T, Gilfillan GD, Maier VK, Becker PB. 2005. The *Drosophila* MSL complex activates the transcription of target genes. *Genes Dev* **19:** 2284–2288.

Straub T, Grimaud C, Gilfillan GD, Mitterweger A, Becker PB. 2008. The chromosomal high-affinity binding sites for the *Drosophila* dosage compensation complex. *PLoS Genet* **4:** e1000302.

Straub T, Zabel A, Gilfillan GD, Feller C, Becker PB. 2013. Different chromatin interfaces of the *Drosophila* dosage compensation complex revealed by high-shear ChIP-seq. *Genome Res* **23:** 473–485.

* Strome S, Kelly WG, Ercan S, Lieb JD. 2014. Regulation of the X chromosomes in *Caenorhabditis elegans*. *Cold Spring Harb Perspect Biol* **6:** a018366.

Strukov YG, Sural TH, Kuroda MI, Sedat JW. 2011. Evidence of activity-specific, radial organization of mitotic chromosomes in *Drosophila*. *PLoS Biol* **9:** e1000574.

Suka N, Luo K, Grunstein M. 2002. Sir2p and Sas2p opposingly regulate acetylation of yeast histone H4 lysine16 and spreading of heterochromatin. *Nat Genet* **32:** 378–383.

Sural TH, Peng S, Li B, Workman JL, Park PJ, Kuroda MI. 2008. The MSL3 chromodomain directs a key targeting step for dosage compensation of the *Drosophila melanogaster* X chromosome. *Nat Struct Mol Biol* **15:** 1318–1325.

Taipale M, Rea S, Richter K, Vilar A, Lichter P, Imhof A, Akhtar A. 2005. hMOF histone acetyltransferase is required for histone H4 lysine 16 acetylation in mammalian cells. *Mol Cell Biol* **25:** 6798–6810.

Turner BM, Birley AJ, Lavender J. 1992. Histone H4 isoforms acetylated at specific lysine residues define individual chromosomes and chromatin domains in *Drosophila* polytene nuclei. *Cell* **69:** 375–384.

Vaquerizas JM, Suyama R, Kind J, Miura K, Luscombe NM, Akhtar A. 2010. Nuclear pore proteins nup153 and megator define transcriptionally active regions in the *Drosophila* genome. *PLoS Genet* **6:** e1000846.

Villa R, Forné I, Müller M, Imhof A, Straub T, Becker PB. 2012. MSL2 combines sensor and effector functions in homeostatic control of the *Drosophila* dosage compensation machinery. *Mol Cell* **48:** 647–654.

Wang Y, Zhang W, Jin Y, Johansen J, Johansen KM. 2001. The JIL-1 tandem kinase mediates histone H3 phosphorylation and is required for maintenance of chromatin structure in *Drosophila*. *Cell* **105:** 433–443.

Wu L, Zee BM, Wang Y, Garcia BA, Dou Y. 2011. The RING finger protein MSL2 in the MOF complex is an E3 ubiquitin ligase for H2B K34 and is involved in crosstalk with H3 K4 and K79 methylation. *Mol Cell* **43:** 132–144.

Zhang Y, Oliver B. 2010. An evolutionary consequence of dosage compensation on *Drosophila melanogaster* female X-chromatin structure? *BMC Genomics* **11:** 6.

Zhang W, Deng H, Bao X, Lerach S, Girton J, Johansen J, Johansen KM. 2006. The JIL-1 histone H3S10 kinase regulates dimethyl H3K9 modifications and heterochromatic spreading in *Drosophila*. *Development* **133:** 229–235.

WWW RESOURCES

www.FlyBase.org A database of *Drosophila* genes and genomes.

C H A P T E R 25

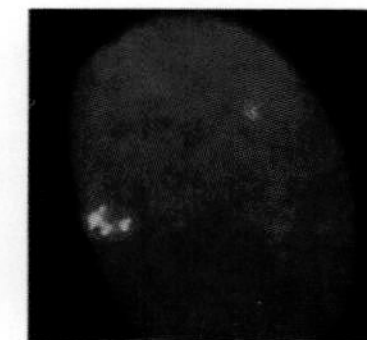

Dosage Compensation in Mammals

Neil Brockdorff[1] and Bryan M. Turner[2]

[1]Department of Biochemistry, University of Oxford Oxford OX1 3QU, United Kingdom; [2]School of Cancer Sciences, Institute of Biomedical Research, University of Birmingham Medical School, Birmingham B15 2TT, United Kingdom

Correspondence: turnerbm@adf.bham.ac.uk

SUMMARY

Many organisms show major chromosomal differences between sexes. In mammals, females have two copies of a large, gene-rich chromosome, the X, whereas males have one X and a small, gene-poor Y. The imbalance in expression of several hundred genes is lethal if not dealt with by dosage compensation. The male–female difference is addressed by silencing of genes on one female X early in development. However, both males and females now have only one active X chromosome. This is compensated by twofold up-regulation of genes on the active X. This complex system continues to provide important insights into mechanisms of epigenetic regulation.

Outline

Cite as *Cold Spring Harb Perspect Biol* doi: 10.1101/cshperspect.a019406

OVERVIEW

In mammals, as in many other organisms, there is a major chromosomal difference between the sexes. For example, humans have 22 pairs of chromosomes known as autosomes that are present in both males and females; one member of each pair is inherited from the mother and one from the father. But there are two other chromosomes, the sex chromosomes, designated X and Y, that differ between the sexes; females have two Xs, whereas males have one X and one Y. This matters because although the X is a medium-sized chromosome with more than 1000 genes, the Y is small and gene poor. A similar situation exists in other mammals, including rodents and marsupials. Chromosomal differences are linked to the mechanism by which sex is determined and seem to have evolved over many millions of years.

Sex chromosome imbalance presents the organism with a problem: The two sexes differ in the copy number of X-linked genes. This can lead to an imbalance in the amount of gene products (RNAs and proteins), which would, in turn, require differences in metabolic control and other cellular processes. To avoid this, dosage compensation mechanisms have evolved that balance the level of X-linked gene products between the sexes. There are three general methods by which this can be performed: first, a twofold up-regulation in the expression of X-linked gene in males; second, a twofold down-regulation of genes on each of the two X chromosomes in females; and finally, the complete inactivation of one of the two X chromosomes in females. The first strategy has been adopted in the fruit fly, *Drosophila* (see Ch. 24 [Lucchesi and Kuroda 2014]), the second in the worm *Caenorhabditis elegans* (see Ch. 23 [Strome et al. 2014]), and, it now seems, both the first and the last in mammals.

Over recent years, studies of dosage compensation in mammals have provided crucial insights into fundamental epigenetic mechanisms and how patterns of gene expression are regulated through development. It can be confidently predicted that they will continue to do so.

Cite as *Cold Spring Harb Perspect Biol* doi: 10.1101/cshperspect.a019406

1 INTRODUCTION

1.1 Sex Determination Creates a Need for Dosage Compensation

Sexual reproduction is common among eukaryotes. Even plants that can replicate perfectly well asexually by sending out shoots or runners often have an alternative sexual mode of reproduction. A possible explanation is that sexual reproduction brings an enormous increase in genetic variability on which natural selection can operate. The reshuffling of alleles that occurs with every sexual generation produces a population better able to cope with environmental shifts compared with a relatively homogeneous population derived from asexual methods of reproduction. But sex is complicated, requiring developmental pathways that lead to male and female sexual organs, as well as the physiological and biochemical apparatus required for meiosis, germ cell maturation, the attraction of partners, and mating (see Marshall Graves and Shetty 2001, and references therein, for further discussion of these issues).

Genetic mechanisms used in defining different sexes vary widely from one organism to another. The simplest system involves a single locus that is homozygous in one sex (the "homogametic" sex) and heterozygous in the other (the "heterogametic" sex; Fig. 1). This system has evolved in different ways to reach varying levels of complexity in different organisms. In some, mechanisms have been put in place that suppress meiotic recombination (crossing-over) of the sex-determining alleles in the heterogametic sex (Fig. 1), a step that helps prevent the generation of mixtures of alleles leading to intersex states. The inability to recombine, in many cases, has spread to include part or all of one chromosome, with an accompanying loss of genetic information. The evolutionary pressures that have driven this chromosome degeneration are still not understood, but the end result in many species is that the two sexes show differences, not just in alleles at one or a few loci, but in complete chromosomes. In mammals, it is the males who carry the degenerate chromosome, whereas in birds, it is the females (Marshall Graves and Shetty 2001).

Sexual differentiation is usually triggered by one or a small number of crucial genes being switched on or off during development. The products of these genes initiate a cascade of gene regulatory events that mediate progression down one or the other pathway of sex determination (see Ch. 23 [Strome et al. 2014] for details in *C. elegans*; see Ch. 24 [Lucchesi and Kuroda 2014] for details in *Drosophila*). In humans, it is the protein product of the *SRY* gene on the Y chromosome that sends the early embryo down the male pathway (reviewed by Quinn and Koopman 2012). A mechanism of this sort does not need major chromosome differences to operate successfully, so why have such differences arisen so often and in such diverse organisms, including mammals, birds, and fruit flies? It may be that they have occurred as a by-product of the suppression of crossing-over required to prevent intersex states (Fig. 1). Mathematical analysis of the factors that influence the spread of alleles through populations shows that suppression of crossing-over will inevitably lead to the gradual accumulation of deleterious mutations around the region of crossover suppression. This is largely because such mutations rarely become homozygous, which is necessary if they are to be

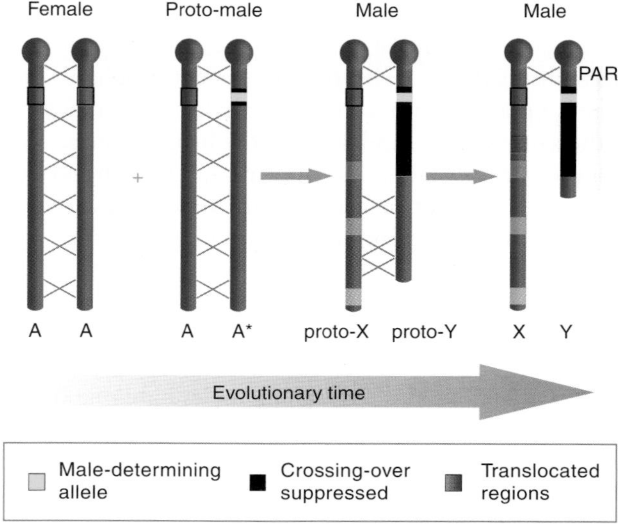

Figure 1. Evolution of the Y chromosome. Early in evolution, the two sexes may have differed at only a single, autosomal locus (marked by a black box); one sex is homozygous at this locus (female) and the other sex (male) is heterozygous (designated proto-male). The "male-determining allele" is shown in yellow. If mating requires one member of each sex, then individuals homozygous for the male-determining allele cannot arise. At this early stage, physiological differences between the sexes will be subtle, comparable to those that distinguish the two mating types in yeast. To prevent the formation of intersex states, crossing-over will be suppressed within and around the male-determining locus (dark shading). Mutations, including deletions and inversions, will accumulate and cause the degenerate region in which crossing-over is suppressed to gradually expand ("Muller's ratchet") until the chromosome has lost most of its active, functional genes. (Mutations accumulate because suppression of crossing-over reduces the probability that they will occur in homozygous form, hence, reducing the selection pressure against them.) A small, active region must remain that is homologous to the X chromosome to allow pairing and crossing-over at meiosis (indicated by a gray ×). This is the pseudoautosomal region (PAR). The autosome, originally homologous to the future X (A in the diagram), will itself evolve sometimes through translocations from other chromosomes (shown as red shaded areas), eventually forming the distinctive X chromosome. The X, like other chromosomes, is a mosaic of DNA fragments put in place at different periods through evolution; some of these are ancient and some are relatively recent. On the human X, the more recent arrivals are enriched in genes that escape X inactivation.

selected against. Mutations, including deletions and translocations, gradually spread beyond the original suppressed region, leading to the progressive degeneration of one of the two originally homologous chromosomes (Fig. 1). This irreversible, degenerative process has been termed "Müller's ratchet," in recognition of the geneticist who first proposed it and modeled it mathematically. There is no selection for this process, it just happens as a consequence of local crossover suppression, which, in turn, was made necessary by the adoption of a two-sex strategy for reproduction (discussed in Charlesworth 1996; Charlesworth and Charlesworth 2005). But whatever the evolutionary drive behind chromosome degeneration, the fact that it has occurred (and is presumably continuing) has required the coevolution of mechanisms to cope with two contingent problems: First, there is a major chromosomal difference between members of the same species, and second, the heterogametic sex is monosomic for a large chromosome and thus monoallelic for a large number of genes. Both these issues must be addressed by mechanisms of dosage compensation.

1.2 Solving the Male Problem: Up-Regulation of X-Linked Genes

In both mammals and *Drosophila*, males have one copy of each sex chromosome, an X and a Y, whereas females have two copies of the X. In both groups of organisms, the Y is gene poor and largely heterochromatic. It contains just a few genes needed for male development or fertility. In contrast, the X is a large, gene-rich chromosome. In all organisms gene products (RNAs and proteins) are made in direct proportion to the number of copies of the gene per cell. Thus, the presence of one or two X chromosomes would cause a twofold difference in the intracellular concentrations of many gene products between the sexes. Furthermore, XY males are monoallelic for the great majority of X-linked genes (only a few genes have homologs on the Y or autosomes). This is surprising, considering higher eukaryotes are usually very intolerant of the loss of even a part of a chromosome; small chromosome deletions generally cause major deformities, whereas autosomal monosomies are always lethal. How, then, do males survive with only one X chromosome?

In attempting to answer this question, it is important to realize that cells and organisms tolerate monoallelism for individual genes rather well. For example, a heterozygous individual with an allele encoding an inactive form of an enzyme, and thus with only half the normal level of active enzyme, is usually perfectly healthy, although homozygosity for the defective allele may be lethal. However, some genes are clearly more dose sensitive than others. Recent studies have suggested that a twofold dosage change in genes encoding components of large multiprotein complexes are particularly likely to exert a phenotypic effect (Pessia et al. 2012). Also, having half the normal level of "multiple" gene products can have a cumulative effect. For example, if several components of a metabolic or signaling pathway are all reduced twofold, the end product of that pathway may be reduced severalfold, with likely phenotypic effects (Fig. 2) (Oliver 2007). Thus, problems will inevitably accumulate for the heterogametic sex as the proto-Y chromosome degenerates over evolutionary time, both through loss of X-linked genes that are individually dosage sensitive and the progressive loss of genes that are individually dosage tolerant, but collectively less so (Fig. 2).

The problem faced by males was recognized more than 40 years ago by the geneticist Susumu Ohno, who speculated that the problem could be solved by up-regulating twofold, expression of genes on the single male X chromosome (Ohno 1967). He also noted that this could occur on a gene-by-gene basis over evolutionary time. Certainly, loss of a dosage-sensitive gene (or a gene in a pathway in which some members had already been lost) would result in a strong selection pressure in favor of up-regulating (i.e., compensating) expression of the remaining gene copy. However, if the male (heterogametic, XY) sex were to adopt this strategy, it would cause problems for the female (homogametic, XX) members of the species, who would then have to cope with increasing numbers of genes whose expression had been "increased" twofold. It may be, therefore, that silencing of one of the two X chromosomes in female mammals, the subject of most of this chapter, is a response to this overexpression.

The general feasibility of Ohno's suggestion was established by work on dosage compensation in the fruit fly *Drosophila melanogaster* in which genetic, biochemical, and recent high-throughput gene expression studies have confirmed that dosage compensation occurs through up-regulation of X-linked genes in XY males (discussed in Ch. 24 [Lucchesi and Kuroda 2014]). Crucially, the mechanism that evolved in the fly is such that up-regulation occurs only in males, circumventing the problem of overexpression in XX females. Detailed exploration of Ohno's hypothesis in mammals has had to wait for the advent of methodologies that can accurately assay the expression of large numbers of genes. The reason is that assaying transcript levels of individual genes, however accurately, cannot determine whether the gene is up-regulated or not. Genes vary enormously, one from another, in the level at which they are expressed, and even the same gene often changes expression depending on the tissue or cell type in which it is housed, the stage of development, or even over time in the same cell. How, then, can we tell whether any particular gene is twofold up-regulated? What is our baseline?

 Cite as *Cold Spring Harb Perspect Biol* doi: 10.1101/cshperspect.a019406

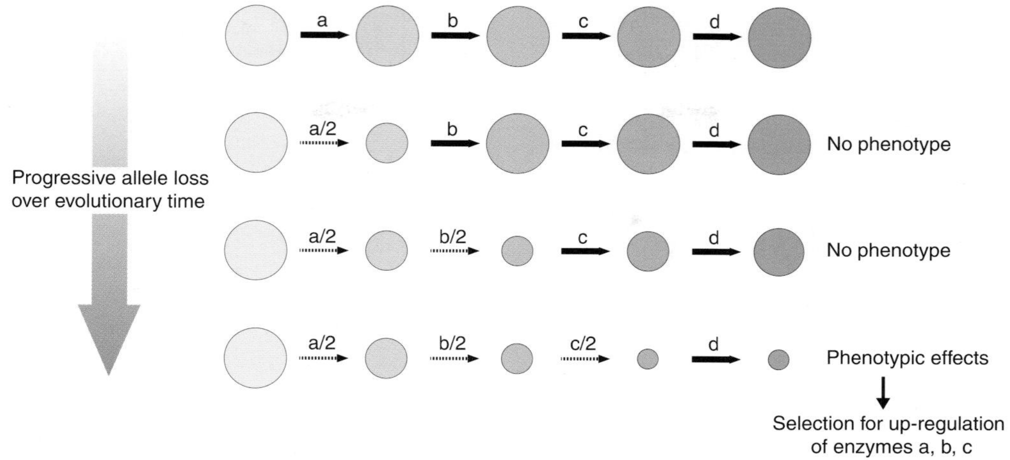

Figure 2. The effects of progressive loss of X-linked alleles on a hypothetical signaling pathway. The diagram shows successive components of the pathway as colored discs, each put in place by the actions of enzymes a, b, c, and d. The size of the discs is proportional to the amount of each component (level 1). The model proposes that enzymes a, b, and c are all encoded by genes on the X chromosome, and early in evolution were encoded by genes on the two proto-X chromosomes. Alleles are lost as the proto-X progressively degenerates, eventually forming the gene-poor Y chromosome (Fig. 1). A twofold reduction in the amount of an enzyme is likely to cause a reduction in its product, although not necessarily twofold. The cell's normal homeostatic mechanisms are likely to correct small disturbances and there is likely to be little or no effect on subsequent steps in the pathway (level 2). Even the loss of two enzymes may be corrected with no physiological effect (level 3). However, a stage will eventually be reached when the cumulative effects of enzyme (gene) depletion cause key components to decrease below a critical level and trigger an effect on phenotype (shown here on level 4). Selection pressure will be exerted to correct the phenotypic effect, most readily by up-regulating expression of one or more of the remaining, single alleles of enzymes a, b, or c.

Recent technologies, first microarrays and now high-throughput RNA sequencing, can assay the expression of large numbers of genes, allowing us to test the proposition that, in any particular tissue or cell type, the "overall" expression of X-linked genes in males (i.e., over the whole, wide range of expression levels) is higher than that of autosomal genes. If the expression ratio of X-linked to autosomal genes in males, with one X and two copies of each autosome, is 0.5, then there is no up-regulation; that is, expression levels reflect gene copy number. On the other hand, if it is 1.0, then there is complete, twofold up-regulation and Ohno's hypothesis is validated (Fig. 3).

Over the past 5 years or so, several studies using either microarray data (Nguyen and Disteche 2006; Lin et al. 2007) or, more recently, RNA sequencing (Deng et al. 2011) have addressed this issue providing strong evidence for the up-regulation of X-linked genes in mammals. However, the data requires careful interpretation. Complications arise from the fact that the X chromosome contains a higher proportion of tissue-specific genes (often involved in sexual development) than autosomes, with the result that the proportion of X-linked genes in any given tissue or cell type that is switched off (silenced) in both sexes is higher than for autosomes (Ellegren and Parsch 2007; Meisel et al. 2012). This must be taken into account. Though the weight of evidence strongly supports the up-regulated expression of X-linked genes in both male and female cells, the situation is complex, as is often the case with dosage compensation. Although the X:autosome expression ratio is consistently >0.5, it does not reach 1.0 (Deng et al. 2011; Lin et al. 2011). The reason for this may be that up-regulation is not applied to all X-linked genes. Perhaps only the more dosage-sensitive genes are up-regulated (Pessia et al. 2012), something that would fit with an evolutionary model in which up-regulation was determined on a gene-by-gene basis (Fig. 1).

There are many questions that remain to be answered concerning expression of X-linked genes, and the issues they raise are relevant to the much more intensively studied question of X-chromosome inactivation, with which the rest of this chapter is concerned. Perhaps most pressing is the need to decide whether genes are up-regulated by a common, chromosome-wide mechanism, whether different genes have adopted different mechanisms, or whether both factors come into play. Similar issues have been addressed in attempting to unravel the dosage compensation mechanism in *D. melanogaster*, and although the male-specific process that is used in flies is unlikely to operate in mammals, there are undoubtedly important lessons to be learned from this widely studied model organism.

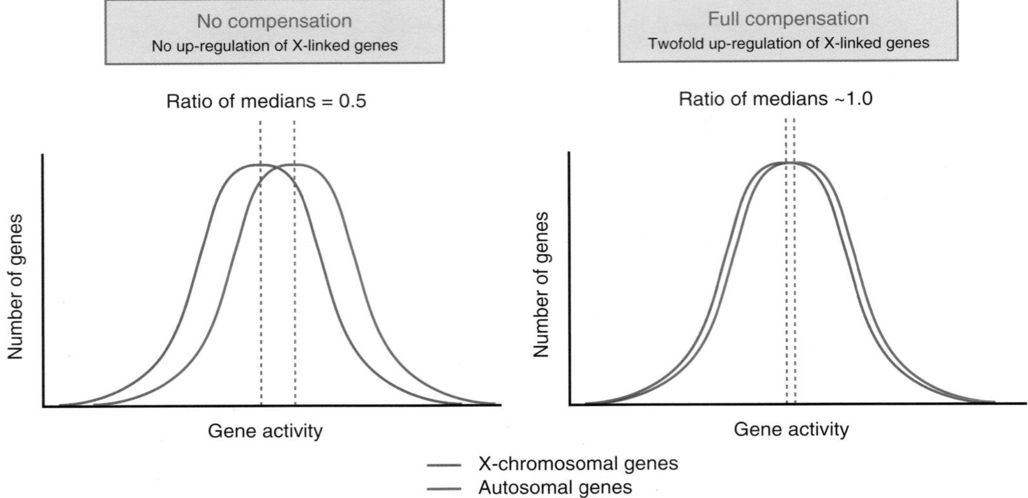

Figure 3. Shift in median expression caused by selective up-regulation of X-linked genes. Measurement of the expression of large numbers of genes on either the X chromosome (blue) or autosomes (red) by microarrays or RNA sequencing shows that transcript levels are normally distributed with a range of expression levels spread over several orders of magnitude. If expression of genes on the single male X (or the single active X in females) is not compensated, then the X:autosome ratio of median expression levels should be 0.5, reflecting the difference in (active) copies (*left* panel). Alternatively, if expression of X-linked genes is up-regulated twofold to compensate for dosage differences, then the ratio of medians should be close to 1.0 (*right* panel).

1.3 Identification of an Inactive X in Mammalian Females

In 1949, Barr and Bertram described the sex chromatin body, a structure visible under the light microscope in the nuclei of only female cells in various mammalian species. The structure proved useful in studies of sexual abnormalities, but it was not until 1959 that Ohno and colleagues showed that this structure was derived from one of the two female X chromosomes (see Ohno 1967). Shortly thereafter, in 1961, Mary Lyon described genetic experiments on the expression of X-linked coat color genes in female mice. To explain the patterns of inheritance for this variable patchwork (mosaic) of coat color in individual female mice, Lyon hypothesized that in each female cell one of the two X chromosomes is stably inactivated early in development (Lyon 1961). The sex chromatin body, now known as the Barr body, is thus the cytological manifestation of the inactive X chromosome. Elegant experiments using skin fibroblasts from females heterozygous for a polymorphism of the X-linked enzyme glucose-6-phosphate dehydrogenase, showed that only one of the two possible alleles was expressed in colonies grown from individual cells (clones), thereby demonstrating the heritability of the inactive state from one cell generation to the next (Davidson et al. 1963), and confirming the occurrence of X inactivation in human females (Beutler et al. 1962). Further studies of X inactivation in human females with multiple copies of the X (with karyotypes such as 47XXX or 48XXXX), showed that all X chromosomes in excess of one were inactivated. This has been generalized as the "$n - 1$ rule," which states that if an individual has n X chromosomes, then $n - 1$ will be inactivated (Ohno 1967). This rule explains the remarkably mild clinical symptoms associated with X-chromosome aneuploidies. The X inactivation hypothesis has continued to provide an explanation for the peculiarities of X-linked gene expression in female cells and has remained essentially unchanged since first proposed. But the past 50 years or so have been spent trying to work out the molecular mechanisms by which it operates.

2 OVERVIEW OF X INACTIVATION

2.1 X Inactivation Is Developmentally Regulated

X inactivation in female mammals is developmentally regulated. Both X chromosomes are active in the early zygote (Epstein et al. 1978), and inactivation then proceeds coincident with cellular differentiation from the pluripotent state. Normally there is an equal probability that cells will inactivate the X chromosome that is either derived maternally (Xm) or paternally (Xp). Exceptions to this are imprinted X inactivation that occurs throughout marsupials and in early preimplantation mouse embryos, in which it is always Xp that is inactivated. In the latter case,

Cite as *Cold Spring Harb Perspect Biol* doi: 10.1101/cshperspect.a019406

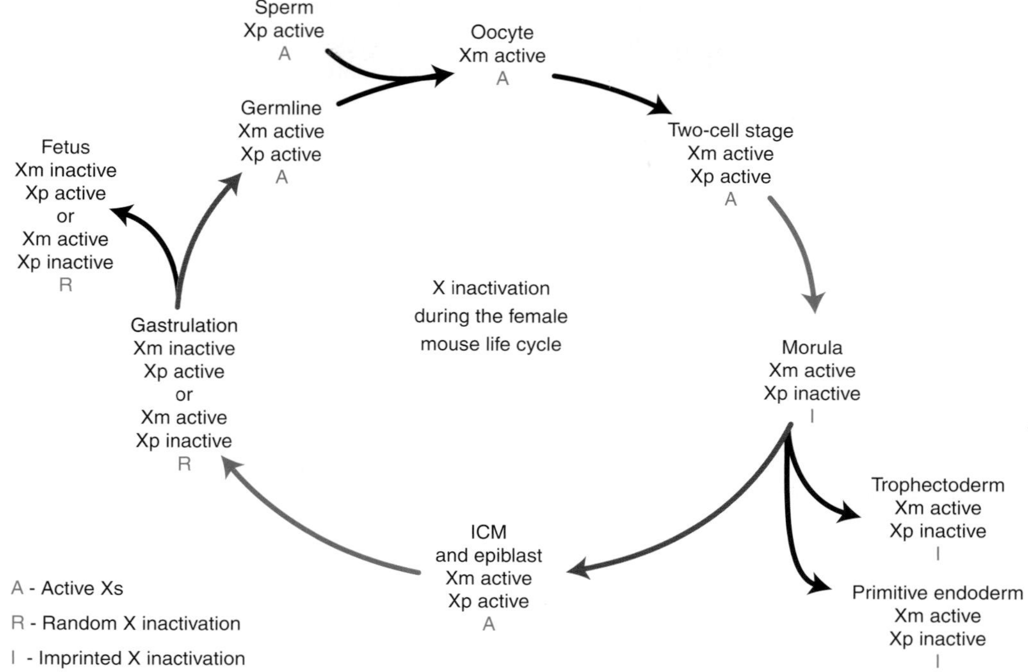

Figure 4. The cycle of X inactivation and reactivation. The X chromosome undergoes a cycle of X inactivation and X reactivation during development. Red arrows indicate X inactivation steps and green arrows indicate X reactivation steps. Inactivation first occurs in early preimplantation embryos (imprinted X inactivation) and subsequently in cells of the epiblast at the time of gastrulation (random X inactivation). The inactive X is reactivated in ICM cells when they are first allocated at the blastocyst stage, and also in the developing germ cells.

imprinted Xp inactivation is maintained in the first differentiating lineages, namely, the extraembryonic trophectoderm (TE) and primitive endoderm (PE) cells, but the inactive X is reactivated in the inner cell mass (ICM) cells that give rise to the embryo. Reversal of X inactivation also occurs in developing primordial germ cells (PGCs), ensuring the X chromosome is again active in the gamete. Figure 4 illustrates the cycle of X inactivation and reactivation in the female mouse.

2.2 Chromosome Silencing Involves Multiple Levels of Chromatin Modification

Silencing of the X chromosome is achieved at the level of chromatin structure by modification of histone tails, incorporation or exclusion of variant histones, DNA methylation of some CpG islands, and reorganization of higher-order chromatin folding, all contributing to a stable facultative heterochromatic structure. There is redundancy built into the system and not all components are essential for all aspects of silencing (e.g., Sado et al. 2004). The layers of chromatin modification are established progressively through ontogeny, as detailed in Section 4.4. Collectively, they ensure stable propagation of the inactive X through multiple rounds of cell division.

2.3 Some Genes Escape X Inactivation

X inactivation affects most of the X chromosome, but some genes escape silencing (Berletch et al. 2011). These include genes within a small region on the X chromosome that pairs with the Y chromosome during male meiosis, referred to as the PAR or XY pairing region (Fig. 1). Genes located in this region do not require dosage compensation as two copies are present in both males and females.

Other genes that escape X inactivation, both with and without Y-linked homologs, have also been characterized. They total ~15% of genes on the human X chromosome (Carrel and Willard 2005). Interestingly, many of these genes lie on the short arm of the chromosome (also referred to as the p arm), which, in evolutionary time, is a recently acquired segment of the X chromosome. Studies in mouse indicate that some escapees can be inactivated in early ontogeny, with progressive reactivation occurring during development (Sec. 4.6). In marsupials, most genes studied have been found to escape X inactivation to some extent. This may reflect a failure to maintain silencing through ontogeny, possibly related to the lack of CpG island methylation on the inactive X (Xi) in these species.

2.4 X Inactivation Is Regulated by a Master Switch Locus: The X Inactivation Center (Xic)

Classical genetic studies showed that X inactivation is mediated by a single *cis*-acting master switch locus, referred to as the X inactivation center (Xic). The Xic was shown to be required both for silencing the X chromosome in *cis*, and ensuring correct and appropriate initiation of random X inactivation. More recent studies have characterized the Xic at the molecular level. The locus produces a large noncoding RNA termed *Xist* (X-inactive-specific transcript) that has the unique property of binding in *cis* and accumulating along the entire length of the chromosome from which it is transcribed (Fig. 5) (Brown et al. 1991; Brockdorff et al. 1992; Brown et al. 1992). Coating of the chromosome with *Xist* RNA provides the trigger for X-chromosome silencing (Lee et al. 1996; Penny et al. 1996; Wutz and Jaenisch 2000). Studies to date indicate that this occurs, at least in part, through *Xist*-mediated recruitment of chromatin modifying complexes (Fig. 5A).

A second noncoding RNA, *Tsix*, is also located in the Xic region (Lee et al. 1999) and plays a key role in regulating *Xist* expression. *Tsix* overlaps with the *Xist* gene, but is transcribed in the antisense direction; hence, its name is *Xist* spelled backwards.

Phylogenetic studies have revealed that the noncoding *Xist* RNA evolved from a protein-coding transcript, *Lnx3* (Duret et al. 2006). The *Lnx3* gene has retained protein-coding capacity in other vertebrate species and also in marsupial mammals. The latter finding was unanticipated as obvious similarities in X inactivation in marsupials led to the supposition that chromosome silencing is mediated by a direct homolog of *Xist*. After many years of searching, a recent study has revealed that marsupials independently evolved a *cis*-acting noncoding RNA locus, *Rsx* (RNA on

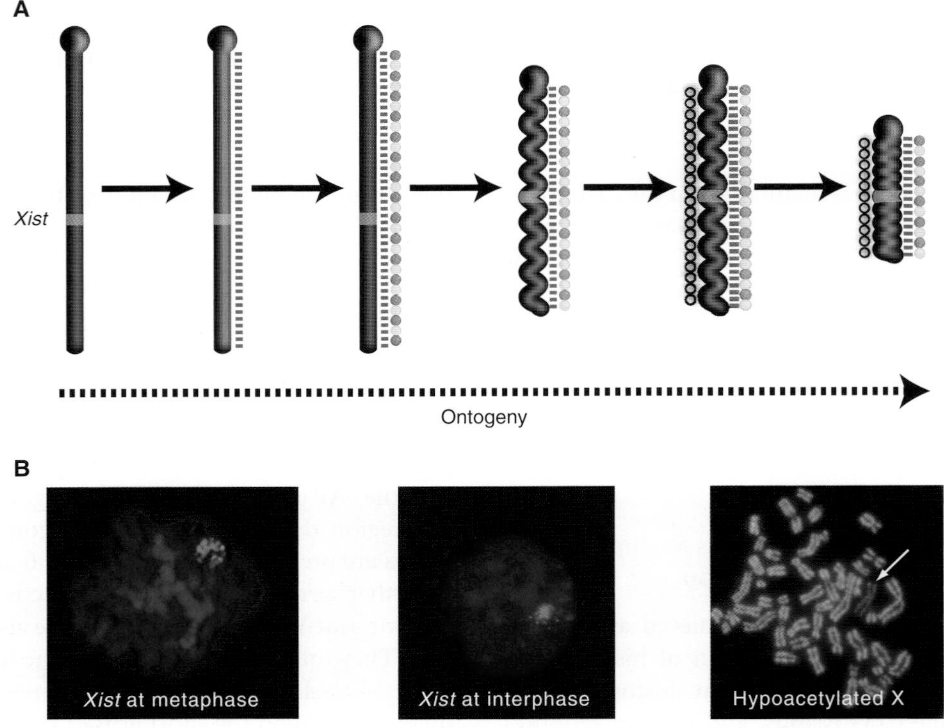

Figure 5. Progressive chromosome-wide heterochromatinization induced by *Xist* RNA. (*A*) When the *Xist* gene is expressed, the RNA binds to and coats the X chromosome from which it is transcribed (green dashed line). *Xist* RNA is thought to trigger silencing of the chromosome by recruiting chromatin modifying activities (red and yellow circles). The initial wave of silencing, in turn, leads to recruitment of additional layers of epigenetic modification (white circles), further stabilizing the heterochromatic structure. Establishment of these different levels of epigenetic silencing is achieved in a stepwise manner through development and ontogeny. (*B*) Localization of *Xist* RNA along the X chromosomes is shown by in situ hybridization in both interphase and metaphase.

 Cite as *Cold Spring Harb Perspect Biol* doi: 10.1101/cshperspect.a019406

the silent X), which fulfills the same function as *Xist* in marsupial species (Grant et al. 2012). Like *Xist* RNA, *Rsx* RNA is transcribed specifically from Xi and coats the length of the X chromosome in *cis*.

3 INITIATION OF X INACTIVATION

3.1 Imprinted versus Random X Inactivation

The decision to inactivate an X chromosome needs to be tightly regulated. Male cells must avoid silencing their single X chromosome, and female cells must avoid silencing both X chromosomes or keeping both X chromosomes active. Two different modes of regulation have been shown to operate. The imprinted mode of X inactivation silences the paternally derived X chromosome whereas the random mode randomly inactivates either the maternal or paternal X chromosome. Metatherian mammals (marsupials) use only the imprinted mode. Some eutherian (placental) mammals (e.g., mouse) use the imprinted mode in extraembryonic lineages and the random mode in the embryo proper (Fig. 2). Other species, notably rabbit and human, show only random X inactivation (Okamoto et al. 2011). This variation may be linked, in part, to differences in the timing of embryonic genome activation, which in mouse occurs relatively early, at the two-cell stage, compared with the four- to eight-cell stages in humans.

Important model systems for studying the initiation of X inactivation are early mouse embryos and embryonic stem (ES) cells derived from the ICM of these embryos. XX ES cells are especially useful as they recapitulate initiation of random X inactivation in vitro when cells are induced to undergo differentiation. There are currently no in vitro models that recapitulate initiation of imprinted X inactivation.

3.2 Regulation of Imprinted X Inactivation

Paternally imprinted X inactivation was first observed in a marsupial (Sharman 1971). Subsequently, imprinted X inactivation was shown to occur in the TE and PE of mouse embryos (Takagi and Sasaki 1975). In imprinted X inactivation, it is the parent of origin from which the X derives that governs its status; that is, paternal, but not maternal, X chromosomes are inactivated regardless of how many X chromosomes or chromosome sets are present. Note that the single X in XY males is always maternally derived and therefore not inactivated in imprinted tissues.

What then is the nature of the imprint? Studies of *Xist* expression indicate that there is a repressive imprint on the Xm allele from morula stage mouse embryos. This imprint prevents *Xist* expression, keeping the X chromosome active (see Fig. 4). Nuclear transfer experiments showed that the repressive *Xist* imprint is established during oocyte maturation (Tada et al. 2000). The molecular basis of the imprint is unknown, but DNA methylation is not required, contrasting with many other imprinted genes (see Ch. 26 [Barlow and Bartolomei 2014] for details on genomic imprinting).

One theory for the preferential inactivation of Xp in the zygote is that there is carryover of silencing of the XY bivalent that is established during the pachytene stage of male meiosis (meiotic sex-chromosome inactivation, MSCI; Huynh and Lee 2003). Recent studies argue against this. First, MSCI has been shown to be a distinct and *Xist*-independent mechanism that is triggered in pachytene by the presence of unpaired chromosomal regions on both sex chromosomes and autosomes (Turner et al. 2006, and references therein). Second, expression analysis of a number of X-linked genes in early zygotes has shown that Xp silencing occurs de novo in response to zygotic Xp *Xist* expression (Okamoto et al. 2005, and references therein).

The paternal *Xist* expression (and resultant Xp silencing) that begins at the onset of zygotic gene activation (at the two- to four-cell stage) indicates that the Xp *Xist* allele is poised to express (Fig. 6). A region-specific demethylation of CpG sites in the *Xist* promoter occurs during spermatogenesis (Norris et al. 1994) and is thought to be important for this.

The *Tsix* gene, an antisense regulator of *Xist*, is required for imprinted X inactivation as deletion of the major promoter results in early embryo lethality when transmitted by the maternal, but not the paternal, gamete (Lee 2000). Lethality appears to be attributable to inappropriate Xm *Xist* expression (i.e., a failure to retain an active X chromosome both in XmY and XmXp embryos). It is not known if expression of *Tsix* RNA is the primary imprint or functions only later to maintain the imprint.

3.3 Regulation of Random X Inactivation: Counting

In the random mode of X inactivation, cells use the $n - 1$ rule described in Section 1.3, in which all X chromosomes except one are inactivated per diploid chromosome set. The process that senses the number of X chromosomes is often referred to as counting. Where more than one X chromosome is present, the selection of active and inactive X chromosome—referred to as choice—is normally random. However, there are factors that can bias this decision resulting in nonrandom or skewed X inactivation. The process of choice is considered separately in Section 3.4, but this division is one of convenience as it is clear that counting and choice must be inextricably linked.

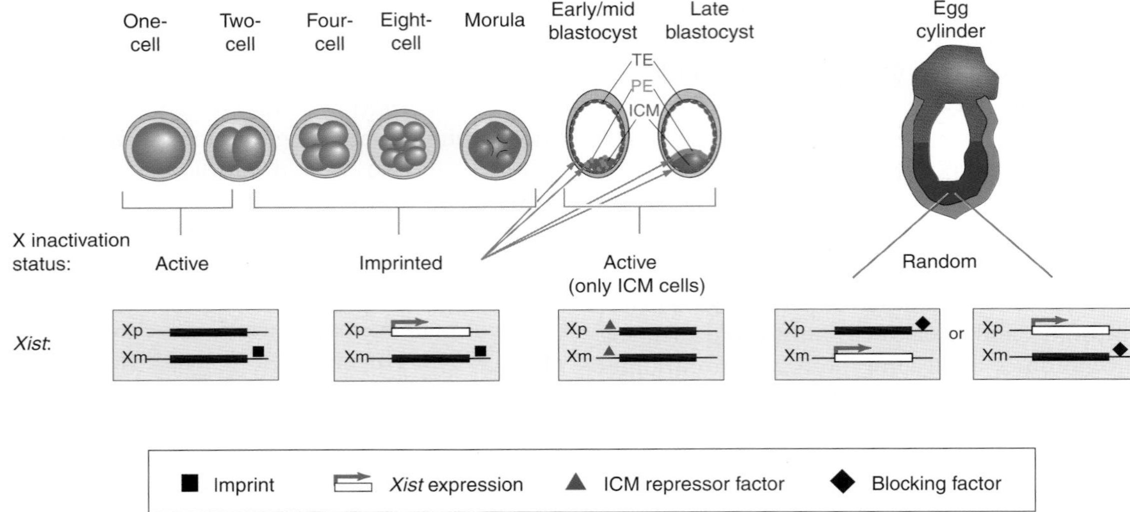

Figure 6. *Xist* gene regulation in early development. The figure illustrates current knowledge and models for imprinted and random *Xist* regulation in early XX mouse embryos. The Xm *Xist* allele arrives in the zygote with a repressive imprint possibly mediated through the antisense *Tsix* locus (black square). The Xp *Xist* allele is primed to be active and is expressed as soon as embryonic gene activation occurs at the two-cell stage. From the two- to four-cell stage up until morula stage, Xp *Xist* is expressed in all cells (expression indicated by open rectangle and arrow at 5′ end). This pattern is maintained at the early blastocyst stage and subsequently in TE and PE cells and their fully differentiated derivative (extraembryonic) tissues. In the late blastocyst, ICM *Xist* expression is extinguished, possibly by an ICM-specific repressor factor (blue triangle). *Xist* expression then commences subsequently at the time of gastrulation. Here, the blocking factor (black diamond) ensures that *Xist* expression cannot occur on one of the two alleles (counting).

A number of different models for counting and choice have been proposed (Fig. 7). Early models, developed before the discovery of *Xist*, invoked a single autosomally encoded blocking factor present in limiting quantities sufficient to bind and repress a single X inactivation center (Rastan 1983). In this model, X inactivation is a default pathway that in diploid cells is inhibited on a single X chromosome (the active X). In cells with more than one X chromosome, choice is determined by the probability of the blocking factor binding to a given X inactivation-center allele.

A related model invokes two factors: an autosomally encoded blocking factor and an X-encoded competence factor (Gartler and Riggs 1983). The blocking factor is suggested to be sufficient to disable the competence factor on a single X inactivation center. Other X inactivation centers can be bound by competence factor leading to the onset of X inactivation. Although initially developed before the discovery of *Xist*, this model has been readopted to help explain experimental observations arising from the deletion of *Tsix*, the antisense regulator of *Xist* (Lee and Lu 1999).

A third and more contemporary model invokes a stochastic process in which autosomal factors promote *Xist* repression, for example, by inducing *Tsix*, whereas the X chromosome produces *Xist* gene activators that compete with the repressors. The resultant competition could create a probability for *Xist* gene activation that is modulated as cells begin to differentiate. The model predicts a stochastic probability of maintaining a single X active in all circumstances, but also invokes checkpoint and feedback mechanisms that ensure incorrect X inactivation patterns, notably inactivation of both or neither X chromosome in XX cells, can be reset at an early stage. In part, this could be attributable to levels of X-linked activators becoming limiting once silencing has spread to the locus encoding the activator on one of the two X chromosomes (Nora and Heard 2009).

These models are not necessarily mutually exclusive and, moreover, they continue to evolve to take into account new experimental findings. As things stand, none of the models can be said to satisfactorily account for all available data, but they nevertheless provide a useful framework both for integrating current data and determining new experimental directions.

Although a complete model for *Xist* regulation in random X inactivation has not yet been found, there is a growing consensus that there is a finely balanced competition

 Cite as *Cold Spring Harb Perspect Biol* doi: 10.1101/cshperspect.a019406

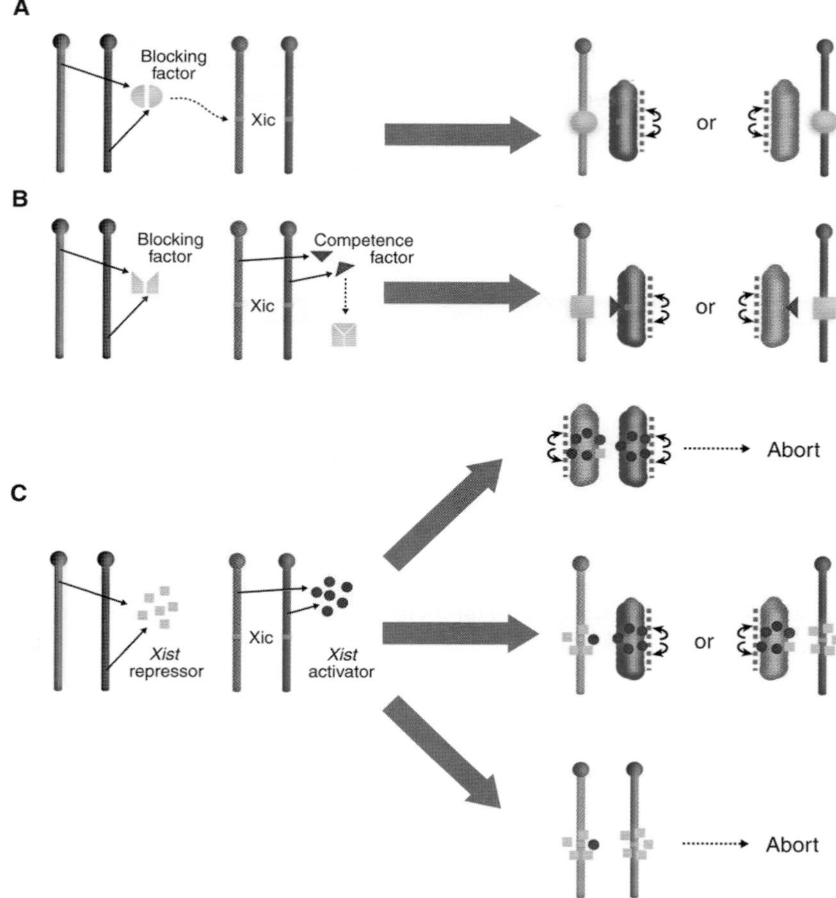

Figure 7. Models for the regulation of random X inactivation. (*A*) Autosomally encoded blocking factor (yellow shapes) is produced in sufficient quantities to occupy a single Xic. Binding of blocking factor to the Xic inhibits *Xist* transcription, thus defining a single active X chromosome. *Xist* transcription occurs on any additional X chromosomes leading to X inactivation (dark green dots). Blocking factor binds to either the maternal (Xm) or paternal (Xp) X chromosome with equal probability and in a cell-autonomous manner. (*B*) The two-factor model invokes an X-encoded competence factor (purple triangle) and an autosomally encoded blocking factor (yellow shapes). Blocking factor titrates away competence factor (purple triangle). In cells with a single X chromosome, there is insufficient available competence factor to activate *Xist*, but in cells with additional X chromosomes, competence factor can activate all X chromosomes except the single X chromosome bound by blocking factor. (*C*) The stochastic model invokes that autosomally encoded repressors (yellow circles) and X-encoded activators (purple circles) compete with one another. All *Xist* alleles have an equal probability of being activated and this is increased in cells with more than one X chromosome (higher levels of *Xist* activators). By chance, some cells with two X chromosomes will initiate inactivation of either both or no X chromosomes. This may be dealt with by checkpoint mechanisms or cell death.

between the pathways that repress *Xist* and those that activate *Xist*. The decision to initiate and then maintain *Xist* up-regulation on a given allele may then be reinforced by feedback and/or feedforward loops.

Genetic studies have shown that the antisense gene, *Tsix*, is an important *Xist* repressor. *Tsix* is transcribed on the X chromosome before and concordant with the onset of *Xist* expression in random X inactivation. *Tsix* promoters lie immediately downstream from the *Xist* locus (Fig. 8). The antisense transcript spans the entire *Xist* locus, terminating immediately upstream of the major *Xist* promoter. Antisense transcription across the *Xist* promoter region is required for *Tsix*-mediated repression (Ohhata et al. 2008). The detailed molecular mechanisms are not fully understood, but are thought to involve a switch in the histone modification state of the *Xist* promoter (Sado et al. 2005; Navarro and Avner 2010), and recruitment of the de novo DNA methyltransferase Dnmt3a (Sun et al. 2006).

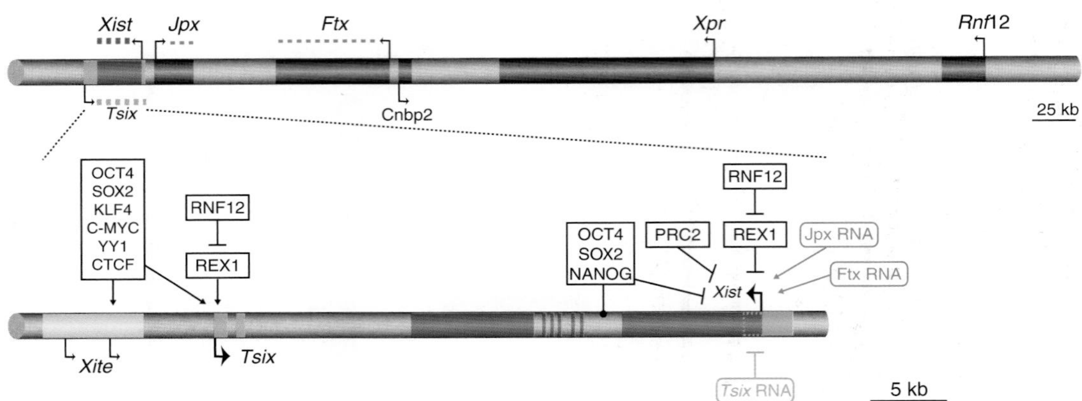

Figure 8. Genes and regulatory elements in the X inactivation-center region. The key region on the mouse X chromosome comprising known elements involved in *Xist* gene regulation is illustrated, showing noncoding RNA (ncRNA) genes and protein-coding genes. The *Xpr* region, several hundred kilobases upstream of *Xist*, has been implicated in *trans*-interaction of Xic alleles in XX cells and as such is thought to be important for initiation of X inactivation. The expanded view illustrates the intron/exon structure of the *Xist* and *Tsix* loci, including the Xite elements that function as *Tsix* enhancers. The network of protein factors (boxes) and ncRNAs (ovals) implicated in *Xist* gene regulation is shown with arrows and bars indicating repressor and activator function, respectively. Note that RNF12 mediates degradation of REX1, which functions both as a *Xist* repressor and a *Tsix* activator.

Tsix is thought to play a role in counting based on the observation that *Xist* is strongly up-regulated in *Tsix* null XY ES cells undergoing differentiation. However, not all *Tsix* null ES cell lines show this effect (the reasons for the discrepancy remain unclear) and, moreover, *Tsix* deletion leads to overt *Xist* gene up-regulation in undifferentiated XY ES cells. These observations indicate that other pathways contribute to *Xist* repression. Recent work has highlighted an important role for transcription factors linked to pluripotent cell circuitry, notably Oct4, Nanog, Sox2, and Rex1 (Donohoe et al. 2009; Navarro et al. 2010). These repressive effects are mediated by promoting *Tsix* expression, directly repressing *Xist*, and also by suppressing levels of the *Xist* activator, Rnf12 (see below). Because levels of pluripotency factors decline rapidly at the onset of differentiation, the regulatory balance at the *Xist* locus will be tipped toward activation, providing a plausible explanation for the link between the onset of *Xist* expression and cell differentiation.

Polycomb family repressors (elaborated in Secs. 4.4 and 4.5) are also implicated in *Xist* repression based on observed synergistic effects in cells with mutations affecting both *Tsix* and the PRC2 Polycomb pathway (Shibata et al. 2008). It is not yet known if this is linked to the pluripotency factor pathway or represents a third independent *Xist* repression pathway.

The existence and location of X chromosome-encoded activators of *Xist* expression was predicted through the analysis of polyploid ES cells and a deletion encompassing known *Xist* regulatory elements on a single X chromosome in XX ES cells (Monkhorst et al. 2008). This study led directly to the identification of the first known *Xist* activator, the ubiquitin E3 ligase Rnf12 (Jonkers et al. 2009). Overexpression of *Rnf12* promotes *Xist* expression in XY ES cells, whereas *Rnf12* deletion delays *Xist* up-regulation in differentiating XX ES cells. The location of the *Rnf12* gene immediately upstream of *Xist* supports the idea of a feedback loop in which the spread of X inactivation on one X chromosome represses *Rnf12*, reduces levels of Rnf12 protein, and thereby serves to reduce the probability of inappropriately inactivating remaining X chromosomes. Rnf12 functions as an E3 ligase for ubiquitin-mediated degradation of the pluripotency factor Rex1, found at the promoter regions of both *Xist* and *Tsix* genes. Thus, Rnf12-mediated degradation of Rex1 may simultaneously activate *Xist* and repress *Tsix* (Gontan et al. 2012).

Although it is clear that Rnf12 plays an important role in *Xist* gene activation, the fact that *Rnf12* null XX ES cells are able to up-regulate *Xist*, albeit with some delay, suggests the presence of additional activators, also possibly encoded on the X chromosome. Such a role has been ascribed to a ncRNA, *Jpx*, which is located immediately upstream of the *Xist* locus (Tian et al. 2010). The mechanism of Jpx RNA function is not yet fully understood although there is recent evidence that it evicts the silencing protein CTCF from the *Xist* locus (Sun et al. 2013). Similarly, another ncRNA produced by the *Ftx* locus, immediately upstream of *Jpx*, is implicated in *Xist* gene activation (Chureau et al. 2011). The loci and pathways known to participate in *Xist* repression and *Xist* activation are summarized in Figure 8.

3.4 Regulation of Random X Inactivation: Choice

In XX cells, the choice of which X chromosome to inactivate in a given cell is normally random. However, animals that are heterozygous for certain Xic alleles can show a bias toward one allele or the other. Before discussing these, it is important to first make the distinction between primary and secondary nonrandom X inactivation. Primary nonrandom X inactivation relates to a bias in the choice step that occurs at the time random X inactivation is initiated. Secondary nonrandom X inactivation, on the other hand, occurs as a consequence of selection against cells that have nominated a given X chromosome as the inactive X. This latter process does not strictly relate to choice, but occurs commonly when mutations on one X chromosome confer a selective disadvantage to cells that inactivate the wild-type allele. Examples include the secondary nonrandom X inactivation seen in carriers of serious genetic disorders such as Duchenne muscular dystrophy. The observed cell selection effects can occur at the level of the whole developing embryo or in specific tissues or cell types, depending on the function of the mutant protein.

The first documented example of primary nonrandom X inactivation came in classical genetic studies that identified variants at the "X controlling element" (*Xce*) in different mouse strains (Cattanach 1974, and references therein). *Xce* was mapped to the approximate location of the Xic on the mouse X chromosome, suggesting the two loci could be synonymous. This has been confirmed in more contemporary studies that locate *Xce* to a region immediately downstream from *Xist*. The molecular basis of the variation underlying nonrandom X inactivation, however, remains to be determined.

Primary nonrandom X inactivation has also been observed in animals heterozygous for various Xic alleles, established in gene targeting experiments aimed at the functional dissection of *Xist* and *Tsix*. A common theme emerging from these studies is that alleles that increase the probability of being selected for X inactivation are associated with mutations that either reduce levels of antisense *Tsix* transcription or increase levels of sense *Xist* transcription (Lee and Lu 1999; Nesterova et al. 2003). There is also recent evidence that choice is linked to an *Xist* RNA splicing switch (Royce-Tolland et al. 2010). In the context of models for regulation of random X inactivation, biased choice can be viewed as tipping the balance between *Xist* repressors and *Xist* activators on a given allele.

What then is the symmetry-breaking event that results in the counting/choice machinery ultimately associating with only a single allele when there is equivalent access to both alleles? One idea that has emerged is that this may involve *trans*-interactions between Xic alleles at the time X inactivation is initiated. Specifically, three-dimensional (3D) fluorescence in situ hybridization (FISH; Xu et al. 2006; Augui et al. 2007) and locus tagging experiments (Masui et al. 2011) have shown frequent associations of Xic alleles mediated by contacts close to the *Tsix* promoter and a region located several hundred kilobases upstream of *Xist*, termed *Xpr*. Current models posit that *Xpr* is required to establish initial *trans*-interactions, thus facilitating contacts in the *Tsix* region. These *trans*-interactions have been proposed to mediate exchange of factors from one allele to another, providing an opportunity to break symmetry (see Fig. 7 of Ch. 19 [Dekker and Misteli 2014]).

3.5 Switching Modes of X Inactivation Regulation in Early Embryogenesis

How do early mouse embryos instigate the switch from the imprinted to the random mode of regulation (see Fig. 4)? Until recently, it was thought that the initiation of imprinted and random X inactivation were both linked to cellular differentiation (Monk and Harper 1979). Thus, trophectoderm and primitive endoderm lineages were thought to inactivate their Xp in response to the parental imprints on *Xist* when they first differentiate. The three germ lineages that give rise to the embryo proper by gastrulation were thought to first erase the *Xist* imprint and then undergo random X inactivation (Fig. 6). More recent data, however, show that Xp inactivation occurs before the onset of cellular differentiation in cleavage stage embryos, and it occurs in all cells, including the precursors of the ICM (Fig. 6) (Mak et al. 2004; Okamoto et al. 2004). Thus, imprinted X inactivation in trophectoderm and primitive endoderm is a relic of the X inactivation pattern established in early cleavage embryos. ICM cells must instigate a program to reverse this initial wave of imprinted Xp inactivation, thereby setting the scene for subsequent random X inactivation. The basis for the reversal of Xp inactivation is unknown, but may involve an ICM-specific program that represses Xp *Xist* expression (see Sec. 5.1).

3.6 Evolutionary Variation in X Inactivation Regulatory Mechanisms

To date, the laboratory mouse has been the primary model system for studies on the initiation of X inactivation. However, recent work suggests that the lessons drawn from such studies may not be as broadly applicable as once thought (Okamoto et al. 2011). Analysis of human preimplantation embryos has shown that *XIST* is up-regulated from the X chromosome in male embryos and both X chromosomes in female embryos during early preimplantation development. This pattern is resolved in late blastocysts, in which males extinguish *XIST* expression and females express it from only one X chromosome. The early *XIST* expression in human embryos is not linked to chromosome silencing, consistent with the requirement for expression of X-linked genes. A further variation was revealed through experiments with rabbits, in which female preimplantation embryos expressed *Xist* from both alleles in ~25% of cells. In this case, *Xist* expression is linked to chromosome silencing, suggesting either a checkpoint mechanism that can reverse and correct inappropriate biallelic *Xist* expression, or cell selection events resulting in the loss of XX cells that silence both X chromosomes (Okamoto et al. 2011). An important goal for the future will be to determine if there are conserved features in these apparently diverse systems as these would likely point to the fundamental underlying mechanisms.

4 PROPAGATION AND MAINTENANCE OF THE INACTIVE STATE

4.1 *Xist* RNA, Gene Silencing, and Heterochromatin Assembly

There is strong evidence that the *Xist* gene and its RNA product provide both the switch that initiates X inactivation in *cis* and the means by which silencing spreads across the chromosome. The evidence comes from experiments, indicating that (1) *Xist* is unique in being expressed only from Xi, (2) *Xist* RNA levels increase dramatically in preimplantation embryos at the time of X inactivation, (3) *Xist* up-regulation precedes X inactivation and appears to be an absolute requirement for it to occur, (4) *Xist* RNA colocalizes with Xi in interphase nuclei and is distributed along one of the two metaphase X chromosomes (see Fig. 5B), and (5) *Xist*-containing transgenes, when inserted into autosomes, can coat the autosome in *cis* with *Xist* RNA and initiate the adoption of a heterochromatin-like, transcriptionally silent chromatin structure. These findings suggest that *Xist* RNA is both necessary and sufficient to trigger heterochromatin formation and transcriptional silencing. However, continuing *Xist* expression is not required for the "maintenance" of X inactivation. For example, silencing of human X-linked genes is maintained in human:rodent somatic cell hybrids, in which *Xist* expression is lost on the retained human Xi chromosome (Brown and Willard 1994). This issue is discussed further in Section 5.

The mechanism(s) by which *Xist* RNA associates with and spreads along the Xi in *cis* and the mechanisms that bring about changes in chromatin structure and gene silencing are still not understood in detail. We do know that different regions of the *Xist* RNA molecule are responsible for gene silencing and spreading along the X chromosome. Experiments with an inducible *Xist* expression system in mouse ES cells, in which the functions of Xist RNA molecules carrying defined deletions could be tested, showed that silencing can be attributed to a conserved repeat sequence, the A-repeat, located at the 5′ end of the molecule, whereas coating of the X is mediated by sequences scattered throughout the rest of the molecule (Wutz et al. 2002). These observations show that *Xist* RNA spreading and chromosome-silencing functions are mechanistically separable.

4.2 Spreading of *Xist* RNA and Chromosome Silencing

The association of *Xist* RNA with Xi is selective. It is not found along the PAR, which remains active and euchromatic, or at constitutive (centric) heterochromatin. Moreover, analysis of metaphase chromosomes shows a banded localization that appears to correlate with gene-rich G-light bands (Duthie et al. 1999). In dividing cells, *Xist* RNA association with Xi is lost when cells enter anaphase. It is then rapidly resynthesized in daughter cells at the onset of G_1. *Xist* RNA is tightly associated with the nuclear matrix in the interphase nucleus and localization is retained following removal of chromatin by micrococcal nuclease, suggesting that *Xist* RNA does not contact underlying DNA sequences directly (Clemson et al. 1996).

There are approximately 2×10^3 molecules of *Xist* RNA in a female somatic cell (Buzin et al. 1994). The transcript is relatively stable with an estimated half-life of 6–8 h in dividing cells, a figure that is consistent with turnover occurring predominantly as a consequence of dissociation through mitosis. However, it may be that *Xist* RNA turnover is inhibited in the presence of transcriptional inhibitors, as a recent study using GFP-tagged *Xist* transcripts in living cells determined a relatively fast turnover rate for *Xist* RNA (Ng et al. 2011).

What then is known of the factors regulating Xist RNA localization? One interesting link has come from analysis of the protein SAF-A/hnRNPU. Originally identified as an hnRNP protein and a major component of the insoluble nuclear scaffold, SAF-A/hnRNPU was first linked to X inactivation through cell imaging studies demonstrating enrichment of the protein over the interphase Xi territory in XX somatic cells (Pullirsch et al. 2010). This observation is consistent with the association of *Xist* RNA with the nuclear matrix. More recent data suggest that genetic disruption of SAF-A/hnRNPU causes dispersal of *Xist* RNA through the nucleoplasm, indicating a role in *Xist* RNA localization (Hasegawa and Nakagawa 2011). Moreover, biochemical analysis suggests that the RNA binding domain of hnRNPU/SAFA interacts directly with *Xist* RNA (Fig. 9A).

Cite as *Cold Spring Harb Perspect Biol* doi: 10.1101/cshperspect.a019406

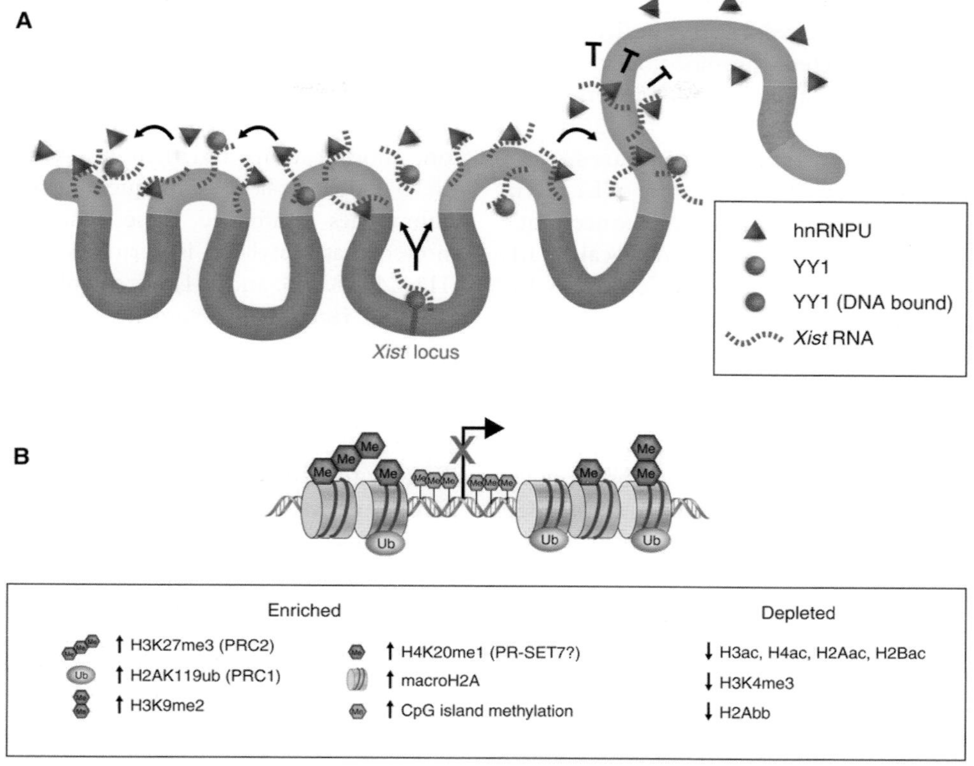

Figure 9. Factors involved in X inactivation spread. (*A*) The organization of LINE-1-rich (gray-shaded) and gene-rich (blue-shaded) domains is important in defining the extent of *Xist* RNA spreading as illustrated by the observation that large gene-rich domains attenuate *Xist* RNA spreading (barred arrows). YY1 is implicated in tethering *Xist* RNA (green wavy line) at the site of synthesis (brown circle) and interacting with *Xist* RNA to facilitate spreading (red circle). hnRNPU/SAFA also plays a role in localization of *Xist* RNA in *cis*, binding to *Xist* RNA directly. (*B*) Summary of characterized DNA methylation, histone modification, and histone changes at a silent gene on Xi.

A second factor implicated in *Xist* RNA localization is the transcription factor YY1, which is proposed to be a bifunctional protein capable of binding both DNA elements at the Xic and *Xist* RNA (Fig. 9A) (Jeon and Lee 2011; Thorvaldsen et al. 2011). Binding elements in the *Xist* gene are suggested to function as nucleation sites that mediate docking of *Xist* RNA particles with Xi.

4.3 Booster Elements Facilitate Spreading of X Inactivation

Although *Xist* RNA transgenes function on autosomes, autosomal silencing is less efficient than silencing of the X chromosome. The relative insensitivity of autosomes to *Xist*-mediated silencing was first reported in classical genetic studies that analyzed gene silencing in X:autosome translocations. Specifically the spreading of silent chromatin along the autosomal chromosome arm was found to be variable between translocations and limited in extent. This is attributable to autosomes resisting the initial spreading of *Xist* RNA and associated gene silencing, at least in some cases (Popova et al. 2006). In other cases, limited silencing appears to result from aberrant long term maintenance of silencing on autosomes following efficient spreading at the onset of X inactivation (Cattanach 1974).

Attenuated silencing on autosomes led to the idea that there are sequences on the X chromosome, originally referred to as "way stations" or "booster elements" (described in Gartler and Riggs 1983), that serve to amplify or enhance the spread/maintenance of X inactivation. More recently, it has been proposed that a common dispersed repeat family, the L1 long-interspersed repeats (LINE-1), are good candidates for the "way station" elements (Lyon 2003). These repeat sequences are common in the human and mouse genomes, but are particularly frequent along the X chromosome. Further, LINE-1 elements are most common in the more condensed, gene-poor, G-banded regions of the human and mouse genomes, suggesting that they may, in some way, favor a chromatin conformation associated with transcriptional silencing. In support of this idea, analysis of gene silencing meditated by *Xist* transgenes located on autosomes and in X autosome translocations indicates that large gene-rich LINE-1 depleted chromosomal domains are relatively refractory to the spreading of X inactivation (Popova et al. 2006; Chow et al. 2010; Tang et al. 2010).

A common idea is that way station elements are synonymous with *Xist* RNA-binding sites. In the case of L1 LINE elements, this seems improbable as *Xist* RNA is concentrated in gene-rich chromosomal domains, reciprocal to L1 LINE distribution. Interestingly, the high density of L1 LINEs on the X chromosome has fragmented gene-rich domains such that they are, on average, much smaller than on autosomes. Thus, one idea is that L1 LINE domains influence the higher-order topological folding of the chromosome in such a way as to favor efficient in *cis* spreading of *Xist* RNA within gene-rich domains (Popova et al. 2006; Tang et al. 2010). It is important to remember that the efficiency of *Xist* RNA spreading from its site of synthesis (i.e., the *Xist* locus) through the X-chromosome territory will depend on both the overall configuration of that territory in three dimensions and how the *Xist* locus is positioned within the territory.

RNA FISH analysis has shown that there is a major burst of L1 LINE transcription maintained specifically on the inactive X chromosome at the time that X inactivation is first established (Chow et al. 2010). L1 transcription has been directly linked to silencing through a mechanism involving overlapping transcripts and the production of short RNAs, but may also play a role in facilitating spread of X inactivation in *cis*.

Knowledge of *Xist* RNA binding sites beyond that provided by cytological analysis is at present lacking, and the development of a method to map binding sites at near nucleotide resolution is an important goal for the future.

4.4 Heterochromatic Structure of the Inactive X: The Link with Chromatin Modifications

Since the very earliest light microscopical studies, it has been realized that Xi shares properties with heterochromatin. Like the constitutive heterochromatin found at and around centromeres, Xi remains visible and apparently condensed throughout interphase (as the Barr body), and its DNA is usually replicated late in S-phase. Xi is said to consist of "facultative" heterochromatin.

Further parallels between Xi and constitutive heterochromatin have come from the use of indirect immunofluorescence microscopy to study the distribution of histone modifications and variants both along metaphase chromosomes and in interphase nuclei. The facultative heterochromatin of the inactive X chromosome in both human and mouse cells is depleted in acetylated histone H4 (Jeppesen and Turner 1993), and in this way resembles constitutive, centric heterochromatin. This was the first demonstration that the inactive X chromosome was marked by a specific type of histone modification. (Histone modifications and their functions are described in Ch. 3 [Allis et al.]). Subsequent experiments in several laboratories confirmed these observations and further showed that acetylated isoforms of all four core histones (H2A, H2B, H3, and H4) were depleted in both constitutive and facultative heterochromatin in interphase and metaphase cells (Fig. 5B). In particular, both centric heterochromatin and Xi are depleted in H3 di- and trimethylated at lysine 4 (H3K4me2 and H3K4me3; O'Neill et al. 2008). Like acetylation, H3K4me2/me3 are generally thought to be markers of transcriptionally active, or potentially active, chromatin.

Other chromatin modifications that are enriched on Xi include the histone modifications H3 trimethylated at lysine 27 (H3K27me3) and H2A monoubiquitinated at lysine 119 (H2AK119ub1; Fig. 9B). These modifications are catalyzed by Polycomb repressor complexes (see Ch. 17 [Grossniklaus and Paro 2014] for further discussion). There is also enrichment of H3K9me3, normally associated with constitutive pericentric heterochromatin, in humans and some other species. In mouse, H3K9me2, and not H3K9me3, enrichment has been observed. Because distinct histone lysine methyltransferases (KMTs) are associated with these modifications, this appears to represent a fundamental difference in Xi heterochromatin in the different species. Elevated H4K20me1, a histone modification associated with chromosome condensation, has also been observed on Xi (Kohlmaier et al. 2004). Finally, in addition to specific histone modifications, a variant histone, macroH2A, is enriched in Xi heterochromatin (Costanzi and Pehrson 1998), and conversely a different variant, H2A.bbd, is specifically depleted on Xi (Chadwick and Willard 2001).

A careful analysis of the distributions of histone modifications across Xi in human cultured cells has provided insights into the complexity of the system (Chadwick and Willard 2003). H3K9me3 and H3K27me3/H2AK119ub1 are enriched at defined, but nonoverlapping, regions across Xi. Thus, unlike loss of histone acetylation, enrichment in these modifications is a regional, not an overall, property of Xi. Intriguingly, those regions enriched in H3K27me3 are also enriched in *Xist* RNA and the variant histone macroH2A1.2. Conversely, those regions of Xi that are enriched in H3K9me3 also show enhanced levels of heterochromatin protein HP1 (known to bind to methylated H3K9) and H4K20me3 (also a mark associated with constitutive, centric heterochromatin). Importantly, immunostaining of the Barr body in interphase and the inactive X in

 Cite as *Cold Spring Harb Perspect Biol* doi: 10.1101/cshperspect.a019406

metaphase cells showed the same costaining patterns, suggesting that the different domains are retained through the cell cycle.

Constitutive centric heterochromatin is enriched in methylated DNA, primarily 5′-methylcytosine at CpG dinucleotides (see Ch. 15 [Li and Zhang 2014]). This is consistent with its low level of transcriptional activity. Perhaps surprisingly, the level of CpG methylation on Xi is not, overall, significantly higher than the rest of the genome. However, CpG islands associated with silenced genes are highly methylated and experimental evidence suggests that DNA methylation plays an important role in the stabilization of the inactive state. The overall reduction in CpG methylation on Xi is apparently attributable to reduced methylation in introns, intergenic regions, and possibly at common repeat elements.

In most cases, chromatin features of the inactive X chromosome have been identified by immunofluorescence analysis of either metaphase chromosomes or the Barr body in interphase cells. However, localized changes in histone modifications may also play important roles in the various stages of the X inactivation process. Such changes can be identified by high-resolution microscopy, or by chromatin immunoprecipitation (ChIP) approaches. ChIP analysis of chromatin surrounding the *Xist* locus has identified a region extending >340 kb 5′ of the *Xist* gene that is characterized by enrichment in methylated H3K9 and H3K27 in undifferentiated ES cells (Rougeulle et al. 2004). Hypermethylation diminishes as the cells differentiate and X inactivation proceeds. Sites within this region are enriched in acetylated H3 and H4 (O'Neill et al. 1999). Ongoing investigations will reveal to what extent these localized histone modifications in the Xic region are early, causative events driving the X inactivation process, or downstream events that are (possibly essential) components of an ongoing chromatin remodeling process.

An important recent innovation—high-throughput sequencing applied to chromatin fragments obtained by ChIP (ChIP-seq)—has been coupled with the use of single-nucleotide polymorphisms (SNPs) in interspecific mouse crosses to determine high-resolution maps of specific modifications/factors on the inactive (Xi) relative to the active (Xa) X chromosome. Using this approach it was shown that H3K27me3 on Xi occurs in large blocks, distributed broadly over promoters, gene bodies, and intergenic regions (Marks et al. 2009).

4.5 The Enzymology of Chromatin Modifications on Xi

The enzymes responsible for the deacetylation of core histones (HDACs) during X inactivation or the demethylation of H3K4 are as yet unknown. We know more about the enzymes responsible for putting histone modifications in place. H2AK119u1 and H3K27me3 are deposited on Xi by the Polycomb repressive complexes PRC1 and PRC2, respectively (Silva et al. 2003; de Napoles et al. 2004; detailed in Ch. 17 [Grossniklaus and Paro 2014]). Recruitment of PRC2 to Xi is *Xist* dependent and it has been suggested that this is mediated by direct interaction of PRC2 components with the A-repeat element of *Xist* RNA (Zhao et al. 2008). Direct interaction of PRC2 with A-repeat elements is unlikely, however, to be the whole story as PcG recruitment to Xi does not occur during early mouse preimplantation development (Okamoto et al. 2004), and transgenic *Xist* RNA from which the A-repeat element is deleted can still recruit PRC2, albeit inefficiently (Kohlmaier et al. 2004).

Recruitment of the PRC1 complex to Xi is, in part, attributable to the interaction of the chromodomain of the core protein CBX2/4/7 with H3K27me3, a histone modification put in place by the PRC2 complex (see Fig. 5 of Ch. 17 [Grossniklaus and Paro 2014]). There is also a PRC2-independent pathway that recruits variant RYBP-PRC1 complexes in which CBX proteins are replaced by the RYBP subunit (Tavares et al. 2012).

Specific KMT enzymes catalyzing H3K9me3 and H3K9me2 on Xi in human and mouse, respectively, have not yet been formally identified, although the enzyme systems that establish these modifications at other sites in the genome are well described (Ch. 6 [Cheng 2014]). DNA methylation of CpG islands on Xi requires the de novo methyltransferase Dnmt3b, whereas Dnmt3a and the accessory protein, Dnmt3L, are dispensable (Gendrel et al. 2012). An additional factor required for DNA methylation at many Xi CpG islands is the protein Smchd1, an atypical member of the SMC (structural maintenance of chromosomes) superfamily that includes components of the condensin and cohesin complexes with roles in chromosome organization and dynamics (Blewitt et al. 2008). Smchd1 protein is enriched on Xi and homozygous mutant mice show female-specific embryo lethality attributable to incomplete silencing of X-linked genes. However, the mechanism by which Smchd1 influences CpG island methylation and silencing remains to be determined.

4.6 Higher-Order Chromatin Structure on Xi

Although Xi chromatin is often described as "condensed," careful microscopic analysis and 3D reconstruction of Xa and Xi chromosomes labeled with X-specific DNA probes suggests that the difference between them is more a matter of shape than the amount of chromatin per unit volume (Eils et al. 1996; Splinter et al. 2011). The position of Xi relative to other nuclear structures may also be important. For example, it has often been observed that the Barr body localizes to the nuclear periphery and/or the periphery of the nucleolus.

Further insights have come from analysis of the 3D organization of genes on the Xi chromosome relative to *Xist* RNA territories in the interphase nucleus (Chaumeil et al. 2006). *Xist* RNA is found to describe a territory that includes common repeat sequences on the X chromosome and in which RNA polymerase II (Pol II) is depleted. During the establishment of X inactivation and in line with gene silencing, X-linked genes are recruited from positions external to the *Xist* RNA territory to sites either at the periphery or within the territory. Silencing deficient *Xist* RNA transgenes in which the A-repeat region is deleted can form an *Xist* RNA territory but fail to recruit genes. That the *Xist* RNA territory has been found to correspond to the location of common repeat elements on the X chromosome is somewhat paradoxical given that on metaphase chromosomes *Xist* RNA shows a reciprocal localization relative to LINE-1 elements (Sec. 4.3). Further studies are needed to resolve this point.

An important role for higher-order chromosome organization in X inactivation is also suggested by the identification of proteins potentially involved in chromosome architecture, notably Smchd1 (see Sec. 4.5) and hnRNPU/SAFA (Sec. 4.2). An additional nuclear scaffold factor, SATB1, has also been implicated in conferring competence for *Xist*-mediated silencing (Agrelo et al. 2009), although its role has recently been questioned because of the lack of an X inactivation defect in SATB1 knockout mice (Nechanitzky et al. 2012).

New methodologies to analyze 3D chromosome topology are shedding further light on higher-order chromosome organization on Xi. The 4C method described in Ch. 19 (Dekker and Misteli 2014) and illustrated in their Fig. 5, which quantifies the frequency of contact between defined positions on the chromosome, has been used in conjunction with SNPs to discriminate Xi and Xa alleles in XX somatic cells demonstrating that preferred long-range contacts involving loci on Xa are lost on Xi (Splinter et al. 2011). These interactions are partially restored by deletion of the *Xist* locus, despite the fact that *Xist* deletion does not result in reactivation of X-linked genes (see Sec. 4.1). Similarly, the 5C method has been applied to study the regulatory landscape of the Xic during the onset of X inactivation demonstrating that *Xist* and *Tsix* lie within separate topological domains (Nora et al. 2012; see Fig. 7 of Ch. 19 [Dekker and Misteli 2014)]. The application of these new methodologies will facilitate important advances in our understanding of Xi structure, in particular, when coupled to advanced microscopy approaches such as 3D structured illumination microscopy that are extending the resolution limit of conventional fluorescence microscopy (Schermelleh et al. 2008).

4.7 The Order of Events That Leads to X Inactivation

Differentiating XX ES cells have provided an invaluable model system for studying the dynamics of X-chromosome inactivation. In undifferentiated cells both X chromosomes are active and *Xist* and *Tsix* are expressed at low levels. Increased levels of *Xist* RNA and its coating of one X chromosome are first detected in a high proportion of cells after 1–2 d of differentiation. This is followed by rapid depletion of RNA Pol II within *Xist* RNA territories and then depletion of H3K4me3 (O'Neill et al. 2008). Recruitment of PcG proteins with associated methylation of H3K27 and monoubiquitination of H2A occur in a similar timeframe, along with deacetylation of H3K9 and loss of H3K4 methylation (Silva et al. 2003; de Napoles et al. 2004; Rougeulle et al. 2004; O'Neill et al. 2008). Global histone deacetylation and the accumulation of H3K9me2 on Xi are established with some delay, occurring at days 3–5 in the majority of cells (Fig. 10) (Keohane et al. 1996). The delayed appearance of some modifications implies that they are likely to be involved in the maintenance/stability of the inactive state, rather than its establishment. This interpretation assumes that patterns of acetylation and methylation at the promoters of individual genes undergoing inactivation reflect those determined by immunofluorescence analysis of the whole chromosome, or large domains. Initial studies by ChIP suggest that this is indeed the case (O'Neill et al. 2008), but further experimentation, using ChIP-seq, for example, is necessary.

Accumulation of the variant macroH2A1.2 histone on Xi occurs much later during XX ES cell differentiation (Mermoud et al. 1999). This variant histone has more than 200 additional amino acids in its carboxy-terminal tail and several amino acid substitutions throughout the molecule. Interestingly, *Xist* RNA expression is required to retain macroH2A on Xi in somatic cells (Csankovszki et al. 1999), but is not sufficient to recruit macroH2A in early differentiation stages (Mermoud et al. 1999; Wutz et al. 2002).

The late recruitment of three other factors to Xi has now been reported: hnRNPU/SAFA, Ash2l (Pullirsch et al. 2010), and Smchd1 (Gendrel et al. 2012). These observations indicate that the establishment of silent chromatin on Xi occurs in a sequential manner with at least two clearly separated phases (Fig. 10).

Selective DNA methylation of Xi CpG islands in ES cells accumulates slowly during differentiation (Gendrel et al. 2012). This is consistent with early studies showing that

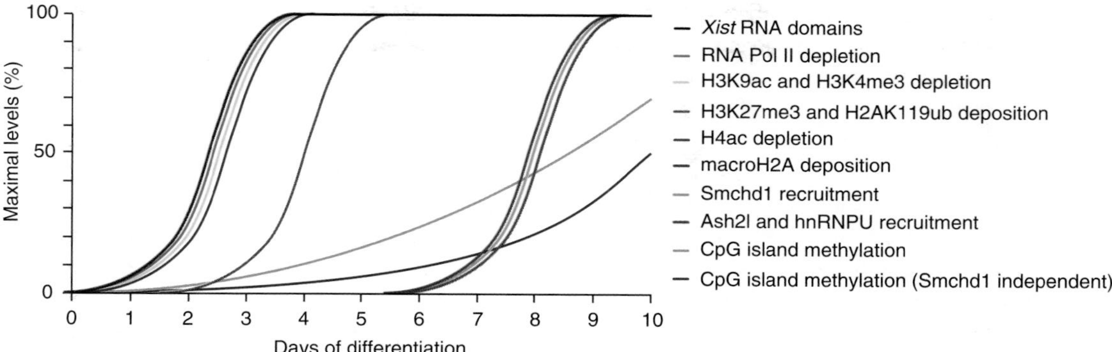

Figure 10. The order of events in differentiating XX ES cells. The diagram summarizes the order in which different silencing pathways are integrated during establishment of X inactivation in differentiating XX ES cells. Early events, depletion of RNA Pol II, loss of H3K4me3/H3K9Ac, and deposition of Polycomb-associated modifications occur coincident with the onset of *Xist* RNA expression. H4 hypoacetylation and a transition to late replication in S-phase occur slightly later. Enrichment for macroH2A, Smchd1 Ash2l, and hnRNPU/SAF-A occur in a defined temporal window relatively late in the differentiation time course. Accumulation of DNA methylation over CpG island occurs slowly following recruitment of Smchd1, although a subset of CpG islands acquire DNA methylation more rapidly and in a Smchd1-independent manner.

methylation of the *Hprt* promoter on Xi occurs relatively late in the developing embryo (Lock et al. 1987), a finding that led to the idea that DNA methylation is responsible for stabilization, or locking, of the inactive state rather than in initiation and spreading. In differentiating XX ES cells, a large proportion of CpG islands acquire little or no DNA methylation before day 7 of differentiation, attributable to the absence of Smchd1 on Xi before this time (see Sec. 4.5). A proportion of CpG islands acquire methylation earlier and at a faster rate, and in these cases methylation is Smchd1 independent (Gendrel et al. 2012). Thus, the picture that emerges is of a coordinated and carefully regulated sequence of events by which chromatin changes on the Xi are put in place as development proceeds (summarized in Fig. 10). It is remarkable that some of these changes, such as histone deacetylation and DNA methylation, take place after the cells have started to progress down various different pathways of differentiation. It seems that the program responsible for the completion of X inactivation proceeds independently of other cell differentiation programs. However, it is important to note that some aspects of random X inactivation can proceed only after differentiation has begun. For example, switching on expression of *Xist* transgenes in "undifferentiated" ES cells triggers various histone modifications associated with heterochromatinization, and also the transition to replication in late S-phase (Wutz and Jaenisch 2000), but there is no detectable incorporation of macroH2A; only after the cells have been induced to differentiate does macroH2A colocalize with *Xist* RNA on the chromosome containing the *Xist* transgene (Rasmussen et al. 2001). Association of macroH2A with *Xist*-coated chromatin is dependent on the continued presence of *Xist* RNA (Csankovszki et al. 1999), but does not require transcriptional silencing, as it is seen also in chromosomes coated with a mutant *Xist* RNA lacking regions necessary for silencing (Wutz et al. 2002). Thus, X inactivation can be seen as the end result of a series of parallel processes, only some of which are interdependent.

It should also be noted that a different order of events may occur during the establishment of imprinted X inactivation in preimplantation embryos. Notably enrichment of H3K27me3 is not detected until the 16-cell stage, considerably later than the onset of *Xist* expression (two- to four-cell stage; Mak et al. 2004; Okamoto et al. 2004). This may indicate a requirement for specific developmentally regulated cofactors to recruit the PRC2 PcG complex to Xi.

The relationship between the various chromosome-wide modifications and gene silencing that occur on the inactivating X in female ES cells is by no means clear. Recent data using microarrays (Lin et al. 2007) or RNA sequencing (Deng et al. 2011) to measure expression of X-linked genes shows that individual genes are inactivated at various times during ES cell differentiation, with some genes escaping inactivation altogether. It seems that, for most X-linked genes, silencing is triggered by conditions that occur at differing stages of ES cell differentiation.

4.8 *Xist*-Mediated Silencing: "Belts and Braces"

Accumulating evidence illustrates multiple pathways contributing to the establishment of gene silencing on the Xi. At the nucleosomal level there is a gain and loss of specific

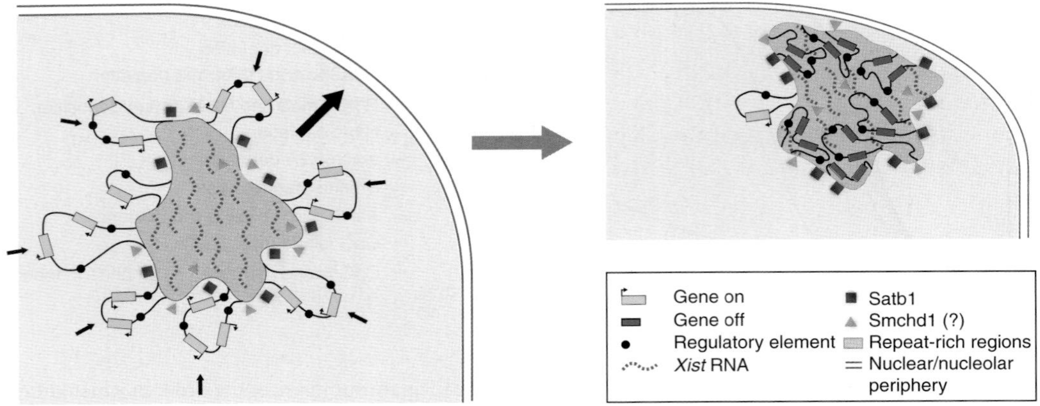

Figure 11. Factors involved in *Xist*-mediated silencing. Depiction of changes in higher-order chromosome architecture during the establishment of X inactivation. *Xist* RNA initially coats repeat rich chromosomal domains; genes and other regulatory elements occupy an external position. As X inactivation proceeds, genes are internalized within the *Xist* territory with consequent restriction in the mobility of chromosome loops. Establishment of X inactivation is also linked to positioning of the chromosome on the nuclear and/or nucleolar periphery. Nuclear scaffold factors (SATB1) and chromosome structure factors (Smchd1) may play a role in the reorganization of chromosome architecture on Xi.

posttranslational modifications, incorporation of different histone variants, and DNA methylation at CpG islands (Fig. 9B). At a higher-order level there are changes in the architecture of chromatin loops and the reorganization of chromosome domains and chromosome position in the nucleus, potentially mediated by chromosomal proteins such as Smchd1, SATB1, and SAFA/hnRNPU (Fig. 11). As such, Xi can be viewed as a "belts and braces" system in which different pathways play overlapping or redundant roles. Within this framework it is likely that different pathways are more or less important at specific times in development, an idea that is supported by the observation that, in contrast to somatic cells, chromosome silencing in cells of the early embryo is dependent on ongoing *Xist* expression. Finally, it is becoming apparent that different pathways contribute differently to silencing of specific genes or subsets of genes on Xi. For example, female Smchd1 null embryos that fail in mid-gestation, show up-regulation of only a small proportion of loci on Xi (Blewitt et al. 2008). In this regard, it is important to note that there is significant variability in the time that individual Xi genes are silenced following the onset of *Xist* expression. Thus, to some degree it may be necessary to consider the contribution of different X inactivation pathways on a gene-by-gene basis.

Although significant progress has been made toward identifying pathways involved in *Xist*-mediated silencing, it is likely that key factors remain to be found. Notably, we do not yet know the critical factors that interact with the A-repeat region of *Xist* RNA to initiate the silencing process. Thus, it remains a possibility that the known modifications of Xi and associated pathways are secondary, occurring in response to silencing established by an as yet uncharacterized primary mechanism.

5 X-CHROMOSOME REACTIVATION AND REPROGRAMMING

5.1 X Reactivation in Normal Development

Multiple layers of epigenetic modification contribute to the silencing of the inactive X chromosome. As a result, the repressed state is generally highly stable and attempts to reverse it experimentally have been consistently unsuccessful. However, there are circumstances in the course of normal development in which the entire X chromosome is reactivated. The best studied example is reversal of X inactivation in developing PGCs. In mouse, PGCs are specified at about 7–8 days of development, shortly after gastrulation. At this time, cells of the embryo have already undergone random X inactivation. Subsequently, the developing PGCs migrate along the hindgut region of the embryo and arrive at the genital ridges, the structures that give rise to the adult gonads. It is at this time that XX PGCs reactivate their Xi (Monk and McLaren 1981). This event occurs coincident with a more general epigenetic reprogramming that includes erasure of parental imprints and genome-wide DNA demethylation (see Fig. 5 of Ch. 26 [Barlow and Bartolomei 2014]; see also Ch. 27 [Reik and Surani 2014] for more detail).

X-chromosome reactivation in PGCs may indicate a specialized mechanism for reversing the multilayered heterochromatic structure. Extinction of *Xist* RNA expression

 Cite as *Cold Spring Harb Perspect Biol* doi: 10.1101/cshperspect.a019406

has been seen to correlate with X reactivation, but given that silencing is *Xist* independent in XX somatic cells, it is not certain that this is causative. It is possible that PGCs fail to establish all of the marks associated with silencing and are, therefore, more susceptible to reactivation. Consistent with this is the evidence that CpG island methylation does not occur on the Xi in developing PGCs in mouse (Grant et al. 1992).

A second example of X reactivation is the reversal of imprinted Xp inactivation in the ICM lineage of blastocyst stage embryos, discussed in Section 3.5, which again is associated with wider genome reprogramming events. This reactivation also correlates with extinction of *Xist* RNA and a loss of epigenetic marks associated with silencing. Again, it is possible that pre-ICM cells fail to establish all of the marks associated with silencing and are, therefore, more susceptible to X reactivation.

5.2 X Reactivation during Experimental Reprogramming

X reactivation has also been observed under specific experimental circumstances. It occurs during nuclear transfer of somatic nuclei to unfertilized oocytes and following fusion of somatic cells with totipotent cell types such as ES, embryonic germ (EG), or embryonal carcinoma (EC) cells (e.g., see Tada et al. 2000). Finally, X reactivation occurs when XX somatic cells are converted to induced pluripotent stem (iPS) cells (Maherali et al. 2007).

Nuclear transfer embryos provide a particularly interesting example. Experiments in mice (Eggan et al. 2000) showed rapid reactivation of a marker gene on Xi in cleavage stage nuclear transfer embryos. Despite this, the nucleus retained some memory of which X had been inactive because in cloned embryos the donor cell Xi was also the Xi in trophectoderm cells of the placenta. In contrast, cells of the embryo proper showed random X inactivation (see Fig. 12). Presumably, X reactivation and reprogramming that occurs in the developing ICM gives the embryo a second chance to reset epigenetic information from the donor nucleus (see Mekhoubad et al. 2012).

Recent studies have shown that ectopic *Xist* activation makes a significant contribution to the inefficiency of reproductive cloning in mice (Inoue et al. 2010). Using XY donor somatic nuclei, the single *Xist* allele is activated in cloned embryos. Similarly, using XX donor somatic nuclei, both *Xist* alleles are activated. Use of donor cells in which *Xist* is deleted dramatically improves cloning efficiency. Presumably, the host oocyte erases *Xist* gene repression present on the active X chromosome in XY and XX somatic cells. It follows that embryos cloned using donor nuclei with an intact *Xist* locus can only survive early preimplantation development if reprogramming of somatic *Xist* repression fails. This could explain why, in surviving embryos cloned from normal XX somatic cells, the donor cell Xi remains inactive in the extraembryonic lineages for which X inactivation patterns are normally determined at the two- to four-cell stage (Fig. 6).

Although X reactivation occurs during iPS reprogramming of mouse somatic cells, the situation in human iPS cells appears more complex. Early iPS cultures retain an Xi, but X reactivation can occur subsequently, dependent on

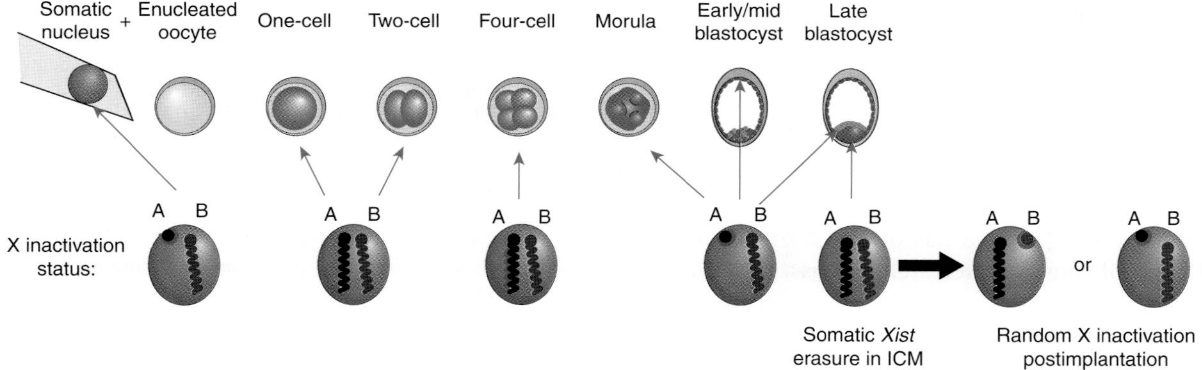

Figure 12. Regulation of X inactivation in cloned mouse embryos. The figure illustrates an XX donor cell with the inactive X chromosome (*A*) coated with *Xist* RNA (green line). In this model, transcription from the donor nucleus, including *Xist* RNA, is repressed by oocyte factors until the two-cell stage, resulting in X reactivation. Recommencement of *Xist* expression then occurs at the two-cell stage. *Xist* is then reexpressed again from the inactive X allele from the donor cell. This would be attributable to retention of a mark such as DNA methylation at the *Xist* promoter. This pattern is maintained in cells allocated to the TE and PE lineages, but not in pluripotent epiblast in which *Xist* expression is again extinguished, leading to a second reactivation event. In the ICM, erasure of the epigenetic marks governing donor *Xist* expression allows for subsequent random X inactivation in the embryo proper.

exact culture conditions. This is likely to relate to different levels of pluripotency as observed also in comparing mouse ES cells and epiblast stem cells (EpiSCs; Bao et al. 2009). Human ES cells (hESCs) are more similar to EpiSCs than to ES cells from mouse, and consistent with this, XX hESCs retain an inactive X chromosome (Tchieu et al. 2010).

5.3 Lessons from Inducible *Xist* Transgenes

A series of experiments using inducible *Xist* transgenes in ES cells has greatly increased our understanding of stability versus reversibility of X inactivation. First, it was shown that *Xist* RNA can establish X inactivation in undifferentiated ES cells and during very early stages of differentiation, but not subsequently (referred to as the "window of opportunity"; Wutz and Jaenisch 2000). Ectopic expression of the nuclear scaffold/matrix protein, SATB1, confers the ability to respond to *Xist* RNA in thymic lymphoma and fibroblast cells (Agrelo et al. 2009), suggesting that this is at least one component that is important for developmental competence (but see also Nechanitzky et al. 2012). The ability of cells to respond to *Xist* RNA broadly correlates with reversibility of X inactivation. Thus, silencing was reversed when the transgene was switched off in ES cells or during early differentiation stages, but not in later differentiation or somatic cells.

Returning to X reactivation and reprogramming, the inducible transgene data imply that in defined cellular environments, namely, undifferentiated ES cells, X reactivation will occur when expression of *Xist* RNA is extinguished. If we consider that those cells in which X reactivation has been documented to occur (i.e., PGCs, ICM cells, iPS, EG, and EC cells) are all similar to ES cells in terms of pluripotency and plasticity, then extinction of *Xist* expression may underlie X reactivation in all cases.

6 SUMMARY AND FUTURE DIRECTIONS

In recent years, there has been significant progress in our understanding of the molecular mechanism of X inactivation. To date, this progress has been fed by advances in related fields of epigenetic research and has, in turn, stimulated advances in other fields. An example of the latter is the growing evidence that some clusters of imprinted genes are regulated by *cis*-acting ncRNAs in much the same way that *Xist* regulates the X chromosome (see Ch. 26 [Barlow and Bartolomei 2014]). Similarly, the independent evolution of ncRNA regulating X inactivation in marsupials further illustrates the potential generality of in *cis* silencing by ncRNA. Conversely, other studies point to distinct types of ncRNA that function in *trans* in gene silencing (e.g., HOTAIR; Rinn et al. 2007; Ch. 2 [Rinn 2014]). It will be interesting to see if mechanistic links with *Xist* and marsupial Rsx ncRNAs emerge in future studies.

There remain many unanswered questions. Although progress has been made in defining the *cis*-acting sequences and *trans*-acting factors that regulate counting and choice, their further elucidation provides an exciting challenge. Similarly, although we now know some of the chromatin-modifying complexes involved in maintaining X inactivation, such as the Polycomb-group complexes, the signal for establishing chromosome-wide silencing, triggered by *Xist* RNA, remains unknown. Similarly, the mechanistic links between chromosome-wide changes in Xi chromatin and the silencing of individual genes remain elusive. Other key questions are to understand how silencing spreads across the chromosome and what role, if any, way stations (perhaps LINE elements) play in this process and in the stabilization/maintenance of the silent state. This may relate to the intriguing question of how X inactivation is reversed in some cell types and stages of development, but is essentially irreversible in others. This latter question relates to the wider and crucially important issue of understanding genome plasticity and reprogramming through development. Finally, the recent validation of an old hypothesis, namely, the up-regulated expression of genes on the single active X in males and females, has given us a new view of the context in which X inactivation operates and a better understanding of the true complexity of dosage compensation in mammals.

REFERENCES

**Reference is also in this book.*

Agrelo R, Souabni A, Novatchkova M, Haslinger C, Leeb M, Komnenovic V, Kishimoto H, Gresh L, Kohwi-Shigematsu T, Kenner L, et al. 2009. *SATB1* defines the developmental context for gene silencing by *Xist* in lymphoma and embryonic cells. *Dev Cell* **16:** 507–516.

Augui S, Filion GJ, Huart S, Nora E, Guggiari M, Maresca M, Stewart AF, Heard E. 2007. Sensing X chromosome pairs before X inactivation via a novel X-pairing region of the Xic. *Science* **318:** 1632–1636.

Bao S, Tang F, Li X, Hayashi K, Gillich A, Lao K, Surani MA. 2009. Epigenetic reversion of post-implantation epiblast to pluripotent embryonic stem cells. *Nature* **461:** 1292–1295.

* Barlow DP, Bartolomei MS. 2014. Genomic imprinting in mammals. *Cold Spring Harb Perspect Biol* **6:** 018382.

Berletch JB, Yang F, Xu J, Carrel L, Disteche CM. 2011. Genes that escape from X inactivation. *Hum Genet* **130:** 237–245.

Beutler E, Yeh M, Fairbanks VF. 1962. The normal human female as a mosaic of X-chromosome activity: Studies using the gene for C-6-PD-deficiency as a marker. *Proc Natl Acad Sci* **48:** 9–16.

Blewitt ME, Gendrel AV, Pang Z, Sparrow DB, Whitelaw N, Craig JM, Apedaile A, Hilton DJ, Dunwoodie SL, Brockdorff N, et al. 2008. SmcHD1, containing a structural-maintenance-of-chromosomes hinge domain, has a critical role in X inactivation. *Nat Genet* **40:** 663–669.

Brockdorff N, Ashworth A, Kay GF, McCabe VM, Norris DP, Cooper PJ, Swift S, Rastan S. 1992. The product of the mouse *Xist* gene is a 15 kb inactive X-specific transcript containing no conserved ORF and located in the nucleus. *Cell* **71:** 515–526.

Brown CJ, Willard HF. 1994. The human X-inactivation centre is not required for maintenance of X-chromosome inactivation. *Nature* **368:** 154–156.

Brown CJ, Ballabio A, Rupert JL, Lafreniere RG, Grompe M, Tonlorenzi R, Willard HF. 1991. A gene from the region of the human X inactivation centre is expressed exclusively from the inactive X chromosome. *Nature* **349:** 38–44.

Brown CJ, Hendrich BD, Rupert JL, Lafreniere RG, Xing Y, Lawrence J, Willard HF. 1992. The human *XIST* gene: Analysis of a 17 kb inactive X-specific RNA that contains conserved repeats and is highly localized within the nucleus. *Cell* **71:** 527–542.

Buzin CH, Mann JR, Singer-Sam J. 1994. Quantitative RT-PCR assays show *Xist* RNA levels are low in mouse female adult tissue, embryos and embryoid bodies. *Development* **120:** 3529–3536.

Carrel L, Willard HF. 2005. X-inactivation profile reveals extensive variability in X-linked gene expression in females. *Nature* **434:** 400–404.

Cattanach BM. 1974. Position effect variegation in the mouse. *Genet Res* **23:** 291–306.

Chadwick BP, Willard HF. 2001. A novel chromatin protein, distantly related to histone H2A, is largely excluded from the inactive X chromosome. *J Cell Biol* **152:** 375–384.

Chadwick BP, Willard HF. 2003. Chromatin of the Barr body: Histone and non-histone proteins associated with or excluded from the inactive X chromosome. *Hum Mol Genet* **12:** 2167–2178.

Charlesworth B. 1996. The evolution of chromosomal sex determination and dosage compensation. *Curr Biol* **6:** 149–162.

Charlesworth D, Charlesworth B. 2005. Sex chromosomes: Evolution of the weird and wonderful. *Curr Biol* **15:** R129–R131.

Chaumeil J, Le Baccon P, Wutz A, Heard E. 2006. A novel role for *Xist* RNA in the formation of a repressive nuclear compartment into which genes are recruited when silenced. *Genes Dev* **20:** 2223–2237.

* Cheng X. 2014. Structural and functional coordination of DNA and histone methylation. *Cold Spring Harb Perspect Biol* **6:** a018747.

Chow JC, Ciaudo C, Fazzari MJ, Mise N, Servant N, Glass JL, Attreed M, Avner P, Wutz A, Barillot E, et al. 2010. LINE-1 activity in facultative heterochromatin formation during X chromosome inactivation. *Cell* **141:** 956–969.

Chureau C, Chantalat S, Romito A, Galvani A, Duret L, Avner P, Rougeulle C. 2011. Ftx is a non-coding RNA which affects *Xist* expression and chromatin structure within the X-inactivation center region. *Hum Mol Genet* **20:** 705–718.

Clemson CM, McNeil JA, Willard HF, Lawrence JB. 1996. *XIST* RNA paints the inactive X chromosome at interphase: Evidence for a novel RNA involved in nuclear/chromosome structure. *J Cell Biol* **132:** 259–275.

Costanzi C, Pehrson JR. 1998. Histone macroH2A1 is concentrated in the inactive X chromosome of female mammals. *Nature* **393:** 599–601.

Csankovszki G, Panning B, Bates B, Pehrson JR, Jaenisch R. 1999. Conditional deletion of *Xist* disrupts histone macroH2A localization but not maintenance of X inactivation. *Nat Genet* **22:** 323–324.

Davidson RG, Nitowsky HM, Childs B. 1963. Demonstration of two populations of cells in the human female heterozygous for glucose-6-phosphate dehydrogenase variants. *Proc Natl Acad Sci* **50:** 481–485.

* Dekker J, Misteli T. 2014. Long-range chromatin interactions. *Cold Spring Harb Perspect Biol* doi: 10.1101/cshperspect.a019356.

de Napoles M, Mermoud JE, Wakao R, Tang YA, Endoh M, Appanah R, Nesterova TB, Silva J, Otte AP, Vidal M, et al. 2004. Polycomb group proteins Ring1A/B link ubiquitylation of histone H2A to heritable gene silencing and X inactivation. *Dev Cell* **7:** 663–676.

Deng X, Hiatt JB, Nguyen DK, Ercan S, Sturgill D, Hillier LW, Schlesinger F, Davis CA, Reinke VJ, Gingeras TR, et al. 2011. Evidence for compensatory upregulation of expressed X-linked genes in mammals, *Caenorhabditis elegans* and *Drosophila melanogaster*. *Nat Genet* **43:** 1179–1185.

Donohoe ME, Silva SS, Pinter SF, Xu N, Lee JT. 2009. The pluripotency factor Oct4 interacts with Ctcf and also controls X-chromosome pairing and counting. *Nature* **460:** 128–132.

Duret L, Chureau C, Samain S, Weissenbach J, Avner P. 2006. The *Xist* RNA gene evolved in eutherians by pseudogenization of a protein-coding gene. *Science* **312:** 1653–1655.

Duthie SM, Nesterova TB, Formstone EJ, Keohane AM, Turner BM, Zakian SM, Brockdorff N. 1999. *Xist* RNA exhibits a banded localization on the inactive X chromosome and is excluded from autosomal material in cis. *Hum Mol Genet* **8:** 195–204.

Eggan K, Akutsu H, Hochedlinger K, Rideout W 3rd, Yanagimachi R, Jaenisch R. 2000. X-chromosome inactivation in cloned mouse embryos. *Science* **290:** 1578–1581.

Eils R, Dietzel S, Bertin E, Schrock E, Speicher MR, Ried T, Robert-Nicoud M, Cremer C, Cremer T. 1996. Three-dimensional reconstruction of painted human interphase chromosomes: Active and inactive X chromosome territories have similar volumes but differ in shape and surface structure. *J Cell Biol* **135:** 1427–1440.

Ellegren H, Parsch J. 2007. The evolution of sex-biased genes and sex-biased gene expression. *Nat Rev Genet* **8:** 689–698.

Epstein CJ, Smith S, Travis B, Tucker G. 1978. Both X chromosomes function before visible X-chromosome inactivation in female mouse embryos. *Nature* **274:** 500–503.

Gartler SM, Riggs AD. 1983. Mammalian X-chromosome inactivation. *Annu Rev Genet* **17:** 155–190.

Gendrel AV, Apedaile A, Coker H, Termanis A, Zvetkova I, Godwin J, Tang YA, Huntley D, Montana G, Taylor S, et al. 2012. Smchd1-dependent and -independent pathways determine developmental dynamics of CpG island methylation on the inactive X chromosome. *Dev Cell* **23:** 265–279.

Gontan C, Achame EM, Demmers J, Barakat TS, Rentmeester E, van Ijcken W, Grootegoed JA, Gribnau J. 2012. RNF12 initiates X-chromosome inactivation by targeting REX1 for degradation. *Nature* **485:** 386–390.

Grant M, Zuccotti M, Monk M. 1992. Methylation of CpG sites of two X-linked genes coincides with X-inactivation in the female mouse embryo but not in the germ line. *Nat Genet* **2:** 161–166.

Grant J, Mahadevaiah SK, Khil P, Sangrithi MN, Royo H, Duckworth J, McCarrey JR, VandeBerg JL, Renfree MB, Taylor W, et al. 2012. *Rsx* is a metatherian RNA with *Xist*-like properties in X-chromosome inactivation. *Nature* **487:** 254–258.

* Grossniklaus U, Paro R. 2014. Transcriptional silencing by *Polycomb*-group proteins. *Cold Spring Harb Perspect Biol* **6:** a019331.

Hasegawa Y, Nakagawa S. 2011. Revisiting the function of nuclear scaffold/matrix binding proteins in X chromosome inactivation. *RNA Biol* **8:** 735–739.

Huynh KD, Lee JT. 2003. Inheritance of a pre-inactivated paternal X chromosome in early mouse embryos. *Nature* **426:** 857–862.

Inoue K, Kohda T, Sugimoto M, Sado T, Ogonuki N, Matoba S, Shiura H, Ikeda R, Mochida K, Fujii T, et al. 2010. Impeding *Xist* expression from the active X chromosome improves mouse somatic cell nuclear transfer. *Science* **330:** 496–499.

Jeon Y, Lee JT. 2011. YY1 tethers *Xist* RNA to the inactive X nucleation center. *Cell* **146:** 119–133.

Jeppesen P, Turner BM. 1993. The inactive X chromosome in female mammals is distinguished by a lack of histone H4 acetylation, a cytogenetic marker for gene expression. *Cell* **74:** 281–289.

Jonkers I, Barakat TS, Achame EM, Monkhorst K, Kenter A, Rentmeester E, Grosveld F, Grootegoed JA, Gribnau J. 2009. RNF12 is an X-encoded dose-dependent activator of X chromosome inactivation. *Cell* **139:** 999–1011.

Keohane AM, O'Neill LP, Belyaev ND, Lavender JS, Turner BM. 1996. X-inactivation and histone H4 acetylation in embryonic stem cells. *Dev Biol* **180:** 618–630.

Kohlmaier A, Savarese F, Lachner M, Martens J, Jenuwein T, Wutz A. 2004. A chromosomal memory triggered by *Xist* regulates histone methylation in X inactivation. *PLoS Biol* **2:** E171.

Lee JT. 2000. Disruption of imprinted X inactivation by parent-of-origin effects at *Tsix*. *Cell* **103:** 17–27.

Lee JT, Davidow LS, Warshawsky D. 1999. *Tsix*, a gene antisense to *Xist* at the X-inactivation centre. *Nat Genet* **21:** 400–404.

Lee JT, Lu N. 1999. Targeted mutagenesis of *Tsix* leads to nonrandom X inactivation. *Cell* **99:** 47–57.

Lee JT, Strauss WM, Dausman JA, Jaenisch R. 1996. A 450 kb transgene displays properties of the mammalian X-inactivation center. *Cell* **86:** 83–94.

* Li E, Zhang Y. 2014. DNA methylation in mammals. *Cold Spring Harb Perspect Biol* **6:** a019133.

Lin H, Gupta V, Vermilyea MD, Falciani F, Lee JT, O'Neill LP, Turner BM. 2007. Dosage compensation in the mouse balances up-regulation and silencing of X-linked genes. *PLoS Biol* **5:** e326.

Lin H, Halsall JA, Antczak P, O'Neill LP, Falciani F, Turner BM. 2011. Relative overexpression of X-linked genes in mouse embryonic stem cells is consistent with Ohno's hypothesis. *Nat Genet* **43:** 1169–1170; author reply 1171–1172.

Lock LF, Takagi N, Martin GR. 1987. Methylation of the *Hprt* gene on the inactive X occurs after chromosome inactivation. *Cell* **48:** 39–46.

* Lucchesi JC, Kuroda MI. 2014. Dosage compensation in *Drosophila*. *Cold Spring Harb Perspect Biol* doi: 10.1101/cshperspect.a019398.

Lyon MF. 1961. Gene action in the X-chromosome of the mouse (*Mus musculus L.*). *Nature* **190:** 372–373.

Lyon MF. 2003. The Lyon and the LINE hypothesis. *Semin Cell Dev Biol* **14:** 313–318.

Maherali N, Sridharan R, Xie W, Utikal J, Eminli S, Arnold K, Stadtfeld M, Yachechko R, Tchieu J, Jaenisch R, et al. 2007. Directly reprogrammed fibroblasts show global epigenetic remodeling and widespread tissue contribution. *Cell Stem Cell* **1:** 55–70.

Mak W, Nesterova TB, de Napoles M, Appanah R, Yamanaka S, Otte AP, Brockdorff N. 2004. Reactivation of the paternal X chromosome in early mouse embryos. *Science* **303:** 666–669.

Marks H, Chow JC, Denissov S, Francoijs KJ, Brockdorff N, Heard E, Stunnenberg HG. 2009. High-resolution analysis of epigenetic changes associated with X inactivation. *Genome Res* **19:** 1361–1373.

Marshall Graves JA, Shetty S. 2001. Sex from W to Z: Evolution of vertebrate sex chromosomes and sex determining genes. *J Exp Zool* **290:** 449–462.

Masui O, Bonnet I, Le Baccon P, Brito I, Pollex T, Murphy N, Hupe P, Barillot E, Belmont AS, Heard E. 2011. Live-cell chromosome dynamics and outcome of X chromosome pairing events during ES cell differentiation. *Cell* **145:** 447–458.

Meisel RP, Malone JH, Clark AG. 2012. Disentangling the relationship between sex-biased gene expression and X-linkage. *Genome Res* **22:** 1255–1265.

Mekhoubad S, Bock C, de Boer AS, Kiskinis E, Meissner A, Eggan K. 2012. Erosion of dosage compensation impacts human iPSC disease modeling. *Cell Stem Cell* **10:** 595–609.

Mermoud JE, Costanzi C, Pehrson JR, Brockdorff N. 1999. Histone macroH2A1.2 relocates to the inactive X chromosome after initiation and propagation of X-inactivation. *J Cell Biol* **147:** 1399–1408.

Monk M, Harper MI. 1979. Sequential X chromosome inactivation coupled with cellular differentiation in early mouse embryos. *Nature* **281:** 311–313.

Monk M, McLaren A. 1981. X-chromosome activity in foetal germ cells of the mouse. *J Embryol Exp Morphol* **63:** 75–84.

Monkhorst K, Jonkers I, Rentmeester E, Grosveld F, Gribnau J. 2008. X inactivation counting and choice is a stochastic process: Evidence for involvement of an X-linked activator. *Cell* **132:** 410–421.

Navarro P, Avner P. 2010. An embryonic story: Analysis of the gene regulative network controlling *Xist* expression in mouse embryonic stem cells. *Bioessays* **32:** 581–588.

Navarro P, Oldfield A, Legoupi J, Festuccia N, Dubois A, Attia M, Schoorlemmer J, Rougeulle C, Chambers I, Avner P. 2010. Molecular coupling of *Tsix* regulation and pluripotency. *Nature* **468:** 457–460.

Nechanitzky R, Davila A, Savarese F, Fietze S, Grosschedl R. 2012. Satb1 and Satb2 are dispensable for X chromosome inactivation in mice. *Dev Cell* **23:** 866–871.

Nesterova TB, Johnston CM, Appanah R, Newall AE, Godwin J, Alexiou M, Brockdorff N. 2003. Skewing X chromosome choice by modulating sense transcription across the *Xist* locus. *Genes Dev* **17:** 2177–2190.

Ng K, Daigle N, Bancaud A, Ohhata T, Humphreys P, Walker R, Ellenberg J, Wutz A. 2011. A system for imaging the regulatory noncoding Xist RNA in living mouse embryonic stem cells. *Mol Biol Cell* **22:** 2634–2645.

Nguyen DK, Disteche CM. 2006. Dosage compensation of the active X chromosome in mammals. *Nat Genet* **38:** 47–53.

Nora EP, Heard E. 2009. X chromosome inactivation: When dosage counts. *Cell* **139:** 865–867.

Nora EP, Lajoie BR, Schulz EG, Giorgetti L, Okamoto I, Servant N, Piolot T, van Berkum NL, Meisig J, Sedat J, et al. 2012. Spatial partitioning of the regulatory landscape of the X-inactivation centre. *Nature* **485:** 381–385.

Norris DP, Patel D, Kay GF, Penny GD, Brockdorff N, Sheardown SA, Rastan S. 1994. Evidence that random and imprinted *Xist* expression is controlled by preemptive methylation. *Cell* **77:** 41–51.

Ohhata T, Hoki Y, Sasaki H, Sado T. 2008. Crucial role of antisense transcription across the *Xist* promoter in *Tsix*-mediated *Xist* chromatin modification. *Development* **135:** 227–235.

Ohno S. 1967. *Sex chromosomes and sex linked genes*. Springer-Verlag, Berlin.

Okamoto I, Otte AP, Allis CD, Reinberg D, Heard E. 2004. Epigenetic dynamics of imprinted X inactivation during early mouse development. *Science* **303:** 644–649.

Okamoto I, Arnaud D, Le Baccon P, Otte AP, Disteche CM, Avner P, Heard E. 2005. Evidence for de novo imprinted X-chromosome inactivation independent of meiotic inactivation in mice. *Nature* **438:** 369–373.

Okamoto I, Patrat C, Thepot D, Peynot N, Fauque P, Daniel N, Diabangouaya P, Wolf JP, Renard JP, Duranthon V, et al. 2011. Eutherian mammals use diverse strategies to initiate X-chromosome inactivation during development. *Nature* **472:** 370–374.

Oliver B. 2007. Sex, dose, and equality. *PLoS Biol* **5:** e340.

O'Neill LP, Keohane AM, Lavender JS, McCabe V, Heard E, Avner P, Brockdorff N, Turner BM. 1999. A developmental switch in H4 acetylation upstream of *Xist* plays a role in X chromosome inactivation. *EMBO J* **18:** 2897–2907.

O'Neill LP, Spotswood HT, Fernando M, Turner BM. 2008. Differential loss of histone H3 isoforms mono-, di- and tri-methylated at lysine 4 during X-inactivation in female embryonic stem cells. *Biol Chem* **389:** 365–370.

Penny GD, Kay GF, Sheardown SA, Rastan S, Brockdorff N. 1996. Requirement for *Xist* in X chromosome inactivation. *Nature* **379:** 131–137.

Pessia E, Makino T, Bailly-Bechet M, McLysaght A, Marais GA. 2012. Mammalian X chromosome inactivation evolved as a dosage-compensation mechanism for dosage-sensitive genes on the X chromosome. *Proc Natl Acad Sci* **109:** 5346–5351.

Popova BC, Tada T, Takagi N, Brockdorff N, Nesterova TB. 2006. Attenuated spread of X-inactivation in an X;autosome translocation. *Proc Natl Acad Sci* **103:** 7706–7711.

Pullirsch D, Hartel R, Kishimoto H, Leeb M, Steiner G, Wutz A. 2010. The Trithorax group protein Ash2l and Saf-A are recruited to the inactive X chromosome at the onset of stable X inactivation. *Development* **137:** 935–943.

Quinn A, Koopman P. 2012. The molecular genetics of sex determination and sex reversal in mammals. *Semin Reprod Med* **30:** 351–363.

Rasmussen TP, Wutz AP, Pehrson JR, Jaenisch RR. 2001. Expression of *Xist* RNA is sufficient to initiate macrochromatin body formation. *Chromosoma* **110:** 411–420.

Rastan S. 1983. Non-random X-chromosome inactivation in mouse X-autosome translocation embryos—Location of the inactivation centre. *J Embryol Exp Morphol* **78:** 1–22.

* Reik W, Surani MA. 2014. Germline and pluripotent stem cells. *Cold Spring Harb Perspect Biol* doi: 10.1101/cshperspect.a019422.

* Rinn JL. 2014. lncRNAs: Linking RNA to chromatin. *Cold Spring Harb Perspect Biol* **6:** a018614.

Rinn JL, Kertesz M, Wang JK, Squazzo SL, Xu X, Brugmann SA, Goodnough LH, Helms JA, Farnham PJ, Segal E, et al. 2007. Functional demarcation of active and silent chromatin domains in human *HOX* loci by noncoding RNAs. *Cell* **129:** 1311–1323.

Rougeulle C, Chaumeil J, Sarma K, Allis CD, Reinberg D, Avner P, Heard E. 2004. Differential histone H3 Lys-9 and Lys-27 methylation profiles on the X chromosome. *Mol Cell Biol* **24:** 5475–5484.

Royce-Tolland ME, Andersen AA, Koyfman HR, Talbot DJ, Wutz A, Tonks ID, Kay GF, Panning B. 2010. The A-repeat links ASF/SF2-dependent *Xist* RNA processing with random choice during X inactivation. *Nat Struct Mol Biol* **17:** 948–954.

Sado T, Okano M, Li E, Sasaki H. 2004. De novo DNA methylation is dispensable for the initiation and propagation of X chromosome inactivation. *Development* **131:** 975–982.

Sado T, Hoki Y, Sasaki H. 2005. *Tsix* silences *Xist* through modification of chromatin structure. *Dev Cell* **9:** 159–165.

Schermelleh L, Carlton PM, Haase S, Shao L, Winoto L, Kner P, Burke B, Cardoso MC, Agard DA, Gustafsson MG, et al. 2008. Subdiffraction multicolor imaging of the nuclear periphery with 3D structured illumination microscopy. *Science* **320:** 1332–1336.

Sharman GB. 1971. Late DNA replication in the paternally derived X chromosome of female kangaroos. *Nature* **230:** 231–232.

Shibata S, Yokota T, Wutz A. 2008. Synergy of Eed and *Tsix* in the repression of *Xist* gene and X-chromosome inactivation. *EMBO J* **27:** 1816–1826.

Silva J, Mak W, Zvetkova I, Appanah R, Nesterova TB, Webster Z, Peters AH, Jenuwein T, Otte AP, Brockdorff N. 2003. Establishment of histone h3 methylation on the inactive X chromosome requires transient recruitment of Eed-Enx1 polycomb group complexes. *Dev Cell* **4:** 481–495.

Splinter E, de Wit E, Nora EP, Klous P, van de Werken HJ, Zhu Y, Kaaij LJ, van Ijcken W, Gribnau J, Heard E, et al. 2011. The inactive X chromosome adopts a unique three-dimensional conformation that is dependent on *Xist* RNA. *Genes Dev* **25:** 1371–1383.

* Strome S, Kelly WG, Ercan S, Lieb JD. 2014. Regulation of the X chromosomes in *Caenorhabditis elegans*. *Cold Spring Harb Perspect Biol* **6:** a018366.

Sun BK, Deaton AM, Lee JT. 2006. A transient heterochromatic state in *Xist* preempts X inactivation choice without RNA stabilization. *Mol Cell* **21:** 617–628.

Sun S, Del Rosario BC, Szanto A, Ogawa Y, Jeon Y, Lee JT. 2013. Jpx RNA activates *Xist* by evicting CTCF. *Cell* **153:** 1537–1551.

Tada T, Obata Y, Tada M, Goto Y, Nakatsuji N, Tan S, Kono T, Takagi N. 2000. Imprint switching for non-random X-chromosome inactivation during mouse oocyte growth. *Development* **127:** 3101–3105.

Takagi N, Sasaki M. 1975. Preferential inactivation of the paternally derived X chromosome in the extraembryonic membranes of the mouse. *Nature* **256:** 640–642.

Tang YA, Huntley D, Montana G, Cerase A, Nesterova TB, Brockdorff N. 2010. Efficiency of *Xist*-mediated silencing on autosomes is linked to chromosomal domain organisation. *Epigenetics Chromatin* **3:** 10.

Tavares L, Dimitrova E, Oxley D, Webster J, Poot R, Demmers J, Bezstarosti K, Taylor S, Ura H, Koide H, et al. 2012. RYBP-PRC1 complexes mediate H2A ubiquitylation at polycomb target sites independently of PRC2 and H3K27me3. *Cell* **148:** 664–678.

Tchieu J, Kuoy E, Chin MH, Trinh H, Patterson M, Sherman SP, Aimiuwu O, Lindgren A, Hakimian S, Zack JA, et al. 2010. Female human iPSCs retain an inactive X chromosome. *Cell Stem Cell* **7:** 329–342.

Thorvaldsen JL, Weaver JR, Bartolomei MS. 2011. A YY1 bridge for X inactivation. *Cell* **146:** 11–13.

Tian D, Sun S, Lee JT. 2010. The long noncoding RNA, Jpx, is a molecular switch for X chromosome inactivation. *Cell* **143:** 390–403.

Turner JM, Mahadevaiah SK, Ellis PJ, Mitchell MJ, Burgoyne PS. 2006. Pachytene asynapsis drives meiotic sex chromosome inactivation and leads to substantial postmeiotic repression in spermatids. *Dev Cell* **10:** 521–529.

Wutz A, Jaenisch R. 2000. A shift from reversible to irreversible X inactivation is triggered during ES cell differentiation. *Mol Cell* **5:** 695–705.

Wutz A, Rasmussen TP, Jaenisch R. 2002. Chromosomal silencing and localization are mediated by different domains of *Xist* RNA. *Nat Genet* **30:** 167–174.

Xu N, Tsai CL, Lee JT. 2006. Transient homologous chromosome pairing marks the onset of X inactivation. *Science* **311:** 1149–1152.

Zhao J, Sun BK, Erwin JA, Song JJ, Lee JT. 2008. Polycomb proteins targeted by a short repeat RNA to the mouse X chromosome. *Science* **322:** 750–756.

C H A P T E R 26

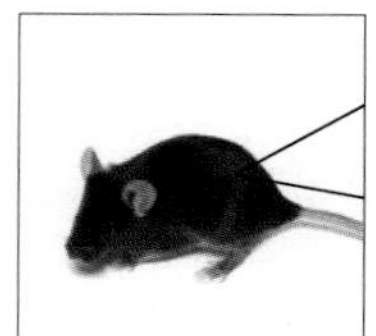

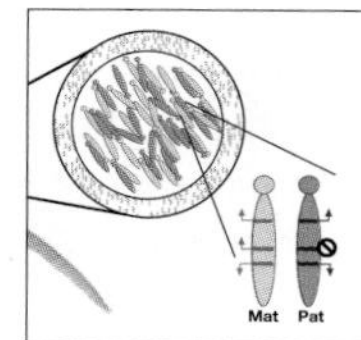

Genomic Imprinting in Mammals

Denise P. Barlow[1] and Marisa S. Bartolomei[2]

[1]CeMM Research Center for Molecular Medicine of the Austrian Academy of Sciences, CeMM, 1090 Vienna, Austria; [2]Department of Cell and Developmental Biology, University of Pennsylvania Perelman School of Medicine, Philadelphia, Pennsylvania 19104-6148

Correspondence: dbarlow@cemm.oeaw.ac.at and bartolom@mail.med.upenn.edu

SUMMARY

Genomic imprinting affects a subset of genes in mammals and results in a monoallelic, parental-specific expression pattern. Most of these genes are located in clusters that are regulated through the use of insulators or long noncoding RNAs (lncRNAs). To distinguish the parental alleles, imprinted genes are epigenetically marked in gametes at imprinting control elements through the use of DNA methylation at the very least. Imprinted gene expression is subsequently conferred through lncRNAs, histone modifications, insulators, and higher-order chromatin structure. Such imprints are maintained after fertilization through these mechanisms despite extensive reprogramming of the mammalian genome. Genomic imprinting is an excellent model for understanding mammalian epigenetic regulation.

Outline

Cite as *Cold Spring Harb Perspect Biol* doi: 10.1101/cshperspect.a018382

OVERVIEW

Mammals are diploid organisms whose cells possess two matched sets of chromosomes, one inherited from the mother and one from the father. Thus, mammals have two copies of every gene. Normally both the maternal and paternal copy of each gene has the same potential to be active in any cell. Genomic imprinting is an epigenetic mechanism that changes this potential because it restricts the expression of a gene to one of the two parental chromosomes. It is a phenomenon displayed by only a few hundred of the approximately 25,000 genes in our genome, the majority being expressed equally when inherited from either parent. Genomic imprinting affects both male and female offspring and is therefore a consequence of parental inheritance, not of sex. As an example of what is meant by this, an imprinted gene that is active on a maternally inherited chromosome will be active on the maternal chromosome and silent on the paternal chromosome in all males and females.

The definition of genomic imprinting is restricted here to "parental-specific gene expression in diploid cells." Thus, diploid cells that contain two parental copies of all genes will express only one parental copy of an imprinted gene and silence the other parental copy. In contrast, nonimprinted genes will be expressed from both parental gene copies in a diploid cell. To understand the concept of genomic imprinting it is important to distinguish between imprinted genes and those showing apparent parental-specific expression because of unequal parental genetic contribution to the embryo. Examples of unequal parental genetic contribution include Y chromosome–linked genes present only in males, genes that escape X inactivation in females (producing a double dose of X-linked gene products compared with males), mitochondrial genes contributed mainly by the maternal parent, and messenger RNAs (mRNAs) and proteins present only in the sperm or egg cytoplasm.

Many features of genomic imprinting in mammals make it a fascinating biological problem in postgenomic times. It is intriguing that the subset of genes subject to genomic imprinting largely code for factors regulating embryonic and neonatal growth. Thus, it is likely that genomic imprinting evolved to play a specific role in mammalian reproduction. It is also providing clues as to a possible evolutionary response to parental conflict, to the adaptation of the maternal parent to an internal reproduction system, and, perhaps, providing a glimpse of the way the mammalian genome protects itself against invading DNA sequences. Genomic imprinting is an intellectually challenging phenomenon, not least because it raises the question of why a diploid organism would evolve a silencing system that forsakes the advantages of the diploid state.

At this stage of our knowledge, genomic imprinting does not appear to be widespread among the four eukaryotic kingdoms that include Protista, Fungi, Plants, and Animals. However, it does exist, in a possibly related form, in two invertebrate arthropods—Coccidae and Sciaridae, and in the endosperm of some seed-bearing plants, such as maize and *Arabidopsis*. This distribution indicates that genomic imprinting arose independently at least three times during the evolution of life. Surprisingly, despite this predicted independent evolution of genomic imprinting, some similarities among the imprinting mechanism are emerging. It is likely that this reflects conservation of basic epigenetic regulatory mechanisms that underlie both genomic imprinting and normal gene regulation.

Cite as *Cold Spring Harb Perspect Biol* doi: 10.1101/cshperspect.a018382

1 HISTORICAL OVERVIEW

The presence of genomic imprinting in mammals has considerable medical, societal, and intellectual implications in terms of (1) the clinical management of genetic traits and diseases, (2) the capacity to control human and animal breeding by assisted reproductive technologies, and (3) the progress of biotechnology and postgenomic medical research. Any modern day discussion of genetic problems, whether in research or medicine, must consider if a gene shows a biparental (i.e., diploid) mode of expression, or, is subject to genomic imprinting and shows parental-specific (i.e., haploid) expression. Despite the importance of genomic imprinting to human health and well-being, it is surprising that widespread acceptance of its existence and significance did not happen until the early nineties after three genes were unequivocally shown to display parental-specific expression in mice.

Parental-specific behavior of whole chromosomes had been observed in cytogenetic studies of chromosomes in Arthropods as early as the 1930s (Chandra and Nanjundiah 1990). Interestingly, the term "chromosome imprinting" was first coined to describe paternal-specific chromosome elimination that plays a role in sex determination in some Arthropod species (Crouse et al. 1971). Chromosomal imprinting of the mammalian X chromosome was also noted, which leads to paternal-specific inactivation of one of the two X chromosomes in all cells of female marsupials and the extraembryonic tissues of the mouse (Cooper et al. 1971). During the same period, classical geneticists were generating mouse mutants carrying chromosomal translocations that laid the foundation for the observation of imprinted gene expression. Some of these "translocation" mice, initially used to map the position of genes on chromosomes, showed a parental-specific phenotype when certain chromosomal regions were inherited as duplications of one parental chromosome in the absence of the other parental chromosome (known as uniparental disomy or UPD; Fig. 1). These results indicated the possibility "that haploid expression of particular maternal or paternal genes is important for normal mouse development" (Searle and Beechey 1978). At the same time, other geneticists used an unusual mouse mutant known as the "hairpin-tail" mouse that carried a large deletion of chromosome 17 to unequivocally set aside a basic tenet of genetics "that organisms heterozygous at a given locus are phenotypically identical irrespective of which gamete contributes which allele to the genotype" (Johnson 1974). Instead, offspring who received the Hairpin-tail deletion from a maternal parent were increased in size and died midway through embryonic development, whereas paternal transmission of the genetically identical chromosome produced viable and fertile mice (Fig. 1). It is notable with hindsight that in spite of the

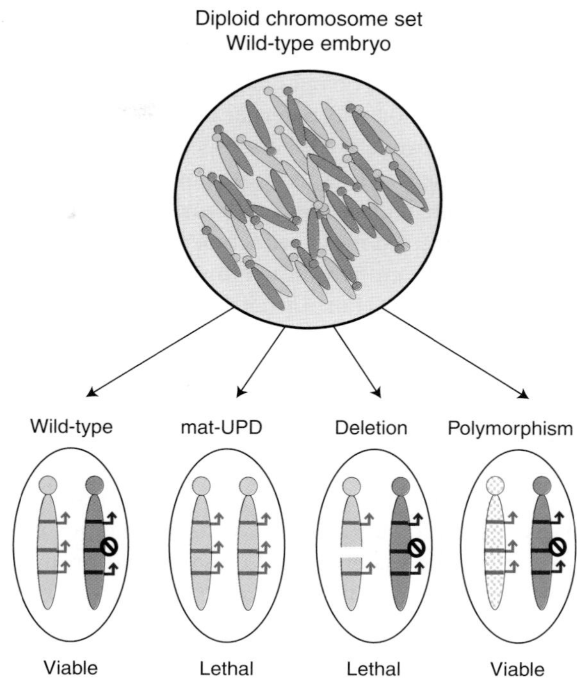

Figure 1. Mouse models to study genomic imprinting that allow the maternal and paternal chromosome to be distinguished. Mammals are diploid and inherit a complete chromosome set from the maternal and paternal parent. However, mice can be generated that (1) inherit two copies of a chromosome pair from one parent and no copy from the other parent (known as UPD), (2) inherit a partial chromosomal deletion from one parent and a wild-type chromosome from the other parent, and (3) inherit chromosomes carrying single-nucleotide polymorphisms (known as SNPs) from one parent and a wild-type chromosome from the other parent. Offspring with UPDs or deletions are likely to display lethal phenotypes, whereas SNPs will allow the production of viable offspring.

previously published description of imprinted X-chromosome inactivation in mammals, the favored interpretation of these genetic translocation and deletion experiments was not that the regions contained imprinted genes, but that genes on these autosomes primarily acted in the haploid egg or sperm to modify proteins used later in embryonic development. Despite this, the concept of differential functioning of the maternal and paternal genome was gaining ground and a suggestion made that "the maternal genome might be normally active at the Hairpin-tail chromosomal region while its paternal counterpart is preferentially inactivated" (McLaren 1979).

A major step forward in establishing the existence of genomic imprinting in mammals came several years later with the development of an improved nuclear transfer technology being used to test the possibility of generating diploid uniparental embryos solely from mouse egg nuclei. The nuclear transfer technique took a donor male or female pronucleus from a newly fertilized egg and used a fine micropipette to place it inside a host fertilized egg from which either the maternal or paternal pronucleus had been removed. This regenerated diploid embryos, but with two maternal or two paternal genomes (known, respectively, as gynogenetic and androgenetic embryos; Fig. 2). The technique was first used to show that nuclei from fertilized Hairpin-tail mutant embryos could not be rescued when transferred into a wild-type host egg. This provided proof that the embryonic genome, and not the oocyte cytoplasm, carried the Hairpin-tail defect. It also confirmed the suggestion that genes on the maternal and paternal copy of chromosome 17 functioned differently during embryonic development (McGrath and Solter 1984b). Subsequently, nuclear transfer was used to show that embryos, reconstructed from two maternal pronuclei (known as gynogenetic embryos) or two paternal pronuclei (androgenetic embryos), failed to survive, whereas only embryos reconstructed from one maternal and one paternal pronucleus produced viable and fertile offspring (McGrath and Solter 1984a; Surani et al. 1984). This work overturned a previous claim that uniparental mice could develop to adulthood (Hoppe and Illmensee 1982). Gynogenetic embryos at the time of death were defective in extraembryonic tissues that contribute to the placenta, whereas androgenetic embryos were defective in embryonic tissue. These outcomes led to the hypothesis that embryonic development required imprinted genes expressed from the maternal genome, whereas the paternal genome expressed imprinted genes required for extraembryonic development (Barton et al. 1984). Subsequent identification of imprinted genes in the mouse did not confirm a bias in the function of imprinted genes, but indicated that the observed differences between gynogenetic and androgenetic embryos may be explained by a dominant effect of one or a few imprinted genes.

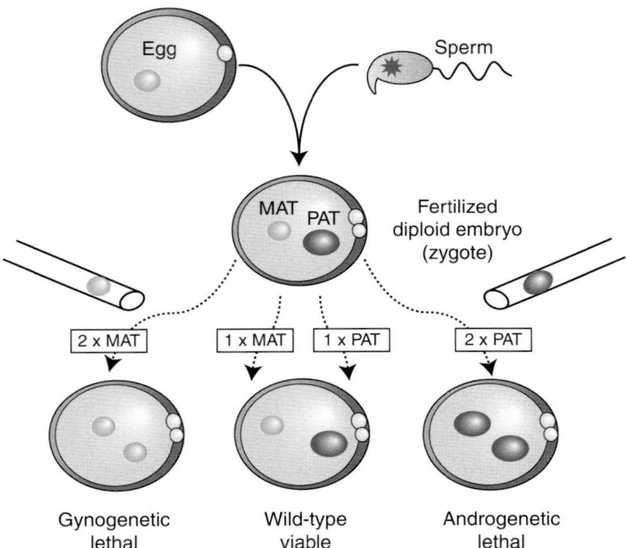

Figure 2. A maternal and paternal genome are needed for mammalian reproduction. The nuclear transfer technique used micropipettes and high-powered microscopes to remove the male or female nuclei from a newly fertilized egg and place them in various combinations into a second "host" fertilized egg that had already been enucleated, thereby generating anew diploid embryos with two maternal (gynogenetic) or two paternal (androgenetic) genomes or a biparental genome (wild-type). Gynogenetic and androgenetic embryos were lethal at early embryonic stages. Only reconstituted embryos that received both a maternal and paternal nucleus (wild-type) survived to produce living young. These experiments show the necessity for both the maternal and paternal genomes in mammalian reproduction, and indicate the two parental genomes express different sets of genes needed for complete embryonic development.

The nuclear transfer experiments, combined with supporting data from mouse genetics, provided convincing evidence that both parental genomes were required for embryogenesis in mice, laying a strong foundation for the existence of genomic imprinting in mammals (Fig. 2). An extensive survey of parental chromosome contribution to embryonic development, using "translocation" mice to create UPD chromosomes (Fig. 1), identified two regions on mouse chromosomes 2 and 11 that showed opposite phenotypes when present either as two maternal or two paternal copies. This further strengthened the argument for parental-specific gene expression in mammals (Cattanach and Kirk 1985). In addition, human data strongly indicated that some genetic conditions, most notably the Prader–Willi syndrome, which appears to arise exclusively by paternal transmission, could best be explained by parental-specific gene expression (Reik 1989). Further clues came from experiments applying the newly developed technology for making transgenic mice by microinjecting gene sequences into a fertilized mouse egg. This was often beset by the problem of DNA methylation unexpectedly inducing silencing of the transgene in somatic tissues. Some transgenes even showed parental-specific differences in their ability to acquire DNA methylation, adding weight to the argument that parental chromosomes behave differently. This normally followed the pattern that maternally transmitted transgenes were methylated, whereas paternally transmitted transgenes were not. However, only in a few cases did DNA methylation differences correlate with parental-specific expression. Although many similarities were later found between "transgene" methylation imprinting and genomic imprinting of endogenous mouse genes, several features distinguish them (Reik et al. 1990). This includes a high susceptibility to strain-specific background effects, an inability to maintain imprinted expression at different chromosomal integration sites, and a requirement for foreign DNA sequences to produce the imprinted effect (Chaillet et al. 1995).

 Cite as *Cold Spring Harb Perspect Biol* doi: 10.1101/cshperspect.a018382

Despite the wealth of supportive data, final proof of the existence of genomic imprinting in mammals depended on the identification of genes showing imprinted parental-specific expression. This occurred in 1991 when three imprinted mouse genes were described. The first of these, *Igf2r* (insulin-like growth factor type 2 receptor that is a "scavenger" receptor for the growth hormone insulin-like growth factor type 2 [*Igf2*]) was identified as a maternally expressed imprinted gene. This gene was later shown to explain the overgrowth phenotype of the Hairpin-tail mutant mouse (Barlow et al. 1991). A few months later, the *Igf2* gene was identified as a paternally expressed imprinted gene (DeChiara et al. 1991; Ferguson-Smith et al. 1991). Finally, the *H19* gene (cDNA clone number 19 isolated from a fetal hepatic library), an unusual long noncoding RNA (lncRNA) was subsequently shown to be a maternally expressed imprinted gene (Bartolomei et al. 1991). Diverse strategies were used to identify these three imprinted genes, each of which depended on emerging technologies in mouse genetics. For *Igf2r*, positional cloning was used to identify genes that mapped to the Hairpin-tail deletion on chromosome 17. Mice then inheriting the deletion from one parent were used to identify those genes showing maternal-specific expression (Fig. 1). For *Igf2*, the physiological role of this growth factor in embryonic development was being tested by gene knockout technology. Surprisingly, mice carrying the mutant nonfunctional allele showed a phenotype following paternal transmission, but no phenotype on maternal transmission. The *H19* lncRNA was identified as an imprinted gene after this gene was mapped close to the *Igf2* locus on chromosome 7, proving the hypothesis that imprinted genes could be clustered together. Although these strategies were to prove useful in subsequent attempts to identify imprinted genes, the demonstration that imprinted genes were closely clustered has proven to be a pivotal discovery in understanding the mechanism controlling genomic imprinting in mammals.

2 GENOMIC IMPRINTING—AN EPIGENETIC GENE REGULATORY SYSTEM

The defining characteristic of genomic imprinting is that it is *cis* acting (see Box 1). Thus, the imprinting mechanism acts only on one chromosome. The two parental chromosomes will normally contain many single base pair differences (known as single-nucleotide polymorphisms [SNPs]) if the population is outbred, but they can be genetically identical if inbred mouse strains are used. Because genomic imprinting occurs in inbred mice that have genetically identical parental chromosomes, it was concluded that the process must use an epigenetic mechanism to modify the information carried by the DNA sequence, yet create an expression difference between the two parental gene copies. These observations also indicate that a *cis*-acting silencing mechanism, which is restricted to one chromosome, is operating so that the silencing factors cannot freely diffuse through the nucleus to reach the active gene copy. Although imprinted genes are repressed on one parental chromosome relative to the other, genomic imprinting is not necessarily a silencing mechanism and has the potential to operate at any level of gene regulation (i.e., at the promoter, enhancers, splicing junctions, or polyadenylation sites) to induce parental-specific differences in expression.

Genomic imprinting must therefore depend on an epigenetic system that modifies or "imprints" one of the two parental chromosomes (Fig. 3). This imprint is subsequently used to attract or repel transcriptional factors or mRNA processing factors, thereby changing expression of the imprinted gene on one parental chromosome. Because inbred mice with genetically identical chromosomes also show genomic imprinting, parental imprints are not likely to be acquired after the embryo becomes diploid because there would be no way for the cells' epigenetic machinery to distinguish between identical parental gene copies. Thus, parental imprints must be acquired when the two parental chromosome sets are separate and this only occurs during

BOX 1. KEY FEATURES OF GENOMIC IMPRINTING IN MAMMALS

- *cis*-acting mechanism
- A consequence of inheritance not sex
- Imprints are epigenetic modifications acquired by one parental gamete
- Imprinted genes are mostly clustered together with a noncoding RNA
- Imprints can modify long-range regulatory elements that act on multiple genes
- Imprinted genes play a role in mammalian development

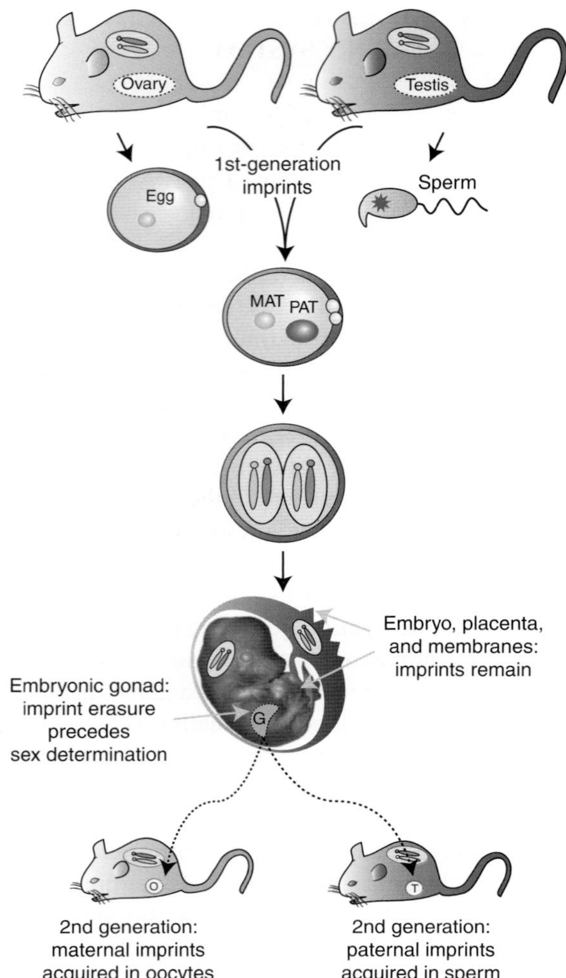

Figure 3. Imprint acquisition and erasure in mammalian development. Imprints are acquired by the gametes; thus, oocytes and sperm already carry imprinted chromosomes (first-generation imprints). After fertilization when the embryo is diploid, the imprint is maintained on the same parental chromosome after each cell division in cells of the embryo, yolk sac, placenta, and also in the adult. The germ cells are formed in the embryonic gonad and the imprints are erased only in these cells before sex determination. As the embryo develops into a male, the gonads differentiate to testes that produce haploid sperm that acquire a paternal imprint on their chromosomes. Similarly, in developing females, chromosomes in the ovaries acquire maternal imprints (second-generation imprints).

gamete formation, and for ~12 h postfertilization (Fig. 3). The most likely scenario is that gametic imprints are placed on paternally imprinted genes during sperm production and on maternally imprinted genes during egg formation. A key feature of the "imprinted" DNA sequence is that it would only be modified in one of the two parental gametes; thus, two types of recognition system are required, one sperm-specific and one oocyte-specific, each directed toward a different DNA sequence. Several other features are required of the imprint. First, once established, it must remain on the same parental chromosome after fertilization when the embryo is diploid. Second, the imprint must be stably inherited through mitosis of the embryo and adult animal. Last, it must be erasable. The latter is necessary because the embryo will follow either a male or female developmental path midway through development and its gonads will need to produce only one type of imprinted haploid parental gamete. Thus, germ cells that have arisen from embryonic diploid cells (Fig. 3) must first lose their inherited maternal and paternal imprints before they gain that of the gamete.

How are gametic imprints identified? An imprint can be defined as the epigenetic modification that distinguishes the two parental copies of a given gene. Once formed, the imprint must also allow the transcription machinery to treat the maternal and paternal gene copy differently within the same nucleus. A gametic imprint is predicted to be continuously present at all developmental stages (Fig. 3), thus imprints can be found by comparing epigenetic modifications on maternal and paternal chromosomes in embryonic or adult tissues (using strategies outlined in Fig. 1) and tracing them back in development to one of the two gametes. Candidates for gametic imprints could be modifications of DNA or histone proteins that package DNA into chromosomes (Ch. 3 [Allis et al.]). There are now two types of epigenetic DNA modification known in mammals; 5-methylcytosine and 5-hydroxymethylcytosine (Ch. 15 [Li and Zhang 2014]). Histones can bear multiple types of modification including methylation, acetylation, phosphorylation, sumoylation, and ubiquitylation (Ch. 3 [Allis et al.]). They can also be replaced by variant histones with specific functions (Ch. 20 [Henikoff and Smith 2014]). Any of these epigenetic modifications could qualify as an imprint. One would predict that enzymes responsible for these epigenetic modifications or an essential cofactor would be exclusively expressed in one of the two gametes, and specifically associate with one parental chromosome to copy the modification when the cell divides. However, as will be described in Section 3 on "key discoveries," only 5-methylcytosine has been clearly shown to function as the gametic imprint for imprinted genes in mammals and, to date, is the only known heritable modification.

How does a gametic imprint control imprinted expression? To understand how the imprint operates, three pieces of information are required: which parental chromosome carries the imprint, which parental chromosome carries the expressed allele of the imprinted gene, and the position of the imprinted sequence relative to the expressed or silenced allele of the imprinted gene. Using this type of approach it has been shown that gametic imprints can act on whole clusters of genes at once. These imprinted clusters contain 3–12

Cite as *Cold Spring Harb Perspect Biol* doi: 10.1101/cshperspect.a018382

imprinted genes and span from 100–3700 kb of genomic DNA (for more details, see http://www.mousebook.org/catalog.php?catalog=imprinting). The majority of genes in any one cluster are imprinted protein-coding mRNA genes; however, at least one is always an imprinted lncRNA.

Because of the arrangement of imprinted genes in clusters, with some genes expressed from one parental chromosome and some from the other, it is not trivial to determine how the imprint operates. It is possible to study the effect of the imprint on single genes in the cluster, but it may prove more informative to study the effects of the imprint on the entire cluster. This will be described in more detail in Section 3. One thing, however, is clear. Nature has not chosen the simplest mechanism whereby the imprint is directed toward a promoter to preemptively silence an imprinted gene in one gamete. Instead, imprints appear, in general, to be directed toward long-range *cis*-acting regulators that influence the expression of multiple genes, and are located a long distance away on the same chromosome.

3 KEY DISCOVERIES IN GENOMIC IMPRINTING

3.1 Imprinted Genes Control Embryonic and Neonatal Growth

What is the function of genomic imprinting in mammals? One way to answer this question is to determine the function of known imprinted genes in vivo. This can be performed by mutating the gene sequence to impair its function using the "homologous recombination" technique. The function of many of the known imprinted genes has been determined in this fashion (for original references, see http://www.mousebook.org/catalog.php?catalog=imprinting). The most significantly represented function among imprinted genes includes genes that affect growth of the embryo, placenta, and neonate. In this category are paternally expressed imprinted genes that function as growth promoters (i.e., *Igf2*, *Peg1*, *Peg3*, *Rasgrf1*, *Dlk1*) and show growth retardation in embryos deficient for the gene. There are also maternally expressed imprinted genes that function as growth repressors (i.e., *Igf2r*, *Gnas*, *Cdkn1c*, *H19*, *Grb10*), as shown by a growth enhancement in embryos deficient for the gene. Another significant category includes genes with behavioral or neurological defects (e.g., *Nesp*, *Ube3a*, *Kcnq1*). These results are, at one level, disappointing because they do not identify one function for all imprinted genes. Nevertheless, these results show that the majority of imprinted genes function as embryonic or neonatal growth regulators. More interestingly, the ability to regulate growth appears to be neatly divided with maternally expressed growth-regulating genes acting to repress growth of the offspring, whereas paternally expressed genes in this category act to increase growth. Moreover, numerous tested imprinted genes are active in neurological processes, some of which affect neonatal growth rate by altering maternal behavior.

3.2 The Function of Genomic Imprinting in Mammals

Can analyses of gene function help us understand why genes are imprinted in mammals? Assessment of genomic imprinting in different types of mammals has been informative. Placental mammals such as mice and humans, and marsupials such as opossum and wallaby, have genomic imprinting. Egg-laying mammals, such as platypus and echidna, appear to lack imprinted genes, although extensive studies have not yet been performed (Renfree et al. 2009). Placental mammals and marsupials are distinguished from egg-laying mammals by a reproductive strategy that allows the embryo to directly influence the amount of maternal resources used for its own growth. In contrast, embryos that develop within eggs are unable to directly influence maternal resources. Most invertebrates and vertebrates use an egg-laying reproductive strategy. Notably, they can also undergo parthenogenesis—a form of reproduction in which the female gamete develops into a new diploid individual without fertilization by a male gamete (note that parthenogenetic embryos arise from the duplication of the same maternal genome, whereas the gynogenetic embryos described in Fig. 2 arise from two different maternal genomes). The ability of organisms to undergo parthenogenesis most likely indicates a complete absence of genomic imprinting as it shows the paternal genome is dispensable. In mammals, however, a direct consequence of imprinted gene expression controlling fetal growth is that parthenogenesis is not possible. Both parents are necessary to produce viable offspring making mammals completely reliant on sexual reproduction to reproduce (Fig. 4). Parthenogenesis has thus not yet been observed in mammals despite claims to the contrary, although manipulating expression of the *Igf2* and *Dlk1* imprinted clusters has generated some rare mice with a diploid maternal genome (Kawahara et al. 2007).

Why should genomic imprinting have evolved only in some mammals, but not in vertebrates in general? Three features of genomic imprinting—the growth regulatory function of many imprinted genes, the restriction of imprinted genes to placental and marsupial mammals, and last, the necessity of the paternal genome for fetal development, provide evidence that can fit two equally attractive hypotheses.

The first hypothesis proposes that genomic imprinting evolved in response to a "parental conflict" situation

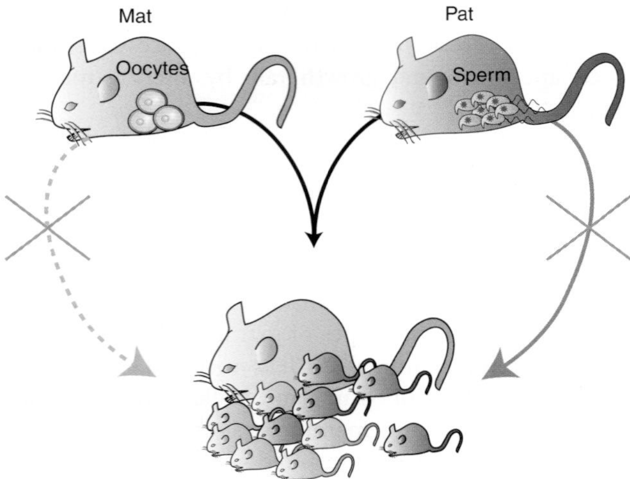

Figure 4. Imprinted genes play a role in mammalian reproduction. Mammals are diploid and reproduction requires fertilization of a haploid female egg by a haploid male sperm to recreate a diploid embryo. Only females are anatomically equipped for reproduction, but they cannot use parthenogenesis to reproduce (the possibility of which is represented by a pink dashed line) because essential imprinted genes needed for fetal growth are imprinted and silenced on maternal chromosomes. These genes are expressed only from paternal chromosomes; thus, both parental genomes are needed for reproduction in mammals. Parthenogenesis is the production of diploid offspring from two copies of the same maternal genome.

(Moore and Haig 1991). This arises from the opposing interests of the maternal and paternal genome: Embryonic growth is dependent on one parent, but influenced by an embryo whose genome comes from two parents. Paternally expressed imprinted genes are proposed to increase embryonic growth, thereby maximizing the fitness of an individual offspring bearing a particular paternal genome. Maternally expressed imprinted genes are proposed to suppress fetal growth. This would allow a more equal distribution of maternal resources to all offspring and increase transmission of the maternal genome to multiple offspring, which may have different paternal genomes.

The second hypothesis is named "trophoblast defense" (Varmuza and Mann 1994). This proposes that the maternal genome is at risk from the consequences of being anatomically equipped for internal reproduction should spontaneous oocyte activation lead to full embryonic development. Because males lack the necessary anatomical equipment for internal reproduction, they do not share the same risks should spontaneous activation of spermatozoa occur. Imprinting is thus proposed to either silence genes on the maternal chromosome that promote placental development or to activate genes that limit this process. The genes necessary for placental invasion of the maternal uterine vasculature would consequently only be expressed from a paternal genome after fertilization has occurred.

Which, if any, of these hypotheses explains the evolution of genomic imprinting in mammals? Both hypotheses indicate a role for imprinted genes in regulating the development and function of the placenta; however, neither the parental conflict nor the trophoblast defense models can provide a full explanation for all the data (Wilkins and Haig 2003). It is interesting to note that imprinted genes have also been identified in the plant endosperm, a tissue that has been compared to the placenta by virtue that it transfers nutrient resources from the parent plant to the embryo (see Ch. 17 [Grossniklaus and Paro 2014]). This finding strengthens arguments that genomic imprinting evolved as a means to regulate nutrient transfer between the parent and offspring, but it does not tell us why.

Fuller or alternative explanations of the function of genomic imprinting in mammals could come from two sources. The first would be to examine the function of "imprinting" across a complete gene cluster in contrast to examining the phenotype of mice lacking a single imprinted gene product. This would require an ability to reverse an imprint and generate biparental gene expression across the whole imprinted cluster. The second approach is to learn exactly how genes are imprinted. It is possible that not all genes in a cluster are deliberate targets of the imprinting mechanism and that some may just be "innocent bystanders" of the process, and their function would not be informative about the role of genomic imprinting. The existence of innocent bystander genes affected by the imprinting mechanism may satisfactorily explain the curious abundance of imprinted genes with no obvious biological function in development.

3.3 Imprinted Genes Are Clustered and Controlled by Imprint Control Elements (ICEs)

To date, about 150 imprinted genes have been mapped to 17 mouse chromosomes including the X chromosome. More than 80% of the identified imprinted genes are clustered into 16 genomic regions that contain two or more genes (Wan and Bartolomei 2008). The discovery of clusters of imprinted genes was a strong indication that a common DNA element may regulate imprinted expression of multiple genes in *cis*. To date, seven of the 16 imprinted clusters have been well characterized, and these are listed in Table 1 by the name of the principle imprinted mRNA gene in the cluster or after a disease association (e.g., the *Pws* cluster for Prader–Willi syndrome; Ch. 33 [Zoghbi and Beaudet 2014]). These seven clusters contain three to 12 (or more) imprinted genes and are spread over 80–3700 kb of DNA.

Table 1. Features of imprinted gene clusters in the mouse genome

Cluster name	Chromosome mouse/human	ICE (gametic methylation imprint)	Cluster size (kb)	Gene number in cluster	Parental expression M/P	lncRNA and expression (M or P)
Igf2r	17/6	Region 2 (M)	490	4	3 M (pc) 1 P (nc)	*Airn* (P)
Kcnq1	7/11	*KvDMR1* (M)	780	12	11 M (pc) 1 P (nc)	*Kcnq1ot1* (P)
Pws	7/15	*Snrpn-CGI* (M)	3700	>8	2 M (pc)/ >7 P (nc and pc)	*Ube3aas* (P)[a] *Ipw* (P)[a] *Zfp127as* (P)[a] *PEC2* (P)[a] *PEC3* (P)[a] *Pwcr1* (P)[a]
Gnas	2/20	*Nespas* DMR (M)	80	7	2 M (pc) 5 P (4 nc and 1 pc)	*Nespas* (P)[b] *Exon1A* (P) *miR-296* (P)[b] *miR-298* (P)[b]
Grb10	11/7	Meg1/Grb10 DMR (M)	780	4	2 M (pc)/ 2 P (pc)	NI
Igf2	7/11	H19-DMD (P)	80	3	1 M (nc)/ 2 P (pc)	*H19* (M)
Dlk1	9/14	IG-DMR (P)	830	>5	>1 M (nc)/ 4 P (pc)	*Gtl2* (M)[c] *Rian* (M)[c] *Rtl1as* (M)[c] *Mirg* (M)[c] miRNAs (M)[c] snoRNAs (M)[c]

Note that cluster size and number of genes in the cluster are provisional and await a genome-wide analysis of imprinted expression. Pws and Dlk1 clusters contain overlapping transcripts in which the number of distinct genes is not yet clear. Details are given in the text.

M, maternal; P, paternal; DMR, differentially methylated region; pc, protein coding; nc, noncoding RNA; NI, none identified; miRNA, micro RNA; snoRNA, small nucleolar RNAs.

[a] May be one long lncRNA.

[b] Part of *Nespas* transcript.

[c] May be part of one or multiple lncRNAs.

A common feature of these seven clusters is the presence of a DNA sequence carrying a gametic methylation imprint that is known as a gametic DMR (differentially DNA-methylated region). A gametic DNA methylation imprint is defined as a methylation imprint established in one gamete and maintained only on one parental chromosome in diploid cells of the embryo. In five clusters (*Igf2r*, *Kcnq1*, *Gnas*, *Grb10*, and *Pws*), the gametic DMR has a maternal methylation imprint acquired in oogenesis, whereas in two clusters (*Igf2* and *Dlk1*), it has a paternal methylation imprint acquired during spermatogenesis. In these examples, the gametic DMR controls imprinted expression of the whole or part of the cluster and is therefore designated as the imprint control element, or ICE, for the cluster (Barlow 2011).

Table 1 shows that each imprinted gene cluster contains multiple mRNAs and, with the exception of *Grb10*, at least one lncRNA. Two trends emerge. First, the imprinted protein-coding genes in each cluster are expressed, for the most part, from the same parental chromosome, whereas the lncRNA is expressed from the opposite parental chromosome (as illustrated in Fig. 5 for a maternal gametic DMR). Second, the ICE deletion causes loss of imprinted expression only when deleted from the parental allele expressing the lncRNA. Table 1 shows that in three clusters (*Igf2r*, *Kcnq1*, and *Gnas*) the lncRNA promoter sits in an intron of one of the imprinted mRNAs, whereas in the remaining clusters the lncRNA promoter is separated, but lies close to the imprinted mRNA genes. This close intermingling of active and silent genes in an imprinted cluster indicates that the silencing and activating mechanisms affecting imprinted genes do not spread and may be restricted to the affected gene. In particular, the fact that the promoter of a silent lncRNA can reside in the intron of an actively tran-

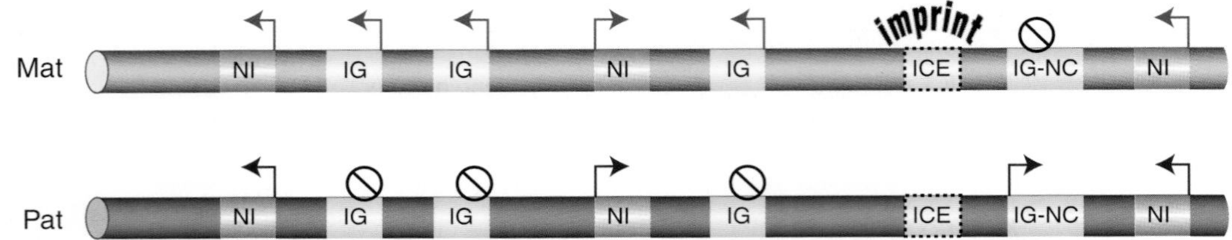

Figure 5. Imprinted genes are expressed from one parental allele and often clustered. Most imprinted genes (yellow) are found in clusters that include multiple protein-coding mRNAs (IG) and at least one noncoding RNA (IG-NC). Nonimprinted genes can also be present (NI in gray). The imprinting mechanism is *cis* acting and imprinted expression is controlled by an imprint control element (ICE) that carries an epigenetic imprint inherited from one parental gamete. One pair of diploid chromosomes is shown: the pink is of maternal origin and the blue of paternal origin. Arrow, expressed gene; slashed circle, repressed gene.

scribed gene indicates that silencing mechanisms may not even spread throughout the length of a gene, but may just be restricted to regulatory elements.

What is the role of the gametic DMR? Despite the fact the gametic DMRs can be maternally or paternally methylated, experiments that deleted these elements have produced broadly similar results albeit with a few interesting exceptions (Fig. 6). For three clusters (*Igf2r, Kcnq1, Dlk1*), experimental deletion of the methylated gametic DMR produced no effect. In contrast, deletion of the unmethylated gametic DMR eliminated parental-specific expression causing a loss of lncRNA expression in *cis* and biallelic mRNA expression (Lin et al. 1995; Zwart et al. 2001; Fitzpatrick et al. 2002). Two clusters (*Gnas* and *Pws*) appear to contain more than one gametic DMR and show a more complex behavior, yet they still share some similarities with the pattern presented in Figure 6 (Williamson et al. 2006). The *Igf2* cluster, however, behaves differently: deletion of both the methylated and unmethylated gametic DMR causes changes in mRNA and lncRNA expression in *cis* (Thorvaldsen et al. 1998).

The results from the above gametic DMR deletion experiments do not at first glance indicate a common function for gametic DMRs. However, an understanding of their exact function depends on knowing the position of the DMR with respect to the imprinted genes in each cluster. In the three clusters with the simplest pattern (*Igf2r, Kcnq1*, and *Dlk1*), the gametic DMR either contains or controls expression of the lncRNA, thus deletion of this element will clearly lead to loss of lncRNA expression. The gametic

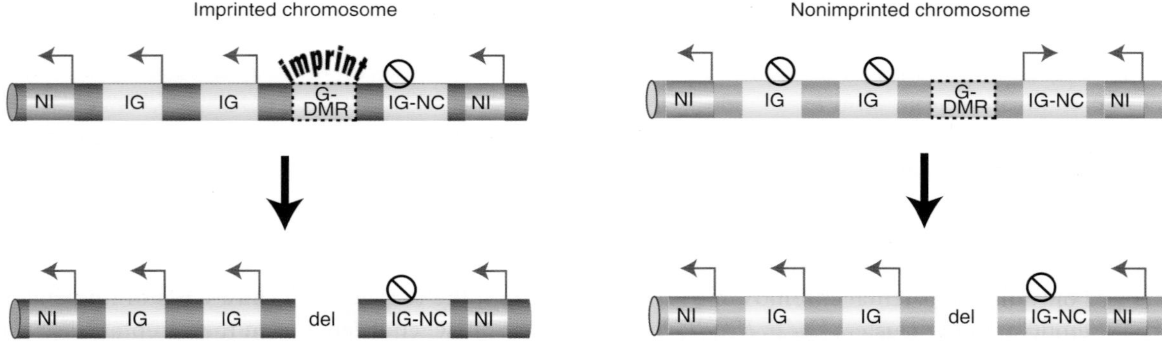

Figure 6. Imprinted expression is regulated by gametic DMRs (G-DMR). (*Left*) The effect of deleting the gametic DMR from the imprinted chromosome (green). (*Right*) The effect of deleting the G-DMR from the nonimprinted chromosome (yellow). In many imprinted clusters (e.g., *Igf2r, Kcnq1*, and *Dlk1*), experimental deletion of the G-DMR only affects the chromosome carrying the nonimprinted G-DMR. This results in a loss of repression of the imprinted protein-coding mRNA genes (IG) and a gain of repression of the imprinted lncRNA gene (IG-NC). Note that in some imprinted clusters (*Igf2* and *Pws*) that are not illustrated here, the methylated G-DMR appears also to be required for expression of some of the imprinted mRNAs in *cis*. del, deleted DNA; G-DMR, gametic differentially DNA-methylated region; NG, nonimprinted gene; arrow, expressed allele; slashed circle, repressed allele; imprint, epigenetic modification leading to a change in gene expression in *cis*.

 Cite as *Cold Spring Harb Perspect Biol* doi: 10.1101/cshperspect.a018382

DMR in the *Igf2* cluster, however, does not directly promote *H19* transcription, but changes the interaction between *Igf2* and *H19* and their shared enhancers, and in this way regulates their expression. Despite these differences, in general, the unmethylated gametic DMR is implicated in all six clusters as a positive regulator of lncRNA expression, and the presence of the DNA methylation imprint is associated with repression of the lncRNA. The conclusion from the data obtained from deletion of gametic DMRs clearly identifies these regions as an ICE, whose activity is regulated by DNA methylation.

3.4 Imprinted Gene Clusters Contain at Least One lncRNA

The majority of imprinted clusters contain an lncRNA, which is currently defined as a noncoding transcript >200 nucleotides (Guttman and Rinn 2012). lncRNAs, with the exception of those involved in RNA processing and translation, were previously thought to be a rarity in the mammalian genome. Now, because of the availability of the mouse and human genome sequence, transcriptome analyses have resulted in the identification of all RNA transcripts in a given cell population. This work has shown that a large part of the mammalian transcriptome is composed of lncRNAs. There are several types of mammalian noncoding RNAs (ncRNAs) that possess gene regulatory functions, including "short" ncRNAs that participate in silencing pathways (Ch. 16 [Martienssen and Moazed 2014]), "longer" processed lncRNAs such as *Xist*, which are involved in X-chromosome activation or inactivation (Ch. 25 [Brockdorff and Turner 2014]; Ch. 24 [Lucchesi and Kuroda 2014]), and lncRNAs that are associated with *cis* or *trans* activation or silencing of protein-coding genes (Ch. 3 [Allis et al.]; Ch. 2 [Rinn 2014]).

What types of lncRNAs are associated with imprinted gene clusters? The analysis of the lncRNAs associated with the well-characterized imprinted clusters shown in Table 1 is still incomplete, highlighting some similarities, but also some differences. Three imprinted lncRNAs are unusually long mature RNAs: *Airn* is 108 kb (Lyle et al. 2000), *Kcnq1ot1* is ~100 kb (Pauler et al. 2012), and *Ube3aas* may be >1000 kb (Landers et al. 2004). The *H19* lncRNA, in contrast, is only 2.3 kb (Brannan et al. 1990). The *Gtl2* lncRNA contains multiple alternatively spliced transcripts; however, downstream intergenic transcription has also been noted, suggesting longer transcription units are likely (Tierling et al. 2005). *Nespas* lncRNA is larger than can be resolved on RNA blots and the full size exceeds 27 kb (Robson et al. 2012). These latter lncRNAs appear to be intron poor with a low intron–exon ratio or are unspliced as mature transcripts (Seidl et al. 2006; Pandey et al. 2008). One further feature is that three imprinted lncRNAs (*H19*, *Ube3aas*, and the *Gtl2* downstream transcripts) act as host transcripts for snoRNAs (small nucleolar RNAs that direct modifications to ribosomal RNA, snRNAs, and possibly mRNAs thereby acting as posttranscription regulators) and miRNAs (microRNAs). The snoRNAs are not directed toward the imprinted mRNA genes in the cluster and it is presently unclear if they play a role in the imprinting mechanism itself (Seitz et al. 2004). Similarly, the miRNAs in the *H19* and *Gtl2* lncRNAs are involved in posttranscriptional repression of mRNA genes, but do not play a direct role in regulating imprinted expression of the cluster (Davis et al. 2005; Keniry et al. 2012).

Two features of imprinted lncRNAs indicate they may play a role in the silencing of the imprinted mRNA (i.e., protein-coding) genes in the cluster. The first is that the lncRNA generally shows reciprocal parental-specific expression compared to the imprinted mRNA genes (Table 1). Second, the DMR that carries the gametic methylation imprint, which controls imprinted expression of the whole cluster, overlaps with the lncRNA promoter in multiple instances (*Airn region2*, *KvDMR1*, *Snrpn-CGI*, and *Nespas-DMR*). This finding could indicate that imprints evolved to regulate the lncRNA in each imprinted cluster. This interpretation is supported by experiments that deleted the unmethylated sequence carrying the gametic DMR causing a loss of lncRNA expression concomitant with a gain of expression of imprinted mRNA genes (Fig. 6) as tested at the *Igf2r*, *Kcnq1*, *Gnas*, *Pws*, and *Dlk1* clusters (Wutz et al. 1997; Bielinska et al. 2000; Fitzpatrick et al. 2002; Lin et al. 2003; Williamson et al. 2006).

Experiments that directly test the role of the lncRNA itself have now been performed for a number of loci (*Airn*, *Nespas*, *Kcnq1ot1*, and *H19* lncRNAs). These lncRNAs were analyzed by genetic manipulation of the endogenous locus. The first three loci were assayed by inserting a polyadenylation signal to truncate the lncRNA. Truncation of the 108-kb *Airn* lncRNA to 3 kb showed that the lncRNA itself is necessary to silence all three mRNA genes in the *Igf2r* cluster, indicating a clear regulatory role for this lncRNA (Sleutels et al. 2002). Truncation of the ~100-kb *Kcnq1ot1* lncRNA to 1.5 kb also showed that this lncRNA was directly needed to silence all 10 mRNA genes in the larger *Kcnq1* cluster (Mancini-DiNardo et al. 2006). And last, truncation of the ~27-kb *Nespas* lncRNA showed it was necessary to silence the overlapped *Nesp* gene in the *Gnas* imprinted cluster (Williamson et al. 2011). In contrast, precise deletion of the *H19* ncRNA had no effect on imprinting in the *Igf2* cluster in endoderm tissues, although some loss of imprinting was seen in mesoderm tissue (Schmidt et al. 1999). Thus, three maternally imprinted clusters (*Igf2r*, *Kcnq1*, and *Gnas*) share a common lncRNA-dependent silencing mechanism, whereas the single paternally imprinted cluster (*Igf2*) so far examined uses a different, insulator-dependent model (see Sec. 3.6).

3.5 The Role of DNA Methylation in Genomic Imprinting

The identification of the first three endogenous imprinted genes in 1991 enabled investigators to study how the cell's epigenetic machinery marked an imprinted gene with its parental identity. The first and most easily testable candidate was DNA methylation, a modification in mammals that covalently adds a methyl group to the cytosine residue in CpG dinucleotides. DNA methylation is acquired through the action of de novo methyltransferases and maintained in situ each time the cell divides by the action of maintenance methyltransferases (see Ch. 15 [Li and Zhang 2014]). Hence, this modification fulfills the criteria outlined in Fig. 3 for a parental identity mark or "imprint" because (1) it can be established in either the sperm or oocyte by de novo methyltransferases that act only in one gamete, (2) it can be stably propagated at each embryonic cell division by a maintenance methyltransferase, and (3) it can be erased in the germline to reset the imprint in the next generation, either by passive demethylation (DNA replication followed by the failure to undergo maintenance methylation) or through the action of a demethylating activity (possibly through conversion of 5-methylcytosine to 5-hydroxymethylcytosine by the ten-eleven translocation family of enzymes or through excision of 5-methylcytosine by the DNA repair machinery; Tan and Shi 2012; see Fig. 6 of Ch. 15 [Li and Zhang 2014]).

DNA methylation could potentially perform two different functions in genomic imprinting. It could act as the imprinting mark by being acquired de novo only by the chromosomes in one gamete. It could also serve to silence one of the parental alleles because DNA methylation is associated with gene repression (Ch. 15 [Li and Zhang 2014]). To determine which function it has, it is first necessary to show that DNA methylation is present only on one parental chromosome (i.e., that it is a DMR). Second, it is necessary to identify which imprinted gene in the cluster and which regulatory sequences are marked by DNA methylation. The location of methylation marks on a promoter, or on distant positive or negative regulatory elements will have different consequences for gene expression. Finally, it is necessary to identify when the DMR forms during development. If it forms during gametogenesis and is continuously maintained in place in somatic cells (known as a gametic DMR), it may serve as the imprinting mark. If, however, it is placed on the gene after the embryo has become diploid when both parental chromosomes are in the same cell (known as a somatic DMR), it is unlikely to serve as the identity mark, but may serve to maintain parental-specific silencing.

Parental allele-specific DNA methylation has been found at most imprinted clusters that have been examined. For example, the *Igf2* cluster has a gametic DMR located 2 kb upstream of the *H19* lncRNA promoter that is methylated only in the paternal gamete and is maintained thereafter in all somatic tissues (Bartolomei et al. 1993). A similar gametic DMR was identified covering the promoter of the *Airn* lncRNA, present only on the silent maternal gene copy, and acquired in the female gamete (Stoger et al. 1993). Surprisingly, gametic DMRs were not identified at the promoters of the principal imprinted protein-coding genes in these clusters (respectively, *Igf2* and *Igf2r*). Instead, the silenced *Igf2* promoter is free of DNA methylation, whereas the silenced *Igf2r* promoter lies within a somatic DMR that is placed after fertilization (Sasaki et al. 1992; Stoger et al. 1993). Similar findings of gametic DMRs methylated on the chromosome carrying the silent copy of the imprinted lncRNA (as illustrated in Fig. 6) have been made for other well-studied imprinted gene clusters, including *Pws*, *Kcnq1*, *Gnas*, *Dlk1*, *and Grb10* (Shemer et al. 1997; Liu et al. 2000; Takada et al. 2002; Yatsuki et al. 2002; Shiura et al. 2009).

Somatic DMRs are relatively rare but have been reported for some imprinted clusters, which suggests that this type of epigenetic modification plays a limited role in maintaining imprinted gene expression (Stoger et al. 1993; Moore et al. 1997; Yatsuki et al. 2002; John and Lefebvre, 2011). Deletions of gametic DMRs in mice result in complete loss of imprinting for multiple genes, thereby proving that this class of DMRs also serves as a major ICE for the whole cluster (Fig. 6) (Wutz et al. 1997; Thorvaldsen et al. 1998; Bielinska et al. 2000; Fitzpatrick et al. 2002; Lin et al. 2003; Williamson et al. 2006). In contrast, deletion of the somatic DMRs affects expression of the adjacent imprinted gene, but imprinted expression is maintained by other genes in the cluster (Constancia et al. 2000; Sleutels et al. 2003).

Genome-wide deficiency in DNA methylation caused by mutations in the *Dnmt* gene family underscores the essential role of DNA methylation in regulating imprinted gene expression. Mutations in the de novo DNA methyltransferase *Dnmt3a*, the DNA methyltransferase stimulatory factor *Dmnt3L*, or the *Dnmt1* maintenance DNA methyltransferase generate DNA methylation deficient embryos that show alterations in imprinted gene expression. The type of perturbations shown for four imprinted clusters (*Igf2*, *Igf2r*, *Kcnq1*, and *Dlk1*) indicates that DNA methylation is generally acting to suppress the action of the gametic DMR. Thus, in the absence of DNA methylation, the gametic DMR cannot function appropriately (i.e., cannot silence the lncRNA). As a consequence, the lncRNA is aberrantly expressed and several imprinted protein-coding genes, including *Igf2*, *Igf2r*, *Kcnq1*, and *Dlk1*, become repressed on both parental chromosomes. This indicates that these mRNA genes require epigenetic modification of a *cis*-regulatory element to be expressed. Notably, the *H19* lncRNA that is normally only expressed on the chromosome carrying the unmethylated gametic DMR becomes expressed on both parental chromosomes. Some exceptions to this general pattern have been reported for genes that show imprinted expression only in the placenta (Lewis et al. 2004).

Are other types of epigenetic modification used as gametic imprints? Given the sheer abundance of epigenetic mechanisms acting to modify genetic information in the mammalian genome, DNA methylation is unlikely to be the only imprinting mechanism. Histone modifications that affect chromatin activity states are also likely candidates for parental imprints because they could fulfill many of the prerequisites shown in Figure 3. In one example, the Polycomb group protein known as EED (part of the PRC2 complex that catalyzes methylation of H3K27, i.e., histone H3 at lysine 27) has been shown to affect a few paternally repressed genes in the placenta. The effects of *Eed* mutation on genomic imprinting, however, are relatively minor compared to that of DNA methylation (Mager et al. 2003). In another example, the EHMT2 histone methyltransferase acting specifically on H3K9 is required to repress a few imprinted genes, but also only in the placenta (Nagano et al. 2008). Thus, evidence to date suggests that histone modifications and modifying enzymes play a minor role in genomic imprinting.

Although much is known about the identity and epigenetic modifications of gametic DMRs, much less is known about how these sequences are chosen for methylation in the gametes. To date, many more maternally than paternally methylated gametic DMRs have been identified (Bartolomei and Ferguson-Smith 2011). The maternally methylated DMRs are methylated during oocyte growth and the paternally methylated DMRs are methylated prenatally in prospermatogonia (Fig. 7) (Lucifero et al. 2002). For maternal gametic DMRs, a sequence comparison of known gametic DMRs reveals no striking sequence conservation although some contain a series of direct repeats that may adopt a secondary structure that attracts DNA methylation (Neumann et al. 1995). The tandem direct repeats in the *Igf2r* cluster gametic DMR have, for instance, been shown to be essential for oocyte-specific DNA methylation (Koerner et al. 2012). Those in the *Kcnq1* cluster gametic DMR, however, are not essential (Mancini-DiNardo et al. 2006). Another feature of maternal DMRs is that they are markedly CpG rich compared to the remainder of the genome. One idea for how these regions are recognized comes from the structural analysis of the complexed carboxy-terminal domains of DNMT3A and DNMT3L, which was obtained by X-ray crystallography (Jia et al. 2007). A tetrameric complex consisting of these two enzymes preferentially methylates a pair of CpGs that are 8–10 base pairs apart (Ch. 6 [Cheng 2014]). Such spacing is found in maternally methylated, but not in paternally methylated imprinted loci. This CpG spacing, however, is widespread in the genome, questioning the specificity of such a mechanism or indicating that additional features are required (Ferguson-Smith and Greally 2007). Additional specificity has been suggested by the demonstration that DNMT3L interacts with the amino terminus of histone H3 if the H3K4 residue is unmethylated, and promotes local DNA methylation (for more detail, see Ooi et al. 2007; Ch. 6 [Cheng 2014]). Another factor contributing to the specificity of de novo DNA methylation at DMRs in the oocyte is transcription across differentially methylated regions (Chotalia et al. 2009). Importantly, only protein-coding transcripts traversing the germline ICEs are thought to be involved in DNA methylation establishment. Although it is, as yet, unclear how this transcription may be attracting the DNA methylation machinery, it has been suggested that transcription across ICEs is required to establish or maintain open chromatin domains that are permissive for the establishment of DNA methylation. To investigate and define the mechanism further, it will be necessary to describe the temporal relationship between transcription and de novo DNA methylation in greater detail. Nevertheless, CpG spacing, posttranslational histone modifications, and transcription in oocytes could provide a starting point for the acquisition of maternal-specific DNA methylation imprints.

There is far less information regarding how paternal-specific DNA methylation imprints are established in the male germline. Nevertheless, early experiments suggest there could be some similarities with the female germline. It has recently been shown that high transcriptional read-through, predominantly from one strand, is detected at two paternal gametic DMRs in primordial germ cells, H19-DMD and IG-DMR, at the time of imprint establishment (Henckel et al. 2011). It also appears that maternal gametic DMRs, which are protected from DNA methylation, are enriched for H3 lysine 4 trimethylation (H3K4me3) in male primordial germ cells.

One of the most mysterious questions in genomic imprinting is how the DNA methylation marks at imprinted genes escape the genome-wide reprogramming that occurs after fertilization, including the DNA demethylation that occurs in the preimplantation embryo and the subsequent wave of de novo DNA methylation (Fig. 7) (see Morgan et al. 2005; Fig. 3 in Ch. 6 [Li and Zhang 2014]). It is likely that a combination of *cis*-acting sequences and *trans*-acting fac-

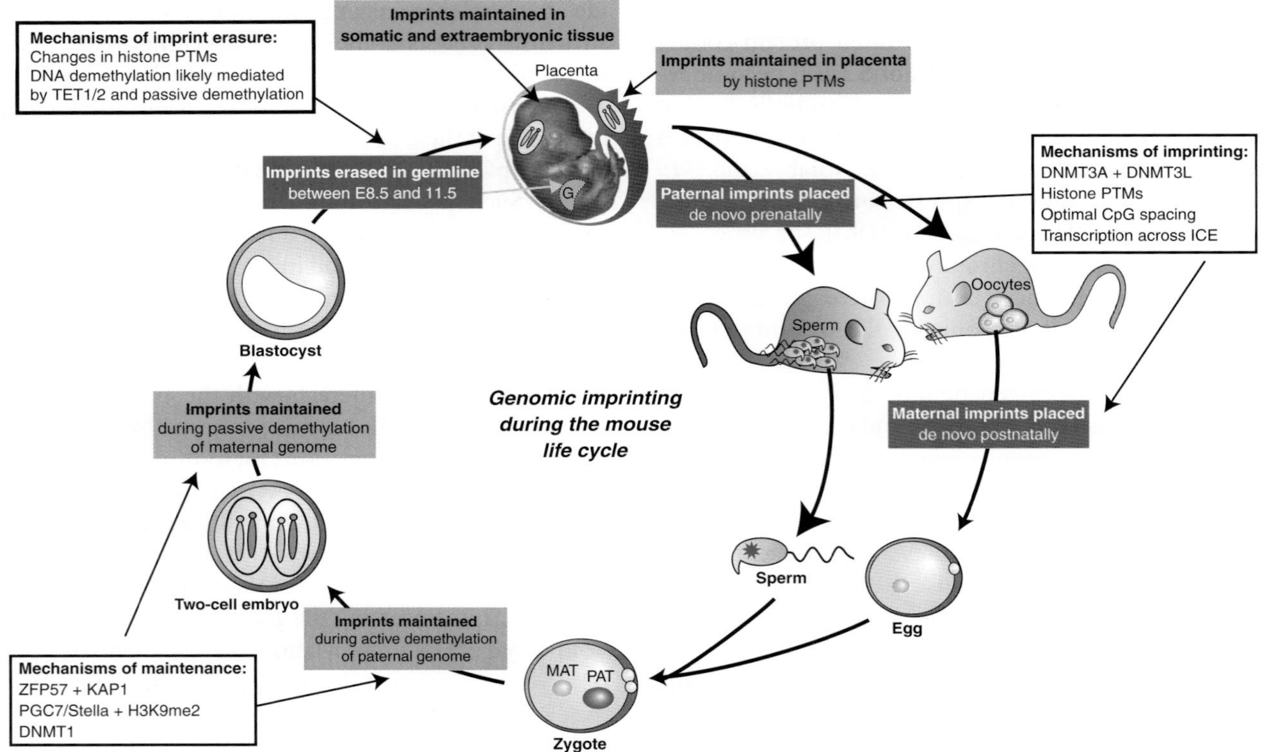

Figure 7. Establishment, maintenance, and erasure of genomic imprints in mouse development. In the germline, primordial germ cells (PGCs) undergo multiple changes in chromatin structure and DNA demethylation during migration into the genital ridge (gonad). Imprints are then acquired in a sex-specific manner in the germline (green shading). DNA methylation is targeted specifically to paternally and maternally DNA-methylated ICEs—prenatally in prospermatogonia and postnatally during oocyte maturation. These imprints are maintained despite global changes in DNA methylation after fertilization (orange shading): active demethylation of the paternal genome in the zygote and passive maternal demethylation in the preimplantation embryo. Candidates for protection of methylation regions include ZFP57 and PGC7/STELLA. De novo DNA methylation of the genome begins at the morula stage, during which time unmethylated alleles of imprinted genes must be protected. These imprints are maintained in somatic cells throughout the lifetime of the organism, whereas imprinting in extraembryonic tissues is thought to be less dependent on maintenance of DNA methylation. In the germline, imprints are erased and reset for the next generation (red shading). PTM, posttranslational modification; MAT, maternal genome; PAT, paternal genome.

tors are mediating the protection. One maternal factor, PGC7/STELLA, appears to have a general role in maintaining DNA methylation in the early mouse embryo through interactions with H3K9me2 (Nakamura et al. 2012). However, a factor that may be more specific for imprinted genes is ZFP57. Studies have shown that *ZFP57* mutations identified in transient neonatal diabetes patients are associated with defects in DNA methylation at multiple imprinted loci (Mackay et al. 2008). Additionally, *Zfp57* null mice show embryonic lethality and loss of imprinting at many (but not all) loci (Li et al. 2008). More recently, it has been shown that ZFP57 binds to cofactor KAP1, which can then recruit other epigenetic regulators (Quenneville et al. 2011). Thus, sequence- and DNA methylation-dependent binding of ZFP57 could act as an anchor to specify allelic binding of KAP1, which would subsequently recruit other major repressive epigenetic regulators such as SETDB1, HP1, DNMT1, DNMT3A, and DNMT3B to the heterochromatic, silenced allele at imprinted loci. It is possible that other yet-to-be-identified proteins also maintain DNA methylation at imprinted loci in the early embryo.

3.6 Two Types of *cis*-Acting Silencing Identified in Imprinted Gene Clusters

Currently, two major classes of *cis*-acting silencing mechanisms are hypothesized to govern imprinting at various clusters: the insulator model applicable to the *Igf2* cluster and the lncRNA-mediated silencing model applicable to the *Igf2r* and *Kcnq1* clusters. Although not yet completely defined, most of the clusters in Table 1 incorporate aspects of one of the two models. The breakthrough that led to the definition of the insulator model at the *Igf2* locus was the deletion of the gametic DMR (H19-DMD) that is located 2 kb upstream of the start of *H19* transcription and 80 kb downstream of *Igf2* (Fig. 8) (Thorvaldsen et al. 1998). When deleted, *H19* and *Igf2* showed a loss of imprinting regardless of whether the deletion was inherited maternally or paternally, identifying this DMR as an ICE. It was subsequently shown that this ICE bound CTCF, a protein shown to mediate insulator activity at the *beta-globin* locus, and that the ICE itself functioned as an insulator (Bell and Felsenfeld 2000; Hark et al. 2000). In this context, an insulator is defined as an element that blocks enhancer and promoter interactions when placed between them. Thus, the model for imprinted gene expression at this locus is as follows: on the maternal allele, CTCF binds to the ICE and blocks the access of *Igf2* and *Ins2* to enhancers shared with

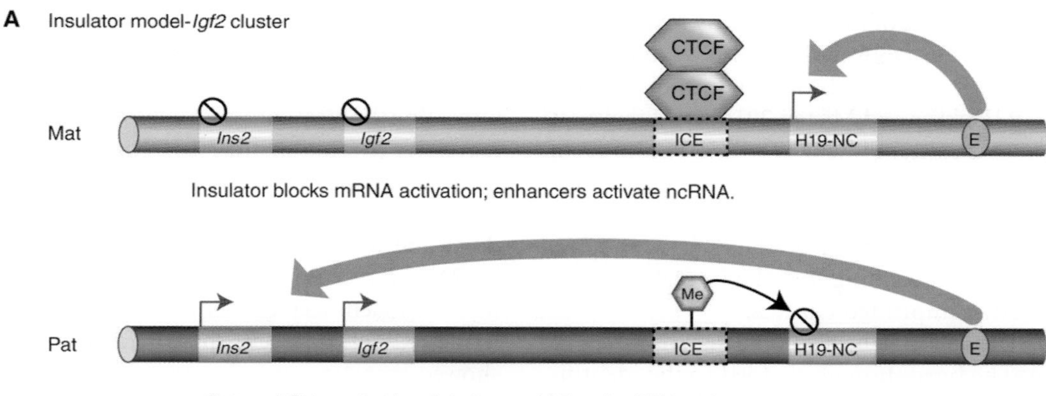

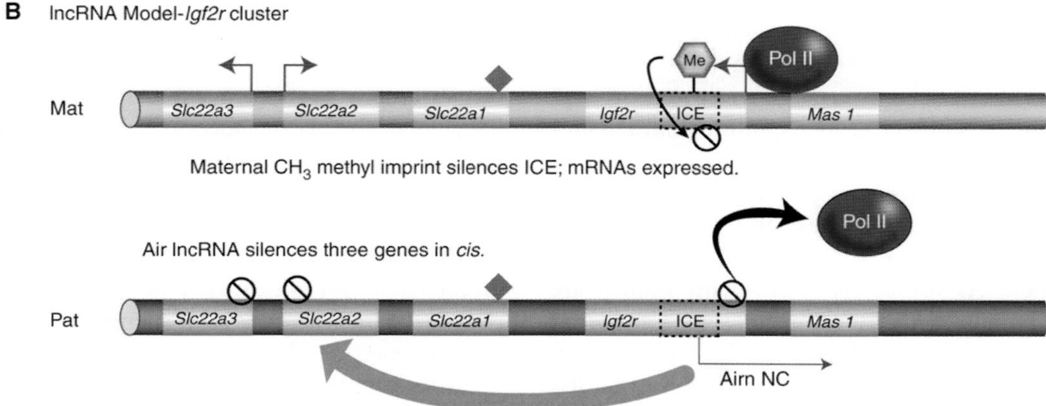

Figure 8. Two *cis*-acting silencing mechanisms at imprinted gene clusters. (*A*) Insulator model for the *Igf2* cluster. The expression pattern for endoderm is shown. On the maternal chromosome, the unmethylated ICE binds the CTCF protein and forms an insulator that prevents the common endoderm enhancers (E) from activating *Igf2* and *Ins2*. Instead the enhancers activate the nearby *H19* lncRNA promoter. On the paternal chromosome, the methylated ICE cannot bind CTCF and an insulator does not form; hence the *Igf2* and *Ins2* mRNA genes are expressed only on this chromosome. The *H19* lncRNA is methylated, most likely because of spreading from the 2-kb distant methylated ICE, and silenced. (*B*) lncRNA model for the *Igf2r* cluster. The expression pattern for placenta is shown. On the maternal chromosome, the methylated ICE contains the *Airn* lncRNA promoter that is directly silenced by the DNA methylation imprint. The *Igf2r*, *Slc22a2*, and *Slc22a3* mRNA genes are expressed only on this chromosome. *Mas1* and *Slc22a1* are not expressed in placenta (filled diamond). On the paternal chromosome, the Airn lncRNA promoter lying in the unmethylated ICE is expressed and silences *Igf2r* (in part by kicking off RNA polymerase II), *Slc22a2*, and *Slc22a3* in *cis*. Note that in both models, the DNA methylation imprint silences the lncRNA and permits mRNA expression. ICE, imprint control element; gray arrow, expressed allele of an imprinted gene; slashed circle, repressed allele of an imprinted gene; thick gray arrows, long distance effect in *cis*.

the *H19* lncRNA that are located downstream of the three genes. This thereby allows *H19* exclusive access to the enhancers (Fig. 8). On the paternal allele, the ICE acquires DNA methylation in the male germline, preventing CTCF from binding to it. Thus, on the paternal chromosome, *Igf2* and *Ins2* interact with the enhancers and are expressed from this chromosome. The presence of DNA methylation on the paternal ICE leads to secondary methylation of the *H19* promoter by an unknown mechanism and it becomes silenced on the paternal chromosome. Although the insulator model is widely accepted, it is unclear how the insulator acts at this locus. One of the most widely held views is that CTCF interacts with DNA molecules in *cis* to insulate genes through the formation of chromatin loops (for more detail, see Ch. 19 [Dekker and Misteli 2014]). Moreover, it has been shown that cohesin interacts with CTCF to form these loops (Nativio et al. 2009). The involvement of CTCF in the insulator model has led to the identification of CTCF binding sites at other imprinted genes such as *Rasgrf1*, *Grb10*, and *Kcnq1ot1*, indicating that the insulator model may operate in other imprinted clusters.

The lncRNA class of imprinting model may, however, be more common. The breakthrough that led to the identification of functional ncRNAs in imprinted clusters was an experiment that truncated the 108-kb *Airn* lncRNA to 3 kb (Sleutels et al. 2002). This shortened lncRNA retained imprinted expression and the *Airn* promoter retained imprinted DNA methylation—yet silencing of all three mRNA genes in the *Igf2r* cluster was lost (Fig. 8). lncRNA-mediated silencing has also now been shown to operate at the *Kcnq1* cluster (Mancini-DiNardo et al. 2006), although in a tissue-specific manner, suggesting that another mechanism such as one that uses insulators may also be involved at this cluster (Shin et al. 2008), and in the *Gnas* imprinted cluster (Williamson et al. 2011). At this time, it is not precisely known how lncRNAs silence genes but many models are possible. Two possibilities arise from the sense–antisense overlap between an mRNA and the lncRNA that occurs in each cluster. The first possibility is that double-stranded RNA can form between the mRNA and lncRNA and induce RNA-interference (RNAi) (described in Ch. 16 [Martienssen and Moazed 2014]). Absence of the RNAi machinery, however, does not affect imprinted expression in the *Kcnq1* cluster (Redrup et al. 2009). Thus, a second possibility is that this sense–antisense overlap causes a form of transcriptional interference of a promoter or an enhancer, which affects transcription from the mRNA promoter (Pauler et al. 2012). In this case, the first event could be silencing of the overlapped promoter or enhancer followed by accumulation of repressive chromatin that can spread and induce transcriptional gene silencing throughout the cluster. Evidence for this model comes from a series of recombinant endogenous chromosomes generated at the *Igf2r*/*Airn* locus in ES (embryonic stem) cells (Latos et al. 2012). The onset of allele-specific expression at this locus in the embryo can be recapitulated by ES cell differentiation, in which *Igf2r* is initially biallelically expressed, but the initiation of *Airn* expression results in *Igf2r* imprinting (Latos et al. 2009). To test whether *Airn* transcription or the lncRNA itself was required for *Igf2*r silencing, *Airn* was shortened to different lengths, with the result that silencing only required *Airn* transcription overlap of the *Igf2r* promoter, which interferes with RNA polymerase II recruitment (Latos et al. 2012). This model suggests that *Airn* acts predominantly through its transcription rather than as an lncRNA.

It is, however, also possible that imprinted lncRNAs act by coating the local chromosomal region and directly recruit repressive chromatin proteins to the imprinted cluster, in a manner similar to that described for the action of the *Xist* lncRNA in X-chromosome inactivation (Ch. 25 [Brockdorff and Turner 2014]). Evidence for a function of the lncRNA in recruitment of histone posttranslational modification machinery comes from experiments in placental tissues. RNA fluorescence in situ hybridization experiments showed that *Airn* and *Kcnq1ot1* form RNA clouds at their site of transcription (Nagano et al. 2008; Pandey et al. 2008; Terranova et al. 2008; Redrup et al. 2009). Terranova and colleagues show that these long ncRNAs are associated with a repressive histone compartment and Polycomb group proteins (Terranova et al. 2008). This nuclear compartment is also devoid of RNA polymerase II and exists in a three-dimensionally contracted state. Other studies on the *Airn* lncRNA go further in suggesting that the lncRNAs actively recruit repressive histone modifications (Nagano et al. 2008), but only in the placenta. In this latter case, *Airn* was shown to actively recruit the EHMT2 H3K9 methyltransferase. This resulted in the paternal-specific silencing of the *Slc22a3* gene but not the *Igf2r* gene. These experiments indicate that lncRNA-mediated silencing of imprinted genes may depend on different downstream mechanisms.

Importantly, other mechanisms of imprinted gene regulation are likely. For example, Wood and colleagues described a new imprinted locus (*H13*) in which alternative polyadenylation sites are used in an allele-specific manner (Wood et al. 2008). The *H13* gene contains a maternally methylated internal CpG island that acquires DNA methylation in oocytes (it has not been tested for ICE activity yet). Hypermethylation of this CpG island ensures synthesis of the full length and functional *H13* gene transcript from the maternal chromosome. Experiments showed that the unmethylated CpG island on the paternal allele allowed transcription from the promoter for the *Mcts2* retrogene. *Mcts2* expression, in turn, correlates with the premature polyadenylation of *H13* and, hence, expression

 Cite as *Cold Spring Harb Perspect Biol* doi: 10.1101/cshperspect.a018382

of truncated *H13* transcripts. This locus raises the possibility that other less widely used mechanisms of genomic imprinting will be identified once the full catalog of imprinted genes is elucidated.

4 GENOMIC IMPRINTING—A MODEL FOR MAMMALIAN EPIGENETIC REGULATION

Studying genomic imprinting has an advantage over other mammalian epigenetic gene regulation models because both the active and inactive parental allele reside in the same nucleus and are exposed to the same transcriptional environment (Bartolomei 2009; Barlow 2011). As a result, any epigenetic difference between the two parental alleles is more likely to correlate to their transcriptional state in contrast to "before and after" epigenetic systems, in which epigenetic changes may also reflect the altered differentiation state of the cell. The presence of both the active and silent parental allele in the same nucleus makes genomic imprinting an ideal system to study epigenetic gene regulation. At the same time, it imposes a difficulty because it is necessary to first distinguish between the parental alleles so that specific features associated with gene activity and silencing can be attributed to the right parental allele. This difficulty has been largely overcome in the mouse by the development of model systems that allow the maternal and paternal chromosomes to be distinguished (Fig. 1). Despite the fact that epigenetic gene regulatory mechanisms are highly conserved in evolution, there are likely to be differences that relate to the type of genome organization for each organism. The mammalian genome shows an unusual organization that intersperses genes with high copy number repeats (also known as transposable elements). This greatly increases the length of most genes as well as the distance between adjacent genes. This contrasts with other model organisms such as yeast, nematodes, plants, and *Drosophila*, whose genomes show a tendency toward remaining repeat-free or, at least to separate repeats from genes (for organismal comparisons, see Rabinowicz et al. 2003; Fig. 19 of Ch. 3 [Allis et al.]). How can genomic imprinting contribute to an understanding of mammalian epigenetics? Although the characterization of imprinted gene clusters is far from complete, they clearly have the potential to provide information about how genes are controlled in local regions or domains. To date, imprinted gene clusters have already provided examples of *cis*-acting DNA sequences that are regulated by DNA methylation, genes that are silenced by default in the mammalian genome and require epigenetic activation to be expressed, long-range regulatory elements that can act as insulators, and unusual lncRNAs that silence large domains of genes in *cis*. Time will tell whether these types of epigenetic regulatory mechanisms are unique to imprinted clusters or whether they can also be found regulating expression of nonimprinted genes in the mammalian genome.

5 FUTURE DIRECTIONS

Genomic imprinting has been the focus of intense interest since the discovery of the first imprinted genes in mammals in 1991. Whereas early experiments relied on molecular and genetic strategies to identify imprinted genes, high-throughput technology on polymorphic individuals is allowing the complete determination of imprinted genes (Deveale et al. 2012) and regions containing parental-specific DNA methylation (Xie et al. 2012). These experiments are indicating that most genes showing ubiquitous imprinted expression have already been identified (http://www.mousebook.org/catalog.php?catalog=imprinting). However, it is possible that some genes showing tissue-specific imprinted expression remain to be identified (Prickett and Oakey 2012). Some questions still await conclusive answers, particularly those concerning why mammals alone among vertebrates use imprinted genes to regulate embryonic and neonatal growth. This lack of knowledge contrasts with the extensive progress during the intervening 20 years on elucidating the epigenetic mechanisms controlling imprinted expression in mammals. From this information, we think we understand the general principles of how the imprinting mechanism operates at imprinted gene clusters, although all the details are still not clear. At this stage, it is clear that genomic imprinting uses the cell's normal epigenetic machinery to regulate parental-specific expression, and that everything is set in motion by restricting this machinery in the gamete to just one parental allele. Although there are general similarities in the mechanism controlling imprinted expression at different gene clusters, it is not yet understood how many variants of this mechanism exist in the mammalian genome. In the future, it will also be of interest to determine to what degree nonimprinted genes are controlled by the epigenetic mechanisms described for imprinted gene clusters. Ultimately, transferring this knowledge for therapeutic use in humans, for example, by inducing re-expression of the silent parental alleles in patients with the Prader–Willi and Angelman syndromes to ameliorate their symptoms would be of great benefit (for further discussion, see Huang et al. 2011; Ch. 33 [Zoghbi and Beaudet 2014]). An understanding of the way the cell controls epigenetic information is of increasing importance, with the realization that epigenetic regulation can also be disturbed in cancers (Ch. 34 [Baylin and Jones 2014]), in assisted reproductive technologies and also in the aging process (Rando and Chang 2012; Ch. 30 [Berger and Sassone-Corsi 2014]).

An improved understanding of genomic imprinting will undoubtedly continue to provide an important model to discover how the mammalian genome uses epigenetic mechanisms to regulate gene expression.

ACKNOWLEDGMENTS

We are grateful to the past and present members of the Barlow and Bartolomei laboratories for discussions and debates on the ideas presented here. We apologize that limitations on the number of references prevented citation of all original data.

REFERENCES

*Reference is also in this book.

Barlow DP. 2011. Genomic imprinting: A mammalian epigenetic discovery model. *Annu Rev Genet* **45:** 379–403.

Barlow DP, Stoger R, Herrmann BG, Saito K, Schweifer N. 1991. The mouse insulin-like growth factor type-2 receptor is imprinted and closely linked to the Tme locus. *Nature* **349:** 84–87.

Bartolomei MS. 2009. Genomic imprinting: Employing and avoiding epigenetic processes. *Genes Dev* **23:** 2124–2133.

Bartolomei MS, Ferguson-Smith AC. 2011. Mammalian genomic imprinting. *Cold Spring Harb Perspect Biol* doi: 10.1101/cshperspect.a002592.

Bartolomei MS, Webber AL, Brunkow ME, Tilghman SM. 1993. Epigenetic mechanisms underlying the imprinting of the mouse *H19* gene. *Genes Dev* **7:** 1663–1673.

Bartolomei MS, Zemel S, Tilghman SM. 1991. Parental imprinting of the mouse H19 gene. *Nature* **351:** 153–155.

Barton SC, Surani MA, Norris ML. 1984. Role of paternal and maternal genomes in mouse development. *Nature* **311:** 374–376.

* Baylin SB, Jones PA. 2014. Epigenetic determinants of cancer. *Cold Spring Harb Perspect Biol* doi: 10.1101/cshperspect.a019505.

Bell AC, Felsenfeld G. 2000. Methylation of a CTCF-dependent boundary controls imprinted expression of the *Igf2* gene. *Nature* **405:** 482–485.

* Berger SL, Sassone-Corsi P. 2014. Metabolic signaling to chromatin. *Cold Spring Harb Perspect Biol* doi: 10.1101/cshperspect.a019463.

Bielinska B, Blaydes SM, Buiting K, Yang T, Krajewska-Walasek M, Horsthemke B, Brannan CI. 2000. De novo deletions of SNRPN exon 1 in early human and mouse embryos result in a paternal to maternal imprint switch. *Nat Genet* **25:** 74–78.

Brannan CI, Dees EC, Ingram RS, Tilghman SM. 1990. The product of the H19 gene may function as an RNA. *Mol Cell Biol* **10:** 28–36.

* Brockdorff N, Turner BM. 2014. Dosage compensation in mammals. *Cold Spring Harb Perspect Biol* doi: 10.1101/cshperspect.a019406.

Cattanach BM, Kirk M. 1985. Differential activity of maternally and paternally derived chromosome regions in mice. *Nature* **315:** 496–498.

Chaillet JR, Bader DS, Leder P. 1995. Regulation of genomic imprinting by gametic and embryonic processes. *Genes Dev* **9:** 1177–1187.

Chandra HS, Nanjundiah V. 1990. The evolution of genomic imprinting. *Dev Suppl* **1990:** 47–53.

* Cheng X. 2014. Structural and functional coordination of DNA and histone methylation. *Cold Spring Harb Perspect Biol* **6:** a018747.

Chotalia M, Smallwood SA, Ruf N, Dawson C, Lucifero D, Frontera M, James K, Dean W, Kelsey G. 2009. Transcription is required for establishment of germline methylation marks at imprinted genes. *Genes Dev* **23:** 105–117.

Constancia M, Dean W, Lopes S, Moore T, Kelsey G, Reik W. 2000. Deletion of a silencer element in Igf2 results in loss of imprinting independent of H19. *Nat Genet* **26:** 203–206.

Cooper DW, VandeBerg JL, Sharman GB, Poole WE. 1971. Phosphoglycerate kinase polymorphism in kangaroos provides further evidence for paternal X inactivation. *Nat New Biol* **230:** 155–157.

Crouse HV, Brown A, Mumford BC. 1971. Chromosome inheritance and the problem of chromosome "imprinting" in Sciara (Sciaridae, Diptera). *Chromosoma* **34:** 324–398.

Davis E, Caiment F, Tordoir X, Cavaille J, Ferguson-Smith A, Cockett N, Georges M, Charlier C. 2005. RNAi-mediated allelic *trans*-interaction at the imprinted *Rtl1/Peg11* locus. *Curr Biol* **15:** 743–749.

DeChiara TM, Robertson EJ, Efstratiadis A. 1991. Parental imprinting of the mouse insulin-like growth factor II gene. *Cell* **64:** 849–859.

* Dekker J, Misteli T. 2014. Long-range chromatin interactions. *Cold Spring Harb Perspect Biol* doi: 10.1101/cshperspect.a019356.

Deveale B, van der Kooy D, Babak T. 2012. Critical evaluation of imprinted gene expression by RNA-seq: A new perspective. *PLoS Genet* **8:** e1002600.

Ferguson-Smith AC, Greally JM. 2007. Epigenetics: Perceptive enzymes. *Nature* **449:** 148–149.

Ferguson-Smith AC, Cattanach BM, Barton SC, Beechey CV, Surani MA. 1991. Embryological and molecular investigations of parental imprinting on mouse chromosome 7. *Nature* **351:** 667–670.

Fitzpatrick GV, Soloway PD, Higgins MJ. 2002. Regional loss of imprinting and growth deficiency in mice with a targeted deletion of KvDMR1. *Nat Genet* **32:** 426–431.

* Grossniklaus U, Paro R. 2014. Transcriptional silencing by *Polycomb*-group proteins. *Cold Spring Harb Perspect Biol* **6:** a019331.

Guttman M, Rinn JL. 2012. Modular regulatory principles of large non-coding RNAs. *Nature* **482:** 339–346.

Hark AT, Schoenherr CJ, Katz DJ, Ingram RS, Levorse JM, Tilghman SM. 2000. CTCF mediates methylation-sensitive enhancer-blocking activity at the H19/Igf2 locus. *Nature* **405:** 486–489.

Henckel A, Chebli K, Kota SK, Arnaud P, Feil R. 2011. Transcription and histone methylation changes correlate with imprint acquisition in male germ cells. *EMBO J* **31:** 606–615.

* Henikoff S, Smith M. 2014. Histone variants and epigenetics. *Cold Spring Harb Perspect Biol* doi: 10.1101/cshperspect.a019364.

Hoppe PC, Illmensee K. 1982. Full-term development after transplantation of parthenogenetic embryonic nuclei into fertilized mouse eggs. *Proc Natl Acad Sci* **79:** 1912–1916.

Huang HS, Allen JA, Mabb AM, King IF, Miriyala J, Taylor-Blake B, Sciaky N, Dutton JW Jr, Lee HM, Chen X, et al. 2011. Topoisomerase inhibitors unsilence the dormant allele of *Ube3a* in neurons. *Nature* **481:** 185–189.

Jia D, Jurkowska RZ, Zhang X, Jeltsch A, Cheng X. 2007. Structure of Dnmt3a bound to Dnmt3L suggests a model for de novo DNA methylation. *Nature* **449:** 248–251.

John RM, Lefebvre L. 2011. Developmental regulation of somatic imprints. *Differentiation* **81:** 270–280.

Johnson DR. 1974. Hairpin-tail: A case of post-reductional gene action in the mouse egg. *Genetics* **76:** 795–805.

Kawahara M, Wu Q, Takahashi N, Morita S, Yamada K, Ito M, Ferguson-Smith AC, Kono T. 2007. High-frequency generation of viable mice from engineered bi-maternal embryos. *Nat Biotechnol* **25:** 1045–1050.

Keniry A, Oxley D, Monnier P, Kyba M, Dandolo L, Smits G, Reik W. 2012. The H19 lincRNA is a developmental reservoir of miR-675 that suppresses growth and Igf2r. *Nat Cell Biol* **14:** 659–665.

Koerner MV, Pauler FM, Hudson QJ, Santoro F, Sawicka A, Guenzi PM, Stricker SH, Schichl YM, Latos PA, Klement RM, et al. 2012. A downstream CpG island controls transcript initiation and elongation and the methylation state of the imprinted Airn macro ncRNA promoter. *PLoS Genet* **8:** e1002540.

Landers M, Bancescu DL, Le Meur E, Rougeulle C, Glatt-Deeley H, Brannan C, Muscatelli F, Lalande M. 2004. Regulation of the large (approximately 1000 kb) imprinted murine *Ube3a* antisense transcript by alternative exons upstream of *Snurf/Snrpn. Nucleic Acids Res* **32:** 3480–3492.

Latos PA, Pauler FM, Koerner MV, Senergin HB, Hudson QJ, Stocsits RR, Allhoff W, Stricker SH, Klement RM, Warczok KE, et al. 2012. Airn transcriptional overlap, but not its lncRNA products, induces imprinted *Igf2r* silencing. *Science* **338:** 1469–1472.

Latos PA, Stricker SH, Steenpass L, Pauler FM, Huang R, Senergin SH, Regha K, Koerner MV, Warczok KE, Unger C, et al. 2009. An in vitro ES cell imprinting model shows that imprinted expression of the *Igf2r* gene arises from an allele-specific expression bias. *Development* **136:** 437–448.

Lewis A, Mitsuya K, Umlauf D, Smith P, Dean W, Walter J, Higgins M, Feil R, Reik W. 2004. Imprinting on distal chromosome 7 in the placenta involves repressive histone methylation independent of DNA methylation. *Nat Genet* **36:** 1291–1295.

Li X, Ito M, Zhou F, Youngson N, Zuo X, Leder P, Ferguson-Smith AC. 2008. A maternal-zygotic effect gene, *Zfp57*, maintains both maternal and paternal imprints. *Dev Cell* **15:** 547–557.

* Li E, Zhang Y. 2014. DNA methylation in mammals. *Cold Spring Harb Perspect Biol* **6:** a019133.

Lin MS, Zhang A, Fujimoto A. 1995. Asynchronous DNA replication between 15q11.2q12 homologs: Cytogenetic evidence for maternal imprinting and delayed replication. *Hum Genet* **96:** 572–576.

Lin SP, Youngson N, Takada S, Seitz H, Reik W, Paulsen M, Cavaille J, Ferguson-Smith AC. 2003. Asymmetric regulation of imprinting on the maternal and paternal chromosomes at the Dlk1-Gtl2 imprinted cluster on mouse chromosome 12. *Nat Genet* **35:** 97–102.

Liu J, Litman D, Rosenberg MJ, Yu S, Biesecker LG, Weinstein LS. 2000. A *GNAS1* imprinting defect in pseudohypoparathyroidism type IB. *J Clin Invest* **106:** 1167–1174.

* Lucchesi JC, Kuroda MI. 2014. Dosage compensation in *Drosophila. Cold Spring Harb Perspect Biol* doi: 10.1101/cshperspect.a019398.

Lucifero D, Mertineit C, Clarke HJ, Bestor TH, Trasler JM. 2002. Methylation dynamics of imprinted genes in mouse germ cells. *Genomics* **79:** 530–538.

Lyle R, Watanabe D, te Vruchte D, Lerchner W, Smrzka OW, Wutz A, Schageman J, Hahner L, Davies C, Barlow DP. 2000. The imprinted antisense RNA at the *Igf2r* locus overlaps but does not imprint Mas1. *Nat Genet* **25:** 19–21.

Mackay DJ, Callaway JL, Marks SM, White HE, Acerini CL, Boonen SE, Dayanikli P, Firth HV, Goodship JA, Haemers AP, et al. 2008. Hypomethylation of multiple imprinted loci in individuals with transient neonatal diabetes is associated with mutations in ZFP57. *Nat Genet* **40:** 949–951.

Mager J, Montgomery ND, de Villena FP, Magnuson T. 2003. Genome imprinting regulated by the mouse Polycomb group protein Eed. *Nat Genet* **33:** 502–527.

Mancini-DiNardo D, Steele JS, Levorse JM, Ingram RS, Tilghman SM. 2006. Elongation of the Kcnqot1 transcript is required for genomic imprinting of neighboring genes. *Genes Dev* **20:** 1268–1282.

* Martienssen R, Moazed D. 2014. RNAi and heterochromatin assembly. *Cold Spring Harb Perspect Biol* doi: 10.1101/cshperspect.a019323.

McGrath J, Solter D. 1984a. Completion of mouse embryogenesis requires both the maternal and paternal genomes. *Cell* **37:** 179–183.

McGrath J, Solter D. 1984b. Maternal Thp lethality in the mouse is a nuclear, not cytoplasmic, defect. *Nature* **308:** 550–551.

McLaren A. 1979. *Maternal effects in development: The fourth symposium of the British Society for Developmental Biology* (ed. Newth DR, Balls M). Cambridge University Press, Cambridge.

Moore T, Haig D. 1991. Genomic imprinting in mammalian development: A parental tug-of-war. *Trends Genet* **7:** 45–49.

Moore T, Constancia M, Zubair M, Bailleul B, Feil R, Sasaki H, Reik W. 1997. Multiple imprinted sense and antisense transcripts, differential methylation and tandem repeats in a putative imprinting control region upstream of mouse Igf2. *Proc Natl Acad Sci* **94:** 12509–12514.

Morgan DH, Santos F, Green K, Dean W, Reik W. 2005. Epigenetic reprogramming in mammals. *Hum Mol Genet* **14:** R47–R48.

Nagano T, Mitchell JA, Sanz LA, Pauler FM, Ferguson-Smith AC, Feil R, Fraser P. 2008. The Air noncoding RNA epigenetically silences transcription by targeting G9a to chromatin. *Science* **322:** 1717–1720.

Nakamura T, Liu YY, Nakashima H, Umehara H, Inoue K, Matoa S, Tachibana M, Ogura A, Shinkai Y, Nakana T. 2012. PGC7 binds histone H3K9me2 to protect against conversion of 5mC to 5hmC in early embryos. *Nature* **486:** 415–419.

Nativio R, Wendt KS, Ito Y, Huddleston JE, Uribe-Lewis S, Woodfine K, Krueger C, Reik W, Peters JM, Murrell A. 2009. Cohesin is required for higher-order chromatin conformation at the imprinted IGF2-H19 locus. *PLOS Genet* **5:** e1000739.

Neumann B, Kubicka P, Barlow DP. 1995. Characteristics of imprinted genes. *Nat Genet* **9:** 12–13.

Ooi SK, Qiu C, Bernstein E, Li K, Jia D, Yang Z, Erdjument-Bromage H, Tempst P, Lin SP, Allis CD, et al. 2007. DNMT3L connects unmethylated lysine 4 of histone H3 to de novo methylation of DNA. *Nature* **448:** 714–717.

Pandey RR, Mondal T, Mohammad F, Enroth S, Redrup L, Komorowski J, Nagano T, Mancini-Dinardo D, Kanduri C. 2008. Kcnq1ot1 antisense noncoding RNA mediates lineage-specific transcriptional silencing through chromatin-level regulation. *Mol Cell* **32:** 232–246.

Pauler FJ, Barlow DP, Hudson QJ. 2012. Mechanisms of long range silencing by imprinted macro non-coding RNAs. *Curr Opin Genet Dev* **22:** 1–7.

Prickett AR, Oakey RJ. 2012. A survey of tissue-specific imprinting in mammals. *Mol Genet Genomics* **287:** 621–630.

Quenneville S, Verde G, Corsinotti A, Kapopoulou A, Jakobsson J, Offner S, Baglivo I, Pedone PV, Grimaldi G, Riccio A, et al. 2011. In embryonic stem cells, ZFP57/KAP1 recognize a methylated hexanucleotide to affect chromatin and DNA methylation of imprinting control regions. *Mol Cell* **44:** 361–377.

Rabinowicz PD, Palmer LE, May BP, Hemann MT, Lowe SW, McCombie WR, Martienssen RA. 2003. Genes and transposons are differentially methylated in plants, but not in mammals. *Genome Res* **13:** 2658–2664.

Rando TA, Chang HY. 2012. Aging, rejuvenation and epigenetic reprogramming: Resetting the aging clock. *Cell* **148:** 46–57.

Redrup L, Branco MR, Perdeaux ER, Krueger C, Lewis A, Santos F, Nagano T, Cobb BS, Fraser P, Reik W. 2009. The long noncoding RNA Kcnq1ot1 organises a lineage-specific nuclear domain for epigenetic gene silencing. *Development* **136:** 525–530.

Reik W. 1989. Genomic imprinting and genetic disorders in man. *Trends Genet* **5:** 331–336.

Reik W, Howlett SK, Surani MA. 1990. Imprinting by DNA methylation: From transgenes to endogenous gene sequences. *Dev Suppl* **1990:** 99–106.

Renfree MB, Hore TA, Shaw G, Graves JA, Pask AJ. 2009. Evolution of genomic imprinting: Insights from marsupials and monotremes. *Annu Rev Genomics Hum Genet* **10:** 214–262.

* Rinn JL. 2014. IncRNAs: Linking RNA to chromatin. *Cold Spring Harb Perspect Biol* **6:** a018614.

Robson JE, Eaton SA, Underhill P, Williams D, Peter J. 2012. MicroRNAs 296 and 298 are imprinted and part of the *GNAS/Gnas* cluster and miR-296 targets IKBKE and *Tmed9*. *RNA* **18:** 135–144.

Sasaki H, Jones PA, Chaillet JR, Ferguson-Smith AC, Barton SC, Reik W, Surani MA. 1992. Parental imprinting: Potentially active chromatin of the repressed maternal allele of the mouse insulin-like growth factor II (*Igf2*) gene. *Genes Dev* **6:** 1843–1856.

Schmidt JV, Levorse JM, Tilghman SM. 1999. Enhancer competition between *H19* and *Igf2* does not mediate their imprinting. *Proc Natl Acad Sci* **96:** 9733–9738.

Searle AG, Beechey CV. 1978. Complementation studies with mouse translocations. *Cytogenet Cell Genet* **20:** 282–303.

Seidl CI, Stricker SH, Barlow DP. 2006. The imprinted Air ncRNA is an atypical RNAPII transcript that evades splicing and escapes nuclear export. *EMBO J* **25:** 3565–3567.

Seitz H, Royo H, Lin SP, Youngson N, Ferguson-Smith AC, Cavaille J. 2004. Imprinted small RNA genes. *Biol Chem* **385:** 905–911.

Shemer R, Birger Y, Riggs AD, Razin A. 1997. Structure of the imprinted mouse Snrpn gene and establishment of its parental-specific methylation pattern. *Proc Natl Acad Sci* **94:** 10267–10272.

Shin JY, Fitzpatrick GV, Higgins MJ. 2008. Two distinct mechanisms of silencing by the KvDMR1 imprinting control region. *EMBO J* **27:** 168–178.

Shiura H, Nakamura K, Hikichi T, Hino T, Oda K, Suzuki-Migishima R, Kohda T, Kaneko-Ishino T, Ishino F. 2009. Paternal deletion of *Meg1/Grb10* DMR causes maternalization of the *Meg1/Grb10* cluster in mouse proximal Chromosome 11 leading to severe pre- and postnatal growth retardation. *Hum Mol Genet* **18:** 1424–1438.

Sleutels F, Tjon G, Ludwig T, Barlow DP. 2003. Imprinted silencing of *Slc22a2* and *Slc22a3* does not need transcriptional overlap between *Igf2r* and *Air*. *EMBO J* **22:** 3696–3704.

Sleutels F, Zwart R, Barlow DP. 2002. The non-coding Air RNA is required for silencing autosomal imprinted genes. *Nature* **415:** 810–813.

Stoger R, Kubicka P, Liu CG, Kafri T, Razin A, Cedar H, Barlow DP. 1993. Maternal-specific methylation of the imprinted mouse *Igf2r* locus identifies the expressed locus as carrying the imprinting signal. *Cell* **73:** 61–71.

Surani MA, Barton SC, Norris ML. 1984. Development of reconstituted mouse eggs suggests imprinting of the genome during gametogenesis. *Nature* **308:** 548–550.

Takada S, Paulsen M, Tevendale M, Tsai CE, Kelsey G, Cattanach BM, Ferguson-Smith AC. 2002. Epigenetic analysis of the *Dlk1-Gtl2* imprinted domain on mouse chromosome 12: Implications for imprinting control from comparison with *Igf2-H19*. *Hum Mol Genet* **11:** 77–86.

Tan L, Shi YG. 2012. Tet family proteins and 5-hydroxymethylcytosine in development and disease. *Development* **139:** 1895–1902.

Terranova R, Yokobayashi S, Stadler MB, Otte AP, van Lohuizen M, Orkin SH, Peters AH. 2008. Polycomb group proteins Ezh2 and Rnf2 direct genomic contraction and imprinted repression in early mouse embryos. *Dev Cell* **15:** 668–679.

Thorvaldsen JL, Duran KL, Bartolomei MS. 1998. Deletion of the H19 differentially methylated domain results in loss of imprinted expression of H19 and Igf2. *Genes Dev* **12:** 3693–3702.

Tierling S, Dalbert S, Schoppenhorst S, Tsai CE, Oliger S, Ferguson-Smith AC, Paulsen M, Walter J. 2005. High-resolution map and imprinting analysis of the Gtl2-Dnchc1 domain on mouse chromosome 12. *Genomics* **87:** 225–235.

Varmuza S, Mann M. 1994. Genomic imprinting–defusing the ovarian time bomb. *Trends Genet* **10:** 118–123.

Wan LB, Bartolomei MS. 2008. Regulation of imprinting in clusters: Noncoding RNAs versus insulators. *Adv Genet* **61:** 207–223.

Wilkins JF, Haig D. 2003. What good is genomic imprinting: The function of parent-specific gene expression. *Nat Rev Genet* **4:** 359–368.

Williamson CM, Turner MD, Ball ST, Nottingham WT, Glenister P, Fray M, Tymowska-Lalanne Z, Plagge A, Powles-Glover N, Kelsey G, et al. 2006. Identification of an imprinting control region affecting the expression of all transcripts in the *Gnas* cluster. *Nat Genet* **38:** 350–355.

Williamson CM, Ball ST, Dawson C, Mehta S, Beechey CV, Fray M, Teboul L, Dear TN, Kelsey G, Peters J. 2011. Uncoupling antisense-mediated silencing and DNA methylation in the imprinted Gnas cluster. *PLoS Genet* **7:** e1001247.

Wood AJ, Schulz R, Woodfine K, Koltowska K, Beechey CV, Peters J, Bourc'his D, Oakey RJ. 2008. Regulation of alternative polyadenylation by genomic imprinting. *Genes Dev* **22:** 1141–1146.

Wutz A, Smrzka OW, Schweifer N, Schellander K, Wagner EF, Barlow DP. 1997. Imprinted expression of the *Igf2r* gene depends on an intronic CpG island. *Nature* **389:** 745–749.

Xie W, Barr CL, Kim A, Yue F, Lee AY, Eubanks J, Dempster EL, Ren B. 2012. Base-resolution analyses of sequence and parent-of-origin dependent DNa methylation in the mouse genome. *Cell* **148:** 816–831.

Yatsuki H, Joh K, Higashimoto K, Soejima H, Arai Y, Wang Y, Hatada I, Obata Y, Morisaki H, Zhang Z, et al. 2002. Domain regulation of imprinting cluster in Kip2/Lit1 subdomain on mouse chromosome 7F4/F5: Large-scale DNA methylation analysis reveals that DMR-Lit1 is a putative imprinting control region. *Genome Res* **12:** 1860–1870.

* Zoghbi HY, Beaudet AL. 2014. Epigenetics and human disease. *Cold Spring Harb Perspect Biol* doi: 10.1101/cshperspect.a019497.

Zwart R, Sleutels F, Wutz A, Schinkel AH, Barlow DP. 2001. Bidirectional action of the Igf2r imprint control element on upstream and downstream imprinted genes. *Genes Dev* **15:** 2361–2366.

WWW RESOURCES

http://www.mousebook.org/catalog.php?catalog=imprinting MouseBook, Medical Research Council.

Cite as *Cold Spring Harb Perspect Biol* doi: 10.1101/cshperspect.a018382

CHAPTER 27

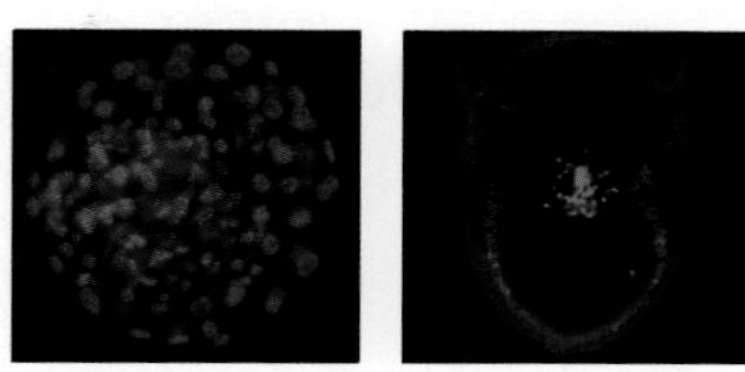

Germline and Pluripotent Stem Cells

Wolf Reik[1,2] and M. Azim Surani[2]

[1]The Babraham Institute, Babraham Research Campus, Cambridge CB2 3EG, United Kingdom; [2]Wellcome Trust Cancer Research UK Gurdon Institute & Wellcome Trust-Medical Research Council Cambridge Stem Cell Institute, University of Cambridge, Cambridge CB2 1QN, United Kingdom

Correspondence: a.surani@gurdon.cam.ac.uk

SUMMARY

Epigenetic mechanisms play an essential role in the germline and imprinting cycle. Germ cells show extensive epigenetic programming in preparation for the generation of the totipotent state, which in turn leads to the establishment of pluripotent cells in blastocysts. The latter are the cells from which pluripotent embryonic stem cells are derived and maintained in culture. Following blastocyst implantation, postimplantation epiblast cells develop, which give rise to all somatic cells as well as primordial germ cells, the precursors of sperm and eggs. Pluripotent stem cells in culture can be induced to undergo differentiation into somatic cells and germ cells in culture. Understanding the natural cycles of epigenetic reprogramming that occur in the germline will allow the generation of better and more versatile stem cells for both therapeutic and research purposes.

Outline

Cite as *Cold Spring Harb Perspect Biol* doi: 10.1101/cshperspect.a019422

OVERVIEW

An egg or oocyte is a most remarkable cell because it is the only cell in the body that is potentially capable of developing into a whole organism. William Harvey was the first to recognize this in 1651 when he remarked *"Ex Ovo Omni"* or "everything comes from an egg." He recognized that an egg probably develops progressively into an organism, and this insight was important for the concept of "epigenesis" or progressive development. This eventually led to the demise of the "preformationist" view of development, a theory proposing that individuals develop from the enlargement of tiny fully formed organisms (the so-called homunculus) contained in the germ cells. Conrad Waddington later depicted this concept in his famous illustration as an "epigenetic landscape," a symbolic representation of sequential development from an egg (Waddington 1956; a variation of which is illustrated in Ch. 2 [Takahashi 2014]). Development of an entire organism from an egg is possible in some organisms without any contribution from a male, which is called "parthenogenesis," but this cannot occur in mammals because of the phenomenon of "genomic imprinting" in which fertilization of an egg by sperm is obligatory for development to adulthood.

In most organisms, development commences following fusion between sperm and eggs to generate a zygote, which gives rise not only to a new individual but, theoretically at least, to an endless series of generations. In this way, germ cells provide the enduring link between all generations. The newly fertilized egg or zygote is therefore unique because no other cell has the potential to develop into an entirely new organism. This property is referred to as "totipotency." Germ cells are unique as transmitters of both genetic and epigenetic information to subsequent generations, and they show many exceptional properties that are required to fulfill this potential. The oocyte also has the striking property of conferring totipotency on cell nuclei from somatic cells, such as a nerve cell when it is transplanted into the egg, a process referred to as cloning or nuclear reprogramming.

During development from a zygote onward, there is a progressive decline in totipotency of the newly dividing cells. In mammals, only the products of very early cell divisions retain totipotency in which each of the cells is, in principle, separately capable of generating a new organism.

Further on in development, the mammalian embryo gives rise to a blastocyst, a structure with an outer group of trophectoderm cells destined to form the placenta, and an inner group of cells that will give rise to the entire fetus and, eventually, a new organism (Gardner 1985). These inner cells will therefore differentiate into all the known 200 or so specialized somatic cells found in adults and they are, therefore, referred to as "pluripotent." Under certain culture conditions, these pluripotent cells can be "rescued" from early embryos and made to grow indefinitely in vitro while still retaining the ability to differentiate into any specific cell type found in embryos and adults, including sperm and eggs themselves (Evans and Kaufman 1981; Martin 1981). Such cells have been derived from human, mouse, and rat embryos and are called pluripotent embryonic stem (ES) cells. The capacity to generate pluripotent stem cells is lost quite rapidly when the embryo implants and commences the program of embryonic development. Our recent understanding of how pluripotency is regulated by transcription factors epigenetically has given rise to the exciting technology of "induced pluripotent (iPS) cells" by which somatic cells can be reprogrammed to iPS cells that are similar to ES cells.

Among the earliest cell types to emerge during embryonic development, after implantation, are the precursors of sperm and eggs called primordial germ cells (PGCs) (McLaren 2003). This early developmental event ensures that PGCs that eventually give rise to subsequent generations are set aside from the remaining cells that form somatic tissues. These are highly specialized cells that eventually develop into mature sperm or eggs in the adult organism, thus repeating the cycle of life, while the rest of the body's cells eventually perish. PGCs are therefore very special cells. PGCs can be isolated to derive pluripotent stem cells called embryonic germ (EG) cells.

Stem cells are also present in adults. For example, adult stem cells generate billions of different blood cells that arise from blood stem cells in the bone marrow. Similarly, our skin cells or the cells in the gut are continually replaced through differentiation of their appropriate stem cells. Adult stem cells normally only have the potential to generate cells of specific tissues and not the diverse cell types that can be made from pluripotent stem cells. One of the key research objectives is to understand the similarities and differences between pluripotent ES and adult stem cells, including the underlying epigenetic mechanisms that regulate their properties. Interestingly, our understanding of the principles behind the differentiation of adult cells has resulted in the ability to reprogram one somatic cell type into another, often termed transdifferentiation, converting, for example, skin cells into cells of the pancreas and fibroblasts into neuronal cells. Understanding the unique epigenetic properties of germ cells and pluripotent stem cells will contribute to enabling us to develop new concepts for therapies, particularly in regenerative medicine.

Cite as *Cold Spring Harb Perspect Biol* doi: 10.1101/cshperspect.a019422

1 THE GENETIC AND EPIGENETIC CONTINUUM OF THE MAMMALIAN LIFE CYCLE

The genetic information encoded in an individual's genome is established at fertilization and does not change during development, with the exception of mutations and some directed sequence changes occurring, for example, during VDJ gene segments recombination in the immune system (elaborated in Sec. 4 of Ch. 29 [Busslinger and Tarakhovsky 2014]). Epigenetic information, on the other hand, is established in the gametes (Fig. 1A); the information stored in the chromatin template undergoes major changes during development and differentiation. DNA methylation, histone modifications, histone variants, and nonhistone chromatin proteins, as well as noncoding RNAs and higher-order chromatin structure encode this information. The key feature of epigenetic marks is that they are usually heritable from one cell generation to the next, and they can regulate gene expression. Epigenetic information is thus thought to be of critical importance for the determination and maintenance of defined and stable gene expression programs that underlie cell fate decisions during development. In totipotent and pluripotent cells, it is imagined that epigenetic marks are less stable and more plastic to allow for these cells to differentiate into any number of cell types. As development proceeds and the potency of cells becomes more and more restricted, epigenetic marks become more rigid and restrictive. Totipotent and pluripotent cells such as germ cells or ES cells (see Fig. 1B) also have the unique property of being able to reprogram the genome and erase existing epigenetic marks, a characteristic that underlies their developmental plasticity.

The interdependence of developmental decisions and epigenetic gene regulation sets up a continuum of genetic and epigenetic events in mammals. This is because developmental events that are, for example, a result of intercell signaling give rise to a specific program of gene expression that can be epigenetically fixed. Such developmental events can also set up new epigenetic events (e.g., the methylation or demethylation of imprinted genes in germ cells). In turn, the setting or erasure of epigenetic marks can determine new gene expression programs and, hence, influence the way individual cells respond to developmental cues. The resulting developmental and epigenetic continuum is particularly fascinating when it includes the germline, as this extends to the next generation and possibly beyond that into the future.

Does life therefore really begin at fertilization? It is true that, genetically, a new individual can be identified from the time of fertilization of an egg by sperm? This is when a haploid set of chromosomes from the mother comes together with a haploid set of chromosomes from the father, and the diploid genome of the offspring is formed. But, epigenetic information that will be transmitted to the offspring is also present in the gametes (Fig. 1A). For example, imprinted genes carry DNA methylation marks that differ between the male and female germ cells and these preexisting patterns are inherited by offspring (see Ch. 26 [Barlow and Bartolomei 2014]). These epigenetic marks are introduced into the germ cell genomes of a parent as early

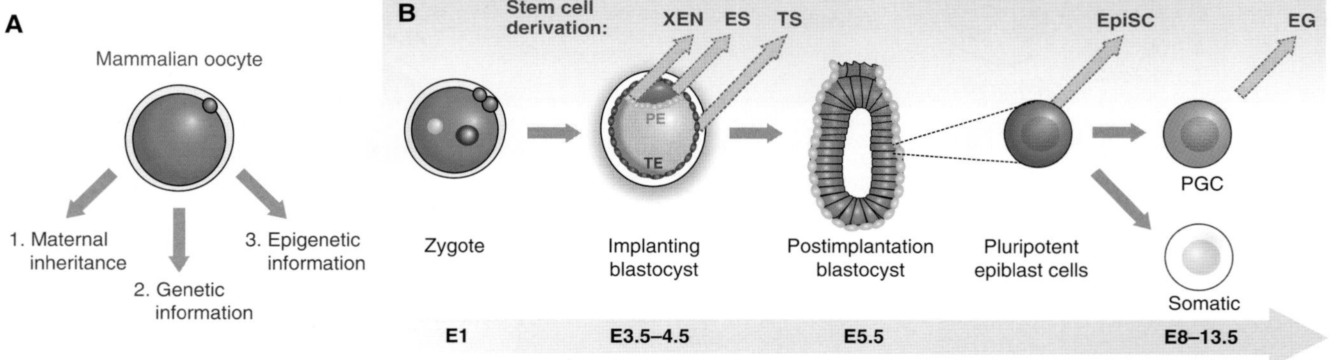

Figure 1. Potential of the mammalian oocyte, zygote, and blastocyst. (*A*) The mammalian oocyte contains maternal RNAs and proteins (maternal inheritance), which can determine early developmental events, genetic information (maternal chromosomes), and epigenetic information (DNA methylation and chromatin marks). (*B*) The zygote gives rise to the blastocyst with its inner cell mass (ICM) cells (blue) giving rise to ES cells in culture. The epiblast derivative of the ICM in the postimplantation blastocyst gives rise to all somatic cells and PGCs. A range of pluripotent stem cells (*top* line) can be derived from the various cell types isolated from early- and late-stage blastocysts and later primitive streak embryos. Types of stem cell include XEN, extraembryonic endoderm; ES, embryonic stem; TS, trophoblast stem; EpiSC, epiblast stem cell; EG, embryonic germ.

as during fetal or early postnatal development. Increasing data are showing that epigenetic marks may also occasionally be transmitted from one generation to the next resulting in transgenerational epigenetic inheritance (discussed in Sec. 2.4 of Ch. 14 [Blewitt and Whitelaw 2013]).

The first lineage allocation decisions occur before implantation, during the morula to blastocyst transition, producing the inner cell mass (ICM), which gives rise to the whole of the adult organism (and pluripotent ES cells) and an outer layer of trophectoderm (TE) cells (Fig. 2). These cell-fate decisions precede specification of PGCs, which occurs after blastocyst implantation. TE cells facilitate implantation and ultimately generate the placenta, whereas the ICM cells differentiate into epiblast and primitive endoderm (PE) cells (this later blastocyst-stage implanting is illustrated in Fig. 1B).

The relative contributions of epigenetic modifications, egg cytoplasmic factors (represented as maternal inheritance in Fig. 1A), and cell–cell interactions in directing genetically driven development are being investigated. This is complementing research into elucidating the details of how different genetic programs first arise in the early embryo. We know that control of development is initially governed by maternally inherited factors. It passes to the embryo itself following the destruction of maternal factors and the activation of embryonic genome, usually at the two- to four-cell stage, when these blastomere cells acquire the totipotent state from which the ICM and TE lineages emerge. We also know that epigenetic information is carried across from the gametes to the embryo at imprinted and perhaps other genes. Other dramatic epigenetic events occur around the time of fertilization: The sperm genome (i.e., the paternal pronucleus) loses DNA methylation rapidly in the fertilized zygote, and then regains DNA and histone modifications over the subsequent cell divisions (Fig. 2). Meanwhile, the maternal genome resists zygotic DNA demethylation, but then becomes demethylated in a more protracted fashion during cleavage

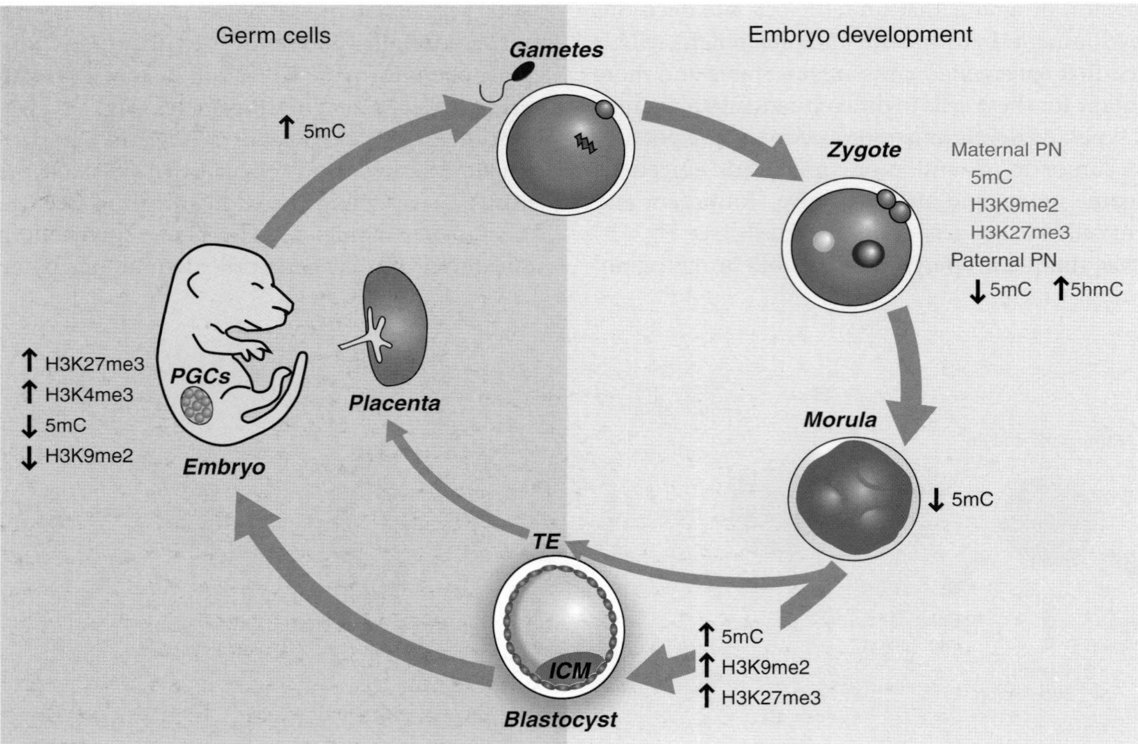

Figure 2. The epigenetic reprogramming cycle in mammalian development. Immediately after fertilization in the zygote, the paternal pronucleus (PN) is packaged with histones that lack H3K9me2 and H3K27me3, whereas the maternal chromatin contains these marks. The paternal PN also rapidly loses 5-methylcytosine (5mC) on a genome-wide scale, whereas the maternal does not. Passive loss of 5mC occurs during preimplantation development until the blastocyst stage when the ICM cells begin to acquire high levels of 5mC, H3K9me2, and H3K27me3. The placenta, which is largely derived from the TE of the blastocyst, remains relatively hypomethylated. PGCs undergo demethylation of 5mC and H3K9me2 progressively as they migrate into the gonads. De novo DNA methylation, including parent-specific imprinting, takes place during gametogenesis.

Cite as *Cold Spring Harb Perspect Biol* doi: 10.1101/cshperspect.a019422

divisions of the early embryo discussed later in Section 3.2 (illustrated in Fig. 3 of Ch. 15 [Li and Zhang 2014]). These reprogramming events may be important to achieve totipotency of the zygote and pluripotency of ICM cells. Yet, these somewhat opposing and distinct epigenetic programs lead to an overall loss of gametic epigenetic information, and it is likely that these dynamic reprogramming events interact with the cellular and genetic processes that determine the earliest processes of cell allocation into ICM and TE lineages.

Epigenetic regulation at the later blastocyst stage differs considerably between the extraembryonic (TE) and embryonic lineages (ICM). For example, the overall levels of DNA methylation are lower in the extraembryonic tissues, and maintenance of imprinting and imprinted X inactivation can be different. Ultimately, however, ICM and TE cells, like the later differentiated PGCs, are largely determined by a genetic program involving transcription factors and, where appropriate, pluripotency genes. Yet, some of these transcription factor genes appear to be epigenetically regulated, which then contributes to the maintenance of cell-fate decisions.

After implantation of blastocysts, the postimplantation epiblast cells start to undergo modifications of DNA and chromatin. These are the cells that give rise to both germ cells and those that will differentiate into diverse cell types of the body. The earliest signs for the onset of germ cell fate are seen among a small group of cells formed in the early postimplantation embryo at E6.25, in response to receiving signaling molecules that come from other parts of the conceptus, primarily the adjacent extraembryonic lineages, which include the cells destined to form the placenta. Germ cell specification entails repression of somatic cell gene products while they begin to show expression of the unique germ-cell-specific genes. This genetically governed function is also responsible for initiating epigenetic reprogramming in early germ cells, leading to the erasure and reestablishment of imprinting, chromosome recombination during meiosis, and reduction divisions to form haploid gametes. These various stages of early development thus far described clearly illustrate that the epigenetic modifications occur progressively and accompany changes in genetic programs.

2 MECHANISMS REGULATING GERM CELL SPECIFICATION

The specification of germ cells in animals is one of the earliest events during development, segregating them from somatic cells (Surani et al. 2004). Germ cells eventually generate the totipotent state. This section focuses on the processes of mouse-germline specification and their subsequent maturation as they migrate into the developing gonads.

The events and mechanisms known to direct the process of germline specification in mice occur in postimplantation epiblast cells that emerge from ICM cells after the blastocyst (which contains just three cell types: ICM, TE, and PE) has implanted. PGCs after specification migrate into the developing gonads and come to reside in their final destination, the male or female gonads (illustrated schematically on the left side of Fig. 2). A description of the events and mechanisms regulating early embryogenesis, from early cleavage and morula stages through to the early blastocyst, are detailed in Section 3 (corresponding to stages of the life cycle shown on the right side of Fig. 2).

2.1 Principles of Germline Development in Different Animal Groups

There are two key modes by which a germ cell lineage can be established; these are referred to as the preformation mode (this is distinct from the old usage of the word as in preformationism) and the epigenesis mode (Extavour and Akam 2003). The first involves the maternal inheritance of preformed germ cell determinants by specific cells, as occurs in *Caenorhabditis elegans* and *Drosophila melanogaster* (Fig. 3) (Leatherman and Jongens 2003; Blackwell 2004). In contrast, the epigenesis mode of germ cell specification is a process in which a group of potentially equivalent pluripotent cells acquire a germ cell fate in response to inductive signals, whereas the remaining cells acquire the somatic fate (Lawson and Hage 1994; McLaren 2003). This mechanism for germ cell specification operates in mice and probably in other mammals, but also in some other vertebrates such as *Axolotls*.

2.2 Early Germline Development in Mammals

PGCs in mice are first detected at E7.5 (the early bud stage), as a cluster of approximately 30 to 40 cells that constitute the founder population of the germ cell lineage (Lawson and Hage 1994; McLaren 2003). They are positive for alkaline phosphatases and located within the extraembryonic mesoderm at the base of the allantois (see Fig. 4). Clonal analysis reveals that the proximal epiblast cells located adjacent to the extraembryonic ectoderm at E6.0–E6.5 (prestreak and early-streak-stage embryos) give rise to both PGCs and tissues of the extraembryonic mesoderm (Lawson and Hage 1994). Before PGC specification, however, the epiblast cells acquire a state of competence in response to signals that include Wnt signaling (Ohinata et al. 2009). Consequently, proximal epiblast cells form PGCs in response to signaling molecules produced by surrounding

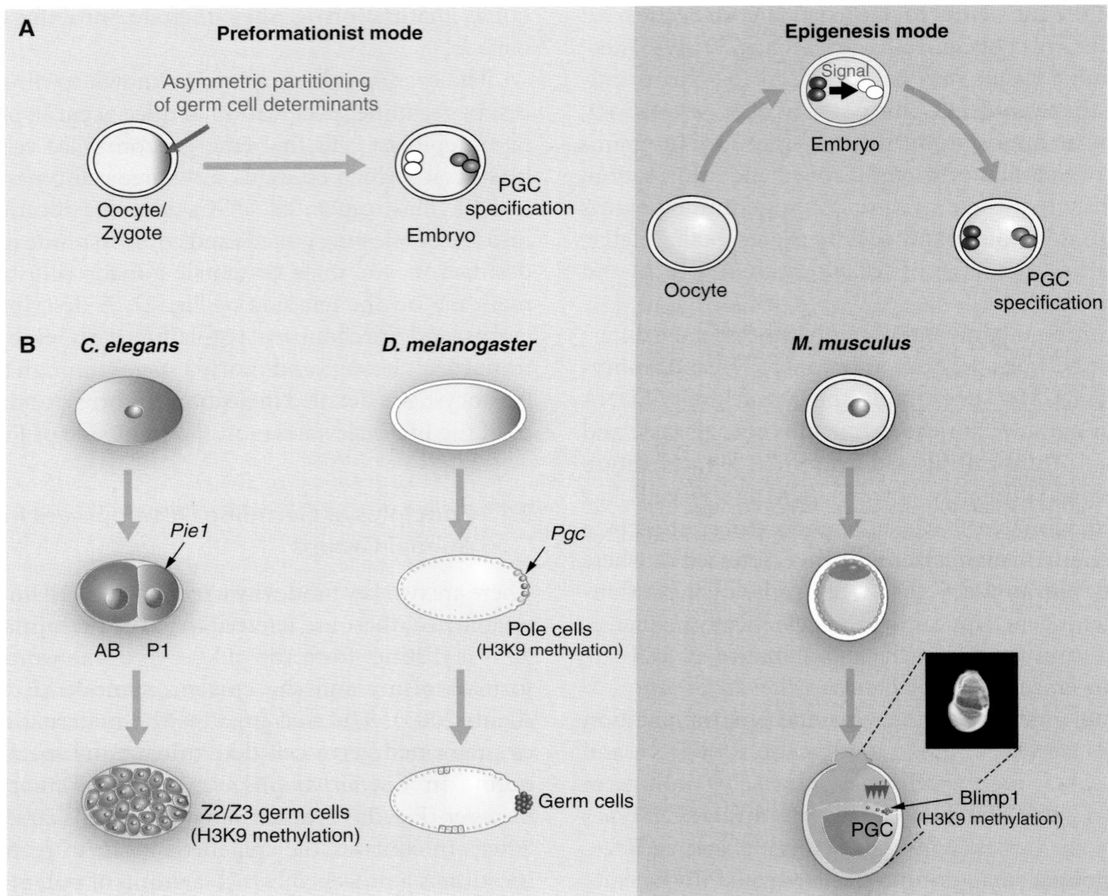

Figure 3. Early germ cell determination in the mouse. (*A*) Two models summarize the mode by which germ cells are determined in various organisms. The preformationist mode assumes one or more localized determinants in the oocyte or early embryo specify progeny cells becoming PGCs. In the epigenesis mode, a signal emanating from a neighboring cell(s) in the early embryo determines the future PGCs. (*B*) This part of the figure highlights the features that contribute to the repression of somatic gene programs during germline specification in various organisms. In *Caenorhabditis elegans*, the germline lineage (red) is specified after the first division of the zygote by expression of *Pie1*, which confers transcriptional quiescence. The other cell (blue) gives rise to somatic tissues. In *Drosophila melanogaster*, the precursors of the germ cells are the so-called pole cells contained on one side of the zygote syncytium (i.e., multinucleated); transcriptional quiescence in these cells depends on localized RNA from the gene *Pgc* and high levels of H3K9 methylation. In *Mus musculus*, the earliest precursors of the germ cells are visible by expression of Blimp1 at the base of the allantois. Blimp1 initiates transcriptional quiescence in these cells.

extraembryonic ectoderm and primary endoderm. Among the most important signals for PGC specification itself are Bmp8b and Bmp4, which are expressed most strongly from the posterior extraembryonic ectoderm, contributing to PGC specification (Fig. 4) (Lawson et al. 1999).

To gain detailed insights into the genetic program of PGC specification, single-cell copy DNAs were generated from the founder PGCs and their neighboring somatic cells (Saitou et al. 2002). A variety of markers were used to distinguish between PGCs and somatic cells. This screen initially identified *fragilis*, a novel member of the interferon-inducible transmembrane protein family implicated in cell aggregation, and *stella* (also known as *PGC7*), a nucleocytoplasmic protein. Further investigations showed that *fragilis* is expressed in the proximal epiblast cells at E6.0 (Fig. 4) when they gain competence to give rise to both PGCs and the neighboring extraembryonic mesoderm cells. The *fragilis*-positive cells move to the posterior proximal region during gastrulation, although the precise role of *fragilis* at this time is unknown. The founder population of PGCs is subsequently detected among this *fragilis*-positive population of cells by expression of *stella*. At the same time, founder PGCs show expression of pluripotency genes including *Sox2*, *Oct4*, and *Nanog* (see Sec. 4.2 for a more

 Cite as *Cold Spring Harb Perspect Biol* doi: 10.1101/cshperspect.a019422

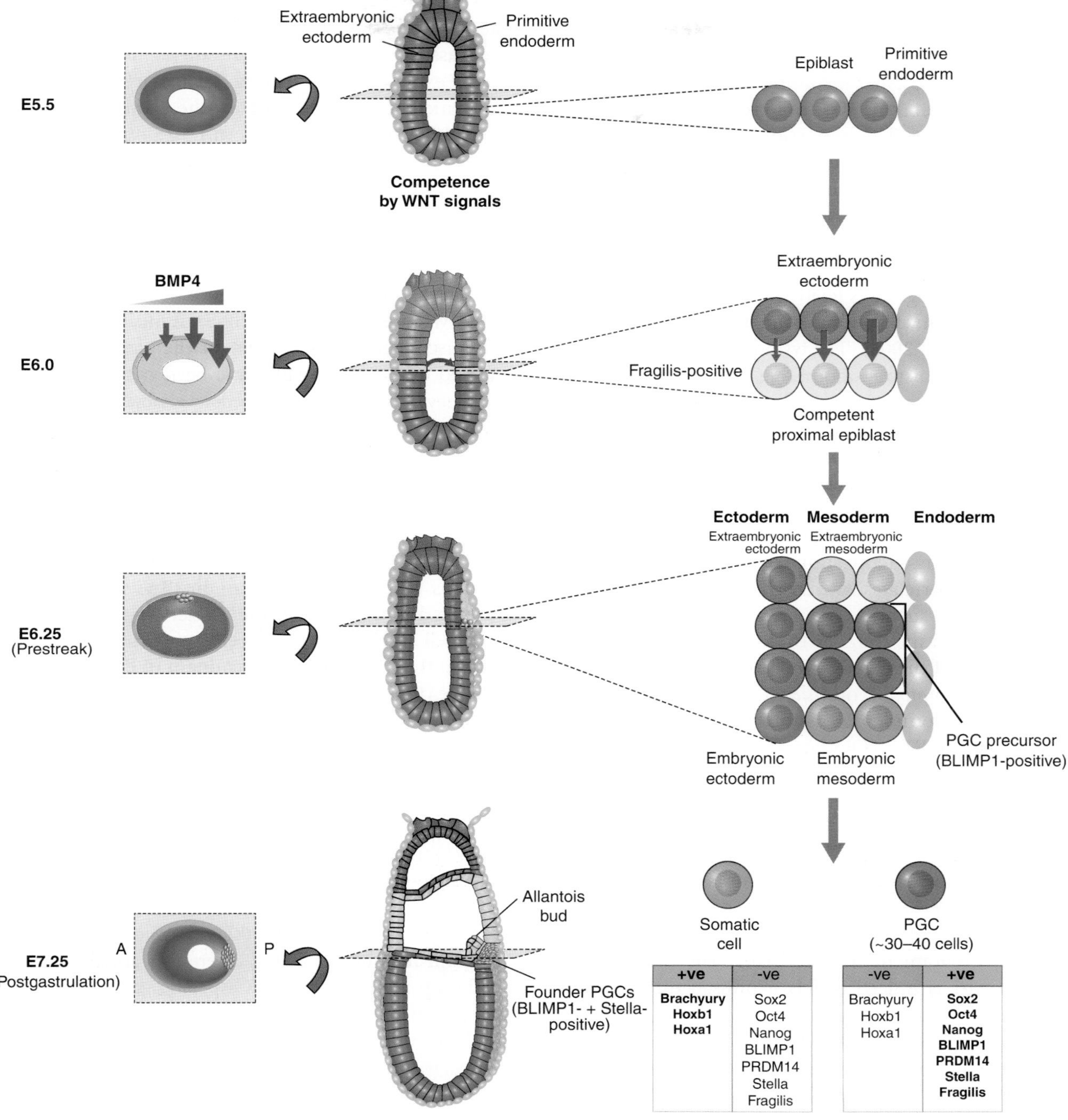

+ve	-ve
Brachyury **Hoxb1** **Hoxa1**	Sox2 Oct4 Nanog BLIMP1 PRDM14 Stella Fragilis

-ve	+ve
Brachyury Hoxb1 Hoxa1	**Sox2** **Oct4** **Nanog** **BLIMP1** **PRDM14** **Stella** **Fragilis**

Figure 4. Germline development in mice. Postimplantation proximal competent epiblast cells at E6 (pink) respond to BMP4 signal from the extraembryonic tissues (pink), which activates BLIMP1. Expression of BLIMP1 marks the onset of commitment to the primordial germ cell fate (red), whereas other cells become somatic cells.

detailed description), suggesting that PGCs reacquire an underlying pluripotency, which is lost in the neighboring somatic cells (Fig. 4). In contrast, the founder PGCs show repression of some genes including *Brachyury*, *Hoxb1*, and *Hoxa1*, which are, at this time, significantly up-regulated in somatic neighbors. The repression of *Hox* genes among others is part of an important mechanism that underlies repression of the somatic cell fate in founder PGCs (elaborated in Sec. 2.4; Saitou et al. 2002).

Based on the analysis of the emergence of founder PGCs, it is evident that, as in other organisms, repression of the somatic program is likely to be a key feature of PGC specification in mice (Seydoux and Strome 1999; Blackwell 2004; Surani et al. 2004). The lysine (K) methyltransferase (KMT) class of histone-modifying enzymes or "writers" (for more detail, see Ch. 3 [Allis et al.]; Ch. 6 [Cheng 2014]) were analyzed for differential expression between PGCs and neighboring somatic cells. KMTs modify histone lysine residues through the addition of one to three methyl groups. Methylation of histones H3 and H4 at positions H3K9 and H4K20 is generally associated with repressive chromatin regions. Methylation at H3K4 and H3K36, however, generally correlates with active transcription. Some of the KMT genes, such as *G9a*, *Pfm1*, *Set1*, and *Ezh2*, were detected both in the founder PGCs and the somatic cells. However, one of these genes, *Blimp1* (*B lymphocyte maturation-induced protein-1* or *prdm1*), showed expression exclusively in the founder PGCs and not in the neighboring somatic cells at E7.5 (Ohinata et al. 2005). *Blimp1* is a known transcriptional repressor with a SET/PR domain (a domain that typically acts as a methyltransferase), a proline-rich region that can recruit Groucho and HDAC2, five C2H2 zinc fingers that can form a complex with Prmt5, and an acidic tail (Gyory et al. 2004; Sciammas and Davis 2004, Ancelin et al. 2006). *Blimp1* was first identified for its role during specification of plasma cells following repression of the B-cell program in the precursor cells (Turner et al. 1994). *Blimp1* is indeed widely expressed during mouse development.

Detailed analysis of *Blimp1* in early mouse embryos led to some unexpected findings. Among them was the discovery that *Blimp1* expression commences in the proximal epiblast cells at E6.25 at the onset of gastrulation, initially in only four to six cells that are in direct contact with the extraembryonic ectoderm cells (Fig. 4) (Ohinata et al. 2005, 2009). *Blimp1* expression is detected at one end of the short anterior–posterior axis in a region that is destined to form the posterior proximal region. The number of *Blimp1*-positive cells increase progressively so that there are approximately 20 cells at the midstreak stage that are seen to form a tight cluster in the posterior proximal region at E6.75. At the E7.5 early bud stage, the number of *Blimp1*-positive cells increases to approximately 40. These cells constitute the founder population of PGCs, and show expression of the classical alkaline phosphatase PGC marker and commence expression of *stella* (Fig. 4). A genetic lineage tracing experiment confirmed that all the *Blimp1*-positive cells originating in the epiblast from E6.25 onward are indeed lineage restricted PGC precursor cells. These data contrast with the previous hypothesis, based on clonal analysis, which suggested that the proximal epiblast cells at E6.0–E6.5 are not lineage restricted to give rise exclusively to PGCs because clonal descendants of individual cells could give rise to both a somatic and germ cell (Lawson and Hage 1994; McLaren and Lawson 2005). A likely explanation for this discrepancy may be that in the clonal analysis, the marked cells may initially have been negative for *Blimp1* and they subsequently divided to generate a positive cell that gave rise to PGCs, whereas the daughter cell produced a somatic descendant. The mechanism that regulates the accretion of *Blimp1*-positive cells is currently unknown.

2.3 The Role of *Blimp1* in Specification of PGCs

Analysis of the role of *Blimp1* in PGC specification has generated insights into the underlying mechanism of germ cell specification in mice. Loss of function of *Blimp1* showed that this is a key determinant of PGC specification in mice (Ohinata et al. 2005; Vincent et al. 2005). At E7.5, *Blimp1* mutant embryos contain an aberrant cluster of approximately 20 PGC-like cells, unlike control embryos in which the PGCs continue to proliferate and commence migration out of the cluster. Furthermore, the number of aberrant PGC-like cells fails to increase when examined at E8.5 (Ohinata et al. 2005).

Single-cell analysis of mutant PGC cells revealed a lack of consistent repression of *Hox* genes. Therefore, it is likely that *Blimp1* has a role in the repression of the somatic program in founder PGCs. There was also inconsistency in the up-regulation of PGC-specific genes such as *stella* and *Nanos3*, and some pluripotency-specific genes such as *Sox2* in mutants. These findings stress that *Blimp1* has a critical role as a transcriptional regulator during PGC specification and in the prevention of these cells from acquiring a somatic cell fate.

Studies on B cells have revealed that *Blimp1* is necessary to induce differentiation into plasma cells through repression of key molecules that maintain B-cell identity (see Sec. 3.2 of Ch. 29 [Busslinger and Tarakhovsky 2014] for details; Turner et al. 1994; Sciammas and Davis 2004). It does this through the formation of a Groucho and HDAC2 repressor complex (Ren et al. 1999). Its zinc fingers also seem important for the formation of a complex with G9a (Gyory et al. 2004), a histone KMT that is required for

Cite as *Cold Spring Harb Perspect Biol* doi: 10.1101/cshperspect.a019422

H3K9me2. *Blimp1*, which itself contains a SET/PR domain that typically acts as a methyltransferase, has no known KMT activity, and how it functions during PGC specification is unknown.

Blimp1 is an evolutionarily conserved gene in both vertebrates and invertebrates and it has a variety of functions. For example, it has a role in the development of several lineages in vertebrates such as the zebrafish and *Xenopus* (de Souza et al. 1999; Roy and Ng 2004; Hernandez-Lagunas et al. 2005), although not specifically in germ cell specification. This implies that the gene has acquired a new role in PGC specification in mice and perhaps in all mammals. For this highly conserved gene, it suggests that additional control elements must have evolved to drive its expression in PGC precursors and founder cells.

Analysis and comparison of *Blimp1* mutant versus control PGC cells led to the discovery of a second key regulator of PGC specification called *Prdm14*, which like *Blimp1* is another PR/SET domain family member (Yamaji et al. 2008). *Blimp1*, together with Prdm14, is involved in the repression of the somatic program and onset of epigenetic reprogramming in PGCs, leading to global DNA demethylation described in Section 2.5.

2.4 Repression of the Somatic Program in Germ Cells: An Evolutionarily Conserved Phenomenon

The mechanism of germ cell specification is not evolutionarily conserved, which is evident when comparing the mechanism in mice with the events in two other well-studied model organisms, *D. melanogaster* and *C. elegans*, as discussed previously in Section 2.1 (Seydoux and Strome 1999). The differences in the mechanism of germ cell specification are primarily accounted for by the differences in the mode of early development in these different organisms. The additional complexities imposed by the phenomenon of genomic imprinting in mammals contribute to these differences. Importantly, however, the repression of the somatic gene expression program during the specification of germ cells is a shared phenomenon in diverse organisms although the molecular mechanisms may differ (Seydoux and Strome 1999; Leatherman and Jongens 2003; Saitou et al. 2003; Blackwell 2004).

In *C. elegans*, the first cell division of the zygote is asymmetric: it establishes a somatic cell (AB), whereas the second cell (P1) is set aside to establish the germ cell lineage (Fig. 3B). Indeed, each of the P1, P2, and P3 cells produce a somatic cell when they divide, and the latter commence transcription and differentiation. The P1–P3 cells destined for the germ cell lineage remain transcriptionally quiescent. Transcriptional quiescence is maintained by a zinc finger protein called PIE-1. This protein competes with the carboxy-terminal domain (CTD) of RNA polymerase II (Pol II) for the phosphorylating protein CDK9. When CDK9 is not available to phosphorylate Ser-2 of the RNA Pol II CTD, RNA Pol II cannot transition into the active holoenzyme necessary for transcriptional elongation (for more detail, see Seydoux and Strome 1999; Zhang et al. 2003; Ch. 3 [Allis et al.]). However, both the somatic and germ cell blastomeres show transcriptionally permissive chromatin states as seen by high levels of genome-wide H3K4 methylation. Later, when the P4 blastomere divides to form two germline cells, Z2 and Z3, a repressive chromatin state becomes evident, with loss of H3K4 methylation and acquisition of high levels of repressive H3K9 methylation (Schaner et al. 2003). Thus, during the establishment of the germ cell lineage in *C. elegans*, despite transcriptional quiescence, the chromatin starts off in a transcriptionally permissive state and changes to an inactive state.

The establishment of the germ cell lineage in *D. melanogaster* is again distinct from what is observed in the mouse and worm. The germline precursors, called the pole cells, are detected before the onset of embryonic development in the fertilized syncytial (multinucleated) egg (Fig. 3B), and these are again transcriptionally quiescent because of a lack of RNA Pol II CTD phosphorylation, as observed in *C. elegans* (Seydoux and Dunn 1997; Van Doren et al. 1998; Schaner et al. 2003). Although this transcriptional silencing is also associated with the repressive chromatin modification H3K9me, pole cells are destined to form only germ cells. These cells are thus equivalent to the Z2 and Z3 cells that divide from the parent P4 cell in *C. elegans* that are destined for the germ cell lineage only. Furthermore, regulation of transcriptional silencing in pole cells is dependent on the *polar granule component* (*pgc*) gene because loss of *pgc* causes a loss of gene repression although the pole cells are still detected. The mutant pole cells show RNA Pol II CTD phosphorylation of Ser 2, indicative of a transcriptionally active elongating RNA Pol II complex (Deshpande et al. 2004; Martinho et al. 2004). This has led to the suggestion that *pgc* might sequester critical components needed for the phosphorylation of RNA Pol II CTD at Ser 2, preventing the transition of RNA Pol II from the preinitiation complex to an active elongation complex.

As discussed in Section 2.3, *Blimp1* not only is a key marker of PGCs, but is key in repressing somatic gene programs. The analysis of specification of germ cells during development in mice, flies, and worms clearly illustrates the fact that transcriptional repression that is presumably essential to repress the somatic cell fate is found in all three organisms, although the precise mechanisms by which this is achieved differ markedly. This is evidently because of the differences in events associated with early development of the different species.

2.5 Regulation of Epigenetic Programming after PGC Specification in Mice

Extensive epigenetic programming and reprogramming occurs in the germ cell lineage following the specification of PGCs (Hajkova et al. 2002; Lee et al. 2002; Seki et al. 2005, 2007; Hajkova et al. 2008, 2010; Popp et al. 2010; Guibert et al. 2012; Hackett et al. 2012, 2013; Kobayashi et al. 2012; Seisenberger et al. 2012; Vincent et al. 2013; Yamaguchi et al. 2013). This period of development is marked by the erasure of some of the repressive epigenetic modifications (Fig. 2), allowing the germ cell lineage to acquire an underlying pluripotent characteristic, which may be a prerequisite for subsequent totipotency.

Among the key changes observed is the erasure of the repressive histone H3K9me2 mark at E8.0, together with a decrease in the levels of HP1α within the euchromatic and pericentric heterochromatic regions by E9.0 (Seki et al. 2005). At the same time, there is also a decline in the overall levels of DNA methylation in PGCs from E8.0 onward. Simultaneous to the decline in H3K9me2 and DNA methylation is the progressive increase in H3K27me3, a repressive modification mediated by the polycomb group protein, Ezh2 (Fig. 5). The loss of DNA methylation is due to the repression of de novo DNA methyltransferases Dnmt3a and Dnmt3b, and a decline in the Np95 protein (also called UHRF1) that targets Dnmt1 (the maintenance DNA methyltransferase) to the replication fork (discussed in Kurimoto et al. 2008; Seisenberger et al. 2012; Kagiwada et al. 2013; Ch. 6 [Cheng 2014]). It is noteworthy that the loss of H3K9me2 and DNA methylation also coincides with the reexpression of a key pluripotency-associated gene, *Nanog* (Yamaguchi et al. 2005). *Nanog* is first expressed in the inner cells of the late morula and ICM cells of blastocysts. However, expression of this gene is down-regulated after implantation, and the gene is only the reexpressed in PGCs and not in the neighboring somatic cells. The collective expression of pluripotency genes including *Nanog*, *Oct4*, *Sox2*, and *Esg1* shows that germ cells acquire characteristics of pluripotency (Fig. 5). One factor that is apparently im-

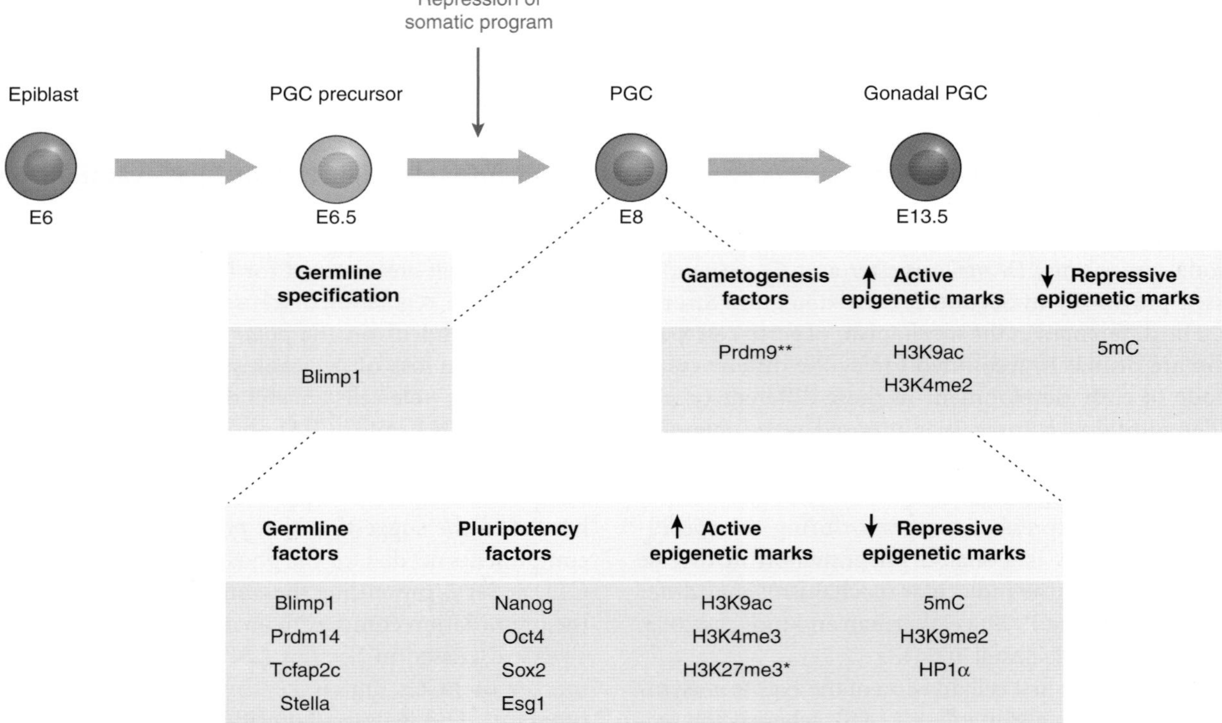

Figure 5. Early epigenetic events during germ cell specification. Expression of Blimp1, Prdm14, and Tcfap2c in descendants of epiblast cells leads to repression of the somatic gene expression program and initiation of the germ cell program (red). This is followed by expression of Stella, Nanog, and Esg1, increase in the H3K4me3 and H3K9ac active marks, as well as the repressive H3K27me3 mark (*), and loss of H3K9me2 and 5mC. The PGCs start to show loss of DNA methylation as they migrate to the developing gonads, with comprehensive loss of DNA methylation and the erasure of imprints occurring shortly after they enter the gonads. PRDM9 is crucial in the later process of gametogenesis, marking the transition from PGCs to gametes. This occurs with the onset of meiosis at E13.5 in females. **In males gametogenesis occurs postnatally.

portant for expression of pluripotency genes in early PGCs is the H3K27me3 demethylase, Utx, likely counteracting the repressive effects of the overall increase in H3K27me3 (Mansour et al. 2012). The expression of the pluripotency network in PGCs is extensive and comparable to ES cells (Seisenberger et al. 2012).

Additional extensive epigenetic programming events ensue when PGCs enter into the developing gonads (Surani et al. 2004). First, there are increases in H3K4me2 methylation and H3K9 acetylation, which are characteristic of permissive chromatin states, excluding H3K9 methylation. In addition, there is very extensive genome-wide DNA demethylation (Fig. 5) that includes the erasure of parental imprints and methylation in single-copy genes. In female embryos, the inactive X chromosome is also reactivated at this time. Genome-wide demethylation is likely to occur by a combination of active and passive mechanisms, including down-regulation of Np95 (discussed in Kurimoto et al. 2008; Ch. 6 [Cheng 2014]), the activation-induced deaminase (AID) and thymine-DNA glycosylase (Popp et al. 2010; Cortellino et al. 2011), and TET1 and TET2, which are potentially coupled to the base excision repair pathway (see Fig. 6 of Ch. 15 [Li and Zhang 2014] and Feng et al. 2010; Hajkova et al. 2010; Hackett et al. 2012; Saitou et al. 2012; Dawlaty et al. 2013; Hackett and Surani 2013; Vincent et al. 2013; Yamaguchi et al. 2013).

Not all epigenetic marks are completely removed during germ cell development, despite effective mechanisms operating to erase "acquired" epigenetic modifications. For example, DNA methylation of the intracisternal A particle (IAP) retrotransposon family and some 200 or so other genomic locations is only partially reprogrammed (Lane et al. 2003; Popp et al. 2010; Guibert et al. 2012; Seisenberger et al. 2012; Hackett and Surani 2013). When some epigenetic marks are incompletely removed during gametogenesis, this can apparently lead to epigenetic inheritance through the germline, of which there are a number of examples now in mammals (e.g., see Fig. 6 in Ch. 14 [Blewitt and Whitelaw 2013] and Chong and Whitelaw 2004), and this could potentially explain the transgenerational epigenetic inheritance of some metabolic phenotypes (Ferguson-Smith and Patti 2011). How widespread this phenomenon is and how many gene loci it involves needs to be established.

2.6 The Germline and Stem Cells: A Reversible Phenotype

Pluripotent stem cells can be derived from ICM and germ cells (Fig. 1B). More specifically, pluripotent EG cells can be made from PGCs when cultured in the presence of FGF2 (Matsui et al. 1992; Resnick et al. 1992). They are, in many respects, similar to pluripotent ES cells (discussed more extensively in Sec. 4) except that EG cells may show erasure of parental imprints during their derivation (Tada et al. 1998; Leitch et al. 2010). Recent studies have shown that EG cells can also be derived from rat PGCs (Leitch et al. 2010).

As PGCs show some characteristics of pluripotency while retaining unipotency in that they are only able to form sperm or eggs, it is probable that mechanisms exist to allow PGCs to retain their distinct lineage-specific characteristics. How this is achieved is as yet unclear, but it is possible that *Blimp1* may have a continuing role following the initial specification of PGCs. During the derivation of EG cells, it is assumed that unipotent restriction is relieved and they acquire an overtly pluripotent character with the ability to differentiate into many distinct cell types, which seldom occurs with germ cells in vivo. It is noteworthy that the derivation of EG cells becomes progressively less efficient, however, when later-stage PGCs from E11.5 and E12.5 are used. This further suggests a change in the characteristics of these cells from E11.5 when they begin their differentiation pathway toward definitive male and female germ cells.

2.6.1 *Development of Germ Cells from Pluripotent ES Cells*

A characteristic of pluripotent stem cells is their ability to differentiate into all types of somatic tissues when introduced into blastocysts, including germ cells (Fig. 6). Increasing efforts are being made to generate different tissues more efficiently from ES cells in culture. It is now possible to generate PGCs, and possibly sperm- and egg-like structures from ES cells in culture (Hubner et al. 2003; Toyooka et al. 2003; Geijsen et al. 2004; Hayashi et al. 2011;

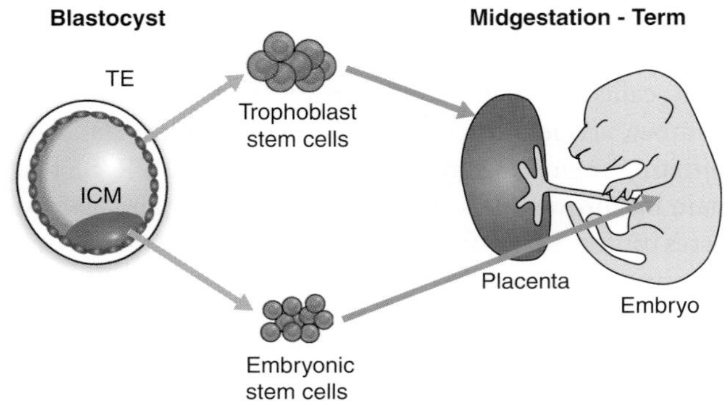

Figure 6. Differentiation of ES cells into different cell types in vitro. ES cells can be differentiated in vitro under suitable culture conditions into many different cell types such as neurons, muscle cells, and even germ cells (oocytes).

Vincent et al. 2011). This has opened up the possibility of studying the process of germ cell specification in vitro, with the hope of being able to determine the precise mechanisms involved, aided by an ever-increasing knowledge base of the genetic programs that govern PGC and gamete cell function. The ability to in vitro differentiate PGCs from ES cells may also provide a model system to examine the regulation of epigenetic reprogramming in this lineage. Such an approach could ultimately advance our understanding of the human germ cell lineage, which, for ethical reasons, has been hard to research to date. Furthermore, if it becomes possible to direct differentiation toward human oocytes from cultured ES cells, it may be possible to use them for "therapeutic" cloning, circumventing the need for donor oocytes that are difficult to obtain. These oocytes could then be used for somatic nuclear transplantation to generate blastocysts, and subsequently to derive ES cells from. This is because somatic nuclei undergo reprogramming to totipotency when transplanted into oocytes (elaborated in Ch. 28 [Hochedlinger and Jaenisch 2014]). This procedure is likely to complement the currently used direct derivation of iPS cells from somatic tissues.

The use of human embryos and ES cells in research and therapy does raise many ethical issues. A variety of guidelines and regulations exist in different countries to monitor research in this area. Within these ethical frameworks, it is hoped that the generation of viable gametes from ES cells, if possible, may lead to advances in reproductive medicine.

2.7 From PGCs to Gametes

The next stage in the development of the germ cell lineage is the initiation of gametogenesis and entry of germ cells into meiosis. The gonadal somatic environment regulates the timing of this event. In females, germ cells arrest in meiotic prophase whereas male germ cells enter into mitotic arrest. A number of environmental signals dictate whether germ cells enter meiosis or not. Recently, a novel gene, *Prdm9* (also called *Meisetz*), another PR/SET domain family member, was identified and shown to play a crucial role in initiating meiosis (Hayashi et al. 2005). The PR/SET domain has proven H3K4me3 catalytic activity and also possesses multiple zinc-fingers. Expression of *Prdm9* is specific to germ cells, and is detected at the time of entry into meiotic prophase in females at E13.5 and in postnatal testis. It is likely that Prdm9/PRDM9 in mouse and human germ cells may introduce epigenetic marks or "hot spots" that constitute narrow segments of the genome as sites for meiotic recombination. Mutation in *Prdm9* results in sterility in both males and females, demonstrating its essential role in germ cells. The mutant germ cells show marked deficiency in the DNA double-strand break repair pathway and pairing of homologous chromosomes during meiosis. These studies suggest a significant role for epigenetic mechanisms in germ cells during meiosis (Baudat et al. 2010; Parvanov et al. 2010).

Extensive chromatin modifications continue during spermatogenesis. Eventually, the somatic linker histones are replaced by testis-specific variants (Kimmins and Sassone-Corsi 2005), followed by the replacement of most histones with protamines. Studies have shown that Suv39, an H3K9 histone methyltransferase, is involved in gene repression and chromosome pairing. Two such SET-domain proteins, Suv39h1 and Suv39h2, have roles in male germ cells, the latter being expressed preferentially in the testis, and accumulating in the chromatin of the sex vesicle (i.e., XY chromosome pair). Mutations in both Suv39h1 and Suv39h2 result in infertility because of the arrest of spermatogenic cells (Peters et al. 2001). In addition, there is also a chromatoid body; this is a cloud-like cytoplasmic structure present in male germ cells. It is an RNA-processing body consisting of Dicer and Argonaute proteins and microRNAs, a germ-cell-specific cytoplasmic organelle that interacts with the nucleus and contains compacted messenger RNA.

Noncoding RNAs, the RNA interference (RNAi) machinery, and histone lysine methyltransferases are implicated in the process of germ cell renewal during spermatogenesis. The members of the Piwi/Argonaute (called *Miwi* in mice) family have been reported to play a role in RNAi pheomena. Loss of Miwi-like proteins (Mili) results in sterility in males (Kuramochi-Miyagawa et al. 2004), causing elevated expression of retrotransposon transcripts, IAP, and Line 1. The involvement of Miwi-like proteins in their repression has been directly demonstrated shown through the piRNA pathway (Siomi et al. 2011).

Interestingly, pluripotent stem cells from spermatogonial stem cells can even be recovered from adult mouse testis (Kanatsu-Shinohara et al. 2004; Guan et al. 2006). These cells may be maintained in culture indefinitely, but unlike ES cells, they have a paternal (androgenetic) imprint. Nonetheless, they can differentiate into a variety of somatic cell types in vitro and in vivo, and can contribute to the germline in vivo. These cells thus provide an important tool to study many aspects of spermatogenesis, including the role of epigenetic mechanisms in regulating their stemness and capability for differentiating into male gametes.

Erasure of imprints in early germ cells leads to epigenetically equivalent parental chromosomes for the first and only time in the life of mammals. Transplantation of such "imprint-free" nuclei directly into oocytes leads to the development of embryos that are aberrant and die at early embryonic stages. This is presumably because without the appropriate epigenetic modifications, there is misexpression of genes that normally undergo imprinting. The experiment also shows that imprints cannot be acquired by imprint-free nuclei if transplanted directly to the oocyte. The initiation of DNA methylation imprints in females begins after birth, during the growth of oocytes. In male germ cells, it occurs at later fetal stages. The de novo methyltransferase Dnmt3a and its cofactor Dnmt3L play critical roles in this process (for details, see Ch. 26 [Barlow and Bartolomei 2014]). Imprinting in mammals is a major barrier to parthenogenetic development. Attempts to manipulate the epigenotype of female gametes have now made it possible, however, to allow the development of mammalian embryos that are of maternal origin only (Kawahara et al. 2007).

3 FROM THE OOCYTE TO THE EARLY EMBRYO

We have seen in Section 1 how the cycle of germline specification involves epigenetic mechanisms, and how mature sperm and oocytes acquire very specific and different epigenetic marks during gametogenesis. Some of these differences, such as parental imprints, are maintained faithfully in the embryo after fertilization (illustrated in Fig. 7 of Ch. 26 [Barlow and Bartolomei 2014]). Many others, as we will see, become dramatically reprogrammed as the embryonic genome attains totipotency. It is important to ask: To what extent do epigenetic marks inherited from the gametes play a role in the earliest differentiation events in the embryo (focusing on the right side of Fig. 2)? If you start life with one cell (the fertilized zygote) containing a complete genome that then divides, how do you ever get differentiation of gene expression and appropriate developmental programs in daughter cells? It is presumably a combination of transcription factors, epigenetic regulators, and external signaling influences that contribute to the earliest lineage decisions in mammalian development (Hemberger et al. 2009).

3.1 Maternal Inheritance and Potential Asymmetric Distribution?

In organisms with a large egg cell (such as *Drosophila*, *Xenopus*, or chicken), some maternally made proteins or RNAs are located asymmetrically in the egg (Fig. 3). They are then only inherited by some of the descendant cells, which subsequently develop a particular fate, whereas others that do not inherit these determinants develop differently (Huynh and St Johnston 2004). Such a strategy is possible with relatively large eggs (e.g., *Drosophila*), but becomes more difficult with smaller mammalian eggs. However, the developmental program may not be dictated simply by the size of the egg but, more importantly, by the necessity to generate a blastocyst in mammals that has to implant and generate a placenta to sustain the embryo. The ICM can thus be considered as developmentally equivalent to a *Drosophila* oocyte in the sense that it undergoes patterning in response to signals from the extraembryonic somatic tissues during early development, akin to the *Drosophila* syncytial cells responding to differentially distributed maternal-effect gene products. There have been some suggestions that link the symmetry of the fertilized zygote to the symmetry of the blastocyst and even the post-implantation embryo (Gardner 1997; Weber et al. 1999), but there is no unequivocal evidence for asymmetrically localized determinants of differentiation in mammalian eggs. Furthermore, mammalian embryos show a remarkable ability to "regulate" development; that is, when cells are removed or perturbed, compensatory growth or cell movements will often be able to keep the embryo developing normally (Kelly 1977). Nevertheless, there may be slightly different propensities of individual cells (blastomeres) to develop along the ICM and TE lineages as early as the four-cell stage (Fujimori et al. 2003; Piotrowska-Nitsche et al. 2005), which could potentially be brought about by epigenetic factors.

3.2 Epigenetic Events at Fertilization

During development and differentiation, somatic cell lineages acquire very specific and specialized DNA methylation and histone modification patterns. These patterns are apparently difficult to erase or reverse when a somatic nucleus is transferred to an oocyte (elaborated in Chan et al. 2012; Ch. 28 [Hochedlinger and Jaenisch 2014]). The epigenetic marks of the oocyte and the sperm are specialized too, but these are reprogrammed efficiently at fertilization so that the embryonic genome can take up its new function, namely, to become totipotent (Reik et al. 2001; Surani 2001). A number of features of the epigenetic makeup of the gametes, and the details of the epigenetic reprogramming that occurs after fertilization are now known (Fig. 2). For instance, both oocyte and sperm genomes have considerable levels of DNA methylation; as an example, the retrotransposon family, IAP, which has a copy number of approximately 1000 per mouse genome, are highly methylated both in the oocyte and sperm genome (Lane et al. 2003). In contrast, certain sequences, particularly differentially DNA-methylated regions (DMRs) in imprinted genes, are methylated only in the oocyte or sperm (see Ch. 26 [Barlow and Bartolomei 2014]). Oocyte and sperm DNA methylation profiles have been determined in detail by reduced representation bisulfite sequencing or bisulfite sequencing methods (Box 1 of Ch. 15 [Li and

Zhang 2014] describes the main DNA methylation assay techniques; Smallwood et al. 2011; Kobayashi et al. 2012; Smith et al. 2012).

The oocyte genome has high levels of histone modifications, both active ones (e.g., H3K9 acetylation and H3K4 methylation) and repressive ones (e.g., H3K9 methylation, and H3K27 methylation; Morgan et al. 2005). At this point before fertilization, the oocyte genome is transcriptionally inactive, but contains maternally inherited transcripts and proteins needed during the first few cleavage divisions, including those required for important reprogramming events (Fig. 1A). The sperm genome, in contrast, is highly specialized: The majority of the histones have been replaced during spermatogenesis by highly basic protamines, which may facilitate the packaging of DNA into the compacted sperm head (McLay and Clarke 2003). However, a minority of the sperm chromatin is organized with histones that carry modifications such as H3K4 or H3K27 methylation that are proposed to be connected with gene transcription or repression, respectively, in the early embryo (Hammoud et al. 2009; Brykczynska et al. 2010).

Shortly after fertilization, a highly regulated sequence of reprogramming events occurs in the sperm genome. Protamines are rapidly removed and replaced by histones. It is likely that being DNA replication–independent involves incorporation of the histone variant H3.3 by the histone chaperone HIRA (illustrated in Fig. 9 of Ch. 20 [Henikoff and Smith 2014]; also see van der Heijden et al. 2005). At the same time, there is genome-wide demethylation of DNA in the male pronucleus involving single-copy and repetitive sequences, but not paternally methylated imprinted genes (Olek and Walter 1997; Oswald et al. 2000; Mayer et al. 2000; Dean et al. 2001; Santos et al. 2002; Lane et al. 2003; Smith et al. 2012).

Before DNA replication, histones in the paternal pronucleus are acetylated (H3 and H4), H3K4 methylated, and rapidly acquire H3K9me1 and H3K27me1 (Arney et al. 2002; Santos et al. 2002; Lepikhov and Walter 2004; Santos et al. 2005). H3K9me2/3 and H3K27me2/3, however, only occur subsequent to DNA replication, likely in conjunction with the incorporation of core histone H3.1 instead of H3.3 (Santos et al. 2005). By the first mitosis, most histone marks analyzed begin to be quite similar on the maternal and paternal chromosomes, at least as determined by low-resolution immunofluorescence staining (Santos et al. 2005). With the advent of chromatin immunoprecipitation-sequencing, in the years to come it should be possible to obtain higher resolution results with smaller starting material (e.g., even single cell) at a genome-wide and individual loci level.

The enzyme activities that are responsible for these early reprogramming steps are all likely to be present in the oocyte, either as protein or RNA molecules that can be rapidly translated. We already mentioned HIRA, but after DNA replication, it is CAF1 that is needed for replication-dependent incorporation of histone H3.1. The su(var) enzymes methylate H3K9, and Ezh2, together with its co-factor Eed, methylates H3K27 (Erhardt et al. 2003; Santos et al. 2005). It is likely that the dramatic DNA demethylation of the paternal genome is caused by a process of "active demethylation," which is in part now explained by oxidation of 5-methylcytosine (5mC) to 5-hydroxymethylcytosine (5hmC) and further to 5-formylcytosine (5fC) and 5-carboxylcytosine (5caC) by the TET hydoxylase enzyme family (Branco et al. 2011; see Figs. 3 and 6 of Ch. 15 [Li and Zhang 2014]). The paternal pronucleus thus rapidly loses 5mC and gains 5hmC (as well as some 5fC and 5caC; Gu et al. 2011; Inoue et al. 2011; Inoue and Zhang 2011; Iqbal et al. 2011; Wossidlo et al. 2011). Indeed, TET3 protein is highly expressed in the zygote, localizes preferentially to the male pronucleus after fertilization, and is responsible for some DNA demethylation and the gain of 5hmC (Gu et al. 2011; Iqbal et al. 2011; Wossidlo et al. 2011). Base excision repair (potentially excising modified methylcytosine, illustrated in Fig. 6 of Ch. 15 [Li and Zhang 2014]), the elongator complex, and DNA replication are also thought to contribute to demethylation of the paternal genome, the biological function of which remains unknown (Hajkova et al. 2010; Okada et al. 2010; Inoue and Zhang 2011; Santos et al. 2013). Why the maternal genome is not demethylated at the same time as the paternal one is a key outstanding question in this field of research. Apart from preferential localization of TET3, the maternal chromatin or pronucleus also possesses a specific protection mechanism in the form of H3K9me2, which is apparently recognized by Stella and protects from TET3-mediated demethylation (Arney et al. 2002; Santos et al. 2002, 2005; Nakamura et al. 2007; Nakamura et al. 2012). A number of other factors, including Trim28, Kap1, and Zfp57, are also involved in "protecting" the imprints (Messerschmidt 2012).

Although the evidence mainly suggests that histone modifications are acquired rather than lost at the global level during this period, it is possible that histone arginine methylation is more dynamic. Indeed, a candidate for erasing histone arginine methylation by "deimination", Padi4, is present in the oocyte (Sarmento et al. 2004).

The main result of the rapid chromatin changes that occur at fertilization seems to be that, at the two-cell stage, the paternal genome is similar to the maternal one. This excludes DNA methylation, which differs considerably between the two genomes largely as a result of the demethylation of the sperm genome. Also, the level of analysis so far has not excluded identifying gene-specific differences in histone modifications that may be established at this stage.

 Cite as *Cold Spring Harb Perspect Biol* doi: 10.1101/cshperspect.a019422

3.3 From the Zygote to the Blastocyst

The general theme of reprogramming, particularly of genome-wide DNA methylation patterns, continues from the two-cell stage through to the cleavage stages during preimplantation development until the embryo reaches the blastocyst stage (Monk et al. 1987; Howlett and Reik 1991; Rougier et al. 1998; Smallwood et al. 2011; Kobayashi et al. 2012; Smith et al. 2012). The precise dynamics of histone modifications are not fully described yet in the mouse, but DNA methylation is reduced stepwise with each nuclear division until the 16-cell morula stage. The reason for this is that Dnmt1, the methyltransferase that maintains methylation at CpG dinucleotides in a semiconservative fashion during DNA replication (see Fig. 2 of Ch. 15 [Li and Zhang 2014]), is excluded from the nucleus (Carlson et al. 1992). Therefore, at each division, 50% of all genomic DNA methylation is lost. However, some sequences, such as DMRs in imprinted genes, maintain their methylation because of a certain amount of Dnmt1 being retained in the nucleus, potentially targeted by the zinc finger protein Zfp57 (Hirasawa et al. 2008; Li et al. 2008). Remarkably, at the eight-cell stage, the Dnmt1 protein appears to enter the nucleus for one replication cycle. If this Dnmt1 protein is removed (by its genetic ablation in the oocyte, which provides most, if not all, of the protein during the cleavage divisions), methylation in DMRs is indeed reduced by 50%, consistent with it being needed for maintenance of methylation for one round of replication only (Howell et al. 2001).

At the eight- to 16-cell stage, the outer cells of the morula flatten and become epithelial (Fig. 2); this is called compaction. This is the first outward sign of differentiation in the mammalian embryo. Over the next two to three divisions, the morula then cavitates (i.e., a cavity forms) and forms a blastocyst, distinguished by its ICM and outer TE cells. The ICM cells go on to form all lineages of the embryo and fetus, whereas the TE cells form most (but not all) lineages of the placenta (extraembryonic lineages). Shortly after this stage, another epithelial layer of cells forms on the surface of the ICM; these are PE cells, which again contribute to the placenta and the yolk sac, but not the embryo. A few genetic determinants of these very early allocation events are known: *Oct4*, *Nanog*, and *Sox2* are important for the determination or maintenance of ICM cells, whereas *Cdx2* is required for the early maintenance of the TE cell fate (Nichols et al. 1998; Avilion et al. 2003; Chambers et al. 2003; Mitsui et al. 2003; Niwa et al. 2005), and *Gata6* is important for the formation of PE cells (Lanner and Rossant 2010). The TE transcription factor *Elf5* becomes rapidly methylated in the epiblast and this is important for maintaining the distinction between embryonic and extraembryonic cell fate (Ng et al. 2008). To what extent maternal proteins present in the oocyte or the epigenetic regulation of these genes contribute to the early cell-fate decisions in early embryos or their maintenance is currently unknown (Dean and Ferguson-Smith 2001; Torres-Padilla et al. 2010).

Major epigenetic programming events do, however, occur right at this developmental stage. The ICM cells begin to acquire DNA methylation, as judged by immunofluorescence, which is brought about by the de novo DNA methyltransferase Dnmt3b (Santos et al. 2002; Borgel et al. 2010; Smith et al. 2012). This is accompanied by increases in histone H3K9 and H3K27 methylation, transduced by G9a and Eset, and Ezh2, respectively (Fig. 2) (Erhardt et al. 2003). Although de novo methylation of DNA is not critical for the initial establishment of ICM cells, histone H3K27 methylation by Ezh2 and Eset is critical; in gene knockouts of either gene, the ICM cells do not develop properly (O'Carroll et al. 2001; Dodge et al. 2004).

In contrast to the increase in epigenetic modifications in the ICM, the TE remains largely DNA hypomethylated, as do most of the cell lineages in the later placenta (Chapman et al. 1984; Santos et al. 2002). It is thought that placental cell types need less epigenetic stability because their lifespan is much more restricted (i.e., lasting only through gestation) than that of the fetus, which continues to develop into the adult organism.

In addition to genome-wide epigenetic events during morula/blastocyst-stage development, considerable change and reprogramming occur in females on the X chromosomes. In female mouse embryos (XX), the paternally inherited X chromosome is always inactivated during the cleavage stages and remains so in the extraembryonic tissues (i.e., the TE and the placenta; Huynh and Lee 2003, Okamoto et al. 2005). But, in the ICM, the inactive X is reactivated and this is followed by random inactivation of one X chromosome after differentiation in the ICM-derived lineages (Mak et al. 2004; for more detail, see Figs. 4 and 6 of Ch. 25 [Brockdorff and Turner 2014]). Mechanistically, imprinted X inactivation (i.e., paternal X inactivation) in the preimplantation embryo involves expression of the noncoding RNA *Xist* from the paternal X chromosome, whose "coating" of the chromosome is thought to lead to gene silencing and the establishment of repressive epigenetic modifications (Heard 2004). *Xist* transcription is then down-regulated in the newly formed ICM cells, and repressive histone modifications are subsequently lost. At this stage, the chromosome becomes reactivated (Mak et al. 2004; Okamoto et al. 2004). *Xist* down-regulation is achieved, at least in part, by the pluripotency transcription factor network (Deuve and Avner 2011). This is followed, shortly afterward, by the initiation of random X inactivation in epiblast cells. We will see in Section 4 that ES cells are "frozen" at the stage after reactivation of the X chromosome such that female ES cells contain two active X chromosomes.

4 FROM PLURIPOTENT STEM CELLS TO SOMATIC CELLS AND BACK TO GERM CELLS

4.1 Derivation of Pluripotent Stem Cells

In Section 3.3, we learned that there are dramatic epigenetic reprogramming events in the zygote, cleavage-stage embryos, and the blastocyst, resulting in different epigenetic patterns in the ICM and TE. We now consider the genetic and epigenetic properties of early stem cells derived into culture from the blastocyst and later lineages (Fig. 1B), such as ES cells (Smith 2001), trophoblast stem (TS) cells (Rossant 2001), extraembryonic endoderm (XEN) stem cells (Kunath et al. 2005), and EG stem cells (Matsui et al. 1992).

The feature that is common to these cell types is that they can be isolated or established from intact embryos and put into culture under certain culturing conditions. Once established, they can be cultured for extended periods of time and show no signs of senescence. They can also be genetically manipulated during culture and then reintroduced into living embryos to participate in the development of the appropriate lineages.

The development of methods to generate pluripotent ES cells from ICMs was one of the most important discoveries in mammalian embryology during the 1980s. The ES cells explanted from mouse blastocysts into culture were maintained for extended culture periods, and when microinjected back into blastocysts, they colonize all embryonic lineages (Fig. 7), thus forming chimeras (Evans and Kaufman 1981; Martin 1981). What was particularly striking was that descendants of the ES cells could colonize the germ cells and give rise to normal offspring, which were derived wholly from the ES cell genotype. This, together with the ability to genetically manipulate the ES cell genome by homologous recombination techniques leading to gene knockouts, revolutionized mouse genetics and, consequently, made the mouse the mammalian genetic model organism of choice. Considerably later, it became possible to derive ES cells from human and rat preimplantation embryos (Thomson et al. 1998; Buehr et al. 2008).

ES cells share properties with ICM/epiblast cells, but also show substantial differences, making it likely that they are a "synthetic" cell type that do not exist in the normal embryo (Smith 2001). The same is likely to apply to the other pluripotent cell lines. For example, whereas the self-renewal of mouse ES cells requires a functional Lif/gp130/Stat3 signaling pathway, embryos with mutations in this pathway still develop a normal ICM (Smith 2001). It is thus likely that epigenetic changes occur, and may be necessary, for the derivation and maintenance of ES cells from ICM cells. Also, outgrowths of ICM cells into culture rapidly lose expression of Oct4, and only a mouse strain from which it is relatively easy to derive ES cells, called 129Sv, retains some Oct4-expressing cells on culture. Another feature of ES cells is the epigenetic change reported in imprinted genes in both mouse and rhesus monkey; in the mouse, this can result in aberrant development of

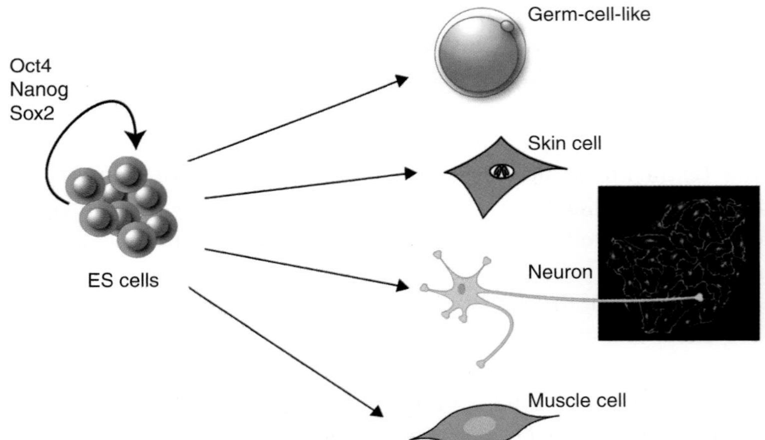

Figure 7. ES and TS cells from the blastocyst. ES cells are derived from ICM cells and can be kept in culture without differentiating. They can be genetically manipulated while in culture. ES cells can be reintroduced into blastocysts and then colonize all tissues in the embryo, including the germline, but excluding the TS cells of the placenta. TS cells can be established similarly into culture from the TE cells of the blastocyst and, when reintroduced into blastocysts, contribute to placental cell types.

 Cite as *Cold Spring Harb Perspect Biol* doi: 10.1101/cshperspect.a019422

the cells when reintroduced into chimeras (Dean et al. 1998; Humpherys et al. 2001). One of the important determinants of ES cells in culture is the signaling system they respond to; it seems that ES cells grown in serum and leukemia inhibitory factor are epigenetically and transcriptionally somewhat heterogeneous, whereas ES cells grown in the presence of inhibitors of ERK (FGF4) and GSK3 are more homogeneous (Hayashi et al. 2008; Lanner and Rossant 2010; Leitch et al. 2010; Marks et al. 2012; Ficz et al. 2013). This may reflect the natural tendency of epiblast cells to progress from a naïve state of pluripotency in the early ICM to a state in which they are primed for differentiation and thus receive increased levels of prodifferentiation FGF signals through ERK or GSK3 (for more detail on epiblast stem cells, see Ch. 28 [Hochedlinger and Jaenisch 2014]).

4.2 Epigenetic Properties of Pluripotent Cell Lines

ES cells can be differentiated in vitro into a number of different cell lineages (Fig. 6). To what extent do epigenetic mechanisms maintain cells in an undifferentiated or differentiated state? Clearly, there are epigenomic differences between undifferentiated ES cells, differentiated ES cells, and somatic cells. Pluripotential cells are particularly characterized by the hyperdynamic plasticity of their chromatin and dynamics of DNA methylation and hydroxymethylation (Meshorer et al. 2006; Branco et al. 2011). Deletion in ES cells of Parp (poly-ADP ribosylase), which is involved in controlling the alteration of histone marks, leads to a higher frequency of transdifferentiation into TS cells, suggesting that epigenetic marks in ES cells are needed to maintain their identity (Hemberger et al. 2009). DNA methylation is also important; it epigenetically silences, for instance, the extraembryonic lineage transcription factor *Elf5*, whereas a lack of DNA methylation at other loci is needed to maintain ES cell fate.

The maintenance of pluripotency of ES cells depends on the transcription factors Oct4, Nanog, Sox2, and others—the so-called pluripotency network (Young 2011). These bind alone or in combination with many gene loci in ES cells, which either need to be expressed or silenced for pluripotency to be maintained (Boyer et al. 2005; Loh et al. 2006). Differentiation of ES cells in vitro is characterized by the transcriptional silencing of pluripotency genes, which then remain repressed in somatic tissues. Epigenetic mechanisms are indeed important for their silencing; the *Oct4* promoter, for example, accumulates repressive histone modifications and DNA methylation during differentiation once it has been silenced (Feldman et al. 2006). Reexpression of *Oct4* in differentiated cells can, however, be triggered by a loss of DNA methylation using *Dnmt1* knockout embryos (Feldman et al. 2006). This typifies the many emerging links between the pluripotency gene network and epigenetic modifiers involved in the epigenetic reprogramming of germ cells or early embryos (Gifford and Meissner 2012).

Insights into pluripotency transcription factors such as Oct4 or Nanog enabled Yamanaka and colleagues to devise their strategy for iPS (see Ch. 28 [Hochedlinger and Jaenisch 2014]). The original protocol used Oct4, Sox2, Klf4, and c-myc to reprogram somatic cells to iPS cells, which are often indistinguishable from ES cells (Fig. 8) (described in Takahashi and Yamanaka 2006; Ch. 2 [Takahashi 2014]). There are now many variants of the iPS protocol, including those using RNA, protein, microRNAs, or small molecule inhibitors of epigenetic modifiers. Nevertheless, the details of the reprogramming process and its kinetics, including importantly epigenetic reprogramming, remain incompletely understood (Gifford and Meissner 2012). There is the possibility that many iPS cell lines retain epigenetic memory of some kind from the tissue of origin, which needs to be taken into account when examining their differentiation potential and contemplating their use in the clinic.

ES cells and their differentiated derivatives have served as a model for the epigenetic regulation of X chromosome inactivation. As introduced in Section 3.3, female ICM and ES cells have a down-regulated *Xist* gene and two active X chromosomes. On differentiation in vitro, *Xist* becomes up-regulated on one of the X chromosomes, *Xist* RNA begins to "coat" this chromosome in *cis*, and silencing of genes on the X concomitant with the accumulation of repressive histone modifications and DNA methylation ensues (Heard 2004).

Other pluripotent cell types can be similarly established in culture, but their epigenetic properties are less well characterized than those of ES cells. It is, however, known from epigenetic studies of X inactivation that in female TS stem cells, there is a paternally inactivated X chromosome, as found in their originating TE cells, containing repressive histone marks (Huynh and Lee 2003). Female XEN cells also have a paternal X chromosome, which is inactive (Kunath et al. 2005).

Pluripotent cell lines can also be established from PGCs either during their period of migration in the embryo (E8–E10.5) or once they have reached the embryonic gonads (E11.5–E13.5; Fig. 1B) (Matsui et al. 1992). PGCs, during the E8.5–11.5 stages of development, undergo extensive epigenetic reprogramming including DNA demethylation of imprinted genes and other sequences in the genome (Hajkova et al. 2002; Lee et al. 2002). Indeed, most EG cells have undergone DNA demethylation erasure of imprinted genes and other sequences and this alters their developmental potential, as shown when introduced into chimeras

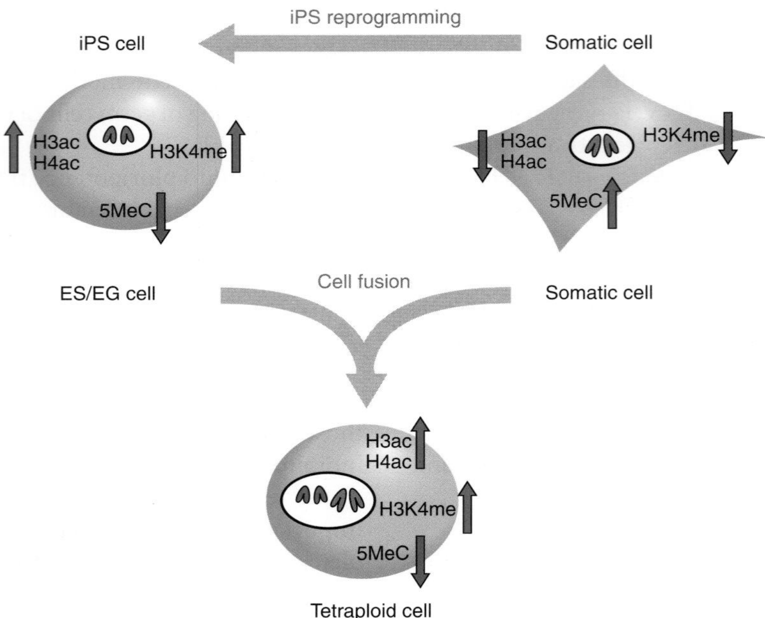

Figure 8. Pluripotent cells have the capacity to reprogram somatic cells. ES or EG cells can be fused with somatic cells, resulting in tetraploid hybrids. This leads to epigenetic reprogramming of the somatic nucleus, with changes in, for example, 5MeC, H3 and H4 acetylation, and H3K4 methylation. The tetraploid cells resulting from this fusion or reprogramming that occurs when producing iPSCs also have a pluripotent phenotype: When injected into blastocysts, they can contribute to many different cell types in the embryo.

(Tada et al. 1998). It is not clear yet if there are epigenetic differences between endogenous PGCs and in vitro cultured EG cells, similar to those suspected to exist between ICM and ES cells.

4.3 Reprogramming Capacity of Stem Cells

The continued state of pluripotency in culture without cell senescence may well require continued epigenetic reprogramming of stem cells. That these cells indeed have reprogramming activities has been shown in fusion experiments in which EG or ES cells were fused to differentiated somatic cells (Tada et al. 1997, 2001; Cowan et al. 2005). In the tetraploid cell lines resulting from fusion, the somatic epigenotype is reprogrammed (Fig. 8). In EG-somatic cell fusions, the somatic genome loses DNA methylation at imprinted genes as well as other sequences in the genome (Tada et al. 1998). In ES-somatic cell fusions, in contrast, imprinted gene DNA methylation is retained, but the inactive X chromosome (in female cells) is reactivated and the promoter of the Oct4 gene becomes DNA demethylated, resulting in Oct4 reexpression (Tada et al. 2001; Cowan et al. 2005; Surani 2005). Both the AID catalyzed deamination and Tet catalyzed hydroxylation pathways for DNA demethylation (see Fig. 6 of Ch. 15 [Li and Zhang 2014]) have been implicated in the reprogramming of somatic cells by ES cell fusion (Piccolo and Fisher 2013).

5 PERSPECTIVE

The next few years will see decisive and exciting advances in our understanding of the genetic and epigenetic factors that are critical for totipotency and pluripotency of germ and stem cells. High-throughput and sensitive methods for determining the various layers of epigenetic information of the genome are becoming widely used together with computational epigenomics developments. Factors that regulate epigenetic information, particularly those that are needed to reprogram epigenetic marks in somatic cells to ones found in pluripotent cells, have been identified. The precise temporal and cell-specific connections between the pluripotency transcription factor network, epigenetic modifiers, and signaling networks still need to be worked out and this may require systems biology-type approaches. Alongside this is the development of better methods to selectively and safely manipulate epigenetic states in vivo, ultimately for therapeutic ends.

Pluripotent stem cells present many exciting opportunities for fundamental studies as well as for their potential applications in biomedicine. In fundamental research, the uniqueness of the pluripotent state promises to provide insights into the mechanisms that regulate cell-fate decisions. The ability to differentiate into diverse cell types also provides a potential to generate replacement cells in the quest to repair diseased tissues. Pluripotent stem cells are also being used to develop disease models to explore how various human diseases originate from the very beginning as a result of specific mutations and epimutations (i.e., mutations caused by alterations of DNA methylation or the chromatin template, but not involving changes to the DNA sequence). Such disease models may, in turn, allow the development of new drugs to cure or even prevent diseases. Transdifferentiation technologies that directly transform one differentiated cell type into another (e.g., skin into neurons) have recently been established and this may offer new routes to therapy, avoiding the use of pluripotent cell types with the potential danger of tumor formation when cells are used for transplantation in patients (discussed further in Ch. 28 [Hochedlinger and Jaenisch 2014]). There is also the prospect of reversing epigenetic effects during aging by "epigenetic rejuvenation," which is reprogramming cells to a younger and healthier cell state without going back to pluripotency (Rando and Chang 2012).

ACKNOWLEDGMENTS

The use of human pluripotent stem cells raises sensitive ethical issues that are being debated by the wider public. Appropriate ethical and regulatory frameworks are being established and continue to be refined in step with scientific progress for the use of stem cells in research and biomedical applications.

REFERENCES

*Reference is also in this book.

Ancelin K, Lange UC, Hajkova P, Schneider R, Bannister AJ, Kouzarides T, Surani MA. 2006. Blimp1 associates with Prmt5 and directs histone arginine methylation in mouse germ cells. *Nat Cell Biol* **8:** 623–630.

Arney KL, Bao S, Bannister AJ, Kouzarides T, Surani MA. 2002. Histone methylation defines epigenetic asymmetry in the mouse zygote. *Int J Dev Biol* **46:** 317–320.

Avilion AA, Nicolis SK, Pevny LH, Perez L, Vivian N, Lovell-Badge R. 2003. Multipotent cell lineages in early mouse development depend on SOX2 function. *Genes Dev* **17:** 126–140.

* Barlow DP, Bartolomei MS. 2014. Genomic imprinting in mammals. *Cold Spring Harb Perspect Biol* **6:** a018382.

Baudat F, Buard J, Grey C, Fledel-Alon A, Ober C, Przeworski M, Coop G, de Massy B. 2010. PRDM9 is a major determinant of meiotic recombination hotspots in humans and mice. *Science* **327:** 836–840.

Blackwell TK. 2004. Germ cells: Finding programs of mass repression. *Curr Biol* **14:** R229–R230.

* Blewitt M, Whitelaw E. 2013. The use of mouse models to study epigenetics. *Cold Spring Harb Perspect Biol* **5:** a017939.

Borgel J, Guibert S, Li Y, Chiba H, Schübeler D, Sasaki H, Forné T, Weber M. 2010. Targets and dynamics of promoter DNA methylation during early mouse development. *Nat Genet* **42:** 1093–1100.

Boyer LA, Lee TI, Cole MF, Johnstone SE, Levine SS, Zucker JP, Guenther MG, Kumar RM, Murray HL, Jenner RG, et al. 2005. Core transcriptional regulatory circuitry in human embryonic stem cells. *Cell* **122:** 947–956.

Branco MR, Ficz G, Reik W. 2011. Uncovering the role of 5-hydroxymethylcytosine in the epigenome. *Nat Rev Genet* **13:** 7–13.

* Brockdorff N, Turner BM. 2014. Dosage compensation in mammals. *Cold Spring Harb Perspect Biol* doi: 10.1101/cshperspect.a019406.

Brykczynska U, Hisano M, Erkek S, Ramos L, Oakeley EJ, Roloff TC, Beisel C, Schübeler D, Stadler MB, Peters AH. 2010. Repressive and active histone methylation mark distinct promoters in human and mouse spermatozoa. *Nat Struct Mol Biol* **17:** 679–687.

Buehr M, Meek S, Blair K, Yang J, Ure J, Silva J, McLay R, Hall J, Ying QL, Smith A. 2008. Capture of authentic embryonic stem cells from rat blastocysts. *Cell* **135:** 1287–1298.

* Busslinger M, Tarakhovsky A. 2014. Epigenetic control of immunity. *Cold Spring Harb Perspect Biol* **6:** a019307.

Carlson LL, Page AW, Bestor TH. 1992. Properties and localization of DNA methyltransferase in preimplantation mouse embryos: Implications for genomic imprinting. *Genes Dev* **6:** 2536–2541.

Chambers I, Colby D, Robertson M, Nichols J, Lee S, Tweedie S, Smith A. 2003. Functional expression cloning of Nanog, a pluripotencysustaining factor in embryonic stem cells. *Cell* **113:** 643–655.

Chan MM, Smith ZD, Egli D, Regev A, Meissner A. 2012. Mouse ooplasm confers context-specific reprogramming capacity. *Nat Genet* **44:** 978–980.

Chapman V, Forrester L, Sanford J, Hastie N, Rossant J. 1984. Cell lineage-specific undermethylation of mouse repetitive DNA. *Nature* **307:** 284–286.

* Cheng X. 2014. Structural and functional coordination of DNA and histone methylation. *Cold Spring Harb Perspect Biol* **6:** a018747.

Chong S, Whitelaw E. 2004. Epigenetic germline inheritance. *Curr Opin Genet Dev* **14:** 692–696.

Cortellino S, Xu J, Sannai M, Moore R, Caretti E, Cigliano A, Le Coz M, Devarajan K, Wessels A, Soprano D, et al. 2011. Thymine DNA glycosylase is essential for active DNA demethylation by linked deamination-base excision repair. *Cell* **146:** 67–79.

Cowan CA, Atienza J, Melton DA, Eggan K. 2005. Nuclear reprogramming of somatic cells after fusion with human embryonic stem cells. *Science* **309:** 1369–1373.

Dawlaty MM, Breiling A, Le T, Raddatz G, Barrasa MI, Cheng AW, Gao Q, Powell BE, Li Z, Xu M, et al. 2013. Moation. *Dev Cell* **24:** 310–323.

Dean W, Bowden L, Aitchison A, Klose J, Moore T, Meneses JJ, Reik W, Feil R. 1998. Altered imprinted gene methylation and expression in completely ES cell-derived mouse fetuses: Association with aberrant phenotypes. *Development* **125:** 2273–2282.

Dean W, Santos F, Stojkovic M, Zakhartchenko V, Walter J, Wolf E, Reik W. 2001. Conservation of methylation reprogramming in mammalian development: Aberrant reprogramming in cloned embryos. *Proc Natl Acad Sci* **98:** 13734–13738.

Deshpande G, Calhoun G, Schedl P. 2004. Overlapping mechanisms function to establish transcriptional quiescence in the embryonic *Drosophila* germline. *Development* **131:** 1247–1257.

de Souza FS, Gawantka V, Gomez AP, Delius H, Ang SL, Niehrs C. 1999. The zinc finger gene Xblimp1 controls anterior endomesodermal cell fate in Spemann's organizer. *EMBO J* **18:** 6062–6072.

Deuve JL, Avner P. 2011. The coupling of X-chromosome inactivation to pluripotency. *Annu Rev Cell Dev Biol* **27:** 611–629.

Dodge JE, Kang YK, Beppu H, Lei H, Li E. 2004. Histone H3-K9 methyltransferase ESET is essential for early development. *Mol Cell Biol* **24:** 2478–2486.

Erhardt S, Su IH, Schneider R, Barton S, Bannister AJ, Perez-Burgos L, Jenuwein T, Kouzarides T, Tarakhovsky A, Surani MA. 2003. Consequences of the depletion of zygotic and embryonic enhancer of zeste 2 during preimplantation mouse development. *Development* **130:** 4235–4248.

Evans MJ, Kaufman MH. 1981. Establishment in culture of pluripotential cells from mouse embryos. *Nature* **292:** 154–156.

Extavour CG, Akam M. 2003. Mechanisms of germ cell specification across the metazoans: Epigenesis and preformation. *Development* **130:** 5869–5884.

Feldman N, Gerson A, Fang J, Li E, Zhang Y, Shinkai Y, Cedar H, Bergman Y. 2006. G9a-mediated irreversible epigenetic inactivation of Oct-3/4 during early embryogenesis. *Nat Cell Biol* **8:** 188–194.

Feng S, Jacobsen SE, Reik W. 2010. Epigenetic reprogramming in plant and animal development. *Science* **330:** 622–627.

Ferguson-Smith AC, Patti ME. 2011. You are what your dad ate. *Cell Metab* **13:** 115–117.

Ficz G, Hore TA, Santos F, Lee HJ, Dean W, Arand J, Krueger F, Oxley D, Paul YL, Walter J, et al. 2013. FGF signaling inhibition in ESCs drives rapid genome-wide demethylation to the epigenetic ground state of pluripotency. *Cell Stem Cell* **13:** 351–359.

Fujimori T, Kurotaki Y, Miyazaki J, Nabeshima Y. 2003. Analysis of cell lineage in two- and four-cell mouse embryos. *Development* **130:** 5113–5122.

Gardner RL. 1985. Clonal analysis of early mammalian development. *Philos Trans R Soc Lond B Biol Sci* **312:** 163–178.

Gardner RL. 1997. The early blastocyst is bilaterally symmetrical and its axis of symmetry is aligned with the animal–vegetal axis of the zygote in the mouse. *Development* **124:** 289–301.

Geijsen N, Horoschak M, Kim K, Gribnau J, Eggan K, Daley GQ. 2004. Derivation of embryonic germ cells and male gametes from embryonic stem cells. *Nature* **427:** 148–154.

Gifford CA, Meissner A. 2012. Epigenetic obstacles encountered by transcription factors: Reprogramming against all odds. *Curr Opin Genet Dev* **22:** 409–415.

Gu TP, Guo F, Yang H, Wu HP, Xu GF, Liu W, Xie ZG, Shi L, He X, Jin SG, et al. 2011. The role of Tet3 DNA dioxygenase in epigenetic reprogramming by oocytes. *Nature* **477:** 606–610.

Guan K, Nayernia K, Maier LS, Wagner S, Dressel R, Lee JH, Nolte J, Wolf F, Li M, Engel W, et al. 2006. Pluripotency of spermatogonial stem cells from adult mouse testis. *Nature* **440:** 1199–1203.

Guibert S, Forné T, Weber M. 2012. Global profiling of DNA methylation erasure in mouse primordial germ cells. *Genome Res* **22:** 633–641.

Gyory I, Wu J, Fejer G, Seto E, Wright KL. 2004. PRDI-BF1 recruits the histone H3 methyltransferase G9a in transcriptional silencing. *Nat Immunol* **5:** 299–308.

Hackett JA, Surani MA. 2013. Beyond DNA: Programming and inheritance of parental methylomes. *Cell* **172:** 5427–5440.

Hackett JA, Zylicz JJ, Surani MA. 2012. Parallel mechanisms of epigenetic reprogramming in the germline. *Trends Genet* **28:** 164–174.

Hackett JA, Sengupta R, Zylicz JJ, Murakami K, Lee C, Down TA, Surani MA. 2013. Germline DNA demethylation dynamics and imprint erasure through 5-hydroxymethylcytosine. *Science* **339:** 448–452.

Hajkova P, Erhardt S, Lane N, Haaf T, El-Maarri O, Reik W, Walter J, Surani MA. 2002. Epigenetic reprogramming in mouse primordial germ cells. *Mech Dev* **117:** 15–23.

Hajkova P, Ancelin K, Waldmann T, Lacoste N, Lange UC, Cesari F, Lee C, Almouzni G, Schneider R, Surani MA. 2008. Chromatin dynamics during epigenetic reprogramming in the mouse germ line. *Nature* **452:** 877–881.

Hajkova P, Jeffries SJ, Lee C, Miller N, Jackson SP, Surani MA. 2010. Genome-wide reprogramming in the mouse germ line entails the base excision repair pathway. *Science* **329:** 78–82.

Hammoud SS, Nix DA, Zhang H, Purwar J, Carrell DT, Cairns BR. 2009. Distinctive chromatin in human sperm packages genes for embryo development. *Nature* **460:** 473–478.

Hayashi K, Yoshida K, Matsui Y. 2005. A histone H3 methyltransferase controls epigenetic events required for meiotic prophase. *Nature* **438:** 374–378.

Hayashi K, Lopes SM, Tang F, Surani MA. 2008. Dynamic equilibrium and heterogeneity of mouse pluripotent stem cells with distinct functional and epigenetic states. *Cell Stem Cell* **3:** 391–401.

Hayashi K, Ohta H, Kurimoto K, Aramaki S, Saitou M. 2011. Reconstitution of the mouse germ cell specification pathway in culture by pluripotent stem cells. *Cell* **146:** 519–532.

Heard E. 2004. Recent advances in X-chromosome inactivation. *Curr Opin Cell Biol* **16:** 247–255.

Hemberger M, Dean W, Reik W. 2009. Epigenetic dynamics of stem cells and cell lineage commitment: Digging Waddington's canal. *Nat Rev Mol Cell Biol* **10:** 526–537.

* Henikoff S, Smith MM. 2014. Histone variants and epigenetics. *Cold Spring Harb Perspect Biol* doi: 10.1101/cshperspect.a019364.

Hernandez-Lagunas L, Choi IF, Kaji T, Simpson P, Hershey C, Zhou Y, Zon L, Mercola M, Artinger KB. 2005. Zebrafish *narrowminded* disrupts the transcription factor *prdm1* and is required for neural crest and sensory neuron specification. *Dev Biol* **278:** 347–357.

Hirasawa R, Chiba H, Kaneda M, Tajima S, Li E, Jaenisch R, Sasaki H. 2008. Maternal and zygotic Dnmt1 are necessary and sufficient for the maintenance of DNA methylation imprints. *Genes Dev* **22:** 1607–1616.

* Hochedlinger K, Jaenisch R. 2014. Induced pluripotency and epigenetic reprogramming. *Cold Spring Harb Perspect Biol* doi: 10.1101/cshperspect.a019448.

Howell CY, Bestor TH, Ding F, Latham KE, Mertineit C, Trasler JM, Chaillet JR. 2001. Genomic imprinting disrupted by a maternal effect mutation in the *Dnmt1* gene. *Cell* **104:** 829–838.

Howlett SK, Reik W. 1991. Methylation levels of maternal and paternal genomes during preimplantation development. *Development* **113:** 119–127.

Hubner K, Fuhrmann G, Christenson LK, Kehler J, Reinbold R, De La Fuente R, Wood J, Strauss JF 3rd, Boiani M, Scholer HR. 2003. Derivation of oocytes from mouse embryonic stem cells. *Science* **300:** 1251–1256.

Humpherys D, Eggan K, Akutsu H, Hochedlinger K, Rideout WM 3rd, Biniszkiewicz D, Yanagimachi R, Jaenisch R. 2001. Epigenetic instability in ES cells and cloned mice. *Science* **293:** 95–97.

Huynh KD, Lee JT. 2003. Inheritance of a pre-inactivated paternal X chromosome in early mouse embryos. *Nature* **426:** 857–862.

Huynh JR, St Johnston D. 2004. The origin of asymmetry: Early polarisation of the *Drosophila* germline cyst and oocyte. *Curr Biol* **14:** R438–R449.

Inoue A, Zhang Y. 2011. Replication-dependent loss of 5-hydroxymethylcytosine in mouse preimplantation embryos. *Science* **334:** 194.

Inoue A, Shen L, Dai Q, He C, Zhang Y. 2011. Generation and replication-dependent dilution of 5fC and 5caC during mouse preimplantation development. *Cell Res* **21:** 1670–1676.

Iqbal K, Jin SG, Pfeifer GP, Szabó PE. 2011. Reprogramming of the paternal genome upon fertilization involves genome-wide oxidation of 5-methylcytosine. *Proc Natl Acad Sci* **108:** 3642–3647.

Kagiwada S, Kurimoto K, Hirota T, Yamaji M, Saitou M. 2013. Replication-coupled passive DNA demethylation for the erasure of genome imprints in mice. *EMBO J* **32:** 340–353.

Kanatsu-Shinohara M, Inoue K, Lee J, Yoshimoto M, Ogonuki N, Miki H, Baba S, Kato T, Kazuki Y, Toyokuni S, et al. 2004. Generation of pluripotent stem cells from neonatal mouse testis. *Cell* **119:** 1001–1012.

Kawahara M, Wu Q, Takahashi N, Morita S, Yamada K, Ito M, Ferguson-Smith AC, Kono T. 2007. High-frequency generation of viable mice from engineered bi-maternal embryos. *Nat Biotechnol* **25:** 1045–1050.

Kelly SJ. 1977. Studies of the developmental potential of 4- and 8-cell stage mouse blastomeres. *J Exp Zool* **200:** 365–376.

Kimmins S, Sassone-Corsi P. 2005. Chromatin remodelling and epigenetic features of germ cells. *Nature* **434:** 583–589.

Kobayashi H, Sakurai T, Imai M, Takahashi N, Fukuda A, Yayoi O, Sato S, Nakabayashi K, Hata K, Sotomaru Y, et al. 2012. Contribution of intragenic DNA methylation in mouse gametic DNA methylomes to establish oocyte-specific heritable marks. *PLoS Genet* **8:** e1002440.

Kunath T, Arnaud D, Uy GD, Okamoto I, Chureau C, Yamanaka Y, Heard E, Gardner RL, Avner P, Rossant J. 2005. Imprinted X-inactivation in extra-embryonic endoderm cell lines from mouse blastocysts. *Development* **132:** 1649–1661.

Kuramochi-Miyagawa S, Kimura T, Ijiri TW, Isobe T, Asada N, Fujita Y, Ikawa M, Iwai N, Okabe M, Deng W, et al. 2004. *Mili*, a mammalian member of *piwi* family gene, is essential for spermatogenesis. *Development* **131:** 839–849.

Kurimoto K, Yabuta Y, Ohinata Y, Shigeta M, Yamanaka K, Saitou M. 2008. Complex genome-wide transcription dynamics orchestrated by Blimp1 for the specification of the germ cell lineage in mice. *Genes Dev* **22:** 1617–1635.

Lane N, Dean W, Erhardt S, Hajkova P, Surani A, Walter J, Reik W. 2003. Resistance of IAPs to methylation reprogramming may provide a mechanism for epigenetic inheritance in the mouse. *Genesis* **35:** 88–93.

Lanner F, Rossant J. 2010. The role of FGF/Erk signaling in pluripotent cells. *Development* **137:** 3351–3360.

Lawson KA, Dunn NR, Roelen BA, Zeinstra LM, Davis AM, Wright CV, Korving JP, Hogan BL. 1999. Bmp4 is required for the generation of primordial germ cells in the mouse embryo. *Genes Dev* **13:** 424–436.

Lawson KA, Hage WJ. 1994. Clonal analysis of the origin of primordial germ cells in the mouse. *Ciba Found Symp* **182:** 68–84.

Leatherman JL, Jongens TA. 2003. Transcriptional silencing and translational control: Key features of early germline development. *Bioessays* **25:** 326–335.

Lee J, Inoue K, Ono R, Ogonuki N, Kohda T, Kaneko-Ishino T, Ogura A, Ishino F. 2002. Erasing genomic imprinting memory in mouse clone embryos produced from day 11.5 primordial germ cells. *Development* **129:** 1807–1817.

Leitch HG, Blair K, Mansfield W, Ayetey H, Humphreys P, Nichols J, Surani MA, Smith A. 2010. Embryonic germ cells from mice and rats exhibit properties consistent with a generic pluripotent ground state. *Development* **137:** 2279–2287.

Lepikhov K, Walter J. 2004. Differential dynamics of histone H3 methylation at positions K4 and K9 in the mouse zygote. *BMC Dev Biol* **4:** 12.

* Li E, Zhang Y. 2014. DNA methylation in mammals. *Cold Spring Harb Perspect Biol* **6:** a019133.

Li X, Ito M, Zhou F, Youngson N, Zuo X, Leder P, Ferguson-Smith AC. 2008. A maternal-zygotic effect gene, *Zfp57*, maintains both maternal and paternal imprints. *Dev Cell* **15:** 547–557.

Loh YH, Wu Q, Chew JL, Vega VB, Zhang W, Chen X, Bourque G, George J, Leong B, Liu J, et al. 2006. The Oct4 and Nanog transcription network regulates pluripotency in mouse embryonic stem cells. *Nat Genet* **38:** 431–440.

Mak W, Nesterova TB, de Napoles M, Appanah R, Yamanaka S, Otte AP, Brockdorff N. 2004. Reactivation of the paternal X chromosome in early mouse embryos. *Science* **303:** 666–669.

Mansour AA, Gafni O, Weinberger L, Zviran A, Ayyash M, Rais Y, Krupalnik V, Zerbib M, Amann-Zalcenstein D, Maza I, et al. 2012. The H3K27 demethylase Utx regulates somatic and germ cell epigenetic reprogramming. *Nature* **488:** 409–413.

Marks H, Kalkan T, Menafra R, Denissov S, Jones K, Hofemeister H, Nichols J, Kranz A, Stewart AF, Smith A, et al. 2012. The transcriptional and epigenomic foundations of ground state pluripotency. *Cell* **149:** 590–604.

Martin GR. 1981. Isolation of a pluripotent cell line from early mouse embryos cultured in medium conditioned by teratocarcinoma stem cells. *Proc Natl Acad Sci* **78:** 7634–7638.

Martinho RG, Kunwar PS, Casanova J, Lehmann R. 2004. A noncoding RNA is required for the repression of RNApolII-dependent transcription in primordial germ cells. *Curr Biol* **14:** 159–165.

Matsui Y, Zsebo K, Hogan BL. 1992. Derivation of pluripotential embryonic stem cells from murine primordial germ cells in culture. *Cell* **70:** 841–847.

Mayer W, Niveleau A, Walter J, Fundele R, Haaf T. 2000. Demethylation of the zygotic paternal genome. *Nature* **403:** 501–502.

McLaren A. 2003. Primordial germ cells in the mouse. *Dev Biol* **262:** 1–15.

McLaren A, Lawson KA. 2005. How is the mouse germ-cell lineage established? *Differentiation* **73:** 435–437.

McLay DW, Clarke HJ. 2003. Remodelling the paternal chromatin at fertilization in mammals. *Reproduction* **125:** 625–633.

Meshorer E, Yellajoshula D, George E, Scambler PJ, Brown DT, Misteli T. 2006. Hyperdynamic plasticity of chromatin proteins in pluripotent embryonic stem cells. *Dev Cell* **10:** 105–116.

Messerschmidt DM. 2012. Should I stay or should I go: Protection and maintenance of DNA methylation at imprinted genes. *Epigenetics* **7:** 969–975.

Mitsui K, Tokuzawa Y, Itoh H, Segawa K, Murakami M, Takahashi K, Maruyama M, Maeda M, Yamanaka S. 2003. The homeoprotein Nanog is required for maintenance of pluripotency in mouse epiblast and ES cells. *Cell* **113:** 631–642.

Monk M, Boubelik M, Lehnert S. 1987. Temporal and regional changes in DNA methylation in the embryonic, extraembryonic and germ cell lineages during mouse embryo development. *Development* **99:** 371–382.

Morgan HD, Santos F, Green K, Dean W, Reik W. 2005. Epigenetic reprogramming in mammals. *Hum Mol Genet* **14:** R47–R58.

Nakamura T, Arai Y, Umehara H, Masuhara M, Kimura T, Taniguchi H, Sekimoto T, Ikawa M, Yoneda Y, Okabe M, et al. 2007. PGC7/Stella protects against DNA demethylation in early embryogenesis. *Nat Cell Biol* **9:** 64–71.

Nakamura T, Liu YJ, Nakashima H, Umehara H, Inoue K, Matoba S, Tachibana M, Ogura A, Shinkai Y, Nakano T. 2012. PGC7 binds histone H3K9me2 to protect against conversion of 5mC to 5hmC in early embryos. *Nature* **486:** 415–419.

Ng RK, Dean W, Dawson C, Lucifero D, Madeja Z, Reik W, Hemberger M. 2008. Epigenetic restriction of embryonic cell lineage fate by methylation of Elf5. *Nat Cell Biol* **10:** 1280–1290.

Nichols J, Zevnik B, Anastassiadis K, Niwa H, Klewe-Nebenius D, Chambers I, Scholer H, Smith A. 1998. Formation of pluripotent stem cells in the mammalian embryo depends on the POU transcription factor Oct4. *Cell* **95:** 379–391.

Niwa H, Toyooka Y, Shimosato D, Strumpf D, Takahashi K, Yagi R, Rossant J. 2005. Interaction between Oct3/4 and Cdx2 determines trophectoderm differentiation. *Cell* **123:** 917–929.

O'Carroll D, Erhardt S, Pagani M, Barton SC, Surani MA, Jenuwein T. 2001. The *Polycomb*-group gene *Ezh2* is required for early mouse development. *Mol Cell Biol* **21:** 4330–4336.

Ohinata Y, Payer B, O'Carroll D, Ancelin K, Ono Y, Sano M, Barton SC, Obukhanych T, Nussenzweig M, Tarakhovsky A, et al. 2005. Blimp1 is a critical determinant of the germ cell lineage in mice. *Nature* **436:** 207–213.

Ohinata Y, Ohta H, Shigeta M, Yamanaka K, Wakayama T, Saitou M. 2009. A signaling principle for the specification of the germ cell lineage in mice. *Cell* **137:** 571–584.

Okada Y, Yamagata K, Hong K, Wakayama T, Zhang Y. 2010. A role for the elongator complex in zygotic paternal genome demethylation. *Nature* **463:** 554–558.

Okamoto I, Otte AP, Allis CD, Reinberg D, Heard E. 2004. Epigenetic dynamics of imprinted X inactivation during early mouse development. *Science* **303:** 644–649.

Okamoto I, Arnaud D, Le Baccon P, Otte AP, Disteche CM, Avner P, Heard E. 2005. Evidence for de novo imprinted X-chromosome inactivation independent of meiotic inactivation in mice. *Nature* **438:** 369–373.

Olek A, Walter J. 1997. The pre-implantation ontogeny of the H19 methylation imprint. *Nat Genet* **17:** 275–276.

Oswald J, Engemann S, Lane N, Mayer W, Olek A, Fundele R, Dean W, Reik W, Walter J. 2000. Active demethylation of the paternal genome in the mouse zygote. *Curr Biol* **10:** 475–478.

Parvanov ED, Petkov PM, Paigen K. 2010. Prdm9 controls activation of mammalian recombination hotspots. *Science* **327:** 835.

Peters AH, O'Carroll D, Scherthan H, Mechtler K, Sauer S, Schofer C, Weipoltshammer K, Pagani M, Lachner M, Kohlmaier A, et al. 2001. Loss of the Suv39h histone methyltransferases impairs mammalian heterochromatin and genome stability. *Cell* **107:** 323–337.

Piccolo FM, Fisher AG. 2013. Getting rid of DNA methylation. *Trends Cell Biol* doi: 10.1016/j.tcb.2013.09.001.

Piotrowska-Nitsche K, Perea-Gomez A, Haraguchi S, Zernicka-Goetz M. 2005. Four-cell stage mouse blastomeres have different developmental properties. *Development* **132:** 479–490.

Popp C, Dean W, Feng S, Cokus SJ, Andrews S, Pellegrini M, Jacobsen SE, Reik W. 2010. Genome-wide erasure of DNA methylation in mouse primordial germ cells is affected by AID deficiency. *Nature* **463:** 1101–1105.

Rando TA, Chang HY. 2012. Aging, rejuvenation and epigenetic reprogramming: Resetting the aging clock. *Cell* **148:** 46–57.

Reik W, Dean W, Walter J. 2001. Epigenetic reprogramming in mammalian development. *Science* **293:** 1089–1093.

Ren B, Chee KJ, Kim TH, Maniatis T. 1999. PRDI-BF1/Blimp-1 repression is mediated by corepressors of the Groucho family of proteins. *Genes Dev* **13:** 125–137.

Resnick JL, Bixler LS, Cheng L, Donovan PJ. 1992. Long-term proliferation of mouse primordial germ cells in culture. *Nature* **359:** 550–551.

Rossant J. 2001. Stem cells from the mammalian blastocyst. *Stem Cells* **19:** 477–482.

Rougier N, Bourc'his D, Gomes DM, Niveleau A, Plachot M, Paldi A, Viegas-Pequignot E. 1998. Chromosome methylation patterns during mammalian preimplantation development. *Genes Dev* **12:** 2108–2113.

Roy S, Ng T. 2004. Blimp-1 specifies neural crest and sensory neuron progenitors in the zebrafish embryo. *Curr Biol* **14:** 1772–1777.

Saitou M, Barton SC, Surani MA. 2002. A molecular programme for the specification of germ cell fate in mice. *Nature* **418:** 293–300.

Saitou M, Payer B, Lange UC, Erhardt S, Barton SC, Surani MA. 2003. Specification of germ cell fate in mice. *Philos Trans R Soc Lond B Biol Sci* **358:** 1363–1370.

Saitou M, Kagiwada S, Kurimoto K. 2012. Epigenetic reprogramming in mouse pre-implantation development and primordial germ cells. *Development* **139:** 15–31.

Santos F, Hendrich B, Reik W, Dean W. 2002. Dynamic reprogramming of DNA methylation in the early mouse embryo. *Dev Biol* **241:** 172–182.

Santos F, Peters AH, Otte AP, Reik W, Dean W. 2005. Dynamic chromatin modifications characterise the first cell cycle in mouse embryos. *Dev Biol* **280:** 225–236.

Santos F, Peat J, Burgess H, Rada C, Reik W, Dean W. 2013. Active demethylation in mouse zygotes involves cytosine deamination and base excision repair. *Epigenetics Chromatin* **6:** 39.

Sarmento OF, Digilio LC, Wang Y, Perlin J, Herr JC, Allis CD, Coonrod SA. 2004. Dynamic alterations of specific histone modifications during early murine development. *J* Cell Sci **117:** 4449–4459.

Schaner CE, Deshpande G, Schedl PD, Kelly WG. 2003. A conserved chromatin architecture marks and maintains the restricted germ cell lineage in worms and flies. *Dev Cell* **5:** 747–757.

Sciammas R, Davis MM. 2004. Modular nature of Blimp-1 in the regulation of gene expression during B cell maturation. *J Immunol* **172:** 5427–5440.

Seisenberger S, Andrews S, Krueger F, Arand J, Walter J, Santos F, Popp C, Thienpont B, Dean W, Reik W. 2012. The dynamics of genome-wide DNA methylation reprogramming in mouse primordial gene cells. *Mol Cell* **48:** 849–862.

Seki Y, Yamaji M, Yabuta Y, Sano M, Shigeta M, Matsui Y, Saga Y, Tachibana M, Seki Y, Hayashi K, Itoh K, Mizugaki M, Saitou M, Matsui Y. 2005. Extensive and orderly reprogramming of genome-wide chromatin modifications associated with specification and early development of germ cells in mice. *Dev Biol* **278:** 440–458.

Seki Y, Yamaji M, Yabuta Y, Sano M, Shigeta M, Matsui Y, Saga Y, Tachibana M, Shinkai Y, Saitou M. 2007. Cellular dynamics associated with the genome-wide epigenetic reprogramming in migrating primordial germ cells in mice. *Development* **13:** 2627–2638.

Seydoux G, Dunn MA. 1997. Transcriptionally repressed germ cells lack a subpopulation of phosphorylated RNA polymerase II in early embryos of *Caenorhabditis elegans* and *Drosophila melanogaster. Development* **124:** 2191–2201.

Seydoux G, Strome S. 1999. Launching the germline in *Caenorhabditis elegans*: Regulation of gene expression in early germ cells. *Development* **126:** 3275–3283.

Siomi MC, Sato K, Pezic D, Aravin AA. 2011. PIWI-interacting small RNAs: The vanguard of genome defence. *Nat Rev Mol Cell Biol* **12:** 246–258.

Smallwood SA, Tomizawa S, Krueger F, Ruf N, Carli N, Segonds-Pichon A, Sato S, Hata K, Andrews SR, Kelsey G. 2011. Dynamic CpG island methylation landscape in oocytes and preimplantation embryos. *Nat Genet* **43:** 811–814.

Smith AG. 2001. Embryo-derived stem cells: Of mice and men. *Annu Rev Cell Dev Biol* **17:** 435–462.

Smith ZD, Chan MM, Mikkelsen TS, Gu H, Gnirke A, Regev A, Meissner A. 2012. A unique regulatory phase of DNA methylation in the early mammalian embryo. *Nature* **484:** 339–344.

Surani MA. 2001. Reprogramming of genome function through epigenetic inheritance. *Nature* **414:** 122–128.

Surani MA, Ancelin K, Hajkova P, Lange UC, Payer B, Western P, Saitou M. 2004. Mechanism of mouse germ cell specification: A genetic program regulating epigenetic reprogramming. *Cold Spring Harb Symp Quant Biol* **69:** 1–9.

Surani MA. 2005. Nuclear reprogramming by human embryonic stem cells. *Cell* **122:** 653–654.

Tada M, Tada T, Lefebvre L, Barton SC, Surani MA. 1997. Embryonic germ cells induce epigenetic reprogramming of somatic nucleus in hybrid cells. *EMBO J* **16:** 6510–6520.

Tada T, Tada M, Hilton K, Barton SC, Sado T, Takagi N, Surani MA. 1998. Epigenotype switching of imprintable loci in embryonic germ cells. *Dev Genes Evol* **207:** 551–561.

Tada M, Takahama Y, Abe K, Nakatsuji N, Tada T. 2001. Nuclear reprogramming of somatic cells by in vitro hybridization with ES cells. *Curr Biol* **11:** 1553–1558.

* Takahashi K. 2014. Cellular reprogramming. *Cold Spring Harb Perspect Biol* **6:** a018606.

Takahashi K, Yamanaka S. 2006. Induction of pluripotent stem cells from mouse embryonic and adult fibroblast cultures by defined factors. *Cell* **126:** 663–676.

Thomson JA, Itskovitz-Eldor J, Shapiro SS, Waknitz MA, Swiergiel JJ, Marshall VS, Jones JM. 1998. Embryonic stem cell lines derived from human blastocysts. *Science* **282:** 1145–1147.

Torres-Padilla ME, Parfitt DE, Kouzarides T, Zernicka-Goetz M. 2010. Histone arginine methylation regulates pluripotency in the early mouse embryo. *Nature* **445:** 214–218.

Toyooka Y, Tsunekawa N, Akasu R, Noce T. 2003. Embryonic stem cells can form germ cells in vitro. *Proc Natl Acad Sci* **100:** 11457–11462.

Turner CA Jr, Mack DH, Davis MM. 1994. Blimp-1, a novel zinc finger-containing protein that can drive the maturation of B lymphocytes into immunoglobulin-secreting cells. *Cell* **77:** 297–306.

van der Heijden GW, Dieker JW, Derijck AA, Muller S, Berden JH, Braat DD, van der Vlag J, de Boer P. 2005. Asymmetry in histone H3 variants and lysine methylation between paternal and maternal chromatin of the early mouse zygote. *Mech Dev* **122:** 1008–1022.

Vincent SD, Dunn NR, Sciammas R, Shapiro-Shalef M, Davis MM, Calame K, Bikoff EK, Robertson EJ. 2005. The zinc finger transcriptional repressor Blimp1/Prdm1 is dispensable for early axis formation but is required for specification of primordial germ cells in the mouse. *Development* **132:** 1315–1325.

Vincent JJ, Li Z, Lee SA, Liu X, Etter MO, Diaz-Perez SV, Taylor SK, Gkountela S, Lindgren AG, Clark AT. 2011. Single cell analysis facilities staging of Blimp1-dependent primordial germ cells derived from mouse embryonic stem cells. *PLoS ONE* **6:** e28960.

Vincent JJ, Huang Y, Chen PY, Feng S, Calvopiña JH, Nee K, Lee SA, Le T, Yoon AJ, Faull K, et al. 2013. Stage-specific roles for Tet1 and Tet2 in DNA demethylation in primordial germ cells. *Cell Stem Cell* **12:** 470–478.

Waddington C. 1956. *Principles of embryology.* Allen and Unwin, London.

Weber RJ, Pedersen RA, Wianny F, Evans MJ, Zernicka-Goetz M. 1999. Polarity of the mouse embryo is anticipated before implantation. *Development* **126:** 5591–5598.

Wossidlo M, Nakamura T, Lepikhov K, Marques CJ, Zakhartchenko V, Boiani M, Arand J, Nakano T, Reik W, Walter J. 2011. 5-Hydroxymethylcytosine in the mammalian zygote is linked with epigenetic reprogramming. *Nat Commun* **2:** 241.

Yamaguchi S, Kimura H, Tada M, Nakatsuji N, Tada T. 2005. Nanog expression in mouse germ cell development. *Gene Expr Patterns* **5:** 639–646.

Yamaguchi S, Hong K, Liu R, Inoue A, Shen L, Zhang K, Zhang Y. 2013. Dynamics of 5-methylcytosine and 5-hydroxymethylcytosine during germ cell reprogramming. *Cell Res* **23:** 329–339.

Yamaji M, Seki Y, Kurimoto K, Yabuta Y, Yuasa M, Shigeta M, Yamanaka K, Ohinata Y, Saitou M. 2008. Critical function of Prdm14 for the establishment of the germ cell lineage in mice. *Nat Genet* **40:** 1016–1022.

Young RA. 2011. Control of the embryonic stem cell state. *Cell* **144:** 940–954.

Zhang F, Barboric M, Blackwell TK, Peterlin BM. 2003. A model of repression: CTD analogs and PIE-1 inhibit transcriptional elongation by P-TEFb. *Genes Dev* **17:** 748–758.

C H A P T E R 28

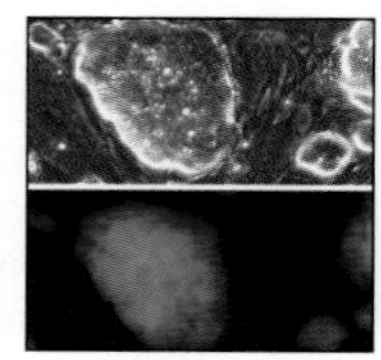
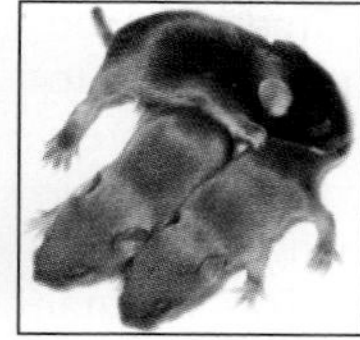

Induced Pluripotency and Epigenetic Reprogramming

Konrad Hochedlinger[1,2] and Rudolf Jaenisch[2]

[1]Howard Hughes Medical Institute at Massachusetts General Hospital, Department of Stem Cell and Regenerative Biology, Harvard University and Harvard Medical School, Boston, Massachusetts 02114; [2]Whitehead Institute and Department of Biology, Massachusetts Institute of Technology, Cambridge, Massachusetts 02142

Correspondence: khochedlinger@helix.mgh.harvard.edu

SUMMARY

Induced pluripotency defines the process by which somatic cells are converted into induced pluripotent stem cells (iPSCs) upon overexpression of a small set of transcription factors. In this chapter, we put transcription factor–induced pluripotency into a historical context, review current methods to generate iPSCs, and discuss mechanistic insights that have been gained into the process of reprogramming. In addition, we focus on potential therapeutic applications of induced pluripotency and emerging technologies to efficiently engineer the genomes of human pluripotent cells for scientific and therapeutic purposes.

Outline

Chapter-opening figures: (*Right*) Reprinted from Stadtfeld et al. (2010).

Cite as *Cold Spring Harb Perspect Biol* doi: 10.1101/cshperspect.a019448

OVERVIEW

Somatic cell nuclear transfer experiments in animals showed that the genome of a differentiated cell remains equivalent to that of a fertilized egg. Consequently, differential gene expression responsible for the formation of the 200 cell types of our body is the result of reversible epigenetic changes that are imposed on the genome during development. This seminal discovery raised fundamental questions about the mechanisms by which a somatic genome is epigenetically reprogrammed to an early embryonic state. In addition, the marriage of cloning and embryonic stem cell technology provided a means to generate custom-tailored cells in potential therapeutic settings. Although ethical, legal, and biological barriers associated with somatic cell nuclear transfer prevented significant progress toward this goal over the past 10 years, it motivated attempts to directly reprogram adult cells into pluripotent cells. Indeed, this concept was realized in 2006 by the isolation of induced pluripotent stem cells (iPSCs) directly from skin cells. iPSCs are generated by activating a handful of embryonic genes in somatic cells, giving rise to cells that closely resemble embryonic stem cells without ever going through development. Studies on the process of induced pluripotency have yielded important insights into the mechanisms by which transcription factors and epigenetic regulators cooperate to establish cell fates during development. They further revealed an unexpected plasticity of the differentiated cell state and led to the successful interconversion of other differentiated cell types by activating alternative sets of genes. Importantly, iPSCs have been derived from human patients, raising the possibility that these cells could be used to study and, perhaps, treat degenerative diseases.

Cite as *Cold Spring Harb Perspect Biol* doi: 10.1101/cshperspect.a019448

1 HISTORY OF CELLULAR REPROGRAMMING

The discovery of induced pluripotency represents the synthesis of scientific principles and technologies that have been developed over the last six decades (Fig. 1) (Stadtfeld and Hochedlinger 2010). These are notably (1) the demonstration by somatic cell nuclear transfer (SCNT) that differentiated cells retain the same genetic information as early embryonic cells; (2) the development of techniques that allowed researchers to derive, culture, and study pluripotent cell lines; and (3) the observation that transcription factors are key determinants of cell fate whose enforced expression can switch one mature cell type into another. In this section, we will briefly summarize these three areas of research and the influence they have had on the generation of iPSCs.

1.1 Nuclear Transfer and the Cloning of Animals

During mammalian development, cells gradually lose potential and become progressively differentiated to fulfill the specialized functions of somatic tissues. For example, only zygotes and blastomeres of early morulae (Kelly 1977) retain the ability to give rise to all embryonic and extraembryonic tissues and are therefore called "totipotent," whereas cells of the inner cell mass (ICM) of the blastocyst give rise to all embryonic, but not to extraembryonic tissues, and are hence coined "pluripotent." Stem cells residing in adult tissues can only give rise to cell types within their lineage and are, depending on the number of cell types they produce, either called "multipotent" or "unipotent" (Table 1). On terminal differentiation, cells entirely lose their developmental potential.

During the 1950s and 1960s, Briggs and King established the technique of SCNT or "cloning" to probe the developmental potential of nuclei isolated from late-stage embryos and tadpoles by transplanting them into enucleated oocytes (Fig. 1) (Briggs and King 1952, 1957). This work, together with seminal experiments by John Gurdon (Gurdon 1962), showed that differentiated amphibian cells

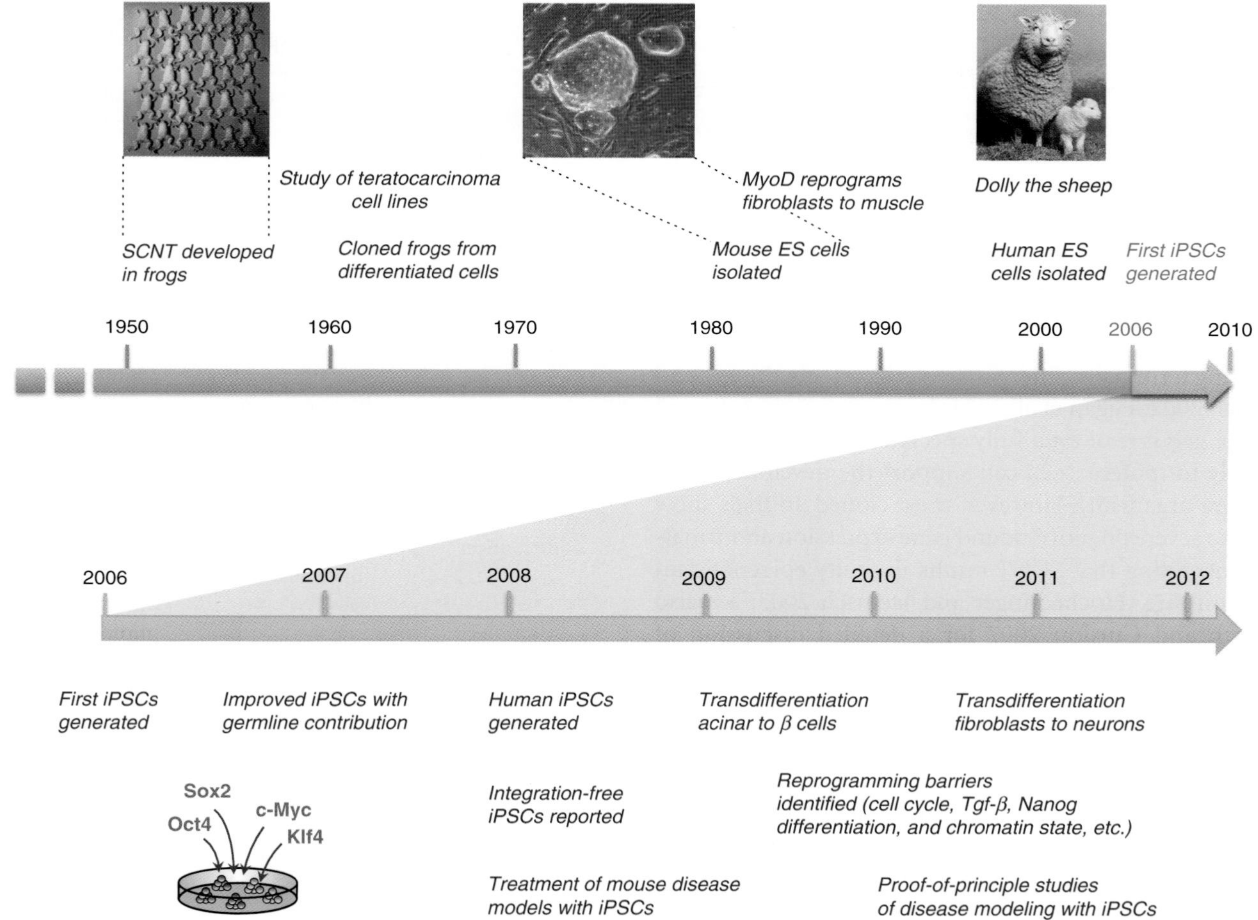

Figure 1. Historic time line of reprogramming research. Shown are seminal discoveries leading to the first generation of iPSCs in 2006, as well as progress in the generation and subsequent application of iPSCs.

Table 1. Definition of some terms

Potency	Sum of developmental options accessible to the cell
Totipotent	Ability to form all lineages of the organism; in mammals, only the zygote and first cleavage blastomeres are totipotent.
Pluripotent	Ability to form all lineages of the body (e.g., embryonic stem cells).
Multipotent	Ability of adult stem cells to form multiple cell types of one lineage (e.g., hematopoietic stem cells).
Unipotent	Cells form one cell type (e.g., spermatogonial stem cells, which can only generate sperm).
Reprogramming	Increase in potency and dedifferentiation; can be induced by nuclear transfer, cell fusion, genetic manipulation.
Transdifferentiation, plasticity	Notion that somatic stem cells have broadened potency and can generate cells of other lineages, a concept that is controversial in mammals. More recently, transdifferentiation also refers to transcription factor–induced lineage conversions among differentiated cell types.

indeed retain the genetic information necessary to support the generation of cloned frogs. The major conclusion from these experiments was that development imposes reversible epigenetic rather than irreversible genetic changes on the genome during cellular differentiation. The cloning of Dolly the sheep (Wilmut et al. 1997) and several other mammals from adult cells (Meissner and Jaenisch 2006) including terminally differentiated cells (Hochedlinger and Jaenisch 2002; Eggan et al. 2004; Inoue et al. 2005) showed that the genome of even fully specialized cells remains genetically totipotent (i.e., can support the development of an entire organism). However, most cloned animals show subtle to severe phenotypic and gene expression abnormalities, suggesting that SCNT results in faulty epigenetic reprogramming (Hochedlinger and Jaenisch 2003; see also Jaenisch and Gurdon 2007 for a detailed discussion of SCNT).

1.2 Pluripotent Cell Lines and Fusion Hybrids

Although SCNT is a powerful tool to probe the developmental potential of a cell, it is technically challenging and not well suited for genetic or biochemical studies. A major advance toward isolating iPSCs was the establishment of immortal pluripotent cell lines that maintained their ability to differentiate into essentially all cell types of the body when reintroduced into early embryos. Pluripotent stem cell lines were initially derived from teratocarcinomas, tumors of germ cell origin, giving rise to so-called embryonal carcinoma (EC) cells (Kleinsmith and Pierce 1964). Although EC cell lines fulfilled some pluripotency criteria (Table 2) such as teratoma formation and chimera contribution, they rarely contributed to the germline because of their tumorigenic origin. These findings motivated attempts to isolate pluripotent cell lines directly from embryos and subsequently led to the derivation of embryonic stem (ES) cells from the ICM of mouse and human blastocysts (Fig. 2) (Evans and Kaufman 1981; Martin 1981; Thomson et al. 1998). Mouse ES cells not only contribute

Table 2. Commonly used functional criteria to assess the developmental potential of cells

Assay	Experimental approach	Limitations
In vitro differentiation	Differentiation is induced in cultured cells, and cells are assayed for the expression of cell type–specific markers.	The expression of differentiation markers is no test for functionality; marker expression can be due to cellular stress response.
Teratoma formation	The induction of tumors shows the potential to generate differentiated cell types of various lineages.	Does not test for the ability of cells to promote normal development.
Chimera formation	Cells injected into a host blastocyst can be assessed for their contribution to normal development.	Host-derived cells in chimera may complement cell nonautonomous defects.
Germline contribution	Ability of test cells to generate functional germ cells.	Excludes genetic, but not epigenetic, defects that could interfere with promoting development.
Tetraploid complementation	Injection of test cells into 4n host blastocyst. Because 4n host cells cannot contribute to somatic lineages, an embryo is exclusively composed of test cells.	Most stringent test for pluripotency; does not test for the ability to form the trophectoderm (placental) lineage.

Cite as *Cold Spring Harb Perspect Biol* doi: 10.1101/cshperspect.a019448

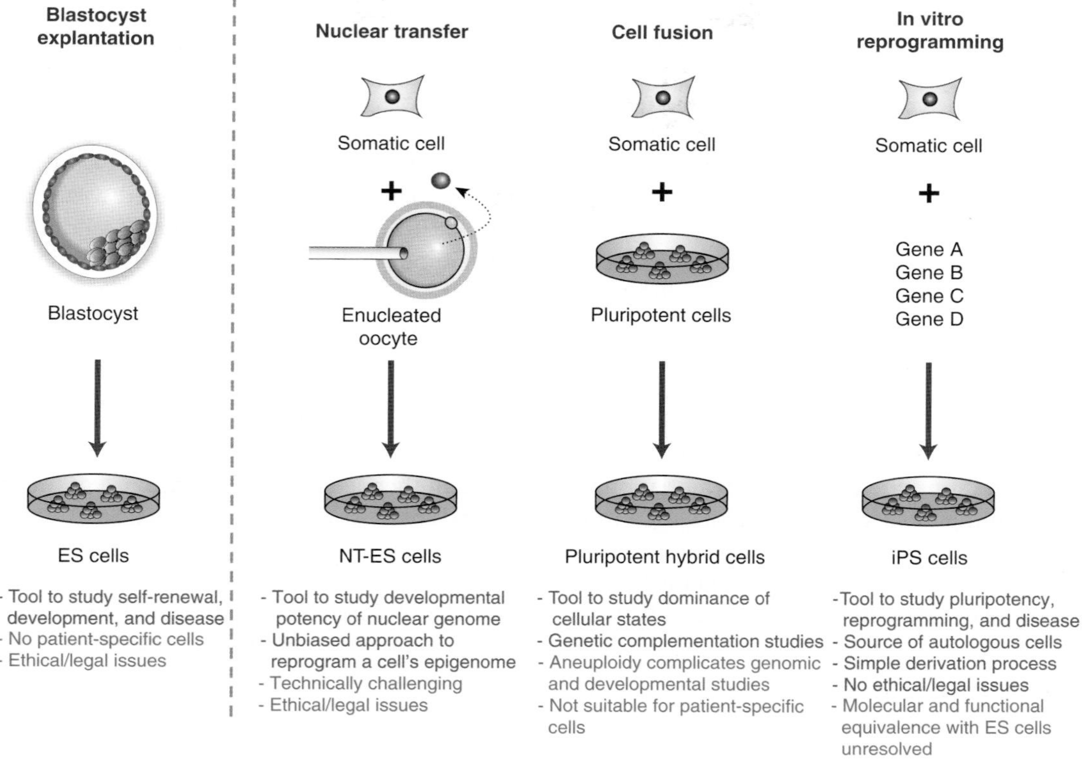

Figure 2. Sources of pluripotent stem cells. Comparison of different strategies used to derive pluripotent stem cell lines; their advantages (in green) or disadvantages (in red) are summarized at the *bottom* of each column. ES cells, embryonic stem cells; NT-ES cells, nuclear transfer-ES cells.

to adult tissues, including germ cells in chimeric mice, but also support the development of entirely ES cell--derived animals after injection into tetraploid blastocysts (Nagy et al. 1990; Eggan et al. 2001). Tetraploid blastocysts are generated by electrofusion of fertilized two-cell embryos; these embryos can only develop into extraembryonic tissues (i.e., the placenta), but fail to give rise to the fetus. This "tetraploid complementation assay" (referred to in Table 2) represents the most stringent developmental assay available in the mouse to test for pluripotency. ES cell lines can also be derived from cloned mouse (Munsie et al. 2000; Wakayama et al. 2001) and human (Tachibana et al. 2013) blastocysts generated by SCNT, generating so-called NT-ES cells (Fig. 2). In contrast to the abnormalities seen in directly cloned animals, NT-ES cells are molecularly and functionally indistinguishable from fertilization-derived ES cells, presumably because of a selection of faithfully reprogrammed cells in culture (Brambrink et al. 2006; Wakayama et al. 2006).

The study of hybrids generated by cell fusion between different cell types has also been instrumental for the identification of factors that could directly induce pluripotency in somatic cells (Yamanaka and Blau 2010). Specifically, when EC or ES cells are fused with somatic cells, the resulting hybrid cells acquire biochemical and developmental properties of pluripotent cells and extinguish features of the somatic fusion partner (Fig. 2) (Miller and Ruddle 1976; Tada et al. 2003). This dominance of the pluripotent state over the somatic state in hybrids suggested that soluble transacting factors must exist within pluripotent cells that can confer a pluripotent state on somatic cells and these factors should be identifiable (Yamanaka and Blau 2010).

1.3 Transcription Factors and Lineage Switching

The third principle that contributed to the discovery of induced pluripotency was the observation that lineage-associated transcription factors can change cell fate when ectopically expressed in certain heterologous cells. Transcription factors help to establish and maintain cellular identity during development by driving the expression of cell type–specific genes while suppressing lineage-inappropriate genes. This principle was first shown by the formation of myofibers in fibroblast cell lines transduced with retroviral vectors expressing the skeletal muscle transcription factor MyoD (Davis et al. 1987). Subsequently, Graf and colleagues discovered that primary B and T cells could be efficiently converted into functional macrophages on overexpression of the myeloid transcription factor C/EBPα (Xie et al. 2004; Laiosa et al. 2006). More recently, researchers have identified sets of transcription factors that induce the conversion of pancreatic acinar cells into insulin-producing β cells by overexpressing the pancreatic factors *MafA*, Pdx,1 and *Ngn3* (Zhou et al. 2008). Similarly, the conversion of fibroblasts into neurons can be achieved by the activation of the neural factors *Ascl1*, *Brn2*, and *Myt1l* (Vierbuchen et al. 2010); fibroblasts can be made into cardiomyocytes by the cardiac factors *Gata4*, *Mef2c*, and *Tbx5* (Ieda et al. 2010); and fibroblasts can be converted into hepatocytes on overexpression of *HNF1alpha*, *Foxa3*, and optionally *Gata4 factors* (Huang et al. 2011). The early muscle and immune cell transdifferentiation experiments provided the intellectual framework for a more systematic search for transcription factors that could induce the conversion of differentiated cells to a pluripotent state as discussed below (see also Ch. 2 [Takahashi 2014]).

2 GENERATION OF iPSCs

2.1 Screen for Reprogramming Factors

To identify transcriptional regulators that are sufficient for reprogramming adult cells into pluripotent cells, Yamanaka and Takahashi devised an elegant screen for factors that could activate a dormant drug resistance allele integrated into the ES cell–specific *Fbxo15* locus (Fig. 3). This selection approach was chosen to ensure that potentially rare reprogrammed cells could be detected and nonreprogrammed colonies and transformed cells would be eliminated. The investigators selected 24 genes that were specifically expressed in pluripotent cells or had previously been implicated in ES cell biology. The combination of all 24 factors, when coexpressed from retroviral vectors in mouse fibroblasts, indeed activated *Fbxo15* and induced the formation of drug-resistant colonies with characteristic ES cell morphology, albeit at extremely low efficiencies (0.01%–0.1%; Fig. 3) (Takahashi and Yamanaka 2006). Successive rounds of elimination of individual factors from the gene cocktail then led to the identification of the minimally required core set of four factors, Klf4, Sox2, c-Myc, and Oct4. iPSCs generated by this selection approach also expressed markers of pluripotent stem cells such as the surface antigen SSEA-1 and Nanog, generated teratomas when injected subcutaneously into immunocompromised mice, and contributed to different tissues of the developing embryo upon blastocyst injection, thereby meeting some criteria of pluripotency (Table 2). However, these iPSCs expressed lower levels of several other key pluripotency genes compared to ES cells, showed incomplete reprogramming of

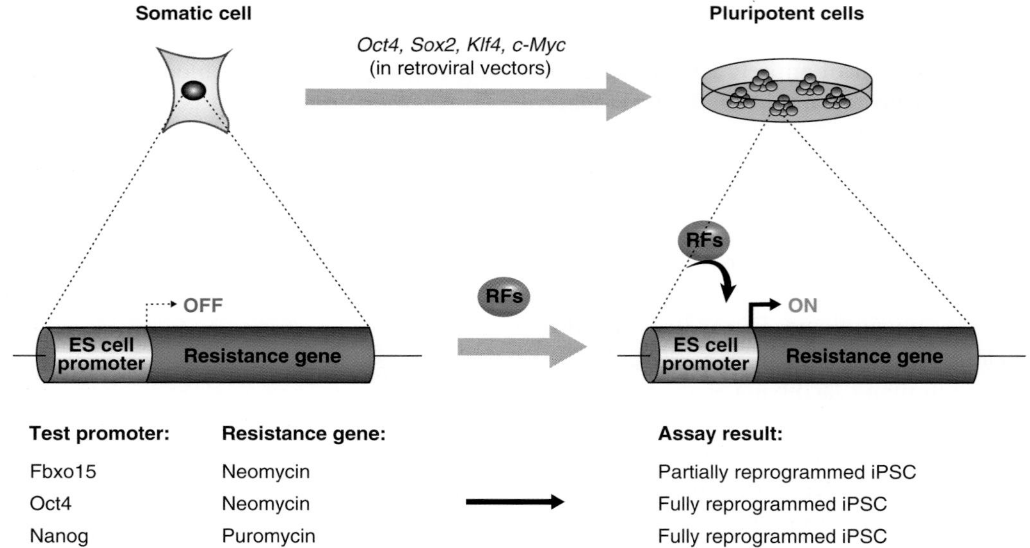

Figure 3. Strategy to derive iPSCs. (*Top*) Schematic representation of the first successful attempt to produce iPSCs by Takahashi and Yamanaka. (*Bottom*) The genetic assay system used to screen for factors that could reprogram to pluripotency (reprogramming factors [RFs]). Partial reprogramming to iPSCs was achieved by viral infection of cells with Oct4, Sox2, Klf4, and c-Myc, followed by drug selection for *Fbxo15*-expressing cells. In contrast, subsequent modifications to the assay selecting for *Oct4*- or *Nanog*-expressing cells gave rise to fully reprogrammed iPSCs. Note that drug selection is not essential for producing high-quality iPSCs, but was used as part of the assay to identify factors that induced embryonic gene expression (see text).

epigenetic marks, and failed to generate postnatal chimeras or contribute to the germline. These initially derived iPSCs therefore appeared to be partially reprogrammed.

Soon after this report, several laboratories including Yamanaka's were able to reproduce and improve upon these findings. For example, by selecting for the reactivation of the essential pluripotency genes *Nanog* or *Oct4* instead of *Fbxo15*, iPSCs were generated that molecularly and functionally more closely resembled ES cells (Fig. 3) (Maherali et al. 2007; Okita et al. 2007; Wernig et al. 2007). More recently, iPSCs have been identified that are even capable of generating "all-iPSC" mice after injection into tetraploid blastocysts (Table 2) (Boland et al. 2009; Kang et al. 2009; Zhao et al. 2009), thus suggesting that at least some iPSC clones have a developmental potency equivalent to that of ES cells.

Importantly, high-quality mouse iPSCs can be derived from genetically unmodified somatic cells without drug selection by simply using morphological criteria (Blelloch et al. 2007; Maherali et al. 2007; Meissner et al. 2007). This discovery was critical for extending induced pluripotency to other species for which transgenic tools were not readily available. For example, iPSCs have been successfully generated from human (Takahashi et al. 2007; Yu et al. 2007; Park et al. 2008), rat (Li et al. 2009b), and rhesus monkey fibroblasts (Liu et al. 2008) by expression of the same four Yamanaka factors, demonstrating that fundamental features of the transcriptional network governing pluripotency remain conserved during evolution. In addition, iPSCs have been derived from other somatic cell populations such as keratinocytes (Aasen et al. 2008; Maherali et al. 2008), neural cells (Eminli et al. 2008; Kim et al. 2008), stomach and liver (Aoi et al. 2008), as well as from genetically labeled pancreatic β cells (Stadtfeld et al. 2008a), melanocytes (Utikal et al. 2009a), and terminally differentiated B and T lymphocytes (Hanna et al. 2008b; Eminli et al. 2009), thus underscoring the generality of induced pluripotency across different cell types.

2.2 Genetically Unmodified iPSCs

Retroviral transgenes used to deliver the reprogramming factors are usually silenced toward the end of reprogramming (Stadtfeld et al. 2008b) by a mechanism that involves DNA (Lei et al. 1996) and histone methylation (Matsui et al. 2010). However, this process is often incomplete, resulting in partially reprogrammed cell lines that fail to activate endogenous pluripotency genes and therefore continue to depend on transgenic reprogramming factor expression for indefinite growth (Takahashi and Yamanaka 2006; Mikkelsen et al. 2008; Sridharan et al. 2009). In addition, residual activity or reactivation of viral transgenes in iPSC-derived somatic cells can interfere with their developmental potential and frequently leads to the formation of tumors in chimeric animals (Okita et al. 2007). These shortcomings spurred efforts to derive iPSCs devoid of viral vector sequences. The first integration-free iPSCs were generated from adult mouse hepatocytes using nonintegrating adenoviral vectors (Stadtfeld et al. 2008c) and from mouse embryonic fibroblasts (MEFs) after transfection with plasmids (Okita et al. 2008) or RNA viruses (Fusaki et al. 2009) that persisted only transiently inside cells. Importantly, chimeric animals produced from integration-free iPSCs were tumor-free. Alhough these methods were extremely inefficient, they led to two important conclusions. First, viral integration and insertional mutagenesis are not required for stable cellular reprogramming. Second, direct reprogramming does not necessarily generate pluripotent cells of poorer quality or compromised safety than ES cells. However, it is important in all of these approaches to completely exclude the possibility that vector fragments or RNA viruses persisted in the resulting iPSCs.

Various new methods have been developed to generate genetically unmodified or "reprogramming factor-free" human iPSCs with alternative technologies. These include transfection with messenger RNA (mRNA) instead of DNA, replacement of factors by small molecules, and delivery of factors as recombinant proteins or extracts. The extremely low efficiency of reprogramming cells with recombinant proteins (Zhou et al. 2009) or extracts (Kim et al. 2009a) makes this an impractical strategy for iPSC generation. Another approach has used modified mRNA constructs to express the reprogramming factors in somatic donor cells, giving rise to human iPSCs (Warren et al. 2010). This method is efficient in yielding factor-free iPSCs and thus may be the preferred method for reprogramming. Last, a variety of small molecules have been shown to replace individual reprogramming factors (Huangfu et al. 2008; Xu et al. 2008; Ichida et al. 2009; Lyssiotis et al. 2009). Remarkably, the combination of certain compounds is sufficient to induce pluripotency in somatic cells according to a recent report (Hou et al. 2013).

In summary, the generation of genetically unmodified vector-free iPSCs has been resolved in principle. It is likely that, for a given application, one of the various strategies outlined above may be more optimal than another.

2.3 Reprogrammable Mice

A technical advance for studying molecular mechanisms of reprogramming has been the development of so-called "secondary reprogramming systems" and "reprogrammable mice" (Fig. 4). This approach entails differentiating "primary" iPSC clones using either in vitro differentiation for

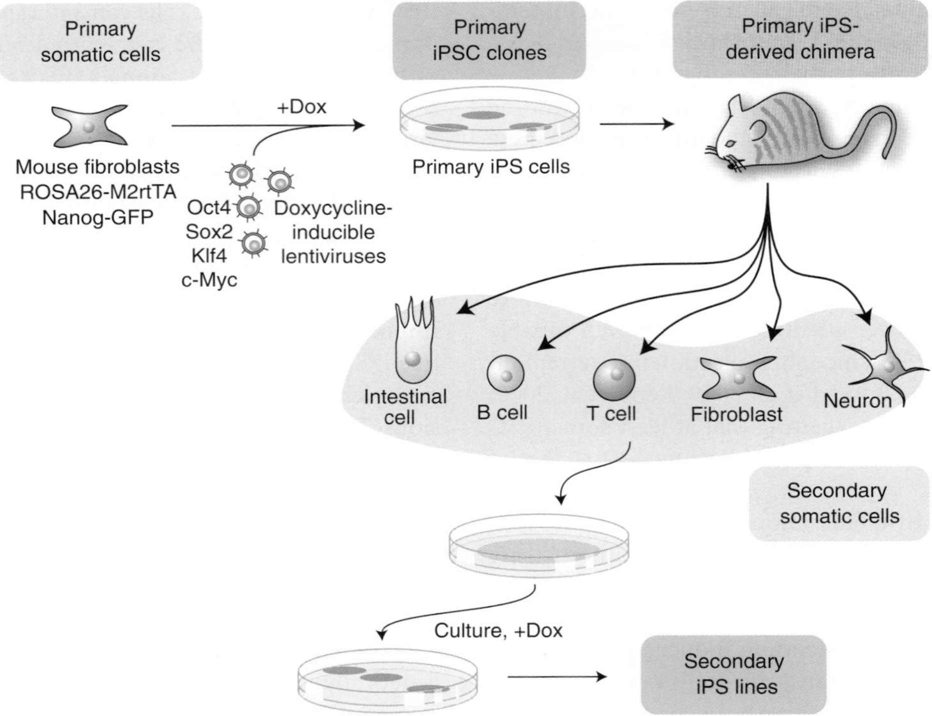

Figure 4. Generation of genetically homogeneous cell cultures for epigenetic reprogramming. Scheme for obtaining genetically homogeneous somatic cells that are more efficient at pluripotency induction. Primary somatic cells with stable integrations of a Nanog-GFP (green fluorescent protein) marker and reverse tetracycline transactivator (M2rtTA) are infected with DOX-inducible lentiviruses encoding the four reprogramming factors. Primary iPSCs are generated by culturing the cells in DOX to activate the factors. After DOX withdrawal, the primary iPSCs are injected into mouse blastocysts and "secondary" somatic cells carrying the DOX-inducible vectors are cultured in the presence of DOX to produce secondary iPSCs. The key advantage of this system is that reprogramming can be induced without new virus infection at a much higher efficiency. (Modified, with permission, from Hanna et al. 2008a.)

human cells (Hockemeyer et al. 2008; Maherali et al. 2008) or blastocyst injection for mice (Wernig et al. 2008; Woltjen et al. 2009). Primary iPSCs were generated by introducing the Yamanaka transcription factor cocktail carried on doxycyline-inducible lentiviral vectors or transposons into somatic cells. These genetically homogeneous differentiated cells are then cultured in doxycycline-containing media, thus triggering the formation of "secondary" iPSCs at efficiencies that are generally several orders of magnitude higher (1%–5%) than the efficiencies obtained after primary infection (0.01%–1%). This observation showed that the low efficiency of reprogramming is not solely the result of ineffective transduction of somatic cells by all four viral vectors as had originally been assumed. It is instead consistent with the notion that other, presumably epigenetic, roadblocks must exist to limit the acquisition of pluripotency.

In a modification of the conventional secondary system, "reprogrammable" mouse strains have been developed that contain a single doxycycline-inducible polycistronic transgene that has been targeted to a defined genomic position by homologous recombination. This system does not depend on viral infection anymore and facilitates the derivation of iPSCs from virtually any cell type of the mouse by simply adding doxycycline to the culture medium (Carey et al. 2010; Stadtfeld et al. 2010b).

3 MECHANISMS UNDERLYING iPSC FORMATION

In the following section, we introduce models that have been developed to explain the low efficiency of reprogramming at a cellular level. We then discuss key molecular events that act as barriers during the reprogramming process and speculate on the role of the individual reprogramming factors.

3.1 Deterministic versus Stochastic Models

Two main models have been put forward to explain the reprogramming process (Fig. 5A) (Yamanaka 2009). The "deterministic" model posits that individual somatic cells synchronously convert into iPSCs with a constant latency

 Cite as *Cold Spring Harb Perspect Biol* doi: 10.1101/cshperspect.a019448

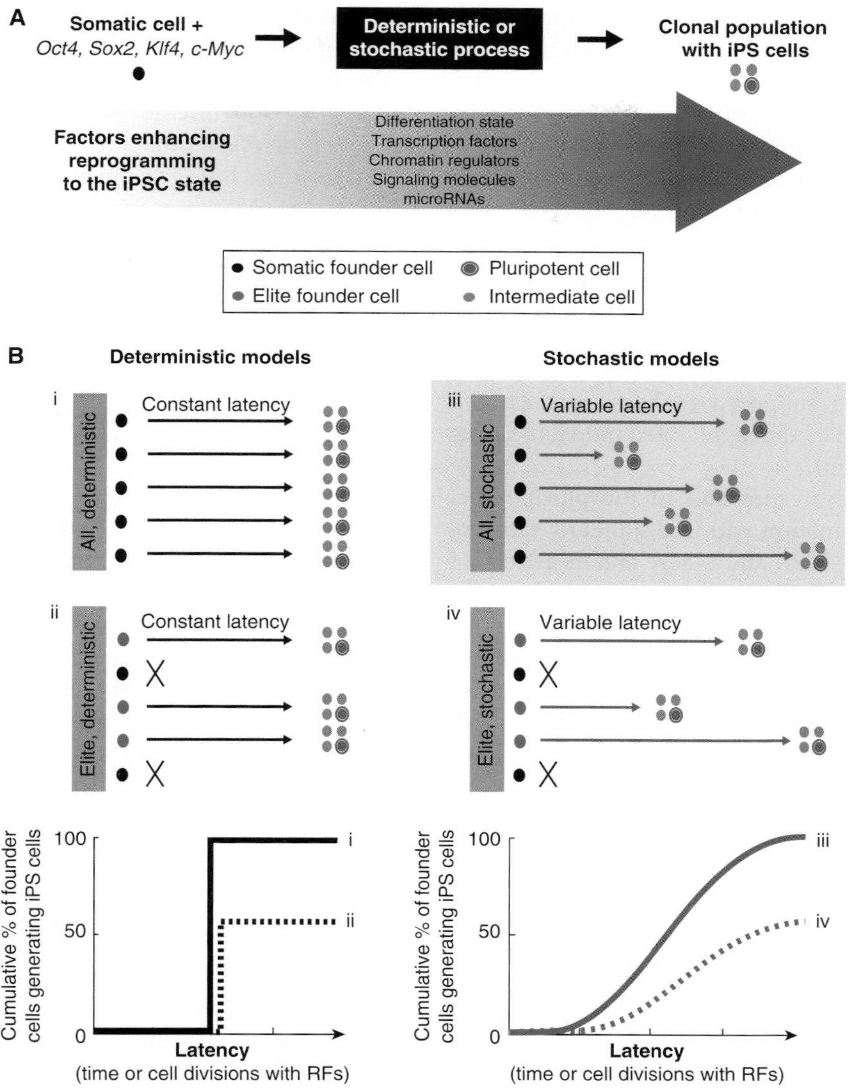

Figure 5. Stochastic and deterministic models of cellular reprogramming into iPSCs. (*A*) Schematic representation of the reprogramming process. (*B*) Representation of four possible models to explain the low efficiency of reprogramming. The deterministic model posits that (i) all somatic cells or (ii) a subset of somatic cells termed elite founder cells gives rise to iPSCs with the same predetermined latency. In contrast, the stochastic model predicts that (iii) all cells or (iv) a subpopulation of "elite" cells produces iPSCs with different latencies. Latency can be measured in elapsed time or number of cell divisions necessary to activate pluripotency genes. Expected outcomes for the individual models are shown at the bottom. Experimental evidence using clonal B cell and monocyte populations supports a stochastic model of type "iii" (highlighted; see text for details). RFs, reprogramming factors. (Modified, with permission, from Hanna et al. 2009b.)

(or number of cell divisions; models i and ii in Fig. 5B) whereas the "stochastic" model predicts that somatic cells give rise to iPSCs with variable latencies or after going through different numbers or cell division (models iii and iv). In addition, one has to consider whether all somatic cells or only a few "elite" cells will yield iPSCs; somatic stem or progenitor cells that are present in most adult tissues, and possibly persist in explanted cell populations, are the most obvious candidate "elite" cells as they are rare and developmentally closer to pluripotent cells than differentiated cells.

Reprogramming is unlikely to follow a purely deterministic process in all cells (model i) as this disagrees with the low efficiency of iPSC formation. Models based exclusively on an elite component (models ii or iv) are also difficult to sustain because iPSCs can be derived from many different somatic cells including fully differentiated B and T lymphocytes (Hanna et al. 2008b; Eminli et al. 2009) as well as pancreatic β cells (Stadtfeld et al. 2008a). Moreover, when following clonal populations of early B cells and monocytes expressing the reprogramming factors, almost all cell clones ultimately give rise to daughter cells that form iPSCs, although this process requires continuous growth over several weeks to months (Hanna et al. 2009b). The latter observation suggested that continuous cell proliferation allows rare cells in almost every clonal cell population to acquire molecular changes that facilitate their conversion into a pluripotent state. These findings, combined with mathematical modeling, support a stochastic model of cellular reprogramming (i.e., the highlighted model iii in Fig. 5B) (Hanna et al. 2010b).

Interestingly, the overexpression of the pluripotency factor Nanog in combination with the standard Yamanaka factors, Oct4, Sox2, Klf4, and c-Myc enhances cellular reprogramming in a cell division–independent manner (Hanna et al. 2009b). This result indicated that cellular reprogramming is amenable to acceleration by additional treatments. Consistently, several other factors have been identified wherein overexpression or depletion during cellular reprogramming increased the generation of iPSCs (Fig. 5A). These molecules include transcription factors (e.g., Tbx3, Sall4, Glis1), chromatin regulators (e.g., UTX, BAF, Dnmt1, Mbd3), microRNAs (e.g., miR-294, miR-302/367), and signaling molecules (e.g., Wnt, Tgf-β, Jak/Stat) (Stadtfeld and Hochedlinger 2010; Maekawa et al. 2011; Orkin and Hochedlinger 2011; Mansour et al. 2012).

Another parameter that has been shown to contribute to the efficiency of reprogramming is the differentiation state of the starting cell. For example, clonally plated hematopoietic stem and progenitor cells give rise to iPSCs with significantly higher efficiencies than mature lymphocytes and myeloid cells (10%–40% vs. 0.01% to 1%; Eminli et al. 2009; Stadtfeld et al. 2010b). Similarly, subpopulations of fibroblasts produce iPSCs sooner and more efficiently than the bulk population when following individual cells with live cell imaging (Smith et al. 2010) or upon sorting of immature cells with surface markers (Nemajerova et al. 2012). In conclusion, cellular reprogramming is most consistent with a stochastic model. However, additional parameters such as differentiation stage, growth factors, and the supplementation of other transcription factors can influence this process.

3.2 Molecular Changes during Cellular Reprogramming

The low efficiency and slow kinetics of iPSC derivation are in stark contrast to somatic cell lineage switching triggered by transcription factor overexpression, such as the conversion of B cells into macrophages induced by C/EBPα, which occurs at efficiencies of up to 100% and within 48 hours (Bussmann et al. 2009). This suggests that the induction of pluripotency by defined factors faces more barriers than lineage conversion, possibly because of a higher degree of transcriptional and epigenetic similarity among mature cell types than between mature cells and pluripotent cells. Thus, what are the major molecular roadblocks a somatic cell faces during reprogramming into an iPSC?

3.2.1 *Silencing of Somatic Genes and Activation of Pluripotency Genes*

Studies in fibroblasts suggest that reprogramming follows an organized sequence of events that begins with the down-regulation of somatic markers (Stadtfeld et al. 2008b) and morphological changes reminiscent of a mesenchymal-to-epithelial transition (MET) (Fig. 6) (Li et al. 2010; Samavarchi-Tehrani et al. 2010). In accordance, interference with genes involved in MET such as the epithelial molecule *E-cadherin* and bone morphogenetic protein (BMP) receptor signaling abrogate reprogramming. These events are followed by the activation of the early pluripotency markers SSEA-1, alkaline phosphatase, and *Fbxo15* before bona fide pluripotency genes such as *Nanog* or *Oct4* become expressed (Fig. 6) (Brambrink et al. 2008; Stadtfeld et al. 2008b). The telomerase enzyme, responsible for extending the shortened telomeres of somatic cells, is reactivated at the same time as endogenous *Nanog* and *Oct4*.

Reprogramming intermediates isolated based on combinations of the aforementioned markers have an increased probability to form iPSC colonies (Stadtfeld et al. 2008b), suggesting that these cells have overcome several transcriptional and epigenetic barriers that normally prevent the induction of pluripotency. Interestingly, the majority of fibroblasts expressing reprogramming factors fail to down-regulate somatic markers and activate pluripotency genes (Wernig et al. 2008; Stadtfeld et al. 2010b), indicating that many cells become refractory to reprogramming. Such "nonresponding" fibroblasts do not give rise to iPSCs even after prolonged culture. Collectively, these results suggest that the extinction of the somatic program and the subsequent activation of endogenous pluripotency genes represent roadblocks during iPSC formation. In support of this conclusion, the down-regulation of genes that stabilizes the differentiated state (e.g., Pax5, Pax7, Gata6) (Hanna et al. 2008b; Mikkelsen et al. 2008) or the ectopic expression of other pluripotency factors in combination with Oct4, Sox2, Klf4, and c-Myc has been shown to enhance iPSC formation (Stadtfeld and Hochedlinger 2010).

Cite as *Cold Spring Harb Perspect Biol* doi: 10.1101/cshperspect.a019448

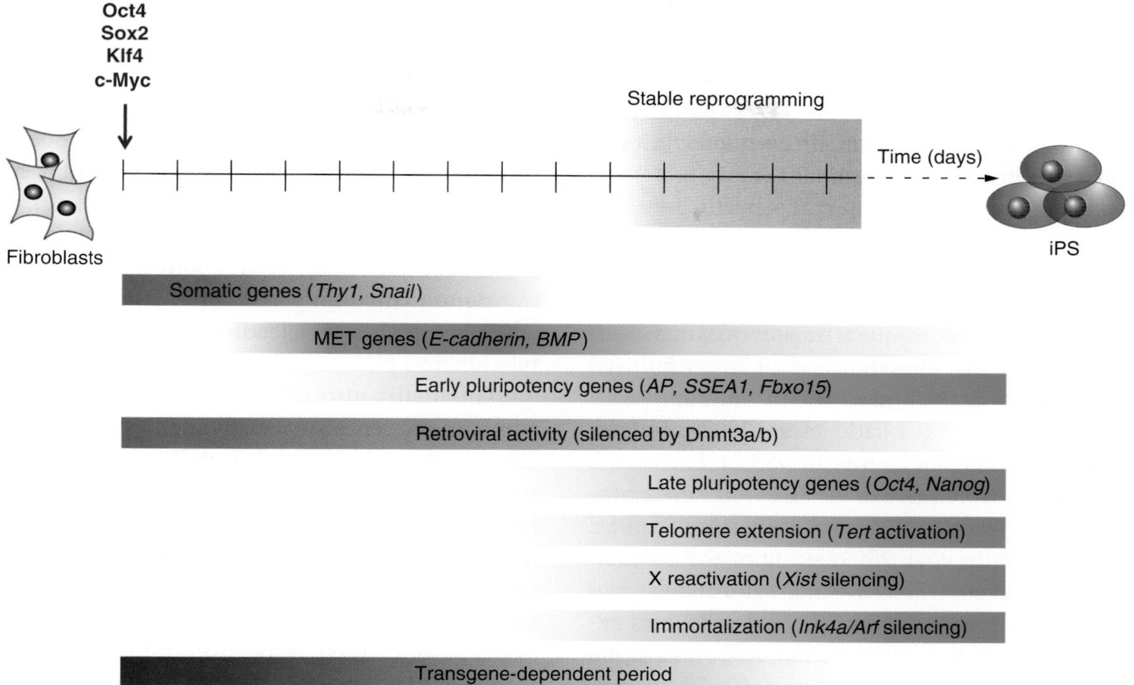

Figure 6. Molecular and cellular cornerstones of cellular reprogramming into iPSCs. Depicted are key events and examples of genes that are regulated during the reprogramming of fibroblasts to pluripotency. "Stable reprogramming" indicates the time window when cells activate endogenous pluripotency loci and become transgene independent. MET, mesenchymal-to-epithelial transition. (Adapted, with permission, from Stadtfeld et al. 2008b.)

3.2.2 Resetting of Global Histone and DNA Methylation Patterns

Gene expression in somatic and differentiated cells is maintained by characteristic patterns of DNA methylation and histone tail modifications. In general, pluripotency genes such as *Oct4* and *Nanog* are silenced in somatic cells by both repressive histone modifications such as histone H3 lysine 27 trimethylation (H3K27me3) and DNA methylation. In contrast, in pluripotent cells, the *Oct4* promoter is devoid of promoter DNA methylation and carries the activating histone mark H3 lysine 4 trimethylation (H3K4me3). Successful reprogramming requires resetting of both of these epigenetic modifications from a somatic to a pluripotent state at a genome-wide level. Genome-wide analyses (chromatin immunoprecipitation, ChIP) combined with deep sequencing (ChIP-seq) of iPSCs revealed that the overall histone modification and DNA methylation landscapes are correctly reprogrammed in most authentic iPSC lines, whereas they are incompletely restored in partially reprogrammed iPSCs (Maherali et al. 2007; Mikkelsen et al. 2008; Sridharan et al. 2009). A number of histone-modifying enzymes have recently been discovered that are involved in this process. For example, the histone lysine (K) demethylase UTX, which removes inhibitory H3K27 methylation marks from silenced pluripotency loci, is critical for efficient iPSC formation (Mansour et al. 2012). Likewise, components regulating the activating H3K4 mark such as WD repeat domain 5 (Wdr5) influence efficient reprogramming by ensuring proper expression of key pluripotency genes (Ang et al. 2011).

DNA methylation patterns are established during mammalian development by the de novo methyltransferases Dnmt3a and Dnmt3b and are maintained throughout adulthood by the maintenance methyltransferase Dnmt1 (Reik et al. 2001). Loss of the DNA maintenance methylation machinery is incompatible with embryonic development (Li et al. 1992). Surprisingly, iPSC formation is not affected in the absence of *Dnmt3a* and *Dnmt3b*, indicating that de novo methylation is dispensable for cellular reprogramming (Pawlak and Jaenisch 2011). It is likely that other repressive mechanisms such as histone modifications compensate for the loss of de novo methylation. In contrast, reducing global genomic methylation levels, by either using short hairpins against *Dnmt1* or treating cells with the demethylating drug 5-aza-cytidine, boosts cellular reprogramming (Mikkelsen et al. 2008). Specifically, these treatments enhance overall colony formation and facilitate the conversion of partially reprogrammed iPSCs into fully reprogrammed iPSCs. Although the underlying mechanisms remain unclear, it is likely that DNA demethylation enhances reprogramming through the derepression of pluripotency genes such as *Oct4* and *Nanog*. Together, these results show that DNA demethylation rather than the acquisition of methylation provides additional barriers to cellular reprogramming.

3.3 Importance of Cell Proliferation

In contrast to ES cells, which grow indefinitely in culture, fibroblasts have a restricted proliferative potential and eventually undergo apoptosis, growth arrest, or stress-induced senescence because of activation of the tumor-suppressor genes p53 and *Ink4a/Arf* (Collado et al. 2007). Indeed, expression of the Yamanaka factors in *p53* or *Ink4a/Arf*-deficient fibroblasts, which fail to senesce and hence proliferate indefinitely, leads to a dramatic increase in iPSC colony numbers (Banito et al. 2009; Hong et al. 2009; Kawamura et al. 2009; Li et al. 2009a; Utikal et al. 2009b). It is important to note, however, that different cell types expressing the four factors elicit different responses upon loss of p53. In fibroblasts, the main effect of p53 loss appears to be the inhibition of senescence and cell death, whereas in blood cells that express the same reprogramming factors, the loss of p53 mainly contributes to reprogramming by accelerating cell cycle (Hanna et al. 2009b). This finding emphasizes the fact that barriers inherent to cellular reprogramming can be cell context–dependent.

The acquisition of pluripotency may not be complete on attaining independence from exogenous Yamanaka factor expression; the full activation of pluripotency genes may require several rounds of cell divisions, as suggested by the finding that early- and late-passage iPSCs show discernible differences in telomere length (Marion et al. 2009) and global changes in transcription and DNA methylation patterns (Chin et al. 2009; Polo et al. 2010). This finding is consistent with the notion that freshly derived iPSC lines show an "epigenetic memory" that is characterized by residual epigenetic marks and gene expression signatures inherited from the somatic cell of origin (Kim et al. 2010; Polo et al. 2010).

3.4 Transcription Factors Maintain the Pluripotent State of ES Cells

Oct4, Nanog, and Sox2 form a core of transcription factors that maintain ES cells in a self-renewing and undifferentiated state that is poised for differentiation. Accordingly, deletion of any of these factors in ES cells abrogates or severely compromises ES cell self-renewal (Chambers and Smith 2004). Studies analyzing the occupancy of these factors across the genome in mouse and human ES cells indicates that they serve two main purposes: their association with a particular genomic region controls either the repression of genes associated with differentiation or the activation of ES cell–specific targets (Fig. 7A) (Jaenisch and Young 2008). Gene suppression by pluripotency factors in ES cells is at least in part mediated by the recruitment to target promoters of repressive chromatin remodeling complexes such as the histone deacetylase-containing NuRD complex (Kaji et al. 2006) and the lysine methyltransferase-containing Polycomb complex 2 (Boyer et al. 2006; Lee et al. 2006), leading to repressive histone deacetylation and H3K27me3, respectively.

This dual control of target gene regulation might explain why somatic genes are usually silenced before pluripotency genes become activated during iPSC formation. Although repressive complexes can immediately form upon binding of individual reprogramming factors to target sites, key components of the more elaborate activating complexes such as Nanog or Dax1 (Wang et al. 2006) may be limiting or absent at early stages of reprogramming and only become available once their respective endogenous genomic loci have been transcriptionally activated (Fig. 7B). This process might be facilitated by nucleosome remodelers such as Chd1 (Gaspar-Maia et al. 2009) and BAF (Singhal et al. 2010); indeed, both of these molecules have been shown to enhance reprogramming efficiencies when overexpressed. Once the majority of core pluripotency factors are expressed, they presumably engage in positive-feedback loops (Jaenisch and Young 2008) to sustain pluripotency in the absence of exogenous factor expression (Fig. 7A).

3.5 Contributions of Individual Factors to Cellular Reprogramming

Studies on partially reprogrammed cells suggest that the inability of Oct4, Sox2, and Klf4 to bind to their targets is a limiting factor for acquiring pluripotency (Sridharan et al. 2009). In contrast, c-Myc efficiently occupies targets associated with proliferation and metabolism, indicating that c-Myc plays a distinct role compared with Oct4, Sox2, and Klf4 (see cell-cycle gene in Fig. 7B). Accordingly, c-Myc expression is only required for the first few days of reprogramming (Sridharan et al. 2009), whereas Sox2 expression is essential only at late stages (Chen et al. 2011). In further agreement with a supportive role of c-Myc during early steps in reprogramming, premature expression of c-Myc and Klf4 in fibroblasts before activation of all four factors increases reprogramming efficiencies and speed, whereas early expression of Sox2 and Oct4 has no effect (Markoulaki et al. 2009). Mechanistically, c-Myc expression might enhance reprogramming by facilitating the binding of Oct4 and Sox2 to cognate targets—for example, by establishing or

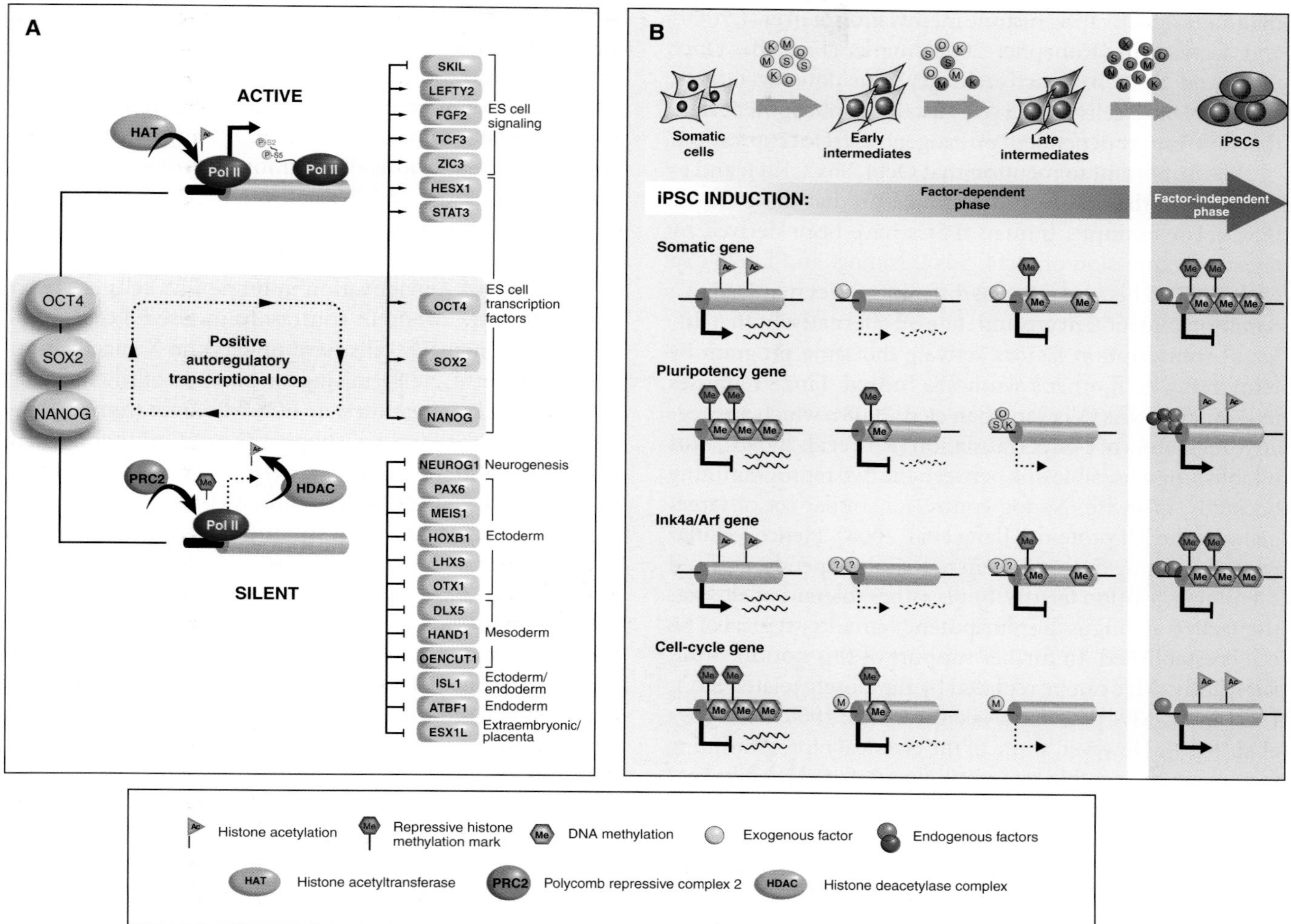

Figure 7. Roles of transcription factors in inducing and maintaining pluripotency. (*A*) The transcription factors Oct4, Sox2, and Nanog maintain the pluripotency of ES cells by activating genes important for self-renewal and suppressing genes that drive differentiation. These factors either collaborate with chromatin activators such as the histone acetyltransferase p300 (blue, HAT) and elongating RNA polymerase II (green, Pol II) to induce genes or with chromatin repressors such as PRC2 and histone deacetylase complexes (HDAC) to inhibit genes through repressive histone methylation and the removal of histone acetylation. Moreover, pluripotency factors positively regulate their own transcription, thus establishing a transcriptional circuitry typical of ES cells and iPSCs. (*B*) Model of how reprogramming factors act during iPSC induction. Somatic cells expressing exogenous reprogramming factors transition from "early intermediates" to "late intermediates" to iPSCs (blue circles: O, Oct4; S, Sox2, K, Klf4, M, c-Myc). To acquire pluripotency, cells must activate endogenous pluripotency genes (dark blue circles) as well as essential cofactors (blue circles labeled N, Nanog; X, other factors) to sustain self-renewal in the absence of exogenous factors. (*Bottom*) Model of how reprogramming factors establish pluripotency in somatic cells via their action at different types of genes. Single factors may suppress somatic genes early, whereas combinations of factors activate pluripotency genes late in reprogramming. Pluripotency is stabilized once suppressive chromatic marks are deposited at somatic genes and removed from silenced pluripotency loci. Successful reprogramming is tightly linked with the acquisition of indefinite self-renewal properties through activation of proliferation genes such as cyclins (mostly targeted by c-Myc) and suppression of cell-cycle inhibitor genes such as *Ink4a/Arf* by unknown factors. (*A*, Adapted, with permission, from Jaenisch and Young 2008; *B*, adapted, with permission, from Stadtfeld and Hochedlinger 2010.)

maintaining activating histone methylation (Lin et al. 2009) and acetylation (Knoepfler 2008) marks (Fig. 7B). Once Oct4 and Sox2 have activated key pluripotency targets, such as *Nanog*, cells enter a self-sustaining pluripotent state that is no longer dependent on exogenous factor expression.

It is important to mention that Oct4, Sox2, Klf4, and c-Myc are not the only factor combination that can generate iPSCs. For example, human iPSCs have been derived by enforced expression of Oct4, Sox2, Nanog, and Lin28 (Yu et al. 2007). This suggests that different routes may lead to a common pluripotent ground state or, alternatively, that different transcription factors activate the same program by reinforcing each other's synthesis. Indeed, Lin28 represses let-7 microRNAs (Viswanathan et al. 2008), which are negative regulators of c-Myc translation (Kim et al. 2009b), thus establishing a possible link between the two reprogramming cocktails. Likewise, Nanog controls a similar set of target genes as the Klf proteins (Jiang et al. 2008). Hence, cellular reprogramming does not seem to strictly depend on a fixed set of transcription factors, but is rather tolerant of alternative factors as long as the pluripotency circuitry typical of ES cell is established. In further support of this notion, Sox2, Klf4, and c-Myc can be replaced by the closely related Sox1, Klf2, and L-Myc proteins (Nakagawa et al. 2008; Nakagawa et al. 2010). However, some of the classical reprogramming factors are replaceable by seemingly unrelated members of the nuclear orphan receptor family. For instance, Klf4 can be replaced by Esrrb (Feng et al. 2009) and Oct4 by Nr5a2 (Heng et al. 2010) during mouse fibroblast reprogramming. The mechanisms by which these alternative proteins operate during reprogramming remain elusive.

3.6 X Chromosome Inactivation

X chromosome inactivation in female mammals ensures balanced gene expression of X-linked genes compared with males (Augui et al. 2011). X inactivation occurs randomly on one of the two female Xs per cell during early postimplantation development and is stably maintained in all somatic daughter cells throughout adulthood. X inactivation is accomplished by a complex interaction of epigenetic mechanisms that involve the noncoding RNA *Xist*, which coats the future inactive X and recruits repressive chromatin regulators, resulting in the acquisition of inhibitory histone and DNA methylation marks that induce stable silencing (summarized in Fig. 10 of Ch. 25 [Brockdorff and Turner 2014]). Although all differentiated female cells show X inactivation, with one active and one inactive X chromosome (XiXa), mouse ICM cells and derivative ES cells are in a preinactivation state and thus carry two active Xs (XaXa). This observation raised the question of whether induced pluripotency entails faithful reactivation of the somatically silenced X chromosome. Analysis of mouse iPSCs derived from female fibroblasts showed that the silenced X chromosome did indeed become reactivated and undergo random inactivation when cells were induced to differentiate (Maherali et al. 2007). This finding is reminiscent of the previous observation that embryos cloned from fibroblasts by SCNT reactivate the silenced X chromosome and undergo random X inactivation during development (Eggan et al. 2000).

The state of X inactivation in human ES cells has been puzzling (Wutz 2012). In contrast to mouse ES cells, conventional human ES cells have undergone X inactivation (XiXa; Shen et al. 2008), raising the question of whether this reflects the X inactivation state of ICM cells in human blastocysts. Two observations support the notion that human ICM cells are in a pre–X inactivation state: (1) Direct observation indicates that cells of human blastocysts have not as yet commenced X inactivation (Okamoto et al. 2011), and (2) human ES cells isolated and propagated under physiological oxygen conditions (5% O_2) displayed a pre–X inactivation status and, similar to mouse ES cells, initiated random inactivation on differentiation (Lengner et al. 2010). The latter observation suggested that suboptimal culture conditions such as oxidative stress may interfere with the in vitro capture of the more immature XaXa state of human ICM cells. More recent data have shown that human ES cells and iPSCs can lose *Xist* expression and gain biallelic expression of some X-linked genes (Mekhoubad et al. 2012), and that prolonged culture may select for overexpression of growth-promoting X-linked genes (Anguera et al. 2012). These observations are consistent with the possibility that the state of X inactivation in pluripotent cells may be less stable than in somatic cells and that continuous propagation of the cells may result in cultures with partially reactivated Xi. It is important to note that the state of pluripotency of ES cells and iPSCs has a profound impact on the state of X inactivation as discussed in Section 3.7.

3.7 Alternative States of Pluripotency: Naïve versus Primed Cells

Pluripotent cell lines exist in two distinct states that are characterized by different growth factor requirements and developmental properties (Fig. 8). Murine ES cells, established from the ICM of preimplantation blastocysts in the presence of leukemia inhibitory factor (LIF) and BMP, exist in a more primitive pluripotent state in contrast to epiblast stem cells (EpiSCs), derived from the implanted embryo in the presence of bFgf and Activin. Nichols and Smith have designated the ICM-like state of ES cells as the "naïve" state and that of the epiblast-derived EpiSCs as the "primed" pluripotent state (Nichols and Smith 2009). This definition

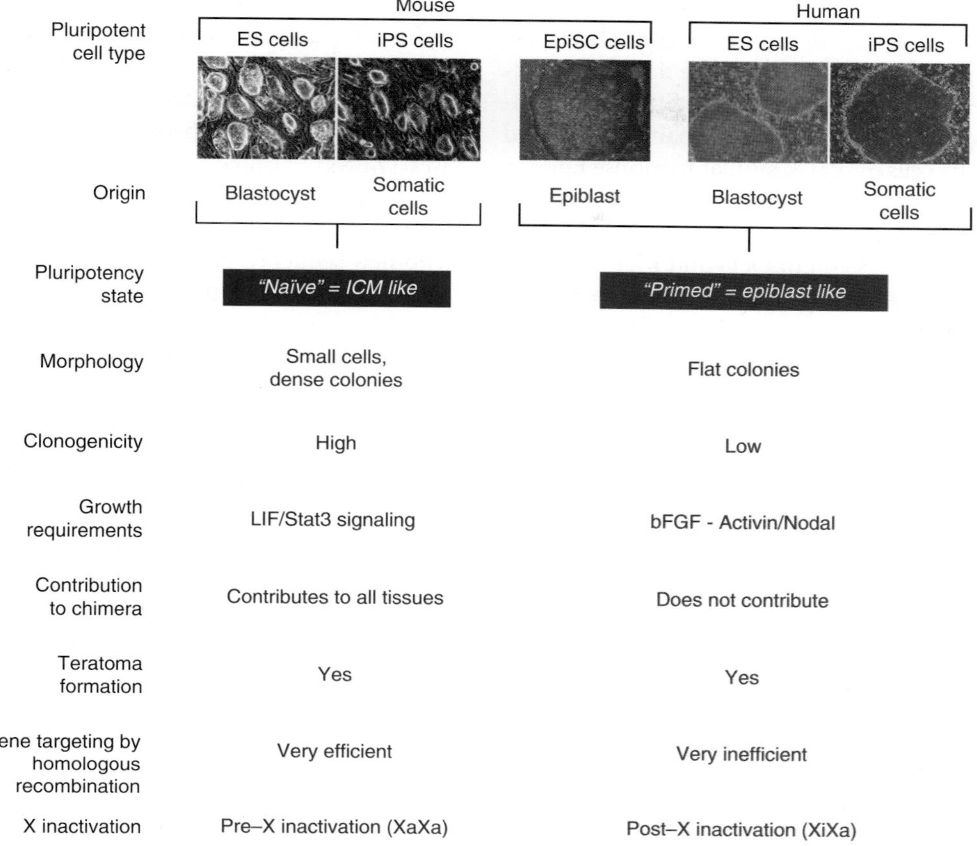

Figure 8. Different states of pluripotency. Classical mouse ES cells are derived from the ICM of the blastocyst and are designated as "naïve." In contrast, EpiSCs are derived from the epiblast of the implanted embryo and are designated as "primed," implying that these cells are less immature and more differentiated than naïve cells. The differences between the two states of pluripotency are reflected in morphology, clonogenicity (ability to form discrete colonies), signal transduction pathways, their pluripotency as assayed by their ability to contribute to tissues in a chimera or form a teratoma, gene targeting by homologous recombination, and the state of X inactivation. Human ES cells, although also derived from the blastocyst, resemble the primed state by many criteria and differ from the naïve state of pluripotency.

implies that the primed state is prone to differentiation whereas the naïve ES cells correspond to a more immature state of pluripotency. The state of X inactivation reflects the different states of pluripotency: Naïve female ES cells are in the XaXa preinactivation state, whereas primed EpiSCs have already undergone X-chromosome inactivation (Fig. 8). Consistent with their advanced developmental state, EpiSCs show some pluripotency criteria such as teratoma formation, but fail to contribute to any tissues in chimeric mice (Brons et al. 2007; Tesar et al. 2007). Of interest is that the two alternative states of pluripotency are metastable and can be interconverted by changes in culture conditions; EpiSCs can be reverted to naïve ES cell–like cells on exposure to LIF/Stat3 signaling, and this conversion can be boosted by transient expression of pluripotency factors including Klf4, Klf2, Nanog, or c-Myc, or by cultivation of cells in LIF and "2i" conditions (2i: GSK3β inhibitor and ERK1/2 inhibitor or Kenpaullone; Guo et al. 2009; Hanna et al. 2009a). Conversely, exposure to bFGF and Activin converts the naïve ES cells into primed EpiSCs. Thus, activating different signaling pathways through different culture conditions can alter and stabilize the two alternative states of pluripotency.

Human ES cells, like mouse ES cells, are isolated from explanted preimplantation blastocysts by a protocol established by Thomson and colleagues (Thomson et al. 1998). These cells share multiple defining features with mouse EpiSCs rather than mouse ES cells, including a flat morphology, signaling dependence on bFGF/Activin, propensity for X-chromosome inactivation, and reduced tolerance to single cell dissociation (Fig. 8). Thus, these molecular and biological similarities with mouse EpiSCs suggest that human ES cells correspond to the primed pluripotent state rather than to the naïve state of mouse ES cells. This raised the question of whether conditions can be devised that allow isolation of human pluripotent cells with defining biological and epigenetic features of mouse naïve ES cells and whether human ES cells or iPSCs, similar to mouse EpiSCs, could be converted to a naïve pluripotent state. Indeed, the propagation of human pluripotent stem cells in LIF/"2i" conditions and overexpression of Oct4 and Klf4 or KLF2/KLF4 induced conversion of conventional human ES cells to a naïve pluripotent state (Hanna et al. 2010a). The naïve human ES cells and iPSCs resembled naïve mouse ES cells by several criteria: They had reactivated the inactive X chromosome resulting in a XaXa pre–X inactivation status, showed high single-cell cloning efficiency (i.e., high clonogenicity), were dependent on LIF/STAT3 instead of bFGF/Activin signaling, could routinely be passaged as single cells, and showed a gene expression pattern that more closely resembled that of naïve mouse ES cells. However, the naïve state cannot be robustly maintained and depends on the continuous expression of transgenes. Thus, it is of major interest to define conditions that would allow the maintenance of genetically unmodified naïve cells and to isolate mouse ES-like cells directly from human blastocysts.

4 APPLICATIONS OF iPSC TECHNOLOGY IN DISEASE RESEARCH

The most exciting application of iPSC technology is the potential of deriving patient-specific pluripotent cells for disease research. One can distinguish between two different applications of patient-derived iPSCs: (1) studying diseases in tissues culture ("diseases in the dish" approach), and (2) cell transplantation therapy. A crucial requirement to fully realize the potential of iPSCs in disease research is efficient gene-targeting methods in human pluripotent cells. We will first summarize the different approaches to genetically manipulate human ES cells and iPSCs, followed by a discussion of iPSC technology in disease modeling and cell therapy.

4.1 Genetic Modification of Human ES Cells and iPSCs

Gene targeting by homologous recombination is efficient in mouse ES cells and has facilitated the generation of thousands of genetically modified mouse models. In contrast, homologous recombination has proven to be difficult in human ES cells and iPSCs, and only few reports have described successful gene targeting since the derivation of the first human ES cells more than 15 years ago. The difficulty of genetic manipulation has been a major obstacle to realizing the full potential of human ES cells and iPSCs in disease research.

Novel tools to facilitate homologous recombination, based on the introduction of DNA double-strand breaks (DSBs) by site-specific nucleases, have been used to target genes in human cells. Two approaches that can introduce site-specific DSBs have been devised: (1) zinc-finger nucleases (ZFNs) (Urnov et al. 2010) and (2) "transcription activator–like effector" (TALE) proteins (Bogdanove and Voytas 2011). In both strategies, DNA-binding domains with new sequence specificities are generated and fused to nucleases that introduce a DSB at a specific nucleotide. A ZFN is generated by fusing the FokI nuclease domain to a DNA recognition domain composed of engineered zinc-finger motifs that specify the genomic DNA-binding site for the chimeric protein. On binding of two such fusion proteins at adjacent genomic sites, the nuclease domains dimerize, become active, and cut the genomic DNA (Fig. 9A). When a donor DNA that is homologous to the target on both sides of the DSB is provided, the genomic site can be repaired by homology-directed repair, allowing the incorporation of exogenous sequences placed between the homologous regions. Although zinc-finger domains recognize nucleotide triplets, the DNA-binding domains of TALEs recognize single nucleotides: multiple ~34-amino-acid units (also referred to as TALE repeats) are arranged in tandem, their sequences being nearly identical except for two highly variable amino acids that establish the base recognition specificity for each unit. Each individual domain determines the specificity of binding to one DNA base pair and, therefore, four different repeat units are sufficient to specify binding to a novel site. As in the ZFN approach, the nuclease fused to the TALE module introduces a specific DSB between the two DNA-binding domains.

The insertion of donor sequences at the DSB is accomplished by cotransfecting the ZFN or TALEN pair together with a donor plasmid designed to carry approximately 500–750 bp of homologous sequence flanking both sides of the recognition site allowing the generation of reporter ES cells or iPSCs that carry a GFP marker in key transcription factor genes (Hockemeyer et al. 2009, 2011). Gene editing using the ZFN or TALEN strategy has been used to introduce disease-relevant mutations into normal ES cells or correct the mutations in patient-derived iPSCs (Soldner et al. 2011). This creates "isogenic" pairs of disease and control cells (Fig. 9B) that, as outlined below, allow a meaningful comparison between experimental and control cells.

4.2 Disease Modeling in the Culture Dish ("Disease in the Dish")

iPSC technology facilitates the generation of genetically identical cells from patients afflicted with a disease of known or unknown etiology. Because the cells are derived

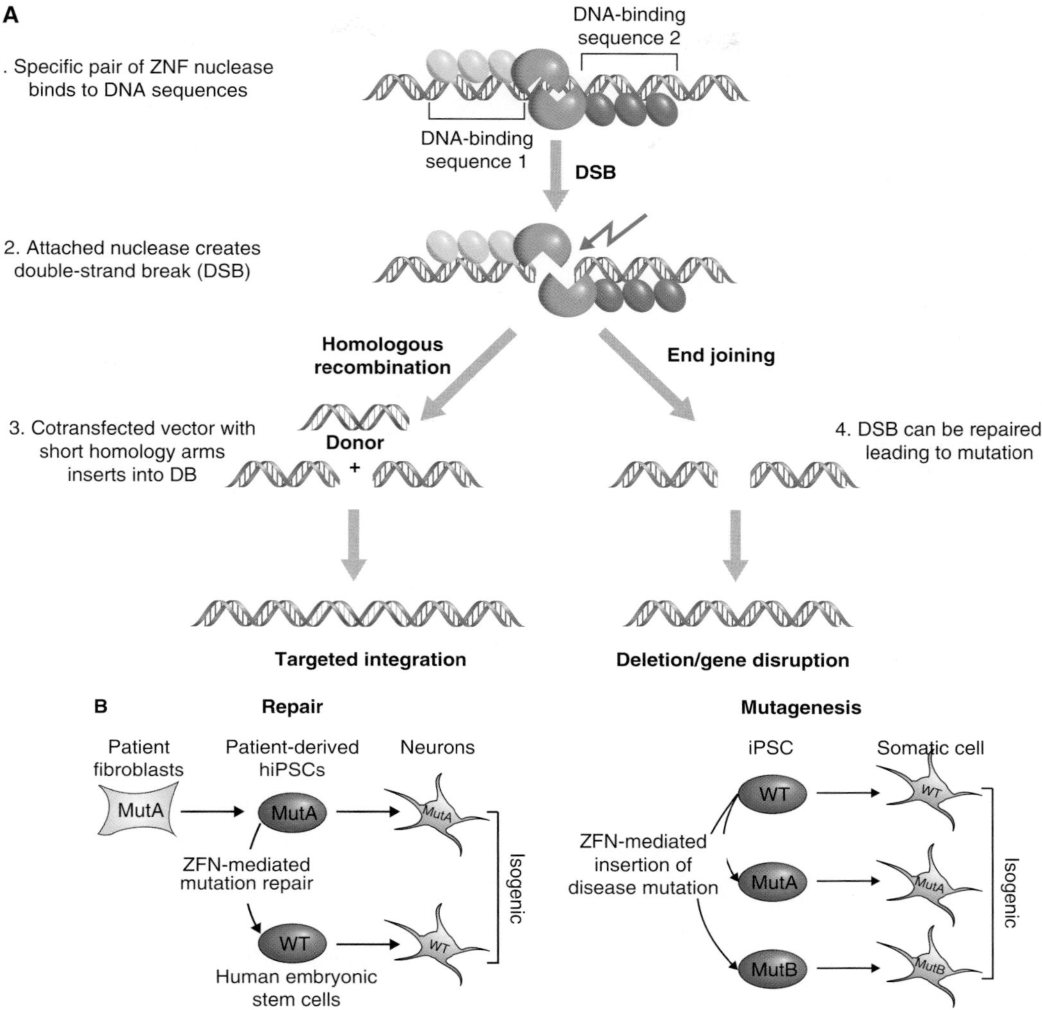

Figure 9. ZFN and TALE nuclease-mediated gene targeting. (*A*) (*1*) DNA-binding proteins—either zinc-finger or TALE proteins in blue fused to a Fok1 restriction nuclease in orange—are designed to specifically recognize two adjacent DNA-binding sequences with a defined spacing. (*2*) On binding of the zinc fingers, the FOK1 nuclease domains dimerize, become active, and cut the DNA. (*3*) If a donor plasmid carrying DNA (red, DNA) homologous to the DSB is ectopically provided to the cell, this can be used to repair the DNA lesion. A donor plasmid can be designed so that it carries additional sequence in between the homology arms. On repair of the DSB with such a donor, the genomic locus will be altered to carry this additional sequence as an insertion at the site of the DSB. (*4*) Alternatively, the DSB is repaired, incurring deletion or sequence alteration that disrupts gene function. (*B*) Using ZFN (or TALEN)-mediated gene targeting, a disease causing mutation is either corrected in a patient-derived iPSC (*left* illustration), or disease-causing mutations are introduced into wild-type (WT) ES cells (*right* illustration). The result of either manipulation will be the generation of isogenic sets of iPSCs, providing a genetically matched control for functional studies. (*B*, Adapted, with permission, from Soldner et al. 2011.)

from the patient, they carry all genetic alterations that may have contributed to the disease manifestation allowing, in principle, the investigation of the genetic basis of the disorder even if the genes that contribute to the disease have not yet been identified.

The basis for modeling diseases in the culture dish is the ability to differentiate iPSCs into the cell type that is affected in the patient (Fig. 10). For example, iPSCs need to be differentiated into dopaminergic neurons to model Parkinson's disease (PD) and they need to be coaxed into motor neurons to study spinal motor atrophy (SMA; depicted in Fig. 11), a fatal disease causing paralysis of the lower body. Importantly, the differentiation of the disease-specific iPSCs into functional somatic cells must show a quantifiable phenotype when compared with proper control cells. In PD, patient-derived dopaminergic neurons can be ana-

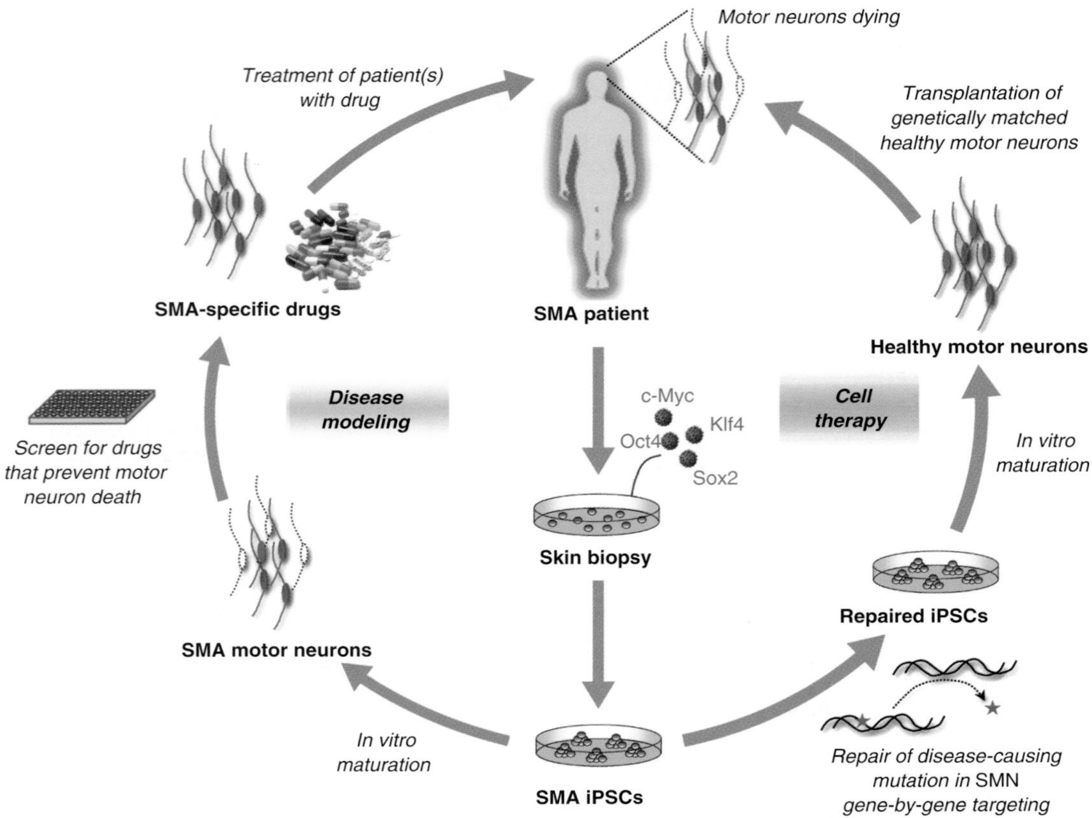

Figure 10. Potential applications of iPSC technology. Shown are the potential applications of iPSC technology for cell therapy and disease modeling using SMA as an example. In SMA patients, motor neurons are afflicted and die causing the devastating symptoms of the disease. SMA-specific iPSCs could be coaxed into motor neurons in vitro to establish a culture model of the disease that may lead to the identification of novel drugs that prevent the abnormal death of motor neurons in patients. Alternatively, the disease-causing mutation could be repaired, if known (in this case, *SMN* gene), in iPSCs by gene targeting before their differentiation into healthy motor neurons and transplantation into the patient's brain. (Adapted, with permission, from Stadtfeld and Hochedlinger 2010.)

lyzed for phenotypic abnormalities when compared to neurons derived from control iPSCs. Similarly, to study SMA, patient-derived motor neurons can be examined for an in vitro phenotype that may correspond to neuronal defects seen in the patient. An important goal of such experiments is to screen for small molecules that can affect the observed in vitro phenotype (Fig. 10) (Rubin 2008). If such compounds could be identified, they may represent promising candidates for drug development to treat the disease.

How far are we from identifying new drugs with iPSCs? In fact, several research laboratories have already derived iPSCs from patients suffering from Huntington's and Parkinson's disease, ALS, juvenile diabetes, muscular dystrophy, Fanconi anemia, Down syndrome, and others (Raya et al. 2009; Soldner et al. 2009), which will facilitate these studies. Moreover, three promising reports showed that iPSCs derived from patients suffering from the devastating disorders SMA (Ebert et al. 2009), familial dysautonomia (Lee et al. 2009), and LEOPARD syndrome (Carvajal-Vergara et al. 2010) recapitulated the cell abnormalities in a Petri dish as they are seen in patients. Remarkably, when the cultured cells were exposed to experimental drugs for these diseases, the "symptoms" were partially alleviated in culture. It is noteworthy that neurons differentiated from iPSCs derived from patients suffering from Rett syndrome (Marchetto et al. 2010) or schizophrenia (Brennand et al. 2011) were shown to display a patient-specific phenotype in the culture dish. These observations suggest that iPSC technology may allow studying even such complex mental disorders such as autism (Rett) or psychoses (schizophrenia) on a cellular level. The hope is that this experimental approach can be applied to many other diseases and cell types for which we currently do not have treatments, and this may result in the development of drugs from which not just one individual, as in cell therapy, but millions of patients may benefit.

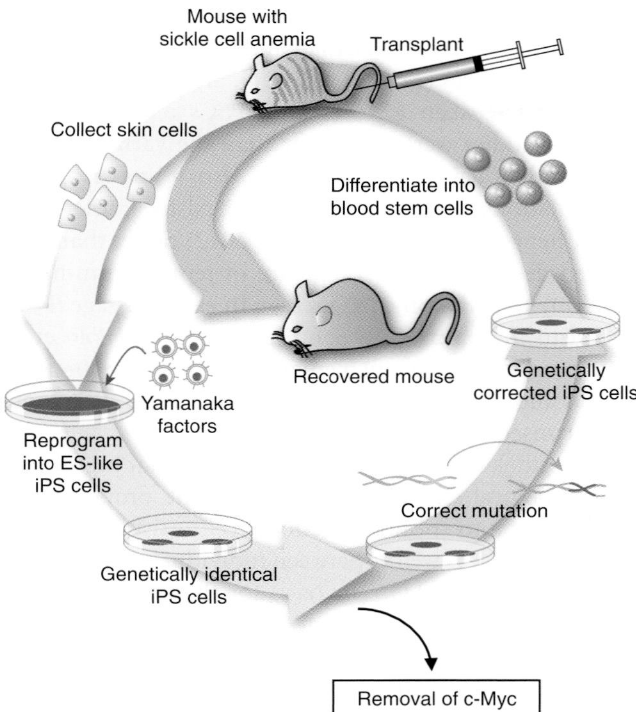

Figure 11. Proof of concept for cell therapy using iPSCs in a humanized model of sickle cell anemia. Transgenic mice carrying the human α-globin gene and the anemia-causing β-globin variant develop disease that resembles human sickle cell anemia. Skin cells were reprogrammed to iPSCs by the four Yamanaka factors and c-Myc was removed by Cre-mediated excision. Homologous recombination was used to correct the mutation in the defective β-globin gene; the corrected iPSCs were differentiated into hematopoietic stem cells and transplanted into the mutant mice. The cells engrafted and generated normal red blood cells curing the anemia.

4.3 Cell Therapy

A widely popularized application of iPSCs is cell replacement therapy. Because iPSCs are genetically matched with the patient donor, this approach eliminates the need for immune-suppressive therapy needed in conventional transplantation settings using cells or organs from unmatched donors (Fig. 10). Indeed, recent experiments have validated this concept using a humanized mouse model of sickle cell anemia (Hanna et al. 2007). Sickle cell anemia is the result of a single-point mutation in the hemoglobin gene, causing red blood cells to adopt a crescent-like shape, which renders them nonfunctional. In this proof-of-concept study, skin cells from the mouse model, which recapitulates the human condition, were first reprogrammed into iPSCs. The disease-causing mutation was subsequently fixed in iPSCs by gene targeting and the repaired cells were then coaxed into blood-forming progenitors (Fig. 11). These now healthy progenitors were transplanted back into anemic mice where they produced normal red blood cells and cured the disease. In principle, this approach could be applied to any disease in humans for which the underlying mutation is known and can be treated by cell transplantation.

However, major challenges need to be overcome before iPSC-based cell therapy can be considered for clinical use (Daley 2012). These include potential tumor (teratoma) formation, the development of robust protocols to derive the cells used for transplantation, and effective delivery of the cells into the patient.

4.3.1 Teratoma Formation

A significant safety consideration of using iPSCs for clinical practice is the risk of tumor formation. Undifferentiated ES cells or iPSCs induce teratomas when injected into immune-compromised animals (see Table 2). A teratoma is a complex tumor consisting of undifferentiated embryonic as well as differentiating cell types. Thus, a crucial challenge for any iPSC-based therapy is to eliminate all undifferentiated cells that may be present in the cell preparation used for transplantation.

4.3.2 Differentiation into Functional Cells

A major issue of using iPSCs or ES cells for transplantation therapy is the derivation of functional differentiated cells from undifferentiated stem cells. Growing evidence indicates that current differentiation protocols yield mostly immature cells (Wu and Hochedlinger 2011). For example, only immature β cells of the pancreas have been derived from ES cells, yielding low levels of insulin, which would be insufficient for replacement therapy in type 1 diabetes patients. Also, differentiation into some cell types such as the hematopoietic lineage is extremely inefficient; so far, no hematopoietic stem cells (HSCs) have been generated that have successfully engrafted over the long term in immune-compromised animals. Likewise, it remains unclear whether neuroblasts or mature neurons are the better donor cell population for transplantation therapy of neurodegenerative diseases such as PD. It will, therefore, be of great importance to generate not only mature functional cells but also the committed, self-renewing stem cells from ES cells/iPSCs as these may be the cells that are most appropriate for cell therapy. Thus, a major challenge for current research is the development of robust protocols that yield homogenous populations of functional cells that could be used for cell replacement therapy.

4.3.3 Delivery of Cells

As with gene therapy, delivery of the therapeutic agent is a crucial issue for regenerative medicine. Depending on the target tissue, cell delivery can be straightforward or complex. For example, HSCs normally colonize the bone marrow when injected into the circulation, a principle that has been successfully used in the clinic for decades. Thus, if HSCs could be successfully derived from iPSCs, their delivery would pose no problem. Similarly, pancreatic β cells derived from cadavers have been shown to regulate glucose levels when transplanted into the livers of type 1 diabetes patients, although long-term survival of the cells was compromised owing to immune rejection. If mature β cells could be differentiated from patient-specific iPSCs, they would still be rejected because of the underlying autoimmune disorder inherent to diabetes and would therefore require repeated transplantation. Some applications, such as cell replacement therapy in PD patients, face yet additional challenges because the transplanted cells would need to be placed into a specific brain region by stereotactic injection.

In summary, many hurdles need to be overcome before the iPSC approach can be considered for clinical application (Daley 2012). However, recent progress in tissue engineering is encouraging and suggests that some of these obstacles are of a technical nature and, hence, should be surmountable. For example, epithelial and endothelial cells seeded on scaffolds of decellularized material have produced artificial tissues that can function for a limited time when transplanted into animals (Wu and Hochedlinger 2011). Nevertheless, it should be emphasized that cell transplantation for tissue regeneration will be extremely challenging for diseases such as Alzheimer's, muscular dystrophy, or cystic fibrosis in which a large fraction of dysfunctional cells in complex organs such as brain, skeletal muscle, or lung and intestine would need to be replaced.

5 AN UNRESOLVED ISSUE: ARE iPSCs EQUIVALENT TO ES CELLS?

Both iPSCs and ES cells are equivalent based on criteria such as expression of pluripotency markers, in vitro differentiation to various cell types, and in vivo differentiation in the teratoma assay (Fig. 8). However, controversy exists as to whether specific epigenetic or genetic differences distinguish iPSCs from ES cells. A number of studies found that iPSCs showed a higher level of mutations (Gore et al. 2011; Hussein et al. 2011) and different global expression patterns (Chin et al. 2009) when compared with ES cells. It was also discovered that iPSCs displayed "epigenetic memory" of the donor cells; iPSCs derived from fibroblasts, blood, or liver cells had a DNA methylation signature reminiscent of the respective donor cells (Kim et al. 2010; Lister et al. 2011; Ohi et al. 2011). However, these issues remain controversial as more extensive studies comparing large numbers of ES cells and iPSCs failed to find significant expression (Guenther et al. 2010; Newman and Cooper 2010) or epigenetic differences (Bock et al. 2011), arguing that the dissimilarities between ES cells and iPSCs reported in the previous studies were not larger than the variations seen within individual ES cells or iPSCs. Also, it was discovered that mutations found in iPSCs cells preexisted in the donor cells (Cheng et al. 2012; Young et al. 2012) rather than were accrued de novo as a consequence of reprogramming. In addition, methylation changes seen in early-passage iPSCs tended to disappear with continuous culture (Polo et al. 2010), suggesting that such epigenetic differences between ES cells and iPSCs are not stable and may be of little functional consequence.

Nevertheless, much evidence indicates that highly variable biological characteristics such as the propensity to differentiate into specific functional cells exist between ES cells and iPSCs, as well as between individual iPSCs and ES cells (Bock et al. 2011). The basis for these profound differences is manifold and includes differences in genetic background, variegation effects, and residual transgene expression of the viral vectors used to induce reprogramming (Soldner et al. 2009). It has further been shown that parameters associated with the iPSC derivation process, such as the stochiometry of the reprogramming factors (Stadtfeld et al. 2010a; Carey et al. 2011) or the particular medium composition used to culture the cells (Stadtfeld et al. 2012), can profoundly affect the quality of the resultant iPSCs.

For any iPSC-based disease research, the variability between individual iPSC clones in their ability to proliferate or differentiate poses a potentially serious limitation. For example, if a subtle phenotype in the survival of dopaminergic neurons from a PD-derived iPSC line was seen in comparison to cells derived from control iPSCs, the question is whether this phenotype truly reflects a disease-specific abnormality rather than variability between genetically unrelated iPSC lines. The use of ZNF-based gene-targeting approaches to create isogenic pairs of experimental and control iPSCs that differ exclusively at a single-disease relevant mutation offers one approach to overcome this potential problem (Fig. 9B) (Soldner et al. 2011). Any cellular change seen in the patient-derived cells, but not in isogenic controls, would give confidence that the respective phenotype is indeed disease-related and not a consequence of variability between different iPSCs. Thus, for long latency diseases such as PD or Alzheimer's, in which the patient-specific phenotype may only be subtle, the generation of isogenic pairs of disease and control cells may be particularly important. However, it should be emphasized that genetically matched control cells cannot be generated for iPSCs from polygenic diseases.

Cite as *Cold Spring Harb Perspect Biol* doi: 10.1101/cshperspect.a019448

6 CONCLUDING REMARKS

The generation of iPSCs eight years ago has provided researchers with a unique platform to dissect the mechanisms of cellular reprogramming, which largely remained elusive for the past six decades. Although many questions remain, interesting insights have been gained into the process of reprogramming, such as the finding that cells undergo defined sequential molecular events in an apparently stochastic manner, which are influenced by the choice and number of transcription factors as well as the starting cell type and environmental cues. The ease with which iPSCs can be generated with improved methodology has facilitated the development of chemical and small interfering RNA screens as well as biochemical studies that should further unravel the mechanisms of this process.

The discovery of iPSCs has also influenced our view of normal development; only a few transcription factors are actually needed to potently change cell fate and, hence, mammalian cells must have developed epigenetic mechanisms to efficiently lock in a cell once it has differentiated. These mechanisms are often broken in cancer cells, which show features of stem cells and signs of dedifferentiation (discussed further in Ch. 34 [Baylin and Jones 2014]). Notably, many signaling pathways mutated in cancer cells have recently been shown to affect the formation of iPSCs, indicating remarkable similarities between tumorigenesis and cellular reprogramming.

The isolation of iPSCs has also sparked new interest in interconverting mature cell types directly into each other, which has already led to a number of remarkable examples for pancreatic, cardiac, hepatic, and neural cell types. It is likely that many other direct cell switches will be achieved in the near future. It remains to be tested, however, whether cells generated by direct transdifferentiation as well as in vitro differentiation of iPSCs are functionally equivalent to their in vivo counterparts.

Despite numerous technical advances in the derivation of human iPSCs, relatively little is known about their molecular and functional equivalence with ES cells, which might ultimately affect their potential clinical utility. Addressing these questions will require a careful analysis of the genomic and epigenomic integrity of human iPSCs, as well as the development of novel differentiation protocols and reliable assays to evaluate the functionality of iPSC-derived specialized cells.

ACKNOWLEDGMENTS

K.H. and R.J. declare that portions of the text and figures have been extracted from previous articles by the authors, which are referenced.

REFERENCES

Reference is also in this book.

Aasen T, Raya A, Barrero MJ, Garreta E, Consiglio A, Gonzalez F, Vassena R, Bilic J, Pekarik V, Tiscornia G, et al. 2008. Efficient and rapid generation of induced pluripotent stem cells from human keratinocytes. *Nat Biotechnol* **26:** 1276–1284.

Ang YS, Tsai SY, Lee DF, Monk J, Su J, Ratnakumar K, Ding J, Ge Y, Darr H, Chang B, et al. 2011. Wdr5 mediates self-renewal and reprogramming via the embryonic stem cell core transcriptional network. *Cell* **145:** 183–197.

Anguera MC, Sadreyev R, Zhang Z, Szanto A, Payer B, Sheridan SD, Kwok S, Haggarty SJ, Sur M, Alvarez J, et al. 2012. Molecular signatures of human induced pluripotent stem cells highlight sex differences and cancer genes. *Cell Stem Cell* **11:** 75–90.

Aoi T, Yae K, Nakagawa M, Ichisaka T, Okita K, Takahashi K, Chiba T, Yamanaka S. 2008. Generation of pluripotent stem cells from adult mouse liver and stomach cells. *Science* **321:** 699–702.

Augui S, Nora EP, Heard E. 2011. Regulation of X-chromosome inactivation by the X-inactivation centre. *Nat Rev Genet* **12:** 429–442.

Banito A, Rashid ST, Acosta JC, Li S, Pereira CF, Geti I, Pinho S, Silva JC, Azuara V, Walsh M, et al. 2009. Senescence impairs successful reprogramming to pluripotent stem cells. *Genes Dev* **23:** 2134–2139.

* Baylin SB, Jones PA. 2014. Epigenetic determinants of cancer. *Cold Spring Harb Perspect Biol* doi: 10.1101/cshperspect.a019505.

Blelloch R, Venere M, Yen J, Ramalho-Santos M. 2007. Generation of induced pluripotent stem cells in the absence of drug selection. *Cell Stem Cell* **1:** 245–247.

Bock C, Kiskinis E, Verstappen G, Gu H, Boulting G, Smith ZD, Ziller M, Croft GF, Amoroso MW, Oakley DH, et al. 2011. Reference maps of human ES and iPS cell variation enable high-throughput characterization of pluripotent cell lines. *Cell* **144:** 439–452.

Bogdanove AJ, Voytas DF. 2011. TAL effectors: Customizable proteins for DNA targeting. *Science* **333:** 1843–1846.

Boland MJ, Hazen JL, Nazor KL, Rodriguez AR, Gifford W, Martin G, Kupriyanov S, Baldwin KK. 2009. Adult mice generated from induced pluripotent stem cells. *Nature* **461:** 91–94.

Boyer LA, Plath K, Zeitlinger J, Brambrink T, Medeiros LA, Lee TI, Levine SS, Wernig M, Tajonar A, Ray MK, et al. 2006. Polycomb complexes repress developmental regulators in murine embryonic stem cells. *Nature* **441:** 349–353.

Brambrink T, Hochedlinger K, Bell G, Jaenisch R. 2006. ES cells derived from cloned and fertilized blastocysts are transcriptionally and functionally indistinguishable. *Proc Natl Acad Sci* **103:** 933–938.

Brambrink T, Foreman R, Welstead GG, Lengner CJ, Wernig M, Suh H, Jaenisch R. 2008. Sequential expression of pluripotency markers during direct reprogramming of mouse somatic cells. *Cell Stem Cell* **2:** 151–159.

Brennand KJ, Simone A, Jou J, Gelboin-Burkhart C, Tran N, Sangar S, Li Y, Mu Y, Chen G, Yu D, et al. 2011. Modelling schizophrenia using human induced pluripotent stem cells. *Nature* **473:** 221–225.

Briggs R, King TJ. 1952. Transplantation of living nuclei from blastula cells into enucleated frogs' eggs. *Proc Natl Acad Sci* **38:** 455–463.

Briggs R, King TJ. 1957. Changes in the nuclei of differentiating endoderm cells as revealed by nuclear transplantation. *J Morphol* **100:** 269–311.

* Brockdorff N, Turner BM. 2014. Dosage compensation in mammals. *Cold Spring Harb Perspect Biol* doi: 10.1101/cshperspect.a019406.

Brons IG, Smithers LE, Trotter MW, Rugg-Gunn P, Sun B, Chuva de Sousa Lopes SM, Howlett SK, Clarkson A, Ahrlund-Richter L, Pedersen RA, et al. 2007. Derivation of pluripotent epiblast stem cells from mammalian embryos. *Nature* **448:** 191–195.

Bussmann LH, Schubert A, Vu Manh TP, De Andres L, Desbordes SC, Parra M, Zimmermann T, Rapino F, Rodriguez-Ubreva J, Ballestar E, et al. 2009. A robust and highly efficient immune cell reprogramming system. *Cell Stem Cell* **5:** 554–566.

Carey BW, Markoulaki S, Beard C, Hanna J, Jaenisch R. 2010. Single-gene transgenic mouse strains for reprogramming adult somatic cells. *Nat Methods* **7:** 56–59.

Carey BW, Markoulaki S, Hanna JH, Faddah DA, Buganim Y, Kim J, Ganz K, Steine EJ, Cassady JP, Creyghton MP, et al. 2011. Reprogramming factor stoichiometry influences the epigenetic state and biological properties of induced pluripotent stem cells. *Cell Stem Cell* **9:** 588–598.

Carvajal-Vergara X, Sevilla A, D'Souza SL, Ang YS, Schaniel C, Lee DF, Yang L, Kaplan AD, Adler ED, Rozov R, et al. 2010. Patient-specific induced pluripotent stem-cell-derived models of LEOPARD syndrome. *Nature* **465:** 808–812.

Chambers I, Smith A. 2004. Self-renewal of teratocarcinoma and embryonic stem cells. *Oncogene* **23:** 7150–7160.

Chen J, Liu J, Yang J, Chen Y, Chen J, Ni S, Song H, Zeng L, Ding K, Pei D. 2011. BMPs functionally replace Klf4 and support efficient reprogramming of mouse fibroblasts by Oct4 alone. *Cell Res* **21:** 205–212.

Cheng L, Hansen NF, Zhao L, Du Y, Zou C, Donovan FX, Chou BK, Zhou G, Li S, Dowey SN, et al. 2012. Low incidence of DNA sequence variation in human induced pluripotent stem cells generated by non-integrating plasmid expression. *Cell Stem Cell* **10:** 337–344.

Chin MH, Mason MJ, Xie W, Volinia S, Singer M, Peterson C, Ambartsumyan G, Aimiuwu O, Richter L, Zhang J, et al. 2009. Induced pluripotent stem cells and embryonic stem cells are distinguished by gene expression signatures. *Cell Stem Cell* **5:** 111–123.

Collado M, Blasco MA, Serrano M. 2007. Cellular senescence in cancer and aging. *Cell* **130:** 223–233.

Daley GQ. 2012. The promise and perils of stem cell therapeutics. *Cell Stem Cell* **10:** 740–749.

Davis RL, Weintraub H, Lassar AB. 1987. Expression of a single transfected cDNA converts fibroblasts to myoblasts. *Cell* **51:** 987–1000.

Ebert AD, Yu J, Rose FF Jr, Mattis VB, Lorson CL, Thomson JA, Svendsen CN. 2009. Induced pluripotent stem cells from a spinal muscular atrophy patient. *Nature* **457:** 277–280.

Eggan K, Akutsu H, Hochedlinger K, Rideout W 3rd, Yanagimachi R, Jaenisch R. 2000. X-chromosome inactivation in cloned mouse embryos. *Science* **290:** 1578–1581.

Eggan K, Akutsu H, Loring J, Jackson-Grusby L, Klemm M, Rideout WM 3rd, Yanagimachi R, Jaenisch R. 2001. Hybrid vigor, fetal overgrowth, and viability of mice derived by nuclear cloning and tetraploid embryo complementation. *Proc Natl Acad Sci* **98:** 6209–6214.

Eggan K, Baldwin K, Tackett M, Osborne J, Gogos J, Chess A, Axel R, Jaenisch R. 2004. Mice cloned from olfactory sensory neurons. *Nature* **428:** 44–49.

Eminli S, Utikal J, Arnold K, Jaenisch R, Hochedlinger K. 2008. Reprogramming of neural progenitor cells into induced pluripotent stem cells in the absence of exogenous Sox2 expression. *Stem Cells* **26:** 2467–2474.

Eminli S, Foudi A, Stadtfeld M, Maherali N, Ahfeldt T, Mostoslavsky G, Hock H, Hochedlinger K. 2009. Differentiation stage determines potential of hematopoietic cells for reprogramming into induced pluripotent stem cells. *Nat Genet* **41:** 968–976.

Evans MJ, Kaufman MH. 1981. Establishment in culture of pluripotential cells from mouse embryos. *Nature* **292:** 154–156.

Feng B, Jiang J, Kraus P, Ng JH, Heng JC, Chan YS, Yaw LP, Zhang W, Loh YH, Han J, et al. 2009. Reprogramming of fibroblasts into induced pluripotent stem cells with orphan nuclear receptor Esrrb. *Nat Cell Biol* **11:** 197–203.

Fusaki N, Ban H, Nishiyama A, Saeki K, Hasegawa M. 2009. Efficient induction of transgene-free human pluripotent stem cells using a vector based on Sendai virus, an RNA virus that does not integrate into the host genome. *Proc Jpn Acad Ser B Phys Biol Sci* **85:** 348–362.

Gaspar-Maia A, Alajem A, Polesso F, Sridharan R, Mason MJ, Heidersbach A, Ramalho-Santos J, McManus MT, Plath K, Meshorer E, et al. 2009. Chd1 regulates open chromatin and pluripotency of embryonic stem cells. *Nature* **460:** 863–868.

Gore A, Li Z, Fung HL, Young JE, Agarwal S, Antosiewicz-Bourget J, Canto I, Giorgetti A, Israel MA, Kiskinis E, et al. 2011. Somatic coding mutations in human induced pluripotent stem cells. *Nature* **471:** 63–67.

Guenther MG, Frampton GM, Soldner F, Hockemeyer D, Mitalipova M, Jaenisch R, Young RA. 2010. Chromatin structure and gene expression programs of human embryonic and induced pluripotent stem cells. *Cell Stem Cell* **7:** 249–257.

Guo G, Yang J, Nichols J, Hall JS, Eyres I, Mansfield W, Smith A. 2009. Klf4 reverts developmentally programmed restriction of ground state pluripotency. *Development* **136:** 1063–1069.

Gurdon J. 1962. The developmental capacity of nuclei taken from intestinal epithelium cells of feeding tadpoles. *J Embryol Exp Morphol* **10:** 622–640.

Hanna J, Wernig M, Markoulaki S, Sun CW, Meissner A, Cassady JP, Beard C, Brambrink T, Wu LC, Townes TM, et al. 2007. Treatment of sickle cell anemia mouse model with iPS cells generated from autologous skin. *Science* **318:** 1920–1923.

Hanna J, Carey BW, Jaenisch R. 2008a. Reprogramming of somatic cell identity. *Cold Spring Harb Symp Quant Biol* **73:** 147–155.

Hanna J, Markoulaki S, Schorderet P, Carey BW, Beard C, Wernig M, Creyghton MP, Steine EJ, Cassady JP, Foreman R, et al. 2008b. Direct reprogramming of terminally differentiated mature B lymphocytes to pluripotency. *Cell* **133:** 250–264.

Hanna J, Markoulaki S, Mitalipova M, Cheng AW, Cassady JP, Staerk J, Carey BW, Lengner CJ, Foreman R, Love J, et al. 2009a. Metastable pluripotent states in NOD-mouse-derived ESCs. *Cell Stem Cell* **4:** 513–524.

Hanna J, Saha K, Pando B, van Zon J, Lengner CJ, Creyghton MP, van Oudenaarden A, Jaenisch R. 2009b. Direct cell reprogramming is a stochastic process amenable to acceleration. *Nature* **462:** 595–601.

Hanna J, Cheng AW, Saha K, Kim J, Lengner CJ, Soldner F, Cassady JP, Muffat J, Carey BW, Jaenisch R. 2010a. Human embryonic stem cells with biological and epigenetic characteristics similar to those of mouse ESCs. *Proc Natl Acad Sci* **107:** 9222–9227.

Hanna JH, Saha K, Jaenisch R. 2010b. Pluripotency and cellular reprogramming: Facts, hypotheses, unresolved issues. *Cell* **143:** 508–525.

Heng JC, Feng B, Han J, Jiang J, Kraus P, Ng JH, Orlov YL, Huss M, Yang L, Lufkin T, et al. 2010. The nuclear receptor Nr5a2 can replace Oct4 in the reprogramming of murine somatic cells to pluripotent cells. *Cell Stem Cell* **6:** 167–174.

Hochedlinger K, Jaenisch R. 2002. Monoclonal mice generated by nuclear transfer from mature B and T donor cells. *Nature* **415:** 1035–1038.

Hochedlinger K, Jaenisch R. 2003. Nuclear transplantation, embryonic stem cells, and the potential for cell therapy. *N Engl J Med* **5:** S114–S117.

Hockemeyer D, Soldner F, Cook EG, Gao Q, Mitalipova M, Jaenisch R. 2008. A drug-inducible system for direct reprogramming of human somatic cells to pluripotency. *Cell Stem Cell* **3:** 346–353.

Hockemeyer D, Soldner F, Beard C, Gao Q, Mitalipova M, DeKelver RC, Katibah GE, Amora R, Boydston EA, Zeitler B, et al. 2009. Efficient targeting of expressed and silent genes in human ESCs and iPSCs using zinc-finger nucleases. *Nat Biotechnol* **27:** 851–857.

Hockemeyer D, Wang H, Kiani S, Lai CS, Gao Q, Cassady JP, Cost GJ, Zhang L, Santiago Y, Miller JC, et al. 2011. Genetic engineering of human pluripotent cells using TALE nucleases. *Nat Biotechnol* **29:** 731–734.

Hong H, Takahashi K, Ichisaka T, Aoi T, Kanagawa O, Nakagawa M, Okita K, Yamanaka S. 2009. Suppression of induced pluripotent stem cell generation by the p53–p21 pathway. *Nature* **460:** 1132–1135.

Hou P, Li Y, Zhang X, Liu C, Guan J, Li H, Zhao T, Ye J, Yang W, Liu K, et al. 2013. Pluripotent stem cells induced from mouse somatic cells by small-molecule compounds. *Science* **341:** 651–654.

Huang P, He Z, Ji S, Sun H, Xiang D, Liu C, Hu Y, Wang X, Hui L. 2011. Induction of functional hepatocyte-like cells from mouse fibroblasts by defined factors. *Nature* **475:** 386–389.

Huangfu D, Maehr R, Guo W, Eijkelenboom A, Snitow M, Chen AE, Melton DA. 2008. Induction of pluripotent stem cells by defined factors is greatly improved by small-molecule compounds. *Nat Biotechnol* **26:** 795–797.

Hussein SM, Batada NN, Vuoristo S, Ching RW, Autio R, Narva E, Ng S, Sourour M, Hamalainen R, Olsson C, et al. 2011. Copy number variation and selection during reprogramming to pluripotency. *Nature* **471:** 58–62.

Ichida JK, Blanchard J, Lam K, Son EY, Chung JE, Egli D, Loh KM, Carter AC, Di Giorgio FP, Koszka K, et al. 2009. A small-molecule inhibitor of tgf-β signaling replaces sox2 in reprogramming by inducing nanog. *Cell Stem Cell* **5:** 491–503.

Ieda M, Fu JD, Delgado-Olguin P, Vedantham V, Hayashi Y, Bruneau BG, Srivastava D. 2010. Direct reprogramming of fibroblasts into functional cardiomyocytes by defined factors. *Cell* **142:** 375–386.

Inoue K, Wakao H, Ogonuki N, Miki H, Seino K, Nambu-Wakao R, Noda S, Miyoshi H, Koseki H, Taniguchi M, et al. 2005. Generation of cloned mice by direct nuclear transfer from natural killer T cells. *Curr Biol* **15:** 1114–1118.

Jaenisch R, Gurdon J. 2007. Nuclear transplantation and the reprogramming of the genome. In *Epigenetics* (ed. Allis CD, et al.), pp. 415–434. Cold Spring Harbor Laboratory Press, Cold Spring Harbor, NY.

Jaenisch R, Young R. 2008. Stem cells, the molecular circuitry of pluripotency and nuclear reprogramming. *Cell* **132:** 567–582.

Jiang J, Chan YS, Loh YH, Cai J, Tong GQ, Lim CA, Robson P, Zhong S, Ng HH. 2008. A core Klf circuitry regulates self-renewal of embryonic stem cells. *Nat Cell Biol* **10:** 353–360.

Kaji K, Caballero IM, MacLeod R, Nichols J, Wilson VA, Hendrich B. 2006. The NuRD component Mbd3 is required for pluripotency of embryonic stem cells. *Nat Cell Biol* **8:** 285–292.

Kang L, Wang J, Zhang Y, Kou Z, Gao S. 2009. iPS cells can support full-term development of tetraploid blastocyst-complemented embryos. *Cell Stem Cell* **5:** 135–138.

Kawamura T, Suzuki J, Wang YV, Menendez S, Morera LB, Raya A, Wahl GM, Izpisua Belmonte JC. 2009. Linking the p53 tumour suppressor pathway to somatic cell reprogramming. *Nature* **460:** 1140–1144.

Kelly SJ. 1977. Studies of the developmental potential of 4- and 8-cell stage mouse blastomeres. *J Exp Zool* **200:** 365–376.

Kim JB, Zaehres H, Wu G, Gentile L, Ko K, Sebastiano V, Arauzo-Bravo MJ, Ruau D, Han DW, Zenke M, et al. 2008. Pluripotent stem cells induced from adult neural stem cells by reprogramming with two factors. *Nature* **454:** 646–650.

Kim D, Kim CH, Moon JI, Chung YG, Chang MY, Han BS, Ko S, Yang E, Cha KY, Lanza R, et al. 2009a. Generation of human induced pluripotent stem cells by direct delivery of reprogramming proteins. *Cell Stem Cell* **4:** 472–476.

Kim HH, Kuwano Y, Srikantan S, Lee EK, Martindale JL, Gorospe M. 2009b. HuR recruits let-7/RISC to repress c-Myc expression. *Genes Dev* **23:** 1743–1748.

Kim K, Doi A, Wen B, Ng K, Zhao R, Cahan P, Kim J, Aryee MJ, Ji H, Ehrlich LI, et al. 2010. Epigenetic memory in induced pluripotent stem cells. *Nature* **467:** 285–290.

Kleinsmith LJ, Pierce GB Jr. 1964. Multipotentiality of single embryonal carcinoma cells. *Cancer Res* **24:** 1544–1551.

Knoepfler PS. 2008. Why myc? An unexpected ingredient in the stem cell cocktail. *Cell Stem Cell* **2:** 18–21.

Laiosa CV, Stadtfeld M, Xie H, de Andres-Aguayo L, Graf T. 2006. Reprogramming of committed T cell progenitors to macrophages and dendritic cells by C/EBP α and PU.1 transcription factors. *Immunity* **25:** 731–744.

Lee TI, Jenner RG, Boyer LA, Guenther MG, Levine SS, Kumar RM, Chevalier B, Johnstone SE, Cole MF, Isono K, et al. 2006. Control of developmental regulators by Polycomb in human embryonic stem cells. *Cell* **125:** 301–313.

Lee G, Papapetrou EP, Kim H, Chambers SM, Tomishima MJ, Fasano CA, Ganat YM, Menon J, Shimizu F, Viale A, et al. 2009. Modelling pathogenesis and treatment of familial dysautonomia using patient-specific iPSCs. *Nature* **461:** 402–406.

Lei H, Oh SP, Okano M, Juttermann R, Goss KA, Jaenisch R, Li E. 1996. De novo DNA cytosine methyltransferase activities in mouse embryonic stem cells. *Development* **122:** 3195–3205.

Lengner CJ, Gimelbrant AA, Erwin JA, Cheng AW, Guenther MG, Welstead GG, Alagappan R, Frampton GM, Xu P, Muffat J, et al. 2010. Derivation of pre-X inactivation human embryonic stem cells under physiological oxygen concentrations. *Cell* **141:** 872–883.

Li E, Bestor TH, Jaenisch R. 1992. Targeted mutation of the DNA methyltransferase gene results in embryonic lethality. *Cell* **69:** 915–926.

Li H, Collado M, Villasante A, Strati K, Ortega S, Canamero M, Blasco MA, Serrano M. 2009a. The Ink4/Arf locus is a barrier for iPS cell reprogramming. *Nature* **460:** 1136–1139.

Li W, Wei W, Zhu S, Zhu J, Shi Y, Lin T, Hao E, Hayek A, Deng H, Ding S. 2009b. Generation of rat and human induced pluripotent stem cells by combining genetic reprogramming and chemical inhibitors. *Cell Stem Cell* **4:** 16–19.

Li R, Liang J, Ni S, Zhou T, Qing X, Li H, He W, Chen J, Li F, Zhuang Q, et al. 2010. A mesenchymal-to-epithelial transition initiates and is required for the nuclear reprogramming of mouse fibroblasts. *Cell Stem Cell* **7:** 51–63.

Lin CH, Lin C, Tanaka H, Fero ML, Eisenman RN. 2009. Gene regulation and epigenetic remodeling in murine embryonic stem cells by c-Myc. *PloS One* **4:** e7839.

Lister R, Pelizzola M, Kida YS, Hawkins RD, Nery JR, Hon G, Antosiewicz-Bourget J, O'Malley R, Castanon R, Klugman S, et al. 2011. Hotspots of aberrant epigenomic reprogramming in human induced pluripotent stem cells. *Nature* **471:** 68–73.

Liu H, Zhu F, Yong J, Zhang P, Hou P, Li H, Jiang W, Cai J, Liu M, Cui K, et al. 2008. Generation of induced pluripotent stem cells from adult rhesus monkey fibroblasts. *Cell Stem Cell* **3:** 587–590.

Lyssiotis CA, Foreman RK, Staerk J, Garcia M, Mathur D, Markoulaki S, Hanna J, Lairson LL, Charette BD, Bouchez LC, et al. 2009. Reprogramming of murine fibroblasts to induced pluripotent stem cells with chemical complementation of Klf4. *Proc Natl Acad Sci* **106:** 8912–8917.

Maekawa M, Yamaguchi K, Nakamura T, Shibukawa R, Kodanaka I, Ichisaka T, Kawamura Y, Mochizuki H, Goshima N, Yamanaka S. 2011. Direct reprogramming of somatic cells is promoted by maternal transcription factor Glis1. *Nature* **474:** 225–229.

Maherali N, Sridharan R, Xie W, Utikal J, Eminli S, Arnold K, Stadtfeld M, Yachechko R, Tchieu J, Jaenisch R, et al. 2007. Directly reprogrammed fibroblasts show global epigenetic remodeling and widespread tissue contribution. *Cell Stem Cell* **1:** 55–70.

Maherali N, Ahfeldt T, Rigamonti A, Utikal J, Cowan C, Hochedlinger K. 2008. A high-efficiency system for the generation and study of human induced pluripotent stem cells. *Cell Stem Cell* **3:** 340–345.

Mansour AA, Gafni O, Weinberger L, Zviran A, Ayyash M, Rais Y, Krupalnik V, Zerbib M, Amann-Zalcenstein D, Maza I, et al. 2012. The H3K27 demethylase Utx regulates somatic and germ cell epigenetic reprogramming. *Nature* **488:** 409–413.

Marchetto MC, Carromeu C, Acab A, Yu D, Yeo GW, Mu Y, Chen G, Gage FH, Muotri AR. 2010. A model for neural development and treatment of Rett syndrome using human induced pluripotent stem cells. *Cell* **143:** 527–539.

Marion RM, Strati K, Li H, Tejera A, Schoeftner S, Ortega S, Serrano M, Blasco MA. 2009. Telomeres acquire embryonic stem cell characteristics in induced pluripotent stem cells. *Cell Stem Cell* **4:** 141–154.

Markoulaki S, Hanna J, Beard C, Carey BW, Cheng AW, Lengner CJ, Dausman JA, Fu D, Gao Q, Wu S, et al. 2009. Transgenic mice with defined combinations of drug-inducible reprogramming factors. *Nat Biotechnol* **27:** 169–171.

Martin GR. 1981. Isolation of a pluripotent cell line from early mouse embryos cultured in medium conditioned by teratocarcinoma stem cells. *Proc Natl Acad Sci* **78:** 7634–7638.

Matsui T, Leung D, Miyashita H, Maksakova IA, Miyachi H, Kimura H, Tachibana M, Lorincz MC, Shinkai Y. 2010. Proviral silencing in embryonic stem cells requires the histone methyltransferase ESET. *Nature* **464:** 927–931.

Meissner A, Jaenisch R. 2006. Mammalian nuclear transfer. *Dev Dyn* **235:** 2460–2469.

Meissner A, Wernig M, Jaenisch R. 2007. Direct reprogramming of genetically unmodified fibroblasts into pluripotent stem cells. *Nat Biotechnol* **25:** 1177–1181.

Mekhoubad S, Bock C, de Boer AS, Kiskinis E, Meissner A, Eggan K. 2012. Erosion of dosage compensation impacts human iPSC disease modeling. *Cell Stem Cell* **10:** 595–609.

Mikkelsen TS, Hanna J, Zhang X, Ku M, Wernig M, Schorderet P, Bernstein BE, Jaenisch R, Lander ES, Meissner A. 2008. Dissecting direct reprogramming through integrative genomic analysis. *Nature* **454:** 49–55.

Miller RA, Ruddle FH. 1976. Pluripotent teratocarcinoma-thymus somatic cell hybrids. *Cell* **9:** 45–55.

Munsie MJ, Michalska AE, O'Brien CM, Trounson AO, Pera MF, Mountford PS. 2000. Isolation of pluripotent embryonic stem cells from reprogrammed adult mouse somatic cell nuclei. *Curr Biol* **10:** 989–992.

Nagy A, Gocza E, Diaz EM, Prideaux VR, Ivanyi E, Markkula M, Rossant J. 1990. Embryonic stem cells alone are able to support fetal development in the mouse. *Development* **110:** 815–821.

Nakagawa M, Koyanagi M, Tanabe K, Takahashi K, Ichisaka T, Aoi T, Okita K, Mochiduki Y, Takizawa N, Yamanaka S. 2008. Generation of induced pluripotent stem cells without Myc from mouse and human fibroblasts. *Nat Biotechnol* **26:** 101–106.

Nakagawa M, Takizawa N, Narita M, Ichisaka T, Yamanaka S. 2010. Promotion of direct reprogramming by transformation-deficient Myc. *Proc Natl Acad Sci* **107:** 14152–14157.

Nemajerova A, Kim SY, Petrenko O, Moll UM. 2012. Two-factor reprogramming of somatic cells to pluripotent stem cells reveals partial functional redundancy of Sox2 and Klf4. *Cell Death Differ* **19:** 1268–1276.

Newman AM, Cooper JB. 2010. Lab-specific gene expression signatures in pluripotent stem cells. *Cell Stem Cell* **7:** 258–262.

Nichols J, Smith A. 2009. Naive and primed pluripotent states. *Cell Stem Cell* **4:** 487–492.

Ohi Y, Qin H, Hong C, Blouin L, Polo JM, Guo T, Qi Z, Downey SL, Manos PD, Rossi DJ, et al. 2011. Incomplete DNA methylation underlies a transcriptional memory of somatic cells in human iPS cells. *Nat Cell Biol* **13:** 541–549.

Okamoto I, Patrat C, Thepot D, Peynot N, Fauque P, Daniel N, Diabangouaya P, Wolf JP, Renard JP, Duranthon V, et al. 2011. Eutherian mammals use diverse strategies to initiate X-chromosome inactivation during development. *Nature* **472:** 370–374.

Okita K, Ichisaka T, Yamanaka S. 2007. Generation of germline-competent induced pluripotent stem cells. *Nature* **448:** 313–317.

Okita K, Nakagawa M, Hyenjong H, Ichisaka T, Yamanaka S. 2008. Generation of mouse induced pluripotent stem cells without viral vectors. *Science* **322:** 949–953.

Orkin SH, Hochedlinger K. 2011. Chromatin connections to pluripotency and cellular reprogramming. *Cell* **145:** 835–850.

Park IH, Zhao R, West JA, Yabuuchi A, Huo H, Ince TA, Lerou PH, Lensch MW, Daley GQ. 2008. Reprogramming of human somatic cells to pluripotency with defined factors. *Nature* **451:** 141–146.

Pawlak M, Jaenisch R. 2011. De novo DNA methylation by Dnmt3a and Dnmt3b is dispensable for nuclear reprogramming of somatic cells to a pluripotent state. *Genes Dev* **25:** 1035–1040.

Polo JM, Liu S, Figueroa ME, Kulalert W, Eminli S, Tan KY, Apostolou E, Stadtfeld M, Li Y, Shioda T, et al. 2010. Cell type of origin influences the molecular and functional properties of mouse induced pluripotent stem cells. *Nat Biotechnol* **28:** 848–855.

Raya A, Rodriguez-Piza I, Guenechea G, Vassena R, Navarro S, Barrero MJ, Consiglio A, Castella M, Rio P, Sleep E, et al. 2009. Disease-corrected haematopoietic progenitors from Fanconi anaemia induced pluripotent stem cells. *Nature* **460:** 53–59.

Reik W, Dean W, Walter J. 2001. Epigenetic reprogramming in mammalian development. *Science* **293:** 1089–1093.

Rubin LL. 2008. Stem cells and drug discovery: The beginning of a new era? *Cell* **132:** 549–552.

Samavarchi-Tehrani P, Golipour A, David L, Sung HK, Beyer TA, Datti A, Woltjen K, Nagy A, Wrana JL. 2010. Functional genomics reveals a BMP-driven mesenchymal-to-epithelial transition in the initiation of somatic cell reprogramming. *Cell Stem Cell* **7:** 64–77.

Shen Y, Matsuno Y, Fouse SD, Rao N, Root S, Xu R, Pellegrini M, Riggs AD, Fan G. 2008. X-inactivation in female human embryonic stem cells is in a nonrandom pattern and prone to epigenetic alterations. *Proc Natl Acad Sci* **105:** 4709–4714.

Singhal N, Graumann J, Wu G, Arauzo-Bravo MJ, Han DW, Greber B, Gentile L, Mann M, Scholer HR. 2010. Chromatin-remodeling components of the BAF complex facilitate reprogramming. *Cell* **141:** 943–955.

Smith ZD, Nachman I, Regev A, Meissner A. 2010. Dynamic single-cell imaging of direct reprogramming reveals an early specifying event. *Nat Biotechnol* **28:** 521–526.

Soldner F, Hockemeyer D, Beard C, Gao Q, Bell GW, Cook EG, Hargus G, Blak A, Cooper O, Mitalipova M, et al. 2009. Parkinson's disease patient-derived induced pluripotent stem cells free of viral reprogramming factors. *Cell* **136:** 964–977.

Soldner F, Laganiere J, Cheng AW, Hockemeyer D, Gao Q, Alagappan R, Khurana V, Golbe LI, Myers RH, Lindquist S, et al. 2011. Generation of isogenic pluripotent stem cells differing exclusively at two early onset Parkinson point mutations. *Cell* **146:** 318–331.

Sridharan R, Tchieu J, Mason MJ, Yachechko R, Kuoy E, Horvath S, Zhou Q, Plath K. 2009. Role of the murine reprogramming factors in the induction of pluripotency. *Cell* **136:** 364–377.

Stadtfeld M, Hochedlinger K. 2010. Induced pluripotency: History, mechanisms, and applications. *Genes Dev* **24:** 2239–2263.

Stadtfeld M, Brennand K, Hochedlinger K. 2008a. Reprogramming of pancreatic β cells into induced pluripotent stem cells. *Curr Biol* **18:** 890–894.

Stadtfeld M, Maherali N, Breault DT, Hochedlinger K. 2008b. Defining molecular cornerstones during fibroblast to iPS cell reprogramming in mouse. *Cell Stem Cell* **2:** 230–240.

Stadtfeld M, Nagaya M, Utikal J, Weir G, Hochedlinger K. 2008c. Induced pluripotent stem cells generated without viral integration. *Science* **322:** 945–949.

Stadtfeld M, Apostolou E, Akutsu H, Fukuda A, Follett P, Natesan S, Kono T, Shioda T, Hochedlinger K. 2010a. Aberrant silencing of imprinted genes on chromosome 12qF1 in mouse induced pluripotent stem cells. *Nature* **465:** 175–181.

Stadtfeld M, Maherali N, Borkent M, Hochedlinger K. 2010b. A reprogrammable mouse strain from gene-targeted embryonic stem cells. *Nat Methods* **7:** 53–55.

Stadtfeld M, Apostolou E, Ferrari F, Choi J, Walsh RM, Chen T, Ooi SS, Kim SY, Bestor TH, Shioda T, et al. 2012. Ascorbic acid prevents loss of Dlk1-Dio3 imprinting and facilitates generation of all-iPS cell mice from terminally differentiated B cells. *Nat Genet* **44:** 398–405, S391–S392.

Tachibana M, Amato P, Sparman M, Gutierrez NM, Tippner-Hedges R, Ma H, Kang E, Fulati A, Lee HS, Sritanaudomchai H, et al. 2013. Human embryonic stem cells derived by somatic cell nuclear transfer. *Cell* **153:** 1228–1238.

Tada M, Morizane A, Kimura H, Kawasaki H, Ainscough JF, Sasai Y, Nakatsuji N, Tada T. 2003. Pluripotency of reprogrammed somatic genomes in embryonic stem hybrid cells. *Dev Dyn* **227:** 504–510.

* Takahashi K. 2014. Cellular reprogramming. *Cold Spring Harb Persp Biol* **6:** a018606.

Takahashi K, Yamanaka S. 2006. Induction of pluripotent stem cells from mouse embryonic and adult fibroblast cultures by defined factors. *Cell* **126:** 663–676.

 Cite as *Cold Spring Harb Perspect Biol* doi: 10.1101/cshperspect.a019448

Takahashi K, Tanabe K, Ohnuki M, Narita M, Ichisaka T, Tomoda K, Yamanaka S. 2007. Induction of pluripotent stem cells from adult human fibroblasts by defined factors. *Cell* **131:** 861–872.

Tesar PJ, Chenoweth JG, Brook FA, Davies TJ, Evans EP, Mack DL, Gardner RL, McKay RD. 2007. New cell lines from mouse epiblast share defining features with human embryonic stem cells. *Nature* **448:** 196–199.

Thomson JA, Itskovitz-Eldor J, Shapiro SS, Waknitz MA, Swiergiel JJ, Marshall VS, Jones JM. 1998. Embryonic stem cell lines derived from human blastocysts. *Science* **282:** 1145–1147.

Urnov FD, Rebar EJ, Holmes MC, Zhang HS, Gregory PD. 2010. Genome editing with engineered zinc finger nucleases. *Nat Rev Genet* **11:** 636–646.

Utikal J, Maherali N, Kulalert W, Hochedlinger K. 2009a. Sox2 is dispensable for the reprogramming of melanocytes and melanoma cells into induced pluripotent stem cells. *J Cell Sci* **122:** 3502–3510.

Utikal J, Polo JM, Stadtfeld M, Maherali N, Kulalert W, Walsh RM, Khalil A, Rheinwald JG, Hochedlinger K. 2009b. Immortalization eliminates a roadblock during cellular reprogramming into iPS cells. *Nature* **460:** 1145–1148.

Vierbuchen T, Ostermeier A, Pang ZP, Kokubu Y, Sudhof TC, Wernig M. 2010. Direct conversion of fibroblasts to functional neurons by defined factors. *Nature* **463:** 1035–1041.

Viswanathan SR, Daley GQ, Gregory RI. 2008. Selective blockade of microRNA processing by Lin28. *Science* **320:** 97–100.

Wakayama T, Tabar V, Rodriguez I, Perry AC, Studer L, Mombaerts P. 2001. Differentiation of embryonic stem cell lines generated from adult somatic cells by nuclear transfer. *Science* **292:** 740–743.

Wakayama S, Jakt ML, Suzuki M, Araki R, Hikichi T, Kishigami S, Ohta H, Van Thuan N, Mizutani E, Sakaide Y, et al. 2006. Equivalency of nuclear transfer-derived embryonic stem cells to those derived from fertilized mouse blastocysts. *Stem Cells* **24:** 2023–2033.

Wang J, Rao S, Chu J, Shen X, Levasseur DN, Theunissen TW, Orkin SH. 2006. A protein interaction network for pluripotency of embryonic stem cells. *Nature* **444:** 364–368.

Warren L, Manos PD, Ahfeldt T, Loh YH, Li H, Lau F, Ebina W, Mandal PK, Smith ZD, Meissner A, et al. 2010. Highly efficient reprogramming to pluripotency and directed differentiation of human cells with synthetic modified mRNA. *Cell Stem Cell* **7:** 618–630.

Wernig M, Meissner A, Foreman R, Brambrink T, Ku M, Hochedlinger K, Bernstein BE, Jaenisch R. 2007. In vitro reprogramming of fibroblasts into a pluripotent ES-cell-like state. *Nature* **448:** 318–324.

Wernig M, Lengner CJ, Hanna J, Lodato MA, Steine E, Foreman R, Staerk J, Markoulaki S, Jaenisch R. 2008. A drug-inducible transgenic system for direct reprogramming of multiple somatic cell types. *Nat Biotechnol* **26:** 916–924.

Wilmut I, Schnieke AE, McWhir J, Kind AJ, Campbell KH. 1997. Viable offspring derived from fetal and adult mammalian cells. *Nature* **385:** 810–813.

Woltjen K, Michael IP, Mohseni P, Desai R, Mileikovsky M, Hamalainen R, Cowling R, Wang W, Liu P, Gertsenstein M, et al. 2009. *piggyBac* transposition reprograms fibroblasts to induced pluripotent stem cells. *Nature* **458:** 766–770.

Wu SM, Hochedlinger K. 2011. Harnessing the potential of induced pluripotent stem cells for regenerative medicine. *Nat Cell Biol* **13:** 497–505.

Wutz A. 2012. Epigenetic alterations in human pluripotent stem cells: A tale of two cultures. *Cell Stem Cell* **11:** 9–15.

Xie H, Ye M, Feng R, Graf T. 2004. Stepwise reprogramming of B cells into macrophages. *Cell* **117:** 663–676.

Xu Y, Shi Y, Ding S. 2008. A chemical approach to stem-cell biology and regenerative medicine. *Nature* **453:** 338–344.

Yamanaka S. 2009. Elite and stochastic models for induced pluripotent stem cell generation. *Nature* **460:** 49–52.

Yamanaka S, Blau HM. 2010. Nuclear reprogramming to a pluripotent state by three approaches. *Nature* **465:** 704–712.

Young MA, Larson DE, Sun CW, George DR, Ding L, Miller CA, Lin L, Pawlik KM, Chen K, Fan X, et al. 2012. Background mutations in parental cells account for most of the genetic heterogeneity of induced pluripotent stem cells. *Cell Stem Cell* **10:** 570–582.

Yu J, Vodyanik MA, Smuga-Otto K, Antosiewicz-Bourget J, Frane JL, Tian S, Nie J, Jonsdottir GA, Ruotti V, Stewart R, et al. 2007. Induced pluripotent stem cell lines derived from human somatic cells. *Science* **318:** 1917–1920.

Zhao XY, Li W, Lv Z, Liu L, Tong M, Hai T, Hao J, Guo CL, Ma QW, Wang L, et al. 2009. iPS cells produce viable mice through tetraploid complementation. *Nature* **461:** 86–90.

Zhou Q, Brown J, Kanarek A, Rajagopal J, Melton DA. 2008. In vivo reprogramming of adult pancreatic exocrine cells to β-cells. *Nature* **455:** 627–632.

Zhou H, Wu S, Joo JY, Zhu S, Han DW, Lin T, Trauger S, Bien G, Yao S, Zhu Y, et al. 2009. Generation of induced pluripotent stem cells using recombinant proteins. *Cell Stem Cell* **4:** 381–384.

CHAPTER 29

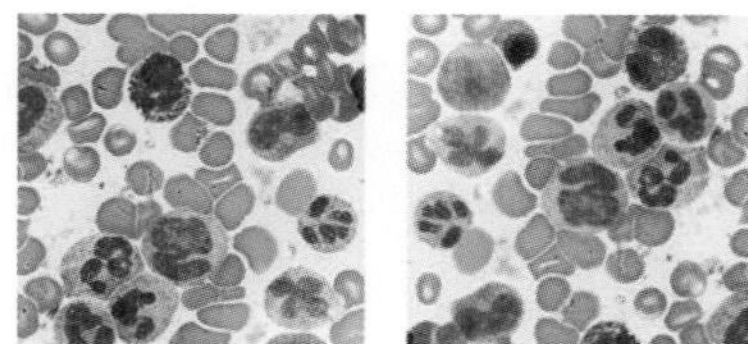

Epigenetic Control of Immunity

Meinrad Busslinger[1] and Alexander Tarakhovsky[2]

[1]Research Institute of Molecular Pathology, Vienna Biocenter, A-1030 Vienna, Austria; [2]Laboratory of Lymphocyte Signaling, The Rockefeller University, New York, New York 10021

Correspondence: busslinger@imp.ac.at and tarakho@mail.rockefeller.edu

SUMMARY

Immunity relies on the heterogeneity of immune cells and their ability to respond to pathogen challenges. In the adaptive immune system, lymphocytes display a highly diverse antigen receptor repertoire that matches the vast diversity of pathogens. In the innate immune system, the cell's heterogeneity and phenotypic plasticity enable flexible responses to changes in tissue homeostasis caused by infection or damage. The immune responses are calibrated by the graded activity of immune cells that can vary from yeast-like proliferation to lifetime dormancy. This chapter describes key epigenetic processes that contribute to the function of immune cells during health and disease.

Outline

Cite as *Cold Spring Harb Perspect Biol* doi: 10.1101/cshperspect.a019307

OVERVIEW

The immune system has a nearly unlimited capacity to respond to environmental triggers. The enormous adaptive potential of the immune system is governed by the mechanisms that enable potential recognition of any "foreign" or "self" triggers, as well as flexibility of immune cell responses to signals of various types and duration. Two types of the immune response, innate and adaptive, provide a comprehensive defense against environmental hazards and also eliminate nonfunctional or malignantly transformed cells.

The innate response involves recognition of nonvariable environmental components, such as pathogen-derived nucleic acids or lipids and noxious substances or venoms. The adaptive immune response relies on a highly diverse immune cell repertoire that is generated through the random arrangement of the antigen receptor genes in B and T lineage cells. The function of the immune system depends also on a highly robust and stringently orchestrated process of hematopoietic cell differentiation, where cells of different types are produced at constant rates and maintain their homeostatic balance through the life of the organism.

A combination of the relatively rigid patterns of immune cell differentiation and "fluid" patterns of responses implies the existence of mechanisms that limit the breadth of phenotypic variation during the generation of immune cells of a defined lineage but increase these variations in mature cells to match the variety and abundance of the environmental triggers. These overreaching models suggest the existence of potentially nonoverlapping epigenetic processes that minimize the variability of the developmental choices within cells of a given lineage, but increase the stochasticity of the differentiated immune cell adaptation to their environment.

The following chapter narrates B-cell development as an example of a highly hierarchical and tightly controlled process governed by well-established transcriptional networks and less understood epigenetic processes. We also discuss the epigenetic regulation of the inflammatory responses that require flexible adaptation to diverse environmental challenges. Finally, we show how the very basic biochemical principles of epigenetic regulation, involving interaction between the histones and effector proteins, could be used for selective interference with immune responses during health and disease.

Cite as *Cold Spring Harb Perspect Biol* doi: 10.1101/cshperspect.a019307

1 INTRODUCTION TO THE ADAPTIVE IMMUNE SYSTEM

The lymphoid system is well suited to study epigenetic mechanisms controlling lineage commitment and differentiation; the developmental pathways from the hematopoietic stem cell (HSC) to mature lymphocytes have been elucidated in great detail (Fig. 1). Commitment of the common lymphoid progenitor (CLP) to the B-lymphoid lineage initiates B-cell development in the bone marrow that is characterized by sequential rearrangements of the immunoglobulin heavy-chain (*IgH*) locus in pro-B cells and immunoglobulin light-chain (*IgL*) genes in pre-B cells. Following elimination of autoreactive B lymphocytes, immature B cells migrate to peripheral lymphoid organs and differentiate into mature B cells that, on antigen encounter, develop into immunoglobulin-secreting plasma cells (Fig. 1). Lymphoid progenitors that flux through the blood stream enter the thymus, where they initiate T-cell development and undergo rearrangements at the T-cell receptor (TCR) β, γ, and δ loci in early T-cell progenitors ($CD4^-CD8^-$ double negative, DN). Successful recombination of *TCRγ* and *TCRδ* genes results in the development of γδ T cells. In-frame *TCRβ* rearrangements activate pre-TCR signaling and subsequent differentiation to $CD4^+$ $CD8^+$ double-positive (DP) thymocytes that undergo recombination at the *TCRα* locus (Fig. 1). Following positive and negative selection, naïve $CD4^+$ and $CD8^+$ single-positive (SP) T cells emerge that further differentiate into distinct $CD4^+$ T helper cell types or $CD8^+$ cytotoxic T cells in peripheral lymphoid organs (Fig. 1). Here, we discuss those epigenetic mechanisms that have been shown to play an important role in the control of lymphopoiesis and immunity.

2 LINEAGE COMMITMENT IN THE LYMPHOID SYSTEM

2.1 Lymphoid Gene Priming in Hematopoietic Progenitors

Early lymphopoiesis is controlled by signaling through the c-Kit, Flt3, and IL-7 receptors as well as by cell-intrinsic transcription factors, such as the zinc-finger transcription factor Ikaros and the Ets-domain protein PU.1, which are essential for the generation of CLPs (Fig. 2A) (Nutt and Kee 2007). The up-regulation of selected lymphoid genes and the simultaneous down-regulation of self-renewal-associated genes constitute the first sign of early lymphopoiesis in multipotent progenitors (MPPs). This priming of lymphoid gene expression is under positive control by the helix-loop-helix protein E2A (Dias et al. 2008) and the transcription factor Ikaros (Ng et al. 2009) that regulates its target genes through interaction with the Mi-2β nucleosome-remodeling and histone-deactylase (NuRD) complex (Zhang et al. 2012b). Lymphoid gene priming is antagonized by the Bmi1-containing Polycomb-repressive complex 1 (PRC1) that promotes HSC renewal and function (Oguro et al. 2010). In addition, both PRC1 and Polycomb-repressive complex 2 (PRC2) are responsible for silencing the B-cell-specific transcription factor genes *Ebf1* and *Pax5* in MPPs and lymphoid progenitors by establishing the repressive histone marks H2AK119ub1 and H3K27me3 at their promoters (Decker et al. 2009; Oguro

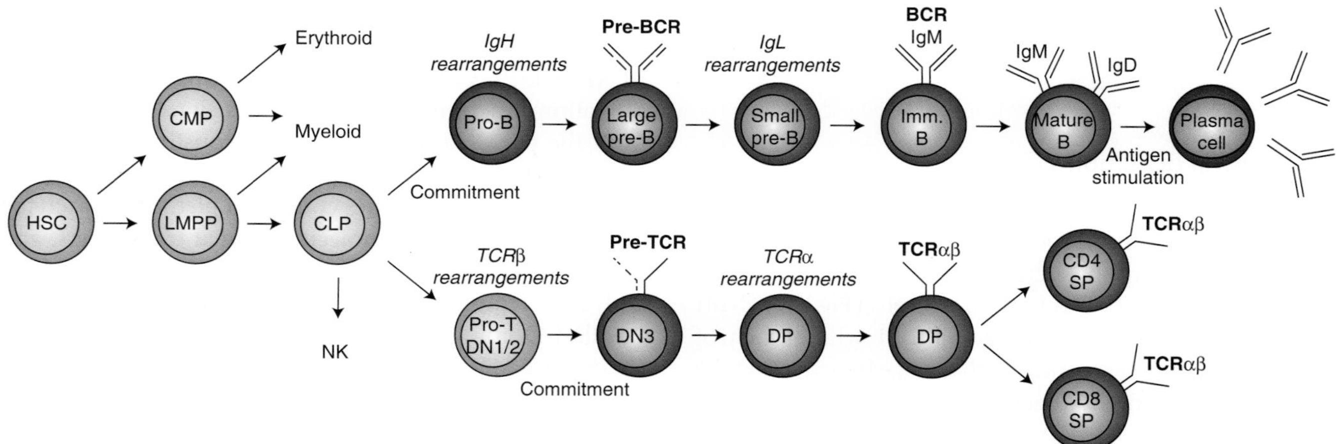

Figure 1. Schematic diagram of B- and T-cell development. Hematopoietic stem cells (HSCs) differentiate via the indicated developmental stages to immunoglobulin-secreting plasma cells or $CD4^+$ helper and $CD8^+$ cytotoxic T cells. LMPP, lymphoid-primed multipotent progenitors; CMP, common myeloid progenitor; CLP, common lymphoid progenitor; DN, double negative; DP, double positive; SP, single positive. Orange, uncommitted progenitors; blue, committed lymphocytes; red, plasma cell.

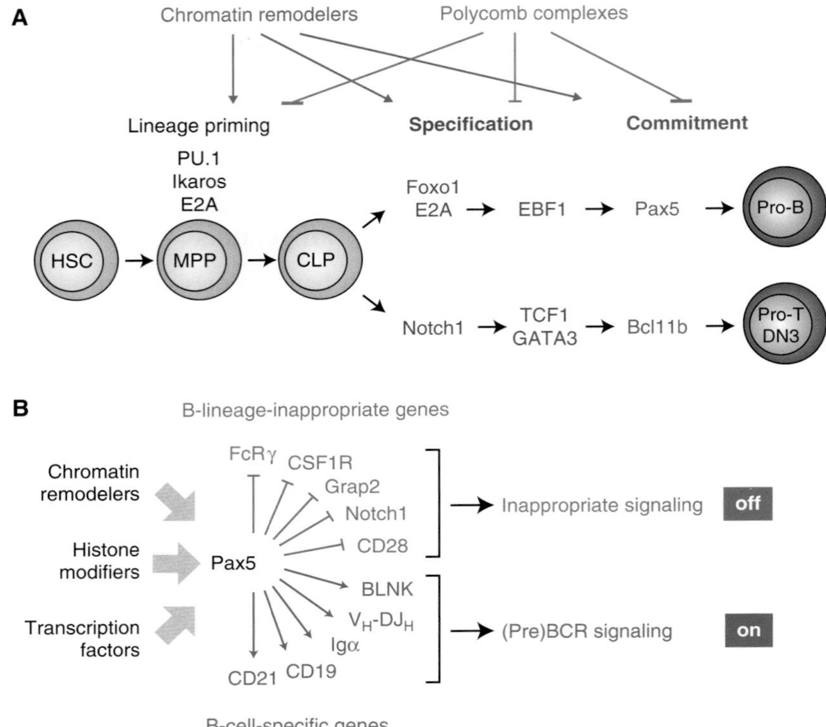

Figure 2. Transcriptional control of early lymphopoiesis. (*A*) Progenitors of the B- and T-cell lineages develop from the HSC under the control of the indicated transcription factors. (*B*) Pax5 activates B-cell-specific genes involved in (pre) B-cell receptor (BCR) signaling, and represses lineage-inappropriate genes of the myeloid (FcRγ, CSF1R) or T-lymphoid (Notch1, CD28, Grap2) pathways. See Revilla-i-Domingo et al. (2012) for a more complete list of regulated Pax5 target genes.

et al. 2010). Remarkably, loss of the Polycomb protein Bmi1 is sufficient to activate the transcription of *Ebf1*, *Pax5*, and their target genes in *Bmi1*$^{-/-}$ MPPs (Oguro et al. 2010). Hence, Polycomb-mediated silencing prevents premature activation of the B-cell gene expression program before the onset of B lymphopoiesis.

2.2 Control of Lymphoid Lineage Commitment

B-cell commitment depends on sequential activity of the instructive transcription factors E2A, EBF1, and Pax5 during the development of CLPs to pro-B cells (Fig. 2A) (Nutt and Kee 2007). The early B-cell factor EBF1, with the help of E2A and Foxo1, specifies the B-cell lineage by activating the expression of B-lymphoid genes in uncommitted pre-pro-B cells (Győry et al. 2012). Pax5 subsequently controls B-cell commitment by restricting the developmental potential of lymphoid progenitors to the B-cell lineage (Nutt et al. 1999; Medvedovic et al. 2011). Pax5 fulfills a dual role at B-cell commitment by repressing a multitude (~230) of B-lineage-inappropriate genes to suppress alternative lineage options and by simultaneously activating many (~120) B-cell-specific genes to promote B-cell development (Fig. 2B) (Nutt et al. 1999; Revilla-i-Domingo et al. 2012). Several of the activated Pax5 target genes code for essential components of (pre)B-cell receptor (BCR) signaling (Igα, BLNK, CD19, CD21; Fig. 2B), indicating that the trans-activation function of Pax5 facilitates signal transduction from the pre-BCR and BCR, which constitute important checkpoints in B-cell development. On the other hand, the repressed Pax5 target genes code for a plethora of secreted proteins, cell adhesion molecules, signal transducers, and nuclear proteins that are expressed in erythroid, myeloid, and/or T-lymphoid cells (Fig. 2B) (Delogu et al. 2006; Revilla-i-Domingo et al. 2012). Among them are the *Csf1r* (M-CSFR) and *Notch1* genes that nicely exemplify how their Pax5-dependent down-regulation renders committed B cells no longer responsive to the myeloid cytokine M-CSF or to T-cell-inducing Notch ligands (Medvedovic et al. 2011). Hence, the repression function of Pax5 is as important as its activation function because it contributes to B-cell commitment by shutting down inappropriate signaling systems.

 Cite as *Cold Spring Harb Perspect Biol* doi: 10.1101/cshperspect.a019307

Pax5 controls gene expression by inducing the formation of active enhancers and promoters at activated target genes and by simultaneously promoting the loss of active DNase I hypersensitive sites at repressed target genes (McManus et al. 2011; Revilla-i-Domingo et al. 2012). Pax5 generates these chromatin and transcription changes by recruiting chromatin-remodeling (BAF), histone-modifying (MLL, CBP, NCoR1), and basal transcription factor (TFIID) complexes to its target genes (McManus et al. 2011). Hence, Pax5 functions as an epigenetic regulator to reprogram gene expression at B-cell commitment.

On entry into the thymus, lymphoid progenitors are exposed to the Notch ligand Delta-like 4 on thymic epithelial cells that activates Notch1 signaling and the T-cell developmental program (Fig. 2A) (Radtke et al. 2010). Notch1 and its downstream transcription factors TCF1 (*Tcf7*) and GATA3 are essential for the formation of the earliest T-cell precursors and subsequent development of pro-T (DN) cells (Ting et al. 1996; Radtke et al. 2010; Weber et al. 2011). Commitment to the T-cell lineage is further controlled by the transcription factor Bcl11b at the transition to the DN3 cell stage (Fig. 2A) (Li et al. 2010). Little is known, however, about how these transcription factors shape the epigenetic landscape and regulate the transcriptional network directing early T-cell development.

3 LINEAGE PLASTICITY IN THE IMMUNE SYSTEM

3.1 Lineage Reprogramming by Ectopic Transcription Factors

The groundbreaking discovery of Takahashi and Yamanaka (2006) showed that most somatic cell types can be epigenetically reprogrammed by ectopic expression of the four transcription factors Oct4, Sox2, Klf4, and c-Myc into pluripotent stem cells with properties similar to embryonic stem cells. Previously, it was already known that forced expression of the myeloid transcription factor C/EBPα induces rapid transdifferentiation of B lymphocytes into functional macrophages (Fig. 3) (Xie et al. 2004). Mecha-

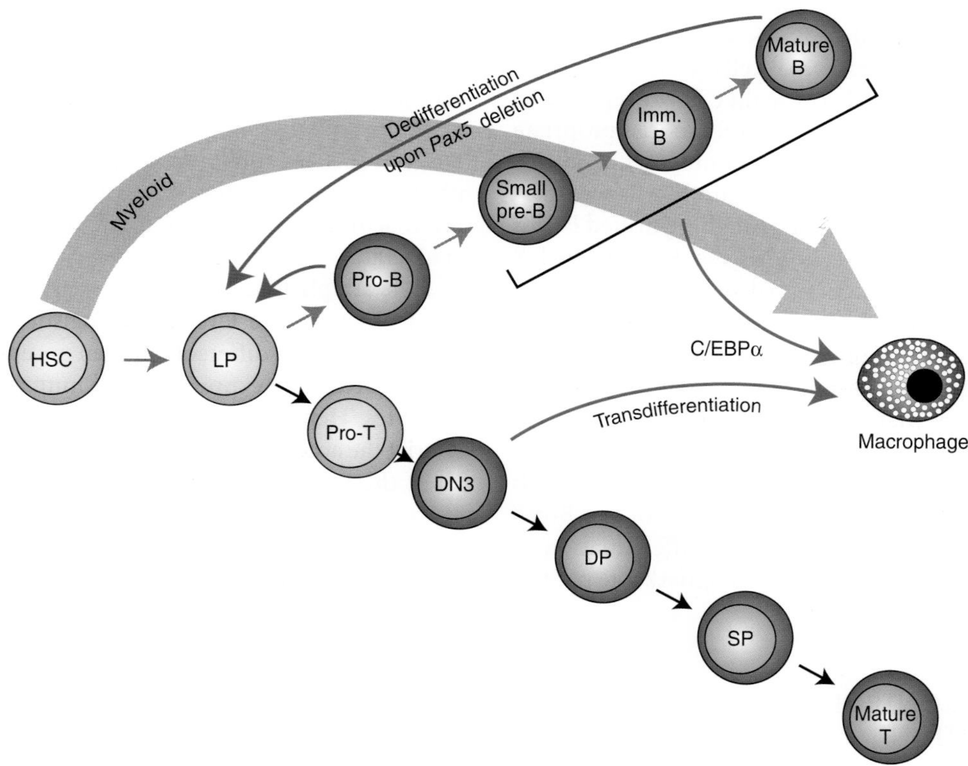

Figure 3. Developmental plasticity of B lymphocytes. Committed $CD19^+$ B lymphocytes and committed DN3 thymocytes undergo rapid transdifferentiation in vitro to macrophages in response to forced C/EBPα expression (red arrows) (Xie et al. 2004). Conditional *Pax5* deletion allows committed pro-B cells and mature B cells to dedifferentiate in vivo to uncommitted lymphoid progenitors (LPs; green arrows) that subsequently develop into other hematopoietic cell types in the bone marrow or T cells in the thymus (black arrows) (Cobaleda et al. 2007). Committed lymphocytes are shown in blue.

nistic analysis of this transdifferentiation process revealed that pre-B cells are converted by C/EBPα activity into macrophage-like cells within 2–3 d by loss of the B-cell expression signature and simultaneous activation of the macrophage-specific gene expression program (Bussmann et al. 2009). Likewise, ectopic C/EBPα expression can efficiently reprogram committed DN3 thymocytes into functional macrophages (Fig. 3) (Laiosa et al. 2006). This myeloid lineage conversion is, however, prevented in the presence of Notch signaling (Laiosa et al. 2006), indicating that lymphoid signals and transcription factors can antagonize the C/EBPα-mediated reprogramming of lymphocytes to macrophages.

3.2 Conversion of Mature B Cells into Functional T Cells

The last differentiation step in the B-cell pathway has characteristic features of a lineage switch because it results in a radical change from a Pax5-dependent expression program in mature B cells to a Blimp1-dominated transcription program in plasma cells (Fig. 1) (Shaffer et al. 2002; Delogu et al. 2006). The developmental potential of mature B cells may thus be plastic rather than restricted to the B-cell fate. Consistent with this idea, mature B cells seem to lose their B-cell identity on Pax5 loss because they down-regulate B-cell-specific genes and reactivate lineage-inappropriate genes (Delogu et al. 2006; Revilla-i-Domingo et al. 2012). The destiny of the Pax5-deficient mature B cells was analyzed by conditional *Pax5* deletion in highly purified B cells followed by intravenous injection of the *Pax5*-deleted B cells into T-cell-deficient mice (Cobaleda et al. 2007). These experiments showed that the loss of Pax5 allows mature B cells from peripheral lymphoid organs to dedifferentiate in vivo back to early uncommitted progenitors that migrate to the bone marrow (Fig. 3) (Cobaleda et al. 2007). These dedifferentiated progenitors are subsequently able to develop into macrophages and functional T cells (Fig. 3) (Cobaleda et al. 2007). Hence, loss of the B-cell identity factor Pax5 is able to induce dedifferentiation of mature B lymphocytes, which reveals a remarkable plasticity of these cells.

4 EPIGENETIC CONTROL OF V(D)J RECOMBINATION

4.1 Developmental Regulation of Antigen Receptor Gene Rearrangements

The guiding principle of the acquired immune system is that every newly generated lymphocyte recognizes a unique antigen and that the overall diversity of lymphocytes is great enough to counteract any possible antigen. To this end, B and T cells express lineage-specific antigen receptors that mediate antibody-dependent humoral or cellular immunity, respectively. The BCR consists of an immunoglobulin heavy-chain (IgH) and an Igκ or Igλ light-chain (IgL). T cells of the αβ lineage, which comprise the majority of T lymphocytes in mouse and man, express the TCRβ polypeptide in association with TCRα, whereas the functionally distinct γδ T cells contain TCRγ paired with TCRδ on their cell surface. These antigen receptor proteins are encoded by large gene loci containing discontinuous variable (V), diversity (D), and joining (J) gene segments that are assembled by V(D)J recombination into a functional gene during lymphocyte development (Fig. 4A). The multiplicity of D, J, and especially V gene segments, combined with the randomness of their recombination, is responsible for the virtually unlimited diversity of the immune repertoire (Bassing et al. 2002).

The mechanics of V(D)J recombination at the DNA level is rather simple. All V, D, and J gene segments are flanked by recombination signal sequences (RSSs) that consist of relatively conserved heptamer and nonamer elements separated by a spacer of either 12 or 23 bp. The lymphoid-specific recombinase proteins RAG1 and RAG2, assisted by high-mobility group proteins, assemble 12- and 23-bp RSSs into a synaptic complex and then generate double-strand DNA breaks between the RSSs and coding segments. These DNA breaks are subsequently processed and relegated by ubiquitous repair factors of the nonhomologous end-joining machinery to form coding and signal joints (Bassing et al. 2002).

The simplicity of the V(D)J recombination process at the DNA template level poses logistic problems for the assembly of the different antigen receptors because the RAG proteins are expressed in all immature B and T lymphocytes. Hence, stringent regulation must be in place to restrict the access of RAG proteins to only specific subsets of all of the recombination substrates (Yancopoulos and Alt 1985; Stanhope-Baker et al. 1996). V(D)J recombination is tightly controlled in a lineage- and stage-specific manner. Within the B-lymphoid lineage, the *IgH* locus is rearranged in pro-B cells before recombination of *Igκ* and *Igλ* genes in pre-B cells, whereas *TCRβ* and *TCRα* genes are rearranged in pro-T (DN) and DP thymocytes, respectively (Fig. 1). Moreover, V(D)J recombination of the *IgH* gene occurs in a defined temporal order with D_H-J_H rearrangements preceding V_H-DJ_H recombination. Rearrangements of the *TCRβ* locus also proceed in the same order ($D_β$-$J_β$ before $V_β$-$DJ_β$) during pro-T-cell development. Control mechanisms must therefore exist to shield all V genes from RAG-mediated cleavage during D-J recombination and to facilitate rearrangement of only one out of 100 V genes

Cite as *Cold Spring Harb Perspect Biol* doi: 10.1101/cshperspect.a019307

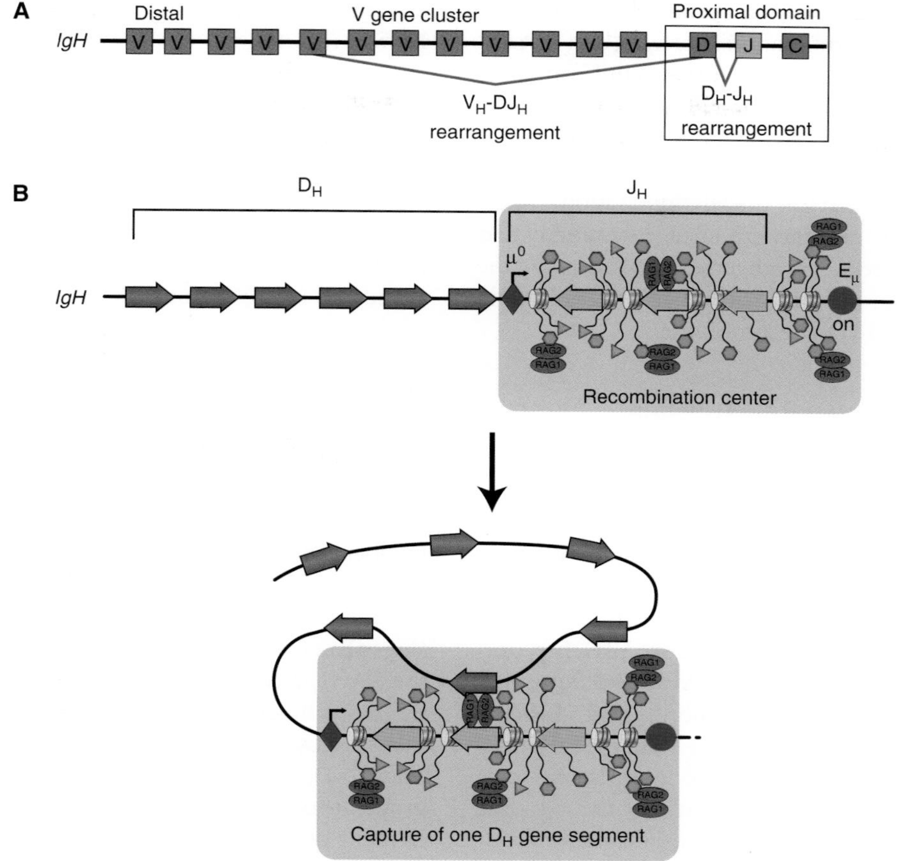

Figure 4. Involvement of RAG1/2 in the formation of recombination centers. (*A*) Structure of the *IgH* locus. The murine *IgH* locus is composed of the 3′ proximal region of 270 kb length consisting of 16 D_H, 4 J_H, and 8 C_H gene segments and of the distal V_H gene cluster extending over a 2.5-Mb region containing 200 V_H genes. (*B*) The recombination center model. In lymphoid progenitors, the proximal J_H gene region of the *IgH* locus is activated as a recombination center under the control of the μ^0 promoter and E_μ enhancer. Binding of the RAG2 PHD finger to the active H3K4me3 modification (green hexagons) in the recombination center recruits the RAG1/2 complex (brown oval), whose binding is further stabilized by the interaction of RAG1 with the J_H RSS element (arrowhead). The tethered RAG1/2 complex captures one of several D_H gene segments followed by D_H-J_H recombination. Blue triangles indicate acetylated lysine residues of histone H3.

during V-DJ recombination. Consequently, the process of antigen receptor generation entirely depends on accurate regulation of the accessibility of RSSs for the RAG1/2 recombinase.

Successful V-DJ recombination of the *IgH* or *TCRβ* gene leads to expression of the Igμ or TCRβ protein as part of the pre-BCR or pre-TCR complex that acts as an important checkpoint to inhibit V-DJ recombination of the second, DJ-rearranged allele and to promote development to pre-B cells or DP thymocytes that initiate *IgL* or *TCRα* gene rearrangements, respectively (Fig. 1). Expression of a signaling-competent BCR or TCR subsequently arrests V(D)J recombination by transcriptional repression of *Rag1/2* genes in immature B or T cells (Jankovic et al. 2004). Signaling of an autoreactive BCR can, however, restart immunoglobulin light-chain gene rearrangement that results in the generation of a BCR with novel antigen specificity (Jankovic et al. 2004). Moreover, signaling of the cytokine IL-7 via the transcription factor STAT5 is essential for promoting recombination of the *TCRγ* gene in pro-T cells (Ye et al. 2001) and for suppressing premature rearrangements at the *Igκ* locus in pro-B cells (Malin et al. 2010). Hence, V(D)J recombination is controlled not only intrinsically by developmental and lineage-specific nuclear mechanisms, but also extrinsically by signals generated at the cell surface.

The developmental and locus-specific constraints on V(D)J recombination are largely imposed at the epigenetic level (Jhunjhunwala et al. 2009). In nonlymphoid cells, the *Ig* and *TCR* genes are present in inaccessible chromatin because exogenously expressed RAG proteins readily cleave transfected episomal recombination substrates but not endogenous antigen receptor genes in kidney cells (Romanow et al. 2000). Moreover, recombinant RAG proteins added to isolated lymphocyte nuclei can only cleave the *Ig* or *TCR* gene that is actively undergoing V(D)J recombination at the developmental stage used for nuclei preparation (Stanhope-Baker et al. 1996). Hence, the lineage specificity and temporal ordering of gene rearrangements is caused by the sequential opening of local chromatin that renders specific RSSs accessible to the V(D)J recombinase.

4.2 Chromatin-Mediated RAG Function at Recombination Centers

Numerous promoters that are associated with V, D, and J segments of antigen receptor loci control the rearrangements of promoter-proximal sequences within relatively short distances (Bassing et al. 2002). These promoters give rise to germline transcription of short sense RNA from an unrearranged gene segment before V(D)J recombination (Yancopoulos and Alt 1985). Moreover, antisense intergenic transcription throughout the V_H gene cluster is known to precede V_H-DJ_H rearrangements at the *IgH* locus in pro-B cells, suggesting that these long antisense transcripts may direct chromatin remodeling of the V_H gene domain (Bolland et al. 2004). An essential role for transcription in controlling locus accessibility is further indicated by the fact that enhancers exert long-range control of V(D)J recombination at antigen receptor loci (Bassing et al. 2002; Perlot et al. 2005). Deletion of endogenous enhancers strongly impairs V(D)J recombination of antigen receptor loci, whereas the insertion of additional lineage-specific enhancers leads to a novel V(D)J recombination pattern (Bassing et al. 2002; Perlot et al. 2005).

The occurrence of active chromatin at antigen receptor loci has been most extensively studied for the *IgH* locus. The intronic Eμ enhancer and adjacent J_H segments are characterized by the abundant presence of the three active histone marks H3K4me2, H3K4me3, and H3K9ac in pro-B cells, where the *IgH* locus undergoes V(D)J recombination (Chakraborty et al. 2009; Malin et al. 2010). Local chromatin accessibility and histone acetylation are thereby established by the core Eμ enhancer itself (Chakraborty et al. 2009). Surprisingly, however, these three active histone modifications are largely absent at V_H genes except for members of the V_H3609 gene family in the distal V_H gene cluster that carry detectable levels of H3K9ac and H3K4me2 (Malin et al. 2010). Hence, most V_H genes do not show the active chromatin signature characteristic of expressed genes but must be accessible at the chromatin level in pro-B cells, where they undergo germline transcription and V_H-DJ_H recombination (Yancopoulos and Alt 1985).

A possible explanation for this asymmetrical distribution of active chromatin at the *IgH* locus and other antigen receptor loci may be provided by the fact that the RAG2 protein contains at its carboxyl terminus a plant homeodomain (PHD) finger that specifically recognizes the active histone mark H3K4me3 (Matthews et al. 2007). Notably, the RAG2 protein can bind via its PHD finger to H3K4me3 islands at active promoters throughout the genome (Ji et al. 2010). More importantly, RAG2 also binds to the H3K4me3 island at the J gene segments of the different antigen receptor loci (*IgH*, *Igκ*, *TCRβ*, and *TCRα/δ*) once they become accessible during lymphocyte development (Fig. 4B) (Ji et al. 2010). The RAG1 protein interacts in a more restrictive manner only with the RSS elements of the accessible J gene segments and, thus, further contributes to the focal and stable recruitment of the RAG1/2 complex to the active chromatin region in the proximal region of antigen receptor loci referred to as the recombination center (Fig. 4B) (Ji et al. 2010). As exemplified for the *IgH* locus, the RAG1/2 complex, once tethered to the J_H segments at the recombination center, can capture a partner RSS element of a D_H segment to undergo synapse formation, which leads to D_H-J_H recombination in uncommitted lymphoid progenitors (Fig. 4B). As a consequence, the rearranged D_H element becomes part of the active recombination center and is bound by the RAG1/2 complex that is now able to capture an RSS element of one of the many V_H genes that are brought into close proximity by long-range contraction of the V_H gene cluster (see Sec. 4.5), thus resulting in V_H-DJ_H rearrangement in pro-B cells. The H3K4me3-mediated focal targeting of the RAG1/2 endonuclease to the proximal recombination center is likely the reason for the fact that active chromatin is asymmetrically distributed at the proximal J_H region and not throughout the D_H and V_H gene clusters in pro-B cells.

4.3 Monoallelic Recombination by Homologous Pairing of Immunoglobin Loci

Allelic exclusion ensures the productive rearrangement of only one of the two *Ig* alleles, which leads to the expression of a single antibody molecule with a unique antigen specificity in B cells. The process of allelic exclusion can be divided into two distinct steps. During the initiation phase, one of the two *Ig* alleles is selected to rearrange first, which precludes simultaneous recombination of the two alleles. Expression of the productively rearranged allele subsequently prevents recombination of the second allele by feedback inhibition, thus maintaining allelic exclusion. Similar to allelic exclusion of antigen receptor loci, X-chromosome inactivation in female cells leads to monoallelic expression of X-linked genes by random silencing of one X chromosome. Transient pairing of the two X chromosomes is known to initiate X inactivation by differentially marking one of the two chromosomes for subsequent epigenetic silencing (Bacher et al. 2006). Notably, the initiation of allelic exclusion also depends on homologous pairing of antigen receptor loci as revealed by three-dimensional (3D) DNA fluorescence in situ hybridization (3D DNA-FISH) (Hewitt et al. 2009). Transient pairing of the two *IgH* alleles is high in pre-pro-B and pro-B cells undergoing D_H-J_H and V_H-DJ_H recombination, respectively, whereas the *Igκ* alleles frequently associate in pre-B and immature B cells during $V_κ$-$J_κ$ recombination (Fig. 5A) (Hewitt et al. 2009). Bind-

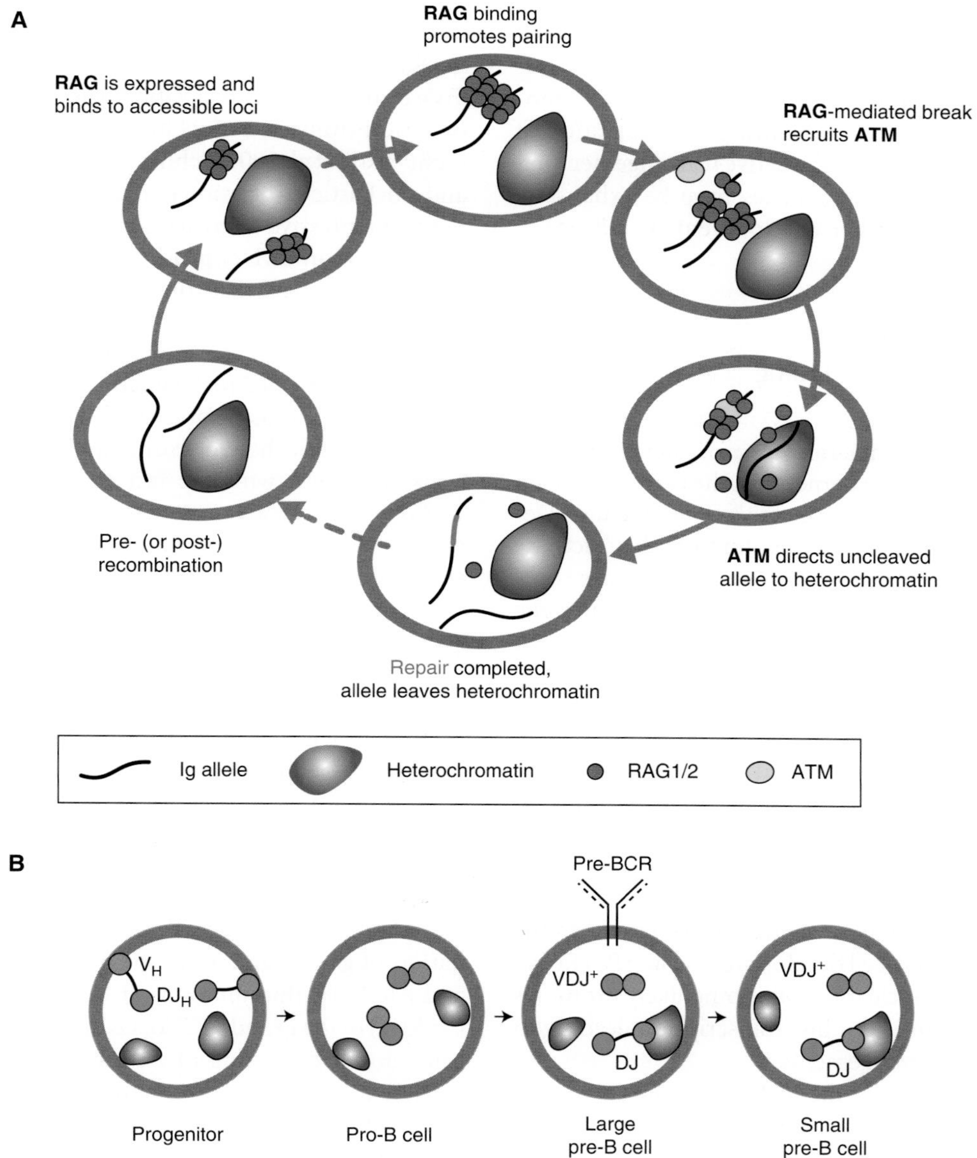

Figure 5. Control of monoallelic V(D)J recombination and subnuclear location of the *IgH* locus in early B-cell development. (*A*) Control of monoallelic V(D)J recombination by pairing of *IgH* alleles. See Section 4.2 for detailed explanation of the model describing the regulation of V(D)J recombination at the *IgH* locus (black line) by the RAG1/2 endonuclease (brown) and repair-checkpoint protein ATM (yellow). (*B*) Subnuclear location of the two *IgH* alleles at different stages of early B-cell development. The distal V_H region (red) and proximal J_H-C_H domain (green) of the *IgH* locus are indicated together with their location relative to the repressive compartments at the nuclear periphery (gray) and pericentromeric heterochromatin (blue). The contraction and decontraction of the *IgH* alleles are schematically shown.

ing of the RAG1/2 complex is essential for homologous pairing, which selects one of the two *Ig* alleles for RAG-mediated cleavage (Fig. 5A). The introduction of a DNA break on the rearranging *Ig* allele activates the repair-checkpoint protein ATM that in turn provides a signal in *trans* for repositioning of the uncleaved *Ig* allele to pericentromeric heterochromatin (Fig. 5A) (Hewitt et al. 2009). The repositioned *Ig* allele is likely protected from RAG1/2 cleavage in this heterochromatic environment during the time at which the rearrangement on the second allele is completed and functionally tested. Hence, V(D)J recombination initiates on paired *Ig* chromosomes and proceeds in a monoallelic manner through cycles of pairing and separation of homologous *Ig* alleles until a productive rearrangement has been generated or the two *Ig* alleles have be exhausted, which leads to either further development or cell death, respectively.

4.4 Subnuclear Relocation of Antigen Receptor Loci in Developing Lymphocytes

The nuclear periphery and pericentromeric heterochromatin are two repressive compartments in the nucleus that are important for propagating the inactive state of genes (Deniaud and Bickmore 2009). Depending on their activity state, genes are repositioned between these repressive compartments and central nuclear positions that facilitate gene transcription. Interestingly, the *IgH* and *Igκ* loci are located in their default state at the nuclear periphery in all non-B cells including uncommitted lymphoid progenitors (Kosak et al. 2002). The *IgH* locus is thereby anchored via distal V_H genes at the nuclear periphery and is oriented with the proximal *IgH* domain toward the center of the nucleus, facilitating D_H-J_H rearrangements in lymphoid progenitors (Fig. 5B) (Fuxa et al. 2004). An important step of *IgH* locus activation consists of relocation of *IgH* and *Igκ* loci from the nuclear periphery to more central positions within the nucleus at the onset of B-cell development (Fig. 5B) (Kosak et al. 2002). This subnuclear repositioning likely promotes chromatin opening and germline transcription, leading to proximal V_H-DJ_H rearrangements.

Although both alleles of the *IgH* and *Igκ* loci are repositioned to central nuclear positions in pro-B cells (Kosak et al. 2002; Fuxa et al. 2004), the two alleles behave differently at the next developmental stage in pre-B cells. Following successful V_H-DJ_H rearrangement, pre-BCR signaling leads to repositioning of the incompletely rearranged *IgH* allele to repressive pericentromeric heterochromatin, whereas the functionally rearranged *IgH* alleles remain in a central nuclear position consistent with its continuous expression in pre-B cells (Fig. 5B) (Roldán et al. 2005). Notably, one of the two *Igκ* alleles is also recruited to pericentromeric heterochromatin in pre-B cells before rearrangement, which favors V_κ-J_κ recombination on the second *Igκ* allele located in euchromatic regions of the nucleus (Hewitt et al. 2008). Interestingly, pericentromeric recruitment of the nonfunctional *IgH* allele occurs via the distal V_H gene region suggesting that the same DNA sequences are involved in the recruitment of the *IgH* locus to either the nuclear periphery or pericentromeric heterochromatin (Fig. 5B) (Roldán et al. 2005).

Surprisingly, the two alleles of the *TCRβ* locus show little change in their association with the nuclear periphery or pericentromeric heterochromatin during T-cell development (Skok et al. 2007; Schlimgen et al. 2008). The frequent and stochastic association of one *TCRβ* allele with either of the two repressive compartments is likely to inhibit V(D)J recombination in DN pro-T cells and may thus promote V_β-DJ_β rearrangements on the second *TCRβ* allele (Skok et al. 2007; Schlimgen et al. 2008), similar to the situation described above for the monoallelic recruitment of *Igκ* to pericentromeric heterochromatin. In contrast, the *TCRα/δ* locus shows little association with the nuclear periphery or pericentromeric heterochromatin in developing thymocytes consistent with the fact that this locus undergoes *TCRδ* or *TCRα* rearrangements on both alleles in DN and DP thymocytes, respectively (Skok et al. 2007; Schlimgen et al. 2008).

4.5 Spatial Control of V(D)J Recombination by Locus Contraction

All four antigen receptor loci have a large size ranging from 0.67 Mb (*TCRβ*) and 1.6 Mb (*TCRα/δ*) to 3 Mb (*IgH* and *Igκ*) and show a complex organization with the D, J, and C gene segments constituting the 3′ domain of each locus (Fig. 4A) (Jhunjhunwala et al. 2009). Enhancers and promoters located within this proximal domain control local chromatin structure, germline transcription, and recombination of gene segments at the 3′ end of the locus (Perlot and Alt 2008; Jhunjhunwala et al. 2009). The largest part (>80%) of the *Ig* and *TCR* loci is, however, devoted to the V gene cluster consisting of an array of 31 (*TCRβ*) to 200 (*IgH*) V genes, whose chromatin accessibility and germline transcription is regulated independently of the proximal domain (Hawwari and Krangel 2005; Jhunjhunwala et al. 2009). Hence, antigen receptor loci can be viewed as consisting of two distinct entities, the proximal domain and the V gene region, that are separated by a large distance on linear DNA. This separation spatially restricts the V(D)J recombination process, because antigen receptor loci are present in an extended conformation in nonlymphoid cells and lymphoid progenitors, as revealed by 3D DNA-FISH

Cite as *Cold Spring Harb Perspect Biol* doi: 10.1101/cshperspect.a019307

analysis (Fig. 6A) (Kosak et al. 2002; Fuxa et al. 2004; Roldán et al. 2005; Skok et al. 2007). As first shown for the *IgH* locus, both alleles undergo long-range contraction in committed pro-B cells, which juxtaposes distal V_H genes next to the rearranged proximal DJ_H domain, thus facilitating V_H-DJ_H rearrangements (Fig. 6A) (Kosak et al. 2002; Fuxa et al. 2004). The V_H gene cluster thereby undergoes looping in such a way that the different V_H genes can undergo recombination at similar frequency, which is essential for the generation of a highly diverse immunoglobulin repertoire (Roldán et al. 2005; Sayegh et al. 2005; Jhunjhunwala et al. 2008; Medvedovic et al. 2013). Following successful V_H-DJ_H recombination, pre-BCR signaling leads to decontraction of the nonfunctional *IgH* allele at the next develop-

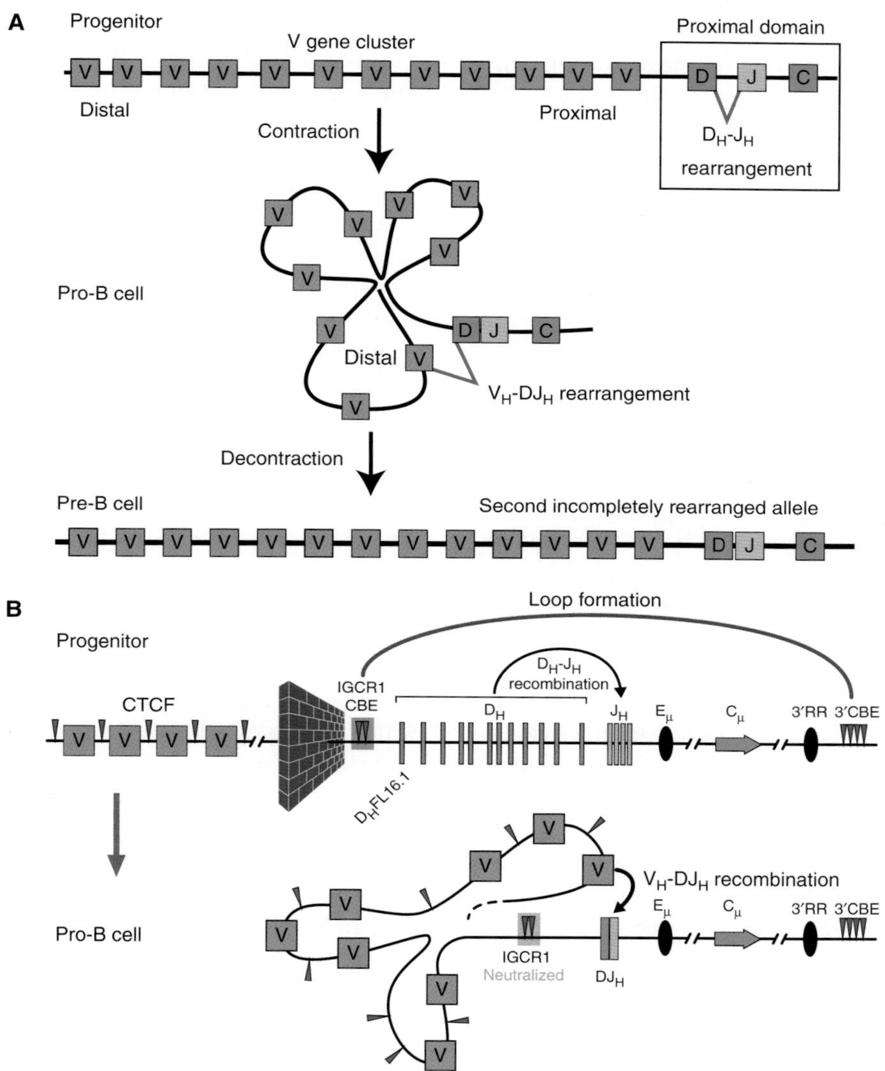

Figure 6. Spatial regulation and IGCR1-mediated control of *IgH* recombination in early B-lymphopoiesis. (*A*) Reversible *IgH* contraction. The *IgH* locus is in an extended configuration in uncommitted progenitors, which allows D_H-J_H recombination to take place in the proximal domain. In pro-B cells, all 200 V_H genes participate in V_H-DJ_H rearrangements because of contraction of the *IgH* locus by looping. The V_H genes of the incompletely rearranged *IgH* allele are no longer available for V_H-DJ_H recombination because of decontraction in pre-B cells. (*B*) Control of *IgH* recombination by the IGCR1 insulator. See Chapter 4.6 for detailed description how the IGCR1 region controls the order of V(D)J recombination at the *IgH* locus. CTCF-binding regions (CBEs) are symbolized by red arrowheads. A brick wall indicates that the IGCR1 insulator restrains loop formation and recombination to the proximal *IgH* domain in uncommitted lymphoid progenitors. In pro-B cells, the IGCR1 element is neutralized by a so-far unknown mechanism, and locus contraction promotes V_H-DJ_H rearrangements. 3′RR, 3′ regulatory region; E_μ, intronic enhancer.

mental stage that physically separates the distal V_H genes from the proximal *IgH* domain, thus preventing further rearrangement of the second *IgH* allele in pre-B cells (Fig. 6A) (Roldán et al. 2005). Decontraction of the nonfunctional *IgH* allele occurs at the pericentromeric heterochromatin, where the *IgH* allele is recruited in pre-B cells by interchromosomal pairing with the pericentromerically located *Igκ* allele in a manner that depends on the 3′ *Igκ* enhancer (Fig. 5B) (Hewitt et al. 2008). All other antigen receptor loci (*Igκ*, *TCRβ*, and *TCRα/δ*) also show reversible contraction at the developmental stage, during which they undergo V(D)J recombination (Roldán et al. 2005; Skok et al. 2007). Hence, reversible locus contraction by looping is a general mechanism that promotes the spatial communication between V genes and the proximal domain and thus allows V genes to participate in V(D)J recombination. On the other hand, decontraction prevents further recombination at the next developmental stage and thereby contributes to the maintenance of allelic exclusion, which ensures the productive rearrangement of only one of the two antigen receptor loci.

To date, only a few *trans*-acting factors and *cis*-regulatory elements have been implicated in the control of locus contraction. Pax5 was identified as the first transcription factor to be involved in the pro-B-cell-specific contraction of the *IgH* locus (Fuxa et al. 2004). In its absence, only the four most proximal of the 200 V_H genes can efficiently undergo V_H-DJ_H recombination, despite the fact that all V_H genes are accessible and give rise to germline transcripts in Pax5-deficient pro-B cells (Hesslein et al. 2003; Roldán et al. 2005). A similar V_H-DJ_H recombination phenotype was observed in pro-B cells lacking the histone methyltransferase Ezh2, an essential component of PRC2 that implies a critical role of this Polycomb complex in the regulation of *IgH* locus contraction (Su et al. 2003; A Ebert et al., pers. comm.). In addition to these two regulators, contraction of the *IgH* locus also depends on the ubiquitous transcription factor YY1 (Liu et al. 2007). Another potential regulator is the CCCTC-binding factor (CTCF) that has been implicated, through its association with cohesin, in long-range chromatin looping at several complex loci (Splinter et al. 2006; Hadjur et al. 2009; Nativio et al. 2009). CTCF and cohesin bind together to multiple sites along the V_H gene cluster of the *IgH* locus (Degner et al. 2011; Ebert et al. 2011), and short hairpin RNA (shRNA) knockdown experiments have suggested a role for the CTCF/cohesin complex in the regulation of looping at the *IgH* locus in pro-B cells (Degner et al. 2011).

A systematic search for intergenic regions with active chromatin has recently identified 14 Pax5-activated intergenic repeat (PAIR) elements in the distal V_H gene region of the *IgH* locus (Ebert et al. 2011). These PAIR elements are bound by Pax5, E2A, CTCF, and cohesin, carry Pax5-dependent active chromatin, and give rise to Pax5-regulated noncoding antisense transcripts exclusively in pro-B cells (Ebert et al. 2011). The pro-B-cell-specific and Pax5-dependent activity implicates the PAIR elements in the regulation of distal V_H-DJ_H recombination, possibly by inducing *IgH* locus contraction (Ebert et al. 2011).

4.6 Control of Ordered V(D)J Recombination by CTCF and Cohesin

The transcription factor CTCF has been implicated not only in chromosome looping but in chromatin boundary formation, transcriptional insulation, activation, or repression (Phillips and Corces 2009). A large array of regularly spaced CTCF-binding sites characterizes the V gene clusters of all antigen receptor loci, whereas the proximal domains contain CTCF-binding sites only at their 5′ and 3′ boundaries (Degner et al. 2011; Ebert et al. 2011; Ribeiro de Almeida et al. 2011). The 100-kb region separating the D_H and V_H genes of the *IgH* locus contains two CTCF-binding elements (CBEs) located 2 kb upstream of the D_HFL16.1 gene segment at the 5′ end of the proximal *IgH* domain (Fig. 6B). Specific mutation of the two CTCF-binding sites within the so-called intergenic control region 1 (IGCR1) revealed that these two sites function as insulator elements to regulate ordered and lineage-specific V(D)J recombination at the *IgH* locus (Fig. 6B) (Guo et al. 2011). In uncommitted lymphoid progenitors, the IGCR1/CBE insulator is thought to restrict the potent E_μ enhancer activity and all other long-range interactions to the proximal domain, which generates a local chromatin architecture that promotes D_H-J_H recombination on both *IgH* alleles (Fig. 6B) (Guo et al. 2011). During normal development, the IGCR1/CBE insulator is only neutralized in committed pro-B cells possibly by Pax5-dependent contraction of the *IgH* locus that facilitates V_H-DJ_H recombination (Fig. 6B). In IGCR1/CBE mutant mice, the activity of the E_μ enhancer is, however, no longer restrained and thus induces active chromatin and germline transcription at the first few proximal V_H genes that undergo premature V_H-DJ_H and even V_H-D_H recombination in lymphoid progenitors and developing thymocytes (Guo et al. 2011). As a consequence of the precocious V_H gene activity in lymphoid progenitors, fewer DJ_H-rearranged *IgH* alleles are available for V_H-DJ_H recombination at the next developmental stage in committed pro-B cells, which leads to a strong increase of proximal V_H gene rearrangements and a corresponding skewing of the immunoglobulin repertoire. Hence, the CTCF-regulated IGCR1 activity and Pax5-mediated locus contraction fulfill opposing roles by preventing or promoting the participation of V_H genes in V(D)J recom-

Cite as *Cold Spring Harb Perspect Biol* doi: 10.1101/cshperspect.a019307

bination in lymphoid progenitors or committed pro-B cells, respectively.

Similar to the *IgH* locus, the intergenic V_κ-J_κ region of the *Igκ* locus also contains a regulatory element (known as Sis) that functions as a silencer of recombination through binding of the transcription factors CTCF and Ikaros (Liu et al. 2006; Ribeiro de Almeida et al. 2011). Deletion of the Sis element or conditional inactivation of CTCF similarly affect V_κ-J_κ rearrangements in pre-B cells by increasing germline transcription and recombination of proximal V_κ genes at the expense of distal V_κ gene recombination (Ribeiro de Almeida et al. 2011; Xiang et al. 2011). The loss of CTCF thereby skews the repertoire of V_κ-J_κ rearrangements by strongly increasing the interactions of the iEκ and 3′Eκ enhancers with proximal V_κ genes (Ribeiro de Almeida et al. 2011). This evidence suggests that the Sis element functions as an insulator to negatively regulate proximal V_κ gene recombination by blocking the long-range activity of *Igκ* enhancers (Ribeiro de Almeida et al. 2011) similar to the role of the IGCR1 element in the control of *IgH* rearrangements (Guo et al. 2011). CTCF binds together with cohesin to both regulatory regions (Ebert et al. 2011), indicating that cohesin may mediate the enhancer-blocking function of CTCF at these elements. In support of such a hypothesis, conditional loss of the cohesin subunit Rad21 impairs *TCRα* rearrangements in DP thymocytes by affecting the histone modification and transcription patterns in the proximal *TCRα* domain due to altered long-range interactions of the E_α enhancer (Seitan et al. 2011).

5 ROLE OF EPIGENETIC REGULATORS IN LYMPHOID MALIGNANCIES

High-resolution genomic analyses of single-nucleotide polymorphisms and genome-wide DNA and RNA sequencing have provided fascinating new insight into the genetic basis of leukemia and lymphoma development. As shown by these global analyses, the majority of mutations detected in lymphoid malignancies affect the function of hematopoietic transcription factors and epigenetic regulators in addition to the signaling of cytokine and antigen receptors (Mullighan et al. 2007; Morin et al. 2011; Zhang et al. 2012a). Among the different classes of epigenetic regulators, several writers (CREBBP, EP300, MLL1, MLL2, PRC2, MMSET, and SETD2) and one eraser (UTX) are frequently mutated in lymphoid malignancies (Fig. 7), whereas genetic alterations affecting readers have so far not been reported. The highly related lysine acetyltransferases CREBBP (CBP) and EP300 (p300) are known to function as coactivators for a large number of DNA-binding transcription factors by modifying histone and nonhistone nuclear proteins (Ogryzko et al. 1996). Monoallelic deletions or point mutations inactivate the *CREBBP* or *EP300* gene in 41% of follicular lymphoma (FL), 39% of diffuse large B-cell lymphoma (DLBCL), and 18% of relapsed B-cell acute lymphoblastic leukemia (B-ALL), suggesting that the two acetyltransferases function as haploinsufficient tumor suppressors in these lymphoid malignancies (Mullighan et al. 2011; Pasqualucci et al. 2011). Notably, the mutant CREBBP and EP300 proteins are un-

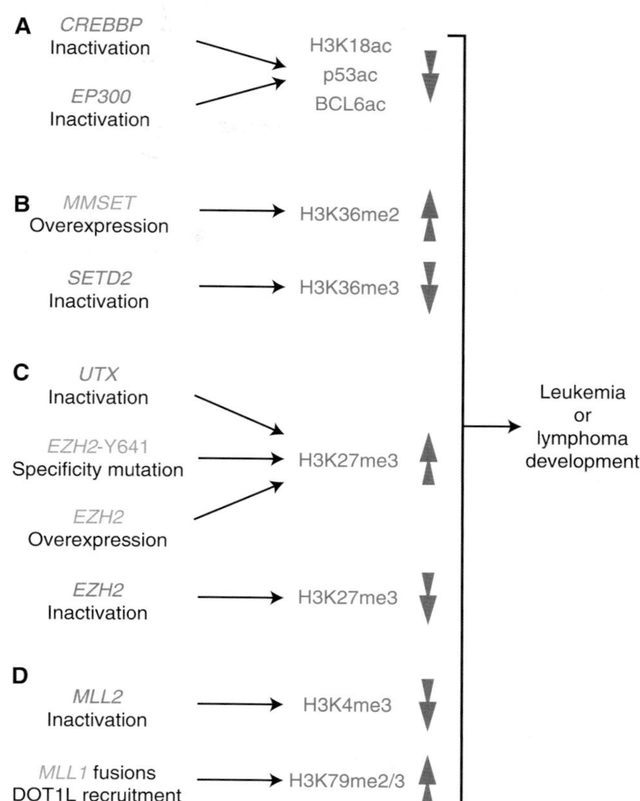

Figure 7. Role of histone-modifying enzymes in leukemia and lymphoma. (*A*) Monoallelic mutation of the *CREBBP* (CBP) or *EP300* (p300) gene results in reduced acetylation of BCL6, p53, and histone H3 in B-cell lymphoma. (*B*) Overexpression of the H3K36 dimethylase MMSET (NSD2, WHSC1) in multiple myeloma and loss of the H3K36 trimethylase SETD2 in early T-cell precursor acute lymphoblastic leukemia (ETP-ALL) contribute to tumorigenesis by altering the methylation state of lysine 36 of histone H3. (*C*) Loss of the H3K27 demethylase UTX as well as an altered specificity mutation and overexpression of the H3K27 methyltransferase EZH2 are implicated in the formation of B-cell malignancies by increasing the abundance of the repressive histone mark H3K27me3, whereas loss of EZH2 is associated with T-ALL. (*D*) Inactivation of the H3K4 methyltransferase MLL2 results in decreased levels of the active modification H3K4me3 in B-cell lymphoma, whereas MLL1 fusion proteins recruit the methyltransferase DOT1L to focally increase H3K79 methylation in *MLL1*-rearranged leukemias. The following color code is used: oncogenes (blue), tumor-suppressor genes (brown), active (green), and repressive (red) protein modifications.

able to acetylate the transcription factors p53 and BCL6, which are both expressed in FL and DLBCL (Pasqualucci et al. 2011). The fact that acetylation activates the tumor suppressor p53 (Lill et al. 1997) and inactivates the oncoprotein BCL6 (Bereshchenko et al. 2002) suggests the following mechanism of action for the *CREBBP* or *EP300* mutations. In addition to a general effect on histone acetylation (H3K18ac) and gene transcription (Mullighan et al. 2011), the decreased activity of CREBBP or EP300 may lead to increased BCL6 and decreased p53 activity, which favors enhanced tolerance of DNA damage (mediated by BCL6) at the expense of diminished apoptosis and cell cycle arrest (mediated by p53ac) in FL and DLBCL (Fig. 7A) (Pasqualucci et al. 2011). It may, therefore, be of clinical benefit to use already approved histone deacetylase inhibitors to restore physiological acetylation levels in tumors with CREBBP and EP300 mutations.

The histone methyltransferase MMSET (NSD2, WHSC1) is overexpressed in 15%–20% of the plasma-cell-derived tumor multiple myeloma (MM) as a consequence of the t(4;14)(p16.3;q32) translocation that places the *MMSET* gene under the control of the E_{μ} enhancer of the *IgH* locus (Keats et al. 2003). MMSET catalyzes H3K36 dimethylation, and its increased expression in t(4;14)-positive myeloma cells alters the genome-wide pattern of H3K36me2, which results in localized chromatin relaxation and expression of normally silent genes that promote plasma cell transformation (Fig. 7B) (Kuo et al. 2011). Consistent with this notion, ectopic expression of a wild-type but not of a catalytically dead MMSET protein in t(4;14)-negative myeloma cells rapidly induces tumor development in a mouse xenograft model, suggesting that MMSET is a potent oncoprotein (Kuo et al. 2011). Interestingly, the nonredundant methyltransferase SETD2, which mediates all H3K36 trimethylation (Edmunds et al. 2008), functions as a tumor suppressor, because its gene is inactivated by biallelic mutations in early T-cell precursor acute lymphoblastic leukemia (ETP-ALL), leading to a reduction of global H3K36me3 levels (Fig. 7B) (Zhang et al. 2012a). Hence, MMSET and SETD2 appear to have antagonistic functions in cancer development, which may reflect distinct functions of the H3K36 di- and trimethylation states in controlling gene expression.

PRC2 consists of the core components EED and SUZ12 and the methyltransferase EZH2 that mediates gene silencing by methylating histone H3 at lysine 27. *EED*, *SUZ12*, or *EZH2* are mutated by deletions or inactivating point mutations in 42% of ETP-ALL and 25% of T-ALL (Fig. 7C) (Ntziachristos et al. 2012; Zhang et al. 2012a). Some of these mutations are even homozygous, further indicating a tumor-suppressing function of PRC2 in T-ALL (Ntziachristos et al. 2012). In support of this finding, T-cell leukemia arises at a high frequency in mice undergoing conditional *Ezh2* inactivation in HSCs (Simon et al. 2012). Notably, genetic alterations of *PRC2* are frequently associated with *NOTCH1* mutations suggesting that the loss of the repressive H3K27me3 mark cooperates with constitutively active NOTCH1 in inducing an oncogenic gene expression program in T-ALL (Ntziachristos et al. 2012).

Mutations of *EZH2* are also found in 22% of germinal-center B-cell-like DLBCL and 7% of FL (Morin et al. 2010). Surprisingly, however, all heterozygous mutations result in the replacement of a single tyrosine residue (Y641) in the catalytic site of the SET domain of EZH2 (Morin et al. 2010). Detailed biochemical analysis revealed that the wild-type EZH2 enzyme displays greatest catalytic activity for the H3K27 monomethylation reaction, but diminished activity for the subsequent di- and trimethylation steps (Sneeringer et al. 2010). Importantly, all amino acid replacements at Y641 result in an altered substrate specificity of the mutant EZH2 protein, which has limited ability to perform the first methylation reaction but displays enhanced catalytic activity for the subsequent methylation steps (Sneeringer et al. 2010). This change in substrate specificity is caused by the enlarged active site of the mutant EZH2 protein, which favors di- and trimethylation but interferes with the first methyl transfer reaction. The coexpression of wild-type and mutant EZH2 proteins consequently leads to coupling of the enzymatic activities between the H3K27-monomethylating wild-type enzyme and the efficiently di- and trimethylating mutant enzyme, which results in substantially increased H3K27me3 levels in lymphoma cells (Fig. 7C) (Sneeringer et al. 2010; Yap et al. 2011). Hence, the Y641 substitutions of EZH2 are oncogenic gain-of-function mutations that increase H3K27 trimethylation and thus enhance gene silencing similar to the previously reported overexpression of EZH2 in B-cell lymphoma (Bracken et al. 2003). Interestingly, the X-linked gene coding for the H3K27 demethylase UTX is frequently mutated in human cancer including multiple myeloma and B-ALL, which leads to increased H3K27me3 levels and cell proliferation (Fig. 7C) (van Haaften et al. 2009; Mar et al. 2012). In summary, the Y641 mutation and overexpression of EZH2, as well as the loss of UTX, affect the same epigenetic pathway by increasing H3K27 trimethylation and gene repression in tumor cells (Fig. 7C).

The Trithorax group protein MLL2 is one of six human methyltransferases that generate the active H3K4me3 mark at promoters of transcribed genes. Notably, *MLL2* is one of the most frequently mutated genes in FL (89%) and DLBCL (32%; Morin et al. 2011). The majority of these mutations are heterozygous and truncate the large MLL2 protein, thus identifying this methyltransferase as a prominent tumor suppressor in B-cell non-Hodgkin's lympho-

Cite as *Cold Spring Harb Perspect Biol* doi: 10.1101/cshperspect.a019307

ma (Morin et al. 2011). Hence, the partial loss of the activating MLL2 protein contributes to tumor development by decreasing the active H3K4me3 mark, thus reducing expression of key genes in B-cell lymphoma (Fig. 7D).

Approximately 50% of infant ALLs and ~10% of adult leukemias are characterized by the presence of *MLL1* translocations that code for fusion proteins with potent leukemia-inducing activity (Liu et al. 2009). MLL1, like MLL2, is also a large multidomain protein containing the H3K4-methylating SET domain at the very carboxyl terminus. *MLL1* translocations generate fusion proteins consisting of the amino-terminal third of MLL1 linked to different carboxy-terminal partner proteins (Liu et al. 2009). Although 71 distinct fusion partners are known to date, the majority of *MLL1*-rearranged leukemias express fusion proteins involving the partners AF4, AF9, AF10, AFF4, ENL, or ELL1, all of which are components of the super elongation complex regulating transcription elongation by RNA polymerase II (RNA Pol II) (Luo et al. 2012). Hence, these MLL fusions proteins have been implicated in leukemia development by deregulating transcription elongation. In addition to this mechanism, the AF4, AF9, AF10, and ENL partner proteins also interact with DOT1L, which methylates lysine 79 in the globular domain of histone H3 (Okada et al. 2005; Luo et al. 2012). These MLL fusion proteins immortalize leukemic cells only in the presence of a functional DOT1L methyltransferase (Okada et al. 2005), leading to increased H3K79 methylation and transcription at gene loci coding for developmental regulators (Fig. 7D) (Guenther et al. 2008).

6 CHROMATIN-MEDIATED CONTROL OF THE INFLAMMATORY RESPONSE

6.1 Transcriptional Dynamics of the Inflammatory Response

Inflammation is the human or animal response to infection or tissue damage. Similar to cognitive and metabolic processes, inflammation is essential for the adaptation of the organism to the environment and maintenance of the body's homeostasis (Medzhitov 2008). The inflammatory response is largely mediated by myeloid lineage cells that either migrate (neutrophils and monocytes) or reside steadily in the tissue (Kuppfer cells in liver or microglia cells in the brain). Epigenetic mechanisms that cooperate with transcriptional control, discussed in this section, play a significant role in effecting and regulating the inflammatory response in macrophages, but are likely to represent a common mechanism that operates in various cell types.

One of the common hallmarks of inflammation is a temporal reprofiling of gene expression patterns in affected cells (Zak and Aderem 2009; Smale 2010b). Indeed, exposure of cells to inflammatory signals modulates transcription of several hundreds of genes, including those that enable the whole body awareness of a local inflammatory state (Gilchrist et al. 2006; Medzhitov and Horng 2009; Smale 2010b; Bhatt et al. 2012). The elaborate dissection of the transcriptional response associated with inflammation has been performed on in vitro cultured macrophages treated with Gram-negative bacteria-derived lipopolysaccharide (LPS) or its active component lipid A (Fig. 8) (Ramirez-Carrozzi et al. 2009; Bhatt et al. 2012). LPS-activated genes can be divided into several major groups based on their timing of expression and the biochemical nature of the encoded protein. Analysis of the chromatin-bound and cytosolic RNA transcripts was performed in lipid A-triggered bone-marrow-derived macrophages. This showed up-regulation of gene transcription as early as 15 min after stimulation (16 genes) followed by activation of hundreds of genes in the course of a 2-h period after stimulation (Bhatt et al. 2012). The degree of gene up-regulation varied significantly among individual loci: 246 genes were induced five- and 10-fold, 247 between 10- and 100-fold, and 67 >100-fold.

A common theme that has emerged from studies of LPS-inducible genes is a connection between timing of gene expression and the epigenetic state of the inducible gene. In particular, the different temporal patterns of gene expression, in response to LPS, appear to be embedded within the CpG content of inducible gene promoters (Ramirez-Carrozzi et al. 2009; Smale 2010b; Natoli et al. 2011; Bhatt et al. 2012; see Ch. 15 [Li and Zhang 2014] for details on CpG islands). In macrophages, promoters rich in CpG islands are highly prevalent among so-called primary and weakly induced secondary response genes. CpG-low promoters, however, are more prevalent among secondary response genes that are induced at a higher level (Fig. 9) (Ramirez-Carrozzi et al. 2009; Bhatt et al. 2012). In both primary and secondary response genes, CpG abundance correlates with the presence of chromatin marks that are commonly associated with either ongoing gene transcription typical of actively transcribed genes or those with a predisposition to rapid gene activation in response to stimuli. Rapidly activated genes with a mostly high CpG content, described as poised, are common among stress-response genes (Adelman and Lis 2012). They are characterized by gene promoters with a relatively high abundance of RNA Pol II and methylated histone H3 at lysine 4 (H3K4me3; see primary and weakly induced secondary response genes in Fig. 9) (Gilchrist and Adelman 2012; Kwak et al. 2013).

Low CpG content at promoters stabilizes nucleosome formation. This explains why CpG-low genes (i.e., strongly

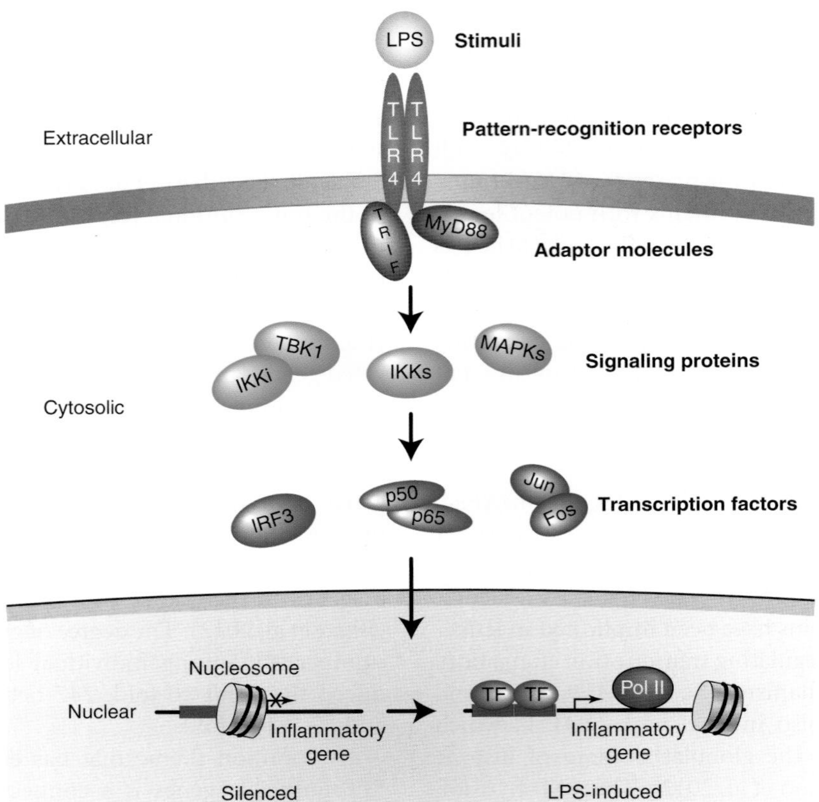

Figure 8. Signaling control of LPS-induced gene expression. The Gram-negative bacteria-derived LPS binds to the surface-expressed Toll-like receptor 4 (TLR4). Binding to TLR4 leads to activation of cytosolic signaling proteins and ensuing activation of diverse transcription factors such as NF-κB (p50/p65) and AP-1 (Jun/Fos). Transcription factors enter the cell nucleus and bind to the promoters of proinflammatory genes. In numerous cases, binding of transcription factors requires prior chromatin remodeling that provides transcription factor access to the otherwise nucleosome-occluded regulatory regions.

induced secondary response genes) strongly depend on chromatin remodeling for their activation (Ramirez-Carrozzi et al. 2009). Thus, LPS-induced inflammatory genes with CpG-low promoters can be viewed as examples of covered promoters (Lam et al. 2008; Cairns 2009; Bai and Morozov 2010). They are considered covered when nucleosomes occlude the transcriptional start site (TSS), the regions flanking the TSS, and most of the binding sites for transcriptional activators (Fig. 9) (Lam et al. 2008; Cairns 2009). At covered promoters, nucleosomes compete effectively with transcription factors for occupancy of key *cis*-regulatory binding sites, rendering covered promoters more reliant on chromatin remodeling and modifying enzymes to help "uncover" *cis*-regulatory sites and allow activity (Bai et al. 2011). In addition to the dependence on chromatin remodelers, the activation of genes controlled by covered promoters may require pioneer factors that bind to linker DNA among nucleosomes or directly to nucleosomes and facilitate accompanying chromatin remodeling (Smale 2010a). In the context of inflammation, one can predict that the combination of high- and low-affinity-binding sites for specific transcriptional regulators as well as the differences in promoter occlusion by nucleosomes help to adjust inflammatory response genes to signals of various types and strengths.

The differences in chromatin architecture between primary and secondary response genes are likely to yield differences in gene dynamics. Studies of gene dynamics in prokaryotic and eukaryotic cells has described the discontinuous nature of gene transcription occurring through bursts of activity (gene "on"), in which many messenger RNAs (mRNAs) are transcribed and interspersed with periods of inactivity (gene "off") (Fig. 10A) (Cai et al. 2008; Pedraza and Paulsson 2008; Chubb and Liverpool 2010; Larson 2011). The duration of transcriptional bursts, the size of the bursts, and the interval separating the bursts vary from a few minutes to many hours (Larson 2011; Yosef and Regev 2011). The "burstiness" of individual genes and duration of

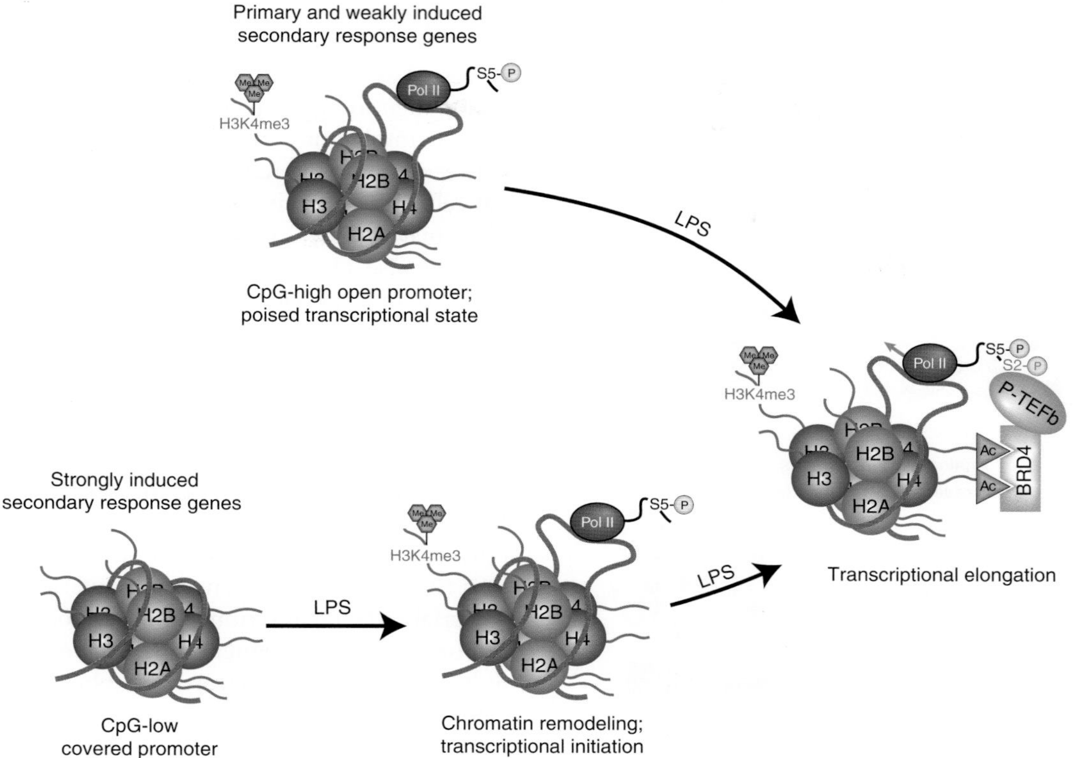

Figure 9. Chromatin-mediated control of the temporal pattern of proinflammatory gene expression. Key features of chromatin associated with LPS-induced genes are shown. The emphasis is placed on the differences in the CpG content of the differentially induced gene promoters as well as the abundance of the chromatin mark H3K4m3 and serine 5-phosphorylated RNA Pol II, which is characteristic for the poised state of gene expression. In both the primary and secondary response genes, RNA elongation is supported by binding of BRD4 to acetylated histone H4. The histone-bound BRD4 recruits P-TEFb, which governs an initial phase of the RNA elongation process.

the refractory phase depend on the state of the chromatin and the local concentration of transcription factors (Raj and van Oudenaarden 2009). In such a model, referred to as the promoter progression model, changes in nucleosome occupancy occur at a slow rate (minutes to hours), but once the promoter is open, the transcription machinery will assemble rapidly (in seconds or less) and trigger the "burst" that will continue as long as RNA Pol II is supplied and used by the RNA elongating machinery (Larson 2011).

According to the promoter progression model (Larson 2011), one can expect a direct correlation between the number of steps that are required for gene activation and the duration of the refractory state (i.e., bursty behavior) during inflammatory responses. Secondary response genes are likely to have a limited number of activation cycles, thus limiting the production of inflammatory triggers that can propagate the inflammatory process even in the absence of the initial stimulus.

The discontinuous pattern of gene regulation may contribute not only to timing of gene expression during initial cell response to the proinflammatory signals, but also during cell responses to repetitive triggering. Some initially activated cells fail to be reactivated by LPS (termed tolerance), as judged by the inability of "tolerant" cells to express numerous inflammatory triggers, including cytokines such as IL-6 or IL-12 (Foster et al. 2007; Medzhitov and Horng 2009). However, certain genes, including those that encode secreted antimicrobial peptides, escape tolerance induction (Foster et al. 2007). Studies by Medzhitov's group show that, after stimulation of naïve macrophages with LPS, RNA Pol II was recruited equally well to promoters of tolerizable and nontolerizable genes (Foster et al. 2007). Tolerance to LPS selectively affects RNA Pol II recruitment to the tolerizable genes after secondary LPS triggering. Although promoters of both tolerizable and nontolerizable genes were initially acetylated at histone H4 after the first instance of macrophage activation, only histones at nontolerizable gene promoters were reacetylated after LPS stimulation of tolerant macrophages. Tolerizable

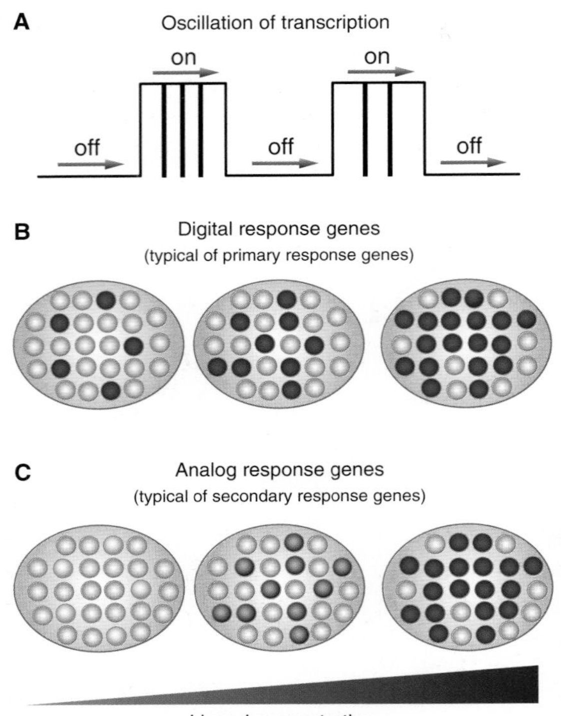

Figure 10. Dynamic control of gene expression. (*A*) Oscillatory dynamics of gene expression. The transition from the silent to active state is shown in a two-dimensional fashion. The rectangle represents the phase of active gene transcription and elongation (transcriptional burst), in which individual vertical bars indicate the number of transcripts generated during the burst. The transcriptional burst is followed by gene silencing. The amount of RNA transcripts produced per burst defines the size of the burst, and the number of bursts per defined time periods (from seconds to hours) corresponds to the burst frequency. (*B*,*C*) Genes can be induced in a "digital" (*B*) or "analog" (*C*) fashion. The responding or nonresponding cells are shown as closed or open circles, respectively. The graded color of the closed circles reflects differences in gene expression levels.

genes appear to maintain the covered state of their promoters in the tolerant state.

The described dynamics of chromatin changes of nontolerizable and tolerizable genes, that do not correlate with primary and secondary gene expression during initial cell triggering by LPS, matches well the predictive behavior of genes with an extended "off" state of the transcriptional cycle. The duration of this cycle can be determined either by intrinsic factors (e.g., local gene environment) or by extrinsic factors (e.g., signaling and metabolic changes). Tolerance in macrophages is associated with expression of proteins that attenuate LPS signaling and, hence, reduce the overall extrinsic signaling input (Foster et al. 2007). The reduced signaling, although generic by nature, may have a selectively stronger impact on genes that depend critically on signaling for repetitive entry into the transcriptional cycle. As a consequence, these genes might display an extended refractory phase to a point in which cells will be perceived as unresponsive or tolerant.

By mentioning perception of a phenotype, we intentionally imply our currently limited ability to distinguish between population response and the response of individual cells. It is conceivable that stable production of antimicrobial peptides, as well as other nontolerizable genes in otherwise LPS-tolerant macrophages, may reflect the existence of a specialized or randomly appearing cell subpopulation in which stochastic arrangement of signaling proteins, metabolic enzymes, and transcription factors generate conditions that do not favor LPS tolerance. Support for intrapopulational diversity during inflammatory responses is offered by studies that show stochastic expression of various proinflammatory cytokines in responding cells. In brief, during viral infection, only a limited fraction of cells are able to activate type I interferon IFNβ gene expression (Zhao et al. 2012). The type I interferon, induced by bacterial and viral infection, plays a key role in antiviral host response. The low frequency of IFNβ-expressing cells, as well as differentially triggered lymphoid cells expressing cytokines such as IL-2, IL-4, IL-10, IL-5, and IL-13 (Guo et al. 2004; Murphy 2005; Paixao et al. 2007), reflects the stochastic intercellular distribution of factors that contribute critically to cytokine expression but are expressed at too low levels to be distributed equally among the cells of a given population. Although the nature of limiting factors that control stochastic gene expression is not fully understood, the underlying mechanism of uneven gene expression is likely to rely on the chromatin architecture of stochastically expressed genes. The covered promoters have a much higher level of transcriptional variation as compared to genes controlled by open promoters (Bai and Morozov 2010). In addition, the presence of suppressive histone marks, such as H3K9me2 and H3K27me3, may create an additional barrier for activation of inflammatory genes and contribute to intra- and interpopulation differences in inflammatory gene expression (Saccani and Natoli 2002; De Santa et al. 2007; Fang et al. 2012). Although chromatin plays an important role in gene expression, there are other sources of variation in gene expression. Gene positioning within the nuclear space or gene "looping" could, for instance, contribute to the variability of gene expression among individual cells (de Wit and van Steensel 2009).

6.2 Transcriptional Noise, Digital versus Analog Gene Regulation, and Diversity of the Immune Response

The notion of individual differences in gene expression between cells involved in inflammation reverberates with

Cite as *Cold Spring Harb Perspect Biol* doi: 10.1101/cshperspect.a019307

a question about the sources and role of transcriptional diversity (transcriptional noise) during inflammation. Transcriptional noise refers to the variability in the expression of individual gene alleles in a cell population held under constant conditions (Blake et al. 2003; Raser and O'Shea 2005; Eldar and Elowitz 2010; Balazsi et al. 2011). The transcriptional noise could be either intrinsic, that is, governed by concentration of transcription factor(s) within a cell, or extrinsic and driven by environmental triggers. The major purpose of transcriptional noise is to adapt to changes in the environment (Balazsi et al. 2011). In yeast that have been engineered to transit randomly but at different switching rates between two phenotypes in response to stochastic fluctuations in gene expression, the environment stability had a strong influence on the dominance of a particular strain (Cairns 2009). When environmental conditions are stable, the strain with slower phenotypic transitions dominates the population, but when the environment fluctuates more quickly, faster transitions are beneficial. Similarly, microbial populations use phenotypic heterogeneity as a strategy to respond to unpredictable changes in the environment (Maheshri and O'Shea 2007; Eldar and Elowitz 2010).

The unpredictability of the environment best describes the situation that awaits migratory inflammatory cells, that is, neutrophils, lymphocytes, or monocytes. Therefore, one would predict a propensity for transcriptional noise in these cells at levels that will enable rapid population adaptation to different environmental settings. Kinetically static tissue cells such as hepatocytes or muscle cells, however, are likely to limit the noise to reduce the cell's responsiveness to random inflammatory signals. A mechanism that would significantly increase the noise of proinflammatory genes in hematopoietic cells but "de-noise" these genes in nonhematopoietic cells has thus provided the basis for investigations. Results in fibroblasts as well as in cardiac myocytes and neurons showed that the IFN-α/β-stimulated genes (ISGs) and NF-κB-inducible genes display enrichment in a suppressive H3K9me2 mark at their promoters, whereas the same genes in macrophages and dendritic cells are largely H3K9me2-depleted (Fang et al. 2012). It is possible that H3K9me2 as well as other suppressive modifications, that is, H3K27me3, at the inflammatory genes may establish the level of transcriptional noise of inflammatory genes and determine the range of cell responsiveness to proinflammatory signals. In support of this model, removal of H3K9me2 from fibroblasts decreased the threshold for cell activation and allowed for virus-induced expression of IFN-α/β and interferon-stimulated genes at levels characteristic for "professional" IFN-α/β-producing dendritic cells (Fang et al. 2012). One can speculate that factors that reduce the noise are likely to play a key role in protecting nonmigratory cells from erroneous inflammatory responses caused by minor tissue damage or metabolic stress. In turn, those factors that increase noise within migratory cells are likely to increase the probability of inflammatory responses during infection.

The probability of an initial transcriptional response to proinflammatory signals determines the scope or degree of population engagement into an inflammatory process. As discussed in Sec. 6.1, the dynamics of the initiation process is gene specific, in which collaboration among a gene sequence, chromatin state, intranuclear gene position, intra- and intergenic interactions, and finally, transcription factors (lineage-specific, signal-induced, and generic) determines the timing of gene expression as well as gene transcriptional dynamics (i.e., transcriptional bursting). The gene will start to transcribe only when all the requirements for transcriptional initiation are accomplished. Therefore, at a single-cell level, a given signal may not gradually increase transcription of an inducible gene locus but rather enable a digital transition from an "off" to an "on" state of gene expression (Fig. 10B) (Stevense et al. 2010). For example, triggering of 3T3 mouse fibroblast cells with the proinflammatory cytokine TNFα results in activation of the transcription factor NF-κB and primary "open" promoter response genes in a digital-like fashion (Covert et al. 2005; Tay et al. 2010). Although the number of cells responding to TNFα is proportional to the TNFα concentration, the activated cells display equal levels of NF-κB and primary gene expression (Fig. 10B). However, at the population level, a transition from an "off" to "on" state will not occur simultaneously, thus giving the impression of gene expression being increased in a gradual, analog fashion. The latter analog response, however, appears to describe the expression of TNFα-induced secondary response genes (Fig. 10C). The expression of secondary response genes (i.e., with "closed" promoters), opposite of the primary response genes, follows the analog-like pattern wherein an increase in concentration increases the level of gene expression within an individual cell (Fig. 10C) (Covert et al. 2005; Tay et al. 2010).

Future studies of individual cell epigenomes combined with comprehensive studies of gene expression at a single-cell level may provide a statistical foundation to support a model that could predict the dynamics of gene expression during inflammation.

6.3 Regulation of Inflammatory Gene Expression by Signal-Induced RNA Elongation

A reason for digital versus analog patterns of primary and secondary TNFα-inducible genes and, perhaps, other inducible gene responses might lie in the differential coordi-

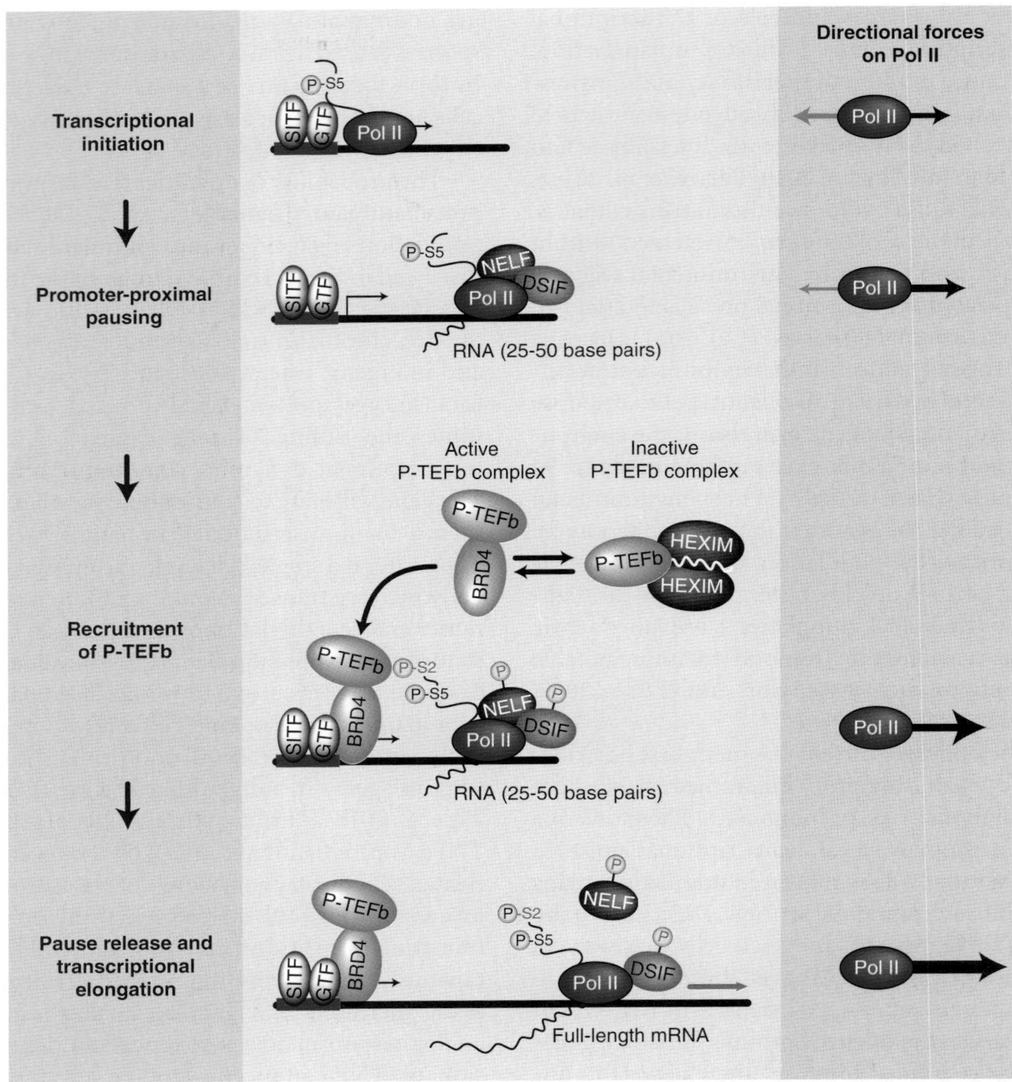

Figure 11. Coupling of transcriptional initiation to elongation. The key biochemical events that control the transition from transcriptional initiation to elongation are shown. The arrows shown to the right illustrate the combinatorial nature of forces acting on the RNA polymerase II (Pol II) at various phases of RNA expression. SITF, signal induced transcription factor; GTF, general transcription factors; P-S5, phosphorylation of serine 5 in the carboxy-terminal domain (CTD) of Pol II; NELF, negative elongation factor; DSIF, DRB sensitivity-inducing factor; P-TEFb, positive transcription-elongation factor-b; BRD4, bromodomain-containing protein 4; HEXIM, hexamethylene bis-acetimide inducible 1; P-S2, phosphorylation of serine 2 in the CTD of Pol II.

nation between transcriptional initiation and elongation at primary and secondary gene loci. In both types of genes, elongation requires the activity of factors that enable RNA Pol II processivity. These factors could be schematically assigned to a few key regulatory hubs that determine the efficiency of elongation (Fig. 11). The first of these regulatory hubs is at gene promoters where RNA Pol II comes under the control of factors that significantly slow or halt elongation (Zhou et al. 2012). Two negative elongation factors, the 5,6-dichloro-1-β-D-ribofuranosylbenzimidazole (DRB) sensitivity-inducing factor (DSIF) and negative elongation factor (NELF), associate with RNA Pol II during initiation, leading to the generation of poised polymerases (Adelman and Lis 2012). The amount of promoter-proximal RNA Pol II determines the size of the polymerase pool potentially available for elongation.

The release of Pol II into elongation depends on the activity and/or amount of gene-bound cyclin-dependent kinase-pausing machinery (i.e., P-TEFb and associated factors), which enables RNA Poll II processivity (Peterlin and

Cite as *Cold Spring Harb Perspect Biol* doi: 10.1101/cshperspect.a019307

Price 2006; Zhou et al. 2012). The amount of P-TEFb available for RNA Pol II release and activation relies on the level of gene activity. This is effectively regulated in a "demand and supply" fashion that is gauged through the amount of promoter-bound BRD4. BRD4 is a "reader" of the acetylation state of histone H4 at gene promoters (Fig. 11) (Zeng and Zhou 2002; Peterlin and Price 2006; Mujtaba et al. 2007; Filippakopoulos et al. 2012). Indeed, the amount of BRD4 bound to chromatin is a key determinant of the gene transition from initiation into elongation. BRD4 plays a dual role in the transition because it not only mediates the release of P-TEFb from an inhibitory macromolecular complex in a graded fashion but also recruits it to chromatin via binding to acetylated histone H4. The inhibitory complex includes 7SK snRNP, the RNA-bound hexamethylene bis-acetamide-inducible (HEXIM) protein, and other proteins that stabilize the complex (Fig. 11) (Zhou et al. 2012). The binding of the CDK9 kinase of P-TEFb to the inhibitory domains of HEXIM1 or HEXIM2 render it inactive (Zhou et al. 2012). Binding of the carboxy-terminal portion of BRD4 to P-TEFb, however, releases this protein from the complex and is followed by the activation of RNA elongation (Fig. 11). In summary, both the amount of RNA Pol II stalled at the promoter and the amount of P-TEFb delivered to release the stalled RNA Pol II will determine the initial efficiency of RNA elongation. At later stages of RNA synthesis, other elongation factors become engaged into the complex and support elongation at a steady 3.8 kb/min rate for up to 2 million base pairs (Luo et al. 2012; Zhou et al. 2012).

mRNA elongation is important for controlling the inducible inflammatory gene expression program in response to Toll-like receptor signaling in macrophages. In the absence of stimulation, RNA Pol II generates low levels of full-length but unspliced and untranslatable transcripts at many of the CpG-rich primary response genes (Hargreaves et al. 2009). Gene induction is accomplished through the signal-dependent recruitment of P-TEFb by BRD4, which itself recognizes the inducibly acquired histone H4 acetylation on lysines 5, 8, and 12 (Hargreaves et al. 2009). This results in robust RNA Pol II phosphorylation on serine 2 by P-TEFb (Fig. 11) and production of high levels of fully spliced mature mRNA transcripts.

Interaction between BRD4 and acetylated histones is mediated by the evolutionary conserved ~110aa-long bromodomain, which is present in two tandemly arranged modules (BDI and BDII) in BRD4. This configuration is also present in the related BRD2 and BRD3 proteins of the BET family (Zeng and Zhou 2002; Mujtaba et al. 2007; Filippakopoulos et al. 2012). Interruption of the bromodomain interaction with acetylated histones by highly selective synthetic antagonists (i.e., JQ1 or I-BET) disrupts

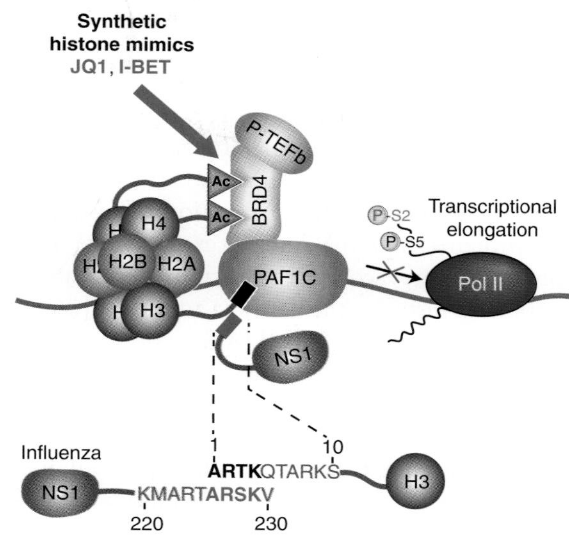

Figure 12. Control of gene expression by "histone mimics." Nucleosomes contribute to the control of signal-induced transcriptional elongation by recruiting BRD4 and its associated P-TEFb complex via the acetylated amino-terminal tail of histone H4 and the PAF1 complex (PAF1C) via the amino-terminal ARTK motif of histone H3. The small molecules JQ1 and I-BET (red arrow) function as synthetic histone mimics by preventing the recruitment of BRD4 to promoters through competitive binding to BRD4 bromodomains. The carboxy-terminal ARSK sequence (red rectangle) of the influenza NS1 protein acts as a histone mimic by competing with the amino-terminal tail of histone 3 for binding to the PAF1 complex.

BET association with chromatin, followed by changes in the expression of numerous proinflammatory genes such as *IL-12* or *IL-6* (Fig. 12) (Filippakopoulos et al. 2010; Nicodeme et al. 2010). Thus, both primary and secondary genes are dependent on BRD4 and perhaps other BET proteins, as suggested by the reduced expression of primary and secondary LPS-induced genes during small interfering RNA (siRNA)-mediated knockdown of BRD4 or multiple BET proteins (Fig. 9) (Hargreaves et al. 2009; Medzhitov and Horng, 2009). However, treatment with I-BET had a highly selective impact on secondary response genes and minimal effect on primary LPS-inducible gene expression. This may be because I-BET predominantly affects a pool of BET proteins that are not yet bound to chromatin, while having a limited impact on histone-bound BET proteins. The chromatin-bound pool of BET proteins at primary gene loci might be sufficient for transcriptional elongation. Contrary to the primary response genes, recruitment of BET proteins to secondary response genes follows signal-induced chromatin remodeling and transcriptional initiation. By that time, a significant fraction of BET proteins are likely to be trapped by I-BET, thus limiting the amount of BET proteins available for RNA synthesis.

7 "HISTONE MIMICRY" AND ITS IMPLICATION FOR REGULATION OF INFLAMMATION

The ability of I-BET to interfere with proinflammatory gene expression suggests a more general possibility of interference with gene expression by using synthetic or natural molecules that could mimic the interaction of histones with effector proteins. Mammalian cells express large numbers of proteins that carry sequences similar to the amino-terminal portions of histones, that is, short amino acid sequences such as ARTK or ARKS (Fig. 12) (A Tarakhovsky, unpubl.). In some cases, histone-like sequences ("histone mimics") can fully recapitulate the protein-binding capacity of its histone H3 counterpart, such as the mimicking motif within the methyltransferase G9a protein shown to contain similarity to histone H3, and results in the binding to self and autocatalytical methylation (Sampath et al. 2007). Histone mimics may also serve as recognition modules that enable posttranslational modification of nonhistone proteins for purposes not directly linked to chromatin function at all (Lee et al. 2012). In any scenario, the proteins carrying histone mimics may compete with histones for histone-modifying enzymes as well as for histone-binding proteins.

Histone mimics are present in numerous bacteria or viral proteins (A Tarakhovsky, unpubl.). The physiological significance of the pathogen-derived histone mimics has been underscored by the effect of a histone H3 mimicking domain present in the carboxy-terminal portion of the influenza protein NS1 (with a sequence of ARSK vs. the amino-terminal endogenous H3 sequence ARTK) (Fig. 12) (Marazzi et al. 2012). In humans, most of the NS1 action probably takes place in the nucleus, where the amount of NS1 protein can reach levels close to the amounts of nucleosomes (5×10^6 vs. 3×10^7, respectively) (Marazzi et al. 2012). The NS1 histone mimic domain binds directly to the PAF1 protein (Marazzi et al. 2012), an essential subunit of the multiprotein elongation complex PAF1C (Fig. 12) (Kim et al. 2009; Jaehning 2010; Smith and Shilatifard 2010). Binding of NS1 to PAF1 inhibits elongation of virus-induced genes, thus contributing to the attenuation of a host's antiviral response (Marazzi et al. 2012). The presumed mode of NS1 action on PAF1C-mediated elongation is likely to reflect a competition between histone H3 and NS1 for PAF1, resulting in a reduction of PAF1 abundance at the transcribed gene loci (Fig. 12) (Marazzi et al. 2012). Parallel experiments showed that siRNA-mediated knockdown of PAF1 expression results in the selective down-regulation of inducible antiviral gene expression, confirming that PAF1 is important in the antiviral transcriptional response (Marazzi et al. 2012).

The discovery of histone mimicry is a relatively novel finding. It has shown that it is not only important as a mechanism of pathogen adaption but may provide novel therapeutic avenues to target the regulation of the immune response.

8 CONCLUDING REMARKS

Healthy immunity thrives on the robustness of immune cell generation, functional diversity of the differentiated immune cells, and a high level of immune cell adaptability to environmental changes. Although operating as a system, immunity relies on the individual cell's preexisting or adaptive ability to tune to the constantly changing environment. This feature of immune cells implies the existence of mechanisms that enhance diversification of the individual cells while maintaining the integrity of the system. In this context, epigenetic mechanisms must capacitate cell adaptation without affecting cell differentiation. Failure to adapt, or an exaggerated response, will cause systemic immune system failures followed by the development of immune disorders. The epigenetic mechanisms that are supposed to control the memory of the environmental impact may also contribute to the persistence of disease-associated phenotypes, even in the absence of the initial trigger. In this context, it would be attractive to consider the possibility of treating chronic inflammatory states by pharmacological "erasure" of the diseased epigenetic landscape, followed by restoration of the healthy gene expression patterns.

There is growing evidence suggesting that pathogens can affect host immunity through the interference with multiple epigenetic processes. Histone mimicry by pathogen-derived proteins offers a mechanistic understanding of the short- and long-term effects of infections on the organism. Accordingly, the identification of proteins that bind to the pathogen-derived histone mimics may guide the identification of novel targets for therapeutic intervention of gene expression. In this context, the rational design of small molecules that mimic histone binding to various effector proteins may lead to the development of drugs that affect gene expression in a highly selective fashion and in accordance with the epigenetic state of individual genes.

REFERENCES

**Reference is also in this book.*

Adelman K, Lis JT. 2012. Promoter-proximal pausing of RNA polymerase II: Emerging roles in metazoans. *Nat Rev Genet* **13:** 720–731.

Bacher CP, Guggiari M, Brors B, Augui S, Clerc P, Avner P, Eils R, Heard E. 2006. Transient colocalization of X-inactivation centres accompanies the initiation of X inactivation. *Nat Cell Biol* **8:** 293–299.

Bai L, Morozov AV. 2010. Gene regulation by nucleosome positioning. *Trends Genet* **26:** 476–483.

Bai L, Ondracka A, Cross FR. 2011. Multiple sequence-specific factors generate the nucleosome-depleted region on CLN2 promoter. *Mol Cell* **42:** 465–476.

Balazsi G, van Oudenaarden A, Collins JJ. 2011. Cellular decision making and biological noise: From microbes to mammals. *Cell* **144:** 910–925.

Bassing CH, Swat W, Alt FW. 2002. The mechanism and regulation of chromosomal V(D)J recombination. *Cell* **109:** S45–S55.

Bereshchenko OR, Gu W, Dalla-Favera R. 2002. Acetylation inactivates the transcriptional repressor BCL6. *Nat Genet* **32:** 606–613.

Bhatt DM, Pandya-Jones A, Tong AJ, Barozzi I, Lissner MM, Natoli G, Black DL, Smale ST. 2012. Transcript dynamics of proinflammatory genes revealed by sequence analysis of subcellular RNA fractions. *Cell* **150:** 279–290.

Blake WJ, Kærn M, Cantor CR, Collins JJ. 2003. Noise in eukaryotic gene expression. *Nature* **422:** 633–637.

Bolland DJ, Wood AL, Johnston CM, Bunting SF, Morgan G, Chakalova L, Fraser PJ, Corcoran AE. 2004. Antisense intergenic transcription in V(D)J recombination. *Nat Immunol* **5:** 630–637.

Bracken AP, Pasini D, Capra M, Prosperini E, Colli E, Helin K. 2003. EZH2 is downstream of the pRB-E2F pathway, essential for proliferation and amplified in cancer. *EMBO J* **22:** 5323–5335.

Bussmann LH, Schubert A, Vu Manh TP, De Andres L, Desbordes SC, Parra M, Zimmermann T, Rapino F, Rodriguez-Ubreva J, Ballestar E, et al. 2009. A robust and highly efficient immune cell reprogramming system. *Cell Stem Cell* **5:** 554–566.

Cai L, Dalal CK, Elowitz MB. 2008. Frequency-modulated nuclear localization bursts coordinate gene regulation. *Nature* **455:** 485–490.

Cairns BR. 2009. The logic of chromatin architecture and remodelling at promoters. *Nature* **461:** 193–198.

Chakraborty T, Perlot T, Subrahmanyam R, Jani A, Goff PH, Zhang Y, Ivanova I, Alt FW, Sen R. 2009. A 220-nucleotide deletion of the intronic enhancer reveals an epigenetic hierarchy in immunoglobulin heavy chain locus activation. *J Exp Med* **206:** 1019–1027.

Chubb JR, Liverpool TB. 2010. Bursts and pulses: Insights from single cell studies into transcriptional mechanisms. *Curr Opin Genet Dev* **20:** 478–484.

Cobaleda C, Jochum W, Busslinger M. 2007. Conversion of mature B cells into T cells by dedifferentiation to uncommitted progenitors. *Nature* **449:** 473–477.

Covert MW, Leung TH, Gaston JE, Baltimore D. 2005. Achieving stability of lipopolysaccharide-induced NF-κB activation. *Science* **309:** 1854–1857.

Decker T, Pasca di Magliano M, McManus S, Sun Q, Bonifer C, Tagoh H, Busslinger M. 2009. Stepwise activation of enhancer and promoter regions of the B cell commitment gene *Pax5* in early lymphopoiesis. *Immunity* **30:** 508–520.

Degner SC, Verma-Gaur J, Wong TP, Bossen C, Iverson GM, Torkamani A, Vettermann C, Lin YC, Ju Z, Schulz D, et al. 2011. CCCTC-binding factor (CTCF) and cohesin influence the genomic architecture of the *Igh* locus and antisense transcription in pro-B cells. *Proc Natl Acad Sci* **108:** 9566–9571.

Delogu A, Schebesta A, Sun Q, Aschenbrenner K, Perlot T, Busslinger M. 2006. Gene repression by Pax5 in B cells is essential for blood cell homeostasis and is reversed in plasma cells. *Immunity* **24:** 269–281.

Deniaud E, Bickmore WA. 2009. Transcription and the nuclear periphery: Edge of darkness? *Curr Opin Genet Dev* **19:** 187–191.

De Santa F, Totaro MG, Prosperini E, Notarbartolo S, Testa G, Natoli G. 2007. The histone H3 lysine-27 demethylase Jmjd3 links inflammation to inhibition of polycomb-mediated gene silencing. *Cell* **130:** 1083–1094.

de Wit E, van Steensel B. 2009. Chromatin domains in higher eukaryotes: Insights from genome-wide mapping studies. *Chromosoma* **118:** 25–36.

Dias S, Mansson R, Gurbuxani S, Sigvardsson M, Kee BL. 2008. E2A proteins promote development of lymphoid-primed multipotent progenitors. *Immunity* **29:** 217–227.

Ebert A, McManus S, Tagoh H, Medvedovic J, Salvagiotto G, Novatchkova M, Tamir I, Sommer A, Jaritz M, Busslinger M. 2011. The distal V_H gene cluster of the *Igh* locus contains distinct regulatory elements with Pax5 transcription factor-dependent activity in pro-B cells. *Immunity* **34:** 175–187.

Edmunds JW, Mahadevan LC, Clayton AL. 2008. Dynamic histone H3 methylation during gene induction: HYPB/Setd2 mediates all H3K36 trimethylation. *EMBO J* **27:** 406–420.

Eldar A, Elowitz MB. 2010. Functional roles for noise in genetic circuits. *Nature* **467:** 167–173.

Fang TC, Schaefer U, Mecklenbrauker I, Stienen A, Dewell S, Chen MS, Rioja I, Parravicini V, Prinjha RK, Chandwani R, et al. 2012. Histone H3 lysine 9 di-methylation as an epigenetic signature of the interferon response. *J Exp Med* **209:** 661–669.

Filippakopoulos P, Qi J, Picaud S, Shen Y, Smith WB, Fedorov O, Morse EM, Keates T, Hickman TT, Felletar I, et al. 2010. Selective inhibition of BET bromodomains. *Nature* **468:** 1067–1073.

Filippakopoulos P, Picaud S, Mangos M, Keates T, Lambert JP, Barsyte-Lovejoy D, Felletar I, Volkmer R, Muller S, Pawson T, et al. 2012. Histone recognition and large-scale structural analysis of the human bromodomain family. *Cell* **149:** 214–231.

Foster SL, Hargreaves DC, Medzhitov R. 2007. Gene-specific control of inflammation by TLR-induced chromatin modifications. *Nature* **447:** 972–978.

Fuxa M, Skok J, Souabni A, Salvagiotto G, Roldán E, Busslinger M. 2004. Pax5 induces *V-to-DJ* rearrangements and locus contraction of the *immunoglobulin heavy-chain* gene. *Genes Dev* **18:** 411–422.

Gilchrist DA, Adelman K. 2012. Coupling polymerase pausing and chromatin landscapes for precise regulation of transcription. *Biochim Biophys Acta* **1819:** 700–706.

Gilchrist M, Thorsson V, Li B, Rust AG, Korb M, Roach JC, Kennedy K, Hai T, Bolouri H, Aderem A. 2006. Systems biology approaches identify ATF3 as a negative regulator of Toll-like receptor 4. *Nature* **441:** 173–178.

Guenther MG, Lawton LN, Rozovskaia T, Frampton GM, Levine SS, Volkert TL, Croce CM, Nakamura T, Canaani E, Young RA. 2008. Aberrant chromatin at genes encoding stem cell regulators in human mixed-lineage leukemia. *Genes Dev* **22:** 3403–3408.

Guo L, Hu-Li J, Paul WE. 2004. Probabilistic regulation of IL-4 production in Th2 cells: Accessibility at the *Il4* locus. *Immunity* **20:** 193–203.

Guo C, Yoon HS, Franklin A, Jain S, Ebert A, Cheng HL, Hansen E, Despo O, Bossen C, Vettermann C, et al. 2011. CTCF-binding elements mediate control of V(D)J recombination. *Nature* **477:** 424–430.

Győry I, Boller S, Nechanitzky R, Mandel E, Pott S, Liu E, Grosschedl R. 2012. Transcription factor Ebf1 regulates differentiation stage-specific signaling, proliferation, and survival of B cells. *Genes Dev* **26:** 668–682.

Hadjur S, Williams LM, Ryan NK, Cobb BS, Sexton T, Fraser P, Fisher AG, Merkenschlager M. 2009. Cohesins form chromosomal *cis*-interactions at the developmentally regulated *IFNG* locus. *Nature* **460:** 410–413.

Hargreaves DC, Horng T, Medzhitov R. 2009. Control of inducible gene expression by signal-dependent transcriptional elongation. *Cell* **138:** 129–145.

Hawwari A, Krangel MS. 2005. Regulation of TCR δ and α repertoires by local and long-distance control of variable gene segment chromatin structure. *J Exp Med* **202:** 467–472.

Hesslein DGT, Pflugh DL, Chowdhury D, Bothwell ALM, Sen R, Schatz DG. 2003. *Pax5* is required for recombination of transcribed, acetylated, 5′ IgH V gene segments. *Genes Dev* **17:** 37–42.

Hewitt SL, Farmer D, Marszalek K, Cadera E, Liang HE, Xu Y, Schlissel MS, Skok JA. 2008. Association between the *Igk* and *Igh* immunoglobulin loci mediated by the 3′ *Igk* enhancer induces 'decontraction' of the *Igh* locus in pre-B cells. *Nat Immunol* **9:** 396–404.

Hewitt SL, Yin B, Ji Y, Chaumeil J, Marszalek K, Tenthorey J, Salvagiotto G, Steinel N, Ramsey LB, Ghysdael J, et al. 2009. RAG-1 and ATM coordinate monoallelic recombination and nuclear positioning of immunoglobulin loci. *Nat Immunol* **10:** 655–664.

Jaehning JA. 2010. The Paf1 complex: Platform or player in RNA polymerase II transcription? *Biochim Biophys Acta* **1799:** 379–388.

Jankovic M, Casellas R, Yannoutsos N, Wardemann H, Nussenzweig MC. 2004. RAGs and regulation of autoantibodies. *Annu Rev Immunol* **22:** 485–501.

Jhunjhunwala S, van Zelm MC, Peak MM, Cutchin S, Riblet R, van Dongen JJM, Grosveld FG, Knoch TA, Murre C. 2008. The 3D structure of the immunoglobulin heavy-chain locus: Implications for long-range genomic interactions. *Cell* **133:** 265–279.

Jhunjhunwala S, van Zelm MC, Peak MM, Murre C. 2009. Chromatin architecture and the generation of antigen receptor diversity. *Cell* **138:** 435–448.

Ji Y, Resch W, Corbett E, Yamane A, Casellas R, Schatz DG. 2010. The in vivo pattern of binding of RAG1 and RAG2 to antigen receptor loci. *Cell* **141:** 419–431.

Keats JJ, Reiman T, Maxwell CA, Taylor BJ, Larratt LM, Mant MJ, Belch AR, Pilarski LM. 2003. In multiple myeloma, t(4;14)(p16;q32) is an adverse prognostic factor irrespective of FGFR3 expression. *Blood* **101:** 1520–1529.

Kim J, Guermah M, McGinty RK, Lee JS, Tang Z, Milne TA, Shilatifard A, Muir TW, Roeder RG. 2009. RAD6-mediated transcription-coupled H2B ubiquitylation directly stimulates H3K4 methylation in human cells. *Cell* **137:** 459–471.

Kosak ST, Skok JA, Medina KL, Riblet R, Le Beau MM, Fisher AG, Singh H. 2002. Subnuclear compartmentalization of immunoglobulin loci during lymphocyte development. *Science* **296:** 158–162.

Kuo AJ, Cheung P, Chen K, Zee BM, Kioi M, Lauring J, Xi Y, Park BH, Shi X, Garcia BA, et al. 2011. NSD2 links dimethylation of histone H3 at lysine 36 to oncogenic programming. *Mol Cell* **44:** 609–620.

Kwak H, Fuda NJ, Core LJ, Lis JT. 2013. Precise maps of RNA polymerase reveal how promoters direct initiation and pausing. *Science* **339:** 950–953.

Laiosa CV, Stadtfeld M, Xie H, de Andres-Aguayo L, Graf T. 2006. Reprogramming of committed T cell progenitors to macrophages and dendritic cells by C/EBPα and PU.1 transcription factors. *Immunity* **25:** 731–744.

Lam FH, Steger DJ, O'Shea EK. 2008. Chromatin decouples promoter threshold from dynamic range. *Nature* **453:** 246–250.

Larson DR. 2011. What do expression dynamics tell us about the mechanism of transcription? *Curr Opin Genet Dev* **21:** 591–599.

Lee JM, Lee JS, Kim H, Kim K, Park H, Kim JY, Lee SH, Kim IS, Kim J, Lee M, et al. 2012. EZH2 generates a methyl degron that is recognized by the DCAF1/DDB1/CUL4 E3 ubiquitin ligase complex. *Mol Cell* **48:** 572–586.

* Li E, Zhang Y. 2014. DNA methylation in mammals. *Cold Spring Harb Perspect Biol* **6:** a019133.

Li L, Leid M, Rothenberg EV. 2010. An early T cell lineage commitment checkpoint dependent on the transcription factor Bcl11b. *Science* **329:** 89–93.

Lill NL, Grossman SR, Ginsberg D, DeCaprio J, Livingston DM. 1997. Binding and modulation of p53 by p300/CBP coactivators. *Nature* **387:** 823–827.

Liu H, Cheng EH, Hsieh JJ. 2009. MLL fusions: Pathways to leukemia. *Cancer Biol Ther* **8:** 1204–1211.

Liu Z, Widlak P, Zou Y, Xiao F, Oh M, Li S, Chang MY, Shay JW, Garrard WT. 2006. A recombination silencer that specifies heterochromatin positioning and Ikaros association in the immunoglobulin κ locus. *Immunity* **24:** 405–415.

Liu H, Schmidt-Supprian M, Shi Y, Hobeika E, Barteneva N, Jumaa H, Pelanda R, Reth M, Skok J, Rajewsky K, et al. 2007. Yin Yang 1 is a critical regulator of B-cell development. *Genes Dev* **21:** 1179–1189.

Luo Z, Lin C, Shilatifard A. 2012. The super elongation complex (SEC) family in transcriptional control. *Nat Rev Mol Cell Biol* **13:** 543–547.

Maheshri N, O'Shea EK. 2007. Living with noisy genes: How cells function reliably with inherent variability in gene expression. *Annu Rev Biophys Biomol Struct* **36:** 413–434.

Malin S, McManus S, Cobaleda C, Novatchkova M, Delogu A, Bouillet P, Strasser A, Busslinger M. 2010. Role of STAT5 in controlling cell survival and immunoglobulin gene recombination during pro-B cell development. *Nat Immunol* **11:** 171–179.

Mar BG, Bullinger L, Basu E, Schlis K, Silverman LB, Döhner K, Armstrong SA. 2012. Sequencing histone-modifying enzymes identifies UTX mutations in acute lymphoblastic leukemia. *Leukemia* **26:** 1881–1883.

Marazzi I, Ho JS, Kim J, Manicassamy B, Dewell S, Albrecht RA, Seibert CW, Schaefer U, Jeffrey KL, Prinjha RK, et al. 2012. Suppression of the antiviral response by an influenza histone mimic. *Nature* **483:** 428–433.

Matthews AG, Kuo AJ, Ramón-Maiques S, Han S, Champagne KS, Ivanov D, Gallardo M, Carney D, Cheung P, Ciccone DN, et al. 2007. RAG2 PHD finger couples histone H3 lysine 4 trimethylation with V(D)J recombination. *Nature* **450:** 1106–1110.

McManus S, Ebert A, Salvagiotto G, Medvedovic J, Sun Q, Tamir I, Jaritz M, Tagoh H, Busslinger M. 2011. The transcription factor Pax5 regulates its target genes by recruiting chromatin-modifying proteins in committed B cells. *EMBO J* **30:** 2388–2404.

Medvedovic J, Ebert A, Tagoh H, Busslinger M. 2011. Pax5: A master regulator of B cell development and leukemogenesis. *Adv Immunol* **111:** 179–206.

Medvedovic J, Ebert A, Tagoh H, Tamir IM, Schwickert TA, Novatchkova M, Sun Q, Huis in 't Veld PJ, Guo C, Yoon HS, et al. 2013. Flexible long-range loops in the V_H gene region of the *Igh* locus facilitate the generation of a diverse antibody repertoire. *Immunity* **39:** 229–244.

Medzhitov R. 2008. Origin and physiological roles of inflammation. *Nature* **454:** 428–435.

Medzhitov R, Horng T. 2009. Transcriptional control of the inflammatory response. *Nat Rev Immunol* **9:** 692–703.

Morin RD, Johnson NA, Severson TM, Mungall AJ, An J, Goya R, Paul JE, Boyle M, Woolcock BW, Kuchenbauer F, et al. 2010. Somatic mutations altering EZH2 (Tyr641) in follicular and diffuse large B-cell lymphomas of germinal-center origin. *Nat Genet* **42:** 181–185.

Morin RD, Mendez-Lago M, Mungall AJ, Goya R, Mungall KL, Corbett RD, Johnson NA, Severson TM, Chiu R, Field M, et al. 2011. Frequent mutation of histone-modifying genes in non-Hodgkin lymphoma. *Nature* **476:** 298–303.

Mujtaba S, Zeng L, Zhou MM. 2007. Structure and acetyl-lysine recognition of the bromodomain. *Oncogene* **26:** 5521–5527.

Mullighan CG, Goorha S, Radtke I, Miller CB, Coustan-Smith E, Dalton JD, Girtman K, Mathew S, Ma J, Pounds SB, et al. 2007. Genome-wide analysis of genetic alterations in acute lymphoblastic leukaemia. *Nature* **446:** 758–764.

Mullighan CG, Zhang J, Kasper LH, Lerach S, Payne-Turner D, Phillips LA, Heatley SL, Holmfeldt L, Collins-Underwood JR, Ma J, et al. 2011. *CREBBP* mutations in relapsed acute lymphoblastic leukaemia. *Nature* **471:** 235–239.

Murphy KM. 2005. Fate vs choice: The immune system reloaded. *Immunol Res* **32:** 193–200.

Nativio R, Wendt KS, Ito Y, Huddleston JE, Uribe-Lewis S, Woodfine K, Krueger C, Reik W, Peters JM, Murrell A. 2009. Cohesin is required for higher-order chromatin conformation at the imprinted *IGF2-H19* locus. *PLoS Genet* **5:** e1000739.

Natoli G, Ghisletti S, Barozzi I. 2011. The genomic landscapes of inflammation. *Genes Dev* **25:** 101–106.

Ng SY-M, Yoshida T, Zhang J, Georgopoulos K. 2009. Genome-wide lineage-specific transcriptional networks underscore Ikaros-dependent lymphoid priming in hematopoietic stem cells. *Immunity* **30:** 493–507.

Nicodeme E, Jeffrey KL, Schaefer U, Beinke S, Dewell S, Chung CW, Chandwani R, Marazzi I, Wilson P, Coste H, et al. 2010. Suppression of inflammation by a synthetic histone mimic. *Nature* **468:** 1119–1123.

Ntziachristos P, Tsirigos A, Van Vlierberghe P, Nedjic J, Trimarchi T, Flaherty MS, Ferres-Marco D, da Ros V, Tang Z, Siegle J, et al. 2012. Genetic inactivation of the Polycomb repressive complex 2 in T cell acute lymphoblastic leukemia. *Nat Med* **18:** 298–301.

Nutt SL, Kee BL. 2007. The transcriptional regulation of B cell lineage commitment. *Immunity* **26:** 715–725.

Nutt SL, Heavey B, Rolink AG, Busslinger M. 1999. Commitment to the B-lymphoid lineage depends on the transcription factor Pax5. *Nature* **401:** 556–562.

Ogryzko VV, Schiltz RL, Russanova V, Howard BH, Nakatani Y. 1996. The transcriptional coactivators p300 and CBP are histone acetyltransferases. *Cell* **87:** 953–959.

Oguro H, Yuan J, Ichikawa H, Ikawa T, Yamazaki S, Kawamoto H, Nakauchi H, Iwama A. 2010. Poised lineage specification in multipotential hematopoietic stem and progenitor cells by the polycomb protein Bmi1. *Cell Stem Cell* **6:** 279–286.

Okada Y, Feng Q, Lin Y, Jiang Q, Li Y, Coffield VM, Su L, Xu G, Zhang Y. 2005. hDOT1L links histone methylation to leukemogenesis. *Cell* **121:** 167–178.

Paixao T, Carvalho TP, Calado DP, Carneiro J. 2007. Quantitative insights into stochastic monoallelic expression of cytokine genes. *Immunol Cell Biol* **85:** 315–322.

Pasqualucci L, Dominguez-Sola D, Chiarenza A, Fabbri G, Grunn A, Trifonov V, Kasper LH, Lerach S, Tang H, Ma J, et al. 2011. Inactivating mutations of acetyltransferase genes in B-cell lymphoma. *Nature* **471:** 189–195.

Pedraza JM, Paulsson J. 2008. Effects of molecular memory and bursting on fluctuations in gene expression. *Science* **319:** 339–343.

Perlot T, Alt FW. 2008. Cis-regulatory elements and epigenetic changes control genomic rearrangements of the *IgH* locus. *Adv Immunol* **99:** 1–32.

Perlot T, Alt FW, Bassing CH, Suh H, Pinaud E. 2005. Elucidation of IgH intronic enhancer functions via germ-line deletion. *Proc Natl Acad Sci* **102:** 14362–14367.

Peterlin BM, Price DH. 2006. Controlling the elongation phase of transcription with P-TEFb. *Mol Cell* **23:** 297–305.

Phillips JE, Corces VG. 2009. CTCF: Master weaver of the genome. *Cell* **137:** 1194–1211.

Radtke F, Fasnacht N, Macdonald HR. 2010. Notch signaling in the immune system. *Immunity* **32:** 14–27.

Raj A, van Oudenaarden A. 2009. Single-molecule approaches to stochastic gene expression. *Annu Rev Biophys* **38:** 255–270.

Ramirez-Carrozzi VR, Braas D, Bhatt DM, Cheng CS, Hong C, Doty KR, Black JC, Hoffmann A, Carey M, Smale ST. 2009. A unifying model for the selective regulation of inducible transcription by CpG islands and nucleosome remodeling. *Cell* **138:** 114–128.

Raser JM, O'Shea EK. 2005. Noise in gene expression: Origins, consequences, and control. *Science* **309:** 2010–2013.

Revilla-i-Domingo R, Bilic I, Vilagos B, Tagoh H, Ebert A, Tamir IM, Smeenk L, Trupke J, Sommer A, Jaritz M, et al. 2012. The B-cell identity factor Pax5 regulates distinct transcriptional programmes in early and late B lymphopoiesis. *EMBO J* **31:** 3130–3146.

Ribeiro de Almeida C, Stadhouders R, de Bruijn MJ, Bergen IM, Thongjuea S, Lenhard B, van Ijcken W, Grosveld F, Galjart N, Soler E, et al. 2011. The DNA-binding protein CTCF limits proximal Vκ recombination and restricts κ enhancer interactions to the immunoglobulin κ light chain locus. *Immunity* **35:** 501–513.

Roldán E, Fuxa M, Chong W, Martinez D, Novatchkova M, Busslinger M, Skok JA. 2005. Locus 'decontraction' and centromeric recruitment contribute to allelic exclusion of the immunoglobulin heavy-chain gene. *Nat Immunol* **6:** 31–41.

Romanow WJ, Langerak AW, Goebel P, Wolvers-Tettero ILM, van Dongen JJM, Feeney AJ, Murre C. 2000. E2A and EBF act in synergy with the V(D)J recombinase to generate a diverse immunoglobulin repertoire in nonlymphoid cells. *Mol Cell* **5:** 343–353.

Saccani S, Natoli G. 2002. Dynamic changes in histone H3 Lys 9 methylation occurring at tightly regulated inducible inflammatory genes. *Genes Dev* **16:** 2219–2224.

Sampath SC, Marazzi I, Yap KL, Sampath SC, Krutchinsky AN, Mecklenbrauker I, Viale A, Rudensky E, Zhou MM, Chait BT, et al. 2007. Methylation of a histone mimic within the histone methyltransferase G9a regulates protein complex assembly. *Mol Cell* **27:** 596–608.

Sayegh C, Jhunjhunwala S, Riblet R, Murre C. 2005. Visualization of looping involving the immunoglobulin heavy-chain locus in developing B cells. *Genes Dev* **19:** 322–327.

Schlimgen RJ, Reddy KL, Singh H, Krangel MS. 2008. Initiation of allelic exclusion by stochastic interaction of *Tcrb* alleles with repressive nuclear compartments. *Nat Immunol* **9:** 802–809.

Seitan VC, Hao B, Tachibana-Konwalski K, Lavagnolli T, Mira-Bontenbal H, Brown KE, Teng G, Carroll T, Terry A, Horan K, et al. 2011. A role for cohesin in T-cell-receptor rearrangement and thymocyte differentiation. *Nature* **476:** 467–471.

Shaffer AL, Lin K-I, Kuo TC, Yu X, Hurt EM, Rosenwald A, Giltnane JM, Yang L, Zhao H, Calame K, et al. 2002. Blimp-1 orchestrates plasma cell differentiation by extinguishing the mature B cell gene expression program. *Immunity* **17:** 51–62.

Simon C, Chagraoui J, Krosl J, Gendron P, Wilhelm B, Lemieux S, Boucher G, Chagnon P, Drouin S, Lambert R, et al. 2012. A key role for EZH2 and associated genes in mouse and human adult T-cell acute leukemia. *Genes Dev* **26:** 651–656.

Skok JA, Gisler R, Novatchkova M, Farmer D, de Laat W, Busslinger M. 2007. Reversible contraction by looping of the *Tcrb* and *Tcra* loci in rearranging thymocytes. *Nat Immunol* **8:** 378–387.

Smale ST. 2010a. Pioneer factors in embryonic stem cells and differentiation. *Curr Opin Genet Dev* **20:** 519–526.

Smale ST. 2010b. Selective transcription in response to an inflammatory stimulus. *Cell* **140:** 833–844.

Smith E, Shilatifard A. 2010. The chromatin signaling pathway: Diverse mechanisms of recruitment of histone-modifying enzymes and varied biological outcomes. *Mol Cell* **40:** 689–701.

Sneeringer CJ, Scott MP, Kuntz KW, Knutson SK, Pollock RM, Richon VM, Copeland RA. 2010. Coordinated activities of wild-type plus mutant EZH2 drive tumor-associated hypertrimethylation of lysine 27 on histone H3 (H3K27) in human B-cell lymphomas. *Proc Natl Acad Sci* **107:** 20980–20985.

Splinter E, Heath H, Kooren J, Palstra R-J, Klous P, Grosveld F, Galjart N, de Laat W. 2006. CTCF mediates long-range chromatin looping and local histone modification in the β-globin locus. *Genes Dev* **20:** 2349–2354.

Stanhope-Baker P, Hudson KM, Shaffer AL, Constantinescu A, Schlissel MS. 1996. Cell type-specific chromatin structure determines the targeting of V(D)J recombinase activity in vitro. *Cell* **85:** 887–897.

Stevense M, Muramoto T, Muller I, Chubb JR. 2010. Digital nature of the immediate-early transcriptional response. *Development* **137:** 579–584.

Su I-H, Basavaraj A, Krutchinsky AN, Hobert O, Ullrich A, Chait BT, Tarakhovsky A. 2003. Ezh2 controls B cell development through histone H3 methylation and *Igh* rearrangement. *Nat Immunol* **4:** 124–131.

Takahashi K, Yamanaka S. 2006. Induction of pluripotent stem cells from mouse embryonic and adult fibroblast cultures by defined factors. *Cell* **126:** 663–676.

Tay S, Hughey JJ, Lee TK, Lipniacki T, Quake SR, Covert MW. 2010. Single-cell NF-κB dynamics reveal digital activation and analogue information processing. *Nature* **466:** 267–271.

Ting CN, Olson MC, Barton KP, Leiden JM. 1996. Transcription factor GATA-3 is required for development of the T-cell lineage. *Nature* **384:** 474–478.

van Haaften G, Dalgliesh GL, Davies H, Chen L, Bignell G, Greenman C, Edkins S, Hardy C, O'Meara S, Teague J, et al. 2009. Somatic mutations of the histone H3K27 demethylase gene *UTX* in human cancer. *Nat Genet* **41:** 521–523.

Weber BN, Chi AW-S, Chavez A, Yashiro-Ohtani Y, Yang Q, Shestova O, Bhandoola A. 2011. A critical role for TCF-1 in T-lineage specification and differentiation. *Nature* **476:** 63–68.

Xiang Y, Zhou X, Hewitt SL, Skok JA, Garrard WT. 2011. A multifunctional element in the mouse *Igκ* locus that specifies repertoire and *Ig* loci subnuclear location. *J Immunol* **186:** 5356–5366.

Xie H, Ye M, Feng R, Graf T. 2004. Stepwise reprogramming of B cells into macrophages. *Cell* **117:** 663–676.

Yancopoulos GD, Alt FW. 1985. Developmentally controlled and tissue-specific expression of unrearranged V_H gene segments. *Cell* **40:** 271–281.

Yap DB, Chu J, Berg T, Schapira M, Cheng SW, Moradian A, Morin RD, Mungall AJ, Meissner B, Boyle M, et al. 2011. Somatic mutations at EZH2 Y641 act dominantly through a mechanism of selectively altered PRC2 catalytic activity, to increase H3K27 trimethylation. *Blood* **117:** 2451–2459.

Ye S-K, Agata Y, Lee H-C, Kurooka H, Kitamura T, Shimizu A, Honjo T, Ikuta K. 2001. The IL-7 receptor controls the accessibility of the TCRγ locus by Stat5 and histone acetylation. *Immunity* **15:** 813–823.

Yosef N, Regev A. 2011. Impulse control: Temporal dynamics in gene transcription. *Cell* **144:** 886–896.

Zak DE, Aderem A. 2009. Systems biology of innate immunity. *Immunol Rev* **227:** 264–282.

Zeng L, Zhou MM. 2002. Bromodomain: An acetyl-lysine binding domain. *FEBS Lett* **513:** 124–128.

Zhang J, Ding L, Holmfeldt L, Wu G, Heatley SL, Payne-Turner D, Easton J, Chen X, Wang J, Rusch M, et al. 2012a. The genetic basis of early T-cell precursor acute lymphoblastic leukaemia. *Nature* **481:** 157–163.

Zhang J, Jackson AF, Naito T, Dose M, Seavitt J, Liu F, Heller EJ, Kashiwagi M, Yoshida T, Gounari F, et al. 2012b. Harnessing of the nucleosome-remodeling-deacetylase complex controls lymphocyte development and prevents leukemogenesis. *Nat Immunol* **13:** 86–94.

Zhao M, Zhang J, Phatnani H, Scheu S, Maniatis T. 2012. Stochastic expression of the interferon-β gene. *PLoS Biol* **10:** e1001249.

Zhou Q, Li T, Price DH. 2012. RNA polymerase II elongation control. *Annu Rev Biochem* **81:** 119–143.

 Cite as *Cold Spring Harb Perspect Biol* doi: 10.1101/cshperspect.a019307

C H A P T E R 30

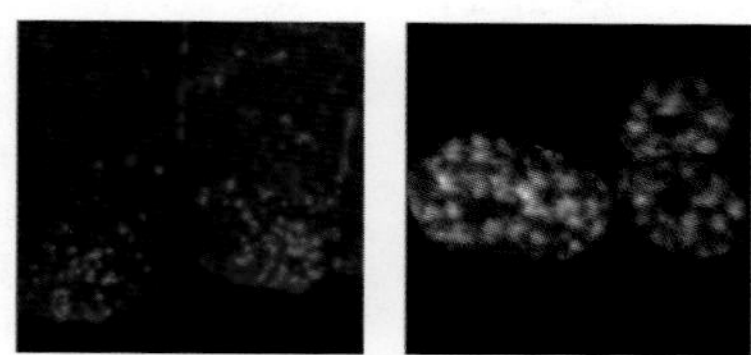

Metabolic Signaling to Chromatin

Shelley L. Berger[1] and Paolo Sassone-Corsi[2]

[1]Department of Cell & Developmental Biology, Department of Biology, and Department of Genetics, Epigenetics Program, University of Pennsylvania, Philadelphia, Pennsylvania 19104-6508; [2]Center for Epigenetics and Metabolism, Department of Biological Chemistry, University of California, Irvine, Irvine, California 92697-4049

Correspondence: bergers@mail.med.upenn.edu; psc@uci.edu

SUMMARY

There is a dynamic interplay between metabolic processes and gene regulation via the remodeling of chromatin. Most chromatin-modifying enzymes use cofactors, which are products of metabolic processes. This chapter explores the biosynthetic pathways of the cofactors nicotinamide adenine dinucleotide (NAD), acetyl coenzyme A (acetyl-CoA), *S*-adenosyl methionine (SAM), α-ketoglutarate, and flavin adenine dinucleotide (FAD), and their role in metabolically regulating chromatin processes. A more detailed look at the interaction between chromatin and the metabolic processes of circadian rhythms and aging is described as a paradigm for this emerging interdisciplinary field.

Outline

Chapter-opening figures: (*Left*) Reproduced from Crosio et al. (2000).

Cite as *Cold Spring Harb Perspect Biol* doi: 10.1101/cshperspect.a019463

OVERVIEW

A prominent area in epigenetic research that has emerged in recent years relates to how cellular metabolism regulates various events of chromatin remodeling. Cells sense changes in the environment and translate them into specific modulations of the epigenome through a variety of signaling components, several of which are proteins with histone- and DNA-modifying enzymatic activity. There are now a myriad of residues on DNA and histone tails that can undergo modification at a given time. The enzymes that elicit these modifications rely critically on the availability of phosphate, acetyl, and methyl groups, to mention a few. This constitutes an intriguing link between cellular metabolism and epigenetic control that has previously been largely unappreciated. Although the specificity and degree of change in the levels of cellular metabolites may influence the epigenome, we are unsure of to what degree. However, a number of remarkable studies discussed in this chapter are revealing a range of responses to the environment.

An example of the intimate connection between metabolism and epigenetic control is epitomized by findings on the *FTO* (fat mass and obesity-associated) gene. The gene encodes an *N*6-methyladenosine (m6A) demethylase, an enzyme that controls the levels of RNA methylation (Gerken et al. 2007). Single-nucleotide polymorphisms of the *FTO* gene, associated with obesity, fundamentally alter its response to food intake and thereby influence the development of obesity. The epigenetic connection came through chromosome capture experiments (described in Ch. 19 [Dekker and Misteli 2014]), showing that certain variants of the *FTO* gene are uniquely correlated with obesity and physically loop to and interact with *IRX3*, a homeobox gene located megabases away from *FTO*. Thus, metabolic dysfunction, such as obesity, manifesting at one gene, directly influences the function of a distant gene that, remarkably, is also implicated in controlling body mass (Smemo et al. 2014).

Chromatin regulation involves enzymes that use cofactors for the reactions that modify DNA or histones. These enzymes either attach small chemical units (i.e., posttranslational modifications or PTMs) or alter nucleosome positioning or composition (i.e., of histone variants). It is assumed that this control depends partly on the variable levels of cellular metabolites acting as enzyme cofactors. For example, acetyltransferases use acetyl-coenzyme A (acetyl-CoA), methyltransferases use *S*-adenosyl methionine, and kinases use ATP as donors of acetyl, methyl, or phospho groups, respectively; deacetylases can use nicotinamide adenine dinucleotide (NAD), and demethylases can use flavin adenine dinucleotide (FAD) or α-ketoglutarate as coenzymes. In addition, another relevant example relates to remodeler complexes that use ATP for moving, ejecting, or restructuring nucleosomes, as discussed in Ch. 21 (Becker and Workman 2013).

In this chapter, we first present a summary of several metabolites that are known to alter the activity of their cognate enzymes and, hence, their potential in regulating chromatin states is discussed. We then present examples of enzymes that directly respond to changes in metabolism, such as the sirtuin enzymes and polyadenosine diphosphate (ADP)-ribose polymerases (PARPs). Finally, we discuss several fields that have been greatly influenced by ideas that altered metabolic signaling could affect epigenetic regulation, such as (1) altered nutrient availability regulating histone and DNA modifications, (2) circadian biological rhythms, and (3) cellular replicative aging and senescence.

 Cite as *Cold Spring Harb Perspect Biol* doi: 10.1101/cshperspect.a019463

1 METABOLITES

Although it is conceptually conceivable that a variety of metabolites are implicated in various aspects of epigenetic control (Katada et al. 2012; Lu and Thompson 2012), to date solid evidence is scarce. The assumption is that levels of metabolites may change in response to various physiological stimuli, and thereby influence enzymes implicated in chromatin remodeling. In addition, the subcellular concentration of specific metabolites may direct the local activation or inhibition of enzymatic activities. Thus, the presence of "niches" of chromatin-associated metabolites has been hypothesized, which could rationalize the locus-specific modification of otherwise undistinguishable histones and DNA (Katada et al. 2012). The metabolites discussed in this section can be functionally linked to enzymatic activities involved in chromatin remodeling and DNA methylation (Fig. 1).

1.1 Acetyl-CoA

Acetylation of both histone and nonhistone proteins has critical biological consequences (Guan and Xiong 2011). The availability of appropriate levels of acetyl-CoA, the metabolite that provides the acetyl group for the acetylation reaction, could modulate the efficacy and specificity of the reaction. Acetyl-CoA appears to exist in two separate pools in the cell: a mitochondrial and nuclear/cytosolic pool. The mitochondrial pool is derived mainly from the action of the enzyme pyruvate dehydrogenase and fatty acid oxidation (Fig. 2). The nuclear/cytosolic pool of acetyl-CoA, responsible for protein acetylation and fatty acid synthesis, is produced by two enzymes in metazoans: the acetyl-CoA synthetase 1 (AceCS1) and ATP-citrate lyase (ACL) (Albaugh et al. 2011). ACL uses citrate (produced during the tricarboxylic acid [TCA] cycle) as a substrate for the production of acetyl-CoA, whereas AceCS1 uses acetate. In

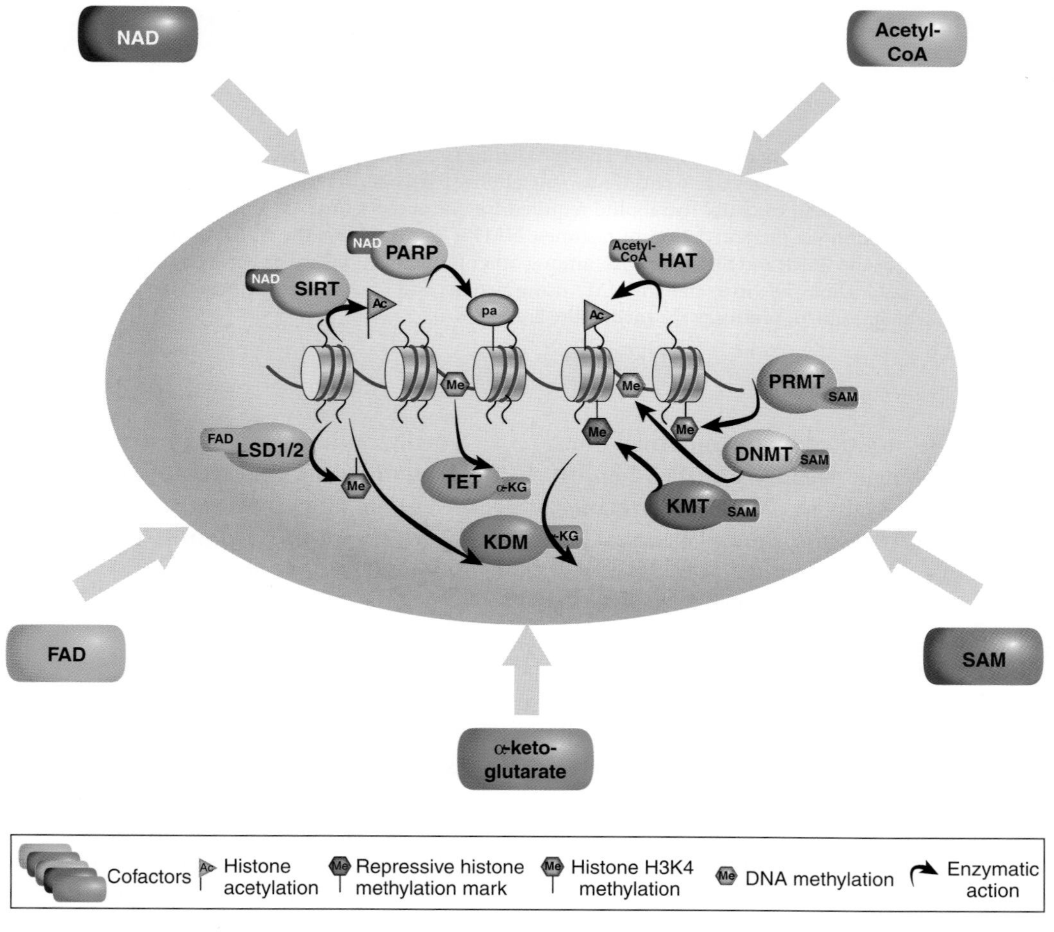

Figure 1. The major cofactors involved in enzyme-mediated DNA or histone posttranslational modification (PTM). NAD, nicotinamide adenine dinucleotide; acetyl-CoA, acetyl coenzyme A; SAM, *S*-adenosyl methionine; FAD, flavin adenine dinucleotide; PARP, poly-ADP-ribose polymerase; HAT, histone acetyltransferase; PRMT, protein arginine methyltransferase; DNMT, DNA methyltransferase; KMT, lysine (K) methyltransferase; KDM, lysine demethylase; TET, ten-eleven translocation protein; LSD, lysine-specific demethylase.

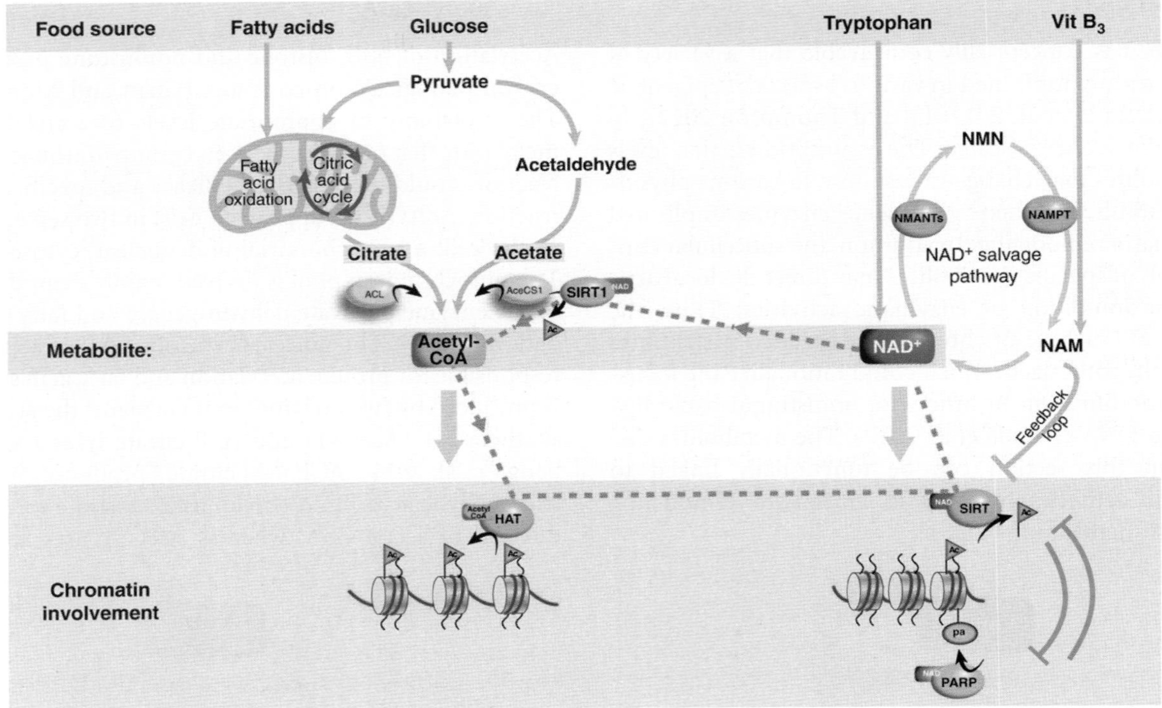

Figure 2. Biosynthesis pathways of acetyl-CoA and NAD cofactors, and their involvement in chromatin-related processes. Acetyl-CoA is produced via two pathways to metabolize pyruvate, involving the key catalytic action of ACL or AceCS1. Acetyl-CoA is an essential metabolite required for the activity of HATs involved in creating an active chromatin conformation via the acetylation of histones. NAD is produced via the NAD salvage pathway. It is an essential cofactor for the PARP and SIRTenzymes, among other proteins. A link exists between the two pathways, as indicated by the dashed lines, by virtue that the NAD-using enzyme, SIRT1, activates the AceCS1 enzyme via protein deacetylation, which, in turn, produces the metabolite acetyl-CoA.

mammals, acetate can be produced by the intestinal flora, alcohol metabolism, prolonged fasting, and class I and II histone deacetylases (HDACs). In *Saccharomyces cerevisiae*, the homolog of AceCS1 is the major source of acetyl-CoA for histone acetylation (Takahashi et al. 2006). ACL and AceCS1 are present in both the cytosol and nucleus of mammalian cells, and the loss of either leads to reduction in global histone acetylation (Wellen et al. 2009; Ariyannur et al. 2010). In addition, reduction in histone acetylation upon loss of ACL can be rescued by supplementing the cells with acetate, suggesting a critical role for AceCS1 in acetyl-CoA biosynthesis, possibly when cells are shifted from glucose- to acetate-driven metabolism. Overall, these studies thereby indicate that changes in the activity of both ACL and AceCS1 and, thus, in the intracellular levels of acetyl-CoA, can influence histone acetylation and, consequently, chromatin remodeling. Additional work has shown that AceCS1 is itself acetylated (Hallows et al. 2006; Sahar et al. 2014), an event controlled in a cyclic manner by the deacetylase SIRT1 (see Sec. 2.1).

Finally, an intriguing twist has linked acetyl-CoA with DNA methylation; the levels of DNA methyltransferase 1 (DNMT1) in adipocytes seem to be controlled in part by ACL, which in turn controls adipocyte differentiation (Londoño Gentile et al. 2013).

1.2 NAD

Nicotinamide adenine dinucleotide (NAD) is a key metabolite involved in a large array of cellular metabolic pathways. It is synthesized from tryptophan or two different forms of vitamin B_3: nicotinic acid and nicotinamide (NAM) (Fig. 2). The tryptophan pathway provides de novo NAD from protein-based nutrients in all life forms. A switch has occurred during evolution, concerning the synthesis of NAD from vitamin B_3, such that lower eukaryotes and invertebrates (including yeast, worms, and flies) use nicotinic acid as the major precursor of NAD, whereas in mammals it is NAM (Magni et al. 2004). The synthesis of NAD in mammals is controlled by an enzymatic loop that is often referred

 Cite as *Cold Spring Harb Perspect Biol* doi: 10.1101/cshperspect.a019463

to as the NAD salvage pathway. NAD is metabolized into NAM by NAD-using enzymes. NAM, subsequently, functions as a potent inhibitor of NAD-dependent enzymes, thereby constituting an enzymatic feedback regulatory mechanism (Magni et al. 2004; Nikiforov et al. 2011). NAM is in turn converted into nicotinamide mononucleotide (NMN) by a single, rate-limiting step enzyme: the NAM-phosphoribosyltransferase (NAMPT) (Garten et al. 2009). The regulation of NAMPT is particularly intriguing and will be discussed later in the context of the circadian clock (Sec. 4.4). Finally, a group of three mononucleotide adenylyltransferases convert the NMN to NAD. Changing levels of NAD may then influence the activity of cellular enzymes in two ways, either because NAD is used as a coenzyme, thereby determining the efficacy of the enzymatic activity, or because the enzymes consume NAD to generate NAM (Magni et al. 2004; Nikiforov et al. 2011).

Two groups of enzymes involved in epigenetic control use NAD as coenzyme or are NAD-consumers (Fig. 2). The first is the class III HDACs. This group of seven proteins, commonly called the sirtuins (pronounced "sir-two-ins") (Finkel et al. 2009), is composed of the mammalian homologs of the yeast Sir2 (silent information regulator 2) (elaborated in Sec. 2.1). The second group of enzymes comprises PARPs, which consume large amounts of NAD to modify proteins by attaching long chains of ADP-ribose (Schreiber et al. 2006; Gibson and Kraus 2012). These two groups of enzymes are discussed in Section 2.2.

1.3 SAM

Methylation of both DNA and histones (as well as nonhistone proteins) requires the metabolite *S*-adenosyl methionine (SAM) as a source of methyl groups. SAM is a common cosubstrate involved in the transfer of methyl groups and is obtained from ATP and methionine by the methionine adenosyltransferases (MATs) enzymes (Fig. 3) (Grillo and Colombatto 2008). The methyl group (CH_3) in SAM is reactive and donated to acceptor substrates by *trans*-methylation. After this reaction, SAM becomes *S*-adenosyl homocysteine (SAH), which is a potent inhibitor of all methyltransferases. A variety of methyltransferases have been characterized and are involved in epigenetic control (Fig. 1), including the DNA methyltransferases (DNMTs), the peptidylarginine methyltransferases (PRMTs), and the lysine (K) methyltransferases (KMTs) (detailed in Ch. 6 [Cheng 2014]). Because ATP is a source for SAM, the intracellular ATP concentration and sublocalization could influence the availability of SAM for methylation reactions. Furthermore, MATII, one of the MATs responsible for SAM synthesis, serves as a transcriptional corepressor in the oxidative stress response. Specifically, MATII directly interacts with the his-

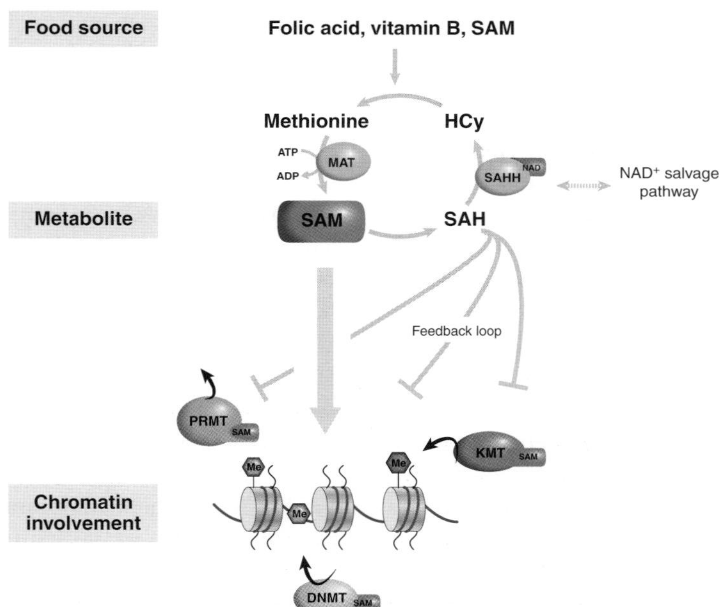

Figure 3. The biosynthesis pathway of *S*-adenosyl methionine (SAM) and its involvement in chromatin-related processes. SAM is an essential cofactor for PRMT, DNMT, and KMT chromatin-modifying enzymes. This pathway is metabolically influenced by the NAD salvage pathway by virtue that the SAH hydrolase (SAHH) enzyme in its biosynthesis pathway uses NAD. The primary product of SAM metabolism is SAH, which has an inhibitory effect on all the SAM-dependent chromatin-modifying complexes. SAH, *S*-adenosyl homocysteine; HCy, homocysteine.

tone H3 at lysine 9 (H3K9) methyltransferase SETDB1, thereby promoting H3K9 trimethylation and repression of the cyclooxygenase 2 gene (Kera et al. 2013), a prostaglandin synthase involved in the inflammatory response.

Various studies have linked nutrition regimes to the pattern of DNA methylation, indicating that food containing folic acid, vitamin B, and SAM can improve a number of pathophysiological conditions. A revealing example relates to the chemical bisphenol A (BPA), used to make polycarbonate plastic, which was reported as unsafe for human health. Pregnant yellow *agouti* mice fed with BPA generate a more yellow, unhealthy progeny that correlates with undermethylation of the *agouti* gene. Feeding pregnant, BPA-treated mothers with methyl-rich foods reverts the negative effect of BPA, and offspring are healthy and brown (Dolinoy et al. 2007). This and additional evidence indicate that diets high in methyl-donating nutrients (such as folic acid, B vitamins, and SAM-e) can alter gene expression, especially during early development when the epigenome is thought to be first established. Conversely, nutritional challenge, such as a high-fat diet regimen, has been shown to induce new oscillations in SAM and SAH levels through a transcriptional and epigenetic reprogramming of the circadian system (Eckel-Mahan et al. 2013), discussed below in Section 4.

1.4 FAD

Riboflavin (vitamin B_2) is an essential component for the enzyme cofactors flavin adenine dinucleotide (FAD) and flavin mononucleotide. As with other B vitamins, riboflavin is required for the metabolism of fats, ketone bodies, carbohydrates, and proteins (Fischer and Bacher 2008). Riboflavin is present in many vegetables and meat. During digestion, the various flavoproteins in food are degraded and riboflavin is resorbed. FAD is a redox cofactor used by several enzymes in the control of metabolic pathways (Macheroux et al. 2011). Amine oxidations are widespread reactions in nature. Flavin-dependent amine oxidases are enzymes that catalyze oxidative cleavage of the C–N bond by two-electron reduction of the FAD coenzyme, which produces an imine intermediate that is then hydrolyzed nonenzymatically. Reduced FAD can be reoxidized by molecular oxygen, which generates hydrogen peroxide and thereby makes the enzyme available for a new catalytic cycle (illustrated in Fig. 1 of Ch. 2 [Shi and Tsukada 2013]).

Different classes of demethylases use different metabolites (Hou and Yu 2010). For example, the JMJC family (e.g., Jumonji C) uses a dioxygenase reaction that is dependent on Fe (II) and α-ketoglutarate for demethylating mono-, di-, and trimethylated residues (see Fig. 1 of Ch. 2 [Shi and Tsukada 2013]). In contrast, the demethylases, lysine-specific demethylase 1 and 2 (LSD1 and LSD2), use a FAD-dependent amine oxidation reaction to demethylate mono- and dimethylated substrates (Fig. 1). LSD1, specifically, removes one or two methyl groups from histone H3 at Lys4 via a FAD-dependent reaction (Metzger et al. 2010). LSD1 is inactive in vitro on H3K4me3, consistent with the flavin-catalyzed amine oxidation reaction that requires a lone pair of electrons on the lysine amino group. Thus, depending on the metabolite, different levels of control for H3K4 methylation are exerted.

LSD1 enzymatic activity depends on protein kinase Cα (PKCα)-dependent phosphorylation. PKCα responds both to calcium and to increasing levels of the lipid second messenger diacylglycerol, possibly linking LSD1-mediated demethylation to these signaling pathways (Metzger et al. 2010). Intriguingly, variable phosphorylation of LSD1 appears to control circadian gene expression by inducing direct interaction of LSD1 with the CLOCK:BMAL1 complex (see also Sec. 4.2; Nam et al. 2014). Finally, the balance of methyl groups toward both histone and DNA methylation may be interdependent on reciprocal modifications, as highlighted by the requirement of LSD1 for the maintenance of global DNA methylation (Wang et al. 2009).

2 ENZYMES

The activity of several classes of enzymes may be controlled by changes in metabolite availability or the use of specific metabolites as coenzymes. Here, we discuss examples of enzymes whose role in chromatin regulation has been explored in greater detail.

2.1 SIRT1 and Other Sirtuins

The class III HDACs are composed of a family of mammalian proteins originally identified by similarity to the yeast gene Sir2. In mammals, the sirtuin family is composed of seven members, some of which are mitochondrial (Sirt3, Sirt4, and Sirt5), others that are principally nuclear (Sirt1, Sirt6, and Sirt7), and others are expressed in more than one cellular compartment (Finkel et al. 2009). Sir2 was initially identified as an NAD^+-dependent deacetylase and linked to the control of longevity (Imai et al. 2000; Schwer and Verdin 2008; Chalkiadaki and Guarente 2012), although the contribution to life span of Sir2 orthologs in various other organisms is debated (Burnett et al. 2011).

The principal mammalian ortholog of Sir2, SIRT1, has emerged as a key component, linking both histone and nonhistone deacetylation to cellular metabolism. SIRT1, as a class III HDAC, differs from the class I and II deacetylases in that it requires NAD^+ as a cofactor for its enzymatic activity, similar to yeast Sir2. SIRT1 breaks down NAD^+

 Cite as *Cold Spring Harb Perspect Biol* doi: 10.1101/cshperspect.a019463

during the process of lysine deacetylation, producing O-acetyl-ADP-ribose. During fasting, levels of NAD^+ are high, and SIRT1 activity is elevated. However, when energy is in excess, NAD^+ is depleted because the rampant flux through the glycolytic cycle promotes the conversion of NAD^+ to NADH. It is generally thought that SIRT1 levels do not change (Nakahata et al. 2008) in response to different physiological states, although its variable expression in some specific physiological settings cannot exclude that possibility. Evidence is clear, however, that enzymatic activity is regulated, in large part, by the availability of the enzyme's cofactor, NAD^+ (Finkel et al. 2009).

SIRT1 HDAC activity is mostly directed to K9ac and K14ac of the H3 tail (Chalkiadaki and Guarente 2012). However, increasing evidence indicates that SIRT1 may be a more efficient deacetylase for nonhistone nuclear proteins rather than histones. Indeed, in the dietary fasted state, SIRT1 affects the activity of numerous target proteins, including many involved directly or indirectly in metabolic homeostasis, such as PGC-1α, FOXO, IRS1/2, LXR, HNF-4α, FXR, RAR, TORC2, BMAL1, eNOS, LKB1, AMPK, and SREBP1 (reviewed in Houtkooper et al. 2012). In most cases, the enzymatic deacetylation reaction has been linked to increased activity of a target protein. So, SIRT1-mediated deacetylation of PGC-1α, for instance, activates the protein promoting gluconeogenic gene transcription and the inhibition of glycolytic gene transcription in the liver (Nemoto et al. 2005; Rodgers et al. 2005).

SIRT1 may also contribute to acetylation of histones and nonhistone proteins, indirectly, by regulating the activity of AceCS1 and thereby acetyl-CoA synthesis (Fig. 2). Indeed, the acetylated form of AceCS1 at Lys-661 is inactive, whereas SIRT1-mediated deacetylation activates the enzyme. Interestingly, SIRT1 is the only sirtuin able to deacetylate AceCS1 (Hallows et al. 2006), and this acetylation is cyclic, controlled by the circadian clock (see more in Sec. 4.3), leading to oscillating levels of acetyl-CoA. Because acetyl-CoA is the carbon source for synthesis and elongation of fatty acids, it is of interest that the clock and AceCS1 contribute to the elongation of long fatty acids into very long fatty acids (Sahar et al. 2014). Thus, there is a direct link between NAD^+, the coenzyme for SIRT1, and acetyl-CoA synthesis to control the acetylation of both histone and nonhistone proteins (Fig. 2, dashed lines). In summary, the genomic function of SIRT1 appears to result from the combination of a variety of converging molecular functions. The first high-throughput transcriptomic analysis in liver-specific *Sirt1*-null mice has shown that SIRT1, together with SIRT6, contributes to genomic partitioning of circadian transcription (Masri et al. 2014).

Another sirtuin that may contribute to epigenetic regulation is SIRT3. This enzyme appears to be the major mitochondrial deacetylase in which it targets the local acetyl-CoA synthetase, lecithin-cholesterol aceyltransferase, and 3-hydroxy-3-methylglutaryl-CoA synthase 2, which controls the levels of ketone body production (Shimazu et al. 2010). SIRT3 may also operate as a critical link between circadian metabolism and aging (Peek et al. 2013). This was intimated by the finding that the majority of clock-controlled acetylation relates to mitochondrial proteins, using an unbiased analysis of the circadian acetylome (Masri et al. 2013). In the liver, SIRT3 appears to generate rhythms in the acetylation and activity of oxidative enzymes and respiration. In hematopoietic stem cells, SIRT3 regulates mitochondrial antioxidative stress required to maintain mitochondrial homeostasis during oxidative stress or aging (Peek et al. 2013). Furthermore, it has been observed that, in conditions of critical oxidative stress, the cell responds by resetting the circadian clock, activating antioxidant pathways through BMAL1, heat shock factor 1, and casein kinase II (Tamaru et al. 2013). Interestingly, SIRT1 also participates in a mitochondrial antiaging function by modulating the nuclear-mitochondrial communication. It appears that the decrease in NAD^+ levels during aging affects SIRT1 activity and thereby the control of nuclear-encoded mitochondrial genes (discussed further in Sec. 5.3). Hence, the decreased levels in NAD^+ might impinge negatively on mitochondrial homeostasis by reducing the activity of sirtuins.

Two other sirtuins that are nuclear and may conceivably participate directly in chromatin remodeling are SIRT6 and SIRT7. Little is known, however, about their deacetylase activity and whether they consume NAD^+ with the same affinity and efficacy as SIRT1. It has been proposed that the three nuclear sirtuins may compete for specific NAD^+ nuclear pools, thus influencing each other's enzymatic activities. Intriguingly, free fatty acids are potent endogenous activators of SIRT6 HDAC activity, but not of SIRT1 (Feldman et al. 2013). Thus, endogenous fatty acids could play a role in activating or sensitizing SIRT6, a notion that is particularly appealing when taking into account that the intranuclear concentration and localization of free fatty acids is likely to be dynamic in response to metabolic and nutritional changes.

SIRT6 is unique in its constitutive localization to chromatin (Mostoslavsky et al. 2006). Its genome-wide occupancy is prominent at the transcriptional start site (TSS) of active genomic loci, coinciding with serine 5 phosphorylated RNA polymerase II binding sites (Ram et al. 2011). SIRT6 binding to chromatin has also been reported to be dynamic in response to stimuli, such as tumor necrosis factor α, subsequently, resulting in the alteration of the transcriptional landscape of aging and stress-related genes (Kawahara et al. 2011). SIRT6, specifically, deacetylates

H3K9 (Michishita et al. 2008; Kawahara et al. 2009) and H3K56 (Michishita et al. 2009; Yang et al. 2009; Toiber et al. 2013) in a nucleosome-dependent manner (Gil et al. 2013), modulating gene expression and affecting telomere maintenance and genomic stability. Also, SIRT6 interacts with transcription factors, such as NF-κB and HIF1α (Kawahara et al. 2009; Zhong et al. 2010), and contributes to their targeting to gene promoters. A full-scale analysis by high-throughput transcriptomics and metabolomics in the liver has revealed that SIRT6 contributes to genomic partitioning of circadian transcription, which leads to segregated control of cellular metabolism (Masri et al. 2014).

Physiologically, SIRT6 has been implicated in metabolic regulation, as *Sirt6*$^{-/-}$ mice die at 2–4 wk of age because of a severe accelerated-aging phenotype and hypoglycemia. Lethality is a result of altered rates of glycolysis, glucose uptake, and mitochondrial respiration (Mostoslavsky et al. 2006; Xiao et al. 2010; Zhong et al. 2010). SIRT6 also controls the acetylation state of PGC-1α in a GCN5-dependent manner, which regulates blood glucose levels (Dominy et al. 2010). Liver-specific *Sirt6*$^{-/-}$ mice develop fatty liver as a result of altered expression of genes involved in fatty acid β-oxidation and triglyceride synthesis (Kim et al. 2010). The SIRT6-targeted deacetylation of H3K9 is an event that has been associated with the down-regulation of cell senescence and apoptotic pathways, further described in Section 5.1.

SIRT7 was first described as being highly concentrated at the level of the nucleolus, involved in ribosomal DNA transcription and interacting with the UBF transcription factor (Grob et al. 2009). This apparently limited function was challenged by the added finding that SIRT7 controls H3K18 acetylation in a NAD-dependent manner, which, in turn, leads to the regulation of genes involved in oncogenic transformation (Barber et al. 2012). Additional findings have confirmed the extranucleolar role of SIRT7. Functional proteomics revealed that SIRT7 interacts with chromatin-remodeling complexes and components of RNA polymerase I machinery (Tsai et al. 2012). Indeed, in the liver, SIRT7 functions at chromatin to suppress endoplasmic reticulum (ER) stress to protect from fatty liver disease. Consistent with this, SIRT7-deficient mice develop chronic hepatosteatosis, in part because SIRT7 acts as a cofactor of MYC-mediated transcriptional repression (Shin et al. 2013).

2.2 PARPs

Poly(ADP-ribosyl)ation is a PTM of proteins, which is catalyzed by PARPs. The founding member of the PARP family, poly(ADP-ribose) polymerase-1 (PARP-1), has been well characterized. The PARP family is composed of at least 18 members, in large part nuclear, which are linked to various genomic functions, including DNA repair and programmed cell death (Schreiber et al. 2006; Gibson and Kraus 2012). PARPs hydrolyze NAD^+ and transfer the ADP-ribose moiety to acceptor proteins. Many cellular proteins, including histones, are acceptors of ADP-ribose and can be mono- or poly(ADP-ribosyl)ated. The use of NAD^+ by the PARP enzymatic process can be quite massive, thereby reducing the availability of NAD^+ for other enzymes; for example, PARP-1 activity depletes the NAD^+ pool, thus reducing SIRT1 activity, which causes cell death (Kolthur-Seetharam et al. 2006; Bai et al. 2011). Owing to its role in DNA repair, PARP inhibitors have become attractive drugs against DNA damage-response defective tumors, such as BRCA1/2 mutant cancers, described in Section 1.3.2 of Ch. 35 (Audia and Campbell 2014). Thus, PARP enzymes may have an indirect role in chromatin remodeling through regulation of sirtuins.

In contrast, PARPs may be directly involved in epigenetic regulation via other mechanisms. For example, PARP-1 may be tethered at CpG regions bound by the insulator protein CTCF, activating PARP's own auto-ADP-ribosylation. The resulting ADP-ribose polymers then interact in a noncovalent way with DNMT1 to inhibit its methylation activity, ensuring the methylation-free status of CTCF bound promoters (Reale et al. 2005; Guastafierro et al. 2008; Zampieri et al. 2012). This mechanism may ensure that a locus' promoter region (e.g., of the Dnmt1 housekeeping gene) (Yu et al. 2004) or imprinted control region (e.g., of the Igf2 gene cluster) (Zampieri et al. 2009) are DNA methylation free.

Evidence also exists for PARP-1-mediated poly(ADP-ribosyl)ation of several core histones (at positions K13 of H2A, K30 of H2B, K27 and K37 of H3, and K16 of H4). Inhibitory cross talk with other histone modifications, such as acetylation, has been suggested (Beneke 2012). In particular, acetylated H4K16 appears to inhibit PARP-1-mediated ADP-ribosylation (Messner et al. 2010). Intriguingly, PARP-1 has also been found to be a structural component of chromatin and to modulate the enzymatic activity of various other regulators, such as the histone demethylase KDM5B (Krishnakumar and Kraus 2010). The complexity of poly(ADP-ribosyl)ation and the multitude of enzymes able to elicit this reaction has impeded a full understanding of the pathways through which PARPs may direct specific and functionally relevant chromatin-remodeling events.

3 ALTERED INTERMEDIARY METABOLISM REGULATES EPIGENETIC STATE

A key concept underlying the role that chromatin plays in regulating cell physiology and maintaining homeostasis is

 Cite as *Cold Spring Harb Perspect Biol* doi: 10.1101/cshperspect.a019463

that nuclear activity requires coordination with, and response to, the overall cellular metabolic state. However, nuclear processes involving DNA-associated transactions on chromatin are segregated in location and regulation from cytoplasm-based intermediary metabolic enzymes. An emerging idea to reconcile this separation is that alterations in metabolism may directly influence acetyl-CoA, SAM, NAD, and FAD levels in the nucleus, which may then, in turn, alter DNA and histone modifications and hence transcription (Katada et al. 2012). As described below, there is abundant evidence indicating that metabolic state alters histone acetylation, in particular, both during normal growth cycles and as a result of altered metabolic state in disease, such as cancer. However, although histone and DNA methylation can be altered in disease via mutations in enzymes occupying metabolic pathways, the question of whether normal growth cycles actually alter the activity of methylation and demethylation enzymes remains.

3.1 Histone Acetylation Is Regulated by Metabolic State

Initial observations showing the agile histone acetylation response to growth were made in the yeast, *S. cerevisiae.* A major pathway of acetyl-CoA generation is via the acetyl-CoA synthetases Acs1 and Acs2 (or AceCS1 in mammals), which are enzymes that hydrolyze ATP to catalyze the ligation of acetyl-CoA from acetate and CoA (Fig. 2). Although long considered to be exclusively cytoplasmic, the enzymes also localize to the nucleus (Takahashi et al. 2006). Moreover, impairment of Acs2 globally lowered levels of acetylated histone H3 and H4 tails, and showed remarkably broad decreases in gene expression occurring in specific pathways (Takahashi et al. 2006). These discoveries provided a direct link between enzymes that produce acetyl-CoA to nuclear chromatin regulation and gene expression.

Observations, again in *S. cerevisiae*, established that the nutrient environment/metabolic state can lead to altered global histone acetylation, namely, that prolonged quiescence during stationary phase leads to decreased histone acetylation (Sandmeier et al. 2002; Ramaswamy et al. 2003). A direct link was shown when elevated glucose levels (the preferred carbon source for yeast) used to stimulate growth were accompanied by the direct metabolic induction of histone acetylation. This acetylation occurred broadly over the genome, catalyzed by the histone acetyltransferases (HATs), Gcn5 and Esa2, in association with their protein complexes, SAGA and picNuA4, respectively, functioning in an untargeted manner (Friis et al. 2009).

A related analysis of intrinsic yeast metabolic cycling provided more insight into histone acetylation fluxes. Yeast metabolic cycling occurs when cells are starved at high cell density and, then, are continuously supplemented with low concentrations of glucose. A synchronous cycle is established in the population, consisting of a growth period accompanied by oxidation, followed by a period of cessation of growth while reductive, building processes occur. Under these conditions, transcription occurs in waves, with up-regulation of growth-related genes followed by induction of stress-related genes, within each respective period (Tu et al. 2005). Interestingly, during the growth portion of the metabolic cycle, a critical metabolite is acetyl-CoA. This transient pulse of acetyl-CoA leads to increased activity of the SAGA-Gcn5 complex and both a transient global increase of acetylation at its substrate sites in histone H3 and a transient targeted increase specifically at growth-related genes (Cai et al. 2011).

These observations of metabolic signaling to directly alter histone acetylation have been extended to mammalian cells. Glucose is the major carbon source for mammalian cells and it can be used by ACL to produce acetyl-CoA (Fig. 2). A small proportion of the ACL enzyme localizes to the nucleus in several mammalian cell types in culture, and small interfering RNA-induced silencing of ACL led to a decrease in global histone acetylation. Indeed, in adipocyte differentiation, ACL is critical for inducing histone acetylation at specific genes and activating transcription (Wellen et al. 2009).

What might be the mechanism by which HATs respond to altered acetyl-CoA concentrations? One possible mechanism is that the binding affinity of the enzymes for acetyl-CoA is in the range of in vivo fluctuations of acetyl-CoA concentrations. Thus, the measured acetyl-CoA concentration during the yeast growth oscillatory cycle (~3–30 mM) (Cai et al. 2011) bracket the in vitro measured K_d and K_m of Gcn5 (~8.5 and ~2.5 mM) (Berndsen and Denu 2008). Further speculation is that nuclear localization of the acetyl-CoA-producing enzymes may be induced during growth phases and may directly provide acetyl-CoA to the histone-acetylating enzymes.

Important future questions include determining how far reaching are these mechanisms of increasing activity or driving acetyl-CoA-producing enzymes to localize to the nucleus for directly influencing chromatin modifications. Is this pathway used only to respond to nutrients, or can this pathway be used to respond to cell differentiation (Kaochar and Tu 2012)? In this respect, a new study showed that nutritional challenge, in the form of a high-fat diet in the mouse, leads to a reprogramming of circadian gene expression, using a shift in chromatin remodeling, specifically in H3K9/K14 acetylation. Remarkably, at the metabolic level, it is clear that the shift in histone modifications is paralleled by changes in the levels of NAD^+ and acetyl-CoA (Eckel-Mahan et al. 2013).

3.2 Connections of Metabolism to Histone Methylation

The profound changes in acetyl-CoA levels, histone acetylation, and transcription lead to the question of whether levels of metabolites associated with methylation, such as SAM, are similarly altered in the nucleus in response to the nutrient state to then alter DNA and histone methylation (Teperino et al. 2010). A massive response to metabolic change via the up-regulation of histone methylation, as described above for histone acetylation, has not been reported. There are, however, interesting examples, cited below, of altered histone methylation caused by more restricted alterations in metabolic regulation.

The high-energy methyl donor, SAM, is generated from methionine by MAT enzymes in an ATP-consuming reaction. Interestingly, in mammalian cultured immune T cells, MAT has been shown to directly associate with repressed genes, leading to chromatin-localized synthesis of SAM, and fine-tuning of histone methylation and gene down-regulation (Katoh et al. 2011).

Mouse stem cells represent an interesting example of altered histone methylation in response to metabolic adaptation. In general, histone methylation appears to have a critical function in transcriptional regulation and self-renewal in stem cells (Ang et al. 2011; Liang and Zhang 2013). Mouse embryonic stem cells (mESCs) have a specific mode of threonine catabolism, which affects SAM metabolism and H3K4me3, and is crucial for maintaining pluripotency (Shyh-Chang et al. 2013). Culturing proliferating mESCs in threonine-depleted media, therefore, leads to a lowering of SAM and H3K4me3 levels, but not other residue methylation, which slows growth and increases differentiation. However, threonine dependency is specific to mESCs; hence, it is unclear whether histone methylation is similarly influenced by metabolic activities in other stem cell populations, such as human ESCs or adult human stem cells. Furthermore, it remains unknown whether, akin to histone acetylation, histone methylation is a broad energy sensing modification that adapts transcription to the overall metabolic state.

3.3 Oncogenic Mutations in Metabolic Pathways Alter Epigenetic Methylation State

Although it is not clear whether histone methylation responds to normal fluctuations in growth and ranges of nutrient levels, the activity of methyltransferase enzymes can be dramatically altered as a result of mutations in key metabolic pathways. There are leukemic and other malignancies that result from oncogenic gain-of-function mutations in isocitrate dehydrogenase enzymes, IDH1 and IDH2 (Figueroa et al. 2010; Lu et al. 2012; Turcan et al. 2012). Large amounts of 2-hydroxyglutarate are produced from α-ketoglutarate in certain isocitrate dehydrogenase (IDH) mutant enzymes. α-ketoglutarate is a cofactor for DNA ten-eleven translocation (TET) enzymes (which convert 5meC to further oxidized forms, as described in Ch. 15 [Li and Zhang 2014]) and histone lysine demethylases (KDMs); thus, high levels of 2-hydroxyglutarate act to repress the TET and KDM enzymes. Consequently, the IDH1/2 mutations lead to increased levels of DNA methylation and histone lysine methylation and altered transcription (Fig. 4). One model to explain the physiology is that growth-promoting genes have reduced levels of repressive methylation, causing increased transcription, which leads to the oncogenic state. Based on these observations, there may be other cancers that result specifically from mutations or altered activity of metabolic enzymes that change epigenetic regulation.

4 THE CIRCADIAN EPIGENOME

There are certain major areas of physiology that are exquisitely reactive to environmental signals and can be gauged by the levels of metabolites and other entities. Circadian rhythms are one area of physiology in which an organism responds to external cycles, including nutrients, via establishing internal cycles. The circadian response largely involves transcriptional regulation and epigenetic mechanisms, as discussed in detail in this section. A second major organismal response to the environment is in the related areas of aging and senescence, discussed in Section 5.

4.1 The Circadian Clock

A wide variety of physiological functions, including sleep–wake cycles, body temperature, hormone secretion, locomotor activity, and feeding behavior depend on the circadian clock—a highly conserved system that enables organisms to adapt to common daily changes, such as the day–night cycle and food availability (Reppert and Weaver 2002; Sahar and Sassone-Corsi 2012). Based on evidence accumulated over several decades, circadian rhythms represent possibly the most prominent area of physiology that has effectively been studied using systems biology.

In mammals, the anatomical structure in the brain that governs circadian rhythms is a small area consisting of approximately 15,000 neurons localized in the anterior hypothalamus, called the suprachiasmatic nucleus (SCN). This "central pacemaker" in the SCN receives signals from the environment and, in turn, coordinates the oscillating activity of peripheral clocks, which are located in almost all tissues (Schibler and Sassone-Corsi 2002). This highly orchestrated network is based on cascades of signal-

Cite as *Cold Spring Harb Perspect Biol* doi: 10.1101/cshperspect.a019463

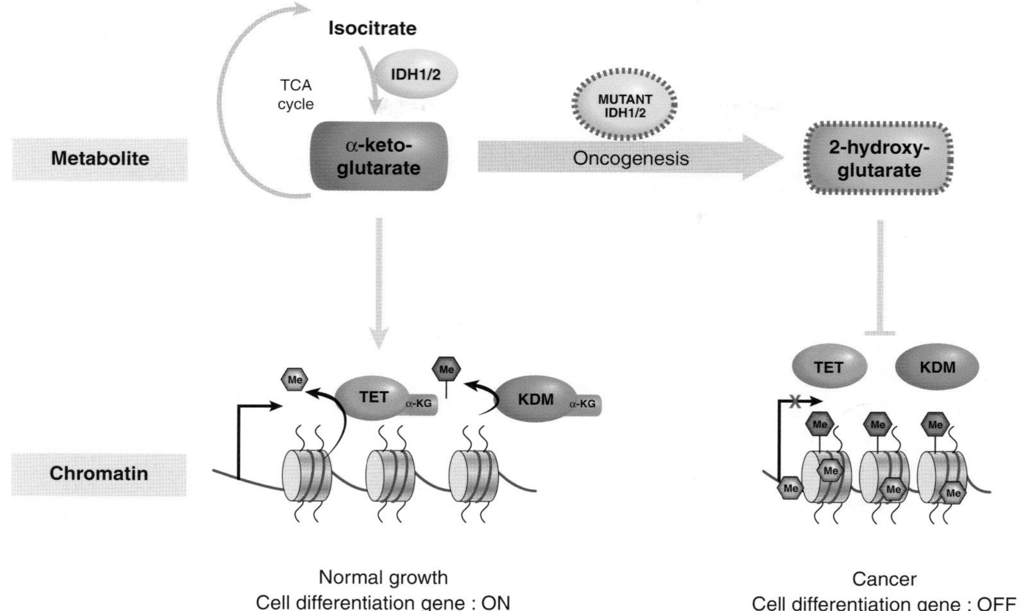

Figure 4. Metabolism of the α-ketoglutarate cofactor regulates the activity of the TET and KDM enzymes in normal and cancer cells. The α-ketoglutarate metabolite is produced by the enzymatic action of the IDH proteins. Some cancer driving dominant-negative mutants of IDH cause the accumulation of an aberrant metabolite, 2-hydroxy-glutarate, which blocks TET and KDM activity.

ing pathways that lead to the activation of transcriptional programs, involving a remarkable fraction of the genome. Transcriptome studies indicate that ~10% of all transcripts oscillate and belong to the category of clock-controlled genes (CCGs). This fraction is actually much higher because different tissues express different CCGs (Panda et al. 2002; Masri and Sassone-Corsi 2010). A study covering 14 mouse tissues identified approximately 10,000 of the total almost 25,000 known protein-coding genes as showing circadian oscillations in at least one tissue. The number of common genes showing circadian oscillation in multiple tissues, however, decreases drastically as the number of tissues included in the comparative analysis is increased; that is, only 41 genes displayed circadian oscillation in at least eight out of 14 tissues (Yan et al. 2008).

The capacity for genes to be regulated in a circadian manner appears to depend, in part, on the metabolic state of a given tissue. For example, nutritional challenge in the form of a high-fat diet reprograms the circadian transcriptome in the liver so that normally noncyclic transcriptional and epigenetic regulatory programs become cyclically activated (Eckel-Mahan et al. 2013). Thus, a portion of genes larger than originally thought is intrinsically capable of becoming regulated in a circadian manner under the appropriate metabolic setting.

The circadian expression program is governed by interlocking transcription–translation feedback loops, a system that is conceptually conserved among species (Reppert and Weaver 2002). At the heart of the circadian clock molecular network are the core transcription factors CLOCK and BMAL1. These two proteins heterodimerize and direct transcriptional activation of CCGs by binding to E-box sites within their promoters (Fig. 5A). Among these CCGs, CLOCK and BMAL1 also direct transcription of their own repressors, period (PER) and cryptochrome (CRY) family members, creating a tightly self-regulated system. During the day, transcription of PER and CRY is high, leading to protein translation of the circadian repressors and resulting in formation of the inhibitory complex with CLOCK and BMAL1 that abolishes transcription of CCGs (Fig. 5B). The degradation of PER and CRY by nighttime alleviates transcriptional repression and allows CLOCK:BMAL1-mediated transcription to again proceed, establishing an oscillatory rhythm in circadian gene expression (Fig. 5C). An additional level of circadian regulation exists via the orphan nuclear receptors RORα and REV-ERBα, which activate and repress transcription of the *Bmal1* gene, respectively (Fig. 5D) (Reppert and Weaver 2002; Sahar and Sassone-Corsi 2012).

The cyclic transcriptional program invokes a similarly cyclic, plastic remodeling of chromatin organization at CCGs, with the need to be strictly timed to ensure the appropriate anticipation of the next cycle of gene expression. The first demonstration that chromatin remodeling is

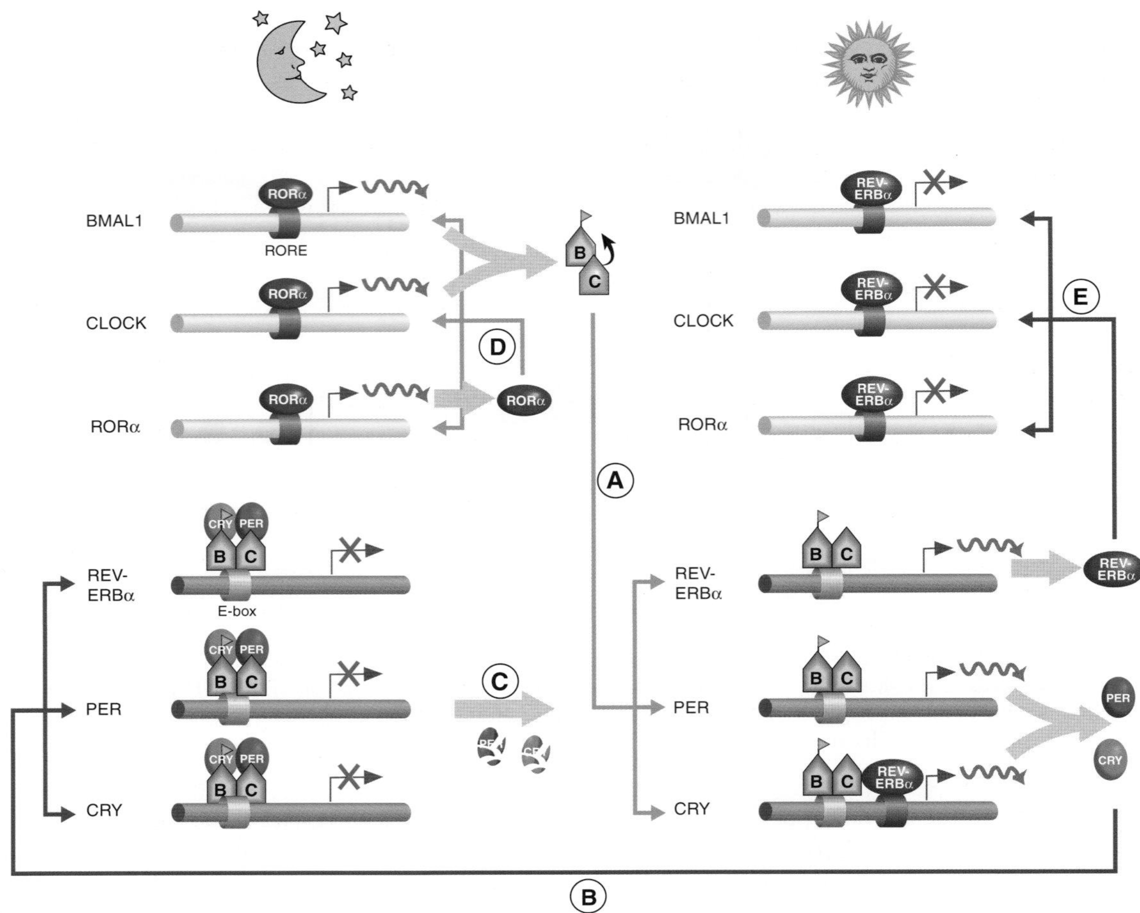

Figure 5. A simplified scheme of the timing of key regulator activity in the molecular circadian clock machinery. (*A*) A maximum accumulation of CLOCK:BMAL1 heterodimers is achieved by daybreak following nighttime transcription and translation. CLOCK:BMAL1 binding to E-box elements of clock-controlled genes (CCGs) leads to chromatin remodeling and activation of genes. (*B*) The daytime expression of the repressors, PER and CRY, partially through the acetylated BMAL1 K537 residue, leads to CCG transcriptional repression at night. (*C*) The nighttime degradation of PER/CRY repression gradually leads to the derepression of CCGs by daybreak. (*D*) At nighttime, the predominance of RORα binding to the RORE (retinoic acid–related orphan receptor response element) lends to transcription of BMAL1, CLOCK, and RORα. (*E*) The predominance of REV-ERBα, as a result of the gene's daytime transcription, has an inhibitory action on BMAL1, CLOCK, and RORα.

indeed implicated in circadian gene expression came from in vivo studies in SCN neurons; circadian gene transcription was associated with the induction of H3S10 phosphorylation in response to light, via the GABA-B pathway (Crosio et al. 2000). Subsequently, other histone modifications were found to display oscillatory profiles at CCG promoters (Etchegaray et al. 2003; Doi et al. 2006; Ripperger and Schibler 2006; Masri and Sassone-Corsi 2010), and genome-wide profiling has revealed coordinated kinetics of circadian transcription with cyclic histone modifications (Koike et al. 2012; Le Martelot et al. 2012). Chromosome conformation capture 4C technology (described in Ch. 19 [Dekker and Misteli 2014]) has further revealed that circadian genes are physically organized in nuclear interactomes. Remarkably, the formation and disassembly of a given interactome correspond to peaks and troughs of cyclic transcription for the genes constituting the interactome (Aguilar-Arnal et al. 2013).

4.2 CLOCK Is a HAT, and Other Regulators

The search for a specific element that could contribute to circadian chromatin remodeling revealed that CLOCK, the master regulator of circadian transcription, has intrinsic HAT activity (Doi et al. 2006). CLOCK has a MYST (Moz, Ybf2/Sas3, SAS2, Tip60)-type of acetyl-CoA binding domain, although its general structure resembles that of ACTR (acetyltransferase amplified in breast cancer) (Chen et al. 1997). The HAT function of CLOCK is linked to circadian transcriptional activation and preferentially targets H3K9 and K14. In addition, it acetylates its own transcriptional partner, BMAL1, at a unique lysine residue in position 537, an event essential for circadian rhythmicity, described in more detail in Section 4.3 (Fig. 5A) (Hirayama et al. 2007). Another nonhistone target of CLOCK is the glucocorticoid receptor (GR), a finding with intriguing implications because GR-dependent transcription has vast effects on cellular metabolism (Nader et al. 2009).

Several other remodelers have been implicated in circadian chromatin transitions (Fig. 6). Both cAMP response element binding-binding protein (CBP) and p300 are thought to participate in rhythmic histone acetylation and associate with the CLOCK:BMAL1 complex (Curtis et al. 2004). Furthermore, cyclic methylation at H3K27 has been linked to the KMT, EzH2, although the contribution of this specific enzyme to circadian transcription is not established (Etchegaray et al. 2006). Also, a search for components of the PER complex has revealed the HP1γ-Suv39h methyltransferase, suggesting that some of the repressive action of the PER proteins is linked to H3K9 methylation (Duong and Weitz 2014).

The critical role of H3K4 methylation in transcriptional activation has pointed to this site as potentially key for circadian gene expression. Remarkably, mono- and di-

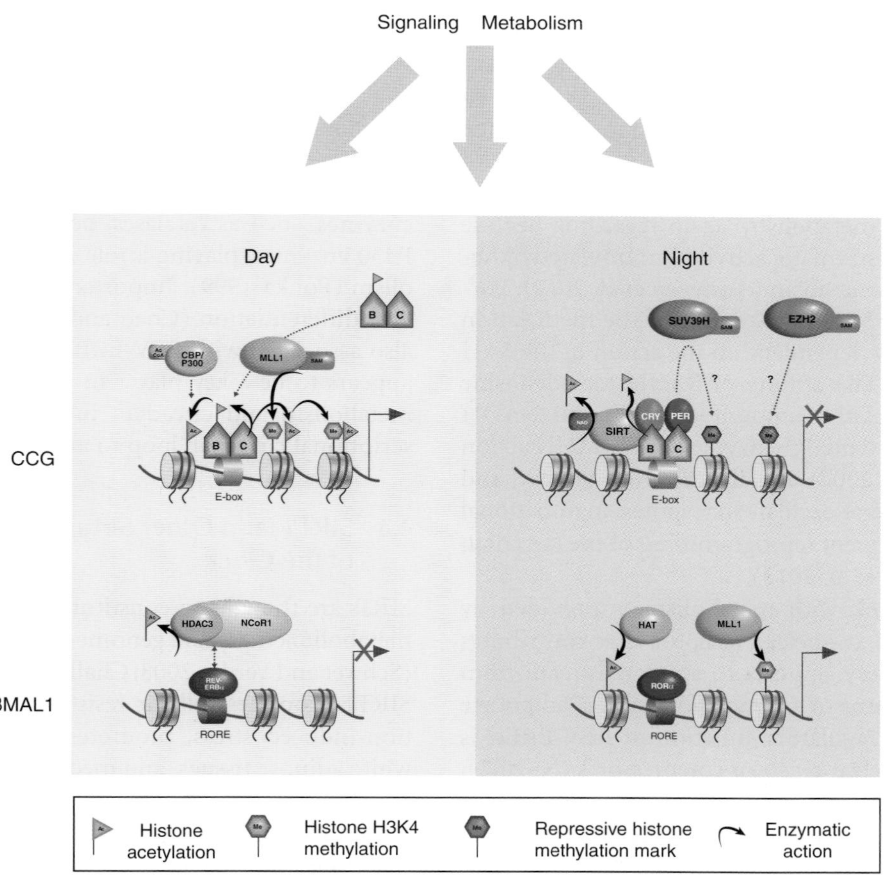

Figure 6. Key chromatin-remodeling factors involved in the activation and repression of CCGs and BMAL1. Recruitment of the CLOCK:BMAL1 heterodimer to CCGs during the day is facilitated by binding to the E-box and via the MLL1 enzyme. A transcriptionally active chromatin conformation is achieved through the combined action of H3K4me3, via the MLL1 lysine methyltransferase enzyme, and the histone acetylating activity of CLOCK, CBP, and P300 (cyan shaded proteins). Repressive chromatin conformations at CCGs may be achieved by the SIRT1-mediated deacetylation of histones at H3K9, K14, and H4K16, and possibly the actions of EZH2-catalyzed H3K27me3 or SUV39H-catalyzed H3K9 methylation. SIRT1 also has an inhibitory effect by deacetylating BMAL1 K537-ac. Silencing of RORE-containing genes, such as BMAL1, by day is, in part, achieved by recruitment to the RORE-bound inhibitor, REV-ERBα, of the NCoR1 complex and associated HDAC3 deacetylase. At night, predominance of RORα-bound ROREs promotes active chromatin structure, typified by histone acetylation and H3K4 methylation.

methylation is not cyclic at CCGs promoters, whereas H3K4 trimethylation cycles (Katada and Sassone-Corsi 2010). H3K4me3 is mediated by MLL1 and favors recruitment of the CLOCK:BMAL1 complex to chromatin to induce subsequent K9/K14 acetylation. Cells lacking the *mll1* gene display a drastically impaired circadian cycle (Katada and Sassone-Corsi 2010). Although MLL1 is a central KMT for the circadian system, more than one demethylase has been implicated. First, the histone KDM, Jarid1A, is involved in circadian control, although not by directly counteracting MLL1 nor by demethylating H3K4 (DiTacchio et al. 2011). Rather, Jarid1A is proposed to block HDAC1 to enhance CLOCK:BMAL1-mediated activation, although the precise role of HDAC1 in clock function is not fully established. A second lysine-specific demethylase, LSD1, associates with CLOCK:BMAL1 in a phosphorylation-dependent manner and participates in their recruitment (Nam et al. 2014). As both Jarid1A and LSD1 target H3K4, it is still unclear whether they act on the circadian epigenome in concert or on distinct genomic subdomains or at different times of the circadian cycle.

Further links have been established between circadian control and cellular metabolism, again regarding histone methylation. MLL1 enzymatic activity is modulated by xenobiotic and lipid metabolism (Austenaa et al. 2012). Furthermore, the SAM/SAH ratio, and thus the methylation potential, is critically dependent on the action of the SAH hydrolase (SAHH). The affinity of SAHH for adenosine is affected by the relative amount of NAD and NADH because NAD is an essential coenzyme for the SAHH action (see Fig. 3) (Li et al. 2007). Finally, the levels of SAM and SAH have been found to oscillate in response to nutritional challenge and consequent reprogramming of the circadian clock (Eckel-Mahan et al. 2013).

An intriguing link with metabolism is provided by HDAC3. REV-ERBα, a nuclear receptor that contributes to the clock machinery, appears to act downstream from PPARγ, a key regulator of fat metabolism and adipocyte differentiation. The regulatory function of REV-ERBα is controlled by the nuclear receptor corepressor 1 (NCoR1). This is a corepressor that recruits the HDAC3 histone deacetylase to mediate transcriptional repression of target genes, such as *Bmal1*. When the NCoR1-HDAC3 association is genetically disrupted in mice, circadian and metabolic defects develop. These mice show a shorter oscillation period, increased energy expenditure, and are resistant to diet-induced obesity. Moreover, there is alteration of cyclic expression in liver lipid metabolism. HDAC3 recruitment to the genome is rhythmic (high during the day and low at night). At these HDAC3-binding sites, REV-ERBα and NCoR1 recruitment are in phase with HDAC3 recruitment, whereas histone acetylation and RNA polymerase II recruitment are antiphasic (Alenghat et al. 2008). Interestingly, genes involved in lipid metabolism in the liver also appear to be the major targets of HDAC3 and REV-ERBα. Depletion of either HDAC3 or REV-ERBα causes fatty liver phenotype, such as increased hepatic lipid and triglyceride content (Alenghat et al. 2008; Feng et al. 2011).

Additional reports show that heme functions as a ligand for REV-ERBα (Raghuram et al. 2007; Yin et al. 2007). Heme binding increases the thermal stability of REV-ERBα and enhances its interaction with the corepressor complex and, hence, is required for its repressor function. Requirement for heme places REV-ERBα in a pivotal position in the regulation of circadian rhythms and metabolism because, importantly, the circadian clock controls cellular heme levels. Aminolevulinate synthase 1 (ALAS1), the rate limiting enzyme in heme biosynthesis, is expressed in a circadian manner and is a specific target gene for the NPAS2 (a paralog of CLOCK)/BMAL1 heterodimer (Yin et al. 2007). Heme, in turn, binds to NPAS2 and inhibits its transactivation ability (Dioum et al. 2002). ALAS1 expression is also regulated by PGC-1α. The metabolic roles attributed to heme are well known; heme is a cofactor for enzymes, such as catalases, peroxidases, and cytochrome P450 enzymes, playing a role in oxygen and drug metabolism (Ponka 1999). Importantly, heme promotes adipocyte differentiation (Chen and London 1981), a function also associated with REV-ERBα and BMAL1. Thus, heme appears to be a key player in the precise coordination of metabolism and circadian function, connecting a transcriptional circadian loop to an enzymatic pathway.

4.3 SIRT1 (and Other Sirtuins) as a Rheostat of the Clock

SIRTs are thought to constitute a functional link between metabolic activity and genome stability and, possibly, aging (Schwer and Verdin 2008; Chalkiadaki and Guarente 2012). SIRT1 promotes cellular resistance to oxidative or radiation-induced stress, promotes mobilization of fat from white adipose tissues, and mediates the metabolism of energy sources in metabolically active tissues (Schwer and Verdin 2008). Thus, the finding that SIRT1 acts as a "rheostat" to modulate CLOCK-mediated acetylase activity and circadian function established an intriguing molecular link (Asher et al. 2008; Nakahata et al. 2008). SIRT1 associates with CLOCK and is recruited to circadian promoters, whereas *Sirt1* genetic ablation or pharmacological inhibition of SIRT1 activity leads to significant disturbances in the circadian cycle. Importantly, although the protein levels of SIRT1 do not oscillate, as analyzed in several tissues and under various experimental conditions (Nakahata et al. 2008; Ramsey et al. 2009), its enzymatic activity oscillates

Cite as *Cold Spring Harb Perspect Biol* doi: 10.1101/cshperspect.a019463

in a circadian manner with a peak that corresponds to the lowest level of H3 acetylation at various CCGs (Nakahata et al. 2008; Bellet et al. 2013). This may also be caused by a role of SIRT1 in controlling the stability of some clock proteins, such as PER2 (Asher et al. 2008). Thus, it is possible that SIRT1 contributes to transducing signals originated by cellular metabolites to the circadian machinery. Further evidence comes from the observed circadian oscillation of NAD^+ levels in serum-entrained mouse embryonic fibroblasts and mice livers (Nakahata et al. 2009; Ramsey et al. 2009).

Sirtuins, and SIRT1 in particular, establish functional interactions with other proteins implicated in cellular metabolism and signaling (Chalkiadaki and Guarente 2012). At the level of chromatin, the SIRT1 enzymatic activity preferentially targets histone H3K9ac, H3K14ac, and H416ac (Nakahata et al. 2008). In addition, a number of nonhistone proteins are regulated by SIRT1-mediated deacetylation, including p53, FOXO3, PGC-1a and LXR, underlining SIRT1's pivotal function in cellular control. SIRT1 also deacetylates BMAL1, contributing to its cyclic acetylation levels (Fig. 6A) (Hirayama et al. 2007). An antibody that recognizes BMAL1 acetylated at the unique Lys-537 (K537) has been used in a variety of settings (Hirayama et al. 2007; Nakahata et al. 2008; Chang and Guarente 2013). The BMAL1 K537 antibody shows that acetylation is dependent on CLOCK and highly specific, as shown using a K537R mutant. Using this antibody, it was shown that BMAL1 deacetylation occurs in the presence of both SIRT1 and NAD^+ and is inhibited by NAM (Nakahata et al. 2008). Also, BMAL1 acetylation is circadian and acetylation controls the efficient repression of BMAL1 by CRY proteins (Hirayama et al. 2007). These finding were corroborated by quantitative analyses of the BMAL1–CRY interaction by fluorescence polarization and isothermal titration calorimetry (Czarna et al. 2011). Also, comparative brain immunostaining between wild-type and brain-specific *Sirt1*-deficient mice showed the dependence on SIRT1 for Lys-537 deacetylation of BMAL1 (Chang and Guarente 2013).

An interdependence had been reported between SIRT1 and the nutrient-responsive adenosine monophosphate-activated protein kinase (AMPK) that contributes to the metabolic adaptation to fasting and exercise in skeletal muscles (Cantó et al. 2010). Importantly, AMPK plays a role in the control of circadian rhythms by phosphorylating and inducing the destabilization of CRY1 in a cyclic manner (Lamia et al. 2009). Stimulation of AMPK signaling alters circadian rhythms and mice in which the AMPK pathway is genetically disrupted show alterations in peripheral clocks. Thus, phosphorylation by AMPK enables CRY1 to transduce nutrient signals to circadian clocks in mammalian peripheral organs. How SIRT1, and possibly NAD^+ oscillations, may be involved in this control system is still unclear.

A comprehensive analysis of the roles played by SIRT1 and SIRT6 in circadian function by high-throughput transcriptomics in individual liver-specific ablation of these sirtuins showed that SIRT1 and SIRT6 control separate subdomains of the circadian epigenome (Masri et al. 2014). Analysis by metabolomic profiling showed that, whereas SIRT1 controlled metabolites associated with peptides and cofactors, SIRT6 directed the oscillation of lipids and carbohydrates (Masri et al. 2014). This corresponds with the finding that SIRT6 responds more efficiently to long-chain fatty acids rather than NAD (Feldman et al. 2013).

4.4 NAD^+, an Oscillating Metabolite

The involvement of SIRT1 in circadian regulation showed a direct link between cyclic rhythms and energy metabolism in the cell (Sahar and Sassone-Corsi 2012). Given that the expression levels of the SIRT1 gene and protein are noncyclic (Nakahata et al. 2008), a major question has been: How can two regulators (CLOCK and SIRT1) that are mostly expressed in a nonoscillatory manner lead to the circadian regulated acetylation of histone H3K9/K14 and BMAL1K537? This apparent discrepancy was solved by subsequent studies, which revealed that NAD^+ levels oscillate in a circadian fashion in all cell types analyzed, and it is through the cyclic availability of its own coenzyme that SIRT1 HDAC activity is circadian (Nakahata et al. 2009; Ramsey et al. 2009). The circadian regulation of NAD^+ synthesis is itself conceptually remarkable as it links the transcriptional feedback loop of the circadian clock to the NAD^+ salvage-dependent enzymatic pathway.

The heavy use of the NAD^+ coenzyme by enzymes, such as SIRT1 and PARP-1 (Fig. 1), risks depleting the intracellular stores, which can lead to cell death. Thus, levels of NAD^+ need to be controlled, even in the absence of de novo biosynthesis, through nutritional pathways. Production of the NAMPT enzyme is the rate-limiting step within the NAD^+ salvage pathway (Nakahata et al. 2009; Ramsey et al. 2009), thus, changes in NAMPT activity directly dictate the levels of intracellular NAD^+ (Fig. 2).

The rhythmicity of NAD^+ levels, which parallels the antiphasic oscillation in NAM, is governed by the circadian clock, as proven by the abolishment in rhythmicity in cells with a mutation in the circadian machinery (Nakahata et al. 2009). Because SIRT1 associates with and modulates CLOCK:BMAL1 (Nakahata et al. 2008), the oscillation of both SIRT1 activity and NAD^+ levels via the NAD^+ salvage pathway was suspected to occur via an enzymatic feedback loop. This was shown to, indeed, be the case. First, the regulatory region of the *Nampt* gene was found to contain two E-box promoter elements that bind CLOCK:BMAL1 and, second, its expression was shown to be controlled by CLOCK:BMAL1 in complex with SIRT1 (Fig. 7). Thus, SIRT1 is present both in the transcriptional regulatory loop of the *Nampt* gene and in the NAD^+ enzymatic salvage pathway. This two-way control results in the circadian expression of the *Nampt* gene, giving a circadian function to the NAD^+ salvage pathway, and thereby regulating the synthesis of NAD^+ in a circadian manner.

Importantly, the use of a highly specific NAMPT pharmacological inhibitor, FK866, abolishes NAD^+ circadian oscillations and, therefore, SIRT1 cyclic activity (Nakahata et al. 2009). This finding is of interest because FK866 is used to control cell death in human cancer tissues. Thus, in addition to revealing a critical enzymatic circadian cycle, these results suggest that a direct molecular coupling exists between the circadian clock, energy metabolism, and cell survival.

The circadian clock is directly implicated in controlling the intracellular levels of critical metabolites, thereby interlocking the transcriptional feedback clock loop with the enzymatic feedback loop of the NAD^+ salvage pathway (Fig. 7). This view has been confirmed by recent results, using mice deficient in a NAD^+ hydrolase (CD38), in which NAD^+ levels become elevated during most of the circadian cycle. These alterations in circadian rhythmicity and CCG expression lead to aberrant metabolism (Sahar et al. 2011). Thus, the oscillation of NAD^+ has important consequences on cellular physiology, including changes in chromatin remodeling and other downstream molecular pathways, providing a direct link between circadian control and metabolic regulation via SIRT1, and controlling the cellular levels of its own coenzyme (Fig. 7).

5 THE AGING AND SENESCENT EPIGENOMES AND CONNECTIONS TO METABOLISM

The question of whether epigenetic alterations are a driving feature of cellular aging is important, given the longstanding known connections between life span, chromatin enzymes, and metabolic changes. In particular, the discovery that yeast Sir2, a key protein in heterochromatic silencing and promoting longevity, is an NAD^+-dependent HDAC, has suggested that chromatin is involved in aging (Imai et al. 2000; Landry et al. 2000; Lin et al. 2000; Schwer and Verdin 2008; Chalkiadaki and Guarente 2012). Further observations, linking Sirtuin pathways with life span extension by calorie/dietary restriction, underscored the role of metabolism in regulating longevity. Cellular senescence is a response to environmental stress and cellular damage whereby cells cease to replicate, but are metabolically active. This process is, in fact, physiologically linked to aging, accelerating the process and involving enormous epigenetic alterations. Thus, as discussed below, numerous findings

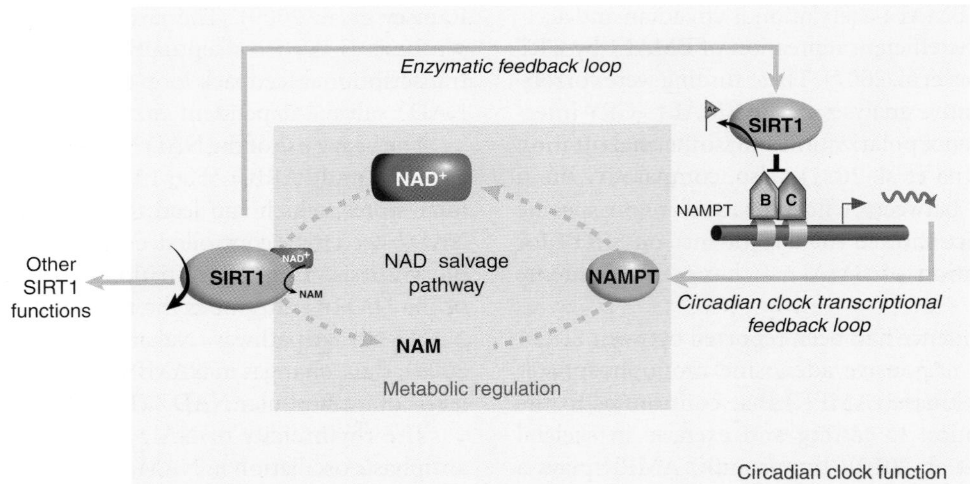

Figure 7. Linking the circadian clock with the NAD^+ salvage pathway. The NAMPTenzyme is the limiting factor in the NAD^+ salvage pathway; thus, because the NAMPT gene is regulated by the circadian CLOCK:BMAL1 machinery, its product, the NAD^+ metabolite, oscillates via this NAMPT transcriptional feedback loop. The HDAC, SIRT1, among other functions, acts as a repressor of the clock machinery by effecting deacetylation of the BMAL1 protein and histones at CCGs (see also Fig. 6). Thus, whereas the enzymatic activity of SIRT1 oscillates in a circadian manner via the circadian controlled supply of its metabolite, NAD^+, its activity also constitutes an enzymatic feedback loop for the circadian clock via its repressive activity at the NAMPT gene.

 Cite as *Cold Spring Harb Perspect Biol* doi: 10.1101/cshperspect.a019463

have been made, establishing the involvement of epigenetic regulation in the aging processes.

5.1 Role of Histone Acetylation and Histone Methylation in Longevity

The Sir (silent information regulator) pathway, which includes the Sir2/Sir3/Sir4 proteins in budding yeast, functions to silence gene expression in heterochromatin (detailed in Ch. 8 [Grunstein and Gasser 2013]). Mutations in *Sir* genes also shorten yeast life span, whereas, conversely, *SIR2* gene duplications lengthen life span (Kaeberlein et al. 1999; Lamming et al. 2005). Hst2 is another regulator of life span in yeast (see Table 1 of Ch. 5 [Seto and Yoshida 2014]) and a member of the large family of catalytically related deacetylases, called Sirtuins. Mammalian Sirtuins have many substrates, both nuclear and outside the nucleus. For example, p53 deacetylation by Sirt1 has been linked to cellular senescence (Chua et al. 2005). The large number of substrates has thus called into question whether chromatin is in fact a crucial substrate for Sirtuin action in the aging process and, more generally, whether life span is directly impacted by epigenetic regulation.

Chromatin structure studies over the last few years have, however, confirmed a direct involvement of epigenetics in regulating aging. In yeast, global levels of histone H4K16 acetylation increase in older cells, particularly at heterochromatin regions, such as telomeres and ribosomal DNA. This was corroborated by experiments substituting glutamine at histone H4K16Q genome-wide, to mimic H4K16 acetylation, which led to premature aging via a Sir2-dependent genetic pathway. Also, the constitutive reduction in acetylation near heterochromatin via loss of Sas2, a heterochromatin-directed HAT, lengthened life span (Dang et al. 2009). Taken together, these results confirm chromatin as an aging-relevant Sir2 substrate.

Histone acetylation is typically induced at promoters during gene activation. This modification is able to generally open chromatin structure, as shown directly for H4K16 acetylation on in vitro assembled chromatin (Shogren-Knaak et al. 2006). Thus, the finding that acetylation increases as cells age, and loss of certain acetylation pathways, such as Sas2, extends life span, suggests that chromatin relaxation and opening may negatively impact longevity (Dang et al. 2009). Furthermore, disruptions in chromatin structure occur as a consequence of the age-related decline in overall core histones, which are needed for the continued Asf1 chaperone-assisted chromatin assembly process (Feser et al. 2010), also accelerating aging. Thus, the global reduction of histone levels leads to increased transcription genome-wide and mobilization of transposable elements, which causes genome instability and chromosomal translocations (Hu et al. 2014). In contrast, elevation of histone expression, either by direct up-regulation or by deleting the Hir repressors of histone transcription, delays aging (Feser et al. 2010). These findings emphasize the importance of maintaining the integrity of chromatin structure for appropriate gene expression and genome stability and, hence, enhancing longevity.

Key hypotheses that have followed on from the *S. cerevisiae* studies are that the modulation of chromatin toward a more relaxed state is relevant to longevity in more complex eukaryotes; there may be a trend toward open chromatin during aging, and these chromatin trends may include alterations in histone methylation as well. In the worm, *Caenorhabditis elegans*, life span extension was observed following a reduction in components of the trithorax ASH-2 complex, consistent with the latter hypothesis. The ASH-2 complex is orthologous to COMPASS, the complex that methylates H3K4 over promoter TSSs during gene activation. Interestingly, the ASH-2 reduction was unable to extend life span if the reverse key demethylase for H3K4me3 was reduced, strongly implicating the actual need for a reduction in H3K4 methylation to observe a reduction in life span. A gene ontology category, showing altered gene expression with age, includes known life span–regulatory genes, emphasizing the fact that specific genes may have a role in extending life span (Greer et al. 2010). Interestingly, this pathway to extend life span requires an intact germline and it persists through the germline to extend life span in a finite number of subsequent generations (i.e., providing an example of transgenerational memory of epigenetic alterations [Greer et al. 2011]).

Turning to studies in the fly, *Drosophila melanogaster*, loss of repressive chromatin occurs during aging. The level of histone H3K9me3, a heterochromatic modification associated with pericentric and telomeric locations, decreases at these locations and broadly increases at euchromatin genome-wide. This imbalance in repressive chromatin observed with age is mirrored by similar changes in the localization of HP1, a histone H3K9me3-binding protein, as well as changes in transcription across the genome (Wood et al. 2010). These observations in the yeast, worm, and fly model systems consistently implicate a reduction of heterochromatin and the relaxation of chromatin-mediated transcriptional control during aging, leading to shortened life span.

In the mouse, among the Sirtuins, *Sirt6* has a direct relationship to aging via maintaining chromatin integrity (introduced in Sec. 2.1). SIRT6 is nuclear localized and chromatin associated (Masri et al. 2014). *Sirt6* gene deletion leads to mice with a short life span and an aging-like degenerative condition because of the impairment of a DNA damage response base-excision repair pathway (Mostoslavsky et al. 2006). In human cultured fibroblasts, the reduction of Sirt6 levels leads to premature senescence (described below in Sec. 5.3), in which cellular replicative life span is dramatically shortened. A key question in all of these correlations with life span is whether chromatin and histones are direct substrates for Sirt6-related rapid aging; SIRT6 was found to specifically target H3K9ac, among numerous major histone acetylation sites, for deacetylation both in vitro and at telomeres in vivo. Further, SIRT6 stabilizes telomeres to antagonize replication-associated dysfunction and chromosome end joining (Michishita et al. 2008). One key SIRT6 pathway, regulating cellular senescence in the mouse, is via NF-κB-regulated genes involved in senescence and apoptosis; these genes are normally maintained in a down-regulated state via Sirt6-dependent deacetylation, but up-regulated on Sirt6 depletion in an NF-κB-dependent manner (Kawahara et al. 2009). Hence, mammalian aging is clearly regulated by histone acetylation, and, consistent with studies in model organisms, low histone acetylation maintains genome integrity, leading to a general model in which compaction of the epigenome and, in particular, the heterochromatic compartment, enhances longevity.

5.2 Calorie Restriction, Life Span Extension, and Epigenetic Regulation

Calorie restriction (CR), an extreme form of altered metabolism, has a beneficial effect on longevity in fungal and most animal models; anecdotal evidence suggests that humans may also benefit from dietary restriction. One pathway to extend life span through lowered nutrients is via the targets of rapamycin (TOR) signaling pathway, which limits protein translation (Kaeberlein et al. 2005). Another well-established CR pathway is via the activation of Sirtuins through elevated NAD^+ levels, a cofactor for Sirtuins (Fig. 2) (Cohen et al. 2004). Given the beneficial role of maintaining Sir-2-mediated H4K16 deacetylation for yeast longevity, chromatin is highly likely to be a CR-relevant target.

The ATP-dependent Isw2 complex mediates another CR pathway for life span regulation. In this case, Isw2-dependent nucleosome remodeling represses stress response genes, whereas deletion of Isw2, which partially mimics the effect of CR through the up-regulation of stress response genes, extends life span. This is a conserved life span regulatory pathway in yeast and worms, and reduction of the orthologous complexes also extends life span (Dang et al. 2014).

5.3 Senescence, Premature Aging, and Profound Epigenome Restructuring

Altered chromatin features in aging mammalian cells include large-scale reorganization of cytological heterochromatic structures and dramatic changes in broad epigenomic domains. These alterations involve the nuclear lamina, a protein structural matrix that associates with the interior face of the nuclear membrane. Normally, broad heterochromatic gene-poor genomic regions associate with the lamina, called lamin-associated domains (LADs). They contain relevant repressive chromatin features (e.g., H3K9me3 and HP1 binding) and cognate enzymes map to these large domains (e.g., H3K27me3/EZH2) (Kind and van Steensel 2010). The premature aging syndrome, HGPS (Hutchinson–Guilford progeria syndrome), is caused by dominant negative mutations in lamin A, correlating with reduced levels of heterochromatic H3K9me3 and lowered HP1, the heterochromatin protein that binds to H3K9me3 (Scaffidi and Misteli 2005). Similarly, aged fibroblasts display reconfiguration of nuclear lamin A and a reduction in H3K9me3 marks and HP1, as well as an increase in foci bearing the phosphorylated H2AX variant (indicating DNA damage; see Ch. 20 [Henikoff and Smith 2014]), which are common hallmarks of HGPS (Scaffidi and Misteli 2006). Thus, HGPS is an accelerated aging syndrome caused by dramatic alterations of the nuclear lamina and concomitant reduced heterochromatin.

As mentioned above, cellular senescence is an irreversible cell cycle arrest in response to stress caused by expression of activated oncogenes or long-term replication, both of which are associated with cellular dysfunction and, thus, the cell cycle arrest is protective against cancer (Narita et al. 2003). However, senescence also correlates with aging in primates (Herbig et al. 2006) and, importantly, was recently causally linked to aging in a mouse progerin premature aging model via demonstration that removal of senescent cells ameliorates age-associated tissue pathologies (Baker et al. 2011). Hence, senescence is a provocative state of protection against cancerous malignancy, but also correlates with cellular damage associated with aging. Aberrant large-scale chromatin foci, called senescent-associated heterochromatin foci (SAHF), have been shown to arise during senescence and are enriched in heterochromatin proteins and chromatin changes indicative of DNA damage, such as HMGA (high mobility group AT hook protein; Narita et al. 2003, 2006), H2AXph (S139 phosphorylated histone variant H2AX protein; Sedelnikova et al. 2004), and macroH2A (H2A variant protein with macro domain; Zhang et al. 2005). These acquired SAHFs repress certain genes required for cell proliferation (Narita et al. 2006).

SAHFs have been thought to represent increased heterochromatin, but they do not correlate with obvious changes in heterochromatic H3K9 methylation or H3K27me3. Rather, they show possible higher-order epigenetic changes (Chandra et al. 2012). A reduction of LaminB1 was shown to lead to premature senescence (Shimi et al. 2011; Shah et al. 2013). Interestingly, extremely large domains of reduced heterochromatin develop in senescent cells, which correspond to LADs in proliferating cells, displaying broad increases in H3K4me3 (Shah et al. 2013) and decreases in DNA methylation (Cruickshanks et al. 2013). Hence, senescent cells appear to undergo complex epigenetic restructuring, with focal regions of increased heterochromatin (SAHFs) and broad regions of decreased heterochromatin over LADs (Fig. 8).

There are challenging questions surrounding alterations of chromatin during the aging process. Much is known about altered concentration of certain metabolites (such as NAD/NADH ratio) during aging and how these changes affect, for example, Sirtuins; however, virtually nothing is known about how these changes might drive the chromatin alterations genome-wide. It remains to be seen whether other metabolites, such as acetyl-CoA and SAM, alter chromatin during aging, and how this is involved in CR conditions to extend life span. Also, what is the precise relationship between large-scale nuclear structural changes and genome-wide changes in chromatin? And, which age-associated chromatin changes reflect degradative processes and which reflect protective processes?

6 CONCLUDING REMARKS

It is evident from the breadth of subjects discussed in this chapter that there are many mechanistic and physiological aspects within the broad scope of metabolism and epigenetics. The mechanisms that transduce metabolic signals into epigenetic changes encompass enzymes typified by acetylases and deacetylases, which respond to profound shifts in the cellular milieu via their use of small molecule metabolites as direct cofactors and substrates that alter their

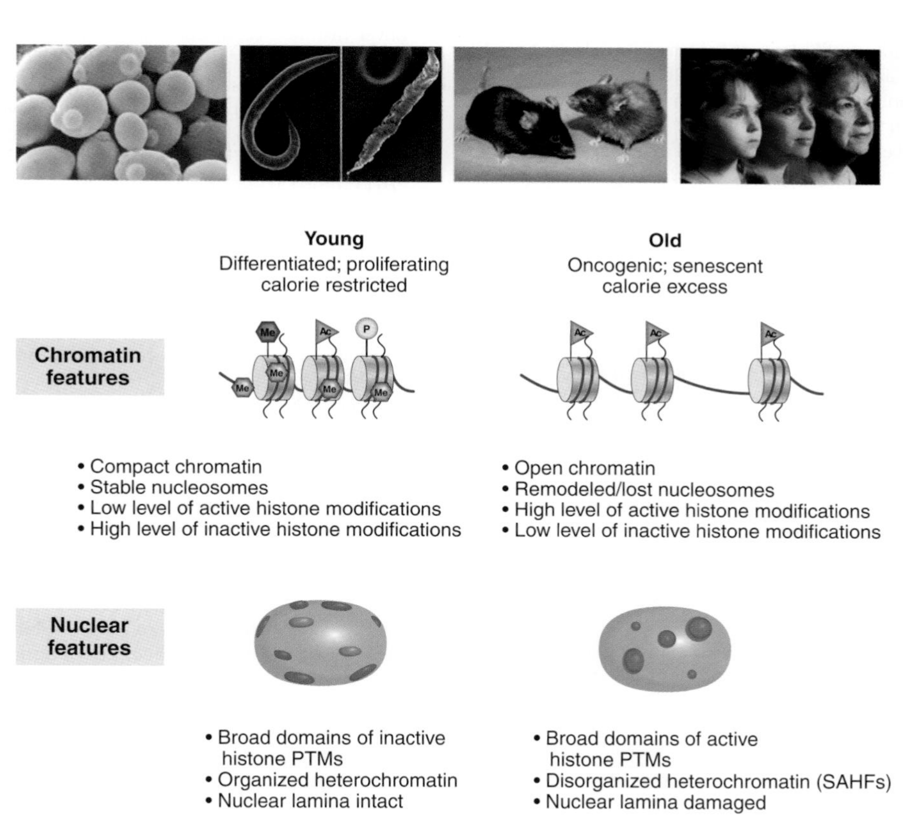

Figure 8. Features of aging and senescent cells. (*Top*) Aging is a biological inevitability for all eukaryotes, from budding yeast and *C. elegans* to mice and humans (depicted). The features of senescence and cellular aging are compared between a normal or "young" cell (*left*) and a diseased or "old" cell (*right*) at three levels: chromatin, nuclear organization, and integrity of the genome. (*Top* images [left to right] from chemistryland.com; from Hazreet Gill, Francis Ghandi, and Arjumand Ghazi, University of Pittsburgh School of Medicine, http://www.chp.edu/CHP/ghazilab; courtesy of Christopher Fisher; from victorymedspa.com.)

enzymatic activity. One key proposal to come from the studies of HATs and HDACs is that many other epigenetic enzymes may also nimbly interact with the environment via their response to changing concentrations of metabolites.

The "circadian epigenome" and the "aging epigenome" represent examples of striking physiological states that are influenced by metabolic changes, which impinge on chromatin. Many more physiological states are altered by metabolic epigenetics, such as numerous types of cancers; many others await elucidation.

REFERENCES

*Reference is also in this book.

Aguilar-Arnal L, Hakim O, Patel VR, Baldi P, Hager GL, Sassone-Corsi P. 2013. Cycles in spatial and temporal chromosomal organization driven by the circadian clock. *Nat Struct Mol Biol* **20:** 1206–1213.

Albaugh BN, Arnold KM, Denu JM. 2011. KAT(ching) metabolism by the tail: Insight into the links between lysine acetyltransferases and metabolism. *Chembiochem* **12:** 290–298.

Alenghat T, Meyers K, Mullican SE, Leitner K, Adeniji-Adele A, Avila J, Bućan M, Ahima RS, Kaestner KH, Lazar MA. 2008. Nuclear receptor corepressor and histone deacetylase 3 govern circadian metabolic physiology. *Nature* **456:** 997–1000.

Ang YS, Gaspar-Maia A, Lemischka IR, Bernstein E. 2011. Stem cells and reprogramming: Breaking the epigenetic barrier? *Trends Pharmacol Sci* **32:** 394–401.

Ariyannur PS, Moffett JR, Madhavarao CN, Arun P, Vishnu N, Jacobowitz DM, Hallows WC, Denu JM, Namboodiri AM. 2010. Nuclear-cytoplasmic localization of acetyl coenzyme a synthetase-1 in the rat brain. *J Comp Neurol* **518:** 2952–2977.

Asher G, Gatfield D, Stratmann M, Reinke H, Dibner C, Kreppel F, Mostoslavsky R, Alt FW, Schibler U. 2008. SIRT1 regulates circadian clock gene expression through PER2 deacetylation. *Cell* **134:** 317–328.

* Audia JE, Campbell RM. 2014. Histone modifications and cancer. *Cold Spring Harb Perspect Biol* doi: 10.1101/cshperspect.a019521.

Austenaa L, Barozzi I, Chronowska A, Termanini A, Ostuni R, Prosperini E, Stewart AF, Testa G, Natoli G. 2012. The histone methyltransferase Wbp7 controls macrophage function through GPI glycolipid anchor synthesis. *Immunity* **36:** 572–585.

Bai P, Cantó C, Oudart H, Brunyánszki A, Cen Y, Thomas C, Yamamoto H, Huber A, Kiss B, Houtkooper RH, et al. 2011. PARP-1 inhibition increases mitochondrial metabolism through SIRT1 activation. *Cell Metab* **13:** 461–468.

Baker DJ, Wijshake T, Tchkonia T, LeBrasseur NK, Childs BG, van de Sluis B, Kirkland JL, van Deursen JM. 2011. Clearance of p16Ink4a-positive senescent cells delays ageing-associated disorders. *Nature* **479:** 232–236.

Barber MF, Michishita-Kioi E, Xi Y, Tasselli L, Kioi M, Moqtaderi Z, Tennen RI, Paredes S, Young NL, Chen K, et al. 2012. SIRT7 links H3K18 deacetylation to maintenance of oncogenic transformation. *Nature* **487:** 114–118.

* Becker PB, Workman JL. 2013. Nucleosome remodeling and epigenetics. *Cold Spring Harb Perspect Biol* **5:** a017905.

Bellet MM, Nakahata Y, Boudjelal M, Watts E, Mossakowska DE, Edwards KA, Cervantes M, Astarita G, Loh C, Ellis JL, et al. 2013. Pharmacological modulation of circadian rhythms by synthetic activators of the deacetylase SIRT1. *Proc Natl Acad Sci* **110:** 3333–3338.

Beneke S. 2012. Regulation of chromatin structure by poly(ADP-ribosyl)-ation. *Front Genet* **3:** 169.

Berndsen CE, Denu JM. 2008. Catalysis and substrate selection by histone/protein lysine acetyltransferases. *Curr Opin Struct Biol* **18:** 682–689.

Burnett C, Valentini S, Cabreiro F, Goss M, Somogyvári M, Piper MD, Hoddinott M, Sutphin GL, Leko V, McElwee JJ, et al. 2011. Absence of effects of Sir2 overexpression on lifespan in *C. elegans* and *Drosophila*. *Nature* **477:** 482–485.

Cai L, Sutter BM, Li B, Tu BP. 2011. Acetyl-CoA induces cell growth and proliferation by promoting the acetylation of histones at growth genes. *Mol Cell* **42:** 426–437.

Cantó C, Jiang LQ, Deshmukh AS, Mataki C, Coste A, Lagouge M, Zierath JR, Auwerx J. 2010. Interdependence of AMPK and SIRT1 for metabolic adaptation to fasting and exercise in skeletal muscle. *Cell Metab* **11:** 213–219.

Chalkiadaki A, Guarente L. 2012. Sirtuins mediate mammalian metabolic responses to nutrient availability. *Nat Rev Endocrinol* **8:** 287–296.

Chandra T, Kirschner K, Thuret JY, Pope BD, Ryba T, Newman S, Ahmed K, Samarajiwa SA, Salama R, Carroll T, et al. 2012. Independence of repressive histone marks and chromatin compaction during senescent heterochromatic layer formation. *Mol Cell* **47:** 203–214.

Chang HC, Guarente L. 2013. SIRT1 mediates central circadian control in the SCN by a mechanism that decays with aging. *Cell* **153:** 1448–1460.

Chen JJ, London IM. 1981. Hemin enhances the differentiation of mouse 3T3 cells to adipocytes. *Cell* **26:** 117–122.

Chen H, Lin RJ, Schiltz RL, Chakravarti D, Nash A, Nagy L, Privalsky ML, Nakatani Y, Evans RM. 1997. Nuclear receptor coactivator ACTR is a novel histone acetyltransferase and forms a multimeric activation complex with P/CAF and CBP/p300. *Cell* **90:** 569–580.

* Cheng X. 2014. Structural and functional coordination of DNA and histone methylation. *Cold Spring Harb Perspect Biol* **6:** a018747.

Chua KF, Mostoslavsky R, Lombard DB, Pang WW, Saito S, Franco S, Kaushal D, Cheng HL, Fischer MR, Stokes N, et al. 2005. Mammalian SIRT1 limits replicative life span in response to chronic genotoxic stress. *Cell Metab* **2:** 67–76.

Cohen HY, Miller C, Bitterman KJ, Wall NR, Hekking B, Kessler B, Howitz KT, Gorospe M, de Cabo R, Sinclair DA. 2004. Calorie restriction promotes mammalian cell survival by inducing the SIRT1 deacetylase. *Science* **305:** 390–392.

Crosio C, Cermakian N, Allis CD, Sassone-Corsi P. 2000. Light induces chromatin modification in cells of the mammalian circadian clock. *Nat Neurosci* **3:** 1241–1247.

Cruickshanks HA, McBryan T, Nelson DM, Vanderkraats ND, Shah PP, van Tuyn J, Singh Rai T, Brock C, Donahue G, Dunican DS, et al. 2013. Senescent cells harbour features of the cancer epigenome. *Nat Cell Biol* **15:** 1495–1506.

Curtis AM, Seo SB, Westgate EJ, Rudic RD, Smyth EM, Chakravarti D, FitzGerald GA, McNamara P. 2004. Histone acetyltransferase-dependent chromatin remodeling and the vascular clock. *J Biol Chem* **279:** 7091–7097.

Czarna A, Breitkreuz H, Mahrenholz CC, Arens J, Strauss HM, Wolf E. 2011. Quantitative analyses of cryptochrome-mBMAL1 interactions: Mechanistic insights into the transcriptional regulation of the mammalian circadian clock. *J Biol Chem* **286:** 22414–22425.

Dang W, Steffen KK, Perry R, Dorsey JA, Johnson FB, Shilatifard A, Kaeberlein M, Kennedy BK, Berger SL. 2009. Histone H4 lysine 16 acetylation regulates cellular lifespan. *Nature* **459:** 802–807.

Dang W, Sutphin GL, Dorsey JA, Otte GL, Cao K, Perry RM, Wanat JJ, Saviolaki D, Murakami CJ, Tsuchiyama S, et al. 2014. Inactivation of yeast Isw2 chromatin remodeling enzyme mimics longevity effect of calorie restriction via induction of genotoxic stress response. *Cell Metab* **19:** 952–966.

* Dekker J, Misteli T. 2014. Dosage compensation in *Drosophila*. *Cold Spring Harb Perspect Biol* doi: 10.1101/cshperspect.a019356.

Dioum EM, Rutter J, Tuckerman JR, Gonzalez G, Gilles-Gonzalez MA, McKnight SL. 2002. NPAS2: A gas-responsive transcription factor. *Science* **298:** 2385–2387.

DiTacchio L, Le HD, Vollmers C, Hatori M, Witcher M, Secombe J, Panda S. 2011. Histone lysine demethylase JARID1a activates CLOCK-BMAL1 and influences the circadian clock. *Science* **333:** 1881–1885.

Doi M, Hirayama J, Sassone-Corsi P. 2006. Circadian regulator CLOCK is a histone acetyltransferase. *Cell* **125:** 497–508.

Dolinoy DC, Huang D, Jirtle RL. 2007. Maternal nutrient supplementation counteracts bisphenol A-induced DNA hypomethylation in early development. *Proc Natl Acad Sci* **104:** 13056–13061.

Dominy JE Jr, Lee Y, Gerhart-Hines Z, Puigserver P. 2010. Nutrient-dependent regulation of PGC-1α's acetylation state and metabolic function through the enzymatic activities of Sirt1/GCN5. *Biochim Biophys Acta* **1804:** 1676–1683.

Duong HA, Weitz CJ. 2014. Temporal orchestration of repressive chromatin modifiers by circadian clock Period complexes. *Nat Struct Mol Biol* **21:** 126–132.

Eckel-Mahan KL, Patel VR, de Mateo S, Orozco-Solis R, Ceglia NJ, Sahar S, Dilag-Penilla SA, Dyar KA, Baldi P, Sassone-Corsi P. 2013. Reprogramming of the circadian clock by nutritional challenge. *Cell* **155:** 1464–1478.

Etchegaray JP, Lee C, Wade PA, Reppert SM. 2003. Rhythmic histone acetylation underlies transcription in the mammalian circadian clock. *Nature* **421:** 177–182.

Etchegaray JP, Yang X, DeBruyne JP, Peters AH, Weaver DR, Jenuwein T, Reppert SM. 2006. The polycomb group protein EZH2 is required for mammalian circadian clock function. *J Biol Chem* **281:** 21209–21215.

Feldman JL, Baeza J, Denu JM. 2013. Activation of the protein deacetylase SIRT6 by long-chain fatty acids and widespread deacylation by mammalian sirtuins. *J Biol Chem* **288:** 31350–31356.

Feng D, Liu T, Sun Z, Bugge A, Mullican SE, Alenghat T, Liu XS, Lazar MA. 2011. A circadian rhythm orchestrated by histone deacetylase 3 controls hepatic lipid metabolism. *Science* **331:** 1315–1319.

Feser J, Truong D, Das C, Carson JJ, Kieft J, Harkness T, Tyler JK. 2010. Elevated histone expression promotes life span extension. *Mol Cell* **39:** 724–735.

Figueroa ME, Abdel-Wahab O, Lu C, Ward PS, Patel J, Shih A, Li Y, Bhagwat N, Vasanthakumar A, Fernandez HF, et al. 2010. Leukemic IDH1 and IDH2 mutations result in a hypermethylation phenotype, disrupt TET2 function, and impair hematopoietic differentiation. *Cancer Cell* **18:** 553–567.

Finkel T, Deng CX, Mostoslavsky R. 2009. Recent progress in the biology and physiology of sirtuins. *Nature* **460:** 587.

Fischer M, Bacher A. 2008. Biosynthesis of vitamin B_2: Structure and mechanism of riboflavin synthase. *Arch Biochem Biophys* **474:** 252.

Friis RM, Wu BP, Reinke SN, Hockman DJ, Sykes BD, Schultz MC. 2009. A glycolytic burst drives glucose induction of global histone acetylation by picNuA4 and SAGA. *Nucleic Acids Res* **37:** 3969–3980.

Garten A, Petzold S, Körner A, Imai S, Kiess W. 2009. Nampt: Linking NAD biology, metabolism and cancer. *Trends Endocrinol Metab* **20:** 130–138.

Gerken T, Girard CA, Tung YC, Webby CJ, Saudek V, Hewitson KS, Yeo GS, McDonough MA, Cunliffe S, McNeill LA, et al. 2007. The obesity-associated *FTO* gene encodes a 2-oxoglutarate-dependent nucleic acid demethylase. *Science* **318:** 1469–1472.

Gibson BA, Kraus WL. 2012. New insights into the molecular and cellular functions of poly(ADP-ribose) and PARPs. *Nat Rev Mol Cell Biol* **13:** 411–424.

Gil R, Barth S, Kanfi Y, Cohen HY. 2013. SIRT6 exhibits nucleosome-dependent deacetylase activity. *Nucleic Acids Res* **41:** 8537–8545.

Greer EL, Maures TJ, Hauswirth AG, Green EM, Leeman DS, Maro GS, Han S, Banko MR, Gozani O, Brunet A. 2010. Members of the H3K4 trimethylation complex regulate lifespan in a germline-dependent manner in *C. elegans*. *Nature* **466:** 383–387.

Greer EL, Maures TJ, Ucar D, Hauswirth AG, Mancini E, Lim JP, Benayoun BA, Shi Y, Brunet A. 2011. Transgenerational epigenetic inheritance of longevity in *Caenorhabditis elegans*. *Nature* **479:** 365–371.

Grillo MA, Colombatto S. 2008. S-adenosylmethionine and its products. *Amino Acids* **34:** 187–193.

Grob A, Roussel P, Wright JE, McStay B, Hernandez-Verdun D, Sirri V. 2009. Involvement of SIRT7 in resumption of rDNA transcription at the exit from mitosis. *J Cell Sci* **122:** 489–498.

* Grunstein M, Gasser SM. 2013. Epigenetics in *Saccharomyces cerevisiae*. *Cold Spring Harb Perspect Biol* **5:** a017491.

Guan KL, Xiong Y. 2011. Regulation of intermediary metabolism by protein acetylation. *Trends Biochem Sci* **36:** 108–116.

Guastafierro T, Cecchinelli B, Zampieri M, Reale A, Riggio G, Sthandier O, Zupi G, Calabrese L, Caiafa P. 2008. CCCTC-binding factor activates PARP-1 affecting DNA methylation machinery. *J Biol Chem* **283:** 21873–21880.

Hallows WC, Lee S, Denu JM. 2006. Sirtuins deacetylate and activate mammalian acetyl-CoA synthetases. *Proc Natl Acad Sci* **103:** 10230–10235.

* Henikoff S, Smith MM. 2014. Histone variants and epigenetics. *Cold Spring Harb Perspect Biol* doi: 10.1101/cshperspect.a019364.

Herbig U, Ferreira M, Condel L, Carey D, Sedivy JM. 2006. Cellular senescence in aging primates. *Science* **311:** 1257.

Hirayama J, Sahar S, Grimaldi B, Tamaru T, Takamatsu K, Nakahata Y, Sassone-Corsi P. 2007. CLOCK-mediated acetylation of BMAL1 controls circadian function. *Nature* **450:** 1086–1090.

Hou H, Yu H. 2010. Structural insights into histone lysine demethylation. *Curr Opin Struct Biol* **20:** 739–748.

Houtkooper RH, Pirinen E, Auwerx J. 2012. Sirtuins as regulators of metabolism and healthspan. *Nat Rev Mol Cell Biol* **13:** 225–238.

Hu Z, Chen K, Xia Z, Chavez M, Pal S, Seol JH, Chen CC, Li W, Tyler JK. 2014. Nucleosome loss leads to global transcriptional up-regulation and genomic instability during yeast aging. *Genes Dev* **28:** 396–408.

Imai S, Armstrong CM, Kaeberlein M, Guarente L. 2000. Transcriptional silencing and longevity protein Sir2 is an NAD-dependent histone deacetylase. *Nature* **403:** 795–800.

Kaeberlein M, McVey M, Guarente L. 1999. The SIR2/3/4 complex and SIR2 alone promote longevity in *Saccharomyces cerevisiae* by two different mechanisms. *Genes Dev* **13:** 2570–2580.

Kaeberlein M, Powers RW 3rd, Steffen KK, Westman EA, Hu D, Dang N, Kerr EO, Kirkland KT, Fields S, Kennedy BK. 2005. Regulation of yeast replicative life span by TOR and Sch9 in response to nutrients. *Science* **310:** 1193–1196.

Kaochar S, Tu BP. 2012. Gatekeepers of chromatin: Small metabolites elicit big changes in gene expression. *Trends Biochem Sci* **37:** 477–483.

Katada S, Sassone-Corsi P. 2010. The histone methyltransferase MLL1 permits the oscillation of circadian gene expression. *Nat Struct Mol Biol* **17:** 1414–1421.

Katada S, Imhof A, Sassone-Corsi P. 2012. Connecting threads: Epigenetics and metabolism. *Cell* **148:** 24–28.

Katoh Y, Ikura T, Hoshikawa Y, Tashiro S, Ito T, Ohta M, Kera Y, Noda T, Igarashi K. 2011. Methionine adenosyltransferase II serves as a transcriptional corepressor of Maf oncoprotein. *Mol Cell* **41:** 554–566.

Kawahara TL, Michishita E, Adler AS, Damian M, Berber E, Lin M, McCord RA, Ongaigui KC, Boxer LD, Chang HY, et al. 2009. SIRT6 links histone H3 lysine 9 deacetylation to NF-κB-dependent gene expression and organismal life span. *Cell* **136:** 62–74.

Kawahara TL, Rapicavoli NA, Wu AR, Qu K, Quake SR, Chang HY. 2011. Dynamic chromatin localization of Sirt6 shapes stress- and aging-related transcriptional networks. *PLoS Genet* **7:** e1002153.

Kera Y, Katoh Y, Ohta M, Matsumoto M, Takano-Yamamoto T, Igarashi K. 2013. Methionine adenosyltransferase II-dependent histone H3K9 methylation at the COX-2 gene locus. *J Biol Chem* **288:** 13592–13601.

Kim HS, Xiao C, Wang RH, Lahusen T, Xu X, Vassilopoulos A, Vazquez-Ortiz G, Jeong WI, Park O, Ki SH, et al. 2010. Hepatic-specific disruption of SIRT6 in mice results in fatty liver formation due to enhanced glycolysis and triglyceride synthesis. *Cell Metab* **12:** 224–236.

Kind J, van Steensel B. 2010. Genome-nuclear lamina interactions and gene regulation. *Curr Opin Cell Biol* **22:** 320–325.

Koike N, Yoo SH, Huang HC, Kumar V, Lee C, Kim TK, Takahashi JS. 2012. Transcriptional architecture and chromatin landscape of the core circadian clock in mammals. *Science* **338:** 349–354.

Kolthur-Seetharam U, Dantzer F, McBurney MW, de Murcia G, Sassone-Corsi P. 2006. Control of AIF-mediated cell death by the functional interplay of SIRT1 and PARP-1 in response to DNA damage. *Cell Cycle* **5:** 873–877.

Krishnakumar R, Kraus WL. 2010. PARP-1 regulates chromatin structure and transcription through a KDM5B-dependent pathway. *Mol Cell* **39:** 736–749.

Lamia KA, Sachdeva UM, DiTacchio L, Williams EC, Alvarez JG, Egan DF, Vasquez DS, Juguilon H, Panda S, Shaw RJ, et al. 2009. AMPK regulates the circadian clock by cryptochrome phosphorylation and degradation. *Science* **326:** 437–440.

Lamming DW, Latorre-Esteves M, Medvedik O, Wong SN, Tsang FA, Wang C, Lin SJ, Sinclair DA. 2005. HST2 mediates SIR2-independent life-span extension by calorie restriction. *Science* **309:** 1861–1864.

Landry J, Sutton A, Tafrov ST, Heller RC, Stebbins J, Pillus L, Sternglanz R. 2000. The silencing protein SIR2 and its homologs are NAD-dependent protein deacetylases. *Proc Natl Acad Sci* **97:** 5807–5811.

Le Martelot G, Canella D, Symul L, Migliavacca E, Gilardi F, Liechti R, Martin O, Harshman K, Delorenzi M, Desvergne B, et al. 2012. Genome-wide RNA polymerase II profiles and RNA accumulation reveal kinetics of transcription and associated epigenetic changes during diurnal cycles. *PLoS Biol* **10:** e1001442.

* Li E, Zhang Y. 2014. DNA methylation in mammals. *Cold Spring Harb Perspect Biol* **6:** a019133.

Li QS, Cai S, Borchardt RT, Fang J, Kuczera K, Middaugh CR, Schowen RL. 2007. Comparative kinetics of cofactor association and dissociation for the human and trypanosomal S-adenosylhomocysteine hydrolases. 1. Basic features of the association and dissociation processes. *Biochemistry* **46:** 5798–5809.

Liang G, Zhang Y. 2013. Embryonic stem cell and induced pluripotent stem cell: An epigenetic perspective. *Cell Res* **23:** 49–69.

Lin SJ, Defossez PA, Guarente L. 2000. Requirement of NAD and SIR2 for life-span extension by calorie restriction in *Saccharomyces cerevisiae*. *Science* **289:** 2126–2128.

Londoño Gentile T, Lu C, Lodato PM, Tse S, Olejniczak SH, Witze ES, Thompson CB, Wellen KE. 2013. DNMT1 is regulated by ATP-citrate lyase and maintains methylation patterns during adipocyte differentiation. *Mol Cell Biol* **33:** 3864–3878.

Lu C, Thompson CB. 2012. Metabolic regulation of epigenetics. *Cell Metab* **16:** 9–17.

Lu C, Ward PS, Kapoor GS, Rohle D, Turcan S, Abdel-Wahab O, Edwards CR, Khanin R, Figueroa ME, Melnick A, et al. 2012. IDH mutation impairs histone demethylation and results in a block to cell differentiation. *Nature* **483:** 474–478.

Macheroux P, Kappes B, Ealick SE. 2011. Flavogenomics—A genomic and structural view of flavin-dependent proteins. *FEBS J* **278:** 2625–2634.

Magni G, Amici A, Emanuelli M, Orsomando G, Raffaelli N, Ruggieri S. 2004. Structure and function of nicotinamide mononucleotide adenylyltransferase. *Curr Med Chem* **11:** 873–885.

Masri S, Sassone-Corsi P. 2010. Plasticity and specificity of the circadian epigenome. *Nat Neurosci* **13:** 1324–1329.

Masri S, Patel VR, Eckel-Mahan KL, Peleg S, Forne I, Ladurner AG, Baldi P, Imhof A, Sassone-Corsi P. 2013. Circadian acetylome reveals regulation of mitochondrial metabolic pathways. *Proc Natl Acad Sci* **110:** 3339–3344.

Masri S, Rigor P, Cervantes M, Ceglia N, Sebastian C, Xiao C, Roqueta-Rivera M, Deng C, Osborne TF, Mostoslavsky R, et al. 2014. Partitioning circadian transcription by SIRT6 leads to segregated control of cellular metabolism. *Cell* **158:** 659–672.

Messner S, Altmeyer M, Zhao H, Pozivil A, Roschitzki B, Gehrig P, Rutishauser D, Huang D, Caflisch A, Hottiger MO. 2010. PARP1 ADP-ribosylates lysine residues of the core histone tails. *Nucleic Acids Res* **38:** 6350–6362.

Metzger E, Imhof A, Patel D, Kahl P, Hoffmeyer K, Friedrichs N, Müller JM, Greschik H, Kirfel J, Ji S, et al. 2010. Phosphorylation of histone H3T6 by PKCβ(I) controls demethylation at histone H3K4. *Nature* **464:** 792–796.

Michishita E, McCord RA, Berber E, Kioi M, Padilla-Nash H, Damian M, Cheung P, Kusumoto R, Kawahara TL, Barrett JC, et al. 2008. SIRT6 is a histone H3 lysine 9 deacetylase that modulates telomeric chromatin. *Nature* **452:** 492–496.

Michishita E, McCord RA, Boxer LD, Barber MF, Hong T, Gozani O, Chua KF. 2009. Cell cycle-dependent deacetylation of telomeric histone H3 lysine K56 by human SIRT6. *Cell Cycle* **8:** 2664–2666.

Mostoslavsky R, Chua KF, Lombard DB, Pang WW, Fischer MR, Gellon L, Liu P, Mostoslavsky G, Franco S, Murphy MM, et al. 2006. Genomic instability and aging-like phenotype in the absence of mammalian SIRT6. *Cell* **124:** 315–329.

Nader N, Chrousos GP, Kino T. 2009. Circadian rhythm transcription factor CLOCK regulates the transcriptional activity of the glucocorticoid receptor by acetylating its hinge region lysine cluster: Potential physiological implications. *FASEB J* **23:** 1572–1583.

Nakahata Y, Kaluzova M, Grimaldi B, Sahar S, Hirayama J, Chen D, Guarente LP, Sassone-Corsi P. 2008. The NAD^+-dependent deacetylase SIRT1 modulates CLOCK-mediated chromatin remodeling and circadian control. *Cell* **134:** 329–340.

Nakahata Y, Sahar S, Astarita G, Kaluzova M, Sassone-Corsi P. 2009. Circadian control of the NAD^+ salvage pathway by CLOCK-SIRT1. *Science* **324:** 654–657.

Nam HJ, Boo K, Kim D, Han DH, Choe HK, Kim CR, Sun W, Kim H, Kim K, Lee H, et al. 2014. Phosphorylation of LSD1 by PKCα is crucial for circadian rhythmicity and phase resetting. *Mol Cell* **53:** 791–805.

Narita M, Nũnez S, Heard E, Narita M, Lin AW, Hearn SA, Spector DL, Hannon GJ, Lowe SW. 2003. Rb-mediated heterochromatin formation and silencing of E2F target genes during cellular senescence. *Cell* **113:** 703–716.

Narita M, Narita M, Krizhanovsky V, Nuñez S, Chicas A, Hearn SA, Myers MP, Lowe SW. 2006. A novel role for high-mobility group a proteins in cellular senescence and heterochromatin formation. *Cell* **126:** 503–514.

Nemoto S, Fergusson MM, Finkel T. 2005. SIRT1 functionally interacts with the metabolic regulator and transcriptional coactivator PGC-1{α}. *J Biol Chem* **280:** 16456–16460.

Nikiforov A, Dölle C, Niere M, Ziegler M. 2011. Pathways and subcellular compartmentation of NAD biosynthesis in human cells: From entry of extracellular precursors to mitochondrial NAD generation. *J Biol Chem* **286:** 21767–21778.

Panda S, Antoch MP, Miller BH, Su AI, Schook AB, Straume M, Schultz PG, Kay SA, Takahashi JS, Hogenesch JB. 2002. Coordinated transcription of key pathways in the mouse by the circadian clock. *Cell* **109:** 307–320.

Peek CB, Affinati AH, Ramsey KM, Kuo HY, Yu W, Sena LA, Ilkayeva O, Marcheva B, Kobayashi Y, Omura C, et al. 2013. Circadian clock NAD^+ cycle drives mitochondrial oxidative metabolism in mice. *Science* **342:** 1243417.

Ponka P. 1999. Cell biology of heme. *Am J Med Sci* **318:** 241–256.

Raghuram S, Stayrook KR, Huang P, Rogers PM, Nosie AK, McClure DB, Burris LL, Khorasanizadeh S, Burris TP, Rastinejad F. 2007. Identification of heme as the ligand for the orphan nuclear receptors REV-ERBα and REV-ERBβ. *Nat Struct Mol Biol* **14:** 1207–1213.

Ram O, Goren A, Amit I, Shoresh N, Yosef N, Ernst J, Kellis M, Gymrek M, Issner R, Coyne M, et al. 2011. Combinatorial patterning of chromatin regulators uncovered by genome-wide location analysis in human cells. *Cell* **147:** 1628–1639.

Ramaswamy V, Williams JS, Robinson KM, Sopko RL, Schultz MC. 2003. Global control of histone modification by the anaphase-promoting complex. *Mol Cell Biol* **23:** 9136–9149.

Ramsey KM, Yoshino J, Brace CS, Abrassart D, Kobayashi Y, Marcheva B, Hong HK, Chong JL, Buhr ED, Lee C, et al. 2009. Circadian clock feedback cycle through NAMPT-mediated NAD^+ biosynthesis. *Science* **324:** 651–654.

 Cite as *Cold Spring Harb Perspect Biol* doi: 10.1101/cshperspect.a019463

Reale A, Matteis GD, Galleazzi G, Zampieri M, Caiafa P. 2005. Modulation of DNMT1 activity by ADP-ribose polymers. *Oncogene* **24:** 13–19.

Reppert SM, Weaver DR. 2002. Coordination of circadian timing in mammals. *Nature* **418:** 935–941.

Ripperger JA, Schibler U. 2006. Rhythmic CLOCK-BMAL1 binding to multiple E-box motifs drives circadian Dbp transcription and chromatin transitions. *Nat Genet* **38:** 369–374.

Rodgers JT, Lerin C, Haas W, Gygi SP, Spiegelman BM, Puigserver P. 2005. Nutrient control of glucose homeostasis through a complex of PGC-1α and SIRT1. *Nature* **434:** 113–118.

Sahar S, Sassone-Corsi P. 2012. Circadian rhythms and memory formation: Regulation by chromatin remodeling. *Front Mol Neurosci* **5:** 37.

Sahar S, Nin V, Barbosa MT, Chini EN, Sassone-Corsi P. 2011. Altered behavioral and metabolic circadian rhythms in mice with disrupted NAD^+ oscillation. *Aging* **3:** 794–802.

Sahar S, Masubuchi S, Eckel-Mahan K, Vollmer S, Galla L, Ceglia N, Masri S, Barth TK, Grimaldi B, Oluyemi O, et al. 2014. Circadian control of fatty acid elongation by SIRT1 protein-mediated deacetylation of acetyl-coenzyme A synthetase 1. *J Biol Chem* **289:** 6091–6097.

Sandmeier JJ, French S, Osheim Y, Cheung WL, Gallo CM, Beyer AL, Smith JS. 2002. RPD3 is required for the inactivation of yeast ribosomal DNA genes in stationary phase. *EMBO J* **21:** 4959–4968.

Scaffidi P, Misteli T. 2005. Reversal of the cellular phenotype in the premature aging disease Hutchinson–Gilford progeria syndrome. *Nat Med* **11:** 440–445.

Scaffidi P, Misteli T. 2006. Lamin A-dependent nuclear defects in human aging. *Science* **312:** 1059–1063.

Schibler U, Sassone-Corsi P. 2002. A web of circadian pacemakers. *Cell* **111:** 919–922.

Schreiber V, Dantzer F, Ame JC, de Murcia G. 2006. Poly(ADP-ribose): Novel functions for an old molecule. *Nat Rev Mol Cell Biol* **7:** 517–528.

Schwer B, Verdin E. 2008. Conserved metabolic regulatory functions of sirtuins. *Cell Metab* **7:** 104.

Sedelnikova OA, Horikawa I, Zimonjic DB, Popescu NC, Bonner WM, Barrett JC. 2004. Senescing human cells and ageing mice accumulate DNA lesions with unrepairable double-strand breaks. *Nat Cell Biol* **6:** 168–170.

* Seto E, Yoshida M. 2014. Erasers of histone acetylation: The histone deacetylase enzymes. *Cold Spring Harb Perspect Biol* **6:** a018713.

Shah PP, Donahue G, Otte GL, Capell BC, Nelson DM, Cao K, Aggarwala V, Cruickshanks HA, Rai TS, McBryan T, et al. 2013. Lamin B1 depletion in senescent cells triggers large-scale changes in gene expression and the chromatin landscape. *Genes Dev* **27:** 1787–1799.

* Shi YG, Tsukuda Y. 2013. The discovery of histone demethylases. *Cold Spring Harb Perspect Biol* **5:** a017947.

Shimazu T, Hirschey MD, Hua L, Dittenhafer-Reed KE, Schwer B, Lombard DB, Li Y, Bunkenborg J, Alt FW, Denu JM, et al. 2010. SIRT3 deacetylates mitochondrial 3-hydroxy-3-methylglutaryl CoA synthase 2 and regulates ketone body production. *Cell Metab* **12:** 654–661.

Shimi T, Butin-Israeli V, Adam SA, Hamanaka RB, Goldman AE, Lucas CA, Shumaker DK, Kosak ST, Chandel NS, Goldman RD. 2011. The role of nuclear lamin B1 in cell proliferation and senescence. *Genes Dev* **25:** 2579–2593.

Shin J, He M, Liu Y, Paredes S, Villanova L, Brown K, Qiu X, Nabavi N, Mohrin M, Wojnoonski K, et al. 2013. SIRT7 represses Myc activity to suppress ER stress and prevent fatty liver disease. *Cell Rep* **5:** 654–665.

Shogren-Knaak M, Ishii H, Sun JM, Pazin MJ, Davie JR, Peterson CL. 2006. Histone H4-K16 acetylation controls chromatin structure and protein interactions. *Science* **311:** 844–847.

Shyh-Chang N, Locasale JW, Lyssiotis CA, Zheng Y, Teo RY, Ratanasirintrawoot S, Zhang J, Onder T, Unternaehrer JJ, Zhu H, et al. 2013. Influence of threonine metabolism on *S*-adenosylmethionine and histone methylation. *Science* **339:** 222–226.

Smemo S, Tena JJ, Kim KH, Gamazon ER, Sakabe NJ, Gómez-Marín C, Aneas I, Credidio FL, Sobreira DR, Wasserman NF, et al. 2014. Obesity-associated variants within FTO form long-range functional connections with IRX3. *Nature* **507:** 371–375.

Takahashi H, McCaffery JM, Irizarry RA, Boeke JD. 2006. Nucleocytosolic acetyl-coenzyme a synthetase is required for histone acetylation and global transcription. *Mol Cell* **23:** 207–217.

Tamaru T, Hattori M, Ninomiya Y, Kawamura G, Varès G, Honda K, Mishra DP, Wang B, Benjamin I, Sassone-Corsi P, et al. 2013. ROS stress resets circadian clocks to coordinate pro-survival signals. *PLoS One* **8:** e82006.

Teperino R, Schoonjans K, Auwerx J. 2010. Histone methyl transferases and demethylases; can they link metabolism and transcription? *Cell Metab* **12:** 321–327.

Toiber D, Erdel F, Bouazoune K, Silberman DM, Zhong L, Mulligan P, Sebastian C, Cosentino C, Martinez-Pastor B, Giacosa S, et al. 2013. SIRT6 recruits SNF2H to DNA break sites, preventing genomic instability through chromatin remodeling. *Mol Cell* **51:** 454–468.

Tsai YC, Greco TM, Boonmee A, Miteva Y, Cristea IM. 2012. Functional proteomics establishes the interaction of SIRT7 with chromatin remodeling complexes and expands its role in regulation of RNA polymerase I transcription. *Mol Cell Proteomics* **11:** 60–76.

Tu BP, Kudlicki A, Rowicka M, McKnight SL. 2005. Logic of the yeast metabolic cycle: Temporal compartmentalization of cellular processes. *Science* **310:** 1152–1158.

Turcan S, Rohle D, Goenka A, Walsh LA, Fang F, Yilmaz E, Campos C, Fabius AW, Lu C, Ward PS, et al. 2012. IDH1 mutation is sufficient to establish the glioma hypermethylator phenotype. *Nature* **483:** 479–483.

Wang J, Hevi S, Kurash JK, Lei H, Gay F, Bajko J, Su H, Sun W, Chang H, Xu G, et al. 2009. The lysine demethylase LSD1 (KDM1) is required for maintenance of global DNA methylation. *Nat Genet* **41:** 125–129.

Wellen KE, Hatzivassiliou G, Sachdeva UM, Bui TV, Cross JR, Thompson CB. 2009. ATP-citrate lyase links cellular metabolism to histone acetylation. *Science* **324:** 1076–1080.

Wood JG, Hillenmeyer S, Lawrence C, Chang C, Hosier S, Lightfoot W, Mukherjee E, Jiang N, Schorl C, Brodsky AS, et al. 2010. Chromatin remodeling in the aging genome of *Drosophila*. *Aging Cell* **9:** 971–978.

Xiao C, Kim HS, Lahusen T, Wang RH, Xu X, Gavrilova O, Jou W, Gius D, Deng CX. 2010. SIRT6 deficiency results in severe hypoglycemia by enhancing both basal and insulin-stimulated glucose uptake in mice. *J Biol Chem* **285:** 36776–36784.

Yan J, Wang H, Liu Y, Shao C. 2008. Analysis of gene regulatory networks in the mammalian circadian rhythm. *PLoS Comput Biol* **4:** e1000193.

Yang B, Zwaans BM, Eckersdorff M, Lombard DB. 2009. The sirtuin SIRT6 deacetylates H3 K56Ac in vivo to promote genomic stability. *Cell Cycle* **8:** 2662–2263.

Yin L, Wu N, Curtin JC, Qatanani M, Szwergold NR, Reid RA, Waitt GM, Parks DJ, Pearce KH, Wisely GB, et al. 2007. Rev-erbα, a heme sensor that coordinates metabolic and circadian pathways. *Science* **318:** 1786–1789.

Yu W, Ginjala V, Pant V, Chernukhin I, Whitehead J, Docquier F, Farrar D, Tavoosidana G, Mukhopadhyay R, Kanduri C, et al. 2004. Poly(ADP-ribosyl)ation regulates CTCF-dependent chromatin insulation. *Nat Genet* **36:** 1105–1110.

Zampieri M, Passananti C, Calabrese R, Perilli M, Corbi N, De Cave F, Guastafierro T, Bacalini MG, Reale A, Amicosante G, et al. 2009. Parp1 localizes within the Dnmt1 promoter and protects its unmethylated state by its enzymatic activity. *PLoS One* **4:** e4717.

Zampieri M, Guastafierro T, Calabrese R, Ciccarone F, Bacalini MG, Reale A, Perilli M, Passananti C, Caiafa P. 2012. ADP-ribose polymers localized on Ctcf-Parp1-Dnmt1 complex prevent methylation of Ctcf target sites. *Biochem J* **441:** 645–652.

Zhang R, Poustovoitov MV, Ye X, Santos HA, Chen W, Daganzo SM, Erzberger JP, Serebriiskii IG, Canutescu AA, Dunbrack RL, et al. 2005. Formation of macroH2A-containing senescence-associated heterochromatin foci and senescence driven by ASF1a and HIRA. *Dev Cell* **8:** 19–30.

Zhong L, D'Urso A, Toiber D, Sebastian C, Henry RE, Vadysirisack DD, Guimaraes A, Marinelli B, Wikstrom JD, Nir T, et al. 2010. The histone deacetylase Sirt6 regulates glucose homeostasis via Hif1α. *Cell* **140:** 280–293.

CHAPTER 31

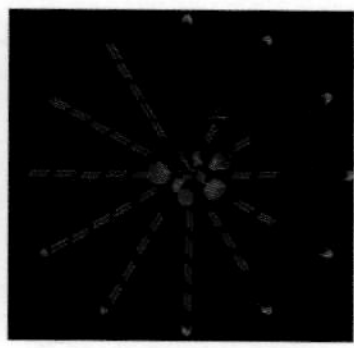

Epigenetic Regulation in Plant Responses to the Environment

David C. Baulcombe[1] and Caroline Dean[2]

[1]Department of Plant Science, University of Cambridge, Cambridge CB2 3EA, United Kingdom; [2]Department of Cell and Developmental Biology, John Innes Centre, Norwich NR4 7UH, United Kingdom

Correspondence: dcb40@cam.ac.uk

SUMMARY

In this chapter, we review environmentally mediated epigenetic regulation in plants using two case histories. One of these, vernalization, mediates adaptation of plants to different environments, and it exemplifies processes that are reset in each generation. The other, virus-induced silencing, involves transgenerationally inherited epigenetic modifications. Heritable epigenetic marks may result in heritable phenotypic variation, influencing fitness, and so be subject to natural selection. However, unlike genetic inheritance, the epigenetic modifications show instability and are influenced by the environment. These two case histories are then compared with other phenomena in plant biology that are likely to represent epigenetic regulation in response to the environment.

Outline

Cite as *Cold Spring Harb Perspect Biol* doi: 10.1101/cshperspect.a019471

OVERVIEW

Epigenetic modification of plant genomes resembles that of mammals in that there is a similar profile of histone marks and the DNA can be methylated at cytosine residues. However, plant epigenomes are more susceptible to environmental influence than those in animals. In this chapter we review environmentally mediated epigenetic regulation in plants using two case histories. One of these involves a repressor of flowering, *FLC*. The basal level of expression of *FLC* is set by opposing processes that either activate or repress its level of expression. The activation involves a transcriptional regulator—FRIGIDA—and the repression involves activities grouped within the so-called "autonomous pathway," which reduce H3K4 and H3K36 methylation modifications associated with active chromatin and increase H3K27 methylation associated with silent chromatin. *FLC* expression is then epigenetically silenced following exposure to prolonged cold through a process known as vernalization. An early step in vernalization is the cold-induced expression of noncoding RNAs at the *FLC* locus. The cold exposure also leads to progressive recruitment of a plant homeodomain (PHD)-Polycomb protein Su(z)12 (PRC2) complex to a localized region within the *FLC* gene. Subsequent spreading of this complex then leads to high levels of H3K27 methylation over the whole locus. The quantitative nature of vernalization is the result of cell-autonomous switching of *FLC* expression off in an increasing proportion of cells. Variation in this epigenetic silencing mechanism underpins the adaptation of plants to climates with different winters. The vernalization requirement is reset during gametogenesis or embryogenesis so that flowering is dependent on exposure to cold in each generation.

The second case history involves virus-induced silencing and epigenetic marks associated with methylation of DNA. The mechanism of virus-induced epigenetic modification is likely to involve a pathway that has also been implicated in transposon silencing in which Dicer generates small interfering RNAs (siRNAs) that are targeted to a chromatin-bound scaffold RNA in association with an Argonaute protein. DNA methyltransferases are then recruited to the chromatin locus where they introduce methyl groups onto C residues in CG, CHG, and CHH contexts. These DNA methylation marks may result in silencing of gene expression if they are in or close to promoter sequences and, in some instances, they are inherited across generations. If there is heritable phenotypic variation due to the gene silencing effect, there also can be an effect on fitness that is subject to natural selection. However, unlike genetic inheritance, the epigenetic modifications are unstable and are influenced by the environment.

These two case histories illustrate mechanisms that are likely to explain phenomena in plant biology in which stresses of various kinds can trigger responses that persist for longer than the inducing stimulus. In some instances, the persistence extends across generations indicating the potential role of epigenetic mechanisms in evolution.

 Cite as *Cold Spring Harb Perspect Biol* doi: 10.1101/cshperspect.a019471

1 EPIGENETIC REGULATION AS ENVIRONMENTAL MEMORY

Plants monitor day/night and seasonal cycles to align their metabolism, growth, and development to changing environmental conditions. This plasticity requires signal perception/integration, changing gene expression in response to those signals, and then maintenance of that response until conditions change again. Epigenetic mechanisms, by definition, allow changed states to persist through cell divisions, even in the absence of the inducing stimulus, and they provide a molecular memory that underpins the maintenance phase of these responses.

Many of the environmentally induced epigenetic changes in plants are reset during gametogenesis, as are most epigenetic marks in animals. However, some persist through gametogenesis and can be stable through many generations. There is, therefore, the definite potential for transgenerational epigenetic change in plants whereas, in animals, this possibility is more controversial.

There are two factors that account for the greater potential for heritable epigenetic regulation in plants versus animals. First, there is the late differentiation of the germline. It is not laid down in embryogenesis as in animals, but it arises from somatic tissue after flowering in the male and female reproductive organs (illustrated in Fig. 1 of Ch. 13 [Pikaard and Mittelsten Scheid 2014]). The plant germline cells are, therefore, descended from somatic cells and they carry epigenetic marks as persistent remnants of earlier environmental stimuli. The second factor to differentiate transgenerational inheritance in plants and animals is related to epigenetic erasure during embryogenesis, which is more complete in animals than in plants (Gutierrez-Marcos and Dickinson 2012).

The first sections of this chapter describe plant case histories in which induced epigenetic changes are well-understood and can be used as a general framework for further analysis of the role of the environment as a trigger for epigenetic changes. Later, in Sections 4–8 of this chapter, we discuss various examples in plant biology that are likely to represent epigenetic responses to environmental stimuli.

2 CASE HISTORY I—VERNALIZATION

Plants are sessile and they have to continually adjust their growth and physiology to changing environmental conditions. This adjustment is particularly important in developmental timing: Plants need to align their seed production to periods with favorable environmental conditions to maximize reproductive success. Environmental cues are therefore monitored and can act to regulate the timing of different developmental switches. One of the earliest characterized processes involving epigenetic regulation in plants is vernalization. This is the ability of plant cells to "remember" they have experienced periods of prolonged cold (measured by the number of weeks at ~5°C) or winter (Purvis and Gregory 1952).

This vernalization memory even persists through tissue culture. Single cells of a vernalized plant can be cultured and regenerated into new plants that flower without prolonged cold (Burn et al. 1993). Each sexual generation of plants, however, needs re-vernalizing because the vernalized state is effectively reset during meiosis or embryo formation (Fig. 1).

The biological function of vernalization is to align flowering with spring and the return of more favorable environmental conditions. This ensures effective flower formation, pollination, and fruit set. Breeding for vernalization has increased the production range of most of our major crops although the process has malign association from the Soviet era. The Soviet scientist and politician, Lysenko, claimed falsely that a vernalized state could be inherited into subsequent generations and increase wheat yields. His anti-Mendelian doctrine gained him power and influence in the political establishment and influenced practice in wheat production. However, when crops failed, it led to mass starvation and the persecution of Vavilov and other geneticists.

2.1 Memory of Winter Involves Polycomb Silencing

Vernalization involves the epigenetic silencing of a floral repressor in response to cold periods. In *Arabidopsis*, the repressor has been identified as FLOWERING LOCUS C (*FLC*; Michaels and Amasino 1999; Sheldon et al. 1999), a MADS box transcriptional regulator that represses genes required to switch the meristem to a floral fate. Prolonged cold progressively silences expression of *FLC* and this is epigenetically maintained during subsequent development in the warm (Fig. 1). Vernalization thus provides a clear example of the separation of the establishment and maintenance phases of epigenetic gene silencing.

In the absence of cold, *FLC* acts as a brake to flowering. The restraint is removed following prolonged cold so that, once plants have detected inductive photoperiods and warm ambient temperatures, the switch to flowering is activated. The repression of *FLC* is epigenetically stable and is maintained for many months after the cold exposure until embryogenesis in the next generation.

Many pathways regulate *FLC* and variation in their activity determines the reproductive habit of the plant. High *FLC* levels cause plants to overwinter before flowering, thereby limiting flowering to once a year. Low *FLC* levels enable plants to flower without the need for cold, opening up the possibility of reproducing more than once a year.

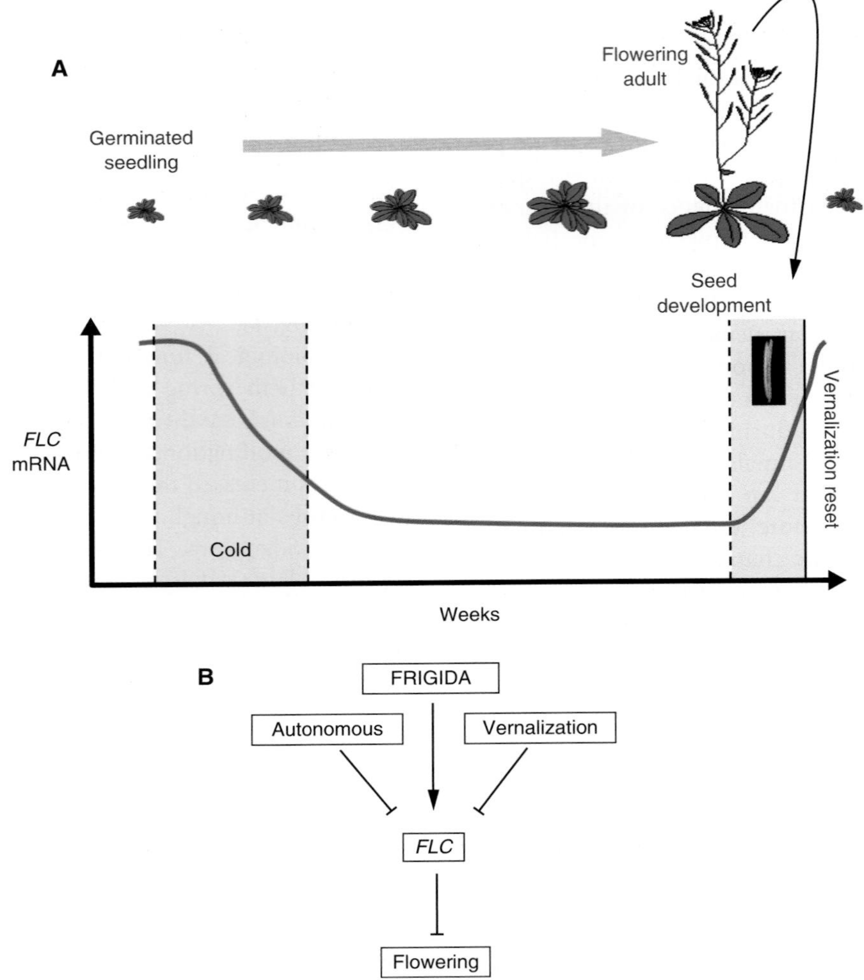

Figure 1. *FLC* expression is epigenetically silenced by cold and reset during embryo development. (*A*) The floral repressor gene, *FLC*, is highly expressed in young seedlings. As plants perceive cold, the expression is quantitatively repressed, dependent on the length of cold experienced. As temperatures warm in spring, the repression is epigenetically maintained until seed development when it is reset. This ensures that each generation of seedlings requires vernalization. (*B*) Epigenetic and transcriptional pathways activate or inhibit *FLC* expression and, hence, contribute to flowering time control. Chromatin modifications and noncoding RNAs contribute in different ways to each pathway.

The level of expression of *FLC* is set very early in development: in the early multicellular embryo for maternally derived *FLC*, and in the sporogenous pollen mother cells or single-celled zygote for the paternal copy (Sheldon et al. 2008; Choi et al. 2009). Many regulators determine this expression level (summarized by pathway in Fig. 1B; Crevillen and Dean 2010) and this is subject to extensive natural variation (Shindo et al. 2005).

FRIGIDA, the main activator of *FLC* expression (Johanson et al. 2000), is a novel protein with coiled-coil domains that directly interacts with the nuclear cap-binding complex (Geraldo et al. 2009). Its function requires the conserved RNA polymerase-associated factor 1 complex (He et al. 2004; Oh et al. 2004; Park et al. 2010; Yu and Michaels 2010) of H3K4 methyltransferases of the Set1-class (*Arabidopsis* TRITHORAX-RELATED7, ATRX7) and Trithorax-class (ATX1/2; Tamada et al. 2009); and a Set2-class methyltransferase, EARLY FLOWERING IN SHORT DAYS, (also called SET domain group 8; Xu et al. 2008). Thus, increased *FLC* expression is linked to trimethylation on histone H3 at lysine 4 (H3K4me3) and lysine 36 (H3K36me3; Xu et al. 2008; Jiang et al. 2009).

Functioning antagonistically to these activators is the autonomous pathway, which reduces H3K4 and H3K36 methylation and increases H3K27 methylation. The autonomous pathway is comprised of a series of activities linking

RNA processing of the *FLC* antisense transcripts with the H3K4 demethylase FLOWERING LOCUS D (Liu et al. 2007, 2010) and Polycomb regulation involving VERNALIZATION2 (VRN2, a Su(z)12 homolog), SWINGER, or CURLY LEAF (SWN, CLF, E(z) HMTase homologs), FERTILIZATION-INDEPENDENT ENDOSPERM (an extra sex combs homolog), and MSI1 (a p55 homolog; Fig. 2) (Wood et al. 2006). The balance of these antagonistic FRIGIDA and autonomous pathways determines levels of *FLC* expression in the young seedlings and establishes whether or not they require vernalization for flowering (i.e., whether or not they need to overwinter before flowering).

FLC expression is progressively silenced by vernalization as plants are exposed to increasing periods of cold. This epigenetic process translates the prolonged exposure to cold into a stable silencing of *FLC* expression. This is maintained throughout the rest of development until it is reset in the embryo (Fig. 1). Early molecular work investigated whether the mitotic memory of vernalization involved changes in DNA methylation, a well-characterized epigenetic mark in plants (reviewed in Finnegan et al. 2000). However, there is little DNA methylation at the *FLC* locus and it does not change with cold (Finnegan et al. 2005). A classic mutagenesis strategy identified H3K27me3, and the Polycomb protein Su(z)12 (PRC2) as required for the memory of *FLC* silencing (Gendall et al. 2001). Reminiscent of Polycomb silencing in mammalian embryonic stem cells, *Arabidopsis* PRC2 is already localized at *FLC* before silencing (De Lucia et al. 2008). This is unlike the situation in *Drosophila* in which PRC2 gener-

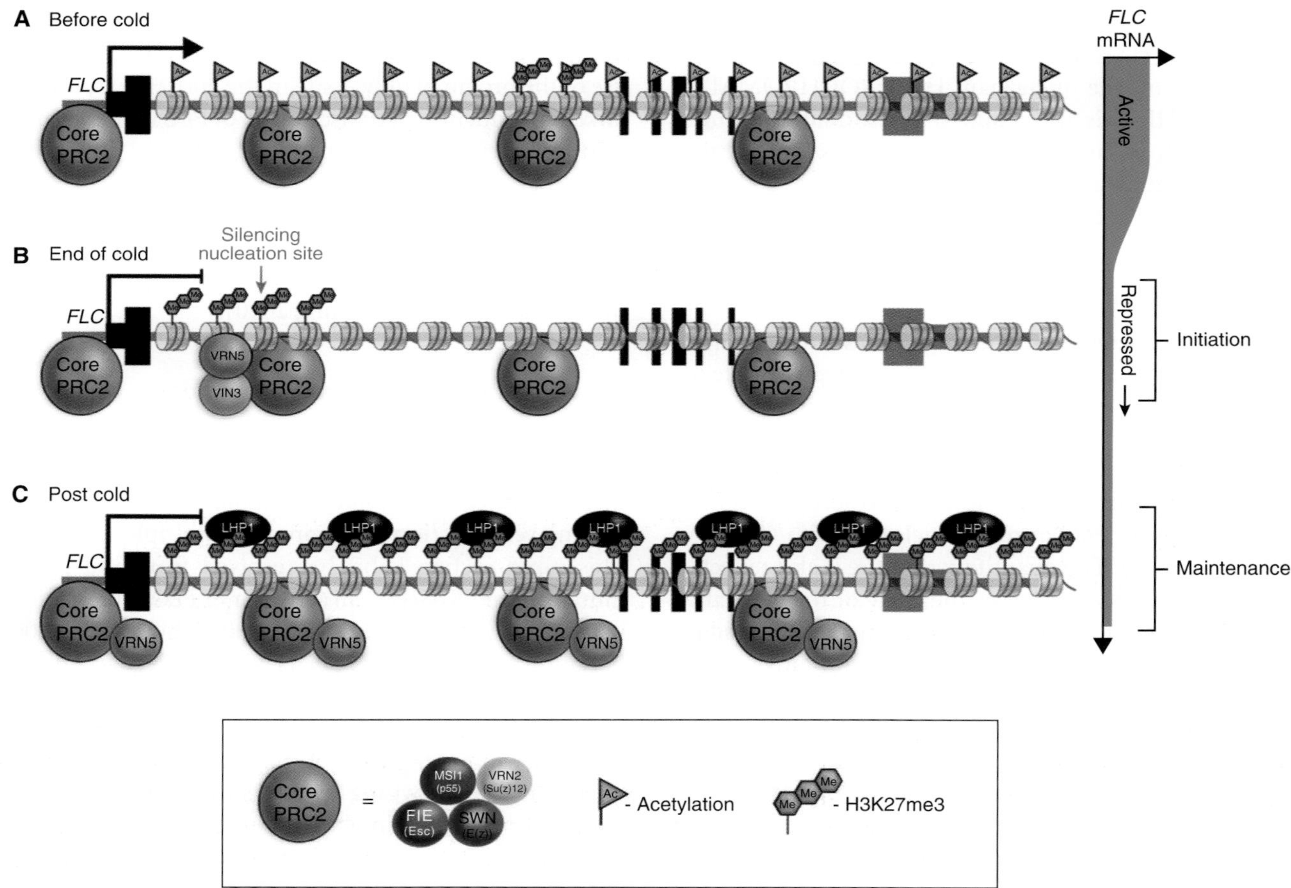

Figure 2. The Polycomb complex composition and localization changes dynamically at *FLC* during different phases of vernalization. (*A*) Before the onset of cold, which triggers vernalization, the PRC2 core complex is already associated with chromatin over the length of the active *FLC* locus. The exon–intron structure is indicated beneath the chromatin fiber as black bars for each exon. (*B*) Prolonged cold leads to the accumulation and nucleation of an alternative Polycomb complex containing plant homeodomain (PHD) proteins (VIN3, VRN5) at a specific intragenic site near the beginning of the first intron. (*C*) In plants returned to warm conditions, the cold-induced VIN3 PHD protein is lost. A modified PHD-PRC2 complex associates across the whole locus, inducing high levels of H3K27me3, which blanket the locus and provide repressive epigenetic stability (maintenance).

ally associates with targets after repression of transcription by other factors (for more detail, see Ch. 17 [Grossniklaus and Paro 2014]). Two proteins, VRN5 and VIN3, containing PHD and FNIII domains then associate with PRC2 to trigger the silencing (Mylne et al. 2006; Sung et al. 2006b; Wood et al. 2006; Greb et al. 2007; De Lucia et al. 2008).

An important part of the cold-induced epigenetic silencing at *FLC* is the cold-induced expression of one of the PHD proteins (VIN3; Sung and Amasino 2004). VIN3 heterodimerizes with the constitutively expressed PHD protein (VRN5; Greb et al. 2007) and they associate with PRC2 at a site within the first intron of *FLC*, causing local accumulation of the histone modification H3K27me3, typical of Polycomb silencing (Fig. 2). On transfer of plants back to warm temperatures, a PHD–PRC2 complex then spreads along the length of *FLC*, generating very high levels of H3K27me3 across the whole locus. These very high levels of H3K27me3 are required for the epigenetic maintenance of the silencing (De Lucia et al. 2008). LHP1, the *Arabidopsis* homolog of the metazoan HETEROCHROMATIN PROTEIN1 (HP1), is also required to maintain the epigenetic silencing, and intriguingly binds to H3K27me3, rather than H3K9me3 as in mammalian cells (Fig. 2; Mylne et al. 2006; Sung et al. 2006a; Zhang et al. 2007).

2.2 Cell-Autonomous Bistable Switching

An unusual characteristic of the epigenetic silencing during vernalization is its quantitative nature: The degree of silencing is dependent on how much cold the plant perceives. This feature ensures the plant can distinguish a cold snap in autumn from a whole winter. Detailed analysis on whole tissue showed a quantitative cold-induced accumulation of H3K27me3 at the *FLC* intronic nucleation site (Fig. 3A) (Angel et al. 2011). The quantitative nature is also reflected in the level of H3K27me3 over the body of the gene after transfer to warm. Mathematical modeling, constrained by the H3K27me3 data, predicted that the quantitative nature of vernalization was generated by a cell-autonomous switch, in which the *FLC* locus flipped into a fully epigenetically silenced state, marked by high levels of H3K27me3 over the whole locus (Fig. 3B) (Angel et al. 2011). Lengthening cold would increase the proportion of cells that have undergone this switch (Fig. 3C). The prediction that vernalization is quantitative because of the degree of cells that flip was confirmed in transgenic plants carrying an *FLC* fusion that could be visualized at the cellular level: A bistable expression pattern was indeed verified in partially vernalized plants (see middle panel of Fig. 3C) (Angel et al. 2011). The quantitative increase is thus a reflection of an increasing proportion of cells with a fully epigenetically silenced *FLC* locus, rather than all cells carrying an increasingly silenced *FLC*.

The cold-induced nucleation of the PHD-PRC2 complex is a key regulatory step in the environmental induction of this process. However, the establishment of silencing still occurs in plants deficient for the PHD proteins (Swiezewski et al. 2009), so there must be multiple cold-dependent steps in the epigenetic silencing of *FLC*—some to establish silencing and others to induce the accumulation of epigenetic memory. These multiple, independent cold steps suggest there is not just one "thermometer" for vernalization but that temperature is perceived by a number of different steps.

A search for cold-induced steps required for establishment of the silencing showed an early and robust induction of antisense transcripts (termed COOLAIR) to *FLC* (Swiezewski et al. 2009). These antisense transcripts encompass the whole length of the sense transcript, are alternatively polyadenylated and alternatively spliced (Fig. 4) and they are an integral part of *FLC* regulation both in the warm and in the cold (Hornyik et al. 2010; Liu et al. 2010). Increased use of the proximal poly A site in the antisense transcript is linked to reduction in transcription of *FLC* in the warm and cold (Swiezewski et al. 2009; Liu et al. 2010), in a mechanism that is independent of the PHD proteins. There is also, somewhat later, cold induction of an *FLC* noncoding sense strand (termed COLDAIR) from a cryptic promoter within *FLC* intron 1 that is required for targeting PRC2 to the *FLC* locus (Heo and Sung 2010).

2.3 Adaptation to Different Climates

Variation in various steps of the vernalization mechanism has played a major role in the evolution of different reproductive strategies. *Arabidopsis thaliana* accessions not requiring vernalization can reproduce multiple times a year. This rapid-cycling habit has evolved from independent mutations in FRIGIDA, an activator of *FLC* expression (Johanson et al. 2000; Gazzani et al. 2003; Shindo et al. 2005). Variation at *Arabidopsis FLC* itself only rarely accounts for the rapid-cycling habit (Werner et al. 2005). *FLC* variation, however, is a major factor contributing to the need for different winter lengths in *A. thaliana* accessions (Shindo et al. 2006). How winter length is determined is still unknown, but studies in *Arabidopsis halleri* growing in its native Japanese habitat have pointed to a temperature/time-averaging mechanism that integrates temperature over a six-week period (Aikawa et al. 2010). The mechanistic basis of adaptation to different winters has been explored through analysis of *Arabidopsis* accessions. Those from the northern limit of its range (e.g., Lov-1 from Northern Sweden [latitude 62.5°N]) initially silence *FLC* within the same time frame, yet need much longer periods of cold for the maintenance of full epigenetic silencing.

Cite as *Cold Spring Harb Perspect Biol* doi: 10.1101/cshperspect.a019471

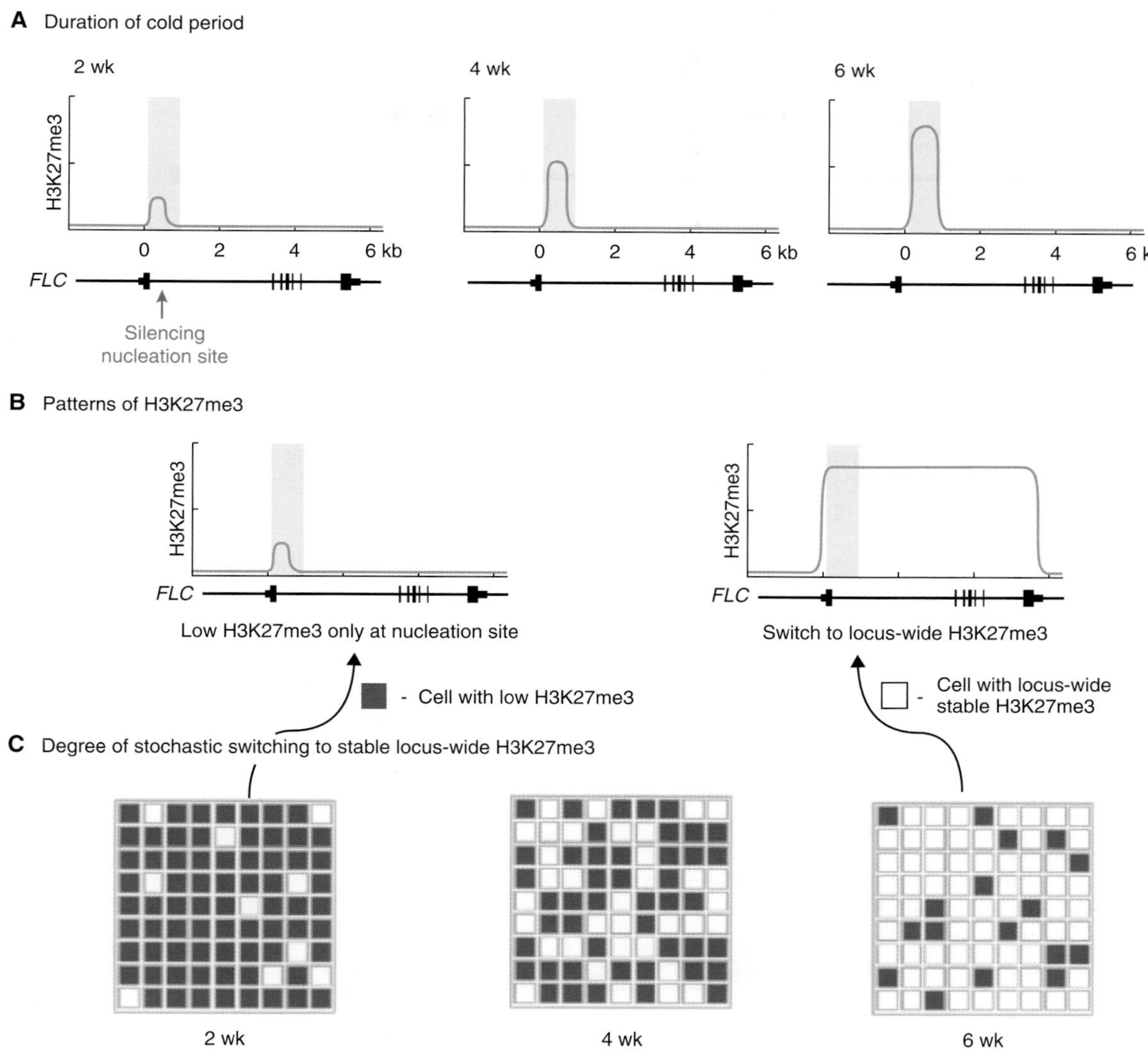

Figure 3. Stochastic switching mechanism underlies the quantitative nature of vernalization. (*A*) During cold, H3K27me3 quantitatively accumulates in the nucleation region of the *FLC* gene, indicated schematically below each graph, with increasing weeks of cold (*top* row of figure). (*B*) After cold, the nucleated H3K27me3 causes some cells to switch to a silenced state with high levels of H3K27me3 blanketing the gene. This epigenetic switch is cell-autonomous. (*C*) The quantitative nature of the vernalization response is due to an increasing number of cells switching to a silenced state after increasing cold exposure. Each cell is indicated by a square. (Figure courtesy of Dr. Jie Song.)

Shorter cold leads to *FLC* expression resuming after transfer back to the warm (Fig. 5). A genetic analysis revealed that a major portion of this variation for winter length mapped to *cis* elements within *FLC* itself, and not the *trans* factors mediating vernalization (Shindo et al. 2006; Coustham et al. 2012). The natural variants are likely to provide important information on the *cis* elements and dynamics of the chromatin complexes that mediate both activation and repression of *FLC*.

This molecular variation in the epigenetic silencing also appears to apply to the molecular basis of perenniality (i.e., the ability to flower year after year). Analysis of *Arabis alpina*, a relative of *Arabidopsis*, revealed that the *FLC* homolog (called *PEP1*) shows interesting regulatory differences with *FLC* in *A. thaliana* (Wang et al. 2009). During prolonged cold, *PEP1* expression in *A. alpina* plants decreases, enabling expression of downstream floral activators as in *A. thaliana*. However, unlike *A. thaliana*, the silencing is not epigenetically stable (Wang et al. 2009). Differences in epigenetic silencing thus seem to have contributed widely to the evolution of reproductive strategies in land plants.

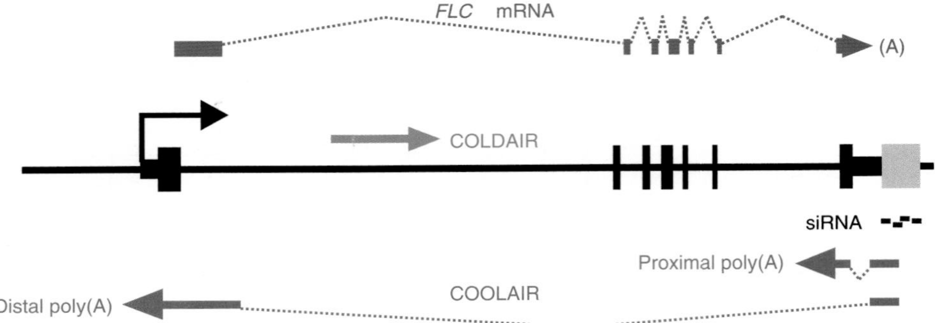

Figure 4. Noncoding transcripts at the *FLC* locus. Many classes of noncoding transcripts have been characterized at the *FLC* locus. A set of antisense transcripts have collectively been called COOLAIR (red). These are alternatively spliced and alternatively polyadenylated, and encompass the whole length of the sense transcript. They are an integral part of *FLC* regulation both in the warm and in the cold. An *FLC* noncoding sense transcript, termed COLDAIR (blue), is transcribed from a cryptic promoter in intron 1. There are also homologous 24- and 30-mer siRNAs (gray) mapping just upstream of the COOLAIR transcription start site.

3 CASE HISTORY II—VIRUS-INDUCED GENE SILENCING AND EPIGENETICS

In a second case history we describe an example in which the external stimulus is a virus and the maintenance mechanism is associated with RNA silencing and DNA methylation. This viral mechanism illustrates the potential for endogenous genome elements to initiate persistent gene silencing if they are activated by stress or other external stimuli.

3.1 RNA Silencing in Virus-Infected Plants

As an introduction to virus-induced epigenetic modification, we first describe the process of RNA silencing that is part of the antiviral defense systems in plants and inverte-

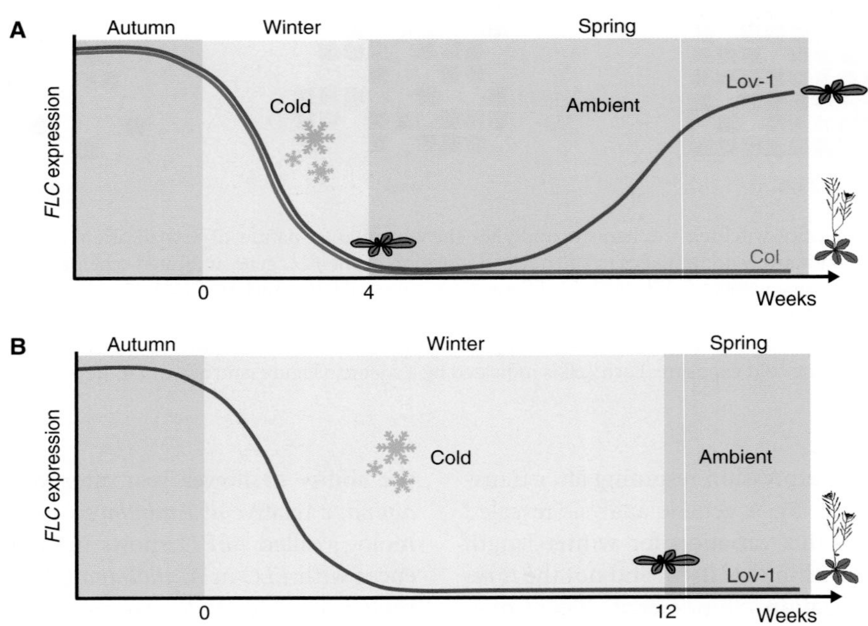

Figure 5. Quantitative variation in the epigenetic silencing of *FLC* in *Arabidopsis* accessions from different climates. (*A*) An *Arabidopsis* accession from Germany (Col, red line) requires only 4 wk of cold to epigenetically silence *FLC*. Lov-1, from the northern limit of its range in Northern Sweden (latitude 62.5°N), sees reactivation of the *FLC* gene if the cold period is so short, resulting in an inability to become vernalized and, hence, does not flower. (*B*) Lov-1 needs a much longer period of cold (12 wk) for full epigenetic silencing. Molecular analysis has shown this difference is the result of a small number of *cis* polymorphisms near the PHD-PRC2 nucleation region in intron 1.

Cite as *Cold Spring Harb Perspect Biol* doi: 10.1101/cshperspect.a019471

brate animals (Baulcombe 2004). RNA silencing operates when plant cells recognize the viral RNAs as foreign and copy it with an RNA-dependent RNA polymerase (RdRP) into a double-stranded (ds) form. The double-stranded RNA (dsRNA) is then processed by a Dicer nuclease into 21–24-nt siRNAs (Dunoyer and Voinnet 2005). In plants, the Dicer proteins are referred to as Dicer-like (DCL) although strictly they are not DCL—they are, in fact, Dicer proteins. The small RNAs are initially in a ds form but eventually a single-stranded (ss) version is then incorporated into an effector ribonucleoprotein complex that also contains an Argonaute (AGO) nuclease (Fig. 6A). AGO has structural similarity to ribonuclease H and the siRNA serves as a guide for AGO: The siRNA is able to base pair with other RNAs in the cell and thereby guide AGO to its target for degradation or to be translationally suppressed. The rule for targeting is to have base-pairing between the siRNA and its target at most of the 21–24 positions. The most frequent target of viral siRNAs would, of course, be the viral RNA itself (Dunoyer and Voinnet 2005).

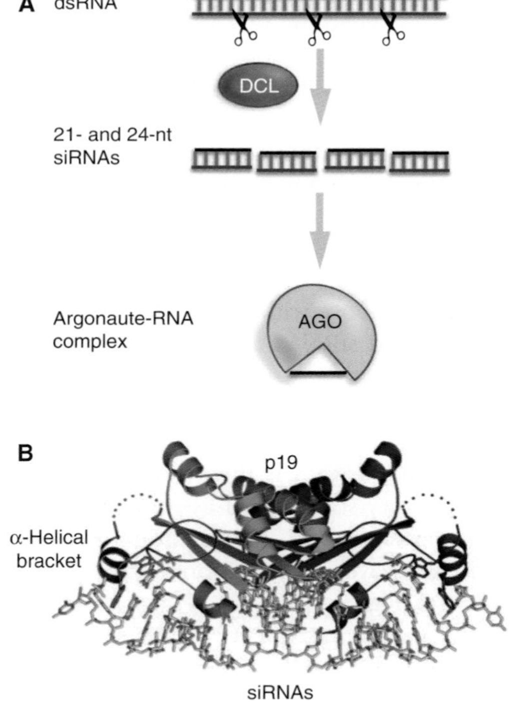

Figure 6. Virus- and transgene-induced RNA silencing in plants. (*A*) The core RNA silencing pathway: dsRNA is processed into 21-nt and 24-nt RNA by Dicer and then bound to the Argonaute slicer protein, to guide the complex to specific target RNA sequences. (*B*) The p19 viral suppressor of RNA silencing (VSR): The siRNA (stick diagram) is bound to a dimeric form of the p19 viral suppressor and held in place by helical brackets formed by the respective amino termini. (*B*, Reproduced from Vargason et al. 2003, with permission from Elsevier.)

In an uninfected cell there are no viral siRNAs, hence AGO proteins have no guides to target viral RNAs. Thus, following infection, the first few rounds of virus replication would be unconstrained by RNA silencing. However, once the viral RNA has accumulated to a substantial level it could prime AGO and the rate of viral RNA accumulation would slow down. The steady state level of virus accumulation would be a reflection of the relative kinetics of silencing and the rate of virus accumulation. The steady state levels are also influenced by the extent to which the viral RNA can evade the silencing machinery by, for example, encapsidation (Baulcombe 2004), or through the action of VSR (Ding and Voinnet 2007).

VSRs can act by binding to siRNA or proteins of the RNA silencing pathway, directly or indirectly, effectively sequestering them to prevent the viral RNA from being targeted. One of the best-understood suppressors is a 19-kDa protein (p19) from tombuviruses that forms a dimeric clamp around the ds form of siRNA (Fig. 6B) (Vargason et al. 2003). The two p19 amino termini are positioned precisely around the two ends of the RNA duplex structure so that the two strands of the duplex cannot separate and release the active ss form. Other well-understood VSRs include P38 from carmoviruses that binds the AGO protein itself via a glycine-tryptophan (GW) hook motif (Azevedo et al. 2010), and P0 from poleroviruses that is an atypical F box protein that directs AGO degradation through an autophagy rather than the normal proteasome pathway (Derrien et al. 2012).

There are probably very few, if any, plant viruses that do not trigger RNA silencing in one form or another although there may be subtle differences. The RdRP step, for example, can be bypassed if the viral RNA has base-paired regions in foldback structures or, in viruses with circular DNA genomes, the dsRNA is formed by annealing of complementary RNAs transcribed from opposite DNA strand. The AGO step can also be dispensed with if the viral RNA exists in a ds form, allowing Dicer to act directly as the antiviral nuclease (Ding and Voinnet 2007).

RNA silencing operates in animals and plants, and the virus defense system described above is probably related to a primitive RNA silencing pathway that existed in primitive eukaryotic cells. Defense against viruses is likely to have been an essential function in this ancient cell type and it is likely that it used RNA silencing in a defense role, as in modern plants. RNA silencing might even be a vestige of the RNA world that is thought to have existed before DNA acquired its central role in inheritance (Salgado et al. 2006).

Since the divergence of animals and plants, it seems that RNA silencing pathways diversified in the different lineages of animals, plants, fungi, and other eukaryotic organisms. The end result is a number of variations on the RNA silencing theme in different organisms (also discussed in Ch. 16 [Martienssen and Moazed 2014]). These variations involve endogenous genetic elements including genes as well as viruses and transposons (Zamore 2006).

3.2 Multiple RNA Silencing Pathways

The diversity of RNA silencing mechanisms is very evident in plants (Eamens et al. 2008). The small RNAs in these pathways are referred to as either microRNAs (miRNAs) or siRNAs depending on the structure of their precursor (Fig. 7). The miRNAs are derived from a precursor RNA containing an inverted repeat structure whose two arms fold back on each other to form a base-paired structure that is the substrate for Dicer. A key feature of these RNAs is the mismatches that occur between the two arms of the foldback region that guide the Dicer nuclease to a single site in the foldback structure. Each precursor RNA of this type therefore generates a single miRNA that associates, as described above, with AGO. Most miRNAs are 21 nt although there are 22-nt and 24-nt forms. They are normally implicated in posttranscriptional silencing; The AGO nuclease either cleaves a target messenger RNA (mRNA) or blocks its translation through a mechanism that is not understood.

The siRNAs are generated by Dicer from precursors that, in contrast to the pre-miRNAs, are perfectly base-paired (Fig. 7). The absence of mismatched structures means that Dicer has no guide. It either cleaves randomly in the base-paired region or processively from one end. If cleavage is initiated from one end, the siRNAs are predominantly in the same register and phased (Chen et al. 2007; 2010). However, irrespective of whether the siRNAs are phased (Chen et al. 2010), multiple siRNAs can be generated from a single precursor. The existence of single or multiple short RNAs from a single precursor is a fundamental difference between miRNAs and siRNAs. Both host and viral genomes can generate siRNAs. siRNA precursors, which are ds, can be generated either by RdRP transcription of an ss template, the annealing of complementary RNAs, or foldback and base-pairing of inverted repeat regions as described above for virus-infected cells (Figs. 7 and 8).

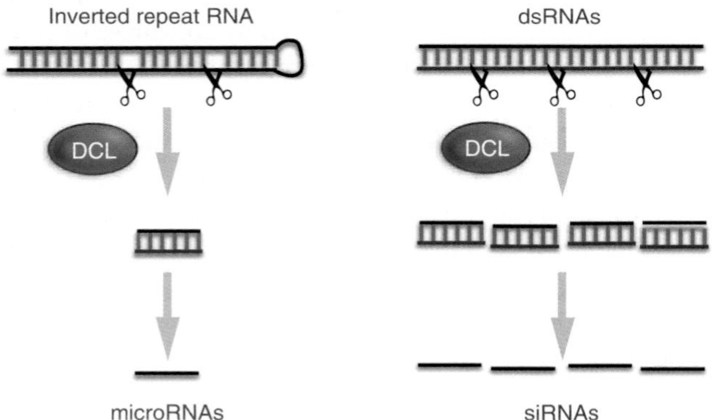

Figure 7. Endogenous RNA silencing in plants. The difference between miRNAs and siRNAs: miRNAs are generated by DCL cleavage of a single RNA molecule in which there is a secondary structure. Mismatches in base-paired regions guide the DCL protein so that it releases a single miRNA from the long precursor. In contrast, the siRNAs are derived from a perfectly base-paired precursor molecule that is cleaved at several sites to release multiple siRNAs.

Although there are several small RNA pathways in animals, the potential for diversity is more pronounced in plants. This difference between plants and animals is probably a consequence of the extreme plasticity of plant genomes that have facilitated the duplication and neofunctionalization of Dicer, AGO, and other RNA silencing factor genes. In flowering plants, genome plasticity accounts for more than a 1000-fold difference in range of haploid genome size between different species. This large range of genome sizes is due to repeated rounds of whole genome duplication, the activity of transposable elements, but also reductions in genome size due to genome deletions. The net result of this genomic flux is that many proteins in modern plants, not only those in RNA silencing, are encoded in multigene families (Van de Peer et al. 2009).

Plant Dicers provide one of the best examples of functional diversification within a gene family. These multigene families encode four or more members, depending on the species, and the different forms of DCL are functionally distinct (Gasciolli et al. 2005). Some Dicers generate 21-nt molecules that are involved in posttranscriptional silencing, as described above with miRNAs. They can produce both miRNAs and siRNAs. Other Dicers generate 22-nt siRNA or miRNAs that somehow influence the structure of the associated AGO so that the targeted RNA is converted into a ds form by an RdRP (Chen et al. 2010; Manavella et al. 2012). This dsRNA is then processed by Dicers to generate secondary siRNAs that associate with AGO proteins and participate in further rounds of targeting through RNA cleavage, translational arrest or RdRP recruitment (see Fig. 7 in Ch. 13 [Pikaard and Mittelsten Scheid 2014]). The consequence of this secondary siRNA pathway is to amplify the RNA silencing effect: A single 22-nt initiator siRNA or miRNA gives rise to many secondary siRNAs and a regulatory cascade (Chen et al. 2007).

A third class of Dicers generates 24-nt siRNAs (Xie et al. 2004). Like their shorter relatives, these 24-nt siRNAs associate with AGOs. However AGOs, like Dicers, are encoded by multigene families and they are functionally diverse (Havecker et al. 2010). The AGOs associating with 24-nt siRNAs, for example, are different from those that bind to 21- or 22-nt siRNAs; they act in the nucleus rather than the

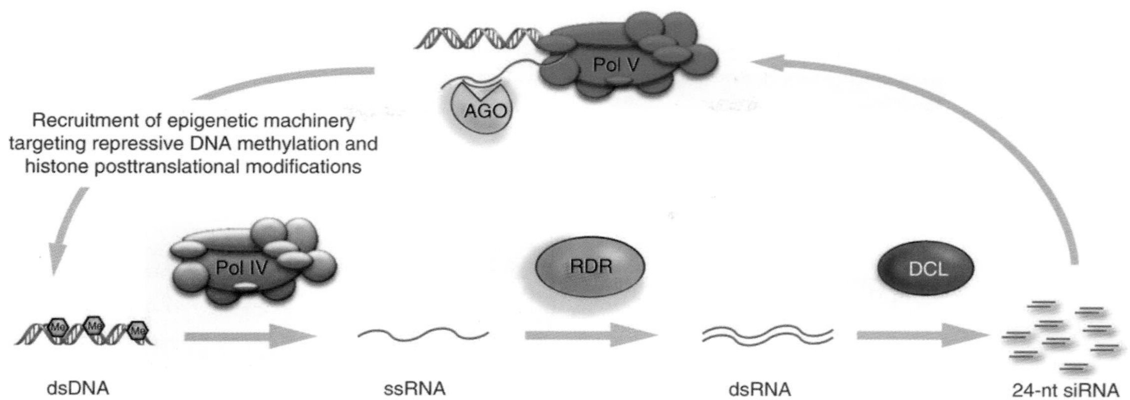

Figure 8. The Pol IV pathway of siRNA biogenesis. The variant form of RNA polymerase II (Pol II), known as Pol IV, generates single-stranded RNA (ssRNA) from a DNA template. The ssRNA is converted to a ds form by RdRP and then processed into 24-nt siRNAs by a DCL protein. The siRNA then binds to an AGO protein and it targets nascent transcripts in noncoding regions of the genome that are transcribed by a second variant form of Pol II known as Pol V. The AGO protein then recruits DNA methyltransferases to introduce methyl groups at cytosine bases (pink hexagon symbols) of the DNA template as well as other histone-modifying enzymes.

cytoplasm and they affect epigenetic modification of DNA or chromatin rather than mRNA (see Fig. 3 in Ch. 13 [Pikaard and Mittelsten Scheid 2014]).

This 24-nt RNA-directed epigenetic silencing mechanism is similar to posttranscriptional silencing because the AGOs are involved, and the specificity of targeting is due to Watson–Crick base-pairing of the siRNA. However, the role of the 24-nt siRNAs is distinct because they target a chromatin-associated RNA rather than an mRNA (Wierzbicki et al. 2008). This chromatin RNA is transcribed by an atypical form of DNA-dependent RNA polymerase II, known as Pol V (Fig. 8). The AGO proteins bound to 24-nt siRNA are distinct from those with 21-nt of 22-nt siRNAs because they guide the DRM2 DNA methyltransferase (Law and Jacobsen 2010). DRM2 introduces methyl groups onto C residues in DNA when it has been targeted to a region of DNA by an AGO complex so that the 24-nt siRNA is, in effect, the determinant of genomic regions that are to be methylated by DRM2. The involvement of DNA methylation gives this pathway an epigenetic property. This is discussed in detail in Section 3.5.2 of Ch. 13 (Pikaard and Mittelsten Scheid 2014) and illustrated in their Figure 6.

3.3 Virus-Induced Silencing of Promoters

The different RNA silencing pathways described in Section 3.2 are relevant to the epigenetic case history of virus-induced promoter silencing because viral siRNAs include the 24-nt-long species (Deleris et al. 2006) that have been implicated in epigenetic silencing, as described above. The smaller 21- or 22-nt species are also produced from viral RNA, but they are associated with posttranscriptional gene silencing. These smaller siRNAs are important in defense, particularly against RNA viruses.

Virus-induced posttranscriptional gene silencing is illustrated very clearly by an experiment with transgenic plants expressing the green fluorescent protein under the control of a strong plant viral promoter, called 35S (Fig. 9). These plants were green fluorescent under UV light. If these plants were then infected with an RNA virus that had been modified to carry a part of the green fluorescent protein (GFP) coding sequence (Fig. 9B), the GFP fluorescence faded and the GFP mRNA was degraded through the siRNA- and AGO-based mechanism described above (Fig. 6). The virus used in this experiment did produce a weak VSR (Martin-Hernandez and Baulcombe 2008), but clearly it could not completely block the virus-induced silencing of the GFP transgene in the infected plants.

Virus-induced posttranscriptional silencing affects the host if, by chance, a virus or virus-associated nucleic acid has sequence similarity to a host mRNA. The viral siRNAs target the host gene and the symptoms of the infected plant reflect a corresponding loss-of-function (Shimura et al. 2011). Viral silencing is also a useful technology in functional genomics (Baulcombe 1999); the basic idea is simply to produce a library of viral clones with inserts corresponding to different host genes. Plants are then infected with individual clones and monitored for symptoms in the expectation that they would reflect the function of the gene insert. This approach has been very successful, for example, in the identification of genes required for disease resistance (Lu et al. 2003).

An epigenetic rather than posttranscriptional effect of the virus-derived 24-nt siRNA is also illustrated by the use of

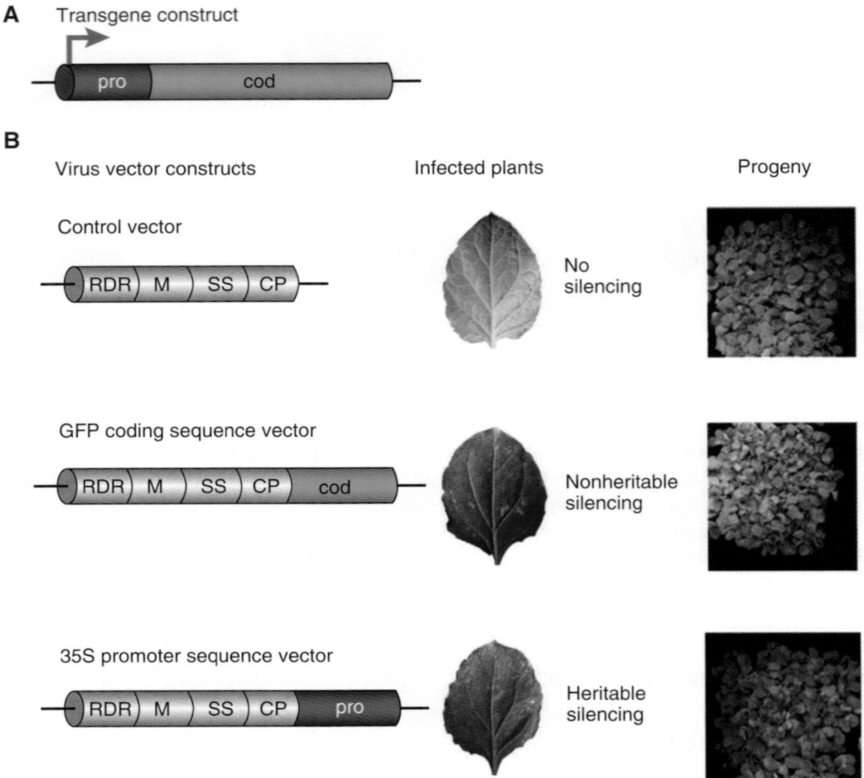

Figure 9. Virus-induced gene silencing—VIGS. A demonstration of VIGS in *Nicotiana benthamiana*, a plant related to tobacco, carrying an endogenous copy of a transgene construct (*A*) containing a 35S promoter (pro) driving GFP coding sequences (cod). (*B*) The tobacco rattle virus vector constructs are replicating RNA molecules that encode several proteins. The proteins include a viral RNA-dependent RNA polymerase (RDR), a movement protein (M), a suppressor of silencing (SS), and a coat protein (CP). The control virus vector (top) has no insert. The experimental constructs carried an insert corresponding either to "pro" or "cod." The loss of green fluorescence in the plants infected with the experimental constructs indicated that there was gene silencing using both constructs (*center*); however, with the pro construct, silencing persisted into the progeny seedlings (*right* side).

GFP transgenic plants. For epigenetic silencing, the insert in the virus vector must correspond to the transgene promoter rather than the coding sequence. The infected plants show loss of GFP with virus constructs carrying promoter or coding sequence inserts (Fig. 9B). However, heritable GFP silencing is caused by targeted methylation of the DNA of the associated gene (in this case, the integrated transgene).

This transcriptional gene silencing (TGS) mechanism can also target endogenous plant genomic elements if there is sequence similarity between the virus and the plant genome. This occurs because C residues in the endogenous promoter become methylated in the infected plant, blocking transcription of its adjacent gene. There was transcriptional silencing and promoter DNA methylation, for example, when the GFP transgenic plants described above were infected with a tobacco rattle RNA virus carrying part of the 35S promoter sequence in an RNA form (Fig. 9) (Jones et al. 2001).

The mechanism of gene silencing has not been explored in the tobacco rattle virus experiments but, based on other epigenetic systems in plants, it is likely to involve the DRM1/2 DNA methyltransferases (Dnmt homologs) and histone-modifying enzymes including histone deacetylases and lysine methyltransferases. Acetylation of lysine residues weakens the binding of histones to DNA so that there is less resistance to the translocation of RNA polymerase. Methylation of lysines, particularly at histone H3 lysine 9, is similarly correlated with silent chromatin (Fig. 8).

3.4 Separate Establishment and Maintenance of Virus-Induced Gene Silencing

Epigenetic mechanisms are classically defined when they persist through cell generations even if the initiation stimulus is transient. This defining feature of epigenetic regulation is clearly illustrated by the persistence of GFP

Cite as *Cold Spring Harb Perspect Biol* doi: 10.1101/cshperspect.a019471

silencing in the progeny of viral infected plants possessing the pro insert (Fig. 9B) (Jones et al. 2001). The viral constructs in these GFP experiments are designed to not transmit the virus through the seed, so the different generations provide clear separation of establishment and maintenance of the silencing phenotype; that is, establishment of silencing in the infected plant is clearly dependent on the virus whereas maintenance in subsequent generations is virus-independent. The transcriptional silencing observed using viral constructs containing 35S promoter homology was truly epigenetic because it persisted in several generations of virus-free progeny. In contrast, the experiments with the GFP coding sequence virus showed no silencing in the next generation (Fig. 9B). The persistence of posttranscriptional silencing is, therefore, dependent on the presence of the initiator virus and the effect is not epigenetic.

The difference between the establishment and maintenance phases arises because DNA methylation in TGS can be maintained, at least partially, in the absence of the initiator virus by maintenance DNA methyltransferase activity that is distinct from the de novo enzymes involved in establishment. The methyltransferase MET1, for example, causes C residue methylation to be replicated if the C is adjacent to a G on its 3′ side (Fig. 10). The daughter strand is methylated on the C residue opposite to this G and, following the same pattern in a second round of replication, the methylation is added to a C in the original position. Such a mechanism does not, of course, maintain methylation at C residues that are not adjacent to G, and so the extent of DNA methylation in the maintenance phase of an epigenetic mechanism is less extensive than in the presence of the initiator.

This distinction of initiation and maintenance phases is central to many epigenetic phenomena including those in which the mechanisms do not involve DNA methylation. In the vernalization case history described above, for example, the initiation phase is the nucleation of the PHD-PRC at the intronic site in *FLC* and maintenance is associated with the complex spread and LHP1 function.

3.5 Mobile Silencing

Maintenance of epigenetic effects in plants is effective if the initiator stimulus is applied to the meristems. The silenced state can then be passed to new cells, tissues, and organs as they arise. In vernalization, for example, the cold stimulus is perceived in meristems and the silenced state of *FLC* is propagated with the replication of chromatin as the cells divide. However, there are also instances in which initiation and maintenance are in separate cells and there are mobile signals that move between cells (the topic of Ch. 2 [Dunoyer et al. 2013]).

In instances of epigenetic RNA silencing, mobile signals were illustrated by experiments in which shoots from one genotype of plant were grafted to another (Fig. 11) (Melnyk et al. 2011). One illustrative experiment was based on plants in which a GFP target transgene was expressed under the control of a meristem-specific promoter and enhancer, and in which a silencer transgene was designed to produce siRNA to target the necessary enhancer component. The target transgene was methylated at C residues and silenced in root meristems if they were grafted to shoots carrying the silencer. However, this promoter DNA methylation and silencing did not occur if the silencer shoots were defective for the 24-nt siRNA-specific Dicer. Correspondingly, the 24-nt siRNAs were present in the grafted root if the shoot was from the wild type but not Dicer mutant shoot. These and other experiments illustrated clearly that 24-nt siRNAs are the

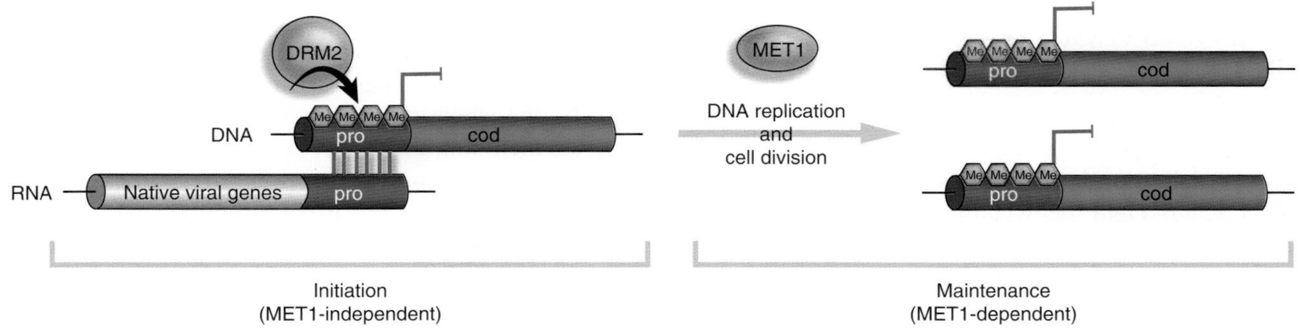

Figure 10. Separate establishment and maintenance of transcriptional silencing induced by a virus. The red vertical lines represent the interaction between the viral RNA containing the promoter sequence insert (pro) and the cognate DNA. The promoter DNA is assumed to be the target of the RNA silencing pathway to which methyl groups are introduced (Fig. 8). TGS initially occurs through de novo DNA methylation (pink hexagon symbols) of the promoter sequence, catalyzed by DRM2. Maintenance of silencing relies on the maintenance DNA methyltransferase, MET1 propagating methylation patterns through DNA replication and cell division.

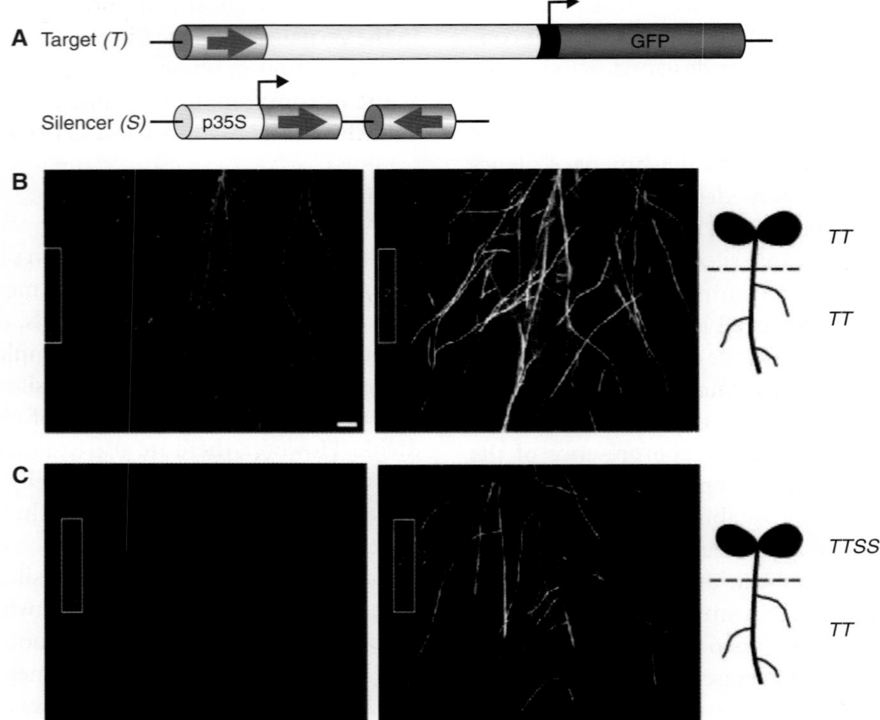

Figure 11. An experiment to show mobile silencing. (*A*) Plants carrying the target *GFP* transgene coupled to a promoter with meristem-specific enhancer region (arrow) are denominated TT, and were hybridized to SS plants carrying a silencer construct with a 35S promoter that directed transcription of an inverted repeat of the enhancer (arrows). The TT (*B*) and TTSS plants (*B*) were grafted as a shoot scion to the TT roots and the expression of GFP was monitored by GFP fluorescence and RNA gel blotting to track spread of silencing from the shoot into the root. In this analysis the panels shown in *B* and *C* are images of the roots under UV (*left*) or white (*right*) light. TTSS shoot grafts cause GFP silencing in the roots, indicating the mobile nature of the silencing signal. (Adapted from Melnyk et al. 2011.)

mobile signals in plants and that they can initiate epigenetic effects in recipient meristems (Molnar et al. 2010; Melnyk et al. 2011; described in Ch. 2 [Dunoyer et al. 2013]).

4 RESETTING VERSUS TRANSGENERATIONAL INHERITANCE

These two case histories illustrate examples of environmentally induced epigenetic change that are fundamentally different. Vernalization involves Polycomb group proteins, is independent of DNA methylation, and is reset in each generation. The resetting is important as an environmental response: Plants with a vernalization requirement have to experience an appropriate cold period (winter) in each generation before they will flower. Virus-induced transcriptional gene silencing, in contrast, is dependent on DNA methylation, is targeted by siRNA, and persists between generations.

Currently, we do not understand the mechanism that ensures that methylation marks at promoter DNA persist through meiosis, gametogenesis and early embryogenesis. In fact, we do not understand the broader question of why some epigenetic marks are heritable whereas others are not. In part the difference could be associated with the different types of epigenetic marks: Polycomb marks could be more prone to resetting whereas DNA methylation is more likely to be transmitted between generations. However, this cannot be the whole story because many DNA methylation marks, including those that are set by viruses, are sometimes lost in the next generation (Kanazawa et al. 2011; Otagaki et al. 2011). The persistent marks may have more or particular types of methylated C residues and be more difficult to erase than those that are reset. Alternatively, it could be that heritable epigenetic marks generate a signal that is transmitted from the maternal plant to the progeny, which guides reestablishment in the early zygote or meristematic cells of the next generation.

Evidence in support of the latter model derives from an analysis of transposon silencing in the pollen of *Arabidopsis*. The transposon marks are lost in the somatic cell of the

 Cite as *Cold Spring Harb Perspect Biol* doi: 10.1101/cshperspect.a019471

pollen so that transposons are activated. RNA from the active transposons serves as a precursor of siRNA production that is then transported into the germ cells of the pollen grain. These cells then fertilize the egg cell and give rise to the next generation. In addition, they carry the mobile siRNA into the fertilized egg so that it can reinforce the establishment of epigenetic marks in the zygote (Slotkin et al. 2009; further discussed in Ch. 2 [Dunoyer et al. 2013]).

5 TRANSIENT EPIGENETIC REGULATION IN RESPONSE TO STRESS

The two case histories of environmentally induced epigenetic regulation are well understood at the molecular level. However, there are many other examples of effects in plants that may have an epigenetic basis including those that are induced by stress.

One of these candidate epigenetic phenomena is well-known to gardeners who "harden off" plants by a mild exposure to low temperature so that they are then protected against later freezing temperatures. Of course, this process needs to be repeated in each generation. The concept of deficit irrigation, in which long-term drought tolerance can be induced by transient or partial drying, could similarly be explained in terms of epigenetic mechanisms that are reset in each generation (Davies et al. 2010). Epigenetics is invoked in these two examples because the response persists for an extended period after the initial induction; that is, there is a molecular memory that is unlikely to involve mutation or other types of genome alteration. Other stresses appear to be less effective than temperature or drought at inducing epigenetic changes, although exposure of *Arabidopsis* to UV irradiation may induce a long-lived increase in somatic recombination (Molinier et al. 2006).

The mechanisms underlying these phenomena are still being elucidated. In principle, self-reinforcing mechanisms such as autoregulatory transcription factors (Ptashne 2007) could play a role. However, there is good evidence that chromatin- and RNA-based mechanisms are involved, as described above. Early work showing a role of temperature in siRNA production hinted at a role for RNA in this environmental memory (Szittya et al. 2003). Other studies revealed changes in nucleosome spacing and transient chromatin decondensation following exposure to different high temperature regimes (Pecinka et al. 2010). With mild increases in temperature, there is evidence that H2A.Z may act as a thermosensor (Kumar and Wigge 2010). This histone H2A variant is normally associated with transcription start points.

Most of these temperature-sensitive chromatin changes and the associated up- or down-regulation of gene expression are transient. However, the transcripts of a COPIA 78 retrotransposon family were enhanced for up to 7 d after exposure to stress. This effect was independent of DNA demethylation or changes to histone H3K9 methylation, but it was affected by siRNA pathways. One interpretation of these results is that, as a result of siRNA action, an epigenetic mark may constitute the way nucleosome loading occurs rather than direct chemical modification of the DNA or chromatin (Ito et al. 2011).

A longer-lived epigenetic "memory" occurs following exposure of plants to various pathogens. The initial exposure to the pathogen may activate defense systems involving hormones including salicylic acid, and genes that interfere with growth of the pathogen. The primary responses are, however, transient, but if the plants are then exposed to a second pathogen, the genes are activated more rapidly than after the first inoculation. A similar defense priming effect is induced by β-amino butyric acid.

This defense priming memory persists for 28 d or even longer. The possibility that pathogen-induced changes are epigenetic is further reinforced by the extensive and dynamic changes to the methylation of genomic DNA in *Arabidopsis* infected with bacteria or treated with the defense hormone salicylic acid (Dowen et al. 2012). The key changes may involve loss of DNA methylation at defense gene loci so that they are more easily activated at the transcriptional level (Yu et al. 2013).

6 TRANSGENERATIONAL EPIGENETIC REGULATION IN RESPONSE TO STRESS

The examples of stress-induced effects discussed in the previous section are mostly transient. However, the virus-induced gene silencing case history illustrates the potential for transgenerational epigenetic effects in plants if a stress induces siRNAs or chromatin changes. There are several intriguing observations that are consistent with this prediction. In *Mimulus*, for example, artificial herbivory induces heritable alterations in expression of a gene controlling trichome production and, correspondingly, transgenerational changes in trichome density (Holeski et al. 2010). In dandelions (*Taraxacum officinale*), the genome-wide pattern of DNA methylation is modified if the parental plants are exposed to environmental stress and the progeny show modifications of root/shoot biomass ratio, P content, leaf morphology, and stress tolerance relative to the control. DNA methylation is implicated because chemical suppression of DNA methyltransferase blocks the transgenerational effects (Verhoeven and Van Gurp 2012).

A third example involves rice plants subjected to the stress involved in regeneration from tissue culture. The regenerated plants had changes to the genome-wide pattern of DNA methylation, including some at the promoters of genes. The changes are predominantly loss rather than gain of DNA methylation and they persist in the regenerated plants and their progeny (Pellegrini et al. 2013).

These miscellaneous examples should be taken as indicators rather than proof of transgenerational epigenetic change affecting an adaptive phenotype. None of these examples provide definitive evidence that changes to gene expression and/or the epigenome are causal of phenotypic changes; there are either no phenotypes or the evidence is correlative. In *T. officinale*, in addition, the seeds are produced apomictically (i.e., asexually, without meiosis) and the mechanisms may not represent those that apply in sexual reproduction.

However, there is one example related to defense priming in which there is good evidence for a transgenerational epigenetic effect that could be adaptive (Luna et al. 2012). The progeny of *Arabidopsis* plants infected with bacteria were more resistant to secondary infection with an oomycete (fungi-like unicellular organisms) than the control progeny of unprimed plants. This effect was evident in the first-generation progeny, as with the phenomena described in the previous paragraph. However, oomycete resistance was also present in the second-generation progeny that were produced on plants that were not themselves primed by infection. This observation is crucial because it rules out inherited resistance being due to the transport of a physiological or biochemical factor from parents to seed—the most likely explanation is therefore epigenetic.

Chromatin analysis of defense genes has reinforced the notion that inherited priming is because of epigenetic mechanisms. Defense genes that are up-regulated following primed resistance were associated with the enrichment of acetylated histones—a known activating epigenetic mark—in the promoter region. In contrast, down-regulated genes following primed resistance had higher levels of the repressive mark H3K27me3. However, this is not a simple mechanism because plants that were defective for DNA methylation at CpHpG sites mimicked the transgenerational priming (Luna and Ton 2012). It is likely, therefore, that transgenerational priming is mediated by hypomethylation of DNA at these CpHpG sites; there may be cascades of epigenetic regulation in which stress leads to a loss of repressive marks that, in turn, trigger activating epigenetic marks.

7 EPIGENETIC EFFECTS ON GENOME STRUCTURE

Epigenetic mechanisms are linked in various ways to genome structure and there are several indications that environmental stimuli induce epigenetic mechanisms with genetic consequences. For example, there is evidence that DNA repair is affected by RNA silencing in plants (Wei et al. 2012). It might be expected, therefore, that there is an increase in mutations in the progeny of stressed plants because of the targeting of the DNA repair machinery by siRNAs. However, this possibility has not yet been tested experimentally and, at present, is hypothetical. Similarly, there is the hypothetical possibility, based on a precedent in mammalian cancer, that methylation of C bases accelerates the transition to T at genomic loci that are hypermethylated in response to stress (Laird and Jaenisch 1994).

Other genetic effects could be related to the epigenetic control of transposons. The transcription of the Onsen family of copia retrotransposons, for example, was increased after extreme temperature shifts (4°C to 37°C) and, consistent with the involvement of epigenetics, this effect persisted for up to seven days. This activation of Onsen results in frequent retrotranspositions in the progeny of the stressed plants if the plants are mutant for siRNA production (Ito et al. 2011) and the new inserts may disrupt the gene or direct the establishment of epigenetic marks. In either scenario, the induced epigenetic effect would have induced a genetic rearrangement leading to an altered pattern of gene expression.

A study of nutrient stress-induced heritable changes in flax may also illustrate the potential of the environment to induce DNA rearrangements that influence gene expression. Flax plants with a "plastic" genotype grew as large plants under high nutrient conditions and were small under low nutrient conditions. Many of the induced differences persisted for several generations in progeny grown under standard nutrient conditions. The inherited differences include changed height and weight, nuclear DNA content and methylation, ribosomal gene number, and seed capsule characters. These phenotypes were not graft-transmissible, but they are stable in subsequent generations (Cullis 1986).

An insertion element (LIS-1) was reconstituted reproducibly and independently in different flax lineages after the nutrient stress (Chen et al. 2005) and is likely to be one of many genomic changes associated with this heritable adaptation. Unfortunately, this phenomenon to nutrient stress is not yet fully characterized by advanced genomic techniques or next-generation sequencing. However, there are striking resonances with the genome rearrangements that occur in protozoans, mediated by a variation on the basic RNA-directed mechanism of epigenetic modification (Mochizuki et al. 2002; see Ch. 11 [Chalker et al. 2013] for more detail). It will be interesting to find out whether this flax system is a plant example of a link between epigenetics and RNA-directed genome rearrangement.

 Cite as *Cold Spring Harb Perspect Biol* doi: 10.1101/cshperspect.a019471

8 THE HERITABLE LEGACY OF INDUCED EPIGENETIC CHANGES

Epigenetic marks show a tendency to undergo spontaneous gain or loss, as indicated by the analysis of a 30-generation pedigree in *Arabidopsis* (Becker et al. 2011; Schmitz et al. 2011). It is not obvious why some loci should be prone to spontaneous epigenetic change, although the presence of overlapping and diverging transcripts may have an effect (Havecker et al. 2012). Perhaps such a configuration affects the chromatin structure so that epigenetic marks are gained and lost more easily than in other regions of the genome. Other changes may have resulted from genome interactions in hybrid ancestors of the modern varieties. In the allotetraploids of *Arabidopsis*, for example, there is a loss of repressive histone marks from the *CCA*1 and *LHY* circadian clock regulators, and a consequent up-regulation of 130 downstream genes that are regulated by these factors (Ni et al. 2009).

There is also evidence in hybrids of cultivated and wild tomato of altered production of siRNAs and, at one locus at least, a change in the methylation status at the associated DNA. In this instance, it seems likely that there is indeed an induced epigenetic change as a result of genome shock in the hybrid plant. However, the epigenetic change was not initiated until the F_2 or later generations. It could be that there is a delayed onset of the epigenetic change or that it was initiated in the F_1 and that its strength increased progressively over several generations (Shivaprasad et al. 2012). A priority for future research will be to find out the relative contribution of the different mechanisms to the epigenomes of wild and cultivated plants.

9 PERSPECTIVE

Epigenetic regulation can be considered as an additional layer in the genetic regulation of complex systems that is subject to environmental influence. However, unlike many other regulatory mechanisms, the epigenetic systems have the potential to store information over time—they are a molecular memory. This memory can be viewed as part of a "soft inheritance" system (Richards 2006). The "soft" descriptor refers to the potential for environmental influence and rapid introduction of heritable phenotypic effects; the "hard" inheritance of genetics, in contrast, is relatively insensitive to these external influences.

A striking illustration of soft inheritance is shown by genotypically identical (doubled haploid) oilseed rape lineages that were selected for either high- or low-respiration rates (Hauben et al. 2009). Four rounds of selection gave rise to lineages with heritable differences in energy use efficiencies and yield potential. It is unlikely that such differences could have arisen through genetic change in such a short time; thus, an epigenetic explanation is most likely.

There is, therefore, great potential for environmentally mediated epigenetic change to generate variation, and thus influence evolution in plants. The ease of connecting environmental and epigenetic regulation in plants is likely to be good for the cross-fertilization of ideas between the plant and animal kingdoms.

REFERENCES

*Reference is also in this book.

Aikawa S, Kobayashi MJ, Satake A, Shimizu KK, Kudoh H. 2010. Robust control of the seasonal expression of the *Arabidopsis FLC* gene in a fluctuating environment. *Proc Natl Acad Sci* **107:** 11632–11637.

Angel A, Song J, Dean C, Howard MA. 2011. A Polycomb-based switch underlying quantitative epigenetic memory. *Nature* **476:** 105–108.

Azevedo J, Garcia D, Pontier D, Ohnesorge S, Yu A, Garcia S, Braun L, Bergdoll M, Hakimi MA, Lagrange T, et al. 2010. Argonaute quenching and global changes in Dicer homeostasis caused by a pathogen-encoded GW repeat protein. *Genes Dev* **24:** 904–915.

Baulcombe DC. 1999. Fast forward genetics based on virus-induced gene silencing. *Curr Opin Plant Biol* **2:** 109–113.

Baulcombe D. 2004. RNA silencing in plants. *Nature* **431:** 356–363.

Becker C, Hagmann J, Muller J, Koenig D, Stegle O, Borgwardt K, Weigel D. 2011. Spontaneous epigenetic variation in the *Arabidopsis thaliana* methylome. *Nature* **480:** 245–249.

Burn JE, Bagnall DJ, Metzger JD, Dennis ES, Peacock WJ. 1993. DNA methylation, vernalization, and the initiation of flowering. *Proc Natl Acad Sci* **90:** 287–291.

* Chalker DL, Meyer E, Mochizuki K. 2013. Epigenetics of ciliates. *Cold Spring Harb Perspect Biol* **5:** a017764.

Chen Y, Schneeberger RG, Cullis CA. 2005. A site-specific insertion sequence in flax genotrophs induced by environment. *New Phytol* **167:** 171–180.

Chen H-M, Li Y-H, Wu S-H. 2007. Bioinformatic prediction and experimental validation of a microRNA-directed tandem trans-acting siRNA cascade in *Arabidopsis*. *Proc Natl Acad Sci* **104:** 3318–3323.

Chen H-M, Chen L-T, Patel K, Li Y-H, Baulcombe DC, Wu S-H. 2010. 22-nucleotide RNAs trigger secondary siRNA biogenesis in plants. *Proc Natl Acad Sci* **107:** 15269–15274.

Choi J, Hyun Y, Kang MJ, In Yun H, Yun JY, Lister C, Dean C, Amasinno RM, Noh B, Noh YS, et al. 2009. Resetting and regulation of *Flowering Locus C* expression during *Arabidopsis* reproductive development. *Plant J* **57:** 918–931.

Cousthain V, Li P, Strange A, Lister C, Song J, Dean C. 2012. Quantitative modulation of Polycomb silencing underlies natural variation in vernalization. *Science* **337:** 584–587.

Crevillén P, Dean C. 2010. Regulation of the floral repressor gene *FLC*: The complexity of transcription in a chromatin context. *Curr Opin Plant Biol* **14:** 38–44.

Cullis CA. 1986. Phenotypic consequences of environmentally induced changes in plant DNA. *Trends Genet* **2:** 307–309.

Davies WJ, Zhang J, Yang J, Dodd IC. 2010. Novel crop science to improve yield and resource use efficiency in water-limited agriculture. *J Agric Sci* **149:** 123–131.

Deleris A, Gallego-Bartolome J, Bao J, Kasschau KD, Carrington JC, Voinnet O. 2006. Hierarchical action and inhibition of plant Dicer-like proteins in antiviral defense. *Science* **313:** 68–71.

De Lucia F, Crevillen P, Jones AM, Greb T, Dean C. 2008. A PHD-polycomb repressive complex 2 triggers the epigenetic silencing of *FLC* during vernalization. *Proc Natl Acad Sci* **105:** 16831–16836.

Derrien B, Baumberger N, Schepetilnikov M, Viotti C, De Cillia J. 2012. Degradation of the antiviral component ARGONAUTE1 by the autophagy pathway. *Proc Natl Acad Sci* **109:** 15942–15946.

Ding S-W, Voinnet O. 2007. Antiviral immunity directed by small RNAs. *Cell* **130:** 413–426.

Dowen RH, Pelizzola M, Schmitz RJ, Lister R, Dowen JM, Nery JR, Dixon JE, Ecker JR. 2012. Widespread dynamic DNA methylation in response to biotic stress. *Proc Natl Acad Sci* **109:** E2183–E2191.

Dunoyer P, Voinnet O. 2005. The complex interplay between plant viruses and host RNA-silencing pathways. *Curr Opin Plant Biol* **8:** 415–423.

* Dunoyer P, Melnyk CW, Molnar A, Slotkin RK. 2013. Plant mobile small RNAs. *Cold Spring Harb Perspect Biol* **5:** a017897.

Eamens A, Wang M-B, Smith NA, Waterhouse PM. 2008. RNA silencing in plants: yesterday, today, and tomorrow. *Plant Physiol* **147:** 456–468.

Finnegan EJ, Peacock WJ, Dennis ES. 2000. DNA methylation, a key regulator of plant development and other processes. *Curr Opin Genet Dev* **2:** 217–223.

Finnegan E, Kovac KA, Jaligot E, Sheldon CC, James Peacock W, Dennis ES. 2005. The downregulation of *FLOWERING LOCUS C* (*FLC*) expression in plants with low levels of DNA methylation and by vernalization occurs by distinct mechanisms. *Plant J* **44:** 420–432.

Gasciolli V, Mallory AC, Bartel DP, Vaucheret H. 2005. Partially redundant functions of *Arabidopsis* DICER-like enzymes and a role for DCL4 in producing trans-acting siRNAs. *Curr Biol* **15:** 1494–1500.

Gazzani S, Gendall AR, Lister C, Dean C. 2003. Analysis of the molecular basis of flowering time variation in *Arabidopsis* accessions. *Plant Physiol* **132:** 1107–1114.

Gendall AR, Levy YY, Wilson A, Dean C. 2001. The *VERNALIZATION 2* gene mediates the epigenetic regulation of vernalization in *Arabidopsis*. *Cell* **107:** 525–535.

Geraldo N, Bäurle I, Kidou S, Hu X, Dean C. 2009. FRIGIDA delays flowering in *Arabidopsis* via a cotranscriptional mechanism involving direct interaction with the nuclear cap-binding complex. *Plant Physiol* **150:** 1611–1618.

Greb T, Mylne JS, Crevillen P, Geraldo N, An H, Gendall AR, Dean C. 2007. The PHD finger protein VRN5 functions in the epigenetic silencing of *Arabidopsis FLC*. *Curr Biol* **17:** 73–78.

* Grossniklaus U, Paro R. 2014. Transcriptional silencing by *Polycomb*-group proteins. *Cold Spring Harb Perspect Biol* **6:** a019331.

Gutierrez-Marcos JF, Dickinson HG. 2012. Epigenetic reprogramming in plant reproductive lineages. *Plant Cell Physiol* **53:** 817–823.

Hauben M, Haesendonckx B, Standaert E, Van Der Kelen K, Azmi A, Van Breusegem F, Guisez Y, Bots M, Lambert B, Laga B, et al. 2009. Energy use efficiency is characterized by an epigenetic component that can be directed. *Proc Natl Acad Sci* **106:** 20109–20114.

Havecker ER, Wallbridge LM, Hardcastle TJ, Bush MS, Kelly KA, Dunn RM, Schwach F, Doonan JH, Baulcombe DC. 2010. The *Arabidopsis* RNA-directed DNA methylation argonautes functionally diverge based on their expression and interaction with target loci. *Plant Cell* **22:** 321–334.

Havecker ER, Wallbridge LM, Fedito P, Hardcastle TJ, Baulcombe DC. 2012. Metastable differentially methylated regions within *Arabidopsis* inbred populations are associated with modified expression of non-coding transcripts. *PloS One* **7:** e45242.

He Y, Doyle MR, Amasino RM. 2004. PAF1-complex-mediated histone methylation of *FLOWERING LOCUS C* chromatin is required for the vernalization-responsive, winter-annual habit in *Arabidopsis*. *Genes Dev* **18:** 2774–2784.

Holeski LM, Chase-Alone R, Kelly JK. 2010. The genetics of phenotypic plasticity in plant defense: Trichome production in *Mimulus guttatus*. *Am Nat* **175:** 391–400.

Hornyik C, Terzi LC, Simpson GG. 2010. The spen family protein FPA controls alternative cleavage and polyadenylation of RNA. *Dev Cell* **18:** 203–213.

Ito H, Gaubert H, Bucher E, Mirouze M, Vaillant I, Paszkowski J. 2011. An siRNA pathway prevents transgenerational retrotransposition in plants subjected to stress. *Nature* **472:** 115–119.

Jiang D, Gu X, He Y. 2009. Establishment of the winter-annual growth habit via *FRIGIDA*-mediated histone methylation at *FLOWERING LOCUS C* in *Arabidopsis*. *Plant Cell* **21:** 1733–1746.

Johanson U, West J, Lister C, Michaels S, Amasino R, Dean C. 2000. Molecular analysis of *FRIGIDA*, a major determinant of natural variation in *Arabidopsis* flowering time. *Science* **290:** 344–347.

Jones L, Ratcliff F, Baulcombe DC. 2001. RNA-directed transcriptional gene silencing in plants can be inherited independently of the RNA trigger and requires Met1 for maintenance. *Curr Biol* **11:** 747–757.

Kanazawa A, Inaba J, Shimura H, Otagaki S, Tsukahara S, Matsuzawa A, Kim BM, Goto K, Masuta C. 2011. Virus-mediated efficient induction of epigenetic modifications of endogenous genes with phenotypic changes in plants. *Plant J* **65:** 156–168.

Kumar SV, Wigge PA. 2010. H2A.Z-containing nucleosomes mediate the thermosensory response in *Arabidopsis*. *Cell* **140:** 136–147.

Laird PW, Jaenisch R. 1994. DNA methylation and cancer. *Hum Mol Genet* **3:** 1487–1495.

Law JA, Jacobsen SE. 2010. Establishing, maintaining and modifying DNA methylation patterns in plants and animals. *Nat Rev Genet* **11:** 204–220.

Liu F, Quesada V, Crevillén P, Bäurle I, Swiezewski S, Dean C. 2007. The *Arabidopsis* RNA-binding protein FCA requires a lysine-specific demethylase 1 homolog to downregulate *FLC*. *Mol Cell* **28:** 398–407.

Liu F, Marquardt S, Lister C, Swiezewski S, Dean C. 2010. Targeted 3′ processing of antisense transcripts triggers *Arabidopsis FLC* chromatin silencing. *Science* **327:** 94–97.

Lu R, Malcuit I, Moffett P, Ruiz MT, Peart J, Wu A-J, Rathjen JP, Bendahmane A, Day L, Baulcombe D. 2003. High throughput virus-induced gene silencing implicates heat shock protein 90 in plant disease resistance. *EMBO J* **22:** 5690–5699.

Luna E, Bruce TJA, Roberts MR, Flors V, Ton J. 2012. Next-generation systemic acquired resistance. *Plant Physiol* **158:** 844–853.

Luna E, Ton J. 2012. The epigenetic machinery controlling transgenerational systemic acquired resistance. *Plant Signal Behav* **7:** 615–618.

Manavella PA, Koenig D, Weigel D. 2012. Plant secondary siRNA production determined by microRNA-duplex structure. *Proc Natl Acad Sci* **109:** 2461–2466.

* Martienssen R, Moazed D. 2014. RNAi and heterochromatin assembly. *Cold Spring Harb Perspect Biol* doi: 10.1101/cshperspect.a019323.

Martin-Hernandez AM, Baulcombe DC. 2008. Tobacco rattle virus 16-kilodalton protein encodes a suppressor of RNA silencing that allows transient viral entry in meristems. *J Virol* **82:** 4064–4071.

Melnyk CW, Molnar A, Bassett A, Baulcombe DC. 2011. Mobile 24 nt small RNAs direct transcriptional gene silencing in the root meristems of *Arabidopsis thaliana*. *Curr Biol* **21:** 1678–1683.

Michaels SD, Amasino RM. 1999. FLOWERING LOCUS C encodes a novel MADS domain protein that acts as a repressor of flowering. *Plant Cell* **11:** 949–956.

Mochizuki K, Fine NA, Fujisawa T, Gorovsky MA. 2002. Analysis of a piwi-related gene implicates small RNAs in genome rearrangement in *Tetrahymena*. *Cell* **110:** 689–699.

Molinier J, Ries G, Zipfel C, Hohn B. 2006. Transgeneration memory of stress in plants. *Nature* **442:** 1046–1049.

Molnar A, Melnyk CW, Bassett A, Hardcastle TJ, Dunn R, Baulcombe DC. 2010. Small silencing RNAs in plants are mobile and direct epigenetic modification in recipient cells. *Science* **328:** 872–875.

Mylne JS, Barrett L, Tessadori F, Mesnage S, Johnson L, Bernatavichute YV, Jacobsen SE, Fransz P, Dean C. 2006. LHP1, the *Arabidopsis* homolog of HETEROCHROMATIN PROTEIN1, is required for epigenetic silencing of *FLC*. *Proc Natl Acad Sci* **103:** 5012–5017.

Ni Z, Kim ED, Ha MS, Lackey E, Liu JX, Zhang YR, Sun QX, Chen ZJ. 2009. Altered circadian rhythms regulate growth vigour in hybrids and allopolyploids. *Nature* **457:** 327–331.

Oh S, Zhang H, Ludwig P, van Nocker S. 2004. A mechanism related to the yeast transcriptional regulator Paf1c is required for expression of the *Arabidopsis FLC/MAF* MADS box gene family. *Plant Cell* **16:** 2940–2953.

Otagaki S, Kawai M, Masuta C, Kanazawa A. 2011. Size and positional effects of promoter RNA segments on virus-induced RNA-directed DNA methylation and transcriptional gene silencing. *Epigenetics* **6:** 681–691.

Park S, Oh S, Ek-Ramos J, van Nocker S. 2010. *PLANT HOMOLOGOUS TO PARAFIBROMIN* is a component of the PAF1 complex and assists in regulating expression of genes within H3K27ME3-enriched chromatin. *Plant Physiol* **153:** 821–831.

Pecinka A, Dinh HQ, Baubec T, Rosa M, Lettner N, Mittelsten Scheid O. 2010. Epigenetic regulation of repetitive elements is attenuated by prolonged heat stress in *Arabidopsis*. *Plant Cell* **22:** 3118–3129.

Pellegrini M, Wang G, Meyers BC, Fellowship DY. 2013. Plants regenerated from tissue culture contain stable epigenome changes in rice. *eLife* **2:** e00354.

* Pikaard CS, Mittelsten Scheid O. 2014. Epigenetic regulation in plants. *Cold Spring Harb Perspect Biol* doi: 10.1101/cshperspect.a019315.

Ptashne M. 2007. On the use of the word "epigenetic." *Bioessays* **17:** R233–R236.

Purvis O, Gregory F. 1952. Studies in vernalisation of cereals. XII. The reversibility by high temperature of the vernalised condition in Petkus winter rye. *Ann Bot* **16:** 1–21.

Richards EJ. 2006. Revisiting soft inheritance. *Nat Rev Genet* **7:** 395–402.

Salgado PS, Koivunen MRL, Makeyev EV, Bamford DH, Stuart DLI, Grimes JM. 2006. The structure of an RNAi polymerase links RNA silencing and transcription. *PLoS Biol* **4:** e434.

Schmitz RJ, Schultz MD, Lewsey MG, O'Malley RC, Urich MA, Libiger O, Schork NJ, Ecker JR. 2011. Transgenerational epigenetic instability is a source of novel methylation variants. *Science* **334:** 369–373.

Sheldon CC, Burn JE, Perez PP, Metzger J, Edwards JA, Peacock WJ, Dennis ES. 1999. The *FLF* MADS box gene: A repressor of flowering in *Arabidopsis* regulated by vernalization and methylation. *Plant Cell* **11:** 445–458.

Sheldon CC, Hills MJ, Lister C, Dean C, Dennis ES, Peacock WJ. 2008. Resetting of *FLOWERING LOCUS C* expression after epigenetic repression by vernalization. *Proc Natl Acad Sci* **105:** 2214–2219.

Shimura H, Pantaleo V, Ishihara T, Myojo N, Inaba J, Sueda K, Burgyán J, Masuta C. 2011. A viral satellite RNA induces yellow symptoms on tobacco by targeting a gene involved in chlorophyll biosynthesis using the RNA silencing machinery. *PLoS Pathog* **7:** e1002021.

Shindo C, Aranzana MJ, Lister C, Baxter C, Nicholls C, Nordborg M, Dean C. 2005. Role of *FRIGIDA* and *FLOWERING LOCUS C* in determining variation in flowering time of *Arabidopsis*. *Plant Physiol* **138:** 1163–1173.

Shindo C, Lister C, Crevillen P, Nordborg M, Dean C. 2006. Variation in the epigenetic silencing of *FLC* contributes to natural variation in *Arabidopsis* vernalization response. *Genes Dev* **20:** 3079–3083.

Shivaprasad PV, Dunn RM, Santos BA, Bassett A, Baulcombe DC. 2012. Extraordinary transgressive phenotypes of hybrid tomato are influenced by epigenetics and small silencing RNAs. *EMBO J* **31:** 257–266.

Slotkin RK, Vaughn M, Borges F, Tanurdzić M, Becker JD, Feijó JA, Martienssen RA, Tanurdzic M, Feijo JA. 2009. Epigenetic reprogramming and small RNA silencing of transposable elements in pollen. *Cell* **136:** 461–472.

Sung S, Amasino RM. 2004. Vernalization in *Arabidopsis thaliana* is mediated by the PHD finger protein VIN3. *Nature* **427:** 159–164.

Sung S, He Y, Eshoo TW, Tamada Y, Johnson L, Nakahigashi K, Goto K, Jacobsen SE, Amasino RM. 2006a. Epigenetic maintenance of the vernalized state in *Arabidopsis thaliana* requires LIKE HETEROCHROMATIN PROTEIN 1. *Nat Genet* **38:** 706–710.

Sung S, Schmitz RJ, Amasino RM. 2006b. A PHD finger protein involved in both the vernalization and photoperiod pathways in *Arabidopsis*. *Genes Dev* **20:** 3244–3248.

Swiezewski S, Liu F, Magusin A, Dean C. 2009, Cold-induced silencing by long antisense transcripts of an *Arabidopsis* Polycomb target. *Nature* **462:** 799–802.

Szittya G, Silhavy D, Molnár A, Havelda Z, Lovas A, Lakatos L, Bánfalvi Z, Burgyán J. 2003. Low temperature inhibits RNA silencing-mediated defence by the control of siRNA generation. *EMBO J* **22:** 633–640.

Tamada Y, Yun JY, Woo SC, Amasino RM. 2009. *ARABIDOPSIS TRITHORAX-RELATED7* is required for methylation of lysine 4 of histone H3 and for transcriptional activation of *FLOWERING LOCUS C*. *Plant Cell* **21:** 3257–3269.

Van de Peer Y, Fawcett JA, Proost S, Sterck L, Vandepoele K. 2009. The flowering world: A tale of duplications. *Trends Plant Sci* **14:** 680–688.

Vargason JM, Szittya G, Burgyan J, Tanaka Hall TM. 2003. Size selective recognition of siRNA by an RNA silencing suppressor. *Cell* **115:** 799–811.

Verhoeven KJF, Van Gurp TP. 2012. Transgenerational effects of stress exposure on offspring phenotypes in apomictic dandelion. *PloS One* **7:** e38605.

Wang R, Farrona S, Vincent C, Joecker A, Schoof H, Turck F, Alonso-Blanco C, Coupland G, Albani MC. 2009. *PEP1* regulates perennial flowering in *Arabis alpina*. *Nature* **459:** 423–427.

Wei W, Ba Z, Gao M, Wu Y, Ma Y, Amiard S, White CI, Rendtlew Danielsen JM, Yang Y-G, Qi Y. 2012. A role for small RNAs in DNA double-strand break repair. *Cell* **149:** 101–112.

Werner JD, Borevitz JO, Uhlenhaut NH, Ecker JR, Chory J, Weigel D. 2005. *FRIGIDA*-independent variation in flowering time of natural *Arabidopsis thaliana* accessions. *Genetics* **170:** 1197–1207.

Wierzbicki AT, Haag JR, Pikaard CS. 2008. Noncoding transcription by RNA polymerase Pol IVb/Pol V mediates transcriptional silencing of overlapping and adjacent genes. *Cell* **135:** 635–648.

Wood CC, Robertson M, Tanner G, Peacock WJ, Dennis ES, Helliwell CA. 2006. The *Arabidopsis thaliana* vernalization response requires a polycomb-like protein complex that also includes VERNALIZATION INSENSITIVE 3. *Proc Natl Acad Sci* **103:** 14631–14636.

Xie Z, Johansen LK, Gustafson AM, Kasschau KD, Lellis AD, Zilberman D, Jacobsen SE, Carrington JC. 2004. Genetic and functional diversification of small RNA pathways in plants. *PLoS Biol* **2:** 642–652.

Xu L, Zhao Z, Dong A, Soubigou-Taconnat L, Renou JP, Steinmetz A, Shen WH. 2008. Di- and tri- but not monomethylation on histone H3 lysine 36 marks active transcription of genes involved in flowering time regulation and other processes in *Arabidopsis thaliana*. *Moll Cell Biol* **28:** 1348–1360.

Yu X, Michaels SD. 2010. The *Arabidopsis* Paf1c complex component CDC73 participates in the modification of *FLOWERING LOCUS C* chromatin. *Plant Physiol* **153:** 1074–1084.

Yu A, Lepère G, Jay F, Wang J, Bapaume L, Wang Y, Abraham A-L, Penterman J, Fischer RL, Voinnet O, et al. 2013. Dynamics and biological relevance of DNA demethylation in *Arabidopsis* antibacterial defense. *Proc Natl Acad Sci* **110:** 2389–2394.

Zamore PD. 2006. RNA interference: Big applause for silencing in Stockholm. *Cell* **127:** 1083–1086.

Zhang X, Germann S, Blus BJ, Khorasanizadeh S, Gaudin V, Jacobsen S. 2007. The *Arabidopsis* LHP1 protein colocalizes with histone H3 Lys27 trimethylation. *Nat Struct Mol Biol* **14:** 869–871.

C H A P T E R 32

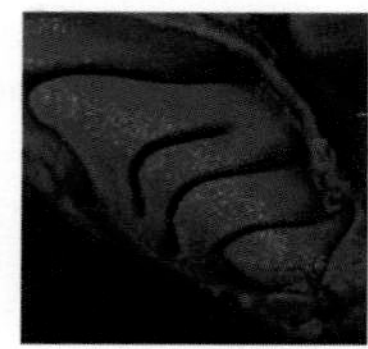
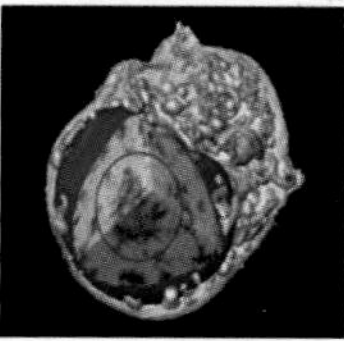

Histone and DNA Modifications as Regulators of Neuronal Development and Function

Stavros Lomvardas and Tom Maniatis

Department of Biochemistry and Molecular Biophysics, Columbia University Medical Center, New York, New York 10032

Correspondence: sl682@columbia.edu; tm2472@columbia.edu

SUMMARY

DNA and histone modifications, together with constraints imposed by nuclear architecture, contribute to the transcriptional regulatory landscape of the nervous system. Here, we provide select examples showing how these regulatory layers, often referred to as epigenetic, contribute to neuronal differentiation and function. We describe the interplay between DNA methylation and Polycomb-mediated repression during neuronal differentiation, the role of DNA methylation and long-range enhancer–promoter interactions in Protocadherin promoter choice, and the contribution of heterochromatic silencing and nuclear organization in singular olfactory receptor expression. Finally, we explain how the activity-dependent expression of a histone variant determines the longevity of olfactory sensory neurons.

Outline

Chapter-opening figures: (*Left*) Adapted from Skykind (2005), by permission of Oxford University Press; (*right*) generated in the National Center of X-Ray Tomography at Lawrence Berkeley National Laboratory in collaboration with Dr. Carolyn Larabell.

Cite as *Cold Spring Harb Perspect Biol* doi: 10.1101/cshperspect.a024208

OVERVIEW

The nervous system of higher organisms is characterized by an enormous diversity of cell types that function in concert to perform a myriad of neuronal functions. In mammals, there are an estimated 10^{11} neurons that form 10^{15} synapses. Differences in connectivity, and subsequent physiology of the connected neurons, are the result of differences in transcriptional programs. The extraordinary complexity of the nervous system requires a complex regulatory system. Neurons must interpret subtle fluctuations in spatiotemporal cues during development in order to commit to specific differentiation programs, but at the same time, they must retain a level of synaptic and transcriptional plasticity. It is well established that transcription factor combinations and the organization of *cis*-regulatory sequences control commitment to differentiation programs and preserve a nuclear plasticity required for neuronal functions. An additional level of regulation is provided by epigenetic regulatory mechanisms that allow stochastic, mutually exclusive transcriptional choices during neuronal differentiation. This affords long-lasting transcriptional changes in response to internal stimuli and external experiences. In fact, the evolution of epigenetic mechanisms that regulates the nervous system may have been a driving force in the expansion and evolution of the large numbers of chromatin- and DNA-modifying enzymes, capable of producing a highly complex, yet plastic nervous system in higher eukaryotes.

This chapter focuses primarily on epigenetic mechanisms involved in neurogenesis, the specification of neuronal fates, and the development of neural circuitry in the brain. We do not address epigenetic studies of neuronal plasticity, neurodegeneration, and psychiatric conditions, such as drug addiction (Fass et al. 2014; Nestler 2014; Rudenko and Tsai 2014). Changes in chromatin and DNA modifications occur during the very earliest stages of neuronal development (i.e., during neurogenesis), and they involve the interplay between DNA methylation and the Polycomb and Trithorax complexes. These changes coincide with the gene expression patterns that govern the developmental decision to adopt neuronal or nonneuronal (glia) cell identity. On commitment to the neuronal fate, developing neurons must make choices between similar, but functionally distinct transcription programs. Because these neurons are postmitotic, inheritance of histone and DNA posttranslational modifications cannot be tested. Thus, we use the word "epigenetic" in neural processes to describe posttranslational modifications of DNA that are associated with changes in gene expression. We describe two examples of this regulation, both of which are examples of the specification of neuronal identity during neural circuitry development. The first is the generation of enormous cell-surface diversity required for neuronal identity by the stochastic expression of the clustered protocadherin (Pcdh) genes. The second is the control of olfactory receptor (OR) gene expression in olfactory neurons, which govern the sense of smell. Both of these examples contribute to the establishment of complex neuronal architectures, as well as functional single-cell diversity, through mechanisms that involve distinct examples of monoallelic expression.

Monoallelic gene expression is an unusual form of gene regulation in which only one of two alleles of a gene is expressed. The challenge to the regulatory system is to discern and differentially express two alleles in the same nucleus. Recent studies have revealed an important role for chromatin modifications in this process, and similar mechanisms have also been identified in the immune system in which enormous single-cell diversity of immunoglobulin and T-cell receptors is generated (see Ch. 29 [Busslinger and Tarakhovsky 2014]). It is not surprising that similar epigenetic mechanisms have evolved to generate cellular diversity in the nervous system. Generating panneuronal single-cell diversity in the central and peripheral nervous system occurs via mechanisms that ensure the stochastic combinatorial expression of cadherin-like Pcdh proteins from a cluster of genes. In the case of the OR genes, all but one of more than 1000 possible receptor isoforms are stochastically silenced. This stochastic process contrasts with genomic imprinting, which is an epigenetic process ensuring the parent of origin determined monoallelic expression of certain genes. These "classic" examples of monoallelic gene expression, however, do bear some similarities to genomic imprinting in that only one of two identical copies of a gene or genomic region are silenced in a single-cell nucleus.

We also describe another "epigenetic peculiarity" in olfactory neurons involving the expression of a histone H2b isoform (or variant) named H2be. This histone variant, which differs by only five amino acids from the canonical H2b protein, appears to be a gauge of the external olfactory sensory environment by being exclusively expressed from understimulated olfactory neurons, signaling the shortening of their life span.

 Cite as *Cold Spring Harb Perspect Biol* doi: 10.1101/cshperspect.a024208

1 EPIGENETIC REGULATION OF NEUROGENESIS

Multipotent neuronal stem and progenitor cells give rise to both neuronal and nonneuronal lineages of the nervous system (Fig. 1A) (reviewed in Olynik and Rastegar 2012). Epigenetic processes, some of which are described in this section, are involved in generating neuronal cells from nonneuronal multipotent precursor cells during neurogenesis. There are two main sites of adult neurogenesis: the subventricular (SVZ) zones in the forebrain and the dentate gyrus of the hippocampus (Alvarez-Buylla and Garcia-Verdugo 2002; Alvarez-Buylla and Lim 2004). These regions, which contain neuronal stem cells (NSCs), give rise to the majority of the neurons present throughout the lives of mice and humans (Doetsch et al. 1999). NSCs can be isolated, cultured, and differentiated into various types of neurons in vitro, thus, making it possible to study epigenetic regulation during neurogenesis in vitro, as well as in vivo. Studies thus far have revealed that DNA- and chromatin-modifying enzymes (particularly, the DNA methyltransferase, Trithorax group, and Polycomb group repressive enzymes) play key roles in this differentiation process (Fig. 1B) (Hirabayashi and Gotoh 2010).

DNA methylation is unequivocally implicated in NSC differentiation, as deletion of the de novo DNA methyltransferase, Dnmt3a, shows an approximately 10-fold reduction in the rate of differentiation of postmitotic cells expressing neuronal markers (Wu et al. 2010). Studies on the role of DNA methylation during neurogenesis (an epigenetic mark traditionally associated with repression; discussed in Ch. 15 [Li and Zhang 2014]) have shown that, atypically, this type of DNA modification contributes to transcriptional activation, as well as repression of neurogenic genes (Wu et al. 2010). A closer look at the genome-wide binding of Dnmt3a revealed a widespread distribution of binding sites across the euchromatic genome with significant enrichment within genes and intergenic regions (Fig. 2). Many of these Dnmt3a-rich regions localize to transcriptionally active genes in the NSCs in which the chromatin is marked by high levels of H3K4me3 on promoters (Wu et al. 2010). A statistically significant number of these Dnmt3a target genes, such as Dlx2 and Sp8, are involved in neurogenesis or neuronal functions. In contrast, in transcriptionally inactive genes, the peak of Dnmt3a-dependent methylation maps to transcription start sites (TSS). Thus, it appeared that the role of DNA methylation in transcriptional activity may depend on the genomic context.

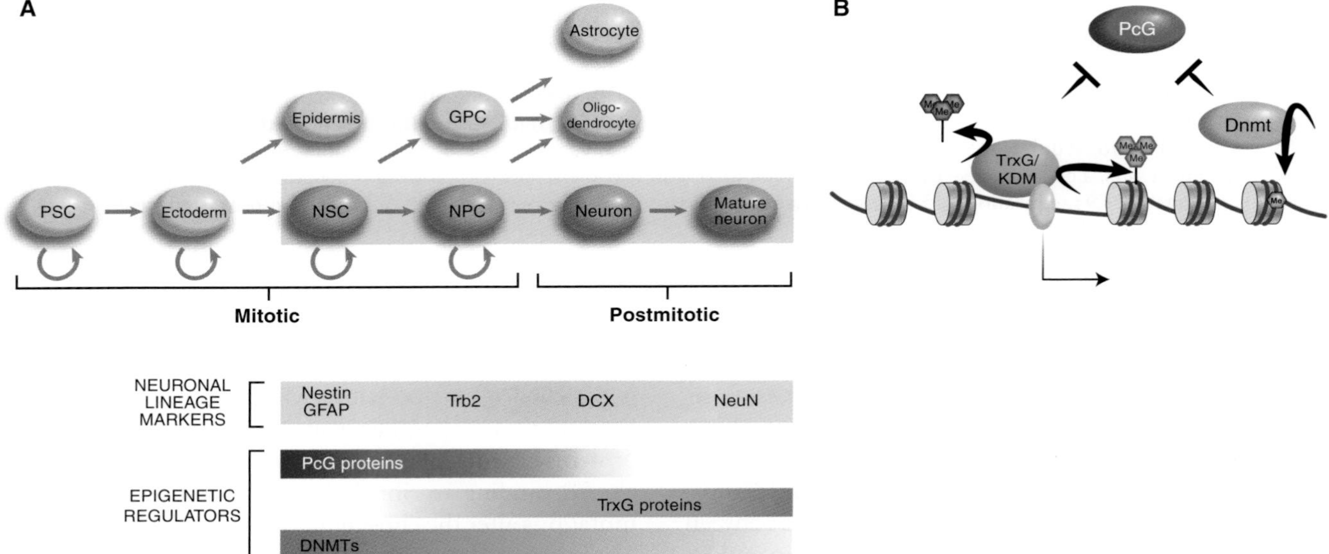

Figure 1. Epigenetic regulation of neurogenesis. (*A*) The temporal involvement of various major epigenetic regulator molecules, acting at key fate-defining genes, such as Dlx2, are indicated alongside the neurogenic differentiation pathway (i.e., cells shaded in pink). PcG, Polycomb group; TrxG, Trithorax group; DNMT, DNA methyltransferase; PSC, pluripotent stem cell; NSC, neuronal stem cell; NPC, neuronal precursor cell; GPC, glial progenitor cell. (*B*) The interplay and modes of action of the three types of epigenetic-regulating molecules involved in the neuronal differentiation process are illustrated. Dnmt3a methylates the gene bodies and upstream regulatory sequences of neurogenic genes, such as Dlx2. This nonpromoter methylation protects these genes from Polycomb-mediated silencing. MLL1, through the trimethylation of H3K4 and demethylation of H3K27me3 (by association of KDM6), also antagonizes Polycomb-mediated silencing of Dlx2 to promote neurogenesis.

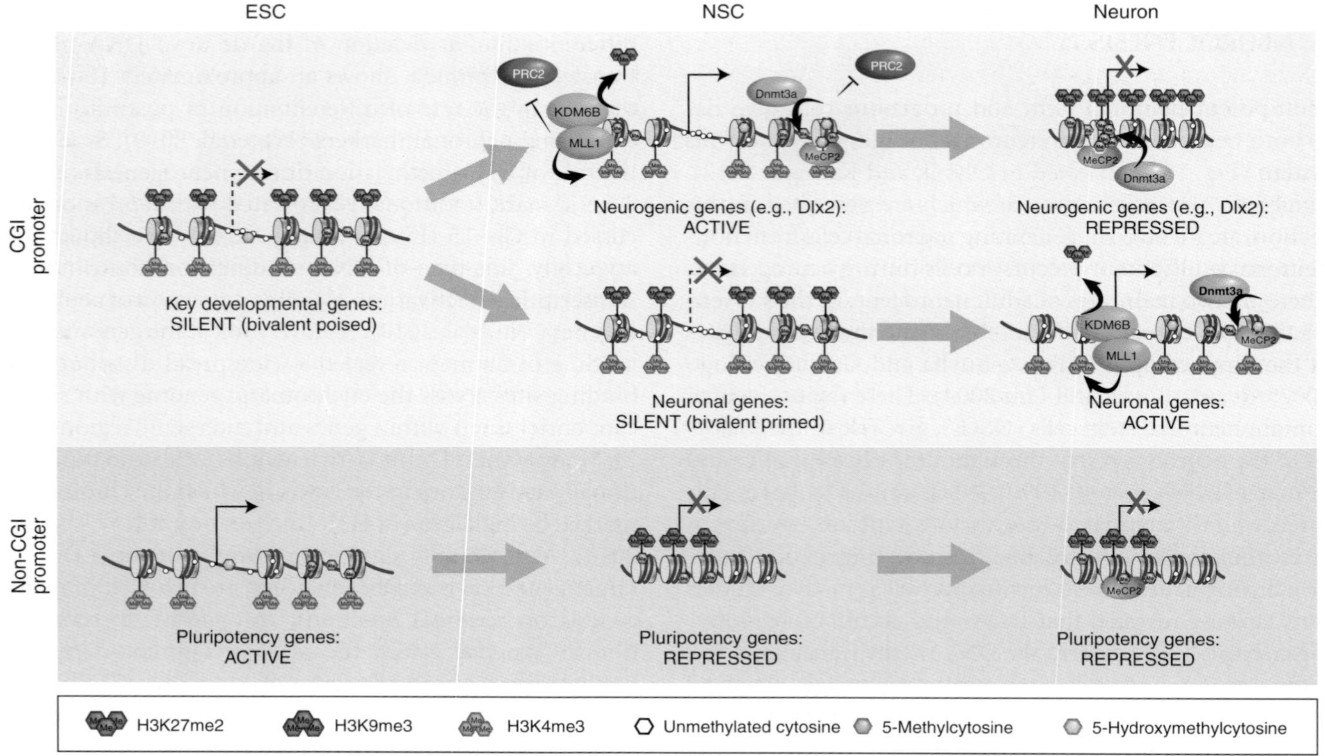

Figure 2. Epigenetic mechanisms involved in the regulation of gene expression during neurogenesis. Typically, CpG island–containing genes are mostly housekeeping genes and, consequently, are ubiquitously active. In early development, just postimplantation, these genes may not yet be active, but are marked by a bivalent chromatin signature, consisting of H3K4me3 and H3K27me3. Most of these genes will be activated during development via H3K27 demethylation (*top* line). A subset of these genes that are tissue-specific neurogenic genes may, subsequently, become silenced in nonneuronal or fully differentiated neuronal cells, through H3K9 methylated heterochromatin formation (*top right*). Other neuronal-specific genes may delay activation, but remain bivalently primed for expression until neuronal differentiation ensues (*middle* line). Pluripotency genes, largely, do not contain CpG islands in their promoters. They are active during early development and become repressed following differentiation by conventional H3K9 methylation and DNA methylation mechanisms. ESC, embryonic stem cell; NSC, neuronal stem cell; CGI, CpG island.

An explanation for this finding may be provided by the discovery that 5-hydroxymethylcytosine (5hmC) is present in the mammalian genome, and is approximately 10-fold more abundant in neurons than in peripheral tissues or embryonic stem (ES) cells (Kriaucionis and Heintz 2009; Munzel et al. 2010; Szulwach et al. 2011). Until recently, it was not possible to distinguish between 5mC and 5hmC. We now know that the distribution of 5hmC varies between distinct regions of the brain and between the brain and ES cells (Szulwach et al. 2011). For example, 5hmC is enriched in gene bodies and depleted from TSS in neurons, whereas the opposite is the case in ES cells. More recent studies revealed that the relationship between the distribution of 5hmC, 5mC, and gene expression is brain cell specific, and the methyl-CpG-binding protein 2 (MeCP2) binds to 5hmC in the brain (Mellen et al. 2012). Additional studies led to the proposal that when MeCP2 binds to 5hmC, it facilitates transcription in neural cell types, but can act as a repressor when bound to 5mC-containing DNA (Fig. 2). This proposal was based on the observation that 5hmC is enriched in euchromatin, whereas 5mC is enriched in heterochromatin (Mellen et al. 2012). Remarkably, MeCP2 proteins bearing a Rett syndrome mutation R133C display altered binding to 5hmC DNA, suggesting that MeCP2 mutations alter the chromatin distribution of the proteins. The investigators of this work speculated that neuronal cell-specific dynamic gene regulation is the consequence of the three-dimensional (3D) chromatin structure, which is determined by the levels of 5mC, 5hmC, and MeCP2 (Mellen et al. 2012).

Dnmt3a-dependent de novo DNA methylation has been shown to inhibit the binding of the Polycomb repressive complex PRC2 (Wu et al. 2010). The neurogenic genes that require Dnmt3a are known targets of PRC2 in nonneuronal lineages (Hirabayashi et al. 2009). For this reason, it is thought that Dnmt3a-dependent DNA methylation prevents the binding of the inhibitor, in this case, PRC2 (Fig. 2; see the active NSC locus). Indeed, deletion of Dnmt3a results in an increase in PRC2 binding to these neurogenic genes in NSCs and, thus, enrichment of H3K27me3 on their promoters. Conversely, partial rescue of the neurogenic Dnmt3a deficit and reexpression of neuronal markers occurs following knockdown of PRC2 components in these Dnmt3a knockout NSCs.

Dnmt3a-dependent DNA methylation is not the only factor involved in the control of Polycomb-mediated silencing in neurogenesis. The Trithorax (trxG) protein complex is the bona fide inhibitor of the Polycomb group, as shown by complementary genetic and biochemical studies first performed in *Drosophila* and, subsequently, extended to other model organisms (described in Ch. 17 [Grossniklaus and Paro 2014]; Ch. 18 [Kingston and Tamkun 2014]). It is important to note that genes subject to PcG regulation typically contain CpG islands at their promoters and are initially bivalently marked by H3K4me3 and H3K27me3 in pluripotent stem cells (Fig. 2). A member of the trx group, Mll1 (mixed lineage leukemia 1), is required for neurogenesis in the postnatal brain, as is Dnmt3a (Fig. 1) (Lim et al. 2009). Mll1 trimethylates H3K4 (H3K4me3), an epigenetic mark that is associated with transcriptional competence. Mll1 also mediates H3K27 demethylation by recruiting the histone demethylase KDM6 (Fig. 1B) (Burgold et al. 2008). Conditional deletion of MLL1 in NSCs results in postnatal atrophy of brain regions, which involves postnatal neurogenesis, ataxia, and death between P25 and P30 (Lim et al. 2009). Expression analysis of the MLL1 knockout NSCs revealed significant down-regulation of neurogenic genes, such as Dlx2. Further analysis of the SVZ showed that, in wild-type mice, nucleosomes in the Dlx2 promoter are marked only by H3K4me3, but in Mll1 knockout mice, it is bivalently marked by H3K4me3 and H3K27me3. The simultaneous presence of these histone marks is typically found in ES cells at bivalent loci that remain poised until cell differentiation signals cause transcriptional activation and H3K27me3 demethylation of certain neural genes (Bernstein et al. 2006). Therefore, a possible role of Mll1 in neurogenesis is to protect Dlx2 from Polycomb-mediated silencing by reversing H3K27me3. Indeed, the neurogenesis deficits observed in the Mll1 knockout mice were partially rescued by Dlx2 overexpression. Thus, the intersection of the three epigenetic pathways on the Dlx2 gene, which encodes a transcription factor with a pivotal role in neurogenesis, determines whether a multipotent cell commits to the neuronal lineage. When Dnmt3a and Mll1 prevent Polycomb-mediated repression of the Dlx2 promoter, the cell differentiates into a neuron (top line of Fig. 2). When an NSC fails to activate Dlx2, then a glial fate ensues (see Fig. 1A). For nonneuronal cells, Dnmt3a and Mll1 are not activated at neuronal-specific loci and, thus, PRC2 acts to trimethylate H3K27, resulting in gene repression.

There are, however, numerous open questions regarding the epigenetic control of early neurogenesis and, without a doubt, Dnmt3a, Mll1, and PRC2 are only a few of the regulatory factors involved in this developmental process. For example, as mentioned earlier, it is not yet clear if further modification of 5-methylcytocine to 5-hydroxymethylcytocine in gene bodies and intergenic regions contributes to the competition between Polycomb and Trithorax, and whether additional DNA- or chromatin-modifying complexes and transcription factors are involved in determining the final balance between neurogenesis or gliogenesis (Riccio 2010). Research is also needed to understand the cross talk between the various epigenetic mechanisms (Jobe et al. 2012). Significant advances are being made into understanding the role of nucleosome remodelers (elaborated in Ch. 21 [Becker and Workman 2013]), such as microRNAs, histone variants, and histone deacetylases (see Ch. 5 [Seto and Yoshida 2014]) in the process of neurogenesis (reviewed in Hirabayashi and Gotoh 2010; Ma et al. 2010; Tyssowski et al. 2014).

2 EPIGENETIC REGULATION OF Pcdh PROMOTER CHOICE

Gene regulation involving DNA and chromatin modifications enables the fine-tuning of the nervous system. This is illustrated by two examples discussed in the following sections: (1) the control of the expression of clustered Pcdh genes in this section, and (2) the OR gene choice described in Section 3. A common feature of the two gene families is that their regulation generates vast diversity among neurons, which allows them to adopt specialized functional identities. However, this is accomplished by distinct mechanisms. In the case of the Pcdh genes, stochastic promoter choice occurs by multiple independent, random choice events across three clusters of genes on two chromosomes. This usually results in monoallelic expression as a result of simple probability, but there is not a mechanism that enforces monoallelism. Promoter choice in the OR system is truly "monoallelic" in that it relies on random activation of a single OR gene coupled with strict monoallelic expression (Chess 2013).

2.1 Clustered Pcdh Genes

Individual neurons engage in multiple and precise synaptic connections with other neurons through an elaborate

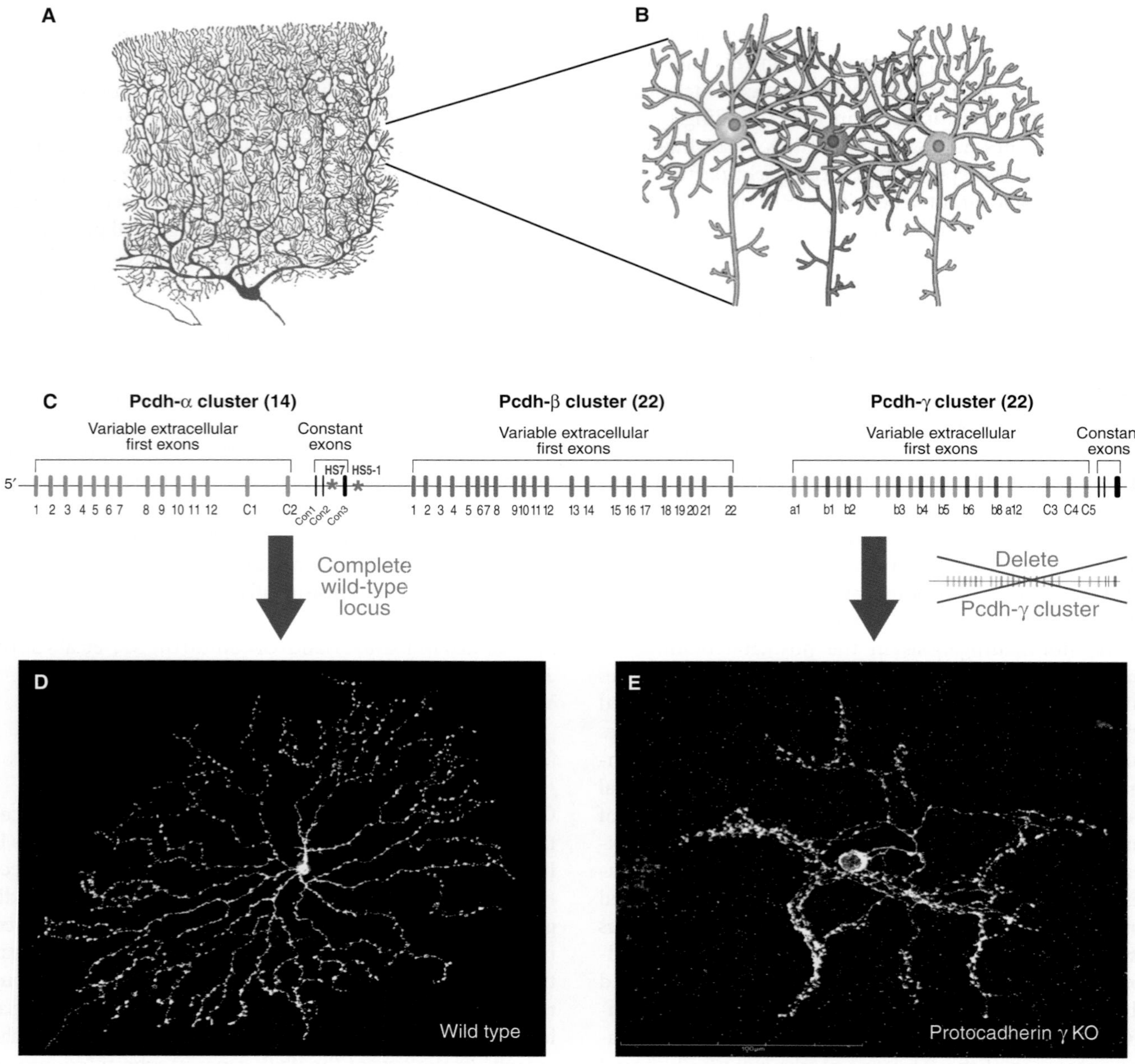

Figure 3. Neurons require a mechanism for self-recognition. (*A*) Drawing of a Purkinje cerebellar neuron emphasizes the elaborate dendritic arborization. (*B*) Arbors of the blue neuron must make connections with the neighboring orange neurons while avoiding connectivity with other blue arbors. This image emphasizes the need for a "barcoding" mechanism for self-avoidance. (*C*) The genomic organization of clustered Pcdh genes. Pcdh-α, -β, and -γ clusters span a genomic region of ~1 Mb. The α cluster has 12 monoallelically variable exons and two biallelically expressed variable exons (C1 and C2). All variable exons splice to the three constant exons that code for the intracellular region. Similarly, the Pcdh-γ cluster has 19 monoallelically expressed variable exons and three variable exons (C3, C4, and C5) that are biallelically expressed. All variable exons splice to the constant exons except at the β cluster. (*D*) Wild-type retinal starburst amacrine cells. (*E*) Deletion of the 22 variable exons of the Pcdh-γ cluster abolishes self-avoidance of retinal starburst amacrine cells resulting in dendritic collapse and luck of arborization.

Cite as *Cold Spring Harb Perspect Biol* doi: 10.1101/cshperspect.a024208

network of dendrites and axons, which collectively form a neural arbor (see Fig. 3A). During development, a growing neurite (i.e., axons and dendrites) must precisely navigate through a sea of hundreds, if not millions, of other developing neurites and find the appropriate target to connect with, while maintaining the ability to distinguish and minimize interactions with neurites emerging from its own cell (i.e., sister neurites; see Fig. 3B). Thus, prevention of synaptic interactions between sister neurites is essential for proper neural connectivity, signal transmission, and computation of neuronal signals. In both vertebrates and higher invertebrates, the solution to this problem is to provide a unique extracellular identity tag to every neuron, similar to a "barcode," in which neurites containing the same barcode recognize and repel each other (Cameron and Rao 2010; Grueber and Sagasti 2010). This mechanism is called self-avoidance, a process that ensures that neuronal processes from the same neuron repel each other during arborization and axon branching to avoid clumping. Self-avoidance also prevents extensive overlapping in the arborization pattern and facilitates the coverage of the neuronal processes across different regions of the nervous system during development.

The mechanisms by which a highly diverse barcode is generated in vertebrates and flies are different. Diversity in *Drosophila* neurons is generated by one of the most complex examples of alternative pre-mRNA splicing known, whereas, in mammals, diversity is achieved through a combination of higher-order chromatin organization, stochastic promoter choice, and alternative splicing. *Drosophila* neurons express the Dscam1 gene, which encodes a transmembrane protein with extracellular immunoglobulin-like domains. Through alternative splicing of a large number of variable exons, an individual *Drosophila* neuron expresses approximately 10 to 15 isoforms, out of about 38,000 available combinations, of which 19,008 have a different extracellular domain (Schmucker et al. 2000). The Dscam1 immunoglobulin domains on the cell surface of one dendrite interact exclusively (i.e., "homophilically" with its counterpart on the opposing dendrite). These homophilic interactions ultimately lead to repulsion of the dendrites through a process requiring the intracellular domains of the Dscam1 proteins.

Single-cell identify of mammalian neurons is thought to be provided by the clustered Pcdhs rather than the mammalian Dscam homologs, as the latter have a simple organization compared with their *Drosophila* counterparts, and therefore cannot generate high levels of cell-surface diversity (Schmucker and Chen 2009). The clustered Pcdh gene family consists of approximately 70 family members in the mouse and 58 in humans (Wu and Maniatis 1999; Wu et al. 2001). The mechanism by which Pcdh diversity is generated is distinct from that which generates Dscam diversity. The organization of the Pcdh gene clusters is, however, remarkably similar to that of the immunoglobulin and T-cell receptor gene clusters discussed in Ch. 29 (Busslinger and Tarakhovsky 2014) and Wu and Maniatis (1999). The extracellular domains of Pcdhs are encoded by "variable" exons, whereas the intracellular domains are encoded by three "constant" exons (see Fig. 3C). The variable extracellular domains of Pcdh proteins function in self-recognition through homophilic interactions (Schreiner and Weiner 2010; Chen and Maniatis 2013; Zipursky and Grueber 2013; Thu et al. 2014), whereas the constant intracellular domain appears to be involved in cell signaling (Wu and Maniatis 2000; Han et al. 2010; Schalm et al. 2010; Suo et al. 2012). The intracellular domain can be released from the membrane by the combined activities of a metalloproteinase and γ-secretase in a process that requires endocytosis (Haas et al. 2005; Hambsch et al. 2005; Reiss et al. 2006; Bonn et al. 2007; Buchanan et al. 2010). The intracellular domain is then thought to translocate to the nucleus, but neither the regulation of this process nor the cytoplasmic or nuclear function of the intracellular domain is understood.

The extracellular domain of Pcdh proteins, like Dscams, appears to mediate homotypic repulsion (Schreiner and Weiner 2010; Zipursky and Sanes 2010; Lefebvre et al. 2012; Yagi 2012, 2013; Chen and Maniatis 2013; Zipursky and Grueber 2013; Thu et al. 2014). This has been shown by genetic manipulation of the Pcdh gene clusters; deletion of the 22 Pcdh-γ genes (Fig. 3C) results in the disruption of dendritic self-avoidance in retinal starbust amacrine cells (a type of retinal interneuron) (see Fig. 3E) and cerebellar Purkinje cells (inhibitory neurons involved in motor coordination) (Lefebvre et al. 2012). Considering the large difference in the number of distinct Dscam and Pcdh protein isoforms (19,000 compared with 58), it is surprising that the Pcdhs could generate the cell-surface diversity sufficient for single-cell identity. However, cell aggregation studies provided evidence that Pcdhs engage in homophilic interactions as multimeric complexes, thus dramatically increasing the potential single-cell diversity provided by the Pcdhs (Schreiner and Weiner 2010; Thu et al. 2014). In fact, theoretical calculations suggest that comparable levels of single-cell diversity can be generated by the invertebrate Dscam gene and vertebrate clustered Pcdhs (Yagi 2012; Thu et al. 2014). Recent studies have shown that all but one of the members of the Pcdh-α, -β, and -γ gene clusters are capable of individually engaging in homophilic interactions as monomers, and that multimers display highly specific interactions. Examination of homophilic interactions between specific combinations of multiple Pcdh isoforms revealed that Pcdh combinatorial recognition specificities depend on the identity of all of the expressed isoforms. However, the nature of the multimer (dimer, trimer, or higher-order multimers) and complexes on the cell surface have yet to be established (Thu et al. 2014).

2.2 Pcdh Gene Regulation

This section describes the genomic organization of the three Pcdh gene clusters, and Section 2.3 describes the chromatin and epigenetic mechanisms that underlie their stochastic expression in individual neurons. Remarkably, the pattern of Pcdh expression in mouse and human cell lines is maintained indefinitely through many rounds of cell division. The timing of promoter choice during neuronal differentiation in vivo and its stability during cell division in vivo (if it occurs) remains to be shown. Nevertheless, the stability of promoter choice in neuroblastoma cell lines suggests the existence of mechanisms that faithfully maintain expression patterns during DNA replication. Such mechanisms could allow a stable pattern of Pcdh expression through the life of an individual neuron, thereby maintaining neuronal self-identity. Continued studies on the mechanisms of Pcdh promoter choice during development, and the epigenetic mechanisms involved in determining whether and how the choice is maintained, should provide important insights into the role of Pcdh diversity in the assembly of neural circuits.

To understand the epigenetic regulation of this gene cluster, it is necessary to describe its genomic organization and expression. The extracellular domains of the mouse Pcdh-α and -γ genes are encoded by 14 and 22 variable exons, respectively (Fig. 3C). Each of the variable exons is transcribed from its own promoter. Transcripts initiated from these promoters read through the downstream variable exons (i.e., C1 and C2 in Fig. 4A). The 5′ splice site nearest the start site of transcription is then *cis*-spliced to the first constant exon (Con1 in Fig. 4B). Two types of variable exons have been identified: the alternately expressed exons (i.e., Pcdh-α 1-12, and Pcdh-γ 1–22) and the biallelic ubiquitously expressed exons (Pcdh-α C1 and C2, and Pcdh-γ C3, C4, C5) (Esumi et al. 2005).

The expression of the alternately expressed exons is determined by random promoter choice. For example, in the case of the Pcdh-α cluster, promoter choice is stochastic and occurs independently on each chromosome in an individual neuron, based on single-cell analyses of Purkinje cells from F1 mouse hybrids (Esumi et al. 2005). Thus, different maternal and paternal alternate promoters are stochastically activated in each neuron. This choice is not strictly limited to a single alternate promoter; sometimes, two or more different alternate Pcdh isoforms from a cluster can be expressed on each chromosome. This mechanism ultimately results in monoallelic promoter choice (i.e., each chromosome expresses one or more unique isoforms) (Fig. 4B). This is different than the monoallelic expression of X-linked genes in females (see Ch. 25 [Brockdorff and Turner 2014]) or immunoglobulin receptors (described in Ch. 29 [Busslinger and Tarakhovsky 2014]); in these cases, a single copy is expressed and chosen at random from only one chromosome in each cell (see Chess 2013 for a comparison of the different monoallelic expression mechanisms). Another distinction from monoallelic immunoglobulin genes is that, in addition to the alternately expressed exons, five "C"-type exons are expressed ubiquitously and biallelically (C1–C5, of which two are expressed from the α cluster and three from the γ cluster) (Esumi et al. 2005; Kaneko et al. 2009). These C1–C5-type exons (Fig. 3B) are also spliced to the constant exons (Con1–3) (Fig. 4B). The Pcdh-β cluster differs in that the variable exons encode complete proteins with only short cytoplasmic domains and, thus, are unlikely to have signaling potential (at least on their own). Taken together, this stochastic, combinatorial (i.e., a combination of 1–2 Pcdh-α, 1–2 Pcdh-β, and 1–2 Pcdh-γ isoforms in a single cell), and partially monoallelic pattern of expression of the three Pcdh gene clusters can potentially generate extraordinary cell-surface diversity (see Yagi 2012 for discussion).

2.3 Epigenetic Control of Pcdh Expression

In this section, we will focus on what is known about the genetic and epigenetic mechanisms that regulate promoter choice in the Pcdh-α cluster as a paradigm, but it is virtually certain that similar mechanisms are used in the Pcdh-β and -γ gene clusters.

The promoters of every alternately expressed exon, and a subset of the biallelic exons, share a conserved sequence element (CSE) required for transcriptional activation (Wu et al. 2001; Tasic et al. 2002). Beyond this common promoter element, long-distance enhancers are required for maximum levels of Pcdh expression. For the Pcdh-α cluster, these include two *cis*-regulatory sequences, hypersensitive sites HS7 and HS5-1, identified by in vivo DNase I sensitivity assays, which function as nervous-system-specific enhancers based on transgenic reporter assays (Ribich et al. 2006).

Deletion of HS5-1 in mice results in a decrease in Pcdh-α gene expression throughout the brain (Kehayova et al. 2011). In particular, expression of exons proximal to HS5-1, which contain a CSE element (i.e., Pcdh-α6–12, in addition to Pcdhα-C1) is most affected by the deletion (Fig. 5A). In nonneuronal tissues, however, deletion of HS5-1 results in up-regulation of Pcdh-α gene expression. Examination of the DNA sequence of HS5-1 revealed that this

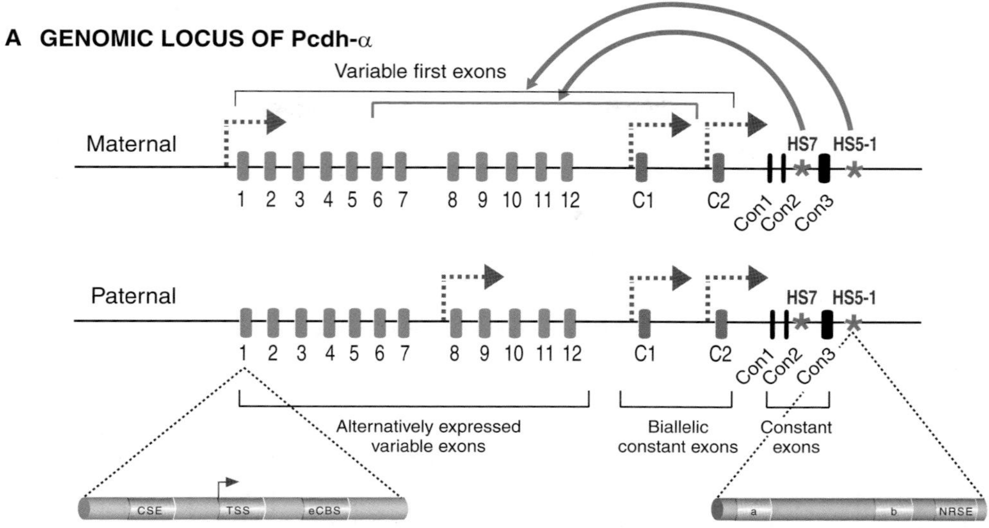

B pre-mRNA TRANSCRIPTS

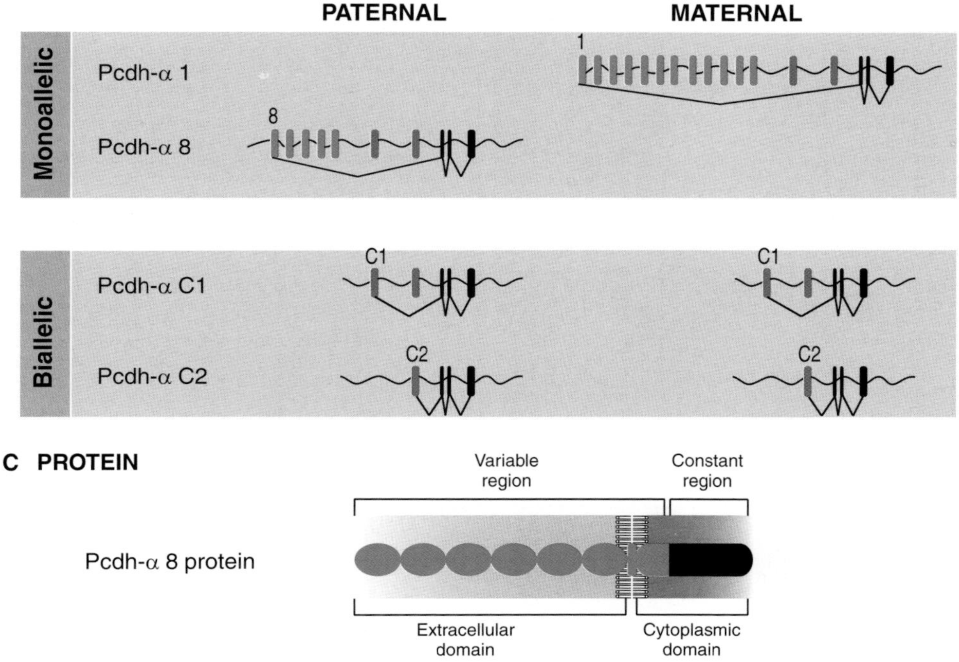

Figure 4. Alternative splicing produces variable Pcdh isoforms. (*A*) Hypothetical example showing the expression pattern of the Pcdh-α cluster in a single neuron, in which one of the alternative promoters is activated in a seemingly stochastic fashion on each chromosome. An enlargement of the exon 1 promoter region shows the two CCCTC-binding factor (CTCF) binding sites (conserved sequence element [CSE] and exonic CTCF binding site [eCBS]). The enlargement of the HS5-1 enhancer shows the two CTCF binding sites (a and b) and the neuron-restrictive silencer factor (NRSF) binding sequence, neuron-restrictive silencer element (NRSE). (*B*) The pre-mRNAs produced the maternal and paternal alleles, showing monoallelic expression of variable exons α1 and α8, but biallelic expression of the C-type exons. The variable pre-mRNAs contain all the variable exons located downstream from the selected promoter, which are subsequently spliced to the Con1–3 exons. Only the exon located immediately downstream from the selected promoter will be spliced to the exons encoding the intracellular constant region. (*C*) Each variable exon encodes for six ecto- (cadherin) domains involved in self-recognition, most probably through homotypic repulsion.

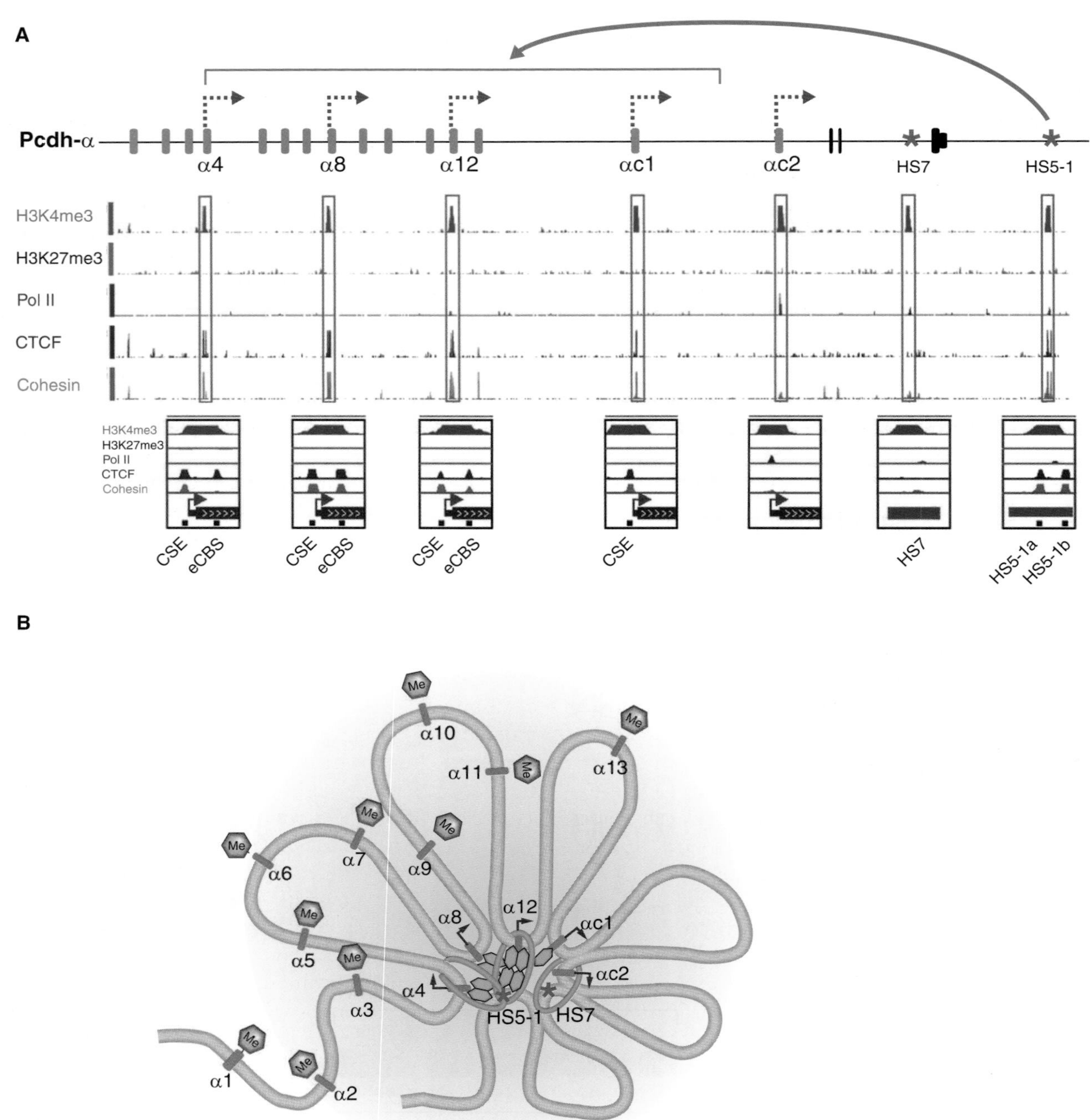

Figure 5. Summary of the epigenetic properties of the Pcdh-α cluster. (*A*) Summary of ChIP-seq (parallel sequencing technologies coupled to chromatin immunoprecipitation) analysis in a human diploid neuroblast cell line (SK-N-SH) that expresses only three variable Pcdh-α exons. There are two CTCF binding sites at each active variable exon (CSE and eCBS sequences) and two at the HS5-1 enhancer (HS5-1a and HS5-1b). (*B*) Summary of the epigenetic properties of active and silenced Pcdh-α exons on forming a 3D transcriptional hub, based on chromosome conformation capture (3C) data. Regions outside the central transcriptional hub are thought to be packaged in a repressive chromatin conformation (blue-shaded region), suggested by the DNA methylation of inactive Pcdh-α promoter regions. Active genes are thought to be concentrated in a transcriptional hub, aided by double-clamp binding of CTCF between the variable exons and the HS5-1 enhancer and cohesin.

Cite as *Cold Spring Harb Perspect Biol* doi: 10.1101/cshperspect.a024208

enhancer contains a sequence element known to function as a silencer (Kehayova et al. 2011). This element, referred to as the neuron-restrictive silencer element (Fig. 4A) (Kehayova et al. 2011), is known to act as a repressor when bound by NRSF, a repressor expressed in nonneuronal tissues that inhibits neurogenic transcription by the recruitment of repressive protein complexes (Lunyak and Rosenfeld 2005).

The HS7 enhancer is also required for maximum levels of Pcdh-α expression in the brain, as its deletion results in a moderate down-regulation of Pcdh-α gene expression (Kehayova et al. 2011). This observation indicates that maximum levels of Pcdh-α gene expression require the combined activity of both the HS5-1 and HS7 enhancers. A simple model to explain the data is that all variable exon promoters have an equal chance of being activated, but a stochastic process selects one for activation (Fig. 4), the determining factor of which may be its interaction with HS7 and HS5-1.

Beyond the DNA sequence elements of the Pcdh gene clusters, recent studies have revealed that the random stochastic promoter choice correlates with a combination of epigenetic marks on DNA (i.e., DNA methylation) and histones (e.g., H3K4me3), and DNA-binding factors that affect higher-order chromatin structure. Because there is a mixed population of cells in the nervous system, and each cell expresses a distinct set of Pcdh isoforms, it is difficult to gain mechanistic insights into the regulation of Pcdh promoter choice in vivo. Progress, however, has been made using certain cell lines that stably express specific sets of Pcdhs to make correlations between DNA methylation, protein binding, chromatin modifications, and Pcdh isoform expression. For example, such studies have shown that the insulator binding protein, CTCF, and the cohesin complex subunit, Rad21, bind to two sites at active Pcdh promoters, the CSE upstream of the TSS and an exonic CTCF binding site (eCBS) downstream from TSS, and to two sites within the HS5-1 enhancer sequences (Golan-Mashiach et al. 2012; Monahan et al. 2012). This observation was corroborated by studies of a diploid human neuroblastoma cell line SK-N-SH, which stably expresses a small subset of Pcdhα isoforms (Fig. 5A) (Guo et al. 2012). The CTCF protein is known, among other functions, to promote long distance *cis* and *trans* genomic interactions (Phillips and Corces 2009), and Rad21 (a cohesin subunit) was recently shown to stabilize such interactions in a fashion similar to the stabilization that occurs for sister chromatid cohesion during mitosis (Kagey et al. 2010). Thus, the simultaneous presence of these two molecules on both the active variable promoters and a distant enhancer suggested that CTCF and cohesin may mediate enhancer/promoter interactions (see Ch. 2 [Kim et al. 2014] for further discussion). This does not explain, however, how only one, or sometimes two, of the 13 variable Pcdh-α promoters are selected for activation in individual cells. It is intriguing, though, that CTCF and the cohesin subunit Rad21 are only bound to promoters that are transcriptionally active, suggesting that initial choice may be accomplished by the selective binding of this complex to one of these competing promoters. In support of this idea, all the nontranscribed variable promoters are heavily methylated on CpG dinucleotides (Tasic et al. 2002; Kawaguchi et al. 2008). CpG methylation of CTCF-binding sites is known to abolish CTCF binding and, specifically, to CSE (Guo et al. 2012), in vitro, providing an epigenetic mechanism for promoter choice. Of course, in this case, the question is only moved a step back because promoter choice then becomes a question of how all but one variable promoter are methylated. However, the initial promoter choice is made, pharmacological inhibition of DNA methylation in cell lines that express only a subset of variable exons leads to reactivation of all the variable promoters that were normally silent in these cells (Kawaguchi et al. 2008). These results are equally consistent with an alternative mechanism, though, in which initially all the variable promoters are unmethylated and CTCF/Rad21 is bound, but only interaction with HS5-1 stabilizes this complex on the chosen promoter. The other variable promoters would eventually lose CTCF/Rad21 binding and they would become methylated, "locking" them into a permanent silent state without basal transcription activity. The silencing of the variable promoters not chosen could then be reinforced by heterochromatinization, as shown in olfactory sensory neurons (OSNs) by the enrichment of H3K9me3 and H4K20me3 marks on the silent variable exons (Magklara et al. 2011). Recently, a conditional knockout of the CTCF gene was used to show that the alternately expressed promoters require CTCF for normal levels of transcription in the cortex and hippocampus, whereas the biallelic promoters, ac2 and gc4, do not (Hirayama et al. 2012). This observation is consistent with the notion that CTCF/cohesin organizes the enhancer/promoter interactions in such a way that the alternately expressed promoters are chosen stochastically, in contrast to the biallelic exons.

The symmetric relationships between the two CTCF/cohesin binding sites in the promoter/exon and the HS5-1 enhancer suggests the interesting possibility that if the HS5-1 enhancer and promoter do interact through DNA looping, the CTCF/cohesin complex simultaneously binds to the enhancer and promoter to form a "double clamp" (i.e., two binding points between each DNA element and CTCF/cohesin complex), which may play a role in the epigenetic stabilization of promoter choice (Fig. 5B) (Guo et al. 2012; Monahan et al. 2012).

Long-range DNA-looping interactions in SK-N-SH cells between enhancers and promoters of the Pcdh-α cluster were detected using quantitative 3C assays (described in the text and Fig. 5 of Ch. 19 [Dekker and Misteli 2014]). Interactions of varying strength were detected between both the HS7 and HS5-1 enhancers and the transcriptionally active promoters. The same experiments performed after small hairpin RNA knockdown of CTCF or the cohesin subunit Rad21 significantly decreased all of the enhancer/promoter DNA-looping interactions. These results, coupled to data indicating the requirement of both enhancers for maximal expression of Pcdh-α alternately expressed exons, suggest a complex mechanism of promoter choice; Guo et al. (2012) proposed that the DNA-looping interactions in the Pcdh-α gene cluster recruit CTCF-bound promoters to enhancers in an active "transcriptional hub" (Fig. 5B). This model is based on extensive and highly specific functional and physical interactions between promoters and enhancers, and the fact that the formaldehyde cross-linking used in the ChIP-seq and 3C studies would be expected to cross-link a large complex containing multiple enhancers and promoters. In this model, the DNA-looping interactions between HS5-1 and the promoters of α8 and α12 are formed by a double-clamping mechanism between the HS5-1a/HS5-1b sites of the enhancer and CSE/eCBS sites of alternate promoters (Fig. 5A). At the same time, HS5-1 must interact with αc1 because HS5-1 is required for its expression (Ribich et al. 2006; Kehayova et al. 2011). Also, HS7 must directly interact with αc2, as well as with the active alternate promoters and ubiquitous promoters, because this enhancer interacts with and is required for the maximum activation of these promoters (Kehayova et al. 2011). Finally, long-range DNA-looping interactions are formed between enhancers and active promoters. All of these observations are consistent with the model illustrated in Fig. 5B, in which the apparent simultaneous interactions between the two enhancers and multiple promoters leads to the formation of a large "transcriptional hub."

In summary, the regulation of Pcdh gene expression constitutes a complex regulatory system that involves unusual transcriptional and posttranscriptional events of paramount importance for the development and wiring of the nervous system. Although much has been learned since the discovery of this remarkable gene family, significant open questions, nonetheless, remain that likely involve epigenetic processes. For example, how does the splicing machinery recognize only the 5′ splice site proximal to active promoters, and ignore all of the downstream 5′-splice sites? This is likely to be a consequence of the organization of chromatin on the active and inactive variable region exons (Magklara et al. 2011; Monahan et al. 2012). Other questions include: At what point during neural differentiation does promoter choice occur, and how stable is the choice once it is made? As mentioned above, the specific pattern of Pcdh isoform expression in neuronal cell lines is stable over many generations, so, it is likely that an epigenetic mechanism is involved. However, given the elaborate enhancer/promoter interactions described above, the mechanism of epigenetic stability must be complex. Another question is: How can stochastic promoter choice and selective expression of individual Pcdh gene cluster occur at the same time? For example, Pcdh-α, but not -β or -γ, is expressed in the dorsal raphe and serotonergic system, but one presumes that stochastic choice still occurs in the Pcdh-α gene cluster (Katori et al. 2009). Finally, it will be of great interest to understand the organization of the proposed Pcdh transcription hub in individual neurons.

3 EPIGENETIC REGULATION OF OR CHOICE

Until recently, the H3K9me3 and H4K20me3 histone marks were considered hallmarks of constitutive heterochromatin described as occurring exclusively at pericentromeric and subtelomeric regions of the mammalian genome. These are primarily repetitive regions of the genome, with little abundance of genic regions. Epigenomic analysis of primary olfactory neurons, however, revealed an unexpected enrichment of H3K9me3 and H4K20me3 on ∼5% of mouse genes, namely, the OR genes (Magklara et al. 2011). Thus, this type of heterochromatin may actually be dynamic and directly involved in gene regulation, rather than solely to preserve genomic integrity and repress retrotransposons and other repetitive elements.

3.1 Anatomy and Genetics of the Olfactory System

A description of the anatomy of the olfactory epithelium and bulb are helpful in understanding how this organ is functionally regulated at the genetic and epigenetic levels. In higher organisms, volatile odorants are detected by OSNs via G-coupled OR proteins, with seven membrane-spanning domains found on the surface (Fig. 6D) (Buck and Axel 1991). In mammals, these neurons reside in a specialized sensory organ called the main olfactory epithelium (MOE), which covers the lumen of the nasal cavity (Fig. 6A,B) (Axel 1995). The olfactory epithelium regenerates throughout life, and is composed of three main cell types organized in intermingling layers: basal cells, which are multipotent stem cells that give rise to all the other cell types of this sensory organ; OSNs, which are postmitotic neurons with odor-detection potential; and sustentacular (supporting) cells, which are nonneuronal cells that line the apical layer of the epithelium. The dendrites of the olfactory neu-

Cite as *Cold Spring Harb Perspect Biol* doi: 10.1101/cshperspect.a024208

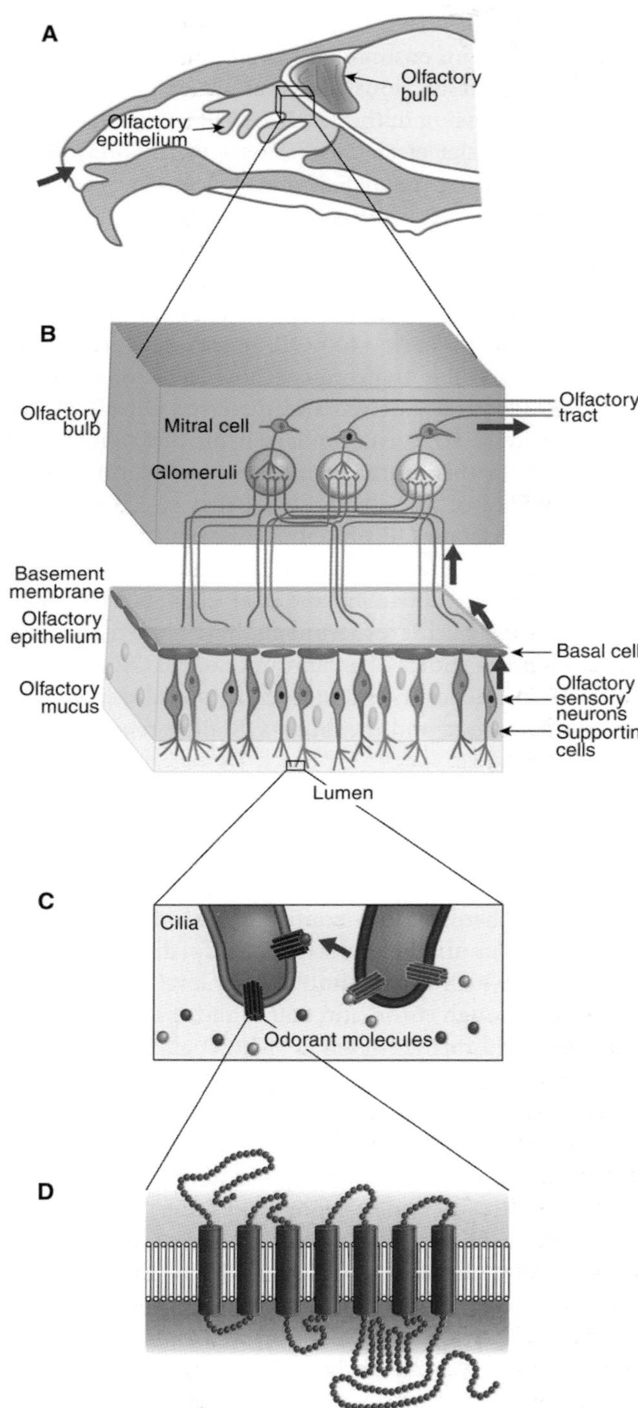

Figure 6. Introduction to olfactory receptors (ORs). The anatomy of the olfactory system is shown in increasing degrees of magnification from *A* to *D*, down to the depiction of ORs, which are G-protein-coupled receptors with seven transmembrane α-helices (*D*). Purple arrows throughout indicate the pathway of olfactory system transduction from the entrance of odorant molecules into the nasal cavity (*A*), their binding to ORs (*C*), and transduction of the signal through the olfactory system (*B*). (*A*) Head cross section in the mouse, indicating the olfactory tissues containing OSNs. The MOE is the olfactory sensory organ containing OSNs, basal pluripotent cells (blue), and sustentacular supporting cells (yellow nuclei). Each OSN expresses one type of OR (three types illustrated for simplicity as green, red, and purple) and the axons of OSNs expressing the same OR project to the same glomerulus in the olfactory bulb in which they synapse with interneuron/mitral cells to transmit the signal to the brain. Glomeruli map spatially into zones based on similarity of OR characteristics. (*C*) Enlargement of OSN cilia, depicting the single type of OR (red or purple), which binds a particular type of odorant molecule. (*D*) Structure of the seven transmembranes spanning the OR.

rons terminate in cilia, which project beyond the apical layer of the MOE and contain ORs, and make contact with volatile odorants in the lumen of the nasal cavity (Fig. 6C).

Most mammals have more than 1000 olfactory or odorant receptor genes, organized in genomic clusters of variable size (from two to about 200 OR genes), distributed across most chromosomes (Glusman et al. 2000; Zhang and Firestein 2002; Nei et al. 2008). The extraordinary number of genes dedicated to odor detection illustrates the importance of this chemosensory system in the survival and reproduction of most animal species. The question of how this myriad of OR genes are regulated in OSNs has intrigued researchers for a number of years. A striking characteristic of OR genes is that they are expressed in a mutually exclusive, monogenic, and monoallelic fashion; that is, each OSN expresses only one OR gene from a choice of more than 1000, from only one parental allele (Fig. 7A) (Chess et al. 1994; Shykind 2005). It is important to note that unlike parental imprinting, which is described in detail elsewhere in this collection (see Ch. 26 [Barlow and Bartolomei 2014]), maternal and paternal OR alleles are expressed in equal frequency overall in the olfactory epithelium, so, expression is termed "monoallelic" because both alleles are never coexpressed in the same neuron (Fig. 7B,C). This expression pattern is distinct from imprinting because the parent of origin does not determine which allele is used, and it differs from the Pcdh promoter choice because only one OR allele is expressed in an individual neuron, rather than one, two, or more isoforms that can be expressed independently from each parental copy from the Pcdh clustered genes. This expression pattern, known as the "one receptor per neuron" rule, is essential for the proper functioning of the olfactory system because the identity of the OR protein not only provides specificity in odorant detection, but also plays an instructive role in the guidance of its axon toward a specific glomerulus within the olfactory bulb (Fig. 7D,E) (Mombaerts et al. 1996; Wang et al. 1998; Barnea et al. 2004; Serizawa et al. 2006).

The glomerulus is a spherical structure located near the surface of the olfactory bulb, and it contains synapses between incoming OSN axons and the dendrites of mitral, tufted, and periglomerular cells residing in the olfactory bulb. If a neuron were to express more than one OR allele, it would respond inappropriately to different odorants, and would send signals to the wrong brain regions, resulting in sensory confusion and an olfactory system with impaired discriminatory power. Therefore, immense evolutionary pressure has resulted in perfecting this unique olfactory system and epigenetic mechanisms have evolved to safeguard the singularity of OR expression, as mentioned above (Fig. 4B). Simply stated, in the case of Pcdhs, the two chromosomes make independent stochastic choices, whereas in OR regulation, only one OR allele from only one cluster and one chromosome is expressed.

Some spatiotemporal specificity of OR gene regulation can be attributed to the promoter sequence signatures that ensure olfactory-restricted gene expression, through the binding of combinations of highly abundant transcription factors, such as Emx2, Lhx2, and Olf/Ebf (O/E) family members (Wang et al. 1997; Hirota and Mombaerts 2004; Rothman et al. 2005; McIntyre et al. 2008). Mapping of the TSS of OR genes revealed that they have extremely AT-rich sequences, with similar transcription factor-binding sites predicted ~1 kb upstream of and ~1 kb downstream from their TSS (Clowney et al. 2011; Plessy et al. 2011). Promoter signature variations ensure that each OR gene is subject to spatiotemporal restrictions visible as, at least, four broad "zones" of expression in the MOE, two of which are visible in Fig. 7D (Ressler et al. 1993; Vassar et al. 1993). This transcription factor–mediated control, however, is not sufficient to restrict OR activation to only one promoter per neuron. Thus, the fundamental question of how only one of the thousands of equivalent promoters is transcriptionally active in each OSN remains unanswered. In the mouse, there are approximately 1400 OR genes, of which about 1100 have detectable transcripts (Clowney et al. 2011) in the MOE. Identical paternal and maternal OR alleles are never coexpressed (i.e., its monoallelic property). A critical mechanism for this mutually exclusive expression pattern is the existence of a feedback signal that is generated by the expression of an OR protein to prevent the transcriptional activation of additional ORs (Serizawa et al. 2003; Lewcock and Reed 2004; Shykind et al. 2004; Nguyen et al. 2007). Lineage tracing experiments have shown that if a neuron makes a choice of a "nonfunctional" OR (because ~20% of OR genes in the mouse are pseudogenes), the choice will be reset and the faulty OR allele replaced with a functional allele.

3.2 Epigenetic Regulation of Monoallelic OR Expression

Another mechanism that contributes to the specificity missing at the promoter level is the action of distant enhancer elements, which usually implicates a level of epigenetic regulation through chromatin marks and spatial organization (described further in Ch. 2 [Kim et al. 2014]). Some distal OR enhancers have been described as being required for the proper expression of some OR genes (Serizawa et al. 2003; Khan et al. 2011). For example, the H enhancer element is necessary for the transcription of three ORs located 75 Kb downstream from its genomic location. This element can also drive the widespread expression of a transgenic OR when inserted proximal to its promoter (Serizawa et al. 2003). Moreover, chromosome conformation capture (3C and 4C), and two-color DNA fluorescence in situ hybridization (FISH) experiments, have shown that the H element interacts both in *cis* and *trans* (i.e., on the same and different chromosomes) with actively transcribed ORs (Lomvardas et al. 2006). However, the interaction of H with OR alleles in *trans* is not necessary for their expression (Fuss et al. 2007) and, based on results presented below, the *trans* interaction is most likely the consequence of the spatial segregation of active and silent OR alleles. A few ORs can, nonetheless, be expressed without distal regulatory elements as mini transgenes (minigenes) with <500 bp of DNA sequence up-

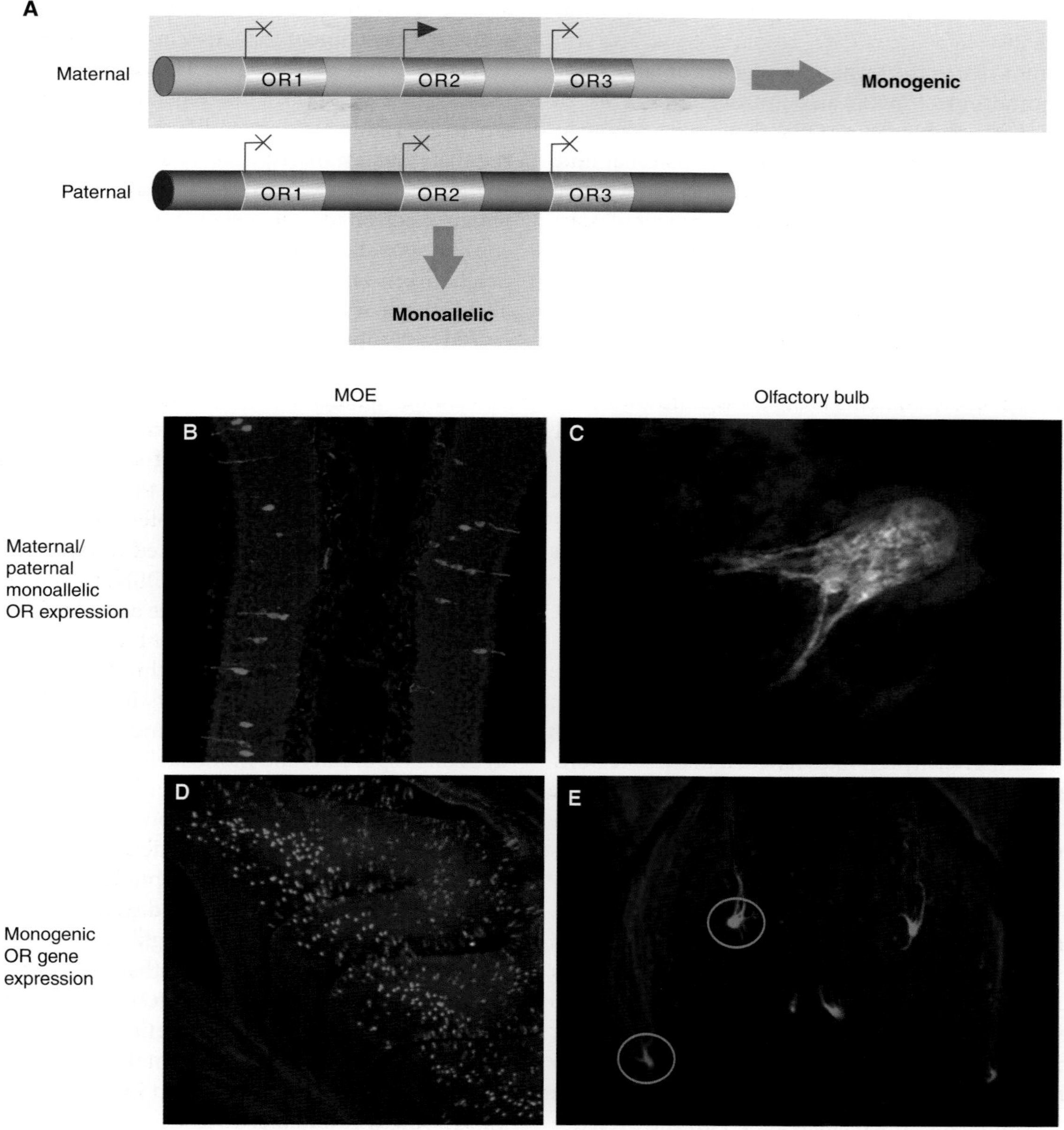

Figure 7. Monogenic and monoallelic OR gene expression in the MOE and olfactory bulb. (*A*) A schematic representation of an OR gene cluster containing three genes, illustrating the monogenic and monoallelic nature of OR expression. (*B,C*) Immunofluorescence hybridization illustrates the monoallelic nature of OR gene expression. Maternal and paternal alleles of an OR gene are tagged with lacZ (red) and green fluorescent protein (GFP, green) by knocking-in two reporter genes. (*B*) A section of the MOE shows that there is a lack of coexpression of both parental alleles in a single OSN nucleus. (*C*) A section of the olfactory bulb shows that neurons expressing an OR from either the paternal or maternal allele coalesce in the same glomerulus. (*D,E*) OR expression is monogenic and expressed in a zonal fashion. Two different OR genes have been tagged with red fluorescent protein (red) and GFP (green), respectively. (*D*) Whole-mount imaging of the MOE shows the zonal distribution of expression between the two genes. (*E*) The axons of the neurons that express different OR genes coalesce in distinct and spatially separated glomeruli in the olfactory bulb. (Images kindly donated by Dr. Thomas Bozza.)

stream of the TSS (Vassalli et al. 2002; Rothman et al. 2005; Vassalli et al. 2011).

Transgenic ORs, in the form of minigenes or large yeast artificial chomosome (YAC)-based transgenes, are never coexpressed with endogenous ORs that share identical regulatory sequences. In addition, identical paternal and maternal OR alleles are never coexpressed. The fact that transgenic and endogenous ORs with the same regulatory sequences (Ebrahimi et al. 2000; Serizawa et al. 2000; Vassalli et al. 2002), or the identical paternal and maternal OR alleles, are never coexpressed in the same neuron (Chess et al. 1994; Feinstein and Mombaerts 2004) argues for a mechanism of gene silencing to achieve monoallelic expression. Based on experiments using transgenic OR minigenes, the coding sequence of ORs, specifically the second exons of these genes, appears to contain the sequence information necessary for the repression of transgenes (Nguyen et al. 2007). Thus, a simple solution for the maintenance of the monogenic and monoallelic OR expression is epigenetic silencing.

3.3 The Chromatin Structure and Nuclear Organization of OR Clusters in OSN Development

OR genes found in many cell lines have led to these regions being viewed as "epigenetic deserts," based on numerous ChIP-on-chip or ChIP-seq experiments (Larson and Yuan 2010). This description is attributed to the lack of enrichment for any known chromatin modifications (Mikkelsen et al. 2007). However, ChIP-on-chip analysis with chromatin prepared from the MOE, the tissue in which OR genes are expressed, revealed the unexpected enrichment for H3K9me3 and H4K20me3 on ORs (Magklara et al. 2011). The presence of these heterochromatin marks is so specific to OR loci (and only a few other monoallelically expressed gene families, such as vomeronasal receptors) that genomic OR clusters can be identified solely based on the presence of H3K9me3 and H4K20me3 (Fig. 8A). Further analysis showed that H3K9 and H4K20 trimethylation of OR chromatin occurs in a differentiation-dependent and cell-type-specific manner. Specifically, in the pluripotent basal cells that support the constant regeneration of this tissue, because olfactory neurons have a finite life span, or in the case of injury, ORs are marked only with H3K9me2. On differentiation and commitment to the neuronal lineage, the trimethyl marks are deposited on ORs (Fig. 8B). Importantly, these marks are not restricted to promoters, as is the case for Polycomb-mediated repression, but spread over the entire gene into the intergenic regions between neighboring ORs, generating continuous genomic blocks of heterochromatin often >1 Mb long. The active allele in each mature OSN appears to be liberated from this epigenetic silencing with high enrichment for activating histone marks, such as H3K4me3, and depletion of H3K9me3 and H4K20me3 (Magklara et al. 2011). This epigenetic switch from repressive to activating histone modifications is restricted only to the allele that is transcriptionally active; ChIP-qPCR (quantitative polymerase chain reaction) experiments from fluorescence-activated cell-sorted neurons, which express a specific OR allele that is tagged by GFP, confirmed that neighboring ORs from the same cluster or the identical allele inherited from the other parent are epigenetically silenced as are the rest of the ORs.

The marking of OR chromatin with H3K9me3 and H4K20me3 coincides with molecular and biochemical manifestations of heterochromatinization, such as recruitment of HP1β, reduced DNase I sensitivity, and altered sedimentation properties. Thus, the epigenetic silencing of OR loci could render OR promoters inaccessible to abundant transcription factors predicted to bind to them, potentially explaining why, in each OSN, the nonchosen OR genes and other parental alleles are completely silent at the transcriptional level. In agreement with this hypothesis, inserting a reporter transgene within the heterochromatic boundaries of an OR locus results in them being marked with H3K9me3 and H4K20me3, and leads to its expression in a zonal, sporadic, and monoallelic fashion that coincides with the neighboring OR gene (Pyrski et al. 2001; Magklara et al. 2011). In a rare showing of bona fide epigenetic regulation, this reporter transgene, which normally is expressed independently of Emx2, a transcription factor that activates OR transcription, becomes Emx2 dependent when inserted within OR heterochromatin (Magklara et al. 2011). Insertion of the same transgene into other parts of the genome results in its expression in most MOE neurons (Pyrski et al. 2001), suggesting that the epigenetic state of the neighboring OR influences the expression of the transgene, in a fashion similar to classic position effect variegation described in Ch. 12 (Elgin and Reuter 2013).

The timing of OR silencing, which occurs before OR gene activation, combined with the fact that the active OR allele is devoid of H3K9-methyl marks, immediately suggests that demethylation of H3K9 is a requirement for OR gene activation. Indeed, genetic experiments revealed that H3K9 demethylase Lsd1 is necessary for the initiation, but not the maintenance, of OR transcription (Fig. 8B) (Lyons et al. 2013). Moreover, timely down-regulation of Lsd1 in response to a signaling pathway triggered by OR expression (Dalton et al. 2013; Lyons et al. 2013) is necessary for the stabilization of OR choice, providing evidence for the paramount role of chromatin-mediated silencing and desilencing in OR gene regulation (Rodriguez 2013; Tan et al. 2013; Ferreira et al. 2014).

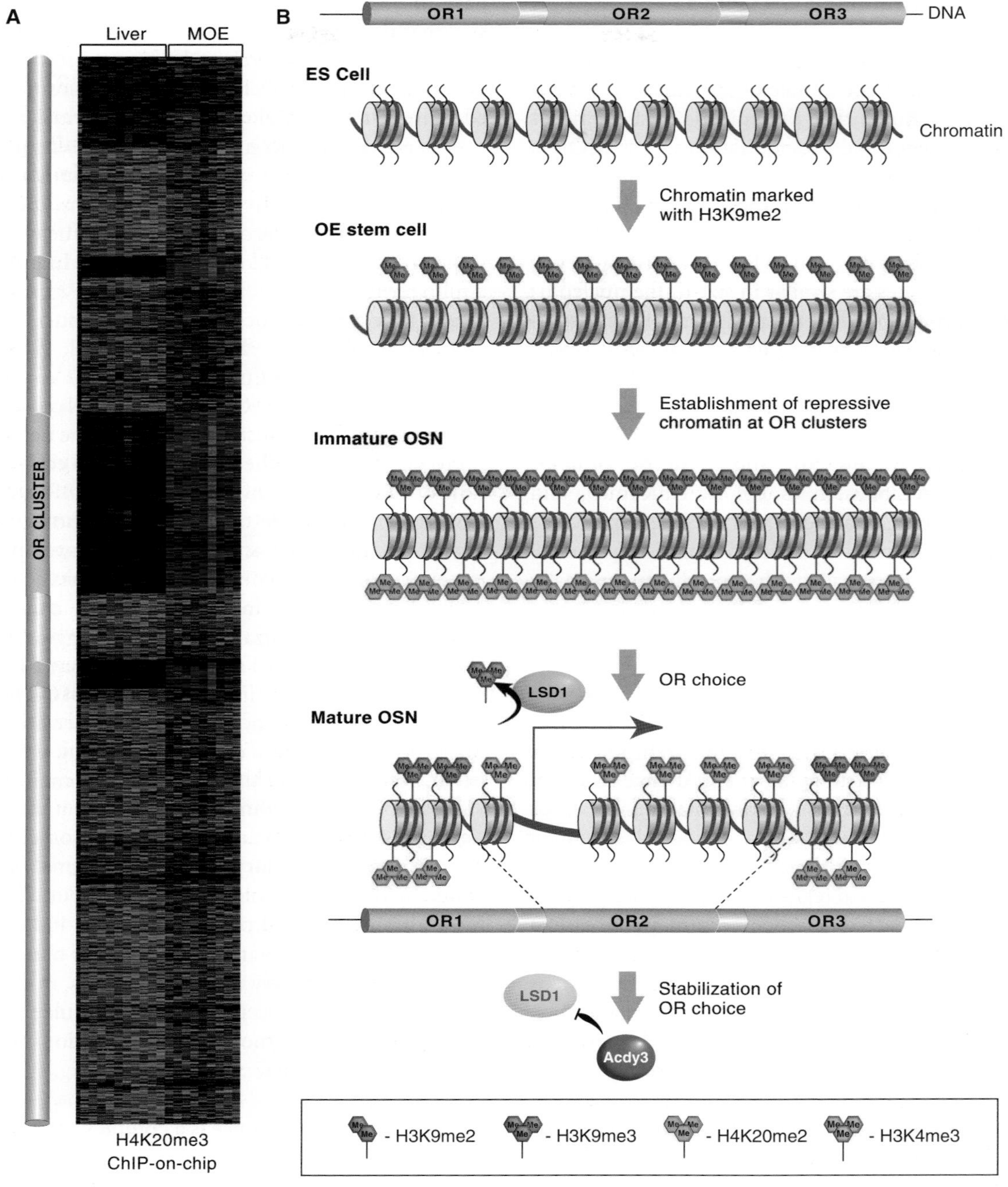

Figure 8. MOE-specific epigenetic regulation of OR gene choice during OSN development. (*A*) A heatmap depicting various enrichment levels for H4K20me3 on all the genes of mouse chromosome 2, from ChIP-on-chip analysis from MOE or liver. Red reflects high enrichment levels and green shows no enrichment. The three "red clusters" of high H4K20me3 levels coincide with the genomic coordinates of the OR clusters, schematically represented on this chromosome in orange to the left of the heatmap. (*B*) OR clusters start in early development (ES cells), devoid of most histone marks. At the onset of olfactory neurogenesis, OR clusters are marked with H3K9me2. Most OR genes then become marked by H3K9me3 and H4K20me3 during the transition from basal pluripotent cells to OSN. The OR allele that is chosen for expression (i.e., OR2) becomes liberated from its repressive chromatin marks and remodeled into an active chromatin configuration, aided by the action of the H3K9 demethylase, LSD1, and becomes positive for H3K4me3. Toward the end of the OSN differentiation pathway, the locking in of a single active OR gene is facilitated by the OR protein-induced expression of adenylyl cyclase III (Adcy3), which inhibits further LSD1 histone demethylation.

3.4 Spatial Organization of OR Genes

At the 3D level, nonexpressed OR genes converge in the nucleus into foci near the center of the nucleus. Recent experiments have shown that the epigenetic silencing of OR genes during neuronal differentiation in the MOE is coincident with widespread nuclear reorganization that culminates in intra- and interchromosomal association of hundreds of OR alleles within a few (approximately five) OR-selective heterochromatic foci (Fig. 9A) (Clowney et al. 2012). These aggregates are similar in size to the nucleolus and are frequently arranged on the periphery of the heterochromatic chromocenter of the olfactory nucleus, which primarily contains pericentromeric and centromeric repeats. Thus, OR foci represent nuclear territories highly enriched for heterochromatic markers, such as H3K9me3, H4K20me3 and HP1β (Fig. 9B), and are devoid of landmarks of euchromatin, such as Pol II, H3K4me3, and H3K27ac. The active OR allele in each OSN escapes these heterochromatic OR foci, but is found in transcriptionally competent nuclear territories with typically euchromatic epigenetic signatures (Clowney et al 2012; Armelin-Correa et al. 2014) that also frequently contain the H enhancer (Lomvardas et al. 2006).

The fact that OR foci are incompatible with transcription or other nuclear processes is also suggested by measurements made with the novel imaging technique SXT. SXT is a high-resolution imaging method that relies on the same principle as medical X-ray imaging (Le Gros et al. 2009); as X rays penetrate a biological specimen (e.g., an olfactory nucleus or your broken hand), they are more efficiently absorbed in regions of concentrated organic material (i.e., compacted heterochromatin, in the case of the nucleus, or bone tissue, in the case of your hand). The fraction of X ray absorbed in different parts of the nucleus can be quantified and is higher in heterochromatic than euchromatic territories. SXT of olfactory nuclei reveals a unique nuclear architecture with high-absorbing heterochromatin in the center of the nucleus and even denser chromatin particles on the periphery of this heterochromatic core (Fig. 9C). Because these dense aggregates are OSN specific and have a nuclear arrangement and number similar to that of OR foci, we hypothesized that they correspond to the previously described aggregates revealed by the pan-OR FISH. Notably, these aggregates have X-ray absorption properties higher than pericentromeric heterochromatin (Clowney et al. 2012). Only sperm nuclei, in which protamines replace histones, have more compacted chromatin than these OSN-specific foci as determined by SXT.

Does the OSN-specific spatial aggregation of OR loci contribute to OR silencing and the monogenic expression of these genes? An unusual feature of OSN nuclei is that they have an "inside-out" nuclear morphology, whereby heterochromatin is aggregated in the center of the nucleus and euchromatin is peripheral, in sharp contrast to the textbook view of nuclear organization with heterochromatin restricted to the nuclear envelope. This inverse nuclear organization is actually found in other types of sensory neurons, such as photoreceptor neurons, in the retinal epithelium (Solovei et al. 2009). In olfactory neurons, the reason for the "collapse" of heterochomatin in the nuclear core is the lack of the nuclear envelope protein, LBR. In most cells, the amino terminus of LBR interacts with HP1 proteins and is shown to recruit heterochromatin to the nuclear periphery (e.g., basal cell in Fig. 9E) (Worman et al. 1988; Pyrpasopoulou et al. 1996; Hoffmann et al. 2002). A loss-of-function mutation in the LBR gene in non-OSN cells of the MOE (Shultz et al. 2003) causes pericentromeric heterochromatin to be localized to the nuclear core and ectopic aggregation of OR genes. Conversely, ectopic LBR expression in OSNs reverts the inside-out nuclear morphology by recruiting heterochromatin back to the nuclear envelop and allowing euchromatin to claim the interior parts of the nucleus as in most mammalian cell types. One consequence of this dramatic reorganization is that OR aggregation is perturbed and the OR foci are dissolved. From a biochemical perspective, this results in decompaction of OR heterochromatin as detected by increased DNase I sensitivity and reduced X-ray absorption using SXT imaging. These changes occur despite the fact that these genes remain marked by H3K9me3 and H4K20me3, suggesting that this epigenetic signature translates to an inaccessible chromatin structure only on proper secondary (intrachromosomal) and tertiary (interchromosomal) folding (Fig. 9D). Functionally (i.e., from a transcriptional perspective), LBR-induced disruption of OR foci results in coexpression of hundreds of OR alleles in each olfactory neuron and, thus, in the dramatic violation of the one receptor per neuron rule that describes the monogenic and monoallelic expression of OR genes (illustrated in the defective OSN cell, Fig. 9E) (Clowney et al. 2012).

In summary, OR choice is a complex process that relies on chromatin-mediated silencing and selective desilencing on one out of thousands of available alleles. Although the aforementioned data provide a conceptual framework for the regulatory principles of OR expression, major questions must be answered for a comprehensive understanding of this unique process. For example, there is no information on the source of "singularity" that activates only one OR allele at a time, or signaling pathway and the molecular targets of the feedback signal elicited by OR expression. Moreover, the identity of the histone methyltransferases and demethylases responsible for the silencing and activation of OR genes in not known and neither are the mech-

Cite as *Cold Spring Harb Perspect Biol* doi: 10.1101/cshperspect.a024208

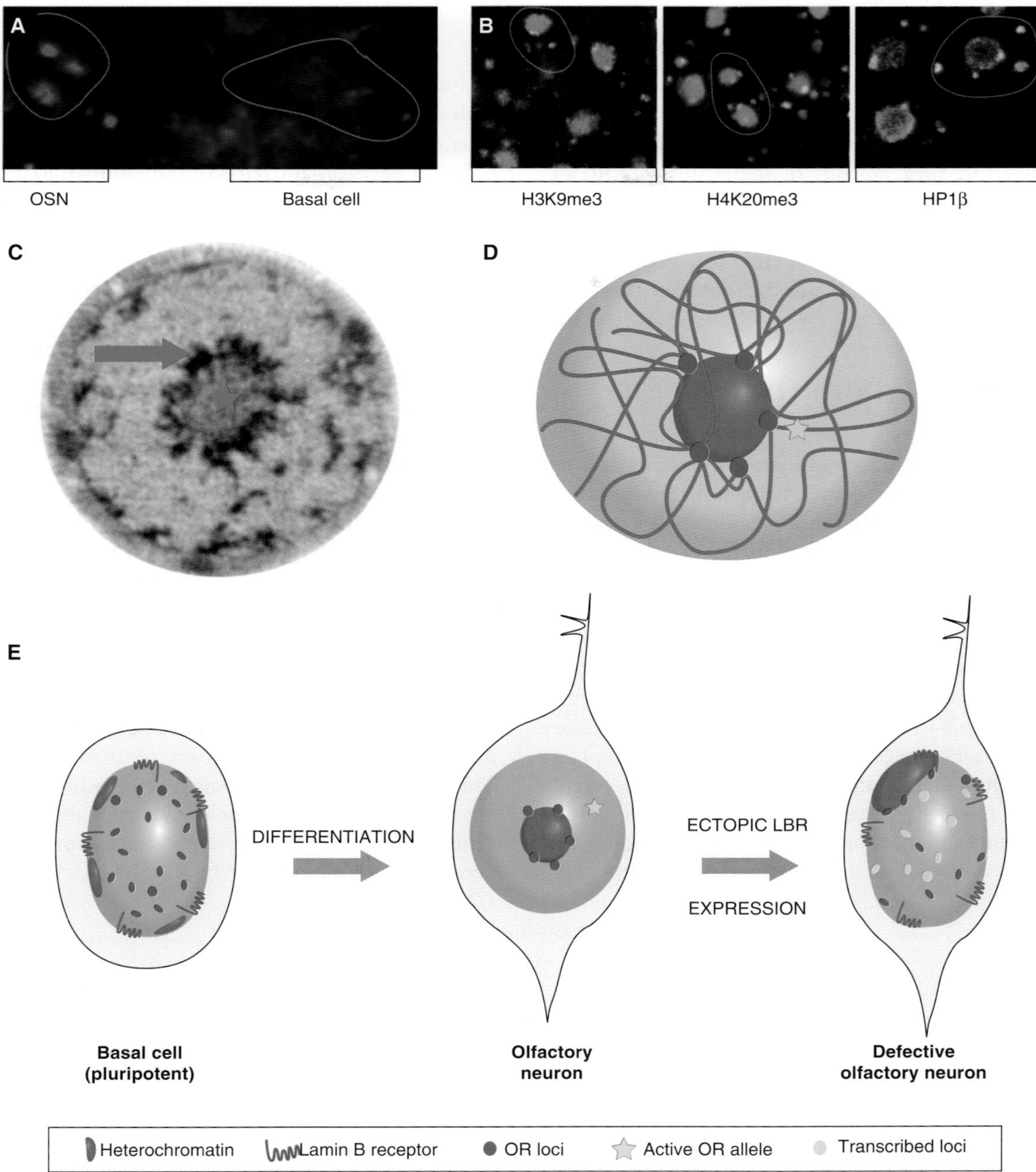

Figure 9. The spatial organization of OR gene clusters during OSN differentiation. (*A*) DNA FISH showing OR gene cluster distribution using a panOR probe (red) on MOE sections. The signal is diffuse at the basal pluripotent cells, but aggregated in neuronal nuclei (the nuclear borders are highlighted). (*B*) PanOR foci (red) are enriched at regions marked with H3K9me3, H4K20me3, and HP1β (green), suggesting that they are heterochromatic aggregates of the silent OR alleles. (*C*) Soft X-ray tomography (SXT) of an olfactory nucleus shows that compacted chromatin (highlighted by a blue star) is aggregated at the center of the nucleus. At the periphery of this heterochromatic core, SXTreveals even more compacted structures (pointed by blue arrow) that are OSN specific. (*D*) A schematic 3D nuclear landscape shows all the heterochromatic OR clusters converging, in *cis* and *trans*, to form silent foci incompatible with transcription. The active OR allele is physically separated from the inactive OR aggregates. This is based on epigenetic data obtained from ChIP-on-chip experiments depicted as heatmaps. (*E*) Prior to OSN differentiation, the precursor basal stem cell has a typical nuclear configuation with regard to heterochromatin with peripheral foci. The OR clusters are randomly distributed in nuclear space. Following differentiation, aggregation of all but one OR allele to heterochromatic foci surrounding the pericentromeric chromocenter of an OSN nucleus occurs. This serves the monogenic and monoallelic expression of these genes and depends on the down-regulation of lamin B receptor (LBR). Ectopic LBR expression in OSNs dissolves the OR foci, reverses the nuclear morphology, and causes the simultaneous expression of hundreds of ORs in a striking violation of the one receptor per neuron rule.

anisms by which they are recruited with such specificity on the OR loci. Finally, the mechanism responsible for the specific aggregation of silenced OR alleles and placement of the active OR allele outside of the repressive OR foci remains unknown. Of great interest will be to determine if the same methyltransferase activities responsible for the aggregation of pericentromeric heterochromatin (Pinheiro et al. 2012) are also responsible for the focal distribution of OR genes.

4 EPIGENETIC REGULATION OF OLFACTORY NEURON LIFE SPAN

Olfactory neurons, unlike most neurons, have a finite life span of ~90 d and are constantly being replaced by neurogenesis that lasts through adulthood. OSNs, however, do not all have the same life span; their longevity is determined by the type of OR they express and abundance of its cognate odorant in the environment (Santoro and Dulac 2012). A key regulator of this process is, surprisingly, a previously uncharacterized histone variant named H2be. This histone variant differs by only five amino acids from the common H2b isoform and is expressed only in mature olfactory and vomeronasal (pheromone-detecting chemoreceptors) neurons. Although the molecular mechanism by which H2be affects longevity is not yet understood, it is striking that this histone variant cannot be acetylated or methylated on lysine 5, two posttranslational modifications associated with transcriptional elongation. As a result, olfactory neurons with high levels of H2be, which are particularly enriched at euchromatic parts of the genome, have low levels of methylated and acetylated lysine 5 of H2b. This is indicative of some kind of replication-independent mode of H2b substitution with H2be.

H2be has a sporadic expression pattern in the olfactory epithelium; some neurons have undetectable levels of this histone and others have moderate to high levels, based on immunofluorescence experiments. Strikingly, the identity of the receptor seems to determine the expression rates of H2be, given that the levels of H2be are the same among neurons that express the same OR gene. In fact, olfactory neurons with ORs that are frequently activated in a specific environment retain low levels of H2be in their nuclei (i.e., the red OSNs in Fig. 10). Loss- and gain-of-function experiments with knockout and knockin mice revealed that H2be promotes the apoptosis of understimulated olfactory neurons and reduces their life span. Thus, H2be acts as a sensor of neuronal stimulation and determines the longevity of the neuron; if a neuron is frequently activated, then H2be levels are low and the life span of that neuron is extended. If the neuron is dormant, then it will die faster. The benefit of this mechanism is obvious: Because ORs are stochastically chosen, many neurons might express receptors that are useless in a specific environment. If these

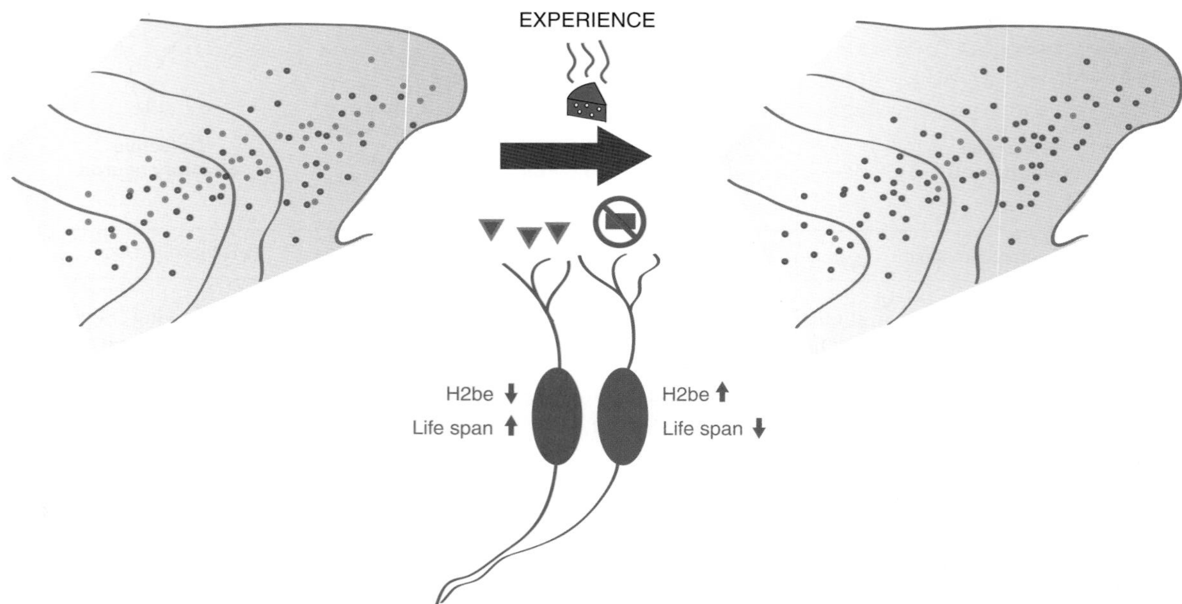

Figure 10. The H2be histone variant in OSNs. The figure summarizes results of experimental manipulations showing that olfactory experience regulates the expression levels of H2be and dictates the longevity of olfactory neurons. Neurons frequently activated (in red) do not accumulate H2be in the chromatin of OR clusters and live longer, eventually dominating the MOE.

Cite as *Cold Spring Harb Perspect Biol* doi: 10.1101/cshperspect.a024208

neurons have a shorter life span, after several rounds of neurogenesis and OR choice, the olfactory epithelium will have a higher percentage of "useful" olfactory neurons and thus operate as a sensory organ that is better "tuned" to a specific odorant environment (Monahan and Lomvardas 2012; Santoro and Dulac 2012).

The unexpected discovery that a histone variant with extreme tissue specificity determines the life span of olfactory neurons in an activity-dependent manner provides an elegant showing of the diversity of epigenetic processes that contribute to the development and function of the nervous system. In addition to the regulation of H2be expression, neuronal activity may directly affect the stability of OR choice because Lsd1 down-regulation requires OR-induced expression of Adcy3 (Lyons et al. 2013), the major contributor of cAMP, in response to OR activation by odorants. Thus, the olfactory system, with its extreme regulatory requirements, provides an excellent model system for the study of the interplay between neuronal activity and epigenetic processes that control nuclear plasticity and regulate neuronal longevity.

5 SUMMARY AND PERSPECTIVE

In this chapter, we provided four specific examples of epigenetic processes involved in the development and wiring of the nervous system. There are, of course, other epigenetic processes involved in the nervous system in both vertebrates and invertebrates. For this, we point you to several excellent reviews and papers therein (Dulac 2010; Zocchi and Sassone-Corsi 2010; Qureshi and Mehler 2012; Russo and Nestler 2013). In three of the cases covered in this chapter, the cell under study has to "make" decisions that will determine its developmental fate and function: choice between neurogenesis or gliogenesis, choice between a combination of Pcdh isoforms, and choice of a single OR allele. In all three cases, a choice is determined by the balance between epigenetic silencing and activation, but the mechanistic details differ depending on specific regulatory needs. In the case of neural precursor cells, in which this choice is determined by extrinsic signals and thus is highly regulated during development, there is a balance between Polycomb and Trithorax complexes modulated by Dnmt3a activity. In the case of Pcdh and olfactory gene choice, however, which are seemingly stochastic and the goal is to achieve maximum diversity of expression programs rather than specific transcriptional outcomes, different, unknown, epigenetic regulators are involved, which assure that only one of multiple promoters will become accessible to the transcription machinery. On the other hand, in the case of the histone variant H2be, the role of this protein is not to determine a developmental decision or make a transcriptional choice, but to affect the longevity of the neuron through mechanisms that are not yet understood. Most certainly, novel epigenetic mechanisms will emerge from the study of other neuronal processes, such as learning and memory, neuronal plasticity, critical periods, neurodegeneration, and replacement of neurons during adult neurogenesis. In all of these cases, a common question that needs to be answered is how external input received by different neuronal populations translates into specific epigenetic changes resulting in transcriptional and physiological output.

REFERENCES

*Reference is also in this book.

Alvarez-Buylla A, Garcia-Verdugo JM. 2002. Neurogenesis in adult subventricular zone. *J Neurosci* **22:** 629–634.

Alvarez-Buylla A, Lim DA. 2004. For the long run: Maintaining germinal niches in the adult brain. *Neuron* **41:** 683–686.

Armelin-Correa LM, Gutiyama LM, Brandt DYC, Malnic B. 2014. Nuclear compartmentalization of odorant receptor genes. *Proc Natl Acad Sci* **111:** 2782–2787.

Axel R. 1995. The molecular logic of smell. *Sci Am* **273:** 154–159.

* Barlow DP, Bartolomei MS. 2014. Genomic imprinting in mammals. *Cold Spring Harb Perspect Biol doi* **6:** a018382.

Barnea G, O'Donnell S, Mancia F, Sun X, Nemes A, Mendelsohn M, Axel R. 2004. Odorant receptors on axon termini in the brain. *Science* **304:** 1468.

* Becker PB, Workman JL. 2013. Nucleosome remodeling and epigenetics. *Cold Spring Harb Perspect Biol* **5:** a017905.

Bernstein BE, Mikkelsen TS, Xie X, Kamal M, Huebert DJ, Cuff J, Fry B, Meissner A, Wernig M, Plath K, et al. 2006. A bivalent chromatin structure marks key developmental genes in embryonic stem cells. *Cell* **125:** 315–326.

Bonn S, Seeburg PH, Schwarz MK. 2007. Combinatorial expression of α- and γ-protocadherins alters their presenilin-dependent processing. *Mol Cell Biol* **27:** 4121–4132.

* Brockdorff N, Turner BM. 2014. Dosage compensation in mammals. *Cold Spring Harb Perspect Biol* doi: 10.1101/cshperspect.a019406.

Buchanan SM, Schalm SS, Maniatis T. 2010. Proteolytic processing of protocadherin proteins requires endocytosis. *Proc Natl Acad Sci* **107:** 17774–17779.

Buck L, Axel R. 1991. A novel multigene family may encode odorant receptors: A molecular basis for odor recognition. *Cell* **65:** 175–187.

Burgold T, Spreafico F, De Santa F, Totaro MG, Prosperini E, Natoli G, Testa G. 2008. The histone H3 lysine 27-specific demethylase Jmjd3 is required for neural commitment. *PLoS ONE* **3:** e3034.

* Busslinger M, Tarakhovsky A. 2014. Epigenetic control of immunity. *Cold Spring Harb Perspect Biol* **6:** a019307.

Cameron S, Rao Y. 2010. Molecular mechanisms of tiling and self-avoidance in neural development. *Mol Brain* **3:** 28.

Chen WV, Maniatis T. 2013. Clustered protocadherins. *Development* **140:** 3297–3302.

Chess A. 2013. Random and non-random monoallelic expression. *Neuropsychopharmacology* **38:** 55–61.

Chess A, Simon I, Cedar H, Axel R. 1994. Allelic inactivation regulates olfactory receptor gene expression. *Cell* **78:** 823–834.

Clowney EJ, Magklara A, Colquitt BM, Pathak N, Lane RP, Lomvardas S. 2011. High-throughput mapping of the promoters of the mouse olfactory receptor genes reveals a new type of mammalian promoter and provides insight into olfactory receptor gene regulation. *Genome Res* **21:** 1249–1259.

Clowney EJ, LeGros MA, Mosley CP, Clowney FG, Markenskoff-Papadimitriou EC, Myllys M, Barnea G, Larabell CA, Lomvardas S. 2012. Nuclear aggregation of olfactory receptor genes governs their monogenic expression. *Cell* **151:** 724–737.

Dalton RP, Lyons DB, Lomvardas S. 2013. Co-opting the unfolded protein response to elicit olfactory receptor feedback. *Cell* **155:** 321–332.

* Dekker J, Misteli T. 2014. Long-range chromatin interactions. *Cold Spring Harb Perspect Biol* doi: 10.1101/cshperspect.a019356.

Doetsch F, Caille I, Lim DA, Garcia-Verdugo JM, Alvarez-Buylla A. 1999. Subventricular zone astrocytes are neural stem cells in the adult mammalian brain. *Cell* **97:** 703–716.

Dulac C. 2010. Brain function and chromatin plasticity. *Nature* **465:** 728–735.

Ebrahimi FA, Edmondson J, Rothstein R, Chess A. 2000. YAC transgene-mediated olfactory receptor gene choice. *Dev Dyn* **217:** 225–231.

* Elgin SCR, Reuter G. 2013. Position-effect variegation, heterochromatin formation, and gene silencing in *Drosophila*. *Cold Spring Harb Perspect Biol* **5:** a017780.

Esumi S, Kakazu N, Taguchi Y, Hirayama T, Sasaki A, Hirabayashi T, Koide T, Kitsukawa T, Hamada S, Yagi T. 2005. Monoallelic yet combinatorial expression of variable exons of the protocadherin-α gene cluster in single neurons. *Nat Genet* **37:** 171–176.

Fass DM, Schroeder FA, Perlis RH, Haggarty SJ. 2014. Epigenetic mechanisms in mood disorders: Targeting neuroplasticity. *Neuroscience* **264C:** 112–130.

Feinstein P, Mombaerts P. 2004. A contextual model for axonal sorting into glomeruli in the mouse olfactory system. *Cell* **117:** 817–831.

Ferreira T, Wilson SR, Choi YG, Risso D, Dudoit S, Speed TP, Ngai J. 2014. Silencing of odorant receptor genes by G protein βγ signaling ensures the expression of one odorant receptor per olfactory sensory neuron. *Neuron* **81:** 847–859.

Fuss SH, Omura M, Mombaerts P. 2007. Local and *cis* effects of the H element on expression of odorant receptor genes in mouse. *Cell* **130:** 373–384.

Glusman G, Bahar A, Sharon D, Pilpel Y, White J, Lancet D. 2000. The olfactory receptor gene superfamily: Data mining, classification, and nomenclature. *Mamm Genome* **11:** 1016–1023.

Golan-Mashiach M, Grunspan M, Emmanuel R, Gibbs-Bar L, Dikstein R, Shapiro E. 2012. Identification of CTCF as a master regulator of the clustered protocadherin genes. *Nucleic Acids Res* **40:** 3378–3391.

* Grossniklaus U, Paro R. 2014. Transcriptional silencing by *Polycomb*-group proteins. *Cold Spring Harb Perspect Biol* **6:** a019331.

Grueber WB, Sagasti A. 2010. Self-avoidance and tiling: Mechanisms of dendrite and axon spacing. *Cold Spring Harb Perspect Biol* **2:** a001750.

Guo Y, Monahan K, Wu H, Gertz J, Varley KE, Li W, Myers RM, Maniatis T, Wu Q. 2012. CTCF/cohesin-mediated DNA looping is required for protocadherin α promoter choice. *Proc Natl Acad Sci* **109:** 21081–21086.

Haas IG, Frank M, Véron N, Kemler R. 2005. Presenilin-dependent processing and nuclear function of γ-protocadherins. *J Biol Chem* **280:** 9313–9319.

Hambsch B, Grinevich V, Seeburg PH, Schwarz MK. 2005. γ-Protocadherins, presenilin-mediated release of C-terminal fragment promotes locus expression. *J Biol Chem* **280:** 15888–15897.

Han MH, Lin C, Meng S, Wang X. 2010. Proteomics analysis reveals overlapping functions of clustered protocadherins. *Mol Cell Proteomics* **9:** 71–83.

Hirabayashi Y, Gotoh Y. 2010. Epigenetic control of neural precursor cell fate during development. *Nat Rev Neurosci* **11:** 377–388.

Hirabayashi Y, Suzki N, Tsuboi M, Endo TA, Toyoda T, Shinga J, Koseki H, Vidal M, Gotoh Y. 2009. Polycomb limits the neurogenic competence of neural precursor cells to promote astrogenic fate transition. *Neuron* **63:** 600–613.

Hirayama T, Tarusawa E, Yoshimura Y, Galjart N, Yagi T. 2012. CTCF is required for neural development and stochastic expression of clustered Pcdh genes in neurons. *Cell Rep* **2:** 345–357.

Hirota J, Mombaerts P. 2004. The LIM-homeodomain protein Lhx2 is required for complete development of mouse olfactory sensory neurons. *Proc Natl Acad Sci* **101:** 8751–8755.

Hoffmann K, Dreger CK, Olins AL, Olins DE, Shultz LD, Lucke B, Karl H, Kaps R, Muller D, Vaya A, et al. 2002. Mutations in the gene encoding the lamin B receptor produce an altered nuclear morphology in granulocytes (Pelger-Huet anomaly). *Nat Genet* **31:** 410–414.

Jobe EM, McQuate AL, Zhao X. 2012. Crosstalk among epigenetic pathways regulates neurogenesis. *Front Neurosci* **6:** 59.

Kagey MH, Newman JJ, Bilodeau S, Zhan Y, Orlando DA, van Berkum NL, Ebmeier CC, Goossens J, Rahl PB, Levine SS, et al. 2010. Mediator and cohesin connect gene expression and chromatin architecture. *Nature* **467:** 430–435.

Kaneko R, Kawaguchi M, Toyama T, Taguchi Y, Yagi T. 2009. Expression levels of Protocadherin-α transcripts are decreased by nonsense-mediated mRNA decay with frameshift mutations and by high DNA methylation in their promoter regions. *Gene* **430:** 86–94.

Katori S, Hamada S, Noguchi Y, Fukuda E, Yamamoto T, Yamamoto H, Hasegawa S, Yagi T. 2009. Protocadherin-α family is required for serotonergic projections to appropriately innervate target brain areas. *J Neurosci* **29:** 9137–9147.

Kawaguchi M, Toyama T, Kaneko R, Hirayama T, Kawamura Y, Yagi T. 2008. Relationship between DNA methylation states and transcription of individual isoforms encoded by the protocadherin-α gene cluster. *J Biol Chem* **283:** 12064–12075.

Kehayova P, Monahan K, Chen W, Maniatis T. 2011. Regulatory elements required for the activation and repression of the protocadherin-α gene cluster. *Proc Natl Acad Sci* **108:** 17195–17200.

Khan M, Vaes E, Mombaerts P. 2011. Regulation of the probability of mouse odorant receptor gene choice. *Cell* **147:** 907–921.

* Kim T-K, Hemberg M, Gray JM. 2014. Enhancer RNAs: A class of long noncoding RNAs synthesized at enhancers. *Cold Spring Harb Perspect Biol* doi: 10.1101/cshperspect.a018622.

* Kingston RE, Tamkun JW. 2014. Transcriptional regulation by trithorax-group proteins. *Cold Spring Harb Perspect Biol* **6:** a019349.

Kriaucionis S, Heintz N. 2009. The nuclear DNA base 5-hydroxymethylcytosine is present in Purkinje neurons and the brain. *Science* **324:** 929–930.

Larson JL, Yuan GC. 2010. Epigenetic domains found in mouse embryonic stem cells via a hidden Markov model. *BMC Bioinformatics* **11:** 557.

Lefebvre JL, Kostadinov D, Chen WV, Maniatis T, Sanes JR. 2012. Protocadherins mediate dendritic self-avoidance in the mammalian nervous system. *Nature* **488:** 517–521.

Le Gros MA, McDermott G, Uchida M, Knoechel CG, Larabell CA. 2009. High-aperture cryogenic light microscopy. *J Microsc* **235:** 1–8.

Lewcock JW, Reed RR. 2004. A feedback mechanism regulates monoallelic odorant receptor expression. *Proc Natl Acad Sci* **101:** 1069–1074.

* Li E, Zhang Y. 2014. DNA methylation in mammals. *Cold Spring Harb Perspect Biol* **6:** a019133.

Lim DA, Huang YC, Swigut T, Mirick AL, Garcia-Verdugo JM, Wysocka J, Ernst P, Alvarez-Buylla A. 2009. Chromatin remodelling factor Mll1 is essential for neurogenesis from postnatal neural stem cells. *Nature* **458:** 529–533.

Lomvardas S, Barnea G, Pisapia DJ, Mendelsohn M, Kirkland J, Axel R. 2006. Interchromosomal interactions and olfactory receptor choice. *Cell* **126:** 403–413.

Lunyak VV, Rosenfeld MG. 2005. No rest for REST: REST/NRSF regulation of neurogenesis. *Cell* **121:** 499–501.

Lyons DB, Allen WE, Goh T, Tsai L, Barnea G, Lomvardas S. 2013. An epigenetic trap stabilizes singular olfactory receptor expression. *Cell* **154:** 325–336.

Ma DK, Marchetto MC, Guo JU, Ming GL, Gage FH, Song H. 2010. Epigenetic choreographers of neurogenesis in the adult mammalian brain. *Nat Neurosci* **13:** 1338–1344.

Magklara A, Yen A, Colquitt BM, Clowney EJ, Allen W, Markenscoff-Papadimitriou E, Evans ZA, Kheradpour P, Mountoufaris G, Carey C, et al. 2011. An epigenetic signature for monoallelic olfactory receptor expression. *Cell* **145:** 555–570.

McIntyre JC, Bose SC, Stromberg AJ, McClintock TS. 2008. Emx2 stimulates odorant receptor gene expression. *Chem Senses* **33:** 825–837.

Mellen M, Ayata P, Dewell S, Kriaucionis S, Heintz N. 2012. MeCP2 binds to 5hmC enriched within active genes and accessible chromatin in the nervous system. *Cell* **151:** 1417–1430.

Mikkelsen TS, Ku M, Jaffe DB, Issac B, Lieberman E, Giannoukos G, Alvarez P, Brockman W, Kim TK, Koche RP, et al. 2007. Genome-wide maps of chromatin state in pluripotent and lineage-committed cells. *Nature* **448:** 553–560.

Mombaerts P, Wang F, Dulac C, Chao SK, Nemes A, Mendelsohn M, Edmondson J, Axel R. 1996. Visualizing an olfactory sensory map. *Cell* **87:** 675–686.

Monahan K, Lomvardas S. 2012. How keeping active pays off in the olfactory system. *eLife* **1:** e00326.

Monahan K, Rudnick ND, Kehayova PD, Pauli F, Newberry KM, Myers RM, Maniatis T. 2012. Role of CCCTC binding factor (CTCF) and cohesin in the generation of single-cell diversity of protocadherin-α gene expression. *Proc Natl Acad Sci* **109:** 9125–9130.

Munzel M, Globisch D, Bruckl T, Wagner M, Welzmiller V, Michalakis S, Muller M, Biel M, Carell T. 2010. Quantification of the sixth DNA base hydroxymethylcytosine in the brain. *Angew Chem Int Ed Engl* **49:** 5375–5377.

Nei M, Niimura Y, Nozawa M. 2008. The evolution of animal chemosensory receptor gene repertoires: Roles of chance and necessity. *Nat Rev* **9:** 951–963.

Nestler EJ. 2014. Epigenetic mechanisms of drug addiction. *Neuropharmacology* **76 Pt B:** 259–268.

Nguyen MQ, Zhou Z, Marks CA, Ryba NJ, Belluscio L. 2007. Prominent roles for odorant receptor coding sequences in allelic exclusion. *Cell* **131:** 1009–1017.

Olynik BM, Rastegar M. 2012. The genetic and epigenetic journey of embryonic stem cells into mature neural cells. *Front Genet* **3:** 81.

Phillips JE, Corces VG. 2009. CTCF: Master weaver of the genome. *Cell* **137:** 1194–1211.

Pinheiro I, Margueron R, Shukeir N, Eisold M, Fritzsch C, Richter FM, Mittler G, Genoud C, Goyama S, Kurokawa M, et al. 2012. Prdm3 and Prdm16 are H3K9me1 methyltransferases required for mammalian heterochromatin integrity. *Cell* **150:** 948–960.

Plessy C, Pascarella G, Bertin N, Akalin A, Carrieri C, Vassalli A, Lazarevic D, Severin J, Vlachouli C, Simone R, et al. 2011. Promoter architecture of mouse olfactory receptor genes. *Genome Res* **22:** 486–497.

Pyrpasopoulou A, Meier J, Maison C, Simos G, Georgatos SD. 1996. The lamin B receptor (LBR) provides essential chromatin docking sites at the nuclear envelope. *EMBO J* **15:** 7108–7119.

Pyrski M, Xu Z, Walters E, Gilbert DJ, Jenkins NA, Copeland NG, Margolis FL. 2001. The OMP-lacZ transgene mimics the unusual expression pattern of OR-Z6, a new odorant receptor gene on mouse chromosome 6: Implication for locus-dependent gene expression. *J Neurosci* **21:** 4637–4648.

Qureshi IA, Mehler MF. 2012. Emerging roles of non-coding RNAs in brain evolution, development, plasticity and disease. *Nat Rev Neurosci* **13:** 528–541.

Reiss K, Maretzky T, Haas IG, Schulte M, Ludwig A, Frank M, Saftig P. 2006. Regulated ADAM10-dependent ectodomain shedding of γ-protocadherin C3 modulates cell–cell adhesion. *J Biol Chem* **281:** 21735–21744.

Ressler KJ, Sullivan SL, Buck LB. 1993. A zonal organization of odorant receptor gene expression in the olfactory epithelium. *Cell* **73:** 597–609.

Ribich S, Tasic B, Maniatis T. 2006. Identification of long-range regulatory elements in the protocadherin-α gene cluster. *Proc Natl Acad Sci* **103:** 19719–19724.

Riccio A. 2010. Dynamic epigenetic regulation in neurons: Enzymes, stimuli and signaling pathways. *Nat Neurosci* **13:** 1330–1337.

Rodriguez I. 2013. Singular expression of olfactory receptor genes. *Cell* **155:** 274–277.

Rothman A, Feinstein P, Hirota J, Mombaerts P. 2005. The promoter of the mouse odorant receptor gene M71. *Mol Cell Neurosci* **28:** 535–546.

Rudenko A, Tsai LH. 2014. Epigenetic regulation in memory and cognitive disorders. *Neuroscience* **264C:** 51–63.

Russo SJ, Nestler EJ. 2013. The brain reward circuitry in mood disorders. *Nat Rev Neurosci* **14:** 609–625.

Santoro SW, Dulac C. 2012. The activity-dependent histone variant H2BE modulates the life span of olfactory neurons. *eLife* **1:** e00070.

Schalm SS, Ballif BA, Buchanan SM, Phillips GR, Maniatis T. 2010. Phosphorylation of protocadherin proteins by the receptor tyrosine kinase Ret. *Proc Natl Acad Sci* **107:** 13894–13899.

Schmucker D, Chen B. 2009. Dscam and DSCAM: Complex genes in simple animals, complex animals yet simple genes. *Genes Dev* **23:** 147–156.

Schmucker D, Clemens JC, Shu H, Worby CA, Xiao J, Muda M, Dixon JE, Zipursky SL. 2000. *Drosophila* Dscam is an axon guidance receptor exhibiting extraordinary molecular diversity. *Cell* **101:** 671–684.

Schreiner D, Weiner JA. 2010. Combinatorial homophilic interaction between γ-protocadherin multimers greatly expands the molecular diversity of cell adhesion. *Proc Natl Acad Sci* **107:** 14893–14898.

Serizawa S, Ishii T, Nakatani H, Tsuboi A, Nagawa F, Asano M, Sudo K, Sakagami J, Sakano H, Ijiri T, et al. 2000. Mutually exclusive expression of odorant receptor transgenes. *Nat Neurosci* **3:** 687–693.

Serizawa S, Miyamichi K, Nakatani H, Suzuki M, Saito M, Yoshihara Y, Sakano H. 2003. Negative feedback regulation ensures the one receptor-one olfactory neuron rule in mouse. *Science* **302:** 2088–2094.

Serizawa S, Miyamichi K, Takeuchi H, Yamagishi Y, Suzuki M, Sakano H. 2006. A neuronal identity code for the odorant receptor-specific and activity-dependent axon sorting. *Cell* **127:** 1057–1069.

* Seto E, Yoshida M. 2014. Erasers of histone acetylation: The histone deacetylase enzymes. *Cold Spring Harb Perspect Biol* **6:** a018713.

Shultz LD, Lyons BL, Burzenski LM, Gott B, Samuels R, Schweitzer PA, Dreger C, Herrmann H, Kalscheuer V, Olins AL, et al. 2003. Mutations at the mouse ichthyosis locus are within the lamin B receptor gene: A single gene model for human Pelger-Huet anomaly. *Hum Mol Genet* **12:** 61–69.

Shykind BM. 2005. Regulation of odorant receptors: One allele at a time. *Hum Mol Genet* **14:** R33–R39.

Shykind BM, Rohani SC, O'Donnell S, Nemes A, Mendelsohn M, Sun Y, Axel R, Barnea G. 2004. Gene switching and the stability of odorant receptor gene choice. *Cell* **117:** 801–815.

Solovei I, Kreysing M, Lanctot C, Kosem S, Peichl L, Cremer T, Guck J, Joffe B. 2009. Nuclear architecture of rod photoreceptor cells adapts to vision in mammalian evolution. *Cell* **137:** 356–368.

Suo L, Lu H, Ying G, Capecchi MR, Wu Q. 2012. Protocadherin clusters and cell adhesion kinase regulate dendrite complexity through Rho GTPase. *J Mol Cell Biol* **4:** 362–376.

Szulwach KE, Li X, Li Y, Song CX, Wu H, Dai Q, Irier H, Upadhyay AK, Gearing M, Levey AI, et al. 2011. 5-hmC-mediated epigenetic dynamics during postnatal neurodevelopment and aging. *Nat Neurosci* **14:** 1607–1616.

Tan L, Zong C, Xie XS. 2013. Rare event of histone demethylation can initiate singular gene expression of olfactory receptors. *Proc Natl Acad Sci* **110:** 21148–21152.

Tasic B, Nabholz CE, Baldwin KK, Kim Y, Rueckert EH, Ribich SA, Cramer P, Wu Q, Axel R, Maniatis T. 2002. Promoter choice determines splice site selection in protocadherin α and γ pre-mRNA splicing. *Mol Cell* **10:** 21–33.

Thu CA, Chen WV, Rubinstein R, Chevee M, Wolcott HN, Felsovalyl KO, Tapia JC, Shapiro L, Honig B, Maniatis T. 2014. Single-cell identity generated by combinatorial homophilic interactions between α, β, γ protocadherins. *Cell* **158:** 1045–1059.

Tyssowski K, Kishi Y, Gotoh Y. 2014. Chromatin regulators of neural development. *Neuroscience* **264C:** 4–16.

Vassalli A, Rothman A, Feinstein P, Zapotocky M, Mombaerts P. 2002. Minigenes impart odorant receptor-specific axon guidance in the olfactory bulb. *Neuron* **35:** 681–696.

Vassalli A, Feinstein P, Mombaerts P. 2011. Homeodomain binding motifs modulate the probability of odorant receptor gene choice in transgenic mice. *Mol Cell Neurosci* **46:** 381–396.

Vassar R, Ngai J, Axel R. 1993. Spatial segregation of odorant receptor expression in the mammalian olfactory epithelium. *Cell* **74:** 309–318.

Wang SS, Tsai RY, Reed RR. 1997. The characterization of the Olf-1/EBF-like HLH transcription factor family: Implications in olfactory gene regulation and neuronal development. *J Neurosci* **17:** 4149–4158.

Wang F, Nemes A, Mendelsohn M, Axel R. 1998. Odorant receptors govern the formation of a precise topographic map. *Cell* **93:** 47–60.

Worman HJ, Yuan J, Blobel G, Georgatos SD. 1988. A lamin B receptor in the nuclear envelope. *Proc Natl Acad Sci* **85:** 8531–8534.

Wu Q, Maniatis T. 1999. A striking organization of a large family of human neural cadherin-like cell adhesion genes. *Cell* **97:** 779–790.

Wu Q, Maniatis T. 2000. Large exons encoding multiple ectodomains are a characteristic feature of protocadherin genes. *Proc Natl Acad Sci* **97:** 3124–3129.

Wu Q, Zhang T, Cheng JF, Kim Y, Grimwood J, Schmutz J, Dickson M, Noonan JP, Zhang MQ, Myers RM, et al. 2001. Comparative DNA sequence analysis of mouse and human protocadherin gene clusters. *Genome Res* **11:** 389–404.

Wu H, Coskun V, Tao J, Xie W, Ge W, Yoshikawa K, Li E, Zhang Y, Sun YE. 2010. Dnmt3a-dependent nonpromoter DNA methylation facilitates transcription of neurogenic genes. *Science* **329:** 444–448.

Yagi T. 2012. Molecular codes for neuronal individuality and cell assembly in the brain. *Front Mol Neurosci* **5:** 45.

Yagi T. 2013. Genetic basis of neuronal individuality in the mammalian brain. *J Neurogenet* **27:** 97–105.

Zhang X, Firestein S. 2002. The olfactory receptor gene superfamily of the mouse. *Nat Neurosci* **5:** 124–133.

Zipursky SL, Grueber WB. 2013. The molecular basis of self-avoidance. *Annu Rev Neurosci* **36:** 547–568.

Zipursky SL, Sanes JR. 2010. Chemoaffinity revisited: Dscams, protocadherins, and neural circuit assembly. *Cell* **143:** 343–353.

Zocchi L, Sassone-Corsi P. 2010. Joining the dots: From chromatin remodeling to neuronal plasticity. *Curr Opin Neurobiol* **20:** 432–440.

Cite as *Cold Spring Harb Perspect Biol* doi: 10.1101/cshperspect.a024208

C H A P T E R 33

Epigenetics and Human Disease

Huda Y. Zoghbi[1,2] and Arthur L. Beaudet[2]

[1]Howard Hughes Medical Institute, Baylor College of Medicine, and Jan and Dan Duncan Neurological Research Institute at Texas Children's Hospital, Houston, Texas 77030; [2]Department of Molecular and Human Genetics, Baylor College of Medicine, Houston, Texas 77030

Correspondence: hzoghbi@bcm.edu

SUMMARY

Genetic causes for human disorders are being discovered at an unprecedented pace. A growing subclass of disease-causing mutations involves changes in the epigenome or in the abundance and activity of proteins that regulate chromatin structure. This chapter focuses on research that has uncovered human diseases that stem from such epigenetic deregulation. Disease may be caused by direct changes in epigenetic marks, such as DNA methylation, commonly found to affect imprinted gene regulation. Also described are disease-causing genetic mutations in epigenetic modifiers that either affect chromatin in *trans* or have a *cis* effect in altering chromatin configuration.

Outline

Cite as *Cold Spring Harb Perspect Biol* doi: 10.1101/cshperspect.a019497

OVERVIEW

The last two decades have witnessed unparalleled success in identifying the genetic bases for hundreds of human disorders and, more recently, via sequencing, the whole exome or genome. Studies of genotype–phenotype relationships have, nonetheless, challenged clinicians and researchers because some observations are not easily explained. For example, monozygotic twins carrying the same disease mutation can be quite different clinically. Or a mutation passed on in a multigenerational family can cause vastly different diseases depending on the sex of the transmitting parent. The study of such unusual cases uncovered the role of the epigenome (altered genetic information without change in DNA sequence) in health and disease. For instance, some studies showed that certain regions of the mammalian genome are not functionally equivalent on the maternal and paternal alleles. Patients who inherit both homologous chromosomes (or segments thereof) from the same parent—termed uniparental disomy or UPD—lose expression of some genes that are only expressed on one parental allele. So, in the case of paternal UPD, expression of maternal alleles is lost with increased levels for paternally expressed genes. On regions of UPD, the altered patterns of DNA modifications (termed epigenetic mutations) quickly became recognized as the molecular bases for a variety of developmental and neurological disorders. It is interesting that, for many of these disorders, either epigenetic or genetic mutations can lead to the same phenotype. This is often because the genetic mutations disrupt the function of a gene, whereas epigenetic defects typically misregulate gene expression through altering the chromatin context of the locus.

In another class of diseases, genetic mutations can cause loss of function of proteins involved in epigenetic processes, such as modifying DNA methylation, chromatin remodeling, or histone posttranslational modifications, with phenotypic consequences resulting from altered epigenetic states at one or more loci. This relationship between the genome and epigenome has broadened our understanding of the types of molecular events that cause human diseases. These could be de novo or inherited, genetic or epigenetic, and, most interestingly, some might be influenced by environmental factors. The finding that environmental factors, such as diet and experience, alter the epigenome (currently gauged predominantly by DNA methylation patterns) is likely to provide mechanistic insight into disorders with genetic predispositions that are highly influenced by the environment. Such disorders include neural tube defects (NTDs) and psychiatric illnesses. Identifying environmental factors that can affect the epigenome provides hope for developing interventions that might decrease the risk or the burden of developmental abnormalities, cancer, and neuropsychiatric disorders that currently have a known epigenetic factor to their etiologies.

Cite as *Cold Spring Harb Perspect Biol* doi: 10.1101/cshperspect.a019497

1 INTRODUCTION

Two genetically identical male monozygotic twins, raised in the same environment, manifested very different neurological functions. Both twins carried the same mutation in the X-linked adrenoleukodystrophy (ALD) gene, yet one developed blindness, balance problems, and loss of myelin in the brain, features typical of the progressive and lethal neurological disease, whereas the other remained healthy. The conclusion of the investigators reporting the unusual occurrence was "some nongenetic factors may be important for different adrenoleukodystrophy phenotypes" (Korenke et al. 1996). That indeed was a valid conclusion in 1996, given the focus of medical genetics on DNA sequence. If the DNA sequence could not explain a phenotypic variation, then environmental factors did. Similar to the case of the ALD-discordant monozygotic twins, many monozygotic twins have been found to be discordant for schizophrenia despite similar environmental rearing conditions (Petronis 2004). More recently, attention has focused on epigenetic changes, which are modifications of the genetic information that do not alter DNA sequence, as a potential explanation for discordant phenotypes in monozygotic twins and individuals who otherwise share similar DNA sequence alterations (Dennis 2003; Fraga et al. 2005).

Epigenetic modifications control gene expression patterns in a cell. These modifications are stable and at least somatically heritable, such that a mother liver cell can give rise to more liver cells with the same (or similar) patterns of gene expression after it divides. In the case of nondividing cells, such as neurons, adaptation of chromosomal regions through chromatin modifications offers an epigenetic mechanism for maintaining and possibly mediating the reproducible response of an individual neuron to specific stimuli (elaborated in Ch. 32 [Lomvardas and Maniatis 2014]). An epigenotype (i.e., the epigenetic state of a locus) is defined as the ensemble of DNA methylation states, histone modifications, histone variant composition, and the yet-to-be elucidated activities of noncoding RNAs.

In mammals, DNA methylation, which is the best-studied, generally repressive epigenetic signal when located at promoters, occurs predominantly at the carbon-5 position of symmetrical CpG (cytosine and guanine separated by a phosphate) dinucleotides (5mC) (see Ch. 15 [Li and Zhang 2014]). The state of DNA methylation is maintained after cell division through the activity of DNA methyltransferase 1 (DNMT1), which methylates hemimethylated CpG dinucleotides in daughter cells. DNA methylation is particularly important in regulating imprinted gene expression, and thus, its modulation has been found to be disease causing for such genomic regions, as described later in Section 3.1. Since the previous edition of this collection, the discovery of 5-hydroxymethylcytosine (5hmc) as an oxidative product of 5mC and the family of catalyzing enzymes, TET1-3, has added a new dimension to the study of DNA methylation (Ch. 2 [Kriaucionis and Tahiliani 2014]). The role of this modified cytosine base, however, is largely unexplored in a clinical setting. Owing to the abundance of 5hmC found in the brain, it would appear to be important for the processes of development, differentiation, and aging; hence, alterations are likely to be implicated in neurological diseases, such as Alzheimer's and Rett syndrome (RTT), described in Section 3.2.

Chromatin modifications involve covalent posttranslational modifications (PTMs) of mostly the protruding amino-terminal histone tails by the addition of acetyl, methyl, phosphate, ubiquitin, or other groups (Fig. 6 of Ch. 3 [Allis et al.] or App. 2 [Zhao and Garcia 2014] for a full listing). Methyl modifications can be mono-, di-, or trimethylation, as in the case of lysine. These modifications contribute to regulatory functions in gene expression through a direct effect on chromatin structure, or the attraction or repulsion of effector binding proteins referred to throughout this collection as PTM writers, readers, and erasers. Because chromatin consists of densely packed DNA strands wrapped around histone octamers, the folding pattern of DNA into chromatin is clearly at the root of gene activity changes. Although histone PTMs and chromatin structures can be stably transmitted from a parent cell to a daughter cell, the mechanisms underlying the replication of such structures are not fully understood (for more detail, see Ch. 22 [Almouzni and Cedar 2014]). The epigenotype shows plasticity during development and postnatally, depending on environmental factors and experiences (see Sec. 3.4); thus, it is not surprising that epigenotypes contribute not only to developmental human disorders, but also to postnatal and even adult diseases. The most recent class of molecules contributing to the epigenetic signal is that of non-protein-coding RNAs (ncRNAs). For years, the class of ncRNA included only transfer RNA, ribosomal RNA, and spliceosomal RNA. More recently, because of the availability of the genome sequences from multiple organisms, together with cross-species molecular genetic studies (from *Escherichia coli* to humans), the list of ncRNAs has expanded and resulted in the identification of hundreds of small ncRNAs, including small nucleolar RNA (snoRNA), microRNA (miRNA), short interfering RNA, and small double-stranded RNA, as well as some longer regulatory ncRNAs (see Fig. 26 of Ch. 3 [Allis et al.]). Some of these RNA molecules regulate chromatin modifications, imprinting, DNA methylation, and transcriptional silencing (as discussed in Ch. 3 [Allis et al.]; Ch. 26 [Barlow and Bartolomei 2014]; Ch. 25 [Brockdorff and Turner 2014]; Ch. 2 [Kim et al. 2014]; Ch. 2 [Rinn 2014]).

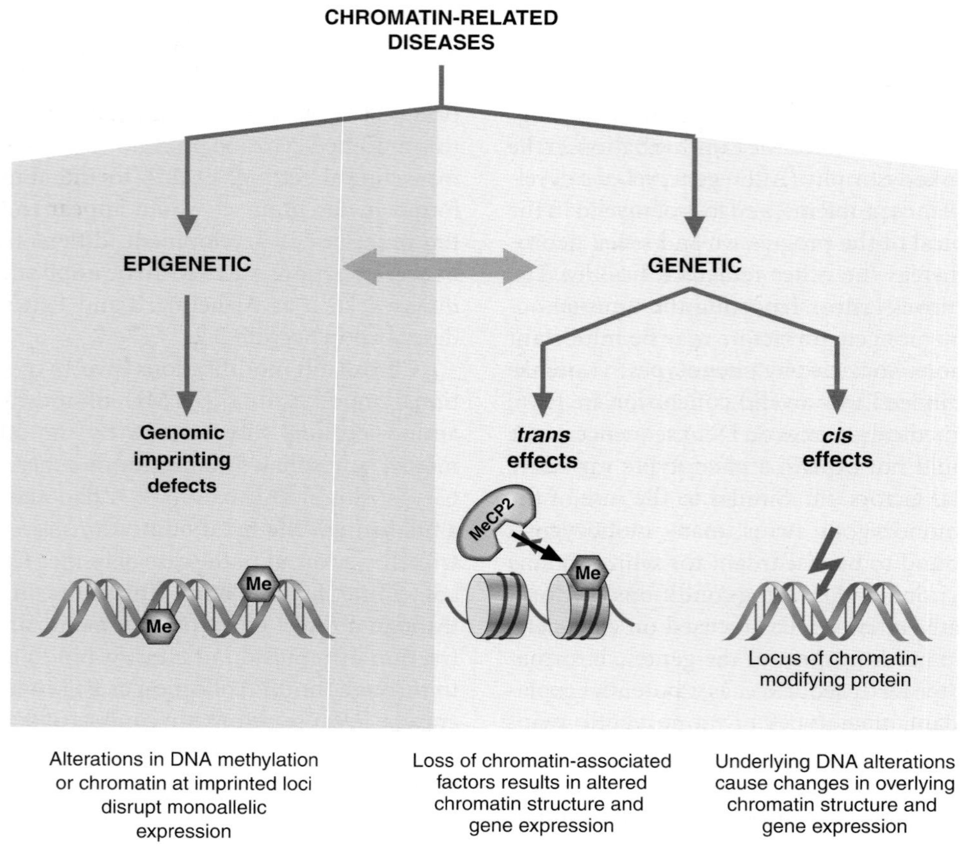

Figure 1. Genetic and epigenetic mechanisms underlying chromatin-related disorders.

Definitive evidence of a role for epigenetics in human disease came about after the understanding of genomic imprinting and finding that several genes are subject to regulation by this mechanism (Reik 1989). Genomic imprinting is a form of epigenetic regulation in which the expression of a gene depends on whether it is inherited from the mother or father. Thus, at an imprinted diploid locus, there is unequal expression of the maternal and paternal alleles. In each generation, the parent-specific imprinting marks have to be erased, reset, and maintained, thus rendering imprinted loci vulnerable to any errors that may occur during this process (discussed in Ch. 26 [Barlow and Bartolomei 2014]). Such errors, as well as mutations in genes encoding proteins involved in methylating DNA, binding to methylated DNA, or binding to histone modifications, all contribute to the fast-growing class of human disorders affecting the epigenome (Fig. 1).

2 STUDIES OF HUMAN CASES UNCOVER THE ROLE OF EPIGENETICS IN BIOLOGY

There is no doubt that the study of model organisms has been crucial for understanding many biological principles, especially in the fields of genetics, developmental biology, and neuroscience. It is often forgotten, however, that humans represent one of the most important model organisms when it comes to all aspects of biology. The characterization of thousands of human diseases represents the largest mutant screen for any species, and, if carefully and systematically studied, these phenotypes are likely to reveal biological insights in addition to ultimately providing clinical benefits. In the process, some genotype–phenotype relationships challenged Mendelian inheritance patterns in the case of "dynamic mutations" (a term coined to describe expansion of unstable repeats), revealed through the study of patients with fragile-X syndrome (FXS) (Pieretti et al. 1991).

Patients with unique features and the observant physicians who study them often break open a new field in biology, revealing novel genetic and molecular mechanisms. This, indeed, proved to be the case in revealing the role of epigenetics in human development and disease. A female patient made medical history for being reported twice by the physicians who saw her over the span of 10 years. At the age of 7 yr, she was reported in the medical literature because she suffered from cystic fibrosis (CF) and growth hormone deficiency, and was very short. During the race to find the CF gene, Beaudet and colleagues sought unusual patients who had CF plus additional features, in the hopes of identifying small deletions or chromosomal rearrangements that might facilitate the mapping and identification of the CF gene. Hence, the aforementioned patient was brought to their attention; she was 16 years of age, measured 130 cm, and had normal intelligence, but clearly had some body asymmetry. Analysis of her DNA revealed that she is homozygous for multiple polymorphic DNA markers on chromosome 7, including the centromeric alphoid repeats (Spence et al. 1988). Spence and colleagues concluded that this patient inherited two identical copies of the centromeric region of chromosome 7 from her maternal grandmother (mother was deceased) after excluding nonpaternity and hemizygosity, which could be the result of a deletion (Spence et al. 1988). Given Engel's theoretical proposal that uniparental disomy (UPD) is a possibility in humans (Engel 1980), Beaudet and colleagues immediately recognized that maternal UPD for chromosome 7 uncovered a recessive mutation in the CF gene and accounted for the additional somatic features. The constellation of clinical features in the patient, together with the laboratory evaluations, not only resulted in the identification of the first human case of UPD, but also illustrated that the maternal and paternal genomes are not equivalent for at least some portion of chromosome 7. This provided a novel mechanism of non-Mendelian inheritance to explain disease and developmental abnormalities (Fig. 2) and preceded the first reports of genomically imprinted genes in the mouse in 1991 (described in Ch. 26 [Barlow and Bartolomei 2014]). Although in 1988, it was thought by some that UPD of a chromosome was a rare event, today, we know that UPD has been reported thus far for almost all human chromosomes.

The study of unusual patients not only identified cases of UPD for additional chromosomes, but in 1989, also led to the proposal that UPD causes disease because of changes in epigenotype and disruption of genomic imprinting (Nicholls et al. 1989). Nicholls et al. (1989) studied a male patient with Prader–Willi syndrome (PWS) who had a balanced Robertsonian translocation t(13;15), but this was also present in his asymptomatic mother and maternal relatives. The fact that the founding patient inherited both copies of chromosome 15 from his mother (whereas all asymptomatic individuals inherited one copy from their fathers), led the authors to conclude that maternal UPD led to the PWS phenotype. After confirming maternal UPD15 in a second PWS patient with an apparently normal karyotype, the authors proposed a role for genomic imprinting in the etiology of PWS. Furthermore, they concluded that either paternal deletions or maternal UPD from 15q11–13 leads to PWS, and they predicted that the converse situation of paternal UPD15 would lead to Angelman syndrome (AS), just as maternal deletions of this region do. All of these predictions proved true (Fig. 3).

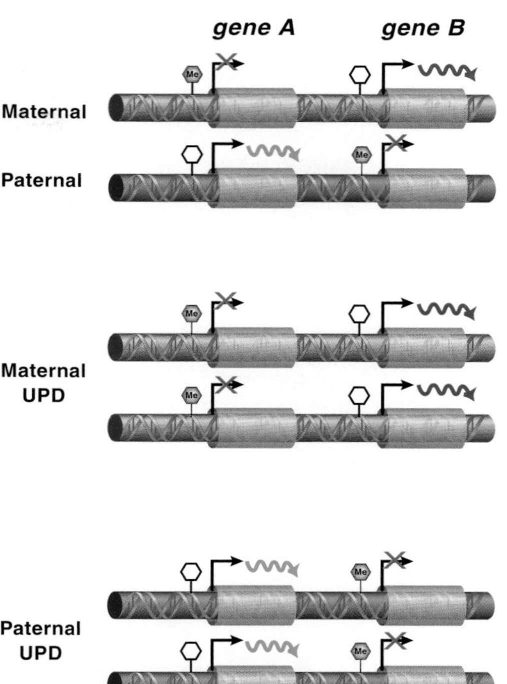

Figure 2. Consequences of uniparental disomy (UPD). In maternal UPD, transcripts expressed from the maternally inherited alleles are doubled, whereas those that are on the paternal alleles are lost. The opposite occurs in paternal UPD.

3 HUMAN DISEASES

3.1 Disorders of Genomic Imprinting

The discovery of UPD was the clinical entry point into disorders of genomic imprinting in humans. Whereas PWS and AS were the first genomic imprinting disorders to be studied, Beckwith–Wiedemann syndrome (BWS), pseudohypoparathyroidism (PHP), and Silver–Russell syndrome (SRS) expanded the list and introduced many intriguing questions about how epigenetic defects lead to the disease phenotype. In the following section, we give a brief review of the clinical features of each disorder, the various mechanisms leading to epigenotypic defects, and the phenotypes and biological insight gained from the study of this class of disorders (see Table 1).

3.1.1 Mutations and Epimutations Causing the Same

For the purposes of this chapter, we will define epigenetic mutations or epimutations as changes in the epigenome

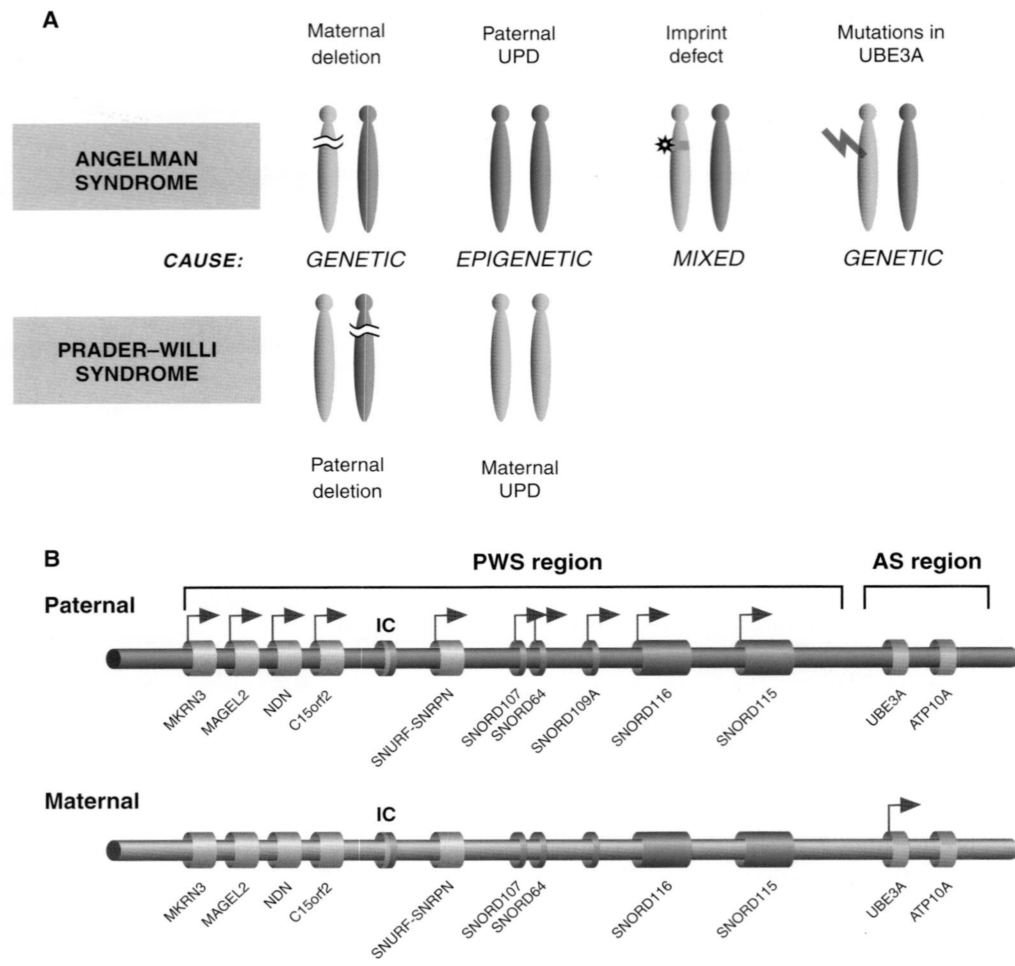

Figure 3. The genetics and epigenetics of Prader–Willi syndrome (PWS) and Angelman syndrome (AS). (*A*) PWS and AS can be caused by genetic, epigenetic, or mixed defects. (*B*) The imprinted gene clusters associated with PWS and AS, indicating genes that are normally maternally or paternally expressed. The bipartite regulatory imprinted control (IC) region is indicated, showing the region critical for imprinted AS cluster control (green) and PWS gene cluster control (purple).

that are different from the consensus epigenome. As with genetic mutations, epimutations may phenotypically display as being benign or disease causing, and they may be common or rare. Most known clinical examples involve altered DNA methylation and/or differences in histone modifications. Genomic regions subject to imprinting are uniquely susceptible to causing clinical disorders, primarily because the genes in the region are already functionally hemizygous in the normal state, and hence a loss of function of the single expressing allele can lead to a complete absence of function for a gene. This is analogous to loss of function for a gene on the X chromosome in a male. The loss of function for an essential imprinted gene may occur by a genetic mechanism, such as gene deletion or point mutation, or an epigenetic mutation, often referred to as an imprinting defect.

3.1.2 Sister Syndromes: Prader–Willi and Angelman

PWS (online Mendelian inheritance in man or OMIM 176270) and AS (OMIM 105830) are caused, in the majority of cases, by the same 5- to 6-Mb deletion in 15q11-q13, but their phenotypes are vastly different. Genomic imprinting in the region accounts for the phenotypic differences, given that PWS is caused by paternally inherited deletions, whereas in AS, the deletion is of maternal origin (Fig. 3A) (Ledbetter et al. 1981; Magenis et al. 1987; Nicholls et al. 1989).

PWS, which occurs in approximately 1/10,000 births, was described about 50 years ago and is characterized by infantile hypotonia, developmental delay, failure to thrive due to poor feeding, and lethargy, followed by hyperphagia,

Table 1. Selected disorders of genomic imprinting

Disorder	Type of mutation (% frequency where known)	Genomic region (cluster name)	Gene(s) involved
Prader–Willi syndrome	Deletion (70%) Maternal UPD (25%) Imprint defect (2%–5%)	15q11-q13 (Pws cluster)	snoRNAs and other (?)
Angelman syndrome	Deletion (70%) Paternal UPD (2%–5%) Epimutation (2%–5%) Point mutation Duplication[a]	15q11-q13 (*Pws* cluster)	*UBE3A*
Beckwith–Wiedemann syndrome	Epimutation Loss of maternal ICR2/Kcnq1 methylation Gain of H19 methylation (5%) Paternal UPD for Igf2 cluster 11p15.5 duplication including Igf2 Translocation at KCNQ1 maternal Point mutation (CDKN1C)	11p15.5 (Kcnq1 and Igf2 cluster)	*IGF2, CDKNIC*
Silver–Russell syndrome	UPD, maternal (10%) Duplication Translocation, inversion	7p11.2 (Grb10 cluster)	Several candidates in the region
	Epimutation, loss of paternal ICR1 methylation (40%)	11p15.5 (Kcnq1 cluster)	Biallelic expression of H19 and decrease of IGF2
Pseudohypoparathyroidism	Point mutation Imprint defect UPD, paternal	20q13.2 (Gnas cluster)	*GNAS1*

snoRNAs, small nucleolar RNAs; *CDKN1C*, cyclin-dependent kinase inhibitor; *UBE3A*, ubiquitin E3 ligase gene.
[a] Maternal duplications, trisomy, and tetrasomy for this region cause autism and other developmental abnormalities.

severe obesity, short stature, secondary hypogonadism with genital hypoplasia, and mild cognitive impairment. PWS patients also have distinct physical characteristics, such as small hands and feet, almond-shaped eyes, and thin upper lip. Most have mild to moderate intellectual disability, and the vast majority displays a variety of obsessive–compulsive behaviors, anxiety, and sometimes a withdrawn unhappy disposition (Fig. 4A). In contrast, patients with AS have a "happy disposition," smile frequently, and have unexplained bouts of laughter. AS patients suffer from severe developmental delay, very minimal (if any) verbal skills, balance problems (ataxia), abnormal hand-flapping movements, microcephaly, seizures, and some dysmorphic features, such as prominent mandible and wide mouth (Fig. 4B). Both can display hypotonia, hypopigmentation of the skin and irises, and occurrence of a squint. The hypopigmentation is caused by a heterozygous deficiency for a nonimprinted albinism gene (*OCA2*) and hence manifests in deleted genotypes.

The majority (~70%) of PWS and AS cases are caused by paternal and maternal deletions of 15q11-q13, respectively. About 25% of PWS cases are caused by maternal UPD of 15q11-q13, whereas paternal UPD of this region accounts for 2%–5% of AS patients (Fig. 3). The difference in frequency of UPD between PWS and AS starts with maternal nondisjunction, as influenced by maternal age, leading to a conceptus with trisomy or monosomy 15. These are then "rescued," leading to either loss of paternal chromosome 15 and resulting in maternal UPD and PWS from the initial trisomic embryo, or rescue by duplication of paternal chromosome 15 (i.e., UPD and AS). The difference in frequency of the two UPDs is presumably related to the frequency of the two abnormal eggs and the probability of rescue for the two circumstances. Translocations within the PWS/AS critical region account for <10% of the cases, but it is of note that such translocations are associated with a high recurrence risk (up to 50%) depending on the sex of the transmitting parent. In fact, PWS and

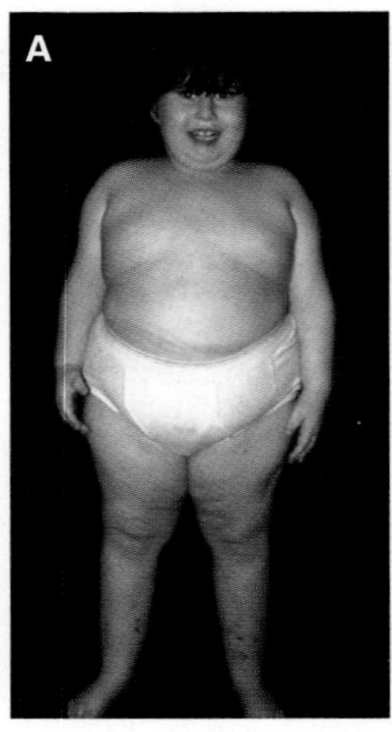

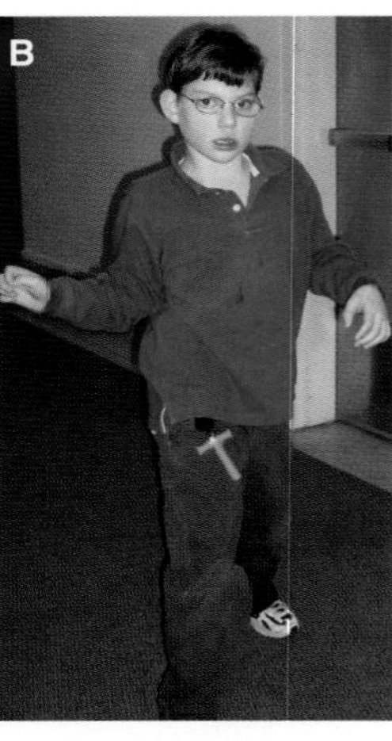

Figure 4. Images of a Prader–Willi syndrome patient (*A*) and Angelman syndrome patient (*B*) illustrate the dramatic differences in the clinical features of the disorders resulting from defects in an imprinted region. (Images kindly provided by Dr. Daniel J. Driscoll and Dr. Carlos A. Bacino, respectively.)

AS co-occurred in some families as a result of translocations or other structural abnormalities of 15q11-q13, and the phenotype was determined by the sex of the transmitting parent (Hasegawa et al. 1984; Smeets et al. 1992).

Imprinting defects represent another class of mutations leading to PWS or AS phenotypes. These defects cause a chromosome of one parental origin to have an altered epigenotype, typically that of the chromosome originating from the opposite parent. During mapping and functional studies of the PWS and AS imprinted regions, a bipartite imprinting center (IC) was characterized within 15q11-q13, found to be necessary for resetting appropriate parental imprints across the whole imprinted cluster (Fig. 3B) (Ohta et al. 1999). Incidentally, in mouse research, an IC is typically called an imprinting control element, as elaborated in Ch. 26 (Barlow and Bartolomei 2014). Imprinting defects often involve deletion of the IC, but there are instances when it appears to be caused by an epigenetic mutation that does not involve the DNA sequence. The phenotypic outcome of either imprinting defect (i.e., IC deletion or changes of epigenotype across the region) is similar to that of broader deletions across the imprinted region or UPD, and at the chromatin level includes alterations in DNA methylation, histone PTMs, chromatin structure, and, ultimately, gene expression patterns. Imprinting defects account for 2%–5% of PWS and AS cases, and the IC deletions are typically associated with 50% recurrence risk, depending on the sex of the transmitting parent, whereas the recurrence risk is low for families without IC deletions.

The identification of imprinting defects in a handful of AS patients who were conceived after intracytoplasmic sperm injection (ICSI) raised the possibility that this approach of in vitro fertilization might cause imprinting defects. The finding of imprinting defects also among AS cases born to subfertile couples who did not receive ICSI (although did receive hormonal stimulation) raises questions about whether there are common mechanisms contributing to infertility and imprinting defects, or assisted reproductive technology (hormones and/or ICSI) has epigenetic consequences (Ludwig et al. 2005). A recent review of epigenetic abnormalities associated with assisted reproductive technologies is available (Dupont and Sifer 2012).

Exactly which gene(s) in the genomically imprinted 15q11-q13 cluster is causal to the AS phenotype is known. About 10%–15% of AS cases are caused by loss-of-function mutations in the ubiquitin E3 ligase gene (*UBE3A*) (Fig. 3A), encoding the E6-associated protein (Kishino et al. 1997; Matsuura et al. 1997). Expression studies showed that *Ube3a* is expressed exclusively from the maternal allele in cerebellar Purkinje cells and hippocampal neurons. Furthermore, *Ube3a*$^{+/-}$ mice lacking the maternal allele reproduce features of AS (Jiang et al. 1998). These results, together with human data, pinpoint the *UBE3A* gene as the causative gene in AS. Paternal UPD or maternal deletions of 15q11-q13 lead to loss of expression of *UBE3A* in Purkinje cells. It appears that expression of an antisense transcript normally expressed from the IC region only from the paternal chromosome is expressed from the maternal chromosome with the imprinting defect, leading to suppression of *UBE3A* (Rougeulle et al. 1998; Meng et al. 2012). In a mouse model, depletion of this antisense transcript leads to activation of maternal UBE3A, showing improvement of some of the phenotypic effects of AS (Meng et al. 2013). It is intriguing that ~10% of AS cases remain without a molecular diagnosis. A subset of these patients appears to have mutations in a chromatin-remodeling protein, methyl-CpG-binding protein 2 (MeCP2), more typically found in RTT individuals, but it is conceivable that Rett and AS can be confused clinically.

In the case of PWS, there are several candidate imprinted genes that are only expressed from the paternal allele; however, careful studies of rare translocation and deletion families support the interpretation that deficiency of PWCR1/HBII-85 snoRNAs causes PWS (Schule et al. 2005). There is more recent evidence that deletion of the HBII-85 snoRNAs transcribed from the *SNORD116* locus can cause most of the phenotypic expression for PWS (Sahoo et al. 2008, and references thereafter). Recently, truncating mutations in *MAGEL2* were identified in patients with features of PWS and autism underscoring the complexity of the genetic basis of this disorder (Schaaf et al. 2013). It is still possible that loss of expression for *SNURF-SNRPN* and/or *Necdin* may contribute to the phenotype

 Cite as *Cold Spring Harb Perspect Biol* doi: 10.1101/cshperspect.a019497

independently or as part of a contiguous gene effect with *SNORD116*. *SNURF-SNRPN* has its major transcriptional start site at the IC, and encodes a small nuclear ribonucleoprotein (SNRPN) that functions in the regulation of splicing. Another gene, a "SNRPN upstream reading frame" or *SNURF*, along with upstream noncoding exons, is thought to be the major site of imprinting defects, because disruption of this gene leads to altered imprinting of *SNRPN* and other 15q11-q13 imprinted genes. It is interesting that mice lacking *Snrpn* appear normal, but mice with deletions spanning *Snrpn* and other genes homologous to those in 15q11-q13 are hypotonic, develop growth retardation and die before weaning (Tsai et al. 1999).

3.1.3 Beckwith–Wiedemann Syndrome

The story of BWS (OMIM 130650) represents an excellent example of how a human disorder uncovered the importance of epigenetics not only in normal development, but also in the regulation of cell growth and tumorigenesis. BWS is characterized by somatic overgrowth, congenital abnormalities, and a predisposition to childhood embryonal malignancies (Weksberg et al. 2003). BWS patients typically manifest gigantism, macroglossia (large tongue), hemihypertrophy, variable degrees of ear and other organ anomalies, and omphalocele (protrusion of abdominal organs through the navel). In addition, many patients suffer from increased size of internal organs, embryonic tumors such as Wilms tumor, hepatoblastoma, or rhabdomyosarcoma, and hyperplasia and hypertrophy of pancreatic islets, often leading to neonatal hypoglycemia.

The majority of BWS cases are sporadic, but a small number of families with an autosomal dominant inheritance pattern suggested a genetic etiology, linking the syndrome to chromosome 11p15 (Ping et al. 1989). The findings, however, that there was preferential loss of maternal alleles in the 11p15 region in BWS-related tumors, an excess of transmitting females in the dominant form of the disease, and paternal UPD of 11p15.5 in some cases of BWS provided evidence that epigenetics and imprinting must play an important role in the etiology of BWS, and the disease might result from a mixture of genetic and epigenetic abnormalities occurring either de novo or inherited. The cluster of imprinted genes implicated in BWS actually contain both the Igf2 and Kcnq1 clusters discussed in the mouse context in Ch. 26 (Barlow and Bartolomei 2014), and map to a ~1-Mb region in 11p15.5, and include at least 12 imprinted genes (Fig. 5A) (Weksberg et al. 2003). The first imprinted cluster includes the reciprocally imprinted *H19* and insulin-like growth factor (*IGF2*) genes interspersed by a differentially methylated region thought to represent one imprinting control region (ICR1) (Joyce et al. 1997; Weksberg et al. 2003). *H19* encodes a maternally expressed ncRNA, and *IGF2* encodes a paternally expressed growth factor. These two genes share a common set of enhancers, access to which is affected by the methylation state of ICR1 and binding of CTCF, a zinc-finger protein (Hark et al. 2000). The second imprinting gene cluster, Kcnq1, contains several maternally expressed genes, including the cyclin-dependent kinase inhibitor (*CDKN1C*, encoding p57*kip2*), a component of the potassium channel (*KCNQ1*), and a putative cation transporter (*SLC22A1L*). The differentially methylated region in ICR2 maps to an intron of *KCNQ1* and is unmethylated on paternal alleles, leading to expression of *KCNQ1OT1* in an antisense direction of *KCNQ1*. Methylation of ICR2 on the maternal allele is believed to silence maternal expression of *KCNQ1OT1*, allowing expression of the maternally expressed *KCNQ1* and *CDKN1C* (Lee et al. 1999; Smilinich et al. 1999).

Various epigenetic, as well as genetic molecular, defects have provided insight into which genes contribute to the BWS phenotype. On unmethylated maternal alleles, CTCF binds ICR1 and establishes a chromatin boundary whereby the *IGF2* promoter is insulated from enhancers. These enhancers can then access the *H19* promoter (proximal to the boundary), permitting transcription of *H19*. Methylation of ICR1 on paternal alleles abrogates the binding of CTCF, permitting expression of *IGF2* and silencing of *H19*. The findings that either duplications in 11p15.5 spanning the *IGF2* locus or paternal UPD of this region (expected to lead to overexpression of *IGF2*), coupled with data showing that transgenic mice overexpressing *IGF2* develop overgrowth and large tongues, has implicated *IGF2* overexpression as one potential cause of the overgrowth phenotype in BWS (Henry et al. 1991; Weksberg et al. 1993; Sun et al. 1997). Alternatively, loss-of-function mutations in *CDKN1C* also give rise to BWS patients who are phenotypically similar to those caused by overexpression of *IGF2*. Intriguingly, mice lacking *Cdkn1c* only develop omphaloceles, but not the overgrowth phenotype, and it is only when loss of *Cdkn1c* is coupled with increased expression of *Igf2* that the animals reproduce many features of BWS (Caspary et al. 1999). The types of molecular lesions that cause BWS are indicated in Table 1. Of note, translocations on the maternal chromosome disrupting *KCNQ1* affect imprinting of *IGF2* but, curiously, not ICR2. Also, the most common mechanism, loss of imprinting for ICR2/*KCNQ1OT1*, again alters imprinting of *IGF2* and suggests some regulatory interactions between ICR1 and ICR2 (Cooper et al. 2005). Some of the epigenetic changes identified in BWS, such as methylation defects at the *H19* ICR1, have also been confirmed in individuals who develop Wilms tumor, but not BWS, suggesting that the timing of the epigenetic defect might dictate whether abnormal growth regulation will

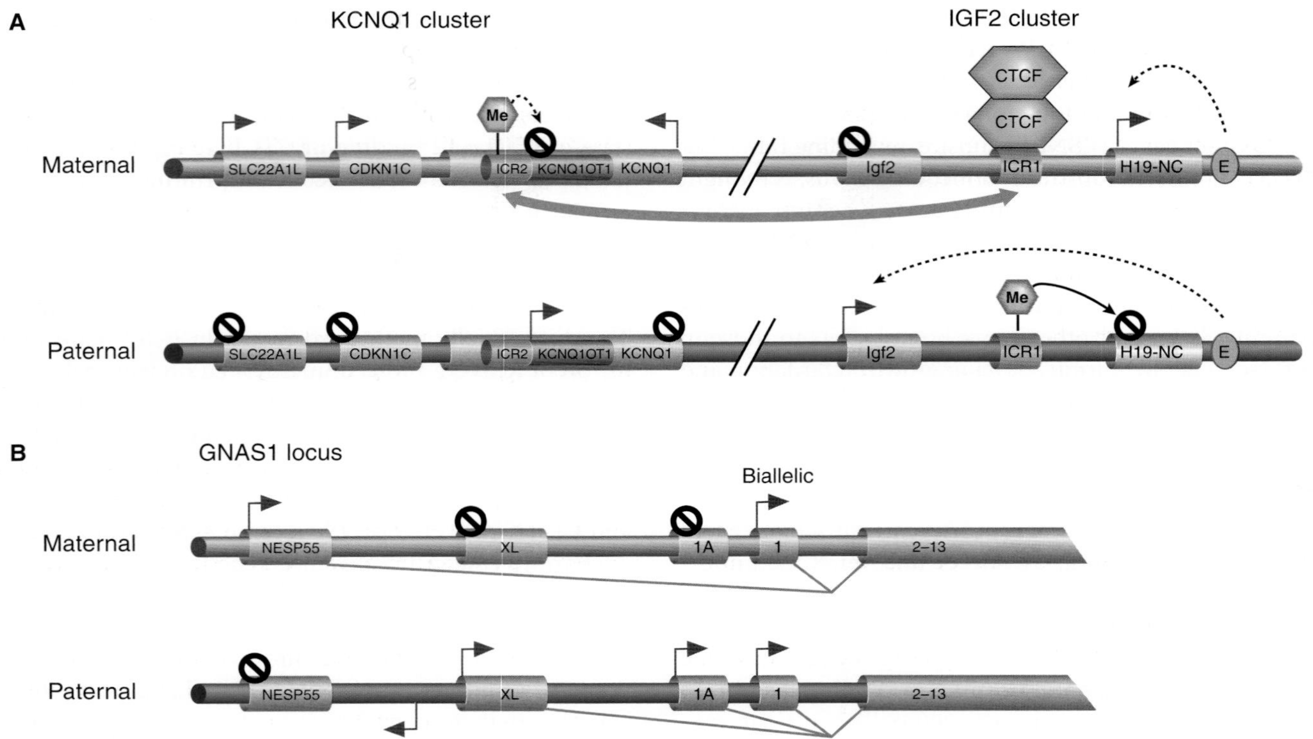

Figure 5. Imprinted clusters associated with the human Beckwith–Wiedemann and pseudohypoparathyroidism imprinting disorders. (*A*) The expression of imprinted genes at adjacent KCNQ1 and IGF2 imprinting clusters associated with BWS is displayed. Expression patterns from both parental chromosomes in control individuals are indicated. ICRs are indicated in green, which when imprinted, are shown to be DNA methylated (pink hexagon). ICR2 and the antisense KCNQ1OT1 lie within the KCNQ1 locus. The gray arrow connecting ICR1 and ICR2 indicates some kind of regulatory interaction that has been postulated. E, enhancer. (*B*) The 5′ region of the GNAS1 locus is illustrated, a gene implicated in PHP, indicating the parental expression in certain tissues of the different transcripts produced from alternative 5′ exons (NESP55, XL, and 1A). The reverse arrow indicates a NESP55 antisense transcript.

affect the whole organism or a specific organ. The fact that aberrant methylation at ICR1 often leads to Wilms tumor, and at ICR2 often leads to rhabdomyosarcoma and hepatoblastoma in BWS, suggests that there is more than one locus in 11p15.5 predisposing to tumorigenesis (Weksberg et al. 2001; DeBaun et al. 2003; Prawitt et al. 2005).

3.1.4 Silver–Russell Syndrome

SRS (OMIM 180860) is a developmental disorder characterized by growth retardation, short stature often with asymmetry, and some dysmorphic facial and cranial features, as well as digit abnormalities. The most prominent feature is the somatic growth abnormality, with other features being highly variable. SRS is genetically heterogeneous, but it is estimated that ~10% of the cases result from maternal UPD for chromosome 7 (Eggermann et al. 1997). It is proposed that loss of function of a paternally expressed gene, possibly one that promotes growth, causes SRS, but an alternate model of overexpression of a maternally expressed growth-suppressing gene cannot be excluded. It is interesting that an epigenetic mutation causing demethylation of ICR1 on chromosome 11p15 has been identified in several individuals with SRS (Fig. 5A). This epigenetic defect causes biallelic expression of H19 and decreased expression of IGF2 (Gicquel et al. 2005), and occurs in ~40% of patients (Eggermann et al. 2012).

3.1.5 Pseudohypoparathyroidism

PHP represents a group of phenotypes that result from functional hypoparathyroidism despite normal parathyroid hormone (PTH) levels. These patients are resistant to PTH. There are several clinical subtypes: Ia, Ib, Ic, II, and Albright hereditary osteodystrophy (OMIM 103580). In addition to the functional hypoparathyroidism and osteodystrophy, these clinical variants may show a variety of developmental and somatic defects. The clinically het-

 Cite as *Cold Spring Harb Perspect Biol* doi: 10.1101/cshperspect.a019497

erogeneous phenotypes result from mutations in the *GNAS1* gene encoding the α-stimulating activity polypeptide 1 (GSα), a guanine nucleotide-binding protein. *GNAS1* maps to chromosome 20q13.2. The *GNAS1* locus has three upstream alternative first exons (exons 1A, XL, and NESP55) that are spliced to exons 2–13 to produce different transcripts and, in the case of NESP55 and XL, this alternative splicing produces unique proteins (Fig. 5B). There are differentially methylated regions near these exons, causing NESP55 to be expressed exclusively from maternal alleles, whereas XL, exon 1A, and an antisense transcript for NESP55 are paternally expressed. Although the transcript encoding the GSα protein is biallelically expressed, the maternal allele is preferentially expressed in some tissues, such as the proximal renal tubule. The combination of genomic and tissue-specific imprinting accounts for the variable phenotypes and parent-of-origin effect even for mutations that have a clear autosomal dominant inheritance pattern (Hayward et al. 1998). Of note is the finding that one patient with paternal UPD of the *GNAS1* region developed PHP type Ib disease (Bastepe et al. 2003).

3.1.6 Other Disorders of Imprinted Regions

There are a few more reports of imprinted gene deregulation causing disease. For example, maternal and paternal UPD, deletions with different phenotypes based on parent of origin, and epimutations cause disorders involving a cluster of imprinted genes at chromosome 14q32. The imprinted genes in the region include paternally expressed *DLK1* and maternally expressed *MEG3*. Maternal UPD 14 is characterized by failure to thrive followed by obesity, learning difficulties, and precocious puberty and can be confused with PWS. Paternal UPD 14 is characterized by thoracic dysplasia and intellectual disability. Deletions and epimutations of 14q32 can give rise to phenotypes that overlap those of maternal and paternal UPD 14 (Kagami et al. 2008).

Paternal UPD 6 is associated with transient neonatal diabetes, which resolves spontaneously but may recur later in life. Paternal duplications or loss of maternal methylation at the 6q24 also can be associated with transient neonatal diabetes (Temple and Shield 2002).

The genotype–phenotype studies of the clinical disorders described in this section show that almost all of the genomic imprinting disorders can be caused by a mixture of genetic or epigenetic abnormalities, either de novo or inherited. It is hard to believe that such a mixed genetic model for disease would remain unique for this small subset of disorders. A little more than a decade ago, UPD was only a theoretical possibility, but now it is established to occur in many chromosomal regions and result in diverse diseases and developmental phenotypes. One challenge in human genetics research is to uncover which genes are responsible for which UPD-associated phenotypes to establish a list of diseases that are likely to result from mixed genetic/epigenetic mechanisms.

3.1.7 Epimutations Outside Imprinted Regions

Somatic epimutations are well documented in cancer as discussed in multiple chapters in this collection, but particularly in Ch. 34 (Baylin and Jones 2014) and Ch. 35 (Audia and Campbell 2014). Here, the focus is on constitutional epimutations that affect most or all of the cells in an individual. Curiously, most constitutional epimutations reported to date affect cancer-related genes. Whether this represents a true biological bias or an ascertainment bias is unclear. Epimutations affecting *MLH1*, for instance, were among the first to be described and remain the most frequently reported (Suter et al. 2004; Pineda et al. 2012). Vertical or transgenerational inheritance has been reported for epimutations in *MLH1* (Crepin et al. 2012). Constitutional epimutations involving *BRCA1* and *BRCA2* have also been reported (Hansmann et al. 2012). Constitutional epimutations are sometimes associated with sequence variants near or in promoters (Ward et al. 2012). The full extent of the role of epimutations in human disease will only become apparent when investigators begin to search for such mutations systemically. For any gene in which heterozygous loss-of-function mutation causes a phenotype, an epimutation silencing a promoter should give rise to the same phenotype.

3.2 Genetic Disorders Affecting Chromatin Structure in *trans*

The rapidly growing list of human diseases caused by mutations in genes encoding proteins essential for chromatin structure and remodeling highlights the importance of finely tuned chromatin structure in human health. These disorders are not caused by epigenetic mutations, but the mutated genes secondarily alter chromatin states that are critical components of the epigenotype. The vast differences in phenotypes, as well as the subtle changes in protein levels or even conserved amino acid substitution for these chromatin-modifying or -remodeling proteins found in human disease, have revealed a lot about their interactions and the need for their tightly controlled expression. Disorders that affect chromatin in *trans* result from the functional disruption of proteins directly involved in posttranslationally modifying histones, such as the histone-acetylating cAMP response element binding (CREB)-binding protein (CBP) or EP300 enzymes, modifiers of DNA cytosines (i.e., the DNMTs), readers of histone or cytosine PTMs, such as MeCP2, or histone remodelers (see Table 2; also Fig. 1 of Ch. 34 [Baylin and Jones 2014]). Disruption of the function of any of these genes causes complex multisystem phenotypes or neoplasia owing to the downstream effects of misregulated expression of a large number of target genes. Although yet to be discovered, there is ample opportunity to reveal that mutations in ncRNA genes known to act on chromatin in *trans* may contribute to certain diseases.

3.2.1 Recurrent Hydatidiform Mole

Complete hydatidiform mole (CHM) or recurrent hydatidiform mole 1 (OMIM 231090) is an abnormal pregnancy in which there is hyperproliferative vesicular-appearing trophoblast (i.e., extraembryonic tissue) but absent embryo development. CHM has the potential to become invasive and malignant, and patients can have other symptoms, such as early-onset preeclampsia, related to the large amount of human chorionic gonadotropin secreted by the molar tissues. Most CHMs are sporadic, but rare familial and recurrent cases have been described. The majority of CHMs is androgenetic (comprised entirely of DNA of paternal origin) and believed to originate from the fertilization of an oocyte lacking a functional nucleus (Kajii and Ohama 1977). This occurs either through fertilization by a single sperm with subsequent duplication of the paternal pronuclear DNA, or dispermic fertilization to generate a diploid genome that only contains paternally inherited DNA. Thus, the disrupted regulation and expression of imprinted genes is likely responsible for the abnormal trophoblast phenotype seen in these pregnancies.

In contrast to more common androgenetic CHMs, there are much rarer cases of recurrent and often familial hydatidiform moles that have normal biparental inheritance of their genome, referred to as "biparental hydatidiform moles" or "BiHM." This led to the hypothesis that the development of the abnormal trophoblast in these BiHM

Table 2. Selected genetic disorders affecting chromatin structure in *trans*

Disorder	Gene	Comments
Coffin–Siris syndrome, intellectual disability	*ARID1A*	Component of the BRG1-associated factor complex
	ARID1B	Component of SWI/SNF complexes
α-thalassemia/mental retardation syndrome	*ATRX*	Helicase, SNF2-like family
CHARGE	*CHD7*	Transcriptional regulator
Autism spectrum disorders	*CHD8*/Duplin	
Rubinstein–Taybi syndrome	*CREBBP*	Histone acetyltransferase
	EP300	Histone acetyltransferase
Neuropathy, hereditary sensory, type IE	*DNMT1*	Maintenance DNA methyltransferase
Immunodeficiency-centromeric instability-facial anomalies syndrome 1 (ICF1)	*DNMT3B*	DNA methyltransferase 3B
Immunodeficiency-centromeric instability-facial anomalies syndrome 2 (ICF2) and intellectual disability	*ZBTB24*	DNA methylation
Intellectual disability, seizures, dysmorphism; Kleefstra syndrome	*EHMT1*/ *KMT1D*	Histone methyltransferase
Intellectual disability, seizures, syndromic, Claes–Jensen type	*KDM5C*/ JARID1C	Histone H3 K4me 3 and K4Me2 demethylase
Kabuki 1 syndrome	*MLL2*	Histone lysine methyltransferase
Kabuki 2 syndrome	*KDM6A*	Histone H3 K27 demethylase
Rett syndrome	*MECP2*	Transcriptional modulator
Sotos syndrome; acromegaly, intellectual disability	*NSD1*/ *KMT3B*	Nuclear receptor-binding Su-var; transcriptional coregulator
Recurrent biparental hydatidiform mole	*NLRP7*	
	KHDC3L/ *C6orf221*	
Intellectual disability, cleft lip/palate Siderius syndrome	*PHF8*	Histone H4K20me1 demethylase
Skeletal malformations, intellectual disability, hearing deficits, Coffin–Lowry syndrome	*RPS6KA3*/*RSK2*	EGF-stimulated phosphorylation of H3
Intellectual disability, seizures, short stature, sparse hair, Nicolaides–Baraitser syndrome	*SMARCA2*	Chromatin regulator
Immune defects, nephritis, skeletal abnormalities, Schimke immuno-osseous dysplasia	*SMARCAL1*	SNF2-like family, DNA-dependent ATPase activity

SWI/SNF, switch/sucrose nonfermentable; KMT, lysine methyltransferase; *CREBBP*, CREB-binding protein gene; *MLL2*, mixed leukemia lineage 2.

was caused by disrupted expression of imprinted genes, as per CHMs. This proved to be the case when several studies showed that there is loss of methylation at the differentially methylated regions of most maternally imprinted, paternally expressed genes, similar to what is seen in androgenetically inherited moles (Fisher et al. 2002), although some maternal imprints were correctly specified.

In 2006, Murdoch et al. and, subsequently, several other studies showed that most women with recurrent BiHM have homozygous mutations in the *NLRP7* gene, encoding a member of the NLRP family of CATERPILLER proteins, which have known functions in innate immunity and apoptosis (Murdoch et al. 2006). NLRP7 is a cytoplasmic protein that is a member of the nucleotide oligomerization domain-like family characterized by an amino-terminal pyrin domain, a NACHT domain found in proteins involved in apoptosis, and a carboxy-terminal leucine-rich repeat region. In 2011, Parry and colleagues identified biallelic mutations in C6orf221 in three families with BiHM, now termed HYDM2 (OMIM 614293) (Parry et al. 2011), implicating the gene as a regulator of genomic imprinting in the oocyte. Studies on hydatidiform moles have clearly reinforced the notion that the correct regulation of imprinted gene expression is essential in human health.

3.2.2 Rubinstein–Taybi Syndrome

Rubinstein–Taybi syndrome (RSTS) (OMIM 180849) is characterized by intellectual disability, broad thumbs and toes, facial abnormalities, congenital heart defects, and increased risk of tumor formation. The high concordance rate in monozygotic twins, together with a few cases of mother-to-child transmission, suggested that this disease has a genetic basis and an autosomal dominant inheritance was most likely. Cytogenetic abnormalities involving 16p13.3 were identified in several RSTS patients (Tommerup et al. 1992) and found to map to the region that contains the CBP (or CREBBP) gene. Heterozygous mutations in *CREBBP* showed that haploinsufficiency of CBP causes RSTS (Petrij et al. 1995). CBP was first described as a coactivator of the cAMP-responsive binding protein, CREB. Once bound, CBP, in turn, activates transcription from a cAMP response element–containing promoter through the acetylation of all four core histones in the adjacent nucleosomes (Ogryzko et al. 1996). CBP also interacts through a region in its carboxyl terminus directly with the basal transcription factor TFIIIB. In vitro functional analysis of one of the CBP missense mutations (Arg-1378 to proline) that cause RSTS revealed that this mutation abolishes the histone acetyltransferase (HAT) activity of CBP (Murata et al. 2001). These data, together with the finding that mice haploinsufficient for CBP have impaired learning and memory, altered synaptic plasticity, and abnormal chromatin acetylation, support the conclusion that the decreased HAT activity of CBP is a key contributor to the RSTS phenotype (Alarcon et al. 2004). Consistent with the role of decreased HAT activity in disease is the recent discovery that mutations in a second gene, *p300*, encoding another potent HAT and transcriptional coactivator, cause some cases of RSTS (Roelfsema et al. 2005). Interestingly, some of the synaptic plasticity defects, as well as learning and memory deficits of the $CBP^{+/-}$ mice, can be reversed by using histone deacetylase (HDAC) inhibitors (Alarcon et al. 2004), raising the question of whether pharmacologic therapy using such reagents can ameliorate some of the mental deficits in RSTS.

3.2.3 Rett Syndrome

3.2.3.1 Clinical Characterization and Discovery of Genetic Cause. RTT (OMIM 312750) is a dominant X-linked, postnatal, neurological disorder characterized by motor abnormalities, ataxia, seizures, replacement of hand use by purposeless hand-wringing, and language regression (Hagberg et al. 1983). RTT shares three main features with autism spectrum disorders (ASD): First, they manifest postnatally, often after a period of apparently normal development; second, they disrupt social and language development; third, they are accompanied by unusual stereotypical hand or arm movements (Fig. 6A). Although RTT is a sporadic disorder in the vast majority of cases ($>99\%$), the discovery of a handful of families in whom the gene was transmitted through maternal lines suggested a genetic basis for this disorder. Such families, together with findings that RTT was typically observed in females and carrier females can be asymptomatic, led to the hypothesis that RTT is an X-linked dominant disorder. An exclusion mapping strategy localized the RTT gene to Xq27-qter, and candidate gene analysis pinpointed the causative gene as encoding MeCP2 (Amir et al. 1999).

3.2.3.2 The Genetics of MECP2 Expression in RTT. The discovery that *MECP2* mutations were the major cause of RTT syndrome also provided molecular evidence for a relationship between RTT and autism. Mutations in *MECP2* are now known to cause an even broader spectrum of phenotypes in females, including learning disabilities, isolated intellectual disability, Angelman-like syndrome, and ASD. X-chromosome inactivation (XCI) patterns are the major molecular determinants for clinical variability because expression from the mutant *MECP2* locus can deviate from the 50% of cells expected if XCI was completely random. Females with *MECP2* mutations and balanced XCI patterns do typically have classic RTT with the exception of a few

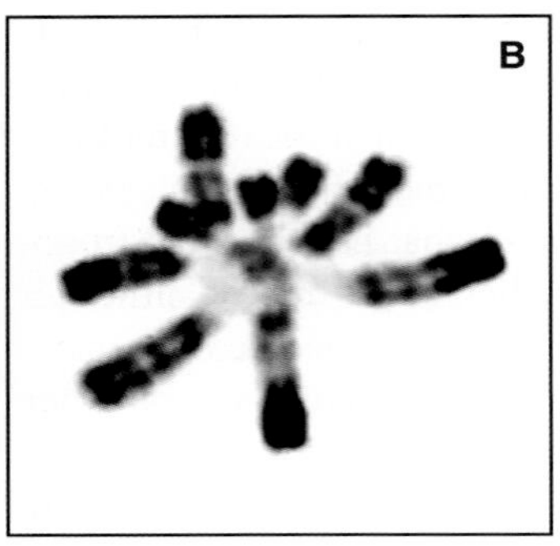

Figure 6. Genetic disorders affecting chromatin in *cis*. (*A*) This photo of a Rett syndrome patient illustrates the unusual stereotyped hand movements, teeth grinding, and abnormal posture. (Photo kindly provided by Dr. Daniel G. Glaze.) (*B*) Micrograph of chromosomes from an immunodeficiency, centromeric region instability, and facial anomalies (ICF) syndrome patient. (Courtesy of Drs. Timothy H. Bestor, Robert A. Rollins, and Deborah Bourc'his.)

hypomorphic alleles. Females with unbalanced XCI patterns favoring the wild-type allele, however, typically have the milder phenotypes (Wan et al. 1999; Carney et al. 2003). Males with *MECP2* mutations display a more severe phenotype than females because of their hemizygosity for the locus (i.e., they only possess a single X chromosome, hence, a single mutant allele). Typically, RTT-causing mutations cause neonatal lethality unless the male is mosaic for the mutations or has an XXY karyotype, in which case, all the phenotypes seen in females are also seen in these males (Zeev et al. 2002). On the other hand, males that have hypomorphic alleles that barely cause a phenotype in females develop any combination of features, including intellectual disability, seizures, tremors, enlarged testes, bipolar disease, or schizophrenia (Meloni et al. 2000; Couvert et al. 2001). It is interesting that doubling the dose of MeCP2 in mice and humans leads to progressive postnatal phenotypes that are quite severe and overlap with some of the loss-of-function phenotypes (Collins et al. 2004; Meins et al. 2005; Van Esch et al. 2005; Ramocki et al. 2009), indicating that the balance of MeCP2 expression is crucial.

3.2.3.3 The Role of MECP2 in RTT. MeCP2 was identified based on its ability to bind to symmetrically methylated CpG dinucleotides (Lewis et al. 1992). It binds methylated DNA through its methyl-CpG-binding domain (MBD) and interacts with corepressors Sin3A and HDACs through its transcription repression domain (Nan et al. 1997; see Fig. 9 of Ch. 15 [Li and Zhang 2014]). MeCP2 localizes to heterochromatin and was initially believed to be a transcriptional repressor (Jones et al. 1998; Nan et al. 1998), but data from animal studies revealed that many genes have decreased expression on its loss, and enhanced expression when MeCP2 levels are doubled or tripled (Yasui et al. 2007; Chahrour et al. 2008), raising questions about the exact mechanisms by which MeCP2 affects gene expression.

Recent data revealed that MeCP2 is highly abundant in neurons and binds widely throughout the genome—and is perhaps as abundant as one molecule for every two nucleosome (Skene et al. 2010). This finding led the investigators to propose that MeCP2 may function as an alternative to a linker histone given the competitive nature of H1 and MeCP2 binding to chromatin. Other possibilities to explain the opposing effects of MeCP2 loss and gain of gene expression include either a direct effect through its interactions with activating factors such as CREB (Chahrour et al. 2008) or the aforementioned silencing via Sin3A/HDAC complexes. It could also have an indirect secondary effect caused by altered neuronal activity on its loss or gain. The finding that MeCP2 may bind to 5hmC as well as 5mC, and a specific *MECP2* mutation that produces RTT-like phenotype reduces binding specifically to 5hmC suggests that gene body 5-hydroxymethylation, concomitant with MeCP2-binding enrichment in these active regions of neuronal lineages, may be important for neuronal function (Mellen et al. 2012). Even more recently, a study suggests that gene body 5-hydroxymethylation plays an essential role in neuronal differentiation, perhaps serving to inhibit repressive Polycomb repressive complex 2 (PRC2)-mediated H3K27 methylation (Hahn et al. 2013). It remains to be determined whether MeCP2 or other MBD proteins interpret the 5hmC signal on active genes. Clearly, we need a better understanding of the exact molecular and biochemical functions of MeCP2 in vivo.

An intriguing feature of RTT is the delayed postnatal onset of phenotypes in the absence of neurodegeneration. Studies on the distribution and abundance of MeCP2 revealed that it is detected in mature neurons, probably after synapse formation (Shahbazian et al. 2002; Kishi and

Macklis 2004; Mullaney et al. 2004). Such a distribution suggests that MeCP2's function is essential after neuronal maturation, once neuronal activity has been established, playing a role in regulating gene expression in response to neuronal activity. It is also expressed in astrocytes and other types of glia, but at lower levels than in neurons (Ballas et al. 2009; Tsujimura et al. 2009). Using mouse studies, neuron-specific deletions revealed that loss of MeCP2 compromises the function of all neurons tested and leads to a decrease in the enzymes, neurotransmitters, and neuropeptides critical for mediating their functions (Gemelli et al. 2006; Fyffe et al. 2008; Samaco et al. 2009; Chao et al. 2010; Ward et al. 2011). These animals display varying degrees of pathology and recapitulate one or more aspects of the RTT phenotype. Interestingly, the deletion of Mecp2 in adult animals still displays a delay of symptom onset, typical of the constitutive deletion that eventually causes RTT. This indicates a lifelong need for MeCP2 to maintain the epigenetic program required for normal brain function (McGraw et al. 2011). Importantly, these data show that the delayed onset in RTT phenotypes is not attributable to functional redundancy and/or lack of requirement in early postnatal life, but rather because of the period it takes for brain cells to succumb functionally as a result of the loss of protein. The finding that reexpression of MeCP2 in adult animals lacking a functional Mecp2 allele can rescue several RTT phenotypes (Giacometti et al. 2007; Guy et al. 2007; Robinson et al. 2012) is quite exciting and provides hope that symptoms of the disorder might be reversible in humans. Furthermore, expression of MeCP2 in either astrocytes or microglia improves breathing, motor function, and survival (Lioy et al. 2011; Derecki et al. 2012). Indeed, Garg et al. (2013) provided proof-of-principle that gene therapy works in reversing RTT symptoms using mouse models. Altogether, these data suggest that several features of RTT are likely to be reversed or subdued when adequate interventions are identified.

3.2.4 α-Thalassemia X-Linked Mental Retardation

Males with α-thalassemia mental retardation syndrome X-linked (ATRX) (OMIM 301040) display α-thalassemia, moderate to severe intellectual disability, dysmorphic facial features, microcephaly, skeletal and genital abnormalities, and, usually, an inability to walk. Heterozygous females are typically asymptomatic. Mutations in the ATRX gene, which maps to Xq13, cause this syndrome, as well as a host of additional phenotypes, including variable degrees of X-linked intellectual disability with or without spastic paraplegia, and acquired α-thalassemia myelodysplastic syndrome owing to somatic mutations (Gibbons et al. 1995; Villard et al. 1996; Yntema et al. 2002; Gibbons et al. 2003). The ATRX protein contains a plant homeodomain zinc-finger motif, as well as a SNF2 (sucrose nonfermentable) family DNA-dependent ATPase motif. This, together with its localization to pericentromeric heterochromatic domains, in association with heterochromatin1α (McDowell et al. 1999), suggests a role as a chromatin-remodeling protein. Mutations in *ATRX* cause down-regulation of the α-globin locus, silencing of the maternal H19 imprinting control regions (i.e., ICR1 in Fig. 5A), and abnormal methylation of several highly repeated sequences, including subtelomeric repeats, Y-specific satellite, and ribosomal DNA arrays. ATRX functions via interaction with other proteins that are key epigenetic regulators, including the methyl-CpG-binding protein, MeCP2, and cohesin. One study showed that ATRX is essential for the survival of cortical neurons, hinting that increased neuronal loss might contribute to the severe intellectual disability and spasticity seen in patients with ATRX mutations (Berube et al. 2005).

It is interesting that the levels of ATRX are tightly regulated and either decreases or increases (similarly to MeCP2 in RTT) cause major neurodevelopmental problems. For example, human patients with mutations that result in 10%–30% of normal ATRX levels display the full ATRX phenotype despite having significant amounts of the normal ATRX protein (Picketts et al. 1996). Too much ATRX seems to be equally devastating. Transgenic mice that overexpress ATRX develop NTDs, have growth retardation, and die during embryogenesis. Those that survive develop craniofacial abnormalities, compulsive facial scratching, and seizures. The features are reminiscent of clinical features in ATRX patients with ATRX loss-of-function mutations, raising the possibility that levels of ATRX are tightly regulated for the functional integrity of the protein complex within which it resides. ATRX clearly plays a variety of key roles in chromatin-related processes for which active research is ongoing (Clynes et al. 2013).

3.2.5 Immunodeficiency, Centromeric Region Instability, and Facial Anomalies Syndrome

The ICF syndrome (OMIM 242860) is a rare autosomal recessive chromosome breakage disorder. ICF patients display two invariant phenotypes, immunodeficiency, and cytogenetic abnormalities. Highly variable and less penetrant phenotypes include craniofacial defects, such as a broad and flat nasal bridge, epicanthal folds, high forehead and low-set ears, psychomotor retardation, and intestinal dysfunction (Smeets et al. 1994). The immunodeficiency is typically severe and often the cause of premature death during childhood due to respiratory or gastrointestinal infections. A decrease in serum IgG levels is the most common immunological defect, but decreased numbers of B or

T cells are also observed (Ehrlich 2003). Cytogenetic abnormalities primarily affect chromosomes 1 and 16, and to a lesser degree 9, and are seen on routine karyotype analysis of blood and cultured cells of ICF patients (Fig. 6B) (Tuck-Muller et al. 2000).

Hypomethylation of juxtacentromeric repeat sequences on chromosomes 1, 9, and 16 was observed well before the identification of the ICF causing gene (Jeanpierre et al. 1993). These chromosomes contain the largest blocks of classic satellite (satellites 2 and 3) tandem repeats near their centromeres. The finding that ICF is caused by loss-of-function mutations in the de novo DNA methyltransferase gene (*DNMT3B*) provided insight into why a decrease in methylation at centromeric satellites 2 and 3 was observed (Hansen et al. 1999; Okano et al. 1999; Xu et al. 1999). However, it remains unclear why loss of function of a widely expressed de novo methyltransferase selectively affects specific repetitive sequences. One possible explanation entails the subcellular distribution and/or context-specific protein interaction of DNMT3B (Bachman et al. 2001). Another possibility is that the catalytic activity of DNMT3B is more essential for methylating sequences that have a high density of CpGs over large genomic regions, as in the case of satellite 2 (Gowher and Jeltsch 2002) or the D4Z4 repetitive sequence implicated in facioscapulohumeral muscular dystrophy (discussed later in Sec. 3.3; Kondo et al. 2000). Using lymphoblastoid cell lines of normal and ICF patients, gene expression studies revealed alterations in the expression of genes involved in maturation, migration, activation, and homing of lymphocytes (Ehrlich et al. 2001). It is not clear from this study, however, whether loss of DNMT3B causes dysregulation of such genes because the methylation patterns at their promoter did not seem to be altered. A more recent study by Jin and colleagues, however, did reveal that methylation patterns are decreased at the promoters of specific genes in cells from ICF patients. These methylation changes were accompanied with altered histone modifications (a decrease in repressive H3K27 trimethylation and increase in activating H4K9 acetylation and H3K4 trimethylation marks), resulting in increased gene expression (Jin et al. 2008). Based on these studies, DNMT3B is clearly important in regulating the methylation of genes involved in development and immune function at gene-specific sites, in addition to heterochromatic regions. A related development has been the finding that loss of DNMT3B leads to an increase in the number of protocadherins (PCDH) expressed in an individual Purkinje cell. This causes abnormal dendritic arborization, akin to neuronal "short-circuiting," which is detrimental for establishing fully functioning neuronal circuitry (Toyoda et al. 2014). Normally, each neuron should only express one or a few Pcdh genes, which acts as a unique cell identification mechanism. This pseudo-monoallelic expression pattern is established through epigenetic processes described in Ch. 32 (Lomvardas and Maniatis 2014) evidently involving DNMT3B.

Using homozygosity mapping and exome sequencing, de Greef and colleagues identified homozygous loss-of-function mutations in *ZBTB24* as the genetic basis for a variant form of ICF, termed ICF2 (OMIM 614069). ZBTB24 is a transcriptional repressor involved with regulating hematopoietic development. Patients with ICF2 present with agammaglobulinemia, facial abnormalities, and intellectual disability, and their cells show hypomethylation of the α-satellite repeat on chromosome 9 (de Greef et al. 2011).

3.2.6 Schimke Immuno-Osseous Dysplasia

Schimke immuno-osseous dysplasia (SIOD) (OMIM 242900) is an autosomal recessive multisystem disorder characterized by dysplasia of the spine and the ends of long bones, growth deficiency, renal function abnormalities due to focal and segmental glomerulosclerosis, hypothyroidism, and defective T-cell-mediated immunity (Schimke et al. 1971; Spranger et al. 1991). SIOD is caused by mutations in *SMARCAL1* (SW1/SNF2-related, matrix-associated, actin-dependent regulator of chromatin, subfamily alike1), which encodes a protein proposed to regulate transcriptional activity through chromatin remodeling (Boerkoel et al. 2002). Nonsense and frameshift mutations cause severe phenotypes, whereas some of the missense mutations cause milder or partial phenotypes (Boerkoel et al. 2002). A patient with B-cell lymphoma and SIOD was found to have mutations in *SMARCAL1*, suggesting that loss of function of this protein can also cause a fatal lymphoproliferative disorder (Taha et al. 2004). The exact mechanism by which loss of SMARCAL1 causes the phenotypes of SIOD remains to be elucidated, although we do know that it is a factor required in DNA damage response pathways (reviewed in Bansbach et al. 2010).

3.2.7 Kabuki Syndrome

Kabuki syndrome (KABUK1) (OMIM 147920) is a congenital intellectual disability disorder that is typically associated with dwarfism, high arched eyebrows, long palpebral fissures, eversion of lower eyelids, large prominent ears, a depressed nasal tip, and various skeletal abnormalities. In addition, autistic features are prominent in a subset of patients with KABUK1 and some patients have other psychiatric features, including aggressive/oppositional behavior, hyperactive/impulsive behavior, anxiety, and obsessions. Using exome sequencing, Ng et al. (2010) discovered that mutations in the mixed leukemia lineage 2 (*MLL2*) gene account for the cause in 56%–76% of cases. Additional studies identified mutation in up to 74% of KABUK1 patients (Hannibal et al. 2011). *MLL2* encodes a *Trithorax*-group histone lysine (K) methyltransferase (KMT) and preferentially mediates trimethylation of histone H3 lysine 4 (H3K4me3). Studies of patients with a Kabuki phenotype, but lacking a mutation in *MLL2*, revealed that mutations in *KDM6A* (often referred to as *UTX*), a gene encoding an MLL2-interacting protein, cause some cases of KABUK1 (Lederer et al. 2012), now referred to as KABUK2 (OMIM 300867). Interestingly, KDM6A is a histone demethylase that removes mono-, di-, and trimethyl marks from H3K27 residues. The KDM6A protein also interacts with the switch/sucrose nonfermentable (SWI/SNF) remodeling complex that contains the transcription activator Brg1, thus, linking it to the control of higher-order chromatin structure (Lederer et al. 2012). Hence, both MLL2 and KDM6A are part of the gene activation machinery, functioning in the addition of active histone marks via MLL2 and the removal of repressive histone marks via KDM6A, relying on the Brg1-containing SWI/SNF complex remodeling capacity for access to chromatin. The role of these crucial chromatin functions are further discussed in Ch. 36 (Pirrotta 2014) and illustrated in its Figure 2.

3.2.8 Other Disorders

Since the first edition of this collection, many additional disorders of chromatin modification have been described, several of which are listed in Table 3, including mutations in *KDM5C* (JARID1C), *EHMT1* (KMT1D), ARID1D, and *NSD1* (KMT3B). All of these cause complex phenotypes, including intellectual disability. Quite different is the case of mutations in *DNMT1* that cause hereditary sensory neuropathy, type IE. Also, many other disorders have been reported to affect helicases, ligases, DNA repair machinery, and the cohesin complex; their gene products all interact with chromatin in *trans* in one way or another.

Table 3. Selected genetic disorders affecting chromatin structure in *cis*

Disorder	Gene	Comments
γδβ- and δβ-thalassemia	Deletion of LCR causes decreased globin expression	
Fragile-X syndrome	Expansion of CCG repeat leads to abnormal methylation and silencing of *FMR1*	Premutation alleles (60–200) cause a neurodegenerative disorder
FSH dystrophy	Contraction of D4Z4 repeats causes less repressive chromatin	
Multiple cancers	Germline epimutation of *MLH1*	

FSH, facioscapulohumeral.

3.3 Genetic Disorders Affecting Chromatin Structure in *cis*

The genes for most Mendelian disorders are usually identified by finding mutations in either exons or splice sites, whereby the gene products, RNA or protein, are altered or not produced. For many of these disorders, however, there is, frequently, a small group of patients in whom mutations cannot be identified after sequencing the coding and noncoding regions of the gene despite linkage to the specific locus. It is becoming increasingly clear that epigenetic abnormalities can also affect gene expression in *cis* and underlie some Mendelian disorders for cases lacking exonic mutations. The following three examples show how *cis*-linked alterations in chromatin structure can result in human disease (see Fig. 1 and Table 3).

3.3.1 γδβ- and δβ-Thalassemia

The thalassemias are the most common single-gene disorders in the world. They are a heterogeneous group of hemoglobin synthesis disorders caused by reduced levels of one or more of the globin chains of hemoglobin. The imbalance in synthesis of various globin chains leads to abnormal erythropoiesis and profound anemia (Weatherall et al. 2001). Hundreds of coding and splicing mutations have been identified, but it was the deletions of the regulatory sequences that pinpointed how changes in chromatin structure can explain some subtypes of thalassemia. In particular, the initial discovery that deletions of ~100 kb upstream of the α-globin gene (while leaving the gene intact) caused γδβ-thalassemia helped identify the locus control region (LCR) that regulates β-globin expression (Kioussis et al. 1983; Forrester et al. 1990). Smaller deletions involving part of the LCR caused δβ-thalessemia (Curtin et al. 1985; Driscoll et al. 1989). This kick-started a new branch of research looking into gene regulatory elements, such as enhancers, that are involved in more long-range gene control achieved in a chromatin context. These initial deletions resulted in an altered chromatin state at the β-globin locus despite being tens of kilobases upstream of the coding region (Grosveld 1999). Continued research has, particularly, brought an understanding of epigenetic gene control from a chromatin topological and nuclear organizational point of view, showing how chromatin looping can bring distal regions together to ensure correct gene regulation, as extensively discussed in Ch. 19 (Dekker and Misteli 2014).

3.3.2 FXS

FXS mental retardation (OMIM 309550) is one of the most common causes of inherited intellectual disability. More than 60 years ago, Martin and Bell (1943) described a family that showed that intellectual disability segregated as an X-linked disorder. In 1969, Lubs reported a constriction on the long arm of the X chromosome in some mentally retarded males and one asymptomatic female (Lubs 1969). Cytogenetic studies, especially those using culture media deficient in folic acid and thymidine, revealed the fragile site in families with X-linked intellectual disability, and they were then diagnosed as having FXS (Sutherland 1977; Richards et al. 1981). This chromosomal variant was mapped to Xq27.3 and dubbed the fragile X chromosome (Harrison et al. 1983). Affected males have moderate to severe intellectual disability, macroorchidism, connective tissue abnormalities, such as hyperextensibility of joints, and large ears (Fig. 7) (Hagerman et al. 1984). The gene responsible for FXS is *FMR1*, which encodes the fragile-X mental retardation protein (FMRP). The most common mutational mechanism is an expansion of an unstable noncoding CGG repeat at the 5′-UTR (untranslated region) of the *FMR1* gene (Warren and Sherman 2001). Normal alleles contain 6–60 repeats, premutation alleles have 60–200, and the full mutation contains more than 200 repeats (Fig. 8A). This repeat expansion provides an excellent example of a genetic disorder that is mediated through altered chromatin structure in *cis*. A CpG island in the 5′-regulatory region of *FMR1* becomes aberrantly methylated on repeat expansion in the case of the full mutation (Verkerk et al. 1991). Decreased histone acetylation at the 5′-end has also been documented in cells from fragile-X patients compared with healthy controls (Coffee et al. 2002). The altered DNA methylation and histone acetylation patterns, in turn, lead to loss of *FMR1* expression and, therefore, loss of FMRP function in patients with FXS. Thus, these patients have a primary genetic mutation (noncoding repeat expansions) and secondary epigenetic mutation (DNA methylation and histone PTMs), causing silencing of the FMR1 gene. Carriers of the fragile-X premutation (60–200 repeats) develop a distinct neurodegenerative syndrome characterized by tremor and ataxia (Hagerman and Hagerman 2004). Interestingly, these premutations may induce this distinctive pathogenesis at the RNA level because the *FMR1* RNA and protein are present. Studies in animal models suggest that the RNA encoded by CGG repeats may bind to and alter the function of some cellular proteins, causing them to accumulate (Jin et al. 2003; Willemsen et al. 2003).

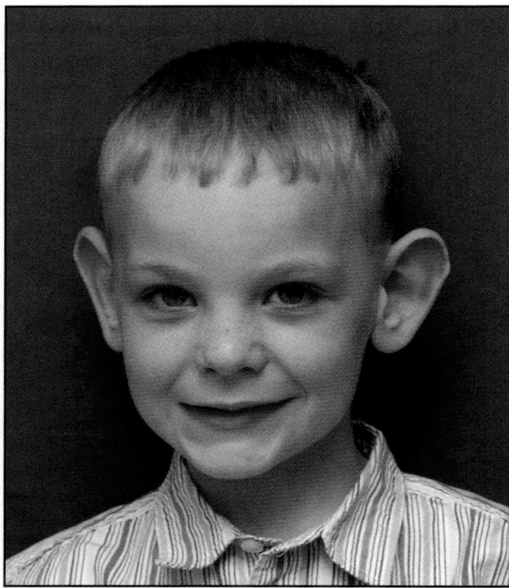

Figure 7. An example of a genetic disorder affecting chromatin in *trans*. The photograph is of a patient with fragile-X syndrome who, in addition to mental retardation, has the typical features of prominent forehead and large ears. (Photograph kindly provided by Dr. Stephen T. Warren.)

An interesting epigenetic mechanism has been proposed to explain how the CGG FMR1 repeat gets methylated and subsequently silenced, involving RNA interference (RNAi). The fact that a premutation CGG repeat can form a single and stable hairpin structure (Handa et al. 2003) and rCGG repeats can be cleaved by Dicer raised the possibility that expanded FMR1-associated CGG repeats (which are unmethylated during early development) could be transcribed, resulting in an RNA transcript that can form a hairpin structure cleavable by Dicer to produce small ncRNAs. These small RNA molecules could be envisaged to associate with a mammalian equivalent of RNA-induced initiator of transcriptional gene silencing and recruit DNA de novo methyltransferases and/or histone methyltransferases to the 5′-UTR of *FMR1*, leading to full methylation of the CGG repeat and transcriptional repression of *FMR1* as development progresses (Jin et al. 2004).

The RNAi model, leading to *FMR1* silencing, remains to be verified, leaving open the question of how an epigenetically controlled switch can occur during development in FXS individuals to transcriptionally silence the *FMR1* gene in the presence of significant repeat expansion. Some progress has been made in generating human models to study this developmentally regulated switch. Eiges et al. (2007) reported generating human embryonic stem cells from in vitro fertilization-generated embryos diagnosed to have the full FMR1 expansion mutation. Subsequently, in vitro experiments pointed toward histone deacetylation

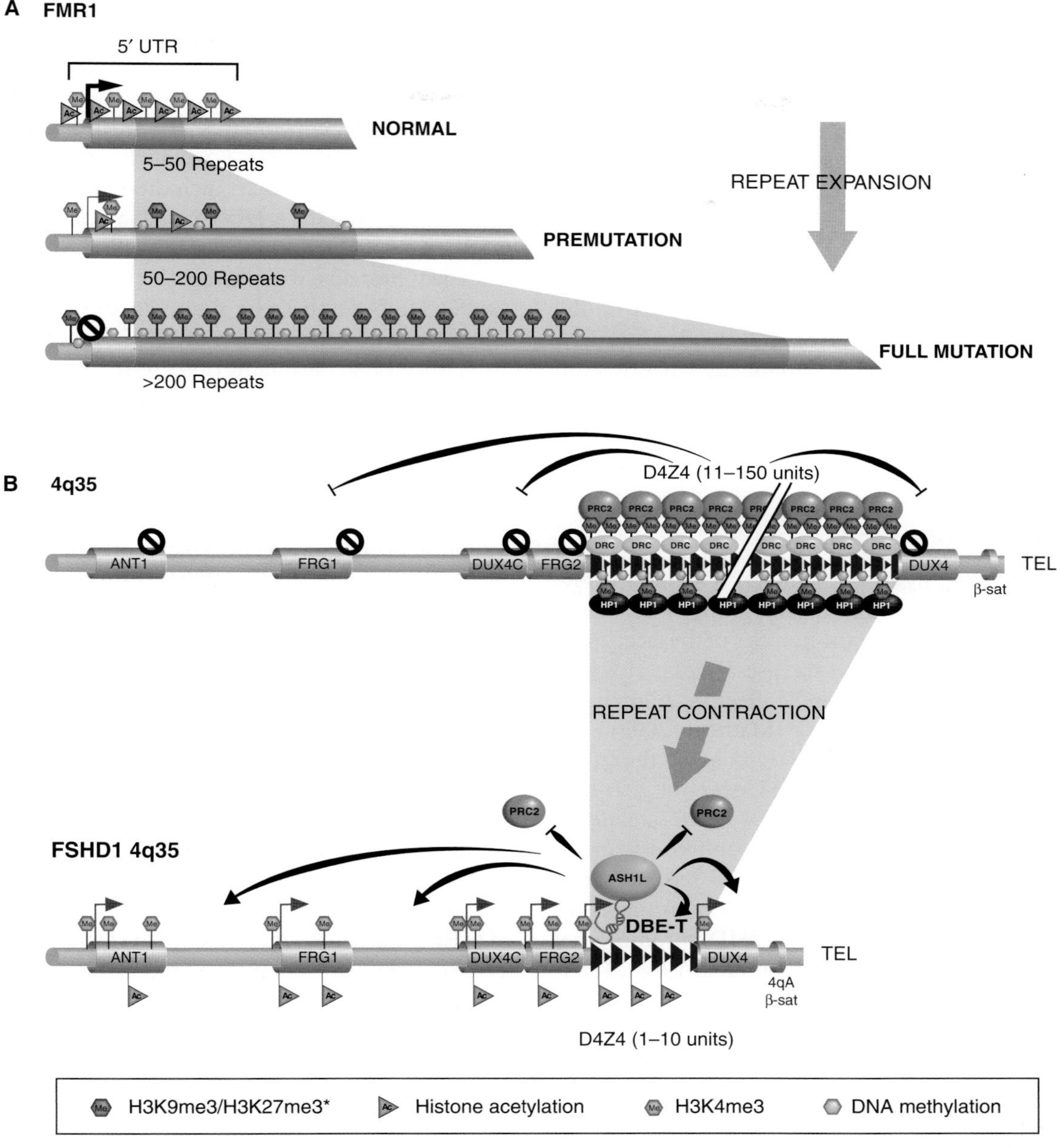

Figure 8. Human diseases showing genetic alterations in regions with triplet repeats, in the form of repeat expansion or contraction, with consequential in *cis* chromatin structural changes. (*A*) The 5′ region of the FMR1 gene is shown with the CGG triplet repeat (blue shading) in normal and fragile-X-affected individuals. The chromatin-associated features of the 5′ FMR1 region are indicated in each case. The normal repeat number range (5–50) typically shows active chromatin features, whereas fully expanded repeat alleles (more than 200) have heterochromatic features. (*B*) The 4q35 region associated with normal and facioscapulohumeral dystrophy individuals (FSHD1) is depicted. In the normal repeat number range (11–150 units), heterochromatinization is presumed to initiate from the D4Z4 repeats (blue triangles) and spread throughout the 4q35 region (chromatin marks not shown throughout), silencing all genes. The 4q35 region in FSHD1 individuals has a contracted number of D4Z4 repeats, which is permissive to transcription of the DBE-T lncRNA (indicated in red), recruiting ASH1L and associated factors to remodel, generating euchromatin and allowing gene expression from 4q35 genes with myopathic potential. PRC2, Polycomb repressive complex 2; DRC, D4Z4-repressing complex; HP1, heterochromatin protein 1; ASH1L, absent small and homeotic disks protein 1; DBE-T, D4Z4-binding element transcript.

and H3K9 methylation as being likely to precede DNA methylation in the establishment of a silent FMR1 locus. More recently, induced pluripotent stem cells have been generated from FXS patients and aberrant neuronal differentiation was recapitulated in vitro (Sheridan et al. 2011). It is hoped that these cell models will be able to elucidate the specifics of how expanded repeat mutants of FMR1 become epigenetically silenced during development.

Another major question in FXS research is how FMR1 silencing leads to the FXS phenotype. The FMRP gene product is an RNA-binding protein thought to inhibit translation of many genes involved in neurodevelopment, particularly proteins that function following synaptic stimulation. It is thought to do this via RNAi pathways, involving miRNAs, the RNA-induced silencing complex, and Argonaute 2 protein (Jin et al. 2004).

3.3.3 Facioscapulohumeral Dystrophy

Facioscapulohumeral dystrophy (FSHD) (OMIM 158900) is an autosomal dominant muscular dystrophy characterized by progressive wasting of the muscles of the face, upper arm, and shoulder. The more severe cases have hearing loss, and a very small subset of severely affected children has intellectual disability and seizures (Mathews 2003). The major locus linked to FSHD (FSHD1) maps to the subtelomeric region of chromosome 4q35 near the germline DUX4 gene. The region contains D4Z4 macrosatellite repeats arranged in tandem arrays, each unit consisting of 3.3 kb of GC-rich sequence (Fig. 8B). This repeated array is polymorphic, containing 11–150 units on normal chromosomes, but only 1–10 units on FSHD chromosomes (Wijmenga et al. 1992; van Deutekom et al. 1993). Interestingly, this decrease in copy number repeats contrasts with the expansion of repeats associated with the FMR1 locus in FXS discussed in the previous section. A second variable satellite repeat sequence (β-68bp *Sau*3A) distal to D4Z4 also appears to play a role in developing FSHD. The 4qA variant at the β-satellite repeat, along with the contraction of the D4Z4 repeat, is necessary for the manifestation of FSHD (Fig. 8B) (Lemmers et al. 2002).

We are beginning to unravel how epigenetic processes in the 4q35 region cause disease with contracted D4Z4 repeats and the 4qA β-satellite variant present. The 4q35 region normally displays features typical of heterochromatin, such as DNA methylation, histone hypoacetylation H3K9 methylation, HP1γ and cohesin binding, but also methylation of H3K27, which is more typical of developmentally regulated genes (reviewed in Casa and Gabellini 2012). It is presumed that D4Z4 repeats are the trigger for heterochromatin formation when the range of repeats is 11–150. When the D4Z4 repeat number decreases below 11, the locus and flanking regions become associated with euchromatic structures, enriched in histone acetylation, H3K4me3, and H3K36me2, binding of CTCF, with a loss of repressive marks (Cabianca et al. 2012).

The 4q35 region contains a number of genes normally silenced in muscle cells, but which are significantly expressed in FSHD1 muscle, including *FRG1* and *-2* (FSHD region genes 1 and 2), DUX4C, adenine nucleotide transporter 1 (ANT1), and DUX4 (Fig. 8B). Among the 4q35 encoded proteins with myopathic potential, FRG1 and DUX4 are believed to be the main proteins causative to FSHD based on the analysis of deletion mutants; however, it is conceivable that all the derepressed genes in 4q35 contribute to the disease's penetrance and severity (reviewed in Cabianca and Gabellini 2010).

Bickmore and van der Maarel, and Gabellini and colleagues, proposed that contraction of the repeats causes a less repressive chromatin state leading to increased transcription of 4q35-qter genes (Bickmore and van der Maarel 2003; Tupler and Gabellini 2004). However, the loss of repressive H3K9 and DNA methylation marks alone was shown to be insufficient for myopathic gene derepression and FSHD onset, and therefore they are likely to be downstream epigenetic mechanisms, locking in a repressive chromatin state (Cabianca et al. 2012). The chromatin actually needs to be remodeled for active transcription. Thus, continued studies have endeavored to determine what provides the epigenetic switch that permits derepression of 4q35 genes in the presence of contracted D4Z4 repeats and the 4qA variant. In vitro and in vivo studies identified a 27-bp sequence in D4Z4, called the D4Z4-binding element (DRE), which show similarities to the *Drosophila* Polycomb response element, known to bind PRCs (Gabellini et al. 2002; Bodega et al. 2009). This DRE element binds a complex termed the D4Z4-repressing complex (DRC) that comprises the transcriptional repressor Ying Yang1, high mobility group box 2, and nucleolin (Gabellini et al. 2002), as well as the EZH2 component of the PRC2 complex (Fig. 8B). Accordingly, chromatin is also marked by PRC1/2-catalyzed H3K27me3 and H2AK119 ubiquitination, and enriched in the macroH2A histone variant (Casa and Gabellini 2012). Recent research showed that, although in normal individuals with a sufficiently high D4Z4 repeat number, the concentration of DRCs provides a platform for stable Polycomb binding and the establishment of repressive chromatin, created primarily through methylation of H3K27; this threshold binding capacity is compromised when D4Z4 repeats number less than 10, leading to the derepression of chromatin. The key finding, however, has been the transcription of a long ncRNA called DBE-T, which recruits the Trithorax group protein, ASH1L, and other chromatin remodeling factors in *cis* (Cabianca et al.

 Cite as *Cold Spring Harb Perspect Biol* doi: 10.1101/cshperspect.a019497

2012). The DBE-T–ASH1L interaction acts in a self-reinforcing feedback loop to remodel chromatin and lock in the transition to an active chromatin conformation, marked by H3K4me3, H3K36me2, and histone acetylation (Cabianca et al. 2012). This remodeled chromatin efficiently inhibits the binding of the repressive PRC machinery. The discovery and functional characterization of the DBE-T long noncoding RNA (lncRNA) is the first documented activating ncRNA associated with human disease and currently represents the best druggable therapeutic target available in the treatment of FSHD. Preliminary DBE-T RNA knockdown experiments have shown promise toward that end (Cabianca et al. 2012).

Another consideration in trying to understand why changes in chromatin at the 4q35 region make it susceptible to the onset of FSHD is looking at its chromatin organization in 3D nuclear space. Studies have been performed that look into the localization of 4q35 in normal and FSHD cells in relation to nuclear subcompartment (e.g., its proximity to the nuclear periphery or direct association with the nuclear lamina) or use of chromosome conformation capture (3C) to determine the higher-order chromatin organization and specific intrachromosomal interactions, the principles of which are discussed in Ch. 19 (Dekker and Misteli 2014). Interestingly, studies have shown that 4q35 in both normal and FSHD1 individuals localize to the nuclear periphery, but using different sequences. It is not yet clear whether they are associated with different nuclear peripheral subdomains and whether this has a functional consequence. Also, based on different regional intrachromosomal interactions ascertained by 3C, topological chromatin domains infer that looping at normal 4q35 alleles enables the spreading of heterochromatin into adjacent genes (reviewed in Cabianca and Gabellini 2010). Proving a functional role for nuclear compartmentalization or chromatin topology is notoriously hard to test, and we await further research to answer these questions in the case of FSHD.

3.4 Epigenetic and Environment Interactions

Data from human studies, as well as animal models, are providing evidence that the environment can affect epigenetic marks and, as a result, gene function. The finding that monozygotic twins have similar epigenotypes during the early years of life, but show remarkable differences in the content and distribution of 5-methylcytosines and acetylated histones in later life, provides strong evidence that the epigenotype is metastable and displays temporal variability (Fraga et al. 2005). It is likely that many environmental factors and stochastic events contribute to the variations in the epigenome (Fig. 9) (Anway et al. 2005), but diet and early experiences are emerging as potential key players.

3.4.1 *Diet and Epigenotypes in Aging and Disease*

Several reports indicate that there is an age-dependent decrease of global DNA methylation, whereas concurrently there might be site-specific hypermethylation (Hoal-van Helden and van Helden 1989; Cooney 1993; Rampersaud et al. 2000). Given the large body of data linking altered DNA methylation to cancer risk or progression (Mays-Hoopes 1989; Issa et al. 1994), such epigenetic changes might contribute to the age-related increase in cancer risk. The role of diet may be a contributing factor in controlling global methylation status and has best been illustrated in adult males suffering from uremia and undergoing hemodialysis. The presence of hyperhomocystinemia in

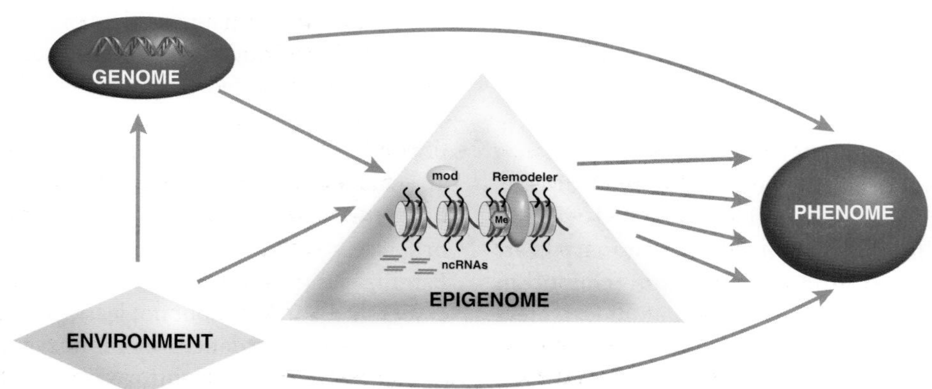

Figure 9. The epigenotype plays a critical role, along with the genotype and environmental factors, in determining phenotypes.

these patients suggested low methionine availability, presumably caused by folate depletion (see Ch. 30 [Berger and Sassone-Corsi 2014]). These males had consequent reduced global and locus-specific DNA methylation thought to be because of the lack of the S-adenosyl L-methionine (SAM) cosubstrate required in the methylation reactions of DNMTs and histone KMTs. The methylation patterns were reversed after the administration of high doses of folic acid (Ingrosso et al. 2003).

Because several of the neuropsychiatric features resulting from folate and B12 deficiencies overlap with those seen with sporadic neuropsychiatric disorders, it was proposed that sporadic onset might be caused by alterations in methylation patterns in the central nervous system (Reynolds et al. 1984). For instance, low levels of the SAM cosubstrate were found in folate-responsive depression; furthermore, SAM supplementation was helpful as an adjunct therapeutic in some forms of depression (Bottiglieri et al. 1994). Turning to early development, we know that increased folic acid intake by childbearing women reduces the risk of neural tube defect, and it is tempting to propose that some epigenetic-mediated effects on DNA or histone methylation are the reason. The finding that supplementing maternal diets with extra folic acid, B12, and betaine alters the epigenotype and phenotype of the offspring of "agouti viable yellow" mice corroborates this presumption. This is likely to be the first of many examples in humans and other mammals to support a mechanistic link between diet and epigenetic processes (Wolff et al. 1998; Waterland and Jirtle 2003).

Another new entry to the field of diet and epigenetics is the finding that vitamin C affects the efficiency of Tet-mediated conversion of 5mC to 5hmC in the early embryo at the time of implantation (Blaschke et al. 2013). There is no doubt that researchers will continue to uncover how dietary components affect the epigenome, which, in the case of absence or overintake, particularly during early development, may contribute to the onset of disease.

3.4.2 *Fetal or Infant Experiences and Epigenotypes*

The Barker or thrifty phenotype hypothesis, which has evolved into the fetal origins hypothesis of adult disease, posits that reduced fetal nutrition is associated with an increased risk of adult disorders including coronary heart disease, stroke, diabetes, and hypertension (Guilloteau et al. 2009; Calkins and Devaskar 2011; Dyer and Rosenfeld 2011). Although widely accepted and studied in some scientific circles, the validity of these hypotheses remains controversial and the molecular basis unknown. There is extensive speculation that epigenetic changes might mediate such a phenomenon, and the term "thrifty epigenotype" has been used (Sebert et al. 2011). Robust demonstration of the phenomenon in animal models, although, is lacking. Moreover, limitations in access to human tissues from fetal and corresponding adult sources have hampered human research. This topic is likely to continue to be debated in science and public health policy until more data can be generated to substantiate the hypothesis or not.

The effect of the exposure of pregnant women to famine is a related topic. It is well documented that poor nutrition increases the risk of NTDs and folic acid supplementation can reduce the incidence of NTDs. Again, this is presumed to be because of insufficient provision of the SAM substrate for DNMTs and KMTs, operating to establish and maintain appropriate patterns of gene expression through DNA and histone methylation.

The best example of how early experiences and maternal behavior might alter the mammalian epigenotype has so far been described only in rats. Frequent licking and grooming by rat mothers altered the DNA methylation status in the promoter region of the glucocorticoid receptor (GR) gene in the hippocampus of their pups. The highly licked and groomed pups have decreased DNA methylation and increased histone acetylation at the GR promoter compared with pups that were raised by low-licking and -grooming mothers (Weaver et al. 2004). The increased levels of GR, secondary to the epigenotype change, affect the regulation of stress hormone levels and lifelong response to stress in the rat pups (Liu et al. 1997; Weaver et al. 2004). This and follow-up studies in humans answer and raise new questions, driving research toward an understanding of the role of early experiences in modulating epigenotypes and risk for psychiatric disorders.

4 LOOKING INTO THE FUTURE

During the next decade, we anticipate that mutations that alter the epigenotype will become increasingly recognized as mutational mechanisms that cause a variety of human disorders. Traditionally, the identification of disease-causing genes has focused on disorders in which familial cases or patients with chromosomal abnormalities facilitated the positional cloning of the responsible gene. At this time, we are challenged as we attempt to discover the mutational bases for some of the most common and devastating disorders, such as schizophrenia, autism, and mood disorders. Familial cases are not very common, genetic heterogeneity is very likely, and, last but not least, genetic data—especially the rate of discordances in monozygotic twins—do not always support a straightforward Mendelian inheritance model. These findings, coupled with the strong environmental effects on the penetrance of some of these disorders, underscore the importance of investigating the epigenomes

Cite as *Cold Spring Harb Perspect Biol* doi: 10.1101/cshperspect.a019497

in such diseases. Even single-gene disorders, such as AS, BWS, and SRS, can be caused either by genomic mutations or mutations that affect the epigenotype, and either inherited or de novo. Such molecular variations will undoubtedly be unearthed for other human disorders. Furthermore, data demonstrating that the levels of several proteins involved in epigenetic regulation are tightly regulated and perturbations of such levels, either through loss-of-function mutations or duplications, cause human disorders suggest that epigenetic mutations that will affect transcription, RNA splicing, or protein modifications are also likely to cause disease.

ACKNOWLEDGMENTS

We thank Dr. Timothy H. Bestor, Dr. Robert A. Rollins, and Dr. Deborah Bourc'his for the image of chromosomes from an ICF syndrome patient; Dr. Daniel J. Driscoll for the image of a PWS patient, Dr. Carlos A. Bacino for the image of an AS patient, Dr. Daniel G. Glaze for the image of an RTT patient, Dr. Stephen T. Warren for the image of an FXS patient, and Dr. Igna Van den Veyver for input on recurrent hydatidiform moles. We also thank past and current laboratory members who have contributed to our work on RTT, PWS, and AS. Last, but not least, we are grateful to our patients with RTT, UPD of chromosome 7, PWS, AS, and autism and to their families for enlightening us about the role of epigenetics in human diseases. Our work has been supported by grants from the National Institutes of Health (5 P01 HD040301-05; 5 P30 HD024064-17; 5 P01 HD37283; 5 R01 NS057819-08), International Rett Syndrome Research Foundation, Rett Syndrome Research Trust, The Simons Foundation, March of Dimes (12-FY03-43), and the Blue Bird Clinic Rett Center. H.Y.Z. is an investigator with the Howard Hughes Medical Institute. We regret that because of space constraints, we had to eliminate many important and relevant citations.

REFERENCES

*Reference is also in this book.

Alarcon JM, Malleret G, Touzani K, Vronskaya S, Ishii S, Kandel ER, Barco A. 2004. Chromatin acetylation, memory, and LTP are impaired in CBP$^{+/-}$ mice: A model for the cognitive deficit in Rubinstein–Taybi syndrome and its amelioration. *Neuron* **42:** 947–959.

* Almouzni G, Cedar H. 2014. Maintenance of epigenetic information. *Cold Spring Harb Perspect Biol* doi: 10.1101/cshperspect.a019372.

Amir RE, Van den Veyver IB, Wan M, Tran CQ, Francke U, Zoghbi HY. 1999. Rett syndrome is caused by mutations in X-linked MECP2, encoding methyl-CpG-binding protein 2. *Nat Genet* **23:** 185–188.

Anway MD, Cupp AS, Uzumcu M, Skinner MK. 2005. Epigenetic transgenerational actions of endocrine disruptors and male fertility. *Science* **308:** 1466–1469.

* Audia JE, Campbell RM. 2014. Histone modifications and cancer. *Cold Spring Harb Perspect Biol* doi: 10.1101/cshperspect.a019521.

Bachman KE, Rountree MR, Baylin SB. 2001. Dnmt3a and Dnmt3b are transcriptional repressors that exhibit unique localization properties to heterochromatin. *J Biol Chem* **276:** 32282–32287.

Ballas N, Lioy DT, Grunseich C, Mandel G. 2009. Non-cell autonomous influence of MeCP2-deficient glia on neuronal dendritic morphology. *Nat Neurosci* **12:** 311–317.

Bansbach CE, Boerkoel CF, Cortez D. 2010. SMARCAL1 and replication stress: An explanation for SIOD? *Nucleus* **1:** 245–248.

* Barlow DP, Bartolomei MS. 2014. Genomic imprinting in mammals. *Cold Spring Harb Perspect Biol* **6:** a018382.

Bastepe M, Frohlich LF, Hendy GN, Indridason OS, Josse RG, Koshiyama H, Korkko J, Nakamoto JM, Rosenbloom AL, Slyper AH, et al. 2003. Autosomal dominant pseudohypoparathyroidism type Ib is associated with a heterozygous microdeletion that likely disrupts a putative imprinting control element of GNAS. *J Clin Invest* **112:** 1255–1263.

* Baylin SB, Jones PA. 2014. Epigenetic determinants of cancer. *Cold Spring Harb Perspect Biol* doi: 10.1101/cshperspect.a019505.

* Berger SL, Sassone-Corsi P. 2014. Metabolic signaling to chromatin. *Cold Spring Harb Perspect Biol* doi: 10.1101/cshperspect.a019463.

Berube NG, Mangelsdorf M, Jagla M, Vanderluit J, Garrick D, Gibbons RJ, Higgs DR, Slack RS, Picketts DJ. 2005. The chromatin-remodeling protein ATRX is critical for neuronal survival during corticogenesis. *J Clin Invest* **115:** 258–267.

Bickmore WA, van der Maarel SM. 2003. Perturbations of chromatin structure in human genetic disease: Recent advances. *Hum Mol Genet* **12:** R207–R213.

Blaschke K, Ebata KT, Karimi MM, Zepeda-Martínez JA, Goyal P, Mahapatra S, Tam A, Laird DJ, Hirst M, Rao A, et al. 2013. Vitamin C induces Tet-dependent DNA demethylation and a blastocyst-like state in ES cells. *Nature* **500:** 222–226.

Bodega I, Ramirez GD, Grasser F, Cheli S, Brunelli S, Mora M, Meneveri R, Marozzi A, Mueller S, Battaglioli E, Ginelli E. 2009. Remodeling of the chromatin structure of the facioscapulahumeral muscular dystrophy (FSHD) locus and upregulation of FSHD-related gene 1 (*FRG1*) expression during human myogenic differentiation. *BMC Biol* **7:** 41.

Boerkoel CF, Takashima H, John J, Yan J, Stankiewicz P, Rosenbarker L, Andre JL, Bogdanovic R, Burguet A, Cockfield S, et al. 2002. Mutant chromatin remodeling protein SMARCAL1 causes Schimke immuno-osseous dysplasia. *Nat Genet* **30:** 215–220.

Bottiglieri T, Hyland K, Reynolds EH. 1994. The clinical potential of ademetionine (*S*-adenosylmethionine) in neurological disorders. *Drugs* **48:** 137–152.

* Brockdorff N, Turner BM. 2014. Dosage compensation in mammals. *Cold Spring Harb Perspect Biol* doi: 10.1101/cshperspect.a019406.

Cabianca DS, Gabellini D. 2010. The cell biology of disease: FSHD: Copy number variations on the theme of muscular dystrophy. *J Cell Biol* **191:** 1049–1060.

Cabianca DS, Casa V, Bodega B, Xynos A, Ginelli E, Tanaka Y, Gabellini D. 2012. A long ncRNA links copy number variation to a polycomb/trithorax epigenetic switch in FSHD muscular dystrophy. *Cell* **149:** 819–831.

Calkins K, Devaskar SU. 2011. Fetal origins of adult disease. *Curr Probl Pediatr Adolesc Health Care* **41:** 158–176.

Carney RM, Wolpert CM, Ravan SA, Shahbazian M, Ashley-Koch A, Cuccaro ML, Vance JM, Pericak-Vance MA. 2003. Identification of MeCP2 mutations in a series of females with autistic disorder. *Pediatr Neurol* **28:** 205–211.

Casa V, Gabellini D. 2012. A repetitive elements perspective in Polycomb epigenetics. *Front Genet* **3:** 199.

Caspary T, Cleary MA, Perlman EJ, Zhang P, Elledge SJ, Tilghman SM. 1999. Oppositely imprinted genes p57(Kip2) and igf2 interact in a mouse model for Beckwith–Wiedemann syndrome. *Genes Dev* **13:** 3115–3124.

Chahrour M, Yung SY, Shaw C, Zhou X, Wong STC, Qin J, Zoghbi HY. 2008. MeCP2, a key contributor to neurological disease, activates and represses transcription. *Science* **320:** 1224–1229.

Chao HT, Chen H, Samaco RC, Xue M, Chahrour M, Yoo J, Neul JL, Gong S, Lu HC, Heintz N, et al. 2010. Dysfunction in GABA signalling mediates autism-like stereotypies and Rett syndrome phenotypes. *Nature* **468:** 263–269.

Clynes D, Higgs DR, Gibbons RJ. 2013. The chromatin remodeller ATRX: A repeat offender in human disease. *Trends Biochem Sci* **38:** 461–466.

Coffee B, Zhang F, Ceman S, Warren ST, Reines D. 2002. Histone modifications depict an aberrantly heterochromatinized *FMR1* gene in fragile X syndrome. *Am J Hum Genet* **71:** 923–932.

Collins AL, Levenson JM, Vilaythong AP, Richman R, Armstrong DL, Noebels JL, Sweatt JD, Zoghbi HY. 2004. Mild overexpression of MeCP2 causes a progressive neurological disorder in mice. *Hum Mol Genet* **13:** 2676–2689.

Cooney CA. 1993. Are somatic cells inherently deficient in methylation metabolism? A proposed mechanism for DNA methylation loss, senescence and aging. *Growth Dev Aging* **57:** 261–273.

Cooper WN, Luharia A, Evans GA, Raza H, Haire AC, Grundy R, Bowdin SC, Riccio A, Sebastio G, Bliek J, et al. 2005. Molecular subtypes and phenotypic expression of Beckwith–Wiedemann syndrome. *Eur J Hum Genet* **13:** 1025–1032.

Couvert P, Bienvenu T, Aquaviva C, Poirier K, Moraine C, Gendrot C, Verloes A, Andres C, Le Fevre AC, Souville I, et al. 2001. MECP2 is highly mutated in X-linked mental retardation. *Hum Mol Genet* **10:** 941–946.

Crepin M, Dieu MC, Lejeune S, Escande F, Boidin D, Porchet N, Morin G, Manouvrier S, Mathieu M, Buisine MP. 2012. Evidence of constitutional MLH1 epimutation associated to transgenerational inheritance of cancer susceptibility. *Hum Mutat* **33:** 180–188.

Curtin P, Pirastu M, Kan YW, Gobert-Jones JA, Stephens AD, Lehmann H. 1985. A distant gene deletion affects β-globin gene function in an atypical γ δβ-thalassemia. *J Clin Invest* **76:** 1554–1558.

DeBaun MR, Niemitz EL, Feinberg AP. 2003. Association of in vitro fertilization with Beckwith–Wiedemann syndrome and epigenetic alterations of LIT1 and H19. *Am J Hum Genet* **72:** 156–160.

de Greef JC, Wang J, Balog J, den Dunnen JT, Frants RR, Straasheijm KR, Aytekin C, van der Burg M, Duprez L, Ferster A, et al. 2011. Mutations in ZBTB24 are associated with immunodeficiency, centromeric instability, and facial anomalies syndrome type 2. *Am J Hum Genet* **88:** 796–804.

* Dekker J, Misteli T. 2014. Long-range chromatin interactions. *Cold Spring Harb Perspect Biol* doi: 10.1101/cshperspect.a019356.

Dennis C. 2003. Epigenetics and disease: Altered states. *Nature* **421:** 686–688.

Derecki NC, Cronk JC, Lu Z, Xu E, Abbott SB, Guyenet PG, Kipnis J. 2012. Wild-type microglia arrest pathology in a mouse model of Rett syndrome. *Nature* **484:** 105–109.

Driscoll MC, Dobkin CS, Alter BP. 1989. γ δ β-thalassemia due to a de novo mutation deleting the 5′ β-globin gene activation-region hypersensitive sites. *Proc Natl Acad Sci* **86:** 7470–7474.

Dupont C, Sifer C. 2012. A review of outcome data concerning children born following assisted reproductive technologies. *ISRN Obstet Gynecol* **2012:** 405382.

Dyer JS, Rosenfeld CR. 2011. Metabolic imprinting by prenatal, perinatal, and postnatal overnutrition: A review. *Semin Reprod Med* **29:** 266–276.

Eggermann T, Wollmann HA, Kuner R, Eggermann K, Enders H, Kaiser P, Ranke MB. 1997. Molecular studies in 37 Silver–Russell syndrome patients: Frequency and etiology of uniparental disomy. *Hum Genet* **100:** 415–419.

Eggermann T, Spengler S, Gogiel M, Begemann M, Elbracht M. 2012. Epigenetic and genetic diagnosis of Silver–Russell syndrome. *Expert Rev Mol Diagn* **12:** 459–471.

Ehrlich M. 2003. The ICF syndrome, a DNA methyltransferase 3B deficiency and immunodeficiency disease. *Clin Immunol* **109:** 17–28.

Ehrlich M, Buchanan KL, Tsien F, Jiang G, Sun B, Uicker W, Weemaes CM, Smeets D, Sperling K, Belohradsky BH, et al. 2001. DNA methyltransferase 3B mutations linked to the ICF syndrome cause dysregulation of lymphogenesis genes. *Hum Mol Genet* **10:** 2917–2931.

Eiges R, Urbach A, Malcov M, Frumkin T, Schwartz T, Amit A, Yaron Y, Eden A, Yanuka O, Benvenisty N, et al. 2007. Developmental study of fragile X syndrome using human embryonic stem cells derived from preimplantation genetically diagnosed embryos. *Cell Stem Cell* **1:** 568–577.

Engel E. 1980. A new genetic concept: Uniparental disomy and its potential effect, isodisomy. *Am J Med Genet* **6:** 137–143.

Fisher RA, Hodges MD, Rees HC, Sebire NJ, Seckl MJ, Newlands ES, Genest DR, Castrillon DH. 2002. The maternally transcribed gene p57(KIP2) (CDNK1C) is abnormally expressed in both androgenetic and biparental complete hydatidiform moles. *Hum Mol Genet* **11:** 3267–3272.

Forrester WC, Epner E, Driscoll MC, Enver T, Brice M, Papayannopoulou T, Groudine M. 1990. A deletion of the human β-globin locus activation region causes a major alteration in chromatin structure and replication across the entire β-globin locus. *Genes Dev* **4:** 1637–1649.

Fraga MF, Ballestar E, Paz MF, Ropero S, Setien F, Ballestar ML, Heine-Suner D, Cigudosa JC, Urioste M, Benitez J, et al. 2005. Epigenetic differences arise during the lifetime of monozygotic twins. *Proc Natl Acad Sci* **102:** 10604–10609.

Fyffe SL, Neul JL, Samaco RC, Chao HT, Ben-Shachar S, Moretti P, McGill BE, Goulding EH, Sullivan E, Tecott LH, et al. 2008. Deletion of Mecp2 in Sim1-expressing neurons reveals a critical role for MeCP2 in feeding behavior, aggression, and the response to stress. *Neuron* **59:** 947–958.

Gabellini D, Green MR, Tupler R. 2002. Inappropriate gene activation in FSHD: A repressor complex binds a chromosomal repeat deleted in dystrophic muscle. *Cell* **110:** 339–348.

Garg SK, Lioy DT, Cheval H, McGann JC, Bissonnette JM, Murtha MJ, Foust KD, Kaspar BK, Bird A, Mandel G. 2013. Systemic delivery of MeCP2 rescues behavioral and cellular deficits in female mouse models of Rett syndrome. *J Neurosci* **33:** 13612–13620.

Gemelli T, Berton O, Nelson ED, Perrotti LI, Jaenisch R, Monteggia LM. 2006. Postnatal loss of methyl-CpG binding protein 2 in the forebrain is sufficient to mediate behavioral aspects of Rett syndrome in mice. *Biol Psychiatry* **59:** 468–476.

Giacometti E, Luikenhuis S, Beard C, Jaenisch R. 2007. Partial rescue of MeCP2 deficiency by postnatal activation of MeCP2. *Proc Natl Acad Sci* **104:** 1931–1936.

Gibbons RJ, Picketts DJ, Villard L, Higgs DR. 1995. Mutations in a putative global transcriptional regulator cause X-linked mental retardation with α-thalassemia (ATR-X syndrome). *Cell* **80:** 837–845.

Gibbons RJ, Pellagatti A, Garrick D, Wood WG, Malik N, Ayyub H, Langford C, Boultwood J, Wainscoat JS, Higgs DR. 2003. Identification of acquired somatic mutations in the gene encoding chromatin-remodeling factor ATRX in the α-thalassemia myelodysplasia syndrome (ATMDS). *Nat Genet* **34:** 446–449.

Gicquel C, Rossignol S, Cabrol S, Houang M, Steunou V, Barbu V, Danton F, Thibaud N, Le Merrer M, Burglen L, et al. 2005. Epimutation of the telomeric imprinting center region on chromosome 11p15 in Silver–Russell syndrome. *Nat Genet* **37:** 1003–1007.

Gowher H, Jeltsch A. 2002. Molecular enzymology of the catalytic domains of the Dnmt3a and Dnmt3b DNA methyltransferases. *J Biol Chem* **277:** 20409–20414.

Grosveld F. 1999. Activation by locus control regions? *Curr Opin Genet Dev* **9:** 152–157.

Guilloteau P, Zabielski R, Hammon HM, Metges CC. 2009. Adverse effects of nutritional programming during prenatal and early postnatal life, some aspects of regulation and potential prevention and treatments. *J Physiol Pharmacol* **60:** 17–35.

Guy J, Gan J, Selfridge J, Cobb S, Bird A. 2007. Reversal of neurological defects in a mouse model of Rett syndrome. *Science* **315:** 1143–1147.

Hagberg B, Aicardi J, Dias K, Ramos O. 1983. A progressive syndrome of autism, dementia, ataxia, and loss of purposeful hand use in girls: Rett's syndrome: Report of 35 cases. *Ann Neurol* **14:** 471–479.

Hagerman PJ, Hagerman RJ. 2004. The fragile-X premutation: A maturing perspective. *Am J Hum Genet* **74:** 805–816.

Hagerman RJ, Van Housen K, Smith AC, McGavran L. 1984. Consideration of connective tissue dysfunction in the fragile X syndrome. *Am J Med Genet* **17:** 111–121.

Hahn MA, Qiu R, Wu X, Li AX, Zhang H, Wang J, Jui J, Jin SG, Jiang Y, Pfeifer GP, Lu Q. 2012. Dynamics of 5-hydroxymethylcytosine and chromatin marks in mammalian neurogenesis. *Cell Rep* **3:** 291–300.

Handa V, Saha T, Usdin K. 2003. The fragile X syndrome repeats form RNA hairpins that do not activate the interferon-inducible protein kinase, PKR, but are cut by Dicer. *Nucleic Acids Res* **31:** 6243–6248.

Hannibal MC, Buckingham KJ, Ng SB, Ming JE, Beck AE, McMillin MJ, Gildersleeve HI, Bigham AW, Tabor HK, Mefford HC, et al. 2011. Spectrum of MLL2 (ALR) mutations in 110 cases of Kabuki syndrome. *Am J Med Genet A* **155A:** 1511–1516.

Hansen RS, Wijmenga C, Luo P, Stanek AM, Canfield TK, Weemaes CM, Gartler SM. 1999. The DNMT3B DNA methyltransferase gene is mutated in the ICF immunodeficiency syndrome. *Proc Natl Acad Sci* **96:** 14412–14417.

Hansmann T, Pliushch G, Leubner M, Kroll P, Endt D, Gehrig A, Preisler-Adams S, Wieacker P, Haaf T. 2012. Constitutive promoter methylation of BRCA1 and RAD51C in patients with familial ovarian cancer and early-onset sporadic breast cancer. *Hum Mol Genet* **21:** 4669–4679.

Hark AT, Schoenherr CJ, Katz DJ, Ingram RS, Levorse JM, Tilghman SM. 2000. CTCF mediates methylation-sensitive enhancer-blocking activity at the H19/Igf2 locus. *Nature* **405:** 486–489.

Harrison CJ, Jack EM, Allen TD, Harris R. 1983. The fragile X: A scanning electron microscope study. *J Med Genet* **20:** 280–285.

Hasegawa T, Hara M, Ando M, Osawa M, Fukuyama Y, Takahashi M, Yamada K. 1984. Cytogenetic studies of familial Prader–Willi syndrome. *Hum Genet* **65:** 325–330.

Hayward BE, Kamiya M, Strain L, Moran V, Campbell R, Hayashizaki Y, Bonthron DT. 1998. The human GNAS1 gene is imprinted and encodes distinct paternally and biallelically expressed G proteins. *Proc Natl Acad Sci* **95:** 10038–10043.

Henry I, Bonaiti-Pellie C, Chehensse V, Beldjord C, Schwartz C, Utermann G, Junien C. 1991. Uniparental paternal disomy in a genetic cancer-predisposing syndrome. *Nature* **351:** 665–667.

Hoal-van Helden EG, van Helden PD. 1989. Age-related methylation changes in DNA may reflect the proliferative potential of organs. *Mutat Res* **219:** 263–266.

Ingrosso D, Cimmino A, Perna AF, Masella L, De Santo NG, De Bonis ML, Vacca M, D'Esposito M, D'Urso M, Galletti P, et al. 2003. Folate treatment and unbalanced methylation and changes of allelic expression induced by hyperhomocysteinaemia in patients with uraemia. *Lancet* **361:** 1693–1699.

Issa JP, Ottaviano YL, Celano P, Hamilton SR, Davidson NE, Baylin SB. 1994. Methylation of the oestrogen receptor CpG island links ageing and neoplasia in human colon. *Nat Genet* **7:** 536–540.

Jeanpierre M, Turleau C, Aurias A, Prieur M, Ledeist F, Fischer A, Viegas-Pequignot E. 1993. An embryonic-like methylation pattern of classical satellite DNA is observed in ICF syndrome. *Hum Mol Genet* **2:** 731–735.

Jiang YH, Armstrong D, Albrecht U, Atkins CM, Noebels JL, Eichele G, Sweatt JD, Beaudet AL. 1998. Mutation of the Angelman ubiquitin ligase in mice causes increased cytoplasmic p53 and deficits of contextual learning and long-term potentiation. *Neuron* **21:** 799–811.

Jin P, Zarnescu DC, Zhang F, Pearson CE, Lucchesi JC, Moses K, Warren ST. 2003. RNA-mediated neurodegeneration caused by the fragile X premutation rCGG repeats in *Drosophila*. *Neuron* **39:** 739–747.

Jin P, Alisch RS, Warren ST. 2004. RNA and microRNAs in fragile X mental retardation. *Nat Cell Biol* **6:** 1048–1053.

Jin B, Tao Q, Peng J, Soo HM, Wu W, Ying J, Fields CR, Delmas AL, Liu X, Qiu J, et al. 2008. DNA methyltransferase 3B (DNMT3B) mutations in ICF syndrome lead to altered epigenetic modifications and aberrant expression of genes regulating development, neurogenesis and immune function. *Hum Mol Genet* **17:** 690–709.

Jones PL, Veenstra GJ, Wade PA, Vermaak D, Kass SU, Landsberger N, Strouboulis J, Wolffe AP. 1998. Methylated DNA and MeCP2 recruit histone deacetylase to repress transcription. *Nat Genet* **19:** 187–191.

Joyce JA, Lam WK, Catchpoole DJ, Jenks P, Reik W, Maher ER, Schofield PN. 1997. Imprinting of IGF2 and H19: Lack of reciprocity in sporadic Beckwith–Wiedemann syndrome. *Hum Mol Genet* **6:** 1543–1548.

Kagami M, Sekita Y, Nishimura G, Irie M, Kato F, Okada M, Yamamori S, Kishimoto H, Nakayama M, Tanaka Y, et al. 2008. Deletions and epimutations affecting the human 14q32.2 imprinted region in individuals with paternal and maternal upd(14)-like phenotypes. *Nat Genet* **40:** 237–242.

Kajii T, Ohama K. 1977. Androgenetic origin of hydatidiform mole. *Nature* **268:** 633–634.

* Kim T-K, Hemberg M, Gray JM. 2014. Enhancer RNAs: A class of long noncoding RNAs synthesized at enhancers. *Cold Spring Harb Perspect Biol* doi: 10.1101/cshperspect.a018622.

Kioussis D, Vanin E, deLange T, Flavell RA, Grosveld FG. 1983. β-globin gene inactivation by DNA translocation in γ β-thalassaemia. *Nature* **306:** 662–666.

Kishi N, Macklis JD. 2004. MECP2 is progressively expressed in post-migratory neurons and is involved in neuronal maturation rather than cell fate decisions. *Mol Cell Neurosci* **27:** 306–321.

Kishino T, Lalande M, Wagstaff J. 1997. UBE3A/E6-AP mutations cause Angelman syndrome. *Nat Genet* **15:** 70–73.

Kondo T, Bobek MP, Kuick R, Lamb B, Zhu X, Narayan A, Bourc'his D, Viegas-Pequignot E, Ehrlich M, Hanash SM. 2000. Whole-genome methylation scan in ICF syndrome: Hypomethylation of non-satellite DNA repeats D4Z4 and NBL2. *Hum Mol Genet* **9:** 597–604.

Korenke GC, Fuchs S, Krasemann E, Doerr HG, Wilichowski E, Hunneman DH, Hanefeld F. 1996. Cerebral adrenoleukodystrophy (ALD) in only one of monozygotic twins with an identical ALD genotype. *Ann Neurol* **40:** 254–257.

* Kriaucionis S, Tahiliani M. 2014. Expanding the epigenetic landscape: Novel modifications of cytosine in genomic DNA. *Cold Spring Harb Perspect Biol* **6:** a018630.

Ledbetter DH, Riccardi VM, Airhart SD, Strobel RJ, Keenan BS, Crawford JD. 1981. Deletions of chromosome 15 as a cause of the Prader–Willi syndrome. *N Engl J Med* **304:** 325–329.

Lederer D, Grisart B, Digilio MC, Benoit V, Crespin M, Ghariani SC, Maystadt I, Dallapiccola B, Verellen-Dumoulin C. 2012. Deletion of KDM6A, a histone demethylase interacting with MLL2, in three patients with Kabuki syndrome. *Am J Hum Genet* **90:** 119–124.

Lee MP, DeBaun MR, Mitsuya K, Galonek HL, Brandenburg S, Oshimura M, Feinberg AP. 1999. Loss of imprinting of a paternally expressed transcript, with antisense orientation to KVLQT1, occurs frequently in Beckwith–Wiedemann syndrome and is independent of insulin-like growth factor II imprinting. *Proc Natl Acad Sci* **96:** 5203–5208.

Lemmers RJ, de Kievit P, Sandkuijl L, Padberg GW, van Ommen GJ, Frants RR, van der Maarel SM. 2002. Facioscapulohumeral muscular dystrophy is uniquely associated with one of the two variants of the 4q subtelomere. *Nat Genet* **32:** 235–236.

Lewis JD, Meehan RR, Henzel WJ, Maurer-Fogy I, Jeppesen P, Klein F, Bird A. 1992. Purification, sequence, and cellular localization of a novel chromosomal protein that binds to methylated DNA. *Cell* **69:** 905–914.

* Li E, Zhang Y. 2014. DNA methylation in mammals. *Cold Spring Harb Perspect Biol* **6:** a01933.

Lioy DT, Garg SK, Monaghan CE, Raber J, Foust KD, Kaspar BK, Hirrlinger PG, Kirchhoff F, Bissonnette JM, Ballas N, et al. 2011. A role for glia in the progression of Rett's syndrome. *Nature* **475:** 497–500.

Liu D, Diorio J, Tannenbaum B, Caldji C, Francis D, Freedman A, Sharma S, Pearson D, Plotsky PM, Meaney MJ. 1997. Maternal care, hippocampal glucocorticoid receptors, and hypothalamic-pituitary-adrenal responses to stress. *Science* **277:** 1659–1662.

⋆ Lomvardas S, Maniatis T. 2014. Histone and DNA modifications as regulators of neuronal development and function. *Cold Spring Harb Perspect Biol* doi: 10.1101/cshperspect. a019489.

Lubs HA. 1969. A marker X chromosome. *Am J Hum Genet* **21:** 231–244.

Ludwig M, Katalinic A, Gross S, Sutcliffe A, Varon R, Horsthemke B. 2005. Increased prevalence of imprinting defects in patients with Angelman syndrome born to subfertile couples. *J Med Genet* **42:** 289–291.

Magenis RE, Brown MG, Lacy DA, Budden S, LaFranchi S. 1987. Is Angelman syndrome an alternate result of del(15)(q11q13)? *Am J Med Genet* **28:** 829–838.

Martin J, Bell J. 1943. A pedigree of mental defect showing sex-linkage. *Arch Neurol Psychiat* **6:** 154–157.

Mathews KD. 2003. Muscular dystrophy overview: Genetics and diagnosis. *Neurol Clin* **21:** 795–816.

Matsuura T, Sutcliffe JS, Fang P, Galjaard RJ, Jiang YH, Benton CS, Rommens JM, Beaudet AL. 1997. De novo truncating mutations in E6-AP ubiquitin-protein ligase gene (UBE3A) in Angelman syndrome. *Nat Genet* **15:** 74–77.

Mays-Hoopes LL. 1989. Age-related changes in DNA methylation: Do they represent continued developmental changes? *Int Rev Cytol* **114:** 181–220.

McDowell TL, Gibbons RJ, Sutherland H, O'Rourke DM, Bickmore WA, Pombo A, Turley H, Gatter K, Picketts DJ, Buckle VJ, et al. 1999. Localization of a putative transcriptional regulator (ATRx) at pericentromeric heterochromatin and the short arms of acrocentric chromosomes. *Proc Natl Acad Sci* **96:** 13983–13988.

McGraw CM, Samaco RC, Zoghbi HY. 2011. Adult neural function requires MeCP2. *Science* **333:** 186.

Meins M, Lehmann J, Gerresheim F, Herchenbach J, Hagedorn M, Hameister K, Epplen JT. 2005. Submicroscopic duplication in Xq28 causes increased expression of the MECP2 gene in a boy with severe mental retardation and features of Rett syndrome. *J Med Genet* **42:** e12.

Mellén M, Ayata P, Dewell S, Kriaucionis S, Heintz N. 2012. MeCP2 binds to 5hmC enriched within active genes and assessible chromatin in the nervous system. *Cell* **151:** 1417–1430.

Meloni I, Bruttini M, Longo I, Mari F, Rizzolio F, D'Adamo P, Denvriendt K, Fryns JP, Toniolo D, Renieri A. 2000. A mutation in the Rett syndrome gene, MECP2, causes X-linked mental retardation and progressive spasticity in males. *Am J Hum Genet* **67:** 982–985.

Meng L, Person RE, Beaudet AL. 2012. Ube3a-ATS is an atypical RNA polymerase II transcript that represses the paternal expression of Ube3a. *Hum Mol Genet* **21:** 3001–3012.

Meng L, Person RE, Huang W, Zhu PJ, Costa-Mattioli M, Beaudet AL. 2013. Truncation of Ube3a-ATS unsilences paternal Ube3a and ameliorates behavioral defects in the Angelman syndrome mouse model. *PLoS Genet* **9:** e1004039.

Mullaney BC, Johnston MV, Blue ME. 2004. Developmental expression of methyl-CpG binding protein 2 is dynamically regulated in the rodent brain. *Neuroscience* **123:** 939–949.

Murata T, Kurokawa R, Krones A, Tatsumi K, Ishii M, Taki T, Masuno M, Ohashi H, Yanagisawa M, Rosenfeld MG, et al. 2001. Defect of histone acetyltransferase activity of the nuclear transcriptional coactivator CBP in Rubinstein–Taybi syndrome. *Hum Mol Genet* **10:** 1071–1076.

Murdoch S, Djuric U, Mazhar B, Seoud M, Khan R, Kuick R, Bagga R, Kircheisen R, Ao A, Ratti B, et al. 2006. Mutations in NALP7 cause recurrent hydatidiform moles and reproductive wastage in humans. *Nat Genet* **38:** 300–302.

Nan X, Campoy FJ, Bird A. 1997. MeCP2 is a transcriptional repressor with abundant binding sites in genomic chromatin. *Cell* **88:** 471–481.

Nan X, Ng HH, Johnson CA, Laherty CD, Turner BM, Eisenman RN, Bird A. 1998. Transcriptional repression by the methyl-CpG-binding protein MeCP2 involves a histone deacetylase complex. *Nature* **393:** 386–389.

Ng SB, Bigham AW, Buckingham KJ, Hannibal MC, McMillin MJ, Gildersleeve HI, Beck AE, Tabor HK, Cooper GM, Mefford HC, et al. 2010. Exome sequencing identifies MLL2 mutations as a cause of Kabuki syndrome. *Nat Genet* **42:** 790–793.

Nicholls RD, Knoll JHM, Butler MG, Karam S, Lalande M. 1989. Genetic imprinting suggested by maternal heterodisomy in non-deletion Prader–Willi syndrome. *Nature* **342:** 281–285.

Ogryzko VV, Schiltz RL, Russanova V, Howard BH, Nakatani Y. 1996. The transcriptional coactivators p300 and CBP are histone acetyltransferases. *Cell* **87:** 953–959.

Ohta T, Gray TA, Rogan PK, Buiting K, Gabriel JM, Saitoh S, Muralidhar B, Bilienska B, Krajewska-Walasek M, Driscoll DJ, et al. 1999. Imprinting-mutation mechanisms in Prader–Willi syndrome. *Am J Hum Genet* **64:** 397–413.

Okano M, Bell DW, Haber DA, Li E. 1999. DNA methyltransferases Dnmt3a and Dnmt3b are essential for de novo methylation and mammalian development. *Cell* **99:** 247–257.

Orstavik KH, Eiklid K, van der Hagen CB, Spetalen S, Kierulf K, Skjeldal O, Buiting K. 2003. Another case of imprinting defect in a girl with Angelman syndrome who was conceived by intracytoplasmic semen injection. *Am J Hum Genet* **72:** 218–219.

Parry DA, Logan CV, Hayward BE, Shires M, Landolsi H, Diggle C, Carr I, Rittore C, Touitou I, Philibert L, et al. 2011. Mutations causing familial biparental hydatidiform mole implicate c6orf221 as a possible regulator of genomic imprinting in the human oocyte. *Am J Hum Genet* **89:** 451–458.

Petrij F, Giles RH, Dauwerse HG, Saris JJ, Hennekam RC, Masuno M, Tommerup N, van Ommen GJ, Goodman RH, Peters DJ, et al. 1995. Rubinstein–Taybi syndrome caused by mutations in the transcriptional co-activator CBP. *Nature* **376:** 348–351.

Petronis A. 2004. The origin of schizophrenia: Genetic thesis, epigenetic antithesis, and resolving synthesis. *Biol Psychiatry* **55:** 965–970.

Picketts DJ, Higgs DR, Bachoo S, Blake DJ, Quarrell OW, Gibbons RJ. 1996. ATRX encodes a novel member of the SNF2 family of proteins: Mutations point to a common mechanism underlying the ATR-X syndrome. *Hum Mol Genet* **5:** 1899–1907.

Pieretti M, Zhang F, Fu Y-H, Warren ST, Oostra BA, Caskey CT, Nelson DL. 1991. Absence of expression of the *FMR-1* gene in fragile X syndrome. *Cell* **66:** 817–822.

Pineda M, Mur P, Iniesta MD, Borras E, Campos O, Vargas G, Iglesias S, Fernandez A, Gruber SB, Lazaro C, et al. 2012. MLH1 methylation screening is effective in identifying epimutation carriers. *Eur J Hum Genet* **20:** 1256–1264.

Ping AJ, Reeve AE, Law DJ, Young MR, Boehnke M, Feinberg AP. 1989. Genetic linkage of Beckwith–Wiedemann syndrome to 11p15. *Am J Hum Genet* **44:** 720–723.

Pirrotta V. 2014. The necessity of chromatin: A view in perspective. *Cold Spring Harb Perspect Biol* doi: 10.1101/cshperspect.a019547.

Prawitt D, Enklaar T, Gartner-Rupprecht B, Spangenberg C, Oswald M, Lausch E, Schmidtke P, Reutzel D, Fees S, Lucito R, et al. 2005. Microdeletion of target sites for insulator protein CTCF in a chromosome 11p15 imprinting center in Beckwith–Wiedemann syndrome and Wilms' tumor. *Proc Natl Acad Sci* **102:** 4085–4090.

Ramocki MB, Peters SU, Tavyev YJ, Zhang F, Carvalho CM, Schaaf CP, Richman R, Fang P, Glaze DG, Lupski JR, et al. 2009. Autism and other neuropsychiatric symptoms are prevalent in individuals with MeCP2 duplication syndrome. *Ann Neurol* **66:** 771–782.

Rampersaud GC, Kauwell GP, Hutson AD, Cerda JJ, Bailey LB. 2000. Genomic DNA methylation decreases in response to moderate folate depletion in elderly women. *Am J Clin Nutr* **72:** 998–1003.

Reik W. 1989. Genomic imprinting and genetic disorders in man. *Trends Genet* **5:** 331–336.

Reynolds EH, Carney MW, Toone BK. 1984. Methylation and mood. *Lancet* **2:** 196–198.

Richards BW, Sylvester PE, Brooker C. 1981. Fragile X-linked mental retardation: The Martin-Bell syndrome. *J Ment Defic Res* **25:** 253–256.

⋆ Rinn JL. 2014. lncRNAs: Linking RNA to chromatin. *Cold Spring Harb Perspect Biol* **6:** a018614.

Robinson L, Guy J, McKay L, Brockett E, Spike RC, Selfridge J, De Sousa D, Merusi C, Riedel G, Bird A, et al. 2012. Morphological and functional reversal of phenotypes in a mouse model of Rett syndrome. *Brain* **135:** 2699–2710.

Roelfsema JH, White SJ, Ariyurek Y, Bartholdi D, Niedrist D, Papadia F, Bacino CA, den Dunnen JT, van Ommen GJ, Breuning MH, et al. 2005. Genetic heterogeneity in Rubinstein–Taybi syndrome: Mutations in both the CBP and EP300 genes cause disease. *Am J Hum Genet* **76:** 572–580.

Rougeulle C, Cardoso C, Fontes M, Colleaux L, Lalande M. 1998. An imprinted antisense RNA overlaps UBE3A and a second maternally expressed transcript. *Nat Genet* **19:** 15–16.

Sahoo T, del Gaudio D, German JR, Shinawi M, Peters SU, Person RE, Garnica A, Cheung SW, Beaudet AL. 2008. Prader–Willi phenotype caused by paternal deficiency for the HBII-85 C/D box small nucleolar RNA cluster. *Nat Genet* **40:** 719–721.

Samaco RC, Mandel-Brehm C, Chao HT, Ward CS, Fyffe-Maricich SL, Ren J, Hyland K, Thaller C, Maricich SM, Humphreys P, et al. 2009. Loss of MeCP2 in aminergic neurons causes cell-autonomous defects in neurotransmitter synthesis and specific behavioral abnormalities. *Proc Natl Acad Sci* **106:** 21966–21971.

Schaaf CP, Gonzalez-Garay ML, Xia F, Potocki L, Gripp KW, Zhang B, Peters BA, McElwain MA, Drmanac R, Beaudet AL, et al. 2013. Truncating mutations of MAGEL2 cause Prader–Willi phenotypes and autism. *Nat Genet* **45:** 1405–1408.

Schimke RN, Horton WA, King CR. 1971. Chondroitin-6-sulphaturia, defective cellular immunity, and nephrotic syndrome. *Lancet* **2:** 1088–1089.

Schule B, Albalwi M, Northrop E, Francis DI, Rowell M, Slater HR, Gardner RJ, Francke U. 2005. Molecular breakpoint cloning and gene expression studies of a novel translocation t(4;15)(q27;q11.2) associated with Prader–Willi syndrome. *BMC Med Genet* **6:** 18.

Sebert S, Sharkey D, Budge H, Symonds ME. 2011. The early programming of metabolic health: Is epigenetic setting the missing link? *Am J Clin Nutr* **94:** 1953S–1958S.

Shahbazian MD, Antalffy B, Armstrong DL, Zoghbi HY. 2002. Insight into Rett syndrome: MeCP2 levels display tissue- and cell-specific differences and correlate with neuronal maturation. *Hum Mol Genet* **11:** 115–124.

Sheridan SD, Theriault KM, Reis SA, Zhou F, Madison JM, Daheron L, Loring JF, Haggarty SJ. 2011. Epigenetic characterization of the FMR1 gene and aberrant neurodevelopment in human induced pluripotent stem cell models of fragile X syndrome. *PLoS One* **6:** e26203.

Skene PJ, Illingworth RS, Webb S, Kerr AR, James KD, Turner DJ, Andrews R, Bird AP. 2010. Neuronal MeCP2 is expressed at near histone-octamer levels and globally alters the chromatin state. *Mol Cell* **37:** 457–468.

Smeets DF, Hamel BC, Nelen MR, Smeets HJ, Bollen JH, Smits AP, Ropers HH, van Oost BA. 1992. Prader–Willi syndrome and Angelman syndrome in cousins from a family with a translocation between chromosomes 6 and 15. *N Engl J Med* **326:** 807–811.

Smeets DF, Moog U, Weemaes CM, Vaes-Peeters G, Merkx GF, Niehof JP, Hamers G. 1994. ICF syndrome: A new case and review of the literature. *Hum Genet* **94:** 240–246.

Smilinich NJ, Day CD, Fitzpatrick GV, Caldwell GM, Lossie AC, Cooper PR, Smallwood AC, Joyce JA, Schofield PN, Reik W, et al. 1999. A maternally methylated CpG island in KvLQT1 is associated with an antisense paternal transcript and loss of imprinting in Beckwith–Wiedemann syndrome. *Proc Natl Acad Sci* **96:** 8064–8069.

Spence JE, Perciaccante RG, Greig GM, Willard HF, Ledbetter DH, Hejtmancik JF, Pollack MS, O'Brien WE, Beaudet AL. 1988. Uniparental disomy as a mechanism for human genetic disease. *Am J Hum Genet* **42:** 217–226.

Spranger J, Hinkel GK, Stoss H, Thoenes W, Wargowski D, Zepp F. 1991. Schimke immuno-osseous dysplasia: A newly recognized multisystem disease. *J Pediatr* **119:** 64–72.

Sun FL, Dean WL, Kelsey G, Allen ND, Reik W. 1997. Transactivation of Igf2 in a mouse model of Beckwith–Wiedemann syndrome. *Nature* **389:** 809–815.

Suter CM, Martin DI, Ward RL. 2004. Germline epimutation of MLH1 in individuals with multiple cancers. *Nat Genet* **36:** 497–501.

Sutherland GR. 1977. Fragile sites on human chromosomes: Demonstration of their dependence on the type of tissue culture medium. *Science* **197:** 265–266.

Taha D, Boerkoel CF, Balfe JW, Khalifah M, Sloan EA, Barbar M, Haider A, Kanaan H. 2004. Fatal lymphoproliferative disorder in a child with Schimke immuno-osseous dysplasia. *Am J Med Genet A* **131:** 194–199.

Temple IK, Shield JP. 2002. Transient neonatal diabetes, a disorder of imprinting. *J Med Genet* **39:** 872–875.

Tommerup N, van der Hagen CB, Heiberg A. 1992. Tentative assignment of a locus for Rubinstein–Taybi syndrome to 16p13.3 by a de novo reciprocal translocation, t(7;16)(q34;p13.3). *Am J Med Genet* **44:** 237–241.

Toyoda S, Kawaguchi M, Kobayashi T, Tarusawa E, Toyama T, Okano M, Oda M, Nakauchi H, Yoshimura Y, Sanbo M, et al. 2014. Developmental epigenetic modification regulates stochastic expression of clustered protocadherin genes, generating single neuron diversity. *Neuron* **82:** 94–108.

Tsai TF, Jiang YH, Bressler J, Armstrong D, Beaudet AL. 1999. Paternal deletion from Snrpn to Ube3a in the mouse causes hypotonia, growth retardation and partial lethality and provides evidence for a gene contributing to Prader–Willi syndrome. *Hum Mol Genet* **8:** 1357–1364.

Tsujimura K, Abematsu M, Kohyama J, Namihira M, Nakashima K. 2009. Neuronal differentiation of neural precursor cells is promoted by the methyl-CpG-binding protein MeCP2. *Exp Neurol* **219:** 104–111.

Tuck-Muller CM, Narayan A, Tsien F, Smeets DF, Sawyer J, Fiala ES, Sohn OS, Ehrlich M. 2000. DNA hypomethylation and unusual chromosome instability in cell lines from ICF syndrome patients. *Cytogenet Cell Genet* **89:** 121–128.

Tupler R, Gabellini D. 2004. Molecular basis of facioscapulohumeral muscular dystrophy. *Cell Mol Life Sci* **61:** 557–566.

van Deutekom JC, Wijmenga C, van Tienhoven EA, Gruter AM, Hewitt JE, Padberg GW, van Ommen GJ, Hofker MH, Frants RR. 1993. FSHD associated DNA rearrangements are due to deletions of integral copies of a 3.2 kb tandemly repeated unit. *Hum Mol Genet* **2:** 2037–2042.

Van Esch H, Bauters M, Ignatius J, Jansen M, Raynaud M, Hollanders K, Lugtenberg D, Bienvenu T, Jensen LR, Gecz J, et al. 2005. Duplication of the MECP2 region is a frequent cause of severe mental retardation and progressive neurological symptoms in males. *Am J Hum Genet* **77:** 442–453.

Verkerk AJMH, Pieretti M, Sutcliffe JS, Fu Y-H, Kuhl DPA, Pizutti A, Reiner O, Richards S, Victoria MF, Zhang R, et al. 1991. Identification of a gene (FMR-1) containing a CGG repeat coincident with a breakpoint cluster region exhibiting length variation in fragile X syndrome. *Cell* **65:** 905–914.

Villard L, Gecz J, Mattei JF, Fontes M, Saugier-Veber P, Munnich A, Lyonnet S. 1996. XNP mutation in a large family with Juberg-Marsidi syndrome. *Nat Genet* **12:** 359–360.

Wan M, Lee SS, Zhang X, Houwink-Manville I, Song HR, Amir RE, Budden S, Naidu S, Pereira JL, Lo IF, et al. 1999. Rett syndrome and beyond: Recurrent spontaneous and familial MECP2 mutations at CpG hotspots. *Am J Hum Genet* **65:** 1520–1529.

Ward CS, Arvide EM, Huang TW, Yoo J, Noebels JL, Neul JL. 2011. MeCP2 is critical within HoxB1-derived tissues of mice for normal lifespan. *J Neurosci* **31:** 10359–10370.

Ward RL, Dobbins T, Lindor NM, Rapkins RW, Hitchins MP. 2012. Identification of constitutional MLH1 epimutations and promoter variants in colorectal cancer patients from the Colon Cancer Family Registry. *Genet Med* **15:** 25–35.

Warren ST, Sherman SL. 2001. *The fragile X syndrome.* McGraw-Hill, New York.

Waterland RA, Jirtle RL. 2003. Transposable elements: Targets for early nutritional effects on epigenetic gene regulation. *Mol Cell Biol* **23:** 5293–5300.

Weatherall DJ, Clegg MB, Higgs DR, Wood WG. 2001. The hemoglobinopathies. In *The metabolic and molecular bases of inherited disease* (ed. Scriver CR, Beaudet AL, Sly WS, et al.), pp. 4571–4636. McGraw-Hill, New York.

Weaver IC, Cervoni N, Champagne FA, D'Alessio AC, Sharma S, Seckl JR, Dymov S, Szyf M, Meaney MJ. 2004. Epigenetic programming by maternal behavior. *Nat Neurosci* **7:** 847–854.

Weksberg R, Teshima I, Williams BR, Greenberg CR, Pueschel SM, Chernos JE, Fowlow SB, Hoyme E, Anderson IJ, Whiteman DA, et al. 1993. Molecular characterization of cytogenetic alterations associated with the Beckwith–Wiedemann syndrome (BWS) phenotype refines the localization and suggests the gene for BWS is imprinted. *Hum Mol Genet* **2:** 549–556.

Weksberg R, Nishikawa J, Caluseriu O, Fei YL, Shuman C, Wei C, Steele L, Cameron J, Smith A, Ambus I, et al. 2001. Tumor development in the Beckwith–Wiedemann syndrome is associated with a variety of constitutional molecular 11p15 alterations including imprinting defects of KCNQ1OT1. *Hum Mol Genet* **10:** 2989–3000.

Weksberg R, Smith AC, Squire J, Sadowski P. 2003. Beckwith–Wiedemann syndrome demonstrates a role for epigenetic control of normal development. *Hum Mol Genet* **12:** R61–R68.

Wijmenga C, Hewitt JE, Sandkuijl LA, Clark LN, Wright TJ, Dauwerse HG, Gruter AM, Hofker MH, Moerer P, Williamson R, et al. 1992. Chromosome 4q DNA rearrangements associated with facioscapulohumeral muscular dystrophy. *Nat Genet* **2:** 26–30.

Willemsen R, Hoogeveen-Westerveld M, Reis S, Holstege J, Severijnen LA, Nieuwenhuizen IM, Schrier M, Van Unen L, Tassone F, Hoogeveen AT, et al. 2003. The FMR1 CGG repeat mouse displays ubiquitin-positive intranuclear neuronal inclusions; implications for the cerebellar tremor/ataxia syndrome. *Hum Mol Genet* **12:** 949–959.

Wolff GL, Kodell RL, Moore SR, Cooney CA. 1998. Maternal epigenetics and methyl supplements affect agouti gene expression in Avy/a mice. *Faseb J* **12:** 949–957.

Xu GL, Bestor TH, Bourc'his D, Hsieh CL, Tommerup N, Bugge M, Hulten M, Qu X, Russo JJ, Viegas-Pequignot E. 1999. Chromosome instability and immunodeficiency syndrome caused by mutations in a DNA methyltransferase gene. *Nature* **402:** 187–191.

Yasui DH, Peddada S, Bieda MC, Vallero RO, Hogart A, Nagarajan RP, Thatcher KN, Farnham PJ, Lasalle JM. 2007. Integrated epigenomic analyses of neuronal MeCP2 reveal a role for long-range interaction with active genes. *Proc Natl Acad Sci* **104:** 19416–19421.

Yntema HG, Poppelaars FA, Derksen E, Oudakker AR, van Roosmalen T, Jacobs A, Obbema H, Brunner HG, Hamel BC, van Bokhoven H. 2002. Expanding phenotype of XNP mutations: Mild to moderate mental retardation. *Am J Med Genet* **110:** 243–247.

Zeev BB, Yaron Y, Schanen NC, Wolf H, Brandt N, Ginot N, Shomrat R, Orr-Urtreger A. 2002. Rett syndrome: Clinical manifestations in males with MECP2 mutations. *J Child Neurol* **17:** 20–24.

* Zhao Y, Garcia BA. 2014. Comprehensive catalog of currently documented histone modifications. *Cold Spring Harb Perspect Biol* doi: 10.1101/cshperspect.a025064.

CHAPTER 34

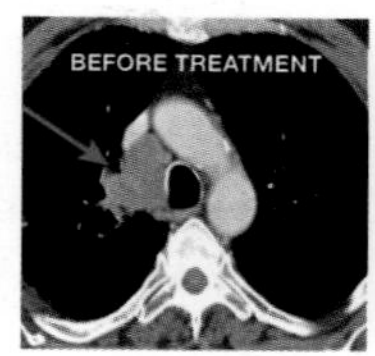

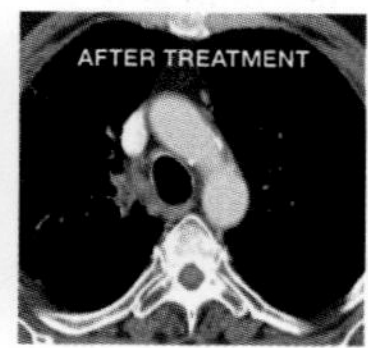

Epigenetic Determinants of Cancer

Stephen B. Baylin[1] and Peter A. Jones[2]

[1]Cancer Biology Program, Johns Hopkins University, School of Medicine, Baltimore, Maryland 21287; [2]Van Andel Research Institute, Grand Rapids, Michigan 49503

Correspondence: sbaylin@jhmi.edu

SUMMARY

Epigenetic changes are present in all human cancers and are now known to cooperate with genetic alterations to drive the cancer phenotype. These changes involve DNA methylation, histone modifiers and readers, chromatin remodelers, microRNAs, and other components of chromatin. Cancer genetics and epigenetics are inextricably linked in generating the malignant phenotype; epigenetic changes can cause mutations in genes, and, conversely, mutations are frequently observed in genes that modify the epigenome. Epigenetic therapies, in which the goal is to reverse these changes, are now one standard of care for a preleukemic disorder and form of lymphoma. The application of epigenetic therapies in the treatment of solid tumors is also emerging as a viable therapeutic route.

Outline

Chapter-opening figures: (*Left, right*) Adapted from Juergens et al. (2011), with permission from American Association of Cancer Research.

Cite as *Cold Spring Harb Perspect Biol* doi: 10.1101/cshperspect.a019505

OVERVIEW

Cancer is caused by the somatically heritable deregulation of genes that control the processes governing when cells divide, die, and move from one part of the body to another. During carcinogenesis, genes can become activated in such a way that enhances division or prevents cell death (oncogene). Alternatively, genes can become inactivated so that they are no longer available to apply the brakes to these processes (tumor-suppressor gene). It is the interplay between these two classes of genes that results in the formation of cancer.

Tumor-suppressor genes (TSGs) can become inactivated by at least three pathways: (1) through mutations, in which their functions become disabled; (2) a gene can be completely lost and thus not be available to work appropriately (loss of heterozygosity); and (3) a gene can be switched off in a somatically heritable fashion by epigenetic changes, rather than by mutation of the DNA sequence. Epigenetic silencing can occur by deregulation of the epigenetic machinery at several different levels; it may involve inappropriate methylation of cytosine (C) residues in CpG sequence motifs that reside within control regions governing gene expression. Also, changes to histone posttranslational modifications (PTMs) or aberrations in the way histone-modifying enzymes function may occur. A change in a protein's ability to read histone marks, and hence bind to chromatin, or alterations in the way nucleosome-remodeling or histone exchange complexes function can result. Finally, changes in regulatory microRNA (miRNA) expression patterns have been noted.

This chapter focuses predominantly on how cancer is affected by this third pathway (i.e., epigenetic mechanisms). The basic molecular mechanisms responsible for maintaining the silenced state are quite well understood, as outlined in this collection. Consequently, we also know that epigenetic silencing has profound implications for cancer prevention, detection, and therapies. We now have drugs approved by the U.S. Food and Drug Administration (FDA) that are used to reverse epigenetic changes and restore gene activity to cancer cells. Also, because changes in DNA methylation can be detected with a high degree of sensitivity, many strategies are able to detect cancer early by finding changes in DNA methylation. The translational opportunities for epigenetics in human cancer research, detection, prevention, and treatment are, therefore, quite extraordinary.

Cite as *Cold Spring Harb Perspect Biol* doi: 10.1101/cshperspect.a019505

1 THE BIOLOGICAL BASIS OF CANCER

Cancer is ultimately a disease of gene expression in which the complex networks governing homeostasis in multicellular organisms become deranged, allowing cells to grow without reference to the needs of the organism as a whole. Great advancements have been made in delineating the subset of cellular control pathways subject to derangement in human cancer (Table 1). The realization that distinct sets of cellular control pathways are affected and heritably disabled in almost all cancers is a key concept that has advanced the field (Hanahan and Weinberg 2011). Historically, research has focused on the genetic basis of cancer, particularly, in terms of how mutational activation of oncogenes or inactivation of tumor-suppressor genes (TSGs) underpins these above pathway changes. However, since the 1990s, a growing research endeavor has centered on the recognition that heritable changes, regulated by epigenetic alterations, may also be critical for the evolution of all human cancer types (Baylin and Jones 2011).

Epigenetic alterations can be observed as abnormal patterns of DNA methylation, disrupted patterns of histone posttranslational modifications (PTMs), and changes in chromatin composition and/or organization. Changes in the epigenome largely occur through disrupting the epigenetic machinery, and Figure 1 illustrates the different elements of the epigenetic machinery that are now known to be perturbed in cancer. These epigenomic changes not only are associated with altered patterns of expression for otherwise wild-type genes, but, in some cases, may also be causal to their changed expression state. The recognition of an epigenetic component in tumorigenesis, or the existence of a cancer "epigenome," has led to new opportunities for the understanding, detection, treatment, and prevention of cancer.

Table 1. Examples of key cellular pathways disrupted in human cancers by genetic and epigenetic mechanisms

Pathway	Example of genetic alteration	Example of epigenetic alteration
Self sufficiency in growth signals	Mutations in *Ras* gene	Methylation of *RASSFIA* gene
Insensitivity to antigrowth signals	Mutation in TGF-β receptors	Down-regulation of TGF-β receptors
Tissue invasion and metastasis	Mutation in *E-cadherin* gene	Methylation of *E-cadherin* promoter
Limitless replicative potential	Mutation in *p16* and pRb genes	Silencing of *p16* or pRb genes by promoter methylation
Sustained angiogenesis		Silencing of thrombospondin-1
Evading apoptosis	Mutation in p53	Methylation of *DAPK*, ASC/TMS1, and HIC1
DNA repair capacity	Mutations in MLH1, MSH2	Methylation of *GST Pi*, *O6-MGMT*, *MLH1*
Monitoring genomic stability	Mutations in *Chfr*	Methylation of *Chfr*
Protein ubiquination functions regulating mitotic control genes	Mutations in *Chfr*	Methylation of *Chfr*

TGF-β, transforming growth factor β; *DAPK*, death-associated protein kinase.

Signaling gene (oncogene) mutations in many human cancers are often dominant and drive the formation of cancers. An example would be *ras*, which when mutated, enhances the activity of the gene product to stimulate growth. Genetic mutations or epigenetic silencing of TSGs, on the other hand, are often recessive, requiring disruptive events in both allelic copies of a gene for the full expression of the transformed phenotype. The idea that both copies of a TSG have to be incapacitated in a malignant cell line was proposed by Knudson (2001) in his "two- or multiple-hit" hypothesis and has found wide acceptance. It is now realized that three classes of "hits" can participate in different combinations to cause a complete loss of activity of TSGs. Direct mutations in the coding sequence may occur, loss of parts or entire copies of genes, or epigenetic silencing, can cooperate with each other to result in the disablement of key control genes. Another growing concept discussed in this chapter is that there is an intense cooperation between genetic and epigenetic abnormalities to drive the initiation and progression of cancer (Fig. 1) (Baylin and Jones 2011; You and Jones 2012; Garraway and Lander 2013; Shen and Laird 2013). Most recently, excitement has centered on the realization that most cancers actually harbor frequent mutations in genes that encode for components of the epigenetic machinery, potentially resulting in abnormalities in the epigenome, which may affect gene expression patterns and genomic stability (Baylin and Jones 2011; You and Jones 2012; Garraway and Lander 2013; Shen and Laird 2013). Some of the growing list of genes frequently mutated in cancer, encoding proteins central to establishing normal control of chromatin and DNA methylation patterns, are illustrated in Figure 1 and more exhaustively, listed in Table 2 or the Appendices of Ch. 35 (Audia and Campbell 2014) (Baylin and Jones 2011; You and Jones 2012; Garraway and Lander 2013; Shen and Laird 2013). Although most of the consequences of these mutations remain to be elucidated, this concept is critical not only for understanding the biology of cancer, but also for implications regarding cancer therapy. Conversely, epigenetic silencing or activation of genes may predispose cells to further mutations (e.g., the epigenetic silencing of the key MLH1 DNA repair protein leads to new mutations because of a lack of efficient DNA repair). Other chapters in this collection provide details concerning our understanding of

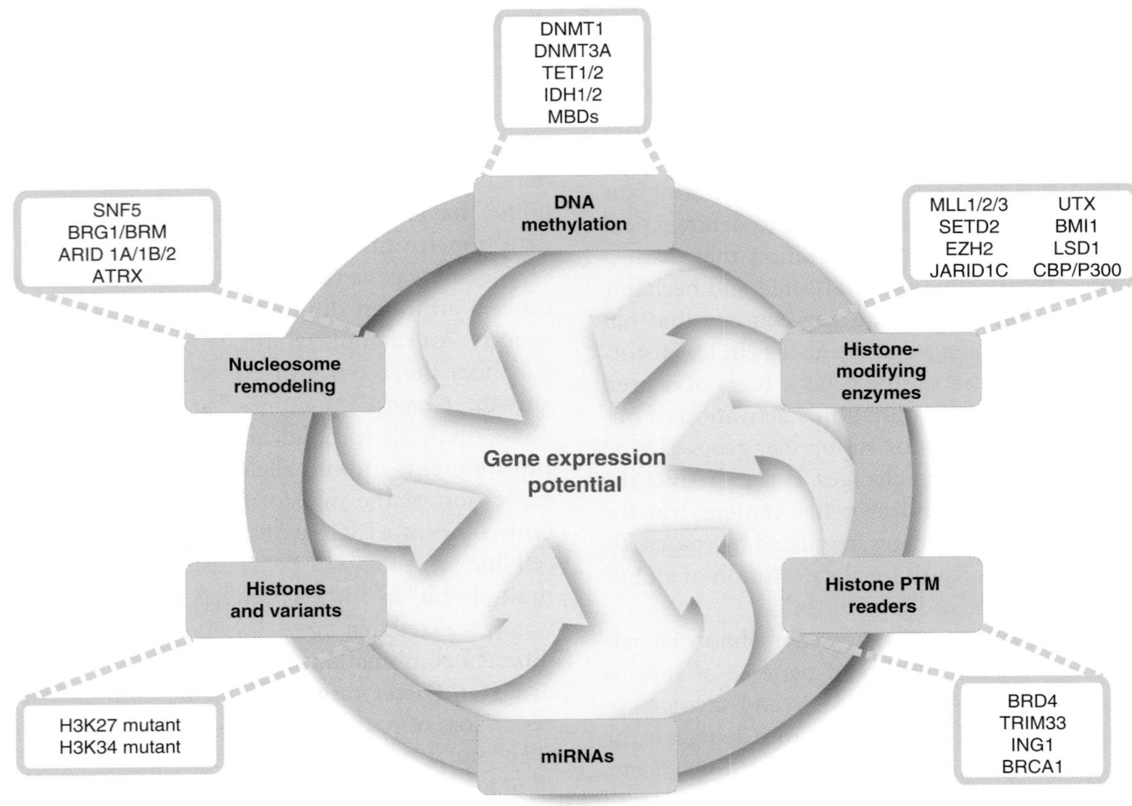

Figure 1. Genetic mutations of epigenetic modifiers in cancer. The drawing shows the input of epigenetic processes in specifying gene expression patterns. Recent whole-exome sequencing studies show that mutations in various classes of epigenetic modifiers are frequently observed in many types of cancers, further highlighting the cross talk between genetics and epigenetics. Examples of some, but not all, of these mutations are illustrated here and listed in Table 2. The mutations of epigenetic modifiers potentially cause genome-wide epigenetic alterations in cancer, but, save for isocitrate dehydrogenase (IDH) mutations as discussed in the text, these have yet to be shown on a genome-wide scale. Understanding the relationship of genetic and epigenetic changes in cancer will offer novel insights for cancer therapies. MBDs, methylcytosine-binding proteins; PTM, posttranslational modification. (Adapted from You and Jones 2012.)

how the various epigenetic processes contribute to regulating the genome and can become deregulated in cancer.

2 THE IMPORTANCE OF CHROMATIN TO CANCER

Despite the major advances in understanding the key molecular lesions in cellular control pathways that contribute to cancer, it is true that microscopic examination of nuclear structure by a pathologist remains a gold standard in cancer diagnosis. The human eye can accurately discern changes in nuclear architecture, which largely involve the state of chromatin configuration, and definitively diagnose the cancer phenotype in a single cell. Foremost in the cues used by pathologists are the size of the nucleus, nuclear outline, a condensed nuclear membrane, prominent nucleoli, dense "hyperchromatic" chromatin, and a high nuclear/cytoplasmic ratio. These structural features, visible under a microscope (Fig. 2), likely correlate with profound alterations in chromatin structure and function, with resultant changes in gene expression states and/or chromosome stability. Linking changes observable at a microscopic level with the molecular marks discussed throughout this collection remains one of the great challenges in cancer research. In this chapter, we review epigenetic marks that are abnormally distributed in cancer cells, typified by changes in DNA cytosine methylation at CpG dinucleotides, changes in histone modifications, nucleosomal composition (i.e.,

Table 2. Mutations in selected epigenetic modifiers in human cancers

Process	Gene	Function	Tumor type	Alteration
DNA methylation	*DNMT1*	DNA methyltransferase	Colorectal, Non–small cell lung, pancreatic, gastric, breast cancer	Mutation (Kanai et al. 2003) Overexpression (Wu et al. 2007)
	DNMT3A	DNA methyltransferase	MDS; AML	Mutation (Ley et al. 2010; Yamashita et al. 2010; Yan et al. 2011)
	DNMT3B	DNA methyltransferase	ICF syndrome, SNPs in breast and lung adenoma	Mutation (Wijmenga et al. 2000) Mutation (Shen et al. 2002)
	MBD1/2	Methyl-binding protein	Lung and breast cancer	Mutation (Sansom et al. 2007)
	TET1	5′-Methylcytosine hydroxylase	AML	Chromosome translocation (De Carvalho et al. 2010; Wu and Zhang 2010)
	TET2	5′-Methylcytosine hydroxylase	MDS, myeloid malignancies, gliomas	Mutation/silencing (Araki et al. 2009)
	IDH1/2	Isocitrate dehydrogenase	Glioma, AML	Mutation (Figueroa et al. 2010; Lu et al. 2012; Turcan et al. 2012)
	AID	5′-Cytidine deaminase	CML	Aberrant expression (De Carvalho et al. 2010)
	MLL1/2/3	Histone methyltransferase H3K4	Bladder TCC, hematopoietic, non-Hodgkin lymphoma, B-cell lymphoma, prostate (primary)	Translocation, mutation, aberrant expression (Gui et al. 2011; Morin et al. 2011)
Histone modification enzymes	*EZH2*	Histone methyltransferase H3K27	Breast, prostate, bladder, colon, pancreas, liver, gastric, uterine tumors, melanoma, lymphoma, myeloma, and Ewing's sarcoma	Mutation, aberrant expression (Chase and Cross 2011; Tsang and Cheng 2011)
	BMI-1	PRC1 subunit	Ovarian, mantle cell lymphomas, and Merkel cell carcinomas	Overexpression (Jiang and Song 2009; Lukacs et al. 2010)
	G9a	Histone methyltransferase H3K9	HCC, cervical, uterine, ovarian, and breast cancer	Aberrant expression (Varier and Timmers 2011)
	PRMT1/5	Protein arginine methyltransferase	Breast/gastric	Aberrant expression (Miremadi et al. 2007)
	LSD1	Histone demethyltransferase H3K4/H3K9	Prostate	Mutation (Rotili and Mai 2011)
	UTX (KDM6A)	Histone demethyltransferase H3K27	Bladder, breast, kidney, lung, pancreas, esophagus, colon, uterus, brain, hematological malignancies	Mutation (Rotili and Mai 2011)
	JARID1B/C (KDM5C)	Histone demethyltransferase H3K4/H3K9	Testicular and breast, RCCC	Overexpression (Rotili and Mai 2011)
	EP300 (P300/ KAT3B)	Histone acetyltransferase	Breast, colorectal, pancreatic cancer	Mutation (Miremadi et al. 2007)
	CREBBP (CBP/ KAT3A)	Histone acetyltransferase	Gastric and colorectal, epithelial, ovarian, lung, esophageal cancer	Mutation, overexpression (Miremadi et al. 2007)
	PCAF	Histone acetyltransferase	Epithelial	Mutation (Miremadi et al. 2007)
	HDAC2	Histone deacetyltransferase	Colonic, gastric, endometrial cancer	Mutation (Ropero et al. 2006)
	SIRT1, HDAC5/7A	Histone deacetyltransferase	Breast, colorectal, prostate cancer	Mutation, aberrant expression (Miremadi et al. 2007)
Chromatin-remodeling enzymes	*SNF5 (SMARCB1, INI1)*	BAF subunit	Kidney malignant rhabdoid tumors, atypical rhabdoid/teratoid tumors (extrarenal), epithelioid sarcomas, small cell hepatoblastomas, extraskeletal myxoid chondrosarcomas, and undifferentiated sarcomas	Mutation, silencing, loss of expression (Wilson and Roberts 2011)

Continued

Table 2. *Continued*

Process	Gene	Function	Tumor type	Alteration
	BRG1 (SMARCA4)	ATPase of BAF	Lung, rhabdoid, medulloblastoma	Mutation, low expression (Wilson and Roberts 2011)
	BRM (SMARCA2)	ATPase of BAF	Prostate, basal cell carcinoma	Mutation, low expression (Sun et al. 2007; de Zwaan and Haass 2010)
	ARID1A (BAF250A)	BAF subunit	Ovarian clear cell carcinomas, 30% of endometrioid carcinomas, endometrial carcinomas	Mutation, genomic rearrangement, low expression (Jones et al. 2010; Guan et al. 2011)
	ARID2 (BAF200)	PBAF subunit	Primary pancreatic adenocarcinomas	Mutation (Li et al. 2011)
	BRD7	PBAF subunit	Bladder TCC	Mutation (Drost et al. 2010)
	PBRM1 (BAF180)	PBAF subunit	Breast tumors	Mutation (Varela et al. 2011)
	SRCAP	ATPase of SWR1	Prostate	Aberrant expression (Balakrishnan et al. 2007)
	P400/Tip60	ATPase of SWR1, acetylase of SWR1	Colon, lymphomas, head and neck, breast	Mutation, aberrant expression (Mattera et al. 2009)
	CHD4/5	ATPase of NuRD	Colorectal and gastric cancer, ovarian, prostate, neuroblastoma, hematopoietic	Mutation (Bagchi et al. 2007; Kim et al. 2011; Wang et al. 2011)
	CHD7	ATP-dependent helicase	Gastric and colorectal	Mutation (Wessels et al. 2010)

Adapted from You and Jones 2012.

MDS, myelodysplastic syndromes; AML, acute myeloid leukemia; ICF, immunodeficiency, centromere instability, and facial anomalies; SNPs, single-nucleotide polymorphisms; TCC, transitional cell carcinoma; HCC, hepatocellular carcinoma; RCCC, renal clear cell carcinoma; TET, ten-eleven translocation; NuRD, nucleosome remodeling and deacetylation.

the incorporation of histone variants), and nucleosome positioning.

Understanding what the pathologist's visible cellular phenotype means will require researchers to link it to the relationship between nuclear organization, chromatin structure, molecular marks, and genome function. This is an exciting new domain of research only touched on in this chapter, but it is likely to yield important contributions to our understanding of cancer initiation and progression, thanks to the continued advances in technologies, such as chromosome conformation capture (see Ch. 19 [Dekker and Misteli 2014]), epigenome-wide mapping studies, massive parallel sequencing, genome tethering techniques, and advanced fluorescence microscopy modeling (Bernstein et al. 2010; Cancer Genome Atlas Research Network 2013b; Garraway and Lander 2013; Reddy and Feinberg 2013).

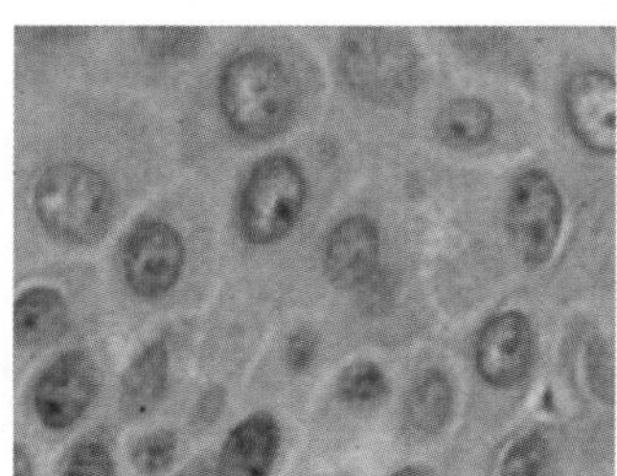
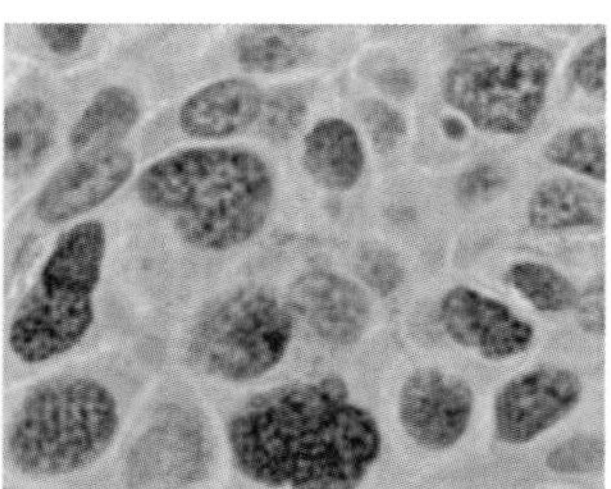

Figure 2. Chromatin structural changes in cancer cells. These two photomicrographs were taken from a patient with a squamous cell carcinoma of the skin. The *left* panel shows normal epidermal cells within one millimeter of the contiguous tumor shown at the same magnification on the *right*. The chromatin, which stains purple as a result of its affinity to hematoxylin, appears much more coarse and granular in the cancer cells than in normal epidermis. Such changes in the staining characteristics of chromatin are used by pathologists as diagnostic criteria for cancer.

One of the most recent exciting developments in the understanding of normal and cancer epigenomes comes from the results of whole-exon sequencing, whole-genome sequencing, genome-wide DNA methylation and chromatin analyses, and RNA expression approaches, which all supersede previous genome-wide analyses (Bernstein et al. 2010; Jones 2012; Cancer Genome Atlas Research Network 2013b; Garraway and Lander 2013; Reddy and Feinberg 2013). We, therefore, now recognize that epigenetic control involves not only canonical coding genes, but also noncoding RNA (ncRNA), microRNAs (miRNAs), and other regions that provide important genome regulatory function (Bernstein et al. 2010; Jones 2012; Cancer Genome Atlas Research Network 2013b; Garraway and Lander 2013; Reddy and Feinberg 2013). Thousands of solid and liquid tumors have been analyzed, showing, as introduced above, that there is an unexpected plethora of mutations in genes that control the function of the epigenome (Fig. 1; Table 2) (Baylin and Jones 2011; Dawson et al. 2011; You and Jones 2012; Garraway and Lander 2013; Shen and Laird 2013; Timp and Feinberg 2013; Ch. 35 [Audia and Campbell

Cite as *Cold Spring Harb Perspect Biol* doi: 10.1101/cshperspect.a019505

2014]). Importantly, many of these mutations occur at high enough frequencies to justify their roles as "driver" mutations in the cancers—that is, the results clearly show that disruption of the epigenome by mutations may lead to the initiation and/or progression of cancer. A major challenge, however, is to understand their precise contribution to cancer-specific alterations in chromatin and DNA methylation, and the exact consequences of these mutations in the key steps of tumorigenesis. It is important to remember that epigenetic changes in cancer may arise independently of mutations in chromatin-modifying factors; the epigenome is also subject to damage and heritable alterations induced by environmental or physiological events inherent to cancer risk states and steps during cancer progression (O'Hagan et al. 2008, 2011; Zheng et al. 2012), as will be discussed.

3 THE ROLE OF DNA METHYLATION IN CANCER

The initial discovery that the cytosine base in DNA can be methylated to become 5-methylcytosine (5mC), sometimes referred to as the 5th base, soon led to the proposal that alterations in DNA methylation may contribute to oncogenesis (Table 3). Over the last 40 years, there have been many studies showing that alterations in the 5mC distribution patterns can distinguish cancer cells from normal cells. At least three major routes have been identified by which CpG methylation can contribute to the oncogenic phenotype. The first is by general hypomethylation of the cancer genome. Second, focal hypermethylation at TSG promoters may occur. Third, direct mutagenesis of 5mC-containing sequences by deamination, UV irradiation, or exposure to other carcinogens is possible (Fig. 3) (Jones and Laird 1999; Jones and Baylin 2002; Herman and Baylin 2003; Baylin and Jones 2011). It is significant that all three of these alterations generally occur simultaneously to contribute to cancer, suggesting that altered homeostasis of epigenetic mechanisms is central to the evolution of human cancer.

Table 3. Time line for elucidating the role of DNA methylation in cancer

Observation	Reference
Hypothesis of "methylases as oncogenic agents"	Srinivasan and Borek 1964
Decreased levels of 5-methylcytosine in animal tumors	Lapeyre and Becker 1979
5-Azacytidine and 5-aza-2′-deoxycytidine inhibit methylation and activate genes	Jones and Taylor 1980
Decreased genomic and gene-specific methylation in human tumors	Ehrlich et al. 1982; Feinberg and Vogelstein 1983; Flatau et al. 1984
Inhibitors of DNA methylation alter tumorigenic phenotype	Frost et al. 1984
Methylation of a CpG island in cancer	Baylin et al. 1987
Hot spots for p53 mutations are methylated CpG sites	Rideout et al. 1990
Allele-specific methylation of the retinoblastoma TSG	Sakai et al. 1991
Loss of imprinting in cancer	Rainier et al. 1993
Hypermethylation of CpG islands is associated with aging	Issa et al. 1994
Mice with decreased methylation develop fewer tumors	Laird et al. 1995
Coupling DNA methylation and HDAC inhibitors leads to rapid isolation of TSGs	Suzuki et al. 2002; Yamashita et al. 2002
DNA repair gene (MLH1) is methylated in somatic cells	Gazzoli et al. 2002
Hypomethylation contributes to cancer	Gaudet et al. 2003
5-Azacytidine is FDA approved for treatment of myelodysplastic syndrome	Kaminskas et al. 2005
Discovery of the 5-hydroxymethylcytosine base and the TET1/2/3 enzymes that catalyze this conversion	Kriaucionis and Heintz 2009; Tahiliani et al. 2009

Adapted from You and Jones 2012.
HDAC, histone deacetylase; FDA, [U.S.] Food and Drug Administration; TSG, tumor-suppressor gene; TET, ten-eleven translocation.

3.1 DNA Hypomethylation in Cancer

The most prominent and earliest recognized change in DNA methylation patterns in cancer cells was regional decreases in this modification (Feinberg and Vogelstein 1983; Ehrlich and Lacey 2013), now recognized by genome-wide analyses as a global DNA hypomethylation (Hansen et al. 2011; Berman et al. 2012; Bert et al. 2013). Although all of the ramifications of these losses still need definition, DNA demethylation potentially contributes to genomic instability and increases in aneuploidy (Ehrlich and Lacey 2013), which are both classic hallmarks of cancer. Indeed, deletion or reduction of the maintenance DNA methyltransferase, Dnmt1, results in increased mutation rates, aneuploidies, and tumor induction, a clear indication that DNA hypomethylation plays an active role in increasing chromosomal fragility (Chen et al. 1998; Narayan et al. 1998; Gaudet et al. 2003; Ehrlich and Lacey 2013). Loss of DNA methylation may be accompanied by the activation of transcription, allowing transcription of repeats, transposable elements (TEs), and oncogenes (Jones and Baylin 2007; Ehrlich and Lacey 2013; Hur et al. 2014). Activation of repeats may predispose the genome of a cell to recombination, as corroborated by the increased frequency of chromosomal recombination at certain genomic regions (hot spots) or

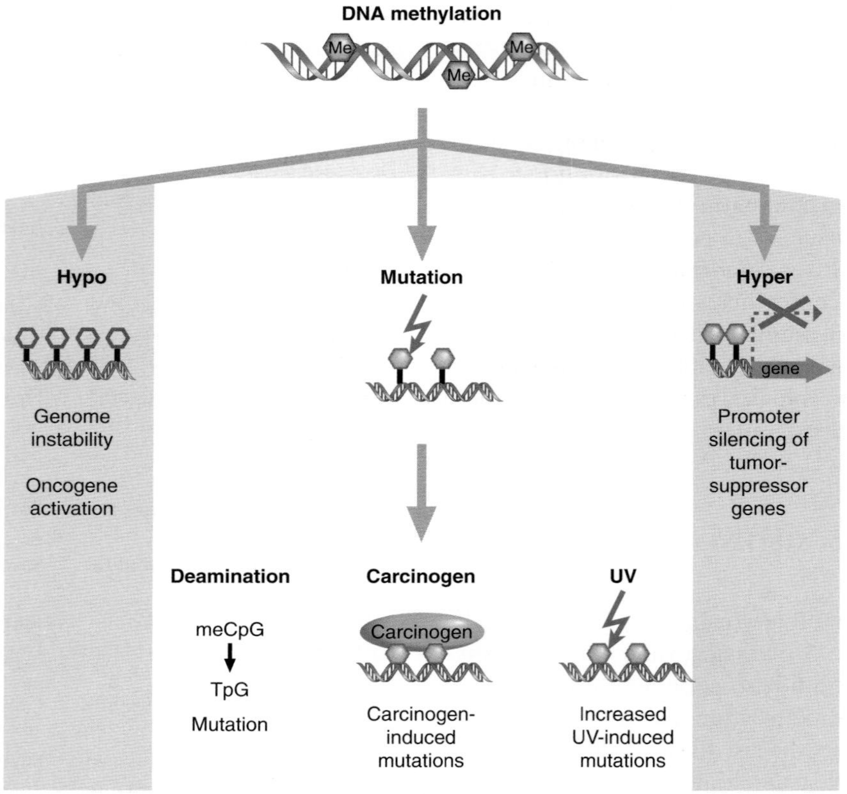

Figure 3. Epigenetic alterations involving DNA methylation can lead to cancer by various mechanisms. Loss of DNA cytosine methylation (white hexagons) illustrated in the hypo column results in genome instability. Focal hypermethylation (pink hexagons) at gene promoters shown in the hyper column causes heritable silencing and, therefore, inactivation of tumor suppressors and other genes. Additionally, methylated CpG sites (pink hexagons) are prone to mutation: They are hot spots for C to T transition mutations caused by spontaneous hydrolytic deamination; or methylation of CpG sites can increase the binding of some chemical carcinogens to DNA; and it increases the rate of UV-induced mutations.

may express nearby proto-oncogenes (Wolffe 2001; Jones and Baylin 2007; Ehrlich and Lacey 2013; Hur et al. 2014). Indeed, the activation of TEs is another potential source of mutations during the transposition process.

We know that most of the CpGs in the genome, apart from CpG-rich regions, are 80% methylated. In cancer, the average CpG methylation levels are 40%–60%. Advances in mapping technologies are allowing researchers to map the patterns more precisely. Such studies have revealed that DNA hypomethylation can be concentrated in blocks of 28 kb–10 Mb, covering about one-third of the genome (Hansen et al. 2011; Berman et al. 2012; Hon et al. 2012; Bert et al. 2013). The exact mechanisms by which DNA methylation is lost from the cancer epigenome and how functional consequences occur are not yet fully understood; however, we are beginning to be able to dissect these mechanisms. For example, a leading possibility is that many regions of DNA hypomethylation could be integrally tied to broad shifts in chromatin organization, typical in cancer (discussed further in Sec. 6). The broad epigenomic changes, in turn, could, in some instances, result from mutations in chromatin regulators that affect DNA methylation homeostasis, such that the active or passive process of removing DNA methylation is promoted. This could occur, for example, as discussed below and in other chapters, by the deregulated activation of ten-eleven translocation (TET) family members or the partial loss of function of the DNA methyltransferase (DNMT) proteins.

3.2 DNA Hypermethylation in Cancer

A well-chronicled DNA methylation change in cancer is abnormal hypermethylation of CpG islands in the 5′ regions of cancer-related genes (i.e., hypermethylation, Fig. 3). This change can be integrally associated with transcriptional silencing, providing an alternative mechanism to mutation for the inactivation of genes with tumor-suppressor function (Jones and Baylin 2007; Baylin and Jones 2011;

Cite as *Cold Spring Harb Perspect Biol* doi: 10.1101/cshperspect.a019505

Shen and Laird 2013). In this regard, 60% of all gene promoters have CpG islands, most of which are not DNA methylated at any time in normal development or in adult cell renewal systems (Jones and Baylin 2007; Baylin and Jones 2011; Shen and Laird 2013). This lack of methylation is fundamental to the more open chromatin states, and active, or ready to be activated, expression status of these genes (Jones and Baylin 2007; Baylin and Jones 2011; Shen and Laird 2013). The fact that methylated CpG island promoters are so prevalent in cancers (~5%–10% of CGI genes) and are known to directly contribute to carcinogenesis has led to new possibilities in the area of epigenetic therapy—that is, where epigenetic changes are targeted for therapeutic reversal, as discussed further in Section 9 (Egger et al. 2004; Spannhoff et al. 2009; Kelly et al. 2010; Bernt et al. 2011; Daigle et al. 2011; Dawson et al. 2012; Azad et al. 2013).

It should be noted that 5mC commonly occurs in the gene body of active genes and functional ramifications in this region may often be opposite to presence of this modification in promoters (Jones 2012; Kulis et al. 2012; Shen and Laird 2013). Thus, rather than being associated with repression of transcription, gene body DNA methylation may facilitate transcriptional elongation and enhance gene expression (Fig. 4) (Jones 2012; Kulis et al. 2012; Shen and Laird 2013). Interestingly, *DNMT3A* somatic muta-

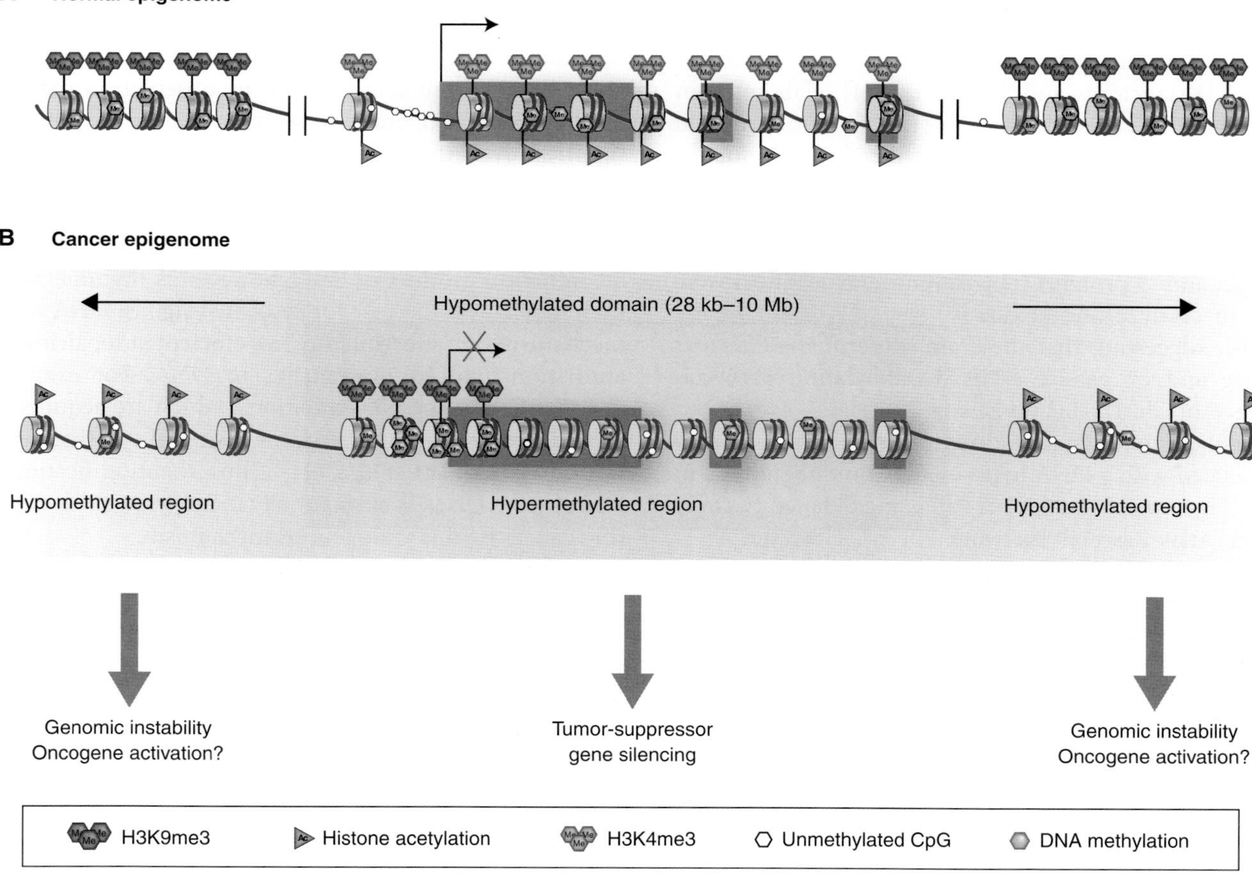

Figure 4. Chromatin structural changes in cancer cells. (*A*) In a typical cell, a CpG-island-containing active gene can be recognized by virtue of a nucleosome-depleted promoter, absence of promoter DNA methylation, but marked by H3K4me3 surrounding the promoter and histone acetylation along the locus. Gene body CpG methylation often can be observed. Nongenic regions flanking an active gene are frequently marked by repressive epigenetic marks, such as H3K9me3 and 5mC. (*B*) The cancer epigenome is characterized by simultaneous global losses in DNA methylation (gray shading), interspersed with silenced genes that have abnormal gains of DNA methylation and repressive histone modifications in CpG island promoter regions. These silenced genes may be hypomethylated in their gene body, similar to surrounding chromatin. The hypomethylated regions can have an abnormally open nucleosome configuration and acetylated histone lysines. Conversely, abnormal DNA hypermethylation in promoter CpG islands of silenced genes is associated with nucleosomes positioned over the transcription start sites.

tions that occur in certain patients with acute myeloid leukemia (AML) may predispose them to a loss of gene body DNA methylation (Cancer Genome Atlas Research Network 2013a), the causal consequences of which are currently unclear.

Our mechanistic understanding of how DNA methylation homeostasis may be disturbed in cancer is continually being enriched by discoveries challenging two key assumptions in epigenetics and cancer: All mammalian DNA methylation is confined to CpG sequences and it is a very stable mark. The first assumption was challenged when DNA methylation at CpHpG sequences was documented in human embryonic stem (ES) cells (Lister et al. 2009). The significance of this remains to be determined and it has not been well documented in cancers. The second assumption was challenged following proof that methylated cytosines can be actively demethylated; this has been extremely significant in both the epigenetic and cancer fields (described in Ch. 2 [Kriaucionis and Tahiliani 2014]; elaborated in Sec. 3 of Ch. 15 [Li and Zhang 2014]). DNA demethylation was first discovered through the identification of oxidative derivatives of 5mC, including 5-hydroxymethylcytosine (Kriaucionis and Heintz 2009; Tahiliani et al. 2009), 5-formylcytosine, and 5-carboxylcytosine. Simultaneously, the TET1, -2, and -3 proteins (ten-eleven translocations) were shown to catalyze these oxidative steps (Wu and Zhang 2011a,b), suggesting that these are some of the effectors of active and/or passive DNA demethylation pathways (see Fig. 6 of Ch. 15 [Li and Zhang 2014]).

The suggestion that mutations in the TET enzymes may be associated with a DNA hypermethylation phenotype in cancer (Figueroa et al. 2010) is still being debated (Cancer Genome Atlas Research Network 2013a). However, TET-mediated DNA demethylation has been linked to altered cellular metabolism and cancer through mutations in the upstream isocitrate dehydrogenase enzymes, IDH1 and IDH2. These enzymes normally produce α-ketoglutarate, an essential cofactor for the TET hydroxylases (elaborated in Sec. 5.2) (Lu et al. 2012, 2013; Shen and Laird 2013; Venneti et al. 2013). Mutations in IDH1/2, however, lead to a marked increase in the formation of an abnormal metabolite, 2-hydroxy-glutarate, formed from α-ketoglutarate (see Fig. 6 of Ch. 30 [Berger and Sassone-Corsi 2014]). In this scenario, an increased frequency of DNA hypermethylation can be observed, as seen with leukemias and brain tumors (Noushmehr et al. 2010; Turcan et al. 2012; Shen and Laird 2013). The fact that TET and IDH mutations in cancer are mutually exclusive underscores the need for constant demethylation in ensuring the correct level of cellular 5mC (Williams et al. 2011). Importantly, IDH mutations in the hematopoietic system (Sasaki et al. 2012a) appear to drive tumorigenesis because it blocks a cell's response to differentiation cues and, hence, skews lineage choice (Borodovsky et al. 2013; Turcan et al. 2013). Importantly, the experimental drug reversal of abnormal DNA methylation patterns associated with IDH mutations appears to restore an element of cellular differentiation responses, showing therapeutic promise for treating these types of cancers (Borodovsky et al. 2013; Turcan et al. 2013).

3.3 Mutation of 5mCs

A third mechanism, which we have known of for some time, by which methylation of cytosine residues (5mC) contributes disproportionately to cancer is its propensity of cytosines to be mutated in this sequence context (Fig. 3). Thus, when looking at the human germline, CpG sites typically methylated in the soma constitute more than a third of all transition mutations. Early examples of such mutations were documented in the cancer-causing p53 gene (Rideout et al. 1990). More surprising is the observation that this mechanism also operates in somatic tissues, contributing significantly to the formation of inactivating mutations in many TSGs. This occurs because methylation of the 5 position of the cytosine ring increases the rate of hydrolytic deamination of the base in double-stranded DNA. The deamination product of 5mC, however, is thymine rather than uracil, as is the case for cytosine (Fig. 3). DNA repair mechanisms are subsequently less efficient at repairing deamination-induced mismatches in DNA. For example, >50% of all of the p53 mutations, which are acquired in sporadic colorectal cancers, occur at sites of cytosine methylation (Greenblatt et al. 1994). Thus, the modification of DNA by the DNMTs substantially increases the risk of getting cancer by this endogenous mechanism.

Methylation of cytosine residues have also been shown to favor the formation of carcinogenic adducts between DNA and carcinogens, such as benzo(a)pyrene in cigarette smoke (Fig. 3). In this case, methylation of the cytosine residue increases the formation of carcinogenic adducts between an adjacent guanine residue and benzo(a)pyrene diol epoxide, resulting in increased mutations at CpG sites in the lungs of cigarette smokers (Greenblatt et al. 1994; Pfeifer et al. 2000).

Interestingly, DNA methylation can also alter the rate of mutations in the p53 gene in sunlight-exposed skin (Greenblatt et al. 1994; Pfeifer et al. 2000). This is because the methyl group changes the absorption spectrum for cytosine into the range of incident sunlight, increasing the formation of pyrimidine dimers in the DNA of skin cells exposed to sunlight. In summary, the 5mC modification of DNA not only increases spontaneous mutagenesis, but can influence the way DNA interacts with carcinogens and UV light (Pfeifer et al. 2000).

4 HYPERMETHYLATED GENE PROMOTERS IN CANCER

A main focus of this chapter, described in this section, is the characterization and role of DNA methylation in cancer and, in particular, its effect on TSGs. We are beginning to see how this intersects with other modes of epigenetic regulation, discussed further in Section 6.

4.1 The Genes Involved

The most well-understood mechanism by which DNA methylation contributes to cancer is through association with the focal hypermethylation of promoters at TSGs. This clearly is a significant pathway by which genes that would normally suppress cancer development are heritably silenced (Jones and Baylin 2002, 2007; Herman and Baylin 2003; Baylin and Jones 2011; Shen and Laird 2013). Usually, DNA hypermethylation occurs at CpG-rich regions or CpG islands that are located in and around the transcriptional start site of abnormally silenced genes in cancer (Fig. 4). Typically, 5%–10% of these CpG island promoters are DNA methylated in cancer (Baylin and Jones 2011). It is important to recognize that cytosine methylation in CpG islands is usually restricted to the vicinity of the gene start site position often spanning the transcription start site, but also occurring in the island at proximal upstream or downstream positions; this same DNA modification occurring within bodies of genes generally has either no correlation to transcription status or, as discussed earlier, can actually accompany increased gene expression, possibly through facilitating the transcriptional elongation process (Jones 2012; Kulis et al. 2012; Shen and Laird 2013).

The list of cancer-related genes affected by transcription disruption through DNA hypermethylation continues to grow and involves genes found at all chromosome locations. In an individual tumor, hundreds of genes can be disrupted by promoter hypermethylation and this mechanism holds true for virtually every type of cancer (Jones and Baylin 2002, 2007; Baylin and Jones 2011; Hammerman et al. 2012; Cancer Genome Atlas Research Network 2013a; Shen and Laird 2013). Indeed, as more deep analyses of DNA methylation are being performed in multiple tumor types, the frequency of this epigenetic change appears to be outnumbering gene mutations in human tumors (Jones and Baylin 2002, 2007; Baylin and Jones 2011; Hammerman et al. 2012; Cancer Genome Atlas Research Network 2013a; Shen and Laird 2013), promoter regions occur in genes involved in virtually every signaling pathway altered in tumorigenesis. Involvement of such a large number of genes has created one of the most important conundrums for the cancer epigenetics field: Why would so many genes be involved in cancer and which silencing events are truly important for the process of tumorigenesis? Clearly, experimentally, it is difficult to test whether each gene is critical for tumor initiation and progression by loss-of-function analyses. However, as well reviewed and mentioned below, some of the genes involved are clearly driver TSGs (Esteller 2007; Jones and Baylin 2007; Baylin and Jones 2011; Shen and Laird 2013). Moreover, just as analyzing signaling pathway participation has been important for understanding the myriad of genetic alterations in cancer, categorizing DNA hypermethylated genes in this manner has great potential to facilitate our understanding of their significance in the process of tumorigenesis (Jones and Baylin 2007; Baylin and Jones 2011; Shen and Laird 2013).

A first group of DNA hypermethylated genes found in cancers constitutes those in which loss of function clearly has a "driver function" for all stages of cancer evolution (Jones and Baylin 2007; Baylin and Jones 2011; Shen and Laird 2013). Typically, true cancer driver mutations involve a relatively limited group of genes. The first examples of epigenetically silenced genes to be characterized were instrumental in defining gene silencing by promoter hypermethylation as an important mechanism for loss of TSG function in cancer (Table 4). The genes were easily recognized as classic TSGs, known to cause inherited forms of cancer when mutated in the germline of families (Jones and Laird 1999; Jones and Baylin 2002, 2007; Esteller 2008; Shen and Laird 2013). They were also often mutated in sporadic forms of cancers and, notably, were also frequently hypermethylated on one or both alleles in such tumors

Table 4. Discovery classes of hypermethylated genes

Class of hypermethylated gene	Examples
Known TSG[a]	*VHL*
	E-cadherin
	P16Ink4a
	MLH1
	APC
	Stk4
	Rb
Candidate TSG	*FHIT*
	Rassf1a
	O6-MGMT
	Gst-Pi
	GATA4/5
	DAP-kinase
Gene discovered through random screens for hypermethylated genes	*HIC-1*
	SFRP1, -2, -4, -5
	BMP-3
	SLC5A8
	SSI1

[a] A classic tumor-suppressor gene (TSG) is known to be mutated in the germline of families with hereditary cancer syndromes.

(Jones and Laird 1999; Jones and Baylin 2002; Herman and Baylin 2003). Also, for these genes, it was noted that promoter hypermethylation sometimes constituted the "second hit" in Knudson's hypothesis, that is, the first hit constituted a germline mutation in familial tumors, whereas the second hit arose from the loss of function through DNA methylation of the second copy of the gene (Grady et al. 2000; Esteller et al. 2001a). In some instances, 5-aza-cytidine (5-aza-CR) treatment in cultured tumor cells induced the reactivation of these genes, restoring the key TSG function lost during tumor progression. This was shown for the mismatch repair gene, MLH1, which is typically silenced in colon cancer cells (Herman et al. 1998).

A second group of epigenetically silenced genes are those previously identified as candidate TSGs by virtue of their function, but were not found to have an appreciable frequency of mutational inactivation, which would indicate that they are true driver mutations (Table 4). Despite the paucity of known cancer-associated mutations in this category of genes, they often reside in chromosome regions frequently suffering deletions in cancers. Examples include *RasFF1a* and *FHIT*, located on chromosome arm 3p, frequently deleted in lung and other types of tumors (Dammann et al. 2000; Burbee et al. 2001). Other candidate TSGs fit into this category because they are known to encode proteins, which subserve functions critical for the prevention of tumor progression, such as the proapoptotic gene, *DAP-kinase* (Katzenellenbogen et al. 1999), families of genes that antagonize WNT signaling (Suzuki et al. 2004; Jones and Baylin 2007; Zhang et al. 2008; Baylin and Jones 2011; Shen and Laird 2013). Yet, others qualify because it is now recognized that promoter CpG island hypermethylation can silence noncoding miRNA genes, which are necessary for modulating signaling networks (Saito and Jones 2006; Saito et al. 2006; Chaffer et al. 2013; Tam and Weinberg 2013; Nickel and Stadler 2014; Sun et al. 2014). These genes present an important challenge for the field of cancer epigenetics because, although they are often hypermethylated in tumors, many of them are not frequently mutated, making it difficult to be sure that they actually contribute to tumorigenesis. Section 4.3 describes the strategies being used to determine whether these are truly TSGs.

The third and largest group of genes (Table 4) continues to be populated as more and more genome-wide screens randomly identify aberrant DNA hypermethylation, involving coding and noncoding regions (Baylin and Jones 2011; Shen and Laird 2013; Taberlay et al. 2014). As compared with genes in the first two groups, it is a challenge to place these genes into a functional context for cancer progression because their precise roles are not yet obvious.

A very important relationship exists between a large number of genes hypermethylated in cancer and their tendency to evolve promoter CpG island methylation with age (Issa 2014; Maegawa et al. 2014). This has been very well shown for genes in the colon in which this increasing DNA methylation virtually parallels the age-related risk for colon cancer (Issa et al. 1994; Toyota et al. 1999; Issa 2014; Maegawa et al. 2014). This relationship has now been well documented for other cancers as well and appears to relate to such increases with age that occur not only in humans, but across mammalian species also (Maegawa et al. 2014). The mechanisms for these changes need further dissection, but clearly, this epigenetic change is closely linked to risk for human cancer.

4.2 Technology Used to Identify DNA Methylation Patterns

Class II hypermethylated genes, shown in Table 4, are categorized on the presumption that any abnormal DNA methylation is potentially a causal mechanism in the loss of TSG function, especially when genetic mutations are lacking, yet expression of the gene is low or absent in tumor versus normal tissue. These characteristics provide the basis for a candidate gene approach. The robust application of global mapping assays of genome-wide DNA methylation patterns is now a mandatory approach to identify new hypermethylated genes involved in cancer (i.e., class III in Table 4). Importantly, these technologies also can now place the promoter changes under discussion here into context for their importance, as compared with other regions in which there are DNA methylation changes in tumorigenesis (Bernstein et al. 2010; Cancer Genome Atlas Research Network 2012a; Taberlay et al. 2014). These newer platforms for genome-wide hybridization and/or next-generation sequencing yield comprehensive genome coverage of the DNA methylation landscape. These assays are being used in consortial efforts to map DNA methylation in both normal and disease cells, and illustrate the power to rapidly identify large numbers of hypermethylated genes and other cancer DNA methylation abnormalities (Cancer Genome Atlas Research Network 2012b, 2013a; ENCODE 2012; Shen and Laird 2013).

Many mapping studies to date have used a high-throughput method, which is very cost-effective, when there is a need to broadly screen DNA methylation in many human samples. Termed the Illumina Infinium 450K microarray platform, the assay involves bisulfite treatment of genomic DNA and subsequent hybridization to approximately 450,000 candidate CpG sites throughout the genome. Bisulfite treatment distinguishes methylated from unmethylated cytosines by virtue of the fact that cytosines are converted to uracil, whereas 5mC is resistant to this modification. The Infinium 450K platform queries sites

Cite as *Cold Spring Harb Perspect Biol* doi: 10.1101/cshperspect.a019505

not only at gene promoters, but also at other candidate sequences, including enhancer and ncRNA promoter regions. However, the coverage, although wide, is often not deep in a given sequence region and valuable as a first screening tool, to be followed by deeper probing of selected samples from among those being studied (Dedeurwaerder et al. 2011). This platform is currently used by the Cancer Genome Atlas project for matching DNA methylation abnormalities to genome-wide screens for gene mutations, copy number alterations, translocations, expression changes, and their integration for delineating signaling pathway abnormalities in cancer (Cancer Genome Atlas Research Network 2012b, 2013a; ENCODE 2012; Shen and Laird 2013). A prime goal is to outline cancer-specific abnormalities that suggest new therapy targets for development and biomarker strategies for cancer detection and prognostic predictions. These studies have already produced large lists of newly defined genes with epigenetic abnormalities for brain, colon, lung, breast, and other (Cancer Genome Atlas Research Network 2012b, 2013a; ENCODE 2012; Shen and Laird 2013).

Other integrative studies, such as the ENCODE project and the Epigenome Roadmap projects (Cancer Genome Atlas Research Network 2012b, 2013a; ENCODE 2012), are also increasing our understanding of the role of DNA methylation and chromatin abnormalities in cancer and, specifically, the hypermethylated genes under discussion. The more costly approaches involve methods, such as capturing differentially DNA methylated sequences by either methylcytosine antibodies, antibodies recognizing methylcytosine-binding proteins or their binding domains or sequences generated by methylation-sensitive restriction enzymes, and then identifying these via next-generation sequencing (Harris et al. 2010; Aryee et al. 2013). Even more extensive information is also being compiled by direct sequencing of virtually all candidate CpG sites following bisulfite treatment of DNA (Lister et al. 2009; Lister et al. 2011; Berman et al. 2012). All of these approaches are providing a detailed view of DNA methylation patterns inherent to normal development in normal mature tissues and primary and cultured tumor samples.

The high-throughput DNA methylation detection approaches can be combined with data obtained by treatment of cultured cells with demethylating agents, such as 5-aza-CR or 5-aza-2′-deoxycytidine (5-aza-CdR). RNA from before and after drug treatment is hybridized to gene microarrays, or subject to RNA-sequencing (RNA-seq) analysis, to detect drug-induced up-regulated genes (Suzuki et al. 2002; Yamashita et al. 2002; Schuebel et al. 2007). It must be recognized, however, that the very low expression levels of many of the induced genes before and after drug treatment challenge the sensitivity of gene expression platforms and reduce the efficiency of these approaches (Suzuki et al. 2002; Schuebel et al. 2007). Use of quantitative RNA-seq assays may provide a more dynamic range of gene expression changes, which enhances the utility of combining induced gene expression with genome-wide DNA methylation assays.

4.3 Determining the Functional Importance of Genes Hypermethylated in Cancer

The large number of genes with hypermethylated DNA at their promoters in cancer presents a formidable research challenge for understanding the functional scope of these changes. Frequent promoter hypermethylation in a given gene does not, in and of itself, guarantee that the silenced gene has a functional significance in cancer, as is often the case for genetic mutations. This is especially the case when the hypermethylated gene is not a known tumor suppressor and there is no evidence that the gene is frequently mutated in cancers. Thus, it is obligatory that the gene in question is studied in such a way as to determine the significance of loss of function, in terms of both the processes controlled by the encoded protein and the implications for tumor progression. In fact, sorting out the driver versus passenger roles for this class of genes is one of the biggest research challenges in cancer epigenetics.

Several initial steps are useful, but do not absolutely confirm the importance of a given gene in cancer (summarized in Table 5). First, of course, is the precise documentation of its cancer-specific hypermethylation profile,

Table 5. Steps in documenting the importance of a hypermethylated gene for tumorigenesis

1. Document CpG island promoter methylation and correlate with transcriptional silencing of the gene and ability to reverse the silencing with demethylating drugs in culture.
2. Document correlation of promoter hypermethylation with specificity for this change in tumor cells (cell culture and primary tumors) versus normal cell counterparts and incidence for the hypermethylation change in primary tumors.
3. Document the position of the hypermethylation change for tumor progression of given cancer types.
4. Document the potential significance for the gene silencing in tumorigenesis through gene reinsertion studies in culture and effects on soft agar cloning, growth of tumor cells in nude mouse explants, etc.
5. Establish function of the protein encoded by the silenced gene, through either known characteristics of the gene or testing for activity of recognized protein motifs in culture systems, etc.
6. Document tumor-suppressor activity and functions of the gene for cell renewal, etc., especially for totally unknown genes, through mouse knockout studies.

including its position in the gene promoter and consequences on the expression state of the gene. This might include assessing the ability of the gene to undergo reexpression following drug-induced promoter demethylation. Second, the incidence of hypermethylation and gene silencing must be well established in primary, as well as cultured, tumor samples. Third, it is often useful to know at what point the silencing of the gene occurs in tumor progression, as exemplified in Figure 5 for colon cancer.

Confirming a gene is a bona fide TSG requires studies that assess its contribution to tumorigenicity following loss of function. The function of the encoded protein is important and can be established through knowledge about the type of protein, aspects of the protein structure, and/or relationships to gene families and signaling pathways. In an age when many known genes have been subjected to genetic knockout studies, the phenotypes produced and attendant biology can be informative in pointing to the potential contribution of gene silencing in tumorigenesis. Candidate TSGs can be assayed for their tumorigenic potential following gene knockout by assessing in cultured cells the effects of their loss on (1) soft agar cloning (to detect any capacity for malignant transformation) and (2) tumorigenicity of the cells when grown as heterotransplants in immunocompromised mice and (3) assessing the cellular properties, such as the induction of apoptosis following gene reinsertion. Ultimately, however, additional transgenic knockout approaches may be needed to establish the role of a gene as a tumor suppressor and to understand the functions of the encoded protein in development, adult cell renewal, etc. Mouse knockout studies documenting the function of the transcription factor and developmental gene, *HIC-1*, provide an example of how this gene was experimentally validated as a TSG (Chen et al. 2003, 2004). It was initially identified through screens of genomic regions that have undergone loss of heterozygosity in cancer cells (Wales et al. 1995). Clearly, discovering genes that are epigenetically silenced in cancer is of great value, yet, the major scope of work lies ahead in definitively trying to show where loss of function for the gene is important in cancer.

5 THE IMPORTANCE OF EPIGENETIC GENE SILENCING IN EARLY TUMOR PROGRESSION

In the classic view of cancer evolution, as articulated by Vogelstein and colleagues (Kinzler and Vogelstein 1997; Vogelstein et al. 2013), a series of genetic changes drives progression from early premalignant stages through the

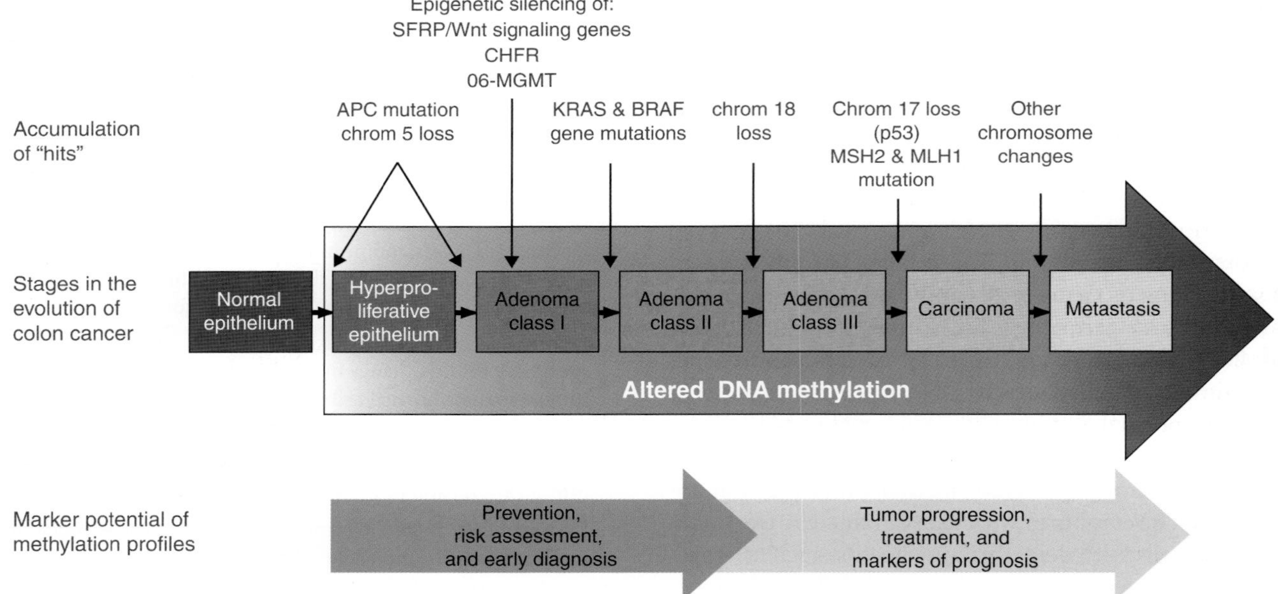

Figure 5. The position of abnormal DNA methylation in tumor progression. This is depicted in the classic model (Kinzler and Vogelstein 1997) for genetic alterations during the evolution of colon cancer. Altered DNA methylation is shown to occur from very early on in tumorigenesis (red arrow), as discussed in the text, during the conversion of normal to hyperplastic epithelium, accruing during the progression from noninvasive to invasive and, ultimately, metastatic tumors. This places it in a strategic position for channeling stem cells into abnormal clonal expansion (illustrated in Fig. 6) by cooperating with key genetic alterations. These epigenetic abnormalities also have connotations for cancer treatment and markers of prognosis.

appearance of invasive cancer to the onset of metastatic disease (Fig. 5), although this progression does not necessarily occur in the same exact linear order from tumor to tumor. We now know that epigenetic changes are occurring throughout this course of events, and this includes the early appearance of widespread loss of normal DNA methylation and more focal gains in gene promoters, discussed in Section 4 (Fig. 4). Other features of the epigenome can also be deregulated, including the altered occurrence and distribution of histone marks, and this may be caused by mutations in components of the epigenetic machinery. Thus, there is the potential for interaction between epigenetic and genetic events to drive progressive cellular abnormalities throughout the entire course of neoplastic progression (Fig. 1). Two epigenetic processes, loss of gene imprinting or LOI (as discussed in Ch. 33 [Zoghbi and Beaudet 2014]) and epigenetic gene silencing, are extremely important mechanisms contributing to the very early stages of cancer development.

5.1 Loss of Imprinting

Loss of imprinting (LOI) and epigenetic gene silencing are the most studied processes involving epigenetic aberrations that affect tumor evolution. LOI is a process in which the silenced allele of an imprinted gene becomes activated during tumorigenesis. This results in biallelic expression of the gene, generating excess gene product (Rainier et al. 1993). The most studied example of LOI occurs at the *IGF2* gene in tumors, such as colon cancer (Kaneda and Feinberg 2005). This occurs because hypermethylation of a regulatory element upstream of the neighboring imprinted H19 gene removes its insulator function (illustrated in Fig. 8 of Ch. 26 [Barlow and Bartolomei 2014]). This insulator, which normally prevents the *IGF2* gene from being activated through interaction with its distal enhancer, allows *IGF2* to become expressed on the maternal copy of chromosome 11p in some cancers (Kaneda and Feinberg 2005). The resultant biallelic *IGF2* expression leads to excess production of the growth-promoting IGF2 protein. Experimental evidence suggests that this could play a role in the very early progression steps of colon cancer (Kaneda and Feinberg 2005; Sakatani et al. 2005). In fact, studies in mouse models suggest that LOI events alone may be sufficient to initiate the tumorigenesis process (Holm et al. 2005).

5.2 IDH Mutations Leading to Epigenetic Deregulation in Oncogenesis

Another compelling story illustrating how epigenetic regulation is central in cancer initiation and progression involves IDH1 and IDH2 oncogenic mutations in brain, colon, and hematologic cancers (Figueroa et al. 2010; Noushmehr et al. 2010; Prensner and Chinnaiyan 2011; Turcan et al. 2012; Cancer Genome Atlas Research Network 2013b; Losman and Kaelin 2013). The IDH1/2 alterations appear to alter the DNA and histone demethylation pathways, resulting in imbalances in histone methylation levels, such as increases in H3K36, H3K9 methylation (Lu et al. 2012; Lu et al. 2013; Venneti et al. 2013). There is also an associated increase in the frequency of promoter region CpG island DNA hypermethylation, which resembles the well-characterized CpG island methylator phenotype (CIMP) in colon and other cancers (Figueroa et al. 2010; Noushmehr et al. 2010; Turcan et al. 2012). The genes heavily involved are those with a history of an embryonic state chromatin pattern and often involved in the regulation of development (discussed further in Sec. 6.3).

Delineating the precise causes for these above chromatin and DNA methylation changes is an ongoing area of investigation. The leading data-driven hypothesis is that the changes result from the accumulation of 2-hydroxy-glutarate from α-ketoglutarate in cancer cells. This abnormal 2-hydroxy-glutarate metabolite, which increases to mM levels in cells with the IDH mutations, constituting a biomarker in and of itself, competes with the necessary α-ketoglutarate metabolite needed by the TET and lysine (K) demethylase (KDM) enzymes, which regulate either chromatin demethylase function or levels of DNA methylation. Interestingly, other Krebs cycle control genes, when mutated in certain tumor types, can also result in decreased α-ketoglutarate levels and similar chromatin and DNA methylation abnormalities (Xiao et al. 2012; Mason and Hornick 2013). Experiments have specifically shown that this leads to the buildup of repressive histone marks in gene promoter regions and, subsequently, DNA hypermethylation in what may constitute a molecular progression (Lu et al. 2012, 2013; Venneti et al. 2013). Mouse models for IDH1 or IDH2 mutagenesis suggest that these mutations are implicated in early tumor progression events (Sasaki et al. 2012b). Engineering the mutations into mice or cells, in vitro, appears to trap stem/progenitor cells in states of abnormal self-renewal and/or diminishes their capacity for lineage commitment and differentiation, as illustrated in Figure 6 (Lu et al. 2012; Turcan et al. 2012; Borodovsky et al. 2013). Then, inducing DNA demethylation can partially restore the capacity of cells with the mutations to respond to differentiation cues (Borodovsky et al. 2013; Turcan et al. 2013).

5.3 Early Event IDH or H3 Mutations Drive Oncogenesis

Beyond the IDH1/2 example in cancer, as discussed earlier, there are an increasing number of common mutations be-

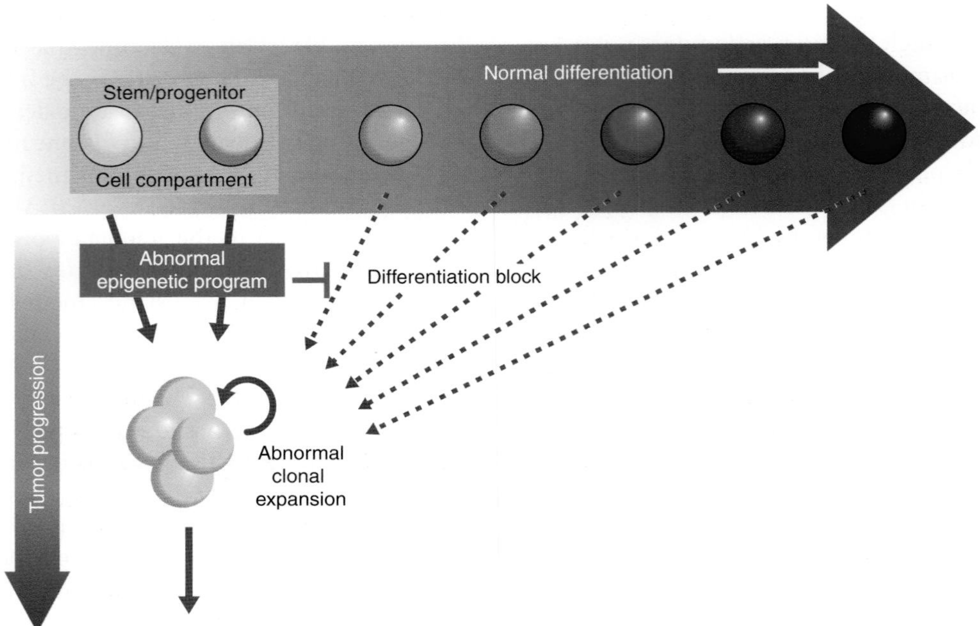

Figure 6. Epigenetic gene-silencing events and tumorigenesis. The earliest steps in tumorigenesis are depicted as abnormal clonal expansion, which evolves during the stress of cell renewal. This is caused by factors, such as aging and chronic injury from, for example, inflammation. These cell clones are those at risk of subsequent genetic and epigenetic events that could drive tumor progression. Abnormal epigenetic events, such as the aberrant gene silencing focused on in this chapter, could be the earliest heritable causes, in many instances, for a potential role in inducing the abnormal clonal expansion from within stem/progenitor cell compartments in a renewing adult cell system. The gene silencing is triggered by chromatin modifications that repress transcription, and the DNA hypermethylation of this chromatin serves as the tight lock to stabilize the heritable silencing. The gene silencing, in turn, disrupts normal homeostasis, preventing stem and progenitor cells from moving properly along the differentiation pathway for a given epithelial cell system (blue arrow) and channels them into the abnormal clonal expansion (red arrow).

ing found in genes coding for proteins that establish and maintain appropriate chromatin configurations (i.e., the normal epigenome [Figs. 1 and 4]). In fact, recent studies looking at the pathways that are activated during typical early cancer cell insults implicates the epigenetic machinery, and this is beginning to explain why epigenetic alterations are a common event in the early stages of cancer and even in precancerous changes preceding frank malignancy. Intriguingly, the timing of key mutations and cell compartments in which they occur may actually dictate and/or accompany the evolution of tumor subtypes. This may involve either a prominent presence of DNA methylation abnormalities or chromatin changes, both of which can play a major driver role.

A dramatic example of contrasting epigenetic patterns in cancer is evident when comparing IDH mutations in a pediatric subtype of brain tumor versus a histone mutation occurring at a key PTM site in another subtype of brain tumor. The IDH mutations are associated with CIMP and confined to low-grade gliomas arising in proneural progenitor cells in younger patients with better survival rates than those with advanced gliomas. These tumors arise in glial cell progenitors (Parsons et al. 2008; Noushmehr et al. 2010). In contrast, in the other subtype, mutations in H3K27 have been recently described and these tumors do not have CIMP. Although these H3K27 mutations are present in only one of the many H3 alleles, they apparently exert a dominant negative effect, which blunts all activity of the EZH2 enzyme catalyzing H3K27 methylation. The result is a dramatic loss of H3K27me3 (Chan et al. 2013; Lewis et al. 2013; Shen and Laird 2013), which probably leads to the activation of many genes that can drive tumorigenesis in a particular progenitor cell within a cell compartment.

5.4 Known Examples of TSG Epigenetic Silencing in Oncogenesis

Evidence for the involvement of specific genes in cancer progression continues to build. p16, for example, is a classic TSG that can be mutated or epigenetically silenced in human cancers. In lung cancer, the epigenetic silencing of $p16^{ink4a}$ (listed in Table 1) occurs very early in populations of premalignant cells before tumor formation (Swafford et al. 1997). In breast cancer, small populations of hyperplastic epithelial cells are also prone to $p16^{ink4a}$ epigenetic silencing (Holst et al. 2003). In fact, in cell culture (on plastic), normal human mammary epithelial cells require this type of *p16* silencing as a prerequisite for the very early steps toward cell transformation (Kiyono et al. 1998; Romanov et al. 2001). This loss of gene function through epigenetic means accompanies a failure of subsets of the mammary cells to reach a mortality checkpoint, allowing these cells to then develop progressive chromosomal abnormalities and the reexpression of telomerase as they continue to proliferate. Furthermore, it also involves the expansion of stem cells, as observed in *p16* mouse knockout models (Janzen et al. 2006).

A second example concerns the mismatch repair gene *MLH1*. This gene is typically mutated in the germline of families predisposed to a type of colon cancer; this form displays multiple genetic alterations and the "microsatellite" instability phenotype (Fishel et al. 1993; Liu et al. 1995). However, 10%–15% of patients with this tumor phenotype have nonfamilial colon cancer, in which the MLH1 gene is epigenetically silenced rather than genetically mutated (Herman et al. 1998; Veigl et al. 1998). It follows that its loss of function in a DNA repair capacity could lead to multiple genetic alterations and microsatellite instability. Indeed, in cell culture, reexpression of epigenetically silenced MLH1 produces a functional protein, which restores a considerable portion of the DNA damage mismatch repair function (Herman et al. 1998). This illustrates the clear link between genetics and epigenetics for these types of colon cancers in which MLH1 is epigenetically silenced. However, we do not have a full understanding of all the mechanisms involved yet; for instance, it is intriguing that virtually all of these colon tumors have the CpG island hypermethylator phenotype (Toyota et al. 1999; Weisenberger et al. 2006; Hinoue et al. 2012) and mutations of the B-RAF oncogene (Weisenberger et al. 2006; Shen and Laird 2013). Recent work by Hitchins et al. (2011) has interestingly shown that a single nucleotide variant in the promoter region of the MLH1 gene, which results in reduced expression of the allele, predisposes it to becoming methylated. Such reduced transcription may bias these alleles to evolve DNA methylation at promoters, which may deepen the silencing and make the gene more difficult to transcribe. It is most important, however, to pursue the underlying mechanisms leading to these outcomes.

Another example of a gene that is subject to early and important epigenetic changes is *Chfr*, a checkpoint-regulating gene that also controls genomic integrity, chromosomal stability, and ploidy (Table 1) (Sanbhnani and Yeong 2012). This gene is infrequently mutated in tumors, although it is often epigenetically silenced in lung and other cancers and, importantly, silenced early in the progression of colon cancer (Fig. 5) (Mizuno et al. 2002). Mouse knockout studies have revealed a tumor-suppressor role for this gene based on its function as an E3 ubiquitin ligase that regulates Aurora A, a control gene for mitosis (Yu et al. 2005). Consequently, embryonic cells from the mice display chromosomal instability and a predisposition to transformation.

5.5 Defining Epigenetic Silencing of TSGs as Drivers or Passengers of Oncogenesis

Many of the hypermethylated genes in cancer can only be defined as candidate TSGs and often only have a history of epigenetic change, but no genetic mutations. Research, as highlighted in Section 4.3, is needed to determine whether these genes are silenced early, which would represent a key event in early tumor progression. For example, the DNA repair gene, O6-MGMT, is silenced early in premalignant stages of colon cancer progression (Fig. 5) (Esteller et al. 2001b) and this loss of function can predispose cells to persistent alkylation damage at guanosines, resulting in G to A point mutations. Indeed, silencing of this gene occurs in premalignant colon polyps before the appearance of a high rate of p53 and RAS gene mutations in later colon tumor progression phases (Esteller et al. 2001b; Wolf et al. 2001). Similarly, the GST-Pi gene is epigenetically silenced via promoter hypermethylation in virtually all premalignant lesions that are predisposing to prostate cancer, putting cells at risk of oxidative damage at adenine (Lee et al. 1994).

The random screening approaches used to identify DNA hypermethylated genes in cancer has uncovered a particularly intriguing scenario in the progression of colon cancer: Epigenetic loss of function seems to occur in a number of components of the Wnt signaling family of genes, as discovered through a microarray approach (Suzuki et al. 2002). Silencing of genes required for regulating signal transduction may, thus, allow the abnormal activation of the *WNT* developmental pathway, driving early cancer progression (Suzuki et al. 2004; Jones and Baylin 2007; Zhang et al. 2008; Baylin and Jones 2011). For instance, frequent mutations (genetic and epigenetic) in another member of the Wnt pathway, the *APC* tumor suppressor, are also known to be universally involved with the initiation and progression of this disease and, hence, can be considered as driver mutations for this type of cancer. Other components of the Wnt signaling pathway were later implicated in colon tumorigenesis and gene silencing, such as the family of secreted frizzled-related protein genes (SFRPs) (Suzuki et al. 2004) and the transcription factor SOX17 (Zhang et al. 2008). The silencing of *SFRPs* relieves repression of the pathway at the level of membrane and cytoplasmic events. Loss of *SOX17*, which normally antagonizes the β-catenin transcription factor, relieves repression of this nuclear step that normally blocks Wnt ligand signal transduction (Finch et al. 1997; Zorn et al. 1999; Zhang et al. 2008), resulting in the up-regulation of downstream cellular β-catenin transcription factor levels. These silencing events occur in very early lesions predisposing to colon cancer, sometimes before common mutations in downstream Wnt pathway proteins (Suzuki et al. 2004; Zhang et al. 2008). Thus, early activation of the Wnt pathway by epigenetic events promotes the early expansion of cells. Persistence of both epigenetic and genetic alterations seems to complement one another in further driving progression of the disease (Suzuki et al. 2004).

The HIC-1 (hypermethylated-in-cancer 1) gene, which encodes a zinc finger transcriptional repressor, provides a final example of how a putative TSG, when its expression is epigenetically altered, can be cancer driving. HIC-1 was discovered by a random screening looking for hypermethylated CpG islands in a hot spot for chromosomal loss in cancer cells (Wales et al. 1995). HIC-1, although not mutated, was epigenetically silenced early on in cancer progression and, through mouse knockout modeling, proven to be a tumor suppressor (Chen et al. 2003). It complements p53 mutations (Chen et al. 2003) leading to up-regulation of SIRT1 (Chen et al. 2005), which contributes to enhanced stem/progenitor cell growth (Howitz et al. 2003; Nemoto et al. 2004; Kuzmichev et al. 2005). In pediatric medulloblastoma tumors, *Hic1* silencing was shown to exert cancer-driving function by depressing the Atoh1 transcription factor required for neuronal cell growth (Briggs et al. 2008).

A key issue for understanding the processes leading to altered DNA methylation and chromatin patterns in early tumorigenesis is elucidating the causative factors that may trigger them. In this regard, some of the environmental factors that induce cellular stress responses, as highlighted in Figure 37 of Ch. 3 (Allis et al.), appear critical. These exposure scenarios are linked to multiple disease states, including cancer. Recent experimental evidence, for instance, has directly linked exposure to cellular stress and the reversion of key cell population to a stem/progenitor state for survival, with the recruitment of protein silencing complexes involving PcG, histone deactylases (HDACs), and DNMTs, to CpG-rich gene promoters, and consequent gene silencing (O'Hagan et al. 2011). The molecular progression to DNA methylation is often then triggered at vulnerable low-expressed genes (see O'Hagan et al. 2011). Examples of stress stimuli, often observed in the cancer risk state of chronic inflammation and injury, are increases in reactive oxygen species (ROS) or DNA double-strand breaks (O'Hagan et al. 2008, 2011). The genes subject to permanent chromatin and DNA methylation changes at promoters after such insults may be those for which loss of function sets the stage for cell survival (Hahn et al. 2008; O'Hagan et al. 2011). Such cells are then poised for clonal expansion as stem/progenitor types and will be predisposed to later genetic and epigenetic events that drive tumor progression (Fig. 6) (Easwaran et al. 2014).

All of the data discussed above support the hypothesis outlined in Figure 5, which suggests that some of the earliest heritable changes in the evolution of tumors are epigenetic ones, particularly the transcriptional silencing of genes maintained by promoter DNA methylation. Although the precise effects of mutations in epigenetic regulators on cellular phenotypes and the epigenome of the cancer cell are not fully understood, these genetic alterations have highlighted how important epigenetic changes are in tumor initiation and progression. The key challenges now, as outlined in Table 6 and discussed more fully in Section 7, are to understand the molecular dynamics of epigenomic changes causal to cancer progression. This, in turn, will feed into discovering molecular strategies aimed at the prevention and early intervention of cancer, as well as providing more markers for improved diagnosis and prognosis of cancers.

6 THE MOLECULAR ANATOMY OF EPIGENETICALLY SILENCED CANCER GENES

Knowing which genes are silenced in neoplastic cells is important for understanding what contributes to the initiation and maintenance of cancer. Silenced loci also serve as excellent models for understanding how gene silencing is initiated and maintained, and how the mammalian genome is packaged to facilitate regions of transcription and repression. An understanding of chromatin function, which is a major emphasis of many of the chapters in this collection, is facilitating our understanding of what may trigger aberrant gene silencing in cancer and how the components of this silencing maintain the attendant transcriptional repression. Also, they are unveiling how (onco)genes and regions can be transcriptionally derepressed and what bearing that has on cancer development.

6.1 Chromatin Characteristics of Active and Repressive Genomic Regions

This chapter has concentrated on DNA methylation in cancer, as well as associated chromatin changes, which may occur in association with or without altered DNA methylation. In particular, we have described the finding and role of aberrant DNA methylation in the gene silencing

Cite as *Cold Spring Harb Perspect Biol* doi: 10.1101/cshperspect.a019505

Table 6. Major research challenges for understanding the molecular events mediating epigenetic gene silencing in cancer

Questions to be addressed	Research required
The cancer methylome	Elucidate links between simultaneous losses and gains of DNA methylation in the same cancer cells.
Chromatin boundaries	Determine the molecular nature of boundaries, and how they change during tumorigenesis, that separate areas of transcriptionally active zones encompassing gene promoters from the transcriptionally repressive areas that surround them and which may prevent the repressive chromatin from spreading through the active zone. Among the candidate mechanisms are roles that may be played by key histone modifications, insulator proteins, chromatin-remodeling proteins, etc.
Hierarchy of epigenetic events leading to gene silencing	What is the order of events for the evolution of gene silencing in cancer with respect to histone modifications, DNA methylation, etc.? Which comes first and what are the key protein complexes that target the processes (DNA methylating enzymes, histone deacetylating and methylation enzymes, cytosine methyl-binding proteins, polycomb silencing complexes, etc.) that determine the events?
Targeting and composition of DNA methylation machinery	Which specific DNA methylating enzymes are required for initiating and/or maintaining the most stable gene silencing and what protein complexes contain them, including their interaction with key histone posttranslational modifications?
Composition of the epigenetic machinery in maintaining silencing	Once established, what are all of the components of chromatin and DNA methylation machinery, and the hierarchy of their involvement, required to maintain the gene silencing and how are they reversible?

of TSGs and our understanding of factors involved in DNA methylation homeostasis (i.e., DNMTs, TET enzymes, and IDH1/2). The fundamental defect in cancers with regard to these abnormalities, particularly in proximal gene promoters, appears to be a disruption of chromatin borders that normally separate transcriptionally repressive from active chromatin. In this regard, several laboratories have highlighted how, in cancer cells, chromatin configuration found at hypermethylated CpG islands near the promoters of aberrantly silenced genes differs from when these genes are basally expressed in normal settings (Kelly et al. 2012; Yang et al. 2012). The promoter CpG islands of active genes in normal (or cancer) cells are characterized by a zone of open chromatin, a lack of DNA methylation, nucleosome depletion (detected by hypersensitive sites), and histone PTMs, which are typical for active genes (Fig. 4A) (Kelly et al. 2012; Yang et al. 2012). Active covalent histone marks at gene promoters, which typically become altered along with abnormal DNA methylation in cancer, include acetylation of H3 at lysines 9 and 14 (H3K9ac and H3K14ac) and methylation of H3K4 (Nguyen et al. 2001; Fahrner et al. 2002; McGarvey et al. 2008; Baylin and Jones 2011; Shen and Laird 2013). In addition, the histone variant H2A.Z is present in the nucleosomes flanking the start sites, and its presence is strongly anticorrelated with DNA methylation (Zilberman et al. 2008; Yang et al. 2012).

Beyond the 5′ and 3′ borders of active genes, there appears to be a stark transition in chromatin structure with characteristics of transcriptionally repressed genomic regions (Fig. 4A). Historically, chromatin characterization has been restricted to analyses of relatively short DNA stretches that are biologically relevant. Such studies, using normal cells, revealed that just upstream of promoter CpG islands, the less frequent CpG sites are mostly methylated (Berman et al. 2005; Hansen et al. 2011). These sites were found to recruit methylcytosine-binding proteins (MBDs) and their partners (e.g., HDACs) (illustrated in Fig. 9 of Ch. 15 [Li and Zhang 2014]) and are accessible to enzymes that catalyze repressive histone methylation marks, particularly H3K9me2, accompanied by deacetylation of key histone residues (Nguyen et al. 2001; Fahrner et al. 2002; Kondo et al. 2003; McGarvey et al. 2008; Baylin and Jones 2011; Shen and Laird 2013).

6.2 The Mistargeting of Epigenetic Machinery Recruited by Oncogenic Translocation Products

Several examples showing that chromatin-modifying activities play a role in human cancer have been known for some time (Wolffe 2001). For example, the use of HDACs is altered by chromosomal translocations in AML and acute promyelocytic leukemia (PML) (Di Croce et al. 2002). Histone acetylation is associated with open transcriptionally active chromatin regions (see Ch. 4 [Marmorstein and Zhou 2014]; Ch. 36 [Pirrotta 2014]; Ch. 5 [Seto and Yoshida 2014]). In PML, the PML gene is fused to the retinoic acid receptor (RAR). The PML part of the fusion receptor recruits HDAC and DNA methylation activity and causes a state of transcriptional silencing at RAR target loci (Di Croce et al. 2002). This ultimately participates in a cellular differentiation block (Di Croce et al. 2002). In AML, the DNA-binding domain of the transcription factor AML-1 is fused to a protein called ETO, which, similar to PML, interacts with a HDAC. The mistargeted HDAC contributes to aberrant gene repression, blocking cellular differentiation and ultimately leading to leukemia (Amann et al. 2001).

Another translocation occurring in infants with a highly aggressive form of acute leukemia involves the mixed lineage leukemia (*MLL*) gene, which codes for a histone K methyltransferase (KMT). The *MLL* gene product normally catalyzes the formation of the histone H3K4me3 active mark, which helps to repel the de novo DNA methylation machinery (Popovic and Licht 2012). The *MLL* translocation, however, inactivates the enzyme, thus losing the ability to generate the active histone mark. This fusion gene product can then associate with DNA hypermethylation at some promoters, which may contribute to the disease phenotype (Stumpel et al. 2009). These are just three examples of the direct involvement of chromatin-modifying factors contributing to the oncogenic phenotype.

6.3 The Makeup and Distribution of Typical Epigenomic Alterations in Cancer

More recently, in-depth analyses of CpG methylation across genomes are providing an exciting and enriched look at chromatin transitions at CpG-island-containing promoters, which are prone to abnormal DNA methylation in cancer. These studies suggest that for both normal and cancer cells, there are important configurations across defined, megabase regions of most chromosomes (~100 kb–10 Mb). In normal ES cell and differentiated cell types, the majority of these megabase domains are not CpG rich, although where they occur, these CpGs are heavily methylated, but in a mosaic fashion, across different tissue types, which have been termed partially methylated domains, that is, ~80% methylated (Fig. 7A) (Hansen et al. 2011; Berman et al. 2012; Bert et al. 2013; Shen and Laird 2013). In cancer, substantial loss of normal DNA methylation is found throughout these regions, creating hypomethylated domains, with only ~40%–60% of CpGs methylated, as documented in colon and other cancers (Hansen et al. 2011; Berman et al. 2012; Bert et al. 2013; Shen and Laird 2013). This creates megabase "islands" of reduction, located frequently throughout the genome, commonly termed "hypomethylated blocks" or "domains." Other epigenomic mapping approaches have termed broadly similar regions in cancer as "large organized chromatin K" domains, corresponding to regions rich in histone lysine methylation, such as H3K9 (Wen et al. 2009; Hansen et al. 2011; Hon et al. 2012). A key question that needs deciphering is whether these broadly defined regions are configured in a repressive chromatin environment (e.g., H3K9me3) or in a more open chromatin environment, as depicted in Figure 4B. Data indicate that both exist in cancer, the significance and consequence of which are actively being investigated (Berman et al. 2012; Hon et al. 2012; Brennan et al. 2013; Reddy and Feinberg 2013; Timp and Feinberg 2013).

Of great interest for the focus of this chapter, and what may be most functionally significant about hypomethylated blocks in cancer, is the occurrence of opposite, focal gains in promoter CpG island, or gains in DNA methylation for genes embedded within these regions (Berman et al. 2012). Although there is some disagreement about the exact positioning of this methylation, it seems to be within the CpG islands of gene promoters that reside within hypomethylated domains. These promoter islands are virtually always protected from methylation in normal cells, even when they reside in partially methylated domains best characterized in differentiated cells (Berman et al. 2012). Thus, these large domains may harbor a much higher than expected percentage of genes that are vulnerable to abnormal CpG island DNA hypermethylation (Ohm et al. 2007; Schlesinger et al. 2007; Widschwendter et al. 2007; Berman et al. 2012). Thus, hypomethylated blocks consist of juxtaposed regions of losses and more focal CpG island gains of DNA methylation in cancer (Fig. 7A).

Several laboratories have now identified that hypermethylated genes are heavily biased to Polycomb repressive complex 2 (PRC2)-regulated, H3K27me3-marked genes, in ES and adult stem cells (Fig. 7B) (Ohm et al. 2007; Schlesinger et al. 2007; Widschwendter et al. 2007). Interestingly, the above partially DNA methylated or hypomethylated domains in which many of these genes reside broadly correspond to late replicating and lamin-associated domains at the nuclear periphery, generally associated with repressive chromatin domains and PcG-marked, often bivalent, genes in ES cells (Peric-Hupkes and van Steensel 2010; Peric-Hupkes et al. 2010; Berman et al. 2012). This PcG-mediated transcriptional repression in a stem-cell setting is most often, in the context of bivalent chromatin (i.e., dually marked with H3K27me3 and H3K4me3), thought to mediate a low, poised transcription state for genes important for cell commitment and/or that must not be highly expressed to preserve states of stem cell self-renewal (Bernstein et al. 2006; Chi and Bernstein 2009). Importantly, these bivalently marked promoters are virtually never associated with the presence of DNA methylation in normal cells at any stage of development (Fig. 7B) (Bernstein et al. 2006; Chi and Bernstein 2009; Baylin and Jones 2011; Shen and Laird 2013). A working model envisions a molecular progression during tumorigenesis during which, in the abnormally expanding adult stem or progenitor cell compartments depicted in Figure 6, the bivalent and/or PcG-repressed chromatin at CpG island promoters is replaced with more stable silencing states associated with

 Cite as *Cold Spring Harb Perspect Biol* doi: 10.1101/cshperspect.a019505

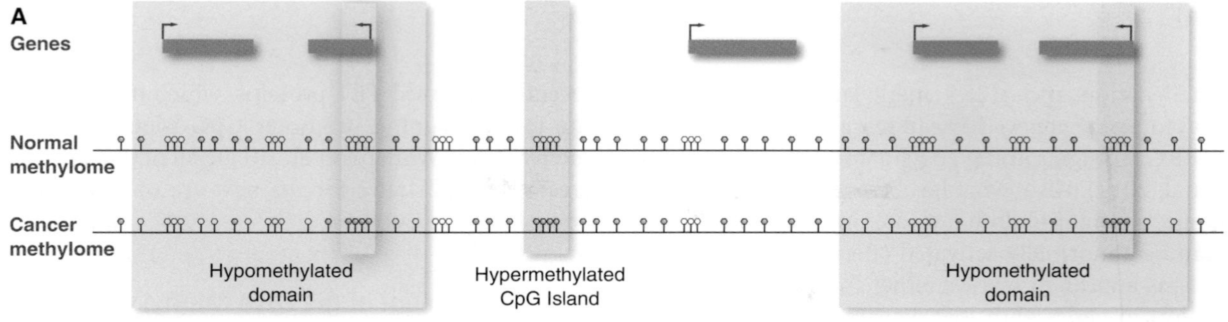

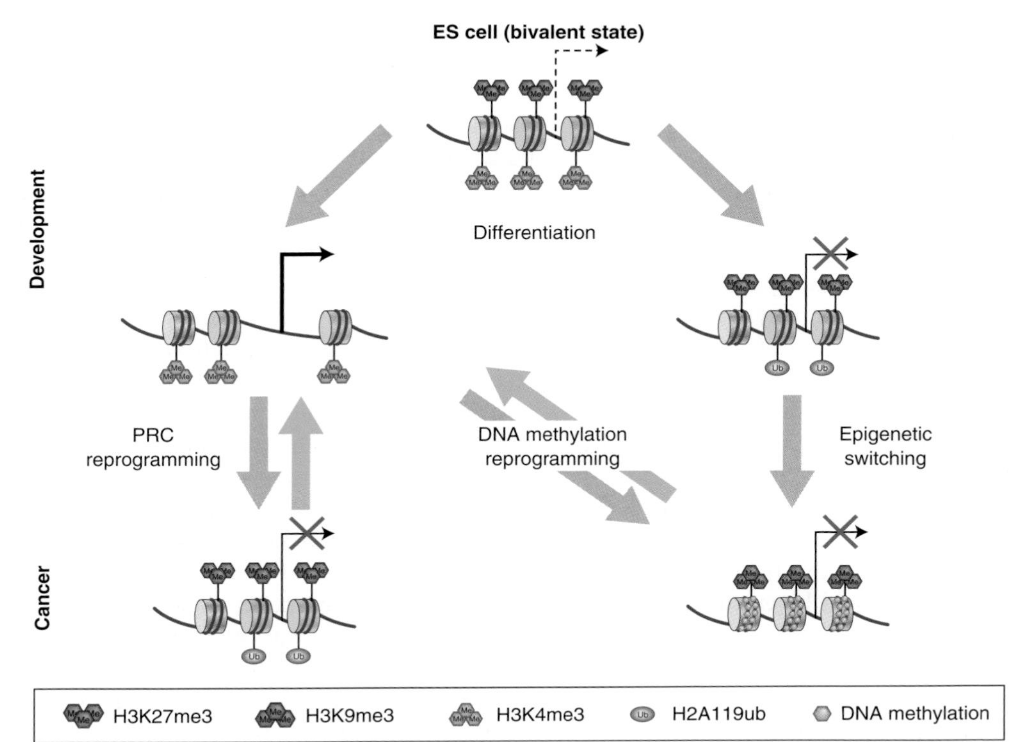

Figure 7. Reprogramming of DNA methylation patterns and abnormal modes of gene silencing in cancer. (*A*) Common DNA methylome changes observable in cancer versus a normal somatic cell are illustrated. This is shown in the context of large hypomethylated blocks (gray shading) of the genome seen in cancer interspersed with focal hypermethylation of promoter region CpG-island-containing genes (pink shading). In normal cells, background DNA methylation is high (pink shaded hexagons) with the exception of CpG islands (densely packed white shaded hexagons). In the cancer methylome, overall genome DNA methylation declines, particularly in the hypomethylated blocks, whereas CpG island promoter genes frequently become methylated (pink shading), most of which are located in the hypomethylated blocks. (*B*) The currently suggested routes to abnormally silenced CpG-island-containing genes in cancer are shown. Genes that are active in cells throughout development and adult cell renewal initially have active promoter chromatin, which is characterized by the presence of the bivalent histone modification pattern consisting of H3K4me, the repressive H3K27me3 mark, and a lack of DNA methylation. Genes that become transcriptionally active lose much of their Polycomb-mediated repressive H3K27 methylation, whereas those that become silenced (indicated by a red X) can do so by the loss of H3K4 methylation and acquisition of, or increases in, Polycomb-mediated repressive chromatin (PRC) mark and H2A119 ubiquitination. During tumor progression, active genes may become silenced through either the aberrant PRC-mediated reprogramming (*bottom left*) or DNA methylation and H3K9me marks (*bottom right*). Some normally silent genes may change the way in which they are transcriptionally repressed from H3K27-methylation-type repression to H3K9-methylation-based silencing and/or DNA hypermethylation (epigenetic switching). The reverse yellow arrows indicate the potential for epigenetic abnormalities in cancer to be corrected by epigenetic therapies. Representative of such therapies are DNMT inhibitors, HDAC inhibitors, KMT inhibitors, and others, as discussed in this and other chapters. These inhibitors can all potentially promote gene activation by producing losses of DNA methylation, or deacetylating lysines, or alleviating silencing mediated by histone methylation PTMs, such as H3K27 methylation. (*A*, Adapted from Reddington et al. 2014; *B*, adapted from Sharma et al. 2010.)

DNA methylation and H3K9 methylation (Ohm et al. 2007), or, for some genes, they can remain in an abnormal state of PRC reprogramming (Baylin and Jones 2011; Easwaran et al. 2012). Also, what has been observed for these long-range chromatin domains are instances in which genes can be abnormally activated (Bert et al. 2013). The mechanisms appear to involve either focal losses of DNA methylation in genes with low-density CpG islands or a switch to alternative transcription start sites, because the canonical sites harbor focal gains in CpG island methylation (Bert et al. 2013).

What is essential for extending the concept of chromatin transitions at bivalently marked genes during tumorigenesis is to unravel the molecular mechanisms underlying this progression. Scenarios can be hypothesized in which abnormal retention of PcG complexes may initially occur and, then, DNA methylation subsequently ensues (i.e., epigenetic switching, Fig. 7B). Once DNA methylation evolves, the PcG complex and accompanying H3K27me3 histone may be completely or quantitatively replaced (Gal-Yam et al. 2008; McGarvey et al. 2008; Bartke et al. 2010). Experimental data corroborating this show that methylated DNA, when in a nucleosomal context, is resistant to the presence of PcG complexes and, hence, the imposition of the silencing H3K27me marks (Schlesinger et al. 2007; Widschwendter et al. 2007; Gal-Yam et al. 2008; Bartke et al. 2010). A scenario in which active bivalent genes are PRC reprogrammed (Fig. 7B) could be explained by surrounding hypomethylation allowing PRC2 access, which could then extend repression to neighboring active genes (reviewed in Reddington et al. 2014). Continued research is needed to understand the interplay between different repressive mechanisms.

6.4 Chromatin Boundaries

We also need to understand that although, in normal cells, CpG islands at promoters of resident genes have a narrow band of protection from the surrounding DNA methylation (O'Hagan et al. 2011; Berman et al. 2012), why does the molecular maintenance of chromatin and DNA methylation boundaries "break down" during tumor progression (Fig. 4B)? One idea is that factors, such as insulator proteins (e.g., CTCF), which separate transcriptionally repressive and active chromatin states, may be altered (Taberlay et al. 2014). Also, the chromatin-modifying machinery may be altered and cause shifts in chromatin boundaries and configuration (O'Hagan et al. 2011). Cancer risk states, such as chronic inflammation and DNA damage, can participate in inducing such shifts (Hahn et al. 2008; O'Hagan et al. 2011). Within these hypomethylated domains, there could also be alterations in function and/or targeting of the recently identified TET proteins, which normally contribute to maintaining promoter CpG islands free of DNA methylation (Williams et al. 2011). All of these possibilities create rich substrates for the next era of defining normal and cancer epigenomes.

6.5 Involvement of the DNA Methylation Machinery in Tumorigenesis

The question of how DNMTs are targeted and then establish and maintain abnormal patterns of DNA methylation in cancer cells needs continued study, most especially the complexes through which these enzymes act cooperatively to target gene promoters and modify DNA methylation patterns. For DNMT1, the protein UHFR1 and proteins associated with it seem to facilitate the targeting of this protein to DNA replication and other sites (Bostick et al. 2007; Nishiyama et al. 2013). Although less is known about the targeting DNMT3A and -B, specific types of DNA configuration (e.g., DNA-RNA triplex structures) may exert targeting effects for DNMT3B (Schmitz et al. 2010). Very importantly, past and present studies indicate that transcriptional repression complexes, which include histone-modifying enzymes, such as methyltransferases (KMTs) and demethylases (KDMs) are key for the recruitment of, or being recruited by, DNMTs, as elaborated in Figure 7 and Section 2.2.2 of Ch. 22 (Almouzni and Cedar 2014) (Di Croce et al. 2002; Fuks et al. 2003; Brenner et al. 2005; O'Hagan et al. 2011). Indeed, some studies suggest that deregulation of the chromatin machinery precedes DNA methylation changes (Bachman et al. 2003; O'Hagan et al. 2011; Sproul et al. 2011, 2012). As noted above, the changes discussed in key cancer risk states, such as chronic inflammation and the buildup of ROS, appear capable of rapidly triggering the assembly of DNMTs with HDAC and MBD protein partners and recruiting them to promoter CpG islands (O'Hagan et al. 2011). In these events, there is a rapid tightening of DNMT1 and SIRT1 (an HDAC) to chromatin. The DNMT1 step seems to be upstream in this process, highlighting the multitasking potential for this protein, in addition to it catalyzing DNA methylation (O'Hagan et al. 2011).

Genetic disruption studies of DNMTs in cultured colon cancer cells indicate that the maintenance of most DNA methylation, including at hypermethylated promoters and its attendant gene silencing, requires both DNMT1 and DNMT3b (Rhee et al. 2000, 2002). Studies in other cancer cell types have produced more variable results (Leu et al. 2003; Jones and Liang 2009). Whatever the mechanism of molecular progression to hypermethylated bivalent genes in cancer, bear in mind that mammalian DNMTs appear to have complex functions, which include not only catalytic DNMT activity at the carboxy-terminal regions, but also

Cite as *Cold Spring Harb Perspect Biol* doi: 10.1101/cshperspect.a019505

direct transcriptional repression activities at their amino-terminal domains (Robertson et al. 2000; Rountree et al. 2000; Fuks et al. 2001; Clements et al. 2012). Thus, a role for DNMTs potentially has many facets in transcriptional silencing, from initiation to maintenance, and is not necessarily restricted to steps involving DNA methylation (see also Ch. 6 [Cheng 2014]; Ch. 7 [Patel 2014]).

7 SUMMARY OF MAJOR RESEARCH ISSUES FOR UNDERSTANDING EPIGENETIC GENE SILENCING IN CANCER

Despite progress in understanding the molecular events that can drive the epigenetic abnormalities typifying the cancer epigenome, Table 6 summarizes some of the most important questions that remain to be resolved through future research. First, molecular events determining the simultaneous appearance of overall DNA hypomethylation and more localized promoter DNA hypermethylation must continue to be elucidated. These juxtaposed states suggest a broad mistargeting of chromatin states in cancer cells. We particularly need to relate DNA methylation patterns in cancer with other chromatin marks, such as H3K9 methylation, histone acetylation, and H3K27 methylation. We also need more research relating how all these epigenetic features are organized in a three-dimensional fashion in normal versus cancer cells, and whether nuclear architecture is a regulating factor in the deregulation that occurs in cells during transformation. What we learn about how the changing cancer epigenome contributes to cancer etiology should prove equally illuminating in understanding how mammalian cells normally package their genomes for proper patterns of gene expression and the maintenance of chromosome integrity.

A second important question will be to identify the determinants and function of chromatin boundaries. This will obviously need to be performed in the context of how DNA methylation patterns around individual gene promoters relate to the general chromatin configuration of other surrounding regions, such as gene enhancers, bodies, and insulators; this will need to be worked out in both normal and abnormal states of transcription. A third consideration is addressing the evolution of chromatin states across the various regulatory regions of cancer relevant loci during the course of tumorigenesis and comparing them to normal developmental scenarios. Fourth, key components of this must be dissected, particularly during specific phases of tumor initiation and progression. This should include assessing the molecular interactions determining the constitution and targeting of DNMTs and other silencing complexes, such as Polycomb repressive complexes (described in Ch. 17 [Grossniklaus and Paro 2014]) and how this relates to gene expression, PTM signatures, and ncRNAs. Also, determining what are truly the causal epigenetic mechanisms that cause TSG silencing must be resolved. Finally, once abnormal heritable gene silencing is established in cancer, what is the precise hierarchy of molecular steps that maintain it? This latter question is not only a key basic question, but also central to the translational implications discussed in Section 8 for using epigenetic abnormalities as cancer biomarkers and, in Section 9, for reversing abnormal gene silencing as a cancer prevention or therapy strategy.

8 DNA METHYLATION ABNORMALITIES AS BIOMARKERS FOR CANCER DETECTION AND MONITORING CANCER PROGNOSIS

The pervasive nature of epigenetic abnormalities being characterized at all stages of cancer development constitutes an ever-increasing pool of potential biomarkers, which can be developed for predicting cancer risk states, the early detection of cancer, and for use as prognostic indicators. Methods that can sensitively detect changes in DNA methylation and chromatin have already been developed and more are being pursued not only for use on tumor and other tissue biopsies, but also those that can be applied to body fluids for noninvasive detection methods.

Focal, promoter region DNA hypermethylation of CpG islands, which is so common in cancer, is currently the most well-studied and developed biomarker. A number of very sensitive polymerase chain reaction (PCR)-based assays have been developed to be used in combination with sodium bisulfite pretreated DNA to detect levels of DNA methylation (Herman et al. 1996; Laird 2003). PCR approaches, such as methylation-specific PCR, now being used quantitatively, and new nano-assay approaches, in which primers are designed to amplify only methylated regions, are very sensitive (Bailey et al. 2010). Other methods to detect methylated DNA include techniques based on real-time PCR, such as "MethyLight" (Campan et al. 2009), in which a fluorescent probe can only bind to methylated DNA. These techniques can detect one methylated allele in a background of about 1000–50,000 alleles, depending on the particular assay design and specific needs of application. Thus, these approaches are applicable to a mixture of cells or even various biological fluids, such as plasma, urine, or sputum (Laird 2003).

Cancer detection by identification of altered cytosine methylation is quite robust because of the inherent stability of DNA compared with RNA or proteins. Also, because altered methylation patterns are often cancer-specific, these approaches may be able to distinguish one type of cancer from another. There are now a host of studies providing "proof of principle" for the use of promoter DNA hypermethylated sequences as an extremely sensitive strategy for predicting cancer risk and/or detection. For example, synchronous detection of abnormal promoter DNA methylation in tumor and chest lymph nodes, which were deemed microscopically free, shows promise for predicting the rapid recurrence of early stage lung cancer (Brock et al. 2008). Similarly, sensitive detection of such abnormalities in DNA from stool may offer a test for predicting the presence of colon tumors (Hong and Ahuja 2013; Imperiale et al. 2014). The detection of both CpG island DNA methylation and specific mutations is even more promising for detecting colon polyps and/or cancer by assaying stool blood DNA (Hong and Ahuja 2013; Imperiale et al. 2014) and this approach is moving toward clinical practice. The clinical value of this approach is being tested in larger studies in which the current hypotheses can be fully validated over the next few years. Likewise, detection of DNA hypermethylated genes in prostate needle biopsies is now being used clinically to augment histological detection of prostate cancer (Van Neste et al. 2012).

Several approaches for using CpG island hypermethylation to predict cancer patients' response to therapies are very promising. Examples include the detection of this change in the promoter of the O^6*MGMT* gene to predict the response to alkylating agents as the main treatment approach in gliomas (Esteller et al. 2000; Hegi et al. 2005). Use of this methylation marker is now becoming standard practice in the management of patients with gliomas. Tumors in which O^6*MGMT* is silenced in association with DNA methylation changes are more sensitive to alkylating therapy because the repair gene is not available to remove guanosine adducts from the genome (Esteller et al. 2000; Hegi et al. 2005). Another recent promising example includes the promoter methylation of *SMAD1* to predict resistance to the chemotherapeutic agent, doxorubicin, in patients with diffuse large B-cell lymphoma (DBCL) (Clozel et al. 2013). The silencing of this gene, when reversed by low doses of DNMT inhibitors, appears to be key in mediating the reversal of this chemoresistance (Clozel et al. 2013). Early findings in a phase I clinical study of patients with DBCL suggests that low doses of azacitidine can prime for increased responses to chemotherapy (Clozel et al. 2013).

9 EPIGENETIC THERAPY

The heritable inactivation of cancer-related genes by altered DNA methylation and chromatin modification has led to the realization that silenced chromatin may represent a viable target for cancer therapy. Thus, a new treatment approach called "epigenetic therapy" has been developed in which drugs that can modify chromatin or DNA methylation patterns are used alone or in combination to affect therapeutic outcomes (Egger et al. 2004; Kelly et al. 2010; Dawson and Kouzarides 2012; Azad et al. 2013; Ahuja et al. 2014).

9.1 DNMT Inhibitors

Powerful mechanism-based inhibitors of DNA cytosine methylation represent the most advanced epigenetic therapeutics currently available for cancer treatment. The nucleoside analogs, 5-aza-CR (Vidaza) and 5-aza-CdR (Dacogen or Decitibine), have been in clinical trials for many years. More recently, a new prodrug-like agent for 5-aza-CdR, called SGI-110, has entered the scene showing promise (Fig. 8) (Chuang et al. 2005; Yoo et al. 2007). These drugs, or the prodrug derivative, are incorporated into the DNA of replicating cells after they have been metabolized to the appropriate deoxynucleoside triphosphate or cleaved by phosphodiesterase in the case of SGI-110 (Chuang et al. 2005; Yoo et al. 2007). Once incorporated into DNA, they interact with all three known DNMTs to form covalent intermediates, which ultimately inhibit DNA methylation in subsequent rounds of DNA synthesis. The mechanism of action of these compounds for blocking the catalytic site of DNMTs is quite well understood and they have been used for some time to reactivate silenced genes in tissue culture or xenograft models (Santi et al. 1984; Ghoshal et al. 2005; Kelly et al. 2010; Tsai and Baylin 2011; Azad et al. 2013). However, it is often overlooked that the above DNA-demethylating agents not only induce the above catalytic block, but also cause degradation of the DNMTs (Ahuja et al. 2014). This latter action is quickly triggered, even by low doses of the drugs, when used in vivo (Tsai and Baylin 2011).This protein loss is potentially very important for the DNMT inhibitors to achieve reexpression of key cancer genes because, experimentally, all three biologically active DNMTs can exert transcriptional repression independent of catalyzing DNA methylation (Fuks et al. 2000; Robertson et al. 2000; Rountree et al. 2000; Bachman et al. 2001; Clements et al. 2012). These latter events are related to the potential scaffolding properties of these proteins with respect to binding key mediators of gene silencing, such as HDAC1 and HDAC2 (Fuks et al. 2000; Robertson et al. 2000; Rountree et al. 2000; Bachman et al. 2001; Clements et al. 2012). Thus, loss of DNMTs as proteins cannot be overlooked as a key event linked to any therapeutic efficacy of the drugs discussed above.

Initially, when used at what we now know in retrospect to be very high doses, the DNMT inhibitors were too toxic to patients to gain any traction for the treatment of cancer. Later, however, as the doses were profoundly lowered, these

Cite as *Cold Spring Harb Perspect Biol* doi: 10.1101/cshperspect.a019505

Compound	Structure	Cancer type	Clinical trial status
DNA methylation inhibitors			
5-Azacytidine 5-Aza-CR (Vidaza)	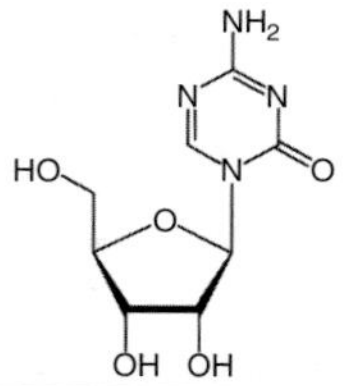	Myelodysplastic syndrome; AML	FDA-approved for MDS in 2004
5-Aza-2′-deoxycytidine 5-Aza-CdR Decitabine (Dacogen)	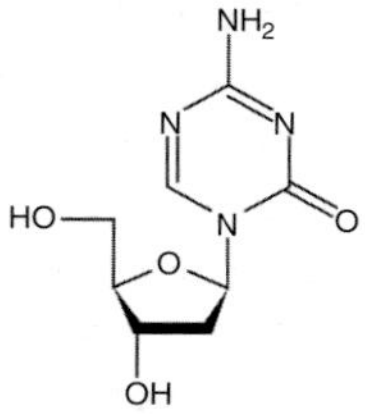	Myelodysplastic syndrome; AML	FDA-approved for MDS in 2006
SGI-110	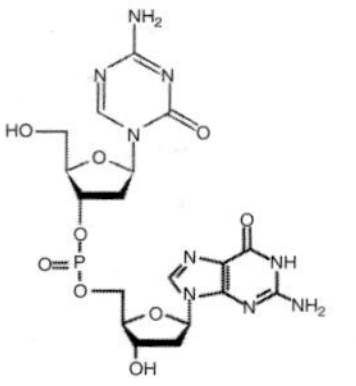	Acute myeloid leukemia; AML	Phase 2
Histone deacetylase inhibitors			
Suberoylanilide hydroxamic acid (SAHA) Vorinostat (Zolinza)	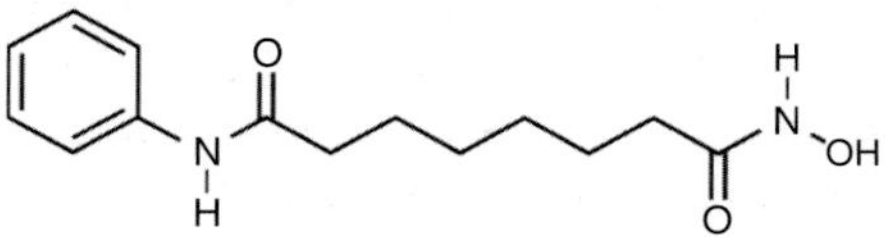	T-cell lymphoma	FDA-approved in 2006
Depsipeptide FK-229 FR901228 Romidepsin (Istodax)	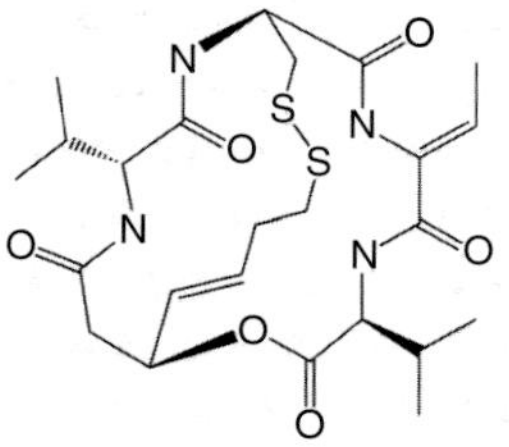	T-cell lymphoma	FDA-approved in 2009

Figure 8. Structures of selected epigenetic drugs. Three nucleoside analogs are known that can inhibit DNA methylation after incorporation into DNA. 5-aza-CR (Vidaza) and 5-aza-CdR (decitabine) have been FDA approved for the treatment of the preleukemic disorder, myelodysplasia. Two HDAC inhibitors are also FDA approved for cutaneous T-cell lymphoma and several others are in clinical trials. Drugs targeting other epigenetic processes are in earlier stages of clinical development (see also Figs. 5 and 6 of Ch. 35 [Audia and Campbell 2014]).

agents have now found application in the treatment of certain hematological malignancies, particularly myelodysplastic syndrome, which is a preleukemic condition occurring mainly in elderly patients (Lubbert 2000; Wijermans et al. 2000; Silverman et al. 2002; Issa et al. 2004). Clinical responses for patients with this disorder, and with leukemias that may have progressed from the preleukemic stage, are becoming increasingly dramatic. Accordingly, drugs with the clinical names Vidaza and Dacogen, for 5-aza-CR and 5-aza-CdR, respectively, have now been approved by the U.S. FDA for the treatment of patients with these disorders (Fig. 8). Although Vidaza and decitabine have been shown to be clinically efficacious, it has been more difficult to establish with clarity whether the targets of drug action are methylated gene promoters. Preliminary experiments suggested that the p15 TSG became demethylated following decitabine treatment (Daskalakis et al. 2002); however, it remains to be shown whether the drugs act by inducing gene expression or some other mechanism, such as triggering an immune response to the tumor. Based on preclinical studies looking at the above responses, and applying the approaches to solid tumor models, it seems likely that at very low nanomolar doses, both Vidaza and Dacogen can "reprogram" cancer cells and cause antitumor responses, which are most likely caused by the specific targeting of DNMTs rather than producing other less off-target effects (Tsai et al. 2012).

Using these concepts, DNA-demethylating agents, poised for new therapeutic applications, may assume a major role in cancer therapy. Newer versions of DNA-demethylating drugs are being developed with these concepts in mind. For example, as noted earlier, SGI-110, which is a dinucleotide prodrug of 5-aza-CdR, is also an inhibitor of DNMTs after cleavage by phosphodiesterases. Also, it has a longer half-life in patients because it is not deaminated by plasma cytidine deaminase, which causes the rapid inactivation of the 5-azanucleosides (Chabot et al. 1983; Qin et al. 2011). To date, effective inhibitors that do not require incorporation into DNA have not been developed, but these might be more desirable in the clinic because they might have fewer side effects. Numerous approaches to synthesize and/or discover such drugs are now ongoing.

9.2 HDAC Inhibitors

Another key set of proteins being targeted for cancer therapy are the HDACs (Dawson and Kouzarides 2012; Bose et al. 2014; West and Johnstone 2014). This large family of enzymes removes acetylation marks from histone tails (as well as other nonhistone proteins), typically acting in the context of larger protein complexes, sometimes associated with DNA methylation, to establish repressive chromatin environments (the topic of Ch. 5 [Seto and Yoshida 2014]). Inhibitors of HDACs (HDACis) have a general transcriptional activating effect and their therapeutic use in cancer treatment has been presumed to be largely through the activation of abnormally silenced TSGs, although this eminently remains to be proven. Two of these inhibitors, suberoylanilide hydroxamic acid (SAHA or Vorinostat) and depsipeptide (Romidepsin), which are more specific inhibitors of HDACs (Fig. 8B), have now been approved by the FDA for the treatment of cutaneous T-cell lymphoma. However, the molecular mechanisms responsible for the unusual sensitivity of this tumor type to these drugs are still unclear. A significantly larger number of drugs are known to cause substantial inhibition of HDACs (see our Fig. 8 and Sec. 7 of Ch. 5 [Seto and Yoshida 2014] for more detail). Some of these, such as 4-phenylbutyrate or valproic acid (VPA), have been in clinical use to treat other conditions for some time (Marks et al. 2001; Richon and O'Brien 2002), whereas newer ones are now in clinical trials. HDACis, used alone, however, have had little success, especially in solid tumors (Azad et al. 2013; Ahuja et al. 2014). Interestingly, preclinical studies have recently suggested that these drugs may be able to reprogram cancer cells in a way that reverses treatment resistance or sensitizes cancers to conventional chemotherapy and newer targeted therapies (Sharma et al. 2010). Pursuant to these concepts, clinical data in patients with advanced non–small cell lung carcinoma (NSCLC) and breast cancer are accruing to corroborate this hypothesis. For example, a newer HDACi called entinostat, combined with the epidermal growth factor receptor inhibitor erlotinib, showed a significant overall survival benefit in patients with recurrent advanced NSCLC (Witta et al. 2012). Also, vorinostat therapy increased response rates significantly in combination with carboplatin and paclitaxel as a front line treatment of patients with metastatic NSCLC, and may extend overall survival (Ramalingam et al. 2010). Moreover, entinostat significantly increased survival of patients with advanced breast cancer in combination with an aromatase inhibitor (Yardley et al. 2013).

9.3 Epigenetic Drug Development

The clinical successes with current epigenetic drugs have led to a big increase in interest from the pharmaceutical industry in developing compounds that will target epigenetic abnormalities in cancers (Kelly et al. 2010; Dawson et al. 2011; Arrowsmith et al. 2012). The challenges and strategies being adopted by the research and industry sector are discussed in Section 3 of Ch. 35 (Audia and Campbell 2014). Examples of epigenetic drugs in development include a potent small-molecule DOT1L inhibitor, which can selectively kill MLL cells (Daigle et al. 2011). Inhibitors of BRD4 represent another class of small-molecule epigenetic therapeutics that has been developed to interfere with their capacity for reading histone acetyllysine marks (Filippakopoulos et al. 2010; Nicodeme et al. 2010; also reviewed in Ch. 2 [Qi 2014; Schaefer 2014]). The BRD4 protein constitutes part of the machinery that activates transcription and, in particular, may be key for multiple gene activation events controlled by the pervasive oncogene, c-MYC (illustrated in Fig. 1 of Ch. 2 [Qi 2014]) (Filippakopoulos et al. 2010; Delmore et al. 2011; Zuber et al. 2011; Dawson and Kouzarides 2012). The BRD4 inhibitors appear very effective in preclinical studies for the

 Cite as *Cold Spring Harb Perspect Biol* doi: 10.1101/cshperspect.a019505

treatment of MLL-fusion leukemias (Dawson et al. 2011) and might be a therapeutic strategy for countering c-MYC overactivity (Delmore et al. 2011).

9.4 Combination Epigenetic Therapy

One of the major concepts emerging from all of the above clinical trial and drug development activities is that of combination epigenetic therapy. This is currently being tested in the clinic for the older drugs, targeting DNA demethylation and inhibiting HDACs. This will surely emerge for the newer drugs either in combination with these above drugs or other novel combinatorial strategies. In terms of the older drugs, the approach has been to exploit preclinical data showing that blocking HDAC activity subsequent to inhibiting DNA methylation can additively lead to the reexpression of DNA hypermethylated genes (Cameron et al. 1999; Suzuki et al. 2002; Cai et al. 2014). This concept exploits the fact that interactions between HDAC-mediated histone deacetylation (especially via HDAC1 and -2), collaborate with DNA methylation for the silencing of these genes (Cameron et al. 1999; Suzuki et al. 2002; Cai et al. 2014), as discussed in Section 6. This treatment paradigm has been applied in the clinical treatment of hematologic malignancies. The first study used Vidaza and the older HDACi, sodium phenyl butyrate, on patients with myelodysplastic syndrome and AML (Gore et al. 2006). This was well tolerated, and clinical responses were frequent, with five of 14 patients achieving complete or partial response. Another pilot study resulted in three out of 10 patients with myelodysplastic syndrome or AML developing a partial response (Maslak et al. 2006). Investigators at M.D. Anderson Cancer Center administered decitabine and VPA, and 12 of 54 patients achieved complete remissions (Garcia-Manero 2008). Subsequently, a study of Vidaza and VPA also suggests increased efficacy in high-risk myelodysplastic syndrome (Voso et al. 2009).

Controversy over the efficacy of DNA-demethylating agents used in combination with HDAC is in myelodysplastic syndrome/AML has arisen in subsequent studies. Thus, the U.S. Leukemia Intergroup undertook a study with a randomized phase using entinostat (HDACi) in combination with Vidaza (DNMTi). The combination did not show increased efficacy and suggested even less efficacy (Prebet et al. 2014). The reasons for the mixed results are not clear, but the approach still bears promise for myelodysplastic syndrome/AML. However, it remains to be established whether combination therapies are more effective than single-agent demethylating therapies, how to best use such agents together, and what molecular mechanisms account for any efficacies seen.

Much less has been performed to test the efficacy of combination therapy in solid tumors. A recent study on a lung cancer model in mice has shown promise that DNA methylation inhibitors (e.g., azacytidine) and HDAC inhibitors (e.g., entinostat) might have strong synergistic, antitumor effects (Belinsky et al. 2011). Closely related to this, recently completed clinical trials in 65 patients with advanced lung cancer, the deadliest of all human cancers, show promise that these approaches can, in a small subset of patients, induce robust, durable responses (Juergens et al. 2011). Moreover, in these same trials, there is an early indication that the epigenetic therapy may lead to sensitization to subsequent therapies in many more (Juergens et al. 2011). These latter include not only standard chemotherapies, but also, intriguingly, an exciting new immunotherapy (Brahmer et al. 2012; Topalian et al. 2012), which targets breaking lymphocyte immune tolerance to render these cells immune competent (Wrangle et al. 2013). This last possibility is backed in the laboratory by work suggesting that, in lung cancer cells and other solid tumor types, DNA-demethylating agents up-regulate a very complex, immune attraction effect with pathways harboring hundreds of genes (Wrangle et al. 2013; Li et al. 2014). Both the possibilities for sensitizing patients with advanced lung cancer to chemotherapy and immunotherapy are now being tested in larger trials that are under way. It is worthwhile noting that others are reporting that DNA-demethylating agents can sensitize patients with advanced ovarian cancer to subsequent chemotherapy (Matei et al. 2012), as well as the aforementioned beneficial effects of HDACi in sensitizing patients to chemotherapy.

Combinations of newer agents targeting additional steps in chromatin assembly are just starting to be explored at the preclinical level for cancer therapy paradigms. Examples include the finding that synergistic antitumor activity is achieved using a BRD4 inhibitor and an HDACi, and when using a LSD1 inhibitor and HDACi in the treatment of human AML cells (Fiskus et al. 2014a, 2014b). The concept underlying the first therapeutic strategy relies on combinatorially activating the histone acetylation pathway via HDAC is to reexpress abnormally repressed TSGs, whereas the BET (double bromodomain proteins) inhibitor interferes with *myc* oncogene-activated genes. The second strategy also activates histone acetylation while boosting the H3K4me3 activating mark to combinatorially target and activate abnormally repressed genes.

In summary, the concept of epigenetic therapy for cancer has a rationale and expanding basis in theory, and clinical efficacies are emerging, which suggest great promise. However, much needs to be performed at a mechanistic and clinical level to realize this promise, especially for the common human cancers. One problem that is broadly discussed is the lack of specificity of some of the older agents being used, like the DNA-demethylating agents. However, most steps in epigenetic regulation control many genes and pathways in normal and cancer cells (Jones and Baylin 2007; Baylin and Jones 2011; Jones 2012). This is the very nature of epigenetic control of a cell program. In cancer, the epigenome is widely altered and drugs that can broadly "reprogram" such cells and blunt many tumor pathways may be the most valuable (Jones and Baylin 2007; Baylin and Jones 2011; Dawson et al. 2012; Azad et al. 2013; Ahuja et al. 2014). These arguments are not to say that targeting individual genes abnormally regulated in cancer would not be a highly desirable goal for personalizing cancer therapy. A second problem is always the possibility of the inadvertent reactivation of normal genes as the result of therapy. This, however, in terms of contribution to therapy related toxicities has not been documented. Nor has increased tumorigencitiy associated with this possibility been noted in scenarios such as MD/AML, in which the older epigenetic therapy drugs have been used the longest. Thus, the cancer epigenome as a target for cancer therapies remains a vital possibility and one for which exciting advances are anticipated in the coming years.

REFERENCES

*Reference is also in this book.

Ahuja N, Easwaran H, Baylin SB. 2014. Harnessing the potential of epigenetic therapy to target solid tumors. *J Clin Invest* **124:** 56–63.

* Almouzni G, Cedar H. 2014. Maintenance of epigenetic information. *Cold Spring Harb Perspect Biol* doi: 10.1101/cshperspect.a019372.

Amann JM, Nip J, Strom DK, Lutterbach B, Harada H, Lenny N, Downing JR, Meyers S, Hiebert SW. 2001. ETO, a target of t(8;21) in acute leukemia, makes distinct contacts with multiple histone deacetylases and binds mSin3A through its oligomerization domain. *Mol Cell Biol* **21:** 6470–6483.

Araki S, Doi H, Sano Y, Tanaka S, Miyake Y. 2009. Preparation and CO(2) adsorption properties of aminopropyl-functionalized mesoporous silica microspheres. *J Colloid Interface Sci* **339:** 382–389.

Arrowsmith CH, Bountra C, Fish PV, Lee K, Schapira M. 2012. Epigenetic protein families: A new frontier for drug discovery. *Nat Rev Drug Discov* **11:** 384–400.

Aryee MJ, Liu W, Engelmann JC, Nuhn P, Gurel M, Haffner MC, Esopi D, Irizarry RA, Getzenberg RH, Nelson WG, et al. 2013. DNA methylation alterations exhibit intraindividual stability and interindividual heterogeneity in prostate cancer metastases. *Sci Transl Med* **5:** 169ra110.

* Audia JE, Campbell RM. 2014. Histone modifications and cancer. *Cold Spring Harb Perspect Biol* doi: 10.1101/cshperspect.a019521.

Azad N, Zahnow CA, Rudin CM, Baylin SB. 2013. The future of epigenetic therapy in solid tumours—Lessons from the past. *Nat Rev Clin Oncol* **10:** 256–266.

Bachman KE, Rountree MR, Baylin SB. 2001. Dnmt3a and Dnmt3b are transcriptional repressors that exhibit unique localization properties to heterochromatin. *J Biol Chem* **276:** 32282–32287.

Bachman KE, Park BH, Rhee I, Rajagopalan H, Herman JG, Baylin SB, Kinzler KW, Vogelstein B. 2003. Histone modifications and silencing prior to DNA methylation of a tumor suppressor gene. *Cancer Cell* **3:** 89–95.

Bagchi A, Papazoglu C, Wu Y, Capurso D, Brodt M, Francis D, Bredel M, Vogel H, Mills AA. 2007. CHD5 is a tumor suppressor at human 1p36. *Cell* **128:** 459–475.

Bailey VJ, Zhang Y, Keeley BP, Yin C, Pelosky KL, Brock M, Baylin SB, Herman JG, Wang TH. 2010. Single-tube analysis of DNA methylation with silica superparamagnetic beads. *Clin Chem* **56:** 1022–1025.

Balakrishnan A, Bleeker FE, Lamba S, Rodolfo M, Daniotti M, Scarpa A, van Tilborg AA, Leenstra S, Zanon C, Bardelli A. 2007. Novel somatic and germline mutations in cancer candidate genes in glioblastoma, melanoma, and pancreatic carcinoma. *Cancer Res* **67:** 3545–3550.

* Barlow DP, Bartolomei MS. 2014. Genomic imprinting in mammals. *Cold Spring Harb Perspect Biol* **6:** a018382.

Bartke T, Vermeulen M, Xhemalce B, Robson SC, Mann M, Kouzarides T. 2010. Nucleosome-interacting proteins regulated by DNA and histone methylation. *Cell* **143:** 470–484.

Baylin SB, Jones PA. 2011. A decade of exploring the cancer epigenome—Biological and translational implications. *Nat Rev Cancer* **11:** 726–734.

Baylin SB, Fearon ER, Vogelstein B, de Bustros A, Sharkis SJ, Burke PJ, Staal SP, Nelkin BD. 1987. Hypermethylation of the 5′ region of the calcitonin gene is a property of human lymphoid and acute myeloid malignancies. *Blood* **70:** 412–417.

Belinsky SA, Grimes MJ, Picchi MA, Mitchell HD, Stidley CA, Tesfaigzi Y, Channell MM, Liu Y, Casero RA Jr, Baylin SB, et al. 2011. Combination therapy with Vidaza and entinostat suppresses tumor growth and reprograms the epigenome in an orthotopic lung cancer model. *Cancer Res* **71:** 454–462.

* Berger SL, Sassone-Corsi P. 2014. Metabolic signaling to chromatin. *Cold Spring Harb Perspect Biol* doi: 10.1101/cshperspect.a019463.

Berman H, Zhang J, Crawford YG, Gauthier ML, Fordyce CA, McDermott KM, Sigaroudinia M, Kozakiewicz K, Tlsty TD. 2005. Genetic and epigenetic changes in mammary epithelial cells identify a subpopulation of cells involved in early carcinogenesis. *Cold Spring Harb Symp Quant Biol* **70:** 317–327.

Berman BP, Weisenberger DJ, Aman JF, Hinoue T, Ramjan Z, Liu Y, Noushmehr H, Lange CP, van Dijk CM, Tollenaar RA, et al. 2012. Regions of focal DNA hypermethylation and long-range hypomethylation in colorectal cancer coincide with nuclear lamina-associated domains. *Nat Genet* **44:** 40–46.

Bernstein BE, Mikkelsen TS, Xie X, Kamal M, Huebert DJ, Cuff J, Fry B, Meissner A, Wernig M, Plath K, et al. 2006. A bivalent chromatin structure marks key developmental genes in embryonic stem cells. *Cell* **125:** 315–326.

Bernstein BE, Stamatoyannopoulos JA, Costello JF, Ren B, Milosavljevic A, Meissner A, Kellis M, Marra MA, Beaudet AL, Ecker JR, et al. 2010. The NIH Roadmap Epigenomics Mapping Consortium. *Nat Biotechnol* **28:** 1045–1048.

Bernt KM, Zhu N, Sinha AU, Vempati S, Faber J, Krivtsov AV, Feng Z, Punt N, Daigle A, Bullinger L, et al. 2011. MLL-rearranged leukemia is dependent on aberrant H3K79 methylation by DOT1L. *Cancer Cell* **20:** 66–78.

Bert SA, Robinson MD, Strbenac D, Statham AL, Song JZ, Hulf T, Sutherland RL, Coolen MW, Stirzaker C, Clark SJ. 2013. Regional activation of the cancer genome by long-range epigenetic remodeling. *Cancer Cell* **23:** 9–22.

Borodovsky A, Salmasi V, Turcan S, Fabius AW, Baia GS, Eberhart CG, Weingart JD, Gallia GL, Baylin SB, Chan TA, et al. 2013. 5-azacytidine reduces methylation, promotes differentiation and induces tumor regression in a patient-derived IDH1 mutant glioma xenograft. *Oncotarget* **4:** 1737–1747.

Bose P, Dai Y, Grant S. 2014. Histone deacetylase inhibitor (HDACI) mechanisms of action: Emerging insights. *Pharmacol Ther* **pii: S0163-7258:** 00092–00098. doi: 10.1016/j.pharmthera.2014.04.004.

Bostick M, Kim JK, Esteve PO, Clark A, Pradhan S, Jacobsen SE. 2007. UHRF1 plays a role in maintaining DNA methylation in mammalian cells. *Science* **317:** 1760–1764.

Brahmer JR, Tykodi SS, Chow LQ, Hwu WJ, Topalian SL, Hwu P, Drake CG, Camacho LH, Kauh J, Odunsi K, et al. 2012. Safety and activity of anti-PD-L1 antibody in patients with advanced cancer. *N Engl J Med* **366:** 2455–2465.

Brennan CW, Verhaak RG, McKenna A, Campos B, Noushmehr H, Salama SR, Zheng S, Chakravarty D, Sanborn JZ, Berman SH, et al. 2013. The somatic genomic landscape of glioblastoma. *Cell* **155:** 462–477.

Brenner C, Deplus R, Didelot C, Loriot A, Vire E, De Smet C, Gutierrez A, Danovi D, Bernard D, Boon T, et al. 2005. Myc represses transcription through recruitment of DNA methyltransferase corepressor. *Embo J* **24:** 336–346.

Briggs KJ, Corcoran-Schwartz IM, Zhang W, Harcke T, Devereux WL, Baylin SB, Eberhart CG, Watkins DN. 2008. Cooperation between the Hic1 and Ptch1 tumor suppressors in medulloblastoma. *Genes Dev* **22:** 770–785.

Brock MV, Hooker CM, Ota-Machida E, Han Y, Guo M, Ames S, Glockner S, Piantadosi S, Gabrielson E, Pridham G, et al. 2008. DNA methylation markers and early recurrence in stage I lung cancer. *N Engl J Med* **358:** 1118–1128.

Burbee D, Forgacs E, Zochbauer-Muller S. 2001. Epigenetic inactivation of RASSF1A in lung and breast cancers and malignant phenotype suppression. *J Natl Cancer Inst* **93:** 691–699.

Cai Y, Geutjes EJ, de Lint K, Roepman P, Bruurs L, Yu LR, Wang W, van Blijswijk J, Mohammad H, de Rink I, et al. 2014. The NuRD complex cooperates with DNMTs to maintain silencing of key colorectal tumor suppressor genes. *Oncogene* **33:** 2157–2168.

Cameron EE, Bachman KE, Myohanen S, Herman JG, Baylin SB. 1999. Synergy of demethylation and histone deacetylase inhibition in the re-expression of genes silenced in cancer. *Nat Genet* **21:** 103–107.

Campan M, Weisenberger DJ, Trinh B, Laird PW. 2009. MethyLight. *Methods Mol Biol* **507:** 325–337.

Cancer Genome Atlas Research Network. 2012a. An integrated encyclopedia of DNA elements in the human genome. *Nature* **489:** 57–74.

Cancer Genome Atlas Research Network. 2012b. Comprehensive genome characterization of squamous cell lung cancer. *Nature* **489:** 519–525.

Cancer Genome Atlas Research Network. 2013a. Comprehensive molecular characterization of clear cell renal cell carcinoma. *Nature* **499:** 43–49.

Cancer Genome Atlas Research Network. 2013b. Genomic and epigenomic landscapes of adult de novo acute myeloid leukemia. *N Engl J Med* **368:** 2059–2074.

Chabot GG, Bouchard J, Momparler RL. 1983. Kinetics of deamination of 5-aza-2′-deoxycytidine and cytosine arabinoside by human liver cytidine deaminase and its inhibition by 3-deazauridine, thymidine or uracil arabinoside. *Biochem Pharmacol* **32:** 1327–1328.

Chaffer CL, Marjanovic ND, Lee T, Bell G, Kleer CG, Reinhardt F, D'Alessio AC, Young RA, Weinberg RA. 2013. Poised chromatin at the ZEB1 promoter enables breast cancer cell plasticity and enhances tumorigenicity. *Cell* **154:** 61–74.

Chan K-M, Fang D, Gan H, Hashizume R, Yu C, Schroeder M, Gupta N, Mueller S, James CD, Jenkins R, et al. 2013. The histone H3.3K27M mutation in pediatric glioma reprograms H3K27 methylation and gene expression. *Genes Dev* **27:** 985–990.

Chase A, Cross NC. 2011. Aberrations of EZH2 in cancer. *Clin Cancer Res* **17:** 2613–2618.

Chen RZ, Pettersson U, Beard C, Jackson-Grusby L, Jaenisch R. 1998. DNA hypomethylation leads to elevated mutation rates. *Nature* **395:** 89–93.

Chen WY, Zeng X, Carter MG, Morrell CN, Chiu Yen RW, Esteller M, Watkins DN, Herman JG, Mankowski JL, Baylin SB. 2003. Heterozygous disruption of Hic1 predisposes mice to a gender-dependent spectrum of malignant tumors. *Nat Genet* **33:** 197–202.

Chen W, Cooper TK, Zahnow CA, Overholtzer M, Zhao Z, Ladanyi M, Karp JE, Gokgoz N, Wunder JS, Andrulis IL, et al. 2004. Epigenetic and genetic loss of Hic1 function accentuates the role of p53 in tumorigenesis. *Cancer Cell* **6:** 387–398.

Chen WY, Wang DH, Yen RC, Luo J, Gu W, Baylin SB. 2005. Tumor suppressor HIC1 directly regulates SIRT1 to modulate p53-dependent DNA-damage responses. *Cell* **123:** 437–448.

* Cheng X. 2014. Structural and functional coordination of DNA and histone methylation. *Cold Spring Harb Perspect Biol* **6:** a018747.

Chi AS, Bernstein BE. 2009. Developmental biology. Pluripotent chromatin state. *Science* **323:** 220–221.

Chuang JC, Yoo CB, Kwan JM, Li TW, Liang G, Yang AS, Jones PA. 2005. Comparison of biological effects of non-nucleoside DNA methylation inhibitors versus 5-aza-2′-deoxycytidine. *Mol Cancer Ther* **4:** 1515–1520.

Clements EG, Mohammad HP, Leadem BR, Easwaran H, Cai Y, Van Neste L, Baylin SB. 2012. DNMT1 modulates gene expression without its catalytic activity partially through its interactions with histone-modifying enzymes. *Nucleic Acids Res* **40:** 4334–4346.

Clozel T, Yang S, Elstrom RL, Tam W, Martin P, Kormaksson M, Banerjee S, Vasanthakumar A, Culjkovic B, Scott DW, et al. 2013. Mechanism-based epigenetic chemosensitization therapy of diffuse large B-cell lymphoma. *Cancer Discov* **3:** 1002–1019.

Daigle SR, Olhava EJ, Therkelsen CA, Majer CR, Sneeringer CJ, Song J, Johnston LD, Scott MP, Smith JJ, Xiao Y, et al. 2011. Selective killing of mixed lineage leukemia cells by a potent small-molecule DOT1L inhibitor. *Cancer Cell* **20:** 53–65.

Dammann R, Li C, Yoon J, Chin P, Bates S, Pfeifer G. 2000. Epigenetic inactivation of a RAS association domain family protein from the lung tumor suppressor locus 3p21.3. *Nat Genet* **25:** 315–319.

Daskalakis M, Nguyen TT, Nguyen C, Guldberg P, Kohler G, Wijermans P, Jones PA, Lubbert M. 2002. Demethylation of a hypermethylated P15/INK4B gene in patients with myelodysplastic syndrome by 5-Aza-2′-deoxycytidine (decitabine) treatment. *Blood* **100:** 2957–2964.

Dawson MA, Kouzarides T. 2012. Cancer epigenetics: From mechanism to therapy. *Cell* **150:** 12–27.

Dawson MA, Prinjha RK, Dittmann A, Giotopoulos G, Bantscheff M, Chan WI, Robson SC, Chung CW, Hopf C, Savitski MM, et al. 2011. Inhibition of BET recruitment to chromatin as an effective treatment for MLL-fusion leukaemia. *Nature* **478:** 529–533.

Dawson MA, Kouzarides T, Huntly BJ. 2012. Targeting epigenetic readers in cancer. *N Engl J Med* **367:** 647–657.

De Carvalho DD, You JS, Jones PA. 2010. DNA methylation and cellular reprogramming. *Trends Cell Biology* **20:** 609–617.

de Zwaan SE, Haass NK. 2010. Genetics of basal cell carcinoma. *Australasian J Dermatol* **51:** 81–92; quiz 93–84.

Dedeurwaerder S, Defrance M, Calonne E, Denis H, Sotiriou C, Fuks F. 2011. Evaluation of the Infinium Methylation 450K technology. *Epigenomics* **3:** 771–784.

* Dekker J, Misteli T. 2014. Long-range chromatin interactions. *Cold Spring Harb Perspect Biol* doi: 10.1101/cshperspect.a019356.

Delmore JE, Issa GC, Lemieux ME, Rahl PB, Shi J, Jacobs HM, Kastritis E, Gilpatrick T, Paranal RM, Qi J, et al. 2011. BET bromodomain inhibition as a therapeutic strategy to target c-Myc. *Cell* **146:** 904–917.

Di Croce L, Raker VA, Corsaro M, Fazi F, Fanelli M, Faretta M, Fuks F, Lo Coco F, Kouzarides T, Nervi C, et al. 2002. Methyltransferase recruitment and DNA hypermethylation of target promoters by an oncogenic transcription factor. *Science* **295:** 1079–1082.

Drost J, Mantovani F, Tocco F, Elkon R, Comel A, Holstege H, Kerkhoven R, Jonkers J, Voorhoeve PM, Agami R, et al. 2010. BRD7 is a candidate tumour suppressor gene required for p53 function. *Nat Cell Biol* **12:** 380–389.

Easwaran H, Johnstone SE, Van Neste L, Ohm J, Mosbruger T, Wang Q, Aryee MJ, Joyce P, Ahuja N, Weisenberger D, et al. 2012. A DNA hypermethylation module for the stem/progenitor cell signature of cancer. *Genome Res* **22:** 837–849.

Easwaran H, Tsai HC, Baylin SB. 2014. Cancer epigenetics: Tumor heterogeneity, plasticity of stem-like states, and drug resistance. *Mol Cell* **54:** 716–727.

Egger G, Liang G, Aparicio A, Jones PA. 2004. Epigenetics in human disease and prospects for epigenetic therapy. *Nature* **429:** 457–463.

Ehrlich M, Lacey M. 2013. DNA hypomethylation and hemimethylation in cancer. *Adv Exp Med Biol* **754:** 31–56.

Ehrlich M, Gama-Sosa MA, Huang LH, Midgett RM, Kuo KC, McCune RA, Gehrke C. 1982. Amount and distribution of 5-methylcytosine in human DNA from different types of tissues of cells. *Nucl Acids Res* **10:** 2709–2721.

ENCODE 2012. Decoding ENCODE. *Nat Chem Biol* **8:** 871.

Esteller M. 2007. Cancer epigenomics: DNA methylomes and histone-modification maps. *Nat Rev Genet* **8:** 286–298.

Esteller M. 2008. Epigenetics in cancer. *N Engl J Med* **358:** 1148–1159.

Esteller M, Garcia-Foncillas J, Andion E, Goodman SN, Hidalgo OF, Vanaclocha V, Baylin SB, Herman JG. 2000. Inactivation of the DNA-repair gene MGMT and the clinical response of gliomas to alkylating agents. *N Engl J Med* **343:** 1350–1354.

Esteller M, Fraga MF, Guo M, Garcia-Foncillas J, Hedenfalk I, Godwin AK, Trojan J, Vaurs-Barriere C, Bignon YJ, Ramus S, et al. 2001a. DNA methylation patterns in hereditary human cancers mimic sporadic tumorigenesis. *Hum Mol Genet* **10:** 3001–3007.

Esteller M, Risques RA, Toyota M, Capella G, Moreno V, Peinado MA, Baylin SB, Herman JG. 2001b. Promoter hypermethylation of the DNA repair gene O(6)-methylguanine-DNA methyltransferase is associated with the presence of G:C to A:T transition mutations in p53 in human colorectal tumorigenesis. *Cancer Res* **61:** 4689–4692.

Fahrner JA, Eguchi S, Herman JG, Baylin SB. 2002. Dependence of histone modifications and gene expression on DNA hypermethylation in cancer. *Cancer Res* **62:** 7213–7218.

Feinberg AP, Vogelstein B. 1983. Hypomethylation distinguishes genes of some human cancers from their normal counterparts. *Nature* **301:** 89–92.

Figueroa ME, Abdel-Wahab O, Lu C, Ward PS, Patel J, Shih A, Li Y, Bhagwat N, Vasanthakumar A, Fernandez HF, et al. 2010. Leukemic IDH1 and IDH2 mutations result in a hypermethylation phenotype, disrupt TET2 function, and impair hematopoietic differentiation. *Cancer Cell* **18:** 553–567.

Filippakopoulos P, Qi J, Picaud S, Shen Y, Smith WB, Fedorov O, Morse EM, Keates T, Hickman TT, Felletar I, et al. 2010. Selective inhibition of BET bromodomains. *Nature* **468:** 1067–1073.

Finch PW, He X, Kelley MJ, Uren A, Schaudies RP, Popescu NC, Rudikoff S, Aaronson SA, Varmus HE, Rubin JS. 1997. Purification and molecular cloning of a secreted, Frizzled-related antagonist of Wnt action. *Proc Natl Acad Sci* **94:** 6770–6775.

Fishel R, Lescoe MK, Rao MR, Copeland NG, Jenkins NA, Garber J, Kane M, Kolodner R. 1993. The human mutator gene homolog MSH2 and its association with hereditary nonpolyposis colon cancer. *Cell* **75:** 1027–1038.

Fiskus W, Sharma S, Qi J, Valenta JA, Schaub LJ, Shah B, Peth K, Portier BP, Rodriguez M, Devaraj SG, et al. 2014a. Highly active combination of BRD4 antagonist and histone deacetylase inhibitor against human acute myelogenous leukemia cells. *Mol Cancer Ther* **13:** 1142–1154.

Fiskus W, Sharma S, Shah B, Portier BP, Devaraj SG, Liu K, Iyer SP, Bearss D, Bhalla KN. 2014b. Highly effective combination of LSD1 (KDM1A) antagonist and pan-histone deacetylase inhibitor against human AML cells. *Leukemia.* doi: 10.1038/leu.2014.119.

Flatau E, Gonzales FA, Michalowsky LA, Jones PA. 1984. DNA methylation in 5-aza-2′-deoxycytidine-resistant variants of C3H 10T1/2 C18 cells. *Mol Cell Biol* **4:** 2098–2102.

Frost JK, Ball WC Jr, Levin ML, Tockman MS, Baker RR, Carter D, Eggleston JC, Erozan YS, Gupta PK, Khouri NF, et al. 1984. Early lung cancer detection: Results of the initial (prevalence) radiologic and cytologic screening in the Johns Hopkins study. *Am Rev Respir Dis* **130:** 549–554.

Fuks F, Burgers WA, Brehm A, Hughes-Davies L, Kouzarides T. 2000. DNA methyltransferase Dnmt1 associates with histone deacetylase activity. *Nat Genet* **24:** 88–91.

Fuks F, Burgers WA, Godin N, Kasai M, Kouzarides T. 2001. Dnmt3a binds deacetylases and is recruited by a sequence-specific repressor to silence transcription. *Embo J* **20:** 2536–2544.

Fuks F, Hurd PJ, Deplus R, Kouzarides T. 2003. The DNA methyltransferases associate with HP1 and the SUV39H1 histone methyltransferase. *Nucleic Acids Res* **31:** 2305–2312.

Gal-Yam EN, Egger G, Iniguez L, Holster H, Einarsson S, Zhang X, Lin JC, Liang G, Jones PA, Tanay A. 2008. Frequent switching of Polycomb repressive marks and DNA hypermethylation in the PC3 prostate cancer cell line. *Proc Natl Acad Sci* **105:** 12979–12984.

Garcia-Manero G. 2008. Demethylating agents in myeloid malignancies. *Curr Opin Oncol* **20:** 705–710.

Garraway LA, Lander ES. 2013. Lessons from the cancer genome. *Cell* **153:** 17–37.

Gaudet F, Hodgson JG, Eden A, Jackson-Grusby L, Dausman J, Gray JW, Leonhardt H, Jaenisch R. 2003. Induction of tumors in mice by genomic hypomethylation. *Science* **300:** 489–492.

Gazzoli I, Loda M, Garber J, Syngal S, Kolodner RD. 2002. A hereditary nonpolyposis colorectal carcinoma case associated with hypermethylation of the *MLH1* gene in normal tissue and loss of heterozygosity of the unmethylated allele in the resulting microsatellite instability-high tumor. *Cancer Res* **62:** 3925–3928.

Ghoshal K, Datta J, Majumder S, Bai S, Kutay H, Motiwala T, Jacob ST. 2005. 5-Aza-deoxycytidine induces selective degradation of DNA methyltransferase 1 by a proteasomal pathway that requires the KEN box, bromo-adjacent homology domain, and nuclear localization signal. *Mol Cell Biol* **25:** 4727–4741.

Gore SD, Baylin S, Sugar E, Carraway H, Miller CB, Carducci M, Grever M, Galm O, Dauses T, Karp JE, et al. 2006. Combined DNA methyltransferase and histone deacetylase inhibition in the treatment of myeloid neoplasms. *Cancer Res* **66:** 6361–6369.

Grady WM, Willis J, Guilford PJ, Dunbier AK, Toro TT, Lynch H, Wiesner G, Ferguson K, Eng C, Park JG, et al. 2000. Methylation of the CDH1 promoter as the second genetic hit in hereditary diffuse gastric cancer. *Nat Genet* **26:** 16–17.

Greenblatt MS, Bennett WP, Hollstein M, Harris CC. 1994. Mutations in the p53 tumor suppressor gene: Clues to cancer etiology and molecular pathogenesis. *Cancer Res* **54:** 4855–4878.

* Grossniklaus U, Paro R. 2014. Transcriptional silencing by *Polycomb*-group proteins. *Cold Spring Harb Perspect Biol* **6:** a019331.

Guan B, Mao TL, Panuganti PK, Kuhn E, Kurman RJ, Maeda D, Chen E, Jeng YM, Wang TL, Shih Ie M. 2011. Mutation and loss of expression of ARID1A in uterine low-grade endometrioid carcinoma. *Am J Surg Pathol* **35:** 625–632.

Gui Y, Guo G, Huang Y, Hu X, Tang A, Gao S, Wu R, Chen C, Li X, Zhou L, et al. 2011. Frequent mutations of chromatin remodeling genes in transitional cell carcinoma of the bladder. *Nat Genet* **43:** 875–878.

Hahn MA, Hahn T, Lee DH, Esworthy RS, Kim BW, Riggs AD, Chu FF, Pfeifer GP. 2008. Methylation of polycomb target genes in intestinal cancer is mediated by inflammation. *Cancer Res* **68:** 10280–10289.

Hammerman PS, Hayes DN, Wilkerson MD, Schultz N, Bose R, Chu A, Collisson EA, Cope L, Creighton CJ, Getz G, et al. 2012. Comprehensive genomic characterization of squamous cell lung cancers. *Nature* **489:** 519–525.

Hanahan D, Weinberg RA. 2011. Hallmarks of cancer: The next generation. *Cell* **144:** 646–674.

Hansen KD, Timp W, Bravo HC, Sabunciyan S, Langmead B, McDonald OG, Wen B, Wu H, Liu Y, Diep D, et al. 2011. Increased methylation variation in epigenetic domains across cancer types. *Nat Genet* **43:** 768–775.

Harris RA, Wang T, Coarfa C, Nagarajan RP, Hong C, Downey SL, Johnson BE, Fouse SD, Delaney A, Zhao Y, et al. 2010. Comparison of sequencing-based methods to profile DNA methylation and identification of monoallelic epigenetic modifications. *Nat Biotechnol* **28:** 1097–1105.

Hegi ME, Diserens AC, Gorlia T, Hamou MF, de Tribolet N, Weller M, Kros JM, Hainfellner JA, Mason W, Mariani L, et al. 2005. MGMT gene silencing and benefit from temozolomide in glioblastoma. *N Engl J Med* **352:** 997–1003.

Herman JG, Baylin SB. 2003. Gene silencing in cancer in association with promoter hypermethylation. *N Engl J Med* **349:** 2042–2054.

Herman JG, Graff JR, Myohanen S, Nelkin BD, Baylin SB. 1996. Methylation-specific PCR: A novel PCR assay for methylation status of CpG islands. *Proc Natl Acad Sci* **93:** 9821–9826.

Herman JG, Umar A, Polyak K, Graff JR, Ahuja N, Issa JP, Markowitz S, Willson JK, Hamilton SR, Kinzler KW, et al. 1998. Incidence and functional consequences of hMLH1 promoter hypermethylation in colorectal carcinoma. *Proc Natl Acad Sci* **95:** 6870–6875.

Hinoue T, Weisenberger DJ, Lange CP, Shen H, Byun HM, Van Den Berg D, Malik S, Pan F, Noushmehr H, van Dijk CM, et al. 2012. Genome-scale analysis of aberrant DNA methylation in colorectal cancer. *Genome Res* **22:** 271–282.

Hitchins MP, Rapkins RW, Kwok CT, Srivastava S, Wong JJ, Khachigian LM, Polly P, Goldblatt J, Ward RL. 2011. Dominantly inherited constitutional epigenetic silencing of MLH1 in a cancer-affected family is linked to a single nucleotide variant within the 5′UTR. *Cancer Cell* **20:** 200–213.

Holm TM, Jackson-Grusby L, Brambrink T, Yamada Y, Rideout WM 3rd, Jaenisch R. 2005. Global loss of imprinting leads to widespread tumorigenesis in adult mice. *Cancer Cell* **8:** 275–285.

Holst CR, Nuovo GJ, Esteller M, Chew K, Baylin SB, Herman JG, Tlsty TD. 2003. Methylation of p16(INK4a) promoters occurs in vivo in histologically normal human mammary epithelia. *Cancer Res* **63:** 1596–1601.

Hon GC, Hawkins RD, Caballero OL, Lo C, Lister R, Pelizzola M, Valsesia A, Ye Z, Kuan S, Edsall LE, et al. 2012. Global DNA hypomethylation coupled to repressive chromatin domain formation and gene silencing in breast cancer. *Genome Res* **22:** 246–258.

Hong L, Ahuja N. 2013. DNA methylation biomarkers of stool and blood for early detection of colon cancer. *Genet Test Mol Biomarkers* **17:** 401–406.

Howitz KT, Bitterman KJ, Cohen HY, Lamming DW, Lavu S, Wood JG, Zipkin RE, Chung P, Kisielewski A, Zhang LL, et al. 2003. Small molecule activators of sirtuins extend *Saccharomyces cerevisiae* lifespan. *Nature* **425:** 191–196.

Hur K, Cejas P, Feliu J, Moreno-Rubio J, Burgos E, Boland CR, Goel A. 2014. Hypomethylation of long interspersed nuclear element-1 (LINE-1) leads to activation of proto-oncogenes in human colorectal cancer metastasis. *Gut* **63:** 635–646.

Imperiale TF, Ransohoff DF, Itzkowitz SH, Levin TR, Lavin P, Lidgard GP, Ahlquist DA, Berger BM. 2014. Multitarget stool DNA testing for colorectal-cancer screening. *N Engl J Med* **370:** 1287–1297.

Issa JP. 2014. Aging and epigenetic drift: A vicious cycle. *J Clin Invest* **124:** 24–29.

Issa JP, Ottaviano YL, Celano P, Hamilton SR, Davidson NE, Baylin SB. 1994. Methylation of the oestrogen receptor CpG island links ageing and neoplasia in human colon. *Nat Genet* **7:** 536–540.

Issa JP, Garcia-Manero G, Giles FJ, Mannari R, Thomas D, Faderl S, Bayar E, Lyons J, Rosenfeld CS, Cortes J, et al. 2004. Phase 1 study of low-dose prolonged exposure schedules of the hypomethylating agent 5-aza-2′-deoxycytidine (decitabine) in hematopoietic malignancies. *Blood* **103:** 1635–1640.

Janzen V, Forkert R, Fleming HE, Saito Y, Waring MT, Dombkowski DM, Cheng T, Depinho RA, Sharpless NE, Scadden DT. 2006. Stem-cell ageing modified by the cyclin-dependent kinase inhibitor p16(INK4a). *Nature* **443:** 421–426.

Jiang L, Li J, Song L. 2009. Bmi-1, stem cells and cancer. *Acta Biochim Biophys Sinica* **41:** 527–534.

Jones PA. 2012. Functions of DNA methylation: Islands, start sites, gene bodies and beyond. *Nat Rev Genet* **13:** 484–492.

Jones PA, Baylin SB. 2002. The fundamental role of epigenetic events in cancer. *Nat Rev Genet* **3:** 415–428.

Jones PA, Baylin SB. 2007. The epigenomics of cancer. *Cell* **128:** 683–692.

Jones PA, Laird PW. 1999. Cancer epigenetics comes of age. *Nat Genet* **21:** 163–167.

Jones PA, Liang G. 2009. Rethinking how DNA methylation patterns are maintained. *Nat Rev Genet* **10:** 805–811.

Jones PA, Taylor SM. 1980. Cellular differentiation, cytidine analogs and DNA methylation. *Cell* **20:** 85–93.

Jones S, Wang TL, Shih Ie M, Mao TL, Nakayama K, Roden R, Glas R, Slamon D, Diaz LA Jr, Vogelstein B, et al. 2010. Frequent mutations of chromatin remodeling gene *ARID1A* in ovarian clear cell carcinoma. *Science* **330:** 228–231.

Juergens RA, Wrangle J, Vendetti FP, Murphy SC, Zhao M, Coleman B, Sebree R, Rodgers K, Hooker CM, Franco N, et al. 2011. Combination epigenetic therapy has efficacy in patients with refractory advanced non-small cell lung cancer. *Cancer Discov* **1:** 598–607.

Kaminskas E, Farrell A, Abraham S, Baird A, Hsieh LS, Lee SL, Leighton JK, Patel H, Rahman A, Sridhara R, et al. 2005. Approval summary: Azacitidine for treatment of myelodysplastic syndrome subtypes. *Clin Cancer Res* **11:** 3604–3608.

Kanai Y, Ushijima S, Nakanishi Y, Sakamoto M, Hirohashi S. 2003. Mutation of the DNA methyltransferase (DNMT) 1 gene in human colorectal cancers. *Cancer Lett* **192:** 75–82.

Kaneda A, Feinberg AP. 2005. Loss of imprinting of IGF2: A common epigenetic modifier of intestinal tumor risk. *Cancer Res* **65:** 11236–11240.

Katzenellenbogen RA, Baylin SB, Herman JG. 1999. Hypermethylation of the DAP-kinase CpG island is a common alteration in B-cell malignancies. *Blood* **93:** 4347–4353.

Kelly TK, De Carvalho DD, Jones PA. 2010. Epigenetic modifications as therapeutic targets. *Nat Biotechnol* **28:** 1069–1078.

Kelly TK, Liu Y, Lay FD, Liang G, Berman BP, Jones PA. 2012. Genome-wide mapping of nucleosome positioning and DNA methylation within individual DNA molecules. *Genome Res* **22:** 2497–2506.

Kim MS, Chung NG, Kang MR, Yoo NJ, Lee SH. 2011. Genetic and expressional alterations of CHD genes in gastric and colorectal cancers. *Histopathology* **58:** 660–668.

Kinzler KW, Vogelstein B. 1997. Cancer-susceptibility genes. Gatekeepers and caretakers. *Nature* **386:** 761–763.

Kiyono T, Foster SA, Koop JI, McDougall JK, Galloway DA, Klingelhutz AJ. 1998. Both Rb/p16INK4a inactivation and telomerase activity are required to immortalize human epithelial cells. *Nature* **396:** 84–88.

Knudson AG. 2001. Two genetic hits (more or less) to cancer. *Nat Rev Cancer* **1:** 157–162.

Kondo Y, Shen L, Issa JP. 2003. Critical role of histone methylation in tumor suppressor gene silencing in colorectal cancer. *Mol Cell Biol* **23:** 206–215.

Kriaucionis S, Heintz N. 2009. The nuclear DNA base 5-hydroxymethylcytosine is present in Purkinje neurons and the brain. *Science* **324:** 929–930.

* Kriaucionis S, Tahiliani M. 2014. Expanding the epigenetic landscape: Novel modificatons of cytosine in genomic DNA. *Cold Spring Harb Perspect Biol* **6:** a018630.

Kulis M, Heath S, Bibikova M, Queiros AC, Navarro A, Clot G, Martinez-Trillos A, Castellano G, Brun-Heath I, Pinyol M, et al. 2012. Epigenomic analysis detects widespread gene-body DNA hypomethylation in chronic lymphocytic leukemia. *Nat Genet* **44:** 1236–1242.

Kuzmichev A, Margueron R, Vaquero A, Preissner TS, Scher M, Kirmizis A, Ouyang X, Brockdorff N, Abate-Shen C, Farnham P, et al. 2005. Composition and histone substrates of polycomb repressive group complexes change during cellular differentiation. *Proc Natl Acad Sci* **102:** 1859–1864.

Laird PW. 2003. The power and the promise of DNA methylation markers. *Nat Rev Cancer* **3:** 253–266.

Laird PW, Jackson-Grusby L, Fazeli A, Dickinson SL, Jung WE, Li E, Weinberg RA, Jaenisch R. 1995. Suppression of intestinal neoplasia by DNA hypomethylation. *Cell* **81:** 197–205.

Lapeyre JN, Becker FF. 1979. 5-Methylcytosine content of nuclear DNA during chemical hepatocarcinogenesis and in carcinomas which result. *Biochem Biophys Res Commun* **87:** 698–705.

Lee WH, Morton RA, Epstein JI, Brooks JD, Campbell PA, Bova GS, Hsieh WS, Isaacs WB, Nelson WG. 1994. Cytidine methylation of regulatory sequences near the pi-class glutathione S-transferase gene accompanies human prostatic carcinogenesis. *Proc Natl Acad Sci* **91:** 11733–11737.

Leu YW, Rahmatpanah F, Shi H, Wei SH, Liu JC, Yan PS, Huang TH. 2003. Double RNA interference of DNMT3b and DNMT1 enhances DNA demethylation and gene reactivation. *Cancer Res* **63:** 6110–6115.

Lewis PW, Muller MM, Koletsky MS, Cordero F, Lin S, Banaszynski LA, Garcia BA, Muir TW, Becher OJ, Allis CD. 2013. Inhibition of PRC2 activity by a gain-of-function H3 mutation found in pediatric glioblastoma. *Science* **340:** 857–861.

Ley TJ, Ding L, Walter MJ, McLellan MD, Lamprecht T, Larson DE, Kandoth C, Payton JE, Baty J, Welch J, et al. 2010. DNMT3A mutations in acute myeloid leukemia. *N Engl J Med* **363:** 2424–2433.

Li H, Chiappinelli KB, Guzzetta AA, Easwaran H, Yen RW, Vatapalli R, Topper MJ, Luo J, Connolly RM, Azad NS, et al. 2014. Immune regulation by low doses of the DNA methyltransferase inhibitor 5-azacitidine in common human epithelial cancers. *Oncotarget* **5:** 587–598.

* Li E, Zhang Y. 2014. DNA methylation in mammals. *Cold Spring Harb Perspect Biol* **6:** a019133.

Li M, Zhao H, Zhang X, Wood LD, Anders RA, Choti MA, Pawlik TM, Daniel HD, Kannangai R, Offerhaus GJ, et al. 2011. Inactivating mutations of the chromatin remodeling gene *ARID2* in hepatocellular carcinoma. *Nat Genet* **43:** 828–829.

Lister R, Pelizzola M, Dowen RH, Hawkins RD, Hon G, Tonti-Filippini J, Nery JR, Lee L, Ye Z, Ngo QM, et al. 2009. Human DNA methylomes at base resolution show widespread epigenomic differences. *Nature* **462:** 315–322.

Lister R, Pelizzola M, Kida YS, Hawkins RD, Nery JR, Hon G, Antosiewicz-Bourget J, O'Malley R, Castanon R, Klugman S, et al. 2011. Hotspots of aberrant epigenomic reprogramming in human induced pluripotent stem cells. *Nature* **471:** 68–73.

Liu B, Nicolaides NC, Markowitz S, Willson JK, Parsons RE, Jen J, Papadopolous N, Peltomaki P, de la Chapelle A, Hamilton SR, et al. 1995. Mismatch repair gene defects in sporadic colorectal cancers with microsatellite instability. *Nat Genet* **9:** 48–55.

Losman JA, Kaelin WG Jr, 2013. What a difference a hydroxyl makes: Mutant IDH, (R)-2-hydroxyglutarate, and cancer. *Genes Dev* **27:** 836–852.

Lu C, Ward PS, Kapoor GS, Rohle D, Turcan S, Abdel-Wahab O, Edwards CR, Khanin R, Figueroa ME, Melnick A, et al. 2012. IDH mutation impairs histone demethylation and results in a block to cell differentiation. *Nature* **483:** 474–478.

Lu C, Venneti S, Akalin A, Fang F, Ward PS, Dematteo RG, Intlekofer AM, Chen C, Ye J, Hameed M, et al. 2013. Induction of sarcomas by mutant IDH2. *Genes Dev* **27:** 1986–1998.

Lubbert M. 2000. DNA methylation inhibitors in the treatment of leukemias, myelodysplastic syndromes and hemoglobinopathies: Clinical results and possible mechanisms of action. *Curr Top Microbiol Immunol* **249:** 135–164.

Lukacs RU, Memarzadeh S, Wu H, Witte ON. 2010. Bmi-1 is a crucial regulator of prostate stem cell self-renewal and malignant transformation. *Cell Stem Cell* **7:** 682–693.

Maegawa S, Gough SM, Watanabe-Okochi N, Lu Y, Zhang N, Castoro RJ, Estecio MR, Jelinek J, Liang S, Kitamura T, et al. 2014. Age-related epigenetic drift in the pathogenesis of MDS and AML. *Genome Res* **24:** 580–591.

Marks P, Rifkind RA, Richon VM, Breslow R, Miller T, Kelly WK. 2001. Histone deacetylases and cancer: Causes and therapies. *Nat Rev Cancer* **1:** 194–202.

* Marmorstein R, Zhou M-M. 2014. Writers and readers of histone acetylation: Structure, mechanism, and inhibition. *Cold Spring Harb Perspect Biol* **6:** a018762.

Maslak P, Chanel S, Camacho LH, Soignet S, Pandolfi PP, Guernah I, Warrell R, Nimer S. 2006. Pilot study of combination transcriptional modulation therapy with sodium phenylbutyrate and 5-azacytidine in patients with acute myeloid leukemia or myelodysplastic syndrome. *Leukemia* **20:** 212–217.

Mason EF, Hornick JL. 2013. Succinate dehydrogenase deficiency is associated with decreased 5-hydroxymethylcytosine production in gastrointestinal stromal tumors: Implications for mechanisms of tumorigenesis. *Mod Pathol* **26:** 1492–1497.

Matei D, Fang F, Shen C, Schilder J, Arnold A, Zeng Y, Berry WA, Huang T, Nephew KP. 2012. Epigenetic resensitization to platinum in ovarian cancer. *Cancer Res* **72:** 2197–2205.

Mattera L, Escaffit F, Pillaire MJ, Selves J, Tyteca S, Hoffmann JS, Gourraud PA, Chevillard-Briet M, Cazaux C, Trouche D. 2009. The p400/Tip60 ratio is critical for colorectal cancer cell proliferation through DNA damage response pathways. *Oncogene* **28:** 1506–1517.

McGarvey KM, Van Neste L, Cope L, Ohm JE, Herman JG, Van Criekinge W, Schuebel KE, Baylin SB. 2008. Defining a chromatin pattern that characterizes DNA-hypermethylated genes in colon cancer cells. *Cancer Res* **68:** 5753–5759.

Miremadi A, Oestergaard MZ, Pharoah PD, Caldas C. 2007. Cancer genetics of epigenetic genes. *Hum Mol Genet* **16:** R28–R49.

Mizuno K, Osada H, Konishi H, Tatematsu Y, Yatabe Y, Mitsudomi T, Fujii Y, Takahashi T. 2002. Aberrant hypermethylation of the CHFR prophase checkpoint gene in human lung cancers. *Oncogene* **21:** 2328–2333.

Morin RD, Mendez-Lago M, Mungall AJ, Goya R, Mungall KL, Corbett RD, Johnson NA, Severson TM, Chiu R, Field M, et al. 2011. Frequent mutation of histone-modifying genes in non-Hodgkin lymphoma. *Nature* **476:** 298–303.

Narayan A, Ji W, Zhang XY, Marrogi A, Graff JR, Baylin SB, Ehrlich M. 1998. Hypomethylation of pericentromeric DNA in breast adenocarcinomas. *Int J Cancer* **77:** 833–838.

Nemoto S, Fergusson MM, Finkel T. 2004. Nutrient availability regulates SIRT1 through a forkhead-dependent pathway. *Science* **306:** 2105–2108.

Nguyen CT, Gonzales FA, Jones PA. 2001. Altered chromatin structure associated with methylation-induced gene silencing in cancer cells: Correlation of accessibility, methylation, MeCP2 binding and acetylation. *Nucleic Acids Res* **29:** 4598–4606.

Nickel A, Stadler SC. 2014. Role of epigenetic mechanisms in epithelial-to-mesenchymal transition of breast cancer cells. *Transl Res* **pii: S1931-524400125**–X doi: 10.1016/j.trsl.2014.04.001.

Nicodeme E, Jeffrey KL, Schaefer U, Beinke S, Dewell S, Chung CW, Chandwani R, Marazzi I, Wilson P, Coste H, et al. 2010. Suppression of inflammation by a synthetic histone mimic. *Nature* **468:** 1119–1123.

Nishiyama A, Yamaguchi L, Sharif J, Johmura Y, Kawamura T, Nakanishi K, Shimamura S, Arita K, Kodama T, Ishikawa F, et al. 2013. Uhrf1-dependent H3K23 ubiquitylation couples maintenance DNA methylation and replication. *Nature* **502:** 249–253.

Noushmehr H, Weisenberger DJ, Diefes K, Phillips HS, Pujara K, Berman BP, Pan F, Pelloski CE, Sulman EP, Bhat KP, et al. 2010. Identification of a CpG island methylator phenotype that defines a distinct subgroup of glioma. *Cancer Cell* **17:** 510–522.

O'Hagan HM, Mohammad HP, Baylin SB. 2008. Double strand breaks can initiate gene silencing and SIRT1-dependent onset of DNA methylation in an exogenous promoter CpG island. *PLoS Genet* **4:** e1000155.

O'Hagan HM, Wang W, Sen S, Destefano Shields C, Lee SS, Zhang YW, Clements EG, Cai Y, Van Neste L, Easwaran H, et al. 2011. Oxidative damage targets complexes containing DNA methyltransferases, SIRT1, and polycomb members to promoter CpG islands. *Cancer Cell* **20:** 606–619.

Ohm JE, McGarvey KM, Yu X, Cheng L, Schuebel KE, Cope L, Mohammad HP, Chen W, Daniel VC, Yu W, et al. 2007. A stem cell-like chromatin pattern may predispose tumor suppressor genes to DNA hypermethylation and heritable silencing. *Nat Genet* **39:** 237–242.

Parsons DW, Jones S, Zhang X, Lin JC, Leary RJ, Angenendt P, Mankoo P, Carter H, Siu IM, Gallia GL, et al. 2008. An integrated genomic analysis of human glioblastoma multiforme. *Science* **321:** 1807–1812.

* Patel DJ. 2014. A structural perspective on readout of epigenetic histone and DNA methylation marks. *Cold Spring Harb Perspect Biol* doi: 10.1101/cshperspect.a018754.

Peric-Hupkes D, van Steensel B. 2010. Role of the nuclear lamina in genome organization and gene expression. *Cold Spring Harb Symp Quant Biol* **75:** 517–524.

Peric-Hupkes D, Meuleman W, Pagie L, Bruggeman SW, Solovei I, Brugman W, Graf S, Flicek P, Kerkhoven RM, van Lohuizen M, et al. 2010. Molecular maps of the reorganization of genome-nuclear lamina interactions during differentiation. *Mol Cell* **38:** 603–613.

Pfeifer GP, Tang M, Denissenko MF. 2000. Mutation hotspots and DNA methylation. *Curr Top Microbiol Immunol* **249:** 1–19.

* Pirrotta V. 2014. The necessity of chromatin: A view in perspective. *Cold Spring Harb Perspect Biol* doi: 10.1101/cshperspect.a019547.

Popovic R, Licht JD. 2012. Emerging epigenetic targets and therapies in cancer medicine. *Cancer Discov* **2:** 405–413.

Prebet T, Sun Z, Figueroa ME, Ketterling R, Melnick A, Greenberg PL, Herman J, Juckett M, Smith MR, Malick L, et al. 2014. Prolonged administration of azacitidine with or without entinostat for myelodysplastic syndrome and acute myeloid leukemia with myelodysplasia-related changes: Results of the US Leukemia intergroup trial E1905. *J Clin Oncol* **32:** 1242–1248.

Prensner JR, Chinnaiyan AM. 2011. The emergence of lncRNAs in cancer biology. *Cancer Discov* **1:** 391–407.

* Qi J. 2014. Bromodomain and extraterminal domain inhibitors (BETi) for cancer therapy: Chemical modulation of chromatin structure. *Cold Spring Harb Perspect Biol* doi: 10.1101/cshperspect.a018663.

Qin T, Castoro R, El Ahdab S, Jelinek J, Wang X, Si J, Shu J, He R, Zhang N, Chung W, et al. 2011. Mechanisms of resistance to decitabine in the myelodysplastic syndrome. *PLoS One* **6:** e23372.

Rainier S, Dobry CJ, Feinberg AP. 1995. Loss of imprinting in hepatoblastoma. *Cancer research* **55:** 1836–1838.

Rainier S, Johnson LA, Dobry CJ, Ping AJ, Grundy PE, Feinberg AP. 1993. Relaxation of imprinted genes in human cancer. *Nature* **362:** 747–749.

Ramalingam SS, Maitland ML, Frankel P, Argiris AE, Koczywas M, Gitlitz B, Thomas S, Espinoza-Delgado I, Vokes EE, Gandara DR, et al. 2010. Carboplatin and Paclitaxel in combination with either vorinostat or placebo for first-line therapy of advanced non-small-cell lung cancer. *J Clin Oncol* **28:** 56–62.

Reddington JP, Sproul D, Meehan RR. 2014. DNA methylation reprogramming in cancer: Does it act by re-configuring the binding landscape of Polycomb repressive complexes? *Bioessays* **36:** 134–140.

Reddy KL, Feinberg AP. 2013. Higher order chromatin organization in cancer. *Semin Cancer Biol* **23:** 109–115.

Rhee I, Jair KW, Yen RW, Lengauer C, Herman JG, Kinzler KW, Vogelstein B, Baylin SB, Schuebel KE. 2000. CpG methylation is maintained in human cancer cells lacking DNMT1. *Nature* **404:** 1003–1007.

Rhee I, Bachman KE, Park BH, Jair KW, Yen RW, Schuebel KE, Cui H, Feinberg AP, Lengauer C, Kinzler KW, et al. 2002. DNMT1 and DNMT3b cooperate to silence genes in human cancer cells. *Nature* **416:** 552–556.

Richon VM, O'Brien JP. 2002. Histone deacetylase inhibitors: A new class of potential therapeutic agents for cancer treatment. *Clin Cancer Res* **8:** 662–664.

Rideout WM 3rd, Coetzee GA, Olumi AF, Jones PA. 1990. 5-Methylcytosine as an endogenous mutagen in the human LDL receptor and p53 genes. *Science* **249:** 1288–1290.

Robertson KD, Ait-Si-Ali S, Yokochi T, Wade PA, Jones PL, Wolffe AP. 2000. DNMT1 forms a complex with Rb, E2F1 and HDAC1 and represses transcription from E2F-responsive promoters. *Nat Genet* **25:** 338–342.

Romanov SR, Kozakiewicz BK, Holst CR, Stampfer MR, Haupt LM, Tlsty TD. 2001. Normal human mammary epithelial cells spontaneously escape senescence and acquire genomic changes. *Nature* **409:** 633–637.

Ropero S, Fraga MF, Ballestar E, Hamelin R, Yamamoto H, Boix-Chornet M, Caballero R, Alaminos M, Setien F, Paz MF, et al. 2006. A truncating mutation of HDAC2 in human cancers confers resistance to histone deacetylase inhibition. *Nat Genet* **38:** 566–569.

Rotili D, Mai A. 2011. Targeting histone demethylases: A new avenue for the fight against cancer. *Genes Cancer* **2:** 663–679.

Rountree MR, Bachman KE, Baylin SB. 2000. DNMT1 binds HDAC2 and a new co-repressor, DMAP1, to form a complex at replication foci. *Nat Genet* **25:** 269–277.

Saito Y, Jones PA. 2006. Epigenetic activation of tumor suppressor microRNAs in human cancer cells. *Cell Cycle* **5:** 2220–2222.

Saito Y, Liang G, Egger G, Friedman JM, Chuang JC, Coetzee GA, Jones PA. 2006. Specific activation of microRNA-127 with downregulation of the proto-oncogene BCL6 by chromatin-modifying drugs in human cancer cells. *Cancer Cell* **9:** 435–443.

Sakai T, Toguchida J, Ohtani N, Yandell DW, Rapaport JM, Dryja TP. 1991. Allele-specific hypermethylation of the retinoblastoma tumor-suppressor gene. *Am J Hum Genet* **48:** 880–888.

Sakatani T, Kaneda A, Iacobuzio-Donahue CA, Carter MG, de Boom Witzel S, Okano H, Ko MS, Ohlsson R, Longo DL, Feinberg AP. 2005. Loss of imprinting of Igf2 alters intestinal maturation and tumorigenesis in mice. *Science* **307:** 1976–1978.

Sanbhnani S, Yeong FM. 2012. CHFR: A key checkpoint component implicated in a wide range of cancers. *Cell Mol Life Sci* **69:** 1669–1687.

Sansom OJ, Maddison K, Clarke AR. 2007. Mechanisms of disease: Methyl-binding domain proteins as potential therapeutic targets in cancer. *Nat Clin Practice Oncol* **4:** 305–315.

Santi DV, Norment A, Garrett CE. 1984. Covalent bond formation between a DNA-cytosine methyltransferase and DNA containing 5-azacytosine. *Proc Natl Acad Sci* **81:** 6993–6997.

Sasaki M, Knobbe CB, Itsumi M, Elia AJ, Harris IS, Chio II, Cairns RA, McCracken S, Wakeham A, Haight J, et al. 2012a. D-2-hydroxyglutarate produced by mutant IDH1 perturbs collagen maturation and basement membrane function. *Genes Dev* **26:** 2038–2049.

Sasaki M, Knobbe CB, Munger JC, Lind EF, Brenner D, Brustle A, Harris IS, Holmes R, Wakeham A, Haight J, et al. 2012b. IDH1(R132H) mutation increases murine haematopoietic progenitors and alters epigenetics. *Nature* **488:** 656–659.

* Schaefer U. 2014. Pharmacological inhibition of bromodomain-containing proteins in inflammation. *Cold Spring Harb Perspect Biol* **6:** a018671.

Schlesinger Y, Straussman R, Keshet I, Farkash S, Hecht M, Zimmerman J, Eden E, Yakhini Z, Ben-Shushan E, Reubinoff BE, et al. 2007. Polycomb-mediated methylation on Lys27 of histone H3 pre-marks genes for de novo methylation in cancer. *Nat Genet* **39:** 232–236.

Schmitz KM, Mayer C, Postepska A, Grummt I. 2010. Interaction of noncoding RNA with the rDNA promoter mediates recruitment of DNMT3b and silencing of rRNA genes. *Genes Dev* **24:** 2264–2269.

Schuebel KE, Chen W, Cope L, Glockner SC, Suzuki H, Yi JM, Chan TA, Van Neste L, Van Criekinge W, van den Bosch S, et al. 2007. Comparing the DNA hypermethylome with gene mutations in human colorectal cancer. *PLoS Genet* **3:** 1709–1723.

* Seto E, Yoshida M. 2014. Erasers of histone acetylation: The histone deacetylase enzymes. *Cold Spring Harb Perspect Biol* **6:** a018713.

Sharma SV, Lee DY, Li B, Quinlan MP, Takahashi F, Maheswaran S, McDermott U, Azizian N, Zou L, Fischbach MA, et al. 2010. A chromatin-mediated reversible drug-tolerant state in cancer cell subpopulations. *Cell* **141:** 69–80.

Shen H, Laird PW. 2013. Interplay between the cancer genome and epigenome. *Cell* **153:** 38–55.

Shen XM, Ohno K, Fukudome T, Tsujino A, Brengman JM, De Vivo DC, Packer RJ, Engel AG. 2002. Congenital myasthenic syndrome caused by low-expressor fast-channel AChR delta subunit mutation. *Neurology* **59:** 1881–1888.

Silverman LR, Demakos EP, Peterson BL, Kornblith AB, Holland JC, Odchimar-Reissig R, Stone RM, Nelson D, Powell BL, DeCastro CM, et al. 2002. Randomized controlled trial of azacitidine in patients with the myelodysplastic syndrome: A study of the cancer and leukemia group B. *J Clin Oncol* **20:** 2429–2440.

Spannhoff A, Hauser AT, Heinke R, Sippl W, Jung M. 2009. The emerging therapeutic potential of histone methyltransferase and demethylase inhibitors. *ChemMedChem* **4:** 1568–1582.

Sproul D, Nestor C, Culley J, Dickson JH, Dixon JM, Harrison DJ, Meehan RR, Sims AH, Ramsahoye BH. 2011. Transcriptionally repressed genes become aberrantly methylated and distinguish tumors of different lineages in breast cancer. *Proc Natl Acad Sci* **108:** 4364–4369.

Sproul D, Kitchen RR, Nestor CE, Dixon JM, Sims AH, Harrison DJ, Ramsahoye BH, Meehan RR. 2012. Tissue of origin determines cancer-associated CpG island promoter hypermethylation patterns. *Genome Biol* **13:** R84.

Srinivasan PR, Borek E. 1964. Enzymatic alteration of nucleic acid structure. *Science* **145:** 548–553.

Stumpel DJ, Schneider P, van Roon EH, Boer JM, de Lorenzo P, Valsecchi MG, de Menezes RX, Pieters R, Stam RW. 2009. Specific promoter methylation identifies different subgroups of MLL-rearranged infant acute lymphoblastic leukemia, influences clinical outcome, and provides therapeutic options. *Blood* **114:** 5490–5498.

Sun A, Tawfik O, Gayed B, Thrasher JB, Hoestje S, Li C, Li B. 2007. Aberrant expression of SWI/SNF catalytic subunits BRG1/BRM is associated with tumor development and increased invasiveness in prostate cancers. *Prostate* **67:** 203–213.

Sun J, Song Y, Wang Z, Wang G, Gao P, Chen X, Gao Z, Xu H. 2014. Clinical significance of promoter region hypermethylation of microRNA-148a in gastrointestinal cancers. *Onco Targets Ther* **7:** 853–863.

Suzuki H, Gabrielson E, Chen W, Anbazhagan R, van Engeland M, Weijenberg MP, Herman JG, Baylin SB. 2002. A genomic screen for genes upregulated by demethylation and histone deacetylase inhibition in human colorectal cancer. *Nat Genet* **31:** 141–149.

Suzuki H, Watkins DN, Jair KW, Schuebel KE, Markowitz SD, Chen WD, Pretlow TP, Yang B, Akiyama Y, Van Engeland M, et al. 2004. Epigenetic inactivation of SFRP genes allows constitutive WNT signaling in colorectal cancer. *Nat Genet* **36:** 417–422.

Swafford DS, Middleton SK, Palmisano WA, Nikula KJ, Tesfaigzi J, Baylin SB, Herman JG, Belinsky SA. 1997. Frequent aberrant methylation of p16INK4a in primary rat lung tumors. *Mol Cell Biol* **17:** 1366–1374.

Taberlay PC, Statham AL, Kelly TK, Clark SJ, Jones PA. 2014. Reconfiguration of nucleosome depleted regions at distal regulatory elements accompanies DNA methylation of enhancers and insulators in cancer. *Genome Res* **pii:** gr.163485.113

Tahiliani M, Koh KP, Shen Y, Pastor WA, Bandukwala H, Brudno Y, Agarwal S, Iyer LM, Liu DR, Aravind L, et al. 2009. Conversion of 5-methylcytosine to 5-hydroxymethylcytosine in mammalian DNA by MLL partner TET1. *Science* **324:** 930–935.

Tam WL, Weinberg RA. 2013. The epigenetics of epithelial-mesenchymal plasticity in cancer. *Nat Med* **19:** 1438–1449.

Timp W, Feinberg AP. 2013. Cancer as a dysregulated epigenome allowing cellular growth advantage at the expense of the host. *Nat Rev Cancer* **13:** 497–510.

Topalian SL, Hodi FS, Brahmer JR, Gettinger SN, Smith DC, McDermott DF, Powderly JD, Carvajal RD, Sosman JA, Atkins MB, et al. 2012. Safety, activity, and immune correlates of anti-PD-1 antibody in cancer. *N Engl J Med* **366:** 2443–2454.

Toyota M, Ahuja N, Ohe-Toyota M, Herman JG, Baylin SB, Issa JP. 1999. CpG island methylator phenotype in colorectal cancer. *Proc Natl Acad Sci* **96:** 8681–8686.

Tsai HC, Baylin SB. 2011. Cancer epigenetics: Linking basic biology to clinical medicine. *Cell Res* **21:** 502–517.

Tsai HC, Li H, Van Neste L, Cai Y, Robert C, Rassool FV, Shin JJ, Harbom KM, Beaty R, Pappou E, et al. 2012. Transient low doses of DNA-demethylating agents exert durable antitumor effects on hematological and epithelial tumor cells. *Cancer Cell* **21:** 430–446.

Tsang DP, Cheng AS. 2011. Epigenetic regulation of signaling pathways in cancer: Role of the histone methyltransferase EZH2. *J Gastroenterol Hepatol* **26:** 19–27.

Turcan S, Rohle D, Goenka A, Walsh LA, Fang F, Yilmaz E, Campos C, Fabius AW, Lu C, Ward PS, et al. 2012. IDH1 mutation is sufficient to establish the glioma hypermethylator phenotype. *Nature* **483:** 479–483.

Turcan S, Fabius AW, Borodovsky A, Pedraza A, Brennan C, Huse J, Viale A, Riggins GJ, Chan TA. 2013. Efficient induction of differentiation and growth inhibition in IDH1 mutant glioma cells by the DNMT Inhibitor Decitabine. *Oncotarget* **10:** 1729–1736.

Van Neste L, Bigley J, Toll A, Otto G, Clark J, Delree P, Van Criekinge W, Epstein JI. 2012. A tissue biopsy-based epigenetic multiplex PCR assay for prostate cancer detection. *BMC Urol* **12:** 16.

Varela I, Tarpey P, Raine K, Huang D, Ong CK, Stephens P, Davies H, Jones D, Lin ML, Teague J, et al. 2011. Exome sequencing identifies frequent mutation of the SWI/SNF complex gene PBRM1 in renal carcinoma. *Nature* **469:** 539–542.

Varier RA, Timmers HT. 2011. Histone lysine methylation and demethylation pathways in cancer. *Biochim Biophys Acta* **1815:** 75–89.

Veigl ML, Kasturi L, Olechnowicz J, Ma A, Lutterbaugh JD, Periyasamy S, Li GM, Drummond J, Modrich PL, Sedwick WD, et al. 1998. Biallelic inactivation of hMLH1 by epigenetic gene silencing, a novel mechanism causing human MSI cancers. *Proc Natl Acad Sci* **95:** 8698–8702.

Venneti S, Felicella MM, Coyne T, Phillips JJ, Gorovets D, Huse JT, Kofler J, Lu C, Tihan T, Sullivan LM, et al. 2013. Histone 3 lysine 9 trimethylation is differentially associated with isocitrate dehydrogenase mutations in oligodendrogliomas and high-grade astrocytomas. *J Neuropathol Exp Neurol* **72:** 298–306.

Vogelstein B, Papadopoulos N, Velculescu VE, Zhou S, Diaz LA Jr, Kinzler KW. 2013. Cancer genome landscapes. *Science* **339:** 1546–1558.

Voso MT, Santini V, Finelli C, Musto P, Pogliani E, Angelucci E, Fioritoni G, Alimena G, Maurillo L, Cortelezzi A, et al. 2009. Valproic acid at therapeutic plasma levels may increase 5-azacytidine efficacy in higher risk myelodysplastic syndromes. *Clin Cancer Res* **15:** 5002–5007.

Wales MM, Biel MA, el Deiry W, Nelkin BD, Issa JP, Cavenee WK, Kuerbitz SJ, Baylin SB. 1995. p53 activates expression of HIC-1, a new candidate tumour suppressor gene on 17p13.3. *Nat Med* **1:** 570–577.

Wang J, Chen H, Fu S, Xu ZM, Sun KL, Fu WN. 2011. The involvement of CHD5 hypermethylation in laryngeal squamous cell carcinoma. *Oral Oncol* **47:** 601–608.

Weisenberger DJ, Siegmund KD, Campan M, Young J, Long TI, Faasse MA, Kang GH, Widschwendter M, Weener D, Buchanan D, et al. 2006. CpG island methylator phenotype underlies sporadic microsatellite instability and is tightly associated with BRAF mutation in colorectal cancer. *Nat Genet* **38:** 787–793.

Wen K, Azevedo MS, Gonzalez A, Zhang W, Saif LJ, Li G, Yousef A, Yuan L. 2009. Toll-like receptor and innate cytokine responses induced by lactobacilli colonization and human rotavirus infection in gnotobiotic pigs. *Vet Immunol Immunopathol* **127:** 304–315.

Wessels K, Bohnhorst B, Luhmer I, Morlot S, Bohring A, Jonasson J, Epplen JT, Gadzicki D, Glaser S, Gohring G, et al. 2010. Novel CHD7 mutations contributing to the mutation spectrum in patients with CHARGE syndrome. *Eur J Med Genet* **53:** 280–285.

West AC, Johnstone RW. 2014. New and emerging HDAC inhibitors for cancer treatment. *J Clin Invest* **124:** 30–39.

Widschwendter M, Fiegl H, Egle D, Mueller-Holzner E, Spizzo G, Marth C, Weisenberger DJ, Campan M, Young J, Jacobs I, et al. 2007. Epigenetic stem cell signature in cancer. *Nat Genet* **39:** 157–158.

Wijermans P, Lubbert M, Verhoef G, Bosly A, Ravoet C, Andre M, Ferrant A. 2000. Low-dose 5-aza-2′-deoxycytidine, a DNA hypomethylating agent, for the treatment of high-risk myelodysplastic syndrome: A multicenter phase II study in elderly patients. *J Clin Oncol* **18:** 956–962.

Wijmenga C, Hansen RS, Gimelli G, Bjorck EJ, Davies EG, Valentine D, Belohradsky BH, van Dongen JJ, Smeets DF, van den Heuvel LP, et al. 2000. Genetic variation in ICF syndrome: Evidence for genetic heterogeneity. *Human Mutat* **16:** 509–517.

Williams K, Christensen J, Pedersen MT, Johansen JV, Cloos PA, Rappsilber J, Helin K. 2011. TET1 and hydroxymethylcytosine in transcription and DNA methylation fidelity. *Nature* **473:** 343–348.

Wilson BG, Roberts CW. 2011. SWI/SNF nucleosome remodellers and cancer. *Nat Rev Cancer* **11:** 481–492.

Witta SE, Jotte RM, Konduri K, Neubauer MA, Spira AI, Ruxer RL, Varella-Garcia M, Bunn PA Jr, Hirsch FR. 2012. Randomized phase II trial of erlotinib with and without entinostat in patients with advanced non-small-cell lung cancer who progressed on prior chemotherapy. *J Clin Oncol* **30:** 2248–2255.

Wolf P, Hu YC, Doffek K, Sidransky D, Ahrendt SA. 2001. O(6)-Methylguanine-DNA methyltransferase promoter hypermethylation shifts the p53 mutational spectrum in non-small cell lung cancer. *Cancer Res* **61:** 8113–8117.

Wolffe AP. 2001. Chromatin remodeling: Why it is important in cancer. *Oncogene* **20:** 2988–2990.

Wrangle J, Wang W, Koch A, Easwaran H, Mohammad HP, Vendetti F, Vancriekinge W, Demeyer T, Du Z, Parsana P, et al. 2013. Alterations of immune response of non-small cell lung cancer with azacytidine. *Oncotarget* **4:** 2067–2079.

Wu SC, Zhang Y. 2010. Active DNA demethylation: Many roads lead to Rome. *Nav Rev Mol Cell Biol* **11:** 607–620.

Wu H, Zhang Y. 2011a. Mechanisms and functions of Tet protein-mediated 5-methylcytosine oxidation. *Genes Dev* **25:** 2436–2452.

Wu H, Zhang Y. 2011b. Tet1 and 5-hydroxymethylation: A genome-wide view in mouse embryonic stem cells. *Cell Cycle* **10:** 2428–2436.

Wu Y, Strawn E, Basir Z, Halverson G, Guo SW. 2007. Aberrant expression of deoxyribonucleic acid and methyltransferases DNMT1, DNMT3A, and DNMY3B in women with endometriosis. *Fertil Steril* **87:** 24–32.

Xiao M, Yang H, Xu W, Ma S, Lin H, Zhu H, Liu L, Liu Y, Yang C, Xu Y, et al. 2012. Inhibition of α-KG-dependent histone and DNA demethylases by fumarate and succinate that are accumulated in mutations of FH and SDH tumor suppressors. *Genes Dev* **26:** 1326–1338.

Yamashita K, Upadhyay S, Osada M, Hoque MO, Xiao Y, Mori M, Sato F, Meltzer SJ, Sidransky D. 2002. Pharmacologic unmasking of epigenetically silenced tumor suppressor genes in esophageal squamous cell carcinoma. *Cancer Cell* **2:** 485–495.

Yamashita Y, Yuan J, Suetake I, Suzuki H, Ishikawa Y, Choi YL, Ueno T, Soda M, Hamada T, Haruta H, et al. 2010. Array-based genomic resequencing of human leukemia. *Oncogene* **29:** 3723–3731.

Yan XJ, Xu J, Gu ZH, Pan CM, Lu G, Shen Y, Shi JY, Zhu YM, Tang L, Zhang XW, et al. 2011. Exome sequencing identifies somatic mutations of DNA methyltransferase gene DNMT3A in acute monocytic leukemia. *Nat Genet* **43:** 309–315.

Yang X, Noushmehr H, Han H, Andreu-Vieyra C, Liang G, Jones PA. 2012. Gene reactivation by 5-aza-2′-deoxycytidine-induced demethylation requires SRCAP-mediated H2A.Z insertion to establish nucleosome depleted regions. *PLoS Genet* **8:** e1002604.

Yardley DA, Ismail-Khan RR, Melichar B, Lichinitser M, Munster PN, Klein PM, Cruickshank S, Miller KD, Lee MJ, Trepel JB. 2013. Randomized phase II, double-blind, placebo-controlled study of exemestane with or without entinostat in postmenopausal women with locally recurrent or metastatic estrogen receptor-positive breast cancer progressing on treatment with a nonsteroidal aromatase inhibitor. *J Clin Oncol* **31:** 2128–2135.

Yoo CB, Jeong S, Egger G, Liang G, Phiasivongsa P, Tang C, Redkar S, Jones PA. 2007. Delivery of 5-aza-2′-deoxycytidine to cells using oligodeoxynucleotides. *Cancer Res* **67:** 6400–6408.

You JS, Jones PA. 2012. Cancer genetics and epigenetics: Two sides of the same coin? *Cancer Cell* **22:** 9–20.

Yu X, Minter-Dykhouse K, Malureanu L, Zhao WM, Zhang D, Merkle CJ, Ward IM, Saya H, Fang G, van Deursen J, et al. 2005. Chfr is required for tumor suppression and Aurora A regulation. *Nat Genet* **37:** 401–406.

Zhang W, Glockner SC, Guo M, Machida EO, Wang DH, Easwaran H, Van Neste L, Herman JG, Schuebel KE, Watkins DN, et al. 2008. Epigenetic inactivation of the canonical Wnt antagonist SRY-box containing gene 17 in colorectal cancer. *Cancer Res* **68:** 2764–2772.

Zheng L, Dai H, Zhou M, Li X, Liu C, Guo Z, Wu X, Wu J, Wang C, Zhong J, et al. 2012. Polyploid cells rewire DNA damage response networks to overcome replication stress-induced barriers for tumour progression. *Nat Commun* **3:** 815.

* Zoghbi HY, Beaudet AL. 2014. Epigenetics and human disease. *Cold Spring Harb Perspect Biol* doi: 10.1101/cshperspect.a019497.

Zilberman D, Coleman-Derr D, Ballinger T, Henikoff S. 2008. Histone H2A.Z and DNA methylation are mutually antagonistic chromatin marks. *Nature* **456:** 125–129.

Zorn AM, Barish GD, Williams BO, Lavender P, Klymkowsky MW, Varmus HE. 1999. Regulation of Wnt signaling by Sox proteins: XSox17 α/β and XSox3 physically interact with β-catenin. *Mol Cell* **4:** 487–498.

Zuber J, Shi J, Wang E, Rappaport AR, Herrmann H, Sison EA, Magoon D, Qi J, Blatt K, Wunderlich M, et al. 2011. RNAi screen identifies Brd4 as a therapeutic target in acute myeloid leukaemia. *Nature* **478:** 524–528.

C H A P T E R 35

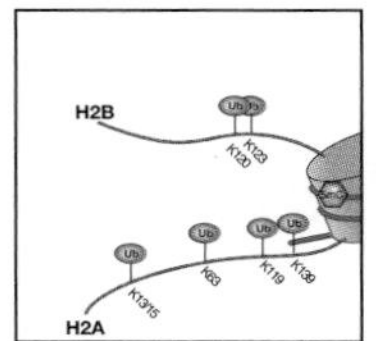

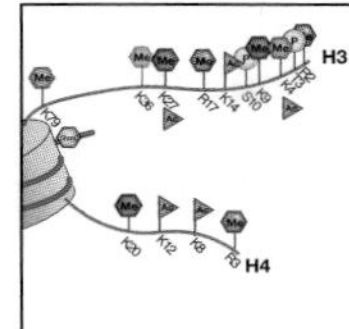

Histone Modifications and Cancer

James E. Audia[1] and Robert M. Campbell[2]

[1]Constellation Pharmaceuticals, Cambridge, Massachusetts 02142; [2]Eli Lilly and Company, Lilly Corporate Center, Indianapolis, Indiana 46285

Correspondence: jim.audia@constellationpharma.com

SUMMARY

Histone posttranslational modifications represent a versatile set of epigenetic marks involved not only in dynamic cellular processes, such as transcription and DNA repair, but also in the stable maintenance of repressive chromatin. In this chapter, we review many of the key and newly identified histone modifications known to be deregulated in cancer and how this impacts function. The latter part of the chapter addresses the challenges and current status of the epigenetic drug development process as it applies to cancer therapeutics.

Outline

Cite as *Cold Spring Harb Perspect Biol* doi: 10.1101/cshperspect.a019521

OVERVIEW

Cancer is a diverse collection of diseases characterized by the dysregulation of important pathways that control normal cellular homeostasis. This escape from normal control mechanisms leads to the six hallmarks of cancer, which include sustaining proliferative signaling, evading growth suppressors, resisting cell death, enabling replicative immortality, inducing angiogenesis, and activating invasion and metastasis (Hanahan and Weinberg 2011). The systematic investigation of the acquired and inherited molecular alterations in the genomes of somatic cells has revealed a great deal about the genetic basis for cancer initiation, progression, and maintenance. This progress has afforded a number of promising and effective opportunities for therapeutic intervention.

The epigenetic mechanisms that govern transcriptional regulation and the corresponding dysregulation in cancer, initially less well studied, have increasingly become the focus of cancer researchers, although the transgenerational effects remain largely unexplored. "Epigenetic" transcriptional control can occur through DNA methylation, covalent histone modification, the reading of these modifications by protein recognition modules, histone exchange, and alteration by ATP-dependent chromatin remodelers and via the effects of noncoding RNA. Because DNA methylation in cancer is addressed elsewhere, this chapter focuses on many of the covalent histone modifications that are altered in cancer, particularly the well-studied acetylation and methylation modifications.

Collectively, the combination of histone marks found in a localized region of chromatin function through multiple mechanisms as part of a "chromatin-based signaling" system (Jenuwein and Allis 2001; Schreiber and Bradley 2002). The landscape of known histone and nonhistone modifications that affect chromatin-based processes continues to expand. The diverse set of observed histone adornments includes phosphorylation, citrullination, sumoylation, adenosine diphosphate (ADP) ribosylation, deimination, and crotonylation. The various known epigenetic mechanisms work in a concerted and interdependent fashion to regulate gene expression. In cancers, their misregulation can result in the inappropriate activation of oncogenes or, conversely, the inappropriate inactivation of tumor suppressors. There is also a growing understanding and appreciation for the genetic basis contributing to epigenetic changes observed in cancer. This adds to the complexity of cancer etiology and, perhaps, offers important general insights into the epigenetic basis of human disease.

This chapter speaks of some of the challenges faced by the epigenetic drug discovery process, searching for molecules targeting epigenetic regulators and also supporting an emerging role of epigenetic alterations that occur during chemotherapy, thought to contribute to drug resistance. Finally, it highlights some of the more recent progress toward developing therapeutic agents in this promising target space.

 Cite as *Cold Spring Harb Perspect Biol* doi: 10.1101/cshperspect.a019521

1 INTRODUCTION

DNA within cells is packaged as chromatin, a dynamic structure composed of nucleosomes as the fundamental building blocks. Histones are the central component of the nucleosomal subunit, forming an octamer containing the four core histone proteins (H3, H4, H2A, H2B) around which is wrapped a 147-base-pair segment of DNA. Each of the largely globular histone proteins possesses a characteristic side chain, or tail, which is densely populated with basic lysine and arginine residues. The histone tails are subject to extensive covalent posttranslational modifications (PTMs) that cooperate to govern the chromatin state. Some PTMs can alter the charge density between histones and DNA, impacting chromatin organization and underlying transcriptional processes, but they can also serve as recognition modules for specific binding proteins that, when bound, may then signal for alterations in chromatin structure or function.

Alterations in the patterns of histone PTMs have been extensively linked to cancer, both at the global level across the genome (Seligson et al. 2005; Bannister and Kouzarides 2011) and at specific gene loci, using ChIP-chip (chromatin immunoprecipitation with DNA microarray analysis) and ChIP-sequencing (parallel sequencing technologies coupled to chromatin immunoprecipitation) technology. These findings have come on the back of the earlier, more established findings linking aberrant DNA methylation to cancer, discovered in the early 1980s (see Ch. 34 [Baylin and Jones 2014]). In addition to recent PTM mapping projects, sequencing efforts have also now identified many of the enzymes responsible for placing ("writers") and removing ("erasers") such epigenetic marks (Fig. 1). Mutations in such enzymes turn out to be among the most frequently mutated targets in cancers (Shen and Laird 2013). Collectively, these findings show interplay between cancer genetics and epigenetics, adding to the complexity in our understanding of the oncogenic process. Advances in whole genome/exome sequencing of patient tumors have allowed for the identification of possible key epigenetic drivers of cancer. These epigenetic drivers may silence one or more tumor-suppressor genes and/or activate oncogenes, thus providing an alternative mechanism by which oncogenic reprogramming of the genome may occur (Shen and Laird 2013). Genomic studies have clearly implicated dysregulation of chromatin modifiers as drivers in many types of cancer (Garraway and Lander 2013) and recurrent mutations occur in the genes that encode the enzymes, which add, remove, and interpret the covalent histone modifications. Intriguingly, certain chromatin

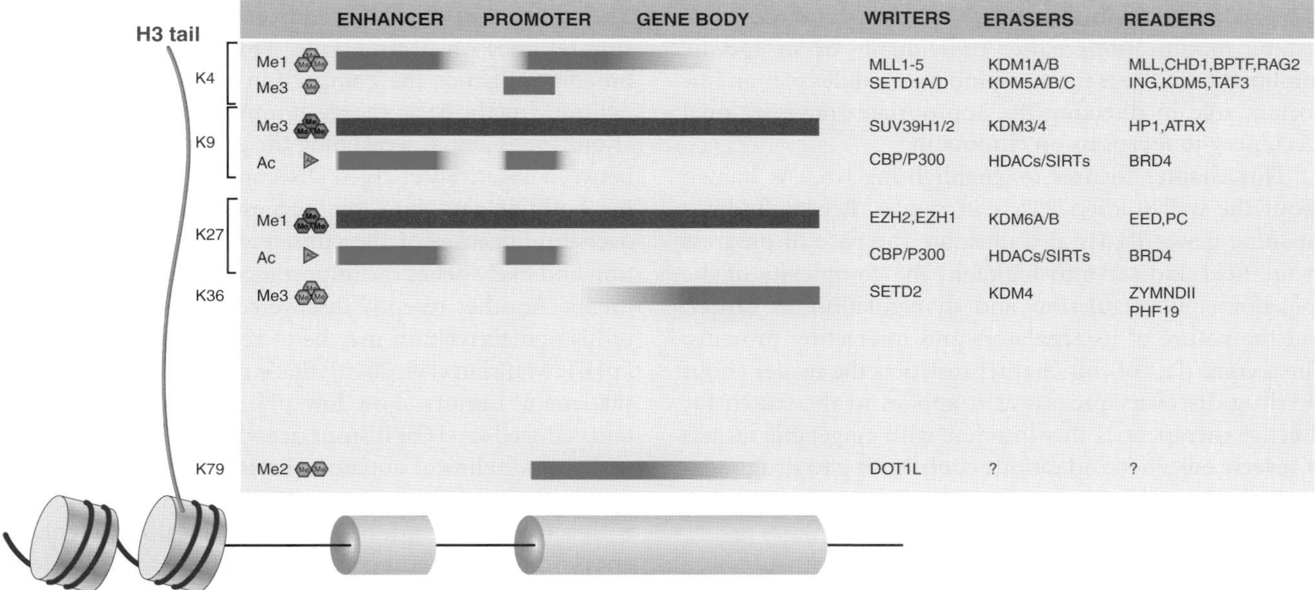

Figure 1. Histone writers, erasers, and readers in cancer. Histone H3 tail lysine residues, frequently subject to posttranslational modifications (PTMs), are indicated along the *left* side. The typical distribution of these H3 PTMs is also indicated along the length of gene loci (including distal enhancers) as shaded blocks. Green (methylation) or cyan (acetylation) indicates histone marks associated with active genes, whereas red shading is indicative of silent genes. A few examples of writers, erasers, and readers that may propagate a mark or act as an effector protein are listed on the *right* side of the figure. For a more complete listing of these proteins, see Appendices A–D at the end of this chapter.

modifiers have been identified in cancer, for both increased and decreased levels of functionality, suggesting they operate both as tumor-suppressor genes and oncogenes. The loss-of-function mutations in affected tumors are often heterozygous, suggesting that haploinsufficiency for these chromatin-modifying enzymes drives the cancer, whereas total absence of function is cell lethal. As a result, this class of enzymes has broad appeal as potential therapeutic targets with both gain-of-function aberrations associated with oncogenic behavior and monoallelic loss-of-function changes rendering tumor cells particularly vulnerable to further inhibition.

An additional level of epigenetic control, and with it additional complexity, is brought by the wide variety of proteins that act at the chromatin level through domains that "read" histone and even DNA PTMs. Direct aberrations in reader proteins have recently been shown to drive oncogenic transformation in some contexts (French et al. 2001, 2008). The discovery of potent and selective small-molecule inhibitors of prototypic members of readers from the bromodomain family, which recognize acetyl-lysine marks, suggests that these may represent attractive points of therapeutic intervention in cancers arising from other genetic and epigenetic drivers (elaborated in Ch. 2 [Qi 2014; Schaefer 2014]; also see Filippakopoulos et al. 2010; Gallenkamp et al. 2014). In summary, the ensemble of chromatin regulators and adaptors possess a diverse set of specialized domains that serve to bind to and recognize histone modifications either individually or in specific combinations. These protein-binding modules play an important role in directing the appropriate transcriptional machinery to locations on chromatin.

This chapter focuses on highlighting what is known about the well-studied PTMs in cancer. Recent findings on some newer PTMs also illustrate the pace of progress in the field and serve to highlight the complexity of the epigenome, its regulation and dysregulation in cancer, and the nature of its regulators and interacting proteins. The second part of this chapter instructs the reader about the drug-discovery process as it applies to the search for effective therapeutics that interfere with epigenetic targets or reverse epigenetic adaptions contributing to drug resistance during cancer treatment.

2 HISTONE MODIFICATIONS

It is well noted that PTM of histones mediates a variety of critical biological processes, generally via chromatin modification that is conducive to the expression or repression of target genes. The bulk of the literature has focused on acetylation, methylation, and phosphorylation. However, in addition to these well-published modifications, histones may also be modified in other ways, some of which are illustrated in Figure 6 of Ch. 3 (Allis et al.) or included in App. 2 (Zhao and Garcia 2014). These include citrullination, ubiquitination, ADP-ribosylation, deamination, formylation, *O*-GlcNAcylation, propionylation, butyrylation, crotonylation, and proline isomerization (Chen et al. 2007; Martin and Zhang 2007; Ruthenburg et al. 2007; Tan et al. 2011; Herranz et al. 2012; Tweedie-Cullen et al. 2012). It is generally accepted that the sum of all these PTMs largely determines the chromatin structure and, hence, biological outcome. The contextual reading and interpretation of the collection of histone modifications is critical, as the same combination of marks may result in distinct biologic outcomes at different genes within the same (or different) cells. This may, at least in part, be due to different reader recognition, DNA-binding proteins, and/or chromatin conformations. Experimentally, the modification and the reader cannot be readily separated for a specific output, and this is an important limitation of our current ability to readily translate combinatorial PTMs. Furthermore, the enzymes that modulate the addition or removal of these moieties are not fully understood (see Ch. 6 [Cheng 2014]; Ch. 4 [Marmorstein and Zhou 2014]; Ch. 5 [Seto and Yoshida 2014]) or, in many cases, yet to be identified.

2.1 Histone Acetylation

The addition of an acetyl group can occur at multiple lysine residues on histone tails. This has the potential to broadly influence the compaction state of chromatin by neutralizing the basic charge at unmodified lysine residues (Kouzarides 2007), weakening the electrostatic interaction between negatively charged DNA and histones. An increasing body of data, however, suggests that this may be an oversimplification of the consequences of this modification and likely belies the importance of specific acetylation events. Another recently discovered, generalized function of histone acetylation may be to regulate intracellular pH (pHi) (McBrian et al. 2013); this is corroborated by the fact that many tumors show low pHi and display concomitant reduced levels of histone acetylation, which correlates with a poor clinical outcome. From a functional perspective, we know that histone acetylation is largely associated with active transcription, particularly localized at enhancers, promoters, and the gene body (Di Cerbo and Schneider 2013). Altered global levels of histone acetylation, particularly acetylation of H4 at lysine (K)16, have been linked to a cancer phenotype in a variety of cancers (Fraga et al. 2005) and have even been found to be of potential prognostic value (Seligson et al. 2009). When hyperacetylation occurs, specifically involving proto-oncogenes, gene expression may be activated, whereas hypoacetylation of tumor sup-

Cite as *Cold Spring Harb Perspect Biol* doi: 10.1101/cshperspect.a019521

pressors often localizes to promoters, co-occurring with DNA methylation, causing the genes to be silenced (see Ch. 34 [Baylin and Jones 2014]).

The enzymes that catalyze the addition of acetyl groups to histone lysine residues are the lysine (K) acetyltransferases (KATs), commonly referred to as histone acetyltransferases (HATs). These enzymes can also acetylate a broad range of nonhistone proteins, including p53, Rb, and MYC. The opposing histone deacetylases (HDACs) are responsible for their removal. Although HATs are generally associated with transcriptional activity, both HAT and HDAC activity together are required for properly regulated gene expression (Struhl 1998). This is partly to do with their activity often being an integral part of multisubunit chromatin-modifying complexes. As a result, genetic or epigenetic aberrations affecting HAT and HDAC expression, translation, protein stability, or domain function can have chromatin regulatory consequences beyond changes in the histone acetylation states. In addition to the structural effects acetylation has on chromatin, it can also function in chromatin as a signal that is recognized by a specific protein module (i.e., "reader"), such as the bromodomain. Epigenetic-based drugs, targeting acetyl readers, are thus likely to be of clinical relevance in the treatment of cancer. Given that acetylation modifications are reversible, the pharmacological intervention of HATs, HDACs, and acetyllysine readers represent known, in the case of HDACs, or viable strategies of therapeutic value in the treatment of cancer.

2.1.1 *Histone Acetylation Writers*

In humans, there are three major families of HATs: the Gcn5-related *N*-acetyltransferase family (GNAT), the MYST family (MOZ, Ybf2, Sas2, TIP60), and the orphan family (CBP/EP300 and nuclear receptors), whose structure and mechanism of action are elaborated in Ch. 4 (Marmorstein and Zhou 2014). A variety of studies have implicated HATs as both oncogenes and tumor suppressors, suggesting that the balance of acetylation is critical. Many of the HAT mutations frequently detected in a variety of cancers are captured in Appendix A at the end of this chapter (Di Cerbo and Schneider 2013). Interestingly, alterations in HAT levels, both upward (Chen et al. 2012b; Hou et al. 2012) and downward (Seligson et al. 2009), often occur without DNA mutation in cancers and are associated with poor outcome.

Somatic mutations in a single allele of the p300 and CBP genes have been identified in multiple cancers. The resulting loss of heterozygosity implicates them as tumor suppressors (Muraoka et al. 1996; Gayther et al. 2000; for more on the characterization of tumor-suppressor genes, see Ch. 34 [Baylin and Jones 2014]). It is increasingly evident that CBP and/or EP300 are important in cancers (e.g., transgenic mice deficient in CBP or EP300 can develop hematologic malignancies [Iyer et al. 2004]). Another facet of HAT function revealed by the many mutations in CBP and EP300 is the protein's ability to acetylate the non-histone transcription factors, p53 and BCL6; a lack of P53 and BCL6 acetylation abrogates their transcriptional activator and repressor functions (Fig. 7A of Ch. 29 [Busslinger and Tarakhovsky 2014]), making the resultant cells more tumorigenic via altered pathways that tolerate DNA damage yet evade apoptosis and cell-cycle arrest (Pasqualucci et al. 2011).

Enhanced HAT activity, conversely, has an overall oncogenic effect in cancers and usually occurs as the result of chromosomal translocations with diverse fusion partners, such as mixed lineage leukemia (MLL)-CBP, MLL-EP300, MOZ-EP300, or MOZ-CBP in hematological malignancies (Krivtsov and Armstrong 2007). The oncogenic effect arises when the translocations generate chimeric oncoproteins. This may allow HATs to abnormally acetylate the genomic targets of its fusion partner. Another mechanism by which HATs, such as EP300/CBP, are oncogenic is when they are recruited by more common fusion proteins, such as acute myloid leukemia (AML)1-ETO, to serve as transcriptional coactivators (Wang et al. 2011). Overall, a HAT's oncogenic or tumor-suppressor effects in cancer are dependent on its dose; overexpression correlates with oncogenic potential, whereas loss of expression results in loss of acetylation capacity. This, therefore, suggests that HATs could be a good drug target, although, to date, progress toward producing a viable HAT inhibitor has lagged behind the development of inhibitors of its counterpart enzyme, HDACs, described in the following section.

2.1.2 *Histone Acetylation Erasers*

Alterations in HDACs, the enzymes that remove acetyl groups from histone lysine residues, have been observed in cancer. There are four major families of HDACs, termed class I, II, III, and IV. Class I, II, and IV are Zn^{2+}-dependent, whereas class III/Sirtuins are nicotinamide adenine dinucleotide (NAD)-dependent (see Fig. 1 in Ch. 5 [Seto and Yoshida 2014]). The contributions of specific HDAC subtypes to individual cancers are currently not fully understood (see Appendix B at the end of the chapter) (Barneda-Zahonero and Parra 2012). This may be, in part, because of the relatively low substrate specificity shown by the HDACs themselves; each individual enzyme is capable of deacetylating multiple divergent histone sites. Interestingly, mutations in HDACs are rare, but overexpression of HDACs is frequently observed in cancer patients (Dell'Aversana et al. 2012).

The class I HDACs that are misregulated, usually by overexpression or mistargeting, are found in a variety of human cancers. The frequency with which they are deregulated in cancer, perhaps, reflects their normal function in such a wide range of tissues, but in cancer, this is often associated with poor prognosis (Nakagawa et al. 2007). The role of HDACs in cancer, as with HATs, may not be confined to histones. The growing list of reported HDAC targets that become deacetylated include α-tubulin, HSP90 and cortactin (HDAC6), p53 (HDAC5), and ERRα (HDAC8); HDACs also directly act on proteins involved in tumor migration, metastasis, and growth (see Appendix B at the end of the chapter). The example of HDAC2 in lung cancer is a case in point; HDAC2 directly deacetylates p53 and the CDKN1B/1C/2A proteins, attenuating a cell's ability to activate the apoptotic machinery or regulate cell cycle (Jung et al. 2012; Reichert et al. 2012). This, therefore, makes it hard to dissect what the exact epigenetic effects of HDAC deregulation are, and thus how HDAC inhibitors interfere with their activity.

Another mode by which HDACs are implicated in cancer is through the abnormal recruitment of HDAC-containing complexes to promoters as a consequence of chromosomal translocation, observed in leukemias (Mercurio et al. 2010). For example, the promyelocytic leukemia (PML)-retinoic acid receptor α (RARα) translocation, which is the driver for many acute promyelocytic leukemia cases, represses many RAR target genes by the aberrant recruitment of the N-CoR/HDAC repressor complexes (Minucci and Pelicci 2006).

The class III HDACs, the Sirtuins (SIRT1-7), are also capable of deacetylating a combination of histones (SIRT1-3, 6, 7) and nonhistone proteins (SIRT1-3, 5, 7), as well as ADP-ribosylating (SIRT4), and desuccinylating various proteins (SIRT5) (see Appendix B at the end of the chapter). They are distinguished from the other classes of HDACs by a distinct catalytic mechanism of action (Ch. 5 [Seto and Yoshida 2014]). The bulk of efforts has focused on studying SIRT1, but has, to date, failed to clarify whether SIRT1 is a tumor suppressor or oncogene, suggesting that these activities may be contextual (Stunkel and Campbell 2011). SIRT1 is overexpressed in many tumors (leukemia, lymphoma, prostate, liver, breast, ovarian, gastric, colorectal, and melanoma) but significantly reduced in others (bladder, colon, glioma). It may be that SIRT1 plays an oncogenic role by inactivating other tumor suppressors (e.g., HIC1) and/or activating tumor promoting genes (e.g., through N-Myc stabilization, or p53) or other proteins (cortactin).

SIRT1 levels are elevated in drug-resistant SK-N-SH neuroblastoma cells, a result that prompted the evaluation of class I/II (Vorinostat) and class III (Cambinol) HDAC inhibitors in animal models of neuroblastoma (Lautz et al. 2012). In neuroblastoma, N-Myc induces SIRT1 transcription, which, in turn, enhances N-Myc stability in a feed-forward loop. It is believed that SIRT1 represses MKP3 phosphatase, leading to elevated ERK phosphorylation/activation and then to N-Myc phosphorylation, a more stable form of N-Myc. The antitumor efficacy of cambinol, a SIRT1/2 inhibitor, in both wild-type and doxorubicin-resistant neuroblastoma is consistent with a tumor-promoting role for SIRT1 and/or SIRT2. Gene silencing and use of pharmacological inhibitors in MCF7 breast tumor cells indicate that blockade of both SIRT1 and SIRT2 may be required to induce apoptosis. Other dual SIRT1/2 inhibitors have shown efficacy in xenograft models (melanoma, Burkitt's lymphoma). In genetically engineered mouse models, SIRT1/PTEN-null transgenic mice spontaneously develop aggressive prostate and thyroid carcinomas. In contrast, $SIRT^{+/-}/p53^{+/-}$ mice develop tumors in multiple organs and SIRT-null mice form prostatic intraepithelial neoplasia.

SIRT3-5 are localized primarily to the mitochondria (although SIRT3 has been reported to deacetylate histones) and can modify a number of substrates involved in energy metabolism via deacetylation, ADP-ribosylation, or desuccinnylation (see Appendix B at the end of the chapter). It is believed that these Sirtuins act as metabolic sensors to adjust mitochondrial energy production during stress or energy deprivation (Haigis and Sinclair 2010). SIRT3, and perhaps also SIRT6, have been identified as tumor suppressors that regulate glycolysis in tumor cell (Haigis et al. 2012; Sebastian et al. 2012), with reduced expression in human breast cancer tissue versus normal. Its loss correlates with hypoxia-induced factor (HIF)α stabilization, up-regulation of HIF1α target genes, increased glycolysis, tumor cell proliferation, and fibroblast transformation (Bell et al. 2011), whereas overexpression has the opposite effect (Finley et al. 2011; Sebastian et al. 2012). Similarly, depletion of SIRT6 leads to tumorigenesis, with transformed SIRT6-deficient MEFs displaying increased glycolysis and tumor growth. Conditional SIRT6 knockout in vivo increases the number, size, and aggressiveness of tumors. These data suggest that SIRT6 plays a role in both the establishment and maintenance of cancer (Sebastian et al. 2012).

SIRT7 deacetylates H3K18 and is highly expressed in many cancers (see Appendix A at the end of the chapter) (Van Damme et al. 2012; Paredes et al. 2014). Hypoacetylation of H3K18, which is indicative of high SIRT7 activity, causes tumor-suppressor gene repression, and is associated with cancer progression and poor prognosis (Seligson et al. 2005; Barber et al. 2012; Paredes et al. 2014).

Broad-spectrum (class I/II) HDAC (e.g., vorinostat, romidepsin) and DNA methylation inhibitors (Dacogen, Vidaza) were the first types of epigenetically targeted therapies approved by the FDA for the treatment of hematological malignancies (discussed in more depth in Ch. 34 [Baylin and Jones 2014]). Data are evolving that suggest that these therapies may be even more effective in combination with standards of care or even with each other (e.g., HDACi + DNA methylation inhibitor), and perhaps be used to ameliorate chemoresistance (discussed further in Sec. 3.1). However, it is abundantly clear that the functions of each specific HDAC isoform are much more complicated and different than anticipated. This is, in part, due to their contextual function in various protein complexes and ability to modify both histone and nonhistone proteins. In the near future, it is expected that more HDAC isoform-specific small molecules (especially inhibitors) will emerge to provide more clarity as to the roles of each HDAC in tumorigenesis and the therapeutic value of more advanced HDAC modulators.

2.1.3 Histone Acetylation Readers

Proteins that read histone acetylated lysines can do so via the bromodomain motif. There are more than 40 bromodomain-containing proteins, which share a high level of sequence homology and structural similarity within the domain. The bromodomain was the first histone-binding module and, consequently, is the most prominent and thoroughly studied histone recognition domain. It is intrinsic to a number of histone-modifying writers (e.g., p300 and MLL) and remodelers (e.g., SMARCA2), as well as other proteins associated with chromatin function and transcriptional control. Its occurrence in such a vast array of chromatin-associating proteins has made it an obvious druggable motif in the search for new epigenetic drugs. The primary structural feature of the bromodomain is a hydrophobic acetyllysine-binding pocket surrounding, most typically, an asparagine residue, which engages in a hydrogen-bonding interaction with the modified histone substrate (for details, see Fig. 6 of Ch. 4 [Marmorstein and Zhou 2014]).

The BET (bromodomain and extraterminal domain) subset of bromodomain proteins consists of four family members, BRD2, BRD3, BRD4, and BRDT, with a common architecture and structural design. BRD4 is a protein that, when bound to acetylated histones, can activate transcription (see Sec. 6 of Ch. 29 [Busslinger and Tarakhovsky 2014]). The reversal of cancer cell phenotype (i.e., the promotion of differentiation and growth impairment using bromodomain-specific inhibitors, such as JQ1 and I-BET) provided the first proof of concept that this histone mark reader could act as a potential therapeutic target for the treatment of cancers, such as leukemias (described in Ch. 2 [Qi 2014; Schaefer 2014]). The inhibitors were effectively shown to phenocopy BRD4 small hairpin RNA knockdown experiments. These approaches were applied to murine MLL-AF9/NRas^{G12D} leukemia models, MLL-fusion cancer cell lines, and patient-derived cells, leading to differentiation and growth impairment (Fig. 2A) (Dawson et al. 2011; Zuber et al. 2011). Further testing of the effects of selective BET inhibitors in other cancer models continues to reinforce the notion that BET proteins are potential therapeutic targets in other cancers, including nuclear protein in testis (NUT)-midline carcinoma, multiple myeloma, lymphoma, lung cancer, and neuroblastoma (Delmore et al. 2011; Mertz et al. 2011; Lockwood et al. 2012; Wyce et al. 2013). The NUT-midline carcinoma, for example, has a translocation, which fuses the BRD3 or 4 protein to the NUT transcriptional regulator. This fusion produces an oncoprotein that, through binding to acetylated histones, is thought to promote transcription of proliferation genes (e.g., MYC). Although work on BET inhibitors continues to be the main focus of research, it is interesting that a dual-drug strategy involving HDAC inhibitors coupled to the JQ1 BET inhibitor has shown positive results as a therapeutic approach for the treatment of AML (Fiskus et al. 2014).

BRD7 and, more recently, BRD9 bromodomain proteins have been identified as components of chromatin-remodeling switch/sucrose nonfermenting (SWI/SNF) complexes, increasingly found to be mutated in a variety of human cancers (Kaeser et al. 2008; Kadoch et al. 2013). BRD7 has specifically been linked to the control of p53 and BRCA1 transcriptional pathways (Drost et al. 2010; Harte et al. 2010). In the case of p53, BRD7 is thought to help recruit the P300 HAT to p53 target genes, activating transcription of senescence genes (Drost et al. 2010). In the case of estrogen receptor (ER)-α response genes, BRD7 is believed to function in the recruitment of the BRCA1 transcriptional activator (Harte et al. 2010).

CBP and EP300, previously described in the context of their HAT activity, also possess bromodomains and transcription factor binding domains. These multifunctional proteins are critical as mediators of histone marks at transcriptional enhancers (Fig. 1) (elaborated in Ch. 36 [Pirrotta 2014]) and required for transcriptional activation. This is illustrated by the loss of expression at MYB target genes, controlling the proliferation and differentiation of hematopoietic stem and progenitor cells, because *MYB* is mutated at its P300 interacting domain and thus is no longer able to activate gene expression (Sandberg et al. 2005). It should be noted that transcriptional activation occurs in a three-dimensional context, in which looping together of enhancer and promoter regions is thought to ensure proper expression. This relies on correct local histone signatures written by HATs and lysine methyltransferases (KMTs) to facilitate an open chromatin structure and

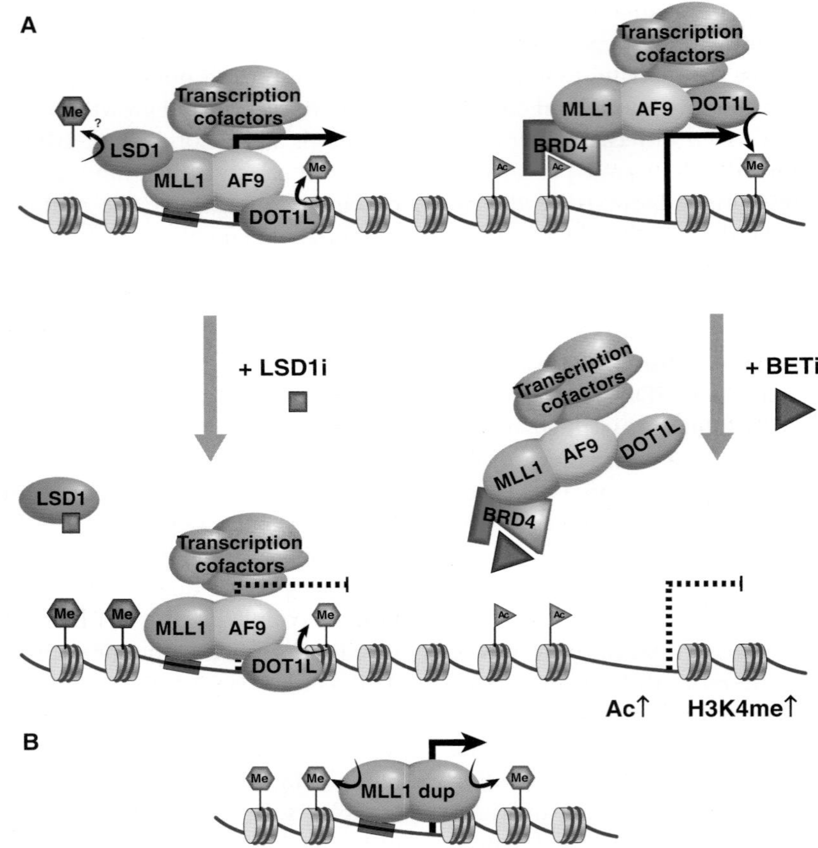

Figure 2. MLL as an oncogene. (*A*) The MLL1-AF9 (or ENL) fusion oncoprotein activates transcription via two possible mechanisms; the *left* mechanism attributes recruitment of MLL1-AF9 to chromatin via the MLL1 portion, and transcription is activated via association with cofactors, including the DOT1 methyltransferase, methylating H3K79 (green hexagon) and the pTEFb complex, which modifies RNA Polymerase II into the active elongating form. (*B*) The partial duplication of an MLL gene can result in duplication of an internal region that includes chromatin binding features and protein–protein interaction domains, providing oncogenic methyltransferase H3K4me3 activity and increased transcriptional activation. LSD1 may be involved via the MLL supercomplex, or a transcription elongation complex contributing to oncogenic activity through H4K4me2 or H3K9me2 demethylase activity. LSD1 inhibition somehow reduces the oncogenic program, promoting differentiation. In the mechanism on the *right*, recruitment of MLL1-AF9 to chromatin is attributed to its association with BRD-2, -3, or -4 via acetylated chromatin. This mechanism of gene activation can therapeutically be targeted via treatment with BET inhibitors.

the recruitment of the full transcription machinery (Fig. 1), often through histone PTM reader proteins, such as BRD4 (reviewed in Ong and Corces 2011; exemplified in Fig. 1 of Ch. 2 [Qi 2014]).

P300 was also found to mark binding sites of leukemic fusion proteins, such as AML1-ETO and PML-RARα, and to directly acetylate AML1-ETO fusion proteins, indicating it may have particular importance in contributing to oncogenic transcriptional programs in leukemia (Wang et al. 2011; Saeed et al. 2012). The specific function of the bromodomains in this context has not been definitively shown, but the recent identification of potent and selective chemical probe molecules should afford a similar opportunity to interrogate the function of these domains as has occurred for the BET proteins (Jennings et al. 2014). One would anticipate that, by virtue of BRD proteins being involved in multiple complexes, their mutation must have far-reaching oncogenic effects.

2.2 Histone Methylation

Methylation of lysine and arginine residues on histone tails represents a complex and more subtle chromatin modification than does acetylation. Multiple methylation states exist for both lysine and arginine residues, but when modified, they retain their fundamental basic character (Bannister and Kouzarides 2011). The suggestion that individual methylation states for each histone lysine or argi-

Cite as *Cold Spring Harb Perspect Biol* doi: 10.1101/cshperspect.a019521

nine can be meaningfully recognized generates enormous functional complexity. This modification is tightly regulated by a number of methyltransferase writers and demethylase eraser enzymes that act in concert to place and remove specific methyl marks critical for gene expression, cell fate, and genomic stability. Individual methyltransferases are highly specific with regard to both the lysine residue on which they operate and the degree of methylation. Appendix C at the end of this chapter captures some of the important information about lysine and arginine methyltransferases in cancer. Appendix D at the end of this chapter summarizes some of the important demethylases implicated in cancer. It is important to note that, whereas many of the methyltransferases and demethylases are amplified, overexpressed, deleted, misregulated, rearranged, or mutated, their direct causality on cancer as a result of changes to lysine methylation has not been broadly shown (Black et al. 2012). Additionally, cancer associations have been observed both for elevations and reductions in histone methylation (illustrated in Fig. 7 of Ch. 29 [Busslinger and Tarakhovsky 2014]). The literature has much more on histone KMTs, as elaborated in this section, partly because of their earlier discovery. Discussion of protein arginine methyltransferases (PRMTs) in cancer is discussed in Section 1.4 on PTM cross talk and further in Ch. 6 (Cheng 2014).

2.2.1 Histone H3K4 Writers

The MLL family of KMTs is implicated in many forms of cancer, either by loss of function or, in the case of MLL1, through dysregulation following translocation or rearrangement (see Appendix C at the end of this chapter). Members of the MLL (KMT2) methyltransferases specifically methylate histone H3 at lysine 4. MLL1 is frequently translocated in myeloid and lymphoid leukemias, in which such rearrangements account for ~80% of infant leukemias and 5%–10% of adult leukemias (Smith et al. 2011; Zhang et al. 2012a; Li et al. 2013a). The MLL1 fusion proteins generally do not retain the catalytic SET methyltransferase domain, but do retain their DNA-binding motifs that target Hox genes. The chimeric proteins inappropriately recruit epigenetic factors to MLL targets, altering the transcriptional control of critical genes, such as *HOXA9*. Constitutive HOXA9 expression prevents differentiation in myeloid lineages, thus, contributing to the maintenance of a multipotent phenotype. The fusion partners also frequently play a role in transcriptional elongation at the target genes. The partial duplication of MLL1's set KMT domain further corroborates a role for MLL1 in cancer, showing in this case, increased HoxA gene expression associated with increased H3K4me3 marks at the promoter, albeit by an unknown mechanism (Fig. 2B) (Dorrance et al. 2006). In the case of MLL2/3 loss-of-function mutations, the molecular consequences are, at present, unclear, although increasing evidence suggests that cancers may alter H3K4 methylation states as a common method of achieving a growth advantage.

2.2.2 Histone Lysine Erasers

LSD1 (KDM1A) was the first reported lysine demethylase (KDM) enzyme specific to H3K4 and H3K9 residues (Ch. 2 [Shi and Tsukada 2013]). It is a classic oncogene, based on the numerous reports of its overexpression in many types of cancer (see Appendix D at the end of this chapter). This overexpression, coupled with the fact that LSD1 is part of the MLL supercomplex and found at MLL target genes in MLL-fusion instances of AML, implicates it in the oncogenic gene expression program of these leukemia stem cells. Promising results have been obtained using a small molecule inhibitor of LSD1 in both human and murine AML cell lines with the MLL-AF9 translocation, in which induction of differentiation of otherwise proliferative undifferentiated tumor cells was observed (Fig. 2A) (Harris et al. 2012). The similar use of an LSD1 inhibitor in PML-RARα translocated AML, but this time in combination with all-*trans*-retinoic acid (ATRA), produced a superior antileukemic effect to the single-agent approach (Schenk et al. 2012). ATRA alone can enhance differentiation of tumor cells through the dissociation of RARα-recruited repressor complexes at RARα target genes. The combination therapy of LSD1 inhibitor plus ATRA caused ATRA-insensitive cancer cells (i.e., a drug-resistant population) to become differentiated. Although these two results show great therapeutic promise, the mechanism by which LSD1 modulates expression is not clear at present.

Apart from KDM1 (i.e., LSD1 and -2), all the other known KDMs are members of the larger JmjC domain family of demethylases. The characterization of many of these proteins is in progress through both functional studies and the development of small molecule inhibitors targeted against them and stand to be invaluable for our understanding and treatment of cancer. One interesting finding made by Sharma et al. (2010) was the discovery of an underlying epigenetic basis for transient and reversible drug resistance that develops in certain cancer cell populations during treatment with cancer drugs. The H3K4-specific JmjC domain demethylase, KDM5A, in particular, was implicated in the development of drug resistance. Further research should be able to confirm whether epigenetic mechanisms, such as histone demethylation truly represent a broad and adaptive mechanism by which cancer cells can avoid eradication during chemotherapy (Sharma et al. 2010). The related KDM5B has also been associated with divergent roles in cancer: in metastatic melanoma, it has a putative tumor-suppressive function (Roesch et al. 2006; Roesch et al. 2008); in breast cancer, a proproliferative role (Mitra et al. 2011); and in prostate cancer, overexpression has been observed.

H3K9 is another histone residue in which methylation is aberrantly regulated in multiple cancers. This could be explained by alterations in H3K9-specific KMTs or opposing KDM enzymes (see Appendices C and D). With respect to the G9a KMT, both deletion or lowered expression and increased expression have been observed in cancer (see Appendix C at the end of this chapter). For H3K9-specific demethylases, KDM3A and KDM4 are often amplified or highly expressed, whereas attenuated expression of KDM4A has been observed in cancer (see Appendix D at the end of this chapter). One mechanism by which KDM3 and KDM4 could become overexpressed is by the targeted gene-activating effect of the HIF, itself induced by tumorigenic hypoxic conditions.

2.2.3 Histone H3K27 Writers and Erasers

EZH2 is the catalytic component of the Polycomb repressive complex 2 (PRC2) responsible for the di- and trimethylation of H3K27 (H3K27me2 and -me3) via its SET domain. EZH2 was among the earliest methyltransferase to have putative oncogenic capabilities because it was shown to be overexpressed or amplified in multiple cancers, including breast, prostate, and bladder (Bracken et al. 2003). More recently, altered EZH2 functionality was reported as oncogenic. Intriguingly, it is also believed to act as a tumor suppressor in some cancers, suggesting the balance of resulting H3K27 methylation is key.

Change-of-function mutations within the SET domain of EZH2 are oncogenic, giving rise to H3K27me3 accumulation. Such mutations modify EZH2's H3K27 substrate preference, increasing catalytic activity at H3K27me2 substrates (not increased for H3K27me1 substrates vs. wild type) (see Appendix C at the end of the chapter). These heterozygous oncogenic gain-of-function mutations strongly implicate the catalytic activity in the malignant transformation process, and support consequent gene silencing as being the driver for genesis or maintenance of tumor cells. Moreover, diffuse large B-cell lymphoma tumors harboring such EZH2 mutations are extremely sensitive to EZH2 inhibition with small molecule inhibitors, suggesting promise for their clinical translation and offering a path toward molecular stratification of patients on the basis of the mutational status. More compelling evidence that accumulation of H3K27me3 is central to EZH2's oncogenic effect in medulloblastoma is that EZH2 overexpression or amplification was found to be mutually exclusive to inactivating mutations in the reversing H3K27 demethylase UTX (KDM6A) (Robinson et al. 2012).

A potential tumor-suppressor role for EZH2, or PRC2, has been shown by loss-of-function mutations in myeloid malignancies (Khan et al. 2013). This was further supported by the observation that a K27M point mutation in the histone H3 gene (H3F3A) in pediatric glioblastoma resulted in lowered or absent H3K27me3 (Venneti et al. 2013). This indicated that a loss of a lysine residue at position 27 of H3 abrogated the capacity for chromatin to appropriately repress PRC2-mediated gene expression. EZH2's repressive effects are not only modulated by the opposing UTX demethylase, but also by the activity of the SWI/SNF chromatin-remodeling complex. SWI/SNF, in fact, critically depends on EZH2 activity, as shown in cell lines and mouse models in which inactivation of EZH2 blocked tumor formation driven by SNF5 loss (Wilson et al. 2010). Collectively, these results indicate that correct genomic and developmental gene repression programs are largely ensured by H3K27 methylation. The writers and erasers of this modification, as well as other epigenetic modulators, are central to maintaining the balance of H3K27 methylation and represent viable drug-discovery targets in cancer. It is interesting to note that continued endeavors to epigenomically map cancers are suggesting that the correct balance of H3K27 methylation may be affected by other chromatin PTMs, namely DNA methylation in cancer, but also the interplay with other histone PTMs, such as acetylation (see Ch. 36 [Pirrotta 2014]).

2.2.4 Other Histone Lysine Writers and Erasers

H3K36-specific methyltransferase writers are implicated in a variety of cancers (see Appendix C at the end of this chapter). These enzymes are members of the NSD (nuclear receptor-binding SET domain) family (i.e., KMT3B/3F/3G). In AML, NSD1 (KMT3B) is fused to nucleoporin 98 (*NUP98*), resulting in enhanced H3K36me3 at critical HOXA gene loci, accompanied by increased transcription (Wang et al. 2007). The loss of gene repression may partly be achieved by NSD1/H3K36me3 preventing PRC2 complex access. Similarly, NSD2 (WHSC1, KMT3G) is translocated in 20% of multiple myelomas and has been reported to act as a potent coactivator of NF-κβ, which plays an important role in cancer progression. The investigators suggest that NSD2 is recruited to NF-κβ target gene promoters on pathway induction, resulting in an elevation of histone H3K36me2 and H3K36me3 marks at their promoters, directly implicating NSD2 methyltransferase activity in gene deregulation (Yang et al. 2012).

Another common site of histone lysine methylation is at H4K20, which when monomethylated (H4K20me1) is associated with transcriptional repression and when dimethylated (H4K20me2) is linked to DNA repair pathways. The presence of H4K20me3, accompanied by the loss of H4K16 acetylation, constitutes a common H4 cancer "epigenetic signature" predominantly at DNA repetitive sequences (Fraga et al. 2005). This suggests that one or more of the known histone H4K20 methyltransferases—that is, SUV420H1 and SUV420H2 (H4K20me2 and H4K20me3) or SETD8 (HK20me1) (KMT5A)—could be targets for disruption in cancer cells.

2.3 "Atypical" Histone Modifications

The growing base of knowledge regarding some of these less prevalent or newer PTMs, namely, histone deamination/citrullination, ubiquitination, ADP-ribosylation, deamination, N6-formylation, and *O*-GlcNAcylation, are discussed below.

2.3.1 Histone Arginine Citrullination

Histone arginine residues may become methylated like lysines, and this modification is catalyzed by the PRMT family of enzymes. Histone arginine residues, unmodified or monomethylated, may also be modified by hydrolysis to Citrulline, a process termed citrullination or deimination. This removes the positive charge, as does histone acetylation, and is enzymatically catalyzed by the partitioning and anchoring domain (PADs or PADIs). Of the PAD family, only PAD4 has a nuclear localization signal and been clearly shown to deiminate/citrullinate histones H3 (Arg residues 2, 8, 17, 26), H2A, H4 (Arg3), and H1R54 (Wang et al. 2004; Tanikawa et al. 2012; Christophorou et al. 2014), although a recent report showed PAD2 as deiminating H3R26 (Zhang et al. 2012b). The citrullination of histone tails is associated with the decondensation of chromatin, but this is tightly regulated to occur only in naïve pluripotent (i.e., embryonic stem or inner cell mass cells) cells, conducive to the transcriptional activation of key stem-cell genes, or in myeloid lineages as part of the immune response to inflammatory stimuli. The H1-specific deimination at stem-cell loci seems to be one mechanism by which PAD4 effects chromatin decondensation through the weakening and eventual displacement of H1 on chromatin in ground-state pluripotent cells (Christophorou et al. 2014). Citrullination of H3R8, in some instances, may also contribute to chromatin decondensation, possibly through the interference of HP1a binding to H3K9me3 (Sharma et al. 2012). Given that PAD4 is highly expressed in many tumor tissues and cell lines (e.g., non-small-cell lung carcinoma [NSCLC], ovarian, breast, and hepatocellular carcinomas) compared with normal tissues or more-benign hyperplastic tissues (Chang et al. 2009), it will be interesting to see whether PAD4 contributes to tumorigenesis. It is conceivable that this may occur by promoting chromatin decondensation, resulting in the activation of pluripotent stem cell genes, or interference of HP1b heterochromatin, or via its gene-repressive effect via involvement in the p53 pathway described below.

Citrullination may have a gene-repressive effect, reported first in conjunction with HDAC1-mediated histone deacetylation of estrogen-regulated genes (Cuthbert et al. 2004; Wang et al. 2004; Denis et al. 2009). Depletion of PAD4 in colorectal tumor cells has an antitumorigenic effect, increasing the expression of p53 target genes (p21, GADD45, PUMA) and inducing cell-cycle arrest and apoptosis (Li et al. 2008). As p53 and PAD4 directly interact, it is not certain whether PAD4 acts in these cells via a scaffolding effect, enzymatic deiminase activity, or a combination of both.

Based on the accumulated evidence, there is interest in developing inhibitors of PADs (especially PAD2 and PAD4) for the treatment of cancer, and possibly also inflammatory diseases. Currently, PAD inhibitors are in early preclinical development, so more time will be required to understand the therapeutic potential of these targeted epigenetic therapies.

2.3.2 Histone Lysine ADP-Ribosylation

During the cell cycle, DNA is being damaged and repaired by an elaborate set of mechanisms that have evolved to conserve genomic integrity. Although cancers may result from improper/incomplete DNA repair, targeted inhibition of DNA repair mechanisms may also be exploited to kill cancer cells. One clinical paradigm is to interfere with the tumor's ability to repair its DNA following treatment with a cytotoxic chemotherapy (i.e., chemosensitization).

ADP-ribosylation of lysine residues is a relatively rare histone modification, occurring in $<1\%$ of all histone proteins, but is observed particularly in instances of single-DNA-strand breaks (Boulikas 1989). Some of the NAD^+-dependent Poly(ADP-ribose) polymerases (PARPs), more recently referred to as ADP-ribosyltransferases (ART), ADP-ribosylate histone and nonhistone proteins, as do some of the Sirtuins (Stunkel and Campbell 2011). PARP1 (ARTD1), the best-characterized histone ADP-ribosylating enzyme, is activated by environmental stresses, such as DNA damage, and has been reported to ADP-ribosylate the amino-terminal tails of all core histones, specifically at H2AK13, H2BK30, H3K27, H3K37, and H4K16 (Messner et al. 2010). This PTM is thought to cause chromatin decondensation and recruit DNA repair machinery.

Interestingly, acetylation of H4K16 inhibits ADP-ribosylation by PARP1 (Messner et al. 2010). This is the first direct piece of evidence pointing to PTM cross talk between lysine ADP-ribosyl and acetylation marks. Interestingly, a group of macrodomain-containing proteins have been shown as mono-ADP-ribosylhydrolases and define a class of enzymes that renders mono-ADP-ribosylation a reversible modification (Rosenthal et al. 2013).

Responding to DNA damage repair is crucial for maintaining genomic stability and, typically, tumors are deficient in a DNA repair pathway. It is still early for deciphering a functional understanding of the biological role of histone ADP-ribosylation and individual PARP enzymes in chromatin function. Nonetheless, observed histone ADP-ribosylation and PARP1 detected during DNA repair has led to the use of PARP1 inhibitors in the treatment of breast and ovarian cancers. PARP1 inhibitors (olaparib) induce synthetic lethality (i.e., when two or more mutations together cause cell lethality) in cancer cells already mutated for BRCA1 or -2 genes. The BRCA proteins are essential components of the double-strand break (DSB) repair pathway through homologous recombination (discussed in Sec. 1.4.3), and in their absence, cancer cells become reliant on PARP1-dependent DNA repair pathways. Thus, the use of PARP1 inhibitors in BRCA mutant cancers results in tumor cell death, whereas normal cells survive owing to their still-intact BRCA1-dependent DNA damage repair pathway.

The positive progression of several PARP inhibitors through clinical trials (breast, colorectal, and ovarian cancers), although yet unapproved, lends credence to a parallel clinical approach with histone ADP-ribosylation inhibitors.

2.3.3 Other Histone PTMs with Implications in Cancer

A relatively new entry to histone PTMs is histone lysine deamination. This PTM already appears to play a known role in cancer progression and metastasis via genetic studies of its writing enzyme, lysyl oxidase–like 2 enzyme (LOXL2). Histone H3K4me3 can be deaminated by LOXL2 and this enzymatic activity is required to silence the tumor-suppressor gene, CDH1, and induce the epithelial to mesenchymal transition, commonly associated with metastasis, in breast tumor cells (Herranz et al. 2012).

N6-formylation of histone lysine residues is one of the rarest histone PTMs, found at very low abundancy (0.04%–0.1% of all lysines in acid-soluble chromatin proteins) (Jiang et al. 2007). The linker histone H1 is *N*-formylated most frequently (13 residues), but core histones also contain multiple sites of Lys formylation (19 sites in total for H2A, H2B, H3, and H4) (Jiang et al. 2007; Wisniewski et al. 2008). Lysine formylation increases during oxidative stress (Jiang et al. 2007), which has been associated with various disease states, including cancer. The biological relevance of N6-lysine histone formylation may lie in its potential to disrupt gene expression mediated by other histone lysine marks. For example, *N*-formylation is observed at H3K79, which could possibly interfere with methylation at that site catalyzed by DOT1L, a target that has been linked to MLL-rearranged leukemias. However, specific cross talk between lysine *N*-formylation and acetylation or methylation remains unverified and there is no clear disease linkage established with histone formylation.

Histone PTM is not restricted to Arg and Lys residues. Histone Ser or Thr phosphorylation has long been known to be an important PTM involved in the cellular response to DNA damage (Rossetto et al. 2012). For example, H2A.X is phosphorylated at Ser 139 shortly after DNA damage, demarcating the region of chromatin around the DNA lesion.

Recently, *O*-GlcNAcylation (addition of β-*N*-acetylglucosamine) of Ser and Thr residues on core histones H2A at Thr101, H2B at Ser36, and H4 at Ser47 was reported, although more residues are thought to be subject to this modification (Sakabe et al. 2010). This PTM increases during heat shock and the mitotic phase of the cell cycle. The known sites of *O*-GlcNAcylation are also modified by phosphorylation, the latter of which are potentially important for H2A-H2B dimerization. This suggests that *O*-GlcNAcylation may interfere with residues that are potentially to be phosphorylated, hence, explaining the observed increase in chromatin condensation when the *O*-GlcNAc transferase (OGT) enzyme is overexpressed or *O*-GlcNAcylation is triggered following heat shock (Hanover 2010). Thus, the increase in histone *O*-GlcNAcylation during heat shock, concurrent with DNA condensation, suggests a possible role in DNA damage.

Recently, OGT was found to operate in a ten eleven translocation (TET)2-dependent manner, particularly at transcriptional start sites. This suggests there is a combinatorial effect of DNA 5-hydroxymethylation and *O*-GlyNAcylation in regulating gene expression (Chen et al. 2013). Collectively, these findings suggest a possible role in transcriptional regulation, DNA damage, and thus cancer, particularly, given that the OGT partner, TET2, is a known tumor suppressor involved in myeloproliferative diseases (see Ch. 34 [Baylin and Jones 2014]).

2.4 Cross Talk among Histone Modifications

As noted above, histones can be decorated with a wide variety of PTMs, presenting numerous combinatorial pat-

Cite as *Cold Spring Harb Perspect Biol* doi: 10.1101/cshperspect.a019521

terns (Jenuwein and Allis 2001). These can occur on any of the four core histones in the histone octamer (two copies each of histones H2A, H2B, H3, and H4) and may even differ between tails of the same histone within a nucleosome or between nucleosomes (Fig. 3) (Ruthenburg et al. 2007). There is no doubt that PTMs are implicated in cancer. Although single marks have been generalized as activating (e.g., H3K9ac, H3K4me3) or repressive (H3K9me3, H3K27me3) (Ruthenburg et al. 2007; Zhu et al. 2013), it is clear that certain PTMs can affect the ability of other marks to either be put down or read (more extensively reviewed by Ruthenburg et al. 2007; Suganuma and Workman 2008). Extracting which are the important modifications that drive tumorigenesis versus which are passengers is a key endeavor for the field. It is complicated by the fact that multiple interactions occur between PTMs, often referred to as cross talk. A few examples that have a bearing on our understanding of the epigenetic role in cancer are described in this section.

2.4.1 Histone Lysine and Arginine Methylation Cross Talk

There are instances in which histone PTMs can act as exclusion marks, preventing the occurrence of other marks and/or binding of chromatin readers (Migliori et al. 2010). For instance, the presence of H3K4me3 appears to inhibit the deposition of the H3R2me2a mark by PRMT6 (Guccione et al. 2007; Hyllus et al. 2007). Conversely, the H3R2me2a mark prevents methylation of H3K4 by an MLL complex, making the coexistence of these marks mutually exclusive (Fig. 3A). Another example of exclusion marks is seen when PRMT7 promotes symmetric dimethylation of H4R3 (H4R3me2s) and inhibits expression of MLL2-dependent target genes (Dhar et al. 2012). PRMT7 knockdown causes an increase in MLL2-catalyzed H3K4 methylation, suggesting an inverse relationship between these methylation sites. It has been hypothesized that the H4R3me2s mark may block binding

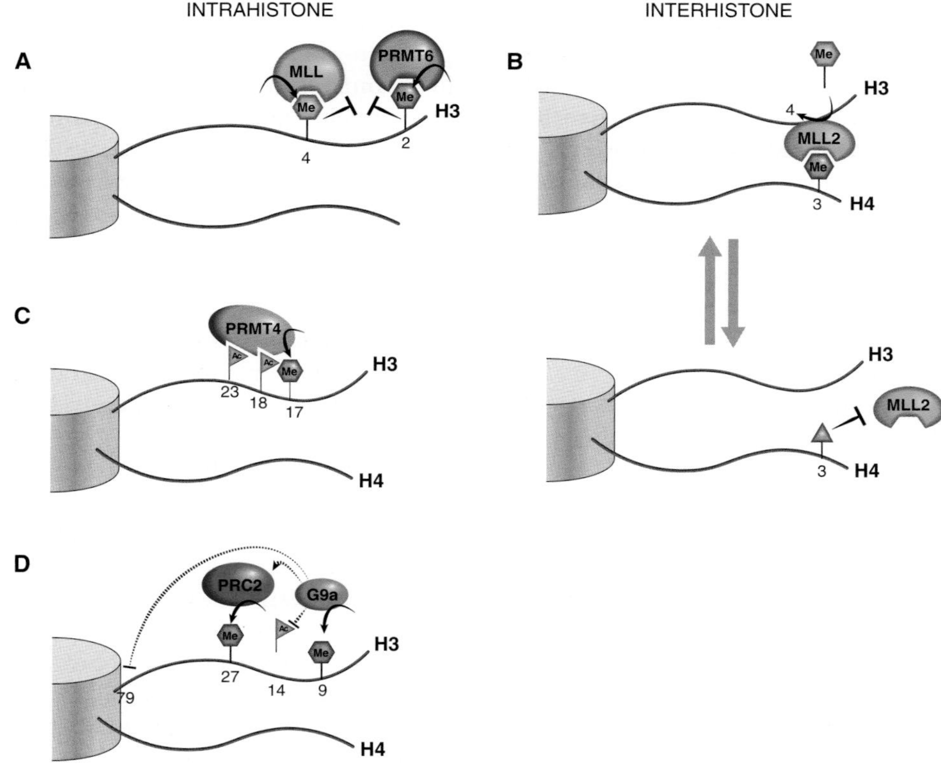

Figure 3. Histone tail cross talk. (*A*) H3R2me2a and H3K4 methylation are examples of mutually exclusive intrahistone H3 PTMs. (*B*) An example of an interhistone cross talk is when H4R3me0 or H4R3me2a marks (purple hexagon) are converted into H4R3me2s (purple triangle), which is thought to block the binding of MLL2 via its PHD4-6 domain, thus preventing the methyltransferase activity of MLL2 at H3K4. (*C*) PRMT4, a histone arginine methyltransferase, is thought to partially rely on H3 acetylation at K18 and K23 for recruitment to H3 and subsequent dimethylation of the nearby R17 residue. (*D*) An illustration of the increasingly complex picture of histone H3 tail cross talk, involving H3K9, H3K27, and H3K79 methylation and H3K14 acetylation. See text for a more detailed explanation.

by tandem PHD4-6 reader domain(s) within MLL2, which normally recognize H4R3me0 or H4R3me2a, and are required for its methyltransferase activity (Fig. 3B) (Dhar et al. 2012). In this way, the tails of two histones, H3 and H4, could interact contextually to dictate target gene expression.

2.4.2 Cis *Histone Cross Talk*

Histone lysine acetylation and arginine methylation can also act cooperatively to localize and activate other methyltransferases. A sequential process has been proposed in which estrogen stimulation, CBP/EP300, first acetylates H3K18, then H3K23, and, finally, PRMT4 is attracted to the histone tail in which it dimethylates H3R17 (Daujat et al. 2002). In vitro histone H3K18/K23 acetylation was shown to tether recombinant PRMT4 to the H3 tail to efficiently catalyze arginine dimethylation (Fig. 3C) and activate estrogen-dependent genes.

Histone methylation and phosphorylation can interact in a regulatory fashion. One key example is when chromodomain recognition of a methylated lysine can be disrupted by an adjacent phosphorylation event. H3K9me3 is important for recruiting heterochromatin protein 1 (HP1) to distinct chromosomal regions, required for heterochromatin formation, and thereby regulating gene expression (Stewart et al. 2005). H3 serine 10 phosphorylation adjacent to the trimethylated H3 (H3K9me3S10ph) by Aurora B causes dissociation of HP1 from heterochromatin during M phase and allows for mitotic progression (Fischle et al. 2005; Hirota et al. 2005). This apparently occurs by a disruption of the chromodomain binding of HP1 to H3K9me3 (see Fig. 12 of Ch. 3 [Allis et al.]).

With the advent of modern mass spectrometry and corresponding analytical tools, the entire panoply of histone marks may be interrogated simultaneously from cell extracts. An example is provided by the mass spectrometric analysis of HEK293 cells when G9a/GLP-1 (EHMT1/EHMT2) is knocked down, revealing not only a reduction in the expected methyltransferase products, H3K9me1 and H3K9me2 (H3K9me2 being predominant), but also an increase in H3K79me2, a mark associated with gene expression, added by DOT1L (Plazas-Mayorca et al. 2010). In addition, the greatest reduction in H3K9me2 was seen on peptides containing H3K14 acetylation. The biological significance of these mark changes is unknown, but it infers cross talk between G9a/GLP-1, H3K9me2, H3K79me2, and H3K14ac (Fig. 3D). It is conceivable that H3K9me2 and H3K14ac are mutually exclusive marks representing a form of conditional epigenetic switch. The situation is further complicated by PRC2 recruitment being shown as dependent on G9a and GLP-1 association, affecting, therefore, the degree of H3K27me3 at target genes, which mediates gene silencing (Mozzetta et al. 2014). This shows that there is additional interplay between H3K27me3 and H3K9 methylation, which are both typically repressive marks.

Ultimately, the challenge of relating changes in primary chromatin definitively to cancer, and then determining how cross talk contributes to cancer are key questions in the field. A clear understanding of these marks is limited both by the technologies available with which to interrogate this, as well as our ability to interpret the role that the readers themselves play on the biologic output of chromatin modifications.

2.4.3 *DNA Damage Repair: Ubiquitination and Other Cross-Talking PTMs*

The repair of DNA damage is an essential cellular function, requiring a rapid dynamic response. It is important to note that repair has to occur in a chromatin context and involves extensive chromatin remodeling. The loss of total repair functionality is cell lethal; however, in cancer, there is often partial loss of function that is accompanied by an increase in genomic instability and mutation rates. PTMs of histone proteins around the DNA damage site and recruited repair proteins is a central means by which the process is regulated. In fact, our current knowledge of mammalian DNA repair in a chromatin context, illustrates the central involvement of histone PTMs as signals in the recruitment or binding inhibition of repair proteins, involving extensive cross talk. Ubiquitination, in particular, represents one of the first key modifications activating the repair pathway. Ubiquitination occurs at histone lysine residues, as do the already discussed acetylation and methylation modifications.

One of the earliest stages of the DNA damage response is when H2AX undergoes monoubiquitination at Lys119/Lys120, mediated by the RNF2-BMI1 complex (components of the PRC1 complex). This is necessary for the recruitment of early sensors of DNA damage. They include the ATM protein, which phosphorylates histone variant H2AX to form γ-H2AX, MRN complex at broken DNA ends, and MDC1 (Fig. 4B) (Ginjala et al. 2011; Pan et al. 2011; reviewed in Panier and Durocher 2013). As RNF2-BMI1 complex-depleted cells have an impaired DNA repair capability and increased sensitivity to ionizing radiation (Pan et al. 2011), it is tempting to speculate that pharmacological inhibitors of H2AX monoubiquitination may act as radio-sensitizing agents in cancer. Monoubiquitination of H2A at Lys119, incidentally, is also involved in transcriptional repression via PRC proteins, such as the RING1A and 1B ubiquitin ligases (de Napoles et al. 2004; Fang

Cite as *Cold Spring Harb Perspect Biol* doi: 10.1101/cshperspect.a019521

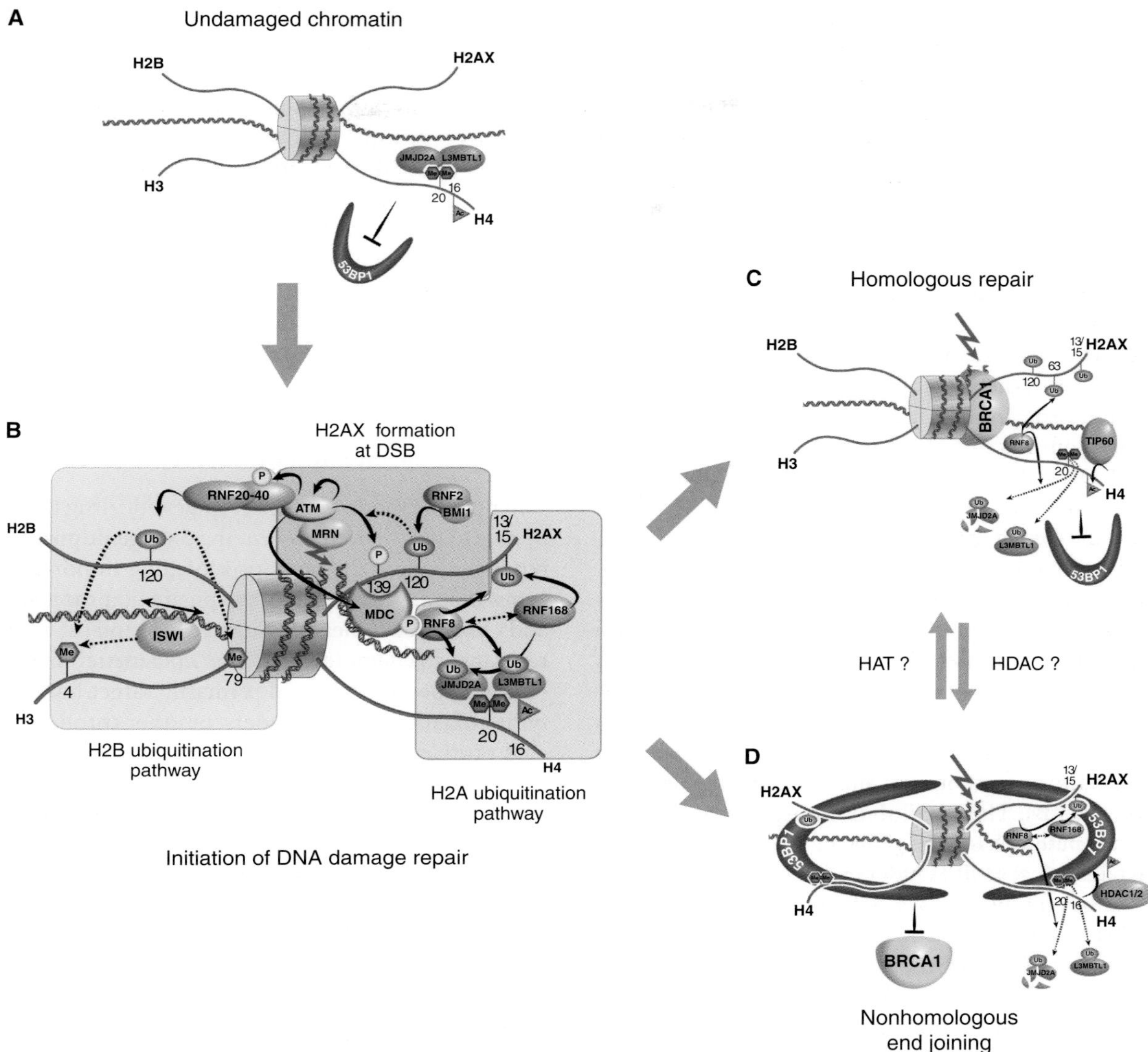

Figure 4. Role of histone PTMs in DNA damage repair. (*A*) Before DNA damage, L3MBTL1 and JMJD2A (possibly also JMJD2B) bind, via their Tudor domains, to H4K20me2 (red hexagons), hindering access by DNA repair proteins, such as 53BP1. (*B*) When double-strand DNA breaks occur (red arrow), the DNA damage–sensing proteins, such as MRN, act to initiate cascades of protein and histone tail PTMs (blue-shaded area). In particular, the phosphorylation of H2A at S139 represents the conversion of H2A into H2A.X. Subsequent chromatin-remodeling events occur as a consequence of H2B ubiquitination (green-shaded area), whereas further ubiquitination events on H2A.X alter local chromatin structure (pink-shaded area). This latter RNF8/RNF168-catalyzed H2A ubiquitination pathway also polyubiquitinates JMJD2A/B for degradation, whereas L3MBTL1 is removed by RNF8-mediated ubiquitination. This then allows DSBs in DNA to be repaired by homologous recombinant repair (HRR), where joining occurs between two similar or identical strands of DNA (*C*); or by nonhomologous end joining (NHEJ), in which the two DNA ends are joined directly, usually with no sequence homology, although, in some cases, regions of microhomology are used (*D*). (*C*) During HRR, TIP60 acetylates H4K16, RNF8 ubiquitinates H2AK63, and selectively allows for BRCA1 binding but not 53BP1. (*D*) If NHEJ is required, H4K16 is deacetylated (presumably by HDAC1,2) and H2AK15 is ubiquitinated. 53BP1 binds to both H4K20me2 and H2AK15ub, whereas BRCA1 is excluded. To illustrate 53BP1 oligmer formation, only histone H4 and H2AX tails are illustrated in this panel.

et al. 2004; Endoh et al. 2008; Trojer et al. 2011). BMI-1 functions as an oncogene in medulloblastoma and other malignancies, although its deregulation seems to mostly impact its function as a stem-cell factor (see Ch. 17 [Grossniklaus and Paro 2014]). Given BMI-1's dual role, it is hard to separate its role when considering using it as a drug target, as both functions must be taken into consideration. RING1B has been proposed to control the balance between early myeloid progenitors and mature myeloid populations (Cales et al. 2008); potential dysfunction could lead to myeloid dysfunction and/or malignancies. Indeed, RING1B deficiency was found to result in an accelerated onset of hematopoietic neoplasia in the absence of the tumor suppressor, *Ink4a* (Cales et al. 2008).

Monoubiquitination of H2B at Lys120 is also induced at sites of DNA damage and required for DNA DSB repair (reviewed in Prinder et al. 2013). The ubiquitin ligase, RNF20 (really interesting new gene [RING]-finger), when phosphorylated by ATM, likely mediates H2B ubiquitination. Ubiquitination of histone H2B, in turn, promotes methylation of H3K4 and H3K79, which is critical for remodeling chromatin to allow access to the DNA repair machinery. The remodeling is partly facilitated by the SNF2h subunit of the ISWI complex, likely through interaction with methylated H3K4, as SNF2h depletion reduces repair, at least by the homologous repair (HR) pathway (Fig. 4B). This H2B ubiquitination branch of the DNA repair pathway is a necessary part of either NHEJ or HR mechanisms at DSB (Pinder et al. 2013).

The HR pathway involves polyubiquitination of H2A or H2AX at Lys63 by the RNF8 and RNF168 ubiquitin ligases at sites of DNA DSBs, acting to recruit DNA repair factors, such as BRCA1 (Doil et al. 2009; Campbell et al. 2012). Repair by NHEJ involves histone lysine PTM cross talk to dictate 53BP1 protein recruitment, which interacts with the nucleosome in an oligovalent *trans*-histone fashion (Fig. 4D). Focal accumulation of 53BP1 at DSBs depends on the specific interaction of its tandem Tudor domain with dimethylated H4K20 (H4K20me2) and ubiquitinylated H2AK15. Rapid induction of H4K16 deacetylation after DSBs facilitates 53BP1-induced DNA damage signaling and DSB repair (Hsiao and Mizzen 2013). In the absence of DNA damage, TIP60 acetyltransferase-mediated H4K16 acetylation inhibits the interaction between 53BP1 and H4K20me2 (Tang et al. 2013). Following the initiation of DSB repair, the acetylation status of histone H4K16, written by TIP60 or removed by HDAC1/2, in fact, may act as a switch between BRCA1 and 53BP1 localization to DSB chromatin by affecting the binding affinity of the 53BP1 Tudor domain for H4K20me2 (Tang et al. 2013). These histone modifications (H4K16ac, H4K20me2) could function to balance 53BP1 DSB chromatin occupancy for NHEJ, versus BRCA1 DSB localization and HRR (Fig. 4).

In summary, DSB repair is a complex process involving many proteins and PTMs. When mutated, many of these factors contribute to cancer: among them are TIP60, BRCA1, and 53BP1, documented to have tumor-suppressor function, thus making these and other epigenetic modifying proteins involved in the DNA repair pathway potential cancer drug targets. Conversely, tumors may be made more susceptible to standard cytotoxic therapies if their DNA damage repair pathways are perturbed by targeted pharmacological agents, such as epigenetic therapies.

3 DRUG-DISCOVERY CHALLENGES FOR HISTONE-MODIFYING TARGETS

Two possible modes by which epigenetic drug therapy may be able to halt or even prevent the oncogenic process are (1) repressing oncogenes and/or activating tumor-suppressor genes that are deregulated by epigenetic processes (Baylin and Ohm 2006; Jones and Baylin 2007) and (2) overcoming resistance to chemotherapy. The "epigenetic" drugs that are currently being developed primarily target histone-modifying enzymes, histone readers, or other chromatin-associated proteins.

3.1 Epigenetics and Drug Resistance

There is evidence to suggest there is an epigenetic basis for resistance to cancer chemotherapy. Because resistance develops to virtually all forms of cancer therapy, overcoming drug resistance remains the biggest obstacle to improving a positive and durable outcome cancer treatment. Examples of epigenetic alterations being attributed to resistance include the aforementioned KDM5A in Section 2.2.2 and overexpression of EZH2, observed during cisplatin chemotherapy; EZH2 overexpression results in increased gene repression and enhanced cell proliferation (Hu et al. 2010; Crea et al. 2011). Clearly, more research is required to understand how epigenetic alterations mechanistically contribute to drug resistance and how to apply epigenetic drug therapy appropriately to patients, depending on an individual's drug and/or (epi)genomic profile.

3.2 Approved Epigenetic Drugs

The earliest clinical entries to epigenetic drugs were actually not developed with a focus on histone-modifying targets, but were discovered to interact with chromatin modifiers and express their biological effects through these targets after the fact. Dimethylsulfoxide and the later agent

Cite as *Cold Spring Harb Perspect Biol* doi: 10.1101/cshperspect.a019521

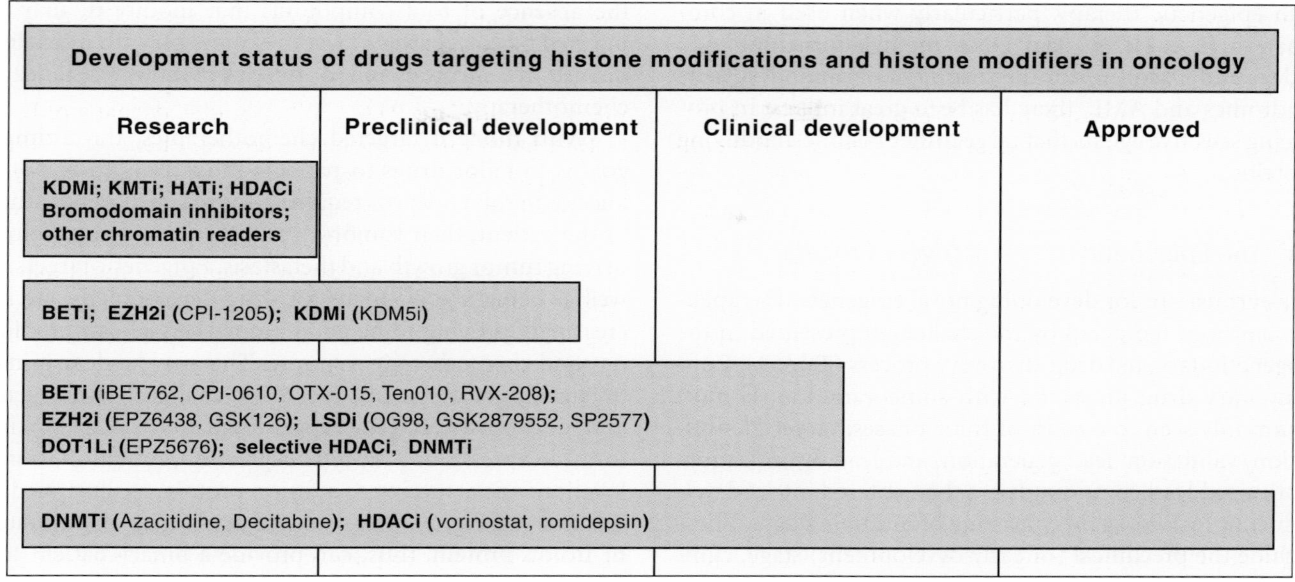

Figure 5. The development status of epigenetic drugs. KDMi, lysine demethylase inhibitor; KMTi, lysine methyltransferase inhibitor; HATi, histone acetyltransferase inhibitor; HDACi, histone deacetylase inhibitor; BETi, bromodomain and extracarboxy terminal domain inhibitor; DNMTi, DNA methyltransferase inhibitor.

HMBA, for example, although explored therapeutically since the 1950s, were only recently shown as inhibitory to a subset of bromodomains (including the BET family).

Among the agents specifically designed against histone-modifying targets that have been clinically approved thus far are the HDAC inhibitors (see Figs. 5 and 6). These drugs (Zolinza, Istodax) broadly inhibit HDACs and show single-agent activity only in a limited set of patients with cutaneous T-cell lymphoma. A large number of clinical trials are now ongoing with these and newer, more subtype-selective HDAC inhibitors (some examples are found in Fig. 8 of Ch. 5 [Seto and Yoshida 2014]) in combination with standard care drugs, in the hopes of expanding the utility of these treatments. Given the early successes

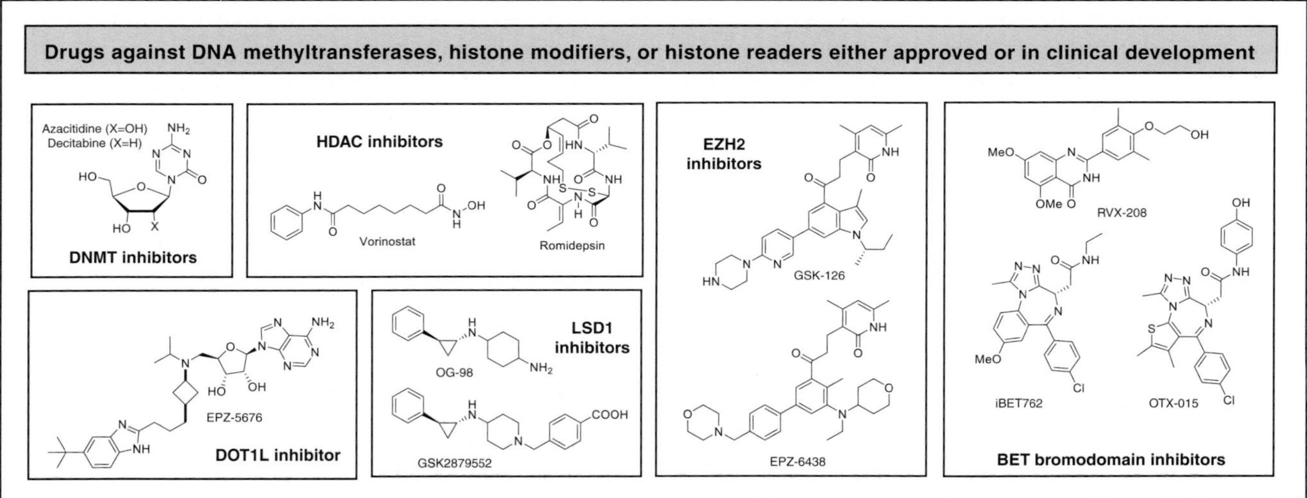

Figure 6. Representative structures of drugs targeting DNA methylation, histone modifications, and histone readers in oncology.

with epigenetic therapy, particularly when used in combination (i.e., HDAC and DNA methylation inhibitors, Vidaza and Dacogen) for the treatment of myelodysplastic syndromes and AML, there has been great interest in producing selective agents that target other histone-modifying proteins.

3.3 The Epigenetic Drug-Discovery Process

The enthusiasm for developing novel epigenetic therapeutics has been tempered by the challenges presented in the epigenetic-targeted drug-discovery process (Table 1). Contemporary drug discovery, with some variation, is most commonly seen to consist of three phases: target identification/validation, lead generation, and lead optimization. Optimized lead compounds are then entered into a development pipeline, again consisting of multiple stages. These include the preclinical (or early development) stage, clinical stage (which is typically further subdivided into Phases 1, 2, and 3), and the registration/approval phase. From beginning to end, this discovery/development process can take more than a decade despite myriad efforts on the part of academics, industry, and the regulatory agencies to streamline the process.

Epigenetic drug discovery presents some unique challenges at each stage of discovery and development. For example, in the target identification/validation stage, we still have yet to fully "map" the epigenetic landscape of all cell types, normal or tumorigenic, on a structural, biochemical, and functional level. This fundamental lack of knowledge introduces a number of pragmatic and scientific issues, which must be considered in the drug-discovery process. For example, challenges include identifying the most disease-relevant targets, understanding the associated biology, developing biologically relevant assays with limited tools (e.g., proteins, antibodies, etc.), and finding suitable small molecule activators or inhibitors with few chemical starting points. Cancer relevance for a given target (e.g., a particular KMT like MLL1) is often determined from genomic analyses of patient tumors; this may reveal mutations, translocations, rearrangements, fusions, amplifications, or copy number changes. Additional experimental data is also used to corroborate their role in cancer, such as testing for synthetic lethality or gene silencing. Experimental data using RNA silencing, although beneficial in these analyses, must be interpreted with care, especially for epigenetic enzymes, as knockdown can disrupt protein complexes and not necessarily be directly associated with the catalytic activity. Use of catalytically inactive "dominant-negative" mutants in concert with selective small molecule activators/inhibitors can be more informative. Clinical relevance of a target in oncology, especially in the absence of tool compounds that modify it, may be inferred by correlating a target's expression with mortality, disease-free survival, and/or development of resistance to chemotherapy.

With modern targeted chemotherapies, the ultimate goal is to tailor drugs to patients most likely to respond. Success in this endeavor requires a thorough understanding of the patient, their tumor(s), and the underlying biology driving tumor growth and metastasis. Epigenetic targets, as well as being a good focus for drug development, are increasingly proving to be good biomarkers of cancer etiology and chemotherapy response. This may be determined by tracking various histone mark changes, target gene, and/or downstream gene expression patterns (e.g., looking for overexpression, wild-type expression, mutation, translocation, or partnering as a fusion protein, as illustrated in scenarios of Fig. 2). The presence or absence of a mutated or fusion protein, thus, can provide a binary patient-tailoring marker. For instance, EZH2-activating mutations are observed in subsets of lymphoma, providing a biomarker for this type of lymphoma (Yap et al. 2011; Majer et al. 2012; McCabe et al. 2012), whereas MLL-fusion proteins in rearranged leukemias have altered DOT1L-dependent H3K79 methylation profiles, which is again a biomarker for MLL-rearranged cancers (Bernt and Armstrong 2011). However, aside from EZH2, very few epigenetic activating mutations or translocations have been found in human tumors; many epigenetic gene alterations are predicted to result in loss of function (i.e., loss of tumor-suppressor function, which is extensively discussed in Ch. 34 [Baylin and Jones 2014]), or alteration of function, the latter of which does not constitute a straightforward overexpression consequence. This makes epigenetic targeting and patient tailoring quite challenging—that is, how does one "reactivate" the target when deleted, truncated, or conformationally unfavorable?

Once a compelling target has been identified and selected, a series of assays are constructed to screen compounds and ultimately optimize drug-like properties. An initial high-throughput screening assay is typically biochemical and -analytical in nature, investigating disruption or enhancement of protein–protein interactions and catalytic activity. It is critical that these biochemical assays reflect (or at least, generally, predict) the cellular context of the active process one wishes to modulate. For example, chromatin-modifying proteins can exist in heterogeneous multimeric complexes (e.g., PRC1, PRC2, COMPASS) such that partners and enzymatic substrates may be recruited or exchanged to elicit specific biological responses. It is therefore conceivable that different results may be obtained when screening with various protein forms, such as the apoprotein lacking its nonpolypeptide moiety versus the

 Cite as *Cold Spring Harb Perspect Biol* doi: 10.1101/cshperspect.a019521

Table 1. Epigenetic drug-discovery challenges

	Issues
Target selection	
	Few activating mutations, translocations, or synthetic lethal relationships known.
	Understanding change-of-function mutations in large multisubunit complexes that represent the most frequent epimutations.
	Limited high quality, specific antibodies to epigenetic proteins and histone marks (e.g., confirm target expression, linkage of target to mark).
	Biology-driving cancer phenotype unknown or poorly understood.
	Roles of PTM of histone versus nonhistone substrates by epigenetic targets unclear.
Chemistry	
	Existing chemical libraries may not have adequate diversity to provide good starting points.
	Few crystal structures solved. Are structures relevant if not reflecting complete complex? Multiple complexes with different functions possible.
	Different enzyme/chaperone/substrate combinations may yield different SARs.
Assay development	
	Few reference compounds to establish assay signal window, sensitivity, reproducibility.
	Assay may not be configured properly to achieve optimal sensitivity (e.g., enzyme in proper complex, with relevant substrate to achieve adequate turnover).
	Producing active enzymes difficult, may require coexpression of multimeric complex and specific substrate (nucleosome, histone, nonhistone).
	Limited high-quality antibodies to epigenetic proteins and histone marks (quantify mark or target gene product).
	Cellular target engagement readout of chromatin adaptors and remodelers.
In vivo biology	
	Histone marks and target genes slow to change, require longer duration studies to assess target engagement (pharmacodynamic [PD] biomarker) and efficacy.
	May necessitate higher-compound requirement to conduct studies, earlier optimization of PK properties than traditional paradigm.
	May require novel models for tumors with mutations, translocations.
Toxicology	
	Acute and/or chronic liabilities of specific isoform-targeted epigenetic therapies currently unknown.
	Knockout animal data limited; inducible knockouts, dominant negatives preferred, but more scarce, and technically challenging.
Clinical	
	Identify and implement appropriate patient selection markers, more challenging if not an activating mutation (overexpression, gene profile?).
	Identify and implement suitable PD marker (PTM, target gene, surrogate tissue, or tumor?).
	Epigenetic changes at metastatic sites can differ from primary tumor; which should be targeted clinically?

Modified, from Campbell and Tummino 2014, with permission of the American Society for Clinical Investigation.

complexed form, or using various substrate forms (i.e., peptide vs. protein/histone, core histones, or nucleosomes, as discussed in Sec. 16.2 of Ch. 7 [Patel 2014]). The choice of enzymatic substrates is further complicated by the fact that many epigenetic enzymes can posttranslationally modify nonhistone targets. Among the many known examples are HDAC6, SIRT1, SIRT2, SET7/9, SETD8, SMYD2, PRMT5, and EHMT2/G9a) (Hubbert et al. 2002; Huang et al. 2006; Shi et al. 2007; Pradhan et al. 2009; Karkhanis et al. 2011; Stunkel and Campbell 2011).

The presence of "tool compounds" or chemical probes to understand the translation from biochemical to cellular assays brings us a step closer to the elucidation of in vivo biological function. This collection of specific epigenetic probes is currently quite limited; however, efforts to expand the collection of such tools continue to progress. A tool compound, such as the *S*-adenosyl methionine (SAM) analog, Sinefungin, for example, can be useful for SAM-dependent enzyme assays on lysine and arginine histone methyltransferase enzymes, but is poorly cell permeable. Other tools that can be used are the recently described selective, cell-active inhibitors of EZH2, DOT1L, SMYD2, G9a/GLP1, PRMT3, JMJD3, LSD1, BET (BRD2-4), L3MBTL1, L3MBTL3, CBP/EP300, PCAF, and BAZ bromodomains (Arrowsmith et al. 2012; Muller and Brown 2012; James et al. 2013; Gallenkamp et al. 2014). It is anticipated that these and future chemical tools will greatly facilitate assay development and our understanding of target biology. However, a cautionary note is warranted: as powerful as probes may be in revealing the mechanistic and phenotypic consequences of inhibiting specific domains of chromatin modifiers and their readers, such data is clearly dependent on the quality of the probes, the thoughtful design of experiments, and the cellular assays designed to monitor their readout, such that the outcome can be most directly linked to the specific target. As more data is generated and more targets are credentialed, probes must continue to be reviewed for specificity and cellular target engagement (i.e., ability to specifically modulate the downstream target-dependent biology, such as a histone mark change) to ensure that the biology observed is properly interpreted.

The challenge of identifying chemical probe molecules of sufficient potency, selectivity, and cell-permeability suggests that traditional chemical libraries may be deficient in the most relevant chemotypes and screening methodologies were not optimal. As such, new chemical diversity is being explored, particularly, using small molecular weight fragment approaches, and generating substrate or cofactor analogs, such as SAM- and Lys/Arg-peptide mimetics for use with histone methyltransferase targets. A growing collection of crystal structures has been solved to facilitate drug design, some of which are covered in the structural chapters of this collection (see Ch. 6 [Cheng 2014]; Ch. 4 [Marmorstein and Zhou 2014]; Ch. 7 [Patel 2014]; Ch. 5 [Seto and Yoshida 2014]). However, still too few structures represent protein complexes or have a small molecule modulator bound, a problem discussed in Section 16.2 of Ch. 7 (Patel 2014).

Epigenetic-targeted drug discovery is in its infancy, but steadily evolving to overcome the unique challenges of prosecuting its targets. To be successful, ideally, a comprehensive "epigenetics toolbox" must be assembled to test and better understand the dominant biology, relevant histone marks (or nonhistone PTM), structural biology, appropriate in vitro assays, pharmacodynamics of lead drug and target interactions, and patient tailoring biomarkers (perhaps, a major challenge with today's pharma-favoring genetic features to reflect a real epigenetic basis for sensitivity). The possibility of "antitarget" action must also be determined (i.e., action with another target), which might contribute to clinically unmanageable toxicities.

To accelerate the production of epigenetic drug development tools, a new paradigm has emerged whereby private-public partnerships, such as the Structural Genomics Consortium, have joined forces to generate necessary epigenetic chemical probes, assays, antibodies, and X-ray crystal structures. Other large group efforts are performing sequencing (exome and next-generation) of patient tumors and performing synthetic lethal screens in genomically characterized tumor cell lines. This is helping to identify epigenetic target opportunities and determine patient populations sensitive to therapeutic intervention. It is anticipated that substantial progress will be made in the coming years as a result of knowledge sharing between these pooled collaborative efforts. Such progress is occurring at an increasing rate and, as reflected in Figure 5, early stage drugs targeting histone modifications and modifiers are now progressing through to clinical phases of development. Beyond the initial DNA methyltransferase and HDAC drugs already approved, new chemical entities targeting the BET bromodomains have been entered into a number of clinical studies in a variety of cancers by multiple sponsors (Figs. 5 and 6). Similarly, small molecule inhibitors of the histone methyltransferases, DOT1L and EZH2, have advanced to the clinic, as well as inhibitors of the histone demethylase LSD1. Data from these and subsequent studies will be critical for refining our understanding of the role of histone modifications and histone-modifying enzymes (and their readers) in cancer initiation, cancer progression, and, ultimately, translating to new and improved cancer therapies.

Cite as *Cold Spring Harb Perspect Biol* doi: 10.1101/cshperspect.a019521

APPENDICES

Appendix A. HAT mutations in cancer

Gene	Common name	Tumor types	Cell type	Tissue type	Mutation types	Fusion proteins
MYST family						
KAT5	TIP60	Colorectal, head and neck, stomach	Somatic	Epithelial	Missense, frameshift nonsense	
KAT7	HBOI	Lung, colorectal, breast, prostate, ovarian, sarcoma	Somatic	Epithelial	Amplification, missense, splice	
KAT6A/ MYST3	MOZ	Colorectal, lung, breast, acute myelogenous leukemia	Somatic	Epithelial, leukemia/ lymphoma	Nonsense, missense amplification, deletion, translocation	MOZ-CBP, MOZ-EP300, MOZ-TIF2, MOZ-NCOA3, MOZ-ASXL2
KAT6B/ MYST4	MORF	Colorectal, glioblastoma, lung, ovarian, acute myelogenous leukemia	Somatic	Epithelial, leukemia/ lymphoma	Nonsense, missense amplification, deletion translocation	MORF-CBP
MYST1	MOF	Lung, colorectal, medulloblastoma	Somatic	Epithelial	Missense, nonsense, deletion	
GNAT family						
KAT2A	GCN5	Breast, colorectal, prostate, lung, kidney, sarcoma	Somatic	Epithelial	Deletion, amplification	
KAT2B	PCAF	Lung, kidney, sarcoma, colorectal	Somatic	Epithelial	Missense, frameshift, deletion, amplification	
Orphan family						
EP300	p300	Colorectal, breast, pancreatic, AML, ALL, DLBCL (10%), NHL (7%), FL (8.7%)	Somatic	Epithelial, leukemia lymphoma	Translocation, nonsense, frameshift, missense, other	p300-MOZ, MLL-p300
CREBBP	CBP	ALL (18.3%), AML, DLBCL (29%), NHL (21%)	Somatic	Leukemia lymphoma	Translocation, non-sense, frameshift, missense, other	CBP-MOZ, CBP-MORF, MLL-CBP
CREBBP	CBP	Hematological (Rubstein- Taybe syndrome)	Germline	Leukemia/ lymphoma	Deletion	
NCOAI/ KATI3A	SRCI	Lung, colorectal	Somatic	Epithelial	Missense, deletion	PAX3-NCOAI
NCOA3/ KATI3B	CRC-3/ ACTR	Colorectal, ovarian, lung	Somatic	Epithelial	Nonsense, missense, amplification, in frame insertion	NCOA3-MOZ
KATI3D	CLOCK	Colorectal, glioblastoma, lung	Somatic	Epithelial	Missense, nonsense, amplification, other	
KAT4	TAFI	Lung, colorectal, breast, glioblastoma, ovarian, kidney	Somatic	Epithelial	Missense, nonsense, splice	

Adapted from Di Cerbo and Schneider 2013, by permission of Oxford University Press.

NHL, B-cell non-Hodgkin lymphoma; AML, acute myelogenous leukemia; ALL, acute lymphocytic leukemia; DLBCL, diffuse large B-cell lymphoma.

Appendix B. Histone deacetylases in cancer

Name	HDAC class	Substrate(s)	Links to cancer
HDAC1	I	H4K16, H3K56	Overexpressed in ALL, CLL, Hodgkin's lymphoma, renal cell cancer, breast, gastric, pancreatic, and colorectal (associated with poor prognosis), prostate (high grade, hormone-resistant), lung, and hepatocellular carcinoma (advanced stage)
HDAC2	I	H4K16, H3K56	Overexpressed in ALL, Hodgkin's lymphoma, renal cell cancer, lung, colorectal (higher in polyps), cervical, gastric, and prostate (associated with advanced stage and/or poor prognosis); HDAC2-inactivating mutation in MSI^{+} colon, gastric, and endometrial tumors
HDAC3	I	H3K9,K14 H4K5 H4K12	Overexpression in CLL, Hodgkin's lymphoma, renal cell cancer, gastric, colorectal, prostate correlated with poor prognosis (coincident with HDAC1, -2 overexpression); higher expression correlated with poor 5-year event-free survival
HDAC8	I	ERRα,	Overexpressed in ALL, pediatric neuroblastoma (associated with advanced stage disease; poor survival)
HDAC4	IIA	Not defined	High expression in breast cancer (relative to renal, bladder, and colorectal tumors); associated with prednisone-poor response in ALL
HDAC5	IIA	p53 (Sen et al. 2013)	High expression in colorectal cancer (relative to bladder, renal, breast cancer) and medulloblastoma (associated with poor overall survival); underexpressed in AML blasts
HDAC7	IIA	Not defined	High expression in CLL, colorectal (relative to bladder, renal, breast cancer), pancreatic cancer, and childhood ALL (associated with poor survival)
HDAC9	IIA	Not defined	Lower expression in higher grade astrocytoma and glioblastoma (vs. lower grade astrocytoma and normal brain), lung tumors (vs. nontumor epithelial cells, although no correlation with disease-free survival) (Okudela et al. 2013), higher expression observed in medulloblastoma, CLL, Philadephia-negative chronic myeloproliferative neoplasms, and childhood ALL associated with poorer overall survival
HDAC6	IIB	α-tubulin, HSP90, cortactin	Overexpressed in CLL, oral squamous cell cancer (higher in advanced stage), breast cancer (but correlated with better survival; sensitivity to endocrine treatment), DLBCL and AML; expression progressively increased with progression of Philadephia-negative chronic myeloproliferative neoplasms
HDAC10	IIB	Not defined	Reduced expression linked to poor prognosis in lung cancer; overexpressed in CLL; low expression correlated to better survival in advanced stage primary neuroblastoma (Oehme et al. 2013)
HDAC11	IV	Not defined	Overexpressed in mixed lobular and ductal breast carcinoma (vs. normal breast tissue) (Deubzer et al. 2013); elevated in Mantle cell lymphoma (Shah et al. 2012) and Philadephia-negative chronic myeloproliferative neoplasms
SIRT1	III	H4K16, H3K9, p53, p73, PTEN, FOX01, FOX03a, FOX04, NICD, MEF2, HIF-1α, HIF-2α, TAF(I)68, SREBP-1c, β-catenin, RelA/p65, PGC1α, BMAL1, Per2, Ku70, XPA, SMAD7, cortactin, IRS-2, APE1, PCAF, TIP60, p300, SUV39H1, AceCS1, PPARγ, ER-α, ERRα, AR, LXR	Overexpressed in AML, CLL, colon carcinoma, prostate, ovarian, gastric, melanoma, and a subset of HCC (Chen et al. 2012a); underexpressed in bladder, colon, glioma, prostate (?), ovarian tumors (?)

Continued

 Cite as *Cold Spring Harb Perspect Biol* doi: 10.1101/cshperspect.a019521

Appendix B. *Continued*

Name	HDAC class	Substrate(s)	Links to cancer
SIRT2	III	H4K16, H3K56, α-tubulin, ATP-citrate lyase (Lin et al. 2013), SETD8 (Serrano et al. 2013), CDK9 (Zhang et al. 2013)	Higher nuclear expression in GBM (vs. astrocytoma or normal by IHC) (Imaoka et al. 2012) and grade 3 ER^{+} breast cancer (associated with reduced disease-free survival and increased frequency of relapse, although the reverse trend was observed in Grade 2 tumors) (McGlynn et al. 2014); however, decreased expression of total SIRT2 in NSCLC (Li et al. 2013b), breast, GBM, HCC as compared with normal tissue (Park et al. 2012)
SIRT3	III	H4K16, H3K56, H3K9, GDH, IDH2, HMGCS2, SOD2, AceCS2, Ku70, LCAD, SdhA, PDHA1, and PDP (Fan et al. 2014)	Overexpressed in CLL (Van Damme et al. 2012); higher levels associated with node-positive breast cancer
SIRT4	III	GDH (ADP-ribosylation)	Underexpressed in AML blasts
SIRT5	III	Cytochrome c, CPS1 HMGCS2, and SOD1 (desuccinylation) (Rardin et al. 2013)	SIRT5 suggested to desuccinylate and activate SOD1, inhibiting lung tumor growth in vitro (Lin et al. 2013)
SIRT6	III	H3K9, H3K56	Overexpressed in CLL (Van Damme et al. 2012)
SIRT7	III	H3K18 , p53	Low levels of H3K18ac predict higher risk of prostate cancer, poor prognosis in pancreatic adenocarcinoma; overexpressed in CLL (Van Damme et al. 2012) and hepatoceullar cancer; higher levels associated with node-positive breast cancer

Adapted from Witt et al. 2009; Bosch-Presegue and Vaquero 2011; Stunkel and Campbell 2011; Barneda-Zahonero and Parra 2012; Houtkooper et al. 2012; Bernt 2013; Gong and Miller 2013; Paredes et al. 2014.

HDAC, histone deacetylase; ALL, acute lymphoblastic leukemia; CLL, chronic lymphocytic leukemia; AML, acute myelogenous leukemia; HCC, hepatocellular carcinoma; ER-α, estrogen receptor α; GBM, glioblastoma multiforme; GDH, glutamate dehydrogenase; IHC, immunohistochemistry; NSCLC, non-small-cell lung carcinoma.

Appendix C. Selected histone methyltransferases in cancer

Name	Synonyms	Histone target(s)	Links to cancer
KMT1A	SUV39H1	H3K9	Overexpressed in colorectal cancer (Kang et al. 2007), associated with transcriptional repression
KMT1C	G9a, EHMT2	H3K9	Overexpressed in lung cancers (Watanabe et al. 2008; Chen et al. 2010, for advanced lung cancer), regulation of centrosome duplication via chromatin structure (Kondo et al. 2008)
KMT1F	SETDB2	H3K9	Involved in chromosome segregation (Falandry et al. 2010)
KMT2A	MLL	H3K4	Rearranged and translocated in leukemias (Zhang et al. 2012a)
KMT2B	MLL2	H3K4	Frequently mutated in non-Hodgkin lymphoma (Morin et al. 2011), critical role in lymphomagenesis (Chung et al. 2012), role in multiple myeloma and Kabuki syndrome
KMT2C	MLL3	H3K4	Germline mutation in colorectal cancer and AML (Li et al. 2013a); mutations in glioblastoma, melanoma, pancreatic carcinoma, and colorectal cancer
KMT2D	MLL4	H3K4	Regulates cell-cycle progression and viability in colon cancer (Ansari et al. 2012)
KMT2E	MLL5	H3K4	Implicated as a tumor suppressor, MLL5 expression positive prognostic in AML (Damm et al. 2011)
KMT3A	SETD2	H3K36	Mutations in high-grade gliomas (Fontebasso et al. 2013), tumor suppressor in breast cancer and renal cell carcinoma (Hakimi et al. 2012)
KMT3B	NSD1	H3K36	Mutated in AML, myeloma, and lung cancers, NUP98-NSD1 translocation linked to tumorogenesis in AML (Wang et al. 2007)
KMT3C	SMYD2	H3K36	Overexpression in esophageal squamous cell carcinoma correlated with poor survival (Komatsu et al. 2009)
KMT3E	SMYD3	H3K4, H4K5	Overexpressed in liver, breast, and rectal carcinomas (Van Aller et al. 2012)
KMT3F	WHSC1L1/ NSD3	H3K36	Overexpressed in CML, bladder, lung, and liver cancers (Kang et al. 2013); amplified in human breast cancer cell lines (Angrand et al. 2001)
KMT3G	WHSC1/NSD2	H3K36	Implicated in constitutive NF-κB signaling for cancer cell proliferation, survival, and tumor growth (Yang et al. 2012), overexpressed in myeloma because of t(4;14) chromosomal translocation, modulates cMYC in myeloma (Min et al. 2012)

Continued

Appendix C. *Continued*

Name	Synonyms	Histone target(s)	Links to cancer
KMT4	DOT1L	H3K79	Implicated in MLL-rearranged leukemias (Krivtsov et al. 2008; Bernt and Armstrong 2011)
KMT5A	SETD8/PR-SET7	H4K20	Overexpressed in bladder cancer, NSCLC, small cell lung cancer, pancreatic cancer, hepatocellular carcinoma, and chronic myelogenous leukemia (Takawa et al. 2012)
KMT6	EZH2	H3K27	Catalytic component of PRC2 complex (Kuzmichev et al. 2002), overexpressed, is a marker for advanced and metastatic breast and prostate cancer (Chase and Cross 2011), essential for glioblastoma cancer stem-cell maintenance (Suva et al. 2009), somatic mutations in follicular and diffuse large B-cell lymphomas (Morin et al. 2010)
KMT7	SET7/SET9/SETD7	H3K4	Regulation of the estrogen receptor (Subramanian et al. 2008)
PRDM14		Unknown	Amplified and overexpressed in breast cancer (Nishikawa et al. 2007; Moelans et al. 2010), overexpressed in lymphoid neoplasms, and implicated in initiation of lymphoblastic leukemia (Dettman et al. 2011)
PRMT4	CARM1	H3R17, H3R26	Deregulated in melanoma (Limm et al. 2013), methylation CBP/P300 required for estrogen-induced targeting to chromatin (Ceschin et al. 2011)
PRMT5		H4R3, H3R8	Essential component of HIF-1 signaling (Lim et al. 2012); silences the tumor suppressor ST7 and is overexpressed in GBM (Yan et al. 2014), non-Hodgkin's lymphoma (Chung et al. 2013), and melanoma (Nicholas et al. 2013)
PRMT6	H3R2, H3R42		Overexpressed in prostate carcinoma (Vieira et al. 2014); PRMT6 silencing reduces PELP1-mediated ER activation, proliferation, and colony formation in breast tumor cells (Mann et al. 2014)

AML, acute myelogenous leukemia; CML, chronic myelogenous leukemia; MLL, mixed lineage leukemia;
NSCLC, non-small-cell lung carcinoma; PRC2, Polycomb repressive complex 2; HIF-1, hypoxia-inducible factor 1; GBM, glioblastoma multiforme; ER, estrogen receptor.

Appendix D. Histone demethylases in cancer

Name	Synonyms	Targets	Links to cancer
KDM1A	LSD1, AOF2	H3K4me2/me1, H3K9me2/me1	Overexpressed in prostate carcinoma, bladder cancer (Kauffman et al. 2011), ER-negative breast cancer (Lim et al. 2010), neuroblastoma (Schulte et al. 2009); inhibition in animal models of engrafted AML (Schenk et al. 2012)
KDM1B	LSD2, AOF1	H3K4me2/me1	
KDM2A	FBXL11A, JHDM1A	H3K36me2/me1	Overexpressed in NSCLC; KDM2A knockdown inhibits NSCLC tumor growth in mouse xenografts (Wagner et al. 2013)
KDM2B	FBXL10B, JHDM1B	H3K36me2/me1, H3K4me3	Required for initiation and maintenance of AML (He et al. 2011)
KDM3A	JMJD1A, JHDM2A	H3K9me2/me1	High expression correlated with bad prognosis in colorectal cancer (Uemura et al. 2010), overexpressed in renal cell carcinoma (Guo et al. 2011)
KDM3B	JMJD1B, JHDM2B	H3K9me2/me1	
KDM4A	JMJD2A, JHDM3A	H3K9me3/me2, H3K36me3/me2	Required for proliferation of breast cancer cells (Lohse et al. 2011), attenuated expression in bladder cancer (Kauffman et al. 2011), required for latency and replication of viruses that cause cancer (Chang et al. 2011)
KDM4B	JMJD2B	H3K9me3/me2, H3K36me3/me2	Overexpressed in gastric cancer (Li et al. 2011), required for proliferation and formation of metastasis in breast cancer cells (Kawazu et al. 2011)
KDM4C	JMJD2C, GASC1	H3K9me3/me2, H3K36me3/me2	Overexpressed in breast cancer (Liu et al. 2009), esophageal cancer (Yang et al. 2000), MALT lymphoma (Vinatzer et al. 2008), AML (Hélias et al. 2008), and lung sarcomatoid carcinoma (Italiano et al. 2006)
KDM4D	JMJD2D	H3K9me3/me2/me1, H3K36me3/me2	Required for cell proliferation and survival in colon carcinoma cells (Kim et al. 2012a; Kim et al. 2012b)
KDM4E	JMJD2E	H3K9me3/me2	

Continued

Appendix D. *Continued*

Name	Synonyms	Targets	Links to cancer
KDM5A	Jarid1A, RBP2	H3K4me3/me2	Involved in drug resistance (Sharma et al. 2010)
KDM5B	Jarid1B, PLU1	H3K4me3/me2	Tumor-suppressive function in metastatic melanoma cells (Roesch et al. 2006, 2008), proproliferative in breast cancer (Mitra et al. 2011), and overexpressed in prostate cancer
KDM5C	Jarid1C, SMCX	H3K4me3/me2	Inactivating mutations found in clear cell renal carcinoma (Dalgliesh et al. 2010)
KDM5D	Jarid1D, SMCY	H3K4me3/me2	
KDM6A	UTX, MGC141941	H3K27me3/me2	Tumor-suppressive function (Tsai et al. 2010)
KDM6B	JMJD3, KIAA0346, PHF8, KIAA1111, ZNF422	H3K27me3/me2, H3K9me2/me1, H4K20me1	Overexpressed in Hodgkin's lymphoma (Anderton et al. 2011)
KDM7	KIAA1718, JHDM1D (KDM7A)	H3K9me2/me1, H3K27me2/me1	Increased expression of KDM7A suppresses tumor growth in HeLa and B16 xenograft models (Osawa et al. 2011)
KDM8	JMJD5, FLJ13798	H3K36me2	Overexpressed in breast tumors; may play a critical role in regulation of cell cycle by activating the cyclin A1 locus via demethylation of H3K36me2 (Hsia et al. 2010)

Adapted, with permission, from Hoffmann et al. 2012, © Elsevier.

From ER, estrogen receptor; AML, acute myloid leukemia; NSCLC, non-small-cell lung carcinoma; KDM2A, lysine demethylase 2A; MALT, mucosa-associated lymphoid tissue.

REFERENCES

**Reference is also in this book.*

Anderton JA, Bose S, Vockerodt M, Vrzalikova K, Wei W, Kuo M, Helin K, Christensen J, Rowe M, Murray PG, et al. 2011. The H3K27me3 demethylase, KDM6B, is induced by Epstein-Barr virus and over-expressed in Hodgkin's Lymphoma. *Oncogene* **30:** 2037–2043.

Angrand PO, Apiou F, Stewart AF, Dutrillaux B, Losson R, Chambon P. 2001. NSD3, a new SET domain-containing gene, maps to 8p12 and is amplified in human breast cancer cell lines. *Genomics* **74:** 79–88.

Ansari KI, Kasiri S, Mishra BP, Mandal SS. 2012. Mixed lineage leukaemia-4 regulates cell-cycle progression and cell viability and its depletion suppresses growth of xenografted tumour in vivo. *Br J Cancer* **107:** 315–324.

Arrowsmith CH, Bountra C, Fish PV, Lee K, Schapira M. 2012. Epigenetic protein families: A new frontier for drug discovery. *Nat Rev Drug Discov* **11:** 384–400.

Bannister AJ, Kouzarides T. 2011. Regulation of chromatin by histone modifications. *Cell Res* **21:** 381–395.

Barber MF, Michishita-Kioi E, Xi Y, Tasselli L, Kioi M, Moqtaderi Z, Tennen RI, Paredes S, Young NL, Chen K, et al. 2012. SIRT7 links H3K18 deacetylation to maintenance of oncogenic transformation. *Nature* **487:** 114–118.

Barneda-Zahonero B, Parra M. 2012. Histone deacetylases and cancer. *Mol Oncol* **6:** 579–589.

* Baylin SB, Jones PA. 2014. Epigenetic determinants of cancer. *Cold Spring Harb Perspect Biol* doi: 10.1101/cshperspect.a019505.

Baylin SB, Ohm JE. 2006. Epigenetic gene silencing in cancer—A mechanism for early oncogenic pathway addiction? *Nat Rev Cancer* **6:** 107–116.

Bell EL, Emerling BM, Ricoult SJ, Guarente L. 2011. SirT3 suppresses hypoxia inducible factor 1α and tumor growth by inhibiting mitochondrial ROS production. *Oncogene* **30:** 2986–2996.

Bernt M. 2013. HDAC expression patterns in pediatric ALL. *Leuk Res* **37:** 1191–1192.

Bernt KM, Armstrong SA. 2011. A role for DOT1L in MLL-rearranged leukemias. *Epigenomics* **3:** 667–670.

Black JC, Van Rechem C, Whetstine JR. 2012. Histone lysine methylation dynamics: Establishment, regulation, and biological impact. *Mol Cell* **48:** 491–507.

Bosch-Presegue L, Vaquero A. 2011. The dual role of sirtuins in cancer. *Genes Cancer* **2:** 648–662.

Boulikas T. 1989. DNA strand breaks alter histone ADP-ribosylation. *Proc Natl Acad Sci* **86:** 3499–3503.

Bracken AP, Pasini D, Capra M, Prosperini E, Colli E, Helin K. 2003. EZH2 is downstream of the pRB-E2F pathway, essential for proliferation and amplified in cancer. *EMBO J* **22:** 5323–5335.

* Busslinger M, Tarakhovsky A. 2014. Epigenetic control of immunity. *Cold Spring Harb Perspect Biol* **6:** a019307.

Cales C, Roman-Trufero M, Pavon L, Serrano I, Melgar T, Endoh M, Perez C, Koseki H, Vidal M. 2008. Inactivation of the polycomb group protein Ring1B unveils an antiproliferative role in hematopoietic cell expansion and cooperation with tumorigenesis associated with Ink4a deletion. *Mol Cell Biol* **28:** 1018–1028.

Campbell RM, Tummino PJ. 2014. Cancer epigenetics drug discovery and development: The challenge of hitting the mark. *J Clin Invest* **124:** 64–69.

Campbell SJ, Edwards RA, Leung CC, Neculai D, Hodge CD, Dhe-Paganon S, Glover JN. 2012. Molecular insights into the function of RING finger (RNF)-containing proteins hRNF8 and hRNF168 in Ubc13/Mms2-dependent ubiquitylation. *J Biol Chem* **287:** 23900–23910.

Ceschin DG, Walia M, Wenk SS, Duboe C, Gaudon C, Xiao Y, Fauquier L, Sankar M, Vandel L, Gronemeyer H. 2011. Methylation specifies distinct estrogen-induced binding site repertoires of CBP to chromatin. *Genes Dev* **25:** 1132–1146.

Chang X, Han J, Pang L, Zhao Y, Yang Y, Shen Z. 2009. Increased PADI4 expression in blood and tissues of patients with malignant tumors. *BMC Cancer* **9:** 40.

Chang PC, Fitzgerald LD, Hsia DA, Izumiya Y, Wu CY, Hsieh WP, Lin SF, Campbell M, Lam KS, Luciw PA, et al. 2011. Histone demethylase JMJD2A regulates Kaposi's sarcoma-associated herpesvirus replication and is targeted by a viral transcriptional factor. *J Virol* **85:** 3283–3293.

Chase A, Cross NC. 2011. Aberrations of EZH2 in cancer. *Clin Cancer Res* **17:** 2613–2618.

Chen Y, Sprung R, Tang Y, Ball H, Sangras B, Kim SC, Falck JR, Peng J, Gu W, Zhao Y. 2007. Lysine propionylation and butyrylation are novel post-translational modifications in histones. *Mol Cell Proteomics* **6:** 812–819.

Chen MW, Hua KT, Kao HJ, Chi CC, Wei LH, Johansson G, Shiah SG, Chen PS, Jeng YM, Cheng TY, et al. 2010. H3K9 histone methyltransferase G9a promotes lung cancer invasion and metastasis by silencing the cell adhesion molecule Ep-CAM. *Cancer Res* **70:** 7830–7840.

Chen HC, Jeng YM, Yuan RH, Hsu HC, Chen YL. 2012a. SIRT1 promotes tumorigenesis and resistance to chemotherapy in hepatocellular carcinoma and its expression predicts poor prognosis. *Ann Surg Oncol* **19:** 2011–2019.

Chen YF, Luo RZ, Li Y, Cui BK, Song M, Yang AK, Chen WK. 2012b. High expression levels of COX-2 and P300 are associated with unfavorable survival in laryngeal squamous cell carcinoma. *Eur Arch Otorhinolaryngol* **270:** 1009–1017.

Chen Q, Chen Y, Bian C, Fujiki R, Yu X. 2013. TET2 promotes histone O-GlcNAcylation during gene transcription. *Nature* **493:** 561–564.

* Cheng X. 2014. Structural and functional coordination of DNA and histone methylation. *Cold Spring Harb Perspect Biol* **6:** a018747.

Christophorou MA, Castelo-Branco G, Halley-Stott RP, Oliveira CS, Loos R, Radzisheuskaya A, Mowen KA, Bertone P, Silva JC, Zernicka-Goetz M, et al. 2014. Citrullination regulates pluripotency and histone H1 binding to chromatin. *Nature* **507:** 104–108.

Chung YR, Schatoff E, Abdel-Wahab O. 2012. Epigenetic alterations in hematopoietic malignancies. *Int J Hematol* **96:** 413–427.

Chung J, Karkhanis V, Tae S, Yan F, Smith P, Ayers LW, Agostinelli C, Pileri S, Denis GV, Baiocchi RA, et al. 2013. Protein arginine methyltransferase 5 (PRMT5) inhibition induces lymphoma cell death through reactivation of the retinoblastoma tumor suppressor pathway and polycomb repressor complex 2 (PRC2) silencing. *J Biol Chem* **288:** 35534–35547.

Crea F, Nobili S, Paolicchi E, Perrone G, Napoli C, Landini I, Danesi R, Mini E. 2011. Epigenetics and chemoresistance in colorectal cancer: An opportunity for treatment tailoring and novel therapeutic strategies. *Drug Resist Updat* **14:** 280–296.

Cuthbert GL, Daujat S, Snowden AW, Erdjument-Bromage H, Hagiwara T, Yamada M, Schneider R, Gregory PD, Tempst P, Bannister AJ, et al. 2004. Histone deimination antagonizes arginine methylation. *Cell* **118:** 545–553.

Dalgliesh GL, Furge K, Greenman C, Chen L, Bignell G, Butler A, Davies H, Edkins S, Hardy C, Latimer C, et al. 2010. Systematic sequencing of renal carcinoma reveals inactivation of histone modifying genes. *Nature* **463:** 360–363.

Damm F, Oberacker T, Thol F, Surdziel E, Wagner K, Chaturvedi A, Morgan M, Bomm K, Gohring G, Lubbert M, et al. 2011. Prognostic importance of histone methyltransferase MLL5 expression in acute myeloid leukemia. *J Clin Oncol* **29:** 682–689.

Daujat S, Bauer UM, Shah V, Turner B, Berger S, Kouzarides T. 2002. Crosstalk between CARM1 methylation and CBP acetylation on histone H3. *Curr Biol* **12:** 2090–2097.

Dawson MA, Prinjha RK, Dittmann A, Giotopoulos G, Bantscheff M, Chan WI, Robson SC, Chung CW, Hopf C, Savitski MM, et al. 2011. Inhibition of BET recruitment to chromatin as an effective treatment for MLL-fusion leukaemia. *Nature* **478:** 529–533.

Dell'Aversana C, Lepore I, Altucci L. 2012. HDAC modulation and cell death in the clinic. *Exp Cell Res* **318:** 1229–1244.

Delmore JE, Issa GC, Lemieux ME, Rahl PB, Shi J, Jacobs HM, Kastritis E, Gilpatrick T, Paranal RM, Qi J, et al. 2011. BET bromodomain inhibition as a therapeutic strategy to target c-Myc. *Cell* **146:** 904–917.

de Napoles M, Mermoud JE, Wakao R, Tang YA, Endoh M, Appanah R, Nesterova TB, Silva J, Otte AP, Vidal M, et al. 2004. Polycomb group proteins Ring1A/B link ubiquitylation of histone H2A to heritable gene silencing and X inactivation. *Dev Cell* **7:** 663–676.

Denis H, Deplus R, Putmans P, Yamada M, Metivier R, Fuks F. 2009. Functional connection between deimination and deacetylation of histones. *Mol Cell Biol* **29:** 4982–4993.

Dettman EJ, Simko SJ, Ayanga B, Carofino BL, Margolin JF, Morse HC 3rd, Justice MJ. 2011. Prdm14 initiates lymphoblastic leukemia after expanding a population of cells resembling common lymphoid progenitors. *Oncogene* **30:** 2859–2873.

Deubzer HE, Schier MC, Oehme I, Lodrini M, Haendler B, Sommer A, Witt O. 2013. HDAC11 is a novel drug target in carcinomas. *Int J Cancer* **132:** 2200–2208.

Dhar SS, Lee SH, Kan PY, Voigt P, Ma L, Shi X, Reinberg D, Lee MG. 2012. Trans-tail regulation of MLL4-catalyzed H3K4 methylation by H4R3 symmetric dimethylation is mediated by a tandem PHD of MLL4. *Genes Dev* **26:** 2749–2762.

Di Cerbo V, Schneider R. 2013. Cancers with wrong HATs: The impact of acetylation. *Brief Funct Genomics* **12:** 231–243.

Doil C, Mailand N, Bekker-Jensen S, Menard P, Larsen DH, Pepperkok R, Ellenberg J, Panier S, Durocher D, Bartek J, et al. 2009. RNF168 binds and amplifies ubiquitin conjugates on damaged chromosomes to allow accumulation of repair proteins. *Cell* **136:** 435–446.

Dorrance AM, Liu S, Yuan W, Becknell B, Arnoczky KJ, Guimond M, Strout MP, Feng L, Nakamura T, Yu L, et al. 2006. Mll partial tandem duplication induces aberrant Hox expression in vivo via specific epigenetic alterations. *J Clin Invest* **116:** 2707–2716.

Drost J, Mantovani F, Tocco F, Elkon R, Comel A, Holstege H, Kerkhoven R, Jonkers J, Voorhoeve PM, Agami R, et al. 2010. BRD7 is a candidate tumour suppressor gene required for p53 function. *Nat Cell Biol* **12:** 380–389.

Endoh M, Endo TA, Endoh T, Fujimura Y, Ohara O, Toyoda T, Otte AP, Okano M, Brockdorff N, Vidal M, et al. 2008. Polycomb group proteins Ring1A/B are functionally linked to the core transcriptional regulatory circuitry to maintain ES cell identity. *Development* **135:** 1513–1524.

Falandry C, Fourel G, Galy V, Ristriani T, Horard B, Bensimon E, Salles G, Gilson E, Magdinier F. 2010. CLLD8/KMT1F is a lysine methyltransferase that is important for chromosome segregation. *J Biol Chem* **285:** 20234–20241.

Fan J, Shan C, Kang HB, Elf S, Xie J, Tucker M, Gu TL, Aguiar M, Lonning S, Chen H, et al. 2014. Tyr phosphorylation of PDP1 toggles recruitment between ACAT1 and SIRT3 to regulate the pyruvate dehydrogenase complex. *Mol Cell* **53:** 534–548.

Fang J, Chen T, Chadwick B, Li E, Zhang Y. 2004. Ring1b-mediated H2A ubiquitination associates with inactive X chromosomes and is involved in initiation of X inactivation. *J Biol Chem* **279:** 52812–52815.

Filippakopoulos P, Qi J, Picaud S, Shen Y, Smith WB, Fedorov O, Morse EM, Keates T, Hickman TT, Felletar I, et al. 2010. Selective inhibition of BET bromodomains. *Nature* **468:** 1067–1073.

Finley LW, Carracedo A, Lee J, Souza A, Egia A, Zhang J, Teruya-Feldstein J, Moreira PI, Cardoso SM, Clish CB, et al. 2011. SIRT3 opposes reprogramming of cancer cell metabolism through HIF1α destabilization. *Cancer Cell* **19:** 416–428.

Fischle W, Tseng BS, Dormann HL, Ueberheide BM, Garcia BA, Shabanowitz J, Hunt DF, Funabiki H, Allis CD. 2005. Regulation of HP1-chromatin binding by histone H3 methylation and phosphorylation. *Nature* **438:** 1116–1122.

Fiskus W, Sharma S, Qi J, Valenta JA, Schaub LJ, Shah B, Peth K, Portier BP, Rodriguez M, Devaraj SG, et al. 2014. Highly active combination of BRD4 antagonist and histone deacetylase inhibitor against human acute myelogenous leukemia cells. *Mol Cancer Ther* **13:** 1142–1154.

Fontebasso AM, Schwartzentruber J, Khuong-Quang DA, Liu XY, Sturm D, Korshunov A, Jones DT, Witt H, Kool M, Albrecht S, et al. 2013. Mutations in SETD2 and genes affecting histone H3K36 methylation target hemispheric high-grade gliomas. *Acta Neuropathol* **125:** 659–669.

Fraga MF, Ballestar E, Villar-Garea A, Boix-Chornet M, Espada J, Schotta G, Bonaldi T, Haydon C, Ropero S, Petrie K, et al. 2005. Loss of acetylation at Lys16 and trimethylation at Lys20 of histone H4 is a common hallmark of human cancer. *Nat Genet* **37:** 391–400.

French CA, Miyoshi I, Aster JC, Kubonishi I, Kroll TG, Dal Cin P, Vargas SO, Perez-Atayde AR, Fletcher JA. 2001. BRD4 Bromodomain gene rearrangement in aggressive carcinoma with translocation t(15;19). *Am J Pathol* **159:** 1987–1992.

French CA, Ramirez CL, Kolmakova J, Hickman TT, Cameron MJ, Thyne ME, Kutok JL, Toretsky JA, Tadavarthy AK, Kees UR, et al. 2008. BRD-

NUT oncoproteins: A family of closely related nuclear proteins that block epithelial differentiation and maintain the growth of carcinoma cells. *Oncogene* **27:** 2237–2242.

Gallenkamp D, Gelato KA, Haendler B, Weinmann H. 2014. Bromodomains and their pharmacological inhibitors. *ChemMedChem* **9:** 438–464.

Garraway LA, Lander ES. 2013. Lessons from the cancer genome. *Cell* **153:** 17–37.

Gayther SA, Batley SJ, Linger L, Bannister A, Thorpe K, Chin SF, Daigo Y, Russell P, Wilson A, Sowter HM, et al. 2000. Mutations truncating the EP300 acetylase in human cancers. *Nat Genet* **24:** 300–303.

Ginjala V, Nacerddine K, Kulkarni A, Oza J, Hill SJ, Yao M, Citterio E, van Lohuizen M, Ganesan S. 2011. BMI1 is recruited to DNA breaks and contributes to DNA damage-induced H2A ubiquitination and repair. *Mol Cell Biol* **31:** 1972–1982.

Gong F, Miller KM. 2013. Mammalian DNA repair: HATs and HDACs make their mark through histone acetylation. *Mutat Res* **750:** 23–30.

* Grossniklaus U, Paro R. 2014. Transcriptional silencing by *Polycomb*-group proteins. *Cold Spring Harb Perspect Biol* **6:** a019331.

Guccione E, Bassi C, Casadio F, Martinato F, Cesaroni M, Schuchlautz H, Luscher B, Amati B. 2007. Methylation of histone H3R2 by PRMT6 and H3K4 by an MLL complex are mutually exclusive. *Nature* **449:** 933–937.

Guo X, Shi M, Sun L, Wang Y, Gui Y, Cai Z, Duan X. 2011. The expression of histone demethylase JMJD1A in renal cell carcinoma. *Neoplasma* **58:** 153–157.

Haigis MC, Sinclair DA. 2010. Mammalian sirtuins: Biological insights and disease relevance. *Annu Rev Pathol* **5:** 253–295.

Haigis MC, Deng CX, Finley LW, Kim HS, Gius D. 2012. SIRT3 is a mitochondrial tumor suppressor: A scientific tale that connects aberrant cellular ROS, the Warburg effect, and carcinogenesis. *Cancer Res* **72:** 2468–2472.

Hakimi AA, Chen YB, Wren J, Gonen M, Abdel-Wahab O, Heguy A, Liu H, Takeda S, Tickoo SK, Reuter VE, et al. 2012. Clinical and pathologic impact of select chromatin-modulating tumor suppressors in clear cell renal cell carcinoma. *Eur Urol* **63:** 848–854.

Hanahan D, Weinberg RA. 2011. Hallmarks of cancer: The next generation. *Cell* **144:** 646–674.

Hanover JA. 2010. Epigenetics gets sweeter: O-GlcNAc joins the 'histone code'. *Chem Biol* **17:** 1272–1274.

Harris WJ, Huang X, Lynch JT, Spencer GJ, Hitchin JR, Li Y, Ciceri F, Blaser JG, Greystoke BF, Jordan AM, et al. 2012. The histone demethylase KDM1A sustains the oncogenic potential of MLL-AF9 leukemia stem cells. *Cancer Cell* **21:** 473–487.

Harte MT, O'Brien GJ, Ryan NM, Gorski JJ, Savage KI, Crawford NT, Mullan PB, Harkin DP. 2010. BRD7, a subunit of SWI/SNF complexes, binds directly to BRCA1 and regulates BRCA1-dependent transcription. *Cancer Res* **70:** 2538–2547.

He J, Nguyen AT, Zhang Y. 2011. KDM2b/JHDM1b, an H3K36me2-specific demethylase, is required for initiation and maintenance of acute myeloid leukemia. *Blood* **117:** 3869–3880.

Hélias C, Struski S, Gervais C, Leymarie V, Mauvieux L, Herbrecht R, Lessard M. 2008. Polycythemia vera transforming to acute myeloid leukemia and complex abnormalities including 9p homogeneously staining region with amplification of MLLT3, JMJD2C, JAK2, and SMARCA2. *Cancer Genet Cytogenet* **180:** 51–55.

Herranz N, Dave N, Millanes-Romero A, Morey L, Diaz VM, Lorenz-Fonfria V, Gutierrez-Gallego R, Jeronimo C, Di Croce L, Garcia de Herreros A, et al. 2012. Lysyl oxidase-like 2 deaminates lysine 4 in histone H3. *Mol Cell* **46:** 369–376.

Hirota T, Lipp JJ, Toh BH, Peters JM. 2005. Histone H3 serine 10 phosphorylation by Aurora B causes HP1 dissociation from heterochromatin. *Nature* **438:** 1176–1180.

Hoffmann I, Roatsch M, Schmitt ML, Carlino L, Pippel M, Sippl W, Jung M. 2012. The role of histone demethylases in cancer therapy. *Mol Oncol* **6:** 683–703.

Hou X, Li Y, Luo RZ, Fu JH, He JH, Zhang LJ, Yang HX. 2012. High expression of the transcriptional co-activator p300 predicts poor survival in resectable non-small cell lung cancers. *Eur J Surg Oncol* **38:** 523–530.

Houtkooper RH, Pirinen E, Auwerx J. 2012. Sirtuins as regulators of metabolism and healthspan. *Nat Rev Mol Cell Biol* **13:** 225–238.

Hsia DA, Tepper CG, Pochampalli MR, Hsia EY, Izumiya C, Huerta SB, Wright ME, Chen HW, Kung HJ, Izumiya Y. 2010. KDM8, a H3K36me2 histone demethylase that acts in the cyclin A1 coding region to regulate cancer cell proliferation. *Proc Natl Acad Sci* **107:** 9671–9676.

Hsiao KY, Mizzen CA. 2013. Histone H4 deacetylation facilitates 53BP1 DNA damage signaling and double-strand break repair. *J Mol Cell Biol* **5:** 157–165.

Hu S, Yu L, Li Z, Shen Y, Wang J, Cai J, Xiao L, Wang Z. 2010. Overexpression of EZH2 contributes to acquired cisplatin resistance in ovarian cancer cells in vitro and in vivo. *Cancer Biol Ther* **10:** 788–795.

Huang J, Perez-Burgos L, Placek BJ, Sengupta R, Richter M, Dorsey JA, Kubicek S, Opravil S, Jenuwein T, Berger SL. 2006. Repression of p53 activity by Smyd2-mediated methylation. *Nature* **444:** 629–632.

Hubbert C, Guardiola A, Shao R, Kawaguchi Y, Ito A, Nixon A, Yoshida M, Wang XF, Yao TP. 2002. HDAC6 is a microtubule-associated deacetylase. *Nature* **417:** 455–458.

Hyllus D, Stein C, Schnabel K, Schiltz E, Imhof A, Dou Y, Hsieh J, Bauer UM. 2007. PRMT6-mediated methylation of R2 in histone H3 antagonizes H3 K4 trimethylation. *Genes Dev* **21:** 3369–3380.

Imaoka N, Hiratsuka M, Osaki M, Kamitani H, Kambe A, Fukuoka J, Kurimoto M, Nagai S, Okada F, Watanabe T, et al. 2012. Prognostic significance of sirtuin 2 protein nuclear localization in glioma: An immunohistochemical study. *Oncol Rep* **28:** 923–930.

Italiano A, Attias R, Aurias A, Pérot G, Burel-Vandenbos F, Otto J, Venissac N, Pedeutour F. 2006. Molecular cytogenetic characterization of a metastatic lung sarcomatoid carcinoma: 9p23 neocentromere and 9p23~p24 amplification including JAK2 and JMJD2C. *Cancer Genet Cytogenet* **167:** 122–130.

Iyer NG, Ozdag H, Caldas C. 2004. p300/CBP and cancer. *Oncogene* **23:** 4225–4231.

James LI, Korboukh VK, Krichevsky L, Baughman BM, Herold JM, Norris JL, Jin J, Kireev DB, Janzen WP, Arrowsmith CH, et al. 2013. Small-molecule ligands of methyl-lysine binding proteins: Optimization of selectivity for L3MBTL3. *J Med Chem* **56:** 7358–7371.

Jennings LE, Measures AR, Wilson BG, Conway SJ. 2014. Phenotypic screening and fragment based approaches to the discovery of small molecule bromodomain ligands. *Future Med Chem* **6:** 179–204.

Jenuwein T, Allis CD. 2001. Translating the histone code. *Science* **293:** 1074–1080.

Jiang T, Zhou X, Taghizadeh K, Dong M, Dedon PC. 2007. N-formylation of lysine in histone proteins as a secondary modification arising from oxidative DNA damage. *Proc Natl Acad Sci* **104:** 60–65.

Jones PA, Baylin SB. 2007. The epigenomics of cancer. *Cell* **128:** 683–692.

Jung KH, Noh JH, Kim JK, Eun JW, Bae HJ, Xie HJ, Chang YG, Kim MG, Park H, Lee JY, et al. 2012. HDAC2 overexpression confers oncogenic potential to human lung cancer cells by deregulating expression of apoptosis and cell cycle proteins. *J Cell Biochem* **113:** 2167–2177.

Kadoch C, Hargreaves DC, Hodges C, Elias L, Ho L, Ranish J, Crabtree GR. 2013. Proteomic and bioinformatic analysis of mammalian SWI/SNF complexes identifies extensive roles in human malignancy. *Nat Genet* **45:** 592–601.

Kaeser MD, Aslanian A, Dong MQ, Yates JR 3rd, Emerson BM. 2008. BRD7, a novel PBAF-specific SWI/SNF subunit, is required for target gene activation and repression in embryonic stem cells. *J Biol Chem* **283:** 32254–32263.

Kang MY, Lee BB, Kim YH, Chang DK, Kyu Park S, Chun HK, Song SY, Park J, Kim DH. 2007. Association of the SUV39H1 histone methyltransferase with the DNA methyltransferase 1 at mRNA expression level in primary colorectal cancer. *Int J Cancer* **121:** 2192–2197.

Kang D, Cho HS, Toyokawa G, Kogure M, Yamane Y, Iwai Y, Hayami S, Tsunoda T, Field HI, Matsuda K, et al. 2013. The histone methyltransferase Wolf-Hirschhorn syndrome candidate 1-like 1 (WHSC1L1) is involved in human carcinogenesis. *Genes Chromosomes Cancer* **52:** 126–139.

Karkhanis V, Hu YJ, Baiocchi RA, Imbalzano AN, Sif S. 2011. Versatility of PRMT5-induced methylation in growth control and development. *Trends Biochem Sci* **36:** 633–641.

Kauffman EC, Robinson BD, Downes MJ, Powell LG, Lee MM, Scherr DS, Gudas LJ, Mongan NP. 2011. Role of androgen receptor and associated lysine-demethylase coregulators, LSD1 and JMJD2A, in localized and advanced human bladder cancer. *Mol Carcinog* **50:** 931–944.

Kawazu M, Saso K, Tong KI, McQuire T, Goto K, Son DO, Wakeham A, Miyagishi M, Mak TW, Okada H. 2011. Histone demethylase JMJD2B functions as a co-factor of estrogen receptor in breast cancer proliferation and mammary gland development. *PLoS One* **6:** e17830.

Khan SN, Jankowska AM, Mahfouz R, Dunbar AJ, Sugimoto Y, Hosono N, Hu Z, Cheriyath V, Vatolin S, Przychodzen B, et al. 2013. Multiple mechanisms deregulate EZH2 and histone H3 lysine 27 epigenetic changes in myeloid malignancies. *Leukemia* **27:** 1301–1309.

Kim TD, Oh S, Shin S, Janknecht R. 2012a. Regulation of tumor suppressor p53 and HCT116 cell physiology by histone demethylase JMJD2D/KDM4D. *PLoS One* **7:** e34618.

Kim TD, Shin S, Berry WL, Oh S, Janknecht R. 2012b. The JMJD2A demethylase regulates apoptosis and proliferation in colon cancer cells. *J Cell Biochem* **113:** 1368–1376.

Komatsu S, Imoto I, Tsuda H, Kozaki KI, Muramatsu T, Shimada Y, Aiko S, Yoshizumi Y, Ichikawa D, Otsuji E, et al. 2009. Overexpression of SMYD2 relates to tumor cell proliferation and malignant outcome of esophageal squamous cell carcinoma. *Carcinogenesis* **30:** 1139–1146.

Kondo Y, Shen L, Ahmed S, Boumber Y, Sekido Y, Haddad BR, Issa JP. 2008. Downregulation of histone H3 lysine 9 methyltransferase G9a induces centrosome disruption and chromosome instability in cancer cells. *PLoS One* **3:** e2037.

Kouzarides T. 2007. Chromatin modifications and their function. *Cell* **128:** 693–705.

Krivtsov AV, Armstrong SA. 2007. MLL translocations, histone modifications and leukaemia stem-cell development. *Nat Rev Cancer* **7:** 823–833.

Krivtsov AV, Feng Z, Lemieux ME, Faber J, Vempati S, Sinha AU, Xia X, Jesneck J, Bracken AP, Silverman LB, et al. 2008. H3K79 methylation profiles define murine and human MLL-AF4 leukemias. *Cancer Cell* **14:** 355–368.

Kuzmichev A, Nishioka K, Erdjument-Bromage H, Tempst P, Reinberg D. 2002. Histone methyltransferase activity associated with a human multiprotein complex containing the Enhancer of Zeste protein. *Genes Dev* **16:** 2893–2905.

Lautz TB, Naiditch JA, Clark S, Chu F, Madonna MB. 2012. Efficacy of class I and II vs class III histone deacetylase inhibitors in neuroblastoma. *J Pediatr Surg* **47:** 1267–1271.

Li P, Yao H, Zhang Z, Li M, Luo Y, Thompson PR, Gilmour DS, Wang Y. 2008. Regulation of p53 target gene expression by peptidylarginine deiminase 4. *Mol Cell Biol* **28:** 4745–4758.

Li W, Zhao L, Zang W, Liu Z, Chen L, Liu T, Xu D, Jia J. 2011. Histone demethylase JMJD2B is required for tumor cell proliferation and survival and is overexpressed in gastric cancer. *Biochem Biophys Res Commun* **416:** 372–378.

Li WD, Li QR, Xu SN, Wei FJ, Ye ZJ, Cheng JK, Chen JP. 2013a. Exome sequencing identifies an MLL3 gene germ line mutation in a pedigree of colorectal cancer and acute myeloid leukemia. *Blood* **121:** 1478–1479.

Li Z, Xie QR, Chen Z, Lu S, Xia W. 2013b. Regulation of SIRT2 levels for human non-small cell lung cancer therapy. *Lung Cancer* **82:** 9–15.

Lim S, Janzer A, Becker A, Zimmer A, Schüle R, Buettner R, Kirfel J. 2010. Lysine-specific demethylase 1 (LSD1) is highly expressed in ER-negative breast cancers and a biomarker predicting aggressive biology. *Carcinogenesis* **31:** 512–520.

Lim JH, Choi YJ, Cho CH, Park JW. 2012. Protein arginine methyltransferase 5 is an essential component of the hypoxia-inducible factor 1 signaling pathway. *Biochem Biophys Res Commun* **418:** 254–259.

Limm K, Ott C, Wallner S, Mueller DW, Oefner P, Hellerbrand C, Bosserhoff AK. 2013. Deregulation of protein methylation in melanoma. *Eur J Cancer* **49:** 1305–1313.

Lin R, Tao R, Gao X, Li T, Zhou X, Guan KL, Xiong Y, Lei QY. 2013. Acetylation stabilizes ATP-citrate lyase to promote lipid biosynthesis and tumor growth. *Mol Cell* **51:** 506–518.

Liu G, Bollig-Fischer A, Kreike B, van de Vijver MJ, Abrams J, Ethier SP, Yang ZQ. 2009. Genomic amplification and oncogenic properties of the GASC1 histone demethylase gene in breast cancer. *Oncogene* **28:** 4491–4500.

Lockwood WW, Zejnullahu K, Bradner JE, Varmus H. 2012. Sensitivity of human lung adenocarcinoma cell lines to targeted inhibition of BET epigenetic signaling proteins. *Proc Natl Acad Sci* **109:** 19408–19413.

Lohse B, Nielsen AL, Kristensen JB, Helgstrand C, Cloos PA, Olsen L, Gajhede M, Clausen RP, Kristensen JL. 2011. Targeting histone lysine demethylases by truncating the histone 3 tail to obtain selective substrate-based inhibitors. *Angew Chem Int Ed Engl* **50:** 9100–9103.

Majer CR, Jin L, Scott MP, Knutson SK, Kuntz KW, Keilhack H, Smith JJ, Moyer MP, Richon VM, Copeland RA, et al. 2012. A687V EZH2 is a gain-of-function mutation found in lymphoma patients. *FEBS Lett* **586:** 3448–3451.

Mann M, Zou Y, Chen Y, Brann D, Vadlamudi R. 2014. PELP1 oncogenic functions involve alternative splicing via PRMT6. *Mol Oncol* **8:** 389–400.

* Marmorstein R, Zhou M-M. 2014. Writers and readers of histone acetylation: Structure, mechanism, and inhibition. *Cold Spring Harb Perspect Biol* **6:** a018762.

Martin C, Zhang Y. 2007. Mechanisms of epigenetic inheritance. *Curr Opin Cell Biol* **19:** 266–272.

McBrian MA, Behbahan IS, Ferrari R, Su T, Huang TW, Li K, Hong CS, Christofk HR, Vogelauer M, Seligson DB, et al. 2013. Histone acetylation regulates intracellular pH. *Mol Cell* **49:** 310–321.

McCabe MT, Graves AP, Ganji G, Diaz E, Halsey WS, Jiang Y, Smitheman KN, Ott HM, Pappalardi MB, Allen KE, et al. 2012. Mutation of A677 in histone methyltransferase EZH2 in human B-cell lymphoma promotes hypertrimethylation of histone H3 on lysine 27 (H3K27). *Proc Natl Acad Sci* **109:** 2989–2994.

McGlynn LM, Zino S, MacDonald AI, Curle J, Reilly JE, Mohammed ZM, McMillan DC, Mallon E, Payne AP, Edwards J, et al. 2014. SIRT2: Tumour suppressor or tumour promoter in operable breast cancer? *Eur J Cancer* **50:** 290–301.

Mercurio C, Minucci S, Pelicci PG. 2010. Histone deacetylases and epigenetic therapies of hematological malignancies. *Pharmacol Res* **62:** 18–34.

Mertz JA, Conery AR, Bryant BM, Sandy P, Balasubramanian S, Mele DA, Bergeron L, Sims RJ 3rd. 2011. Targeting MYC dependence in cancer by inhibiting BET bromodomains. *Proc Natl Acad Sci* **108:** 16669–16674.

Messner S, Altmeyer M, Zhao H, Pozivil A, Roschitzki B, Gehrig P, Rutishauser D, Huang D, Caflisch A, Hottiger MO. 2010. PARP1 ADP-ribosylates lysine residues of the core histone tails. *Nucleic Acids Res* **38:** 6350–6362.

Migliori V, Phalke S, Bezzi M, Guccione E. 2010. Arginine/lysine-methyl/methyl switches: Biochemical role of histone arginine methylation in transcriptional regulation. *Epigenomics* **2:** 119–137.

Min DJ, Ezponda T, Kim MK, Will CM, Martinez-Garcia E, Popovic R, Basrur V, Elenitoba-Johnson KS, Licht JD. 2012. MMSET stimulates myeloma cell growth through microRNA-mediated modulation of c-MYC. *Leukemia* **27:** 686–694.

 Cite as *Cold Spring Harb Perspect Biol* doi: 10.1101/cshperspect.a019521

Minucci S, Pelicci PG. 2006. Histone deacetylase inhibitors and the promise of epigenetic (and more) treatments for cancer. *Nat Rev Cancer* **6:** 38–51.

Mitra D, Das PM, Huynh FC, Jones FE. 2011. Jumonji/ARID1 B (JARID1B) protein promotes breast tumor cell cycle progression through epigenetic repression of microRNA let-7e. *J Biol Chem* **286:** 40531–40535.

Moelans CB, de Weger RA, Monsuur HN, Vijzelaar R, van Diest PJ. 2010. Molecular profiling of invasive breast cancer by multiplex ligation-dependent probe amplification-based copy number analysis of tumor suppressor and oncogenes. *Mod Pathol* **23:** 1029–1039.

Morin RD, Johnson NA, Severson TM, Mungall AJ, An J, Goya R, Paul JE, Boyle M, Woolcock BW, Kuchenbauer F, et al. 2010. Somatic mutations altering EZH2 (Tyr641) in follicular and diffuse large B-cell lymphomas of germinal-center origin. *Nat Genet* **42:** 181–185.

Morin RD, Mendez-Lago M, Mungall AJ, Goya R, Mungall KL, Corbett RD, Johnson NA, Severson TM, Chiu R, Field M, et al. 2011. Frequent mutation of histone-modifying genes in non-Hodgkin lymphoma. *Nature* **476:** 298–303.

Mozzetta C, Pontis J, Fritsch L, Robin P, Portoso M, Proux C, Margueron R, Ait-Si-Ali S. 2014. The histone H3 lysine 9 methyltransferases G9a and GLP regulate polycomb repressive complex 2-mediated gene silencing. *Mol Cell* **53:** 277–289.

Muller S, Brown PJ. 2012. Epigenetic chemical probes. *Clin Pharmacol Ther* **92:** 689–693.

Muraoka M, Konishi M, Kikuchi-Yanoshita R, Tanaka K, Shitara N, Chong JM, Iwama T, Miyaki M. 1996. p300 gene alterations in colorectal and gastric carcinomas. *Oncogene* **12:** 1565–1569.

Nakagawa M, Oda Y, Eguchi T, Aishima S, Yao T, Hosoi F, Basaki Y, Ono M, Kuwano M, Tanaka M, et al. 2007. Expression profile of class I HDACs in human cancer tissues. *Oncology Rep* **18:** 769–774.

Nicholas C, Yang J, Peters SB, Bill MA, Baiocchi RA, Yan F, Sif S, Tae S, Gaudio E, Wu X, et al. 2013. PRMT5 is upregulated in malignant and metastatic melanoma and regulates expression of MITF and p27(Kip1.). *PLoS One* **8:** e74710.

Nishikawa N, Toyota M, Suzuki H, Honma T, Fujikane T, Ohmura T, Nishidate T, Ohe-Toyota M, Maruyama R, Sonoda T, et al. 2007. Gene amplification and overexpression of PRDM14 in breast cancers. *Cancer Res* **67:** 9649–9657.

Oehme I, Linke JP, Bock BC, Milde T, Lodrini M, Hartenstein B, Wiegand I, Eckert C, Roth W, Kool M, et al. 2013. Histone deacetylase 10 promotes autophagy-mediated cell survival. *Proc Natl Acad Sci* **110:** E2592–E2601.

Okudela K, Mitsui H, Suzuki T, Woo T, Tateishi Y, Umeda S, Saito Y, Tajiri M, Masuda M, Ohashi K. 2013. Expression of HDAC9 in lung cancer—Potential role in lung carcinogenesis. *Int J Clin Exp Pathol* **7:** 213–220.

Ong CT, Corces VG. 2011. Enhancer function: New insights into the regulation of tissue-specific gene expression. *Nat Rev Genet* **12:** 283–293.

Osawa T, Muramatsu M, Wang F, Tsuchida R, Kodama T, Minami T, Shibuya M. 2011. Increased expression of histone demethylase JHDM1D under nutrient starvation suppresses tumor growth via down-regulating angiogenesis. *Proc Natl Acad Sci* **108:** 20725–20729.

Pan MR, Peng G, Hung WC, Lin SY. 2011. Monoubiquitination of H2AX protein regulates DNA damage response signaling. *J Biol Chem* **286:** 28599–28607.

Panier S, Durocher D. 2013. Push back to respond better: Regulatory inhibition of the DNA double-strand break response. *Nat Rev Mol Cell Biol* **14:** 661–672.

Paredes S, Villanova L, Chua KF. 2014. Molecular pathways: Emerging roles of mammalian Sirtuin SIRT7 in cancer. *Clin Cancer Res* **20:** 1741–1746.

Park SH, Zhu Y, Ozden O, Kim HS, Jiang H, Deng CX, Gius D, Vassilopoulos A. 2012. SIRT2 is a tumor suppressor that connects aging, acetylome, cell cycle signaling, and carcinogenesis. *Transl Cancer Res* **1:** 15–21.

Pasqualucci L, Dominguez-Sola D, Chiarenza A, Fabbri G, Grunn A, Trifonov V, Kasper LH, Lerach S, Tang H, Ma J, et al. 2011. Inactivating mutations of acetyltransferase genes in B-cell lymphoma. *Nature* **471:** 189–195.

* Patel DJ. 2014. A structural perspective on readout of epigenetic histone and DNA methylation marks. *Cold Spring Harb Perspect Biol* doi: 10.1101/cshperspect.a018754.

Pinder JB, Attwood KM, Dellaire G. 2013. Reading, writing, and repair: The role of ubiquitin and the ubiquitin-like proteins in DNA damage signaling and repair. *Front Genet* **4:** 45.

* Pirrotta V. 2014. The necessity of chromatin: A view in perspective. *Cold Spring Harb Perspect Biol* doi: 10.1101/cshperspect.a019547.

Plazas-Mayorca MD, Bloom JS, Zeissler U, Leroy G, Young NL, DiMaggio PA, Krugylak L, Schneider R, Garcia BA. 2010. Quantitative proteomics reveals direct and indirect alterations in the histone code following methyltransferase knockdown. *Mol Biosyst* **6:** 1719–1729.

Pradhan S, Chin HG, Esteve PO, Jacobsen SE. 2009. SET7/9 mediated methylation of non-histone proteins in mammalian cells. *Epigenetics* **4:** 383–387.

* Qi J. 2014. Bromodomain and extraterminal domain inhibitors (BETi) for cancer therapy: Chemical modulation of chromatin structure. *Cold Spring Harb Perspect Biol* doi: 10.1101/cshperspect.a018663.

Rardin MJ, He W, Nishida Y, Newman JC, Carrico C, Danielson SR, Guo A, Gut P, Sahu AK, Li B, et al. 2013. SIRT5 regulates the mitochondrial lysine succinylome and metabolic networks. *Cell Metab* **18:** 920–933.

Reichert N, Choukrallah MA, Matthias P. 2012. Multiple roles of class I HDACs in proliferation, differentiation, and development. *Cell Mol Life Sci* **69:** 2173–2187.

Robinson G, Parker M, Kranenburg TA, Lu C, Chen X, Ding L, Phoenix TN, Hedlund E, Wei L, Zhu X, et al. 2012. Novel mutations target distinct subgroups of medulloblastoma. *Nature* **488:** 43–48.

Roesch A, Becker B, Schneider-Brachert W, Hagen I, Landthaler M, Vogt T. 2006. Re-expression of the retinoblastoma-binding protein 2-homolog 1 reveals tumor-suppressive functions in highly metastatic melanoma cells. *J Invest Dermatol* **126:** 1850–1859.

Roesch A, Mueller AM, Stempfl T, Moehle C, Landthaler M, Vogt T. 2008. RBP2-H1/JARID1B is a transcriptional regulator with a tumor suppressive potential in melanoma cells. *Int J Cancer* **122:** 1047–1057.

Rosenthal F, Feijs KL, Frugier E, Bonalli M, Forst AH, Imhof R, Winkler HC, Fischer D, Caflisch A, Hassa PO, et al. 2013. Macrodomain-containing proteins are new mono-ADP-ribosylhydrolases. *Nat Struct Mol Biol* **20:** 502–507.

Rossetto D, Avvakumov N, Côté J. 2012. Histone phosphorylation: A chromatin modification involved in diverse nuclear events. *Epigenetics* **7:** 1098–1108.

Ruthenburg AJ, Li H, Patel DJ, Allis CD. 2007. Multivalent engagement of chromatin modifications by linked binding modules. *Nat Rev Mol Cell Biol* **8:** 983–994.

Saeed S, Logie C, Francoijs KJ, Frige G, Romanenghi M, Nielsen FG, Raats L, Shahhoseini M, Huynen M, Altucci L, et al. 2012. Chromatin accessibility, p300, and histone acetylation define PML-RARα and AML1-ETO binding sites in acute myeloid leukemia. *Blood* **120:** 3058–3068.

Sakabe K, Wang Z, Hart GW. 2010. β-N-acetylglucosamine (O-GlcNAc) is part of the histone code. *Proc Natl Acad Sci* **107:** 19915–19920.

Sandberg ML, Sutton SE, Pletcher MT, Wiltshire T, Tarantino LM, Hogenesch JB, Cooke MP. 2005. c-Myb and p300 regulate hematopoietic stem cell proliferation and differentiation. *Dev Cell* **8:** 153–166.

* Schaefer U. 2014. Pharmacological inhibition of bromodomain-containing proteins in inflammation. *Cold Spring Harb Perspect Biol* **6:** a018671.

Schenk T, Chen WC, Göllner S, Howell L, Jin L, Hebestreit K, Klein HU, Popescu AC, Burnett A, Mills K, et al. 2012. Inhibition of the LSD1 (KDM1A) demethylase reactivates the all-trans-retinoic acid differentiation pathway in acute myeloid leukemia. *Nat Med* **18:** 605–611.

Schreiber SLB, Bradley E. 2002. Signaling network model of chromatin. *Cell Cycle* **111:** 771–778.

Schulte JH, Lim S, Schramm A, Friedrichs N, Koster J, Versteeg R, Ora I, Pajtler K, Klein-Hitpass L, Kuhfittig-Kulle S, et al. 2009. Lysine-specific demethylase 1 is strongly expressed in poorly differentiated neuroblastoma: Implications for therapy. *Cancer Res* **69:** 2065–2071.

Sebastian C, Zwaans BM, Silberman DM, Gymrek M, Goren A, Zhong L, Ram O, Truelove J, Guimaraes AR, Toiber D, et al. 2012. The histone deacetylase SIRT6 is a tumor suppressor that controls cancer metabolism. *Cell* **151:** 1185–1199.

Seligson DB, Horvath S, Shi T, Yu H, Tze S, Grunstein M, Kurdistani SK. 2005. Global histone modification patterns predict risk of prostate cancer recurrence. *Nature* **435:** 1262–1266.

Seligson DB, Horvath S, McBrian MA, Mah V, Yu H, Tze S, Wang Q, Chia D, Goodglick L, Kurdistani SK. 2009. Global levels of histone modifications predict prognosis in different cancers. *Am J Pathol* **174:** 1619–1628.

Sen N, Kumari R, Singh MI, Das S. 2013. HDAC5, a key component in temporal regulation of p53-mediated transactivation in response to genotoxic stress. *Mol Cell* **52:** 406–420.

Serrano L, Martinez-Redondo P, Marazuela-Duque A, Vazquez BN, Dooley SJ, Voigt P, Beck DB, Kane-Goldsmith N, Tong Q, Rabanal RM, et al. 2013. The tumor suppressor SirT2 regulates cell cycle progression and genome stability by modulating the mitotic deposition of H4K20 methylation. *Genes Dev* **27:** 639–653.

* Seto E, Yoshida M. 2014. Erasers of histone acetylation: The histone deacetylase enzymes. *Cold Spring Harb Perspect Biol* **6:** a018713.

Shah BD, Martin P, Sotomayor EM. 2012. Mantle cell lymphoma: A clinically heterogeneous disease in need of tailored approaches. *Cancer Control* **19:** 227–235.

Sharma SV, Lee DY, Li B, Quinlan MP, Takahashi F, Maheswaran S, McDermott U, Azizian N, Zou L, Fischbach MA, et al. 2010. A chromatin-mediated reversible drug-tolerant state in cancer cell subpopulations. *Cell* **141:** 69–80.

Sharma P, Azebi S, England P, Christensen T, Møller-Larsen A, Petersen T, Batsché E, Muchardt C. 2012. Citrullination of histone H3 interferes with HP1-mediated transcriptional repression. *PLoS Genet* **8:** e1002934.

Shen H, Laird PW. 2013. Interplay between the cancer genome and epigenome. *Cell* **153:** 38–55.

* Shi YG, Tsukada Y. 2013. The discovery of histone demethylases. *Cold Spring Harb Perspect Biol* **5:** a017947.

Shi X, Kachirskaia I, Yamaguchi H, West LE, Wen H, Wang EW, Dutta S, Appella E, Gozani O. 2007. Modulation of p53 function by SET8-mediated methylation at lysine 382. *Mol Cell* **27:** 636–646.

Smith E, Lin C, Shilatifard A. 2011. The super elongation complex (SEC) and MLL in development and disease. *Genes Dev* **25:** 661–672.

Stewart MD, Li J, Wong J. 2005. Relationship between histone H3 lysine 9 methylation, transcription repression, and heterochromatin protein 1 recruitment. *Mol Cell Biol* **25:** 2525–2538.

Struhl K. 1998. Histone acetylation and transcriptional regulatory mechanisms. *Genes Dev* **12:** 599–606.

Stunkel W, Campbell RM. 2011. Sirtuin 1 (SIRT1): The misunderstood HDAC. *J Biomol Screen* **16:** 1153–1169.

Subramanian K, Jia D, Kapoor-Vazirani P, Powell DR, Collins RE, Sharma D, Peng J, Cheng X, Vertino PM. 2008. Regulation of estrogen receptor α by the SET7 lysine methyltransferase. *Mol Cell* **30:** 336–347.

Suganuma T, Workman JL. 2008. Crosstalk among histone modifications. *Cell* **135:** 604–607.

Suva ML, Riggi N, Janiszewska M, Radovanovic I, Provero P, Stehle JC, Baumer K, Le Bitoux MA, Marino D, Cironi L, et al. 2009. EZH2 is essential for glioblastoma cancer stem cell maintenance. *Cancer Res* **69:** 9211–9218.

Takawa M, Cho HS, Hayami S, Toyokawa G, Kogure M, Yamane Y, Iwai Y, Maejima K, Ueda K, Masuda A, et al. 2012. Histone lysine methyltransferase SETD8 promotes carcinogenesis by deregulating PCNA expression. *Cancer Res* **72:** 3217–3227.

Tan M, Luo H, Lee S, Jin F, Yang JS, Montellier E, Buchou T, Cheng Z, Rousseaux S, Rajagopal N, et al. 2011. Identification of 67 histone marks and histone lysine crotonylation as a new type of histone modification. *Cell* **146:** 1016–1028.

Tang J, Cho NW, Cui G, Manion EM, Shanbhag NM, Botuyan MV, Mer G, Greenberg RA. 2013. Acetylation limits 53BP1 association with damaged chromatin to promote homologous recombination. *Nat Struct Mol Biol* **20:** 317–325.

Tanikawa C, Espinosa M, Suzuki A, Masuda K, Yamamoto K, Tsuchiya E, Ueda K, Daigo Y, Nakamura Y, Matsuda K. 2012. Regulation of histone modification and chromatin structure by the p53-PADI4 pathway. *Nat Commun* **3:** 676.

Trojer P, Cao AR, Gao Z, Li Y, Zhang J, Xu X, Li G, Losson R, Erdjument-Bromage H, Tempst P, et al. 2011. L3MBTL2 protein acts in concert with PcG protein-mediated monoubiquitination of H2A to establish a repressive chromatin structure. *Mol Cell* **42:** 438–450.

Tsai MC, Wang JK, Chang HY. 2010. Tumor suppression by the histone demethylase UTX. *Cell Cycle* **9:** 2043–2044.

Tweedie-Cullen RY, Brunner AM, Grossmann J, Mohanna S, Sichau D, Nanni P, Panse C, Mansuy IM. 2012. Identification of combinatorial patterns of post-translational modifications on individual histones in the mouse brain. *PLoS One* **7:** e36980.

Uemura M, Yamamoto H, Takemasa I, Mimori K, Hemmi H, Mizushima T, Ikeda M, Sekimoto M, Matsuura N, Doki Y, et al. 2010. Jumonji domain containing 1A is a novel prognostic marker for colorectal cancer: In vivo identification from hypoxic tumor cells. *Clin Cancer Res* **16:** 4636–4646.

Van Aller GS, Reynoird N, Barbash O, Huddleston M, Liu S, Zmoos AF, McDevitt P, Sinnamon R, Le B, Mas G, et al. 2012. Smyd3 regulates cancer cell phenotypes and catalyzes histone H4 lysine 5 methylation. *Epigenetics* **7:** 340–343.

Van Damme M, Crompot E, Meuleman N, Mineur P, Bron D, Lagneaux L, Stamatopoulos B. 2012. HDAC isoenzyme expression is deregulated in chronic lymphocytic leukemia B-cells and has a complex prognostic significance. *Epigenetics* **7:** 1403–1412.

Venneti S, Garimella MT, Sullivan LM, Martinez D, Huse JT, Heguy A, Santi M, Thompson CB, Judkins AR. 2013. Evaluation of histone 3 lysine 27 trimethylation (H3K27me3) and enhancer of Zest 2 (EZH2) in pediatric glial and glioneuronal tumors shows decreased H3K27me3 in H3F3A K27M mutant glioblastomas. *Brain Pathol* **23:** 558–564.

Vieira FQ, Costa-Pinheiro P, Ramalho-Carvalho J, Pereira A, Menezes FD, Antunes L, Carneiro I, Oliveira J, Henrique R, Jeronimo C. 2014. Deregulated expression of selected histone methylases and demethylases in prostate carcinoma. *Endocr Relat Cancer* **21:** 51–61.

Vinatzer U, Gollinger M, Müllauer L, Raderer M, Chott A, Streubel B. 2008. Mucosa-associated lymphoid tissue lymphoma: Novel translocations including rearrangements of ODZ2, JMJD2C, and CNN3. *Clin Cancer Res* **14:** 6426–6431.

Wagner KW, Alam H, Dhar SS, Giri U, Li N, Wei Y, Giri D, Cascone T, Kim JH, Ye Y, et al. 2013. KDM2A promotes lung tumorigenesis by epigenetically enhancing ERK1/2 signaling. *J Clin Invest* **123:** 5231–5246.

Wang Y, Wysocka J, Sayegh J, Lee YH, Perlin JR, Leonelli L, Sonbuchner LS, McDonald CH, Cook RG, Dou Y, et al. 2004. Human PAD4 regulates histone arginine methylation levels via demethylimination. *Science* **306:** 279–283.

Wang GG, Cai L, Pasillas MP, Kamps MP. 2007. NUP98-NSD1 links H3K36 methylation to Hox-A gene activation and leukaemogenesis. *Nat Cell Biol* **9:** 804–812.

Wang L, Gural A, Sun XJ, Zhao X, Perna F, Huang G, Hatlen MA, Vu L, Liu F, Xu H, et al. 2011. The leukemogenicity of AML1-ETO is dependent on site-specific lysine acetylation. *Science* **333:** 765–769.

Watanabe H, Soejima K, Yasuda H, Kawada I, Nakachi I, Yoda S, Naoki K, Ishizaka A. 2008. Deregulation of histone lysine methyltransferases contributes to oncogenic transformation of human bronchoepithelial cells. *Cancer Cell Int* **8:** 15.

Cite as *Cold Spring Harb Perspect Biol* doi: 10.1101/cshperspect.a019521

Wilson BG, Wang X, Shen X, McKenna ES, Lemieux ME, Cho YJ, Koellhoffer EC, Pomeroy SL, Orkin SH, Roberts CW. 2010. Epigenetic antagonism between polycomb and SWI/SNF complexes during oncogenic transformation. *Cancer Cell* **18:** 316–328.

Wisniewski JR, Zougman A, Mann M. 2008. Nepsilon-formylation of lysine is a widespread post-translational modification of nuclear proteins occurring at residues involved in regulation of chromatin function. *Nucleic Acids Res* **36:** 570–577.

Witt O, Deubzer HE, Lodrini M, Milde T, Oehme I. 2009. Targeting histone deacetylases in neuroblastoma. *Curr Pharm Des* **15:** 436–447.

Wyce A, Ganji G, Smitheman KN, Chung CW, Korenchuk S, Bai Y, Barbash O, Le B, Craggs PD, McCabe MT, et al. 2013. BET inhibition silences expression of MYCN and BCL2 and induces cytotoxicity in neuroblastoma tumor models. *PloS One* **8:** e72967.

Yan F, Alinari L, Lustberg ME, Martin LK, Cordero-Nieves HM, Banasavadi-Siddegowda Y, Virk S, Barnholtz-Sloan J, Bell EH, Wojton J, et al. 2014. Genetic validation of the protein arginine methyltransferase PRMT5 as a candidate therapeutic target in glioblastoma. *Cancer Res* **74:** 1752–1765.

Yang ZQ, Imoto I, Fukuda Y, Pimkhaokham A, Shimada Y, Imamura M, Sugano S, Nakamura Y, Inazawa J. 2000. Identification of a novel gene, GASC1, within an amplicon at 9p23-24 frequently detected in esophageal cancer cell lines. *Cancer Res* **60:** 4735–4739.

Yang P, Guo L, Duan ZJ, Tepper CG, Xue L, Chen X, Kung HJ, Gao AC, Zou JX, Chen HW. 2012. Histone methyltransferase NSD2/MMSET mediates constitutive NF-κB signaling for cancer cell proliferation, survival, and tumor growth via a feed-forward loop. *Mol Cell Biol* **32:** 3121–3131.

Yap DB, Chu J, Berg T, Schapira M, Cheng SW, Moradian A, Morin RD, Mungall AJ, Meissner B, Boyle M, et al. 2011. Somatic mutations at EZH2 Y641 act dominantly through a mechanism of selectively altered PRC2 catalytic activity, to increase H3K27 trimethylation. *Blood* **117:** 2451–2459.

Zhang H, Park SH, Pantazides BG, Karpiuk O, Warren MD, Hardy CW, Duong DM, Park SJ, Kim HS, Vassilopoulos A, et al. 2013. SIRT2 directs the replication stress response through CDK9 deacetylation. *Proc Natl Acad Sci* **110:** 13546–13551.

Zhang Y, Chen A, Yan XM, Huang G. 2012a. Disordered epigenetic regulation in MLL-related leukemia. *Int J Hematol* **96:** 428–437.

Zhang X, Bolt M, Guertin MJ, Chen W, Zhang S, Cherrington BD, Slade DJ, Dreyton CJ, Subramanian V, Bicker KL, et al. 2012b. Peptidylarginine deiminase 2-catalyzed histone H3 arginine 26 citrullination facilitates estrogen receptor α target gene activation. *Proc Natl Acad Sci* **109:** 13331–13336.

* Zhao Y, Garcia BA. 2014. Comprehensive catalog of currently documented histone modifications. *Cold Spring Harb Persp Biol* doi: 10.1101/cshperspect.a025064.

Zhu J, Adli M, Zou JY, Verstappen G, Coyne M, Zhang X, Durham T, Miri M, Deshpande V, De Jager PL, et al. 2013. Genome-wide chromatin state transitions associated with developmental and environmental cues. *Cell* **152:** 642–654.

Zuber J, Shi J, Wang E, Rappaport AR, Herrmann H, Sison EA, Magoon D, Qi J, Blatt K, Wunderlich M, et al. 2011. RNAi screen identifies Brd4 as a therapeutic target in acute myeloid leukaemia. *Nature* **478:** 524–528.

C H A P T E R 36

The Necessity of Chromatin: A View in Perspective

Vincenzo Pirrotta

Department of Molecular Biology and Biochemistry, Rutgers University, Piscataway, New Jersey 08854

Correspondence: pirrotta@biology.rutgers.edu

SUMMARY

Epigenomics has grown exponentially, providing a better understanding of the mechanistic aspects of new and old phenomena originally described through genetics, as well as providing unexpected insights into the way chromatin modulates the genomic information. In this overview, some of the advances are selected for discussion and comment under six topics: (1) histone modifications, (2) weak interactions, (3) interplay with external inputs, (4) the role of RNA molecules, (5) chromatin folding and architecture, and, finally, (6) a view of the essential role of chromatin transactions in regulating the access to genomic DNA.

Outline

Cite as *Cold Spring Harb Perspect Biol* doi: 10.1101/cshperspect.a019547

OVERVIEW

The size and scope of this new Cold Spring Harbor Laboratory Press Epigenetics collection gives a tangible sense of the vast expansion of the field in the 7 years since the first edition. Not all of it goes under the tag of epigenetic in the strict sense of involving long-lasting, heritable responses in the gene-expression machinery. Epigenomics has revealed that chromatin is not just a way to package genomic DNA—it is a structure that provides the genome with a large range of variables that can be exploited to tune and articulate the function of the underlying DNA. Thus, in chromatin a set of complex interactions intervenes between the genetic material and the transcriptional readout.

The expansion has been both qualitative and quantitative, driven by the emergence of technologies for genome-wide analysis: first with the broad availability of genomic tiling microarrays and then of the next-generation parallel-sequencing technologies. We can now view the distribution of chromatin proteins, nucleosomes, histone modifications, and transcriptional activity, not just at an individual site but over the entire genome. It is true that the results are generally averages over a large number of cells, but, in some cases, the analysis of chromatin activities in single cells has become possible. Massively parallel sequencing has also made possible modes of analysis that were previously unthinkable. We can evaluate how often a given sequence is represented in the transcriptional output of the genome, and we can determine how much time a given genomic sequence spends in the vicinity of any other sequence using one of the currently available chromatin conformation capture (3C) approaches.

The broad availability of high-throughput technologies applied to the analysis of genomic structures and their transactions has sent colossal volumes of data pouring into computer storage. The DNA sequence data stored at the European Bioinformatics Institute has grown from a few terabases in 2008 to >200 terabases in 2012. The genomic data now exceeds 2 petabytes and is set to double in a year's time (Marx 2013). What has this deluge of information produced? A word of caution first: The vast mass of chromatin-related data is highly heterogeneous in format, structure, degree of documentation, ease of access, and, above all, I suspect, quality. Given the vertiginous pace of research and publications, there are likely many incorrect results scattered in the literature owing to unexamined assumptions, inappropriate applications, insufficient controls, and hasty conclusions. Antibodies, an indispensable resource, are both extraordinary enablers and treacherous tools. Most researchers are aware of the need for rigorous testing for specificity. But criteria that are sufficient for one application are not necessarily adequate for another. This is particularly true when the antibodies are used for chromatin immunoprecipitation. The sheer volume of data means that the resulting inconsistencies may often go unnoticed for a long time because investigators analyze their results primarily with respect to the specific concerns that prompted their experiments. Although sophisticated bioinformatics analysis has been applied, in some cases, to extract features and patterns out of the data, the majority of the information is far from fully used and will likely remain so. New technological advances will probably mean that new data sets will supersede those of today in the databases of the future.

In attempting to assess the advances brought on by this vast accumulation of data, I have highlighted a few developments that give food for thought, and I end with the synthesis of a global chromatin accessibility hypothesis. Topics discussed include the following.

1. Histone modifications, the chromatin complexes that write, read, and erase them, and the interplay between intra- and internucleosomal interactions can produce long-lasting chromatin states.
2. Weak interactions along the epigenome provide an opportunistic or searching mode by which chromatin associations can occur.
3. A context is provided for chromatin regulation—the interplay with external inputs.
4. The pervasive role of RNA molecules is discussed: The dark energy of the nucleus is beginning to be not so dark.
5. The importance of chromatin folding and nuclear localization is discussed.
6. Finally, I will spend several pages discussing what seems to me to be the antithesis that underlies so many aspects of chromatin transactions: the conflict between the need to prevent and the imperative of allowing access to the underlying DNA.

These topics do not, by any means, exhaust what can be said about progress in chromatin research and epigenomics. For that, a whole collection would be needed. And, indeed, a whole collection is what you have in your hands.

Cite as *Cold Spring Harb Perspect Biol* doi: 10.1101/cshperspect.a019547

1 CONTEXTUALIZING SOME ADVANCES

1.1 Chromatin

1.1.1 Histone Posttranslational Modifications: Instructors or Bystanders?

The effort to catalog, classify, and interpret the significance of the histone modifications, chromatin complexes, and individual factors with which chromatin is thickly encrusted has fascinated a large number of researchers and led to a new era in molecular genetics. The task now is to understand their role in modulating the multiple activities that operate on the genomic material.

Many questions have been raised about the relationship between the molecular machinery that manipulates chromatin and histone modifications that are associated with the various kinds of chromatin transactions. Do these modifications carry information that extends the genetically encoded information? Do they instruct activities or simply accompany or facilitate these activities? Are they self-maintaining or do they fade more or less rapidly after the events with which they are associated? In essence, are they causes or consequences of mechanisms that act on the underlying DNA (Henikoff and Shilatifard 2011)? There is probably no single answer; each modification may play somewhat different roles depending on the context.

If histone modifications carry information, it is clearly not the same kind as genetic information. Unlike sequence-based information, this "epigenetic" information is not meant to be stable over the long term. Chromatin marks such as histone or DNA methylation are the products of activities operating on the chromatin. They carry information about the activities that have resulted in the placement of the methylation marks, but also affect subsequent activities. Many characteristic epigenetic states such as those found in heterochromatic regions, Polycomb-repressed regions, or genes silenced by DNA methylation are reset during the course of differentiation. It came, nevertheless, as something of a shock to many researchers to discover that there existed demethylases that could remove even the most treasured marks of epigenetic repression, histone H3K9 methylation, associated with heterochromatin as well as histone H3K27 methylation associated with Polycomb silencing. This was just unfolding at the time of the first edition of this collection. Now, even cytosine methylation on the DNA has a means of being actively erased via the Tet enzymes (Kohli and Zhang 2013). The consequent implication is that epigenetic silencing is not forever; it could be and is constantly tinkered with both at the level of the histone marks and complexes that write and read them. This does not mean that epigenetic states cannot be maintained for long periods, through many rounds of cell division and, in some cases, transgenerationally. It means, however, that an epigenetic state cannot simply be equated to the presence of a particular mark. The mechanisms that put chromatin marks in place, maintain them, and respond to them are more complex and dynamic than had previously been thought.

The fact that histone modifications can endure through multiple cell cycles means that they can serve as a mark of the antecedent events that produced them and, therefore, influence the activities of a chromatin region as a function of its history. We now know that various complexes that interact with chromatin (chromatin complexes) include structural domains that recognize specific histone marks (commonly referred to as readers) and that these interactions govern the function of many molecular machines that operate on chromatin. In some cases, the presence of a histone mark is recognized by the same chromatin complex that deposits that mark and this can stimulate further deposition of the same mark on neighboring nucleosomes. This feed-forward effect means that histone marks such as H3K9 and H3K27 methylation can promote and help to maintain their own presence, thereby explaining their "epigenetic" persistence from one cell cycle to the next.

In the case of H3K9 methylation, the protein that binds to the H3K9me3 mark, HP1, is associated with the H3K9 methyltransferase, together helping to maintain the histone modification and spread heterochromatin. The complex containing both partners not only recognizes H3K9 methylation in surrounding chromatin but is also stimulated by its presence (Al-Sady et al. 2013). Interestingly, heterochromatin-associated H3K9 trimethylation appears in some circumstances to be able to maintain itself once it is generated de novo. For instance, once a domain of H3K9 trimethylation was generated at a GAL4 binding site, through the expression of a transgenic HP1 fused with a GAL4 DNA-binding domain (GAL4-HP1), the methylated and repressed state were then maintained even after the GAL4-HP1 was no longer produced (Hathaway et al. 2012). This implies that H3K9 methylation set by GAL4-HP1 is sufficient to recruit endogenous HP1 and ensures further methylation, a key feature of Epigenetics. Thus, HP1 binding is tightly linked to methylation activity in heterochromatin formation, more so than the corresponding activities of Polycomb complexes PRC1 and PRC2; a Polycomb-repressed domain cannot maintain itself, at least in *Drosophila*, when the Polycomb-response element (PRE) is deleted (Busturia et al. 1997). The extensive H3K27 trimethylation and H2A ubiquitination associated with it are important contributors, but not sufficient to continue recruiting the Polycomb complexes. The long-term persistence of the Polycomb repressed state requires a recruiting element, in addition to the histone modifications, and the absence of strong antagonizing activities (see, however, Secs. 1.2 and 2.5).

A surprise of the last few years has been the discovery that many nonhistone chromatin proteins are targets of posttranslational modifications—in some cases, placed by the same chromatin factors that modify histones. For instance, p53 is acetylated by CBP; it is also methylated by SMYD at K370 to prevent its binding to DNA or methyltransferase SET7/9 at K372, which prevents methylation at K370 in response to DNA damage (Ivanov et al. 2007). Do such modifications convey information? Of course they do; they switch the function of the target protein. Another surprising case is retinoic acid receptor α, which can be methylated by EZH2, targeting it for degradation by the proteasome. EZH2 acts, in this case, without the participation of the other components of the PRC2 complex that are normally required for methyltransferase activity (Lee et al. 2012). These effects remind us that some of the well-known chromatin factors can have additional effects that are entirely independent of chromatin, including roles outside of the nucleus (Su et al. 2005).

1.1.2 Nucleosome Marks and Chromatin Differentiation

Nucleosomes can carry a bewildering number of histone modifications simultaneously. Unfortunately, there is no evidence that histone marks have simple instructive combinatorial properties. In general, chromatin complexes distinguish nucleosomes by the presence or absence of one particular histone mark, and each mark has its proper interacting protein or "reader." There are some very interesting exceptions. For example, phosphorylation of H3S10 or H3S28 prevents the binding of chromodomains to adjacent methylated H3K9 or H3K27, respectively, and may constitute a toggle or binary switch to rapidly neutralize the role of these methylations. Another example of the combinatorial action of histone marks is the way certain "reader" proteins require a specific combination of marks with the right spacing and position. Double acetylation marks recognized by proteins containing two bromodomains provide such an example. Another interesting example is the reading of multiple histone marks by the DNA repair protein 53BP, which binds to nucleosomes containing both H4K20me1/2 and H2AK15ub marks through a combination of its Tudor domain and a carboxy-terminal ubiquitin-dependent recruitment domain (Fradet-Turcotte et al. 2013). The question these examples raise is do they represent ways to extend the repertoire of "meaningful" marks on a nucleosome? Or are they ways to integrate multiple inputs that control a chromatin process?

We tend to focus on the functional consequences that a nucleosome and interacting protein reader can have. Ultimately, however, what a given stretch of chromatin can do also depends on how it folds locally to allow interactions between the nucleosomes, and proteins bound to them and between *trans*-acting factors and the DNA. This appears to be the role of the HMG (high-mobility group) proteins, small proteins with a highly negatively charged carboxy-terminal tail. HMG proteins are abundant components of chromatin that are generally thought to have accessory "architectural" roles, binding to DNA and helping it to fold in particular structures important for the function of chromatin regulators. But, HMG proteins are also targets of posttranslational modifications including acetylation, methylation, phosphorylation, and probably others. Is there a whole new world of chromatin marks hidden there? In the case of HMGA1, for example, acetylation at K64 by CBP is said to destabilize the interferon-β enhanceosome, leading to transcriptional turnoff, whereas acetylation of K70 by PCAF/GCN5 (CBP-associated factor) has the opposite effect (see Zhang and Wang 2010). Furthermore, HMG proteins, with or without posttranslational modifications, could have an important role in modulating modifications of the histones. If so, the complexities of chromatin modifications through posttranslational modifications are enormously multiplied. So far, the differential distribution of HMGs and their modifications in the genome, and the relationship between the histone modifications involved in HMG association and resultant gene activity, have not been studied systematically.

1.2 The Power of Weak Interactions and Their Role in Chromatin Scanning

Classically, strong interactions between DNA-binding proteins and their cognate DNA are viewed as the efficient way to target and regulate specific genomic sites predictably, given sufficient nuclear concentrations. However, efficient use of specificity for genome regulation requires a strategy that relies on weak interactions. Weak interactions do not imply lack of interactions; on the contrary, potential binding sites are frequently visited (or perhaps "sampled" is a better word). But, in the absence of other events that might stabilize the interaction, the binding is weak and transient. This kind of interaction is typical of a specific DNA-binding protein with nonspecific DNA sequences and, indeed, constitutes the searching phase during which the protein spends much time weakly bound to nonspecific sequences. When the protein finds a consensus or near-consensus sequence, the interaction then becomes stronger and takes longer to dissociate. These dynamics were recognized a long time ago by Peter von Hippel (von Hip-

 Cite as *Cold Spring Harb Perspect Biol* doi: 10.1101/cshperspect.a019547

pel et al. 1974), who showed that weak interactions reduce the three-dimensional volume that a protein needs to explore and find its specific binding sites.

The binding of chromatin complexes to chromatin is, in general, of the weak and transient variety because interactions with histones or histone modifications are, at best, in the micromolar range of affinities. However, in the nucleus, all nucleosomes, except for those in highly condensed regions, are, in principle, accessible for interaction. In most cases, such transient interactions do not necessarily leave a trace detectable by chromatin immunoprecipitation (ChIP) or similar approaches unless the complex has an enzymatic activity such as a methyltransferase or deacetylase activity, which can mark the genome with a change in histone modifications. But when a factor or complex reaches an environment in which additional interactions are possible, such as binding to other histone marks, interactions with a specific DNA-binding protein, or cooperative interactions with other factors, the interactions stabilize a longer residence time and we call that binding to a specific site.

Weak binding is essential for all regulated processes and is enshrined in the concept of mass-action equilibria. Any molecule in the nucleus can potentially interact with any other. We tend to lose track of this fact when we think, for example, that RNA polymerase binds only at promoters. It also binds, with greater or lesser affinity, to any other accessible site. If it binds long enough or frequently enough, it will produce some kind of transcript (see the argument in Sec. 2). Thus, when a chromatin region becomes accessible, this can produce unwanted transcripts. Various mechanisms are available to control this, from exosome surveillance of the transcripts, to H3K36 methylation and recruitment of repressive complexes in regions that become too accessible because of active transcription (Li et al. 2007).

In addition to strong protein binding, cooperative binding, or binding via "readers" of histone marks, an important emerging issue is that relatively weak or transient interactions produce an opportunistic surveillance strategy for targeting chromatin complexes. A variant PRC1 complex illustrates the point: We know it binds specifically via the CXXC zinc finger motif of its KDM2B component to mediate repression at a subset of unmethylated CpG (cytosine and guanine separated by a phosphate) island target genes, rather than via chromobox homolog protein (CBX), the more conventional PRC1 component that reads H3K27me3 marks. But, interestingly, Farcas et al. (2012) also detected a much weaker presence of the complex at most CpG islands. Klose et al. (2013) argue that this is part of an opportunistic strategy of recruitment through weak interactions at all such sites, rather than targeting to a specific subset of sites. At any given site, if any cooperating factors are present, they can stabilize encounters for which weak interactions have created the opportunity. I will argue in the second half of this essay that such genome-wide scanning activities may be an underlying prerequisite for many basic chromatin functions.

1.3 The Interplay of Epigenetic Mechanisms and the Environment

We have seen that histone modifications alone cannot be responsible for long-lived epigenetic states. These modifications can come and go and are, in many cases, the target of activities that remove them. It is rather the whole mechanism with which they are associated that can generate long-lived states. We are beginning to get glimpses of the fact that epigenetic mechanisms are far more complex and likely to involve the interplay of the genome, physiological state of the organism, nervous system, and other environmental input.

One of the most remarkable stories to emerge in the past decade is the role that chromatin modifications play in extending the effects of early newborn experiences in rats to later functions in the adult via epigenetic programming in the hypothalamus. Specifically, Weaver, Szyf, and collaborators (Weaver et al. 2004, 2005; McGowan et al. 2009) found that maternal licking behavior is important to reduce alarm in rat pups. Inadequate maternal licking resulted in stressed and insecure rats that grew up to be stressed, insecure, and aggressive. This was traced to chromatin alterations in the genes associated with glucocorticoid response, a major stress response pathway in rats and man. Thus, early experience caused long-lasting phenotypes (epigenetic changes) affecting the expression of a pathway important for stress behavior. How the chromatin changes resulting from early stress are maintained over a period of years in the lifetime of an individual is not entirely clear. This pathway also has a remarkable transgenerational effect: The stressed and aggressive adult female behavior, resulting from the epigenomic programming occurring after birth as a consequence of negligent maternal nurturing, propagates to the next generation through their inadequate maternal licking behavior to their offspring. This stress syndrome is thus transmitted over multiple generations by a combination of epigenetic memory, behavior, and what might be called cultural inheritance. It is not hard to see how this interplay might be relevant to a variety of behavioral patterns such as autism spectrum disorders (see, e.g., LaSalle 2013). This is just a taste of what research on the interplay between behavior and chromatin modifications in the central nervous system is likely to reveal in many other aspects of behavior. And, in principle, a similar interplay can be expected between chromatin modifications and other inputs such as metabolic, nutritional, and physiological states, environmental conditions, and chemical or immunological challenges.

1.4 The Pervasive Roles of Nuclear RNA Molecules

Another major theme in the past decade of research is the new-found ascendancy of RNA molecules. RNAs are far from the subservient intermediary between genomic DNA and proteins envisioned by Francis Crick's central dogma. Deep sequencing of the transcriptome has revealed virtually ubiquitous transcription. Far from being mere genomic deserts, many intergenic regions are actively transcribed to produce RNA molecules that lack appreciable protein-coding potential (noncoding RNAs or ncRNAs) and whose functions are still, for the most part, unknown. Indeed, the expansion in the field has led to the emergence of a variety of new names to refer to the different types of ncRNAs with different functions (see Fig. 26 of Ch. 3 [Allis et al.]).

I will not attempt to summarize the well-known roles of small RNAs in directing cleavage of RNA molecules, modulating translation, heterochromatin formation, and transcriptional silencing. I will focus, instead, on the role of long noncoding RNAs (lncRNAs) that has generated much excitement in the epigenetic field. Some lncRNAs have been found to be critical components of chromatin complexes that regulate chromatin states and, therefore, transcription itself. These discoveries have now led to the surmise that many, if not all, lncRNAs have similar roles.

Enough is known at this time to show that the role of lncRNAs such as Xist, Tsix, HOTAIR, Air, and enhancer RNAs (eRNAs) described in this collection are varied. Some act in *cis*, perhaps by being captured as they are being transcribed, and are important in X chromosome inactivation, allele-specific regulation, and imprinting. Others act in *trans*. They may recognize DNA sequence motifs in some way, possibly by forming DNA–RNA hybrids, and serve a targeting role for RNA-binding proteins. Or they may act as scaffolds to assemble multiple chromatin-modifying complexes to activate or silence target genes. Overall, they seem to be highly versatile and powerful components of the nuclear arsenal of tools that regulate transcriptional activity, even though they themselves are the direct products of transcription.

Are all genomic sequences, thus, "functional" in some way? Although it is true that some ncRNAs are potentially involved in important chromatin processes, this has not been shown for the vast majority. Ubiquitous ncRNA functionality is not supported by the fact that many sequences are not evolutionarily conserved. However, sequence conservation is not necessarily required—often it is the transcriptional activity itself that is important. In other cases, the functional properties of an ncRNA depend on very short motifs and secondary structure rather than on extensive sequence conservation. Recent work has shown that sequence specificity is not involved in the binding of the PRC2 complex to lncRNAs such as HOTAIR (Davidovich et al. 2013; Kaneko et al. 2013). Rather, PRC2 has high nonspecific, but size-dependent, affinities for all RNA including nascent transcripts. This, in turn, implies that we need a different way of thinking about the role of such lncRNAs and suggests that the local concentration of RNA molecules is important in controlling the availability of complexes such as PRC2.

A particularly surprising role of RNA molecules and an application of the concept of small RNAs as guides for sequence-specific targeting of chromatin complexes is the clustered regularly interspaced short palindromic repeats (CRISPR)-mediated mechanism recently discovered in bacteria (see the review by Brouns 2012) and now increasingly applied to eukaryotic genomic engineering. The CRISPR system stores snippets of DNA sequence from previous invading DNA in a context that can produce corresponding short CRISPR RNAs (scRNAs). These act as guides for CRISPR-associated (Cas) proteins to target and cleave homologous DNA. Although widespread in bacteria, where they are used primarily for host defense, CRISPR-like mechanisms have not yet been discovered in mammals. However, the ease and versatility of the CRISPR targeting mechanism has attracted many researchers seeking ways to direct sequence-specific activities of all sorts to a desired genomic target. Modifications or chimeric fusions of a Cas protein can be used to direct, not only nucleolytic cleavage, but transcriptional activation, repression, chromatin modifications, fluorescent tags, or any other desired activity, including just plain binding, to a specific genomic sequence using an appropriate short scRNA. Although the major application of the CRISPR/Cas technique so far has been to edit (insert or delete) sequences into the genome, its versatility is just beginning to be exploited. It would not be surprising if this powerful technique will find varieties of additional uses in combination with the rich armamentarium of micro RNA- and RNA interference (RNAi)-related mechanisms. Potentially, even more powerful, given the ability to store a true memory of earlier DNA-based events, is the mechanism by which new CRISPR spacer sequences are acquired from foreign DNA. Little is known about the molecular details of this process, but could it be harnessed to "teach" a genome?

1.5 Getting Around in the Nucleus

Another important development in the past 10 years has been the growing awareness that larger-scale chromatin domains and nuclear localization play crucial roles in the

Cite as *Cold Spring Harb Perspect Biol* doi: 10.1101/cshperspect.a019547

regulation of the genome. The nucleus is not just an amorphous bag of chromatin: There are chromosomes, centromeres, and telomeres; ribosomal RNA genes assemble a nucleolus; and certain chromatin regions are more likely to be in contact with the nuclear envelope.

Through numerous genome-wide and locus-specific studies, mostly using 3C approaches (explained in Gibcus and Dekker 2013; Ch. 19 [Dekker and Misteli 2014]), we have seen now that different parts of the genome tend to associate with one another in long-range interactions on different scales of distance along the chromatin fiber. There are local interactions that form loops allowing enhancers to contact one another and their cognate promoters on a scale of tens of kilobases. The function of these interactions seems evident: to bring different regulatory regions controlling one or a few genes into contact with the corresponding promoters. Larger chromatin domains, of the order of a few megabases, contain many genes among which there is much internal looping. The function of this kind of interaction is much less clear in most cases other than in special regions such as the mammalian X-chromosome inactivation center or immunoglobulin genes. On a larger scale, chromatin from a given chromosome tends to occupy a distinct chromosomal territory, but individual regions clearly loop out of the territory into a common space and often associate with other regions with similar properties on the same or different chromosomes. These are structures such as transcription factories, with which active genes are thought to associate for a few rounds of transcription at a time, although their existence is still debated (Zhang et al. 2013; Zhao et al. 2014), or Polycomb bodies, in which chromatin domains repressed by Polycomb mechanisms often associate. Chromatin domains such as promyelocytic leukemia bodies, regions associated with the nuclear lamina, and certain heterochromatin regions may be other examples.

Surprisingly, two types of factors have turned out to be involved in many, perhaps all, such interactions: insulator proteins and cohesin complexes. Although the designation insulator protein is now very likely irrevocably affixed, these proteins have, in fact, turned out to be organizers of chromatin architecture, which, in some cases, prevents an enhancer or silencer from acting on neighboring regions, but more frequently functions to juxtapose enhancers and promoters. The role of cohesin complexes may be related to that of insulator proteins in some cases, but is much less well understood although surprising connections continue to be reported. How transcription factories or Polycomb bodies assemble is still very unclear, but it is likely that insulator proteins and cohesin play a role there as well. It is surprising, in this respect, that mammalian genomes contain only one reported insulator protein, CTCF, whereas *Drosophila* contains several. And, although a close link is apparent between mammalian CTCF and cohesin, such a link, if it exists, is much less direct in *Drosophila*. Mechanistically, we will need much more work to understand how insulator proteins function and what controls their activities. In particular, it remains unclear how a given insulator-binding site distinguishes which other sites to interact with and which not, particularly if, as appears to be the case in mammals, there really is just one DNA-binding insulator protein: CTCF. Do insulator protein modifications such as small ubiquitin-related modifier (SUMO)ylation (MacPherson et al. 2009), or perhaps interactions with ncRNA as suggested by the reported involvement of Argonaute proteins (Moshkovich et al. 2011), "customize" CTCF according to specific functions?

2 BUILDING A GLOBAL CHROMATIN ACCESSIBILITY HYPOTHESIS

2.1 The Accessibility Problem: The Difficulty of Reading Genomic DNA

Ultimately, the distinction of one bit of chromatin from another depends on the local DNA sequence. Wrapping the DNA into nucleosomes makes the DNA sequence difficult to access, and if it cannot be accessed, one nucleosome cannot be distinguished from another. Histone modifications help to restore a functional differentiation. The presence of nucleosomes and nature of the histone modifications they carry produce chromatin states that determine whether, when, or how the sequence is read by DNA-binding proteins. Unless DNA-binding proteins have access to the DNA sequence, the apparatus that binds to chromatin and acts on it would not know where to go and, most importantly, the genes and other sequence information could not be read.

In the second part of this essay, I want to consider the question of the accessibility of genomic DNA, apart from cooperative binding or the remodeling that follows the binding of pioneer factors (elaborated in Sec. 2.3); this is a problem that underlies many of the questions concerning histone modifications—their multiplicities and dynamics. I propose an argument that begins to integrate the mass of information about histone modifications from the vantage point of DNA accessibility.

Textbook accounts explain the wrapping of the DNA into nucleosomes as a means of packaging eukaryotic genomes into the nucleus. The wrapping of 147 bp of DNA into a nucleosome clearly reduces the volume occupied, although more by reducing the degrees of freedom than by actually decreasing the space occupied by the double helix. DNA can be compacted much more than it generally is in a eukaryotic nucleus, as shown by the packing of a bacteriophage genome into a capsid or of the human genome in a sperm nucleus. The difference is that the bacteriophage DNA is not available except by releasing the entire content of the capsid, and neither is much, or all, of the sperm genome. The eukaryotic genome in a somatic cell, instead, is available or specifically addressable in part (i.e., euchromatin), and the heterochromatic complement could be said to be accessible at least part of the time. What makes the difference, of course, is that the eukaryotic genome is partitioned into myriads of small packets (i.e., nucleosomes). These can be further packaged hierarchically into higher-order structures that, potentially, can be folded or unfolded locally. In other words, an arbitrary fraction of the genome down to single nucleosomes can, in principle, be opened up individually, allowing access to its DNA content. To do this, however, a nucleosome has to be differentiated from its neighbors or its DNA content has to be searchable at least intermittently.

2.2 Chromatin as a Response to DNA Concentration in the Nucleus

Although space is limiting in the nucleus and compaction of at least part of the genomic DNA, an imperative, far more important tension exists: a tension between the need for controlled genomic accessibility to transcription and the need to reduce the accessibility to most of the genomic DNA. Packaging of a large genome in the nucleus is a necessity, but perhaps far more important is the imperative to hide most of the DNA so that it is not easily accessible to the protein machinery that needs to act on it. All mechanisms that distinguish one genomic site from another, except those based on bulk physical or topological properties, must be able to recognize specific nucleotide sequence motifs, short enough to be readable and bindable by a single protein with sufficient discrimination to minimize the noise resulting from inappropriate binding. The ratio between signal and noise (i.e., specific versus nonspecific binding) is critical for carrying out any kind of gene regulation. This discrimination is lost if the concentration of nonspecific DNA is so great that any DNA-binding protein would spend its time bound inappropriately to the wrong sequences. The key to specificity in the nucleus is therefore intrinsically tied to the need to mask most of the genomic DNA so as to make it unavailable for binding to the regulatory and transcriptional machinery. However, the differentiation of chromatin and specific activities of the genetic information depend ultimately on the local DNA sequence. Packaging into nucleosomes not only prevents access to the DNA sequence, but, by itself, it would reduce the genome to an undifferentiated collection of more or less structurally identical nucleosomes. Viewed in this light, it is clear then that a large part of eukaryotic gene regulation must consist of ways to (1) remove or remodel nucleosomes so as to make the underlying DNA available for DNA-binding proteins; (2) do this in a sequence-specific way or at least produce sequence-specific results; and (3) develop a way to mark nucleosomes or nucleosomal domains so as to restore some specificity for the action of regulatory proteins (e.g., acetylating histone tails to make certain nucleosomes easier to displace or remodel).

2.3 Nucleosome Density

It is well known that the density of nucleosomes is important for retaining regulatory specificity. If insufficient histones are produced, nucleosomal density is reduced. In *Drosophila*, this causes loss of heterochromatic silencing and suppression of position-effect variegation (Moore et al. 1983). In yeast and mammalian cells, histone insufficiency causes derepression of many conditionally expressed genes (Han and Grunstein 1988; Lenfant et al. 1996; Wyrick et al. 1999; Celona et al. 2011; Gossett and Lieb 2012). Interestingly, a reduced nucleosome density changes the occupancy (the frequency with which a position is occupied) rather than the distribution of nucleosomes. This is because certain DNA sequences favor the formation of nucleosomes by more easily wrapping around the histone core than others. In addition, the more DNA becomes accessible, the more DNA-binding proteins can bind to their preferred sequences and compete with nucleosome formation.

As shown long ago, initially by *Drosophila* histone gene deletion experiments, nucleosome density is a necessary precondition for heterochromatic silencing. If the nucleosome density is too low, the DNA becomes too accessible to DNA-binding proteins, in particular, RNA polymerase, and increasing evidence shows that indiscriminate access results in indiscriminate transcription. Transcriptional activity is associated with many other nucleosome-modifying activities, particularly histone acetylation, which prevents the establishment of the heterochromatic state and promotes further accessibility. A similar effect is obtained in the presence of a normal histone gene complement if the concentration of a gene activator is increased (Ahmad and Henikoff 2001). These effects show that without nucleosomes the ability to repress transcription is lost and the ability of nucleosomes to prevent access is in competition with the concentration-dependent binding of the transcriptional machinery. What is lost is not only the ability to repress, but also the control of transcriptional activation in that the requirement for activators to produce transcription is at least partly absolved if RNA polymerase no longer needs the help of various remodeling activities to access the DNA sequence. Furthermore, although the control of ac-

Cite as *Cold Spring Harb Perspect Biol* doi: 10.1101/cshperspect.a019547

cess is a major component of the repressive activities of heterochromatin, in normal cells there are windows of opportunity to access even the heterochromatic DNA sequence, and a sufficiently high concentration of a DNA-binding activator can exploit these to bind to its sequence and produce local derepression. In many cases, then, a major limiting factor in the control of transcription is access of RNA polymerase to DNA. Many genes, particularly in higher eukaryotes, have developed ways to ensure that RNA polymerase is preloaded, often transcriptionally initiated, but arrested (paused polymerase) and ready to respond to transcriptional signals that allow it to elongate. In most cases, this requires the access of DNA-binding proteins that configure the nucleosomes around the promoter site. Here too, however, the pausing is dependent on the need of additional factors to overcome nucleosomal obstacles to elongation.

When the DNA is fully occupied by nucleosomes, or at least when histone levels do not limit the nucleosome density, most of the DNA sequence is not directly accessible to DNA-binding proteins. It has been shown that some transcription factors are better able than others to bind to nucleosomal DNA sequences, at least when the binding sites are close to one edge of the nucleosome (Zaret and Carroll 2011). Such "pioneer" factors can gain a toehold by binding to the DNA entering the nucleosome even in a compacted chromatin structure, evict linker histone H1, and invoke nucleosome remodeling machines to unravel the DNA and expose it for the binding of other enhancer binding factors in a multistage process (Li et al. 2010).

2.4 Roaming Activities

Special features may allow certain sequence-specific binding proteins to find their binding sites, possibly by profiting opportunistically from transient opening of the chromatin structure. In general, however, access to DNA requires the help of remodeling machines. Thus, to allow access to nucleosomal DNA without prior sequence information, we would need to hypothesize roaming activities that survey the genomic chromatin and periodically turn it over, so to speak, temporarily opening access to the underlying sequence. At the same time, to prevent such access and ensure regulated opening, we might expect an opposing activity. Are there any known chromatin marks or chromatin activities that might support this hypothesis?

Two features have been identified that are characteristic of sites in which DNA has to be maintained in an accessible state: One is the binding of the CBP histone acetylase or its close relative p300, whether or not accompanied by steady-state enrichment in histone H3K27 acetylation. The other is the chromatin mark H3K4me1, whose role in accessibility is not well understood. These features are characteristically found at enhancer sites, in which CBP is thought to be recruited by most enhancer-binding factors (Heintzman et al. 2007; Xi et al. 2007; Visel et al. 2009). They are also found at promoters and wherever DNA-binding proteins find access to the genomic DNA. These sites have also been found to be hot spots of active nucleosome turnover, detectable by the deposition of nucleosomes containing the histone variant H3.3, often together with the histone variant H2A.Z. This combination of variants is less stable than normal, and easier to remodel or turn over (Jin et al. 2009). The fact that acetylation is not always detected at enhancer sites, despite the presence of CBP, suggests that the acetylated nucleosomes are those that have been displaced to create the nucleosome-free region, which is occupied by DNA-binding proteins.

CBP is often associated with a histone-remodeling activity (e.g., *Drosophila Brahma*, ortholog of human SNF2L2) and UTX, one of two known histone H3K27 demethylases, but the only one found in *Drosophila* (Tie et al. 2012). UTX is an essential component of the Trithorax-related (TRR) H3K4 methyltransferase (or MLL3 and MLL4 in mammals), which is the source of H3K4me1 and sometimes H3K4me2, found at enhancers, promoters, and other protein-bound DNA-binding sites (Herz et al. 2012). We might wonder what a H3K27 demethylase might be doing at these sites, but there is a strong connection; CBP is responsible for H3K27 acetylation and this activity is blocked by the simple presence of preexisting H3K27 methylation, without need to recruit repressive complexes of any kind.

2.5 Ubiquitous H3K27 Methylation

H3K27 methylation is, in fact, ubiquitous in the genome. It is produced by the Polycomb repressive complex 2 (PRC2), whose methyltransferase subunit in *Drosophila* is E(z) (ortholog of mammalian Ezh1 and Ezh2). PRC2 is responsible for mono-, di-, trimethylated H3K27. The trimethylated state is the one that has received the most attention because it is the one associated with Polycomb-repressed genes. However, its most abundant product is not H3K27me3, which in somatic cells constitutes some 5%–10% of total histone H3, but H3K27me2, which is found in a staggering 50%–60% of all H3 (Peters et al. 2003; Ebert et al. 2004; Jung et al. 2010; Voigt et al. 2012). Dimethylation is therefore the major activity of PRC2 in flies as in man. Kinetic studies (McCabe et al. 2012) show, in fact, that although PRC2 monomethylates and dimethylates rapidly, trimethylation is enzymatically more difficult and probably occurs in vivo primarily where PRC2 is stably bound. H3K27me2 is also the most abundant and broadly distributed type of histone modification, found everywhere except in regions that are enriched in H3K27me3 or undergoing transcriptional activity. The reason for the former is obvious. The reason for the latter (i.e., its depletion in transcribed regions) is most likely twofold: (1) nucleosomes are less densely distributed in such regions because of increased instability and turnover, and (2) transcriptionally active regions are targeted by the UTX H3K27 demethylase, producing H3K27me1 and H3K27me0. Very similar results have been recently reported for mouse embryonic stem cells (Ferrari et al. 2014). In these cells, H3K27me3, me2, and me1 constitute, respectively, 7%, 70%, and 4% of total H3, whereas H3K27ac is 2% and unmodified H3K27 is 16%. H3K27me2 is confined to transcriptionally inactive regions, except those that have stably bound PRC2, whereas H3K27me1 is found only in transcriptionally active regions. Most likely then, H3K27me2 is actively removed by demethylation, with H3K27me1 as an intermediate to complete demethylation. Not surprisingly, transcriptionally active regions are also enriched in UTX, the only known H3K27 demethylase in *Drosophila*. Removal of H3K27 methylation in active regions is required for H3K27 acetylation, which is generally found in the 5′ region of active transcription units and enhancers. But this leaves the twin questions: Why is there H3K27 methylation in the first place? And why is H3K27 acetylation specifically needed?

Unlike H3K27me3, which is found mainly at genomic sites that can stably recruit the PRC2 complex, the activity that produces H3K27me2 must target the whole genome. Although it might be associated with the replication fork, it is most likely explained by the transient interaction of free PRC2 with nucleosomes by a hit-and-run mechanism. It is not a completely random mechanism, however. The methylation activity of PRC2 is modulated by several inputs from the surrounding chromatin. One of these is dependent on a hydrophobic pocket in the PRC2 subunit extra sex combs (ESC)/Eed, which binds methylated H3K27 (Margueron et al. 2009). When this binding occurs, it effects a conformational shift in the catalytic subunit E(z) that greatly stimulates its methylation activity. Although H3K27me3 binds more strongly, H3K27me2 also binds to the aromatic pocket. Therefore, the presence of H3K27me2 or H3K27me3 in surrounding nucleosomes promotes the methylation of newly deposited nucleosomes. Mutations in the ESC hydrophobic pocket drastically reduce the global level of both H3K27me3 and H3K27me2. Other mechanisms that modulate PRC2 activity probably contribute, although they have not been tested in vivo. Thus, the nucleosome density surrounding a target nucleosome appears also to stimulate methylation activity (Yuan et al. 2012), whereas the presence of H3K4me3 or H3K36me2/me3 on the target nucleosome reduces the methylation activity of PRC2 (Schmitges et al. 2011; Yuan et al. 2011). As a consequence, regions that already contain H3K27 methylation are better targets for PRC2, whereas regions that have a lower nucleosome density or nucleosomes that bear H3K4me3 or H3K36me2/me3, all marks of transcriptional activity, are poor targets. The discovery of these several devices in PRC2 and other methyltransferases has made it clear that feedback and feed-forward mechanisms can be incorporated in chromatin-modifying machines to both self-renew a chromatin mark and avoid regions marked with certain other histone modifications. It is worth pointing out here that these mechanisms not only help to maintain Polycomb repression from one cell cycle to the next through the maintenance of H3K27me3, but, by modulating the deposition of H3K27me2, they also provide a memory of transcriptional activity. Regions that have been recently transcribed have lower nucleosome density and are enriched in the H3K4me3 and H3K36me2/me3 marks that favor renewed transcriptional activity and, at the same time, inhibit H3K27 methylation. In other words, in the case of a globally distributed histone mark such as H3K27me2, the absence of the mark is itself a mark that carries information of previous transcriptional activity (Fig. 1).

PRC2, therefore, provides a global mechanism to mark chromatin according to its recent usage. But what does the H3K27me2 mark do and how is it interpreted? We have become accustomed to thinking of histone modifications as marks that are "read" by chromatin proteins that possess appropriate binding domains. This is possible, but unlikely for H3K27me2. The "reader" approach is suitable for marks that distinguish a region from the rest of the chromatin. A global mark such as H3K27me2 would bind the "reader" virtually everywhere. A more economical interpretation is that, rather than being "read," the presence of the mark provides both the reading and response at the same time; H3K27 methylation preempts the lysine so that it cannot be acetylated. H3K27ac is a mark associated with the 5' region of active genes. As we have seen, it is also potentially present at all sites containing CBP—that is, sites such as enhancers that involve the access of DNA-binding proteins to DNA. In principle, monomethylation of H3K27 would serve the same purpose. H3K27me1 has been often considered to be associated, not with repression, but with transcriptional activity. H3K27me1 is found in transcriptionally active genes most likely because these are sites in which H3K27me2 is demethylated by UTX. The monomethylated state is most likely a stage in demethylation or remethylation.

2.6 The Accessibility Hypothesis

A hypothesis that would integrate these various findings is that access to the DNA content of nucleosomes is a major limiting factor in controlling all sequence-specific activity

Cite as *Cold Spring Harb Perspect Biol* doi: 10.1101/cshperspect.a019547

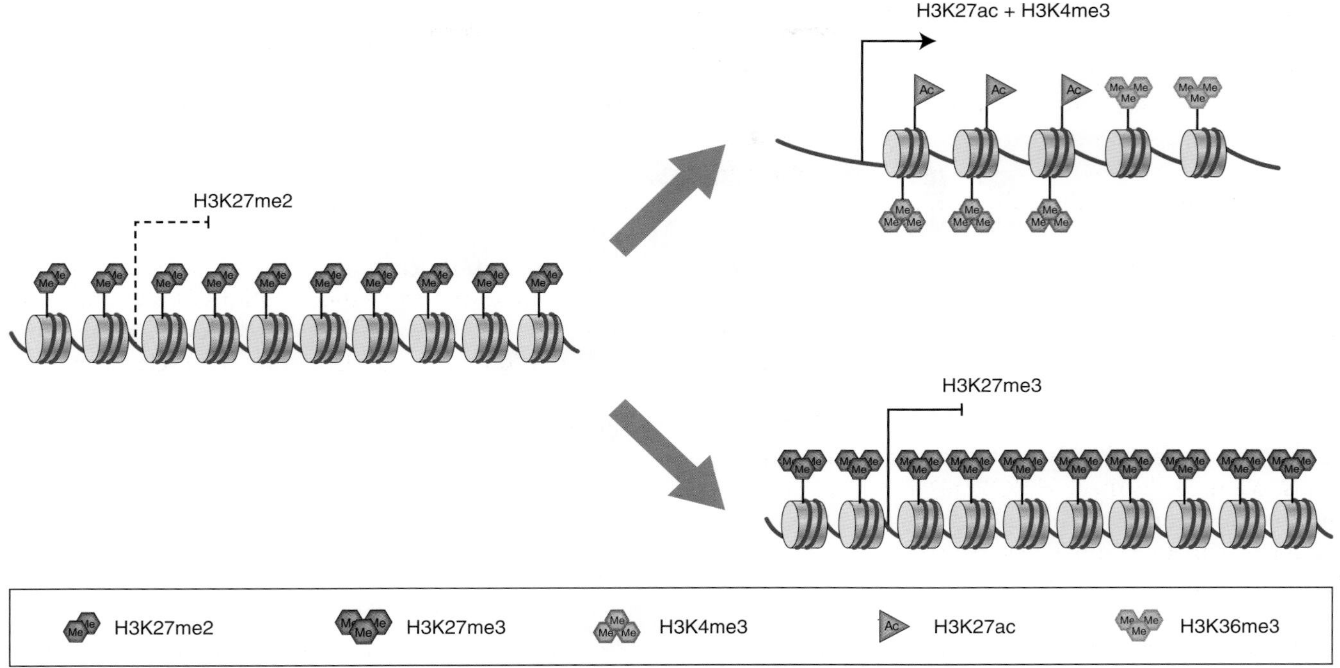

Figure 1. Transcriptional memory of the chromatin state. The schematic drawing illustrates some key changes in the chromatin marks associated with a chromatin region that has recently been transcribed or becomes stably repressed by Polycomb mechanisms. A region that has not been recently transcribed is marked by heavy H3K27me2. A region recently transcribed has lost H3K27me2, but instead gained H3K27ac and H3K4me3 marks in the promoter-proximal part and H3K36me3 (which, in turn, recruits deacetylating complexes) to control the excessive access allowed by the loss of H3K27me2. Regions that can recruit stable binding of Polycomb complexes PRC1 and PRC2 acquire H3K27me3. For simplicity, other histone marks are not shown.

in the genome. Control of the access to DNA is therefore a key regulatory principle. According to the hypothesis (Fig. 2), access is provided intermittently by a roaming nucleosome remodeling activity that "turns over" nucleosomes, temporarily making their DNA accessible. For some reason that is not clear at this stage, this activity involves H3K27 acetylation, perhaps transiently, and is blocked by preventing H3K27 acetylation. In general, this acetylation is constantly removed by roaming histone deacetylases, as has been shown in yeast (Vogelauer et al. 2000). The remodeling that provides accessibility is counteracted by a roaming PRC2 activity that dimethylates H3K27 genome-wide; this PRC2-mediated H3K27 dimethylation thereby preempts this position and blocks its acetylation. The H3K27 dimethylation can be removed by the UTX demethylase, whose main activity is therefore to remove the block to acetylation and allow more stable access to the DNA. This is needed at sites such as enhancers, promoters, PREs, and others in which multiple DNA-binding proteins need to see the nucleotide sequence. The binding of these factors recruits stable CBP and is associated with remodeling activity whose longer-term presence displaces nucleosomes, producing a nucleosome-depleted region. Such regions are typically hypersensitive to DNaseI treatment and found associated with enhancers, promoters, PREs, sites whose edges are also enriched for histone H3.3 relative to their surroundings. This, as analyzed by Mito et al. (2007), is caused by nucleosome replacement, which means that nucleosomes found there are not formed as part of the replicative process, but are due to continuous turnover.

2.7 Transcription of Nucleosome-Free Regions and the RNAi Response

The accessibility hypothesis requires that sites of nucleosomal depletion or remodeling, or any region that is not densely populated by nucleosomes, will have a high probability of binding RNA polymerase on an opportunistic basis and producing some transcriptional products. The amount and length of such transcripts is likely to be very variable and dependent on the sequence, the vicinity to some enhancer-like activity, and probably many other factors. There should be little strand specificity for these transcriptional starts, which, therefore, are likely to result in production of RNA from both strands and thus would be targeted by RNAi mechanisms.

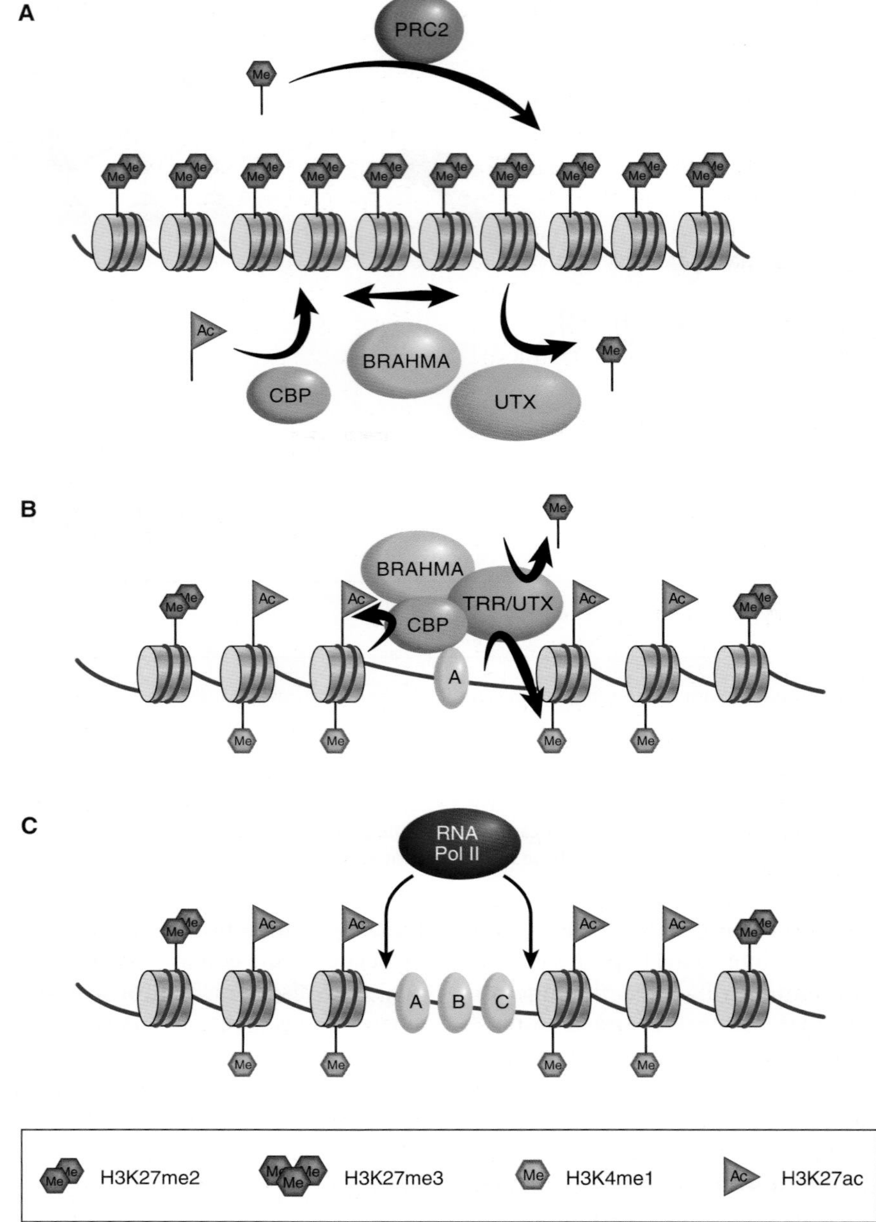

Figure 2. A model for the control of DNA accessibility in chromatin. (*A*) The model proposes that antagonistic roaming activities transiently interact with genomic chromatin: one, caused by PRC2, deposits the H3K27me2 mark. Another removes this methylation mark and remodels nucleosomes, allowing transient access to the DNA sequence. These activities are attributed to UTX, CBP, and BRAHMA. (*B*) A DNA-binding factor A binds to its cognate binding motif in the DNA, transiently made accessible, and recruits stable binding of CBP together with a remodeling activity (BRAHMA) and the TRR/MLL3,4 complex containing UTX. These activities remove H3K27 methylation, depositing instead the H3K27ac and H3K4me1 marks. (*C*) The remodeling activity provides stable access to the DNA, leading to the binding of additional factors B and C to an enhancer region (or other regulatory element on the DNA). The region of DNA made accessible can also be opportunistically targeted by RNA polymerase, which may produce short transcripts from both DNA strands.

Cite as *Cold Spring Harb Perspect Biol* doi: 10.1101/cshperspect.a019547

Promoters, which have a short nucleosome-depleted region, are well known to produce short transcripts from both strands within a region of a few hundred nucleotides surrounding the transcription start called TSSa-RNAs (Seila et al. 2008; Affymetrix/Cold Spring Harbor Laboratory ENCODE Transcriptome Project 2009; Taft et al. 2009). Promoter regions have evolved ways to minimize the production of RNA from both strands by selecting a high frequency of polyadenylation signals such that productive elongation occurs predominantly in the direction downstream from the gene TSS (Almada et al. 2013; Ntini et al. 2013). Enhancers are also the source of transcripts, the so-called enhancer RNAs or eRNAs, from both strands (De Santa et al. 2010; Kim et al. 2010; Ørom and Shiekhattar 2013). DNA damage sites, where nucleosomes are removed for a considerable length surrounding a double-strand break, also produce transcripts from both strands. These RNAs are now known to be processed by Dicer and Drosha and required for the binding of ATM, a kinase that phosphorylates the histone variant H2AX to initiate the formation of DNA damage repair foci (Francia et al. 2012). RNAs produced from all such nucleosome-depleted regions are by-products of the nucleosome remodeling processes occurring at these sites. They need not have particular function, but it should not be surprising to find that they have acquired a function at certain sites.

Regions that are partially depleted of nucleosomes or become too easily accessible to RNA polymerase are prone to initiate transcription, which is not strand-specific. In the general case, therefore, most such accessible regions would produce RNA transcripts from both strands. One possible consequence of bidirectional transcription is the recruitment of the RNAi machinery. The bidirectional transcripts produced from DNA damage sites clearly recruit components of the RNAi machinery (Francia et al. 2012). RNAi proteins such as Dicer2 and AGO2 are associated with active promoters that produce small bidirectional RNAs (Cernilogar et al. 2011). It has been claimed that the RNAi protein, AGO2, associates with a variety of sites that are expected to be depleted of nucleosomes, including CTCF-binding sites, promoters, and PREs (Moshkovich et al. 2011). It is not clear what the function of AGO2 might be in these cases, but its loss leads to a decrease in insulator activity or Polycomb repression (Grimaud et al. 2006; Lei and Corces 2006).

RNAi mechanisms are often thought to be protectors of genome integrity against attacks by viruses or proliferating transposons. In the nucleus, they result in the recruitment of histone H3K9 methylation, the binding of heterochromatin proteins such as HP1 and histone deacetylases, and the stabilization of nucleosomes, in essence, the opposite of the process that opened up the chromatin and produced the bidirectional transcripts at enhancers, promoters, etc. The connection between DNA accessibility and the RNAi response is, I suggest, not accidental. Regions that are partially depleted of nucleosomes or become too easily accessible to RNA polymerase are prone to initiate transcription, which is not strand-specific. In the general case, therefore, most such accessible regions would produce RNA transcripts from both strands. If these RNAs recruit the RNAi response, this response is endemic and inseparable from the fundamental necessity of gaining access to the genomic DNA. It could be argued, therefore, that the RNAi response might be, in its basic form, a way to recruit proteins that stabilize nucleosomes (HP1, linker histone, histone deacetylases), restore nucleosomal occupation, or restore nucleosomal stability to sites that, for whatever reason, might have transiently become open. The fact that the RNAi response has become a valuable protection against invading genetic elements would not be incompatible with the even more basic function of keeping the genomic DNA covered and ensuring that transiently opened regions do not get out of hand.

2.8 PRC2 and Heterochromatin

RNAi mechanisms are thought to be important for the establishment of heterochromatin. This has been worked out in detail for the fission yeast *Schizosaccharomyces pombe*, but many aspects of this relationship apply to *Drosophila* and mammalian heterochromatin formation. The arguments presented above help to understand why E(z) has been found to play a role in the efficient establishment of heterochromatin and, in fact, is known as a suppressor of position-effect variegation in *Drosophila* (Laible et al. 1997). This role has been a puzzle for many years because there is no specific presence of E(z) or H3K27me3 in heterochromatin. This role is better understood in terms of accessibility. In *Drosophila*, the early embryonic stages are a time of extremely rapid and synchronous nuclear divisions. These slow down by the 14th cycle (3 h postfertilization), but the chromatin produced must now be the target of a massive H3K27 methylation effort. This is accomplished thanks to correspondingly massive amounts of PRC2 components that are deposited in the egg during oogenesis. By the time nuclear proliferation slows down and heterochromatin first becomes detectable, global dimethylation of H3K27 must be in place. At this stage, it is important to suppress H3K27 acetylation, remodeling, and adventitious transcriptional activity to allow the RNAi and other mechanisms to initiate and maintain heterochromatin. Access to DNA is never completely prevented even in heterochromatin, as shown by the fact that strong activators can prevent heterochromatic silencing of a reporter gene (Ahmad and Henikoff 2001), but the absence of H3K27me2 would certainly result in a level of access to activators and RNA polymerase that would interfere with the establishment of heterochromatic silencing.

2.9 Effects of Loss of PRC2

If H3K27 methylation plays such a global genomic role, loss of PRC2 function would surely have major consequences—increasing pervasive transcription, among others. Unfortunately, it has not been possible yet to separate the global H3K27 dimethylation function from the Polycomb-related and more specific H3K27 trimethylation. Loss of PRC2 is an early embryonic lethal both in mammals and *Drosophila*, and it produces embryos with classical homeotic derepression phenotypes (Struhl and Brower 1982). Loss of Polycomb repression of *Hox* genes and many other developmentally important genes would certainly be sufficient to account for lethality. In addition, it would make it difficult to determine whether any other effects should be attributed to indirect consequences of derepression or tloss of H3K27 dimethylation. Nevertheless, loss of PRC2 activity is not cell lethal. Mammalian embryonic stem cells with knockouts of Ezh2 or Eed are viable, although unable to differentiate. In mouse embryonic stem cells lacking PRC2 function, H3K27 acetylation appears at new sites together with H3K4me1, forming a signature typical of poised enhancer regions (Ferrari et al. 2014). This suggests that many normally silent regions become accessible and transcriptionally active. Activation of new transcription sites not normally associated with H3K27me3 was also observed. An increase in the accessibility of chromatin to RNA polymerase was also observed in *Drosophila* embryos lacking maternal and zygotic ESC, an essential component of PRC2 (Chopra et al. 2011). Promoters of thousands of genes became occupied by RNA polymerase II, whether or not they were transcriptionally activated.

Mutations affecting EZH2 and PRC2 activity are associated with a variety of aggressive cancers but, strangely, both hyperactivity and loss of activity appear to be oncogenic. The interpretation has generally been that these effects are mediated by the hyperrepression or derepression of Polycomb target genes, and this is undoubtedly true, at least in part. For example, genes that block cell cycle progression, such as INK4A/B, are regulated by Polycomb mechanisms and hyperrepression would remove brakes to cell proliferation. Because much evidence supports a cancer-promoting role of PRC2 activity, the discovery that loss of PRC2 function can also promote cancers, such as myeloid leukemia, has been puzzling (Hock 2012; Simon et al. 2012; Tamagawa et al. 2013).

A particularly interesting case is that of the recently characterized mutations converting K27 to methionine in histone H3 or H3.3 genes. This mutation was found associated with a particularly malignant glioblastoma, in which it has a dominant effect totally out of proportion with the relatively small fraction of total histone H3 that is produced by the mutated histone H3 gene copy (Chan et al. 2013; Lewis et al. 2013). The methionine at position 27 mimics, in part, the K27 methylation, but lacks the moderating positive charge that the amine nitrogen retains even when trimethylated. As a consequence, the EZH2 catalytic domain binds, but does not readily release, the H3K27M peptide. Although this has not been directly shown, one consequence might be that the PRC2 complex becomes effectively sequestered and unavailable, causing H3K27 to become undermethylated genome-wide. Attention has been focused on the partial loss of H3K27me3 and consequent derepression of Polycomb target genes. I suggest a somewhat different interpretation. Genome-wide H3K27 dimethylation would be even more powerfully affected because it is strongly dependent on a hit-and-run mechanism and, therefore, on the pool of free PRC2. The loss of H3K27 methylation would be expected to derepress a large number of genes whose silencing depends on the inability of activators and RNA polymerase II to access the promoter. More important, perhaps, is that transcription may start anywhere, including within gene bodies, producing partial proteins that would have unexpected neomorphic effects.

These observations support the global accessibility hypothesis, at least in part, but are a long way from providing substantial proof. They suggest, nevertheless, that the abundance and ubiquitousness of H3K27 dimethylation are not without a significance that could help to understand the way in which different chromatin modifications jostle with one another, the interplay that has provided the raw material from which evolution has shaped the chromatin landscape and its functions.

REFERENCES

Reference is also in this book.

Affymetrix/Cold Spring Harbor Laboratory ENCODE Transcriptome Project (Fejes-Toth K, Sotirova V, Sachidanandam R, Assaf G, Hannon GJ, Kapranov P, Foissac S, Willingham AT, Duttagupta R, Dumais E, Gingeras TR). 2009. Post-transcriptional processing generates a diversity of 5′-modified long and short RNAs. *Nature* **457:** 1028–1032.

Ahmad K, Henikoff S. 2001. Modulation of a transcription factor counteracts heterochromatic gene silencing in *Drosophila*. *Cell* **104:** 839–847.

Almada AE, Wu X, Kriz AJ, Burge CB, Sharp PA. 2013. Promoter directionality is controlled by U1 snRNP and polyadenylation signals. *Nature* **499:** 360–363.

Al-Sady B, Madhani HD, Narlikar GJ. 2013. Division of labor between the chromodomains of HP1 and Suv39 methylase enables coordination of heterochromatin spread. *Mol Cell* **51:** 80–91.

Brouns SJ. 2012. A Swiss army knife of immunity. *Science* **337:** 808–809.

Cite as *Cold Spring Harb Perspect Biol* doi: 10.1101/cshperspect.a019547

Busturia A, Wightman CD, Sakonju S. 1997. A silencer is required for maintenance of transcriptional repression throughout *Drosophila* development. *Development* **124:** 4343–4350.

Celona B, Weiner A, Di Felice F, Mancuso FM, Cesarini E, Rossi RL, Gregory L, Baban D, Rossetti G, Grianti P, et al. 2011. Substantial histone reduction modulates genomewide nucleosomal occupancy and global transcriptional output. *PLoS Biol* **9:** e1001086.

Cernilogar FM, Onorati MC, Kothe GO, Burroughs AM, Parsi KM, Breiling A, Sardo FL, Saxena A, Miyoshi K, Siomi H, et al. 2011. Chromatin-associated RNA interference components contribute to transcriptional regulation in *Drosophila*. *Nature* **480:** 391–395.

Chan K-M, Fang D, Gan H, Hashizume R, Yu C, Schroeder M, Gupta N, Mueller S, James CD, Jenkins R, et al. 2013. The histone H3.3K27M mutation in pediatric glioma reprograms H3K27 methylation and gene expression. *Genes Dev* **27:** 985–990.

Chopra VS, Hendrix DA, Core LJ, Tsui C, Lis JT, Levine M. 2011. The polycomb group mutant esc leads to augmented levels of paused Pol II in the *Drosophila* embryo. *Mol Cell* **42:** 837–844.

Davidovich C, Zheng L, Goodrich KJ, Cech TR. 2013. Promiscuous RNA binding by Polycomb repressive complex 2. *Nat Struct Mol Biol* **20:** 1250–1257.

* Dekker J, Misteli T. 2014. Long-range chromatin interactions. *Cold Spring Harb Perspect Biol* doi: 10.1101/cshperspect.a017905.

De Santa F, Barozzi I, Mietton F, Ghisletti S, Polletti S, Tusi BK, Muller H, Ragoussis J, Wei C-L, Natoli G. 2010. A large fraction of extragenic RNA Pol II transcription sites overlap enhancers. *PLoS Biol* **8:** e1000384.

Ebert A, Schotta G, Lein S, Kubicek S, Krauss V, Jenuwein T, Reuter G. 2004. Su(var) genes regulate the balance between euchromatin and heterochromatin in *Drosophila*. *Genes Dev* **18:** 2973–2983.

Farcas AM, Blackledge NP, Sudbery I, Long HK, McGouran JF, Rose NR, Lee S, Sims D, Cerase A, Sheahan TW, et al. 2012. KDM2B links the Polycomb repressive complex 1 (PRC1) to recognition of CpG islands. *eLife* **1:** e00205.

Ferrari KJ, Scelfo A, Jammula S, Cuomo A, Barozzi I, Stützer A, Fischle W, Bonaldi T, Pasini D. 2014. Polycomb-dependent H3K27me1 and H3K27me2 regulate active transcription and enhancer fidelity. *Mol Cell* **53:** 49–62.

Fradet-Turcotte A, Canny MD, Escribano-Diaz C, Orthwein A, Leung CCY, Huang H, Landry M-C, Kitevski-LeBlanc J, Noordermeer SM, Sicheri F, et al. 2013. 53BP1 is a reader of the DNA-damage-induced H2A Lys 15 ubiquitin mark. *Nature* **499:** 50–54.

Francia S, Michelini F, Saxena A, Tang D, de Hoon M, Anelli V, Mione M, Carninci P, d'Adda di Fagagna F. 2012. Site-specific DICER and DROSHA RNA products control the DNA-damage response. *Nature* **488:** 231–235.

Gibcus JH, Dekker J. 2013. The hierarchy of the 3D genome. *Mol Cell* **49:** 773–782.

Gossett AJ, Lieb JD. 2012. In vivo effects of histone H3 depletion on nucleosome occupancy and position in *Saccharomyces cerevisiae*. *PLoS Genet* **8:** e1002771.

Grimaud C, Bantignies F, Pal-Bhadra M, Ghana P, Bhadra U, Cavalli G. 2006. RNAi components are required for nuclear clustering of Polycomb group response elements. *Cell* **124:** 957–971.

Han M, Grunstein M. 1988. Nucleosome loss activates yeast downstream promoters in vivo. *Cell* **55:** 1137–1145.

Hathaway NA, Bell O, Hodges C, Miller EL, Neel DS, Crabtree GR. 2012. Dynamics and memory of heterochromatin in living cells. *Cell* **149:** 1447–1460.

Heintzman ND, Stuart RK, Hon G, Fu YT, Ching CW, Hawkins RD, Barrera LO, Van Calcar S, Qu C, Ching KA, et al. 2007. Distinctive and predictive chromain signatures of transcriptional promoters and enhancers in the human genome. *Nat Genet* **39:** 311–318.

Henikoff S, Shilatifard A. 2011. Histone modification: Cause or cog? *Trends Genet* **27:** 389–396.

Herz H-M, Mohan M, Garruss AS, Liang K, Takahashi Y-H, Mickey K, Voets O, Verrijzer CP, Shilatifard A. 2012. Enhancer-associated H3K4 monomethylation by Trithorax-related, the *Drosophila* homolog of mammalian Mll3/Mll4. *Genes Dev* **26:** 2604–2620.

Hock H. 2012. A complex Polycomb issue: The two faces of EZH2 in cancer. *Genes Dev* **26:** 751–755.

Ivanov GS, Ivanova T, Kurash J, Ivanov A, Chuikov S, Gizatullin F, Herrera-Medina EM, Rauscher F III, Reinberg D, Barlev NA. 2007. Methylation-acetylation interplay activates p53 in response to DNA damage. *Mol Cell Biol* **27:** 6756–6769.

Jin C, Zang C, Wei G, Cui K, Peng W, Zhao K, Felsenfeld G. 2009. H3.3/H2A.Z double variant-containing nucleosomes mark 'nucleosome-free regions' of active promoters and other regulatory regions. *Nat Genet* **41:** 941–945.

Jung HR, Pasini D, Helin K, Jensen ON. 2010. Quantitative mass spectrometry of histones H3.2 and H3.3 in Suz12-deficient mouse embryonic stem cells reveals distinct, dynamic post-translational modifications at Lys-27 and Lys-36. *Mol Cell Proteomics* **9:** 838–850.

Kaneko S, Son J, Shen SS, Reinberg D, Bonasio R. 2013. PRC2 binds active promoters and contacts nascent RNAs in embryonic stem cells. *Nat Struct Mol Biol* **20:** 1258–1264.

Kim T-K, Hemberg M, Gray JM, Costa AM, Bear DM, Wu J, Harmin DA, Laptewicz M, Barbara-Haley K, Kuersten S, et al. 2010. Widespread transcription at neuronal activity-regulated enhancers. *Nature* **465:** 182–187.

Klose RJ, Cooper S, Farcas AM, Blackledge NP, Brockdorff N. 2013. Chromatin sampling—An emerging perspective on targeting polycomb repressor proteins. *PLoS Genet* **9:** e1003717.

Kohli RM, Zhang Y. 2013. TETenzymes, TDG and the dynamics of DNA demethylation. *Nature* **502:** 472–479.

Laible G, Wolf A, Dorn R, Reuter G, Nislow C, Lebersorger A, Popkin D, Pillus L, Jenuwein T. 1997. Mammalian homologues of the *Polycomb*-group gene *Enhancer of zeste* mediate gene silencing in *Drosophila* heterochromatin and at *S. cerevisiae* telomeres. *EMBO J* **16:** 3219–3232.

LaSalle JM. 2013. Epigenomic strategies at the interface of genetic and environmental risk factors for autism. *J Hum Genet* **58:** 396–401.

Lee JM, Lee JS, Kim H, Kim K, Park H, Kim J-Y, Lee SH, Kim IS, Kim J, Lee M, et al. 2012. EZH2 generates a methyl degron that is recognized by the DCAF1/DDB1/CUL4 E3 ubiquitin ligase vomplex. *Mol Cell* **48:** 572–586.

Lei EP, Corces VG. 2006. RNA interference machinery influences the nuclear organization of a chromatin insulator. *Nat Genet* **38:** 936–941.

Lenfant F, Mann RK, Thomsen B, Ling X, Grunstein M. 1996. All four core histone N-termini contain sequences required for the repression of basal transcription in yeast. *EMBO J* **15:** 3974–3985.

Lewis PW, Müller MM, Koletsky MS, Cordero F, Lin S, Banaszynski LA, Garcia BA, Muir TW, Becher OJ, Allis CD. 2013. Inhibition of PRC2 activity by a gain-of-function H3 mutation found in pediatric glioblastoma. *Science* **340:** 857–861.

Li B, Gogol M, Carey M, Pattenden SG, Seidel C, Workman JL. 2007. Infrequently transcribed long genes depend on the Set2/Rpd3S pathway for accurate transcription. *Genes Dev* **21:** 1422–1430.

Li G, Margueron R, Hu G, Stokes D, Wang Y-H, Reinberg D. 2010. Highly compacted chromatin formed in vitro reflects the dynamics of transcription activation in vivo. *Mol Cell* **38:** 41–53.

MacPherson MJ, Beatty LG, Zhou W, Du M, Sadowski PD. 2009. The CTCF insulator protein is posttranslationally modified by SUMO. *Mol Cell Biol* **29:** 714–725.

Margueron R, Justin N, Ohno K, Sharpe ML, Son J, Drury WJ III, Voigt P, Martin SR, Taylor WR, De Marco V, et al. 2009. Role of the Polycomb protein EED in the propagation of repressive histone marks. *Nature* **461:** 762–767.

Marx V. 2013. Biology: The big challenges of big data. *Nature* **498:** 255–260.

McCabe MT, Ott HM, Ganji G, Korenchuk S, Thompson C, Van Aller GS, Liu Y, Graves AP, Della Pietra A III, Diaz E, et al. 2012. EZH2 inhibition as a therapeutic strategy for lymphoma with EZH2-activating mutations. *Nature* **492:** 108–112.

McGowan PO, Sasaki A, D'Alessio AC, Dymov S, Labonte B, Szyf M, Turecki G, Meaney MJ. 2009. Epigenetic regulation of the glucocorticoid receptor in human brain associates with childhood abuse. *Nat Neurosci* **12:** 342–348.

Mito Y, Henikoff JG, Henikoff S. 2007. Histone replacement marks the boundaries of *cis*-regulatory domains. *Science* **315:** 1408–1411.

Moore GD, Sinclair DA, Grigliatti TA. 1983. Histone gene multiplicity and position effect variegation in *Drosophila melanogaster. Genetics* **105:** 327–344.

Moshkovich N, Nisha P, Boyle PJ, Thompson BA, Dale RK, Lei EP. 2011. RNAi-independent role for Argonaute2 in CTCF/CP190 chromatin insulator function. *Genes Dev* **25:** 1686–1701.

Ntini E, Järvelin AI, Bornholdt J, Chen Y, Boyd M, Jørgensen M, Andersson R, Hoof I, Schein A, Andersen PR, et al. 2013. Polyadenylation site-induced decay of upstream transcripts enforces promoter directionality. *Nat Struct Mol Biol* **20:** 923–928.

Ørom UA, Shiekhattar R. 2013. Long noncoding RNAs usher in a new era in the biology of enhancers. *Cell* **154:** 1190–1193.

Peters AH, Kubicek S, Mechtler K, O'Sullivan RJ, Derijck AA, Perez-Burgos L, Kohlmaier A, Opravil S, Tachibana M, Shinkai Y, et al. 2003. Partitioning and plasticity of repressive histone methylation states in mammalian chromatin. *Mol Cell* **12:** 1577–1589.

Schmitges FW, Prusty AB, Faty M, Stützer A, Lingaraju GM, Aiwazian J, Sack R, Hess D, Li L, Zhou S, et al. 2011. Histone methylation by PRC2 is inhibited by active chromatin marks. *Mol Cell* **42:** 330–341.

Seila AC, Calabrese JM, Levine SS, Yeo GW, Rahl PB, Flynn RA, Young RA, Sharp PA. 2008. Divergent transcription from active promoters. *Science* **322:** 1849–1851.

Simon C, Chagraoui J, Krosl J, Gendron P, Wilhelm B, Lemieux SB, Boucher GV, Chagnon P, Drouin S, Lambert RL, et al. 2012. A key role for EZH2 and associated genes in mouse and human adult T-cell acute leukemia. *Genes Dev* **26:** 651–656.

Struhl G, Brower D. 1982. Early role of the esc+ gene product in the determination of segments in *Drosophila. Cell* **31:** 285–292.

Su I-H, Dobenecker M-W, Dickinson E, Oser M, Basavaraj A, Marqueron R, Viale A, Reinberg D, Wülfing C, Tarakhovsky A. 2005. Polycomb group protein Ezh2 controls actin polymerization and cell signaling. *Cell* **121:** 425–436.

Taft RJ, Glazov EA, Cloonan N, Simons C, Stephen S, Faulkner GJ, Lassmann T, Forrest ARR, Grimmond SM, Schroder K, et al. 2009. Tiny RNAs associated with transcription start sites in animals. *Nat Genet* **41:** 572–578.

Tamagawa H, Oshima T, Numata M, Yamamoto N, Shiozawa M, Morinaga S, Nakamura Y, Yoshihara M, Sakuma Y, Kameda Y, et al. 2013. Global histone modification of H3K27 correlates with the outcomes in patients with metachronous liver metastasis of colorectal cancer. *Eur J Surg Oncol* **39:** 655–661.

Tie F, Banerjee R, Conrad PA, Scacheri PC, Harte PJ. 2012. Histone demethylase UTX and chromatin remodeler BRM bind directly to CBP and modulate acetylation of histone H3 lysine 27. *Mol Cell Biol* **32:** 2323–2334.

Visel A, Blow MJ, Li Z, Zhang T, Akiyama JA, Holt A, Plajzer-Frick I, Shoukry M, Wright C, Chen F, et al. 2009. ChIP-seq accurately predicts tissue-specific activity of enhancers. *Nature* **457:** 854–858.

Vogelauer M, Wu J, Suka N, Grunstein M. 2000. Global histone acetylation and deacetylation in yeast. *Nature* **408:** 495–498.

Voigt P, LeRoy G, Drury WJ III, Zee BM, Son J, Beck DB, Young NL, Garcia BA, Reinberg D. 2012. Asymmetrically modified nucleosomes. *Cell* **151:** 181–193.

von Hippel P, Revzin A, Gross C, Wang A. 1974. Nonspecific DNA binding of genome regulating proteins as a biological control mechanism: 1. The lac operon: Equilibrium aspects. *Proc Natl Acad Sci* **71:** 4808–4812.

Weaver IC, Cervoni N, Champagne FA, D'Alessio AC, Sharma S, Seckl JR, Dymov S, Szyf M, Meaney MJ. 2004. Epigenetic programming by maternal behavior. *Nat Neurosci* **7:** 847–854.

Weaver IC, Champagne FA, Brown SE, Dymov S, Sharma S, Meaney MJ, Szyf M. 2005. Reversal of maternal programming of stress responses in adult offspring through methyl supplementation: Altering epigenetic marking later in life. *J Neurosci* **25:** 11045–11054.

Wyrick JJ, Holstege FC, Jennings EG, Causton HC, Shore D, Grunstein M, Lander ES, Young RA. 1999. Chromosomal landscape of nucleosome-dependent gene expression and silencing in yeast. *Nature* **402:** 418–421.

Xi S, Zhu H, Xu H, Schmidtmann A, Geiman TM, Muegge K. 2007. Lsh controls Hox gene silencing during development. *Proc Natl Acad Sci* **104:** 14366–14371.

Yuan W, Xu M, Huang C, Liu N, Chen S, Zhu B. 2011. H3K36 methylation antagonizes PRC2-mediated H3K27 methylation. *J Biol Chem* **286:** 7983–7989.

Yuan W, Wu T, Fu H, Dai C, Wu H, Liu N, Li X, Xu M, Zhang Z, Niu T, et al. 2012. Dense chromatin activates Polycomb repressive complex 2 to regulate H3 lysine 27 methylation. *Science* **337:** 971–975.

Zaret KS, Carroll JS. 2011. Pioneer transcription factors: Establishing competence for gene expression. *Genes Dev* **25:** 2227–2241.

Zhang Q, Wang Y. 2010. HMG modifications and nuclear function. *Biochim Biophys Acta* **1799:** 28–36.

Zhang Y, Wong C-H, Birnbaum RY, Li G, Favaro R, Ngan CY, Lim J, Tai E, Poh HM, Wong E, et al. 2013. Chromatin connectivity maps reveal dynamic promoter-enhancer long-range associations. *Nature* **504:** 306–310.

Zhao ZW, Roy R, Gebhardt JC, Suter DM, Chapman AR, Xie XS. 2014. Spatial organization of RNA polymerase II inside a mammalian cell nucleus revealed by reflected light-sheet superresolution microscopy. *Proc Natl Acad Sci* **111:** 681–686.

 Cite as *Cold Spring Harb Perspect Biol* doi: 10.1101/cshperspect.a019547

APPENDIX 1

WWW Resource Pages

WWW Resource pages

Resource	Description	URL
Teaching material, news, or videos on epigenetics		
An Introduction to Epigenetics video	Cell Signaling Technology's introductory teaching animation on chromatin and histone modifications	http://www.cellsignal.com/common/content/content.jsp?id=resources-tutorials-epigenetics
Epigenie	An epigenetics-focused "news" website	http://epigenie.com
Learn.Genetics pages on Epigenetics	Epigenetics teaching module at the Genetic Science Learning Center	http://learn.genetics.utah.edu/content/epigenetics/
RNAi video by Nature video	Teaching video on RNAi	http://www.youtube.com/watch?v=cK-OGB1_ELE&feature=relmfu
Scitable pages on Epigenetics	Introductory material by Nature Education	http://www.nature.com/scitable/spotlight/epigenetics-26097411
SciShow video on Epigenetics	Explaining epigenetics by the SciShow	http://www.youtube.com/watch?v=kp1bZEUgqVI
Chromatin and epigenetics-related websites		
NIH Roadmap Epigenomics Project	An epigenomic mapping consortium	http://www.roadmapepigenomics.org
ENCODE	Encyclopedia of DNA Elements: identifying functional elements in human	http://www.genome.gov/12513456
ChromDB	The plant chromatin database	http://www.chromdb.org/
modENCODE	A comprehensive encyclopedia of genomic functional elements in *C. elegans* and *D. melanogaster*	http://www.modencode.org
Blueprint	An EU-funded FP7 project: "blueprint of haematopoietic epigenomes"	http://www.blueprint-epigenome.eu
EpiGeneSys	An EU-funded research initiative on epigenetics advancing toward systems biology	http://www.epigenesys.eu/en/homepage
International Human Epigenome Consortium	An international research initiative aiming to map 1000 epigenomes	http://ihec-epigenomes.org
Histone resources		
Antibody Validation Database	Providing information on antibody performance in the public research domain	http://compbio.med.harvard.edu/antibodies
The NHGRI Histone Database	Contains histone sequence information	http://research.nhgri.nih.gov/histones/
Abcam histone page	Contains downloads of histone modification maps	http://www.abcam.com/chromatin
Millipore	Histone modification app generated by Millipore	http://www.millipore.com/antibodies/flx4/histone_mobile_app
Imprinting		
Geneimprint	A website cataloging recent publications and listing genomically imprinted genes	http://www.geneimprint.com/site/home

Continued

WWW Resource pages *Continued*

Resource	Description	URL
RNA		
LNCipedia	A comprehensive compendium of long noncoding RNAs	http://www.lncipedia.org
miRBase	microRNA data resource	http://microrna.sanger.ac.uk/
Modomics	A database of RNA modification pathways	http://modomics.genesilico.pl
NONCODE	A database of noncoding RNAs, except tRNA and rRNAs, and related websites	http://www.noncode.org/index7.htm
ncRNA Databases Resource	Data resource accessing more than 100 ncRNA databases	http://www.ime.usp.br/~durham/ncrnadatabases/index.php
Commercial epigenetic resource sites		
Abcam	Suppliers of histone antibodies	http://www.abcam.com/
Active Motif	Suppliers of histone antibodies and chromatin-related molecular biology reagents	http://www.activemotif.com
Upstate	Suppliers of histone antibodies	http://www.millipore.com/antibodies/flx4/epigenetics
Epigenomics	DNA methylation diagnostic screening	http://www.epigenomics.com/
Organismal genome resources and databases		
All eukaryotes		
NCBI	Portal to multiple genome analysis and reference sites	http://www.ncbi.nlm.nih.gov
Ensembl	Eukaryotic genome browser	http://www.ensembl.org
UCSC Genome Bioinformatics	Genome sequence and resource portal	http://genome.ucsc.edu
EBI	Portal to various genomic and proteomic computational analysis resources	http://www.ebi.ac.uk/services
Sanger Institute	Portal to sequence, bioinformatics, and proteomic resources	http://www.sanger.ac.uk/
zPicture	Comparative sequence analysis tool	http://zpicture.dcode.org
iHOP	Information hyperlinked over proteins—protein resource	http://www.ihop-net.org/UniPub/iHOP/
RepeatMasker	Repeat sequence algorithm to identify repetitive DNA sequences	http://repeatmasker.org
Individual organisms		
Saccharomyces cerevisiae	SGD (*Saccharomyces* Genome Database)	http://www.yeastgenome.org
	Ensembl *S. cerevisiae* genome analysis portal	http://www.ensembl.org/Saccharomyces_cerevisiae/index.html
Schizosaccharomyces pombe	PomBase (portal to *S. pombe*–related genomic sequence and analysis sites)	http://www.sanger.ac.uk/Projects/S_pombe/
Neospora crassa	*Neurospora crassa* Database	http://www.broadinstitute.org/annotation/genome/neurospora/MultiHome.html
Tetrahymena thermophila	TGD (*Tetrahymena* Genome Database)	http://www.ciliate.org/
	Tetrahymena Macronuclear Genome Sequencing Project	http://www.genome.gov/12512294
Plants	TAIR (The *Arabidopsis* Information Resource)	http://www.arabidopsis.org
	Gramene: A Comparative Resource for Plants	http://www.gramene.org/
Caenorhabditis elegans	Portal to *C. elegans*–related resources	http://www.wormbase.org/
	Ensembl *C. elegans* genome analysis portal	http://www.ensembl.org/Caenorhabditis_elegans/index.html
	Portal to *C. elegans*–related genomic sequence and analysis sites	http://www.sanger.ac.uk/Projects/C_elegans/
Drosophila	Database of the *Drosophila* genome	http://www.flybase.net
	Ensembl *D. melanogaster* genome analysis portal	http://www.ensembl.org/Drosophila_melanogaster/index.html

Continued

WWW Resource pages *Continued*

Resource	Description	URL
Xenopus laevis	Information database	http://www.xenbase.org/
Mouse	Mouse Genome Informatics	http://www.informatics.jax.org
	Mouse strain resource	http://jaxmice.jax.org/index.html
	Mouse Ensembl resource	http://www.ensembl.org/Mus_musculus/index.html
Human	Human Genome Resources portal	http://www.ncbi.nlm.nih.gov/genome/guide/human/
	Ensembl human genome analysis portal	http://www.ensembl.org/Homo_sapiens/index.html
	The Human Gene Mutation Database	http://www.hgmd.org/
	1000 Genomes—a Deep Catalog of Human Genetic Variation	http://www.1000genomes.org/home
	International HapMap Project	http://hapmap.ncbi.nlm.nih.gov
	The Cancer Genome Atlas (TCGA)—understanding genomics to improve cancer care	http://cancergenome.nih.gov

NIH, The National Institutes of Health; EU, European Union; NHGRI, National Human Genome Research Institute; tRNA, transfer RNA; rRNA, ribosomal RNA; ncRNA, noncoding RNA; NCBI, National Center for Biotechnology Information; UCSC, University of California, Santa Cruz; EBI, European Bioinformatics Institute; iHOP, Information hyperlinked over proteins.

APPENDIX 2

Comprehensive Catalog of Currently Documented Histone Modifications

Yingming Zhao[1] and Benjamin A. Garcia[2]

[1]Ben May Department for Cancer Research, The University of Chicago, Chicago, Illinois 60637; [2]Epigenetics Program, Department of Biochemistry and Biophysics, Perelman School of Medicine, University of Pennsylvania, Philadelphia, Pennsylvania 19104

Correspondence: yingming.zhao@uchicago.edu; bgarci@mail.med.upenn.edu

SUMMARY

Modern techniques in molecular biology, genomics, and mass spectrometry–based proteomics have identified a large number of novel histone posttranslational modifications (PTMs), many of whose functions are still under intense investigation. Here, we catalog histone PTMs under two classes: first, those whose functions have been fairly well studied and, second, those PTMs that have been more recently identified but whose functions remain unclear. We hope that this will be a useful resource for researchers from all biological or technical backgrounds, aiding in their chromatin and epigenetic pursuits.

Outline

Cite as *Cold Spring Harb Perspect Biol* doi: 10.1101/cshperspect.a025064

Histone posttranslational modifications (PTMs) have been linked to a variety of processes, including transcription, DNA replication, and DNA damage (Kouzarides 2007; Murr 2010; for a list of reviews, see Suggested Reviews). The growing list of histone PTMs has exploded in the last several years as a consequence of the considerable advances in available antibody reagents, peptide and protein array technology, and mass spectrometry–based proteomics (Karch et al. 2013). These approaches allow for identification and quantification of histone PTMs from global or more local chromatin states, especially when combined with chromatin immunoprecipitation experiments (Han and Garcia 2013). Here, we attempt to catalog the ever-growing number of histone PTMs that have been studied over the last several years. The tables are broken down into two groups: Section 1 (Tables 1–8) lists the histone PTMs that have been studied more and thus some functional aspects are known concerning the marks; Section 2 (Tables 9–16) lists the histone PTMs that have been detected, but in which rather limited function has been determined. The latter category of histone marks, being in their observed infancy, is very intriguing and represents a large opportunity for the chromatin biology and epigenetics communities to decipher their biological consequences and outcomes in the years to come. Histone acetylation and methylation were first discovered about 50 years ago (Allfrey et al. 1964); however, only in the last decade or so has its physiological function been uncovered. We anticipate that this will serve as a useful resource for those already working in the field, but also as motivation for those newer scientists beginning their research work in this area to continue to push the boundaries of knowledge.

Model organism key

An	*Aspergillus nidulans*
At	*Arabidopsis thaliana*
Bt	*Bos torus*
Ce	*Caenorhabditis elegans*
Dm	*Drosophila melanogaster*
Hs	*Homo sapiens*
Mm	*Mus musculus*
Nc	*Neurospora crassa*
Rn	*Rattus norvegicus*
Sc	*Saccharomyces cerevisiae*
Sp	*Schizosaccharomyces pombe*
Tt	*Tetrahymena thermophila*
Xl	*Xenopus laevis*

Modification key

ac	acetylation	mal	malonylation
ar1	mono-ADP-ribosylation	me1	monomethylation
bio	biotinylation	me2	dimethylation
but	butyrlyation	me3	trimethylation
cit	citrullination	og	*O*-GlcNAcylation
cr	crotonylation	oh*	hydroxylation
for	formylation	ox*	oxidation
gt*	glutathionylation	ph	phosphorylation
hib	2-hydroxyisobutyrylation	su	SUMOylation
iso	isomerization	ub	ubiquitination

* XYZ modifications are known to occur on histone proteins but have not been included in these tables at individual amino acid residues.

Table heading key

Site	A known histone PTM is indicated, based on the Brno nomenclature (Turner 2005), by the numbered amino acid residue at which it occurs followed by the abbrevation for its chemical modification (see Modification key).
Model	Refers to the model organism in which a PTM was characterized.
Enzyme	Where known, the histone-modifying enzyme (writer) that transduces a PTM is indicated. *Italics* specify the valency of modification that an enzyme can catalyze.
Function	The associated biological function of a histone PTM is indicated where known.
Reference	The key primary references where a PTM and/or its function are described are listed.

Cite as *Cold Spring Harb Perspect Biol* doi: 10.1101/cshperspect.a025064

1 MORE-STUDIED HISTONE PTMS

The tables in this section list histone modifications with known functions and modifying enzymes, and primary references are indicated wherever possible (up until 2014). Distinct modification states are indicated in *italics* in the Enzyme column.

Additional modifications with currently unknown function are listed in Section 2. These modifications were obtained from a combination of sources.

The tables that constitute Section 1 have been extended from Appendix 2 in Allis et al. (2007) by Ben Garcia and Yingming Zhao, with additional input from Le Hehuang, Monika Lachner, and Marie-Laure Caparros. That appendix was based on an original setup from Lachner et al. (2003) and significantly extended by Roopsha Sengupta, Mario Richter, and Marie-Laure Caparros and verified by Patrick Trojer.

The histone modifications follow the nomenclature as proposed by Turner (2005).

Table 1. Histone H2A

Site	Model	Enzyme	Function	Reference(s)
K5ac	Hs, Sc	Tip60, p300/CBP, Hat1	Transcriptional activation	Yamamoto and Horikoshi 1997; Kimura and Horikoshi 1998; Verreault et al. 1998
K9bio		HCS Biotinidase	Acetylation and methylation dependent Involved in cell proliferation, gene silencing, and cellular response to DNA damage	Stanley et al. 2001; Kothapalli et al. 2005a; Chew et al. 2006
K7ac	Sc	Hat1, Esa1	Transcriptional activation	Suka et al. 2001
K13bio		HCS Biotinidase	Acetylation and methylation dependent Involved in cell proliferation, gene silencing, and cellular response to DNA damage	Stanley et al. 2001; Kothapalli et al. 2005a; Chew et al. 2006
K13ub	Mm	Rnf168	Part of the DNA damage response to double-stranded DNA breaks	Mattiroli et al. 2012; Gatti et al. 2012
K15ub	Mm	Rnf168	Part of the DNA damage response to double-stranded DNA breaks	Mattiroli et al. 2012; Gatti et al. 2012
K63ub	Mm	Rnf8	Part of the DNA damage response to double-stranded DNA breaks	Huen et al. 2007; Mailand et al. 2007
Q105me	Sc, Hs	Nop1, fibrillarin: *me1*	Ribosomal gene expression	Tessarz et al. 2014
K119ub	Dm, Hs	dRing, RING1B	Polycomb silencing UV damage response	Wang et al. 2004; Kapetanaki et al. 2006
S121ph **(S122ph)**	Sc Sp	Mec1 PIKK Bub1	DNA damage response Telomere silencing Chromosomal stability	Wyatt et al. 2003; Harvey et al. 2005 Kawashima et al. 2010
T125ph	Sc	Mec1 PIKK	DNA damage response Telomere silencing	Wyatt et al. 2003
K126bio	Hs	HCS Biotinidase	Acetylation and methylation dependent Involved in cell proliferation, gene silencing, and cellular response to DNA damage.	Stanley et al. 2001; Kothapalli et al. 2005a; Chew et al. 2006
K126su	Sc		Transcriptional repression Blocks histone acetylation and histone ubiquitination	Nathan et al. 2006
K127bio	Hs	HCS Biotinidase	Acetylation and methylation dependent Involved in cell proliferation, gene silencing, and cellular response to DNA damage	Stanley et al. 2001; Kothapalli et al. 2005a; Chew et al. 2006
S128ph **(S129ph)**	Sc	Mec1 PIKK	DNA damage response Telomere silencing	Downs et al. 2000; Redon et al. 2003; Wyatt et al. 2003; Downs et al. 2004
K130bio	Hs	HCS Biotinidase	Acetylation and methylation dependent Involved in cell proliferation, gene silencing, and cellular response to DNA damage	Stanley et al. 2001; Kothapalli et al. 2005a; Chew et al. 2006

Additional H2A modifications: K4ac, K21ac, K74me (Pantazis and Bonner 1981; Song et al. 2003; Aihara et al. 2004).
HCS, holocarboxylase synthetase; PIKK, phosphatidylinositol 3-kinase-related kinase.

Table 2. **Histone H2AX**

Site	Model	Enzyme	Function	Reference(s)
K13ub	Mm	Rnf168	Part of the DNA damage response to double-stranded DNA breaks	Mattiroli et al. 2012; Gatti et al. 2012; Panier and Durocher 2013
K15ub	Mm	Rnf168	Part of the DNA damage response to double-stranded DNA breaks	Mattiroli et al. 2012; Gatti et al. 2012; Panier and Durocher 2013
K63ub	Mm	Rnf8	Part of the DNA damage response to double-stranded DNA breaks	Huen et al. 2007; Mailand et al. 2007; Panier and Durocher 2013
S139ph	Hs, Sc, Dm, Xl	ATM DNA-PK ATR	DNA repair M-phase related Also known as γH2AX	Rogakou et al. 1998; Rogakou et al. 1999; Burma et al. 2001; Stiff et al. 2004; Ichijima et al. 2005; Mukherjee et al. 2006; Ward and Chen 2001
Y142ph	Hs, Mm	WSTF	DNA damage	Xiao et al. 2009

ATM, ataxia telangiectasia mutated; PK, protein kinase; ATR, ataxia telangiectasia and Rad3-related; WSTF, Williams–Beuren syndrome transcription factor.

Table 3. **Histone H2B**

Site	Model	Enzyme	Function	Reference(s)
K5ac	Hs		Transcriptional activation	Puerta et al. 1995; Galasinski et al. 2002
S10ph	Sc	Ste20	Apoptosis	Ahn et al. 2005
S14ph	Hs, Mm	Mst1/krs2 kinase	Apoptosis Somatic hypermutation and class switch recombination	Ajiro 2000; Cheung et al. 2003; Odegard et al. 2005
K16su	Sc		Gene repression	Nathan et al. 2006
K17su	Sc		Gene repression	Nathan et al. 2006
S33ph	Dm	CTK TAF1	Transcriptional activation	Maile et al. 2004
K34ub	Sc	MSL2	Transcriptional activation	Wu et al. 2011
K120ub	Hs	RNF20/40	Cell-cycle progression in concert with SAGA for transcriptional activation through H3 methylation, DNA damage response, meiosis	Robzyk et al. 2000; Sun and Allis 2002; Kao et al. 2004; Zhu et al. 2005
K123ub	Sc	Rad6(E2) Bre1(E3); *ub1*	Telomeric silencing by lowering histone methylation at H3K4 and H3K79	Emre et al. 2005

Table 4. **Histone H3**

Site	Model	Enzyme	Function	Reference(s)
R2me	Hs Mm	CARM1; *me1*, *me2a* PRMT5; *me1*, *me2s* PRMT6; *me1*, *me2a* PRMT7; *me1*, *me2s*	Gene expression	Chen et al. 1999; Schurter et al. 2001; Greer and Shi 2012
T3ph	Hs At	Haspin	Centromere mitotic spindle function	Polioudaki et al. 2004; Dai et al. 2005
K4ac	Sc	GCN5, RTT109, Sir2, Hst1	Transcription activation at some promoters	Guillemette et al. 2011
K4me	Sc Ce Ds Hs	Set1; *me3* Set-2; *me1–3* Set1; *me2/3* SETD1A; *me1–3* SETD1B	rDNA/telomeric silencing (Sc) Germ cell maintenance Transcriptional activation (All)	Briggs et al. 2001; Roguev et al. 2001; Nagy et al. 2002; Bryk et al. 2002; Bernstein et al. 2002; Santos-Rosa et al. 2002; Lee and Skalnik 2005; Lee et al. 2007; Xiao et al. 2011
	Tt		Transcriptional activation	Strahl et al. 1999
	Ds	Trx	Trithorax activation	Milne et al. 2002; Nakamura et al. 2002; Greer and Shi 2012
	Hs	MLL; *me1–3* MLL2	Gene activation	

Continued

Cite as *Cold Spring Harb Perspect Biol* doi: 10.1101/cshperspect.a025064

Table 4. *Continued*

Site	Model	Enzyme	Function	Reference(s)
	Ds	Trr	Enhancer function	Herz et al. 2013
	Hs	MLL3; *me1–3* MLL4		
	Ce	Ash-2; *me1–3*	Germ cell specification	Beisel et al. 2002; Xiao et al. 2011
	Ds	Ash1; *me3*	Trithorax activation	
	Hs	ASH1L; *me1/3*	Gene activation	
	Hs	SETD7; *me1*	Transcriptional activation	Wang et al. 2001a; Nishioka et al. 2002a; Wilson et al. 2002; Zegerman et al. 2002
	Hs	SMYD3-*me2/3*	Transcriptional activation	Hamamoto et al. 2004
	Mm	Meisetz-*me3*	Meiotic prophase progression	Hayashi et al. 2005
T6ph	Hs	PKCβ	Inhibits AR-dependent transcription	Metzger et al. 2010
R8me	Hs	PRMT5; *me1*, *me2s*	Transcriptional repression	Pal et al. 2004
K9ac	Sc	SAGA GCN5	Transcriptional activation	Grant et al. 1999
	Hs	SRC1	Nuclear receptor coactivator	Spencer et al. 1997; Schubeler et al. 2000; Vaquero et al. 2004
	Dm		Transcriptional activation	Nowak and Corces 2000
K9me	Sp	Clr4; *me1*, *me2*	Centromeric and mating-type silencing	Bannister et al. 2001; Nakayama et al. 2001
	Nc	Dim5; *me3*	DNA methylation	Tamaru and Selker 2001
	Ce	Met-2; *me3* Mes-2; *me3*	Germ cells	Bessler et al. 2010
	Dm	Su(var)3–9; *me2/3*	Dominant PEV modifier	Czermin et al. 2001; Schotta et al. 2002; Ebert et al. 2004
	At	KRYPTONITE; *me2*	DNA methylation	Jackson et al. 2002; Jackson et al. 2004
	Mm	Suv39h1; *me2/3* Suv39h2; *me2/3*	Pericentric heterochromatin	O'Carroll et al. 2000; Rea et al. 2000; Lachner et al. 2001; Peters et al. 2001
	Hs	SUV39H1; *me3*	Rb-mediated silencing	Nielsen et al. 2001; Vandel et al. 2001
	Hs, Mm	ESET; *me2/me3* (SETDB1)	Transcriptional repression	Schultz et al. 2002; Yang et al. 2002; Dodge et al. 2004; Wang et al. 2004
	Mm, Hs	G9a; *me1/me2*	Transcriptional repression Imprinting	Tachibana et al. 2001, 2002; Ogawa et al. 2002; Xin et al. 2003
	Hs	EHMT1/GLP; *me1/me2*	Transcriptional repression	Ogawa et al. 2002; Tachibana et al. 2005
	Hs	PRDM2/RIZ1; *me2*	Tumor suppression and response to female sex hormones	Kim et al. 2003; Carling et al. 2004
S10ph	Sc	Snf1	Transcriptional activation	Lo et al. 2001
	Dm	Jil-1	Transcriptional up-regulation of male X-chromosome	Jin et al. 1999; Wang et al. 2001c
	Hs	Rsk2 Msk1 Msk2	Transcriptional activation of immediate early genes (in concert with H3-K14 acetylation)	Sassone-Corsi et al. 1999; Thomson et al. 1999; Cheung et al. 2000; Clayton et al. 2000
	Hs	IKKα	Transcriptional up-regulation	Anest et al. 2003; Yamamoto et al. 2003
	Sc, Ce	Ip11/AuroraB	Mitotic chromosome condensation	Hendzel et al. 1997; Wei et al. 1999; Hsu et al. 2000
	An	NIMA	Mitotic chromosome condensation	De Souza et al. 2000
	Hs, Ce	Fyn kinase	UVB-induced MAP kinase pathway	He et al. 2005
T11ph	Hs	Dlk/ZIP	Mitosis-specific phosphorylation	Preuss et al. 2003
K14ac	Sc, Tt, Mm	Gcn5	Transcriptional activation	Brownell et al. 1996; Kuo et al. 1996
	Hs, Dm	$TAF_{II}230$ $TAF_{II}250$	Transcriptional activation	Mizzen et al. 1996
	Hs	p300	Transcriptional activation	Schiltz et al. 1999
	Hs	PCAF	Transcriptional activation	Schiltz et al. 1999
	Mm	SRC1	Nuclear receptor coactivator	Spencer et al. 1997
R17me	Hs, Mm	CARM1; *me1*, *me2a*	Transcriptional activation (in concert with H3-K18/23 acetylation)	Chen et al. 1999; Schurter et al. 2001; Bauer et al. 2002; Daujat et al. 2002

Continued

Table 4. *Continued*

Site	Model	Enzyme	Function	Reference(s)
K18ac	Sc	SAGA Ada GCN5	Transcriptional activation	Grant et al. 1999
	Hs	p300	Transcriptional activation	Schiltz et al. 1999
	Hs	CBP	Transcriptional activation (in concert with H3-R17 methylation)	Daujat et al. 2002
K23ac	Sc	SAGA	Transcriptional activation	Grant et al. 1999
	Hs	CBP	Transcriptional activation (in concert with H3-R17 methylation)	Daujat et al. 2002
R26me	Hs	CARM1; *me1*, *me2a*	In vitro methylation site	Chen et al. 1999; Schurter et al. 2001
K27ac	Sc, Dm	CBP, P300, GCN5	Enhancer function, gene expression	Tie et al. 2009; Suka et al. 2001; Creyghton et al. 2010
K27me	Hs, Dm	E(z)/EZH2; *me3*	Polycomb repression Early B-cell development X-chromosome inactivation	Cao et al. 2002; Czermin et al. 2002; Kuzmichev et al. 2002; Muller et al. 2002; Su et al. 2003
S28ph	Hs	Aurora-B	Mitotic chromosome condensation	Goto et al. 1999; Goto et al. 2002
	Hs	MSK1	UVB-induced phosphorylation	Zhong et al. 2001
K36me	Sc	Set2; *me2*	Gene repression	Strahl et al. 2002; Kizer et al. 2005; Sun et al. 2005
	Nc	Set2; *me2*	Transcription activation	Adhvaryu et al. 2005
	Sp	Set2; *me2*	Transcription elongation	Morris et al. 2005
	Ce	MES-4; *me2* MET-1; *me3*	Dosage compensation in germline Meiosis	Bender et al. 2006; Andersen and Horvitz 2007
	Dm	MES4; *me3* SET2; *me3*	Transcription elongation	Bell et al. 2007
	Hs, Mm	SETD2; *me1–3* NSD1–3; *me1*, *me2*	Transcription activation	Edmunds et al. 2008 Wang et al. 2007
K36ac	Sc, Mm, Hs	GCN5	Promoter mark on active genes	Morris et al. 2007
P38iso	Sc	Fpr4	Gene expression	Nelson et al. 2006
Y41ph	Hs	JAK2	Gene expression	Dawson et al. 2009
R43me	Hs	CARM1, PRMT6; *me2a*	Transcriptional activation	Casadio et al. 2013
T45ph	Sc, Hs	Cdc7, PKC	DNA replication; apoptosis	Baker et al. 2010; Hurd et al. 2009
K56ac	Sc	SPT10	Transcriptional activation; DNA damage	Xu et al. 2005; Ozdemir et al. 2005; Masumoto et al. 2005
K56me	Hs	G9a; *me1*	DNA replication	Yu et al. 2012
	Hs	Suv39h; *me3*	Heterochromatin	Jack et al. 2013
K64ac	Hs/Mm	p300	Nucleosome dynamics and transcription	Di Cerbo et al. 2014
K64me	Mm	*me3*	Pericentric heterochromatin	Daujat et al. 2009
K79me	Sc, Hs	Dot1/DOT1L; *me1–3*	Telomeric silencing, pachytene checkpoint DNA damage response	Feng et al. 2002; Lacoste et al. 2002; Ng et al. 2002; van Leeuwen et al. 2002; Greer and Shi 2012
T80ph	Hs		Mitosis	Hammond et al. 2014

Table 5. **Histone H3.3**

Site	Model	Enzyme	Function	Reference(s)
K4me	Dm	*me1, me2, me3*	Transcriptional activation	McKittrick et al. 2004
K9me	Dm	*me1, me2*	Transcriptional repression	McKittrick et al. 2004
K9ac	Dm, Hs		Transcriptional activation	McKittrick et al. 2004; Hake et al. 2006
K14me	Dm	*me1, me2*		McKittrick et al. 2004
K14ac	Dm, Hs		Transcriptional activation	McKittrick et al. 2004; Hake et al. 2006
K18ac	Hs		Transcriptional activation	Hake et al. 2006
K23ac	Hs		Transcriptional activation	Hake et al. 2006
K27me	Dm	*me1, me2, me3*	Transcriptional repression	McKittrick et al. 2004
S31ph	Mammals		Mitosis-specific phosphorylation	Hake et al. 2005
K36me	Dm, Hs	*me1, me2, me3*	Transcriptional activation	McKittrick et al. 2004; Hake et al. 2006
K37me	Dm	*me1, me2*		McKittrick et al. 2004
K79me	Dm, Hs	*me1, me2*	Transcriptional activation	McKittrick et al. 2004; Hake et al. 2006

Table 6. **CEN-H3/CENP-A**

Site	Model	Enzyme	Function	Reference
G1me3	Hs	RCC1	Mitosis	Bailey et al. 2013
S7ph	Hs		Mitosis	Zeitlin et al. 2001
S16ph	Hs		Chromosome segregation during mitosis	Bailey et al. 2013
S18ph	Hs		Chromosome segregation during mitosis	Bailey et al. 2013

Table 7. **Histone H4**

Site	Model	Enzyme	Function	Reference(s)
S1ph	Hs, Sc	Casein kinase II	DNA damage response	Ruiz-Carrillo et al. 1975; Cheung et al. 2005; van Attikum and Gasser 2005
R3me	Hs, Sc	PRMT1; *me1, me2a* PRMT5; *me1, me2s* PRMT6; *me1, me2a* PRMT7; *me1, me2s*	Transcriptional activation	Wang et al. 2001b; Strahl et al. 2001; Greer and Shi 2012
K5ac	*Tt, Dm, Hs*	Hat1	Histone deposition	Sobel et al. 1995; Parthun et al. 1996; Taplick et al. 1998; Kruhlak et al. 2001
	Sc	Esa1/NuA4	Cell-cycle progression	Smith et al. 1998; Allard et al. 1999; Clarke et al. 1999; Miranda et al. 2006; Bird et al. 2002
	Hs, Mm	ATF2	Sequence-specific transcription factor	Kawasaki et al. 2000a
	Hs	p300	Transcriptional activation	Schiltz et al. 1999; Turner and Fellows 1989
K5me	Hs	Smyd3-*me1*	Contributes to cancer phenotype	Van Aller et al. 2012
K8ac	Hs, Mm	Y-ATF2	Excluded from Xi Sequence-specific transcription factor	Jeppesen and Turner 1993; Choy et al. 2001; Kruhlak et al. 2001; Kawasaki et al. 2000b
	Hs	PCAF/ p300	Transcriptional activation	Schiltz et al. 1999; Turner and Fellows 1989
K8me	Sc	SET5; *me1*	Stress response	Green et al. 2012
K12ac	Sc, Hs	Hat1	Excluded from Xi Histone deposition	Jeppesen and Turner 1993; Kleff et al. 1995; Sobel et al. 1995; Parthun et al. 1996; Chang et al. 1997; Kruhlak et al. 2001; Turner and Fellows 1989
	Sc	NuA4	Mitotic and meiotic progression	Choy et al. 2001
K12me	Sc	SET5; *me1*	Stress response	Green et al. 2012

Continued

Table 7. *Continued*

Site	Model	Enzyme	Function	Reference(s)
K12bio	Hs	HCS Biotinidase	Decrease in response to DNA double-strand breaks Effects on cell proliferation	Stanley et al. 2001; Kothapalli et al. 2005a,b
K16ac	Mm		Excluded from Xi Cell-cycle-dependent acetylation	Jeppesen and Turner 1993; Taplick et al. 1998
	Dm	MOF	Transcriptional up-regulation of male X chromosome	Akhtar and Becker 2000; Hsu et al. 2000
	Hs, Mm	ATF2	Sequence-specific transcription factor	Kawasaki et al. 2000a; Turner 2000; Kruhlak et al. 2001; Turner and Fellows 1989; Vaquero et al. 2004
K20me	Mm, Dm	Suv4–20h1; *me2, me3* Suv4–20h2; *me2, me3*	Gene silencing	Schotta et al. 2004
	Hs, Dm	SETD8/Pr-SET7; *me1*	Transcriptional silencing Mitotic condensation	Fang et al. 2002; Nishioka et al. 2002b; Rice et al. 2002
	Dm	Ash1; *me2*	Trithorax activation in concert with H3K4 and H3K9 methylation	Beisel et al. 2002
K59me	Sc		Silent chromatin formation	Zhang et al. 2003
K59su	Hs	SUMO-1 SUMO-3	Transcriptional repression	Shiio and Eisenman 2003

Table 8. **Histone H1**

Site	Model	Enzyme	Function	Reference(s)
E2arn	Rn	PARP-1; *ar1*	Involved in neurotrophic activity	Ogata et al. 1980b; Visochek et al. 2005
T10ph	Hs		Mitosis specific Transcriptional activation H1b	Chadee et al. 1995; Garcia et al. 2004; Sarg et al. 2006
E14arn	Rn	PARP-1; *ar1*	Involved in neurotrophic activity	Ogata et al. 1980b; Visochek et al. 2005
S17ph	Hs		Interphase specific Transcriptional activation H1b	Chadee et al. 1995; Garcia et al. 2004; Sarg et al. 2006
K26me	Hs	EZH2; *me2*	Mediates HP1 binding	Kuzmichev et al. 2004; Daujat et al. 2005
S27ph	Hs	EZH2; *me2*	Blocks HP1 binding	Garcia et al. 2004; Daujat et al. 2005
R54cit	Mm	PADI4	Cellular reprogramming/nucleosome binding	Christophorou et al. 2014
T137ph	Hs		Mitosis specific Transcriptional activation H1b	Chadee et al. 1995; Garcia et al. 2004; Sarg et al. 2006
T154ph	Hs		Mitosis specific Transcriptional activation H1b	Chadee et al. 1995; Garcia et al. 2004; Sarg et al. 2006
S172ph	Hs		Interphase specific Transcriptional activation H1b	Chadee et al. 1995; Garcia et al. 2004; Sarg et al. 2006
S188ph	Hs		Interphase specific Transcriptional activation H1b	Chadee et al. 1995; Garcia et al. 2004; Sarg et al. 2006
K213ar	Rn	PARP-1; *ar1*	Involved in neurotrophic activity	Ogata et al. 1980b; Visochek et al. 2005

Cite as *Cold Spring Harb Perspect Biol* doi: 10.1101/cshperspect.a025064

2 LESS-STUDIED HISTONE PTMS

Methods key

The method by which novel histone modifications have been identified is indicated by the following abbreviations:

Ab	antibody
Au	autoradiography
MS	mass spectrometry

These tables reflect modification sites that have been detected, but where no function has been assigned.

Table 9. **Histone H2A[a,b]**

Site	Model	Method	Reference(s)
S1ph	Mm	Au	Pantazis and Bonner 1981
R3me3	Mm	MS	Tweedie-Cullen et al. 2012
K5hib	Mm	MS	Dai et al. 2014
K9me1; me2; suc; hib	Hs; Mm; Hs; Hs	MS	Tan et al. 2011; Tweedie-Cullen et al. 2012; Xie et al. 2012; Dai et al. 2014
R11me1, me2	Hs	MS	Waldmann et al. 2011
K13me1, ac; suc	Bt; Sc	MS	Zhang et al. 2003; Xie et al. 2012
K15ac	Bt	MS	Zhang et al. 2003
K21suc	Sc	MS	Xie et al. 2012
R29me1, me2	Hs	MS/Ab	Waldmann et al. 2011
K36ac, suc; for; hib; cr	Hs/Dm; Hs/Mm; Mm; Hs/Mm	MS	Xie et al. 2012; Wisniewski et al. 2008; Dai et al. 2014; Tan et al. 2011
Y39oh	Hs	MS	Tan et al. 2011
R42me1	Hs; Hs	MS/Ab	Tan et al. 2011
R71me1	Mm	MS	Tweedie-Cullen et al. 2012
K74ac, me1; hib	Mm; Mm	MS	Tweedie-Cullen et al. 2012; Dai et al. 2014
K75me1; hib	Bt; Mm	MS	Zhang et al. 2003; Dai et al. 2014
R77me1	Bt	MS	Zhang et al. 2003
T79ac	Mm	MS	Tweedie-Cullen et al. 2012
R88me1	Hs	MS	Tan et al. 2011
K95cr, but, pr, me1, me2; for; ub; suc; hib	Mm; Hs/Mm; Mm; Hs; Hs/Mm	MS	Tweedie-Cullen et al. 2012; Wisniewski et al. 2008; Tweedie-Cullen et al. 2009; Xie et al. 2012; Dai et al. 2014
K99me1; me2	Mm; Mm	MS	Tweedie-Cullen et al. 2012; Tweedie-Cullen et al. 2009
T101og	Hs	MS/Ab	Sakabe et al. 2010
K118for, me1, cr; for; ub, me2; hib	Hs/Mm; Hs/Mm; Mm; Mm	MS	Tan et al. 2011; Wisniewski et al. 2008; Tweedie-Cullen et al. 2009; Dai et al. 2014
K119cr; mal	Hs; Sc	MS	Tan et al. 2011; Xie et al. 2012
T120ph	Dm	Ab	Aihara et al. 2004
K125me1, cr; me2, pr; ub	Hs; Mm; Mm	MS	Tan et al. 2011; Tweedie-Cullen et al. 2012; Tweedie-Cullen et al. 2009
K127ac	Mm	MS	Tweedie-Cullen et al. 2012
K129ac	Mm	MS	Tweedie-Cullen et al. 2012

[a]Carbonylation of H2A has been detected in Rn indirectly by Sharma et al. (2006).

[b]5-Hydroxylation of lysine residues has been reported for canonical histones (H2A, H2B, H3, and H4) by Unoki et al. (2013).

Table 10. **Histone H2AX**

Site	Model	Method	Reference
K118ub	Mm	MS	Tweedie-Cullen et al. 2009
K119ub	Mm	MS	Tweedie-Cullen et al. 2009

Table 11. Histone H2A.Z

Site	Model	Method	Reference(s)
K4ac; me1	Hs; Hs/Mm	MS; MS/Ab	Tweedie-Cullen et al. 2009; Binda et al. 2013
K7ac; me1	Hs; Hs/Mm	MS; MS/Ab	Bonenfant et al. 2006; Binda et al. 2013
K11ac	Hs	MS	Bonenfant et al. 2006
K13ac	Mm	MS	Tweedie-Cullen et al. 2009
K120ub	Mm	MS	Ku et al. 2012
K121ub	Mm	MS	Ku et al. 2012
K125ub	Mm	MS	Ku et al. 2012

Table 12. Histone macroH2A

Site	Model	Method	Reference(s)
K17me1	Hs	MS	Chu et al. 2006
K115ub	Hs	MS	Ogawa et al. 2005; Chu et al. 2006
K122me2	Hs	MS	Chu et al. 2006
T128ph	Hs	MS/Ab	Chu et al. 2006; Bernstein et al. 2008
K238me1	Hs	MS	Chu et al. 2006
K238me2	Hs	MS	Chu et al. 2006

Table 13. Histone H2B

Site	Model	Method	Reference(s)
E2arn	Rn	Au	Ogata et al. 1980a
K5me1; cr; for; hib; suc	Bt; Hs; Hs/Mm; Hs/Mm; Hs	MS	Zhang et al. 2003; Tan et al. 2011; Wisniewski et al. 2008; Dai et al. 2014; Weinert et al. 2013
S6ph	Mm	MS	Tweedie-Cullen et al. 2009
K11ac; cr	Sc; Hs/Mm	MS	Jiang et al. 2007; Tan et al. 2011
K12me1, cr; me3; hib	Hs; Mm; Mm	MS	Tan et al. 2011; Tweedie-Cullen et al. 2012; Dai et al. 2014
K15ac; me1, cr	Hs	MS	Tan et al. 2011
K16ac; cr	Sc; Hs	MS	Jiang et al. 2007; Tan et al. 2011
T19ac	Mm	MS	Tweedie-Cullen et al. 2012
K20me1, cr; hib	Hs; Mm	MS	Tan et al. 2011; Dai et al. 2014
K21but	Sc	MS	Zhang et al. 2009
K23me1, cr; me2; hib	Hs; Bt; Mm	MS	Tan et al. 2011; Zhang et al. 2003; Dai et al. 2014
K24hib	Mm	MS	Dai et al. 2014
K34for; cr; suc; me1; hib	Hs/Mm; Hs/Mm; Sc/Hs; Mm; Mm	MS	Wisniewski et al. 2008; Tan et al. 2011; Xie et al. 2012; Tweedie-Cullen et al. 2012; Dai et al. 2014
S36og	Hs	MS/Ab	Sakabe et al. 2010
K37me1	Sc	MS	Zhang et al. 2009
E38me2	Sc	MS	Zhang et al. 2009
Y37oh	Hs	MS	Tan et al. 2011
K43me1; for; hib; suc	Bt; Hs/Mm; Mm; Hs	MS	Zhang et al. 2003; Wisniewski et al. 2008; Dai et al. 2014; Weinert et al. 2013
K46for; suc; hib	Hs/Mm; Sc/Dm; Hs/Mm	MS	Wisniewski et al. 2008; Xie et al. 2012; Dai et al. 2014
K57me1; ac; hib	Hs; Mm; Mm	MS	Tan et al. 2011; Tweedie-Cullen et al. 2012; Dai et al. 2014
E64me2	Sc	MS	Zhang et al. 2009
S76ph	Mm	MS	Tweedie-Cullen et al. 2009
K79me1	Hs	MS	Tan et al. 2011
K85ac; me1; suc; hib	Bt; Hs; Hs; Hs/Mm	MS	Zhang et al. 2003; Tan et al. 2011; Weinert et al. 2013; Dai et al. 2014
S88ph	Mm	MS	Tweedie-Cullen et al. 2009
T89ph	Mm	MS	Tweedie-Cullen et al. 2009
S92ph	Mm	MS	Tweedie-Cullen et al. 2009
K99me1	Hs	MS	Tan et al. 2011
K108for; cr; ub, ac; hib; suc	Hs/Mm; Mm; Mm; Hs/Mm; Mm	MS	Wisniewski et al. 2008; Tweedie-Cullen et al. 2012; Tweedie-Cullen et al. 2009; Dai et al. 2014; Park et al. 2013
S113ph	Mm	MS	Tweedie-Cullen et al. 2009
K116for, me1; suc, mal; ac; hib; cr	Hs; Hs/Dm; Mm; Hs/Mm; Mm	MS	Tan et al. 2011; Xie et al. 2012; Tweedie-Cullen et al. 2012; Dai et al. 2014; Montellier et al. 2013
K120for; suc; ac, ub; hib	Hs; Hs/Dm/Mm; Mm; Hs/Mm	MS	Tan et al. 2011; Xie et al. 2012; Tweedie-Cullen et al. 2009; Dai et al. 2014
K125ac	Mm	MS	Tweedie-Cullen et al. 2012

Also, 5-hydroxylation of lysine residues has been reported (Unoki et al. 2013).

Table 14. **Histone H3**

Site	Model	Method	Reference(s)
K4cr; hib; ac	Hs/Mm; Mm; Hs	MS	Tan et al. 2011; Dai et al. 2014; Garcia et al. 2007
T6ac	Tt	MS	Britton et al. 2013
K9cr; hib	Hs/Mm; Mm	MS	Tan et al. 2011; Dai et al. 2014
S10ac; og	Sc/Mm/Hs	MS; Ab; Ab	Britton et al. 2013; Zhang et al. 2011
K14suc; but; hib	Hs; Sc; Mm	MS	Xie et al. 2012; Zhang et al. 2009; Dai et al. 2014
K18cr; for; me1; hib	Hs/Mm; Hs/Mm; Mm; Mm	MS	Tan et al. 2011; Wisniewski et al. 2008; Garcia et al. 2005; Dai et al. 2014
T22ac	Sc/Dm/Hs	MS	Britton et al. 2013
K23cr; for; pr; hib; suc	Hs/Mm; Hs/Mm; Sc; Hs/Mm; Mm	MS	Tan et al. 2011; Wisniewski et al. 2008; Zhang et al. 2009; Dai et al. 2014; Park et al. 2013
K27cr; but; hib; suc	Hs/Mm; Sc; Mm; Mm	MS	Tan et al. 2011; Zhang et al. 2009; Dai et al. 2014; Park et al. 2013
S28ac	Mm	MS	Britton et al. 2013
T32og	Hs	MS	Fong et al. 2012
K36hib	Mm	MS	Dai et al. 2014
R52me1	Sc/Bt	MS	Hyland et al. 2005
R53me1	Sc/Bt	MS	Hyland et al. 2005
Y54ac	Tet	MS	Britton et al. 2013
K56cr, for; suc, mal; pr; ub; hib	Hs/Mm; Dm/Mm/Hs; Sc; Mm; Hs/Mm	MS	Tan et al. 2011; Xie et al. 2012; Zhang et al. 2009; Tweedie-Cullen et al. 2009; Dai et al. 2014
E59me2	Sc	MS	Zhang et al. 2009
R63me1	Hs/Mm	MS	Tan et al. 2011
K64for; hib; ac, me1	Hs/Mm; Mm; Sc, Hs	MS	Wisniewski et al. 2008; Dai et al. 2014; Garcia et al. 2007
K79suc; for; cr; ub; hib; ac	Sc/Dm/Mm/Hs; Hs/Mm; Mm; Mm; Hs/Mm; Hs	MS	Xie et al. 2012; Wisniewski et al. 2008; Tweedie-Cullen et al. 2012; Tweedie-Cullen et al. 2009; Dai et al. 2014; Garcia et al. 2007
T80ac	Mm	MS	Tweedie-Cullen et al. 2012
R83me1, me2	Mm	MS	Tweedie-Cullen et al. 2012
S86ph	Mm	MS	Tweedie-Cullen et al. 2012
T107ph	Mm	MS	Tweedie-Cullen et al. 2012
C110gt	Hs/Mm	Indirect chemical labeling	Garcia-Gimenez et al. 2013
K115ac; but	Sc/Bt; Mm	MS	Hyland et al. 2005; Tweedie-Cullen et al. 2012
T118ph	Sc/Bt	MS	Hyland et al. 2005
R128me1	Hs/Mm	MS	Tan et al. 2011
K122suc; for; me2; hib; cr	Hs; Hs/Mm; Mm; Hs/Mm; Mm	MS	Tan et al. 2011; Wisniewski et al. 2008; Tweedie-Cullen et al. 2009; Dai et al. 2014; Montellier et al. 2013
K134me1	Mm	MS	Tweedie-Cullen et al. 2012

Also, 5-hydroxylation of lysine residues has been reported (Unoki et al. 2013).

Table 15. **Histone H4**

Site	Model	Method	Reference(s)
R3me3	Mm	MS	Tweedie-Cullen et al. 2012
K5cr; me3; pr, but; hib	Mm/Hs; Mm; Hs; Mm	MS	Tan et al. 2011; Tweedie-Cullen et al. 2012; Chen et al. 2007; Dai et al. 2014
K8cr; pr, but; hib	Mm/Hs; Hs; Mm	MS; MS; MS/Ab	Tan et al. 2011; Chen et al. 2007; Dai et al. 2014
K12cr; for; suc; pr, but; hib	Mm/Hs; Hs; Hs; Mm	MS	Tan et al. 2011; Wisniewski et al. 2008; Xie et al. 2012; Chen et al. 2007; Dai et al. 2014
K16cr, me1; pr; pr, but; hib	Mm/Hs; Mm; Hs; Mm	MS	Tan et al. 2011; Tweedie-Cullen et al. 2012; Chen et al. 2007; Dai et al. 2014
R17me1, me2	Mm	MS	Tweedie-Cullen et al. 2012
R17me1, me2, me3	Mm	MS	Tweedie-Cullen et al. 2012
K20ac	Sc	MS	Garcia et al. 2007

Continued

Table 15. *Continued*

Site	Model	Method	Reference(s)
K31for; suc; hib; me1; pr	Hs; Sc/Dm/Mm/Hs; Hs/Mm; Hs/Mm/ Sc; Hs	MS	Tan et al. 2011; Xie et al. 2012; Dai et al. 2014; Garcia et al. 2007; Chen et al. 2007
R23me3	Mm	MS	Tweedie-Cullen et al. 2012
R35me1	Hs	MS	Tan et al. 2011
K44pr; hib	Hs; Mm	MS	Chen et al. 2007; Dai et al. 2014
S47og; ph	Hs; Sc/Bt	MS/Ab; MS	Sakabe et al. 2010; Hyland et al. 2005
Y51oh	Hs	MS	Tan et al. 2011
R55me1	Hs	MS	Tan et al. 2011
K59me1, for; hib	Hs; Mm/Hs; Mm	MS	Tan et al. 2011; Wisniewski et al. 2008; Dai et al. 2014
R67me1	Hs	MS	Tan et al. 2011
K77me1; suc; for; hib; ac; pr; cr	Hs; Sc/Dm/Mm/Hs; Mm/Hs; Hs/Mm; Sc/Bt; Hs; Mm	MS	Tan et al. 2011; Xie et al. 2012; Wisniewski et al. 2008; Dai et al. 2014; Hyland et al. 2005; Chen et al. 2007; Montellier et al. 2013
K79for; suc, ac; hib; pr	Hs; Dm/Mm; Mm; Hs/Mm; Hs	MS	Wisniewski et al. 2008; Xie et al. 2012; Tweedie-Cullen et al. 2012; Dai et al. 2014; Chen et al. 2007
Y88ox; ph	Hs; Mm	MS	Tan et al. 2011; Tweedie-Cullen et al. 2009
K91ac; for; suc; cr; hib; pr	Hs; Hs; Dm/Mm/Hs; Mm; Hs/Mm; Hs	MS	Tan et al. 2011; Wisniewski et al. 2008; Xie et al. 2012; Tweedie-Cullen et al. 2012; Dai et al. 2014; Chen et al. 2007
R92me1	Sc/Bt	MS	Hyland et al. 2005

Also, 5-hydroxylation of lysine residues has been reported (Unoki et al. 2013).

Table 16. **Histone H1**

Site	Model	Method	Reference(s)
S1ph	Hs	MS	Garcia et al. 2004
E2arn	Rn	Au	Ogata et al. 1980b
T3ph	Hs	MS	Garcia et al. 2004
K12me1	Hs	MS	Lu et al. 2009
E14arn	Rn	Au	Ogata et al. 1980b
K16ac; me1, me2; for	Hs/Mm; Hs/Mm	MS	Wisniewski et al. 2007 Wisniewski et al. 2008
T17ph	Hs	MS	Garcia et al. 2004
K21ac; me1	Hs; Mm	MS	Wisniewski et al. 2007; Tweedie-Cullen et al. 2012
K22hib	Mm	MS	Dai et al. 2014
K25hib	Mm	MS	Dai et al. 2014
K26hib	Mm	MS	Dai et al. 2014
S30ph	Hs	MS	Garcia et al. 2004
K33ac; me2, for; cr; ub; hib	Hs; Mm; Hs/Mm; Mm; Mm	MS	Wisniewski et al. 2007; Tweedie-Cullen et al. 2012; Tan et al. 2011; Tweedie-Cullen et al. 2009
S35ph; ac	Hs; Mm	MS	Garcia et al. 2004; Tweedie-Cullen et al. 2012
K45ac,ub; hib; for; suc	Hs; Hs; Hs/Mm; Mm	MS	Wisniewski et al. 2007; Dai et al. 2014; Wisniewski et al. 2008; Park et al. 2013
K48ac	Hs	MS	Wisniewski et al. 2007
S50ac	Mm	MS	Tweedie-Cullen et al. 2012
K51ac; me1; hib	Hs; Mm; Hs/Mm	MS	Wisniewski et al. 2007; Tweedie-Cullen et al. 2012; Dai et al. 2014
K54me1	Hs	MS	Tan et al. 2011
K62ac; for; hib; suc	Hs; Hs/Mm; Hs/Mm; Hs	MS	Garcia et al. 2004; Wisniewski et al. 2008; Dai et al. 2014; Weinert et al. 2013
K63ac; me1; for; cr; hib	Hs; Hs; Hs/Mm; Hs/ Mm; Hs	MS	Wisniewski et al. 2007; Lu et al. 2009; Wisniewski et al. 2008; Tan et al. 2011; Dai et al. 2014
K64ac	Hs	MS	Tan et al. 2011
K66for	Hs/Mm	MS	Wisniewski et al. 2008

Continued

Table 16. *Continued*

Site	Model	Method	Reference(s)
S72ph	Hs	MS	Garcia et al. 2004
Y73oh	Hs	MS	Tan et al. 2011
K74for; hib	Hs/Mm; Mm	MS	Wisniewski et al. 2008; Dai et al. 2014
K80hib	Mm	MS	Dai et al. 2014
K81me1	Hs	MS	Lu et al. 2009
K83for	Hs	MS	Tan et al. 2011
K84for; cr; hib	Hs/Mm; Hs/Mm; Hs/Mm	MS	Wisniewski et al. 2008; Tan et al. 2011; Dai et al. 2014
K85ac	Hs	MS	Wisniewski et al. 2007
S87ph[a]	Hs	MS	Garcia et al. 2004
K87for[a]	Hs	MS	Wisniewski et al. 2007
K89ac, for; cr; suc; hib	Hs; Hs; Mm; Hs	MS	Wisniewski et al. 2007; Tan et al. 2011; Park et al. 2013; Dai et al. 2014
K92me1	Hs	MS	Tan et al. 2011
K96ac; me1; me2; for; cr; hib; suc	Hs; Hs; Mm; Hs/Mm; Hs; Hs/Mm; Mm	MS	Wisniewski et al. 2007; Lu et al. 2009; Tweedie-Cullen et al. 2012; Wisniewski et al. 2008; Tan et al. 2011; Dai et al. 2014; Park et al. 2013
K101me1	Hs	MS	Lu et al. 2009
K105me1; suc	Hs; Mm	MS	Lu et al. 2009; Park et al. 2013
K107me1	Hs	MS	Lu et al. 2009
K109me2; for; hib	Mm; Hs/Mm; Mm	MS	Tweedie-Cullen et al. 2012; Wisniewski et al. 2008; Dai et al. 2014
S112ac	Mm	MS	Tweedie-Cullen et al. 2012
K118me1	Hs	MS	Lu et al. 2009
K120hib; suc	Mm; Mm	MS	Dai et al. 2014; Park et al. 2013
K128hib	Mm	MS	Dai et al. 2014
K131me1	Hs	MS	Tan et al. 2011
K135hib	Mm	MS	Dai et al. 2014
K140for	Hs/Mm	MS	Wisniewski et al. 2008
T145ph	Hs	MS	Garcia et al. 2004
T146ph	Hs	MS	Garcia et al. 2004
K147me1; hib	Hs; Mm	MS	Lu et al. 2009; Dai et al. 2014
K150me1	Hs	MS	Tan et al. 2011
K158cr; hib	Hs/Mm; Mm	MS	Tan et al. 2011; Dai et al. 2014
K159for	Hs/Mm	MS	Wisniewski et al. 2008
T164ph	Hs	MS	Wisniewski et al. 2007
K167cr; hib	Hs; Mm	MS	Tan et al. 2011; Dai et al. 2014
T179ph	Hs	MS	Garcia et al. 2004
K187me1	Hs	MS/Ab	Weiss et al. 2010
K201me1	Hs	MS	Tan et al. 2011
K212hib	Mm	MS	Dai et al. 2014
K226me1	Hs	MS	Tan et al. 2011

Pham (2000) found an enzyme that ubiquitinates histone H1
[a]H1S87 and H1K87 represent different histone H1 variants.

REFERENCES

Adhvaryu KK, Morris SA, Strahl BD, Selker EU. 2005. Methylation of histone H3 lysine 36 is required for normal development in *Neurospora crassa*. *Eukaryot Cell* **4:** 1455–1464.

Ahn SH, Cheung WL, Hsu JY, Diaz RL, Smith MM, Allis CD. 2005. Sterile 20 kinase phosphorylates histone H2B at serine 10 during hydrogen peroxide-induced apoptosis in *S. cerevisiae*. *Cell* **120:** 25–36.

Aihara H, Nakagawa T, Yasui K, Ohta T, Hirose S, Dhomae N, Takio K, Kaneko M, Takeshima Y, Muramatsu M, et al. 2004. Nucleosomal histone kinase-1 phosphorylates H2A Thr 119 during mitosis in the early *Drosophila* embryo. *Genes Dev* **18:** 877–888.

Ajiro K. 2000. Histone H2B phosphorylation in mammalian apoptotic cells. An association with DNA fragmentation. *J Biol Chem* **275:** 439–443.

Akhtar A, Becker PB. 2000. Activation of transcription through histone H4 acetylation by MOF, an acetyltransferase essential for dosage compensation in *Drosophila*. *Mol Cell* **5:** 367–375.

Allard S, Utley RT, Savard J, Clarke A, Grant P, Brandl CJ, Pillus L, Workman JL, Cote J. 1999. NuA4, an essential transcription adaptor/histone H4 acetyltransferase complex containing Esa1p and the ATM-related cofactor Tra1p. *EMBO J* **18:** 5108–5119.

Allfrey VG, Faulkner R, Mirsky AE. 1964. Acetylation and methylation of histones and their possible role in the regulation of RNA synthesis. *Proc Natl Acad Sci* **51:** 786–794.

Allis CD, Jenuwein T, Reinberg D. eds. 2007. *Epigenetics*. Cold Spring Harbor Laboratory Press, Cold Spring Harbor, NY.

Andersen EC, Horvitz HR. 2007. Two *C. elegans* histone methyltransferases repress lin-3 EGF transcription to inhibit vulval development. *Development* **134:** 2991–2999.

Anest V, Hanson JL, Cogswell PC, Steinbrecher KA, Strahl BD, Baldwin AS. 2003. A nucleosomal function for IκB kinase-α in NF-κB-dependent gene expression. *Nature* **423:** 659–663.

Bailey AO, Panchenko T, Sathyan KM, Petkowski JJ, Pai PJ, Bai DL, Russell DH, Macara IG, Shabanowitz J, Hunt DF, et al. 2013. Posttranslational modification of CENP-A influences the conformation of centromeric chromatin. *Proc Natl Acad Sci* **110:** 11827–11832.

Baker SP, Phillips J, Anderson S, Qiu Q, Shabanowitz J, Smith MM, Yates JR 3rd, Hunt DF, Grant PA. 2010. Histone H3 Thr 45 phosphorylation is a replication-associated post-translational modification in *S. cerevisiae*. *Nat Cell Biol* **12:** 294–298.

Bannister AJ, Zegerman P, Partridge JF, Miska EA, Thomas JO, Allshire RC, Kouzarides T. 2001. Selective recognition of methylated lysine 9 on histone H3 by the HP1 chromo domain. *Nature* **410:** 120–124.

Bauer UM, Daujat S, Nielsen SJ, Nightingale K, Kouzarides T. 2002. Methylation at arginine 17 of histone H3 is linked to gene activation. *EMBO Rep* **3:** 39–44.

Beisel C, Imhof A, Greene J, Kremmer E, Sauer F. 2002. Histone methylation by the *Drosophila* epigenetic transcriptional regulator Ash1. *Nature* **419:** 857–862.

Bell O, Wirbelauer C, Hild M, Scharf AN, Schwaiger M, MacAlpine DM, Zilbermann F, van Leeuwen F, Bell SP, Imhof A, et al. 2007. Localized H3K36 methylation states define histone H4K16 acetylation during transcriptional elongation in *Drosophila*. *EMBO J* **26:** 4974–4984.

Bender LB, Suh J, Carroll CR, Fong Y, Fingerman IM, Briggs SD, Cao R, Zhang Y, Reinke V, Strome S. 2006. MES-4: An autosome-associated histone methyltransferase that participates in silencing the X chromosomes in the *C. elegans* germ line. *Development* **133:** 3907–3917.

Bernstein BE, Humphrey EL, Erlich RL, Schneider R, Bouman P, Liu JS, Kouzarides T, Schreiber SL. 2002. Methylation of histone H3 Lys 4 in coding regions of active genes. *Proc Natl Acad Sci* **99:** 8695–8700.

Bernstein E, Muratore-Schroeder TL, Diaz RL, Chow JC, Changolkar LN, Shabanowitz J, Heard E, Pehrson JR, Hunt DF, Allis CD. 2008. A phosphorylated subpopulation of the histone variant macroH2A1 is excluded from the inactive X chromosome and enriched during mitosis. *Proc Natl Acad Sci* **105:** 1533–1538.

Bessler JB, Andersen EC, Villeneuve AM. 2010. Differential localization and independent acquisition of the H3K9me2 and H3K9me3 chromatin modifications in the *Caenorhabditis elegans* adult germ line. *PLoS Genet* **6:** e1000830.

Binda O, Sevilla A, LeRoy G, Lemischka IR, Garcia BA, Richard S. 2013. SETD6 monomethylates H2AZ on lysine 7 and is required for the maintenance of embryonic stem cell self-renewal. *Epigenetics* **8:** 177–183.

Bird AW, Yu DY, Pray-Grant MG, Qiu Q, Harmon KE, Megee PC, Grant PA, Smith MM, Christman MF. 2002. Acetylation of histone H4 by Esa1 is required for DNA double-strand break repair. *Nature* **419:** 411–415.

Bonenfant D, Coulot M, Towbin H, Schindler P, van Oostrum J. 2006. Characterization of histone H2A and H2B variants and their post-translational modifications by mass spectrometry. *Mol Cell Proteomics* **5:** 541–552.

Briggs SD, Bryk M, Strahl BD, Cheung WL, Davie JK, Dent SY, Winston F, Allis CD. 2001. Histone H3 lysine 4 methylation is mediated by Set1 and required for cell growth and rDNA silencing in *Saccharomyces cerevisiae*. *Genes Dev* **15:** 3286–3295.

Britton LM, Newhart A, Bhanu NV, Sridharan R, Gonzales-Cope M, Plath K, Janicki SM, Garcia BA. 2013. Initial characterization of histone H3 serine 10 O-acetylation. *Epigenetics* **8:** 1101–1113.

Brownell JE, Zhou J, Ranalli T, Kobayashi R, Edmondson DG, Roth SY, Allis CD. 1996. *Tetrahymena* histone acetyltransferase A: A homolog to yeast Gcn5p linking histone acetylation to gene activation. *Cell* **84:** 843–851.

Bryk M, Briggs SD, Strahl BD, Curcio MJ, Allis CD, Winston F. 2002. Evidence that Set1, a factor required for methylation of histone H3, regulates rDNA silencing in *S. cerevisiae* by a Sir2-dependent mechanism. *Curr Biol* **12:** 165–170.

Burma S, Chen BP, Murphy M, Kurimasa A, Chen DJ. 2001. ATM phosphorylates histone H2AX in response to DNA double-strand breaks. *J Biol Chem* **276:** 42462–42467.

Cao R, Wang L, Wang H, Xia L, Erdjument-Bromage H, Tempst P, Jones RS, Zhang Y. 2002. Role of histone H3 lysine 27 methylation in Polycomb-group silencing. *Science* **298:** 1039–1043.

Carling T, Kim KC, Yang XH, Gu J, Zhang XK, Huang S. 2004. A histone methyltransferase is required for maximal response to female sex hormones. *Mol Cell Biol* **24:** 7032–7042.

Casadio F, Lu X, Pollock SB, LeRoy G, Garcia BA, Muir TW, Roeder RG, Allis CD. 2013. H3R42me2a is a histone modification with positive transcriptional effects. *Proc Natl Acad Sci* **110:** 14894–14899.

Chadee DN, Taylor WR, Hurta RA, Allis CD, Wright JA, Davie JR. 1995. Increased phosphorylation of histone H1 in mouse fibroblasts transformed with oncogenes or constitutively active mitogen-activated protein kinase kinase. *J Biol Chem* **270:** 20098–20105.

Chang L, Loranger SS, Mizzen C, Ernst SG, Allis CD, Annunziato AT. 1997. Histones in transit: Cytosolic histone complexes and diacetylation of H4 during nucleosome assembly in human cells. *Biochemistry* **36:** 469–480.

Chen D, Ma H, Hong H, Koh SS, Huang SM, Schurter BT, Aswad DW, Stallcup MR. 1999. Regulation of transcription by a protein methyltransferase. *Science* **284:** 2174–2177.

Chen Y, Sprung R, Tang Y, Ball H, Sangras B, Kim SC, Falck JR, Peng J, Gu W, Zhao Y. 2007. Lysine propionylation and butyrylation are novel post-translational modifications in histones. *Mol Cell Proteomics* **6:** 812–819.

Cheung P, Tanner KG, Cheung WL, Sassone-Corsi P, Denu JM, Allis CD. 2000. Synergistic coupling of histone H3 phosphorylation and acetylation in response to epidermal growth factor stimulation. *Mol Cel* **5:** 905–915.

Cheung WL, Ajiro K, Samejima K, Kloc M, Cheung P, Mizzen CA, Beeser A, Etkin LD, Chernoff J, Earnshaw WC, et al. 2003. Apoptotic phosphorylation of histone H2B is mediated by mammalian sterile twenty kinase. *Cell* **113:** 507–517.

Cheung WL, Turner FB, Krishnamoorthy T, Wolner B, Ahn SH, Foley M, Dorsey JA, Peterson CL, Berger SL, Allis CD. 2005. Phosphorylation of histone H4 serine 1 during DNA damage requires casein kinase II in *S. cerevisiae*. *Curr Biol* **15:** 656–660.

Chew YC, Camporeale G, Kothapalli N, Sarath G, Zempleni J. 2006. Lysine residues in N-terminal and C-terminal regions of human histone H2A are targets for biotinylation by biotinidase. *J Nutr Biochem* **17:** 225–233.

Choy JS, Tobe BT, Huh JH, Kron SJ. 2001. Yng2p-dependent NuA4 histone H4 acetylation activity is required for mitotic and meiotic progression. *J Biol Chem* **276:** 43653–43662.

Christophorou MA, Castelo-Branco G, Halley-Stott RP, Oliveira CS, Loos R, Radzisheuskaya A, Mowen KA, Bertone P, Silva JC, Zernicka-Goetz M, et al. 2014. Citrullination regulates pluripotency and histone H1 binding to chromatin. *Nature* **507:** 104–108.

Chu F, Nusinow DA, Chalkley RJ, Plath K, Panning B, Burlingame AL. 2006. Mapping post-translational modifications of the histone variant MacroH2A1 using tandem mass spectrometry. *Mol Cell Proteomics* **5:** 194–203.

Clarke AS, Lowell JE, Jacobson SJ, Pillus L. 1999. Esa1p is an essential histone acetyltransferase required for cell cycle progression. *Mol Cell Biol* **19:** 2515–2526.

Clayton AL, Rose S, Barratt MJ, Mahadevan LC. 2000. Phosphoacetylation of histone H3 on c-*fos*- and c-*jun*-associated nucleosomes upon gene activation. *EMBO J* **19:** 3714–3726.

Creyghton MP, Cheng AW, Welstead GG, Kooistra T, Carey BW, Steine EJ, Hanna J, Lodato MA, Frampton GM, Sharp PA, et al. 2010. Histone H3K27ac separates active from poised enhancers and predicts developmental state. *Proc Natl Acad Sci* **107:** 21931–21936.

Czermin B, Schotta G, Hulsmann BB, Brehm A, Becker PB, Reuter G, Imhof A. 2001. Physical and functional association of SU(VAR)3–9 and HDAC1 in *Drosophila*. *EMBO Rep* **2:** 915–919.

Czermin B, Melfi R, McCabe D, Seitz V, Imhof A, Pirrotta V. 2002. *Drosophila* enhancer of Zeste/ESC complexes have a histone H3 methyltransferase activity that marks chromosomal Polycomb sites. *Cell* **111:** 185–196.

Dai J, Sultan S, Taylor SS, Higgins JM. 2005. The kinase haspin is required for mitotic histone H3 Thr 3 phosphorylation and normal metaphase chromosome alignment. *Genes Dev* **19:** 472–488.

Dai L, Peng C, Montellier E, Lu Z, Chen Y, Ishii H, Debernardi A, Buchou T, Rousseaux S, Jin F, et al. 2014. Lysine 2-hydroxyisobutyrylation is a widely distributed active histone mark. *Nat Chem Biol* **10:** 365–370.

Daujat S, Bauer UM, Shah V, Turner B, Berger S, Kouzarides T. 2002. Crosstalk between CARM1 methylation and CBP acetylation on histone H3. *Curr Biol* **12:** 2090–2097.

Daujat S, Zeissler U, Waldmann T, Happel N, Schneider R. 2005. HP1 binds specifically to Lys26-methylated histone H14, whereas simultaneous Ser27 phosphorylation blocks HP1 binding. *J Biol Chem* **280:** 38090–38095.

Daujat S, Weiss T, Mohn F, Lange UC, Ziegler-Birling C, Zeissler U, Lappe M, Schubeler D, Torres-Padilla ME, Schneider R. 2009. H3K64 trimethylation marks heterochromatin and is dynamically remodeled during developmental reprogramming. *Nat Struct Mol Biol* **16:** 777–781.

Dawson MA, Bannister AJ, Gottgens B, Foster SD, Bartke T, Green AR, Kouzarides T. 2009. JAK2 phosphorylates histone H3Y41 and excludes HP1α from chromatin. *Nature* **461:** 819–822.

De Souza CP, Osmani AH, Wu LP, Spotts JL, Osmani SA. 2000. Mitotic histone H3 phosphorylation by the NIMA kinase in *Aspergillus nidulans*. *Cell* **102:** 293–302.

Di Cerbo V, Mohn F, Ryan DP, Montellier E, Kacem S, Tropberger P, Kallis E, Holzner M, Hoerner L, Feldmann A, et al. 2014. Acetylation of histone H3 at lysine 64 regulates nucleosome dynamics and facilitates transcription. *eLife* **3:** e01632.

Dodge JE, Kang YK, Beppu H, Lei H, Li E. 2004. Histone H3-K9 methyltransferase ESET is essential for early development. *Mol Cell Biol* **24:** 2478–2486.

Downs JA, Lowndes NF, Jackson SP. 2000. A role for *Saccharomyces cerevisiae* histone H2A in DNA repair. *Nature* **408:** 1001–1004.

Downs JA, Allard S, Jobin-Robitaille O, Javaheri A, Auger A, Bouchard N, Kron SJ, Jackson SP, Cote J. 2004. Binding of chromatin-modifying activities to phosphorylated histone H2A at DNA damage sites. *Mol Cell* **16:** 979–990.

Ebert A, Schotta G, Lein S, Kubicek S, Krauss V, Jenuwein T, Reuter G. 2004. Su(var) genes regulate the balance between euchromatin and heterochromatin in *Drosophila*. *Genes Dev* **18:** 2973–2983.

Edmunds JW, Mahadevan LC, Clayton AL. 2008. Dynamic histone H3 methylation during gene induction: HYPB/Setd2 mediates all H3K36 trimethylation. *EMBO J* **27:** 406–420.

Emre NC, Ingvarsdottir K, Wyce A, Wood A, Krogan NJ, Henry KW, Li K, Marmorstein R, Greenblatt JF, Shilatifard A, et al. 2005. Maintenance of low histone ubiquitylation by Ubp10 correlates with telomere-proximal Sir2 association and gene silencing. *Mol Cell* **17:** 585–594.

Fang J, Feng Q, Ketel CS, Wang H, Cao R, Xia L, Erdjument-Bromage H, Tempst P, Simon JA, Zhang Y. 2002. Purification and functional characterization of SET8, a nucleosomal histone H4-lysine 20-specific methyltransferase. *Curr Biol* **12:** 1086–1099.

Feng Q, Wang H, Ng HH, Erdjument-Bromage H, Tempst P, Struhl K, Zhang Y. 2002. Methylation of H3-lysine 79 is mediated by a new family of HMTases without a SET domain. *Curr Biol* **12:** 1052–1058.

Fong JJ, Nguyen BL, Bridger R, Medrano EE, Wells L, Pan S, Sifers RN. 2012. β-N-Acetylglucosamine O-GlcNAc is a novel regulator of mitosis-specific phosphorylations on histone H3. *J Biol Chem* **287:** 12195–12203.

Galasinski SC, Louie DF, Gloor KK, Resing KA, Ahn NG. 2002. Global regulation of post-translational modifications on core histones. *J Biol Chem* **277:** 2579–2588.

Garcia BA, Busby SA, Barber CM, Shabanowitz J, Allis CD, Hunt DF. 2004. Characterization of phosphorylation sites on histone H1 isoforms by tandem mass spectrometry. *J Proteome Res* **3:** 1219–1227.

Garcia BA, Busby SA, Shabanowitz J, Hunt DF, Mishra N. 2005. Resetting the epigenetic histone code in the MRL-lpr/lpr mouse model of lupus by histone deacetylase inhibition. *J Proteome Res* **4:** 2032–2042.

Garcia BA, Hake SB, Diaz RL, Kauer M, Morris SA, Recht J, Shabanowitz J, Mishra N, Strahl BD, Allis CD, et al. 2007. Organismal differences in post-translational modifications in histones H3 and H4. *J Biol Chem* **282:** 7641–7655.

Garcia-Gimenez JL, Olaso G, Hake SB, Bonisch C, Wiedemann SM, Markovic J, Dasi F, Gimeno A, Perez-Quilis C, Palacios O, et al. 2013. Histone h3 glutathionylation in proliferating mammalian cells destabilizes nucleosomal structure. *Antioxid Redox Signal* **19:** 1305–1320.

Gatti M, Pinato S, Maspero E, Soffientini P, Polo S, Penengo L. 2012. A novel ubiquitin mark at the N-terminal tail of histone H2As targeted by RNF168 ubiquitin ligase. *Cell Cycle* **11:** 2538–2544.

Goto H, Tomono Y, Ajiro K, Kosako H, Fujita M, Sakurai M, Okawa K, Iwamatsu A, Okigaki T, Takahashi T, et al. 1999. Identification of a novel phosphorylation site on histone H3 coupled with mitotic chromosome condensation. *J Biol Chem* **274:** 25543–25549.

Goto H, Yasui Y, Nigg EA, Inagaki M. 2002. Aurora-B phosphorylates histone H3 at serine28 with regard to the mitotic chromosome condensation. *Genes Cells* **7:** 11–17.

Grant PA, Eberharter A, John S, Cook RG, Turner BM, Workman JL. 1999. Expanded lysine acetylation specificity of Gcn5 in native complexes. *J Biol Chem* **274:** 5895–5900.

Green EM, Mas G, Young NL, Garcia BA, Gozani O. 2012. Methylation of H4 lysines 5, 8 and 12 by yeast Set5 calibrates chromatin stress responses. *Nat Struct Mol Biol* **19:** 361–363.

Greer EL, Shi Y. 2012. Histone methylation: A dynamic mark in health, disease and inheritance. *Nat Rev Genet* **13:** 343–357.

Guillemette B, Drogaris P, Lin HH, Armstrong H, Hiragami-Hamada K, Imhof A, Bonneil E, Thibault P, Verreault A, Festenstein RJ. 2011. H3 lysine 4 is acetylated at active gene promoters and is regulated by H3 lysine 4 methylation. *PLoS Genet* **7:** e1001354.

Hake SB, Garcia BA, Kauer M, Baker SP, Shabanowitz J, Hunt DF, Allis CD. 2005. Serine 31 phosphorylation of histone variant H33 is specific to regions bordering centromeres in metaphase chromosomes. *Proc Natl Acad Sci* **102:** 6344–6349.

Hake SB, Garcia BA, Duncan EM, Kauer M, Dellaire G, Shabanowitz J, Bazett-Jones DP, Allis CD, Hunt DF. 2006. Expression patterns and post-translational modifications associated with mammalian histone H3 variants. *J Biol Chem* **281:** 559–568.

Hamamoto R, Furukawa Y, Morita M, Iimura Y, Silva FP, Li M, Yagyu R, Nakamura Y. 2004. SMYD3 encodes a histone methyltransferase involved in the proliferation of cancer cells. *Nat Cell Biol* **6:** 731–740.

Hammond SL, Byrum SD, Namjoshi S, Graves HK, Dennehey BK, Tackett AJ, Tyler JK. 2014. Mitotic phosphorylation of histone H3 threonine 80. *Cell Cycle* **13:** 440–452.

Han Y, Garcia BA. 2013. Combining genomic and proteomic approaches for epigenetics research. *Epigenomics* **5:** 439–452.

Harvey AC, Jackson SP, Downs JA. 2005. *Saccharomyces cerevisiae* histone H2A Ser122 facilitates DNA repair. *Genetics* **170:** 543–553.

Hayashi K, Yoshida K, Matsui Y. 2005. A histone H3 methyltransferase controls epigenetic events required for meiotic prophase. *Nature* **438:** 374–378.

He Z, Cho YY, Ma WY, Choi HS, Bode AM, Dong Z. 2005. Regulation of ultraviolet B-induced phosphorylation of histone H3 at serine 10 by Fyn kinase. *J Biol Chem* **280:** 2446–2454.

Hendzel MJ, Wei Y, Mancini MA, Van Hooser A, Ranalli T, Brinkley BR, Bazett-Jones DP, Allis CD. 1997. Mitosis-specific phosphorylation of histone H3 initiates primarily within pericentromeric heterochromatin during G2 and spreads in an ordered fashion coincident with mitotic chromosome condensation. *Chromosoma* **106:** 348–360.

Hsu JY, Sun ZW, Li X, Reuben M, Tatchell K, Bishop DK, Grushcow JM, Brame CJ, Caldwell JA, Hunt DF, et al. 2000. Mitotic phosphorylation of histone H3 is governed by Ipl1/aurora kinase and Glc7/PP1 phosphatase in budding yeast and nematodes. *Cell* **102:** 279–291.

Huen MS, Grant R, Manke I, Minn K, Yu X, Yaffe MB, Chen J. 2007. RNF8 transduces the DNA-damage signal via histone ubiquitylation and checkpoint protein assembly. *Cell* **131:** 901–914.

Hurd PJ, Bannister AJ, Halls K, Dawson MA, Vermeulen M, Olsen JV, Ismail H, Somers J, Mann M, Owen-Hughes T, et al. 2009. Phosphorylation of histone H3 Thr-45 is linked to apoptosis. *J Biol Chem* **284:** 16575–16583.

Hyland EM, Cosgrove MS, Molina H, Wang D, Pandey A, Cottee RJ, Boeke JD. 2005. Insights into the role of histone H3 and histone H4 core modifiable residues in *Saccharomyces cerevisiae*. *Mol Cell Biol* **25:** 10060–10070.

Ichijima Y, Sakasai R, Okita N, Asahina K, Mizutani S, Teraoka H. 2005. Phosphorylation of histone H2AX at M phase in human cells without DNA damage response. *Biochem Biophys Res Commun* **336:** 807–812.

Jack AP, Bussemer S, Hahn M, Punzeler S, Snyder M, Wells M, Csankovszki G, Solovei I, Schotta G, Hake SB. 2013. H3K56me3 is a novel, conserved heterochromatic mark that largely but not completely overlaps with H3K9me3 in both regulation and localization. *PloS One* **8:** e51765

Jackson JP, Lindroth AM, Cao X, Jacobsen SE. 2002. Control of CpNpG DNA methylation by the KRYPTONITE histone H3 methyltransferase. *Nature* **416:** 556–560.

Jackson JP, Johnson L, Jasencakova Z, Zhang X, PerezBurgos L, Singh PB, Cheng X, Schubert I, Jenuwein T, Jacobsen SE. 2004. Dimethylation of histone H3 lysine 9 is a critical mark for DNA methylation and gene silencing in *Arabidopsis thaliana*. *Chromosoma* **112:** 308–315.

Jeppesen P, Turner BM. 1993. The inactive X chromosome in female mammals is distinguished by a lack of histone H4 acetylation, a cytogenetic marker for gene expression. *Cell* **74:** 281–289.

Jiang L, Smith JN, Anderson SL, Ma P, Mizzen CA, Kelleher NL. 2007. Global assessment of combinatorial post-translational modification of core histones in yeast using contemporary mass spectrometry. LYS4 trimethylation correlates with degree of acetylation on the same H3 tail. *J Biol Chem* **282:** 27923–27934.

Jin Y, Wang Y, Walker DL, Dong H, Conley C, Johansen J, Johansen KM. 1999. JIL-1: A novel chromosomal tandem kinase implicated in transcriptional regulation in *Drosophila*. *Mol Cell* **4:** 129–135.

Kao CF, Hillyer C, Tsukuda T, Henry K, Berger S, Osley MA. 2004. Rad6 plays a role in transcriptional activation through ubiquitylation of histone H2B. *Genes Dev* **18:** 184–195.

Kapetanaki MG, Guerrero-Santoro J, Bisi DC, Hsieh CL, Rapic-Otrin V, Levine AS. 2006. The DDB1-CUL4A^{DDB2} ubiquitin ligase is deficient in xeroderma pigmentosum group E and targets histone H2A at UV-damaged DNA sites. *Proc Natl Acad Sci* **103:** 2588–2593.

Karch KR, Denizio JE, Black BE, Garcia BA. 2013. Identification and interrogation of combinatorial histone modifications. *Front Genet* **4:** 264.

Kawasaki H, Taira K, Yokoyama K. 2000a. Histone acetyltransferase (HAT) activity of ATF-2 is necessary for the CRE-dependent transcription. *Nucleic Acids Symp Ser* **2000:** 259–260.

Kawasaki H, Schiltz L, Chiu R, Itakura K, Taira K, Nakatani Y, Yokoyama KK. 2000b. ATF-2 has intrinsic histone acetyltransferase activity which is modulated by phosphorylation. *Nature* **405:** 195–200.

Kawashima SA, Yamagishi Y, Honda T, Ishiguro K, Watanabe Y. 2010. Phosphorylation of H2A by Bub1 prevents chromosomal instability through localizing shugoshin. *Science* **327:** 172–177.

Kim KC, Geng L, Huang S. 2003. Inactivation of a histone methyltransferase by mutations in human cancers. *Cancer Res* **63:** 7619–7623.

Kimura A, Horikoshi M. 1998. Tip60 acetylates six lysines of a specific class in core histones in vitro. *Genes Cells* **3:** 789–800.

Kizer KO, Phatnani HP, Shibata Y, Hall H, Greenleaf AL, Strahl BD. 2005. A novel domain in Set2 mediates RNA polymerase II interaction and couples histone H3 K36 methylation with transcript elongation. *Mol Cell Biol* **25:** 3305–3316.

Kleff S, Andrulis ED, Anderson CW, Sternglanz R. 1995. Identification of a gene encoding a yeast histone H4 acetyltransferase. *J Biol Chem* **270:** 24674–24677.

Kothapalli N, Camporeale G, Kueh A, Chew YC, Oommen AM, Griffin JB, Zempleni J. 2005a. Biological functions of biotinylated histones. *J Nutr Biochem* **16:** 446–448.

Kothapalli N, Sarath G, Zempleni J. 2005b. Biotinylation of K12 in histone H4 decreases in response to DNA double-strand breaks in human JAr choriocarcinoma cells. *J Nutr* **135:** 2337–2342.

Kouzarides T. 2007. Chromatin modifications and their function. *Cell* **128:** 693–705.

Kruhlak MJ, Hendzel MJ, Fischle W, Bertos NR, Hameed S, Yang XJ, Verdin E, Bazett-Jones DP. 2001. Regulation of global acetylation in mitosis through loss of histone acetyltransferases and deacetylases from chromatin. *J Biol Chem* **276:** 38307–38319.

Ku M, Jaffe JD, Koche RP, Rheinbay E, Endoh M, Koseki H, Carr SA, Bernstein BE. 2012. H2A.Z landscapes and dual modifications in pluripotent and multipotent stem cells underlie complex genome regulatory functions. *Genome Biol* **13:** R85.

Kuo MH, Brownell JE, Sobel RE, Ranalli TA, Cook RG, Edmondson DG, Roth SY, Allis CD. 1996. Transcription-linked acetylation by Gcn5p of histones H3 and H4 at specific lysines. *Nature* **383:** 269–272.

Kuzmichev A, Nishioka K, Erdjument-Bromage H, Tempst P, Reinberg D. 2002. Histone methyltransferase activity associated with a human multiprotein complex containing the Enhancer of Zeste protein. *Genes Dev* **16:** 2893–2905.

Kuzmichev A, Jenuwein T, Tempst P, Reinberg D. 2004. Different EZH2-containing complexes target methylation of histone H1 or nucleosomal histone H3. *Mol Cell* **14:** 183–193.

Lachner M, O'Carroll D, Rea S, Mechtler K, Jenuwein T. 2001. Methylation of histone H3 lysine 9 creates a binding site for HP1 proteins. *Nature* **410:** 116–120.

Lachner M, O'Sullivan RJ, Jenuwein T. 2003. An epigenetic road map for histine lysine methylation. *J Cell Sci* **116:** 2117–2124.

Lacoste N, Utley RT, Hunter JM, Poirier GG, Cote J. 2002. Disruptor of telomeric silencing-1 is a chromatin-specific histone H3 methyltransferase. *J Biol Chem* **277:** 30421–30424.

Lee JH, Skalnik DG. 2005. CpG-binding protein (CXXC finger protein 1) is a component of the mammalian Set1 histone H3-Lys4 methyltransferase complex, the analogue of the yeast Set1/COMPASS complex. *J Biol Chem* **280:** 41725–41731.

Lee JH, Tate CM, You JS, Skalnik DG. 2007. Identification and characterization of the human Set1B histone H3-Lys4 methyltransferase complex. *J Biol Chem* **282:** 13419–13428.

Lo WS, Duggan L, Emre NC, Belotserkovskya R, Lane WS, Shiekhattar R, Berger SL. 2001. Snf1—A histone kinase that works in concert with the histone acetyltransferase Gcn5 to regulate transcription. *Science* **293:** 1142–1146.

Lu A, Zougman A, Pudelko M, Bebenek M, Ziolkowski P, Mann M, Wisniewski JR. 2009. Mapping of lysine monomethylation of linker histones in human breast and its cancer. *J Proteome Res* **8:** 4207–4215.

Mailand N, Bekker-Jensen S, Faustrup H, Melander F, Bartek J, Lukas C, Lukas J. 2007. RNF8 ubiquitylates histones at DNA double-strand breaks and promotes assembly of repair proteins. *Cell* **131:** 887–900.

Maile T, Kwoczynski S, Katzenberger RJ, Wassarman DA, Sauer F. 2004. TAF1 activates transcription by phosphorylation of serine 33 in histone H2B. *Science* **304:** 1010–1014.

Masumoto H, Hawke D, Kobayashi R, Verreault A. 2005. A role for cell-cycle-regulated histone H3 lysine 56 acetylation in the DNA damage response. *Nature* **436:** 294–298.

Mattiroli F, Vissers JH, van Dijk WJ, Ikpa P, Citterio E, Vermeulen W, Marteijn JA, Sixma TK. 2012. RNF168 ubiquitinates K13–15 on H2A/H2AX to drive DNA damage signaling. *Cell* **150:** 1182–1195.

McKittrick E, Gafken PR, Ahmad K, Henikoff S. 2004. Histone H33 is enriched in covalent modifications associated with active chromatin. *Proc Natl Acad Sci* **101:** 1525–1530.

Metzger E, Imhof A, Patel D, Kahl P, Hoffmeyer K, Friedrichs N, Muller JM, Greschik H, Kirfel J, Ji S, et al. 2010. Phosphorylation of histone H3T6 by PKCβ(I) controls demethylation at histone H3K4. *Nature* **464:** 792–796.

Milne TA, Briggs SD, Brock HW, Martin ME, Gibbs D, Allis CD, Hess JL. 2002. MLL targets SET domain methyltransferase activity to *Hox* gene promoters. *Mol Cell* **10:** 1107–1117.

Miranda TB, Sayegh J, Frankel A, Katz JE, Miranda M, Clarke S. 2006. Yeast Hsl7 (histone synthetic lethal 7) catalyses the in vitro formation of ω-N^G-monomethylarginine in calf thymus histone H2A. *Biochem J* **395:** 563–570.

Mizzen CA, Yang XJ, Kokubo T, Brownell JE, Bannister AJ, Owen-Hughes T, Workman J, Wang L, Berger SL, Kouzarides T, et al. 1996. The $TAF_{II}250$ subunit of TFIID has histone acetyltransferase activity. *Cell* **87:** 1261–1270.

Montellier E, Boussovar F, Rousseaux S, Zhang K, Buchou T, Fenaille F, Shiota H, Debernardi A, Hery P, Jamshidikia M, et al. 2013. Chromotin-to-nucleoprotamine transition is controlled by the histone H2B variant TH2B. *Genes Dev* **27:** 1680–1692.

Morris SA, Shibata Y, Noma K, Tsukamoto Y, Warren E, Temple B, Grewal SI, Strahl BD. 2005. Histone H3 K36 methylation is associated with transcription elongation in *Schizosaccharomyces pombe*. *Eukaryot Cell* **4:** 1446–1454.

Morris SA, Rao B, Garcia BA, Hake SB, Diaz RL, Shabanowitz J, Hunt DF, Allis CD, Lieb JD, Strahl BD. 2007. Identification of histone H3 lysine 36 acetylation as a highly conserved histone modification. *J Biol Chem* **282:** 7632–7640.

Mukherjee B, Kessinger C, Kobayashi J, Chen BP, Chen DJ, Chatterjee A, Burma S. 2006. DNA-PK phosphorylates histone H2AX during apoptotic DNA fragmentation in mammalian cells. *DNA Repair* **5:** 575–590.

Muller J, Hart CM, Francis NJ, Vargas ML, Sengupta A, Wild B, Miller EL, O'Connor MB, Kingston RE, Simon JA. 2002. Histone methyltransferase activity of a *Drosophila* Polycomb group repressor complex. *Cell* **111:** 197–208.

Murr R. 2010. Interplay between different epigenetic modifications and mechanisms. *Adv Genet* **70:** 101–141.

Nagy PL, Griesenbeck J, Kornberg RD, Cleary ML. 2002. A trithorax-group complex purified from *Saccharomyces cerevisiae* is required for methylation of histone H3. *Proc Natl Acad Sci* **99:** 90–94.

Nakayama J, Rice JC, Strahl BD, Allis CD, Grewal SI. 2001. Role of histone H3 lysine 9 methylation in epigenetic control of heterochromatin assembly. *Science* **292:** 110–113.

Nakamura T, Mori T, Tada S, Krajewski W, Rozovskaia T, Wassell R, Dubois G, Mazo A, Croce CM, Canaani E. 2002. ALL-1 is a histone methyltransferase that assembles a supercomplex of proteins involved in transcriptional regulation. *Mol Cell* **10:** 1119–1128.

Nathan D, Ingvarsdottir K, Sterner DE, Bylebyl GR, Dokmanovic M, Dorsey JA, Whelan KA, Krsmanovic M, Lane WS, Meluh PB, et al. 2006. Histone sumoylation is a negative regulator in *Saccharomyces cerevisiae* and shows dynamic interplay with positive-acting histone modifications. *Genes Dev* **20:** 966–976.

Nelson CJ, Santos-Rosa H, Kouzarides T. 2006. Proline isomerization of histone H3 regulates lysine methylation and gene expression. *Cell* **126:** 905–916.

Ng HH, Xu RM, Zhang Y, Struhl K. 2002. Ubiquitination of histone H2B by Rad6 is required for efficient Dot1-mediated methylation of histone H3 lysine 79. *J Biol Chem* **277:** 34655–34657.

Nielsen SJ, Schneider R, Bauer UM, Bannister AJ, Morrison A, O'Carroll D, Firestein R, Cleary M, Jenuwein T, Herrera RE, et al. 2001. Rb targets histone H3 methylation and HP1 to promoters. *Nature* **412:** 561–565.

Nishioka K, Chuikov S, Sarma K, Erdjument-Bromage H, Allis CD, Tempst P, Reinberg D. 2002a. Set9, a novel histone H3 methyltransferase that facilitates transcription by precluding histone tail modifications required for heterochromatin formation. *Genes Dev* **16:** 479–489.

Nishioka K, Rice JC, Sarma K, Erdjument-Bromage H, Werner J, Wang Y, Chuikov S, Valenzuela P, Tempst P, Steward R, et al. 2002b. PR-Set7 is a nucleosome-specific methyltransferase that modifies lysine 20 of histone H4 and is associated with silent chromatin. *Mol Cell* **9:** 1201–1213.

Nowak SJ, Corces VG. 2000. Phosphorylation of histone H3 correlates with transcriptionally active loci. *Genes Dev* **14:** 3003–3013.

O'Carroll D, Scherthan H, Peters AH, Opravil S, Haynes AR, Laible G, Rea S, Schmid M, Lebersorger A, Jerratsch M, et al. 2000. Isolation and characterization of *Suv39h2*, a second histone H3 methyltransferase gene that displays testis-specific expression. *Mol Cell Biol* **20:** 9423–9433.

Odegard VH, Kim ST, Anderson SM, Shlomchik MJ, Schatz DG. 2005. Histone modifications associated with somatic hypermutation. *Immunity* **23:** 101–110.

Ogata N, Ueda K, Hayaishi O. 1980a. ADP-ribosylation of histone H2B. Identification of glutamic acid residue 2 as the modification site. *J Biol Chem* **255:** 7610–7615.

Ogata N, Ueda K, Kagamiyama H, Hayaishi O. 1980b. ADP-ribosylation of histone H1 identification of glutamic acid residues 2, 14, the COOH-terminal lysine residue as modification sites. *J Biol Chem* **255:** 7616–7620.

Ogawa H, Ishiguro K, Gaubatz S, Livingston DM, Nakatani Y. 2002. A complex with chromatin modifiers that occupies E2F- and Myc-responsive genes in G_0 cells. *Science* **296:** 1132–1136.

Ogawa Y, Ono T, Wakata Y, Okawa K, Tagami H, Shibahara KI. 2005. Histone variant macroH2A12 is mono-ubiquitinated at its histone domain. *Biochem Biophys Res Commun* **336:** 204–209.

Ozdemir A, Spicuglia S, Lasonder E, Vermeulen M, Campsteijn C, Stunnenberg HG, Logie C. 2005. Characterization of lysine 56 of histone H3 as an acetylation site in *Saccharomyces cerevisiae*. *J Biol Chem* **280:** 25949–25952.

Pal S, Vishwanath SN, Erdjument-Bromage H, Tempst P, Sif S. 2004. Human SWI/SNF-associated PRMT5 methylates histone H3 arginine 8 and negatively regulates expression of ST7 and NM23 tumor suppressor genes. *Mol Cell Biol* **24:** 9630–9645.

Pantazis P, Bonner WM. 1981. Quantitative determination of histone modification H2A acetylation and phosphorylation. *J Biol Chem* **256:** 4669–4675.

Park J, Chen Y, Tishkoff DX, Peng C, Tan M, Dai L, Xie Z, Zhang Y, Zwaans BM, Skinner ME, et al. 2013. SIRTS-mediated lysine desuccinylation impacts diverse metabolic pathways. *Mol Cell* **50:** 919–930.

Parthun MR, Widom J, Gottschling DE. 1996. The major cytoplasmic histone acetyltransferase in yeast: Links to chromatin replication and histone metabolism. *Cell* **87:** 85–94.

Peters AH, O'Carroll D, Scherthan H, Mechtler K, Sauer S, Schofer C, Weipoltshammer K, Pagani M, Lachner M, Kohlmaier A, et al. 2001. Loss of the *Suv39h* histone methyltransferases impairs mammalian heterochromatin and genome stability. *Cell* **107:** 323–337.

Pham AD, Sauer F. 2000. Ubiquitin-activating/conjugating activity of $TAF_{II}250$, a mediator of activation of gene expression in *Drosophila*. *Science* **289:** 2357–2360.

Polioudaki H, Markaki Y, Kourmouli N, Dialynas G, Theodoropoulos PA, Singh PB, Georgatos SD. 2004. Mitotic phosphorylation of histone H3 at threonine 3. *FEBS Lett* **560:** 39–44.

Preuss U, Landsberg G, Scheidtmann KH. 2003. Novel mitosis-specific phosphorylation of histone H3 at Thr11 mediated by Dlk/ZIP kinase. *Nucleic Acids Res* **31:** 878–885.

Puerta C, Hernandez F, Lopez-Alarcon L, Palacian E. 1995. Acetylation of histone H2AH2B dimers facilitates transcription. *Biochem Biophys Res Commun* **210:** 409–416.

Rea S, Eisenhaber F, O'Carroll D, Strahl BD, Sun ZW, Schmid M, Opravil S, Mechtler K, Ponting CP, Allis CD, et al. 2000. Regulation of chromatin structure by site-specific histone H3 methyltransferases. *Nature* **406:** 593–599.

Redon C, Pilch DR, Rogakou EP, Orr AH, Lowndes NF, Bonner WM. 2003. Yeast histone 2A serine 129 is essential for the efficient repair of checkpoint-blind DNA damage. *EMBO Rep* **4:** 678–684.

Rice JC, Nishioka K, Sarma K, Steward R, Reinberg D, Allis CD. 2002. Mitotic-specific methylation of histone H4 Lys 20 follows increased PR-Set7 expression and its localization to mitotic chromosomes. *Genes Dev* **16:** 2225–2230.

Robzyk K, Recht J, Osley MA. 2000. Rad6-dependent ubiquitination of histone H2B in yeast. *Science* **287:** 501–504.

Rogakou EP, Pilch DR, Orr AH, Ivanova VS, Bonner WM. 1998. DNA double-stranded breaks induce histone H2AX phosphorylation on serine 139. *J Biol Chem* **273:** 5858–5868.

Rogakou EP, Boon C, Redon C, Bonner WM. 1999. Megabase chromatin domains involved in DNA double-strand breaks in vivo. *J Cell Biol* **146:** 905–916.

Roguev A, Schaft D, Shevchenko A, Pijnappel WW, Wilm M, Aasland R, Stewart AF. 2001. The *Saccharomyces cerevisiae* Set1 complex includes an Ash2 homologue and methylates histone 3 lysine 4. *EMBO J* **20:** 7137–7148.

Ruiz-Carrillo A, Wangh LJ, Allfrey VG. 1975. Processing of newly synthesized histone molecules. *Science* **190:** 117–128.

Sakabe K, Wang Z, Hart GW. 2010. β-*N*-acetylglucosamine O-GlcNAc is part of the histone code. *Proc Natl Acad Sci* **107:** 19915–19920.

Santos-Rosa H, Schneider R, Bannister AJ, Sherriff J, Bernstein BE, Emre NC, Schreiber SL, Mellor J, Kouzarides T. 2002. Active genes are tri-methylated at K4 of histone H3. *Nature* **419:** 407–411.

Sarg B, Helliger W, Talasz H, Forg B, Lindner HH. 2006. Histone H1 phosphorylation occurs site-specifically during interphase and mitosis: Identification of a novel phosphorylation site on histone H1. *J Biol Chem* **281:** 6573–6580.

Sassone-Corsi P, Mizzen CA, Cheung P, Crosio C, Monaco L, Jacquot S, Hanauer A, Allis CD. 1999. Requirement of Rsk-2 for epidermal growth factor-activated phosphorylation of histone H3. *Science* **285:** 886–891.

Schiltz RL, Mizzen CA, Vassilev A, Cook RG, Allis CD, Nakatani Y. 1999. Overlapping but distinct patterns of histone acetylation by the human coactivators p300 and PCAF within nucleosomal substrates. *J Biol Chem* **274:** 1189–1192.

Schotta G, Ebert A, Krauss V, Fischer A, Hoffmann J, Rea S, Jenuwein T, Dorn R, Reuter G. 2002. Central role of *Drosophila* SU(VAR)3–9 in histone H3-K9 methylation and heterochromatic gene silencing. *EMBO J* **21:** 1121–1131.

Schotta G, Lachner M, Sarma K, Ebert A, Sengupta R, Reuter G, Reinberg D, Jenuwein T. 2004. A silencing pathway to induce H3-K9 and H4-K20 trimethylation at constitutive heterochromatin. *Genes Dev* **18:** 1251–1262.

Schubeler D, Francastel C, Cimbora DM, Reik A, Martin DI, Groudine M. 2000. Nuclear localization and histone acetylation: A pathway for chromatin opening and transcriptional activation of the human β-globin locus. *Genes Dev* **14:** 940–950.

Schultz DC, Ayyanathan K, Negorev D, Maul GG, Rauscher FJ III. 2002. SETDB1: A novel KAP-1-associated histone H3, lysine 9-specific methyltransferase that contributes to HP1-mediated silencing of euchromatic genes by KRAB zinc-finger proteins. *Genes Dev* **16:** 919–932.

Schurter BT, Koh SS, Chen D, Bunick GJ, Harp JM, Hanson BL, Henschen-Edman A, Mackay DR, Stallcup MR, Aswad DW. 2001. Methylation of histone H3 by coactivator-associated arginine methyltransferase 1. *Biochemistry* **40:** 5747–5756.

Sharma R, Nakamura A, Takahashi R, Nakamoto H, Goto S. 2006. Carbonyl modification in rat liver histones: Decrease with age and increase by dietary restriction. *Free Radic Biol Med* **40:** 1179–1184.

Shiio Y, Eisenman RN. 2003. Histone sumoylation is associated with transcriptional repression. *Proc Natl Acad Sci* **100:** 13225–13230.

Smith ER, Eisen A, Gu W, Sattah M, Pannuti A, Zhou J, Cook RG, Lucchesi JC, Allis CD. 1998. ESA1 is a histone acetyltransferase that is essential for growth in yeast. *Proc Natl Acad Sci* **95:** 3561–3565.

Sobel RE, Cook RG, Perry CA, Annunziato AT, Allis CD. 1995. Conservation of deposition-related acetylation sites in newly synthesized histones H3 and H4. *Proc Natl Acad Sci* **92:** 1237–1241.

Song OK, Wang X, Waterborg JH, Sternglanz R. 2003. An N^{α}-acetyltransferase responsible for acetylation of the N-terminal residues of histones H4 and H2A. *J Biol Chem* **278:** 38109–38112.

Spencer TE, Jenster G, Burcin MM, Allis CD, Zhou J, Mizzen CA, McKenna NJ, Onate SA, Tsai SY, Tsai MJ, et al. 1997. Steroid receptor coactivator-1 is a histone acetyltransferase. *Nature* **389:** 194–198.

Stanley JS, Griffin JB, Zempleni J. 2001. Biotinylation of histones in human cells. Effects of cell proliferation. *Eur J Biochem* **268:** 5424–5429.

Stiff T, O'Driscoll M, Rief N, Iwabuchi K, Lobrich M, Jeggo PA. 2004. ATM and DNA-PK function redundantly to phosphorylate H2AX after exposure to ionizing radiation. *Cancer Res* **64:** 2390–2396.

Strahl BD, Ohba R, Cook RG, Allis CD. 1999. Methylation of histone H3 at lysine 4 is highly conserved and correlates with transcriptionally active nuclei in *Tetrahymena*. *Proc Natl Acad Sci* **96:** 14967–14972.

Strahl BD, Briggs SD, Brame CJ, Caldwell JA, Koh SS, Ma H, Cook RG, Shabanowitz J, Hunt DF, Stallcup MR, et al. 2001. Methylation of histone H4 at arginine 3 occurs in vivo and is mediated by the nuclear receptor coactivator PRMT1. *Curr Biol* **11:** 996–1000.

Strahl BD, Grant PA, Briggs SD, Sun ZW, Bone JR, Caldwell JA, Mollah S, Cook RG, Shabanowitz J, Hunt DF, et al. 2002. Set2 is a nucleosomal histone H3-selective methyltransferase that mediates transcriptional repression. *Mol Cell Biol* **22:** 1298–1306.

Su IH, Basavaraj A, Krutchinsky AN, Hobert O, Ullrich A, Chait BT, Tarakhovsky A. 2003. Ezh2 controls B cell development through histone H3 methylation and *Igh* rearrangement. *Nat Immunol* **4:** 124–131.

Suka N, Suka Y, Carmen AA, Wu J, Grunstein M. 2001. Highly specific antibodies determine histone acetylation site usage in yeast heterochromatin and euchromatin. *Mol Cell* **8:** 473–479.

Sun ZW, Allis CD. 2002. Ubiquitination of histone H2B regulates H3 methylation and gene silencing in yeast. *Nature* **418:** 104–108.

Sun XJ, Wei J, Wu XY, Hu M, Wang L, Wang HH, Zhang QH, Chen SJ, Huang QH, Chen Z. 2005. Identification and characterization of a novel human histone H3 lysine 36-specific methyltransferase. *J Biol Chem* **280:** 35261–35271.

Tachibana M, Sugimoto K, Fukushima T, Shinkai Y. 2001. Set domain-containing protein, G9a, is a novel lysine-preferring mammalian histone methyltransferase with hyperactivity and specific selectivity to lysines 9 and 27 of histone H3. *J Biol Chem* **276:** 25309–25317.

Tachibana M, Sugimoto K, Nozaki M, Ueda J, Ohta T, Ohki M, Fukuda M, Takeda N, Niida H, Kato H, et al. 2002. G9a histone methyltransferase plays a dominant role in euchromatic histone H3 lysine 9 methylation and is essential for early embryogenesis. *Genes Dev* **16:** 1779–1791.

Tachibana M, Ueda J, Fukuda M, Takeda N, Ohta T, Iwanari H, Sakihama T, Kodama T, Hamakubo T, Shinkai Y. 2005. Histone methyltransferases G9a and GLP form heteromeric complexes and are both crucial for methylation of euchromatin at H3-K9. *Genes Dev* **19:** 815–826.

Tamaru H, Selker EU. 2001. A histone H3 methyltransferase controls DNA methylation in *Neurospora crassa*. *Nature* **414:** 277–283.

Tan M, Luo H, Lee S, Jin F, Yang JS, Montellier E, Buchou T, Cheng Z, Rousseaux S, Rajagopal N, et al. 2011. Identification of 67 histone marks and histone lysine crotonylation as a new type of histone modification. *Cell* **146:** 1016–1028.

Taplick J, Kurtev V, Lagger G, Seiser C. 1998. Histone H4 acetylation during interleukin-2 stimulation of mouse T cells. *FEBS Lett* **436:** 349–352.

Tessarz P, Santos-Rosa H, Robson SC, Sylvestersen KB, Nelson CJ, Nielsen ML, Kouzarides T. 2014. Glutamine methylation in histone H2A is an RNA-polymerase-I-dedicated modification. *Nature* **505:** 564–568.

Thomson S, Clayton AL, Hazzalin CA, Rose S, Barratt MJ, Mahadevan LC. 1999. The nucleosomal response associated with immediate-early gene induction is mediated via alternative MAP kinase cascades: MSK1 as a potential histone H3/HMG-14 kinase. *EMBO J* **18:** 4779–4793.

Tie F, Banerjee R, Stratton CA, Prasad-Sinha J, Stepanik V, Zlobin A, Diaz MO, Scacheri PC, Harte PJ. 2009. CBP-mediated acetylation of histone H3 lysine 27 antagonizes *Drosophila* Polycomb silencing. *Development* **136:** 3131–3141.

Turner BM. 2005. Reading signals on the nucleosome with a new nomenclature for modified histones. *Nat Struct Mol Biol* **12:** 110–112.

Turner BM, Fellows G. 1989. Specific antibodies reveal ordered and cell-cycle-related use of histone-H4 acetylation sites in mammalian cells. *Eur J Biochem* **179:** 131–139.

Tweedie-Cullen RY, Reck JM, Mansuy IM. 2009. Comprehensive mapping of post-translational modifications on synaptic, nuclear, and histone proteins in the adult mouse brain. *J Proteome Res* **8:** 4966–4982.

Tweedie-Cullen RY, Brunner AM, Grossmann J, Mohanna S, Sichau D, Nanni P, Panse C, Mansuy IM. 2012. Identification of combinatorial patterns of post-translational modifications on individual histones in the mouse brain. *PloS One* **7:** e36980.

Unoki M, Masuda A, Dohmae N, Arita K, Yoshimatsu M, Iwai Y, Fukui Y, Ueda K, Hamamoto R, Shirakawa M, et al. 2013. Lysyl 5-hydroxylation, a novel histone modification, by Jumonji domain containing 6 JMJD6. *J Biol Chem* **288:** 6053–6062.

Van Aller GS, Reynoird N, Barbash O, Huddleston M, Liu S, Zmoos AF, McDevitt P, Sinnamon R, Le B, Mas G, et al. 2012. Smyd3 regulates cancer cell phenotypes and catalyzes histone H4 lysine 5 methylation. *Epigenetics* **7:** 340–343.

van Attikum H, Gasser SM. 2005. The histone code at DNA breaks: A guide to repair? *Nat Rev Mol Cell Biol* **6:** 757–765.

Vandel L, Nicolas E, Vaute O, Ferreira R, Ait-Si-Ali S, Trouche D. 2001. Transcriptional repression by the retinoblastoma protein through the recruitment of a histone methyltransferase. *Mol Cell Biol* **21:** 6484–6494.

van Leeuwen F, Gafken PR, Gottschling DE. 2002. Dot1p modulates silencing in yeast by methylation of the nucleosome core. *Cell* **109:** 745–756.

Vaquero A, Scher M, Lee D, Erdjument-Bromage H, Tempst P, Reinberg D. 2004. Human SirT1 interacts with histone H1 and promotes formation of facultative heterochromatin. *Mol Cell* **16:** 93–105.

Verreault A, Kaufman PD, Kobayashi R, Stillman B. 1998. Nucleosomal DNA regulates the core-histone-binding subunit of the human Hat1 acetyltransferase. *Curr Biol* **8:** 96–108.

Visochek L, Steingart RA, Vulih-Shultzman I, Klein R, Priel E, Gozes I, Cohen-Armon M. 2005. PolyADP-ribosylation is involved in neurotrophic activity. *J Neurosci* **25:** 7420–7428.

Waldmann T, Izzo A, Kamieniarz K, Richter F, Vogler C, Sarg B, Lindner H, Young NL, Mittler G, Garcia BA, et al. 2011. Methylation of H2AR29 is a novel repressive PRMT6 target. *Epigenetics Chromatin* **4:** 11.

Wang H, Cao R, Xia L, Erdjument-Bromage H, Borchers C, Tempst P, Zhang Y. 2001a. Purification and functional characterization of a histone H3-lysine 4-specific methyltransferase. *Mol Cell* **8:** 1207–1217.

Wang H, Huang ZQ, Xia L, Feng Q, Erdjument-Bromage H, Strahl BD, Briggs SD, Allis CD, Wong J, Tempst P, et al. 2001b. Methylation of histone H4 at arginine 3 facilitating transcriptional activation by nuclear hormone receptor. *Science* **293:** 853–857.

Wang Y, Zhang W, Jin Y, Johansen J, Johansen KM. 2001c. The JIL-1 tandem kinase mediates histone H3 phosphorylation and is required for maintenance of chromatin structure in *Drosophila*. *Cell* **105:** 433–443.

Wang H, Wang L, Erdjument-Bromage H, Vidal M, Tempst P, Jones RS, Zhang Y. 2004. Role of histone H2A ubiquitination in Polycomb silencing. *Nature* **431:** 873–878.

Wang GG, Cai L, Pasillas MP, Kamps MP. 2007. NUP98-NSD1 links H3K36 methylation to Hox-A gene activation and leukaemogenesis. *Nat Cell Biol* **9:** 804–812.

Ward IM, Chen J. 2001. Histone H2AX is phosphorylated in an ATR-dependent manner in response to replicational stress. *J Biol Chem* **276:** 47759–47762.

Wei Y, Yu L, Bowen J, Gorovsky MA, Allis CD. 1999. Phosphorylation of histone H3 is required for proper chromosome condensation and segregation. *Cell* **97:** 99–109.

Weinert BT, Schölz C, Wagner SA, Iesmantavicius V, Su D, Daniel JA, Choudhary C. 2013. Lysine succinylation is a frequently occurring modification in prokaryotes and eukaryotes and extensively overlaps with acetylation. *Cell Rep* **4:** 842–851.

Weiss T, Hergeth S, Zeissler U, Izzo A, Tropberger P, Zee BM, Dundr M, Garcia BA, Daujat S, Schneider R. 2010. Histone H1 variant-specific lysine methylation by G9a/KMT1C and Glp1/KMT1D. *Epigenetics Chromatin* **3:** 7.

Wilson JR, Jing C, Walker PA, Martin SR, Howell SA, Blackburn GM, Gamblin SJ, Xiao B. 2002. Crystal structure and functional analysis of the histone methyltransferase SET7/9. *Cell* **111:** 105–115.

Wisniewski JR, Zougman A, Kruger S, Mann M. 2007. Mass spectrometric mapping of linker histone H1 variants reveals multiple acetylations, methylations, and phosphorylation as well as differences between cell culture and tissue. *Mol Cell Proteomics* **6:** 72–87.

Wisniewski JR, Zougman A, Mann M. 2008. N^{ε}-formylation of lysine is a widespread post-translational modification of nuclear proteins occurring at residues involved in regulation of chromatin function. *Nucleic Acids Res* **36:** 570–577.

Wu L, Zee BM, Wang Y, Garcia BA, Dou Y. 2011. The RING finger protein MSL2 in the MOF complex is an E3 ubiquitin ligase for H2B K34 and is involved in crosstalk with H3 K4 and K79 methylation. *Mol Cell* **43:** 132–144.

Wyatt HR, Liaw H, Green GR, Lustig AJ. 2003. Multiple roles for *Saccharomyces cerevisiae* histone H2A in telomere position effect, Spt phenotypes and double-strand-break repair. *Genetics* **164:** 47–64.

Xiao A, Li H, Shechter D, Ahn SH, Fabrizio LA, Erdjument-Bromage H, Ishibe-Murakami S, Wang B, Tempst P, Hofmann K, et al. 2009. WSTF regulates the H2AX DNA damage response via a novel tyrosine kinase activity. *Nature* **457:** 57–62.

Xiao Y, Bedet C, Robert VJ, Simonet T, Dunkelbarger S, Rakotomalala C, Soete G, Korswagen HC, Strome S, Palladino F. 2011. *Caenorhabditis elegans* chromatin-associated proteins SET-2 and ASH-2 are differentially required for histone H3 Lys 4 methylation in embryos and adult germ cells. *Proc Natl Acad Sci* **108:** 8305–8310.

Xie Z, Dai J, Dai L, Tan M, Cheng Z, Wu Y, Boeke JD, Zhao Y. 2012. Lysine succinylation and lysine malonylation in histones. *Mol Cell Proteomics* **11:** 100–107.

Xin Z, Tachibana M, Guggiari M, Heard E, Shinkai Y, Wagstaff J. 2003. Role of histone methyltransferase G9a in CpG methylation of the Prader-Willi syndrome imprinting center. *J Biol Chem* **278:** 14996–15000.

Xu F, Zhang K, Grunstein M. 2005. Acetylation in histone H3 globular domain regulates gene expression in yeast. *Cell* **121:** 375–385.

Yamamoto T, Horikoshi M. 1997. Novel substrate specificity of the histone acetyltransferase activity of HIV-1-Tat interactive protein Tip60. *J Biol Chem* **272:** 30595–30598.

Yamamoto Y, Verma UN, Prajapati S, Kwak YT, Gaynor RB. 2003. Histone H3 phosphorylation by IKK-α is critical for cytokine-induced gene expression. *Nature* **423:** 655–659.

Yang L, Xia L, Wu DY, Wang H, Chansky HA, Schubach WH, Hickstein DD, Zhang Y. 2002. Molecular cloning of ESET, a novel histone H3-

specific methyltransferase that interacts with ERG transcription factor. *Oncogene* **21:** 148–152.

Yu Y, Song C, Zhang Q, DiMaggio PA, Garcia BA, York A, Carey MF, Grunstein M. 2012. Histone H3 lysine 56 methylation regulates DNA replication through its interaction with PCNA. *Mol Cell* **46:** 7–17.

Zegerman P, Canas B, Pappin D, Kouzarides T. 2002. Histone H3 lysine 4 methylation disrupts binding of nucleosome remodeling and deacetylase (NuRD) repressor complex. *J Biol Chem* **277:** 11621–11624.

Zeitlin SG, Barber CM, Allis CD, Sullivan KF. 2001. Differential regulation of CENP-A and histone H3 phosphorylation in G_2/M. *J Cell Sci* **114:** 653–661.

Zhang L, Eugeni EE, Parthun MR, Freitas MA. 2003. Identification of novel histone post-translational modifications by peptide mass fingerprinting. *Chromosoma* **112:** 77–86.

Zhang K, Chen Y, Zhang Z, Zhao Y. 2009. Identification and verification of lysine propionylation and butyrylation in yeast core histones using PTMap software. *J Proteome Res* **8:** 900–906.

Zhang S, Roche K, Nasheuer HP, Lowndes NF. 2011. Modification of histones by sugar β-*N*-acetylglucosamine GlcNAc occurs on multiple residues, including histone H3 serine 10, and is cell cycle-regulated. *J Biol Chem* **286:** 37483–37495.

Zhong S, Jansen C, She QB, Goto H, Inagaki M, Bode AM, Ma WY, Dong Z. 2001. Ultraviolet B-induced phosphorylation of histone H3 at serine 28 is mediated by MSK1. *J Biol Chem* **276:** 33213–33219.

Zhu B, Zheng Y, Pham AD, Mandal SS, Erdjument-Bromage H, Tempst P, Reinberg D. 2005. Monoubiquitination of human histone H2B: The factors involved and their roles in HOX gene regulation. *Mol Cell* **20:** 601–611.

SUGGESTED REVIEWS

Bannister AJ, Kouzarides T. 2005. Reversing histone methylation. *Nature* **436:** 1103–1106.

Berger SL. 2002. Histone modifications in transcriptional regulation. *Curr Opin Genet Dev* **12:** 42–148.

Cosgrove MS, Boeke JD, Wolberger C. 2004. Regulated nucleosome mobility and the histone code. *Nat Struct Mol Biol* **11:** 1037–1043.

Davie JR, Spencer VA. 2001. Signal transduction pathways and the modification of chromatin structure. *Prog Nucleic Acid Res Mol Biol* **65:** 299–340.

Dobosy JR, Selker EU. 2001. Emerging connections between DNA methylation and histone acetylation. *Cell Mol Life Sci* **58:** 721–727.

Dunleavy E, Pidoux A, Allshire R. 2005. Centromeric chromatin makes its mark. *Trends Biochem Sci* **30:** 172–175.

Elgin SC, Grewal SI. 2003. Heterochromatin: Silence is golden. *Curr Biol* **13:** R895–R898.

Emre NC, Berger SL. 2004. Histone H2B ubiquitylation and deubiquitylation in genomic regulation. *Cold Spring Harbor Symp Quant Biol* **69:** 289–299.

Esteller M. 2006. Epigenetics provides a new generation of oncogenes and tumour-suppressor genes. *Br J Cancer* **94:** 179–183.

Feinberg AP, Tycko B. 2004. The history of cancer epigenetics. *Nat Rev Cancer* **4:** 143–153.

Fischer A, Hofmann I, Naumann K, Reuter G. 2006. Heterochromatin proteins and the control of heterochromatic gene silencing in *Arabidopsis*. *J Plant Physiol* **163:** 358–368.

Fischle W, Wang Y, Allis CD. 2003. Binary switches and modification cassettes in histone biology and beyond. *Nature* **425:** 475–479.

Fischle W, Wang Y, Allis CD. 2003. Histone and chromatin cross-talk. *Curr Opin Cell Biol* **15:** 172–183.

Grewal SI, Elgin SC. 2002. Heterochromatin: New possibilities for the inheritance of structure. *Curr Opin Genet Dev* **12:** 178–187.

Grunstein M. 1997. Histone acetylation in chromatin structure and transcription. *Nature* **389:** 349–352.

Grunstein M. 1997. Molecular model for telomeric heterochromatin in yeast. *Curr Opin Cell Biol* **9:** 383–387.

Grunstein M. 1998. Yeast heterochromatin: Regulation of its assembly and inheritance by histones. *Cell* **93:** 325–328.

Henikoff S, Ahmad K. 2005. Assembly of variant histones into chromatin. *Annu Rev Cell Dev Biol* **21:** 133–153.

Herz HM, Garruss A, Shilatifard A. 2013. SET for life: Biochemical activities and biological functions of SET domain-containing proteins. *Trends Biochem Sci* **38:** 621–639.

Hild M, Paro R. 2003. Anti-silencing from the core: A histone H2A variant protects euchromatin. *Nat Cell Biol* **5:** 278–280.

Jenuwein T, Allis CD. 2001. Translating the histone code. *Science* **293:** 1074–1080.

Kimmins S, Sassone-Corsi P. 2005. Chromatin remodelling and epigenetic features of germ cells. *Nature* **434:** 583–589.

Kurdistani SK, Grunstein M. 2003. Histone acetylation and deacetylation in yeast. *Nat Rev Mol Cell Biol* **4:** 276–284.

Lachner M, O'Sullivan RJ, Jenuwein T. 2003. An epigenetic road map for histone lysine methylation. *J Cell Sci* **116:** 2117–2124.

Luger K, Richmond TJ. 1998. The histone tails of the nucleosome. *Curr Opin Genet Dev* **8:** 140–146.

Mellone BG, Allshire RC. 2003. Stretching it: Putting the CEN(P-A) in centromere. *Curr Opin Genet Dev* **13:** 191–198.

Millar CB, Kurdistani SK, Grunstein M. 2004. Acetylation of yeast histone H4 lysine 16: A switch for protein interactions in heterochromatin and euchromatin. *Cold Spring Harbor Symp Quant Biol* **69:** 193–200.

Nightingale KP, O'Neill LP, Turner BM. 2006. Histone modifications: Signalling receptors and potential elements of a heritable epigenetic code. *Curr Opin Genet Dev* **16:** 125–136.

Panier S, Durocher D. 2013. Push back to respond better: Regulatory inhibition of the DNA double-strand break response. *Nat Rev Mol Cell Biol* **14:** 661–772.

Peterson CL, Laniel MA. 2004. Histones and histone modifications. *Curr Biol* **14:** R546–R551.

Pinder JB, Attwood KM, Dellaire G. 2013. Reading, writing, repair: The role of ubiquitin and the ubiquitin-like proteins in DNA damage signaling and repair. *Front Genet* **4:** 45.

Reinberg D, Chuikov S, Farnham P, Karachentsev D, Kirmizis A, Kuzmichev A, Margueron R, Nishioka K, Preissner TS, Sarma K, et al. 2004. Steps toward understanding the inheritance of repressive methyl-lysine marks in histones. *Cold Spring Harbor Symp Quant Biol* **69:** 171–182.

Sarma K, Reinberg D. 2005. Histone variants meet their match. *Nat Rev Mol Cell Biol* **6:** 139–149.

Spencer VA, Davie JR. 2000. Signal transduction pathways and chromatin structure in cancer cells. *J Cell Biochem Suppl* **35:** 27–35.

Sternglanz R. 1996. Histone acetylation: A gateway to transcriptional activation. *Trends Biochem Sci* **21:** 357–358.

Turner BM. 2000. Histone acetylation and an epigenetic code. *Bioessays* **22:** 836–845.

van Attikum H, Gasser SM. 2005. The histone code at DNA breaks: A guide to repair? *Nat Rev Mol Cell Biol* **6:** 757–765.

Vaughn MW, Tanurdzic M, Martienssen R. 2005. Replication, repair, reactivation. *Dev Cell* **9:** 724–725.

Wade PA, Wolffe AP. 1997. Histone acetyltransferases in control. *Curr Biol* **7:** R82–R84.

Wade PA, Pruss D, Wolffe AP. 1997. Histone acetylation: Chromatin in action. *Trends Biochem Sci* **22:** 128–132.

Zilberman D, Henikoff S. 2005. Epigenetic inheritance in *Arabidopsis*: Selective silence. *Curr Opin Genet Dev* **15:** 557–562.

Cite as *Cold Spring Harb Perspect Biol* doi: 10.1101/cshperspect.a025064

Index

Page references followed by b, f, or t denote boxes, figures, or tables respectively.

E

I

N

Q

R

S